Textbook of
Medical Biochemistry

FOREWORD

The discipline of biochemistry has expanded by leaps and bounds. It is an independent subject with a separate examination in almost all medical colleges. Though there are a few books on biochemistry written by Indian authors, there is need for a comprehensive yet simplified textbook. I am happy that the *Textbook of Medical Biochemistry*, written by Dr (Brig) MN Chatterjea and Dr Rana Shinde, fulfils this need. Both authors have long experience in teaching biochemistry to undergraduate and postgraduates. They have tried their best to make the text simple and lucid and at the same time have also incorporated the recent developments in the subject. The textbook is useful not only for undergraduates but also for postgraduates in biochemistry and others registering for diplomate examination of the National Board of Examinations.

I hope the book will receive appreciation as well as encouragement from all biochemists.

Dr C Sita Devi MD FAMS FIMSA
Vice President-Lab Services
Retd. Principal and Professor
Department of Biochemistry
Andhra Medical College
Vishakhapatnam, Andhra Pradesh, India

FOREWORD

I am pleased to go through the *Textbook of Medical Biochemistry* written by Dr (Brig) MN Chatterjea and Dr Rana Shinde. The book makes a lucid reading, is full of necessary facts for medical students, and the text is clinically oriented.

Both the authors are well-known teachers of repute and have really tried to make the book simple as well as useful for medical students at undergraduate and postgraduate levels.

The book though similar to many other such textbooks, is unique for its clarity and comprehension. I am sure students and teachers will gladly accept the book.

Dr KP Sinha MD PhD (Lond)
Ex-Professor and Head
Department of Biochemistry
Patna Medical College
Patna, Bihar, India

Textbook of Medical Biochemistry

Eighth Edition

Dr (Brig) MN Chatterjea
BSc MBBS DCP MD (Biochemistry)
Ex-Professor and Head, Department of Biochemistry
Armed Forces Medical College, Pune, Maharashtra, India
Christian Medical College and Hospital, Ludhiana, Punjab, India
MGM Medical College, Aurangabad, Maharashtra, India
(Specialist in Pathology and Ex-Reader in Pathology)

Rana Shinde
PhD FACB MRC Path (Chemical Pathology)
Professor and Head and Chief, Chemical Pathologist
Department of Biochemistry, SSR Medical College
Belle Rive, Mauritius

Ex-Reader and Consultant, Department of Biochemistry
Christian Medical College and Hospital, Ludhiana, Punjab, India

Ex-Professor, Department of Biochemistry
JN Medical College, Belgaum, Karnataka, India

Formerly
Wanless Hospital
Miraj and Christian Medical College and Hospital, Ludhiana, Punjab, India

Forewords
T Venkatesh
Vijaykant B Kambli
C Sita Devi
KP Sinha

JAYPEE BROTHERS MEDICAL PUBLISHERS (P) LTD

New Delhi • Panama City • London

Jaypee Brothers Medical Publishers (P) Ltd

Headquarter

Jaypee Brothers Medical Publishers (P) Ltd
4838/24, Ansari Road, Daryaganj
New Delhi 110 002, India
Phone: +91-11-43574357
Fax: +91-11-43574314
Email: jaypee@jaypeebrothers.com

Overseas Offices

JP Medical Ltd
83 Victoria Street, London
SW1H 0HW (UK)
Phone: +44-2031708910
Fax: +02-03-0086180
Email: info@jpmedpub.com

Jaypee-Highlights Medical Publishers Inc.
City of Knowledge, Bld. 237, Clayton
Panama City, Panama
Phone: 507-317-0160
Fax: +50-73-010499
Email: cservice@jphmedical.com

Website: www.jaypeebrothers.com
Website: www.jaypeedigital.com

Inquiries for bulk sales may be solicited at: jaypee@jaypeebrothers.com

Publisher: Jitendar P Vij
Publishing Director: Tarun Duneja
Editor: Richa Saxena
Cover Design: Seema Dogra, Sumit Kumar

Textbook of Medical Biochemistry

First Edition	:	1993
Second Edition	:	1995
Third Edition	:	1998
Fourth Edition	:	2000
Fifth Edition	:	2002
Sixth Edition	:	2005
Seventh Edition	:	2007
Eighth Edition	:	**2012**

ISBN 978-93-5025-484-4

Printed at Sanat Printers, Kundli.

Foreword

"Today's Biochemistry is Tomorrow's Medicine" rightly said by many is truly justified by Dr (Brig) MN Chatterjea and Dr Rana Shinde in their *Textbook of Medical Biochemistry*. Authors' sincere efforts are appreciated both by students and faculty who have been using this book, which is meeting most of the requirements of MCI regulations. I have found lot of appreciation and greater acceptance in India and abroad for which authors have put in special efforts to cover wide spectrum of current topics in medical biochemistry through constant review and changes from time-to-time. This popular book itself is a good guide and a tool to both undergraduate students and teachers in medical biochemistry.

Prof T Venkatesh PhD FACBI
Director
National Referral Centre for Lead Poisoning in India
Professor, Department of Biochemistry and Biophysics
Past President, ACBI
St John's National Academy of Health Sciences
Bengaluru, Karnataka, India

Foreword

I am very happy to write the foreword for this edition of the *Textbook of Medical Biochemistry*. In keeping with its prior editions, the authors have given full attention to include the advancements occurred in the field of clinical biochemistry. I am a practicing clinical chemist in community hospital who used tools in basic concepts of the subject, when teaching students the art and science of clinical biochemistry. I encourage the teachers in medical schools in India and abroad to use the book for teaching the subject of clinical biochemistry. I am a strong advocate of teaching the fundamental concepts in biochemistry to medical students, clinical laboratory technologists and clinical pathologists to advance their understanding of the subject.

The authors are well-known teachers and have spent decades in this field. This edition is enriched with relevant subject information for those who are seeking to understand medical biochemistry. I recommend the book for medical students at undergraduate as well as postgraduate levels.

I am sure the medical college community in India and abroad will welcome this edition of *Textbook of Medical Biochemistry*.

Dr Vijaykant B Kambli PhD MBA FACB DABCC
Director
Quality Assurance and Point of Care Testing
Former Director, Clinical Biochemistry
Department of Pathology
Norwalk Hospital
Norwalk, Connecticut, USA

Preface to the Eighth Edition

It is a great pleasure to present the Eighth Edition of *Textbook of Medical Biochemistry* to our beloved teachers and students.

The main framework of the book has been retained, without much changes, new additions have been made, some portions rewritten in certain chapters.

Few teachers wanted that the university question types and MCQs be added at the end of each chapter which were in existence in first three editions but later removed to reduce pages. To quote, "I suggest that short type of questions and multiple choice questions should be included at the end of each chapter in future edition of the book." We have included the same in the topics pertaining to MBBS.

We feel confident that this edition will fulfil the requirement of the undergraduate students as per MCI recommendations and also meet the needs and expectations of postgraduate students of Biochemistry. As the book is clinically oriented, it will be of much help to PG students of other disciplines, viz. Pathology/medicine/pediatrics, etc.

Colour printing has been used to make the book more attractive, easy recording and highlighting the important portions like Clinical Aspects/Significance and Biochemistry Importance.

The overall objective of the book has been to provide concise yet authoritative coverage on the basics of Biochemistry with clinical approach to understand the disease processes. The important points in the text to be remembered by the students have been highlighted in bold and italicised prints.

We do not claim to the perfect. Errors/mistakes may creep-up due to oversight/or printing errors. We shall look forward for any valuable comments and useful suggestions from teachers and students so that errors are rectified and suggestions are taken into account for future.

We sincerely thanks Shri Jitendar P Vij (Chairman and Managing Director), Mr PG Bandhu (Director-Sales), and Mr Tarun Duneja (Director-Publishing) for their untiring work and keen efforts to bring out the new revised edition of the book.

Dr (Brig) MN Chatterjea
Rana Shinde

Preface to the First Edition

Biochemistry holds a key position in the curricula of medical colleges under Indian universities, and is one of the basic preclinical science subjects for first professional MBBS students.

Biochemistry is being transformed with astonishing rapidity and current efflorescence in the knowledge in this subject has necessitated that it should be learnt separately from physiology. The three basic science subjects make a plinth for the house of medicine. A sound and comprehensive learning of biochemistry will help a medical student understand medicine and pathology more clearly.

A large number of books on biochemistry for medical students are available in the market—both international and Indian. Many of the international books are voluminous and too difficult for our students of medicine to handle and comprehend.

The *Textbook of Medical Biochemistry* for the medical students is the outcome of the joint efforts of a medical and a nonmedical biochemist, who possess considerable experience in teaching biochemistry to undergraduate and postgraduate medical students of Indian universities.

We have tried our utmost to ensure that the language used is lucid and simple, makes an easy-reading, and the text provides an intelligent and comprehensive study. At the same time, we have attempted to maintain a high standard after incorporating the recent developments and concepts.

Though the book is primarily meant for the first professional MBBS students, certain chapters have been dealt with in greater detail to meet the requirements of postgraduates, viz. MSc, MD (Biochemistry) students and those preparing for Diplomate in NBE. It meets the requirements of students of medical, dental science, agricultural science, home science, and others who have to take a basic course in biochemistry.

The text of the book is spread over 40 chapters and special mention has been made to introduce to the reader some recent topics such as cyclic nucleotides—cyclic AMP and cyclic GMP, prostaglandins, prostacyclins, thromboxanes and leukotrienes, immunoglobulins, recombinant DNA technology, clinical significance of enzymes and isoenzymes, radioisotopes and their clinical and therapeutic uses, etc.

Recently, much importance has been given to self-study by students in small groups and to avoid or to restrict to the minimum the traditional way of learning based on "didactic" lectures. Keeping this in view, we have included in the beginning of each chapter the "major concepts" and "specific objectives" of the chapter for the information of the readers so that medical students know what to study and learn. The text of each chapter has been written keeping in view the objectives so that students can make a self-study. We wish to emphasise that these are "behavioural objectives" and are self-explanatory.

In our several years of teaching experience, we have observed that medical students have a "fear-complex" that biochemistry involves too many structural formulae and chemical equations. Though this is unavoidable to some extent for proper understanding of the subject, we have tried to restrict the chemical formulae to the minimum and used them to explain certain reactions.

At the end of each chapter we have given "essay type" or "short notes type" questions which we have compiled from the examination papers of different Indian universities. It will be seen that there is repetition of some questions, but the framing of the question and language is different.

We have also tried to give some 15 to 30 MCQs (with answers at the end) in each chapter which may be useful for the medical students for their homework.

While writing the chapters and compiling the questions, we have consulted syllabi of several Indian universities to cover all the topics prescribed for undergraduate and postgraduate medical students. We have included a large number of tables and comparative discussions wherever possible to meet the needs of the students. Our aim has been to provide a comprehensive, self-contained textbook of biochemistry to effectively satisfy the curricular requirements of medical students of Indian universities.

In addition, our main target has been to make the book clinically-oriented. We have given the clinical significance and biomedical importance wherever it is applicable. Biochemical aspects of certain pathological conditions, specially those due to abnormal metabolism, have been discussed in detail. We earnestly hope that the book will be of help to both the undergraduate and postgraduate medical students and their teachers.

No one can be perfect, and there could have been some flaws or shortcomings in the book. We will welcome constructive criticism and comments, if any, along with fruitful suggestions to improve the text in its future editions.

In writing a textbook of this nature, one has to take help from others and this book is no exception. We are highly indebted to our colleagues and friends, and other authors whom we have consulted in compiling this book.

Our thanks to Mrs KN Valsa and Mrs Gracy for their untiring efforts and forbearance in typing the manuscript.

We are also grateful to Shri Jitendar P Vij (Chairman and Managing Director), and Mr YN Arjuna, Editorial and Publishing Consultant, of Jaypee Brothers Medical Publishers (P) Ltd. for their sincere and untiring efforts in transforming the typed manuscript to printed form.

MN Chatterjea
Rana Shinde

Contents

Section One: Cell Biology

Section Two: Chemistry of Biomolecules

SECTION THREE: Molecular Biology

SECTION FOUR: Metabolism

Section Five: Clinical Biochemistry

Section Six: Miscellaneous

SECTION

1

Cell Biology

1. Cell and Cell Organelles: Chemistry and Functions
2. Biological Membranes: Structure and Function

CHAPTER 1

CELL AND CELL ORGANELLES: CHEMISTRY AND FUNCTIONS

Major Concept

To study the molecular and functional organisation of a cell and its subcellular organelles.

Specific Objectives

1. To know importance of cell, and the types: Prokaryotic and eukaryotic cell.
2. Learn the essential differences of a prokaryotic cell and eukaryotic cell.
3. Draw a diagram of an eukaryotic cell showing different cell organelles.
4. Study the following cellular organelles:
   - Nucleus—its structure and functions.
   - Mitochondrion, the **power house** of a cell. Learn its structure and functions.
   - Study endoplasmic reticulum, its types, structure and functions.
   - Learn structure and functions of golgi complexes.
   - Study about lysosomes, their functions, inherited disorder—**I cell disease**.
   - Learn about peroxisomes: Their structure and functions.
   - Study the structure and functions of cytoskeleton.

All organisms are built from cells. All animal tissues including human are also organised from collections of cells. **Thus cell is the fundamental unit of life**. If cell dies, tissue dies and it cannot function.

Modern cell theory can be divided into the following fundamental statements:

- Cells make up all living matter
- All cells arise from other cells
- The genetic information required during the maintenance of existing cells and the production of new cells passes from one generation to the other next generation
- The chemical reactions of an organism that is its metabolism, both anabolism and catabolism, takes place in the cells.

Types of Cells

In general **two types** of cells exist in nature. They are:

1. Prokaryotic cells
2. Eukaryotic cells

1. Prokaryotic Cells

Typical prokaryotic cells (Greek: *Pro*-before and *karyon*-nucleus) include the bacteria and cyanobacteria. Most studied prokaryotic cell is *Escherichia coli (E. coli).*

Characteristics

- It has a minimum of internal organisation and smaller in size
- It does not have any membrane bound organelles.
- Its genetic material is not enclosed by a nuclear membrane
- Its DNA is not complexed with histones. **Histones are not found in prokaryotic cells**
- Its respiratory system is closely associated with its plasma membrane and
- Its sexual reproduction does not involve mitosis or meiosis.

2. Eukaryotic Cells

The eukaryotic cells (Greek: *Eu*-true and *karyon*-nucleus) include the protists, fungi, plants and animals including humans. Cells are larger in size (Fig. 1.1).

Characteristics

- It has considerable degree of internal structure with a large number of distinctive membrane enclosed having specific functions
- Nucleus is the site for informational components collectively called **chromatin**

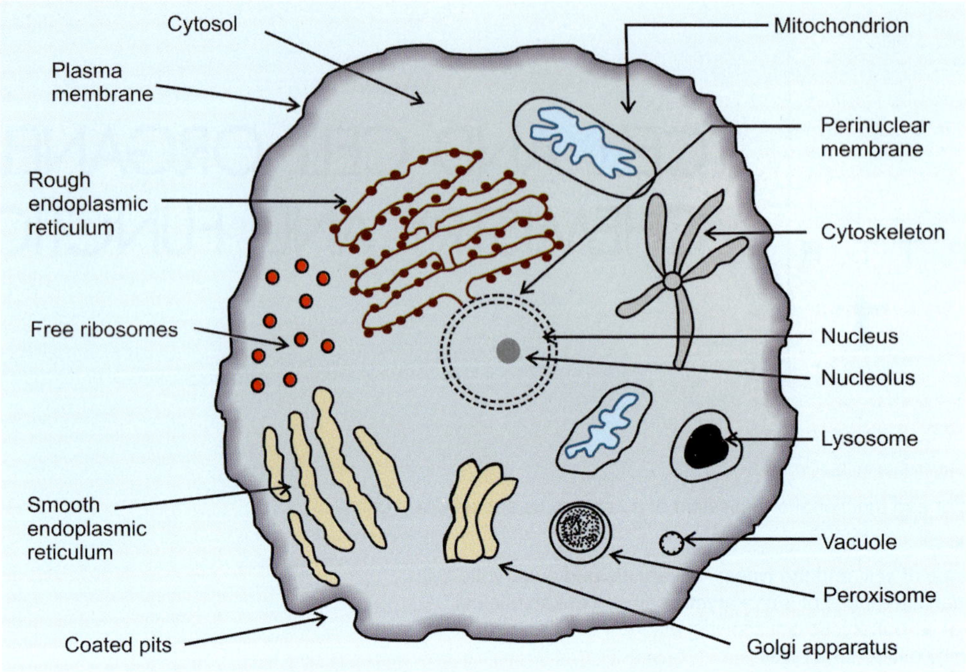

Fig. 1.1: Schematic representation of an eukaryotic cell with cell organelles

- Sexual reproduction involves both mitosis and meiosis
- The respiratory site is the mitochondria
- In the plant cells, the site of the conversion of radiant energy to chemical energy is the highly structural chloroplasts.

Essential differences of prokaryotic and eukaryotic cells are given in Table 1.1.

Table 1.1: Essential differences between prokaryotic and eukaryotic cells

Prokaryotic cell	*Eukaryotic cell*
1. Smaller in size 1 to 10 µm	1. Larger in size 10 to 100 µm or more
2. Mainly unicellular	2. Mainly multicellular (with few exceptions). Several different types present
3. Single membrane, surrounded by rigid cell wall	3. Lipid bilayer membrane with proteins
4. Anaerobic or aerobic	4. Aerobic
5. Not well defined nucleus, only a nuclear zone with DNA **Histones absent**	5. Nucleus well defined, 4 to 6 µm in diameter, contains DNA and surrounded by a perinuclear membrane **Histones present**
6. No nuclei	6. Nucleolus present, rich in RNA
7. Cytoplasm contains no cell organelles	7. Membrane bound cell organelles are present
8. Ribosomes present free in cytoplasm	8. Ribosomes studded on outer surface of endoplasmic reticulum present
9. Mitochondria absent. Enzymes of energy metabolism bound to membrane	9. Mitochondria present **Power house** of the cell. Enzymes of energy metabolism are located in mitochondria
10. Golgi apparatus absent. Storage granules with polysaccharides	10. Golgi apparatus present—flattened single membrane vesicles
11. Lysosomes—absent	11. **Lysosomes present**—single membrane vesicle containing packets of hydrolytic enzymes
12. Cell division usually by fission, no mitosis	12. Cell division—by mitosis
13. Cytoskeleton—absent	13. Cytoskeleton—present
14. RNA and protein synthesis in same compartment	14. RNA synthesised and processed in nucleus. Proteins synthesised in cytoplasm
15. **Examples** are bacteria, cyanobacteria, rickettsia	15. **Examples**: Protists, fungi, plants and animal cells

A. Cell Organelles

Eukaryotic cells contain many membrane-bound organelles that carryout specific cellular processes. Chief organelles and their functions are as follows:

1. ***Nucleus:*** The nucleus contains more than 95 per cent of the cell's DNA and is the control centre of the eukaryotic cell.

- ***Nuclear envelope:*** A double membrane structure called the nuclear envelope separates the nucleus from the cytosol.
- ***Nuclear pore complexes:*** These are embedded in the nuclear envelope. These complex structures control the movement of proteins and the nucleic acid ribonucleic acids (RNAs) across the nuclear envelope.
- ***Chromatin:*** DNA in the nucleus is coiled into a dense mass called chromatin, so named because it is stained darkly with certain dyes.
- ***Nucleolus:*** A second dense mass closely associated with the inner nuclear envelope is called nucleolus.
- ***Nucleoplasm:*** Nucleoplasm of nucleus contain various enzymes such as ***DNA polymerases***, and ***RNA polymerases***, for m-RNA and t-RNA synthesis.

Functions

- DNA replication and RNA transcription of DNA occur in the nucleus. Transcription is the first step in the expression of genetic information and is the major metabolic activity of the nucleus.
- The nucleolus is nonmembranous and contains ***RNA polymerase, RNAase, ATPase*** and other enzymes **but no DNA polymerase**. Nucleolus is the site of synthesis of ribosomal RNA (r-RNA).
- Nucleolus is also the major site where ribosome subunits are assembled.

2. ***Mitochondrion:*** Mitochondrion is the **power house** of cell (**Figs 1.2A and B**).

- ***Number:*** The number of mitochondria in a cell varies dramatically. Some algae contain only one mitochondrion, whereas the protozoan **Chaos** contain half a million. A mammalian liver cell contains from 800 to 2500 mitochondria.
- ***Size:*** They vary greatly in size. A typical mammalian mitochondrion has a diameter of 0.2 to 0.8 μ and a length of 0.5 to 1.0 μm.
- ***Shape:*** The ***shape of mitochondrion is not static.*** Mitochondria assume many different shapes under different metabolic conditions.

Structure and Functions

The mitochondrion is bounded by **two concentric membranes** that have markedly different properties and biological functions.

Mitochondrial Membranes

(a) Outer mitochondrial membrane: The outer mitochondrial membrane consists mostly of phospholipids and contains a considerable amount of cholesterol. The outer membrane also contains many copies of the protein called **Porin**.

Functions of Porin and other Proteins

(i) ***These proteins form channels*** that permit substances with molecular weights of less than < 10,000 to diffuse freely across the outer mitochondrial membrane.

(ii) Other proteins in the outer membrane carry out various reactions in fatty acid and phospholipid biosynthesis and are responsible for some oxidation reactions.

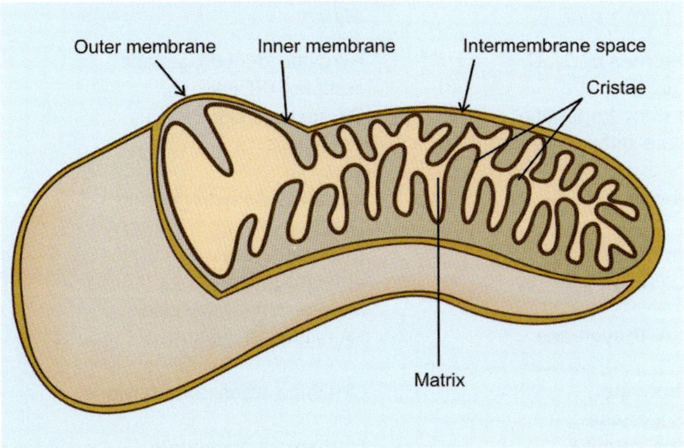

Fig. 1.2A: A mitochondrion—shows half split to show the inner membrane with cristae

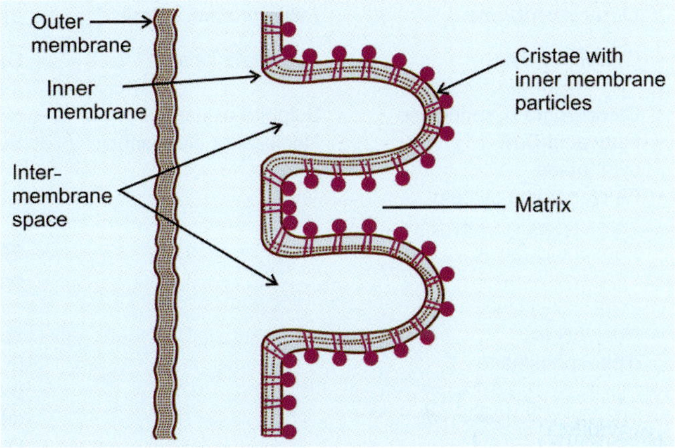

Fig. 1.2B: Cross-section of a mitochondrion—showing various layers and cristae

(b) Inner mitochondrial membrane: The inner mitochondrial membrane is very rich in proteins and the ratio of lipid to proteins is only 0.27:1 by weight. ***It contains high proportion of the phospholipid cardiolipin.*** **In contrast to outer membrane, the inner membrane is virtually impermeable to polar and ionic substances.** These substances enter the mitochondrion only through the mediation of specific transport proteins.

- *Cristae:* The inner mitochondrial membrane is highly folded. The tightly packed inward folds are called *"cristae"*.

Functional changes: It is now known that mitochondria undergo dramatic changes when they switch over from resting state to a respiring state. **In the respiring state, the inner membrane is *not* folded into cristae,** rather it seems to shrink leaving a much more voluminous inter membrane space.

(c) Intermembrane space: The space between the outer and inner membranes is known as the ***intermembrane space.*** Since the outer membrane is freely permeable to small molecules, the intermembrane space has about the same ionic composition as the cytosol.

(d) Mitochondrial matrix: The region enclosed by the inner membrane is known as the mitochondrial matrix.

Composition of matrix: The **enzymes responsible for citric acid cycle and fatty acid oxidation are located in the matrix.** The matrix also contains several strands of circular DNA, ribosomes and enzymes required for the biosynthesis of the proteins coded in the mitochondrial genome. The mitochondrion is not, however, genetically autonomous, and the genes encoding most mitochondrial proteins are present in nuclear DNA.

Functions

- Many enzymes associated with carbohydrates, fatty acids and nitrogen metabolism are located within the mitochondrion. Enzymes of electron transport and oxidative phosphorylation are also located in different areas of this cell organelle.
  Table 1.2 gives the names of some of the important enzymes and their location.
- The mitochondrion is specialised for the rapid oxidation of NADH (reduced NAD) and FAD. H_2 (reduced FAD) produced in the reactions of glycolysis, the citric acid cycle and the oxidation of fatty acids. The ***energy produced is trapped and stored as ATP,*** for future use of energy in the body.

CLINICAL ASPECT

A disease known as **Luft's disease** involving mitochondrial energy transduction has been reported. Further mitochondrial DNA can be damaged by free radicals. ***Age related degenerative disorders such as Parkinson's disease, cardiomyopathy may have a component of mitochondrial damage.***

3. Endoplasmic reticulum (ER): Eukaryotic cells are characterised by several membrane complexes that are interconnected by separate organelles. These organelles are involved in protein synthesis, transport, modification, storage and secretion.

Varying in shape, size and amount, the endoplasmic reticulum (ER) extends from the cell membrane, coats the nucleus, surrounds the mitochondria and appears to connect directly to the Golgi apparatus. These membranes and the aqueous channels they enclose are called ***cisternae.***

Table 1.2: Location of some of the important enzymes in mitochondrion

Outer membrane	*Intermediate space*	*Inner membrane*	*Matrix*
• Cytochrome b_5	• Adenylate kinase	• Cytochromes b, C_1, C, a and a_3	• Pyruvate dehydrogenase complex (PDH)
• Cytochrome b_5 reductase	• Sulphite oxidase	• NADH dehydrogenase	• Citrate synthase
• Fatty acid CoA synthetase	• Nucleoside diphosphokinase	• Succinate dehydrogenase	• Aconitase
• FA elongation system		• Ubiquinone	• Isocitrate dehydrogenase (ICD)
• Phospholipase A		• Electron-transferring flavoproteins (ETF)	• α-oxoglutarate dehydrogenase
			• Malate dehydrogenase
• Nucleoside diphosphokinase		• Vector ATP synthase (F_0F_1)	• FA oxidation system
		• β-OH-butyrate dehydrogenase	• Ornithine transcarbamoylase
		• Carnitine palmityl transferases	• Carbamoyl phosphate synthetase I
		• All translocases	

Types: There are **two kinds** of endoplasmic reticulum (ER):

(i) Rough surfaced ER, also known as ***ergastoplasm***. They are coated with ribosomes. Near the nucleus, this type of ER merges with the outer membrane of the nuclear envelope.

(ii) Smooth surfaced ER: They do not have attached ribosomes.

Functions

(a) Function of rough ER: Rough ER synthesises membrane lipids, and secretory proteins. These proteins are inserted through the ER membrane into the lumen of the cisternae where they are modified and transported through the cell.

(b) Function of smooth ER: Smooth endoplasmic reticulum is involved:

(i) In lipid synthesis and

(ii) Modification and transport of proteins synthesised in the rough ER

Note: A number of important enzymes are associated with the endoplasmic reticulum of mammalian liver cells. These include the enzymes responsible for the synthesis of sterol, triacylglycerol (TG), Phospholipids (PL) and the enzymes involved in detoxification of drugs. Cytochrome P_{450} which participates in drug hydroxylation reside in the ER.

4. Golgi complexes (or Golgi apparatus): They are also called ***Dictyosomes.*** Each eukaryotic cell contains a unique stack of smooth surfaced compartments or cisternae that make up the Golgi complex. The ER is usually closely associated with the Golgi complexes, which contain flattened, fluid filled golgi sacs.

The Golgi complex has a ***Proximal*** or ***Cis*** compartment, a ***medial*** compartment and a ***distal*** or ***trans*** compartment.

Recent evidence suggests strongly that the complex serves as a unique sorting device that receives newly synthesized proteins, all containing signal or transit peptides from the ER. ***It is interesting to note that those proteins with no signal or transit peptides regions are rejected*** by the Golgi apparatus without processing it further and remain as cytoplasmic protein.

Functions

(i) On the proximal or cis side, the Golgi complexes receive the newly synthesised proteins by ER via transfer vesicles.

(ii) The post-translational modifications take place in the golgi lumen (median part) where the carbo hydrates and lipid precursors are added to proteins to form glycoproteins and lipoproteins respectively.

(iii) On the distal or trans side they release proteins via modified membranes called **secretory vesicles.** These secretory vesicles move to and fuse with the plasma membrane where the contents may be expelled by a process called ***exocytosis.***

5. Lysosomes: Lysosomes are cell organelles found in cells which contain packet of enzymes. Lysosome word derived from Greek word *Gree,* meaning lysis (loosening). Discovered and described for the first time as a new organelle by the **Belgian Biochemist de Duve in 1955.**

- *Size:* Mean diameter is approximately 0.4 μ (varies in between that of microsomes and mitochondria). They are surrounded by a lipoprotein membrane.
- Lysosomes are found in all animal cells, ***except erythrocytes,*** in varying numbers and types.
- *pH:* pH inside the lysosomes is lower than that of cytosol. The lysosomal enzymes have an optimal pH around 5. ***Acid phosphatase is used as a marker enzyme for this organelle.***

Enzyme Groups Present in Lysosomes

Essentially the enzymes about 30 to 40, are hydrolytic in nature. They can be grouped as follows:

Lysosomal Enzymes	
1. ***Proteolytic enzymes***	• *Cathepsins (Proteinase)* • *Collagenase* • *Elastase*
2. ***Nucleic acid hydrolysing enzymes***	• *Ribonucleases* • *Deoxyribonucleases*
3. ***Lipid hydrolysing enzymes***	• *Lipases* • *Phospholipases* • *Fatty acyl esterases*
4. ***Carbohydrate splitting enzymes***	• *α-glucosidase* • *β-galactosidase* • *Hyaluronidase* • *Aryl sulphatase, etc.*
5. ***Other enzymes***	• *Acid phosphatase* • *Catalase, etc.*

- As long as the lysosomal membrane is intact, the encapsulated enzymes can act only locally. But when the membrane is ruptured, the enzymes are released into the cytoplasm and can hydrolyse external substrates (biopolymers).

Biomedical Importance

- In autophagic processes, cellular organelles such as mitochondria and the endoplasmic reticulum undergo digestion within the lysosome. The enzymes are active at **postmortem autolysis.**
- In the death of a cell, lysosomal bodies disintegrate, releasing hydrolytic enzymes into the cytoplasm with the result that the cell undergoes autolysis.

- There are good evidences that the metamorphosis of tadpoles to frogs, the regression of the tadpole's tail is accomplished by the lysosomal digestion of the tail cells.
- Bacteria are digested by white blood cells by engulfment of the bacteria and lysosomal action.
- The acrosome, located at the head of the spermatozoa, is a specialised lysosome and is probably involved in some way in the penetration of ovum by the sperm.

CLINICAL ASPECT

1. ***Allergic responses and arthritic conditions:*** Released enzymes from ruptured lysosomal membrane can hydrolyse external biopolymers (substrates) leading to tissue damage in many types of allergic responses and arthritic conditions.

In Gout: Urate crystals are deposited around joints. These crystals when phagocytosed cause physical damage to lysosomes and release of enzymes producing inflammation and arthritis.

2. ***Inherited disorders:*** A number of hereditary diseases involving the abnormal accumulation of complex lipids or polysaccharides in cells of the afflicted individual have now been traced to the absence of key acid hydrolases in the lysosomes of these individuals.

3. ***I-Cell disease:*** I-cell disease is a rare condition in which lysosomes lack all of the normal lysosomal enzymes.
   - The disease is characterised by severe progressive psychomotor retardation and a variety of physical signs, with death often occurring in the first decade.
   - Cultured cells from patients with I-cell disease was found to lack almost all of the normal lysosomal enzymes. The lysosomes thus accumulate many different types of undegraded molecules forming **inclusion bodies**.
   - Samples of plasma from patients with the disease were observed to contain very high activities of lysosomal enzymes; this suggested that the enzymes were being synthesised but failed to reach their proper intracellular **locations and were instead** being secreted.
   - **Mannose-6-P is the marker.** Studies have shown that lysosomal enzymes from patients with I-cell disease lack a recognition marker. Cultured cells from patients with I-cell disease found to be deficient in the enzyme ***GlcNAc phosphotransferase***, leading to lack of normal transfer of GlcNAc-1-P in specific mannose residues of certain lysosomal enzymes. Hence they can not be targeted to lysosomes. **Sequence of events in genesis of I-cell disease:**

Mutations in DNA

↓

Mutant GlcNAc phosphotransferase

↓

Lack of normal transfer of GlcNAc-1-P to specific mannose residues of certain enzymes destined for lysosomes

↓

Hence these enzymes lack Mannose-6-P (the marker) and are secreted into plasma leading to high plasma level. They are not targeted to lysosomes

↓

Lysosomes are deficient in certain hydrolases, and do not function properly

↓

They accumulate partly digested cellular material, manifesting as **"inclusion bodies"**

6. *Peroxisomes:* Peroxisomes are small organelles also called ***Microbodies,*** present in eukaryotic cell. The particles are approximately 0.5 μ in diameter. These subcellular respiratory organelles have no energy-coupled electron transport systems and ***are probably formed by budding from smooth endoplasmic reticulum (ER).***

Functions

(*i*) They carryout oxidation reactions in which toxic hydrogen peroxide (H_2O_2) is produced, which is destroyed by the enzyme ***catalase***.

(*ii*) Recently it has been shown that liver peroxisomes have an unusually active β-oxidative system capable of oxidising long chain fatty acids (C 16 to 18 or > C 18)

The β-oxidation enzymes of peroxisomes are rather unique in that the first step of the oxidation is catalysed by a flavoprotein, an *"acyl Co-A oxidase"*

$$\text{Acyl-CoA} + O_2 \rightarrow \alpha,\ \beta \text{ unsaturated acyl-CoA} + H_2O_2$$

H_2O_2 produced is destroyed by catalase.

Peroxisomes may be absent in inherited disorder ***Zellweger's syndrome*** (Refer to Chapter on fatty acid oxidation).

7. *Cytoskeleton:* For many years, biochemists have considered the cytosol a compartment containing soluble enzymes, metabolites and salts in an aqueous but gel like environment.

Studies now support the idea that this compartment contains actually a complex network of fine structures called **(a) microtubules, (b) microfilaments and (c) microtrabeculae.**

(a) Microtubules: They are long unbranched slender cylindrical structures with an average diameter of about 25 nm. The structures are made primarily by the self-assembly of the heterodimer, **tubulin** having molecular weight 50,000.

Functions

- An important function of microtubules is ***their role in the assembly and disassembly of the spindle structures during mitosis.***

- They also provide internal structure to the cell and helps in maintenance of shape of the eukaryotic cell.
- As they seem to associate with the inner face of plasma membrane, ***they may be involved in transmembrane signals.***

(b) Microfilaments: They are more slender cylinder like structures made up of the contractile protein **actin**. They are linked to the inner face of the plasma membrane.

Function

These structures may be involved in the generation of forces for internal cell motion.

(c) Microtrabeculae: They appear to be very fragile tubes that form a transient network in the cytosol.

Function

It is not yet clearly understood and established fully whether or not soluble enzymes are associated or clustered with these structures to form unstable multi-enzyme complexes.

B. Cytoplasm (Cytosol)

This is the simplest structure of the cell. ***Organelles free sap is called as cytosol*** Many metabolic reactions take place in cytosol where substrates and cofactors interact with various enzymes. There is no specific structure for cytosol. It has a high protein contents. The actual physio-chemical state of cytosol is poorly understood. ***A major role of cytosol is to support synthesis of proteins on the rough endoplasmic reticulum by supplying cofactors and energy.***

Cytosol also contains free ribosomes often in the polysome form. They contain many different types of proteins and ribosomal RNA or r-RNA. They exist as 2 subunits and act as the site of protein synthesis.

Summary

Functions of Various Subcellular organelles

- **Cytosol:** Involved in protein synthesis, purine synthesis, carbohydrate metabolism, HMP shunt. Lipid metabolism-FA synthesis, cholesterol synthesis, partly heme synthesis, urea formation and pyrimidine synthesis.
- **Mitochondria: *Power house* of the cell,** ETC and ATP synthesis, TCA cycle, β-oxidation of fatty acids, ketone body formation, partly heme synthesis, urea synthesis, gluconeogenesis, pyrimidine synthesis.
- **Nucleus**: DNA replication and transcription.
- **Endoplasmic reticulum:** Biogenesis of proteins, lipoproteins, drug metabolism, ethanol oxidation, synthesis of cholesterol (partly).
- **Golgy body:** Maturation of synthesised proteins, protein sorting, packaging and secretion.
- **Lysosomes:** Degradation of proteins carbohydrates, lipids and nucleotides.

CHAPTER 2

BIOLOGICAL MEMBRANES: STRUCTURE AND FUNCTION

Major Concept

To study the structure and function of cell membrane **(Plasma membrane as a prototype)**

Specific Objectives

A. 1. Learn the chemical composition of the membrane—Lipids and its types.
   2. Learn about proteins present and types:
      (i) Integral membrane proteins
      (ii) Peripheral membrane proteins
      (iii) Transmembrane proteins
   3. Learn about nature of carbohydrates
   4. Learn how lipid bilayer is formed

B. Study the fluid mosaic model structure of membrane and additional structures—Lipid rafts and caveolae

C. Learn about special structural characteristics of red cell membrane
   1. Integral protein: Glycophorins, and Band-3-Protein
   2. Peripheral proteins: Spectrin, Actin, Ankyrin and Band 4, 1 Protein

D. Function:
   (a) Ion channels, Ionophores, water channels, gap junctions
   (b) Transport mechanisms
      1. Passive or simple diffusion
      2. Facilitated diffusion
      3. Active transport—Learn about uniport system, and co-transport system—symport and antiport
   (c) Mechanisms of transport of macromolecules
      1. Exocytosis
      2. Endocytosis: Learn about
         (i) Phagocytosis
         (ii) Pinocytosis—Mechanism of receptor mediated absorptive pinocytosis
      3. Inherited disorders
         (i) Leber's hereditary optic neuropathy (LHON)
         (ii) Cystic fibrosis.

The plasma membrane, a prototype cell membrane, studied extensively. It separates the cell contents from the outer environment. Such a membrane barrier that separates cellular contents from the environment is an absolute necessity for life. Plasma membranes have selective permeability that mediate the flow of molecules and ions into and out of the cell. They also contain molecules at their surfaces that provide for cellular recognition and communication.

Eukaryotic cells contain many internal membrane system that surround the cell organelles. Each internal membrane system is specialised to assist in the function of organelle it surrounds.

I. Chemical Composition of the Membranes

Membranes are composed of lipids, proteins and carbohydrates. The relative content of these components varies widely from one type of membrane to another, but typically it contains, 40 per cent of the dry weight is lipids, about 60 per cent proteins and 1 to 10 per cent carbohydrates. All membrane carbohydrate is covalently attached to proteins or lipids.

Table 2.1 shows the composition of different membranes—content of lipids, proteins and carbohydrates as per centage of dry weight.

Table 2.1: Composition of different membranes: Content of lipid, protein and carbohydrates as percentage of dry weight

Type of membranes	*Lipid*	*Protein*	*Carbohydrate*
• Plasma membrane (mammals)	43	49	8
• Nuclear membrane	35	59	3
• Outer mitochondrial membrane	48	52	Trace
• Inner mitochondrial membrane	24	76	Trace
• Endoplasmic reticulum	44	54	2
• Myelin	75	22	3

Table 2.2: Composition of different membranes: Content of various lipids as percentage of total lipids

Type of membranes	*Various types of lipids*					
	Cholesterol	*Lecithin*	*Cephalin*	*Phosphatidyl-serine*	*Sphingo-myelin*	*Glycolipid*
• Plasma membrane (mammals)	20	19	12	7	12	10
• Nuclear membrane	3	45	20	3	2	0
• Outer mitochondrial membrane	8	45	20	2	4	0
• Inner mitochondrial membrane	0	35	25	0	3	0
• Endoplasmic reticulum	5	48	19	4	5	0
• Myelin	28	11	17	6	7	29

Table 2.2 shows the content of lipids as percentage of total lipid.

(a) Lipids: Lipids are the basic structural components of cell membranes. (Refer to chapter on Chemistry of Lipids for details of lipids).

Lipid molecules have a 'polar' or ionic head hence hydrophilic and the other end is a 'nonpolar' and hydrophobic tail. Hence they **are amphipathic** (Fig. 2.1).

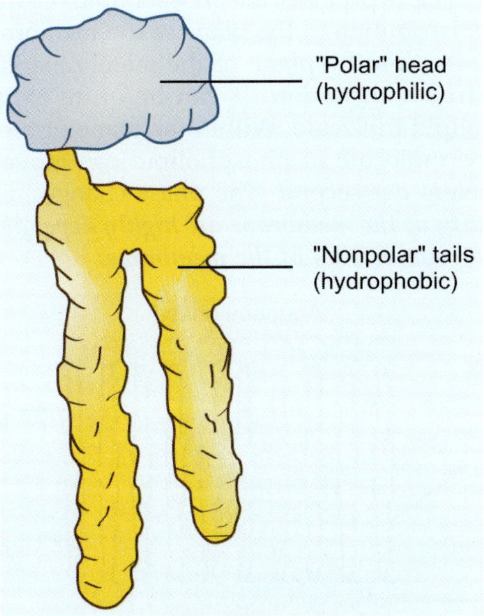

Fig. 2.1: Basic lipid structure—polar head and nonpolar tails

Types of Lipids Present in Biomembranes

1. Fatty acids: They are major components of most membrane lipids. The nonpolar tails of most membrane lipids are long chain fatty acids attached to polar head groups, such as glycerol-3-P.

About 50 per cent of the fatty acid groups are saturated, i.e. they contain no double bond. The most common saturated fatty acid groups in membrane lipids in animals contain 16 to 18 carbon atoms. The other half of fatty acid molecules contain one or more double bonds, i.e. unsaturated or polyunsaturated fatty acids. **Oleic acid is the most abundant unsaturated fatty acid in animal membrane lipids;** others are arachidonic acid, linoleic and linolenic acids. ***The degree of unsaturation determines the fluidity of the membranes.***

2. Glycerophospholipids: They are another group of major components of biomembranes. Phosphatidyl-ethanol amine (cephalin), phosphatidylcholine (Lecithin) and phosphatidylserine are among the most of common glycerophospholipids.

3. Sphingolipids: They comprise another group of lipids found in biological membranes ***specially in the tissues of nervous system.*** There are three types of sphingolipids sphingomyelin, cerebrosides and gangliosides. **About 6 per cent of the membrane lipids of grey matter cells in the brain are gangliosides.**

4. Cholesterol: Cholesterol is another common component of the biomembranes of animals but ***not*** of plants and prokaryotes. It is oriented with its hydrophilic polar heads exposed to water and its hydrophobic fused ring system and attached hydrocarbon groups buried in the interior. ***Cholesterol helps to regulate fluidity of animal membranes.***

(b) Proteins: Main types of membrane proteins are (Fig. 2.3):

1. ***Integral membrane proteins*** (also called ***intrinsic*** membrane proteins): These proteins are deeply embedded in the membrane. Thus portions of these proteins are in Van der Waals contact with the hydrophobic region of the membrane.

2. ***Peripheral membrane proteins*** (also called ***extrinsic*** proteins): These may be weakly bound to the surface of the membrane by ionic interactions or by hydrogen bonds that form between the proteins and the 'polar' heads of the membrane lipids. They may also interact with integral membrane proteins. They can be removed without disrupting the membrane.

3. ***Transmembrane proteins:*** Some of the integral proteins span the whole breadth of the membrane and are called as ***transmembrane proteins***. The hydrophobic side chains of the amino acids are embedded in the hydrophobic central core of the membrane. ***These proteins can serve as receptors for hormones, neurotransmitters, tissue specific antigens, growth factors,*** etc.

(c) Carbohydrates: Occurs in cell membranes and in Lipoproteins. Approximately 5 per cent of the weight of cell membranes is carbohydrate, in the form of glycoproteins and glycolipids.

Their presence on the outer surface of the plasma membrane, the glycocalyx, has been shown with the use of plant lections, protein agglutinins that bind specific glucosyl residues.

Example: ***Concanavalin***—A binds α-glucosyl and α-mannosyl residues.

Glycophorin is a major integral membrane glycoprotein of human erythrocytes. It has 130 amino acid residues and spans the lipid membrane, with polypeptide regions outside both the external and internal (cytoplasmic) surfaces. Carbohydrate chains are attached to the amino terminal portion outside the external surface.

Carbohydrates are also present in apoprotein B of plasma lipoproteins.

The carbohydrate chains of many glycoproteins show structural variation from one molecule to another, a phenomenon known as **microheterogeneity**.

Formation of Lipid Bilayer

Membrane glycerophospholipids and ***sphingolipids spontaneously form bilayers,*** which is the basis of living biological membranes.

Lipid bilayers are oriented with their hydrophobic tails inside the bilayer while hydrophilic 'polar' heads are in contact with the aqueous solution on each side. ***Not all lipids can form bilayers. A lipid bilayer can form only when the cross-sectional areas of the hydrophobic tail and hydrophilic polar head are about equal.*** Glycerophospholipids and sphingolipids fulfil this criteria and hence can form bilayer. The lysophospholipids have only one fatty acyl group, it cannot form the bilayer as the polar heads are too large, similarly cholesterol also cannot form bilayers as the rigid fused ring systems and additional nonpolar tails are too large.

The hydrophobic effect and the solvent entropy provide the driving force for the formation of lipid bilayer. A lipid bilayer is about 6 nm across and this is so thin that it may be regarded as a two-dimensional fluid. Lipid molecules in a bilayer are highly oriented (Fig. 2.2).

Fluid Mosaic Model of Membrane Structure

The fluid mosaic model of membrane structure proposed by **Singer** and **Nicholson in 1972** is now accepted widely. The membrane proteins, intrinsic proteins (integral) deeply embedded and peripheral proteins loosely attached, float in an environment of fluid phospholipid bilayers. It can be compared like icebergs floating in sea water (Figs 2.3 and 2.4).

Early evidences for the model point to rapid and random redistribution of species—specific integral proteins in the plasma membrane of an interspecies hybrid cell formed by the artificially induced fusion of two different parent cells. It has subsequently been shown that it is not only the integral proteins, the phospholipids also undergo rapid redistribution in the plane of the membrane. This diffusion within the plane of the membrane is termed ***translational diffusion.*** It can be quite rapid for a phospholipid molecule. Within the plane of the membrane, one molecule of phospholipid can move several micrometers per second. The phase changes, and ***thus the fluidity of the membrane are highly dependant upon the lipid composition of the membrane.***

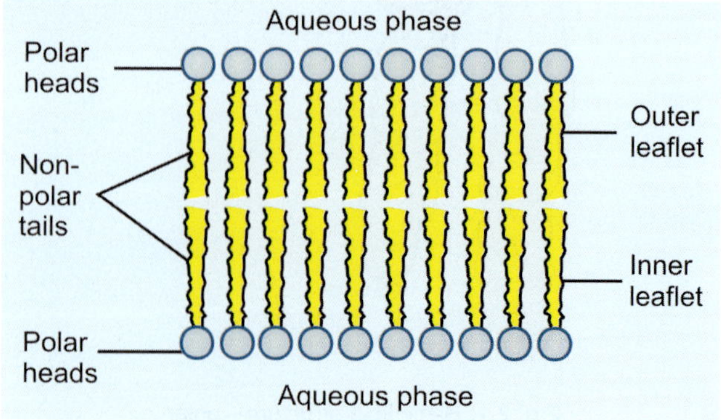

Fig. 2.2: Lipid bilayer

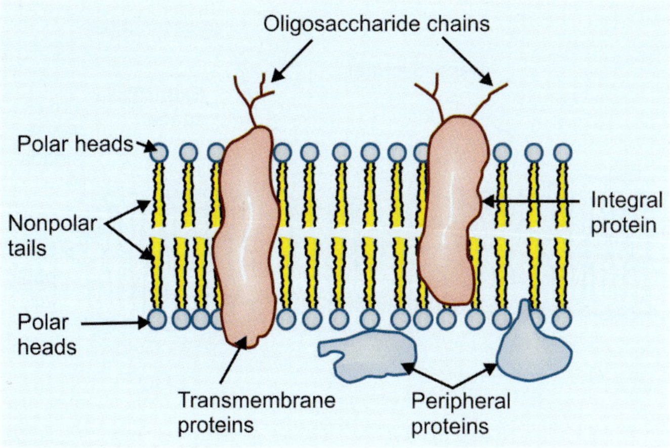

Fig. 2.3: Proteins in fluid bilayer

Effect of temperature: In a lipid bilayer, the hydrophobic chains of the fatty acids can be highly aligned or ordered to provide a rather stiff structure.

- As the temperature increases, the hydrophobic side chains undergo a transition from the ordered state which is more gel like or crystalline to a disordered state, taking on a more liquid like or fluid arrangement. The temperature at which the structure undergoes the transition from ordered to disordered state, i.e. melts, is called the ***transition temperature (Tm)***. The longer and more saturated fatty acid chains interact more strongly with each other via their longer hydrocarbon chains and thus cause higher values of Tm. Hence, higher temperatures are required to increase the fluidity of the bilayer. On the other hand, unsaturated bonds that exist in the "Cis" configuration tend to increase the fluidity of a bilayer by decreasing the compactness of the side chains packing without diminishing the hydrophobicity. The phospholipids of cellular membranes generally contain at least one unsaturated fatty acid with at least one `Cis' double bond.

Effects of Fluidity of Membrane

The fluidity of membrane significantly affects its functions:

- As membrane fluidity increases, its permeability to water and other small hydrophilic molecules also increases.
- As fluidity increases, the lateral mobility of integral proteins also increases.

Additional Special Features of Some Membranes

In addition to fluid mosaic model, some additional features of membrane structures and functions have recently come up. The following two structures which currently drawn attention are:

(a) Lipid rafts: They are dynamic areas of the exoplasmic leaflet of the lipid bilayers enriched in cholesterol and sphingolipids.

Function: They are involved in signal transduction and possibly other processes.

(b) Caveolae: They are probably derived from lipid rafts.

Many of the caveolae contain a special protein called **caveolin-1**, which probably may be involved in their formation from lipid rafts. By electron microscope they look like flask-shaped indentations of the cell membranes.

Functions: They also take part in signal transduction. Proteins detected in caveolae include various components of the signal transduction system, e.g. the insulin receptor and some G-proteins, the folate receptor, and endothelial nitric oxide synthase (eNOS).

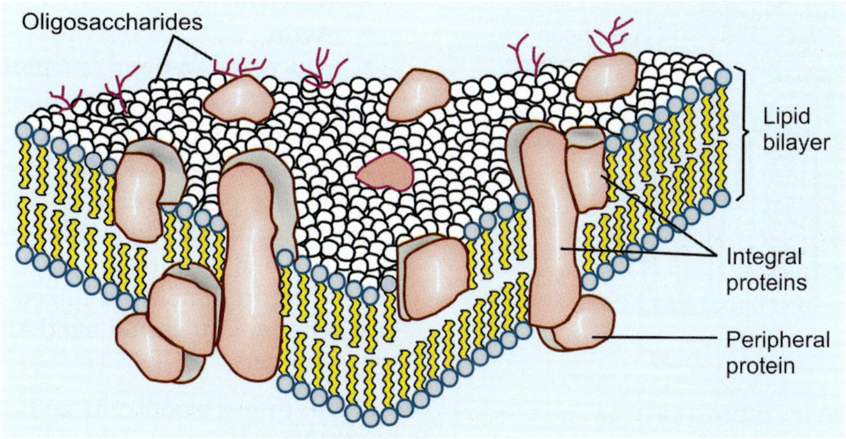

Fig. 2.4: Fluid mosaic model of biomembrane

Special Structural Characteristics of Red Cells Membranes

The same integral proteins and peripheral proteins, as discussed above, are present in cell membrane of nearly all vertebrate erythrocytes. The nature and function of these proteins require special mention.

1. Integral proteins: Two major integral proteins are found in red cells membrane. They are:

(a) **Glycophorin** and
(b) **Band-3-Protein.**

(a) Glycophorins (Fig.2.5): Glycophorins are glycoproteins. It contains 60 per cent carbohydrates by weight. The oligosaccharides bound to glycophorin are linked to serine, threonine and asparagine residues.

Red blood cells membrane contains about 6×10^5 glycophorin molecules. The polypeptide chain of glycophorin contains 130 amino acid residues. A sequence of 23 hydrophobic amino acid residues lies within the nonpolar hydrocarbon phase of the phospholipid bilayer, tightly associated with phospholipids and cholesterol. This 23 amino acid residue sequence has an α-helical conformation.

Function

- Some of the oligosaccharides of glycophorin are ***the M and N blood group antigens***.
- Other carbohydrates bound to glycophorin are ***sites through which the influenza virus becomes attached to red blood cells.***

(b) Band-3-Protein (Fig. 2.6): It is another major integral protein found in red cell membrane. It is *dimeric* having molecular weight of 93,000. ***The polypeptide chain of the dimer is thought to traverse the membrane about a dozen time.*** Both the C and N terminals of band-3-protein are on the cytosolic side of the membrane. The N-terminal residues extend into the cytosol and interact with components of the cytoskeleton.

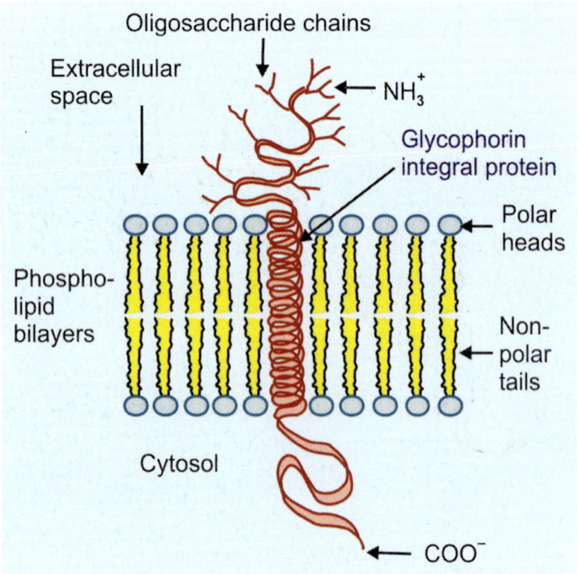

Fig. 2.5: Glycophorin integral protein

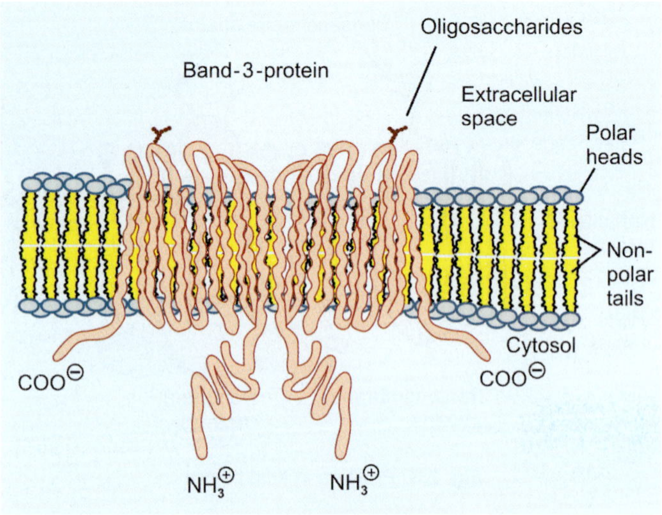

Fig. 2.6: Band-3-protein

Function: Band-3-protein plays an important role in the function of red blood cells. As red blood cells flow through the capillaries of the lungs, they exchange bicarbonate anions (HCO_3^-.) produced, by the reaction of CO_2 and H_2O, for chloride (Cl^-) ions. This exchange occurs by way of a channel in band-3-protein, which forms a **Pore** through the membrane. ***Thus band-3-protein is an example of a membrane transport protein.***

2. Peripheral proteins: The inner face of the red blood cells membrane is laced with a ***network of proteins called cytoskeletons*** that stabilises the membrane and is responsible for the biconcave shape of the cells:

The special peripheral membrane proteins participate in this stability of red cells are:

- **Spectrin**
- **Actin**
- **Ankyrin** and **Band 4, 1 protein.**

- **Spectrin:** Spectrin consists of an α-chain, having molecular weight 240,000 and a β-chain having molecular weight 220,000. It is a **fibrous protein** in which the polypeptide chains are thought to coil around each other to give an α-β dimer, 100 nm long and 5 nm in breadth.

  Spectrin dimers are linked through short chains of actin molecules and band 4, 1 proteins to form $\alpha_2 \beta_2$ tetramers.

- **Actin:** In red blood cells and other nonmuscle cells, actin is a component of the cytoskeleton. An erythrocyte contains 5×10^5 actin molecules. About 20 actin molecules polymerise to form short *actin* **filaments**.

- **Ankyrin:** The network of spectrin, actin and band 4, 1 protein forms the skeleton of the red blood cell, but none of these proteins is attached directly to the membrane. The network of proteins is instead attached to another peripheral protein called **ankyrin**.

  Ankyrin has a molecular weight 200,000. The protein has 2 domains: One binds to spectrin, and the other to the N-terminal region of band-3-protein that extends into the cytoskeleton.

  It is now known that the protein network can also be bound directly to glycophorin (integral protein) or to band-3-protein.

CLINICAL ASPECT

Hereditary spherocytosis and ***hereditary elliptocytosis*** are inherited genetic abnormalities of red cells in which red cells are of abnormal shape. In hereditary spherocytosis the red cells are spherical and in hereditary elliptocytosis they are ellipsoidal. These defects in shape of red blood cells lead to increased haemolysis, anaemia and jaundice. These abnormally shaped red blood cells are fragile and have shorter life than normal erythrocytes.

Defect: They result from mutations in the genes coding for proteins of the membrane. The abnormality may be from defective spectrin that is unable to bind either ankyrin or band 4, 1 protein and in some cases band 4, 1 protein is absent.

II. Functions: Transport Systems

An essential role of biomembranes is to allow movement of all compounds necessary for the normal function of a cell across the membrane barrier. These compounds include a vast array of substances like sugars, amino acids, fatty acids, steroids, cations and anions to mention a few. These compounds must enter or leave the cells in an orderly manner for normal functioning of the cell.

A. 1. Ion Channels

Ion channels are transmembrane channels, ***pore like structures composed of proteins.*** Specific channels for Na^+, K^+, Ca^{++}, and Cl^- have been identified.

Cation conductive channels are negatively charged within the channel and have an average diameter of about 5 to 8 nm.

All ion channels are basically made up of transmembrane subunits that come together to form a central pore through which ions pass selectively.

All channels have *gates*, and are controlled by opening and closing.

Types of Gates

Two types of gated channels. They are:

a. Ligand gated channels: In this a specific molecule binds to a receptor and opens the channel.

Example: Acetylcholine receptor is present in postsynaptic membrane. It is a **complex of five subunits,** having a binding site for acetylcholine.

Acetylcholine released from the presynaptic region binds with the binding site of postsynaptic region, which triggers the opening of the channel and influx of Na^+.

b. Voltage gated channels: These channels open or close in response to changes in membrane potential.

Some properties of ion channels

- Composed of transmembrane protein subunits.
- Highly selective.
- Well regulated by presence of "gates".
- Two main types of gates: Ligand-gated and voltage gated.
- Activities are affected by certain drugs.
- **Mutations of genes encoding transmembrane proteins can cause specific diseases.**

2. Ionophores

Certain microorganisms can synthesise small organic molecules, called **ionophores**, which function as shuttles for the movement of ions across the membrane.

Structure: These ionophores contain hydrophilic centres that bind specific ions and are surrounded by peripheral hydrophobic regions.

Types: Two types:
(a) Mobile ion carriers: Like valinomycin (Refer uncouplers of oxidative phosphorylation).
(b) Channel formers: Like gramicidin.

3. Water Channels (Aquaporins)

In certain cells, e.g. in red blood cells, and cells of the collecting ductules of the kidney, the movement of water by simple diffusion is enhanced by movements of water through **water channels**, composed of tetrameric transmembrane proteins called **aquaporins**. About five distinct types of aquaporins have been recognised.

CLINICAL ASPECT

Recently mutations in the gene encoding AP-2 (Aquaporin 2) protein, have been shown to be the cause of one type of nephrogenic diabetes insipidus.

4. Gap Junction

Certain cells develop specialised regions on their membranes for intercellular communications which are in close proximity.

Function: They mediate and regulate the passage of ions and small molecules upto 1000 to 2000 mol wt, through a narrow hydrophilic core connecting the cytosol of adjacent cells.

Structure: They are primarily composed of protein, called **connexon** which contains four membrane spanning α-helices.

B. Types of Transport Mechanisms

The following are **three important mechanisms** for transport of various compounds across the bio-membrane

(a) Passive or simple diffusion
(b) Facilitated diffusion and
(c) Active transport

(a) **Passive or simple diffusion:** It depends on the concentration gradient of a particular substance across the membrane. The solute passes from higher concentration to lower concentration till equilibrium is reached. **The process neither requires any carrier protein nor energy. It operates unidirectionally.**

Initial rate (v) at which a solute (s) diffuses across a phospholipid bilayer is directly proportional to the concentration gradient across the membrane ([s] out – [s] in) and inversely proportional to the thickness (t) of the membrane. Thus,

$$v = \frac{D\,([s]\text{ out} - [s]\text{ in})}{t}$$

D is the **diffusion coefficient**, which is expressed in terms of area divided by time.

The diffusion of molecules across a bilayer is described by a "Permeability coefficient", which is equal to the diffusion coefficient (D) divided by the thickness of the membrane (t).

Examples: water, gases, pentose sugars.

Factors affecting net diffusion:

- *Concentration gradient:* The solutes move from high to low concentration.
- *Electrical potential:* Solutes move toward the solution that has the opposite charge. The inside of the cell usually has a negative charge.
- *Hydrostatic pressure gradient:* Increased pressure will increase the rate and force of the collision between the molecules.
- *Temperature:* Increased temperature will increase particle motion and thus increase the frequency of collisions between external particles and the membrane.
- *Permeability coefficient:* Net diffusion also depends on the permeability coefficient for the membrane.

(b) **Facilitated diffusion:** It is similar to passive diffusion in that solutes move along the concentration gradient. But it differs from passive diffusion in that it requires a ***carrier or transport protein. Hence the rate of diffusion is faster than simple diffusion.*** The process ***does not require any energy*** and ***can operate bidirectionally.***

Mechanism of facilitated diffusion has been explained by **ping-pong model**. (For details of facilitated diffusion and 'ping-pong' model—refer to chapter on Digestion and Absorption of Carbohydrates).

Example: D-fructose is absorbed from intestine by facilitated diffusion.

(c) **Active transport:** Active transport occurs against a concentration gradient and electrical gradient. Hence it **requires energy. *About 40 per cent of the total energy requirement in a cell is utilised for active transport system. It requires the mediation of specific*** **carrier or transport proteins.**

Types of transport system: Transport systems can be classified as follows:

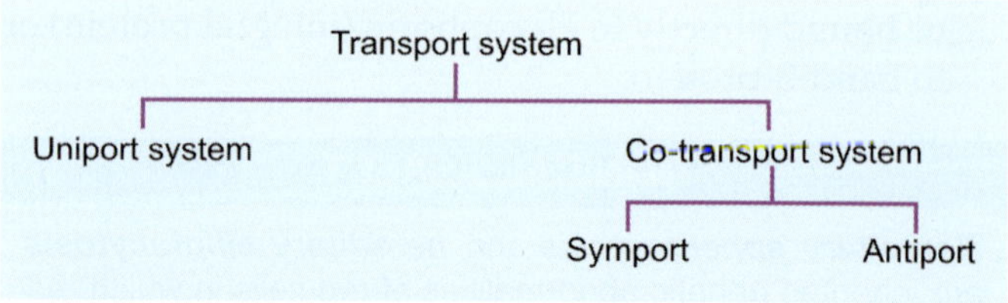

1. *Uniport system:* This system involves the transport of a single solute molecule through the membrane. *Example:* Glucose transporters in various cells.
2. *Co-transport system:* D-Glucose, D-Galactose and L-amino acids are transported into the cells by Na^+-dependant co-transport system. Na^+ is not allowed to accumulate in the cells and it is pumped out by "sodium pump".
   (i) *Symport system (Fig. 2.7)*: It is a co-transport system in which the transporter carries the two solutes in the same direction across the membrane.
   (ii) *Antiport system (Fig. 2.8)*: It is a type of co-transport system in which two solutes or ions are transported simultaneously in opposite directions. *Example:* Chloride and bicarbonate ion exchange in lungs in red blood cells.

C. Transport of Macromolecules

The mechanism of transport of macromolecules such as proteins, hormones, immunoglobulins, low density lipoproteins (LDL) and even viruses takes place across the membrane by two independant mechanisms.

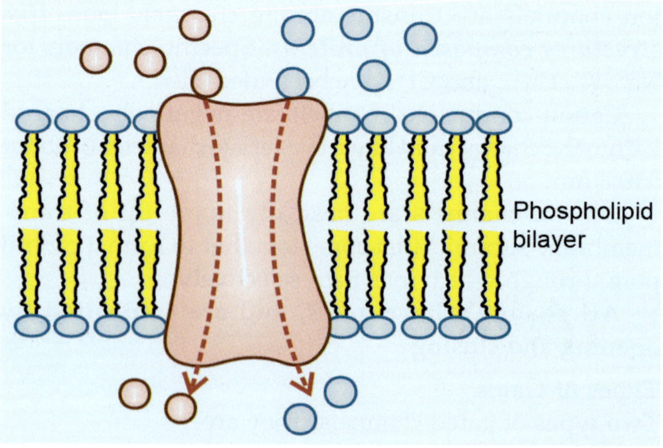

Fig. 2.7: Symport: Transport of two different molecules (or ions) in same direction

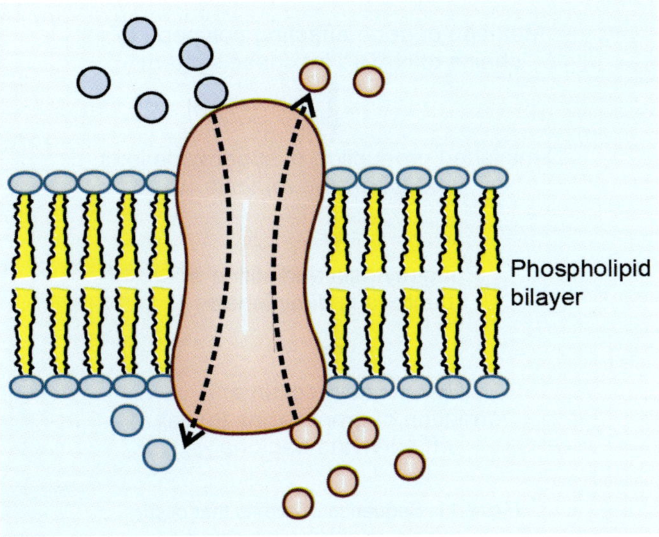

Fig. 2.8: Antiport: Transport of two different molecules (or ions) in opposite direction

1. **Exocytosis**
2. **Endocytosis.**

1. **Exocytosis (Fig. 2.9):** Most cells release macromolecules to the exterior by the process called exocytosis. This process is also involved in membrane remodelling when the components synthesised in the Golgi apparatus are carried in vesicles to the plasma membrane. The movement of the vesicle is carried out by cytoplasmic contractile elements in the microtubular system.

   Mechanism: The inner membrane of the vesicle fuses with the outer plasma membrane, while cytoplasmic side of vesicle fuses with the cytoplasmic side of plasma membrane. Thus, ***the contents of vesicles are externalised. The process is also called* reverse pinocytosis.** ***The process induces a local and transient change in Ca^{++} concentration which triggers exocytosis.***

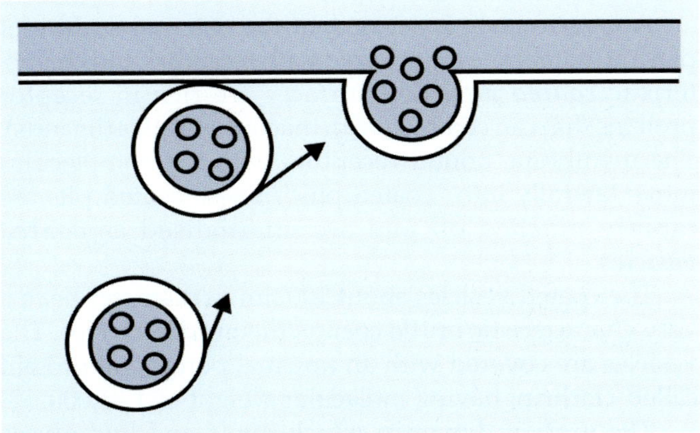

Fig. 2.9: Exocytosis—involves the contact of two inside surface (cytoplasmic side) monolayers

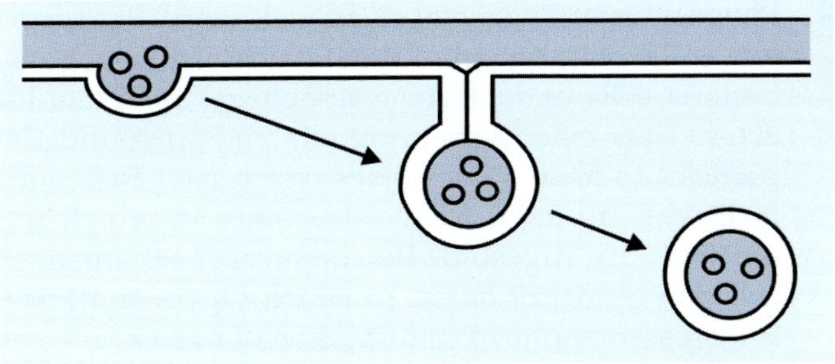

Fig. 2.10: Endocytosis—results from the contact of two outer surfaces monolayers

Types of macromolecules released by exocytosis: They fall into **3 categories**.

(i) They can attach to the cell surface and become peripheral proteins, e.g. antigens.
(ii) They can become part of extracellular matrix, e.g. collagen and glycosaminoglycans (GAGs)
(iii) Hormones like insulin, parathormone (PTH) and catecholamines are all packaged in granules, processed within cells to be released upon appropriate stimuli.

2. **Endocytosis (Fig. 2.10):** All eukaryotic cells are continuously ingesting parts of their plasma membrane. Endocytic vesicles are formed when segments of plasma membrane invaginates enclosing a minute volume of extracellular fluid (ECF) and its contents. The vesicle then pinches off as the fusion of plasma membranes seal the neck of the vesicle at the original site of invagination. The vesicle fuses with other membrane structures and thus transports of its contents to other cellular compartments.

   Factors required for endocytosis: Endocytosis requires the following:
   - Energy: Usually derived from ATP hydrolysis.
   - Ca^{++}
   - Contractile element in the cell-probably the microfilament system.

Fate: Most endocytic vesicles fuse with primary lysosomes to form secondary lysosomes which contain hydrolyic enzymes, and are, therefore, specialised organelles for intracellular disposal. Vesicular contents are digested liberating simple sugars, amino acids, etc. which diffuse out of the vesicles to be reutilised in the cytoplasm.

Types of endocytosis: The endocytosis is of following types:

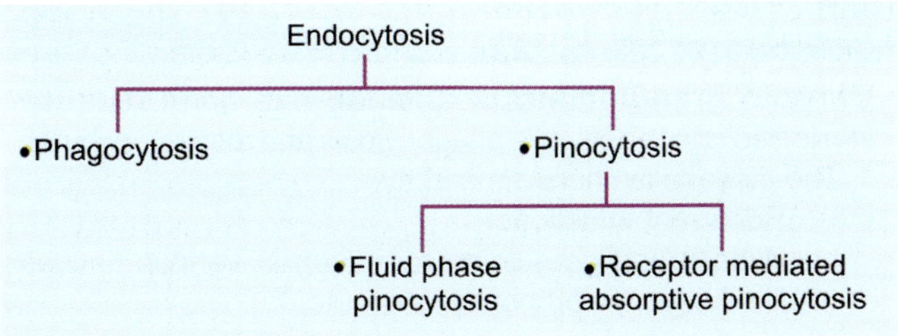

1. **Phagocytosis:** Phagocytosis (Greek word-*Phagein*-to eat) is the engulfment of large particles like viruses, bacteria, cells, or debris by macrophages and granulocytes. They extend pseudopodia and surround the particles to form ***phagosomes*** which later fuse with lysosomes to form ***Phagolysosomes*** in which the particles are digested. Biochemical mechanism is called ***respiratory burst,*** in which O_2 consumption is increased and lead to formation of superoxide ion O_2^-.

Superoxide anion O_2^- is converted to H_2O_2 and other free radicals $OH^\bullet$ and OCl^-, etc. which are potent microbial agent.

$$O_2^- + O_2^- + 2H^+ \rightarrow H_2O_2 + O_2$$

The electron transport chain system responsible for the "respiratory burst" is *"NADPH oxidase"*. In resting phagocyte it is in an **inactive** form, consisting of cytochrome b 558 + two polypeptides (heterodimer).

The NADPH oxidase system is activated by recruitment in plasma membrane by two more cytoplasmic polypeptides. Thus:

$$\text{Cytb}_{558} + 2\text{ Polypeptides} \xrightarrow{\text{2 Cytoplasmic polypeptides}} \text{Cytb}_{588} + 4\text{ Polypeptides}$$

(Inactive NADPH oxidase) → **(Active NADPH oxidase)**

NADPH oxidase is activated upon contact with various ligands like complement fragment C5a, chemotactic peptides, etc.

Events resulting in activation of the NADPH oxidase system involve G proteins, activation of phospholipase C and generation of inositol-1, 4, 5-triphosphate (P_3). The P_3 mediates a transient increase in the level of cytosolic Ca^{++}, which is essential for the induction of the respiratory burst.

Killing of bacteria within phagolysosomes appears to depend on the combined action of elevated pH, superoxide ions or other "free radicals" like H_2O_2, $OH^\bullet$, and HOCl (hypochlorous acid) and on the action of certain bactericidal peptides, called ***defensins*** and other proteins, e.g. **cathepsin G** and certain cationic proteins present in phagocytic cells.

Macrophages are extremely active and may ingest 25 per cent of their volume per hour. In such a process, a macrophage may internalize 3 per cent of its plasma membrane each minute or the entire membrane every 1/2 hour.

CLINICAL ASPECT

Chronic granulomatous disease has been recently implicated due to defective phagocytosis and respiratory burst.

The disease is characterised by:

- Recurrent infections
- Widespread granuloma formation in various tissues like lungs, lymph nodes, skin, etc.

Mutation of gene affecting polypeptide chains of NADPH-oxidase system

↓

Diminished production of superoxide ions and other free radicals

↓

Phagocytosis and killing of bacteria is diminished

↓

Recurrent infection and formation of granulomas to wall off surviving bacteria

Fig. 2.11: Sequence of events that occur in chronic granulomatous disease

Defect: The disorder is attributed to **mutations in the genes encoding the four polypeptides** that constitute the active NADPH oxidase system.

The granulomas are formed as attempts to wall off bacteria that have not been killed due to genetic deficiencies in the NADPH oxidase system (Fig. 2.11).

2. **Pinocytosis:** It is a property of all cells and leads to the cellular uptake of fluid and fluid contents.
   (a) Fluid phase pinocytosis: It is a nonselective process in which uptake of a solute by formation of small vesicles is simply proportionate to its concentration in the surrounding extracellular fluid (ECF). The formation of these vesicles is an extremely ***active process.***
   (b) Receptor mediated absorptive pinocytosis (Fig. 2.12): By coated vesicles and endosomes

All eukaryotic cells have transient structures like **coated vesicles** and **endosomes** that are involved in the transport of macromolecules from the exterior of the cells to its interior.

Approximately 2 per cent of the external surface of plasma membrane are covered with ***receptors*** and characteristic ***coated pits.*** Cell surfaces are rich in receptor proteins that can combine with macromolecules (ligands). The membrane bound receptors with macromolecules move laterally into "coated pits". These coated pits are rapidly pinched off and are internalised as ***coated vesicles.***

The coated vesicles about 100 nm in diameter have a very characteristic brittle coat on their outer surface. The vesicles are covered with an unusual peripheral protein called **clathrin,** having molecular weight of 1,85,000.

The protein ***dynamin*** *which binds and hydrolyses GTP, is necessary for the pinching off of clathrin-coated vesicles from the cell surface.*

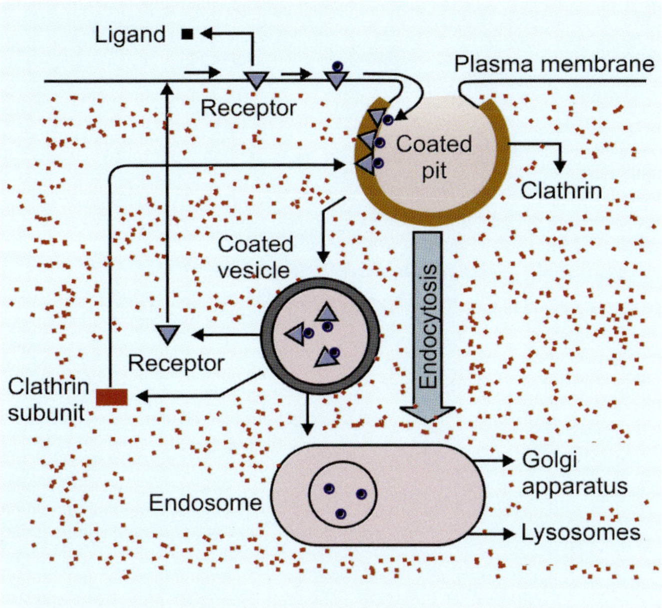

Fig. 2.12: Receptor mediated absorptive pinocytosis

Near the periphery of the cell's interior, another structure called **endosome (also called receptosome)**, having diameter 0.3 to 1 μ found. They do not contain hydrolytic enzymes, are less dense than lysosomes and have an internal pH of 5.0.

The internalised coated vesicles fuse with the endosomes and discharge their macromolecules into the interior of the endosomes. The low pH breaks the linkage between receptor-macromolecule, with a simultaneous release of **clathrin**, macromolecule, free receptors and membrane fragments, most of which **recycle back** to the plasma membrane to replenish the population of receptors and coated pits.

The macromolecules containing endosomes now move, by the help of microtubule to further interior of the cells where they fuse with lysosomes or become associated with vesicles derived from the Golgi apparatus (Fig. 2.12).

Example: The low density lipoproteins (LDL) molecule bound to receptors are internalised by means of coated pits.

CLINICAL ASPECT

Receptor mediated endocytosis with viruses are responsible for many diseases, viz.

- **Hepatitis virus affecting liver cells**
- **Poliovirus affecting motor neurons**
- **AIDS affecting T cells.**

Iron toxicity also occurs with excessive uptake due to endocytosis.

DISEASES DUE TO GENETIC MUTATIONS

1. Leber's Hereditary Optic Neuropathy (LHON)

In this disease mutations in genes encoding mitochondrial membrane proteins involved in oxidative phosphorylation can produce neurologic and vision problems.

2. Cystic Fibrosis

Inheritence: A recessive genetic disorder, prevalent among whites in N America and certain parts of Northern Europe.

Clinical Features: The disease is characterised by:

- Chronic bacterial infections of the respiratory tract and sinuses
- Fat maldigestion due to pancreatic exocrine insufficiency
- Infertility in males due to abnormal development of the vas deferens, and
- **Elevated levels of chloride in sweat**, greater than > 60 mmol/L.

Defect

Cystic fibrosis transmembrane protein (CFTR) is a cyclic AMP dependant regulatory protein for **chloride channel**.

Gene for CFTR has been identified on chromosome 7. This gene is responsible for encoding CFTR protein, a polypeptide of 1480 amino acids which regulates chloride channel.

Genetic mutation produces an abnormal CFTR, which produces an abnormality of membrane Cl^- permeability resulting to increased viscosity of many bodily secretions.

The commonest mutation found involves deletion of three bases resulting to ***loss of phenylalanine in 508 position***.

Prognosis: It is bad, life threatening and serious complication is recurrent lung infections due to overgrowth of bacteria in viscous secretions. Efforts are in progress to use gene therapy to restore the activity of CFTR protein.

SECTION

2

Chemistry of Biomolecules

CHAPTER 3

CHEMISTRY OF CARBOHYDRATES

Major Concepts

A. What are carbohydrates? Their general properties and biomedical importance.
B. List the monosaccharides of biological importance and learn their properties.
C. List the disaccharides of biological importance and learn their properties.
D. Study the chemistry and properties of various polysaccharides.
E. Study the chemistry and functions of proteoglycans.

Specific Objectives

A. 1. Define carbohydrates in chemical terms.
   2. Classify carbohydrates into four major groups with examples of each group.
   3. Describe the biomedical importance of carbohydrates.
   4. Learn the general properties of carbohydrates with reference to glucose.

B. 1. List and describe the monosaccharides of biological importance, viz. trioses, tetroses, pentoses, hexoses, etc. Example of both aldoderivatives and ketoderivatives.
   2. Study important properties of monosaccharides.
   3. Study the sugar derivatives of biological importance.
      • Deoxy sugars • Amino sugars • Amino sugar acids—Neuraminic acid • *Glycosides*—Learn the chemistry and biological/medical importance, viz. cardiac *glycosides, ouabain, phlorizin, etc.* • Define "aglycone".

C. 1. List the disaccharides of biological importance.
   2. Study the chemistry and properties of three important disaccharides.
   3. What are invert sugars and what is "inversion"? Learn about lactalose, a Keso disaccharide and its clinical importance.
   4. Differentiate sucrose from either lactose or maltose.

D. Learn the chemistry and properties of polysaccharides of biological importance.
   a. Homopolysaccharides (homoglycans)
      1. *Starch:* Chemistry and properties. Differentiate between amylose and amylopectin (in tabular form)
      2. Glycogen (animal starch)—Chemistry and properties.
      3. Inulin—Chemistry and physiological importance.
      4. Cellulose —'Roughage' value.
      5. *Dextrins* and *Dextran—Differentiate*. Use of dextran as plasma expander.
   b. Heteropolysaccharides (heteroglycans). Example—Mucopolysaccharides (glycosaminoglycans)
      1. Sulphate free acid MPS—Hyaluronic acid and chondroitin
      2. Sulphate containing MPS
         • Chondroitin sulphate—A, B, C and D
         • Keratan sulphate
         • Heparin and Heparitin sulphate
      3. Neutral MPS—Blood group substances. Learn chemistry of each MPS, distribution in body, and its biological importance. Relation of MPS to mucopolysaccharidoses.

E. Learn chemistry and functions of proteoglycans.

CARBOHYDRATES

Definition: Carbohydrates are defined chemically as aldehyde or ketone derivatives of the higher polyhydric alcohols, or compounds which yield these derivatives on hydrolysis.

CLASSIFICATION

Carbohydrates are divided into **four** major groups—monosaccharides, disaccharides, oligosaccharides and polysaccharides.

1. ***Monosaccharides:*** (also called **'simple' sugars**) are those which cannot be hydrolysed further into simpler forms.

General formula : $C_nH_{2n}O_n$

They can be subdivided further:

(a) Depending upon the number of carbon atoms they possess, as trioses, tetroses, pentoses, hexoses, etc.

(b) Depending upon whether aldehyde (– CHO) or ketone (– CO) groups are present as aldoses or ketoses.

General formula	Aldosugars	Ketosugars
• **Trioses** ($C_3H_6O_3$)	Glyceraldehyde	Dihydroxyacetone
• **Tetroses** ($C_4H_8O_4$)	Erythrose	Erythrulose
• **Pentoses** ($C_5H_{10}O_5$)	Ribose	Ribulose
• **Hexoses** ($C_6H_{12}O_6$)	Glucose	Fructose

2. ***Disaccharides:*** Those sugars which yield two molecules of the same or different molecules of monosaccharide on hydrolysis.

General formula : $C_n(H_2O)_{n-1}$

Examples

- **Maltose** yields 2 molecules of glucose on hydrolysis.
- **Lactose** yields one molecule of glucose and one molecule of galactose on hydrolysis.
- **Sucrose** yields one molecule of glucose and one molecule of fructose on hydrolysis.
- **Lactulose** a ketodisaccharide

3. ***Oligosaccharides:*** Those sugars which yield 3 to 10 monosaccharide units on hydrolysis, e.g. Maltotriose.

4. ***Polysaccharides (Glycans):*** Those sugars which yield more than ten molecules of monosaccharides on hydrolysis.

General formula: $(C_6H_{10}O_5)_n$

Polysaccharides are further divided into **two** groups:

a. ***Homopolysaccharides (homoglycans):*** Polymer of same monosaccharide units.
   Examples—Starch, glycogen, inulin, cellulose, dextrins, dextrans.

b. ***Heteropolysaccharides (heteroglycans):*** Polymer of different monosaccharide units or their derivatives.
   Example—Mucopolysaccharides (glycosaminoglycans).

Biomedical Importance of Carbohydrates

- Chief source of energy.
- Constituents of compound lipids and conjugated proteins.
- Degradation products act as "promoters" or 'catalysts'.
- Certain carbohydrate derivatives are used as drugs like cardiac glycosides/antibiotics.
- Lactose principal sugar of milk—in lactating mammary gland.
- Degradation products utilised for synthesis of other substances such as fatty acids, cholesterol, amino acid, etc.
- Constituents of mucopolysaccharides which form the ground substance of mesenchymal tissues.
- Inherited deficiency of certain enzymes in metabolic pathways of different carbohydrates can cause diseases, e.g. galactosemia, glycogen storage diseases (GSDs), lactose intolerance, etc.
- Derangement of glucose metabolism is seen in diabetes mellitus.

General Properties in Reference to Glucose

Asymmetric carbon: A carbon atom to which four different atoms or groups of atoms are attached is said to be asymmetric (Fig. 3.1).

Van't Hoff's rule of 'n': The number of possible isomers of any given compound depends upon the number of asymmetric carbon atoms the molecule possesses.

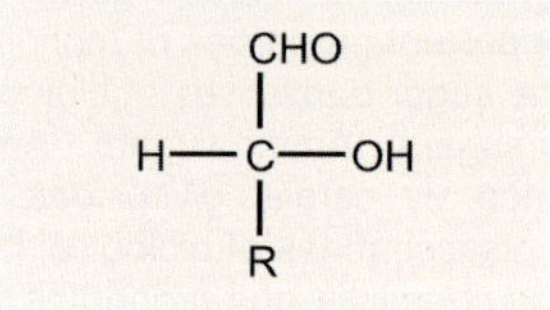

Fig. 3.1: Asymmetric carbon

According to **Van't Hoff's rule of 'n'**; 2^n equals the possible isomers of that compound, where, n = represents the number of asymmetric carbon atoms in a compound.

Stereoisomerism: The presence of asymmetric carbon atoms in a compound gives rise to the formation of isomers of that compound. Such compounds which are identical in composition and differs only in spatial configuration are called **stereoisomers.**

Two such isomers of glucose—D-Glucose and L-Glucose are **mirror image of each other** (Fig. 3.2).

D-Series and L-Series: The orientation of the H and OH groups around the carbon atom just adjacent to the terminal primary alcohol carbon, e.g. C-atom 5 in glucose determines the series. **When the – OH group on this carbon is on the right, it belongs to D-series, when the – OH group is on the left, it is a member of L-series** (Fig. 3.3).

Most of the monosaccharides occurring in mammals are D-sugars, and the enzymes responsible for their metabolism are specific for this configuration.

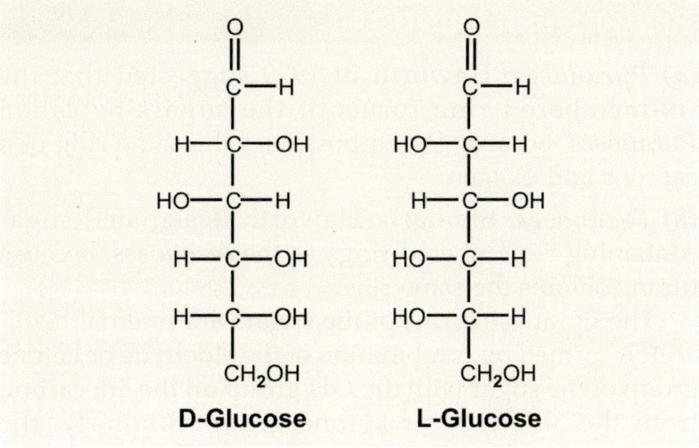

Fig. 3.2: Stereoisomers of glucose

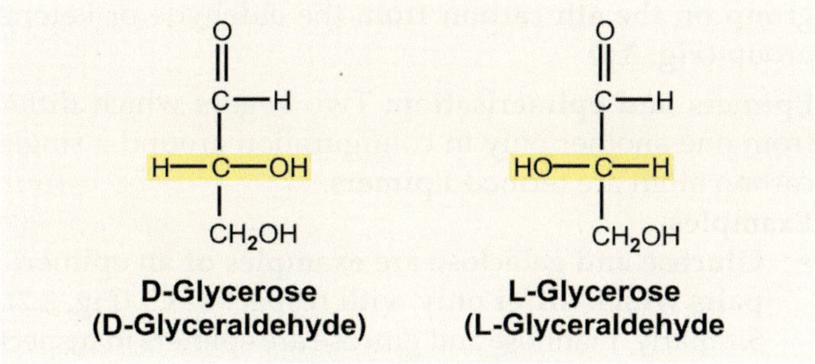

Fig. 3.3: D- and L-series

Optical activity: Presence of asymmetric carbon atoms also confers optical activity on the compound. When a beam of ***plane-polarised light*** is passed through a solution exhibiting optical activity, it will be rotated to the right or left in accordance with the type of compound, i.e. the ***optical isomers*** or ***enantiomorphs***; when it is rotated to right, the compound is called

- **Dextrorotatory (D or + sign),**
  when rotated to left, the compound is called
- **Laevorotatory (I or – sign).**

Racemic: When equal amounts of dextrorotatory and laevorotatory isomers are present, the resulting mixture has no optical activity, since the activities of each isomer cancels each other. Such a mixture is said to be **Racemic.**

Resolution: The separation of optically active isomers from a racemic mixture is called resolution.

CYCLIC STRUCTURES

As the two reacting groups aldehyde and alcoholic group belong to the same molecule, a cyclic structure takes place.

If the open-chain form of D-Glucose, which may be called as **Aldehydo-D-Glucose** is taken, and condense the aldehyde group on carbon-1, with the alcoholic-OH group on carbon-5, two different forms of glucose are formed. **When the OH group extends to right, it is *α-D-Glucose* and it extends to *left*, it is *β-D-Glucose*** **(Fig. 3.4)**.

Anomers and anomeric carbon: Carbon-1, after cyclization has four different groups attached to it and thus it becomes now **asymmetric.** The two cyclic compounds, α and β have different optical rotations, but they will not be same because the compounds as a whole are not mirror-images of each other. Compounds related in this way are called **anomers** and carbon-1, after cyclisation becomes asymmetric is called now **anomeric carbon atom** **(Fig. 3.4)**.

MUTAROTATION

Definition: When an aldohexose is first dissolved in water and the solution is put in optical path so that plane polarized light is passed, the initial optical rotation shown by the sugar gradually changes until a constant *fixed* rotation characteristic of the sugar is reached. This phenomenon of change of rotation is called as **mutarotation.**

Explanation: **Ordinary crystalline glucose happens to be in the α-form.** The above change in optical rotation represents a conversion from α-Glucose to an equilibrium mixture of α and β-forms. The **mechanism of mutarotation** probably involves **opening of the hemiacetal ring** to form traces of the aldehyde form, and then

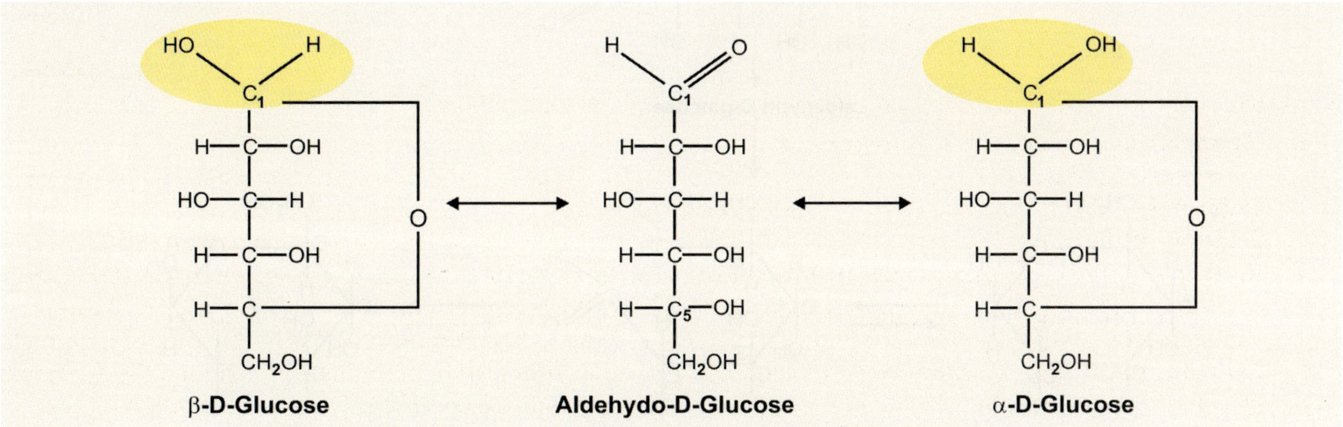

Fig. 3.4: C_1 after cyclisation becomes asymmetric—it is called "anomeric" carbon and α-D-glucose and β-D-glucose are "anomers"

recondensation to the cyclic forms. The aldehyde form is extremely ***unstable*** and exists only as a transient intermediate.

Experimental evidence: **Tanret** provided the correct interpretation. He prepared two isomeric forms of D-Glucose by crystallisation under different conditions:

- When D-Glucose is crystallised from water or dilute alcohol at room temperature, α-D-Glucose separates, having initial specific rotation of +112° which changes to fixed rotation +52.5°.
- On the other hand, when crystallisation takes place from water at temperature above 98°C, a different β-form of glucose separates, having initial specific rotation of +19° which changes to fixed rotation +52.5°.

α-D-Glucose + 112° ⟶ + 52.5° ⟵ + 19° β-D-Glucose

This work of **Tanret** showed that glucose exists in isomeric forms which in solution changes into the same equilibrium mixture regardless of which form is dissolved. ***In glucose solutions—approx 2/3 of the sugar exists as the β-form and 1/3 as α-form, at equilibrium*** *(Fig. 3.5).*

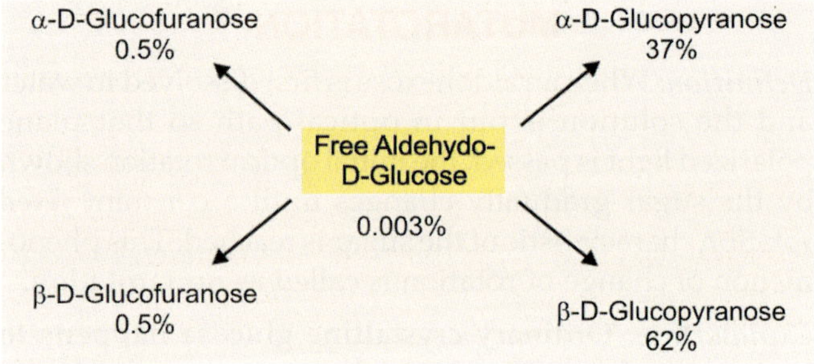

Fig. 3.5: Pyranose and furanose forms of glucose in solution

HAWORTH PROJECTION

(a) Pyranoses: **Haworth** in 1929 suggested that the six-membered ring forms of the sugars be called ***Pyranoses,*** because Pyran possesses the same ring of 5 carbons and oxygen.

(b) Furanoses: Similarly Haworth designated sugar containing 5-membered rings as the ***furanoses,*** because furan contains the same ring.

The Pyranose forms of the sugars are internal hemiacetals formed by combination of the aldehyde or ketone group of the sugar with the **OH group on the 5th carbon** from the aldehyde or ketone group. Similarly, the furanose forms of the sugars are formed by reaction between the aldehyde or ketone group with the **OH group on the 4th carbon from** the aldehyde or ketone group **(Fig. 3.6)**.

Epimers and epimerisation: Two sugars which differ from one another only in configuration around a single carbon atom are termed **Epimers**.

Examples

- **Glucose and galactose are examples of an epimeric pairs which differ only with respect of C_4 (Fig. 3.7).** Similarly, mannose and glucose are epimers in respect of C_2.

Epimerisation: Process by which one epimer is converted to other is called epimerisation and it requires the ***enzyme epimerase,*** e.g. conversion of galactose to glucose in liver.

MONOSACCHARIDES

MONOSACCHARIDES OF BIOLOGICAL IMPORTANCE

(a) Trioses: Both D-glyceraldehyde and dihydroxy-acetone occur in the form of phosphate *esters,* as

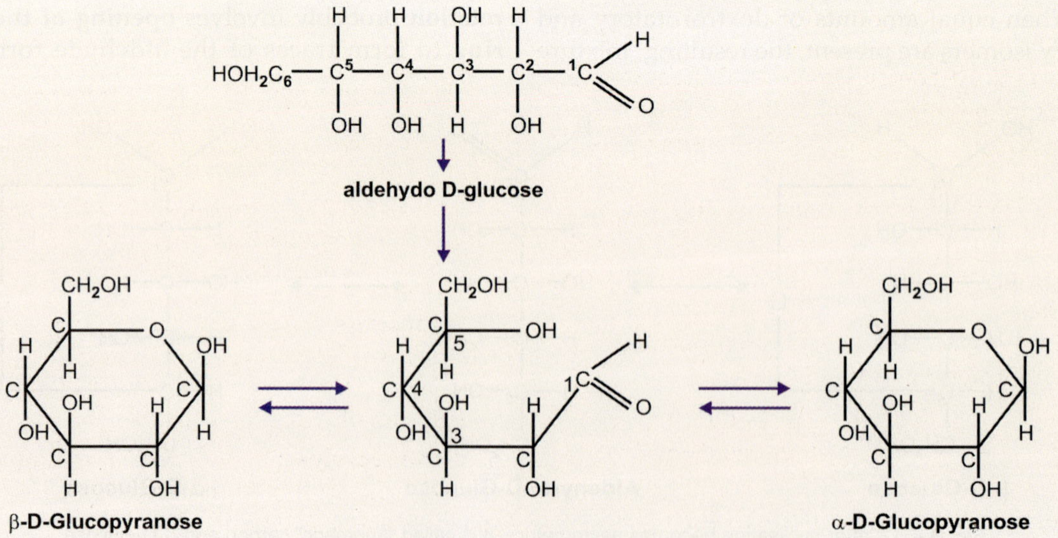

Fig. 3.6: Haworth projection

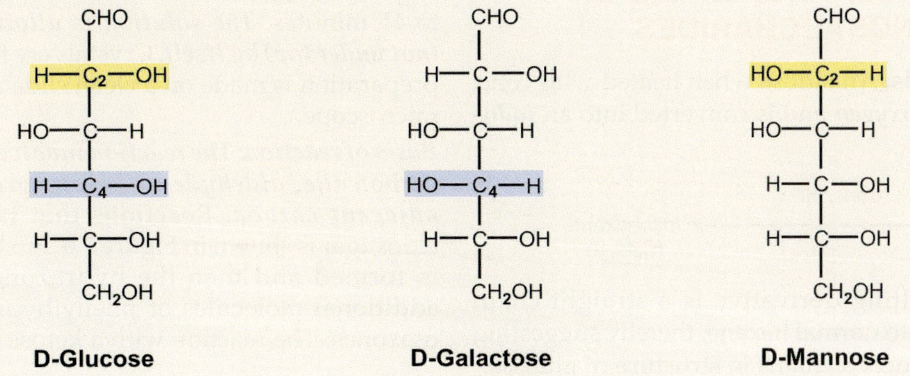

Fig. 3.7: Epimers

intermediates in glycolysis. ***They are also the precursors of glycerol,*** which the organism synthesises and incorporates into various types of lipids.

(b) Tetroses: Erythrose-4-P occurs as an intermediate in hexosemonophosphate shunt which is an alternative pathway for glucose oxidation.

(c) Pentoses

- D-ribose is a constituent of nucleic acid *RNA;* also as a constituent of certain coenzymes, e.g. FAD, NAD, coenzyme A.
- D-2-deoxyribose is a constituent of *DNA.*
- Phosphate esters of ketopentoses—D-ribulose and D-xylulose occur as intermediates in HMP shunt.
- L-xylulose is a metabolite of D-glucuronic acid and is excreted in urine of humans afflicted with a hereditary abnormality in metabolism called ***pentosuria.***
- L-fucose (methyl pentose): occurs in glycoproteins.
- ***D-Lyxose: It forms a constituent of lyxoflavin*** isolated from human heart muscle whose function is not clear.

(d) Hexoses

1. D-Glucose: (Synonyms: Dextrose, Grape Sugar)

- It is the **chief physiological sugar** present in normal blood continually and at fairly constant level, i.e. about 0.1 per cent.
- All tissues utilise glucose for energy. ***Erythrocytes and Brain cells utilise glucose solely for energy purposes.***
- Occurs as a constituent of disaccharide and polysaccharides.
- Stored as glycogen in liver and muscles mainly.
- Shows mutarotation.

α-form	Constant rotation	β-form
+ 112° →	+ 52.5°	← + 19°

2. D-galactose: Seldom found free in nature. In combination it occurs both in plants and animals.

- ***Occurs as a constituent of milk sugar lactose*** and also in tissues as a constituent of galactolipid and glycoproteins.
- It is an **epimer of glucose** and differs in orientation of H and OH on carbon-4.
- It is less sweet than glucose and less soluble in water.
- It is dextrorotatory and shows mutarotation.

α-form	Specific rotation	β-form
+ 150.7° →	+ 80°	← + 43°

- On oxidation with hot HNO_3, it yields dicarboxylic acid, **mucic acid;** which helps in its identification, since the crystals of mucic acid are not difficult to produce and have characteristic shape.

3. D-fructose: It is a ketohexose and commonly called as **fruit sugar**, as it occurs free in fruits.

- It is very sweet sugar, much sweeter than sucrose and more reactive than glucose. It occurs as a constituent of sucrose and also of the ***polysaccharide inulin. It is laevorotatory and hence is also called laevulose.***
- Exhibits mutarotation.

α-form	Specific rotation	β-form
– 21° →	– 92°	← – 133.5°

Biomedical Importance

Seminal fluid is rich in fructose and sperms utilise fructose for energy. Fructose is formed in the seminiferous tubular epithelial cells from glucose.

4. D-mannose: It does not occur free in nature but is widely distributed in combination as the polysaccharide mannan, e.g. in ivory nut. In the body, it is found as a constituent of glycoproteins.

5. Sedoheptulose: It is a ketoheptose found in plants of the sedum family. Its phosphate is important as an intermediate in the HMP-shunt and has been identified as a product of photosynthesis.

IMPORTANT PROPERTIES OF MONOSACCHARIDES

1. **Iodocompounds:** An aldose when heated with conc. HI *loses all of its oxygen* and is converted into an *iodo-compound.*

$$\text{Glucose} \xrightarrow{\text{Conc. HI}} \text{Iodohexane } (C_6H_{13}I)$$

Since the resulting derivative is a straight chain compound related to normal hexane, thereby suggesting the lack of any branched chains in structure of glucose.

2. **Acetylation or ester formation:** The ability to form sugar esters, e.g. acetylation with acetyl chloride (CH_3–COCl) indicates the presence of alcohol groups. Due to alcoholic –OH groups, it can react with anhydrides and chlorides of many organic and inorganic acids, like acetic acid, phosphoric acid, sulphuric and benzoic acids to form esters of corresponding acids.

3. **Osazone formation:** It is a useful means of preparing crystalline derivatives of sugars.

Osazones have characteristic
- **Melting points**
- **Crystal structures,** and
- **Precipitation time** and thus are valuable in identification of sugars.

Preparation: They are obtained by adding a mixture of phenylhydrazinehydrochloride and sodium acetate to the sugar solution and heating in a boiling water bath for 30 to 45 minutes. ***The solution is allowed to cool slowly (not under tap) by itself.*** Crystals are formed. A coverslip preparation is made on a clean slide and seen under the microscope.

Basis of reaction: ***The reaction involves only the carbonyl carbon (i.e. aldehyde or ketone group) and the next adjacent carbon.*** Reactions that take place with an aldosugar is shown in Figure 3.8. First phenyl hydrazone is formed and then the hydrazone reacts with two additional molecules of phenylhydrazine to form the osazones. The reaction with a ketose is similar.

Types of Crystals (Fig. 3.9)

- ***Glucosazone crystals:*** These are fine, yellow needles in fan-shaped aggregates or sheaves or crosses, typically described as **Bundle of Hay**. Melting point = 204 to 205°C.

  Note: Glucose, mannose and fructose due to similarities of structures form the same osazones. ***But since the structure of galactose differs on C-4, that part of the molecule unaffected in osazone formation, it would form a different osazone.***
- ***Lactosazone crystals:*** These are irregular clusters of fine needles and look like a **Powder puff**.
- ***Maltosazone:*** These are star-shaped and compared to ***Sunflower*** petals.

4. **Interconversion of sugars:** Glucose, fructose and mannose are ***interconvertible in solutions of weak***

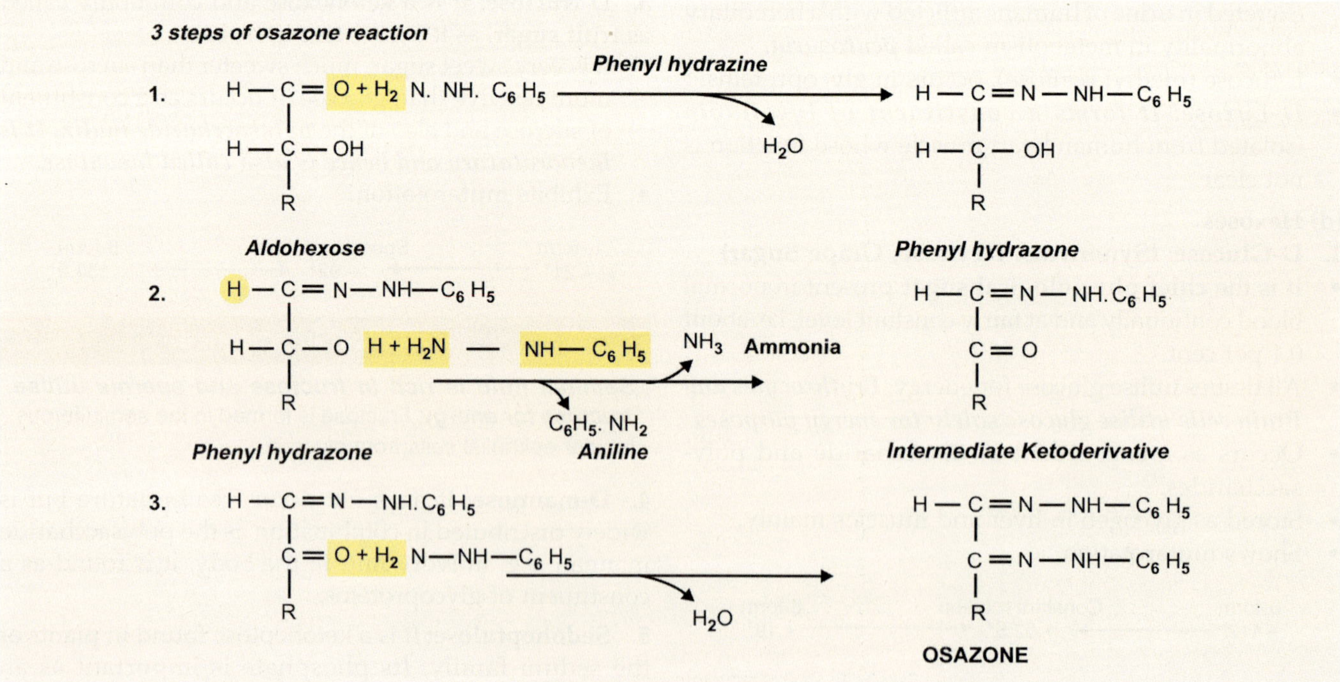

Fig. 3.8: Osazone formation

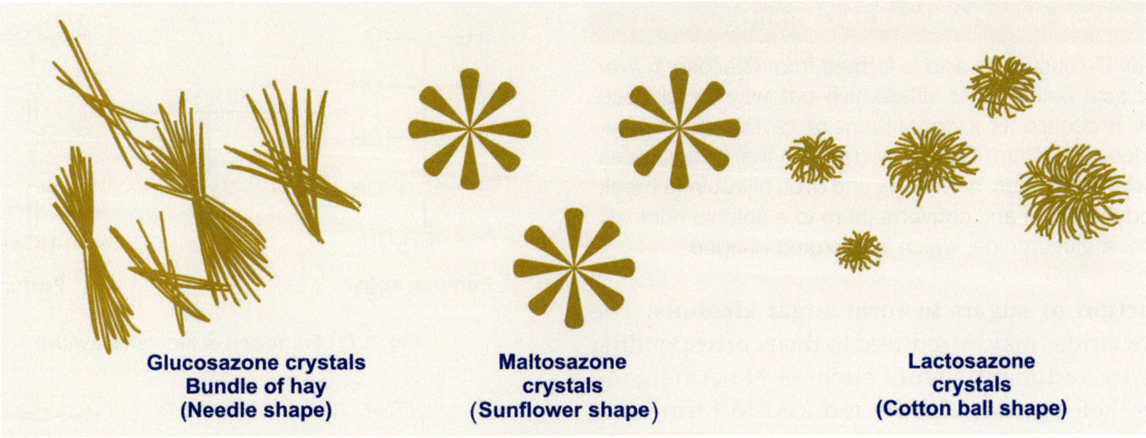

Fig. 3.9: Osazone crystals

alkalinity such as $Ba(OH)_2$ or $Ca(OH)_2$ **(Fig. 3.10)**. These interconversions are due to the fact that all give the same **Enediol form**, which tautomerizes to all three sugars. This interconversion of related sugars by the action of dilute alkali is referred to as ***Lobry de Bruyn-Van Ekenstein reaction.***

5. Oxidation to produce sugar acids: When oxidised under different conditions, the aldoses may form:

- **Monobasic Aldonic acids or**
- **Dibasic Saccharic acids or**
- **Monobasic uronic acids containing aldehyde groups thus possessing reducing properties.**

1. Aldonic acids: ***Oxidation*** of an aldose with Br_2— water converts the aldehyde group to a – COOH group ***aldonic acid*** Br_2 reacts with water to form hypobromous acid, HOBr, which acts as the oxidising agent.

Example

$$\text{D-Glucose} \xrightarrow[H_2O]{Br_2} \text{D-Gluconic acid}$$

2. Saccharic or aldaric acid: Oxidation of aldoses with conc. HNO_3 under proper conditions converts both aldehyde and primary alcohol groups to –COOH groups, forming dibasic sugar acids, the ***saccharic or aldaric acids***.

Examples

D-Glucose ⟶ D-Glucaric acid

D-Galactose ⟶ Mucic acid

3. Uronic acids: When an aldose is oxidised in such a way that the primary alcohol group is converted to – COOH group, **without oxidation of aldehyde group**, a uronic acid is formed. They ***exert reducing action due to presence of free –CHO group.***

Examples

D-Glucose ⟶ D-Glucuronic acid

D-Galactose ⟶ D-Galacturonic acid

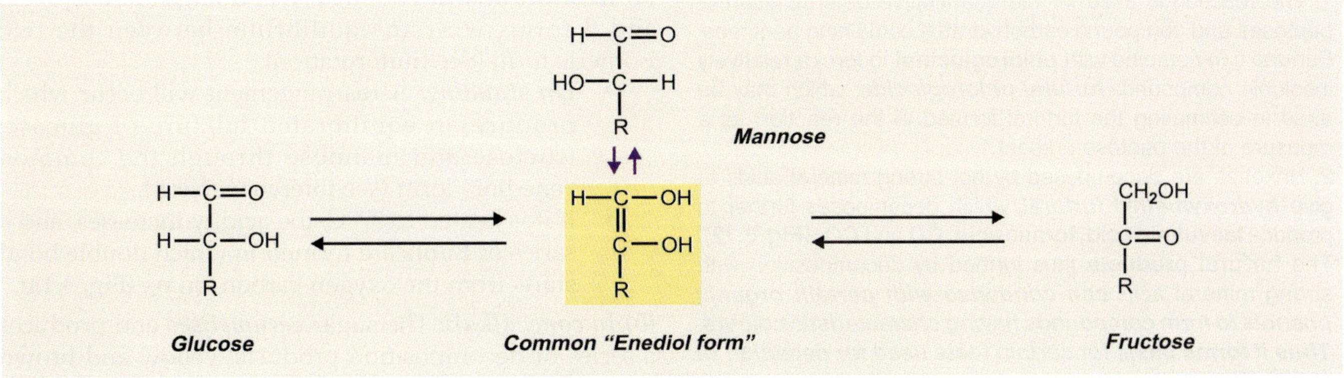

Fig. 3.10: Interconversions of sugars in weak alkalinity- Lobry de Bruyn-Van Ekenstein reaction

Biomedical importance of D-Glucuronic acid

In the body D-Glucuronic acid is formed from Glucose *in liver* by uronic acid pathway, an alternative pathway for glucose oxidation. It occurs as a constituent of certain mucopolysaccharides. In addition, it is of importance in that it conjugates toxic substances, drugs, hormones and even bilirubin (a break down product of Hb) and converts them to a soluble nontoxic substance, a glucuronide, which is excreted in urine.

6. Reduction of sugars to form sugar alcohols: The monosaccharides may be reduced to their corresponding alcohols by reducing agents such as Na-Amalgam. Similarly, ketoses may also be reduced to form keto-alcohol.

Examples

- *D-Glucose yields D-Sorbitol.*
- *D-Galactose yields D-Dulcitol.*
- *D-Mannose yields D-Mannitol.*
- *Ketosugar D-Fructose yields D-Mannitol and D-Sorbitol.*

Practical Application

In microbiology sugar alcohols have been used to identify type of bacteria. Different bacteria gives different pattern.

7. Action of acids on carbohydrates: Polysaccharides and the compound carbohydrates in general are hydrolyzed into their constituent monosaccharides by boiling with dilute mineral acids (0.5 to 1.0 N) such as HCl or H_2SO_4.

- With conc. mineral acids the monosaccharides are decomposed.
- *Pentoses* yield the cyclic aldehyde *"furfural"* (Fig. 3.11). Twelve percent (12%) HCl has been found most satisfactory for decomposition.

Practical Application

1. The reaction is used for the quantitative determination of pentoses and compound carbohydrates containing pentoses. Furfural can combine with **phloroglucinol** to form a relatively insoluble compound, ***furfural phloroglucide***, which may be used in estimating the furfural formed in the reaction as a measure of the pentose present.
2. Hexoses are decomposed by hot strong mineral acids to give ***hydroxymethyl furfural,*** which decomposes further to produce **laevulinic acid, formic acid, CO** and **CO_2** (Fig. 3.12). The **furfural products** thus formed by decomposition with strong mineral acid ***can condense with certain organic phenols to form compounds having characteristic colours. Thus it forms basis for certain tests used for detection of sugars.***

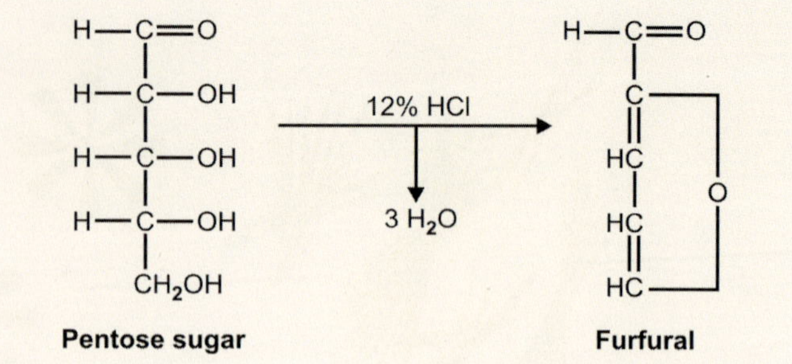

Fig. 3.11: Formation of furfural derivative

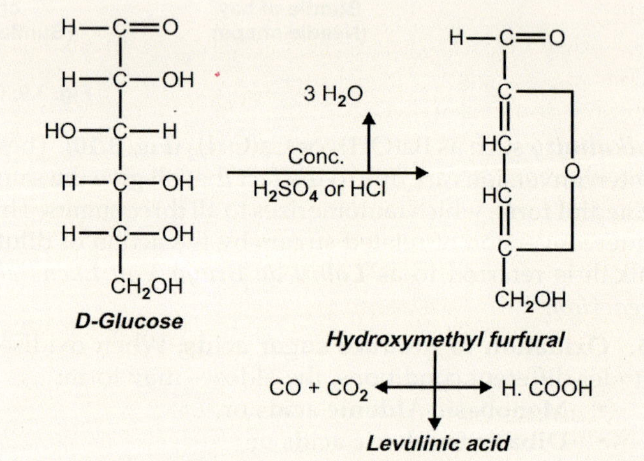

Fig. 3.12: Hydroxymethyl furfural formation

Examples

- ***Molisch's test:*** With α-naphthol (in alcoholic solution) gives red-violet ring. A sensitive reaction but non-specific, given by all sugars.
- ***Seliwanoff's test:*** With resorcinol, a cherry-red colour is produced. It is characteristic of D-fructose.

Other tests are Anthrone test, Bial-orcinol test, etc.

8. Action with alkalies: With alkalies, monosaccharides react in various ways:

(a) ***In dilute alkali:*** The sugar will change to the cyclic α and β forms with an equilibrium between the two isomeric form (See mutarotation).

- *On standing:* A rearrangement will occur which produce an equilibrated mixture of glucose, fructose and mannose through the common "enediol" form (see interconversion).
- If it is heated to 37°C, the acidity increases, and a series of **Enols** are formed in which double bond shifts from the oxygen-carbon atoms (Fig. 3.13).

(b) In conc. alkali: The sugar *caramelises* and produces a series of decomposition products, yellow and brown pigments develop, salts may form, many double bonds

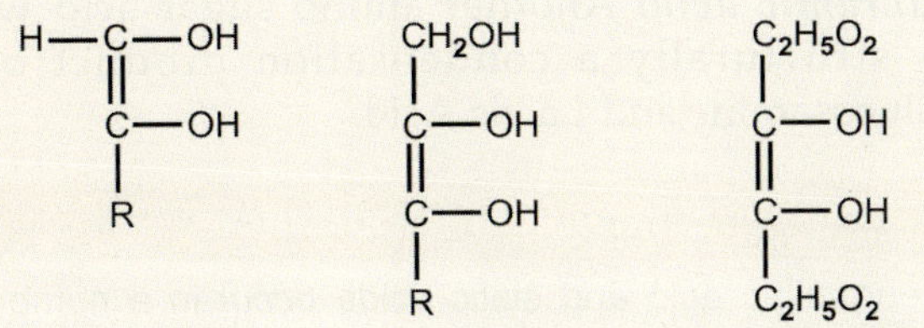

1, 2-Enediol 2, 3-Enediol 3, 4-Enediol

Fig. 3.13: Enediols formation

between C-atoms are formed, and C bond C bonds may rupture.

9. Reducing action of sugars in alkaline solution: All the sugars that contain free sugar group undergo enolisation and various other changes when placed in alkaline solution. ***The* enediol *forms of the sugars are highly reactive*** and are easily oxidised by O_2 and other oxidising agents and forms sugar acids. As a consequence they readily reduce oxidising ions such as Ag^+. Hg^+, Bi^{+++}, Cu^{++} (cupric) and $Fe(CN)_6^{---}$.

Practical Application

This reducing action of sugars in alkaline solution is utilised for both qualitative and quantitative determinations of sugars. Reagents containing Cu^{++} (ic) ions are most commonly used. These are generally alkaline solution of cupric sulphate containing

- Sodium potassium tartarate ***(Rochelle salt)*** and strong alkali NaOH/KOH as in **Fehling's solution (not used now).**
- Sodium citrate and weak alkali sodium carbonate as in **Benedict's Qualitative reagent.**

Functions of Ingredients

- *Sodium citrate/Rochelle salt in the reagents prevent precipitation of cupric hydroxide or cupric carbonate* by forming soluble, slightly dissociable complexes which dissociate sufficiently to provide supply of readily available Cu^{++} (cupric) ions for oxidation.
- The alkali of the reagents enolises the sugars and thereby causes them to be strong reducing agents. ***Enolization is better in weak alkali than strong alkali.***

Reaction: When a solution of reducing sugar is heated with one of the alkaline copper reagents, the following reactions occur (given in box below).

The Cu^{++} (ic) ions take electrons from the enediols and oxidize them to sugar acids, and are, in turn reduced to Cu^+ (ous) ions. The Cu^+ (ous) ions combine with $-OH^-$ ions to form **yellow cuprous hydroxide**, which upon heating is converted to **red cuprous oxide**. The appearance of a yellow to red precipitate indicates reduction and the quantity of sugar present can be roughly estimated from colour and amount of precipitate.

OTHER SUGAR DERIVATIVES OF BIOMEDICAL IMPORTANCE

1. Deoxy sugars: Deoxy sugars represent sugars in which the *oxygen of a –OH gr. has been removed*, leaving the hydrogen. Thus, –CHOH or $-CH_2OH$ becomes $-CH_2$ or $-CH_3$. Several of the deoxy sugars have been synthesised and others are natural products.

Deoxy sugars of biological importance are:

- **2-deoxy-D-Ribose** is found in nucleic acid *(DNA).*
- **6-deoxy-L-Galactose** is found as a constituent of glycoproteins, blood group substances and bacterial polysaccharides.

Practical Application

Feulgen-staining reaction for 2-deoxy-sugar derivatives in tissues is based upon the reaction of 2-deoxy-sugars with Schiff's reagent.

BENEDICT'S QUALITATIVE TEST: BASIS

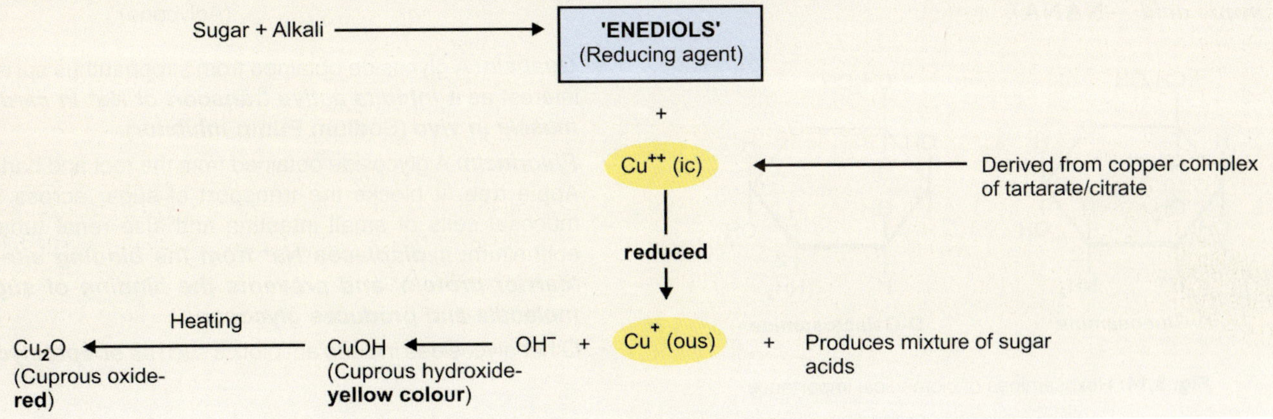

2. Amino sugars (hexosamines): Sugars containing an $-NH_2$ group in their structure are called ***amino sugars***.

Types: Two types of amino sugars of physiological importance are:

- *Glycosylamine:* The ***anomeric –OH group*** is replaced by an $-NH_2$ group.

Example: A compound belonging to this group is ***Ribosylamine***, a derivative of which is involved in the synthesis of purines.

- *Glycosamine (Glycamine):* In this type, the alcoholic – OH group of the sugar molecule is replaced by $- NH_2$ group. Two naturally occurring members of this type are derived from glucose and galactose, in which – OH group on carbon 2 is replaced by $- NH_2$ group, and forms respectively ***Glucosamine*** and ***Galactosamine*** (Fig. 3.14).

Biomedical Importance

- N-acetyl derivative of D-Glucosamine occur as a constituent of certain mucopolysaccharides *(MPS)*.
- Glucosamine is the chief organic constituent of cell wall of fungi, and a constituent of shells of crustaceae (crabs, Lobsters, etc), where it occurs as ***Chitin,*** which is made of repeating units of N-acetylated glucosamine. ***Hence Glucosamine is often called as* Chitosamine.**
- Galactosamine occurs as N-acetyl-Galactosamine in chondroitin sulphates which are present in cartilages, bones, tendons and heart valves. ***Hence Galactosamine is also known as* Chondrosamine.**
- ***Antibiotics:*** Certain antibiotics, such as **Erythromycin, carbomycin,** contain amino sugars. Erythromycin contains dimethyl amino sugar and carbomycin 3-amino-D-Ribose. It is believed that amino sugars are related to the antibiotic activity of these drugs.

3. Amino Sugar Acids

- ***Neuraminic acid:*** It is an amino sugar acid and structurally an aldol condensation product of pyruvic acid and D-Mannosamine. Neuraminic acid is unstable and found in nature in the form of acylated derivatives known as ***Sialic acids (N-acetyl Neuraminic acid* —NANA*)***.

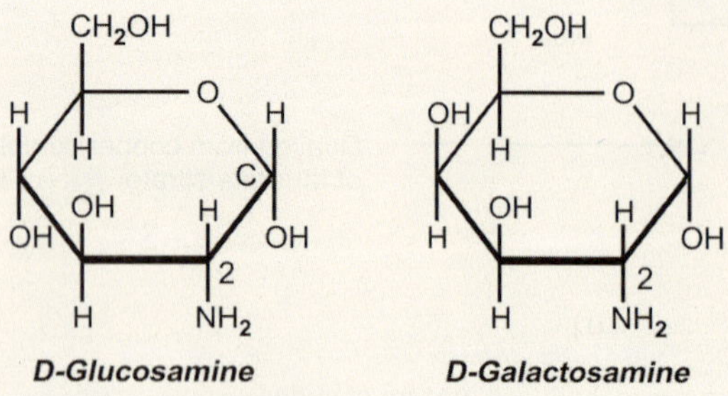

Fig. 3.14: Hexosamines of biomedical importance

- ***Muramic acid:*** Another amino sugar acid which is structurally a condensation product of D-Glucosamine and Lactic Acid.

Biomedical Importance

- Neuraminic acid and sialic acids *occur* in a number of *mucopolysaccharides* and in glycolipids like ***gangliosides***.
- A number of nitrogenous oligosaccharides which contain neuraminic acid are found in human milk.
- Certain bacterial cell walls contain ***muramic acid.***
- ***Neuraminidase*** is the enzyme which hydrolyses to split "NANA" from the compound.

4. Glycosides

Definition: Glycosides are compounds ***containing a carbohydrate* and *a noncarbohydrate residue in the same molecule.*** In these compounds the carbohydrate residue is attached by an ***acetal linkage*** of carbon-I to the noncarbohydrate residue. **The noncarbohydrate residue present in the glycoside is called as Aglycone.** The aglycones present in glycosides vary in complexity from simple substances as methyl alcohol, glycerol, phenol or a base such as adenine to complex substances like sterols, hydroquinones and anthraquinones. The glycosides are named according to the carbohydrate they contain. If it contains glucose, forms ***glucoside.*** If galactose, it forms ***galactoside*** and so on.

Biomedical Importance

Glycosides are found in many drugs, spices and in the constituents of animal tissues. They are widely distributed in plant kingdom.

- ***Cardiac glycosides:*** It is important in medicine because of their action on heart and thus **used in cardiac insufficiency**. They all contain steroids as aglycone component in combination with sugar molecules. They are derivatives of digitalis, strophanthus and squill plants, e.g.

Digitonin ⟶ 4 Galactose + Xylose + Digitogenin (Aglycone)

- ***Ouabain:*** A glycoside obtained from strophanthus sp. is of interest as it ***inhibits active transport of Na^+ in cardiac muscle in vivo*** **(Sodium Pump *inhibitor*).**
- ***Phloridzin:*** A glycoside obtained from the root and bark of Apple tree. It blocks the transport of sugar across the mucosal cells of small intestine and also renal tubular epithelium; it ***displaces Na^+ from the binding site of 'carrier protein' and prevents the binding of sugar molecule and produces glycosuria.***
- Other glycosides include antibiotics such as ***streptomycin***.

DISACCHARIDES

Three most common disaccharides of biological importance are: ***Maltose, Lactose and Sucrose.*** Their general molecular formula is $C_{12}H_{22}O_{11}$ and they are hydrolysed by hot acids or corresponding enzymes as follows:

$$C_{12}H_{22}O_{11} + H_2O \rightarrow C_6H_{12}O_6 + C_6H_{12}O_6$$

Thus on hydrolysis:

Maltose $\xrightarrow{H_2O}$ D-Glucose + D-Glucose

Lactose $\xrightarrow{H_2O}$ D-Glucose + D-Galactose

Sucrose $\xrightarrow{H_2O}$ D-Glucose + D-Fructose

The disaccharides are formed by the union of two constituent monosaccharides with the elimination of one molecule of water. The points of linkage, the **glycosidic linkage** varies, as does the manner of linking ***and the properties of the disaccharides depend to a great extent on the type of the linkage.*** If both of the two potential aldehyde/or ketone groups are involved in the linkage the sugar will not exhibit reducing properties and will not be able to form osazones, e.g. ***sucrose***. But if one of them is not bound in this way, it will permit reduction and osazone formation by the sugars, e.g. ***Lactose and Maltose.***

PROPERTIES OF DISACCHARIDES

1. Maltose: ***Maltose or malt sugar*** is an intermediary in acid hydrolysis of starch and can also be obtained by enzyme hydrolysis of starch. In the body, dietary starch digestion ***by Amylase*** in gut yields maltose, which requires a specific enzyme ***maltase*** to form glucose. It is a rather sweet sugar and is very soluble in water. ***Since it has one aldehyde 'free' or potentially free (Fig. 3.15) it has reducing properties, and forms characteristic osazones,*** which has characteristic appearance ***'Sunflower' like***. As anomeric carbon of one glucose is free, can form α and β forms and exhibit mutarotation. ***On hydrolysis Maltose yields two molecules of glucose.***

2. Lactose: *Lactose* is ***milk sugar*** and found in appreciable quantities in milk to the extent of about 5 per cent and occurs at body temperature as an equilibrium mixture of the α and β forms in 2:3 ratio. It is not very soluble and is not so sweet. It is ***dextrorotatory***. Specific enzyme which hydrolyses is ***lactase*** present in intestinal juice. ***On hydrolysis it yields one molecule of D-Glucose and one molecule of D-Galactose.*** Because it contains galactose as one of its constituents, it yields ***mucic acid*** on being treated with Conc HNO_3 after hydrolysis.

As one of the aldehyde group is free or potentially free (Fig. 3.16), it has reducing properties and can form osazones. Lactosazone crystals have typical ***hedgehog shape or powder puff appearance***. As anomeric carbon of glucose is **free**, can form α and β forms and exhibits mutarotation.

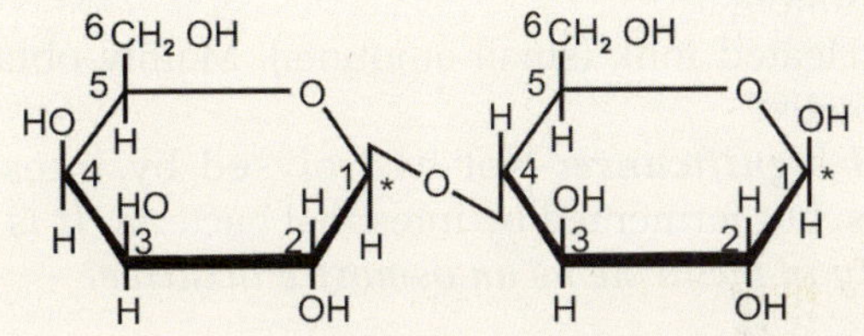

O-β-D-Galactopyranosyl-(1 → 4)-β-D-glucopyranoside

Fig. 3.16: Lactose (β form)

Fearon's Test: Serves to distinguish lactose simultaneously from sucrose and monosaccharides like glucose, galactose and fructose. A mixture of lactose and methylamine hydrochloride solution + NaOH solution, when heated in water bath at 56°C for ½ hour and then cooled by standing at room temperature, **an intense red colour develops.**

3. Sucrose: ***Ordinary table sugar is sucrose.*** It is also called as 'Cane sugar', as it can be obtained from sugarcane. Also obtained from sugar beet, and sugar maple. Also occurs free in most fruits and vegetables, e.g. pineapples, and carrots. It is very soluble and very sweet and ***on hydrolysis yields one molecule of D-Glucose and one molecule of D-Fructose***. The specific enzyme which hydrolyses sucrose is ***sucrase*** present in intestinal juice. ***As both aldehyde and ketone groups are linked together (α 1 → 2), it does not have reducing properties, and cannot form osazones.*** As both anomeric carbons are involved in 'linkage', ***it does not exhibit mutarotation* (Fig. 3.17).**

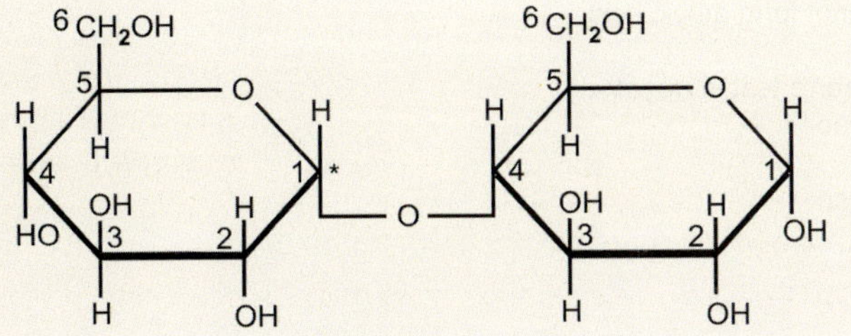

Fig. 3.15: Maltose (α form)

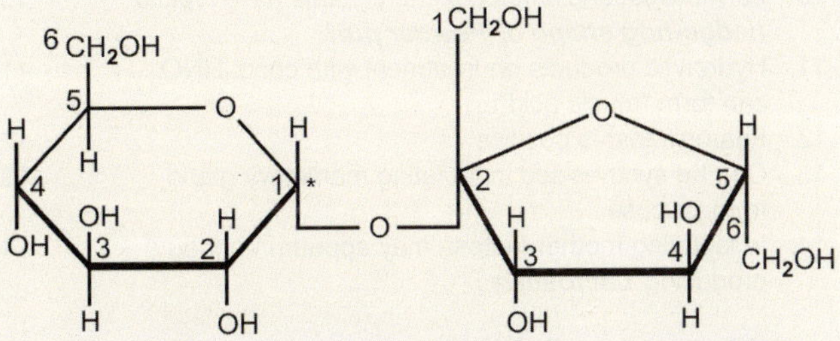

O-α-D-Glucopyranosyl-(1 → 2)-β-D-fructofuranoside

Fig. 3.17: Sucrose

Invert Sugars and 'Inversion'

Sucrose is dextrorotatory (+62.5°) but its hydrolytic products are laevorotatory because fructose has a greater specific laevorotation than the dextrorotation of glucose. As the hydrolytic products invert the rotation, the resulting mixtures of glucose and fructose (hydrolytic products) is called as ***Invert Sugar*** and the process is called as ***Inversion***. Honey is largely 'invert sugar' and the presence of fructose accounts for the greater sweetness of honey.

4. Lactulose: A ketodisaccharide

Structure: It is O-α-D-galctopyranosyl-(1→4)-β-D-fructofuranose

Source: Heated milk (small amounts). Mainly obtained synthetically.

Clinical significance: Not hydrolysed by intestinal enzymes, but fermented by intestinal bacteria. ***It is used clinically in medicine as an osmotice laxative.***

Differences Between Sucrose and Lactose

Detailed differentiation of lactose from sucrose has been given in **Table 3.1**.

Biomedical Importance of Disaccharides

- Various food preparations, such as baby and invalid foods available, are produced by hydrolysis of grains and contain large amounts of maltose. From nutritional point of view they are thus easily digestible.
- In lactating mammary gland, the ***lactose is synthesised from glucose by the duct epithelium*** and lactose present in breast milk is a good source of energy for the newborn baby.
- Lactose is fermented by ***'Coliform' bacilli (E. coli)*** which is usually non-pathogenic ***(lactose fermenter)*** and not by ***Typhoid bacillus*** which is pathogenic ***(lactose non-fermenter). This test is used to distinguish these two microorganisms.***
- ***'Souring' of milk:*** Many organisms that are found in milk, e.g. *E. coli, A. aerogenes,* and *Str. lactis* convert lactose of milk to lactic acid (LA) thus causing souring of milk.
- ***Sucrose if introduced parenterally cannot be utilised,*** but it can change the osmotic condition of the blood and causes a flow of water from the tissues into the blood. Thus ***clinicians use it in oedema like cerebral oedema.*** If sucrose or some other disaccharides are not hydrolysed in the gut, due to deficiency of the appropriate enzyme, diarrhoea is likely to occur.

Table 3.1: Differentiation of lactose from sucrose

Lactose	Sucrose
1. Known also as 'milk sugar'	1. Common table sugar (cane sugar)
2. Structurally one molecule of D-Glucose and one molecule of D-Galactose are joined together by glycosidic linkage (β 1 → 4)	2. Structurally one molecule of D-Glucose and one molecule of D-Fructose joined together (α 1→ 2)
3. Hydrolysed to give one molecule of glucose and one molecule of galactose	3. Hydrolysed to give one molecule of glucose and one molecule of fructose
4. Specific enzyme which hydrolyses is called ***lactase,*** which is present in intestinal juice	4. Specific enzyme which hydrolyses is called ***Sucrase (Invertase)*** which is present in intestinal juice
5. Dextrorotatory disaccharide	5. Also dextrorotatory (+66.5°), but hydrolytic products are laevorotatory (–19.5°) Hydrolytic products are called ***invert sugars*** and process is called ***Inversion.***
6. As anomenic carbon is free, can form α and β forms and exhibits mutarotation	6. As both anomenic carbons are involved in linkage, **cannot form α and β-forms**
7. Specific rotation of the solution is +55.2°	7. **Does not exhibit mutarotation**
8. Can reduce alkaline copper sulphate solution like Benedict's qualitative reagent, Fehling's solution	8. Does not reduce alkaline copper sulphate solution
9. Does not reduce Berfoed's solution	9. Does not reduce Berfoed's solution
10. Forms Osazone. Lactosazone crystals have typical ***hedge-hog shape or Powder puff***	10. **Cannot form osazones**
11. Hydrolytic products on treatment with conc. HNO_3 can form "mucic acid"	11. Cannot form mucic acid
12. Fearon's test is positive	12. Fearon's test is *negative*
13. Can be synthesised in lactating mammary gland from glucose	13. Not so
14. In lactating mother lactose may appear in urine, producing ***Lactosuria***	14. Not so

OLIGOSACCHARIDES

Biomedical Importance: Integral membrane proteins contain covalently attached carbohydrate units, oligosaccharides, on their extracellular face. Many secreted proteins, such as antibodies and coagulation factors also contain oligosaccharide units. These carbohydrates are attached to either the side-chain O_2 atom of serine or threonine residues by O-glycosidic linkages or to the side chain nitrogen of Asparagine residues by N-glycosidic linkages. N-linked oligosaccharides contain a ***common pentasaccharide core*** consisting of three mannose and two N-acetyl glycosamine residues. Additional sugars are attached to this common core in many different ways to form the great variety of oligosaccharide patterns found in glycoproteins.

The diversity and complexity of the carbohydrate oligosaccharide units of glycoprotein suggest that they are rich in information and are functionally important. Carbohydrates participate in molecular targeting and cell-cell recognition. The removal of glycoproteins from the blood is accomplished by ***Surface Protein Receptors*** on Liver cells, e.g. ***Asiulo-glycoprotein receptor.*** Many newly synthesised glycoproteins such as, immunoglobulins (antibodies) and peptide hormones, contain oligosaccharide carbohydrate units with terminal sialic acid residues. When the function of particular protein is over, in hours or days, terminal sialic acid residues are removed by ***Sialyses*** on the surface of blood vessels. The exposed galactose residues of these trimmed proteins are detected by the asialoglycoprotein receptors on liver cell membrane. The complex of the asialoglycoprotein and its receptor is then internalised by the liver, by "endocytosis", to remove the trimmed glycoprotein from the circulating blood. ***The oligosaccharide units actually mark the passage of time and determines when the proteins carrying them should be taken out of circulation.*** The rate of removal of sialic acid from glycoproteins is controlled by the structure of the protein itself. Thus, ***proteins can be designed to have life times ranging from a few hours to many weeks depending on the physiological and biological needs.***

POLYSACCHARIDES

- Polysaccharides are more complex substances. Some are polymers of a single monosaccharide and are termed as ***Homopolysaccharides (Homoglycans),*** e.g. starch, glycogen, etc.
- Some contain other groups other than carbohydrates such as hexuronic acid and are called as ***Heteropolysaccharides (heteroglycans),*** e.g. Mucopolysaccharides.

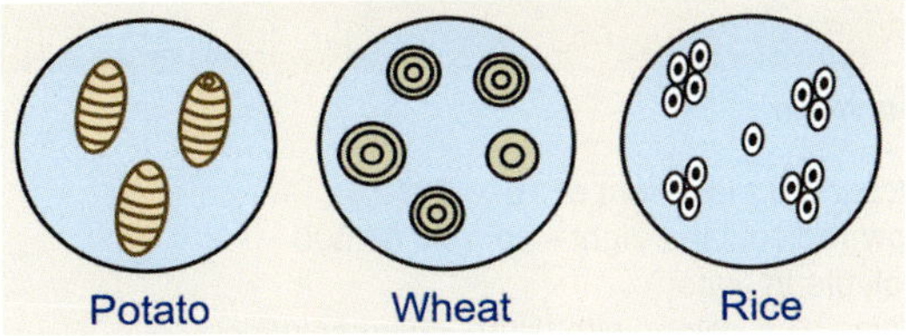

Fig. 3.18: Starch grains under microscope

HOMOPOLYSACCHARIDES (HOMOGLYCANS)

1. Starch

Starch is a **polymer of glucose**, and occurs in many plants as storage foods. It may be found in the leaves, and stem, as well as in roots, fruits and seeds where it is usually present in greater concentration.

- ***Starch granules:*** Appear under microscope as particles made up of concentric layers of material. They differ in shape, size and markings according to the source (Fig. 3.18). Starchy foods are mainstay of our diet.
- ***Composition of starch granule:*** It consists of **two polymeric units** of glucose called (i) ***Amylose*** and (ii) ***Amylopectin,*** but they differ in molecular architecture and in certain properties (Table 3.2).
- ***Solubility:*** Starch granules are insoluble in cold water, but when their suspension is heated, water is taken up and swelling occurs, viscosity increases and ***starch gels or pastes*** are formed.
- ***Reaction with I_2:*** Both the granules and the colloidal solutions react with Iodine to give a blue colour. This is chiefly due to amylose, which forms a deep-blue complex, which dissociates on heating. Amylopectin solutions are coloured blue-violet or purple.
- ***Ester Formation:*** Starches are capable of forming esters with either organic or inorganic acids.
- ***Hydrolysis of starch:*** It yields succession of polysaccharides of gradually diminishing molecular size.

Course of Hydrolysis	*Reaction with Iodine*
Starch ↓	Blue
Soluble starch ↓	Blue
Amylodextrin ↓	Purple
Erythrodextrin ↓	Red
Achroodextrin ↓	Colourless
Maltose	

Table 3.2: Differentiation of amylose and amylopectin

Amylose	*Amylopectin*
1. Occurs to the extent of 15 to 20%	1. Occurs 80 to 85%
2. Low molecular weight—approx. 60,000	2. High molecular weight—approx. 5,00,000
3. Soluble in water	3. Insoluble in water, can absorb water and swells up
4. Gives blue colour with dilute iodine solution	4. Gives reddish-violet colour with I_2 solution
5. ***Structure*** • Unbranched • Straight chain • 250 to 300 D-Glucose units linked by $\alpha 1 \rightarrow 4$ linkages • Twists into a helix, **with six glucose units per turn**	5. • Highly branched structure • More D-Glucose • Units joined together • Structure similar to glycogen • Main stem has α-1 $\rightarrow$ 4 glycoside bonds **At branch point $\alpha 1 \rightarrow 6$ linkage**, Approx 80 branches. One branch after every 24 to 30 D-Glucose units

Amylase / Maltase

• Starch ⟶ Maltose ⟶ Glucose
+
Dextrins

• Starch ⟶ Glucose (acid, Hydrolysis)

Enzyme (amylase) hydrolysis ends at maltose. It is not quantitative conversion, traces of dextrins are also formed. For formation of glucose, it requires the enzyme *maltose*. ***But if the hydrolysis is accomplished by acids much of the starch will be converted into glucose.***

- ***Types of amylases:*** ***Two broad classes of amylases exist as:***
  - ***α-amylase*** is present in **saliva** and **pancreatic juice**,
  - ***β-amylase*** is present in sprouted grains and malts.

Both of them hydrolyse only α-glycosidic linkage. α-amylase produces a random cleavage of glycosidic bonds well inside the starch molecule yielding a mixture of maltose and ***some fragments larger than maltose (dextrins)***, whereas, ***β-amylase splits off*** maltose moietes liberating successive maltose units commencing at the non-reducing end of the starch molecule and ***ends in Limit dextrin.***

2. Glycogen

Glycogen is the reserve carbohydrate of the animal, hence it is called as ***animal starch.*** It has been shown to be present in plants which have no chlorophyll systems, e.g. in fungi and yeasts. It is also found in large amounts in oysters and other shell fish. **In higher animals, it is deposited in the liver and muscle as storage material which are readily available as immediate source of energy.** It is **dextrorotatory** with an [α] D 20° = +196° to +197°. Formation of glycogen from glucose is called as ***Glycogenesis*** and breakdown of glycogen to form glucose is called as ***glycogenolysis.*** Postmortem glycogenolysis is very rapid but ceases when the pH falls to 5.5 due to lactic acid formed from glucose.

- ***Molecular weight:*** The molecular weight varies from 1,000,000 to 4,000,000.
- ***Solubility:*** Glycogen is not readily soluble in water and it forms an opalascent solution. It can be precipitated from opalascent solution by ethyl alcohol, and in drying, it forms a pure white powder.
- ***Action of alkali:*** Glycogen is not destroyed by a hot strong KOH or NaOH solution. ***This property is made use of in the method for determining it quantitatively in tissues.***
- ***Action with iodine:*** Glycogen gives a deep-red colour. In this respect it resembles erythrodextrin.

Structure: Glycogens have a complex structure of highly branched chains. ***It is a polymer of D-Glucose units and resemble amylopectin.*** Glucose units in main stem are joined by α1 → 4 glucosidic linkages and branching occurs at branch points by α1 → 6 glucosidic linkage. A branch point occurs for every 12 to 18 glucose units **(Fig. 3.19)**.

3. Inulin

It is a **polymer of D-fructose** and has a low molecular weight (MW = 5000). It occurs in tubers of the Dehlia, in the roots of the Jerusalem artichoke, dandelion and in the bulbs of onion and garlic. It is a white, tasteless

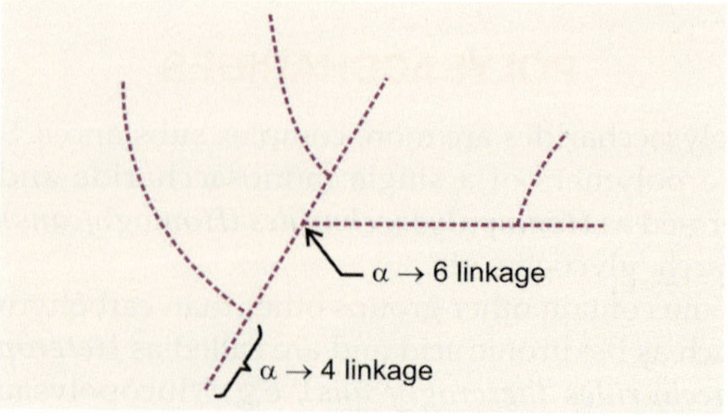

Fig. 3.19: Branched structure of glycogen—showing linkages

powder. It is ***levorotatory*** and gives no colour with iodine. Acids hydrolyse it to D-fructose; similarly it is also hydrolysed by the enzyme ***inulinase***, which accompanies it in plants. ***It has no dietary importance in human beings as inulinase is absent in human.***

Biomedical Importance

- It is **used** in physiological investigation for determination of the **rate of glomerular filtration rate (GFR).**
- It has been also **used** for **estimation of body water (ECF)** volume.

4. Cellulose

Cellulose is a polymer of glucose. It is not hydrolyzed readily by dilute acids, but heating with fairly high concentrations of acids yields, the disaccharide ***Cellobiose*** and D-Glucose. **Cellobiose is made up of two molecules of D-Glucose linked together by *β-Glucosidic linkage* between C_1 and C_4 of adjacent glucose units.**

Biomedical Importance

Cellulose is a very stable insoluble compound. Since, it is the main constituent of the supporting tissues of plants, it forms a considerable part of our vegetable food. Herbivorous animals, with the help of bacteria, can utilise a considerable proportion of the cellulose ingested, ***but in human beings no cellulose splitting enzyme is secreted by GI mucosa, hence it is not of any nutritional value.*** But it is of considerable human dietetic value that it adds bulk to the intestinal contents ***(roughage)*** thereby **stimulating peristalsis and elimination of indigestible food residues.**

5. Dextrins

When starch is partially hydrolysed by the action of acids or enzymes, it is broken down into a number of products of lower molecular weight known as ***dextrins*** (see hydrolysis of starch). They resemble starch by being precipitable by alcohol, forming sticky, gummy masses.

Biomedical Importance

- Dextrin solutions are often **used as *mucilages*** (mucilages on the back of the postage stamp)
- Starch hydrolysates consisting largely of dextrins and maltose are widely **used in infant feeding.**

Limit Dextrin: It is a well defined dextrin. This is the product remaining after the ***β-Amylase*** has acted upon starch until no further action is observed.

6. Dextrans

It is a ***polymer of D-Glucose.*** It is synthesised by the action of ***Leuconostoc mesenteroides,*** a non-pathogenic gram +ve cocci in a sucrose medium. Exocellular enzyme produced by the organisms bring about polymerisation of glucose moiety of sucrose molecule, and forms the polysaccharide known as ***Dextrans.*** They differ from dextrins in structure. They are made up of units of a number of D-Glucose molecules, having $\alpha 1 \rightarrow 6$, $\alpha 1 \rightarrow 4$ or $\alpha 1 \rightarrow 3$ glycosidic linkages, within each unit and the units are joined together to form a network.

Clinical Aspect

Dextran solution, having molecular wt approx. 75,000 have been **used as *Plasma Expander.*** When given IV, in cases of blood loss (haemorrhage), it increases the blood volume. Because of their high viscosity, low osmotic pressure, slow disintegration and utilisation, and slow elimination from the body they remain in blood for many hours to exert its effect.

Disadvantage: Only disadvantage is that ***it can interfere with grouping and cross-matching,*** as it forms false agglutination (Roleux formfation). Hence ***blood sample for grouping and cross-matching should be collected before administration of dextran in a case of haemorrhage and blood loss, where blood transfusion may be required.***

7. Agar

It is a homopolysaccharide. Made up of repeated units of galactose which is sulphated. Present in seaweed. It is obtained from them.

Biomedical Importance

- ***In human:*** **Used as laxative** in constipation. Like cellulose, it is not digested, hence add bulk to the faeces ("roughage" value) and helps in its propulsion.
- ***In microbiology:*** Agar is available in purified form. It dissolves in hot water and on cooling it sets like gel. It is **used in agar plate for culture of bacteria.**

HETEROPOLYSACCHARIDES (HETEROGLYCANS)—MUCOPOLYSACCHARIDES (MPS)

Jeanloz has suggested the name **Glycosaminoglycans** *(GAG)* to describe this group of substances. They are usually composed of ***amino sugar*** and ***uronic acid*** units as the principal components, though some are chiefly made up of amino sugar and monosaccharide units ***without the presence of uronic acid.*** The hexosamine present is generally acetylated. They are essential components of tissues, where they are generally present either in free form or in combination with proteins. Carbohydrate content varies. ***When carbohydrate content is > 4 per cent, they are called* Mucoproteins *and when < 4 per cent they are called as* Glycoproteins.**

SECTION TWO

CLASSIFICATION

Although there is no agreement on classification, the nitrogenous heteropolysaccharides (mucopolysaccharides) are classified as follows:

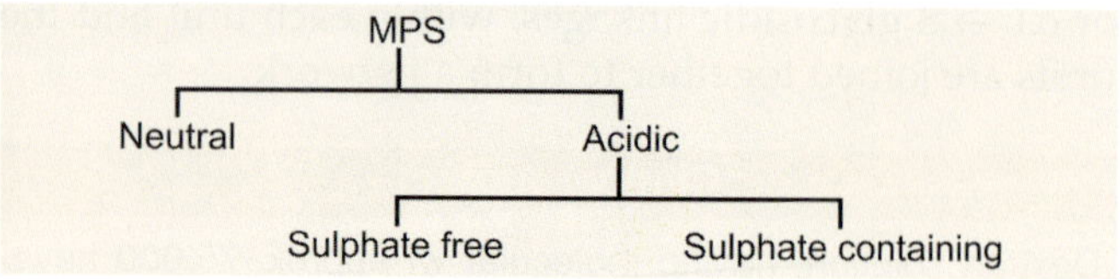

I. Acidic Sulphate Free MPS

1. Hyaluronic Acid

A sulphate free mucopolysaccharide. It was first isolated from vitreous humour of eye. Later it was found to be present in synovial fluid, skin, umbilical cord, haemolytic streptococci and in rheumatic nodule. It occurs both free and salt-like combination with ***proteins and forms so-called ground substance of mesenchyme,*** an integral part of gel-like ground substance of connective and other tissues.

Composition: It is composed of repeating units of ***N-acetyl glucosamine and D-Glucuronic acid. On hydrolysis,*** it yields equimolecular quantities of D-Glucosamine, D-Glucoronic acid and acetic acid (Fig. 3.20).

Functions: (See below under proteoglycans)

Hyaluronidase: An enzyme present in certain tissues, notably testicular tissue and spleen, as well as in several types of pneumococci and haemolytic streptococci. ***The enzyme catalyses the depolymerisation of hyaluronic acid*** and by reducing its viscosity facilitates diffusion of materials into tissue spaces. Hence the enzyme, sometimes, is designated as ***spreading factor.***

Biomedical Importance

- The invasive power of some pathogenic organisms may be increased because they secrete ***hyaluronidase.*** In the testicular secretions, it may dissolve the viscid substances surrounding the ova to permit penetration of spermatozoa.
- ***Clinically*** the enzyme is used to increase the efficiency of absorption of solutions administered by clysis.

2. Chondroitin

Another ***sulphate free*** acid mucopolysaccharide. Found in ***cornea*** and has been isolated from cranial cartilages. It differs from hyaluronic acid only in that it contains N-acetyl galactosamine instead of N-acetyl glucosamine.

II. Sulphate Containing Acid MPS

1. Keratan Sulphate (Kerato Sulphate)

A sulphate containing acid MPS. Found in costal cartilage, and cornea has been isolated from bovine cornea. It has been reported to be present in ***Nucleus pulposus*** and the wall of aorta.

Composition: It is composed of repeating disaccharide unit consisting of N-acetyl glucosamine and galactose.

N-acetyl Glucosamine → Galactose → N-Acetyl Glucosamine $_n$

There are no uronic acids in the molecule. Total sulphate content varies, but ester SO_4 is present at C_6 of both N-acetyl glucosamine and galactose.

Types: **Two-types** have been described. They are found in tissues combined with proteins.

- ***Keratan SO_4 I:*** It occurs in cornea. In this type, linkage is between N-acetyl glucosamine and Asparagine residue to form the N-glycosidic bonding.
- ***Keratan SO_4 II:*** It occurs in skeletal tissues. In this type, the linkage to protein is by way of -OH groups on serine and threonine residues of the protein.

2. Chondroitin Sulphates

They are principal MPS in the ground substance of mammalian tissues and cartilage. They occur in combination with proteins and are called as ***Chondroproteins.***

Four chondroitin sulphates have been isolated so far. They are named as chondroitin SO_4 A, B, C and D.

a. Chondroitin SO_4 A: It is present chiefly in cartilages, adult bone and cornea.

Structure: It consists of repeating units of N-acetyl-D-Galactosamine and D-Glucuronic acid. N-Acetyl galactosamine is esterified with SO_4 in ***position 4*** of galactosamine (Fig. 3.21).

b. Chondroitin SO_4 B: It is present in skin, cardiac valves and tendons. Also isolated from aortic wall and lung

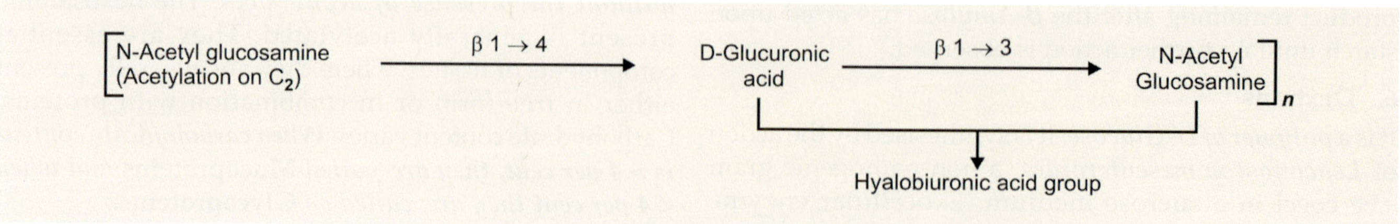

Fig. 3.20: Structure of hyaluronic acid

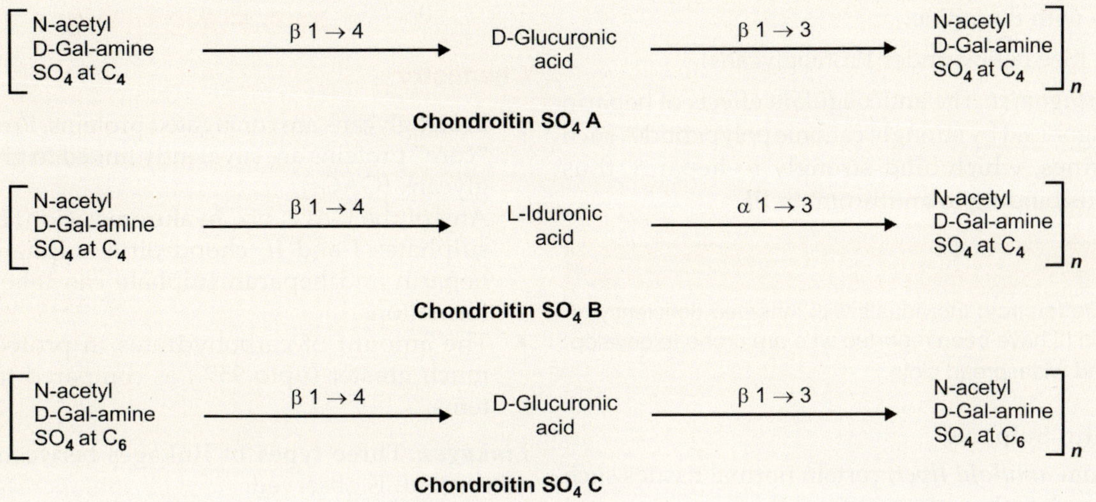

Fig. 3.21: Structure of chondroitin SO_4

parenchyma. ***It has L-iduronic acid in place of glucuronic acid which is found in other chondroitin sulphates.*** It has a ***weak anticoagulant property,*** hence sometimes it is called as **β-Heparin.** As it is found in *skin,* it is also called as ***Dermatan sulphate.***

Structure: It consists of repeating units of ***L-iduronic acid*** and ***N-acetyl galactosamine.*** Sulphate moiety is present at C_4 of N-acetyl galactosamine molecule.

L-Iduronic acid: It is an **epimer of D-Glucuronic acid.** Metabolically it is formed in the liver from D-Glucose.

c. *Chondroitin SO_4 C:* It is found in cartilage and tendons. Structure of chondroitin SO_4 C is the same as that of chondroitin SO_4 A except that the SO_4 group is ***at position 6 of galactosamine*** molecule instead of position 4.

d. *Chondroitin SO_4 D:* It has been isolated from the cartilage of shark. It resembles in structure to chondroitin SO_4 C except that it has a second SO_4 attached probably at carbon 2 or 3 of uronic acid moiety.

3. Heparin

It is also called ***α-Heparin. It is an anticoagulant present in liver*** and it is produced mainly by ***mast cells*** of liver (Originally isolated from liver). In addition, it is also found in lungs, thymus, spleen, walls of large arteries, skin and in small quantities in blood.

Structure: **It is a polymer of repeating disaccharide units of D-Glucosamine (Glc N) and either of the two uronic acids-D-Glucuronic acid (Glc UA) and L-Iduronic acid (IDUA)** (Fig. 3.22). The -NH_2 group at C_2 and OH group at C_6 of D-Glucosamine (Glc N) are sulphated. A few may contain acetyl group on C_2 of D-Glucosamine. In addition, the OH group of C_2 of uronic acids, D-Glucuronic acid and/or L-Iduronic acid, are sulphated. Initially, all of the uronic acids are D-Glucuronic acid (Glc UA), but ***"5-epimerase"*** enzyme converts approximately 90 per cent of the D-Glucuronic acid residues to L-Iduronic acid (IDUA) after the polysaccharide chain is fully formed. ***Hence, in fully formed Heparin molecule 90 per cent or more of uronic acid residues are L-Iduronic acid.***

Properties: It is ***strongly acidic*** due to sulphuric acid groups and readily forms salts. Molecular weight of Heparin varies from 17,000 to 20,000. It occurs in combination with proteins as proteoglycans. The protein molecule of heparin proteoglycan is unique, consisting chiefly ***Serine*** and ***Glycine*** residues. Approximately 2/3 of the serine residue contain GAG chains. Linkage with

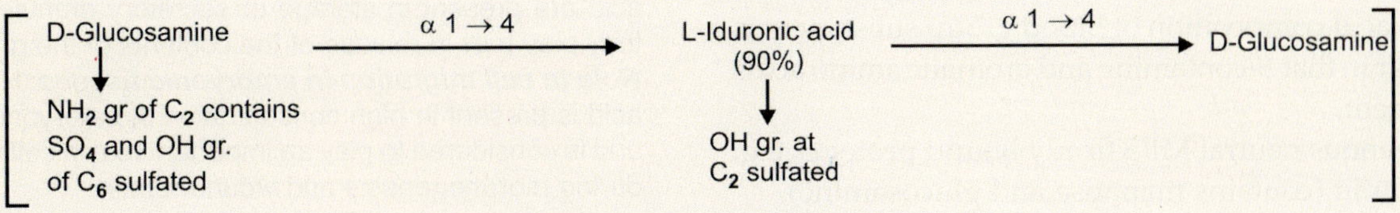

Fig. 3.22: Structure of heparin

protein molecule is usually with GalN and serine/ sometimes with threonine.

Functions: (See below under Proteoglycans).

Heparin antagonist: The anticoagulant effects of heparin can be antagonised by strongly cationic polypeptides such as **protamines,** which bind strongly to heparin, thus inhibiting its binding to antithrombin III.

CLINICAL ASPECT

Inherited Deficiency: Individuals with inherited deficiency of antithrombin III have been reported who are prone to develop frequent and widespread clots.

4. Heparitin Sulphate

Isolated from ***amyloid liver,*** certain normal tissues such as human and cattle aorta, and from the urine, liver and spleen of patients with gargoylism (Hurler's syndrome). ***This compound has negligible anticoagulant activity.*** It seems to be structurally similar to heparin, but has a:

- Lower molecular weight,
- Some of the amino groups carry acetyl groups and percentage of $-SO_4$ groups are smaller.
- Unlike heparin, its predominant uronic acid is D-Glucuronic acid (Glc UA).

Recently it has been shown that it is present on cell surfaces as proteoglycan and is extracellular.

III. Neutral MPS

- Many of the neutral nitrogenous polysaccharides of various types are found in pneumococci capsule. Type specificity of pneumococci resides on specific polysaccharides present on capsule ("hapten"). Preparations of capsular polysaccharides from Type-1 pneumococci yield on hydrolysis glucosamine and Glucuronic acids.
- *Blood group substances:* These contain peptides or amino acids as well as carbohydrates. ***Four monosaccharides*** are found in all types of blood group substances regardless of source: ***Galactose, fucose, Galactosamine (acetylated) and acetylated glucosamine.*** Non-reducing end groups of acetyl glucosamine, galactose and **fucose** are associated with blood group specificities of A, B and H respectively (Fig. 3.23). The amino acid composition of blood group substances is peculiar in that *S*-containing and aromatic amino acids are absent.
- Nitrogenous neutral MPS firmly bound proteins, e.g. ovalbumin (contains mannose and glucosamine).

PROTEOGLYCANS—CHEMISTRY AND FUNCTIONS

Chemistry

- Proteoglycans are conjugated proteins. Proteins called "core" proteins are covalently linked to glycosaminoglycans (GAGs).
- Any of the GAGs viz. hyaluronic acid (HA); keratan sulphates I and II, chondroitin sulphates A, B, C, heparin and heparan sulphate can take part in its formation.
- The amount of carbohydrates in proteoglycans is much greater (upto 95%) as compared to glycoproteins.

Linkages: **Three types of linkages** between GAG and coreprotein is observed.

- *O-glycosidic linkage:* Formed between N-acetyl galactosamine (GalNAc) and serine or threonine of the core protein.

  Example: Typically seen in keratan SO_4 II.
- *N-glycosylamine linkage:* Formed between N-acetyl glucosamine (GlcNAc) and amide N of asparagine (ASn) of core protein.

  Example: Typically seen in keratan SO_4I and N-linked glycoproteins.
- *O-glycosidic linkage:* Formed between xylose (Xyl) and serine of the protein. This **bond is unique to proteoglycans.**

Functions of Proteoglycans

- ***As a constituent of extracellular matrix or ground substance:*** Interacts with collagen and elastin
- ***Acts as polyanions:*** GAG_S present in proteoglycans are polyanions and hence bind to polycations and cations such as Na and K. Thus attracts water by osmotic pressure into extracellular matrix contributing to its turgor.
- ***Acts as a barrier in tissue:*** Hyaluronic acid in tissues acts as a cementing substance and contributes to tissue barrier which permit metabolites to pass through but **resist penetration by bacteria and other infective agents.**
- ***Acts as lubricant in joints:*** Hyaluronic acid in joints acts as a lubricant and shock absorbant. Intraarticular injection of hyaluronic acid in knee joints is used to alleviate pain in chronic osteoarthritis of knee joints.
- ***Role in release of hormone:*** Proteoglycans like hyaluronic acid are present in storage or secretory granules, where they play part in release of the contents of the granules.
- ***Role in cell migration in embryonic tissues:*** Hyaluronic acid is present in high concentration in embryonic tissues and is considered to play an important role in cell migration during morphogenesis and wound repair.

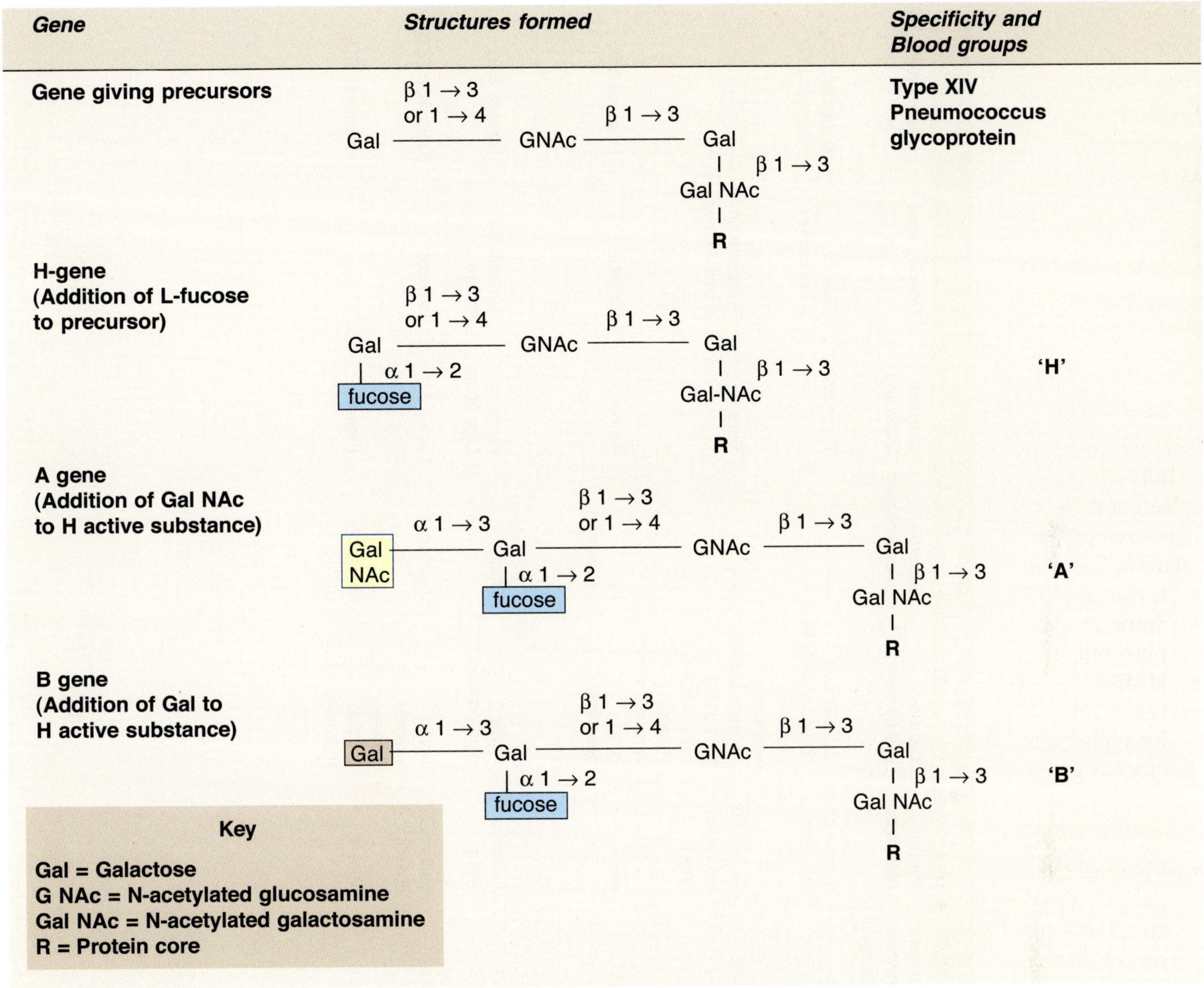

Fig. 3.23: Structure of blood groups—ABO system

- ***Role in glomerular filtration:*** Proteoglycans like hyaluronic acid is present in basement membrane (BM) of glomerulus of kidney where it plays important role in charge-selectiveness of glomerular filtration.
- ***Role as anticoagulant in vitro and in vivo:***
  - ***In vitro,*** heparin is used as an anticoagulant. 2 mg/10 ml of blood is used. Most satisfactory anticoagulant as it does not produce a change in red cell volume or interfere with its subsequent determinations.
  - ***In vivo,*** heparin is an important anticoagulant. It binds with *factor IX and XI*, but its most important action is with ***plasma antithrombin III***. Binding of heparin to lysine residues in antithrombin III ***produces conformational change*** which promotes the binding of the latter to serine protease ***thrombin*** which is inhibited, thus fibrinogen is ***not*** converted to fibrin.

  Four naturally occurring thrombin inhibitors in plasma are:
  (i) Antithrombin III (75% of the activity)
  (ii) α_2-macroglobulin contributes remainder
  (iii) Heparin cofactor II
  (iv) α_1-antitrypsin
  The last two shows minor activity.
- ***Role as a coenzyme:*** Heparin acts in the body to increase the activity of the enzyme ***Lipoprotein lipase***. Heparin binds specifically to the enzyme present in capillary walls, causing a release of the enzyme into the circulation. Hence heparin is called as **Clearing factor.**
- ***As a receptor of cell:*** Proteoglycans like heparan sulphate are components of plasma membrane of cells, where they may act as **receptors** and can participate in cell adhesion and cell-cell interactions.

Table 3.3: Types of mucopolysaccharidoses

Types	*Inheritance*	*Enzyme defect*	*Somatic skeletal changes*	*Mental retardation*	*Cardio-pulmonary*	*Hepato-splenomegaly*	*Corneal clouding*	*Hearing loss*	*Urinary MPS*
MPS-I (Hurler's syndrome)	Autosomal recessive	***α-L-Iduronidase (A-Lysosomal hydrolase)***	+++	Severe after one year	Valvular and coronary disease, Impaired ventilation	+++	Progressive	Present *(Conductive)*	Dermatan SO_4 *Heparan SO_4*
MPS-II (Hunter's syndrome)	Sex linked recessive	***Iduronate sulfatase***	++ to +++	Severe but gradual in onset	Valvular disease, pulmonary hypertension, impaired ventilation	+++	Rare	Present (early onset) *Perceptive*	—Do—
MPS-III SAN Filipos syndrome A, B and C	Autosomal recessive	***A-sulfamidase B-α-N-acetyl Glucosaminidase C-Acetyl transferase***	Mild	+++	Not described	++ (Moderate)	Absent	Present	Heparan SO_4
MPS-IV (Morquio syndrome)	" "	***N-Acetyl galactosamine 6-sulphatase***	+++	Absent or slight	Aortic regurgitation	Slight	Present, Late onset	Present, not severe	Keratan SO_4
MPS-V (Scheie syndrome)	Autosomal recessive	***α-L-Iduro-nidase***	Mild	Essentially absent	Aortic valvular disease	Variable	+++	Variable	Dermatan SO_4
MPS-VI (Maroteaux-Lamy syndrome)	" "	***N-acetyl galactosamine 4-sulphatase (Aryl sulfatase B)***	+++	Absent	Cardiac murmurs	Moderate ++	Present	"	Dermatan SO_4

- ***Role in compressibility of cartilages:*** Chondroitin sulphates and hyaluronic acid are present in high concentration in cartilages and have a role in compressibility of cartilage in weight bearing.
- ***Role in sclera of eye:*** Dermatan sulphate is present in sclera of the eye where it has an important function in maintaining overall shape of the eye.
- ***Role in corneal transparency:*** Keratan sulphate I is present in cornea of the eye and lie between the collagen fibrils. It plays an important role in maintaining corneal transparency.

Biomedical Importance/Clinical Aspect

Mucopolysaccharidoses: The mucopolysaccharidoses are a group of related disorders, ***due to inherited enzyme defect,*** in which skeletal changes, mental retardation, visceral involvement and corneal clouding are manifested to varying degrees.

Defect/defects in these disorders result in:

- Widespread deposits in tissues of a particular MPS
- In excessive excretion of MPS in urine.

At least **six types** of mucopolysaccharidoses have been described. The enzyme deficiency and the clinical findings are tabulated **(Table 3.3).**

Detection of MPS in urine: Cetyl trimethyl ammonium bromide test: Take 5 ml of fresh urine from the suspected case in a test tube. Add 1.0 ml of 5 per cent cetyl trimethyl ammonium bromide (Cetavlon) in 1 M citrate buffer (pH–6.0). Mix and allow to stand at room temperature for ½ hour. A heavy precipitate is given in gargoylism.

SECTION TWO

I. Questions (Essay type and short notes)

1. What are carbohydrates? Classify them giving suitable examples.
2. Classify carbohydrates and give their importance.
3. Describe the chemistry and functions of glycogen.
4. Mention five characteristic reactions of an aldose hexose which occurs in our body.
5. Explain the mutarotation of carbohydrates. Describe in brief the structure and important properties of the following.
   a. Glycogen b. Starch
6. Describe the chemistry and functions of mucopolysaccharides.

Differentiate

a. Amylose and amylopectin,
b. Sucrose and Iactose,
c. Starch and glycogen
d. Hyaluronic acid and heparin,
e. Dextrins and dextran

Short Notes

a. Mutarotation
b. Mucopolysaccharides or glycosaminoglycans (GAG)
c. Glycosides

II. MCQs (Give one correct answer)

1. Milk sugar is:
   a. Mannose b. Galactose
   c. Glucose d. Lactose
   e. Fructose
2. Glucose and galactose are epimers and they differ structurally in orientation of H and OH on:
   a. C_1 b. C_3
   c. C_2 d. C_5
   e. C_4
3. Cane sugar is:
   a. Glucose b. Sucrose
   c. Fructose d. maltose
   e. Galactose
4. Which of the following is not a reducing sugar?
   a. Lactose b. Maltose
   c. Sucrose d. Fructose
   e. Galactose
5. α-D-glucose and β-D-glucose are related by:
   a. Epimers b. Anomers
   c. Mutarotation d. Ketoaldo pair
   e. Ketoenol pair
6. The stable ring formation in D-glucose involves:
   a. C_1 and C_4 b. C_1 and C_2)
   c. C_1 and C_5 d. C_2 and C5
   e. C_3 and C_4
7. Table sugar which is used in making tea is:
   a. Glucose b. Lactose
   c. Mannose d. Sucrose
   e. Maltose
8. Reduction of glucose with Ca in water produces:
   a. Sorbitol b. Dulcitol
   c. Mannitol d. Glucuronic acid
   e. Fructose
9. Starch and glycogen are polymers of:
   a. Frutose b. Mannose
   c. β-D-glucose d. α-D-glucose
   e. Galactose
10. The reducing ability of a carbohydrate is due to the presence/or formation of:
    a. A free carboxyl group b. A free hydroxyl group
    c. Enediol formation d. Presence of a 1-4 linkage
    e. Formation of ring structure
11. Which of the following is not a polymer of glucose:
    a. Amylose b. Inulin
    c. Cellulose d. Dextrins
    e. Glycogen

12. Invert sugar is:
    a. Lactose b. Sucrose
    c. Mannose d. Fructose
    e. Hydrolytic products of sucrose
13. The carbohydrate reserve in human body is:
    a. Starch b. Cellulose
    c. Glucose d. Glycogen
    e. Inulin
14. A disaccharide linked by α 1-4 glycosidic linkage is:
    a. Lactose b. Sucrose
    c. Cellulose d. Maltose
    e. Cellobiose
15. A polymeric unit of starch which has a branched structure is:
    a. Glucose b. Amylopectin
    c. Isomaltose d. Amylose
    e. Cellobiose
16. The repeating unit in hyaluronic acid is:
    a. Glucuronic acid and galactose
    b. Glucuronic acid and galactosamine
    c. Glucuronic acid and glucosamine
    d. Glucuronic acid and N-acetyl glucosamine
    e. Glucuronic acid and N-acetyl galactosamine
17. The repeating disaccharide unit in cellulose is:
    a. Dextrin b. Sucrose
    c. Maltose d. Dextrose
    e. Cellobiose
18. Inversion is related to:
    a. Dextrose b. Fructose
    c. Sucrose d. Maltose
    e. Lactose
19. Seliwanoff's test is answered by:
    a. Glucose b. Fructose
    c. Galactose d. Mannose
    e. Lactose
20. The sugar residues of amylase are:
    a. Fructose units only
    b. In β1→4 linkage
    c. In α1→ 4 linkages
    d. Galactose units only
    e. Both galactose and fructose units
21. Mucic acid is produced by oxidation of:
    a. Glucose b. Galactose
    c. Fructose d. Mannose
    e. Sucrose
22. Spermatozoa in seminal fluid utilises the following sugar for its metabolism:
    a. Galactose b. Glucose
    c. Sucrose d. Fructose
    e. Mannose
23. Match the following:
    A. Fructose and mannose B. Furfural,
    C. Galactosamine D. Sialic acid
    1. Acid catalysed dehydration of pentose _________
    2. A constituent of the heteropolysaccharide of cartilage______
    3. A derivative of neuraminic acid
    4. Mild alkaline treatment of glucose _________

Answers to MCQs

1. d	2. e	3. b	4. c	5. b	6. c
7. d	8. a	9. d	10. c	11 b	12. e
13. d	14. d	15. b	16. d	17. e	18. e
19. b	20. c	21. b	22. d	23. 1-B, 2-C, 3-D, 4-A).	

CHAPTER 4

CHEMISTRY OF LIPIDS

Major Concepts

A. What are lipids, their classification and biomedical importance.
B. List the derived lipids of biological importance and learn their chemistry and properties.
C. List the simple lipids of biological importance and learn their chemistry and properties.
D. List the compound lipids of biological importance and learn their chemistry and properties.

Specific Objectives

A. 1. Define lipids. What are "Bloor's" criteria to call a compound as lipid?
   2. Classify lipids. Learn the three major groups with examples of each group.
   3. Study the biomedical importance of lipids in general.

B. 1. List the derived lipids of biological importance.
   2. • Learn about fatty acid—definition and classification.
      • Study nomenclature of fatty acids and isomerism in unsaturated fatty acids.
      • Learn about essential fatty acids—their chemistry, functions and deficiency manifestations. Learn docosahexaenoic acid, (DHA)
   3. Learn the chemistry, properties and biomedical importance of glycerol.
   4. • Learn about chemistry, properties, occurrence, distribution and biomedical importance of cholesterol.
      • Study other sterols of biological importance.

C. 1. List the simple lipids of biological importance
   2. • Study in detail the simple lipid triacylglycerol (TG) (neutral fat). Learn the chemistry and important physical and chemical properties of TG.
      • Learn how fats and oils can be identified: Saponification number, acid number, Polenske number, Reichert-Meissl number, Iodine number, acetyl number.

D. 1. List the compound lipids of biological importance.
   2. Study in detail about chemistry and functions of phospholipids.
      • Define phospholipid. • Classify the various phospholipids • Learn the chemistry and properties of phospholipids • Study the various types of phospholipases and their site of action and products formed • Study the inherited disorder "Niemann-Pick disease" • Learn the functions of phospholipids and clinical importance of dipalmitoyl lecithin (DPL).
   3. List important glycolipids of biological importance.
      a. Study the chemistry, types and properties of cerebrosides. Learn about the inherited disorder "Gaucher's disease".
      b. Study the chemistry, types and properties of gangliosides. Learn about the inherited disorder "Tay-Sach's disease".
   4. Differentiate in a tabular form cerebrosides and gangliosides.
   5. Study the chemistry of sulfatides and learn about inherited disorders metachromatic leukodystrophy, Fabry's disease and Krabbe's disease.

LIPIDS

What are Lipids?

The lipids constitute a very important heterogenous group of organic substances in plant and animal tissues, and ***related either actually or potentially to the fatty acids.*** Chemically they are various types of esters of different alcohols. **In addition to alcohol and fatty acids, some of the lipids may contain phosphoric acid, nitrogenous base and carbohydrates.**

Bloor's Criteria

According to **Bloor**, lipids are compounds having the following characteristics:

- They are ***insoluble in water.***
- Solubility in one or more organic solvents, such as ether, chloroform, benzene, acetone, etc, so called ***fat solvents.***
- Some relationship to the fatty acids as esters either actual or potential.
- Possibility of utilisation by living organisms.

Thus, lipids include fats, oils, waxes and related compounds. *An oil* is a lipid which is liquid at ordinary temperature. Distinction between fats and oils is a purely physical one. Chemically they are all esters of glycerol with higher fatty acids.

Biomedical Importance

- Lipids are important dietary constituent and acts as ***fuel*** in the body. In some respects lipid is even superior to carbohydrates as a raw material for combustion, since, it yields more energy per gm (9.5 C/gm as compared to carbohydrates 4.0 C/gm)
- Can be stored in the body in almost unlimited amount in contrast to carbohydrates.
- Some deposits of lipids may exert an insulating effect in the body, while lipids around internal organs like kidney, etc. may provide padding and protect the organs.
- ***Building materials:*** Breakdown products of fats can be utilised for building biological active materials.
- Lipids supply so called ***Essential fatty acids (EFA)***, which cannot be synthesised in the body and are essential in the diet for normal health and growth.
- The nervous system is particularly rich in lipids especially certain types and are essential for proper functioning.
- Some vitamins like, A, D, E and K are fat soluble, hence lipid is necessary for these vitamins.
- Lipoproteins and phospholipids are important constituents of many natural membranes such as cell walls and cell organelles like mitochondrion, etc.
- Lipoproteins are also 'carriers' of triglycerides, cholesterol and PL in the body.

CLASSIFICATION OF LIPIDS

Bloor's Classification is generally adopted with a few modifications as follows:

I. Simple Lipids

Esters of fatty acids with various alcohols:

(a) Neutral fats (Triacylglycerol, TG): These are triesters of fatty acids with glycerol.

(b) Waxes are esters of fatty acids with higher monohydroxy aliphatic alcohols.

- *True waxes* are esters of higher fatty acids with cetyl alcohol ($C_{16}H_{33}OH$) or other higher straight chain alcohols.
- *Cholesterol esters* are esters of fatty acid with cholesterol.
- *Vit A and Vit D esters* are palmitic or stearic acids esters of Vit A (Retinol) or Vit D respectively.

II. Compound Lipids

Esters of fatty acids containing groups, other than, and in addition, to an alcohol and fatty acids.

(a) Phospholipids: They are substituted fats containing in addition to fatty acid and glycerol, a **phosphoric acid residue, a nitrogenous base and other substituents.** *Examples: phosphatidyl choline (Lecithin), phosphatidyl ethanolamine (Cephalin), phosphatidyl inositols (Lipositols), phosphatidyl serine, plasmalogens, sphingomyelins,* etc.

(b) Glycolipids: Lipids containing carbohydrate moiety are called glycolipids. ***They contain a special alcohol called sphingosine or sphingol*** and nitrogenous base in addition to fatty acids ***but does not contain phosphoric acid or glycerol***. These are of **two types:**

- *Cerebrosides*
- *Gangliosides*

(c) Sulpholipids: Lipids characterised by possessing sulphate groups.

(d) Aminolipids *(Proteolipids)*

(e) Lipoproteins: Lipids as prosthetic group to proteins.

III. Derived Lipids

Derivatives obtained by hydrolysis of those given in group I and II, which still possess the general characteristics of lipids.

(a) Fatty acids may be saturated, unsaturated or cyclic.

(b) Monoglycerides (Monoacylglycerol) and Diglycerides (Diacylglycerol).

(c) Alcohols

- Straight chain alcohols are water insoluble alcohols of higher molecular weight obtained on hydrolysis of waxes.
- Cholesterol and other steroids including Vit D.
- Alcohols containing the β-ionone ring include Vit A and certain carotenoids.
- Glycerol.

IV. Miscellaneous

- Aliphatic hydrocarbons include isooctadecane found in liver fat and certain hydrocarbons found in bees wax and plant waxes.
- Carotenoids
- Squalene is a hydrocarbon found in shark and mammalian liver and in human sebum.
- Vitamins E and K.

DERIVED LIPIDS

FATTY ACIDS

Definition: A fatty acid (FA) may be defined as an organic acid that occurs in a natural triglyceride and is a monocarboxylic acid ranging in chain length from C4 to about 24 carbon atoms. FA are obtained from hydrolysis of fats.

TYPES OF FATTY ACIDS

Straight chain FA: These may be:

- ***Saturated FA:*** Those which contain **no double bonds.**
- ***Unsaturated FA:*** Those which ***contain one or more double bonds.***

(a) Saturated FA: Their general formula is C_nH_{2n+1} COOH

Examples:

- Acetic acid CH_3COOH
- Propionic acid C_2H_5COOH
- Butyric acid C_3H_7COOH
- Caproic acid $C_5H_{11}COOH$
- Palmitic acid C_{15} $H_{31}COOH$
- Stearic acid C_{17} $H_{35}COOH$ and so on.

Saturated fatty acids having 10 carbon or less number of carbon atoms are called as ***lower fatty acids,*** e.g. acetic acid, butyric acid, etc. Saturated fatty acids having more than 10 carbon atoms are called ***higher fatty acids,*** e.g. palmitic acid, stearic acid, etc. Milk contains significant amount of lower fatty acids.

(b) Unsaturated FA: They are classified further according to degree of unsaturation.

(1) Mono unsaturated (Monoethenoid) fatty acids: They **contain one double bond.**

Their general formula is C_n H_{2n-1} COOH

Example: **Oleic acid** $C_{17}H_{33}$ COOH is found in nearly all fats (formula 18 : 1; 9).

(2) Polyunsaturated (Polyethenoid) fatty acids: There are three polyunsaturated fatty acids of biological importance.

- ***Linoleic acid series (18 : 2; 9, 12):*** It **contains two double bonds** between C9 and C10; and between C12 and C13. Their general formula is C_nH_{2n-3} COOH.
  Dietary sources: Linoleic acid is present in sufficient amounts in peanut oil, corn oil, cottonseed oil, soyabean oil and egg yolk.
- ***Linolenic acid series (18 : 3; 9, 12, 15):*** It **contains three double bonds** between 9 and 10; 12 and 13; and 15 and 16. Their general formula is C_nH_{2n-5} COOH.
  Dietary Source: Found frequently with linoleic acid, but particularly present in linseed oil, rapeseed oil, soybean oil, fish visceras and liver oil (cod liver oil).
- ***Arachidonic acid series (20 : 4; 5, 8, 11, 14):*** It **contains four double bonds.** Their general formula: C_nH_{2n-7} COOH
  Dietary source: Found in small quantities with linoleic acid and linolenic acid but particularly found in peanut oil. Also found in animal fats including Liver fats.

Note: ***These three polyunsaturated fatty acids, viz. linoleic acid, linolenic acid and arachidonic acid are called as "Essential fatty acids (EFA)".*** They have to be ***provided in the diet,*** as they cannot be synthesised in the body.

(c) Branched chain FA: Odd and even carbon branched chain fatty acids occur in animal and plant lipids, e.g.

- ***Sebaceous glands: Sebum*** contain branched chain FA
- ***Branched chain FA*** is present in certain foods, e.g. ***phytanic acid in butter.***

CLINICAL ASPECT

Virulent strains of tubercle bacilli, *Mycobacterium tuberculosis*, show a "cord-like growth" and lipid factor termed ***"cord factor"*** has been isolated from them. It is chemically "trehalose-6-6'-dimycolate" and is present in extremely small quantity on bacterial surface and ***appears to be responsible for virulence of the organism.***

(d) Substituted fatty acids: In hydroxy fatty acid and methyl fatty acid, one or more of the hydrogen atoms have been replaced by – OH group or – CH_3 group respectively. Both saturated and unsaturated hydroxy fatty acids, particularly with long chains, are found in nature, e.g. ***cerebronic acid*** of brain glycolipids, ***Ricinoleic acid*** in castor oil.

(e) Cyclic fatty acids: Fatty acids bearing cyclic groups are present in some seeds, e.g.

- ***Chaulmoogric acid*** obtained from chaulmoogra seeds,
- ***Hydnocarpic acid.***

Biomedical Importance

Both of them have been used earlier for long time **for treatment of leprosy.**

(f) Eicosanoids: These are derived from eicosapolyenoic FA.

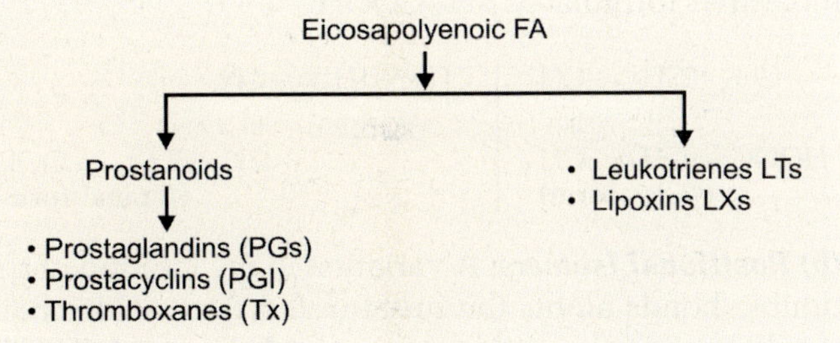

Nomenclature of Fatty Acids

(a) The most frequently used systemic nomenclature is based on naming the fatty acids after the hydrocarbon with the same number of carbon atoms, "oic" being substituted for the final "e" in the name of the hydrocarbon.

- Saturated acids end in ***"anoic"*** e.g. octanoic acid and
- Unsaturated acids with double bonds end in ***enoic,*** e.g. octadecenoic acid (oleic acid).

(b) Carbon atoms are numbered from the – COOH carbon (carbon No. 1). The carbon adjacent to –COOH gr. i.e. carbon number 2 is known as α-carbon, carbon atom 3 is β-carbon and the end $-CH_3$ carbon is known as the ω-carbon ('Omega' carbon).

(c) Various conventions are used for indicating the number and position of the double bonds, e.g. 9 indicates a double bond between carbon atoms 9 and 10 of the fatty acids, for this Δ sign is used, e.g. Δ9 denotes double bond between 9 and 10 C.

(d) A widely used convention is to *express the fatty acids by formula* to indicate:

- **The number of carbon atoms**
- **The number of double bonds** and
- **The positions of the double bonds.**

For example oleic Acid (mol. formula $C_{17}H_{33}$ COOH) has one double bond between C_9 and C_{10}, thus:

$$\overset{10}{} \quad \overset{9}{}$$
$$CH_3\ (CH_2)_7 -CH = CH\ (CH_2)_7 - COOH$$

According to above criteria, it is **expressed as 18 : 1; 9**, [18 indicates the number of carbon atoms, 1 indicates the number of double bond and 9 indicates the position of the double bond].

Isomerism

Two types of isomers can occur in an unsaturated fatty acid.

(a) Geometric Isomers: They depend on the orientation of the radicals around the axis of the double bonds. If the radicals are on the same side of the bond, it is called as ***'cis' form.*** If the radicals are on the opposite side, a ***'trans' form*** is produced. 'Cis' form is comparatively unstable and is more reactive.

Example:* *Oleic acid and elaidic acid both have same molecular formula $-C_{17}H_{33}COOH$

$CH_3 -(CH_2)_7 -CH$	$CH_3- (CH_2)_7 - CH$
‖	‖
$HOOC - (CH_2)_7 -CH$	$CH-(CH_2)_7 -COOH$
("Cis" form)	**('Trans' form)**

(b) Positional Isomers: A variation in the location of the double bonds along the unsaturated fatty acids chain produces isomer of that compound. Thus, **oleic acid could have 15 different positional isomers.**

ESSENTIAL FATTY ACIDS

Three polyunsaturated fatty acids, linoleic acid, linolenic acid and arachidonic acid are called "essential fatty acids" (EFA). They cannot be synthesised in the body and must be provided in the diet. Lack of EFA in the diet can produce growth retardation and other deficiency manifestation symptoms.

Which EFA is Important?

Linoleic acid is most important as, arachidonic acid can be synthesised from linoleic acid by ***a three stage reaction*** by addition of acetyl-CoA. Pyridoxal phosphate is necessary for this conversion. ***Biologically arachidonic acid is very important as it is precursor from which prostaglandins and leukotrienes are synthesised in the body.***

Why EFA cannot be Synthesised?

Introduction of additional double bonds in unsaturated fatty acid is limited to the area between – COOH group and the existing double bond and that it is not possible to introduce a double bond between the – CH_3 group at the opposite end of the molecule and the first unsaturated linkage. This would explain body's inability to synthesise an EFA from oleic acid.

Functions of EFA: (Biomedical Importance)

- ***Structural elements of tissues:*** Polyunsaturated fatty acids occur in higher concentration in lipids associated with structural elements of tissues.
- ***Structural element of gonads:*** Lipids of gonads also contain a high concentration of polyunsaturated fatty acids, which suggests importance of these compounds in reproductive function.
- ***Synthesis of prostaglandins and other compounds:*** Prostaglandins are synthesised from Arachidonic acid by **cyclooxygenase enzyme system**. Leucotrienes are conjugated trienes formed from arachidonic acid in leucocytes by the **Lipoxygenase pathway.**
- ***Structural element of mitochondrial membrane:*** A deficiency of EFA causes swelling of mitochondrial membrane and reduction in efficiency of oxidative phosphorylation. This may explain for increased heat production noted in EFA deficient animals.
- ***Serum level of cholesterol:*** Fats with high content of polyunsaturated fatty acids tends to lower serum level of cholesterol.
- ***Effect on clotting time:*** Prolongation of clotting time is noted in ingestion of fats rich in EFA.
- ***Effect on fibrinolytic activity:*** An increase in fibrinolytic activity follows the ingestion of fats rich in EFA.
- ***Role of EFA in fatty liver:*** Deficiency of EFA produces fatty liver.
- ***Role in vision:* Docosahexaenoic acid (22:6n-3)** is the most abundant polyenoic fatty acids **present in retinal photoreceptor membranes.** Docosahexaenoic acid is formed from dietary linolenic acid. It enhances the electrical response of the photoreceptors to illumination. ***Hence linolenic acid is necessary in the diet for optimal vision.***

Deficiency manifestations: A deficiency of EFA has not yet been unequivocally demonstrated in humans. In weaning animals, symptoms of EFA deficiency are readily produced. They are:

- Cessation of growth.
- **Skin lesions:** Acanthosis (hypertrophy of prickle cells) and hyperkeratosis (hypertrophy of stratum corneum). Skin becomes abnormally permeable to water. Increased loss of water increases BMR.
- Abnormalities of pregnancy and lactation in adult females.
- **Fatty liver** accompanied by increased rates of fatty acids synthesis, lessened resistance to stress.
- Kidney damage.

CLINICAL ASPECT

Human deficiency: Some cases of
- Eczema like dermatitis,
- Degenerative changes in arterial wall and
- Fatty liver in man may be due to EFA deficiency.
  There are also some reports that administration of EFA in such cases may produce:
  - Some improvement of eczema in children kept on skimmed milk,
  - Prevent fatty liver (some cases) and
  - Lowering of cholesterol levels.
  Infants and babies with low fat diet develop typical skin lesions which has shown to be improved with EFA (linoleic acid).

Fate of EFA: EFA undergoes β-oxidation after necessary isomerisation and epimerisation, like other unsaturated fatty acids.

CLINICAL ASPECT
Abnormal Metabolism of EFA

Abnormal metabolism of EFA, which may be concerned with dietary insufficiency, has been noted in a number of disease like cystic fibrosis, hepatorenal syndrome, Crohn's disease, acrodermatitis enteropathica, Sjögren's syndrome, Cirrhosis and Reye's syndrome.

Docosahexaenoic acid: DHA (W3, 22:6)

Docosahexaenoic acid (DHA) is a ***polyunsaturated fatty acid*** which is synthesized from α-linolenic or obtained directly from dietary fish oil. This fatty acid is ***present in high concentrations in retina, cerebral cortex, tastes and sperms.***

DHA is particularly needed for development of the brain and retina and is supplied via the placenta and milk. In EFA deficiency, nonessential polyenoic acids of the W9 family replace the essential fatty acids in phospholipids (PL), other complex lipids and membranes.

Clinical Aspect

Low blood levels of DHA have been ***reported*** in patients with ***Retinitis Pigmentosa.***

Triene: Tetraene Ratio

Triene: tetraene ratio in plasma lipids can be used to diagnose the extent of essential fatty acid deficiency.

ALCOHOLS

Alcohols contained in the lipid molecule includes glycerol, cholesterol and the higher alcohols, e.g. cetylalcohol, $C_{16}H_{33}COOH$ (usually found in waxes).

1. Glycerol: Glycerol is commonly called as ***"glycerin"*** it is the simplest trihydric alcohol as it contains three hydroxyl groups in the molecule (Fig. 4.1).

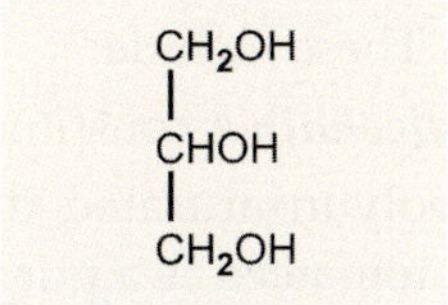

Fig. 4.1: Glycerol

- It is colourless oily fluid with a sweetish taste.
- It is miscible with water and alcohol in all proportions but is almost insoluble in ether.

Source

Industrial
- It is obtained as a by-product of soap manufacture.
- It is also obtainable in the fermentation of glucose by changing conditions in such a way as to decrease the formation of CO_2 and alcohol.

Physiological
- ***Endogenous source:*** Main source is from ***lipolysis*** of fats in adipose tissue.
- ***Exogenous source: Dietary*** Approx. 22 per cent of glycerol directly absorbed to portal blood from the gut.

Test

Acrolein test: The presence of glycerol is detected by acrolein test. Glycerol when dehydrated with heat and $KHSO_4$ produces ***"acryl aldehyde"*** which has pungent or acrid odour (Fig. 4.2).

Uses of Glycerol

(a) ***Industrial:*** Glycerol finds many uses in industry, as a result of its solubility, its solvent action and its

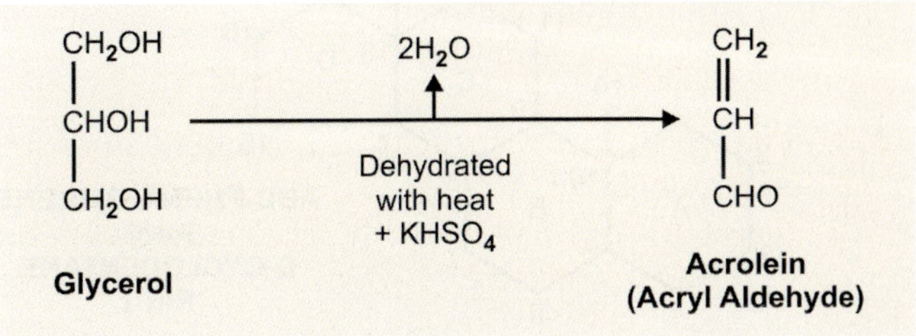

Fig. 4.2: Acrolein test

hygroscopic nature. Many pharmaceuticals and cosmetic preparations have glycerol in their formulas.

(b) In medicine: Nitroglycerine is used as a vasodilator.

- ***Glycerol therapy*** in cerebrovascular (CV) diseases reduces cerebral oedema.

(c) Physiological: In body, glycerol has ***a definite nutritive value.*** It can be converted to glucose/and glycogen, the process called as gluconeogenesis.

Unsaturated Alcohols

Among the unsaturated alcohols found in fats, many of them are pigments. These include:

(a) Phytol (Phytyl alcohol): A constituent of chlorophyll.

(b) Lycophyll: A polyunsaturated dihydroxy alcohol which occurs in tomatoes as a purple pigment.

(c) Carotene: Easily split in the body at the central point of the chain to give two molecules of alcohol, vitamin A.

(d) Sphingosine or sphingol: An ***unsaturated amino alcohol*** present in body as a constituent of phospholipid, sphingomyelin and various glycolipids.

Steroids and Sterols

1. The steroids are often found in association with fat.
2. They may be separated from the fat, after the fat is saponified, since they ***occur in unsaponifiable residue.***
3. All of the steroids have a similar cyclic nucleus resembling ***phenanthrene (ring A, B and C)*** to which a ***cyclopentane ring (ring D)*** is attached. It is designated as **cyclopentano perhydro-phenanthrene nucleus** (Fig. 4.3).
4. Methyl side chains occur typically at positions 10 and 13 (constituting carbon atoms 19 and 18 respectively).
5. A side chain at position 17 is usual. ***If the compound has one or more –OH groups and no carbonyl or carboxyl groups, it is called a 'sterol' and the name terminates in –"ol". Most important sterol in human body is cholesterol.***

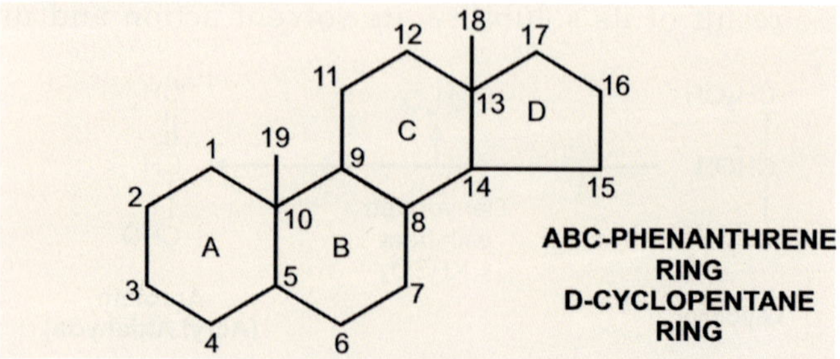

Fig. 4.3: Cyclopentanoperhydrophenanthrene nucleus

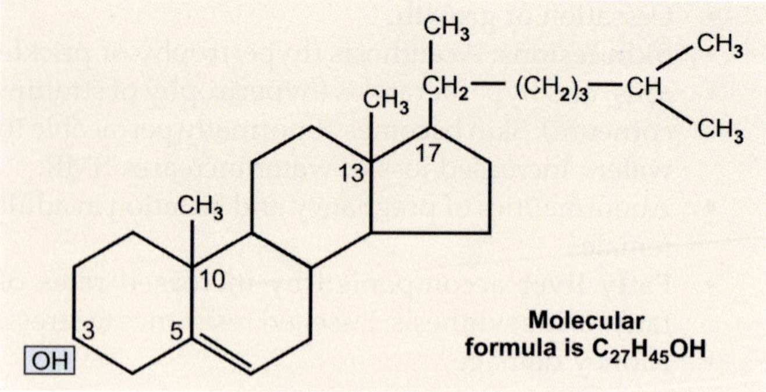

Fig. 4.4: Cholesterol

CHOLESTEROL

Structure: Cholesterol is the most important sterol in human body. Its molecular formula is **$C_{27}H_{45}OH$.** Its structural formula is given in Figure 4.4.

- **It possesses "cyclopentanoperhydrophenanthrene nucleus".**
- It has an –OH group at C_3.
- It has an unsaturated double bond between C_5 and C_6.
- It has two $-CH_3$ groups at C_{10} and C_{13}.
- It has an eight carbon side chain attached to C_{17}.

Properties: The name cholesterol is derived from the Greek word meaning **solid bile**. It occurs as a white or faintly yellow, almost odourless, pearly leaflets or granules. It is insoluble in water, sparingly soluble in alcohol and soluble in ether, chloroform, hot alcohol, ethyl acetate and vegetable oils. It easily crystallises from such solutions in colourless, rhombic plates with one or more characteristic ***notches in the corner*** (Fig. 4.5). It is not saponifiable. Its melting point is 147 to 150° C. Since it has an unsaturated bond, it can take up two halogen atoms.

Source

- *Exogenous:* Dietary cholesterol, approximately 0.3 gm/day. Diet rich in cholesterol are butter, cream, milk, egg yolk, meat, etc. ***A hen's egg weighing 2 oz gives 250 mg cholesterol***
- ***Endogenous: Synthesised in the body from acetyl CoA,*** approximately 1.0 gm/day.

Occurrence: It is widely present in body tissues. Cholesterol is found in largest amounts in normal human adults

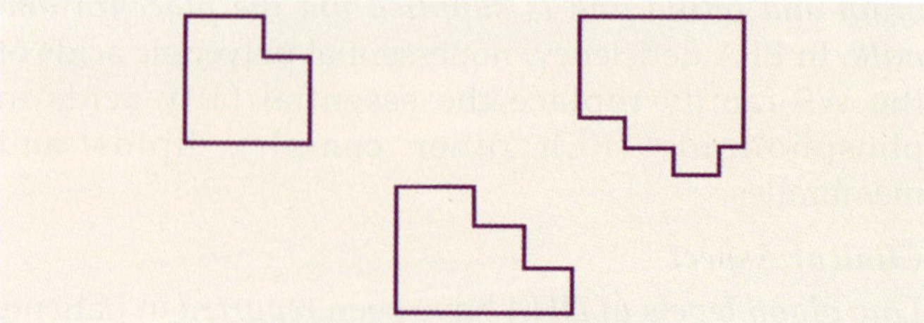

Fig. 4.5: Cholesterol crystals

brain and nervous tissue 2 per cent, in the liver about 0.3 per cent, skin 0.3 per cent and intestinal mucosa 0.2 per cent, certain endocrine glands, viz. ***adrenal cortex contain some 10 per cent or more,*** corpus luteum is also rich in cholesterol. The relatively high content of cholesterol in skin may be related to vit D formation by UV rays and that in the adrenal gland and gonads to steroid hormone synthesis. Cholesterol is present in blood and bile and is usually a major constituent of gallstones.

Forms of cholesterol: Cholesterol occurs both in ***free form*** and in ***ester form,*** in which it is esterified with fatty acids at –OH group at C_3 position. The ester form of cholesterol is also referred as ***bound form.*** The various fatty acids, which form cholesterol esters, are as follows:

- Linoleic acid 50%
- Oleic acid 18%
- Palmitic acid 11%
- Arachidonic acid 5%
- And other fatty acids 16%

Free cholesterol is equally distributed between plasma and red blood cells, but the latter do not contain esters. ***In brain and nervous tissue, free form predominates, whereas in adrenal cortex it occurs mainly as esterified form.***

Esterification of cholesterol: Esterification occurs as follows:

- Some cholesterol esters are formed in tissues by the transfer of acyl groups from acyl-CoA to cholesterol by ***acyl transferases.***
- But most of the plasma cholesterol esters are produced in the plasma itself by the transfer of an acyl group (mostly unsaturated acyl group) from the β-position of lecithin to cholesterol with the help of the enzyme ***lecithin cholesterol acyl transferase*** **(LCAT).**

Lecithin + Cholesterol —(LCAT)→ Lysolecithin + Cholesterol ester

Biomedical Importance of Cholesterol

1. Norum's disease: A genetic deficiency of LCAT produces Norum's disease due to the failure of esterification of cholesterol at the cost of lecithin. The disease is characterised by:

- Rise in free cholesterol ↑
- Rise in lecithin in plasma ↑ and
- Fall in cholesterol ester, ↓ lysolecithin ↓ and α-lipoproteins ↓ in plasma.

2. Normal level and physiological variations: Normal level of serum total cholesterol in an adult varies from 150 to 250 mg per cent. About 40 to 50 mg% occurs as ***free*** cholesterol (approx 30% of total) and 110-200 mg% as cholesterol ***esters*** (approximately 70%).

3. Variations of cholesterol level

- ***Age:*** Blood cholesterol level is low at birth (50-70% of the values in normal adults) and it gradually increases with age. After 55 years there is a tendency to decrease again.
- ***Sex and race:*** Sex and race have little effect, but in case of women, the level is increased just before and decreased during the menstrual period.
- ***Pregnancy:*** The level is also increased during pregnancy, when a progressive rise in free cholesterol and a fall in 'ester' fraction is observed.

4. Pathological variations: Refer to chapter on "Cholesterol metabolism".

Colour Reactions of Sterols

(a) Liebermann-Burchard reaction: A chloroform solution of a sterol, when treated with acetic anhydride and conc. H_2SO_4 gives ***a grass-green colour.*** The usefulness of this reaction is limited by the fact that various sterols give the same or similar colour. This reaction forms the basis for a colorimetric estimation of cholesterol by Sackett's method (Basis of the reaction—see above).

(b) Salkowski test: When a chloroform solution of the sterol is treated with an equal volume of conc. H_2SO_4 develops a **red to purple colour**. The heavier acid, which forms a layer below assumes a yellowish colour with a green fluorescence, whereas the upper chloroform layer becomes bluish red first, and gradually turns violet-red.

(c) Zak's reaction: When glacial acetic acid, ***(aldehyde free),*** solution of cholesterol is treated with ferric chloride and conc. H_2SO_4, ***produces a red colour.*** This reaction forms a basis for the colorimetric estimation of cholesterol (Zak's method).

*Basis of Liebermann-Burchard Reaction

Cholesterol —(conc H_2SO_4)→ 3,5–Cholestadiene

H_2O ↓ (Loses a molecule of water at C_3 and gets oxidised)

3,5–Cholestadiene
↓
Converted to polyenes (acts as chromophore)
+
Sulphonic acid
↓
[Cholesta-Polyene-sulphonic acid] (gives green colour)

OTHER STEROLS OF BIOLOGICAL IMPORTANCE

1. 7-Dehydrocholesterol

It is an important sterol present in the skin. This differs from cholesterol only in having a second double bond, between C_7 and C_8 (Fig. 4.6).

Source: In man, 7-dehydrocholesterol may be obtained partly by synthesis from cholesterol in skin and/or intestinal wall.

Biomedical Importance

In the epidermis of skin, UV rays of sunshine change 7-dehydrocholesterol (pre-cholecalciferol) to cholecalciferol (vitamin D_3). ***Hence 7-dehydrocholesterol is called as provitamin-D_3***. This explains the value of sunshine in preventing rickets, a disease produced from vitamin D deficiency.

2. Ergosterol

It is a **plant sterol**, first isolated from ergot, a fungus of rye and later from yeast and certain mushrooms. Structurally this sterol has the same nucleus as 7-dehydrocholesterol but differs slightly in its side chain (Fig. 4.7).

Biomedical Importance

When irradiated with UV rays (long wave 265 µm) ergosterol is changed to vitamin D_2 by the opening of the ring B of the sterol. ***Hence ergosterol is called as Provitamin-D_2***, over-irradiation may produce toxic products (Fig. 4.8).

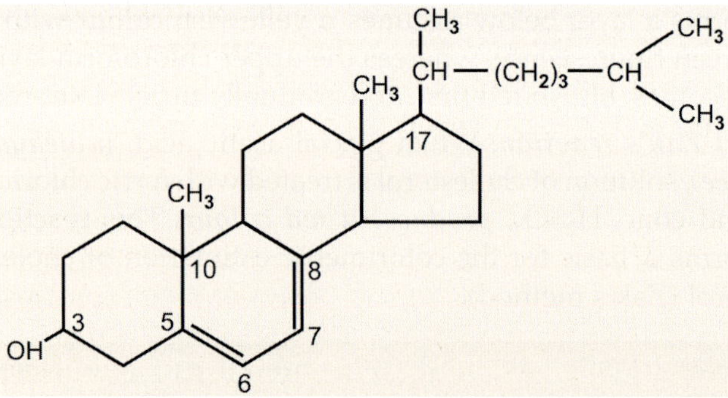

Fig. 4.6: 7-Dehydrocholesterol (Provitamin D_3)

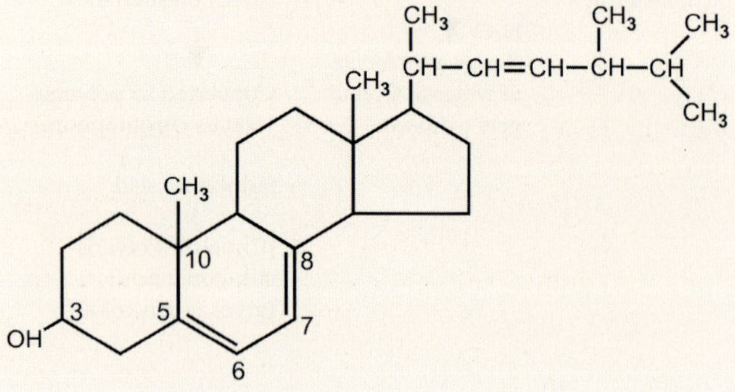

Fig. 4.7: Ergosterol (Provitamin-D_2)

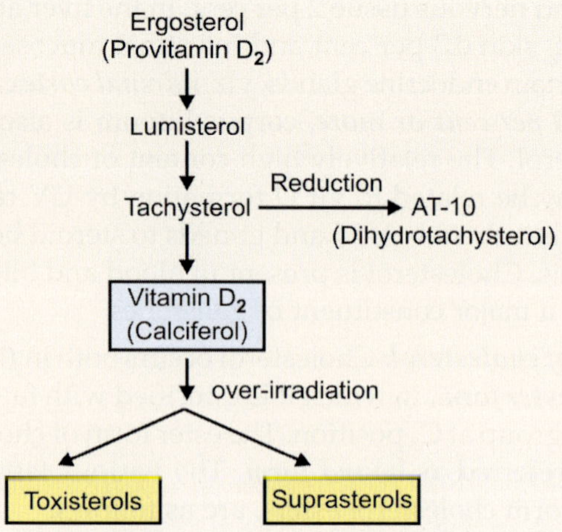

Fig. 4.8: Irradiation of ergosterol

3. Stigmasterol and Sitosterol

They are plant sterols, occurring in higher plants. They have ***no nutritional value for human beings***.

Biomedical Importance

Sitosterol appears to decrease the intestinal absorption of both exogenous and endogenous cholesterol, thus lowering the blood cholesterol level.

4. Coprosterol (Coprostanol)

Occurs in faeces as a result of the reduction of cholesterol (by hydrogenation of double bond). This is brought about by bacterial action, double bond between C_5 and C_6 is saturated. Rings A and B (between C atoms 5 and 10) is "cis" (cf. cholesterol, it is "trans").

5. Other important steroids of biomedical importance

These include the bile acids, adrenocortical hormones, gonadal hormones, D vitamins, cardiac glycosides and some alkaloids. In ***cardiac glycosides, sterols form an important part, i.e. "aglycone" of the structure***.

SIMPLE LIPIDS

NEUTRAL FATS (TRIGLYCERIDES OR TRIACYLGLYCEROL)

Neutral fats (TG) are all triesters of the trihydric alcohol, glycerol with various fatty acids. The type formula for a neutral fat (TG) is given in (Fig. 4.9), in which R_1 R_2 R_3 represent fatty acid chains which may or may not all be the same. Naturally occurring fats have apparently the D-structural configuration.

Physical Properties

1. Neutral fats are colourless, odourless and tasteless substances. The colour and taste of some of the naturally occurring fats is due to extraneous substances.

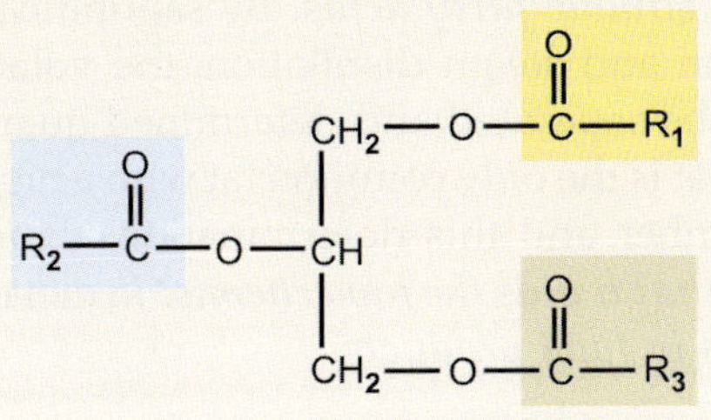

Fig. 4.9: Type formula for neutral fat

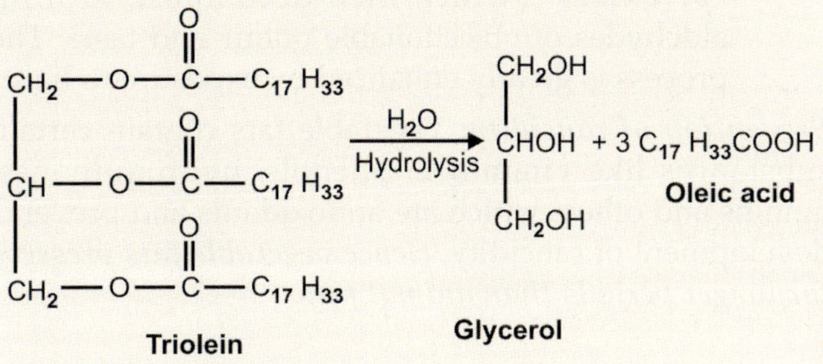

Fig. 4.10: Hydrolysis of neutral fat (NF)

2. ***Solubility:*** They are insoluble in water but soluble in organic fat solvents.

3. ***Specific gravity:*** The specific gravity of all fats is less than 1.0, consequently all fats float in water.

4. ***Emulsification:*** Emulsions of fat may be made by shaking vigorously in water and by emulsifying agents such as gums, soaps and proteins which produce more stable emulsions.

Biomedical Importance of Emulsification

The emulsification of dietary fats in intestinal canal, brought about by bile salts, is a prerequisite for digestion and absorption of fats.

5. ***Melting point and consistency:*** The hardness or consistency of fats is related to their MP glycerides of lower FA melt at lower temperature than those of the higher fatty acids, and the unsaturated fatty acids glycerides at still lower temperature.

Chemical Properties

1. Hydrolysis

The fats may be hydrolysed with:

- **Super heated steam,**
- **By acids, or alkalies,** or
- **By the specific fat spliting enzymes** ***lipases*** **(Fig. 4.10).**

LIPASES

Lipases are enzymes which hydrolyse a triglyceride yielding fatty acids and glycerol.

Sites: Lipases are found in human body in following places.

(a) ***Lingual lipase*** in saliva, **(b)** ***gastric lipase*** in gastric juice, **(c)** ***pancreatic lipase*** in pancreatic juice, **(d)** ***intestinal lipase*** in intestinal epithelial cell, **(e)** ***adipolytic lipase*** in adipose tissue, and **(f)** ***serum lipase.***

Pancreatic lipase is peculiar in that it can hydrolyse ester bonds in positions 1 and 3 preferentially, than in position 2 of TG molecule.

Saponification: **Hydrolysis of a fat by an alkali** is called saponification. The resultant products are glycerol and the alkali salts of the fatty acids, which are called **"soaps".**

$$\text{Triolein} + 3\text{NaOH} \xrightarrow{H_2O} \text{Glycerol} + 3C_{17}H_{33}COO\ Na$$

(Sodium oleate) (soap)

Soaps are cleansing agents because of their emulsifying action. Some soaps of high molecular weight and a considerable degree of unsaturation are selective germicides, others such as Na-ricinoleate, have detoxifying activity against diphtheria and tetanus toxins.

2. Additive Reactions

The unsaturated fatty acids present in neutral fat exhibits all the additive reactions, i.e. ***hydrogenation, halogenation,*** etc. Oils which are liquid at ordinary room temperature, on hydrogenation become solidified. This is the basis of vanaspati (Dalda) manufacture, where inedible and cheap oils like cotton seed oil are hydrogenated and converted to edible solid fat.

3. Oxidation

Fats very rich in unsaturated fatty acids such as linseed oil undergo spontaneous oxidation at the double bond forming aldehydes, ketones and resins which **form transparent coating** on the surfaces to which the oil is applied. These are called **drying oils** and are **used in the manufacture of paints and varnishes.**

4. Rancidity

The unpleasant odour and taste developed by most natural fats on aging is referred to as rancidity.

Cause of rancidity: Rancidity may be caused by the following:

- Hydrolysis of fat yields free fatty acids and glycerol and/or mono and diglycerides. Process is enhanced by presence of lipolytic enzymes *lipases*, which in the presence of moisture and warm temperature bring about hydrolysis rapidly.
- By various oxidative processes, oxidation of double bonds of unsaturated glycerides may form

"peroxides", which then decompose to form aldehydes of objectionable odour and taste. The process is greatly enhanced by exposure to light.

Prevention of rancidity: Vegetable fats contain certain substances like vitamin E, phenols, hydroquinones, tannins and others which are antioxidants and prevents development of rancidity. ***Hence vegetable fats preserve for longer periods than animal fats.***

IDENTIFICATION OF FATS AND OILS

Sometimes it becomes necessary to

- Identify a pure fat.
- Assess the degree of adulteration.
- ***Determine the proportions of different types of fat in a mixture.***

Besides the characteristic melting point and congealing point several other chemical values ("chemical constants") have been used.

1. Saponification Number

Definition: The number of mgms of KOH required to saponify the free and combined FA in one gram of a given fat is called its saponification number.

Basis: The amount of alkali needed to saponify a given quantity of fat will depend upon the number of –COOH group present. ***Thus fats containing short chain fatty acids will have more –COOH groups per gram than long-chain fatty acids and this will take up more alkali and hence will have higher saponification number.***

Examples: **Butter** containing a larger proportion of short-chain fatty acids, such as butyric and caproic acids, has relatively high saponification number from 220 to 230. **Oleomargarine**, with more long chain fatty acids, has a saponification number of 195 or less.

2. Acid Number

Definition: Number of mgms of KOH required to neutralize the fatty acids in a gm of fat is known as the acid number.

Significance: The acid number indicates the degree of rancidity of the given fat.

3. Polenske Number

Definition: It is the number of millilitre of 0.1 normal KOH required to neutralise the insoluble fatty acids (those not volatile with steam distillation) from 5 gram of fat.

4. Reichert-Meissl Number

Definition: It is the number of millilitres of 0.1 (N) alkali required to neutralise the soluble volatile fatty acids distilled from 5 gm of fat.

Significance: The Reichert-Meissl ***measures the amount of volatile soluble fatty acids.*** By saponification of fat, acidification and steam distillation, the volatile soluble acids may be separated and determined quantitatively.

Butter fat is the only common fat with a high Reichert-Meissl number and this determination, therefore, is of interest in ***that it aids the food chemist in detecting butter substitutes in food products.***

5. Iodine Number

Definition: Iodine number is defined as the number of grams of iodine absorbed by 100 gm of fat.

Significance and use: Iodine number is ***a measure of the degree of unsaturation of a fat***. The more the iodine number, the greater the degree of unsaturation. The determination of iodine number ***is useful to the chemist in determining the quality of an oil or its freedom from adulteration.***

Example: Iodine number of **cotton seed** oil varies from 103 to 111, that of **olive oil** from 79 to 88, and that of **linseed oil** from 175 to 202. ***A commercial lot of olive oil which has an iodine number higher than 88 might have been adulterated with cotton seed oil.*** Again a batch of linseed oil with iodine number lower than 175 might also have been adulterated with the cotton seed oil.

6. Acetyl Number

Some of the fatty acid residues in fats contain –OH groups. In order to determine the proportion of these, they are acetylated by means of acetic anhydride. Thus an acetyl group is introduced wherever a free –OH group is present. After washing out the excess acetic anhydride and acetic acid liberated, the acetylated fat can be dried and weighed and the acetic acid in combination determined by titration with standard alkali after it has been set free. ***The acetyl number is thus a measure of the number of –OH group present.***

Definition: The number of mgms of KOH required to neutralise the acetic acid obtained by saponification of 1 gm of fat after it has been acetylated.

Examples: **Castor oil** because of its high content of ricinoleic acid has a high acetyl number.

Acetyl numbers of some oils:

• Castor oil	146-150
• Cod liver oil	1.1
• Cotton seed oil	21-25
• Olive oil	10.5, etc.

It can be **used to detect adulteration.**

COMPOUND LIPIDS

PHOSPHOLIPIDS: CHEMISTRY AND FUNCTIONS

Definition: Phospholipids are compound lipids, they contain in addition to fatty acids and glycerol/or other alcohol, a phosphoric acid residue, nitrogen containing base and other substituents.

Classification: Classification given by **Celmer** and **Carter** is used, which is ***based on the type of alcohol*** present in the phospholipids. Thus they are classified mainly into following **three groups:**

Classification of Phospholipids

- ***Glycerophosphatides*: *In this glycerol is the alcohol group. Examples:*** Phosphatidylethanolamine (cephalin), phosphatidylcholine (Lecithin), phospatidylserine, plasmalogens, phosphatidic acid, cardiolipins and phosphatides.
- ***Phosphoinositides:*** In this group, ***inositol is the alcohol,*** e.g. phosphatidyl inositol (lipositol).
- ***Phosphosphingosides:*** *Alcohol present is sphingosine* (also called as sphingol), an unsaturated amino alcohol, e.g. sphingomyelin.

Following are the chemical and other properties of phospholipids in reference to phosphatidyl choline (lecithin).

PHOSPHATIDYL CHOLINE (LECITHIN)

It is widely distributed in animals in liver, brain, nerve tissues, sperm and egg-yolk, having both metabolic and structural functions. In plants, particularly abundant in seeds and sprouts. Lecithin has been prepared synthetically also (Fig. 4.11). **On hydrolysis,** lecithin yields: **(a) *glycerol,* (b) *fatty acids,* (c) *phosphoric acid and* (d) *nitrogenous base choline.*** Depending on the position of phosphoric acid-choline complex, on α or β carbon, **α-lecithin** and **β-lecithin** can form.

Properties

1. Physical properties: When purified it is a waxy, white substance, becomes brown when exposed to air **(autooxidation)**, is hygroscopic, mixes well with water to form cloudy, colloidal solution, and is soluble in ordinary fat solvents ***except acetone.***

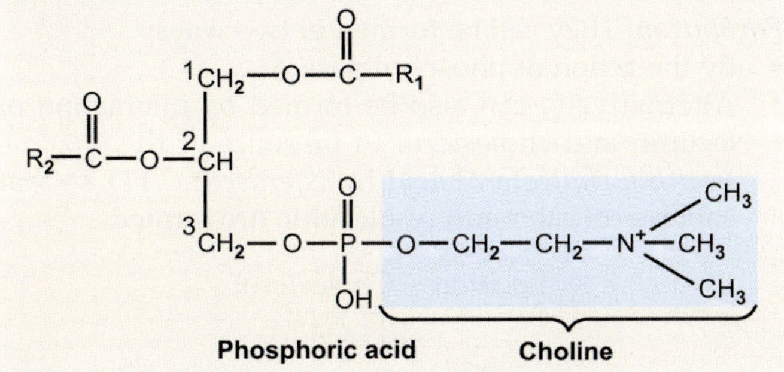

Fig. 4.11: Lecithin (Phosphatidylcholine)

2. Chemical properties

- When aqueous solution of lecithins are shaken with H_2SO_4, choline is split off, forming ***phosphatidic acid.***
- When lecithins are boiled with alkalies or mineral acids, not only choline is split off, phosphatidic acid is further hydrolysed to glycerophosphoric acid and two molecules of fatty acids.

$$\text{Lecithin} \xrightarrow{H_2O} \text{Phosphatidic acid + choline}$$

$$\text{Phosphatidic acid} \xrightarrow{H_2O} \text{Glycerophosphoric acid + 2 fatty acids}$$

PHOSPHOLIPASES (DAWSON)

Dawson found several types of phospholipases. They hydrolyse phospholipids (lecithin) in a characteristic way.

Types of Phospholipases

- *Phospholipase 'A' (A_1):* This is found in human and other mammalian tissues. The enzyme produces degradation of lecithin. Also isolated from cobra venom.
- *Phospholipase 'B':* This is found in association with 'A' and found in mammalian tissues. Also it is isolated from certain fungus like *Aspergillus sp.* and *penicillium notatum.*
- *Phospholipase 'C':* This is found in plant kingdom mainly. *Clostridium welchii* produces phospholipase 'C' and has been isolated from certain venoms. Recently isolated from mammalian brain.
- *Phospholipase 'D':* This is found only in plants, e.g. cabbage, cotton seed, carrots, etc. Not important for human body.
- *Phospholipase, A_2:* This has recently been described. ***Responsible for degradation of PL in humans.***

Specific Site of Action of Phospholipases (Fig. 4.12)

- *Phospholipase A_2:* Attacks 'β' position and form Lysolecithin + one mol. fatty acid.
- *Phospholipase A_1:* Attacks ester bond in position 1 of phospholipid.
- *Phospholipase B:* (also called ***lysophospholipase***) Attacks lysolecithin and hydrolyses ester bond in α position and forms glyceryl phosphoryl choline + 1 mol. fatty acid.
- *Phospholipase C:* Hydrolyses phosphate ester bond and produces α, β-diacyl glycerol + phosphoryl choline.
- *Phospholipase D:* Splits off choline and phosphatidic acid is formed.

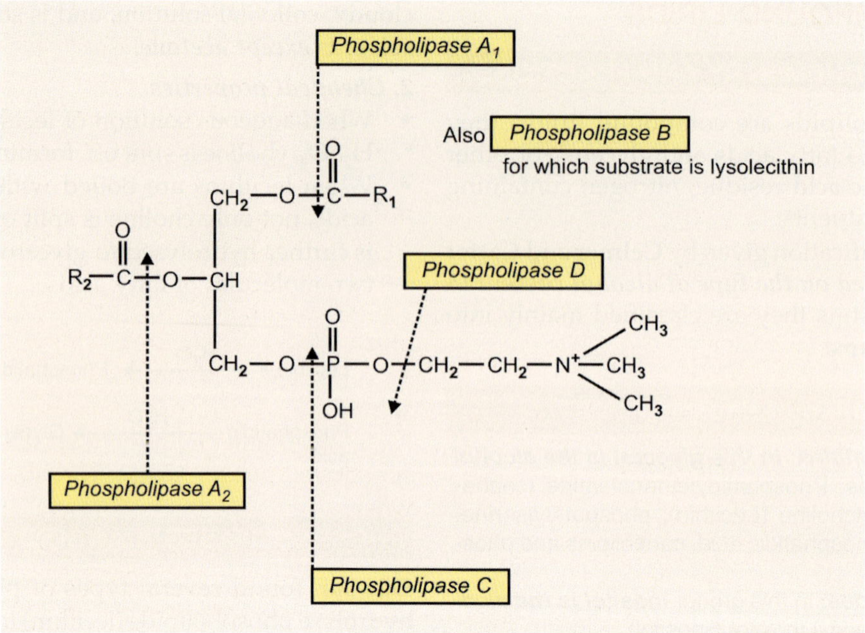

Fig. 4.12: Site of action of phospholipases

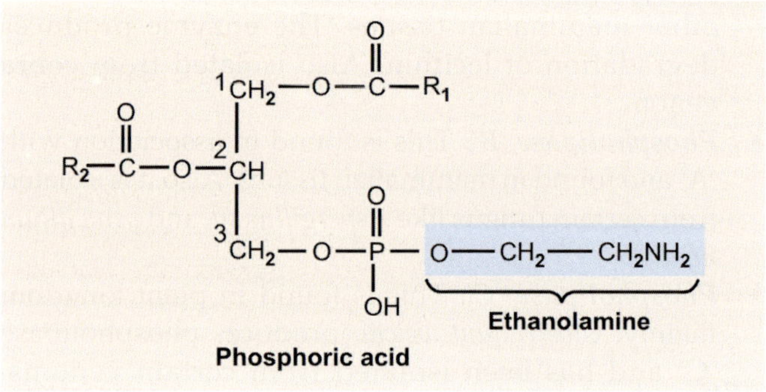

Fig. 4.13: Cephalin

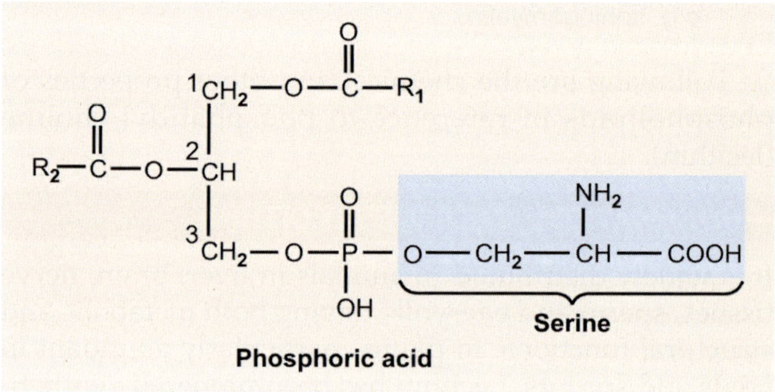

Fig. 4.14: Phosphatidylserine

OTHER PHOSPHOLIPIDS OF BIOLOGICAL IMPORTANCE

1. Phosphatidyl Ethanolamine (Cephalins) (Fig. 4.13)

Cephalins are structurally identical with Lecithins, with the exception that the ***base ethanolamine replaces choline.*** Both α and β cephalins are known. They occur with lecithin, **particularly rich in brain and nervous tissues.**

2. Phosphatidyl Inositol (Lipositols)

Inositol is an alcohol, a cyclic compound hexa hydroxy cyclohexane with molecular formula $C_6H_{12}O_6$. ***It replaces the base choline of lecithin.*** Inositol as a constituent of phospholipids was first discovered in acid fast bacilli. Later, it was **found to occur in brain and nervous tissues,** moderately in soybeans, and also occurs in plant phospholipids.

3. Phosphatidyl Serine

A cephalin like phospholipid (Fig. 4.14) ***contains amino acid serine in place of ethanolamine*** **found in brain and nervous tissues** and small amount in other tissues. Also found in blood.

4. Lysophosphatides

These are phosphoglycerides containing only one acyl radical in α position, e.g. lysolecithin.

Formation: They can be formed in two ways:

- By the action of phospholipase A_2.
- Alternatively, can also be formed by interaction of lecithin and cholesterol in presence of the enzyme ***Lecithin cholesterol acyl transferase (LCAT),*** so that cholesterol ester and lysolecithin are formed.

Lecithin + Cholesterol
↓ LCAT
Lysolecithin + Cholesterol ester

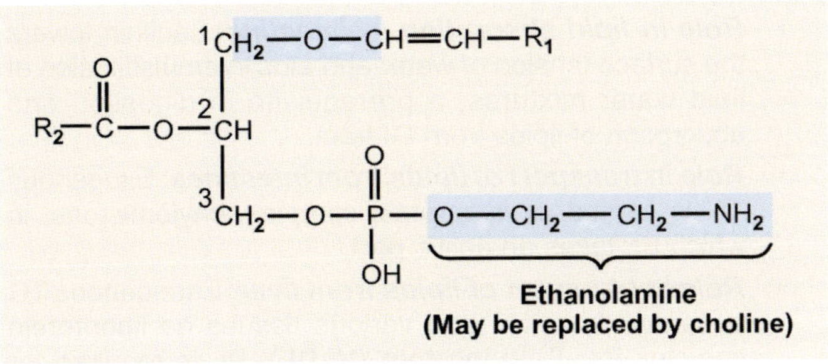

Fig. 4.15: Plasmalogens

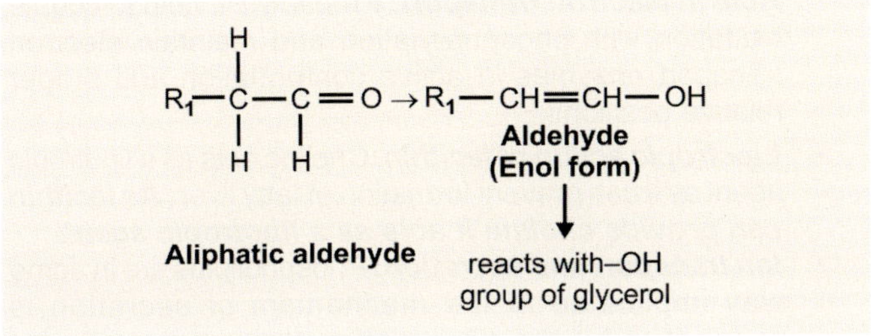

Fig. 4.16: Condensation of enol form of aldehyde with glycerol

5. Plasmalogens (Fig. 4.15)

The plasmalogens make up an appreciable amount, about 10 per cent of total phospholipids of **brain and nervous tissue, muscle and mitochondria.** These compounds yield *on hydrolysis* (a) one molecule each of *long chain aliphatic aldehyde,* (b) a fatty acid, (c) glycerol $-PO_4$, and (d) a nitrogenous base which is usually ethanolamine, but may be sometimes choline.

When hydrolysed, the unsaturated ether group at C_1 position yields a saturated aldehyde such as palmitic or stearic aldehyde, and thus **gives a +ve reaction when tested for aldehydes with Schiff's reagent** (fuchsin-sulphurous acid), after pretreatment of the phospholipid with mercuric chloride.

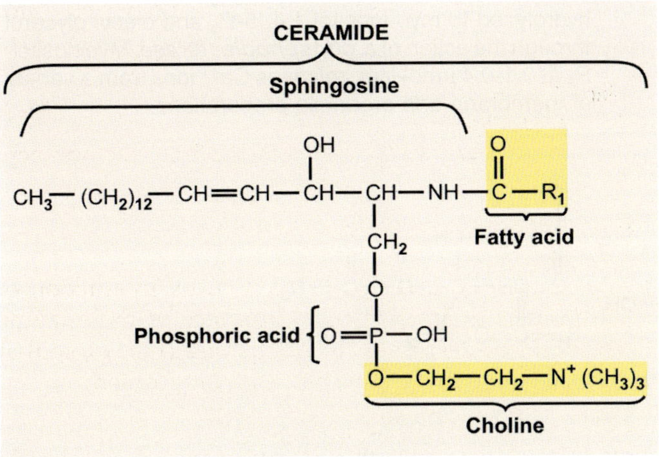

Fig. 4.17: Sphingomyelin

The above group may be considered in effect, to represent the *condensation of the enol form of an aliphatic aldehyde with a glycerol – OH group* (Fig. 4.16).

6. Sphingomyelins (phosphatidyl sphingosides) (Fig. 4.17):

Found in large quantities in brain and nervous tissues, and very small amount in other tissues. **It does not contain glycerol.** In place of glycerol, it contains an 18 carbon unsaturated amino alcohol called **'sphingosine' (sphingol).**

On hydrolysis: Sphingomyelin yields; one molecule of fatty acid + phosphoric acid + nitrogenous base choline + one molecule of complex unsaturated amino alcohol sphingosine (sphingol).

Sphingosine molecule in which a fatty acyl group is substituted on the $-NH_2$ group is called as **ceramide** and when a phosphate group is attached to ceramide, it is called **ceramide phosphate**. When choline is split off from sphingomyelin, ceramide phosphate is left. ***Sphingomyelinase*** is the enzyme which hydrolyses sphingomyelin to form ceramide and phosphoryl choline.

CLINICAL ASPECT OF SPHINGOMYELINS—NIEMANN-PICK DISEASE

Large accumulations of sphingomyelins may occur in brain, liver and spleen of some persons suffering from Niemann-Pick disease.

It is an inherited disorder of sphingomyelin metabolism in which sphingomyelin is not degraded, as a result sphingomyelin accumulates. It is a lipid-storage disease (lipidoses).

Inheritance: Autosomal recessive.

Enzyme defect: Deficiency of the enzyme ***sphingo-myelinase***, a lysosomal enzyme.

Clinical features: It affects children, arises at birth or infancy. The child presents with gradual enlargement of abdomen, enlargement of liver ***(hepatomegaly)***, enlargement of spleen ***(splenomegaly)***, and manifests ***progressive mental deterioration*** (due to accumulation of sphingomyelins in brain). Other lipids and cholesterol are usually normal or may be slightly elevated.

Prognosis: Usually fatal, progressive downhill course. Over 80 per cent of infants die within 2 years.

7. Phosphatidic Acid and Phosphatidyl Glycerol

Phosphatidic acid is important as an intermediate in the synthesis of TG and phospholipids, but is not found in any great quantities in tissues (Fig. 4.18).

8. Cardiolipin

A phospholipid found in mitochondria (inner membrane) and bacterial wall. It is formed from phosphatidyl glycerol. Chemically, it is *diphosphatidyl glycerol. On*

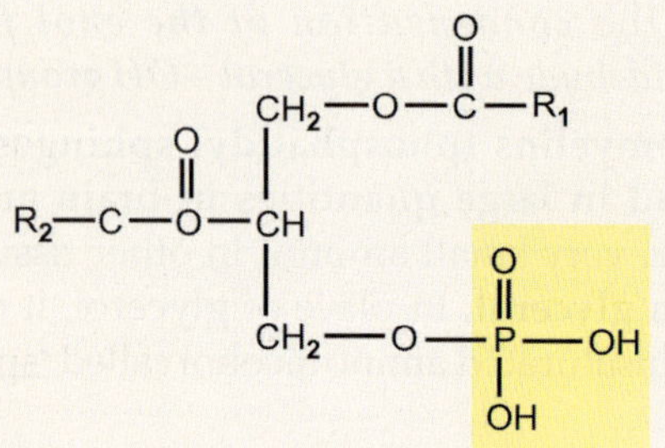

Fig. 4.18: Phosphatidic acid

Hydrolysis: Cardiolipin yields: 4 mols of fatty acids + 2 mols of phosphoric acid + 3 mols of glycerol (Fig. 4.19). **This the only phosphoglyceride that possesses antigenic properties.**

Functions of Phospholipids

- ***Structural:*** Phospholipids participate in the lipoprotein complexes which are thought to constitute the matrix of cell walls and membranes, the myelin sheath, and of such structures as mitochondria and microsomes.
  In this role, they impart certain physical characteristics
  - Unexpectedly high permeability towards certain non-polar (hydrophobic) molecules,
  - Lysis by surface active agents—detergents, bile salts, etc.
- ***Role in enzyme action:*** Certain enzymes require tightly bound phospholipids for their actions, e.g. mitochondrial enzyme system involved in oxidative phosphorylation.
- ***Role in blood coagulation:*** Phospholipids play an essential part in the blood coagulation process. Required at the stage of:
  - Conversion of prothrombin to thrombin by active factor X, and
  - Possibly also in the activation of factor VIII by activated factor IX.

  Platelets provide the chief source of PL and that part of total lipid content of the platelets which contribute to intrinsic blood coagulation process is called ***platelet factor 3.***
- ***Role in lipid absorption in intestine:*** Lecithin lowers the surface tension of water and aids in emulsification of lipid water mixtures, a prerequisite in digestion and absorption of lipids from GI tract.
- ***Role in transport of lipids from intestines:*** Exogenous TG is carried as lipoprotein complex, chylomicrons, in which PL takes an active part.
- ***Role in transport of lipids from liver:*** Endogenous TG is carried from Liver to various tissues as lipoprotein complex **Pre-β-lipoprotein (VLDL),** PL is required for the formation of the lipoprotein complex.
- ***Role in electron transport:*** Probably PL help to couple oxidation with phosphorylation and maintain electron transport enzymes in active conformation and proper relative positions.
- ***Lipotropic action of lecithin:*** Choline acts as a lipotropic agent as it can prevent formation of fatty liver. ***As lecithin can provide choline it acts as a lipotropic agent.***
- ***Ion transport and secretion:*** Phospholipids are in some way implicated in the mechanism of secretion is suggested by the observation that phospholipids, specially phosphatidic acid and phosphoinositides turnover is proportional to the rate of secretion of cells liberating such products as hormones, enzymes, mucins and other proteins.
- ***Membrane phospholipids as source of arachidonic acid:*** Phospholipids of membrane are ***hydrolysed*** by phospholipase A_2 and provide the unsaturated fatty acid arachidonic acid, which is utilised for synthesis of Prostaglandins and leukotrienes.
- ***Insulation:*** Phospholipids of myelin sheaths provide the insulation around the nerve fibres.
- ***Cofactor:*** Phospholipids are required as a cofactor for the activity of the enzyme *lipoprotein lipase* and *triacylglycerol lipase.*
- ***Role of phosphatidyl inositides metabolite in Ca^{++} dependent hormone action:*** Some signal must provide communication between the hormone receptor on the plasma membrane and intracellular Ca^{++} reservoirs. The best candidate appears to be the products of phosphatidyl inositides metabolism. Phosphatidyl inositol, 4,5-P_2 is hydrolysed to myo-inositol 1,4, 5-P_3 and diacyl glycerol through the action of a ***phosphodiesterase.*** Myoinositol-P_3 at 0.1-0.4 µmol/litre, releases Ca^{++} ions from a variety of membrane and organelle preparations.

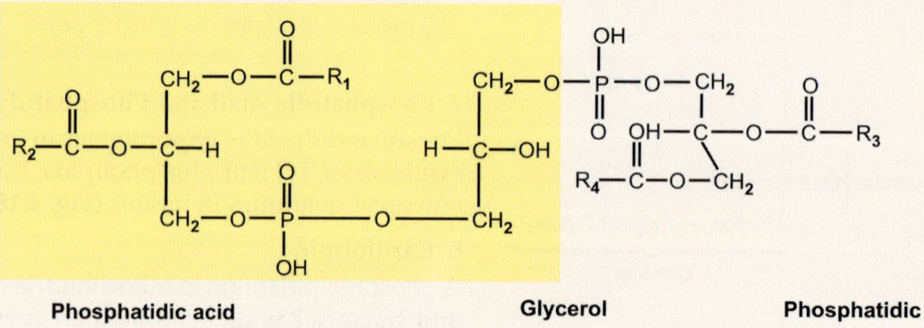

Fig. 4.19: Cardiolipin (diphosphatidyl glycerol)

CLINICAL IMPORTANCE

(a) Dipalmityl lecithin (DPL): DPL acts as a ***surfactant*** and lowers the surface tension in lung alveoli. The lung alveoli contains 2 types of cells.

- **Type-I:** These are thin cells which line the much of alveolar surface, and
- **Type-II:** These are **granular pneumocytes**, round cells and contain lamellar inclusions. These inclusions contain surfactant, DPL, which is secreted from these cells by exocytosis.

Surface activity is that phenomenon whereby the surface tension of the air-alveolar lining interface is lowered with expiration due to presence of DPL. In absence of normal surface activity, **if DPL is absent, the alveolar radius becomes smaller with expirations, the wall tension rises and the alveoli collapse.** ***Absence of DPL, in premature foetus, produces collapse of lung alveoli, which produce respiratory distress syndrome (hyaline-membrane disease).***

(b) Lecithin-Sphingomyelin Ratio (L/S ratio): L/S ratio in Amniotic fluid has been used for the evaluation of fetal lung maturity. Prior to 34 weeks gestation, the amniotic fluid lecithin and sphingomyelin concentrations are approximately equal. After this time, there is a marked increase in lecithin and L/S ratio increases to greater than 5 at term.

- ***A L/S ratio of > 2 or > 5 indicate adequate foetal lung maturity*** and suggests that respiratory distress after delivery is not likely to develop.
- Delivery of a premature low weight fetus, ***with L/S ratio approximately 1 or <1, indicate that the infant will probably develop respiratory distress or hyaline membrane disease.***

(c) Estimation of lecithin: Estimation of lecithin phosphorus in amniotic fluid has been considered to be clinically more useful. A lecithin phosphorus value of 0.100 mg/100 dl indicates adequate foetal lung maturity.

GLYCOLIPIDS

1. CEREBROSIDES (GLYCOSPHINGOSIDES)

Cerebrosides occur in large amounts in the white matter of brain and in the myelin sheaths of nerve. They are not found in embryonic brain but develops as medullation progresses. In smaller amounts they appear to be very widely distributed in animal tissues. In medullated nerves the concentration of cerebrosides are much higher than in non-medullated nerve fibres.

Structure: A cerebroside is considered to be built on the following:

Sphingosine
- FA of high molecular weight
- Usually galactose but sometimes may be glucose.

There is no glycerol, no phosphoric acid and no nitrogenous base. Thus, a cerebroside, ***on hydrolysis, yields:***

- **A sugar, usually galactose, but sometimes glucose**
- **A high molecular weight fatty acid and**
- **The alcohol, sphingosine or dihydrosphingosine.**

Thus they contain nitrogens though there is no nitrogenous base.

Types of cerebrosides: ***Individual cerebrosides are differentiated by the kinds of fatty acids in the molecule.*** **Four types** of cerebrosides have been isolated and their fatty acids have been identified. They are:

- **Kerasin:** Contains normal **lignoceric acid**, $C_{23}H_{47}$ COOH as fatty acid. **Lignoceric acid** is synthesized from 'acetate', by repeated condensation of C_2 units.
- **Cerebron (Phrenosin):** Contains hydroxy lignoceric acid, also called ***cerebronic acid***. This fatty acid is directly formed by hydroxylation (at 2 position) of lignoceric acid.
- **Nervon:** Contains an unsaturated homologue of lignoceric acid called ***nervonic acid*** ($C_{23}H_{45}$ COOH). Nervonic acid has one double bond, appears to be formed by elongation of oleic acid, i.e. from a C_{18} to a C_{24} monounsaturated fatty acid.
- **Oxynervon:** Contains hydroxy derivative of nervonic acid.

Psychosin

By prolonged hydrolysis of any cerebroside with Ba $(OH)_2$, FA is removed and it yields ***psychosin*** (sphingosine + sugar). Psychosin can be further hydrolysed to yield sphingosine and galactose.

CLINICAL ASPECT OF CEREBROSIDES

Gaucher's Disease

An inherited disorder of cerebrosides metabolism (lipidosis).

- ***Inheritance:*** It is autosomal recessive.
- ***Enzyme defect:*** Deficiency of the enzyme ***β-Glucocerebrosidase***, a lysosomal enzyme. Normally this enzyme hydrolyses glucocerebrosides to form ceramide and glucose. In absence of the enzyme, the cerebrosides cannot be degraded in the body, as a result large amounts of glucocerebrosides, usually ***kerasin*** accumulate in RE cells viz., liver, spleen, bone marrow and also brain. Complex lipids appear to collect within mitochondria of the RE cells. ***Biochemically, there is characteristically elevation in serum acid phosphatase level.***

Clinical features: Adults as well as infants are affected.

(a) In infancy and childhood: Fairly acute onset, with rapid course and death in several years. The infant loses weight, fails to grow, ***progressive mental retardation***. Initially ther* e is spasticity, later on followed by flaccidity.

(b) In adult: Progressive enlargement of spleen ***(splenomegaly)*** which may reach to umbilicus or below. Characteristic **bone pain** due to marrow cells replaced by histiocytes loaded with the lipids. As a result leads to ***progressive anaemia, leucopenia and thrombocytopenia***, tendency to get secondary infections and bleeding tendency.

2. GANGLIOSIDES

Klenk, in 1942, isolated from beef brain, a new class of carbohydrate rich glycolipids which he **called as gangliosides**. Gangliosides have been isolated from ganglion cells, neuronal bodies and dendrites, spleen and RBC stroma. The ***highest concentrations are found in gray matter of brain***. Gangliosides are the most complex of the glycosphingolipids. They are large complex lipids, their molecular weights varies from 180,000 to 250,000.

Structure: Although the exact structures of the gangliosides are not definitely established. ***On hydrolysis,*** gangliosides yield the following:

- **A long chain FA (usually C_{18} to C_{24}).**
- **Alcohol-sphingosine.**
- **A carbohydrate moiety which usually contains:**
  - **Glucose and/or galactose,**
  - **One molecule of N-acetyl galactosamine, and**
  - **At least one molecule of N-acetyl neuraminic acid (NANA) (also called 'sialic acid').**

Brain gangliosides are known to be complex and mono-, di-, trisialogangliosides ***containing 1 to 3 sialic acid residues*** have been described.

Types of gangliosides: Over 30 types of gangliosides have been isolated from brain tissue.

Four important types are:

GM-1, GM-2, GM-3 and **GD-3**. Their structures are shown in Figure 4.20.

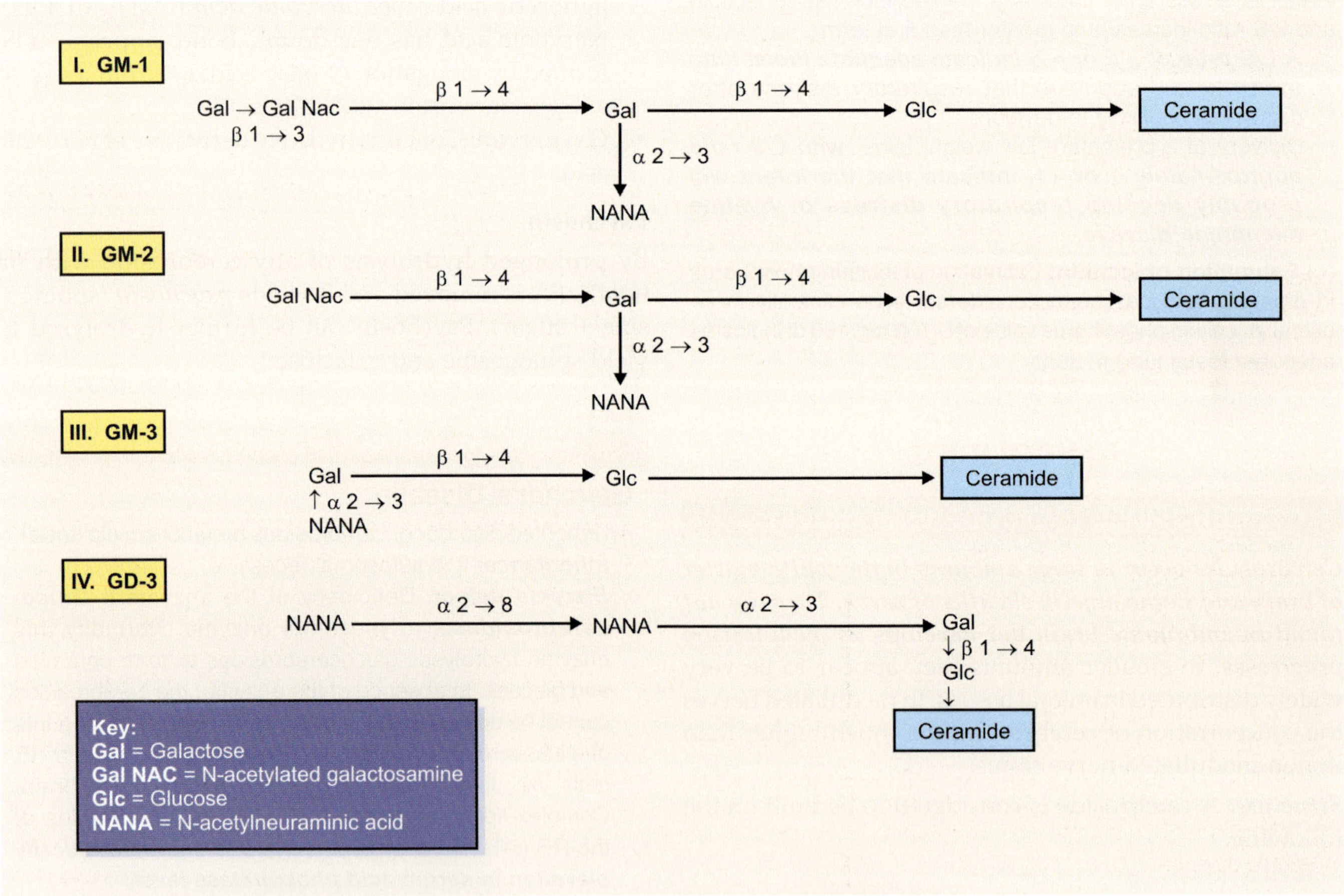

Fig. 4.20: Structure of some common gangliosides

CLINICAL IMPORTANCE

The simplest and common ganglioside found in tissues is **GM_3**, which contains ceramide, one molecule of glucose, one molecule of galactose and one molecule of neuraminic acid (NeuAc).

Gm_1 is a more complex ganglioside derived from Gm_3 is of considerable biologic interest, as it is now known to be ***the receptor in human intestine for cholera toxin.***

Biomedical Importance

Gangliosides are mainly components of 'membranes'. The sugar units and sialic acid sections of the molecule are "water-soluble" (i.e. "hydrophilic") and –vely charged, whereas the ceramide portion is 'lipid soluble' (i.e. "hydrophobic"). The latter appears to be embedded in the membrane lipids, whereas the hydrophilic sections, with its charged units protrudes externally towards the medium. ***The gangliosides, therefore, can serve as specific membrane binding sites (receptor sites)*** for circulating hormones and thereby influence various biochemical processes in the cell.

CLINICAL ASPECT

(a) Tay-Sachs Disease (GM_2 Gangliosidosis)

Accumulation of gangliosides in brain and nervous tissues takes place. The affected ganglioside is GM_2. ***The Enzyme deficiency is hexosaminidase A.***

Inheritance: Autosomal recessive.

Normal degradation of GM_2 requires the action of a specific hydrolyzing enzyme ***hexosaminidase A***, which removes the terminal Gal-NAc. Subsequently the other components are hydrolysed by other specific enzymes. In absence of the enzyme *Hexosaminidase A*, GM_2 cannot be degraded and accumulates.

This rare inherited disorder is associated with:

- Progressive development of ***idiocy and blindness in infants*** soon after birth. This is due to widespread injury to ganglion cells in brain (cerebral cortex) and retina.
- ***A cherry-red spot about the macula, seen ophthalmoscopically, is pathognomonic*** and is caused by destruction of retinal ganglion cells, exposing the underlying vasculature.
- There may be seizures and association of macrocephaly.

Prognosis is bad, usually death follows:

(b) GM_1 Gangliosidosis

It is due to a deficiency of the enzyme **β-galactosidase,** leading to accumulation of GM_1 gangliosides, glycoproteins and the mucopolysaccharide karatan sulphate.

The inheritance pattern and symptoms are similar to Tay-Sach's disease.

Table 4.1 gives the differentiation of cerebrosides and gangliosides.

3. SULPHOLIPIDS

Lipids material containing sulphur has long been known to be present in various tissues and has been found in liver, kidney, testes, brains and certain tumours.

Most abundant in white matter of brain. Several types of sulfur containing lipids have been isolated from brain and other tissues. In general, they appear to be sulfate esters of glycolipids, the sulphate group is esterified with OH gr. of hexose moiety of the molecule.

CLINICAL ASPECT

1. Metachromatic Leucodystrophy (MLD): It is an inherited disorder in which sulfatide accumulates in various tissues. Sulphatide is formed from 'galactocerebroside' through esterification of OH group on C_3 of galactose with H_2SO_4 (SO_4 at C_3 of Gal). ***Ratio of cerebro side: to sulphatide in brain normally is 3:1. In this disorder, it is altered to 1:4.***

Enzyme deficiency: Deficiency of enzyme ***sulphatase*** called as ***Aryl sulfatase A.***

$$\text{Sulphatides} \xrightarrow[H_2O]{\text{Sulphatase}} \text{Galactocerebroside} + H_2SO_4$$

Types: **Two clinical types** are seen.

(a) Late infantile type: Usually manifests before 3 years, gross involvement of CNS:

- Defects in locomotion, weakness, ataxia, hypotonus, and paralysis
- Difficulties in speech
- There may be optic atrophy.

(b) Adult type: Initially associated with psychiatric manifestations but subsequently progressive dementia.

2. Fabry's disease: An inherited disorder, a lipid storage disease (lipidosis).

- ***Inheritance:*** X-linked dominant, full symptoms only in males.
- ***Enzyme deficiency:* α-galactosidase.** The enzyme is found normally in liver, spleen, kidney, brain and small intestine.
- ***Nature of lipid*** that accumulates ***ceramide trihexoside*** (globotriosyl ceramide)
- ***Clinical manifestations: Skin rash*** (reddish purple), Pain in lower extremities ***(painful neuropathy),*** Lipid accumulates in the endothelial lining of blood vessels, may produce ***vascular thrombosis***. Progressive ***renal failure***–due to extensive deposition of lipids in glomeruli. Occasionally manifestations of cardiac enlargement. ***Eye involvement:*** Corneal opacities, cataracts, vascular dilatation.

Table 4.1: Similarities and differentiation of cerebrosides and gangliosides

Cerebrosides	***Gangliosides***
I. Similarities	
1. Both are compound lipids	
2. Both are glycolipids and contain carbohydrates	
3. Both contain sphingol (sphingosine) as an alcohol	
4. Both are found in large quantities in brain and nervous tissue	
5. Both do not have glycerol, phosphoric acid, and nitrogenous base.	
II. Dissimilarities	
1. Occur in white matter of brain and in myelin sheaths	1. Highest concentration in **grey matter of brain**
2. Not found in embryonic brain but develops as medullation progresses	2. Found in ganglion cells, neuronal bodies and dendrites
3. ***Function: Nerve conduction in myelin sheath***	3. ***Transfers biogenic amines***
4. Carbohydrate content—usually galactose sometimes glucose	4. Carbohydrate content is more. In addition to glucose/galactose, contains One mole of Gal-NAC, One or more of N-acetyl neuraminic acid (NANA)-(sialic acid)
5. Long chain fatty acids are—Lignoceric acid and nervonic acid/and their OH-derivatives	5. Contains long chain fatty acids C 18 to C 24
6. On basis of FA content 4 types of cerebrosides viz. kerasin, cerebron (phrenosin), nervon, and oxynervon	6. More than 30 types have been isolated. Four important types are: GM-1, GM-2, GM-3, GD-3
7. Cerebrosides on prolonged hydrolysis with $Ba(OH)_2$, FA is removed and yield ***psychosin***	7. Not so
8. Cerebrosides are degraded by the enzyme ***glucocerebrosidase***, a lysosomal enzyme	8. GM-2 gangliosides are degraded by the enzyme ***hexose aminidase A***, GM-1 gangliosides are degraded by ***β-galactosidase***.
9. Inherited deficiency of "enzyme" *β-glucocerebrosidase* produces the disease ***Gaucher's disease***	9. Inherited deficiency of **a.** Enzyme "Hexosaminidase A" produces ***Tay-Sach's disease*** (GM-2 gangliosidosis) **b.** Enzyme *β-galactosidase* produces ***GM-1 gangliosidosis***
10. Do not act as cell membrane receptors	10. Gangliosides can serve as a 'specific membrane binding site' on cell membrane for circulating hormones and thereby influences various metabolic processes in cells

3. Krabbe's Disease

- An inherited disorder of lipid metabolism, a lipid storage disease (lipidosis)
- *Enzyme deficiency:* ***Galactocerebrosidase* (β-*galactosidase*).** The enzyme normally catalyzes the hydrolysis of galactocerebrosides and it splits the linkage between ceramide and galactose

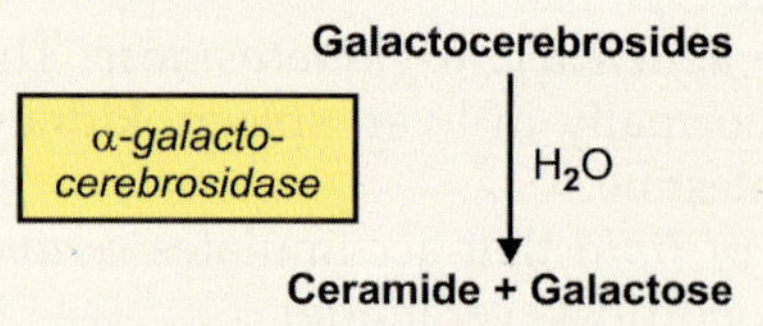

- *Nature of lipid* accumulating: ***Galactosyl ceramide***
- *Clinical manifestations:*
  - Severe mental retardation in infants
  - Total absence of myelin in central nervous system
  - Globoid bodies found in white matter of brain.

Note: Galactocerebroside is an important component of myelin

- *Diagnosis:* Depends on the determination of galactocerebrosidase activity in leucocytes and cultured skin fibroblasts.
- *Prognosis:* Fatal.

Table 4.2 summarises some of the sphingolipidosis with enzymes involved nature of lipid accumulating and clinical symptoms.

AMPHIPATHIC LIPIDS

Lipids as such are insoluble in water, since they contain a predominance of "nonpolar" hydrocarbon groups. But fatty acids, phospholipids (PL), bile salts, and to a lesser extent cholesterol contain "Polar" groups. Hence, the part of the molecule is ***hydrophobic*** or water insoluble and part is ***hydrophilic*** or water soluble. Such molecules are called ***amphipathic.***

Orientation of amphipathic lipids: Amphipathic lipids get oriented at oil-water interfaces with the polar groups in the water phase and the non-polar groups in the oil phase **(Fig. 4.21A)**.

- *Membrane bilayers:* Orientation of amphipathic lipids as above forms the basic structure of biological membranes **(Fig. 4.21B)**.

Table 4.2: Some examples of sphingolipidoses

Disease	*Enzyme deficiency*	*Lipid accumulating (see key below)*	*Clinical symptoms*
• **Niemann-Pick disease**	***Sphingomyelinase***	Cer + p-choline sphingomyelin	Enlarged liver and spleen, mental retardation, fatal in early life
• **Gaucher's disease**	***β-glucosidase***	Cer + Glc glycosylceramide	Enlarged liver, massive splenomegaly, erosion of long bones, mental retardation in infants
• **Tay-Sachs disease**	***Hexosaminidase B***	Cer – Glc – Gal(NeuAc) + Gal NAC GM_2 ganglioside	Mental retardation, muscle weakness, blindness
• **Metachromatic leukodystrophy**	***Arylsulphatase A***	Cer – Gal – Gal + O SO_3 3 sulfogalactosyl ceramide	Mental retardation, demyelination, psychologic disturbances in adults
• **Fabry's disease**	***α-Galactosidase***	Cer – Glc – Gal + Gal Globotriaosyl ceramide	Skin rash, renal failure (full symptoms only in males X-linked recessive)
• **Krabbe's disease**	***β-Galactosidase***	Cer + Gal galactosyl ceramide	Mental retardation myelin almost absent

Key : Cer – Ceramide
Glc – Glucose
Gal – Galactase
Neu-Ac – N-acetyl neuraminic acid
\+ Denotes site of deficient enzyme reaction

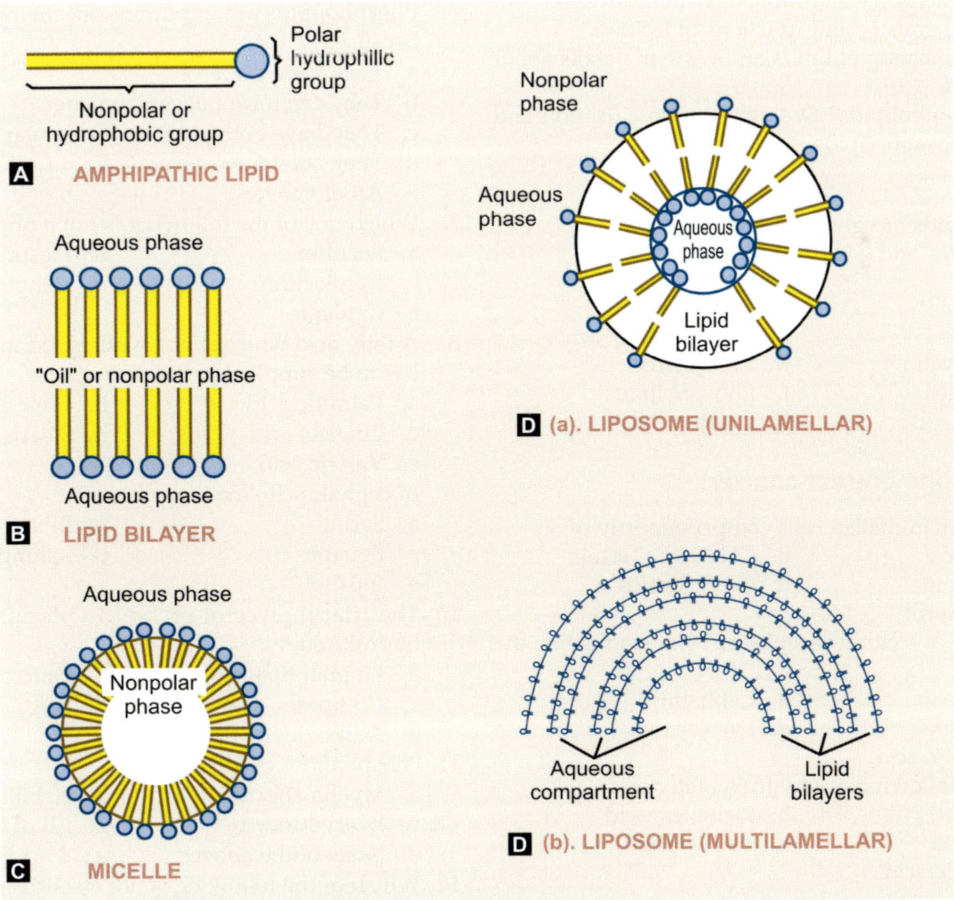

Figs 4.21A to D: Formation of lipid bilayer membrane, micelle, and liposomes from amphipathic lipids

- ***Micelles:*** When a critical concentration of these amphipathic lipids is present in an aqueous medium, they form ***micelles.*** Micelles formation, facilitated by bile salts, is prerequisite for fat digestion and absorption from the intestine **(Fig. 4.21C)**.
- ***Liposomes:*** Liposomes are formed by sonicating an amphipathic lipid in an aqueous medium.
  Characteristic of liposomes: They consist of spheres of lipid bilayers that enclose part of the aqueous medium **(Fig. 4.21D)**.

Uses of Liposomes

i. They are of potential clinical use, particularly when combined with tissue-specific **antibodies**, as carriers of **drugs** in the circulation, targeted to specific organs, e.g. in cancer therapy,
ii. They are being used for ***gene transfer into vascular cells,*** and
iii. As carriers for topical and transdermal delivery of drugs and cosmetics.

- ***Emulsions:*** They are larger in size and formed usually by non-polar lipids (e.g. T-G) are mixed with water (aqueous medium). They are stabilised by emulsifying agents such as amphipathic lipids (e.g. phosphatidyl choline) which form a surface layer separating the main bulk of non-polar material from the water.

I. Questions (Essay type and short notes)

1. What are lipids? Classify them giving suitable example.
2. Describe the chemistry and functions of phospholipids.
3. Describe the chemistry and functions of cholesterol.
4. Describe the chemistry and functions of lecithin.
5. Describe classification of lipids and write in details about phospholipids.
6. What are phospholipids? Describe the classification and functions of phospholipids.

Differentiate

a. Phospholipids are glycolipids
b. Cerebrosides and gaugliosides

Short Notes

a. Glycolipids b. Essential fatty acids
c. Sphingomyelin d. Cholesterol
e. Cerebrosides f. Phospholipids
g. Reacidity h. Lipoproteins

II. MCQs (Give one correct answer)

1. Depot fats of mammalian cells comprise mostly of:
   a. Cholesterol b. Cholesterol esters
   c. Phospholipids d. Cerebrosides
   e. Triacylglycerol
2. When choline of lecithin is replaced by ethanolamine the product is
   a. Spingomyelin b. Cephalin
   c. Plasmalogens d. Lysolecithin
   e Phosphatidic acid.
3. Which of the following is a hydroxyl fatty acid:
   a. Oleic acid b. Ricinoleic acid
   c. Caproic acid d. Stearic acid
   e. Arachidonic acid.
4. Acrolein test is answered by:
   a. Cholesterol b. Glycerol
   c. Glycosides d Sphingol
   e. Saturated fatty acid.
5. The smell of the fat turned rancid is due to:
   a. Presence of vit E b. Presence of quiniones
   c. Phenols d. Volatile fatty acids
   e. Cholesterol.
6. Phospholipids are important cell membrane components because:
   a. They have glycerol
   b. They can form bilayers in water
   c. They have both polar and nonpolar portions
   d. They combine covalently with proteins
   e. All of above.
7. Which one of the following is not a phospholipid?
   a. Lecithin b. Plasmalogen
   c. Lysolecithin d. Gangliosides
   e. Cephalin
8. A fatty acid which is not synthesised in human body and has to be supplied in the diet:
   a. Palmitic acid b. Oleic acid
   c. Linoleic acid d. Stearic acid
   e. Valeric acid.
9. In cephalin choline is replaced by:
   a. Serine b. Ethanolamine
   c. Betaine d. Sphingosine
   e. Inositol
10. The triacyl glycerol present in plasma lipoproteins are hydrolysed by:
    a. Lingual lipase b. Pancreatic lipase
    c. Co-lipase d. Lipoprotein lipase
    e. Adipokinetic lipase.
11. Phosphatides occur in:
    a. Myelin sheath b. Stablilises chylomicrons
    c. In erythrocyte d. All of the above
    e None of the above.
12. Which of the following is not essential fatty acid?
    a. Oleic acid b. Linoleic acid
    c. Arachidonic acid d. Linolenic acid
    e. All of the above.

13. The calorific value of lipid is:
    a. 4.0 Kcal/gm b. 6.0 Kcal/gm
    c. 9.0 Kcal/gm d. 15 Kcal/gm
    e 12.0 kcal/gm.
14. Rancidity of butter is prevented by the addition of:
    a. Vitamin D b. Tocopherols
    c. Presence of biotin d. Presence of copper
    e. Presence of lead
15. Sphingolyelins on hydrolysis yield:
    a. Glycerol, fatty acids, phosphoric acid, choline
    b. Glycerol, Sphingosine choline and fatty acids
    c. Sphingosine, phosphoric acid, glycerol and inositol
    d. Sphingosine, fatty acids, phosphoric acid and cholien
    e. Glycerol, fatty acids, phosphoric acid and ethanol amine.
16. Inherited deficiency of enzyme cerebrosidase produces:
    a. Fabry's disease b. Niemann-Pick disease
    c. Gaucher's disease d. Tay-Sach's disease
    e. Krabbe's disease.
17. Phosphatidic acid on hydrolysis yields:
    a. Glycerol, fatty acids, phosphoric acid, cholien
    b. Glycerol, fatty acids, phosphoric acid
    c. Glycerol, fatty acids, phosphoric acid, glucose
    d. Sphingol, fatty acids, phosphoric acid
    e. Sphingol, fatty acids, choline
18. The maximum number of double bonds present in essential fatty acids is:
    a. 1 b. 2
    c. 3 d. 4
    e. 5
19. Cerebrosides are composed of:
    a. Sphingosine, fatty acids, glycerol, phosphoric acid
    b. Sphingosine, fatty acids galactose
    c. Glycerol, fatty acids, galactose
    d. Glycerol, fatty acids, galactose, sphingol
    e. Sphingosine, glucose, fatty acids, glycerol.
20. All the following are constituents of ganglioside molecule except:
    a. Glycerol b. Sialic acid
    c. Hexose sugar d. Sphingosine
    e. Long chain fatty acid
21. An alcoholic amine residue is present in which of the following lipids?
    a. Phosphatidic acid b. Cholesterol
    c. Glucocerebroside d. Ganglioside
    e. Sphingomyelin
22. Sphingosine is the backbone of all the following except:
    a. Cerebroside b. Ceramide
    c. Sphingomyelin d. Lecithin
    e. Ganglioside
23. Match the following:
    a. Neutral fat b. Plasmalogens
    c. Phosphatidic acid d. Wax
    e. Cardiolipin
    1. Hydrolysis results in an aliphatic aldehyde________
    2. Hydrolysis yield 4 mols of fatty acids, 3 mols glycerol and 2 mols of phosphates ________
    3. Hydrolysis yields 2 mols of fatty acids, one mole of glycerol and one mole of phosphoric acid ________
    4. Hydrolysis yields one mole of fatty acid, and mole of long chain aliphatic alcohol________
    5. Hydrolysis yields 3 mols of fatty acids and one mole of glycerol________.

Answers to MCQs

1. e	2. b	3. b	4. b	5. d	6. c
7. d	8. c	9. b	10. d	11. d	12. a
13. c	14. b	15. d	16. c	17. b	18. d
19. b	20. a	21. e	22. d		

23. 1-B, 2-E, 3-C, 4-D, 5-1.

SECTION TWO

CHAPTER 5

PROSTAGLANDINS—CHEMISTRY AND FUNCTIONS

Major Concepts

A. What are eicosanoids? Study the classification, chemistry, biosynthesis and catabolism of prostaglandins.
B. Study the important functions of PGs.
C. What are prostacyclins and thromboxanes, their important role in the thrombus formation? Study the chemistry and functions of leucotrienes and lipoxins.

Specific Objectives

A. 1. Study what is prostanoic acid?
   2. How prostaglandins are classified, the major four groups and their structural features.
   3. Study how PGs are synthesised and catabolised in the body. Also study the important inhibitors-stimulants of PG synthesis.
   4. Study the occurrence and distribution of PGs in the body and its mechanism of action.
B. 1. Study the important functions of PGs and,
   2. The limitations of PGs as drugs.
C. 1. Study the chemistry and functions of prostacyclins (PG-I_2), thromboxanes (Tx), leucotrienes (LTs) and Lipoxins (LXs)
   2. Differentiate in a tabular form PG-I_2 from Tx.

PROSTAGLANDINS

A generic term for a family of closely related biologically 'active' lipids, now called as **eicosanoids**.

Prostaglandins (PGs)

- Have been detected in almost every mammalian tissue and body fluids;
- Their production increases or decreases in response to diverse stimuli or drugs;
- They are produced in minute amounts;
- Broad spectrum and diverse biological effects;
- Not stored in body; and
- They have also been found to modulate cyclic AMP activity in cells either by activating or inhibiting *adenyl cyclase* activity.

The term is now used to describe a family of closely related derivatives of hypothetical C_{20} molecule named ***Prostanoic acid*** (Fig. 5.1).

Classification: Prostaglandins and related compounds are now classified under the heading **Eicosanoids** as these compounds are derived from **Eicosa (20 C) Polyenoic FA** (Fig. 5.2).

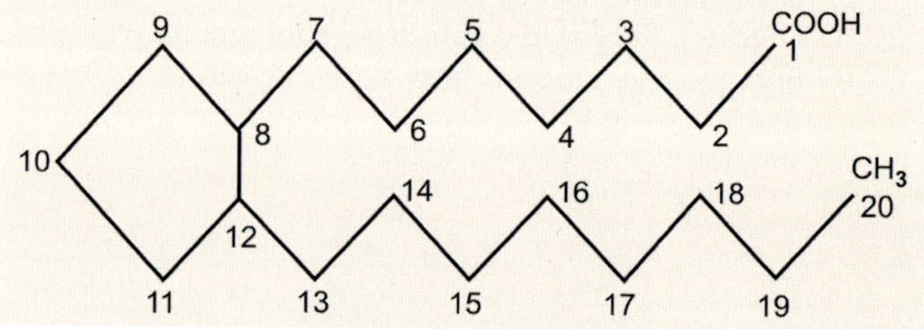

Fig. 5.1: Prostanoic acid

They are classified mainly in **two groups:**
(a) Prostanoids (PGs), and
(b) Leucotrienes (LT's) and Lipoxins (Lxs)

Prostanoids are further subdivided into **three groups** as follows:

1. **Prostaglandins (PGs)**
2. **Prostacyclins (PGI)**
3. **Thromboxanes (Tx)**

CHEMISTRY OF PROSTAGLANDINS

According to structures, PGs can be divided in **four** main groups

(a) **PG-E group:** PGE-1, PGE-2 and PGE 3
(b) **PG-F group:** $PGF_{1\alpha}$, $PGF_{2\alpha}$ and $PGF_{3\alpha}$

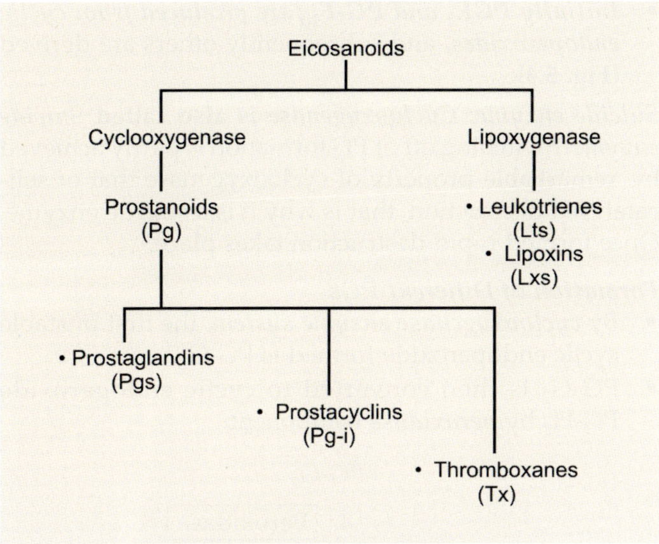

Fig. 5.2: Classification of eicosanoids

(c) **PG-A group:** PG-A_1, PG-A_2, 19-OH PG-A_1, 19-OH PG-A_2

(d) **PG-B group:** PG-B_1, PG-B_2, 19-OH PG-B_1 19-OH PG-B_2

Besides above 14 PGs, PG-C and PG-D group have also been recognised.

Characteristic Features of Structures

1. All naturally occurring PGs are 20C fatty acids containing a ***cyclopentane ring*** (Fig. 5.3). Structures are based on ***parent saturated acid called 'prostanoic acid.'***
2. All PGs have the following salient structural features
   - –OH group at 15 position,
   - *Trans* double bond at 13 position
3. Differences in the four main groups is due to difference in structure of cyclopentane ring.

Primary Prostaglandins: **Six PGs of the E and F series are referred to as primary PGs** because none is precursor of the other. A new type of Prostaglandin has been isolated from human seminal plasma which has been designated as **PG x.**

Occurrence and Distribution: PGs were first discovered in seminal plasma and vesicular glands. They are ***ubiquitous in mammalian tissues*** and have been detected and isolated from pancreas, kidney, brain, thymus, iris, synovial fluid, CS fluid, etc. Recently identified in human amniotic fluid and umbilical cord vessels. **Thirteen** different PGs have been isolated and identified from seminal fluid and together they amount to 300 $\mu g/ml$ of semen and has been found to contain PG-E series, PG-F series **(except PG-$F_{3\alpha}$)**, PG-A and PG-B series with 19-OH derivatives.

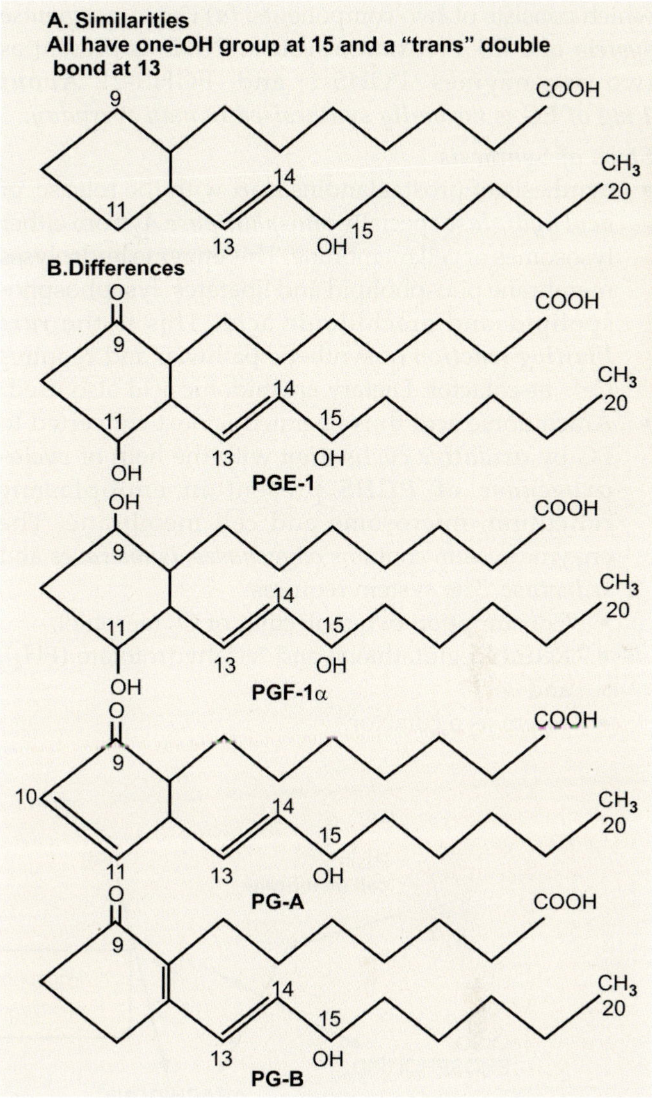

Fig. 5.3: Showing the structural similarities and differences of PGs

Release: Apart from the presence of PGs in tissues, ***spontaneous release*** of these substances from many sites have been demonstrated. These include:

- Cat superfused somatosensory cortex and cerebellar cortex,
- Frog intestine,
- ***Human medullary carcinoma of thyroid,***
- ***Pheochromocytoma,*** and
- ***Kaposi's sarcoma.***

METABOLISM OF PROSTAGLANDINS

Biosynthesis: PGs are synthesised ***aerobically*** from polyunsaturated fatty acid arachidonic acid (5, 8, 11, 14–eicosatetraenoic acid) with the help of a ***multienzyme complex*** now called ***Prostaglandin H Synthase (PGHS)***

which consists of two components: ***(a) Cyclo-oxygenase system*** and ***(b) Peroxidase system.*** PGHS is present as two isoenzymes PGHS-1 and PGHS-2. ***About 1 mg of PG is normally synthesised in man everyday.***

Steps of Synthesis

- Synthesis of prostaglandins start with the release of ***acyl hydrolase*** specially ***phospholipase*** A_2 from either lysosomes or cell membrane. This enzyme hydrolyses membrane phospholipid and liberates lysophosphospolipid and arachidonic acid. This is the ***rate limiting reaction*** of synthetic pathway and requires Ca^{++} as cofactor. Dietary arachidonic acid also used.
- Arachidonic acid thus liberated is next converted to PG by ***oxidative cyclisation*** with the help of ***cyclo-oxygenase*** **of** ***PGHS*** present in endoplasmic reticulum, microsome and cell membrane. The enzyme system contains ***oxygenases, isomerases*** and ***reductase***. The system requires:
  - Consumption two molecules of O_2 (aerobic),
  - Reduced glutathione and tetrahydrofolate (FH_4), and
  - Haem as a cofactor.
- ***Initially*** PGE_2 ***and*** $PG\text{-}F_{2\alpha}$ ***are produced from cyclic endoperoxides,*** and subsequently others are derived (Fig. 5.4).

Suicide enzyme: Cyclooxygenase **is also called** ***Suicide enzyme.*** Switching off of PG formation is partly achieved by remarkable property of cyclooxygenase that of self-catalysed destruction, that is why it is a suicide enzyme. Once formed rapid destruction takes place.

Formation of Different PGs

- By ***cyclooxygenase enzyme system,*** the first unstable cyclic endoperoxide formed is $PG\text{-}G_2$.
- PG G_2 is then converted to cyclic endoperoxide $PG\text{-}H_2$ by ***peroxidase*** component.

$PG\text{-}G_2$
↓ Peroxidase
$PG\text{-}H_2$

- $PG\text{-}H_2$ is the precursor for "prostanoids"—PGs, $PG\text{-}I_2$ and Tx
- ***First PG to form is usually*** $PG\text{-}E_2$, sometimes it may be $PG\text{-}D_2/PGF_{2\alpha}$

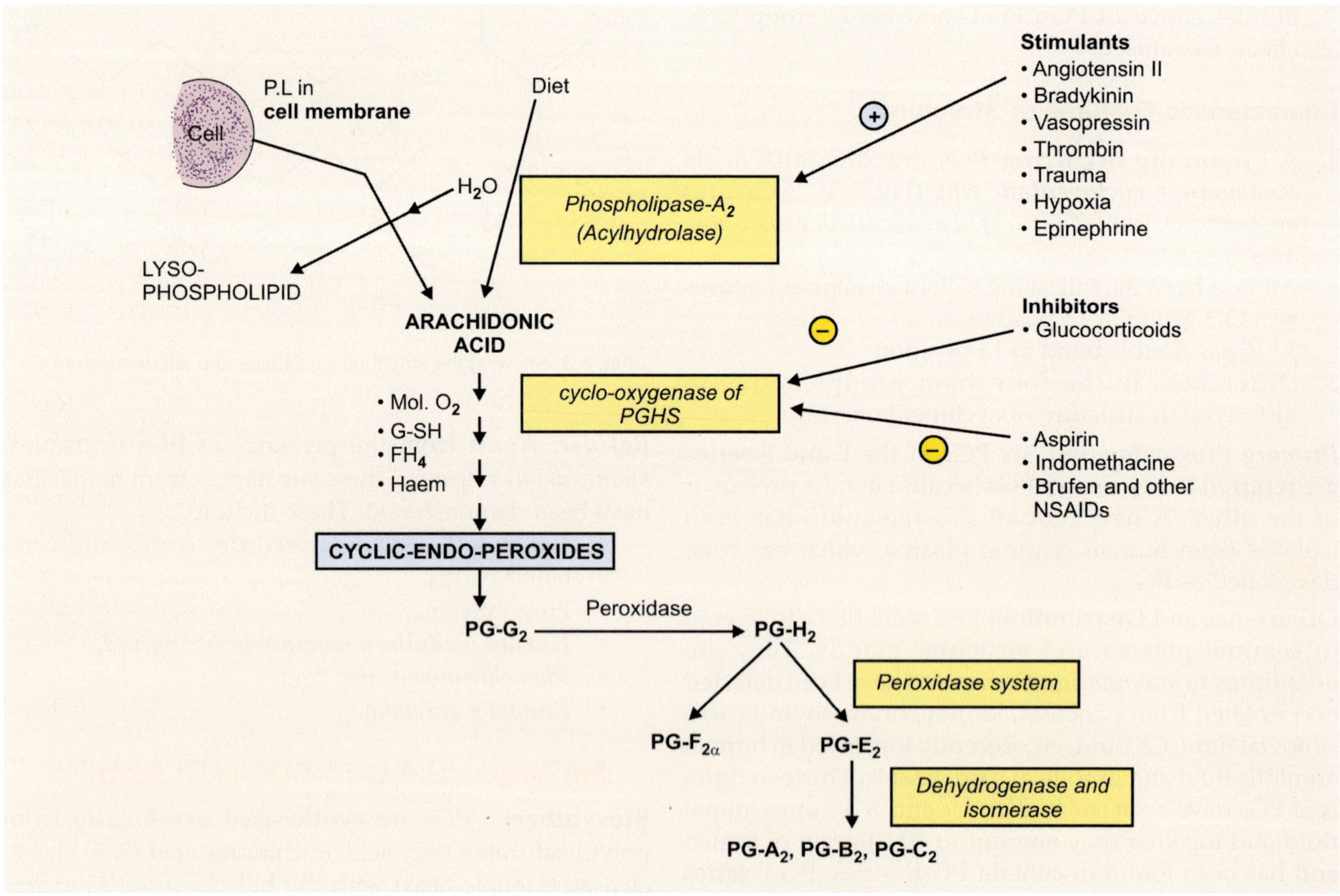

Fig. 5.4: Showing biosynthetic pathway and regulation of PGs

Formation of different PGs are shown below:

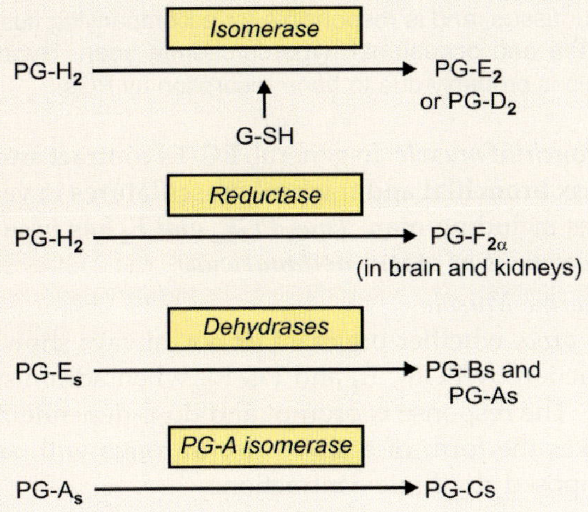

- **Group 1 PGs** can be formed from dietary linolenic acid:

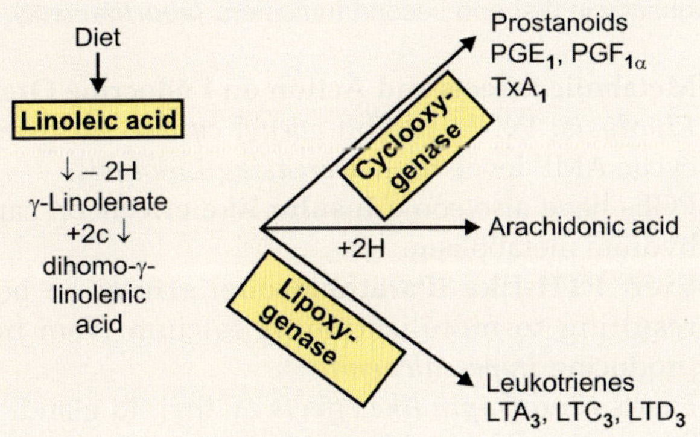

- **Group 2 PGs** can be formed from dietary α-linolenic acid:

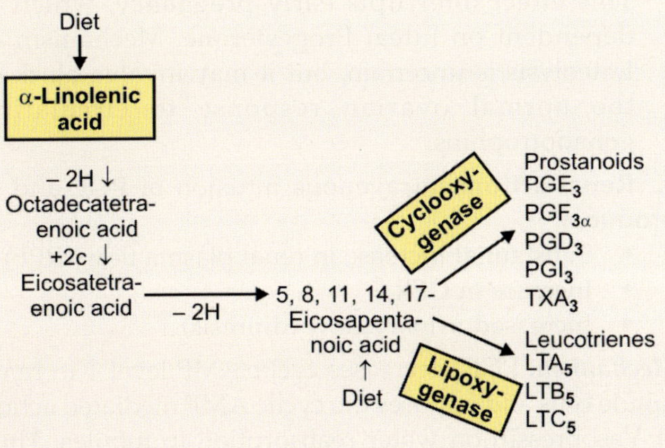

Inhibitors and Stimulants of PG-Synthesis

Inhibitors

- **"False" substrate,** e.g. 5, 8, 11, 14–eicosatetraynoic acid (TYA)
- **Mepacrine:** inhibits phospholipase A_2
- **Glucocorticoids:** Completely inhibits the transcription of PGHS-2 (but not PGHS-1)
- Aspirin, indomethacin, ibuprofen (brufen), phenylbutazone, fenclozic acid and other nonsteroidal antiinflammatory agents (NSAIDS): Aspirin-inhibits cyclooxygenase of both PGHS-1 and PGHS-2 by acetylation. Other NSAIDs inhibits cyclooxygenase by competing with arachidonic acid.
- Cu^{++} and dihydrolipoamide inhibits PG-E formation, but increases that of PG-F.

Stimulants

- Trauma, hypoxia, angiotensin II, bradykinin, vasopressin increase PG synthesis by activating *phospholipase A_2*.
- Catecholamines: They also enhance PG synthesis by changing inactive cyclooxygenase to its active form.
- Addition of G-SH stimulates the formation of PGE at the expense of PG-F.
- Addition of TSH to bovine thyroid cells *in vitro* has been reported to increase PG synthesis, probably it stimulates the *phospholipase A_2* and release more of arachidonate.

Catabolism

PGs are very rapidly removed from circulation and metabolised in lungs, brain, liver and other tissues. ***Some 80 to 90 per cent or more is destroyed during a single passage through the liver/or the lungs.***

The initial and major steps for both PG-E and PG-Fs is:

- ***Oxidation*** of secondary alcohol group at C_{15}. This is achieved by a widely distributed PG-Specific dehydrogenase called, ***15-OH-PG-dehydrogenase* (PGDH).** This is the ***rate limiting step*** in catabolic pathway.
- The above is followed by ***reduction*** of the Δ^{13} double bond. The resulting dihydroderivatives have little or no biological activity.
- It is followed by β-oxidation and ω-hydroxylation and oxidation of the side chains.

MECHANISM OF ACTION OF PROSTANOIDS (PGs, PG-I_2 AND Tx)

Prostanoids bind to specific receptors on the Plasma membrane of target cells and brings about changes in concentration of "Second messengers" which may be cyclic AMP, Ca^{++} or even cyclic GMP, which then mediate the biological effects.

Examples:

- ***PGEs:*** They mostly act through second messenger "Cyclic AMP".

SECTION TWO

- ***PG-Fs and Tx-A_2:*** These may use Ca^{++} as second messenger in some tissues.
- ***PG-$F_{2\alpha}$:*** Action is probably mediated through Cyclic-GMP as second messenger in some tissues.

FUNCTIONS OF PROSTAGLANDINS

Prostaglandins have numerous and diverse effects. ***Diversity is awesome and bewildering.*** Not only is the spectrum of actions broad, but also different PGs show different activities both qualitatively and quantitatively.

1. C_2V_2 System

Antihypertensive action: In most species and in most vascular beds, PG-Es and PG-As are potent vasodilators.

BP: Systemic BP generally falls in response to PGE and PG-As.

2. Haematological Response

- Capillary permeability is increased by PGs-E_1, E_2, $F_{1\alpha}$ and $F_{2\alpha}$. Intradermal injection of PGs in man causes ***wheal and flare*** similar to histamine.
- ***Platelets:*** PGE_1 is a potent inhibitor of human platelet aggregation. **Thus PGE_1 has proved useful for harvesting and storage of blood platelets for therapeutic transfusion.**

3. Action on GI Secretions

(a) Gastric Secretion: PGs E_1, E_2 and A_1 **(Not $F_{2\alpha}$)** ***inhibit gastric secretion,*** whether basal or stimulated by feeding, histamine or pentagastrin. There is decrease in volume, acid and pepsin content. Action is believed to be exerted directly on secretory cells through c-AMP. Based on this, **PGs have been used for preventing or alleviating gastric ulcers.**

(b) Pancreatic Secretion: Its action is opposite. There is increase in volume, bicarbonate and enzyme content of pancreatic juice.

(c) Intestinal Secretion

Mucus secretion is increased. There is substantial movement of water and electrolytes into intestinal lumen. ***PG-E_1 given orally in human volunteers produces watery diarrhoea.***

4. Effects on Smooth Muscles

(a) G_2I_2 Musculature

- ***In vitro*** responses vary widely with species, segment, type of muscle and type of PGs. In human beings PG-E and F produce contraction of longitudinal muscle from stomach to colon.
- ***In vivo*** effects are variable in man. Diarrhea, cramps and reflux of bile have been noted in human volunteers given PG-E orally ***(Purgative action).***

> **CLINICAL ASPECT**
>
> In **medullary carcinoma of thyroid**, PGs are released by the tumour tissue, and is responsible for accompanying flushing, diarrhea and occasional hypercalcaemia seen. Hypercalcaemia is probably due to bone resorption by PGs.

(b) Bronchial muscle: In general, **PG-Fs contract and PG-Es relax bronchial and tracheal musculatures** in various species including man. ***Thus PGE_1 and E_2 has been used for treatment of status asthmaticus.***

(c) Uterine Muscle

- ***In vivo,*** whether pregnant or not, always show contraction by PGE_1, E_2 and PG-$F_{2\alpha}$ when administered IV. The response is prompt and dose-dependent and takes the form of a sharp rise in tonus with superimposed rhythmic contractions.

> **CLINICAL APPLICATION**
>
> Thus PGE_2 at a rate of 0.5 μg/ml has been used for induction of labour at or near term. Same PG in higher doses 5 μg/ml has been reported to be effective in therapeutic termination of pregnancy in first and second trimesters ***(abortifacient).***

5. Metabolic Effects and Action on Endocrine Organs

- ***Lipolysis:*** PG-Es inhibit *adenyl cyclase* and lowers cyclic AMP level, thus ***decreasing Lipolysis.***
- PGEs have also some **Insulin like effects** on carbohydrate metabolism.
- Exert **PTH-like (Parathormone) effects** on bone, resulting to mobilisation of calcium from bone producing ***hypercalcaemia***
- Exerts ***thyrotropin like effects*** on thyroid gland.
- Stimulation of steroid production by adrenal cortex **(Steroidogenesis)**
- ***Luteolysis:*** Prompt subsidence of progesterone secretion and regression of corpus luteum follows parenteral injection of $PGF_{2\alpha}$ in a wide variety of mammals. This effect interrupts early pregnancy, which is dependent on luteal Progesterone. Mechanism of Luteolysis is uncertain, but it may involve block of the normal ovarian response to circulating gonadotrophins.

6. Renal action:

Intravenous infusion of PGE and A produces:

- Substantial increase in renal plasma flow (RPF) ↑
- Increase in GFR ↑
- Increased urinary flow (diuresis) ↑

Mechanism: PGE_2 decreases cyclic AMP level ↓ in renal tubule cells and opposes the cyclic AMP mediated action of Vasopressin on water reabsorption in tubules. Thus

$PG-E_2$ reduces the resorption of water in distal tubules and collecting ducts and produce increased urinary flow (dilute and hypotonic urine) ↑

- Output of Na^+ and K^+ is increased (natriuresis and kaliuresis) ↑
- PGEs stimulate renin secretion from **JG cells**.

For relation of kinins with PGs in renal tissue: (Refer chapter on Water and Electrolyte Balance and Imbalance)

APPLIED ASPECT

Inhibitors of PGE synthesis may be used in treating diabetes insipidus resulting from vasopressin (ADH) insufficiency.

7. Role of PGs in inflammation: PGEs and $PG-D_2$ released at the site of trauma or burn, produce vasodilatation, increases capillary permeability, erythema and **wheal** formation.

CLINICAL ASPECTS

Role of $PG-D_2$ in anaphylaxis: When $PG-D_2$ is injected in nanomole amounts into human skin it causes:

- Increased vasodilatation, vasopermeability resulting ***wheal and flare*** (it is nonpruritic), and
- Influx of polymorphs locally.

When inhaled, $PG-D_2$ produces bronchoconstriction.

- $PG-D_2$ has been suggested as an ***important mediator of anaphylaxis.***
- Treatment with Aspirin has eliminated such attacks suggesting that cyclooxygenase product $PG-D_2$ is directly involved in such cases.

8. Immunological response: PGEs secreted by macrophages may modulate or decrease the functions of B and T lymphocytes. They may also reduce the proliferation of lymphocytes in response to lymphocyte mitogenic factors.

Functions of PGs (Summary)

- Antihypertensive: Lowers BP↓
- Inhibits platelets aggregation
- Inhibits gastric secretion
- Stimulates GI musculature *(purgative* action)
- Bronchodilatation, used in treatment of bronchial asthma
- Increases uterine contraction, can be used as *abortifacient*
- Renal action: Increases RPF ↑, increase in GFR ↑, diuresis, natriuresis, and kaliuresis
- Stimulates renin secretion from JG cells.
- Metabolic effects:
  - Decreases lipolysis ↓
  - Insulin like effect
  - PTH like effect—produces ***hypercalcaemia***
  - TSH like effect
  - Steroidogenesis
  - Luteolysis

LIMITATIONS OF USE OF PGs AS DRUGS

Though PGs may be useful as drugs in certain conditions, as discussed under actions of PGs above, but use of PGs as drugs is **limited** due to following factors:

1. *Short Duration of Action:* PGs are metabolised very rapidly in tissues and they have a short duration of action.
2. *Lack of Tissue Specificity*

- $PG-E_2$ in addition to causing uterine smooth muscle to contract, when induction of labour is desired, it causes GI smooth muscle to contract as well, leading to cramping and diarrhoea.
- Same compound when inhaled into the nostrils dilates bronchi and alleviates the attack of asthma but simultaneously it irritates mucosa of throat causing pain and coughing.

Analogues of PGs: Analogues of PGs by introducing CH_3 group specially at 15 position are being tried,

- Which by inhibiting the enzyme can prolong the action,
- Secondly, can change the tissue specificity.

CHEMISTRY AND FUNCTIONS OF PROSTACYCLINS AND THROMBOXANES

Prostacyclins Vs. Thromboxanes: Both prostacyclins ($PG-I_2$) and thromboxanes (Tx) are produced from cyclic endoperoxide $PG-H_2$ (**Fig. 5.5 and 5.6**). Cyclic endoperoxide first formed in PG synthesis is $PG-G_2$, which is converted to $PG-H_2$, an immediate precursor of PGI_2 and Tx. ***Both cyclic endoperoxides have a very short ½ life (t½ = 5 minutes at 37°C) but they are biologically very active,*** and have powerful effect on contraction of GI smooth muscles, bronchial muscles and umbilical cord vessels. Main differentiating points between prostacyclins (PGI_2) and thromboxane (Tx) are shown in **Table 5.1**.

Applied Aspect of $PG-I_2$ and Tx

1. Prevention of thrombus formation in health: Platelets attempting to stick to blood vessel wall release **endoperoxides**, which is converted to prostacyclin ($PG-I_2$) by endothelial cells of blood vessel wall. $PH-I_2$ by its vasodilatation effects and inhibition of platelet aggregation, repel the platelets and prevent them from sticking and forming a "nidus" and thus opposes thromboxane activity. **A balance between these two biochemical processes is critical for the thrombus formation.**

2. Injury to blood vessel wall: Injury to blood vessel wall decreases $PG-I_2$ formation and thus reduces the antiaggregatory action of $PG-I_2$. Unopposed action of $TX-A_2$ in such cases cause platelet aggregation and thrombus formation.

SECTION TWO

Table 5.1: Differentiation of prostacyclins and thromboxanes

	Prostacyclins (PGI_2)	***Thromboxanes (Tx)***
1.	***Structure*** Contains cyclopentane ring.	No cyclopentane ring is present but, contains an oxane ring.
2.	***Site of formation:*** Principally formed in vascular endothelium; other sites are heart, kidneys.	Principally formed in platelets; other sites are neutrophils, lungs, brain, kidney, spleen.
3.	***Synthesis*** Synthesised from cyclic endoperoxide PG-H_2 by the enzyme ***Prostacyclin synthase***.	Also synthesised from cyclic enoperoxide PG-H_2 by the action of the enzyme ***Thromboxane synthase.***
4.	***Mechanism of action*** Principally by increasing cyclic AMP level ↑ in target cells.	Principally by decreasing cyclic AMP level ↓ in target cells, Can also use Ca^{++} as second messenger in some tissues specially in muscle fibres.
5.	***Functions*** Principally two: • **Inhibits platelet aggregation** • **Produces vasodilatation** The above two prevent thrombus formation. It opposes thromboxane action.	Has opposite action: • **Enhance platelets aggregation** • **Produces vasoconstriction** The above two favors thrombus formation.
	Other functions: • Inhibits gastric secretion • Produces renal vasodilatation and GFR↑ • Increases 'renin' production from JG Cells • Relaxes smooth muscles • **Inhibitors:** Ageing, Hyperlipaemia, Vit. E deficiency. Exposure to ionising radiations.	• Produces contraction of smooth muscles • Also induces release of serotonin, Ca^{++} and ADP. • Does not relax smooth muscles. • Imidazoles and dipyridamole: which inhibit 'thromboxane synthase' and stops synthesis.

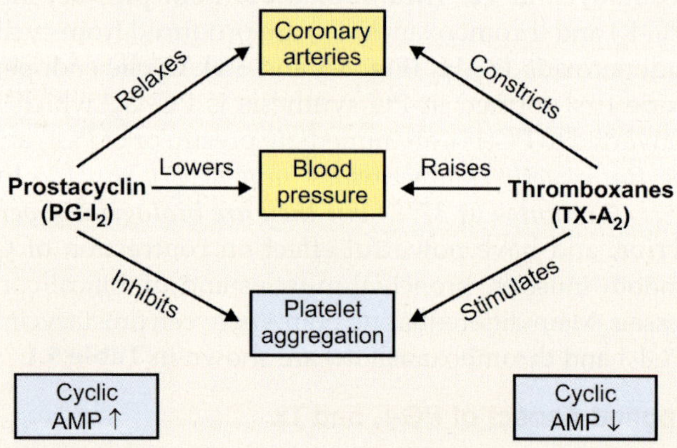

Fig. 5.5: Showing physiological antagonism between prostacyclin (PG-I_2) and thromboxane (TX-A_2)

CLINICAL ASPECTS

1. ***Aspirin as effective antiplatelet aggregator:*** Aspirin (acetyl salicylic acid) has been found to be a most effective drug which prevents platelet aggregation.

Platelets are extremely sensitive to aspirin and a very small amount of aspirin (30 mg) can act as anti-platelet aggregator for 24 hours.

Mechanism

- Aspirin irreversibly acetylates the platelets cyclooxygenase system and inhibits the enzyme so that thromboxane A_2 (TxA_2) a potent aggregator of platelets and vasoconstrictor is not formed.
- At the sametime aspirin also inhibits production of prostacyclin (PGI_2) by endothelial cells which opposes platelet aggregation and is a vasodilator.
- But unlike platelets, the endothelial cells ***regenerate cyclo-oxygenase*** within a few hours. ***Thus, the overall balance shifts towards formation of PGI_2 which opposes platelets aggregation.***

Clinical Uses

- Treatment and management of angina and evolving myocardial infarction.
- For prevention of stroke and death in patients with transient cerebral ischaemic attacks.

2. *Role of marine fish lipids:* Most predominant UFA in fish foods is 5, 8, 11, 14, 17-Eicosapentaenoic acid (EPA), which ***decreases plasma cholesterol and tri-acylglycerol; inhibits the synthesis of TxA_2.*** Thus, low levels of TxA_2 reduce platelet aggregation and thrombosis reducing the risk of myocardial infarction.

LEUCOTRIENES-LTs

A newly discovered family of conjugated trienes formed from eicosanoic acids in leucocytes, mast cells, and macrophages by the ***lipoxygenase pathway***, in response

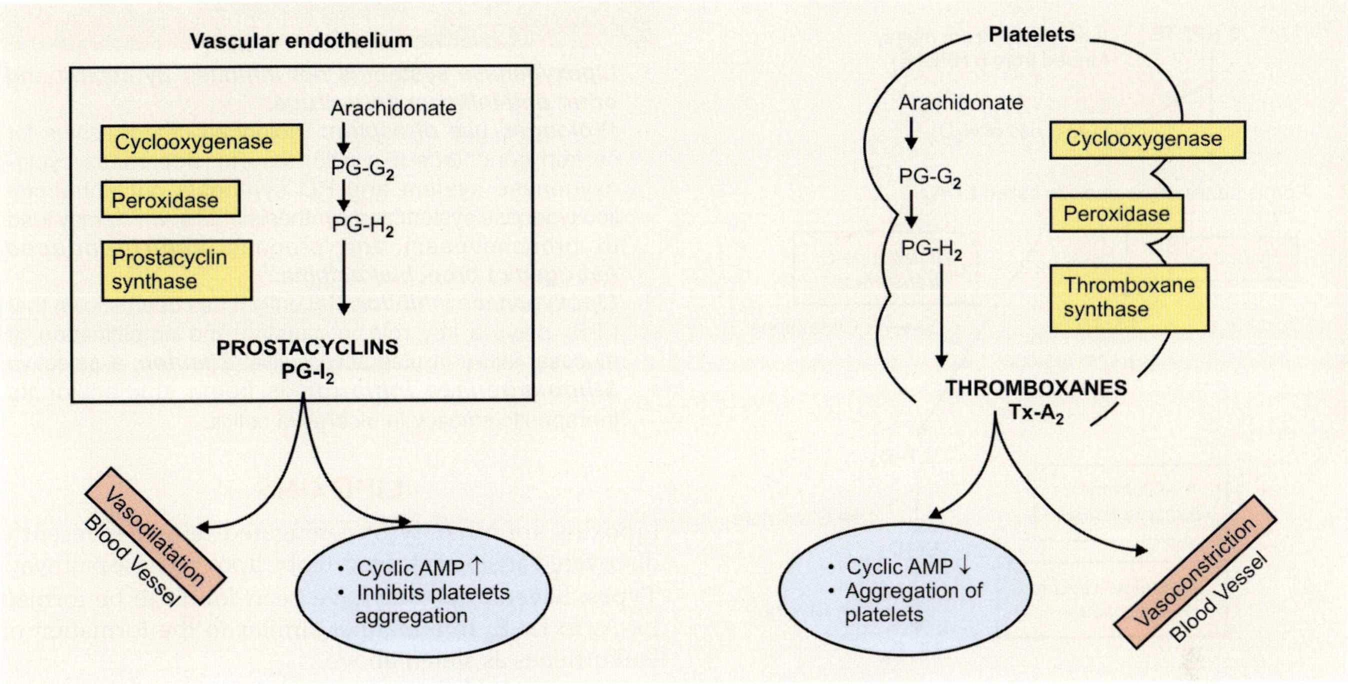

Fig. 5.6: PG-I_2 and Tx-A_2—Formation and actions

to both immunologic and noninflammatory stimuli. LTs possess no ring in its structure but have three characteristic conjugated double bonds.

Synthesis: LTs are synthesised from ***arachidonate*** by the addition of hydroxyperoxy groups to arachidonic acid and produces ***hydroperoxy eicosa-tetraenoates (HPETE).***

Different Types of HPETE: Depending on the position of addition, ***three types*** of HPETE have been found.

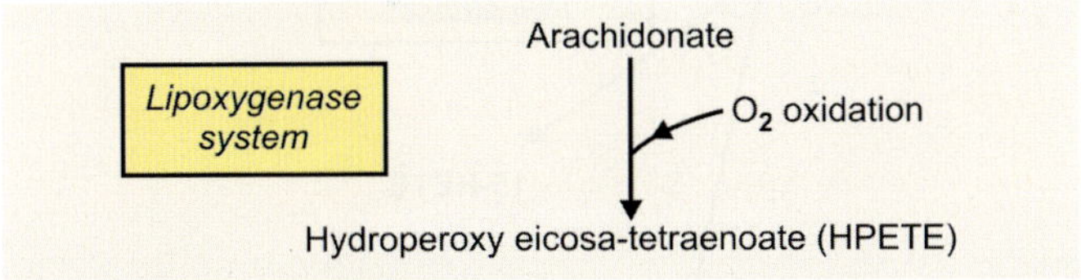

Three types of HPETE are:

- **5-HPETE:** Most common and the major product of 5-Lipoxygenase reaction in polymorphs, basophils, mast cells and macrophages.
- **12-HPETE:** Product of 12-lipooxygenase and occurs in platelets, pancreatic endocrine islet cells and glomerular cells of kidney.
- **15-HPETE:** Occurs in reticulocytes, eosinophils and T-cells.

Functions of LTs

LTs in general appear to act as mediators in inflammation and anaphylaxis.

- ***Capillary Dilatation and Vascular Permeability:*** As little as one n-mol of LT-C_4, D_4 or E_4 elicits erythema and wheal formation like histamine (*wheal and flare* response), also increases vascular permeability.
- ***Action on Bronchial Muscles:*** Inhalation of LTs (C_4, D_4 or E_4) causes bronchospasm.
- ***Mucus Secretion:*** LTs-C_4 and D_4 have been shown to be remarkably potent stimulators of mucus secretion from human airway tissue.
- ***Chemotaxis and Chemokinetic Action:*** LTs-B_4 has been found to stimulate chemotaxis and chemokinesis of neutrophils and eosinophils, which are found in large numbers at the site of inflammation.
- ***SRS-A (slow-reacting substance of anaphylaxis):*** SRS-A which is produced by mast cells during anaphylactic reaction, has now been shown to be mixture of LTs-like LT-C_4, D_4 and E_4. ***These leukotrienes are responsible for intense vasoconstriction of bronchial muscles, vasodilatation and increased vascular permeability seen in anaphylactic/allergic reactions.***

Formation of Different Types of LTs is shown on next page diagrammatically:

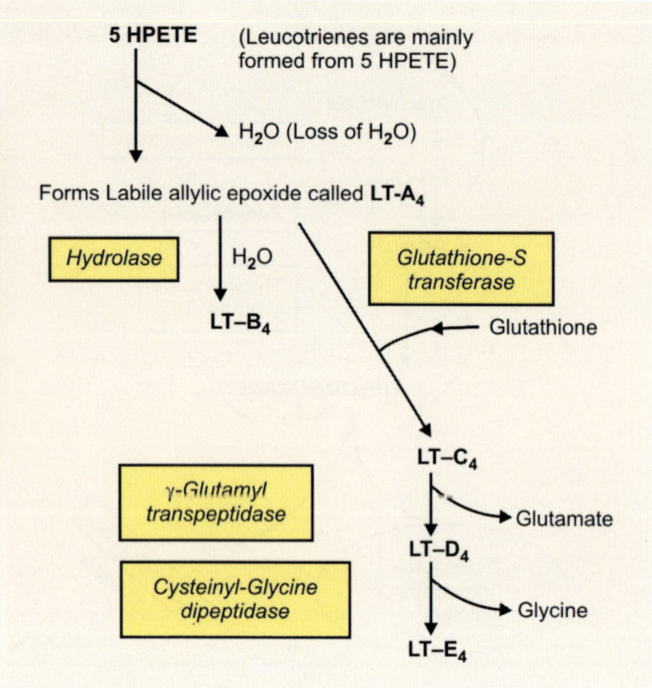

Formation of Lipoxin Lx-A_4 is shown below in the box.

Applied Aspects

- ***Lipoxygenase system is not inhibited by aspirin and other anti-inflammatory drugs.***
- ***Prolonged use of aspirin:*** Prolonged use of aspirin for certain conditions like arthritis, etc. depresses cyclo-oxygenase system and PG synthesis but enhances lipoxygenase system and synthesis of LTs, which may lead to bronchospasm and produce ***aspirin-induced (iatrogenic) bronchial asthma.***
- ***Lipoxygenase inhibitor:*** Recently it has been shown that LT-B_4 plays a key role in causing and amplification of mucosal injury in ulcerative colitis. ***Zileuton,*** a selective ***5-lipoxygenase inhibitor***, is being studied for its therapeutic efficacy in ulcerative colitis.

LIPOXINS

Lipoxins are a family of conjugated tetraenes recently discovered arising in leucocytes by lipoxygenase pathway.

Types: Several lipoxins have been found to be formed Lx-A_4 to Lx-E_4 in a manner similar to the formation of leukotrienes as stated above.

Formation: They are formed by the combined action of more than one lipoxygenase introducing more oxygen into the molecule from arachidonic acid.

Function of Lipoxins: Evidences support a role of lipoxins in vasoactive and immunoregulatory function, e.g. as counter-regulatory compounds (chalones) of the immune response.

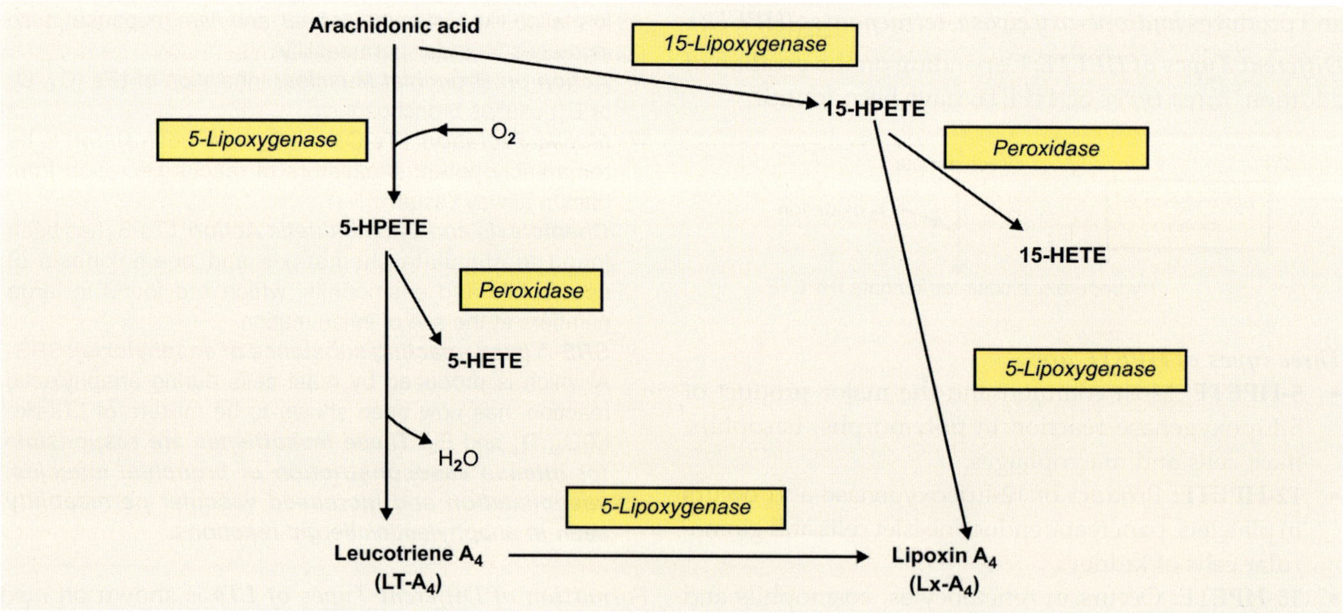

I. Questions (Essay type and short notes)

1. Describe the chemistry and functions of prostaglandins.

Short Notes

a. Prostaglandins b. Prostacyclins
c. Thromboxanes d. Leucotrienes

Differentiate

a. Prostacyclins and Thromboxanes.

II. MCQs (Give one correct answer)

1. PG-Es lower cyclic AMP level in:
   a. Lungs b. Adipose tissue
   c. Thyroids d. Platelets
   e. Spleen
2. All PGs have a structure based on prostanoic acid which contains a cyclopentane ring and has carbon atoms:
   a. 10 b. 15
   c. 18 d. 20
   e. 25
3. Regarding arachidonic acid the following statements are all true except:
   a. Can be synthesised from linoleic acid.
   b. It is a 20 carbon fatty acid with five double bonds.
   c. It is a precursor of PG-I_2
   d. It can form cyclic endoperoxide PG-H_2
   e. It is a precursor of Tx-A_2
4. PG-E_2 when infused intravenously has the following action on kidneys:
   a. Decrease the renal plasma flow (RFP)
   b. Decrease in GFR
   c. Increases reabsorption of water
   d. Increases excretion of sodium
   e. Decreases rennin secretion from JG Cells
5. The release of arachidonic acid from membrane phospholipids is inhibited by:
   a. A specific protein induced by glucocorticoids
   b. By administration of aspirin
   c. Administration of non-steroid anti-inflammatory (NSIDS)
   d. Catecholamines
   e. All of the above.
6. PG Synthesis is increased by activating phospholipases by:
   a. Mepacrine b. Angiotensin II
   c. Glucocorticoids d Indomethacin
   e. Aspirin
7. All of the following functions typifies either PG-I_2 or Tx-A_2, except:
   a. PG-I_2 contracts coronary arteries
   b. PG-I_2 dilates coronary arteries
   c. Tx-A_2 contracts coronary arteries
   d. Tx-A_2 Promotes platelets aggregation
   e. PG-I_2 inhibits platelets aggregation
8. PG-E is catabolised by oxidation of secondary hydroxyl group at
   a. C_9 position b. C_{11} position
   c. C_{15} position d. all of the above
   e. None of the above
9. The leucotrienes are characterised by all of the following except:
   a. Are formed from arachidonate
   b. Enzyme responsible for conversion of arachidonate to leucotrienes is lipoxygenase system
   c. One of LTs stimulates chemotaxis and chemokinesis of neutrophils
   d. All of them contain cyclopentane ring
   e. SRS-A is a mixture of LT-C_4, LT-D_4 and LT-E_4
10. PG-Synthesis is aerobic process and requires two molecules of reduced:
    a. FAD b. NAD^+
    c. FMN d. Glutathione
    e. $NADP^+$
11. All prostaglandins have in common in addition to cyclopentane ring one double bond between positions:
    a. C_5 and C_6 b. C_7 and C_8
    c. C_9 and C_{10} d. C_{11} and C_{12}
    e. C_{13} and C_{14}

Answers to MCQs

1. b. 2. d 3. b 4. d 5. a 6. b
7. a 8. c 9. d 10. d 11. e.

CHAPTER

6

CHEMISTRY OF PROTEINS AND AMINO ACIDS

Major Concepts

A. To know what are proteins and their biomedical importance.
B. To learn what are amino acids, their classification and properties.
C. To learn the classification and properties of proteins.
D. Learn the structure of protein.

Specific Objective

A. 1. Define protein.
   2. Describe the biomedical importance of protein and learn composition of proteins.

B. **Basic monomeric unit of protein is amino acid.**
   1. What are amino acids? Learn the basic structure of amino acid.
   2. Classify amino acids.
   3. Learn the nonstandard amino acids.
   4. Learn the occurrence of amino acids.
   5. List essential amino acids, semiessential amino acids and non-essential amino acids and why they are called so.
   6. Learn the general functions of amino acids.
   7. Learn the physical and chemical properties of amino acids.

C. 1. Classify proteins.
   - Based on size and shape, • Based on functions, • Based on solubility, structure and physical properties—Most commonly employed classification. According to this, proteins are classified as simple, conjugated and derived proteins.

   2. • Learn the physical and chemical properties of proteins.
   - Learn precipitation reaction of proteins and its application.
   - Learn various colour reactions of protein due to specific amino acid.
   - Learn the peptide linkage in a protein molecule and learn few biologically important peptides.

D. • Study the primary structure of protein.
   - Study the secondary structure of protein, linkages and types such as α-helix, β-pleated sheet structure, Triple helix, and Random coil.
   - Learn the tertiary structure, bonds involved in tertiary structure formation.
   - Learn the quaternary structure, bonds that make it and examples.
   - What is denaturation of protein? Learn various factors that cause denaturation, its application and the changes a protein molecule undergoes after denaturation.
   - Study the criteria of purity of protein.

INTRODUCTION

In 1839 Dutch chemist **GJ Mulder** while investigating substances such as those found in milk, egg found that they could be coagulated on heating and were nitrogenous compounds. Swedish scientist **JJ Berzelius** suggested to **Mulder** that these substances should be called proteins. The term is derived from Greek word ***Proteios*** means "primary", or "holding first place" or "pre-eminent" because Berzelius thought them to be most important of biological substances. And now we know that proteins are fundamental structural components of the body. They are ***nitrogenous "macromolecules" composed of many amino acids.***

BIOMEDICAL IMPORTANCE OF PROTEINS

- Proteins are the main structural components of the cytoskeleton. They are the sole source to replace Nitrogen of the body.
- Biochemical catalysts known as ***enzymes*** are proteins.
- Proteins known as ***immunoglobulins*** serve as the first line of defence against bacterial and viral infections.
- Several ***hormones*** are protein in nature.
- Structural proteins furnish ***mechanical support*** and some of them like actin and myosin are contractile proteins and help in the movement of muscle fibre, microvilli, etc.
- Some proteins present in cell membrane, cytoplasm and nucleus of the cell act as ***receptors.***

- The ***transport proteins*** carry out the function of transporting specific substances either across the membrane or in the body fluids.
- ***Storage proteins*** bind with specific substances and store them, e.g. iron is stored as ferritin.
- Few proteins are constituents of ***respiratory pigments*** and occur in electron transport chain or respiratory chain, e.g. cytochromes, hemoglobin, myoglobin.
- Under certain conditions proteins can be ***catabolised to supply energy***.
- Proteins by means of exerting osmotic pressure help in maintenance of electrolyte and water balance in body.

COMPOSITION OF PROTEINS

In addition to C, H, and O which are present in carbohydrates and lipids, proteins also contain N. ***The nitrogen content is around 16 per cent of the molecular weight of proteins.*** Small amounts of S and P are also present. Few proteins contain other elements such as I, Cu, Mn, Zn and Fe, etc.

Amino acids: Protein molecules are very large molecules with a high molecular weight ranging from 5000 to 25,00,000. Protein can be broken down into smaller units by hydrolysis. ***These small units the monomers of proteins are called as amino acids.*** Proteins are made up from, 20 such standard amino acids in different sequences and numbers. So an indefinite number of proteins can be formed and do occur in nature. ***Thus proteins are the unbranched polymers of L- α-amino acids.***

The *L*- *α*-amino acid has a general formula as shown below:

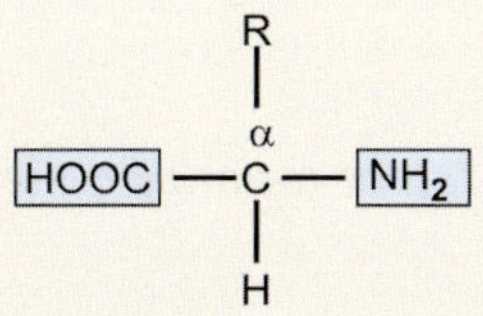

R is called a side chain and can be a hydrogen, aliphatic, aromatic or heterocyclic group. ***Each amino acid has an amino group –NH_2, a carboxylic acid group – COOH and a hydrogen atom each attached to carbon located next to the – COOH group. Thus the side chain varies from one amino acid to the other.***

AMINO ACIDS

CLASSIFICATION AND STRUCTURE OF AMINO ACIDS

Amino acids can be classified into **3 groups** depending on their reaction in solution.

A. Neutral
B. Acidic and
C. Basic.

A. Neutral amino acids: This is the largest group of amino acids and can be further subdivided into aliphatic, aromatic, heterocyclic and S-containing amino acids.

(a) Aliphatic Amino Acids:

1. Glycine (Gly) or α-amino acetic acid.

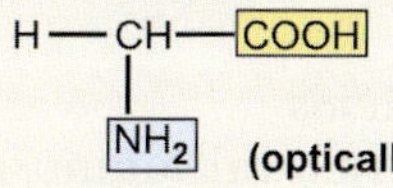

(optically inactive)

2. Alanine (Ala) or α-amino propionic acid.

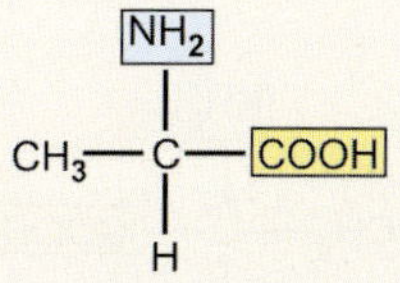

3. Valine (Val) or α-amino-isovaleric acid.

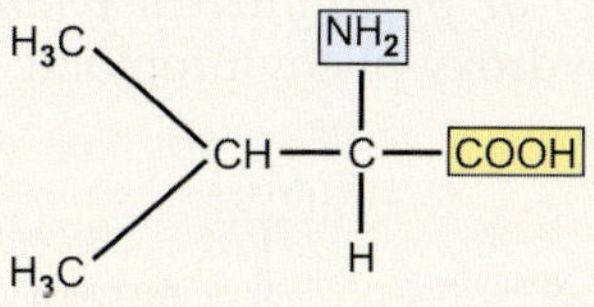

4. Leucine (Leu) or α-amino-isocaproic acid.

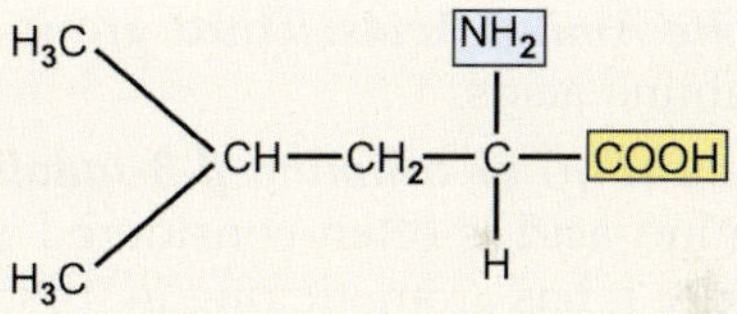

5. Isoleucine (Ile) or α-amino-β-methyl valeric acid.

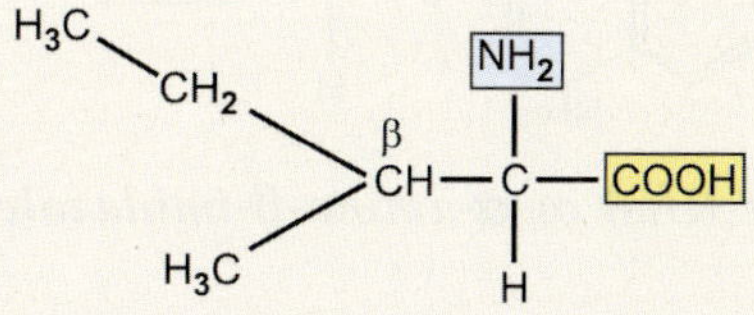

All of the above are ***simple monoamino monocarboxylic acids.*** The next from the neutral group of amino acids are ***hydroxy amino acids***. Since they contain – OH group in their side chains.

6. Serine (Ser) or α-amino-β-hydroxy propionic acid.

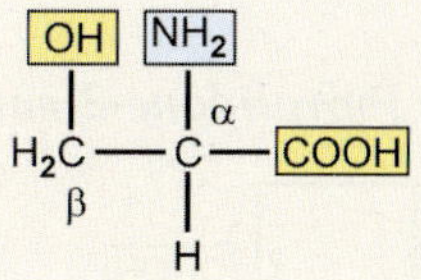

7. Threonine (Thr)* or *α-amino-β-hydroxybutyric acid.

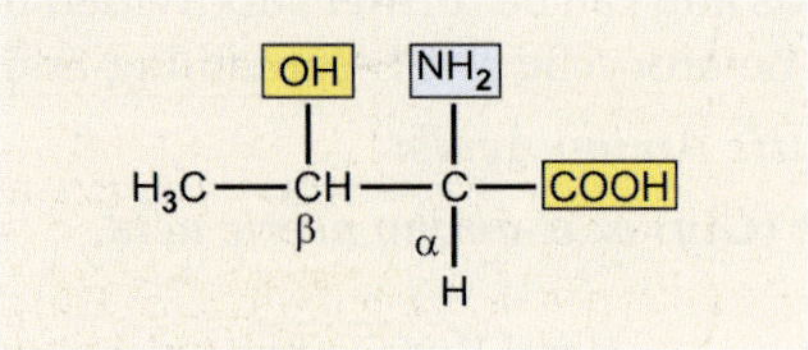

(b) Aromatic Amino Acids

Second subgroup of neutral amino acids consists of aromatic amino acids.

8. Phenylalanine (Phe) or α-amino-β-phenyl propionic acid.

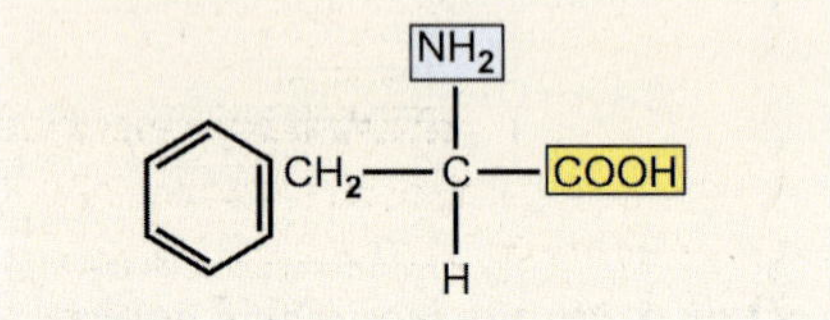

9. Tyrosine (Tyr) or parahydroxy phenylalanine or α-amino-β-parahydroxy phenylpropionic acid.

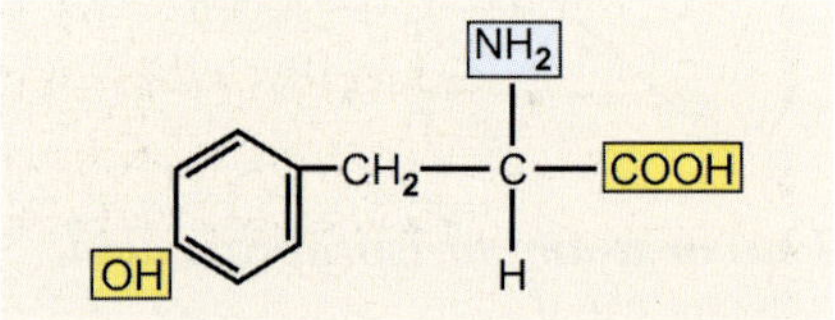

(c) Heterocyclic Amino Acids: Third group belongs to heterocyclic amino acids.

10. Tryptophan (Trp) or α-amino-β-3-indole propionic acid. This amino acid is often considered as aromatic amino acid since it has aromatic ring in its structure.

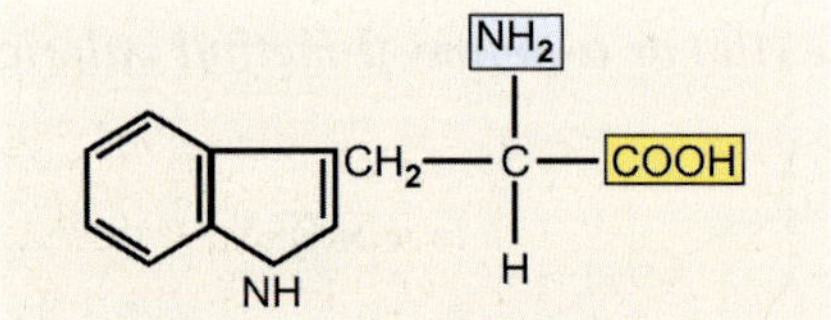

11. Histidine (His) or α-amino-β-imidazole propionic acid.

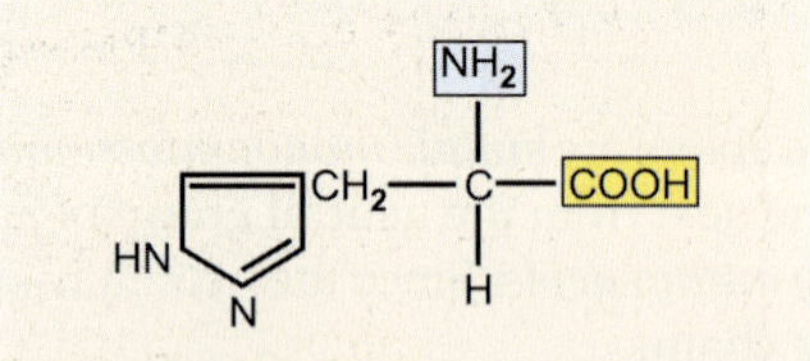

Histidine is basic in solution on account of the imidazole ring and often considered as Basic Amino acid.

(d) Imino Acids

12. Proline (Pro) or Pyrrolidone-2-carboxylic acid.

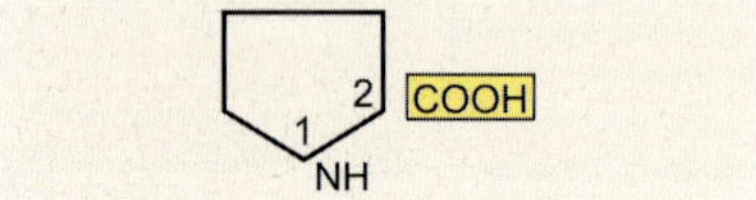

13. Hydroxyproline (Hyp) or 4 hydroxy pyrrolidone-2 carboxylic acid.

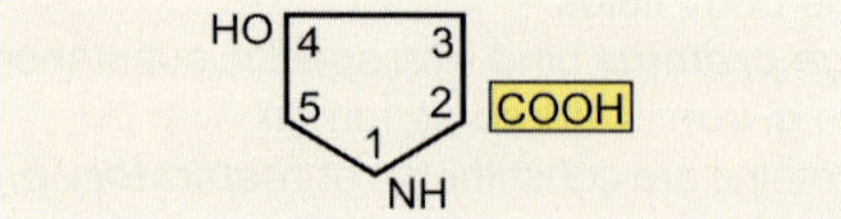

Proline and Hydroxyproline ***do not have a free –NH_2 group*** but ***only a basic pyrrolidone ring*** in which the Nitrogen of the Imino group is in a ring but can still function in the formation of peptides. These amino acids are therefore called as ***imino acids.***

(e) 'S' containing amino acids: The fourth subgroup of neutral amino acids contains two sulphur containing amino acids.

14. Cysteine (Cys) or α-amino-β-mercaptopropionic acid.

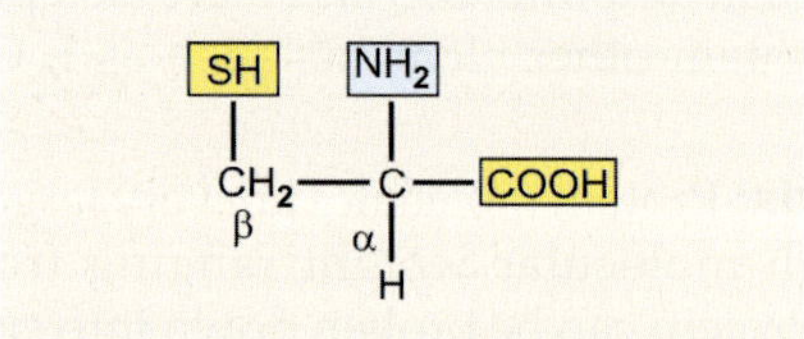

Two molecules of cysteine make cystine (cys-cys) or dithio β, β-α aminopropionic acid. ***The S–S linkage is called as disulphide bridge.***

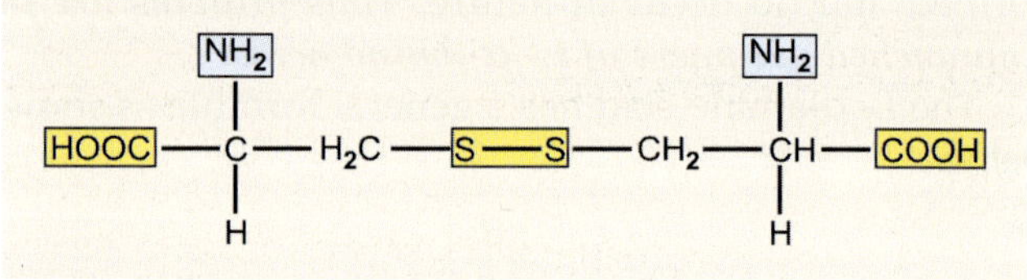

15. Methionine (Met) or α-amino γ-methylthio-η-butyric acid.

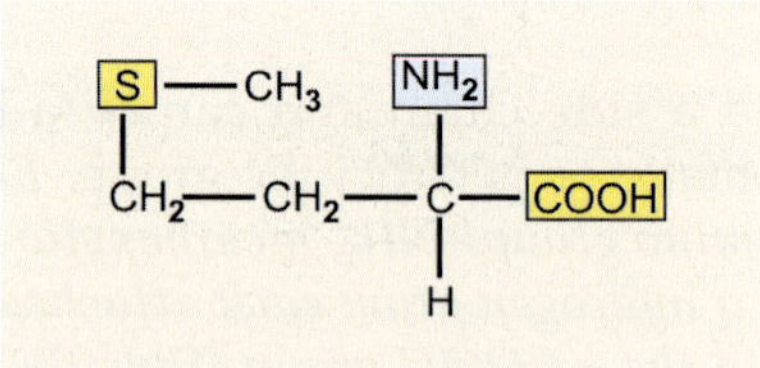

B. Acidic amino acids: These amino acids have two –COOH groups and one – NH_2 group. They are therefore ***monoaminodicarboxylic acids.***

16. Aspartic acid (Asp) or α-amino succinic acid.

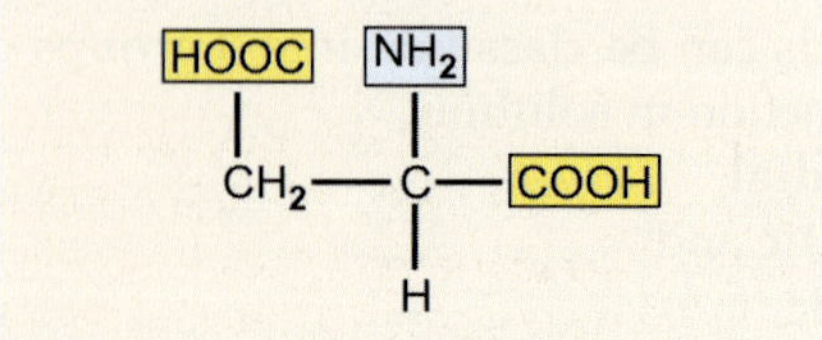

Asparagine (Asn) or γ-amide of α-aminosuccinic acid.

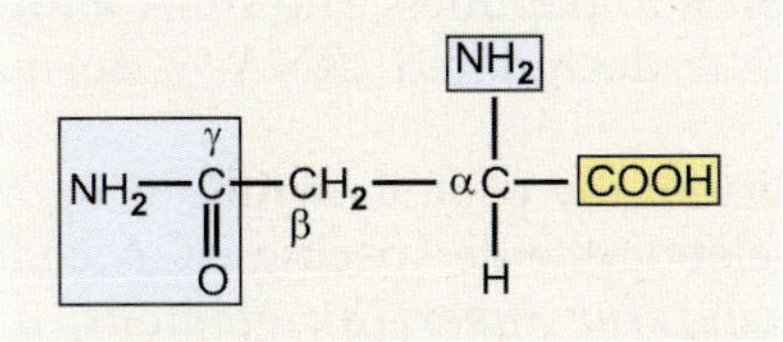

17. Glutamic Acid (Glu) or α-aminoglutaric acid.

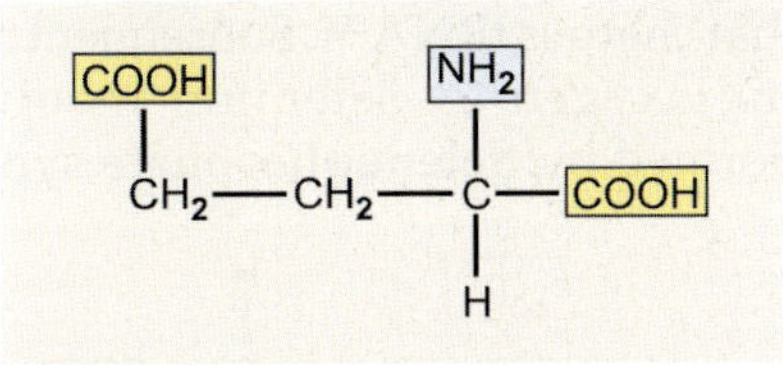

Glutamine (Gln)-Amide of Glutamic Acid or δ-amide of α-aminoglutaric acid.

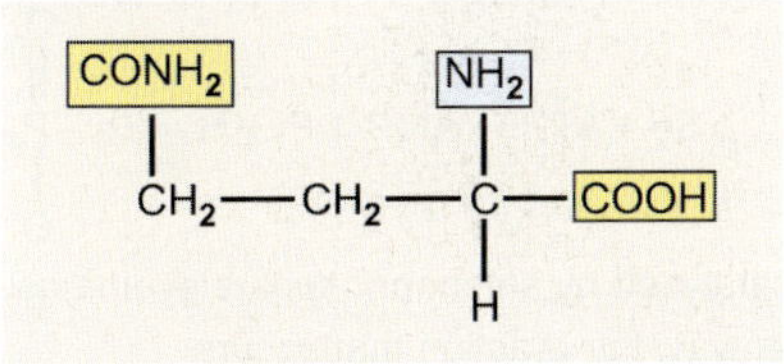

C. Basic amino acids: This class of amino acids consists of those amino acids which have one – COOH group and two $-NH_2$ groups. Thus they are ***diamino monocarboxylic acids.*** Arginine, lysine and hydroxylysine are included in this group.

18. Arginine (Arg) or α-amino- δ-guanidino-n-valeric acid.

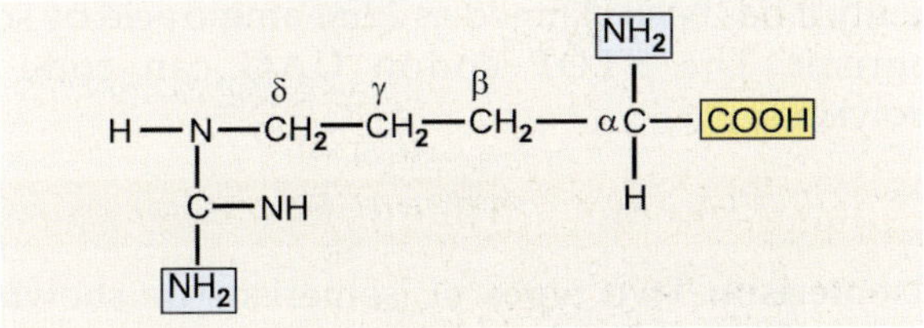

19. Lysine (Lys) or α- ε-diamino caproic acid.

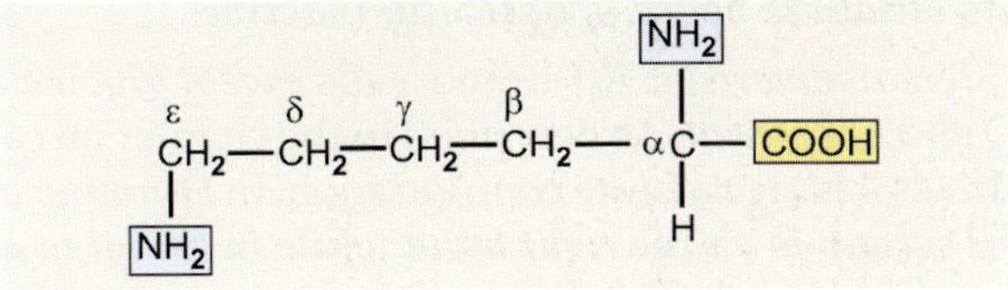

20. Hydroxylysine (Hyl) or α-ε-diamino-δ-hydroxy-n-valeric acid.

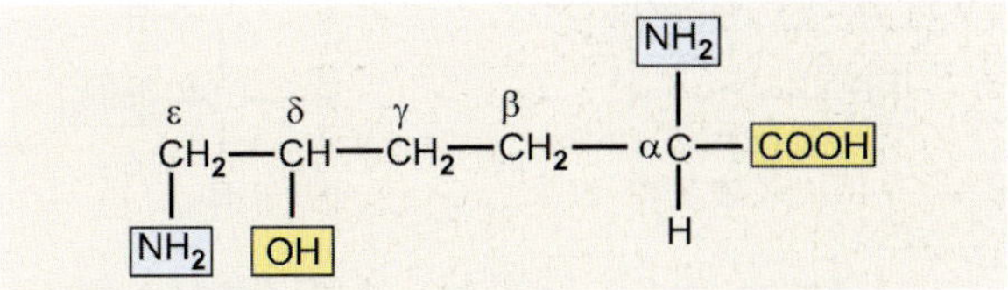

As already mentioned histidine is also classified as basic amino acid.

Non-standard Amino Acids

A. The compounds similar to basic structure of amino acids but do not occur in proteins. Examples of some of those are:

- **β-alanine:** They found in coenzyme A.
- **Taurine:** They found in bile acids
- **Ornithine and citrulline:** They are intermediates in urea cycle
- **Thyroxine (T_4) and Tri-iodo Thyronine (T_3):** Thyroid hormones synthesised from tyrosine.
- **γ-aminobutyric acid (GABA):** A neurotransmitter produced from glutamic acid.
- **β-amino isobutyric acid:** These are end product of pyrimidine metabolism.
- **δ-aminolaevulinic acid (δ-ALA):** These are intermediate in haem synthesis.
- **S-adenosyl methionine (SAM):** These are methyl donor formed from L-methionine
- **3, 4-dihydroxy phenyl alanine (DOPA):** A precursor of mela nine pigment.

B. **D-amino acids:** These are non-standard amino acids—Amino acids normally isolated from animal and plants are L-amino acids. But certain D-amino acids are found in bacteria and antibiotics and in brain tissues of animals.

- **D-glutamic acid** and **D-Alanine** are constituents of bacterial cell walls.
- **D-amino acids** are found in **certain antibiotics,** e.g. gramicidin-S, Actinomycin-D.
- Animal tissues contain L-amino acids which are deaminated by L-amino acid oxidase. But there is also present D-amino acid oxidase the function of which was not known. Now D-amino acids like D-aspartate and D-serine have been found in brain tissue. This explains the existence of D-amino acid oxidase.

Functions of Amino Acids

Apart from being the monomeric constituents of proteins and peptides, amino acids serve variety of functions.

(a) Some amino acids are converted to carbohydrates and are called as ***glucogenic amino acids***.

(b) Specific amino acids give rise to specialised products, e.g.

- **Tyrosine** forms hormones such as ***thyroid hormones***, (T_3, T_4), ***epinephrine*** and ***norepinephrine*** and a pigment called ***melanin.***
- **Tryptophan** can synthesise a vitamin called ***niacin.***
- Glycine, arginine and methionine synthesise ***creatine.***
- Glycine and cysteine help in **synthesis of Bile salts**.
- Glutamate, cysteine and glycine synthesise ***glutathione.***
- **Histidine** changes to ***histamine*** on decarboxylation.
- **Serotonin** is formed from tryptophan.
- Glycine is used for the synthesis of ***haem.***
- Pyrimidines and purines use several amino acids for their synthesis such as aspartate and glutamine for pyrimidines and glycine, aspartic acid, Glutamine and serine for purine synthesis.

(c) Some amino acids such as glycine and cysteine are used as detoxicants of specific substances.
(d) Methionine acts as "active" methionine (S-adenosyl-methionine) and transfers methyl group to various substances by transmethylation.
(e) Cystine and methionine are sources of sulphur.

Essential Amino Acids

Nutritionally, amino acids are of **two types: (a) Essential and (b) Non-essential**. *(c)* There is also a third group of *semi-essential amino acids.*

(a) Essential amino acids: These are the ones which are not synthesised by the body and must be taken in diet. They include ***valine, leucine, isoleucine, phenylalanine, threonine, tryptophan, methionine and lysine. For remembering the following formula is used—MATT VIL PHLY.***

(b) Non-essential amino acids: They can be synthesised by the body and may not be the requisite components of the diet.

(c) Semi-essential amino acids: These are ***growth promoting factors*** since they are not synthesised in sufficient quantity during growth. They include ***arginine*** and ***histidine***. They ***become essential in growing children, pregnancy and lactating women.***

Occurrence of amino acids: All the standard amino acids mentioned above occur in almost all proteins. Cereals are rich in acidic amino acids Asp and Glu while collagen is rich in basic amino acids and also proline and hydroxy-proline.

New Amino Acids

In addition to 20 L-amino acids that take part in protein synthesis, recently two more new amino acids described. They are:

A. Selenocysteine - 21st amino acids
B. Pyrrolysine - 22nd amino acid

A. Selenocysteine

Selenocysteine is recently introduced as 21st amino acid. Selenocysteine ***occurs at the "active site" of several enzymes.***

Examples include:
- ***Thioredoxin reductase***
- ***Glutathione peroxidase*** which scavenges peroxides,
- ***De-iodinase*** that converts thyroxine to tri-iodothyronine
- ***Glycine reductase***
- ***Selenoprotein P***, a glycoprotein containing 10 selenocysteine residues, found in mammalian blood. It has an ***antioxidant function*** and its concentration falls in selenium deficiency.

Selenocysteine arises co-translationally during its incorporation into peptides. The **UGA anticodon** of the unusual tRNA designated tRNA[sec], normally signals **STOP**.

The ability of the protein synthesising apparatus to identify a selenocysteine specific UGA codon ***involves the selenocysteine insertion element, a stem-loop structure*** in the untranslated region of the m-RNA.

Selenocysteine-tRNA[sec] is first charged with **serine** by the **Ligase** that charges tRNA[sec]. Subsequent replacement of the serine oxygen by selenium involves seleno-phosphate formed by **Selenophosphate synthetase**.

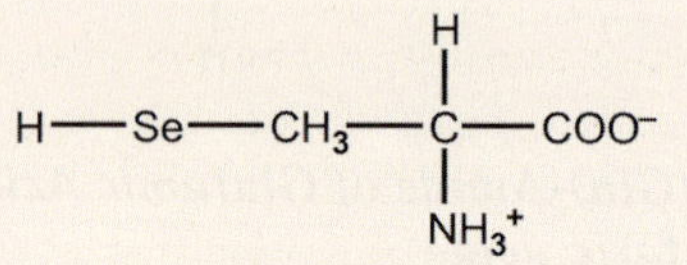

(Selenocysteine)

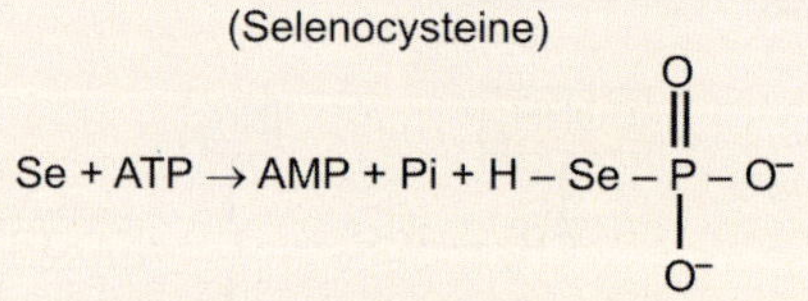

Reaction catalysed by selenophosphate synthetase

In a simpler way, the reaction that occurs

$$Se + ATP \xrightarrow{\text{Selenophosphate synthetase}} Se-P + AMP + Pi$$

$$Serine + Se\text{–}P \longrightarrow Secys + Pi$$

(Selenocysteine)

B. Pyrrolysine—the 22nd Amino Acid

Recently it has been claimed as 22nd amino acid by some scientists. The **STOP** codon **UAG** can code for pyrrolysine.

PROPERTIES OF AMINO ACIDS

A. Isomerism: Two types of isomerism are shown by amino acids basically ***due to the presence of asymmetric carbon atom. Glycine has no asymmetric carbon atom in its structure hence is optically inactive.***

(a) Stereoisomerism: All amino acids ***except glycine*** exist in D and L isomers. As described in the chapter on carbo-hydrates it is an absolute configuration. In ***D-amino acids –NH_2 group is on the right hand while in L-amino acids it is oriented to the left***. It is the same orientation of –OH group of the central carbon of glyceraldehyde **(Fig. 6.1)**.

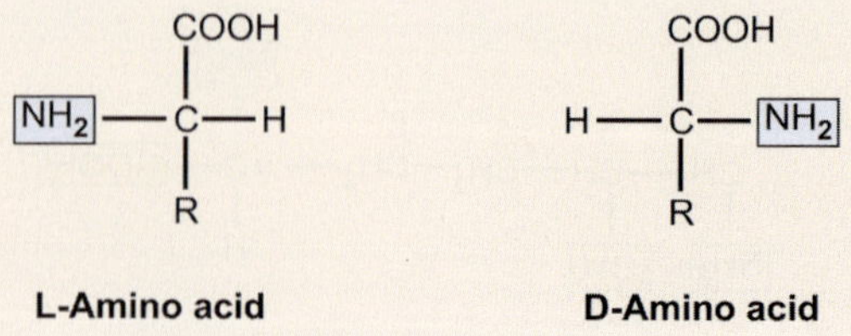

Fig. 6.1: L and D-forms of amino acid

Natural proteins of animals and plants generally contain L-amino acids. D-amino acids occur in bacteria.

(b) Optical Isomerism: All amino acids *except glycine* have asymmetric carbon atom. Few amino acids like isoleucine and threonine have an additional asymmetric carbon in their structures. ***Consequently all but glycine exhibit 'optical' activities and rotate the plane of plane polarised light and exist as dextrorotatory (d) or laevorotatory (l) isomers.*** Optical activity depends on the pH and side chain.

B. Amphoteric Nature and Isoelectric pH: ***The -NH$_2$ and -COOH groups of amino acids are ionizable groups.*** Further, charged polar side chains of few amino acids also ionise. **Depending on the pH of the solution these groups act as proton donors (acids) or proton acceptors (bases).** This property is called as amphoteric and therefore amino acids are called as **ampholytes**. ***At a specific pH the ami11no acid carries both the charges in equal number and exists as dipolar ion or "Zwitterion".*** At this point the net charge on it is zero, i.e. positive charges and negative charges on the protein/amino acid molecule equalizes. ***The pH at which it occurs without any charge on it is called pI or isoelectric pH.*** On the acidic side of its pI amino acids exist as a cation by accepting a proton and on alkaline as anion by donating a proton.

C. Physical Properties: They are colourless, crystalline substances, more soluble in water than in polar solvents. Tyrosine is soluble in hot water. They have high melting point usually more than 200°C. They have a high dielectric constant. They possess a large dipole moment.

D. Chemical Properties

I. Due to Carboxylic (—COOH) Group

1. Formation of esters: They can form esters with alcohols. The COOH group can be esterified with alcohol. Treatment with Na_2CO_3 solution in cold releases the free ester from ester hydrochloride.

2. Reduction to amino alcohol: This is achieved in presence of lithium aluminium hydride.

3. Formation of amines by decarboxylation: Action of specific amino acid decarboxylases, dry distillation or heating with $Ba(OH)_2$ or with diphenylamine evolves CO_2 from the —COOH group and changes the amino acid into its amine (Fig. 6.2).

In vivo, the amino acids can be decarboxylated by the enzyme ***decarboxylase*** and forms the corresponding amines.

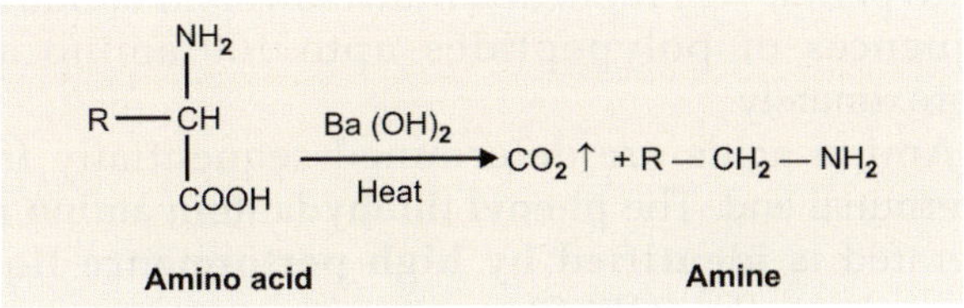

Fig. 6.2: Formation of amine (decarboxylation)

4. Formation of amides: Anhydrous NH_3 may replace alcohol from its combination with an amino acid in an amino acid ester so that an amide of amino acid and a molecule of free alcohol is produced (Fig. 6.3).

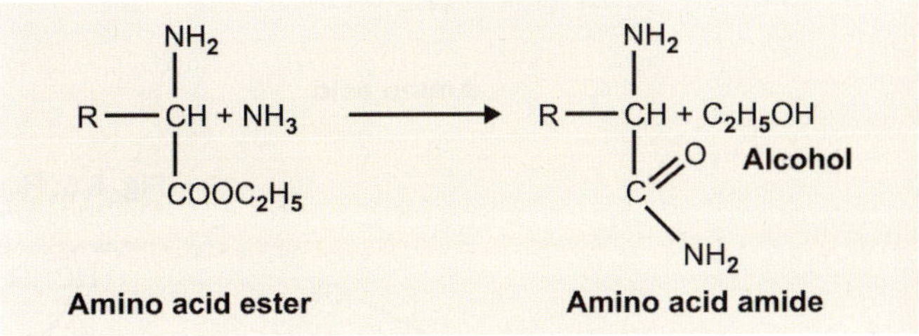

Fig. 6.3: Formation of amide

II. Properties Due to Amino (–NH$_2$) Group

1. Salt formation with acids: The basic amino group reacts with mineral acids such as HCl to form salts like hydrochlorides (Fig. 6.4).

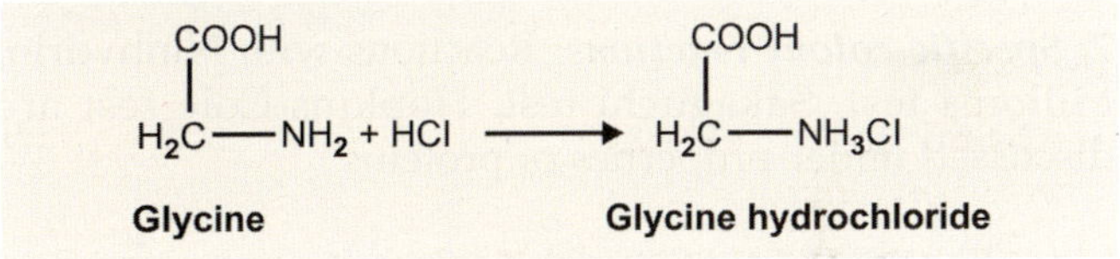

Fig. 6.4: Salt formation of amino acid

2. Formation of acyl derivatives: Amino group reacts with acyl anhydride or acyl halides such as benzoyl chloride and give acyl amino acids like benzoyl glycine (hippuric acid). Incidentally, ***this is one of the mechanisms of detoxication in which glycine is used and this also forms the basis of one of the liver function tests.***

3. Oxidation: Potassium permanganate or H_2O_2 oxidises the NH_2 group and converts the amino acid into imino acid which reacts with water to form NH_3 and α-keto-acid.

4. Reaction with HNO$_2$: Like other primary amines, the amino acids **except proline and hydroxyproline** react with HNO_2 (nitrous acid) libering N_2 from NH_2 group. **This forms the basis of Van Slyke's method for determining -NH$_2$ group (Nitrogen)** (Fig. 6.5).

5. Reaction with CO$_2$: The amino acid anion present in an alkaline solution may react with CO_2 through NH_2 group to form a carbaminoacid anion.

6. Reaction with formaldehyde: Formaldehyde reacts with the-NH_2 group to form a methylene compound.

Application: Because of the presence of free basic amino group in the amino acid molecule its amount cannot be estimated directly by titration with a **standard alkali.** On addition of neutral formaldehyde it combines with the amino group to form either methylene amino acid or dimethylol amino acid. Both these products are strong acids and may be estimated

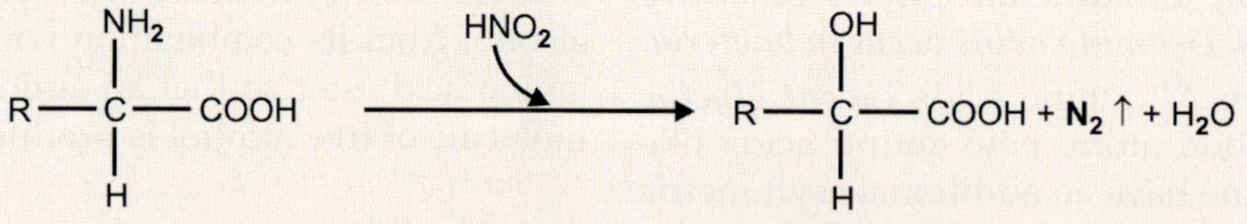

α-Amino acid　　α-Hydroxy acid

Fig. 6.5: Reaction with HNO_2

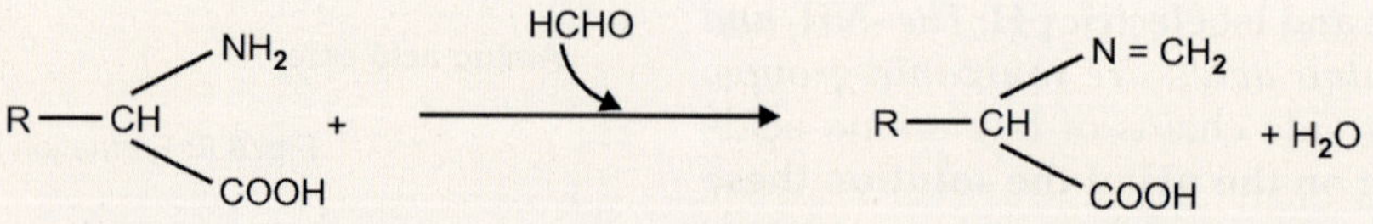

Fig. 6.6: Reaction of glycine with formaldehyde-Basis of Sorensen's formol titration

by titration with a standard alkali. This is known as **"Sorensen's" formol titration method** (Fig. 6.6).

7. Specific colour reactions: Reactions with Ninhydrin, Millon's test, Sakaguchi test, Hopkins-Cole test are discussed under properties of proteins.

Identification of N-terminal Residue

(a) N-terminal residue can be identified by using a reagent that bonds covalently with its α-NH_2 group. Because the bond is stable to hot acid hydrolysis, the derivative of the N-terminal residue can be identified by chromatographic procedures after the protein has been hydrolysed.

Two reagents are commonly used

1. *Sanger's reagent:* The reagent contains 1-fluoro-2, 4-dinitrobenzene (FDNB). It reacts with free $-NH_2$ group in an alkaline medium

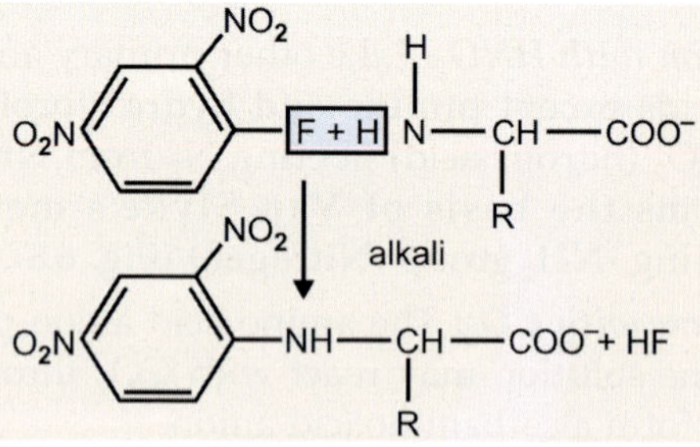

The reaction can also take place with the N-terminal $-NH_2$ group of the polypeptide chain. The compound so formed can be isolated after protein hydrolysis and identified. **Sanger** was first to sequence a polypeptide. He determined the complete primary structure of the hormone insulin.

2. *Reaction with Dansyl Chloride:* The N-terminal $-NH_2$ group can also combine with Dansyl chloride (1-dimethyl aminonaphthalene-5-sulphonyl chloride) to form a fluorescent dansyl derivative which can be isolated and identified.

(b) Edman reaction: A similar reaction with $-NH_2$ group can occur with the reagent phenyl isothiocyanate and thus enables the identification of the N-terminal amino acid (Refer box in right side).

Edman Reaction

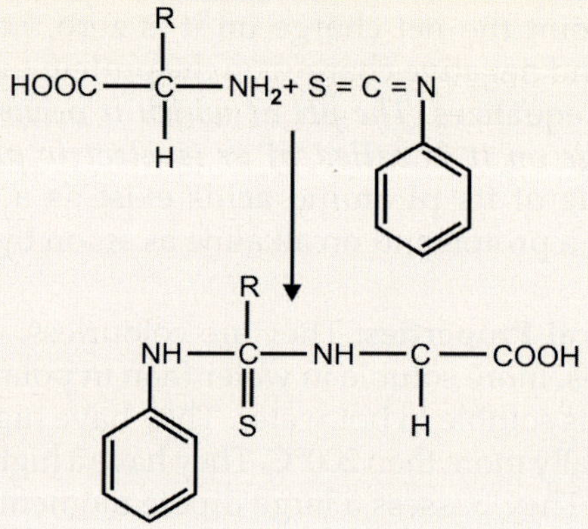

Sequenator

Edman and **G. Begg** have perfected an automated amino acid **sequenator** for carrying out sequential degradation of peptides by the phenylisothiocyanate procedure (Edman's reaction).

Automated amino acid sequencers now widely used, which permit very rapid determination of the amino acid sequences of polypeptides upto 100 amino acid approximately.

Amino acids are determined sequentially from N-terminal end. The **phenyl thiohydantoin amino acid** liberated is **identified by high performance liquid chromatography (HPLC).**

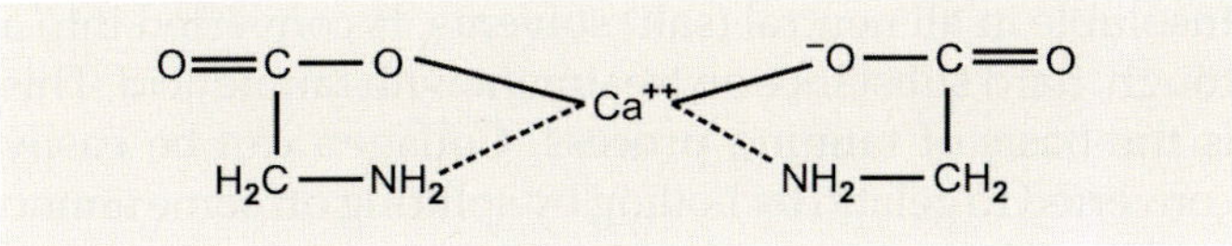

Fig. 6.7: Chelation of glycine with Ca^{++}. Name of compound is calcium diglycinate

III. Properties of Amino acids Due to Both NH_2 and COOH Groups

In addition to the property of reacting with both cation and anion, the amino acids form chelated, co-ordination complexes with certain heavy metals and other ions. These include Cu^{++}, Co^{++}, Mn^{++} and Ca^{++}. An example of chelated complex of Ca and glycine is given in Figure 6.7.

CLINICAL APPLICATION

Chelates are non-ionic and therefore amino acids may be used to remove calcium from bones and teeth. It is possible that the amino acids resulting from the breakdown of enamel and dentine could in this way form soluble calcium complexes thereby causing a loss of calcium and the development of **caries.**

PROTEINS

CLASSIFICATION OF PROTEINS

Proteins are classified:

I. **On the basis of shape and size**
II. **On the basis of functional properties**
III. **On the basis of solubility and physical properties.**

I. On the basis of shape and size

- ***Fibrous proteins:*** When the axial ratio of length: width of a protein molecule is more than 10, it is called a ***fibrous protein.***
  Examples: α-keratin from hair, collagen.
- ***Globular protein:*** When the axial ratio of length: width of a protein molecule is less than 10, it is called as ***globular protein.***
  Examples: Myoglobin, haemoglobin, ribonuclease, etc.

II. On the basis of functional properties: The second way of classifying proteins makes use of their **functional properties,** such as:

- ***Defence proteins:*** Immunoglobulins involved in defence mechanisms.
- ***Contractile proteins:*** Proteins of skeletal muscle involved in muscle contraction and relaxation.
- ***Respiratory proteins:*** Involved in the function of respiration, like haemoglobin, myoglobin, cytochromes.
- ***Structural proteins:*** Proteins of skin, cartilage, nail.
- ***Enzymes:*** Proteins acting as enzymes.
- ***Hormones:*** Proteins acting as hormones.

III. On the basis of solubility and physical properties: However, both the above classification schemes have many overlapping features. Therefore a third most acceptable scheme of classification of proteins is adopted. According to this scheme ***proteins are classified on the basis of their solubility and physical properties and are divided in three different classes.***

A. Simple proteins: These are proteins which on complete hydrolysis yield only amino acids.

B. Conjugated proteins: These are proteins which in addition to amino acids ***contain a non-protein group called prosthetic group*** in their structure.

C. Derived proteins: These are the proteins formed from native protein by the action of heat, physical forces or chemical factors.

A. Simple Proteins

These are further ***subclassified based on their solubilities and heat coagulabilities.*** These properties depend on the size and shape of the protein molecule. Major subclasses of simple proteins are as follows:

1. Protamines

These are small molecules and are soluble in water, dilute acids and alkalies and dilute ammonia and ***non-coagulable by heat.*** They do not contain cysteine, tryptophan and tyrosine but **are rich in arginine.** Their isoelectric pH is around 7.4 and they exist as basic proteins in the body. They combine with nucleic acids to form nucleoproteins.

Examples: Salmine, sardinine and cyprinine of fish (sperms) and testes.

2. Histones

These are basic proteins, ***rich in arginine and histidine,*** with alkaline isoelectric pH. They are soluble in water, dilute acids and salt solutions but insoluble in ammonia. They ***do not readily coagulate on heating.*** They form conjugated proteins with nucleic acids (DNA) and porphyrins. They act as repressors of template activity of DNA in the synthesis of RNA. The protein part of hemoglobin, **globin is an atypical histone having a predominance of histidine and lysine instead of arginine.**

Examples: Nucleohistones, chromosomal nucleoproteins and globin of haemoglobin.

For details of Histone case refer to chaper on chemistry of Nucleic acids.

3. Albumins

These are proteins which are soluble in water and in dilute salt solutions. They are ***coagulable by heat*** and are

SECTION TWO

changed to products that are insoluble in water and solutions of salt. The albumins may be ***precipitated (salted out) of solution by saturating the solution with ammonium sulphate.*** Albumins have low **isoelectric pH of pI 4.7** and therefore they are acidic proteins at the pH 7.4. They are generally deficient in glycine.

Examples: ***Plant albumins:*** Legumelin in legumes, leucosin in cereals.

Animal source: Ovalbumin in egg, lactalbumin in milk.

4. Globulins

Globulins are insoluble in water but soluble in dilute neutral salt solutions. They are also ***heat coagulable.*** Vegetable globulins coagulate rather completely. ***They are precipitated (salted out) by half saturation with ammonium sulphate or by full saturation with sodium chloride.*** Globulins bind with heme, e.g. hemopexin, with metals, e.g. transferrin, ceruloplasmin and with carbohydrates, e.g. immunoglobulins.

Examples: In addition to above, ovoglobulin in eggs, lactoglobulin in milk, legumin from legumes.

5. Gliadins (Prolamines)

Alcohol soluble plant proteins, insoluble in water or salt solutions and absolute alcohol, but they dissolve in 50 to 80 per cent ethanol. They are **very rich in proline, but poor in lysine.**

Examples: Gliadin of wheat and hordein of barley.

6. Glutelins

These are plant proteins, insoluble in water or neutral salt solutions, but soluble in dilute acids or alkalies. They are **rich in glutamic acid**. They are large molecules and can be coagulated by heat.

Examples: Oryzenin of rice and glutelin of wheat.

7. Scleroproteins or Albuminoids

These are **fibrous proteins** with great stability and very low solubility and form supporting structures of animals. In this group are found ***keratins, collagens and elastins.***

(a) Keratins: These are characteristic constituents of chidermal tissue such as horn, hair, nails, wool, hoofs and feathers. All **hard keratins** on hydrolysis yield as part of their amino acids, histidine, lysine and arginine in the ratio of 1:4:12. The **soft or pseudokeratins** such as those occurring in the outermost layers of the skin do not have these amino acids in the same ratio. ***In neurokeratin the ratio is 1:2:2. Human hair has a higher content of cysteine*** than that of other species it is called ***α-keratin. β-keratins are deficient in cysteine and, rich in glycine and alanine.*** They are present in spider's web, silk and reptilian scales.

(b) Collagen: A protein found in connective tissue and bone as long, thin, partially crystalline substance. Insoluble in all neutral (salt) solvents. Is converted into a tough, hard substance on treatment with tannic acid. This is the basis of tanning process. Collagen can be easily converted to gelatin by boiling by splitting off some amino acids.

Gelatin is highly soluble and easily digestible. It forms a gel on cooling and is provided as diet for invalids and convalescents. ***It is not a complete protein as it lacks an amino acid tryptophan which is an essential amino acid.***

(c) Elastins: These are the proteins present in yellow elastic fibre of the connective tissue, ligaments and tendons. They are rich in non-polar amino acids such as alanine, leucine, valine and proline. They do not contain cysteine, methionine, 5-hydroxylysine and histidine. They are ***formed in large amount in the uterus during pregnancy***. Elastins are hydrolysed by pancreatic *elastase* enzyme.

B. Conjugated Proteins

Conjugated proteins are simple proteins combined with a non-protein group called prosthetic group. Protein part is called apoprotein, and entire molecule is called holoprotein.

1. Nucleoproteins

The nucleoproteins are compounds made up of simple basic proteins such as protamine or histone with Nucleic Acids as the prosthetic group. They are proteins of cell nuclei and apparently are the chief constituents of **chromatin**. They are most abundant in tissues having large proportion of nuclear material such as yeast, asparagus tips in plants, thymus, other glandular organs and sperm.

Deoxyribonucleoproteins: They contain DNA as prosthetic group, are found in nuclei, mitochondria and chloroplasts.

Ribonucleoproteins: They occur in nucleoli and ribosome granules. They have *RNA* as prosthetic group.

Examples: Nucleohistone and nucleoprotamine.

2. Mucoproteins or Mucoids

Mucoproteins are the simple proteins combined with **mucopolysaccharides (MPS)** such as hyaluronic acid and the chondroitin sulphate. ***They contain large quantities of N-acetylated hexosamine*** (>4%) and in addition substances such as uronic acid, sialic acid and mucopolysaccharides are also present. Water soluble mucoproteins have been obtained from serum, egg white (α-Ovomucoid) and human urine. These water soluble mucoproteins are not easily denatured by heat or readily precipitated by picric acid or trichloroacetic acid. They have hexosamine and hexose sugars as the prosthetic groups. Mucoproteins are present in large amounts in umbilical cord. They are also present in all kinds of ***mucins*** and blood group substances. Several gonadotropic hormones such as FSH, LH and HCG are mucoproteins. Insoluble mucoproteins are found in egg white (β-ovomucoid), vitreous humour and submaxillary glands.

3. Glycoproteins

Glycoproteins are the proteins with carbohydrate moiety as the prosthetic group. **Karl Meyer** suggested that ***these proteins carry a small amount of carbohydrates <4 per cent such as serum albumin and globulin.*** Carbohydrate is bound much more firmly in the glycoproteins than the mucoprotein. Glycoproteins include mucins, immunoglobulins, complements and many enzymes. They carry mannose, galactose, fucose, xylose, arabinose in their oligosaccharide chains.

4. Chromoproteins

These are proteins that contain coloured substance as the prosthetic group.

(a) Haemoproteins: All hemoproteins are chromoproteins which carry ***haem as the prosthetic group*** which is a red coloured pigment found in these proteins.

- *Haemoglobin:* Respiratory protein found in RB Cells (*See* chapter on haemoglobin for details).
- *Cytochromes:* These are the mitochondrial enzymes of the respiratory chain.
- *Catalase:* This is the enzyme that decomposes H_2O_2 to water and O_2.
- *Peroxidase:* It is an oxidative enzyme.

(b) Others

- *Flavoprotein:* It is a cellular oxidation-reduction protein which has ***riboflavin*** a constituent of B-complex vitamin as its prosthetic group. This is yellow in colour.
- *Visual purple:* It is a protein of the retina in which the prosthetic group is a carotenoid pigment which is purple in colour.

5. Phosphoproteins

These are the proteins with phosphoric acid as organic phosphate but not the phosphate containing substances such as nucleic acids and phospholipids. **(a) Casein and (b) Ovovitellin** are the two important groups of phosphoproteins found in milk, egg-yolk respectively. They ***contain about 1 per cent of phosphorus***. Similar proteins are stated to be present in fish eggs. They are sparingly soluble in water, and very dilute acid in cold but readily soluble in very dilute alkali. The phosphoric acid which is esterified through the –OH groups of serine and threonine is liberated from organic combination by warming with NaOH and can only be detected by Ammonium Molybdate.

6. Lipoproteins

The lipoproteins are formed in combination with lipids as their prosthetic group (Refer chapter on Metabolism of Lipids).

7. Metalloproteins

As the name indicates, they contain a metal ion as their prosthetic group. Several enzymes contain metallic elements such as Fe, Co, Mn, Zn, Cu, Mg, etc.

*Examples: **Ferritin:*** Contains Fe, ***Carbonic Anhydrase:*** Contains Zn, ***Ceruloplasmin:*** Contains Cu.

C. Derived Proteins: This class of proteins includes those protein products formed from the simple and conjugated proteins. It is not a well defined class of proteins. These are ***produced by various physical and chemical factors*** and are divided in ***two major groups.***

(a) Primary derived proteins: Denatured or coagulated proteins are placed in this group. Their molecular weight is the same as native protein, but they differ in solubility, precipitation and crystallisation. Heat, X-ray, UV rays, vigorous shaking, acid, alkali cause denaturation and give rise to primary derived proteins. ***There is an intramolecular rearrangement*** leading to changes in their properties such as solubility. ***Primary derived proteins are synonymous with denatured proteins in which peptide bonds remain intact.***

1. Proteans: These are insoluble products formed by the action of water, very dilute acids and enzymes. They are predominantly formed from certain globulins.

*Example: **Myosan:*** From myosin, ***Edestan:*** From elastin and ***Fibrin:*** From fibrinogen.

2. Metaproteins: They are formed from further action of acids and alkalies on proteins. They are generally soluble in dilute acids and alkalies but insoluble in neutral solvents, e.g. acid and alkali metaproteins.

3. Coagulated proteins: The coagulated proteins are insoluble products formed by the action of heat or alcohol on native proteins.

Examples: include cooked meat, cooked egg albumin and alcohol precipitated proteins.

(b) Secondary derived proteins: These are the proteins ***formed by the progressive hydrolysis of proteins at their peptide linkages.*** They represent a great complexity with respect to their size and amino acid composition. They are roughly called as proteoses, peptones and peptides according to relative average molecular size.

1. Proteoses or albumoses: These are the hydrolytic products of proteins which are ***soluble in water*** and are ***coagulated by heat*** and are precipitated from their solution by saturation with Ammonium Sulphate.

SECTION TWO

2. Peptones: These are the hydrolytic products of proteoses. They are soluble in water, ***not coagulated by heat*** and not precipitated by saturation with Ammonium sulphate. They can be precipitated by phosphotungstic acid.

Examples: Protein products obtained by the enzymatic digestion of proteins.

3. Peptides: Peptides are composed of only a small number of amino acids joined as **peptide bonds**. They are named according to the number of amino acids present in them.

- **Dipeptides**—made up of two amino acids,
- **Tripeptides**—made of three amino acid, etc. Peptides are ***water soluble*** and are ***not coagulated by heat***, are not salted out of solution and can be precipitated by phosphotungstic acid.

Hydrolysis: The complete hydrolytic decomposition of a protein generally follows the stages given below:

Protein→ Protean→ Metaprotein→ Proteose
Aminoacids← Peptides← Peptone←

Note: The products from protein to peptone give a positive Biuret reaction and are relatively large molecules. ***The dipeptide and amino acids do not give Biuret positive reaction and therefore are called as Abiuret products.***

GENERAL PROPERTIES OF PROTEINS

- **Taste:** They are tasteless. However, the hydrolytic products (derived proteins) are bitter in taste.
- **Odour:** They are odourless. When heated to dryness they turn brown and give off the odour of burning feather.
- **Molecular weight:** The proteins in general have a large molecular weight. Proteins therefore are **macromolecules**. Molecular weight is determined by physical methods such as osmotic pressure measurement, depression in freezing point, light scattering effect, X-ray diffraction, turbidity measurement and now by methods such as analytical ultracentrifugation, molecular sieving by gel filtration and SDS-polyacrylamide gel electrophoresis. Mol. Wt. of some common proteins is shown in Table 6.1.

Table 6.1: Molecular weight of some proteins

• Serum Albumin	69,000
• Serum γ-globulin	176,000
• Fibrinogen	330,000
• Haemoglobin	67,000
• Cytochrome-C	15,600
• Pepsin	35,500
• Catalase	250,000

- **Viscosity of protein solutions:** The viscosity of protein varies widely with the kind of protein and its concentration in solution. ***The viscosity is closely related to molecular shape, long molecules (fibrous proteins) being more viscous than globular proteins. Thus fibrinogen can form a more viscous solution than albumin.***
- **Hydration of proteins:** Polar groups of proteins such as $-NH_2$ and -COOH become hydrated in presence of water and swell up when electrolytes, alcohol or sugars that form complexes with water are added to protein solutions. There is competition for water and the degree of hydration of protein is decreased. ***They dehydrate protein and precipitate it from solution.***
- **Heat coagulation of proteins:** Several proteins coagulate forming an insoluble coagulum. ***Coagulation is maximum at the isoelectric pH of the protein.*** During coagulation, protein undergoes a change called as denaturation. ***Denatured proteins are soluble in extremes of pH and maximum precipitation occurs at isoelectric pH (pI) of the protein.***
- **Amphoteric nature of proteins:** In any protein molecule there are amino acids which carry -COOH or $-NH_2$ groups in their side chains. These groups can undergo ionisation in solution producing both anions and cations. In addition to the side chains of polar amino acids, N-terminal $-NH_2$ group and C-terminal -COOH group may also ionise. Depending on the pH few groups act as proton donors while few as proton acceptors. ***Therefore proteins are ampholytes and act both as acids and bases. At a specific pH called an isoelectric pH (pI) a protein exists as a dipolar ion or*** **"Zwitterion"** ***or*** **"Hybrid" ion,** ***carrying equal number of positive and negative charges on its ionizable groups. So the net charge on protein molecule at its isoelectric pH is zero.***

  On the acidic side of its isoelectric pH, a protein exists as a ***cation*** by accepting a proton and migrates towards anode in an electrical field; while **on the alkaline side** of its pI a protein exists ***as anion*** by donating a proton and migrates towards cathode. This property is made use of in electrophoresis to separate different proteins depending on the charge present in them at a particular pH.
- **Precipitation of proteins:** Proteins can be precipitated from solutions by a variety of +ve and –ve ions. Such precipitation is of importance in the isolation of protein, in the deproteinisation of blood and other biological fluids and extracts for analysis and in the preparation of useful protein derivatives.

1. ***Precipitation By +ve Ions:*** The +ve ions most commonly used are those of **heavy metals** such as Zn^{+2}, Ca^{+2}, Hg^{+2}, Fe^{+3}, Cu^{+2}, and Pb^{+2}. ***These***

metals precipitate protein at the pH alkaline to its isoelectric pH. At this pH, protein is dissociated as an anion-proteinate. The metal ions combine with the –COO⁻ group to give insoluble precipitate of metal proteinate. On acidification the metal ions can be removed from the protein or by precipitation of metals by sulphuric acid. Metal protein precipitates are dissolved by addition of strong alkali.

Biomedical Application

- The ***use of $AgNO_3$ in cauteries*** is based on this property. It precipitates the proteins of tissues as Ag-salts.
- Another application of this property is the ***use of proteins as antidotes to metallic poisons.*** Egg white, milk and other proteins can be used to precipitate metal ions. The metallic protein precipitate must be removed from the stomach by an emetic or by stomach-tube to prevent the liberation and absorption of the poisonous metal.

2. *Negative ion precipitation:* ***Negative ions combine with proteins when the pH of the medium is on the acidic side of its isoelectric pH.*** Acidic pH makes the protein to exist as $Protein^+$ and forms precipitate with –ve ions. NH_2-group is the reacting group in this case. Among the more common precipitants involving –ve ion precipitation are Tungstic acid, phosphotungstic acid, trichloroacetic acid, picric acid, tannic acid, ferrocyanic acid and sulphosalicylic acid. When these agents are added to protein solution at proper pH, the protein precipitates as its salt. The precipitate is found to be soluble in alkali.

CLINICAL APPLICATION

- Tricholoroaceticacid, Tungstic acid, are commonly used for the preparation of protein-free filtrate of blood and other biological materials prior to analysis of few constituents such as sugar, urea by specific methods.

COLOUR REACTIONS OF PROTEINS

Proteins produce colour in certain reactions. These reactions are not quite specific for a protein molecule as such but are due to characteristic groups of particular amino acids present in it.

1. Xanthoproteic reaction: The aromatic aminoacids such as *phenylalanine, tyrosine* and *tryptophan* present in the protein give **yellow precipitate** when heated with conc. HNO_3. On addition of alkali, the precipitate turns orange due to nitration of the aromatic ring. ***Collagen and gelatin do not give a positive reaction.***

2. Millon's test: This is ***a specific test for tyrosine of protein.*** Protein gives a **white ppt** with Millon's reagent (10% mercurous chloride in H_2SO_4) on heating. On addition of $NaNO_2$ the precipitate turns pink-red.

3. Sakaguchi test: This is ***a specific test for arginine of*** the protein. Sakaguchi reagent consists of alcoholic α-naphthol and a drop of sodium hypobromite. Guanidine group of arginine (HN=C-NH_2) is responsible for the formation of **red colour**.

4. Hopkins-Cole reaction (Glyoxylic acid reaction): Reagent containing glyoxylic acid which may be prepared by reduction of oxalic acid with sodium amalgam. ***Protein containing tryptophan gives this test positive.*** The ***reaction is characteristic of tryptophan. Gelatin, collagen do not contain tryptophan and hence do not give this test positive.*** Nitrates, chlorates, nitrites and excess chlorides prevent the reaction. A number of aldehydes other than glyoxylic acid give similar colour reactions with tryptophan.

5. Nitroprusside reaction: Proteins with free -SH group of ***cysteine*** give ***reddish colour*** with sodium nitroprusside in ammoniacal solution. Many proteins give this test positive after heat coagulation or denaturation indicating the liberation of free -SH groups.

6. Sullivan reaction: For the determination of ***cysteine*** and ***cystine. Red colour*** is produced when cysteine containing protein is heated with sodium 1, 2 naphthoquinone -4-sulfonate in the alkaline medium in presence of $Na_2S_2O_4$.

7. Lead acetate test (unoxidised sulphur test): This test is ***specific for sulfur containing amino acid.*** The protein containing S-containing amino acids is boiled with strong alkali to split out sulphur as sodium sulphide which reacts with Lead acetate to give **black precipitate of Pb S.**

8. Biuret reaction: When urea is heated it forms biuret.

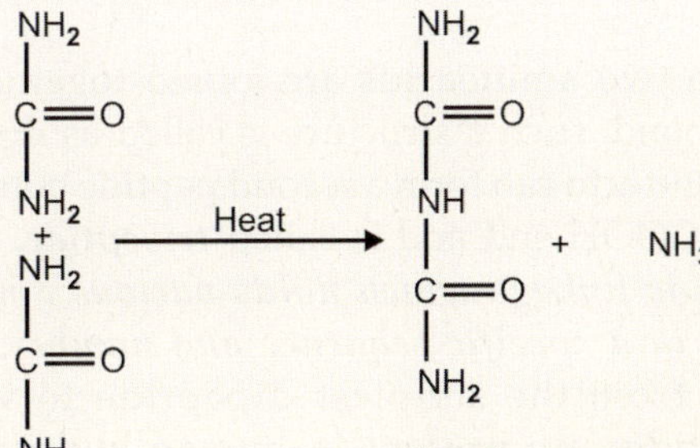

Biuret

Urea (2 moles)

If a strongly alkaline solution of biuret is heated with very dilute copper sulphate a **purple-violet colour** is obtained. ***The colour depends upon the presence of 2 or more peptide linkages.***

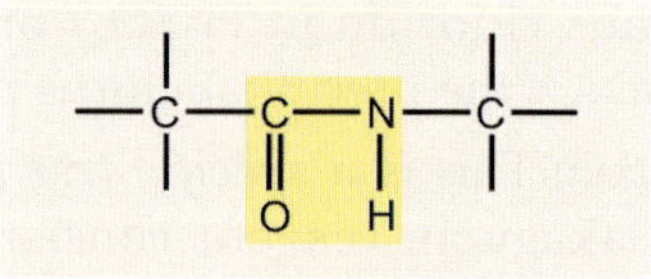

Thus ***dipeptides and free amino acids do not give the biuret test. Only Histidine can give a positive reaction.*** Biuret test is due to co-ordination of cupric ions with the unshared electron pairs of peptide nitrogen and the oxygen of water to form a coloured co-ordination complex.

9. Ninhydrin reaction: Ninhydrin is a powerful oxidising agent and causes oxidative decarboxylation of α-amino acids producing an aldehyde with one carbon less than the parent amino acid. The reduced Ninhydrin Hydrindantin then reacts with ammonia which has been liberated and one molecule of ninhydrin forming **a blue-coloured compound**. ***A molecule of CO_2 is evolved indicating the presence of α-amino acid.***

Peptide Linkage and Peptides

The –COOH group of one amino acid can be joined to the $-NH_2$ group of another by a covalent bond called as ***peptide bond.*** In the process of formation of a peptide bond a molecule of water is eliminated.

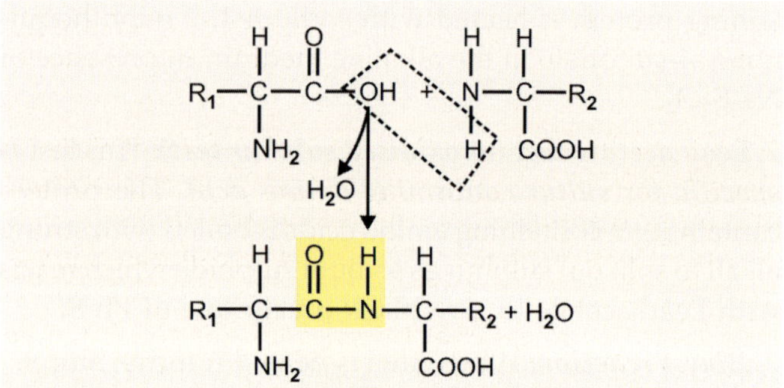

Peptide bond

When two aminoacids are joined together by one peptide bond, such a structure is called as ***dipeptide.*** A third amino acid can form a second peptide bond through its free –COOH end and is called ***tripeptide. Thus it is the peptide linkage which holds various amino acids together in a specific sequence and number.*** Peptides varying from the simplest dipeptide to very long polypeptides are present in human body and serve specific functions. ***Chains that contain fewer than 50 amino acid residues are called peptides or oligopeptides.***

Characteristics of Peptide Bond (Fig. 6.8)

- The **C–N backbone** bonds in each peptide group have some ***double-bond character*** and ***do not rotate.***

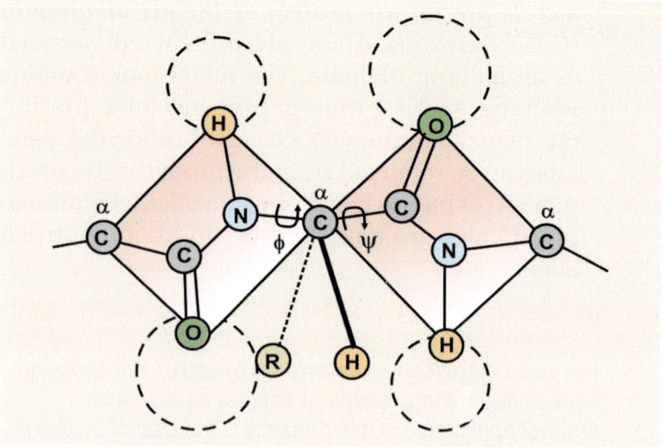

Fig. 6.8: Ball and stick model showing the ϕ and ψ angles between two adjacent peptide groups (Ramachandra angles)

- The other backbone bonds designated $\mathbf{C_\alpha–N}$ and $\mathbf{C_\alpha–C}$, are theoretically ***free to rotate,*** since they are ***true single bonds.***
- But if the R group attached to the central α carbon alone is large enough, it will prevent complete rotation around the C_α–N and C_α–C bonds.
- Moreover, if these bonds are rotated in relation to each other, ***angles will be found where the two H atoms (O atoms) of the peptide bonds would overlap each other and obstruct free rotation.***
- Thus each pair of successive peptide bonds has two kinds of constraints on the freedom of rotation of the C_α–C and C_α–N single bonds.
- The angle of rotation of the C_α–N bond is called the ϕ (Phi) angle and that of C_α–C bond the **ψ (Psi) angle.** These angles are called **Ramachandran angles.**
- For any given pair of ϕ and ψ angles one can ***predict whether the conformation is allowed or not (from Ramachandran plot).***

Biologically Important Peptides

Some small peptides which have significant biological activity are formed as a result of hydrolysis of large proteins while some are formed by synthesis.

- ***Glutathione:*** This is a tripeptide consisting of ***glutamic acid, cysteine and glycine***. By virtue of easy dehydrogenation it gets converted to disulphide form and function in oxidation-reduction systems (Refer Glutathione in chapter of Protein Metabolism).
- ***Carnosine:*** This is a dipeptide of β *alanine* and *histidine.* ***Anserine is methyl carnosine.*** They are water soluble dipeptides of voluntary muscle (Refer to Metabolism of Histidine).
- ***Bradykinin:*** Bradykinin (9 amino acids) or Kallidin I and Kallidin II (10 amino acids) have relaxant effects on smooth muscle.

- ***Oxytocin and Vasopressin:*** Found in pituitary gland, these are cyclic peptide hormones made up of 9 amino acids. (Refer Pituitary Hormones).
- ***Angiotensins:*** The enzyme Renin is released from kidneys and acts on globulin fraction of plasma to release a peptide ***Angiotensin I (10 amino acids)*** which has only a slight effect on blood pressure, Angiotensin I is then converted to ***Angiotensin II*** by splitting off two amino acids which has eight amino acid, has greater effect on BP. Removal of one residue aspartic acid from Angiotensin II results in formation of ***Angiotensin III (7 amino acids)*** which plays role in pathology of hypertension (Refer Renin-angiotensin system).
- ***Gastrin, Secretin and Pancreozymin:*** Are gastrointestinal peptides which act as hormones which stimulate secretion of bile and other enzymes of digestive juices.
- ***β-Corticotropin (ACTH), and β MSH:*** Are peptides which are hormones.
- ***Antibiotics:*** Penicillin, gramicidin, polymyxins, bacitracins, actinomycin, chloramphenicol are all peptides which act as antibiotics.
- Certain microbial peptides are toxic. The cyanobacterial peptides, ***microcystin*** and ***nodularin*** are lethal in large doses. On the other hand, **small quantities promote the formation of hepatic tumours.**
- Peptide synthesised by microbial agents, like ***bleomycin*** can be ***used therapeutically as antitumour agent.***
- ***Brain Peptides:*** Certain brain cells have receptors that bind opiates like morphine and have been termed ***endorphins*** (endogeneous morphine). ***Dynorphin*** is a peptide of 13 amino acids which is called *superopioate* since it is significantly potent. Peptide fragments from brain that reduce intestinal motility are ***met-enkephalins*** and ***leuenkephalin*** both pentapeptides.

STRUCTURAL ORGANISATION OF PROTEINS

Protein structure is normally described at *four levels of organisation.*

A. Primary Structure: Primary structure is the linear sequence of amino acids held together by peptide bonds in its peptide chain. The peptide bonds form the backbone and side chains of amino acid residues project outside the peptide backbone. The free -NH_2 group of the terminal amino acid is called as **N-terminal end** and the free -COOH end is called as **C-terminal end.** ***It is a tradition to number the amino acids from N-terminal end as No. 1 towards the C-terminal end. Presence of specific amino acids at a specific number is very significant for a particular function of a protein. Any change in the sequence is abnormal and may affect the function and properties of protein.***

B. Secondary Structure: The peptide chain thus formed assumes a three dimensional secondary structure by way of folding or coiling consisting of a helically-coiled, zig-zag, linear or mixed form. It results from the steric relationship between amino acids located relatively near each other in the peptide chain. ***The linkages or bonds involved in the secondary structure formation are hydrogen bonds and disulfide bonds.***

- *Hydrogen bond:* These are weak, low energy non-covalent bonds sharing a single hydrogen by two electronegative atoms such as O and N. Hydrogen bonds are formed in secondary structure by sharing H-atoms between oxygen of CO and nitrogen of -NH of different peptide bonds. The hydrogen bonds in secondary structure may form either an ***α-helix or β-pleated sheet structure.***
- *Disulphide bonds:* These are ***formed between two cysteine residues.*** They are strong, high energy covalent bonds.

1. *α-Helix:* ***A peptide chain forms regular helical coils called α-helix.*** These coils are stabilised by hydrogen bonds between carbonyl O of 1st amino and amide N of 4th amino acid residues. ***Thus in α-helix intra chain hydrogen bonding is present. The α-helices can be either right handed or left handed.*** **Left handed α-helix is less stable than right handed α-helix** because of the steric interference between the C = O and the side chains. Only the right handed α-helix has been found in protein structure.

Each amino acid residue advances by 0.15 nm along the helix, and ***3.6 amino acid residues are present in one complete turn.*** The distance between two equivalent points on turn is 0.54 nm and is called ***a pitch.***

Small or uncharged amino acid residues such as ***alanine, leucine and phenylalanine*** are often found in α-helix. More polar residues such as arginine, glutamate and serine may repel and destabilise α-helix. ***Proline is never found in α-helix.*** The proteins of hair, nail, skin contain a group of proteins called ***keratins rich in α-helical structure*** (Fig. 6.9).

α-Helices May be Amphipathic

Normally on the surface of proteins, α-helices may be wholly or partially buried in the interior of a protein. The amphipathic helix, a special case in which residues switch between hydrophobic and hydrophilic about every 3 or 4 residues. It occurs where α-helices interface with both a polar and a non-polar environment.

Occurrence: Amphipathic α-helices occur in plasma lipoproteins, in certain polypeptide hormones, in certain antibiotics, human immunodeficiency virus glycoproteins, certain venoms, and calmodulin-regulated protein kinases.

2. ***β-Pleated Sheet Structure:*** β-Keratins present in spider's web, reptilian claw, fibres of silk form almost

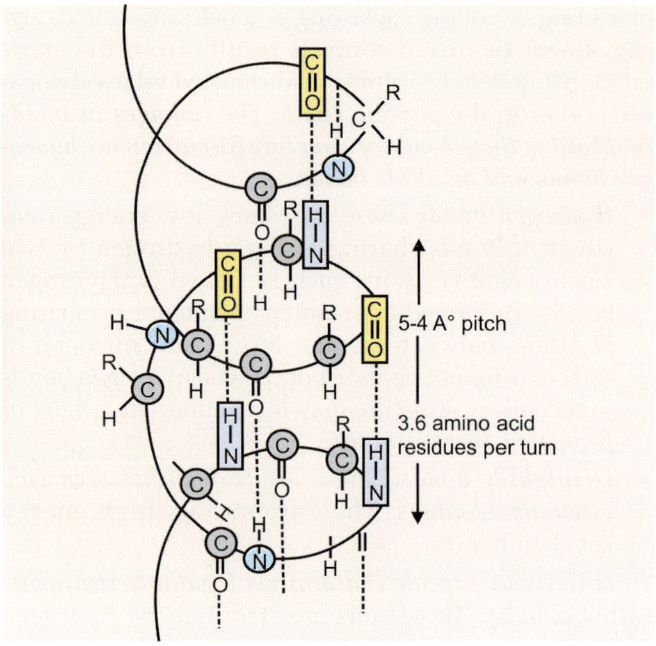

Fig. 6.9: α-Helix

fully extended chain. A conformation called ***β-pleated sheet structure*** is thus formed when hydrogen bonds are formed between the carbonyl oxygens and amide hydrogens of two or more adjacent extended polypeptide chains. ***Thus the hydrogen bonding in β-pleated sheet structure is interchain.*** The structure is not absolutely planar but is slightly pleated due to the angles of bonds. The adjacent chains in β-pleated sheet structure are ***either parallel or antiparallel,*** depending on whether the amino to carbonyl peptide linkage of the chains runs in the same or opposite direction. In both parallel and antiparallel β-pleated sheet structures, the side chains are on opposite sides of the sheet. Generally ***glycine, serine and alanine are most common to form β-pleated sheet. Proline occurs in β-pleated sheet although it tends to disrupt the sheets by producing kinks. Silk fibroin, a protein of silkworm is rich in β-pleated sheets*** (Fig. 6.10).

3. ***Triple Helix:*** ***Collagen is rich in proline and hydroxyproline and cannot form α-helix or β-pleated sheet. It forms a triple helix.*** The triple helix is stabilised by both noncovalent as well as covalent bonds. Interchain hydrogen bonds between different peptide chains are formed which are almost perpendicular to the long axis of α-helix. In addition, interchain additional cross links, secondary amide bonds of peptide bonds also are responsible for triple helix.

Note: Many globular proteins have mixed secondary structure of α-helix, β-pleated sheet and non-helical, non-pleated structures called ***random coil.***

4. ***Reverse Turns or β-bends:*** Since the polypeptide chain of a globular protein changes direction two or more times when it folds, the conformations known as ***reverse turns or β-bends*** are important elements of secondary structure. Reverse turns usually occur on the surfaces of globular proteins where there is little steric hindrance to resist a change in the direction of the polypeptide chain.

Types of Reverse Turns: Two major types, Type I and Type II of reverse turns, each spanning 4 amino acid residues, are particularly common. In each type the carbonyl oxygen of the first residue is hydrogen bonded to the amide hydrogen of the fourth residue, stabilising a loop of 10 atoms. Type I and Type II reverse turns differ by a 180° rotation of the central amide plane of the loop. The third residue of a Type-II reverse turn can only be glycine because the side chains of all other amino acid residues are too large to fit into the restricted space. Because the five-membered ring of ***proline has little***

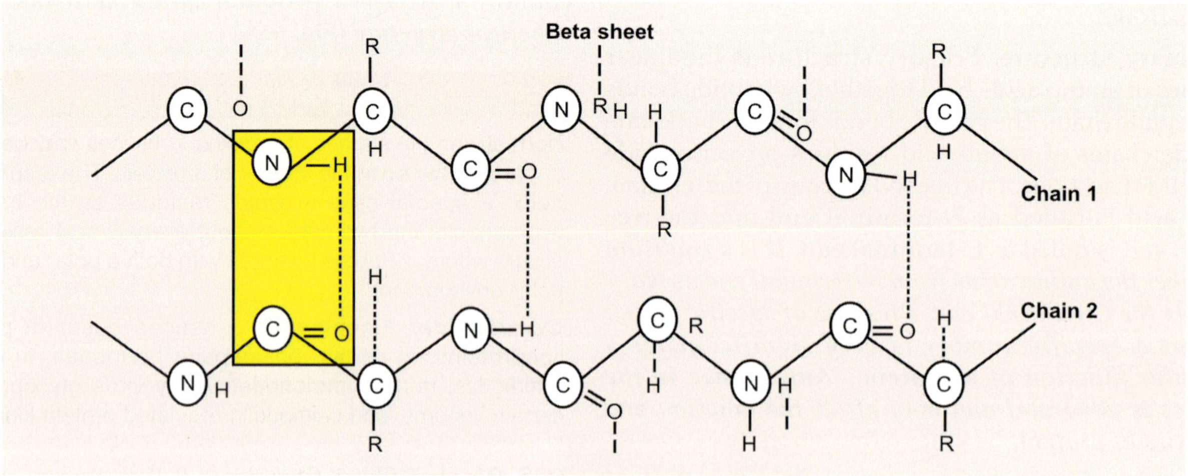

Fig. 6.10: (H --- O) Hydrogen bond—β-pleated sheet

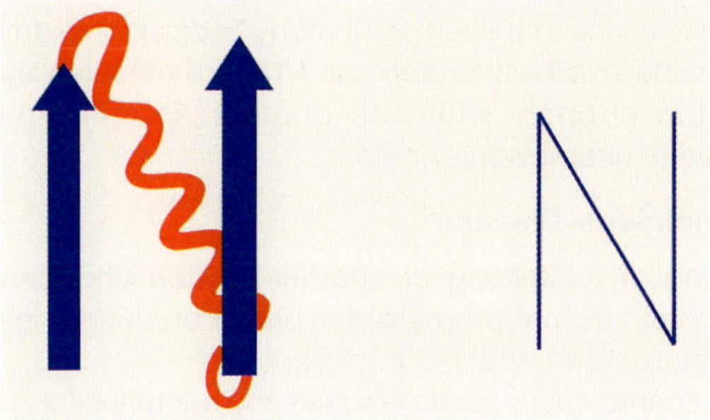

Fig. 6.11: β-α-β unit

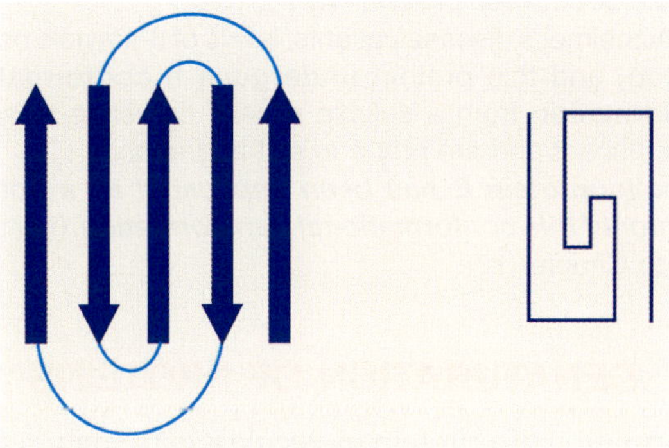

Fig. 6.12: Greek key

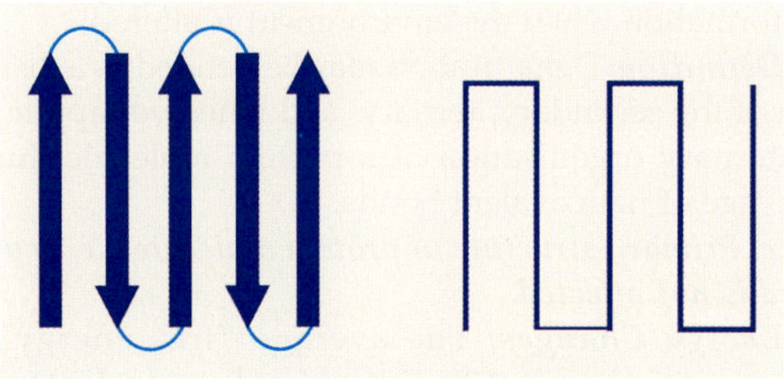

Fig. 6.13: β-meander

conformational mobility, it is ideally suited for the second residue in a reverse turn.

5. *Super Secondary Structures:* Various combinations of secondary structure, called super secondary structure, are commonly found in globular proteins.

These are:

- **β-α-β unit** (Fig. 6.11)
- **Greek key** (Fig. 6.12)
- **β-meander** (Fig. 6.13)

1. *β-α-β Unit:* The β-α-β unit consists of two parallel β-pleated sheets connected by an intervening strand of α-helix (Fig. 6.11).

2. *Greek Key:* Another common super secondary structure is called a **Greek Key**, a conformation that takes its name from a design often found on classical Greek pottery (Fig. 6.12).

3. *β-meander:* The β-meander consists of ***five β-pleated sheets connected by reverse turns*** (Fig. 6.13). The β-meander contains nearly as many hydrogen bonds as an α-helix, and its common occurrence probably reflects the stability conferred by this extensive hydrogen bonding.

C. Tertiary structure: The ***polypeptide chain with secondary structure mentioned above may be further folded, superfolded twisted about itself forming many sizes.*** Such a ***structural conformation is called tertiary structure***. It is only one such conformation which is biologically active and protein in this conformation is called as native protein. Thus the ***tertiary structure is constituted by steric relationship between the amino acids located far apart but brought closer by folding.*** The bonds responsible for interaction between groups of amino acids are as follows:

- *Hydrophobic interactions:* Normally occur between nonpolar side chains of amino acids such as ***alanine, leucine, methionine, isoleucine and phenylalanine.*** They constitute the major stabilising forces for tertiary structure forming a compact three-dimensional structure.
- *Hydrogen bonds:* Normally formed by the polar side chains of the amino acids.
- *Ionic or electrostatic interactions:* These are formed between oppositely charged polar side chains of amino acids, such as basic and acidic amino acids.
- *Van der Waal Forces:* Occur between nonpolar side chains.
- *Disulfide bonds:* These are the S–S bonds between -SH groups of distant cysteine residues.

D. Quaternary Structure: Many proteins are made up of only one peptide chain. However, ***when a protein consists of two or more peptide chains held together by non-covalent interactions or by covalent cross-links, it is referred to as the quaternary structure***. The assembly is often called as oligomer and each constituent peptide chain is called as monomer or subunit. The monomers of oligomeric protein can be identical or quite different in primary, secondary or tertiary structure.

Examples: Protein with **two monomers (dimer)** is an enzyme called creatine phosphokinase (CPK).

Haemoglobin and lactate dehydrogenase (LDH) are **tertramers** consisting of four monomers. **Apoferritin**, an apoprotein of ferritin, an iron binding and storage protein contains 24 identical subunits. An enzyme **aspartate transcarbamoylase** has 72 subunits in its structure.

SECTION TWO

CLINICAL ASPECT

Diseases Resulting from Altered Protein Conformation

A. Prions and Prion Diseases

Prions: Prions are infectious proteins that contain no nucleic acid. Earlier thought to be an infectious agent or a virus. This infectious protein-prions was discovered in ***1982 by Stanley Prusiner.***

Abnormal or pathological prions cause several fatal neurodegenerative disorders known as ***"transmissible spongiform encephalopathies" (TSEs)*** or ***Prion Diseases***.

Types of Prions: There are two isoforms:

(a) Normal or physiologic PrP—called ***(PrPc)*** or ***PrP-Sen.***

(b) Abnormal or pathologic Prp—called ***PrPsc*** or ***PrP-res.***

(a) Normal or physiologic PrPc: It consists of 253aa found in leucocytes and nerve cells. **Gene of this PrP is located in short arm of chromosome 20.** This protein is heat sensitive and protease sensitive.

(b) Abnormal or pathologic PrP: The pathologic isoform is heat resistant and protease resistant. This form is associated with TSEs or prion diseases.

Defect: The basic defect involves alteration of α-helical structure into β-pleated sheet.

Both PrPc and PrP-Sc have identical primary structural and post-translational modification but different tertiary and quarternary structures.

PrPc is rich in α-helix but PrP-Sc consists predominantly of β-sheet. This structural change occurs when PrP-c interacts with the pathologic isoform Pr-Psc.

Pr-Psc serves as a template upon which the α-helical structure of Pr-Pc becomes the β-sheet structure characteristic of Pr-Psc.

Prion Diseases: May manifest themselves as genetic, infectious or sporadic disorders. The disease can occur in humans and also in animals. Prion diseases can be transmitted by the protein alone ***without involvement of DNA or RNA.***

Clinically rapidly progressive dementia sets in with neurological defects and ataxia.

(a) In humans: The disease is called ***"Creutzfeldt-Jakob Disease" (CJD).***

Other human forms of the disease are:

- Gerstmann-Straüssler-Scheinker Disease (GSSD), and
- Fatal familial insomnia (FFI)—rare.

(b) In animals: It produces:

- ***Scrapie in sheep***—From this the term 'Pr-Psc' derived.
- ***Bovine spongiform encephalopathy (BSE) in cattle. Also known as 'Mad Cow Disease'.***

Pathological Changes

Each of the above disease is characterised by ***spongiform changes, astrocytic gliosis*** and ***neuronal loss*** resulting from deposition of insoluble proteins in stable amyloid fibrils.

The protofilaments of amyloid fibrils contain pairs of β-sheets in a helical form that are continuously hydrogen bonded all along the fibrils.

Pr Psc is rich in β sheet with many hydrophobic aminoacyl side chains exposed to solvent. PrPsc molecules therefore associate strongly with one another, forming insoluble protease-resistant aggregates.

B. Alzheimer's Disease

Refolding or misfolding of another protein endogenous to human brain tissue, **β-amyloid** is also a prominent feature of Alzheimer's disease found in old age.

The characteristic senile plaques and neurofibrillary bundles contain aggregates of the protein, **β-Amyloid**, a polypeptide produced by proteolytic cleavage of a larger protein known as **Amyloid precursor protein (APP).**

In Alzheimer's disease patients, levels of β-amyloid become elevated, and this protein **undergoes a conformational transformation** from a soluble α-helix rich state to a state rich in β-sheet and are prone to self-aggregation.

Apolipoprotein E has been implicated as a potential mediator of this conformational transformation. (For details Refer to Chapter 17).

DENATURATION OF PROTEINS

Conformation of a protein molecule is extremely sensitive to changes in their environment. In denaturation, there is a disruption of native or biologically active protein conformation when the environment is altered.

1. *Definition:* Denaturation may be defined as a disruption of the secondary, tertiary, and wherever applicable quaternary organisation of a protein molecule due to cleavage of noncovalent bonds.

Note: ***Primary structure of protein molecule, i.e. peptide bond is not affected.***

2. *Energy Changes:* The average "free energy" of denaturation is only 12 ± 5 Kcal/mol. equivalent to the disruption of three or four hydrogen bonds. Proteins, therefore, are characterised by a narrow range of thermodynamic stability.

3. *Agents that cause denaturation:* Various agents which can disrupt the conformation are:

- *Physical agents:* Heat, UV light, ultrasound, and high pressure. Even violent shaking can denature the protein.
- *Chemical agents:* Organic solvents, acids/alkalies, urea and various detergents.

The disruption/disorganisation of the protein molecules results in alteration of the chemical, physical and biological characteristics of the protein.

Alterations in Protein Molecules after Denaturation

(a) Chemical alterations

- Greatly decreased solubility specially at pI of the protein. Maximum precipitation as floccules occur at pI of the protein.

Behaviour of Denatured Albumin

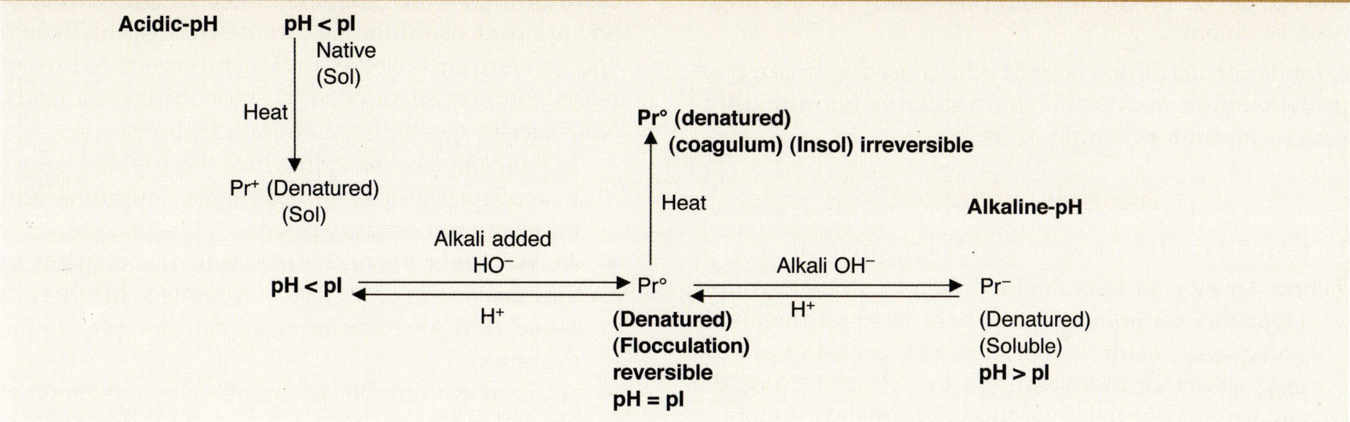

- Many chemical groups which were rather inactive become exposed, e.g. -SH group.
- Denaturation can be ***reversible.***

(b) Physical alterations

- Confers increased viscosity of the solution.
- Rate of diffusion of the protein molecules decreases.

(c) Biological alterations

- Increased digestibility by proteolytic enzymes has been found in the case of certain denatured proteins.
- Denaturation destroys enzymal and hormonal activity.
- **Biologically becomes inactive.**

4. *Relationship of denaturation, flocculation and coagulum formation:* The relationship is explained above schematically in case of albumin. The box above shows the behaviour of denatured albumin.

- ***Denatured albumin is soluble in extremes of pH.***
  - Maximum precipitation occurs at pI of the protein, i.e. albumin, as floccules **(flocculation).**
  - ***Denatured protein as floccules, is reversible and soluble in extremes of pH.***
  - When the floccules at pI is heated further, it becomes dense **coagulum which is irreversible,** and is not soluble in extremes of pH.

5. *Reversibility and irreversibility of denatured protein*

- *Reversibility:* The example of reversibility or controlled denaturation is shown by treatment of protein *ribonuclease* with urea and β-mercaptoethanol, when the disulphide bridges are broken and the polypeptide chain uncoils. On slow re-oxidation of the denatured protein without urea, the protein is converted into original native protein with original tertiary structure.
- *Irreversibility:* An example of irreversible or uncontrolled denaturation occurs in the boiling of an egg during which the tertiary structure of the protein is irreversibly destroyed with the formation of a disorganised mass of polypeptide chains.

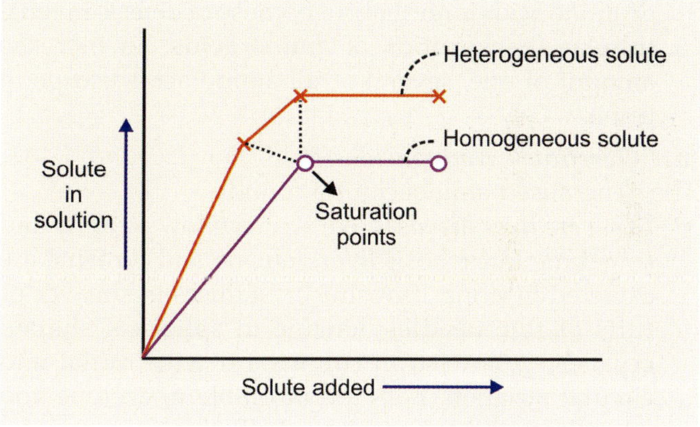

Fig. 6.14: Purity of proteins by solubility

Criteria of Purity of Proteins

The purity or homogeneity of an isolated protein can be tested by employing various methods.

1. *Solubility curve:* When the solute is added repeatedly in increasing quantity to the solvent, the solute in solution rises proportionately and produces a sloping straight line in its solubility curve. When the solution gets saturated, it fails to hold the solute in solution any further. Thus a single sharp break occurs in straight line at saturation point. ***More than one break in the line or a non-linear curve indicates non-homogeneity of the sample*** (Fig. 6.14).

2. *Molecular weight:* Molecular weight determined by electrophoresis, gel filtration or ultracentrifugation indicates the purity of protein.

3. *Ultracentrifugation:* Analytical ultracentrifugation gives a sharp moving boundary between the pure solvent and solute containing layer.

4. *Electrophoresis:* Electrophoretic mobility and a homogeneous sharp level indicates that the protein is pure.

5. Chromatography: The particles of a pure protein emerge out of the chromatography column as one single peak of eluent.

6. Immunoreactivity: The antibodies raised against a pure protein will show only one sharp spike on immunodiffusion or immunoelectrophoresis.

AMINOACIDURIAS

Increased excretion of amino acids than normal.

Types: Aminoacidurias can be divided into **two groups:**

- *Overflow aminoaciduria:* There is some metabolic defect, as a result there occurs an increase in plasma level of one or more amino acids which exceeds the capacity of normal renal tubules to reabsorb them.
- *Renal aminoaciduria:* In this the plasma concentration of amino acids is normal but because of defects in renal tubular reabsorption of amino acids, an increase amount of one, several or all amino acids escape in urine.

(a) "Overflow" Aminoacidurias

The generalised aminoaciduria found:

- ***In severe liver diseases*** like acute yellow atrophy and sometime in cirrhosis liver, leucine and tyrosine are excreted in urine. Tyrosine crystallises in sheaves or tufts of fine needles; leucine in spherical shaped crystals, yellowish in colour and with radial and circular striation. Both are insoluble in acetone and ether; but soluble in acids and alkalies.
- Similarly aminoacidurias occur ***in wasting diseases.***
- ***Metabolic defects*** affecting a single amino acid or a small group include inherited disorders like phenyl ketonuria, maple-syrup disease, histidinaemia, and histidinuria, etc.

(b) Renal Aminoacidurias

The defect may be:

- ***Specific*** to one reabsorption mechanism as in cystinuria, in which there is failure to reabsorb cyst, lys, arg and ornithine (a common transport defect) and in Hartnup disease, in which there is failure of reabsorption of all monoamino-monocarboxylic acids.
- *Nonspecific:* Generalised aminoaciduria seen:
  - *In Fanconi syndrome:* In which there is also failure to reabsorb glucose, phosphates, ammonia and other organic acids, e.g. LA
  - *In Wilson's disease:* In which in addition to aminoaciduria (Ala, Asp, Glutamic acid) there is associated glycosuria, uric acid and phosphate excretion.
  - An increase number of amino acids are excreted by patients suffering from ***muscular dystrophies*** and by their mothers and siblings. Among these are Methionine or valine, Isoleucine or leucine, methionine sulfoxide or sarcosine methyl histidine.
  - A generalised hyperaminoaciduria commonly occurs in ***lead intoxication*** and in persons exposed to other heavy metals like Hg, Cd and Uranium. Aminoaciduria in these cases appear to be due to impaired tubular reabsorption.
  - More than 1.0 gm of peptides is excreted daily by the average normal human adult, which accounts for about 1/2 the urinary amino N_2 and about 2 per cent of total N.
  - Large quantities of ***β-amino isobutyric acid* (BAIB)** are excreted in the urine of a small proportion (about 5% of otherwise normal people). It has also been found in various pathologic states, sometimes with other amino acids. It is apparently due to disturbance in the metabolism of thymine and dihydrothymine, which are precursors of this amino acid.

I. Questions (Essay type and short notes)

1. Describe the classification of proteins with suitable examples.
2. What are proteins? Classify them giving suitable examples in each group.
3. What is the primary structure of proteins? Discuss the biochemical functions of proteins.
4. Classify proteins, give their biological importance.
5. What are proteins? Classify them and give important tests for it. Describe briefly biological importance of proteins.
6. Describe the structure of proteins. Discuss how proteins are precipitated.
7. What are proteins? Classify proteins.
   1. Denaturation of proteins
   2. Structure of proteins

Short Notes

a. Essential amino acids
b. Properties of amino acids
c. Thin layer chromatography
d. Two colour reactions of proteins
e. Amphoteric property of amino acids proteins
f. Conjugated proteins
g. Electrophoresis
h. Isoelectric point
i. Separation and identification of proteins
j. Sclero proteins (Albuminoids)
k. Biuret reaction
l. Paper chromatography.

Differentiate

a. Denaturation and coagulation
b. Primary and secondary structure of proteins
c. Primary derived proteins and secondary derived proteins
d. Tertiary and quarternary structure of proteins
e. Protamines and prolamines.

II. MCQs (Give one correct answer)

1. The number of helices present in a collagen molecule is:
   a. 1 b. 2
   c. 3 d. 4
   e. None of the above
2. Which bond is present in the primary structure of protein?
   a. Ester bond b. Disulfide bond
   c. Hydrogen bond d. Ionic bond
   e. Peptide bond
3. The Phosphoprotein present in the egg yolk is:
   a. Ovalbumin b. Avidin
   c. Ovo mucoid d. Ovovitellin
   e. Ovoglobulin
4. All of the following are sulfur containing amino acids found in proteins except:
   a. Cysteine b. Threonine
   c. Cystine d. Methionine
   e. Homocysteine
5. Which of the following amino acid is optically inactive
   a. Aminoacetic acid b. Serine
   c. Methionine d. Lysine
   e. Proline
6. With the exception of glycine, all amino acids found in proteins are:
   a. Optically inactive b. Dextrorotatory
   c. Of L-configuration d. Laevorotatory
   e. Of D-configuration
7. In protein structure, the α-helix and β-pleated sheet are examples of:
   a. Primary structure b. Secondary structure
   c. Tertiary structure d. Quaternary structure
   e. Subunit structure
8. Glutathione is a:
   a. Dipeptide b. Polypeptide
   c. Oligopeptide d. Tripeptide
   e. None of the above
9. Match the following:
   a. Acidic b. Aromatic
   c. Basic d. Imino
   e. Optically inactive
   1. Argine ________
   2. Glutamic acid ________
   3. Glycine ________
   4. Proline ________
   5. Tyrosine ________.
10. An essential amino acid in man is:
    a. Proline b. Threonine
    c. Asparate d. Tyrosine
    e. Serine
11. An amino acid that does not form an α-helix is:
    a. Asparagine b. Tyrosine
    c. Tryptophan d. Proline
    e. Valine
12. Denaturation of proteins involves breakdown of:
    a. Secondary structure
    b. Tertiary structure
    c. Quaternary structure
    d. None of the above
    e. All of the above
13. In denaturation of proteins, the bond which is not broken:
    a. Disulfide bond b. Peptide bond
    c. Hydrogen bond d. Ionic bond
    e. None of the above
14. The protein present in hair is:
    a. Elastin b. Prolamine
    c. Keratin d. Gliadin
    e. None of the above
15. The pl of albumin is:
    a. 3.6 b. 4.7
    c. 5.0 d. 6.1
    c. None of the above
16. Some amino acids are termed non-essential as they:
    a. Are not component of tissue proteins
    b. Have no role in metabolism
    c. May be synthesised in the body from 'essential' amino acids
    d. May be synthesised in body in prolonged fasting
    e. None of the above
17. Which of the following is a dipeptide:
    a. Glutathione b. Oxytocin
    c. Homocysteine d. Anserine
    e. Protamine
18. In which category keratin belong:
    a. Protamines b. Globulins
    c. Conjugated proteins d. Histones
    e. Albuminoids
19. Ninhydrin reaction gives a purple colour and evolves CO_2 with:
    a. Peptide bonds b. Histamine
    c. Egothioneine d. Asparagine
    e. Serotonin
20. Histones are proteins which:
    a. Are indential to protamiens
    b. Are proteins rich in lysine and arginine
    c. Are covalently bound to DNA
    d. Have high molecular weights
    e. None of the above
21. Match the following:
    A. Covalent bond B. α-Helix
    C. Hydrogen bond D. Hydrophobic bond
    E. S-S bond

1. Formed between side chains of alanine, valine and phenyl alanine______
2. Bond formed between two cysteine mols ______
3. Peptide bond ______
4. Prominent in myosin ______
5 Sharing of hydrogen between nitrogen and oxygen____.

Answers to MCQs

1. c	2. e	3. d	4. b	5. a	6. c
7. b	8. d	9. I-C, 2-A, 3-E, 4-D, 5-B			10. b
11. d	12. e	13. b	14. c	15. b	16. c
17. d	18. e	19. d	20. b		

21. 1-D, 2-E, 3-A, 4-B, 5-C.

CHAPTER 7

PLASMA PROTEINS— CHEMISTRY AND FUNCTIONS

Major Concepts

Learn principal proteins present in plasma, their properties and functions in the body.

Specific Objectives

1. What are plasma proteins? What is the normal level in blood?
2. Learn briefly the different methods by which plasma proteins can be separated. • Precipitation by "Salting out", including Cohn's fractionation. • Electrophoresis.
3. Learn the normal values of both the methods.
4. Learn characteristics and properties of individual plasma proteins. • **Albumin:** Site of synthesis and properties • **Globulins:** Different types α, β and γ-globulins • Learn site of synthesis of each and their properties. • **Fibrinogen:** Site of synthesis and properties.
5. Study the important functions of plasma proteins in the body.
6. Learn briefly about • orosomucoid • α-fetoprotein • Haptoglobin • C-reactive protein • Bence-Jones' Proteins, • Transferrin, • Caeruloplasmin.
7. Study briefly variations of plasma proteins, increase/and decrease and their causes.

PLASMA PROTEINS

The chief solids of plasma are the proteins, which are **about 7.0 to 9.0 gm%.** The human plasma proteins are a mixture of simple proteins such as albumins and conjugated proteins such as glycoproteins, lipoproteins, etc.

SEPARATION OF PLASMA PROTEINS

Various methods have been used to separate out the individual proteins in plasma. A brief outline is given as follows:

(a) Precipitation by salting out: By this method, different concentrations of salt solutions are used to precipitate the various fractions of proteins which are then separated. Salt solutions which have been used are

- **Ammonium sulphate solution** and
- **Mixture of sodium sulphate and sodium sulphite solution.**

(b) Fractionation of plasma proteins by ethanol (Cohn's fractionation)

(c) More recent methods include: Separation of plasma proteins by paper electrophoresis (**Tselius,** 1937), gel electrophoresis, Immunoelectrophoresis, ultra centrifugation, gel filtration and column chromatography, etc.

(a) Salting Out Methods: Using the "salting out" techniques, **three proteins** have been separated and they are: ***albumin, globulins,*** and ***fibrinogen.***

Globulins are precipitated by half-saturation with ammonium sulphate, whereas albumins are precipitated only on full saturation. Fibrinogen is best precipitated by 1/5th saturation with ammonium sulphate. Among the globulins, there is a fraction which can be precipitated by 1/3rd saturation of ammonium sulphate and is termed as **euglobulins** ('true' globulins) and the rest is called as **pseudoglobulins** ('false' globulins). Euglobulins are insoluble in distilled water but soluble in dilute salt solutions (say NaCl), but Pseudoglobulin is soluble even in distilled water.

(b) Cohn's Fractionation: Cohn used varying concentrations of ethanol at low temperature to separate out fractions of proteins which are called fraction I, II, etc. Each fraction is itself a mixture of proteins but contained one of the proteins predominantly.

Cohn's Fractions

- **Fraction I** is rich in fibrinogen
- **Fraction II** is γ-globulins
- **Fraction III** contains α and β-globulins including iso-agglutinins and Prothrombin
- **Fraction IV** contains α and β-globulins
- **Fraction V** contains predominantly albumin

Advantage of this Fractionation

- Solvent used in the procedure can be readily removed by evaporation.
- Mild procedures adopted in fractionation of the proteins do not cause denaturation of the proteins.

CLINICAL USE

Cohn's method is useful for obtaining purified proteins on a large scale for therapeutic purposes.

(c) Electrophoresis: Original paper electrophoresis was followed by more sensitive agar gel electrophoresis, cellulose acetate membrane electrophoresis, starchgel electrophoresis and immunoelectrophoresis. By paper electrophoresis, the serum can be separated into a number of fractions, viz, albumin, globulins α_1, α_2, β and γ-globulins. Depending on sensitivity of the method β-globulins can be resolved into β_1 and β_2.

Note

> ***If plasma is used instead of serum, a band of fibrinogen fraction is seen between β and γ-globulins.*** Using modern analytical methods, more than 80 proteins have been identified in plasma. Many of them occur only in trace amount and are either enzymes or transport proteins.

Values of different proteins as obtained by standard precipitation methods and by paper electrophoresis are given in Table 7.1.

Table 7.1: Normal values of plasma proteins

Total Proteins = 7.0 to 7.5 Gm%

	By Precipitation (Gm %)	*By Paper Electrophoresis (% of total proteins)*
• Albumin	3.7 to 5.2	50 to 70%
• Globulins	1.8 to 3.6	29.5 to 54%
• α_1-globulins	0.1 to 0.4	2.0 to 6.0%
• α_2-globulins	0.4 to 0.8	5.0 to 11.0%
• β-globulins	0.5 to 1.2	7.0 to 16.0%
• γ-globulins	0.7 to 1.5	11.0 to 22.0%
• Fibrinogen (200 to 400 mg%)	0.2 to 0.4	

A:G ratio = 2.5 to 1.0 to about 1.2 to 1.0 most frequently 2:1

Electrophoretic pattern in some common diseases is shown in Figure. 7.1.

Electrophoretic Pattern in different diseases

Diseases	***Electrophoretic pattern***
• ***Nephrosis***	Albumin ↓, α_2-globulin increased markedly ↑, γ-globulin ↓
• ***Chronic liver disease***	γ-globulin increased ↑ diffuse in nature (polyclonal)
• ***Infectious hepatitis***	γ-globulin ↑ α_1 and α_2-globulins ↓
• ***Diabetes mellitus***	α_2-globulins, small *increase* ↑
• ***Rheumatoid arthritis (RA)***	γ-globulin increase ↑ slight to moderate, α_2-globulin ↑
• ***Systemic lupus erythematosus (SLE)***	γ-globulin ↑ α_2-globulin ↑
• ***Sarcoidosis***	γ-globulin ↑, α_2-globulin ↑ β-globulin ↑
• ***Lymphatic leukaemia***	γ-globulin ↓
• ***Myelogenous and monocytic leukaemia***	γ-globulin ↑
• ***Multiple myeloma***	Sharp paraprotein band in β to γ-region (M band-monoclonal)

CHARACTERISTICS OF INDIVIDUAL PLASMA PROTEINS

A. Albumin: It is the most abundant and fairly homogenous protein of plasma, with a molecular weight of 69,000. Approximately half of the total proteins of plasma is albumin. It has a low isoelectric pH (pI = 4.7). The protein migrates fastest in electrophoresis at alkaline pH and precipitates last in "salting out" or alcohol precipitation methods. It is a simple protein, **consisting of a single polypeptide chain, having 585 amino acids,** having 17 interchain disulphide (S–S) bonds. ***The protein is precipitated with full saturation of ammonium sulphate.*** The molecule is ellipsoidal in shape, measuring 150Å × 30Å.

Site of Synthesis: Albumin is **mainly synthesised in liver.** Rate of synthesis is approx. 14.0 gms/day. Albumin is initially synthesized as a "Preproprotein". Its signal peptide is removed as it passes into the cisternae of the rough endoplasmic reticulum, and a hexapeptide at the resulting aminoterminal is subsequently cleaved off further along the secretory pathway.

FUNCTIONS OF ALBUMIN

- **It exerts low viscosity.**
- **Contributes 70 to 80% of osmotic pressure** and plays an important role in exchange of water between tissue fluid and blood.
- Also undergoes constant exchange with the albumin present in extracellular spaces of muscles, skin and intestines.
- Helps in transport of several substances viz. FFA (NEFA/UFA), unconjugated bilirubin, Ca^{++} and steroid hormones.
- Certain drugs also bind to albumin, e.g. sulphonamides, aspirin, penicillin, etc. and are transported to target tissue.
- ***Nutritive function:*** Albumin in plasma is in a dynamic state with a rapid turnover with a specific half-life. It is delivered to cells where it is hydrolysed and cellular proteins are synthesised.

The pattern of serum proteins on electrophoresis may be used in the diagnosis of diseases.

Normal Values	% of Total Proteins
Albumin	50-70%
α_1-Globulin	2-6%
α_2-Globulin	5-11%
β-Globulin	7-16%
γ-Globulin	11-22%

	Total Grams/100 ml
Albumin	3.6-5.2
α_1-Globulin	0.1-0.4
α_2-Globulin	0.4-0.8
β-Globulin	0.5-1.2
γ-Globulin	0.7-1.5

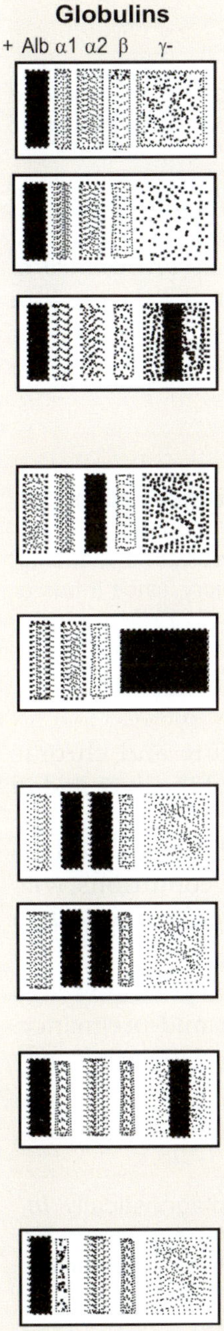

Normal pattern	
Primary immune deficiency	Impaired synthesis of immunoglobulins. Usually familial.
Multiple myeloma	Paraprotein band between α_2 and end of γ region. Normal γ-globulin often decreased. Paraproteins are also found in other diseases.
Nephrotic syndrome	Albumin lost into urine and sometimes γ-globulin, increase in α_2-globulin
Cirrhosis of liver	Decreased albumin, increased production of other unidentified proteins which migrate in βγ region, causing impaired βγ resolution
Infection	Elevated α_2 proteins and α_1 protein, usually decreased albumin
Chronic lymphatic leukaemia	Quite often accompained by decreased γ-Globulin
Plasma should not be used	**If plasma is used instead of serum, fibrinogen band gives the appearance of a paraprotein, leading to misleading diagnosis**
α_1-antitrypsin deficiency	α_1 (antitrypsin) deficiency associated with emphysema of the lung in adults, and juvenile cirrhosis

Fig. 7.1: Electrophoresis pattern of serum proteins

CLINICAL IMPORTANCE

Decrease in albumin concentration: Concentration of albumin decreases in: severe protein calorie malnutrition (PCM), liver diseases like cirrhosis of liver, (albumin synthesis impaired), nephrotic syndrome (albumin is lost in urine).

Decrease in albumin concentration leads to oedema formation. ***Oedema occurs when total proteins fall below about 5.0 gm% and albumin level below approx. 2.5 gm%.***

B. Globulins: *Globulins are separated by half-saturation with ammonium sulphate,* molecular weight ranges from 90,000 to 13,00,000. By electrophoresis, globulins can be separated into different fractions, viz. α_1-globulins, α_2-globulins, β-globulins and γ-globulins.

Site of synthesis: α-and β-globulins are synthesised in the liver. But γ-globulins are synthesised by plasma cells and B-cells of lymphoid tissues (RE system).

NATURE OF DIFFERENT FRACTIONS OF GLOBULINS

I. α-GLOBULINS: They are glycoproteins and are further classified into α_1 and α_2 depending on their electrophoretic mobility.

Some of the α_1-globulins of clinical importance are discussed below:

1. ***α_1-acid glycoprotein:*** Also called as ***orosomucoid.*** Concentration in normal plasma is 0.6 to 1.4 gm per litre (average = 0.9). Its carbohydrate content is about 41 per cent. **Orosomucoid is considered to be a reliable indicator of acute inflammation.**

Functions of Orosomucoid

- Orosomucoid **binds** the hormone **progesterone** and functions as a transport protein for this hormone.
- Probably serves to carry needed carbohydrate constituents to the sites of tissue repair following injury, since it has a high carbohydrate content (41%).

Clinical Importance

Increase: Consistently rises in acute and chronic inflammatory diseases, in cirrhosis of the liver and in many malignant conditions.

Decrease: Low concentrations (abnormally low) occurs in generalised hypoproteinaemic conditions, viz. hepatic diseases, cachexia, malnutrition and in nephrotic syndrome.

2. ***α_1-fetoglobulin (α_1-fetoprotein):*** It is present in high concentrations in foetal blood during mid-pregnancy. Normal adult blood has less than 1 μg/100 ml. It may increase during pregnancy.

CLINICAL IMPORTANCE

Presence of α_1-fetoprotein is useful diagnostically ***in determining presence of hepatocellular carcinoma or teratoblastomas*** **(tumour marker)**. (Refer Chapter on Tumour markers).

3. ***α_1-globulins-inhibitors:*** Some of the α_1-globulins act as inhibitors of coagulation and also inhibit some digestive enzymes like *trypsin* and *chymotrypsin.*

4. ***α_1-antitrypsin (α_1-AT):*** It is α_1-antiproteinase. Molecular weight: approx. 45,000 to 54,000. Isoelectric point pI is 4.0. Normal value in adults: 2 to 4 gm/Litre (approx. 290 mg/ml). The protein is highly polymorphic, the multiple forms can be separated by electrophoresis. It is synthesised by liver and **it is the principal protease inhibitor (Pi) of human plasma**. It inhibits *trypsin, elastase* and certain other proteases by forming complexes with them.

Note

A very low or absent α_1-globulin band in electrophoresis suggests α_1-antitrypsin deficiency (α_1-AT).

Phenotypes: Several phenotypes of α_1-antitrypsin have been identified. The most common phenotype is MM (allele Pi^M) associated with normal antitrypsin activity. Other alleles are Pi^S, Pi^Z, Pi^F, and Pi^- (null). ***The homozygous phenotype ZZ suffers from severe deficiency of α_1-antitrypsin and susceptible to lung disease (emphysema) and cirrhosis of the liver.***

Increases of serum α_1-antitrypsin: In response to inflammation; acute, subacute and chronic. It is considered as one of the ***acute phase reactant*** and increased in trauma, burns, infarction, malignancy, liver disease, etc. Chronic hepatocellular diseases and biliary obstruction, either normal or increased. In pregnancy and also during contraceptive medication. Neonates have serum concentrations much below adult values, but show gradual increase with age and achievement of adult levels is rapid.

Decreases in serum α_1-antitrypsin: In protein losing disorders, e.g. nephrotic syndrome, diffuse hypoproteinaemias, in emphysema of lungs and in juvenile cirrhosis liver.

CLINICAL SIGNIFICANCE

1. ***Role in Emphysema Lung:*** A deficiency of α_1-AT has a role in certain cases, approx. 5 per cent, of emphysema of lung. This occurs mainly with ZZ phenotype who synthesise Pi^Z.

Biochemical mechanism

- **Normally** α_1-AT protects the lung tissues from injurious effects by binding with the proteases, viz. *active elastase.* A particular methionine (358 residue) is involved in binding with the protease.

Thus,

Active elastase + α_1-AT
↓
Inactive elastase: α_1-AT complex
↓
No proteolysis of lung ⟶ **No tissue damage**

- **When α_1-AT is deficient or absent** the above complex with active elastase does not take place and active elastase brings about proteolysis of lung and tissue damage.

Active elastase + No or ↓ α_1-AT
↓
Active elastase ⟶ proteolysis of lung
(complex does not form) ↓
Tissue damage

2. ***Relation of smoking with emphysema:*** Smoking ***oxidises the methionine (358 residue)*** of α_1-AT and thus ***inactivates the protein,*** hence such α_1-AT molecule cannot bind to the protease *'active elastase'* and thus proteolysis of lung and tissue damage occurs accelerating the development of emphysema.

Note

- IV administration of α_1-AT has been proposed as an adjunct in the treatment of patients with emphysema due to α_1-AT deficiency.
- Recently, by protein engineering, attempts are being made to replace methionine at 358 by "another residue" that would not get oxidised by smoking. The resulting mutant new α_1-AT can afford protection against the protease by inactivating it.

Genetherapy in α_1-AT Deficiency

Attempts are being made to develop ***gene therapy* for emphysema due to α-AT deficiency.**

Use of a modified adenovirus (a pathogen of the respiratory tract) into which the gene for α_1-AT has been inserted. The virus would then be introduced into the respiratory tract by an aerosol, hoping that pulmonary epithelial cells would express the gene and secrete α_1-AT locally. This has been successful in experimental animal.

3. ***Role in cirrhosis:*** Juvenile hepatic cirrhosis has also been correlated with α_1-AT deficiency. In this condition molecules of Pi^Z (ZZ phenotype) accumulate and aggregate in the cells of the cisternae of the endoplasmic reticulum of hepatocytes. The hepatocytes cannot secrete this particular type of α_1-AT. Thus Pi^Z protein of α_1-AT is synthesised but not released from the hepatocytes. Aggregation is due to formation of polymers of mutant α_1-antitrypsin, the polymers forming via a strong interaction between a specific loop in one molecule and a prominent β-pleated sheet in another, called **loop-sheet polymerisation** . The aggregates lead to damage to liver cells leading to hepatitis and cirrhosis—accumulation of massive amount of collagen resulting to fibrosis.
4. ***Role as a tumour-marker:*** α_1-AT has been used as a ***tumour marker.*** It is increased in germ cell tumours of testes and ovary.
5. ***As an Inhibitor of Fibrinolysis:*** α_1-AT is one of the most important inhibitors to fibrinolysis along with α_2-antiplasmin and α_2-macroglobulin. All these inhibitors block the action of plasmin on fibrinogen.

II. α_2-GLOBULINS

1. Caeruloplasmin

It is a **copper containing α_2-globulin**, a glycoprotein with enzyme activities. Molecular weight is ≈ 151,000. It has ***eight sites*** for binding copper-contains about eight atoms of copper per molecule-½ as cuprous (Cu^+) and ½ as cupric (Cu^{++}). It carries 0.35 per cent Cu by weight. Normal plasma contains approx. 30 mg/100 ml and about 75 to 100 μg of Cu may be present in 100 ml of plasma. It has enzyme activities, ***e.g. copper oxidase, histaminase and ferrous oxidase.***

Site of Synthesis: It is synthesised in liver, where eight copper atoms are attached to a protein, ***apocaeruloplasmin.***

Level of caeruloplasmin with age and sex: There is low concentrations at birth, gradually increases to adult levels, and slowly continues to rise with age thereafter. Adult females have higher concentrations than males.

SECTION TWO

Functions of Caeruloplasmin

- Although caeruloplasmin is *not* involved in copper transport, 90 per cent or more of total serum copper is contained in caeruloplasmin.
- It mainly functions ***as a ferroxidase*** and helps in oxidation (conversion) of Fe^{++} to Fe^{+++} which can be incorporated into transferrin.

Clinical Importance

Increase: It is found in pregnancy, inflammatory processes, malignancies, oral oestrogen therapy and contraceptive pills.

Decrease: In Wilson's disease and in Menke's disease.

2. *Haptoglobin*

This is another α_2-globulin present in plasma of clinical importance. It is composed of two kinds of polypeptide chains, **two α-chains (possibly three)** and **only one form of β-chain.**

Site of formation: It is **synthesised principally in liver** by hepatocytes and to a very small extent, in cells of RE cells.

Phenotypes: Three phenotypes have been described, viz. 1-1, 2-1 and 2-2.

Clinical variation in concentration: Haptoglobin (Hp) increases from a mean concentration of 0.02 gm/L at birth to adult levels (10×) within the first year of life. As old age approaches, haptoglobin level increases, with a more marked increase being seen in males.

Functions of Haptoglobin

- The function of haptoglobin is to bind ***free Hb*** by its α-chain and minimises urinary loss of Hb. Abnormal Hb such as Bart's-Hb and Hb-H have no α-chains and hence cannot be bound.
- Combining power of Hp with free Hb varies with different phenotypes. **Average binding capacity** of Hp, irrespective of phenotype, can be taken approx. as **100 mg/dl.**
- After binding, Hp-Hb complex circulates in the blood, which cannot pass through glomerular filter and ultimately the complex is destroyed by RE cells.

Fates of free Hb and Hp-Hb complex shown below

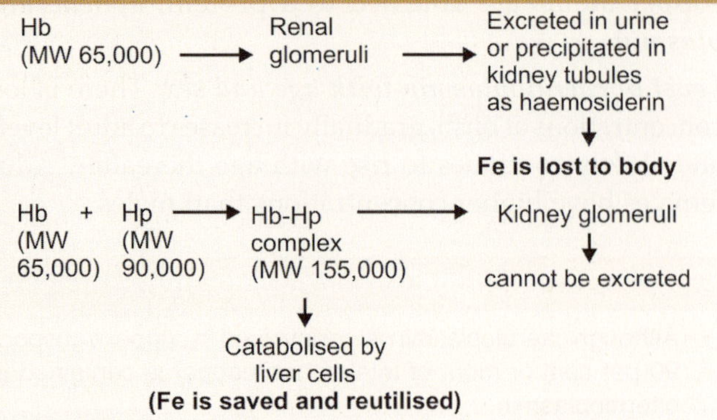

CLINICAL IMPORTANCE

- Serum Hp concentration is found to be **increased in inflammatory conditions.**
- Determination of free Hp (or Hp-binding capacity) has been used ***to evaluate the degree of intravascular haemolysis*** which can occur in mismatched blood transfusion reactions and in haemolytic disease of the newborn (HDN).
- Used also to evaluate the rheumatic diseases.

III. β-GLOBULINS

1. β-Lipoproteins (LDL)

For this refer chapter on *Lipoproteins*.

2. Transferrin

It is non-heme iron-containing protein and formerly used to be called as ***siderophilin***. It exists in plasma as β_1-globulin, a glycoprotein with molecular weight 70,000 approximately and it is a true carrier of Fe. Protein part is **apotransferrin**, a single chain polypeptide about the size of albumin and can bind two atoms of Fe to form **transferrin**. Its concentration in normal plasma is 1.8 to 2.7 gm/litre (average 2.3 gm/litre). It has capacity to bind 3.4 mg of Fe per litre or 1.46 mg of Fe^{+++} per Gm of transferrin. Nor mally transferrin is about 33 per cent saturated with Fe.

Site of synthesis: The protein is **synthesised principally** and **largely in liver**. There is also evidence of its synthesis in bone marrow, spleen, and lymph nodes, perhaps by lymphocytes.

Functions of Transferrin

- Sole function is the transport of Fe between intestine and site of synthesis of Hb and other Fe containing proteins.
- Recently it has also been seen that unsaturated transferrin has a ***bacteriostatic function*** which is attributed to sequestration of Fe required by microorganisms.

Clinical Importance

Increase: Serum transferrin levels are seen increased in Iron deficiency anaemia and in last months of pregnancy.

Decrease: Parallels to albumin in conditions like protein calorie malnutrition (PCM), cirrhosis of the liver, nephrotic syndrome, acute illness such as trauma, myocardial infarction and malignancies or other wasting diseases. (For details refer to Chapter on Iron Metabolism).

3. *C-reactive Protein*

It is also a β-globulin. It is present in concentration less than 1 mg/100 ml in the adult male. It precipitates with group C polysaccharide of pneumococci, in the presence of Ca^{++}, hence it is named as "C-reactive protein". **Two electrophoretic forms** appear on immunoelectrophoresis:

- **CRP alone in the γ-band,**
- **a CRP complex with an acidic mucopolysaccharide termed m-CRP in the β-region.**

Clinical Significance and Functions

- It can bind heme and bears some chemical and antigenic relation to liver catalase.
- It can bind to T-lymphocytes and can activate complement.
- A role in the formation of heme proteins or as an 'opsonin' in augmenting immunity has been proposed.
- It is a sensitive, if not non-specific, indicator of the early phase of an inflammatory process.
- It remains increased in presence of solid tumours, but not in conditions which impair immune response such as leukaemias or Hodgkin's disease.

4. *Haemopexin*

It is a β-globulin and migrates electrophoretically in the globulin region in the electrophoretogram. Molecular weight: 57,000 to 80,000. Normal value in adults = 0.5 to 1.0 gm/L. The level of haemopexin is very low at birth but reaches adult values within the first year of life. It is **synthesised** by parenchymal cells of **liver.**

Functions of Haemopexin

The principal function of haemopexin is to bind and remove circulating haem which is formed in the body from breakdown of Hb, myoglobin or catalase. It binds heme (ferroprotoporphyrin IX) and several other porphyrins in 1: 1 ratio. The haem-haemopexin complex is removed by the parenchymal cells of liver.

Clinical Significance

Decrease: (i) In haemolytic disorders, serum hemopexin is decreased, (ii) At birth in newborns, (iii) Administration of diphenylhydantoin.

Increase: (i) Pregnant mothers have increased serum levels, (ii) In diabetes mellitus, (iii) Duchenne muscular dystrophy, (iv) Some malignancies, especially melanomas.

5. **Complement C1q**

   Complement is a collective term for several plasma proteins that are precursors for certain active proteins circulating in blood. These proteins participate in immune reaction in the body. After the formation of immune complexes, C1q is the first complement factor that is bound. The binding takes place at the 'Fc' or constant part of the IgG or IgM molecule. The binding triggers the classical complement pathway.

Properties: It is thermolabile and is destroyed by heating at 56°C × ½ hour, normal value in adult 0.15 gm/L, Mol. wt 400,000, it can bind heparin and bivalent ions such as Ca^{++}.

Clinical Significance

Decrease: Decrease in plasma C1q concentration is used as an indicator of circulating antigen-antibody complexes. Thus levels are decreased: In active immune disease such as SLE, (i) In disorders of protein synthesis, (ii) In increased protein loss in intestinal diseases.

Increase: Levels are increased (i) In certain chronic infections (ii) In rheumatoid arthritis.

IV. β_2-MICROGLOBULINS

It is a low molecular weight peptide containing 100 amino acid residues and is excreted in urine. It is present in urine to the extent of only 0.01 mg/100 ml. It has also close structural resemblance to immunoglobulins.

CLINICAL SIGNIFICANCE

- **Peterson *et al*** in 1969 first reported that the level of serum or urine β_2-microglobulin (β_2-M) is ***increased in renal diseases and it is a reflection of impairment of function of glomerular membrane or renal tubules.*** Since then much attention has been focussed on the clinical significance of elevated serum or urine β_2-M level and renal diseases, inflammatory disorders and malignant tumours. **Hence β_2-M is now receiving much attention as a *tumour marker* also.**
- Furthermore, it was reported that the measurement of β_2-M is useful to discover renal damage caused by administration of antibiotics like gentamycin and also to check the advance of renal diseases before administration of antibiotics.
- Recently it has been stressed that increase in serum β_2-M in hepatic diseases indicates active inflammatory processes occurring in liver. This requires further evaluation.

Variations of β_2-M in Various Diseases

- ***Urinary β_2-M level is elevated in*** Wilson's disease (Hepatolenticular degeneration), renal diseases like chronic nephritis, chronic cadmium poisoning, Fanconi syndrome.
- ***Serum β_2-M level is elevated in*** Chronic nephritis, myelomatosis, systemic lupus erythematosus (SLE), Malignant tumours, abnormal pregnancies, chronic rheumatism, and Behcet syndrome.

Acute Phase Proteins (or Reactants)

Levels of certain proteins in plasma increase during acute inflammatory states or secondary to certain types of tissue damage. These proteins are called ***Acute phase proteins or reactants.***

They include:

- **C-reactive protein (CRP)**
- **Haptoglobin (Hp)**
- **α_1-antitrypsin**
- **α_1-acid glycoprotein (orosomucoid)** and
- **fibrinogen**

These proteins (except fibrinogen) discussed above.

Role of Interleukins

Interleukin-1 (IL-1), a polypeptide released from mononuclear phagocytic cells (macrophages) seems to be the principal stimulator of the synthesis of the majority of acute phase reactants by liver cells.

Interleukin-6 (IL-6) also plays a role and in conjunction with IL-1 appear to work at the level of ***gene transcription***.

V. γ-GLOBULINS

These are immunoglobulins having antibody activity. (Refer chapter on Immunoglobulins).

OTHER PROTEINS OF CLINICAL INTEREST

1. **Bence-Jones' Protein**

   An abnormal protein occurs in blood and urine of people suffering from a disease called ***multiple myeloma*** (a plasma cell tumour). Defined as **monoclonal** ***light chains*** present in the urine of patients with paraproteinaemic states. Either monoclonal 'κ' or 'λ' light chains are excreted in significant amounts in about 50 per cent cases of multiple myeloma. It has a molecular weight 45000, and has sedimentation coefficient of 3.5 S. Sometimes the chains excreted may be a 'dimer' of L-chains.

Identification—Heat Test

- The protein is identified easily in urine by a simple **Heat Test.** On heating the urine 50 to 60°C,

SECTION TWO

Bence-Jones' proteins are precipitated, but when heated further it dissolves again. Reverse occurs on cooling.

- Best detected by zone electrophoresis and immuno-electrophoresis of concentrated urine.

Note

Normally only very small quantities of 'κ' and 'λ' immunoglobulin L-chains are excreted in urine. They pass through glomerular filter but are reabsorbed by kidney tubules and broken down in the lining cells and thus not found in urine.

2. **Cryoglobulins**

These are proteins which are coagulated when plasma or serum is cooled to very low temperature (2 to 4°C). Most commonly they are monoclonal IgG or IgM or a mixture of two. Traces are present even in normal individuals. Their molecular weights vary from 1,65,000 to 6,00,000.

Increase: They are increased in rheumatoid arthritis, lymphocytic leukaemia, multiple myeloma, lymphosarcomas and systemic lupus erythematosus (SLE).

C. Fibrinogen

A soluble **glycoprotein**, also called as **clotting factor I**, as it takes part in coagulation of blood and it is the precursor of fibrin, the substance required for clotting. Normally constitutes 4 to 6 per cent of total proteins of blood. Like globulins, ***it is also precipitated with 1/5th saturation with ammonium sulphate***. Molecular weight ranges between 3,50,000 and 4,50,000. ***It has a large asymmetrical molecule, which is highly elongated having an axial ratio of about 20:1. Being asymmetrical and large, is important for viscosity of blood.***

Site of synthesis: All the three chains are ***synthesised in the liver;*** **three structural genes** involved in the synthesis are on the same chromosome and their expression is regulated co-ordinately in humans.

Structure: It is made up of **six polypeptide chains**, 2Aα, 2Bβ and 2γ-chains thus formula is $A_{\alpha 2}B_{\beta 2}\gamma_2$. Chains are linked together lengthwise by S-S linkages.

The $-NH_2$ terminal regions of the six chains are held in close proximity by S-S linkages, while –COOH terminal regions are spread apart, giving rise to a highly asymmetric, elongated molecule.

The A and B portions of the Aα and Bβ chains, designated as **fibrinopeptides A (FPA)** and **fibrinopeptides B (FPB) respectively, the amino terminal ends of the chains, bear excess of negative charge** due to presence of ***aspartic acids*** and ***glutamic acids*** residues. Also contains an unusual ***tyrosine-O-SO4*** residue on **FPB**.

These negative charges contribute to **(i)** the **solubility** of fibrinogen in plasma and also **(ii) prevent aggregation** due to electrostatic repulsion between the fibrinogen molecules (Fig. 7.2).

D. α_2-Macroglobulin

- It is a large plasma ***glycoprotein*** (720 KDa).
- It is a ***tetramer*** and it is made up of four identical subunits of 180 KDa.
- It comprises ***8 to 10 per cent of total*** plasma protein in humans.

Synthesis

The protein is synthesised by a variety of cell types, including ***monocytes, hepatocytes*** and ***astrocytes***.

Structure

It belongs to a group of proteins that include complement protein C3 and C4.

These proteins have a unique internal cyclic thiol ester bond, which is formed between a cysteine and glutamine residue. Hence this group of plasma proteins are designated as a **thiol ester plasma protein family**.

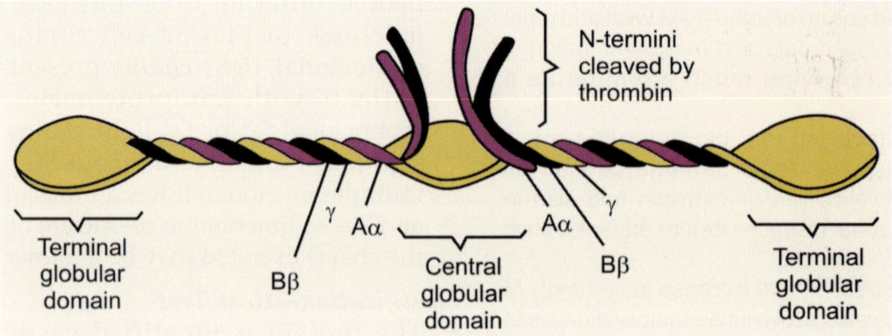

Fig. 7.2: Structure of fibrinogen

Functions

- **Pan proteinase inhibitor:** α_2-macroglobulin binds ***many proteinases.***

The α_2-macroglobulin-proteinase complex thus formed are rapidly cleared from the plasma by a **receptor** located on many cell types.

Cytokines Delivery to Target Tissues

α_2-macroglobulin also ***binds many cytokines*** viz. PDGF, TGF-β, etc. and appears to be involved in targeting them toward particular tissue or cells. Once delivered, the cells take up α_2- macroglobulin-cytokines complex and inside the cells cytokines dissociate from α_2-macroglobulin and ***exerts its effects on cell growth and functions.***

Transportation of Zinc

Approximately 10 per cent of the zinc in plasma is transported by α_2-macroglobulin and the remainder being transported by albumin.

FUNCTIONS OF PLASMA PROTEINS

1. ***Nutritive:*** They are simple proteins and a good source of protein, thus largely involved in the nutritive functions. It is useful in hypoproteinaemic states. These contribute amino acids for tissue protein synthesis.
2. ***Fluid exchange:*** The colloid osmotic pressure of plasma proteins plays an important role in the distribution of water between the blood and tissues (Fig. 7.3).
   - At the ***arterial end*** of capillary: ***Hydrostatic pressure exerted greater than the osmotic pressure. Net filtration pressure is 7 mmHg*** which drives the fluid out from vessels to tissue spaces.
   - On the other hand in the ***venous end*** of capillary loop, ***osmotic pressure is greater than the hydrostatic pressure*** and ***net absorption pressure is 8 mmHg*** which draws fluid from tissue spaces into the vessels. This explanation of the mechanism of exchange of fluids and dissolved materials between the blood and tissue spaces is called the **Starling hypothesis.**
3. ***Buffering action:*** The serum proteins, like other proteins, are ***amphoteric,*** and thus can combine with acids or bases. ***In acidic pH,*** NH_2 group acts as base and can accept a proton and thus is converted to NH_3^+ and ***in alkaline pH,*** COOH group acts as acid and can donate a 'proton' and thus have COO^-. At the normal pH of blood, the proteins can act as an acid and combines with cations (mainly sodium).
4. ***Binding and transport function:*** Already discussed above. See Table 7.2 for important proteins which bind and transport various substances.
5. ***Viscosity of blood:*** Due to presence of proteins, blood is a viscous fluid. ***Globulins and fibrinogen, which are large in size and asymmetrical account for the viscosity of blood.*** The viscosity of blood provides resistance to flow of blood in the blood vessels to maintain blood pressure in the normal range.
6. ***Reserve proteins:*** The amino acids from plasma proteins can be taken up by tissues and used for building up new tissue proteins and *vice versa.*
7. ***Role in blood coagulation and fibrinolysis:*** In addition to prothrombin and fibrinogen, plasma contains a number of other components, enzymes and clotting factors, which participate in the process of coagulation of blood. Intravascular clot, known as ***thrombus,***

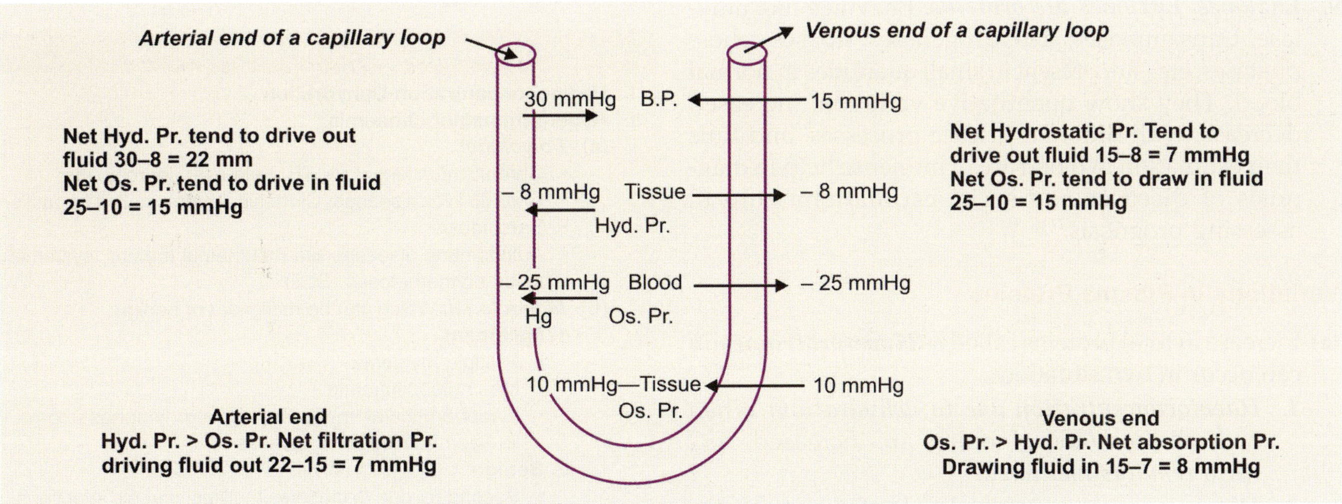

Fig. 7.3: Showing fluid exchange in capillary blood (Hyd. Pr. = Hydrostatic pressure, Os Pr. = Osmotic pressure, BP = Blood pressure)

Table 7.2: Various proteins which have binding and transport functions

Proteins	Mol. Wt.	Normal value in adults mg/100 ml	Biological functions
• ***Transthyretin (Prealbumin)***	61,000	10-40 (Mean 25)	Binds and transports thyroxine and retinol
• ***Albumin***	69,000	3500-5500 (4400)	Osmotic, reserve protein, binding and transport of ions, pigments, drugs, etc.
• ***α_1-lipoprotein (HDL)***	200,000	290-770 (360)	Transport of various lipid fractions, fat soluble vitamins, hormones, etc. removes cholesterol from tissues to liver ***(Scavenging action)***
• ***Transcortin***	45,000	(~ 7)	Binding and transport of cortisol
• ***TBG (thyroxine binding globulin)***	~45,000	(1-2)	Binding and transport of thyroxine
• ***Retinol binding protein***	21,000	(~ 4.5)	Binding and transport of retinol
• ***Caeruloplasmin***	160,000	20-60 (35)	Binding of copper, ***ferroxidase activity***
• ***Haptoglobin phenotypes 1-1***		100-220 (170)	Binding and transport of free Hb
2-1	100,000	160-300 (235)	Peroxidase activity
2-2		120-260 (190)	
• ***β-Lipoprotein*** *(LDL) βLP*	3,200,000	250-800 (500)	Transport of various fractions of Lipids (rich in cholesterol), hormones, fat soluble vitamins. ***Carries cholesterol to tissues***
• ***Hemopexin, (HPX)*** *(β-B-globulin)*	80,000	70-130 (100)	Heme binding
• ***Transferrin, (Tf)*** *(siderophillin)*	80,000 to 90,000	200-400 (295)	Binding of Fe and transport
• ***Transcobalamine*** *I, II and III (α_2-β mobility)*	—	I-20 μg/L II-60 μg/L	Binding and transport of Vit B_{12}

whenever it is formed is digested by the enzymes of fibrinolytic system present in the plasma, which saves from the disastrous effects of thrombosis.

8. *Immunological function (Body defence):* γ-globulins are present in plasma, which are antibodies and protect the body against microbial infections.
9. *Enzymes: Enzymes are proteins.* Enzymes like amylase, transaminases, dehydrogenases, lipases, phosphatases, etc. are present in small quantities in normal blood. They show quantitative variation, increase/decrease, with different disease processes, and thus their estimation in blood are of immense help in diagnosis of diseases and serial estimations help in assessing prognosis.

Variations in Plasma Proteins

(a) Increase in total proteins called as *Hyperproteinaemia* can occur in two situations:
  1. *Haemoconcentration due to dehydration:* When both albumin and globulins are increased. ***A:G ratio remains unaltered.***
  2. *Diseases resulting in high levels of plasma globulins:* Mainly γ-globulins (hypergammaglobulinaemias)-Albumin remains either normal or reduced.

  If albumin is reduced grossly, ***the A:G ratio is reversed.***

- Hypergammaglobulinaemia may be due to:
  - 'Polyclonal' gammopathies.
  - 'Monoclonal' gammopathies (See Table 7.3 for the causes).

Table 7.3: Causes of hyperproteinaemia

I. Haemoconcentration-Dehydration

II. Hypergammaglobulinaemia

(a) ***Polyclonal***
- Chronic infections: Like TB, Kala-azar, leprosy, etc.
- Chronic liver diseases: Cirrhosis, chronic active hepatitis
- Sarcoidosis
- Autoimmune diseases like rheumatoid arthritis, systemic lupus erythematosus (SLE)

(b) ***Monoclonal:*** Which can be malignant or benign:

1. Malignant
- Multiple myeloma
- Macroglobulinaemia
- Lymphoreticular malignancies, e.g. lymphosarcoma, leukaemia, Hodgkin's disease

2. Benign
- Secondary due to diabetes mellitus, chronic infections, etc.
- Idiopathic

Table 7.4: Causes of hypoproteinaemia

I. Haemodilution
- Water intoxication (overload)
- Sample from IV infusion

II. Hypoalbuminaemia

*(a) **Loss from the body***
- ***Renal: Loss of albumin*** in urine in nephrotic syndrome
- ***GI Tract: Protein losing enteropathy***
- Skin: Burns and other exudative skin lesions

*(b) **Decreased synthesis of albumin***
- Severe liver diseases: Chronic hepatitis, cirrhosis liver
- Non-availability of the precursors: Malabsorption syndrome, protein calorie malnutrition
- Genetic deficiency: Analbuminaemia

*(c) **Miscellaneous***

Acute/chronic illness, infections, malignancy, pregnancy

III. Hypogammaglobulinaemia

*(a) **Protein loss:*** Same as above II (a)

*(b) **Decreased synthesis***
1. Transient: Neonates/infants
2. Primary: Genetic deficiency
3. Secondary
   - Certain toxins/and drugs: uraemia, cytotoxic therapy, corticosteroid therapy, etc.
   - Certain haematological disorders: leukaemias, lymphosarcoma
   - Acquired Immune Deficiency Syndrome (AIDS)

*(c) **Miscellaneous:*** Pregnancy

(b) Decrease in total proteins called as ***Hypoproteinaemia*** can occur in two situations:
- *Haemodilution:* Where both albumin and globulins are decreased, ***A:G ratio remains unaltered.***
- *Conditions resulting in low albumin level:* It is more common accompanied either by no increase in globulin or by an increase which is less than the fall in albumin. ***A:G ratio is decreased.*** Hypoproteinaemia may also take place due to decrease in γ-globulin (Hypogammaglobulinaemia).

Various causes of hypoproteinaemia are listed in Table 7.4.

GENETIC DEFICIENCIES OF PLASMA PROTEINS

1. **Analbuminaemia:** Inherited disorder in which albumin is very low or completely absent. **Defect is in the albumin synthesis.** There is usually associated raised levels of plasma lipids and lipoproteins, which is probably secondary defect in lipid transport. All globulin fractions occur in increased concentration.
2. **Bisalbuminaemia:** Another genetic variant in which two albumin peaks are present. It appears to have no clinical importance.
3. **Bruton's agammaglobulinaemia**
   - An inherited disorder, X-linked recessive traits,
   - Differentiation of B-lymphocytes to plasma cells is defective leading **to lack of plasma cells in circulating blood.**
   - There is absence of γ-globulins or γ-globulins level are very low, and lacks humoral immunity and susceptible to bacterial infections.
4. **Afibrinogenaemia:** An inherited disorder characterized by genetic defect in fibrinogen formation. Inherited as a non-X-linked recessive trait characterized by absence of fibrinogen or very low level of fibrinogen. Blood clotting mechanism is hampered and there may be uncontrollable haemorrhages.

PROTEINURIA

When proteins appear in urine in detectable quantities, it is called as ***albuminuria.*** Actually albuminuria term is a misnomer, as seldom albumin found alone, hence the term **proteinuria** ***is to be preferred.*** The proteins that are found in the urine in kidney conditions are commonly believed to be plasma proteins that pass through the damaged renal epithelium. The albumin having smallest molecules pass most easily; globulins next and fibrinogen least readily.

Types of proteinurias: Classified under **two major heads:**

A. Functional proteinurias

B. Organic proteinurias

A. Functional Proteinurias: Those conditions that are not related to a diseased organ. The amount of protein excreted is usually small, majority of cases show < than 0.2 per cent and the condition is usually temporary.

Causes
- *Violent exercise:* Soldiers after long marches, atheletes after strenuous contests can have such temporary proteinurias. Here, there may be slight kidney damage to account for it, but the condition almost always clear up.
- *Cold bathing:* Leading to constriction of renal blood vessels and producing temporary anoxia.
- *Alimentary proteinuria:* Occasionally proteinuria may occur after excessive protein ingestion.
- *Pregnancy:* Proteinuria is frequently, associated with pregnancy, probably as a result of pressure interfering with the return of blood in renal veins.
- *Orthostatic/or postural proteinuria:* This occurs chiefly in children or in adolescents, usually in age group of 14 to 18 years. In these young individuals, the urine contains protein when they are in upright position only. When they are lying down it is free from proteins.

Cause: This is not due to kidney disease but is probably due to some disturbance in the blood supply to the kidneys, leading to venous stasis and temporary anoxia.

B. **Organic Proteinurias:** There are many pathologic conditions that cause organic proteinuria, which may be classified in **three major groups.**
   (a) *Pre-renal,*
   (b) *Renal,* and
   (c) *Post-renal.*

(a) Pre-renal

Conditions causing proteinuria of this group are those that are primarily not related to kidney. In most cases, they affect the kidneys in such a way as to render it more permeable to the protein molecule.

Causes

- *Cardiac diseases:* By affecting the circulation of kidneys leads to proteinuria.
- Any ***abdominal tumors,*** or mass of fluid in the abdomen does the same by exerting pressure on the renal veins.
- *Fevers,* covulsions, anaemias and other blood diseases, liver diseases and many other pathologic states can affect in similar manners as stated above.
- *Cancers:* An increased amount of urinary mucoproteins generally accompanies elevated serum mucoprotein levels. Such has been observed in patients with cancers, with highest values when carcinomatous invasion is widespread.
- *Collagen diseases* and inflammatory conditions also have high mucoprotein levels.

(b) Renal

Proteinurias are found in various types of kidney diseases and are called as ***Renal proteinurias.***

Causes

- *Acute glomerulonephritis:* Always associated with proteinuria.
- *Chronic glomerulonephritis:* Proteinuria is seen in early stages, but may disappear later as the kidney becomes more and more impaired.
- *In nephrosclerosis, TB of kidney and in carcinoma of kidney:* Proteinuria is frequently found but it is not always.
- *In nephrotic syndrome* (Type II): **Large quantities of albumin is lost in the urine** and there may be gross hypoalbuminaemia in blood.
- Polypeptides, the so-called, proteoses and peptones, sometimes are excreted in urine. This may happen in pneumonias, diphtheria, carcinoma and other conditions and is due to some protein containing materials, e.g. an exudate or a tissue mass/pus undergoing autolysis.

(c) Post-renal

These are sometimes called as **false *proteinurias,*** whereas the above two are *true,* because in these conditions (postrenal) ***proteins do not pass through the kidneys.***

Causes

- May be due to inflammatory, degenerative or traumatic lesions of the pelvis of the kidney, ureter, bladder, prostate or urethra.
- Bleeding in genitourinary tract also will account for proteinuria.
- Urine containing pus also contains proteins, since the exudate that accompanies the pus is rich in proteins.

I. Questions (Essay type and short notes)

1. What are plasma proteins? Describe methods of their separation.
2. Enumerate plasma proteins. What are their functions?
3. Describe the chemistry and functions of plasma proteins.
4. Describe in brief different electrophoretic variations seen in various disorders.

Short Notes

a. Functions of plasma proteins
b. Fractionation of plasma proteins
c. Separation of plasma proteins
d. Caeruloplasmin
e. Transferrin
f. C-reactive protein
g. Bence-Jones' protein
h. Fibrinogen
i. Haptoglobin
j. Orosomucoid
k. β_2-microglobulin.

II. MCQs (Give one correct answer)

1. The number of amino acids present in one molecule of albumin are:
   a. 510 b. 590
   c. 600 d. 610
   e. 650
2. One caeruloplasmin molecule can bind ______ copper ions:
   a. 4 b. 6
   c. 8 d. 10
   e. 12
3. Plasma proteins can be separated by:
   a. Electrophoresis
   b. Salting out with ammonium sulphate

c. Ultracentrifugation
d. Immunoelectrophoresis
e. All of the above

4. β-globulin fraction contains:
a. Haptoglobin b. α-fetoprotein
c. Caeruloplasmin d. Transferrin
e. Thyroxine binding proteins

5. Plasma proteins which contain more than 4 per cent hexosamine are designated as:
a. Microglobulins b. Cryoglobulins
c. Mucoproteins d. Glycoproteins
e. Orosomucoids

6. Plasma proteins which are precipitated best by 1/5th saturation of ammonium sulphate:
a. Albumin b. α_2-globulins
c. β-globulins d. Euglobulins
e. Fibrinogen

7. The human plasma proteins are a mixture of:
a. Glycoproteins b. Simple proteins
c. Conjugated proteins d. Lipoproteins
e. All of the above

8. Transferrin can bind _______ atoms of Fe^{+++} per molecule:
a. 1 b. 2
c. 3 d. 4
e. 5

Answers to MCQs

1. d 2. c 3. e 4. d 5. c 6. e
7. e 8. b.

CHAPTER 8

IMMUNOGLOBULINS—CHEMISTRY AND FUNCTIONS

Major Concepts

A. What are immunoglobulins? Study the chemistry and functions of immunoglobulins.

Specific Objectives

1 Define immunoglobulin.
2. Classify immunoglobulins, WHO classification, list the five classes and note the synonyms of each class.
3. Learn the important properties of each class of immunoglobulins and study the differences of immunoglobulin classes as related to structure, properties and antibody functions in a tabular form.
4. Study the antibody function of each class of immunoglobulins.
5. • Note the difference between serum IgA and non-vascular IgA in secretions.
   • Learn what is "secretory component" (T-Piece) and its functions.
6. Study the structure of immunoglobulins
   • What are the Heavy chains (H-chains) and Light chains (L-chains)?, their types and functions.
   • Note the molecular formula of each class.
   • Study the degradation of IgG by proteolytic enzymes: • by papain and • by pepsin.
7. Learn the site of synthesis of Igs and genetic control.
8. Study monoclonal and polyclonal antibodies. What is 'hybridoma'? Learn the uses of monoclonal antibodies.

INTRODUCTION

When foreign substances called antigens are introduced into the body they trigger the appearance of immunoglobulins in the serum and in other body fluids. Immunoglobulins (Igs) are freely circulating proteins. Antibodies are those immunoglobulins that possess a demonstrable specificity for a given antigen.

The term antigen encompasses a wide variety of organic molecules that elicit one or both of the two known types of specific immune response:

Almost any class of organic compound may be antigenic, including many foreign proteins and macromolecular carbohydrates. Nucleic acids may function as antigens in certain experimental situations and in human disease.

Hapten: Certain lipids (e.g. steroids) and other relatively simple organic structures (e.g. dinitrophenyl group) can function as antigens if coupled with a suitably large protein molecule. In such a situation, the organic residue is called as a ***Hapten*** and the protein, a carrier. Both are required to produce an antibody response.

The immune system of the body consists of two major components:

1. **B lymphocytes** and
2. **T lymphocytes**

1. The B lymphocytes are mainly derived from bone marrow cells in higher animals and from the bursa of Fabricius in birds. They are ***responsible for the synthesis of circulating humoral antibodies*** *known* ***as immunoglobulins***. They are synthesised mainly in **plasma cells**, which are specialised cells of B cell lineage that synthesise and secrete immunoglobulins into the plasma in response to exposure to a variety of antigens.
2. The T lymphocytes are of **thymic origin**. They are involved in a variety of important **cell-mediated immunologic processes** such as graft rejection, delayed hypersensitivity reactions, and defense against malignant cells and many viruses. (This will not be discussed in this chapter).

Immunoglobulins normally constitute approximately 20 per cent of total serum proteins. All Igs are definable antibodies. They **are glycoproteins** composed of 82 to 96 per cent polypeptide and 4 to 18 per cent carbohydrates. The polypeptide component possesses almost all of the biologic properties associated with antibody molecules.

Definition: The immunoglobulins constitute a heterogenous family of serum proteins, which either function as antibodies or are chemically related to antibodies. On electrophoresis, they mainly occupy the γ-globulin position but also occur in the β and α_2-regions.

Classification

Immunoglobulins (Igs) are divided into **five main classes** according to their molecular weight, electrophoretic mobility, ultracentrifugal sedimentation and other properties.

WHO Classification

Classes	Corresponding old nomenclatures
1. **IgG** (or γG)	7 s γ, γ_2, γ ss, 6.6 Sγ
2. **IgA** (or γA)	γ_1A, β_2A, 7Sγ^1
3. **IgM** (or γM)	19 Sγ, γ_1M, β_2M
4. **IgD** (or γD)	—
5. **IgE** (γE)	—

PROPERTIES OF INDIVIDUAL IMMUNOGLOBULINS

A. General Properties

The properties of individual immunoglobulins of five classes are given in tabular form in Table 8.1.

B. Special Features of Each Class

I. IgG (γG)

Subclasses of IgGs: **Four** subclasses have been described and they are approximately as follows:
- IgG 1-60 to 70 per cent,
- IgG 2-14 to 20 per cent,
- IgG 3-4 to 8 per cent and
- IgG 4-2 to 6 per cent.

These figures vary somewhat from individual to individual. The capacity of a given individual to produce antibodies of one or another IgG subclass may be under genetic control. Subclasses are not endowed equally with transfer across the placenta, ***IgG_2 being transferred more slowly than the other subclasses.***

Antibodies contained in IgG

IgG class represents the most of antibacterial and antiviral antibodies, thus **majority of acquired antibodies are in this fraction.**

The **following antibodies have been identified** in this class:
- Immune anti-A and anti-B
- Anti-Rh antibodies, incomplete type
- Antistreptolysin
- Opsonins, bacteriocidins and bacteriolysins
- Anti-H flagellar antibodies
- Antitoxins (Diphtheria, tetanus, streptococcal)
- Antiviral antibodies
- Complement fixing antibodies
- Antipneumococcal (capsular)
- Antispirochaetal
- Autoantibodies to thyroid

*Evolution: **In terms of evolution IgG seems to have evolved later than IgM.*** In keeping with this view, the sequence in the organism's response to antigenic stimulation usually consists of **IgM being produced initially, followed later and ultimately replaced by IgG.**

Catabolism of IgG
- Catabolic rate bears a direct relationship to its concentration in the plasma, i.e. the rate of catabolism declines as the concentration falls. (cf. IgA, IgM, fibrinogen, caeruloplasmin, catabolic rates are independent of plasma level).
- It has been suggested that ***Protector sites*** specific for IgG may exist within the body. Accordingly, when IgG molecules are bound to these sites, they are protected from catabolism.

II. IgM (γ M)

Polymerisation

Normally IgM molecule exists as a pentamer with a molecular weight of about 900,000 (19s). Quantitative measurements indicate that there is a single J-chain in each IgM pentamer or in Polymeric IgA molecule.

J-chain

The J-chain is a **small glycopeptide** with an unusually high content of ***aspartic acid (Asp) and glutamic acid.*** The J-chain has a fast electrophoretic mobility on alkaline gels owing to its highly acidic nature. Physiochemical studies indicate that the J-chain molecule is very elongated, with an axial ratio of 18 approximately. Studies have indicated ***that J-chain is not an absolute requirement for polymerisation of Ig basic units. Nevertheless, the presence of J-chain does facilitate the polymerisation of basic units of IgM and IgA molecules into their appropriate polymeric forms.***

Antibody activity of IgM

IgM antibody is prominent in early immune responses to most antigens and predominates in certain antibody responses such as "natural" blood group antibodies.

IgM (with IgD) is the major Ig expressed on the surface of B cells.

Chief antibodies contained in IgM are:
- Naturally occurring anti-A and anti-B
- Some immune anti-A and anti-B (saline type)
- Anti-Rh antibodies (saline type)
- Cold antibodies (anti-i type)
- Heterophil antibodies and Forssman antibodies
- Some opsonins, bacteriocidins, and lysins
- Somatic 'O' antibody of *Salmonella typhi*
- Rheumatoid factor (RA factor)

Table 8.1: The differences in Igs classes as related to structure, properties and functions

Class	*IgG*	*IgA*	*IgM*	*IgD*	*IgE*
• **Alternate names**	γG	γA	γM	γD	γE
• **'H' chains names**	γ	α	μ	δ	ε
• **'H' chains subclasses**	4 (γ_1, γ_2, γ_3 and γ_4)	2 (α_1, α_2)	2 (μ_1, μ_2)	—	—
• **'L' chain types**	K or λ	'K' or λ	'K' or λ	'K' or λ	'K' or λ
• **Molecular formula**	γ_2K_2 $\gamma_2\lambda_2$	α_2K_2 $\alpha_2\lambda_2$	$(\mu_2 K_2)$ 5 $(\mu_2 \lambda_2)$ 5	δ_2K_2 $\delta_2\lambda_2$	ε_2K_2 $\varepsilon_2\lambda_2$
• **Approximate molecular weight**	145,000 Earlier 160,000 to 165,000	150,000 to 500,000	900,000 to 1,000,000	180,000	190,000 to 200,000
• **Sf**	6-7S (7S fraction)	7S-13S	19S	7S to 8S	8S
• **Carbohydrate content**	2.5%	8%	10%	15%	11%
• **Amount**	70 to 80% of total	10 to 20% of total	7% (3 to 10%) of total	0.5 to 2% of total	0.004% of total
• **Proportion of body total in serum**	50%	40%	65-95%	65 to 85%	50%
• **Normal level in serum**	1200 mg% (600-1600)	200 mg% (150-250)	120 mg% (60 to 170)	3 mg%	10 to 70 μg%
• **Half-life**	19 to 24 days	6 days	5 days	2 to 8 days	1 to 6 days
• **Catabolic rate**	6.9% per day (5.6 to 8.4%)	12% per day	14% per day	—	—
• **Synthetic rate**	36 mg/kg/day (24-50)	30 mg/kg/day	7 mg/kg/day	—	—
• **Electrophoretic mobility**	Slow γ_2 to fast γ	Fast $\gamma \rightarrow$ to β	Fast $\gamma \rightarrow$ to β	Fast γ	Fast γ
• **Presence in external secretions**	+	++++	+	±	±
• **Placental transfer**	+++	—	—	—	—
• **Heat stability**	—	+	++	++++ (most labile)	+++ (most labile)
• **Complement fixation**	+	0	+	?	0
• **Antibody activity (ABO blood group and Rh)**	Immune anti-A and anti-B anti-Rh (incomplete type)	—	Naturally occurring anti-A and anti-B Anti-Rh antibodies (saline type)	— —	++++ —
• **Reaginic antibody**	?	—	—	—	++++
• **Skin sensitisation (Passive transfer)**	0	0	0	0	++
• **Antibacterial lysins**	++	+	+++	?	?
• **Antiviral activity**	+	+++	+	?	?

- LE factor and antibodies
- Some antithyroid autoantibodies
- Antitrypanosomal antibodies

CLINICAL ASPECT

1. ***Infectious mononucleosis:*** It has been known for long that haemolytic anaemia may occasionally complicate Infectious mononucleosis. It is of interest in this connection is the finding by **Calvo *et al*** (1965) 19s cold agglutinins of anti-i type often present in this disease.
2. **Waldenström's macroglobulinaemia:** Excessive production of IgM occurs in this condition due to malignant Proliferation of Lymphocytoid cells.

SIA Test: Sia test is often +ve in this condition.

Method

A drop of serum is allowed to fall into about 100 ml of deionized water in a measuring cylinder. **A white precipitate forming at once constitute a +ve test.** Although useful as a simple screening test, both false –ve and false +ve reactions can be obtained in other hyperglobulinaemias. It has been observed that in other hyperglobulinaemic conditions, in which the test is +ve, the *precipitate* falls slowly if at all, in Waldenström's macroglobulinaemia it falls quickly.

Modified SIA Test (Martin, 1960): **Martin** used a modified Sia test in which he diluted the serum 20 times with distilled water at 20°C (pH 6.5 to 7.0). If the test is +ve, a precipitate is formed within 5 minutes and after centrifuging and pouring off the supernate, the precipitate is readily soluble in 0.15 per cent NaCl solution. The modified test was claimed to be +ve in 18 cases of Waldenström's macroglobulinaemia in a series and a false +ve was never encountered.

III. IgA (γA)

IgA in Secretions (Non-vascular IgA): In addition to its antibody function in serum, IgA is the predominant Immunoglobulin class in body secretions. **It is found in external secretions such as colostrum, saliva, tears, GI fluids, prostatic secretion, nasal and bronchial secretions.**

Secretory IgA: IgA present in these secretions is in the form of:

- *Higher polymer:* Each secretory IgA molecule consists of two 4 chain basic units and one molecule of J-chain. Molecular weight of secretory IgA is approximately 400,000.

Table 8.2: Distribution of Igs in non-vascular fluids

	Internal secretion	*External secretions*	*Pathological fluids*
Source	CS fluid, synovial fluids, aqueous humour	Tears, saliva, GI fluids, nasal and bronchial fluids, etc.	Inflammation and irritation —
Igs	Same as serum	Mainly IgA (Predominant)	IgG, IgM, and auto-antibodies, e.g. RF and ANF

- Antigenically slightly different from serum IgA. This dissimilarity between vascular IgA and non-vascular IgA is due to presence of another small protein, called as **transport piece (T-Protein or secretory protein or transport protein)** (Table 8.2).

T-Protein (Transport Piece)

South et al. (1966) demonstrated T-Protein. T-Piece is a single polypeptide chain of approximately molecular weight of 70,000. The carbohydrate content is high but not precisely known. Its amino acid composition differs appreciably from that of every other Igs molecules including J-chain. It has an electrophoretic mobility in the fast β-range. It is ***produced by the ductular epithelial cells of breast,*** salivary glands, mucosal glands of GI tract/bronchial tree and prostate gland. ***Secretory IgA molecule is a*** **dimer** ***and T-Protein is attached to the L-chains (Fc region) of the two IgA molecules.***

Note: Secretory Protein can be found free in secretions of individuals who lack measurable IgA in their serum/or secretions.

Functions of T-Piece

This protein is responsible for two important biological properties of non-vascular IgA:

- Its **selective transport**, from serum to secretions or to facilitate the transport of IgA molecule synthesised in lymphoid cells beneath the epithelium.
- And probably **protection of the IgA molecule** against digestion by proteolytic enzymes like those found in GI tract.

Antibody Functions of IgA Molecule

1. Secretory IgA **provides the primary defence mechanism against some local infections** owing to its abundance in saliva, tears, bronchial secretions, the nasal mucosa, prostatic fluid, vaginal secretions and mucous secretions of small intestine. Thus, ***IgA appears to be essential to ward off*** **sinobronchial** ***infections.***
2. Claims that isolated human IgA globulins have shown isohaemagglutinin and antibacterial activity against Diphtheria bacilli and Brucella organisms have been recently claimed.
3. It has been suggested that IgA levels may be decreased in thymectomised animals.

IV. IgD (γD)

Antibody Functions: The main function and the role is not yet determined. But following has been reported:

1. There are isolated reports of IgD with antibody activity toward certain antigens, including penicillin, milk proteins, insulin, diphtheria toxoid, nuclear antigens and certain thyroid antigens.
2. IgD along with IgM is the predominant immunoglobulin on the surface of human B Lymphocytes and it has been suggested that IgD may be involved in the differentiation of these cells.

Figures 8.1A to C show the presence of J-chain/and T-piece.

V. IgE (γE): The identification of IgE antibodies as **reagins *(or Reaginic antibodies)*** and characterization of this Ig class marked a major breakthrough in the study of the mechanisms involved in allergic diseases like hay fever, asthma, etc. **Ishizakas** (1966) first raised the possibility that reaginic antibody which gives **P-K reaction** is associated with a separate class of Igs which he tentatively called as IgE.

Antibody Function of IgE

1. IgE antibodies provide a striking example of the ***bi-functional*** nature of antibody molecules. IgE antibodies bind **allergens** through the ***Fab*** **protein**, but the binding of IgE antibodies to tissue cells like mast cells, it binds to ***Fc*** **portion** (for 'Fab' and 'Fc' portion see structure of Igs, discussed on page 110).

   Note: Allergen is an alternative term used for any antigen that stimulates IgE production.
2. *Allergic response:* Upon combination of IgE ('reagin' or reaginic antibody) with certain specific antigens, called as allergens, IgE triggers the release from mast cells the pharmacologic mediators responsible for the characteristic *wheal* and *flare* skin reactions evoked by the exposure of the skin of allergic individuals to allergens.

   Prausnitz-Küstner reaction (P-K reaction): Injected into the skin of a non-allergic person, this antibody

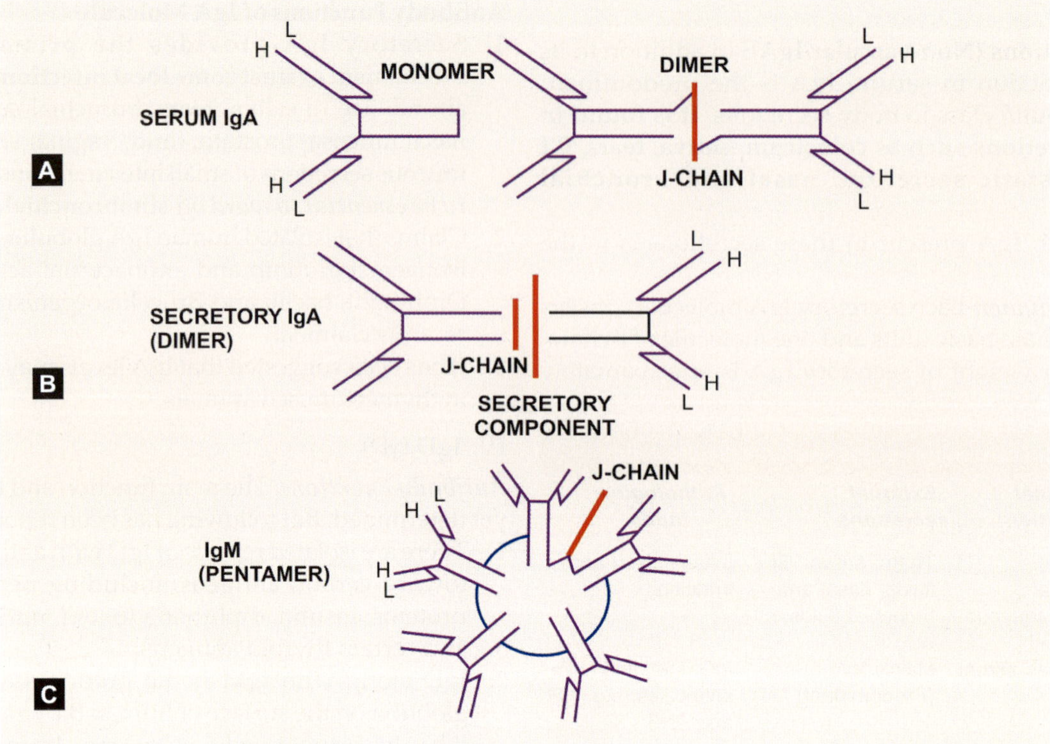

Figs 8.1A to C: J-chain and T-piece secretory component

specifically sensitises the injection site so that the re-exposure to the allergen reproduces the skin reaction. This phenomenon is known as **passive transfer** *or the* **Prausnitz-Küstner** *reaction (P-K reaction).*

3. ***Role in inflammation:* Amman and coworkers** (1969) have demonstrated that IgE not only fixes to skin sites but can also fix to leucocytes and other cells. Thus can release mediators of inflammation upon exposure to allergens or antigens. These may be important protective mechanisms. Histamine release from leucocytes by white cell bound IgE upon antigen contact has been extensively studied.
4. ***Defense against worm infections***
IgE defends against ***worm infections*** by causing release of enzymes from eosinophils. It does not fix complement. **IgE offers main host defense against helminthic infections.**

CLINICAL ASPECT

- IgE deficiency has been demonstrated with chronic sinopulmonary infection (Cain *et al*, 1969).
- IgE deficiency also demonstrated in 11/16 patients with ***Ataxia telangiectasia***, a familial disorder of progressive cerebellar ataxia, oculocutaneous telangiectasia and frequent sinopulmonary infection.

Summary

Igs	*Major Functions*
IgM	• Produced in the primary response to an antigen • Does not cross the placenta hence does not give foetal immunity • Fixes complement • Antigen receptor on the surface of B cells
IgG	• Principal antibody of secondary response • Crosses placenta and confer foetal immunity • Opsonises bacteria, making them easier to phagocytose • Neutralises bacterial toxins viruses • Fixes complement which increases bacterial killing
IgA	• Secretory IgA prevents attachment of bacteria and viruses to mucous membrane • Hence essential to ward off sinobronchial infections • Does not fix complement
IgD	• Uncertain—main function and role yet to be determined • IgD alongwith IgM is predominant Ig on the surface of human B lymphocytes • IgD with antibody activity found towards certain antigens viz. milk proteins, penicillin, insulin, etc.
IgE	• Mediates immediate hypersensitivity by causing release of mediators from mast cells and basophils upon exposure to antigen (allergen) – reagins or reaginic antibodies

- Does not fix complement
- Defends against worm infections by causing release of enzymes from eosinophils
- Defensive action against helminthic infections.

STRUCTURE AND CHEMISTRY OF IMMUNOGLOBULINS—MODEL OF Ig MOLECULE

Edelman-Gally model: Most accepted model. Each ½ of the molecule is considered to be composed of 2 units. Each of these is a polypeptide chain, probably envisaged as folded on itself. The larger of the 2 chains, called as **'H' or Heavy chains** and smaller chain called as **'L' or Light chain**. The major bond between the 'L' and 'H' chains and also between the two halves of the molecule consist of "disulphide" linkages between 'cysteine' residues.

Structural Details (Fig. 8.2)

- Ig molecule has a V-shape. Each molecule is composed of equal numbers of ***two Heavy ('H'-chains)*** and ***two light (L) Polypeptide chains*** which can be represented by the general formula H_2L_2. The chains are held together by non-covalent forces and usually covalent interchain disulfide bridges to form a bilaterally symmetric structure.
- Each polypeptide chain is made up of a number of loops or ***domains*** of rather constant size (100 to 110 amino acid residues) formed by the intrachain disulfide bonds. The N-terminal domain of each chain shows much more variation in amino acid sequence than the others and is designated as ***variable region ('V'-region)*** to distinguish it from the other relatively constant domains collectively called as ***constant region*** in each chain.
- The zone where the variable and constant regions join is termed the ***Switch region (Hinge region).***

1. *Heavy Chain Classes (H-chains):* **Five classes** of H-chains have been identified in human Igs based on structural differences in the constant regions by serologic and chemical methods. The different forms of H-chains are designated as: **γ (gamma), α (alpha), μ (mu), δ (Delta) and ε (epsilon).** They vary in molecular weight from 50,000 to 70,000; the μ and ε-chains possessing 5 domains (one V and 4 C), while γ and α-chains have 4. ***The class of the H-chain determines the class of Immunoglobulins***. Thus, there are **5 classes** of Igs as discussed above: IgG, IgA, IgM, IgD and IgE, having 'H' chains γ, α, μ, δ and ε respectively.
2. *Light Chain Types:* All L-chains have a molecular weight of approximately 23,000, and are **classified into**

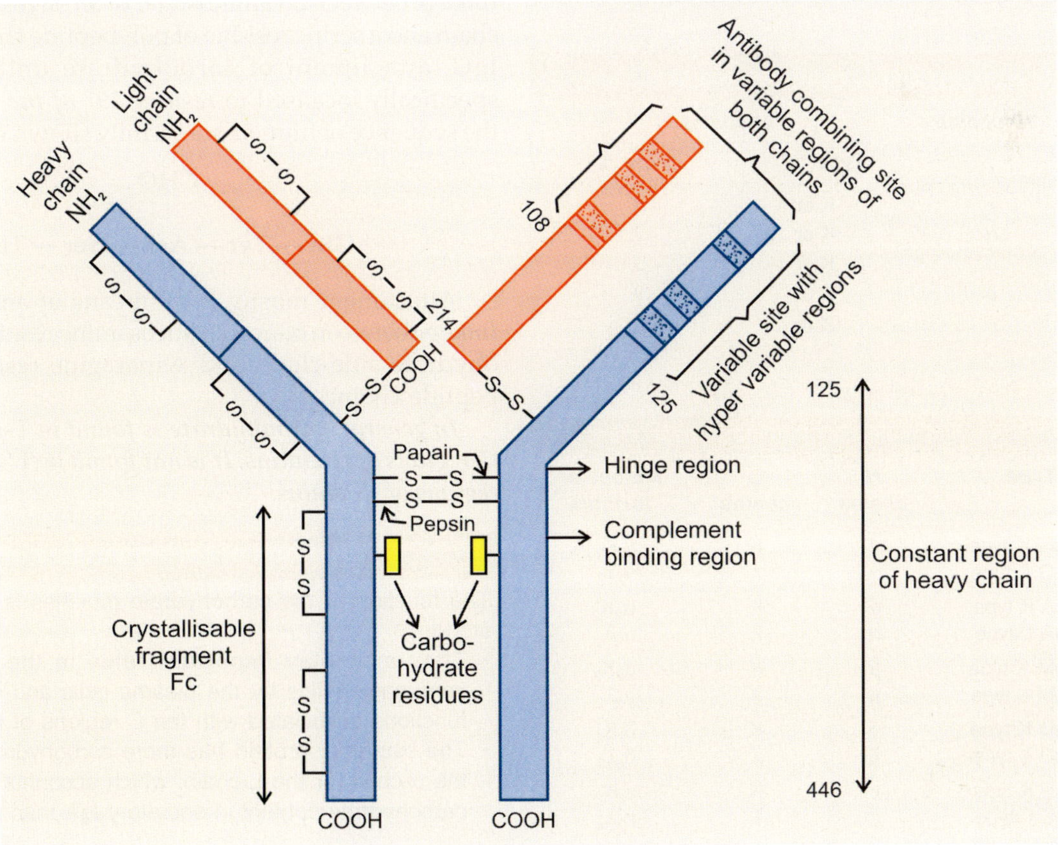

Fig. 8.2: Structure of immunoglobulin

two types: Kappa (K) and **Lambda (λ)** on the basis of multiple structural differences in the 'constant' regions, which are reflected in antigenic differences. The proportions of K to λ chains in Ig molecule varies from species to species, being about 2:1 in humans (Table 8.3). ***A given Ig molecule always contains identical* K *or* λ *chains but never both*.**

Molecular Formula: Each Ig may be written as a formula which expresses both its 'H' and 'L' chain constitutions as in Table 8.4.

Degradation of Igs by Proteolytic Enzymes

Igs are rather insensitive to proteolytic digestion but are most easily cleaved about midway in the H-chains in an area between the first and second constant region domains (CH_1 and CH_2).

(a) *Papain:* The enzyme Papain splits the molecule on the N-terminal side of the inter-H chain disulphide bonds into **three fragments** of similar size.
- **Two "Fab" fragments**, which include an entire 'L' chain and the V_H and CH-1 domains of a heavy chain, and
- **One 'Fc' fragment**, composed of C-terminal halves of the H-chains.

(b) *Pepsin:* If pepsin is used, cleavage occurs on the C-terminal side of the inter-H chain disulfide bonds, yielding a large **F (ab)$_2$** fragment composed of about two "Fab" fragments. The **'Fc'** fragment is extensively degraded by Pepsin.

'Hinge' region: The region of the H-chain susceptible to proteolytic attack is more flexible and exposed to the environment than the more compact globular domains and is known as the ***Hinge* region.**

Antigen Binding Site: ***Antigen binding activity is associated with the* Fab *fragments, or, more specifically with the V_H and V_L domains.*** On the other hand, most of the secondary biologic activities of Igs, e.g. complement fixation are associated with the 'Fc' fragment. Because there are two Fab regions, IgG molecules bind two molecules of antigen and are termed divalent. The site on the antigen to which an antibody binds is termed an **antigenic determinant** or **epitope**.

Table 8.3: 'H' chains and 'L' chains of each of Ig class

Ig class	'H' chains	'L' chains	
		Type 'K'	Type 'L'
• IgG	γ	K or	λ
• IgA	α	K or	λ
• IgM	μ	K or	λ
• IgD	δ	K or	λ
• IgE	ε	K or	λ

Table 8.4: Type and molecular formula of each class

Ig Class	Type	H chains	L chains	Molecular formula
IgG	IgG K type	γ	K	γ_2K_2
	IgG L type	γ	λ	$\gamma_2\lambda_2$
IgA	IgA K type	α	K	α_2K_2
	IgA L type	α	λ	$\alpha_2\lambda_2$
IgM	IgM K type	μ	K	μ_2K_2
	IgM L type	μ	λ	$\mu_2\lambda_2$
IgD	IgD K type	δ	K	δ_2K_2
	IgD L type	δ	λ	$\delta_2\lambda_2$
IgE	IgE K type	ε	K	ε_2K_2
	IgE L type	ε	λ	$\varepsilon_2\lambda_2$

Carbohydrate Content of Igs

All Igs have been shown to be **glycoproteins**. Carbohydrate content of these Igs vary substantially (Table 8.5). The carbohydrate units are primarily ***Asparagine-linked*** heteropolysaccharides. In some IgA and rabbit IgG, ***serine/threonine linked*** units have been demonstrated. Other linkages have also been demonstrated, ***O-glycosidic linkage*** between an aminosugar of an oligosaccharide side chain and a serine residue of polypeptide chain. In human IgG, attachment of carbohydrate units have been specifically localised to ***residue 297 of the 'H' chain*** and the sequence of immediate vicinity shown to be:

$$\begin{array}{c} \quad\quad\quad\quad\;\; \text{CHO} \\ \quad\quad\quad\quad\;\; | \\ \text{Gln — Tyr — Asn — Ser — Thr} \end{array}$$

Attachment mostly is by means of an ***N-glycosidic linkage*** between N-acetyl glucosamine residue of the carbohydrate side-chain and Asparagine residue of polypeptide chain.

In general, carbohydrate is found in T-piece, J-chain, C-regions of H-chains. It is not found in 'L' chains and 'V' regions of H-chains.

Functions of carbohydrates in Igs

The function of the carbohydrate moieties is poorly understood.
- They may play important roles in the secretion of immunoglobulins by the plasma cells and in the biologic functions associated with the C-regions of H-chains.
- The secretory protein has more carbohydrate than either the α-chain or the L-chain, which accounts for the higher carbohydrate content in secretory IgA than serum IgA.

Table 8.5: Carbohydrate content of three major classes of Igs

Class	Galactose	Mannose	N-acetyl glucosamine	Sialic acid	Fucose	Total carb.
IgG	0.4	0.6	1.3	0.2	0.3	2.8
IgA	1.2	1.7	1.6	0.9	0.2	6.4
IgM	1.6	3.3	3.3	1.3	0.7	10.2

Transport of Igs Across the Placenta

- **Only Ig which is transported across the placenta is IgG.** ***This accounts for the immunity to the newborn babies.*** It is to be noted that proteins smaller than the IgG class, e.g. Acid glycoprotein (mol. wt = 35,000), albumin (mol. wt = 65,000) and transferrin (mol. wt = 90,000) are not transferred to any significant amount. ***The importance of molecular structure and composition rather than size for transplacental transport is more important.***
- By splitting the IgG molecule into 2 'Fab' fragments and one 'Fc' fragment each having a molecular weight of 50,000 and by injecting these, after I^{131} labelling, into pregnant women, ***it was observed that the 'Fc' fragment is readily transported across the placental barrier, but the 'Fab' fragments are not*** **(Gitlin and coworkers).** This important work explains why serum IgA globulins are absent from the newborn's blood, although 80 per cent of serum IgA are similar to IgG in molecular size and molecular weight. ***There is a considerable difference in the 'Fc' fragments of both molecules,*** e.g. 8 to 10.5 per cent carbohydrates in IgA against 2.5 per cent in IgG and over 90 per cent of carbohydrates is in the 'Fc' fragment.

Electron Microscopy of Igs

Ig molecules were studied by electron microscope by **Svehag et al** (1967).

1. **IgG molecule:** Spindle-shaped molecule, 250-300 Å long × 40 Å wide; Two combining sites located at opposite ends of the spindle.
2. **IgM molecule:** A 'spider-like' configuration with central relatively rigid portion about 150 × 170 Å, and 5 flexible legs of variable length having a total span of some 350 Å.

Site of Synthesis of Igs

- The principal Ig producing cell is the **"Plasma cell".** Mature plasma cells produce IgG and IgA in their abundant rough surfaced endoplasmic reticulum (ER). Plasma cell is not a normal body cell but appears to develop specifically in response to antigenic stimulation. Plasma cells have a strongly basophilic cytoplasm due to a high concentration of RNA; because of this, they stain intensely with 'Pyronin', which combines specifically with RNA and hence known as ***pyroninophilic cells.*** During Ig production, accumulations of Ig can be seen as hyaline amorphous mass in pockets of endoplasmic reticulum (ER) in HE stained preparations. These collections known as **Russell bodies** can be seen also in EM
- IgM-Cell types that produce IgM globulin is probably **intermediate** in type, between large lymphocytes and plasma cells, and resemble ***Lymphocytoid cells*** of Waldenström's macroglobulinaemia.

Genetic Control of Synthesis of Ig Chains

Both 'H' chains and 'L' chains are **products of multiple genes.**

- Synthesis of each heavy chain (H-chain) is the product of at least **four** different genes:
  - A variable region (VH) gene,
  - A diversity region (D) gene
  - A joining region (J) gene, and
  - A constant region (CH) gene.
- Synthesis of immunoglobulin light chains (L-chains) is the product of at least **three** separate structural genes:
  - A variable region (V_1) gene,
  - A joining region (J) gene (it is separate from joining region (J) gene of H-chains)
  - A constant region (C_L) gene.

The H-chains and L-chains are ***synthesised as separate molecules*** and are subsequently assembled within the B cell or plasma cell into mature immunoglobulin molecules, all of which are ***glycoproteins*** in nature.

Class Switching (Isotypes)

In most humoral responses, antibodies with identical specificity but of different classes are generated in a specific chronologic order in response to the immunising antigen. Antibodies of the IgM class normally precede molecules of the IgG class. ***The switch from one class to another is called class or isotype switching.***

A single type of immunoglobulin light chain can combine with an antigen specific μ chain to produce a specific IgM molecule.

Subsequently, the ***same antigen specific light chain combines with a γ-chain*** with an identical V_H region to form an IgG molecule with antigen specificity identical to that of the original IgM molecule.

Again the same light chain can also combine with an α-heavy chain containing the identical V_H region to form an IgA molecule with identical antigen specificity.

These three classes of immunoglobulins IgM, IgG and IgA molecules against the **same antigen** have identical variable domains of both their light V_L chains and heavy V_H chains and are said to share an **idiotype.** ***Idiotypes are***

SECTION TWO

the antigenic determinants formed by the specific amino acids in the hyper variable regions.

POLYCLONAL VS MONOCLONAL ANTIBODY: HYBRIDOMA

1. **Polyclonal antibody:** In response to an antigenic challenge the body produces different types of antibodies against various antigenic determinants (epitopes) of the antigen. The antibodies thus produced are called polyclonal antibodies. Under such a situation, the different "clones" of antibody forming cells simultaneously synthesises the antibody. Different molecules will have different specificities and affinities.

Example: Body produces polyclonal antibodies in response to all types of microbial infections ***(polyclonal gammopathy).***

2. **Monoclonal antibody:** When one "clone" of antibody producing cells secrete a particular type of antibody against a particular antigenic determinant (epitope), it is called monoclonal antibody.

Example: **Monoclonal antibodies are produced in multiple myeloma,** a plasma cell tumour. In this paraprotein is produced, which gives a sharp paraprotein band in β to γ-region (M-band) in electrophoresis ***(monoclonal gammopathy).***

- *In vitro* in the laboratory monoclonal antibodies can be produced by the hybridoma technology.

3. **Hybridoma:** ***Hybridoma is a hybrid cell capable of producing monoclonal antibodies.*** It has been possible to fuse two different types of cultured animal cells, the resulting "hybrid" cell contained the chromosome of both the parent cells. **Kohler** and **Milstein,** in 1975, first produced monoclonal antibodies from hybridoma cells. They showed that cultured splenic cells from mouse immunized with specific antigen can be fused with that of cultured mouse myeloma cells. The hybrid cells so produced will remain ***immortal in culture*** like the myeloma cell and produce monoclonal antibodies like the immunised splenic cells.

Technique: In principle, the technique of hybridoma cell production is rather simple. It is shown schematically next page top.

Steps

- *Preparation of immunised spleen cells:* The antigen against which monoclonal antibodies are required is injected into a mouse, so that the mouse is immunised. Dose, route and frequency of antigen administration for optimal yield of monoclonal antibodies is standardised. After the immunisation, the mouse is killed and the spleen is removed. Spleen lymphocytes are separated.

Properties of splenic cells

- **Lack proliferation**, hence cannot be maintained in culture for long periods
- Can produce Igs against which the animal has been immunised
- *HGPRTase* +
- HAT resistant

Properties of Myeloma Cells

- ***Proliferation +, cells are self-propagating*** in tissue culture and can be maintained indefinitely.
- Cannot produce Igs (antibody).
- Lacks the enzyme *HGPRTase*, hence they cannot synthesize DNA by salvage pathway.

- *Preparation of mouse myeloma cells:* Usually Sp 2/0 myeloma cell line derived from Bal b/c mouse is used for preparation of hybridoma cell.
- *Preparation of cell mixture:* Immunised splenic lymphocytes are mixed with mouse myeloma cells in the ratio of $10^8 : 2 \times 10^7$.
- *Fusion of the two cells:* Fusion of the splenic cells with myeloma cells is brought about by the addition of **polyethylene glycol** (PEG-1500). Fused cell mixture is maintained in tissue culture in "HAT" medium.

HAT medium: contains

- **Hypoxanthine,**
- **Aminopterin,** and
- **Thymidine.**

Aminopterin, a folic acid antagonist will inhibit the *"de Novo"* synthesis of purines.

- *Hybridisation (Hybridoma):* Result of fusion will be production of hybridoma cells.
- Hybrid of normal + normal cells lack proliferation, hence the ***normal hybrids*** die in the culture medium in 5 to 6 days.
- Unfused myeloma cells also die in HAT medium as they lack *HGPRTase*.
- Only cells that survive are the hybrid cells formed by fusion of normal immunised splenic cells and mouse myeloma cells. ***Such hybridoma cells survive in the culture medium because they can use hypoxanthine and thymidine through salvage pathway.***

Properties of Hybridoma Cell

- Can propagate for indefinite period **(Immortal).**
- **Can secrete Igs.**
- Has *HGPRTase* enzyme, hence salvage pathway of purnie synthesis can operate.

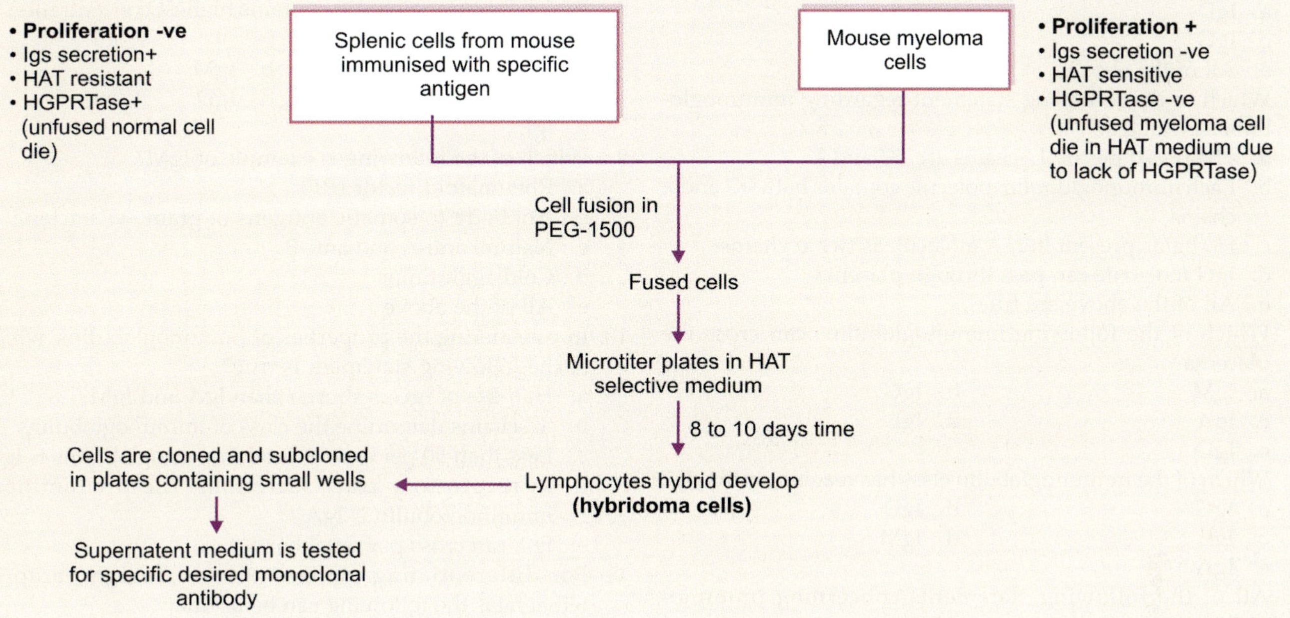

- *Propagation of Hybridoma Cells*
  - The monoclonal antibody producing hybrid cells are cloned and subcloned (separated into individual cells) in plates containing small wells.
  - Supernatent medium is tested for specific antibody. The cells producing desired antibody are selected.
  - These cells are propagated in culture bottles or injected into mice peritoneum where they are grown in ascitic fluid.

Uses of Monoclonal Antibodies – Clinical Aspect

(a) Diagnostic Uses

- Monoclonal antibodies have been raised for the diagnosis of many bacterial, viral and parasitic diseases.
- Monoclonal antibodies have been used for blood grouping.
- Also being used for standardisation and leucocyte identification through the cluster differentiation (CD antigen).
- Recently used against HLA antigens for phenotype screening purposes.

(b) Therapeutic uses: Recently, monoclonal antibodies are being used for treatment viz,

- Antitumour therapy
- Immunosuppression in organ transplantation
- And in autoimmune diseases, e.g. recently in rheumatoid arthritis monoclonal CD_4 antibody tried.

I. Questions (Essay type and short notes)

1. Describe the chemistry and functions of immunoglobulins.

Short Notes

a. Structure of immunoglobulins
b. IgG vs IgM
c. Secretory component or "T" piece
d. IgE
e. Reaginic antibody
f. Hybridoma

Differentiate

a. IgG and IgA
b. IgG and IgE

II. MCQs (Give one correct answer)

1. Which of the following immunoglobulin is a pentamer?
   a. IgG b. IgM
   c. IgA d. IgE
   e. IgD
2. Immunoglobulin classes are differentiated from each other on the basis of their:
   a. Electrophoretic mobility
   b. Molecular weight
   c. By their light chains
   d. By their heavy chains
   e. By Sedimentation co-efficient
3. The serum antibody responsible for fighting gram +ve pyogenic bacteria is:

a. IgD b. IgE
c. IgG d. IgA
e. All of the above

4. Which of the following statement regarding immunoglobulins is not true?
   a. The two types of 'L' chains are 'K' and λ
   b. Each immunoglobulin molecule contains both 'K' and λ chains
   c. H-chains present in IgA molecule is two α chains
   d. IgG molecule can pass through placenta
   e. All of the above are true
5. Which of the following immunoglobulins can cross the placenta?
   a. IgM b. IgG
   c. IgA d. IgE
   e. IgD
6. Which of the immunoglobulin class has reaginic antibody?
   a. IgG b. IgD
   c. IgE d. IgM
   e. IgA
7. All of the following statements concerning immunoglobulins are true except:
   a. The constant regions of 'H' chains are the same in each class of Igs
   b. IgG is the principal antibody in the serum
   c. The 'L' chains are similar in each class of immunoglobulins
   d. IgA is the major immunoglobulin found in external secretions
   e. Ig classes are differentiated from each other by the 'H' chains
8. Which immunoglobulin contains highest concentration of carbohydrates?
   a. IgG b. IgM
   c. IgA d. IgD
   e. IgE
9. Which of the following is example of IgM?
   a. Rheumatoid factor (RF)
   b. Antibody to somatic antigens of gram –ve bacteria
   c. Natural anti-A and anti-B
   d. Cold agglutinins
   e. All of the above
10. In considering the properties of immunoglobulins, which of the following statement is true?
    a. Half-life of IgG is shorter than IgA and IgM
    b. 'L'-chains determine the class of immunoglobulins
    c. Less than 50 per cent of total immunoglobulins is IgG
    d. In respiratory tract secretions, the predominant immunoglobulin is IgA
    e. IgA can cross placental barrier
11. For differentiating malignant from benign paraproteinaemias the following can be useful:
    a. Measurement of serum proteins
    b. Measurement of cryoglobulins
    c. Measurement of serum immunoglobulins
    d. Detection of polyclonal light chains in urine
    e. None of the above

Answers to MCQs

1. b	2. d	3. c	4. b	5. b	6. c
7. a	8. d	9. e	10. d	11. c.	

CHAPTER 9

CHEMISTRY OF ENZYMES

Major Concepts

A. To study what are enzymes, their general properties and classification.
B. To learn the mechanisms of enzyme catalysed reactions and various factors affecting enzyme activity.
C. To learn various types of enzyme inhibition and how the enzyme activity is regulated.

Specific Objectives

A. 1. Define enzyme.
   2. What is meant by catalytic activity of enzymes?
   3. Note that enzymes are protein in nature.
   4. Learn what are coenzymes.
   5. Study the role of metal ions in enzymes.
   6. Study the nomenclature and classification of enzyme as approved by International Union of Biochemistry (IUB). Learn at least two examples from each class.

B. 1. Know what is enzyme catalysed reaction and how an enzyme functions by lowering the energy of activation.
   2. Define specificity of enzyme and learn different types of specificity.
   3. Study Lock-and-Key theory and induced fit theory of mechanism of action of enzymes.
   4. Learn various factors that affect the activity of enzyme, such as, pH, temperature, substrate concentration, enzyme concentration, product concentration, presence of inhibitors or activators.
   5. Know the Michaelis-Menten equation and significance of each term.
   6. Know the importance and application of double reciprocal or Lineweaver-Burk plot and calculate enzyme velocity when $S >> Km$, $S = Km$ and $S << Km$.

C. Learn what is enzyme inhibition and various types of inhibition.
   1. Nonspecific inhibition: List the various agents responsible for it.
   2. Competitive inhibition.
   3. Noncompetitive reversible inhibition, and noncompetitive irreversible inhibition.
   4. Make a tabular form to show the difference between competitive and noncompetitive inhibition.
   5. Learn examples of competitive inhibition in biological system: Clinically used drugs.
   6. Study the various mechanisms by which enzyme activity is regulated, study allosteric enzyme.
   7. Learn about control of enzyme degradation: Ubiquitin-protease pathway.

INTRODUCTION

Enzymes are another important group of biomolecules synthesized by the living cells. They are ***catalysts of biological systems (hence are called as biocatalysts), colloidal, thermolabile and protein in nature.*** They are remarkable molecular devices that determine the pattern of chemical transformations. They also mediate the transformation of different forms of energy. The striking characteristics of enzymes are their catalytic power and specificity. Actions of most enzymes are under strict regulation in a variety of ways. ***Substances on which enzymes act to convert them into products are called*** **substrates.**

Catalytic Activity of Enzymes: Enzymes have immense catalytic power and accelerate reactions at least a million times, by ***reducing the energy of activation***. Before a chemical reaction can occur, the reacting molecules are required to gain a minimum amount of energy, this is called the **energy of activation**. It can be decreased by increasing the temperature of the reaction medium. But in human body which maintains a normal body temperature fairly constant, it is achieved by enzymes.

Protein Nature of Enzymes: In general with the exception of *ribozymes* which are few RNA molecules with enzymatic activity, ***all the enzymes are protein in nature with large mol. wt.*** Few enzymes are simple proteins while some are conjugated proteins. In such enzymes the

SECTION TWO

non-protein part is called prosthetic group or coenzyme and the protein part is called as ***apoenzyme***. ***The complete structure of apoenzyme and prosthetic group is called as holoenzyme.***

Holoenzyme = Apoenzyme + Coenzyme
(Protein part) (Prosthetic group)

Certain enzymes with only one polypeptide chain in their structure are called as ***monomeric enzymes***, e.g. *ribonuclease*. Several enzymes possess more than one polypeptide chain and are called as ***oligomeric enzymes***, e.g. *lactate dehydrogenase, hexokinase*, etc. Each single polypeptide chain of oligomeric enzymes is called as ***subunit***. When many different enzyme catalysing reaction sites are located at different sites of the same macromolecule, it is called as ***multienzyme complex***. The complex becomes inactive when it is fractionated into smaller units each bearing individual enzyme activity, e.g. *fatty acid synthetase, carbamoyl phosphate synthetase II, pyruvate dehydrogenase, prostaglandin synthase*, etc.

Coenzymes

Certain enzymes require a specific, ***thermostable, low mol. wt, non-protein*** organic substance called as ***coenzyme***. A coenzyme may bind covalently or non-covalently to the *apoenzyme*. ***The term prosthetic group denotes a covalently bonded enzyme.*** It is generally observed that reactions involving oxidoreductions, group transfers, isomerisation and covalent bond formation require coenzyme.

Since the involvement of coenzyme in a given reaction on a substrate is so intimate that coenzyme is often called as ***cosubstrate or second substrate.***

Many coenzymes are derived as the physiologically active forms from the constituents of vitamin B-complex viz, **Pantothenic acid:** ***CoASH,*** **Vitamin B_{12}:** ***Cobamide,*** **Folic Acid:** ***Tetrahydrofolate ($F.H_4$),*** **Niacin:** ***NAD^+, NADP,*** **Riboflavin:** ***FMN, FAD,*** **Pyridoxine:** ***Pyridoxal phosphate,*** **Thiamine:** ***TPP.***

Classification of Coenzymes

Coenzymes can be **classified according to the group whose transfer they facilitate**. Based on this concept we may classify coenzymes as follows:

(a) For transfer of groups other than hydrogen
- Sugar phosphates,
- CoASH
- Thiamine pyrophosphate (TPP)
- Pyridoxal phosphate
- Folate coenzymes
- Biotin
- Cobamide coenzyme
- Lipoic acid

(b) For transfer of hydrogen
- NAD^+, $NADP^+$
- FMN, FAD
- Lipoic acid
- Coenzyme Q.

In addition heme acts as coenzyme in cytochromes, peroxidases and PG synthase complex. Many coenzymes contain adenine, ribose and phosphate and are derivatives of adenosine monophosphate (AMP) such as NAD, FAD.

Role of Metal Ions in Enzymes

The activity of many enzymes depends on the presence of certain metal ions such as K^+, Mg^{++}, Ca^{++}, Zn^{++}, Cu^{++}.

- **Metal activated enzymes:** In certain enzymes the metals ***form a loose and easily dissociable complex.*** Such enzymes are called ***metal-activated enzymes.*** The metal ions can be removed by dialysis or any other such method from the enzyme without causing any denaturation of apoenzyme.
- ***Metalloenzymes:*** The second category of metal enzymes is called as ***metalloenzymes*** (Table 9.1). In this case ***metal ion is bound tightly to the enzyme and is not dissociated*** even after several extensive steps of purification.

Metals play variety of roles such as:

- They help in either maintaining or producing (or both), active structural conformation of the enzyme,
- Formation of enzyme-substrate complex,
- Making structural changes in substrate molecule,
- Accept or donate electrons,
- Activating or functioning as nucleophiles, and
- Formation of ternary complexes with enzyme or substrate.

Table 9.1: Examples of some metalloenzymes

Metal	Enzymes
Copper	Superoxide dismutase, cytochrome oxidase, tyrosinase, lysyl oxidase.
Calcium	Lipase, lecithinase
Iron	Catalase, xanthine oxidase, peroxidase cytochrome oxidase.
Manganese	Hexokinase, enolase, phosphoglucomutase, glycosyl transferase
Magnesium	Hexokinase, enolase, glucose-6-phosphatase, phosphofructokinase
Molybdenum	Xanthine oxidase
Zinc	Carbonic anhydrase, alcohol dehydrogenase, carboxy peptidase, alkaline phosphatase (ALP), lactate dehydrogenase (LDH).

NOMENCLATURE AND CLASSIFICATION OF ENZYMES

Enzymes are generally named after adding the suffix ***'ase'*** to the name of the substrate, e.g. enzymes acting on nucleic acids are known as *nucleases,* enzymes hydrolysing dipeptides are called *dipeptidases*. Eventhough few exceptions such as trypsin, pepsin, and chymotrypsin are still in use. Further, **few enzymes exist in their inactive forms** and are called as **proenzymes or zymogens,** e.g. *pepsin* has *pepsinogen* as its zymogen. The zymogens ***become active after undergoing some prior modification*** in its structure by certain agents. ***Many times the active form of enzyme acts on zymogen and catalyses its conversion into active form and this process is called as autocatalysis.***

In order to have a uniformity and unambiguity in identification of enzymes, ***International Union of Biochemistry (IUB)*** adopted a nomenclature system ***based on chemical reaction type and reaction mechanism.*** According to this system, enzymes are grouped in **six main classes.**

- Each enzyme is characterised by a code number (enzyme code No. or E C No) comprising four figures (digits) separated by points, **the first being that of the main class (one of the six).**
- ***The second figure*** indicates the type of group involved in the reaction.
- ***Third figure*** denotes the reaction more precisely indicating substrate on which the group acts.
- ***The fourth figure*** is the serial number of the enzyme. Briefly, the four digits characterise class, sub-class, sub-sub-class and serial number of a particular enzyme.

Six classes are:

1. ***Oxidoreductase:*** Enzymes involved in oxidations and reductions of their substrates, e.g. *alcohol dehydrogenase, lactate dehydrogenase, xanthine oxidase, glutathione reductase, glucose-6-phosphate dehydrogenase.*
2. ***Transferases:*** Enzymes that catalyse transfer of a particular group from one substrate to another, e.g. *aspartate and alanine transaminase (AST/ALT), hexokinase, phosphoglucomutase, hexose-1-phosphate uridyltransferase, ornithine carbamoyl transferase,* etc.
3. ***Hydrolases:*** Enzymes that bring about hydrolysis, e.g. *glucose-6-phosphatase, pepsin, trypsin, esterases, glycoside hydrolases,* etc.
4. ***Lyases:*** Enzymes that facilitate removal of small molecule from a large substrate, e.g. *fumarase, arginosuccinase, histidine decarboxylase.*
5. ***Isomerases:*** Enzymes involved in isomerisation of substrate, e.g. *UDP-glucose, epimerase, retinal isomerase, racemases, triosephosphate isomerase.*
6. ***Ligases:*** Enzymes involved in joining together two substrates, e.g. *alanyl-t. RNA synthetase, glutamine synthetase, DNA ligases.*

Many times the word **'OTHLIL'** is used to remember the six classes.

SPECIFICITY OF ENZYMES

Another important property of enzymes is their specificity. The specificity is of ***three different types*** namely:

1. ***Stereochemical specificity,***
2. ***Reaction specificity,*** and
3. ***Substrate specificity.***

Stereospecificity

1. Optical Specificity

There can be many optical isomers of a substrate. However, it is only one of the isomers which acts as a substrate for an enzyme action, e.g. for the oxidation of *D*- and *L*-amino acids, there are two types of enzyme which will act on *D*- and *L* isomers of amino acids. Secondly there can be a product of enzyme action which can have isomers. However, it is only one kind of isomer which will be produced as a product, e.g. *Succinic dehydrogenase* while acting on succinic acid will give only fumaric acid and not malic acid which is its isomer.

2. Reaction Specificity

A substrate can undergo many reactions but in a reaction specificity ***one enzyme can catalyse only one of the various reactions***. For example, oxaloacetic acid can undergo several reactions but each reaction is catalysed by its own separate enzyme which catalyses only that reaction and none of the others.

3. Substrate Specificity

The extent of substrate specificity varies from enzyme to enzyme. There are two types of substrate specificity viz, absolute specificity and relative specificity

- **Absolute specificity** is comparatively rare such as *urease* which catalyses hydrolysis of urea.
- **Relative substrate specificity** is further divided as:
  - **Group dependent** or
  - **Bond dependent.**

  Examples of group specificity are trypsin, chymotrypsin. *Trypsin* hydrolyses the residues of only *lysine* and *arginine,* while chymotrypsin hydrolyses residues of only aromatic amino acids.

4. Bond Specificity

Bond specificity is observed in case of proteolytic enzymes, *glycosidases* and *lipases* which act on peptide bonds, glycosidic bonds and ester bonds respectively.

SECTION TWO

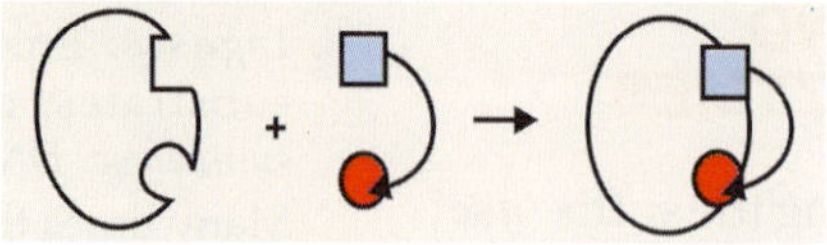

Fig. 9.1A: Template or lock-and-key model

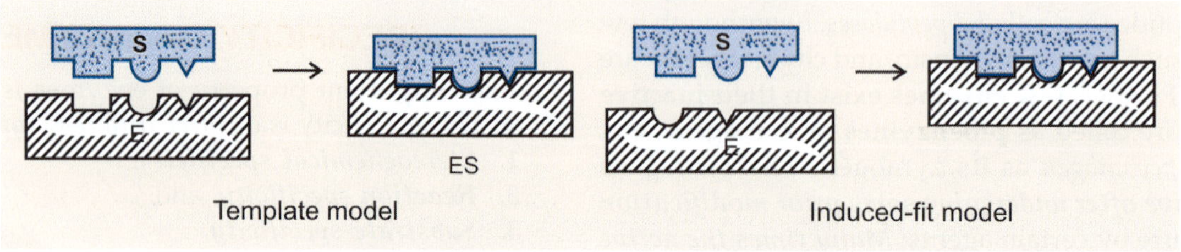

Fig. 9.1B: Models for enzyme-substrate interaction

MECHANISM OF ENZYME ACTION

Michaelis and Menten have proposed a hypothesis for enzyme action, which is most acceptable. According to their hypothesis, ***the enzyme molecule (E) first combines with a substrate molecule (S) to form an enzyme-substrate (ES) complex which further dissociates to form product (P) and enzyme (E) back.*** Enzyme once dissociated from the complex is free to combine with another molecule of substrate and form product in a similar way.

The ***ES complex is an intermediate or transient complex*** and the bonds involved are weak non-covalent bonds, such as H-bonds, Van der Waals forces, hydrophobic interactions. Sometimes two substrates can bind to an enzyme molecule and such reactions are called as **bisubstrate reactions**. The site to which a substrate can bind to the enzyme molecule is extremely specific and is called as ***active site or catalytic site***. Normally the molecular size and shape of the substrate molecule is extremely small compared to that of an enzyme molecule. The active site is made up of several amino acid residues that come together as a result of foldings of secondary and tertiary structures of the enzyme. So, the active site possesses a complex three dimensional form and shape, provides a predominantly non-polar cleft or crevice to accept and bind the substrate. Few groups of active site amino acids are bound to substrate while few groups bring about change in the substrate molecule.

MODELS OF ENZYME-SUBSTRATE COMPLEX FORMATION

These interactions have been described basically of two types.

1. **Template or Lock-and-Key Model**

   This model was originally proposed by **Fischer** which states that the active site already exists in proper conformation even in absence of substrate. Thus the ***active site by itself provides a rigid, pre-shaped template*** fitting with the size and shape of the substrate molecule. **Substrate fits into active site of an enzyme** as the key fits into the lock and hence it is called the **lock-and-key** model. This model proposes that substrate binds with rigid pre-existing template of the active site, provides additional groups for binding other ligands. But this cannot explain change in enzymatic activity in presence of allosteric modulators (Figs 9.1A and B).

2. **Induced-Fit or Koshland Model**

   Because of the restrictive nature of lock-and-key model, another model was proposed by **Koshland** in 1963 ***which is known as induced-fit model. The important feature of this model is the flexibility of the region of active site.*** According to this, active site does not possess a rigid, preformed structure on enzyme to fit the substrate. On the contrary, ***the substrate during its binding induces conformational changes in the active site to attain the final catalytic shape and form*** (Fig. 9.2). This explains several matters related to enzyme action such as:

   - Enzymes become inactive on denaturation,
   - Saturation kinetics,
   - Competitive inhibition, and
   - Allosteric modulation.

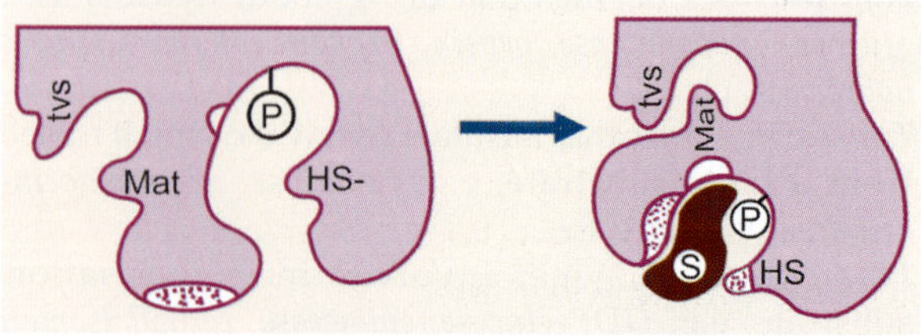

Fig. 9.2: Induced-fit model

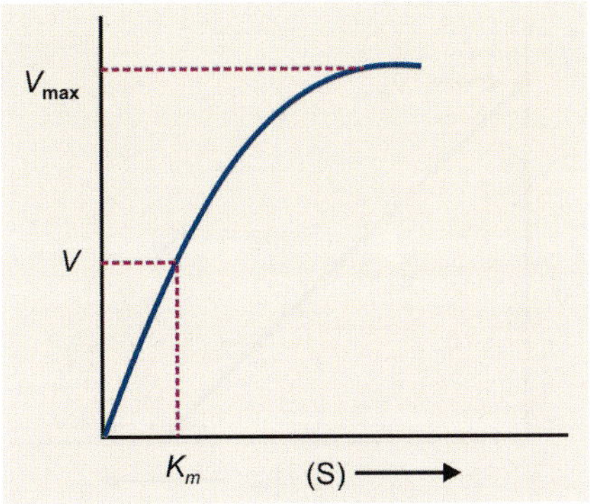

Fig. 9.3: The hyperbolic curve of reaction velocity (V) against substrate concentration (S). K_m is the substrate concentration at ½ V_{max}

KINETIC PROPERTIES OF ENZYMES

Kinetic analysis of enzymes was used for characterization of enzyme-catalysed reactions even before enzymes had been isolated in pure form.

One of the first things that is measured in kinetic analysis is the variation in rate of reaction with substrate concentration. For this purpose a fixed low concentration of enzyme is used in a series of parallel experiments in which only the substrate concentration is varied. Under these conditions initial velocity increases until it reaches a substrate independent maximum velocity at substrate concentration (Fig. 9.3).

The saturation effect is believed to reflect the fact that all the enzyme binding sites are occupied with substrate. This interpretation of the substrate saturation curve led **Hensi, Michaelis and Menten** to develop a general treatment of kinetic analysis of enzyme catalyzed reactions. As already mentioned:

$$E + S \underset{K_2}{\overset{K_1}{\rightleftarrows}} E\text{-}S \underset{K_4}{\overset{K_3}{\rightleftarrows}} E + P \qquad ...(1)$$

Where E is the free enzyme, S is the substrate, ES is enzyme-substrate complex, P is the product, K_1 is the rate constant for the formation of ES, K_2 is the rate constant for the dissociation of ES to E and S and K_3 is the rate constant for the dissociation of ES complex into E and P.

Rate of formation of ES

Rate of formation = K_1 [Et] –[ES] [S] ... (2)

Rate of dissociation of ES

Rate of dissociation = K_2 [ES] + K_3 [ES] ... (3)

Steady state is attained when rate of formation of ES is equal to rate of dissociation,

$$K_1\,[Et] - [ES]\,[S] = K_2\,[ES] + K_3\,[ES] \qquad ...(4)$$

Separation of rate constants

The left side of equation (4) is multiplied to give:

$$K_1\,[Et]\,[S] - K_1\,[ES]\,[S]$$

and right side is simplified to give

$$[K_2 + K_3]\,[ES].$$

We then have,

$$K_1\,[Et]\,[S] — K_1\,[ES]\,[S] = [K_2 + K_3]\,[ES].$$

On transposing and changing sign we get,

$$K_1\,[Et]\,[S] = K_1\,[ES]\,[S] + [K_2 + K_3]\,[ES].$$

On further simplifying and rearranging,

$$[ES] = \frac{K_1\,[Et]\,[S]}{K_1\,[S] + K_2 + K_3} \qquad ...(5)$$

Definition of initial velocity V_o in terms of [ES]:

The initial velocity, according to Michaelis and Menten theory is determined by the rate of dissociation of [ES] in reaction (1) whose rate constant is K_3. So we get,

$$V_o = K_2\,[ES]$$

Substituting value of ES from equation (S) we get

$$V_o = \frac{K_2\,[E]\,[S]}{[S] + [K_3 + K_2]/K_1} \qquad ...(6)$$

Now let us simplify further by defining K_m (the Michaelis-Menten Constant) as $\frac{K_3 + K_2}{K_1}$ and by defining V_{max} as K_2 [Et], i.e. the rate when all the available E is present as ES. On substituting these terms in equation (6).

$$\mathbf{V_o = \frac{V_{max}\,[S]}{[S] + K_m.}}$$

This is called the **Michaelis-Menten equation,** the rate equation for one substrate-enzyme catalysed reaction. It is a statement of the quantitative relationship between the initial velocity V_o, the maximum velocity V_{max} and the initial substrate concentration, all related through the Michaelis-Menten constant K_m.

- An important relationship is observed when the initial reaction rate is exactly one-half the V_{max}. Then,

$$\frac{V_{max}}{2} = \frac{V_{max}\,[S]}{K_m + [S]}$$

Divide by V_{max}

$$½ = \frac{[S]}{K_m + [S]}$$

Solving for K_m, we get

$$K_m + [S] = 2[S]$$
$$K_m = [S]$$

The Michaelis-Menten equation can be algebraically transformed into equivalent equations that are useful in the practical determination of K_m and V_{max}. Therefore, ***K_m is equal to substrate concentration at which the velocity is half the maximum.***

The initial velocity V_o is directly proportional to the molar concentration [S] of the substrate when substrate concentration is very low as compared to K_m. In this stage, a single substrate enzyme reaction is a first order reaction, its rate depending on conc. of single reactant.

- When

$$S << K_m.$$
$$\therefore K_m + [S] \cong K_m$$
$$\therefore V_o = \frac{V_{max}[S]}{K_m + [S]} = \frac{V_{max}[S]}{K_m} K[S]$$

Where K is a new constant equalling $\frac{V_{max}}{K_m}$ because both V_{max} and K_m are constants for a particular enzyme.

- When $S >> K_m$, the initial velocity attains its V_{max} and becomes independent of [S]. The reaction now turns into a zero-order reaction.

Lineweaver-Burk Double-Reciprocal Plot

It is difficult to estimate V_{max} from the position of an asymplote, as in the rectangular hyperbola (Fig. 9.4), linear transforms of the Michaelis-Menten equation are often used.

$$\frac{1}{V} = \frac{1}{V_{max}} + \frac{K_m}{V_{max}} = \frac{1}{[S]}$$

($Y = b + mx$) gives a straight line where m is the slope and (b) is y intercept of the regression of y on x. Figure 9.4 shows the straight line graph obtained by plotting 1/V against 1/[S]. Where y intercept = $1/V_{max,}$ the x intercept = $-1/K_m$ and the slope = $K_m/V_{max.}$

Eadie-Hofstee Transform: This is used to avoid the bunching of values that occurs about the lower end of the double-reciprocal plot. The Eadie-Hofstee transform can be written as:

$$V = V_{max} - K_m \frac{V}{[S]}$$
$$Y = b + mx.$$

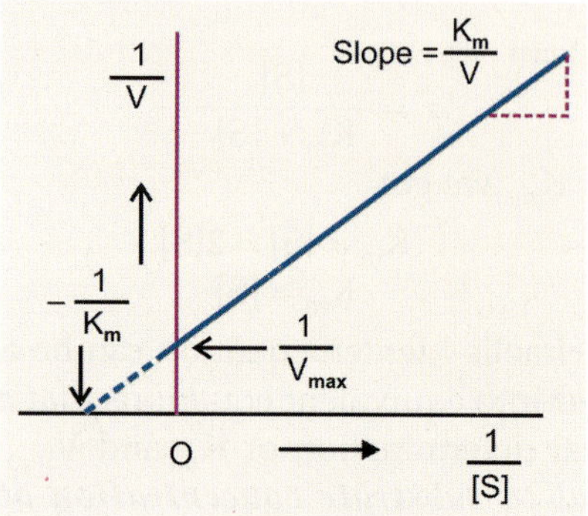

Fig. 9.4: Lineweaver-Burk plot

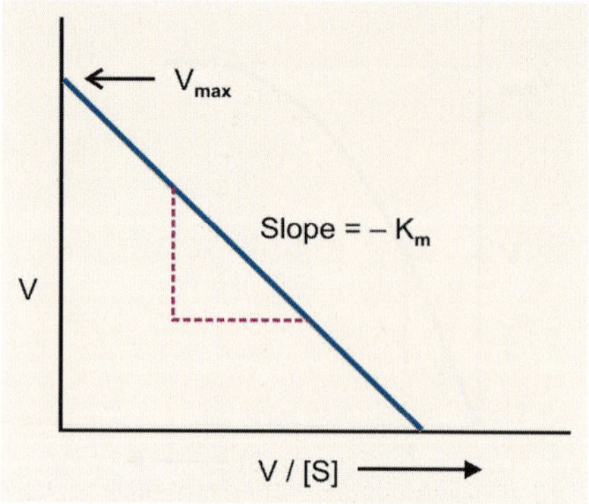

Fig. 9.5: Eadie-Hofstee transform

Figure 9.5 shows the straight line graph obtained by plotting V against V/[S], where the Y intercept = V_{max}, the X intercept = V_{max}/K_m and the slope = $-K_m$.

FACTORS AFFECTING ENZYME ACTION

Activity of enzymes is markedly affected by several factors such as temperature, pH, conc. of other substances, presence of activators or inhibitors, etc.

1. **Effect of Temperature**

   Each enzyme is most active at a specific temperature which is called its ***optimum temperature.*** Temperature increases the total energy of the chemical system with the result the activation energy is increased. The exact ratio by which the velocity changes of 10°C temperature rise is the **Q_{10} or temperature coefficient. Reactions velocity almost doubles with 10°C rise (Q_{10} = 2) in many enzymes.** Activity of enzyme progressively decreases when the temperature of reaction is below or above the optimum temperature. However, increase in temperature also causes denaturation of enzyme.

Note

The shape of the curve is ***bell-shape.*** Most of the enzymes of human system have an optimum temperature within the range of 35–40°C. Thus, **the optimum temperature is that temperature at which the activity of the enzyme is maximum** (Fig. 9.6).

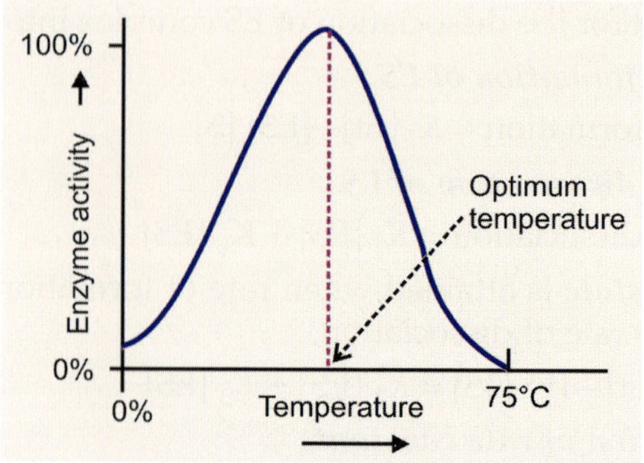

Fig. 9.6: Effect of temperature on enzymatic reaction

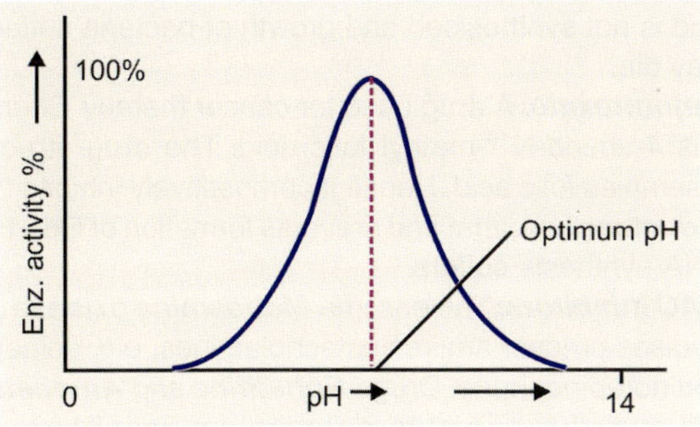

Fig. 9.7: Effect of pH on enzymatic 'reaction'

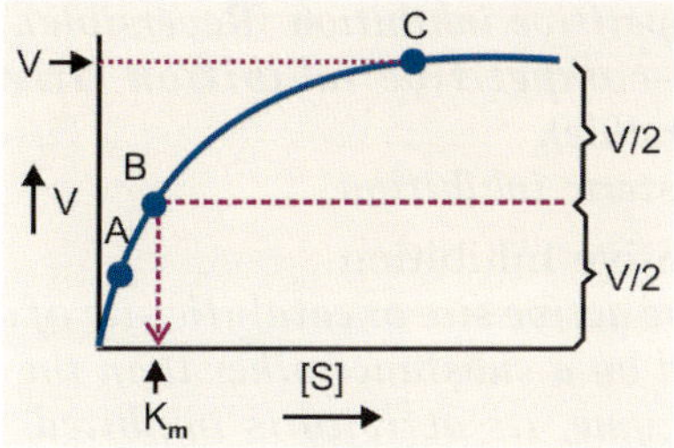

Fig. 9.9: Effect of substrate concentration on enzymatic reaction

2. **Effect of pH**

   The rate of enzymatic reaction also depends on pH of the medium. The enzymatic activity is maximum at a particular pH which is called its ***optimum pH.*** The optimum pH of most enzymes lies in the range of 4–9 **(Fig. 9.7)**.

- Hydrogen ions in the medium may alter the ionisation of active site or substrates. Ionisation is a requirement for ES complex formation and
- pH may influence the separation of coenzyme from holoenzyme complex. At a very low or high pH the H-bonds may be inactivated in the protein structure, destroying its 3D structure. The optimum pH may vary from substrate to substrate for an enzyme acting on a number of substrates because of the ES complex formation and ionisation will vary from substrate to substrate.

3. **Effect of Enzyme Concentration**

   In the beginning velocity of the enzymatic reaction is directly proportional to the enzyme concentration. When the substrate conc. is in large excess exceeding that of V_{max}, because enzyme is the limiting factor in the enzyme-substrate reaction and providing more enzyme molecules enables the conversion of progressively larger numbers of substrate molecules **(Fig. 9.8)**.

4. **Effect of Product Concentration**

   Products formed as a result of enzymatic reaction may accumulate and this excess of product may lower the enzymatic reaction by occupying the active site of the enzyme. It is also possible that under certain conditions of high concentration of products a reverse reaction may be favoured forming back the substrate.

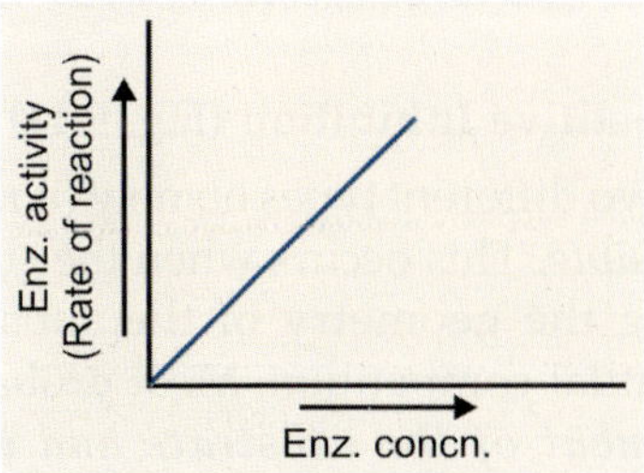

Fig. 9.8: Effect of enzyme concentration on enzymatic reaction

5. **Effect of Substrate Concentration**

   As already described a known quantity of enzyme, the reaction is directly proportional to the substrate concentration. However, this is true only up to a certain concentration after which the increasing concentration of substrate does not further increase the velocity of reaction **(Fig. 9.9)**.

6. **Effect of Activators and Coenzymes**

   The activity of certain enzymes is greatly dependent of metal ion activators and coenzymes. The role of metal ions and coenzymes is already discussed.

7. **Effect of Modulators and Inhibitors**

   Whenever the active site is not available for the binding of the substrate the enzyme activity may be reduced. The substances which stop or modify the enzymatic reaction are called ***inhibitors or modulators.*** Presence of these substances in reaction medium can adversely affect the rate of enzymatic reaction.

8. **Effect of Time**

   The time required for completion of an enzyme reaction increases with decreases in temperature from its optimum. However under the optimum conditions of pH and temperature time required for enzymatic reaction is less.

ENZYME INHIBITION

Enzymes are protein and they can be inactivated by the agents that denature them. The chemical substances which inactivate the enzymes are called as ***inhibitors*** and the process is called as ***enzyme inhibition.*** Inhibitors are sometimes referred to as ***negative modifier***. They may be small inorganic ions, or organic substances. Enzyme inhibition is classified under **three major groups:**

- *Competitive inhibition (Reversible).*
- *Non-competitive inhibition (Irreversible or reversible).*
- *Allosteric inhibition.*

1. Competitive Inhibition

When the active site or catalytic site of an enzyme is occupied by a substance other than the substrate of that enzyme, its activity is inhibited. The type of inhibition of this kind is known as competitive inhibition. This is a type of ***reversible inhibition.*** In such inhibition both the ES and EI (Enzyme-Inhibitor) complexes are formed during the reaction. However, the actual amounts of ES and EI will depend on:

- Affinity between enzyme and substrate/inhibitor,
- Actual concentrations (amounts) of substrate and inhibitor present, and
- Time of preincubation of enzyme with the substrate or inhibitor.

So the affinity of the substrate for the enzyme is progressively decreased with the increase in conc. of inhibitor lowering the rate of enzymatic reaction. Thus, ***the K_m is high,*** but ***V_{max} is the same*** in competitive inhibition. However, when the concentrated substrate is increased, the effect of inhibitor can be reversed forcing it out from EI complex.

Following are few examples of competitive inhibitors.

Enzyme	Substrate	Competitive inhibitor
• ***Lactate Dehydrogenase***	Lactate	Oxamate
• ***Aconitase***	Cisaconitate	Transaconitate
• ***Succinate Dehydrogenase***	Succinate	Malonate
• ***HMG-CoA Reductase***	HMG-CoA	HMG
• ***Dihydrofolate Reductase***	7,8 Dihydrofolate	Amethopterin

Diagrammatic presentation of competitive inhibition is given in Figure 9.10.

EXAMPLES OF COMPETITIVE INHIBITION IN BIOLOGICAL SYSTEM—DRUGS USED CLINICALLY

- ***Allopurinol:*** A drug used for treatment of Gout. Uric acid is formed in the body by oxidation of hypoxanthine by the enzyme ***Xanthine oxidase.*** **Allopurinol structurally resembles hypoxanthine** and thus by competitive inhibition, the drug inhibits the enzyme ***xanthine oxidase*** thus reducing uric acid formation.
- ***Sulphonamides:*** A very commonly used antibacterial agent. Para-aminobenzoic acid (PABA) is essential for synthesis of folic acid by the enzyme action. Folic acid is needed for bacterial growth and survival. Bacterial wall is impermeable to folic acid. Sulphonamide drugs are structurally similar to PABA and competitively inhibit enzyme action. Thus, folic acid is not synthesised and growth of bacteria suffers and they die.
- ***Methotrexate:*** A drug used for cancer therapy. Chemically it is 4-amino-N^{10}-methyl folic acid. The drug structurally resembles folic acid. Hence it competitively inhibits ***"folate reductase"*** enzyme and prevents formation of FH_4. Hence, DNA synthesis suffers.
- ***MAO inhibitors:*** The enzyme ***Monoamine oxidase*** (MAO) oxidises pressor amines catecholamines, e.g. epinephrine and norepinephrine. Drugs ***Ephedrine*** and ***Amphetamine*** structurally resemble catecholamines. Thus, when administered they can competitively inhibit the enzyme "MAO" and prolong the action of pressor amines.
- ***Physostigmine:*** *"Acetylcholinesterase"* is the enzyme which hydrolyses acetylcholine to form choline and acetate. Physostigmine is a drug which competitively inhibits *acetylcholinesterase* and prevents destruction of acetylcholine. Thus, continued presence of acetylcholine in post-synaptic region prolongs the neural impulse.
- ***Dicoumarol:*** Used as an anticoagulant. It is structurally similar to vitamin K and can act as an anticoagulant by competitively inhibiting vitamin K.
- ***Succinylcholine:*** It is used as a muscle relaxant. Succinylcholine is structurally similar to acetylcholine. It competitively fixes on post-synaptic receptors. As it is not hydrolysed easily by the enzyme ***acetylcholinesterase,*** produces continued depolarisation with consequent muscle relaxation.

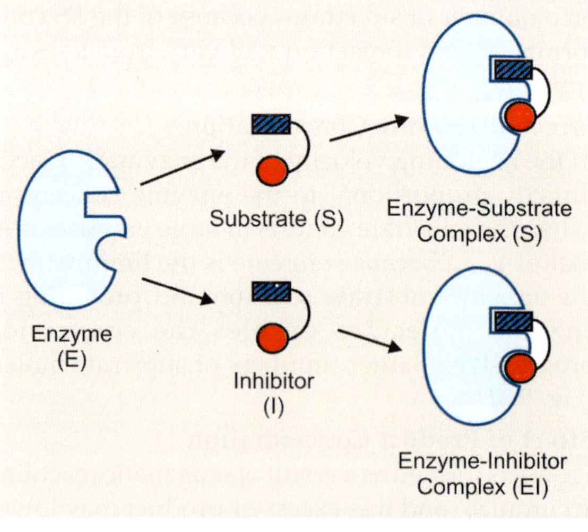

Fig. 9.10: Competitive inhibition

2. Non-competitive Inhibition (Fig. 9.11)

This is of two different types namely (i) ***reversible*** and (ii) ***irreversible.*** This occurs when the substances not resembling the geometry of the substrate do not exhibit mutual competition. Most probably the ***sites of attachment of the substrate and inhibitor are different.*** The inhibitor binds reversibly with a site on enzyme other than the active site. So the inhibitor may

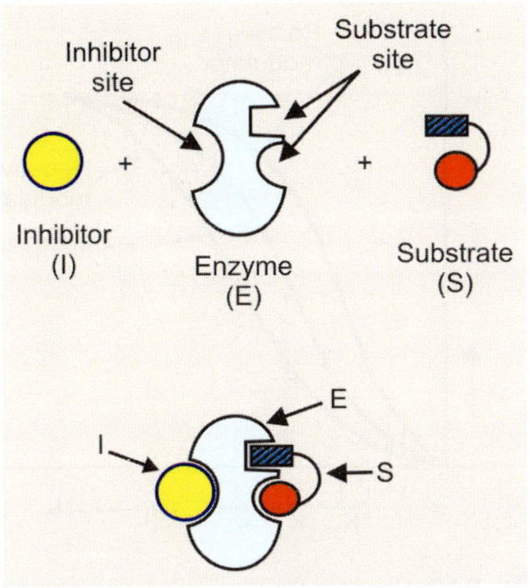

Fig. 9.11: Non-competitive inhibition

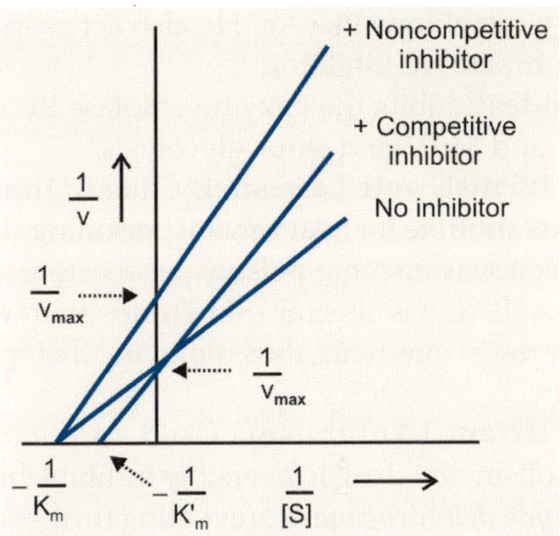

Fig. 9.12: Lineweaver-Burk plot for normal and competitive and non-competitive inhibition

combine with both free enzyme and ES complex. This probably brings about the changes in 3D structure of the enzyme inactivating it catalytically. In non-competitive inhibition ***V_{max}* is lowered**, but ***K_m* is kept constant**. If the inhibitor can be removed from its site of binding without affecting the activity of the enzyme, it is called as ***Reversible-Non-competitive Inhibition***. However, if the inhibitor can be removed only at the loss of enzymatic activity, it is known as ***Irreversible Non-competitive Inhibition***. However, the kinetic properties in case of both are the same.

Figure 9.12 gives graphical presentation of Lineweaver-Burk double reciprocal plot in case of competitive and non-competitive inhibition.

Table 9.2 gives the differences that are observed between competitive and non-competitive inhibition.

Examples of Non-competitive Irreversible Inhibitors

- **Iodoacetate:** An irreversible inhibitor of enzymes like glyceraldehyde-3-p dehydrogenase and papain. It **combines with–SH group** at the active site of the enzyme inactivating the enzyme.

Table 9.2: Differentiation of competitive and non-competitive inhibitions

Competitive inhibition	*Non-competitive inhibition*
1. Reversible	1. Reversible or Irreversible
2. Inhibitor and substrate resemble each other in structure	2. Does not resemble
3. Inhibitor binds the active site	3. Inhibitor does not bind the active site
4. V_{max} is same	4. V_{max} lowered
5. K_m increased	5. K_m unaltered
6. Inhibitor cannot bind with ES complex	6. Inhibitor can bind with ES complex
7. Lowers the substrate affinity to enzyme	7. Does not change substrate affinity for the enzyme
8. Complex is E-I	8. Complex is E-S-I or E-I
9. Michaelis-Menten equation changed to $V = \dfrac{V_{max}\,[S]}{K_m \dfrac{1+(I)}{K_i} + S}$	9. Michaelis-Menten equation changed to: $V = \dfrac{V_{max}\,[S]}{K_m \left[\dfrac{1+(I)}{K_i}\right] [K_m] + [S]}$
10. Lineweaver-Burk plot: $\dfrac{1}{V} = \dfrac{K_m}{V_{max}}[1 + \dfrac{(I)}{K_i}] \times \dfrac{1}{[S]} + \dfrac{1}{V_{max}}$	10. Lineweaver-Burk plot: $\dfrac{1}{V} = \dfrac{K_m}{V_{max}}[1 + \dfrac{(I)}{K_i}] \times \dfrac{1}{[S]} + \dfrac{1}{V_{max}} + [1 + \dfrac{(I)}{K_i}]$
11. Eadie-Hofstee plot: No change in Y intercept (V_{max}) but possesses steeper slope and smaller X intercept	11. Eadie-Hofstee plot: No change in slope ($-K_m$) but Y intercept is lowered and X intercept declines in value

- **Heavy metal ions** like Ag, Hg also act as irreversible noncompetitive inhibitor.
- **Fluoride:** Inhibits the enzyme emolase by removing Mg^{++} and Mn^{++} and stops glycolysis.
- **BAL (British anti Lewesite):** Called **Dimercaprol**, used as antidote for heavy metal poisoning. The heavy metals act as enzyme poisons by reacting with –SH groups. BAL has several –SH groups with which the heavy metal ions react, thus removing their poisonous effects.
- **Disulfiram (Antabuse):** Used in treatment of alcoholism, the drug irreversibly inhibits the enzyme ***aldehyde dehydrogenase*** preventing further oxidation of acetaldehyde which accumulates and produces sickening effect leading to aversion to alcohol.
- **Di-isopropyl fluorophosphate (DFP):** Inhibits enzymes with serine in their active site e.g. acetylcholine esterase.

Suicide Inhibition

It is a ***special type of irreversible noncompetitive inhibition.*** In this type of inhibition, ***substrate analogue is converted to a more effective inhibitor*** with the help of the enzyme to be inhibited. The so formed new inhibitor binds irreversibly with the enzyme.

Examples

- **Allopurinol**
  The best example of suicide inhibition. The drug is used in treatment of gout, as it inhibits the enzyme xanthine oxidase thus decreasing the uric acid formation. But ***allopurinol gets oxidised by the enzyme xanthine oxidase itself to form "alloxanthine" a more potent effective and stronger inhibitor*** of xanthine oxidase thus potentiating the action of allopurinol.
- **Aspirin**
  Most commonly used drug for relieving pain. Anti-inflammatory action of aspirin is also based on the suicide inhibition. Aspirin acetylates a serine residue in the active centre of cyclooxygenase thus inhibiting the PG synthesis and the inflammation.
- **5-fluorouracil**
  Used in cancer therapy, 5-fluorouracil (5-Fu) is converted to fluorodeoxyuridylate (Fdump) by the enzymes of the salvage pathway. Fdump so formed **inhibits the enzyme *thymidylate synthase*** thus inhibiting nucleotide synthesis.

3. Allosteric Inhibition and Allosteric Enzymes

There is a mixed kind of inhibition when the inhibitor binds to the enzyme at a site other than the active site

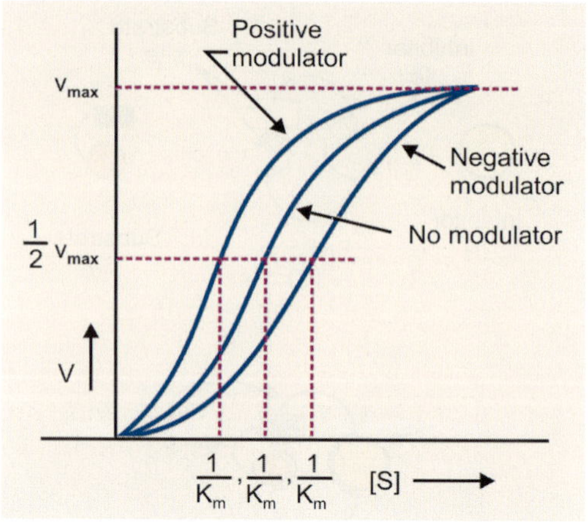

Fig. 9.13: Sigmoid kinetics, allosteric inhibition

but on a different region in the enzyme molecule called ***allosteric site. Allosteric inhibition does not follow the Michaelis-Menten hyperbolic kinetics. Instead it gives a sigmoid kinetics*** (Fig. 9.13). Allosteric inhibitors shift the substrate saturation curve to the right. However as opposite to inhibitors, the presence of activators shifts the curve to the left.

Types: Allosteric enzymes are of *K* and *M* series according to their kinetics.

- In ***K-enzymes***, e.g. *aspartate carbamoylase* and *phosphofructokinase*, the allosteric inhibitor lowers the substrate affinity to raise the K_m of the enzyme; but the V_{max} is unchanged.
- ***In M-enzymes***, e.g. ***acetyl-CoA carboxylase***, the allosteric inhibitor reduces the maximum velocity but no change in K_m or substrate affinity. Allosteric activators produce a fall in K enzymes and a rise in V_{max} in *M* enzymes.
- When the final product allosterically inhibits the enzyme, it is called as feedback allosteric inhibition.

Allosteric Enzyme

Aspartate transcarbamoylase is a model allosteric enzyme

Aspartate transcarbamoylase (ATCase) catalyses the first reaction unique to pyrimidine biosynthesis. ATCase is feedback inhibited by cytidine triphosphate (CTP). Following treatment with mercurials, ATCase loses its sensitivity to inhibition by CTP but retains its full activity for carbamoyl aspartate synthesis. This suggests that CTP is bound at a different (allosteric) site from either substrate. ATCase consists of multiple catalytic and regulatory protomers. Each catalytic protomer contains four aspartate (substrate) sites and each regulatory protomer atleast two CTP (regulatory sites).

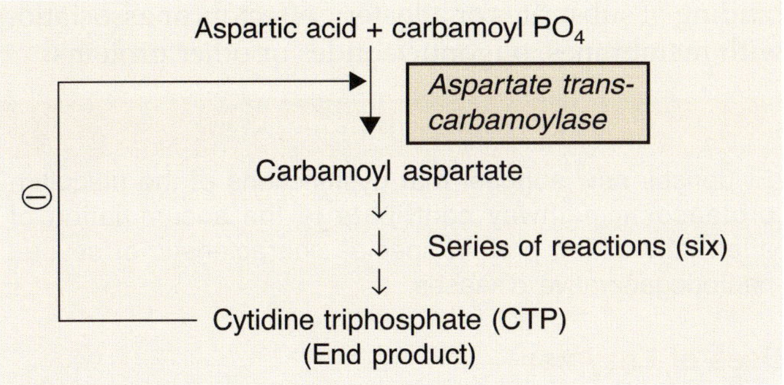

Another example of allosteric enzyme and inhibition: ***Synthesis of isoleucine from threonine involves at least 5 steps*** of enzymatic reactions. Isoleucine, the end product, inhibits the first enzyme ***threonine deaminase*** and stops its own synthesis.

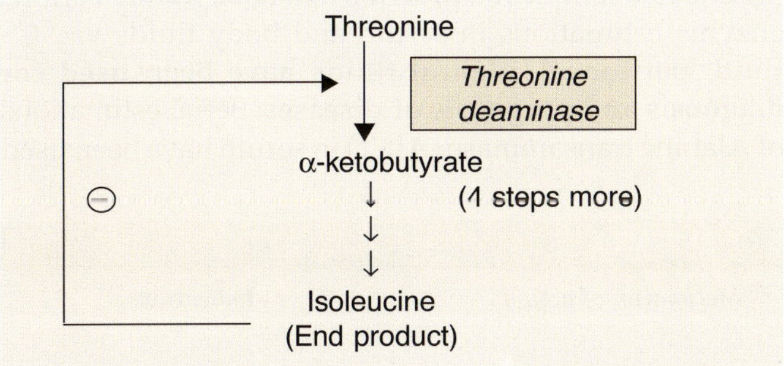

- A metabolite may also cause feed-forward allosteric activation of an enzyme for a subsequent step of its metabolism, e.g. F 1,6-biphosphate allosterically activates *pyruvate kinase* catalysing subsequent step.
- An allosteric effector oppositely influences two allosteric enzymes catalysing reverse reactions. For example, AMP allosterically activates *phosphofructokinase* and allosterically inhibits *FDpase.*

Following box gives some examples of allosteric modulation.

Name of enzyme	*Allosteric activator*	*Allosteric inhibitor*
• Glutamate Dehydrogenase	ADP	ATP, NADH
• Hexokinase, ICD	ADP	G-6-P, ATP
• Protein kinases	c-AMP	—
• Pyruvate carboxylase	Acetyl-CoA	ADP

- **In oligomeric enzymes**, the allosteric site and active site are located on different subunits. ***Changes in the enzyme-substrate interaction due to the allosteric effects of regulatory molecules other than the substrate are called heterotropic allosteric modulations.*** Allosteric activators and inhibitors exhibit respectively positive and negative cooperativities with the substrates. Binding of substrate to one protomer enhances the binding of the same to another protomer or another substrate binding site on the same enzyme molecule. ***When the binding of a substrate enhances the interaction between the allosteric enzyme and more molecules of the same substrate it is homotropic allosteric effect.***

FEEDBACK REGULATION VS FEEDBACK INHIBITION

Feedback regulation and feeback inhibition are not synonymous and they are different.

In both mammalian and bacterial cells, end-products "feedback" and control their own synthesis. In many instances, this involves feedback inhibition of an early biosynthetic enzyme. It is necessary to distinguish between "feedback regulation" and feedback inhibition, a mechanism for regulation of many bacterial and mammalian enzymes.

Example

- Dietary cholesterol restricts the synthesis of cholesterol from acetate in mammalian tissues. This is feedback regulation.

  This feedback regulation, however, does not appear to involve "feedback inhibition" of an early enzyme of cholesterol biosynthesis. An early enzyme *'HMG-CoA reductase'* is affected, but the mechanism involves curtailment by cholesterol or a cholesterol metabolite of the expression of the gene that encodes *'HMG-CoA reductase'*, i.e. enzyme repression. Cholesterol added directly to *'HMG-CoA reductase'* has no effect on its catalytic activity.

CONTROL OF ENZYME DEGRADATION: UBIQUITIN-PROTEASE PATHWAY

In animals, many proteins are degraded by the ubiquitin-protease pathway. This pathway was discovered by Aaron Ciechanover, A Hershko and Irwin Roose who got Nobel Prize.

Proteosome 26s Protein

- Comprises of more than 30 polypeptide subunits, which are arranged in the form of a hollow cylinder.
- Active sites of its proteolytic subunits face the interior of the cylinder, thus preventing the indiscriminate degradation of cellular proteins.
- Proteins requiring degradation are targeted to the proteosome by **"Ubiquitination"**—the process in which there is covalent attachment of one or more ubiquitin molecules.

Ubiquitin: It is a small, approximately 7s residue protein that is highly conserved among eukaryotes.

Ubiquitination: It is catalysed by a large family of enzymes called **"E_3 Ligases"**, which attach ubiquitin to the side chain amino group of lysyl residues.

Ubiquitin-proteasome pathway is responsible both for:

- The regulated degradation of selected cellular proteins, for example cyclins, in response to specific intra or cellular signals and
- For the removal of defective or aberrant protein species

Mechanism

The key to the versatility and selectivity of the ubiquitin-proteasome system resides in both the variety of intracellular E3 ligases and their ability to discriminate between different physical or conformational states of a target protein.

Thus the ubiquitin-proteasome pathway can selectively degrade proteins whose physical integrity and functional competency has been compromised:

- By the loss of or damage to a prosthetic group,
- Oxidation of cysteine or histidine residues or
- Deamination of asparagines or glutamine residues

Recognition by proteolytic enzymes also can be regulated by covalent modifications such as phosphorylation, binding of substrates or allosteric effectors or association with membranes, oligonucleotides or other proteins.

Clinical Significance

Evidences now suggest that dysfunctions of the ubiquitin-proteasome pathway contribute to the accumulation of aberrantly folded protein species characteristic of several neurodegenerative diseases.

Uses of Enzymes

Enzymes are used as follows:

(a) Enzymes estimation in serum and body fluids for diagnosis and prognosis

(b) Enzyme used as laboratory reagent

(c) Therapeutic uses of enzymes

(*a*) ***Enzyme estimation in serum and body fluids:*** Various enzyme estimations in serum and body fluids viz. CS fluid, peritoneal/pleural fluids have been **used for diagnosis and prognosis of diseases.** Serial estimations of Alanine transaminase (AL-T) in serum have been used

THERAPEUTIC USES OF ENZYMES

Name of the enzyme	*Availability*	*Mechanism of action*	*Indications*
A. Enzymes used systemically			
• ***Streptokinase and*** • ***Urokinase***	Pure stabilised • Streptokinase available 750,000 to 15,00,000 IU vial • Urokinase—50,000 to 500,000 IU vial	Increases amounts of proteolytic enzyme ***"plasmin"*** by either • Increasing the circulating level of its precursor ***"plasminogen"*** or • Increasing the conversion of plasminogen to plasmin. Plasmin acts directly on "fibrin" breaking it down to achieve thrombolysis (Fig. 9.14)	• Acute myocardial infarction • Acute thrombosis of arteries • Deep vein thrombosis (DVT) • Pulmonary embolism
• ***L-Asparaginase***	Available as ***"Leunase"***. 10,000 KU of L-Asparaginase per vial	Certain tumour cells require L-Asparagine for growth L-Asparaginase hydrolyses L-Asparagine and growth of tumour cell suffer	• Acute leukaemia • Malignant lymphomas
• ***Digestive enzymes Amylase, lipase and protease***	Available as tablets and syrup	Replacement therapy in pancreatic insufficiency	• Cystic fibrosis • Chronic pancreatitis • Following pancreatectomy
• ***α-chymotrypsin***	5.775 mg sublingual tablets	Mucolytic and proteolytic activity	**Used as adjunct therapy** • In management of inflammatory oedema due to injury, Postsurgical infections and dental procedures
• ***Serratopeptidase***	5 mg tablet	Fibrinolytic activity, high bradykinin decomposing activity, and potent caseinolytic activity	• Effective adjunct in inflammation after traumatic injury and after operation • Subconjunctival bleeding
B. Enzymes used locally			
• Hyaluronidase	Available as ***"Hyalase"*** 1500 IU per ml.	Brings about depolymerisation of ground substance and helps in absorption of fluids	• Promotes diffusion of fluids given subcutaneously (SC) • Intra-articular injection in joints to alleviate pain in osteoarthritis

for prognosis of viral hepatitis (for details—refer to chapter on "Enzymes and Isoenzymes of Clinical Importance").

(b) Enzyme used as laboratory reagent: Some enzymes are used for estimation of biomolecules in serum.

Examples

- ***"Glucose oxidase"*** enzyme is used for estimation of ***"true glucose"*** in blood and body fluids.
- Enzyme ***"uricase"*** is used for estimation of serum uric acid.
- Enzyme ***"urease"*** is used for estimation of urea in blood and body fluids.

(c) Therapeutic uses of enzymes: Enzymes have been used for treatment purposes. Some of the enzymes used therapeutically are given in the box.

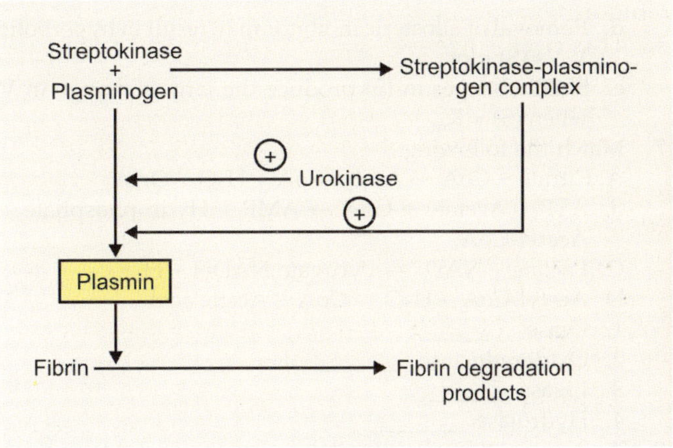

Fig. 9.14: Showing mechanism of action of streptokinase and urokinase

I. Questions (Essay type and short notes)

1. What are enzymes? Classify them giving suitable example.
2. Describe the factors which influence the rate of enzyme action.
3. Describe various factors affecting enzyme action.
4. What are enzymes? Describe various processes of inhibition of enzymes activity.
5. Classify enzymes with suitable examples. Explain the mechanism of enzyme action.
6. Give the classification of enzymes.
7. Classify enzymes giving a few examples in each group. Add a note on enzyme inhibitors.

Short Notes

a. Competitive inhibition of enzymes.
b. Allosteric inhibition
c. Enzyme inhibition
d. Enzyme specificity
e. Enzyme inhibitors
f. Classification of enzymes
g. Enzymes—their classification and mechanism of action.
h. Allosteric enzymes
i. Therapeutic uses of enzymes

Differentiate

a. Competitive and non-competitive inhibition.
b. Coenzymes and isoenzymes

II. MCQs (Give one correct answer)

1. In non-competitive enzyme action:
   a. V_{max} is increased
   b. Apparent K_m is increased
   c. Apparent K_m is decreased
   d. Concentration of active enzyme molecule is reduced
   e. None of the above
2. An allosteric enzyme influences the enzymes activity by:
   a. Competing for the catalytic site with the substrate
   b. Changing the specificity of the enzyme for the substrate
   c. Changing the conformation of the enzyme binding to a site other than catalytic site
   d. Changing the nature of the products formed
   e. All of the above
3. Which of the following regulatory actions involves a reversible covalent modification of an enzyme?
   a. Phosphorylation of serine –OH on the enzyme
   b. Allosteric modulation
   c. Competitive inhibition
   d. Non-competitive inhibition
   e. None of the above
4. A competitive inhibitor of an enzyme has which of the following properties?
   a. It is frequently a feedback inhibitor
   b. It becomes covalently attached to an enzyme
   c. It decreases the V_{max}
   d. It interferes with substrate binding to the enzyme
   e. It causes irreversible inactivation of the enzyme
5. When (S) is equal to K_m, which of the following conditions exist?
   a. Half the enzyme molecules are bound to substrate
   b. The velocity of the reaction is equal to V_{max}
   c. The velocity of the reaction is independent of substrate concentration
   d. Enzyme is completely saturated with substrate
   e. The reaction has reached equilibrium
6. Which of the following statements about an enzyme exhibiting allosteric kinetics with cooperative interaction is false?
   a. A plot of V-Vs (S) has a sigmoidal shape
   b. An inhibitor may increase the apparent K_m
   c. Lineweaver-Burk plot is useful for determining K_m and V_{max}

d. Removal of allosteric inhibitor may result in hyperbolic V-Vs (S) plot
e. Small changes in (S) produce the largest change in V when (S) = K_m

7. Match the following:
A. Citrate + CoA → Acetyl-CoA + H_2O + OAA
B. ATP + Acetate + CoA → AMP + Hydrophosphate + Acetyl-CoA.
C. Lactate + NAD^+ → Pyruvate NADH + H^+
D. Acetyl-CoA + H_2O → CoA + Acetic acid
1. Lyase ________
2. Oxidoreductase ________
3. Ligase ________
4. Hydrolase ________

8. An enzyme that catalyses the coversion of an aldose sugar to a ketose sugar would be classified as one of the:
a. Transferases b. Isomerases
c. Oxidoreductases d. Hydrolases
e. Lyases

9. In which of the following types of enzymes an inducer is not required?
a. Inhibited enzyme b. Cooperative enzyme
c. Allosteric enzyme d. Constitutive enzyme
e. None of the above

10. In which of the following types of enzyme water may be added to a C=C double bond without breaking the bond?
a. Hydrolase b. Hydratase
c. Hydroxylase d. Esterase
e. Oxygenase

11. Lock-and-key" model of enzyme action proposed by Fisher implies that:
a. The active site is flexible and adjust to substrate
b. The active site requires removal of PO_4 group
c. The active site is complementary in shape to that of the substrate
d. Substrates change conformation prior to active site interaction
e. None of the above

12. In competitive inhibition which of the following kinetic effect is true?
a. Decreases both K_m and V_{max}
b. Increases both K_m and V_{max}
c. Decreases K_m without affecting V_{max}
d. Increases K_m without affecting V_{max}
e. Increases V_{max} without affecting K_m

13. In competitive inhibition of enzymes action
a. The apparent K_m is decreased
b. The apparent K_m is increased
c. V_{max} is decreased
d. Apparent concentration of enzyme molecules decreased
e. V_{max} is increased

14. Enzymes increase the rates of reaction by
a. Increasing the free energy of activation
b. Decreasing the engery of activation
c. Changing the equilibrium constant of the reaction
d. Increasing the free energy change of the reaction.
e. Decreasing the free energy change of the reaction

Answers to MCQs

1. d 2. c 3. a 4. d 5. a 6. d
7. 1-A, 2-C, 3-B, 4-D 8. b 9. d 10. b
11. c 12. d 13. b 14. b.

C H A P T E R

10

BIOLOGICAL OXIDATION

Major Concepts

A. Define biological oxidation and enumerate and describe various enzyme systems which carry out this.
B. Describe the respiratory chain and oxidative phosphorylation.

Specific Objectives

A. 1. Define oxidation.
2. List and explain principles of biological oxidation.
3. Learn various processes of oxidation and enzymes involved with it.
a. oxidase b. oxygenase c. aerobic dehydrogenase d. anaerobic dehydrogenases.
4. Study various cytochromes.

B. 1. Learn the terms redox potential and free energy and how they are related to biological oxidation.
2. Define and study in detail mitochondrial electron transport chain.
3. Study each redox reaction and free energy changes.
4. Learn the sites of ATP synthesis.
5. Study the inhibitors of electron transport chain.
6. Learn about mitochondrial shuttles.
7. Define oxidative phosphorylation.
8. Study mechanism of oxidative phosphorylation with special emphasis on Mitchell's chemiosmotic hypothesis.
9. Learn about ATP synthase and mechanism of ATP synthesis (Boyer's hypothesis).
10. Study inhibitors and uncouplers of oxidative phosphorylation.

INTRODUCTION

Oxidation is a reaction with oxygen directly or indirectly or to lose hydrogen and/or electrons. Biologically it is carried out by the enzymes.

The biological oxidations and reductions are restricted to the following three simple classes:

- Loss of one or more electrons, e.g.

$$Fe^{++} \xrightarrow{-e^-} Fe^{+++}$$

- Loss of one or more hydrogen atoms, e.g.

$$CH_3\,CH_2OH \xrightarrow[\text{or } 2H^+,\ 2e^-]{-e^-} CH_3\,CHO$$

- Addition of one or more oxygen atoms

$$CH_3\,CHO \xrightarrow{+O} CH_3\,COOH$$

Thus biological oxidations and reductions can be represented as below:

$$A_{red} \xrightarrow{-ne^-} A_{ox}\ \text{(Oxidation)}$$

$$A_{ox} \xrightarrow{+ne^-} A_{red}\ \text{(Reduction)}$$

Where n is the number of electrons involved.

Since there is no involvement of free electrons or atoms, biological oxidations and reductions can be represented in the following way:

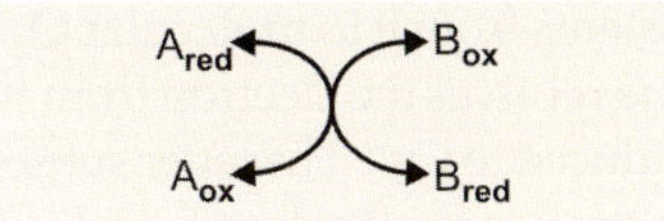

In biological oxidations the terms exothermic and endothermic are replaced by ***exergonic*** and ***endergonic***, e.g. suppose a substance A is oxidized to B with the release of energy and the oxidation is coupled to another reaction in which C is being converted to D. Now, some of the energy liberated in oxidative step A → B is transferred to the synthetic step C → D, in the form other than heat. This is **free energy.** Thus ***in exergonic and endergonic reactions, free energy is released or absorbed respectively***.

The principles of biological oxidations of carbohydrates, proteins and fats may be summarised as follows:

- First of all complex organic molecules are degraded into 2-C compound.
- The 2-C fragments are then broken down by a series of steps. In each step one CO_2 and 2H are removed.
- Decarboxylation of organic acids removes CO_2 without any considerable change in energy.
- The second end-product, water arises from reduced coenzymes of respiratory chain and molecular O_2 of atmosphere with production of some energy.

1. Oxidation by Direct Action of Oxygen

Number of enzymes catalyse direct interaction of substrates with molecular oxygen. Depending upon the fashion in which molecular O_2 is used, these enzymes can be further classified as:

- *Oxidases,*
- *Oxygenases,*
- *Hydroxylases* and
- *Hydroperoxidases*

(a) Oxidases

These are electron transferring oxidases and ***catalyse removal of hydrogen from a substrate by directly using O_2 as hydrogen acceptor.***

$$O_2 + 4e^- \longrightarrow 2O^- \underset{}{\overset{-4H^+}{\rightleftharpoons}} 2H_2O$$

$$\text{or } O_2 + 2e^- \longrightarrow O^-_2 \overset{2H^+}{\rightleftharpoons} H_2O_2$$

Thus ***the product of oxidase action is either H_2O or H_2O_2***
Following are the examples of oxidases:

- **With H_2O as the product:** *Cytochrome Oxidase, Ascorbate oxidase, Catechol oxidase.*
- **With H_2O_2 as the product:** *Urate oxidase, Amino acid oxidase, Xanthine oxidase, Aldehyde oxidase, Glucose oxidase.*

  Some of them are copper-containing enzymes and oxidise the substrate by transferring reducing equivalents from it to molecular O_2. The Cu^{+2} of the enzyme receives the electron from the substrate and gets reduced to Cu^+. The latter subsequently donates the electron to molecular O_2 and gets re-oxidised to Cu^{+2}.

(b) Oxygenases

These enzymes ***incorporate O_2 into their substrates,*** but are not concerned with energy production. They have **two subclasses:**

- ***Dioxygenases:*** These catalyse the incorporation of both the atoms of O_2 into the substrate, e.g. *carotene 15-15′ dioxygenases, Tryptophan 2, 3 dioxygenases.*
- ***Mono-oxygenases or hydroxylases:*** These incorporate one oxygen atom into substrate to form hydroxyl group on it, e.g. *Microsomal cyt-D5 mono-oxygenase, Mitochondrial cyt P_{450} mono-oxygenase,* etc. These enzymes transfer reducing equivalents from NADPH or NADH.

  There is another group of enzymes called *hydroxylase* also fall under this. They are sometimes called as mixed function oxidases, e.g. *Tyrosinase, Phenylalanine hydroxylase,* etc.
- **Hydroxyperoxidases:** They catalyse oxidation ***in which H_2O_2 acts as hydrogen acceptor*** and is reduced to water as:

$$AH_2 + H_2O_2 \rightarrow A + 2H_2O$$
$$A + H_2O_2 \rightarrow AO + H_2O$$

All ***peroxidases*** found in plants and milk and ***catalases*** found in animals and plants are the examples.

$$H_2O_2 + H_2O_2 \xrightarrow{\text{catalase}} 2H_2O + O_2$$

Catalase Vs Peroxidase

Similarities
- Both contain haem
- Both decompose H_2O_2® to H_2O and O_2

Dissimilarities
- Catalase can react directly with H_2O_2, but glutathione peroxidase requires reduced glutathione (G-SH)
- ***K_m of catalase for H_2O_2 is much greater than glutathione peroxidase.***
- Catalase can act in tissues where H_2O_2 is formed by L-amino acid oxidase, but glutathione peroxidase is the active enzyme to remove small amounts of H_2O_2 formed in cells, e.g. RB cells and lens of the eye.

2. Oxidation as a Result of Loss of Hydrogen

Enzymes that remove hydrogen from the substrate fall under this class and are called as ***dehydrogenases.*** When the Hydrogen removed from the substrate is passed on to O_2 directly, it is called as *aerobic dehydrogenase.*

- ***Aerobic dehydrogenases:*** These are ***flavoproteins bearing FMN or FAD as the prosthetic group.*** They accept 2 hydrogens ($2H^+$ and $2e^-$) from it on the FMN or FAD which is thereby reduced to $FMNH_2$ or $FADH_2$. These can be further reoxidised by donating the hydrogen to molecular oxygen forming H_2O_2.

$$AH_2 + FAD \rightleftharpoons FADH_2 + A$$
$$FADH_2 + O_2 \longrightarrow H_2O_2 + FAD$$

These enzymes can also donate hydrogens to artificial electron-acceptors like methylene blue, e.g. *L. Amino acid oxidase, urate oxidase, xanthine oxidase.*

- ***Anaerobic Dehydrogenases:*** In this group of enzymes there is ***direct transfer of electrons to molecular oxygen.*** They make use of intermediate electron-acceptors. The latter reduced thereby transfers the electrons to some other electron acceptor.

- ***Pyridine-linked dehydrogenases:*** These oxidise the substrate by ***transferring a hydride ion (H^-) from the substrate to NAD^+ or NADP. The second hydrogen removed from the substrate is released as free H^+.*** In the process the NAD and NADP get reduced to NADH and NADPH. The chain of reaction continues with another dehydrogenase enzyme.
- ***Flavin-linked dehydrogenases:*** FMN and FAD are the two flavin containing coenzymes that remain linked to specific dehydrogenase enzymes. Some also carry either haem or one or more iron-sulfur clusters. ***They oxidise the substrate by removing $2H^+$ and $2e^-$ from it and transferring them to the flavin coenzymes.*** In the process, the flavin coenzymes get reduced to either $FMNH_2$ and $FADH_2$ depending on the enzyme. The reduced flavin nucleotides then transfer the reducing equivalents to an electron acceptor other than molecular O_2. Most of these enzymes get **reoxidised by transferring reducing equivalents to coenzyme Q.**
  Examples:
  - **FAD linked enzymes:** *D-Amino acid oxidase, Glycine oxidase, Succinic dehydrogenase, Diaphorase,*
  - **FMN-linked enzymes:** *NADP-Cyt-c reductase, Cyt b_2.*

Differences between Aerobic dehydrogenases and Anaerobic dehydrogenases

Aerobic dehydrogenase	*Anaerobic dehydrogenase*
• Can react directly with O_2	• Cannot react directly with molecular O_2
• Transfers hydrogen/ electrons to Fp which is autooxidisable	• Transfers hydrogen/ electron to NAD^+ or Fp which is oxidised in ETC
• **H_2O_2 is produced** which is catabolised by *catalase*	• **H_2O_2 never produced** NADH + H^+ or $Fp.H_2$ are produced
• **ATP is never** produced	• **ATP is produced** by oxidation of NADH + H^+/ $Fp.H_2$ in ETC

3. **Iron-Sulphur Proteins**

Ubiquinones or Coenzyme Q: A group of quinones has been found to be present in the mitochondria. Following types of non-haem iron-sulphur clusters are normally present:
- **FeS:** It has single Fe coordinated to the side chain –SH groups of four cysteine residues.
- **Fe_2S_2:** It contains two iron atoms, two inorganic sulfides and four –SH groups. Each iron is linked to 2-SH and 2 sulphur groups.
- **Fe_4S_4:** It consists of four iron atoms and four cysteine –SH groups and four inorganic sulfides, each iron remains linked to one –SH, 3 inorganic sulphides while each sulphide is coordinated to three iron atoms.
- **Fe_3S_4:** comprises 3 Fe, 4-SH and 4 inorganic sulphides.

The enzymes may have one or more of the combinations of the clusters mentioned above. Fe^{+2} of a reduced iron-sulfur protein gets subsequently re-oxidised by donating its electron to an electron acceptor such as CoQ or Cyt c_1. **Each iron-sulphur protein transfers only one electron at a time.** It is also believed that vit. E, vit. D and plastoquinones in the plant tissues participate in the process of electron transfer.

4. **Cytochromes**

These are very important enzymes which contain haem and are involved in cellular oxidation. The oxidised form of cytochrome possesses a single Fe^{+3} ion and is called *ferricytochrome*. It is reduced to *ferrocytochrome* having a Fe^{+2} ion on accepting an electron.
- Cytochromes are identified by their characteristic ***absorption spectra.*** Ferricytochromes show diffuse and non-characteristic absorption band in visible spectrum. Ferrocytochromes exhibit characteristic absorption bands called **α, β and γ-soret bands.**
- Cytochromes are categorised into different groups according to the light wavelength at which the α-band shows its peak (α absorption maxima). They also differ with respect to haem prosthetic group, the apoprotein and binding between the haem and apoprotein.

Types of Cytochromes

(a) Cytochrome-c

Since it is available in large quantities, it is the best studied of the cytochromes. It is water soluble and easily extractable. It shows characteristic absorption spectra in the reduced form at 550, 521 and 416 mμ. Oxidised form gives two diffuse bands at 530 and 400 mμ. The iron content of Cyt-c is 0.38 per cent. Haem is attached with protein by means of two thioester linkages involving sulphur of two cysteine and apoprotein. ***Cyt-c is incapable of combining with O_2 or CO.*** It is a basic protein with one polypeptide chain with 104 amino acids having Mol. wt. of 12400 to 13000. An enzyme *NADPH- Cyt-c reductase* can readily reduce Cyt-c.

(b) Cytochrome c_1

Like Cyt-c, Cyt-c_1 also possesses an iron-protoporphyrin IX complex-haem-c. It has absorption maxima at 554, 524 and 418 mμ. ***This also is incapable of combining with O_2, CO, CN^-.***

(c) Cytochrome b

These also contain the same protoporphyrin IX complex (haem-b). The apoprotein is a however different. It is found to be tightly bound to flavoproteins and ubiquinones in the mitochondria. The ferrocyt-b has an absorption maxima at 563 mμ, 530 mμ and 430 mμ. It is thermostable and not easily extractable. ***It also does not react with O_2, CO or CN^-***. In normal course its oxidation requires the presence of Cyt-c, a and a_3. Cyt-b is reduced by accepting an electron from reduced CoQ.

(d) Cytochromes a and a_3 (Cytochrome oxidases)

This constitutes the complex IV (cytochrome oxidase) of the mitochondrial electron transport chain. Both possess an identical type of iron-porphyrin complex called ***haem-a***. In spite of identical haem groups, cytochromes a and a_3 differ in their electron affinity and biological activity because of their location of their haem groups at different sites of the apoprotein.

- One haem group is located along with one copper ion. This haem is called as ***haem-a***. This Cyt-a functions as the anaerobic oxidising unit.
- The other "haem a" called ***haem a_3*** is located along with the second copper ion at the binding site for molecular O_2 on subunit I and functions as aerobic reducing unit of the enzyme complex.

Cytochrome a absorbs at 605, 517 and 414 mμ whereas Cyt a_3 absorbs at 600 and 445 mμ. Cyt-a does not react with O_2, CO, or CN^-, whereas ***Cyt-a_3 is autooxidisable and forms compounds with CO and CN^-.***

REDOX POTENTIAL AND FREE ENERGY

In the discussion so far, we have made it clear that an oxidising or reducing agent may exist in two forms, (a) the oxidised form or oxidant which can accept electrons and (b) the reductant which can donate its electrons to a substrate. ***Each oxidising or reducing agent exists as a conjugate pair of electron acceptor oxidant and electron donor reductant forms.***

$$\text{Oxidant} + ne^- \longrightarrow \text{Reductant}$$

Where 'n' is the number of electrons.

The pair consisting of the oxidant and reductant forms of an oxidising or reducing agent is known as a redox couple or conjugate redox pair, e.g. NAD^+/NADH, FMN/$FMNH_2$, Cyt-c Fe^{++}/Cyt-c Fe^{+++}. **Oxidising agents differ in their electron affinity.** The standard redox potential Eo is a measure of the tendency of a redox couple to donate or accept electrons under standard conditions. Redox potential of a given system is intimately related to its free energy change. If redox-potential of a given system is known, the corresponding free energy change which might occur in the system on oxidation or reduction may be found as follows:

$$\Delta F = -nf\Delta E'o \text{ Coulomb Joules}$$

Where F = Standard free energy
n = no. of electrons
f = Faraday's constant (96,500 Coulombs)
ΔE'o = Difference between redox potentials of the couple.

$$\text{or } \Delta F = \frac{nf\Delta E'o}{4.18} \text{ calories}$$

If the sign of free-energy thus calculated is negative, it indicates free energy release and if positive, it indicates free energy consumption.

The redox potentials of many biologically important redox couples are known and are as shown in Table 10.1.

Table 10.1: Redox potentials of redox couples

	System	*E'o volts*
α-Hydroxybutyrate	⇔ acetoacetate	–0.346
Isocitrate	⇔ α-ketoglutarate	–0.36
NADH	⇔ NAD^+	–3.20
$FMNH_2$	⇔ FMN	–0.30
Ubiquinol	⇔ Ubiquinone	+0.10
Cyt-b (Fe^{+2})	⇔ Cyt b Fe^{+3}	+0.08
Cyt c_1 Fe^{+2}	⇔ Cyt c_1 Fe^{+3}	+0.22
Cyt c Fe^{+2}	⇔ Cyt c Fe^{+3}	+0.254
Cyt a Fe^{+2}	⇔ Cyt a Fe^{+3}	+0.29
Cyt a_3 Fe^{+2}	⇔ Cyt a_3 Fe^{+3}	+0.386
O_2	⇔ H_2O	+0.816

A system having relatively more positive redox potential can oxidise another system having low redox potential. ***In the mitochondria, the hydrogen/electrons pass through different carriers in the sequence of increasing order of their positive potential.***

MITOCHONDRIAL ELECTRON TRANSPORT CHAIN

Definition: This is the final common pathway in aerobic cells by which electrons derived from various substrates are transferred to oxygen. Electron transport chain (ETC) is a series of highly organised oxidation-reduction enzymes whose reactions can be represented as:

$$\text{Reduced A} + \text{Oxidised B} \Leftrightarrow \text{Oxidised A} + \text{Reduced B.}$$

Localisation: The ETC is localised in the mitochondria. The outer membrane of mitochondria is permeable to most of the small molecules. There is an intermediate space which presents no barrier to passage of intermediates. **The inner membrane shows a highly selective permeability. It has transport systems only for specific substances such** as ATP, ADP, pyruvate, succinate, α-ketoglutarate, malate and citrate etc **(Refer Fig. 10.1)**. ***The enzymes of the electron transport chain are embedded in the inner membrane*** in association with the enzymes of oxidative phosphorylation.

The most accepted sequence of electron carriers in the mitochondria is as follows:

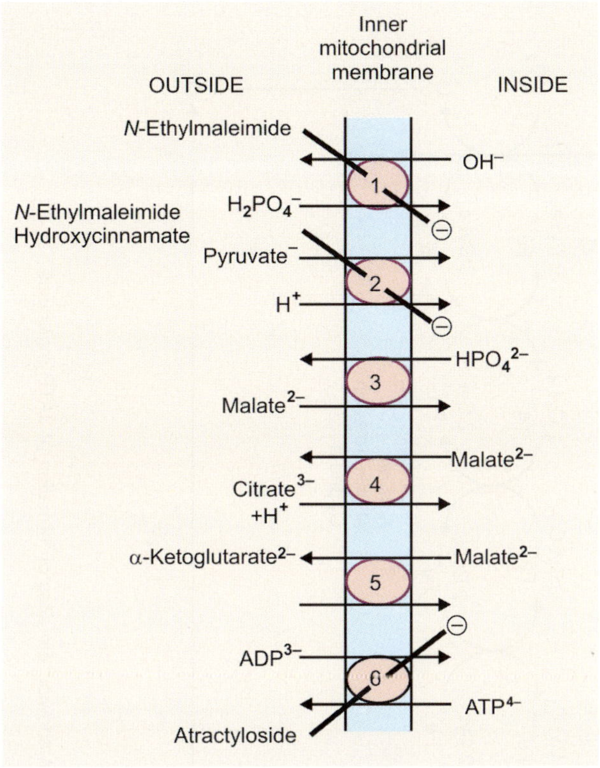

Fig. 10.1: Some of the transporter system in inner mitochondrial membrane and their inhibitors. Transporter systems in the inner mitochondrial membrane. (1) Phosphate transporter; (2) pyruvate symport; (3) dicarboxylate transporter; (4) tricarboxylate transporter; (5) α-ketoglutarate transporter; (6) adenine nucleotide transporter. N-Ethyl-maleimide, hydroxycinnamate, and atractyloside inhibit (–) the indicated systems

$$\text{Substrate} \rightarrow NAD^+ \rightarrow FAD \rightarrow CoQ \rightarrow 2\ \text{Cyt-b}$$
$$\downarrow$$
$$O_2 \leftarrow 2\ \text{Cyt}\ (a_3 + a) \leftarrow 2\ \text{Cyt-c} \leftarrow 2\ \text{Cyt-}c_1$$

As already mentioned ***the redox potentials in the ETC are in increasing order except in the case of ubiquinone.*** The chain actually consists of a series of redox couples, at each step electrons flow from the reductant of a redox couple with more negative redox potential to the oxidant of the next redox couple having a more positive redox potential.

Dehydrogenation of substrate is the first step in the process of respiratory chain oxidation. Most of the dehydrogenases require NAD which can accept a hydride ion (H^-) which is formed by one hydrogen atom and an electron. The electron is received from the second hydrogen atom releasing the second hydrogen atom in the form of a proton (H^+). The second type of dehydrogenase reaction makes use of FAD as the coenzyme. Table 10.2 gives the various reactions and the enzymes that take place in the mitochondrial matrix. In these reactions the oxidants are reduced to NADH + H^+ or $FADH_2$. Fig. 10.2. shows the path of electron transport.

DETAIL STRUCTURE AND FUNCTIONS OF ELECTRON TRANSPORT CHAIN (ETC)

The electron transport chain in the mitochondrial membrane has been separated in **4 (four) complexes** or components as follows:

- **Complex I:** ***NADH-CoQ reductase***
- **Complex II:** ***Succinate-CoQ reductase***
- **Complex III:** ***CoQ-cytochrome C reductase***
- **Complex IV:** ***Cytochrome C oxidase.***

The composition of these four complexes are given in Table 10.3.

When arranged as shown in Figure 10.3 the chain accomplishes the transfer of electrons from NADH or succinic acid to molecular O_2.

CoQ and cytochrome C are not included in any of the complexes because these two components can be removed with relative ease from the membrane.

Cytochrome C is a small, peripheral protein, having Mol. wt 12,000 that is readily extracted by treatment of the membrane with aqueous salt solutions; CoQ can be extracted with butanol.

Complex I: *NADH-CoQ Reductase*

This system has **two functions:**

- *Electron transfer*
- *Acts as a proton pump.*

Table 10.2: Reactions and enzymes

Reaction	Enzyme	Oxidant
• α-Glycerophosphate → Dihydroxyacetone	α-Glycerophosphate dehydrogenase	FAD
• Acyl CoA → Unsaturated acyl-CoA	Acyl-CoA dehydrogenase	FAD
• Glutamate → Ketoglutarate	Glutamate dehydrogenase	NAD⁺
• Pyruvate → Acetyl-CoA	Pyruvate dehydrogenase	NAD⁺
• Succinate → Fumanate	Succinate dehydrogenase	FAD
• Malate → Oxaloacetate	Malate Dehydrogenase	NAD⁺
• α-Ketoglutarate → Succinyl-CoA	α-Ketoglutarate dehydrogenase	NAD⁺

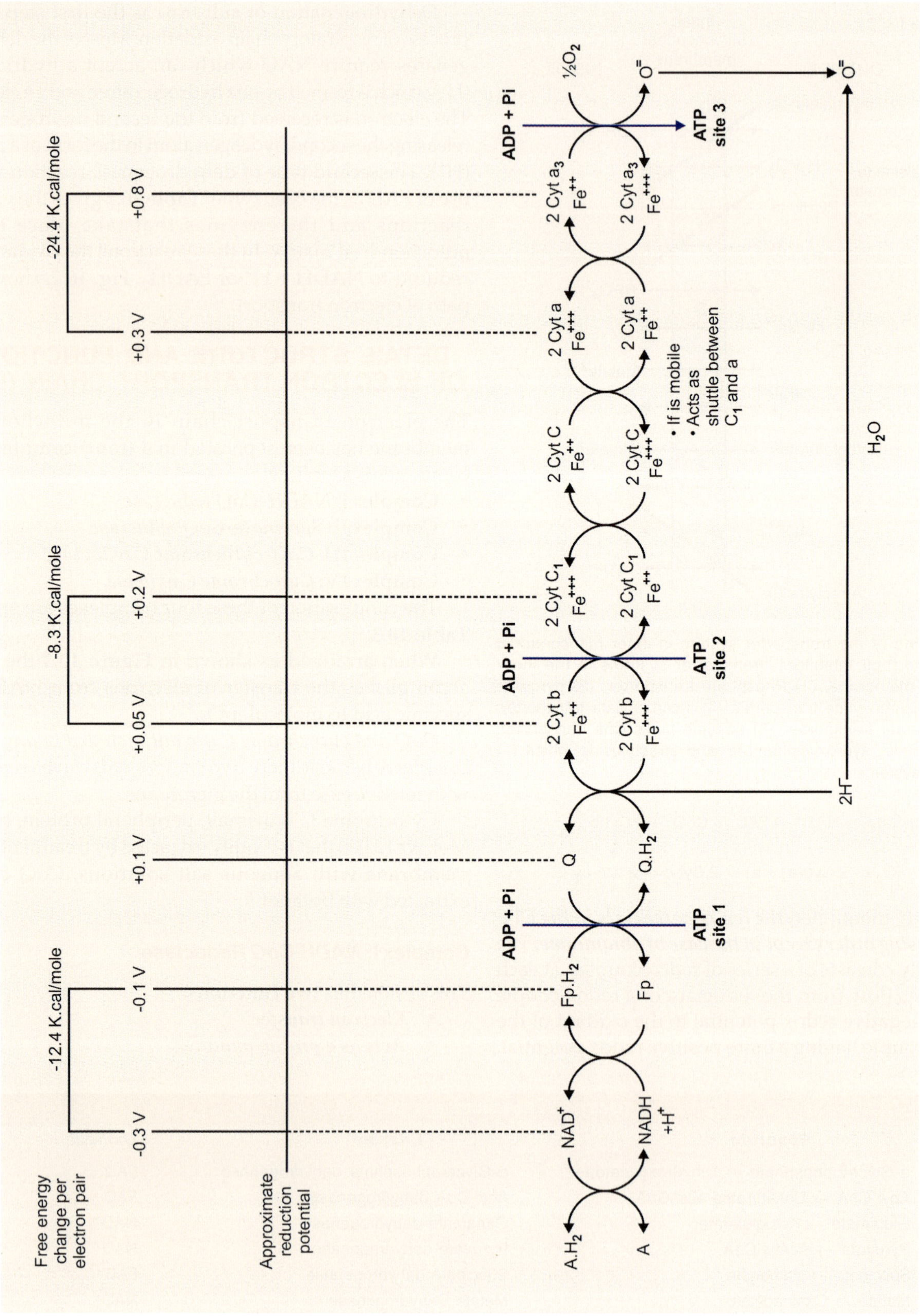

Fig. 10.2: The electron transport system of the respiratory chain showing the sites of formation of 3ATP molecules

Table 10.3: Composition of the four complexes

Complexes	*Mol wt*	*Subunits (polypeptide chains)*	*Cofactors present*
• ***Complex I: NADH-CoQ reductase***	850,000	**16-25** (Minimum 16)	• FMN = 1 • Fes = 22 to 24, in 5 to 8 clusters • 16 to 24 non-haem Fe atoms
• ***Complex II: Succinate-CoQ reductase***	125,000	**4**	• FAD = 1 • Fes = 8, in 3 clusters
• ***Complex III: CoQ-Cyt C reductase***	250,000	**8**	• Fes = 2 clusters • Cyt-b_{562} • Cyt-b_{566} • Cyt-C_1
• ***Complex IV: Cytochrome C oxidase***	300,000	**7**	• Cyt-a • Cyt-a_3 • Two Cu ions

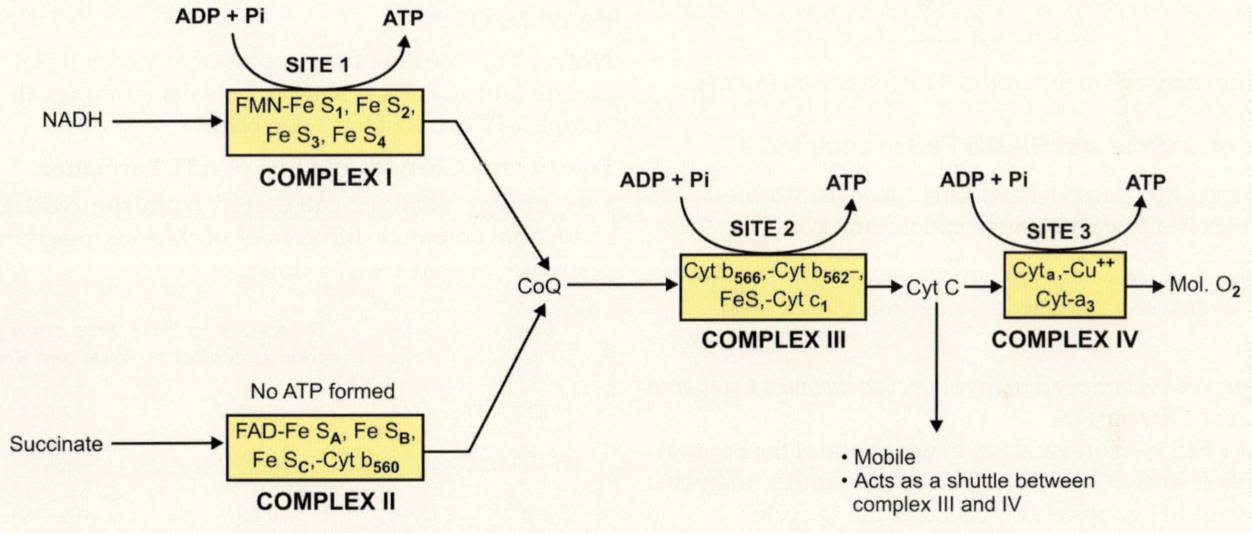

Fig. 10.3: Four complexes of electron transport chain

The system catalyses transfer of two electrons from NADH to small lipid soluble CoQ via FMN and Fes clusters.

$$NADH + H^+ + FMN \rightarrow FMN.H_2 + NAD^+$$

From FMN.H_2 electrons are transferred to a group of Fes proteins. Fe atoms of FeS protein oscillate between Fe^{++} and Fe^{+++}. The electrons are then transferred to CoQ.

$$CoQ \xrightarrow{e^-} CoQ.H \text{ (Semi-Quinone)}$$

$$CoQ.H \xrightarrow{e^-} CoQ.H_2 \text{ (Quinol)}$$

The process is accompanied ***by pumping of protons from mitochondrial matrix into intermembrane space.***

- ***Permits one ATP Formation (Site I).***

Note: Upto CoQ, H is transferred. But from CoQ onward only e^- is transferred, $2H^+$ goes into the medium.

Complex II: *Succinate-CoQ Reductase*

Flow of electrons from succinate to CoQ occurs via FAD.H_2.

$$Succinate + CoQ \longrightarrow Fumarate + CoQ.H_2$$

Standard reduction potential for transfer of electrons from FAD.H_2 to CoQ is + 0.113V (much lower than +0.420 V energy change for the reaction of complex I). The small energy change does not allow *"succinate-CoQ reductase"* system to pump protons across the mitochondrial membrane, hence this protein complex does not contribute to proton gradient. ***Hence no ATP is formed.***

Note: CoQ is the electron acceptor in the reaction catalysed by *NADH-CoQ reductase* (Complex I) and *succinate-CoQ reductase* (Complex II). The electrons received are subsequently transferred from CoQ.H_2, a lipid soluble mobile electron carrier to *CoQ-Cyt.C reductase* (Complex III).

Complex III: *CoQ-Cyt.C Reductase*

Functions as

- *Proton pump,* and
- *Catalyses transfer of electrons*

This system catalyses transfer of electrons from $CoQ.H_2$ to Cyt-c via Cyt-b and Cyt-c_1.

The electrons from $CoQ.H_2$ is first accepted by Cyt-b_{566} and then transferred to Cyt-b_{562}, which reduces

$$Co.Q.H_2 \longrightarrow CoQ$$

Fe^{+++} accepts electron and is oxidised to Fe^{++}.

The system also **acts as a proton pump**. It is believed that 4 (four) protons are pumped across the mitochondrial membrane during the oxidation.

$$CoQ.H_2 + 2\ Cyt\text{-}c\ (Fe^{+++})$$
$$\rightarrow Co.Q + 2\ Cyt.c\ (Fe^{++}) + 2\ H^+$$

The energy change permits ATP formation (Site II).

Role of Q cycle and Rieske FeS in complex III

Electrons are passed from Q.H2 to cytochrome C via complex III-Q cytochrome C oxidoreductase:

$Q.H_2$	+	2 Cyt C	+	$2H^+$	→	Q	+	2 CytC	+	$4H^+$
		Oxidised		matrix		reduced				Inter membrane space

The process is believed to involve *cytochromes C_1, b_1 and bH_H and a Rieske FeS.*

Rieske FeS is an unusual FeS in which one of the Fe atoms is linked to two histidine –SH groups rather than two cysteine –SH goups and is called as the **Q cycle**.

Q may exist in **three forms:**

- Oxidised quinone,
- Reduced quinol, or
- Semiquinone.

The semiquinone is formed transiently during the cycle, one turn of which results in the oxidation of $2Q.H_2$ to Q, releasing $4H^+$ into the intermembrne space, and the reduction of one Q to QH_2, causing $2H^+$ to be taken up from the matrix.

Note

It is to be noted that while Q carries two electrons, the cytochromes carry only one, *thus the oxidation of one $Q.H_2$ is coupled to the reduction of two molecules of cytochrome C via the Q. cycle.*

Complex IV: *Cyt-c Oxidase*

The system functions:

- *As proton pump*
- *Catalyses transfer of electrons to molecular O_2 to form H_2O.*

This is the ***terminal component of ETC. It catalyses the transfer of electrons from Cyt-c to molecular O_2 via Cyt-a, Cu^{++} ions and Cyt-a_3.***

$$4\ Cyt\text{-}c\ (Fe^{++}) + 4H^+ + O_2 \rightarrow 4\ Cyt\text{-}c\ (Fe^{+++}) + 2H_2O$$

The flow of electrons is as follows:

$$Cyt\text{-}c \rightarrow Cyt\text{-}a \rightarrow Cu{++} \rightarrow Cyt\text{-}a_3 \rightarrow O_2$$

Role of Cu Ions: Cu atom adjacent to cyt. haem a is **Cu-A (Sub unit II)** and Cyt-haem a_3 is close to **CuB (Sub unit I).** From Cyt-c, the electrons are transferred to haem-a-CuA cluster and then haem a_3-CuB cluster.

The system **also acts as a proton pump**, it pumps two protons into intermembrane space. *The energy change permits ATP formation (Site III) between cyt a_3 and molecular O_2.*

Note: Cyt c does not form a part of any complexes. It is *mobile* and *acts as a shuttle* between complex-III and complex-IV to transfer e^- (electron).

Free Energy Changes and Site of ATP Formation

Free energy changes calculated from the oxidation-reduction potential differences of various reactions of respiratory chain are as follows:

Step	*Difference in redox potential*	*Free energy changes Kcal*
• NAD $\xrightarrow{2H}$ FP	+0.26 V	–12.004
• FP $\xrightarrow{2e^-}$ 2 Cyt-b	+0.10 V	–4.617
• Cyt-b $\xrightarrow{2e^-}$ 2 Cyt-c_1	+0.23 V	–10.619
• Cyt-c_1 $\xrightarrow{2e^-}$ Cyt-a_1	–0.03 V	–0.923
• Cyt-a_1 $\xrightarrow{2e^-}$ Cyt-a_3	+0.21 V	–9.695
• 2 Cyt-a_3 $\xrightarrow{2e^-}$ O_2	+0.32 V	–14.774
Total	1.14 V	52.63 Kcal

Thus the span of the respiratory chain is 1.14 V which corresponds to 52.6 KCal/mole. There is decline in free energy as electrons flow down the electron transport chain. ***At three sites free energy released per electron pair transferred is sufficient to support the phosphorylation of ADP to ATP*** which requires about 8 KCal/mole.

The sites where ATP is produced in respiratory chain are three (refer to oxidative phosphorylation).

ADP:O or P:O Ratio: The NAD-dependant dehydrogenases such as malate, pyruvate, α-ketoglutarate, iso citrate, etc. produce three high energy phosphate bonds for each pair of electrons transferred to O_2 because they have P : O ratio of 3. ***Thus P : O ratio is a measure of how many moles of ATP are formed from ADP by phosphorylation per gram atom of oxygen used.*** This is usually measured as the number of moles of ADP (or Pi) that disappear per gram atom of oxygen used.

$$\text{P:O ratio} = \frac{\text{Phosphate group esterified}}{\text{Electron pairs transferred}}$$

$$\text{Efficiency} = \frac{\text{P:O ratio} \times 7.3}{51 \text{ K.Cal}} \times 100$$

However, P:O ratio in case of $FADH_2$ is 2 and therefore efficiency is lower in that case.

INHIBITORS OF ELECTRON TRANSPORT CHAIN

Transfer of electrons is selectively inhibited at various components of the electron transport chain by a variety of substances. Some of these are used as poisons (e.g. insecticides) and some of which are used as drugs.

Inhibitors of ETC

Site-I (Complex-I)

- ***Rotenone:*** A fish poison and also insecticide. Inhibits transfer of electrons through complex-I-*NADH-Q-reductase.*
- ***Amobarbital (Amytal) and Secobarbital:*** Inhibits electron transfer through NADH-Q reductase.
- ***Piericidin A:*** An antibiotic. Blocks electron transfer by competing with CoQ.
- ***Drugs:*** Chlorpromazine and hypotensive drug like guanethidine.

Site-II (Complex III)

- ***Antimycin A***
- ***BAL (Dimer-Caprol)***
- ***Hypoglycaemic drugs: like Phenformin***

} Blocks electron transfer from cyt b to c_1

Site-III (Complex IV)

- ***Cyanide***
- ***H_2S***
- ***Azide***

} Inhibits terminal transfer of electrons to molecular O_2

- ***Co (Carbon monoxide):*** Inhibits Cyt. oxidase by combining with O_2 binding site. It can be reversed by illumination with light.

Complex II: *Succinate dehydrogenase FAD*

- ***Carboxin***
- **TTFA**

} Specifically inhibit transfer of reducing equivalent from *succinate dehydrogenase*

- ***Malonate:*** A competitive inhibitor of succinate dehydrogenase.

Mitochondrial Shuttle Systems: Glycolysis produces NADH in the cytoplasm which cannot enter mitochondria. Shuttles between cytoplasm and mitochondria operate. **Two such shuttles** are of considerable importance:

(a) Glycerophosphate Shuttle
(b) Malate Shuttle

(For details—Refer chapter on Carbohydrate Metabolism).

OXIDATIVE PHOSPHORYLATION

The process of oxidative phosphorylation is closely associated with the functioning of the electron transport chain. This was studied by fragmentation of mitochondria. In the first fragmentation step, the outer membrane is removed by treatment with various detergents such as saponin, digitonin. The two particulate fractions that result are:

1. The outer membrane, either in the form of vesicles or completely solubilised.
2. The inner membrane and the mitochondrial matrix enzymes. This fraction is found to contain the enzymes of:
    - **The electron transport chain**
    - **Oxidative phosphorylation**
    - **The TCA cycle.**

In oxidative phosphorylation ATP is produced by combining ADP and Pi with the energy generated by the flow of electrons from NADH to molecular oxygen in the electron transport chain. There are **three sites** in the respiratory chain where ATP is formed by oxidative phosphorylation. These sites have been proved by the free energy changes of the various redox couples. Since hydrolysis of ATP to ADP + Pi releases around 7.3 K. Cal/mole, ***the formation ATP from ADP + Pi requires a minimum of around 8 KCal/mole.*** The formation of ATP is therefore not possible at the sites where free energy released is less than 8 KCal/mole.

Whenever two systems or redox couple of the respiratory chain differ from each other by 0.22 volts in standard redox potential (E'o), the free energy is sufficient to form ATP.

$$G^o = -n \text{ F Eo}$$
$$G^o = -2 \times 23.06 \times 0.22 = -10.15 \text{ KCal}$$

Sites of ATP formation: There are **three sites** in the respiratory chain where ATP can be formed.

- *Site I:* This involves the transfer of electrons from NADH –CoQ. Obviously this step is omitted by *succinic dehydrogenase* whose $FADH_2$ prosthetic group transfers its electrons directly to CoQ bypassing NADH. ***This step is blocked by piericidin, rotenone, amobarbital, certain drugs like chlorpromazine, guanethidine.***
- *Site II:* This involves the transfer of electrons from Cyt-b–Cyt-c_1. This step as well as the previous one is bypassed in oxidation of L-ascorbate whose electrons are directly transferred to Cyt-c. ***This step is blocked by BAL, Antimycin A, Hypoglycaemic drug like phenformin.***

SECTION TWO

- ***Site III:*** Transfer of electrons from Cyt-a_3 to molecular oxygen ***which is blocked by CO, CN, H_2S, and azide.***

Mechanism of Oxidative Phosphorylation

Three major proposals for the mechanism of oxidative phosphorylation have been considered. The synthesis of ATP is carried out by a molecular assembly in the inner mitochondrial membrane. ***This enzyme complex is called Mitochondrial ATPase or H^+-ATPase. It is also called ATP-synthase.***

Theories

The three hypothesis do make use of the information available on *ATP synthase.*

A. ***The chemical coupling hypothesis:*** This is developed from the concept of a high energy intermediate common to both electron transport and phosphorylation of ADP. However, such intermediate has not been identified so far.

B. ***The conformational coupling hypothesis:*** According to this hypothesis the mitochondrial cristae undergo conformational changes and these changes in architecture of the mitochondrial cristae reflect the changes in the different components of the electron chain to one another. It is believed that these conformational change represents the formation of high energy state.

C. ***Chemiosmotic theory:*** This is the most accepted view of oxidative phosphorylation postulated by **Peter Mitchell** in **1961**. Mitchell's chemiosmotic theory postulates that the energy from oxidation of components in the respiration chain is coupled to the translocation of hydrogen ions (Protons, H^+) from the inside to the outside of the inner mitochondrial membrane. **Each of the respiratory chain complexes I, III and IV acts as a proton pump.** The inner membrane is impermeable to ions in general but particularly to protons, which ***accumulate outside the membrane, creating an electrochemical potential*** difference across the membrane ($\Delta\mu H^+$). This consists of a chemical potential **(difference in pH)** and an electrical potential.

The electrochemical potential resulting from the asymmetric distribution of the hydrogen ion is used to ***drive the mechanism responsible for the formation of ATP*** **(Fig. 10.4)**.

Experimental Evidences to Support the Chemiosmotic Hypothesis

- Addition of protons H^+ (acid) to the external medium of intact mitochondria leads to the generation of ATP.
- Oxidative phosphorylation does not take place in soluble system where there is no possibility of a vectorial ATP synthase.
- A closed membrane is a must to achieve oxidative phosphorylation.

ATP SYNTHASE

Much information is now available regarding ATP synthase and its role in ATP formation.

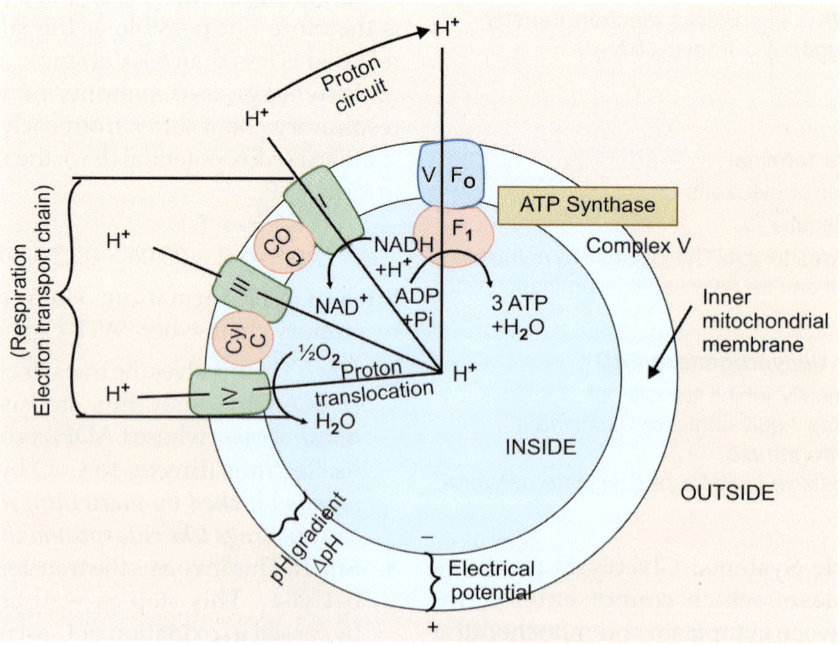

Fig. 10.4: Principles of chemiosmotic theory

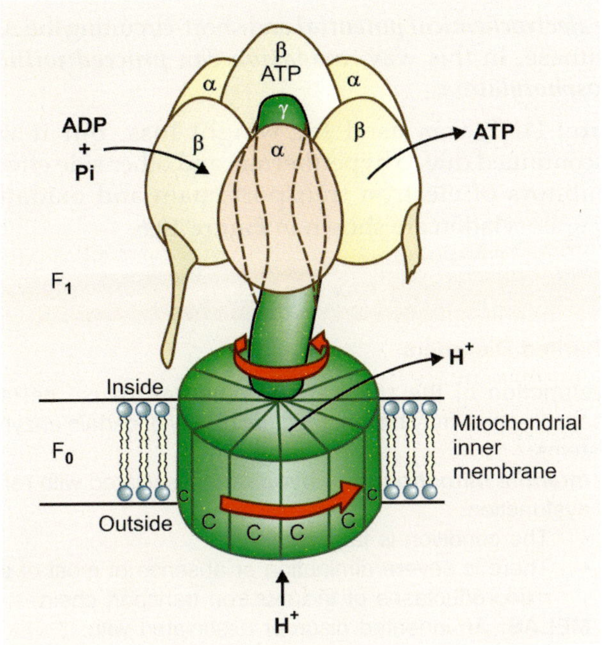

Fig. 10.5: Schematic representation of ATP synthase

Structure: (Fig. 10.5)

It is an *enzyme complex present in the inner mitochondrial* membrane. It is now referred as **COMPLEX V** the enzyme complex has **two subunits – F_0 and F_1.**

- **F_0 Unit or Subcomplex**
  It spans inner mitochondrial membrane and **serves as a proton channel** through which protons enter into mitochondria.
  It is a **disk of C-subunits**. Attached to it is a **γ-subunit** in the form of a ***bent axle. The γ-subunit fits inside the F_1 subcomplex.***
- **F_1 Unit or Subcomplex**
  This projects into the mitochondrial matrix. It catalyses the ATP synthesis. F_1 subcomplex consists of **3β chains ($β_3$)** and **3α chains ($α_3$).**

γ-subunit fits inside the F_1 subcomplex of 3α and 3β subunits which are fixed to the membrane.

In addition, the complex has **sigma** and **epsilon subunits**, the function of which are not known.

Mechanism of ATP Synthesis (Boyer's hypothesis)

Paul Boyer originally proposed a **binding change mechanism.**

According to this hypothesis 3β subunits (catalytic sites) though structurally similar, but functionally not same at a particular time.

It is envisaged that **β-subunits** occur in **3 forms:**

- **'O' form (Open form):** It has low affinity for substrates ADP + Pi.
- **'L' form (Loose form):** Can bind substrates ADP and Pi with more affinity but ***catalytically it is inactive.***
- **'T' form (Tight form):** Binds substrates ADP and Pi tightly and ***catalyses ATP synthesis.***

Protons entering the system cause conformational changes in the β-subunits.

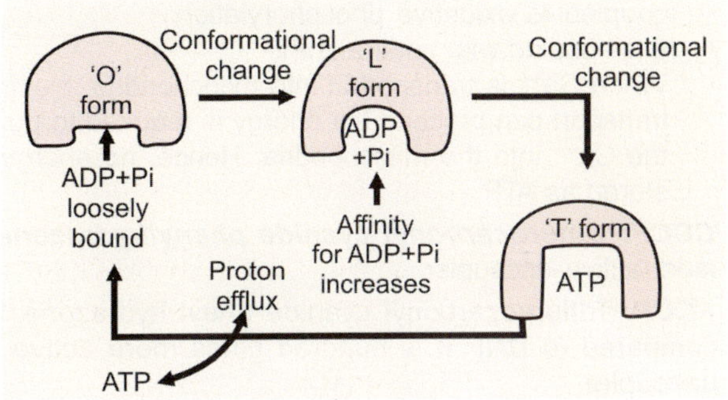

Rotary or Engine Driving Model

Original Boyer's hypothesis is now modified. It is now widely accepted that protons passing through the disk of 'C' subunits of F_0 subcomplex cause it and the attached γ-subunit **to rotate**. The ***β-subunits which are fixed to membrane do not rotate.***

ADP and Pi are taken up sequentially by the β-subunits which undergo conformational changes.

'O' form → 'L' form → 'T' form
↑_______________↑

and forms ATP, which is expelled as the rotating γ-subunit squeezes each β-subunit inturn.

Thus 3 ATP molecules are generated per revolution.

Inhibitors of Oxidative Phosphorylation

- ***Oligomycin:*** It ***binds with the enzyme ATP synthase and blocks the proton channels.*** It thus prevents the translocation of H^+ into the mitochondrial matrix, this leads to accumulation of H^+ at higher concentration in intermembrane space. Since protons cannot be pumped out against steep proton gradients, electron transport stops (respiration stops).
- ***Atractyloside:*** It is a **glycoside, it blocks the translocase that is responsible for movement of ATP and ADP,** across the inner mitochondrial membrane. Adequate supply to ADP is blocked thus preventing phosphoglation and ATP formation (Refer Fig. 10.1).
- ***Bongregate:*** Toxin produced by Pseudomonads. It acts similarly to atractyloside.

UNCOUPLERS OF OXIDATIVE PHOSPHORYLATION

These are the compounds that allow mitochondria to use oxygen regardless of whether or not there is any phosphate (ADP) available. When an uncoupler is added, there is marked increase in O_2 uptake.

SECTION TWO

Uncouplers of Oxidative Phosphorylation

- ***2, 4 Dinitrophenol:*** A classic uncoupler of oxidative phosphorylation (mechanism see below).
- ***Dicoumarol (Vitamin K analogue):*** Used as anticoagulant
- ***Calcium:*** Transport of Ca^{++} ion into mitochondria can cause uncoupling.
  - Mitochondrial transport of Ca^{++} is energetically coupled to oxidative phosphorylation.
  - It is coupled with uptake of Pi.
  - When Ca^{++} is transported into mitochondria, electron transport can proceed but energy is required to pump the Ca^{++} into the mitochondria. Hence, no energy is stored as ATP.
- ***CCCP: Chlorocarbonyl cyanide phenylhydrazone***—most active uncoupler.
- ***FCCP:*** Trifluorocarbonyl cyanide pheyl hydrazone. As compared to DNP it is hundred times more active as uncoupler.
- ***Valinomycin:*** It produced by a type of Streptomyces. Transports K^+ from the cytosol into matrix and H^+ from matrix to cytosol, thereby decreasing the proton gradient.
- ***Physiological Uncouplers:***
  - Excessive thyroxine hormone
  - EFA deficiency
  - Long chain FA in Brown adipose tissue
  - Unconjugated hyperbilirubinaemia

Mechanism of Action of DNP as uncoupler

Dinitrophenol (DNP) a potent uncoupler is ampipathic and increase the permeability of the lipoid inner mitochondrial membrane to protons (H^+), thus *reducing the electrochemical potential* and short-circuiting the ATP synthase. In this way, ***oxidation can proceed without phosphorylation.***

Note: DNP was used for weight loss. But it was discontinued due to hyperthermia and other side effects. Inhibitors of electron transport chain and oxidative phosphorylation are shown in Figure 10.6.

CLINICAL ASPECTS

Inherited Disorders

Dysfunction of the respiratory chain can cause certain diseases which may be inherited deficiency of certain enzyme systems.

1. ***Infantile mitochondrial myopathy*** associated with renal dysfunction:
   - The condition is fatal
   - There is severe diminution or absence of most of the oxidoreductases of the electron transport chain.
2. **MELAS:** An inherited disorder associated with: Mitochondrial myopathy, Lactic acidosis, Encephalopathy and Stroke
   Enzyme deficiency: NADH: Ubiquinone oxidoreductase (complex 1) or deficiency of cytochrome oxidase.

SUMMARY: Various mechanisms of ATP formation are given below:

1. *ATP Formation in Electron Transport Chain (ETC):* The main mechanism by which ATP is formed in the body

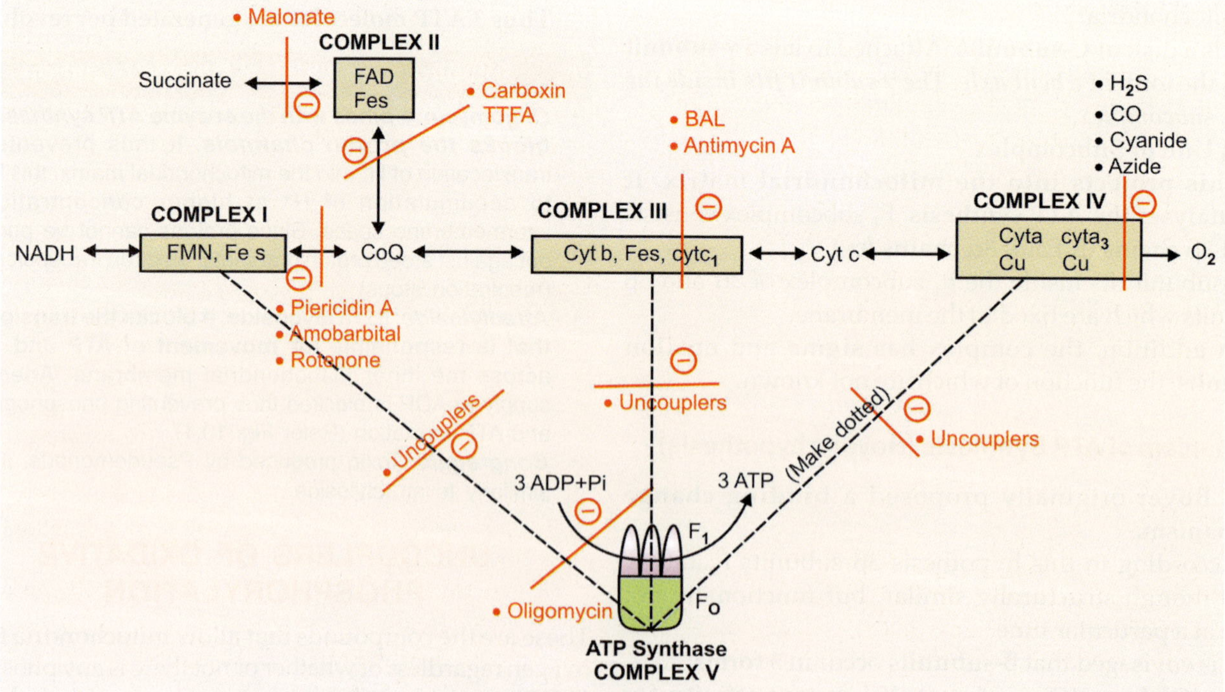

Fig. 10.6: Inhibitors of election transport chain and oxidative phosphorylation

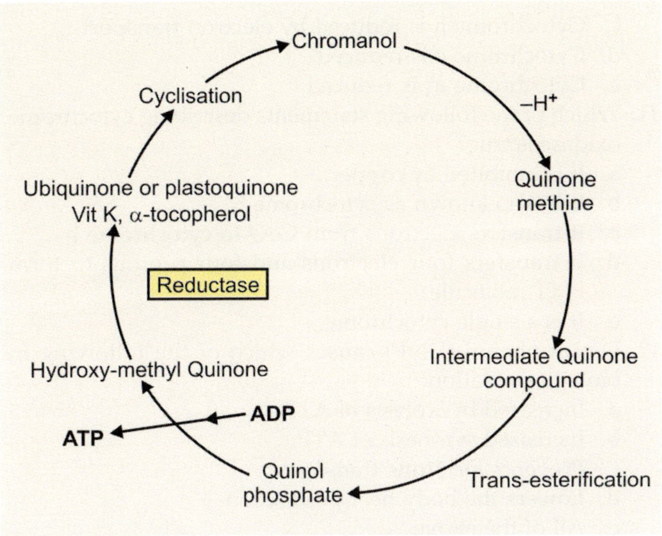

Fig. 10.7: Reaction mechanism involved in ATP formation by quinones

is by oxidative phosphorylation in the electron transport chain (ETC).

2. ***Role of quinones in ATP formation:*** Quinol-p as intermediates in oxidative phosphorylation have been implicated for formation of ATP. Reaction mechanisms proposed by **Nilkas** and **Lederer** are as follows:

   Pi reacts with the 6-chromanol derivatives of ubiquinones, or corresponding derivatives of vit K or α-tocopherol.

Steps of the Reactions

- Cyclisation of the quinones to chromanol, generating a quinone methine.
- Pi is added to the quinone-methine and the quinol-p is generated by trans-esterification.
- The quinol-p is then oxidised to a hydroxy-methyl quinone and transference of phosphate to ADP to form one ATP molecule.
- The final step in the reaction cycle is the reduction of the hydroxy methyl quinone to the original quinone structure (Fig. 10.7).

3. ***Substrate Level Phosphorylation:*** ATP can be produced at substrate level without mediation of electron transport chain.

Examples

a. In glycolytic cycle

1.

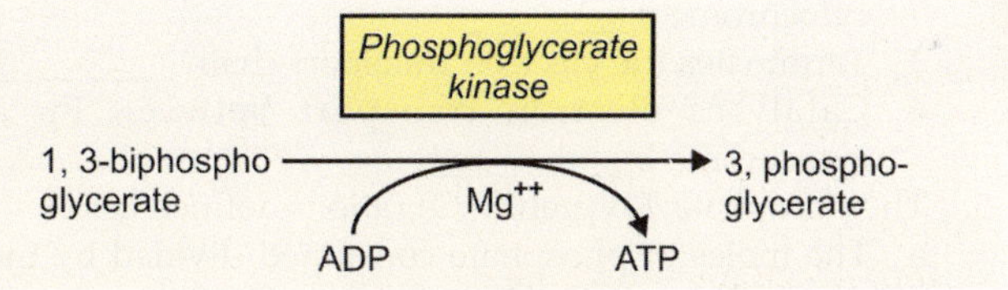

2.

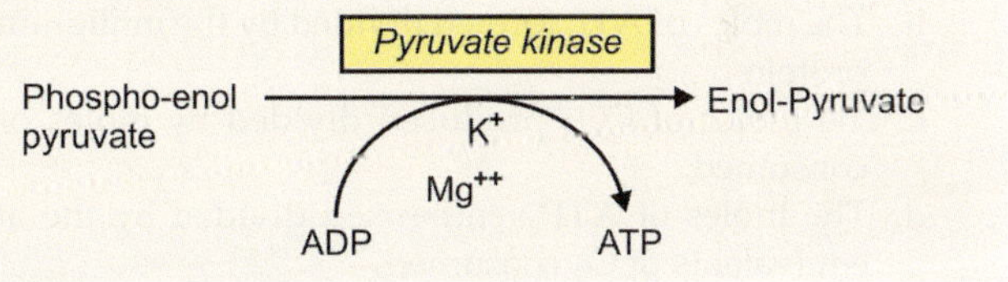

b. In TCA cycle: In conversion of succinyl CoA to succinic acid.

4. ***In Löhman reaction:*** ATP can be formed from creatine ~ P in muscles. The high energy phosphate is transferred to ADP and ATP is formed, called as **Löhmann reaction.**

- ***Myokinase reaction:*** In muscle, two ADP molecules can react to produce one molecule of ATP and AMP, the reaction being catalysed by the enzyme ***myokinase (adenylate kinase).***

I. Questions (Essay type and short notes)

1. What is ATP? Describe the formation, fate and importance with reference to electron transport chain.
2. Describe the electron transport chain reactions, and give its importance in oxidative phosphorylation

Short Notes

a. Respiratory chain,
b. Oxidative phosphorylation,
c. Cytochromes,
d. Cytochrome oxidase,
e. Electron transport chain— Role and importance in biologic oxidation,
f. Cytochromes and its role in biologic oxidation.
g. Redox potential, and
h. Electron transport chain

II. MCQs (Give one correct answer)

1. All of the following electron carriers are components of the mitochondrial electron transport chain except:
   a. FMN b. FAD
   c. NAD^+ d. $NADP^+$
   e. Co-enzyme Q
2. Electron transport chain carriers are located in the mitochondria:
   a. In mitochondrial matrix
   b. On the inner surface of the outer mitochondrial membrane
   c. On the outer surface of the outer mitochondrial membrane
   d. In the intermembrane space
   e. In the inner mitochondrial membrane.

3. Electrons from pyruvic acid enter the mintochondrial electron transport chain at:
   a. Nadh – Q reductase
   b. Coenzyme Q
   c. QH_2– cytochrome c-reductase
   d. Cytochrome c oxidase
   e. Between CoQ and cytochrome b
4. Match the following:
   a. Coenzyme Q b. Flavoprotein
   c. Cytochrome oxidase d. Cytochrome b
   1. Catalyses electron transport between NADH and coenzyme Q ________.
   2. Catalyses electron transport between coenzyme Q and cytochrome c ________.
   3. Terminates the electron transport chain ________.
   4. Catalyses electron transport between Fp and cytochrome b ________.
5. The phosphate: Oxygen (P:O) ratio is defined as:
   a. The moles of phosphate consumed divided by moles of oxygen consumed,
   b. The moles of ATP formed divided by the milligrams of protein,
   c. The moles of CO_2 produced divided by moles of O_2 consumed,
   d. The moles of ATP synthesised divided by the atom equivalents of O_2 consumed,
   e. The moles of O_2 consumed in presence of ADP divided by the moles of O_2 consumed in absence of ADP
6. Which of the following oxidation-reduction systems has highest redox-potential?
   a. Fe^{3+} cytochrome b/Fe^{2+}
   b. Fe^{3+} cytochrome a/Fe^{2+}
   c. Fumarate/succinate
   d. Ubiquinone ox/red
   e. NAD^+/NADH
7. Cyanide is poisonous as it stops respiration because of its:
   a. Inhibition of TCA cycle
   b. Combination with RB cell membrane
   c. Inhibition of myoglobin
   d. Formation of complex with Hb.
   e. Inhibition of cytochrome oxidase
8. In oxidative phosphorylation the oxidation of one molecule NADPH produces how many ATPs?
   a. Zero b. 2
   c. 3 d. 4
   e. 5
9. In oxidative phosphorylation, one molecule of reduced flavoprotein produces how many ATPs:
   a. Zero b. 2
   c. 3 d. 4
   e. 5
10. In the presence of rotenone:
    a. NADH is oxidised by electron transport,
    b. FAD H_2 is oxidised by electron transport
    c. Cytochrome a is reduced by electron transport
    d. Cytochrome c is reduced
    e. Cytochrome a_3 is reduced
11. Which of the following statements describing cytochrome oxidase is true?
    a. It is inhibited by copper
    b. It is also known as cytochrome b
    c. It transfers electrons from CoQ to cytochrome b
    d. It transfers four electrons and four protons to form H_2O molecule
    e. It is a single cytochrome
12. Dinitrophenol (DNP) causes which of the following in biologic oxidation:
    a. Increased bydrolysis of ATP
    b. Increased synthesis of ATP
    c. Prevents electrons transfer
    d. Lowers the body heat production
    e. All of the above
13. The chemical energy required for the synthetic processes is provided by:
    a. Phosphorylation of ATP
    b. Dephosphorylation of ATP
    c. Phosphorylation of ADP
    d. Dephosphorylation of ADP
    e. All of the above processes
14. Which of the following vitamins of not a component of electron transport chain?
    a. Nicotinamide b. Ubiquinone
    c. Biotin d. Riboflavin
    e. None of the above
15. Match the following:
    A. Inhibitors of site I B. Inhibitors of site II
    C. Inhibitors of site III
    1. Antimycin A ________.
    2. Piericidin A________.
    3. H_2S ________.
    4. Amobarbital ________.
    5. BAL ________.
16. Aerobic dehydrogenases have the prosthetic group:
    a. NAD^+ b. $NADP^+$
    c. ATP d. FAD^+
    e. ADP
17. The oxidation and phosphorylation in intact mitochondria is blocked by:
    a. Puromycin b. Oligomycin
    c. Gentamicin d. Streptomycin
    e. Septran

Answers to MCQs

1. d	2. e	3. a	4. 1-B, 2-D, 3-C, 4-A	5. d
6. b	7. e.	8. a.	9. b	10. b 11. d
12. a	13. b	14. c	15. 1-B, 2-A, 3-C, 4-A, 5-C	
16. d	17. b.			

CHAPTER 11

CHEMISTRY OF HAEMOGLOBIN AND HAEMOGLOBINOPATHIES

Major Concepts

A. To study the structure and types/varieties of normal human Hb.
B. To study the properties of Hb: Its physical and chemical properties, combination with various gases and formation of Hb-derivatives.
C. To study the abnormal haemoglobins and haemoglobinopathies.

Specific Objectives

A. Haemoglobin is a conjugated protein, present in RB cells, containing haem as the prosthetic group and globin as the protein part. Study the • Structure of haem • Structure of globin—the four polypeptide chains. • Different varieties of normal human Hb: Hb-A_1, Hb-F, Hb-A_2, Hb-A_3 and embryonic Hb. • Differentiate Hb-A from Hb-F and • Study Hb-A_{1C} (glycosylated Hb) and its clinical significance.
B. Study the properties of haemoglobin
   1. Physical properties: Form and shape, size, molecular weight.
   2. Chemical properties: • Study the role of Hb in acid base balance. • Study the action of acids and alkalies on Hb. • Differentiate acid haematin from alkaline haematin.
   3. Learn about the formation of following Hb derivatives, viz. formation of haemin crystal, haemochromogen, haematoporphyrin, haemopyrrole and haematoidin.
   4. Learn about methaemoglobin and methaemoglobinaemia, its formation, toxic effects and treatment. Learn how methaemoglobin can be converted to Hb.
   5. Study about methaemalbumin and its clinical importance. Learn about Schumm's test.
   6. Learn about combination of Hb with various gases.
      (a) Combination with O_2 to form oxyhaemoglobin: • Learn how O_2 combines with Hb and nature of the combination. • Study what is '***haem-haem interaction***' and the ***"haem-linked groups"***. • Study about two states of Hb: **'R' state** (relaxed) and **'T' state** (taut/tense).
      (b) Combination with CO to form carboxy-Hb: • Learn how CO combines with Hb to form carboxy-Hb and how does it differ from combination with O_2.• Study CO-poisoning: Causes, clinical symptoms, chemical test to demonstrate carboxy-Hb, absorption spectrum.
      (c) Combination with CO_2 to form carbaminocompound.
      (d) Study about formation of sulfhaemoglobin and its clinical significance.
      (e) Study about formation of cyanmethaemoglobin, learn about clinical importance and biochemical basis of treatment of cyanide poisoning.
   7. Study the absorption spectra given by Hb and its derivatives.
C. Study the different abnormal haemoglobins and haemoglobinopathies. The abnormal haemoglobins can result from mutations affecting:
   (a) **Structural gene** coding for amino acid sequences, e.g. Hb-S, Hb-M, Hb-C, Hb-D (Panjab), etc.
   (b) **Regulator gene** affecting the rate of synthesis of the chains, the amino acid sequences remaining unaffected producing thalassaemias.
   (c) List the different types of abnormal haemoglobins produced by mutation in a gene, in a tabular form showing the replacement of a single amino acid and its consequence in brief.
      • Learn in brief Hb-M, Hb-Chesapeake, Hb-Sabin, Hb-C, Hb-D, Hb-E, etc.
      • Study in detail about Hb-S and sickle-cell anaemia. Learn the mechanism of "sickling" in Hb-S and its effects.
   (d) Study the different types of thalassaemia, where the rate of synthesis of different chains is affected.
      1. Study about α-chain thalassaemia: (a) Hb-H, (b) Hb-Barts (c) Hb-Portland, Learn their clinical significance.
      2. Study about β-chain thalassaemia (thalassaemia major): Salient clinical features and biochemical findings.

SECTION TWO

HAEMOGLOBIN

INTRODUCTION

The red colouring matter of the blood is a conjugated protein, Haemoglobin, a chromoprotein, containing ***haem*** as prosthetic group and ***globin*** as the protein part-apoprotein. ***Haem-containing proteins are characteristic of the aerobic organisms,*** and are altogether ***absent in anaerobic forms of life.***

The normal concentration of Hb in an adult male varies from 14.0 to 16.0 gms% and there are about 750 gms of Hb in the total circulating blood of a 70 kg man.* Approximately 6.25 Gm (90 mg/kg) of Hb are produced and destroyed in the body each day.** The structure of *Haem* remains the same in Hb from any animal source, the **basic protein *globin* varies from species to species** in its amino acid composition and sequence and thus responsible for the ***species-specificity. The polypeptide chains of globin of adult Hb, are characterised by a relatively high content of ***histidine (His)*** and ***lysine (Lys)*** and a small amount of ***isoleuciene (Ile)***. By means of isotopic 'tracer' techniques, it has been shown that the ***polypeptide chains are formed from amino acids derived from dietary proteins.***

BIOMEDICAL IMPORTANCE

- Haem-proteins are characteristic of aerobic life. Hb is important in O_2-binding and its transport and delivery to tissues which is required for metabolism.
- Part of CO_2, a waste-product of metabolism, is also carried by the globin part of Hb.
- **2,3-biphosphoglycerate (BPG)** produced in RB cells by Rapoport-Luebering shunt is necessary for stabilisation of Hb-conformation at quaternary level by holding salt bridges and is important for understanding of high-altitude sickness and changes that take place in adaptation at high altitudes.
- ***Cyanide poisoning*** and ***carbon monoxide poisoning*** are fatal because they combine and inhibit heme protein *cytochrome oxidase* in electron transport chain and stops cellular respiration.
- ***Conversion of Hb to methaemoglobin*** by $NaNO_2$/or sodium thiosulphate forms the basis of treatment of cyanide poisoning, as cyanide combines readily with methaemoglobin to form cyanmethaemoglobin which is not toxic and thus spares the cytochrome oxidase.
- Study of Hb chemistry provides an insight into the molecular basis of genetic diseases such as haemoglobinopathies and abnormal Haemoglobins, viz. sickle cell disease is produced due to mutation at structural gene which produces an abnormality in one amino acid sequence or α- and β-chain thalassaemias due to mutation at regulatory genes.

STRUCTURE OF Hb

The structure of Hb molecule has been extensively studied. **It can be discussed under two headings:**

- **Structure of haem, the prosthetic group,** and
- **Structure of globin, the protein part—apoprotein.**

1. Haem

- It is a Fe-porphyrin compound. The porphyrins are complex compounds with a *tetrapyrrole* structure, each pyrrole ring having the following structure:

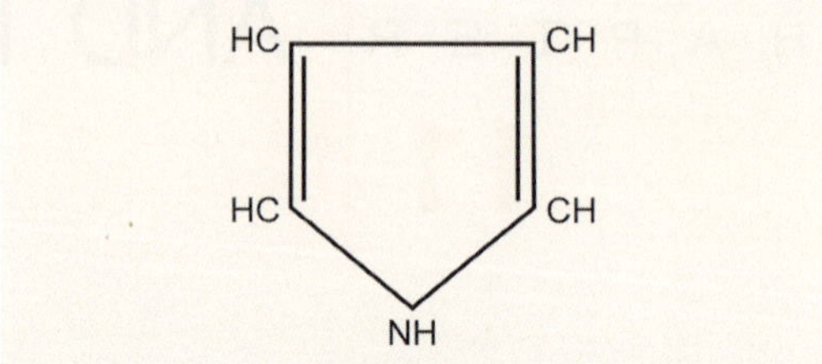

Pyrrole ring

- Four such pyrroles called **I to IV**, are combined through –CH= bridges, called as ***methyne*** or ***methylidene*** bridges to form a porphyrin nucleus.
  Haem may be represented by following structural formula schematically, with its attachment to globin (Fig. 11.1).

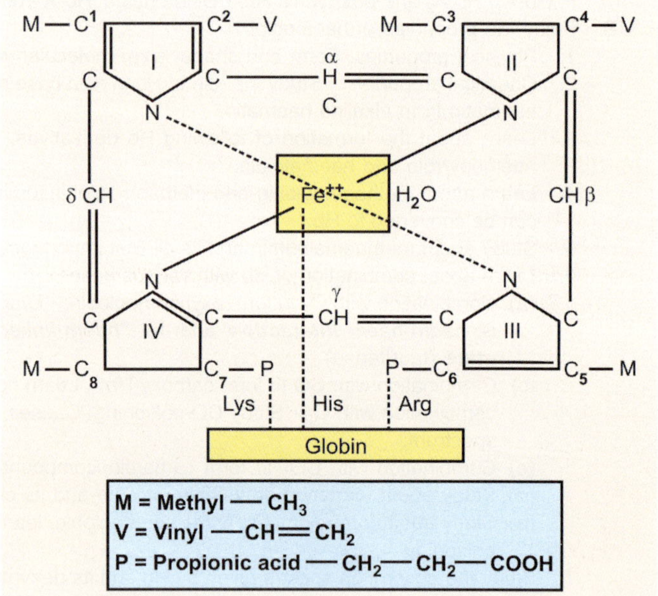

Fig. 11.1: Structure of haem

- The outer carbons of the four pyrrole rings, which are not linked with the methylidene bridges, are numbered 1 to 8.
- The methylidene bridges are referred to as α, β, γ and δ respectively.
- The two hydrogen atoms in the –NH groups of pyrrole rings (II and IV) are replaced by ferrous iron (Fe^{++}) which occupy the centre of the compound ring structure and establish linkages with all the four nitrogens of all the pyrrole rings.

- The Fe, besides its linkages to four nitrogens of the pyrrole rings, is also linked internally (5th linkage) to the nitrogen of the imidazole ring of histidine (His) of the polypeptide chains ***(haem-linked group)***. It is considered to have a valence of six as in ferrocyanide $H_4Fe(CN)_6$ and the sixth valence is directed outwards from the molecule and is linked to a molecule of H_2O in deoxygenated Hb. When Hb is oxygenated, the H_2O is displaced by O_2.

$$\textbf{Hb.}\ H_2O + O_2 \rightleftharpoons \textbf{Hb.}\ O_2 + H_2O$$

- The propionic acid COOH groups of 6 and 7 positions of haem, of III and IV pyrroles, are also linked to the basic groups of amino acids *Arg* and *Lys* of the polypeptide chains.
- If the central Fe is oxidised and converted to Fe(ic) state (Fe^{+++}), ***it will carry a surplus +ve charge*** which is balanced by taking an –OH group from the medium. It may also be balanced by other anions like Cl^- or $SO_4^=$, etc if available (see haematin formation).
- In addition, the hydrogens at positions 1 to 8 are substituted by different groups in different compounds. In the protoporphyrin IX, which forms parent compound of haem, the positions 1 to 8 in pyrroles are **substituted by methyl ($-CH_3$), vinyl ($-CH=CH_2$), methyl, vinyl, methyl, propionic acid ($-CH_2-CH_2-COOH$), propionic acid and methyl groups** in that order respectively.

2. Globin

Studies of globin, the protein part of Hb, have revealed that it is composed of ***four polypeptide chains***, two identical α-chains and two identical β-chains, in normal adult Hb, arranged in the configuration of a ***'tetrahedron'***. Thus there are six edges of contact.

Polypeptide Chains: Each polypeptide chain contains a 'haem', in the so-called ***haem-pocket. Thus one Hb molecule contains four haem units.***

In other normal haemoglobins, in place of two β-chains are replaced by either two δ-chains or two γ-chains or two ε-chains, forming different types.

Formation of haem-pockets: The Hb molecule and its subunits contain mostly hydrophobic amino acids internally and hydrophilic amino acids on their surfaces. Thus, the Hb molecule is waxy inside and soapy outside making it soluble in water, but impermeable to water. Each subunit contains one *haem* moiety hidden within a waxy pocket of the subunit. ***The haem pockets of α-subunits are of size just adequate for entry of an O_2 molecule, but the entry of O_2 into the haem-pockets of the β-subunits is blocked by a valine residue.***

VARIETIES OF NORMAL HUMAN HAEMOGLOBIN

Normal human Hb is of several types, containing four subunits made up of various combinations of possibly 5 different yet related polypeptide chains, designated as α, β, γ, δ and ε-chains. Most normal human Hb contains two α-chains + 2 other chains, which may be β, γ, δ, or ε depending on the type.

1. Hb-A1

Normal adult Hb, commonly called Hb-A, consists of two α- and two β-chains and designated as $\alpha^{A_2}\beta^{A_2}$ (or simply $\alpha_2\beta_2$). 90 to 95 per cent of Hb of normal adult is of this type.

Differences Between α-Chains and β-Chains of Adult Normal Hb-A

	α-Subunit	*β-Subunit*
• Molecular wt	15126	15866
• Total amino acids	141	146
• C-terminal amino acid	Arginine	Histidine
• N-terminal amino acid	*Val-Leu*	*Val-His-Leu*
• α-helices	7	8
• Haem-pocket	Adequate for entry of one molecule of O_2	Entry of O_2 in haem-pocket is blocked by a valine residue

2. Hb-F

Human foetal Hb is designated as Hb-F and it is $\alpha_2\gamma_2$. Hb-F differs in many respects from adult Hb-A_1. Differentiation of Hb-F from Hb-A has been given in Table 11.1.

Table 11.1: Differentiation of Hb-A from Hb-F

	Hb-A	*Hb-F*
1.	***Polypeptide chains:***	
	$\alpha_2\beta_2$	$\alpha_2\gamma_2$
2.	***Behaviour with alkali:***	
	Denatured by alkali	Resistant to alkali denaturation
3.	***Electrophoresis:***	
	At pH 8.9 Hb-A moves ahead of Hb-F.	Hb-F moves behind Hb-A
4.	***BPG-content:***	
	BPG ↑	BPG ↓
5.	***Affinity to O_2:***	
	Affinity of O_2 less	Affinity to O_2 ↑
6.	***Delivery of O_2:***	
	Delivery power of O_2 ↑ (unloading)	Delivery power of O_2↓ decreased
7.	***Concentration:***	Present in foetal life
	At birth	Disappears after one year
	Hb-A_1-85%	
	Hb-F-15%	
	Hb-F disappears by end of first year	
	Persistence of Hb-F after one year is pathological	
	In adult	
	Hb-A_1-90 to 95%	

3. **Hb-A_2**

   A minor component of normal adult Hb, present usually to the extent of 2.5 per cent. Electrophoretically it is a slowly migrating fraction. It contains two α-chains and two δ-chains and thus it is $\boldsymbol{\alpha_2\,\delta_2}$.

4. **Embryonic Hb**

   Another form which is found in first three months of intrauterine life of the baby. It contains two α-chains and two ε-chains and thus it is $\boldsymbol{\alpha_2\,\varepsilon_2}$. Embryonic Hbs occur in human erythrocytes in first three months of embryonic life. They include **Gower I** and **Gower II** Hbs with the quaternary structures of $\xi_2\,\varepsilon_2$, $\xi_2\,\gamma_2$ respectively.

5. **Hb-A_3**

   An electrophoretically fast fraction also appears, amounting to between 3 and 10 per cent of the total. This is designated as A_3 and it ***appears to be an altered form of*** Hb-A, **found chiefly in old red cells.**

6. **Hb-A_{1C} (Glycosylated Hb)**

   In addition to Hb-A_1 the major form of normal adult Hb, a minor glycosylated form is also found in adult red blood cells. ***In normal individuals, it is present in concentration of 3 to 5 per cent of total Hb. However, in patients with diabetes mellitus it may be increased to as much as 6 to 15 per cent of total Hb.*** This minor glycosylated Hb is designated as Hb A_{1C}.

Chemistry: Amino acid sequence of Hb-A_{1C} is exactly the same as that of Hb-A_1.

- The chemical difference consists of attachment of 1-amino-1-deoxyfructose to the $-NH_2$ terminal of valine of the β-chain of Hb-A_1.
- ***Addition of sugar moiety to valine occurs non-enzymatically,*** either by addition of glucose directly to the protein or by formation by G-6-P of an adult with Hb-A_1 (Hb-A_{1b}) which is subsequently dephosphorylated to form **Hb-A_{1C}**.

CLINICAL IMPORTANCE

The level of glycosylated Hb appears to be an ***index of the levels of blood sugar for a period of several weeks prior to the time of sampling. Once the RB cells get glycosylated, it remains for the life span of the RB cells.*** Its measurement would be a more reliable indicator of the adequacy of control of the diabetic state as compared to occasional measurement of blood and urine glucose.

Genetic Control of Chain Synthesis

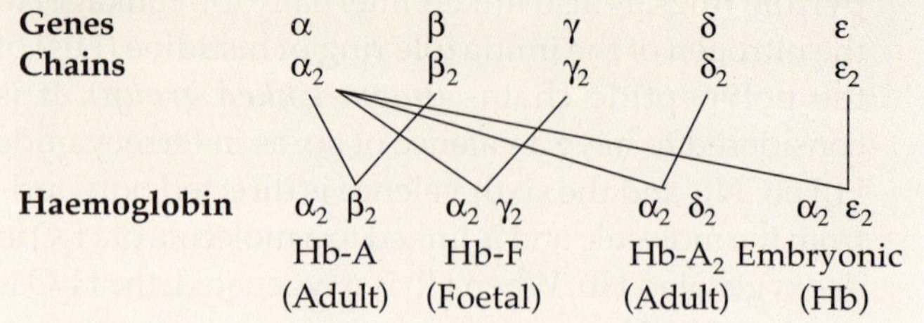

Soon after the foetus begins to develop, there is a rapid production of α and ε-chains. The α-chains persists, whereas the ε-chains disappear and a new polypeptide, the γ-chains make the appearance. During the later stage of foetal growth, and after birth, the β-chains increase in quantity as the γ-chains decrease correspondingly. The pattern is shown graphically in Figure .2.

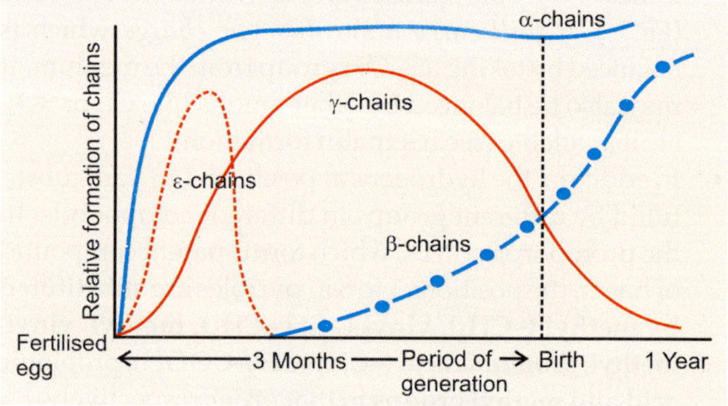

Fig. 11.2: Chains synthesis

Properties

- ***Crystallisable protein:*** Each species has its own crystalline form, which is related to variations in the amino acids of the globin part.
- ***Molecular wt:*** Approximately 65,000 as determined by osmotic pressure and ultracentrifugation measurements.
- ***Shape:*** The overall shape resembles a spheroid, having a length of 64Å, a width of 55Å and a height of 50Å
- ***Acid-base properties:*** (See chapter on Acid-base balance and imbalance).

DERIVATIVES OF HAEMOGLOBIN

1. **Action of Acids and Alkalies**

   Acids and alkalies act upon Hb to form the acid haematin and alkali haematin. The essential differences and similarities between the two are tabulated—refer box next page.

2. **Haemin**

   It is chemically haematin hydrochloride. It is prepared by boiling oxy-Hb with NaCl and glacial acetic acid. **It is a ferric compound**. Haemin crystallises in characteristic brown crystals, which can be seen under the microscope.

3. **Haemochromogen**

Haem and the ferrous porphyrin complexes react readily with basic substances such as hydrazines, primary amines, pyridines, or an imidazole such as the amino acid histidine, resulting compound is called a **haemochromogen (haemochrome).**

On spectroscopy, this compound shows:

(a) *Soret band:* A sharp absorption band near 400 mμ.

Note: It is a distinguishing feature of porphin ring and is characteristic of all porphyrins.

(b) In addition it shows two additional absorption bands:
   (i) *α-band:* Narrow band at 559 mμ, and
   (ii) *β-band:* Broader band at 527 mμ. Both bands are in green part of the spectrum nearer to 'D' line.

4. **Haematoporphyrin**

It is a iron-free derivative. It can exist in two forms acid and alkaline. The derivative is prepared by mixing blood with sulphonic acid.

CLINICAL SIGNIFICANCE

Normal urine may contain trace amounts. It is found in the blood and urine in ***sulphonal poisoning*** and in certain cases of ***liver diseases.***

5. **Haematoidin**

This compound is produced by the breakdown of Hb in the body. It is found as yellowish-red crystals in the region of old blood extravasation. Some authors believe that it is identical with bilirubin.

6. **Methaemoglobin**

It is a derivative in which ***Fe is in the ferric state and it is a true oxidation product of Hb.*** Increased amount of methaemoglobin in blood above normal is called as *"methaemoglobinaemia"*.

Causes: Methaemoglobin can be produced as follows:

- ***In vitro*** by treating blood with potassium ferricyanide
- ***In vivo*** it is produced by certain **oxidant drugs** or exposure to certain poisons, e.g. chlorates, acetanilid, nitrites, nitrobenzene, antipyrin, phenacetin, sulphonal, and perhaps most important, the sulphonamide drugs.

CLINICAL IMPORTANCE

It is interesting to note that earlier treatments like use of $AgNO_3$ solution in burn patients may produce the formation of methaemoglobin and toxic methaemoglobinaemia. Mechanism involves conversion of nitrate anion to "nitrite" by skin bacteria, which in turn converts Hb to methaemoglobin.

- *Industrial hazard*
  - Nitrobenzene is used in manufacture of shoe dyes, floor polishes, cosmetics and explosives. Workers in these industries may be acutely or chronically poisoned if nitrobenzene is absorbed in sufficient amounts.
  - Fumes from carbon arcs contain nitrous oxide which reacts with atmospheric O_2 to form nitrogen dioxide. If this gas is breathed in high concentration methaemoglobin may be produced.
- **Familial methaemoglobinaemia**
  - An ***inherited disorder*** due to lack or absence of the enzyme *methaemoglobin reductase*, which is responsible for conversion of methaemoglobin to normal Hb (Fe^{++}). In absence of the enzyme, methaemoglobin accumulates.
- **Hb-M:** Methaemoglobinaemia may also be found in individuals with abnormal haemoglobin as Hb-M.

Mechanism of Methaemoglobin Formation

1. In methaemoglobin formation, **iron is oxidised to ferric state.**
   - As such it cannot bind O_2
   - The additional +ve charge per haem molecule is balanced with negative group, presumably –OH gr.

SECTION TWO

Differences Between Acid and Alkaline Haematin

Acid Haematin	*Alkaline Haematin*
• Dilute HCl/Other acids split Hb into haem and globin, haem is **"ferrohaem"**	• Alkalies also split Hb into globin and **"ferrohaem"**
• In presence of O_2 haem is quickly oxidised to **"ferri-haem"**	• In presence of O_2 **"ferrohaem"** is converted to **"ferri-haem"**
• Additional +ve charge is balanced by Cl^- ion in case of HCl and forms **'acid haematin'** which is chemically, **"Ferri-haem chloride"**	• Additional +ve charge is balanced by –OH ion in case of NaOH or KOH and forms **alkaline haematin**, which is chemically ***"ferri-haem hydroxide"***
Hb + HCl → globin + ferrohaem 2 ferro haem + 1/2 O_2 + 2 HCl → **2 ferri-haem chloride (acid haematin)**	Hb + NaOH → globin + ferrohaem 2 ferrohaem + ½ O_2 + –OH ions → **2 ferri-haem hydroxide (alkali haematin)**
• Absorption Bands By spectroscopy: It shows a thinner band at 650 mμ between C and D line, but nearer to C line.	• By spectroscopy: It shows a broader band at 600 mμ between C and D line, but near to D line.

2. During methaemoglobin formation, ***oxidation of Hb, mol O_2 is changed to superoxide radical O'_2 by univalent reduction.*** The superoxide is immediately destroyed by Cu-Zn containing enzyme ***superoxide dismutase***. This enzyme helps to transfer electron from superoxide radical to H^+ to produce H_2O_2. H_2O_2 is then either broken down to O_2 and H_2O by ***catalase*** or, it is reduced to H_2O by Se-containing enzyme ***glutathione peroxidase*** and reduced glutathione (G-S H).

Metabolism of Superoxide Radical

Reactions involved in metabolism of superoxide radical is shown below in the box.

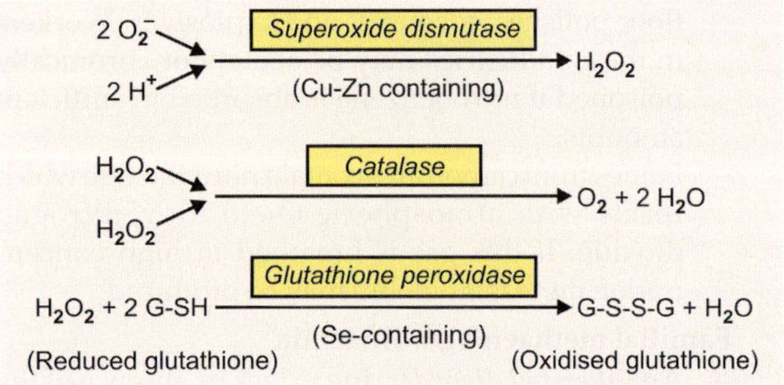

Mechanism of reconversion of methaemoglobin to Hb in normal health: In normal healthy adult, small amount of methaemoglobin may be present approx 0.3 gm/100 ml (about 1.7% of total Hb). This is converted to normal Hb by an enzyme called ***methaemoglobin reductase*** which requires NADH as coenzyme (Refer box below). An additional enzyme ***diaphorase I,*** has been found out, which is NADPH-dependant, can also perform the same function. **Glutathione** and **ascorbic acid**, reducing substances present in significant amounts in erythrocytes, may also be involved in reduction of Met-Hb.

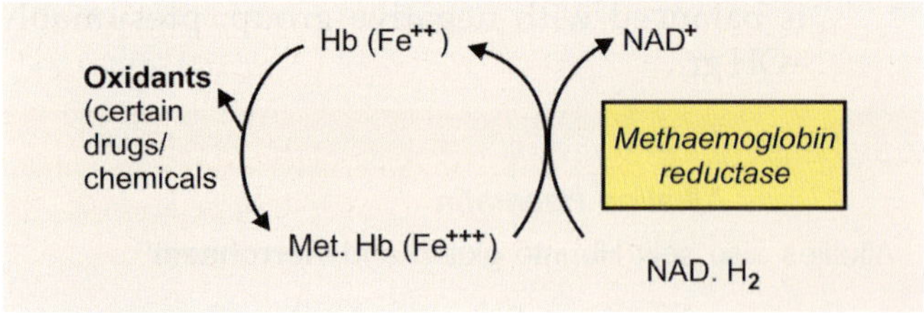

Treatment

Principle: Injection of intravenous **glucose** or **methylene blue,** which helps to reduce Met-Hb (Fe^{+++}) to Hb (Fe^{++}), so that Hb is available again for O_2 binding and transport. Methylene blue activates NADH or NADPH dependant ***methaemoglobin reductase/diaphorase I.*** Administration of ascorbic acid also helps in reduction.

Toxic Effects of Met-Hb

Concentration in blood	*Clinical features*
10 to 20%	Mild cyanosis
20 to 40%	Visible cyanosis, fatigue and dyspnoea with activity
40 to 60%	Produces severe cyanosis, severe cardiopulmonary symptoms, tachycardia, tachypnoea, and depression
> 60%	Causes ataxia, severe cyanosis and dyspnoea, loss of consciousness and death.

Absorption spectra: By spectroscopy methaemoglobin gives characteristic absorption spectra.

1. Dilute neutral Met-Hb gives **four** absorption bands:
   - (i) One **broad band** at 490 mμ in green part nearer to 'F' line
   - (ii) A **narrow band** at 540 mμ in green part
   - (iii) Another **narrow band** at 575 mμ in yellow part of the spectrum
   - (iv) A band at 634 mμ at red part of the spectrum. ***This band is the characteristic one and is used for detection of Met-Hb.***
2. Alkaline Met-Hb gives only **three** bands: Narrow band at 490 mμ is missing.

7. Methaemalbumin

Methaemalbumin is formed by combination of free haematin (ferri-haem) with albumin. ***Normally methaemalbumin is not present in adult blood.*** It may be present at birth and can be detected in umbilical cord blood.

CLINICAL IMPORTANCE

Detection of methaemalbumin is an evidence of IV haemolysis as may occur in mismatched (incompatible) blood transfusion. Methaemalbumin can be detected by a very sensitive test called Schumm's test.

Schumm's Test

- A volume of plasma is covered with layer of ether
- Add 1/10th volume or slightly more of conc. ammonium sulphide (add from the side), mixed by shaking
- The aqueous layer is taken out and examined spectroscopically.

Inference: If methaemalbumin is present an absorption band is seen at 558 mμ.

COMBINATION OF HAEMOGLOBIN WITH GASES

1. **Oxyhaemoglobin**

Combination with Oxygen (Oxy-Hb): The physiological importance lies in the fact that Hb can readily combine with O_2. The combination is ***loose*** and ***reversible.*** The gas is taken up readily at high partial pressures (e.g. in the lungs) and is released as readily at low O_2 pressures (e.g. in tissues), **thus providing an effective system for transport of O_2 from the atmosphere to the cells of the body.** ***At O_2 tensions of 100 mmHg or more, Hb is virtually 100 per cent saturated, approx. 1.34 ml of O_2 is then combined with each gram of Hb.***

For an average Hb concentration of 14.5 gms%, total O_2 which would be carried as oxy-Hb would be $14.5 \times 1.34 = 19.45$ ml%. To this an amount of 0.393 ml of physically dissolved O_2 is to be added taking it to 20 volume%.

Deoxygenated Hb can loosely combine with O_2 forming Oxy-Hb, the attachment with O_2 occurs with Fe in the haem portion. ***Fe remains in the* ferrous *state both in deoxygenated Hb and oxy-Hb.*** O_2 remains attached with the unpaired electrons of Fe.

Each haem can bind only one mol. of O_2. Since each molecule of Hb contains 4 mols. of haem; hence **one mol. of Hb can maximally combine with four mols of O_2.**

Factors: The combination is ***loose*** and ***reversible*** and governed by the following factors:

- Partial pressure of O_2 (Po_2) favours oxygenation
- Partial pressure of CO_2 (Pco_2) favours dissociation
- pH of the medium-acidosis favours liberations of O_2.

Haem-haem Interactions: Combination of O_2 with one haem situated on α-chain, which can permit a molecule of O_2 to enter, ***brings about conformational changes,*** so that O_2 can enter into other haem groups situated on β-chains, ***the valine residue is removed*** permitting O_2 to enter. The increase in O_2 affinity by haem group on β-subunit, after combination of O_2 with α-subunit is called **"haem-haem interactions".**

Haem-linked groups: The "affinity" of Hb for O_2 decreases with decrease in pH, in the physiological range. The converse effect occurs with increase in pH. On the other hand, oxygenation of Hb results in liberation of H^+ ions into the medium. Thus **oxy-Hb is a stronger acid than deoxygenated Hb**. The interrelation of oxygenation and ionization suggests the presence in Hb of ionizable group located in sufficient proximity to the haem ring to be influenced by the state of oxygenation. These have been identified as ***"imidazole" rings of histidine* residues, No-87 on α-subunit and No-92 on β-subunit.**

Two states of Hb: Haemoglobin can occur in two states:

(i) ***'R' state:*** This is a relaxed state and it is oxygenated Hb.

(ii) ***'T' state:*** This is a tense/taut state and it is de-oxygenated Hb.

These two forms are interconvertible and each form has its own equilibrium constant (K_R & K_T) for the binding of O_2. De-oxygenated 'T' form is stabilised by salt bridges and BPG in a central pocket.

Salt Bridges

These are **defined as ionic bonds between +vely charged nitrogen atoms and –vely charged O_2 atoms.** On oxygenation of 'haem' of α-subunit which can easily admit one mole of O_2, the following conformational changes take place in the Hb molecule **("T" form → 'R' form):**

- Rotation of one pair of rigid subunits α_2 β_2 through 15° along the long axis, to the other rigid pair of subunit α_1 β_1 occurs.
- As a result of rotation, salt bridges are broken.
- BPG cannot be held in central pocket as it cannot form salt bridges.
- Valine residue in haem pocket of β-chain is removed and this can now admit a molecule of O_2 (Table 11.2).

Table 11.2: Differentiation of 'T' form and 'R' form

'T' form (Taut form) or (Tense form)	*'R' form (Relaxed form)*
1. **Deoxygenated Hb**	1. **Oxygenated Hb**
2. The rigid subunit $\alpha_1\beta_1$ and $\alpha_2\beta_2$ are close to the long axis	2. Rotation of 15° along the axis
3. Salt bridges are numerous and plenty and intact	3. Due to rotation the salt bridges are broken
4. Valine residue covers the haem pocket of β-subunit and does not allow entry of O_2	4. Valine residue in haem pocket of β-subunit is removed and O_2 can enter
5. BPG can bind and helps in retaining the salt bridges	5. BPG cannot bind as the salt bridges are broken
6. Fe^{++} is 0.07 nm out of the plane of porphyrin ring	6. Fe^{++} comes in plane of porphyrin ring
7. **Affinity of O_2 is low**	7. **Affinity to O_2 becomes high** (Haem-haem interaction)
8. β-chain Histidine residues are protonated (H^+ added)	8. Histidines of β-chains release protons (H^+)

Absorption spectrum of Oxy-Hb: On spectroscopy, oxy-Hb shows two characteristic bands.

- *α-band:* **Narrow-band** in yellow portion of the spectrum at 597 mμ nearer to D-Line
- *β-band:* **Wider broad-band**, nearer to E-Line in green part of spectrum at 542 mμ.

2. Carboxy-Hb: Combination with CO

CO **combines with haem portion** of Hb to form carbon monoxide Hb (also called as carboxy-Hb or carbonyl Hb).

Characteristics

- It is a **much firmer combination**, as compared to oxy-Hb and ***not reversible***
- ***Affinity of Hb to CO is 210 times more than O_2***
- Lethal action is due to ***inhibition of cytochrome oxidase*** of electron transport chain and thus stops cellular respiration.

Poisoning by CO is a common danger of modern life. Carbon monoxide is particularly dangerous as:

- It is **colourless** and **odourless** and hence cannot readily be detected if present in atmosphere, and
- Its action is insidious and rapid. The victim frequently becomes unconscious in a few minutes and death often follows quickly.

Causes

- Incomplete combustion of carbonaceous materials
- Automobile exhaust gas (4–7%)
- Chimney gases and smoke
- Found as a constituent of illuminating gas (derived from coal or oil) in which its presence varies from 4 to 40 percent.

An individual inhaling CO from above mentioned sources can become a victim.

Clinical features: Depends on concentration of carboxy-Hb in blood (% of Hb saturated with CO).

Toxic Effects of Carboxy-Hb

% of Hb saturated with CO	Symptoms
0 to 10%	None
10 to 30%	Possibly slight headache, throbbing in temples.
30 to 50%	Severe headache, weakness and dizziness, nausea and vomiting, dim vision and possibly collapse.
50 to 60%	Above + unconsciousness, coma with intermittent convulsions, Cheyne-Stokes respiration.
60 to 80%	Above + depressed heart action and respiration, respiratory failure and death.

Tests to Identify Carboxy-Hb

- *Physical examination of blood:* It shows a **cherry-red colour.**
- *Chemical test: Dilution test:* To dilute the suspected blood, after treating it with a little NaOH and compare the colour with normal blood similarly treated. Normal blood shows a greenish hue after such treatment ***whereas blood containing carboxy-Hb remains pink.***
- *Absorption spectra:* These resemble like oxy-Hb, but the two bands are slightly nearer to violet end of the spectrum. **On treatment of blood sample with sodium hydrosulphite, there is no change of absorption bands (two bands persist) in case of carboxy-Hb** while the two bands of oxy-Hb are changed to one broad β-band (α-band disappears) as Oxy-Hb is reduced to deoxy-Hb.

Figure 11.3 gives absorption spectra of different derivatives.

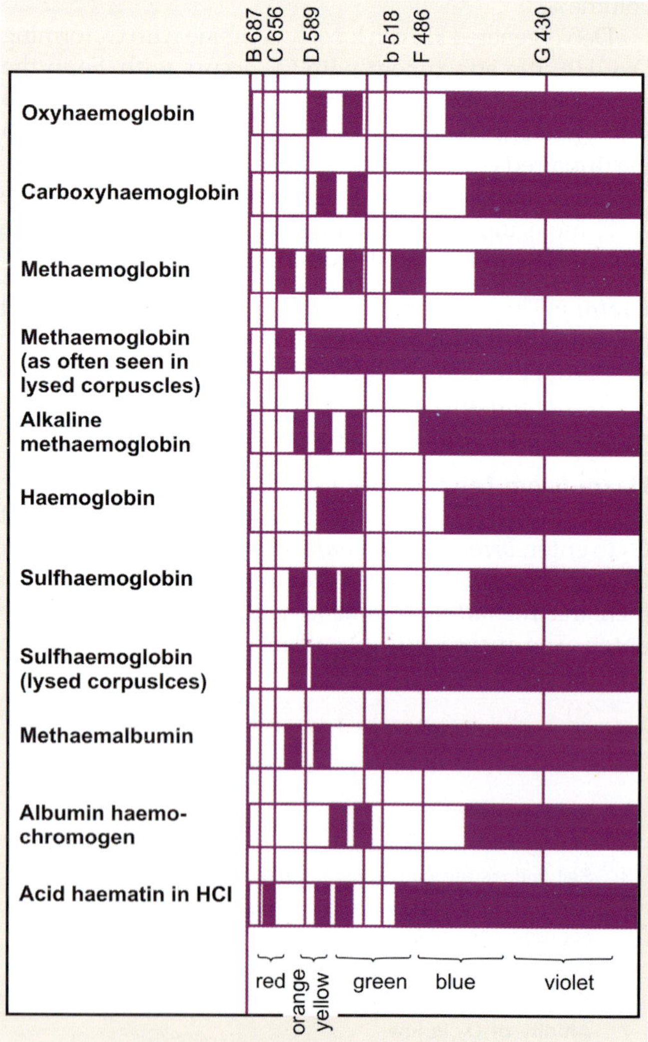

Fig. 11.3: Absorption spectra of different HB-derivatives

3. **Combination with CO_2**

A different type of combination is that of Hb with CO_2 to form *carbaminohaemoglobin*. In this case, the combination is with the globin rather than with the haem. CO_2 combines with NH_2 group.

- $Hb.\ NH_2 + CO_2 \rightleftharpoons Hb.\ NH\ COOH$
$$Hb.\ NH\ COOH \rightleftharpoons Hb.\ NH\ COO^- + H^+$$

It is ***reversible process.*** This is a normal and constant physiologic reaction and accounts for 2 to 10 per cent of CO_2 transported by the blood (see CO_2 transport).

4. **Sulfhaemoglobin**

This is a greenish pigment of uncertain structure **formed by the action of H_2S on oxy-Hb.** It is a stable compound.

Causes
- After administration of aromatic amines or sulphur,
- In severe constipation, and
- In certain types of bacteraemia.

The formation of sulf-Hb involves the production of H_2S in the intestine by bacterial action on proteins. H_2S such formed is normally excreted or oxidised after absorption. Sulfhaemoglobin results from the presence in the blood either: (a) excessive quantities of H_2S; or (b) compounds like aromatic amines which catalyse the formation of sulf-Hb from the normal traces of H_2S in the body.

5. **Action of HCN and Cyanides**

HCN and cyanides ***do not react directly with haemoglobin but they react with methaemoglobin to form cyanmethaemoglobin, which is not toxic.*** The lethal and toxic action of cyanides resides in the fact, that it **inhibits cytochrome oxidase a_3** of electron transport chain and stops cellular respiration.

CLINICAL ASPECT

Biochemical Basis of Treatment of Cyanide Poisoning

1. To enhance formation of met-Hb in the body so that the cyanides combine with met-Hb and spares cytochrome oxidase a_3. ***This is achieved by administering sodium nitrite intravenously to the patient.*** It induces the production of methaemoglobin which quickly combines with cyanide to **produce cyanmethaemoglobin which is non-toxic** and slowly converted subsequently to Hb and cyanate which is non-toxic, is excreted.
2. To administer a drug which will combine with cyanides and thus spares cytochrome oxidase a_3. ***This is achieved by giving Sodium thiosulphate.***

IV *Sodium thiosulphate* reacts with cyanides, **yielding thiocyanate**, an inocuous substance, which is readily excreted out.

ABNORMAL HAEMOGLOBINS AND HAEMOGLOBINOPATHIES

In addition to normal adult Hb and other varieties discussed earlier, more than 30 abnormal types have been described.

SECTION TWO

ABNORMAL HAEMOGLOBINS

The occurrence of these abnormal haemoglobins which are best differentiated by their characteristic electrophoretic mobilities, has given rise to the **concept of "molecular disease"**. These abnormalities are genetically transmitted and are each due to a **single mutant gene**.

- *Types of Abnormal Hb:* They are of **two types:**

1. If the mutation affects ***structural gene***, it results in replacement of a single amino acid residue of Hb-A_1 by some other amino acid resulting into abnormal Hb.

   Examples: Hb-S, Hb-M, Hb-C, Hb-D (Panjab) and others.
2. If the mutation affects the ***regulator gene***, which affect the rate of synthesis of the peptide chains, the amino-acid sequence remains unaffected. This produces ***thalassaemias.*** Depending on the chains affected, it can be ***(a) α-chain thalassaemias and (b) β-chain thalassaemias.***

Detection of abnormal haemoglobins: Usually detected by "fingerprinting technique" or "Hybridisation".

Effects of abnormal haemoglobins: The presence of abnormal Hb in the blood is often associated with the following:
- Abnormalities in red cell morphology
- Definite clinical manifestation like haemolytic anaemia and/or jaundice and other features pertaining to the property of that abnormal Hb.

If the genetic defect is **heterogygous** *the patient will have so-called* **trait** *(e.g. sickle-cell trait)* and may be free from clinical manifestations although the presence of abnormality can be detected electrophoretically/ biochemically by certain tests. On the other hand, *if the defect is "homozygous" the patient will have full-blown disease (e.g. sickle cell disease).*

HAEMOGLOBINOPATHIES

1. *Hb-S*

In Hb-S, both the α-chains have the same amino acid sequence as those of normal Hb-A, but ***in both β-chains glutamic acid in 6th position is replaced by valine.*** Thus its formula will be—$\alpha_2\,\beta_2^{A6val}$.

Note

One single amino acid substitution valine for glutamic acid at 6th position is the only chemical difference between

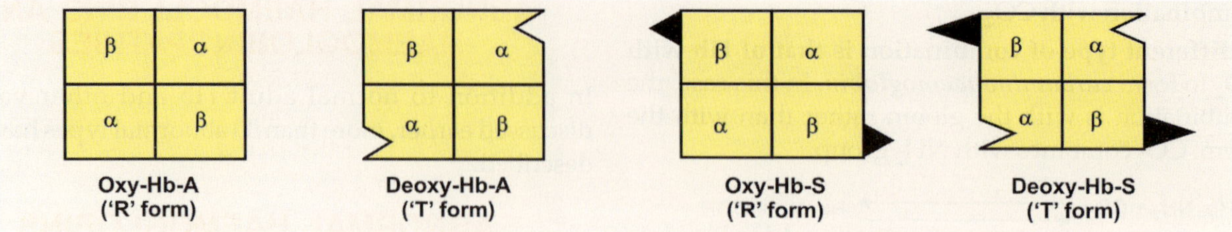

Fig. 11.4: Diagrammatic representation of 'Sticky Patch'

Hb-A and Hb-S among the total of 574 amino acids, and that is responsible for such a dreadful disease.

Solubility: The solubility of Hb-S in the oxygenated state is not very different from that of Hb-A; but in the deoxygenated state, ***deoxygenated Hb-S is only 2 per cent as soluble as deoxygenated Hb-A and about 1 per cent as soluble as its own oxy-form.*** It is likely that the loss of the glutamate residue with its two carboxyl groups alters the distribution of +ve and –ve charges on the protein surface and thus changes its solubility.

Biochemical basis of sickling of RB cells in sickle cell disease: Replacement of a "non-polar" residue valine in β-chain to "polar residue" glutamic acid generates the formation of ***sticky patch*** on the outside of the β-chains. The sticky patch develops on the oxygenated ("R" form) of Hb-S and deoxygenated ('T' form) of HbS, ***but never present on oxy-HbA ("R" form)*** (Fig. 11.4).

On the surface of deoxygenated Hb-S ('T' form), there exists a ***complementary site*** to the sticky patch, but in the oxygenated Hb-S ('R' form) the complementary site is masked and not present.

When Hb-S is deoxygenated, the sticky patch can bind to the complementary patch or site on another deoxygenated Hb-S molecule. This binding causes ***polymerisation of deoxy-Hb-S, forming long fibrous precipitates*** these extend throughout the RB cells and mechanically distort it, causing lysis and multiple secondary clinical effects.

Thus, if Hb-S can be maintained in an oxygenated state, or if the concentration of deoxygenated Hb-S can be minimised, then formation of these polymers will not occur and "sickling" can be prevented. ***Because it is the 'T' form of Hb-S that polymerises, low oxygen tension exacerbates sickling of RB cells.***

Although the deoxy Hb-A contain the receptor sites for the sticky patch present on oxygenated or deoxygenated Hb-S, the binding of sticky Hb-S to deoxy-Hb-A cannot extend the polymer, since the latter (deoxy-Hb-A) does not have a sticky patch to promote binding to another Hb molecule. Hence, ***the binding of deoxy Hb-A to either the 'R' or 'T' form of Hb-S will terminate the polymerisation.***

Nature of polymer: The polymer forms a twisted helical fibre whose cross-section contains 14 Hb-S molecules. These tubular fibres distort the RB cells so that they take the shape of a **sickle** and are vulnerable to lysis as they penetrate the interstices of the splenic sinusoids.

Sickle-cell anaemia/and *Trait*: A hereditary haemolytic anaemia characterised by presence of Hb-S.

In homozygous: Full blown picture of sickle cell anaemia seen. They have 80 to 100 per cent Hb-S and 0 to 20 per cent Hb-A.

In hyterozygous: Manifests *trait* without clinical manifestation. The patients usually have 20 to 40 per cent Hb-S and 60 to 80 per cent Hb-A. Some patients may inherit more than one haemoglobinopathy and produce not only Hb-S but Hb-C, Hb-D or some other abnormal Hb as well.

The **genetic defect is virtually limited to Negroes**. In USA, 8 to 11 per cent of Negroes carry the hereditary abnormality, only one in 40 of these is homozygous. In native populations in Africa, incidence of genetic abnormality may be 45 per cent.

Effects of Sickle Erythrocytes

- ***Sickle cells are more fragile*** due to their shape and predisposes to haemolysis leading to **haemolytic anaemia** and jaundice.
- Packing of these abnormal red cells into vessels, specially capillaries of organs, cause vascular stasis and anoxic damage to the tissues, giving rise to clinical manifestations of the particular organs affected.

Note: "Sickle cells" can be demonstrated *in vitro* from patient's blood.

Relation with Other Diseases: Clinical Aspect

- ***Protection from malaria:*** Persons suffering from sickle cell anaemia/trait **show an increased resistance to malaria.** It has been shown that world distribution of sickle cell anaemia closely parallel the distribution of malarial mosquitoes.
- ***Increased incidence to salmonella infections:*** On the other hand, incidence of salmonella infections has been found to be more in sickle cell anaemia.

ABNORMAL HAEMOGLOBINS WHICH PRODUCE METHAEMOGLOBINEMIA

2. *Hb-M*

In this one amino acid sequence is altered either in α or β-chains. Thus there are different types of Hb-M.

- ***Hb-M (Iwate):*** Histidine of position 87 of α-chain is replaced by Tyrosine.

The formula:

$\alpha_2^{87\ Tyr}\ \beta_2^A$

Other varieties:

$\alpha_2^{58\ Tyr}\ \beta_2^A$

$\alpha_2^A\ \beta_2^{63\ Tyr}$

The O_2-affinity of Hb-M (Iwate) is much lower than that of normal Hb-A. In all of the Hb-M variants, the haem Fe in the Hb of the variant chains is **spontaneously oxidised to the ferric (Fe^{+++}) state, thus producing methaemoglobin.**

3. *Hb-Sabine*

In this, there is substitution in β-chain of ***proline*** for ***Leucine*** at position 91. Formula will be: $\alpha_2^A\ \beta_2^{91Proline}$.

This makes possible for formation of Methaemoglobin. Globin moiety not attached to haem gets precipitated in the erythrocytes and is visible as ***Heinz bodies***, which are inclusion bodies. Attachment of the inclusion bodies to the cell membrane alters the permeability of the cell, as a result osmotic damage (haemolysis) occurs.

ABNORMAL HAEMOGLOBINS ASSOCIATED WITH HIGH O_2-AFFINITY

4. *Hb-Chesapeake*

In this, there is substitution of ***Arginine*** for ***Lysine*** in 92 position of α-chains. Thus, formula is $\alpha_2^{92Arg}\ \beta_2^A$. The change brings about the stabilisation of 'R' form and thus has a high O_2-affinity with decreased release of O_2 to the tissues, resulting to **tissue hypoxia** and compensatory **polycythaemia**.

5. *Hb-Rainier*

Substitution of ***tyrosine*** at 145 is replaced by ***Histidine,*** which modifies the salt bridges of the deoxy form, and this favours stabilisation of 'R' form. Hence O_2 affinity more with decreased release of O_2 to the tissues, which results in **tissue hypoxia** and compensatory **polycythaemia**.

ABNORMAL Hb WHICH INTERFERES WITH m-RNA FORMATION

6. *Hb-Constant Spring*

It produces an unstable m-RNA. In this the **UAA** stop-codon has mutated to a **CAA** which codes for ***Glutamine.*** Hence α-chains are 31 residues longer than normal (i.e., in place of 141 amino acids, it has 172 amino acids). The longer m-RNA is unstable and gets degraded readily.

Other abnormal haemoglobins: Some of the other abnormal haemoglobins are Hb-C, Hb-D (Panjab), Hb-E, etc. Substitution of amino acid and molecular formulae are given below in the box.

Hb-C is confined amongst Negroes in West Africa. All of them bring about change in red cell morphology, affect solubility and may produce mild haemolytic anaemia. Hb-C and Hb-D can occur with Hb-S.

THALASSAEMIAS

Impairment of synthesis of a single kind of Hb-chain result in ***thalassaemias,*** so called because the conditions are more common in Mediterranean countries ("thalasa" = derived from Greek word meaning "sea"). The α-chain thalassaemia is also prevalent in SE Asia, with an occurrence of 1 in 100 in Thailand. They occur due to *mutation of* **Regulator gene.**

1. α-chain Thalassaemias

Synthesis of α-chains are repressed and there occurs a compensatory increase in synthesis of other chains of which the cell is capable either β-chains or γ-chains.

(a) ***Hb-H (β_4):*** In this there is ***inhibition of α-chain synthesis and rate of β-chain production increases.*** The excess of

Other abnormal Hbs

Type	Chain affected	Change in amino acid sequence	Formula
• **Hb-C**	β-chain	Glutamic acid at 6 replaced by lysine	$\alpha_2^A\ \beta_2^{6\,Lys}$
• **Hb-D (Panjab)**	β-chain	Glutamic acid at position 121 is replaced by Glutamine	$\alpha_2^A\ \beta_2^{121\,Gln}$
• **Hb-E**	β-chain	Glutamic acid at 26 is replaced by lysine	$\alpha_2^A\ \beta_2^{26\,Lys}$

β-chains form a large intracellular pool and presumably aggregate to *form β_4 molecules (tetramer) and called Hb-H.*

HB-H disease is characterised by
- Moderate degree of ***haemolytic anaemia***
- Red cell morphological appearance of thalassaemia (see β-chain thalassaemia)
- Variable amount of Hb-H (usually 10–20%).

Inclusion bodies: After incubation of blood sample with brilliant cresyl blue and then drawing a smear on the slide, when seen under microscope, numerous ***inclusion bodies*** can be seen which represent denatured Hb-H.

(b) *Hb-Barts (γ_4):* In this the gross **defect is in repression of α-chain synthesis resulting in great excess of γ-chains, which aggregate to γ_4 molecules (tetramer).**

Clinical Significance

Pregnant ladies suffering from this haemoglobinopathy delivers stillborn infants with clinical picture of ***Hydrops foetalis.*** This disorder is found in SE Asia. 80 per cent of stillborn infants are due to Hb-Barts.

(c) *Hb-Portland ($\varepsilon_2 \gamma_2$):* In this ε chains and γ-chains are produced in excess and can form $\varepsilon_2 \gamma_2$. Foetuses survive for some time by making increased amounts of embryonic Hb, but they commonly die before term or shortly after delivery ***(Hydrops foetalis).***

2. **β-chain Thalassaemia (Thalassaemia Major)**

When the thalassaemia gene represses β-chain synthesis, an excess of α-chains occur which can combine with δ-chains producing **an increase in Hb-A_2** or with γ-chains producing **an increase in Hb-F.** This is the group of thalassaemias associated with **severe anaemias of infancy or early childhood** which was first described by **Cooley,** called ***Cooley's anaemia.*** Infants suffering from this disease have **Mongoloid** features and have stunted growth. They suffer from severe haemolytic anaemia.

On examination: Marked pallor is seen; icterus is variable; enlargement of spleen (splenomegaly) is found.

Blood: Hb very low may be 3 to 5 Gm%; hypochromic microcytic anaemia; osmotic fragility increased.

Blood smear: Shows hypochromasia, polychromasia, basophilic stippling, Target cells ++, nucleated cells +.

Radiological examination: A lateral view of bones of skull shows ***hair-on-end*** appearance, a characteristic radiological finding.

Biochemically: Serum Fe level is normal. **Rise in Hb-F-** 5 to 80 per cent. **Hb-A_2 is significantly increased.**

Note: β-thalassaemia is allelic with Hb-S or Hb-C, so that by interaction with the S or C gene there results a mixture of haemoglobins wherein S or C may comprise as much as 80 to 90 per cent of total Hb.

I. Questions (Essay type and short notes)

1. Write in detail the structure of Haemoglobin.
2. Describe mechanisms of CO_2 transport in blood and the role of Hb.
3. Give a brief account of chemistry of Hb and its properties.
4. What are different abnormal Haemoglobin and how they differ from the normal haemoglobins?

Short Notes

a. Abnormal Haemoglobins
b. Functions of Hb
c. Role of Hb as a buffering agent
d. Methaemoglobin

Differentiate

a. Carboxy-Hb and Carbamino-Hb
b. Oxy-Hb and Carboxy-Hb
c. Hb-A and Hb-F
d. Oxy-Hb and Methaemoglobin

II. MCQs (Give one correct answer)

1. Which is not a part of normal adult haemoglobin molecule?
   a. Vinyl group b. Ferric iron
   c. Propionyl group d. Pyrrole rings
   e. Methylidene bridges
2. All of the following are associated with formation of methaemoglobinaemia, except:
   a. Primaquin b. Acetanilid
   c. Hb-M d. Methylene blue
   e. Deficiency of methaemoglobin reductase
3. In sickle-cell Hb the position 6 of glutamyl residue of β-chain of normal adult Hb is replaced by:
   a. Lysine b. Tyrosine
   c. Valine d. Leucine
   e. Glutamine
4. Which of the following statements about O_2 binding and release by Hb is correct?
   a. O_2 binding of any one of the four haemgroups occurs independently of the other three
   b. Lowering the Ph accelerates the release of O_2 from Oxy-Hb
   c. A high concentration of 2:3 BPG in the erythrocytes the binding of O_2 by Hb
   d. On binding O_2 the iron of the haem is oxidised to ferric state
   e. None of the above statement is correct
5. The most useful diagnostic test for CO poisoning is:
   a. Estimation of CO percentage in the environment
   b. Estimation of CO in expired air
   c. Spectroscopy of blood of patient
   d. Measurement of respiratory rate
   e. Estimation of glucose in urine

6. The protective effect against malaria is related to which of the following?
   a. Hb-F b. Hb-H
   c. Bart's Hb d. Hb-S
   e. None othe above
7. One gram of Hb can carry how much oxygen?
   a. 13.4 ml b. 1.34 ml
   c. 11.34 ml d. 1.84 ml
   e. 3.45 ml
8. Sickle cell anaemia is characterised by all of the following, except:
   a. Presence of Hb-S
   b. Hyperbilrubinaemia
   c. Normocytic red blood cells
   d. Increased osmotic fragility
   e. Sickling of erythrocytes *in vitro*
9. Carbon monoxide inhibits mitochondrial electron transport by:
   a. Inhibiting the electron transfer of NADH-Q reductase
   b. Binding to Hb in RB cells and so blocking the transport of O_2 to tissues
   c. Binding to the O_2 binding site of cytochrome oxidase
   d. Blocking electron transport at the level of the cyt. b. $\rightarrow$ cyt.c_1
   e. Combining with coenzy-Q
10. Sickle cell anaemia is the clinical manifestation of homozygous genes for an abnormal Hb molecule. The mutational event responsible for mutation in the β-chain is:
    a. Deletion b. Insertion
    c. Point mutation d. Crossing over
    e. Non-dysjunction
11. The type of polypeptide chain present in globin of Hb-A_2 are
    a. Two α-chains and two β-chains
    b. Two α-chains and two γ-chains
    c. Two β-chains and two γ-chains
    d. Two α-chains and two δ-chains
    e. Two β-chains and two δ-chains
12. The incidence of *salmonella* infections is most often increased in patients with:
    a. Sickle cell anaemia
    b. β-chain thalassaemia
    c. G-6 PD deficiency of RB cells
    d. Hereditary spherocytosis
    e. Hexokinase deficiency of RB cells
13. Which of the following statement is correct?
    a. In Oxy-Hb formation Fe in haem is converted to ferric form
    b. CO_2 combines with haem to form carboxy-Hb
    c. Co combines with NH_2 group of Hb to form carbaminocompound
    d. H_2S combines with oxy-Hb to form sulfhaemoglobin
    e. Fe in haem remains in ferrous form in methaemoglobin formation
14. In methaemoglobin formation, the iron in haem is in:
    a. Ferric form
    b. Ferrous form
    c. Both ferric and ferrous form
    d. All of the above
    e. None of the above
15. The iron in haem is linked to the globin through:
    a. Lysine b. Arginine
    c. Valine d. Histidine
    e. Glycine
16. Match the following:
    a. Ferroprotoporphyrin b. Protoporphyrin IX
    c. Chlorophyll a d. Porphin
    e. Ferriprotoporphyrin
    1. Hemin ____________
    2. A magnesium porphyrin derivative in which one pyrrole ring is partially reduced ________
    3. Haem ____________
    4. Derivatives are most important porphyrins in nature __________
    5. Basic structure from which Haemoglobin, cytochromes and chlorophyll are derived __________
17. The methylidene bridges of Fe porphyrin ring of haem molecule of Hb arises from:
    a. The α-carbon of succinyl-CoA
    b. The carboxyl carbon of serine
    c. The β-carbon of serine
    d. The α-carbon of serine
    e. The carbon from CO_2
18. Which of the following proteins associated with O_2 is not a haem-protein?
    a. Peroxidase b. Superoxide dismutase
    c. Catalase d. Cytochrome oxidase
    e. Tryptophan pyrrolase
19. Cyanide is poisonous and lethal due to the fact, that it:
    a. Inhibits myoglobin of muscle
    b. Combines with Rb cells membrane
    c. Forms a complex with Hb molecule
    d. Inhibits cytochrome oxidase of electron transport chain
    e. Damages the brain cells
20. Match the following:
    A. βs-chains B. γ-chains C. δ-chains D. ε-chains.
    1. Haemoglobin A_2 ______________
    2. Foetal Hb ______________
    3. Hemoglobin Gower ________________
    4. Sickle cell Hb ______________
21. Methaemoglobin contains:
    a. A denatured globin
    b. A ferroprotoporphyrin
    c. A ferriprotoporphyrin
    d. A porphyrin with reduced methyne bridges
    e. None of the above

Answers to MCQs

1. b	2. d	3. c	4. b	5. c	6. d
7. b	8. d	9. c	10. c	11. d	12. a
13. d	14. a	15. d	16. 1-E, 2-C, 3-A, 4-B, 5-D		
17. d	18. b	19. d	20. 1-C, 2-B, 3-D, 4-A		
21. c					

CHAPTER

12

VITAMINS

Major Concepts

A. Define and classify vitamins. List all vitamins. Study fat soluble vitamins, their chemistry, functions, and deficiency disorders.
B. Study the dietary sources, important metabolites, metabolic functions and deficiency disorders of water soluble vitamins.

Specific Objectives

A. I. 1. Define and classify vitamins. There are four fat soluble vitamins A, D, E and K.
2. Study different forms of vitamin A, dietary sources and daily requirement.
3. Learn the various functions of vitamin A specially the visual cycle and deficiency manifestations.
II. 1. Study different forms of vitamin D and their synthesis specially calcitriol, the active form.
2. Learn the various functions of vitamin D and deficiency diseases.
III. 1. Study different forms of tocopherols (vitamin E), its absorption and transport.
2. Learn the various functions specially antioxidant property of vitamin E and deficiency manifestations.
IV. 1. Study different forms of vitamin K, list the dietary sources and daily requirement.
2. Study various functions of vitamin K specially in coagulation process and its deficiencies.

B. I. 1. There are two water-soluble vitamins—vitamin C and vitamin B complex groups.
2. Study the chemistry, dietary sources and absorption of vitamin C. Learn the various metabolic functions and fate of vitamin C, specially its role in collagen synthesis.
3. Study deficiency manifestations of vitamin C.
II. 1. List all the vitamin B complex constituents.
2. Study the chemistry and "biological active" coenzyme form of each one of them.
3. List the dietary sources, daily requirement of each.
4. Study the metabolic role of each one of them and the deficiency manifestations.
III. Many vitamins manifest certain symptoms in case of excess intake. List and study them.

DEFINITION

Vitamins have been defined as organic compounds occurring in natural foods either as such or as ***utilisable precursors,*** which are required in minute amounts for normal growth, maintenance and reproduction, i.e. for normal nutrition and health.

1. *They differ from other organic food stuffs in that:*
   - They do not enter into tissue structures, unlike proteins.
   - Do not undergo degradation for providing energy unlike carbohydrates and lipids.
   - Several B complex vitamins play an important role as **coenzymes** in several energy transformation reactions in the body.
2. *They differ from hormones:* In not being produced within the organism, and most of them have to be provided in the diet.

The protective substances present in milk were named as **accessory factors** *by Hopkins.* Almost in the same year, **Funk** (1911–12) isolated from rice polishings a crystalline substance which could prevent or cure polyneuritis in pigeons. Chemically it was found to be an ***amine*** and as it was ***vital*** to life, he named it as ***vitamine.*** Further developments in isolation and chemistry of vitamin have shown that only a few are amines. The term ***vitamin*** was retained, omitting the terminal "e" in its spelling.

Classification

All vitamins are broadly divided into ***two groups*** according to solubility.

1. **Fat-soluble Vitamins**
   - **Vitamin A**
   - **Vitamin D**

- **Vitamin E**, and
- **Vitamin K.**

2. **Water-soluble Vitamins**
   (a) **Vitamin C (ascorbic acid),**
   (b) **Vitamin B complex group includes:**
   - **Vitamin B_1 (thiamine)**
   - **Vitamin B_2 (riboflavin)**
   - **Niacin (nicotinic acid)**
   - **Vitamin B_6 (pyridoxine)**
   - **Pantothenic acid**
   - **α-Lipoic acid**
   - **Biotin**
   - **Folic acid group**
   - **Vitamin B_{12} (cyanocobalamine).**

Other water-soluble vitamins included in this group are:

- Inositol
- Para-amino benzoic acid (PABA)
- Choline.

FAT-SOLUBLE VITAMINS

VITAMIN A

Chemistry

In general, the term vitamin A is now used when reference is made to the biological activity of more than one vitamin A active substance. Three important forms of vitamin are shown in the box:

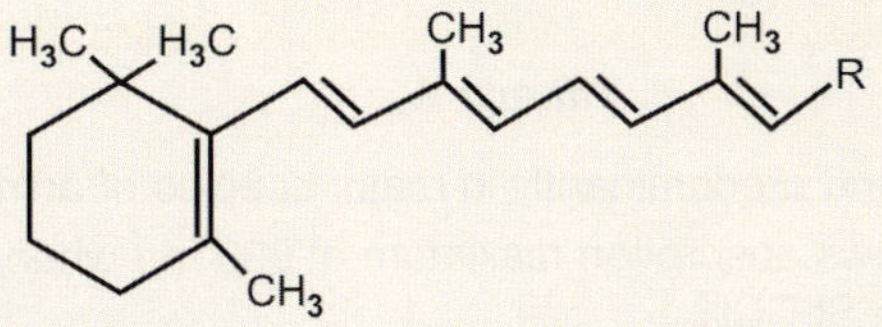

When
R = $-CH_2OH$ Retinol or vitamin A alcohol
R = $-CHO$ Retinal or vitamin A aldehyde
R = $-COOH$ Retinoic acid or vitamin A acid

All three compounds contain as common structural unit:

1. A trimethyl cyclohexenyl ring ***(β-ionone)*** and
2. An all ***trans*** configurated polyene chain, (isoprenoid chain) with four double bonds.
3. They are crystalline substances with limited stability. As already shown above, these three forms are (i) vitamin A alcohol or **retinol**, (ii) vitamin A aldehyde or **retinal** (also called **retinene**) and (iii) vitamin A acid or **retinoic acid**. These forms are sometimes referred to as ***retinoids***. Vitamin A is a derivative of certain carotenoids which are hydrocarbon (polyene) pigments (yellow, red). These are widely distributed in the nature. These are called as ***"Provitamins A"*** and are **α, β and γ-carotenes. Carotenes are $C_{40}H_{56}$ hydrocarbons.** The structure of β-carotene is shown in Figure 12.1. The interconversion of various forms of vitamins are also shown.

   As shown **β-carotene is a symmetrical molecule** containing two characteristic terminal β-ionone rings

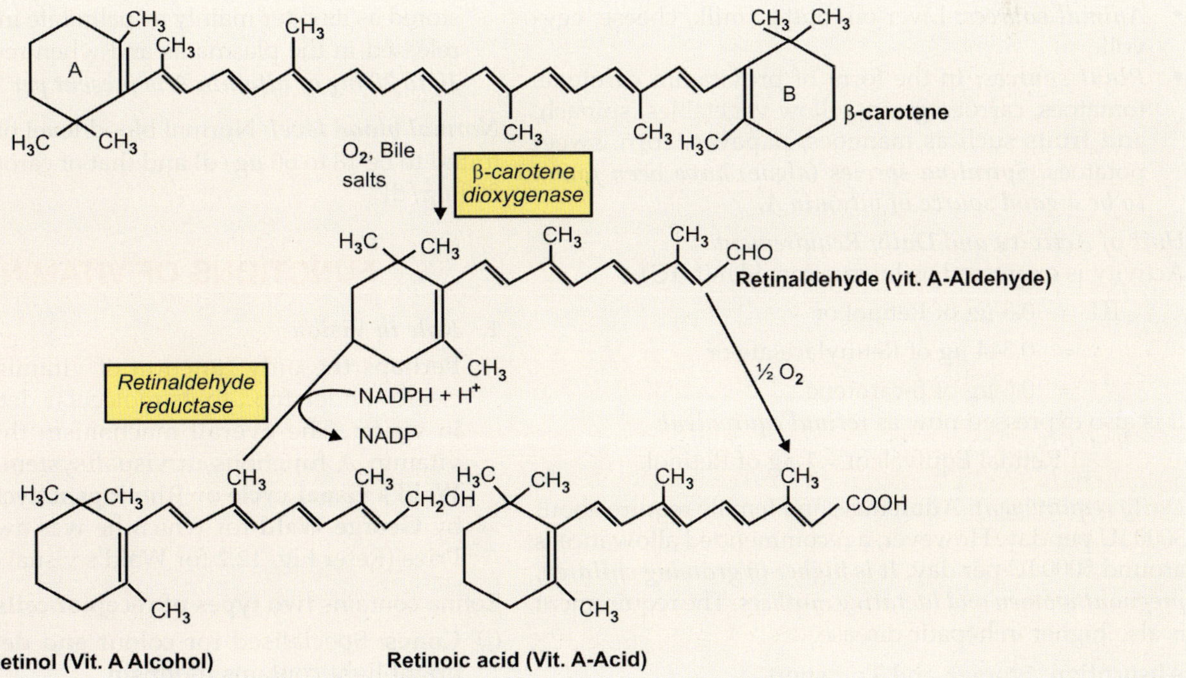

Fig. 12.1: Interconversion of vitamin A metabolites

Differences of vitamin A_1 and vitamin A_2

Vitamin A_1	*Vitamin A_2*
1. Found predominantly in major species of animal.	1. Found in fresh water fish liver and other tissues.
2. Shows absorption maximum at 693 mμ when treated with $SbCCl_3$.	2. Shows absorption maximum at 620 mμ on treatment with $SbCCl_3$.
3. Only one double bond present in β-ionone ring.	3. Two double bonds in β-ionone ring (additional double bond between (C_3-C_4)
4. More potent in its activity than vitamin A_2.	4. Less potent, biological activity is approximately 40% that of A_1.
5. Can be obtained from carotenes.	5. Carotenes cannot give rise to vitamin A_2.

connected by 18-carbon hydrocarbon chain with 2 conjugated double bonds. The three other provitamins A namely α, γ-carotenes and cryptoxanthines differ from β-carotene only in the nature of β ring.

Two mols of vitamin A are formed by symmetrical oxidative scission of β-carotene while only one mole of vitamin A is obtained from α and γ carotenes or cryptoxanthine. Carnivores cannot use carotene.

Forms of vitamin A: Vitamin A occurs in nature in different forms. The usual form vitamin A_1 described above predominates except in fresh water fish, in them another form namely vitamin A_2 is present. Differences between vitamin A_1 and A_2 are shown in a tabular form **(see box above)**.

Neovitamin A: It is a stereoisomer of A_1 and it has about 70 to 80 per cent of the biological activity of vitamin A_1.

Dietary Sources

- *Animal sources:* Liver oil, butter, milk, cheese, egg-yolk.
- *Plant sources:* In the form of provitamin carotene, tomatoes, carrots, green-yellow vegetables, spinach, and fruits such as mangoes, papayas, corn, sweet potatoes. ***Spirulina species (algae) have been found to be a good source of vitamin A.***

Unit of Activity and Daily Requirement

Activity is expressed as International Unit (IU)

IU = 0.3 μg of Retinol or
= 0.344 μg of Retinylacetate or
= 0.6 μg of β-carotene.

It is also expressed now as ***retinal equivalent***

1 Retinal Equivalent = 1 μg of Retinol.

Daily requirement: Adult male and female require about 3000 IU per day. However, a recommended allowance is around 5000 IU per day. ***It is higher in growing children, pregnant women and lactating mothers.*** The requirement is also higher in hepatic disease.

Absorption, Storage and Transport

- Vitamin A and its carotene precursors are absorbed in the small intestine. It is believed that the presence of tocopherols (vitamin E) and other antioxidants protect them against oxidation and destruction in intestinal lumen. Dietary vitamin A is chiefly in ester form which is hydrolysed by ***cholesterol esterase*** into fatty acid and free vitamin A. Free retinol is absorbed and undergoes reesterification in the intestinal epithelial cells. ***It is stored in the liver as retinyl ester, normally as retinyl palmitate.***
- ***Conversion of carotenes to retinol occurs in intestinal epithelial cells*** in few animals, although ***liver also participates in conversion.*** However, ***in man liver is the only organ where carotenes are converted to vitamin A.***
- ***Retinol is transported in the blood in association with a specific retinol binding protein (RBP). There is another Retinoic acid binding protein (RABP) specific for retinoic acid.*** About 95 per cent of vitamin A is stored as its ester mainly as palmitate in the liver. It is released in the plasma as and when required. ***About 10 to 20 mg of vitamin A is present per 100 g of liver.***

Normal blood level: Normal blood level of vitamin A is found to be 18 to 60 μg/dl and that of carotenoids 100 to 300 μg/dl.

FUNCTIONS OF VITAMIN A

1. *Role in Vision*

Perhaps the only function of vitamin A which is clearly understood to its molecular details is its role in vision. The overall machanism through which vitamin A functions in visual system is known as **Wald's visual cycle** or **Rhodopsin cycle** discovered by **George Wald** for which he was awarded Nobel Prize (Refer Fig. 12.2 for Wald's visual cycle).

Retina contains **two types of receptor cells:**

(i) **Cones:** Specialised for colour and detail vision in bright light contains **iodopsin.**

(ii) **Rods:** Specialised for visual activity in dim light (night vision), contains **rhodopsin.**

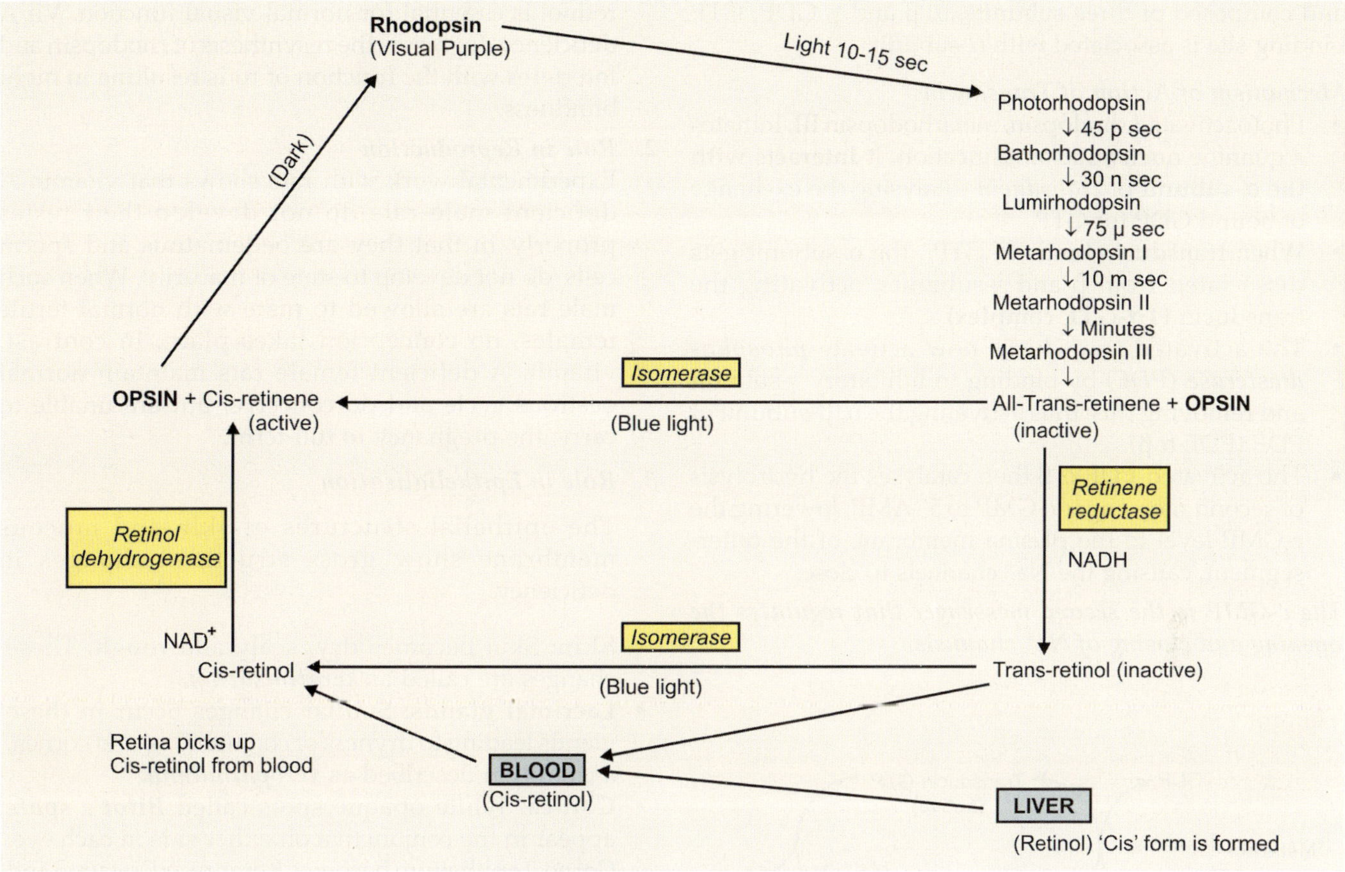

Fig. 12.2: Wald's visual cycle (Rhodopsin cycle)

Light waves striking these receptors produce chemical changes which in turn give rise to nerve impulses. ***Vitamin A plays significant role in the photochemical phase of this process.*** Visual activity of rod cells is dependent on their content of photosensitive pigment called ***rhodopsin*** or ***visual purple*** which is a conjugated protein with a molecular weight of 40,000. It contains ***Opsin*** as its apoprotein and ***retinene*** as its prosthetic group. Retinene or retinal or retinaldehyde present in rhodopsin is ***11-cis-retinal. The aldehyde group of 11-cis-retinal is bound to ε–NH_2 group of lysine of opsin.*** Rhodopsin has a light absorbing property due to polyene group of 11-*cis*-retinal. Even dim light can break rhodopsin.

When the light falls on rhodopsin it is split into ***opsin*** and ***all-trans-retinal*** in a series of events. ***It first forms photorhodopsin, bathorhodopsin, then lumirhodopsin, then metarhodopsin I, II and III. Finally metarhodopsin gets split into opsin and all-trans-retinal.***

At this stage the eye becomes less sensitive to light. ***All-trans-retinal is inactive in synthesis of rhodopsin, it has to be converted to 11-cis-retinal.*** It can take place in the following ways:

- All-*trans*-retinal may be isomerised to its 11-*cis*-isomer in ***presence of blue light—but in the eye this isomerisation is not significant.***
- The all-*trans*-retinal can be converted to all-*trans*-retinol by ***retinene reductase*** by making use of NADH and all-*trans*-retinol then can be isomerised to its *cis* isomer.
- All-*trans*-retinol from blood can be first isomerised to 11-*cis*-retinol. All 11-*cis*-retinol then can be converted to 11-'cis'-retinal by ***retinol dehydrogenase,*** in presence of NAD^+.
- ***Now 11-cis-retinene (retinal) which is active, can combine with opsin to form back rhodopsin in dark.*** Thus the visual process involves continual removal of the active cis-retinol from blood into retina.
- ***Role of cyclic GMP in retinal light-dark adaptation*** (Refer to chapter on Chemistry of Nucleotide).

The closing of Na^+ channels occur by a light-regulated enzymatic reactions in which the original signal-absorption of a single photon by ***rhodopsin*** is amplified manifold. The regulatory protein ***transducin*** is involved in this process. Transducin is a guanine-nucleotide binding protein (G protein), the structure and function of which is similar to those of G-proteins that take part in hormonal signals across biological membranes.

Transducin binds to GDP when it is "inactive", and in "active state" it binds to GTP. Transducin is "trimeric" and composed of three subunits, α, β and γ. GDP/GTP binding site is associated with α-subunit.

Mechanism of Action of Transducin

- Photoactivated rhodopsin, metarhodopsin III, initiates a guanine nucleotide amplification. It **interacts with the α-subunit** of ***transducin*** catalysing the exchange of bound GDP for GTP.
- When transducin binds to GTP, the α-subunit gets dissociated from β and γ-subunits, activating the transducin **(Tα-GTP complex)**.
- The activated transducin now activate ***phosphodiesterase (PDE)*** by binding to inhibitory γ-subunit and removing this; thus activating the α, β-subunit of PDE (PDE α β).
- The activated PDE α β then catalyses the hydrolysis of second messenger c-GMP to 5′-AMP, lowering the c-GMP level to the plasma membrane of the outer-segment, causing the Na^+ channels to close.

The c-GMP is the second messenger that regulates the opening and closing of Na^+ channels.

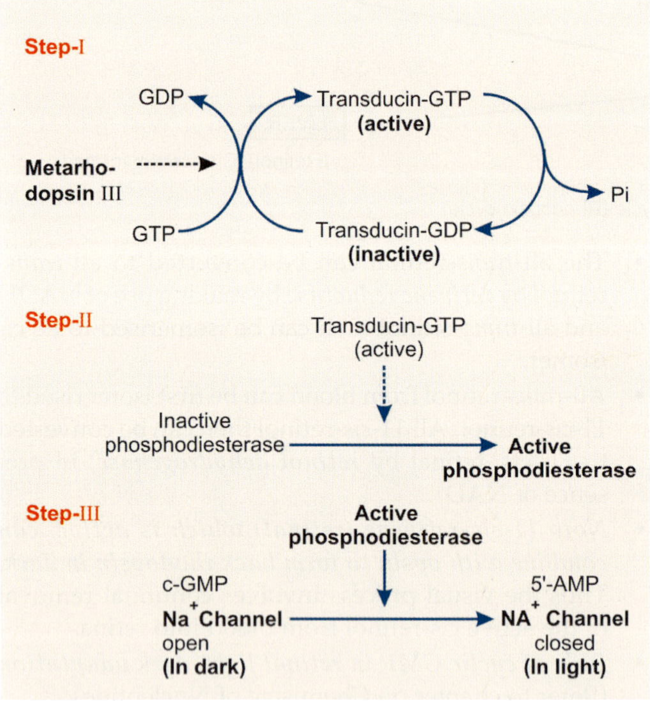

Conclusion

- In the dark, there are high level of c-GMP, which binds to the Na^+ channels, causing them to open.
- In the light, photoactivated rhodopsin, through transducin and phosphodiesterase, *lowers the levels* of c-GMP, thus closing the most of Na^+ channels.

- ***Night blindness (nyctalopia):*** This is ***one of the earliest signals of vitamin A deficiency*** which is impairment of dark adaptation. Therefore continual supply of retinol is essential for normal visual function. Vit A deficiency depresses the resynthesis of rhodopsin and interferes with the function of rods resulting in night blindness.

2. *Role in Reproduction*
   Experimental work with rats shows that vitamin A deficient male rats do not develop their testes properly in that they are oedematous and sperm cells do not develop to state of maturity. When such male rats are allowed to mate with normal fertile females, no conception takes place. In contrast, vitamin A deficient female rats maintain normal oestrous cycle and do conceive, but are unable to carry the pregnancy to full-term.

3. *Role in Epithelialisation*
   The epithelial structures of skin and mucous membrane show gross structural changes in deficiency.

- **Skin:** Skin becomes dry, scaly and rough. These changes are called as ***keratinisation.***
- **Lacrimal glands:** Similar changes occur in these glands leading to dryness of conjunctivae and cornea, a condition described as ***xerophthalmia.***
- **Cornea:** White opaque spots called ***Bitot's spots*** appear in the conjunctiva on either side in each eye. Corneal epithelium becomes Keratinised, opaque and may become softened and ulcerated, condition described as ***keratomalacia.***
- **Respiratory tract:** Keratinisation occurring in the mucous membrane of respiratory tract leads to ***increased susceptibility to infection and lowered resistance to disease.***
- **Urinary tract:** Keratinisation of UT leads to ***calculi 'formation'.***

4. *Role in Bone and Teeth Formation*
   It plays role in construction of normal bone. Deficiency results in slowing of endochondral bone formation and decreased osteoblastic activity, the bone becomes cancellous losing their fine structural details. Mechanical damage to the brain and cord due to arrested limits of bony frame work and cranium and vertebral column in which it has to grow. Teeth become unhealthy due to thinning of enamel and chalky deposits on surface.

5. *Growth*
   Vitamin A alongwith other vitamin is principally involved in growth. Its role in cell differentiation and cell division has been proved beyond doubt.

6. *Metabolism*
   It may be involved in protein synthesis and may play a role in metabolism of DNA.

β-carotene as an Antioxidant and Anticancer

In addition to its antioxidant activity, β-carotene has recently been shown to have anticancer action. β-carotene increases the number of receptors on white blood cells for a molecule known as ***major histocompatibility complex I (MCH I).*** Cancerous cells have different proteins on their surfaces from normal cells. Immune system cells known as killer T-cells or CD 8 cells use such surface proteins to identify foreign invaders or cancer cells. Other immune system cells known as monocytes help direct the CD 8 cells and they use MCH II to do this.

Retinoic Acid: Functions and Therapeutic Uses

Functions

- ***Role in gene expression and tissue differentiation:*** All trans-retinoic acid and 9-cis-retinoic acid are involved. They act on nucleus—binds to specific nuclear receptors that bind to response elements of DNA and regulate the transcription of specific genes.

  Two types of receptors described:
  - **RARs *(Retinoic acid receptors)*** binds all-trans-retinoic acid or 9-cis-retinoic acid.
  - **RXRs *(retinoid X-receptors)*** bind 9-cis-retinoic acid only.
- ***Role in glycoprotein synthesis:*** Retinoic acid is found to be involved in glycoprotein synthesis, probably forming its phosphate which acts as carrier of oligosaccharides chains to glycoprotein molecule.
- ***Role in mucopolysaccharide synthesis:*** Retinoic acid is ***considered to be essential for sulfation of the mucopolysaccharides*** in the matrix of cartilages and bones.
- ***Collagen breakdown:*** Retinoic acid ***inhibits the enzyme "collagenase"*** and thus prevents breakdown of collagen.
- ***Role in keratinisation:*** Retinoic acid prevents keratinisation of epithelial cells. Thus it keeps the mucous membranes healthy and moist.

Therapeutic uses of retinoic acid

- Retinoic acid has been used in ***oral leukoplakia,*** a ***precancerous condition***. It has been claimed to improve the condition by reverting the cells to normal epithelium.
- All trans-retinoic acid has also been found to be useful in treatment of ***promyelocytic leukaemia,*** and brings remission by differentiation of cells.

Effects of excess of vitamin A (hypervitaminosis A): Excess of vitamin A induces series of toxic effects known under the name of the *hypervitaminosis A syndrome.* In man, the main symptoms are (a) alterations of the skin and mucous membrane, (b) hepatic dysfunction and (c) headache, drowsiness, (d) peeling of skin about the mouth and elsewhere. These syndromes were recognised by Eskimos as occurring after eating the livers of polar bears and arctic foxes which are extremely rich in vitamin A.

Chronic effects of continued intake of excessive amounts of vitamin A produces roughening of skin, irritability, coarsening and falling of hair, anorexia, loss of wt. Sometimes seen in children due to excessive feeding of vit. A by parents for prolonged periods.

VITAMIN D

Chemistry

Vitamin D_3 or cholecalciferol occurs in fish liver and also produced in human skin by ultraviolet light. The inactive natural precursors of the vitamin D are the **'provitamins'.** Only two of these have been found in nature.

- **Ergosterol:** ***Provitamin D_2 found in plants.***
- **7-dehydrocholesterol:** ***Provitamin D_3 found in the skin.***

  All the provitamins D possess a certain essential structural characteristics.
- OH group at C_3
- Two conjugated double-bonds between C_5-C_6 and between C_7-C_8.
- A hydrocarbon chain at C_{17}.

Transformation from inactive provitamin to the active vitamin is accomplished by the ***ultraviolet rays***. The photochemical activation, photolysis results only in intramolecular rearrangement (Fig. 12.3).

Dietary Sources

Fish liver oil is the richest source of vitamin D. Egg-yolk, margarine, lard, also contain considerable quantity of vitamin D. Some quantity is also present in butter, cheese, etc.

Ergosterol is widely distributed in plants. It is not absorbed well hence is not of nutritional importance. Calciferol is readily absorbed. 7-dehydrocholesterol is formed from cholesterol in the intestinal mucosa, and principally liver, passed on to the skin where it undergoes activation to vitamin D_3 by the action of solar UV rays.

Daily Requirement

1 USP unit = IU = 0.025 μg of vitamin D_3. About 100 IU or 2.5 μg of vitamin D_3 is the daily requirement in adult man. Pregnant and lactating mother as well as infants and children require about 220 IU per day. Vitamin is easily supplied by cutaneous synthesis in sunlight in tropical countries.

Absorption and Transport

Like most other fat-soluble vitamins, bile salts help in absorption of vitamin D from duodenum and jejunum. After absorption, it is carried in chylomicron droplets of the lymph in combination with serum globulin in blood plasma.

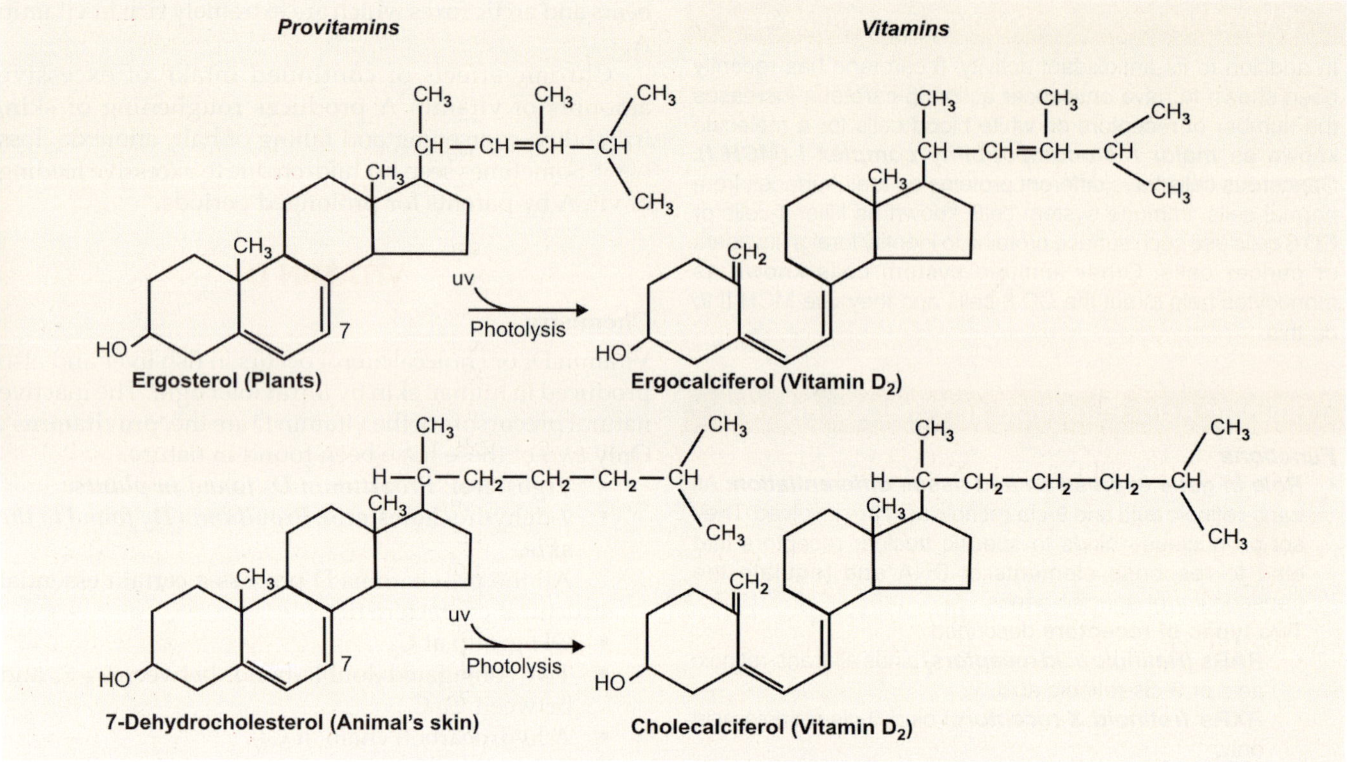

Fig. 12.3: Conversion of provitamin D_2 and D_3 to vitamin D_2 and D_3

Biologically *Active* Form of Vitamin D (Calcitriol)

Formation of Calcitriol

The biologically active form of vitamin D is called **calcitriol,** which is synthesised in **liver** and **kidneys.**

(a) Synthesis of 25-OH-D_3 in Liver (Calcidiol)

- Vitamin D_2 and/or D_3 binds to specific D binding protein and is transported to liver.
- It undergoes hydroxylation at 25 position, by the enzyme ***25-hydroxylase,*** in the endoplasmic reticulum of the mitochondria of liver cells.
- Coenzyme/cofactors required are:
  - **Mg^{++}**
  - **NADPH**, and
  - **Molecular O_2**

A cytoplasmic factor is also required, the exact nature not known. Two enzymes, an NADPH-dependant *cytochrome P_{450} reductase* and *a cytochrome P_{450}* are involved also.

- 25-OH-D_3 (calcidiol) is the ***major storage form of vitamin D in liver*** and found in appreciable amount in circulation. The blood level of 25-OH-D_3 exerts *feedback* inhibition on the enzyme *25-hydroxylase.*

(b) Synthesis of 1, 25-di –OH-D_3 (Calcitriol) in Kidneys

- 25-OH-D_3 is bound to a specific vitamin D binding protein and is carried to kidneys.
- It undergoes hydroxylation at 1-position, by the enzyme ***1 α-hydroxylase,*** in the endoplasmic reticulum of mitochondria of proximal convoluted tubules of kidney.
- The reaction is a complex three component mono-oxygenase reaction requiring Mg^{++}, NADPH and molecular O_2 as coenzymes/cofactors.
- In addition, at least three more enzymes are required. They are:
  - ***Ferrodoxin reductase***
  - **Ferrodoxin**, and
  - **Cytochrome P_{450}**

This system produces 1,25-di-OH-D_3 (calcitriol) which is the most potent metabolite of vitamin D **(Fig. 12.4).**

Note: Renal hydroxylation at position C_1 is the most significant, but recently similar hydroxylation has been found to occur in placenta and bone also.

Regulation: Regulation of calcitriol synthesis is done by:

- Its own concentration—by *feedback* inhibition of *1 α-hydroxylase.*
- Parathyroid hormone (PTH)
- Serum phosphate level.

Hypocalcaemia leads to marked increase in *1 α-hydroxylase* activity, the effect requires PTH. As stated above, calcitriol regulates its own concentration since high

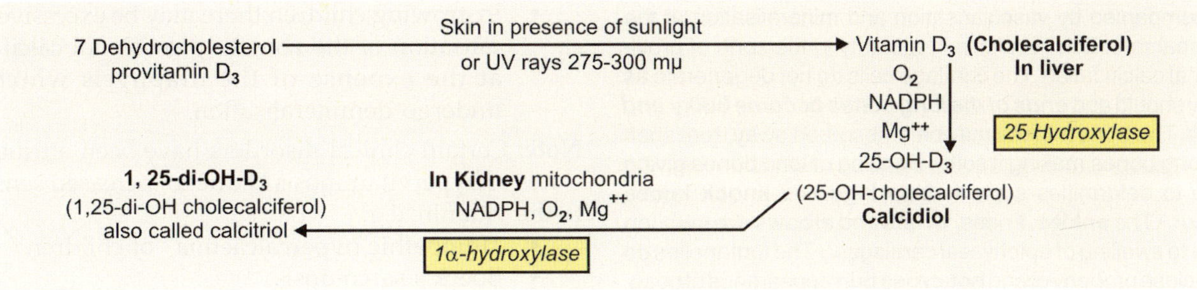

Fig. 12.4: Synthesis of calcitriol

levels of calcitriol inhibit *"1 α-hydroxylase"* and stimulates the formation of 24,25-di-OH-D_3 which is not potent as calcitriol and now supposed to be a storage form.

Mode of action of calcitriol: Calcitriol acts in a similar way as the steroid hormone receptors. This binding is specific and reversible. The receptor has a specific binding site on DNA that appears to contain Zinc-finger motif characteristic of other steroid receptors.

Vitamin D as prohormone: Since calcitriol is synthesised in the body and acts like steroid hormone and has a basic sterol nucleus in its structure, ***it is now regarded*** as a hormone and vitamin D as a prohormone.

Vitamin D is considered as a "Prohormone" and calcitriol (1,25-di-OH-D_3) as a hormone

Points in favour of above statement are:

- Structurally both have *"cyclo-pentanoperhydrophenanthrene"* nucleus like a steroid hormone.
- Vitamin D_3 (cholecalciferol) is synthesised in human skin by UV irradiation from its precursor Provitamin D_3 (7-dehydrocholesterol).
- Vitamin D_3 as such is inactive and is only the storage form. It is converted in liver to 25-OH-D_3 (calcidiol) and biological active 1,25 {$(OH)_2$-D_3 (calcitriol)} in kidney.
- Like hormones, the formation of both the biological active forms 25-OH-D_3 and 1,25-di-OH-D_3 are subject to *"feed-back"* inhibition.
- Like hormones, calcitriol has definite *"target" organs* like small intestine, bones and kidneys to act upon. ***Thus, it is produced in one organ and acts upon distant target organs for its activity (property of hormone).***
- Calcitriol resembles steroid hormone in its mode of action, ***i.e. nuclear action*** (See above).
- Calcitriol maintains calcium homeostasis alongwith—two other protein hormones parathormone (PTH) and calcitonin. Parathormone (PTH) is considered as a "tropic" hormone for calcitriol, it increases the calcitriol production by stimulating the enzyme *"1 α-hydroxylase"* in kidney tubules.

FUNCTIONS OF VITAMIN D

Vitamin D is found to act on target organs like bones, kidneys, intestinal mucosa to regulate calcium and phosphate metabolisms.

- ***Intestinal absorption of calcium and phosphate:*** It binds to the chromatin of target tissue and expresses the genes for calcium binding protein as well as Ca^{++} ATPase in intestinal cells. This increases the Ca^{++} absorption by actively transporting Ca^{++} across the plasma membrane against electrochemical gradients.
- ***Mineralisation of bones:*** Mineralisation of bones is promoted by 1, 25, $(OH)_2D_3$ as well as 24, 25$(OH)_2D_3$. It is believed that the synthesis of Ca^{++}-binding proteins like ***osteocalcin*** and ***alkaline phosphatase*** is promoted which increases calcium and phosphate ions in the bone. These ions enhance the mineral deposition in the bone. 24, 25$(OH)_2D_3$ helps the deposition of hydroxyapatite in bone. Vitamin D is also believed to promote bone resorption and calcium mobilization to raise the levels of Ca and P in blood in association with PTH.

Other Functions

- Renal reabsorption of calcium and phosphorus is also done by 1,25$(OH)_2D_3$ in similar way.
- It lowers the pH in certain parts of the gut such as colon and produces increase in urinary pH.
- It counteracts the inhibitory effect of calcium ions on the hydrolysis of phytate. In adequate amounts and in case of high calcium intake, it suppresses the anticalcifying and rachitogenic effect of phytate.
- In physiologically compatible intake it is found to increase the citrate content of bone, blood, tissues and urinary level.

Deficiency of Vitamin D—Clinical Aspect

1. Rickets

Deficiency produces rickets in growing children and osteomalacia in adults. Vitamin D is required for the normal growth and mineralisation of bones. In its absence, instead of growth occurring normally, the osteoblasts proliferation does not take place in an orderly fashion and is not accompanied by vascularisation and mineralisation at the normal rate. This results in irregularity in the zone of provisional calcification. The cartilage cells do not degenerate as they should and ***ends of the long bones become bulky and soft.*** The bone mineral may be reabsorbed away from shaft of long bones making it soft. • Bending of long bones giving rise to deformities such as **bow legs** and **knock knees** occur. • The **ankles, knees, wrists and elbows are swollen** due to swelling of epiphyseal cartilages. • The fontanelles do not close properly giving **hot-cross bun** appearance of head. • Ribs give beaded appearance and chest gives a **pigeon breast** appearance. Teeth erupt late and are deformed.

Types: There are two types of vitamin D deficiency rickets.

(i) **Type I:** It is inherited as autosomal recessive trait characterised by defect in conversion of 25-OH-D_3 to calcitriol, i.e. 1-α-*Hydroxylase* deficiency.

(ii) **Type II:** It is an autosomal recessive disorder in which there is a single amino acid change in one of the zinc fingers of the DNA binding sites for receptors. This makes the receptors nonfunctional.

2. **Osteomalacia**

The deficiency of vitamin D in adults is osteomalacia which is rare. It can occur:

- ***In pregnancy and lactation:*** Where there is additional requirement of this vitamin and drainage of it in milk.
- ***In women who observe purdah*** or in climates where sunshine is scanty, calcium and phosphorus absorption is decreased. Consequently mineralisation of osteoid to form bone is impaired. Such bones become soft. This particularly affects pelvic bones.

3. **Renal Osteodystrophy**

When renal parenchyma is lost or diseased quite significantly, it is unable to form calcitriol and calcium absorption is impaired. Hypocalcemia leads to increase in PTH which acts on bone to increase Ca^{++}. Consequently there is excessive bone turnover and structural changes. This condition is known as renal osteodystrophy.

HYPERVITAMINOSIS D

Normally vit D is well tolerated if taken in large doses but serious deleterious effects may be produced if taken in extremely large doses, 500 to 1000 times of normal requirement for prolonged periods.

Effects are mainly due to induced hypercalcaemia.

- **Immediate effects** and
- **Delayed effects.**

1. ***Immediate effects:*** Include anorexia, thirst, lassitude, constipation and polyuria. Followed later on by nausea, vomiting and diarrhoea.
2. ***Delayed effects:*** Persistent hypercalcaemia and hyperphosphataemia may produce:
   - *Urinary lithiasis*
   - *Metastatic calcification* which may affect kidneys, bronchi, pulmonary alveoli, muscles, arteries and gastric mucosa. Renal failure may develop and can lead to death.
   - In growing children there may be excessive mineralisation of the zone of provisional calcification at the expense of the diaphysis which may undergo demineralisation.

Note: Certain clinical disorders have been attributed to states of hypervitaminosis D due to increased sensitivity to the vitamin, e.g.

- "Idiopathic hypercalcaemia" of children
- Boeck's sarcoidosis.

CLINICAL ASPECTS

Recently low vitamin D level has been correlated with certain diseases:

- **Heart ailments:** The researchers reported that those individuals whose vitamin D levels were below a certain level (15 ng/ml) had double the risk of developing heart problems in the form of either heart attack/or strokes/or heart failure. It has been surmised from ten year follow-up study that ***vitamin D deficiency is a far better indicator of future heart attack than cholesterol. Most organs and tissues of human body have vitamin D receptors.*** Such receptors are also present in the smooth muscles of the heart and the blood vessels. The presence of vitamin D causes these smooth muscles through vitamin D receptors to relax and this in turn prevents heart attacks/strokes. This is also the reason for the ***recently discovered role of vitamin D in alleviating hypertension.***
- **Diabetes:** Amongst the many recent studies highlighting several new benefits of vitamin D there are some that establish its role in preventing diabetes. This is not a surprising revelation as the ***ubiquitous vitamin D receptors are located in the β-cells of Islet of Langerhans of pancreas.*** The first evidence of the vit D and diabetes link came when it was noticed, mostly through population studies that ***insulin dependent Type-1 diabetes (IDDM-1) occurred commonly in areas which receive less sunlight.*** Observational studies seem to show a link between deficient vit D status of infants and the risk of developing diabetes about 10 to 15 years later. It has ***been proposed that vitamin D probably stimulates not only the release of insulin but also the expression of insulin receptors.*** Researchers from Harvard Medical School have shown that low vitamin D levels may be a particular problem for children and teenagers with Type-1 diabetes.
- **Breast cancer:** Recently a link has been suggested between the occurrence of breast cancer in women and vitamin D deficiency. One study suggests that maintaining optimal vitamin D levels may actually prevent certain types of cancers particularly that of breast cancer. ***Breast cancer cells have vit D receptors. When these receptors are activated with vit D, it brings about some molecular reactions that can make the cancer cells slow down their growth rate or quite often even die or atleast become less aggressive.***
- **Peripheral artery disease (PAD):** Vitamin D may protect against an arterial disease in which fatty deposits restrict blood flow to the limbs. Scientists at the Albert Einstein

College of Medicine, New York, found that ***people with low levels of vitamin D experience an increase risk of peripheral artery disease (PAD),*** most often reduces blood flow to the legs causing pains, numbness, impairing the ability to work and in some cases leading to amputation.

VITAMIN E (TOCOPHEROLS)

Chemistry: The tocopherols differ from each other in the number or position of methyl groups.

- **α-tocopherol:** 5, 7, 8 trimethyl tocol
- **β-tocopherol:** 5, 8 dimethyl tocol
- **γ-tocopherol:** 7, 8 dimethyl tocol
- **δ-tocopherol:** 8 methyl tocol

The α-tocopherol is the most active in vitamin E activity.

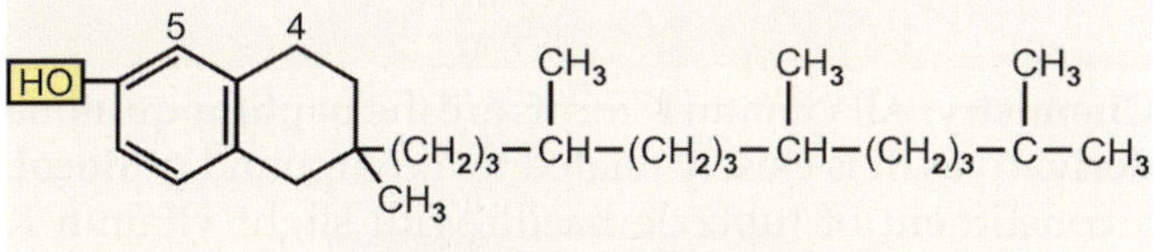

Structure of vitamin E

The *presence of the phenolic –OH group on 6th carbon of the chromane ring is the most important group for its antioxidant activity.*

Dietary Sources and Recommended Allowance

Cottonseed oil, corn oil, sunflower oil, wheat germ oil and margarine are the richest sources of vitamin E. It is also found in fair quantities in dry soyabeans, cabbage, yeast, lettuce, apple seeds, peanuts.

Units : 1 mg of d-α-tocopherol = 1.49 IU
1 mg of dl-α-tocopherol acetate = 1.0 IU

Normal blood level = 1.2 mg/dl.

Recommended Allowance

- **Children:** 10–15 IU/day
- **Adults:** 20–25 IU/day

Special attention has to be given to the dietary intake of unsaturated fatty acids in which case the daily requirement is increased. Requirement is also more in pregnancy and lactation.

Absorption, Distribution and Excretion: Free tocopherols and their esters are readily absorbed in small intestine with the help of bile acids. Absorbed vitamin E is transported to liver where it gets incorporated into lipoproteins and carried by blood to muscle tissues and to adipose tissue for storage. ***The normal value of blood level is around 1 mg/dl and it is transported chiefly in the α-lipoprotein fraction.*** Under normal dietary conditions, there is no significant excretion of tocopherols in urine or faeces as it rapidly and extensively undergoes destruction in the GI tract and in tissues. Placental transfer of vitamin E is limited; mammary transfer is much more extensive. Thus, ***the serum α-tocopherol level of breast-fed infants increases more rapidly than that of bottle-fed infants.***

FUNCTIONS OF VITAMIN E

1. *Antioxidant Property*
   This is the **most important functional aspect of vitamin E.**

- *Removal of free radicals:* Vitamin E is involved in removal of free radicals and prevents their peroxidative effects on unsaturated lipids of membranes and thus helps maintain the integrity of cell membrane. *Vitamin E prevents peroxidation.* Vit E (α-tocopherol) reacts with the lipid peroxide radicals formed by peroxidation of polyunsaturated fatty acids before they can establish a chain reaction, **acting as free radical trapping antioxidant.**

  The tocopheroxy-free radical (Toc.O•) product, formed in the process, is relatively unreactive and ultimately forms non-radical compounds. Usually the tocopheroxyl radical is reduced back to α-tocopherol again by reaction with vitamin C from plasma or reduced glutathione (G-SH) (Fig. 12.5).

2. *Role in Reproduction in Rats*
   Vitamin E helps in maintaining seminiferous epithelium intact. However, its deficiency leads to

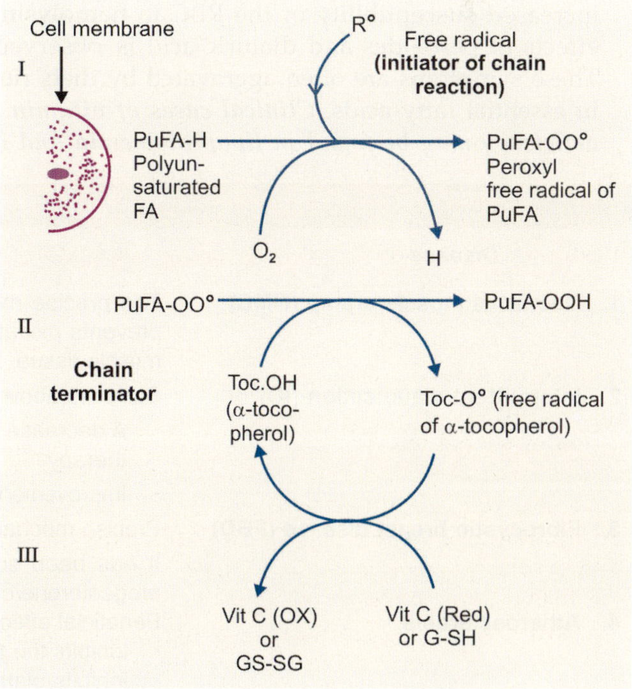

Fig. 12.5: Role of vit E as antioxidant

irreversible degenerative changes leading to permanent sterility. Motility of sperms is lost and spermatogenesis is impaired.

In female rats the ovary is unaffected by vitamine E deficiency, but the foetus does not develop normally, dying *in utero* undergoing resorption.

3. Other Functions

- Tocopherol derivative tocopheranolactone may be involved **in synthesis of coenzyme Q or ubiquinone.**
- Vitamin E may have **some role in nucleic acid synthesis.**

Deficiency of Vitamin E

- *Muscular dystrophy:* Vitamin E deficiency leads to the increased oxidation of polyunsaturated fatty acids in the muscle with a consequent rise in O_2 consumption and peroxide production, peroxides may then cause an increase in intracellular hydrolase activity by affecting the lysosomal membranes. Those hydrolases may then catalyse such breakdowns in muscle and produce muscular dystrophy. The muscle creatine is low and creatinuria occurs.
- *Hemolytic anemia:* Low tocopherol diet may produce low plasma tocopherol, increased susceptibility to hemolysis due to **peroxides** and **dialuric acid**. This could be the reason of ***hemolytic*** or ***macrocytic anemia***. Extensive oedema, reticulocytosis, thrombocytosis and thrombus formation in blood vessels, increased susceptibility of the RBC to hemolysing effects of peroxides and dialuric acid is observed. These symptoms are often aggravated by diets rich in essential fatty acids. ***Clinical cases of vitamin E deficiency may be found in lipoproteinemia and in diseases like sprue, obstructive jaundice, pancreatitis, and steatorrhoea.***
- **Dietary hepatic necrosis:** Diets low in cystine and rich in polyunsaturated fatty acids can cause hepatic necrosis. Fall in acetate utilisation and in respiration of necrotic liver is more effectively cured or prevented by tocopherols. Vitamin E and Factor 3, a selenite compound are complementary to one another in preventing hepatic necrosis or muscular dystrophies (Refer Selenium metabolism).

Clinical and therapeutic uses: Recently, vitamin E has been used in following diseases (Refer box below):

- *Nocturnal muscle cramps (NMC)*
- *Intermittent claudication (IC)*
- *Fibrocystic breast disease (FBD)*
- *Atherosclerosis*

VITAMIN K

Chemistry: All vitamin K forms are the **naphthoquinone derivatives**. It is closely related to a compound **pthiocol,** a constituent of tubercle bacilli with slight vitamin K activity.

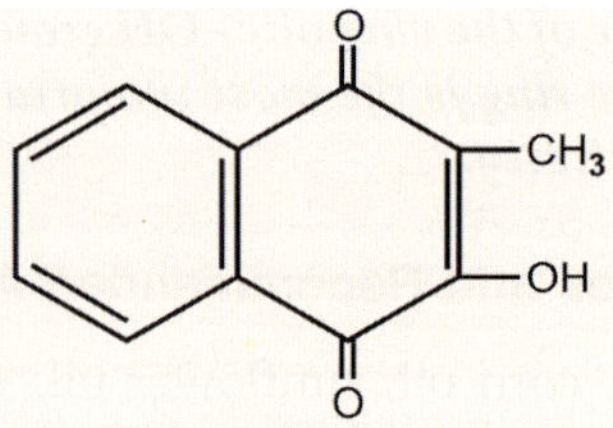

Pthiocol
(2-methyl, 3-hydroxy, 1,4-naphthoquinone)

Vitamins K_1 and K_2 are the two naturally occurring forms of vitamin K that have been identified. The third form vitamin K_3 is the synthetic analogue.

Therapeutic Uses of vitamin E

Disease	*Mechanism of action of vitamin E*
1. **Nocturnal muscle cramp (NMC)**	The precise mechanism not known. By virtue of its antioxidant property, vit. E prevents oxidation of certain radicals and ensures better utilisation of oxygen in muscle tissue, thereby improving muscle metabolism.
2. **Intermittent claudication (IC)**	Same as above. In addition: • A decrease in circulating lactate level and increase in pyruvate level noted after therapy • Improvement in blood supply due to opening of new vessels, improving circulation
3. **Fibrocystic breast disease (FBD)**	Precise mechanism of action in FBD remains obscure It has been suggested that vit. E probably acts by correcting the deranged progesterone/estrogen ratio in women of FBD
4. **Atherosclerosis**	Beneficial effects of vit. E in atherosclerosis are due to: • Inhibits the formation of lipid peroxides and restores PG-I_2 synthesis • Inhibits platelets aggregation • Elevates HDL-cholesterol level ↑ (increased scavenging action).

Types of Vitamin K

1. **Vitamin K_1**

   It is ***phylloquinone*** or phytonadione ***isolated from alfalfa leaves***. Also called ***Mephyton.***

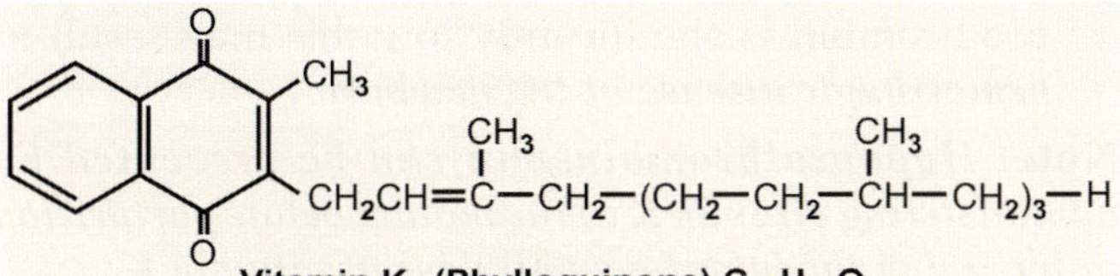

Vitamin K_1 (Phylloquinone) $C_{31}H_{46}O_2$

Thus vitamin K_1 is 2 methyl, 3 phytyl-1,4 naphthoquinone. It is a light yellow oil.

2. **Vitamin K_2**

   Also known as ***farnoquinone,*** it was ***isolated from putrid fish meal*** synthesised by bacteria and has a longer difarnesyl chain attached at position 3. Vitamin K_2 (farnoquinone) is 2 methyl-3-difarnesyl-1,4 naphthoquinone. It is also a yellow oil.

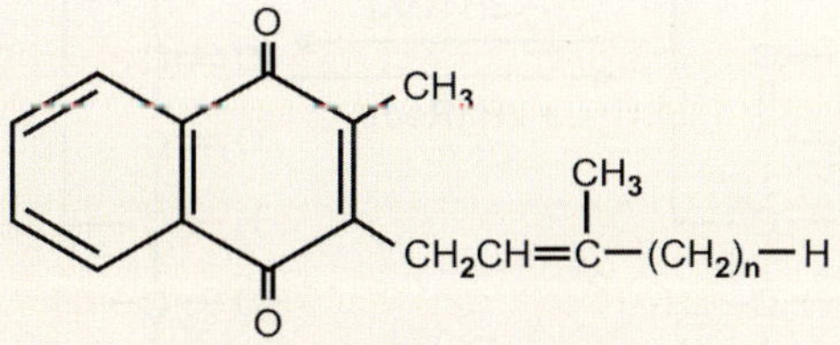

Vitamin K_2 (Menaquinones)

3. **Vitamin K_3**

   Vitamin K_3 is 2 methyl, 1, 4 naphthoquinone without any side chain or OH group (Also known as ***menadione***), is the ***synthetic analogue*** of vitamin K. It is ***three times more potent*** than natural varieties. It is **water-soluble** and can be given parenterally. ***Its activity is related to the presence of methyl group at position 2. Other two forms are Menadiol and Menadioldiacetate.***

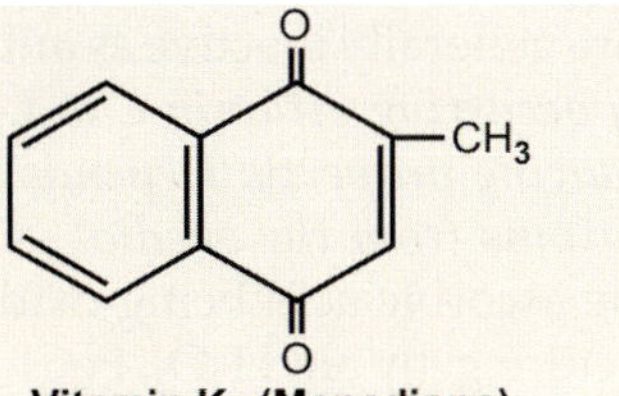

Vitamin K_3 (Menadione)

Dietary sources and daily requirement: Both vitamin K_1 and K_2 are mainly found in plants and synthesised by bacteria respectively. Vitamin K_1 is present chiefly in green leafy vegetables, such as alfalfa, spinach, cauliflower, cabbage, soyabeans, tomatoes.

Vitamin K_2 also called Menaquinones is a product of metabolism of most bacteria ***including the normal intestinal bacteria*** of most higher animal species. Menaquinones (K_2) are absorbed from gut to some extent but it is not clear to what extent they are biologically active as it is possible to induce signs of vit K deficiency simply by feeding a phylloquinone (K_1) deficient diet, without inhibiting intestinal bacterial action.

Absorption and excretion: It is absorbed from the small intestine in presence of bile salts. It is not stored to any appreciable extent. It can cross the placental barrier and is available to the foetus. Vitamin K is not excreted in the urine or bile. Faeces contain large quantities. This may be of bacterial origin. It may also represent actual excretion by the intestinal mucosa.

FUNCTIONS OF VITAMIN K

1. ***Blood Coagulation***

   The main function of vitamin K is the promotion of blood coagulation by helping in the post-transcriptional modifications of blood factors such as prothrombin, and factors II, VII, IX, X. Vitamin K is first converted to its hydroquinone form in liver microsomes by dehydrogenase using NADPH. It then functions as coenzyme for carboxylase. It uses CO_2 to be incorporated as an additional –COOH group at the γ-C of a specific glutamate of these coagulation proteins. ***This converts the glutamate residues into γ-carboxyglutamate.*** Hydroquinone may change to 2, 3 epoxide which is reduced back to quinone by microsomal ***epoxide reductase. Dicumarol is found to inhibit epoxide reductase.***

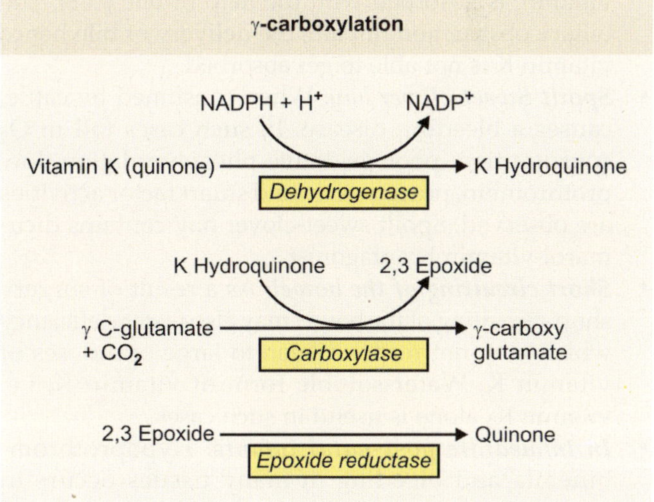

All the first ten glutamate residues of prothrombin are first carboxylated by vitamin K. The γ-carboxyglutamate residues now provide calcium binding sites in the N-terminal portion. This brings together activated factor and accelerin close to the phospholipid membrane of platelets. This enhances blood coagulation manifold.

2. ***Calcium Binding Proteins***
   Vitamin K similarly is found to carboxylate specific glutamate residues of ***calcium binding proteins*** of bones, spleen, placenta and kidneys. This enhances the capacity of these proteins to deposit calcium in the tissues concerned.
3. ***Role in Oxidative Phosphorylation***
   Vitamin K is a necessary cofactor in oxidative phosphorylation being associated with mitochondrial lipids. UV irradiation of isolated mitochondria destroys their vitamin K content and ultimately their ability for oxidative phosphorylation. The normal process of oxidative phosporylation is restored when vitamin K is added to them. Further ***dicumarol, an antagonist of vitamin K, is known to act as uncoupler of oxidative phosphorylation.***

DEFICIENCY OF VITAMIN K

Deficiency of vitamin K is very rare, since most common foods contain this vitamin. In addition, intestinal flora of microorganisms also synthesise vitamin K. However, a deficiency may occur as a result of:

- *Prolonged use of antibiotics and sulfa drugs:* This suppresses the growth of vitamin K_2 producing bacteria thus making vitamin K_2 not available.
- *Malabsorption and biliary tract obstruction:* Sprue, steatorrhoea and coeliac disease can lead sometimes to vitamin K deficiency. Vitamin K being a fat soluble vitamin, is absorbed with the help of bile salts. The biliary obstruction impairs the delivery of bile hence vitamin K is not able to get absorbed.
- *Spoilt Sweet-clover hay:* When consumed by cattle, causes a bleeding disease. In such cases fall in O_2 consumption, poor oxidative phosphorylation, low prothrombin, proconvertin and stuart factor activities are observed. Spoilt sweet-clover hay contains dicumarol-vitamin K antagonist.
- *Short circuiting of the bowel:* As a result of surgery short-circuiting of the bowel may also foster deficiency which may not respond even to large oral doses of vitamin K. Water-soluble form of vitamin K, i.e. vitamin K_3 alone is useful in such cases.
- *In immediate post-natal infants:* Hypoprothrombinemia and bleeding in many tissues occurs in vitamin K deficiency. Relatively small amounts of vitamin K are obtained from the mother through placental membranes and also because the intestinal microflora has not yet been established, this leads to vitamin K deficiency and its consequent effects. If prothrombin is significantly low this may result in ***hemorrhagic disease of the newborn.***

Note: ***Hypoprothrombinemia can be prevented by administering vitamin K to the mother before parturition or by giving the infant a small dose of vitamin K.***

WATER-SOLUBLE VITAMINS

VITAMIN C (ASCORBIC ACID)

Synonyms: Antiscorbutic vitamin.

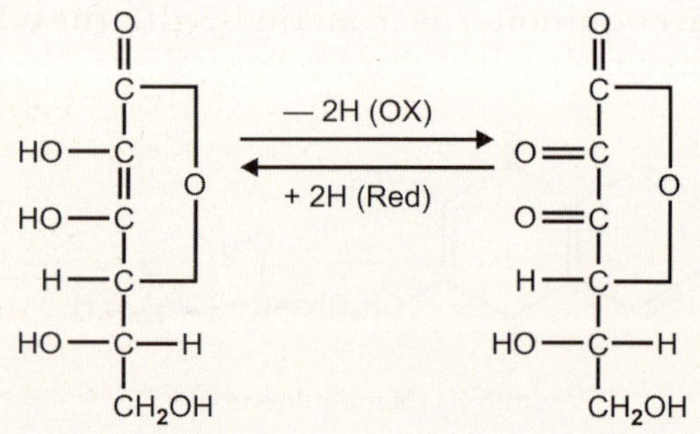

Chemistry

- Ascorbic acid is an ***enediol-lactone*** of an acid with a configuration similar to that of the sugar L-glucose.
- It is a comparatively ***strong acid,*** stronger than acetic acid, owing to dissociation of enolic H at C_2 and C_3.
- D-forms are generally inactive as anti-scorbutic agent. ***Naturally occurring vitamin C is L-Ascorbic acid.***
- ***Strong reducing property: Depends on the liberation of the H-atoms from the enediol –OH groups, on C_2 and C_3;*** the ascorbic acid being oxidised to dehydroascorbic acid, e.g. by air, H_2O_2, $FeCl_3$, methylene blue, ferricyanide, 2:6-dichlorophenol indophenol, etc. The reaction is readily reversible by reducing agents

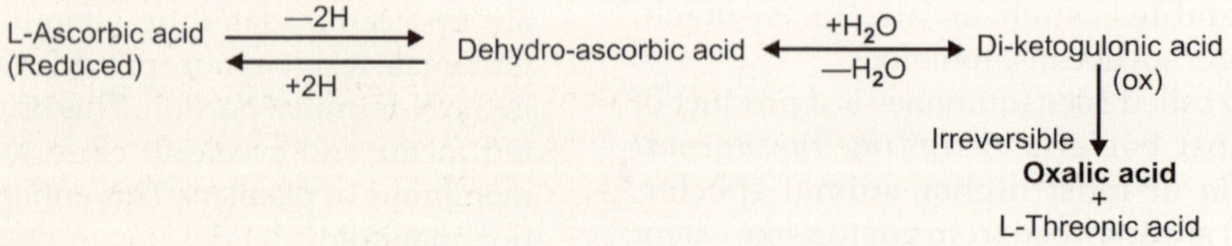

in vitro by H_2S and *in vivo* by –SH compounds, such as 'Glutathione'.

- It is stable in solid form and in acidic solutions, but is rapidly destroyed in alkaline solutions. Oxidative destruction of ascorbic acid is accelerated by increasing pH. Silver (Ag^{++}) and cupric (Cu^{+++}) ions accelerate process.

Biosynthesis

1. Some lower mammals like rats can synthesise the vitamin from glucose by the uronic acid pathway.
2. ***Man, monkey*** and ***guinea pigs lack the enzymes necessary for the synthesis.*** They cannot convert keto-gulonolactone to ascorbic acid. Hence the entire human requirement ***must consequently be supplied by the diet.***

Metabolism: Absorption, distribution and excretion

- It is absorbed readily from the small intestine, peritoneum and subcutaneous tissues. It is widely distributed throughout the body. Some tissues contain high concentrations as compared to others. ***Local concentration roughly parallels the metabolic activity,*** found in descending order as follows: Pituitary gland, adrenal cortex, corpus luteum, Liver, brain, gonads, thymus, spleen, kidney, heart, skeletal muscle, etc. ***From maternal blood, it can cross the placental barrier and supplies the foetus.***
- ***Normal human blood plasma:*** It contains approx. 0.6 to 1.5 mg of ascorbic acid per 100 ml.

The vitamin exists in the body largely in the 'reduced' form, with reversible equilibrium with a relatively small amount of "dehydroascorbic acid" (oxidised form). ***Both forms are physiologically and metabolically active.***

- Under normal dietary intake (of 75 to 100 mg)

50 to 75% are converted to inactive compounds	25 to 50% is excreted in urine as such.

- It is also secreted in milk.
- ***Metabolites***
  - *(a)* Chief terminal metabolites in the rat and guinea pig are ***CO_2*** and ***oxalic acid.***
  - *(b)* In human beings, decarboxylation of Ascorbic acid does not occur, the chief terminal metabolites being, ***oxalic acid*** and diketogulonic acid, which are excreted in urine. ***Conversion of ascorbic acid to oxalate in man may account for the major part of the endogenous urinary oxalate.***

Occurrence and food sources: Widely distributed in plants and animal tissues. In animal tissues, no storage, contains small amount, but highest concentration in metabolically highly 'active' organs, e.g. adrenal cortex, corpus luteum, liver, etc.

Dietary sources: These are chiefly vegetable sources. Good sources are **citrous fruits**—orange/lemon/lime, etc; other fruits like papaya, pineapple, banana, strawberry. Amongst vegetables—**leafy vegetables** like cabbage and cauliflower, germinating seeds, Green peas and beans, potatoes, and tomatoes. ***Amla is the richest source.*** Considerable amount of vitamin C activity is lost during cooking, processing and storage, because of its water-solubility and its irreversible oxidative degradation to inactive compounds.

METABOLIC ROLE AND FUNCTIONS

1. *Role in Cellular Oxidation-Reduction*

The fact that vitamin C is very sensitive to reversible oxidation, Ascorbic acid $\rightleftarrows$ Dehydroascorbic acid, suggests that it may be involved in cellular oxidation-reduction reactions, perhaps serving as hydrogen transport agent.

2. *Role in Collagen Synthesis*

Hydroxyproline and hydroxylysine are important constituents of mature collagen fibres. Precollagen molecules contain the amino acids proline and lysine. They are hydroxylated by corresponding ***hydroxylases*** in presence of vitamin C, Fe^{++} and molecular O_2. Thus:

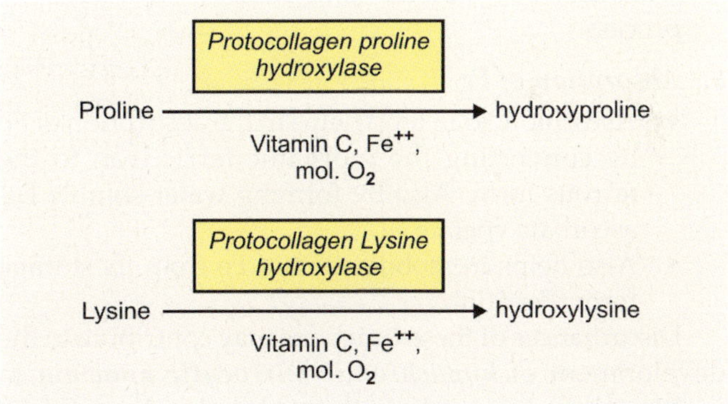

In scurvy, failure of conversion of procollagen to collagen due to the failure of hydroxylation may lead to a rapid destruction of the collagen intermediates.

3. *Functional Activity of Fibroblasts/Osteoblasts*

Ascorbic acid is required for functional activities of fibroblasts, and osteoblasts, and consequently ***for formation of MPS of connective tissues, osteoid tissues, dentine and intercellular cement substance of capillaries.***

4. *Role in Tryptophan Metabolism*

Vitamin C is required as a cofactor for hydroxylation of tryptophan to form 5–OH derivative, in the pathway of biosynthesis of serotonin (5–HT).

5. ***Role in Tyrosine Metabolism***

Required as a cofactor with the enzyme *p-OH phenyl pyruvate hydroxylase,* which is necessary for hydroxylation and conversion of p-OH phenyl pyruvate to Homogentisic acid.

Note

- Scorbutic guinea pigs and premature infants given a high protein and ascorbic acid deficient diet excrete increased amounts of p-OH-phenyl pyruvate and p-OH phenyl lactic acid. Administration of vitamin C corrects this condition.
- Vitamin C deficiency in premature infants may increase urinary excretion of 'homogentisic acid' and ***resemble the inherited disorder "Alkaptonuria".***

6. ***Formation Active FH_4 (Tetrahydrofolate)***

Ascorbic acid in combination with folic acid helps the maturation of the RB cells. Vitamin C regulates the conversion of folic acid to folinic acid (so-called "citrovorum factor"). It has been suggested that vitamin C by maintaining the *folic acid reductase* in its "active" form keeps the folic acid in the reduced tetrahydrofolate FH_4 form.

7. ***Formation of Ferritin***

Ascorbic acid is necessary for the formation of tissue "ferritin". ATP, NAD^+ and $NADP^+$ stimulate the process.

8. ***Absorption of Fe***

- Ascorbic acid in food helps in the absorption of Fe by converting the inorganic ferric iron to the ferrous form. Also by forming water-soluble Fe-ascorbate chelate.
- Also helps in mobilisation of Fe from its storage form 'Ferritin'.

Disturbances of these functions may contribute to the development of ***hypochromic microcytic anaemia*** in scurvy. Absorption of Fe both in normal or Fe-deficient patients is increased by over 10 per cent after administration of vitamin C.

9. ***Role in Electron Transport Systems***

Ascorbic acid seems to take part in electron transport system of mammalian 'microsomes'.The detailed mechanism of role of vitamin C is not definitely known but it has been suggested that the reaction is coupled with hydroxylation.

10. ***Action on Certain Enzymes-Activation/Inhibition***

Vitamin C is capable of both activating and inhibiting different groups of enzymes. *Arginase* and *papain* are activated, whereas, activity of the enzymes like *urease* and β-*amylase* from plants is inhibited.

11. ***Role in Formation of Catecholamines***

Vitamin C is required as a coenzyme with the enzyme *dopamine hydroxylase* which catalyses the conversion of dopamine to norepinephrine.

12. ***Role in Formation of Carnitine***

Formation of **'carnitine'** in liver by hydroxylation of γ-butyrobetaine is helped by vitamin C, α-ketoglutarate, Fe^{++} and a dioxygenase.

13. ***Role in α-Oxidation of FA***

Vitamin C helps in the action of the enzyme *α-hydroxylase (a mono-oxygenase)* which catalyzes the α-oxidation of long-chain F.A. to form α-OH-FA.

14. ***Effect on Cholesterol Level***

Relation of ascorbic acid with hypocholesterolaemia in man and guinea pigs has been reported.

15. ***Role in Stress***

The adrenal cortex contains a large quantity of vitamin C and this is rapidly depleted when the gland is stimulated by ACTH. A similar depletion of adrenocortical vitamin C activity is noted when experimental guinea pigs are injected with large quantities of diphtheria 'toxin'. Increased losses of the vitamin accompany infection and fever. Circulating vitamin C level has been found to be low in acute infectious diseases, congestive heart failure, in renal and hepatic diseases and malignancies. **All of the above suggest that the vitamin C may play an important role in the reaction of the body to *stress.***

16. ***Vitamin C*** has also been reported to act as coenzyme for *cathepsins* and liver *esterases.*
17. Ascorbic acid in both leucocytes and platelets found to be lowered significantly in women taking oral contraceptive pills.

CLINICAL ASPECT

DEFICIENCY MANIFESTATIONS: SCURVY

In the humans, its deficiency produces a disease called **Scurvy**. ***The main defect is a failure to deposit intercellular cement substance.***

- ***Capillaries are fragile*** and there is tendency to haemorrhages: Petechial, subcutaneous, subperiosteal and even internal haemorrhages can occur.
- ***Wound healing is delayed*** due to deficient formation of collagen.
- ***Poor dentine formation*** in children, leading to poor teeth formation.
- ***Gums are swollen*** and becomes spongy and bleeds on slightest pressure–Hyperaemia, swelling, sponginess and bleeding of gums are seen.
- In severe scurvy, may lead to secondary infection and loosening and falling of teeth.

- ***Osteoid of bone is poorly laid*** and mineralisation of bone is poor. The bones are weak and readily fractures. Haemorrhages occurring below the periosteum and into the joints may cause extremely painful swellings of bones and joints.
- Anaemia may be associated which is ***hypochromic microcytic type.***

Bachelor *Scurvy:* Elderly bachelors and widowers who may prepare their own foods are particularly prone to development of vitamin C deficiency.

Detection of Deficiency in Man

- Prompt improvement following administration of vitamin C.
- Determination of concentration of ascorbic acid in blood.
- *Urine ascorbic acid "Saturation" test:* By administering a test dose of 5 mg per lb (pound) body wt. If 50 per cent or more excreted in next 24 hours, the individual has no deficiency of the vitamin.
- *Intradermal test:* Consists of intradermal injection of 2, 6-dichlorophenol indophenol and determination of the time required for decolorisation i.e., reduction of the dye. ***Abnormally long persistence of blue colour in cutaneous wheal indicates subsaturation of ascorbic acid.***
- *Tourniquet test (capillary resistance or fragility test):* A sphygmomanometer cuff is applied around the arm and inflated so that it compresses the venous flow. In a short time, appearance of several petechial haemorrhages on the forearm skin indicates deficiency.

Requirement: ***A daily intake of about 100 mg is quite adequate in normal adults.*** Official recommended minimal daily intakes are:

- Infants: 30 mg per day
- Adults: 75 mg per day
- Adolescence: 80 mg per day
- Pregnant women: 100 mg per day
- Lactating women: 150 mg per day

Requirement is increased in presence of infections.

Hypervitaminosis (Effects of Excess Ascorbic Acid)

Administration of large amounts of ascorbic acid are not known to produce any effects in humans. But in rats, dehydroascorbic acid in enormous doses (1.5 gm/kg body wt) ***produces permanent diabetes, similar to that produced by the glycoside alloxan; it produces probably destruction of β-cells of islets of Langerhans.*** This action is prevented by immediate antecedant IV injection of –SH compounds like cysteine, Glutathione as in the case of alloxan, which resembles dehydroascorbic acid in chemical structure.

Empirical Uses of Vitamin C

Apart from treatment of scurvy, Ascorbic acid has been used ***empirically*** in many other conditions viz. in the control and treatment of infectious diseases. Has been found to help wound healing, in ulcer, trauma and burns, in allergic conditions, common cold and coryza, during labour: vitamin C given in doses of 150 to 250 mg produces an oxytocic action, increasing both the frequency and intensity of uterine contractions, in methaemoglobinaemia may be used for its reducing property.

SECTION TWO

B-COMPLEX VITAMINS

THIAMINE (VITAMIN B_1)

Synonyms: Antiberiberi factor, antineuritic vitamin, aneurin.

Chemistry

Free thiamine is a basic substance and contains ***(a) a pyrimidine, and (b) a thiazole ring. It contains sulphur (sulphur containing vitamin).*** Generally prepared as a chloride-hydrochloride.

- ***Solubility:*** Soluble in water (1 gm/ml) and 95 per cent alcohol (1 gm/100 ml). Not soluble in fat solvents.
- ***Stability:*** Resistant to heat (boiling/autoclaving) in solution < pH 3.5 but loses activity at pH > 5.5.
- Thiamine content of vegetables well preserved by freezing and by storage below 0°C. Rapidly destroyed in alkaline medium.

Chemically it is 2,5-dimethyl, 4-methyl-6-amino pyrimidine 5-OH ethyl thiazole, Thiamine pyrophosphate (thiamine diphosphate).

Biosynthesis: Synthesised by plants, yeasts and bacteria. ***Not synthesised by human beings, hence should be supplied in diet.*** Intestinal bacterial flora can synthesise the vitamin.

Metabolism

Absorption: Free thiamine is absorbed readily from the small intestine, but the pyrophosphate (ester-form) is not. Bulk of the dietary vegetable thiamine is in the "free" form. ***In tissues, it is actively phosphorylated to form Thiamine pyrophosphate (TPP) in Liver, and to a lesser extent in other tissues like muscle, brain and nucleated RB Cells.***

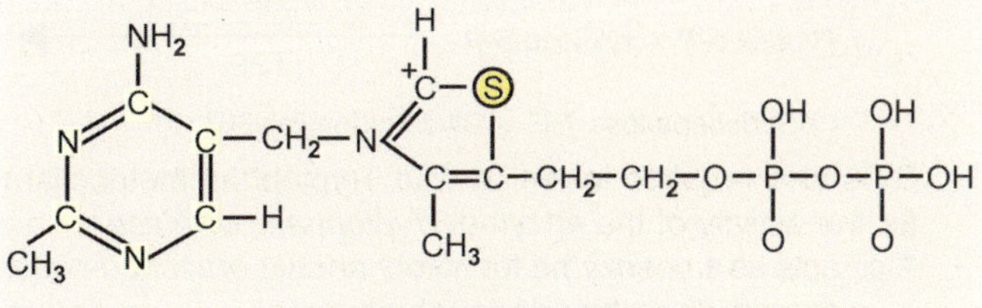

Structure of thiamine

Plasma/blood level: Present in plasma and CS fluid in the "free" form, approx. 1 μg/100 ml. Blood cells contain 6 to 12 μg/100 ml where occurs as TPP.

Storage: Capacity to store is limited. It is present in both free and combined forms in heart (highest concentration), Liver and kidneys. In lower concentration in skeletal muscle and brain. **Total amount of Thiamine in body is approx. 25 mg.**

Excretion: If normal amount of thiamine is taken in the diet:

- About 10 per cent is excreted in the urine
- The remainder is (a) Partly phosphorylated and is used as coenzyme, and (b) Partly degraded to neutral sulphur compounds and inorganic SO_4 which are excreted in urine.

Occurrence and Food Sources

- ***Plant source:*** Widely distributed in plant kingdom. In cereal grains, ***it is concentrated in outer germ/bran layers (e.g. rice polishings)*** **(Richest source).** Other good sources are peas, beans, whole cereal grains, bran, nuts, prunes, etc. ***Whole white bread is a good source.***
- ***Animal source:*** Thiamine is present in most animal tissues. Liver, meat and eggs supply considerable amounts. Ham/pork meats are particularly rich. Milk has low concentration, but a good source as large quantities are consumed.

METABOLIC ROLE

Biological *active* form is Thiamine pyrophosphate (TPP). Acts as a coenzyme in several metabolic reactions.

- Acts as coenzyme to the enzyme *pyruvate dehydrogenase* complex (PDH) which converts pyruvic acid to acetyl-CoA (oxidative decarboxylation)

Pyruvate —(*PDH*, TPP)→ **Acetyl-CoA**

- Similarly acts as a coenzyme to *α-oxoglutarate dehydrogenase complex* and converts α-oxoglutarate to succinyl-CoA (oxidative decarboxylation).

α-Oxoglutarate —(*α-oxoglutarate dehydrogenase*, TPP)→ **Succinyl-CoA**

- TPP also acts as a coenzyme with the enzyme *Transketolase* in transketolation reaction in HMP pathway of glucose metabolism.

Ribose-5-P + xylulose-5-P —(*Transketolase*, TPP)→ Sedoheptulose-7-P + Glyceraldehyde-3-P

- B_1 is also required in amino acid Tryptophan metabolism for the activity of the enzyme *Tryptophan pyrrolase.*
- Also acts as a coenzyme for mitochondrial *branched-chain α-ketoacid decarboxylase* which catalyses oxidative decarboxylations of branched-chain α-ketoacids formed in the catabolism of valine, Leucine and Iso-leucine. TPP binds with and decarboxylates these branched-chain α-ketoacids and transfers the resulting activated –CHO groups to α-lipoic acid.
- TPP acts as the coenzyme *(Co-carboxylase)* of *pyruvate carboxylase* in yeasts for the non-oxidative decarboxylation of pyruvate to acetaldehyde.

CLINICAL ASPECT

DEFICIENCY MANIFESTATIONS: Beriberi

The deficiency of thiamine produces a condition called **beriberi.** It is characterised by the following manifestations.

- ***CV manifestations:*** These include palpitation, dyspnoea, cardiac hypertrophy and dilatation, which may progress to congestive cardiac failure.
- ***Neurological manifestations:*** These are predominantly those of ascending, symmetrical, peripheral polyneuritis. These neurological features may be accompanied occasionally by an acute haemorrhagic polioencephalitis which is then called as **Wernicke's encephalopathy.**
- ***GI symptoms:*** Amongst these, anorexia is an early symptom. There may be gastric atony, with diminished gastric motility and nausea; fever and vomiting occur in advanced stages.

Dry beriberi: When it is not associated with oedema.

Wet beriberi: Oedema is associated. It is probably in part to congestive cardiac failure and in part to protein malnutrition (Low plasma albumin).

Deficiency in Animals: Deficiency produces symptoms resembling beriberi as described above, with certain important additional features in certain species:

- ***In rats:*** Associated with marked bradycardia.
- ***Pigeons:*** Develop a characteristic rigid retraction of head-**opisthotonos.**

The above two are utilised for bioassays.

- **Chastek paralysis:** Observed in foxes eating raw fish. It is characterized by extreme, board-like rigidity with retraction of head. **Reason:** Raw fish contains heat-labile thiamine-splitting enzyme *thiaminase* which destroys thiamine.
- **Bracken disease:** A similar condition which has been reported in grazing animals feeding on ferns and related species of plants which contain the enzyme *thiaminase.*

Biochemical Features in Thiamine Deficiency

- Decreased level of thiamine and cocarboxylase TPP in blood and urine. Determination of amount of thiamine excreted in 4 hours urine is used.
- Accumulation of pentose sugars in RB cells due to retardation of transketolation reaction.

- Increased level of pyruvic acid and lactic acid in blood, due to retardation of oxidative decarboxylation of pyruvic acid.
  - *LA/PA ratio:* Abnormal blood LA / PA ratio is said to be more specific indicator of B_1 deficiency.
- ***Catatorulin effect:*** Decreased uptake of O_2 by thiamine-deficient brain *in vitro;* reversible by addition of thiamine.
- ***Saturation test (thiamine loading test):*** A lower urinary excretion of thiamine and TPP after administration of a test-dose occurring in thiamine-deficient as compared to normal subjects.

Thiamine Antagonist

- **Pyrithiamine:** Thiazole ring is replaced by a pyridine ring, it is a potent antagonist to thiamine.
- **Oxy-thiamine** and **2-n-butyl thiamine:** They are milder antagonists.

Daily Requirements

- **Adult:** 0.5 mg for each 1000 calories; 1.0 to 1.5 mg for diets providing 2000 to 3000 C. Minimum requirement is 1.0 mg. **Actual requirement is related more directly to carbohydrates content of diet than to calorie value of diet.**
- **Children:** Ranges from 0.4 mg for infants to 1.3 mg for preadolescents (10 to 12 years of age).

Requirements Increases in

Anoxia-shock and haemorrhage, Serious illness and injury, During prolonged administration of broad-spectrum oral antibiotics, Increased calorie expenditure like fever, hyperthyroidism, ***Increased carbohydrate intake***, Increased alcohol intake, and pregnancy and in lactation.

RIBOFLAVIN (VITAMIN B_2)

Synonyms: Lactoflavin, ovoflavin, hepatoflavin.

Chemistry

- It is an orange-yellow compound containing,
  - A ribose alcohol: D-Ribitol
  - A heterocyclic parent ring structure **Isoalloxazine (*Flavin* nucleus).** 1-Carbon of ribityl group is attached at the 9 position of isoalloxazine nucleus.

  Ribityl is an alcohol derived from pentose sugar D-ribose.
- ***Stability:*** It is stable to heat in neutral acid solution but not in alkaline solutions. Aqueous solutions are unstable to visible and UV light. The reactions are irreversible.
- Riboflavin undergoes reversible reduction readily in presence of a catalyst to a colourless substance **Leuco-riboflavin.**

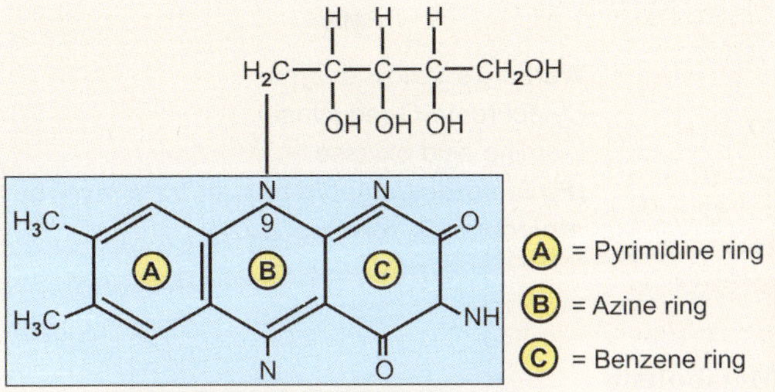

6,7-dimethyl-9 D-ribityl isoalloxazine (Riboflavin (Vitamin B_2))

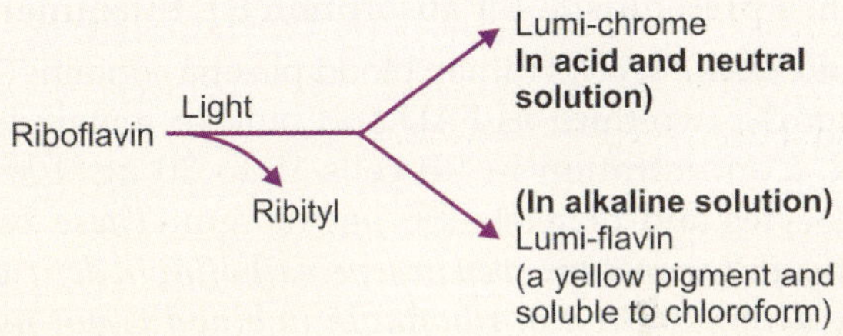

Biological Active Forms

The biological *active* forms, in which riboflavin serves as the prosthetic group (as coenzyme) of a number of enzymes are the phosphorylated derivatives.

Two main derivatives are:

- **FMN (Flavin mononucleotide):** In this the phosphoric acid is attached to ribityl alcoholic group in position 5.

  Flavin-Ribityl-PO_4
- **FAD (Flavin adenine nucleotide):** It may be linked to an adenine nucleotide through a pyrophosphate linkage to form FAD.

  Flavin-ribityl-P-P-ribose-Adenine

Thus, ***FMN and FAD are two coenzymes of this vitamin.***

The acidic properties given by phosphoric acid group influence their capacity for combining with proteins apo-enzyme-forming **flavoproteins** (Holoenzyme). Thus,

F_P (holoenzyme) = FMN/FAD (coenzyme) + Protein (Apoenzyme)

F_P may also unite with metals like Fe and Mo thus forming **Metalloflavoproteins.**

Biosynthesis

All higher plants can synthesise riboflavin. In nature, it occurs both as "free form" and also as "nucleotide" form or as flavoproteins.

Human beings and animals cannot synthesise and hence solely dependant on dietary supply. ***In man, considerable amounts can be synthesised by intestinal bacteria,*** but the quantity absorbed is not adequate to maintain normal nutrition.

FMN	FAD
• Warburg's yellow enzyme • Cytochrome-C-reductase • L-amino acid oxidase ***(Fp is autooxidisable at substrate level by molecular O_2 forming H_2O_2)***	• Xanthine oxidase (Xanthine → uric acid) • D-amino acid oxidase • Aldehyde oxidase • Fumarate dehydrogenase (Succinate → Fumarate) • Glycine oxidase • Acyl-CoA dehydrogenase • Diaphorase

SECTION TWO

Metabolism

Absorption: Flavin nucleotides are readily absorbed in small intestine. **Free riboflavin undergoes phosphorylation, a prerequisite for absorption (cf. thiamine).**

Blood/Plasma level: Human blood plasma contains 2.5 to 4.0 μgm%, two-third as FAD and bulk of remainder as FMN. Concentration in RB cells-15 to 30 μg/100 gm. Leucocytes and platelets-250 μg/100 gm. ***These values remain quite constant even in severe riboflavin deficiency, hence determination of riboflavin in blood is not useful.*** Riboflavin present in all tissues as nucleotides bound to proteins (FP), highest concentration in liver and kidney.

Excretion: Mainly in free form, up to 50 per cent as nucleotides in urine. Daily urinary excretion 0.1 to 0.4 mg (10 to 20% of intake).

- **Milk:** Riboflavin is secreted in milk, 40 to 80 per cent in 'free' form.
- **Faeces:** Free and nucleotides tend to remain quite constant, 500 to 750 μg daily, largely from the unabsorbed bacterial synthesis.

Occurrences and Food Sources

Widely distributed in nature, present in all plant and animal cells.

- *Plant sources:* High concentration occurs in yeasts. Appreciable amount present in whole grains, dry beans and peas, nuts, green vegetable. Germinating seeds, e.g. grams/Dals are very good source.
- *Animal source:* Liver (2–3 mg/100 gm), kidney, milk, eggs, Crab meat has high content.

METABOLIC ROLE

FMN and FAD act as coenzymes in various H-transfer reactions in metabolism. The hydrogen is transported by reversible reduction of the coenzyme by two hydrogen atoms added to the 'N' at positions 1 and 10, thus forming dihydro or leucoriboflavin. The principal enzyme reactions catalysed are as shown above in box.

Antagonists

- **Dichlororiboflavin:** By replacing two $-CH_3$ groups in riboflavin with chlorine atoms.
- **Isoriboflavin:** When $-CH_3$ group is shifted to another position.

CLINICAL ASPECT

Deficiency Manifestations

There is **no definite disease entity**. Deficiency is usually associated with deficiencies in other B-vitamins. In human beings lesions of the mouth, tongue, nose, skin and eyes with weakness, and lassitude reported. They include:

- **Lips:** Redness and shiny appearance of lips.
- **Cheilosis:** Lesions at the mucocutaneous junction at the angles of the mouth leading to painful fissures are characteristic.
- **Tongue:** Painful glossitis, the tongue assuming a red-purple (magenta) colour.
- **Seborrhoeic dermatitis:** Scaly, greasy, desquamation chiefly about the ears, nose and naso-labial folds.
- **Eyes:** May lead to corneal vascularisation and inflammation with cloudiness of cornea, watering, burning of eyes, photophobia, scleral congestion and cataract has also been reported.
- **Protein synthesis:** This is impaired in severe riboflavin deficiency; since protein malnutrition interferes with utilisation and retention of riboflavin.

Demonstration of Deficiency

- Recognition of characteristic deficiency symptoms as above and response to therapy.
- Blood level of riboflavin is not useful as discussed above.
- *Saturation tests:* It may be useful if adequately standardized. Normal subjects excrete at least 20 per cent of a test dose (3 mg) during the subsequent 24 hours.

Daily Requirement

Exact human requirement is not known. ***Related to degree of protein utilisation (cf. thiamine).***

Recommended Daily Intake	
• Adults:	1.5 to 1.8 mg
• Women in later half of pregnancy:	2.0 mg
• During lactation:	2.5 mg
• Infants:	0.6 mg
• Children:	1.0 to 1.8 mg
• Adolescence:	2.0 to 2.5 mg

Requirement increases

After severe injury/burns, etc. during acute illness and during convalescence, during increased protein utilisation, in pregnancy and lactation, during oral broad spectrum antibiotic therapy.

NIACIN (VITAMIN B_3)

Synonyms: Nicotinic acid, P-P factor, Pellagra-preventing factor of Goldberger.

Chemistry: Nicotinic acid (niacin) is chemically Pyridine-3-carboxylic acid.

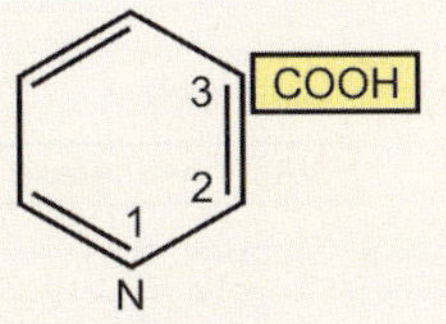

Niacin

In tissues: Occurs principally as the **amide** (nicotinamide, niacinamide). In this form it enters into physiological active combination.

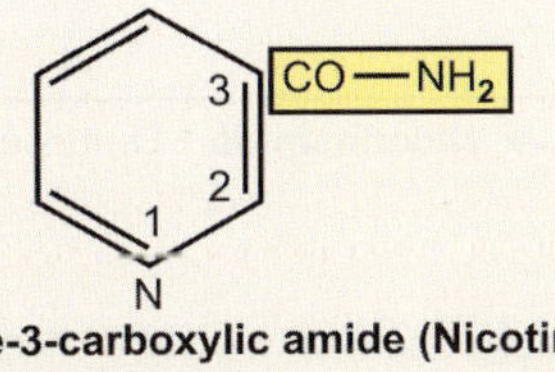

Pyridine-3-carboxylic amide (Nicotinamide)

Biological "Active" Forms

In tissues, nicotinamide is present largely as a "dinucleotide", the pyridine 'N' being linked to a D-ribose residue.

Two such neucleotide active forms are known:

- **Nicotinamide adenine dinucleotide (NAD^+)** Other names are: DPN^+, *coenzyme-I, cozymase*, or *codehydrogenase.* The compound contains:
  - One molecule of nicotinamide,
  - Two molecules of D-ribose,
  - Two molecules of phosphoric acid, and
  - One molecule of adenine. Structure may be shown schematically as follows:

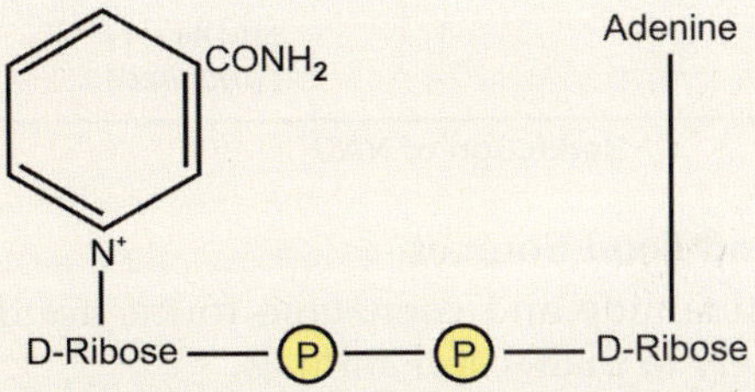

- ***Nicothinamide adenine dinucleotide phosphate ($NADP^+$)*** Other names are TPN^+, *co-enzyme II.*

This compound differs from NAD^+ in that **it contains an additional molecule of phosphoric acid attached to 2-position of D-ribose attached to N-9 of Adenine.** The reduced form of either coenzymes is designated by the prefix "dihydro-nicotinamide adenine dinucleotide (NADH).

Biosynthesis

- ***Amino acid tryptophan is a precursor of nicotinic acid*** in many plants, and animal species including human beings. ***60 mg of tryptophan can give rise 1 mg of Niacin.*** Pyridoxal-P is required as a coenzyme in this synthesis (Refer Tryptophan metabolism).
- ***It can be synthesised also by intestinal bacteria.*** Bacteria in addition to synthesis from tryptophan, can also synthesise from other amino acids, e.g. glutamic acid, proline, ornithine and glycine.
- **In human beings:**
  - In addition to dietary source,
  - It is synthesised in tissues from amino acid tryptophan, and
  - To a limited extent supplemented by bacterial synthesis in intestine.

Applied Aspect

In high corn diet, requirement of dietary niacin increases, as synthesis from tryptophan cannot take place. The reason is the maize protein **Zein** lacks the amino acid tryptophan. ***Hence pellagra is more common in persons whose staple diet is maize.***

Synthesis of NAD^+ and formation of NADP: Synthesis of NAD^+/and $NADP^+$ is shown diagrammatically in the box on next page.

Formation of Nicotinamide: Nicotinamide is not formed directly from nicotinic acid. It is **formed by degradation of NAD^+ and $NADP^+$**. Formation is shown schematically in the box on next page.

Metabolism

Absorption: Nicotinic acid and its amide are absorbed from the small intestine.

Blood/plasma level

- Whole blood: 0.2 to 0.9 mg/100 ml (average 0.6 mg%)
- RB cells: 1.3 mg%
- Plasma-total activity: 0.025 to 0.15 mg% (average 0.075 mg%)

Note

1. Most of the nicotinic acid and its amide in the blood is in RB cells, presumably as coenzyme.
2. ***Values in the blood are not altered significantly even in severe Niacin deficiency. Hence its determination is of no value in the detection of clinical deficiency states.***

Excretion

In urine, it is excreted as follows:

- ***As nicotinic acid and nicotinamide:*** Normal adults on normal diet excretes both nicotinic acid and its amide in urine.

Formation of NAD and NADP

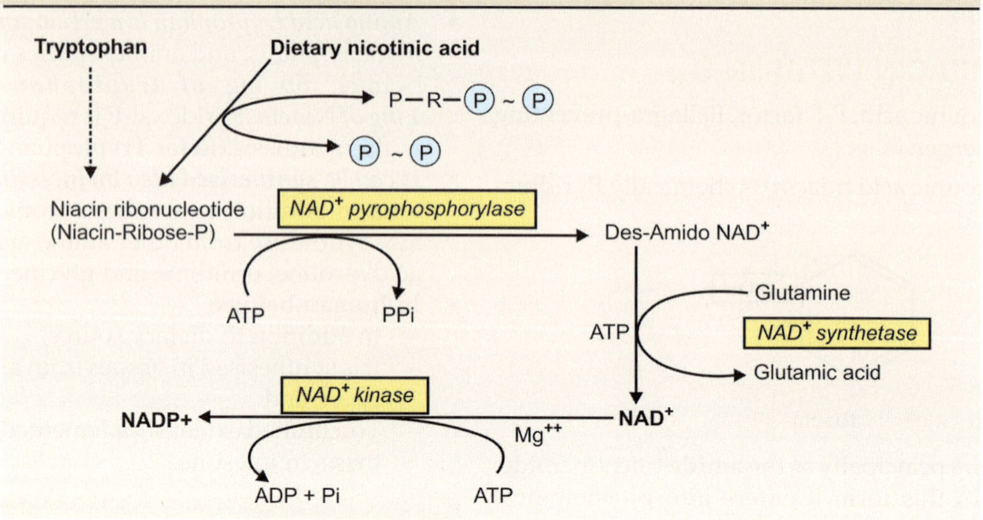

Degradation of NAD

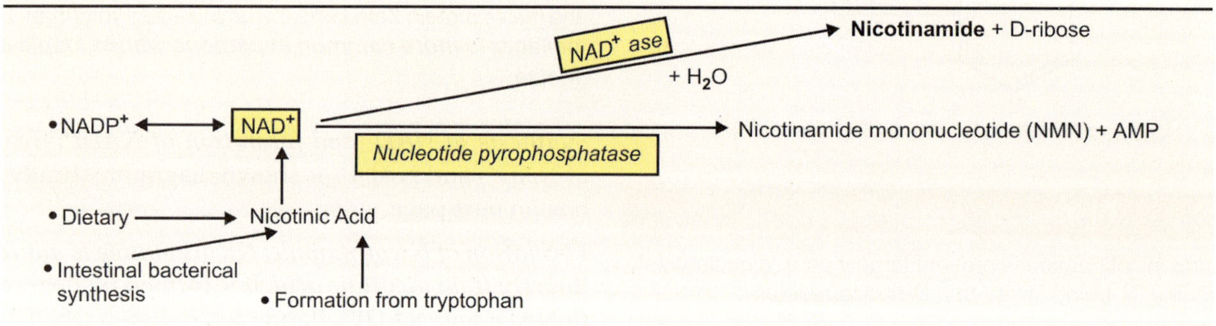

Nicotinic acid: 0.25 to 1.25 mg daily.
Nicotinamide: 0.5 to 4 mg daily.

- ***As N′-methyl nicotinamide:*** Major urinary metabolite is a methylated derivative-N′-methyl nicotinamide. The methylation occurs in liver, by the enzyme *niacinamide methyl transferase*. CH_3 group is given by **Active methionine**.

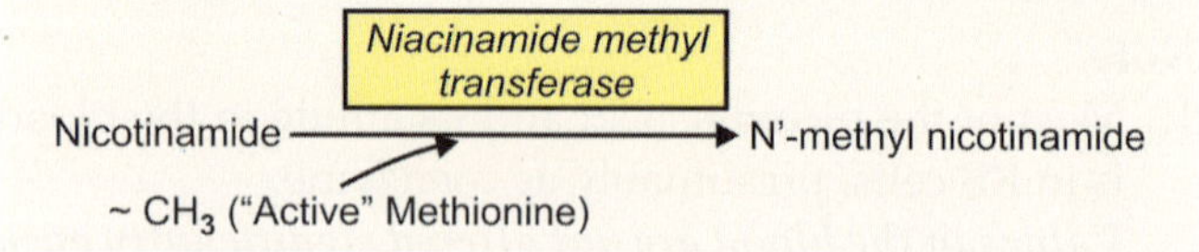

- Oxidation products of N′-methyl-Nicotinamide viz., 6-pyridone and 4-pyridine of N′-methyl Nicotinamide are also excreted in urine.
- Glycine conjugates of methyl derivatives are also excreted. Methylation, oxidation and conjugation take place in liver.

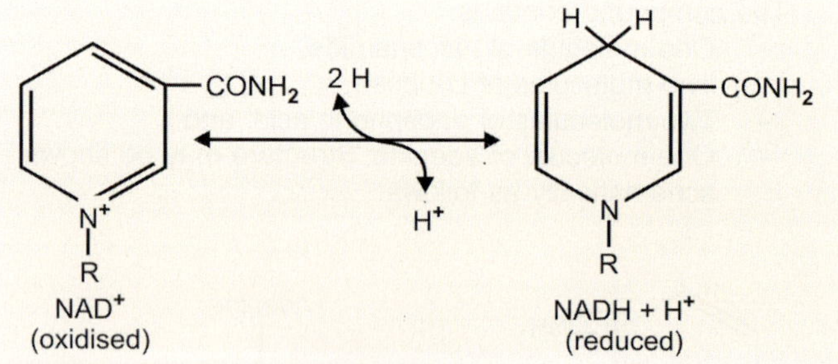

Reduction of NAD$^+$

Occurrence and Food Sources

1. Both nicotinamide and coenzyme forms are distributed widely in plants and animals.
2. Important food sources are:
   - ***Animal source:*** Liver, kidney, meat, fish
   - ***Vegetable source:*** Legumes (peas, beans, lentils), nuts, certain green vegetables, coffee and tea. Nicotinamide is present in highest concentration

in germ and pericarp (bran) in cereal grains. Yeast also particularly rich. Poor sources are: Fruits, milk and eggs.

Metabolic Role

- The coenzymes NAD^+ and $NADP^+$ operate as hydrogen and electron transfer agents by virtue of reversible oxidation and reduction. The mechanism of the transfer of Hydrogen from a metabolite to oxidised NAD^+, thus completing the oxidation of the metabolite and the formation of reduced NAD ($NADH + H^+$) is shown in box. Reduction of NAD^+ occurs in para position; one H loses an electron and enters the medium as H^+.
- Function of $NADP^+$ is similar to that of NAD^+ in hydrogen and electron transport. **The two coenzymes are interconvertible.** The important enzymes to which NAD^+ and $NADP^+$ act as coenzyme are given below in the box.

Other Role of NAD^+

In addition to its coenzyme role, NAD^+ is the source of ADP-ribose for the **ADP-ribosylation** of proteins and poly-ADP ribosylation of nucleoproteins involved in the ***DNA-repair mechanism.***

Niacin antagonists are: (a) Pyridine-3-sulfonic acid (b) 3-Acetyl pyridine.

CLINICAL ASPECT

Deficiency Manifestations

Pellagra:

Nicotinic acid deficiency produces a disease called **pellagra** (Pelle = skin; agra = rough)

Cardinal features described as **"3 D's"** are

- ***Dermatitis,***
- ***Diarrhoea***, and
- ***Dementia.***

Precipitating factors are: (a) High-corn diet and (b) Alcoholism

Clinical Features

(a) Skin lesions: Typically involves areas of skin exposed to sunlight and subjected to pressure, heat or other types of trauma or/irritation. These include face, neck, dorsal surfaces of the wrist, forearms, elbows, breasts and perineum. The skin becomes reddened, later brown, thickened and scaly.

(b) GI manifestations: Include anorexia, nausea, vomiting, abdominal pain, with alternating constipation/ diarrhoea.
- ***Diarrhoea*** becomes intractable later.
- ***Gingivitis*** and ***stomatitis*** with reddening of the tip and margin of the tongue, which become swollen and cracked.
- ***Achlorhydria*** present in about 40% cases.
- Thickening and inflammation of the colon, with cystic lesions of the mucosa, which later becomes atrophic and ulcerated.

(c) Cerebral manifestations: These include headache, insomnia, depression and other mental symptoms ranging from mild psychoneuroses to severe psychosis.

(d) General effects: These include: Inadequate growth, loss of weight and strength, anaemia which may be due to associated deficiency of other vitamins, dehydration and its consequences resulting from diarrhoea.

Additional Factors

- Simultaneous ***deficiency of riboflavin (Vit B_2) or pyridoxine (B_6)*** can contribute to the etiology of pellagra. Both of these are required for synthesis of niacin from tryptophan.
- Incidence of pellagra is more in women than men, reason is ***oestrogen metabolites can inhibit tryptophan metabolism*** and prevents synthesis of niacin from amino acid tryptophan.

Association of Pellagra in other Diseases

- ***In carcinoid syndrome***: Pellagra can occur. There is overproduction of 5-OH tryptamine (5-HT) which may use more than 60 per cent of tryptophan producing deficiency of Niacin synthesis.
- ***In Hartnup disease***: Pellagra may be associated. Tryptophan deficiency occurs due to genetic defect in

NAD^+	*$NADP^+$*
• *Alcohol dehydrogenase* (Ethanol → Acetaldehyde) • *Lactate dehydrogenase (LDH)* (PA ↔ LA) • *Malate dehydrogenase* (Malate → OAA) • *Glyceraldehyde-3-P-dehydrogenase* (Gly-3-P → 1, 3-di-phosphoglycerate) • α-Glycero-P-dehydrogenase • *Pyruvate dehydrogenase complex (PDH)* (PA → Acetyl-CoA) • α-Ketoglutarate dehydrogenase complex (α-ketoglutarate → succinyl-CoA)	• *Glucose-6-P-dehydrogenase (G-6-PD)* (*G-6-P → 6-Phosphogluconate*) • *Glutathione reductase* **Either NAD^+ or $NADP^+$** • *Glutamate dehydrogenase* (Glutamate → α-ketoglutarate + NH_3) • *Isocitrate-dehydrogenase* (I C D) (Isocitrate → Oxalosuccinate)

SECTION TWO

membrane transport mechanism for tryptophan resulting to great **loss of the amino acid. Reason:** intestinal malabsorption and failure of renal reabsorption mechanism, due to genetic defect.

Daily Requirement

- In adult: 17 to 21 mg daily
- Infants: 6 mg
- Pre-adolescence: 17 mg

Requirement Increases in:

Increased calorie intake or expenditure, acute illness or early convalescence, after severe injury, infection and burns, **high corn or Maize diet,** pregnancy and lactation.

OTHER CLINICAL ASPECTS

- ***Development of fatty Liver:*** In rats, administration of large amounts of nicotinic acid or amide produced fatty Liver, which is prevented by simultaneous administration of methionine, choline or betaine.

  Explanation: There occurs diversion of $-CH_3$ group for the formation of excessive amount of N'-methyl nicotinamide producing **relative deficiency of choline.**
- ***Effect on plasma Lipids: Nicotinic acid and NOT amide have been found to reduce the plasma lipid concentration in certain cases of hyperlipidaemia.*** Large doses of Nicotinic acid from 3 to 6 Grams per day have been found to reduce the levels of cholesterol, β-lipoproteins and TG in blood.
- **Niacin toxicity:** Excessive dosage can produce toxic effects:
  - Dilatation of blood vessels and flushing.
  - Skin irritation
  - Can produce liver damage.

PYRIDOXINE (VITAMIN B_6)

Synonyms: Rat antidermatitis factor.

Chemistry

- **Pyridoxol (Pyridoxine):** also called as *Adermin* is chemically 2-methyl-3-OH-4, 5-di (hydroxymethyl) pyridine.

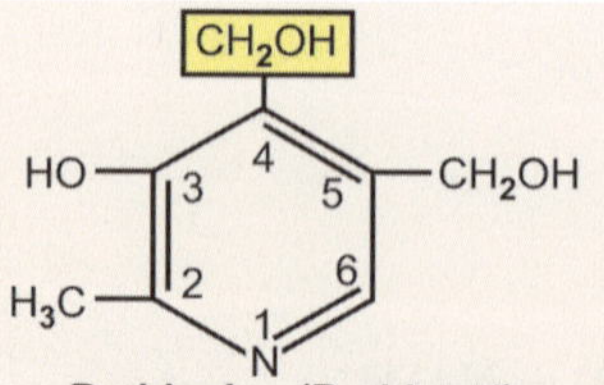

Pyridoxine (Pyridoxol)

- It occurs in association, perhaps in equilibrium, with an aldehyde-**Pyridoxal** and an amine **Pyridoxamine** form. **All three forms exhibit vitamin B_6 activity.**

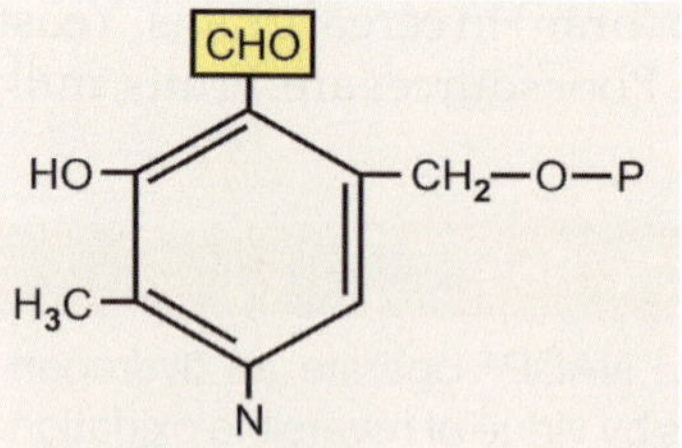

Pyridoxal-P

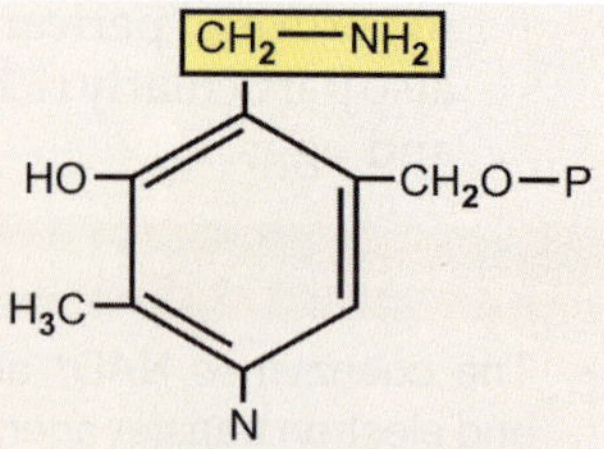

Pyridoxamine-P

Biological 'Active' Forms

Biological 'active' forms of the vitamin are:

- ***Pyridoxal-PO$_4$,*** and
- ***Pyridoxamine-PO$_4$***

The active forms are the phosphorylated derivatives, phosphorylation involves the hydroxymethyl group $-CH_2OH$ at position 5 in the pyridine ring. These forms occur in nature largely in combination with protein (apoenzyme).

Formation of pyridoxal-P: Phosphorylation takes place in Liver, Brain and other tissues with the help of ATP, Zn^{++} and an enzyme *Pyridoxal kinase.*

Two pathways shown below:

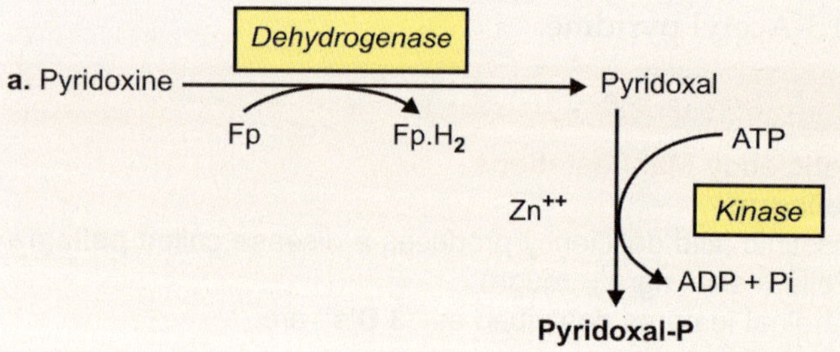

b. **Alternatively,** Pyridoxine first undergoes phosphorylation with the help of ATP-dependant *kinase* to produce pyridoxal-P and then oxidised by FP-dehydrogenase.

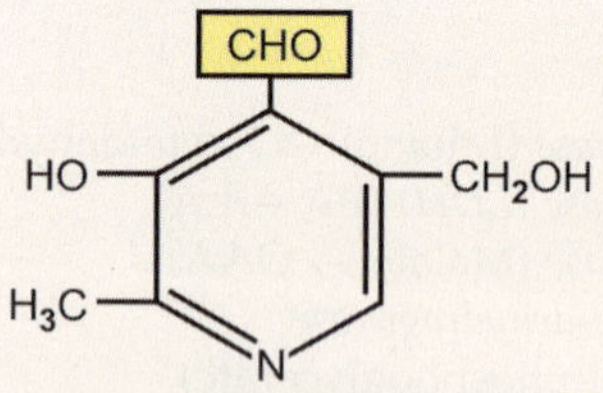

Pyridoxal

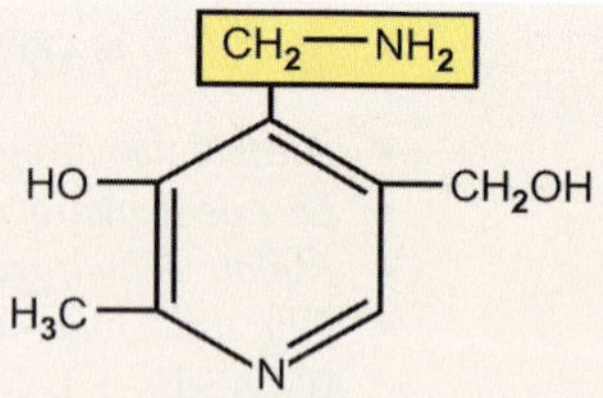

Pyridoxamine

Biosynthesis: Vitamin B_6 can be formed by many micro-organisms and probably also by plants. ***Human beings cannot synthesise the vitamin, hence has to be provided in the diet. Intestinal bacteria can synthesise the vitamin.***

Metabolism

Absorption: Dietary vitamin B_6 is readily absorbed by the intestine.

Excretion

- Pyridoxal and pyridoxamine are excreted in urine in small amounts 0.5 to 0.7 mg daily.
- Major urinary metabolite, about 3 mg daily is the biologically inactive form 4-pyridoxic acid.

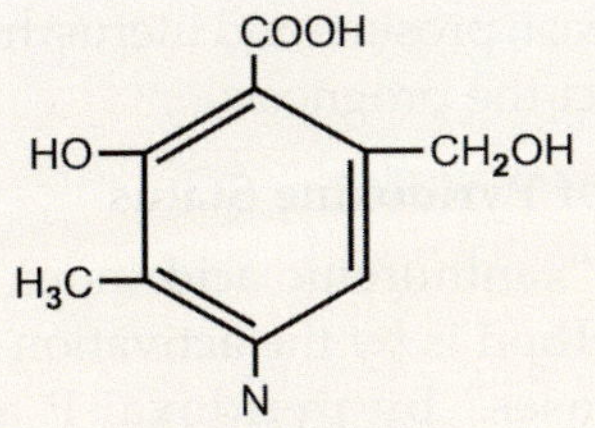

Pyridoxic acid

Occurrence and food sources: The vitamin is distributed widely in animal and plant tissues. Rich sources of the vitamin are yeast, rice polishings, germinal portion of various seeds and cereal grains and egg-yolk. Moderate amounts are present in liver, kidney, muscle, fish. ***Milk is a poor source.*** Highest concentration occurs in royal jelly (bee).

Metabolic Role

Pyridoxal P acts as a coenzyme, it is ***principally involved with metabolism of amino acids.***

- ***Cotransaminase:*** It acts as a coenzyme for the enzyme *transaminases* (aminotransferases) in transamination reaction.
- ***Codecarboxylase:*** It acts as coenzyme for the enzyme *decarboxylases* in decarboxylation reaction. Amino acids are decarboxylated to form corresponding amines.

  Examples:
  - Tyrosine → Tyramine + CO_2
  - Histidine → Histamine + CO_2
  - Glutamic acid → G A B A + CO_2
- ***Acts as coenzyme for deaminases (dehydrases):*** Catalyses non-oxidative deamination of OH-amino acids viz., serine, threonine, etc.
- ***Coenzyme for kynureninase:*** In tryptophan metabolism, pyridoxal-P acts as a coenzyme for the enzyme *kynureninase* which converts 3-OH-kynurenine to 3-OH-anthranilic acid which ultimately forms nicotinic acid. ***Thus in B_6-deficiency niacin synthesis from tryptophan does not take place.*** **In B_6 deficiency,** kynurenine and 3-OH kynurenine levels increases and they are converted to ***xanthurenic** acid* in extrahepatic tissues, which is excreted in urine (See box above). ***xanthurenic acid index is a reliable criterion for B_6 deficiency.*** Examination of urine for xanthurenic acid after the feeding of a test dose of tryptophan has been used to diagnose B_6-deficiency.
- **Transulfuration:** It takes part in transulfuration reactions involving transfer of –SH group, e.g.
  Homocysteine + Serine → homoserine + cysteine.
- ***As coenzyme for desulfhydrases:*** It catalyses non-oxidative deamination of cysteine in which H_2S is liberated.

Formation of Xanthurenic acid

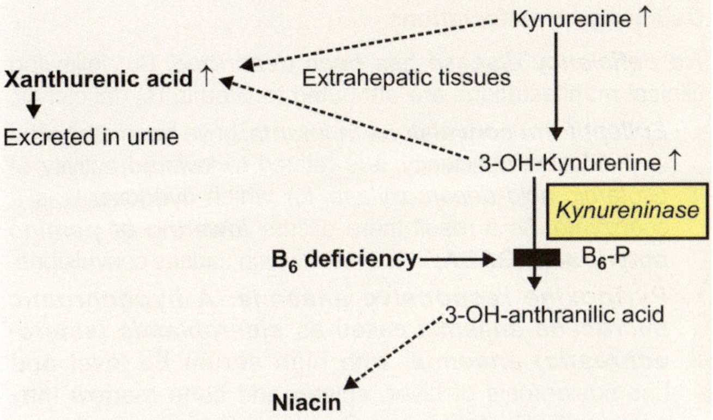

- ***In interconversion of glycine and serine*** by *serine hydroxy methyl transferase:* In this both $F.H_4$ and B_6 are required as coenzymes.
- Pyridoxal-P is required as a coenzyme in the ***biosynthesis of arachidonic acid from linoleic acid.***
- ***Synthesis of Sphingomyelin:*** Pyridoxal-P is required as a coenzyme for activation of serine which is required for synthesis of sphingomyelin.
- Required as a coenzyme for *amino acid racemases:*
  D-Glutamic acid → L-Glutamic acid
  D-Alanine → L-Alanine
- ***Intramitochondrial FA synthesis:*** Required as a coenzyme with *condensing enzyme* for chain elongation of FA in intramitochondrial FA synthesis.
- ***Required for "active transport" of amino acids*** through cell membrane and intestinal absorption of amino acids.
- ***Muscle phosphorylase:*** As a constituent of *muscle phosphorylase:* 4 molecules of pyridoxal-(P) per molecule of enzyme (tetramer).
- ***Transport of K^+:*** Vitamin B_6 has been reported to promote transport of K^+ across the membrane from exterior to interior.
- **As coenzyme for *aminoacetone synthetase*** which is required for formation of aminoacetone from acetyl-CoA and glycine.
- ***Synthesis of CoA-SH (Coenzyme A):*** Vitamin B_6 is involved in synthesis of coenzyme A from pantothenic acid. In B_6 deficiency, coenzyme A level in Liver is reduced.
- ***In porphyrin synthesis:*** Pyridoxal-P is required for conversion of α-amino-β-ketoadipic acid to δ-ALA, an important step in haem synthesis. ***In B_6-deficiency haem synthesis suffers and leads to anaemia.***
- ***Hypercholesterolaemia:*** Relationship of B_6-deficiency, hypercholesterolaemia and atherosclerosis has received considerable attention, although the exact role of vitamin B_6 is not clear.
- ***Immune response:*** In vitamin B_6 deficiency, immune response is impaired.
- ***Oxaluria:*** Vitamin B_6 deficiency has been observed to produce oxaluria in experimental animals.

CLINICAL ASPECT

Deficiency Manifestations

No deficiency disease has been described. But following clinical manifestations are attributed to vitamin B_6 deficiency.

- ***Epileptiform convulsions in infants:*** have been attributed to pyridoxine deficiency. It is related to lowered activity of *Glutamic acid decarboxylase,* for which pyridoxal P is a coenzyme. As a result there occurs ***lowering of γ-amino butyric acid*** **(GABA)** in the brain which causes convulsions.
- ***Pyridoxine responsive anaemia: A hypochromic microcytic anaemia*** called as ***sideroblastic (sideroachrestic) anaemia***, with high serum Fe level and haemosiderosis of Liver, spleen and bone marrow may occur with B_6-deficiency. Pyridoxal-P is required as a coenzyme in the reaction by which α-amino-β-ketoadipic acid is decarboxylated to form δ-ALA in heme synthesis. ***In B_6-deficiency heme synthesis suffers and Fe cannot be utilised.***
- ***Isonicotinic acid hydrazide treatment in tuberculosis:*** A syndrome resembling vitamin B_6 deficiency has been observed in humans during the treatment of tuberculosis with high doses of tuberculostatic drug ***Isonicotinic acid hydrazide*** or **Isoniazid (INH).**
  - 2 to 3 per cent of patients receiving conventional doses of isoniazid, 2 to 3 mg/kg, ***developed neuritis***.
  - 40 per cent receiving 20 mg/kg ***developed neuropathy***.
  - Tryptophan metabolism was also altered, there was increased Xanthurenic acid excretion in urine.

  Signs and symptoms were alleviated by administration of pyridoxine to these patients. 50 mg of pyridoxine per day completely prevented the development of neuritis and neuropathies.

  Mechanism of action: It is believed that isoniazid forms a ***'hydrazone complex'*** with pyridoxine, resulting in incomplete activation of the vitamin.
- Like INH, the drug **penicillamine (β-dimethyl cysteine)** has been incriminated to produce deficiency of Vit B_6. The drug is used in treatment of Wilson's disease to chelate copper. The drug reacts with pyridoxal-P to form inactive thiazolidine derivative. During penicillamine treatment B_6 should be given as a supplement to prevent its deficiency.

  Therapeutic uses: Vitamin B_6 has been found empirically to be of value in treatment of:

  Nausea and vomiting of pregnancy ("morning sickness"), Radiation sickness, Muscular dystrophies, Treatment of hyperoxaluria, and recurring oxalate stones of kidney, and Mild forms of pyridoxine deficiency have been reported to occur sometimes in women taking oral contraceptives containing oestradiol.

Vitamin B_6 may cut risk of Parkinson's Disease

Recently it has been claimed by Dutch researchers and reported in neurology that a **higher intake of vitamin B_6 may decrease the risk of Parkinson's disease.** It could lower Parkinson's disease risk by protecting brain cells from damage caused by free radicals.

Pyridoxine Status Vs Hormone Dependant Cancer

Increased sensitivity to steroid hormones action may be important in the development of hormone-dependant cancer of the breast, prostate and uterus; hence pyridoxine status may affect the prognosis.

Measurement of Pyridoxine Status

In addition of "xanthurenic acid index", **most widely used recent method** is by the activation of "erythrocyte aminotransferases" by pyridoxal-P added *in vitro*, expressed as the activation coefficient.

Daily requirement: It has been difficult to establish definitely the human requirement of vitamin B_6 due to the fact that (i) quantity needed is not large, and (ii) bacterial synthesis in intestine provides a portion of the requirement.

There is evidence that ***requirement of Vitamin B_6 is related to dietary protein intake,*** as it is involved as coenzyme in many metabolic reactions of amino acid metabolism.

- **Adult** : 2 mg/day
- **Infants** : 0.3 to 0.4 mg/day

During second half of pregnancy: 2.5 mg/day

In patients receiving antitubercular treatment with INH, requirement of vitamin B_6 increases much.

LIPOIC ACID (THIOCTIC ACID)

Synonyms: Protogen, Acetate replacement factor (ARF), Pyruvate oxidation factor (POF).

Chemistry

It is a ***sulphur containing*** fatty acid called 6, 8-dithio-octanoic acid (α-lipoic acid or thioctic acid). It contains eight carbon and two sulphur atoms. Oxidised and reduced forms of the compound is shown below.

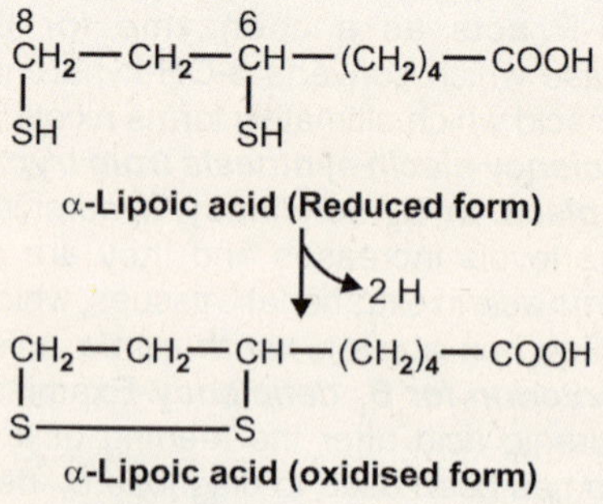

Deficiency manifestations: Not known. Lipoic acid occurs in a wide variety of natural materials. Attempts to induce lipoic acid deficiency in animals have so far been unsuccessful.

Metabolic Role of Lipoic Acid

It is recognised as an essential component in metabolism although it is active in extremely minute amounts.

- ***As a coenzyme of pyruvate dehydrogenase complex (PDH):*** It is required alongwith other coenzymes in oxidative decarboxylation of pyruvic acid to acetyl-CoA.
- ***As a coenzyme of α-oxoglutarate dehydrogenase complex:*** Required alongwith other coenzymes in oxidative decarboxylation of α-oxoglutarate to succinyl CoA.
- ***Lipoic acid is also required for the action of the enzyme sulphite oxidase:*** Required for conversion of SO_2^- to $SO_4^{=}$. Hypoxanthine is also required for the action.

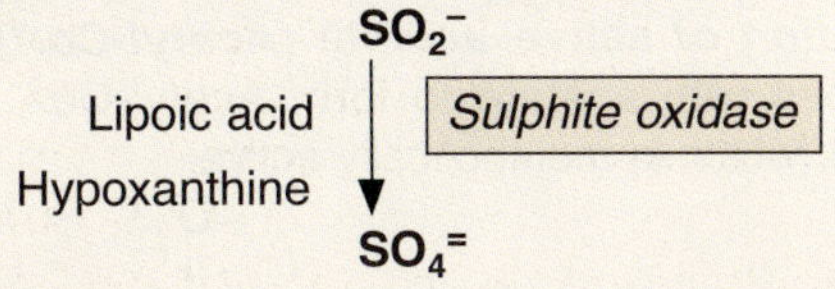

CLINICAL ASPECT

Antioxidant Property:

Recently it has been shown that lipoic acid/or dihydrolipoic acid in large dosage 100 to 500 mg/day can act as an **antioxidant.**

This property has been utilised in treatment of certain diseases (therapeutic uses).

- May be useful in prevention of myocardial infarction and stroke.
- Can mop up "free" radicals in brain tissue and thus can prevent conditions like multiple sclerosis, Alzheimer's disease, etc.
- Helps in reducing the plasma low density lipoproteins (LDL).
- Stimulates production of glutathione (G-SH).

PANTOTHENIC ACID (VITAMIN B_5)

Synonyms: Filtrate factor, Chick antidermatitis factor.

Chemistry

- Pantothenic acid consists of β-alanine in peptide linkage with a dihydroxy dimethyl butyric acid ('Pantoic' acid).

β-alanine + Pantoic acid → Pantothenic acid

- The free acid is soluble in water and is hydrolysed by acids/or alkalies. It is thermolabile and destroyed by heat.

Biological "Active" Form

Active form is coenzyme A. In tissues, this vitamin is present almost entirely in the form of the coenzyme (coenzyme A is also known as *Coacetylase*) and largely bound to proteins (apoenzyme). It may be released from this combination by certain proteolytic enzymes, certain phosphatases, and a liver enzyme system.

Structure of Coenzyme A: Structure of coenzyme A has been delineated and can be represented schematically as follows (Refer to box shown below):

- Pantothenic acid is joined in one hand to adenosine-3'-P by a pyrophosphate bridge, and on the other hand,
- Joined by peptide linkage to β-mercaptoethanol amine, which is obtained from amino acid cysteine. ***The terminal –SH group (thiol group) of β-mercaptoethanolamine is the reactive site of the coenzyme molecule ("Active site" or group).***

Note: For convenience co-enzyme A is represented as CoA – SH. The naturally occurring forms of the coenzyme probably include:

- The reduced –SH form,
- The oxidised –S-S-forms, and
- Combinations of the –SH group with various metabolites, e.g. Acetate, and succinate to form acetyl-CoA and succinyl-CoA respectively.

Biosynthesis and Metabolism

I. Biosynthesis Pantothenic acid

(a) In many microorganisms, including yeast pantothenic acid is synthesized by direct coupling of β-alanine and pantoic acid. ***β-Alanine** is **formed from decarboxylation of Aspartic acid*** and pantoic acid from α-keto-isovalerate.

(b) ***Human tissues cannot synthesise pantothenic acid hence it has to be obtained from diet.*** In addition to

Structure of Coenzyme A

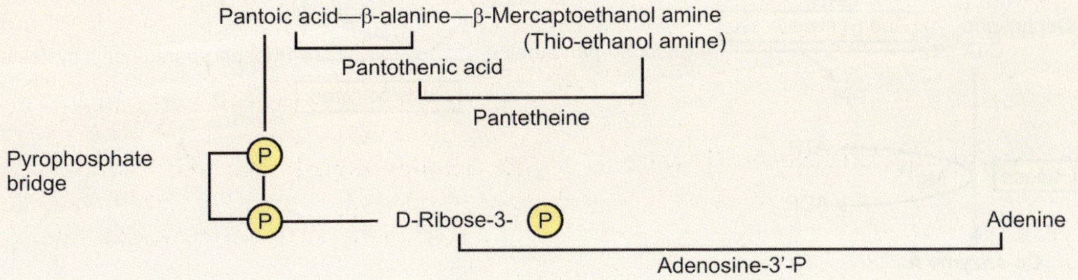

dietary source, ***synthesis by intestinal bacteria*** supply fair amount of pantothenic acid.

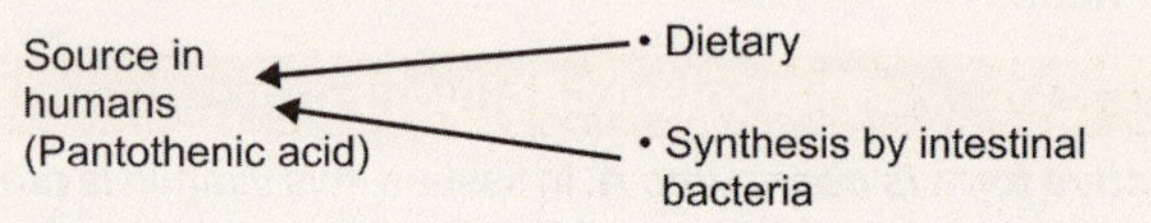

II. Synthesis of Coenzyme A: Complete synthesis of coenzyme A was described by **Khorana** in 1959. Human tissues as well as plants and bacteria can synthesise CoA-SH. Synthesis of coenzyme A is shown schematically in box below.

Whole blood level: The concentration of pantothenic acid in whole blood is 15–45 μg/100 ml (average 30 μg%). It is present in all tissues in small amounts, the highest concentration occurring in liver (40 μg per gram wt) and kidney (30 μg/gm).

Excretion: Catabolic products of pantothenic acid are not known.

- **Urine:** Under ordinary dietary conditions about 2.5 to 5 mg are excreted daily in the urine.
- **Sweat:** 3–4 μg/100 ml.
- **Milk:** Secretes 200 to 300 μg/100 ml.

Occurrence and Food Sources

It is widely distributed in plants, animal tissues and food materials.

- ***Excellent food sources*** (100 to 200 μg/gm of dry materials): include kidney, Liver, egg-yolk and yeasts, cereals and legumes.
- Fair sources (35 to 100 μg/gm): Include skimmed milk, chicken, certain fishes, sweet potatoes, molasses.
- Most vegetables and fruits are rather poor source.
- Richest known source of pantothenic acid is Royal Jelly (also rich in Biotin and pyridoxine).

Note: A 57 per cent loss of pantothenic acid in wheat may occur during the manufacture of patent flour and about 33 per cent is lost during the cooking of meat.

Metabolic Role

Only demonstrated metabolic function of pantothenic acid is as a constituent of coenzyme A. As a constituent of CoA, pantothenic acid is essential to several fundamental metabolic reactions.

- ***Formation of active acetate (Acetyl-CoA):*** It readily combines with acetate to form Acetyl-CoA or "Active" acetate, which is metabolically active.

  Acetyl-CoA chemically is $CH_3-\overset{O}{\overset{||}{C}} \sim S.\,CoA$, the **sulphur bond** of acetyl-CoA is a **high energy bond** equivalent to that of the high energy PO_4 bond of ATP. In the form of ***active* acetate**, it participates in a number of important metabolic reactions, e.g.
  - Utilised directly by combination with oxaloacetate (OAA) to form citric acid, which initiates TCA cycle.
  - Acetylcholine formation.
  - For acetylation reactions.
  - Synthesis of cholesterol.
  - Formation of ketone bodies.
  - Acetyl-CoA and Malonyl-CoA are used in the synthesis and elongation of fatty acids.
- ***Formation of active succinate (Succinyl-CoA):*** Product of oxidative decarboxylation of α-oxoglutarate in TCA cycle is a coenzyme derivative called **Active *succinate***

Synthesis of Coenzyme A

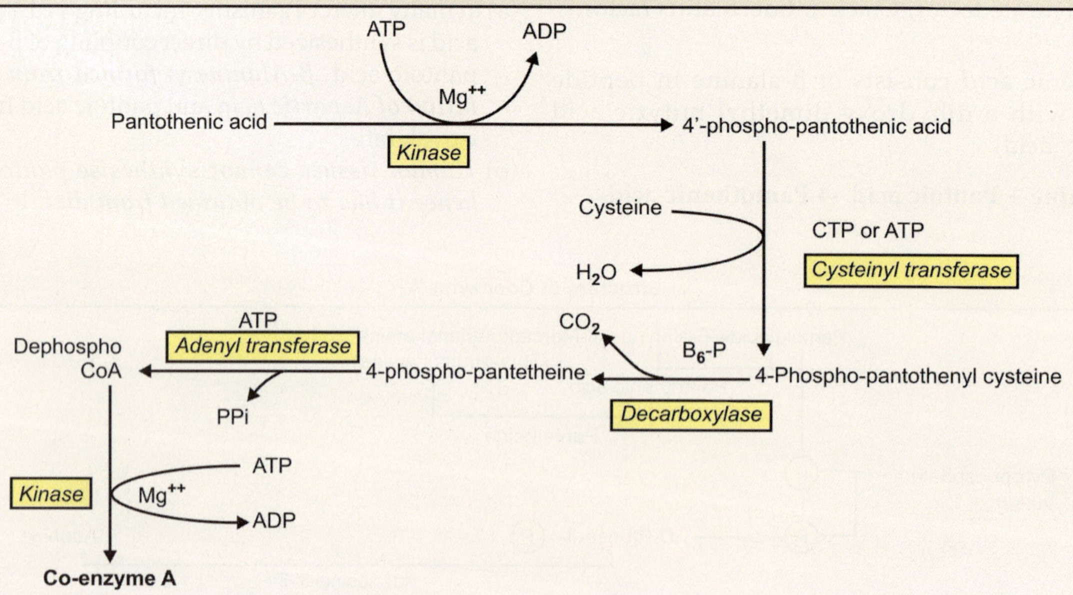

(Succinyl-CoA), Succinyl-CoA is involved in certain important metabolic reactions as follows:

- ***Haem synthesis:*** In haem synthesis, "active" succinate and glycine combines to form δ-ALA, the first step in the pathway of heme formation.
  Applied aspects: Anaemia may occur in pantothenic acid deficiency probably due to deficiency in formation of succinyl-CoA. Due to nonavailability of the substrate the heme synthesis suffers.
- ***Degradation of ketone bodies by extrahepatic tissues:*** Refer to Ketolysis.

- ***Role in lipid metabolism:***
  – ***Oxidation of FA (β-oxidation):*** First step in oxidation of FA catalysed by *Thiokinases* (acyl synthases) involves the activation of the FA by formation of CoA derivatives. (Refer β-oxidation).
  – ***Biosynthesis of FA:*** Pantothenic acid is a constituent of a compound called as "Acyl-carrier protein" (ACP) and also a constituent of "multienzyme complex" in mammals, which is used in the extramitochondrial ***de Novo*** fatty acid synthesis.
- ***Role in Adrenocortical function:*** Pantothenic acid appears to be involved in adrenocortical activity, being essential to the formation of adrenocortical hormones from "active" acetate and cholesterol. Pantothenic acid deficient animals have reduced levels of adrenal cholesterol.
- ***Activation of some amino acids may also involve CoA-SH:*** Occur among the branched chain amino acids such as valine and leucine.

Deficiency Manifestations: No deficiency disease has been recognised in man. This may be due to: Its widespread distribution in food stuffs and supply from synthesis by bacterial flora of intestines.

Deficiency manifestations observed in experimental animals are: Dermatitis, Loss of hair (alopecia): **circum-ocular "spectacle" alopecia** and graying of hairs, **GI manifestations:** Include gastritis and enteritis with ulceration and haemorrhagic diarrhoea, **fatty Liver** develops in dogs and rats, anaemia develops in certain species and in severe cases, hypoplasia of bone marrow.

Nervous system manifestations include: Myelin degeneration of peripheral nerves and degenerative changes in posterior root ganglia.

Human volunteers: ***Experimental deficiency*** in human volunteers is manifested by, gastrointestinal symptoms, easy fatigue, changed sleep patterns, staggering gait, and mental symptoms. Human deficiency can be produced by administration of antagonists like:

- Pantoyltaurine
- ω-methyl Pantothenate

Daily requirement: The human requirement of pantothenic acid is not known due to its widespread distribution.

- **For adults** it is recommended, a daily intake of 5 to 12 mg per 2500 cal.
- **In infants** : 1 to 2 mg
- **In children :** 4 to 5 mg.

Requirement Increases in:
Presence of severe stress, e.g. acute illness, burns, severe injury, etc, oral administration of broad spectrum antibiotics, in pregnancy and lactation, in growing children, and in convalescence.

BIOTIN (VITAMIN B_7)

Synonyms: Bios, vitamin H, Co-enzyme R, Antiegg-white injury factor.

Chemistry

Biotin is a heterocyclic monocarboxylic acid, it is a ***sulphur-containing*** water-soluble B-vitamin.

Structure of the compound was worked out by **du Vigneud,** which is as follows:

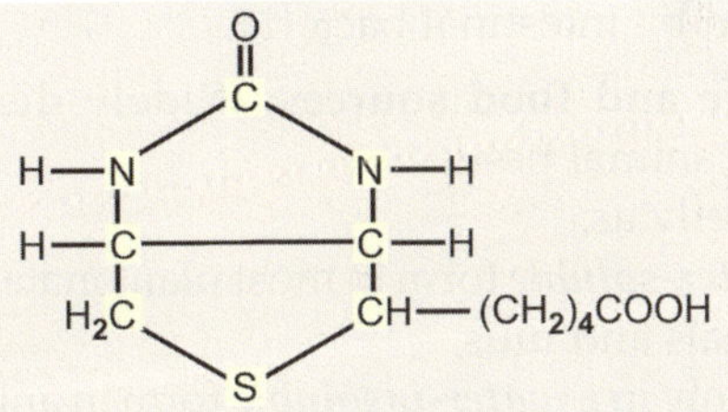

Biotin

It consists of two fused rings, one **imidazole** and the other **thiophene** derivative. Biotin ($C_{10}\ H_{16}\ O_3\ N_2S$) is chemically, Hexahydro-2-oxo-1-thieno-3,4-imidazole-4 valeric acid. Two forms with essentially identical biological activities: **α-Biotin** (egg-yolk) and **β-Biotin** (Liver)-differing only in the nature of the side chain.

Biotin is said to occur both as* free *form and* bound *form in tissues and foods.

Bound forms: Biocytin, desthiobiotin and **oxybiotin** are the bound forms.

- *Biocytin (ε-N-biotinyl-Lysine):* Biotin was found to remain bound with 'Lysine' residues of tissue proteins by amide bonds. Biocytin (ε-N-biotinyl-Lysine) is released on hydrolysing the peptide bonds between the biotin-bound Lysine and the peptide chain.
- *Desthiobiotin and oxybiotin:* These are biologically active in certain strains of yeast and bacteria.

Egg-White Injury

Boas (1927) described egg-white injury in rats, who were fed with diets containing ***raw egg-white***. Cooked egg-white was not found to be toxic. **William and coworkers (1940)** demonstrated that egg-white injury was actually due to ***an antivitamin present in egg-white***. This

substance is ***Avidin***, a basic protein, its ability to inactivate Biotin was confirmed in 1941.

Biosynthesis and Metabolism

- Biotin can be synthesised by many bacteria, yeast, and fungi. In green plants, it may be formed in leaf and root.
- **Coenzyme-R** is a growth essential for the nitrogen-fixing organisms, ***Rhizobium,*** in the root nodules of Leguminous plants. Coenzyme R had been proved to be Biotin. ***Pimelic acid is possible precursor and desthiobiotin, a probable intermediate.***

Storage: Biotin may be stored to a limited extent in the liver and kidneys. Studies with labelled-biotin has shown that a maximum of 14 per cent of administered dose was stored in liver.

Excretion: Excreted in urine, faeces and milk. Normal adult on an adequate diet excretes 10 to 180 μg daily in the urine and 15 to 200 μg daily in the faeces. Faecal excretion probably represents unabsorbed biotin synthesized by intestinal bacteria.

Occurrence and food sources: Widely distributed in plants and animal tissues.

Occurs chiefly as:

- ***Water-soluble*** form in most plant materials, except cereals and nuts,
- Mainly in a ***water-insoluble*** form in animal tissues.

Foods rich in biotin include:

- *Animal sources:* They are liver, kidney, milk and milk products and egg-yolk.
- *Vegetable sources:* Include vegetables, legumes, and grains which are good sources. Molasses contain good amount of biotin. Exceptionally large amounts are present in royal jelly (bee).

Human beings cannot synthesise the vitamin and hence it has to be supplied in diet. But **bacterial flora in intestine can synthesise** the vitamin and is a good source.

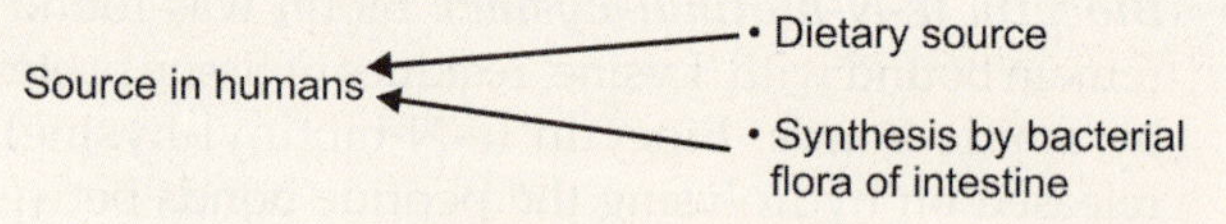

Metabolic Role

Biotin is the prosthetic group of certain enzymes that catalyse ***CO_2-transfer reaction (CO_2-fixation reaction).*** In biologic system, ***biotin functions as the coenzyme for the enzyme called carboxylases, which catalyse the CO_2-fixation (Carboxylation).***

In this process, Biotin is first converted to ***carboxybiotin complex*** by reaction with HCO_3^- and ATP. "CO_2-biotin complex" is the source of "active" CO_2 which is transferred to the substrate, CO_2 becomes attached to the biotin coenzyme as shown below.

Examples of carboxylation or "CO_2-fixation" reactions in biologic systems are given in box.

CO_2 fixation reactions

- ***Conversion of acetyl-CoA to Malonyl-CoA:*** In the first step of extra mitochondrial 'de Novo' FA synthesis, the acetyl-CoA is converted to Malonyl-CoA, the reaction is catalysed by the enzyme ***acetyl-CoA carboxylase.***
- ***Conversion of propionyl-CoA to methylmalonyl-CoA:*** The enzyme catalysing the reaction is ***propionyl-CoA carboxylase.***
- ***Conversion of pyruvic acid to oxalo-acetate:*** The enzyme that catalyzes the reaction is ***Pyruvate carboxylase.***

Other reactions where Biotin has been incriminated are:

- ***Conversion of β-methyl crotonyl-CoA to β-methyl glutaconyl-CoA:*** In their conversion in leucine metabolism, the reaction is catalysed by the enzyme *β-methyl-erotonyl-CoA carboxylase.*

Conversion of PA to OAA

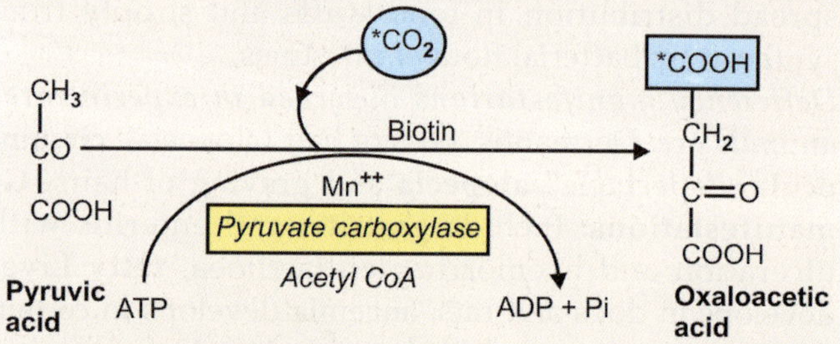

- ***Other enzyme systems:*** A number of other enzyme systems are reportedly influenced by biotin. These include *succinic acid dehydrogenase* and *decarboxylase*, and the *deaminases* of the amino acids aspartic acid, serine and threonine.

CO_2-Biotin Complex

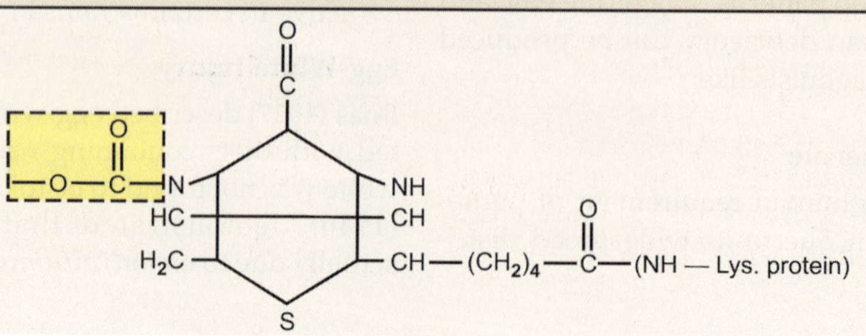

Conversion of Acetyl-CoA to Malonyl-CoA

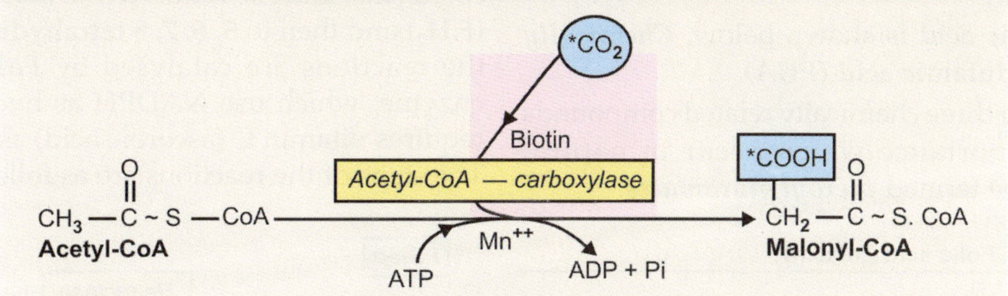

Conversion of Propionyl-CoA to Methyl Malonyl-CoA

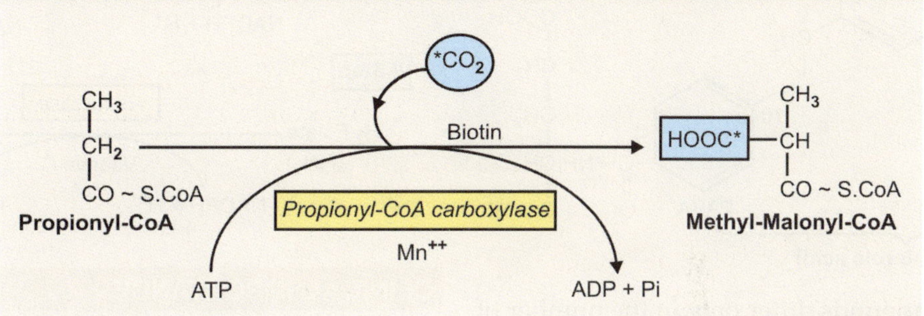

Deficiency manifestations: Biotin deficiency may be induced ***in experimental animals:***

- By inclusion of large amounts of raw-egg white in the diet.
- By using sulphonamide drugs or broad spectrum oral antibiotics for prolonged periods.

The features include: Dermatitis, ***Spectacle-eyed*** appearance, due to circumocular alopecia, thinning or loss of fur/and hairs, graying of hairs/fur of black or brown colours, paralysis of hind legs.

Human volunteers: Deficiency has been produced by excluding dietary biotin and feeding large amounts of raw egg-white (30% of total calories). Such individuals developed following symptoms beginning after 5 to 7 weeks:

Dermatitis of the extremities, pallor of skin and mucous membranes, anorexia and nausea, muscle pains and hyperaesthesia, depression, Lassitude and somnolence, anaemia, and hypercholesterolaemia.

Prompt relief of symptoms occurred when biotin concentrates was given.

DEFICIENCY DISEASES: CLINICAL ASPECTS

There is ***no definite deficiency disease***. But following two conditions have been found to be related to biotin deficiency:

1. **Congenital:** A rare genetic deficiency of ***holocarboxylase synthase*** has been described. The enzyme helps to utilise biotin in metabolic role. The affected child cannot utilise biotin and develops biotin deficiency which is manifested as dermatitis, graying of hair, loss of hair (alopecia) and incoordination of movements. Urine shows high urinary lactate, β-OH-propionate and β-methyl crotonate due to the failure of corresponding enzyme activities.
2. **Acquired (Leiner's disease):** Leiner's disease (erythroderma desquamativum or exfoliative dermatitis) in young infants has been reported by several workers. The disease often occurs in breastfed infants, frequently in association with persistent diarrhoea.
   Mechanism: The low biotin content of human milk together with the poor absorption of biotin due to diarrhoea, appears to cause biotin deficiency. Administration of biotin has been reported to cause a cure in many of these cases.

Daily requirement: It is difficult to arrive at a quantitative requirement of this vitamin as it is ubiquitous and as intestinal bacteria synthesize and supply the vitamin.

- Human adults : 25 to 50 μg daily
- Infants : 10 to 15 μg daily
- Children : 20 to 40 μg daily

Requirement Increases in

- Pregnancy and lactation,
- Oral antibiotic therapy for prolonged periods.

FOLIC ACID GROUPS (VITAMIN B_9)

Synonyms: Liver lactobacillus Caseifactor, vitamin M, Streptococcus lactis R (SLR) factor, vitamin Bc, Fermentation residue factor, pteroyl glutamic acid (PGA).

Chemistry:

- The designation "folic acid" is applied to a number of compounds which contain the following groups:
  - A ***pteridine*** **nucleus** (**pyrimidine** and **pyrazine rings**)

- Para-aminobenzoic acid ("PABA") and
- Glutamic acid.

- Structure of folic acid is shown below. ***Chemically called*** **pteroyl glutamic acid *(PGA).***

There are at least three chemically related compounds of nutritional importance which occur in natural products, all may be termed ***pteroyl glutamates.***

Folic acid (folacin)

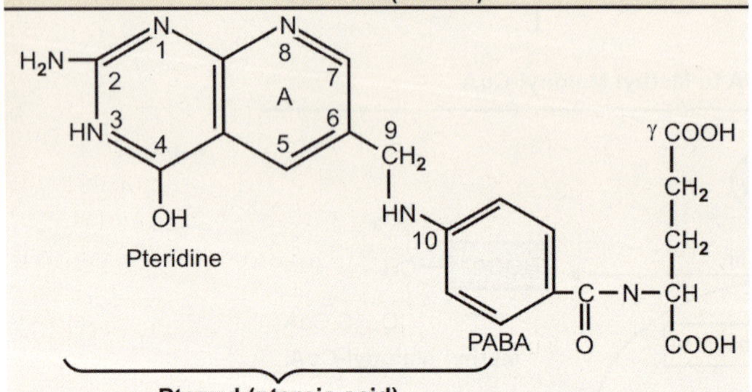

These three compounds differ only in the number of glutamic acid residues attached to **pteridine PABA complex (pteroic acid).**

- ***Monoglutamate:*** Having one glutamic acid. It is synonymous with vitamin Bc.
- ***Triglutamate:*** Having three glutamic acid residues. This substance once designated as "fermentation factor".
- ***Heptaglutamate:*** Having seven glutamate residues, synonymous with vitamin Bc conjugate of yeast.

Pteroyl glutamic acid is liberated from these conjugates by enzymes called ***conjugases.***

Biological "Active" Forms

Active "coenzyme" form of the vitamin is the reduced tetrahydroderivative, **Tetrahydrofolate F.H$_4$**, obtained by addition of four hydrogens to the pteridine moiety at 5, 6, 7 and 8 position.

Structure of tetrahydrofolate is shown below.

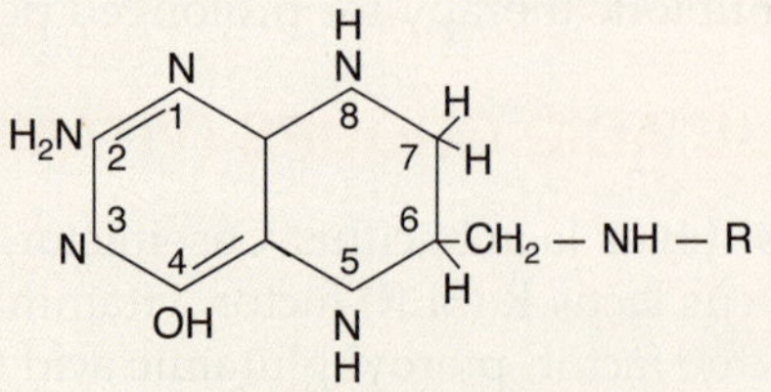

5,6,7,8-tetrahydrofolate (F.H$_4$)

Because of their lability, these occur naturally only in small quantities, being present mainly in the form of N^5-formyl or N^5-methyl derivatives.

Formation of F.H$_4$: Folic acid, before functioning as a coenzyme, must be reduced first to **7, 8-dihydrofolic acid (F.H$_2$)** and then to **5, 6, 7, 8 tetrahydrofolate (F.H$_4$).** Both the reactions are catalysed by ***Folic acid reductases*** enzyme, which use NADPH as hydrogen donor. Also requires vitamin C (ascorbic acid) as cofactor.
The steps of the reactions are as follows:

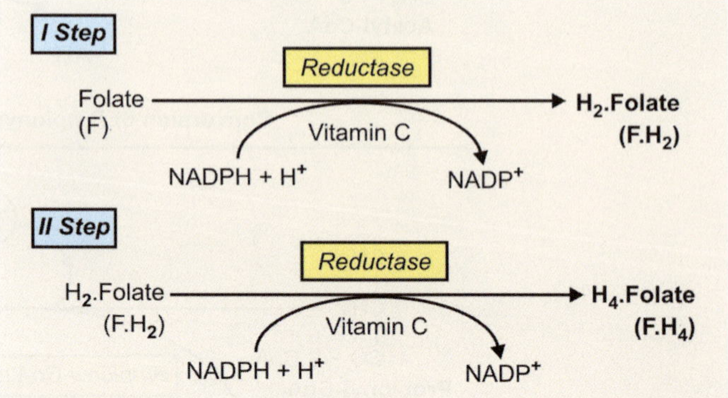

CLINICAL IMPORTANCE

Some of clinically described cases of folic acid deficiency anaemias may actually be due to inherited deficiency of *Folic acid reductase.*

Folinic acid: This is one of the active form of folic acid, a ***formyl* derivative**. It is reduced tetrahydrofolate (F.H$_4$) with a ***formyl* group on position 5 (f$^{5.}$ F-H$_4$).** Folic acid when added to Liver slices is converted to the 'formyl' derivative in presence of NADPH. Ascorbic acid enhances the activity of the liver in this reaction. This form of folic acid was first discovered in liver extracts when it was found to supply an essential growth factor for a *Lactobacillus* called *Leuconostoc citrovorum* and it was termed as ***citrovorum factor*** and when its chemical structure was determined the name ***folinic acid*** was applied.

Structure of folinic acid is similar to folic acid, except

- It is the reduced tetrahydroform (F.H$_4$) and
- With a 'formyl group' at position-5. It is also called as ***leucovorin*** earlier and a similar synthetic form ***Folinic acid-SF*** (synthetic factor) has same structure and function.

Rhizopterin: A similar compound but with **formyl *group at position* 10** has been recovered from liver. Rhizopterin or Streptococcus lactis 'R' factor, is a naturally occurring compound which is a 10-formyl derivative of pteroic acid.

Conversion of f^5 to f^{10}

- The f^5 can be converted to f^{10} by the action of an enzyme system ***formyltetrahydrofolate isomerase*** in presence of ATP, first f$^{5–10}$ FH$_4$ is formed, which by the action of ***cyclohydrase*** is converted to f^{10} FH$_4$ (Refer box on next page top).

Formation of $f^{10}F.H_4$

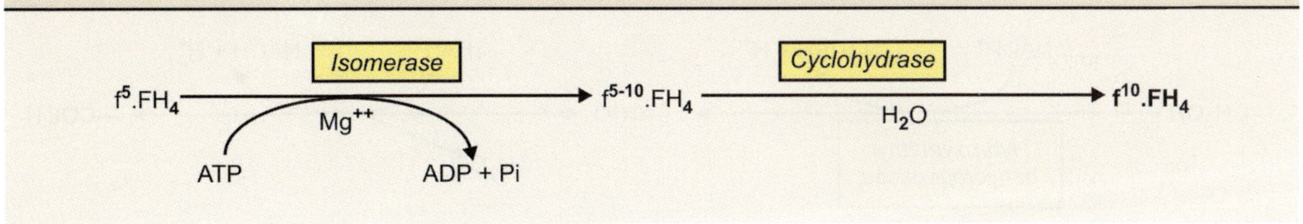

- There is also present in a variety of natural materials, e.g. pigeon liver, many bacteria, an enzyme ***formyl tetrahydrofolate synthetase,*** catalyzes the direct addition of "formate" (H-COOH) to $F.H_4$. The enzyme is specific for formate.

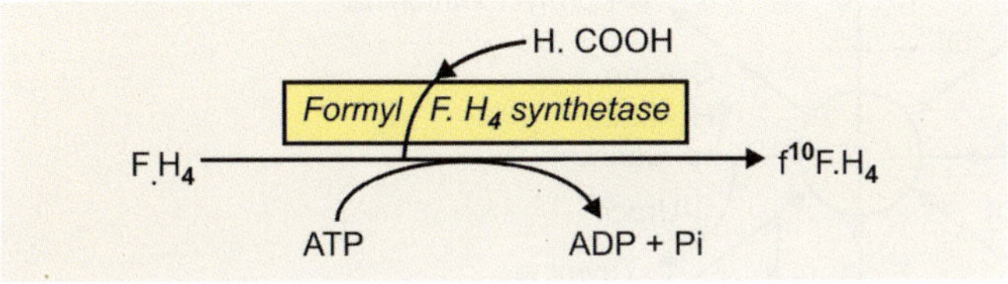

Formation of N^5-methyl $F.H_4$ from f^{5-10} $F.H_4$: f^{5-10} FH_4 can be converted to N_5-methyl $F.H_4$ by an *NAD^+ dependant reductase* and this $-CH_3$ group is then transferable to "De oxyadenosyl B_{12}" (Cobamide coenzyme) to form methyl B_{12}, an important donor of $-CH_3$ group as occurs in methylation of homocysteine to form methionine.

Biosynthesis and Metabolism

- Many microorganisms including those inhabiting the intestinal tract can synthesise folic acid. Some of them cannot synthesise PABA, which has to be supplied. In presence of ATP and CoA-SH, PABA reacts with glutamic acid to form ***"p-amino-benzoyl glutamic acid". The latter then reacts with a "Pterin" to produce "pteroyl monoglutamic acid, PGA" (folic acid); pterin moiety is probably derived from Guanosine.***

Effect of drugs: ***Sulphonamide drugs and antibiotics inhibit their growth by blocking the incorporation of PABA in the synthetic pathway (by competitive inhibition).***

- Higher animals including human beings cannot synthesize folic acid and it has to be supplied in diet. In human beings, intestinal bacteria can synthesise and is a good source.

Absorption: Occurs along the whole length of mucosa of small intestine. Polyglutamates ingested in diet are converted to monoglutamates and dihydrofolates are reduced to tetrahydrofolates by *folate reductase*. Tetrahydrofolates are then converted to methyl tetrahydrofolates, which enter the portal blood and then carried to liver.

Transport: Transported in blood as methyl tetrahydrofolate **bound to a specific protein.**

Plasma level: In normal individuals, it varies from 3 to 21 ng/ml.

Excretion

- **Urine:** 2 to 5 μg/day. This is much increased after an oral dose of folate if the tissues are saturated.
- **Faeces:** 20 per cent of the ingested folates that remains unabsorbed + 60 to 90 μg in the bile that is not reabsorbed + some unabsorbed synthesis of folate by bacterial flora of intestine.

Tissue folate: About 70 mg in the whole body, of which about 1/3 (5 to 15 μg/g) is in the liver.

RB cells folate: Folate is incorporated into the RB cells during erythropoiesis and is retained there during their entire life span except for only a slight fall in concentration. ***Red cell folate is a reliable indicator of the folate status of the body. Average level is 300 ng/ml of whole blood on a PCV of 45 per cent (range 160–640 ng/ml).***

Occurrence and Food Sources

Widely distributed in nature being present in many animal and plant tissues and in microorganisms. Particularly abundant in liver, yeast, kidney and green leafy vegetables. Spinach and cauliflower are also good sources. Other good sources are: meat, fish, wheat. Fair sources: milk, fruits.

Metabolic Role ("one-carbon" Metabolism)

- The folic acid coenzymes are specifically concerned with metabolic reactions involving the transfer and utilisation of the one carbon moiety (C_1).
- ***One carbon moiety (C_1)*** may be either ***Methyl ($-CH_3$), formyl ($-CHO$), formate ($H.COOH$), formimino group ($-CH{=}NH$) or hydroxymethyl ($-CH_2OH$).*** Most of them are metabolically "interconvertible" and catalysed by an NADP-dependant *hydroxymethyl dehydrogenases* (see in box ahead in next page).
- As discussed above, folinic acid is 5-formyl $F.H_4$ ($f^5.FH_4$). However except for the formylation of glutamic acid in the course of the metabolic degradation of Histidine, the F^5 compound is metabolically inert.
- On the other hand, the f^{10} tetrahydrofolate (f^{10} FH_4) or f^{5-10}. FH_4 are the active forms of the folic acid coenzymes in metabolism.
- The "one-carbon" moiety can be derived from several sources and can be utilised to form several compounds (Fig. 12.6).

Interconversion of C_1 moieties

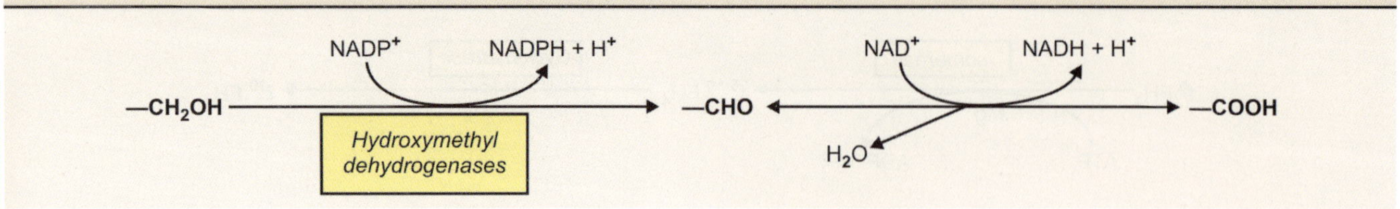

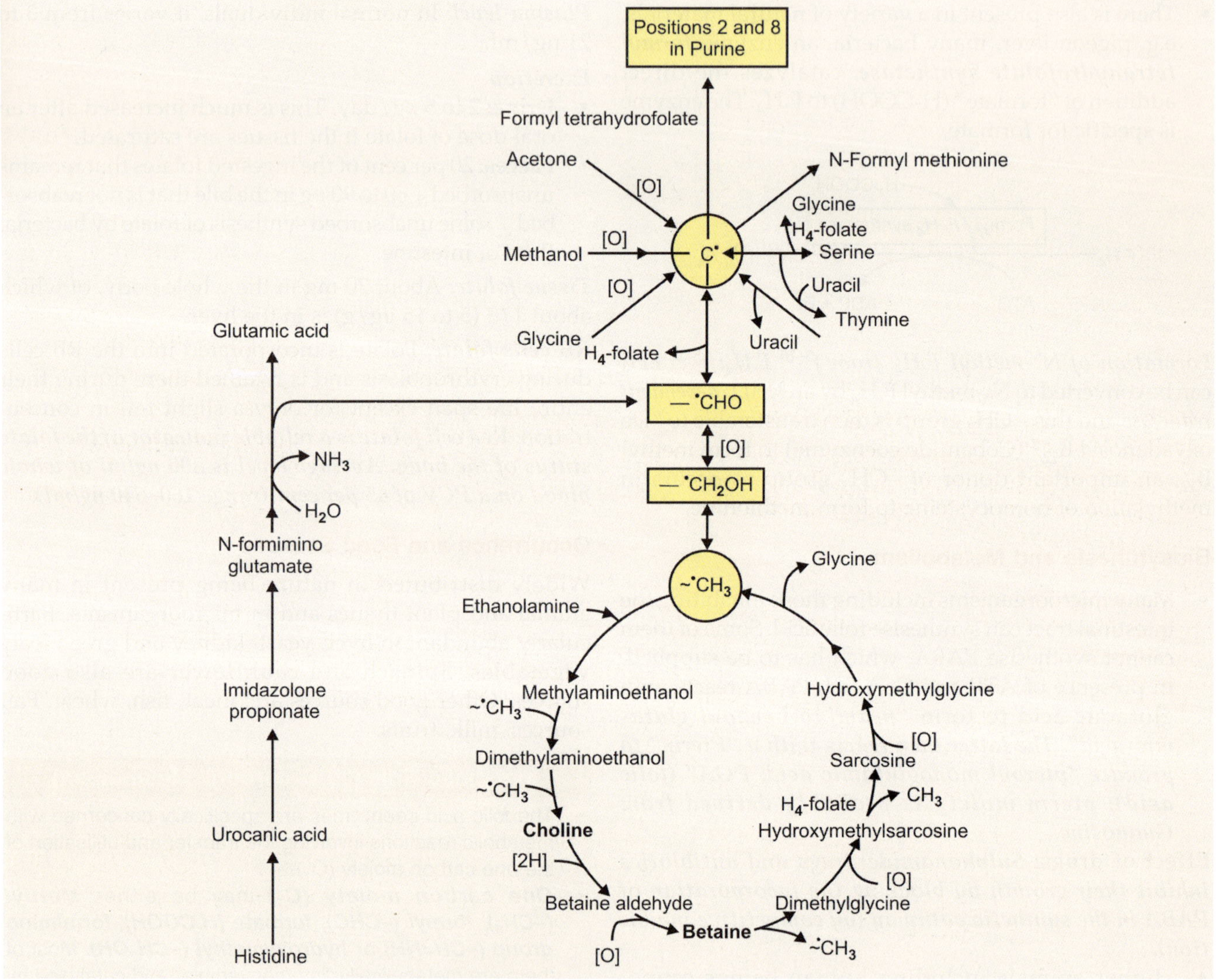

Fig. 12.6: Sources and utilisation of the one carbon moiety

CLINICAL SIGNIFICANCE

Folic acid (antagonists): Several antagonists to this vitamin has been found out. They are of much clinical interest:

On account of their ability to inhibit cell division and multiplication, **they have been used in treatment of conditions where there is unrestricted cell growth,** e.g.

- **In Leukaemias,**
- **In Erythraemias,** and
- **In malignant growths.**

Two folic acid antagonists are important:

1. ***Aminopterin:*** –NH_2 group is substituted for the –OH group in position 4 of the pteridine nucleus. Chemically it is 4-aminofolate. It has maximal inhibitory action.
2. ***Amethopterin or methotrexate:*** It is 4-amino-10 methyl folate. Methotrexate inhibits "*dihydrofolate reductase*" and has been used as anticancer drug.
3. ***Trimethoprim or septran:*** Inhibits *dihydrofolate reductase* and formation FH_4 is decreased. The drug has been used as antibacterial agent.
4. ***Pyrimethamine:*** It is used as antimalarial drug. It also inhibits dihydrofolate reductase.

Deficiency manifestations: Deficiencies have been produced and studied in experimental animals as well as in human volunteers.

In experimental animals deficiency produced most readily in **two ways:**

1. By feeding sulphonamides and broad spectrum antibiotics for prolonged periods—to inhibit growth of intestinal bacteria.
2. By administration of folic acid antagonists.

Outstanding feature is abnormalities of blood formation. Other manifestations include—growth retardation, weakness, lethargy, reproduction difficulties (infertility in females) and inadequate lactation.

CLINICAL ASPECT

Macrocytic Type of Anaemia

- ***Bone marrow*** shows: Arrested development of all elements: erythroid, myeloid and thrombocytes. Megaloblasts and myeloblasts accumulate at the expense of more mature cells viz. erythroblasts, normoblasts and myeloblasts. The number of megakaryocytes decreases.
- ***The peripheral blood picture:*** Reflects these production defects, being characterised by one or more of the following, depending mainly on the degree of deficiency.
  - ***A macrocytic type of anaemia*** at times with normoblasts, erythroblasts, and megaloblasts.
  - ***Granulocytopenia***, occasionally with myelocytes, and
  - ***Thrombocytopenia.***

Clinical conditions: On the basis, chiefly of prompt response to specific replacement therapy, the following clinical conditions have been attributed to folic acid deficiency:

- ***Nutritional macrocytic anaemia*** (cause dietary deficiency of folic acid).
- ***Megaloblastic anaemia of infancy*** (also mainly due to dietary deficiency).
- ***A congenital (inherited) type*** may be due to *reductase* deficiency ("Familial" type).
- ***Megaloblastic anaemia of pregnancy,*** mechanism not clear, may be relative deficiency.
- ***Macrocytic anaemia in liver diseases*** (may be due to inadequate storage/conversion).
- ***Megaloblastic anaemia in caeliac disease and sprue*** (inadequate absorption). In this there may be B_{12} deficiency also, associated with neurological manifestations.
- ***Macrocytic anaemia after extensive intestinal resection*** (inadequate absorption).

'FIGLU' test

The test is **used to detect folate deficiency**. In the metabolism of the amino acid histidine there is folic acid dependent step at the point where formimino-glutamic acid ("Figlu") is converted to glutamic acid (GA)

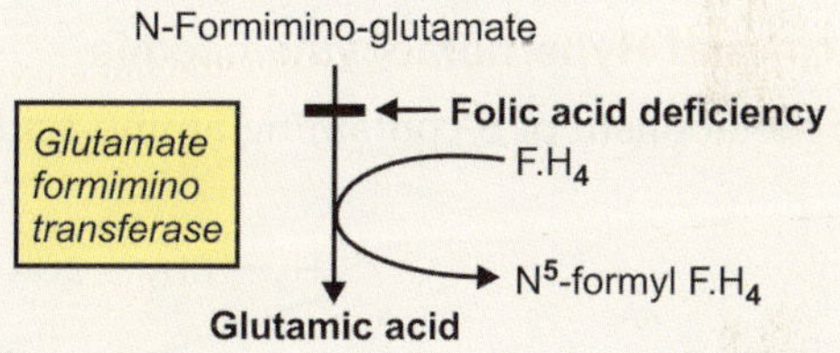

In folic acid deficient patients, this reaction cannot be carried out, as a result, "figlu" accumulates in the blood and excreted in urine. **Figlu *excretion in urine is an index of folic acid deficiency.*** When a "loading dose" of histidine is given, the excretion of 'Figlu' in urine is increased further **(Histidine loading test)**.

One Carbon-Donor and Acceptor

C_1-moiety (Donors)	*C_1-moiety (Acceptors)*
1. Formimino group (–CH=NH) of formimino glutamic acid (formed from Histidine)	**1.** Position 2 and 8 of purine ring.
2. Methyl group ($-CH_3$) of Methionine, choline, Betaine and Thymine. All of which are oxidised to hydroxymethyl ($-CH_2OH$) group and carried as such on $f^{5-10}.FH_4$ The hydroxymethyl group is then oxidized in an NADP-dependant reaction to a 'formyl' group. $h^{5-10}\ FH_4 \xrightarrow{NADP^+ \rightarrow NADPH + H^+} f^{5-10}.FH_4$	**2.** N-formyl methionine of t-RNA (Given by $f^{10}.FH_4$)
	3. Glycine → to serine conversion; formation of β-carbon of serine
	4. Homocysteine → to form methionine
	5. Uracil → to form thymine
	6. Ethanolamine → to form choline
3. β-carbon of serine as a hydroxymethyl group may contribute single carbon moiety.	**7.** Histidine synthesis

Daily requirement: Exact requirement difficult to ascertain due to two reasons:
1. Available in nature-ubiquitous.
2. Most of human need supplied by bacterial synthesis in intestine.

Recommended dietary daily allowance:
- Adults: 400 to 500 μg daily
- Infants: 50 μg
- Children: 100 to 300 μg

Requirement increases in pregnancy and lactation.
- Pregnant women: 800 μg
- Lactating women: 600 μg

Folic Acid and Inositol: Prevents Birth Defects

Based on experiments with "curly tail" mice, researchers have found that B-complex vitamins viz., folic acid and Inositol when given in pregnancy may **prevent formation of *neural tube defects* (NTDs)**, such as **spina bifida**. About 70% of NTDs can be prevented by taking folic acid. Folic acid resistant cases respond to inositol. Exact mechanism is not clear.

Folic Acid and Hyperhomocysteinaemia

Refer to metabolism of S-containing amino acids.

Risks of Excess Folic Acid

- Dosage over 1 mg may cause aggravation of vitamin B_{12} deficiency and may precipitate ***irreversible nerve damage***.
- ***Antagonism*** between folic acid and the ***anticonvulsants*** used in the treatment of epilepsy observed.
- Solubility of folic acid is low, hence large doses of folic acid if given parenterally there is risk of crystallisation in kidney tubules leading to ***renal damage***.

VITAMIN B_{12} (CYANOCOBALAMINE)

Synonyms: Antipernicious anaemia factor, extrinsic factor of Castle, animal protein factor.

Chemistry

(a) Structure of vitamin B_{12}: It has been delineated (Fig. 12.7)
1. Central portion of the molecule consists of four reduced and extensively substituted **pyrrole rings**, surrounding a **single cobalt atom (Co)**. This central structure is called as **Corrin Ring *system.***
2. The above system is similar to porphyrins, but differ in that two of pyrrole rings. Rings I and IV are joined directly.

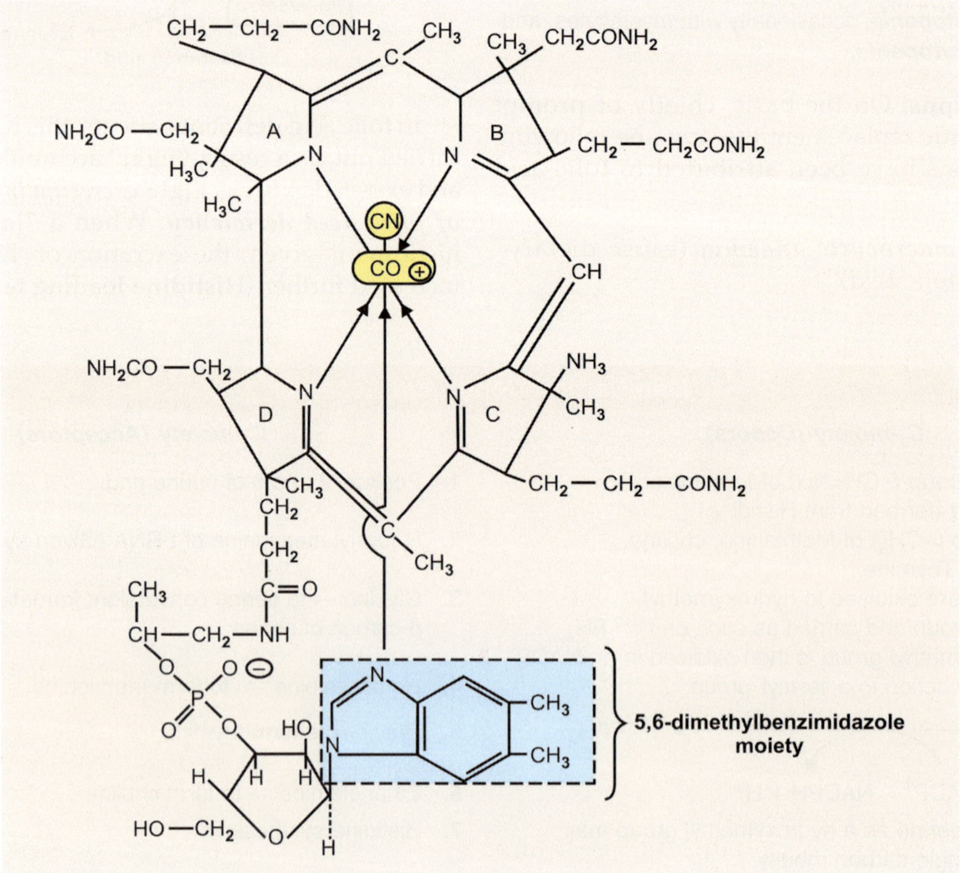

Fig. 12.7: Structure of vitamin B_{12}

3. Below the corrin ring system, is **DBI ring** –5, 6-dimethyl Benzimidazole riboside which is connected:
   - At one end, to central cobalt atom, and
   - At the other end from the riboside moiety to the ring IV of corrin ring system.
4. One PO_4 group connects ribose moiety to **aminopropanol (esterified)**, which in turn is attached to propionic acid side chain of ring IV.
5. A *cyanide* group is coordinately bound to the cobalt atom and then is called as *cyanocobalamine.*

(b) Varieties of Vitamin B_{12}

1. When cyanide is bound to cobalt atom it is called as *cyanocobalamine*, but if cyanide group is removed, then it is called as *cobalamine.* Cyanocobalamine is identical with originally isolated vitamin B_{12}.

   B_{12} which occurs in natural materials does not contain cyanide group. In the original isolation, cyanide group was added only to promote crystallization.
2. –OH group, NO_2, Cl^- and $SO_4^=$ may replace cyanide group, in which case, it is called respectively as:
   - **Hydroxycobalamine (B_{12a}), (Hydroxocobalamine)**
   - **Nitritocobalamine (B_{12c}),**
   - **Chlorocobalamine, and**
   - **Sulphatocobalamine, in that order.**
3. Biologic actions of these derivatives are similar to cobalamine, but *Hydroxycobalamine (B_{12a}) is superior as:*
   - It is more active in enzyme systems
   - It is retained longer in the body when given orally.

   Hence, *B_{12a} is more useful for therapeutic administration of B_{12} by mouth.*

Metabolism

Absorption and Excretion

1. **Vitamin B_{12} is absorbed from Ileum;** for its proper absorption it requires:
   - Presence of HCl, and
   - *Intrinsic factor* (IF) of Castle, a constituent of normal gastric juice.
2. **Intrinsic factor (IF)**

   It is secreted by parietal cells, it is a **glycoprotein**, a constituent of gastric mucoproteins. In addition to amino acids, contain hexoses, hexosamines and sialic acid. It is **non-dialyzable** and **thermolabile**, destroyed by heating at 70° to 80°C for ½ hour. Inactivated by prolonged digestion with pepsin or trypsin. Found in 'Cardiac' end and fundus of stomach, but *not in the pylorus*.

Note: *Atrophy of fundus of stomach and a lack of free HCl (achlorhydria) is usually associated with pernicious anaemia, caused by B_{12} deficiency.*

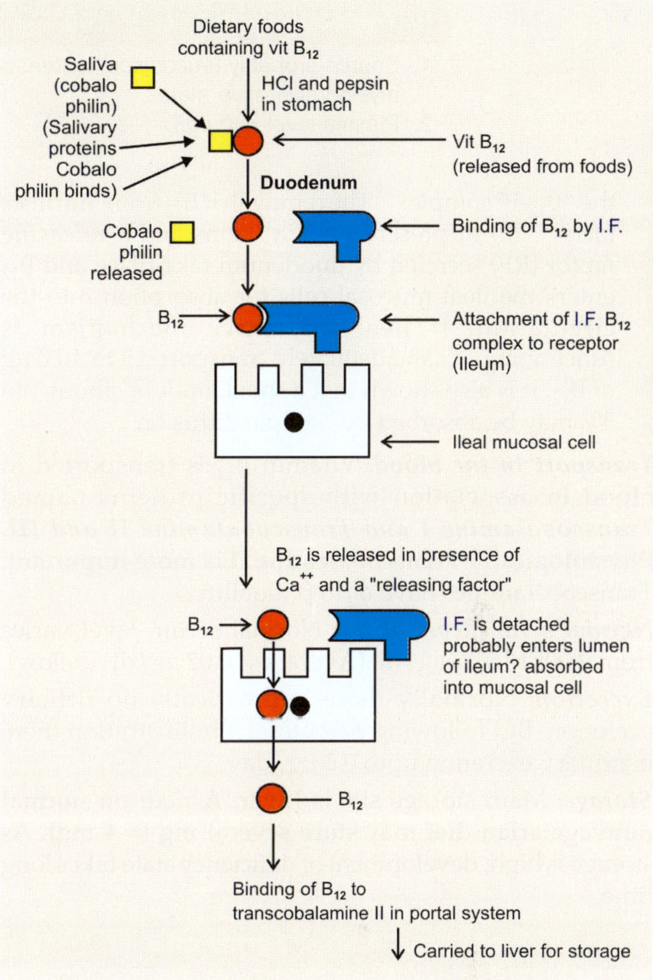

Fig. 12.8: Mechanism of absorption of vitamin B_{12}

3. *Mechanism of Absorption* (Fig. 12.8)

   Recently, it has been shown that *two binding proteins* are required for absorption of Vit B_{12}. They are:
   - **Cobalophilin:** A binding protein secreted in the saliva.
   - **Intrinsic factor (IF)**, a glycoprotein secreted by parietal cells of gastric mucosa.
   - Gastric acid (HCl) and pepsin release the vit B_{12} from protein binding in food and make it available to bind to salivary protein, cobalophilin.
   - In the duodenum, cobalophilin is hydrolysed, releasing the vitamin for binding to "Intrinsic factor" (IF).

CLINICAL ASPECT

In pancreatic insufficiency, the ***cobalophilin-bound vitamin B_{12}*** may not be split and the complex is excreted in faeces resulting to development of vitamin B_{12} deficiency.

- Vitamin B_{12} is absorbed from the distal third of the ileum via *specific binding site* **(receptors)** that binds

Transcobalamine I	*Transcobalamine II*
1. Source-probably leucocytes. Increased in myeloproliferative states. 2. Plasma level =60 $\mu g/L$	1. Source is liver 2. Plasma level = 20 $\mu g/L$

the "B_{12}-IF complex". The removal of B_{12} from 'intrinsic factor' (IF) in presence of Ca^{++} ions and a ***releasing factor*** (RF) secreted by duodenum take place and B_{12} enters the ileal mucosal cells for absorption into the circulation. If ileal absorptive mechanism is functioning, it can adequately transport 0.5 to 10.0 μg of B_{12}. It is also shown that a small amount, about 1 to 3% may be absorbed by "simple diffusion".

Transport in the blood: Vitamin B_{12} is transported in blood in association with specific proteins named ***Transcobalamine I and Transcobalamine II and III.*** **Physiologically Transcobalamine II is more important.** Transcobalamines have α_2 to β mobility.

Normal serum level of B_{12}: Normal serum level varies from 0.008 to 0.42 $\mu g/dl$. (Average = 0.02 $\mu g/dl$) (below).

Excretion: Normally there is practically no urinary excretion. But following parenteral administration there is urinary excretion upto 0.3 μg/day.

Storage: Main storage site is **Liver. A man on normal nonvegetarian diet may store several mg (= 4 mg).** As storage is high, development of deficiency state takes long time.

Biological "Active" Forms of B_{12}

- Biologically active forms are ***cobamide coenzymes,*** act as coenzyme with various enzymes.
- Cobamide coenzyme do not contain the "cyano" group attached to cobalt but instead there is an ***Adenine Nucleoside*** (5'– deoxyadenosine) which is linked to cobalt by a C →CO bond (Fig. 12.9).
- ***5'–adenosyl moiety is derived from ATP,*** which after donating the 'adenosyl' group, releases all three PO_4 groups as inorganic "tripolyphosphates".

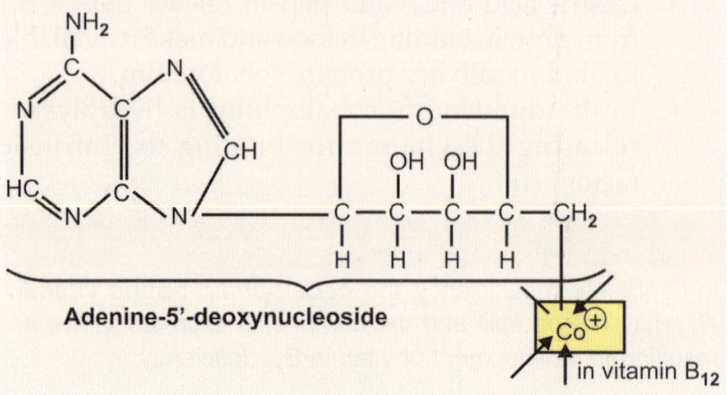

Fig. 12.9: Attachment of adenosyl moiety to vitamin B_{12} through 5'–C to cobalt in the cobamide coenzymes

In formation of "Adenosyl coenzyme" cobalt undergoes successive reduction in a series of steps catalysed by the enzyme B_{12a} *reductase,* which requires NADH and FAD.

- **B_{12} a** — red coloured (Co^{+++})
  ↓
- **B_{12} γ** — orange (Co^{++})
  ↓
- **B_{12} γ** — gray-green (Co^{+})

B_{12} (Co^{+}) reacts with ATP to form **Adenosyl coenzyme.**

Varieties of cobamide coenzymes

At least **four varieties** have been isolated:

- **DBC:** Contains 5,6-dimethyl Benzimidazole (called Dimethyl-Benzimidazole cobamide, DBC).
- **BC:** Benzimidazole cobamide; this contains an unsubstituted methyl free Benzimidazole.
- **AC:** Adenyl cobamide, which contains an adenyl group.
- **MC:** Methyl cobamide. CH_3 group is attached to cobalt atom rather than adenosyl moiety.

These coenzymes donot contain the cyanide group and hence called as ***Corrinoid coenzymes.***

Metabolic Role of Cobamide Coenzymes

- ***Methyl malonyl-CoA to succinyl-CoA conversion:*** Vitamin B_{12} is required as a coenzyme for the conversion of L-methyl malonyl CoA to succinyl-CoA. The reaction is catalyzed by the enzyme *isomerase.*

$$\text{L-Methyl Malonyl-CoA} \xrightarrow[\textit{Isomerase}]{B_{12}} \text{Succinyl-CoA}$$

Normal healthy individuals excrete less than <2 mg/day which is not detectable.

Note: In B_{12} deficiency: Methyl malonic acid accumulates and excretion of methyl malonic acid in urine is increased. ***Methyl malonic aciduria is a sensitive index for B_{12} deficiency.***

- ***Methylation of Homocysteine to Methionine:*** This requires tetrahydrofolate ($F.H_4$) as a $-CH_3$ carrier.
- ***Methylation of pyrimidine ring to form thymine.***
- ***Conversion of Ribonucleotides to deoxyribonucleotides: It is of importance in DNA synthesis.***

$$\text{Ribonucleotides} \xrightarrow[\textit{Reductase}]{B_{12}} \text{Deoxyribonucleotides}$$

Cobamide coenzymes play an essential role as a "H-transferring agent"

- ***Required for metabolism of* Diols*:***

$CH_2.OH$ | $CH_2.OH$ (Ethylene glycol) $\xrightarrow{B_{12}}$ $CH_3.CHO + H_2O$ (Acetaldehyde)

- ***In bacteria*** Interconversion of glutamate and β-methyl aspartate:

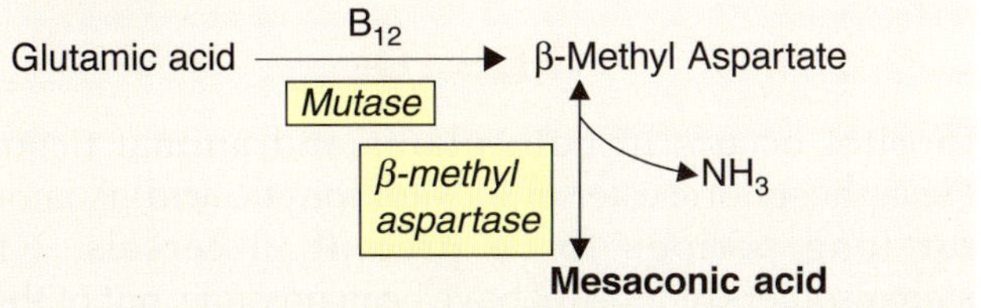

FOLATE TRAP (METHYL TRAP HYPOTHESIS)

- B_{12} is necessary as coenzyme for conversion of N^5–methyl $F.H_4$ to form methyl-B_{12} which is required for conversion of homocysteine to methionine.

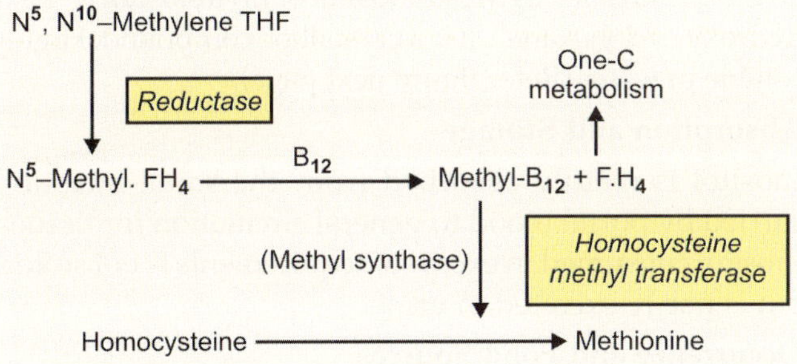

- **In B_{12} deficiency,** the above reactions cannot take place and folate is permanently trapped as N^5–methyl. FH_4 and is therefore not available for C_1-transfer. It is called as **folate trap**. *This results in diminished synthesis of thymidylate and DNA. Increased folate levels are observed in plasma and the activity of the enzyme* **homocysteine methyl transferase** *(also called as Methyl synthase) is low.*

CLINICAL ASPECT

Deficiency Manifestations

I. Adult pernicious anaemia: Vit. B_{12} deficiency produces macrocytic anaemia. In addition to haematological manifestation, it is combined with ***neurological features (subacute combined degeneration of the cord).*** It is an auto-immune disease, antibodies to IF and parietal cells found.

In addition to haematological changes, the following changes are seen:

- **Mucosal atrophy of stomach** and inflammation of tongue ***(glossitis),*** inflammation of mouth ***(stomatitis),*** and pharynx ***(pharyngitis).***
- Absence of HCl ***(Achlorhydria).***
- ***Degenerative changes of posterior and lateral columns of the spinal cord,*** resulting in peripheral sensory disturbances, hyperactive reflexes, ataxia, paralysis.

Cause: Increased concentration of methyl malonic acid, which competes with malonyl-CoA leading to impairment of FA synthesis.

Haematological changes: Peripheral blood picture shows:

- Macrocytic type of anaemia with megaloblasts, erythroblasts, and normoblasts.
- Granulocytopenia with occasional myelocytes.
- Thrombocytopenia.
- Reduction in RB cells count and Hb content.

Other biochemical changes are:

- Serum B_{12} level is decreased↓
- Urinary B_{12} level is decreased↓
- Rise in faecal B_{12} excretion ↑
- **Schilling test** shows decrease in labelled B_{12} absorption. (Refer Chapter on Radioactivity)
- Urinary excretion of methyl malonic acid is increased↑
- **Auto-antibodies** to IF (intrinsic factor) and parietal cells (canalicular lipoproteins). **2 types of IF antibody**
  Type I: Blocks attachment of B_{12}
  Type II: Attaches to complex

Bone marrow:

- Shows evidences of arrested development of all elements: erythroid, myeloid and thrombocytes.
- Megaloblasts and myeloblasts accumulate at the expense of more mature cells, viz. erythroblasts, normoblasts and myelocytes.
- Number of megakaryocytes decrease.

DNA synthesis: In pernicious anaemia, DNA synthesis in maturing red blood cells is depressed, due to:

- Failure in conversion of ribonucleotides to deoxyribonucleotides, and
- Partly owing to the failure in forming the thymidylic acid by the methylation of deoxyuridylic acid.

The fall in DNA synthesis results in prolongation of resting phases between successive mitoses of maturing RB cells as also a rise in the RNA levels ↑ in those cells so that the resulting ***macrocytic RB cells has a much higher RNA: DNA ratio*** ↑ than the normal RB cells. ***The lengthening of resting phase may be the ultimate reason for the formation of Megaloblasts in pernicious anaemia.***

II. Congenital pernicious anaemia: Occurs in postnatal period, before 2½ years of age:

- Due to lack of Intrinsic factor (IF).
- It is not accompanied by gastric acid secretion/or abnormalities of gastric mucosa.

III. Juvenile pernicious anaemia:

- Failure to secrete Intrinsic factor (IF).
- Associated achlorhydria and atrophic gastritis.
- Antibodies to IF/and/or parietal cells have been demonstrated (***autoimmune* in nature).**

IV. Following surgical operations: Pernicious anaemia may occur in following:

- Total gastrectomy
- Extensive resections of small intestine.

*V. **Tapeworm infection of GI tract:*** A megaloblastic anaemia, responsive to vitamin B_{12} therapy occurs with infestation with "Fish tapeworm", ***Diphyllobothrium latum,*** which eats up unusually large amounts of vitamin B_{12} from the gut, creating B_{12} deficiency to the host.

Methyl Malonic Aciduria: This is of **two types:**

1. ***Due to vitamin B_{12} deficiency:*** Vitamin B_{12} is required as a coenzyme for *"isomerase"* which converts methyl malonyl CoA to succinyl CoA. In vitamin B_{12} deficiency, this conversion cannot take place, as a result methyl malonic acid ↑accumulates in blood and tissues and excreted in urine. ***Excretion of methyl malonic acid in urine is a sensitive index for vitamin B_{12} deficiency.***
2. ***Inherited deficiency of the enzyme isomerase:*** In this vitamin B_{12} deficiency is not there. It is observed as inherited disorder in infants and young children, with severe metabolic acidosis.

Occurrence and Sources of Vitamin B_{12}

It is present in foods of animal origin only and is not present in foods of vegetable sources. In nature it is obtained via synthesis by bacteria in soil, water and animal intestine. Good and rich animal sources are: liver, eggs, fish, meat, kidney. Fair sources are; milk, and dairy products.

Note: ***Those people who are purely vegetarians should take enough milk/and milk products and one egg twice a week so that they do not develop B_{12} deficiency in the long run.***

Daily Requirements

- In normal adults: 3 μg/day
- Infants: 0.3 μg/day
- Children: 1 to 2 μg/day
- In pregnancy and lactation: Requirement is increased approx. 4 μg/day.
- In pernicious anaemia: 0.5 to 1.0 μgm/day given parenterally will maintain in complete haematologic and neurologic remissions.

Nieweg's Hypothesis

Deficiency of folic acid is not accompanied by neurological lesion but B_{12} deficiency is associated with neurological lesions, e.g. subacute combined degeneration of the cord, both produces megaloblastic anaemia, Why?

This can be explained by Nieweg's hypothesis:

Nieweg's hypothesis postulates that folic acid is concerned with DNA metabolism, would explain how a deficiency of folic acid could produce GI and bonemarrow abnormalities unaccompanied by neurologic lesions. On the other hand, a deficiency of B_{12} could cause neurological lesions as a result of RNA deficiency as well as changes in the bone marrow and GI tract from DNA deficiency. (Refer box below).

INOSITOL

Inositol occurs in both plants and animal tissues. Hexaphosphoric ester of inositol (phytic acid) is known for long periods to be present in cereals. Nine stereoisomers of inositol have been possible, out of those seven forms are optically inactive and two forms are optically active. Vitamin activity is associated with only one optically inactive form, known as **Meso-inositol** (or **Myo-inositol**).

Chemistry

Inositol, also called as **Meso-inositol** or **myoinositol** is "hexa hydroxy cyclohexane". It is a crystalline compound. Highly soluble in water (Refer figure next page).

Absorption and Storage

Inositol is readily absorbed from the small intestine, carried by portal blood to general circulation for tissues. Inositol consumed in excess of requirements is not stored, but is partly excreted in urine.

Occurrence and Food Sources

Animal tissues such as muscles, brain, red blood cells, and eye contain inositol. It is also known as **muscle sugar** due to its presence in muscles. It is also widely distributed in plants, e.g. in fruits, vegetables, whole grains and yeast.

Types of Inositol

Can occur in **free** and **bound forms.**

Although inositol occurs in the **free state** in muscles, and other tissues, a greater part of it exists in the **bound form.** In animal tissues, it exists mostly as "phospholipids" phosphoinositides, while in plant tissues, it exists mostly in the form of PL, phytic acid and phytin. Inositol containing PL, 'phosphoinositides' is present in brain and liver mostly. Mitochondria and microsomes also contain large quantities of phosphoinositides.

Nieweg's hypothesis

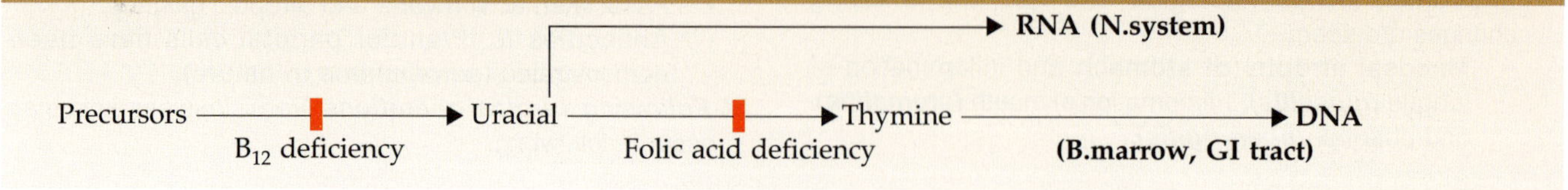

Phytic acid: Cereals, legumes, nuts, etc. contain large amounts of phytic acid and phytin (Ca and Mg salts of phytic acid)

- *Phytic acid is* **inositol-hexaphosphoric acid.** This can be hydrolysed into inositol and phosphoric acid by the enzyme *phytase,* which is present in plants.
- Phytic acid is not absorbed from the intestine unless it is hydrolysed by the enzyme *phytase*. Phytase activity in the intestinal juice is very low and hence a very small percentage of phytic acid present in the diet is hydrolysed and absorbed.
- Phytic acid interferes with absorption of calcium, etc by forming insoluble phytates.

Daily requirements: Daily requirements not yet reported.

Metabolic Role and Functions

- ***Lipotropic action:*** Like choline, inositol exerts a lipotropic effect. It has been shown to exert a curative effect on fatty liver in rats.
- ***Cholesterol level:*** In its presence, blood cholesterol level is not increased as expected on high cholesterol diets in the experimental animals.
- ***Conversion to glucose:*** Isotopically labelled-Inositol when given to animals have been shown to give rise to labelled glucose in the body.
- ***Antiketogenic:*** It has been shown to act as anti-ketogenic compound.

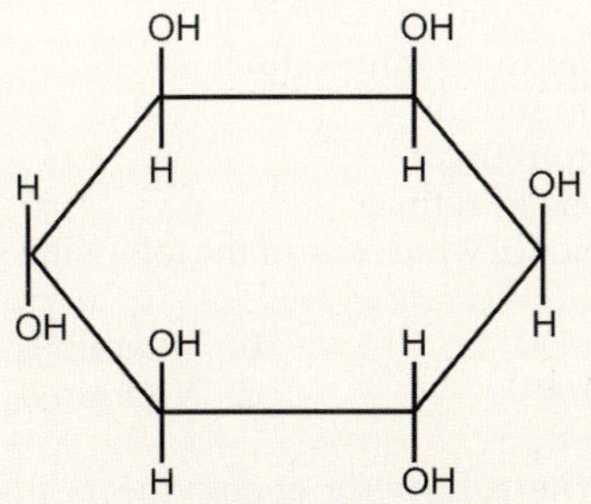

Molecular formula = $C_6H_{12}O_6$

- ***Conversion of glucose to inositol:*** When isotopically labelled glucose is fed to rats, labelled inositol has been isolated from the tissues indicating that some inositol is synthesized in the tissues.
- ***Oxidation:*** Inositol is oxidised to glucuronic acid in the liver by an enzyme *oxygenase*.
- ***Action of cyclase enzyme:*** An enzyme *cyclase* can catalyze the formation of inositol-1-(P) from glucose-6-(P) in the liver in presence of NAD^+.
- ***Growth of fibroblasts:*** Studies in tissue culture have revealed that inositol is required for growth of fibroblasts.
- ***Role in hormone action:*** Phosphatidyl inositol-4, 5-P_2 is hydrolysed to myoinositol -1, 4, 5 -P_3 and diacyl-glycerol through the action of *phosphodiesterase* enzyme. This reaction occurs within seconds after the addition of either vasopressin or epinephrine to hepatocytes. Myoinositol-P_3 at 0.1–0.4 μmol/L releases Ca^{++}. Phosphatidyl inositides products act as the "second messenger" in hormone action and Ca^{++} actually acts as "tertiary messenger".
- ***Action on heart muscle:*** Inositol occurs in large quantities in mammalian heart muscle. Inositol in concentrations found in heart muscle increases the amplitude and rate of contraction.
- ***Action on nerve:*** Inositol causes an increase in nerve chronaxie in rats and it abolishes the reduction in chronaxie caused by 'adrenochrome'.
- Young mice fed on inositol deficient diet lose hairs ("alopecia").
- It stimulates the growth of yeast and fungi.
- In animals, it has been shown that inositol increases the peristalsis of the small intestines.

Deficiency manifestations: No deficiency disease has been reported.

Deficiency manifestations in mice fed on inositol-deficient diets include:

Retardation of growth, failure of lactation, loss of hair over the body ("alopecia"), "Spectacled" eye: loss of hair around the eyes, in chicks: encephalomalacia has been reported.

I. Questions (Essay types and short notes)

1. Describe the chemistry and functions of ascorbic acid in the body.
2. Describe chemistry, functions and deficiency manifestations of Vitamin C.
3. Describe the chemistry, functions and deficiency manifestations of retinol.
4. Describe the chemistry, functions and deficiency manifestations of niacin.
5. Name the coenzymes taking part in metabolic reactions giving one example for each.
6. Name the water-soluble vitamins. Describe the chemistry and role of thiamine.
7. Describe the sources, daily requirements, functions and deficiency symptoms of vitamin A.
8. What are B-complex vitamin? Describe the role of any one.
9. Name the fat-soluble vitamins. Describe the chemistry and functions of vitamin D.
10. What are the "active" forms of (a) Thiamine, (b) Pyridoxine, (c) B_{12}? Describe their metabolic role.
11. Name the fat-soluble vitamins. Describe the chemistry and functions of vitamin A.

SECTION TWO

12. Name the fat-soluble vitamins and describe their physiological functions.
13. Describe the chemistry and functions of pyridoxine.
14. Discuss the sources and metabolic role of vitamin D. Comment on why vitamin D is considered to be a prohormone. Write a note on vitamin D deficiency.
15. Write the sources, requirement and metabolic functions of vitamin D.

Short Notes

a. Thiamine
b. Ascorbic acid (vitamin C)
c. Formation of nicotinic acid in the body
d. Niacin
e. Vitamin K
f. 1, 25-dihydroxy cholecaliferol and its significance
g. Calciferol
h. Hypervitaminosis
i. Retionol
j. Vitamin A
k. Vitamin B_{12}
l. Vitamin B_1
m. Metabolic role of vitamin B_{12}
n. Deficiency of vitamin C
o. Metabolic role of pantothenic acid
p. Biochemical role of vitamin C
q. Metabolic role of folic acid
r. Riboflavin
s. Retinoic acid and its functions.

Differentiate

Coenzyme and isoenzyme

II. MCQs (Give one correct answer)

1. Which coenzyme is not involved in oxidative decarboxylation of pyruvic acid?
   a. TPP b. Lipoic acid
   c. Mg^{++} d. Biotin
   e. CoA-SH
2. FAD is a coenzyme in a reaction catalysed by which of the following enzymes?
   a. β-ketothiolase
   b. Acyl-CoA synthetase
   c. Acy-CoA dehydrogenase
   d. Hydroxy acyl-CoA dehydrogenase
   e. Δ 2-enoyl-CoA dehydrogenase
3. Retinal is reduced to retional by the enzyme retinene reductase in presence of which coenzyme:
   a. $NADP^+$ b. $NADPH + H^+$
   c. NAD^+ d. $NADH + H^+$
   e. FAD^+
4. Retinol is transported in the blood as attached to:
   a. Albumin b. α_1-Globulin
   c. α_2-globulin d. β-globulin
   e. γ-gobulin
5. Vitamin K_2 was originally isolated from:
   a. Soyabean b. Putrid fish meal
   c. Alfalfa d. Oysters
   e. Wheat germs
6. The following forms of vitamin A is used in the visual cycle of Wald:
   a. Retinol b. Retinoic acid
   c. Retinaldehyde d. Retinyl palmitate
   e. Retinyl acetate
7. Increased carbohydrate consumption increases the dietary requirement for:
   a. Thiamine b. Riboflavine
   c. Pyridoxine d. Folic acid
   e. Niacin
8. Pyridoxal-P is a cofactor for which of the following enzymatic reactions:
   a. Fixation of CO_2
   b. Decarboxylation reaction
   c. Phosphate group transfer
   d. Transmethylation reaction
   e. Dehydrogenase reaction
9. The disease pellagra is due to a deficiency of:
   a. Vitamin B_6 b. Nicotinic acid
   c. Pantothenic acid d. Folic acid
   e. Biotin
10. Increased protein intake is accompanied by an increased dietary requirement for:
    a. Ascorbic acid b. Thiamine
    c. Riboflavine d. Folic acid
    e. Nicotinic acid
11. The "xanthurenic index" can be used to measure pyridoxine deficiency, it involves the metabolism of:
    a. Tryptophan b. Tyrosine
    c. Histidine d. Glycine
    e. Methionine
12. The absorption of light by cells in the retina of the eye results in the conversion of:
    a. All *trans*-retinal to *cis*-retinal
    b. *Cis*-retinal to all *trans*-retinal
    c. Retinal to retinol
    d. Retinol to retinal
    e. β-carotene to retinal
13. The deficiency of which one of the following vitamin causes creatinuria:
    a. Vitamin E b. Vitamin K
    c. Vitamin A d. Vitamin B_6
    e. Vitamin B_1
14. Which of the following coenzyme is involved in the formation of hydorxy-proline during collagen synthesis?
    a. Pyridoxal-P b. TPP
    c. Cobamide coenzyme d. Biotin
    e. Ascorbic acid
15. A biochemical indication of vitamine B_{12} deficiency can be obtained by measuring the urinary excretion of:
    a. Pyruvic acid b. Lactic acid
    c. Malic acid d. Methyl malonic acid
    e. Urocanic acid
16. Which of the following characteristic would be seen in a patient with severe deficiency of thiamine?
    a. A decreased level of pyruvic acid
    b. A decreased level of lactic acid
    c. A decreased red cells transminase activity
    d. A decreased red cells transaldolase activity
    e. A decreased red cells transketolase activity

17. To prevent rickets in case of chronic renal disorders, which of the following substances should be administered?
a. Cholecalciferol b. Ergocalciferol
c. 25-OH-calciferol d. 1,25-di-OH-cholecalciferol
e. High dietary calcium
18. 'Chastek paralysis" described in foxes is due to eating of
a. Raw polar Bear's liver b. Raw fish
c. Cereals d. Leguminous plants
e. None of the above
19. Which of the following vitamins is associated with synthesis of coagulation factor prothrombin?
a. Vitamin A b. Vitamin C
c. Vitamin K d. Vitamin E
e. Vitamin D
20. The visual impulse is associated with the:
a. Transport of cis-retinal from liver to retina,
b. Dissociation of rhodopsin into trans-retinal and opsin
c. Dissociation of rhodopsin into cis-retianl and opsin
d. Condensation of opsin with trans-retinal
e. Reduction of $NADP^+$
21. 'Bracken' disease describe in grazing animals is due to feeding of:
a. Cereals b. Grass
c. Leguminous plants d. Ferns
e. Soyabeans
22. Sulphadrugs interfere with bacterial synthesis of:
a. Vitamin E b. Coenzyme A
c. Lipoate d. Vitamin D
e. Tetrahydrofolate ($F.H_4$)
23. In biotin-containing enzymes, the biotin is tightly bound to the enzyme via:
a. An amide linkage to the α-amino group of the protein
b. An amide linkage to the ε-amino group of lysine
c. An amide linkage to the α-carboxyl group of the protein
d. An amide linkage to the γ-carboxyl group of glutamate
e. A covalent bond with CO_2
24. Match the following:
A. Tetrahydrofolate ($F.H_4$)
B. NAD^+
C. Pyridoxal-(P)
D. Thiamine Pyrophosphate (TPP).
1. Required as coenzyme in decarboxylation reaction __________
2. As a coenzyme in oxidative decarboxylation of pyruvic acid __________.
3. As a coenzyme in dehydrogenation __________
4. In formyl transferase __________
25. Match the following:
A. Increase 'Figlu" excretion in urine
B. Cheilosis, glossitis, and dermatitis
C. Increased excretion of methyl malonic acid in urine
D. Increased CSF pressure and periosteal bone formation
E. Increased xanthurenic acid excretion in urine.
1. Hypervitaminosis A __________.
2. Vitamin B_{12} deficiency __________.
3. Vitamin B_6 deficiency __________.
4. Folic acid deficiency __________.
5. Riboflavine deficiency __________.
26. Which of the following statements regarding vitamins is correct?
a. All coenzymes contain vitamins or are vitamins.
b. All vitamins can act as coenzymes.
c. Prostaglandins can be derived from fat soluble vitamins
d. All water-soluble vitamins act as coenzymes or coenzyme precursors.
e. Ergosterol is provitamin A.
27. Which of the following vitamins would most likely become deficient whose staple diet is maize?
a. Ribolfavin b. Thiamine
c. Niacin d. Pantothenic acid
e. Ascorbic acid
28. Pantothenic acid is a constituent of the coenzyme involved in:
a. Dehydrogenation b. Acetylation
c. Decarboxylation d. Reduction
e. Oxidation
29. Biotin is involved in which of the following types of reactions?
a. Deaminations b. Hydroxylation
c. Oxidation d. Reduction
e. Carboxylation

Answers to MCQs

1. d	2. c	3. d	4. b	5. b	6. c
7. a	8. b	9. b	10. c	11. a	12. b
13. a	14. e	15. d	16. e	17. d	18. b
19. c	20. b	21. d	22. e	23. b	

24. 1-C, 2-D, 3-B, 4-A
25. 1-D, 2-C, 3-E, 4-A, 5-B,
26. d 27. c 28. b 29. e.

SECTION TWO

CHAPTER 13

CHEMISTRY OF RESPIRATION AND FREE RADICALS

Major Concepts

A. Learn in detail how O_2 is transported from lungs to tissues and factors affecting its transport.
B. Learn in detail how CO_2 is transported from tissues to lungs and factors affecting its transport.
C. Study about oxygen toxicity and free radicals.

Specific Objectives

A. 1. Define respiration, partial pressure of gases.
 2. Know the composition of atmospheric air, alveolar air and expired air with partial pressure of gases.
 3. Study steps of respiration.
 4. Study various modes of O_2 transport.
 5. Study in detail oxyhaemoglobin formation and oxygen-dissociation curves with various factors affecting O_2 delivery: (i) pO_2 (ii) pCO_2 (iii) 2, 3,-BPG (iv) Temp. (v) Electrolytes.
 6. Study Böhr effect and its importance.
 7. Learn in details the clinical importance of O_2.

B. 1. Study various modes of CO_2 transport.
 2. Study details of events occurring at lung site and tissue site.
 3. Know what is chloride shift and its importance.
 4. Study factors affecting CO_2 transport.
 5. Study the role of N_2 in plasma.

C. 1. Learn how "free" radicals are produced? Why oxygen is more prone to produce superoxide radical O_2?
 2. Learn about scavengers of free radicals.
 3. Study the effects of free radicals on biomembranes.
 4. Learn the various anti-oxidants.
 5. Clinical significance: Role of free radicals in disease processes.

INTRODUCTION

The term respiration is normally employed to indicate an interchange of two gases, Oxygen and Carbon dioxide. It is divided into **four major steps.**

- ***Pulmonary ventilation:*** The inflow and outflow of air between the atmosphere and alveoli.
- ***Diffusion:*** The O_2 and CO_2 undergo diffusion in alveoli and blood.
- ***Transport:*** O_2 and CO_2 are transported to and from cells.
- ***Regulation:*** The whole process of ventilation is strictly regulated.

THE COMPOSITION OF AIR AND PARTIAL PRESSURE OF GASES

On a clear, dry day atmospheric air contains 78.62 per cent N_2, 20.84 per cent O_2, 0.04 per cent CO_2 and 0.5 per cent water vapour. In a mixture, each gas exerts its own partial pressure. At the sea level the atmospheric pressure is 760 mm of Hg. The partial pressure of O_2 (with 20.84% O_2) at sea level would be 159 mmHg. Thus, the pressure exerted by a gas at a given temp depends on number of moles of gas in a given volume. Therefore, the sum total of partial pressures of all gases in the atmosphere will be equal to the atmospheric or barometric pressure (BP). For a dry air the partial pressure of H_2O is subtracted from atmospheric pressure $BP - pH_2O = pO_2 + pCO_2 + pN_2$.

Similarly the partial pressure of each gas can be calculated separately from the percentage as,

$$pO_2 = \frac{BP - pH_2O \times \%O_2}{100}$$

Same is true for other gases.

The partial pressures of all these gases are different in atmospheric, expired and alveolar air and are given as follows.

	pN_2	pO_2	pCO_2	pH_2O
• **Atmospheric Air (dry)**	597	159	0.3	0
• **Expired Air**	566	120	27	47
• **Alveolar Air**	569	104	40	47

Exchange of gases between two compartments takes place by the process of **simple diffusion**, i.e. gas flows from a higher to lower partial pressure. Oxygen passes from atmosphere to lungs, then from lungs to venous blood which becomes oxygenated, hence arterial. Oxygen present in arterial blood then diffuses to tissue cells. The CO_2 produced in the tissue cells is then removed in a reverse manner. Thus, the process of respiration involves the transport of O_2 and CO_2. Both these gases are transported via blood.

TRANSPORT OF OXYGEN

Oxygen is continuously supplied to the tissue cells. The total requirement of O_2 is around 250 ml/minute in the resting state and more than ten times during vigorous exercise. The requirement of O_2 to the tissue is fulfilled in **two ways**:

- ***O_2 in physical solution and***
- ***By oxyhaemoglobin.***

1. **Oxygen in Solution**
   A small amount of O_2 can be dissolved to form a solution. However, a very little amount is carried in this form. It also depends on the pO_2. At a pO_2 of 100 mm of Hg (lung alveoli) ***about 0.3 ml of O_2 remains dissolved in every 100 ml blood (i.e. 0.3 ml vol%).*** Since the maximum blood flow is 25 litres/minute, about 50-60 ml of O_2 can be supplied this way which is far too less to meet the requirement of 250 ml.
2. **Oxygen in Oxyhaemoglobin**
   Most of the O_2 is supplied to the tissues as oxyhaemoglobin. Oxygen combines with Hb to form oxyhaemoglobin, HbO_2. The O_2 thus is carried more efficiently, when fully saturated, **one gram of Hb can carry 1.34 ml of O_2.** A person with normal Hb conc. of 14.5 g/dl will thus carry 14.5 × 1.34 = 19.43 ml of

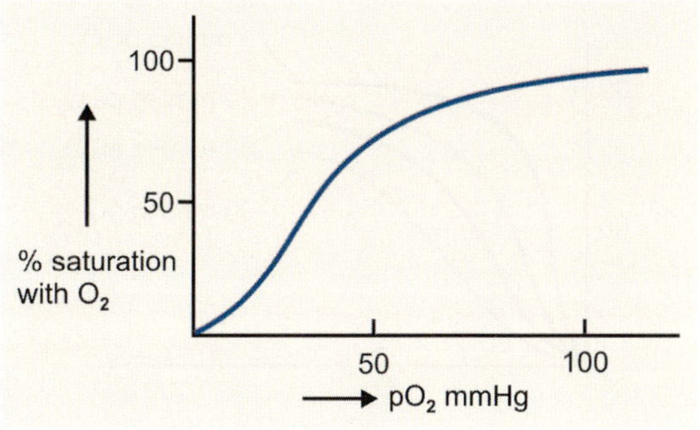

Fig. 13.1: Oxygen-saturation curve

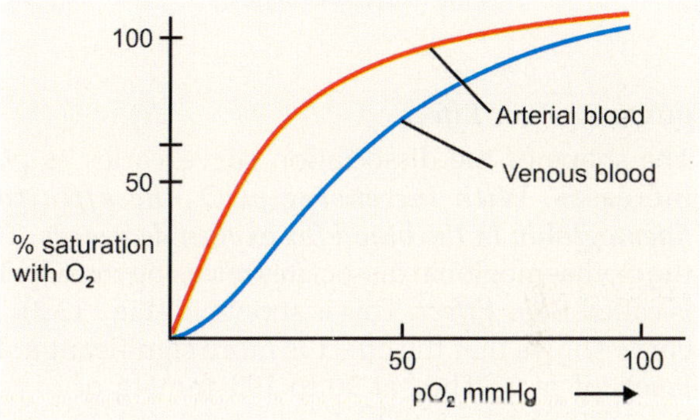

Fig. 13.2: pO_2 of arterial and venous blood

O_2. The O_2 carried in physical form is around 0.3 ml/dl. So the total amount of O_2 carried by 100 ml of blood in a normal individual is about 20 vol%. ***This 20 vol% is known as 100% saturation of haemoglobin or total O_2 carrying capacity of blood.*** Following is the figure which shows the graphical presentation of the saturation curve (**Fig. 13.1**). This curve shows the normal oxygenation up to the point of saturation. There are several factors which bring about the dissociation of oxyhaemoglobin thereby releasing or delivering oxygen.

FACTORS OF DISSOCIATION OF OXYHAEMOGLOBIN—OXYGEN-DISSOCIATION CURVES

1. ***pO_2***
   The actual amount of O_2 which a given amount of haemoglobin takes depends on pO_2 of the medium. The pO_2 is different in arterial and venous blood. It is shown in the (**Fig. 13.2**).

   The curve shows that ***greater the pO_2 higher is the saturation and dissociates as the pO_2 is decreased.*** However, the dissociation is directly proportional to pO_2 indefinitely. Although there is little reduction in content of oxygen taken by Hb up to 50 mmHg of O_2, below this, the content of O_2 taken by haemoglobin decreases very greatly.

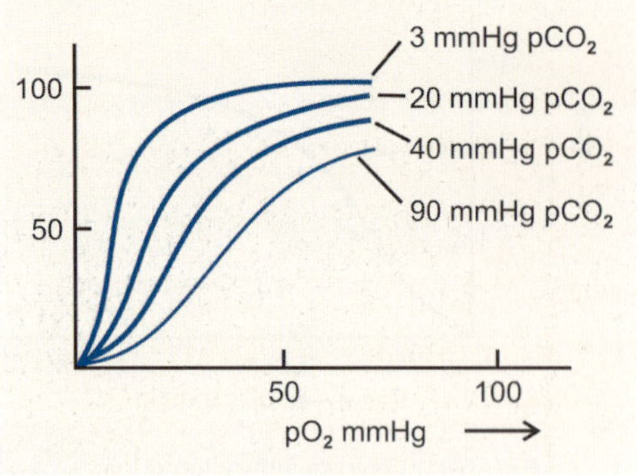

Fig. 13.3: Effect of pCO_2

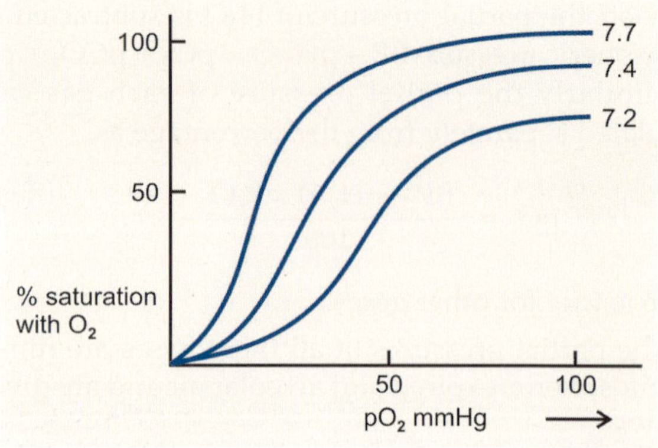

Fig. 13.4: Effect of pH value on oxygen-dissociation curve

2. *pCO_2: or Böhr Effect*

The shape of the dissociation curve varies as pCO_2 increases. ***With increasing pCO_2 the affinity of haemoglobin in the blood for oxygen decreases.*** Thus the oxyhaemoglobin dissociates releasing free O_2. This is called ***Böhr Effect***. This is shown in (Fig. 13.3). The curve shows that the effect is more significant at low levels of pO_2. Thus at 90 to 100 mmHg of oxygen, the pCO_2 is found to be between 20 and 30 mmHg. On the other hand at pO_2 = 20 mmHg the pCO_2 is found to be higher. This Böhr effect has many biological advantages. As already mentioned pCO_2 is low in lung alveoli compared to tissues. Therefore the dissociation of oxyhaemoglobin is more pronounced in tissues. Therefore, the oxygenation of haemoglobin is sure when the alveolar pO_2 is around 80 mmHg. On the other hand, the release of O_2 from blood is sure to take place in the tissues where pO_2 is 30 mmHg. Only about 5 ml of O_2 per 100 ml of blood is given by the blood to tissues and thus a large amount oxygenated blood remains as 'storage' for O_2 in the event of insufficient supply of O_2. This effect can also be attributed to the lowering of pH due to high pCO_2.

$$H_2O + CO_2 \underset{}{\overset{CA}{\rightleftharpoons}} H_2CO_3 \rightleftharpoons H^+ + HCO_3^-$$

Since the decrease in pH has a similar effect on dissociation of oxyhaemoglobin, it is quite clear that pCO_2 levels when high can produce same effect.

The direct effect of CO_2 on the oxy Hb dissociation arises from its combination with the amino terminals of the haemoglobin subunits to form carbaminohaemoglobin which has a decreased affinity for oxygen thus its formation shifts the curve to right.

3. *pH*

As briefly mentioned above, the pH of the medium has a profound effect on oxygen-dissociation curve. ***Increase in pH increases dissociation of oxyhaemoglobin, shifting the oxygen-dissociation curve to the right.*** Apparently formation of HbO_2 is decreased at high pH. As in the case of pCO_2, this effect is more obvious at low pO_2 levels. ***The curve shifts to the left when pH of the medium is increased as shown in*** (Fig. 13.4).

4. *Temperature*

Like the increase in pH, ***the increase in temperature shifts the dissociation curve to the left.*** Since the normal human beings have a constant body temperature this is not of much significance. However, in warm blooded animals oxygen dissociates more readily than in cold blooded animals.

5. *Electrolytes*

It appears that oxyhaemoglobin dissociates more readily in presence of electrolytes at low pO_2 levels.

6. *2,3-Biphosphoglycerate (2,3-BPG)*

2,3-Biphosphoglycerate found in RBCs is produced by Rapoport-Leubering Shunt. ***The presence of 2, 3-BPG facilitates the delivery of O_2 shifting the curve to the right.*** Thus, 2,3-BPG destabilises the HbO_2.

In conclusion it can be summarised that O_2 supply to tissue cells is facilitated by high pO_2 levels in lungs, it is enhanced by the relatively high pCO_2, high acidity (low pH), high temperature in metabolically active tissues. In presence of these factors the supply of O_2 is carried out from the lungs to the tissues cells.

CLINICAL IMPORTANCE OF OXYGEN

Hypoxia: When the supply of O_2 to the cells is inadequate a condition called ***Hypoxia*** results. When there is absolutely no supply of O_2 to the tissues it is a condition known as ***Anoxia.*** There are **four types** of hypoxia known:

Types

1. *Hypoxic Hypoxia*

   Insufficient O_2 supply attributed to:
   - Low pO_2
   - Thickening of lung alveoli hampering diffusion
   - Blood bypasses the arterial alveoli and does not come in contact with O_2 as in congenital heart disease and lobar pneumonia.

2. *Anaemic Hypoxia*

   In anaemia the haemoglobin level in blood is reduced. Since the haemoglobin is the major transporter of O_2, its reduced levels adversely affect the supply of O_2. The supply is similarly affected by poisoning.

3. *Stagnant Hypoxia*

   Blood has to flow at a specific speed to ensure sufficient supply of O_2. However, when the speed is slowed down, O_2 supply is reduced. This condition is known as ***stagnant hypoxia.***

4. *Histotoxic Hypoxia*

   This kind of hypoxia has nothing to do with the mechanism of transport and delivery of O_2. However when the tissue cells are poisoned, they are unable to utilise O_2. These poisons normally inhibit or uncouple the respiratory chain enzymes *cytochrome oxidases.*

TRANSPORT OF CARBON DIOXIDE

Oxygen that is delivered to the tissue cells is used for various oxidative functions and as a result CO_2 is produced. This CO_2 needs to be removed from the body. ***In the resting state, about 200 ml of CO_2 is produced per minute.*** The CO_2 is removed by lungs via corpuscles and also it is transported in plasma. The pCO_2 in tissues is around 60 to 70 mmHg and its level in arterial blood is around 40 mmHg. ***Thus as per the diffusion gradient, CO_2 flows from tissues to the blood.*** It is observed that 4 ml of CO_2 per 100 ml of blood is given off by the tissues and carried to the lungs. Similarly the venous pCO_2 is 46 mmHg and in lungs it is 40 mmHg. Therefore CO_2 diffuses from venous blood to the lungs. It is important to note that diffusion constant of CO_2 is higher than O_2, because CO_2 is more soluble in body fluids such as plasma than O_2.

On the basis of solubility of CO_2 it is observed that ***100 ml of blood can dissolve only 2.9 ml of CO_2 at 37°C at pCO_2 of 40 mmHg.*** Since the CO_2 content of venous and arterial blood is 55.60 ml% and 50.55 ml% respectively, only a small amount of CO_2 is transported in the form of physical solution of CO_2.

The CO_2 is thus transported as follows:
- *In Physical solution—small amount*
- **As Carbamino compound**
- **As Bicarbonates** (Fig. 13.5).

1. **Transport as Physical Solution**

   CO_2 is carried in physical solution by plasma as well as RBCs. ***CO_2 is more soluble in plasma than O_2*** and is directly proportional to the partial pressure. A small fraction of about 1 per cent is present as H_2CO_3. In venous blood at pCO_2 of 46 mmHg, ***3.5 ml of CO_2 is in physical solution per 100 ml of blood.***

2. **Transport as Carbamino Compounds**

   Haemoglobin and certain other plasma proteins carry some CO_2. ***In arterial blood it is around 3 ml/100 of blood while it is 3.7 ml in venous blood.*** The N-terminal NH_2 groups of the polypeptide chains form carbamino haemoglobin. ***Reduced haemoglobin and oxy-Hb carry 8 ml and 3 ml of CO_2 as carbamino compound per 100 ml of venous and arterial blood respectively.***

3. **Transport as Bicarbonate and Chloride Shift**

   This is by far the most important means of CO_2 transport in the blood amounting to 42 ml of CO_2 per 100 ml of arterial and 44.8 ml per 100 ml of venous blood. CO_2 produced in cells finds its way into the RBCs where ***carbonic anhydrase*** catalyses the formation of H_2CO_3. However, at physiological pH of 7.4, carbonic acid is further dissociated into H^+ and HCO_3^-.

$$H_2O + CO_2 \overset{CA}{\rightleftharpoons} H_2CO_3 \rightleftharpoons H^+ + HCO_3^-$$

Some of HCO_3^- combines with intracellular K^+ and rest diffuses out into plasma following the concentration gradient. ***Since HCO_3^- is an electrolyte, its diffusion may disturb the electrolyte balance. This is overcome by shifting equal number of chloride ions into the RBCs. This phenomenon is called as chloride shift or***

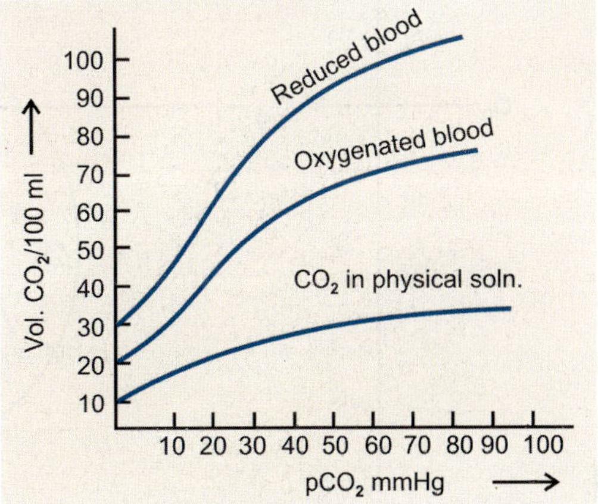

Fig. 13.5: CO_2 carrying capacity of blood

Hamburger's process or phenomenon. The HCO_3^- in plasma combines with Na^+. Thus, the $KHCO_3$ formed in the cell or $NaHCO_3$ formed in plasma is carried by blood to the lungs where it undergoes dissociation (Figs. 13.6A and B).

CO_2 in all the forms ultimately, diffuses into alveolar air. The chemical reactions taking place at the site of the lung are **exactly reverse** of those involved in HCO_3^- formation at the tissue site. **In the lungs,** O_2 diffuses from alveolar air into the red cell where it forms H. Hb O_2 which in turn **donates a proton**. The H^+ combines with HCO_3^- which comes from tissue site to form H_2CO_3. Carbonic anhydrase catalyses reversible reaction to split H_2CO_3 into CO_2 and H_2O. The CO_2 thus formed now diffuses on account of pressure gradient into alveolar air via the plasma. When $KHCO_3$ is decomposed to CO_2 in sufficient amount, the bicarbonate ions formed at the tissue site diffuse into RBC. This leads to disturbance of electrolyte balance and hence ***equal number of chloride ions diffuse out.***

FACTORS AFFECTING CARBON DIOXIDE TRANSPORT

The only important factor which affects the carriage of CO_2 is the partial pressure. However, changes in pH, presence of O_2 do have some effects. The amount of CO_2 carried by the blood is directly proportional to its partial pressure in blood. Increase in pH increases CO_2 carrying capacity. Presence of O_2, however, inversely affects the transport of CO_2. ***Higher the pO_2 lower the amount of CO_2 carried.***

FUNCTION OF NITROGEN IN PLASMA

A large percentage of air contains nitrogen. The major function of nitrogen in respiration is to act as ***an inert diluent of oxygen.*** The percentage of N_2 seen is similar in atmospheric, alveolar and expired air indicating its no role in metabolic processes. ***However, nitrogen is always found in plasma at a conc. of 2.5-3 ml/100 ml of blood.***

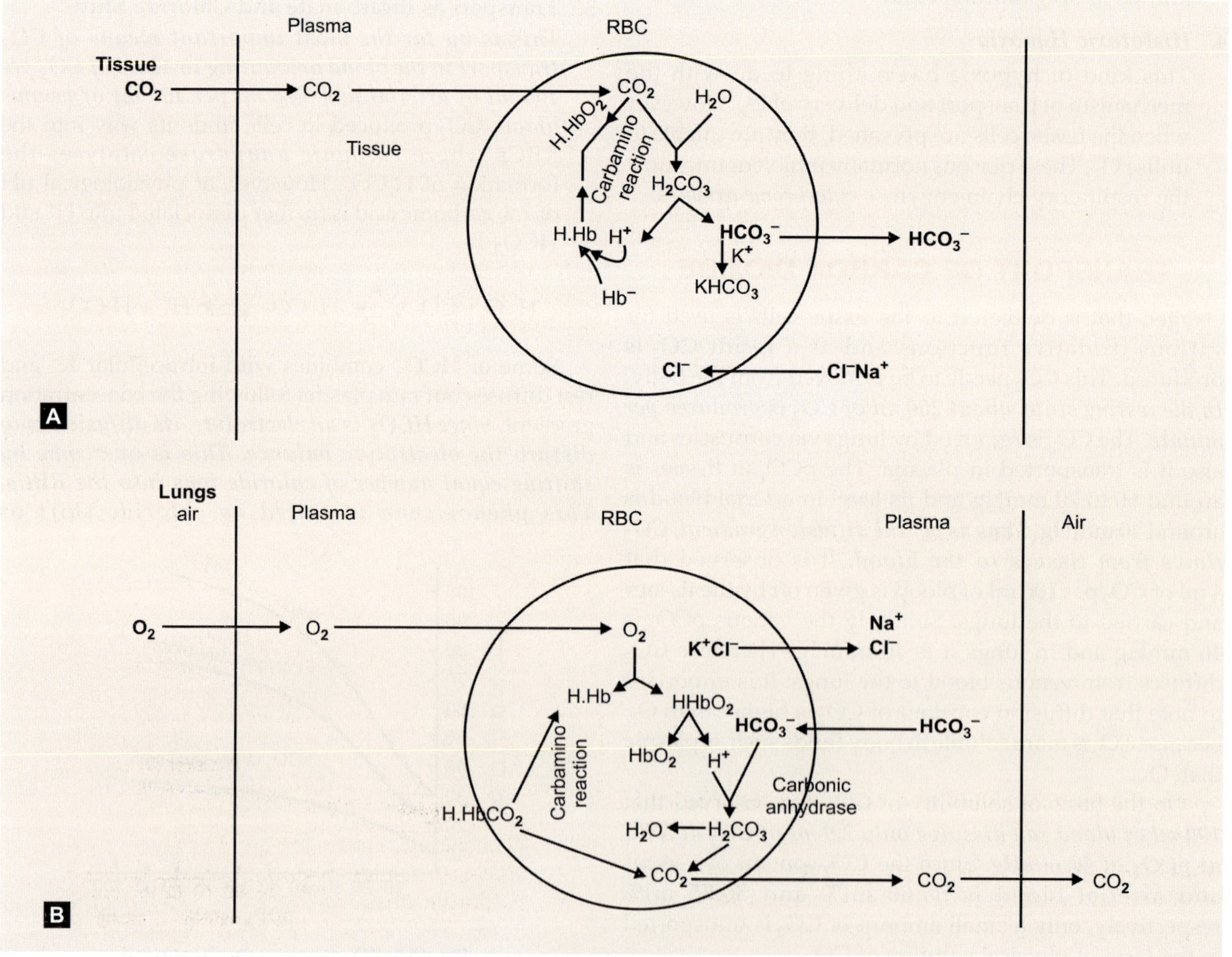

Figs. 13.6A and B: Chemical changes: (A) tissue (B) lung

CLINICAL ASPECT

When a person goes to a high altitude or comes up from deep sea abruptly, the nitrogen in plasma can cause problems. In deep seas due to high pressure excessive amount of N_2 gets dissolved in plasma; when a person comes up, the ***N_2 is quickly released by plasma which may collect in the form of bubbles at the joints, adipose tissue and other tissues which then cause intense pain.*** Sometimes the tissues are damaged impairing their functions. Intravascular clotting is also possible. Similar effects may be observed when a person goes to high altitudes where pressure is low. This effect is called as **Bends or Caisson disease.** Use of pressurisation device is the solution to avoid such abnormalities.

FREE RADICALS

Oxygen is required for all living organisms for their survival. But, at the same time, one has to remember that ***oxygen is potentially toxic.*** **Salvemini** has described oxygen as a **double-edged sword**: it is vital to life but leads to formation of by products that are toxic such as formation of superoxide (O_2^-) anions. Dissolved oxygen at high concentration is toxic to animals. Rats when subjected to breathe pure oxygen at 2 atmospheric pressure, undergo convulsions in 5 to 6 hours and may die.

Formation of *Free* radicals: The univalent reduction of molecular oxygen in tissues gives rise to so-called ***superoxide radical*** O_2^-. It is one of the "free radicals" produced in the body.

Why oxygen is more prone to produce superoxide radical O_2^-?: Molecular oxygen is ***paramagnetic*** and contains two unpaired electrons with parallel spins. These unpaired electrons reside in separate orbitals unless their spins are opposed. Reduction of O_2 by direct insertion of a pair of electrons, e^-, into its partially filled orbitals is not possible without inversion of one electronic spin and such an inversion of spin is a slow change. Hence electrons are added to molecular oxygen as single electron successively.

When oxygen molecule takes up one electron, by univalent reduction, it becomes ***Superoxide*** anion O_2^-. Thus,

$$O_2 + e^- \longrightarrow O_2^-$$

Superoxide anion is highly reactive and toxic to cell membranes.

Other *Free Radicals* in the Body: Sequence of events that can occur in formation of other 'free' radicals are:

1. Superoxide anion O_2^- can capture further electrons to form ***hydrogen peroxide***, H_2O_2, which is toxic and injurious.

$$O_2^- \xrightarrow{e^-,\ 2H^+} H_2O_2 \xrightarrow{2e^-,\ 2H^+} 2H_2O$$

Hydrogen Peroxide

2. H_2O_2 can further react with "Superoxide" anion, in presence of Fe^{++} (ferrous) to form ***Hydroxyl*** radical, and ***Singlet oxygen.***

$$O_2^- + H_2O_2 \xrightarrow{Fe^{++}} OH^\bullet \text{ (free hydroxy radical)} + {}^1O_2 \text{ (Singlet oxygen)} + OH^-$$

The above reaction is called **Haber's reaction** (or **Haber-Weiss-Fenten's reaction**).

Role of Caeruloplasmin: Caeruloplasmin, which acts as *"ferroxidase"*, can serve as antioxidant. ***It can convert $Fe^{++} \rightarrow Fe^{+++}$ (ferric)*** and ***thus it can halt the Haber's reaction*** preventing further formation of highly reactive hydroxyl free radicals.

3. Superoxide anion can accept a H^+ and form **hydroperoxy** radical.

$$O_2^- \xrightarrow{H^+} HOO^\bullet \text{ (Hydroperoxy radical)} \xrightarrow{HOO^\bullet} H_2O_2 + O_2$$

Note: Whenever superoxide anion, O_2^-, is formed in tissues, it will lead to the formation of other "free radicals" like hydroperoxy radical, hydroxyl free radical and hydrogen peroxide. ***All these free radicals are very reactive*** and ***toxic to biological membranes.***

4. Nitric oxide produced in the body from Arginine by the action of ***Nitrogen oxide synthase*** has a short half life 3 to 4 seconds because it reacts with oxygen and superoxide (O_2^-). The reaction with superoxide produces **peroxynitrite** ($ONOO^-$), which decomposes to form the highly reactive $OH^\bullet$ radical.
5. Other toxic 'free' radicals produced in the body are 'free' radical CCl_3^- and 'free' halogen radical like Cl^- formed from CCl_4.

Formation of Superoxide Anion in Metabolic Pathways (Cytosolic Oxidations)

1. Cytosolic oxidations by autooxidisable FP dependent enzymes, e.g.
   - ***Xanthine oxidase***
   - ***Aldehyde dehydrogenase***
   - Oxidative deamination by *L-amino acid oxidase*

   Superoxide anion, O_2^-, may be formed when reduced flavins are reoxidised univalently by molecular O_2.

2. It is also formed during univalent oxidations with molecular oxygen in the respiratory chain.

$$EnZ.\ H_2 + O_2 \rightarrow EnZ.\ H + O_2^- + H^+$$

3. Superoxide anion, O_2, can be formed during methaemoglobin formation.

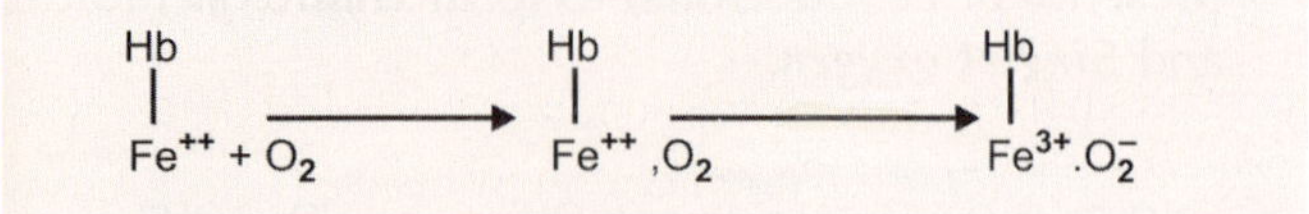

4. Superoxide anion, O_2^-, may also be formed during cytosolic hydroxylations of steroids, drugs, and xenobiotics, by Cyt P_{450} or Cyt P_{448} system. The hydrogens for these hydroxylations are donated by NADH (or NADPH) routed through Fp and Cyt P_{450}.

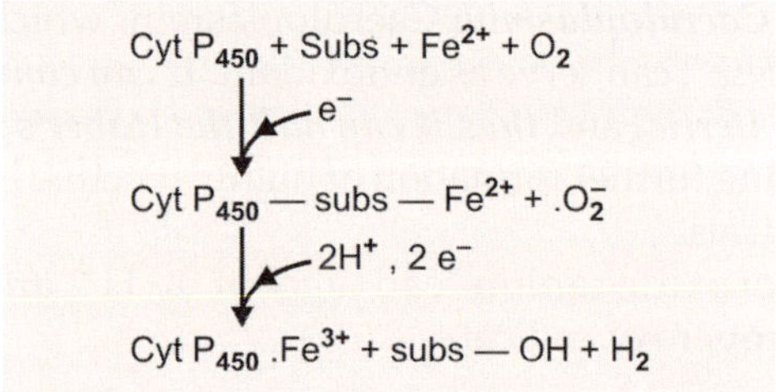

5. Free radicals are also produced in tissues when exposed to ionising radiations.
6. Superoxide anion, O_2^-, is also produced *during phagocytosis*, by NADPH oxidase system during **respiratory burst** when O_2^- consumption is increased.

SCAVENGERS OF FREE RADICALS

- ***Superoxide Dismutase***
  In both cytosol and in mitochondria an enzyme is present called *super-oxide dismutase* which can destroy the superoxide anion, O_2^-.

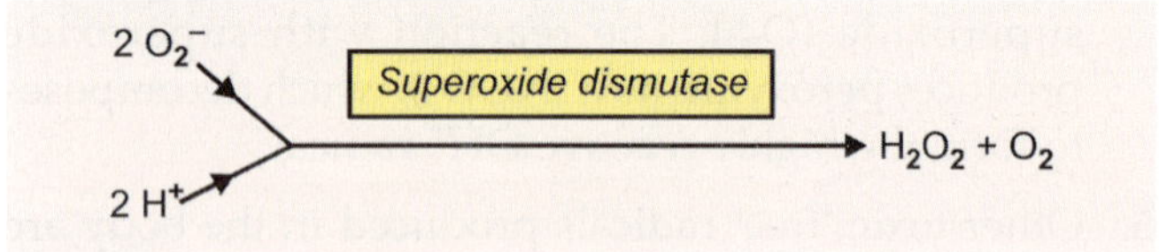

The enzyme is present in all major aerobic tissues. The function of the enzyme seems to be that of protecting aerobic organisms against the potential deleterious effects of superoxide anion, O_2^-.

- ***Catalase***
  Catalase having high Km value which is situated close to aerobic dehydrogenases, like liver peroxisomes, can destroy H_2O_2 formed in the tissues to O_2.

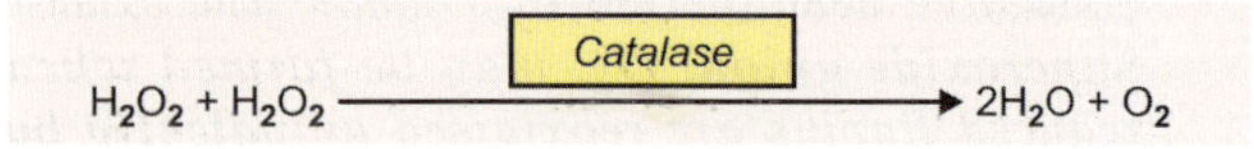

Glutathione Peroxidase

If the concentration of H_2O_2 is less than the optimum required for hydroperoxidation by catalase, then the Selenium-containing enzyme ***glutathione peroxidase*** can destroy H_2O_2 with the help of reduced glutathione (G-SH), **having low K_m**. It is present in cytosol and mitochondria.

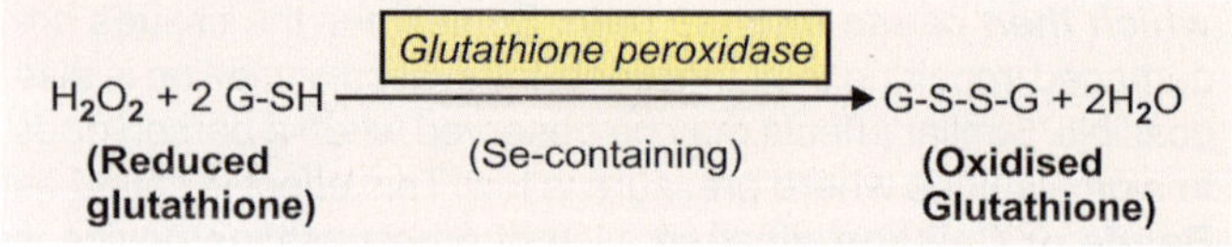

- ***Ferricytochrome***
  Superoxide anion, O_2^- can also be oxidised to O_2 by ferricytochrome.

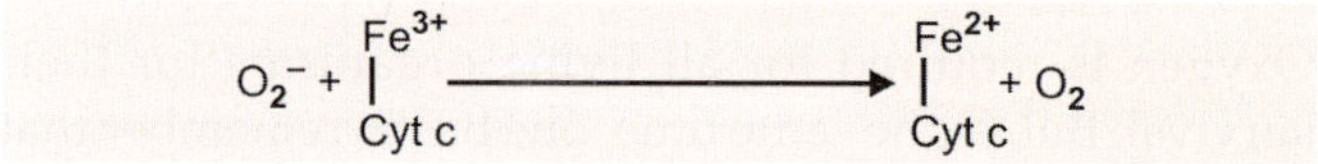

- ***Endogenous Caeruloplasmin can halt Haber's Reaction (see above)***

Effect of 'Free' Radicals on Biomembranes

1. Free radicals are ***highly reactive so can initiate chain reaction and brings about lipid peroxidation producing lipid peroxides and lipoxides.*** These radicals constitute a threat to the integrity of biomembranes which could be oxidised. ***The* free hydroxyl *radical is most reactive and can also be mutagenic. It is an extraordinarily potent oxidant.***
2. These oxidants can oxidise:
   - –SH groups containing membrane proteins in the cells and biomembranes to S–S group.
   - Methionine sulphur is oxidised to its sulphoxide, and
   - Membrane lipids, unsaturated FA to lipid peroxides and lipoxides, called lipid peroxidation. The above will affect the optimum fluidity of the membrane causing ***membranopathy.***
3. ***Lipid peroxidation:*** It is a chain reaction initiated by 'free' radicals which provides a continuous supply of other 'free' radicals formed from unsaturated FA which initiate further peroxidation.

Stages of Lipid Peroxidation

Whole process consists of **three stages:**

- **Initiation:** Production of $R^\bullet$ from a precursor

$$ROOH + metal^{(n)+} \text{———} ROO^\bullet + metal^{(n-1)+} + H^+$$

$$X^\bullet + RH \text{———} R^\bullet + XH$$

- **Propagation:**

$$R^\bullet + O_2 \text{———} ROO^\bullet$$

$$ROO^\bullet + RH \text{———} ROOH + R^\bullet \text{ and so on.}$$

- **Termination**

$$ROO^{\bullet} + ROO^{\bullet} \text{ ——— } ROOR + O_2$$

$$ROO^{\bullet} + R \text{ ——— } ROOR$$

$$R^{\bullet} + R^{\bullet} \text{ ——— } RR$$

Since the molecular precursor for the initiation process is generally the hydroperoxide product ROOH, ***lipid peroxidation is a branching chain reaction with potentially devastating effects.***

4. ***Effect on biomembranes:*** ***$OH^{\bullet}$ free radical is very reactive. O_2^- and $OH^{\bullet}$ can initiate chain reaction and bring about oxidation of polyunsaturated FA of membranes.*** A "free" radical with unpaired electrons may take away hydrogen from "methylene group" of polyunsaturated FA and convert it into a free FA radical, which binds with O_2 to give FA **Peroxy radical**, the latter then changes to FA **hydroperoxide** by accepting hydrogen of the methylene group of another polyunsaturated FA and converting into another free FA radical and so on.

ANTIOXIDANTS

To control and reduce lipid peroxidation both in humans and in nature, antioxidants are used.

A. ***Antioxidants used in vitro:*** These are used to prevent lipid peroxidation in foods, examples are:
   - **Propyl gallate (PG)**
   - **Butylated hydroxy anisole (BHA)**
   - **Butylated hydroxy toluene (BHT)**

B. Naturally occurring antioxidants–Major sources

Antioxidants	*Major sources*
• **β-carotene**	Carrots, yellow papaya, spinach
• **Vit C (Ascorbic acid)**	Citrous fruits, amla richest source, spinach, cabbage
• **Vit E (Tocopherols)**	Vegetable oils, legumes
• **Lycopene**	Red tomatoes
• **Lutein and zeaxanthin**	Egg yolk
• **Polyphenols**	Apple
• **Querectin (a flavonoid)**	Apple, onions
• **Epigallocatectin gallate (ECGg)**	Green tea leaves
• **Selenium**	Sea foods
• **α-Lipoic acid**	Yeast, liver
• **Proanthocyanidines**	Grape seeds
• **Clorogenic and melanic acids**	Coffee beans
• **Curcuminoids**	Turmeric
• **Zinc (trace element)**	Liver, egg, spinach

Antioxidants can be classified into **two classes according to their mode of action.**

1. **Preventive antioxidants:** These reduce the rate of chain initiation, e.g:
   - Catalases
   - Other peroxidases that react with ROOH
   - Natural endogenous caeruloplasmin, and
   - Chelators of metal ions such as:
     - DTPA (diethylene triamine penta acetate), and
     - EDTA (Ethylene diamine tetraacetate)
2. **Chain breaking antioxidants:** Which interfere with chain propagation, examples are:
   - Phenols or aromatic amines.
   - *In vivo* the principal chain breaking antioxidants are:
     - ***Superoxide dismutase*** both cytosolic and mitochondrial
     - Vitamin E and Selenium-containing ***glutathione peroxidase***
     - Urates
   - Peroxidation is also catalysed *in vivo* by heme-containing compounds and by lipoxygenases found in platelets and leucocytes.

Vitamin E action as antioxidants (Refer chapter on Vitamins). Role of Selenium and relation with vitamin E (Refer chapter on Mineral Metabolism).

Synzyme: Recently researchers have prepared a synthetic enzyme that works just like the body's own scavengers like superoxide dismutase (SOD), to mop up "free" radicals.

They have named it as ***synzyme*** and one of its first user may be in treating stroke. The enzyme goes by the experimental name of M40403, based on the metal manganese. The compound is more powerful than traditional antioxidants such as vitamin E or C.

CLINICAL SIGNIFICANCE

1. ***Role in ageing:*** Free radicals play an important role in ageing and aggravate certain disease processes like diabetes mellitus, atherosclerosis, cancer, etc.
2. ***Neonatal oxygen radical diseases:*** ***The preterm baby may be specially vulnerable to free oxygen radicals***, because it is exposed more liberally to oxygen radical generation and its defence against oxygen radicals is low. It has, therefore, been postulated that a **neonatal oxygen radical disease** does exist. Diseases like:
   - ***Bronchopulmonary dysplasia,***
   - ***Retinopathy of prematurity,***
   - ***Necrotising enterocolitis,***
   - ***Periventricular leukomalacia,***
   - ***Patent ductus arteriosus,*** and
   - Perhaps ***intracranial haemorrhage*** represent different facets of this syndrome. The symptoms are determined by which organs are principally affected, but basic pathogenetic mechanisms may be identical.

SECTION TWO

3. ***Rheumatoid arthritis:*** Diseases like rheumatoid arthritis is ***self-propagated by the* free radicals *released by the neutrophils.*** This is further accentuated by decrease of 'scavengers' in joint cavity which mop up the "free radicals." Drugs like corticosteroids and NSAIDs interfere with the formation of "free radicals" and thus provides relief.
4. ***Atherosclerosis Vs Thrombosis*—Role of free radicals*:*** 'Free radicals' released by the endothelial cells of blood vessels oxidises the LDL deposited under the endothelial cells increasing the level of lipid hydroperoxides which in turn ***increases the* thromboxane *production. This decreases the prostacyclin/thromboxane ratio and leads to thrombosis.***
5. ***Chronic granulomatous diseases:*** In patients with ***chronic granulomatous diseases,*** phagocytic function of macrophages have been found to be defective and there is increased susceptibility to microbial and fungal infections (Refer Chapter 2).
6. ***Role in Phagocytosis:*** H_2O_2 is used in killing microorganisms in phagocytosis. In this process, O_2 uptake is increased greatly leading to respiratory burst, which produces O_2^- by NADPH oxidase system. O_2^- subsequently produce H_2O_2.

$$NADPH + 2\,O_2 \longrightarrow NADP^+ + 2\,O_2^- + H^+$$

$$2\,O_2^- + H^+ \longrightarrow H_2O_2 + O_2$$

7. ***Role in Malaria:*** G-6-PD deficient patients are more resistant to infection with *Plasmodium falciparum.* Normally, the parasite generates H_2O_2 by oxidising NADPH, produced by HMP-shunt pathway, by a plasma membrane bound oxidase for their survival. In G-6-PD deficiency, the parasites fail to thrive as they cannot produce H_2O_2 for their survival as NADPH production is lacking.

I. Questions (Essay type and short notes)

1. Describe in detail how O_2 is transported in blood. What are the factors that affect the O_2 transport.
2. Describe the details mechanism of CO_2 transport and factors affecting the transport of CO_2.
3. Describe mechanisms of CO_2 transport in blood and the role of Hb.
4. Give an account of carbon dioxide transport in blood.
5. Give an account of oxygen transport in the body

Short Notes

a. Hypoxia
b. Chloride Shift (Hamburger Phenomenon)
c. Oxygen-dissociation curve
d. Bohr effect
e. Role of N_2 in respiration
f. Chloride bicarbonate shift
g. Transport of oxygen in the body
h. Transport of carbon dioxide (CO_2) in the body
i. Oxygen carrying capacity of blood
j. Superoxide dismutase
k. Antioxidant
l. Haber's reaction
m. Scavengers of free radicals
n. Lipid peroxidation

Differentiate

Oxygen transport and carbon dioxide transport in the blood.

II. MCQs (Give one correct answer)

1. Partial pressure exerted by the gas depends on:
   a. Temperature b. Atmospheric pressure
   c. Volume of gas d. Composition of gas
   e. Moles of gas in a given volume
2. Exchange of gases between two compartments is achieved by:
   a. Active transport b. Passive transport
   c. Diffusion d. Patial pressure
   e. Osmosis
3. In blood 20 vol% of oxygen is regarded as how many percentage saturation of Hb?
   a. 20 b. 50
   c. 80 d. 100
   e. Any one of (a, b, c, d)
4. Hypoxic hypoxia could be due to:
   a. High pO_2
   b. Low pO_2
   c. Low pCO_2
   d. Defective oxygen in atmosphere
   e. All
5. The chloride shift involves:
   a. H^+ leaving the RB cells for each CO_2 that enters
   b. HCO_3^- leaving the erythrocytes in exchange for Cl_3^-
   c. H_2CO_3 leaving
   d. K^+ accompanying each HCO_3^- ion
   e. When CO_2 is discharged in lungs NaCl enters the erythrocytes
6. In the CO_2 transport by blood,
   a. Hb is not a buffering agent
   b. Oxy-Hb is a weaker acid than deoxy-Hb
   c. At physiological pH values the ratio of HCO_3^- to H_2CO_3 is 1:20
   d. Combines with NH_2 group of Hb to form carbamino compound
   e. None of the above

Answers to MCQs

1. e 2. c 3. d 4. b 5. b 6. d.

SECTION

3

Molecular Biology

CHAPTER 14

CHEMISTRY OF NUCLEOTIDES

Major Concepts

A. To study the chemistry of nucleotides.
B. To learn the different biologically important nucleotides that occur in tissues and their functions.
C. Learn what is cyclic AMP and cyclic GMP and their role in the body.

Specific Objectives

A. 1. Study and list the different types of purine and pyrimidine bases that occur in a nucleotide.
   2. Study and draw the purine nucleus and pyrimidine nucleus and number the positions of C and N atoms present in the nucleus.
   3. Study the structure of two major purines and three major pyrimidines and also study their chemical names. Study the other bases, purine and pyrimidine, that can occur in small quantities. Study the *lactam (keto)* and *lactim (enol)* forms of each wherever applicable.
   4. Define a *'nucleoside'* and learn how a nucleoside is formed. Study the *'glycosidic linkage'* by which the pentose sugar ribose/or deoxyribose is linked to the purine and pyrimidine bases. Study the form in which a ribose/deoxyribose sugar is present in a nucleoside.
   5. Study how a *'nucleotide'* is formed by esterification of sugar molecule of nucleoside with phosphoric acid group. Study the possible sites where the esterification with phosphoric acid can take place on the sugar molecule-ribose/ and deoxyribose.
   6. Clearly now define a *"nucleoside"* and *"nucleotide"*. Differentiate the two in a tabular form.
   7. Make a table naming the different types of nucleosides and nucleotides that can occur showing the respective bases/sugar/and phosphoric acid.

B. Study and list the various biologically important nucleosides and nucleotides present in tissues of human beings and study some of their important functions in the body.
   - Learn *"purine nucleotide cycle"*. **IMP** can be formed by the deamination of **AMP**, a reaction which occurs particularly in muscle as part of "purine nucleotide cycle".
   - Study the synthetic derivatives. Certain synthetic nucleobases, nucleosides and nucleotides are widely used in the medical sciences and in clinical medicine.
   - List the synthetic derivatives and their use in clinical medicine.

C. 1. Study the structure of 3'-5' cyclic AMP.
   2. Learn how cyclic AMP is formed in the body from ATP and how it is degraded.
   3. List the important functions of cyclic AMP.
   4. Study how cyclic GMP is formed and degraded.
   5. List the important functions of cyclic GMP.

NUCLEOPROTEINS

One of the groups of conjugated proteins, **characterised by presence of nonprotein prosthetic group, nucleic acid and attached to one or more molecules of a simple protein, a basic protein histone or protamine is called nucleoprotein.** Nucleoproteins are so named because they constitute a large part of nuclear material. Chromatin is largely composed of nucleoproteins, which indicates that these compounds are involved in cell-division, and transmission of hereditary factors. Nucleoproteins are found in all animal and plant tissues and have been extracted from a variety of tissues. Most easily isolated from yeast cells or from tissues with large nuclei where the cells are densely packed such as thymus gland. Extracting agents used include water, dilute alkali, NaCl solutions and buffers ranging from pH of 4.0 to 11.0. In each case extraction is followed by precipitation with acid, saturated $(NH_4)_2SO_4$ or dil. $CaCl_2$. When the purified nucleoprotein is hydrolysed with acid or by the use of enzymes, various components as shown in the box are obtained.

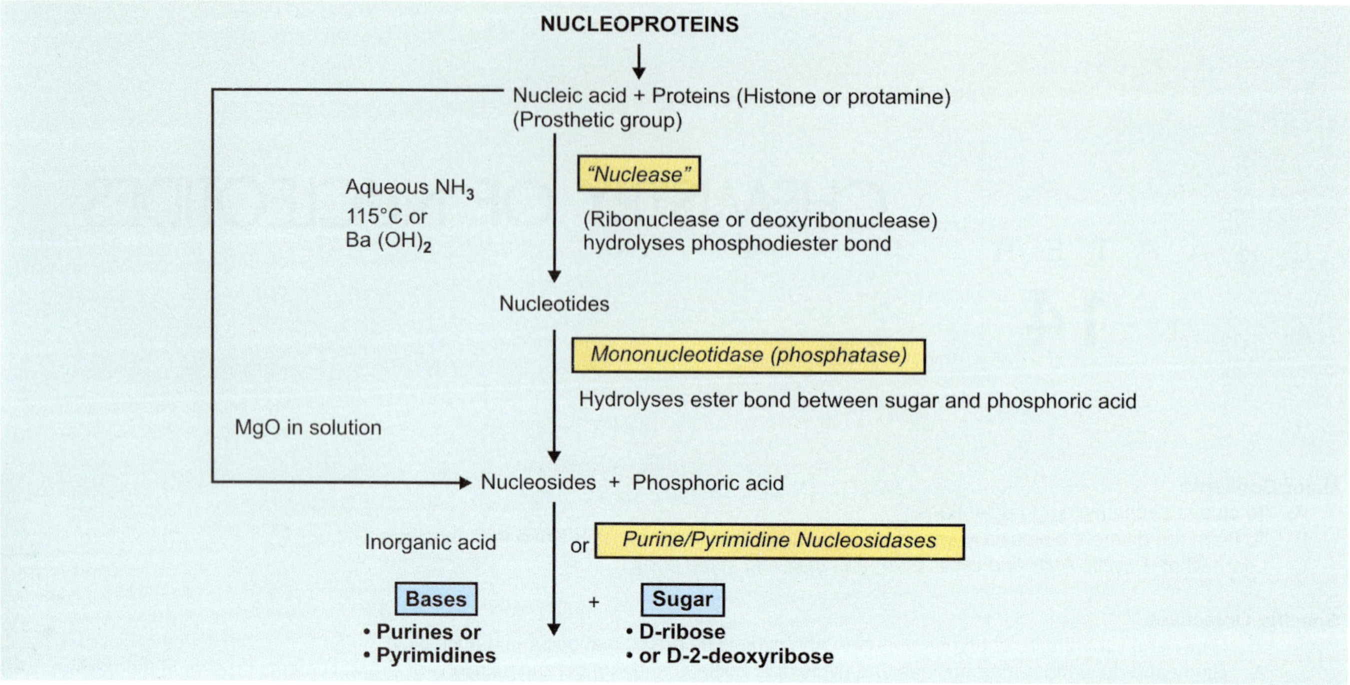

SUGARS

D-ribose and D-2-deoxyribose are the only sugars so far found in the nucleic acids from which the sugars have been isolated and identified, and they are assumed to be the sugars universally present in nucleic acids.

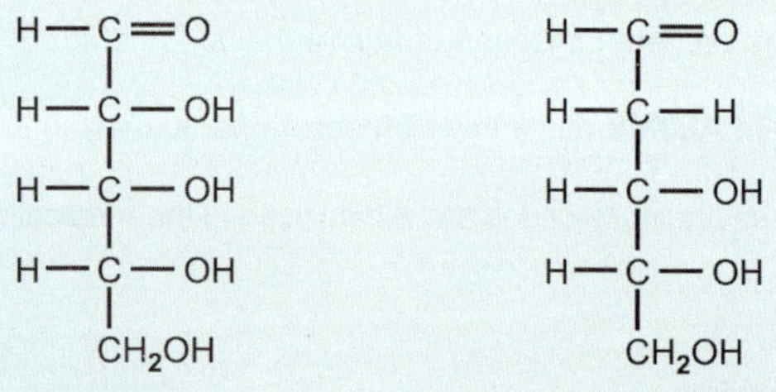

D-Ribose D-2-deoxyribose

Both sugars are present in nucleic acids as the β-Furanoside ring structures as depicted below.

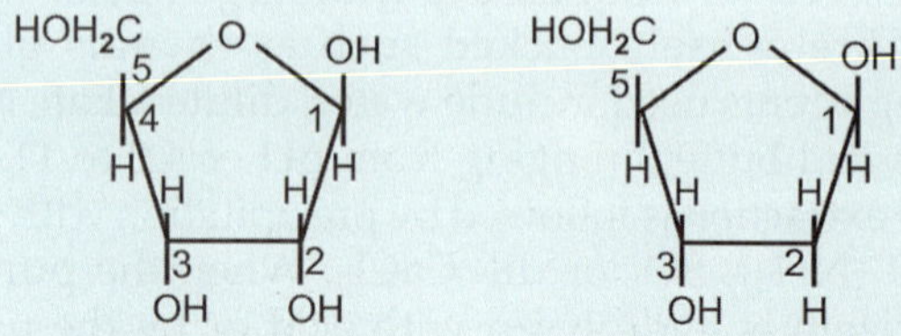

β-D-Ribofuranose β-D-2-deoxyribofuranose

PYRIMIDINE BASES

Pyrimidine bases found in nucleic acids are mainly **three:**

- ***Cytosine*** is found both in DNA and RNA
- ***Thymine*** is found in DNA only
- ***Uracil*** is found in RNA only.

Pyrimidine nucleus is represented below.

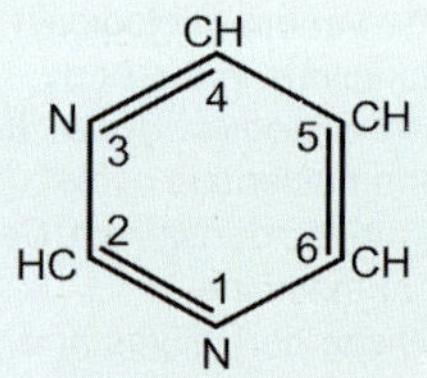

All the pyrimidine bases can exist in ***lactam*** form and ***Lactim*** form. If the group is –HN–CO–, it is called the *Lactam* type *(keto)*, while the same if isomerises to –N = C – OH, it is called *Lactim* form *(enol)*. At the physiological pH, the *lactam* (keto) forms are predominant.

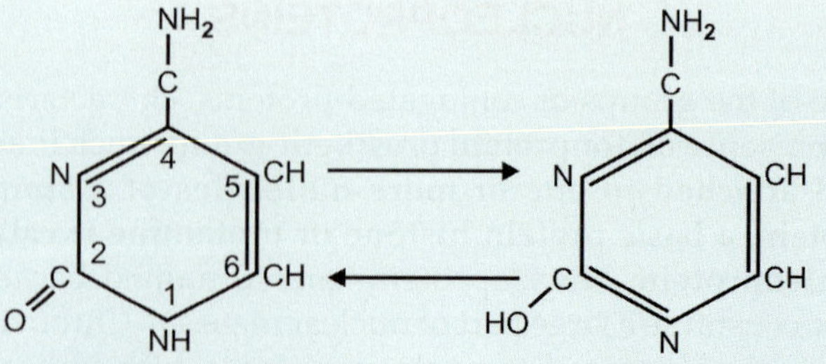

Lactam form of cytosine (keto) **Lactim form of cytosine (enol)**

1. Cytosine

Chemically it is **2-deoxy-4-amino pyrimidine**, it can exist in both *lactam* or *lactim* forms.

Cytosine is found in all nucleic acids except DNA of certain viruses.

2. Thymine (5-methyl uracil)

Chemically it is **2,4-deoxy-5-methyl pyrimidine.** ***Thymine occurs only in DNA***, in nucleic acids which contain deoxyribose as Sugar. Minor amounts have recently been found in t-RNA.

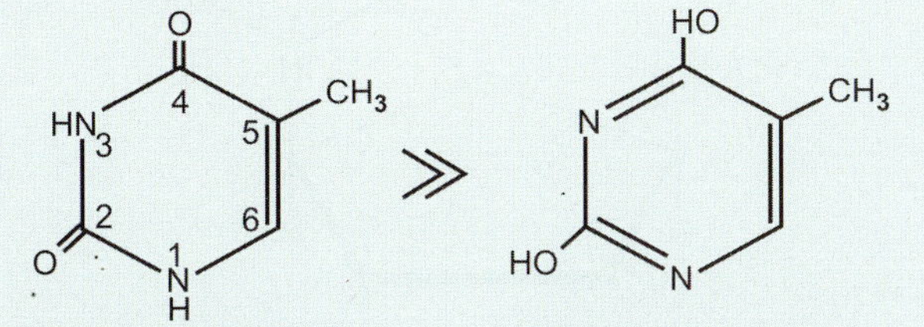

Lactam form of thymine **Lactim form of thymine**

3. Uracil

Chemically it is **2,4-dioxy pyrimidine.**

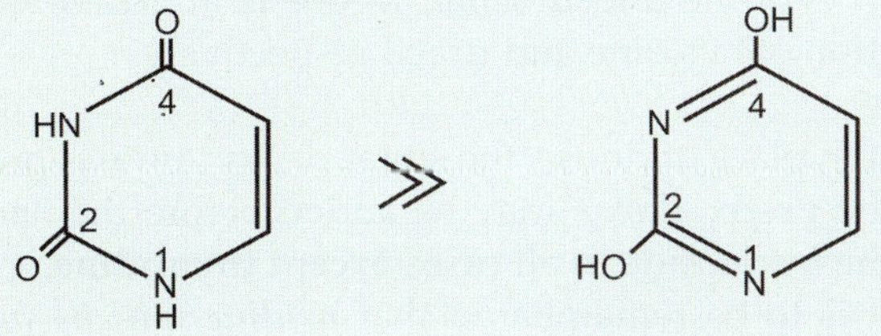

Lactam form of uracil **Lactim form of uracil**

Uracil is confined to RNA only, not found in DNA. In addition to three major pyrimidine bases, there occurs in small quantities bases like 5–OH-methyl cytosine, methylated derivatives and reduced uracil compounds.

PURINE BASES

The Purine ring is more complex than the Pyrimidine ring. ***It can be considered the product of fusion of a pyrimidine ring with an imidazole ring.***

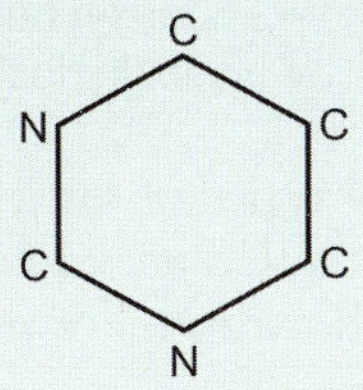

Pyrimidine

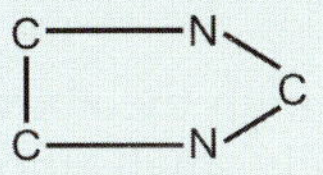

Imidazole

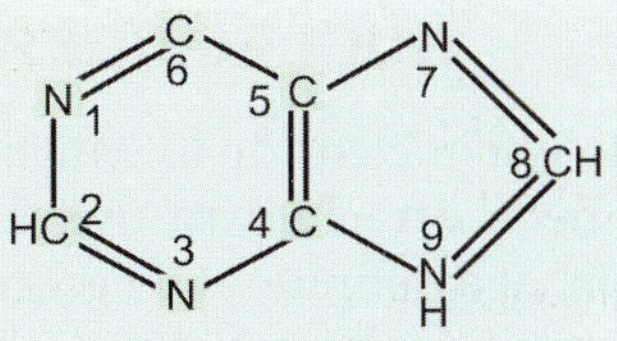

Purine nucleus

Adenine and guanine are the two principal purines found in both DNA and RNA.

1. Adenine

Chemically it is **6-aminopurine.**

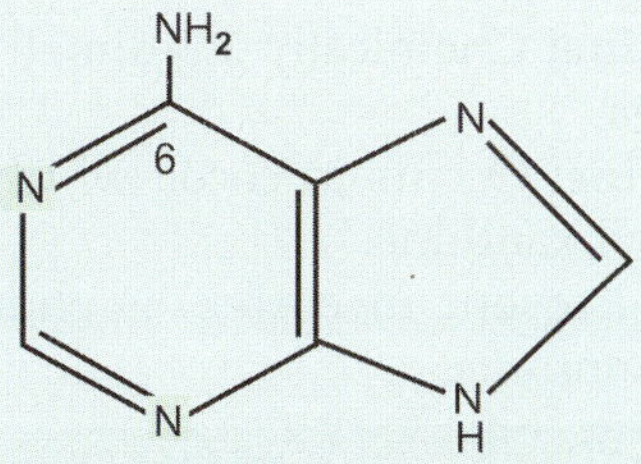

Adenine (6-aminopurine)

2. Guanine

Chemically it is **2-amino-6-oxypurine.** Guanine can be present as lactam and lactim forms.

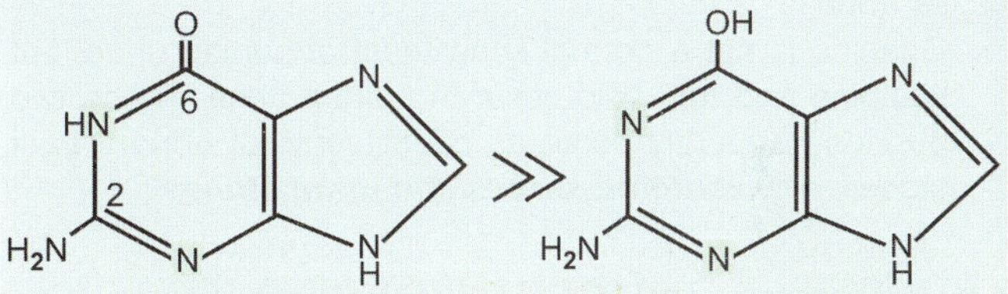

Guanine (Lactam form) **Guanine (Lactim form)**

In addition to above, small amount of **methylated purines** have been shown to be present in nucleic acids. **Two other purine bases, hypoxanthine** and **xanthine** occur as intermediates in the metabolism of adenine and guanine. In human beings, a completely oxidised form of purine base uric acid occurs, which is of great biomedical importance. ***Uric acid is the catabolic end product of purines in human beings.***

Occurrence

A. Nature: In nature, numerous unusual minor bases can exist in addition to the five major bases discussed above. Some of these unusual substituted bases are present only in the nucleic acids of bacteria and viruses. But many are also found in DNA and transfer RNAs of both prokaryotes and eukaryotes.

Examples

- Bacteriophages contain *5-hydroxy methyl cytosine.*
- Certain unusual bases are found in m-RNA molecules of mammalian cells like N^6 – *methyl adenine,* N^7 – *methyl guanine* and N^6 – N^6 – *dimethyl adenine.*
- Both bacterial and human DNA have been found to contain significant quantities of *5-methyl cytosine.*
- A uracil modified at N_3 position by attachment of a propyl group has also been detected in bacteria.

The functions of such substituted Purine and Pyrimidine bases are not clear.

B. *In plants:* Quite a number of purine bases containing methyl substituents occur in plants and plant products and many of them have definite pharmacologic properties. Some of them which occur in foodstuffs having biomedical importance are:

- **Theophylline:** Chemically 1,3-dimethyl xanthine is found in tea.
- **Theobromine:** Occurs in cocoa which is chemically 3,7-dimethyl xanthine.
- **Caffeine:** Present in coffee, is chemically 1,3,7-trimethyl xanthine.

Biomedical Importance

Solubility of the bases

- At neutral pH, guanine is the least soluble of the bases, followed in this respect by xanthine.
- Although uric acid, as urate is relatively soluble at a neutral pH, it is highly insoluble in solutions with a lower pH, such as urine.
- Guanine is not a normal constituent of human urine, but xanthine and uric acid occur in human urine. These two purines frequently occur as constituents of urinary tract stones, e.g. ***xanthine stones/and urate stones***.

NUCLEOSIDES

1. The nucleosides are composed of purine or pyrimidine base linked to either D-ribose (in RNA) or D-2-deoxyribose (in DNA).

Adenosine (Adenine-9-riboside)
(Adenine purine base + ribose sugar)

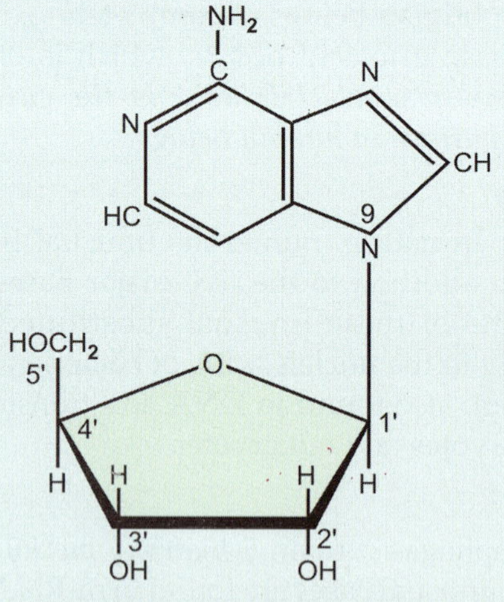

β – N – glycosidic linkage with position 9 of Purine base-adenine and 1' carbon of ribose sugar

2. They are joined by **β-N-glycosidic linkage.**
   a. This linkage in purine nucleosides is at position –9 of the purine base and carbon 1′ of sugar or deoxy sugar.
   Example: Adenosine (adenine-9-riboside)
   b. In pyrimidine nucleosides, β-N-glycosidic linkage is formed at position –1 of the pyrimidine base linked to carbon –1′ of ribose or deoxyribose sugar.
   Example: Uridine (uracil-1-riboside)

Uridine (Uracil-1-riboside)
(Uracil pyrimidine base + ribose sugar)

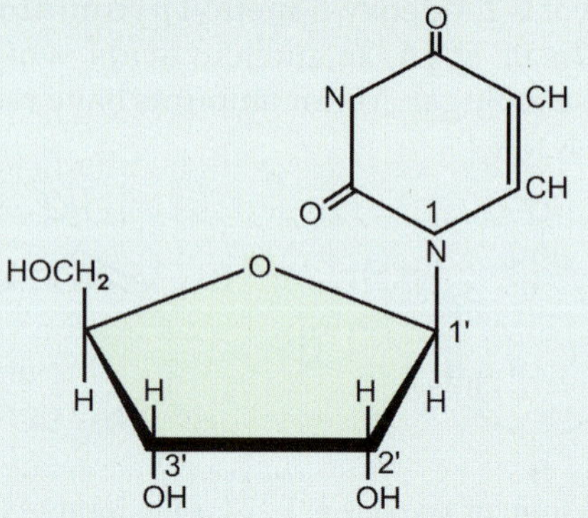

β–N–glycosidic linkage with position 1 of Pyrimidine base-Uracil and 1' carbon of ribose sugar

3. In cytidine and uridine, ribose is attached to N_1 – Position of cytosine and uracil respectively.

Note

1. If in place of ribose, the sugar deoxyribose is present, the prefix *'deoxy'* may be added before the name of the nucleoside in all cases **except thymidine.**
2. It is to be remembered that ***uridine will be present only in RNA but absent in DNA.***

Following shows the different nucleosides and their corresponding base (purine/pyrimidine) and Sugar:

Nucleoside	Base		Sugar
• Adenosine	Adenine	+	Ribose
• Deoxyadenosine	Adenine	+	deoxyribose
• Guanosine	Guanine	+	Ribose
• Deoxyguanosine	Guanine	+	deoxyribose
• Uridine	Uracil	+	Ribose
• Cytidine	Cytosine	+	Ribose
• Deoxycytidine	Cytosine	+	deoxyribose
• Thymidine	Thymine	+	deoxyribose

Though the above are the usual types of nucleoside, relatively small amounts of what is called ***pseudouridine* (ψ)** is also present in RNA in which carbon-5 of pyrimidine is linked to C-1′ of sugar.

NUCLEOTIDES

A nucleotide is a nucleoside to which a phosphoric acid group has been attached to the sugar molecule by 'esterification' at a definite – OH group and thus has the general composition **base – sugar – PO_4.**

Thus nucleotides are nucleoside-P.

- In ribose nucleosides, there are three possible positions for phosphate esterification namely 2′, 3′ and 5′.
- In deoxynucleosides, there are free –OH groups only at the 3′ and 5′ positions in the deoxyribose nucleosides. PO_4 can be attached only at these positions. The name of each nucleotide may be derived from that of its constituent nitrogenous base.

The box below shows the different nucleotides and their corresponding base (purine / pyrimidine) and sugar.

Exceptions and Variations to Above Structures

1. In t-RNA, the ribose is occasionally attached to uracil at the C_5 – Position, forming a C to C linkage instead of usual N to C linkage. This unusual compound present in t-RNA is called ***pseudouridine*** (ψ).
2. Another unusual nucleotide structure, thymine attached to ribose monophosphate may be present in t-RNA. This compound is formed subsequent to the synthesis of the t-RNA by methylation of the UMP residue by S-adenosyl methionine ("active" methionine).

Internucleotide bonds: The bond between the nucleotides is the ***Phosphoric acid di-ester bond***. The phosphate di-ester bond between the nucleotides is ***formed mainly by 3′ OH group of sugar of one nucleotide to 5′ – OH group of sugar of another nucleotide.*** This **3′, 5′-linkage** is to be expected in DNA since these are only sugar – OH groups available in deoxyribose for the formation of phosphate di-ester bond. In RNA–3′–5′ linkages predominate but 2′–3′ linkages are also possible.

Hypoxanthine Derivatives of Biomedical Importance

Inosinic Acid (IMP)

1. Hypoxanthine ribonucleotide, usually called **inosinic acid *(IMP) is a precursor of all purine nucleotides synthesised de novo*** (See Synthesis of Purines)
2. Inosinate can also be formed by the ***deamination* of AMP,** a reaction which occurs particularly in muscle as part of so-called ***"purine nucleotide cycle"***. Inosinate derived from AMP as above, when reconverted to AMP again, results in the net production of NH_3 from asparate (Fig. 14.1).

Synthetic Analogues of Biomedical Importance

Synthetic analogues of nucleobases, nucleosides and nucleotides are recently of wide use in medical sciences and clinical medicine. Synthetic analogues have become the chief fighting weapons in the hands of oncologists for cancer chemotherapy (See box on next page).

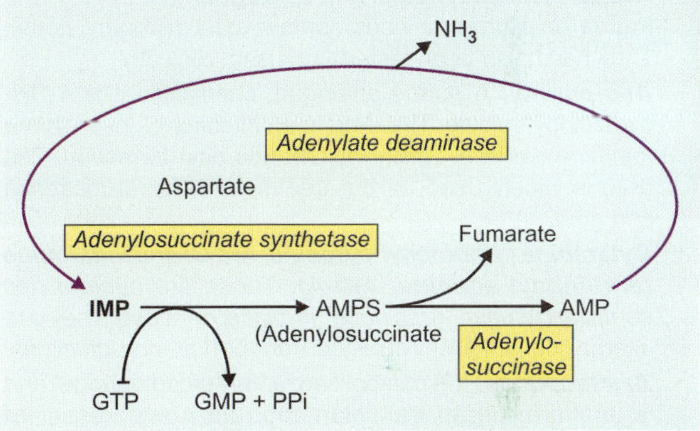

Fig. 14.1: Purine-nucleotide cycle

SECTION THREE

Neucleotide	Base	Sugar	Phosphoric acid
(a) Present in RNA			
• ***Adenylic acid or Adenylate (AMP)***	Adenine	+ Ribose	+ Phosphoric acid
• ***Guanylic acid or Guanylate (GMP)***	Guanine	+ Ribose	+ Phosphoric acid
• ***Cytidylic acid or Cytidylate (CMP)***	Cytosine	+ Ribose	+ Phosphoric acid
• ***Uridylic acid or Uridylate (UMP)***	Uracil	+ Ribose	+ Phosphoric acid
(b) Present in DNA			
• ***Deoxy adenylic acid or deoxy Adenylate (dAMP)***	Adenine	+ Deoxyribose	+ Phosphoric acid
• ***Deoxy guanylic acid or deoxy guanylate (d GMP)***	Guanine	+ Deoxyribose	+ Phosphoric acid
• ***Deoxy cytidylic acid or deoxy cytidylate (d CMP)***	Cytosine	+ Deoxyribose	+ Phosphoric acid
• ***Thymydylic acid or thymidylate (TMP)***	Thymine	+ Deoxyribose	+ Phosphoric acid

Note

1. Uridylic acid occurs in RNA only, hence there will be only ribose. **There is no deoxy uridylic acid.**
2. **There is no thymidylic acid in RNA.**

Basis of Chemotherapy

1. The heterocyclic ring structure or the sugar moiety is altered in such a way as to induce toxic effects when the analogues get incorporated into cellular constituents of the body.
2. Effects result from either:
   - Inhibition by the drug of specific enzyme activities necessary for the nucleic acid synthesis of the cells.
   - From incorporation of metabolites of the drug into the nucleic acids, where they adversely affect the base pairing essential for accurate transferring of information.

Synthetic Derivatives

Some synthetic derivatives are as follows:

- ***6 – thio-guanine and 6 – Mercaptopurine:*** In both of these naturally occurring purines, –OH groups are replaced by "thiol" (–SH) group at 6 – position of purine, widely used in clinical medicine.
- ***Azapurine, Azacytidine, and 8-Azaguanine:*** In these the Purine or pyrimidine rings contain extra nitrogen atoms. They have also been tested and used clinically.
- ***Allo-purinol:*** A purine analogue, chemically it is 4-OH-pyrazolopyrimidine. The drug is an inhibitor of the enzyme *xanthine oxidase*, which inhibits uric acid formation. The drug is widely used for the treatment of hyperuricaemia and gout.
- ***Cytarabine (arabinosyl-cytosine, Ara C) and vidarabine (arabinosyl adenine, Ara-A):*** *These are nucleosides containing sugar "Arabinose" in place of ribose.* They are used in chemotherapy of cancers and certain viral infections.
- ***Azathiopurine:*** It is catabolised to "6-Mercaptopurine" and is useful in organ transplantation as a suppressor of immunologic rejection of grafts.
- ***5-ododeoxyuridine:*** Recently, a series of nucleoside analogues, possessing antiviral activities have been studied. 5-iododeoxyuridine has been demonstrated to have promising result and found to be highly effective in the local treatment of ***Herpetic Keratitis***, an infection of cornea by *Herpes* simplex virus.
- ***Aminophylline and theophylline:*** Recently, both are being used clinically to inhibit the catabolism of cyclic AMP, and thus increasing the cyclic AMP level in the cells.

NUCLEOTIDES AND NUCLEOSIDES OF BIOLOGICAL IMPORTANCE

Besides the nucleotides which occur as integral part of DNA and RNA, there are many biologically important nucleotides present in tissues and cells, where they have diverse biochemical functions.

CLASSIFICATION

A. ***Adenosine nucleotides:*** ATP, ADP, AMP and Cyclic AMP.

B. ***Guanosine nucleotides:*** GTP, GDP, GMP and cyclic GMP.

C. ***Uridine nucleotides:*** UTP, UDP, UMP, UDP-G.

D. ***Cytidine nucleotides:*** CTP, CDP, CMP and certain deoxy CDP derivatives of glucose, choline, ethanolamine.

E. ***Miscellaneous:*** **PAPS ('active' sulphate), active methionine (S-adenosyl methiomine),** certain coenzymes like NAD^+ and $NADP^+$, FAD and FMN, Cobamide coenzyme, CoA.

ADENOSINE CONTAINING NUCLEOTIDES

(a) Adenosine triphosphate (ATP)

It is called as ***storage battery*** of the tissues. It is storehouse of energy.

Formation of ATP: See the chapter of Biologic Oxidation.

Functions: Two of the three phosphate residues are high energy "phosphates (~P)" and on hydrolysis each releases energy (7.6 Kcal); energy is utilised for ***endergonic processes***.

- Many synthetic reactions require energy, e.g. arginino succinate synthetase reaction in the urea cycle.
- ATP is also required in the synthesis of Phosphocreatine from creatine, synthesis of FA from acetyl CoA, synthesis of peptides and proteins from amino acids, formation of glucose from pyruvic acid, synthesis of glutamine, etc.
- ATP is an important source of energy for muscle contraction, transmission of nerve impulses, transport of nutrients across cell membranes, motility of spermatozoa.
- ATP is required for **formation of *active methionine*,** which is required for methylation reactions.
- ATP donates Phosphate for a variety of *phosphotransferase* reactions, e.g. hexokinase reaction.
- ATP is required for formation of 'active' sulphate which is necessary for incorporation of SO_4 in compounds like formation of chondroitin SO_4.
- ***In vivo*** ATP is converted to ADP, AMP and cyclic nucleotides like 3′,5′-cyclic AMP which have important role to play in many biochemical processes.

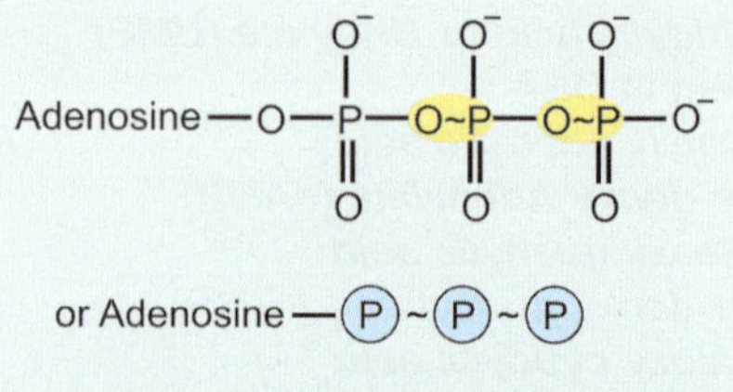

Adenosine triphosphate (ATP)

(b) Adenosine diphosphate (ADP)

- ADP plays an important role as a primary PO_4 acceptor in oxidative phosphorylation and photophosphorylation in addition to its effect on control of cellular respiration, muscle contraction, etc.

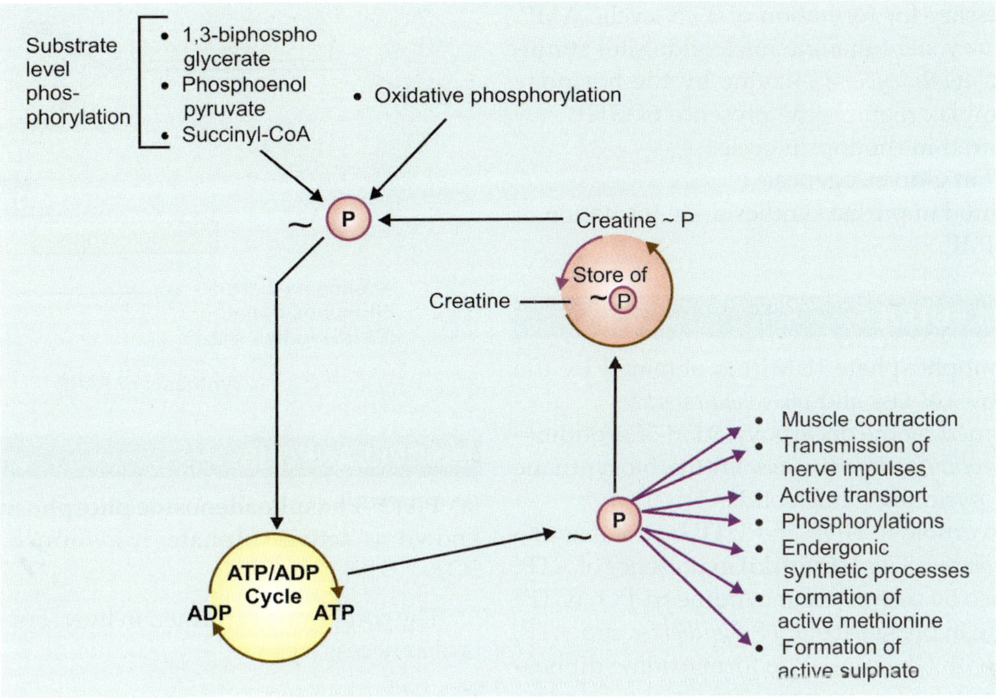

Fig. 14.2: Role of ATP/ADP cycle in transfer of high energy phosphate

- ADP is also important as an activator of the enzyme *glutamate dehydrogenase.*

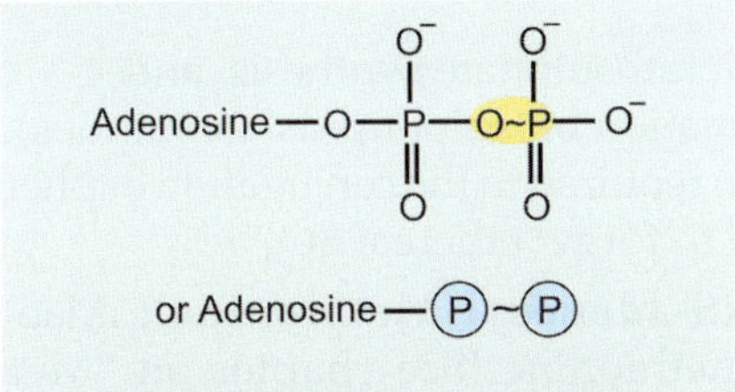

Adenosine diphosphate (ADP)

Figure 14.2 shows the role of ATP/ADP cycle in transfer of high energy phosphate.

(c) Adenosine Monophosphate (AMP)

AMP was the first naturally occurring mononucleotide discovered by **Embden** in 1927. Adenylic acid or Adenosine monophosphate (AMP) obtained by the action of *RNAase,* and *phosphodiesterase* on RNA is the same as the adenylic acid found in muscle.

- AMP acts as an activator of several enzymes in the tissues.

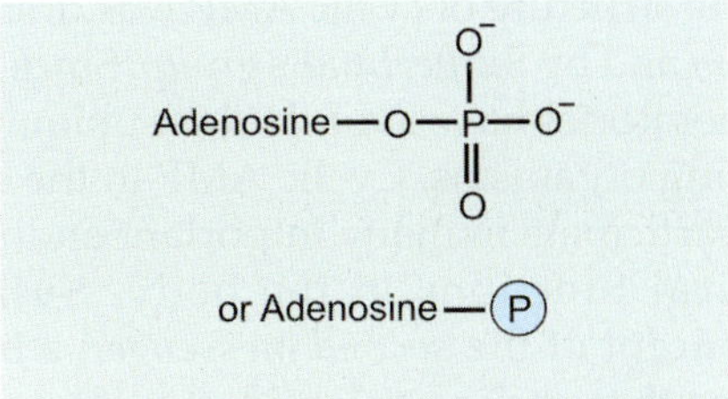

Adenosine monophosphate (AMP)

- In the glycolytic pathway, the enzyme ***phosphofructokinase*** is inhibited by ATP but the inhibition is reversed by AMP, the deciding factor for the reaction being ratio of ATP and AMP.
- AMP is also formed from IMP through adenylo succinate (See purine nucleotide cycle explained above).
- AMP is produced by degradation of 3′, 5′ cyclic AMP by the enzyme ***phosphodiesterase.***
- AMP also can act as an "inhibitor" of certain enzymes like *fructose 1, 6-diphosphatase* in glycolytic cycle and *adenylosuccinate synthetase.*
- In resting muscle, AMP is formed from ADP, by the enzyme *adenylate kinase* (*myokinase*) reaction.

$$2\,ADP \xrightarrow{\text{Myokinase}} ATP + AMP$$

(Adenylate kinase)

The AMP produced activates the *phosphorylase (phosphorylase "b")* enzyme of muscle and increases the breakdown of glycogen.

GUANOSINE NUCLEOTIDES

Guanine analogue of ATP is GTP which is involved in certain important metabolic reactions.

(a) GTP: Guanosine Triphosphate

- The oxidation of succinyl-CoA in the citric acid cycle involves phosphorylation of GDP to form GTP.
- GTP is also required for protein synthesis.

- GTP is necessary for formation of 3′, 5′-cyclic AMP. GTP regulatory site (guanine nucleotide site) stimulation of *adenylate cyclase* enzyme by the hormone receptor complex requires the presence of GTP.
- GTP is required in rhodopsin cycle.
- Role of GTP in gluconeogenesis.
- GTP is required in purine synthesis—in formation of AMP from IMP.

URIDINE NUCLEOTIDES

- Uridine monophosphate (UMP) is obtained by the hydrolysis by *RNAase* and *phosphodiesterase*.
- It is also formed by the decarboxylation of orotidine-5′-PO_4 by *orotidylic decarboxylase* in the biosynthetic pathway of pyrimidine nucleotides.
- UMP is convertible to UDP and UTP by the enzyme *nucleoside diphosphokinase* **(nudiki)** in presence of ATP.
- UTP may also be converted to cytidine tri-PO_4 (CTP) by glutamine in presence of *CTP Synthetase* and ATP.
- UTP reacts with Glucose-1-P to form uridine diphosphoglucose (UDPG), in presence of the enzyme *UDPG-Pyrophosphorylase*. UDPG can donate glucose to a "Primer" molecule of glycogen and is involved in glycogen synthesis.
- UDPG is also required for UDP-Glucose-Galactose *epimerase* for the interconversion of glucose and galactose in the liver.
- Galactose-1-P and UTP similarly can form uridine-diphosphogalactose (UDPGal). Both UDPG and UDP-Gal can be oxidised to UDP-glucuronic acid and UDP-Galacturonic acid respectively by NAD^+-dependant dehydrogenase.

UDP-Glucuronic acid is used for:

- Conjugation and detoxication of bilirubin, benzoic acid, sterols, oestrogens and drugs.
- Also used in the biosynthesis of hyaluronic acid, heparin, and several other MPS.

UDP–Galacturonic acid and UDP–L-iduronic acids are used for the biosynthesis of chondroitin–SO_4.

CYTIDINE NUCLEOTIDES

These include CMP, CDP, and deoxy CDP—derivatives of glucose, choline, ribitol, glycerol, sialic acid, etc.

- CDP-choline, CDP-glycerol, and CDP-ethanolamine are involved in the biosynthesis of phospholipids.
- CMP-acetyl neuraminic acid is an important precursor of cell wall polysaccharides in bacteria.
- CMP-Sialic acid is present in salivary glands and may be concerned with the biosynthesis of salivary mucin.

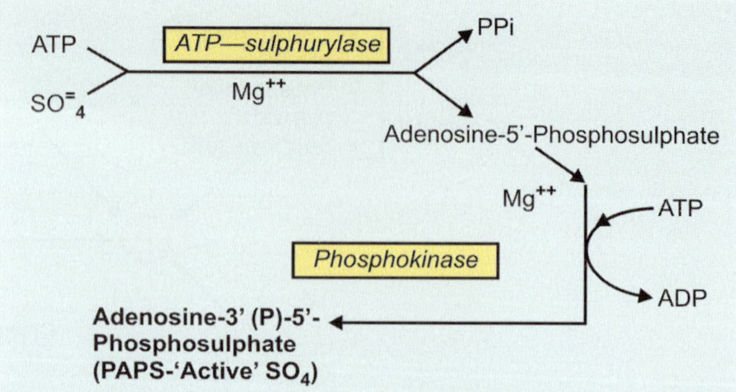

Synthesis of PAPS

MISCELLANEOUS

(a) PAPS-Phosphoadenosine phosphosulphate: It is also known as **active sulphate.** It is formed from ATP and $SO_4^=$.

The compound is formed in liver. For the reaction Vit. is also necessary.

Functions

1. The *sulphatases* are enzymes which catalyse the introduction of a SO_4 group to various biomolecules, e.g.
   - In the biosynthesis of heparin,
   - In the biosynthesis of chondroitin SO_4 A, B, C and D,
   - In Keratosulphate synthesis, and
   - Formation of 'sulpolipids' (sulfatides).
2. It is also required in the conjugation of phenols, indole, skatole to form "ethereal SO_4".

(b) SAM (S-adenosyl Methionine): Also known as "active" methionine (See chapter on "Metabolism of Methionine").

(c) Coenzymes: For these, see chapter on "Vitamins".

CYCLIC NUCLEOTIDES

1. CYCLIC AMP

Cyclic adenosine monophosphate 3′, 5′-cyclic AMP was first discovered as a mediator in hepatic glycogenolysis. ***Sutherland*** and ***Rall*** discovered this factor in the cell while they were studying the mechanism by which epinephrine and glucagon promoted glucose release in the liver. The structure of cyclic AMP was first described by ***Markham*** and by ***Sutherland's group.*** Since then work in many laboratories has established the ubiquity of cyclic AMP in living organisms. Cyclic AMP in the cell is now known to participate in many important endocrinal and physiological functions in the body. **Sutherland's brilliant concept of the *second messenger*, a hypothesis which states that cyclic nucleotides mediate the effect**

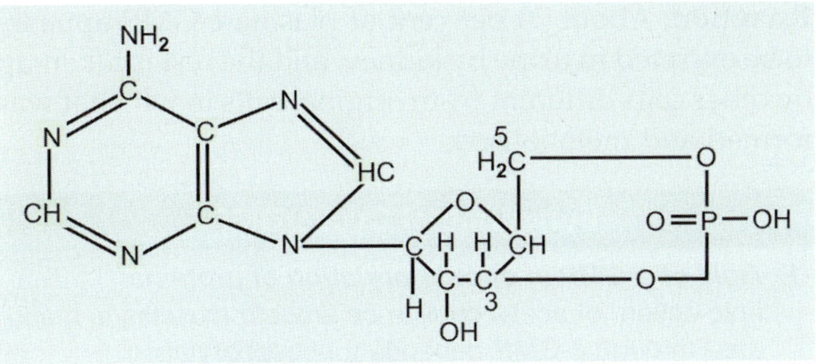

Fig. 14.3: Structure of cyclic AMP

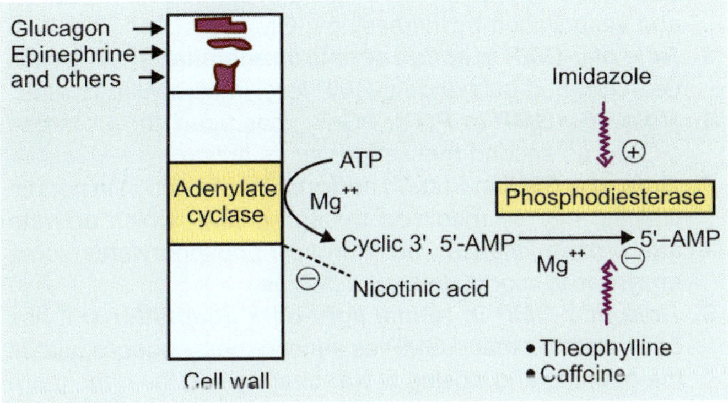

Fig. 14.4: Formation and degradation of cyclic AMP

of a variety of hormones and other biologically active agents has found ample confirmation. ***EW Sutherland (Jr) was awarded the Nobel Prize in 1971 in physiology and medicine*** for his outstanding "discoveries" concerning the mechanisms of the action of hormones.

FORMATION AND DEGRADATION OF CYCLIC AMP

Cyclic AMP is a cyclic nucleotide and chemically it is 3′, 5′-adenosine monophosphate **(Fig. 14.3)**.

Cyclic AMP is synthesised in the tissues from ATP under the influence of an enzyme ***adenyl cyclase*** in the presence of Mg^{++} ions **(Fig. 14.4)**. The activity of the enzyme is regulated by a series of complex interactions many of which involve hormone receptors. c-AMP is degraded in the tissues by its conversion to 5′-AMP in a reaction catalysed by the enzyme ***phosphodiesterase.***

Adenyl Cyclase

The enzyme is widely distributed in nature and has been identified in every mammalian tissue studied with the ***exception of mature mammalian erythrocytes***. Adenyl cyclase has been found to be associated with either the cytoplasmic, mitochondrial or endoplasmic membrane. The profound physiological and metabolic importance of adenyl cyclase resides in the fact that its activity responds to a wide variety of hormones and other pharmacologically active agents, viz histamine, 5-HT (serotonin), ouabain, etc. The activity of the enzyme is inhibited by insulin and prostaglandins. It has been confirmed that insulin inhibits the activation of adenyl-cyclase in liver and fat cells and $PG\text{-}E_1$ that of fat cells.

'Adenyl Cyclase' System—Stimulation and Inhibition

Interaction of the hormone with its receptor results in the activation or inactivation of adenyl cyclase.

Process is mediated by at least two GTP-dependent regulatory proteins.

- **Gs (stimulatory)—also called as Ns.**
- **Gi (inhibitory)—Ni.**

Each of the regulatory protein is composed of 3 subunits—**α, β and γ.**

Two parallel system, a stimulatory (s) one and an inhibitory (i) one, converge upon a single catalytic molecule G. Each consists of a receptor Rs or Ri and regulatory complex—Gs and Gi.

Gs and Gi are each trimers composed of α, β and γ-subunits. β and γ-subunits in Gs appear to be identical to their respective counterparts Gi. α subunit in Gs differs from that of Gi, designated as α_s (MW = 45000) and αi (MW = 41,000).

The binding of a peptide hormone to Rs or Ri results in a receptor-mediated activation of G, which entails Mg^{++}-dependent binding of GTP by α and concomitant dissociation of β and γ from α.

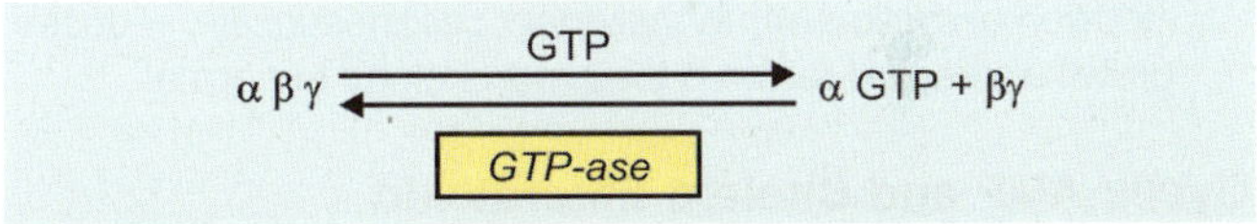

α_s has intrinsic *GTPase* activity and the active form of α_s-GTP is inactivated upon hydrolysis of GTP to → GDP and the trimeric Gs complex (α β γ) is reformed.

Recently, it has also been shown that *adenylate cyclase* activity is also modulated by a heatstable calcium-dependent regulatory protein ('CDR') which appears to be major calcium binding protein within the cell.

Phosphodiesterase: Cyclic AMP is rapidly inactivated by an enzyme, a cyclic nucleotide *Phosphodiesterase* which opens the 3′, 5′-phosphate bond at the 3′ position leaving ordinary 5′-AMP as the product and thus inactivating cyclic AMP.

- Certain activators and inhibitors of the enzyme are known.
- ***Activators*** promote degradation of c-AMP and, thus reduce its level.
- ***Inhibitors*** prevent degradation and, thus increase cyclic-AMP level.
- Known activators and inhibitors of the enzyme are listed below.

Activators	Inhibitors
• Imidazoles • Mg^{++} ions • NH_4 ions	• Methylxanthines • Theophylline > caffeine > theobromine • Papaverine • Bromolysergic acid diethylamide • Reserpine

Functions of Cyclic AMP

- Mediator of hormone action—acts as "Second messenger" in the cell.
- Regulates glycogen metabolism—increased cyclic AMP produces breakdown of glycogen (glycogenolysis).
- Regulates TG metabolism—increased cyclic AMP produces lipolysis (breakdown of TG).
- Cholesterol biosynthesis is inhibited by cyclic AMP.
- Cyclic AMP stimulates protein kinases so that inactive protein kinase is converted to active protein kinase.
- Cyclic AMP modulates both transcription and translation in protein biosynthesis.
- Cyclic AMP activates different steps of steroid biosynthesis.
- Cyclic AMP also regulates permeability of cell membranes to water, sodium, potassium and calcium.
- Plays an important role in regulation of insulin secretion, catecholamine biosynthesis, and melatonin synthesis.
- Histamine increases cyclic AMP production in parietal cells which in turn increases gastric secretion.
- Decrease in cyclic AMP level is involved in the excitation of bitter taste receptors in tongue.
- Cyclic AMP plays an important role in cell differentiation. Addition of cyclic AMP to malignant cell lines *in vitro* reduces growth rates and restores their morphology to normal.

Cyclic AMP and Cholera Enterotoxin

Refer to chapter on Biochemistry of Cholera.

2. CYCLIC GMP

Formation: Cyclic GMP is made from GTP by the enzyme ***guanylate cyclase*** which exists in soluble and membrane-bound form. The enzyme requires Mn^{++} as a cofactor. Guanylate cyclase is also reported to be stimulated by Ca^{++}.

Fate: Cyclic-GMP is hydrolysed by cyclic nucleotide *phosphodiesterase*. The cyclic nucleotide phosphodiesterase is trimeric consisting of three subunits, **α, β** and **γ**. γ-subunit is inhibitory and binding of γ-subunit activates the *phosphodiesterase* (α, β).

The activated phosphodiesterase (α, β) then catalyses the hydrolysis of c-GMP to 5′-AMP.

Phosphodiesterase (α, β, γ)
("inactive")
↓
Phosphodiesterase (α, β)
("active")

Excretion: About 20 per cent of plasma c-GMP appears to be excreted in urine by kidney, and the rest is taken up by other cells different from original cells in which it was formed and metabolised.

Functions of Cyclic GMP

1. ***Role of c-GMP in phosphorylation of proteins:*** "Muscarinic action" of acetyl choline on smooth muscles is mediated through c-GMP dependant phosphorylation.
2. ***Role of c-GMP in vasodilation:*** Compounds like nitroglycerine, sodium nitrite, etc. cause smooth muscle relaxation and vasodilation by increasing c-GMP level.
3. ***Role of c-GMP in action of neurotransmitters:*** GABA has been claimed to change c-GMP level in cerebellar tissues.
4. ***Role of c-GMP in PG's:*** PG-$F_{2\alpha}$ has been shown to use c-GMP as second messenger for its action.
5. ***Role of c-GMP in insulin actions:*** Insulin action in certain tissues may be mediated through c-GMP which activate the *'protein kinases'*, which in turn phosphorylates some enzymes to modulate their activities.
6. ***Role of c-GMP in retinal light-dark adaptations:*** It has been claimed that c-GMP as second messenger regulates the opening and closing of Na^+ channels. In the dark, there are high level of c-GMP which binds to Na^+ channels causing them to open. Reverse occurs in the light (Refer vitamin A).
7. ***Role of c-GMP in vasodilatation produced by nitric oxide.*** Nitric oxide (NO) produced in the tissues by the action of the enzyme *'nitric oxide synthase* on Arginine. It produces vasodilatation and lowering of BP by increasing the c-GMP level.

CLINICAL APPLICATION OF CYCLIC AMP ANALYSIS

- **In Diagnosis of Pseudohypoparathyroidism (PHP):** The discovery that PHP patients had an absent or greatly diminished urinary cyclic AMP increment following bovine PTH challenge has provided the basis for a useful diagnostic procedure. Normal subjects and patients with Surgical or idiopathic hypoparathyroidism increase their urinary cyclic AMP output by as much as 200 fold after the IV injection of 200 MRC units of bovine PTH, ***while the PHP patients show less than a 4-fold rise.***
- **In Liver Diseases:** ***Differentiation of extrahepatic obstruction from intrahepatic cholestasis:***
  In complete "extrahepatic" large duct obstruction an exaggerated plasma cyclic AMP response following the IV injection of 1 mg glucagon, is found, whereas the plasma cyclic AMP elevations in patients with "intrahepatic" cholestasis or with cirrhosis were not significantly different from normal control group. ***Thus plasma cyclic AMP measurements following glucagon stimulation have a possible role in the differentiation of extrahepatic obstructive jaundice and intrahepatic cholestasis*** in those situations where other biochemical findings are inconclusive.

SECTION THREE

- **In Thyroid Disorders:** Thyrotoxicosis in man has been found to have increased plasma levels of cyclic AMP, while in hypothyroidism plasma cyclic AMP levels tend to be low.
- **In Nephrogenous Diabetes Insipidus:** It has been shown that there is a deficient urinary cyclic AMP response to ADH in patients with nephrogenic diabetes insipidus relative to control subjects and to patients with primary ADH deficiency. Nephrogenous diabetes insipidus is also associated with an impairment of the urinary cyclic AMP response to PTH challenge.
- **In Malignant Diseases:** Recent cell culture studies have established that inhibition of cell growth is often associated with increased intracellular cyclic AMP concentrations, whereas increases in cyclic GMP levels are associated with cell proliferation. **Cyclic AMP and cyclic GMP, therefore, probably exert opposing influences on cell growth rates. The addition of cyclic AMP or its derivatives to malignant cell lines reduces growth rates and restores their morphology to normal.**

I. Questions (Essay type and short notes)

1. What is 3′, 5′-cyclic AMP? How it formed and degraded in the body? Describe its role in the body.
2. Discuss the role of cyclic AMP in regulation metabolism. How is it synthesised? Which hormones induce its formation?

Short Notes

a. Cyclic AMP b. Cyclic GMP

II. MCQs (Give one correct answer)

1. Cyclic AMP is a cyclic nucleotide formed from:
   a. ADP b. Adenylic acid
   c. CMP d. ATP
   e. GTP
2. Cholera toxin produces diarrhoea due to:
   a. Inhibits the Na^+ - K^+ stimulated ATPase
   b. By decreasing C-AMP level
   c. By increasing synthesis of C-AMP
   d. Prevents axonal ion transport
   e. Stimulate cholinergic receptors
3. The synthesis of adenyl cyclase is increased by:
   a. Epinephrine b. Parathormone
   c. ACTH d. Thyroid hormones
   e. Glucagons
4. In formation of cyclic AMP the stimulation of adinyl cyclase by the hormone-receptor complex requires the presence of:
   a. CTP b. GTP
   c. UTP d. CDP
   e. UMP
5. Cyclic GMP is a cyclic nucleotide formed form:
   a. ATP b. CTP
   c. GTP d. UTP
   e. Adenylic acid
6. The enzyme cyclic nucleotide "phosphodiesterase" is inhibited by following as a result increasing C-AMP level in cells:
   a. Mg^{++} b. Theophylline
   c. Ammonium ions d. Imidazoles
   e. All of the above
7. Plasma level of cyclic nucleotide in thyroid disorders is associated with
   a. An increase in C-AMP level
   b. A decrease in C-AMP level
   c. An increase in C-GMP level
   d. An increase in activity of cyclic nucleotide phosphodiesterase
   e. None of the above
8. Cyclic AMP shows decrease level in a cell-free cytoplasmic suspension due to addition of:
   a. Aminophylline
   b. Caffeine
   c. Adenylate cyclase
   d. Cyclic AMP phosphodiesterase
   e. Dibutyryl cyclic AMP

Answers to MCQs

1. d 2. c 3. d 4. b 5. c 6. b
7. a 8. d.

CHAPTER 15

METABOLISM OF PURINES AND PYRIMIDINES

Major Concepts

A. Learn the details of pyrimidine synthesis and its catabolism.
B. Learn the details of purine synthesis and its catabolism.
C. Learn various disorders associated with purine and pyrimidine metabolism.

Specific Objectives

A. 1. Draw the structure of pyrimidine base and sources of C and N atoms of pyrimidine.
   2. Synthesis of pyrimidine begins with the formation of carbamoyl phosphate. Learn the details of this reaction.
   3. The synthesis of a nucleotide CTP is the final step. Learn the intermediate steps with special reference to the role of PRPP.
   4. Learn how other derivatives of pyrimidine are synthesised.
   5. Learn how deoxypyrimidine nucleotides are synthesised.
   6. Study in detail how various pyrimidine nucleotides are catabolised.

B. 1. Draw the structure of purine base and sources of C and N atoms of purine.
   2. Study all the steps of the purine biosynthesis.
   3. Formation of IMP is the final step. Study how AMP and GMP are further synthesised.
   4. Study in detail 'salvage pathways' of purine bases
   5. Learn 'Purine-salvage cycle'.
   6. Study how purine synthesis is regulated.
   7. Study the details of how guanosine and adenosine are catabolised to form uric acid, in liver and skeletal muscle.
   8. Study how uric acid is further catabolised in non-primates.

C. 1. Study uric acid metabolism.
   2. Study in detail gout classification, clinical importance and treatment.
   3. Study other inherited disorders of purine and pyrimidine metabolism.

INTRODUCTION

Mammals and most of the lower vertebrates can synthesise purines and pyrimidines and are said to be prototrophic.

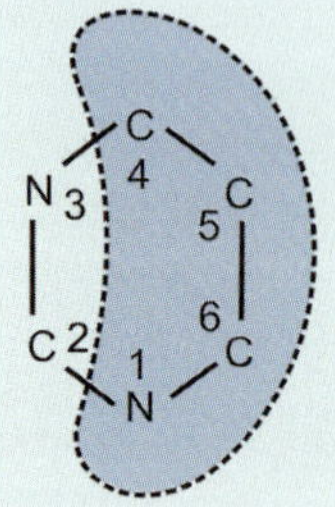

N1, C4, C5 and
C_6 - from Aspartic acid
C_2 - from CO_2
N_3 - from Glutamine amide-N

Fig. 15.1: Sources of C and N atoms of pyrimidine

SYNTHESIS OF PYRIMIDINES

Pyrimidine is a heterocyclic ring **(Fig. 15.1)**.

Materials required for synthesis of pyrimidines

- ***Carbamoyl phosphate:*** Synthesised from CO_2 and glutamine
- ***PRPP:*** 5-phosphoribosyl-1-pyrophosphate
- ***Various enzymes:***
  - *Carbamoyl phosphate synthetase II*
  - *Transcarbamoylase*
  - *Dihydro-orotase*
  - *Dehydrogenase*
  - *Transferase and*
  - *Decarboxylase*

- ***ATP:*** For energy
- ***Amino acid:*** Aspartic acid
- ***Cofactors:*** FAD^+, NAD^+, Mg^{++}

Steps of Synthesis

1. *Synthesis of carbamoyl phosphate:* The synthesis of pyrimidine ring begins with the formation of carbamoyl phosphate from glutamine, CO_2 and ATP, catalysed by the enzyme ***carbamoyl phosphate synthetase II.*** The enzyme is present in cytosol and ***does not require N-acetyl glutamate (NAG).*** The reaction is controlled by the *feed-back* inhibitory effect of pyrimidine nucleotide UTP.

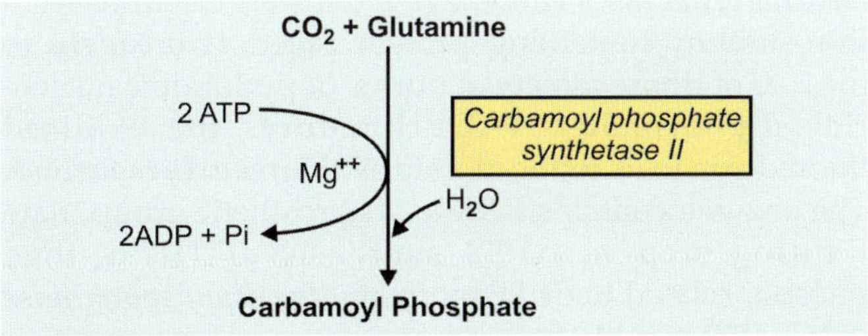

2. *Formation of carbamoyl aspartic acid (CAA):* The enzyme ***Aspartate transcarbamoylase*** then transfers **carbamoyl** group from carbamoyl phosphate to aspartic acid to form carbamoyl aspartic acid (CAA).

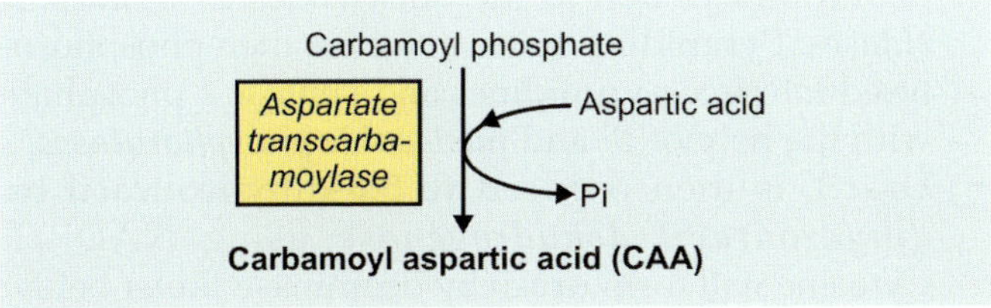

Remarks

- ***Committed*** and ***rate limiting*** step.
- Enzyme is oligomeric.
- Enzyme is inhibited ("feedback" allosteric inhibition) by CTP and UTP.
- Inhibition can be reversed by ATP.

3. *Formation of dihydro-orotic acid:* The reaction is catalysed by the enzyme ***dihydro-orotase,*** which removes a molecule of water (dehydration) from carbamoyl aspartic acid and brings about the closure of the ring to produce dihydro-orotic acid.

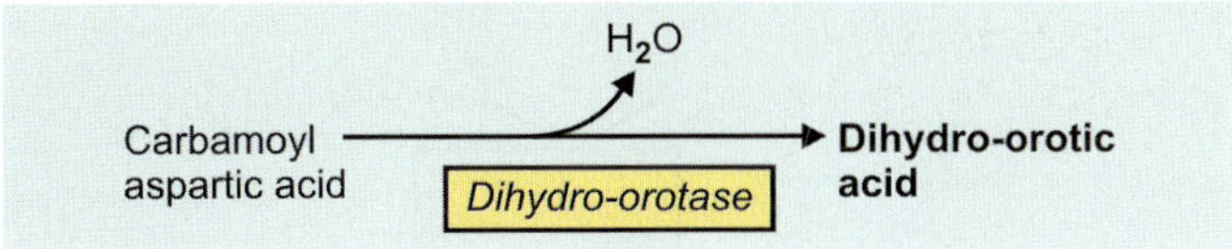

4. *Formation of orotic acid:* The next step is oxidation of dihydro-orotate which is brought about by the enzyme ***dihydro-orotate dehydrogenase.*** The enzyme carries Fe, S, FMN and FAD in its prosthetic group. NAD^+ is required as a coenzyme.

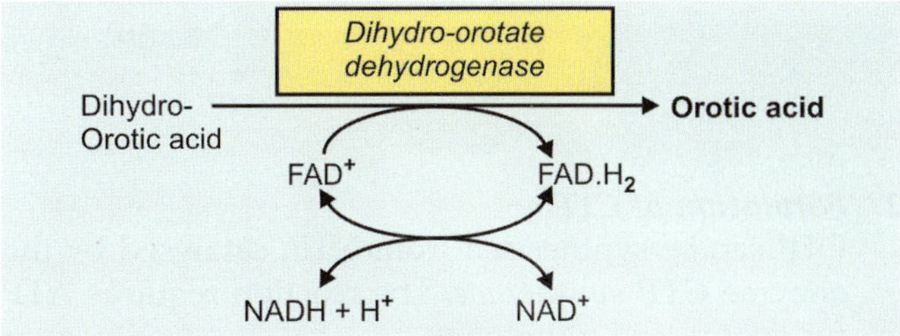

5. *Formation of orotidylic acid:* Under the influence of the enzyme ***Orotate phosphoribosyl trans-ferase,*** 5-phosphoribosyl group from 5-phospho-ribosyl-1-pyrophosphate (PRPP) is transferred to orotic acid to produce orotidylic acid. Mg^{++} ions are required in the reaction.

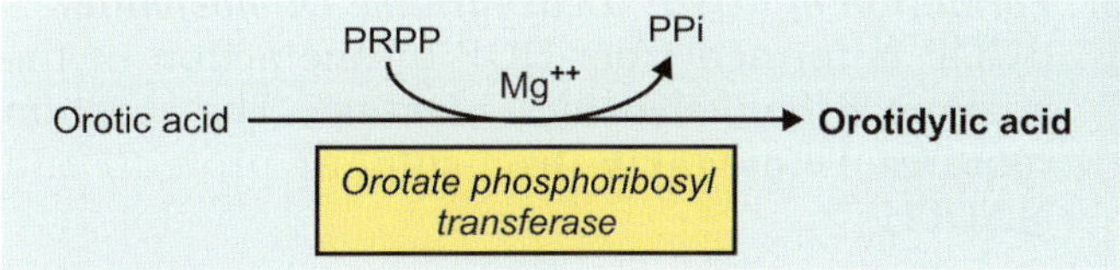

6. *Formation of uridylic acid (UMP):* Orotidylic acid undergoes decarboxylation, catalysed by the enzyme ***Orotidylate decarboxylase,*** and forms uridylic acid (UMP).

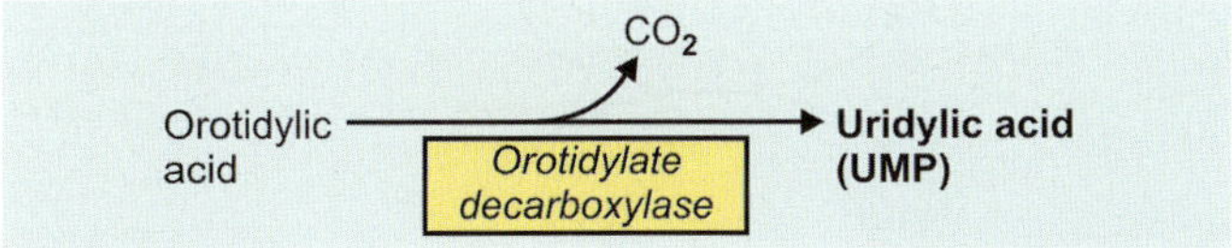

Note

- ***UMP is the first pyrimidine nucleotide to be formed.***
- Other pyrimidine nucleotides viz. UDP, UTP, CTP and d-UDP are synthesised from UMP.
- UMP is the "feedback" inhibitor of the *decarboxylase* enzyme.

Formation of Other Pyrimidine Nucleotides

1. *Formation of UDP and UTP*
   - UMP is phosphorylated by ATP to form UDP, catalysed by the enzyme ***nucleoside monophosphokinase***
   - UDP can be further phosphorylated by ATP to form UTP

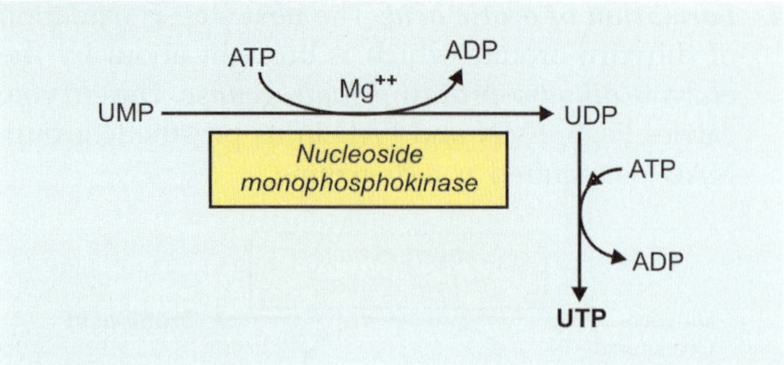

2. ***Formation of CTP***

CTP can be synthesised from UTP, catalysed by the enzyme ***CTP synthetase.*** The reaction requires ATP for energy and glutamine.

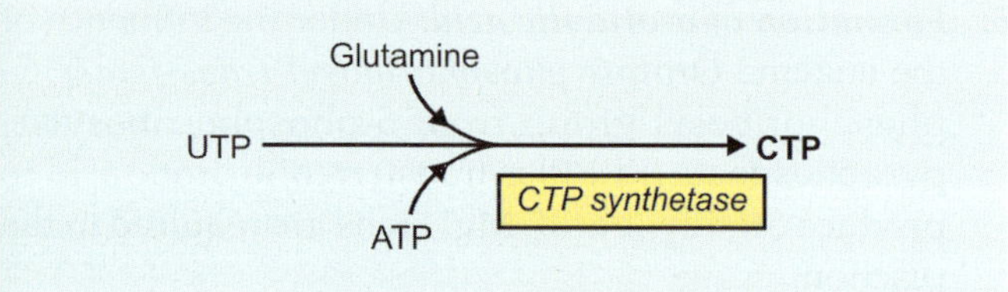

3. ***Formation of dUDP (Deoxyuridine Diphosphate)***

dUDP is formed from UDP by the action of the enzyme ***Ribonucleoside reductase.*** The reaction requires ***Thioredoxin*** (Iron-sulphur protein) and NADPH.

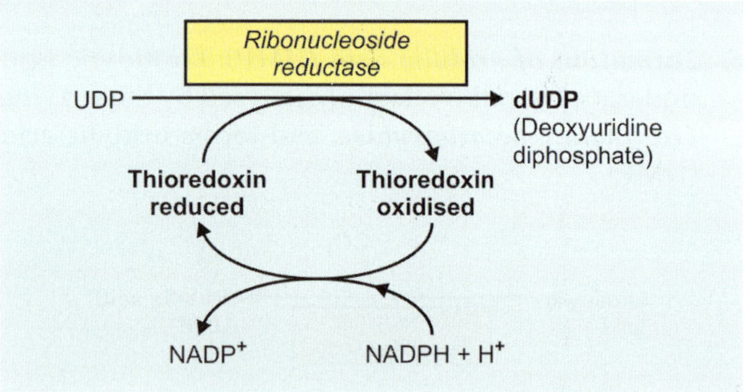

Synthesis of Thymine Deoxyribonucleotides

- dTMP is synthesised from dUMP. dUMP may arise from the hydrolysis of dUDP by ***phosphatase.*** Alternatively, dCMP may be produced by phosphorylation of circulating deoxycytidine with the help of ATP. The enzyme is ***deoxycytidine kinase.***

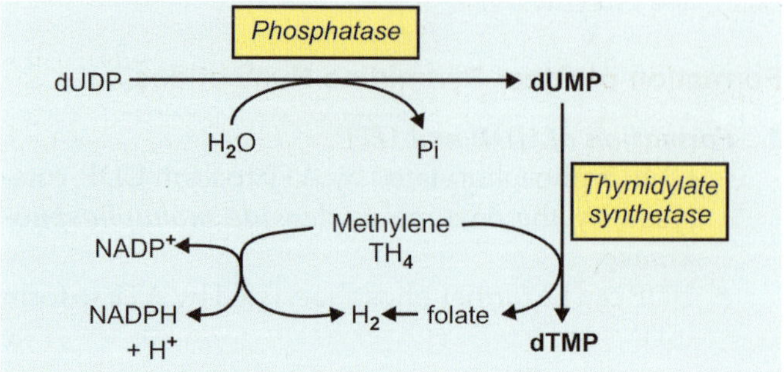

- ***Thymidylate synthetase*** methylates dUMP into dTMP by transferring methyl group from N^5, N^{10} methylene H^4 folate, the C^5 of the uracil moiety. Thymidylate synthetase is covalently inhibited by folate antagonists like ***aminopterin*** and ***amethopterin*** which competitively ***inhibits*** DH_2***-folate reductase.***
- dTMP can be changed to dTDP by transphosphorylation with the help of ATP, Mg^{++} using ***thymidylate kinase.*** This is the synthesis of thymine and pseudouridine ribonucleotides. This occurs in TψU loop of each t-RNA molecule. UMP residue is precursor.

Synthesis of Pyrimidine Deoxyribonucleotides

Reduction of ribose to deoxyribose is achieved by ***ribonucleotide reductase.*** This enzyme transfers electrons from iron-sulphur containing protein called ***thioredoxin*** to the 2′-C of ribose moiety of purine or pyrimidine nucleotide diphosphate. NADPH reduces the oxidised thioredoxin to its reduced state by ***thioredoxin reductase.*** The enzyme contains FAD as its prosthetic group. Rate limiting step in dTTP synthesis. dTDP is then phosphorylated to dTTP by ***thymidine diphosphokinase*** using ATP and Mg^{++}.

CATABOLISM OF PYRIMIDINES

1. **Cytosine and Uracil**

- The first step of the catabolism of pyrimidines is dephosphorylation to the nucleosides by ***5′-nucleotidases.*** Pyrimidine nucleosides are then phosphorolysed into free pyrimidines and pentose 1 phosphate with the help of Pi and ***nucleoside phosphorylases.***
- Uracil, is then reduced to 5,6-dihydrouracil by ***dihydrouracil dehydrogenase*** using NADPH. Cytosine will form uracil by deaminase (Refer below in box).
- The hydrolysis of 5,6-dihydrouracil is the next step. This is done hydrolytically by ***hydropyrimidine hydrase*** to produce β-ureidopropionic acid.

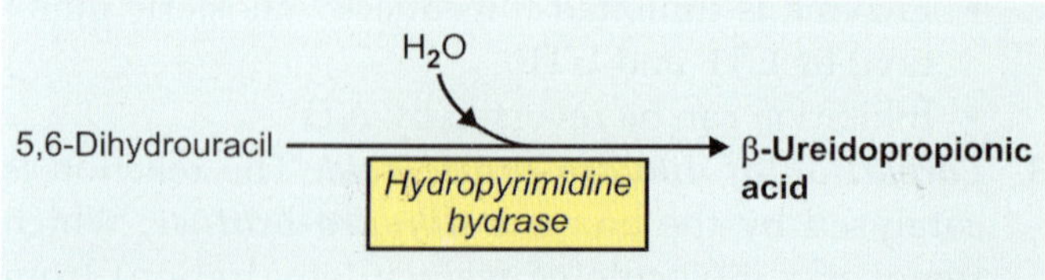

- The next step is further hydrolysis by ***β-ureidopropionase*** into CO_2, NH_3 and β-alanine.

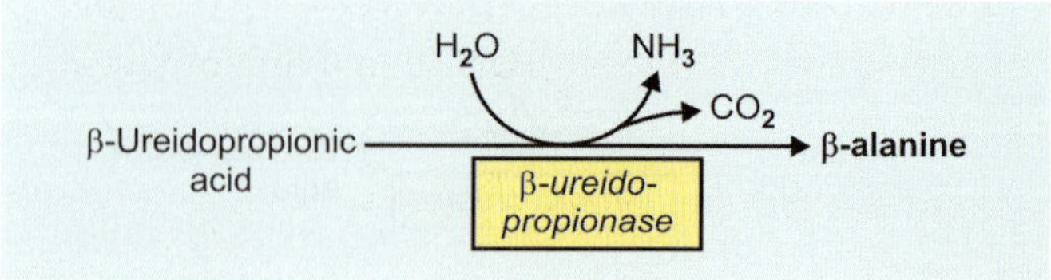

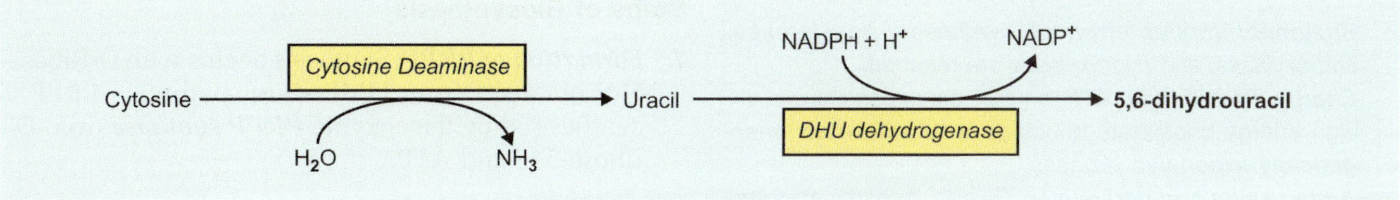

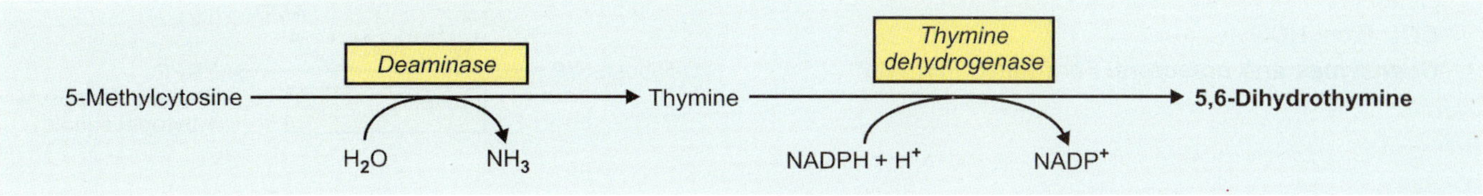

- The β-alanine can either be used in the synthesis of Anserine, carnosine or CoA or can be oxidised to acetate, NH_3 and CO_2.

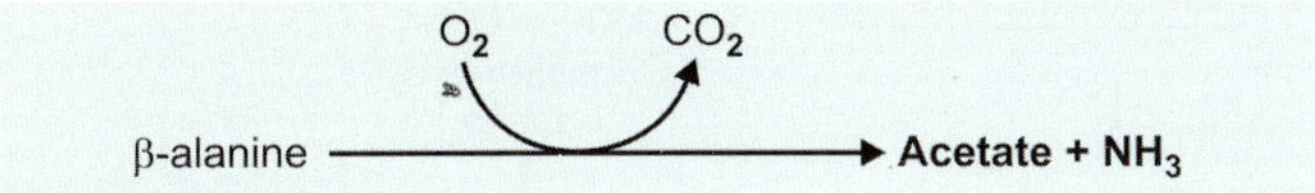

2. Thymine

- Thymine released from thymidine or produced by the deamination of 5-methylcytosine is reduced to dihydrothymine by an NADH dependent ***dehydrogenase*** (Refer box above).
- Next is the hydrase that brings about the hydrolysis of dihydrothymine to give β-ureidoisobutyric acid.
- The β-ureidoisobutyric acid is hydrolysed by ***β-ureidoisobutyrase*** into CO_2, NH_3 and β-amino isobutyrate.

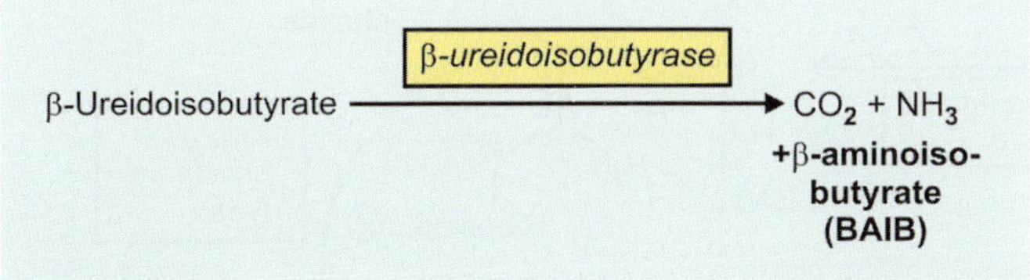

CLINICAL ASPECT

Note: *β-aminoisobutyrate (BAIB) is excreted in large quantities in leukaemia, after body is subjected to X-ray radiation.*

METABOLISM OF PURINES

Adenine and guanine are the two purines either with a ribose or deoxyribose sugar, found in living organisms.

Biosynthesis of Purines

Points to Note: Points to Remember

- **Purines are built up as nucleotides**
- There is no dietary requirement of purines. Body can synthesise them.
- Various enzymes that catalyse specific reactions in the synthetic pathway have been studied. The pathway of synthesis is essentially the same in all the living organisms. Synthesis takes place in the **liver.**
- Purines are ***first synthesised as nucleotide inosinic acid***–(Hypoxanthine ribose 5'-phosphate) which is then converted into the adenine and guanine nucleotides.
  - Purine ring is built on ribose-5'-P.

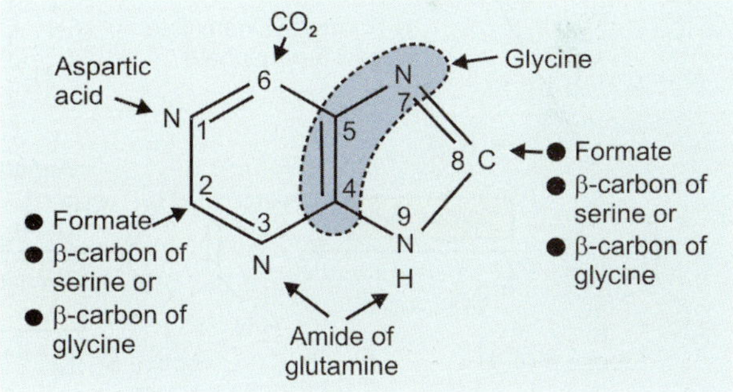

Fig. 15.2: Sources of C and N of purine

- Sources of C and N in purine nucleus are shown in Figure 15.2.

Figure 15.3 gives the reactions of purine synthesis. **Inosonic acid is the first purine formed.**

Materials required for Purine Synthesis

- ***PRPP*** (5-phosphoribosyl-1-pyrophosphate): ***It is the starting material.*** It is formed from D-ribose -5'-P obtained from HMP shunt.

SECTION THREE

- ***Enzymes:*** Various enzymes *synthases, transferases, carboxylases,* and *hydroxylases* are required.
- ***Energy: Provided by ATP*** requires the expenditure of six high energy phosphate bonds. **It is therefore an *energetically expensive process.***
- ***Amino acids and derivatives:*** Glycine, aspartic acid and glutamine
- ***CO_2:*** From HCO_3^-
- ***Coenzymes and cofactors:*** Formylated FH_4, Mg^{++}.

Steps of Biosynthesis

1. ***Formation of PRPP:*** Synthesis begins with D-Ribose-5'-P obtained from HMP-shunt pathway. PRPP is synthesised by the enzyme ***PRPP synthase*** from D'-ribose-5'-P and ATP.

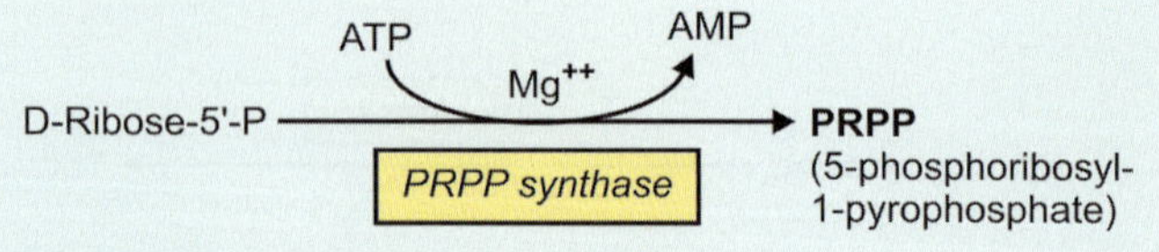

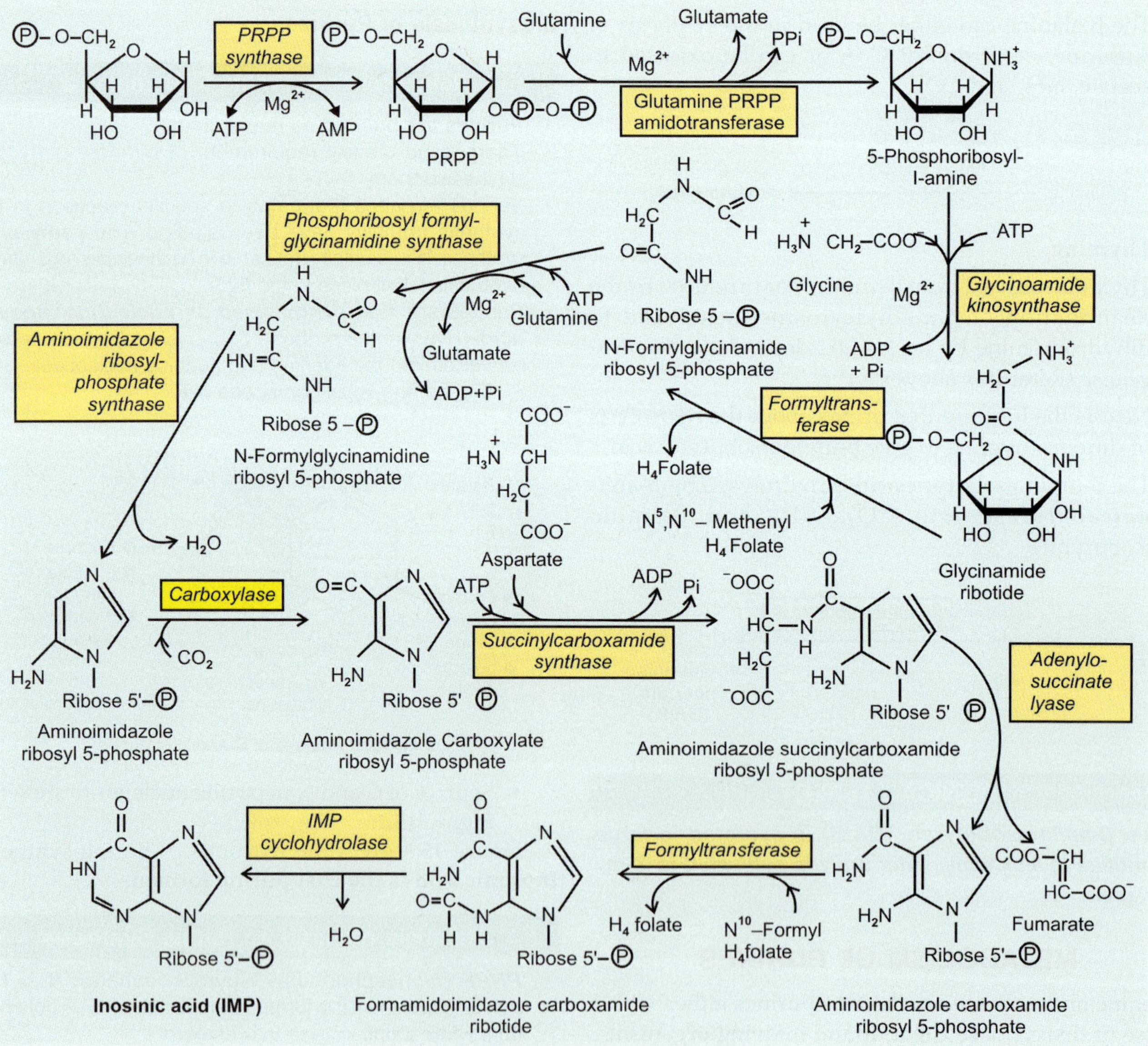

Fig. 15.3: Synthesis of inosinic acid

2. ***Formation of 5′-phosphoribosyl-1 amine (PRA):*** Amide group of glutamine is transferred to C_1 **of PRPP** by the enzyme ***Glutamine PRPP amido-transferase.***
Phosphate group is replaced by $-NH_2$ group. ***This gives the N-9 of purine ring.***

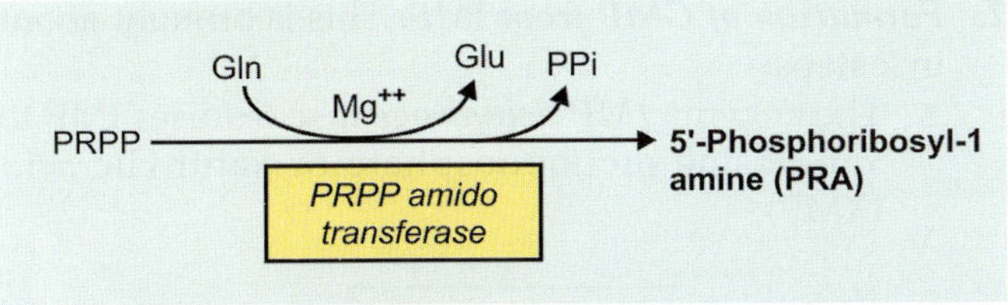

3. ***Formation of glycinamide ribotide, GAR (also called 5′-phosphoribosylglycinamide):*** Glycine condenses with PRA using ATP as energy source to form glycinamide ribotide (GAR), the reaction is catalysed by the enzyme ***Glycinamide kinosynthase. This provides C-4, C-5 and N-7 of the Purine ring.***
4. ***Formation of formylglycinamide ribotide, FGAR (also called 5′-phosphoribosyl-N-formyl-glycinamide):*** The amino nitrogen of glycinamide is formylated by N^{10}-formyl H_4-folate catalysed by the enzyme ***formyl transferase.*** The ***formyl carbon becomes C-8 of the purine ring.***

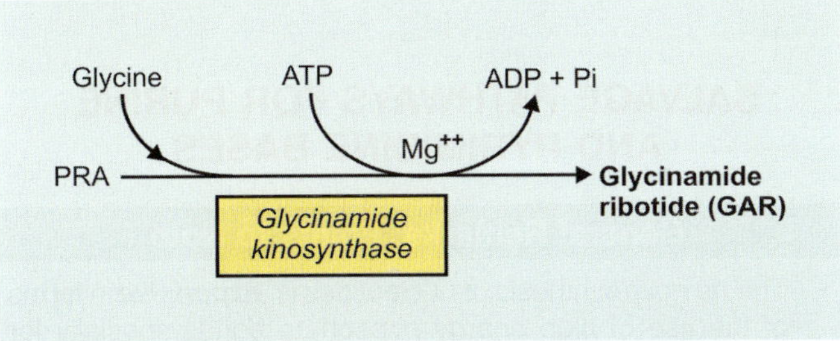

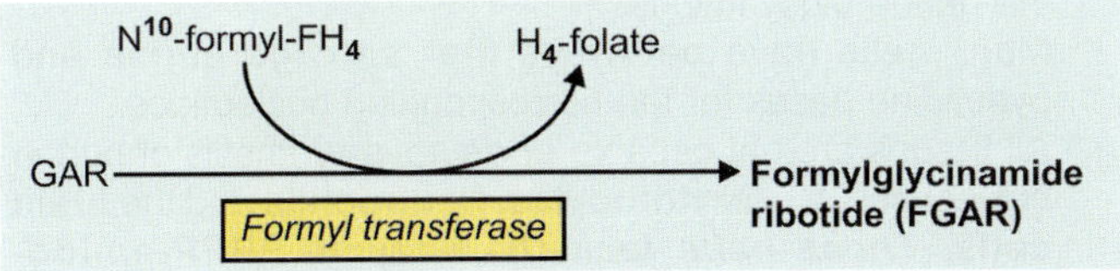

5. ***Formation of a-N-formylglycinamidine ribotide, FGAM: (also called 5′-phosphoribosyl-N-formyl-glycinamidine):*** Another of amide group of glutamine is transferred to FGAR to form FGAM. ***Phosphoribosylglycinamidine synthase*** is the enzyme that catalyses the reaction and ATP provides the energy. The ***reaction contributes N-3 of Purine ring.***

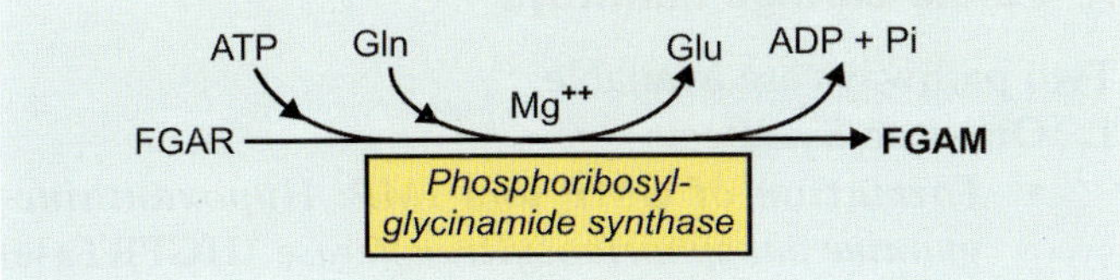

6. ***Formation of 5-Aminoimidazole riboside, AIR (also called 5′-phosphoribosyl-5-aminoimidazole):*** This reaction is catalysed by the enzyme ***aminoimidazole ribosyl phosphate synthetase (AIR-Psynthetase),*** which brings about the dehydratative closure of the ring, by removal of a molecule of H_2O. ATP is required for the reaction.

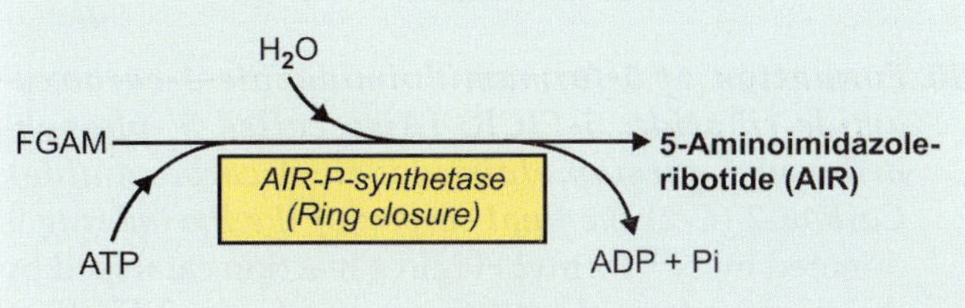

7. ***Formation of 5-aminoimidazole-4-carboxylic acid ribotide, C-AIR (also called 5′-phosphoribosyl-5-aminoimidazole-4 carboxylate):*** This reaction uses CO_2 to carboxylate AIR. ***It contributes to C-6 of the purine nucleus. This reaction is somewhat unusual since neither Biotin nor ATP is required for carboxylation.***

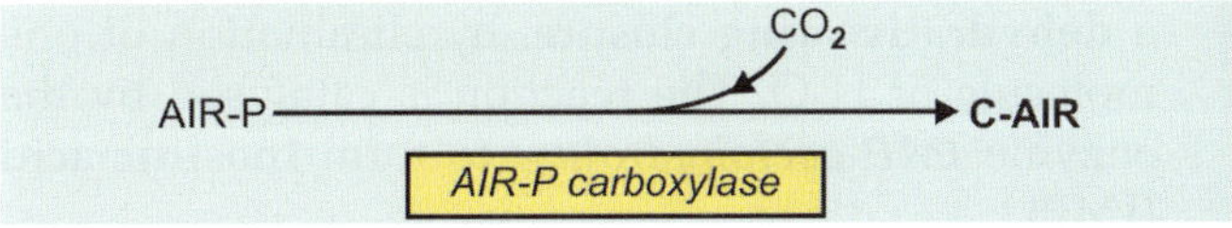

Mechanism: The imidazole ring acts as a nucleophile in this reaction. The atoms from C-4 to the amino group at C-5 constitutes an* enamine*. Since enamines are already activated nucleophiles, hence no further activation is necessary. The reaction occurs by attack of the nucleophilic C-4 atom of the enamine upon the electrophilic carbon of carbon dioxide (CO_2).

8. ***Formation of 5-aminoimidazole-N-succinyl carboxamide ribotide, 5-AISCR. (Also called 5-′phosphoribosyl-5-aminoimidazole-4-N succino carboxamide):*** This reaction is catalysed by the enzyme ***succinyl carboxamide synthetase.*** This uses ATP to condense Aspartic acid with aminoimidazole carboxylate-5-P. ***This contributes to N-1 of the purine nucleus.***

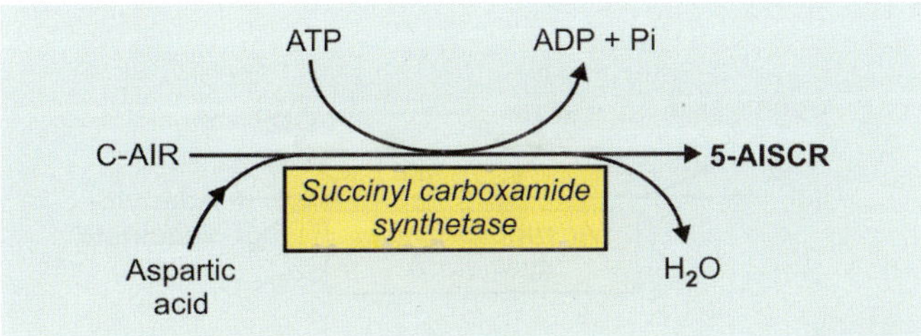

9. ***Formation of 5-aminoimidazole-4-carboxamide ribotide, 5-AICAR. (Also called 5′-Phosphoribosyl 5- aminoimidazole-4-carboxamide):*** 5-AISCR under-

goes cleavage by the cleaving enzyme ***adenylosuccinate lyase*** to form 5-AICAR and fumarate.

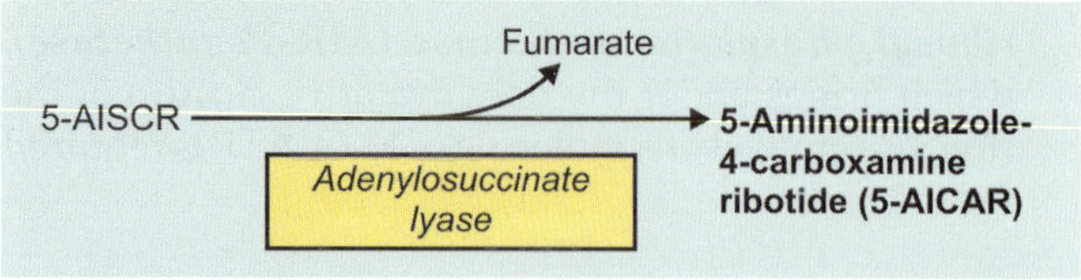

10. *Formation of 5-formamidoimidazole-4-carboxyamide ribotide, 5-FICR.* *(Also called 5′-phosphoribosyl-5-formamidoimidazole-4 carboxamide):* ***Carbon-2 (C-2) the final carbon of the Purine ring*** is donated by N^{10}-formyl FH_4 in a reaction catalysed by the enzyme ***formyl transferase*** and forms 5-FICR.

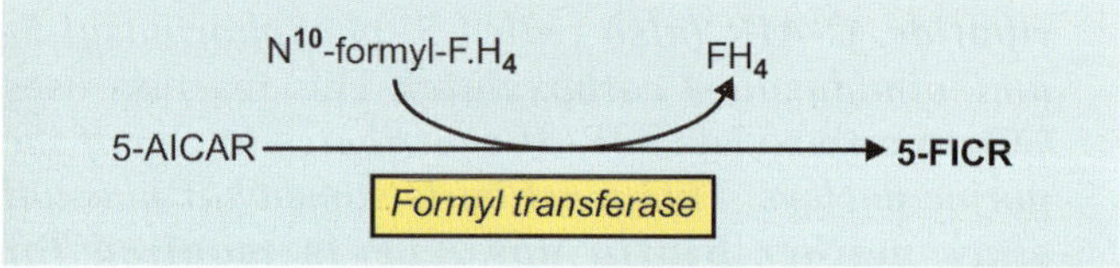

11. *Formation of inosinic acid (IMP):* 5-FICR undergoes a **dehydrative ring closure**, by elimination of one molecule of H_2O. The reaction is catalysed by the enzyme ***IMP cyclohydrolase*** to form **Inosinic acid (IMP).**

Note: ***Hypoxanthine nucleotide (inosinic acid) is the first purine that is synthesised in the body.*** Adenine and guanine nucleotides are then synthesised from the hypoxanthine or xanthine nucleotides respectively by amination.

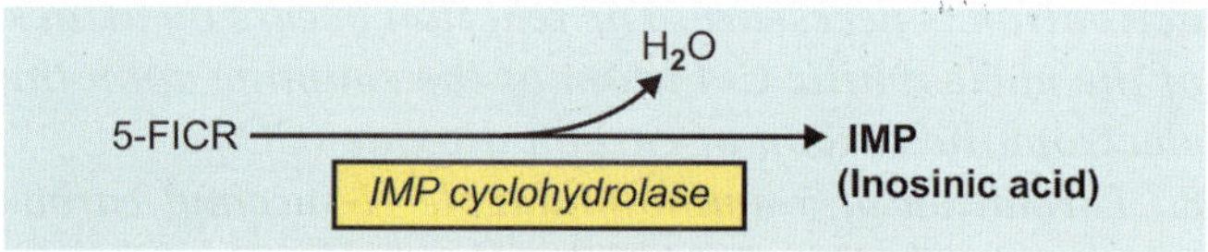

Formation of Other Purine Nucleotides

1. *Formation of AMP from IMP*

This is brought about in **2 steps:**

- The enzyme ***adenylosuccinate synthetase*** catalyses the condensation of Aspartic acid with IMP to form adenylosuccinate. ***GTP provides the required energy.***

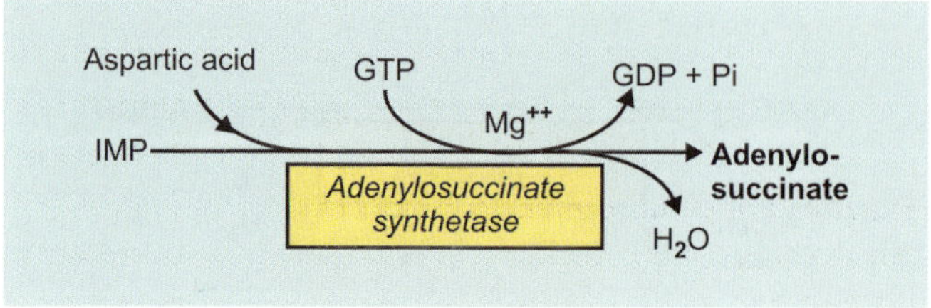

- Adenylosuccinate is then cleaved by the cleaving enzyme ***adenylosuccinate lyase*** to form AMP and fumarate.

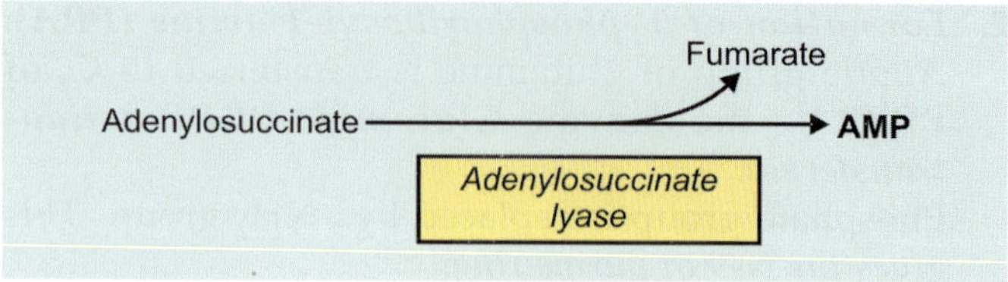

2. *Formation of GMP from IMP:* This is brought about in **2 steps:**

- The enzyme ***IMP dehydrogenase*** oxidises IMP to xanthosine monophosphate or xanthylic acid (XMP).

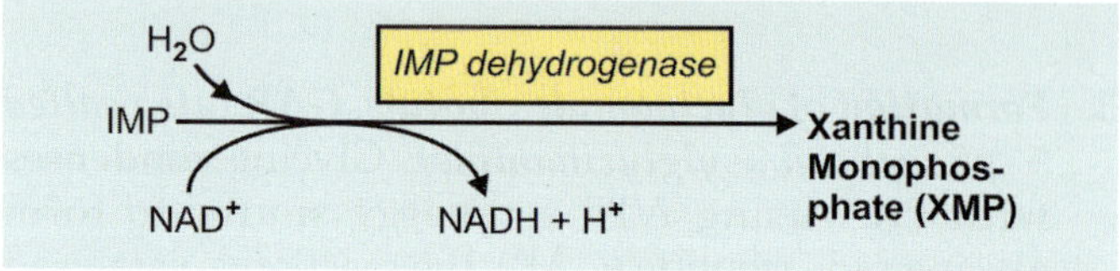

- In the second stage, the amide group of glutamine is transferred to C-2 of XMP catalysed by the enzyme ***GMP synthetase*** to form GMP.

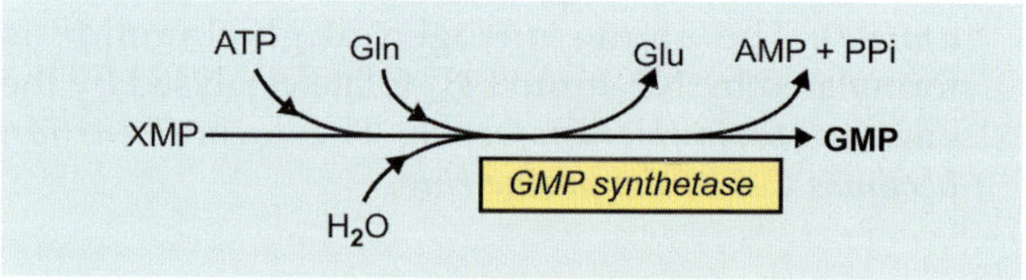

SALVAGE PATHWAYS FOR PURINE AND PYRIMIDINE BASES

General Remarks: Points to Note

- The *de novo* synthesis of nucleotides is ***expensive*** in terms of the use of high energy phosphate bonds specially for the purine biosynthesis.
- Many cells have pathways that ***salvage*** purine and pyrimidine bases for the corresponding nucleotides.
- All tissues are not capable of *de novo* synthesis of purine nucleotides, e.g. **erythrocytes, Neutrophils** and the **brain cells.** ***These cells lack the enzyme*** **PRPP-amidotransferase.**
- Nucleotides do not enter cells directly but are converted to ***nucleosides*** by cell membrane ***nucleosidases.***
  - After entering the cells the nucleoside is either
  - converted to the nucleotide again by a ***kinase*** enzyme or
  - Degraded to corresponding base by the enzyme ***nucleoside phosphorylase.***

A. Purine Salvage Pathways

Two pathways are available.

1. One-step Synthesis

- ***Formation of GMP and IMP:*** *Hypoxanthine-guanine phosphoribosyltransferase* **(HGPRTase)** catalyses the one-step formation of the nucleotides from either guanine or hypoxanthine, using PRPP as the donor of the ribosyl moiety.

Thus,

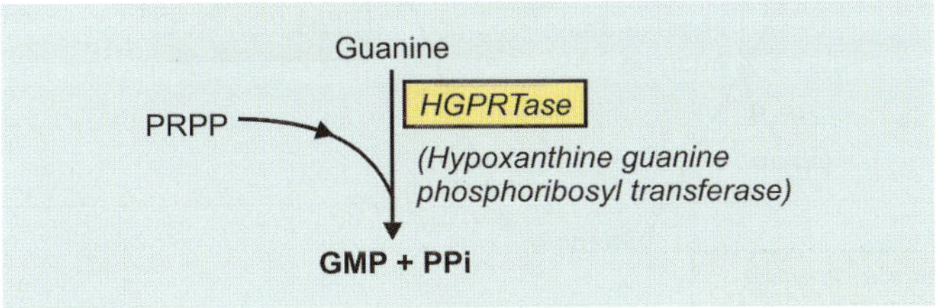

Similarly,

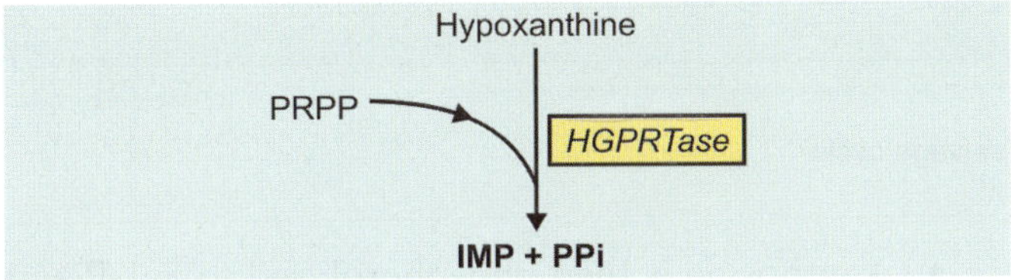

Regulation: The enzyme HGPRTase is regulated by the competitive inhibition of GMP and IMP respectively.

- ***Formation of AMP:*** The enzyme ***Adenine phosphoribosyl transferase*** **(APRTase)** in a similar way catalyses the formation of AMP from adenine, ribosyl moiety is donated by PRPP.

Thus,

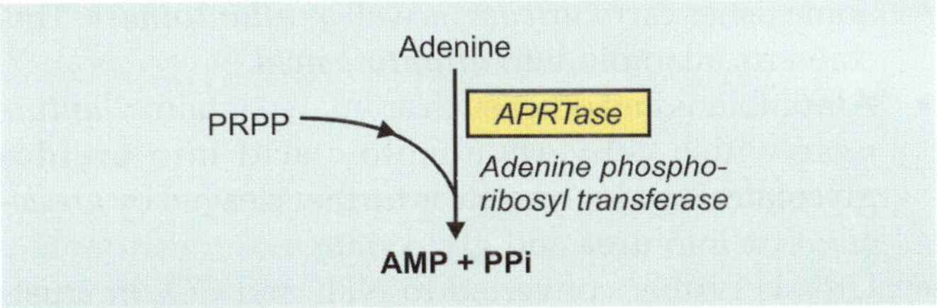

Regulation: The enzyme APRTase is regulated by the competitive inhibition of AMP.

2. Two-Step Synthesis

(Nucleoside phosphorylase-nucleoside kinase pathway) Under some conditions, it is possible for purine bases to be salvaged by a two-step process as under:

Nucleoside Phosphorylase is an enzyme that brings about nucleoside breakdown. But the reaction is readily ***reversible*** and can form back 'nucleoside' which is rather a favourable pathway. Once the nucleoside is formed, a *kinase* enzyme may phosphorylate it to the 5'-nucleotide.

- *Formation of AMP*

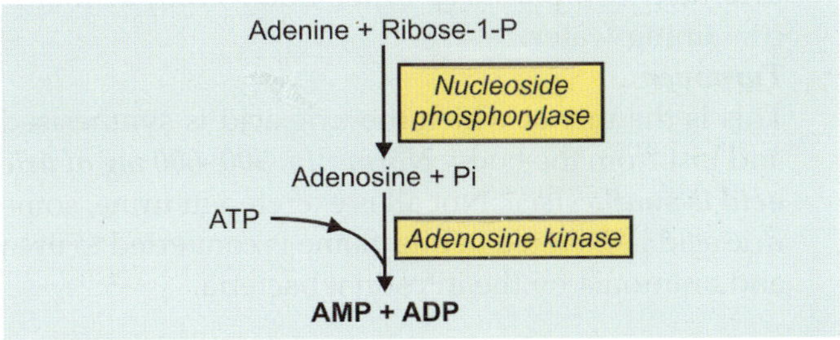

Note: Neither *Guanosine nor Inosine Kinases* have been detected in mammalian cells. ***So Adenine is the only purine that may be salvaged by the two-step pathway.***

Purine Salvage Cycle

In addition to above, there is a cycle in which GMP and IMP as well as their deoxyribonucleotides are converted into their respective 'nucleosides' by a ***purine 5'-nucleotidase*** enzyme. The nucleosides so formed can be hydrolytically cleaved producing the corresponding sugar phosphates and setting free the N-bases. The guanine and hypoxanthine then can be phosphoribosylated again to complete the cycle, called ***Purine salvage cycle*** (Fig. 15.4).

B. Pyrimidine Base Salvage

The enzyme ***Pyrimidine Phosphoribosyl transferase*** catalyzes the formation of pyrimidine nucleotide, using PRPP as the donor of the ribosyl moiety.

Thus,

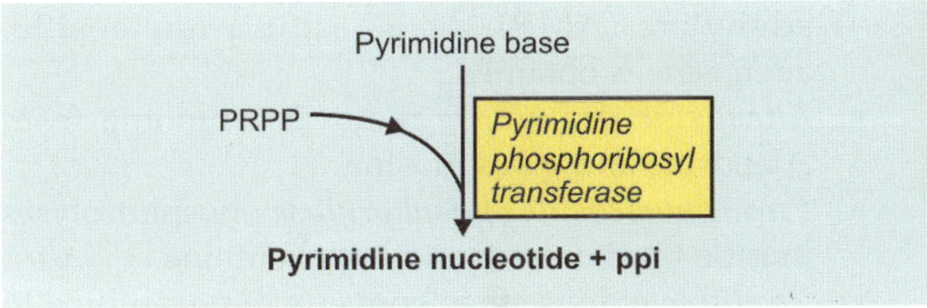

Note: The enzyme obtained from human red cells can use orotate, uracil and thymine as substrates.

Regulation of Purine Synthesis

- ***PRPP synthetase*** is an important enzyme that regulates purine synthesis. It is allosterically inhibited by the feedback effects of PRPP and number of purine nucleotides such as AMP, GMP, ADP, GDP, NAD and FAD.
- ***Glutamine PRPP amidotransferase*** is the enzyme for the rate-limiting step of purine synthesis. It is regulated by feedback inhibitory effects of AMP and GMP.
- A proper balance between the Adenine and guanine concentration is maintained by ***adenylosuccinate synthetase and IMP dehydrogenase*** respectively.

CATABOLISM OF PURINES

Purines are Catabolised to Uric Acid

An average of 600 to 800 mg of uric acid is excreted by human beings most of it is found in urine. Guanine and Adenine nucleotides have their separate enzymes until the formation of a common product—xanthine. The final reaction is the conversion of xanthine to uric acid by ***xanthine oxidase.***

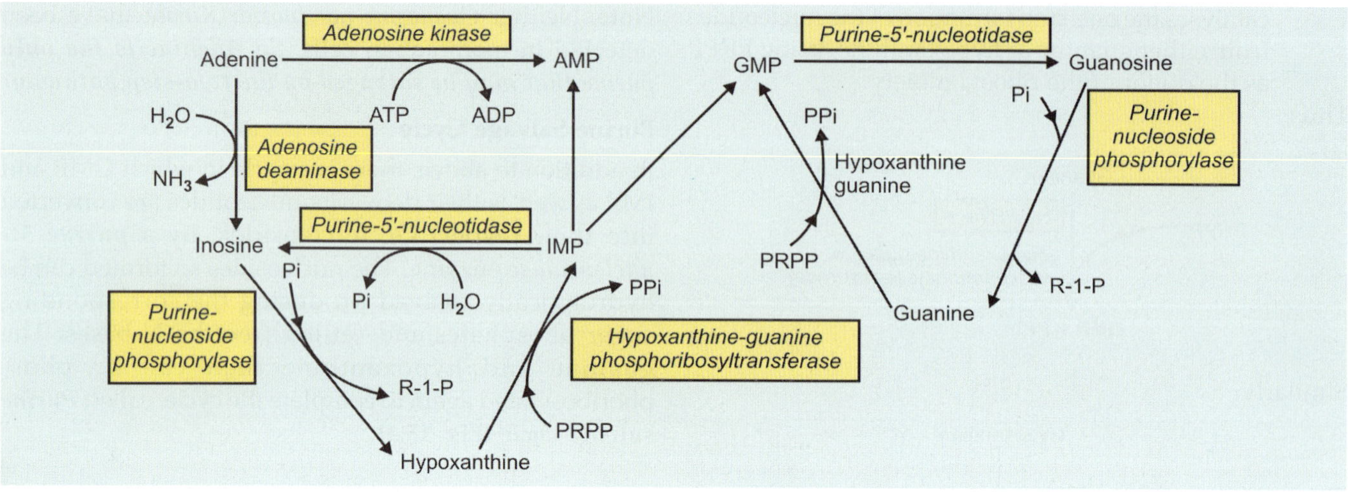

Fig. 15.4: Purine-salvage cycle

A. Adenine Nucleotide Catabolism

1. *In Liver and Heart Muscle*

- An enzyme ***purine-5′-nucleotidase*** hydrolyses adenylate (AMP). As a result a nucleoside, adenosine is obtained.
- ***Adenosine deaminase*** removes ammonia from adenosine and gives inosine.
- ***Purine nucleoside phosphorylase*** phosphorolyses inosine to ribose-1-P and hypoxanthine.
- ***Xanthine oxidase*** then converts hypoxanthine to xanthine and xanthine to uric acid. ***Molecular oxygen is reduced at each stage to the superoxide (O_2^-)*** which is converted to H_2O_2 by ***superoxide dismutase.***

2. *In Skeletal Muscle*

- ***Adenylate deaminase*** converts AMP into inosine monophosphate (IMP).
- Inosine monophosphate is hydrolysed by ***purine-5-nucleotidase*** to inosine.
- Inosine is changed to uric acid as mentioned above. GMP inhibits ***adenylate deaminase*** to reduce the catabolism of AMP.

B. Guanine Nucleotide Catabolism

1. *In Liver, Spleen, Kidneys, Pancreas*

- GMP is hydrolysed by ***purine-5′-nucleotidase*** into guanosine.
- ***Purine nucleoside phosphorylase*** phosphorolyses guanosine into R-1-P and guanine.
- ***Guanine deaminase*** deaminates guanine to xanthine and produces NH_3.
- Oxidation of xanthine to uric acid is brought about by xanthine oxidase.

2. *In Liver Mainly*

- ***Guanosine deaminase*** deaminates guanosine into xanthosine.
- Xanthosine is then phosphorolysed to R-1-P and xanthine by ***purine nucleoside phosphorylase.***
- Xanthine is then oxidised to uric acid by ***xanthine oxidase.*** (Fig. 15.5).

Further Catabolism of Uric Acid (Nonprimates)

- In many nonprimate animals uric acid may be oxidized and decarboxylated by ***uricase,*** a hepatic copper containing enzyme to ***allantoin.***
- Some fishes carry ***uricase*** as well as ***allantoinase.*** This converts allantoin into ***allantoic acid.***
- Amphibians and other such animals contain ***allantoinase*** which converts allantoic acid into **ureidoglycolate.** Ureidoglycolate is further cleaved by *ureidoglycolase* into **urea** and **glyoxylate.**
- Urea is further converted to NH_3 and CO_2 in crustaceans by an enzyme ***urease*** found in their liver.

URIC ACID METABOLISM AND CLINICAL DISORDERS OF PURINE AND PYRIMIDINE METABOLISM

The main site of uric acid formation is **liver** from where it is carried to kidneys.

- ***Miscible Pool***
  It is the quantity of uric acid present in body water. ***In normal subjects an average of 1130 mg of uric acid is present.*** Plasma contains higher concentration of uric acid compared to other body compartments containing water.
- ***Turnover***
  This is the rate at which the uric acid is synthesised and lost from the body. ***Normally, 500–600 mg of uric acid is synthesised.*** Not all is excreted in urine, some uric acid is excreted in bile. Some is converted to urea and ammonia by the intestinal bacteria.

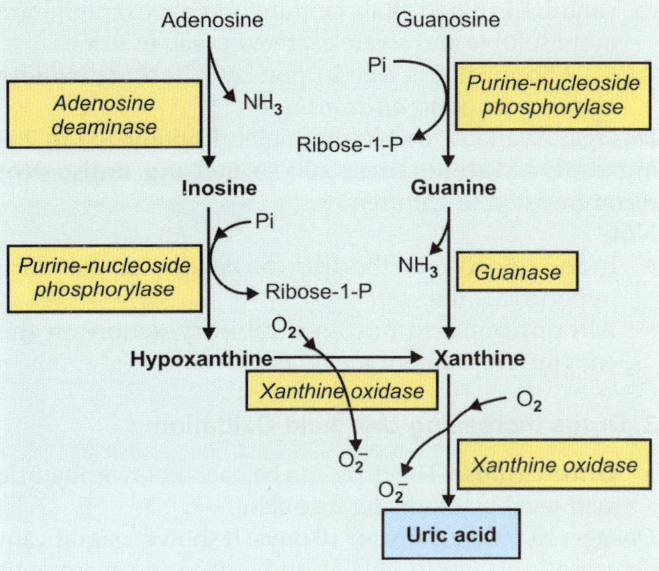

Fig. 15.5: Catabolism of purines

- ***Distribution:*** It is very irregularly distributed in the body. ***Serum contains 3 to 7 mg/dl.*** Average values are slightly higher in males. Red cells contain half as much uric acid as serum. Muscles also contain less amount compared to blood.
- ***Dietary effects:*** Uric acid excretion continues at a rather steady rate during starvation and during a purine-free diet owing to the so-called endogenous purine metabolism. The ingestion of foods high in nucleoproteins such as glandular organs produces a marked increase in urinary uric acid. Diets like milk, eggs and cheese, with low purine contents causes practically no increase in urinary uric acid.
- ***Effect of hormones:*** Administration of the glucocorticoid hormones and ACTH increases the excretion of uric acid in urine.
- ***Excretion of uric acid:*** Uric acid in the plasma is filtered by the glomeruli but is later partially reabsorbed by the renal tubules; **glycine is believed to compete with uric acid for tubular reabsorption.** Certain uricosuric drugs such as **salicylates**, block reabsorption of uric acid. **Lactic acid competes with uric acid in its excretion.** Hence in lactic acidosis uric acid is retained, and can produce gout.

There is now conclusive evidence for tubular secretion of uric acid by kidney. Thus uric acid is cleared both by (a) glomerular filtration; (b) by tubular secretion.

CLINICAL DISORDERS: CLINICAL ASPECTS

1. ***Gout:*** Gout is a chronic disorder characterised by:
   - Excess of uric acid in blood ***(Hyperuricemia)***
   - Deposition of sodium monourate in alveolar and non alveolar structures producing so called ***tophi.***
   - ***Recurring attacks of acute arthritis.*** These are due to deposition of monosodium urate in and around the structures of the affected joints.

Types: There are **two main types** of gout:
1. Primary gout,
2. Secondary gout.

1. Primary Gout

Here the hyperuricaemia is not due to increased destruction of nucleic acid. The essential abnormality is increased formation of uric acid from simple carbon and nitrogen compounds without intermediary incorporation into nucleic acids.

a. ***Primary metabolic gout:*** It is due to inherited metabolic defect in purine metabolism leading to ***excessive rate of conversion of glycine to uric acid.*** X-linked recessive defects enhancing the *de novo* synthesis of purines and their catabolism can also lead to hyperuricaemia. For example, such defects of PRPP synthetase may make it feedback resistant. X-linked recessive defects of hypoxanthine guanine phosphoribosyl transferase may reduce utilisation of PRPP in the salvage pathway. Increased intracellular PRPP enhances *de novo* purine synthesis.

b. ***Primary renal gout:*** It is due to failure in uric acid excretion.

2. Secondary Gout

a. ***Secondary metabolic gout:*** It is due to secondary increase in purine catabolism in conditions like leukaemia, prolonged fasting and polycythemia.

b. ***Secondary renal gout:*** Due to defective glomerular filtration of urate due to generalised renal failure.

c. ***In von-Gierke's disease:*** Deficiency of ***G-6-phosphatase*** leads to elevated rate of pentose formation in HMP. This acts as a good substrate for *PRPP synthetase* and enhances the synthesis of purines followed by their catabolism to uric acid. ***Increase lactic acid competes with uric acid excretion resulting to retention of uric acid*** (Refer, Glycogen Storage Diseases).

Treatment of Gout: Consist of:

(a) Palliative Treatment

(b) Specific Treatment

(a) Palliative treatment: Bed rest in acute stage, Diet—Purine free diet, Restricting alcohol consumption.

- *Anti-inflammatory Drugs*
  1. *Colchicine:* One of the nonspecific anti-inflammatory drug. It has no effect on urate metabolism or excretion. Colchicine therapy is instituted during acute attack.

     Mechanism: Suppresses the synthesis and secretion of the chemotactic factor that is produced in urate crystal-induced inflammation.

     Dosage: Available as 0.5 mg tablet. In acute gout: One tablet hourly till symptoms are relieved or diarrhoea occurs.

     Long-term management: One tab 3 to 4 times a week.

SECTION THREE

2. ***NSAIDS:*** Drugs like: Indomethacin, Diclofenac, Naproxen, Piroxicam, Fenoprofen, Flurbiprofen, Ibuprofen, Rofecoxabin, etc.

These drugs inhibit the synthesis of PGs which are important mediator of the inflammatory response. These drugs have been found effective in treating patients having recurrent attacks of acute gout and also for terminating acute attack of gout.

(b) Specific Treatment

Aim: To lower the uric acid level of blood.

Methods: The above can be achieved in **three ways:**

1. By increasing the renal excretion of uric acid *(uricosuric drugs).*
2. ***By decreasing the synthesis of uric acid*** using enzyme inhibitor
3. ***By increasing oxidation of uric acid.***

1. Uricosuric Drugs

Definition: A uricosuric agent is one that enhances the renal excretion of uric acid probably by specific inhibition of its tubular reabsorption or secretion.

Drugs used are:

- ***Salicylates***

  Effects vary with dosage. In low dosage of 1 to 2 gm/day, salicylates cause uric acid retention but in higher dosage 5 to 6 gm/day it has uricosuric effect. Long-term therapy with high dosage is not desirable due to the side effects.
- ***Probenecid (Benemide)***
  - It is an efficient and harmless uricosuric drug.
  - Lowers the uric acid level—fall is immediate and sustained.

Dose: Available as 500 mg tablet. ½ tablet twice daily for the first week and then one tablet twice daily. Not recommended for children. Therapy is continued for 10 to 12 weeks and patients can return to normal activities.

- ***Halofenate:*** The drug has good uricosuric effect. Also has a hypolipaemic effect. It liberates urates from urate binding sites of proteins of plasma and removes uric acid by normal excretion. The drug can be safely used for short-term and long-term therapy.

Note: Uricosuric drugs are effective provided renal function is normal.

2. Enzyme Inhibitor

- **Allopurinol (zyloprin)** ***drug of choice:*** It has similar structure like hypoxanthine. ***Acts by competitive inhibition on "xanthine oxidase"*** and ***thus uric acid synthesis is impaired.***

  The durg causes a rapid fall in serum uric acid level and an increase in concentration of hypoxanthine and xanthine in blood. Both xanthine and hypoxanthine are more soluble and so are excreted easily in urine.

  Allopurinol is acted upon by ***Xanthine oxidase*** and converted to ***alloxanthine.***

Dosage: Available as 100 mg tablet. Initially 100 to 200 mg daily. Maintenance: 200 to 600 mg daily. Not recommended in children.

Note

- In addition to gout, the drug can be used in secondary hyperuricaemia.
- Allopurinol also has an inhibitory action on the enzyme ***tryptophan pyrrolase.***

3. Drugs Increasing Uric Acid Oxidation

- ***Urate oxidase:*** The drug can be used in lowering uric acid level by oxidising uric acid.

Dosage: 10,000 IU daily for 10 days. It shows a significant decrease in uric acid level. Can be used in severe gout with renal involvement and secondary hyperuricaemia.

2. **Lesch-Nyhan Syndrome**

   Only males are affected by this. It is **X-linked recessive defect of *hypoxanthine-guanine phosphori bosyltrans-ferase.*** The enzyme is almost absent and leads to increased purine salvage pathway from PRPP. This can result in severe gout, renal failure, poor growth, spasticity and ***tendency for self-mutilation***.
3. **Xanthinuria:** An autosomal recessive deficiency of ***xanthine oxidase***, blocks the oxidation of hypoxanthine and xanthine to uric acid. It can cause xanthine lithiasis and **hypouricaemia.**
4. **Adenosine deaminase deficiency: An autosomal recessive deficiency** of ***adenosine deaminase***. It is associated with severe immunodeficiency and both **T cells** and **B cells** (lymphocytes) are deficient. There occurs an accumulation of deoxyribonucleotides which inhibit further production of precursors of DNA synthesis especially d CTP. **Hypouricaemia** occurs which is due to defective breakdown of purine nucleotides. Recently **Gene replacement therapy has been used successfully in a few cases.**
5. **Nucleoside phosphorylase deficiency: An autosomal recessive deficiency** of ***purine nucleoside phosphorylase***, causes the urinary excretion of guanine and hypoxanthine nucleosides. There is reduced production of uric acid. There is severe deficiency of cell-mediated and humoral immunity.
6. ***Orotic Aciduria:*** It is of **2 types:**
   - ***Type I orotic aciduria:*** It is an ***autosomal recessive*** genetic disorder of a protein acting as both ***orotate phosphoribosyltransferase*** and ***OMP decarboxylase***. Orotate fails to be converted to uridylate. This results in accumulation of orotate in blood elevating its level, ***growth retardation*** and ***megaloblastic anaemia.***

SECTION THREE

- ***Type II orotic aciduria:*** It is ***autosomal recessive***, affecting ***OMP decarboxylase*** and is characterised by megaloblastic anaemia and the urinary excretion of asididine in higher concentrations than orotate.

Other Causes of Orotic Aciduria

1. **Orotic aciduria that accompanies *Reye Syndrome*:** Probably is a consequence of the inability of severely damaged mitochondria to utilise carbamoyl phosphate, which then becomes available for ***cytosolic overproduction of orotic acid.***
2. **Associated with deficiency of urea cycle enzyme:** Increased excretion of orotic acid, uracil and uridine, sometimes accompanies a deficiency of liver mitochondrial ***ornithine transcarbamoylase.*** Excess carbamoyl-P ***passes to cytosol*** where it stimulates pyrimidine nucleotide biosynthesis resulting to mild orotic aciduria, which may get aggravated by high N_2 foods.
3. **Drug Related Orotic Aciduria**
   - ***Allopurinol:*** Allopurinol an alternative substrate for **orotate phosphoribosyl transferase** *competes with orotic acid. The resulting nucleotide product also inhibits* **orotidylate decarboxylase** *resulting in orotic aciduria and orotidinuria.*
   - ***6-Azauridine:*** 6-Azauridine, following its conversion to 6-azauridylate, also competively inhibits ***orotidylate decarboxylase*** resulting to increased excretion of orotic acid (orotic aciduria) and orotidine (orotidinuria).

SECTION THREE

I. Questions (Essay type and short notes)

1. Describe briefly the metabolism of purines.
2. Discribe the metabolism of nucleoproteins.
3. How are Purines catabolised?

Short Notes

a. Gout
b. Uric acid
c. Catabolism of purines
d. Salvage pathway for purines
e. Lesch-Nyhan syndrome
f. Orotic acid-uria

II. MCQs (Give one correct answer)

1. The four nitrogen atoms of purines are derived from:
   a. Urea and ammonia
   b. Ammonia, glycine and glutamate
   c. Ammonia, aspartate and glutamate
   d. Aspartate, glutamine and glycine
   e. Glycine, ammonia and aspartate
2. A drug which prevents uric acid synthesis by inhibiting the enzyme xanthine oxidase is:
   a. Aspirin b. Allopurinol
   c. Colchicine d. Probenecid
   e. Phenyl butazone
3. Glycine contributes to the following C and N of purine nucleus:
   a. C-1, C-2, and N-7 b. C-8, C-6 and N-9
   c. C-4, C-5 and N-7 d. C-3, C-4 and N-1
   e. C-4, C-5 and N-9
4. Inosinic acid is the biological precursor of:
   a. Cytosine and uric acid
   b. Adenylic acid and guanylic acid
   c. Orotic acid and uridylic acid
   d. Adenosine and thymidine
   e. Uracil and thymidine
5. The probable metabolic defect in gout is:
   a. A defect in excretion of uric acid by kidney
   b. An overproduction of pyrimidines
   c. An overproductin of uric acid
   d. An underproduction of purine
   e. Rise in calcium leading to deposition of calcium urate
6. Match the following:
   a. Purine b. Pyrimidine
   c. Nucleoside d. Nucleotide
   1. Inosinic acid ________
   2. Caffeine ________
   3. Adenosine ________
   4. Cytosine ________
7. In humans, the principal breakdown product of purines is:
   a. Ammonia b. Allantoin
   c. Alanine d. Urea
   e. Uric acid
8. A key substance in the committed step of pyrimidine biosynthesis is:
   a. Ribose-5′-phosphate b. Carbamoyl phosphate
   c. ATP d. Glutamine
   e. Thiouracil
9. In humans, the principal catabolic product of pyrimidines is:
   a. Uric acid b. Allantoin
   c. Hypoxanthine d. β-Alanine
   e. Urea
10. In most mammals, except primates, uric acid is metabolised by:
    a. Oxidation to allantoin
    b. Reduction to NH_3
    c. Hydrolysis to allantoin
    d. Hydrolysis to ammonia
    e. None of the above
11. Match the following:
    a. Azaserine
    b. 6-Mercaptopurine

c. 6-Azauracil
d. 5-Fluorodeoxyuridine
e. Methotrexate
1. Inhibition of THFA synthesis from DHFA ______
2. Inhibition of dTTP synthesis from dUMP ______
3. Inhibition of UMP synthesis from OMP ______
4. Inhibition of AMP synthesis from IMP ______
5. Inhibition of 5-phosphoribosyl-1-amine synthesis from PRPP ______.

12. Two Nitrogen of the pyrimidine ring are obtained from:
a. Glutamine and carbamoly-P
b. Aspartate and carbamoly-P
c. Glutamate and ammonia
d. Glutamine and ammonia
e. Aspartate and ammonia

13. All are true about Lesch-Nyhan syndrome except:
a. Produces self-mutilation
b. Genetic deficiency of the enzyme HGPRT
c. Elevated levels of uric acid in blood
d. Inheritence is autosomal recessive
e. Mental retardation

14. Synthesis of GMP from IMP requires the following:
a. Ammonia, NAD^+, ATP
b. Glutamine, NAD^+, ATP,
c. Ammonia, GTP, $NADP^+$
d. Glutamine, GTP, $NADP^+$
e. Glutamine, UTP, $NADP^+$

15. Which pathway is correct for catabolism of purines to form uric acid?
a. Guanylate → Adenylate → Xanthine → hypoxanthine → uric acid.
b. Guanylate → inosinate → xanthine → hypoxanthine → uric acid
c. Adenylate → Inosinate → xanthine → hypoxanthine → uric acid
d. Adenylate → inosinate → hypoxanthine → uric acid
e. Inosinate → adenylate → xanthine → hypoxanthine → uric acid.

Answers to MCQs

1. d 2. b 3. c 4. b 5. c
6. 1-d, 2-a, 3-c, 4-b 7. e 8. b
9. d 10. a 11. 1-e, 2-d, 3-c, 4-b, 5-a
12. b 13. d 14. b 15. d.

SECTION THREE

CHAPTER 16

CHEMISTRY OF NUCLEIC ACIDS, DNA REPLICATION AND DNA REPAIR

PART I: Nucleic Acids

Major Concepts

A. To study the chemistry of Nucleic acids.
B. To learn their Biologic significance.

Specific Objectives

A. 1. Nucleotide is the basic monomeric unit of nucleic acids. Study chemisty of basic Protein histone.
2. Study the two types of nucleic acids: Polydeoxyribonucleotides (DNA) and Polyribonucleotides (RNA). Learn about the phosphodiester linkage formed between the nucleotides, i.e. "internucleotide bonds" – 3' – 5' Polarity. (3' – OH group of sugar of one nucleotide is joined to 5' – OH group of another nucleotide).
3. DNA is the genetic material. Genes of eukaryotes are present in 'chromatin' which is made up of protein and nucleic acid—DNA is the carrier of genetic information.
4. Study in detail the structural characteristics of DNA—"Watson and Crick Model of Double Helix". Salient features to be noted and studied: • Two antiparallel Polydeoxyribonucleotide chains. • Stacking of the hydrophobic rings. • Hydrogen bonding occurs between pairs of bases. One base on one strand pairing with another base on the antiparallel strand. • Adenine always pairs with Thymine (A = T) and Guanine always pairs with Cytosine (G≡C). • Define Chargaff's rule • The helix is 20 Å in diameter and bases are separated by 3.4 Å along the axis of the helix.
5. Study three types of DNA—B-DNA, A-DNA, Z-DNA.
6. Draw a diagram of DNA double helical structure showing two strands and connections of bases.
7. Study Dissociation and Reassociation of the double helical chain of DNA. (a) Study denaturation of DNA and the factors affecting denaturation and (b) Renaturation (annealing).
8. Learn briefly the structural organisation of Eukaryotic genome. Learn about Junk DNA; its role and function.

B. 1. Learn the classes of RNA molecules: Three different types of RNAs m-RNA, t-RNA and r-RNA. A precursor of m-RNA is hn-RNA.
2. Study the salient features of structure of three RNAs and their functions.
3. Study about small RNAs: Sn RNAs, Micro RNAs (mi RNAs), Si-RNAs.

NUCLEIC ACIDS

Among all the properties of living organisms, one is absolutely important for continuance of life: a living system must be able to replicate itself. To do so, an organism must possess a complete description of itself. In living organisms, this description is stored in the substances called ***nucleic acids***. These are non-protein nitrogenous substances made up of a monomeric unit called a nucleotide. Two different types of nucleic acids exist in living organisms, namely deoxyribonucleic acid or DNA and ribonucleic acid RNA. The monomeric unit of DNA is deoxyribonucleotide while that of RNA is ribonucleotide. The details of these nucleotides have already been discussed.

DEOXYRIBONUCLEIC ACID (DNA)

DNA is a polymer of deoxyribonucleotides and is found in chromosomes, mitochondria and chloroplasts. The ***nuclear DNA is found bound to basic proteins called histones.*** DNA is present in every nucleated cell and carries the genetic information. It is conveniently isolated from viruses, thymus gland, spleen, leucocytes, etc. Briefly, the tissue is homogenised at neutral pH and centrifuged first at low speed to get the nuclear pellet. The pellet is then extracted with 2 M NaCl which extracts DNA from proteins. The DNA in solution is further precipitated with ethanol. To destroy any RNA present in the solution, it is treated with a *ribonuclease* which destroys only the RNA leaving pure DNA in the precipitate.

SECTION THREE

HISTONES

Histones are a small family of closely related ***basic proteins*** in chromatin.

H1 histones: Easily detectable. Least tightly bound, can be removed by salt solution. Organisational unit of this soluble chromatin is nucleosome.

TYPES OF HISTONES

Nucleosomes: Contain four types of histones. They are—**H2A, H2B, H3** and **H4**, called **core histones.**

The structure of all four types of histones are highly conserved between species. This extreme conservation implies that the function of histones is identical in all eukaryotes.

Structure: The –COOH terminal 2/3rd of the molecules have a typical random amino acid composition, ***but 1/3rd of the –NH_2 terminal are rich in basic amino acids.***

Interaction of histones: The four types of histones interact with each other in very specific ways as follows:

- H3 and H4 form a tetramer containing 2 molecules of each $(H3H4)_2$
- H2A and H2B form dimers (H2A–H2B).

Under physiological conditions, these histone oligomers associate to form the **histone octamer** of the composition $\mathbf{(H3H4)_2\ (H2A–H2B)_2}$

In the nucleosome, the DNA is supercolied in a left hand helix over the surface of the disc-shaped histone octamer. The 146 base pairs of DNA consisting of 1.75 super helical turns, are in contact with the "histone octamer".

Function: This association of histone octamer with ***DNA protects the DNA from digestion by a*** **nuclease.**

(a) Primary Structure of DNA

- Chromosomal DNA consists of very long DNA molecules (MW 1.6×10^6 to 2×10^9).
- Each DNA is a polymer of about 10^{10} deoxyribonucleotides.
- Normally there are only ***four different types*** of deoxyribonucleotides that are found in DNA molecule, namely, ***1. adenine deoxyribonucleotide (dA), 2. thymine deoxyribonucleotide (dT), 3. guanine deoxyribonucleotide (dG),*** and ***4. cytosine deoxyribonucleotide (dC).***
- Nucleotides of each of the two helical strands are bound to each other by covalent ***3′-5′ phosphodiester linkage***. Each such bond is formed by the ester linkages of a single phosphate residue with the 3′ –OH (i.e. C – 3′ –OH group of the ribose sugar) of one nucleotide with the C – 5′ –OH group of ribose of the next nucleotide. This kind of bonding gives rise to a linear polydeoxyribonucleotide strand with 2 free ends on both sides (Fig. 16.1).

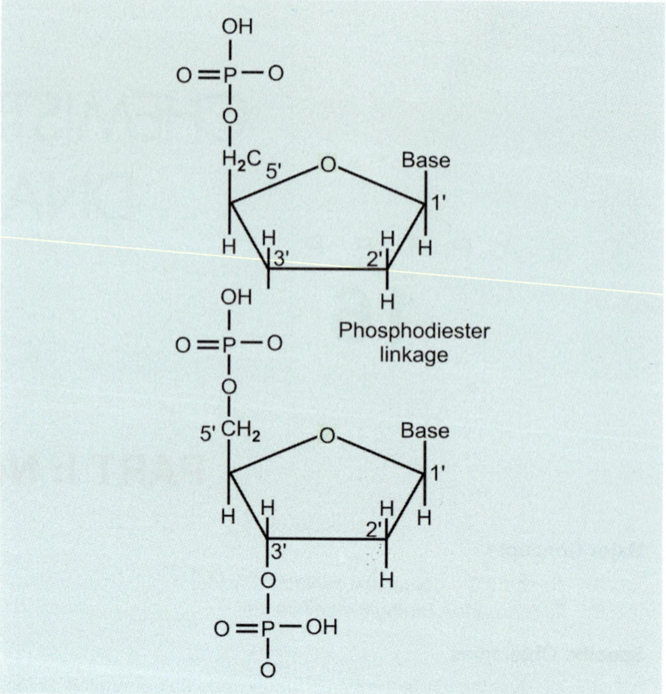

Fig. 16.1: Formation of phosphodiester bond

- That end of the strand which bears a free 5′ phosphate group without phosphodiester linkage is **called the 5′-end.** The opposite end bears a free 3′ – hydroxyl or 3′ phosphate group and is **called the 3′ end.**
- The primary structure is the number and sequence of different deoxyribonucleotides in its strands joined together by phosphodiester linkages.
- The backbone of the primary structure is the linear strand of interconnected sugar phosphate residues while the purine or pyrimidine connected with the sugar residue projects laterally from the backbone.

(b) Secondary Structure of DNA

- This consists of a ***double stranded helix*** formed by the two polydeoxyribonucleotide strands around a central axis. This type of model was first proposed by **Watson and Crick** in their paper in Nature in 1953. Later on they were awarded the Nobel Prize in 1962 alongwith **Maurice Wilkins.**
- ***DNA is a double helix.*** Each of its two strands is coiled about a central axis, ***usually a right handed helix.*** The two sugar phosphate backbones wind around the outside of the bases like the banisters of a spiral staircase and are exposed to the aqueous solution. ***The phosphodiester bonds in the two interwoven strands run in opposite directions.*** Therefore ***the strands are called antiparallel.*** Thus the polarity of the two strands will be 3′ – 5′ and

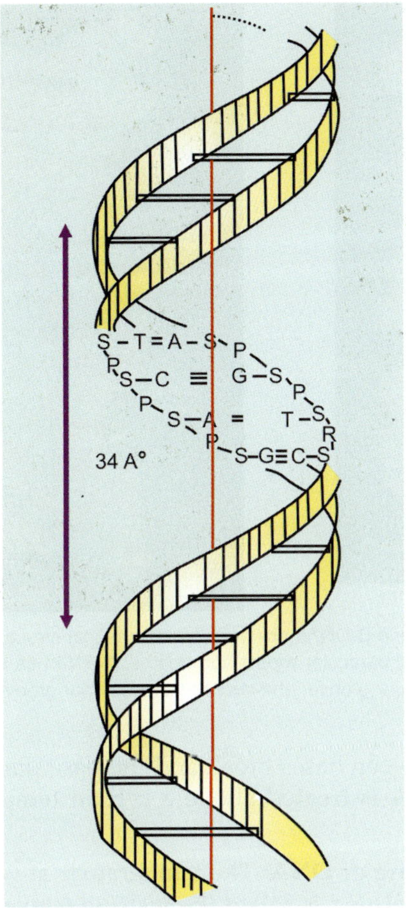

Fig. 16.2: DNA-double helix

5′ – 3′. The **3′ – 5′ strand** is called coding or **"template strand"** and **5' – 3' strand** is called **non-coding 'strand'** (Fig. 16.2).

- The aromatic rings of bases are hydrophobic and they are stacked in the interior, nearly perpendicular to the long axis of the helix.
- ***Adenine base of one strand of DNA is hydrogen bonded to a thymine in the opposite strand; while the guanine is hydrogen bonded to a cytosine.***
- The hydrogen atoms in the bases of DNA can shift from one ring nitrogen or oxygen atom to another. These proton shifts called ***tautomerisation reactions***—interconvert the positions that can serve as hydrogen-bond donors and acceptors in base pairs. ***There are two hydrogen bonds present between Adenine and Thymine while three hydrogen bonds are present between Guanine and Cytosine*** (Figs. 16.3A and B).
- The ratio of purine to pyrimidine bases in the DNA molecule is always around 1 (i.e. G + A/T + C ≈ 1). This is known as ***Chargaff's rule***.
- In DNA, the glycosidic bonds between sugar and bases are not directly opposite each other and two grooves of unequal width form around the double

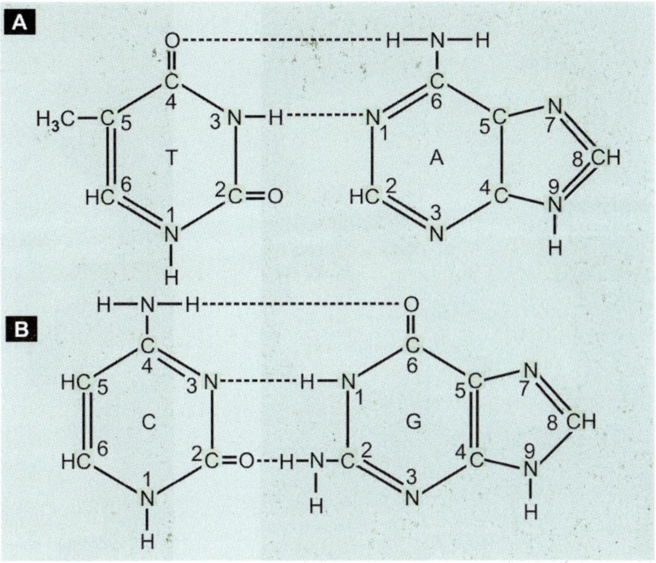

Figs. 16.3A and B: Hydrogen bonds between (A) Adenine and thymine and (B) Between cystosine and guanine

helix. ***The edge of the helix that measures more than 180° from glycosidic bond to glycosidic bond is called the major groove and if it is less than 180° it is called minor groove.***

Types of DNA

- DNA can exist in several conformations depending upon the base composition and under different physical conditions. In all these conformations, the same base pairing rules apply, changes do not alter the information content of the DNA.
- The conformations of DNA have been determined by X-ray crystallography. By far the **most common conformation is B-DNA** (Fig. 16.4).

1. **B-DNA**

   The double helix of B-DNA has the following characteristics:
   - Adjacent nucleotides in each chain are rotated by 34.6° relative to each other. Double helix completes one turn approximately every 10.4 base pairs.
   - One turn of the double helix spans a distance of 3.4 nm. ***This distance is the pitch of the helix.*** Each base pair therefore increases the length of the double helix by 0.33 nm.
   - Diameter of the double helix is 2.37 nm.

2. **A-DNA**
   - When B-DNA crystals are dried or when salt content of the crystal is lowered, the long thin B-DNA molecule becomes short, stubby molecule and is called as A-DNA.
   - The pitch of the helix of A-DNA is 2.46 nm, and number of base pairs per turn is about 11.
   - A-DNA is not found under physiological conditions.

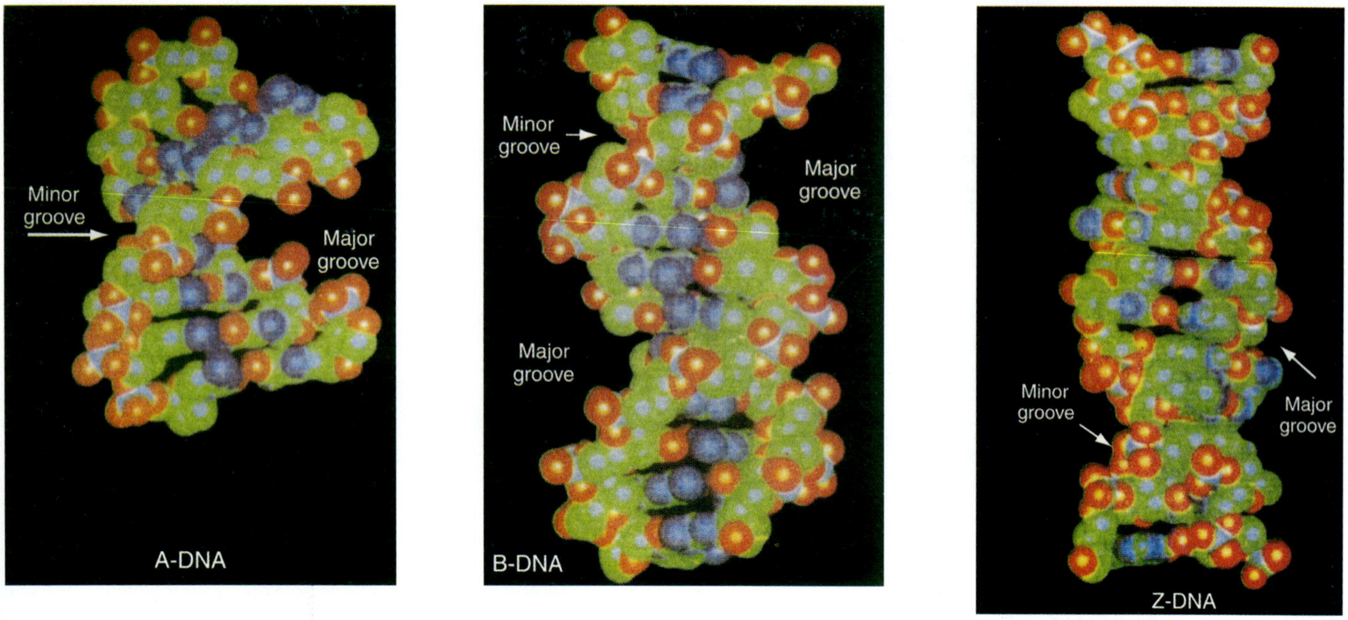

Fig. 16.4: The three general families of DNA double helices are shown here. Both A- and B-DNAs are right-handed helical structures, whereas Z-DNA spirals in a left-handed sense. B-DNA has distinctive major and minor grooves, of particular widths and depths. A-DNA has a very shallow and wide minor groove. The Z helix, long and slender, has a shallow, almost convex major groove, and very narrowed minor groove

Note: In both A and B-DNA, the sugar group and the base are on opposite sides of the glycosidic bond or in the ***anti*** **conformation**. However, in the presence of high concentrations of cations some nucleotides will rotate into *syn* conformations. Under these conditions, a strikingly different DNA conformations can exist. ***The chain zig-zags between syn and anti conformation, and results in the Z-DNA (Z for Zig-zag).***

3. **Z-DNA**
   - Z-DNA is longer and thinner than B-DNA. It has ***left handed helix.***
   - One complete turn of Z-DNA has 12 base pairs, and the pitch of the double helix is 4.56 nm.
   - The diameter of the double helix is 1.84 nm.
   - The major groove in Z-DNA is no more a groove but a convex surface.
   - The minor groove is a deep cleft that spirals around the structure **(Table 16.1)**.

Note: Within the cells, most of the DNA is B-DNA although regions rich in guanine and cytosine base pairs may assume Z-conformation.

Denaturation of DNA

Two strands of DNA double helix can separate or unwind during processes such as DNA replication, RNA transcription and genetic recombination. Complete unwinding of DNA can take place *in vitro* and is called denaturation of DNA or it is also known as ***a helix to coil transition***. Denaturation occurs when the hydrogen bonds between bases break and the base pairs separate when DNA is treated above a certain temperature or melted.

Temperature of DNA: The temperature at which DNA is half denatured is called the ***melting temperature (Tm of DNA)***. At this temperature, the absorbance of DNA at 260 nm is increased by 18.5 per cent, i.e. half the 37 per cent increase in absorbance where DNA is completely denatured. This phenomenon is called ***hyperchromicity***

Table 16.1: The major structural features of A, B and Z DNA

Property	*A-DNA*	*B-DNA*	*Z-DNA*
• **Helix handedness**	right	right	left
• **Repeating unit base pairs**	1 base pair	1 base pair	2 base pair
• **Per turn**	10	11	12
• **Rotation/base pair**	32.7°	34.6°	30°
• **Inclination of basepair to helix axis**	19°	1.2°	9°
• **Rise per base pair along helix axis**	0.23 nm	0.33 nm	0.38 nm
• **Pitch**	2.46 nm	3.40 nm	4.56 nm
• **Diameter**	2.55 nm	2.37 nm	1.84 nm
• **Conformation of glycosidic bond**	anti	anti	anti at C syn at G
• **Major groove**	present	present	non-existent or convex shaped
• **Minor groove**	present	present	deep cleft

or *hyperchromic effect*. The melting temperature of DNA is determined by its base composition. Since there are two hydrogen bonds between A and T while three between G and C, increasing G-C base pairs raises Tm. ***Tm is strongly influenced by the base composition of the DNA.***

- ***DNA rich in G-C pairs has a higher Tm than DNA with high proportion of A-T pairs.***
- Mammalian DNA, which has about 40 per cent G-C pairs, has a Tm of about 87°C.
- The Tm of DNA extracted from different species and measured at pH 7 in an isotonic salt solution varies linearly with G-C content; with synthetic Poly-A-T having a Tm of about 65°C and synthetic Poly G-C having a Tm of 105°C.

Annealing: Once the strands are separated, they can be renatured. If a melted sample of DNA is slowly cooled, the absorbance of the solution decreases. This is indicative of complementary strands being paired again. This process is called as *annealing*. ***Annealing can occur only at a temperature below Tm of DNA*** which is about 70ºC. It is fastest at 20ºC below Tm or 50ºC.

- *Mitochondrial DNA:* The proteins synthesised in cytoplasm by protein synthesising apparatus cannot pass through the mitochondrial membrane. Hence for synthesising the proteins of mitochondrial matrix and membrane, there is a second genetic system with DNA and RNAs in the mitochondrion itself. The ***mitochondrial DNA is also double stranded like nuclear DNA but it is circular.***
- *Satellite DNAs:* The highly repetitive sequence of chromosomal DNA when separated after isopycnic centrifugation in CsCl after shearing the DNA into segments, the DNA distributes into a main band and a set of smaller bands termed **Satellite bands** of DNA. ***In humans, four satellite DNAs constitute 6 per cent of the chromosomal DNA.***

JUNK DNA

In molecular biology, "junk DNA" was a collective label attributed to the portions of the DNA sequence of a chromosome or genome for which no function has been identified. ***About 80 per cent to 90 per cent or more of the human genome had been designated as "junk", including most sequences within introns and intergenic DNA.*** While muech of this sequence ***may be evolutionary artefact*** that serves no present day purpose, some is believed to function in ways that is not currently understood. The conservation of "junk DNA" over millions of years of evolution may imply an essential function. Some authorities consider the "junk" label as something of a misnomer, but others consider it apposite as junk is stored away for possible new uses rather than thrown out, others prefer the term "non-coding DNA" (although junk DNA often includes transposons that encode proteins, with no clear value to their host genome). However, it now appears that although protein-coding DNA genes makes up barely less than 2 per cent of the human genome, the remaining is considered as "junk".

Four years after scientists finished mapping the human genome, they find themselves confronted by a ***biological jungle deeper, denser*** and ***more difficult to penetrate than previously imagined—the so-called* junk *DNA now considered to be crucial*** and ***essential to life*** and ***can trigger several disease processes.***

Role and Function of Junk DNA

Recent studies on junk DNA by scientists have shown that it has some essential and useful functions.

- A study from the University of Michigan in 2002 showed that ***segments of junk DNA called Line-1 elements,*** once thought to be "leftovers from the distant evolutionary past" ***now* deserve more respect *because they are capable of repairing broken strands of DNA.***
- Junk DNA might provide a ***reservoir of sequences from which potentially advantageous new genes may emerge.*** In this way, it may be ***an important genetic basis for evolution.***
- Researchers at the University of Illinois society for Experimental Biology found an **antifreeze protein gene *in a species of fish which evolved from junk DNA.***
- A study from the Cell Press in 2004, suggests that more than one-third of the mouse and human genomes, previously thought to be non-functional, ***may play some role in the regulation of gene expression and promotion of gene diversity.***
- A study from the National Institute of Health in 2005 found that ***social behaviour in rodents and possibly in humans was affected by portions of the genetic code once thought to be* junk.**
- A study from University of California, San Diego in 2005, suggests that junk DNA is critically important to the evolutionary survival of an organism.
- Findings from Purdue University in 2005, stated that many DNA sequences previously believed to have no function–"junk DNA"–***may actually play specialised roles in cell behaviour.***
- Rsearchers from University of Iowa in 2006, documented ***segments of RNA (previously considered junk) that requlated protein production and could generate* micro RNAs.**
- More and more scientists now believe that in fact regulatory layers(s) in the junk DNA, such as through non-coding RNAs, altogether contain genetic programming at least on par with, and possibly much

SECTION THREE

more important that protein coding genes. How much of the 98 per cent would be involved in such activity is not properly known.

- Recently it has been suggested that junk DNA may act as a **protective buffer** ***against genetic damage and harmful mutations.*** For example, a high proportion of nonfunctional sequence makes it unlilkely that a functional element will be destroyed in a chromosomal "crossover" event, possibly making a species more tolerant to this important mechanism of genetic recombination.
- A study from Stanford University, School of Medicine found that ***large swaths of garbled human DNA once dismissed as junk appear to contain some valuable sections which require further studies.***

RIBONUCLEIC ACID (RNA)

Ribonucleic acid is ***a polymer of ribonucleotides*** of Adenine, Uracil, Guanine and Cytosine, joined together by 3′ – 5′ phosphodiester bonds. ***Thymine is absent in RNA.*** RNA is found in the nucleolus, Nissl granules, ribosomes, mitochondria and cytoplasm. **The pentose sugar of the nucleotide is D-ribose.**

Structures of RNA

(a) Primary Structure of RNA

- The primary structure of RNA is defined as the number and sequence of ribonucleotides in the chain.
- Each linear strand is held together by the ribonucleotides bound to each other by 3′,–5′ phosphodiester bonds joining 3′–OH of one nucleotide with the 5′–OH of the next.

(b) Secondary Structure of RNA

- The secondary structure of RNA involves various coil formation of the polyribonucleotide chain.
- These coil structures are stabilised by hydrophobic interactions between the purine and pyrimidine bases.
- There are intrachain hydrogen bonds between G-C and A–U. The hydrogen bonds are the same as in DNA for G–C while N^3 as well as C^4 oxo group of uracil (or dihydrouracil) which pairs with adenine.

(c) Tertiary Structure of RNA

- The tertiary structure of RNA involves the folding of the molecule into three-dimensional structure.
- The crosslinking also occurs at various sites stabilised by hydrophobic and hydrogen bonds producing a compactly coiled globular structure.

Types of RNA

There are mainly **three types of RNA** found in human beings. These are:

1. **Messenger RNA or m-RNA,**
2. **Transfer or soluble RNA or t-RNA,** and
3. **Ribosomal RNA or r-RNA.**

The main function of each of these RNA is protein synthesis. In human cells there are small nuclear RNA or Sn-RNA which are not involved in protein biosynthesis directly. They may have some role in processing of RNA and cellular architecture. They are found in nucleoplasm, nucleolus, perichromatic granules, and cytoplasm and vary in size from 90 nucleotides to 300 nucleotides. A large precursor of m-RNA called as heterogeneous nuclear RNA or hn-RNA is also found in the nucleus.

1. **Messenger RNA (m-RNA)**

   This is the most heterogeneous class of RNA with respect to its size and stability. The molecular weight varies from 3×10^4 to 2×10^6. They consist of 10^3 to 10^4 ribonucleotides. It carries mainly **adenine, guanine, cytosine** and **uracil** as the major bases and methylpurines and methylpyrimidines as minor bases. ***The m-RNA molecules are formed with the help of DNA template strand (3′ – 5′) during the process called* transcription.** The m-RNA carries a specific sequence of nucleotides in "triplets" called ***codons,*** responsible for the synthesis of a specific protein molecule. The 3′–OH end of most m-RNA molecules carries a polymer of adenylate ribonucleotides consisting of 20 to 250 residues in length. This is called as ***"Poly A tail"***, the function of which is not yet fully understood but it seems to maintain the intracellular stability of the specific m-RNA by preventing the attack of *3′ – exonucleases*. On the other hand, the 5' – OH end of the m-RNA ***carries a cap structure consisting of 7 methylguanosine triphosphate.*** The cap is probably involved in recognition of protein biosynthetic machinery and it helps in stabilising the m-RNA by preventing the attack of 5' – *exonucleases.* The ***protein synthesis begins at 5′ end of the capped structure of m-RNA.***

hn-RNA

In mammalian cells, the m-RNA that comes to cytoplasm is the product of processing of a precursor called, heterogeneous nuclear or hn-RNA.

Characteristics

- It is synthesised in the nucleus.
- Has a half life of 23 minutes.
- Has 400 to 4000 nucleotides and it is 10 to 100 times bigger than m-RNA.

- Is bound to macromolecular proteins called *informofers* and exists as "heterogeneous ribonuclear proteins" *(hn-RNP)*.
- 75% of hn-RNA is degraded in the nucleus. Only 25% of the hn-RNA forms a precursor of m-RNA (Pre-mRNA).

PRE-mRNA

- Pre-mRNA is converted to m-RNA.
- Pre-mRNA has regions called as **introns** transcripts—the sequences not required (inactive) and **exons** transcripts (active portion required for translation).
- *Approximately 80 per cent of length of pre-mRNA is removed as* **introns transcripts** *and only 20 per cent of Pre-mRNA, the* **exons transcripts** *are spliced to form the m-RNA.*

2. **Transfer RNA or t-RNA**

These are also called as soluble or s-RNA. ***They remain largely in cytoplasm.*** The t-RNAs are relatively small, ***single-stranded,*** globular molecules with molecular weight of 2 to 3 × 10^4. There are at least **20 different t-RNA molecules.**

(a) ***Primary structure of t-RNA:*** t-RNA molecules consist of approximately 75 nucleotides. Their bases include adenine, guanine, cytosine, uracil, pseudouridine (ψ) or uracil 5-ribofuranoside and thymine are present in one loop.

(b) ***Secondary structure of t-RNA:*** Each single stranded t-RNA molecule remains folded to form ***a clover-leaf secondary structure.*** These folds of the secondary structure are stabilized by H-bonds between complementary bases in different portions of the same strand. These double stranded helical structures are called as ***stems***.

(c) All t-RNA molecules contain **4 main arms or loops:**

1. *Acceptor Arm*

This consists of unpaired sequences of cytosine-cytosine-adenine at the 3′ end also called as ***acceptor end***. The 3′ – OH terminal of adenine may bind with the α-COOH of a specific amino acid and carry the latter as an ***aminoacyl-t-RNA complex*** to ribosomes for protein synthesis. The acceptor arm is borne by a base-paired acceptor stem whose bases are hydrogen bonded with the last few bases at the 5′ end of t-RNA.

2. *Anticodon Arm*

This is another unpaired and non-bonded loop ***carrying specific sequences of three bases constituting the anticodon.*** The bases of anticodon are hydrogen bonded with three complementary bases of codon of m-RNA. The base pair stem leading to anticodon loop is called the ***anticodon stem***.

3. *D arm:* The third is the D-arm because it contains the base dihydrouridine.
4. ***TψC arm:*** Contains thymine, pseudouridine and cytosine.
5. ***Variable arm or extra arm:*** Extra arm is **most variable** structure of t-RNA and it forms the basis of its classification (Fig. 16.5).

CLASSES OF t-RNA

(a) ***Class I t-RNA:*** About 45 per cent of all t-RNA belong to this class and have 3-5 base pairs in its extra arm, e.g. Ala-t-RNA.

(b) ***Class II t-RNA***: This forms about 25 per cent of total t-RNA and has 13-21 base pairs in a long chain, e.g. Phe-t-RNA. They also have a stem loop structure.

The secondary structure of t-RNA is mainly maintained by the base pairing in these arms and this is a consistent feature. The number of base pairs in these arms remains fixed or varies within a specified range as follows:

• TψC arm	7 bases
• T stem	5 base pairs
• D arm	7-11 bases
• D stem	4 base pairs
• Anticodon arm	7 bases
• Anticodon stem	5 base pairs
• Acceptor arm	4 bases
• Acceptor stem	7 base pairs
• Variable, extra arm	4-21 bases depending on the class

- ***L-shaped tertiary structure of t-RNA:*** The L-shaped tertiary structure is formed by further

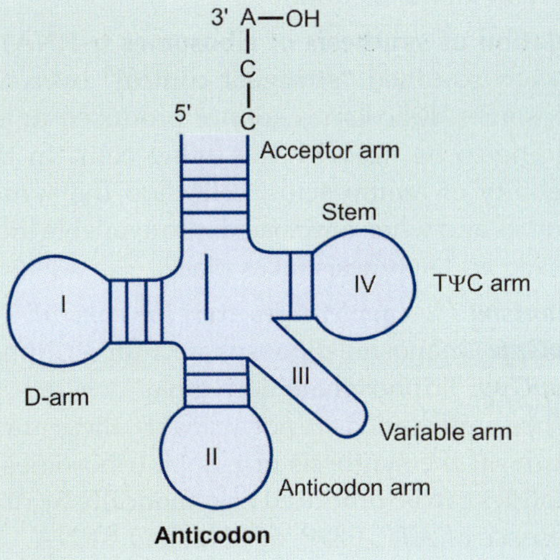

Fig. 16.5: Primary and secondary structure of t-RNA

folding of the cloverleaf due to hydrogen bonds between T and D arms. The base paired double helical stems get arranged into two double helical columns, continuous with and perpendicular to one another. The tertiary structure locates the amino acid acceptor end and the anticodon at the farthest ends of the two columns.

3. **Ribosomal or r-RNA**

A ribosome is present in the cytoplasm and is a nucleoprotein. ***It is on the ribosome that the m-RNA and r-RNA interact during the process of protein biosynthesis.*** Ribosomes contain the third type of RNA known as r-RNA. The r-RNA forms 80 per cent of the total cellular RNA.

- Ribosomes possess a sedimentation coefficient of 80s with a mol. wt. of 4.2×10^6. Mammalian ribosomes are made up of two subunits, larger one with 60s and mol. wt. 2.8×10^6 and smaller subunit with 40s and mol. wt. of 1.4×10^6. The **60S subunit** carries 60 per cent of r-RNA and is a combination of 5S r-RNA, 5.8S r-RNA and 28S r-RNA. The **40S subunit** carries 18S r-RNA and 33 different proteins. All the r-RNA molecules except 5S r-RNA are processed from a precursor of 45S r-RNA.
- Mammalian 5S, 5.8S, 18S, and 28S r-RNAs are made up of about 120, 160, 1900 and 4700 bases respectively. The bases consist mainly of adenine, guanine, cytosine and uracil and a few pseudouridines.
- The function of r-RNA in ribosome particle is still not clearly understood. However, they are necessary for ribosomal assembly and seem to play key roles in the binding of m-RNA to ribosomes and its translation.

Regulation of synthesis of ribosomes (r-RNA): Living cells exert so-called **"stringent control"** over synthesis of ribosomes. Ribosomes are not produced unless they are going to be used by the living cells. In the non-availability of amino acids, cells stop the synthesis of ribosomes and when aminoacids are available the control is relaxed and synthesis takes place.

Mechanism: ***Two nucleotides exert the control.*** They are:

- *ppGpp:* Guanosine diphosphate 2' (or 3') diphosphate
- *pppGpp:* Triphosphate derivative.

When amino acids are not available, these nucleotides can turn off the synthesis of r-RNA (ribosomes). These nucleotides can be produced enzymatically by ribosomes in presence of ATP, GDP, m-RNA and t-RNA.

Evidence: These two nucleotides have been shown to be present as two spots in electrophoretogram, when cell extracts of normal cell under 'stringent control' is subjected to paper electrophoresis. These two spots have been named as ***magic spot I*** and ***magic spot II*** respectively.

SMALL RNAs

Recently a large number of discrete, highly conserved and small stable RNA species have been found in eukaryotes. Most of these molecules are **complexed with proteins to form ribonucleoproteins and are distributed in the nucleus,** in size from 20 to 300 nucleotides and are present about 100,000 to 1,00,000 per cell.

TYPES

A. SnRNAs (Small nuclear RNAs)

SnRNAs are significantly involved in m-RNA processing and gene regulation.

- u_1, u_2, u_4, u_5, and u_6 are ***involved in intron removal*** and the processing of **hn-RNA into m-RNA.**
- u_7 *SnRNA* is involved ***in production of the correct 3' ends*** of histone m-RNA which lacks a poly (A) tail.
- *7SK RNA* associates with several proteins to from a ribonucleoprotein complex that ***modulates m-RNA gene transcription,*** elongation by ***RNA Polymerase II.***

B. Micro RNAs (miRNAs)

Functions

miRNAs play important roles in ***gene regulation.*** All known miRNAs cause ***inhibition of gene expression*** by ***decreasing specific protein production*** via distinct mechanisms.

Structure

miRNAs are ***single stranded RNA molecules*** of about 21 to 23 nucleotides in length. They were ***first described*** in ***1993 by Lee and colleagues.*** But the term micro RNA was introduced only in 2001.

Formation and Processing

The genes encoding miRNAs are much longer than the processed mature miRNA molecules.

miRNAs are first ***transcribed as primary transcripts or* Pre-miRNAs** with a cap and a poly-A tail and processed to short 70 nucleotides stem loop structure known as **Pre-miRNA** in the cell nucleus.

This processng is performed in animals by a protein complex known as the **microprocessor complex,** consisting of the **Nuclease Drosha** and the double stranded RNA binding protein **Pasha.** These Pre-miRNAs are then processed to mature miRNAs in the cytoplasm by interaction with the **Endonuclease dicer,** which also initiates the formation of the **RNA induced silencing complex (RISC).** This complex is responsible for the gene silencing observed due to miRNA expression and RNA interference.

The small processed mature miRNAs typically hybridize via the formation of **imperfect RNA-RNA duplexes** within the 3′-untranslated regions of specific target m-RNAs leading via unknown mechanisms to translation arrest.

Until now hundreds of distinct miRNAs have been described in humans.

C. Si-RNAs (Small Interfering RNAs)

- Si-RNAs also play an ***important role in gene regulation.***
- Si-RNAs are derived by the specific nucleolytic cleavage of larger double stranded RNAs to produce small (21 to 23 nucleotides in length) products from the precursors.
- These short Si-RNAs usually form perfect ***RNA-RNA hybrids*** with their distinct targets potentially anywhere within the length of the m-RNA where the complementary sequence exists.
- Formation of such duplexes Si-RNA and m-RNA results in reduced specific protein production because the Si-RNA-m-RNA complexes are degraded by specific nucleolytic machinery; some or all of these m-RNA degradation occurs in specific cytoplasmic organelles termed "p-bodies".

CLINICAL ASPECT

1. *Therapeutic Drug Development*
   Both miRNAs and Si-RNAs represent exciting new potential targets for therapeutic drug development in humans.
2. *To decrease or "Knock Down" Specific Protein Levels*
   Si-RNAs are frequently now ***being used to decrease or "knock down" specific protein levels*** via Si-RNA homology-directed m-RNA degradation in experimental contexts in the laboratory. It is an extremely useful and powerful alternative to gene knockout technology.
3. *miRNA and Cancer*
   miRNA has been found to have links with some types of cancer.
   - A study of mice altered to produce excess *C-mye*—a protein implicated in several cancers—shows that miRNA has an effect on the development of cancer.
   - Another study found that two types of miRNAs inhibit the E_2F_1 protein, which regulates cell proliferation. miRNA appears to bind to m-RNA before it can be translated to proteins that switch genes on and off.
4. *miRNA and Heart Disease*
   miRNA has been shown to be related to heart disease. Mice were developed which were deficient in a muscle-specific miRNA and these mice had a high rate of the most common congenital heart disease—***ventricular septal defects*** characterised by the holes between the left and right ventricles of the heart. Such mice also ***show hyperplasia, an increase in the number of cardiac muscle cells that leads to heart enlargement and abnormalities in cardiac conduction.***
5. *Si-RNA Used for Killing Cancer Cells*
   Scientists have recently used Si-RNA in experimental mice to kill cancer cells. They used **mini cells called EDVs (EnGenelC Delivery Vehicle)**
   - The first wave of mini cells release ***Si-RNA, which switch off the production of proteins that make the cancer cells esistant to chemotherapy.***
   - A second wave of EDV cells then ***accepted by the cancer cells, releases the chemotherapy drugs killing the cancer cells. This therapy has the potential to directly target cancer cells with chemotherapeutic drugs rather the current treatment where both-healthy as well as cancer cells are targeted indiscriminately.***

Differentiating features of DNA and RNA are given in a tabular form in Table 16.2 and differentiation of m-RNA and t-RNA is given in Table 16.3.

SECTION THREE

SECTION THREE

Table 16.2: Differentiation of DNA and RNA

DNA	RNA
Similarities	
1. Both have adenine, guanine, cytosine.	
2. The nucleotides are linked together by phosphodiester bonds.	
3. The bonding is in 3'– 5' direction.	
4. Main function involves protein biosynthesis.	
Differences	
1. In addition to A, G, C, the fourth base is **T—Uracil absent**	1. In addition to A, G, C, the fourth base is **U—Thymine absent**
2. ***Pentose sugar is deoxyribose***	2. ***Pentose sugar is ribose***
3. Present in nucleus, mitochondria but never in cytoplasm	3. In addition to nucleus, RNA is found in cytoplasm
4. They consist of 2 helical strands	4. ***Single stranded***
5. There are A, B, C, D and E forms of DNA	5. There are t-RNA, m-RNA, r-RNA, hn-RNA and SnRNA
6. Large molecules	6. Only hn-, m- and r-RNA are large molecules
7. One strand 3' – 5' carries genetic information	7. m-RNA transcribed from DNA carries genetic information.
8. DNA can form RNA by the process of "transcription"	8. RNA cannot give rise to DNA under normal conditions, but it can under special experimental conditions using *reverse transcriptase*
9. Purine and pyrimidine contents are almost equal.	9. Not equal
10. Alkali hydrolysis does not give 2' – 3' cyclic diesters.	10. Alkali hydrolysis gives 2'-3' cyclic mononucleotides.

Table 16.3: Differentiation of m-RNA and t-RNA

m-RNA	t-RNA
1. Large mol. wt.	1. Low mol. wt.
2. Most heterogeneous	2. Only about 20 different forms—less heterogeneous
3. ***Acts as a template for protein synthesis***	3. ***Acts as carrier of amino acid***
4. ***Carries codons***	4. ***Carries anticodon***
5. Shape and size is not constant	5. Shape and size is constant for all t-RNAs "cloverleaf"
6. The cap structure is found on 5' OH end	6. No such structure
7. Poly A tail is found on 3' OH end	7. 3' OH end carries CCA sequence where specific amino acid is bound
8. Precursor is hn-RNA and pre-m-RNA	8. No such precursor
9. Unusual bases are not found	9. Unusual bases such as pseudouridine, thymine, etc. are found
10. Stem and loop structure is not found.	10. Stem and loop structure is a consistent feature.

I. Questions (Essay type and short notes)

1. Discuss RNA and DNA in terms of their chemical composition, structures and functions.
2. Describe the chemistry and functions of ribonucleic acid.
3. Describe the chemistry and functions of nucleic acids.
4. Write in brief the chemistry of nucleic acids.
5. What are nucleic acids? Describe the main points of difference between DNA and RNA. Mention the biological importance of nucleic acids.
6. Describe what you know of DNA and its biological role.

Short Notes

a. RNA b. DNA
c. Nucleic acid d. t-RNA
e. m-RNA f. Adenylic acid
g. hn-RNA h. m-RNAs

Differentiate

a. DNA and RNA
b. RNA and DNA
c. m-RNA and t-RNA
d. Nucleotides and nucleosides

II. MCQs (Give one correct answer)

1. The anticodon region is an important part of the structure of:
   a. r-RNA b. m-RNA
   c. hn-RNA d. t-RNA
   e. Z-DNA
2. RNA does not contain:
   a. Uracil b. Adenine
   c. Hydroxymethyl cytosine d. Ribose
   e. Phosphate

3. If the percentage concentration of thymine in DNA is 40%, the concentration of cytosine will be:
   a. 10% b. 20%
   c. 30% d. 40%
   e. 50%
4. Which pyrimidine nucleotide acts as the high energy intermediate?
   a. ATP b. UTP
   c. CMP d. UDPG
   e. Cyclic AMP
5. In a DNA molecule, guanosine nucleotide is held by the cytosine nucleotide by the number of hydrogen bonds:
   a. 1 b. 2
   c. 3 d. 4
   e. 5
6. In t-RNA molecules, there is a 'loop' which contains a minor base:
   a. Cytosine b. Dihydrocytosine
   c. Thymine d. Dihydrouracil
   e. Uracil
7. If the cytosine content of double straded DNA is 20% of the total bases, the adenine content will be:
   a. 10% b. 20%
   c. 30% d. 40%
   e. 50%
8. Match the following:
   A. DNA B. RNA
   C. DNA and RNA both D. Neither DNA nor RNA
   a. Contains ribose in furanose form _______
   b. Hydrolysed by acid _______
   c. Pentose attached to Purine N_9 _______
   d. Not attacked by mild alkali _______
9. The melting temperature of DNA (Tm) is:
   a. Directly proportional to A-T content
   b. Directly proportional to G-C content
   c. Not related to base composition at all
   d. Directly proportional to the length of DNA
   e. None of the above
10. A t-RNA molecule has the following:
   a. A clover-leaf structure
   b. A loop containing minor base dihydrouracil
   c. An anticodon arm
   d. None of the above
   e. All of the above
11. The type of RNA that characteristically contain methylated purines and pyrimidines is:
   a. m-RNA b. hn-RNA
   c. r-RNA d. t-RNA
   e. None of the above
12. In mammalian cells, ribosomal RNA is produced mainly in the:
   a. Nucleus b. Nucleolus
   c. Ribosome d. Golgi apparatus
   e. Endoplasmic reticulum
13. m-RNA is a complementary copy of:
   a. A single strand of DNA b. Ribosomal DNA
   c. Ribosomal RNA d. Transfer RNA
   e. None of the above
14. t-RNA specifically reacts with:
   a. Ribosomal RNA
   b. Ribosomal DNA
   c. Specific aminoacyl adenylates
   d. Nuclear DNA
   e. None of the above
15. Complete hydrolysis of nucleic acids will not yield:
   a. Guanosine b. Ribose
   c. Deoxyribose d. Phosphoric acid
   e. All of the above
16. Match the following:
   A. DNA B. m-RNA
   C. Both D. Neither
   a. Made in nucleus of cell _______
   b. Produced from a structural gene _______
   c. Possess property of self-replication _______
   d. Capable of forming a double helix with DNA _______.

Answers to MCQs

1. d 2. c 3. a 4. d 5. c 6. d
7. c 8. a-B, b-C, c-C, d-A 9. b 10. e
11. d 12. b 13. a 14. c 15. a
16. a-C, b-B, c-A, d-C.

SECTION THREE

PART II: Replication of DNA

Major Concepts

A. Understand in detail the process of DNA replication and its importance.
B. Differentiate between prokaryotic and eukaryotic DNA replication.
C. Study DNA repair mechanisms.

Specific Objectives

A. I 1. Define replication and its importance to living organisms.
2. Learn about semiconservative and conservative DNA replication and Meselson-Stahl experiment.
3. Learn about Theta replication.

II. 1. List various enzymes and proteins that participate in DNA replication with their functions.
2. Describe briefly the sequential process of DNA replication
- Initiation: Define ori.
- Unwinding of DNA helix.
- Clearly understand how 'nick' is formed, the enzymes that form and reseal it.
- Learn Polymerisation process.
- Understand how replication fork and replication bubbles are formed and the enzymatic machinery associated with it.
- Learn how the synthesis of primer takes place and its importance.
- Learn how the strands are joined together.
- Understand the relationship between cell cycle and replication.

B. Make a tabular form to differentiate prokaryotic and eukaryotic DNA replication.

C. • Learn the different mechanisms of DNA repair
• Learn about the disease xeroderma pigmentosum.

REPLICATION AND ITS IMPORTANCE

The model of DNA as proposed by **Watson** and **Crick** suggests that because one strand of DNA is the complement of the other, upon unwinding of the double helix (dsDNA), ***each strand (ssDNA) acts as a template for the formation of a new strand. This process is called as DNA replication.***

TYPES OF REPLICATION

Replication is of **two types:**

1. **Conservative,** and
2. **Semi-conservative.**

1. ***Conservative replication:*** In conservative replication the parental strands never completely separate. Thus, after one round of replication, one daughter duplex contains only parental strands and the other only daughter strands.
2. ***Semiconservative replication:*** The process of unwinding of the double-helical daughter molecules, each of which is composed of a parental strand and a newly synthesized strand formed from the complementary strand. This is called semiconservative replication. **Meselson and Stahl** demonstrated experimentally that the process of replication in *E. coli* was semiconservative. They carried out the following experiment:

   Bacteria were grown in a medium containing the heavy isotope of nitrogen ^{15}N, when all the DNA was labelled with heavy nitrogen. These cells were allowed to divide in a medium containing normal nitrogen, ^{14}N. In the first generation, all DNA molecules were half-labelled. In the second generation, half-labelled and completely unlabelled molecules were present in equal numbers. From this Pioneer experiment, it was proved that DNA replication *in vivo* is semiconservative.
3. ***Theta replication:*** Most prokaryotic DNA replicates when it is in a circular form; such replication is called as Theta replication.

DNA in eukaryotic chromosomes probably exists in looped domains. Thus, the primary function of DNA replication is understood to be the provision of progeny with the genetic information processed by the parent. Hence, ***the replication of DNA must be complete and carried out with high fidelity to maintain genetic stability within the organism and the species.***

The process of DNA replication is complex and involves many cellular functions and several verification procedures to ensure the fidelity in replication. The first important observation was made by **Arthur Kornberg** in *E. coli*, an enzyme called DNA polymerase now also called Kornberg's enzyme. This enzyme has multiple catalytic sites—a complex structure and requires dATP, dCTP, dGTP and dTTP.

DNA Polymerase I is primarily a repair enzyme and brings about deoxyribonucleotide polymerisation. It has both 5′→ 3′ and 3′ → 5′ exonuclease activities. ***When 5′ – 3′ exonuclease domain is removed, the remaining enzyme molecule retains the polymerisation and***

proofreading activities. Such a fragment is called **Klenow fragment** and is used widely in recombinant DNA technology.

MAJOR ENZYMES AND OTHER PROTEINS INVOLVED IN DNA REPLICATION

- *DNA polymerases:* Chief enzyme mainly involved in repair and deoxynucleotide polymerisation.
  In prokaryotes: Three types of DNA polymerases found. They are *DNA polymerase I, DNA polymerase II* and *DNA polymerase III.*
  In eukaryotes: There are **five types** of DNA polymerases and they are called as:
  α, ε, β, γ and δ.
- *DNA helicases:* Required for unwinding of dsDNA.
- *DNA primase:* Required for synthesis of RNA primer.
- *Nick sealing* enzymes:
  Two enzymes:
  - *Topoisomerases*
  - *DNA ligase.*
- **Single strand binding proteins (SSB proteins).**

SOME SALIENT FEATURES

- The primary function—provision of progeny with the genetic make up possessed by the parents.
- *DNA replication must be complete,* to be carried out in such a way as to maintain genetic stability within the organism and the species.
- To *ensure fidelity in replications,* it involves many cellular functions. Several proofreading procedures, and requires the formation of a number of protein-protein and protein-DNA interactions.
- *In all cells, replication can occur only from a single-stranded DNA (ssDNA) template,* which occurs after unwinding double-stranded DNA (dsDNA).
- Mechanisms must exist to target the site of initiation of replication and to unwind the double-stranded DNA (dsDNA) in that region, where the replication complex is formed.
- After replication is complete in an area, the parent and daughter strand must reform dsDNA.
- **In Eukaryotic cells, an additional step is must, in that domain structure must reform,** including nucleosomes, that existed prior to the onset of replication.

Although this entire process has been precisely delineated in prokaryotes, it is not well understood in Eukaryotes, but general principles appear to be same in both.

Classes of proteins involved in replication and their functions are shown in Table 16.4.

The differences between DNA polymerases of prokaryotes and eukaryotes, in types and functions, are shown in Table 16.5.

SEQUENTIAL EVENTS OF DNA REPLICATION

The steps involved in DNA replication in Eukaryotes can be arbitrarily divided into **five steps** for better understanding. They are:

A. *Identification of sites of origin of replication (ori)*
B. *Unwinding of dsDNA to provide ssDNA which can act as template*
C. *Formation of the replication fork*
D. *Initiation and chain elongation*
E. *Formation of replication bubbles and ligation of the newly synthesised DNA segments.*

A. Identification of Site of the Origin of Replication

There are specific sites, called *origin of replication* **(ori)**; where replication starts. At the origin of replication, there is an association of sequence-specific DNA binding proteins with a series of direct repeat DNA sequences. Adjacent to 'ori' is **A + T rich region**.

Table 16.4: Major classes of proteins and their functions

Proteins	*Functions*
• ***DNA polymerases***	Polymerisation of deoxynucleotides **(repair enzyme).**
• ***DNA helicase***	Unwinding of DNA.
• ***DNA primase***	Initiates synthesis of RNA primer.
• ***Topoisomerases***	"Nick" sealing enzyme—relieves torsional strain resulting from helicase induced unwinding. ***Does not require energy.***
• ***DNA ligase***	Nick sealing enzyme—Nick formed between the nascent chain and Okazaki fragments on lagging strand. ***Requires energy.***
• ***Single strand binding proteins (SSB proteins)***	Prevents premature reannealing of dsDNA

Table 16.5: Differences of prokaryotic and eukaryotic DNA polymerases

Types		*Functions*
Prokaryotes	**Eukaryotes**	
I	α	Gap filling and synthesis of lagging strand
II	ε	Proofreading of DNA and repair
	β	DNA repair
	γ	Required for mitochondrial DNA synthesis
III	δ	Helps in leading strand synthesis

SECTION THREE

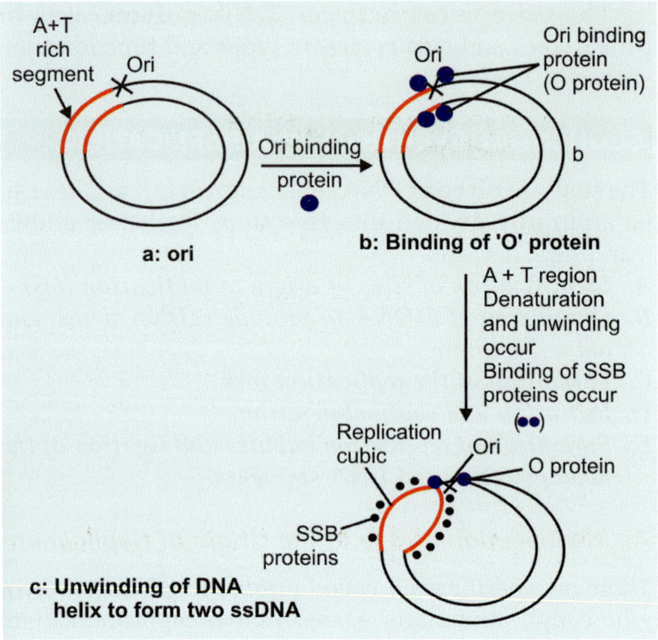

Fig. 16.6: Specific site ori and unwinding of ds DNA

A specific interaction of a protein, say the **O protein** to the origin of replication site leads to local denaturation and unwinding of the adjacent A + T rich region of DNA (Fig. 16.6).

B. Unwinding of DNA to Form ssDNA Which Act as Template

As studied above, the interaction of proteins with 'ori' delineates the start site of replication and provides a short *region of ssDNA* which acts as template for initiation of synthesis of nascent DNA strand. *Main critical enzyme which helps in the unwinding is DNA helicase which allows for processive unwinding of DNA*.

Single-strand binding proteins (SSB proteins) binds to each ssDNA strand and stabilise the complex and prevents re-annealing.

Torsional strain by DNA helicase produces **Nicks** in one strand of unwinding double helix (dsDNA) thereby allowing the unwinding process to proceed. The "nicks" are quickly resealed by the nick-sealing enzyme, called *DNA topoisomerases* which **does not require any energy** because of the formation of a high energy covalent bond (Fig. 16.10A).

C. Formation of the Replication Fork

Formation of a replication fork requires the following **four components** that form in the following sequence:

- DNA helicase unwinds a short segment of the parental dsDNA as discussed above.
- A ***primase initiates synthesis of RNA molecule*** **(Primer)** that is essential for priming DNA synthesis.
- The ***DNA polymerase*** initiates nascent, daughter strand synthesis (ssDNA) and,
- ***SSB proteins bind to ssDNA and prevent premature reannealing*** of ssDNA to dsDNA.

DNA polymerases only ***synthesise DNA in the 5'→ 3' direction*** and only one of the several different types of polymerases is involved in formation of replication fork. ***An enzyme capable of polymerising DNA in 3' → 5' direction does not exist in any organism, so that both of the newly replicated DNA strands cannot grow in the same direction simultaneously.*** Again the same enzyme does not replicate both strands at the same time.

As the DNA strands are antiparallel, the polymerase functions asymmetrically.

- **Leading strand** ***(forward strand):*** The DNA is ***synthesized continuously*** in 5' → 3' direction with same over-all forward direction.
- **Lagging strand** (***Retrograde strand):*** The DNA is ***synthesised*** in a ***discontinuous manner***. In this strand, the DNA is synthesised in short spurts, 1-5 kb fragments, the so-called **Okazaki fragments.**

Several Okazaki fragments, approximately 250 must be synthesized, in sequence, for each replication forks (Fig. 16.7).

The ***DNA helicase*** associates with ***primase*** and a mobile complex is formed which is called as **Primosome**. This association of helicase with primase ***allows the primase to get access to template; which makes the RNA primer*** and in turn allows polymerase to begin the DNA replication. This is an important step as ***DNA polymerases cannot initiate DNA synthesis de novo.***

When synthesis of one Okazaki fragment is completed, the polymerase is released and a new primer is synthesized. The same polymerase molecule remains associated with the replication fork and forms a new Okazaki fragments. ***This process is repeated to form several Okazaki fragments, on the Lagging strands.***

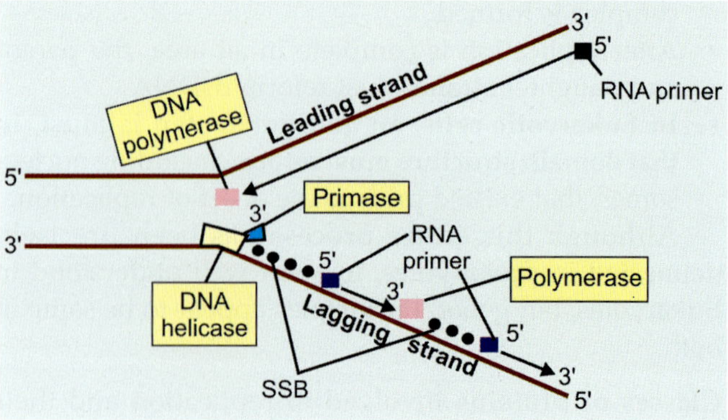

Fig. 16.7: Replication fork-polarity of DNA synthesis

D. Initiation and Elongation of DNA

As discussed above, it will be seen that the initiation of DNA synthesis requires priming by short length of RNA, about 10 to 200 nucleotides.

Priming process: Involves the nucleophilic attack by the 3′–OH group of the RNA primer on the α-phosphate of the first entering deoxynucleotide triphosphate with release of pyrophosphate.

Elongation: Now the 3′–OH group of the newly attached deoxyribonucleotide monophosphate is then free to carry out a nucleophilic attack on the next entering deoxyribonucleotide triphosphate, again at its α-phosphate moiety with release of PPi.

Note

- ***The Nascent DNA is always synthesised in the 5′ → 3′ direction because DNA polymerases can add a nucleotide only to the 3′ end of a DNA strand.***
- The selection of the new nucleotide entrant is governed by proper **base pairing rule.**

Maintenance of Polarity

DNA molecules are double-stranded and the ***two strands*** are ***antiparallel***, running in opposite directions. The replication in both prokaryote and eukaryotes occur on ***both strands simultaneously.***

DNA polymerase enzyme is capable of polymerising DNA in the 3′ → 5′ direction does not exist in any organism, so that both the newly replicated DNA strands cannot grow in the same direction simultaneously. But the same enzyme replicates both strands at the same time.

The single enzyme replicates one strand—the ***Leading strand*** in a ***continuous*** manner in the 5′ → 3′ direction with the same forward direction.

But in the other strand—***Lagging strand***, the same enzyme replicates ***discontinuously***, while polymerising the nucleotides in "short spurts" of 150 to 250 nucleotides again in the 5′ → 3′ direction but at the same time it faces towards the back end of the preceding RNA primer rather than toward the unreplicated portion (Fig. 16.8). In mammals, most of the RNA primers are removed as part of replication process.

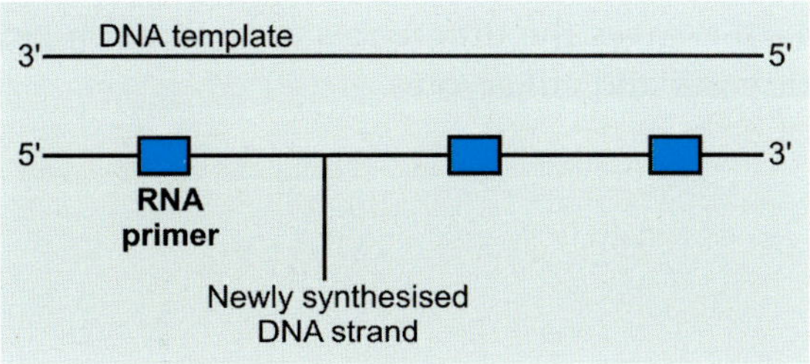

Fig. 16.8: Discontinuous DNA synthesis on lagging strand

E. Formation of Replication Bubbles and Ligation of the Newly Synthesised DNA Segments

The entire mammalian genome replicates in approximately **9 hours**, the average period required for formation of a tetraploid genome from a diploid genome in a replicating cell. If this is done from a single 'ori', it would take approximately 150 hours. **This problem is circumvented by two ways:**

1. Replication occurs **bidirectionally**, and
2. Replication proceeds from **multiple origins** (approx. 100 in humans) in each chromosome. Thus replication occurs in both directions along all of the chromosomes, and both strands are replicated simultaneously resulting in formation of multiple **replication bubbles** (Figs. 16.9A and B).

Ligation of newly synthesised DNA fragments: In mammals, after many Okazaki fragments are formed, the replication complex begins to remove the RNA primers, to fill in the gaps left by their removal with the proper base-paired deoxynucleotides, and then to seal the fragments of newly synthesised DNA by a ***nick–*** **sealing enzyme** called ***DNA ligases, which requires energy from ATP*** (Figs. 16.10B).

DNA Synthesis and Cell Cycle

DNA synthesis occurs during the **S phase** of the cell cycle. This period is referred to as the synthetic S phase. This is usually temporarily separated from the mitotic phase by nonsynthetic period referred to as gap I (G1) and gap-2

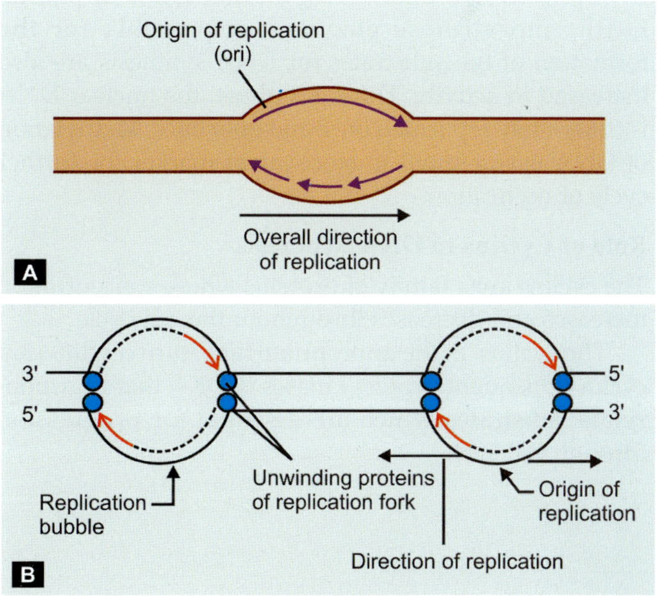

Figs 16.9A and B: (A) Overall direction of replication, (B) Formation of replication bubbles

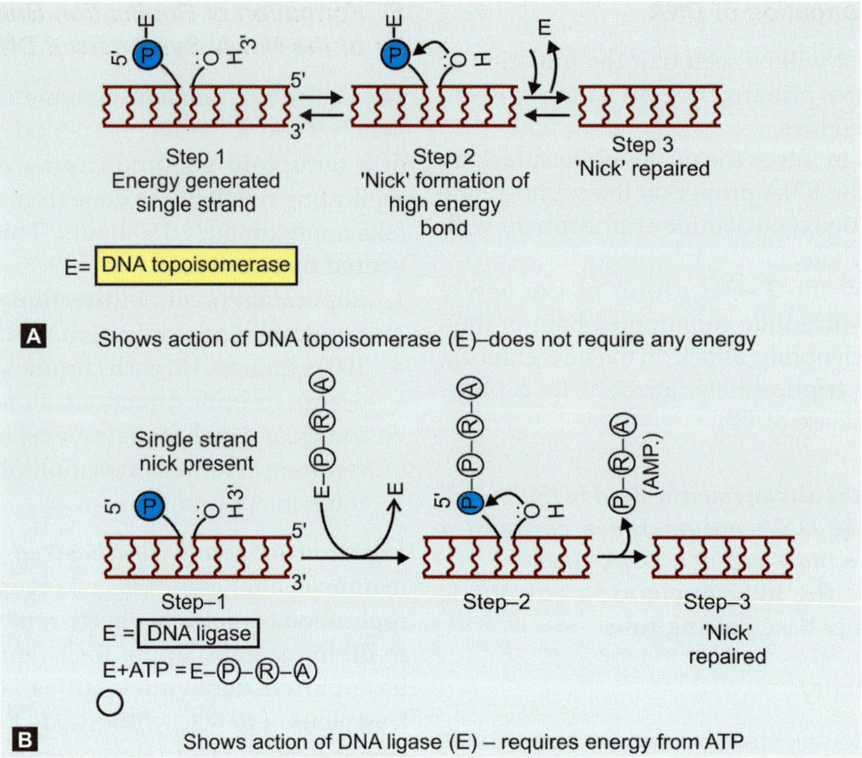

Figs 16.10A and B: Comparison of two types of nick-sealing reactions in DNA

(G2) occurring before and after the S phase respectively. The cell regulates its DNA synthesis grossly by allowing it to occur only at specific times. During S phase of the cell cycle, mammalian cell contains greater quantities of *DNA polymerase* α than during the non-synthetic phases. Furthermore, those enzymes responsible for the formation of the substrates for DNA synthesis are also increased in activity. During S phase, the nuclear DNA is completely replicated once and only once. Methylation of DNA is suggested to be covalent marker for further cycle of replication.

Role of Cyclins in DNA Synthesis

The cyclins are a family of proteins whose concentration increases and decreases throughout the cell cycle.

The cyclins, at the appropriate time, turn on different cyclin-dependent protein kinases (CDKs) that phosphorylate substrates which are essential for progression through cell cycle.

D cyclins activate ***proteinkinases*** **CDK-4 and CDK-6.** These two kinases are also synthesized in G-1 phase, when cell undergoing active division.

Other cyclins like **cyclin E** and **cyclin A** and ***kinases are involved*** in different aspects of cell-cycle progression. **Cyclin-E and CDK2** ***form a complex in late G 1 phase.*** Then **cyclin E** is ***rapidly degraded***, and the ***released CDK 2*** then ***forms a complex with cyclin-A***. This sequence is **necessary** for the ***initiation of DNA synthesis*** in S phase.

Note: Excessive production of cyclins, or production at an inappropriate time may result in abnormal or unrestrained cell division.

PROKARYOTIC AND EUKARYOTIC REPLICATION

Table 16.6 gives the differences of replication between prokaryotes and eukaryotes.

Table 16.6: Differences of replication between prokaryotes and eukaryotes

Prokaryotes	*Eukaryotes*
1. Initiation point specific (ori)	1. ***Initiation point specific*** but other than found in prokaryotes
2. *DNA Polymerases* are of three types, namely I, II and III	2. *DNA Polymerases* are of five types, namely α, ε, β, γ and δ
3. Diverse functional variety specially that of DNA polymerase I	3. Functional variety of DNA polymerases is specific
4. Not applicable	4. γ DNA polymerase is found in mitochondria
5. No repair function	5. β-Polymerase functions as repair enzyme
6. Replication with few replication forks	6. Many replication forks
7. Theta structure observed	7. Theta structure not observed
8. Accessory proteins few with limited functions	8. Many accessory proteins with diverse functions
9. Only unwinding takes place in prokaryotes	9. Histone separation from DNA as well as unwinding takes place
10. Not at all or few replication bubbles	10. Many replication bubbles
11. RNA as primer.	11. RNA/DNA as primer.

Summary of DNA Replication

Steps of DNA replication are as follows:
- Identification of site of the origin of replication (ori)
- Unwinding of parental DNA (dsDNA → ssDNA)
- Formation of ***replication fork***
- Synthesis of RNA ***Primer***, complementary to DNA template, the enzyme required is ***primase***
- Leading strand is synthesised in the 5' to 3' direction by the enzyme ***DNA polymerase***
- Lagging strand is synthesised as ***Okazaki*** fragments.
- RNA pieces are removed, when polymerisation is complete
- The gaps are filled by deoxynucleotides and the pieces are joined by the nick sealing enzyme ***DNA ligase***, which requires energy from ATP.

I. Questions (Essay type and short notes)

1. Describe the process of DNA replication in detail.

Short Notes

a. DNA replication
b. DNA polymerases
c. Replication bubbles
d. Replication fork

Differentiate

a. Prokaryotic and eukaryotic replication
b. Prokaryotic DNA polymerase and eukaryotic DNA polymerase
c. DNA polymerase and RNA polymerase

II. MCQs (Give one correct answer)

1. All of the following statements characterise DNA replication except:
   a. RNA and DNA chains are linked covalently
   b. Growth of the new DNA chains occurs in the 5′ to 3′ direction.
   c. Growth of DNA chain is discontinuous
   d. Overall growth of DNA chain is bidirectional
   e. DNA is synthesised in a 3′ to 5′ direction on one parental strand and in a 5′ to 3′ direction on the other
2. Which of the following statements about DNA polymerase I is correct:
   a. It functions as a DNA repair enzyme, but is not involved in the DNA replication process
   b. It requires a template and a primer to polymerise deoxyribonucleoside triphoshates
   c. It joins together Okazaki fragments to complete the lagging strand during DNA replication
   d. It produces Okazaki fragments linked to RNA primer.
   e. Its presence is not necessary for normal growth in some bacterial mutants.
3. DNA ligase of *E. coli* requires which of the following cofactors
   a. FAD
   b. $NADP^+$
   c. NADH
   d. NAD^+
   e. NADPH
4. Short chains of nucleic acid can be isolated from cells in which DNA is undergoing replication. These segments, known as Okazaki fragments, have which of the following properties:
   a. They are double-stranded
   b. They contain convalently linked RNA and DNA
   c. They are DNA-RNA hybrids
   d. They arise from the nicking of the sugar-phosphate backbone of the parental DNA chain
   e. They are removed by nuclease activity.

5. Which of the sequences listed below best describes the order in which the following enzymes participate in the replication of DNA in bacteria?
   I. *DNA polymerase I* II. 5′ exonuclease
   III. *DNA polymerase III* IV. *Ligase*
   V. *Primase*
   a. *V, III, I, II, IV*
   b. III, II, I, V, IV
   c. *V, III, IV, II, I*
   d. *V, III, II, I, IV*
   e. *III, II, V, I, IV*
6. Which of the following activities is typical of histones during replication of eukaryotic DNA?
   a. They remain bound to the DNA throughout replication.
   b. They undergo proteolysis and are replaced by newly synthesised protein on each daughter strand.
   c. They are both conserved and associated with the leading daughter strand and newly synthesised and associated with the lagging daughter strand.
   d. They are not synthesised for cell division but are retained from generation to generation
   e. They do not become reassociated with newly synthesised DNA until metaphase
7. The question below contains four suggested answers of which, one or more is correct; choose the answer.
   a. if 1, 2 and 3 are correct
   b. if 1 and 3 are correct
   c. if 2 and 4 are correct
   d. if 4 is correct
   e. if 1, 2, 3 and 4 are correct

   Correct statements describing DNA polymerase I include which of the following?
   1. It can add deoxyribonucleoside triphosphates to the 3′–OH end of a DNA primer chain.
   2. It can act as an endonuclease in DNA repair
   3. It can act as a 5′ to 3′ exonuclease
   4. It is a major polymerising enzyme in the replication process

Answers to MCQs

1. e 2. b 3. d 4. b
5. d 6. c 7. b.

SECTION THREE

PART III: DNA Repair Mechanisms

All cells possess machinery by which damage to DNA can be eliminated and the original form of the DNA double helix is restored.

Causes of DNA Damage

DNA damage during DNA replication can occur through:

- Misincorporation of deoxynucleotides during replication
- By spontaneous deamination of bases during normal genetic functions
- From x-radiation that cause "nicks" in the DNA
- From UV irradiation that causes thymine dimer formation
- From various chemicals that interact with DNA.

The rapid repair of DNA damages are necessary since they may be lethal to the cell or cause mutations that may result in abnormal cell growth.

Mechanisms of Repair

Several methods are available. They are:

A. Excision repair
B. Photo reactivation
C. Recombinational repair
D. Mismatch repair.

A. *Excision repair:* Can be of ***two types:***

1. *Repair of Thymine Dimers*
   - A UV specific *"endonuclease"* makes a ***"nick"*** in the affected DNA strand, usually on the 5' end of the dimer, and the defective segment comes out.
   - The enzyme ***"DNA polymerase I" (DNA pol I)*** synthesizes new DNA strand in the 5' to 3' direction with the 3' end of the ***"nicked"*** strand acting as the ***"primer"*** and the intact complementary strand serving as the template.
   - ***5' to 3'-exonuclease*** activity of DNA pol I then removes the damaged sequence.
   - Lastly ***"DNA ligase"*** enzyme seals the gap between the newly synthesized segment and the main chain.
2. *Spontaneous Deamination of Cytosine to Uracil*
   - The deamination of cytosine to form uracil ***can take place due to inherent instability of cytosine.*** On replication in such an event, an 'A' will be inserted to pair with the 'U' formed thus forming a mutation.

The above defect can be repaired as follows:

- A specific enzyme ***"Uracil N-glycosylase"*** removes "U" by cleaving N-glycosidic bond, leaving the deoxyribose phosphate bond intact.
- Another specific ***endonuclease*** cuts the phosphodiester bond on the 5' end.
- ***DNA pol I*** then fills the gap with correct "C" and also removes the deoxyribose residue left over from the excised "U".
- ***DNA ligase*** enzyme finally seals the break and repair is completed.

B. *Photo reactivation:* Also called ***light induced repair***. The enzyme ***photo reactivating (PR) enzyme*** brings about an enzymatic cleavage of thymine dimers activated by the visible light (300-600 mμ) leading to a restoration of the monomeric condition.

Characteristics of the Enzyme

- Enzyme is found in all organisms.
- Binds to **"NN"** dimers, including cytosine dimers, in the DNA.
- Activated by visible light energy and brings about cleavage of **"C-C"** bonds of the dimer.

C. *Recombinational Repair*

- Also called as ***sister-strand*** exchange.
- In this process, the unmutated single stranded segment from homologous DNA is excised from the "good" strand and inserted into the "gap" opposite the dimer.
- It occurs after the first round of DNA replication.

D. *Mismatch Repair of DNA*

- Mismatch repair corrects errors made when DNA is copied.
  Mechanism: This mechanism corrects a single mismatch base pair, e.g. C to A rather than T to A or a short region of unpaired DNA.
- The defective region is recognised by an ***endonuclease*** that makes a single-strand cut at an adjacent methylated GATC sequence. The DNA strand is removed through the mutation, replaced and religated.

SECTION THREE

CLINICAL ASPECT

1. Xeroderma Pigmentosum

- Transmitted as autosomal recessive
- ***Genetic defect:*** DNA repair mechanisms are defective. DNA damage produced by UV irradiation specially thymine dimers, cannot be incised. Results from inborn deficiency of the enzyme ***"nicking endonuclease".***

Clinical Manifestations

- Increased cutaneous sensitivity to UV rays of sunlight.
- Produces blisters on the skin.
- Dry keratosis, hyperpigmentation and atrophy of skin.
- May produce corneal ulcers.

Prognosis: Fatal, death takes place due to formation of ***squamous cell carcinoma of skin.***

2. Ataxia telangiectasia

- A familial disorder
- ***Inheritence:*** Autosomal recessive
- Increased sensitivity to X-rays and UV rays is seen.

Clinical manifestations

- Progressive cerebellar ataxia.
- Oculocutaneous telangiectasia.
- Frequent sinopulmonary infections.
- Lymphoreticular neoplasms are common in this condition.
- IgE deficiency has been demonstrated in 67 per cent of cases.

3. Bloom's Syndrome

Chromosomal breaks and rearrangements are seen in this condition.

- ***Genetic defect:*** Probably defective ***DNA-ligase.***
- ***Clinical Manifestations***
  - Facial erythema
  - Photosensitivity
  - Telangiectasia.

4. Fanconi's Anaemia

- An autosomal recessive anemia. Defective gene is located in **chromosomes 20q** and **9q.**
- ***Defect:*** Defective repair of cross-linking damage.
- ***Characterised by:*** An increased frequency of cancer and by chromosomal instability.

5. Hereditary Nonpolyposis Colon Cancer (HNPCC)

- Most common inherited cancer.
- ***Defect:*** Faulty mismatch repair.

  Genetic defect has been located in ***chromosome 2***, the located gene is called ***hMSH 2***. Mutations of hMSH 2 account for 50 to 60 per cent of HNPCC cases. Another gene also found associated called hMLH 1, probably responsible for remaining cases.

SECTION THREE

CHAPTER 17

PROTEIN SYNTHESIS, GENE EXPRESSION AND RECOMBINANT DNA

Major Concepts

A. Learn the details of the process of transcription.
B. Study the genetic code and its characteristics.
C. Study in detail the process of translation (or protein biosynthesis) and their inhibitors.
D. Study gene expression in prokaryotes and in eukaryotes.
E. Study formation and applications of recombinant DNA technology.

Specific Objectives

A. 1. Define transcription.
   2. Study in detail RNA polymerase and its role in the process of transcription.
   3. Learn various stages of transcription and various details associated with each stage.
      a. Formation of transcription complex b. Initiation c. Elongation d. Termination
   4. Study post-transcriptional modifications of RNA.
   5. Describe inhibitors of transcription.

B. 1. What is genetic code?
   2. Study in detail various characteristics of genetic code.

C. 1. Study the materials required for protein biosynthesis.
   2. Study the details of ribosomes and formation of aminoacyl tRNA.
   3. Study the details of initiation, and various stages associated with it with special reference to initiation factors.
   4. Study the details of the process of elongation and various stages in which it takes place with special reference to role of elongation factors.
   5. Study how the protein biosynthesis is terminated.
   6. Study the role of chaperones: Proteins that prevent faulty folding.
   7. Various antibiotics and other chemicals inhibit the process of protein synthesis. Enumerate them and study their mechanism of action.

D. 1. Study about gene expression in prokaryotes and in eukaryotes.
   2. Study the types of gene expression.
   3. Learn about one cistron-one subunit concept.
   4. Study about constitutive genes and inducible genes.
   5. Learn in details about operon model specially 'Lac' operon.
   6. Learn about repression and derepression of 'Lac' operon.
   7. Study about the role of catabolite (gene) activator protein (CAP).
   8. Learn in details the various modifications of gene expression in eukaryotes, viz.
      - RNA processing
      - Gene amplification
      - Gene rearrangement
      - Class switching and others.
   9. - Learn in details about mutation and mutagens.
      - Study about point mutation, viz. transitions and transversions. Learn the effects with examples.
      - Study about frame shift mutations—deletion type and insertion type.

E. 1. Define recombinant DNA.
   2. What are restriction endonucleases? Study their role in recombinant DNA technology.
   3. Learn the details of cDNA synthesis.

4. What are plasmids? Study their characteristics and their role in recombinant DNA technology. Learn about other vectors, viz. phages and cosmids.
5. Study in detail how chimeric DNA molecule is produced.
6. Study the process of cloning.
7. Learn about gene library.
8. Learn how DNA and RNA are analysed by Southern Blot and Northern Blot tests. What is Western Blot test?
9. Describe few applications of recombinant DNA technology.

PART I

INTRODUCTION

The central dogma defines the most important basis of molecular biology that genes are units perpetuating themselves and functioning through their expression in the form of proteins. ***Genetic information is carried by the sequence of DNA.*** The information is perpetuated by replication. This information is then expressed by a two-step process. (1) ***Transcription*** generates a single stranded RNA identical in sequence with one of the strands of duplex DNA. (2) ***Translation*** converts the nucleotide sequence of the RNA into the sequence of amino acids comprising a protein. ***Central dogma originally stated that this flow of information is irreversible, but now it is proved to be untrue. RNA can now be converted to DNA by an enzyme reverse transcriptase.***

TRANSCRIPTION

Transcription is the process by which the synthesis of RNA molecules is initiated and terminated representing one strand of DNA duplex. By 'representing' means that the RNA is ***identical*** in sequence with one strand of the DNA, it is ***complementary*** to the other strand, which provides the ***template*** for its synthesis. It takes place by the usual process of complementary base pairing, catalysed by the enzyme ***RNA polymerase***. The reaction can be divided into **four stages**.

RNA polymerase: ***RNA polymerase being the key enzyme in transcription,*** it is worthwhile to study the details of its structure and mode of its action. A single type of RNA polymerase is responsible for synthesis of m-RNA, r-RNA and t-RNA in bacteria. However, **in eukaryotes** several different enzymes are required to synthesise the different types of RNA. They are called as:

- *RNA polymerase I,*
- *RNA polymerase II,* **and**
- *RNA polymerase III.*

The complete enzyme or holoenzyme has a mol. wt. of 480,000. Its subunit composition is as follows (Table 17.1):

Table 17.1: Subunit composition of RNA polymerase of *E. coli*

Subunit	*No.*	*Mass (daltons)*	*Location*	*Function*
α	2	40,000 each	Core enzyme	Promoter binding
β	1	185,000	Core enzyme	Nucleotide binding
β'	1	160,000	Core enzyme	Template binding
σ	1	85,000	Sigma factor	Initiation

The holoenzyme ($\alpha_2 \beta\beta' \sigma$) can be separated into two components: **(1)** ***the core enzyme*** **($\alpha_2 \beta\beta'$)** and **(2)** ***sigma factor (σ polypeptide).*** It requires template of double stranded DNA or occasionally a single stranded DNA, four ribonucleotide triphosphates GTP, ATP, UTP, CTP and Mg^{++} or Mn^{++}.

Note: It must be noted that ***only the holoenzyme can initiate transcription,*** but then the sigma factor is released, leaving the core enzyme to undertake elongation. Thus the core enzyme has the ability to synthesise RNA on a DNA template, but cannot initiate transcription at the proper sites.

Stages of Transcription

The process of transcription can be divided into **four stages:**

1. ***Formation of transcription complex (of DNA and RNA polymerase)***
2. ***Initiation***
3. ***Elongation*** **and**
4. ***Termination.***

1. ***Formation of transcription complex:*** The enzyme RNA polymerase needs to bind with specific sequences on a DNA. These sequences recognised by *RNA*

polymerase are called as ***promoter.*** The size of the promoter region is variable. In prokaryotes, it varies from 20-200 bases. As already mentioned, core enzyme cannot recognise the promoter region, sigma factor is required for recognition and formation of the complex. Following four steps occur:

i. Sigma factor recognises the promoter sequences.
ii. *RNA polymerase* attaches to promoter region.
iii. *RNA polymerase* melts the helical structure and separates 2 strands of DNA locally.
iv. *RNA polymerase* initiates RNA synthesis. The site at which the first nucleotide is incorporated is called the ***start site* or *start point.***

Characteristics of promoter sequence: There are few specific characteristics of promoter sequence:

- The ***Pribnow box*** is a sequence contained with the promoter region. It is located 5-10 bases to the left, i.e. upstream the first four bases that will be copied into RNA. It orients *RNA polymerase* as to the direction and start of synthesis.
- All Pribnow boxes are variants of TATAATG sequences and sometimes referred to as **TATA box**.
- The T at position 6 (conserved T) is present in every promoter region of box.
- The –"35" sequence is a second recognition site in many promoter regions upstream from Pribnow box. It is thought to be the initial site of σ subunit binding. Typically it contains nine bases.
- Since the *RNA polymerase* has a huge size, it comes into contact with Pribnow box.
- Once bound to the Pribnow box, *RNA polymerase* dissociates from the initial recognition site.
- This complex is active intermediate in RNA chain initiation. ***The important event is the melting of DNA duplex*** that takes place about 10 base pairs upstream of Pribnow box and extending to the first transcribed base at the start point (Figs. 17.1A to C).

2. ***Initiation***
   - Core enzyme starts transcription at the separated DNA strands of an initiation complex. As the enzyme moves along, the unwound region moves with it.
   - The first base copied is always within six to nine bases of the conserved T of the Pribnow box on the unwound portion of 3′-5′ strand of DNA.
   - It is observed that the subunit of *RNA polymerase* has two specific binding sites for the binding of nucleotide triphosphates (NTP).
   - Formation of hydrogen bonds is always as per the base-pairing rules. The first incoming NTP binds to *RNA polymerase* at the *start point* of initiation site and H-bonds to the complementary base on the DNA within the complex. This site binds only purine NTP—either A or G. ***The binding is with 3′ end of the NTP leaving 5′ end to be free.***
   - The second incoming NTP ***binds to the elongation site on the polymerase***. The NTP is selected as per base-pair rule which can H-bond with complementary base on DNA. This dinucleoside tetraphosphate has either PPPA or PPPG as the 5′ terminal nucleotide. After this phosphodiester bond formation the σ factor is released.
   - First base is then dissociated from initiation site and that marks the completion of initiation.

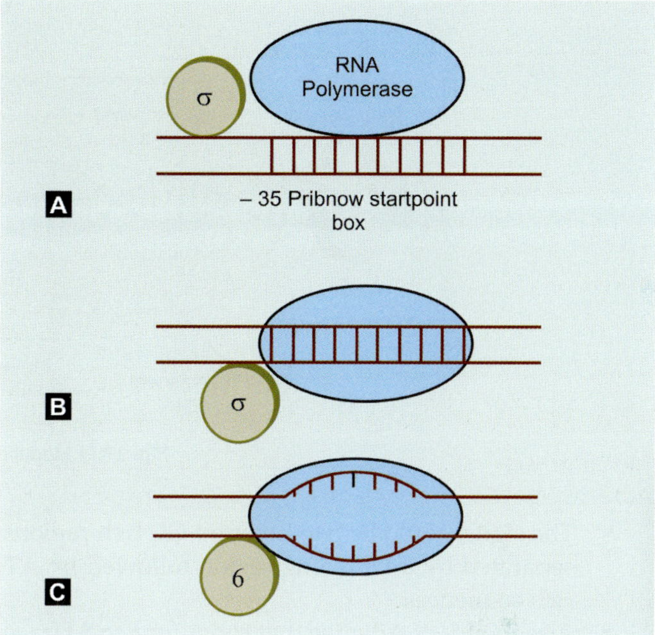

Figs. 17.1A to C: Formation of initiation-complex: (A) Template binding, (B) Dissociation of σ subunit, (C) Unwinding of duplex—open complex formation

3. ***Elongation***
   - The core enzyme *polymerase* moves in 3′-5′ direction of the coding strand and it adds successive NTPs at the 3′-OH end of the ribonucleotide chain already laid down in 5′-3′ direction.
   - The incoming NTP forms a phosphodiester bond with 3′-OH end of the preceding ribonucleotide.
   - The bases are determined by the sense strand by base-pair rules.
   - The DNA helix recloses after *RNA polymerase* transcribes through it and growing RNA chain dissociates from the DNA (Fig. 17.2).
4. ***Termination:*** Specific sequences on the DNA molecule function as the signal for termination of the transcription process.

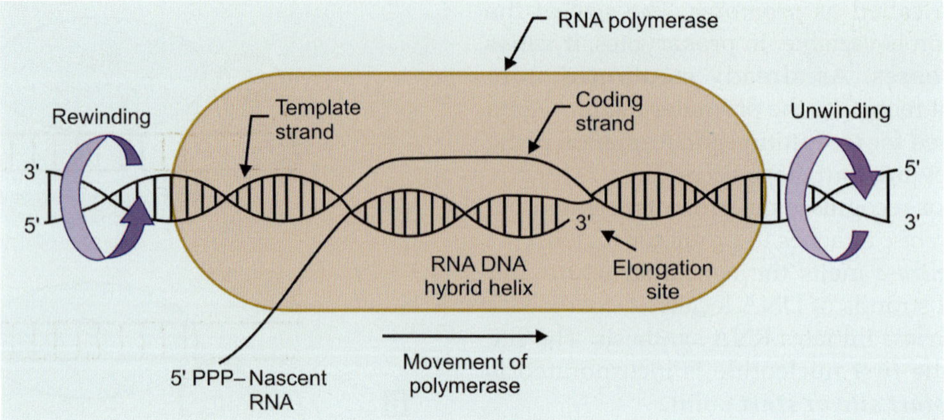

Fig. 17.2: Model of transcription bubble

- The signal could be two inverted GC rich regions separated by intervening region followed by AT rich sequences.
- A sequence of Adenine that codes for 6 to 8 Uracil residues. The Uracil residues are followed by one Adenine.
- There is no unique base for the termination of transcription, e.g. for a given promoter, RNA might end with 5 U's or 6 U's + 1 A.
- ***Rho (ρ) protein and the sequences mentioned, together bring about termination.*** At specific termination sites, the new RNA chain may be released. The rho protein binds very tightly to the RNA (not to polymerase) and in this bound state it acts like *ATPase.*
- Rho factor then dissociates RNA and RNA polymerase from the DNA.

Summary of the transcription is given in Figure 17.3.

Post-transcriptional Modification of RNA

All RNAs, i.e. m-RNA, r-RNA and t-RNA are obtained by transcription. However, the required modifications take place after they are released from polysomes.

- ***Modification in nucleoside:*** *Methylferases, deaminases* and *dehydrogenases* may methylate, deaminate or reduce the bases into the 'minor' bases, e.g. 5 methylcytosine, N^6-methyladenine, hypoxanthine, dihydrouracil, etc. Uridine may be converted into pseudouridine.
- ***Ligations and cleavages of nucleotides:*** The gene contains ***exons*** and ***introns***. ***The introns need to be separated out and exons must be joined as they are actual amino acid coding sequences***. Specific *nucleases* and *ligases* bring about this function. These changed an RNA into functional m-RNA.
- ***Additional nucleotides may be added at the end of RNA transcript, e.g. 7 methyl GTP cap is added at 5′-end while poly A-tail is added at the 3′-end.***

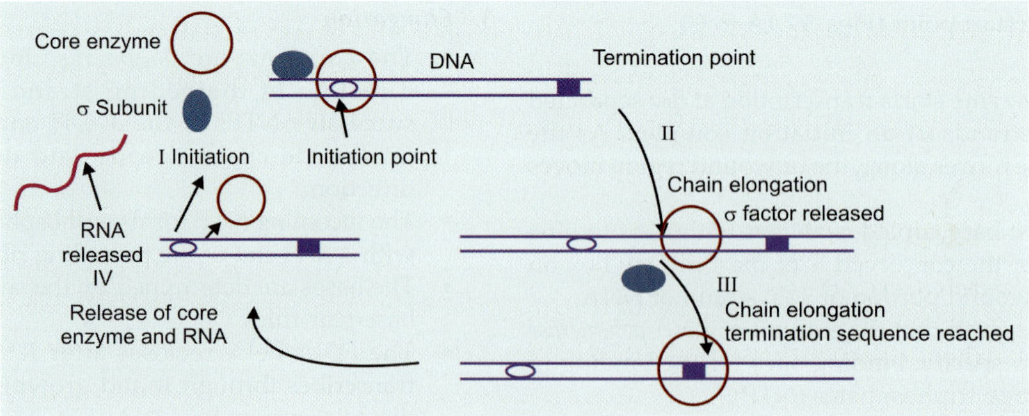

Fig. 17.3: Summary of transcription

Inhibitors of Transcription

Several antibiotics have been found to inhibit the process of transcription.

- ***Rifamycin:*** Rifampicin and streptovaricin bind with β-subunit of the polymerase to block the initiation of transcription.
- ***Actinomycin D:*** It forms a complex with double stranded DNA and prevents the movement of core enzyme and as a result inhibits the process of chain elongation.
- ***Streptoglydigin:*** It binds with the β-subunit of prokaryotic polymerase and thus inhibits the elongation.
- ***Heparin:*** It is a polyanion that binds to the β' subunit and inhibits transcription *in vitro*. The α subunit has no known role in the process.

- Different subspecies of r-RNA such as 5.8S, 18S, 28S r-RNA are made after transcription from a precursor RNA.

GENETIC CODE

Transcription makes the base sequence in the form of m-RNA available to be translated into specific proteins. There are four kinds of base (A, U, G, C) present in RNA while there are 20 different kinds of amino acids found in proteins. Therefore neither one nor two bases can specify all amino acids. However, ***64 kinds of amino acids can be specified by a three-base code.*** Genetic experiments have proved that ***an amino acid is coded by a group of three bases called as codon.***

The 64 combinations of three bases responsible for coding amino acids, initiating protein synthesis and stopping the protein synthesis are arranged in the form of table which is generally known as ***Genetic code.***

Determination of all sequences of bases in codons was carried out by **HG Khorana** and **Marshall Nirenberg.** Later experiments by **Francis Crick, S Brenner** and others fully uncovered the meaning of 64 codons of the genetic code. The sequence of coding strand of DNA read in the direction from 5′ to 3′ consists of triplets corresponding to the amino acid sequence of the protein read from N-terminus to C-terminus. The Genetic code is summarised in Figure 17.4.

Characteristics of Genetic Code

1. Degeneracy

- The striking feature is the degeneracy of the code. 61 codons represent 20 amino acids.
- Every amino acid ***except methionine*** is represented by several codons.
- Codons that represent same amino acid are called as synonyms.
- Codons tend to be clustered in groups representing a single amino acid.
- ***Often the base of the third position is insignificant,*** because the four codons differing only in the third base represent the same amino acid. Sometimes distinction is made only between a purine versus a pyrimidine in this position.
- ***The reduced specificity at the last position is known as third base degeneracy*** or ***Wobbling phenomenon.*** This feature, together with a tendency for similar amino acids to be represented by related codons, minimizes the effects of mutations. It increases the probability that a single random base change will result in no amino acid substitution or in one involving amino acids of similar character.

First Base	*Second Base*								*Third Base*
	U		*C*		*A*		*G*		
U	UUU	Phe	UCU	Ser	UAU	Tyr	UGU	Cys	**U**
	UUC	Phe	UCC	Ser	UAC	Tyr	UGC	Cys	**C**
	UUA	Leu	UCA	Ser	**UAA**	**Stop**	**UGA**	**Stop**	**A**
	UUG	Leu	UCG	Ser	**UAG**	**Stop**	UGG	Tyr	**G**
C	CUU	Leu	CCU	Pro	CAU	His	CGU	Arg	**U**
	CUC	Leu	CCC	Pro	CAC	His	CGC	Arg	**C**
	CUA	Leu	CCA	Pro	CAA	Gln	CGA	Arg	**A**
	CUG	Leu	CCG	Pro	CAG	Gln	CGG	Arg	**G**
A	AUU	lle	ACU	Thr	AAU	Asn	AGU	Ser	**U**
	AUC	lle	ACC	Thr	AAC	Asn	AGC	Ser	**C**
	AUA	lle	ACA	Thr	AAA	Lys	AGA	Arg	**A**
	AUG	**Met**	ACG	Thr	AAG	Lys	AGG	Arg	**G**
G	GUU	Val	GCU	Ala	GAU	Asp	GGU	Gly	**U**
	GUC	Val	GCC	Ala	GAC	Asp	GGC	Gly	**C**
	GUA	Val	GCA	Ala	GAA	Glu	GGA	Gly	**A**
	GUG	Val	GCG	Ala	GAG	Glu	GGG	Gly	**G**

Fig. 17.4: The genetic code

2. *Unambiguity*

A given codon designates only one single specific amino acid and does not incorporate any unspecified amino acid into the peptide chain.

3. *Universality*

- ***In all the living organisms the genetic code is the same***. This phenomenon is called as universality of the code.
- The exception to universality is found in mitochondrial genome where AUA codes for methionine and UGA for tryptophan instead of isoleucine and termination or stop respectively.
- AGA and AGG code for Arginine in normal condition but in mitochondria it terminates protein synthesis.
- Studies of mutations in viruses, bacteria and higher organisms have established the universality of the genetic code.
- Most of the amino acid substitutions in proteins can be accounted for by a change of a single DNA base.

4. *Colinearity of Gene and Product*

- The product of the gene is a protein specified by base sequences.
- High resolution genetic mapping techniques have established that there is a linear correspondence in base sequence in gene and amino acid sequence in protein.

5. *Non-overlapping*

- ***All codons are independent sets of 3 bases.***
- ***There is no overlapping***, i.e. no base functions as a common member of two consecutive codons.

6. *Commalessness*

- Codons are arranged as a continuous structure. There is not one or more nucleotides between consecutive codons.
- The last nucleotide of preceding codon is immediately followed by the first nucleotide of succeeding nucleotide.

Note: All the 64 codons are grouped in 16 families each characterised by first two bases. A mixed codon family codes for more than one amino acid, depending on the third base in its codons. Unmixed codon family codes for the same amino acid irrespective of the third base.

- When the first base in the anticodon is C or A, the pairing with the third base in codon is regular (i.e. G or A).
- When the first base in the anticodon is U then the third base in the codon can be either of the purines (i.e. G or A).
- When the first base in the anticodon is G, then the third base in the codon can be either of the pyrimidines.
- When the first base in the anticodon is ionosine (I), the third base in the codon can be A, C or U.

TRANSLATION OF m-RNA (PROTEIN SYNTHESIS)

Protein is a polymer of amino acids joined together by peptide bonds. In the process of protein synthesis also known as translation of m-RNA, the amino acids are added sequentially in a specific number and sequence, determined by the sequence of codons in the genetic code of the relevant m-RNA.

Materials Required for Protein Synthesis

- Amino acids—at least 20 amino acids
- DNA and three RNAs—m-RNA, t-RNA and r-RNA
- Polyribosomes (Polysomes)
- ***Enzymes***
  - *Amino-acyl-t-RNA synthetase*: enzyme required for activation of amino acids.
  - *Peptide synthetase* (Peptidyl transferase)
- ***Factors***
  - Initiation factors— eIF-1, eIF-2, eIF-3
    eIF-4A, eIF-4B, eIF-4G
    eIF-4E, eIF-5
  - Elongation factors—EF_1 and EF_2
  - Release factors R_1 and R_2
- ***Coenzymes and Cofactors***
  - ***$F.H_4$—required in prokaryotes only*** for formylation of methionine
  - Mg^{++}
- ***Energy:*** ATP and GTP.

Ribosomes

- ***Protein synthesis takes place on ribosomes*** which is a nucleoprotein and contains 65 per cent r-RNA and 35 per cent proteins. It is a large particle and in prokaryotes it has 70S as its sedimentation coefficient while 80S in eukaryotes.
- It can be split into two unequal portions by EDTA which decreases the conc. of Mg^{++} by chelate formation. Mg^{++} is required to hold the two subunits together.
- The two subunits in prokaryotes are 50S large subunit and 30S small subunit while **in eukaryotes they are 60S and 40S.**
- The 50S subunit has been found to contain about 34 proteins (L-proteins) and 2 moles of 23S and 5S r-RNA.
- The 30S subunit contains about 21 proteins (S-proteins) and a 16S RNA molecule.
- Most of the ribosomal proteins (L- and S-) are low molecular weight basic proteins. Due to their basic charge they can easily interact with RNA which is negatively charged.
- The RNAs in ribosomal subunits have a specific well defined secondary structure and they interact with ribosomal proteins in a well defined manner.

SECTION THREE

- The **eukaryotic 60S subunit** of ribosome has 45 proteins and 28S, 5.8S and 5S r-RNA and **40S subunit** contains 30 proteins and 18S RNA.
- The mitochondrial ribosomes are similar to those of prokaryotes.
- A polysome or polyribosome is a beaded string-like linear cluster of 5-8 ribosomes on an m-RNA.
- ***Each ribosome has peptidyl (P) and aminoacyl (A) site.***

STEPS OF PROTEIN SYNTHESIS

The process of protein synthesis (after transcription has taken place) can be divided in **following steps:**

1. **Activation of amino acids,**
2. **Initiation,**
3. **Elongation, and**
4. **Termination.**

1. Activation of amino acids: *(formation of aminoacyl t-RNA)*

The amino acids need to be activated before they can be incorporated into the peptide chain. ***The key enzyme in this process is aminoacyl t-RNA synthetase.*** These are specific for a particular L-amino acid and also for t-RNA. Obviously there are at least ***20 different t-RNAs and 20 different aminoacyl t-RNA synthetases in a protein synthesising system.*** The enzymes vary in molecular size, subunit and amino acid composition. There is at least one and sometimes two specific enzymes for each amino acid. Very high specificity is the most significant feature, because once amino acid is attached to t-RNA, recognition of a specific codon on m-RNA is due entirely to the t-RNA, not to the amino acid. This very high specificity of aminoacyl t-RNA synthetases is due to:

- High selectivity in terms of the acceptance of the amino acid to be activated.
- High selectivity toward the t-RNA to which the activated amino acid will be transferred.

Incorrectly activated amino acids are hydrolysed and do not get incorporated.

Steps of Activation of Amino Acids

- The reaction requires amino acid, t-RNA and ATP. First an intermediate is formed.

$$\text{Amino acid} + \text{ATP} \xrightarrow{Mg^{+2}/Mn^{+2}} \text{(Amino Acyl AMP)} + \text{PPi}$$

Aminoacyl-adenylate complex

- Transfer of aminoacyl group to t-RNA constitutes the next step.

$$\text{Aminoacyl–AMP} + \text{t-RNA} \rightarrow \text{Aminoacyl t-RNA} + \text{AMP}$$

- **Overall reaction is:**

$$\text{Amino acid} + \text{ATP} + \text{t-RNA} \rightarrow \text{Aminoacyl t-RNA} + \text{AMP} + \text{PPi}$$

In the aminoacyl t-RNA, the α-carboxyl group of the amino acid remains esterified with the 3′-OH of the 3′-terminal adenosine on acceptor arm of t-RNA. The hydrolysis of pyrophosphate which renders the reaction virtually ***irreversible,*** drives the reaction to completion. ***Thus 2 high energy bonds of ATP are used in the formation of an aminoacyl t-RNA.***

As each t-RNA forms an aminoacyl t-RNA with specific amino acid, the codon-anticodon reactions bring the amino acid to ribosomes (or polysomes) in a particular sequence depending on the codon sequence in m-RNA.

2. Initiation

The initiation may be divided arbitrarily into following **4 steps:**

A. **Dissociation of the ribosome 80S into 60S and 40S subunits.**
B. **Formation of 43S preinitiation complex**
C. **Formation of initiation complex**
D. **Formation of 80S initiation complex.**

A. *Dissociation of Ribosome*

Before initiation process starts, **80S** ribosome **dissociates** into **40S** and **60S** subunits. Two initiation factors, **elF-3** and **elF-1A** binds to the newly dissociated 40S subunit.

Function

- This binding of initiation factors prevent reassociation of 60S and 40S.
- It allows the other translation initiation factors to associate with 40S subunit and prepares it for formation of 80S initiation complex **(Fig. 17.5)**.

B. *Formation of the 43S Pre-initiation complex*

The process involves the **binding of GTP** with **elF-2**, and forms a binary complex, which then binds to Met-RNA, and forms a ternary complex. Methionine having anticodon UAC is the first aminoacid required to be involved in the binding to the initiation codon AUG on m-RNA.

Formation of N-f-met-t-RNA$_f$

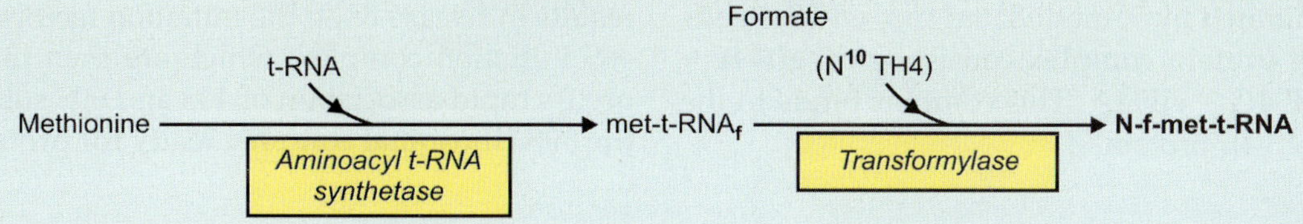

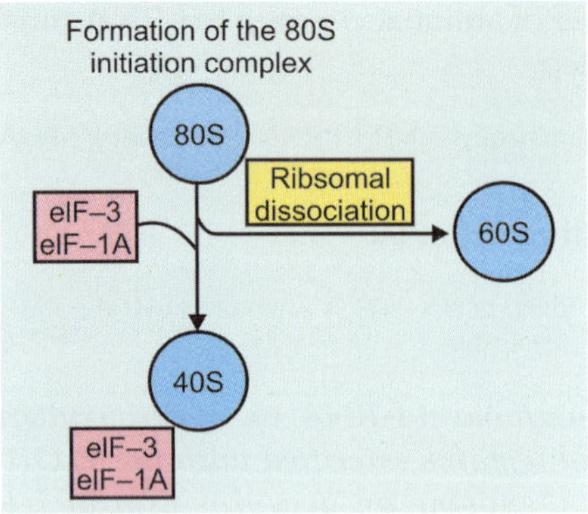

Fig. 17.5: Dissociation of 80S ribosome

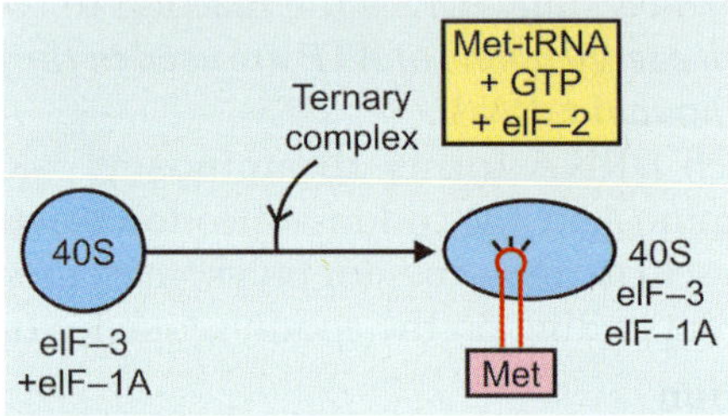

Fig. 17.6: Formation of 43S pre-initiation complex

This ternary complex binds to 40S ribosomal subunit and forms the **43S preinitiation complex**. This complex is stabilised by association with **elF-3** and **elF-1A** (Fig. 17.6).

In eukaryotes, the **elF-2 is the controlling factor in protein synthesis** initiation. Structurally elF-2 is a heterotrimer and consists of 3 subunits α, β, and γ. ***Subunit α is important***, elF-2α is phosphorylated by atleast four different **protein kinases (HCR, PKR, PERK,** and **GCN2)**. The kinases are activated when a cell is under stress, e.g. virus infection, heat shock, carbohydrate and protein deprivation, etc.

CLINICAL IMPORTANCE

PKR protein kinase is important. The kinase is activated by viruses and provides a host defense mechanism that decreases protein synthesis, thereby inhibiting viral replication.

C. *Formation of Initiation Complex (Fig. 17.7)*

Binding of m-RNA with pre-initiation complex is necessary to form the ***Initiation complex.*** In all eukaryotic cells, 5′ terminals of m-RNA are "**capped** which is methyl guanosyl triphosphate (Refer Structure of RNAs). The binding is facilitated by 5′ methylated cap, which needs a **cap binding protein complex** consisting of, **elF-4F** = **elF-4E** and **elF-4G** + **elF-4A**. This complex binds to the cap through elF-4E protein.

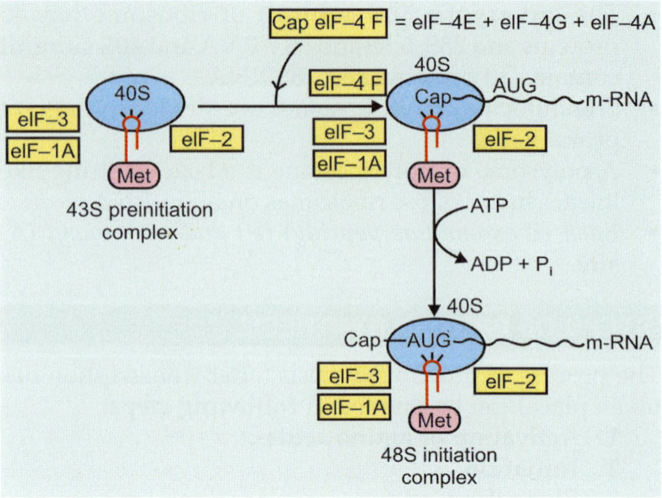

Fig. 17.7: Formation of initiation complex

Then elF-4A and elF-4B bind and reduce the complex secondary structure of the 5′-end of m-RNA through hydrolysis of **ATP** by ***helicase activities*** providing energy. ***The association of m-RNA with the 43S preinitiation complex produces the 48S initiation complex.*** After formation of 48S initiation complex, it searches for precise **initiation codon AUG** for methionine. This is determined by **Kozak consensus sequence** in eukaryotes and in bacteria, it is **Shine-dalgarno sequence**.

elF-4E is the most important factor because it recognises the cap of 5′-methylated end of m-RNA and is the **rate limiting step** in protein synthesis. Insulin and mitogenic factors, e.g. IGF-1, PDGF, inter-leukin-2 and angiotensin II, phosphorylate elF-4E and increases protein synthesis.

Role of 3′-Poly(A) Tail in Initiation

3′-Poly (A) tail has a *binding protein,* "Pablp". This complex helps in the initiation acting synergistically with 'cap'.

Pablp bound to the poly (A) tail interacts with **elF-4G**, which in turn binds to **elF-4E** that is bound to the 'cap' structure, which probably **help to direct the 40S ribosomal subunit** to the 5′-end of the m-RNA.

D. *Formation of 80S Initiation Complex (Fig. 17.8)*

The 48S initiation complex now binds to 60S which was free after ribosomal dissociation, forms **80S initiation complex. This requires hydrolysis of GTP for energy.** The *elF-2 carries GTP and elF-5 GTPase activity*, both interact to bring about hydrolysis. This reaction then results in release of all the initiation factors bound to the 48S initiation complex, which are then recycled. There occurs rapid association of 40S and 60S subunits to **form the 80S ribosomal complex ready for protein synthesis.**

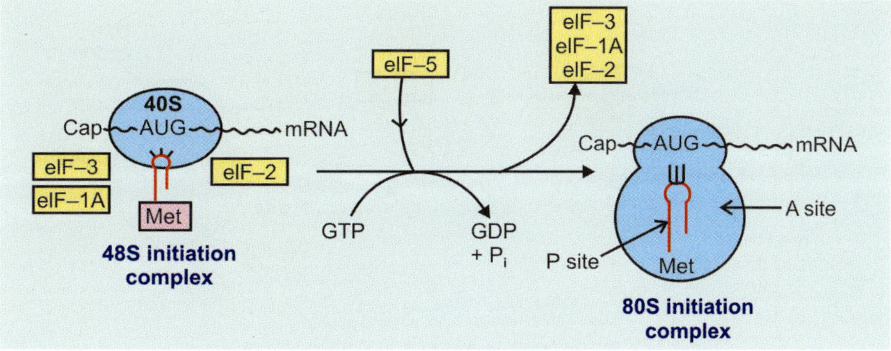

Fig. 17.8: Formation of 80S ribosomal initiation complex

The 80S complex has *two receptor sites*:

- **'P' site or peptidyl site**: *At this point, the met-t-RNA is on the 'P' site*. **On this site, the growing peptide chain will grow.**
- **'A' site** or **aminoacyl site**: *At this point it is free,* the new incoming t-RNA with the amino acid to be added next is taken up, at this site.

The **t-RNA binds with ribosome through the pseudo-uridine arm**.

3. Elongation

- *Elongation is a cyclic process on the ribosome* in which one amino acid is added to the nascent peptide chain.
- The peptide sequence is determined by the **codons** present in the m-RNA.
- It requires *elongation factors*—**EF-IA, EF-2.**

Steps involved in elongation:

The steps are mainly **three:**

A. The binding of new aminoacyl-tRNA to 'A' site
B. Peptide bond formation
C. Translocation process.

A. *Binding of Aminoacyl-tRNA to the A Site* (Fig. 17.9)

In the 80S ribosomal initiation complex, the 'P' site is occupied by met-tRNA and **'A' site is free**. *The fidelity of protein synthesis depends on having correct aminoacyl-tRNA in the 'A' site as per codon reading.*

Elongation factor EF-IA forms a ternary complex with **GTP** and the entering aminoacyl-tRNA (A_1). This complex allows the amino acyl-tRNA to enter the 'A' site. GTP is hydrolysed to give energy and this is catalysed by an active site on the ribosome. This releases the EF-IA-GDP and Pi. The EF-IA-GDP is converted again to EF-IA-GTP by other soluble protein factors and GTP. It is further recycled.

B. *Peptide Bond Formation* (Fig. 17.10)

The α-NH_2 group of the new aminoacyl-tRNA (A_1) in the 'A' site combines with the –COOH group of Met-tRNA (m) occupying the 'P' site. The reaction is catalysed by the enzyme ***Peptidyl transferase***, a component of the 28S RNA of the 60S ribosomal subunit. ***Because the amino acid on the aminoacyl-tRNA is already "activated", the reaction does not require any further energy***. The reaction results in formation of a peptide to the t-RNA in the 'A' site. ***Now the growing peptide chain is occupying 'A' site and naked free t-RNA at 'P' site.***

Note: Peptidyl transferase is an example of a ribozyme where RNA acts as the enzyme (direct role of RNA in protein synthesis).

C. *Translocation*

The t-RNA is fixed at the 'P' site, attached by its anticodon and having no amino acid (free and naked t-RNA) ***by the open CCA tail*** it is bound to an exit site (E site) on the large ribosomal subunit.

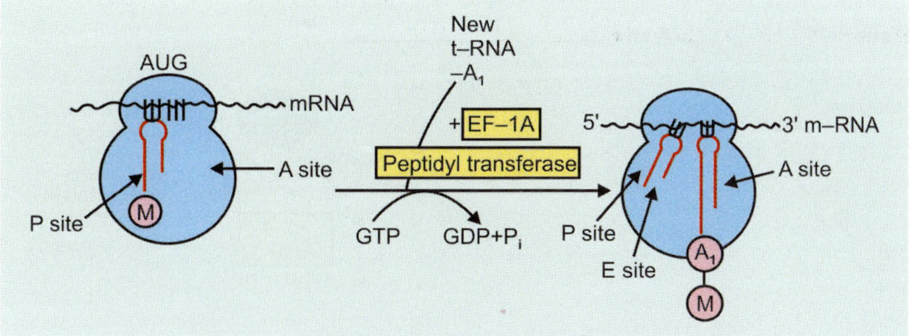

Fig. 17.9: Binding of aminoacyl-tRNA to 'A' site and peptide bond formation

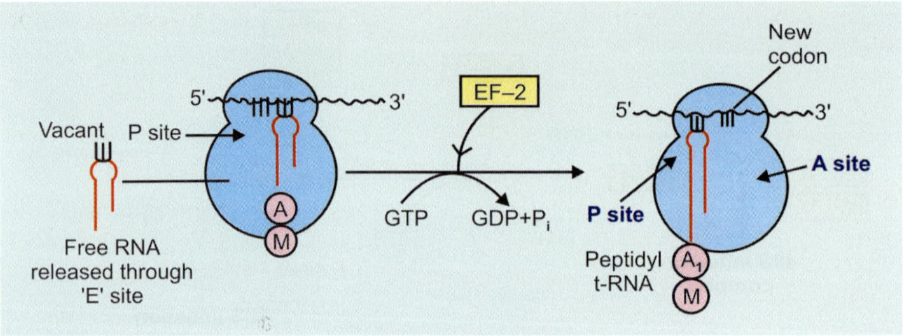

Fig. 17.10: Elongation

At this point *elongation factor 2 (EF-2)* binds to and displaces the peptide-tRNA from the 'A' site to the 'P' site and at the same time the deacylated free *t-RNA* is on the 'E-site' from which the ***t-RNA leaves the ribosome***.

Now, **EF-2-GTP complex** is hydrolysed to **EF-2-GDP**, the energy from hydrolysis ***moves the m-RNA forward by one codon***, leaving the ***'A' site free*** to receive another ternary complex of a new amino acid as per codon and ***repeat the cycle of elongation***.

4. *Termination Process (Fig. 17.11)*

After multiple cycles of elongation process, it results to formation of polypeptide chain. When the desired protein molecule is synthesised, a **stop codon** or **terminating codon** appears in the **A site** of m-RNA. The stop codons are**: UAA, UAG**, or **UGA**. There is no t-RNA with an anticodon capable of recognising such a termination signal.

Releasing factor RF-1 ***recognises that a stop codon has come in the 'A' site***. This ***protein factor RF-1 is a complex consisting of another releasing factor RF-3 with bound GTP***. This complex with the help of *"peptidyl transferase"* brings about hydrolysis of the bond between the peptide and the t-RNA occupying the 'P' site. The energy is provided by GTP → GDP + Pi conversion. ***The hydrolysis releases the synthesised peptide chain, m-RNA and t-RNA from the 'P' site***.

The 80S ribosome now dissociates into 60S and 40S subunits which are then recycled.

Polyribosomes

In eukaryotic cell, a single ribosome is capable of synthesising 400 peptide bonds each minute. Many ribosomes can work on the same m-RNA molecule simultaneously and these aggregate are called **polyribosomes** or **polysomes.** In such cases, each ribosome may be 80 to 100 nucleotides apart on the m-RNA (minimum 35 nucleotides).

Polyribosomes actively synthesising proteins can exist as:

- ***Free cellular particles in cytoplasm*** or
- ***May be attached to the endoplasmic reticulum (ER)***.

Attachment of the particular polyribosomes to the ER makes the:

- **Rough** appearance viewed by electron microscopy and called **Rough endoplasmic reticulum.**
- **Free ribosomes in cytoplasm synthesises cytoplasmic proteins** required for cellular function.

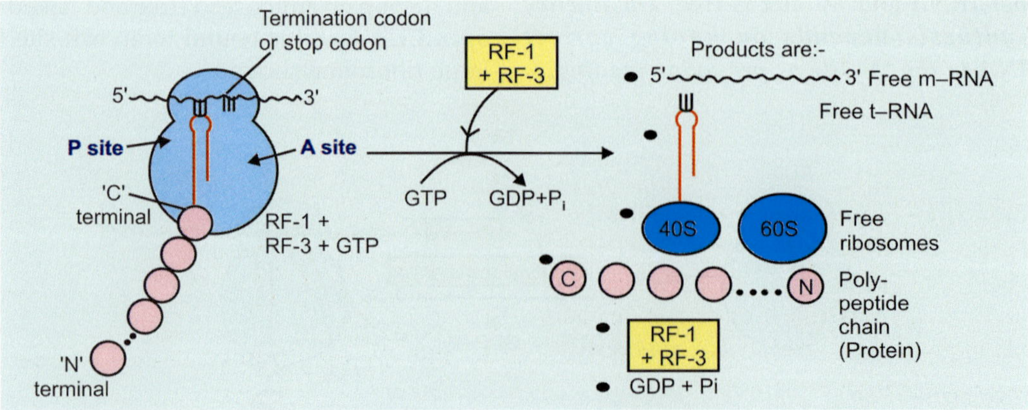

Fig. 17.11: Termination process

- **Bound polyribosomes** of ER synthesise proteins that are transported through cisternal space to Golgi apparatus, where it is packaged and **stored** for eventual export.

Energy requirements in Protein Synthesis

Mainly
- ATP and
- GTP are used

- **ATP:** It is required for the activation of amino acids and formation of t-RNA-amino acid complex. In this reaction, as one molecule of ATP is converted to AMP, it can be considered equivalent to utilisation of 2 ATPs (two high energy PO_4 bonds used).
- One ATP is required for formation of initiation complex.
- One ATP required in formation of 48S initiation complex. The hydrolysis of ATP by helicase activity for associating 5′-end of m-RNA, with binding of eIF-4A and eIF-4B.
- **GTP:** It is required for the following:
  - GTP is required for ***binding with eIF-2*** for forming a binary complex required in formation of 43S preinitiation complex.
  - GTP hydrolysis provides energy ***for formation of 80S initiation complex***. eIF-2 carries GTP and eIF-5 GTPase activity.
  - GTP is required in binding of amino-acyl-tRNA to 'A' site.
  - EF-2 (elongation factor)—GTP complex is hydrolysed to give energy—***for translocation***, movement of m-RNA forward by one codon.
  - RF-3 with bound GTP is required for ***termination process***.

Note: Total energy requirement:
- **ATP = 4**
- **GTP = 5.**

CHAPERONES : PROTEINS THAT PREVENT FAULTY FOLDING

The exit of a protein after synthesis from the endoplasmic reticulum may be the ***rate-limiting step. Certain proteins called chaperones have been found to play a role in the assembly and proper folding of the synthesised proteins so that it has biological activity***. The word chaperones as per dictionary literally means "older woman in charge of young unmarried woman on certain social occasions".

Heat shock proteins (HSPs): The chaperones belong to a large family of proteins called ***heat shock proteins*** **(HSPs)** which were first identified in response to heat shock. Any stress to the cell like radiations, heavy metals, free radicals, toxins, etc, would cause increased production of HSPs and hence also called as ***stress proteins.***

Mechanism of Action

- Most chaperones exhibit ATPase activity and bind ADP and ATP. This activity is important in their effect on folding of proteins.
- The ADP-chaperone complex has a high affinity for the unfolded protein which when bound, stimulates release of ADP with replacement of ATP.
- The ATP-chaperone complex in turn, releases segments of the protein that have folded properly, and the cycle involving ADP and ATP binding is repeated until the folded protein is released.

Properties of chaperone proteins: Some properties of chaperone proteins are listed in the box below.

Properties of Chaperones

- Present in a wide range of species from bacteria to humans.
- Also called as heat shock proteins (HSPs) as stated above.
- Inducible by conditions that cause unfolding of newly synthesised proteins, e.g. elevated temperature, exposure to various chemicals (stress proteins).
- They bind unfolded and aggregated proteins.
- Show ATPase activity and bind ADP and ATP which has important role in effecting folding (see mechanism of action above).
- Found in various cellular compartments, e.g. cytosol, mitochondria and the lumen of endoplasmic reticulum.

Examples of Chaperones

- **Bip *(immunoglobulin heavy chain binding protein):*** It is located in the lumen of the endoplasmic reticulum.
  Function: Chaperone Bip binds to abnormally folded immunoglobulin heavy chains and prevent them from the endoplasmic reticulum in which they are degraded.
- **Calnexin:** It is located in the endoplasmic reticulum membrane.
  Function: It binds the monoglucosylated species of glycoproteins that occur during processing of glycoproteins, retaining them in the endoplasmic reticulum until the glycoprotein has folded properly.
- **PDI *(Protein disulfide isomerase):*** An enzyme chaperone which promotes rapid reshuffling of disulphide (-S-S-) bonds until the correct folding is achieved.
- **PPI *(Peptidyl prolyl cis-trans-isomerase):*** Another enzyme chaperone which accelerates folding of proline containing proteins.

CLINICAL ASPECT

Abnormal folding of proteins, unassisted by chaperones may lead to Alzheimer's disease.

Alzheimer's Disease

Alzheimer's disease occurs in old age and it is the fourth leading cause of death among aged people after myocardial infarction, stroke and cancer.

Clinical features: The disease is characterised by slow progression of memory loss, hallucinations, personality changes, confusion, dementia and finally the patient enters into a vegetative state with no comprehension of outside world.

Pathological changes: The disease is characterised pathologically:

- ***By neurofibril tangles in CNS,***
- ***Formation of senile neuritic plaques and cerebral amyloid deposits.***

Biochemical alterations: The biochemical changes include:

- The neurofibrillary tangles are paired helical filaments made up of abnormal **Tau** proteins. ***Normal* Tau *is soluble and catabolised easily but abnormal* Tau *protein found in this condition is insoluble and cannot be degraded by tissue cathepsins and are deposited around neurons.***
- ***Role of Chaperone Enzyme***
  Pin I: Recently an enzyme ***Pin I*** has been found which is necessary for formation of normal ***Tau*** protein. Absence of ***Pin I*** in this condition produces the abnormal *Tau* protein.
- Deposition of abnormal insoluble "Tau" leads to loss of microtubule, thus damaging the communication channels in nerve fibres. **The synthesis of acetylcholine is reduced leading to memory loss.**
- ***Neuritic plaques have been found to be due to deposition of an abnormal amyloid protein* β *APP.*** This is derived from normal APP (Amyloid Precursor Protein). Mutation in APP gene leads to formation of β-APP from wrong cleavage of APP. **β-APP *is precipitated around neurons as β-amyloid*.**

INHIBITORS OF PROTEIN SYNTHESIS

Many antibiotics inhibit the protein synthesis at some specific steps:

- ***Streptomycin:*** It is a highly basic trisaccharide.
  - It interferes with the binding of f-met t-RNA to ribosomes and thereby inhibits the initiation process.
  - It also leads to misreading of m-RNA.
- ***Puromycin:*** This inhibits protein synthesis by releasing nascent polypeptide chains before their synthesis is complete. ***It binds to the A site on ribosome and inhibits the entry of aminoacyl-t RNA***. It acts both in bacterial and mammalian cells.
- ***Tetracycline:*** It binds to the 30S subunit and inhibits binding of aminoacyl t-RNA, thus inhibits the initiation process.
- ***Chloramphenicol:*** It inhibits the *peptidyl transferase* activity of 50S subunit. Thus it inhibits the process of elongation.
- ***Cycloheximide:*** This inhibits *peptidyl transferase* activity of 60S ribosomal subunit in eukaryotes. It also inhibits elongation.
- ***Erythromycin:*** It binds to the 50S subunit and inhibits translocation.
- ***Diphtheria toxin:*** *Corynebacterium diphtheriae* produces a lethal protein toxin. ***It binds with EF-2 in eukaryotes and blocks its capacity to carry out translocation***.
- ***Ricin and abrin:*** These are toxic lectins. They inhibit protein synthesis by unknown mechanisms, preventing aminoacyl t-RNA from binding to the A site. This operates in eukaryotes, inactivates eukaryotic 28S ribosomal RNA.
- ***Sparsomycin:*** This inhibits *peptidyl transferase* and release factor-dependent termination. It does not prevent termination codon-dependent binding of release factor to the ribosome.
- ***α-Sarcin:*** It is a toxic RNAse that prevents aminoacyl-t-RNA binding by cleaving a single phosphodiester bond in 28S r-RNA.

PART II
GENE EXPRESSION AND REGULATION

Organisms adapt to environmental changes by altering gene expression. The regulation of the expression of genes is necessary for the growth, development, differentiation and the very existence of the organism.

The process of alteration of gene expression has been studied in detail in prokaryotes. It generally involves the interaction of specific binding proteins with various regions of DNA in the immediate vicinity of the transcription site and this produces either a positive or negative effect on transcription.

In eukaryotes, studies are not extensive and not well understood. Cells in eukaryotes utilise the same basic principle but uses other mechanisms to regulate transcription.

Types of Gene Expression

There are **mainly two types** of gene expression and regulation:

a. **Positive regulation, and**
b. **Negative regulation.**

a. ***Positive regulation:*** When the expression of genetic information is quantitatively ***increased*** by the presence of specific regulatory element, it is called as positive regulation. The element or molecule mediating positive regulation is called **positive regulator.**

b. ***Negative regulation:*** When the expression of genetic information is ***decreased*** by the presence of a specific regulatory element, it is called as negative regulation. The element or molecule mediating the negative regulation is called a **negative regulator.**

Note: A ***double negative*** has the effect of acting as a positive. An effector that inhibits the function of a negative regulatory appears to bring about a positive regulation.

One Cistron-One Subunit Concept

Earlier hypothesis proposed that one gene produces one enzyme or protein and "one gene-one enzyme" concept was introduced.

It is now known that some enzymes and protein molecules are composed of two or more nonidentical subunits, which cannot be explained by "one gene-one enzyme" theory and so it is not valid.

The* cistron *is now considered as the genetic unit coding for the structure of the subunit of an enzyme or protein molecule, acting as it does as the smallest unit of genetic expression. Hence, the "one gene-one enzyme" idea might be more accurately regarded as "one cistron-one subunit concept".

Gene Expression in Prokaryotes

Operon

The concept of operon was introduced by **Jacob** and **Monod** in 1961.

Operon is defined as a segment of a DNA strand consisting of:

- ***Structural genes:*** A cluster of several structural genes, which carries the codons which can be translated into protein.
- ***Operator gene:*** One operator gene which has an overall control over the process of translation.
- ***Regulator gene:*** A third gene called regulator gene is located sometimes at a distance from the operator gene on the same DNA strand. Regulator gene transcribes m-RNA which synthesises ***repressor protein*** molecules which regulate the transcription.
- ***P site (promoter site):*** It is situated between operator gene and regulator gene.

Based on data from genetic studies, **Jacob** and **Monod**, proposed the "operon" model.

Type of genes: It is based on the functions carried out, they proposed **two type of genes**:

1. ***Inducible genes:*** The expression of the inducible gene increases in response to an inducer. Inducers are small molecules.
   Some proteins produced by *E. coli*, e.g. β-galactosidase are said to be inducible because they are only produced in significant amounts when a specific inducer "lactose" is present.
2. ***Constitutive genes:*** The constitutive genes are expressed at more or less *constant rate* in almost all the cells and they are not subjected to regulation. ***The products of these genes are required all the time in a cell.***

"Lac-operon" Structure (Fig. 17.12)

The "lac operon" is an inducible catabolic operon of *E. coli*. It consists of:

- ***Structural genes:*** It carries ***three structural genes: 'Z', 'Y'*** and ***'A'***. They code respectively for ***β-galactosidase, galactoside permease*** and ***thiogalactoside transacetylase***.

SECTION THREE

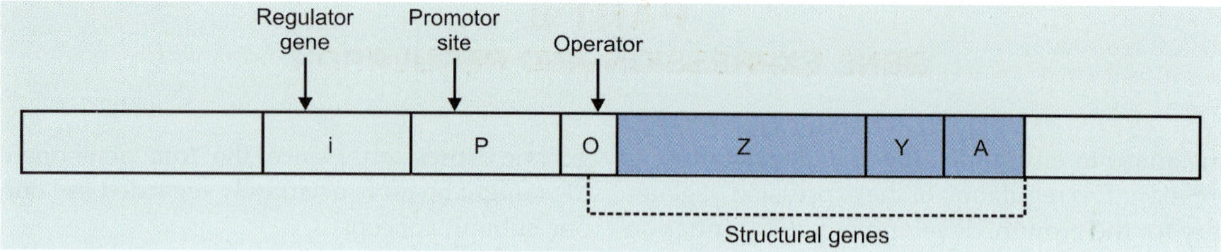

Fig. 17.12: Structure of 'Lac' operon

Functions of these Enzymes

- ***β-galactosidase*** hydrolyses lactose (β-galactoside) to galactose and glucose.
- ***Permease*** is responsible for the transport of lactose into the cell.
- The function of ***acetylase*** coded by 'A' gene is not known properly.

The structural genes Z, Y and A transcribe to form a single large m-RNA with three independent translation units for the synthesis of the three distinct enzymes. Such an m-RNA coding for more than one protein is called ***poly-cistronic m-RNA*** which is a characteristic in prokaryotes.

- ***Operator gene or "O" gene:*** It is located immediately on the 3' or upstream side of the 'Z' gene.
- ***A regulator or "i" gene:*** It is located on the same DNA strand on the upstream side of the 'O' gene and transcribes an m-RNA which codes for ***repressor proteins***, a tetramer.
- ***Promoter site (P site):*** In addition to the above three types of genes, there is a promoter site (P) situated next of the operator gene where the enzyme ***RNA polymerase*** binds and it makes the structural genes Z, Y and A to transcript.

Repression and Derepression of 'Lac' Operon

a. Repression (Fig. 17.13)

- ***Regulator gene (i gene) is a constitutive gene and is independent of inducer.*** It is expressed at a constant rate leading to synthesis of subunits of the 'Lac' repressors. These subunits assemble into a 'Lac' repressor molecule, a protein which is tetramer. Each monomer has a molecular weight of 40,000 (total mol. wt. approx. 160,000).
- This repressor protein molecule has a high affinity for "operator" gene (O). It specifically binds to the operator gene (O).
- ***This binding of* repressors *with 'O' gene prevents the binding of* RNA polymerase *to the promoter site (P) thereby preventing the transcription of structural genes Z, Y and A which are inducible enzymes.***
- This is repression of 'Lac' operon which ***occurs in E. coli in absence of lactose and transcription of structural genes do not occur.*** The repressor molecules act as negative regulator of gene expression.

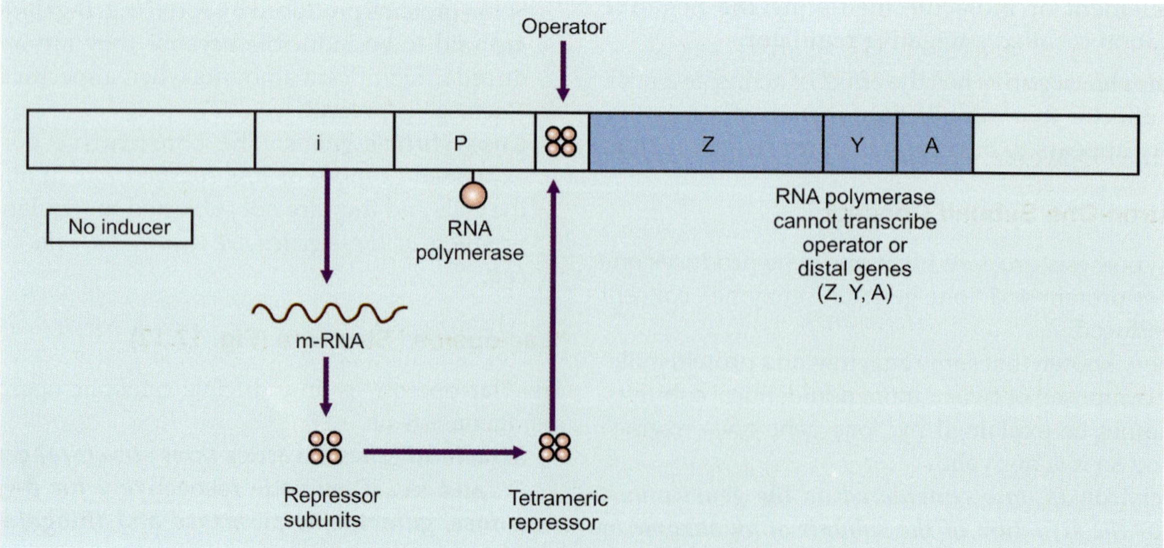

Fig. 17.13: Repression of 'Lac' operon

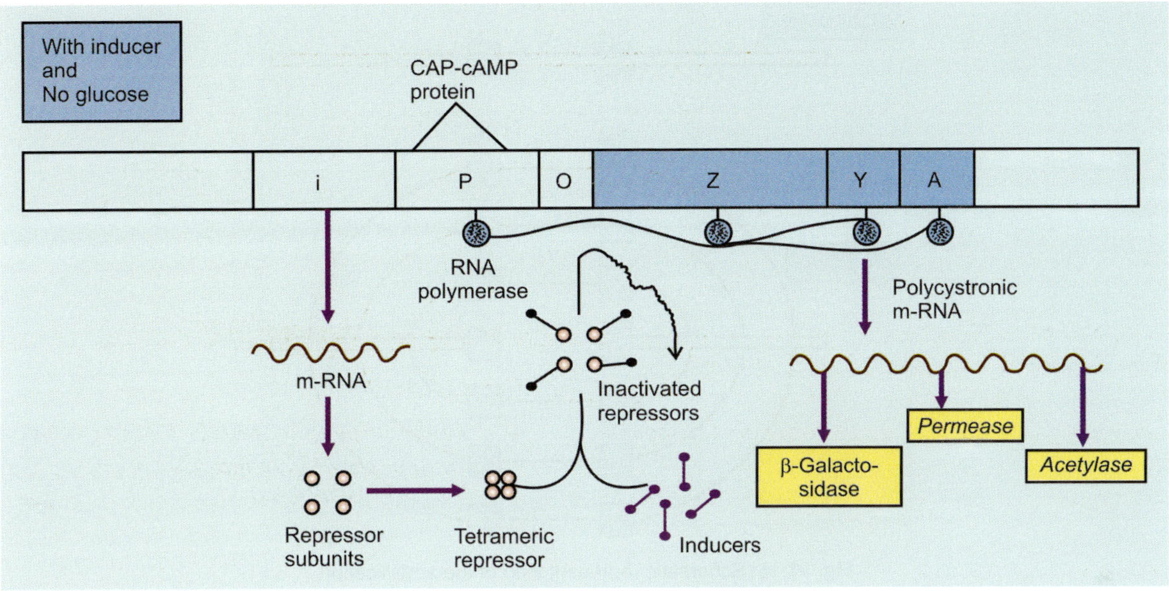

Fig. 17.14: Derepression of 'Lac' operon

b. Derepression (Fig. 17.14)

- The addition of lactose or a gratuitious inducer (see below) results in prompt derepression and induction of 'Lac' operon.
- The inducer molecules react with the repressor molecules attached to the operator locus as well as in cytoplasm, and brings about inactivation of repressor molecules by changing the conformation of the molecules.
- The DNA-dependent *RNA polymerase* attaches to the DNA at the promoter site (P) and permeates to operator locus which is now free and transcription of structural genes takes place leading to formation of polycystronic m-RNA and finally the three enzymes.

Lactose acts as inducer, ***it inactivates the repressor molecules,*** hence the process is called as derepression of 'Lac" operon.

Gratuitious inducers: Certain substances like isopropyl thiogalactoside (IPTG), a structural analogue of lactose, can induce the 'Lac' operon by inactivating repressor molecules. IPTG differs from Lactose in that it cannot be hydrolysed by the enzyme β-galactosidase. Such lactose analogues are called ***gratuitious inducers***.

Role of Catabolite Gene Activator Protein (CAP)

- Catabolite gene activator protein is a dimer and acts as a positive regulator of many catabolic operons like the "Lac operon" in *E. coli*.
- ***Attachment of* RNA polymerase *to the promoter site requires the presence of CAP bound to cAMP.***
- Absence of glucose in the cell activates *adenylate cyclase* which catalyses the synthesis of cAMP from ATP. The latter binds to CAP to form a ***CAP-cAMP complex***.
- Unlike free CAP, this complex binds to promoter site immediately. This stimulates the initiation of transcription of the "Lac operon" structural genes in the absence of repressors.
- ***The binding of* CAP-cAMP complex *to promoter enhances* RNA polymerase *activity.***
- On the other hand, presence of glucose lowers the intracellular cAMP. Due to decreased levels of cAMP, the formation of "CAP-cAMP complex" is lowered. Hence the binding of *RNA polymerase* to P-site is negligible and the transcription does not occur.

Thus, glucose interferes with the expression of* Lac operon *by depleting cAMP level.

Gene Expression in Eukaryotes

Gene expression in the eukaryotes, cannot be the same as that in the prokaryotes because of the following:

- Genome of the eukaryotes is much larger than that of the prokaryotes.
- Human genome consists of 23 pairs of diploid chromosomes containing 10^9 base pairs of DNA with an estimated one lakh genes approximately.
- There is a great deal of redundancy in the DNA of eukaryotes.
- Proteins like the repressor proteins influence gene expression at the DNA level as they are synthesised outside the nucleus and cannot get into the nucleus through the nuclear membrane.

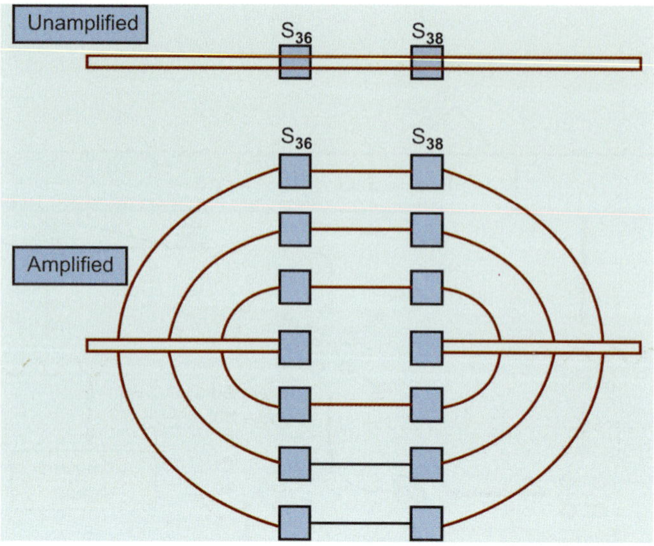

Fig. 17.15: Schematic representation of the amplification of chorion protein genes S_{36} and S_{38}

Various Methods of Regulation of Gene Expression in Eukaryotes

In addition to transcription, eukaryotic cells employ a variety of mechanisms to regulate gene expression. Some of the methods are discussed below:

1. *RNA processing:* RNA processing steps in eukaryotes include:
   - "Capping" of the 5'-end of the primary transcript
   - Addition of a polyadenylate "tail" to the 3'-end of transcription
   - And excision of "intron" regions to generate spliced "exons" in the mature m-RNA molecule.
2. *Gene amplification:* In this mechanism, the expression of a gene is increased several-fold. This is commonly observed during the developmental stages of eukaryotic organisms.

   Examples

   i. In the fruitfly, Drosophila, there occurs during oogenesis an amplification of a few pre-existing genes such as those for the chorion egg shell proteins S_{36} and S_{38} (Fig. 17.15).

   ii. *Methotrexate therapy in cancer:* It has been demonstrated in patients receiving methotrexate for treatment of cancer that malignant cells can develop **drug resistance** by increasing the number of genes for the enzyme ***dihydrofolate reductase***.
3. *Gene rearrangement:* Gene rearrangement can be studied from the mechanism of the synthesis of light chains of immunoglobulins (Igs).
   - ***Each light chain is encoded by three distinct segments:***

     i. ***The variable V_L***

     ii. ***The joining J_L, and***

     iii. ***The constant C_L segments.***

   The mammalian haploid genome contains over 500 V_L segments, 5 or 6 J_L segments and perhaps 10 or 20 C_L segments.
   - During the differentiation of a lymphoid B-cell, a V_L segment is brought from a distant site on the same chromosome to a position in the same chromosome close to the region of the genome containing the J_L and C_L segments.
   - This DNA rearrangement then allows the V_L, J_L and C_L segments to be transcribed as a single m-RNA precursor and later the m-RNA proper to produce a specific antibody light chain.

     The DNA rearrangement is referred to as the VJ joining of the light chain (Fig. 17.16).
4. *Gene regulation by histones and nonhistone proteins:* It has been suggested that the histones of nucleosomes can also regulate gene expression. Though the molecular species of histones is too small in number to regulate 1.25 to 3.0 million genes in an eukaryotic genome, it has been found that during S-phase of cell cycle there are many post-translational modifications of the different histones. Such modified histones can regulate gene expression.
5. *Class switching:* In this process, one gene is switched off and a closely related gene takes up the function.

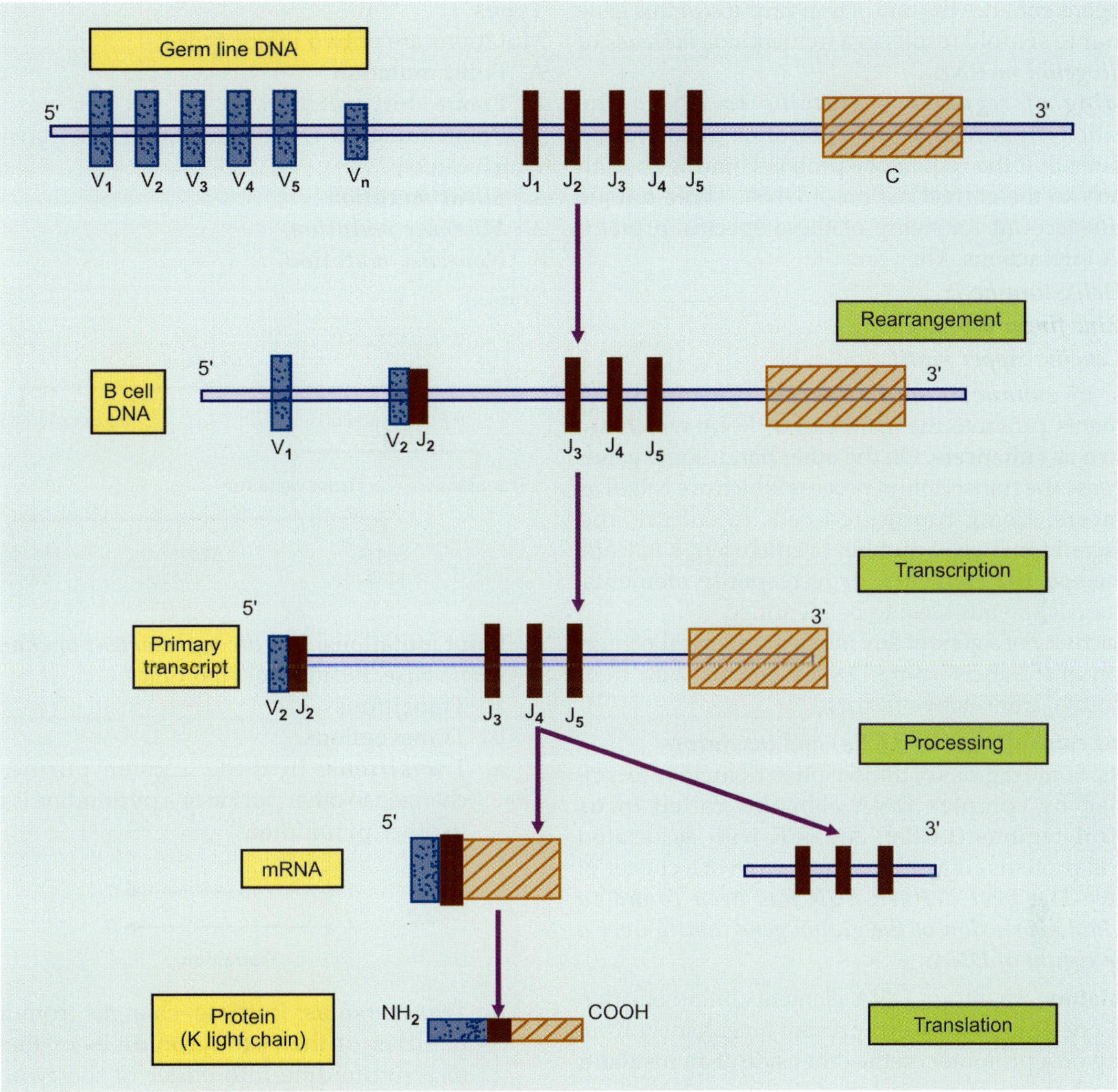

Fig. 17.16: Gene rearrangement. Recombination events leading to a V_2 J_2 K light chain

Examples: Class switching is best illustrated by Hb synthesis and immunoglobulin synthesis.

a. In intrauterine life, first embryonic Hb is produced having two 'zeta' and two 'eta' chains. By the sixth month of intrauterine life, embryonic Hb is replaced by Hb-F consisting of α_2 γ_2 chains. After birth Hb-F is replaced by adult types of HbA, 97 per cent HbA, ($\alpha_2\beta_2$) and 3 per cent Hb A_2 (α_2 δ_2). Thus the gene expression is shifted from:

Zeta-eta $\rightarrow \alpha\,\gamma \rightarrow$ *then* $\alpha\,\beta$ *and* $\alpha\,\delta$

(Refer chapter on Chemistry of Haemoglobin and Haemoglobinopathies).

b. Gene switching is also observed in the formation of immunoglobulins. The primary immune response is the formation of IgM immunoglobulins. This is followed by secondary immune response in formation of IgG immunoglobulins.

6. ***Stability of m-RNA:*** The stability of m-RNA molecules in the cytoplasm can clearly affect the level of gene expression in a positive or negative direction.

It is a potential control site that can be influenced by hormones and other effectors. Some hormones influence the synthesis and degradation of specific m-RNAs. For example, estradiol prolongs the half-life of vitellogenin m-RNA from a few hours to more than 200 hours. This coupled with the fact that estrogens enhance the rate of transcription of this gene by four to six-fold results in a tremendous *increase of vitellogenin m-RNA.*

7. ***Binding of regulatory proteins to DNA:*** The specificity involved in the control of transcription requires that the regulatory proteins bind with high affinity to the correct region of DNA. Three unique motifs-account for many of these specific protein-DNA interactions. They are:
   - *Helix-turn helix*
   - *Zinc finger motif, and*
   - *Leucine zipper motif.*
8. ***Role of enhancers and silencers:*** Certain DNA segments promote the synthesis of RNA which are known as **enhancers**. On the other hand, some genes decrease the transcription process which are called as **silencers**. Using transfected cells in culture and transgenic animals, a number of enhancers, silencers, tissue-specific elements, drug-response elements, hormones/metals have to be identified.
   The activity of a gene at any moment reflects the interaction of these cis-acting DNA elements with their respective trans-active factors.
9. ***Locus control regions (LCRs) and insulators***
   LCRs: Some regions of transcription domains are controlled by complex DNA elements called **locus control regions (LCRs)**. An LCR with associated bound proteins controls the expression of a cluster of genes. The ***best defined LCR has been found to regulate expression of the globin gene family over a large region of DNA.***

   Insulators: These are DNA elements, in association with one or more proteins prevent an enhancer from acting on a promoter on the other side of an insulator in another transcription domain.

MUTATION

DNA is highly stable with regard to its base composition. Mutations result when changes occur in the nucleotide sequence. Although the initial change may not occur in the coding strand of the double-stranded DNA molecule of that gene; but after replication, daughter DNA molecules with mutations in the coding strand would segregate and appear in the population of the organisms.

Definition: Mutation is replacement of nitrogen base with another in one or both the strands or addition or deletion of a base pair in a DNA molecule.

Mutagens: The substances which can induce mutations are collectively known as mutagens. These can be chemicals, radiations or viruses.

Types

Mutations are of **two major types:**

A. Point mutation

B. Frame shift mutation.

Point mutation can be transitions or transversions which can be:

1. ***Silent mutation***
2. ***Missense mutation***
3. ***Nonsense mutation.***

Thus—

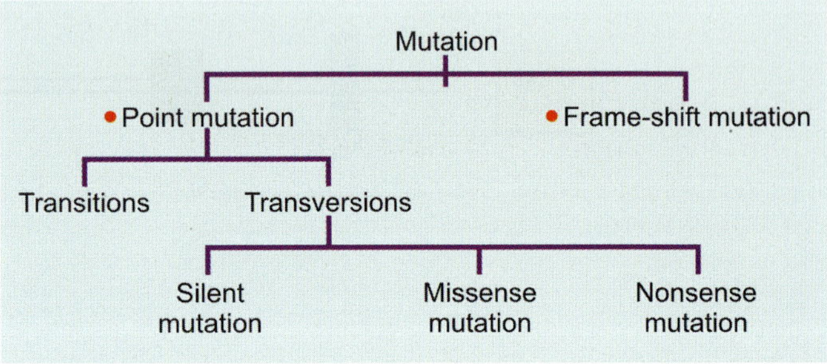

A. Point mutations: It is ***the replacement or changes in a single base***. It can be of ***two types:***

a. **Transitions**

b. **Transversions.**

a. ***Transitions:*** In these a given purine base is changed to other purine or a pyrimidine is changed to other pyrimidine.

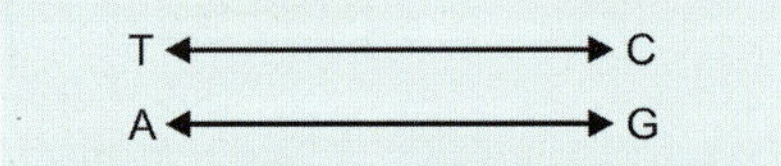

Transition

b. ***Transversions:*** In these, changes from a purine to either of the two pyrimidines or the change of a pyrimidine into either of the two purines takes place.

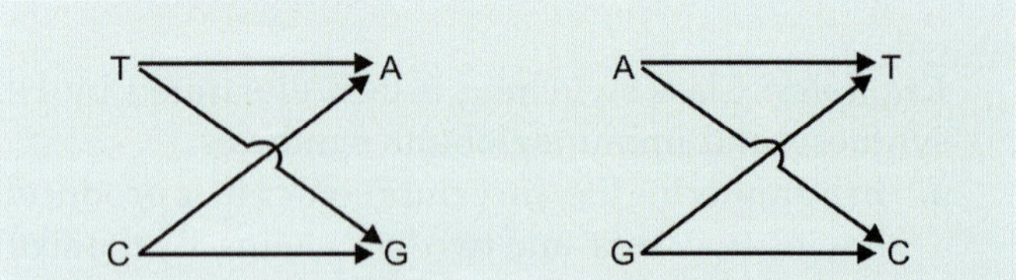

Transversions

Consequences: Single base changes in the m-RNA molecules may have one of several effects as stated below when translated into proteins. These changes may be:

1. ***Silent mutation (No detectable effect).***
2. ***Missense mutation (Missense effect).***
3. ***Nonsense mutation (Nonsense effect).***

1. ***Silent mutation:*** There may be no detectable effect due to the degeneracy of the code. This is likely to occur if

the changed base in the m-RNA molecule occur at the third nucleotide of a codon. ***Due to Wobble phenomenon, the translation of a codon is least sensitive to a change in the third position.***

Examples

- **Hb-A:** If valine at B_{67} is replaced by alanine, no change occurs in the function of Hb (Hb sydney).
- **Leucine** is coded by CUU, CUC, CUA and CUG. Thus a mutation affecting the third nucleotide may have no effect on translated protein.
- **UCU is codon for amino acid serine**, if third U is changed to A, UCA still codes for serine. Hence there will be no detectable effect on such silent mutation.

2. *Missense mutation (missense effect):* A missense effect occurs when a different amino acid is incorporated at the corresponding site in the protein molecule. This mistaken amino acid or missense, depending upon its location in the specific protein molecule might be acceptable, partially acceptable or unacceptable to functioning of that protein molecule.

Thus, missense mutation can have:

i. ***Acceptable missense effect***
ii. ***A partially acceptable missense effect or***
iii. ***An unacceptable missense effect.***

The above three effects can be illustrated in terms of Hb molecule as follows:

i. Acceptable missense	
HbA β chain 61^{Lys} ↓ Hb (Hikari) β chain 61^{Aspn}	*Codons affected* AAA or AAG ↓ Either AAU or AAC (by transversion)

The replacement of a specific lysine at 61 position in β-chain by Asparagine does not alter the normal function of Hb in these individuals, hence it is acceptable missense mutation.

ii. Partially acceptable missense mutations	
Best example for this type is HbS	
HbA-β chain 6^{Gla} ↓ Hb-S β chain 6^{Val}	*Codons involved* GAA or GAG (Glutamic acid) ↓ Changed to GUA or GUG (Valine)

This missense mutation producing HbS interferes with normal function of Hb and results in sickle cell anaemia in homozygous.

But it is classified as partially acceptable because HbS can bind and release O_2 although abnormally.

iii. Unacceptable missense mutation	
Example: HbA, α chain 58^{His} ↓ HbM (Boston) α chain 58^{Tyr}	*Codons involved* CAU or CAC ↓ UAU or UAC

An unacceptable missense mutation in a Hb molecule produces a nonfunctional Hb.

3. *A nonsense Mutation or Effect*

Sometimes the codons with the altered base may become one of the three termination codon called as **"nonsense codon"**. This altered codon acts as a stop signal and causes termination of the protein synthesis at that point, results in a nonfunctioning protein molecule.

B. Frame Shift Mutations

Frame shift mutations can be of **two types**:

1. **Deletion type**
2. **Insertion type**

1. *Deletion type:* Effects of deletion

i. The deletion of a single nucleotide from the coding strand of a gene results in an altered reading frame in the m-RNA (hence called as frame-shift-mutation). The machinery translating the m-RNA does not recognise that a base is missing since there is no punctuation in the reading of codons. Thus a major alteration in the sequence of the amino acids in the protein molecules occur. Such an alteration in the reading frame results in a ***garbled*** translation of the m-RNA distal to the single nucleotide deletion (Fig. 17.17).

ii. Not only the sequence of amino acids distal to the deletion is garbled, there may appear a ***nonsense*** (chain terminating) codon on the way terminating the protein synthesis. ***Thus in such a situation, the polypeptide chain produced is not only garbled but prematurely terminated nonfunctional protein is produced.***

2. *Insertions Type: Effects*

Insertions of one or two nucleotides into a gene results in a m-RNA in which the reading frame is distorted and same type of effects as noted with deletion can occur which ***may result in garbled amino acid sequences*** distal to the insertion or generation of a ***nonsense*** (chain terminating) codon at or distal to insertion ***can lead to termination of the polypeptide chain,*** which may be nonfunctional prematurely terminated protein (Fig. 17.18).

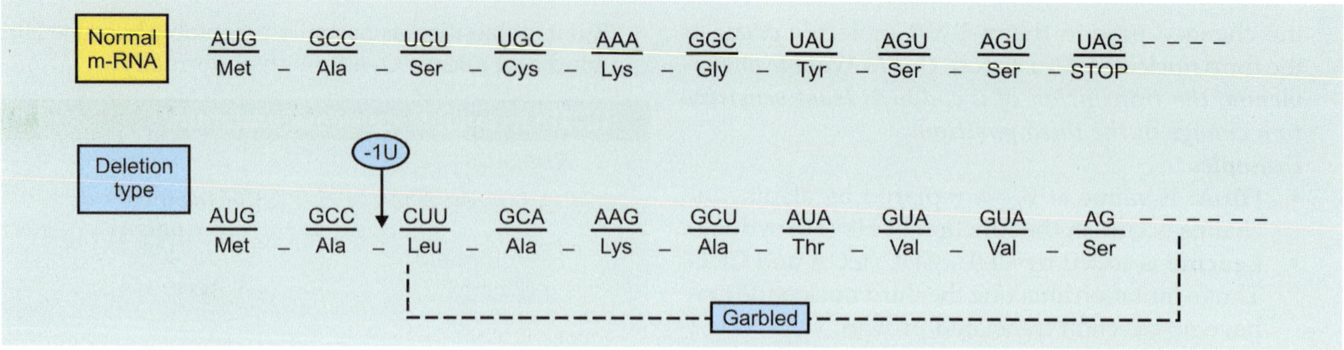

Fig. 17.17: Frame shift mutation—deletion type

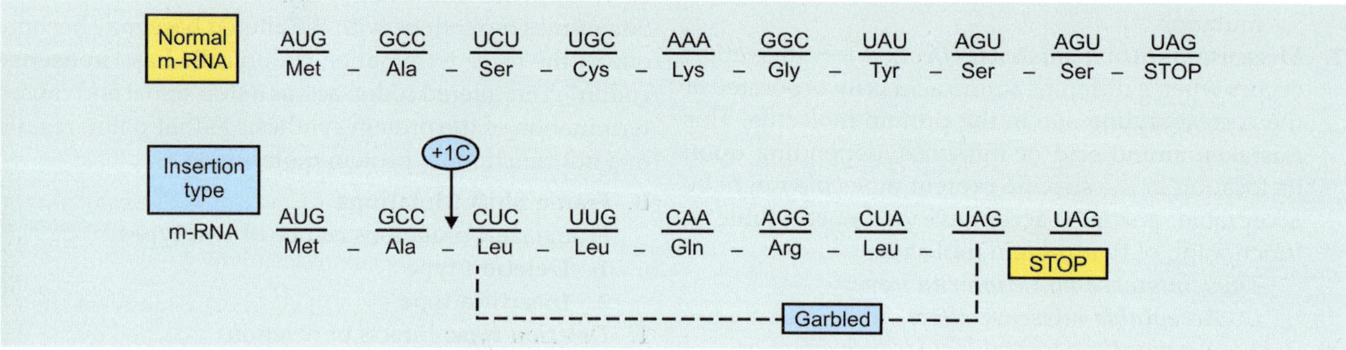

Fig. 17.18: Frame shift mutation—insertion type

Role of Suppressor t-RNA Molecules in Mutations

In prokaryote and lower eukaryote organisms, abnormally functioning t-RNA molecules have been found. Such abnormal t-RNA molecules are formed as the result of alterations in their "anticodon" regions.

Some of these abnormal t-RNA molecules have been found to suppress the effects of mutations in distant structural genes.

Mutation and Cancer

The mutations may be nonlethal, but it may alter the regulatory controls. Such a mutation in somatic cell altering the regulatory control may result to uncontrolled cell divisions leading to cancer.

As some of the mutations may lead to cancer, any substance causing increased mutation can also increase the probability of cancer. Thus all mutagens may be considered as carcinogenic (Refer chapter on Biochemistry of Cancer).

PART III
RECOMBINANT DNA TECHNOLOGY

Recombinant DNA technology is genetic engineering which affects artificial modification of the genetic constitution of a living cell by introduction of foreign DNA through experimental techniques. The technique involves the splicing of DNA by restriction endonucleases, preparation of chimeric DNA molecule, followed by cloning for the production of large number of identical target DNA molecules.

Tools of Recombinant DNA Technology

The various **biological tools** used to bring about genetic manipulations are:
A. ***Enzymes***
B. ***Passenger DNA:*** Foreign DNA (insert DNA fragment) which is passively transferred from one cell to another cell or organ is known as **passenger DNA (Foreign DNA)**
C. ***Vector or vehicle DNA:*** The DNA which acts as the carrier is known as the vector or vehicle DNA

A. Enzymes: The various enzymes which may be required to be used are:
- ***Restriction endonucleases:*** To cut DNA chains at specific locations (called as "chemical knife")
- ***Exonucleases:*** To cut DNA at 5' terminus
- ***Endonucleases:*** To cut in the interior to produce "nicks"
- ***Reverse transcriptase***
- ***DNA polymerases***
- ***DNA ligase (T_4 ligase).***

Table 17.2: Some enzymes used in recombinant DNA technology with their functions other than restriction endonucleases

Name	Reaction	Function
• ***Reverse transcriptase***	Synthesises DNA from RNA template	Synthesis of cDNA from m-RNA
• ***S_1 nuclease***	Degrades single stranded DNA	Removal of "hairpin" in synthesis of cDNA
• ***DNA ligase***	Catalyses bonds between DNA molecules	Joining DNA molecules
• ***DNA polymerase I***	Synthesises double-stranded DNA from single stranded DNA	Synthesis of double-stranded cDNA, 'nick' translation
• ***BAL3 1 nuclease***	Degrades, both 3' and 5' ends of DNA	Progressive shortening of DNA molecule
• ***Alkaline phosphatase***	Dephosphorylates 5' ends of RNA and DNA	Removal of 5'-PO_4 groups prior to kinase labelling to prevent self ligation

- ***S_1 nuclease*** and
- ***Alkaline phosphatase*** (Table 17.2).

B. Passenger DNA (foreign DNA): DNA insert to be introduced into the vector DNA. They are:
a. ***cDNA (complementary DNA)***
b. ***Synthetic DNA***
c. ***Random DNA***

a. Synthesis of cDNA

An enzyme called ***reverse transcriptase*** is used for this purpose. A double stranded DNA molecule, complementary in base sequence to a m-RNA molecule, can be prepared by using this enzyme.

It is prepared from RNA tumour viruses, can use RNA as a template to synthesise on RNA-DNA hybrid molecule. It requires a "primer" which is provided by hybridising a short chain of oligo dT to the 3'-Poly-A tail of m-RNA.

Under ***in vitro*** conditions, the ***reverse transcriptase*** appears to turn the corner at the end of the newly formed DNA chain and starts to form a complementary strand. This forms the **hairpin-bend.** The 3'-end of the hairpin-bend forms a "Primer" for ***DNA polymerase I*** to complete the synthesis of the complementary strand.

The hairpin-bend is then removed by the action of the enzyme ***S_1 nuclease*** which is specific for single stranded DNA.

The cDNA does not contain *introns*, which are present in the genomic DNA sequence, as these have been spliced out during m-RNA processing as explained in the process of transcription.

It is important to note that the source of m-RNA is quite critical. The tissue that is extracted to obtain the m-RNA strand should express adequately the gene that is required to be expressed.

b. Synthetic DNA

Synthetic DNA can be produced purely by chemical means. Short segments (10 to 15 nucleotides) with sticky ends are synthesised chemically, with *T_4 ligases*. These are connected to give synthetic DNA with "Sticky" ends.

c. Random DNA

If it is not possible to use the cDNA or synthetic DNA, a shot-gun experiment is carried out to produce random pieces of DNA (insert DNA) using the enzyme restriction endonucleases.

C. Vector or vehicle DNA: Following types of DNA may be used as vector or vehicle DNA

1. ***Bacterial plasmids***
2. ***Bacteriophages***
3. ***Cosmids.***

1. *Bacterial Plasmids*

They are small circular, duplex DNA molecules whose natural function is to confer antibiotic resistance to the host cell.

Plasmid DNAs replicate independently and they can be easily separated from host bacteria. The DNA sequences and restriction maps of many plasmids are known, hence the precise location of restriction enzyme cleavage sites for inserting the foreign DNA (insert DNA) is available.

The ***plasmids are the most commonly used vectors*** and ***can accept short DNA pieces about 6 to 10 kb long.***

Types of plasmids: Three main types of plasmids have been studied. However, many have been found in a variety of strains of *E. coli.*

- ***F plasmids (Sex plasmids):*** This plasmid transfers a replica of the plasmid from a donor (F+) cell to a recipient (F–) cell without the F+ cell losing its plasmid. The F plasmid DNA can integrate into the chromosome of the recipient cell to produce Hfr cell. When an imperfect excision of F occurs, it produces a plasmid containing chromosomal gene. This is called F'-plasmid.
- ***R plasmids, drug resistance plasmids:*** These plasmids carry genes conferring resistance to one or more antibiotics and usually can transfer this resistance to an R-free recipient cell.
- ***Col plasmids or colicinogenic factor plasmids:*** They carry genes for the synthesis of a protein known as ***colicins***. These proteins can kill related strains of bacteria that lack the *col* plasmid. The *col plasmids* are non-self-transmissible. They can prepare their DNA for transfer but do not have the genes necessary for determining effective contact between donor and recipient cells.
- ***Ent plasmids:*** Plasmids called *Ent* are responsible for traveller's diarrhoea and some types of dysentery. These plasmids contain genes that code for *enterotoxin* which is an intestinal irritant. Some strains of *Bacillus thuringiensis,* which has genes that code for a product that is toxic to gypsy moths and tentworms.

Note: ***Plasmid vector PBR 322*** has both tetracycline *(tet)* and ampicillin *(amp)* resistance genes. A single Pst I site within the ampicillin resistance gene is commonly used as the insertion site for a piece of foreign DNA. In addition to having sticky ends, the DNA inserted at this site disrupts the ampicillin resistance gene and makes the bacterium carrying this plasmid ampicillin-sensitive.

2. *Bacteriophages*

They usually have linear DNA molecules into which foreign DNA can be inserted at several restriction enzyme sites.

The chimeric or hybrid DNA is collected after the phage proceeds through its lytic cycle and produces mature, infective phage particles.

A major advantage of phage vector is that ***they can accept DNA fragments 10 to 20 kb long*** (one kb = 1000 nucleotide long base sequence).

3. *Cosmids:* These are specialised plasmids that contain DNA sequences, so-called **COS sites** required for packaging λ (lambda) DNA into the phage particle.

Larger fragments of DNA can be inserted in cosmids which combine with best features of plasmids and phages. ***Cosmids can accept very large DNA fragments 35 to 50 kb.***

Table 17.3 shows the common vectors and DNA insert size.

Table 17.3: Common vectors and DNA insert size

Vector	*DNA insert size*
• Plasmid PBR 322	0.01 to 10 kb
• Bacteriophage (Lambda charon 4A)	10 to 20 kb
• Cosmids	35 to 50 kb

Stages

I. Isolation of Specific DNA (Insert DNA)

The genomic DNA is very large in size. Isolation of a specific fragment of DNA can be achieved by splicing or cleaving brought about by a group of key enzymes called ***restriction endonucleases.***

Restriction endonucleases: (They are compared to ***chemical knife***). Certain endonucleases, enzymes that cut DNA at specific DNA sequences within the molecule (cf. exonucleases which digest from the ends of DNA molecules), are a key biological tool in recombinant DNA technology (Genetic engineering). These enzymes were originally called *restriction enzymes* because their presence in a given bacterium restricted the growth of certain bacterial viruses called bacteriophages.

Restriction enzymes cut DNA of any source into short pieces in a sequence-specific manner in contrast to other methods, e.g. enzymatic, physical or chemical that break DNA in a random fashion.

Nomenclature of restriction enzymes: Large number of restriction enzymes have been found. They have been named after the bacterium from which they are isolated.

Examples

- **Eco R I and Eco R II** isolated from *Escherichia coli*
- **Bam HI** is from bacillus amyloliquifaciens.

The first three letters of the restriction enzymes name consist of the first letter of the genus (E) and the first two

Table 17.4: A few selected restriction endonuclease and their sequence specificities

Endonuclease	Sequence cleaved	Bacterial source
• **Eco R I**	G|AATTC CTTAA|G	*Escherichia coli* Ry 13
• **Bam H I**	G|GATCC CCTAG|G	Bacillus amyloliquifaciens H
• **Eco R II**	|CCTGG GGACC|	*Escherichia coli* R 245
• **H Pa I**	GTT|AAC CAA|TTG	Haemophilus parainflu-enzae
• **Pst I**	CTGCA|G G|ACGTC	Providencia stuartii 164

Table 17.5: DNA splicing by restriction endonucleases

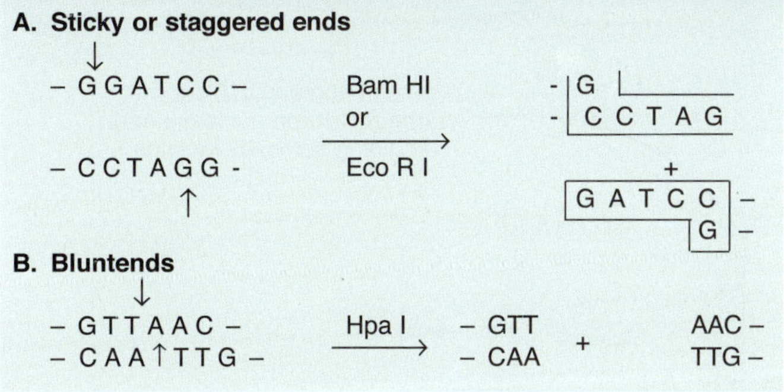

letters of the species (Co). These may be followed by a strain designation (R) and a Roman numeral I or II to indicate the order of discovery, e.g. Eco R I, Eco R II etc.

Table 17.4 shows a few selected restriction endonuclease and their sequence specificities.

Mechanism of action: Each restriction endonuclease enzyme recognises and cleaves a specific double-stranded DNA sequence that is 4 to 7 bp long. These DNA cuts result in:

- ***Blunt ends (by Hpa I)***
- ***Sticky ends (also called staggered or cohesive ends)*** by Bam HI or Eco R I.

Sticky ends are particularly useful in preparation of chimeric or hybrid DNA molecule.

By using different restriction endonucleases for the digestion of a particular DNA, a *restriction map* with characteristic sites of action can be prepared (Table 17.5).

When DNA is digested with a given enzyme, the ends of all the fragments would have the same DNA sequence. The fragments produced can be isolated by agarose electrophoresis or polyacrylamide gel electrophoresis. This is an essential step in cloning and a major use of these enzymes.

II. Chimeric or Hybrid DNA

The main aim of genetic engineering is to insert a DNA of interest (foreign DNA) into a vector DNA so that the DNA fragment replicates along with the vector after annealing. This hybrid combination of two fragments of DNA is referred to as chimeric DNA or hybrid DNA or recombinant DNA.

STEPS OF PREPARATION

- A circular plasmid vector DNA is first cut with a specific ***restriction endonuclease.*** If Eco R I is used, ***sticky ends*** with TTAA sequence on one DNA strand and AATT sequence on the other strand are produced.
- The human DNA (foreign or insert DNA) is also cut with the same *restriction endonuclease*, so that the same sequences are produced on the sticky ends of the cut piece.
- Next, the vector DNA and human cut piece DNA are incubated together so that ***annealing*** occurs. The sticky ends of both vector and human DNA have complementary sequences, hence they come into contact with each other.
- The enzyme ***DNA Ligase*** is allowed to act on the hybrid or chimeric DNA molecule. The enzyme joins the two fragments by covalent phosphodiester linkages between the vector and insert molecules and finally the chimeric DNA molecule is produced (Fig. 17.19).

Problems in Preparation of Chimeric DNA

Although sticky ends ligation is technically easy, but sometimes problems may come up as follows and some special techniques are required to overcome these:

a. Sticky ends of a vector may recombine with themselves without taking up the "insert" molecule.
b. Sticky ends of insert fragments similarly also can anneal without joining the vector.
c. Lastly, sometimes sticky end sites may not be available.

Homopolymer Tailing

To circumvent the above mentioned problems, a reaction enzyme like Hpa I can be used. It produces ***blunt ends.*** The new ends are joined using the enzyme terminal *transferase*. If poly d (G) is added to the 3' ends of the vector and poly d (C) is added to the 3' ends of the insert foreign DNA molecule, the two molecules can only anneal to each other circumventing the problems mentioned above. This procedure is called ***homopolymer tailing***.

Other Methods

1. ***Use of synthetic DNA* Linker:** Sometimes a synthetic DNA called a **linker** is used. The linker helps for the recognition of sequence and effective binding between the human DNA (insert DNA) and vector DNA having "blunt ends".
2. ***Use of T_4 DNA ligase:*** Direct blunt end ligation can be done by using the enzyme bacteriophage *T_4 DNA ligase.* This technique though more difficult than sticky ends ligation has the advantage of joining together any pairs of "blunt ends".

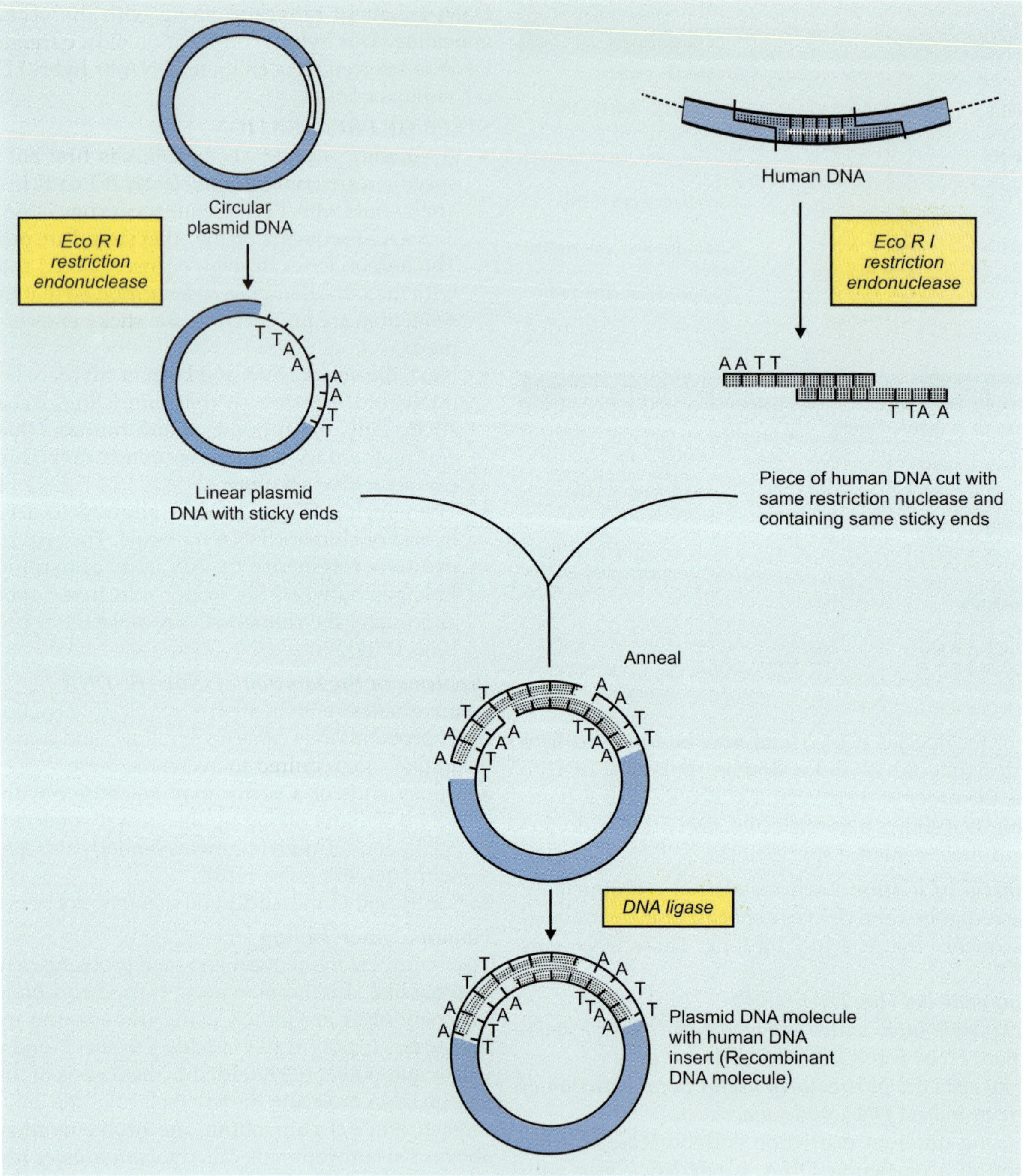

Fig. 17.19: Preparation of chimeric DNA molecule

III. Cloning of Chimeric DNA

What is a Clone?

A clone is a large population of identical molecules, bacteria or cells that arise from a common ancestor.

This cloning allows for the production of a large number of identical DNA molecules which can then be characterised or used for various purposes.

The chimeric DNA contained in a plasmid vector or phages or cosmids can be introduced into bacterial cells *E. coli* strain C 101 by a process called ***transfection.*** The replicating bacterial cell (host cell) permits the amplification of the chimeric DNA of the vector.

In this way, cloning results in the production of large number of identical target DNA molecules. The cloned target DNA is released from its vector by cleavage using appropriate restriction endonucleases, isolated, characterised and used for various purposes.

Gene Library

By the combination of restriction enzymes and various cloning vectors, the entire DNA of an organism can be packed into a suitable vector.

A collection of these recombinant clones is called a gene library. A DNA gene library can be of *two types:*

a. **Genomic library**
b. **cDNA library**

a. *Genomic Library*

- ***A genomic library is prepared from the total DNA of a cell line or tissue.***
- It is prepared by performing partial digestion of total DNA with restriction enzyme, e.g. Saul III A that cuts the DNA frequently to produce larger fragments so that intact genes can be obtained.
- ***Phage vectors are ideal and preferred*** for this as they accept large pieces of DNA up to 20 kb.

b. *cDNA Library*

- ***A cDNA library represents the population of m-RNAs in a tissue.***
- cDNA libraries are prepared by first isolating all the m-RNAs in a tissue.
- m-RNA serves as a template to prepare the cDNA using the enzymes *reverse transcriptase* and *DNA polymerase.* Full length cDNA copies are usually not obtained and smaller DNA fragments are cloned.
- ***Plasmids are the ideal and preferred vectors for*** cDNA libraries as they are workable with smaller fragments.
- cDNA libraries contain relatively smaller DNA fragments compared to genomic libraries.

DNA Probes

Probes are used to search libraries for specific genes or cDNA molecules. Gene libraries contain large numbers of DNA fragments and it is extremely difficult to find out or choose a specific DNA sequence of interest from these.

- A variety of molecules can be used to "Probe" libraries to search for specific genes or cDNAs.
- Probes are generally pieces of DNA or RNA labelled with ^{32}P-containing nucleotide. To be effective, the probes must recognise a complementary sequence.
- A cDNA synthesised from a specific m-RNA can be used to screen either a cDNA library for a longer cDNA or a genomic library for a complementary sequence in the coding region of a gene.
- cDNA probes are used to detect DNA fragments on Southern Blot transfers and to detect and quantitate RNA on Northern Blot transfers (See below).
- Specific antibodies can also be used as probes.

Blotting and Hybridisation

Visualisation and identification of a specific DNA or RNA fragment or protein among the many thousands of molecules requires the convergence of a number of techniques which are called collectively as Blot transfer techniques.

Analysis of DNA, RNA and Proteins

1. **Analysis of chromosomal DNA (Southern Blot):** Following are the steps that are normally employed in Southern Blot Analysis of DNA.
   - ***Cleavage:*** DNA is cleaved with the help of *restriction endonuclease* at specific sites.
   - ***Electrophoresis:*** The DNA thus obtained will be in fragments. These DNA fragments are separated by agarose gel electrophoresis or polyacrylamide gel (Slab) electrophoresis.
   - ***Blotting:*** The separated DNA fragments are transferred to a sheet of nitrocellulose by a flow of buffer. These fragments bind to the nitrocellulose, creating a replica of the pattern of DNA fragments.
   - ***Hybridisation:*** The next step is the hybridisation of fragments with a labelled probe. This probe has a homology with the gene of interest. The probe is hybridised to filter. The probe can be
     - Purified RNA
     - cDNA
     - A segment of cloned DNA.
   - ***Autoradiography:*** This is the final step. The pattern of bands that contain the DNA fragments, or fragments that contain the gene are visualised by virtue of radiation from the probe.
2. **Analysis of RNA (Northern Blot):**
   - ***Electrophoresis:*** Total cellular RNA, or isolated m-RNA is subjected to agarose gel electrophoresis in the presence of a denaturing agent to remove secondary structure constraints.
   - ***Blotting:*** The gel is blotted as above with the difference that, paper has been treated chemically so that it will covalently bind RNA.
   - ***Hybridisation:*** A labelled probe is used. This allows visualisation of the RNA species that are complementary to the probe.
3. **Analysis of Proteins (Western Blot):** By Western blot test, proteins are identified. Proteins are first isolated from the tissues.
   - ***Electrophoresis and fixation:*** Electrophoresis of the whole protein is done and transferred on to a nitrocellulose membrane and fixed.
   - ***Probing:*** After fixation, the protein is probed with radioactive antibody.
   - ***Autoradiography:*** This is the final step. The pattern of bands that contain protein are visualised by virtue of radiation from the probe.

Note: The Western blot is very useful to identify the production of a specific protein in a tissue **(Fig. 17.20)**.

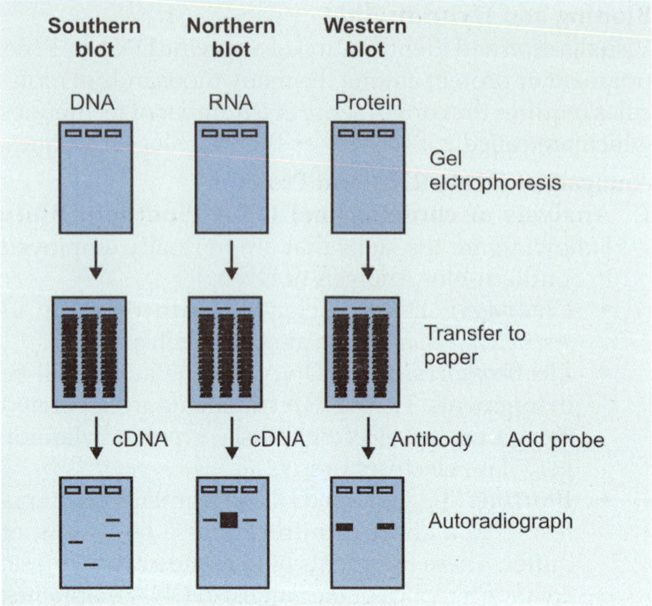

Fig. 17.20: Blot transfer techniques

Applications of Recombinant DNA Technology

Recombinant DNA technology (genetic engineering) has revolutionised the application of molecular biology to medical/agricultural sciences that has immensely benefitted the mankind.

A few of the useful practical applications of recombinant DNA technology are listed below:

1. ***Manufacture of proteins/hormones:*** A practical goal of recombinant DNA research is the production of materials for biomedical application. This technology has two distinct merits:
   - **i.** It can supply large amounts of materials that could not be obtained by conventional purification methods, e.g. interferon, plasminogen activating factor, other blood clotting factors.
   - **ii.** It can provide human materials, e.g. insulin (refer Chapter on Hormones), growth hormone, etc.
2. ***AIDS test:*** By using recombinant DNA techniques, the diagnosis of diseases like AIDS by laboratory has become simple and rapid.
3. ***Diagnosis of molecular diseases:*** Many genetic diseases that yield developmental abnormalities can be detected by characteristic patterns in DNA primary structure. Such mutational changes in DNA sequences are identified by restriction fragments analysis and Southern blotting, using appropriate DNA probes. Analysis of this type could be done in understanding the molecular basis of diseases like sickle cell anaemia, thalassaemias, familial hypercholesterolaemia, cystic fibrosis, etc.
4. ***Prenatal diagnosis:*** If the genetic lesion is understood and a specific probe is available, prenatal diagnosis is possible. DNA from cells collected from as little as 10 ml of amniotic fluid or by chorionic villi biopsy can be analysed by Southern blot transfer.
5. ***Gene therapy:*** Diseases caused by deficiency of a gene product are amenable to replacement therapy. The strategy is to clone a gene into a vector that will readily be taken up and incorporated into genome of a host cell. In 1990, a patient with *adenosine deaminase* deficiency was treated successfully with gene replacement therapy. In future, probably many inherited disorders like sickle cell anaemia, thalassaemias, various enzyme deficiencies, etc. may be treated by gene replacement therapy.
6. ***Applications in agriculture:*** Genetically engineered plants have been developed to resist draught and diseases. Good quality of food and increased yield of crops could be possible by applying this technology. Incorporation of ***nif genes*** to cereals has given higher yield of the crops.
7. ***Industrial application:*** Enzymes synthesised by this technology are in use to produce sugars, cheese and detergents. Certain protein products produced by this technology are used as *food additives* to increase the nutritive value, besides imparting flavour. Ethylene glycol is in great demand for industry. Preparation of ethylene glycol from ethylene is made possible by this technology.
8. ***Application in forensic medicine:*** Advances in genetic engineering have greatly helped to specifically identify criminals and settle the disputes of parenthood of children. The restriction analysis pattern of DNA of one individual will be very specific (DNA finger printing), but the pattern will be different from person to person.
9. ***Transgenesis:*** The somatic gene replacement therapy will not pass on to the offspring. Transgenesis refers to the transfer of genes into fertilised ovum which will be found in somatic as well as germ cells and passed on to the successive generations.

Note: The above depicts the brighter aspects of genetic engineering. But it is to be remembered that there are likely "hazards" also, e.g. by any chance, in the course of cloning using bacterial vectors, some drug-resistant or harmful bacteria are produced and released, the new varieties may cause incurable diseases to the humanity and the live-stock. Scientists should use the technology for useful purposes only like atomic energy and not for harmful purposes to humanity.

SECTION THREE

I. Questions (Essay type and short notes)

1. Write in detail the mode of protein biosynthesis with schematic representations.
2. Describe the process of protein synthesis in our body.
3. Describe peptide formation at the ribosomal level

Differentiate

a. Transcription and translation
b. m-RNA and t-RNA

Short Notes

a. Translation
b. Genetic code
c. 'Lac' operon
d. Restriction endonucleases
e. t-RNA
f. m-RNA
g. Southern Blot test

II. MCQs (Give one correct answer)

1. Polysomes do not contain:
   a. Protein b. DNA
   c. m-RNA d. t-RNA
   e. r-RNA
2. The formation of a peptide bond during the elongation step of protein synthesis results in the splitting of how many high energy bonds?
   a. 1 b. 2
   c. 3 d. 4
   e. 5
3. Translocase is an enzyme required in the process of:
   a. DNA replication
   b. RNA synthesis
   c. Initiation of protein synthesis
   d. Elongation of peptides
   e. Termination of protein synthesis
4. Nonsense codons bring about:
   a. Amino acid activation
   b. Initiation of protein synthesis
   c. Termination of protein synthesis
   d. Elongation of polypeptide chains
   e. Post-translational modification of proteins
5. Which of the following genes of the *E. coli* "Lac operon" codes for a constitutive protein?
   a. The a gene b. The i gene
   c. The c gene d. The z gene
   e. The p gene
6. In the process of transcription, the flow of genetic information is from:
   a. DNA to DNA b. DNA to protein
   c. RNA to protein d. DNA to RNA
   e. Protein to RNA
7. The anticodon region is an important part of the structure of:
   a. r-RNA b. t-RNA
   c. m-RNA d. hn-RNA
   e. DNA
8. The region of the *Lac* operon which must be free (unbound) for structural gene transcription to occur is:
   a. The operator locus b. The promoter site
   c. The 'a' gene d. The 'i' gene
   e. The 'z' gene
9. Another name for reverse transcriptase is:
   a. DNA dependent DNA polymerase
   b. DNA dependent RNA polymerase
   c. RNA dependent DNA polymerase
   d. RNA dependent RNA polymerase
   e. RNA dependent protein synthetase
10. In the "Lac operon" concept, which of the following is a protein?
    a. Operator b. Repressor
    c. Inducer d. Vector
    e. Promoter
11. Degeneracy of the genetic code denotes the existence of:
    a. Base triplets that do not code for any amino acids
    b. Codons consisting of only two bases
    c. Codons that include one or more unusual bases
    d. Different protein synthesising systems in which a given triplet codes for different amino acids
    e. Multiple codons for a single amino acid
12. The normal function of restriction endonucleases is to:
    a. Excise introns from hn-RNA
    b. Polymerise nucleotides to form RNA
    c. Remove primer from Okazaki fragments
    d. Protect bacteria from foreign DNA
    e. Repair damaged DNA
13. In contrast to eukaryotic m-RNA, prokaryotic m-RNA is characterised by:
    a. Having 7-methyl guanosine triphosphate at the 5′-end
    b. Being polycystronic
    c. Being only monocystronic
    d. Being synthesised with introns
    e. Having a poly-A tail
14. DNA ligase of *E. coli* requires which of the following co-factors?
    a. FAD b. NAD^+
    c. $NADP^+$ d. $NADH + H^+$
    e. NADPH
15. Which of the following is transcribed during repression?
    a. Structural gene b. Promoter gene
    c. Regulator gene d. Operator gene
    e. None of the above
16. m-RNA is a complementary copy of:
    a. 5′ – 3′-strand of DNA b. 3′ – 5′-strand of DNA
    c. Antisense strand of DNA d. t-RNA
    e. Ribosomal RNA
17. Synthesis of RNA molecule is terminated by a signal which is recognised by:
    a. α-factor b. β-factor
    c. δ-factor d. ρ-(rho)-factor
    e. None of the above
18. The binding of prokaryotic DNA dependent RNA polymerase to promoter sites of genes is inhibited by the antibiotic:

SECTION THREE

a. Streptomycin
b. Rifampicin
c. Aureomycin
d. Terramycin
e. Puromycin

19. In *E. coli,* the chain intitiating amino acid in protein synthesis is:
a. N-formyl methionine
b. Methionine
c. Serine
d. Cysteine
e. Glycine

20. Match the following:
A. Tetracyclines
B. Puromycin
C. Erythromycin
D. Chloramphenicol
1. Inhibits the peptidyl transferase activity of the 50S ribosomal subunit (Prokaryocytes) _______
2. Binds to 30S subunit and inhibits binding of aminoacyl t-RNAs _______
3. Causes premature chain termination by acting as an analogue of aminoacyl t-RNA _______
4. Binds to 50S subunit and inhibits translocation _______

21. Amanitin the mushroom poison inhibits:
a. Glycoprotein synthesis
b. ATP synthesis
c. DNA synthesis
d. m-RNA synthesis
e. Synthesis of peptides at ribosomal level

22. How many hight-energy phosphate bond equivalents are required for amino acid activation in protein synthesis?
a. One
b. Two
c. Three
d. Four
e. Five

23. Translation results in formation of:
a. m-RNA
b. t-RNA
c. DNA
d. r-RNA
e. A protein molecule

24. Elongation of a peptide chain involves all the following, except:
a. m-RNA
b. GTP
c. Formyl-met-t-RNA
d. Tu, Ts and G factors
e. Peptidyl transferase

25. The rho (ρ) factor is involved:
a. To increase the rate of RNA synthesis
b. In binding catabolite represssor to the promoter region
c. In proper termination of transcription
d. To allow proper initiation of transcription
e. To eliminate binding of RNA polymerase to the promoter

26. In the biosynthesis of c-DNA, the joining enzyme ligase requires:
a. GTP
b. ATP
c. CTP
d. UTP
e. GMP

27. Match the following:
A. Northern blot test
B. Southern blot test
C. Western blot test
D. Polymerase chain reaction
1. Detects the presence of a specific DNA fragment _______
2. Detects the presence of a specific protein _______
3. Synthesises a specific DNA sequence _______
4. Detects the presence of a specific RNA fragment _______.

Answers to MCQs

1. b	2. b	3. d	4. c	5. b	6. d
7. b	8. a	9. c	10. b	11. e	12. d
13. a	14. b	15. c	16. b	17. d	18. b
19. a	20. 1-D, 2-A, 3-B, 4-C.			21. d	22. b
23. e	24. c	25. c	26. b	27. 1-B, 2-C, 3-D, 4-A.	

SECTION THREE

CHAPTER 18

POLYMERASE CHAIN REACTION (PCR) AND REAL-TIME PCR (For Detection and Quantification of Nucleic Acid)

Lt Col AK Sahni, MD DNB PhD, Virology division, Defence R&D establishment, Gwalior, **Lt Col RM Gupta** MD DNB PhD Associate Professor, Department of Microbiology, Armed Forces Medical College, Pune

Molecular techniques based on genomic sequence detection have assumed significance on the rapid diagnosis and identification of viruses and have been gradually accepted as new standard over virus isolation and serodiagnosis. **Polymerase Chain Reaction (PCR)** is a molecular biology technique that ***allows quick replication of DNA. With PCR, minute quantities of genetic material can be amplified millions of times*** within a few hours allowing for the rapid and reliable detection of genetic markers of infectious diseases, cancer and genetic disorders. PCR is a ***highly sensitive technique*** and ***starting material could be a single molecule of rRNA, mRNA of DNA.*** It can be manipulated to make it specific to any desired limits. With PCR, pure DNA fragments from complex genomes could be generated in a matter of hours rather than week and months required with cloning.

Polymerase chain reaction (PCR) is a technique, which allows *in vitro* amplification of a defined target genomic sequence of interest by using a set of primers (oligonucleotides), which anneals to the flanking nucleotide sequence of the target to be amplified. It is a method that allows logarithmic amplification of short DNA sequences (usually 100 to 600 bases) within a longer double stranded DNA molecule. PCR entails the use of a pair of primers, each about 20 nucleotides in length that are complementary to a defined sequence on each of the two strands of the DNA. These primers are extended by a DNA polymerase so that, a copy is made of the designated sequence. After making this copy, the same primers can be used again, not only to make another copy of the input DNA strand, but also of the short copy made in the first round of synthesis. This leads to logarithmic amplification.

PCR is a method of in vitro synthesis of specific DNA sequences. Double-stranded DNA can be disrupted by heat or high pH, giving rise to single stranded DNA. ***The single stranded DNA serves as a template for synthesis of a complementary strand by replicating enzymes,*** **DNA polymerases.** Most polymerases require short regions of double stranded nucleic acid for initiation of synthesis. This can be provided by synthetic oligonucleo-tides (21 to 25 bp) that are complementary to the 5′ region of the single stranded DNA. In PCR, two synthetic oligonucleotides that flank a region of interest are used; one primer is complementary to the negative strand DNA, and the second strand is complementary to the positive strand DNA. In the presence of deoxynucleotide triphosphates (dNTP) and DNA polymerase, a complimentary sequence is synthesised. The primers must be so oriented that DNA synthesis proceeds across the regions defined by the primers. In this way, the product resulting from the extension of one primer will serve as the template for the second primer.

The ***PCR is a cyclical process*** as depicted in the Figure 18.1. Each cycle contains ***three steps that involves:***

- ***Denaturation of the DNA duplex (94-98°C)***
- ***Annealing of the primers (37-60°C) and***
- ***Extension of the primer with a polymerase in the presence of dNTP (about 72°C).*** After 20 cycles of amplification, a million copies of a DNA can be generated from a single copy.

Since it is necessary to raise the temperature to separate the two strands of the double-strand DNA in each round of the amplification process, a major step forward was the discovery of a ***thermostable DNA polymerase (Taq polymerase) that was isolated from Thermus aquaticus,*** a bacterium that grows in hot pools; as a result it is not necessary to add new polymerase in every round of amplification. After several (often about 40) rounds of amplification, the PCR product is analysed on an agarose gel and is abundant enough to be detected with an ethidium bromide stain.

The ***use of Polymerase Chain Reaction*** technology has ***greatly increased scientists' ability to study genetic material.*** The ability to quickly produce large quantities

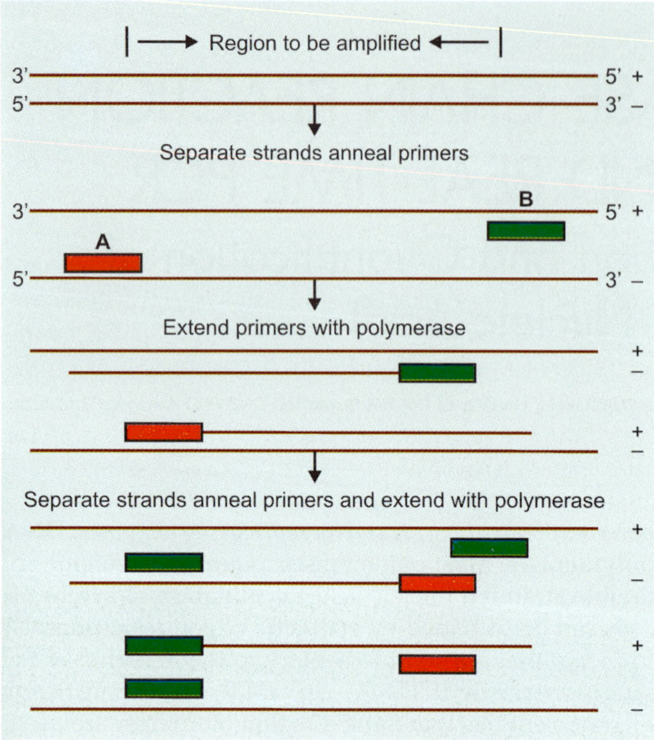

Fig. 18.1: PCR: First and second cycles

of genetic material has enabled significant scientific advances, including ***DNA fingerprinting*** and ***sequencing of the human genome.*** Additionally, PCR technology has heavily influenced the fields of ***disease diagnosis*** and ***patient management***, particularly in the areas of AIDS and Hepatitis C virus. PCR technology has ***become an essential research and diagnostic tool*** for improving human health and quality of life. PCR technology allows scientists to take a specimen of genetic material, even from just one cell, copy its genetic sequence over and over, and generate a test sample sufficient to detect the presence or absence of a specific virus, bacterium or any particular sequence of genetic material. Medical research and clinical medicine are benefiting from PCR technology mainly in **two areas:**

i. **Detection of infectious organisms**, including the viruses that cause AIDS and hepatitis, and other microorganisms;
ii. **Detection of genetic variations, including mutations in human genes.**

PCR technology facilitates the detection of DNA or RNA of pathogenic organisms and, as such, is the basis for a broad range of clinical diagnostic tests for various infectious agents, including viruses and bacteria. These PCR-based tests have several advantages over traditional antibody-based diagnostic methods that measure the body's immune response to a pathogen. In particular, **PCR-based tests are able to detect the presence of pathogenic agents earlier than serologically-based methods**, as patients can take weeks to develop antibodies against an infectious agent. Earlier detection of infection can mean earlier treatment and an earlier return to good health. Capitalising on its exquisite sensitivity, scientists have also developed PCR-based tests designed **to quantify the amount of virus** in a person's blood **(viral load)** thereby allowing physicians to monitor their patients' disease progression and response to therapy. ***Viral load assessment before, during and after therapy has tremendous potential for improving the clinical management of diseases caused by viral infections including AIDS and Hepatitis.***

Various Types of PCR

a. **Conventional DNA based PCR:** These are the classical and conventional PCR assays. The primers target sequences on DNA and amplification follows the usual steps of denaturation, annealing and elongation. Most of the PCR techniques developed for various organisms belong to this category.
b. **Reverse transcription-PCR:** m-RNA, r-RNA can be the starting material in such types of PCR. First step is reverse transcription of RNA to cDNA which could be total or specific RNA by using specific primers. This transcription is useful for diagnostic, viability determination, studies on m-RNA such as for generation of cDNA, etc.
c. **Asymmetric PCR:** This strategy has been employed for generation of single strands for sequencing experiments. By adjusting primer concentrations to favor one strand, after 10 to 15 cycles the second primer is used up and only strand complementary to first strand continues to be copied up.
d. **Inverse PCR:** Inverse PCR amplifies the stretches on either side of a known sequence. Circularising a piece of DNA and hybridising primers to the ends of known sequence in an orientation opposite to the ends of known sequence in an orientation opposite to customary one, results in amplification of primer flanking regions.
e. **Nested PCR:** In this PCR assay, by using, a set of primers (such as genus specific), a fragment is amplified. Using another set (may be changing one more primer, specific to some internal sequence second amplification follows. ***Sensitivity and specificity is considerable improved in nested PCR*** and prove hybridisation may not be necessary. PCR techniques employing this strategy have been developed for detection of many organisms.
f. **Anchored PCR:** When one end of a sequence is known, an anchored PCR can be employed. To unknown end a segment of guanine is covalently

attached which serves as a template with polycytosine primer. Together with primer recognising the known sequences on opposite end of DNA, amplification can occur normally.

g. **PCR systems using other DNA polymerases:** (i) T7 RNA polymerase in transcription based systems, (ii) DNA ligase in ligation amplification reactions and (iii) Qβ replicase from bacteriophage

h. ***In situ* PCR amplification and detection:** Techniques have been developed for the correlation of molecular results with cytological or histological features. In these techniques, the ***amplification can be done directly on the sections. Paraffin embedded or cytospin coated on coated glass slides are digested with protease and amplification solution is added. Taq polymerase is then added at 60°C cycles are done. Product could be detected by in situ hybridisation or by direct incorporation of biotin/digoxigenin labeled nucleotides.***

Analysis of PCR products: PCR products can be analysed by **different methods:**

- ***Gel electrophoresis*** and ***ethidium bromide staining*** is a common method of analysis of PCR products
- With PCR, specific fragments can be amplified and ***RFLP/Sequencing*** undertaken without isolating the organism
- The product could also be ***analysed by blotting/transfer to nylon membrane and hybridisation with specific probes***
- ***Incorporation of biotin/digoxigenin labeled dUTP during amplification*** and detection by colorimetric/luminescent techniques are other approaches for detection and quantitation of PCR products.

In the last decade, PCR technology has ***emerged as a major tool on molecular biological research*** and PCR techniques have already been developed/modified to suit different needs of application. The assays can be tailor made to meet the specific needs. The equipment is not that expensive and the costs of reagents are decreasing. The technique is quite easy to use and it is hoped that in future, the technology would be available as one of easy routine laboratory techniques both for common and sophisticated applications. Well planned laboratory design and laboratory practices are necessary to overcome the problems of contamination. Similarly, technical issues of optimisation, etc. have to be specifically resolved. In addition, the existing RT-PCR test systems are less sensitive, time consuming (3-4 hr), and much more complicated with several steps of amplification (cDNA-PCR-Nested PCR) requiring a high precision thermal-cycler. More sensitive and real time based assays are therefore needed to complement the existing PCR based assay systems.

REAL-TIME PCR

The real-time PCR assay has ***many advantages over conventional*** RT-PCR methods, including ***rapidity, quantitative measurement, lower contamination rate, higher sensitivity, higher specificity***, and ***easy standardisation***. Thus, nucleic acid-based assays or real-time quantitative assay might eventually replace virus isolation and conventional RT-PCR as the new gold standard for ***the rapid diagnosis of virus infection in the acute-phase samples. Real-time PCR has enhanced wider acceptance of the PCR due to its improved rapidity, sensitivity, reproducibility and the reduced risk of carry-over contamination.*** Real-time PCR assays used for quantitative RT-PCR combine the best attributes of both relative and competitive (end-point) RT-PCR in that they are accurate, precise, capable of high throughput, and relatively easy to perform. ***Real-time PCR automates the laborious process of amplification by quantitating reaction products for each sample in every cycle.*** The result is an amasingly broad 10^7-fold dynamic range, with no user intervention or replicates required. Data analysis, including standard curve generation and copy number calculation, is performed automatically. As more labs and core facilities acquire the instrumentation required for real-time analysis, this technique may become the dominant RT-PCR-based quantitation technique.

Advantages of Real-time PCR

- Rapidity due to reduced cycle times and removal of post PCR detection procedures
- Very sensitive
- Reproducible
- Reduced risk of carry-over contamination (sealed reactions)
- High sample throughput (~200 samples/day)
- Easy to perform
- The detection of amplicon could be visualised as the amplification progressed
- Allows for quantitation of results
- Software driven operation.

Real-time Reporters

- ***SYBR® Green,***
- ***TaqMan®, and***
- ***Molecular Beacons.***

All real-time PCR systems rely upon the **detection** and **quantitation of a fluorescent reporter, the signal of which increases in direct proportion to the amount of PCR product in a reaction.**

SECTION THREE

SYBR® Green

In the simplest and most economical format, that reporter is the double-strand DNA-specific dye **SYBR® Green (Molecular Probes).** SYBR Green binds double-stranded DNA, and upon excitation emits light. ***Thus, as a PCR product accumulates, fluorescence increases.***

The **advantages of SYBR Green** are that it's inexpensive, easy to use, and sensitive. The **disadvantage** is that SYBR Green will bind to any double-stranded DNA in the reaction, including primer-dimers and other non-specific reaction products, which results in an overestimation of the target concentration. For single PCR product reactions with well-designed primers, SYBR Green can work extremely well, with spurious non-specific background only showing up in very late cycles.

The **two most popular alternatives** to SYBR Green are **TaqMan®** and **molecular beacons,** both of which are hybridisation probes relying on **fluorescence resonance energy transfer (FRET)** for quantitation.

TaqMan Probes: These are oligonucleotides that **contain a fluorescent dye, typically on the 5' base,** and a **quenching dye, typically located on the 3' base.** When irradiated, the excited fluorescent dye transfers energy to the nearby quenching dye molecule rather than fluorescing, resulting in a non-fluorescent substrate. TaqMan probes are designed to hybridise to an internal region of a PCR product. During PCR, when the polymerase replicates a template on which a TaqMan probe is bound, the 5' exonuclease activity of the polymerase cleaves the probe. This separates the fluorescent and quenching dyes and FRET no longer occurs. **Fluorescence increases in each cycle, proportional to the rate of probe cleavage (Fig. 18.2).**

Molecular beacons: ***Also contain fluorescent and quenching dyes, but FRET only occurs when the quenching dye is directly adjacent to the fluorescent dye.*** Molecular beacons are designed to adopt a hairpin structure while free in solution, bringing the fluorescent dye and quencher in close proximity. When a molecular beacon hybridises to a target, the fluorescent dye and quencher are separated, FRET does not occur, and the fluorescent dye emits light upon irradiation. Unlike TaqMan probes, molecular beacons are designed to remain intact during the amplification reaction, and must rebind to target in every cycle for signal measurement **(Fig. 18.3).**

Real-time Reporters for Multiplex PCR

TaqMan probes and molecular beacons allow multiple DNA species to be measured in the same sample (multiplex PCR), since fluorescent dyes with different emission spectra may be attached to the different probes. Multiplex PCR allows internal controls to be co-amplified

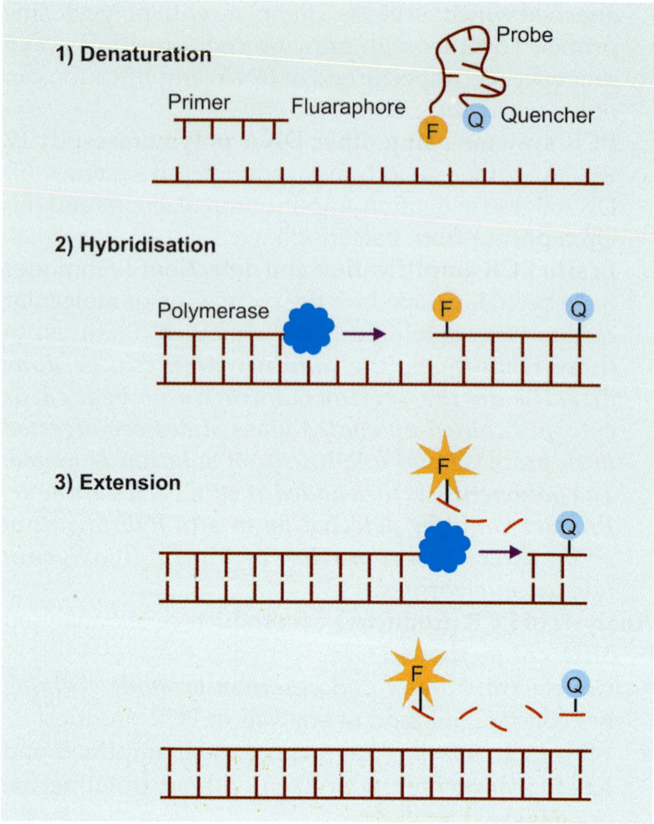

Fig. 18.2: Principle of TaqMan probe system

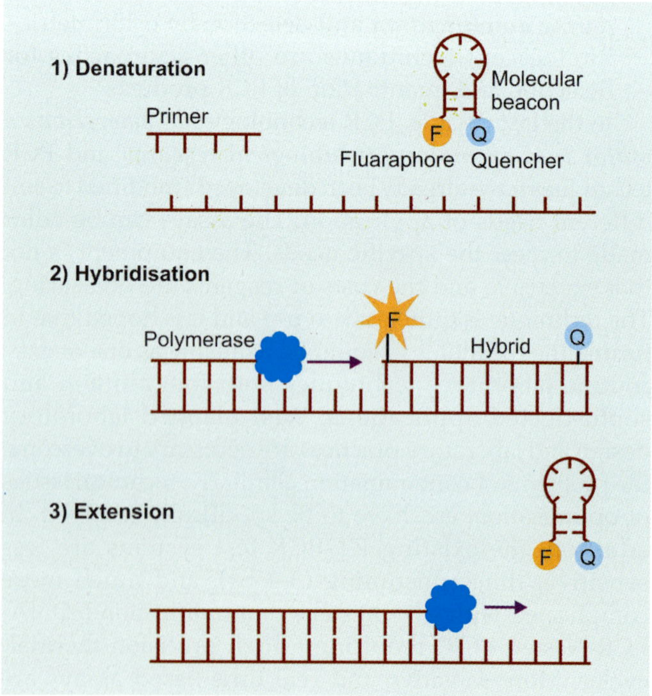

Fig. 18.3: Principle of Beacon probe system

and permits allele discrimination in single-tube, homogeneous assays. These hybridisation probes afford a level of discrimination impossible to obtain with SYBR Green, since they will only hybridise to true targets in a PCR and not to primer-dimers or other spurious products.

Investing in the Real-Time Technique

Real-time PCR requires an instrumentation platform that consists of a thermal cycler, computer, optics for fluorescence excitation and emission collection, and data acquisition and analysis software. These machines, available from several manufacturers, differ in sample capacity (some are 96-well standard format, others process fewer samples or require specialised glass capillary tubes), method of excitation (some use lasers, others broad spectrum light sources with tunable filters), and overall sensitivity. There are also platform-specific differences in how the software processes data. Real-time PCR machines are not cheap, currently about $60-$95, but are well within purchasing reach of core facilities or labs that have the need for high throughput quantitative analysis.

Viral Quantitation

The majority of diagnostic PCR assays reported to date have been used in a qualitative, or 'yes/no' format. The ***development of real-time PCR has brought true quantitation of target nucleic acids*** out of the pure research laboratory and into the diagnostic laboratory. Determining the amount of template by PCR can be performed in two ways:

- As **relative quantitation** and
- As **absolute quantitation**
- **Relative quantitation:** Describes changes in the amount of a target sequence compared with its level in a related matrix.
- **Absolute quantitation:** States the exact number of nucleic acid targets present in the sample in relation to a specific unit.

Generally, relative quantitation provides sufficient information and is simpler to develop. However, ***when monitoring the progress of an infection, absolute quantitation is useful in order to express the results in units*** that are common to both scientists and clinicians and across different platforms. Absolute quantitation may also be necessary when there is a lack of sequential specimens to demonstrate changes in virus levels, no suitably standardised reference reagent is available or when the viral load is used to differentiate active versus persistent infection.

A very accurate approach to absolute quantitation by PCR is the ***use of competitive co-amplification of an internal control nucleic acid of known concentration and a wild-type target nucleic acid of unknown concentration,*** with the former designed or chosen to amplify with an equal efficiency to the latter. However, while conventional competitive PCR is relatively inexpensive, real-time PCR is far more convenient, reliable and better suited to quick decision making in a clinical situation. This is because conventional, quantitative, competitive PCR (qcPCR) requires significant development and optimisation to ensure reproducible performance and a predetermined dynamic range for both the amplification and detection components.

Limitations of Real-time PCR

- Inability to monitor amplicon size without opening the system.
- Incompatibility of some platforms with some fluorogenic chemistries.
- The relatively restricted multiplex capabilities of current applications.
- The start-up expense of real time PCR is prohibitive when used in low-throughput laboratory.

Table 18.1 shows in brief some applications of PCR in clinical diagnosis.

Table 18.1: PCR in clinical diagnosis

1. ***In detection of bacterial infections:*** PCR has been used for the detection of bacterial diseases, e.g. tuberculosis (mycobacterium tubercle)
2. ***In diagnosis of retroviral infection:*** PCR from cDNA is a valuable tool for diagnosis and monitoring of retroviral infections, e.g. HIV infection.
3. ***In detection of infectious agents:*** Specially PCR has been used to detect latent viruses in tissues.
4. ***Application in forensic medicine:*** PCR allows DNA in a single cell, hair follicle, or sperm to be amplified enormously and analysed.
5. ***In diagnosis of inherited disorders:*** PCR technology is being widely used now to amplify gene segments that contain known mutations for diagnosis of inherited diseases viz. sickle cell anaemia, β-thalassaemia, cystic fibrosis, etc.
6. ***In prenatal diagnosis:*** PCR is being used in the prenatal diagnosis of inherited diseases by using chorionic villus samples or cells from amniocentesis.
7. ***In cancer detection:*** Specific chromosomal translocation, e.g. chronic myeloid leukaemia (CML) could be identified by PCR. Several viral induced cancers, e.g. cervical cancer caused by human papilloma virus can be detected by PCR.
8. ***In transplantation:*** PCR has been used to establish precise tissue types for transplants.
9. ***In DNA sequencing:*** As the PCR technique is much simpler and quicker to amplify the DNA, it has conveniently been used for sequencing. For this purpose single strands of DNA are required.

10. ***To study evolution:*** PCR has revolutionised the studies in archeology and in palenteology. By this technique even minute quantities of DNA from any source like fossils, mummified tissues, hair/bones, can be studied.
11. ***In sex determination of embryos:*** Sex of human or live stock embryos fertilised *in vitro* can be determined by PCR. Also useful to detect sex-linked disorders in fertilised embryos.
12. ***To determine viral load:*** By reality-PCR, the "viral load" can be quantitated, which can be useful for diagnosis and for monitoring the progress of an infection.

CONCLUSIONS

Advances in the development of fluorophores, nucleotide labeling chemistries, and the novel applications of oligoprobe hybridisation have provided real-time PCR technology with a broad enough base to ensure its acceptance. Recently, instrumentation has come up that is capable of incredibly short cycling times combined with the ability to detect and differentiate multiple amplicons. New instruments are also flexible enough to allow the use of any of the chemistries making real-time nucleic acid amplification an increasingly attractive and viable proposition for the routine diagnostic laboratory. In many cases these laboratories perform tissue culture to isolate virus and serological methods to confirm the identity of the isolate, which may take a considerable, and clinically relevant, amount of time.

CHAPTER 19

HUMAN GENOME PROJECT

Lt Col RM Gupta MD DNB PhD Associate Professor, Department of Microbiology, Armed Forces Medical College, Pune-40

INTRODUCTION

Started formally in 1990, the US Human Genome Project was a 13-year effort coordinated by the US Department of Energy and the National Institutes of Health.

Project goals were to:

- *Identify* all the approximately 20,000-25,000 genes in human DNA,
- *Determine* the sequences of the 3 billion chemical base pairs that make up human DNA
- *Store* this information in databases
- *Improve* tools for data analysis
- *Transfer* related technologies to the private sector, and
- *Address* the ethical, legal, and social issues (ELSI) that may arise from the project.

To help achieve these goals, researchers also studied the genetic makeup of several nonhuman organisms. These include the common human intestinal pathogen *Escherichia coli*, the fruit fly, and the laboratory mouse.

The project was marked by accelerated progress. Sequencing and analysis of the human genome working draft was published in February 2001 and April 2003 issues of *Nature* and *Science.*

The completion of the human DNA sequence in 2003 coincided with the 50th anniversary of Watson and Crick's description of the fundamental structure of DNA. The analytical power arising from the reference DNA sequences of entire genomes and other genomics resources has jump-started what some call the "biology century."

Historical Background

Human Genome Project (HGP) started officially in 1990, though the project was conceived as early as 1984. It was an international effort whose principal goals were to **sequence the entire human genome.** The project also planned to sequence the genomes of several other model organisms that have been basic to the study of genetics viz.

- *Escherichia coli* (a bacteria),
- *Saccharomyces cerevisae* (yeast)
- *Drosophila melanogaster* (the fruit fly)
- *Caenorhabditis elegans* (the roundworm), and
- *Mus musculus* (the common house mouse)

  Most of the above goals have been accomplished.
- **NCHGR:** The national centre for Human Genome research was established in **1989 in the USA**. The centre was directed by **James D Watson** who delinected the structure of DNA. The NCHGR played a leading role in directing the united states effort in the HGP.
- **NHGRI:** In **1997, NCHGR** became the **National Human Genome Research Institute** (**NHGRI**). The international collaboration involved groups from USA, UK, France, Germany, Japan and China.
- **IHGSC:** The above international collaborating venture came to be known as the **International Human Genome Sequencing Consortium (IHGSC).** The consortium was headed by **Francis collins**.

By the end of **1998**, the above group announced completion of **approximately 60 percent of Human Genome Sequence** and laid down foundations of future work.

Celera Genomics: A second group, a private company Celera Genomies led by **Craig Venter** came up in 1998 and announced that their group had undertaken the objective of sequencing of the Human Genome. ***Venter and colleagues*** already published in 1995, the entire genome sequences of Haemophilus influenzae and Mycoplasma genitalium. **Participation of Venter and colleagues accelerated the progress of human genome sequences.**

Announcement of Draft Sequence of Human Genome

Francis collins and **Crais Venter,** the leaders of the two human genome projects jointly announced the working drafts of human genome sequence covering **more than 90 per cent of the work**, in the presence of US President on **26th June 2000.** This date is to be remembered by the scientists and mankind as a most important and memorable date.

The principal findings of the two groups were published separately in February 2001. IHGSC Published in special issues of Nature and Celera Genomics in "Science".

Published draft involved 10 years of hard work by the six countries of IHGSC and 3 years or less work by Celera associates. The combined achievements has been hailed with applause all over the world as it unfold the:

- **Mysteries of Life** which was hidden in a magic box. Later on many descriptions have been given for the achievement as
- **Supplying a Periodic table of life, Providing Library of life** and finally
- **The Holy Grail of Human Genetics.**

Approaches of Genome Sequencing by Two Groups

The two HGP groups used different approaches in sequencing human genome:

- **IHGSC** basically employed a **map first, sequence later approach**. This approach was because **sequencing was a slow process** and time consuming. The overall approach was referred to as **hierarchical shot gun sequencing**.
  It consisted of fragmenting the entire genome into pieces of approximately 100 to 200 kb and inserting them into **Bacterial artificial chromosomes (BACs),** the vectors. The BACs were then positioned on individual chromosomes by looking for "marker" sequences known as **sequence-tagged sites (STSs)**, whose locations had been already determined. Clones of the BACs then are broken into small fragments (shot gunning). Each cloned fragment was then sequenced, and computer algorithms were used that recognised matching sequence information from overlapping fragments to piece together to complete sequence.
- **Celera and his colleagues** used the **whole genome shot gun approach,** and it **by passed the mapping stap** and **saved time**.

Moreover celera group had **high through-put sequenators, powerful computer programmes** which helped them having **rapid progress**. Thus they made early completion of genome sequence. The combined efforts of the two groups completed the human genome sequencing by 2003.

Genome Mapping

IHGSC employed construction of a series of maps to each chromosome. An outline of different maps used is given below:

- ***Genetic map:*** For the genetic map, the positions of several hypothetical genetic markers are made alongwith the genetic distances in centimorgans between them.
- ***Cytogenetic map:*** For the cytogenetic map, the classic banding pattern of hypothetical chromosome is made.
- ***Physical map:*** For the physical map, the approximate physical positions of the above genetic markers are made, alongwith the relative physical distances in megabase pairs.
- ***Restriction fragment map:*** This consists of the random DNA fragment that have been sequenced.

Principal methods used to identify and isolate normal and disease genes:

Some of the methods used by the two groups are given below:

Method used	*Comments*
• Polymerase chain reaction (PCR)	• Used to amplify fragments of the gene
• DNA sequencing	• Physical map of DNA can be identified with highest resolution. Method can sequence millions of base pairs per day.
• Databases	• Can facilitate gene identification. Compares, DNA and protein sequences obtained from unknown gene with known sequences in database.
• Detection of specific cytogenetic abnormality	• Certain specific genetic disease can be identified by cloning the affected gene.
• Fluorescence *in situ* hybridisation	• Permits localisation of a gene to one chromosome band
• Radiation hybrid mapping	• Most rapid method of localising a gene or DNA fragment. Makes use of a panel of somatic cell hybrids.
• Use of probes to define marker loci	• Probes identify STSs (sequence tagged site), RFLPs (Restriction fragment linked polymorphism) and SNPs (Single nucleotide polymorphism)

Contd...

Contd...

• Use of pulsed field gel electrophoresis (PFGE)	• For the separation and isolation of large DNA fragments. Cutting of the fragments done by use of restriction endonucleases.
• Cloning in vectors like plasmids, phages, cosmids YAC* and BACs**	• Permits isolation of fragments of varying length.

* YAC—Yeast artificial chromosome
** BACs—Bacterial artificial chromosomes

Informations obtained from human genome sequencing.

Determination of the sequence of the human genome has produced a wealth of new findings.

Some of the findings reported in the drafts of human genome are highlighted below:

- The draft represents about 90 per cent of the entire human genome sequenced in draft report of 2000. Gaps large and small remain to be filled in
- The number of protein coding genes varies from 30,000 to 40,000. This number is approximately twice that found in roundworm which is approximately 19,099 and three times that of fruit fly (about 13,060).
- ***Only 1.1 to 1.5 per cent of the genome codes for proteins (exons)***
- About 24 per cent consists of introns and 75 per cent of sequences lying between genes (intragenic) which is of no importance (called **junk DNA**)
- Comparison with roundworm and fruitfly have delineated that exon size across the three species is relatively constant. (Mean size in humans is 145 bp)
- Intron size in humans is much variable, mean being 3300 bp resulting in great variation in gene size.
- Human genes do more work than those of the roundworm or fruitfly.
- Human proteome is more complex than that found in invertebrates.
- ***Repeated sequences probably constitute more than 50 per cent of the human genome.***
- Marked differences noticed among individual chromosome such as gene number per megabase, density of single nucleotide polymorphisms (SNPs), GC content, number of transposable elements and recombination rate.
- ***Chromosome 19 has the richest gene content (23 genes per megabase)***
- Chromocome 13 and γ chromosome have the sparest content (5 genes per megabase)
- Between the humans, the DNA differ only by 0.2 per cent or one in 500 bases.
- Approximately 100 coding regions have been copied and moved by RNA based transposons.
- More than 3 million SNPs (single nucleotide polymorphisms) have been identified.
- A surprising finding is that over 200 genes may be derived from bacteria by lateral transfer. None of these genes are present in nonvertebrate eukaryotes.

Benefits of the Project

Rapid progress in genome science and a glimpse into its potential applications have spurred observers to predict that biology will be the foremost science of the 21st century. Technology and resources generated by the Human Genome Project and other genomics research are already having a major impact on research across the life sciences.

Some current and potential applications of genome research include:

a. ***Molecular medicine***
b. ***Energy sources and environmental applications***
c. ***Risk assessment***
d. ***Bioarchaeology, anthropology, evolution, and human migration***
e. ***DNA forensics (identification)***
f. ***Agriculture, livestock breeding, and bioprocessing.***

a. Molecular Medicine

- *Improved diagnosis of disease*
- *Earlier detection of genetic predispositions to disease*
- *Rational drug design*
- *Gene therapy and control systems for drugs*
- *Pharmacogenomics "custom drugs"*

b. Energy Sources and Environmental Applications

- *Use microbial genomics research to create new energy sources (biofuels)*
- *Use microbial genomics research to develop environmental monitoring techniques to detect pollutants*
- *Use microbial genomics research for safe, efficient environmental remediation*
- *Use microbial genomics research for carbon sequestration*

c. Risk Assessment

- *Assess health damage and risks caused by radiation exposure, including low-dose exposures*
- *Assess health damage and risks caused by exposure to mutagenic chemicals and cancer-causing toxins*
- *Reduce the likelihood of heritable mutations*

d. Bioarchaeology, Anthropology, Evolution, and Human Migration

- *Study evolution through germline mutations in lineages*
- *Study migration of different population groups based on female genetic inheritance*

- *Study mutations on the Y chromosome to trace lineage and migration of males*
- *Compare breakpoints in the evolution of mutations with ages of populations and historical events*

e. **DNA Forensics (Identification)**
- *Identify potential suspects whose DNA may match evidence left at crime scenes*
- *Exonerate persons wrongly accused of crimes*
- *Identify crime and catastrophe victims*
- *Establish paternity and other family relationships*
- *Identify endangered and protected species as an aid to wildlife officials (could be used for prosecuting poachers)*
- *Detect bacteria and other organisms that may pollute air, water, soil, and food*
- *Match organ donors with recipients in transplant programs*
- *Determine pedigree for seed or livestock breeds*
- *Authenticate consumables such as caviar and wine*

To identify individuals, forensic scientists scan about 10 DNA regions that vary from person to person and use the data to create a DNA profile of that individual (sometimes called a DNA fingerprint). There is an extremely small chance that another person has the same DNA profile for a particular set of regions.

f. **Agriculture, Livestock Breeding, and Bioprocessing**
- *Disease-, insect-, and drought-resistant crops*
- *Healthier, more productive, disease-resistant farm animals*
- *More nutritious produce*
- *Biopesticides*
- *Edible vaccines incorporated into food products*
- *New environmental cleanup uses for plants like tobacco*

Additional Benefits

- Gaining a deeper understanding of the microbial world also will provide insights into the strategies and limits of life on this planet.
- Data generated in this young programme have helped scientists identify the minimum number of genes necessary for life and confirm the existence of a third major kingdom of life.
- Additionally, the new genetic techniques now allow us to establish more precisely the diversity of microorganisms and identify those critical to maintaining or restoring the function and integrity of large and small ecosystems; this knowledge also can be useful in monitoring and predicting environmental change.
- Finally, studies on microbial communities provide models for understanding biological interactions and evolutionary history.

SECTION THREE

Project Goals and Completion Dates

Area	*HGP goal*	*Standard achieved*	*Date achieved*
Genetic map	2- to 5-cM resolution map (600-1,500 markers)	1-cM resolution map (3,000 markers)	September 1994
Physical map	30,000 STSs	52,000 STSs	October 1998
DNA sequence	95% of gene-containing part of human sequence finished to 99.99% accuracy	99% of gene-containing part of human sequence finished to 99.99% accuracy	April 2003
Capacity and cost of Finished sequence	Sequence 500 Mb/year at <$0.25 per finished base	Sequence >1,400 Mb/year at <$0.09 per finished base	November 2002
Human sequence variation	100,000 mapped human SNPs	3.7 million mapped human SNPs	February 2003
Gene identification	Full-length human cDNAs	15,000 full-length human cDNAs	March 2003
Model organisms	Complete genome sequences of *E. coli, S. cerevisiae, C. elegans, D. melanogaster*	Finished genome sequences of *E. coli, S. cerevisiae, C. elegans, D. melanogaster,* plus whole-genome drafts of several others, including *C. briggsae, D. pseudoobscura,* mouse and rat	April 2003
Functional analysis	Develop genomic-scale technologies	High-throughput oligonucleotide synthesis	1994
		DNA microarrays	1996
		Eukaryotic, whole-genome knockouts (yeast)	1999
		Scale-up of two-hybrid system for protein-protein interaction	2002

CONCLUSION

The Next Step: Functional Genomics

The words of Winston Churchill, spoken in 1942 after 3 years of war, capture well the HGP era:

"Now this is not the end. It is not even the beginning of the end. But it is, perhaps, the end of the beginning."

The avalanche of genome data grows daily. The new challenge will be to use this vast reservoir of data to explore how DNA and proteins work with each other and the environment to create complex, dynamic living systems. Systematic studies of function on a grand scale-functional genomics—are the focus of biological explorations in this century and beyond. These explorations will encompass studies in transcriptomics, proteomics, structural genomics, new experimental methodologies, and comparative genomics. This enormous task will require the expertise and creativity of ten's of thousands of scientists from varied disciplines.

CHAPTER 20

GENE THERAPY

Lt Col AK Sahni, MD DNB PhD, Virology division, Defence R&D establishment, Gwalior, **Lt Col RM Gupta** MD DNB PhD Associate Professor, Department of Microbiology, Armed Forces Medical College, Pune

Diseases of Genetic Origin

Most of us do not suffer any harmful effects from our defective genes because we carry two copies of nearly all genes, one derived from our mother and the other from our father. The only exceptions to this rule are the genes found on the male sex chromosomes. Males have one X and one Y chromosome, the former from the mother and the latter from the father, so each cell has only one copy of the genes on these chromosomes. In the majority of cases, one normal gene is sufficient to avoid all the symptoms of disease. If the potentially harmful gene is recessive, then its normal counterpart will carry out all the tasks assigned to both. Only if we inherit from our parents two copies of the same recessive gene will a disease develop **(Fig 20.1A)**.

You look a little like your mother and a little like your father because of the genes they gave to you. Genes, those conceptual units composed of deoxyribonucleic acid—DNA, carry the information needed to make proteins, the building blocks of our bodies. The body buries genes deep in the heart of every cell, the nucleus, and organises them in the chromosomes that hold the DNA. But when your DNA is damaged, it no longer makes all the needed proteins and disease results.

To reverse disease caused by genetic damage, researchers isolate normal DNA and package it into a vector, a molecular delivery truck usually made from a disabled virus. Doctors then infect a target cell—usually from a tissue affected by the illness, such as liver or lung cells—with the vector. The vector unloads its DNA cargo, which then begins producing the missing protein and restores the cell to normal **(Figs 20.1B and 20.2)**.

On the other hand, if the gene is dominant, it alone can produce the disease, even if its counterpart is normal. Clearly only the children of a parent with the disease can be affected, and then on average only half the children will be affected. Huntington's chorea, a severe disease of the nervous system, which becomes apparent only in adulthood, is an example of a dominant genetic disease.

Finally, there are the X chromosome-linked genetic diseases. As males have only one copy of the genes from

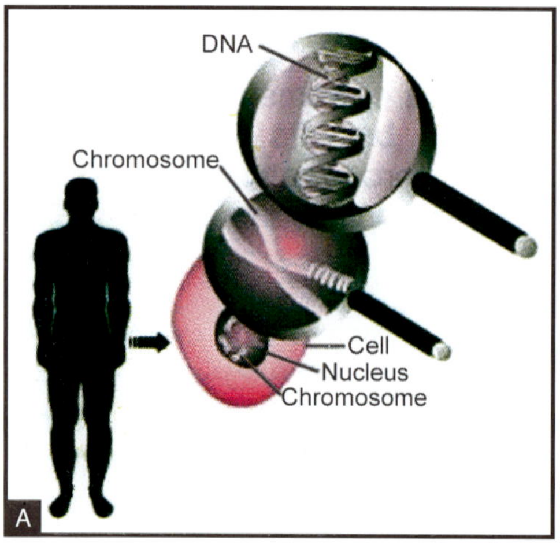

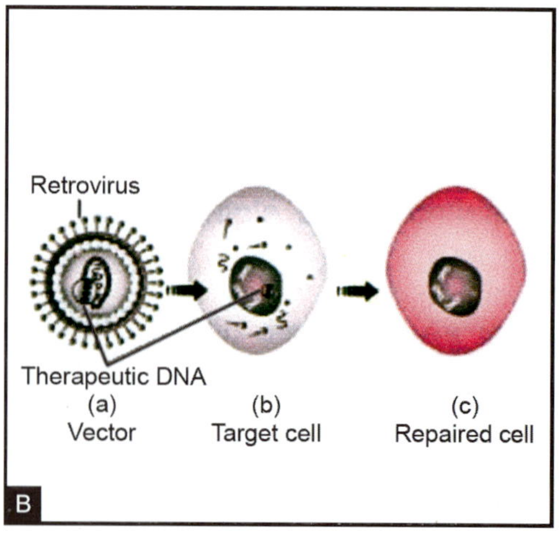

Figs 20.1A and B: Somatic gene therapy close-up view of human cells

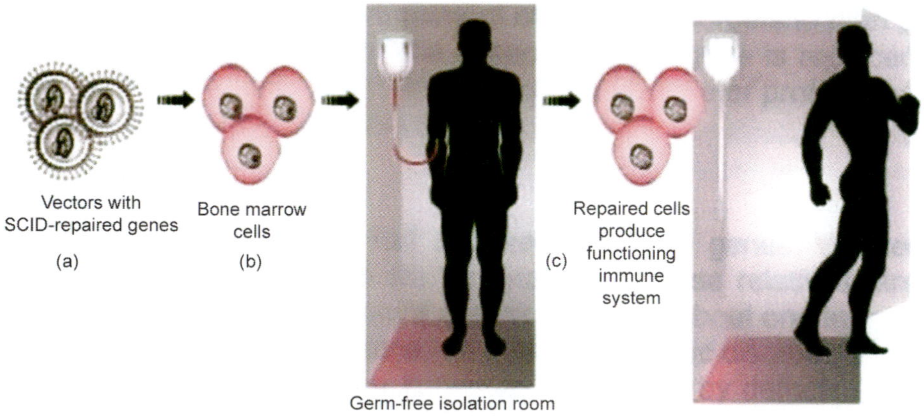

Fig. 20.2: Vectors containing the altered genes are being inserted directly into the human body

this chromosome, there are no others available to fulfil the defective gene's function. Examples of such diseases are **Duchenne** muscular dystrophy and, perhaps most well known of all, **hemophilia**.

Not all defective genes necessarily produce detrimental effects, since the environment in which the gene operates is also of importance. ***A classic example of a genetic disease having a beneficial effect on survival is illustrated by the relationship between sickle-cell anaemia and malaria.*** Only individuals having two copies of the sickle-cell gene, which produces a defective blood protein, suffer from the disease. Those with one sickle-cell gene and one normal gene are unaffected and, more importantly, are able to resist infection by malarial parasites. The clear advantage, in this case, of having one defective gene explains why this gene is common in populations in those areas of the world where malaria is endemic. ***Much attention has been focused on the so-called genetic metabolic diseases in which a defective gene causes an enzyme to be either absent or ineffective in catalysing a particular metabolic reaction effectively.***

A potential approach to the treatment of genetic disorders in man is gene therapy. This is a technique ***whereby the absent or faulty gene is replaced by a working gene,*** so that the body can make the correct enzyme or protein and consequently eliminate the root cause of the disease.

Altered Genes

Each of us carries about half a dozen defective genes. We remain blissfully unaware of this fact unless we, or one of our close relatives, are amongst the many millions who suffer from a genetic disease. About one in ten people has, or will develop at some later stage, an inherited genetic disorder, and approximately 2,800 specific conditions are known to be caused by defects (mutations) in just one of the patient's genes. Some single gene disorders are quite common—cystic fibrosis is found in one out of every 2,500 babies born in the Western World—and in total, diseases that can be traced to single gene defects account for about 5 per cent of all admissions to children's hospitals.

What is Gene Therapy?

One of the most amazing genetic applications in medicine is gene therapy. Also known as somatic gene therapy, ***this procedure involves inserting (or sometimes deleting) portions of the genes in diseased patients so that they can be cured and live healthier lives.*** Gene therapy changes the expression of some genes in an attempt to treat, cure, or ultimately prevent disease. Current gene therapy is primarily experiment based, with a few early human clinical trials under way. Theoretically, gene therapy can be targeted to somatic (body) or germ (egg and sperm) cells.

- In **somatic gene therapy** the recipient's genome is changed, but the change is not passed along to next generation.
- This form of gene therapy is contrasted with **germ line gene therapy**, in which a goal is to pass the change on to offspring. Germ line gene therapy is not being actively investigated, at least in larger animals and humans, the although alot of discussion is being conducted about its value and desirability.

Gene therapy should not be confused with cloning, which has been in the news so much in the past year. Cloning, which is creating another individual with essentially the same genetic makeup, is very different from gene therapy.

Genes, which are carried on chromosomes, are the basic physical and functional units of heredity. Genes are specific sequences of bases that encode instructions on how to make proteins. Although genes get a lot of attention, it's the proteins that perform most life functions and even make up the majority of cellular structures. When genes are altered so that the encoded proteins are unable to carry out their normal functions, genetic disorders can result.

Gene therapy is a technique for correcting defective genes responsible for disease development. Researchers may use one of several approaches for correcting faulty genes:

- A normal gene may be inserted into a nonspecific location within the genome to replace a nonfunctional gene. This approach is most common.
- An abnormal gene could be swapped for a normal gene through homologous recombination.
- The abnormal gene could be repaired through selective reverse mutation, which returns the gene to its normal function.
- The regulation (the degree to which a gene is turned on or off) of a particular gene could be altered.

How Does Gene Therapy Work?

In most gene therapy studies, a "normal" gene is inserted into the genome to replace an "abnormal," disease-causing gene. A **carrier molecule called a vector must be used to deliver the therapeutic gene to the patient's target cells.** Currently, the **most common vector is a virus** that has been genetically altered to carry normal human DNA. Viruses have evolved a way of encapsulating and delivering their genes to human cells in a pathogenic manner. Scientists have tried to take advantage of this capability and manipulate the virus genome to remove disease-causing genes and insert therapeutic genes.

- **Target cells such as the patient's liver or lung cells are infected with the viral vector.** The vector then unloads its genetic material containing the therapeutic human gene into the target cell. **The generation of a functional protein product from the therapeutic gene restores the target cell to a normal state.**
- In gene therapy trials, ***scientist have used a variety of different ways to deliver the genes*** for VEGF-1, VEGF-2 and FGF 4 into the hearts of patients with advanced myocardial ischaemia, after gene therapy, patients had less severe angina (chest pain) and their hearts worked better. Similarly, after gene delivery of VEGF to patients with limb ischaemia, the blood supply improved and leg sores healed better. Gene therapy has prevented below-knee amputation in some patients for whom amputation had been recommended.
- Gene therapy has also been successful in preventing reocclusion, or reblockage, of coronary artery bypass grafts and in keeping arteries open after angioplasty surgery.

Methods of Inserting Genetic Material into Human Chromosomes

Two methods exist for inserting genetic material into human chromosomes.

- **The first, method called the *ex vivo* technique,** involves surgically removing cells from the affected tissue area, injecting or splicing the new DNA (the DNA that will correct the disease) into the cells and letting them divide in cultures. The new tissues are placed back into the affected area of the patient. Often, doctors need only culture the patient's bone marrow because it produces the blood that will eventually travel throughout the body. This type of surgery, however, is especially painful, and then again to replace it—because the culturing time takes many hours to complete.
- The ***second method*** is called the ***in vivo* technique** and requires no surgery or even anaesthesia. In this process, the therapeutic DNA is injected directly into the body cells, usually via one of two types of viruses. The **most frequently used type** is the very simple **retrovirus. Dr Richard Mulligan** of MIT has **synthetically created the perfect retrovirus: It has no reproduction sequence and exists solely to deliver therapeutic DNA during gene therapy.** It has no viral DNA (DNA that would make the cell-and you-sick) whatsoever and only carries the new DNA that has been spliced into it. After injecting the diseased cell with the new therapeutic DNA, it then dies. ***Using retroviruses is very safe and provides long-lasting effects***. unfortunately, the new DNA it injects will only help the new daughter cells and not those that already exist. The **second type of virus used for** the *in vivo* technique is called an **adenovirus,** the equivalent of the common cold virus. Although this virus will also die after injecting its spliced therapeutic DNA, it will be attacked by the immune system and the patient will suffer from a temporary sore throat and runny nose. ***The adenovirus works the same way the retrovirus does, but it effects are much more immediate-within 48 hours***. Unlike the retrovirus, though, the new DNA's effects wear of within weeks. Scientists like the fact that only a few millimetres of altered adenovirus solution is needed to cure the patient, whereas several litres of retrovirus are needed to obtain a much slower result.

There are ***other gene therapy techniques***, although they aren't as frequently used:

- **One method involves inserting therapeutic DNA into cultured endothelium tissue** (endothelium is the membrane that lines all of the blood vessels) and then grafting it into the patient.
- Another technique requires the patient to recieve an electric shock while submerged in a bath of a therapeutic DNA solution. The shock opens the skin pores, allowing the DNA to enter.
- Still other options include skin grafts, connective tissue grafts, and injecting the liver with the therapeutic DNA.

Chemicals called **restriction enzymes** act as the scissors to cut the DNA. Thousands of varieties of restriction enzymes exist, each recognising only a single nucleotide sequence. Once it finds that sequence in a strand of DNA, it attacks it and splits the base pairs apart, leaving single helix strands at the end of two double helixes. Scientists are then free to add any genetic-sequences they wish into the broken chain and afterwards, the chain is repaired (as a longer chain with the added DNA) with another enzyme called ligase. Hence, **any form of genetic material can be spliced together; bacteria and chicken DNA can , and have been, combined.** More often, though splicing is used for important efforts such as the production of insulin and growth hormone to cure human maladies. In the past, insulin was only obtainable from the pancreas of cadavers (and it required 50 cadavers to yield one dose !). With modern splicing techniques, enough insulin can be produced for all diabetics. The insulin-producing genes from human DNA are spliced into plasmid DNA; the plasmids are then allowed to infect bacteria, and, as the bacteria multiply, large amounts of harvestable insulin are produced.

Viruses have evolved a way of encapsulating and delivering their genes to human cells in a pathogenic manner. Scientists have tried to take advantage of the virus's biology and manipulate its genome to remove the disease causing genes and insert therapeutic genes. In the mid-1980 s, the focus of gene therapy was entirely on treating disease caused by such single-gene defects as hemophilia, Duchenne's muscular dystrophy, and sickle cell anemia. In the late 1980s and early 1990s, the concept of gene therapy expanded into a number of acquired disease. When human testing of first-generation vectors began in 1990, scientists learned that the vectors didn't transfer genes efficiently and that they were not sufficiently weakend. Expression and use of the therapeutic genes did not last very long.

Gene Transfer Vehicle

Some of the different types of viruses used as gene therapy vectors **(Table 20.1)**:

- **Retroviruses:** A class of viruses that can create double-stranded DNA copies of their RNA genomes. These copies of its genome can be integrated into the chromosomes of host cells. Human immunodeficiency virus (HIV) is a retrovirus.
- **Adenoviruses:** A class of viruses with double-stranded DNA genomes that cause respiratory, intestinal, and eye infections in humans. The virus that causes the common cold is an adenovirus.
- **Adenoassociated viruses:** A class of small, single-stranded DNA viruses that can insert their genetic material at a specific site on chromosome 19.
- **Herps simples viruses:** A class of double-stranded DNA viruses that infect a particular cell type, neurons. Herpes simplex virus type 1 is a common human pathogen that causes cold sores.

Viral Vectors for Gene Therapy

Three different classes of viral vectors have been used in clinical trials. The first relates to **recombinant retroviruses. To date, retroviruses based on the Mouse Moloney Leukemia virus have been used most frequently in clinical trials.** These vectors are packaged into viral particles, have all viral genes removed but contain some of the viral regulatory sequences, and will only transduce dividing cells. Although efficient at transduction into cells in culture, most cells *in vivo* are

SECTION THREE

Table 20.1: Gene transfer vehicles

Vector	Advantages	Disadvantages
I. Viral		
• **Retrovirus**	Integration into host DNA	Semi-random integration
	All viral genes removed	Transduction requires cell division
	Relatively safe	Relatively low titer
• **Adenovirus**	Higher titer	Toxicity
	Efficient transduction of nondividing cells	Immunological response
	in vitro and *in vivo*	Prior exposure
• **Adenoassociated virus**	All viral genes removed	Small genome limits size of foreign DNA
	Safe	Labor-intensive production
	Transduction of nondividing cells	Status of genome not fully elucidated
	Stable expression	
II. Nonviral		
• **Liposomes**	Absence of viral components	Inefficient gene transfer into the nucleus
	Lack of previous immune recognition	Lack of persistence of DNA
		Lack of tissue targetting

quiescent at any point in time, making this vector less useful for *in vivo* therapies unless the cells in a target organ are stimulated to cycle. A second disadvantage of retroviruses is the relatively low concentration of virus that can be easily produced. **More recently, a chimeric Moloney-Human lentiviral (HIV) vector** has been constructed that can transduce at least some quiescent cells *in vivo* including neurons in the brain of rodents. This promising advance will require further studies to determine the vector's application in the clinic.

Recombinant adenoassociated virus (AAV) vectors contain small, single-stranded DNA genomes and have recently been shown to transduce brain, skeletal muscle and liver by injection into quiescent tissue or vasculature, feeding the tissue in animals. In fact, rAAV has been used to achieve therapeutic or, in some cases, curative concentrations of clotting factor IX in mice without toxicity for at least 9 months by *in vivo* delivery. Unfortunately, these vectors have a disadvantage in that there is a limit in the amount of DNA that can be packaged. Thus, larger cDNAs, genes, or complex regulatory cis elements cannot be used with this vector.

Regulation of gene expression may be important for treating some diseases, and recently, several different approaches to regulate gene expression have been used in animal models of gene expression. These approaches include the addition or subtraction of small molecules that interact with cis DNA elements and turn genes on or off. Moreover, tissue-specific regulation can be achieved by using cell-type-specific promoters or by designing vectors that specifically target an organ. Altering the tropism of the vector by constructing new ligands for receptor-specific targeting will certainly be important for future gene therapy applications.

There are a number of additional viral vectors based on Epstein-Barr virus, herpes, simian virus 40, papilloma, nonhuman lentiviruses, and hepatitis viruses that are currently being derived in the laboratory. Perhaps these chimeric or as yet undiscovered viruses will have properties that offer advantages to clinical gene therapy that are not yet realised.

Non-Viral Vectors: Liposomes

Compared with viral vectors, cationic lipid-based delivery systems have several advantages. Unlike viral vectors, DNA/lipid complexes are easy to prepare and there is no limit to the size of genes that can be delivered. Because carrier systems lack proteins, they may evoke much less immunogenic responses, More importantly, the cationic lipid systems have much less risk of generating the infectious form or inducing tumorigenic mutations because genes delivered have low integration frequency and cannot replicate or recombine.

During the last few years, **two classes of cationic lipids** have been synthesised and show good transfection activity, which is mostly *in vitro* with established cell lines. **The first class has two alkyl chains in each cationic lipid molecule,** and the **other type uses cholesterol as the backbone. Both types of lipid contain either mono- or multiple-amino groups** as the cationic function group to form complexes with DNA via electrostatic interactions. Each type of cationic lipid appears to have its preferred cell lines for an optimal transfection activity, even though both types of cationic lipids may show similar level of transfection activity in a given cell type. With a few exceptions, ***the transfection activity of these cationic lipids is improved when a helper lipid, dioleoylphosphatidlylethanolamine, is included as part of the liposome composition.*** In general, the transfection activity of these cationic liposomes *in vitro* is optimal with slight excess of cationic lipid in the DNA/lipid complexes.

Although additional experiments are needed to show that gene expression *in vivo* by an intravenous administration will be useful for therapeutic purposes, these *in vivo* results reconfirm the potential of the lipid system as a carrier for gene therapy. As with viral vectors, the next challenges are to achieve targeted gene delivery, to control the level of transgene expression, and to devise methods for long-term expression when needed.

Gene Therapy for Haematopoietic Derived Diseases

So far, most clinical trials on gene therapy **focus on gene transfer into haematopoietic (blood cells) and cancer cells. All haematopoietic cells arise from a single cell type** designated as **pluripotent haematopoietic stem cells (PHSCs). Therefore, successful stem cell gene therapy can be applied to a large variety of congenital and acquired blood cell diseases.** In the last few years, the cell fraction that includes these PHSCs is being identified using molecules that are present on the cell surface, e.g. it has been found that PHSCs carry the CD34 antigen (CD34+; this is present in approximatley 1 per cent of the bone marrow cells but are negative for other markers which is the case in approximately 1 per cent of the CD34+ cells). By cell separation techniques, the CD34+lin^{neg} cells can be isolated and used for gene transfer studies. In these stem cell gene transfer studies, the cells are harvested from the patient, gene transfer is performed *ex vivo*, and the transduced cells are subsequently reinfused into the patients. The majority of studies use retroviral vectors as vehicles to mediate gene transfer.

Besides virus-mediated gene-delivery systems, there are several non-viral options for gene delivery. The simplest method is the direct introduction of therapeutic DNA into target cells. This approach is limited in its application because it can be used only with certain tissues and requires large amounts of DNA.

Another non-viral approach involves the creation of an artificial lipid sphere with an aqueous core. This liposome, which carries the therapeutic DNA, is capable of passing the DNA through the target cell's membrane.

Therapeutic DNA also can get inside target cells by chemically linking the DNA to a molecule that will bind to special cell receptors. Once bound to these receptors, the therapeutic DNA constructs are engulfed by the cell membrane and passed into the interior of the target cell. This delivery system tends to be less effective than other options.

Gene Silencing

Another aspect of gene therapy is gene silencing, also called **antisense technology.** With this method, geneticists can **inactivate a gene** that may cause disease or be defective.

When DNA replicates, RNA bonds to half of the split double helix, making a mold of sorts. The RNA (messenger RNA or mRNA) is then used to create an identical DNA strand. To silence a gene on a chromosome, scientists, therefore, simply make an RNA strand 15 to 20 bases in length complementary to the mRNA. The synthesised RNA will attach itself to the mRNA and **prevent that portion of the mRNA from creating the gene on the duplicate DNA strand. This method is highly specific.**

RNA Interference

RNA interference (RNAi) is emerging as a **new strategy to selectively shut off the post-transcriptional expression of mRNA.** This evolutionarily maintained mechanism has been detected from yeast to mammalian cells. Rapid, inexpensive and selective silencing of a gene product in complex biological systems obviously has opened new avenues in fields of virology, cancer research, genetic disorders, drugs designing, etc.

Mechanism of RNA interference: Double stranded RNA (dsRNA) when introduced or synthesised inside the mammalian cells undergoes processing by an enzyme called **Dicer**. This results in formation of dsRNAs, about 21 nucleotides in length—the small interfering RNAs (siRNAs). Coupled with cellular proteins siRNA forms RNA interfering silencing complex (RISC) and heads for target recognition on intracellular mRNAs. RISC binds to complementary sequences on mRNA and effectively leads to the cleavage of mRNA **(Fig. 20.3)**.

Short, synthetic, double stranded siRNA can be easily introduced into mammalian cells using transfecting agents, several of these are available commercially, or by electroporation. The same can also be expressed intracellularly when introduced through plasmid or viral vectors.

The siRNA should be about 21 nucleotides long with 3′ overhang of 2 nucleotides, preferably should have GC content less than 50 per cent and complementary to 50 to 100 nucleotides downstream of initiation codon

SECTION THREE

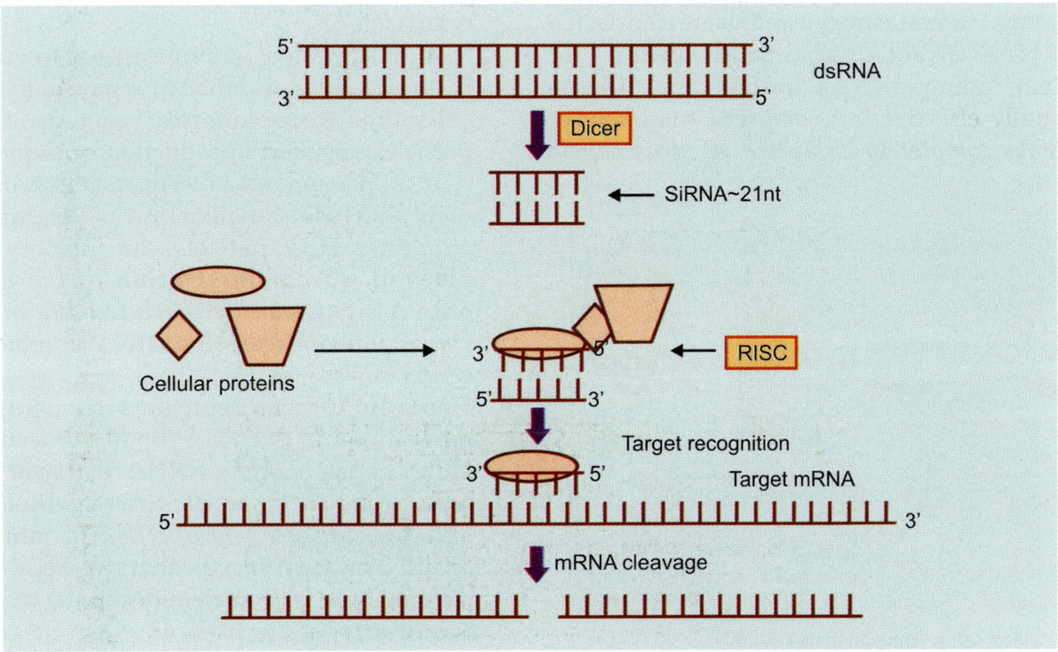

Fig. 20.3: Mechanism of RNA interference (RNAi)

of target mRNA. It is also necessary that the transfection efficiency and siRNA concentration be sufficiently high to introduce siRNA in maximum number of cells. Some cell lines show poor transfection efficiency. Introduction of siRNA in such cell can be achieved by using plasmid or viral vectors.

Antisense

Antisense gene therapy aims to turn off a mutated gene in a cell by targeting the mRNA transcripts copied from the gene.

Genes are made up of two paired DNA strands. During transcription, the sequence of one strand is copied into a single strand of mRNA. This mRNA is called the "sense" strand because it contains the code that will be read by the cell as it makes a protein. The opposite strand is the "antisense" strand.

Antisense gene therapy involves the following steps:
1. Delivery of an RNA strand containing the antisense code of a mutated gene
2. Binding of the antisense RNA strands to the mutated sense mRNA strands, preventing the mRNA from being translated into a mutated protein (Fig. 20.4).

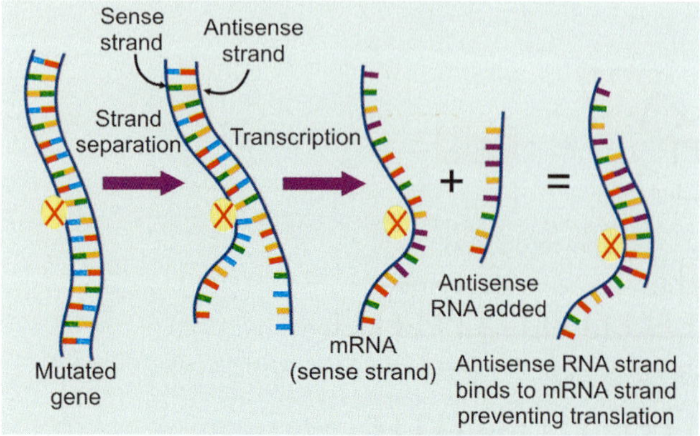

Fig. 20.4: Preventing translation using antisense technology

siRNA as Antiviral Agent

The inhibitory action of siRNAs has been documented for numerous viruses. Some of the examples are highlighted below.

- **JC virus**, a member of the genus polyomavius, causes progressive multifocal leucoencephalopathy. It is also activated in immunocompromised hosts, such as AIDS patients. Human astrocyte cells were transfected with siRNAs directed against T antigen or agnoprotein coding mRNAs. Individually, siRNAs were partially effective but ***combined treatment of both siRNAs completely abolished JC virus capsid protein production.*** Similarly, siRNAs mediated inhibition of viral capsid protein VP1 and agnoprotein resulted in marked inhibition of JC virus production in human glial cells.
- **Influenza** is a global health problem. Estimated 3 to 5 million severe influenza cases and 2,50,000 to 5,00,000 deaths due to influenza occur annually. In order to investigate new strategies for control of virus replication, siRNAs specific to influenza genomic regions were used in cell culture, eggs and in infant mice. ***Of the several siRNAs studied, those specific for nucleoprotein (NP) and polymerase acidic (PA) genes of influenza virus were able to inhibit both, PR8 and WSN strains of influenza virus.*** The same siRNAs were also able to inhibit growth of influenza virus in embryonated eggs.
- **Dengue** is a serious health problem in most tropical countries. The major vector for transmission of dengue viruses (types 1-4) is *Aedes aegypti* mosquito. Dengue virus replication and transmission was significantly reduced by expressing RNAi in *Aedes aegypti* mosquitoes. Attempts to develop transgenic mosquitoes that express dengue virus specific siRNA in targeted tissues such as midgut or salivary gland, are ongoing with ultimate aim to reduce or block vector competence for dengue virus transmission.
- **Hepatitis A virus (HAV)** ***is a major cause of acute liver failure in children in India. HAV RNA replication in HuhT7 cells was inhibited by siRNA.*** Combinations of siRNAs directed against two different genes were more effective and these treatments did not affect expression of endogenous cellular genes.

  Transfection of siRNA targeted towards hepatitis B surface antigen (HBsAg) region into HepG2.2. 15 cells that were constitutively producing HBV particles, resulted in reduction of more than 80 per cent of HBsAg and HBeAg secretion in the culture medium. Additionally, in a mouse model, which generates HBV particles on injection with HBV plasmid, ***co-administration of the plasmid and siRNA significantly inhibited the virus specific transcripts, antigens and DNA in mouse liver and sera.***
- **Hepatitis C virus** replicon system supports HCV replication but does not produce infectious virus in Huh-7 hepatoma cells. ***siRNA treatment of these cells reduced the HCV specific RNA synthesis by 80-fold and cured >98 per cent of cells.*** The inhibitory action of siRNA was so specific that two HCV variants that differed only by 3 nucleotides in the target sequence required specific homologous siRNAs for their inhibitory effects.

SECTION THREE

These examples **suggest that sequence-specific siRNAs interfere with virus replication. However, it is also possible to direct siRNAs to selectively degrade cellular mRNAs, protein products of which interact with viral proteins or in some ways are essential for virus replication.**

Future prospects: There are numerous viral diseases but unfortunately few antiviral drugs are available, viz., acyclovir and its derivatives for herpes group of viruses, alpha-interferon for hepatitis C and B viruses and anti-retrovirals for HIV. The very idea that siRNA can be a therapeutic drug against many diseases is now becoming acceptable. The major question that has always perturbed scientists working on gene therapy is the issue of 'delivery'. How do we deliver siRNAs to target organs? How long would its effect last? Will there be side effects of treatment? The siRNA technology is in its infancy. The questions raised will certainly be addressed and rapid technological advances may even resolve some of these perpetual and burning issues. Surely, ***siRNA as antiviral agent has opened new vistas for possible control of many viral diseases.***

Germline Gene Therapy

Two categories of cells make up the mammalian body, **germline** and **somatic cells.** In genetic engineering, changes only to the germline can pass to the next generation.

The **germline starts** as the fertilised egg, or zygote. This cell divides into a cluster of physiologically identical blastomeres which form a hollow ball called the blastocyst. Several days later the dividing cells become differentially committed to form particular parts of the embryo. One group of cells, is set aside very early from the embryonic yolk sac to continue the germline, migrating to the gonads (testes and ovaries) and later form the gametes (sperm and eggs, which pass their genes to the next generation).

Somatic Cells make up the rest of the body. Changes to their genes do not pass to the next generation. Indeed, a central tenet of Mendelian genetics, the Weissman boundary, asserts that nothing that happens to the somatic cells or tissues of a mammal will have any effect whatsoever on the genetic information transmitted to its offspring. Many research laboratories are currently able to inject DNA successfully into fertilised eggs of frogs, mice and other mammals, but at this time all gene therapy in humans introduces DNA only into somatic cells. If germline gene therapy is ever perfected it will have certain advantages over somatic therapy because changes could be incorporated into every cell of the body, even those inaccessible to somatic techniques. Of course, with germline therapy geneticists would have to incorporate effective genetic control sequences along with the transgenes to ensure that they were expressed only in the intended cell types at the intended times. ***Germline engineering would have the distinct disadvantage, however, that its genetic modifications would be applied to the first cell of the embryo and hence (unlike somatic engineering) could not possibly respond to unforeseen problems arising in an adult.***

Two stages of the germline are suitable for genetic engineering, the released egg, (before or after fertilisation with sperm (when it is known as a Zygote) and ***cells at the* developmental stage of blastomeres,** which are the cells into which the egg divides during cleavage. Both the egg and blastomeres have attractive features for genetic engineering:

1. **The egg** is a very large cell, relatively easy to manipulate and inject with DNA. Remarkably, if DNA is injected into an egg, it will often stably integrate into one of the chromosomes, and therefore be incorporated into all subsequent cells of the body. But, complications can arise. Sometimes a segment of DNA introduced into a fertilised egg will not become integrated into a chromosome until after the egg divides. Then only some of the cells of the embryo will contain a copy of that DNA. That resulting animal will be a mosaic of modified and unmodified cell types. Also, several copies of the introduced DNA can become integrated into different chromosomes. In this case different genotypes will segregate during subsequent generations. In order to obtain a stable new strain, several generations of descendants must be examined and animals with correct genotype selected.
2. **The blastomere stage is especially convenient for genetic engineering because the cells can be grown and manipulated in the test tube.** Lines of cells at this developmental stage, called **embryonal stem (ES) cells**, can be propagated indefinitely in cell culture. One or several ES cells can be injected into a blastocyst (obtained by culturing a fertilised egg for several division cycles in a Petri dish) and then implanted in a surrogate mother. Both types of cells in the hybrid blastocyst can contribute patches of cells to the final adult animal. This holds for both somatic and germline tissues.

The **resulting animal is therefore a mosaic** or **chimera**; an ***individual composed of a mixture of genetically different cells.*** The genotypes of the two types of cells remain distinct, they do not blend. Thus, individual sperm or eggs may be derived from either the ES cells or those of the fertilised egg. In the former case, the resulting progeny has the genetic constitution expected if the test tube of cultured ES cells had been one of its parents.

Embryonic stem cells have many advantages for genetic engineering: A culture of millions of identical cells can be treated with a DNA preparation. Once a suitable selection scheme is devised, a single altered cell can be plucked out from the huge excess of unaltered cells, and grown into a large homogeneous population whose genotype can be analysed in detail. Conceivably this culture can be subjected to a second round of genetic engineering, a third, fourth and so on. In addition, viable samples of the cultures can be preserved for long periods of time by freezing, and any number of mosaic embryos can be constructed from these engineered cell lines. Finally, the source of the egg used to obtain the blastocysts into which the ES cells are placed will not affect the final engineered animal. Eggs can come from "any old breeding stock".

Many of these same advantages also apply to technological extension that **uses nuclear transfers**. Instead of introducing the cultured cells into a blastocyst, the nucleus of one cultured cell is substituted for the nucleus of a fertilised egg. This technique has not been used extensively by genetic engineers but its spectacular recent use in producing clones of animals with identical genotype has sharply increased interest in it. One advantage is that an animal with the genetic constitution of a cultured cell can be obtained in one generation. Another is that cultures of a wider variety of cell types probably can be used. Some selection procedures may be easier to carry out on cultures of cells more differentiated than ES cells. ***The technique of nuclear transplantation pioneered to produce genetically identical clones may find its greatest importance in engineering changes in genomes.***

Germline genetic engineering is still in its infancy. Already, however, it has been important for producing a variety of types of specially altered animals. Some examples are:

- Cows with elevated milk production.
- Sheep which synthesise a valuable hormone or enzyme in the udder and secrete it into the milk. This is an especially convenient source for purifying large scale amounts of medical products.
- "Knock-out" mice with specific genes inactivated for use in analysing their function.
- Mice with a genetically engineered deficiency which mimics some human disease. These mice have provided an important approach for developing treatments of diseases.

Geneticists are rapidly expanding their bag of techniques for gentic engineering and their knowledge about the genetic basis of disease. Considering this unprecedented growth in genetic understanding, any serious consideration of future options in medicine should not exclude the possibilities of human germline engineering.

What is the Current Status of Gene Therapy Research?

The Food and Drug Administration (FDA) has not yet approved any human gene therapy product for sale. Current gene therapy is experimental and has not proven very successful in clinical trials. Little progress has been made since the first gene therapy clinical trial began in 1990.

- On 14th September, 1990, a girl suffering from ***Adenosine deaminase deficiency*** (severe immunodeficiency) was treated successfully by transferring the normal gene for adenosine deaminase.
- In 1999, gene therapy suffered a major setback with the death of 18-year-old Jesse Gelsinger. Jesse was participating in a gene therapy trial for ***ornithine transcarboxylase deficiency (OTCD).*** He died from multiple organ failures 4 days after starting the treatment. His death is believed to have been triggered by a severe immune response to the adenovirus carrier.
- Another major blow came in January 2003, when the FDA placed a temporary halt on all gene therapy trials using retroviral vectors in blood stem cells. FDA took this action after it learned that a second child treated in a French gene therapy trial had developed a leukaemia-like condition. Both this child and another who had developed a similar condition in August 2002 had been successfully treated by gene therapy for X-linked severe combined immunodeficiency disease (X-SCID), also known as **bubble baby syndrome**.

FDA's Biological Response Modifiers Advisory Committee (BRMAC) met at the end of February 2003 to discuss possible measures that could allow a number of retroviral gene therapy trials for treatment of life-threatening diseases to proceed with appropriate safeguards. FDA has yet to make a decision based on the discussions and advice of the BRMAC meeting.

What Factors have kept Gene Therapy from becoming an effective Treatment for Genetic Disease?

- **Short-lived nature of gene therapy:** ***Before gene therapy can become a permanent cure for any condition, the therapeutic DNA introduced into target cells must remain functional and the cells containing the therapeutic DNA must be long-lived and stable.*** Problems with integrating therapeutic DNA into the genome and the rapidly dividing nature of many cells prevent gene therapy from achieving any long-term benefits. Patients will have to undergo multiple rounds of gene therapy.

- **Immune response:** Anytime a foreign object is introduced into human tissues, the immune system is designed to attack the invader. The risk of stimulating the immune system in a way that reduces gene therapy effectiveness is always a potential risk. Furthermore, the immune system's enhanced response to invaders it has seen before makes it difficult for gene therapy to be repeated in patients.
- **Problems with viral vectors:** Viruses, while the carrier of choice in most gene therapy studies, present a variety of potential problems to the patient-toxicity, immune and inflammatory responses, and gene control and targeting issues. In addition, there is always the fear that the viral vector, once inside the patient, may recover its ability to cause disease.
- **Multigene disorders:** Conditions or disorders that arise from mutations in a single gene are the best candidates for gene therapy. Unfortunately, some of the most commonly occurring disorders, such as heart disease, high blood pressure, Alzheimer's disease, arthritis, and diabetes, are caused by the combined effects of variations in many genes. Multigene or multifactorial disorders such as these would be especially difficult to treat effectively using gene therapy.

What are some of the Ethical Considerations for using Gene Therapy?

What is normal and what is a disability or disorder, and who decides?

- Are disabilities diseases? Do they need to be cured or prevented?
- Does searching for a cure demean the lives of individuals presently affected by disabilities?
- Is somatic gene therapy (which is done in the adult cells of persons known to have the disease) more or less ethical than germline gene therapy (which is done in egg and sperm cells and prevents the trait from being passed on to further generations)? In cases of somatic gene therapy, the procedure may have to be repeated in future generations.
- Preliminary attempts at gene therapy are exorbitantly expensive. Who will have access to these therapies? Who will pay for their use?

Future Gene Therapy

The most likely candidates for future gene therapy trials will be rare disease such as **Lesh-Nyhan syndrome**, a distressing disease in which the patients are unable to manufacture a particular enzyme. This ***leads to a bizarre impulse for self-mutilation, including very severe biting*** of the lips and fingers. The normal version of the defective gene in this disease has now been cloned.

If gene therapy dose become practicable, the biggest impact would be on the treatment of disease where the normal gene needs to be introduced into only one organ. One such disease is **phenylketonuria (PKU).** PKU affects about one in 12,000 white children, and if not treated early can result in severe mental retardation. The disease is caused by a defect in a gene producing a liver enzyme. If detected early enough, the child can be placed on a special diet for their first few years, but this is very unpleasant and can lead to many problems within the family.

The types of gene therapy described thus far all have one factor in common: That is, that the tissue being treated are somatic (somatic cells include all the cells of the body, excluding sperm cells and egg cells). In contrast to this is the replacement of defective genes in the germline cells (which contribute to the genetic heritage of the offspring). **Gene therapy in germline cells has the potential to affect not only the individual being treated, but also his or her children as well.** Germline therapy would change the genetic pool of the entire human species, and future generations would have to live with that change. In addition to these ethical problems, a number of technical difficulties would make it unlikely that germline therapy would be tried on human in the near future.

Before treatment for a genetic disease can begin, an accurate diagnosis of the genetic defect needs to be made. It is here that biotechnology is also likely to have a great impact in the near future. Genetic engineering research has produced a powerful tool for pinpointing specific disease rapidly and accurately. Short pieces of DNA called **DNA probes** can be designed to stick very specifically to certain other pieces of DNA. The technique relies upon the fact that complementary pieces of DNA stick together. DNA probes are more specific and have the potential to be more sensitive than conventional diagnostic methods, and it should be possible in the near future to distinguish between defective genes and their normal counterparts, an important development. Already, the genes for duchenne muscular dystrophy, cystic fibrosis, and retinoblastoma have been identified, and more such information is emmerging all the time.

Protein therapeutics currently is manufactured by placing genes in laboratory-cultured organisms that produce the proteins coded by those genes. Examples of such manufactured proteins include insulin, growth hormone, and erythropoietin, all of which must be injected frequently into the patient.

Recent gene therapy approaches promise to avoid these repeated injections, which can be painful, impractical, and extremely expensive. One method uses a new vector called adeno-associated virus, an organism that causes no known disease and doesn't trigger patient immune response. The vector takes up residence in the cells, which ***then express the corrected gene to*** *manufacture the protein. In haemophilia treatments, for example, a gene-carrying vector could be injected into a muscle, prompting the muscle cells to produce Factor IX and thus prevent bleeding. This method would end the need for injections of Factor IX—a derivative of pooled blood products and a potential source of HIV and hepatitis infection. In studies by Wilson and Kathy High (University of Pennsylvania),* patients have not needed Factor IX injections for more than a year. In gene therapies such as those described above, the introduced gene is always "on" so the protein is always being expressed, possibly even in instances when it isn't needed. Wilson described a newer permutation in which the vector contains both the protein-producing gene expression. This may prove to be one of gene therapy's most useful applications as scientists begin to consider it in many other contexts, he said. Wilson's group is conducting experiments with ARIAD Pharmaceuticals to study the modulation of gene expression.

New Approaches to Gene Therapy

There are times, though, when adding a "good" copy of the gene won't solve the problem. For example, if the mutated gene encodes a protein that prevents the normal protein from doing its job, adding back the normal gene won't help. ***Mutated genes that function this way are called* dominant negative** *(Fig. 20.5)*.

How to Deal with a Dominant Negative?

To address this situation, you could either repair the mutated gene's product, or you could get rid of it altogether. Here are *some of the newest methods that scientists are developing as potential approaches to gene therapy.*

Each of these techniques also requires a specific and efficient means of delivering the gene to the target cell.

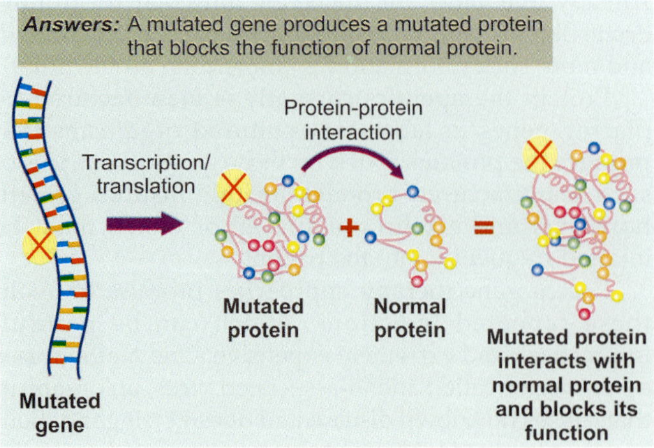

Fig. 20.5: What is a dominant negative mutation?

a. A Technique for Repairing Mutations

- **SMaRT ™**

The term SMaRT™ stands for **Spliceosome-Mediated RNA Transplicing.** *This technique targets and repairs the messenger RNA (mRNA) trancripts copied from the mutated gene.* Instead of attempting to replace the entire gene, this technique *repairs just the section of the mRNA transcript* that contains the mutation.

The sequence of a human gene contains regions that encode the protein (called exons) and regions that don't encode the protein (called intron).

After a gene is copied into mRNA, the cell uses RNA_based machinery called spliceosomes to cut out the non-coding introns and splice the exons together.

SMaRT™ involves **three steps:**

1. Delivery of an RNA strand that pairs specifically with the intron next to the mutated segment of mRNA. Once bound, this RNA strand prevents spliceosomes from including the mutated segment in the final, spliced RNA product.
2. Simultaneous delivery of a correct version of the segment to replace the mutated piece in the final mRNA product.
3. Translation of the repaired mRNA to produce the normal, functional protein

SMaRT™ is a trademark of intronn, Inc.

b. Techniques to Prevent the Production of a Mutated Protein: Triple-helix-Forming Oligonucleotides (Fig. 20.6)

This technique *involves the delivery of short, single-stranded pieces of DNA, called oligonucleotides, that bind specifically in the groove between the double strands*

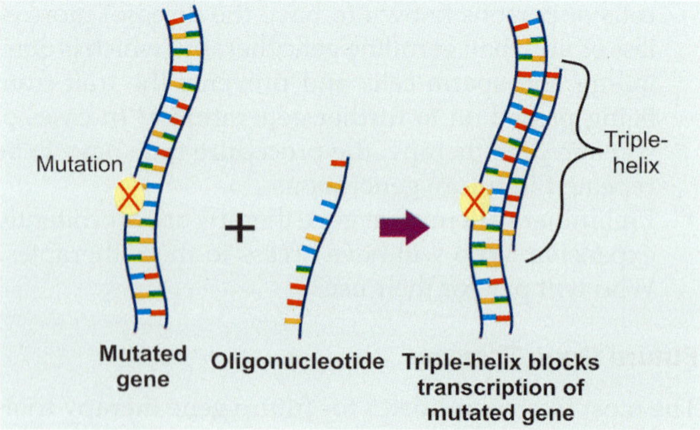

Fig. 20.6: Preventing transcription of a mutated gene using triple-helix-forming oligonucleotides

Triple-helix-forming oligonucleotide (pronounced AHL-ih-go-Nook-leotide) gene therapy targets the DNA sequence of a mutated gene to prevent its transcription.

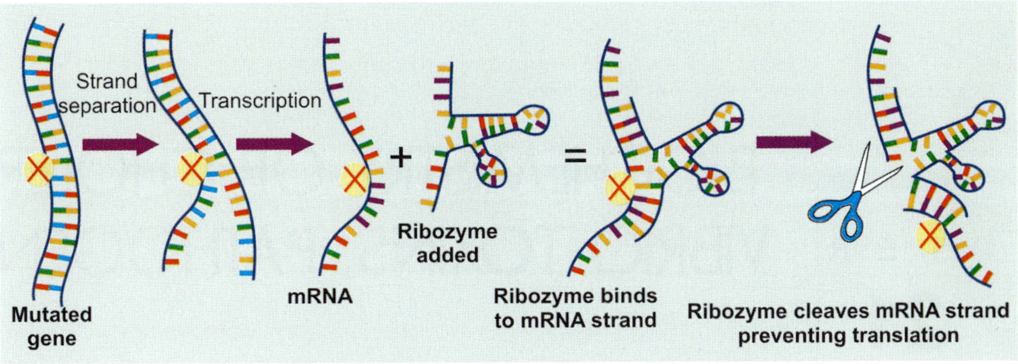

Fig. 20.7: Preventing translation using ribozyme technology

of the mutated gene's DNA. Binding produces a triple-helix structure that prevents that segment of DNA from being transcribed into mRNA.

c. Use of Ribozymes Technology

Like antisense, ribozyme (pronounced RYE-bo-ZYME) gene therapy aims to ***turn off a mutated gene in a cell by targeting the mRNA transcripts copied from the gene. This approach prevents the production of the mutated protein.***

Ribozymes are RNA molecules that act as enzymes. Most often, they act as molecular scissors that cut RNA. For example, spliceosomes (described above) are believed to be a type of ribozyme.

Ribozyme gene therapy involves the following steps:

1. Delivery of RNA strands engineered to function as ribozymes.
2. Specific binding of the ribozyme RNA to mRNA encoded by the mutated gene.
3. Cleavage of the target mRNA, preventing it from being translated into a protein **(Fig. 20.7)**.

CHAPTER 21

BIOCHEMISTRY OF CHOLERA–VIBRIO TOXINS, PATHOGENESIS

Lt Col AK Sahni, MD DNB PhD, Virology Division, Defence R&D establishment, Gwalior, **Lt Col RM Gupta** MD DNB PhD Associate Professor, Department of Microbiology, Armed Forces Medical College, Pune

INTRODUCTION

The genus ***Vibrio*** consists of gram-negative straight or curved rods, motile by means of a single polar flagellum. Vibrios are capable of both respiratory and fermentative metabolism. O_2 is a universal electron acceptor; they do not denitrify. Most species are **oxidase-positive**. In most ways vibrios are related to enteric bacteria, but they share some properties with pseudomonads as well. The family **Vibrionaceae** is found in the "Facultatively Anaerobic gram-negative rods" in Bergey's manual (1986), on the level with the family **Enterobacteriaceae**. In the revisionist taxonomy of 2001 (Bergey's manual), based on phylogenetic analysis, **Vibrionaceae**, **Pseudomonadaceae** and **Enterobacteriaceae** are all landed in the **Gammaproteobacteria.** Vibrios are distinguished from enterics by being oxidase-positive and motile by means of polar flagella. Vibrios are distinguished from pseudomonads by being fermentative as well as oxidative in their metabolism. ***Of the vibrios that are clinically significant to humans, Vibrio cholerae, the agent of cholera, is the most important.***

Most vibrios have relatively simple growth factor requirements and will grow in synthetic media with glucose as a sole source of carbon and energy. However, since vibrios are typically marine organisms, most species require 2-3 per cent NaCl or a sea water base for optimal growth. Vibrios vary in their nutritional versatility, but some species will grow on more than 150 different organic compounds as carbon and energy sources, occupying the same level of metabolic versatility as ***Pseudomonas***. In liquid media vibrios are motile by polar flagella that are enclosed in a sheath continuous with the outer membrane of the cell wall. On solid media they may synthesise numerous lateral flagella which are not sheathed.

Vibrios are one of the most common organisms in surface waters of the world. They occur in both marine and freshwater habitats and in associations with aquatic animals. Some species are bioluminescent and live in mutualistic associations with fish and other marine life. Other species are pathogenic for fish, eels, and frogs, as well as other vertebrates and invertebrates.

V. cholerae and ***V. parahaemolyticus*** are pathogens of humans. Both produce diarrhoea, but in ways that are entirely different. *V. parahaemolyticus* is an invasive organism affecting primarily the colon; ***V. cholerae is noninvasive, affecting the small intestine through secretion of an enterotoxin. Vibrio vulnificus*** is an emerging pathogen of humans. This organism causes **wound infections,** *gastroenteritis*, or a **syndrome known** as **primary septicemia.**

Campylobacter jejuni (formerly ***Vibrio foetus***), is now moved to the class **Epsilonproteobacteria** in the family **Campylobacteriaceae**. ***Campylobacter jejuni*** has been associated with dysentery-like gastroenteritis, as well as with other types of infection, including bacteremic and central nervous system infections in humans. Another vibrio-like organism, ***Helicobacter pylori*** causes duodenal and gastric ulcers and gastric cancer. It is also reclassified into the class **Epsilonproteobacteria** family ***Helicobacteriaceae.***

Cholera

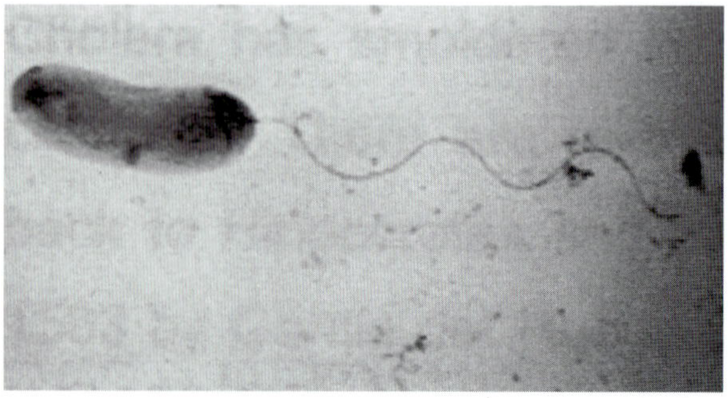

Fig. 21.1: *Vibrio cholerae*

Cholera (frequently called **Asiatic cholera** or **epidemic cholera**) is a severe diarrhoeal disease caused by the bacterium ***Vibrio cholerae*** (Fig. 21.1). ***Transmission to humans is by water or food.*** The natural reservoir of the organism is not known. It was long assumed to be humans, but some evidence suggests that it is the aquatic environment.

V. cholerae produces cholera toxin, the model for enterotoxins, whose action on the mucosal epithelium is responsible for the characteristic diarrhoea of the disease cholera. In its extreme manifestation, cholera is one of the most rapidly fatal illnesses known. A healthy person may become hypotensive within an hour of the onset of symptoms and may die within 2-3 hours if no treatment is provided. More commonly, the disease progresses from the first liquid stool to shock in 4-12 hours, with death following in 18 hours to several days.

The **clinical description** of cholera begins with ***sudden onset of massive diarrhoea***. The patient may ***lose gallons of protein-free fluid*** and ***associated electrolytes, bicarbonates and ions within a day or two.***

This results from the ***activity of the cholera enterotoxin which activates the adenylate cyclase enzyme in the intestinal cells,*** converting them into pumps which extract water and electrolytes from blood and tissues and pump it into the lumen of the intestine.

This loss of fluid **leads to dehydration, anuria, acidosis** and **shock.** The watery diarrhoea is speckled with flakes of mucus and epithelial cells **(rice-water stool)** and contains enormous numbers of vibrios.

The ***loss of potassium ions*** may result in cardiac complications and circulatory failure. Untreated cholera frequently results in high (50-60%) mortality rates.

History and Spread of Epidemic Cholera

Cholera has smoldered in an endemic fashion on the Indian subcontinent for centuries. There are references to deaths due to dehydrating diarrhoea dating back to Hippocrates and Sanskrit writings. **Epidemic cholera was described in 1563 by Garcia del Huerto**, a Portuguese physician at Goa, India. The mode of transmission of cholera by water was proven in 1849 by **John Snow**, a London physician. ***In 1883, Robert Koch successfully isolated the cholera vibrio from the intestinal discharges of cholera patients*** and proved conclusively that it was the agent of the disease.

The first long-distance spread of cholera to Europe and Americas began in 1817, such that by the early 20th century, six waves of cholera had spread across the world in devastating epidemic fashion. Since then, until the 1960s, the disease contracted, remaining present only in southern Asia. **In 1961**, the **"El Tor" biotype *(distinguished from classic biotypes by the production of hemolysins)*** reemerged and produced a major epidemic in the Philippines to initiate a **seventh global pandemic** (Fig. 21.2). Since then, this biotype has spread across Asia, the Middle East, Africa, and parts of Europe.

There are several characteristics of the El Tor strain that confer upon it a high degree of "epidemic virulence" allowing it to spread across the world as previous strains have done. ***First, the ratio of cases to carriers is much less than in cholera*** due to classic biotypes (1:30-100 for El Tor vs. 1:2-4 for "classic" biotypes). **Second,** ***the duration of carriage after infection is longer for the El Tor strain than the classic strains.*** **Third,** the El Tor strain survives for longer periods in the extraintestinal

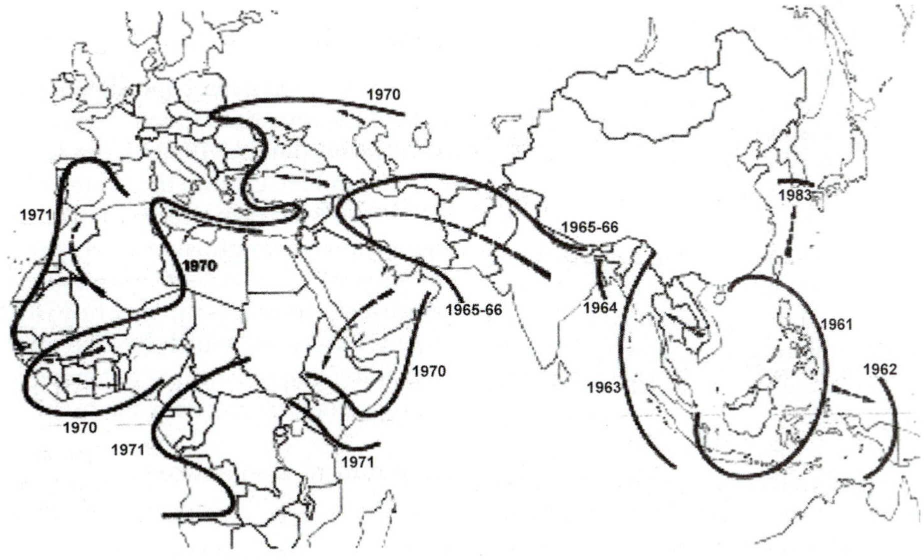

Fig. 21.2: The global spread of cholera during the seventh pandemic, 1961-1971. (CDC)

SECTION THREE

environment. ***Between 1969 and 1974, El Tor replaced the classic strains in the heartland of endemic cholera, the Ganges River Delta of India.***

El Tor broke out explosively in Peru in 1991 (after an absence of cholera there for 100 years), and spread rapidly in Central and South America, with recurrent epidemics in 1992 and 1993. From the onset of the epidemic in January 1991 through September 1, 1994, a total of 1,041,422 cases and 9,642 deaths (overall case-fatality rate: 0.9%) were reported from countries in the Western Hemisphere to the Pan American Health Organisation. In 1993, the numbers of reported cases and deaths were 204,543 and 2362, respectively.

In 1982, in Bangladesh, a classic biotype resurfaced with a new capacity to produce more severe illness, and it rapidly replaced the El Tor strain which was thought to be well-entrenched. This classic strain has not yet produced a major outbreak in any other country.

In **December, 1992, a large epidemic of cholera began in Bangladesh,** and large numbers of people have been involved. The organism has been characterised as ***V. cholerae*** **O139 "Bengal"**. ***It is derived genetically from the El Tor pandemic strain but it has changed its antigenic structure*** such that there is no existing immunity and all ages, even in endemic areas, are susceptible. The epidemic has continued to spread. and *V. cholerae* **O139** has affected at least 11 countries in southern Asia. Specific totals for numbers of *V. cholerae* O139 cases are unknown because affected countries do not report infections caused by O1 and O139 separately.

In April 1997, a cholera outbreak occurred among 90,000 Rwandan refugees residing in temporary camps in the ***Democratic Republic of Congo***. During the first 22 days of the outbreak, 1521 deaths were recorded, most of which occurred outside of health-care facilities.

In the United States, cholera was prevalent in the 1800s but **has been virtually eliminated *by modern sewage and water treatment systems.*** However, as a result of improved transportation, more persons from the United States travel to parts of Latin America, Africa, or Asia where epidemic cholera is occurring. US travellers to areas with epidemic cholera may be exposed to the bacterium. In addition, travellers may bring contaminated seafood back to the United States. A few foodborne outbreaks have been caused by contaminated seafood brought into this country by travellers. Greater than 90 per cent of the cases of cholera in the US have been associated with foreign travel.

V. cholerae may also live in the environment in brackish rivers and coastal waters. **Shellfish eaten raw have been a source of cholera**, and a few persons in the United States have contracted cholera after eating raw or undercooked shellfish from the Gulf of Mexico.

Antigenic Variation and LPS Structure in *Vibrio cholerae*

Antigenic variation plays an important role in the epidemiology and virulence of cholera. The emergence of the Bengal strain, mentioned above, is an example. The flagellar antigens of *V. cholerae* are shared with many water vibrios and therefore are of no use in distinguishing strains causing epidemic cholera. **O-antigens, however, do distinguish strains of *V. cholerae* into 139 known serotypes.** Almost all of these strains of *V. cholerae* are nonvirulent. Until the emergence of the Bengal strain (which is "non-O1") a single serotype, designated O1, has been responsible for epidemic cholera. However, there are **three distinct O1 biotypes**, named **Ogawa, Inaba** and **Hikojima**, and each biotype may display the "classical" or El Tor phenotype. ***The Bengal strain (O139) is a new serological strain with a unique O-antigen which partly explains the lack of residual immunity.***

Antigenic Determinants of *Vibrio cholerae*

Serotype	*O-Antigens*
• **Ogawa**	A, B
• **Inaba**	A, C
• **Hikojima**	A, B, C

Endotoxin is present in *Vibrio cholerae* as in other gram-negative bacteria. Fewer details of the chemical structure of *Vibrio cholerae* LPS are known than in the case of *E. coli* and *Salmonella,* but some unique properties have been described. Most importantly, variations in LPS occur *in vivo* and *in vitro*, which may be correlated with reversion in nature of nonepidemic strains to classic epidemic strains and vice versa.

Cholera Toxin

Cholera toxin **activates the *adenylate cyclase* enzyme in cells of the intestinal mucosa leading to increased levels of intracellular cAMP, and the secretion of H_2O, Na^+, K^+, Cl^-, and HCO_3^- into the lumen of the small intestine.** The effect is dependent on a **specific receptor, monosialosyl ganglioside (GM1 ganglioside)** present on ***the surface of intestinal mucosal cells.*** The bacterium produces an invasin, **neuraminidase**, during the colonisation stage which has the interesting property of degrading gangliosides to the monosialosyl form, which is the specific receptor for the toxin.

Structure of Choleratoxin

The toxin has been characterised and contains:

- **Five binding (B) subunits** of 11,500 daltons
- **An active (A1) subunit of 23,500 daltons**

- **A bridging piece (A2) of 5,500 daltons** that links A1 to the 5B subunits.

Once it has entered the cell, the A1 subunit enzymatically transfers ADP ribose from NAD to a protein (called Gs or Ns), that regulates the adenylate cyclase system which is located on the inside of the plasma membrane of mammalian cells.

Enzymatic Reaction

Enzymatically, fragment A1 catalyses the transfer of the ADP-ribosyl moiety of NAD to a component of the adenylate cyclase system. The process is complex. Adenylate cyclase (AC) is activated normally by a regulatory protein GS and GTP; however activation is normally brief because another regulatory protein (Gi), hydrolyses GTP. The normal situation is described as follows (Fig. 21.3).

The A1 fragment catalyses the attachment of ADP-Ribose (ADPR) to the regulatory protein forming Gs-ADPR from which GTP cannot be hydrolysed. Since GTP hydrolysis is the event that inactivates the **adenylate cyclase, the enzyme remains continually activated.** This situation can be illustrated as follows (Fig. 21.4).

Thus, the **net effect** of the toxin is to **cause cAMP to be produced at an abnormally high rate which stimulates mucosal cells to pump large amounts of Cl^-** into the intestinal contents. H_2O, Na^+ and other electrolytes follow due to the osmotic and electrical gradients caused by the loss of Cl^-. The lost H_2O and electrolytes in mucosal cells are replaced from the blood. Thus, ***the toxin-damaged cells become pumps for water and electrolytes causing the diarrhoea, loss of electrolytes, and dehydration that are characteristic of cholera.***

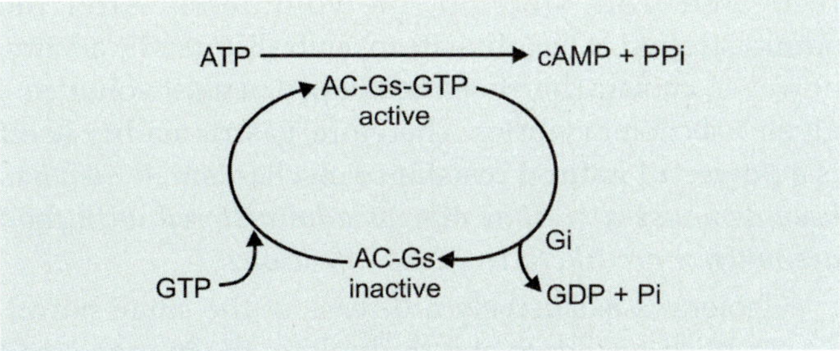

Fig. 21.3: Enzymatic reaction (active and inactive)

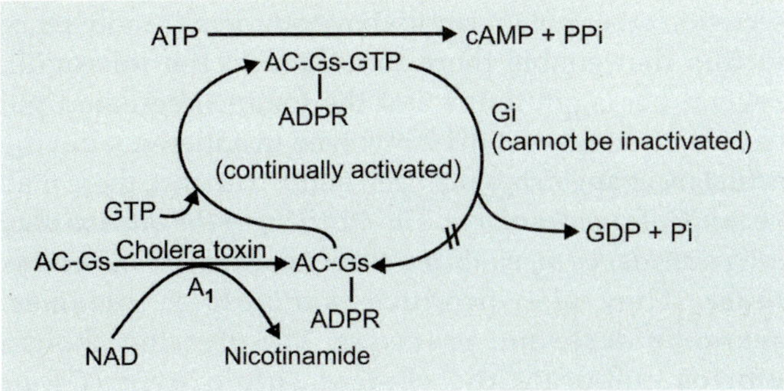

Fig. 21.4: Enzymatic reaction

Chemistry of Colonisation of the Small Intestine

There are several characteristics of pathogenic *V. cholerae* that are important **determinants of the colonisation** process. These include **(i) adhesins, (ii) neuraminidase, (iii) motility, (iv)** ***chemotaxis*** and **toxin production**. If the bacteria are able to survive the gastric secretions and low pH of the stomach, they are well adapted to survival in the small intestine. *V. cholerae* is resistant to bile salts and can penetrate the mucus layer of the small intestine, possibly aided by secretion of neuraminidase and proteases (mucinases). They withstand propulsive gut motility by their own swimming ability and chemotaxis directed against the gut mucosa.

Specific adherence of V. cholerae to the intestinal mucosa is probably mediated by long filamentous fimbriae that form bundles at the poles of the cells. These fimbriae have been termed **Tcp pili (for *toxin coregulated pili*)**, because **expression of these pili genes is coregulated with expression of the cholera toxin genes.** Not much is known about the interaction of Tcp pili with host cells, and the host cell receptor for these fimbriae has not been identified. ***Tcp pili share amino acid sequence similarity with N-methylphenylalanine pili of Pseudomonas and Neisseria.***

Two other possible adhesins in *V. cholerae* are:

- **A surface protein that agglutinates red blood cells (hemagglutinin)**
- **A group of outer membrane proteins which are products of the acf (accessory colonisation factor) genes. acf mutants** have been shown to **have reduced ability to colonise** the intestinal tract. It has been suggested that *V. cholerae* might use these nonfimbrial adhesins to mediate a tighter binding to host cells than is attainable with fimbriae alone.

V. cholerae produces a protease originally called **mucinase** that degrades different types of protein including **fibronectin, lactoferrin** and **cholera toxin** itself. Its role in virulence is not known but it probably is not involved in colonisation since mutations in the mucinase gene (***designated*** **hap** for **hemagglutinin protease**) do not exhibit reduced virulence. ***It has been suggested that the mucinase might contribute to detachment rather than attachment.*** Possibly the vibrios would need to detach from cells that are being sloughed off of the mucosa in order to reattach to newly formed mucosal cells.

SECTION THREE

Genetic Organisation and Regulation of Virulence Factors in *Vibrio cholerae*

In *Vibrio cholerae*, **the production of virulence factors is regulated at several levels.** Regulation of genes at the transcriptional level, especially the genes for toxin production and fimbrial synthesis, has been studied in the greatest detail.

- *V. cholerae* **enterotoxin** is a product of *ctx* genes
- *ctxA encodes the A subunit of the toxin*
- *ctx***B** *encodes the B subunit.*

The genes are part of the same operon. The transcript (mRNA) of the *ctx* operon has two ribosome binding sites (rbs), one upstream of the A coding region and another upstream of the B coding region. ***The rbs upstream of the B coding region is at least seven-times stronger than the rbs of the A coding region. In this way the organism is able to translate more B proteins than A proteins, which is required to assemble the toxin in the appropriate 1A: 5B proportion.*** The components are assembled in the periplasm after translation. Any extra B subunits can be excreted by the cell, but ***A must be attached to 5B in order to exit the cell. Intact A subunit is not enzymatically active, but must be nicked to produce fragments A1 and A2 which are linked by a disulfide bond. Once the cholera toxin has bound to the GM1 receptor on host cells, the A1 subunit is released from the toxin by reduction of the disulfide bond that links it to A2, and enters the cell by an unknown translocation mechanism. One hypothesis is that the 5B subunits form a pore in the host cell membrane through which the A1 unit passes.***

Transcription of the *ctx*AB operon is regulated by a number of environmental signals, including temperature, **pH, osmolarity,** and **certain amino acids**. Several other *V. cholerae* genes are coregulated in the same manner ***including the tcp operon, which is concerned with fimbrial synthesis and assembly. Thus the ctx operon and the tcp operon are part of a regulon, the expression of which is controlled by the same environmental signals.***

The proteins involved in control of this regulon expression have been identified as **ToxR**, **ToxS** and **ToxT**.

- **ToxR** is a ***transmembranous protein*** with about two-thirds of its amino terminal part exposed to the cytoplasm. ToxR dimers, but not ToxR monomers, will bind to the operator region of *ctx*AB operon and activate its transcription.
- **ToxS** is a ***periplasmic protein***. It is thought that ToxS can respond to environmental signals, change conformation, and somehow influence dimerisation of ToxR which activates transcription of the operon. ToxR and ToxS appear to form a standard two-component regulatory system with **ToxS functioning as a sensor protein that phosphorylates and thus converts ToxR to its active DNA binding form.**
- **ToxT** is a *cytoplasmic protein* that is ***a transcriptional activator of the tcp operon.*** Expression of ToxT is activated by ToxR, while ToxT, in turn, activates transcription of tcp genes for synthesis of tcp pili.

Thus, ***the ToxR protein is a regulatory protein which functions as an inducer in a system of positive control. Tox R is thought to interact with ToxS in order to sense some change in the environment and transmit a molecular signal to the chromosome which induces the transcription of genes for attachment (pili formation) and toxin production.*** It is reasonable to expect that the environmental conditions that exist in the GI tract (i.e. 37°C temperature, low pH, high osmolarity, etc.), as opposed to conditions in the extraintestinal (aquatic) environment of the vibrios, are those that are necessary to induce formation of the virulence factors necessary to infect. However, there is conflicting experimental evidence in this regard, which leads to speculation of the ecological function of the toxin during human infection.

Pathogenesis

Laboratory animal models and human volunteers have provided a detailed understanding of the pathogenesis of cholera. Initial attempts to infect healthy American volunteers with cholera vibrios revealed that the oral administration of up to 1011 living cholera vibrios rarely had an effect; in fact, the organisms usually could not be recovered from stools of the volunteers. After the administration of bicarbonate to neutralise gastric acidity, however, cholera diarrhoea developed in most volunteers given 104 cholera vibrios. Therefore, **gastric acidity itself is a powerful natural resistance mechanism.** It ***also has been demonstrated that vibrios administered with food are much more likely to cause infection.***

Cholera is exclusively a disease of the small bowel. To establish residence and multiply in the human small bowel (normally relatively free of bacteria because of the effective clearance mechanisms of peristalsis and mucus secretion), the cholera vibrios have one or more adherence factors that enable them to adhere to the microvilli. Several hemagglutinins and the toxin-coregulated pili have been suggested to be involved in adherence but the actual mechanism has not been defined. In fact, there may be ***multiple mechanisms. The motility of the vibrios may affect virulence by enabling them to penetrate the mucus layer.*** They also produce ***mucinolytic enzymes, neuraminidase,*** and ***proteases.*** The growing cholera vibrios elaborate the ***cholera enterotoxin (CT or choleragen),*** a ***polymeric protein*** (Mr 84,000) consisting of ***two major domains or regions.***

The A region (Mr 28,000), responsible for biologic activity of the enterotoxin, is linked by noncovalent interactions with the B region (Mr 56,000), which is composed of five identical noncovalently associated peptide chains of Mr 11,500.

The B region, also known as choleragenoid, binds the toxin to its receptors on host cell membranes. It is also the immunologically dominant portion of the holotoxin. The structural genes that encode the synthesis of CT reside on a ***transposon-like element in the V cholerae chromosome,*** in contrast to those for the heat-labile enterotoxins (LTs) of *E coli*, which are encoded by plasmids. The amino acid sequences of these structurally, functionally, and immunologically related enterotoxins are very similar. Their differences account for the differences in physicochemical behaviour and the antigenic distinctions that have been noted. There are at least **two antigenically related but distinct forms of cholera enterotoxin**, called **CT-1** and **CT-2**.

Classical O1 V cholerae and the Gulf Coast El Tor strains produce CT-1 whereas most other El Tor strains and O139 produce CT-2. *Vibrio cholerae* exports its enterotoxin, whereas the *E. coli* LTs occur primarily in the periplasmic space. This may account for the reported

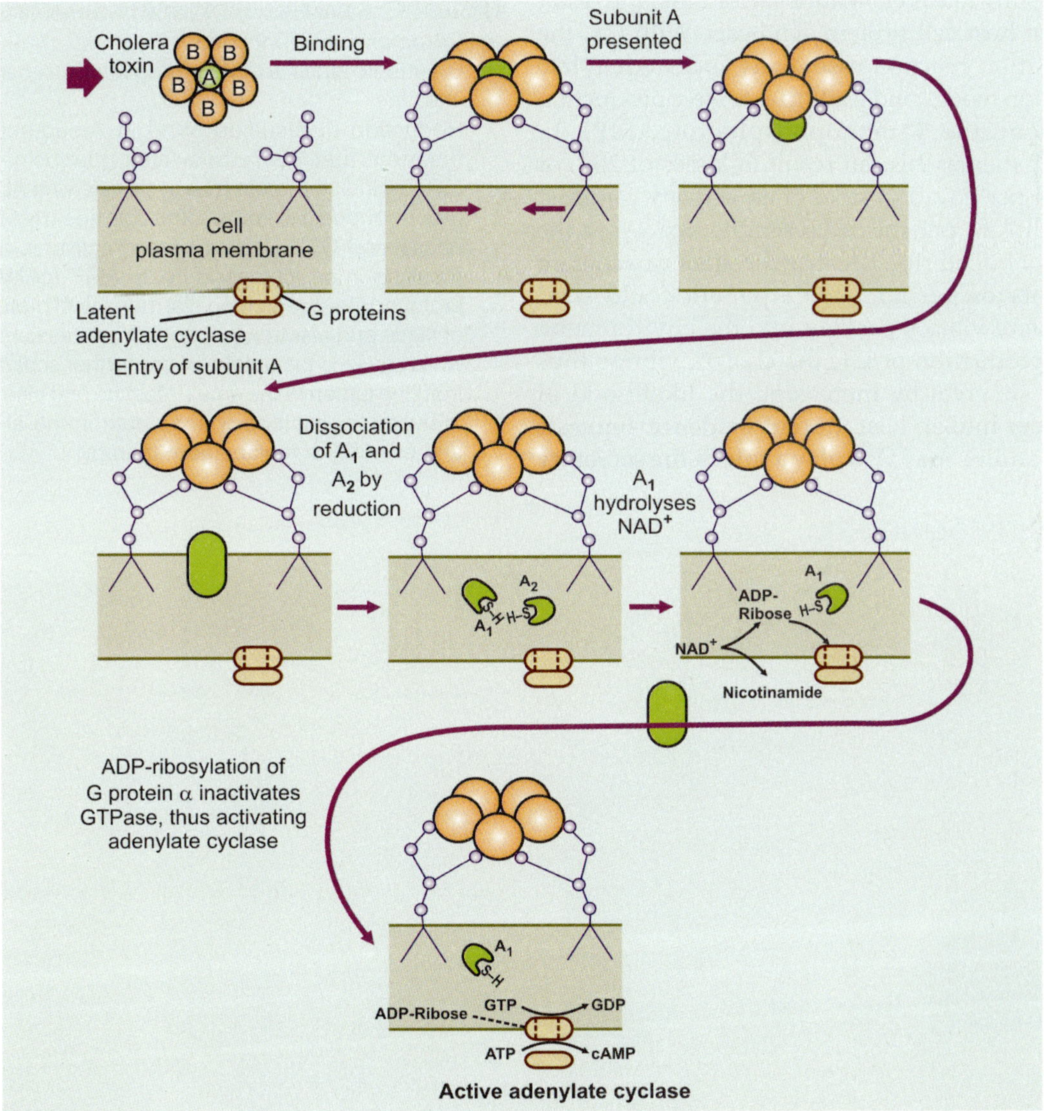

Fig. 21.5: Mechanism of action of cholera enterotoxin according to Finkelstein. Cholera toxin approaches target cell surface. B-subunits bind to oligosaccharide of GM1 ganglioside. Conformational alteration of holotoxin occurs, allowing the presentation of the A-subunit to cell surface. The A subunit enters the cell. The disulfide bond of the A-subunit is reduced by intracellular glutathione, freeing A1 and A2. NAD is hydrolysed by A1, yielding ADP-ribose and nicotinamide. One of the G proteins of adenylate cyclase is ADP-ribosylated, inhibiting the action of GTPase and locking adenylate cyclase in the "on" mode

SECTION THREE

differences in severity of the diarrhoeas caused by these organisms.

The toxins bind through region B to a glycolipid, the GM1 ganglioside, which is practically ubiquitous in eukaryotic cell membranes. Following this binding, the A region, or a major portion of it known as the A1 peptide (Mr 21,000), penetrates the host cell and enzymatically transfers ADP-ribose from nicotinamide adenine dinucleotide (NAD) to a target protein, the guanosine 5'-triphosphate (GTP)-binding regulatory protein associated with membrane-bound adenylate cyclase. Thus, CT (and LT) resembles diphtheria toxin in causing transfer of ADP-ribose to a substrate. With diphtheria toxin, however, the substrate is elongation factor 2 and the result is cessation of host cell protein synthesis. With CT, the ADP-ribosylation reaction essentially locks adenylate cyclase in its "on mode" and leads to excessive production of cyclic adenosine 5′-monophosphate (cAMP). In hospitalised patients, this can result in losses of 20 L or more of fluid per day. The stool of an actively purging, severely ill cholera patient can resemble rice water the supernatant of boiled rice. Because the stool can contain 108 viable vibrios per ml, such a patient could shed 2 × 1012 cholera vibrios per day into the environment. Perhaps by production of CT, the cholera vibrios thus ensure their survival by increasing the likelihood of finding another human host. **Recent evidence suggests that prostaglandins may also play a role in the secretory effects of cholera enterotoxin.** Vibrios have putative mechanisms in addition to CT for causing (milder) diarrheal disease. These include **Zot** (for Zonula occludens toxin) and **Ace** (for **accessory cholera enterotoxin**), and perhaps others, but their role has not been established conclusively. Certainly **CT is the major virulence factor** and the act of colonisation of the small bowel may itself elicit an altered host response (e.g. mild diarrhoea), perhaps by a transmembrane signaling mechanism (Fig. 21.5).

Summary

Cholera toxin (A-5B) (*Vibrio cholerae*):

- Chromosomally-encoded
- B-subunit binds to GM_1 ganglioside receptors in small intestine
- Reduction of disulfide bond in A-subunit activates A_1 fragment that ADP-ribosylates guanosine triphosphate (GTP)-binding protein (G_s) by transferring ADP-ribose from nicotinamide adenine dinucleotide (NAD); the ADP-ribosylated GTP-binding protein activates adenyl cyclase resulting in an increased cyclic AMP (cAMP) level and a profound life-threatening diarrhoea with profuse outpouring of fluids and electrolytes (sodium, potassium, bicarbonate) while blocking the uptake of any further sodium and chloride from the lumen of the small intestine and ultimately resulting in hypovolaemic shock and death in the absence of fluid and electrolyte replacement therapy.

SECTION

4

Metabolism

CHAPTER 22

DIGESTION AND ABSORPTION OF CARBOHYDRATES

Major Concepts

A. To study how the complex carbohydrates present in foodstuffs are broken down to simple sugars in the GI tract.

B. To study how the simple sugars are absorbed from the GI tract, into portal blood.

Specific Objectives

A. 1. Study the principal carbohydrates present in the foodstuffs which we take. List them.

2. Digestion in mouth
   - Liquid food materials, like milk, soup, fruit juice escape digestion in mouth, solid foodstuffs are masticated before they are swallowed.
   - Study the biochemical composition of saliva, with special stress to pH range, activating factors, and action of carbohydrate splitting enzymes which is α-*amylase*.
   - Learn the characteristics of α-*amylase* and its mode of action on starch, and glycogen and the products.

3. Digestion in stomach (gastric digestion): Study the biochemical composition of gastric juice, with special stress to pH ranges and enzymes present.

4. Digestion in duodenum and small intestine
   (a) ***Pancreatic juice:***
   - Study the composition of pancreatic juice and learn the role of carbohydrate splitting enzyme—pancreatic amylase

   (b) ***Intestinal juice:***
   - List the carbohydrate splitting enzymes present. Make in tabular form their pH range of action, mode of action on substrate and products.

B. Absorption

Dietary complex carbohydrates are hydrolysed in GI tract to monosaccharides. Foodstuffs also contain small amount of pentoses and mannoses.

1. Study the site and rate of absorption of monosaccharides from GI tract.
2. Learn the experimental study made by **Cori** on rate of absorption of different sugars and observations made.
3. Name process of absorption of sugars
   • Simple diffusion • Active transport • Facilitated transport.
4. Study **Wilson** and **Craine's** hypothesis. List the criteria laid down by them for sugars which are 'actively' transported.
5. Learn in detail mechanism of ***active*** transport.
   • List the sugars which are actively transported. • Characteristics of the ***receptor or carrier protein*** molecule, i.e.
   • Number of binding sites • Mobility • **Sodium dependency** • Specific nature • **Energy dependent, etc.**
   • Learn about **glucose transporters (GluT),** prepare a table showing tissue location and function.
6. Study the different experimental evidences to support the active transport hypothesis.
7. Learn what is *facilitated transport.* Name the sugar absorbed by this process, and how does it differ from active transport.
8. Study the factors that influence rate of absorption of sugars:
   - State of mucosa of GI tract • Length of contact in GI tract • Hormones • Role of vitamins

C. 1. Study lactase deficiency
   Three types: • Inherited lactase deficiency • Primary low lactase activity • Secondary low lactase activity

2. Sucrase deficiency
3. Disacchariduria
4. Monosaccharide malabsorption.

DIGESTION OF CARBOHYDRATES

Dietary carbohydrates principally consist of the *polysaccharides:* Starch and glycogen. It also contains *disaccharides:* Sucrose (cane sugar), lactose (milk sugar) and maltose and in small amounts ***monosaccharides*** like fructose and pentoses. Liquid food materials like milk, soup, fruit juice escape digestion in mouth as they are swallowed, but solid foodstuffs are masticated thoroughly before they are swallowed.

1. **Digestion in mouth:** Digestion of carbohydrates starts at the mouth, where they come in contact with saliva during mastication. Saliva contains a carbohydrate splitting enzyme called ***salivary amylase (ptyalin).***

Action of Ptyalin (Salivary Amylase)

It is ***α-amylase,*** **requires Cl^- ion for activation** and optimum pH 6.7 (range 6.6 to 6.8). ***The enzyme hydrolyzes α-1 → 4 glycosidic linkage at random deep inside polysaccharide*** molecule like starch, glycogen and dextrins, producing smaller molecules maltose, glucose and trisaccharide maltotriose. Ptyalin action stops in stomach when pH falls to 3.0.

Starch, Glycogen and Dextrins —(*α-Amylase*)→ Glucose, Maltose and Maltotriose

2. **Digestion in stomach:** Practically no action. ***No carbohydrate splitting enzymes available in gastric juice.*** Some dietary sucrose may be hydrolysed to equimolar amounts of glucose and fructose by HCl.
3. **Digestion in duodenum:** Food bolus reaches the duodenum from stomach where it meets the pancreatic juice. Pancreatic juice contains a carbohydrate-splitting enzyme ***pancreatic amylase*** (also called ***amylopsin)*** similar to salivary amylase.

Action of Pancreatic Amylase

It is also an *α-amylase,* optimum pH 7.1. Like *ptyalin* ***it also requires Cl^- for activity.*** The enzyme hydrolyses α-1→4 glycosidic linkage situated well inside polysaccharide molecule. Other criteria and end products of action similar to ptyalin.

4. **Digestion in Small Intestine**

Action of Intestinal Juice

- ***Intestinal amylase:*** This hydrolyses terminal α-1→4, glycosidic linkage in polysaccharides and oligosaccharide molecules ***liberating free glucose molecule.***
- ***Lactase:*** It is a ***β-galactosidase,*** its pH range is 5.4 to 6.0. Lactose is hydrolysed to equimolar amounts of glucose and galactose.

Lactose —(*Lactase*)→ Glucose + Galactose

- ***Isomaltase:*** It ***catalyses hydrolysis of α-1→ 6 glycosidic linkage,*** thus splitting ***α-limit dextrin*** at the branching points and producing **maltose** and **glucose**.
- ***Maltase:*** The enzyme hydrolyses the α-1→4 glycosidic linkage between glucose units in maltose molecule liberating equimolar quantities of two glucose molecules. Its pH range is 5.8 to 6.2.

Maltose —(*Maltase*)→ Glucose + Glucose

Five maltases have been identified in intestinal epithelial cells. ***Maltase V*** can act as *isomaltase* over and above its action on maltose.

- ***Sucrase:*** pH range 5.0 to 7.0. It hydrolyses sucrose molecule to form equimolar quantities of glucose and fructose. ***Maltase III and maltase IV also have sucrase activity***

Sucrose —(*Sucrase*)→ Glucose + Fructose

ABSORPTION OF CARBOHYDRATES

It is observed from above that carbohydrate digestion is complete when the food materials reach small intestine and all complex dietary carbohydrates like starch and glycogen and the disaccharides are ultimately converted to simpler monosaccharides. All monosaccharides, products of digestion of dietary carbohydrates, are practically completely absorbed almost entirely from the small intestine.

Rate of absorption diminishes from above downwards; proximal jejunum three times greater than that of distal ileum. It is also proved that some disaccharides, which escape digestion, may enter the cells lining the intestinal lumen may be by ***pinocytosis;*** and are hydrolysed within these cells. No carbohydrates higher than the monosaccharides can be absorbed directly into the bloodstream in normal health and if administered parenterally, they are eliminated as foreign bodies.

- **Cori** studied the rate of absorption of different sugars from small intestine in rat. Taking glucose absorption as 100, comparative rate of absorption of other sugars were found as follows:

Galactose >	**Glucose >**	**Fructose >**	**Mannose >**
110	**100**	**43**	**19**
Xylose >	**Arabinose**		
15	**9**		

The above study proves that ***glucose*** and ***galactose are absorbed very fast;*** fructose and mannose intermediate rate and the pentoses are absorbed slowly. ***Galactose is absorbed more rapidly than glucose.***

MECHANISMS OF ABSORPTION

Two mechanisms are suggested:

1. **Simple diffusion:** This is dependent on sugar concentration gradients between the intestinal lumen, mucosal cells and blood plasma. All the monosaccharides are probably absorbed to some extent by simple 'passive' diffusion.
2. **"Active" Transport Mechanisms**
   - Glucose and galactose are absorbed very rapidly and hence it has been suggested that they are ***absorbed actively and it requires energy.***
   - Fructose absorption is also rapid but not so much as compared to glucose and galactose, but it is definitely faster than pentoses. Hence fructose is not absorbed by simple diffusion alone and it is suggested that some mechanism facilitates its transport, called as ***facilitated transport.***

Wilson and Crane's Hypothesis of Active Transport

Wilson and Crane have shown that sugars which are 'actively' transported have several chemical features in common. They suggested that to be actively transported sugar must have the following:

- They must have ***a six-membered ring,***
- Secondly, they must have ***one or more carbon atoms attached to C 5,*** and
- Thirdly, they ***must have a –OH group at C-2*** with the same stereoconfiguration as occurs in D-glucose. ***–OH group and 5 hydroxymethyl or methyl group on the pyranose ring appear to be essential structural requirements for the active transport mechanism.***

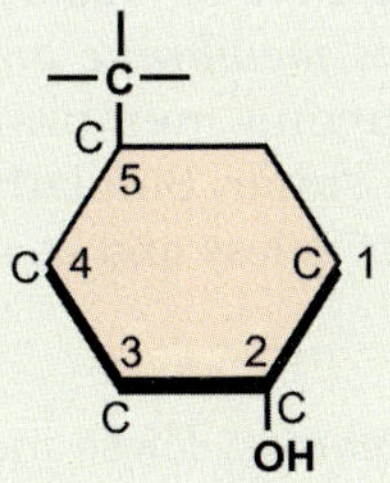

- ***Crane and his collaborators*** explain active transport by envisaging the presence of a **Carrier protein (transport protein)** in the brush border of intestinal epithelial cell.

The 'carrier protein' has the following characteristics:

- It has ***two binding sites one for sodium and another for the glucose.***
- The carrier protein is specific for sugar.
- It is mobile.
- It is ***sodium-dependent***
- It is ***energy-dependent.***

Energy: It is provided by ATP, by the interaction of the sodium dependent sugar carrier and the sodium pumps, actively transported sugars are concentrated within the cell without any back leakage of the sugar into the lumen. ***It is believed that sodium binding by the carrier-protein is pre-requisite for glucose binding.*** Sodium binding changes the conformation of the protein molecule, enabling the binding of glucose to take place and thus the absorption to occur. It is presumed that analogous "carrier protein" exists for D-galactose also. This is a ***cotransport system*** (Fig. 22.1).

Glucose Transporters (GLUT)

Several glucose transporters GLUT-1 to 7 have been described in various tissues (Table 22.1).

Evidences in favour of the cotransport system of glucose absorption:

- The dependence of the active transport of glucose upon the presence of sodium ions has been demonstrated in isolated loops of rat intestine ***by replacing the sodium of bathing fluid by K^+ and lithium. Under these circumstances, the rate of glucose transport is markedly reduced and ultimately stopped.***
- Drugs such as ***strophanthin*** and ***ouabain*** which inhibit sodium pump also inhibit active transport of sugars.
- Substances preventing the liberation of metabolic energy, such as **dinitrophenol (DNP),** also inhibits active transport of sugars.
- **Phloridzin,** a glycoside inhibits glucose transport by probably ***displacing sodium from its binding site,*** as a result glucose cannot be bound and transported.

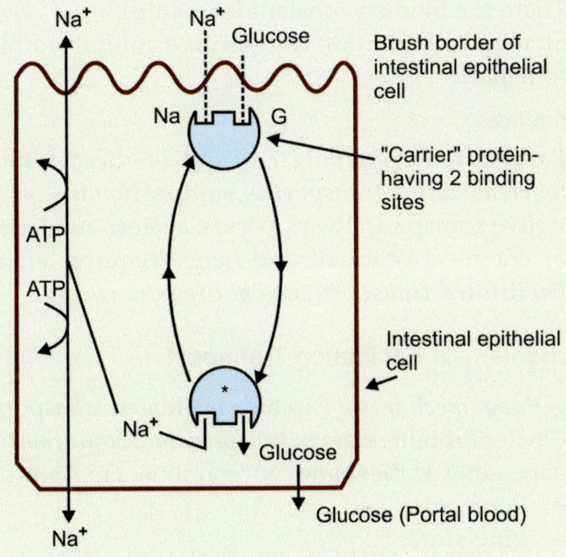

Fig. 22.1: Carrier protein and transport of glucose

SECTION FOUR

	Type of glucose transporter	Tissue locations	Functions
	Table 22.1: Major glucose transporters		
a.	**Sodium dependent transporter SGLT1**	Small intestine and kidney	Active uptake of glucose against a concentration gradient. Requires ATS for energy and Na^+
b.	**Facilitative bidirectional transporters:**		
	GLUT-1	Red blood cells, brain, kidney, colon, placenta	Glucose uptake independent of insulin.
	GLUT-2	Liver, kidney, small intestine, β-cells of pancreas	Rapid uptake or release of glucose.
	GLUT-3	Brain, kidney, placenta	Glucose uptake
	GLUT-4	Skeletal muscle, adipose tissue, heart	Insulin stimulated glucose uptake.
	GULT-5	Small intestine, testes sperms	Poor ability to transport glucose. Fructose transporter
	GLUT-7	Liver, endoplasmic reticulum	Transport glucose from ER to cytoplasm

Absorption of Other Sugars

- Sugars like D-fructose and D-mannose are probably absorbed by ***facilitated transport*** which ***requires the presence of carrier protein but does not require energy.***
- Other sugars like ***pentoses*** and ***L-isomers*** of glucose and galactose are ***absorbed passively by simple diffusion.***

Facilitated Transport Vs Active Transport

Similarities

- Both appear to involve carrier proteins.
- Both show specificity.
- Both resemble a substrate-enzyme type of reaction.
- Both have specific binding sites for solutes.
- 'Carrier' is saturable so it has maximum rate of transport.
- There is a binding constant for solute.
- Structurally similar competitive inhibitors block transport.

Differences

- ***Facilitated transport can act bi-directionally, whereas active transport is unidirectional.***
- Active transport always occurs against an electrical or chemical gradient and hence requires energy. ***Facilitated transport does not require energy.***

Mechanism of Facilitated Transport

Ping-Pong' mechanism explains facilitated transport.

- **Carrier Protein** *exists in two principal conformations* depending on the solute concentration. *Two forms are:*
  - **Pong** *state,* and
  - **Ping** *state.*
- In the **Pong** *state,* it is exposed to high concentrations of solutes, and molecules of solutes bind to specific sites on the 'carrier protein'. This occurs in lipid bilayer of the cell with high solute concentration.
- In inner side, a conformational change occurs to **Ping** *state* and the solute is discharged to the side favouring new equilibrium.
- The empty carrier protein then reverts to the original conformation "Pong" state to complete the cycle.

Factors determining facilitated transport: Rate at which solutes enter a cell by facilitated transport is determined by the following factors:

- Concentration gradient across the membrane.
- The amount of "Carrier protein" available (key control system).
- Rapidity of solute-carrier interaction.
- Rapidity of conversion of conformation state from 'Pong' to 'Ping' and *vice versa.*

Factors Influencing Rate of Absorption

1. ***State of mucous membrane and length of time of contact:*** If mucous membrane is not healthy, absorption will suffer. Similarly in hurried bowel, length of contact is less and as such absorption will be less.
2. ***Hormones***
   - ***Thyroid hormones:*** These increase absorption of hexoses and act directly on intestinal mucosa.
   - ***Adrenal cortex: Absorption decreases in adreno-cortical deficiency,*** mainly due to decreased concentration of sodium in body fluids
   - ***Anterior pituitary:*** This affects absorption mainly through its influence on thyroids. Hyperpituitarism induces thyroid overactivity and *vice versa.*
   - ***Insulin:*** This ***has no effect on absorption of glucose.***
3. ***Vitamins:*** Absorption is diminished in states of deficiency of B-vitamins, viz, thiamine, pyridoxine and pantothenic acid.

SECTION FOUR

4. ***Inherited enzyme deficiencies:*** Inherited enzyme deficiencies like sucrase and lactase can interfere with hydrolysis of corresponding disaccharides and their absorption.

CLINICAL ASPECT

DEFECTS IN DIGESTION AND ABSORPTION OF CARBOHYDRATES (including inherited disorders)

1. Lactase Deficiency

Some infants may have deficiency of the enzyme *lactase* and they show intolerance to lactose, the sugar of milk. Symptoms and signs seen in affected infants include:
- Diarrhoea and flatulence
- Abdominal cramps
- Distension.

Explanation: The above features are explained as follows:
- As lactose of milk cannot be hydrolysed due to deficiency of *lactase* enzyme, there occurs accumulation of lactose in intestinal tract, which is **osmotically active** and holds water, producing diarrhoea.
- Accumulated lactose is also fermented by intestinal bacteria which produce gas and other products, producing flatulence, distension and abdominal cramps.

Types: *Lactase* deficiency can be of **3 types.**

TYPES: Lactase deficiency
- Inherited
- Low lactase activity
  - Primary
  - Secondary

Inherited Lactase Deficiency
- Rare disorder
- Symptoms of intolerance to milk such as diarrhoea and wasting incident to fluid and electrolyte disturbances as well as inadequate nutrition, all develop very soon after birth
- ***Urine:*** Presence of lactose in urine is a prominent feature (lactosuria)
- ***Treatment:*** Feeding of lactose-free diet results in disappearance of the symptoms and marked improvement.

Low Lactase Activity

a. ***Primary low lactase activity:*** It is relatively a common syndrome. It is seen particularly among non-white population in USA as well as other parts of the world specially South East Asia including India. Intolerance to lactose is not a feature in early life and appears later in life. It is presumed to represent a gradual decline in the activity of the enzyme *lactase* in susceptible individuals.

b. ***Secondary low lactase deficiency:*** This is secondary to many GI conditions prevalent in tropics and non-tropical countries like:
   - Tropical and nontropical sprue (Celiac disease)
   - Kwashiorkor
   - Colitis and chronic gastroenteritis

Also can occur after surgery of peptic ulcer.

2. Sucrase Deficiency

Inherited deficiency of ***sucrase*** and ***isomaltase*** have been reported. Symptoms occur in early childhood with ingestion of sugars (cane sugar and table sugar) sucrose, a disaccharide. Symptoms and signs as in lactase deficiency.

3. Disacchariduria

An increase in the excretion of disaccharides may be observed in some patients with ***disaccharidase*** deficiency. As much as 300 mg or more of disaccharides may be excreted in those people and in patients with intestinal damage (e.g. sprue and celiac disease).

4. Monosaccharide Malabsorption

Inherited disorders in which glucose and/or galactose are absorbed very slowly have been reported. The reason probably is absence of "carrier protein" necessary for absorption of glucose/galactose.

I. Questions (Essay type and short notes)

1. Describe the digestion and absorption of carbohydrates from GI tract

Short Notes

a. Digestion and absorption of carbohydrates
b. Absorption of glucose
c. Absorption of monosaccharides
d. Absorption of carbohydrates

II. MCQs (Give one correct answer)

1. Which of the following sugars is absorbed by facilitated transport?
   a. Galactose b. Fructose
   c. Xylose d. Glucose
   e. Arabinose
2. Amylase present in saliva is:
   a. α-amylase b. β-amylase
   c. γ-amylase d. All of the above
   e. None of the above

3. The absorption of glucose is decreased by the deficiency of:
   a. Vitamin A
   b. Vitamin D
   c. Thiamine
   d. Vitamin B_{12}
   e. Folic acid
4. For activity of salivary amylase. Which of the following is required as activator:
   a. HCO_3^- ion
   b. Presence of Na^+
   c. Presence of K^+
   d. Chloride ion
   e. None of the above
5. Which of the following hormone increases the absorption of glucose from GI tract?
   a. Insulin
   b. Thyroid hormones
   c. Glucagon
   d. FSH
   e. LH

Answers to MCQs

1. b 2. a 3. c 4. d 5. b.

CHAPTER 23

METABOLISM OF CARBOHYDRATES

PART I

Major Concepts

A. Utilisation of glucose in the body: General outline.
B. To study catabolism of glucose to CO_2 and H_2O.

Specific Objectives

A. Various monosaccharides after absorption are carried through portal blood to liver.
- Learn the various mechanisms that operate in liver, some utilises monosaccharides and others release the glucose.
- Glucose released in systemic circulation is utilised by extrahepatic tissues.
- Learn the various mechanisms and fate of glucose in the body.

B. I. ***Glycolysis and oxidative decarboxylation of Pyruvic acid.*** There is minimal requirement for glucose in all tissues except erythrocytes and nervous tissues which require substantial amount. Glycolysis is the major pathway for utilisation of glucose and is carried out in all cells. It is a unique pathway in that it can utilise O_2 if available ***(aerobic)*** or it can function in absence of O_2 ***(anaerobic)***.

1. What is glycolysis? Define. What is meant by aerobic/and anaerobic glycolysis? Learn the other synonyms and the biomedical importance of glycolytic pathway.
2. Study the sequence of reactions involved in glycolytic pathway, enzymes and coenzymes required for each reaction.
3. Differentiate in a tabular form the role of enzyme *'hexokinase'* and *'glucokinase'* in phosphorylation of glucose.
4. Note the end-product of glycolytic pathway (a) In presence of O_2 (aerobic) and (b) In absence of O_2 (anaerobic).
5. Learn how regulation of glycolysis is done by substrate, end-products and hormones.
6. Study and list the chemicals that inhibit a particular enzyme of this pathway.
7. Learn how glycolytic pathway differs in RB cells as compared to other tissues.
   - Study ***Rapoport-Luebering shunt*** that operates in RB Cells.
   - What is ***2,3-biphosphoglycerate*** **(2,3-BPG)**? Learn its importance in RB Cells.
8. Learn the energetics (stoichiometry) of glycolysis-ATP formation in presence of O_2 and in absence of O_2.
   - What is ***substrate-level phosphorylation***? Give examples from glycolytic pathway.
   - Learn the other sources of PA and other fates, learn what is meant by ***anaplerotic reactions ('anaplerosis)***.

II. **Fate of pyruvic acid formed from glucose in tissues depends upon the *redox state* of the tissues.**

(a) If O_2 is available, the pyruvic acid is converted to 2-carbon unit "acetyl-CoA" ("active acetate") by the process of oxidative decarboxylation.

(b) If O_2 is not available/or in relative anaerobiosis, pyruvic acid is converted to lactic acid. Learn the details of the reactions involved, enzyme and coenzyme required.

III. ***TCA cycle***

- Learn TCA cycle: A common and final pathway for breakdown of "Active acetate" obtained from carb, lipids and proteins to CO_2 and H_2O ***(Third phase of catabolism)***.
- Study the reactions of TCA cycle, enzymes and coenzymes required.
- Study the energetics of TCA cycle: ATP formation in ETC and reactions which produce ATP at "Substrate level".
- Learn the chemicals that inhibit particular enzyme.
- Learn in detail the conversion of α-oxoglutarate* to succinyl-CoA ("Active succinate") by oxidative decarboxylation.
- Learn the other sources of succinyl-CoA and its fate.
- Study why TCA cycle is called ***amphibolic*** in nature?
- Learn how TCA cycle is regulated.
- Study the 'over-all' bioenergetics of glycolytic-TCA cycle and its efficiency.

*α-ketoglutarate and α-oxoglutarate are synonymous.

INTRODUCTION

Utilisation of Glucose in the Body

General Outline

After absorption of monosaccharides into the portal blood, it passes through the liver *(the first 'filter')* before entering the systemic circulation, a fact of considerable physiological and biochemical importance. ***In liver two mechanisms operate:***

- ***Withdrawal*** of carbohydrates from blood and
- ***Release*** of glucose by liver to the blood.

These two mechanisms are shown below in a tabular form in the box.

All the above processes are finely regulated in the Liver cells, control exerted at substrate level, by the end products and by hormones. ***The amount of glucose reaching the systemic circulation at any instant, will be the resultant of operation of these two groups of opposing forces.*** Once glucose is in systemic circulation, it becomes available for its utilisation by "extrahepatic tissues". Thus extrahepatic tissues are presented with carbohydrates which have already been "picked over" by the liver in a selective manner.

Hence functional state of the liver will be of prime importance and will have a profound influence on the carbohydrate metabolism on the entire organism.

Glucose is taken up by intestinal mucosal cells and kidney tubule cells by "active" transport. ***Hepatic cells are freely permeable to glucose.*** Insulin increases uptake of glucose by many extrahepatic tissues as skeletal muscle, heart muscle, diaphragm, adipose tissue, lactating mammary gland, etc.

SECTION FOUR

Utilisation of Glucose

1. **Oxidation**
   - *For Provision of Energy*

     In response to physiological needs, human body requires energy. ***Oxidation of glucose or glycogen to pyruvate and lactate by EM pathway is called glycolysis.*** Glucose is degraded by glycolysis to pyruvate, which in presence of O_2 is completely oxidized to CO_2 and H_2O. ***Glycolysis occurs in all tissues.***
   - *HMP Shunt*

     An alternative pathway for oxidation of glucose. It is ***not meant for energy.*** The pathway provides (i) NADPH which is used for reductive synthesis and (ii) Pentoses which is used for nucleic acids synthesis. ***This pathway operates only in certain special tissues and not all tissues.***
   - *Uronic Acid Pathway*

     This is another alternative pathway for oxidation of glucose. It ***provides D-glucuronic acid*** which is used for synthesis of mucopolysaccharides and conjugation reactions.
2. *Storage*

   Excess of glucose taken is converted to glycogen in various tissues (glycogenesis) specially liver and skeletal muscle and stored there for future needs. Amount of glycogen storage in liver and muscles is limited. ***Liver can store approx. 72 to 108 gm (4 to 6% of the weight of liver) and muscles can store approx. 245 gm (0.7% of total weight).***
3. *Conversion to Fats*

   As mentioned above, since the amount of glycogen that can be stored is limited, excess of glucose is converted to FA and **stored as *triacyl glycerol* (TG)** in fat depots **(lipogenesis)**. There is no fixed amount for storage of fats as is evidenced from everyday observations on human beings.
4. *Conversion to Other Carbohydrates*

   Small amounts of glucose are used directly or indirectly, in the synthesis of certain other carbohydrates or derivatives, which play important role in the body.
   - ***Formation of ribose and deoxyribose:*** This is required for synthesis of nucleic acids. It is formed by HMP Shunt.

"Withdrawal" of carbohydrates from blood	"Release" of glucose by liver to the blood
• Uptake of hexoses by liver cells such as galactose, and fructose and their conversion to glucose by liver cells.	• Formation of blood glucose from hexoses other than glucose by liver and its release from liver cells.
• Conversion of glucose to glycogen for storage in liver ***(glycogenesis)***.	• Conversion of liver glycogen to blood glucose ***(glycogenolysis)***.
• Utilisation of glucose, by oxidation (glycolysis) for energy production.	• Formation of blood glucose by the liver from non-carbohydrate sources, viz. amino acids (glucogenic), pyruvates and lactates, glycerol and propionyl-CoA ***(gluconeogenesis)***.
• Utilisation of glucose for synthesis of other compounds, viz. FA and certain amino acids.	

Fate of Glucose and its Utilisation

Glucose

- **Oxidation**
  1. Glycolysis (EM pathway)
  2. HMP-Shunt pathway
  3. Uronic acid pathway
- **Storage as glycogen** (glycogenesis)
- **Conversion to fats** (lipogenesis)
- **Conversion to amino acids**
- **Conversion to other carbohydrates**
  1. Ribose and deoxyribose
  2. Fructose
  3. Mannose, fucose, glucosamine, and neuraminic acid
  4. Galactose
  5. D-Glucuronic acid

- ***Formation of fructose from glucose:*** Seminal fluid is rich in fructose and it is required for the metabolism of spermatozoa. ***Fructose is formed from glucose in seminiferous tubular epithelial cells by 'Sorbitol' (polyol) pathway.***
- ***Mannose, fucose, glucosamine and neuraminic acid:*** Form parts of mucopolysaccharides (MPS) and glycoproteins.
- ***Galactose***
  - A component part of glycolipids;
  - Galactose required for synthesis of lactose (milk sugar) in lactating mammary gland is synthesised from glucose.
- ***D-Glucuronic acid:*** Required in the formation of mucopolysaccharides (MPS) and in conjugation reaction for detoxication. It is produced in the body from glucose by uronic acid pathway.

5. ***Conversion to Amino acids***

Certain amino acids are not required in the diet, although they occur in tissue proteins. These amino acids are synthesised in the body. This group is called as ***dispensable*** or ***non-essential amino acids***. The C-skeletons of such amino acids are derived from glucose or its metabolites.

GLYCOLYSIS

Definition: Oxidation of glucose or glycogen to pyruvate and lactate is called glycolysis.

This was described by **Embden, Meyerhof** and **Parnas.** Hence, it is also called as **Embden Meyerhof pathway**. Process of fermentation in yeast cells was similar to breakdown of glycogen in muscles.

It occurs virtually in all tissues. Erythrocytes and nervous tissues derive its energy mainly from glycolysis. This pathway is unique in the sense that it can utilise O_2 if available (aerobic) and it can function in absence of O_2 also (anaerobic).

Two Phases of Glycolysis

- ***Aerobic phase:*** Oxidation is carried out by dehydrogenation and reducing equivalent is transferred to NAD^+. Reduced NAD in presence of O_2 is oxidised in electron-transport chain producing ATP.
- ***Anaerobic phase:*** NADH cannot be oxidised in electron transport chain, so ***no ATP is produced in electron transport chain.*** But the NADH is oxidised to NAD^+ by conversion of pyruvate to lactate, without producing ATP. ***Anaerobic phase limits the amount of energy per mol. of glucose oxidised. Hence, to provide a given amount of energy, more glucose must undergo glycolysis under anaerobic as compared to aerobic.***

Enzymes: Enzymes involved in glycolysis are ***extramitochondrial.***

Biomedical Importance

- This pathway is meant for ***provision of energy.***
- It has importance in skeletal muscle as glycolysis provides ATP even in absence of O_2. ***Muscles can survive anoxic episodes.***
- ***Heart muscle:*** As compared to skeletal muscle, heart muscle is adapted for aerobic performance. It has relatively poor glycolytic activity and poor survival under conditions of ischaemia.
- ***Role in cancer therapy:*** **In fast-growing cancer cells, rate of glycolysis is very high.** Produces more pyruvic acid (PA) than TCA cycle can handle. Accumulation of pyruvic acid leads to excessive formation of lactic acid producing ***local lactic acidosis. Local acid environment may be congenital for certain cancer therapy.***
- ***Haemolytic anaemias:*** Inherited enzyme deficiencies like hexokinase deficiency and pyruvate kinase deficiency in glycolytic pathway enzymes, can produce haemolytic anaemia.

REACTIONS OF GLYCOLYTIC PATHWAY (FIG. 23.1)

Series of reactions of glycolytic pathway which degrades glucose / glycogen to pyruvate / lactate are discussed below. For discussion and proper understanding, the various reactions can be ***arbitrarily*** divided into ***four stages.***

Stage I

This is a ***preparatory stage.*** Before the glucose molecule can be split, the ***rather asymmetric glucose molecule is converted to almost symmetrical form*** fructose 1,6-biphosphate by donation of 2 PO_4 groups from ATP.

1. ***Uptake of glucose by cells and its phosphorylation:*** Glucose is freely permeable to Liver cells. Insulin facilitates the uptake of glucose in skeletal muscles, cardiac muscle, diaphragm and adipose tissue. Glucose is then phosphorylated to form glucose-6-P. The reaction is catalysed by the specific enzyme ***glucokinase*** in liver cells and by non-specific

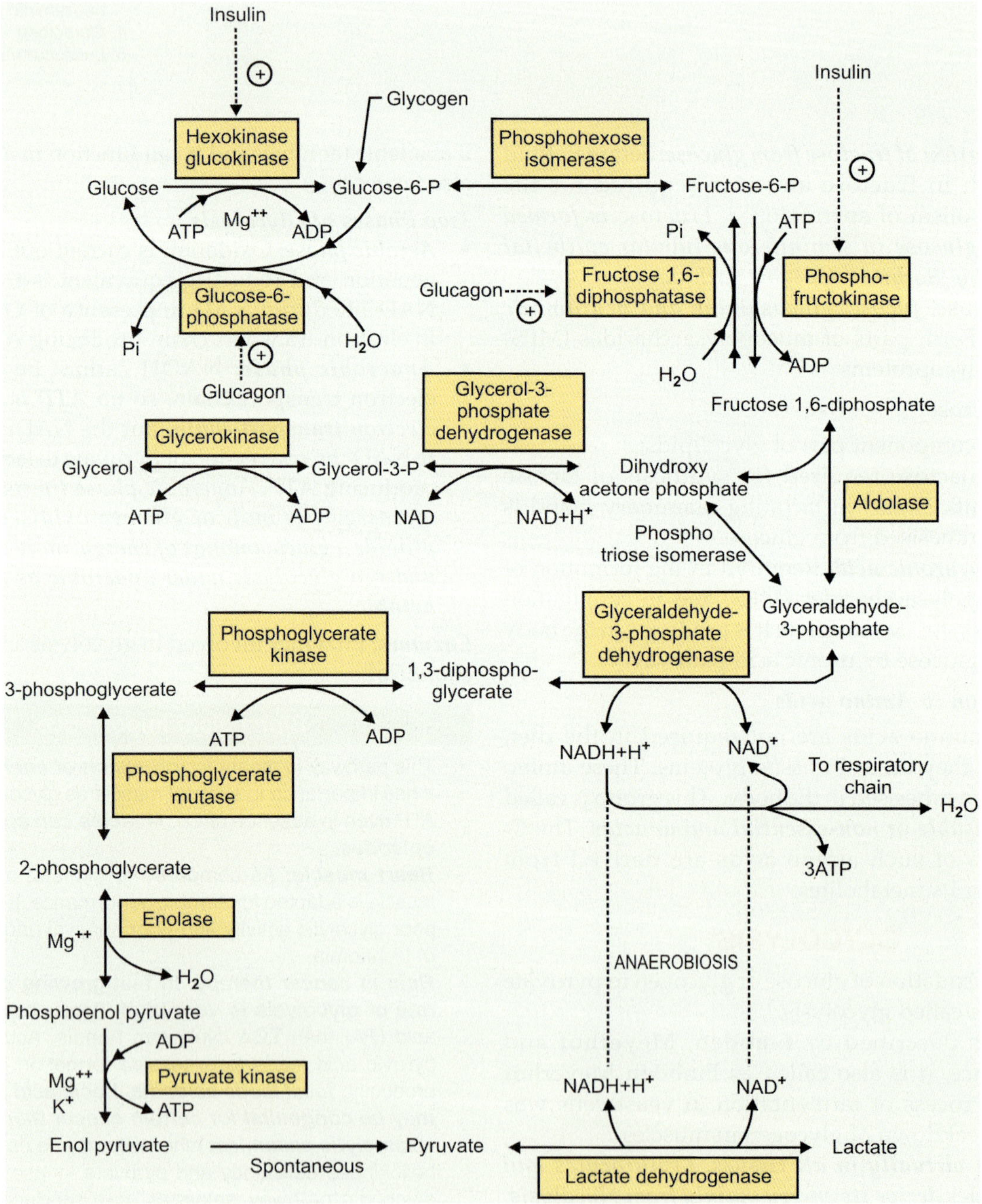

Fig. 23.1: Embden-Meyerhof pathway of glycolysis

hexokinase in liver and extrahepatic tissues *(Refer second box in right hand side this page).*

Note

- **Reaction is *irreversible***
  - ATP acts as PO_4 donor and it reacts as Mg-ATP complex. One high energy PO_4 bond is utilised and ADP is produced. The reaction is accompanied by considerable loss of free energy as heat, and hence under physiologic conditions is regarded as irreversible.
  - Glucose-6-P formed is an important compound at the junction of several metabolic pathways like glycolysis, glycogenesis, glycogenolysis, gluconeogenesis, HMP-Shunt, uronic acid pathway. Thus it is a ***committed step*** in metabolic pathways.

2. ***Conversion of G-6-P to fructose-6-P:*** G-6-P after formation is converted to fructose-6-P by ***phosphohexose isomerase,*** which involves an aldose-ketose isomerisastion. The enzyme can act only on α-anomer of G-6-P.

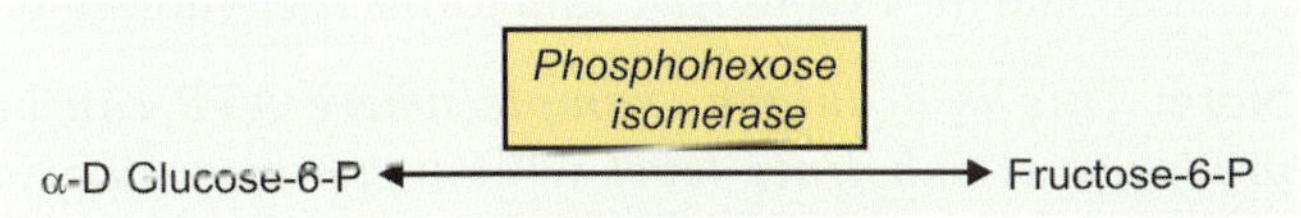

3. ***Conversion of fructose-6-P to fructose-1, 6-bi-P:*** The above reaction is followed by another phosphorylation. Fructose-6-P is phosphorylated with ATP at 1-position catalysed by the enzyme ***phosphofructokinase-1*** to produce the **symmetrical molecule** fructose-1,6-bi-phosphate.

Note

- The reaction is ***irreversible.***
- One ATP is utilised for phosphorylation.
- ***Phosphofructokinase-1*** is the ***key enzyme*** in glycolysis which regulates breakdown of glucose. The enzyme is ***inducible,*** as well as ***allosterically modified.***
- *Phosphofructokinase-2* is an isoenzyme which catalyses the reaction to form fructose-2,6-bi-phosphate.

Energetic: Note that in this stage glucose oxidation does not yield any useful energy rather **there is expenditure of 2 ATP molecules for two phosphorylations (–2 ATP).**

Stage II

Actual Splitting of Symmetrical Fructose-1-6-bi-P. Fructose-1,6-bi-P is split by the enzyme ***aldolase*** into two molecules of triose-phosphates, an aldotriose–glyceraldehyde-3-P and one ketotriose, Dihydroxy acetone-P.

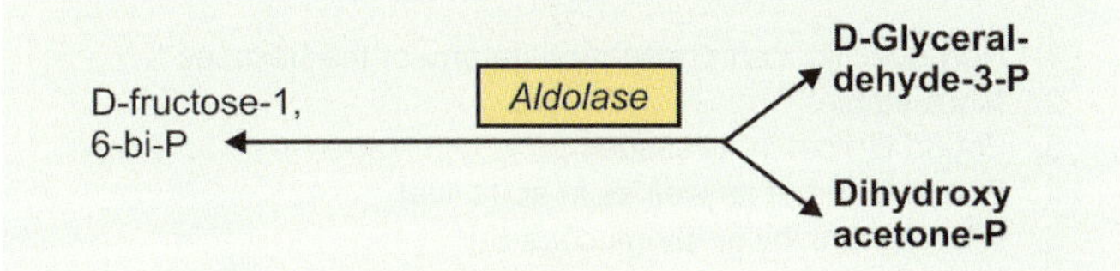

Note

- The reaction is ***reversible*** and there is ***neither expenditure of energy nor formation of ATP***
- ***Aldolases*** are tetramers, containing 4 subunits. Two isoenzymes: ***Aldolase A:*** Occurs in most tissues, ***Aldolase B:*** Occurs in liver and kidney

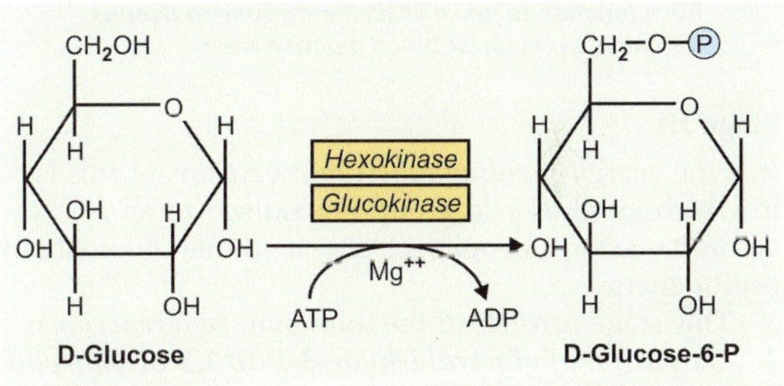

- The fructose-6-P exists in the cells in ***furanose*** form but they react with isomerase, phosphofructokinase-1 and aldolase in the open-chain configuration
- Both triose phosphates are interconvertible.

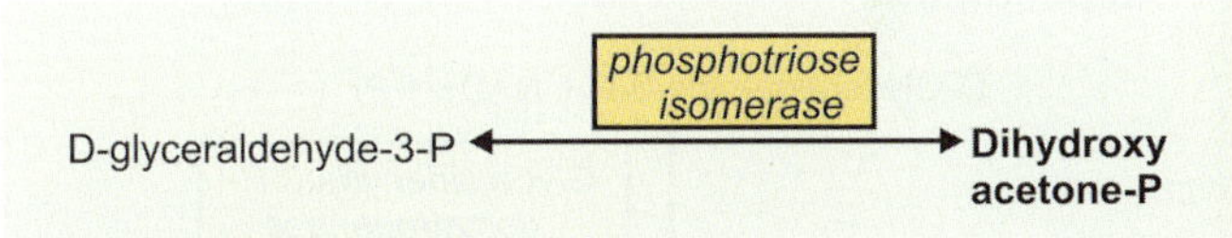

Inhibitors

- ***Bromohydroxyacetone-P:*** It **resembles structurally to dihydroxyacetone-P.** Hence it binds covalently with the γ-COOH group of a glutamate residue of the enzyme ***phosphotriose isomerase*** at the active site of the enzyme molecule. Thus the enzyme becomes inactive and cannot catalyse the reaction. It blocks glycolysis at the stage of dihydroxyacetone-P and leads to accumulation of dihydroxyacetone-P and fructose-1,6-bi-phosphate.

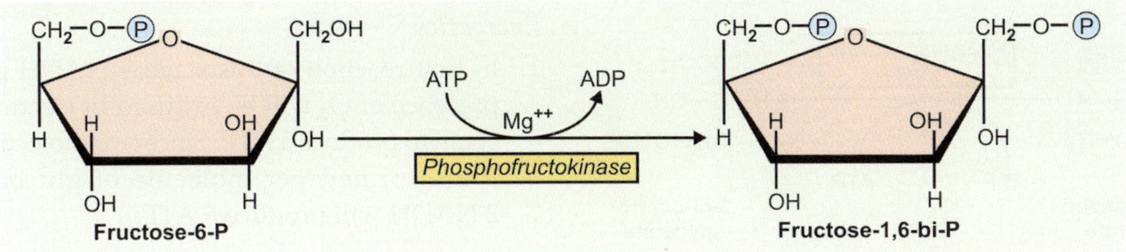

Differences between *hexokinase* and *glucokinase*

Hexokinase	*Glucokinase*
1. Non-specific, can phosphorylate any of the hexoses	1. Specific, can phosphorylate glucose only
2. More stable	2. Physiologically more labile
3. Found almost in all tissues	3. Found only in liver
4. Found in foetal as well as in adult liver	4. Found in adult liver, not in foetal liver
5. Allosteric inhibition by glucose-6-P	5. Not inhibited by Glucose-6-P
6. K_m is low = 0.1 mM, hence high affinity for glucose	6. K_m is high = 10 mM, low affinity for glucose
7. Not very much influenced by diabetic state/or fasting	7. Depressed in fasting and in diabetes. Glucokinase is deficient in patients of DM, changes according to nutritional status
8. No change with glucose feeding	8. Increased by feeding of glucose after fasting
9. Inhibited by glucocorticoids and GH; insulin does not have effect on hexokinase	9. Inhibited by glucocorticoids and GH; glucose and insulin stimulates. Synthesis is induced by insulin, an ***inducible enzyme***
10. Hexokinase activity of liver found in three enzyme proteins (isoenzymes)	10. Not known
11. ***Main function to make available glucose to tissues for oxidation at lower blood glucose level***	11. ***Main function to clear glucose from blood after meals and at blood levels greater than 100 mg/dl***

Stage III

It is the ***energy-yielding reaction.*** Reactions of this type in which an aldehyde group is oxidised to an acid are accompanied by liberation of large amounts of potentially useful energy.

This stage consists of the following ***two reactions:***

1. ***Oxidation of glyceraldehyde-3-P to 1,3-bi-phosphoglycerate:*** Glycolysis proceeds by the oxidation of glyceraldehyde-3-P to form 1,3-bi-phosphoglycerate. Dihydroxyacetone-P also form 1,3-bi-phosphoglycerate via glyceraldehyde-3-P. Enzyme responsible is ***Glyceraldehyde-3-P dehydrogenase*** which is NAD^+ dependant.

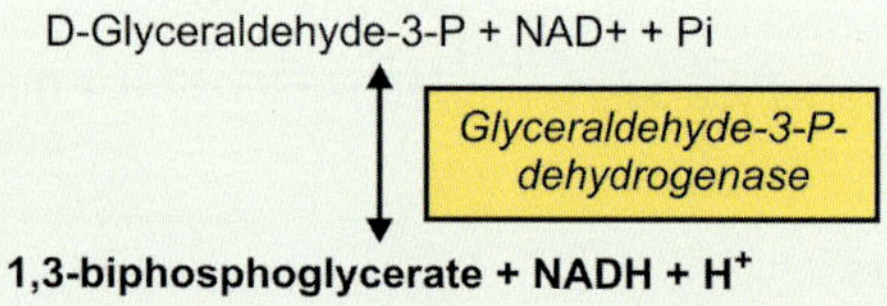

Characteristics of the Enzyme

- The enzyme is a ***tetramer,*** consisting of four identical polypeptides.
- Four –SH groups are present on each polypeptide derived from cysteine residue in the chain.
- One of the –SH group forms the "active site" of the enzyme molecule.

2. ***Conversion of 1,3-Biphosphoglycerate to 3-Phosphoglycerate***

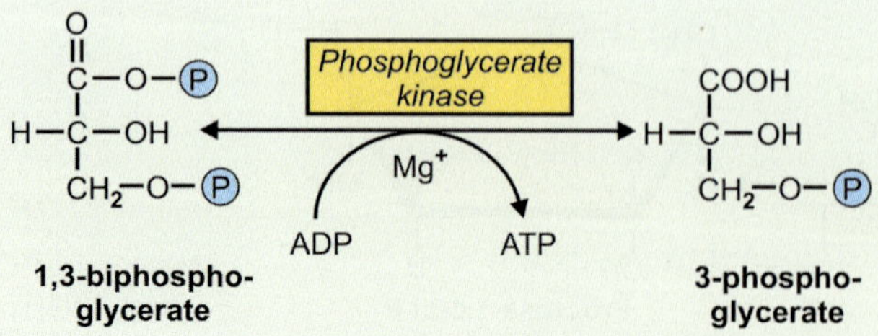

The reaction is catalysed by the enzyme ***phosphoglycerate kinase.*** The high energy PO_4 bond at position-1 can donate the PO_4 to ADP and forms ATP molecule.

Note: ***This is a unique example where ATP can be produced at substrate level*** without participating in electron transport chain. This type of reaction where ATP is formed at substrate level is called as ***Substrate level phosphorylation.***

Inhibitors

- ***Arsenite***

  If present, it competes with inorganic Pi in the reaction of conversion of glyceraldehyde-3-P to 1,3-biphosphoglycerate and produces 1-arseno-3-phosphoglycerate, which hydrolyses spontaneously to yield 3-phosphoglycerate and heat. ***Thus in the next step no ATP is produced.*** This is an important example of the ability of arsenate to uncouple oxidation and phosphorylation.

- ***Iodoacetate and Iodoacetamide***

  They bind covalently with –SH group and alkylate the –SH group of the enzyme *glyceraldehyde-3-P dehydrogenase.* They bind irreversibly with the enzyme and inhibits glycolysis. **This leads to accumulation of glyceraldehyde-3-P.**

Energetics

1. In first reaction of this stage —NADH produced in presence of O_2 will be oxidised in electron transport chain to produce 3 ATP. Since two molecules of triose-P are formed per molecule of glucose oxidised 2 NADH will produce 6 ATP.

+ 6 ATP

2. The second reaction will produce one ATP. Two molecules of substrate will produce 2 ATP.

+ 2 ATP

Net gain at this stage per molecule of glucose oxidised = + 8 ATP.

Stage IV

It is the ***recovery of the PO_4 group*** from 3-Phosphoglycerate. The two molecules of 3-phosphoglycerate, the end-product of the previous stage, still retains the PO_4 group originally derived from ATP in stage 1. Body wants back the two ATP spent in first stage for two phosphorylations. This is achieved by the following ***three reactions:***

1. ***Conversion of 3-Phosphoglycerate to 2-Phosphoglycerate***

   3-phosphoglycerate formed by the above reaction is converted to 2-phosphoglycerate, catalysed by the enzyme ***Phosphoglycerate mutase***. It is likely that 2, 3-bi-phosphoglycerate is an intermediate in the reaction and probably acts catalytically.

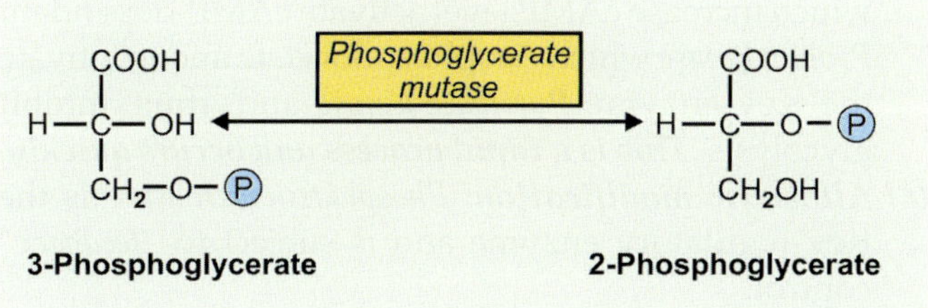

2. ***Conversion of 2-Phosphoglycerate to Phosphoenol Pyruvate***

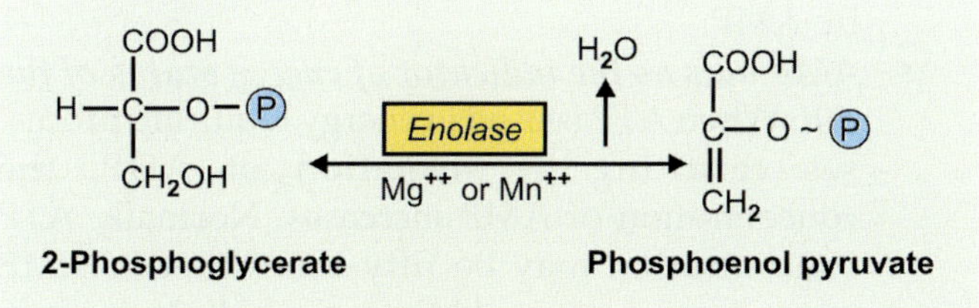

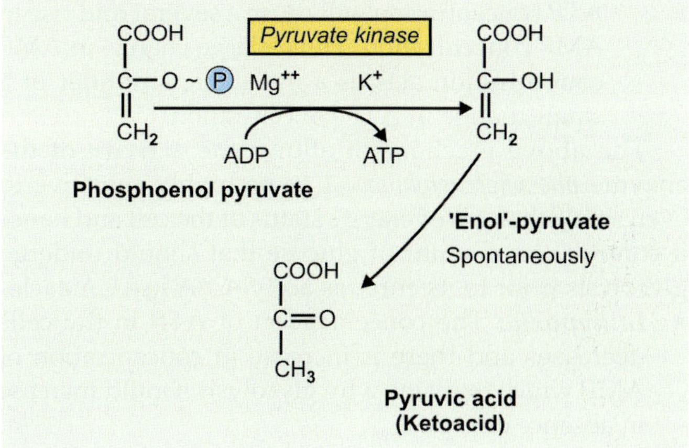

The reaction is catalysed by the enzyme ***Enolase,*** the enzyme requires the presence of either Mg^{++} or Mn^{++} for activity. The reaction ***involves dehydration and redistribution of energy*** within the molecule raising the PO_4 in position 2 to a "high-energy state".

3. ***Conversion of Phosphoenol Pyruvate to Pyruvate***

   Phosphoenol pyruvate is converted to 'Enol' pyruvate, the reaction is catalysed by the enzyme ***Pyruvate kinase.*** The high energy PO_4 of phosphoenol pyruvate is directly transferred to ADP producing ATP (Refer box).

Note

- Reaction is ***irreversible.***
- ATP is formed at the substrate level without electron transport chain. This is another example of ***substrate level phosphorylation*** in glycolytic pathway
- "Enol" pyruvate is converted to 'keto' pyruvate spontaneously.

Inhibitors

Fluoride inhibits the enzyme enolase.

CLINICAL IMPORTANCE

Sodium fluoride is used along with K-oxalate for collection of blood for glucose estimation. If K-oxalate is used alone, then ***in vitro*** glycolysis will reduce the glucose value in the sample.

Functions of Fluoride

- Inhibits ***in vitro*** glycolysis by inhibiting enzyme *enolase*
- Also acts as anticoagulant, and
- Act. as an antiseptic.

Energetics: **In this stage, 2 molecules of ATP are produced, per molecule of glucose oxidised.**

+ 2 ATP

ENERGY YIELD PER GLUCOSE MOLECULE OXIDATION

A. In Glycolysis in Presence of O_2 (Aerobic Phase)

Reaction catalysed by	*ATP Production*
Stage I	
1. Hexokinase/Glucokinase reaction (for phosphorylation)	– 1 ATP
2. Phosphofructokinase-1 (for phosphorylation)	– 1 ATP
Stage III	
3. Glyceraldehyde-3-P dehydrogenase (oxidation of 2 NADH in electron transport chain)	+ 6 ATP

4. Phosphoglycerate kinase (substrate level phosphorylation) + 2 ATP

Stage IV

5. Pyruvate kinase (substrate level phosphorylation) + 2 ATP

Net gain = 10–2
= **8 ATP**

B. In Glycolysis—in Absence of O_2 (Anaerobic Phase)

- In absence of O_2, reoxidation of NADH at glyceraldehyde-3-P-dehydrogenase stage cannot take place in electron-transport chain.
- But the cells have limited coenzyme. Hence to continue the glycolytic cycle NADH must be oxidised to NAD^+. This is achieved by reoxidation of NADH by conversion of pyruvate to lactate **(without producing ATP)** by the enzyme lactate dehydrogenase.

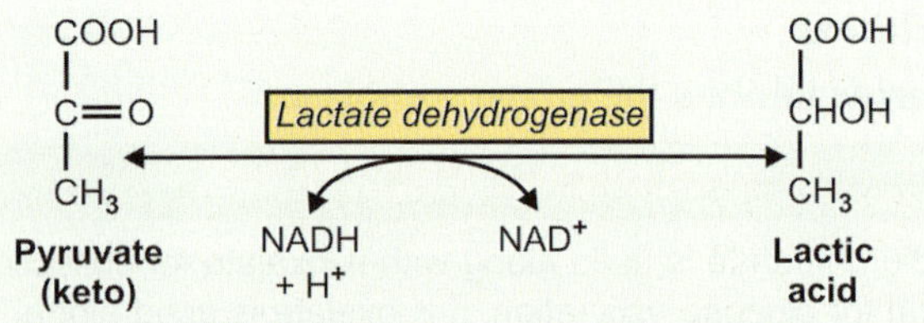

- It is to be noted that in the reaction catalysed by glyceraldehyde-3-P-dehydrogenase, therefore, no ATP is produced.

In anaerobic phase per molecule of glucose oxidation 4 – 2 = 2 ATP will be produced.

+ 2 ATP

CLINICAL IMPORTANCE

- Tissues that function under hypoxic circumstances will produce lactic acid from glucose oxidation, producing ***local acidosis.*** If lactate production is more it can produce metabolic acidosis.
- Vigorously contracting skeletal muscle will produce relative anaerobiosis and glycolysis will produce lactic acid.
- Whether O_2 is present or not, glycolysis in erythrocytes always terminate in pyruvate and lactate.
- When there is relative anaerobiosis, glycolysis will stop as cells will exhaust NAD^+.
- ***Inhibitor of Lactate Dehydrogenase (LDH)*** is ***Oxamate:*** It competitively inhibits lactate dehydrogenase and prevents the reoxidation of NADH.

REGULATION OF GLYCOLYSIS

Regulation of glycolysis achieved by **three types of mechanisms:**

(a) Changes in the rate of enzyme synthesis, Induction/repression.
(b) Covalent modification by reversible phosphorylation.
(c) Allosteric modification.

(a) Induction and repression of key enzymes: This is not rapid and takes several hours to come into operation.

- ***Glucose:*** When there is increased substrate, i.e. glucose, the enzymes involved in utilisation of glucose are activated. On the other hand, enzymes responsible for producing glucose (gluconeogenesis) are inhibited. Glucose also increases the activity of the key enzymes *glucokinase, phosphofructokinase-1* and *pyruvate kinase.*
- ***Insulin:*** The secretion of insulin which is responsive to blood glucose concentration enhances the synthesis of the key enzymes responsible for glycolysis. On the other hand, it antagonises the effects of glucocorticoids and glucagon-stimulated c-AMP in stimulating the key enzymes responsible for gluconeogenesis.

(b) Covalent modification by reversible phosphorylation: Hormones like epinephrine and glucagon which increase cAMP level activate cAMP-dependant *Protein kinase* which can phosphorylate and inactivate the Key enzyme *Pyruvate kinase* and, thus, inhibit glycolysis. ***This is a rapid process and occurs quickly.***

(c) Allosteric modification: *Phosphofructokinase-1* is the Key regulatory enzyme and is subject to *"feedback"* control.

- ***Inhibition of the enzyme:*** The enzyme is inhibited by citrate and by ATP.
- ***Activator of the enzyme:*** The enzyme is activated by AMP.
- ***AMP acts as the indicator of energy status of the cell:*** When ATP is used in energy requiring processes resulting in formation of ADP, the concentration of AMP increases. Normally ATP concentration may be fifty times that of AMP concentration at equilibrium, a small decrease in ATP concentration will cause a several fold rise in AMP concentration. Thus a large change in AMP concentration acts as a metabolic amplifier of a small change in ATP concentration.

The above mechanism allows the activity of the enzyme *phosphofructokinase-1* to be highly sensitive to even small changes of energy status of the cell and hence it controls the amount of glucose that should undergo glycolysis prior to its entry as acetyl-CoA in TCA cycle.

- ***In hypoxia:*** The concentration of ATP in the cells decreases and there is increase in concentration of AMP which explains why glycolysis should increase in absence of O_2.

PECULIARITIES OF GLUCOSE OXIDATION BY RB CELLS AND RAPOPORT-LUEBERING SHUNT

RB cells are structurally and metabolically unique as compared to other cells.

A. ***Structural Peculiarities:*** Structurally mature erythrocytes ***do not possess*** **nucleus nor cytoplasmic subcellular structures.**

B. ***Metabolic peculiarities:*** Metabolically mature erythrocytes:
   - Entirely ***depends on glucose for its energy***, i.e. glycolysis. More than > 90 per cent of total energy is met by glycolysis.
   - Glucose is ***freely permeable*** to erythrocytes like Liver cells.
   - Glucose oxidation always ends in formation of pyruvic acid (PA) and lactic acid (LA), whether O_2 is available or not.
   - The enzyme ***pyruvate dehydrogenase complex is absent*** hence Pyruvic acid is not converted to "acetyl-CoA".

RAPOPORT-LUEBERING SHUNT OR CYCLE

Maintains a high steady state concentration of **2,3-biphosphoglycerate (2,3-BPG)**, produced by a diversion in glycolytic pathway. This diversion is called as **Rapoport-Luebering cycle or shunt (RLC or RLS).** A supplementary to glycolysis.

Functional Significance of this Shunt Pathway

(a) ***Factors which waste energy are not present in RB Cells***
   - Energy demanding ***endergonic*** reactions utilising ATP is not present in mature human red blood cells.
   - *ATPase* activity which controls ATP/ADP ratio is not active in mature RB Cells.

RB cells utilise more glucose than it requires to maintain cellular integrity, resulting in accumulation of ATP and 1,3-BPG, causing cessation of glycolysis. ***RLC or RLS provides a mechanism to dissipate the excess energy.*** RL shunt/cycle is shown schematically in the box below.

(b) *Role in Hb*
   - **Adult Hb-A_1:** 2,3-BPG concentration is high, affinity to O_2 less and unloading/dissociation is more.
   - **Hb-F:** 2,3-BPG concentration is low, affinity to O_2 is more, and unloading/dissociation is less.

(c) ***Role in hypoxia:*** Tissue hypoxia has an important effect on the level of Red cells BPG. Pulmonary hypoxic hypoxia, stagnant hypoxia either as a result of CV failure or shock, and anaemic hypoxia, as in a deficit of red cells mass, ***all favour an increase in red cells BPG level, thus enhancing unloading of O_2 in tissues.*** When normal individuals ascend to a height of 450 metres, there is a rise in red cells BPG level. The maximum rise occurs by 48 hours and returns to normal within a similar period after descent to sea-level.

(d) ***Inherited enzyme deficiency:*** Several hereditary defects in enzyme of red-cell glycolysis that affect red cell BPG concentration such as rare ***hexokinase deficiency*** and much more commonly occurring ***pyruvate kinase (PK) deficiency,*** also exhibit alterations of red cells BPG concentration. In a patient with red cell **hexokinase** ***deficiency*** a decrease in BPG concentration to about 2/3 of normal has been reported. In ***PK deficiency*** (pyruvate kinase) BPG is more than twice normal. As a result, affinity for oxygen of Hb is greater than normal in 'hexokinase' deficiency and less than normal in 'pyruvate kinase' deficiency.

	Hexokinase deficient red cells	*Pyruvate kinase deficient red cells*
• **2,3-BPG**	↓	↑
• **Affinity to O_2**	↑	↓
• **Unloading of O_2**	↓	↑

SECTION FOUR

RAPOPORT-LUEBERING SHUNT OR CYCLE

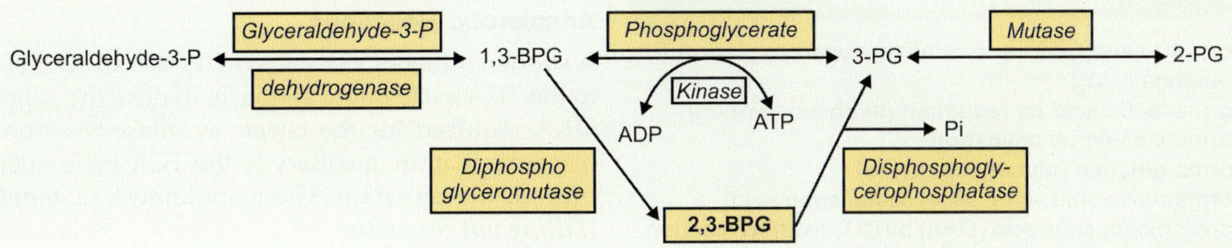

FORMATION AND FATE OF PYRUVIC ACID

Formation of Pyruvic Acid (PA) in the Body

- **From oxidation of glucose (Glycolysis)**
- **From lactic acid by oxidation**
- **Deamination of Alanine**
- **Glucogenic amino acids-pyruvate forming**
- **Decarboxylation of oxaloacetic acid (OAA)**

Pyruvic acid is a key substance in phase-II metabolism.

1. Principally it is formed from **oxidation of glucose** (glycolysis) by EM Pathway.

   In addition to that pyruvic acid can be formed in the body from various other sources. They are:

2. Conversion of lactic acid to pyruvic acid (see below).
3. Also formed from **deamination** of amino acid **alanine**.
4. Certain other amino acids during their catabolism produces pyruvic acid, e.g. ***glycine, serine, cysteine/ and cystine*** and ***threonine (Glucogenic a-a).***
5. Pyruvic acid can also be formed from ***decarboxylation*** of dicarboxylic ketoacid **oxaloacetic acid**, which can be ***spontaneous decarboxylation*** or can be ***catalysed by the enzyme*** *oxaloacetate decarboxylase.*

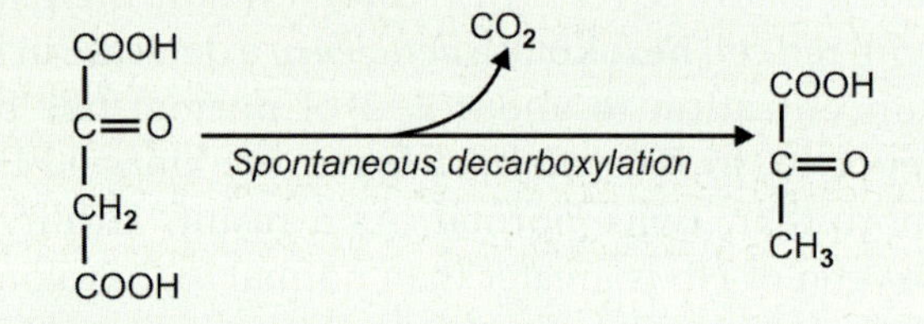

Oxaloacetic acid **Pyruvic acid**

6. Lastly pyruvic acid can be formed in the body from **malic acid** by *malic enzyme.*

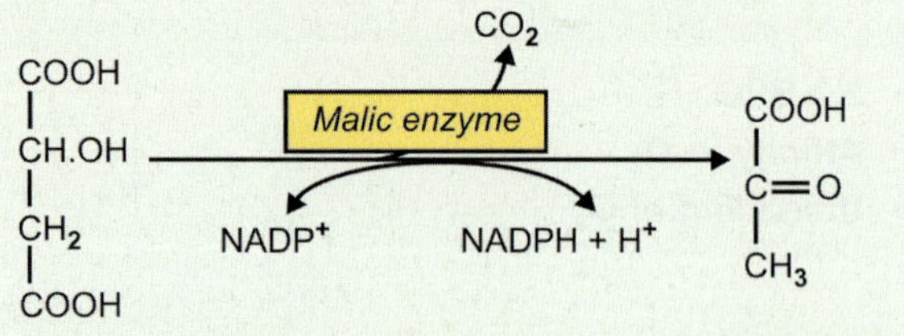

Malic acid **Pyruvic acid**

Fate of Pyruvic Acid (PA)

- **Forms acetyl-CoA by oxidative decarboxylation (in presence of O_2)**
- **Forms lactic acid by reduction (in absence of O_2)**
- **Forms alanine by amination**
- **Forms glucose (gluconeogenesis)**
- **Forms malic acid → to OAA (oxaloacetic acid)**
- **Forms oxaloacetic acid (OAA) by CO_2-fixation reaction.**

Fate of pyruvic acid depends on the redox state of the tissues:

- ***In Presence of O_2:*** Pyruvic acid is oxidatively decarboxylated to two-carbon unit "Acetyl-CoA"
- ***In absence of O_2:*** Pyruvic acid is converted to Lactic acid (LA)

Other fates of Pyruvic acid can be summed up as follows:

- Pyruvic acid can be **aminated** to form the amino acid alanine
- Pyruvic acid can be **converted to form glucose** in the body
- Pyruvic acid can be converted to malic acid, which in turn can form oxaloacetic acid (OAA)
- Pyruvic acid can be converted directly to oxaloacetic acid in the body by ***CO_2-fixation (CO_2-assimilation) reaction.***

Last two reactions are important in the body as OAA can be formed and supplied to TCA cycle in case of relative deficiency (see "anaplerotic" reactions below).

1. *Conversion of Pyruvic Acid to Lactic Acid*

It is an important reaction, because it occurs in skeletal muscles working under conditions of absolute or relative lack of O_2. In anaerobic glycolysis, Pyruvate acts as a temporary H-store. It dehydrogenates (oxidises), the reduced NADH + H^+ back to oxidised NAD^+, so that glycolysis can continue even in absence of O_2. ***Pyruvate is thus reduced to Lactic acid. In presence of O_2, Lactic acid can be oxidised to pyruvic acid again.***

Characteristics of this reaction:

- ***Reversible*** reaction
- Oxidation-reduction
- Same enzyme and co-enzyme required.

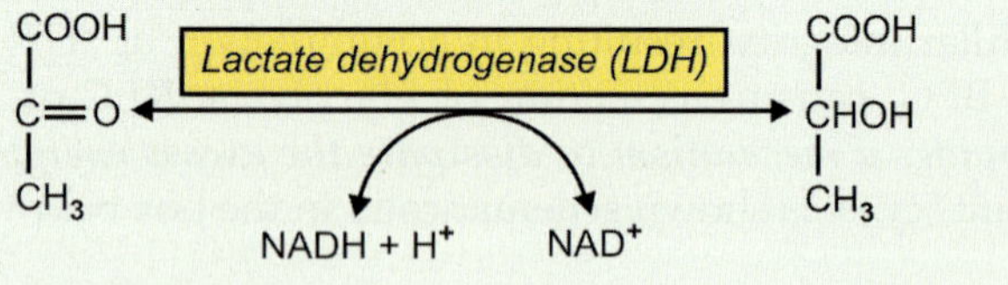

Pyruvic acid (PA) **Lactic acid (LA)**

Anaplerotic Reactions

A sudden influx of Pyruvic acid (PA) or of "acetyl-CoA" to the TCA cycle might seriously deplete the supplies of OAA required for the citrate synthase reaction. Two reactions that are auxiliary to the TCA cycle operate to prevent this situation. These are known as ***anaplerotic (Filling up)*** reactions.

These **two reactions** are:

2. **Conversion of PA to OAA (by CO_2-Fixation Reaction)**

Pyruvic acid can be converted to oxaloacetate by the enzyme ***Pyruvate carboxylase***. The enzyme requires:

- 'Biotin' as a prosthetic group which brings CO_2
- ATP and Mg^{++}
- Requires 'acetyl-CoA'

Acetyl-CoA does not enter into the reaction but may by combination with the enzyme maintains it in "active" conformation ***(+ve modifier)***. The generation of "acetyl-CoA" in metabolic reactions activates the enzyme and promote the formation of oxaloacetic acid (OAA) required for oxidation of acetyl-CoA in the TCA Cycle.

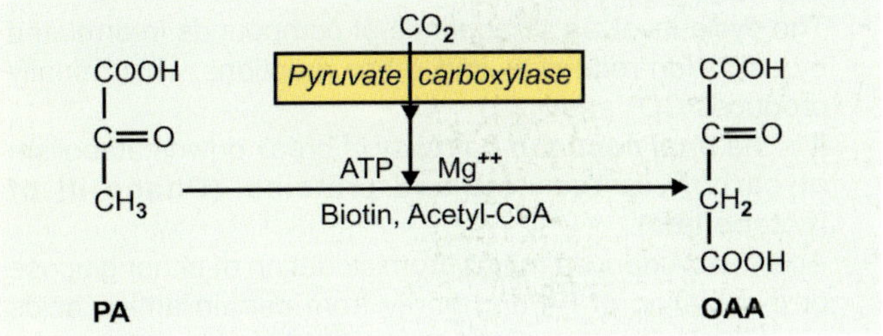

3. ***Conversion of PA to OAA Through Malic Acid Formation***

The other anaplerotic reaction is formation of malic acid by *Malic enzyme* in presence of CO_2 and NADPH. The 'malate' is converted to OAA by dehydrogenation by the enzyme ***Malate dehydrogenase*** in presence of NAD^+.

4. ***Conversion of PA to Acetyl-CoA***

In presence of O_2, Pyruvate undergoes oxidative decarboxylation to form 2-C compound 'acetyl-CoA' ***Pyruvate formed in cytosol is transported to mitochondrion by a 'transport' protein. Since the overall reaction involves both oxidation and loss of CO_2 (decarboxylation) it is termed oxidative decarboxylation.*** The mechanism of the reaction is one of the most complex involved in metabolism of carbohydrates. The reaction is catalysed by a ***multi-enzyme complex*** called ***pyruvate dehydrogenase complex***, which can exist both as "inactive form" and the "active" form (see regulation below):

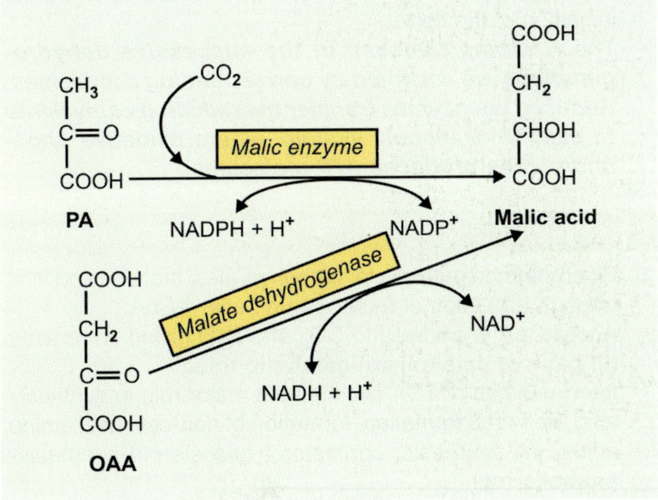

The enzyme complex consists of:

29 molecules of Pyruvate dehydrogenase (PD)
+ 8 molecules of Flavoprotein containing dihydrolipoyl dehydrogenase, and
+ 1 molecule of dihydrolipoyl transacetylase.

The enzyme complex for its activity requires at least six coenzymes/cofactors:

- **Thiamine pyrophosphate (TPP)**
- **Lipoic acid**
- **CoA–SH**
- **FAD**
- **NAD^+ and**
- **Mg^{++}**

The overall reaction can be represented as follows:

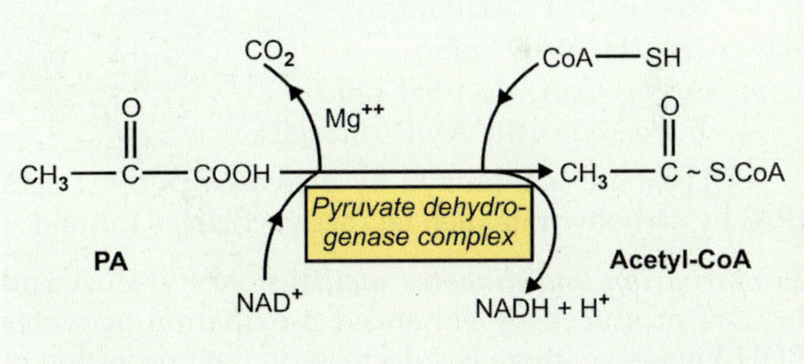

1. 'Acetyl' moiety of PA is transferred to CoA –SH.
2. Carbon of COOH group is liberated as CO_2 (decarboxylation).
3. Remaining two H atoms: One from –COOH group of PA and another from CoA –SH (–SH group) are transferred to NAD^+, by way of a mechanism involving Lipoic acid and FAD.

Details of reactions: Reaction is not so simple as shown in left hand side column. The reaction is catalysed by several different enzymes working sequentially in multienzyme complex.

- Pyruvate is decarboxylated to a hydroxyethyl derivative of the thiazole ring of enzyme bound TPP.
- Which in turn, reacts with oxidised lipoamide to form *acetyl lipoamide*.

- In the presence of ***dihydrolipoyl transacetylase***, acetyl lipoamide reacts with CoA-SH to form "Acetyl-CoA" and reduced lipoamide.
- The latter is reoxidised by a FP in the presence of ***dihydrolipoyl dehydrogenase.***
- Finally, the reduced FP is oxidised by NAD^+, which in turn transfers the reducing equivalent to the 'electron transport chain'.

The sequence of events that occur is shown schematically in Figure 23.2.

Energetics

One molecule of glucose produces two molecules of PA which inturn by oxidative decarboxylation produces 2 molecules of Acetyl-CoA and 2 NADH. Two molecules of NADH will be oxidized to 2 molecules of NAD^+ producing 6 ATP molecules in respiratory chain.

+ 6 A T P

Regulation: Activation and inactivation of Pyruvate dehydrogenase complex (PDH) (See figure next page).

- ***Conversion of inactive to active PDH:***

Insulin: stimulates *phosphatase* and converts 'inactive' to 'active' form by dephosphorylation.

- ***Conversion of active to inactive PDH:*** PDH kinase enzyme is activated by:
  - Rise in ATP/ADP ratio↑
  - NADH/NAD^+ ratio↑
  - Acetyl-CoA/CoA-SH ratio↑
  - Increased cyclic AMP ↑ in cells.

PDH-kinase is inhibited by increased pyruvic acid (PA) by carbohydrate diet, i.e. 'active' PDH is formed.

In starvation and diabetes mellitus: Acetyl-CoA and NADH produced by enhanced β-oxidation activates PDH-kinase, i.e. there is a decrease in the proportion of 'active' form.

Thus, it will be seen that pyruvate dehydrogenase complex and hence glycolysis is inhibited not only by a high energy potential, but also under conditions of FA oxidation, which leads to increases in the ratios stated above.

CLINICAL/APPLIED ASPECTS

- ***Arsenite or Hg(ic)* ions:** Can complex with the –SH group of α-Lipoic acid and inhibit pyruvate dehydrogenase complex.
- ***Dietary deficiencies of vitamin B_1 (Thiamine):*** Has similar effect. Lack of TPP inhibits PDH, leading to accumulation of PA and LA.
- **Alcoholism:** Chronic alcoholics suffer from nutritional deficiency of vitamin B_1 (thiamine) which will have similar effect, resulting to accumulation of PA and LA. If a glucose load is given, exhibit rapid accumulation of PA and LA. Lactic acidosis is seen (specially after glucose load).
- ***Inherited deficiency of pyruvate dehydrogenase:*** This has been reported.

Formation of pyruvate and its fate shown diagrammatically in Figure 23.3.

CITRIC ACID CYCLE

Synonyms: TCA cycle (tricarboxylic acid cycle), Krebs' cycle, Krebs' citric acid cycle.

Points to Remember

- It is a ***cyclic process.***
- The cycle involves a sequence of compounds interrelated by oxidation-reduction and other reactions which finally produces CO_2 and H_2O.
- It is the ***final common pathway*** of break down/catabolism of carbohydrates, fats and proteins. **(Phase III of metabolism)**.
- Acetyl-CoA derived mainly from oxidation of either glucose or β-oxidation of FA and partly from certain amino acids combines with oxaloacetic acid (OAA) to **form *citrate*** the first reaction of citric acid cycle. In this reaction acetyl-CoA transfers its 'acetyl-group' (2-C) to OAA.
- By stepwise dehydrogenations and loss of two molecules of CO_2, accompanied by internal re-arrangements, the ***citric acid is reconverted to OAA***, which again starts the cycle by taking up another acetyl group from acetyl-CoA.
- ***A very small catalytic amount of OAA can bring about the complete oxidation of active-acetate.***
- ***Enzymes are located in mitochondrial matrix,*** either free or attached to the inner surface of the inner mitochondrial membrane, which facilitates the transfer of reducing equivalents to the adjacent enzymes of the respiratory chain.
- The ***whole process is aerobic,*** requiring O_2 as the final oxidant of the reducing equivalents. Absence of O_2 (anoxia) or partial deficiency of O_2 (hypoxia) causes total or partial inhibition of the cycle.
- The ***H atoms removed in the successive dehydrogenations are accepted by corresponding coenzymes. Reduced coenzymes transfer the reducing equivalents to electron-transport system, where oxidative phosphorylation produces ATP molecules.***

BIOMEDICAL IMPORTANCE OF CITRIC ACID CYCLE

- Final common pathway for carbohydrates, proteins and fats, through formation of 2-carbon unit acetyl-CoA.
- Acetyl-CoA is oxidised to CO_2 and H_2O giving out energy **(III pase of catabolism)-catabolic role.**
- Intermediates of TCA cycle play a major role in synthesis also like heme formation, formation of non-essential amino acids, FA synthesis, cholesterol and steroid synthesis-**anabolic role**.

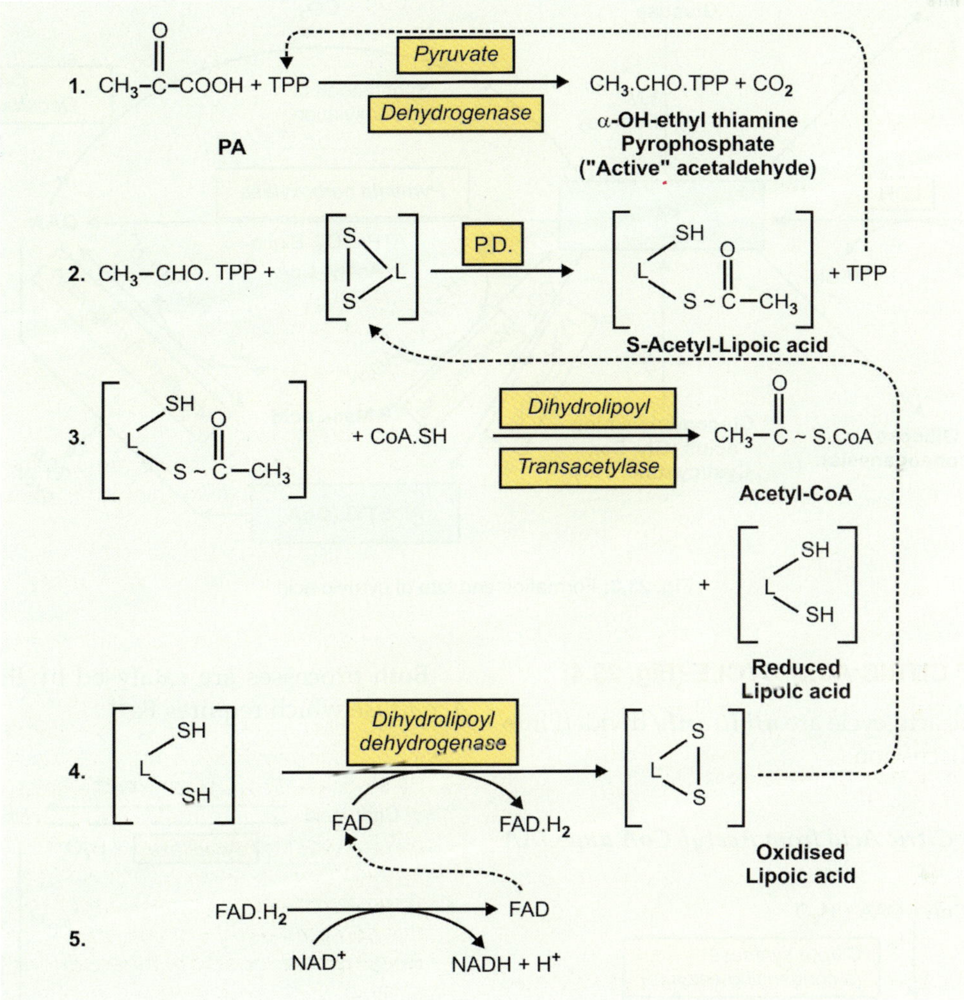

Fig. 23.2: Oxidative decarboxylation of pyruvic acid

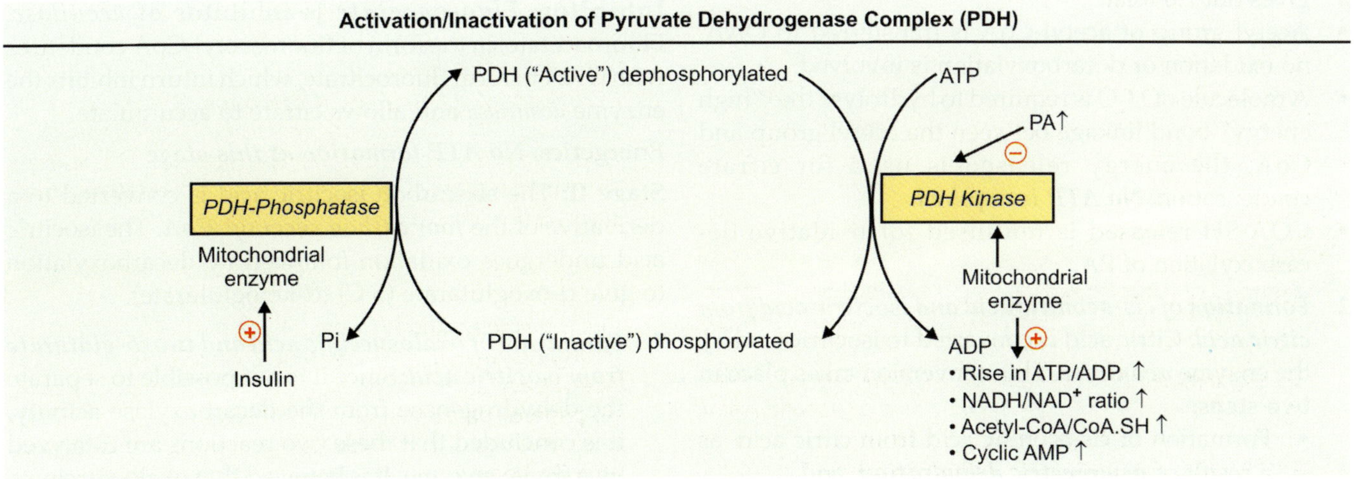

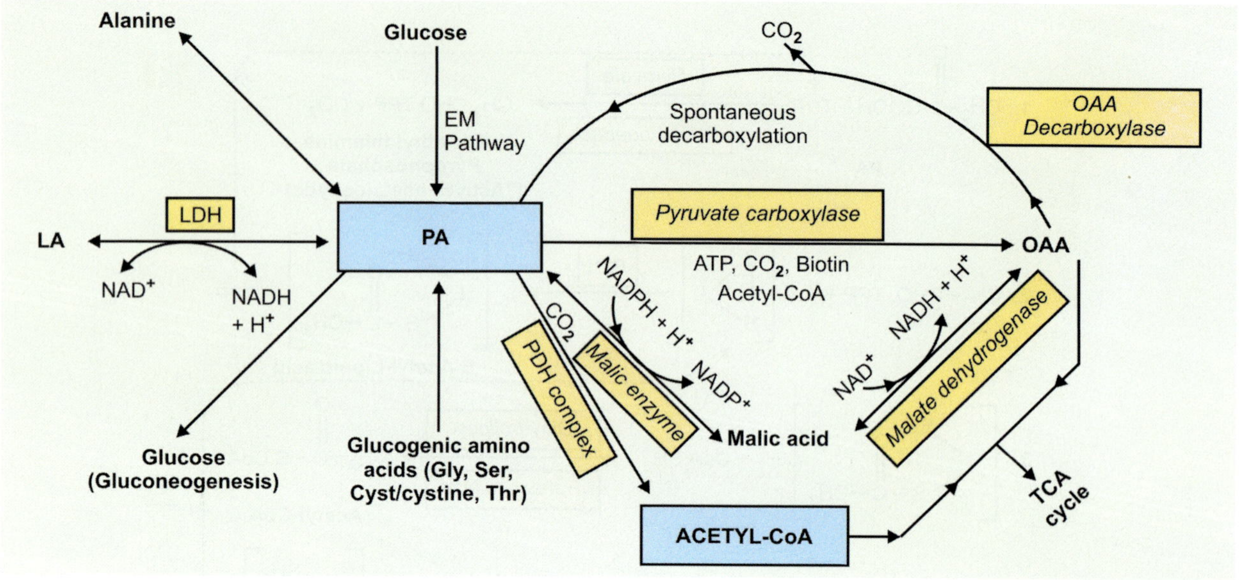

Fig. 23.3: Formation and fate of pyruvic acid

REACTIONS OF CITRIC ACID CYCLE (Fig. 23.4)

Reactions of citric acid cycle are ***arbitrarily*** divided into **four stages** for discussion:

Stage I

1. *Formation of Citric Acid from Acetyl-CoA and OAA*

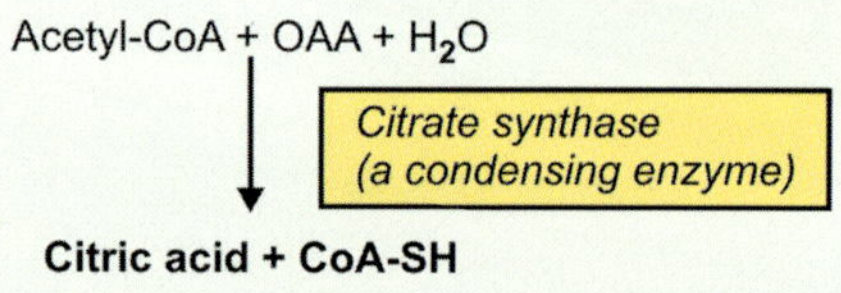

- An ***irreversible*** reaction and an exergonic reaction-gives out 7.8 Kcal.
- Acetyl group of acetyl-CoA is transferred to OAA, no oxidation or decarboxylation is involved.
- A molecule of H_2O is required to hydrolyse the "high energy" bond linkage between the acetyl group and CoA, the energy released is used for citrate condensation. ***No ATP is required.***
- COA-SH released is reutilised for oxidative decarboxylation of PA.

2. *Formation of cis-aconitic acid and isocitric acid from citric acid:* Citric acid is converted to isocitric acid by the enzyme ***aconitase***. This conversion takes place in **two steps:**
   - Formation of cis-aconitic acid from citric acid as a result of ***asymmetric dehydration,*** and
   - Formation of isocitric acid from cis-aconitic acid as a result of ***stereospecific rehydration.***

Both processes are catalysed by the same enzyme ***Aconitase*** which requires Fe^{++}.

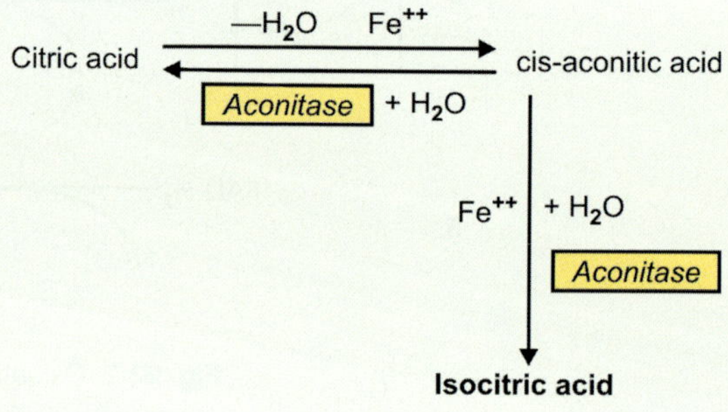

Inhibitor: *Fluoroacetate* is inhibitor of *aconitase*. Fluoroacetate, in the form of fluoroacetyl-CoA condenses with OAA to form fluorocitrate, which inturn inhibits the enzyme *aconitase* and allows citrate to accumulate.

Energetics: *No ATP formation at this stage*

Stage II: The six-carbon isocitric acid is converted to a derivative of the ***four carbon succinyl-CoA***. The isocitric acid undergoes oxidation followed by decarboxylation to give α-oxoglutarate (5 C) (α-ketoglutarate).

1. ***Formation of oxalosuccinic acid and α-oxo-glutarate from isocitric acid:*** Since it is not possible to separate the dehydrogenase from the decarboxylase activity, it is concluded that these two reactions are catalyzed by a single enzyme. It is believed that oxalo-succinate is not a free intermediate but rather exists bound to the enzyme.

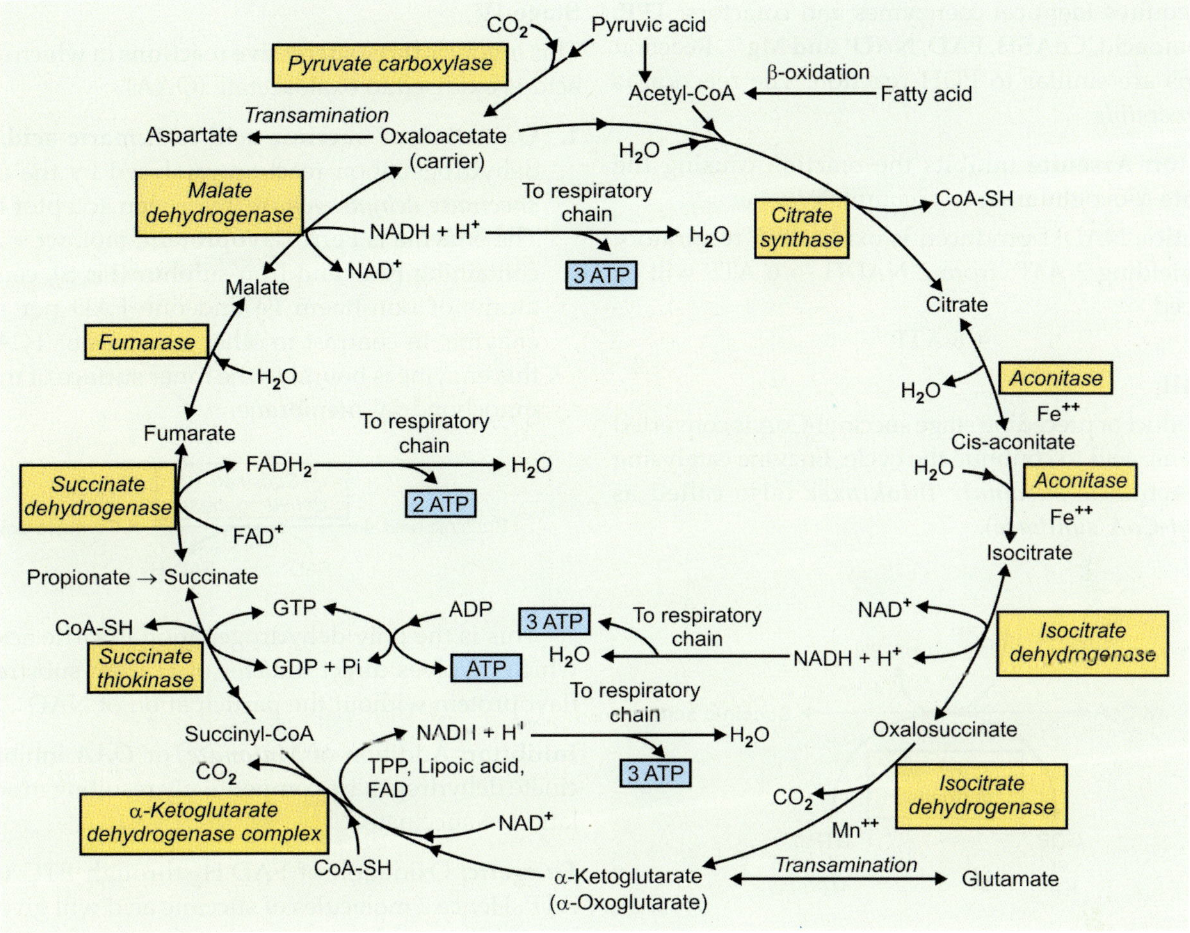

Fig. 23.4: Citric acid cycle

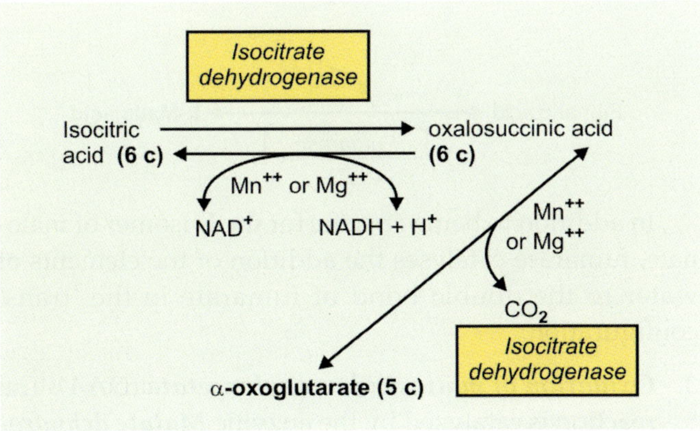

Isocitrate dehydrogenase (ICD) enzyme: Three types described:

- One NADP dependant ICD found in cytosol.
- Another NADP dependant ICD exists in mitochondria, greater activity and more widely distributed.
- One NAD^+ dependant ICD, found only in mitochondria.

Respiratory chain-linked oxidation of isocitrate proceeds almost completely through the NAD^+ dependant ICD in mitochondrion.

2. ***Oxidative decarboxylation of α-oxoglutarate to succinyl-CoA:*** This reaction is analogous to oxidative decarboxylation of Pyruvic acid to acetyl-CoA. Enzyme is α-Ketoglutarate dehydrogenase complex,

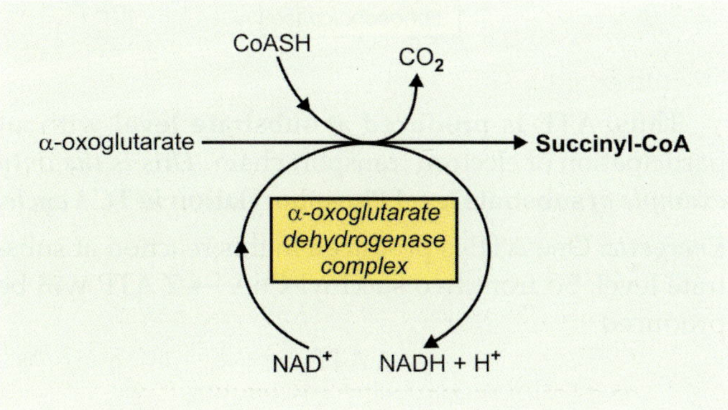

It requires identical coenzymes and cofactors: **TPP, Lipoic acid, CoASH, FAD, NAD^+ and Mg^{++}.** Reaction steps are similar to PDH reaction. The reaction is ***irreversible***

Inhibitor: Arsenite inhibits the reaction causing the substrate α-oxoglutarate to accumulate.

Energetics: NADH produced is oxidised in respiratory chain yielding 3 ATP, from 2 NADH → 6 ATP will be produced

+ 6 ATP

Stage III

The product of preceding stage succinyl-CoA is converted to succinic acid to continue the cycle. Enzyme catalysing this reaction is ***succinate thiokinase*** (also called as ***succinyl-CoA synthase***).

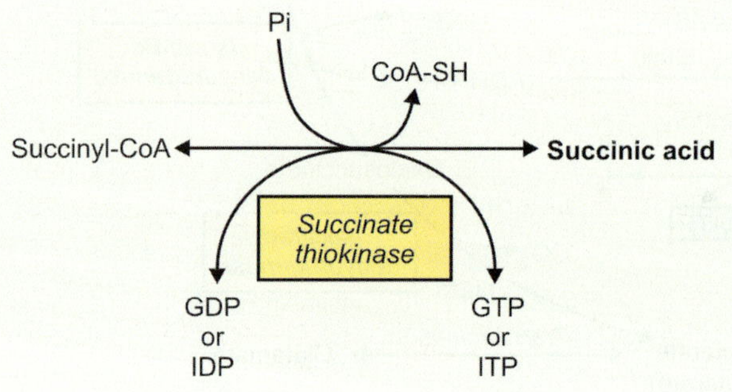

- Reaction requires GDP or IDP, which is converted in presence of Pi to either GTP or ITP.
- The release of free energy from oxidative decarboxylation of α-oxoglutarate is sufficient to generate a high energy bond in addition to the formation of NADH.
- In presence of enzyme ***nucleoside diphosphate kinase,*** ATP is produced either from GTP or ITP.

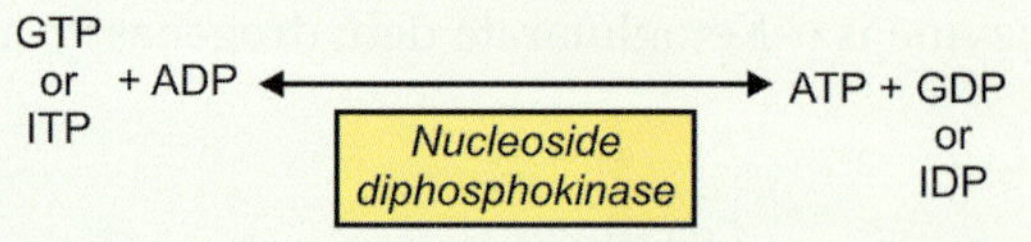

Thus, **ATP is produced at substrate level** without participation of electron transport chain. ***This is the only example of* substrate level Phosphorylation *in TCA cycle.***

Energetic: One ATP is produced in this reaction at substrate level. So from two succinyl-CoA → 2 ATP will be produced

+ 2 ATP

Stage IV

This involves **three successive reactions** in which succinic acid is oxidised to oxaloacetate (OAA).

1. **Oxidation of succinic acid to fumaric acid:** It is a dehydrogenation reaction catalysed by the enzyme ***succinate dehydrogenase,*** hydrogen acceptor is FAD. The enzyme is **Ferri-flavoprotein**, mol. wt = 200,000 containing FAD and Iron-sulphur (Fe: S), contains 4 atoms of non-haem Fe and one FAD per mol. of enzyme. In contrast to other enzymes of TCA cycle, this enzyme is bound to the inner surface of the inner mitochondrial membrane.

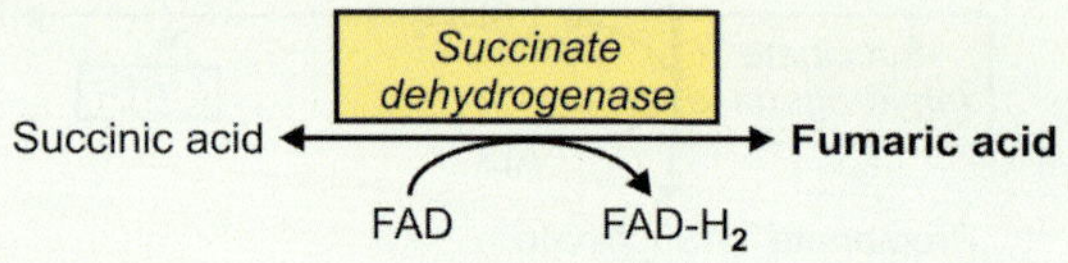

This is the only dehydrogenation in citric acid cycle which involves direct transfer of H from substrate to a flavoprotein without the participation of NAD^+.

Inhibitor: Addition of ***Malonate***/or ***OAA*** inhibits succinate dehydrogenase competitively resulting in accumulation of succinate.

Energetic: Oxidation of $FAD.H_2$ through ETC yields 2 ATP. Hence 2 molecules of succinic acid will give 4 ATP

+ 4 ATP

2. ***Formation of Malic Acid from Fumaric Acid***

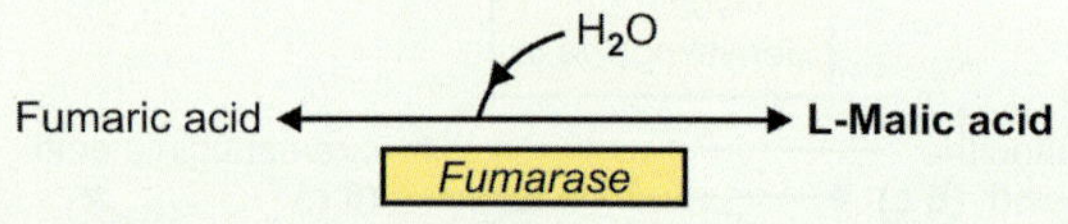

In addition to being specific for the L-isomer of malonate, fumarase catalyses the addition of the elements of water to the double bond of fumarate in the 'trans' configuration.

3. ***Oxidation of malic acid to oxaloacetate (OAA):*** The reaction is catalyzed by the enzyme ***Malate dehydrogenase*** which requires NAD^+ as H-acceptor.

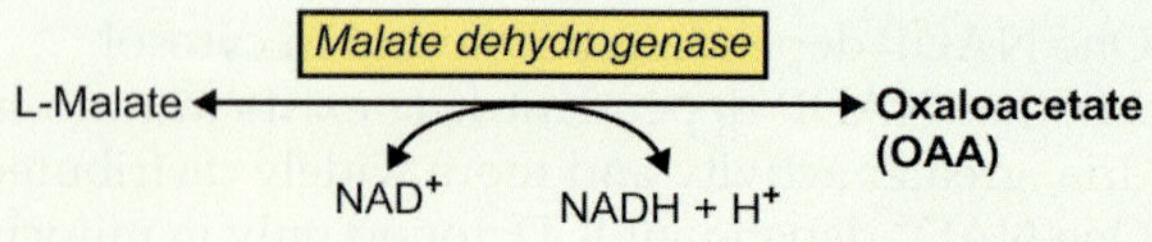

OAA produced acts 'Catalytically', combines with a fresh molecule of acetyl-CoA and the whole process is repeated.

Note: Although the equilibrium of this reaction strongly favours L-malate the net reaction is toward the formation of OAA as this compound together with the other product of reaction like NADH and $FADH_2$ are removed continuously.

Energetic: Oxidation of $NADH_2$ through ETC produces 3 ATP. Hence two molecules of $NADH^+$ will give 6 ATP

+ 6 ATP

ROLE OF VITAMINS IN TCA CYCLE

Five B vitamins are associated with TCA cycle essential for ***yielding energy***.

- Riboflavin: In the form of ***flavin adenine dinuleotide (FAD)***—a cofactor for ***succinate dehydrogenase*** enzyme.
- **Niacin:** In the form of ***nicotinamide adenine dinucleotide (NAD)*** the electron acceptor for ***isocitrate dehydrogenase, α-ketoglutarate dehydrogenase, and malate dehydrogenase.***
- **Thiamine:** As "thiamine diphosphate"—required as coenzyme for decarboxylation in the α-ketoglutarate dehydrogenase reaction.
- **Lipoic acid:** It is required as coenzyme for α-ketoglutarate dehydrogenase reaction.
- **Pantothenic acid:** As part of coenzyme A, the cofactor attached to "active" carboxylic acid residues such as acety CoA and succinyl-CoA.

TCA CYCLE IS CALLED AMPHIBOLIC IN NATURE—WHY?

TCA cycle has dual role

- **catabolic,** and
- anabolic.

(a) ***Catabolic role:*** The two carbon compound acetyl-CoA produced from metabolism of carbohydrates, Lipids and Proteins are oxidised in this cycle to produce CO_2, H_2O and energy as ATP.

(b) ***Anabolic or synthetic role:*** Intermediates of TCA cycle are utilised for synthesis of various compounds.

Examples

1. ***Transamination: Synthesis of non-essential amino acids:*** Transaminase (aminotransferase) reactions produce Ketoacids, PA, OAA and α-ketoglutarate, from alanine, asparate and glutamate respectively. Because these reactions are reversible, TCA cycle serves as a source of C-skeletons for the synthesis of nonessential amino acids.

Asparate + PA ⟷ OAA + Alanine
Glutamate + PA ⟷ α-ketoglutarate + Alanine

2. ***Formation of glucose: (Gluconeogenesis):*** Other amino acids contribute to gluconeogenesis because all or part of their C-Skeletons enter TCA cycle after deamination or transamination.

- ***Pyruvate forming amino acids:*** Glycine, alanine, serine, cysteine/cystine, threonine, Hydroxy-Proline and tryptophan.
- ***α-Ketoglutarate forming amino acids:*** Arginine, histidine, glutamine and proline.
- ***Fumarate forming amino acids:*** Phenylalanine and tyrosine
- ***Succinyl-CoA forming amino acids:*** Valine, Methionine and Isoleucine.

3. ***Fatty acid synthesis:*** Acetyl-CoA formed from PA by the action of PDH complex is the starting material for long chain FA synthesis (palmitic acid). ***But this synthesis is extramitochondrial,*** whereas acetyl-CoA

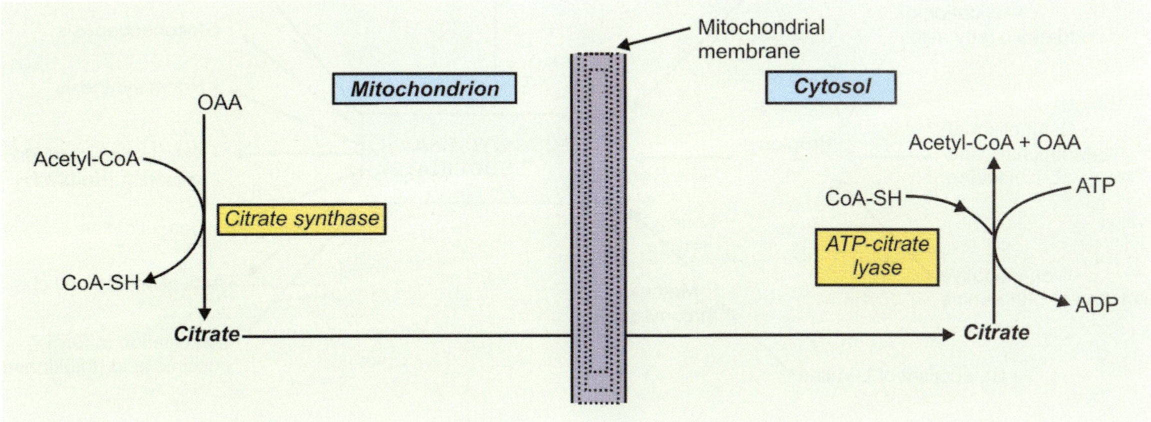

is formed in mitochondria. Acetyl-CoA is impermeable to mitochondrial membrane and hence it has to be transported out. This is achieved by "citric acid", an intermediate of TCA cycle, which is permeable to mitochondrial membrane. Acetyl-CoA is made available in cytosol for synthesis of FA, by cleavage of 'citrate' by the enzyme ***citrate-cleavage enzyme (ATP-citrate lyase)*** present in cytosol. This is shown schematically in the box.

4. ***Synthesis of cholesterol and steroids:*** Acetyl-CoA is used for synthesis of cholesterol, which in turn is required for synthesis of steroids.
5. ***Haem synthesis:*** Succinyl-CoA produced in TCA cycle takes part in heme synthesis.
6. ***Formation of acetoacetyl-CoA:*** Succinyl-CoA is utilised for formation of 'acetoacetyl-CoA' from acetoacetate in extrahepatic tissues (Refer ketolysis).

Formation and fate of succinyl-CoA (Active succinate) is shown schematically in Figure 23.5.

Regulation of TCA Cycle

1. As the primary function of TCA cycle is to provide energy, respiratory control via the ETC and oxidative phosphorylation exerts the main control.
2. In addition to this overall and coarse control, several enzymes of TCA cycle are also important in the regulation.

Three Key enzymes are:

- ***Citrate synthase***
- ***Isocitrate dehydrogenase (ICD)***
- ***α-oxoglutarate dehydrogenase***

These enzymes are responsive to the energy status as expressed by the ATP/ADP ratio and NADH/NAD^+ ratio

- Citrate synthase enzyme is allosterically inhibited by ATP and long-chain acyl-CoA.
- NAD^+-dependant mitochondrial *isocitrate dehydrogenase* (ICD) is activated allosterically by ADP and is inhibited by ATP and NADH.
- α-oxoglutarate dehydrogenase regulation is analogous to PDH complex.

3. In addition to above, *succinate dehydrogenase* enzyme is inhibited by OAA and the availability of OAA is controlled by *malate dehydrogenase*, which depends on NADH/NAD^+ ratio.

Bioenergetics: Overall energy production in glycolysis cum TCA cycle in presence of O_2 is summarised in box an next page.

Efficiency

1. Complete oxidation of glucose to CO_2 and H_2O in a 'Bomb calorimeter' yields 686,000 calories which is liberated as heat.
2. When oxidation occurs in tissues, some of this energy is not lost immediately as 'heat' but captured as "high energy PO_4 bonds". At least 38 high energy PO_4 bonds are generated per molecule of glucose oxidised to CO_2 and H_2O.
3. Assuming each high energy bond to be equivalent to 7600 calories. Total energy captured in ATP per mol. of glucose oxidised:

$$= 7{,}600 \times 38$$
$$= 288{,}800 \text{ calories}$$

Hence, **efficiency** $= \dfrac{288{,}800}{686{,}000} \times 100$

$= $ ***42% of energy of combustion.***

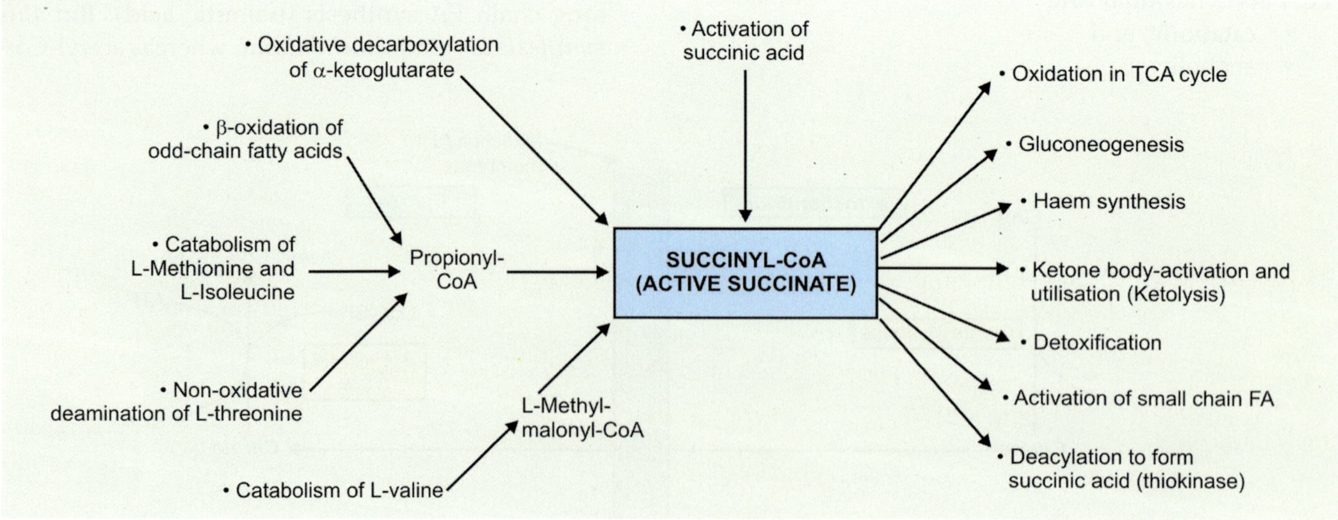

Fig. 23.5: Succinyl-CoA (Active succinate): Formation and fate

4. Most of ATP is formed as a result of oxidative phosphorylation resulting from re-oxidation of reduced coenzymes, viz., NADH and $FADH_2$ by the respiratory chain.

The remainder is generated by Phosphorylation at ***substrate level.***

PASTEUR EFFECT

Anaerobic oxidative reactions (glycolysis) may be decreased by aeration. This was first observed by **Pasteur** in studies of the fermentation of glucose by yeast cells. Pasteur also noted that in the presence of O_2 less glucose was broken down by the yeast cells and less alcohol was formed, whereas under anaerobic conditions more alcohol was formed and more glucose was fermented. ***The phenomenon of inhibition of glycolysis by O_2 is termed the Pasteur effect.***

Explanation: Various views have been offered for the explanation of the phenomenon:

- For a given amount of energy, less glucose is required to be metabolised in presence of O_2, hence less glycolysis.
- Anaerobically more Pi and ADP are available which will stimulate glycolysis. In presence of O_2, due to operation of TCA cycle, more ATP is formed and there is less Pi and ADP available. Increased ATP will inhibit glycolysis.
- In presence of O_2, as the TCA cycle operates, more ATP is produced which inhibits the key enzyme *phosphofructokinase-1* and thus decreasing glycolysis.

Crabtree effect: This is opposite of Pasteur effect, which represents decreased respiration of cellular systems caused by high concentration of glucose.

A. Glycolysis		***ATP yield per hexose unit***
1. Glycogen → Fructose-1, 6 bi — P		–1 ATP
2. Glucose → Fructose-1, 6 bi — P		–2 ATP
3. Glyceraldehyde-3P-dehydrogenase (2 NADH → 2 NAD^+)		+ 6 ATP
4. Substrate level Phosphorylation		
(a) 2-phosphoglycerate kinase		+ 2 ATP
(b) Pyruvate kinase		+ 2 ATP
Net gain		
• For glucose	=	**+ 8 ATP**
• For glycogen	=	+ 9 ATP
B. Oxidative decarboxylation of PA		
1. Pyruvate dehydrogenase complex (2 NADH → 2 NAD^+)		**+ 6 ATP**
C. TCA Cycle		
1. Isocitrate dehydrogenase 2 (NADH → NAD^+)		+ 6 ATP
2. α-Ketoglutarate dehydrogenase complex 2 (NADH → NAD^+)		+ 6 ATP
3. Substrate level phosphorylation succinate thiokinase 2 GTP or 2 ITP → 2 ATP		+ 2 ATP
4. Succinate dehydrogenase 2(FAD.H_2 → FAD)		+ 4 ATP
5. Malate dehydrogenase 2(NADH → NAD^+)		+ 6 ATP
Total per mol. of glucose	=	30 + 8
(under aerobic condition)	=	**38 ATP**
Total per mol. of glycogen (glycogenolysis provides G-1-P		30 + 9
(G-1-P → G-6-P)	=	**39 ATP**
Total per mol. of glucose (under anaerobic condition)	=	**+2ATP** (only in anaerobic glycolysis)

SHUTTLE SYSTEMS

NADH produced in the glycolysis is extramitochondrial, whereas the electron transport chain, where NADH has to be oxidised to NAD^+ is in the mitochondrion. ***NADH is not permeable to mitochondrial membrane.*** It is envisaged that NADH produced in cytosol ***transfer the reducing equivalents through the mitochondrial membrane via substrate pairs,*** linked by suitable dehydrogenases by shuttle systems. It is important that the specific dehydrogenases which act as "shuttle" be present on both sides of mitochondrial membrane. **Two** such shuttle systems are there:

1. Glycerophosphate shuttle
2. Malate shuttle

1. Glycerophosphate shuttle: The α-glycerophosphate shuttle is shown on next page diagrammatically in box.

Note

- α-glycero-P-dehydrogenase enzyme present **in cytosol is NAD-linked**, whereas the enzyme present **in mitochondrion is a FP-dependant.** *So oxidation of reduced FAD in respiratory chain will produce only 2 ATP and not 3 ATP.*
- Although it might be important in Liver and probably in heart muscle, but mitochondrial-FP-dependant α-glycero-P-dehydrogenase enzyme is deficient in other tissues.

SECTION FOUR

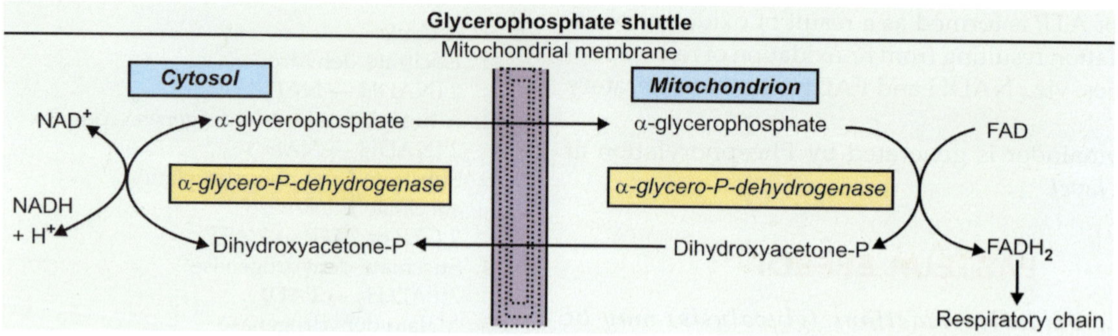

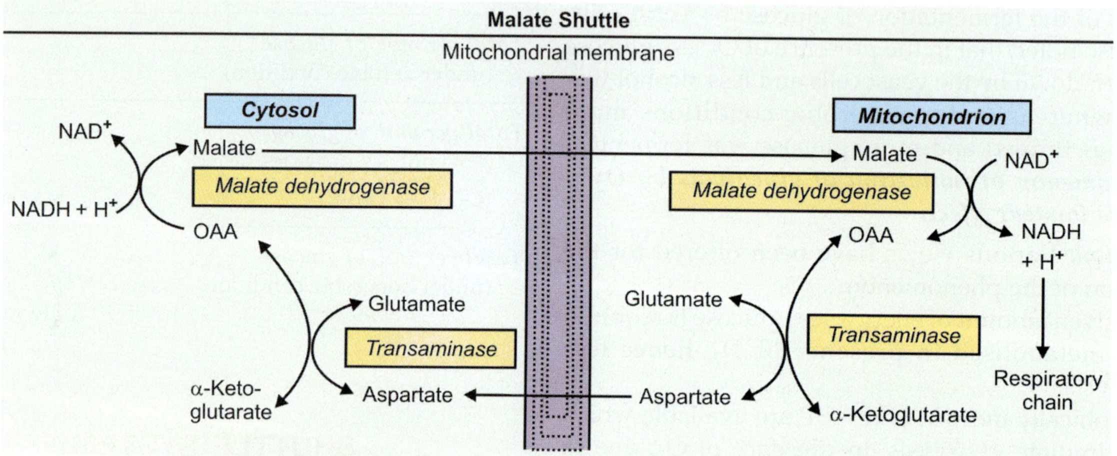

- This shuttle system is not much common to be used in humans. It is present in insect flight muscle and in white muscle.
- FAD-linked mitochondrial α-glycerophosphate dehydrogenase decreases after thyroidectomy and increases after administration of thyroxine in some species.

2. **Malate shuttle:** Malate shuttle is shown diagrammatically above in the box.

Note

- ***This shuttle system is more common and universal.*** Reduced NADH + H^+ is reformed in mitochondrion, which being oxidised in respiratory chain produces 3 ATP.
- The system is rather a little complex as OAA is impermeable to mitochondrial membrane; whereas malate, aspartate, glutamate and α-ketoglutarate are permeable to mitochondrial membrane. Hence OAA reformed in mitochondrion has to be transaminated to form aspartate. In the cytosol again OAA is regenerated by transamination.

Energetics

- **When body utilises α-glycero-P-shuttle, net ATP produced by glycolysis—TCA cycle per molecule glucose oxidised will be 36 ATP (2 ATP less) and NOT 38 ATP.**
- **Use of Malate shuttle will form 38 ATP.**

PART II

Major Concepts

A. Metabolism of Glycogen—to study formation of glycogen from glucose (glycogenesis) and its breakdown (glycogenolysis) and inherited disorders.
B. Other alternative pathways for oxidation of glucose— • Hexose monophosphate pathway • Uronic acid pathway.
C. Formation of glucose from non-carbohydrate sources (gluconeogenesis).
D. Metabolism of other carbohydrates and inherited disorders associated with them. • Metabolism of galactose • Metabolism of fructose.
E. Regulation of Blood Sugar Level
  • Autoregulation • Hormonal control • Learn the normal blood glucose level and clinical significance of its variations.
F. Glycosurias
G. Diabetes mellitus: Clinical and study how experimental diabetes can be produced in brief.
H. Glucose tolerance test (GTT).

Specific Objectives

A. Glycogen Metabolism: Glycogen is the major storage form of carbohydrates in human beings and correspond to starch in plants. It occurs mainly in liver (upto 6%) and muscle, where it rarely exceeds 0.7 per cent.
  1. Glycogen formation (glycogenesis):
     • Learn the reactions by which glucose is converted to glycogen, enzymes and coenzymes required for each.
     • Study specially *Glycogen synthetase,* the key enzyme, active and inactive forms, and how glycogen synthesis is regulated.
  2. Glycogen breakdown (glycogenolysis): *It is* ***Not a reversal of synthetic pathway.***
     • Learn the reactions of glycogen breakdown with enzymes and coenzymes.
     • Study specially ***phosphorylase*** enzyme, active and inactive forms.
     • ***End-product of glycogen breakdown is not glucose but mainly glucose1-P.*** Glucose is only formed on cleavage of α-1-6, linkage.
  3. Regulation of synthetic and catabolic pathway.
  4. Inherited disorders associated with glycogen synthesis/and breakdown. Learn the six major types, pinpoint the enzyme deficiency and salient features of each type.

B. (a) Hexose monophosphate pathway (HMP):
     • An alternate pathway for oxidation of glucose. Its purpose is different. It does not produce ATP. Learn the synonyms.
     • Learn the reactions involved in the pathway, with enzymes and coenzymes involved.
     • How does this pathway differ from EM pathway? Tabulate the similarities and differences.
     • Why this pathway is necessary? Learn the metabolic importance of this pathway.
  (b) Uronic acid pathway:
     • Another alternative pathway for glucose oxidation.
     • Learn the reactions involved with enzymes and coenzymes required.
     • Formation of D-Glucuronic acid in this pathway: Study the biological importance of this compound and its role in the body.
     • Vitamin C (ascorbic acid) can be synthesised by this pathway in lower animals, not in human beings. Why?
     • How it is connected with HMP pathway?
     • Inherited disorders associated with this pathway.

C. Gluconeogenesis: Formation of glucose from *non-carbohydrate sources.*
  • Define gluconeogenesis. What are the non-carbohydrate sources? List them. Why gluconeogenesis is necessary to the body?
  • Learn the metabolic pathways, enzymes and coenzymes required. How it is regulated?
  • Learn how pyruvic acid is converted to glucose? Is it a simple reversal of glycolysis? If not, what are the barriers to simple reversal and how the barriers are circumvented?
  • Study how other non-carbohydrate substances are converted to glucose?
  • Learn the regulation of the pathway.
  • Learn the other fates of lactic acid. Study what is "Cori cycle".

D. (a) Metabolism of Galactose:
     • Learn the reactions involved in metabolism of galactose with name of enzymes and coenzymes.
     • Study how lactose is synthesised in lactating mammary glands? (biosynthesis of lactose) and its regulation.
     • Inherited disorder associated with galactose metabolism. Name the specific enzyme which is deficient and salient features.
  (b) Metabolism of Fructose:
     • Learn the different pathways of fructose metabolism, enzymes and coenzymes.
     • Semen is rich in fructose as sperms utilise for its metabolism. Learn the pathway by which glucose is converted to fructose in seminiferous tubules (sorbitol pathway).
     • List the various inherited disorders associated with fructose metabolism. Name the specific enzyme deficient with the disorders and salient features of the disorders.

E. Regulation of Blood Sugar Level:
  • What is the normal level of blood glucose? What is meant by the term *"true" glucose?*
  • Study the various factors which regulate the blood glucose level. Learn what is autoregulation? and study the hormonal regulation of blood sugar.

SECTION FOUR

- Study whether there is any difference in blood sugar level in arterial vs. venous blood. Explain difference if any.
- List the conditions which can cause increase of blood glucose level (hyperglycaemia) and those which decrease blood glucose level (hypoglycaemia).

F. Glycosurias and other Disorders:

a. Glycosurias: Presence of glucose in urine in detectable amount which can be detected by Benedict's qualitative test. Learn types of glycosurias.
   - Hyperglycaemic glycosuria: List the causes
   - Renal glycosurias: List the causes
   - Experimental glycosurias:
     - Claude Bernad's "picsquare glycosuria" or 'punctured' diabetes.
     - Alloxan diabetes
     - Phloridzin diabetes.

b. Study Diabetes Mellitus:
   - Definition and Types of diabetes mellitus: Type 1 and Type 2
   - Metabolic derangements that occur in diabetes mellitus
   - Study the different methods by which diabetes can be produced in experimental animal.

c. Glucose Tolerance Test:
   Study under following heads:
   - Aim and indications.
   - Precautions to be taken before GTT is done.
   - Procedure, and Interpretation of different types of curves.
   - WHO recommendation for diagnosis of DM.

SECTION FOUR

METABOLISM OF GLYCOGEN

Glycogen is the storage form of glucose, which is stored in animal body specially in liver and muscles. It is mobilised as glucose whenever body tissues require. Students should revise their knowledge regarding chemistry of glycogen.

Why body stores glucose as glycogen and not as glucose itself?

Possible Reasons

1. Being insoluble it exerts no osmotic pressure, and so does not disturb the intracellular fluid content and does not diffuse from its storage sites.
2. It has a higher energy level than a corresponding weight of glucose (though energy has to be expended to make it from glucose).
3. It is readily broken down under the influence of hormones and enzyme:
   - Into glucose in liver (to maintain blood glucose level).
   - Into lower intermediates in skeletal muscle and other tissues for energy.

Role of Liver Glycogen

- ***It is the only immediately available reserve store of blood glucose.***
- A high liver glycogen level protects the liver cells against the harmful effects of many poisons and chemicals, e.g. CCl_4, ethyl alcohol, arsenic, various bacterial toxins.
- Certain forms of detoxication, e.g. conjugation with glucuronic acid; and acetylation reactions, are directly influenced by the liver glycogen level.
- The rate of deamination of amino acids in the liver is depressed as the glycogen level rises, so that amino acids are preserved longer in that form and so remain available for protein synthesis in the tissues.
- Similarly, a high level of liver glycogen depresses the rate of ketone bodies formation.

Biomedical Importance

- Liver glycogen is largely concerned with storage and supply of glucose-1-P, which is converted to glucose, for maintenance of blood glucose, particularly in between meals.
- Muscle glycogen on the other hand, is to act as readily available source of intermediates of glycolysis for provision of energy within the muscle itself. ***Muscle glycogen cannot directly contribute to blood glucose level.***
- Inherited deficiency of enzymes in the pathway of glycogen metabolism produces certain inherited disorders called as ***Glycogen storage diseases (GSDs).***

Metabolism of glycogen can be discussed **under two headings:**

A. **Synthetic phase:** Formation of glycogen
B. **Catabolic phase:** Breakdown of glycogen

Note

- The above two pathways are different from each other. ***Breakdown pathway is not a 'reversal' of synthetic pathway.***
- ***Glycogen in any tissue is not static.*** It is being constantly used up and resynthesised. At any time, the glycogen of the tissue should be considered ***as a balance between constant production and loss.***

- Both the processes are finely controlled at substrate level, by end-products and hormones.
- ***When glycogenesis occurs, glycogenolysis does not take place and vice versa.***

GLYCOGENESIS

Definition: It is the formation of glycogen from glucose.

Sites: Principally it occurs in liver and skeletal muscles, but it can occur in every tissue to some extent.

Limitations of storage: In humans, the liver may contain as much as 4 to 6 per cent of glycogen as per weight of the organ, when analysed shortly after a meal, high in carbohydrate. After 12 to 18 hours of fasting, the liver becomes almost totally depleted of glycogen.

Storage Capacity of Glycogen in a Normal Adult Man (70 Kg)

Organ	*Amount*	*Normal weight of the organ*
• Liver glycogen	4 to 6% = 72 to 108 gm	1800 gm (liver)
• Skeletal muscles	0.7% = 245 gm	35 kg (muscle)
• Extra-cellular glucose	0.1% = 10 gm	Total volume = 10 litres
	Total = 327 to 363 gm	

Reactions of Synthetic Pathway

1. Glucose is first phosphorylated to glucose-6-P by ***Glucokinase (or hexokinase),*** common to first reaction of glycolytic pathway.

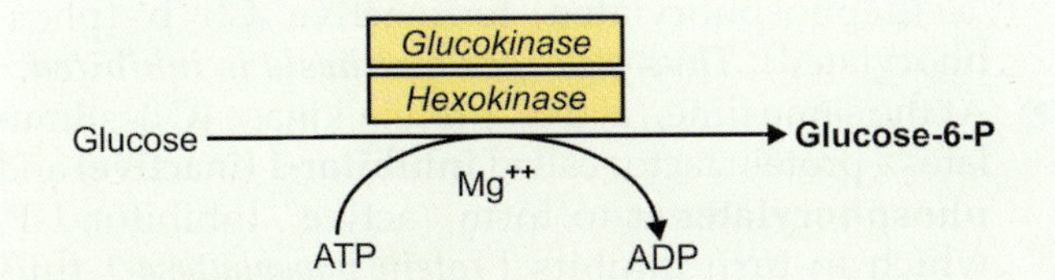

2. Glucose-6-P is converted to glucose-1-P by the enzyme ***phosphoglucomutase.***

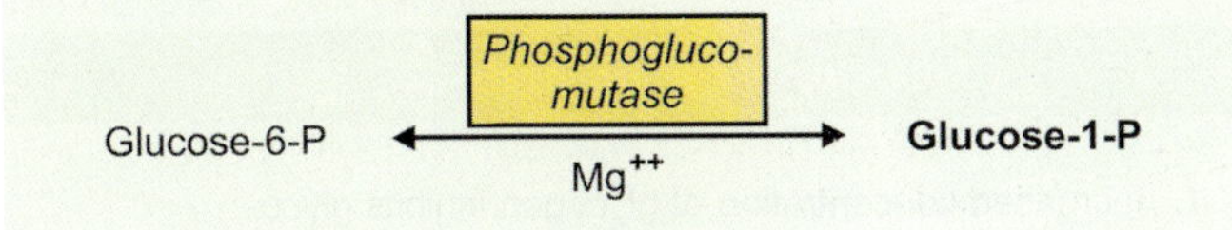

The reaction takes place as follows:

$$\text{Enz-P} + \text{Glucose-6-P} \rightarrow \text{Enz} + \text{Glucose-1,6-bi-P}$$
$$\downarrow \; \uparrow$$
$$\text{Enz-P} + \text{Glucose-1-P}$$

An intermediate glucose-1-6-bi-P is formed temporarily.

3. Glucose-1-P then reacts with one molecule of uridine-triphosphate (UTP) to form the "active" nucleotide **uridine-diphosphate-glucose (UDP-Glc)**, under the influence of the enzyme ***UDP-Glc pyrophosphorylase*** with elimination of a molecule of "pyrophosphate" (PPi). The subsequent hydrolysis of inorganic pyrophosphate (PPi) by *inorganic pyrophosphatase* drives the reaction to the right.

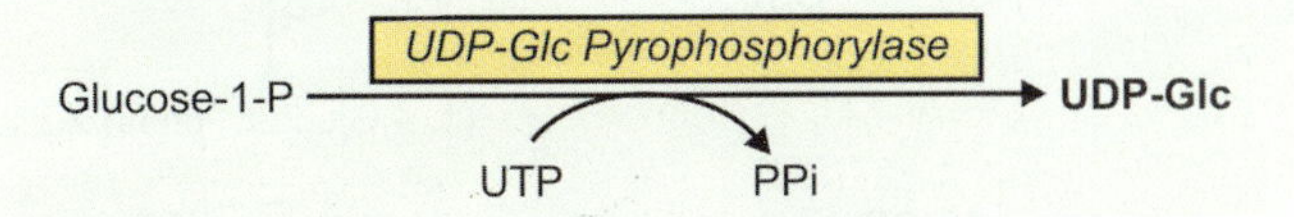

4. By the action of the enzyme ***glycogen synthase,*** C-1 of the activated glucose of UDP-Glc forms a glycosidic bond with the C-4 of a terminal glucose residue of glycogen **primer**, liberating UDP. ***A pre-existing glycogen molecule, or "primer" must be present to initiate this reaction.***

Glycogenin

The ***glycogen 'primer' is formed on a protein primer known as* glycogenin**. Glycogenin is a 37 kDa protein that is glycosylated on a specific tyrosine residue by UDPGlc. Further glucose residues are attached in the 1→ 4 position to form a short chain forming a substrate for ***glycogen synthase***. In skeletal muscle, ***glycogenin*** remains attached in the centre of the glycogen molecule; in liver the number of glycogen molecules is greater than the number of glycogenin molecules.

An existing glycogen chain can be repeatedly extended by one glucose unit.

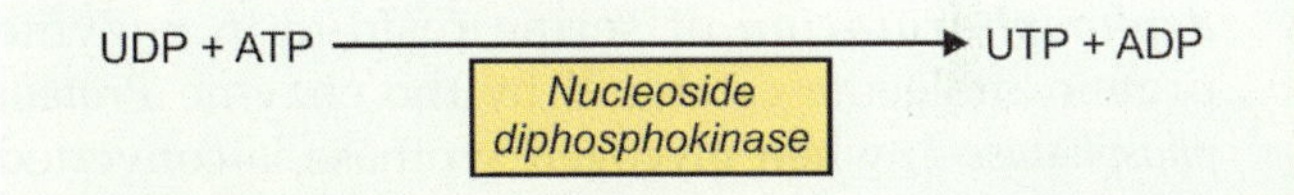

An existing glycogen chain can be repeatedly extended by one glucose unit at a time.

In each extension, 2 ATP molecules are expended:

- ***One in Phosphorylation of glucose to form G-6-P.***
- ***Another in the regeneration of UTP.***

Glycogen synthase requires glucose-6-P as an activator. The addition of a glucose residue to a pre-existing glycogen chain occurs at the "non-reducing" outer end of the molecule, so that the particular branch of the glycogen tree becomes elongated as successive α 1→ 4 linkages occur. ***Glycogen synthase is the principal key enzyme which regulates the glycogen formation*** (See regulation).

5. ***When the chain has been lengthened to minimum of 11 glucose residues,*** a second enzyme ***Branching enzyme***

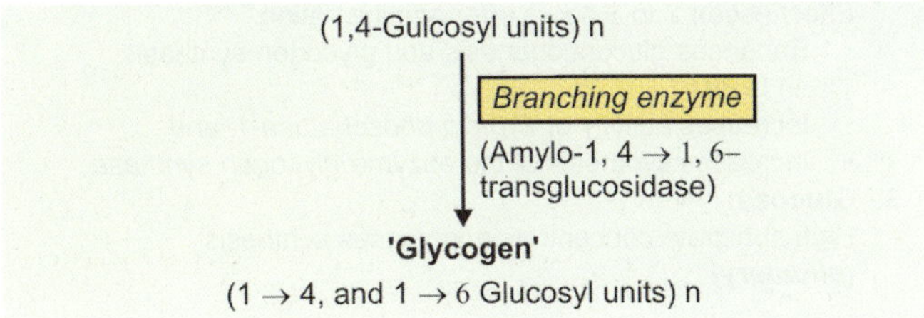

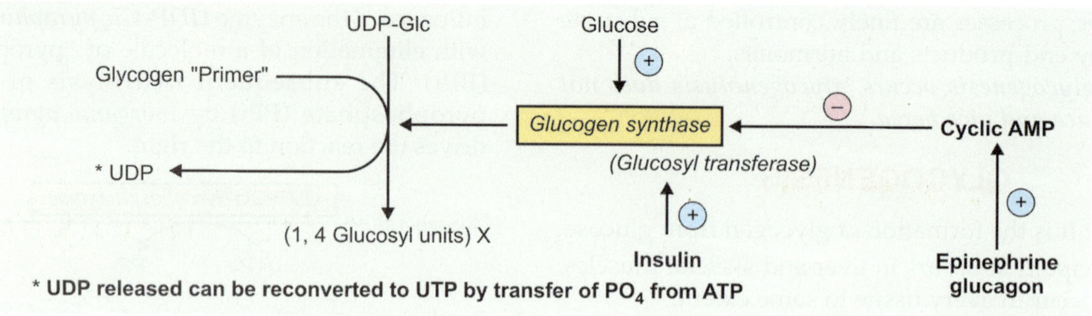

(*Amylo-1, 4 → 1, 6-transglucosidase*) comes into play and transfers a part of α 1 → 4 chain, ***minimum length 6 glucose residues,*** to a neighbouring chain **to form α 1 → 6 linkage thus establishing a branch** point in the molecule. The branches grow by further additions of α 1 → 4 glucosyl units and further branching.

Regulation of glycogenesis: ***Glycogen synthase is the key enzyme*** which regulates the process of glycogenesis. It is present as 'active' and 'inactive' forms, which are interconvertible.

- **'Active' form is: GS 'a'** (previously called as GS-I)
- **'Inactive' form is: GS 'b'** (previously called as GS-D)

1. ***'Active' GS-'a' is converted to 'inactive' GS-'b'* by phosphorylation**, which is modulated by Cyclic—AMP dependant *protein kinase*. When glycogen synthase is converted to 'inactive' form the glycogenesis is inhibited (role of *protein kinase* is discussed below).
2. ***'Inactive' GS-b is converted to 'active' GS 'a' by dephosphorylation*** of serine residue in enzyme protein molecule catalysed by the enzyme *Protein phosphatase-1;* when glycogen synthase is converted to 'active' form glycogenesis occurs. ***The interconversion of active to inactive and vice-versa is controlled by substrate level, end-product and hormones.*** The interconversion of "Glycogen synthase" 'a' and 'b' and its regulation is shown schematically in Figure 23.6 (Refer next page).

Role of cyclic-AMP dependant protein kinases: Protein kinases exist in cells, the enzyme is of wide specificity and c-AMP dependant.

Activation and Inactivation of Protein Kinases: Protein kinases are "tetramer", consisting of 2 pairs of subunits. Each pair consists of ***regulatory* subunit (R)** and a ***'catalytic'* subunit (C)**, the latter contains the "active site". **Inactive *Protein kinase* is R_2C_2.**

When c-AMP level in cells increases, 2 mols of c-AMP bind to each of the regulatory (R) units, and thus releases the 'active' catalytic (C) units. ***Active Proteinkinase is C_2*** thus,

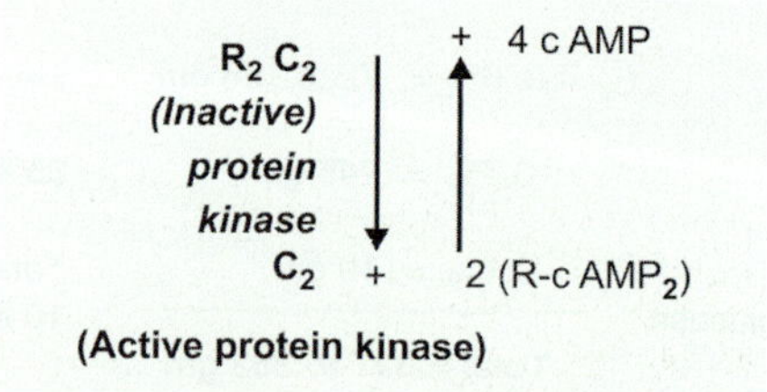

'Active' Protein kinase (C_2) thus formed due to increased cyclic AMP in the cell has ***two effects*** on glycogenesis:

- It brings about **phosphorylation** of *glycogen synthase* with the help of ATP and thus converts "active" GS 'a' (dephosphorylated) to 'inactive' GS 'b' (phosphorylated). ***Thus glycogen synthesis is inhibited.***
- At the same time, 'active' protein kinase (C_2), stimulates a protein factor called **inhibitor-1 (inactive)** and **phosphorylates** it to form "active" Inhibitor-1-P, which in turn inhibits *Protein phosphatase-1* thus conversion of inactive GS-'b' to active GS-'a' does not take place. ***Thus, glycogenesis in the cells is inhibited.***

Stimulation of Glycogenesis	*Inhibition of Glycogenesis*
1. **Insulin:** Insulin increases the *Protein-phosphatase-1* activity 2. **Glucocorticoids:** Effects seen 2 to 3 hours after administration. • Enhances gluconeogenesis and glycogen synthesis in Liver. • Increases activity of *Protein phosphatase-1*, and • Increases synthesis of the enzyme *glycogen synthase* 3. **Glucose:** High substrate concentration increases synthesis ***(allostery)***	1. Increased concentration of glycogen inhibits glycogenesis, "Feedback" inhibition 2. Increased concentration of cyclic-AMP ↑ stimulates **inhibitor-1**, to form 'active' inhibitor-1-P, which in turn inhibits *Protein Phosphatase-1*.

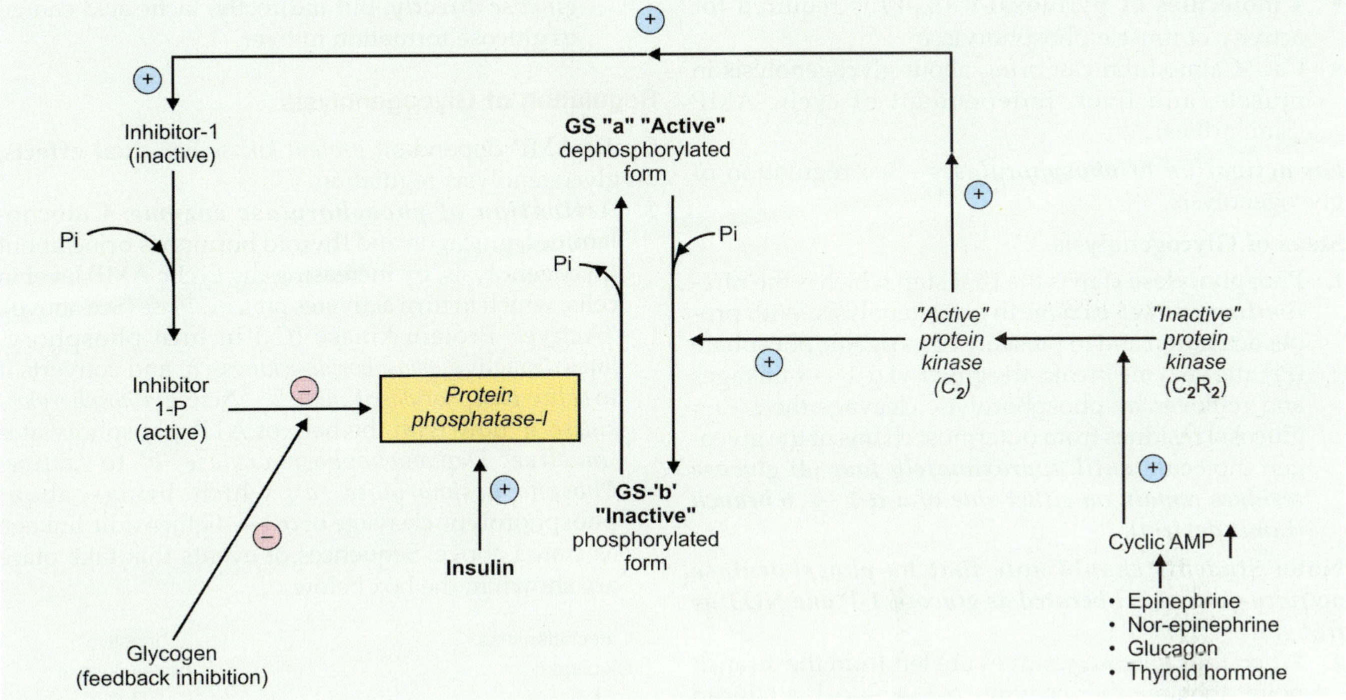

Fig. 23.6: Activation and inactivation of glycogen synthase

CLINICAL ASPECT

Relation of glycogenesis with K^+ influx into the cell:

- Clinical importance in treatment of hyperkalaemia, when insulin and glucose are administered.
- Importance in treatment of diabetic ketoacidosis, with insulin and glucose, danger of "hypokalaemia" to precipitate. ***The patient should be monitored for potassium level in blood while treating diabetic ketoacidosis with insulin and glucose infusion.***

GLYCOGENOLYSIS

Definition: Breakdown of glycogen to glucose is called as glycogenolysis.

"Phosphorylase" Enzyme

It is initiated by the action of a specific enzyme ***phosphorylase***, which ***brings about*** **Phosphorolytic** ***cleavage of*** α ***1→ 4 linkage to yield glucose-1-P.*** Liver and muscle both contain the enzyme ***phosphorylase***. In both, the enzyme can be in 'active' and "inactive" forms and both are interconvertible.

(a) Liver Phosphorylase

- **'Active' form** is the phosphorylated form *Phospho-phosphorylase* (cf. glycogen synthase), having mol. wt 240,000.
- **"Inactive" form** is the dephosphorylated form *Dephosphophosphorylase.*

(b) Muscle Phosphorylase

It can also exist in 'active' and 'inactive' forms and are interconvertible. 'Active' form is *Phosphorylase 'a'* and 'inactive' form is phosphorylase *'b'*. Muscle phosphorylase is immunologically distinct from liver phosphorylase.

Characteristics of muscle phosphorylases	
• ***Muscle phosphorylase 'a'*** **(active)**	***'Active' form*** • Tetramer • Molecular wt = 500,000 • Contains 4 molecules of Pyridoxal-P per molecule of enzyme
• ***Muscle phosphorylase 'b'*** **(inactive)**	***'Inactive' form*** • A 'dimer' • Molecular wt = 250,000 • Contains two molecules of Pyridoxal-P per molecule of the enzyme

Differences of Muscle Phosphorylase from Liver Phosphorylase

- There is ***no cleavage of structure*** with liver phosphorylase as compared to muscle phosphorylase.
- *Muscle phosphorylase* is not affected by glucagon **(no receptor on muscle).**

SECTION FOUR

- **4 molecules of pyridoxal-P (B_6-P)** is required for activity of muscle phosphorylase
- Ca^{++}-Calmodulin can bring about glycogenolysis in muscle/and liver, independent of cyclic-AMP (Similarities).

For activation of phosphorylases—See regulation of glycogenolysis.

Steps of Glycogenolysis

1. **Phosphorylase step** is the **first step** which is the ***rate-limiting and key enzyme*** in glycogenolysis. With proper activation and in presence of inorganic phosphate (Pi), the enzyme breaks the glucosyl α-1 → 4 linkages and removes by phosphorolytic cleavage the 1 → 4 glucosyl residues from outermost chains of the glycogen molecule ***until approximately four (4) glucose residues remain on either side of a α-1 → 6 branch (Limit dextrin).***

Note: ***Students should note that by phosphorylase activity glucose is liberated as glucose-1-P and NOT as free glucose.***

2. When four glucose residues are left from the branch point, then another enzyme, α-1, 4 → α-1, 4 Glucan transferase ***transfers a*** **trisaccharide** ***unit*** from one side to the other thus exposing α 1 → 6 branch point.
3. The hydrolytic splitting of α-1 → 6 glucosidic linkage requires the action of a specific ***debranching enzyme (Amylo-1 → 6-glucosidase).*** As the α-1→ 6 linkage is hydrolytically split, ***one molecule of free glucose is produced,*** rather than one molecule of glucose-1-P as is the case with phosphorolytic cleavage with the enzyme phosphorylase.

Clinical Significance

Note: In this way it is possible for some rise in blood glucose to take place even in absence of glucose-6-phosphatase, in Von Gierke's disease, after epinephrine or Glucagon is administered.

4. ***Fate of Glucose-1-P***
   The combined action of phosphorylase and other enzymes convert glycogen to glucose-1-P mostly. By the action of ***Phosphoglucomutase*** enzyme, glucose-1-P is easily converted to glucose-6-P, as the reaction is reversible.
   - In **liver and kidney,** a specific enzyme ***glucose-6-phosphatase*** is present, that removes PO_4 from glucose-6-P, enabling "free glucose" to form and diffuse from the cells to extracellular spaces including blood. ***This is the final step in hepatic glycogenolysis, which is reflected by a rise in blood glucose.***
   - ***In muscles, enzyme glucose-6-phosphatase is absent.*** Hence glucose-6-P enters into glycolytic cycle and forms pyruvate and LA. ***Muscle glycogenolysis does not contribute to blood glucose directly.*** But indirectly, lactic acid can go to glucose formation in liver.

Regulation of Glycogenolysis

Cyclic-AMP dependant *protein kinase* has **dual effects,** on glycogenolysis regulation.

1. ***Activation of phosphorylase enzyme:*** Catecholamines, glucagon and thyroid hormones bring about glycogenolysis, by increasing the cyclic AMP level in cells, which in turn activates *protein kinase* (See above). "Active" Protein kinase (C_2) in turn phosphorylates 'inactive' *phosphorylase kinase 'b'* and converts it to *active phosphorylase Kinase 'a'*. 'Active' *Phosphorylase kinase 'a'* now with the help of ATP, phosphorylates *'inactive' Dephosphophosphorylase 'b'* to *'active' Phosphophosphorylase 'a'*, which brings about phosphorolytic cleavage of α-1→ 4 glucosidic linkage as stated above. Sequences of events that take place are shown in the box below.

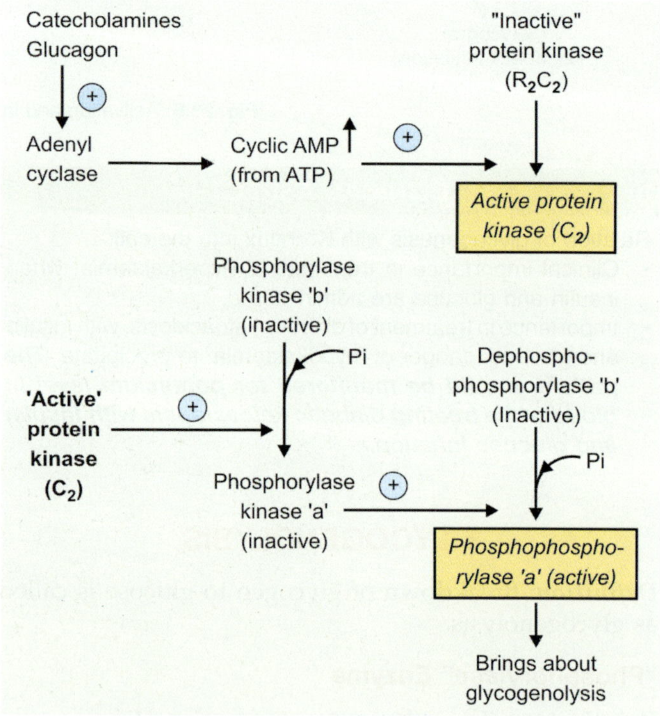

2. ***Activation of inhibitor-1:*** At the same time, *'active' protein kinase* (C_2), phosphorylates the protein factor called "inhibitor-1" (inactive) and converts it to 'active' inhibitor-1-p, which inhibits *Protein phosphatase-1* which in turn ***inhibits conversion of phosphorylase kinase 'a' to 'b'.*** Phosphorylase kinase 'a' in turn phosphorylates "dephosphophosphorylase 'b' to 'active' phosphophosphorylase 'a'.

Activation and Inactivation of Muscle Phosphorylase: In muscle, Phosphorylase can exist in ***two inter convertible forms*** 'active' and 'inactive'.

1. **Conversion of phosphorylase 'a' (active) to 'b' (inactive).** Phosphorylase 'a' contains 4 mols of B_6-PO_4. Phosphorylase 'b' contains 2 mols of B_6-PO_4.

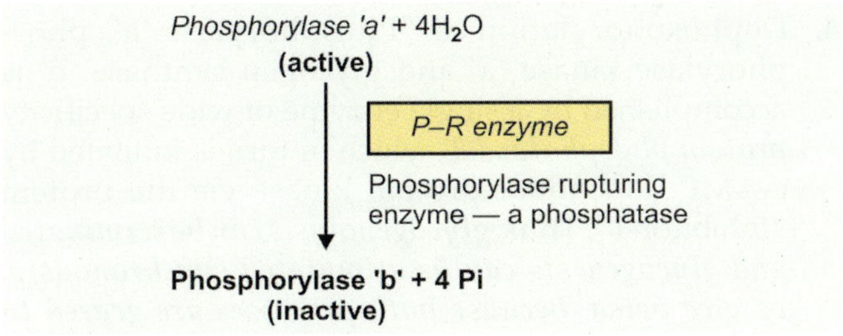

2. **Conversion of Phosphorylase 'b' (inactive) to 'a' (active):**

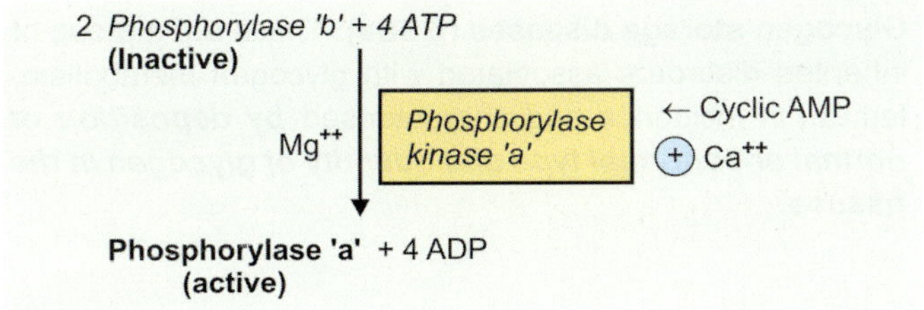

Note

- Catecholamines cause breakdown of liver and muscle glycogen, acting through β-adrenergic cell receptors on cell membrane.
- ***Glucagon breakdowns liver glycogen only,*** and not muscle glycogen as β-adrenergic cell receptors for glucagon are not present on muscle cell membrane.

Glycogenolysis of Muscle—Role of Ca^{++}

Activation by Ca^{++} and its synchronisation with muscle contraction.

1. Glycogenolysis increases in muscles several hundred fold immediately after the onset of the contraction. This involves the rapid activation of *Phosphorylase* owing to activation of *Phosphorylase kinase* by Ca^{++}, the same signal that initiates contraction.
2. Muscle *Phosphorylase kinase* has 4 types of subunits **(tetramer)**—α, β, γ and δ represented as $(\alpha, \beta, \gamma, \delta)_4$. The α and β-subunits contain serine residues that are phosphorylated by cyclic–AMP–dependant ***Protein kinase***. β-subunit binds 4 Ca^{++} ions and is identical to the Ca^{++} binding protein **calmodulin** (See below). **The binding of Ca^{++} activates the "Catalytic site" of the γ-subunit while** the molecule remains in the dephosphorylated 'b' configuration. **However, the phosphorylated 'a' form is fully activated in presence of Ca^{++}.**

   A second molecule of "calmodulin" can interact with the *phosphorylase kinase*, causing further activation. Thus, activation of muscle contraction and glycogenolysis are carried out by the same Ca^{++} binding protein.

Calmodulin

- A ***Ca^{++} dependant regulatory protein,*** which is specific for calcium, having a molecular wt = 17000 daltons. Widely distributed in all nucleated eukaryotic cells.
- ***A heat stable protein,*** containing 148 amino acids. High content of glutamic acid and aspartic acid, low level of aromatic amino acids. Lacks tryptophan, cysteine and OH-Proline.
- It is a ***flexible protein,*** with 4 binding sites for Ca^{++}, distributed in four domains. Homologous to muscle protein, "Troponin C" in structure and function. Binding of 4 molecules of Ca^{++} in four binding sites results in the conformational change linked to activation/inactivation of enzymes. Different enzymes require different conformational changes.
- The efflux of Ca^{++} is associated with a Na-Ca exchange Protein exchanging one Ca^{++} ion for two Na^+ ions.
- Evidences suggest that Ca-Pump, *Ca^{++}-ATPase* is probably identical with calmodulin.
- Some of enzymes regulated directly or indirectly by Ca^{++}, probably through calmodulin are: *Adenylate cyclase, Ca^{++}-dependant protein kinase, Ca^{++}/Mg^{++} ATPase, Ca^{++}/P L dependant protein kinase, Phosphodiesterase, phosphorylase kinase, myosin kinase,* etc.

Two Other Pathways of Glycogenolysis Independant of Phosphorylase

Two other pathways for breakdown of tissue glycogen other than phosphorolytic cleavage have been demonstrated though their physiological significance is uncertain.

1. ***α-Amylolysis***

   Due to presence in tissues of *α-amylase,* which hydrolyses glycogen to oligosaccharides; which are then broken down to glucose by the action of *glucosidases* (maltases), one of which ***α-1→4, glucosidase,* (*acid maltase*)** of human liver, hydrolyzes not only maltose to glucose but even the external branches of oligosaccharides and glycogen to glucose.

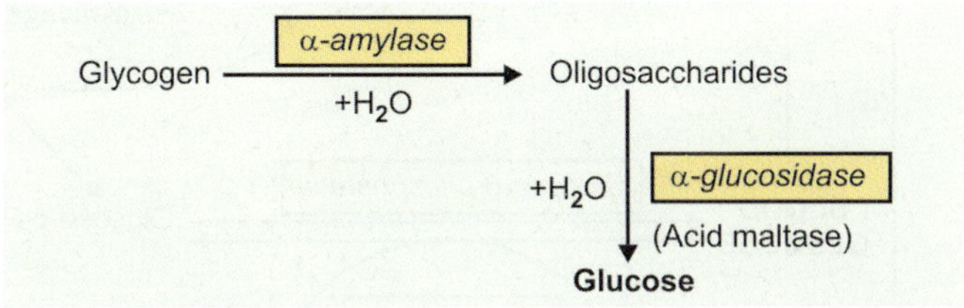

2. ***γ-Amylolysis***

   The other pathway of glycogen breakdown, γ-amylolysis, is affected through ***γ-amylase,*** which occurs generally in tissues. This enzyme splits glucose from glycogen by hydrolyzing the α-1→4 linkages of glycogen branches, forming so-called ***γ-dextrins***. Thus glycogen is only partially degraded to glucose.

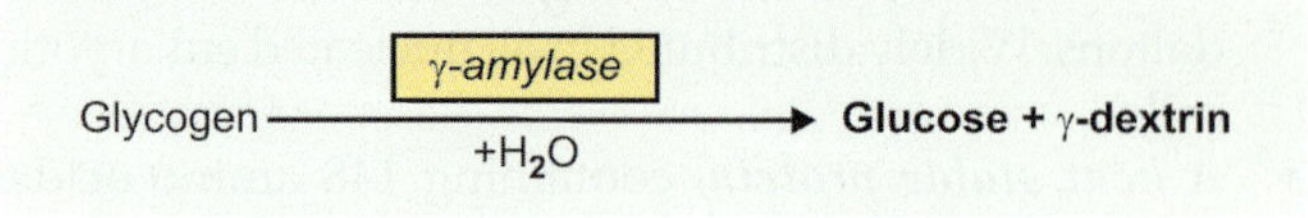

Regulation of Glycogen Metabolism (Fig. 23.7)

1. ***Regulation of glycogen metabolism is achieved by a balance in activities between Glycogen synthase and Phosphorylase***, which are as follows:
   - Substrate control (through ***allostery***) as well as
   - Hormonal control and by
   - End products.
2. Not only *Phosphorylase* enzyme is activated by a rise in concentration of c-AMP↑via phosphorylase kinase, but *Glycogen synthase* enzyme is at the same time converted to "inactive" form, both the effects are mediated via "c-AMP-dependant protein-kinase".
3. Thus inhibition of glycogenolysis increases net glycogenesis and inhibition of glycogenesis increases net glycogenolysis. ***Both processes cannot occur simultaneously together.***
4. Dephosphorylation of "Phosphorylase 'a', phosphorylase kinase 'a' and glycogen synthase 'b' is accomplished by a single enzyme of wide specificity ***protein phosphatase-1,*** which in turn is inhibited by c-AMP dependant protein kinase via the protein "Inhibitor-1". Thus, ***glycogenolysis can be terminated and glycogenesis can be stimulated synchronously, or vice versa, because both processes are geared to the activity of c-AMP dependant proteinkinase.***

INHERITED DISORDERS — CLINICAL ASPECT

Glycogen storage diseases (GSDs): These are a group of inherited disorders associated with glycogen metabolism, familial in incidence and characterised ***by deposition of normal or abnormal type and quantity of glycogen in the tissues.***

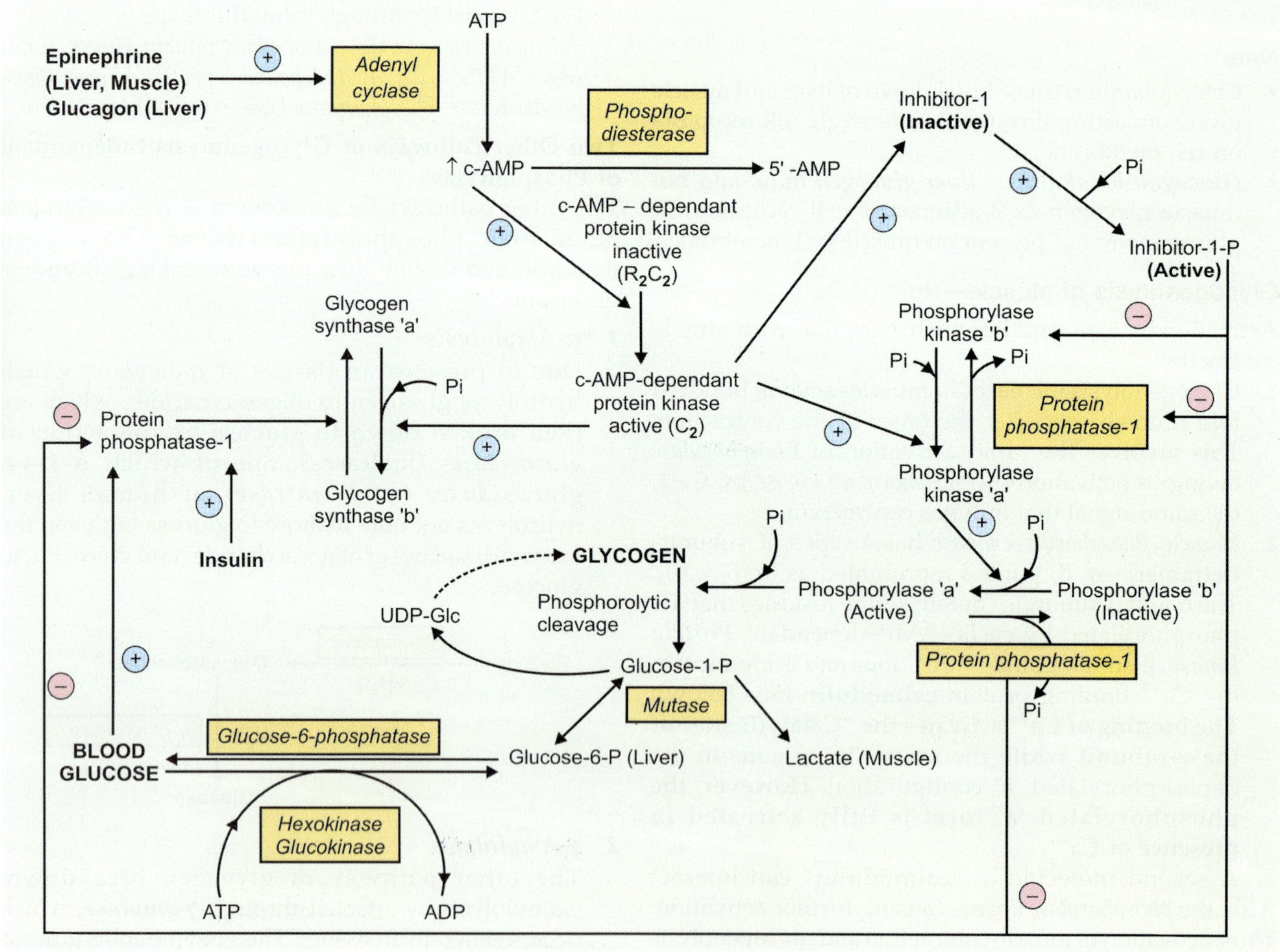

Fig. 23.7: Regulation of glycogen metabolism

Six classical types of GSDs will be considered, though there are quite a number of additions in the list.

Type I: Von Gierke's Disease

- ***Enzyme deficiency: Glucose-6-phosphatase.*** The enzyme is absent in liver cells and also in intestinal mucosa.
- ***Inheritence:*** *Autosomal recessive*
- Liver cells, intestinal mucosa and cells of renal tubular epithelial cells ***are loaded with glycogen which is normal in structure but metabolically not available.***

CLINICAL AND BIOCHEMICAL FEATURES

1. Since very little glucose is derived from the Liver, children with this disease tend to develop ***hypoglycaemia***. Glucose-1-P cannot be converted to glucose-6-P due to deficiency of the enzyme and it is "locked" in the cells.
2. **Fat is utilised** as **energy source**, which leads to ***Lipaemia, Acidaemia and Ketosis.***
3. **Excess acetyl-CoA** is diverted for ***cholesterol synthesis*** resulting to increase in cholesterol↑ level, which may produce ***Xanthomas.***
4. Increased FA synthesis can produce ***fatty infiltration*** of liver.
5. Persistent hypoglycaemia can have **two effects:**
   - Hypoglycaemia **inhibits insulin↓** secretion which in turn inhibits protein synthesis which causes ***stunted growth (dwarfism).***
   - Hypoglycaemia **stimulates secretion of catecholamines**, which cause muscle glycogen to breakdown producing Lactic acid and ***lactic acidosis.***
6. ***Increased blood Lactic acid competes with urate excretion by kidneys leading to increased blood uric acid↑ level.*** There is also evidence that there is increase in uric acid synthesis↑ in those children who develop symptoms of ***gout.***
   - ***Prognosis:*** Although many of these children die young, a number of them have survived for adolescence, when for some unknown reason much improvement can occur.

   Biochemical change in type-1 Von Gierke's disease and its correlation with clinical manifestations are shown schematically in Figure 23.8.

Type-II: Pompe's Disease

- ***Enzyme deficiency: Acid Maltase.*** Enzyme is present in lysosome and catalyses break down of oligosaccharides.

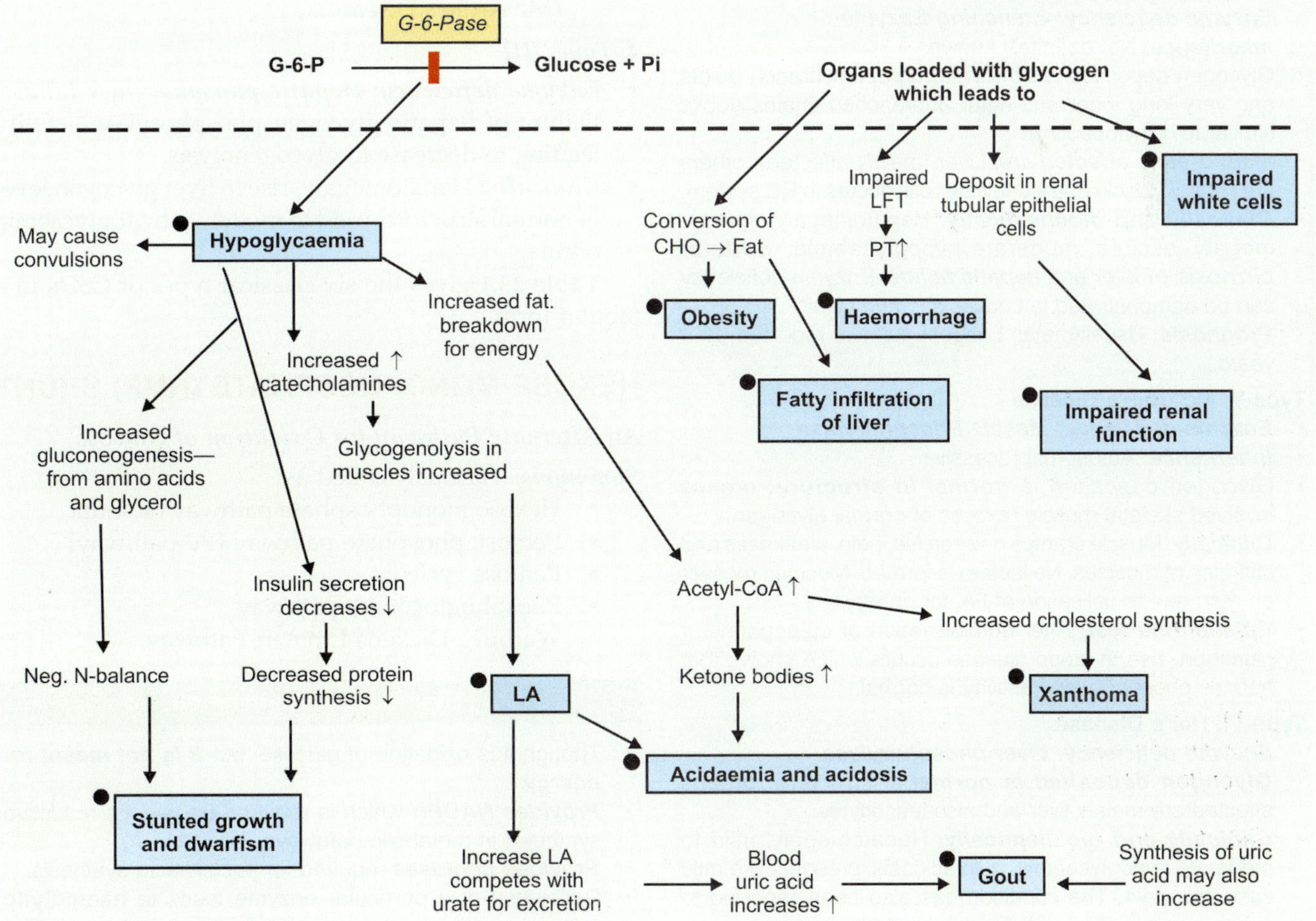

Fig. 23.8: Biochemical changes and correlation of clinical finding in type-I

SECTION FOUR

- ***Inheritance:*** Autosomal recessive.
- ***Glycogen structure is normal***: Generalised involvement of organs seen including heart, liver, smooth and striated muscles. Nearly all tissues contain excessive amount of normal glycogen.
- ***Clinically:*** Enlargement of heart ***(cardiomegaly)*** seen. Muscle hypotonia leading to muscle weakness. ***No hypoglycaemia.***
- ***Prognosis:*** Infants usually die of cardiac failure and bronchopneumonia. Death usually occurs before 9 months. A few cases, milder form survived up to 2½ years.

Type-III: Limit Dextrinosis (Forbe's Disease)

- ***Enzyme deficiency: Debranching Enzyme***
- ***Inheritance:*** Autosomal recessive.
- ***Glycogen structure: Limit dextrin type***, abnormal, short or missing outer chains. Organs involved are liver (18%), heart, and muscle (6%).

Clinically and biochemically: Hepatomegaly, moderate hypoglycaemia, acidosis, progressive myopathy. Similar to type-1, ***but glycogen is abnormal*** and runs a milder course. Enzyme deficiency can be demonstrated in leucocytes (diagnostic).

- ***Prognosis:*** Survives well to adult life.

Type-IV: Amylopectinosis (Andersen's Disease)

- ***Enzyme deficiency: Branching Enzyme***
- ***Inheritence:*** Not definitely known.
- Glycogen deposited is ***abnormal type,*** few branch points and very long inner and outer unbranched chains (looks similar to "amylopectin").
- ***Main organs affected are:*** Liver (mainly affected), others are Heart, Muscle and kidney. Deposit occurs in RE system.
- ***Clinically and biochemically:*** Hepatomegaly, splenomegaly, ascites, moderate hypoglycaemia, ***nodular cirrhosis of liver*** and ***hepatic failure***. Enzyme deficiency can be demonstrated in Leucocytes and Liver.
- ***Prognosis:*** Usually fatal. Longest survival reported as 4 years.

Type-V: McArdle's Disease

- ***Enzyme deficiency: Muscle Phosphorylase***
- ***Inheritence:*** Autosomal recessive.
- Glycogen deposited is ***normal in structure;*** organs involved skeletal muscle (excess of normal glycogen).
- ***Clinically:*** Muscle cramps on exercise, pain, weakness and stiffness of muscles. No lactate is formed. Muscles recover on rest, due to utilisation of FA. for energy.
- ***Epinephrine test:*** After administration of epinephrine/or glucagon, rise in blood glucose occurs which shows that hepatic phosphorylase activity is normal.

Type-VI: Her's Disease

- ***Enzyme deficiency: Liver phosphorylase***
- ***Glycogen deposited is normal*** in structure, organs affected are mainly liver and also leucocytes.
- ***Clinically and biochemically:*** Hepatomegaly, mild to moderate hypoglycaemia, mild acidosis; presents like mild case of type-1. The condition has also been reported to occur in association with Fanconi's syndrome.

Other GSDs: In addition to six classical types as discussed above, many other types of GSD's (total at least XIV) have been described. A few of them are:

1. *Lewis et al (1963): Described Type IX*
   - *Enzyme deficiency: Glycogen synthase*
   - Organs involved: Liver and erythrocytes. Liver glycogen is reduced very much (very low)
   - *Clinically and biochemically:* Hepatomegaly due to fatty infiltration, hypoglycaemia (fasting).
2. *Thompson et al (1963)*

Described deficiency of the enzyme ***phosphoglucomutase***. Glycogen deposited is normal type. Organs affected are: Liver, Muscle. Clinically: Moderate weakness of muscles, wasting of skeletal muscles.

3. *Tarui et al (1965): Described Type VII*
   - *Enzyme deficiency: Phosphofructokinase.*
   - Organs: Principally muscles. Moderate accumulation of glycogen in skeletal muscles. Also there is marked accumulation of G-6-P and fructose-6-P and marked diminution of fructose-1,6-bi-P.
   - *Clinical manifestations:* Similar to type V (Mc Ardle's Disease).
4. *Type VIII*

- ***Enzyme deficiency: Hepatic phosphorylase kinase.*** Failure of hepatic glycogen phosphorylase activity leading to decrease in glycogenolysis.
- *Clinically*: Hepatomegaly, rise in liver glycogen level of normal structure, mild to moderate hypoglycaemia occurs.

Table 23.1 gives the six classical types of GSDs in a tabular form.

HEXOSE MONOPHOSPHATE (HMP) SHUNT

An Alternate Pathway for Oxidation of Glucose

Synonyms: Variously called as:
- Hexose monophosphate pathway or shunt
- Pentose phosphate pathway (PP-pathway)
- Pentose cycle
- Phosphogluconate pathway
- Warburg-Dickens-Lipman Pathway.

Biomedical Importance

- Though it is oxidation of glucose, but ***it is not meant for energy.***
- ***Provides NADPH*** which is required for various reductive synthesis in metabolic pathways.
- ***Provides pentoses*** required for nucleic acid synthesis.
- Deficiency of a particular enzyme leads to **haemolytic anaemia**, which is of great clinical importance.

SECTION FOUR

Table 23.1: Glycogen storage diseases (GSDs)

Type	Name	Deficient Enzyme	Inheritance	Structure of Glycogen	Organs Affected	Clinical Features
I.	**Von Gierke's Disease**	***G-6-Pase***	Autosomal recessive	Normal (Metabollically NOT available)	Liver, Kidney, Intestine	Hypoglycaemia, Ketosis and Acidosis LA ↑, Uric Acid ↑ Failure to thrive, Hepatomegaly
II.	**Pompe's Disease**	***Acid Maltase*** (Present in Lysosomes. Catalyses breakdown of oligosaccharides)	Autosomal recessive	Normal	Liver, Heart, Smooth and Striated muscles	Cardiomegaly, Muscle hypotonia, No Hypoglycaemia
Note: Infants die of cardiac failure and Bronchopneumonia. Death usually before 9 months. A few survive 2½ years.						
III.	**Limit Dextrinosis (Forbe's Disease)**	***Debranching enzyme***	Autosomal recessive	***Abnormal,*** 'Limit Dextrin' type Short missing Outer branches	Liver (18%), Heart and Muscle (6%)	Moderate Hypoglycaemia, Acidosis, Progressive myopathy, Hepatomegaly
Note: Patients survive well to adult life.						
IV.	**Amylopectinosis (Andersen's Disease)**	***Branching enzyme***	Not known	• ***Abnormal*** ("Amylopectin" type) • Very long inner and outer unbranched chains, very few branch point	Liver, Heart, Muscle, RE System	Moderate hypoglycaemia Hepatosplenomegaly, Ascites, Nodular Cirrhosis of Liver, Hepatic failure
Note: Prognosis Fatal, Longest survival reported as 4 years.						
V.	**Mac Ardle's Disease**	***Muscle Phosphorylase***	Autosomal recessive	Normal	Skeletal muscle (Excess normal glycogen muscles)	Muscle cramps on exercise, Pain in muscle, weakness and stiffness of muscle
Note: Affects children and adults, Muscle recovers with rest—Due to utilisation of FA for energy, after inj. epinephrine/glucagon → Blood sugar ↑ ***(shows Liver Phosphorylase not affected)***						
VI.	**Her's Disease**	***Liver Phosphorylase***	Autosomal Dominant	Normal	Liver, Leucocytes	Hypoglycaemia, Mild to moderate acidosis, Hepatomegaly
Note: Presents like a mild case of Type-I, the condition has also been reported to occur in association with Fanconi Syndrome.						

MAJOR DIFFERENCES WITH EM PATHWAY

1. This pathway ***occurs in certain specialised tissues*** only ***to serve specific functions***, e.g. liver, adipose tissue, RB cells, testes and ovary, adrenal cortex and lactating mammary gland. Also lens and cornea of eye. It is unimportant for skeletal muscle and does not operate in non-lactating mammary gland.
2. It is a ***multicyclic process***, 3 molecules of glucose-6-P enter the cycle, producing 3 mols of CO_2 and 3 mols. of 5-C residues, which rearrange to give 2 mols of Glucose-6-P and one mol of Glyceraldehyde-3-P.
3. Oxidation is achieved by dehydrogenation but ***$NADP^+$ is used as hydrogen acceptor and NOT NAD^+.***
4. ***CO_2 is produced in this pathway which is never produced in EM Pathway.***

Similarity: The only similarity is that enzymes are ***extramitochondrial (cytosolic)*** and it operates in cytosol.

Summary of the Reactions

$$3\text{-Glucose-6-P} + NADP^+ \rightarrow 3CO_2\uparrow + \text{Glucose-6-P} + 1\text{-Glyceraldehyde 3-P} + 6\ NADPH + 6H^+$$

METABOLIC PATHWAYS

Reactions of this pathway can be considered **arbitrarily in 2 stages** (Fig. 23.9).

Stage I: (Oxidative Phase)

Oxidation of Glucose and formation of Pentose Phosphates (Oxidative Phase).

Stage II: (Non-oxidation Phase)

- Pentose phosphate reactions (Interconversion of Pentoses)
- Conversion of pentose phosphates to hexose phosphates.

SECTION FOUR

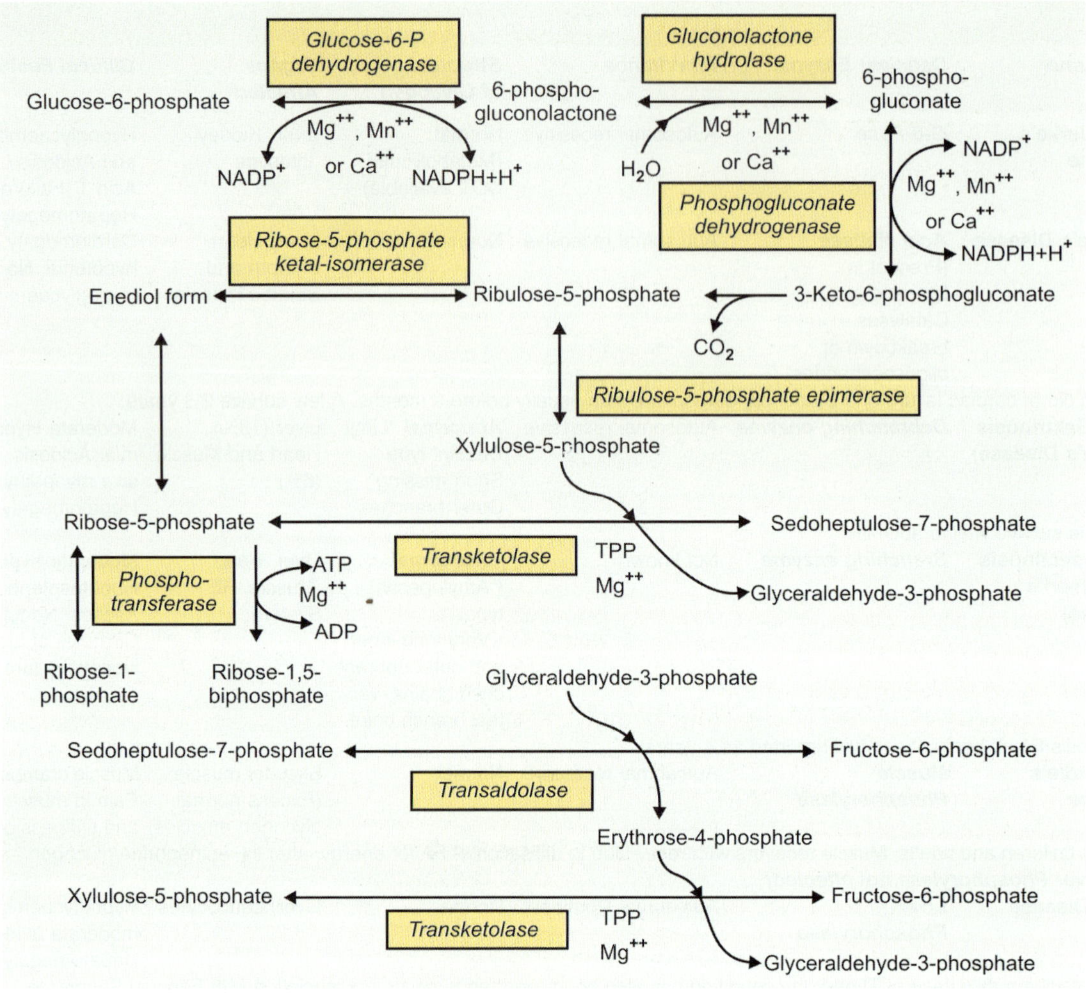

Fig. 23.9: The hexose monophosphate shunt or pentose phosphate pathway

Stage I: Oxidation of Glucose and Formation of Pentose phosphates (Oxidative phase):

1. ***First oxidation:*** Dehydrogenation of Glucose-6-P to form 6-phosphogluconate via the formation of intermediate unstable compound 6-phosphoglucono-lactone, catalysed by the first enzyme in this pathway ***Glucose-6-P-dehydrogenase (G-6-PD),*** an NADP-dependent enzyme. Hydrolysis of 6-phosphoglucono-lactone is accomplished by the enzyme *gluconolactone hydrolase.*

G-6-PD enzyme is a dimer/or tetramer, the first enzyme in the pathway which is ***rate-limiting***. Subunit is 51,000 daltons. The enzyme exhibits ***negative co-operativity,*** being inhibited readily by the product NADPH. ***Glucono-lactone is an unstable compound,*** and ***hydrolyses spontaneously.*** But the reaction is accelerated by a specific *Lactonase* (hydrolase) and equilibrium of the reaction lies to the right.

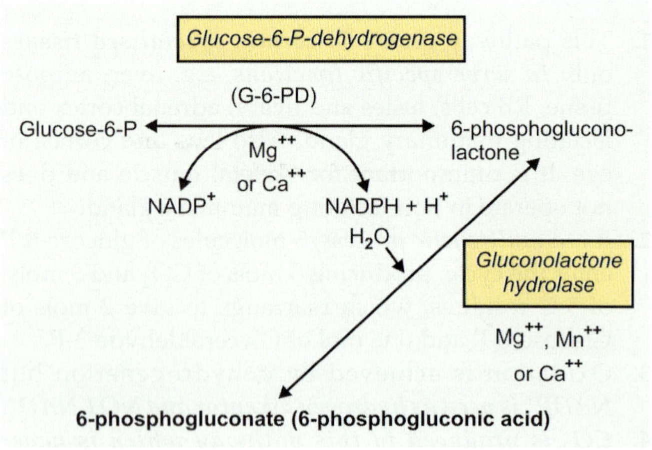

2. ***Second oxidative step:*** The second oxidative reaction takes place in **2 steps:**
- **Firstly** 6-phosphogluconate is converted to 3-keto-6-phosphogluconate, catalysed by the enzyme

Difference of EM Pathway and HMP Shunt

EM Pathway	*HM Pathway*
1. Occurs in all tissues	1. Occurs in certain special tissues for special function
2. Not a multicyclic process	2. Multicyclic process
3. Oxidation by dehydrogenation but NAD^+ is H-acceptor	3. Oxidation achieved by dehydrogenation but $NADP^+$ is used as H-acceptor
4. ATP is required and ATP is produced	4. **Not meant for energy; ATP is not produced,** ATP is required for glucose to glucose-6-P. (for phosphorylation) and for interconversion of Pentoses
5. CO_2 is never formed	5. CO_2 is produced

6-phosphogluconate dehydrogenase, in which $NADP^+$ acts as H-acceptor. Liver dehydrogenase is "dimer", consists of two identical subunits of 51,000 daltons.

- **Secondly,** 3-keto-6-phosphogluconate undergoes decarboxylation, first carbon is removed as CO_2 and form ketopentose, **D-ribulose-5-P**.

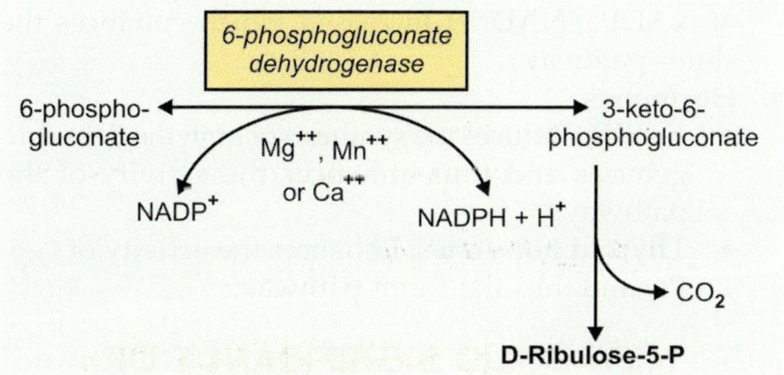

Stage II: Non-Oxidative Phase

(a) Pentose-P reactions (interconversions of pentoses): D-Ribulose-5-P is readily converted into a variety of pentoses.

- D-Ribulose-5-P is acted upon by a *ketoisomerase* and forms *D-ribose-5-P* (aldopentose) which can be converted to D-Ribose-1,5-di-P and D-Ribose-1-P.

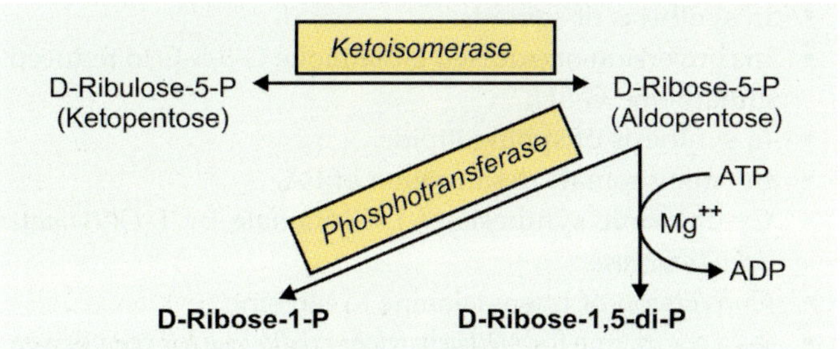

- D-ribulose-5-P can be converted to D-xylulose-5-P, another ketopentose, the reaction being catalysed by an *epimerase*, by altering the configuration at C-3, a transitory intermediate "2-3-enediol" formation may occur.

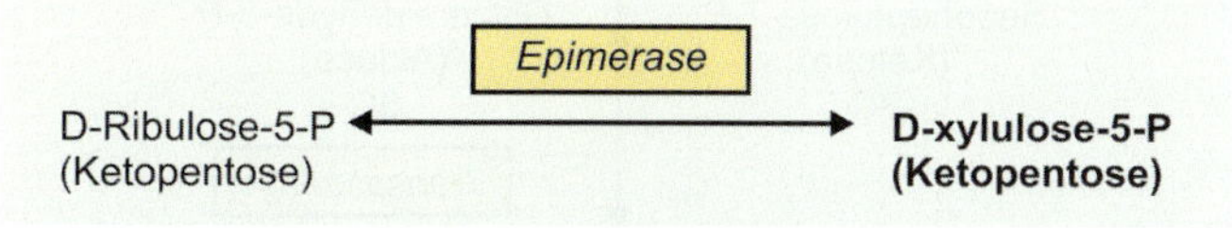

- Some Ribose-5-P may be epimerised by *Phosphopentose-2-epimerase* to form **Arabinose-5-P.**

(b) Conversion of pentose phosphates to hexose-phosphates: Reactions in this stage involves the conversion of pentose phosphates to fructose-6-P which is isomerised to glucose-6-P to begin the cycle over again.

The above is achieved by two peculiar reactions dependant on two enzymes: ***transketolase*** and ***transaldolase*** and reactions are called **transketolation** and **transaldolation** respectively.

(a) Transketolase: It **transfers a 2-carbon unit**, a **ketol group**

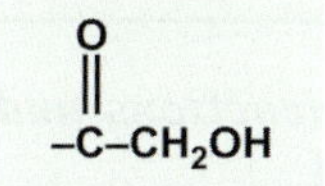

comprising carbons 1 and 2 of a ketopentose to an aldose. Thiamine pyrophosphate (TPP) is required as a coenzyme and serves in the transfer of 2-carbon unit as ***active glycolaldehyde.***

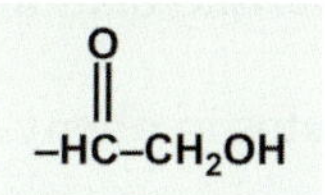

Thus, ***it converts ketosugar into an aldose with 2 carbons less and simultaneously converts an aldose sugar to a ketose with 2 carbons more.***

Transketolase enzyme is a "dimer" having two identical subunits of 70,000 daltons. The "active glycolaldehyde" moiety is α, β-dihydroxyethyl thiamine pyrophosphate.

(b) Transaldolase: It transfers a 3-carbon moiety ***dihydroxyacetone***

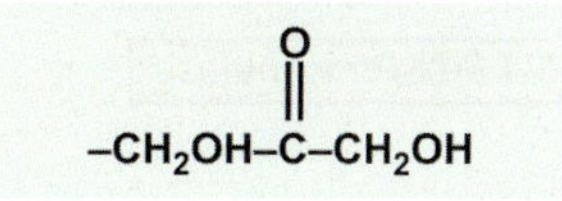

from the ketose to the aldose to form another ketose and aldose.

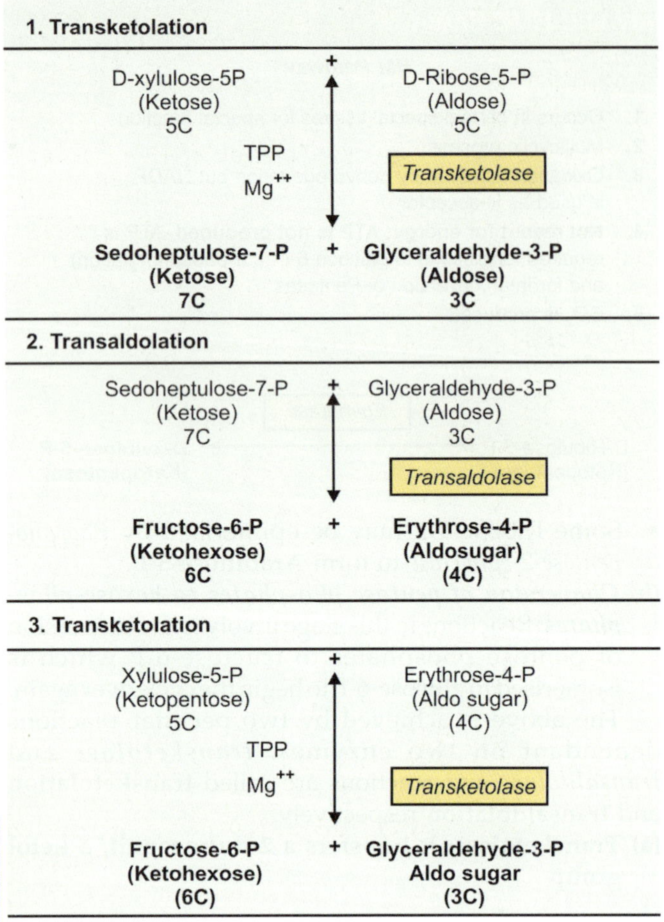

In the HMP-shunt, two transketolation reactions and one transaldolation reaction are involved.
The reactions are shown above in box:

Fate of Erythrose-4-P

- Formation of fructose-6-P by another transketolation as above.
- In plants and microorganisms-participates in phenylalanine synthesis.

Fate of Fructose-6-P and Glyceraldehyde-3P

- Two molecules of Glyceraldehyde-3-P can form one molecule of fructose-1,6-bi-P.

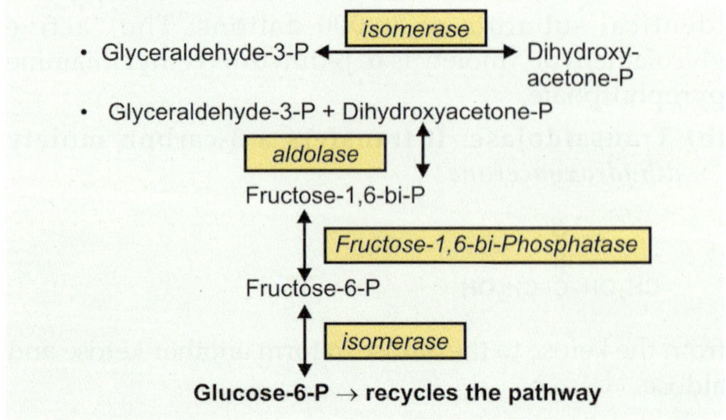

REGULATION OF HMP SHUNT

1. Reaction catalysed by G-6-PD, the first reaction of the pathway constitutes the ***rate-limiting*** step. It is primarily regulated by cytoplasmic levels of $NADP^+$ and NADPH, thus the ratio of the two, i.e. $[NADP^+]/[NADPH]$. If the cytoplasmic ***ratio is high,*** i.e. a rise in $NADP^+$, enhances the "rate-limiting" reaction as well as the shunt pathway. A **decrease in the ratio,** i.e. a rise in NADPH level, inhibits both *G-6-PD* and *6-phosphogluconate dehydrogenase* by making less $NADP^+$ available for their catalytic reactions and also by competing with $NADP^+$ to occupy the enzyme binding site of G-6-PD.
2. Activities of both dehydrogenases and the rate of the pathway are enhanced on feeding high carbohydrate diets and are reduced in starvation and dibetes mellitus.
3. Increase in FA synthesis and steroid synthesis re-oxidises NADPH to $NADP^+$ and cytoplasmic ratio of $NADP^+/NADPH$ increased which enhances the shunt pathway.
4. **Hormones**
   - **Insulin:** Induces the synthesis of both the dehydrogenases and thus enhances the activity of the pathway.
   - **Thyroid hormones:** Enhances the activity of G-6-PD and thus the shunt pathway.

METABOLIC SIGNIFICANCE OF HMP SHUNT

1. Formation of NADPH

NADPH is produced in this pathway which is **used as electron donor** in many reductive synthesis in the body. ***Examples of such reactions where NADPH is used:***

- Extramitochondrial *de novo* fatty acid synthesis,
- In synthesis of cholesterol,
- In synthesis of steroids,
- In conversion of oxidised glutathione G-S-S-G to reduced glutathione G-SH,
- In synthesis of sphingolipids,
- In "microsomal" desaturation of FA,
- Cytoplasmic synthesis of L-Glutamate by *L-Glutamate dehydrogenase,*
- Conversion of phenylalanine to tyrosine,
- As a coenzyme for *methaemoglobin reductase* for conversion of methaemoglobin to Hb,
- In uronic acid pathway.

2. Provision of Pentoses

Provision of pentoses for nucleotide and nucleic acid synthesis; the source of the D-Ribose is the D-ribose-5-P, an intermediate in this pathway. This pathway is not active in skeletal muscles. Muscle tissues contain very small amounts of the *dehydrogenases*, but still skeletal muscle is capable of synthesising ribose. ***Probably this is accomplished by "reversal of shunt pathway", utilising fructose-6-P, glyceraldehyde-3-P and the enzymes transketolase and transaldolase (by non-oxidative pathway).***

3. **Supply of Arabinose-5-P**

   Arabinose-5-P can be produced, as discussed above, and is used for incorporating arabinose in glycoproteins.

4. **Role in RB Cells Fragility**

   HMP-shunt in erythrocytes provides NADPH for:
   - Reduction of oxidised glutathione (G-S-S-G) to reduced glutathione (2 G-SH) catalysed by the enzyme ***glutathione reductase.***
   - Reduced glutathione (G-SH) thus formed in turn removes H_2O_2 from the erythrocytes in a reaction catalyzed by ***glutathione peroxidase*** (Se-containing enzyme).

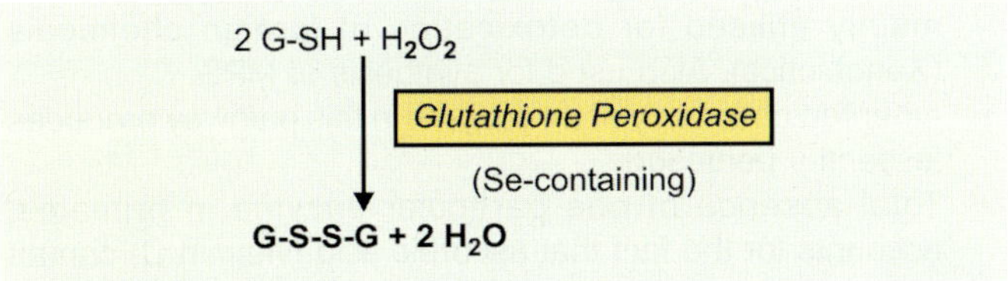

This reaction is important, since accumulation of H_2O_2 may decrease the life span of RB Cells by increasing the rate of oxidation of Hb to methaemoglobin. An inverse correlation exists between the activity of 'G-6-PD' enzyme and the fragility of red cells (Susceptibility to haemolysis) (See G-6-PD deficient haemolytic anaemia below).

5. **Role in Lens Metabolism**

   In studying lens metabolism, it has been observed, at least 10 per cent of Glucose is metabolised by shunt pathway and provides NADPH, which is necessary to convert oxidised glutathione to reduced glutathione, which is necessary for maintenance of lens proteins.

6. **Role in Phagocytosis**

   Reactions of this pathway are increased manifold in Leucocytes during 'phagocytosis'. NADPH generated in this pathway is utilised by "NADPH *oxidase*" in producing **Superoxide anions** (O_2^-) for destroying Phagocytosed materials.

7. **Oxidation**

   Oxidation in this pathway is different from EM Pathway. Oxidation occurs in the first reaction and third reaction, by specific *dehydrogenase* and NADPH is produced. ***Carbon-1 of glucose is removed as CO_2 by decarboxylation, which is never formed in EM Pathway.***

8. **Energy Production**

   Primarily it is NOT the function of this pathway. But if all NADPH produced by this pathway by complete oxidation of one molecule of glucose is converted to NADH in the body by ***transhydrogenase*** reaction requiring **oestrogen** as coenzyme, and NADH thus produced when oxidised in respiratory chain would provide theoretically 36 molecules of ATP per molecule of glucose. The energy yield is comparable to glycolysis-TCA Cycle.

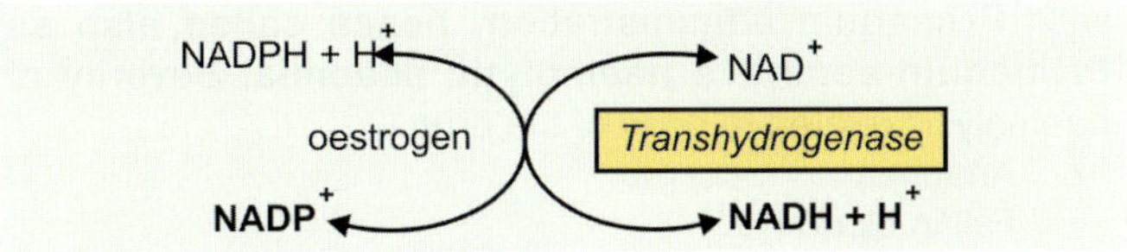

9. **Role in Tissue Anoxia**

   Tissues subjected to extended periods of anoxia develop **fatty infiltration.**

   Explanation: Increased amounts of glucose may be metabolised by way of HMP-shunt, in situations resulting from tissue anoxia. The mechanism involved appears to be that the lack of tissue O_2 decreases the metabolism of pyruvate by way of TCA cycle. Intermediates of anaerobic glycolysis accumulate resulting in a diversion of G-6-P into HMP-shunt pathway, resulting to increased amount of NADPH (lowering the NADP/NADPH ratio). ***The excess NADPH is diverted to increased FA synthesis, thus accounting for "fatty infiltration".***

10. **Production of CO_2**

    Carbon dioxide produced in HMP-shunt ***may be used in "CO_2-fixation" reactions.*** In plants, CO_2 produced in HMP-shunt is utilised for glucose formation by photosynthesis.

11. L-phenylalanine and its metabolite L-phenyl pyruvate accumulate in Phenyl ketonuria. They inhibit hexokinase, pyruvate kinase and 6-phosphogluconate dehydrogenase (6-p-GD) of differentiating brain tissue.

CLINICAL ASPECTS

(A) Haemolytic Anaemia Due to G-6-PD Deficiency: A mutation present in some populations causes a deficiency of the enzyme-G-6-PD with ***impairment of generation of NADPH*** that is manifested as ***red cells haemolysis*** when the susceptible individuals are subjected to ***oxidant group of drugs*** such as antimalarials, like primaquin and aspirin, etc. ***Inheritence:*** Generally transmitted by a **sex-linked** gene of intermediate dominance. Full expression of the trait occurs in hemizygous and homozygous females. Intermediate expression is found in heterozygous females. Female heterozygotes have been shown to have two populations of red cells: One with normal enzyme activity and one with markedly deficient activity.

Race: Defect has a **high incidence in Negroes**. Also found in non-Negro races and seen most frequently in Mediterranean peoples, namely Italians particularly **Sardinians** and some **Greeks**. Other races known to be affected include: Indians, Chinese, Malayan and Thais. In India, the genetic disorder is relatively frequent in Parsis, Punjabis, Sindhis and Kutchi-Lohans. In general, ***the deficiency of the enzyme is more marked in non-Negroes.***

Clinically: Develops acute haemolytic episode after administration of oxidant group of drugs. First described with Primaquin administration, hence called also as **Primaquin-sensitive haemolytic anaemia.** Severity of haemolysis related to dose of the drug.

- Anaemia
- Fall in blood Hb↓
- May be jaundice and
- Blackenning of urine. This type of haemolysis is "self-limiting"; older cells are destroyed whereas the younger cells are resistant.

Mechanism of haemolysis: Discussed above

Red cells abnormalities associated are as follows:

- G-SH-low or low normal
- G-SH stability-low
- Catalase-low
- O_2 consumption-low
- Methaemoglobin reductase-low

SECTION FOUR

Oxidant drugs which may cause haemolysis in G-6-PD deficient subject

- **Antimalarials:** Primaquin, Pamaquin
- **Analgesics:** Acetyl salicylic acid, phenacetin
- **Sulphonamides** group of drugs
- **Sulphones:** Dapsone, sulphoxone, etc.
- **Nitrofurans:** Furadantin, furoxone
- **Miscellaneous:** Vit K, Probenecid, PAS, Phenylhydrazine, methylene blue, etc.

Relationship with Other Diseases

1. **Relation of *P. falciparum* infection:** There is evidence to suggest that **the defect confers some protection against *P. falciparum* infection**, where this genetic disorder is prevalent, it may lessen the severity of malarial infections in young children and infants.
2. **Relation with colour blindness:** It has been shown that the disorder shows a close linkage with colour blindness.

(B) Favism

In G-6-PD deficient subjects, a disorder characterized by an acute haemolytic anaemia of sudden onset, often with haemoglobinuria, occurs in persons sensitive to the ***fava beans (vicia fava)*** either:

- On ingestion of uncooked or lightly cooked beans, or
- On inhalation of pollens from the blossom of the plant.

Although favism occurs typically with Mediterranean peoples, it may also occur in certain non-mediterranean subjects with G-6-PD deficiency including Chinese and Jews.

(C) Wernicke-Korsakoff Syndrome

A genetically 'variant' form of *transketolase*. The enzyme cannot bind TPP, affecting the 'transketolation' reaction. The patient suffering from this disorder shows severe neuro-psychiatric symptoms.

It is characterised by lesions and haemorrhages near the III cerebral ventricles, deranged mental functions, depression, disorientation, loss of memory and mental confusions. There may be abnormal gait or stance, and paralysis of eye movements.

URONIC ACID PATHWAY

It is an alternate pathway for oxidation of glucose.

Biomedical Importance

- In this pathway ***energy is not produced.***
- Major function is **to produce D-Glucuronic acid** which is mainly utilised for detoxication of foreign chemicals (Xenobiotics). Also used for synthesis of MPS.
- Inherited deficiency of an enzyme in this pathway produces 'essential pentosuria'.
- Total absence of one particular enzyme in primates, accounts for the fact that ascorbic acid (vitamin C) cannot be synthesised by humans, and requires to be provided in the diet.

METABOLIC PATHWAY (Fig. 23.10)

I (a) Formation of UDP-G from glucose: The steps are similar to glycogenesis:

- Glucose is phosphorylated to form glucose-6-P.
- Glucose-6-P is converted to glucose-1-P, in presence of the enzyme *Phosphoglucomutase*
- Glucose-1-P then reacts with UTP to form the ***"active nucleotide", uridine-diphosphate glucose (UDP-G).*** The reaction is catalysed by the enzyme ***UDP-G Pyrophosphorylase.***

All the above steps are similar to those already described under glycogenesis.

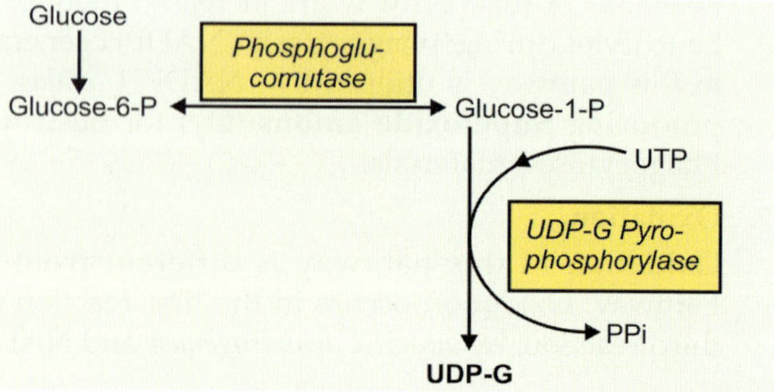

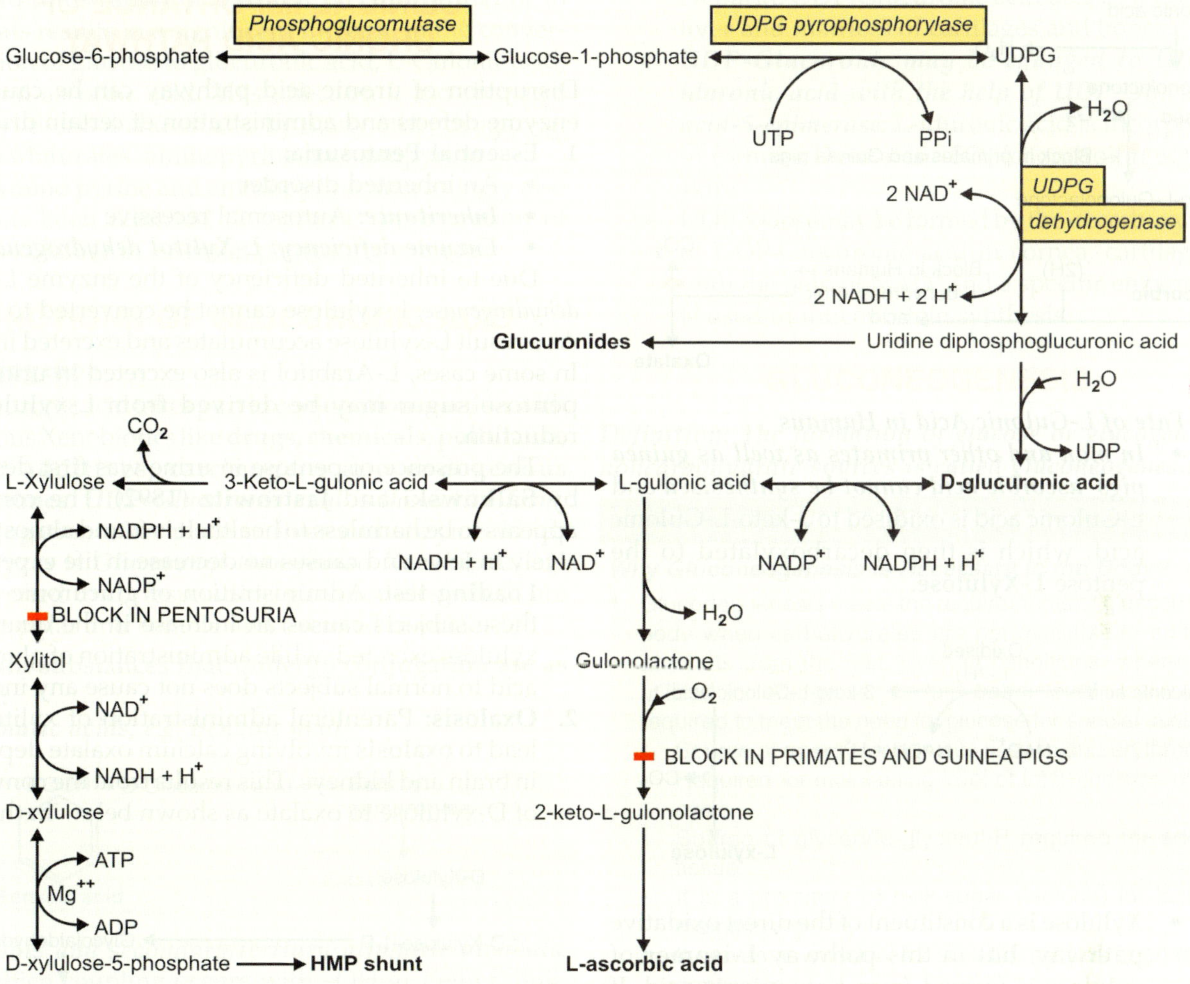

Fig. 23.10: Uronic acid pathway

(b) Formation of D-glucuronic acid: UDP-G is now oxidised at carbon-6 by a ***two-step process*** to "D-Glucuronic acid".

- UDP-G is oxidised by an enzyme *UDP-G dehydrogenase* and forms **UDP-Glucuronic acid.** The enzyme requires NAD^+ as hydrogen acceptor.
- UDP-Glucuronic acid is hydrolysed to form D-Glucuronic acid.

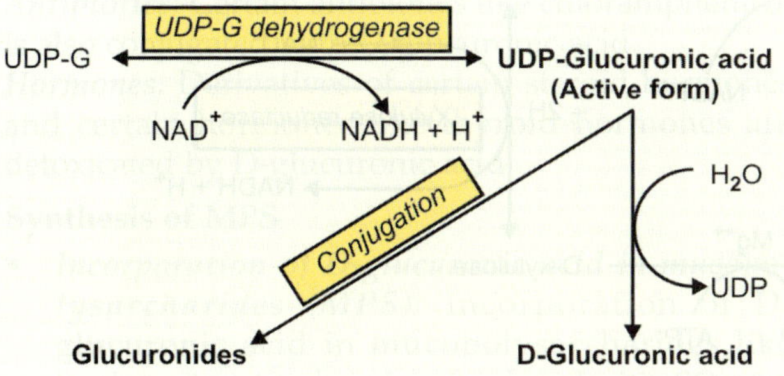

UDP-Glucuronic acid is the "active" form which takes part in conjugation to form **Glucuronides** or in formation of proteoglycans.

II. Formation of L-gulonic acid: In an NADPH-dependant reaction D-Glucuronic acid is first reduced to "L-Gulonic acid".

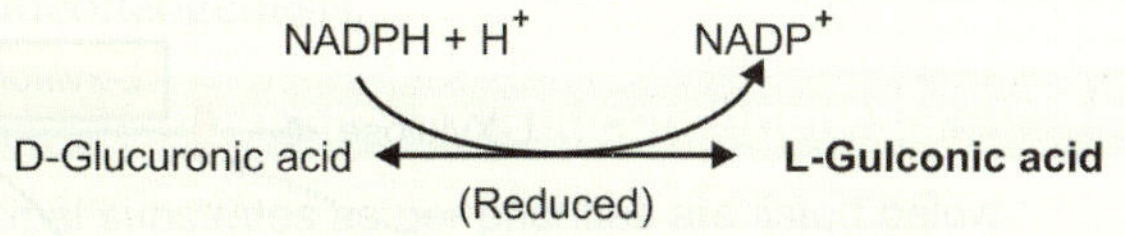

III. Fate of L-gulonic acid: Fate of L-Gulonic acid is different according to the animals.

(a) Synthesis of ascorbic acid: L-Gulonic acid is the direct precursor of Ascorbic acid, in those animals which are capable of synthesising this vitamin. In those animals, synthesis of vitamin C (Ascorbic acid) takes place as follows:

METABOLIC PATHWAYS INVOLVED IN GLUCONEOGENESIS

Gluconeogenesis involves glycolysis, the citric acid cycle and some special reactions.

1. Main Pathway of Gluconeogenesis (Fig. 23.11). *Main pathway is a reversal of glycolytic pathway but thermodynamic "barriers" prevent a simple reversal of glycolysis*. Certain modifications and adaptations are necessary in the EM Pathway. **Krebs** showed that energy barriers obstruct a simple reversal of glycolysis. There are **four such energy barriers which are irreversible:**

- *Between pyruvate and phosphoenol pyruvate,*
- *Between fructose-1, 6-bi-phosphate and fructose-6-P,*
- *Between glucose-6-P and glucose,* and
- *Between glucose-1-P and glycogen.*

The above energy barriers and irreversible reactions are circumvented by special adaptations which are discussed below:

(a) Between pyruvate and phosphoenol pyruvate: The conversion of pyruvic acid (thus also Lactic acid which can be converted to PA) to phosphoenolpyruvate is achieved by **two** important enzymes:

- *Pyruvate carboxylase* a mitochondrial enzyme, and
- *Phosphoenol pyruvate carboxykinase* an enzyme present principally in cytosol, but also small amount present in mitochondrion.

Reactions

1. *Pyruvate carboxylase* in mitochondrion, in presence of ATP, Biotin and CO_2 converts pyruvate to OAA (Refer anaplerotic reactions).

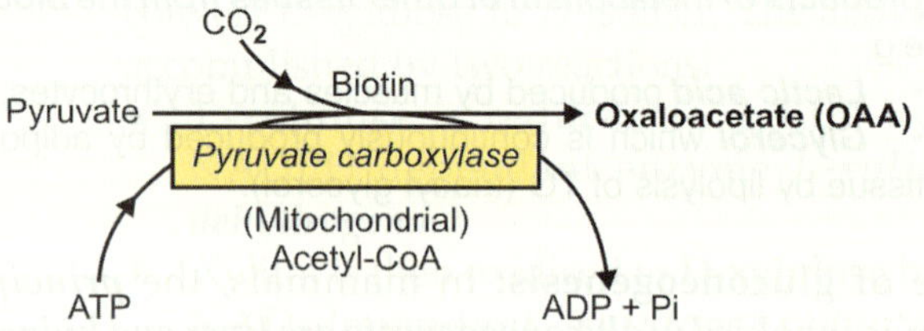

2. *Phosphoenol pyruvate carboxykinase* enzyme present chiefly in cytosol, catalyses the conversion of OAA to phosphoenol pyruvate. *High energy phosphate* in the form of **GTP** or **ITP** is required in this reaction and CO_2 is liberated.

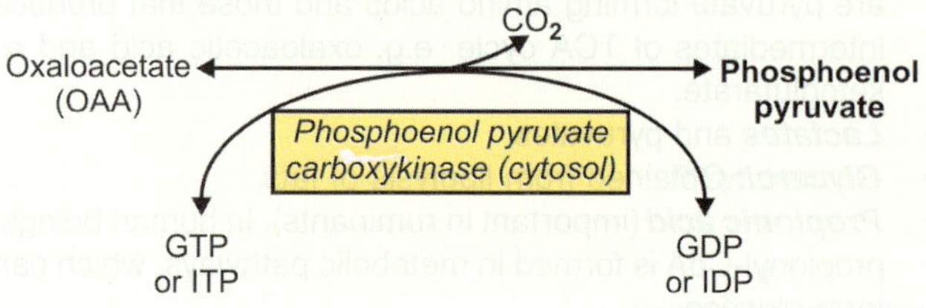

Once Phosphoenol pyruvate is formed from pyruvic acid by the action of above two enzymes it can go into reverse glycolytic pathway.

Note: It is to be noted that OAA is formed inside the mitochondrion but the other reaction occurs in the cytosol. ***OAA is not permeable to mitochondrial membrane.*** OAA is transferred to cytosol by **two mechanisms:**

- **Mainly as malate** which then in cytosol is converted to OAA again.

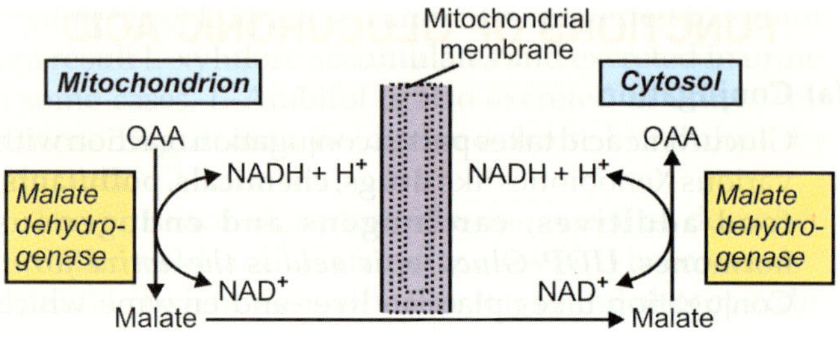

- Secondly, OAA can combine with acetyl-CoA to form **citrate**, which is permeable to mitochondrial membrane. Citric acid is cleaved by ***citrate-cleavage enzyme*** and reforms OAA.
- In addition to Pyruvate carboxylase, PA can be converted to OAA via the formation of malate (See anaplerotic reactions). Thus, conversion of PA to phosphoenol-(P) can be summarised as follows diagrammatically (Fig. 23.12) called **Dicarboxylic acid shuttle**.

(b) Between fructose-1,6-biphosphate to fructose-6-P: The conversion of fructose-1,6-biphosphate to fructose 6-Phosphate, the next **energy barrier;** is circumvented as follows to achieve a reversal of glycolysis.

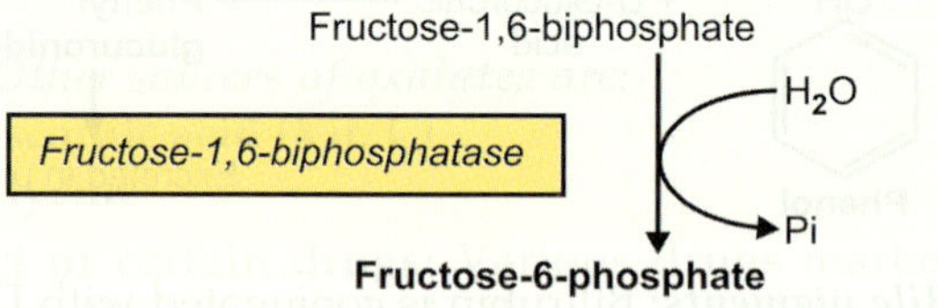

The above reaction is catalysed by a specific enzyme ***fructose-1,6-biphosphatase.*** This is a ***key enzyme*** in gluconeogenic pathway and its presence determines whether or not a tissue is capable of forming glucose/and glycogen not only from pyruvate/lactate but also from triosephosphates.

Site: The enzyme is present in liver, kidneys and intestine. Recently it has been shown to be present also in striated muscles. ***The enzyme is absent in adipose tissue, cardiac muscle and smooth muscles.***

(c) Between glucose-6-P and glucose: The conversion of Glucose-6-P to glucose, the **third energy barrier** in the glycolytic pathway is circumvented as follows:

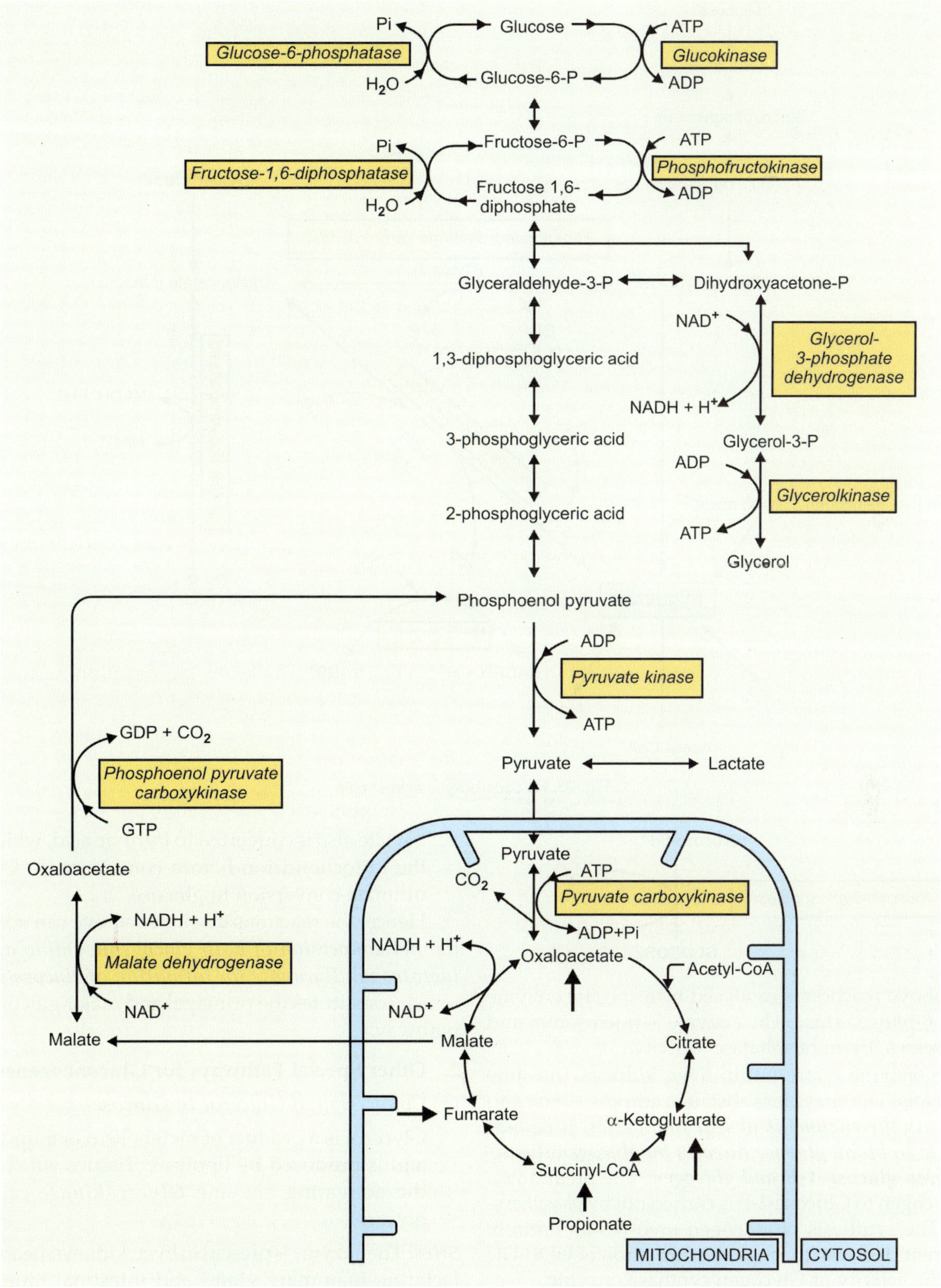

Fig. 23.11: Major pathway of gluconeogenesis in the liver. Entry points of glucogenic amino acids are shown by arrows (bold arrows)

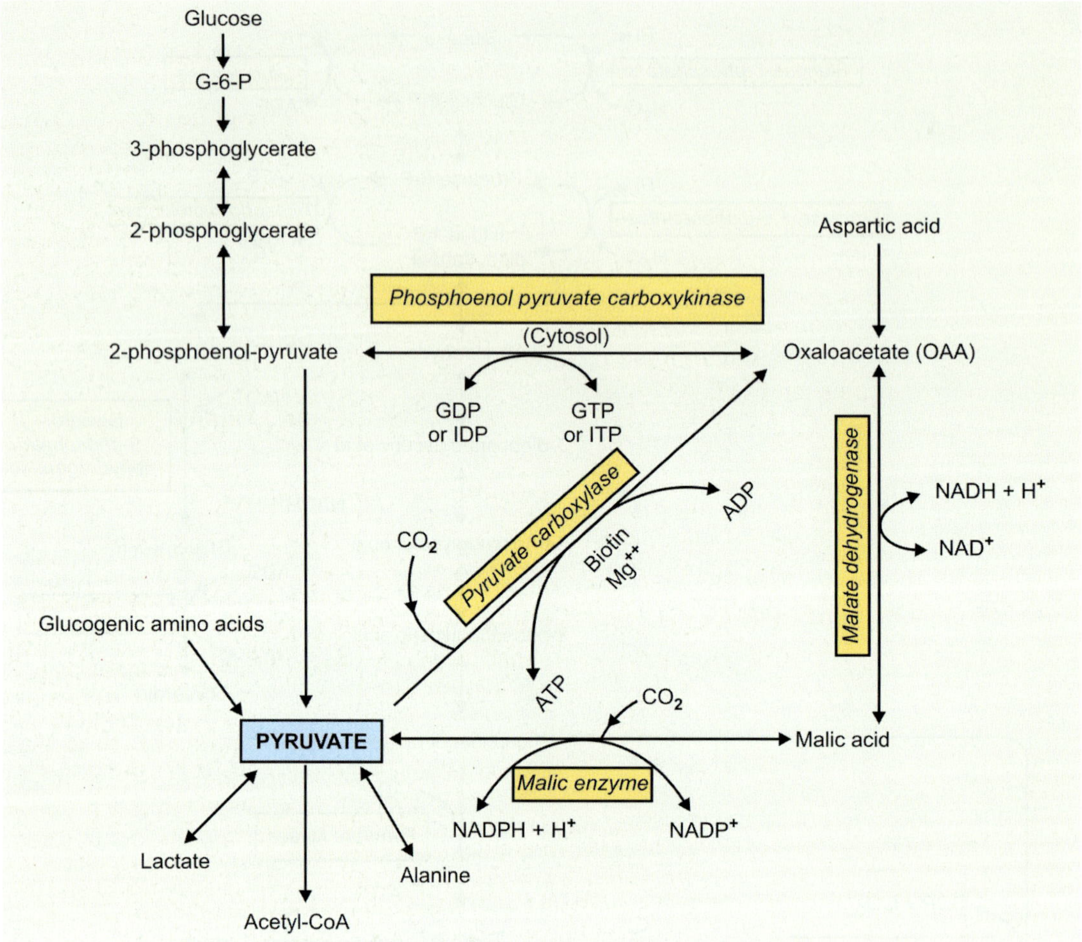

Fig. 23.12: Dicarboxylic acid shuttle

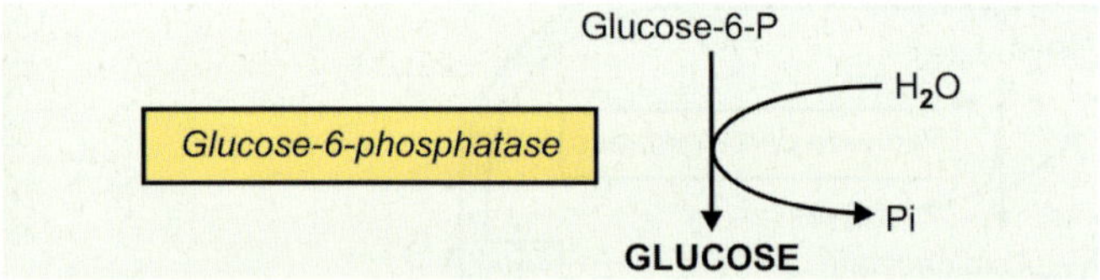

The above reaction is catalysed by a specific enzyme ***Glucose-6-phosphatase.*** The enzyme is microsomal and also possesses 'Pyrophosphatase' activity.

Site: The enzyme is present in liver, kidneys, intestine and platelets. The enzyme is absent in adipose tissue and muscles. ***As the enzyme is absent in muscles, it cannot contribute to blood glucose directly by glycogenolysis.***

(d) Between glucose-1-P and glycogen: The breakdown of glycogen to Glucose-1-P is carried out by *phosphorylase.* The synthesis of glycogen involves an entirely different pathway through the formation of UDP-Glc and the activity of Glycogen synthase enzyme.

Note

- It is to be noted that after transamination or deamination, glucogenic amino acids form either Pyruvic acid or intermediates of TCA cycle as stated above.
- Lactate also is converted to Pyruvic acid, which enters the mitochondrion before conversion to OAA and ultimate conversion to glucose.

Hence, the reactions discussed above can account for the ***net conversion of both glucogenic amino acids and lactates via Pyruvate for formation of glucose/glycogen*** which constitutes the principal and major gluconeogenic pathway.

2. Other Special Pathways for Gluconeogenesis

(a) Conversion of Glycerol to Glucose

Glycerol is a product of metabolism of adipose tissue and is produced by lipolysis. Tissues which possess the activating enzyme ***Glycerokinase*** can utilise glycerol.

Sites: The enzyme is present in liver, kidneys, heart muscle, lactating mammary gland and intestinal mucosa. ***The enzyme is absent in adipose tissue.***

Reactions of the pathway

- ***Glycerol as such cannot enter metabolic pathway. It is activated to α-glycero-P.*** The reaction is catalysed

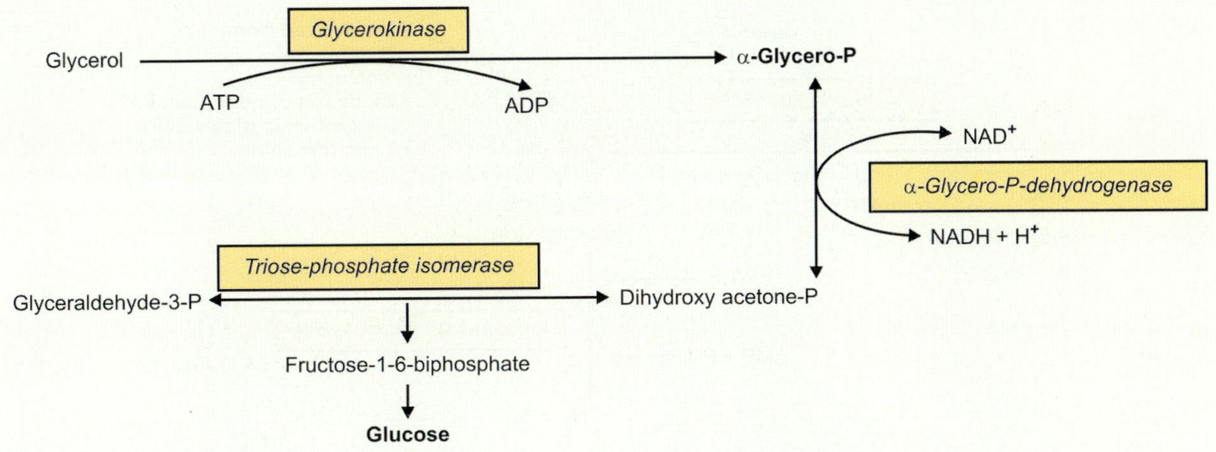

by the enzyme ***Glycerokinase*** which requires ATP. Principally occurs in Liver and Kidneys.

- α-Glycero-P thus formed is oxidised to dihydroxy-acetone-P, in presence of the enzyme *Glycerol-3-P-dehydrogenase* and NAD^+.
- Dihydroxyacetone-P can be converted to glyceraldehyde-3-P by *triose phosphate isomerase*.

This pathway thus connects with the triose-phosphate stages of the glycolytic pathway and triose-P can enter 'reverse' glycolysis after conversion to fructose-1, 6-biphosphate. Then it can form glucose as discussed above. Conversion of glycerol to triose-phosphates is shown above in box.

(b) Conversion of Propionic Acid to Glucose

- Propionic acid is a major source of glucose in ruminants, and enters the main gluconeogenic pathway via TCA cycle after conversion to succinyl-CoA.
- Propionate is first activated with ATP and CoA-SH, the reaction is catalysed by the enzyme ***Acyl-CoA synthase*** (earlier called as *thiokinase*) and converted to 'Propionyl-CoA'
- In humans, Propionic acid is not formed, but propionyl-CoA is formed as a metabolic product.

Chief sources of Propionyl-CoA in humans are:

- Catabolism of **L-methionine** via α-ketobutyrate
- Formed from catabolism of amino acid **isoleucine**
- Formed from β-oxidation of **odd-chain fatty acids**
- Formed in **biosynthesis of bile acids**
- Also formed from non-oxidative deamination of **threonine**.

- Propionyl-CoA undergoes a **CO_2-fixation reaction**, catalysed by the enzyme, ***Propionyl-CoA carboxylase*** and forms D-methylmalonyl-CoA. The reaction requires CO_2, ATP and Biotin.
- D-methylmalonyl-CoA is next converted to its stereoisomer **L-methylmalonyl-CoA** before it can further be metabolised.

Note: L-methylmalonyl-CoA is also produced in the body by the catabolism of amino acid valine.

- L-methylmalonyl-CoA is finally isomerised to succinyl-CoA by the enzyme ***methylmalonyl-CoA isomerase*** which requires **Vit B_{12} as a coenzyme.**

Note

1. Vitamin B_{12} deficiency in humans and animals results in the excretion of large amounts of methylmalonic acid ***(Methylmalonic aciduria).***
2. Though the main fate of propionyl-CoA is its conversion to succinyl-CoA, as an intermediate of TCA cycle, ***propionyl-CoA is also used as the primary molecule for synthesis of odd-chain fatty acids C_{15} to C_{17} in adipose tissue and in mammary gland.***

Propionic acid metabolism is shown schematically in Figure 23.13.

Hormones in Gluconeogenesis

- ***Glucagon:*** It increases gluconeogenesis from Lactic acid and amino acids.
- ***Glucocorticoids:*** They stimulate gluconeogenesis by increasing protein catabolism in the peripheral tissues and increasing hepatic uptake of amino acids and increases activity of transaminases and other enzymes concerned in gluconeogenesis.

List of amino acids which are glucogenic:

- ***Pyruvate forming amino acids are:*** Glycine, alanine, serine, cysteine, cystine, and threonine.
- ***Oxaloacetate forming amino acids:*** Aspartic acid.
- ***α-Oxoglutarate forming amino acids are:*** Glutamate, Glutamine, Proline, Arginine, Histidine, Lysine.

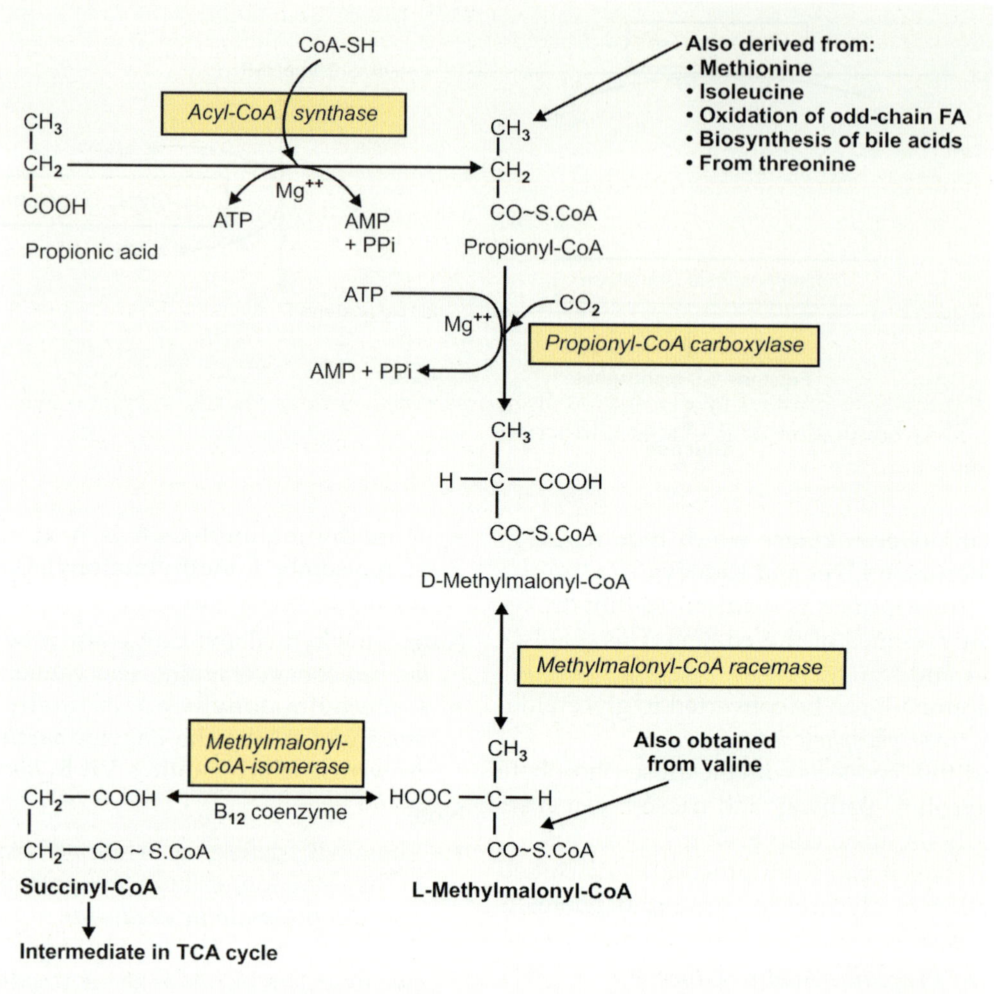

Fig. 23.13: Metabolism of propionic acid

REGULATION OF GLUCONEOGENESIS

Key enzymes which regulate gluconeogenesis are: *Pyruvate carboxylase, Phospho-enol-Pyruvate carboxykinase* (PEPCK), *Fructose-1,6-biphosphatase* and *Glucose-6-phosphatase.*

1. ***High carbohydrate diets:*** These **reduce** gluconeogenesis by ***increasing the insulin/glucagon ratio*** and thereby reducing the activities of all four key gluconeogenic enzymes.
2. ***Glucose-6-pase:*** This enzyme is **induced** by the hormones **glucagon** and **glucocorticoids**, which are secreted during starvation thus enhancing gluconeogenesis. Insulin represses the enzyme.
3. ***Fructose-1,6-Biphosphatase***
   - This enzyme is strongly and allosterically inhibited by AMP, but is activated by citrates. Hence gluconeogenesis is increased when there is increased ATP and citrate levels.
   - Gluconeogenesis is decreased by inhibition of this enzyme when liver cells are rich in AMP and low in citrate concentration.
4. ***Role of fructose-2,6-biphosphate:*** Fructose-2,6-bi-P is formed by phosphorylation of fructose-6-P by the enzyme *phosphofructokinase-2*. The same enzyme protein is also responsible for its breakdown since it also contains *fructose-2,6-biphosphatase* activity. ***This bifunctional enzyme is under the allosteric control of fructose-6-P.***
   - Under conditions of glucose shortage, gluconeogenesis is stimulated by glucagon by decreasing the concentration of fructose-2-6,bi-P, which in turn inhibits *phosphofructokinase-1* and activates the enzyme *fructose-1,6-biphosphatase.*
5. ***Phosphoenol pyruvate carboxykinase:*** The enzyme is **induced** by **Glucagon**, during starvation, thus increasing gluconeogenesis.

   Insulin reduces gluconeogenesis as it represses the enzyme.

6. ***Pyruvate carboxylase:*** This is the key enzyme in gluconeogenetic pathway. ***The enzyme is activated allosterically by acetyl-CoA. It binds with the allosteric site of the enzyme, brings about conformational change at tertiary level, so that the affinity of the enzyme for CO_2 increases (+ve modifier).***

7. **Other Factors**

(a) Fatty acid oxidation promotes gluconeogenesis

- It provides acetyl-CoA which acts as +ve allosteric modifier of the enzyme *Pyruvate carboxylase.*
- Increased acetyl-CoA and NADH from β-oxidation inhibits *Pyruvate dehydrogenase complex* so that oxidative decarboxylation of PA does not occur thus sparing its conversion to OAA
- It provides energy thus it spares PA from degradation.

(b) ***Increased ADP allosterically inhibits Pyruvate carboxylase*** and this decreases gluconeogenesis.

(c) ***Insulin represses the enzyme pyruvate carboxylase*** and reduces gluconeogenesis.

(d) On the other hand, hormones like glucagon, adrenaline and glucocorticoids induce the synthesis of the enzyme *Pyruvate carboxylase* and enhances gluconeogenesis.

FATES OF LACTIC ACID IN THE BODY

1. ***Chief fate is conversion to Pyruvate and its utilisation as pyruvate*** which either undergoes oxidative decarboxylation to form Acetyl-CoA or it can be glucogenic.

2. **What is "Cori Cycle"?**

Once formed Lactic acid can be further metabolised only by its reconversion to pyruvate as stated above. In contrast to the phosphorylated intermediates of glycolysis which are locked in the cells, ***lactates and pyruvates can readily diffuse out from the cells in which they are produced and pass into the circulation.*** From circulation, they are removed by the **liver** and in liver cells they are reconverted to form glucose and glycogen by gluconeogenesis. ***Muscles cannot convert Glucose-6-P to glucose due to lack of the enzyme Glucose-6-Phosphatase.*** This cycle is referred to as **Cori cycle**. It is shown diagrammatically in Figure 23.14.

3. **Lactate-Propanediol Pathway**

Miller and **Olson** observed that lactic acid utilisation by rat ventricle slices *in vitro* could not be completely inhibited by sodium fluoride and by anaerobic conditions. Hence, they postulated that Lactic acid can be metabolised by an alternate pathway not involving its conversion to pyruvate. They proposed a propanediol pathway.

Steps of the Reactions

- Lactic acid is first reduced to form Lactaldehyde.
- It is followed by conversion to Acetol.
- Acetol in turn is reduced, in presence of NADH, to form 1,2-propanediol either "free" form or "Phosphorylated" form. Refer to box on next page.

METABOLISM OF GALACTOSE

Galactose is derived from the hydrolysis of the disaccharide "lactose" (sugar of milk) in the intestine by

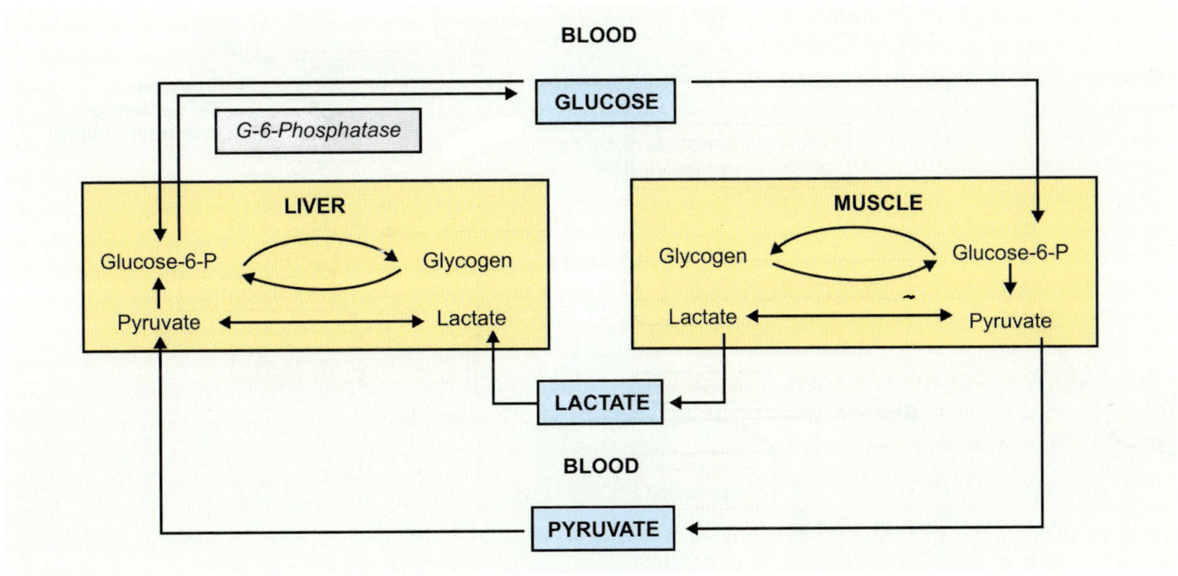

Fig. 23.14: "Cori" cycle

SECTION FOUR

the enzyme *"lactase"*. ***Galactose is "actively" transported, to portal blood. Galactose absorption is faster than glucose.*** Galactose reaches Liver by portal blood where it is readily converted to glucose. This property is used as a test of liver function called ***galactose tolerance test.***

Most of the dietary galactose is converted to glucose and goes to systemic circulation as glucose. Very little galactose as such goes to systemic circulation. ***Tissues requiring galactose synthesises it from glucose,*** taken up by tissues from systemic circulation.

LACTATE-PROPANEDIOL PATHWAY

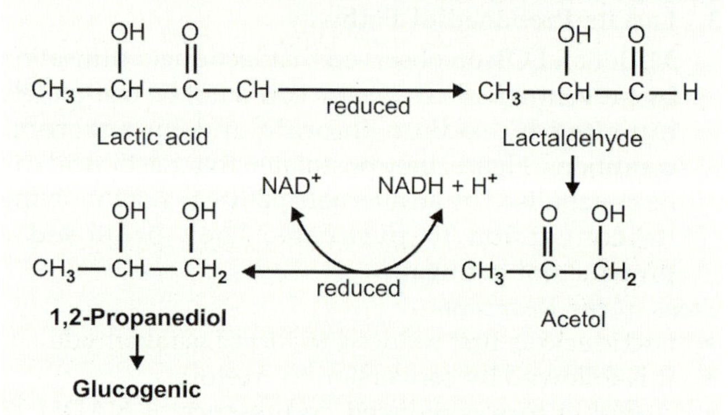

Biomedical Importance

- Galactose is required in lactating mammary gland for synthesis of lactose for breast milk.
- Galactose is utilised in Brain and nervous tissues for synthesis of glycolipids-cerebrosides and gangliosides.
- Galactose is required for synthesis of chondromucoids and mucoproteins.
- Inherited deficiency of certain enzymes in pathway of galactose metabolism produces inherited disorder ***"Galactosemia"***.

METABOLIC PATHWAY (Fig. 23.15)

The pathway by which galactose is converted to glucose and Lactose is synthesised is shown in **Figure 23.15.**

1. ***In reaction-1:*** Galactose is phosphorylated with the enzyme ***galactokinase,*** an **adaptive enzyme**, using ATP as phosphate donor, and produces galactose-1-P. ***The reaction is irreversible.***
2. ***In reaction-2:*** Galactose-1-P reacts with UDP-Glucose to form UDP-galactose and glucose-1-P, in presence of the enzyme ***Galactose-1-P-uridyl transferase.***
3. ***In reaction-3:*** UDP-Gal and UDP-Glucose are interconvertible and catalysed by the enzyme ***epimerase,*** which requires NAD^+. The reaction is freely **reversible**.

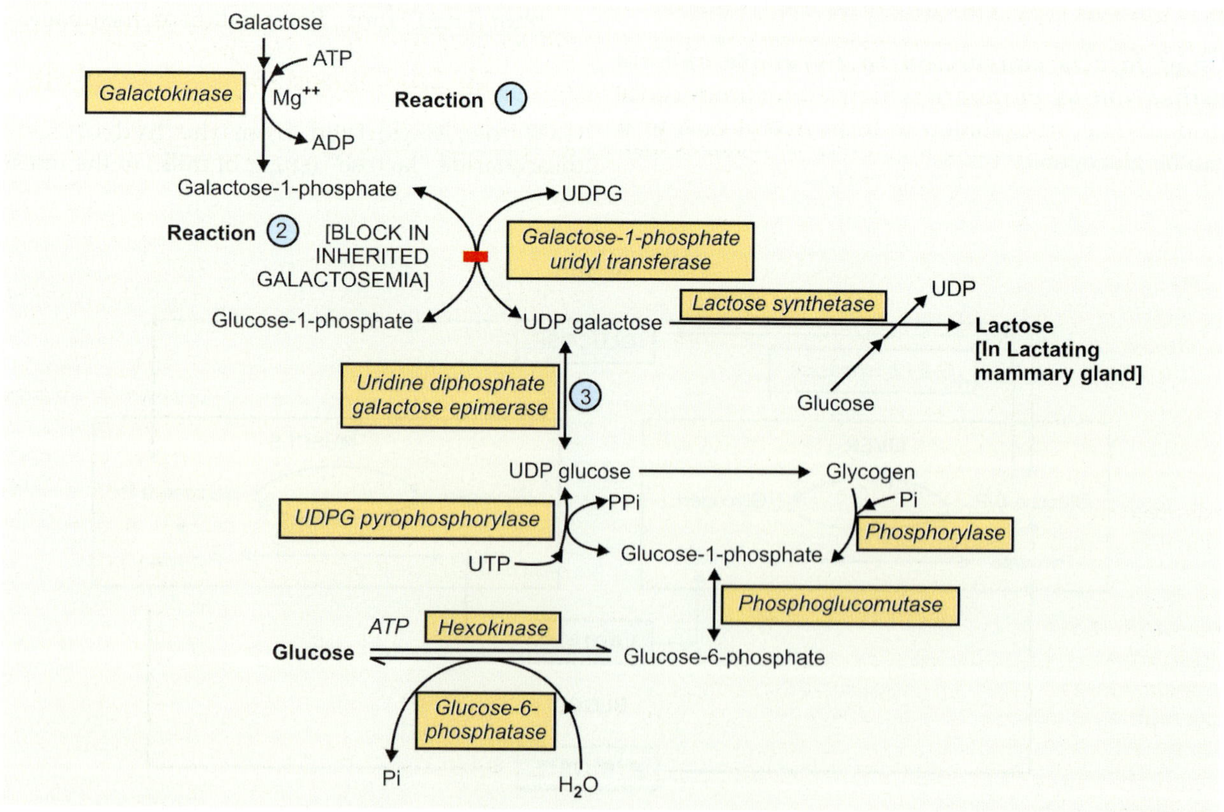

Fig. 23.15: Metabolism of galactose

Epimerisation probably involves an oxidation and reduction at carbon-4. In this manner, glucose can be converted to galactose, so that preformed galactose is not essential in the diet.

Note: Formation of UDP-glucose from glucose is similar to steps in glycogenesis.

BIOSYNTHESIS OF LACTOSE

In synthesis of Lactose in lactating mammary gland, UDP glucose is converted to UDP-galactose by the enzyme ***epimerase***. UDP-Galactose condenses with one molecule of glucose to form **Lactose**, the reaction is catalysed by the enzyme ***Lactose synthetase***. Lactose synthatase also called as ***galactosyl transferase***.

Regulation of Biosynthesis of Lactose

1. *Lactose synthetase* enzyme has **two subunits**:
   - **Catalytic subunit:** Called *Galactosyl transferase* and
   - **Modifier subunit:** Called α-Lactalbumin.
2. *Galactosyl transferase* does not use D-Glucose as one of its substrates in tissues other than mammary gland. In other tissues, it is involved in transfer of "Galactosyl" moiety from UDP-gal to N-acetyl glucosamine to form N-acetyl Galactosamine.
3. In the lactating mammary gland, **binding of α-lactalbumin modifier** to the **catalytic subunit** brings about the **transfer of *galactosyl* moiety to D-Glucose to form Lactose.**
4. **Hormones**
   - α-lactalbumin level in mammary tissues is under hormonal control and the hormonal changes that follow parturition increases the cellular level of modifier subunit↑ "α-lactalbumin".
   - ***Prolactin:*** Increases the rate of synthesis of both subunits.
   - ***Progesterone:*** Inhibits synthesis of α-lactalbumin. At parturition, level of Progesterone decreases↓ and α-lactalbumin synthesis increases↑.

CLINICAL ASPECT INHERITED DISORDER OF GALACTOSE METABOLISM

Galactosaemia:

An inherited disorder, in which there is inability to convert galactose to glucose in normal manner.

Incidence: 1 in 18,000 live births.

Enzyme Defects:

- ***Galactose-1-P-uridyl transferase*** enzyme is deficient ***(classical type)***. In this both galactose and galactose-1-P accumulates in blood and tissues like RB Cells, Liver, Spleen, Kidney, Heart, Lens of eye and cerebral cortex.
- ***Galactokinase*** deficiency ***(minor type):*** In this galactose accumulates in blood and tissues.
- Rarely it is claimed that there may be ***Epimerase*** deficiency.

Inheritence: Autosomal recessive.

Clinically

(a) Infants appear normal at birth but later:

- Failure to thrive, becomes lethargic, may vomit, **hypoglycaemia.**

  Reasons of hypoglycaemia:
  - Due to enzyme deficiency galactose cannot be converted to glucose.
  - Increased galactose level increases insulin secretion which lowers blood glucose.
  - Galactose-1-P inhibits *phosphoglucomutase* enzyme.
- **May manifest jaundice**, which may be prolonged in neonatal period.

(b) After 2 to 3 months:

- **Liver:** May show ***fatty infiltration*** and lead to ***cirrhosis liver***.
- **Mental retardation:** Due to accumulation of galactose and galactose-1-P in cerebral cortex.
- ***Development of cataracts***

  Causes:
  - Excess of galactose in lens is reduced to **"galactitol" *(Dulcitol)*** by the enzyme ***aldose reductase.***
  - Galactitol cannot escape from lens cells. Osmotic effect of the sugar alcohol contributes to injury of lens proteins and development of cataracts.
  - ***Excess of galactose inhibits Glucose-6P-dehydrogenase (G-6-PD)*** leading to ***less NADPH*** which results to less of reduced glutathione (G-SH).

Biochemical and urinary findings

- Increase blood galactose level ↑, Blood Sugar level decreases ↓ (Hypoglycaemia), ***Inorganic P decreases ↓, due to utilisation of PO_4 for gal-1-P.***

Urine:

- Increased excretion of galactose in urine↑, ***(galactosuria)***
- **Albuminuria**, and **aminoaciduria**: amino acids excreted usually are ***serine, alanine*** and ***glycine***.

Prognosis: Not fatal, can survive to puberty and adulthood.

Reasons:

- ***Epimerase reaction*** (provided there is no associated epimerase deficiency) is freely reversible, hence even with 'galactose-free diets', galactosemic children can meet the body requirements of galactose from glucose.
- By the time puberty is reached, **alternate Pathway** for Galactose may develop by which galactose can be converted to D-xylulose-5p and can join HMP pathway.

Galactose
↓
Galactonolactone
↓
Galactonic acid
↓
β-Ketogalactonic acid ⟶ (→ 2H) D-xylulose
↓
D-xylulose-5-P
↓
HMP–shunt pathway

METABOLISM OF FRUCTOSE

Dietary Sources of Fructose: Fructose is present in fruit juices and honey. **Chief dietary source is sucrose**, a diasaccharide, taken as table sugar (cane sugar). Sucrose is hydrolysed in the intestine to one mol. of glucose and one mol. of fructose by the enzyme ***Sucrase.*** Fructose is absorbed by *facilitated transport* and taken by portal blood to liver, where it is mostly converted to glucose.

Biomedical Importance

- Fructose is easily metabolised and a **good source of energy**.
- **Seminal fluid is rich in fructose and spermatozoa utilises fructose for energy.**
- Excess dietary fructose is harmful—leads to increased synthesis of TG.
- In diabetics, fructose metabolism through *'sorbitol'* pathway may account for the development of cataract.
- **Inherited disorder: *"Hereditary fructose intolerance" occurs due to inherited deficiency of the enzyme aldolase-B.***

METABOLIC PATHWAYS OF FRUCTOSE

1. ***Fructose may be phosphorylated*** to form fructose-6-P, catalysed by the enzyme *hexokinase.* But this is not the major pathway, ***as affinity of the enzyme hexokinase for fructose is very low.***

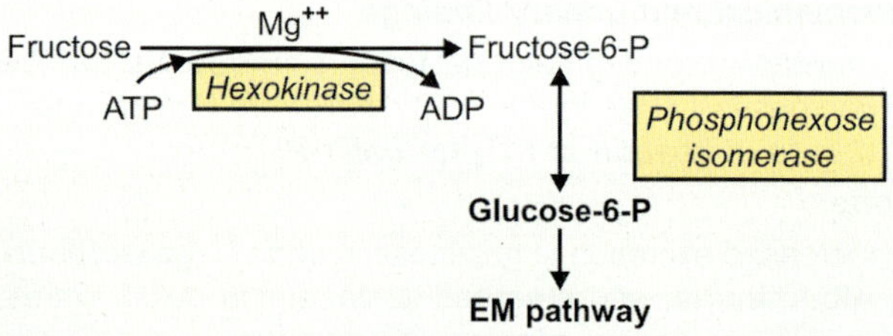

II. (a) ***A specific enzyme fructokinase*** is present in Liver and Muscle. Also present in kidney and intestine. ***This enzyme phosphorylates fructose only*** and will not phosphorylate glucose. This appears to be the ***major pathway*** for phosphorylation of fructose. Its activity is not affected by insulin. ***This explains why fructose disappears from the blood of diabetic patients at a normal rate.***

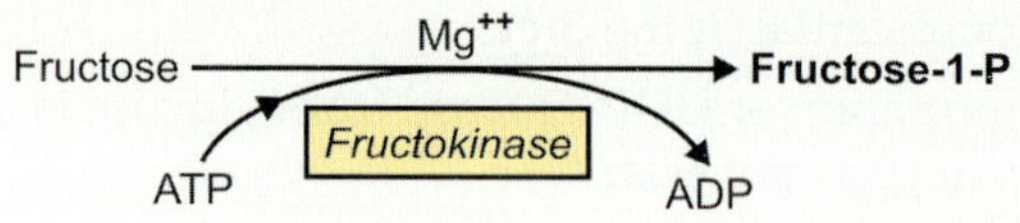

(b) Conversion of fructose-1-P to D-glyceraldehyde: The reaction is catalysed by the enzyme ***aldolase B.***

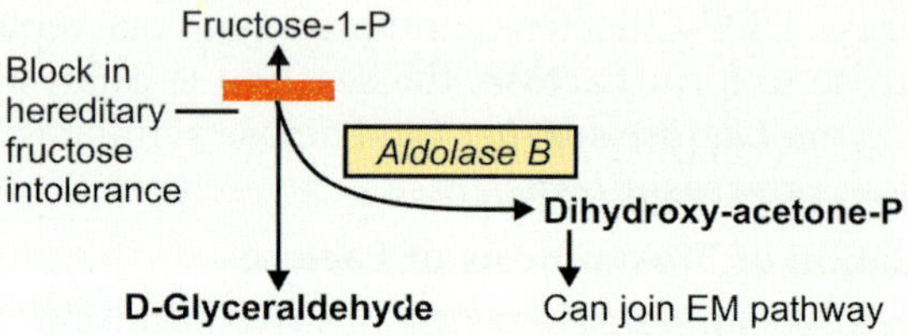

(c) Fate of D-glyceraldehyde: D-glyceraldehyde can enter glycolytic pathway when converted to either to D-Glyceraldehyde-3-P or to some other metabolites of glycolytic pathway.

Three possible mechanisms are available

1. ***Conversion of D-glyceraldehyde to dihydroxy acetone-P:*** This can be achieved as shown in box (Refer to box at bottom of the page).
2. ***Conversion of D-Glyceraldehyde to 2-phosphoglycerate:*** (Refer to box below).

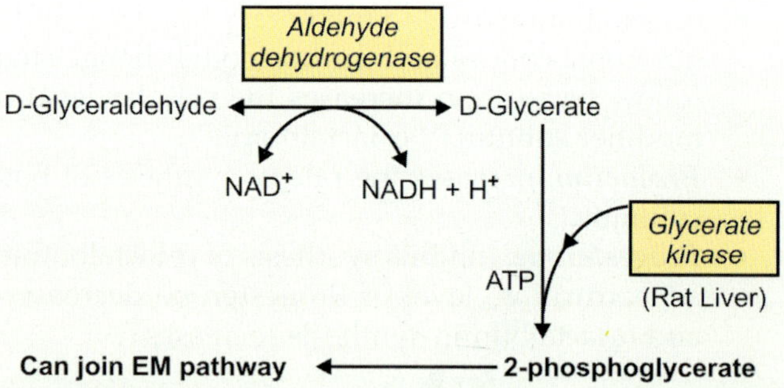

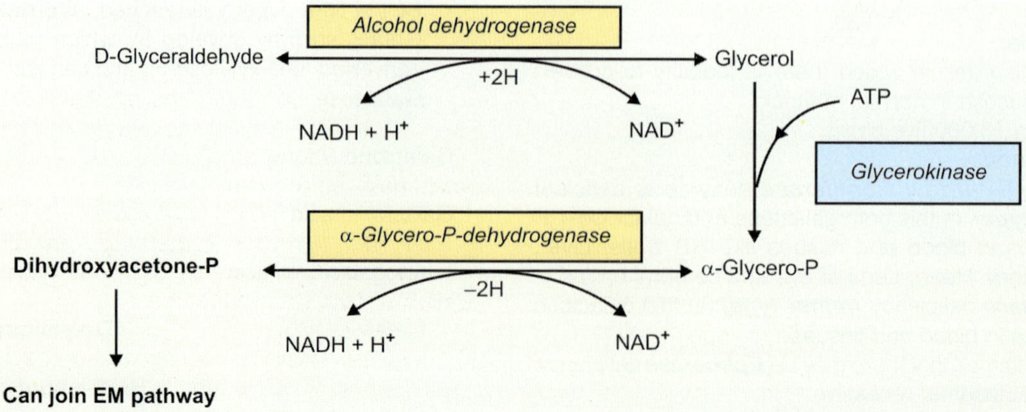

3. ***Conversion of D-glyceraldehyde to glyceraldehyde-3-P:*** This appears to be the ***principal and major pathway.*** An enzyme ***triokinase*** present in Liver catalyzes this phosphorylation reaction with ATP.

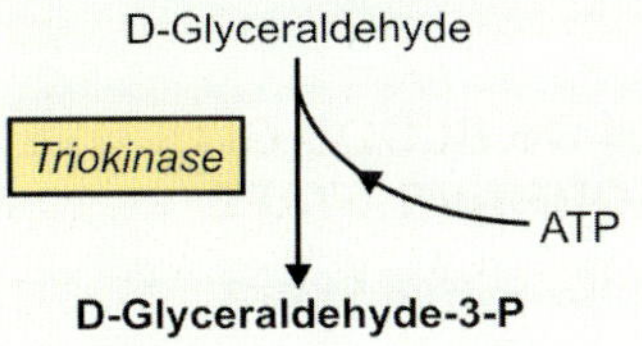

Two triose-P: Dihydroxy acetone-P and glyceraldehyde-3-P may be:
- Degraded via EM Pathway or
- They may combine under the influence of enzyme *aldolase* and be converted to glucose.

4. Another possibility is conversion of fructose-1-P to fructose-1,6-biphosphate.

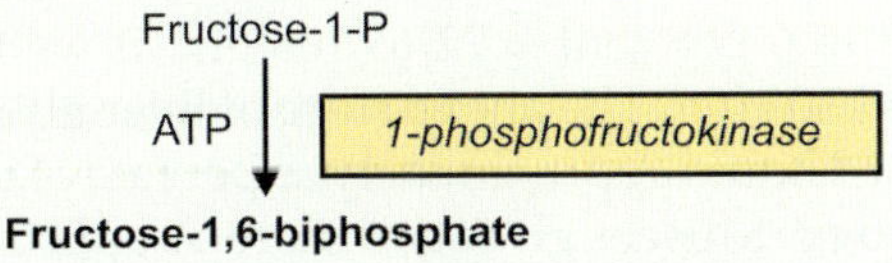

It seems to be not a major pathway, otherwise "hereditary fructose intolerance" probably would not occur.

Note:
- Brain and muscle can utilise singificant quantities of fructose after its conversion to glucose in Liver.
- Fructose is also metabolised actively by adipose tissue.

CLINICAL AND PHYSIOLOGICAL ASPECTS

I. Sorbitol Pathway of Fructose Metabolism
- Seminal fluid is very rich in fructose. Fructose is the principal and major energy source for spermatozoa. ***It is formed from glucose in the seminal vesicle.***
- ***Metabolic Pathways involved are as follows:***
  - Reduction of D-Glucose to D-Sorbitol
  - Oxidation of sorbitol to form D-fructose.

  Reactions involved along with enzymes and coenzymes required are shown below:

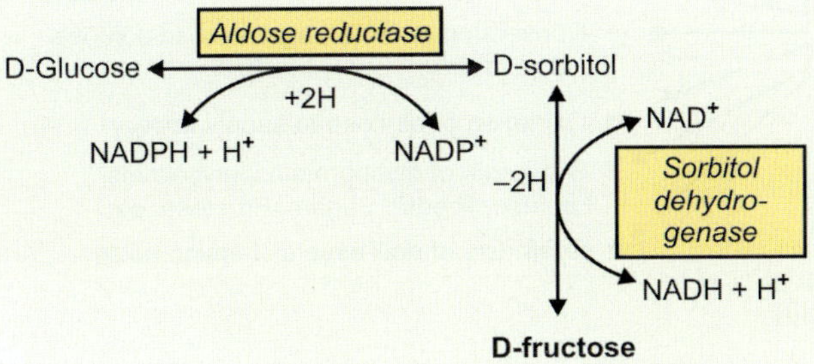

- The fructose concentration in semen may reach 10 mM. Most of this is available for the sperms because fructose is used sparingly by other tissues that come in contact with the seminal fluid.
- ***The mitochondria of sperms are the only such organelles to contain the enzyme LDH (Lactate dehydrogenase).*** Due to presence of LDH, the lactate that is formed by fructolysis can be completely oxidised to CO_2 and H_2O, without the need for a shuttle system to transport reducing equivalents into the mitochondria.

II. Clinical Significance of Sorbitol Metabolism in Diabetes Mellitus: Formation of sorbitol from glucose proceeds rapidly in the lens of the eye and the Schwann cells of the nervous system.
- Sorbitol cannot pass through the cell membrane and in Diabetic individuals, as the rate of oxidation of sorbitol to fructose decreases↓, ***sorbitol level in these cells greatly increases.***
- Elevated sorbitol concentration in these cells increase the osmotic pressure which may be ***responsible for the development of the cataracts of Lens of the eye and Diabetic neuropathies.***

III. Effect of Excessive Fructose on Lipid Metabolism: Fructose is more rapidly metabolised by the Liver than glucose. This is due to the fact that it bypasses the step in glucose metabolism catalysed by *Phosphofructokinase-I*, at which point metabolic control is exerted on the rate of glucose oxidation. This allows fructose to flood the pathways in the Liver, leading to enhanced FA Synthesis, increased esterification and VLDL secretion, which increases serum TG Level.

IV. Liver Cell Necrosis: Fructose when used parenterally for nutrition can ***cause depletion of adenine nucleotides in Liver and can cause liver cell necrosis.***

V. Sorbitol Intolerance: When sorbitol is administered IV, it is converted to fructose rather than to glucose, although if given by mouth, much escapes absorption from the gut and is fermented in the colon by intestinal bacteria to products such as acetate and H_2. Abdominal pain may be caused by "Sugar-free" Sweeteners containing sorbitol ***(Sorbitol intolerance).***

VI. Inherited Disorder—Hereditary Fructose Intolerance: An inherited disorder manifesting with severe clinical features.

***Enzyme deficiency:* Aldolase-B deficiency**

Administration of fructose in these patients leads to:
- Excessive and prolonged rise of fructose - and fructose-1-p ↑ in blood
- Blood glucose falls ↓ **(hypoglycaemia)**. Accompanied by:
  - Nausea and vomiting (may be haemorrhagic),
  - Profuse sweating.

After cessation of symptoms,
- Slight icterus, albuminuria, and aminoaciduria. There may be rapid ***decrease in serum inorganic PO_4↓***, which is attributed to binding of PO_4 with fructose.

Cause of hypoglycaemia: Probably due to:
- Excessive insulin secretion
- Inhibition of *phosphoglucomutase* by fructose-1-P.

SECTION FOUR

REGULATION OF BLOOD GLUCOSE (Homeostasis)

Blood glucose level is maintained within physiological limits 60 to 100 mg% ("true" glucose) in fasting state and 100 to 140 mg% following ingestion of a carbohydrate containing meal, by **a balance between two sets of factors:**

(A) Rate of glucose entrance into the blood stream, and
(B) Rate of its removal from the blood stream.

(A) RATE OF SUPPLY OF GLUCOSE TO BLOOD:

Except for a possible minor contribution by the kidney, which probably does not occur under physiological conditions, the blood glucose may be derived directly from the following sources:

- By absorption from the intestine
- Breakdown of glycogen of Liver (Hepatic glycogenolysis)
- By gluconeogenesis in Liver, source being glucogenic amino acids, lactate and pyruvate, glycerol and propionyl-CoA
- Glucose obtained from other carbohydrates, e.g. fructose, galactose, etc.

(B) RATE OF REMOVAL OF GLUCOSE FROM BLOOD

- Oxidation of glucose by the tissues to supply energy
- Glycogen formation from glucose in Liver (Hepatic glycogenesis)
- Glycogen formation from glucose in muscles (Muscle glycogenesis)
- Conversion of glucose to fats (lipogenesis) specially in adipose tissue
- Synthesis of compounds containing carbohydrates—blood glucose is utilised, e.g.
  - Formation of fructose in seminal fluid, formation of lactose (sugar of milk) in lactating mammary gland, synthesis of glycoproteins and glycolipids,
  - Formation of ribose sugars from glucose required for nucleic acid synthesis.
- Excretion of glucose in urine (glycosuria), when blood glucose level exceeds the renal threshold.

All the above processes are under substrate, end-product, nervous and hormonal control.

The above is shown diagrammatically in the box given below.

1. CONDITION OF BLOOD GLUCOSE IN POST-ABSORPTIVE STATE

What is Post-Absorptive State?

This is fasting state, approx. 12 to 14 hours after last meal. There is practically no intestinal absorption. It is not prolonged starvation, as there are no metabolic abnormalities. It is the condition of a subject between 8 to 10 A.M, if he had his dinner previous evening about 8 P.M and had taken nothing thereafter. ***Under Such a Situation only Source of Glucose is Liver Glycogen.*** At rest, tissues utilise approximately 200 mg of glucose per minute from blood. Glycogen stores of Liver is limited approx. 4 to 6 per cent = 72 to 108 gm, if we take liver weight as 1800 gm. The above store of liver glycogen can supply for 8 hours approximately as per rate of utilisation stated above. Muscle glycogen store is approx. 0.7 per cent of the weight of muscle mass. If the muscle mass of an adult is approx. 35 kg the muscle glycogen amounts to 245 gm. ***Muscle glycogen cannot provide blood glucose by glycogenolysis due to lack of the enzyme Glucose-6-phosphatase.*** It can supply indirectly by gluconeogenesis from the pyruvates and lactates, the products of muscle glycogenolysis. Thus muscle glycogen can provide indirectly blood glucose approx. less than 25 hours.

2. CONDITION OF BLOOD GLUCOSE IN POSTPRANDIAL STATE

Condition following ingestion of food is called ***Post-prandial*** state. ***Absorbed monosaccharides are utilised for oxidation to provide energy. Remaining in excess is stored as glycogen in Liver and muscles.*** In well-nourished

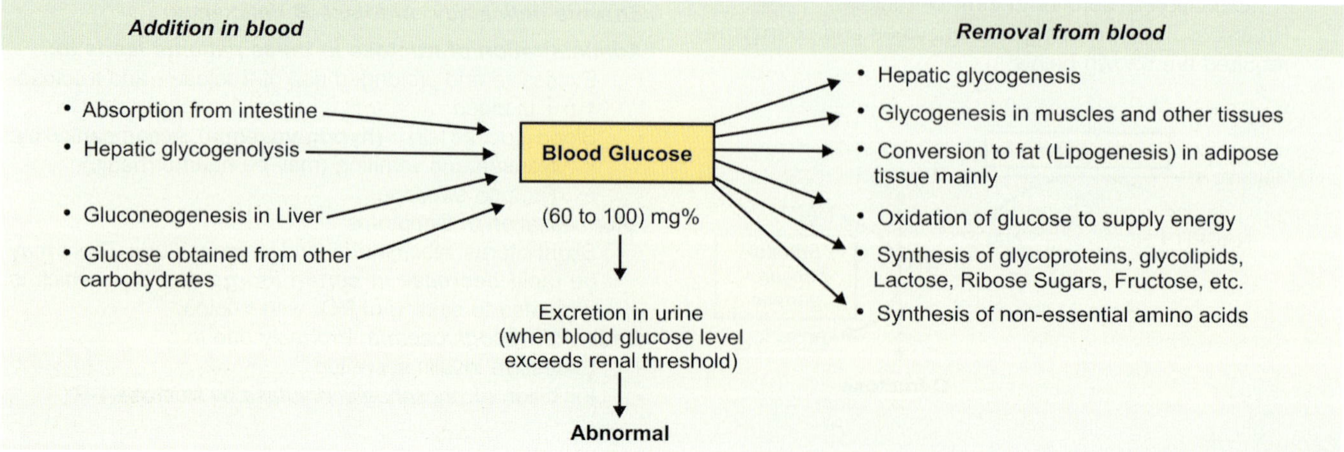

individuals, glycogen stores are fairly saturated. About 40 per cent of absorbed glucose is used for lipogenesis. Some are used for synthesis of glycoproteins, glycolipids etc, when load of glucose is very high renal mechanisms operate. **Tm G is 250 to 350 mg/mt, which is maximum renal tubular reabsorption.** ***When blood glucose rises more than 160 to 180 mg% i.e. the renal threshold, glucose appears in urine (glycosuria).*** **This is an abnormal state.** In normal intestinal absorption such situation does not occur. It can take place with an IV load or disease processes.

AUTOREGULATION (Fundamental Regulatory Mechanisms)

Process of hepatic glycogenesis, glycogenolysis and tissue utilisation of glucose are sensitive to relatively slight deviation from the normal blood sugar concentration.

1. As blood sugar tends to increase ↑:

- Glycogenesis is accelerated, and
- Utilisation of glucose by tissues is increased, resulting to fall in blood glucose level.

The reverse occurs as the blood glucose level tends to fall↓.

Normal balance between production and utilization of blood glucose, at a mean level of circulating glucose of approximately 80 mg % is, therefore, dependant upon the sensitivity of these processes to variations above and below this concentration. ***This level of sensitivity is determined to a considerable extent by the balance between:***

- ***Insulin in one hand, and***
- ***Hormones of adrenal cortex and anterior pituitary on the other hand.*** Overall effect of Insulin is to lower the blood glucose level and adrenocortical/and growth hormone to raise it.

In as much as these two sets of factors are mutually antagonistic to each other. ***It is the ratio between them rather than their absolute amounts that is of Prime importance in this connection.***

The processes of hepatic glycogenesis, glycogenolysis, and glucose utilisation and also the blood glucose concentration are continually exposed to disturbing influences under physiological conditions. These include absorption of glucose from intestine, physical and mental activity, emotional states, etc. Primary effect of majority of these is to cause a rise in blood glucose. This results in:

- Decrease in delivery of glucose by Liver, and
- Acceleration of utilisation by tissues.

Simultaneous increase in insulin secretion, stimulated by elevated blood glucose concentration results in increase in ratio of insulin/adrenocortical hormones and growth hormone. This change in ratio and hormonal balance results in:

- Increased hepatic glycogenesis↑
- Decreased gluconeogenesis↓
- Decreased output of glucose from Liver↓ and
- Increased utilisation of glucose↑.

As a result of above, the blood glucose concentration tends to fall.

2. As blood sugar tends to decrease ↓:

A drop in blood glucose concentration below the normal resting level causes:

- Decrease in secretion of insulin, ↓
- Resulting to decrease in ratio of insulin/glucocorticoids and GH, ↓
- Increased production of blood glucose mainly by gluconeogenesis ↑, and
- Decreased glucose utilisation ↓.

Due to the above actions, blood glucose tends to rise.

If the blood glucose falls below to hypoglycaemic levels, additional emergency mechanisms come into play:

- ***Stimulation of secretion of catecholamines*** by hypoglycaemia resulting in hepatic glycogenolysis and rise in blood glucose.
- The increase in catecholamines may ***also stimulate production of ACTH, hence of adrenocortical hormones,*** causing increased gluconeogenesis.

The blood glucose concentration in normal health-regulates itself. Efficient operation of this autoregulation at physiological levels, however, requires a normal balance between *(a)* insulin and *(b)* the carbohydrate active, adrenocorticoids and anterior pituitary hormones, and also to normal responsiveness of the pancreatic islet cells to variation in blood glucose concentration. ***This constitutes the "autoregulation" or central regulatory mechanism.***

HORMONAL INFLUENCES: (ENDOCRINE INFLUENCES) ON CARBOHYDRATE METABOLISM

Endocrine organs play an important key role in this homeostatic mechanism.

There are **two categories** of endocrine influences:

(a) Those which exert a fundamental regulatory influence, their normal function being essential for normal carbohydrate metabolism, for example, hormones of pancreatic islet cells specially Insulin and hormones of adrenal cortex and anterior pituitary as stated above.

(b) Those which influence carbohydrate metabolism, but are not essential for its autoregulation under normal physiological conditions, e.g. hormones of adrenal medulla and hormones of thyroid gland.

1. Insulin

Administration of insulin is followed by a **fall** ↓ in blood glucose concentration to hypoglycaemic levels, if adequate amounts are given. This results from:

SECTION FOUR

(A) Net decrease of delivery of glucose to systemic blood by Liver, and
(B) Increase in the rate of utilisation of glucose by tissue cells.

(A) Diminished supply of glucose to blood is due to:

- Decreased hepatic glycogenolysis
- Increased hepatic glycogenesis by its direct action on *Protein phosphatase-1*, thus converting glycogen synthase 'b' to glycogen synthase 'a'
- Decreased gluconeogenesis
- The liver glycogen tends to increase although this may be obscured by the hypoglycaemia, which itself tends to accelerate hepatic glycogenolysis.

(B) Increase in the rate of utilisation of glucose by tissue cells: Glucose is removed from the blood more readily and is utilised more actively for:

- Oxidation for energy production
- Increases lipogenesis, and
- For glycogenesis.

The overall effect of insulin is antagonistic to that of adrenal glucocorticoids and growth hormone. Its primary action in extrahepatic tissues is to facilitate entrance of glucose into the cells. In the liver, which is freely permeable to glucose, insulin exerts a regulatory influence upon the activity of *glucokinase*.

2. **Adrenocortical hormones:** Adrenal cortex produces a number of steroid hormones, of which the "glucocorticoids" are important in carbohydrate metabolism. The predominant glucocorticoids in man is **cortisol.**

SECTION FOUR

Glucocorticoids

(a) Increases blood glucose level: By gluconeogenesis, as a result of:

- Increased protein catabolism in the peripheral tissues, so that more amino acids are available.
- Increased hepatic uptake of amino acids and increasing the activity of *transaminases* and all the enzymes concerned with gluconeogenesis, e.g. *Pyruvate carboxylase, PEP-carboxykinase, fructose-1, 6-biphosphatase* and *glucose-6-phosphatase.*
- Diminishing peripheral uptake and utilisation of glucose.

(b) Increases liver glycogen: Attributable in part to increased activity of glycogen synthase, 'b' to 'a' conversion. ***Glucocorticoids are catabolic to peripheral tissues but anabolic to Liver.***

3. **Anterior Pituitary Gland**

Secretes hormones that tend to elevate the blood glucose level and therefore, antagonise the effect of insulin. These are **growth hormone** and **ACTH (corticotropin)** and possibly other "diabetogenic" principles. Growth hormone secretion is stimulated by hypoglycaemia.

- Growth hormone decreases glucose uptake in certain tissues, e.g. muscles. Some of this effect may not be direct since it mobilises FFA from adipose tissue and long-chain fatty acids inhibit glucose utilisation.
- **Liver:** There is increase in liver glycogen due to increased gluconeogenesis.

Chronic administration of GH leads to Diabetes. By producing hyperglycaemia, leads to stimulation of secretion of insulin, which eventually ***produces exhaustion of β-cells.***

4. **Catecholamines**

These are hormones produced by adrenal medulla:

- Produces an increase in blood glucose level and also blood Lactic acid level↑. It stimulates glycogen breakdown (glycogenolysis) in Liver as well as in muscle and is accompanied by a decrease in glycogen content. The action in liver is mediated through cyclic AMP dependant protein kinase.
- In muscle due to absence of *glucose-6-pase,* glycogenolysis does not directly contribute to blood glucose. It increases the pyruvate and lactate. Pyruvate and Lactates diffuse into the blood and in liver are converted to glucose and glycogen.
- Catecholamines also stimulate ACTH formation, which increase glucocorticoids, enhancing gluconeogenesis.
- Epinephrine has direct inhibitory action on Insulin release from β-cells of islets of Langerhans of pancreas. ***In pancreas, α-adrenergic response predominates, which decreases cyclic AMP level↓ and insulin release is inhibited.***

5. **Glucagon**

Glucagon is a protein hormone, a Polypeptide containing 29 amino acids. It is produced by α-cells of islets of Langerhans. It is also known as HGF (hyperglycaemic glycogenolytic factor).

- In response to hypoglycaemia, α-cells produce glucagon which produces rapid glycogenolysis in Liver. It activates phosphorylase enzyme, by increasing cyclic AMP level, which mediates its action through cyclic AMP mediated protein kinase. Active protein kinase, activates *phosphorylase kinase* which in turn activates the enzyme phosphorylase.

Note: Glucagon cannot produce glycogenolysis in muscle as it lacks the receptor.

- Glucagon also enhances "gluconeogenesis" from amino acids, pyruvates and lactates.

6. **Thyroid Hormones**

Thyroxine accelerates hepatic glycogenolysis, with consequent rise in blood glucose. This may be due in part to:

- Increased sensitivity of the tissues to catecholamines, and
- In part to accelerated destruction of insulin.
- Thyroid hormones may also increase the rate of absorption of hexoses from the intestine.
- Increased hepatic *glucose-6-phosphatase* activity.
- Rate of protein catabolism is increased by excessive thyroid hormones and thus increases gluconeogenesis from amino acids.

There is experimental evidences that thyroxine has a diabetogenic action and that thyroidectomy inhibits the development of diabetes.

BLOOD SUGAR LEVEL AND ITS CLINICAL SIGNIFICANCE

1. **Normal values:** The range for normal fasting or Post-absorptive blood glucose taken at least three hours after the last meal is:
   - As per *glucose-oxidase* method ("true" glucose) 60 to 100 mg%, some authorities give as 60 to 95 mg%.
   - As per "Folin and Wu's method": 80 to 120 mg%.
2. **Abnormalities in Blood Glucose Level**
   - Increase in blood glucose level above normal is called **hyperglycaemia**.
   - Decrease in blood glucose level below normal is called **hypoglycaemia**.

(a) Hyperglycaemia

Causes of Hyperglycaemia

- Most common cause is ***Diabetes mellitus*** in which the highest values for fasting blood glucose is obtained, in which it may vary from normal to 500 mg% and over, depending on the severity of the disease.
- ***Hyperactivity of the thyroids, pituitary, and adrenal glands.*** Except in DM, fasting blood glucose rarely exceeds 200 mg%. There may be increased incidence of DM in hyperthyroidism and hyperpituitarism.
- Emotional ***'stress'*** can increase the blood glucose level.
- ***In diffuse diseases of pancreas,*** e.g. in pancreatitis and carcinoma of pancreas some increase in fasting blood glucose may occur.
- Increase in blood glucose in appreciable amount may be seen ***in sepsis*** and in a number of infectious diseases.
- A moderate hyperglycaemia may also be found in some ***intracranial diseases*** such as meningitis, encephalitis, intracranial tumors and haemorrhage.
- ***Anaesthesia:*** Can also increase blood glucose, depending on the degree and duration of anaesthesia.
- ***Asphyxia*** may also increase blood sugar level.
- Increase in blood sugar, rarely exceeding 150 to 180 mg% may be seen ***in convulsions*** and in the terminal stages of many diseases.

(b) Hypoglycaemia

Causes of hypoglycaemia: Hypoglycaemia may be considered to be present when the ***blood glucose is below 40 mg% ("true" glucose value by glucose oxidase method).***

- Most common cause and clinically important to be considered first is ***overdosage of Insulin*** in treatment of diabetes mellitus.
- ***Insulin-secreting tumor (Insulinoma)*** of pancreas produces a severe hypoglycaemia in which blood glucose is very low or may be almost completely absent. It is extremely rare.
- Fasting blood glucose may be reduced in ***hypoactivity of thyroids*** (Myxoedema, and cretinism), ***hypopituitarism*** (Simmond's disease) and ***hypoadrenalism*** (Addison's disease).
- ***Severe liver diseases:*** Low blood glucose levels are often found.
- In childhood, an ***idiopathic hypoglycaemia,*** due to sensitivity to the amino acid leucine has been recognized ***(Leucine-sensitive hypoglycaemia).***
- An acquired Leucine-sensitivity has also been stated to exist.
- ***Spontaneous hypoglycaemia*** in childhood may be due to deficiency of glucagon production.
- ***Severe exercise*** may produce hypoglycaemia due to depletion of liver glycogen.
- Hypoglycaemia is also found in some of the ***Glycogen storage diseases*** (GSDs), e.g. in Von Gierke's disease, Liver phosphorylase deficiency—Due to impaired ability to produce glucose from glycogen.
- Impaired absorption of glucose in some types of ***steatorrhoea.*** The blood glucose may be in the lower part of the normal range, it is rare to be subnormal.
- Transient post-prandial hypoglycaemia ***("reactive" hypoglycaemia),*** may occur in an occasional case, some one and a half to three hours after taking food and more commonly in patients with partial gastrectomy.
- Hypoglycaemia has also been found to be associated with ***alcohol ingestion.***
- ***It has been seen that a variety of tumors of non-endocrine origin, particularly, retroperitoneal fibrosarcoma may produce hypoglycaemia by secreting insulin-like hormones.***

GLYCOSURIA

Under ordinary dietary conditions, glucose is the only sugar present in the free state in blood plasma in demonstrable amounts. Although normal urine contains virtually no sugar; under certain circumstances, glucose or other sugars may be excreted in the urine. This condition is called ***melituria (excretion of sugar in urine).***

The terms glycosuria, fructosuria, galactosuria, lactosuria and pentosuria are applied specially to the urinary excretion of glucose, fructose, galactose, lactose, and pentose respectively.

RENAL THRESHOLD FOR GLUCOSE AND MECHANISM OF GLYCOSURIA

Glucose is present in the glomerular filtrate in the same concentration as in the blood plasma. Under normal conditions, it undergoes practically complete reabsorption by the renal tubular epithelial cells and is returned to the blood stream. **In normal subjects, a very small amount less than 0.5 gm of glucose may escape reabsorption by tubules and be excreted by urine. But this amount is not detected by Benedict's qualitative test**. The reabsorption of glucose by tubular epithelial cells is an "active" process, by "carrier Protein" which is Na^+-dependant and requires energy (similar to absorption of glucose by intestinal epithelial cells). Rate of glucose absorption is expressed as ***TmG (tubular maximum for glucose) which is 350 mg/mt.*** When the blood levels of glucose are elevated, the glomerular filtrate may contain more glucose than can be reabsorbed, the excess passes in urine to produce ***"glycosuria". In normal individuals, glycosuria occurs when the venous blood glucose exceeds 170 to 180 mg/100 ml.*** This level of the venous blood glucose is termed as the **renal threshold** for glucose. Since the maximal rate of reabsorption of glucose by the tubule (TmG—the tubular maximum for glucose) is a constant, it is a more accurate measurement than the renal threshold, which varies with changes in the GFR.

Definition: Glycosuria is defined as the excretion of glucose in urine **which is detectable by Benedict's Qualitative test.**

MECHANISM OF GLYCOSURIA

Excretion of abnormal amounts of glucose in the urine may be due to **two types** of abnormalities:

(a) ***Increase in the amount of glucose entering in the tubule/mt.***
(b) ***Decrease in the glucose reabsorption capacity of the renal tubular epithelium.***

(a) The quantity of glucose entering the tubules is the product of:

- The minute volume of glomerular filtrate, and
- The concentration of glucose in the filtrate, i.e. in the arterial blood plasma.

In as much as glomerular filtration is rarely increased markedly, glycosuria of this type is due almost invariably to an increase in the blood glucose concentration above the "renal threshold level" and called as **hyperglycaemic glycosuria.**

(b) Reabsorption of glucose by renal tubular epithelium: It is accomplished mainly by an "active transport" mechanism, by transport carrier protein. The capacity for reabsorption may be diminished by:

- *'Hereditary'* cause: ***Absence of 'carrier protein' or defective carrier protein.***
- ***Acquired: Due to certain types of kidney diseases*** specially involving the tubules or damage of tubules by chemicals/poisons.
- *Induced:* Experimental glycosuria, e.g. by administration of glycoside "phloridzin".

The above type of glycosuria is called as **Renal glycosuria.** Blood glucose level in this type is normal or even may be subnormal.

TYPES OF GLYCOSURIAS

From above the glycosuria can be divided into **two main groups:**

A. *Hyperglycaemic glycosuria*
B. *Renal glycosuria*

A. Hyperglycaemic Glycosuria

1. *Alimentary glycosuria:* When a large carbohydrate diet is taken, blood sugar rises and may cross renal threshold in occasional case and may produce glycosuria. This condition does not seem to be a normal process, as homeostatic control is so efficient in normal healthy person that such glycosuria should not occur. Alimentary glycosuria, therefore, is only possible in those subjects in whom the power of glucose utilisation is impaired and ***such people should be kept under observation and should be screened regularly for diabetes.***
2. *Nervous or "emotional" glycosuria:* Stimulation of the sympathetic nerves to the Liver or of the splanchnic nerves, breakdown of liver glycogen occurs and produces hyperglycaemia and glycosuria.
   Nervous stimulation mentioned above causes:
   - Glycogenolysis directly, and
   - Also by increased secretion of catecholamines, producing glycogenolysis.

   Thus anything that stimulates sympathetic system such as excitement, stress, etc. may produce glycosuria. ***In one study, college students going for examination 1.6 per cent showed glycosuria.***
3. *Glycosuria due to endocrine disorders:* Deranged function of a number of endocrine glands produces hyperglycaemia which may result in glycosuria.

Examples are:

- *Diabetes mellitus (clinical):* In this case β-cells of islets of Langerhans fail to secrete adequate amount of insulin, ***producing absolute or relative deficiency of insulin.*** Lack of insulin produces hyperglycaemia and glycosuria.

- ***Hyperthyroidism:*** Hyperactivity of thyroid is always attended with low sugar tolerance, hyperglycaemia and may be glycosuria. ***In 25 to 35 per cent of cases, hyperthyroidism and diabetes mellitus can coexist.***
- ***Epinephrine:*** Increased secretion of epinephrine or prolonged administration through subcutaneous route can increase the breakdown of liver glycogen leading to hyperglycaemia and glycosuria.
- ***Hyperactivity of anterior pituitary:*** Hyperactivity of anterior pituitary as in Acromegaly is attended with hyperglycaemia and glycosuria (20 to 30% cases), due to increased secretion of GH and adrenocortical hormones.
- ***Adrenal cortex:*** Hyperactivity of adrenal cortex as in Cushing's syndrome/disease, may cause hyperglycaemia and glycosuria. Glucocorticoids stimulate gluconeogenesis and increased resistance to insulin (glucose uptake by peripheral tissues inhibited).
- ***Glucagon:*** Increased secretion of glucagon by α-cells of islets of Langerhans can cause glycogenolysis producing hyperglycaemia and glycosuria.

4. ***Experimental Hyperglycaemic Glycosurias***

(a) ***"Piqure" glycosuria:*** Certain injuries to the nervous system can cause hyperglycaemia and glycosuria. **Claude Bernard** found that Puncture of a particular spot in the floor of the IV ventricle of rabbits produces hyperglycaemia and glycosuria. ***(Puncture Diabetes).*** This glycosuria persists for 24 hours or more and is accompanied by marked hyperglycaemia.

Mechanism: It is suggested that experimental procedure ***stimulates a group of nerve cells at the floor of IV ventricle*** which sends impulses through the splanchnic nerves to adrenal medulla and liver, increasing glycogenolysis by increased secretion of catecholamines.

(b) ***'Alloxan' diabetes and glycosuria:*** Injection of **alloxan** to an experimental animal like dog, a substance related chemically to "pyrimidine" bases, produces permanent diabetes. ***Diabetes is due to degeneration, necrosis and resorption of β-cells of islets of Langerhans.*** The α-cells and acinar cells are unaffected. ***The alloxan acts directly, promptly and specifically on β-cells.***

Its effect can be prevented by administration of cysteine, glutathione, BAL, or thioglycolic acid immediately before or within a few minutes after injection of the alloxan. ***This protective action is due apparently to the –SH content of these compounds, the alloxan being probably reduced to an inactive substance.***

B. Renal Glycosuria

1. ***Hereditary:*** A milder glycosuria occurs spontaneously, as hereditary familial traits, persisting throughout life, due to absence of "carrier Protein" or altered kinetics of the carrier system due to failure of development.
2. ***Acquired***
   - ***Diseases of renal tubules:*** In some cases of kidney diseases, the renal tubules may be grossly damaged, thus the tubules fail to reabsorb glucose producing glycosuria.
   - ***Due to heavy metal poisoning:*** The heavy metals like Lead (Pb), Cadmium (Cd), Mercury (Hg), etc. can damage the renal tubules thus interfering with the reabsorption of glucose, resulting to glycosuria.
3. ***Lowering of renal threshold: 15 to 20 per cent cases of Pregnancy may be associated with the physiological glycosuria*** with advancement of pregnancy, ***due to lowering of renal threshold. But Pregnancy may be associated with Diabetes mellitus*** in which there will be hyperglycaemic glycosuria. ***These two can be differentiated by performing a fasting blood sugar level.***
4. ***Renal glycosuria:*** It may also occur in association with evidences of other renal tubular transport defects, e.g. aminoacidurias, Renal tubular acidosis, hyperphosphaturia as in Fanconi syndrome.
5. ***'Experimental' renal glycosuria: "phloridzin" glycosuria:*** Phloridzin is a glycoside found in roots of apple tree; when hydrolysed it gives glucose and **aglycone Phloretin**. When phloridzin is administered subcutaneously, it gives rise to intense glycosuria. The dose given to dogs is 1 gm/day, in oil, SC. **Certain other glycosides, such as 'Arbutin' have similar effects.**

Mechanism: Phloridzin displaces sodium from the sodium binding site from "Carrier protein" and hence glucose cannot be bound to the glucose binding site, thus inhibiting glucose reabsorption.

DIABETES MELLITUS

Diabetes mellitus is a common disease in man. A predisposition to the disease is probably ***inherited as an autosomal recessive trait.*** About 25 per cent of relatives of diabetics show abnormal glucose tolerance curves as compared to 1 per cent in the general population.

Definition: A chronic disease due primarily to a disorder of carbohydrate metabolism, **cause of which is deficiency or diminished effectiveness of insulin,** resulting in hyperglycaemia and glycosuria. Secondary changes may occur in the metabolism of proteins, fats, water and electrolytes and in tissues/organs sometimes with grave consequences.

STAGES OF DIABETES MELLITUS

Since "overt" diabetes is not usually seen till after the age of 40, there must be a stage of ***Pre-diabetes*** which dates from the time of conception. American Diabetes Association has divided into **Four stages.** The four stages and findings are shown below in tabular form in the box.

CLINICAL TYPES AND CAUSES

These are *two main groups:*

(a) *Primary (Idiopathic):* constitute major group. Exact cause is not known; ***metabolic defect is insufficient insulin which may be absolute or relative.***

(b) *Secondary:* constitute minor group where it can be secondary to some disease process.

(a) Primary (Idiopathic)

Two clinical types:

- *"Juvenile"-onset diabetes:* Now called as **Type-I-** (Insulin dependent)—**IDDM.**
- *"Maturity" onset diabetes:* **Type-II NIDDM**— (Non-Insulin Dependent).

Differences between the two clinical types are listed in a box in next page.

Other Factors

1. *Heredity:* In both types, familial tendency noted. Genetic factors more important in those who develop after 40. In younger, "Juvenile" type, ***susceptibility is associated with particular HLA phenotype. RISK*** is two to three times more in those **who are HLA phenotype B_8 or BW_{15}.**
2. *Autoimmunity:* ***Insulin-dependent juvenile type may be an autoimmune disorder*** and has been found to co-exist with other autoimmune disorders.
   Evidences in favour of autoimmunity:
   - Lymphocytic and plasma cells infiltrations in pancreas
   - Detection of autoantibodies by immunofluorescence.
3. *Infections: **Certain viral infections may precipitate Juvenile type.*** Experimentally it has been shown that certain viruses can induce diabetes. **Incidence is high after mumps. Antibodies to coxsackie B_4 virus** have been found in young Juvenile type.
4. *Obesity:* Majority of middle aged maturity-onset diabetics are obese, "stress" like pregnancy may precipitate.
5. *Diet:* ***Over-eating*** and ***under activity*** are also predisposing factors in elderly middle aged maturity onset diabetes.
6. *Insulin antagonism:* In ***"maturity onset" diabetes, the deficiency of insulin is relative*** and glucose induced insulin secretion may be greater and more prolonged than normal. This relative deficiency may be due to ***insulin antagonism,*** exact cause for the same is not known but various factors have been incriminated from time to time. They are:
   - **Synalbumin** of **Vallence-Owen** in plasma, dialyzable, thermostable substance.
   - **β_1-lipoprotein factor:** Another similar factor found in β_1-lipoprotein fraction of plasma in diabetics.
   - **Insulin "antibodies".**
   - Secretion of **abnormal** and **less active** insulin or **'altered' insulin.**
   - ***A "tissue barrier" to the transport of insulin to the cells, probably "receptor" deficiency.***
   - ***Lack of cellular response to insulin.***

(b) Secondary: This forms a minor group. Diabetes is secondary to some other diseases.

1. *Pancreatic diabetes:*
   - Pancreatitis
   - Haemochromatosis
   - Malignancy of Pancreas.
2. *Abnormal concentrations of antagonistic hormones:*
   - Hyperthyroidism
   - Hypercorticism: like Cushing's disease and syndrome

Stages	GTT	Fasting blood glucose	Plasma insulin	Symptoms	Angiopathies
• ***Pre-diabetes***	*Normal*	*Normal*	*Normal*	*None*	*+*
• ***'Suspected' Diabetes***	*Abnormal*	*May be normal*	*Normal*	*Symptoms after 'stress'*	*+*
• ***'Chemical'/Latent Diabetes***	*Abnormal*	*Normal or raised*	*Normal or raised*	*Unusual*	*++*
• ***"Overt" Diabetes***	*Abnormal*	*Raised*	*Normal or low*	*Usual*	*+++ to ++++*

Note: Thickening of capillary basement membranes occur in 'Pre-diabetes' before manifestation of insulin lack appear. Degenerative vascular changes become greater in the stages of chemical and 'overt' diabetes.

SECTION FOUR

"Juvenile onset" Diabetes—Type-I	"Maturity onset" Diabetes—Type II
1. Frequency—less	1. Frequency—more common.
2. Commences usually before 15 yrs of age. Males > than Females.	2. Occurs in middle aged individuals. Women are more.
3. Onset—rapid and abrupt	3. Onset—is insidious
4. ***Speedy Progression to Ketoacidosis and coma***	4. Usually mild. ***Ketoacidosis is rare.***
5. Usually patients are thin and underweight	5. Associated with obesity in 2/3 of cases. ***Usually detected during routine check-up of urine.***
6. Deficient Insulin: At first Juvenile diabetics produce more insulin than normal, but the β-cells soon become exhausted and patient becomes "overt" diabetics with atrophied β-cells and practically no insulin	6. β-Cells respond normally. Relative deficiency of insulin, which may be due to "insulin" antagonism.
7. Plasma insulin– It is almost absent. No insulin response is shown to glucose load.	7. Plasma insulin levels may be normal or even raised.
8. ***Insulin therapy–is necessary for control of these cases.***	8. ***Oral hypoglycaemic agents and dietary control are useful in treatment.***

- Hyperpituitarism: Like acromegaly
- Increased glucagon activity.

3. *Iatrogenic:* In genetically susceptibles, may be precipitated by therapy like corticosteroids, thiazide diuretics.

Recent Advances—Excerpts

1. ***Immune "markers" in type IDM (IDDM):*** The recent area of interest is the role of *"glutamic acid decarboxylase"* (GAD) as antigen of potential significance.

Recently anti-GAD antibodies have been deomonstrated in most newly diagnosed IDDM (Type I) patients and in predictable first degree relatives.

In adults presence of GAD antibody is a "marker" for *slow onset* Type I DM (IDDM) and helps to differentiate IDDM with age of onset > greater than 35 years and Type II (NIDDM). ***The term "Latent Autoimmune DM" in adults or "LADA" is now being used for such patients.***

2. ***Genes of DM discovered:*** As seen above, some people may be genetically susceptible to Non-insulin-dependent (NIDDM) or adult onset diabetes. Recently researchers have discovered 2 (two) genes, called **MODY 1** and **MODY 3**, that appear to contribute to the 2 to 5 per cent of Diabetes cases that are clearly inheritable.

MODY 3 gene, located on *chromosome 12,* produces ***hepatocyte nuclear factor-1α (HNF-1α),*** a protein found in the Liver and in the β-cells of the pancreas. Pancreatic β-cells produce insulin, the hormone that regulates blood sugar levels.

"MODY 1 gene", located on *chromosome 20,* makes ***hepatocyte nuclear factor 4α (HNF 4α),*** a cell receptor that plays a role in HNF-1α production.

The biological effects of mutant forms of HNF-1α and HNF-4α are still not known clearly.

Presentation Diabetes Mellitus

The disease has a varied presentation.

- ***Glycosuria*** may be detected during routine examination of urine like annual check-up or when doing routine examination due to some other diseases. There may not be any symptoms/signs.
- Some may present with all classical symptoms like thirst, ***polydypsia, Polyuria, Polyphagia,*** loss of weight etc ("Overt" diabetics).
- Some women present during pregnancy (stress)
- ***A few specially Type-1 cases may present as fulminant ketoacidosis and a few with complications.***

CLINICAL FEATURES AND BIOCHEMICAL CORRELATIONS

- Large amounts of glucose may be excreted in urine (may be 90 to 100 G/day in some cases). Loss of solute produces osmotic diuresis thus large volume of urine ***(polyuria).***
- Loss of fluid leads to thirst and ***polydypsia.***
- ***Polyphagia:*** Eats more frequently. More fond of sweets. The above symptoms may persist for many months in maturity-onset diabetes. In juvenile onset type-1, further symptoms develop if treatment is not started.
- Tissues including muscles received liberal supply of glucose but cannot use glucose due to absolute or relative deficiency of insulin/ or transport defect to cells. This causes ***weakness and tiredness.***
- As glucose cannot be used for fuel, fat is mobilised leading to increase FFA- in blood and liver.
- Increased acetyl-CoA is diverted for cholesterol synthesis—***Hypercholesterolaemia*** and ***atherosclerosis.*** Xanthomas may develop.

- Increased ketone bodies leads to **acidosis**, which leads to hyperventilation ("air-hunger").
- If ketosis is severe, acetone will be breathed out, giving characteristic "fruity" smell in breath (due to acetone).
- Alongwith above, there may be excessive breakdown of tissue proteins. Deaminated amino acids are catabolised to provide energy, which accounts for *Loss of weight.*
- Due to ketosis, develops anorexia, nausea, and vomiting. Continued loss of water and electrolytes increases *dehydration.*
- ***Ketoacidosis produces increasing drowsiness, leading to diabetic coma in untreated cases.***

METABOLIC CHANGES IN DIABETES MELLITUS (Fig. 23.16)

1. **Hyperglycaemia:** Occurs as a result of:
   - Decreased and impaired transport and uptake of glucose into muscles and adipose tissues.
   - Repression of key glycolytic enzymes like Glucokinase, phosphofructokinase and pyruvate kinase takes place.
   - Derepression of key gluconeogenic enzymes like Pyruvate carboxylase, phosphoenol pyruvate carboxykinase, fructose biphosphatase and glucose-6-phosphatase occur, promoting gluconeogenesis in Liver. This further contributes to hyperglycaemia.
   - Elevated amino acid level in the blood particularly alanine provides fuel for gluconeogenesis in Liver.
2. **Amino Acids Level**
   - Transport and uptake of amino acids in peripheral tissues is also depressed causing an elevated circulating level of amino acids, particularly alanine. Glucocorticoid activity predominate having catabolic action on peripheral tissue proteins, releasing more amino acids in blood.
   - Amino acids breakdown in Liver results in increased production of urea N ↑.
3. **Protein synthesis:** Protein synthesis is decreased in all tissues due to:
   - Decreased production of ATP↓
   - Absolute or relative deficiency of Insulin.
4. **Fat Metabolism**
   - Decrease extramitochondrial *de Novo* synthesis of FA and also TG synthesis due to decrease in acetyl-CoA from carbohydrates, ATP, NADPH and α-glycero-(p) in all tissues.
   - Stored lipids are hydrolysed by increased Lipolysis liberating free fatty acids (FFA)↑. Increased FFA interferes at several steps of carbohydrate phosphorylation in muscles, further contributing to hyperglycaemia.

Effects of Increased FFA Level

- FFA reaching the Liver in high concentration inhibits further FA synthesis by a feedback inhibition at the *acetyl-CoA carboxylase* step.
- Fats are mobilised for energy; increased fatty acid oxidation increases acetyl-CoA level, which in turn activates *Pyruvate carboxylase*, stimulating the gluconeogenic pathway required for conversion of amino acids C-skeletons to glucose.
- FA also stimulates gluconeogenesis by entering TCA cycle and increasing production of citrates↑. Citrate in turn inhibits glycolysis at *phosphofructokinase* level.
- Eventually FA inhibits TCA cycle at the level of *citrate synthase* and possibly pyruvate dehydrogenase complex and Isocitrate dehydrogenase level.

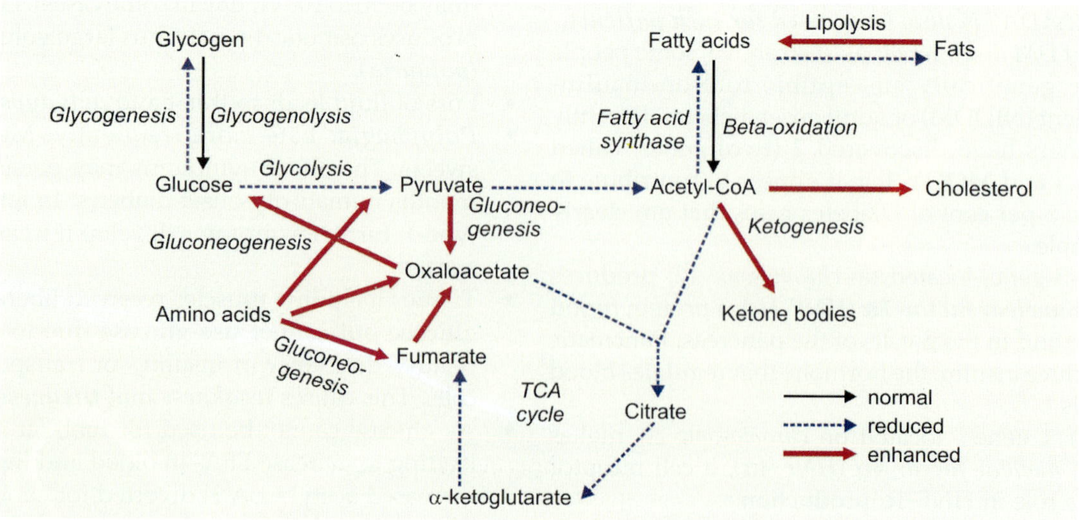

Fig. 23.16: Metabolic changes in diabetes mellitus

SECTION FOUR

- Acetyl-CoA which no longer can be channelised to TCA cycle or be used for FA synthesis are diverted to:
    - ***Cholesterol synthesis,*** and
    - ***Ketone bodies formation.*** Excessive production of ketone bodies increases the concentration of ketone bodies in blood (ketonaemia) and excretion of ketone bodies in urine (ketonuria) and leads to acidosis.

5. **Effect on glycogen synthesis:** Glycogen synthesis is depressed as a result of:
    - Decreased *glycogen synthase* activity due to deficiency of insulin.
    - By activation of phosphorylase producing glycogenolysis through the action of epinephrine and/or glucagon (antagonistic) hormones.
    - By increased ADP: ATP ratio.

Note: The insulin-deficient animal is in a state of hormonal imbalance favouring the action (Preponderance) of glucocorticoids, growth hormone and glucagon, all of which add to the stimulation of gluconeogenesis, lipolysis and decreased intracellular metabolism.

6. **Other Effects of Hyperglycaemia**

(a) Glycosylation of Hb and formation of glycosylated Hb (HbA_{1C}): Glycosylated haemoglobins particularly HbA_{1C} rises in prolonged and uncontrolled diabetes 3 to 4 times than the normal level.

(b) Non-enzymatic glycosylation of other proteins as plasma albumin, collagenous tissues and the lens protein α-crystallin: Such glycosylations of colla-genous tissues bring about thickening and morphological changes of vessel walls and also glomerular basement membrane thickening. Glycosylation of lens protein may also account for diabetic cataract. ***Diabetic cataract biochemically may be due to:***

- Glycosylation of Lens Proteins, i.e. α-crystallin.
- Accumulation of 'sorbitol' which produces osmotic damage.

7. **Sorbitol (polyol) pathway and phosphoinositide metabolism:** Over the past few years, quest to discrete pathogenetic mechanisms for the long-term complications of Diabetes has gradually **focussed on three promising targets** for specific therapeutic intervention:
    - **Non-enzymatic glycosylation of Proteins** (discussed above)
    - **Altered 'microvascular' haemodynamics,** and
    - **Abnormal sorbitol (polyol)—inositol metabolism.**

Hyperglycaemia and its effects on metabolism of glucose through sorbitol pathway and myoinositol metabolism and their associated effects on phosphoinositide metabolism, protein kinase-C, and *Na^+–K^+–ATPase* are shown schematically in Figure 23.17.

1. Striking structural similarity between glucose and Myoinositol (MI) results in competition for high affinity myoinositol transporters which explains the ***selective tissue depletion of myoinositol that accompanies hyperglycaemia.*** Glucose interferes with "carrier-mediated" myoinositol transport in various

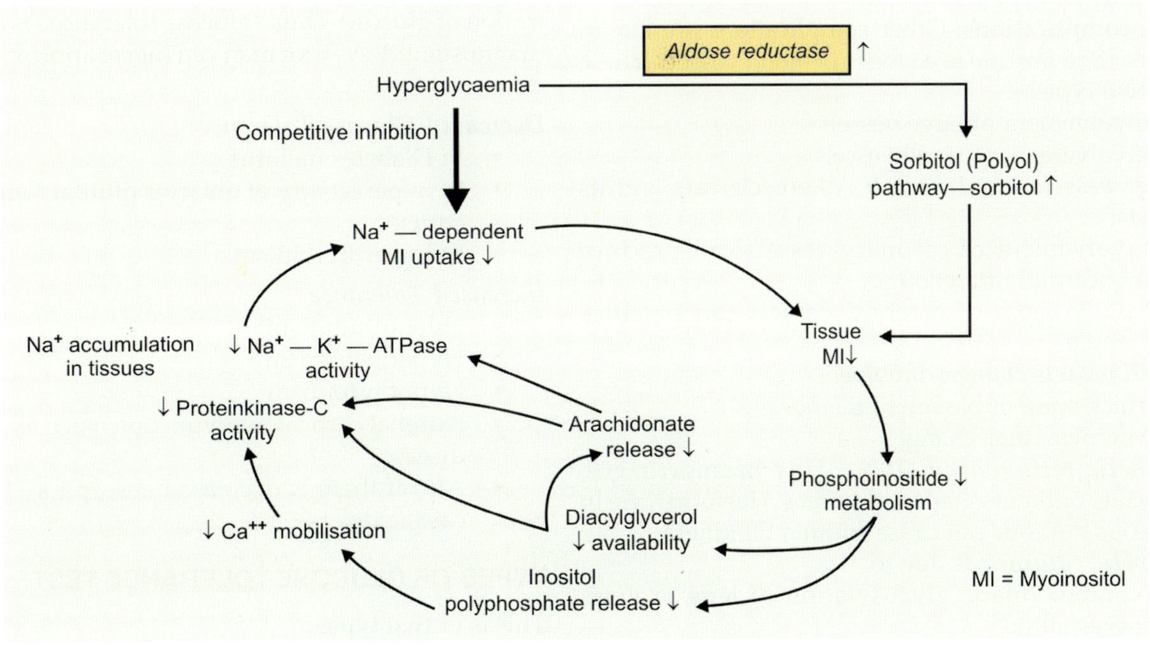

Fig. 23.17: Sorbitol (Polyol)—inositol pathway

tissues specially intestine, renal brushborder membranes, peripheral nerves and renal glomerulus. Thus, ***hyperglycaemia may selectively deplete myoinositol from tissues.***

2. When tissues that exhibit diabetic complications are exposed to hyperglycaemia, sorbitol (Polyol—alcohol), accumulates because of an increase in the conversion of intracellular glucose to sorbitol by the high K_m, i.e. low affinity enzyme *aldose reductase* and myoinositol level falls.
3. High milli-molar levels of sorbitol that accumulate in Lens of eye **cause osmotic damage**, which has been linked to the development of Diabetic cataract.
4. Inositol-1,4,5-triphosphate is thought to mobilise intracellular Ca^{++} sequestered in the endoplasmic reticulum. Due to decrease in Inositol Polyphosphate release, Ca^{++} mobilisation is decreased.
5. Due to decreased availability of Diacylglycerol and lower Ca^{++} mobilisation *proteinkinase C* activity is decreased, resulting in lowered Na^+-K^+ ATPase activity, resulting to Na accumulation in tissues.

Clinical Aspect: Several inhibitors like **Sorbinil**, an *aldose reductase* inhibitor, the first enzyme in the metabolic cascade shown above can prevent or reverse early diabetic complications in laboratory animals and are being extensively undergoing clinical trials in patients with Diabetes mellitus.

COMPLICATIONS OF DIABETES MELLITUS

I. Immediate: Diabetic ketoacidosis and coma is one of the most important and dreaded complication specially in Type-I.

II. Late complications: Other complications are late to appear and are due to changes in blood vessels. These are **two types:**

- Involvement of large vessels
- Involvement of small vessels.

(a) Large vessels involvement: Atherosclerosis and its effects:

- Involvement of coronary vessels can produce myocardial infarction.
- Involvement of cerebral vessels can produce "stroke".

(b) Small vessels changes involve:

- Thickening of basement membrane
- Microvascular changes.

1. ***Diabetic retinopathy (70%):*** Tiny haemorrhages, punctate or flame-shaped, exudates. Haemorrhage in vitreous humour can cause sudden blindness.
2. ***Diabetic cataract:*** Is due to:
   - Non-enzymatic glycosylation of lens protein, α-crystallin;
   - Osmotic damage to lens protein due to accumulation of sorbitol.
3. ***Diabetic nephropathy (50% cases):*** Characterised by (a) Proteinuria, (b) Hypertension and (c) Oedema. The triad is called as ***Kimmelsteil-Wilson syndrome.*** Microscopic lesions are called as ***'Kimmelsteil-Wilson lesions/disease'.*** Lesions are often present when syndrome is not developed. Sometimes kidney lesions may be shown as:
   - ***Papillary necrosis:*** A dangerous complication.
   - ***Pyelonephritis:*** When secondary infections occur.
4. ***Peripheral neuritis (neuropathy):*** Manifestated by loss of sensation and tingling. Biochemically probably the cause is myoinositol deficiency. Sometimes there may be associated myopathies, weakness of muscles.
5. ***Diabetic gangrene:*** Cause is due to diminished blood supply due to atherosclerotic changes in blood vessels. Also associated tissue hypoxia due to formation of HbA_{1C} (glycosylated Hb), less oxygen carrying capacity.
6. ***Skin lesions:*** Prone to infections: boils/ulcers and carbuncles. There may be necrosis of skin, ***Necrobiosis diabeticorum.***
   - May be punctate depigmented atrophy
   - Wound healing is delayed.
7. ***Pulmonary tuberculosis:*** Susceptible to pulmonary tuberculosis.

GLUCOSE TOLERANCE TEST (GTT)

What is "Carbohydrate Tolerance"?: The ability of the body to utilise carbohydrates may be ascertained by measuring its carbohydrate tolerance. It is indicated by the nature of blood glucose curve following the administration of glucose. Thus "glucose tolerance" is a valuable diagnostic aid. A 70 kg man can ingest approx. 1500 gm/day.

Decreased Glucose Tolerance

- In Diabetes mellitus
- In hyperactivity of anterior pituitary and adrenal cortex
- In hyperthyroidism.

Increased Tolerance

- Hypopituitarism
- Hyperinsulinism
- Hypothyroidism
- Adrenal cortical hypofunction (such as Addison's disease)
- Also if there is decreased absorption, like sprue, caeliac disease.

TYPES OF GLUCOSE TOLERANCE TEST

This is of **two types:**

(A) Standard oral glucose tolerance test

(B) IV glucose tolerance test.

(A) Standard Oral GTT

Indications

- In patients with transient or sustained glycosuria, who have no clinical symptoms of Diabetes with normal fasting and PP blood glucose.
- In patients with symptoms of Diabetes but with no glycosuria and normal fasting blood glucose level.
- In persons with strong family history but no overt symptoms.
- In patients with glycosuria associated with thyrotoxicosis, infections / sepsis, Liver diseases, Pregnancy, etc.
- In women with characteristically large babies 9 lbs or individuals who were large babies at birth.
- In patients with neuropathies or retinopathies of undetermined origin.
- In patients with or without symptoms of DM, showing one abnormal value.

Pre-requisites: Precautions to be taken on the day prior to the test:

- The individual takes usual supper at about 2000 hours and does not eat or drink anything after that. Early morning if so desires, a cup of tea / or coffee may be given without sugar or milk. No other food or drink is permitted till the test is over.
- ***Should be on normal carbohydrate diets at least for three days prior to test*** (approx 300 G daily), otherwise 'false' high curve may be obtained.
- Complete mental / and physical rest.
- No smoking is permitted.
- All samples of blood should be venous preferably. If capillary blood from 'finger prick' is used, all samples should be capillary blood.

Procedure

1. A fasting sample of venous blood is collected in **flouride bottle *(fasting sample)***
2. The bladder is emptied completely and urine is collected for qualitative test for glucose and ketone bodies ***(fasting urine).***
3. The individual is given 75 Gm of glucose dissolved in water about 250 ml to drink. Lemon can be added to make it palatable and to prevent nausea / vomiting. Time of oral glucose administration is noted.
4. A total of **five specimens** of venous blood and urine are collected every ½ hour after the oral glucose viz. ½ hour, 1 hour, 1½ hour, 2 hour and 2½ hour.
5. Glucose content of all the **six** (including fasting sample) samples of blood are estimated and corresponding urine samples are tested qualitatively for presence of glucose and ketone bodies. A curve is plotted which is called as **Glucose tolerance curve.**

Explanation and Significance of a Normal Curve

1. A sharp rise to a peak, averaging about 50 per cent above the fasting level within 30 to 60 minutes. Extent of the rise varies considerably from person to person, ***but maximum should not exceed 160 to 180 mg% in normal subjects.***

   Reason
   - Rise is due directly to the glucose absorbed from the intestine, which temporarily exceeds the capacity of the Liver and tissues to remove it.
   - As the blood glucose concentration increases, regulatory mechanisms come into play:
     - Increased insulin secretion due to hyperglycaemia,
     - Hepatic glycogenesis is increased,
     - Hepatic glycogenolysis is decreased, and
     - Glucose uptake and utilization in tissues increase.
2. A sharp fall to approximately the fasting level at the end of 1½ to 2 hours.

   Reason: Glucose now leaves the circulation faster than it is entering. This is due to:
   - Continuing stimulation of the mechanisms stated above, i.e. increased utilisation and hepatic glycogenesis, and
   - To slowing or completion of glucose absorption from the intestines.
3. ***Hypoglycaemic "dip":*** Continued fall to a slightly subfasting (10 to 15 mg lower than fasting value) at 2 hours and subsequent rise to fasting level at 2½ to 3 hours.

   Reason: ***The hypoglycaemic 'dip' is due to "inertia" of the regulatory mechanisms.*** The decreased output of glucose by liver and increased utilisation induced by the rising blood glucose are not reversed as rapidly as the blood sugar falls.

Characteristics of Different Types of GTC (Fig. 23.18)

(a) A Normal GTC

1. Fasting blood glucose within normal limits of 60 to 100 mg% ("True" glucose)
2. The highest peak value is reached within one hour.
3. The highest value does not exceed the renal threshold, i.e. 160 to 180 mg%
4. The fasting level is again reached by 2½ hour
5. No glucose or Ketone bodies are detected in any specimens of urine.

A **typical normal GTC** is given below:

	Glucose—75 G					
	F ↓	*½ hr*	*1 hr*	*1½ hr*	*2 hr*	*2½ hr*
Blood glucose	75	130	150	100	65	76
Urine glucose	–	–	–	–	–	–

(b) Diabetic Type of GTC

1. Fasting blood glucose is definitely raised 110 mg% or more ("True" Glucose).

SECTION FOUR

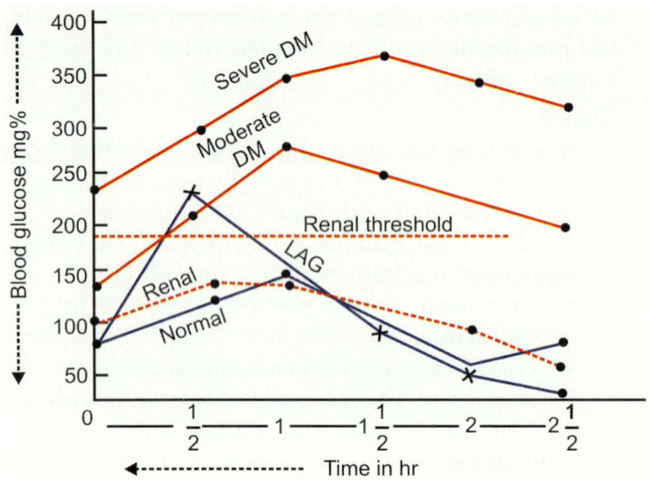

Fig. 23.18: Different glucose tolerance curves

2. The highest value is usually reached after 1 to 1½ hour.
3. The highest value exceeds the normal renal threshold.
4. Urine samples always contain glucose ***except in some chronic diabetics or nephritis who may have raised renal threshold (Dangerous type),*** hyperglycaemia but no glycosuria. Urine may or may not contain ketone bodies depending on the type of Diabetes and severity.
5. ***The blood glucose does not return to the fasting level within 2½ hours.*** **This is the most characteristic feature of true DM.**

According to severity, it may be:
(a) **Mild Diabetic curve,**
(b) **Moderately severe Diabetic curve,** and
(c) **Severe Diabetic curve** (see box below).

Glucose—75 G	F ↓	½ hr	1 hr	1½ hr	2 hr	2½ hr
Moderate Diabetic curve						
Blood glucose	130	200	280	260	220	170
Urine glucose	—	++	++	++	++	±

Glucose—75 G	F ↓	½ hr	1 hr	1½ hr	2 hr	2½ hr
↓ 'Severe' Diabetic curve						
Blood glucose	230	300	345	365	350	330
Urine glucose	++	+++	+++	++++	+++	+++

(c) Renal glycosuria curve: Glucose appears in the urine at levels of blood glucose much below 170 mg%.

Patients who show no glycosuria when fasting may have glycosuria when the blood glucose is raised. The condition may be:

- Idiopathic without any pathological significance
- Occasionally occurs in certain renal diseases and in pregnancy (when there may be lowering of renal threshold)
- May be found in case of "early" Diabetes with low renal threshold
- It has been reported in children of diabetic parents. These cases should be reviewed from time to time (every six months).

Renal glycosuria curve is shown below in the box.

Glucose—75 G	F ↓	½ hr	1 hr	1½ hr	2 hr	2½ hr
Renal Glycosuria						
Blood glucose	90	130	139	110	90	85
Urine glucose	± (or trace)	+	+	+	± (or trace)	± (or trace)

(d) 'Lag' Curve (or Oxyhyperglycaemic Curve):

1. Fasting blood glucose is normal but it rises rapidly in the ½ to 1 hour and exceeds the renal threshold so that the corresponding urine specimens show glucose.
2. The return to normal value is rapid and complete.

This type of GTC may be obtained in:

- **Hyperthyroidism**
- **After gastroenterostomy**
- **During pregnancy**
- **Also in "early" diabetes.**

A patient showing "lag curve" should be reviewed from time to time after every six months. A 'Lag type' of GTC is shown below in the box:

Glucose—75 G	F ↓	½ hr	1 hr	1½ hr	2 hr	2½ hr
Lag curve						
Blood glucose	80	220	190	110	80	65
Urine glucose	—	++	+	—	—	—

GTC of arterial (capillary) blood differs from GTC of venous blood as follows:

1. Rise begins somewhat earlier.
2. Peak is usually reached at 30 to 45 mts.
3. The level in capillary blood may be 20 to 70 mg (average 30) higher than the venous blood.
4. The return to the fasting level at 1½ to 3 hours is not as rapid as in the case of venous blood.

Value of GTT

- Most valuable in investigating a case of "symptomless" glycosuria, such as renal glycosuria, and lag type glycosuria.
- Helpful in recognising milder cases of DM and "early DM."
- Rarely necessary for diagnosis of DM of moderately severe or severe intensity, where characteristic symptoms, fasting high blood glucose and glycosuria is present.
- But helpful in following course and treatment in established DM. In such cases modified GTT can be done. Two specimens: One fasting and the other, postprandial (PP) -2 hours sample.
- May be of use in certain endocrine dysfunction and patients with steatorrhoeas.

Criteria Laid Down to Label "Renal Glycosuria"

Incidence of renal glycosuria is approx 9 to 25 per cent, but with strict criteria, as given below incidence is < 1 per cent.

Criteria laid down for the diagnosis of true renal glycosuria

1. A fasting blood sugar within normal limits.
2. A normal or supernormal 'flat' GTC.
3. Glucose should be detected in every specimen of urine, whether voided in fasting state or after meal.
4. Carbohydrate utilisation should be normal, as evidenced by determination of RQ and normal serum inorganic PO_4 after glucose ingestion.
5. No disturbance of fat metabolism and no ketosis.
6. Moderate doses of Insulin should have little or no effect on glycosuria.

B. Intravenous GTT

Preferred where there are abnormalities in absorption of glucose. Thus **IV GTT is indicated:**

- **In Hypothyroidism**
- **In sprue and caeliac disease**

Dose: 1/3 gm-of glucose/kg body wt., given as 50 per cent solution IV within 3 to 5 minutes.

Procedure: Similar to oral GTT

Blood glucose estimations are done on "fasting" and ½ hourly intervals for 2 hours after IV injection of glucose.

Observations

1. All normal cases required less than 60 minutes for the blood glucose to return to normal.
2. In Diabetes mellitus: Even mild cases take more than 120 minutes to return to initial level.

C. CORTISONE STRESSED GTT

Used for detecting 'Latent' diabetes or Pre-diabetes.

Basis: It is based on the fact that while a large dose of ACTH or a suitable corticosteroid will produce a raised "glucose tolerance curve" and glycosuria in normal persons, smaller doses will do so only in "pre-diabetic" persons.

Dose

1. *Cortisone:* Two doses of 50 mg cortisone orally. First dose to be given 8½ hours before GTT, and second dose 2 hours before GTT.
2. *Prednisolone:* 0.4 mg/kg body wt. ½ the dose at mid-night and ½ at 6 AM, before carrying out GTT at 8 AM.

D. EXTENDED GTT

Instead of ending at 2½ hours after taking glucose, ½ hourly blood sugars are done for periods upto 4 to 5 hours. Partial gastrectomy cases and patients with islet cell tumors may have attacks suggesting hypoglycaemia some 2 to 3 hours after food.

Extended GT are sometimes used:

- To differentiate transient attacks of hypoglycaemia from those due to insulin secreting tumors of islet of Langerhans in the pancreas, and other abnormal endocrine conditions such as Simmond's disease, which cause hypoglycaemia.
- In the endocrine disorders, fall in blood glucose at the end of tolerance test tends to be progressive. However, the fasting blood glucose particularly early morning sample before breakfast may be low enough for diagnosis.

RECOMMENDATIONS OF WHO FOR DIAGNOSIS OF DM BY GTT: In 1980, WHO expert committee on DM has proposed raising the degrees of hyperglycaemia necessary for a diagnosis of DM and they created a new category: ***"Impaired glucose tolerance" (IGT),*** which will not be regarded as diabetic.

(a) New proposals state that ***in patients with symptoms***
- A fasting venous plasma concentration of 8 mMol/L (144 mg/dl) or greater is diagnostic of DM.
- If the concentration is below 6 m Mol/L (108 mg/dl) the dignosis of DM is excluded.
- **Patients with results in intermediate Zone,** i.e. 6 m Mol/L to 8 mMol/L (108 to 144 mg/dl) should be given a 75 gm of oral glucose load.
  1. If the 2 hour venous plasma concentration is greater than 11 mMol/L (198 mg/dl), the test is diagnostic of DM.
  2. If it is less than below 11 mMol/L (198 mg/dl) but greater than 8 mMol/L (144 mg/dl), ***the diagnosis is IGT.***

(b) In ***patients without symptoms,*** the criteria require an additional abnormal value after 75 gm glucose load e.g an one-hour plasma concentration of 11 mMol/L (198 mg/dl) or greater. Should subsequent tests confirm either a raised fasting > than 144 mg/dl or 2 hrs. > value 198 mg/dl, these patients too may be classified as Diabetic.

Note

1. ***A dignosis of DM should not be made on the basis of one abnormal glucose concentration.***
2. 75 gm Glucose load is also a compromise between the American 100 gm load, which often produces nausea, and the 50 gm UK load, which may not be sufficiently rigorous stimulus.
3. The new category of ***Impaired glucose tolerance (IGT)*** will replace the earlier stages and terms like "chemical", latent, 'suspected', 'borderline' and 'subclinical' diabetes.

IGT Group: Must be recognised as at **Risk** of large vessel disease and probably of coronary heart disease in association with Risk factors, as obesity, serum lipoprotein abnormalities and other factors. Though the rate of progression from IGT to frank DM is low, it may be amenable to intervention.

1. Without advice or any treatment, the rate is about 3 per cent a year.
2. With dietary advice, the rate was found to be reduced to 1.3 per cent yearly.
3. In one small study, it has been claimed that with the further addition of tolbutamide (hypoglycaemic drug) the rate was found to be NIL.
4. Another study, addition of "chlorpropamide" in a slightly younger age group reduced the rate of progression to 'overt' DM to 0.1 per cent in a year.

IGT and Pregnancy: In view of implications of IGT in pregnancy, patients diagnosed for the first time in pregnancy should be said to be ***Gestational Diabetes.*** They have an increased incidence of foetal malformation, but treatment can prevent much of the associated "perinatal morbidity" and mortality.

I. Questions (Essay type and short notes)

1. Describe the HMP shunt pathway and state its significance.
2. Describe Embden Meyerhof pathway (glycolysis) in the body.
3. Why is the Krebs' cycle as "final common pathway of metabolism". Explain.
4. Describe the formation and fate of pyruvic acid in the body.
5. Describe the various mechanisms which regulate the blood sugar level.
6. Describe how blood glucose level is maintained. Mention its hormonal controls.
7. Define glycosuria and explain in brief the different types of glycosuria.
8. What is "Glucose tolerance"? Describe the Glucose tolerance test with its features, precautions and significance.
9. Describe the formation, fate and role of Lactic acid in the body.
10. How is blood glucose level regulated?
11. Define glycogenesis, glycogenolysis, and gluconeogenesis. Describe the role of epinephrine in glycogenolysis.
12. Describe the fate of pyruvic acid in the body.
13. Describe how glucose tolerance curve is obtained. Explain its significance.
14. Describe the hormonal regulation of blood sugar level.
15. What is "Active" succinate? Describe the formation and fate in the body.
16. Describe the formation and fate of "Active acetate" (Acetyl-CoA) in the body.
17. What is gluconeogenesis? Describe briefly the metabolic pathways involved in gluconeogenesis.
18. What is the normal blood sugar level? How is it maintained in the body?
19. "Krebs" citric acid cycle is the common pathway of metabolism". Justify the statement.
20. Describe the process of glycogen formation and breakdown in the Liver. Highlight the points of difference and their significance.
21. Write about various steps involved in gluconeogenesis?
22. What is gluconeogenesis? Give an outline of reactions. How are these reactions controlled? What is biological importance of these reactions?
23. Name the various pathways for glucose metabolism. Give in detail the tricarboxylic acid cycle and mention its metabolic importance.
24. Describe reactions of glycogenolysis. Discuss the hormonal control of these reactions.
25. "TCA cycle is called amphibolic in nature"—Justify.

Short Notes

a. Cori's cycle
b. Acetyl-CoA
c. Gluconeogenesis
d. Glycogenolysis
e. Glucuronic acid
f. Hexose Monophosphate shunt (HMP)
g. Glucose tolerance test/curve
h. Metabolic significance of HMP shunt
i. Galactosemia
j. Endocrine regulation of blood sugar
k. Glycogenesis
l. Biosynthesis of Lactose
m. D-Glucuronic acid formation and function
n. Glycosuria
o. Glycogen storage diseases
p. Role of Insulin and Adrenaline in carbohydrate metabolism.

II. MCQs (Give one correct answer)

1. UDP–Glucose is converted to UDP–glucuronic acid by:
   a. ATP b. GTP
   c. $NADP^+$ d. NAD^+
   e. FAD
2. All of the following compounds are intermediates of TCA cycle *except:*
   a. Malate b. Pyruvate
   c. Oxaloacetate d. Fumarate
   e. Succinate
3. In conversion of lactic acid to glucose, three reactions of glycolytic pathway are circumvented, which of the following enzymes do not participate?
   a. Pyruvate carboxylase
   b. Phosphoenolpyruvate carboxykinase
   c. Pyruvate kinase
   d. Glucose-6-phosphatase
   e. Fructose-1, 6-biphosphatase
4. MacArdle's disease involves a deficiency of the which enzyme?
   a. Acid maltase b. Glucose-6-phosphatase
   c. Hepatic phosphorylase d. Muscle phosphorylase
   e. Branching enzyme
5. In the normal resting state of humans, most of the blood glucose brunt as 'fuel' is consumed by:
   a. Liver b. Brain
   c. Kidneys d. Adipose tissue
   e. Muscles
6. A regulator of the *enzyme glycogen synthase* is:
   a. Citric acid
   b. 2,3-biphosphoglycerate (2,3-BPG)
   c. Pyruvate
   d. GTP
   e. Glucose-6-PO_4
7. Which of the following compound is a positive allosteric modifier of the enzyme pyruvate carboxylase?
   a. Biotin b. Acetyl-CoA
   c. Oxaloacetate d. ATP
   e. Fructose-6-P
8. A specific inhibitor for *succinate dehydrogenase* is:
   a. Arsenite b. Malonate
   c. Citrate d. Cyanide
   e. Fluoride
9. Most of metabolic pathways are either anabolic (synthetic) or catabolic (degradation). Which of the following pathways is considered as "amphibolic" in nature?
   a. Glycogenesis b. Glycolytic pathway
   c. Lipolysis d. Citric acid cycle
   e. Glycogenolysis
10. Transketolase activity is affected in:
    a. Biotin deficiency b. Pyridoxine deficiency
    c. PABA deficiency d. Thiamine deficiency
    e. Manganese deficiency
11. The following metabolic abnormalities occur in Diabetes mellitus, except:
    a. Increase plasma FFA
    b. Increased pyruvate carboxylase activity
    c. Decrease lipogenesis
    d. Decrease gluconeogenesis
    e. Increase PEP–carboxykinase activity
12. A substance that is not an intermediate in the formation of D-glucuronic acid from glucose is:
    a. Glucose-1-P b. 6-phosphogluconate
    c. Glucose-6-P d. UDP–Glucose
    e. UDP–glucuronate
13. Von Gierke's disease is characterised by a deficiency of which enzyme?
    a. Glucokinase b. Glucose-6-phosphatase
    c. α-1, 6-glucosidase d. Glycogen synthase
    e. De-branching enzyme
14. The hydrolysis of glucose-6-P is catalysed by a phosphatase that is *not* found in which of the following?
    a. Liver b. Kidney
    c. Muscle d. Small intestine
    e. None of the above
15. An essential for converting glucose to glycogen in Liver is
    a. Lactic acid b. GTP
    c. CTP d. UTP
    e. Pyruvic acid
16. Which of the following is a substrate for aldolase activity in glycolytic pathway?
    a. Glyceraldehyde-3-P b. Glucose-6-P
    c. Fructose-6-P d. 1,3-diphosphoglycerate
    e. Fructose-1,6-bi-P
17. The ratio that approximates the number of net molecule of ATP formed per mole of glucose oxidised in presence of O_2 to the net number formed in absence of O_2 is:
    a. 4:1 b. 10:2
    c. 12:1 d. 18:1
    e. 24:1
18. The "Primaquin sensitivity type" of haemolytic anaemia has been found to be related to reduced RB cells activity of which enzyme?
    a. Pyruvate kinase deficiency
    b. Glucose-6-phosphatase deficiency
    c. Glucose-6-P-dehydrogenase deficiency
    d. Hexokinase deficiency
    e. Phosphogluconate dehydrogenase deficiency.
19. Which of the following hormones is not involved in carbohydrate metabolism:
    a. Cortisol b. ACTH
    c. Glucagon d. Vasopressin
    e. Growth hormone
20. Dehydrogenases involved in HMP shunt are specific for:
    a. $NADP^+$ b. NAD^+
    c. FAD d. FMN
    e. None of the above
21. Which of the following enzymes in glycolytic pathway is inhibited by fluoride?
    a. Glyceraldehyde 3-P-dehydrogenase
    b. Phosphoglycreate kinase
    c. Pyruvate kinase
    d. Enolase
    e. Aldolase

22. Out of 24 mols of ATP formed in citric acid cycle, two molecules of ATP can be formed at "substrate level", by which of the following reaction?
    a. Citric acid → to isocitric acid
    b. Isocitrate → to oxalosuccinate
    c. Malate → to oxaloacetate
    d. Succinic acid → to fumarate
    e. Succinyl-CoA → to Succinic acid
23. The fructokinase reaction in fructose metabolism produces which of the following intermediates?
    a. Fructose-6-P b. Fructose-1-P
    c. Fructose-1, 6-bi-P d. Glyceraldehyde-3-P
    e. Pyruvate
24. Which of the following statements regarding TCA cycle is true?
    a. It is an anaerobic process
    b. It occurs in cytosol
    c. It contains no intermediates
    d. It is amphibolic in nature
    e. It generates 10 molecules of ATP per cycle
25. An allosteric enzyme responsible for controlling the rate of TCA cycle is:
    a. Malate dehydrogenase b. Isocitrate dehydrogenase
    c. Fumarase d. Aconitase
    e. Succinate thiokinase
26. A liver biopsy from an infant with hepatomegaly, stunted growth, hypoglycaemia, lactic acidosis, hyperlipidaemia revealed accumulation of glycogen having ***normal structure.*** A possible diagnosis would be:
    a. Branching enzyme deficiency
    b. Acid maltase deficiency
    c. Liver phosphorylase deficiency
    d. Debranching enzyme deficiency
    e. Glucose-6-phosphatase deficiency
27. How may ATP molecules will be required for conversion of two molecules of lactic acid to glucose?
    a. Two b. Four
    c. Six d. Eight
    e. Ten
28. In Rapaport-Leubering shunt in erythrocytes, 2,3-biphosphoglycerate (2,3-BPG) is produced from which intermediate in glycolytic pathway?
    a. 3-phosphoglycerate b. 2-phosphoglycerate
    c. 1,3-biphosphoglycerate d. Glyceraldehyde-3-P
    e. Dihydoxyacetone-P
29. Which of the following enzyme is not involved in HMP shunt?
    a. Glyceraldehyde-3-P-dehydrogenase
    b. Glucose-6-P-dehydrogenase
    c. Transketolase
    d. Phosphogluconate dehydrogenase
    e. Transaldolase
30. Match the following:
    a. Gulucose-6-P-dehydrogenase
    b. Phosphopentose epimerase
    c. Transketolase
    d. Transaldolase
    e. Phosphopentose isomerase
    1. Transfers C-1, C-2 and C-3 of a ketopentose to C-1 of an aldopentose _______
    2. Convert xylulose-5-P to ribulose-5-P _______
    3. Converts ribulose-5-P to ribose-5-P _______
    4. Generates NADPH _______
    5. Catalyses formation of Sedoheptulose-7-p from P entose-P _______

Answers to MCQs

1. d	2. b	3. c	4. d	5. b	6. e
7. b	8. b	9. d	10. d	11. d	12. b
13. b	14. c	15. d	16. e	17. d	18. c
19. d	20. a	21. d	22. e	23. b	24. d
25. b	26. e	27. c	28. a	29. a	

30. 1-d; 2-b; 3-e; 4-a; 5-c.

SECTION FOUR

CHAPTER 24

DIGESTION AND ABSORPTION OF LIPIDS

Major Concepts

A. To study how the complex lipids present in foodstuffs are broken down to simpler forms in the GI tract.
B. To study how the simpler forms are absorbed from the GI tract.
C. To study the defects in digestion/and absorption of lipids.

Specific Objectives

A. 1. Study the principal lipids present in the foodstuffs, which we take in normal diet. List them.
2. Problems faced in digestion of lipids in GI tract and how it differs from carbohydrates.
3. Digestion of dietary lipids in mouth and stomach:
   - Study the role of *lingual lipase,* and *gastric lipase.*
   - Role of fat in stomach, delays the rate of emptying of stomach. What is the role of "enterogastrone", a GI hormone?
   - "Fats have a high satiety value"—Explain.
4. Digestion in duodenum and small intestine
   Major site of fat digestion is small intestine:
   - Due to presence of a powerful *lipase (steapsin)* in the pancreatic juice and
   - Presence of "bile salts"—that act as effective emulsifying agent for fats
   - Revise your knowledge regarding composition of pancreatic juice and learn the role of *secretin,* 'CCK-PZ' and hepatocrinin-GI hormones.
   - Revise knowledge regarding the composition of Bile. What are Bile Salts? Name them and learn the functions of Bile Salts.
   - Learn in detail about **emulsification of fats**, what are **Micelles**? How they are formed?
   - Study the lipolytic enzymes present in the pancreatic juice. List them and learn their pH range, optimum pH range, optimum pH for activity, activators if any, mode of action, substrates and products formed.

B. Absorption of Lipids
1. Study of action of pancreatic lipase on TG (triacyl glycerol) and what products are formed by hydrolysis of TG in intestinal lumen.
   - What is the function of *isomerase* present in intestinal juice?
   - List the products formed from TG in the small intestine by the action of pancreatic *lipase* and *isomerase* and precentage of absorption of these hydrolytic products.
   - Learn the mechanism of absorption of these products.
2. Study in details what happens to these hydrolytic products in intestinal epithelial cell after absorption from the gut lumen and how TG is again re-synthesised in intestinal epithelial cell.
   - Draw a diagram to illustrate what happens to TG in intestinal lumen.
   - Draw a diagram to illustrate how TG is resynthesised in the intestinal epithelial cell.
3. Resynthesised TG in intestinal epithelial cells being insoluble in water is first converted into soluble form by layering of PL, cholesterol and cholesterol esters and **addition of a specific apo-protein (apo-B_{48}) forming a lipoprotein complex called "chylomicrons",** which are carried through lacteals to thoracic duct then to systemic circulation.
   - Study the composition and size of chylomicrons.
   - How "nascent" chylomicrons differ from "circulating" chylomicrons?
   - Study the fate of chylomicrons in systemic circulation.
4. Learn what happens to dietary cholesterol and PL.

C. Study the defects in digestion and absorption of lipids
- Steatorrhoea
- Chyluria
- Chylothorax

Learn role of medium chain triglycerides in chyluria and chylothorax.

DIGESTION OF LIPIDS

INTRODUCTION

The digestion of fats and other lipids **poses a special problem because of *(a)* the insolubility of fats in water, and *(b)* because lipolytic enzymes, like other enzymes, are soluble in an aqueous medium.** The above problem is solved in the gut by emulsification of fats, particularly by bile salts, present in bile and PL. The breaking of large fat particles or oil globules, into smaller fine particles by emulsification increases the surface exposed to interaction with *Lipases* and thus, the rate of digestion is proportionally increased.

Phases of Digestion and Absorption

The whole process of digestion of dietary lipids and its subsequent absorption may be ***arbitrarily divided*** into ***three phases:***

A. ***Preparatory phase:*** Includes the digestion of lipids in the intestine. The large lipid particles are broken down into smaller particles with the help of lipolytic enzymes.
B. ***Transport phase:*** Includes the transport of digested fats across the membrane of intestinal villous layer into intestinal epithelial cells.
C. ***Transportation phase:*** Includes the events of action that take place inside intestinal epithelial cells and its passage through lacteals to lymph/or in portal blood.

Dietary Sources of Lipids

The chief dietary sources of lipids in human beings:

- ***Animal source:*** Dairy products like milk, butter, ghee, etc. meat and fish, especially pork, eggs.
- ***Vegetable source:*** Various cooking oils from various seeds, viz. sunflower oil, groundnut oil, cotton seed oil, mustard oil, etc. and fats from other vegetable sources.

However, unlike proteins, **vegetable fats are superior to animal fats because:**

- They contain more of polyunsaturated fatty acids
- Vegetable fats are less likely to go rancid due to presence of antioxidants.

A. PREPARATORY PHASE

1. **Digestion in mouth and stomach:** It was believed earlier that little or no fat digestion takes place in the mouth. Recently a *lipase* has been detected called ***lingual lipase*** which is secreted by the dorsal surface of the tongue (Ebner's gland).

Lingual lipase: The pH of activity is 2.0 to 7.5 (optimal pH value is 4.0 to 4.5). *Lingual lipase* activity is continued in the stomach also where the pH value is low. Due to retention of food bolus for 2 to 3 hours, about 30 per cent of dietary triacyl glycerol (TG) may be digested. *Lingual lipase* is more active on TG having shorter FA chains and is found to be more specific for ester linkage at 3-position rather than position-1. Milk fats contain short and medium chain FA which tend to be esterified in the 3-position. Hence, ***milk fat appears to be the best substrate for this enzyme.*** The released short chain fatty acids are relatively more soluble and hydrophilic and can be absorbed directly from the stomach wall and enter the portal vein.

Gastric lipase: There is evidence of presence of small amounts of *gastric lipase* in gastric secretion. The overall digestion of fats, brought about by gastric lipase is negligible because:

- No emulsification of fats takes place in stomach
- The enzyme secreted in small quantity
- pH of gastric juice is not conducive which is highly acidic, whereas gastric lipase activity is more effective at relatively alkaline pH (average pH 7.8). ***Gastric lipase activity requires presence of Ca^{++}.*** Activity of gastric lipase is seen when intestinal contents are regurgitated into the gastric lumen. Recent studies have shown that gastric lipase is not capable of hydrolysing fats containing long-chain FA. Whatever minimal action of gastric lipase is there, it is confined to highly emulsified fats viz. those of milk fats or fats present in egg-yolk or fats with short-chain fatty acids, as these are somewhat relatively more soluble and hydrophilic.

Role of fat in stomach: Fats do play one important role in the stomach in that they ***delay the rate of emptying of stomach***, presumably by way of the hormone ***enterogastrone, which inhibits gastric motility and retards the discharge of bolus of food from the stomach.*** Thus fats have a high **satiety value.**

2. **Digestion in small intestine:** ***The major site of fat digestion is the small intestine.*** This is due to the presence of a powerful lipase ***(steapsin)*** in the pancreatic juice and presence of bile salts, which acts as an effective emulsifying agent for fats. Pancreatic juice and bile enter the upper small intestine, the duodenum, by way of the pancreatic and bile ducts respectively. Secretion of pancreatic juice is stimulated by the:

- Passage of an acid gastric contents (acid chyme) into the duodenum
- By secretion of the GI hormones, ***secretin*** and ***CCK-PZ.***
  – ***Secretin:*** Increases the secretion of electrolytes and fluid components of pancreatic juice.

SECTION FOUR

- *Pancreozymin* of CCK-PZ: Stimulates the secretion of the pancreatic enzymes.
- *Cholecystokinin* of CCK-PZ: Causes contraction of the gallbladder and discharge the bile into the duodenum. Discharge of bile is also stimulated by *secretin* and bile salts themselves.
- *Hepatocrinin:* Released by the intestinal mucosa stimulates more bile formation which is relatively poor in the bile salt content.

The above sequence of events prepares the small intestine for the digestion of fats.

Pancreatic juice has been shown to contain a number of lipolytic enzymes:

1. ***Pancreatic lipase (steapsin)***
2. ***Phospholipase A_2 (Lecithinase)***
3. ***Cholesterol esterase.***

The pancreatic lipase is the most important which hydrolyses TG containing short-chain FA as well as long-chain FA. Other two enzymes are required for phospholipids and cholesterol respectively.

Pancreatic lipase (steapsin): It is an *esterase* with optimum pH value of 6.

1. *Role of bile salts in pancreatic lipase activity:* Bile salts are required for proper functioning of the enzyme.

(a) ***Bile salts help in combination of 'lipase' with two molecules of a small protein called as colipase*** (mol wt = 10,000) in the intestinal lumen. **This combination of *lipase* with colipase has two effects:**

- Enhances the lipase activity of the intestinal pH.
- Also protects the enzyme against inhibitory effects of bile salts and against surface denaturation.

(b) ***Bile salts also help in emulsification of fats.***

2. *Role of Ca^{++}:* In the presence of Ca^{++} in the intestine, theFFA are immediately precipitated as 'soaps' (insoluble Ca-soaps) and are thereby prevented from inhibiting further lipase action. Thus, **Ca^{++} facilitates lipase action.**

Mode of action of pancreatic lipase: The complete hydrolysis of fats (TG) produces glycerol and FA. ***Pancreatic lipase is virtually specific for the hydrolysis of "primary ester linkage". It cannot readily hydrolyse the ester linkage of position-2 (β),*** if it does so it occurs at a very slow rate. Digestion of TG molcule by pancreatic lipase proceeds

- First by removal of a terminal FA to produce an **α, β-diglyceride**
- The other terminal FA is then removed to produce a **β-monoglyceride.**

Since the last FA at position (Sn-2) is linked by a secondary ester group and as it cannot be hydrolysed easily by pancreatic lipase, ***the β-monoglyceride is first converted to α-monoglyceride by isomerisation by an "isomerase" enzyme.*** Then the α-monoglyceride is hydrolysed by pancreatic *lipase.*

Sequence of events that occurs in the intestinal lumen is shown schematically in Figure 24.1.

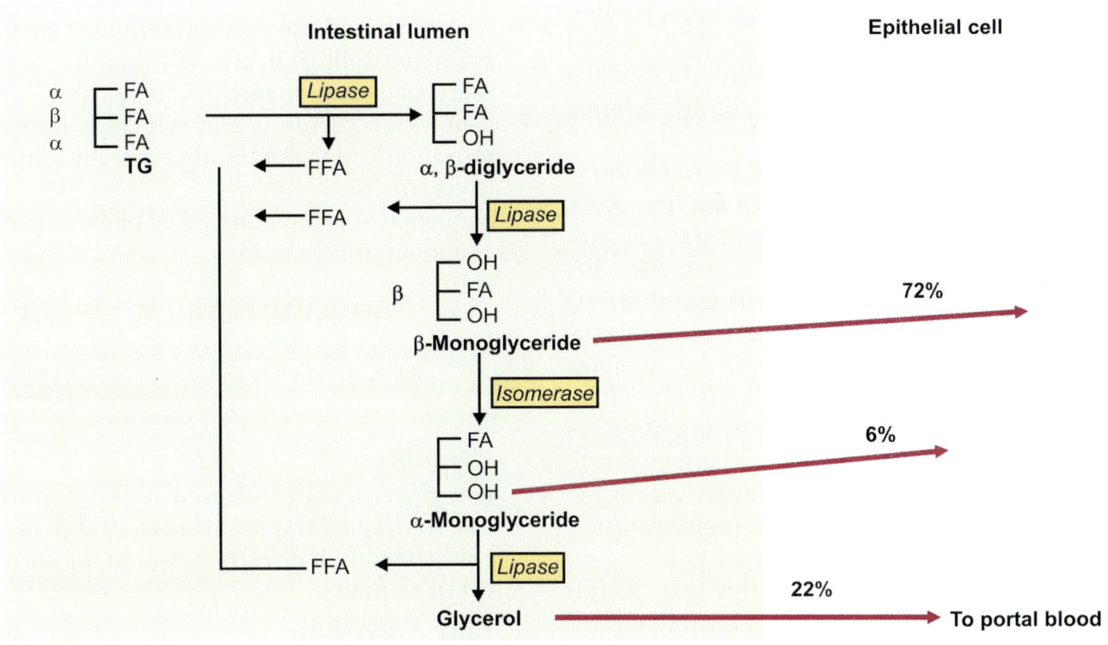

Fig. 24.1: Digestion in intestinal lumen

As it is seen from above, by the action of pancreatic lipase and isomerase, α- and β-monoglycerides are the major end-products of fat digestion, and less than 1/4 of the ingested fat is (25% or less) completely broken down to glycerol and FA.

Note: Recently an enzyme that can hydrolyse FA esterified at β-position (2-position) of a TG, has been found in rat pancreatic juice. This enzyme is not pancreatic lipase, it is a ***sterol-ester hydrolase.***

"Micelle" formation: Mono and diglycerides, along with bile salts, play an important role in stabilising and further increasing the emulsification of fat in the small intestine. Higher FA, mono and diglycerides are relatively insoluble in water and their absorption is largely aided by the **hydrotropic action** of bile salts. Bile salts and soaps formed in the intestinal lumen and bicarbonates of pancreatic and intestinal juices, collect the molecules of higher FA, mono and diglycerides, lecithins, cholesterol, etc. in the form of **"water-soluble molecular aggregates" called mixed "micelles",** which are much smaller than the droplets of emulsified fats (size 0.1 to 0.5 μ in diameter) and are absorbed mainly from duodenum and jejunum.

Bile salts of the "micelles" are not absorbed at this point but are redissolved in other emulsoid particles. They are reabsorbed later in the lower part of the small intestine and return to the liver via the portal vein for resecretion into the bile. This is known as ***enterohepatic circulation of bile salts.***

B. TRANSPORT PHASE

Current evidences, based on electron microscopic studies, indicate that the products of fat digestion, FFA, α- and β-monoglycerides mainly enter the microvilli and the apical pole of absorptive epithelial cells by ***"simple diffusion" through the cell-membrane.***

Short and medium chain FA (6 to 10 C) and unsaturated FA are more readily absorbed than the ling-chain FA (12 to 18 C). Also the short chain FA appear to enhance the absorption of fats in general, whereas long-chain FA tend to impair the process. The products of digestion next appear to be taken up by the smooth endoplasmic reticulum and ***resynthesised into TG again by enzymes present in the membrane and/or cavities of the reticulum.*** The rapid removal of FA and α and β-monoglycerides and their synthesis into TG in intestinal epithelial cell, maintains a sharp gradient of concentration within the mucosal cell that favours the continued rapid diffusion into the cell from the intestinal lumen. There is a merging of the smooth endoplasmic reticulum into rough endoplasmic reticulum, in which probably enzymes for TG resynthesis are formed as well as the protein component (apo-B_{48}) of lipoprotein complex, **chylomicron.**

Note: Pinocytosis does not appear to play a significant role in fat absorption as was formerly believed; probably less than 5 per cent absorbed in emulsified forms may be absorbed by pinocytosis.

C. TRANSPORTATION PHASE

Sequence of events that occurs inside the intestinal mucosal cell are:

Within the intestinal epithelial cell, α-monoglycerides (6%) are further hydrolysed by *intestinal lipase* to produce free FA and glycerol.

Intestinal lipase: A lipase distinct from that of the pancreatic lipase is present in the intestinal mucosal cell. Principal action of this enzyme is confined within the epithelial cell.

- FA absorbed from intestinal lumen and FA formed from hydrolysis of α-monoglycerides, are activated to "Acyl-CoA". An ATP-dependant *thiokinase* has been shown to be present in the mucosal cells of the intestine.

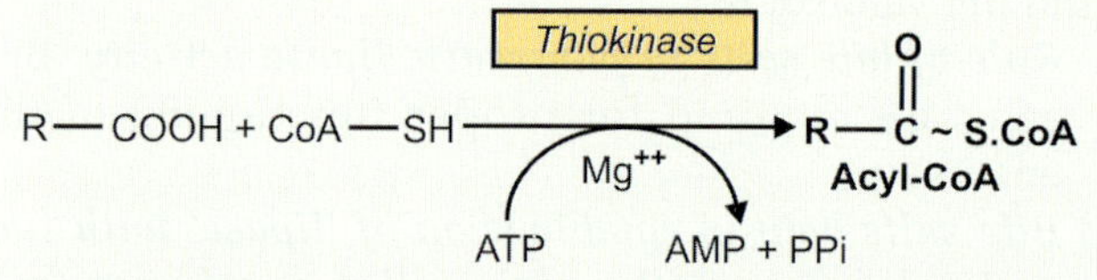

- ***Note that glycerol released within the intestinal wall cells are reutilised for TG resynthesis.*** Glycerol is converted to α-Glycerol-(P) by *glycerokinase* in presence of ATP. Some amount of α-Glycero-(P) can be contributed from glycolysis operating in intestinal epithelial cell. α-glycero (P) thus formed combines with 'acyl-CoA' to form TG molecule.
- β-monoglyceride (72%) which is absorbed from intestinal lumen can combine directly with 'acyl-CoA' to reform TG.

Sequence of events that takes place in resynthesis of TG is shown in Figure 24.2.

ABSORPTION OF LIPIDS

ABSORPTION OF RESYNTHESISED TG AND OTHER PRODUCTS

1. **Glycerol**

   Free glycerol (22%) released in the intestinal lumen is ***not utilised for resynthesis of TG*** in intestinal epithelial cell. ***It directly passes to the portal vein and taken to liver.***

2. **Fatty Acids (FA)**

   Behaviour of absorption of FA differs according to the carbon contents.

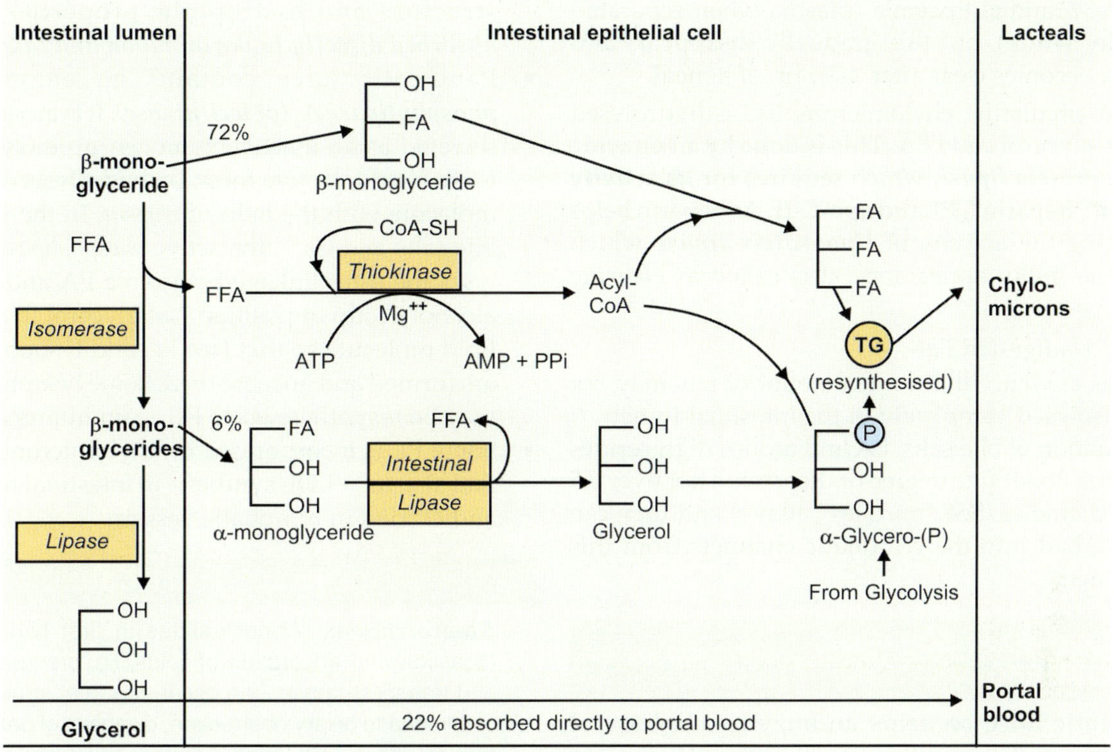

Fig. 24.2: Resynthesis of TG in intestinal epithelial cells

- Short-chain FA and medium-chain FA (less than 8 to 10 C) and unsaturated FA are ***absorbed to portal blood directly*** and taken to Liver.
- Normally, all FFA present in the intestinal wall are ultimately reincorporated in TG after activation.
- Some of absorbed FA, more than 10 C atoms in length, irrespective of the form, in which they are absorbed, are found as "esterified FA" ***in the lymph of the lacteals, which passes through thoracic duct to systemic circulation.***

3. ***Fate of resynthesised TG:*** Re-esterification to form TG molecule takes place in the cisternae of endoplasmic reticulum of the mucosal cells. Resynthesised TG cannot pass to lymphatics (lacteals) nor to portal blood as ***it is insoluble in water (hydrophobic).*** Hence, **it is converted to lipoprotein complex called *chylomicrons.*** Each droplet of *hydrophobic* and water insoluble TG gets covered with a layer of *hydrophilic* PL, cholesterol/cholesterol esters and an apoprotein called **apo-B_{48}.** Addition of the 'polar' ions make it relatively soluble and hydrophilic.
   - ***Chylomicrons:*** They are synthesised in the intestinal wall. Size of chylomicrons range from 0.075 to 1 mμ (average 0.5 mμ in diameter). It is composed largely of TG, cholesterol (both 'free' and esterified), PL and a small per cent less than 2 per cent, of specific protein, called **apo-B_{48}.**
   - ***Average composition of chylomicron molecule*** is:

     - **TG: 87 to 88 per cent**
     - **PL: Approximately 8 per cent**
     - **Free and esterified cholesterol: Approximately 3 per cent and**
     - **Specific "apo-β_{48}" protein: 0.05 to 2 per cent.**

Administration of inhibitor of protein synthesis, e.g. puromycin to rats prevents the formation of chylomicrons and results in accumulation of fats in intestinal epithelial cells.

Chylomicrons pass out through the cell membrane of bases and lateral walls of intestinal epithelial cells and moves through extracellular spaces between those cells to enter lymphatic vessels of abdominal region and later goes to systemic circulation through the thoracic duct.

- ***Difference between nascent chylomicrons and "circulating" chylomicrons in systemic circulation:*** 'Nascent', chylomicrons have 'apo-B_{48}' only while "circulating" chylomicrons contain in addition **apo-C, *which is derived from circulating HDL.***

- ***Fate of circulating chylomicrons:*** Presence of chylomicrons in circulating blood after a meal accounts for the post prandial lipaemia (plasma when separated is **milky white**), but this gradually disappears and plasma becomes clear after 2-2½ hr. of a meal.

Reason: In circulating chylomicrons, TG is hydrolysed to produce glycerol and FFA. **This is done by an enzyme called *lipoprotein lipase,* which requires for its activity presence of "heparin", PL and apo-C-II.** As heparin helps in increasing the activity of ***lipoprotein lipase*** which removes the milky appearance, it is called as ***clearing factor.***

4. ***Fate of Undigested Fat***

There is evidence that some amount of fats may not be hydrolysed completely in the intestinal lumen. A combination of bile salts, FA and mono/diglycerides can bring about fine degree of dispersion not over 0.5 mμ in diameter (See "micelles" above) and they can be absorbed into the lymphatic channels from this dispersion.

DIGESTION AND ABSORPTION OF CHOLESTEROL

- Pancreatic juice contains an enzyme ***cholesterol esterase,*** which may either catalyse the esterification of free cholesterol with FA or it may also catalyse the opposite reaction, i.e. hydrolysis of cholesterol esters. In the intestinal lumen, depending on the equilibrium, the "cholesterol-esters" are hydrolysed by this enzyme.
- Thus, cholesterol appears to be ***absorbed from the intestine almost entirely in "free" (unesterified) form.***
- Nevertheless, 85 to 90 per cent of the cholesterol in the lymph is found to be in esterified form, indicating that esterification of cholesterol, like that of FFA, must take place within the intestinal mucosal cell.
- Absorption of cholesterol has been reported to be facilitated by presence of unsaturated FA and bile salts are necessary for the absorption of cholesterol.

Note: Certain plant sterols like sitosterol and stigmasterol are not absorbed, rather their presence can inhibit cholesterol absorption.

DIGESTION AND ABSORPTION OF PHOSPHOLIPIDS

- Dietary phospholipids may be absorbed from intestine without any digestion. Due to its polar structure and hydrophilic properties, they are ***absorbed directly to portal blood and taken to Liver.***
- Pancreatic juice contains an enzyme called ***phospholipase A_2 (or lecithinase).*** It is an ***esterase,*** and secreted as an inactive zymogen **proenzyme**, which is changed to active form, by hydrolysis of a peptide molecule with the help of *trypsin.* In the presence of bile salts and Ca^{++}, the active *phospholipase A_2* hydrolyses the ester linkage between a FA and secondary alcohol group of position 2 of glycerol in a phospholipid molecule so that free FA and lysophospholipid are formed and are absorbed. Some lysophospholipid may be resynthesised to PL again in mucosal cell.
- Some PL is incorporated in 'chylomicrons' synthesis and also for VLDL synthesis in intestinal mucosal cell and carried in lymphatic vessels.

CLINICAL SIGNIFICANCE

- ***Steatorrhoeas:*** Abnormalities in first two phases of digestion and absorption of lipids, i.e. preparatory phase and transport phase, can occur due to obstruction to flow of bile due to biliary obstruction, diseases of pancreas, and tropical sprue. This leads to increased fat content of faeces causing "steatorrhoea". Normally less than 5 per cent of ingested fat is excreted in faeces.
- ***Obstruction to Transportation Phase in Lacteals:*** It may lead to chylous abdomen, chylothorax and chyluria.

1. ***Chyluria***

It is an abnormality in which the patient excretes ***milky urine*** because of the presence of abnormal connection between urinary tract and lymphatic drainage system of the intestine, forming a so-called **chylous fistula.**

2. ***Chylothorax***

A similar abnormality, an abnormal connection between pleural space and the lymphatic drainage of small intestine results in accumulation of lymph in pleural cavity giving ***milky pleural effusion.***

Hashim and colleagues studied the effects of feeding short and medium chain triglycerides less than 10 C and they found remarkable improvement in that the milky appearance disappeared, chyluria/chylothorax improved. This is due to the fact that short and medium chain TG can be absorbed directly to portal blood system, and thus bypassing the lacteals.

- ***Defect in Chylomicrons Synthesis from Resynthesised TG:*** This defect due to lack of 'apo-B_{48}', TG may accumulate in intestinal epithelial cells.

SECTION FOUR

I. Questions (Essay type and short notes)

1. Describe the digestion and absorption of lipids with special reference to neutral fats.
2. Describe the mode of absorption, transport and utilisation of fats (triglycerides)

Short Notes

a. Absorption of lipids
b. Absorption of fats from GI tract
c. Absorption of fats
d. Lingual lipase
e. Chylomicrons

II. MCQs (Give one correct answer)

1. "Micelles" formed in the intestine have which of the following properties?
   a. They are absorbed intact in the duodenun
   b. They contain mostly triacylglycerol (TG)
   c. They facilitate transfer of hydrolytic products of TG to intestinal mucosal cell
   d. They are secreted by cells lining the intestine
   e. They are mainly absorbed from stomach wall
2. Dietary TGs (triacylglycerols) are absorbed from the intestinal lumen after hydrolysis mainly as:
   a. Fatty acids and glycerol
   b. Fatty acids and two monoacylglycerol
   c. Fatty acids and α, β-diglyceride
   d. Acyl-CoA and glycerol
   e. Glycerol-phosphate and fatty acids
3. The gastric lipase is inactive because of the following reasons except:
   a. No emulsifications of fats take place in stomach
   b. The enzyme is secreted in small quantity
   c. pH of gastric juice is not conducive for its action
   d. The gastric juice is highly acidic
   e. The conc. of bile salts is high in gastric juice
4. Delay in the rate of gastric emptying is due to the action of:
   a. Pancreozymin b. Enterokinase
   c. Enterogastrone d. CCK
   e. Gastric lipase
5. Ca^{++} facilitates action of lipase:
   a. By emulsification
   b. By inhibiting emulsification
   c. By soap formation
   d. By formation of micelle
   e. By formation of tiny droplets

Answers to MCQs

1. c 2. b 3. e 4. b 5. c.

CHAPTER 25

METABOLISM OF LIPIDS

PART I

Major Concepts

A. Study the lipids present in plasma.
B. Study the Metabolism of adipose tissue and role of 'Brown' adipose tissue.

Specific Objectives

A. Plasma Lipids and Lipoproteins
   - Study the composition of plasma lipids
   - Study how the plasma lipids are transported in blood
   - Learn the methods by which plasma lipids can be separated: (1) Ultracentrifugation and (2) Electrophoresis
   - Define lipoproteins. Learn different types, their sources and composition
   - Study how free fatty acids are carried.

B. Metabolism of Adipose Tissue
   1. Lipids in the body physiologically exist in two forms (a) Element 'constant' or 'structural' lipids and (b) Element 'variable', stored lipid (Depot fats)
      - Learn the composition of element constant and its distribution
      - Learn the composition of element variable and its distribution, source and functions.
   2. TG stores continually undergo (a) Synthesis (re-esterification) and (b) Breakdown (lipolysis).
      Both processes are different pathways. Nutritional, metabolic and hormonal factors regulate either of these two mechanisms. Resultant of these two processes determines the level of free FA in plasma.
      (a)
      - Learn the process of re-esterification (Synthesis of TG)
      - Name the substrates required for the synthesis, study their sources
      - Learn the source of Acyl-CoA required for the synthesis.
      - Learn the source of α-Glycerophosphate. Is there any role of "Glycerokinase"? If not, what is the source of α-Glycero-P in adipose tissue.

      (b)
      - Learn the process of breakdown (Lipolysis). What are the products of lipolysis?
      - Study the fate of FA and glycerol after lipolysis
      - Study the enzymes involved in lipolysis: 'Hormone sensitive' 'triacyl glycerol lipase', "Hormone independent" diacyl glycerol lipase and monoacyl glycerol lipase, and Lipoprotein lipase.

      (c) Effect of Glucose
      - Study the effect of glucose on adipose tissue metabolism
      - Study the adipose tissue metabolism in diabetes mellitus and in starvation.
   3. Study the influence of various hormones on adipose tissue
      - List the hormones that help re-esterification and study their mechanism of action in detail
      - List the hormones that help lipolysis and study their mechanism of action in detail
      - Draw a neat diagram showing the mechanism of action of these hormones
      - Study the role of fat mobilising substance isolated from pituitary gland.
   4. Assimilation of TG FA by adipose tissue:
      Major chemical form in which plasma lipids interact with adipose tissue is TG of circulating chylomicrons and 'VLDL'
      - Study in detail the role of ***Lipoprotein lipase,*** its location, its action on TG of chylomicrons and VLDL, cofactors required for the activity of the enzyme.
   5. Brown adipose tissue and its role in thermogenesis
      - Study the location of brown adipose tissue and its purpose
      - What are the salient characteristics and how does it differ from normal "White adipose tissue"?
      - Study its role in heat generation (thermogenesis) and its mechanism.

PLASMA LIPIDS

1. **In mammals, principal Lipids that have metabolic significance** are as follows:
   - **Triacyl glycerol (TG):** Also called **Neutral fats (NF)**
   - **Phospholipids** and
   - **Steroids: Chief of which is cholesterol.**

   Plasma lipids also constitute the products of the metabolism:
   - **Fatty acids:** Long-chain and short-chain (free FA) and
   - **Glycerol.**
2. Extraction of plasma lipids with a suitable Lipid solvent and subsequent separation of the extract into various classes of lipids shows the presence of:
   - Triacyl glycerol (TG), Phospholipids (PL), Cholesterol and } approximately in equal quantities
   - A much smaller fraction of non-esterified long-chain fatty acid (NEFA) or free-fatty acid (FFA), which constitutes less than 5 per cent of total FA present in plasma.
3. *NEFA is now known to be metabolically most active of the plasma lipids* and 1/2 life being approximately 2 to 3 minutes.
4. Plasma lipids at any time may be considered to represent the net balance between production, utilisation and storage.
5. Lipids of the blood plasma in humans are as follows (Table 25.1):

Table 25.1: Lipid fractions in plasma

Lipid fraction	*Plasma level in mg/100 ml*	
	Range	*Mean*
Total lipids	360-820	560
Triacyl glycerol (TG)	80-180	150
Total phospholipids (PL)	125-390	210
• **Phosphatidyl choline (lecithin)**	50-200	
• **Phosphatidyl ethanolamine (cephalin)**	50-130	
• **Sphingomyelins**	15-35	
Total cholesterol	150-250	200
Free cholesterol (nonesterified)	25-105	55
Free fatty acids or nonesterified FA (NEFA)	6-16	10

TRANSPORTATION OF PLASMA LIPIDS

Principal lipid, triacyl glycerol (TG), is hydrophobic material. To transport them in blood in an aqueous medium poses a problem, which is solved by associating the more insoluble lipids with more "Polar" ones, such as phospholipids, cholesterol and combining with a specific, protein molecule (called as 'apo-proteins'). Thus, the hydrophobic and insoluble triacyl glycerol (TG) is converted by above combination into a hydrophilic and "soluble" Lipoprotein "complex".

Thus:

- TG derived from intestinal absorption of fats are transported in the blood as a lipoprotein complex called *chylomicrons. Chylomicrons are small microscopic particles of fats, about 1μ in diameter and are responsible for transport of exogenous (TG) in the blood.*
- Similarly, TG that are synthesised in Liver cells are converted to lipoprotein particles, called **very low density lipoproteins** (VLDL) and thrown into the circulation. *VLDL is mainly concerned with transport of endogenous TG.*

In addition to above:

- Fatty acids released from adipose tissue by hydrolysis of TG are thrown in the circulation as free fatty acid (FFA). They are carried in non-esterified state in plasma, hence also called NEFA. In circulation, FFA/NEFA combines with albumin and are carried as **albumin-FFA complex**. Some 25 to 30 mols of FFA are present in combination with one mol. of albumin.

SEPARATION OF PLASMA LIPIDS

(a) Ultracentrifugation

Pure fat is less dense than water. As the proportion of lipid to protein in lipoprotein complex increases, the density of the molecule decreases. This property has been utilised in separation of plasma lipids, the various lipoprotein fractions, by ultracentrifugation.

Sf unit: The rate at which each lipoprotein fraction floats up through a solution of NaCl (Sp. gr = 1.063) is expressed in **Svedberg units** (Sf units). *One Sf unit is equal to 10^{-13} cm/sec/dyne/gm at 26°C.*

Because of the low densities, the chylomicrons and VLDL rise to the surface most rapidly upon ultracentrifugation and hence have relatively high Sf values. Composition of various Lipoprotein fractions separated by ultracentrifugation is given in Table 25.2.

(b) Electrophoresis

Lipoproteins may be separated also according to their electrophoretic properties and identified more accurately using immunoelectrophoresis. **Fredrickson and others (1967)** identified lipoproteins into **4 groups** by electrophoresis as follows:

- **HDL:** Moves fastest and occupies position of α globulin-called α *lipoproteins*
- **LDL:** β-**lipoproteins**
- **VLDL:** *(Pre-β or α_2 lipoproteins)* and
- **Chylomicrons:** Slowest moving and remains near the origin.

SECTION FOUR

Table 25.2: Plasma lipoproteins

Name of Lipoproteins (ultracentrifugation)	Source	Electrophoretic fraction	Density	Sf value	Diameter (nm)	Proteins %	Total Lipids %	Percentage composition % of total lipids: TG	PL	Ch-ester	Free cholesterol
• Chylomicrons	Intestine	origin	< 0.96	> 400	100-1000	1 to 2	98 to 99	88	8	3	1
• VLDL	Liver (Main source) Also intestine	pre-β	0.96 to 1.006	20 to 400	30-90	7 to 10	90 to 93	56	20	15	8
• LDL	Degradation of VLDL	β	1.019 to 1.063	2 to 12	20-25	21	79	13	28	48	10
• HDL	Liver (Main source) Also intestine	α	1.063 to 1.210	—	10-20	33 to 57	43 to 67	13 to 16	43 to 46	29 to 31	6 to10

VLDL	=	Very low density lipoproteins.
LDL	=	Low density lipoproteins.
HDL	=	High density lipoproteins.
TG	=	Triacyl glycerol.
PL	=	Phospholipids.

Composition of lipoprotein complexes and apoproteins (Refer Lipoproteins Metabolism).

Table 25.3 shows the normal value of Lipoprotein fractions in health.

METABOLISM OF ADIPOSE TISSUE

The lipids in the body physiologically exist in **two forms:**

- *"Element constant" or structural lipids* and
- *"Element variable": stored lipids (Depot fats).*

Although a sharpline of demarcation cannot be made between the two, it has been generally observed that the value of the former remains constant even under extremes of starvation, whereas the latter varies.

Composition of Element Constant

Cytoplasm and cell membranes of all organs are composed of ***element constant***, so that their ***fat content does not diminish in starvation.*** Element constant is composed chiefly of Phospholipids (PL), along with smaller amounts of other lipids, including cholesterol. It is independent of previous feeding. It remains an integral part of cell protoplasm and is essential for its life.

Composition of Element Variable

The lipids which are stored in the body in excess of above. The amount fluctuates and it is composed mainly of triacyl glycerol (TG), also called as neutral fats (NF). Thus, depot fat is chiefly composed of glycerides of various fatty acids and usually contains 75 per cent of oleic acid, 20 per cent of palmitic acid and 5 per cent of stearic acid. Traces of lecithin and cholesterol as well as a little amount of Polyunsaturated FA are also present. The depot fat is called "adipose tissue", they are intracellular fats which remain inside the cells of adipose tissue.

Table 25.3: Normal values of lipoproteins (Normal lipid profile)

Lipid fraction	Normal values
• **Total cholesterol**	150 to 240 mg/dl
• **Serum HDL—cholesterol**	Males—35 to 60 mg/dl Female—40 to 70 gm/dl
• **Serum TG (Triacyl glycerol)**	Males—60 to 165 mg/dl Females—40 to 140 mg/dl
• **Serum chylomicrons**	Up to 28 mg/dl (14 hrs. Post-absorptive state)
• **Serum pre-β-lipoproteins (VLDL)**	Males—up to 240 mg/dl Females—up to 210 mg/dl
• **Serum β-lipoproteins (LDL)**	Up to 550 mg/dl
• ***Serum LDL—cholesterol**	Up to 190 mg/dl

***Serum LDL—cholesterol can be calculated by the Friedewald formula:**

- **LDL—cholesterol in mg/dl =**
  $$\text{Total cholesterol} - \text{HDL cholesterol} - \frac{TG}{5}$$
- **LDL—cholesterol in mmol/L =**
  $$\text{Total cholesterol} - \text{HDL cholesterol} - \frac{TG}{2.2}$$

Note: The formula is not much reliable at TG concentration >4.5 mmol/L (> 400 mg/dl).

Dynamic state of adipose tissue: Adipose tissue is not just a static lump of fats; it is in ***dynamic state;*** breakdown of fats and synthesis take place all the time.

METABOLISM

TG stores in the body is continually undergoing ***(a) Esterification (synthesis)*** and ***(b) Lipolysis (breakdown).***

These two processes are not the forward and reverse processes of the same reaction. They are entirely different

pathways involving different reactants and enzymes. Many of the nutritional, metabolic and hormonal factors regulate either of these two mechanisms, i.e. esterification and lipolysis. ***Resultant of these two processes determine the magnitude of free fatty acid pool in adipose tissue and this, in turn, will determine the level of free fatty acid (FFA) circulating in the blood.***

I. Esterification (Synthesis of TG)

In adipose tissue, for TG synthesis, ***two substrates are required:***

- **Acyl-CoA**
- **α-Glycero-P**

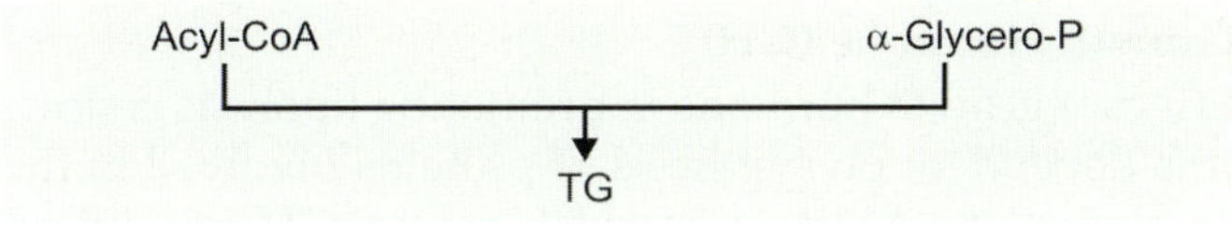

For detailed steps, see biosynthesis of TG.

1. Sources of Acyl-CoA

Sources of FFA in blood are:

- Dietary,
- Synthesis of FA (palmitic acid) from acetyl CoA *'de novo'* synthesis (extramitochondrial). Further elongation to form other fatty acids in microsomes.
- Acyl-CoA obtained from lipolysis taking place in adipose tissue **(FFA-Pool No. 1)**.
- FFA obtained from lipolysis of TG of circulating chylomicrons and VLDL by *lipoprotein lipase* enzyme present in capillary wall **(FFA-Pool No. 2)**, which are taken up by adipose tissue.

2. Source of α-Glycerol-P

Mainly two:

- Conversion of glycerol to α-Glycero-P by the enzyme *Glycerokinase* in presence of ATP.
- The other source is from glucose oxidation. Dihydroxyacetone-P is converted to α-Glycero-P.

The enzyme glycerokinase is practically absent in adipose tissue. If any glycerokinase is present, it has very low activity. **Hence, glycerol produced by lipolysis in adipose tissue cannot be utilised for provision of α-Glycero-P** and thus, **glycerol passes into the blood,** from where it is taken up by liver, kidney and other tissues which possess ***glycerokinase*** and is utilised for gluconeogenesis. ***Thus, for provision of α-Glycero-P in adipose tissue for TG synthesis, the tissue is dependent on a supply of glucose and glycolysis.***

II. Lipolysis (Breakdown of TG)

TG in adipose tissue undergoes hydrolysis by a ***hormone-sensitive TG lipase enzyme*** to form free fatty acids and glycerol.

Adipolytic lipases are three:

1. **Hormone sensitive** ***triacyl glycerol lipase:*** Key regulating enzyme.
2. **Two others are not hormone-sensitive:**
   - ***Diacyl glycerol lipase***
   - ***Monoacyl glycerol lipase.***

These lipases are distinct from *lipoprotein lipase* that catalyses lipoprotein TG (Present in chylomicrons and VLDL) hydrolysis before it is taken up by extrahepatic tissues.

The free fatty acids formed by lipolysis can be reconverted in the tissue to acyl-CoA by *Acyl-CoA synthase* and re-esterified with α-Glycero-P to form TG. Thus, there is a continuous cycle of lipolysis and reesterification within the tissue.

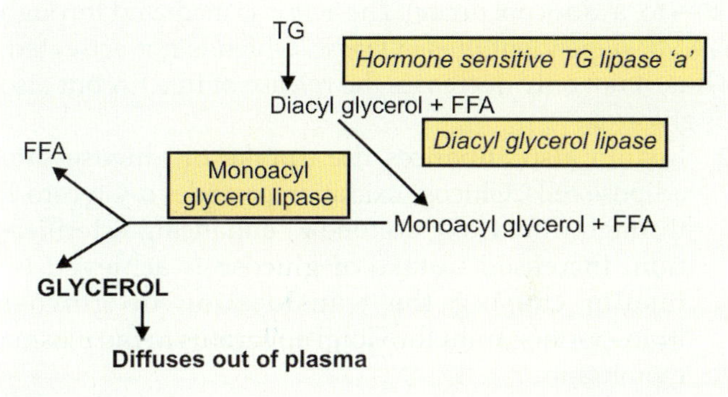

Note

When the rate of re-esterification is less than rate of lipolysis, FFA accumulates and diffuses into the plasma where it raises the level of FFA ↑ in plasma.

Effect of glucose: Under conditions of adequate nutritional intake or when utilisation of glucose by adipose tissue is increased, then more α-Glycero-P will be available. Re-esterification will be greater than lipolysis; as a result FFA outflow decreases and plasma FFA↓ level falls. *In vitro* studies have shown that release of glycerol continues; that means lipolysis continues. Hence, effect of glucose in reducing plasma FFA level is not mediated by decreasing the rate of lipolysis. ***It proves that the effect is due to provision of α-Glycero-P from glycolysis, which enhances esterification.***

Adipose tissue metabolism in diabetes mellitus and in starvation: In diabetes mellitus and in starvation, availability of glucose in adipose tissue is grossly reduced, resulting to lack of α-Glycero-P. Thus, rate of re-esterification is decreased↓. Lipolysis is greater than re-esterification, resulting to accumulation of FFA and increase in plasma FFA level.

INFLUENCE OF HORMONES ON ADIPOSE TISSUE

Rate of release of FFA from adipose tissue, is affected by many hormones which influence either the *(a)* rate of esterification or *(b)* the rate of lipolysis.

I. List of hormones that increase the rate of esterification:

- **Insulin** is the principal hormone
- **Prolactin** effective in large doses.

Insulin

Net result of insulin on adipose tissue is to inhibit the release of free FA from adipose tissue, which ***results in fall of circulating plasma FFA↓.***

1. The above is brought about by decreasing the level of cyclic AMP↓ in the cells. This is achieved by:
   - Inhibiting *adenyl cyclase* and
   - Increasing the *phosphodiesterase* activity.

   Lowered cyclic AMP level in the cell inhibits the activity of *hormonesensitive-TG lipase* (conversion from 'b'→ to 'a' does not occur). The action is mediated through c-AMP-dependent *protein kinase,* which is not activated. Thus, it not only decreases the release of free FA but also of glycerol.
2. Insulin also enhances the uptake of glucose into adipose cells. Glucose oxidation provides α-Glycero-P through dihydroxy-acetone-P, enhancing esterification. Increased uptake of glucose is achieved by insulin causing the translocation of glucose 'transporters' from the Golgi apparatus to the plasma membrane.
3. There is also increased FA synthesis as glucose is oxidised by HMP-shunt pathway and provides NADPH.
4. ***Effect on enzymes:*** Insulin increases the activity of *Pyruvate dehydrogenase, acetyl-CoA carboxylase,* and *Glycerol*-P-acyl transferase, which reinforce the effects arising from increased glucose uptake on the enhancement of FA synthesis and TG synthesis.
5. Insulin has also been shown to inhibit *Hormone-sensitive TG lipase 'a'* independent of c-AMP pathway.

Prolactin

Effect of prolactin is similar to insulin, provided it is given in larger doses.

II. *List of hormones that increases the rate of lipolysis:* **Catecholamines** (epinephrine and norepinephrine) are the principal hormones. Other lipolytic hormones are:

- **Glucagon,**
- **Growth hormone,**
- **Glucocorticoids**
- **ACTH, α and β MSH, TSH** and **Vasopressin.**

These hormones accelerate the release of FFA from adipose tissue and ***raise the plasma FFA ↑*** level by increasing the rate of lipolysis of TG stores. Most of them act by activating *adenyl cyclase,* thus increasing the cyclic AMP level in cells. Cyclic AMP in turn converts "inactive" cyclic AMP-dependent *Protein kinase* to the "active" form, which phosphorylates inactive hormone-sensitive *TG lipase* 'b' to active form 'a' and brings about lipolysis.

Note

- ***For an optimal effect most of these lipolytic process require the presence of glucocorticoids (GC) and "thyroid hormones" in minimal amounts.*** On their own, these hormones, i.e. GC and thyroid hormones do not increase the lipolysis markedly but act in a ***"facilitatory" or "permissive"*** capacity with other lipolytic endocrine hormones.
- Hormones that act rapidly in promoting lipolysis are the catecholamines, they stimulate the activity of *adenyl cyclase* and increases cyclic-AMP level (see above). ***Thyroid hormones in minimal amount are necessary for its full lipolytic activity.***

Growth Hormone (GH)

Effect of growth hormone in promoting lipolysis is slow. It is dependent on synthesis of proteins involved in the formation of c-AMP, i.e. *adenylate cyclase* (GTP is required for the synthesis).

Glucocorticoids (GC)

Net result is increase in plasma FFA ↑ level. This is achieved by:

- Glucocorticoids depresses uptake of glucose and there is less α-Glycero-P↓ available. Thus, decreases rate of esterification↓.
- Stimulates synthesis of *adenylate cyclase* thus increasing c-AMP level in the cells.
- Facilitates adipokinetic property of growth hormones, and
- Increases synthesis of new *lipase* protein by a c-AMP independent pathway.

All the above four mechanisms increase lipolysis, thus increasing the level of plasma FFA.

Conclusion: Human adipose tissue is unresponsive to most of the lipolytic hormones apart from the catecholamines. On consideration of the profound derangement of metabolism in diabetes mellitus, which is due mainly to increased release of FFA from the depots and the fact that insulin to large extent corrects the condition, it must be concluded that ***Insulin plays a prominent role in the regulation of adipose tissue metabolism.***

EFFECT OF DRUGS LIKE METHYL XANTHINES

Drugs like caffeine and theophylline:

- Inhibit *phosphodiesterase* enzyme thus decreases the catabolism of c-AMP and increases its cellular level. Increased c-AMP level in turn increases lipolysis ↑.
- Also inhibits *adenylate cyclase.*

Note: It is significant to note that drinking of coffee, containing caffeine causes marked and prolonged elevation of plasma FFA in humans.

Mechanism of action of the hormones is shown diagrammatically in Figure 25.1.

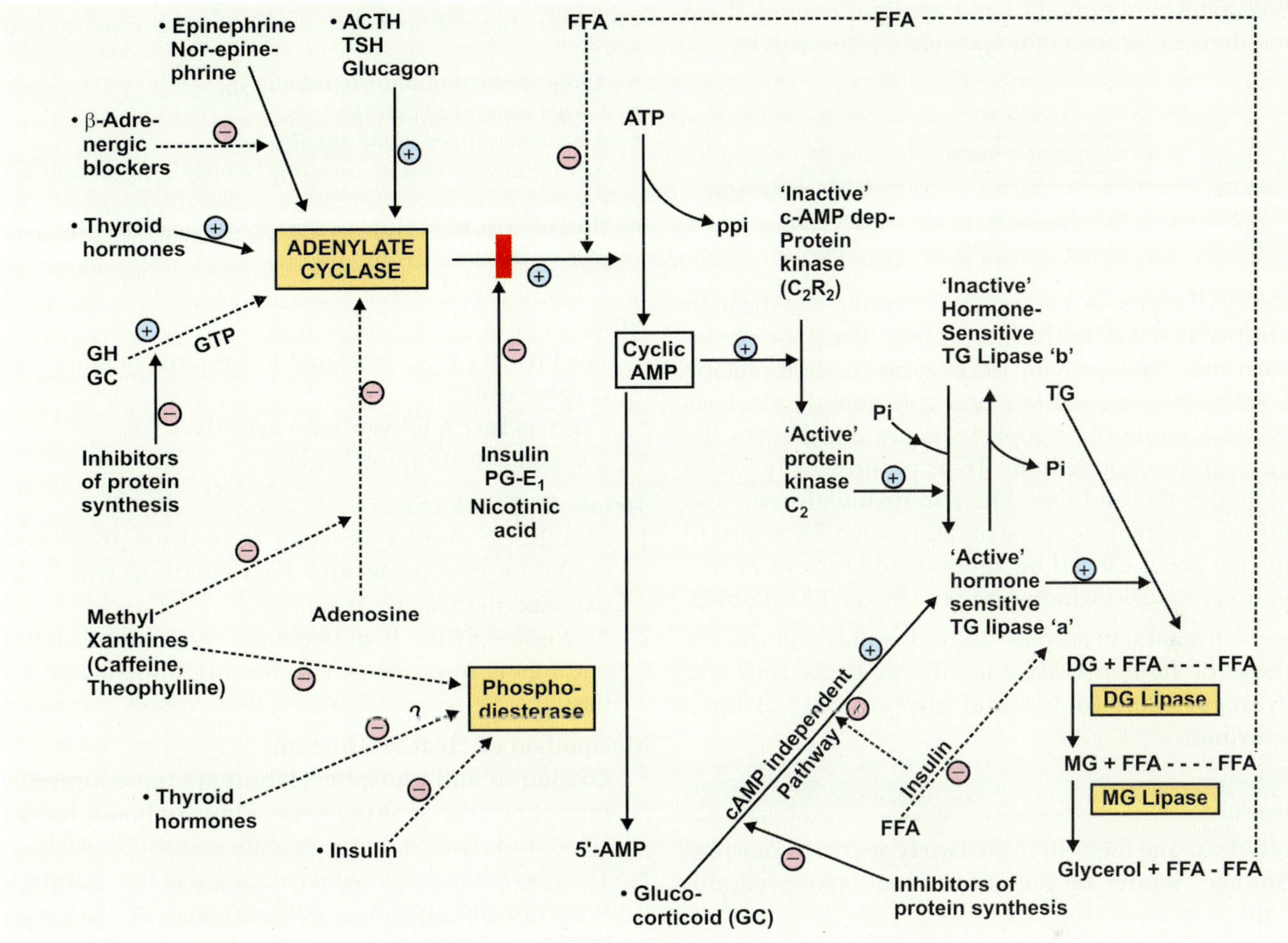

Fig. 25.1: Influence of hormones on adipose tissue metabolism

FAT MOBILISING SUBSTANCE: Besides the recognised hormones, certain other adipokinetic principles have been isolated from pituitary gland. A ***fat mobilising substance*** has been isolated from the urine of several fasting species including humans, provided the pituitary gland is intact. Exact nature of the substance yet to be elucidated.

ASSIMILATION OF TGFA BY ADIPOSE TISSUE: Major chemical forms in which plasma lipids interact with adipose tissue is TG.

- As 'chylomicrons' derived from intestinal absorption of fats.
- As "very low-density lipoprotein" complex (VLDL) by liver.

TG of circulating chylomicrons and VLDL is acted upon by the enzyme ***Lipoprotein lipase*** to hydrolyse TG to form FFA and glycerol. Significant correlation between ability of adipose tissue to incorporate TG FA and the activity of the enzyme *Lipoprotein lipase* has been found and the activity varies with the nutritional and hormonal state. ***Activity of lipoprotein lipase in adipose tissue is high in the fed state and 'Low' in starvation and diabetes mellitus.***

Characteristics of Lipoprotein Lipase

- The enzyme *lipoprotein lipase* is ***located in walls of blood capillaries***. The enzyme remains bound to wall by proteoglycan chains of Heparan-SO_4.
- Has been found in extracts of heart, adipose tissue, spleen, lungs, renal medulla, aorta, diaphragm and lactating mammary gland.
- Normal blood does not contain appreciable quantities of the enzyme.
- However, following injection of Heparin *lipoprotein lipase* is released from its binding with Heparan-SO_4 into the circulation and is accompanied by clearing of lipaemia (hence called as ***clearing factor***). ***Another lipase is also released from liver, a hepatic lipase*** by large quantities of heparin, but this enzyme has properties different from those of *lipoprotein lipase* and does not react readily with TG of chylomicrons and VLDL.

- Both phospholipids (PL) and apolipoprotein-CII are required as cofactors for *lipoprotein lipase* activity.

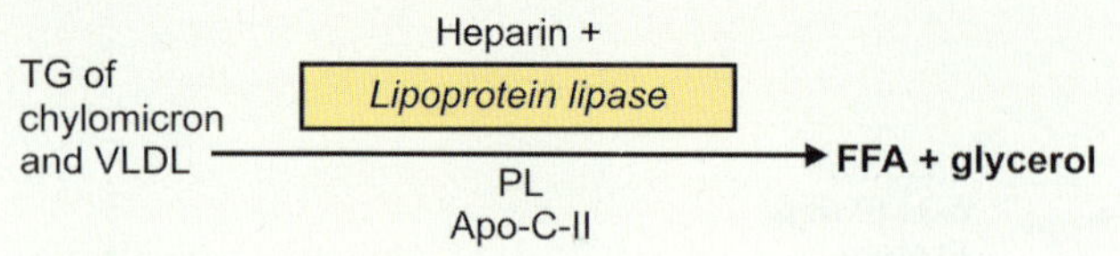

Apo-C-II contains a specific PL binding site through which it is attached to the lipoprotein. Thus, the chylomicrons and VLDL provide the enzyme for their metabolism with both its substrate and cofactors. Hydrolysis takes place while the lipoproteins are attached to the enzyme on the endothelium. TG is progressively hydrolysed to give DG and then MG and finally glycerol and FFA (three molecules). Some of hydrolysed FA return to circulation being carried by albumin and bulk of FFA is taken up by tissues including adipose tissue **(FFA-Pool-2).**

Role of hormones: In adipose tissue, insulin enhances the synthesis of *lipoprotein lipase* in adipose tissue cells and its translocation to luminal surface of capillary endothelium.

BROWN ADIPOSE TISSUE

Types of storage fats: There are **two types** of storage fats:

- **Storage "white" fat** present in depot fats—predominant
- In addition to usual white storage fat, another type of **"pigmented" brown fat** is stored in some species including humans.

Role in Thermogenesis

1. Brown adipose tissue is involved in metabolism particularly at times when a heat generation is necessary.
   Thus, **the tissue is extremely active**
   - ***In arousal from hibernation,***
   - ***In animals exposed to cold,*** and
   - ***In heat production in newborn animals.***

   It is present in rats, throughout the life.
2. Though not a prominent tissue in humans, recently it has been shown to be active in normal individuals, where it appears to be responsible for ***diet-induced thermogenesis***, which may account for how some persons can ***eat and do not get fat***.

Note: It is to be noted that brown adipose tissue is reduced or may be absent in obese persons.

Location: It is located and present particulary in the thoracic region.

Characteristics of Brown Adipose Tissue: It is characterised by:

- ***A high content of mitochondria,***
- ***A high content of cytochromes,***
- ***A well-developed blood supply,***
- Also relatively ***rich in carnitine***, which is significant for FA oxidation,
- Unlike white adipose tissue, it ***has the enzyme glycerokinase.***

Note

- The brown colour is related to a relatively high cytochrome content.
- There is **low ATP-synthase activity.**
- Oxygen consumption is high.

Metabolic Peculiarities

1. Metabolic emphasis is on oxidative processes-oxidation of glucose and FA oxidation, which are oxidised to CO_2 and H_2O.
2. Nor-epinephrine liberated from sympathetic nerve endings is important in increasing lipolysis in this tissue.

Mechanism of Heat Production

- **Oxidation and phosphorylation are not coupled in mitochondria of this tissue.** Dinitrophenol has no effect and there is no respiratory control by ADP.
- The phosphorylation which occurs is at the "substrate level" in the glycolytic pathway and in "succinate thiokinase" step of TCA cycle. Thus, oxidation produces much heat and very little free energy is trapped as ATP due to decreased coupling of oxidation and phosphorylation.
- In terms of chemi-osmotic theory, it appears that the proton gradient, normally present across the inner mitochondrial membrane of coupled mitochondria, is ***continually dissipated in brown adipose tissue by a thermogenic protein, called*** **thermogenin**, which acts as a proton conductance pathway through the membrane. This explains the apparent lack of effects of uncouplers.

Function: Brown storage fat has a somewhat higher temperature than other tissues. It plays a role in heat production for vital organs, serving as a sort of ***heating pad*** or ***furnace*** for the local application of its heat to the vital organs of the thorax, the upper spinal cord, and the autonomic sympathetic chain. The amount of brown fat increases in animals when subjected to "cold stress". Direct measurements of heat production during cold stress in rats showed that brown adipose tissue accounts for 82 per cent of total heat production.

Recent-Advances—An Excerpt

***Brown fat cells vs. White fat cells:* Genesis of Brown fat cells—An experimental study**

As discussed above, brown fat cells have a very different role from the white adipose tissue cells. ***White adipose tissue stores fat, but brown adipose tissue burn it off as heat.***

Babies have lots of brown fat to help them keep warm. ***Until April 2009, scientists believed that the brown fat gradually disappeared and is not found in adults.***

Dr Sven Enerback of the University of Goteborg in Sweden and others reported that ***some brown tissue persisted in adults,*** raising the possibility that if the cells could be made more active, a person might burn off more fat.

Spiegelman of Harvard Medical School, in 1994, found the body's master regulator of fat cells. Turning to brown fat cells, ***he followed the general assumption that they were derived from white fat cells.***

A key element in making brown fat cells seems to be a kind of protein called a **zinc finger.** Spiegelman assumed that if he inactivated all the relevant zinc finger proteins in brown fat cells they should turn back into their precursors, the white fat cells. His experiment worked but to his surprise he found that the ***brown fat cells revert not to white fat cells but to muscle cells.***

Thus he concluded that muscle cells and not white fat cells are the natural precursors of brown fat cells.

Further studies showed that ***zinc finger protein in combination with a second protein produced in the muscle cells is the "master switch" for brown fat cells and it can also convert human skin cells into brown fat cells.*** Spiegelman surmised that further discoveries might produce the natural protein for turning on the zinc-finger switch and this protein might make a useful drug for converting skin cells to brown fat cells.

Summary

- Brown fat cells make white fat cells breakdown fat into fatty acids.
- ***Babis have lots of brown fats. In April 2009, it was found that adults also have them.***
- Previously it was thought that brown fat cells come from white fat cells, but ***now it appears that they are derived from muscle cells.***
- A protein called zinc finger plays a role in the formation of brown fat from muscle cells.
- Zinc finger proteins in combination with a second protein in muscle cells is the master-switch for brown fat cells and also convert skin cells into brown fat.
- Researchers are assuming if this protein can be activated in adults, the body will produce more brown fats and thereby help burn fat.

PART II

Major Concepts

A. To study how fatty acids are oxidised in the body to give energy.
B. Learn the synthesis of Fatty acids, Triacyl glycerol and Phospholipids.
C. To learn how Ketone bodies are produced and utilised in the body.
D. To study how cholesterol is metabolised in the body.

Specific Objectives

A.
- Enumerate and list the various methods by which fatty acids are oxidised in body.
- What is β-oxidation? Study the experiment carried out by **Knoop**, who proposed the β-oxidation.
- Study the tissues in which β-oxidation is carried out. Is it extramitochondrial/or mitochondrial?
- Study what is 'Carnitine' and its role in transportation of long-chain FA from cytosol to mitochondrion.
- Study how fatty acid is activated? Learn the various reactions, involved in β-oxidation, along with enzymes/and coenzymes required.
- What is the end-product of β-oxidation? ***How many acetyl-CoA are produced from β-oxidation of Palmitic acid?***
- What are the end products of β-oxidation of odd number long chain fatty acids?
- Learn the bioenergetics of β-oxidation of palmitic acid and its efficiency as compared to glucose oxidation.
- Make in tabular form to differentiate α-oxidation and ω-oxidation. Learn about Refsum's disease.

B. (a) FA Synthesis
- Enumerate and list the various methods by which FA can be synthesised in the body.
- What is *de novo* extramitochondrial FA synthesis? Study and note the starting material required in this synthesis and the product formed in this pathway.
- Learn in detail the conversion of Acetyl-CoA to Malonyl-CoA, the enzyme and co-enzymes required for this reaction. List the sources of Acetyl-CoA in the cytosol.
- Study in detail the multienzyme complex *Fatty Acid Synthase* system involved in *de novo* synthesis. How does it differ from enzyme which operates in bacteria, plants and lower animals?
- Study in detail the various steps involved in *de novo* synthesis; the enzyme and co-enzymes required for the same. NADPH is required for the reductive synthesis. Note the steps where NADPH is required and sources of supply of NADPH for this process.
- Note the requirement of "primer" for the *de novo* synthesis of long chain FA having odd number of c-atoms.
- Make in tabular form to differentiate the mitochondrial FA synthesis, and microsomal FA synthesis. What is the purpose of these two pathways?

(b) Synthesis of TG and PL
- List the substrates required for the synthesis of TG.
- What are the sources of α-Glycero-P? Learn how Glycerol can be converted into α-Glycero-P.
- In tissues which lack the enzyme *Glycerokinase*, what is the source of α-Glycero-P?
- Study the reactions involved in synthesis of TG, the enzymes and coenzymes, if any, required for the synthesis.
- Study how different phospholipids can be synthesised from TG.
- List the substrates required for the synthesis of sphingosine. Learn the steps by which the unsaturated amino alcohol, sphingosine is synthesised in the body; the enzymes and co-enzymes required for the same. Study how sphingomyelin is formed from sphingosine.
- Learn how Lecithin is catabolised in the body.

C.
- Learn what are ketone bodies? Enumerate the ketone bodies.
- Define the terms: ketonemia, ketonuria and ketosis.
- Study the causes for ketone bodies formation and site of production of ketone bodies.
- Learn the reactions of Ketone bodies formation in Liver (Ketogenesis), enzymes and co-enzymes required.
- Study how the ketone bodies are utilised by extrahepatic tissues (ketolysis).
- What is the fate of acetone? Can acetone be converted to 1,2-Propanediol-P? And thus may be glucogenic.
- Study the factors that determine the magnitude of Ketogenesis.
- Learn how ketosis can be prevented, and the antiketogenic mechanisms.
- Study the different tests used for detection of Ketone bodies in urine.

D.
- Revise your knowledge of chemistry of cholesterol.
- Learn how cholesterol is synthesised in the body? Note the starting material required for cholesterol biosynthesis.
- List the tissues in which cholesterol biosynthesis occurs.
- Learn in detail the various steps in cholesterol biosynthesis, the enzymes and coenzymes required.
- What is HMG-CoA? How it is formed? Learn the fates of HMG-CoA—why it is called "committed step"?
- What is the "rate-limiting" step in biosynthetic pathway and study how the cholesterol biosynthesis is regulated?
- Make in tabular form the various factors involved in:
  - Increasing cholesterol biosynthesis and
  - Decreasing the synthesis of cholesterol. Study the various drugs which have been used to lower the blood cholesterol level and their mechanism of action.

- Study the metabolic fate of cholesterol in the body.
- What are bile acids? Learn how they are formed in the body and their functions.
- Study the role of cholesterol and TG in atherosclerosis and IHD.

SECTION FOUR

OXIDATION OF FATTY ACIDS

Sources of Plasma FFA

Plasma free fatty acids are derived:
- Mainly from lipolysis in adipose tissue **(Pool-1)**.
- Portion of FFA is derived from degradation of circulating chylomicrons and VLDL by the action of the enzyme *lipoprotein lipase* **(Pool-2)**.
- A small portion of plasma FFA is derived from absorption of dietary source specially small chain and medium chain fatty acids.
- Also FFA is obtained from synthesis from acetyl-CoA in liver cells, which are incorporated in TG.

Fatty Acid Uptake

Free fatty acids uptake by tissues is ***related directly to the plasma free fatty acids concentration (FFA)***, which in turn is determined by the rate of lipolysis in adipose tissue. In post absorptive state, plasma contains 10 to 30 mg% of FFA, ***most of which is transported in plasma as a loose complex with albumin*** known as ***"albumin-FFA complex"***. After dissociation of the "FFA-albumin complex" at the plasma membrane, fatty acids ***bind to a membrane fatty acid transport protein that acts as a transmembrane co-transporter with Na^+***. On entering the cytosol, the free fatty ***acids are bound by intracellular fatty acid-binding proteins or Z-proteins***. The role of these proteins in intracellular transport is thought to be similar to that of serum albumin in extracellular transport of long chain fatty acids. A small amount of FA is also associated with HDL. Shorter chain FA are more water soluble and exist as the unionised acid or as a FA anion.

Fatty acids exhibit a very rapid turnover rate with a 1/2 life of only 1 to 3 minutes, they are rapidly taken up by tissues and metabolised. Plasma FFA is decreased by Insulin and Glucose administration. Plasma FFA is increased by catecholamines, Growth hormone, Glucocorticoids and thyroid hormones. Also increased in Diabetes mellitus, starvation and with high fat diets.

Methods by which fatty acids are oxidised in the body are as follows:
- **β-oxidation:** Principal method of oxidation of FA. Other ancillary and specialised methods are:
- **α-oxidation,**
- **ω-oxidation,** and
- **Peroxisomal FA oxidation.**

A. β-OXIDATION

Principal method by which FA is oxidised is called β-oxidation. Several theories have been proposed to explain the mechanism of the oxidation of FA chains. The classical theory of β-oxidation was the outcome of the work of **Knoop.**

Knoop's Experiment: Tagged the $-CH_3$ end of FA by substitution of a phenyl radical, this prevented the complete oxidation. It resulted in urinary excretion of phenyl derivatives:
- On feeding dogs with FA with even carbon atoms, he observed that phenyl acetic acid is always excreted. It is conjugated with glycine and excreted as phenyl aceturic acid.
- When FA with odd number of carbon atoms were fed, Benzoic acid was excreted as glycine conjugate **hippuric acid**.

Conclusion: **Knoop** proposed the β-oxidation theory. According to this mechanism FA chains are oxidised by the removal of 2 carbon atoms at a time. The carbon atom in the β-position to COOH group is assumed to be attacked with the formation of the corresponding β-keto acid; then the two terminal C-atoms are split off as "acetyl-CoA". A new –COOH group is formed at the site of the keto (=CO) grouping, so that ***a fatty acid remains with 2 carbon atoms less than the original.*** Again the new β-carbon atom is attacked and two more carbon atoms are split off as acetyl-CoA. In this way, the FA is degraded by the removal of 2 carbon atoms at a time, until finally the stage of acetoacetic acid is reached.

Tissues in which β-Oxidation is carried out: The circulating FA are taken up by various tissues and oxidised. Tissues like **liver, heart, kidney, muscle, brain, lungs, testes and adipose tissue** have the ability to oxidise long chain FA. In cardiac muscle, fatty acids are an important fuel of respiration (80% of energy derived from FA oxidation).

Enzymes Involved in β-Oxidation

β-oxidation takes place in mitochondrion. Several enzymes known collectively as ***FA-oxidase system*** are found in the mitochondrial matrix, adjacent to the respiratory chain, which is found in the inner membrane. These enzymes catalyse the oxidation of FA to acetyl-CoA.

Activation of FA: Fatty acids are in cytosol of the cell ***(extramitochondrial).*** Fatty acids must be first activated so that they participate in metabolic pathway. The activation requires energy which is provided by ATP. In presence of ATP, and coenzyme A, the enzyme ***acyl-CoA synthetase*** (previously called as *thiokinases*) catalyses the conversion of a free fatty acid to an 'active' FA (acyl-CoA).

The presence of inorganic *pyrophosphatase* ensures that activation goes to completion by facilitating the loss of additional high energy ~ P bond of PPi.

Thus, in effect 2 ~ P bonds are expended during activation of each FA molecule. Not only saturated FA but unsaturated FA and –OH fatty acids are also activated by these acyl-CoA synthetases.

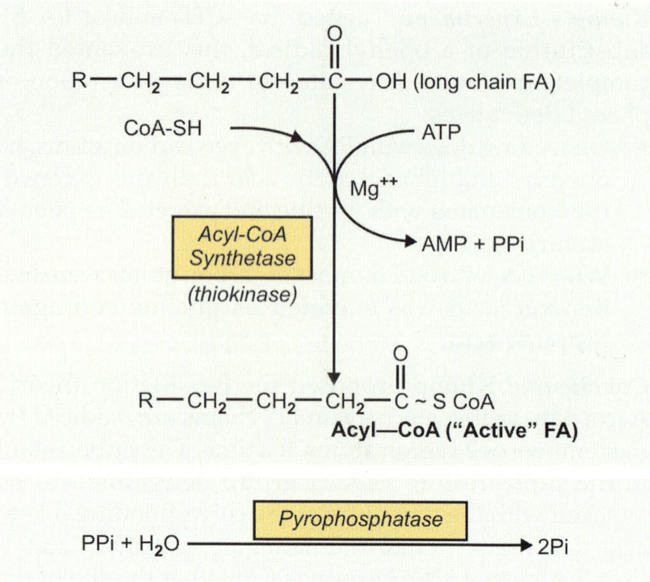

Location and types of Acyl-CoA synthetases: The enzymes are found in the endoplasmic reticulum and inside (for short-chain FA) and outside (for long-chain FA) of the mitochondria. Several varieties of the enzyme have been described, each specific for FA of different chain lengths.

- ***Acetyl-CoA synthetase*** → acts on acetic acid and butyric acid
- ***Second medium chain synthetase*** → acts on FA with chain length C_4 to C_{12}
- ***Long chain acyl-CoA synthetase*** → Acts on FA with chain length C_8 to C_{22}
- Recently, a GTP-specific mitochondrial acyl-CoA synthetase described which forms GDP + Pi.

CARNITINE AND ITS ROLE IN FA METABOLISM

"Active" FA (acyl-CoA) are formed in cytosol, whereas β-oxidation of FA occurs in mitochondrial matrix. ***Acyl-CoA are impermeable to mitochondrial membrane.*** Long-chain activated FA penetrate the inner mitochondrial membrane only in combination with carnitine.

Carnitine: Chemistry and functions

Carnitine is chemically "β–OH–γ–trimethyl ammonium butyrate"

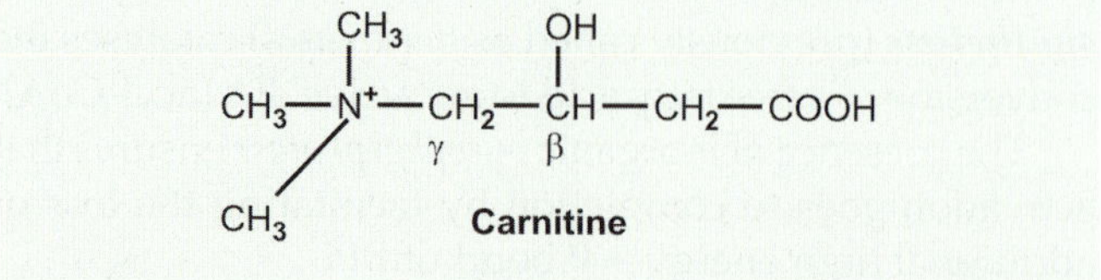

Historical background: **Fraenkel's** vitamin B_T is same as carnitine. It was found to be required as a nutritional factor in meal-worm (Tenebrio molitor). If the meal worms are fed on a synthetic diet, deficient in vitamin B_T, they die in 4 to 5 weeks.

Distribution: Carnitine is widely distributed in yeast, milk, liver and particularly large quantities in muscles and in meat extracts.

Concentration: Fraenkel used bioassay technique based on the rate of growth of larvae of *Tenebrio molitor* and found that in mammals, carnitine content of:

- Skeletal muscle: 1 mg/gm dry weight
- Heart muscle: 560 mcg/gm
- Kidneys: 412 mcg/gm
- Liver: 280 mcg/gm

Blood: Small amounts in blood 7-14 mcg/ml.

Excretion in 24 hrs urine: 50 to 100 mcg/ml.

Biosynthesis of carnitine: It is synthesised from lysine and methionine in liver principally, also in kidneys. Synthesis of carnitine is shown below in the box.

Biosynthesis of Carnitine

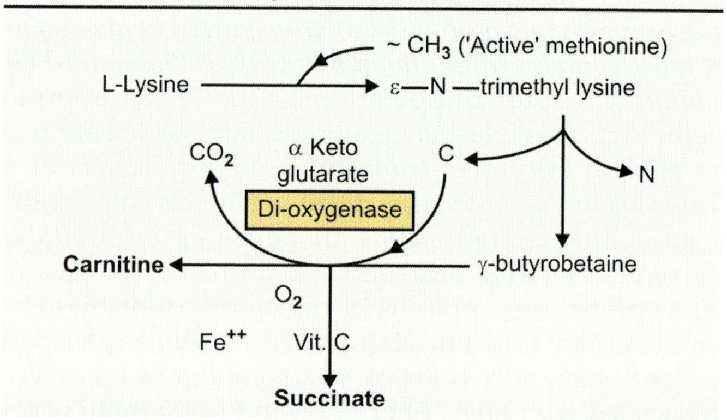

Functions: Carnitine is considered as a **"carrier molecule"; it *acts like a ferry-boat.*** It transports long-chain acyl-CoA across mitochondrial membrane which is impermeable to acyl-CoA.

- Facilitates transport of long-chain acyl-CoA for oxidation in mitochondria.
- Facilitates exit of acetyl-CoA and acetoacetyl-CoA from within mitochondria to cytosol, where FA synthesis takes place.
- ***Methionine-sparing action:*** **Khairallah** and **Wolf** (1965) found that in rats, carnitine has a methionine-sparing action and may thus be considered a food factor required in marginal diets.

Mechanism of transport of long-chain acyl-CoA: Activation of lower FA and their oxidation may occur within the mitochondria, independently, of carnitine; but long-chain acyl-CoA (or FFA) will not Penetrate mitochondria and become oxidised unless they form **acyl carnitines.**

- An enzyme ***carnitine-palmitoyl transferase I***, present on the inner side of the outer mitochondrial membrane, converts long-chain acyl-CoA to

SECTION FOUR

acyl-carnitines; which is able to penetrate mitochondria and gain access to the β-oxidation systems of the enzymes.

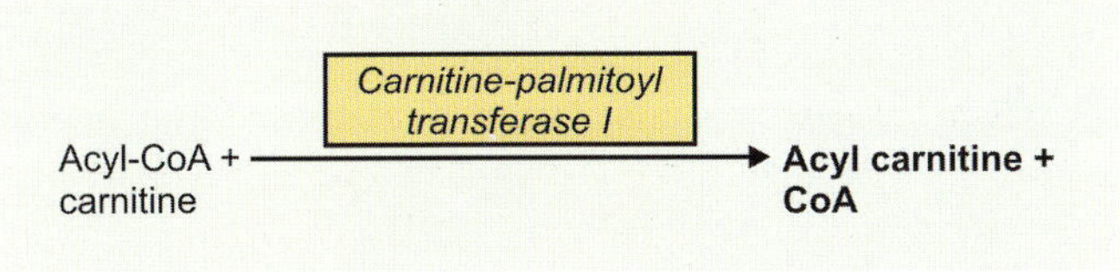

- Another enzyme ***carnitine-acyl-carnitine translocase*** acts as a membrane-carnitine exchange transporter. Acyl-carnitine is transported in, coupled with the transport out of one molecule of carnitine.
- The acyl-carnitine then reacts with CoA-SH, catalysed by ***carnitine-palmitoyl transferase II,*** attached to the inside of the inner membrane. Acyl-CoA is reformed in the mitochondrial matrix and carnitine is liberated.

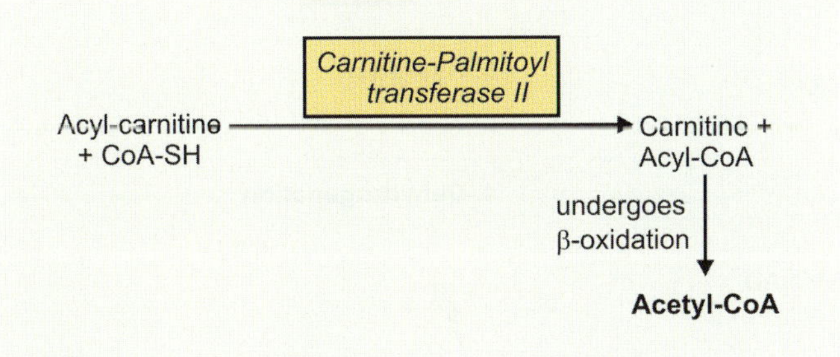

Figure 25.2 shows the mechanism of action of carnitine.

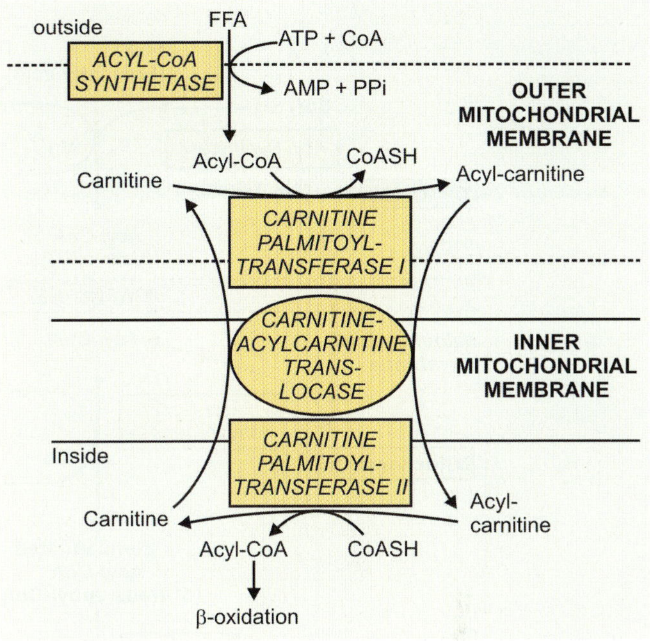

Fig. 25.2: Role of carnitine in the transport of long-chain FA

STEPS OF β-OXIDATION (FIG. 25.3)

Once acyl-CoA is transported by carnitine in the mitochondrial matrix, it undergoes β-oxidation by *Fatty acid oxidase complex*. The successive steps are as follows:

1. ***Dehydrogenation: Removal of 2 H atoms:*** Removal of two hydrogen atoms from the 2 (α) and 3 (β) carbon atoms is catalysed by the enzyme ***acyl-CoA dehydrogenase,*** resulting in formation of Δ^2 - transenoyl-CoA (also called α, β-unsaturated acyl-CoA).

   Hydrogen acceptor, i.e the co-enzyme for this dehydrogenase is a ***flavo-protein,*** containing FAD as prosthetic group, whose re-oxidation in the respiratory chain requires the mediation of another flavoprotein, called **electron transferring flavoprotein (ETF)** (see Biologic oxidation chapter).

+ 2 ATP

Types of acyl-CoA dehydrogenases: At least three acyl-CoA dehydrogenases have been described:

- **"G"-green-coloured,** cu-containing, catalyses oxidation of fatty acids having chain length–C_4 to C_8

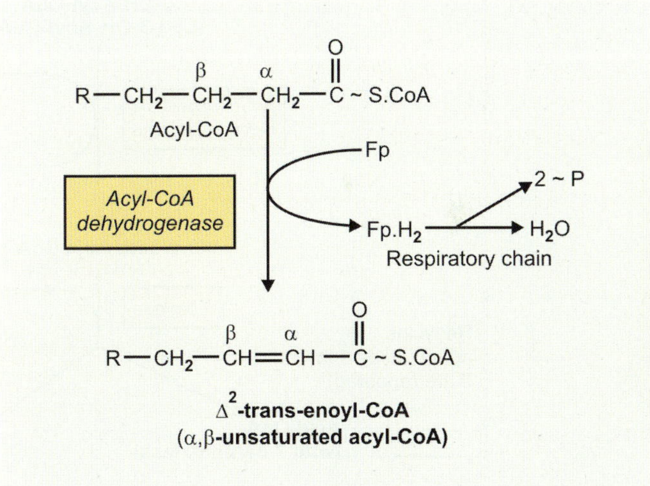

- **'Y' or 'y$_1$'-Yellow** flavo protein, catalyses oxidation of FA having chain length –C_4 to C_{18}, more specific for C_6 ***("hexonyl-CoA dehydrogenase")***
- **'y' or 'Y$_2$'**-more active on FA having chain-length C_6 to C_{18}. Maximum activity on FA having chain-length C_{16} ***("hexa-decanoyl dehydrogenase").***

2. **Hydration: Addition of one molecule of H_2O:** One molecule of water is added to saturate the double bond to form 3-OH acyl-CoA (called also as β-OH acyl-CoA), the reaction is catalysed by the enzyme *"Δ^2 - enoyl - CoA hydratase"* (also called as *Enoyl hydrolase*; earlier called also as *crotonase*).

3. **Dehydrogenation: Removal of 2 hydrogen atoms:** The 3 – OH – Acyl-CoA undergoes further dehydro-

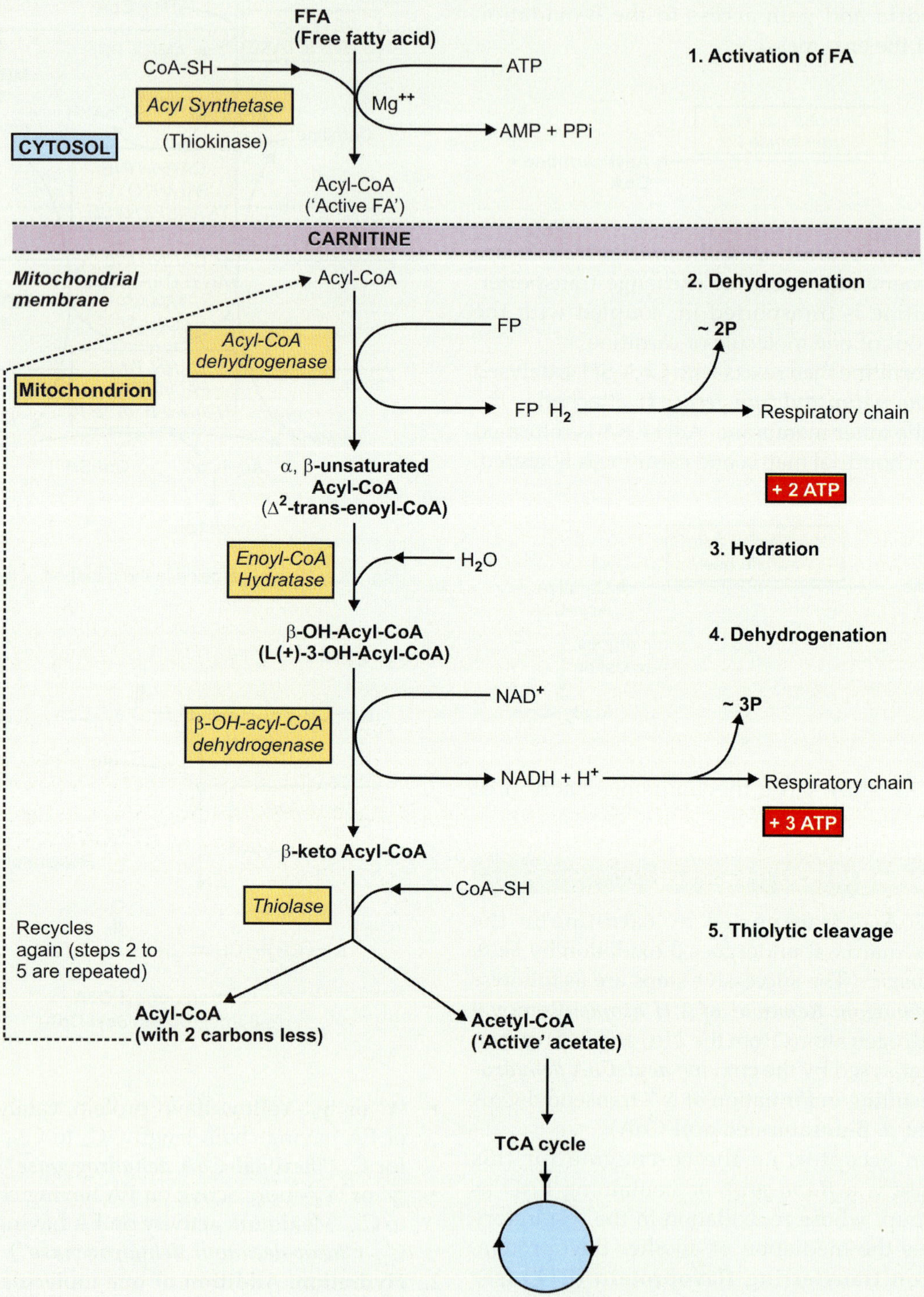

Fig. 25.3: β-oxidation of fatty acids

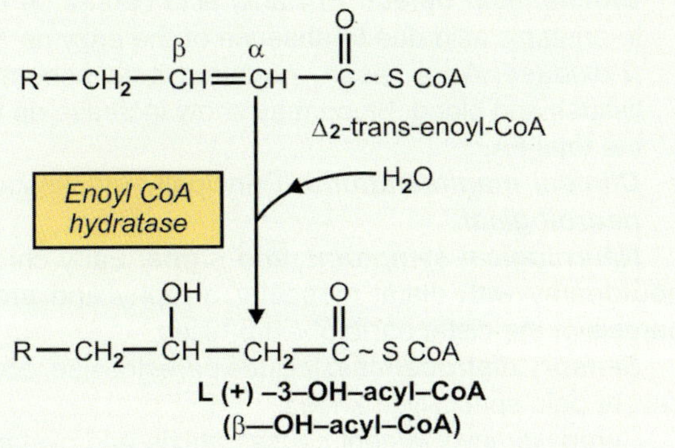

genation on the 3 carbon, catalysed by the enzyme **3–OH–acyl-CoA** ***dehydrogenase,*** to form the corresponding 3–ketoacyl-CoA (β-ketoacyl-CoA).

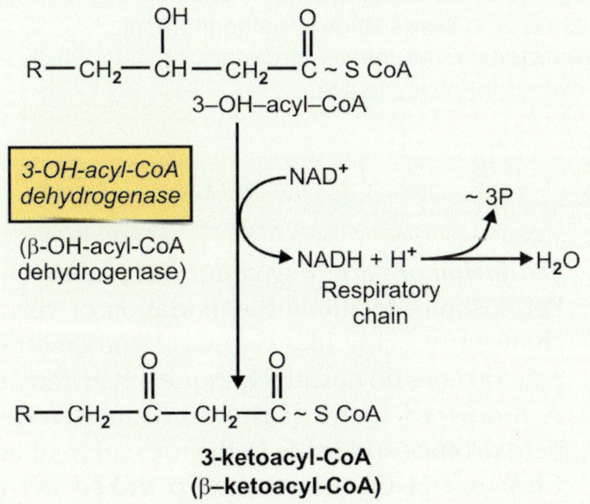

Hydrogen acceptor, i.e. coenzyme of this dehydrogenase is NAD^+. Reduced NAD when oxidised in respiratory chain produces 3 ATP.

+ 3 ATP

4. ***Thiolytic cleavage:*** Finally, 3-keto-acyl-CoA is split at the 2,3 position by ***thiolase*** ("3 – keto acyl thiolase" or "acetyl–CoA acyl transferase"), which catalyses a thiolytic cleavage involving another molecule of CoA.

End-products of this reaction: The thiolytic cleavage results in formation of:

- ***One molecule of acetyl-CoA*** and
- ***An acyl-CoA molecule containing 2-carbons less than the original acyl-CoA*** molecule, which enters for oxidation by the enzyme *acyl-CoA dehydrogenase* (re-enters at step 1).

In this way, a long-chain FA may be degraded completely to "acetyl-CoA" (C-2 units). Acetyl-CoA can be oxidised to CO_2 and H_2O and thus complete oxidation of FA is achieved. Thus, end-product of β-oxidation of a long-chain FA will produce acetyl-CoA molecules (C-2 units).

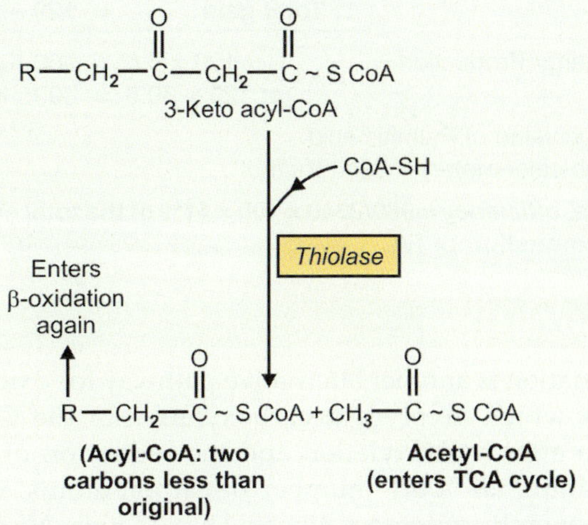

How many acetyl-CoA are produced from β-oxidation of palmitic acid?

Palmitic acid is $C_{15}H_{31}$ COOH. In β-oxidation, it will ***undergo 7 (seven) cycles,*** producing 7 acetyl-CoA (in 7 cycles) + 1 acetyl-CoA (last cycle-one extra). ∴ ***Total acetyl-CoA produced by β-oxidation of one molecule of palmitic acid – 8 acetyl-CoA.***

β-Oxidation of FA with an odd number of carbon Atoms: Fatty acids with an odd number of carbon atoms are oxidised by β-oxidation pathway to produce acetyl-CoA until a 3-carbon residue **propionyl-CoA** is left. Propionyl-CoA is metabolised to succinyl-CoA through methyl malonyl-CoA.

Note:

- ***Propionyl-CoA formed from an odd-chain FA is the only part of the FA which is glucogenic,*** as it is converted to succinyl-CoA.
- Sources of propionyl-CoA in the body (Refer to Gluconeogenesis).

Bioenergetics of β-Oxidation and its Efficiency

Palmitic acid, $C_{15}H_{31}COOH$, on complete oxidation (β-oxidation) **produces 8 acetyl-CoA** (Refer discussion above). Transport of electrons in respiratory chain from reduced Fp and NAD in each cycle produces 5 (five) high energy phosphate bonds.

Hence, 7 cycles (7 × 5)		= 35 ~ P
Total 8 molecules of acetyl-CoA, When oxidised in TCA cycle will produce =	12 × 8	= 96 ~ P
Total high energy phosphate bonds produced		= 131 ~ P
	Total	**= 131 ~ P**

In initial activation of FA~ P bond utilised		=	–2 ~ P
	∴ Total gain	**=**	**129 ~ P**
∴ Energy Production =	129 × 7.6	=	980 Kc
	(or 129 × 30.5	=	3935 Kj)

Caloric value of Palmitic acid
(Bomb calorimeter) = 2340 Kc/mol

Hence, ***efficiency* = 980/2340 × 100 = 41% of the total energy of combustion of FA.**

B. α-OXIDATION

α-oxidation is another alternative pathway for oxidation of FA which involves decarboxylation of the COOH group after hydroxylation and the formation of a FA containing an "odd" number of carbon atoms, which subsequently undergoes repeated β-oxidation. ***No initial activation of FA is necessary in this process.***

C. ω-OXIDATION (VERKADE)

In ω-oxidation, Fatty acids undergo oxidation at the carbon atom farthest removed from the carboxyl group (ω-carbon) producing a ***dicarboxylic acid,*** which is then subjected to β-oxidation and cleavage to form successively smaller dicarboxylic acids.

Both processes occur principally in brain microsomes but are negligible in extent as compared to β-oxidation.

Essential differences and similarities between α-oxidation and ω-oxidation are shown in Table 25.4.

Purpose of α-oxidation: In the body α-oxidation serves:

- To synthesise α-OH fatty acids like cerebronic acid of brain cerebrosides and sulphatides.
- To form odd C-long-chain FA required in brain sphingolipids.
- Also helps to oxidise phytanic acid produced from dietary phytols a constituent of chlorophyll of plant food stuffs. Phytanic acid is oxidised by α-oxidation with ***Phytate α-oxidase*** (an α-hydroxylase enzyme) to yield CO_2 and odd –C chain FA **pristanic acid** which is then completely oxidised by β-oxidation.

CLINICAL ASPECT

Inherited Disorders

Refsum's disease:

A rare genetic disorder

- ***Enzyme deficiency:*** *"Phytanate α-oxidase".*
- ***Inheritance:*** Autosomal recessive.
- ***Age:*** The disease may become manifest at any age from childhood to adult life. In some affected families there was consanguinity of the parents.
- ***Biochemical defect:*** Phytanic acid cannot be converted to pristanic acid due to absence of the enzyme *Phytanate α-oxidase*. As a result, phytanic acid accumulates in tissues and blood. Blood may show increase up to 20% of the total FA.
- ***Clinical manifestations:*** Principally manifestations are ***neurological.***

Neurological symptoms and signs: Early chronic polyneuropathy with distal muscular atrophy and progressive paresis of the distal parts of extremities.

Sensory distrubances: Include paresthesiae, occasionally severe pain specially in knees.

Cerebellar involvement causes ataxia and nystagmus.

- ***Eye manifestations:*** Typical pigmentary retinitis, night blindness, and concentric narrowing of visual fields.
- ***Mental development:*** Usually normal.
- ***CS fluid:*** CSF protein is always considerably increased, whilst the cell count is usually normal.
- ***Diagnosis:*** Demonstration of increased phytanic acid in plasma or in tissue lipids is pathognomonic.
- ***Treatment:*** Omit intake of dietary phytols which is the precursor of phytanic acid.

D. ROLE OF PEROXISOMES: IN OXIDATION OF FATTY ACIDS

- ***Oxidation of very long chain fatty acids:*** Peroxisomes facilitate the oxidation of very long chain fatty acids like C_{20}, C_{22}. The enzymes in peroxisomes do not attack shorter chain fatty acids. A modified form of β-oxidation is found in peroxisomes and leads to the formation of **acetyl-CoA** and **H_2O_2**—from the Fp linked dehydrogenase step. H_2O_2 formed in the process is broken down by the enzyme **Catalase**. The dehydrogenation in peroxisomes is not linked to phosphorylations and thus there is **no ATP formation.** The enzymes of peroxisomes are **induced by high fat diets** and in some speices by hypolipidaemic drugs like clofibrate.
- Another role of peroxisomal β-oxidation is to shorten the sidechain of cholesterol in bile acid formation.
- Peroxisomes also take part in the synthesis of:
  - *Glycolipids*
  - *Cholesterol*
  - *Dolichol.*

Dolichol

A polyisoprenoid.

It is formed from farnesyl diphosphate by further addition of upto 16 isopentenyl residues.

End-products: Are octanoyl-CoA and acetyl-CoA, which are removed from peroxisomes to mitochondria with the help of carnitine for further oxidation.

Table 25.4: Differences in α-and ω-oxidation

α-oxidation	*ω-oxidation*
1. **Substrate:** Even carbon long-chain FA (some of them)	1. Some medium- and long-chain FA
2. **Presence of O_2**—oxidised aerobically in presence of O_2	2. Occurs aerobically in presence of O_2
3. **Sites:** Microsomes of brain and liver	3. Liver microsomes
4. **Enzyme:** *α-hydroxylase*—a monooxygenase	4. *FA ω-hydroxylase* also a monooxygenase
5. **Cofactors required**—Fe^{++}, Vitamin C/FH_4	5. Cytochrome P_{450}, flavoprotein reductase and $NADP^+$
6. ***No initial activation of FA is required***	6. ***No initial activation of FA is necessary.***
7. **Steps:**	7. **Steps:**

α-oxidation steps

(i) **Step-1:** Formation of α-OH-FA

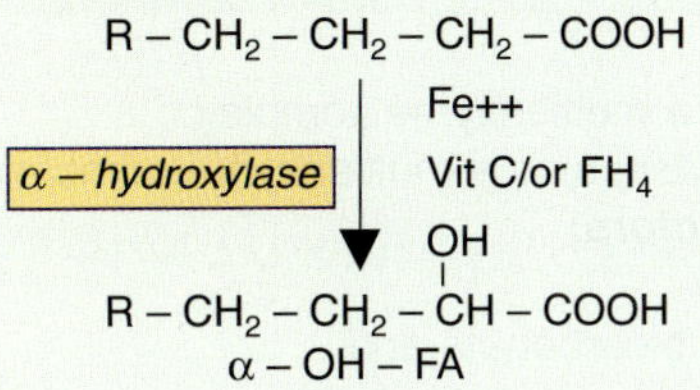

(ii) **Step-2:** Decarboxylation to produce an odd-C long-chain FA

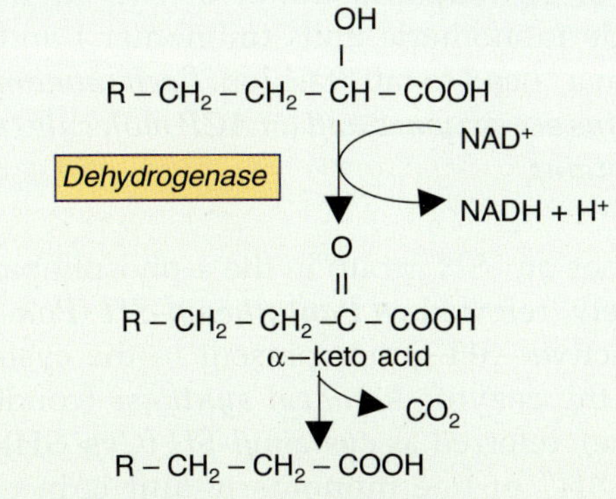

Odd chain FA with one carbon less

(iii) **Step-3:** Odd chain FA undergoes repeated β-oxidation to produce $(Acetyl\text{-}CoA)_n$ + "Propionyl-CoA"

↓

Succinyl-CoA

↓

TCA Cycle

ω-oxidation steps

(i) **Step-1:** Hydroxylation of ω carbon.

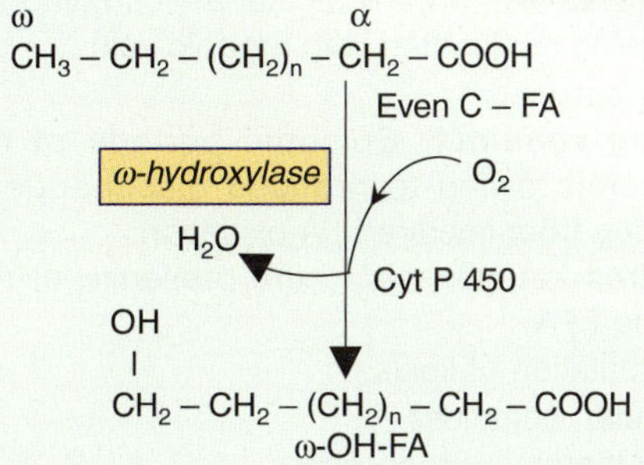

(ii) **Step-2:** ω-OH FA is oxidised with the help of an NADP-dependent enzyme to produce- ***α-ω dicarboxylic acid***

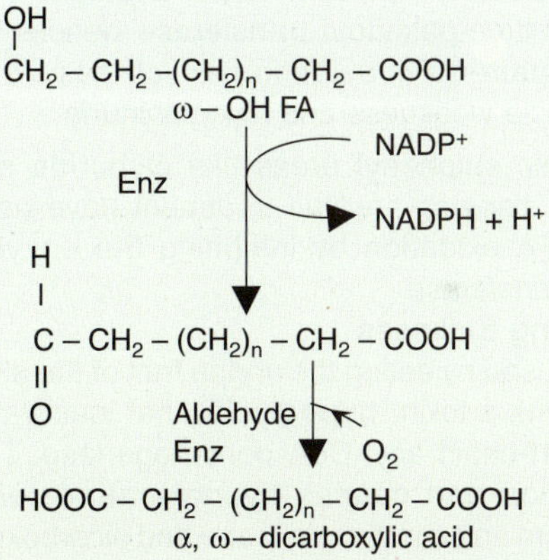

(iii) **Step-3:** Even carbon dicarboxylic acid then undergoes repeated β-oxidation to yield $(Acetyl\text{-}CoA)_n$ and one molecule of succinyl-CoA (intermediate in TCA Cycle).

↓

TCA cycle

CLINICAL ASPECT

Inherited Disorders

1. **Zellweger's syndrome** (Hepato-renal syndrome): Rare inherited disorder. There is ***inherited absence of peroxisomes*** in all tissues. Due to the absence of peroxisomes and its enzymes, fail to oxidise long-chain FA in peroxisomes. As a result there is accumulation of FA C_{26}-C_{38} chain-length in brain tissue and other tissues like liver/kidney.

SECTION FOUR

2. **Brown–Schilder's Disease:**
Also called Adreno leukodystrophy (ALD).
Inheritance: Autosomal recessive.
Defect: Main defect is insufficient oxidation of very long chain fatty acids by peroxisomes due to deficiency of peroxisomal matrix proteins. ***VLCFAS are accumulated and myelin sheaths are destroyed.***
Clinical Feature
Characterised by progressive degeneration of liver, kidney and brain.
3. **Carnitine deficiency:** Deficiency of carnitine can occur:
   (a) In newborns: specially premature infants, owing to inadequate synthesis or renal leakage.
   (b) In adults:
   - Losses can occur in hemodialysis
   - In patients with organic acidurias, carnitine is lost in urine being conjugated with organic acid.

   Clinical Features
   - ***Hypoglycaemia:*** Episodic periods of hypoglycaemia owing to reduced gluconeogenesis resulting from impaired FA oxidation
   - Impaired ketogenesis in the presence of raised plasma FFA
   - Accumulation of lipids
   - Muscular weakness.

Treatment: Oral therapy with carnitine.
4. **Carnitine-Palmitoyl Transferase Deficiency**
   (a) Hepatic deficiency of the enzyme results in hypoglycaemia and low plasma ketone bodies.
   (b) Muscular carnitine-palmitoyl transferase deficiency: Produces impaired FA oxidation which results in recurrent muscle weakness and myoglobinuria.

Note: Hypoglycaemic sulphonyl ureas like glyburide and tolbutamide used in diabetes mellitus treatment have been reported to inhibit FA oxidation by inhibiting the enzyme *carnitine-palmitoyl transferase.*
5. **Jamaican Vomiting Sickness**
The disease is caused by eating the unripe fruit of the akee tree, which contains a toxin, **hypoglycin**, that inactivates medium and short-chain acyl-CoA dehydrogenase. This inhibits β-oxidation and causes hypoglycaemia with excretion of medium and short chain mono and dicarboxylic acids.

FATTY ACID SYNTHESIS

Earlier it was believed that fatty acid synthesis was reversal of fatty acid oxidation. But now it is clear that there are *three systems* for fatty acid synthesis.

A. **Extramitochondrial system:** This is a radically different and highly active system responsible for *de novo* synthesis of palmitic acid from 2-carbon unit acetyl-CoA.

B. **Chain Elongation System**
   1. **Microsomal system:** A system present in microsomes which can lengthen existing fatty acid chains. The palmitic acid formed in cytosol is lengthened to stearic acid and arachidonic acids.
   2. **Mitochondrial system:** This system is mostly restricted to lengthening of an existing fatty acid of moderate chain-length. It operates under *anaerobiosis* and is favoured by a high NADH/NAD^+ ratio.

A. Extramitochondrial (Cytoplasmic) Synthesis of Fatty Acids: (*De Novo* Synthesis)

The synthesis takes place in cytosol. Starting material is acetyl-CoA and synthesis always ends in formation of palmitic acid.

Materials Required for the Synthesis

- ***Enzymes***
  – *Fatty acid synthase,* a multienzyme complex
  – *Acetyl-CoA carboxylase,* also a multienzyme complex
- ***Coenzymes and cofactors:***
  Biotin, NADPH, Mn^{++}
- ***CO_2:*** Source of CO_2 is bicarbonate and
- ***ATP:*** For energy.

Details of Enzymes

- ***Fatty acid synthase:*** In yeast, mammals and birds, the synthetase enzyme system is called the *fatty acid synthase* complex—it is a **multienzyme complex.** It is made up of an ellipsoid dimer of two identical polypeptide monomeric units (monomer I and II), arranged in a "head to tail" fashion. ***Each monomeric unit contains six enzymes and an ACP molecule (Acyl carrier protein).***

Active Site

- The ACP has an –SH group in the 4-phosphopantothene moiety, referred as ***Pantothenyl-SH (Pan-SH)***
- Another active –SH group present in the cysteine moiety of the enzyme ***ketoacyl synthase*** (condensing enzyme), referred as ***cysteinyl-SH*** **(Cys–SH).**

The "Pan-SH" of one monomeric unit is in close proximity to the "Cys-SH" group of other monomeric unit and vice-versa.

Following is the order of enzymes from end to end in each monomeric unit.

3-keto-acyl synthase, transacylase, enoyl reductase, 3-OH-acyl dehydratase, 3-keto-acyl reductase, ACP and thio-esterase (deacylase) (Fig. 25.4).

It is found that complex is functional only when the two monomeric units are in association with each other. The functional activity is lost when they are dissociated. In a dimer form, the complex jointly synthesises 2 molecules of palmitic acid simultaneously.

Note: In bacteria, plants and lower forms of life, the individual enzymes are separate and it is the *Acyl carrier*

SECTION FOUR

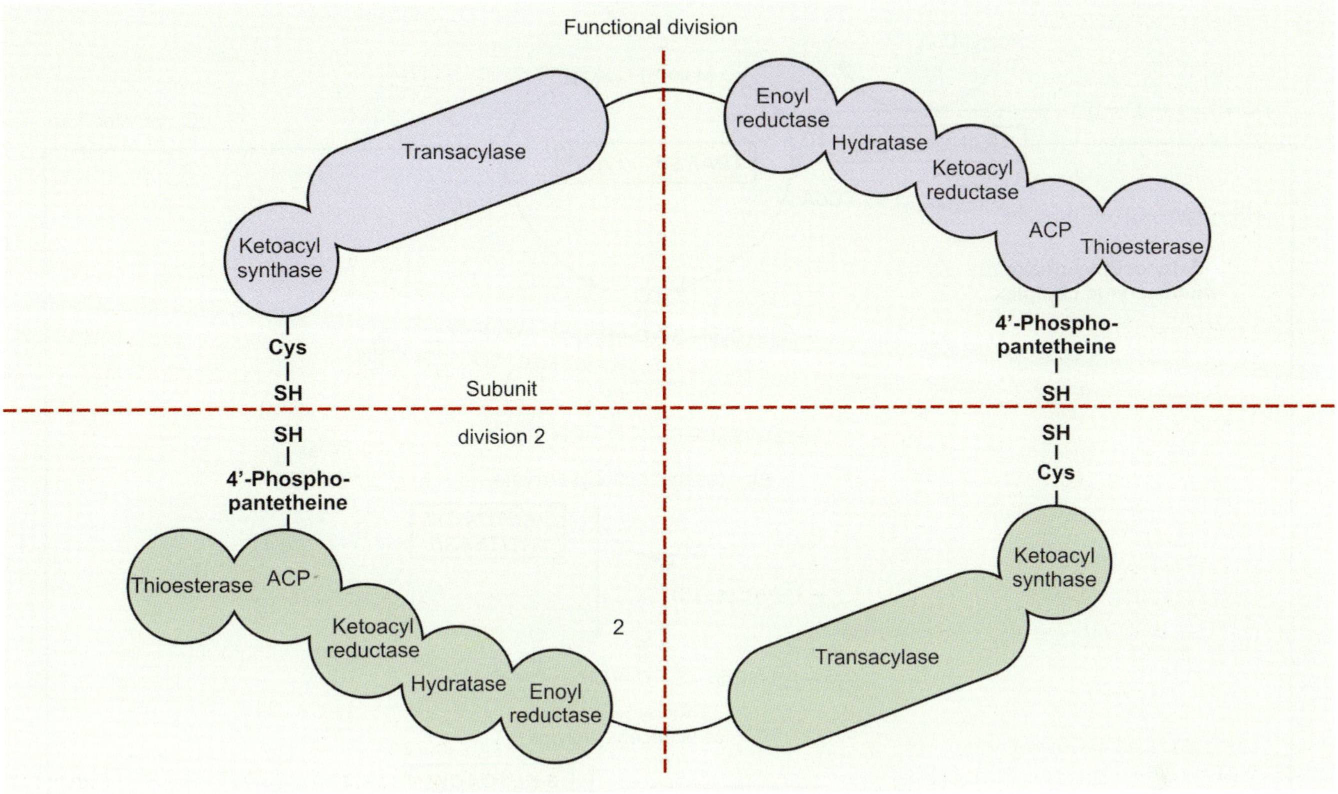

Fig. 25.4: Fatty acid synthase multienzyme complex

Protein (ACP) which binds the acyl radicals. ACP is a single polypeptide chain of 77 amino acids, a serine moiety of this peptide chain is in combination with phosphopantothene. The –SH group of pantothene moiety takes active part in synthesis of fatty acid.

- ***Acetyl-CoA carboxylase:*** Also a multienzyme complex containing:
  - ***Biotin***
  - ***Biotin carboxylase***
  - ***Biotin carboxyl carrier protein***
  - ***Transcarboxylase and***
  - ***A regulatory allosteric site.***

Steps of FA Synthesis (Fig. 25.5)

- ***The starting material for the synthesis is acetyl-CoA.*** Acetyl-CoA is formed in mitochondrion but synthesis occurs in cytosol. ***Acetyl-CoA is impermeable to mitochondrial membrane.*** The various means by which it is made available is discussed later (Refer sources of acetyl-CoA and NADPH).
  1. ***Formation of malonyl CoA from acetyl-CoA:*** In presence of the enzyme ***"acetyl-CoA carboxylase"***, the acetyl-CoA is converted to malonyl-CoA by ***"CO_2-fixation reaction"***. Mn^{++} is required as a cofactor and ATP provides the energy.

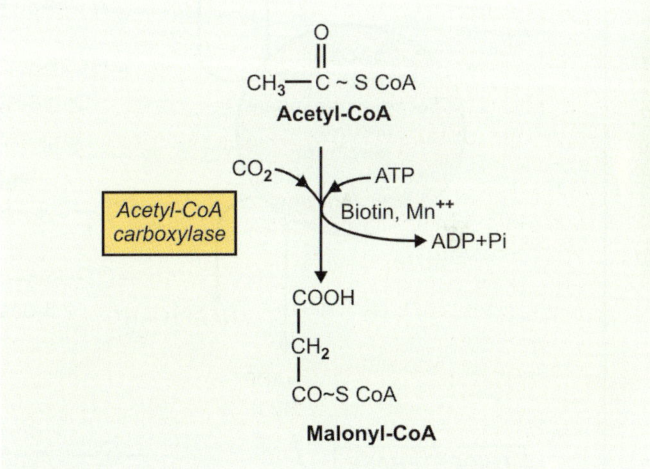

Reaction occurs in two steps:

- Biotin-enzyme + ATP + HCO_3^-

  ↓

  Carboxy-biotin-enzyme + ADP + Pi
- Carboxy-biotin-enzyme + Acetyl-CoA

  ↓

  Malonyl CoA + biotin-enzyme

Characteristics

- The reaction is ***irreversible.***
- CO_2 is provided by HCO^-_3.
- One high energy bond of ATP is utilised.

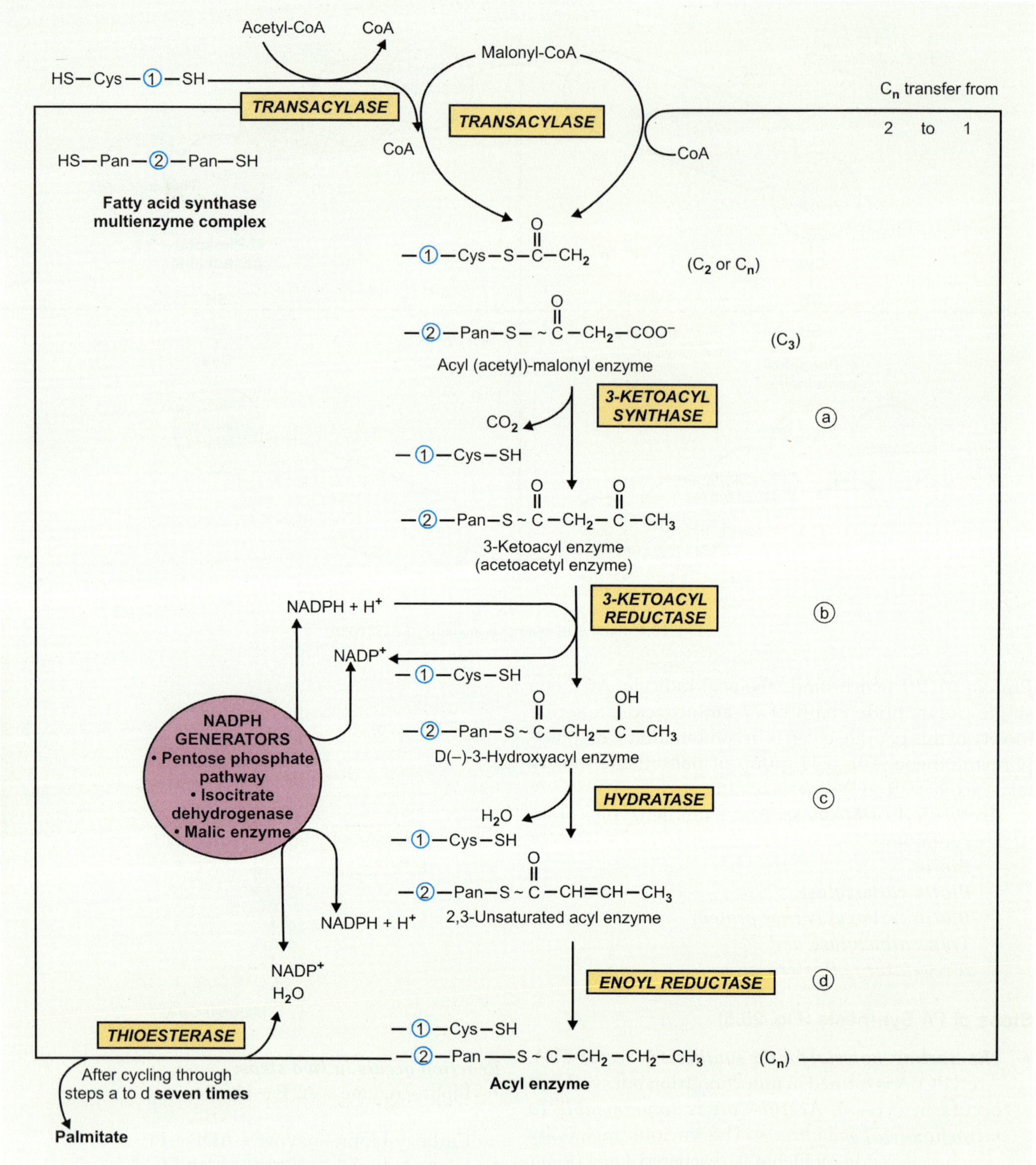

Fig. 25.5: Biosynthesis of long-chain fatty acids—extramitochondrial *de novo* synthesis

- ***Acetyl-CoA carboxylase is a rate-limiting enzyme.*** Citrate is an activator of the enzyme and palmityl-CoA is inhibitor.

2. ***Subsequent steps:*** Once malonyl-CoA is synthesised, rest of fatty acid synthesis reactions take place with FA synthase complex.

"Cys-SH" and "Pan-SH" may be considered as two arms of the enzyme complex. *"Cys-SH" is the acceptor of Acetyl-CoA whereas "Pan-SH" takes up malonyl-CoA.*

- Initially, a molecule of acetyl-CoA combines with the "Cys-SH" of *"keto acyl-synthase"* of one monomeric unit **(monomer I)**. The coenzyme A is removed, the reaction is catalysed by the enzyme *"transacylase"*,
- In a similar manner, a molecule of malonyl-CoA (formed as above) combines with the adjacent "Pan-SH" of ACP of opposite monomeric unit **(Monomer II)**, to form *"Malonyl-ACP-enzyme"*. The coenzyme A of Malonyl-CoA is also removed in this step and the reaction is catalysed by the same *"transacylase"* enzyme.

3. ***Condensation reaction:*** Now, the acetate attacks malonate to form *"aceto-acetyl-ACP"*. The reaction is catalysed by the enzyme *"Keto-acyl synthase" (condensing enzyme)* and ***there is loss of one molecule of CO_2*** (decarboxylation). ***The decarboxylation provides the extra thermodynamic push to make the reaction highly favourable.*** It also makes the central carbon a better nucleo-philic agent for attacking the carbonyl carbon of the acetyl group.

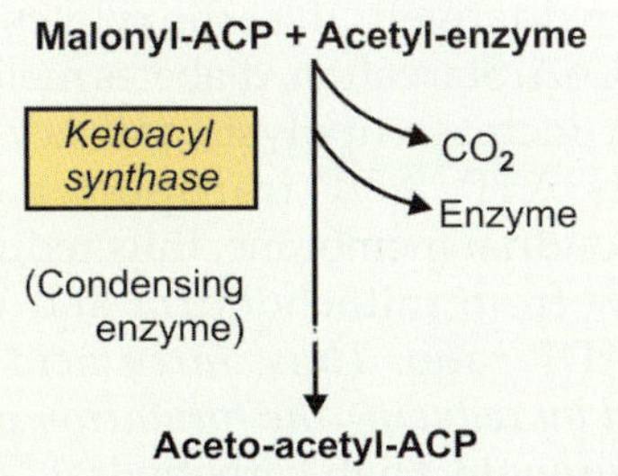

The aceto-acetate remains attached to Pan-SH of monomer II, ***the cys-SH of monomer I becomes free.***

- While aceto-acetate remains attached to "Pan-SH", ***three reactions*** take place, viz. ***reduction, dehydration,*** followed by another ***reduction***.
- ***First reaction (reduction):*** The keto-acyl group is reduced to hydroxy group (–OH) to form *"β-OH butyryl-ACP"* catalysed by the enzyme *"keto-acyl reductase"*.

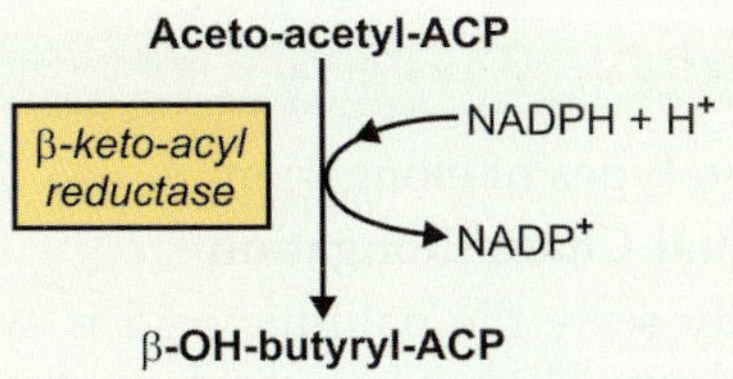

- ***Second reaction (dehydration):*** A molecule of H_2O is removed from *"β-OH-butyryl-ACP"* to form *"α, β-unsaturated butyryl-ACP"* (also called *crotonyl-ACP*), catalysed by the enzyme *"β-OH-acyl dehydratase"*.

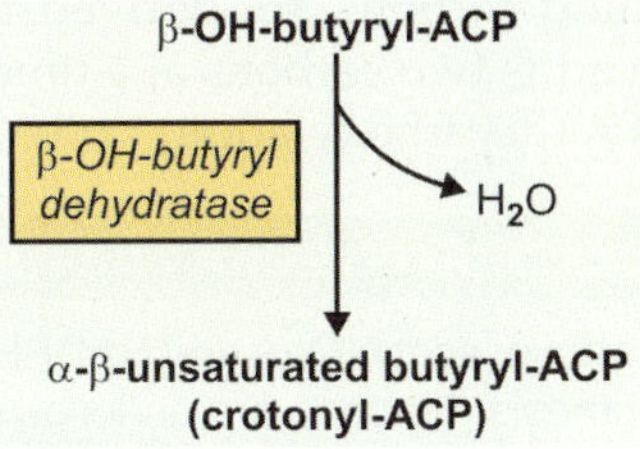

- ***Third reaction (reduction):*** The third and final reduction is catalysed by *"enoyl-reductase"* using NADPH + H^+, as a result the double bond is saturated to form *"butyryl-ACP"* **(4 carbon).**

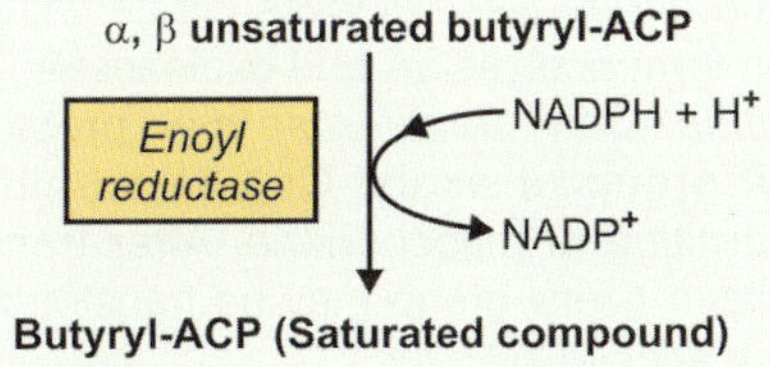

All the above three reactions occur on "Pan-SH" of monomer II. Once saturated butyric acid is formed; it is now transferred to the "Cys-SH" of monomer I which is free to accommodate.

4. ***Continuation reaction:*** Now a fresh molecule of Malonyl-CoA is taken up on to the free "Pan-SH" group of monomer II and the sequence of events is repeated ***to form a saturated six carbon fatty acid.*** Once formed this is again transferred to "Cys-SH" of monomer I.

The set of reactions on each of the monomer is repeated till a ***16 carbon palmityl-ACP*** is formed on "Pan-SH" of monomer II.

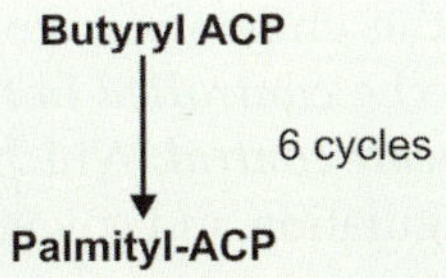

Note: The lengthening of the acyl group by each two carbon units at a time is brought about by one ATP molecule which is used for formation of malonyl-CoA from acetyl-CoA.

5. ***Termination reaction:*** Palmityl-ACP is released as palmitic acid from the enzyme complex by the enzyme *"thioesterase" (deacylase).*

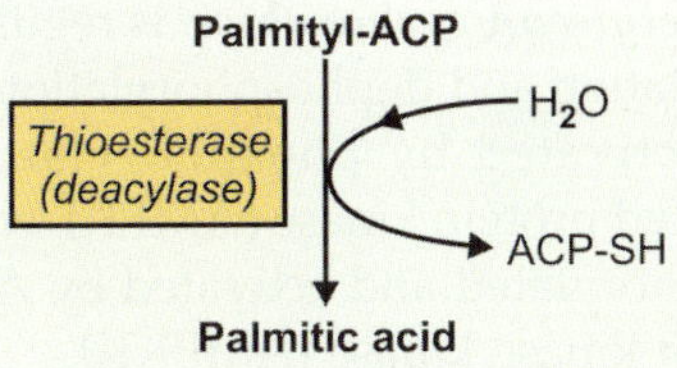

Note

- The two carbons farthest away from the –COOH group are derived directly from "acetate" (acetyl-CoA).
- The remaining carbons are derived from Malonyl-CoA which adds two carbons at a time and the third is lost as CO_2 (decarboxylation).

Sources of Acetyl-CoA and NADPH

As is clear from above, acetyl-CoA and NADPH are substrates for fatty acid synthesis. Sources are as follows:

1. Sources of Acetyl-CoA

- Acetyl-CoA is mainly found in mitochondria which cannot pass out. It forms citrate by condensing with oxaloacetate. Citrate is transported out by a transporter protein in exchange of malate. Once in cytoplasm, an enzyme *citrate lyase* cleaves citrate with the help of ATP to form acetyl-CoA and oxaloacetate.
- *Carnitine-acetyl transferase* may probably transfer acetyl group of acetyl-CoA to carnitine to form acylcarnitine in mitochondria. After translocation to cytoplasm acetyl group may be transferred to CoA to make it acetyl-CoA.

2. Sources of NADPH

- ***Hexose monophosphate shunt or HMP pathway is the main source of NADPH.***
- Oxaloacetate produced by *citrate lyase* in the cytoplasm can be reduced to malate by NADH + H^+ and *malate dehydrogenase.* There is another cytoplasmic enzyme called *malic enzyme* (NADP-malate dehydrogenase). Malate is oxidatively decarboxylated to pyruvate and NADPH is produced.
- An enzyme called *isocitrate dehydrogenase* is present in the cytoplasm. It generates NADPH mainly in ruminants. It uses NADP as the coenzyme.

SECTION FOUR

Regulation of Fatty Acid Synthesis

The activities of the enzymes involved in fatty acid synthesis appear to be ***controlled in two ways:***

(a) ***Short-term or acute control:*** Which involves allosteric or metabolic regulation and covalent modification of enzymes.

(b) ***Long-term control:*** Involving changes in the amounts of the enzymes brought about by changes in the rates of synthesis and degradation.

1. ***Acetyl-CoA carboxylase*** catalyses the rate limiting step in the *de novo* synthesis of fatty acids and provides the earliest unique point at which control can be exerted.

   Acetyl-coA carboxylase is regulated by phosphorylation and dephosphorylation. The enzyme is inactivated by phosphorylation by AMP-activated protein kinase (AMPK), which in turn is phosphorylated and activated by AMP-activated protein kinase kinase (AMPKK).

   Glucagon (and epinephrine), after increasing c-AMP, activate this latter enzyme via cAMP-dependent protein kinase. The kinase kinase enzyme is also believed to be activated by acyl-CoA.

   Insulin activates acetyl-CoA carboxylase, probably through an "activator" protein and insulin-stimulated protein kinase.
2. Glucagon and dibutyryl cAMP which inhibit fatty acid synthesis markedly decrease cytosolic citrate concentration. They also inhibit glycolysis at the level of *phosphofructokinase*, resulting in decreased glycolytic flux into pyruvate which in turn decreases the mitochondrial synthesis of oxaloacetate and citrate. ***Decrease in citrate conc. decreases acetyl-CoA carboxylase activity.***
3. Fatty acyl-CoAs regulate fatty acid synthesis by inhibiting *acetyl-CoA carboxylase* as a result of its depolymerisation.
4. Guanine nucleotides and CoA have been implicated in regulation of *acetyl-CoA carboxylase* by stimulating its activity.
5. It is also believed that regulation of *acetyl-CoA carboxylase* is done by phosphorylation-dephosphorylation of the protein.
6. Some cyclic AMP dependent *kinases* regulate the activity of *carboxylase* enzyme.
7. Fatty acid synthesis may also be regulated by altering the activity of *pyruvate dehydrogenase* which oxidatively decarboxylates pyruvate to acetyl-CoA. Starvation, diabetes mellitus and high fat diet increase lipolysis and acyl-CoA which inhibits ATP/ADP transporter of the inner mitochondrial membrane, thus reducing the ATP outflow from mitochondria and thus, raising ATP/ADP ratio. ***Thus, fatty acid synthesis is reduced by reducing the formation of acetyl-CoA from pyruvate.*** High carbohydrate diet increases fatty acid synthesis.
8. Long-term regulation is done by stimulation of *acetyl-CoA carboxylase, fatty acid synthase, ATP-citrate lyase.*
9. Insulin increases fatty acid synthesis in several ways such as decreasing lipolysis, bringing about activation of *protein phosphatase,* stimulating synthesis of *citratelyase,* enhancing formation of acetyl-CoA from pyruvate, increasing glycolysis which leads to increased pyruvate and thus acetyl-CoA.

ELONGATION OF FATTY ACIDS

There are **two types** of elongation:

I. Microsomal Chain Elongation

As already seen, the palmitic acid is synthesised in the cytoplasm. Higher fatty acids such as stearic acid

Table 25.5: Similarities and differences between FA synthesis by mitochondrial system and microsomal system

Mitochondrial system	*Microsomal system*
1. Not a common pathway.	1. Usual common pathway for chain elongation.
2. Operates in mitochondria.	2. Operates in "microsomal" system. Chain elongation of FA takes place in endoplasmic reticulum (ER).
3. Palmityl-CoA is usually the starting material and converted to stearyl-CoA. Other long-chain FA may be elongated.	3. Acyl group that may act as "Primer" molecule; May be saturated FA series from C_{10} to C_{16} and some unsaturated-C_{18} FA. ***End-product is next higher homologue of the 'Primer' acyl-CoA molecule.***
4. Operates under ***anaerobic*** conditions. It is ***favoured by a high NADH/NAD⁺ ratio*** in cells. Also in presence of excessive ethanol oxidation in liver.	4. Requires presence of O_2 **(aerobic)**.
5. ***Acetyl-CoA*** **(two carbon unit) is** ***directly incorporated*** **into the palmityl-CoA molecule.**	5. ***Acetyl moiety (two carbon) is added through malonyl CoA*** **and** ***not directly*** **by incorporating C-2 units.**
6. NADPH is required which is provided by HMP shunt.	6. NADPH is required as a reductant provided by HMP shunt.
7. Pyridoxal-P is required as a coenzyme of the "condensing enzyme" in the first reaction to incorporate C-2 unit.	7. Pyridoxal-P is not required.

(C_{18}) and others are formed from palmitate by enzymes of the microsomal elongase of chain elongation system in the smooth endoplasmic reticulum. ***It makes use of malonyl CoA and NADPH to add 2-C at a time.*** The enzymes of the elongase system are separate and not clustered like fatty acid synthase complex. ***There is a cycle of four reactions which adds 2-C per cycle.*** The cycle may be repeated to add desirable number of carbon atoms. Acyl-CoA group that acts as Primer molecule may be saturated FA series C_{10} to C_{16} and some unsaturated C_{18} FA are converted to next higher homologue. ***Molecular O_2 is necessary.***

II. Mitochondrial Chain Elongation

There is another elongation system found in mitochondria which is called as mitochondrial elongase. ***This makes use of acetyl-CoA, NADH, NADPH and ATP.*** Neither ACP nor malonyl-CoA is used in this case. Enzymes are separately located with individual activity. ***Usually palmityl-CoA is the starting material*** and converted to stearyl-CoA. But other long-chain FA can also act as substrate. This system requires ***anaerobiosis*** **and a high ratio of NADH/NAD⁺ is favourable.** Mitochondrial chain elongation is meant for producing long-chain fatty acids for incorporation into mitochondrial lipids. Table 25.5 shows similarities and differences between FA synthesis by mitochon-drial system and microsomal system.

METABOLISM OF ACYL GLYCEROLS AND SPHINGOLIPIDS

INTRODUCTION

Acyl glycerols are the major lipids in the body. Triacyl glycerol (TG) is the principal lipid of fat depots in the body and also constitute the major dietary lipids. Phospholipids are major constituents of plasma and other membranes. Glycosphingolipids account for 5 to 10 per cent of the lipids of plasma membrane and are also present in brain tissues.

BIOSYNTHESIS OF TRIACYL GLYCEROLS (TG) (FIG. 25.6)

Substrates required for synthesis of TG are:

- **α-Glycero-P (sn-glycerol-3-P)** and
- **FA (Acyl-CoA)**

Sources of α-Glycero-P

- α-Glycero-P is formed from glycerol by the action of the enzyme *glycerokinase* in presence of ATP.

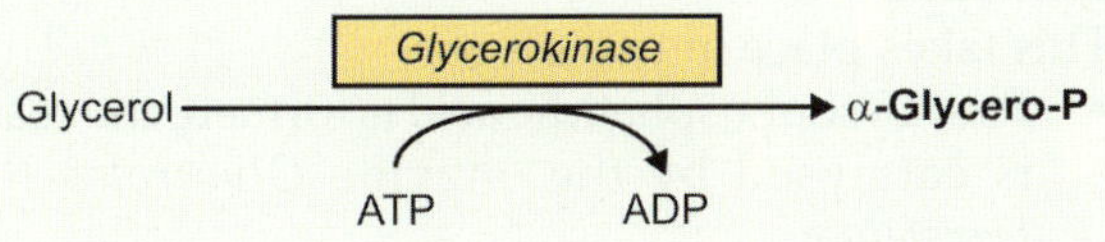

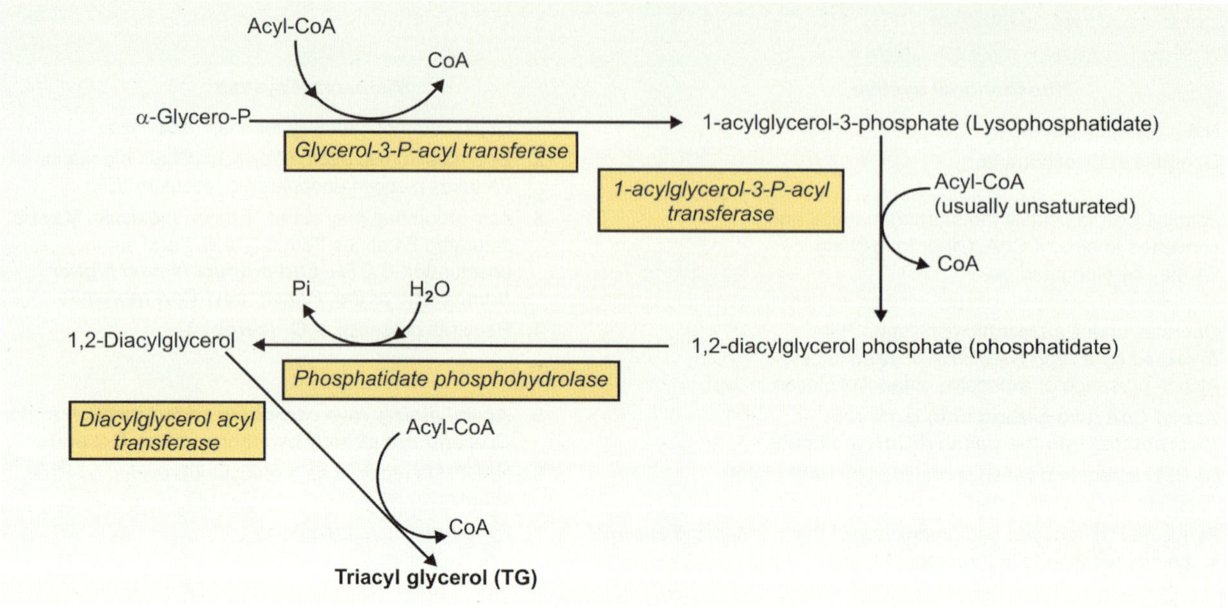

Fig. 25.6: TG synthesis

Tissues which possess the enzyme can form α-Glycero-P. ***Glycerokinase enzyme is absent or very low in activity in muscle and adipose tissue.***

- Alternative source of α-Glycero-P in tissues like muscle/adipose tissue, where *glycerokinase* is lacking is derived from an intermediate of the glycolytic system, dihydroxyacetone-P. α-Glycero-P is formed from dihydroxy-acetone-P by reduction with NADH catalysed by the enzyme *Glycerol-3-P-dehydrogenase.*

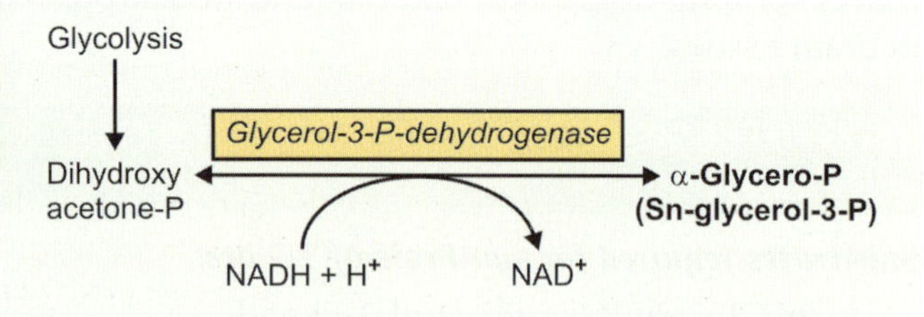

Steps of Synthesis

1. Fatty acids are activated by *Acyl-CoA synthetase* to form acyl-CoA in presence of ATP and CoA-SH.
2. Two molecules of acyl-CoA then combine with α-Glycero-P to form **Phosphatidic acid** (1,2-diacyl glycerol-P).
   This takes place in two stages:
   - First Lysophosphatidic acid is formed, the reaction is catalysed by the enzyme *Glycerol-3-P-acyl transferase.*
   - Next, phosphatidic acid is formed by the enzyme *1-acyl glycerol-3-P-acyl transferase (lysophosphatidate acyl transferase).*
3. Phosphatidic acid is now converted to 1,2-diacyl glycerol by the enzyme *phosphatidate phosphohydrolase.*
4. A further molecule of acyl-CoA is esterified with the diacyl glycerol to form one molecule of **tri-acyl glycerol (TG)**. The reaction is catalysed by the enzyme *diacylglycerol acyltransferase.*

Site of Synthesis

1. Most of the activity of these enzymes resides in endoplasmic reticulum of the cell, but some is found in mitochondria, e.g. *Glycerol-3-P-acyl transferase.*
2. *Phosphatidate Posphohydrolase* activity is found mainly in the particle-free supernatant fraction but also is membrane bound.

Note: *Monoacyl glycerol pathway:* In intestinal mucosa, a monoacyl glycerol pathway exists, whereby monoacyl glycerol can be directly converted to "diacyl glycerol" as a result of the presence of the enzyme *monoacylglycerol acyltransferase.*

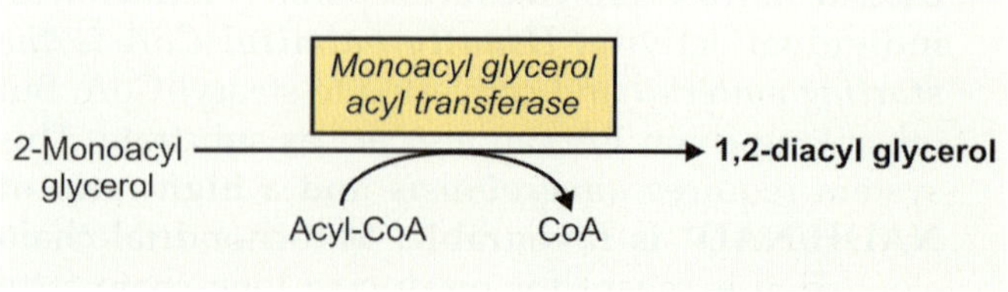

BIOSYNTHESIS OF PHOSPHOLIPIDS

A. Biosynthesis of Phosphatidyl Choline (Lecithin) and Phosphatidyl Ethanol Amine (Cephalin):

Substrates required are:

- **Choline or ethanolamine**
- **1, 2-Diacyl glycerol**

SECTION FOUR

1. Choline (and ethanolamine) has to be activated first so that it can be transferred to 1,2-diacyl glycerol. Activation takes place as shown in the box below.

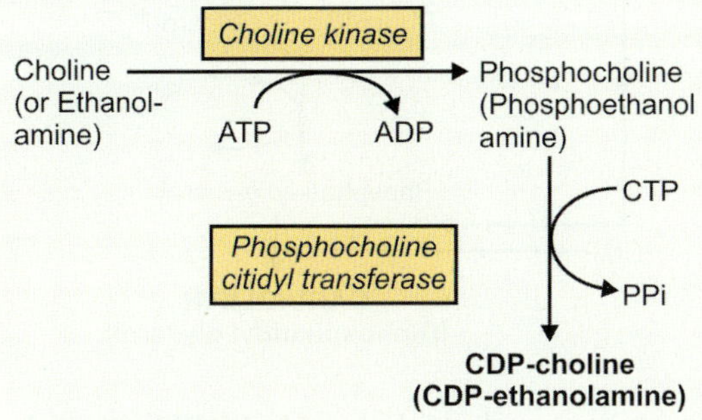

- Choline or ethanolamine reacts with ATP first, catalysed by the enzyme *choline kinase* to form corresponding monophosphate: phosphocholine (phosphoethanolamine).
- It is further activated with a molecule of CTP, catalysed by the enzyme *phosphocholine cytidyl transferase*, to form either cytidine-di-phospho-choline (CDP-choline) or cytidine-di-phospho ethanolamine (CDP-ethanolamine).

2. CDP-choline (or CDP-ethanolamine) then reacts with 1,2-diacyl glycerol catalysed by the enzyme, *phosphocholine diacyl glycerol transferase*, the phosphorylated nitrogenous base either phosphocholine/or phosphoethanolamine is transferred to 1, 2-diacyl glycerol to form either phosphatidyl choline (lecithin) or phosphatidyl ethanolamine (cephalin).

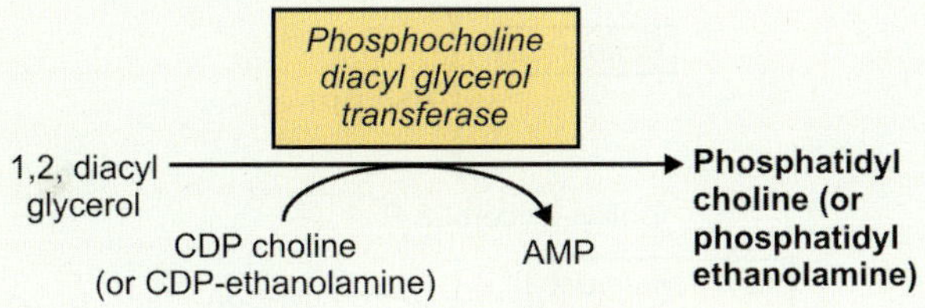

B. Synthesis of Phosphatidyl Serine

Phosphatidyl serine is formed from phosphatidyl ethanolamine directly by reaction with serine. Phosphatidyl serine may reform phosphatidyl ethanolamine by decarboxylation.

An alternative pathway in liver, but not in brain, by progressive methylation of ethanolamine can form phosphatidyl choline from phosphatidyl ethanolamine. ~CH_3 group is donated by 'active' methionine (S-adenosyl-methionine).

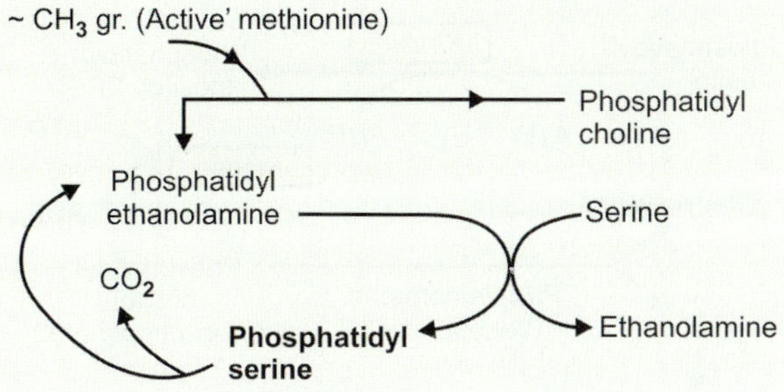

C. Biosynthesis of Phosphatidyl Inositol (lipositol)

Substrates required are:

- **1,2-diacyl glycerol-phosphate (Phosphatidate)**
- **CTP and inositol**

1. CTP reacts with phosphatidate, catalysed by the enzyme *CTP-Phosphatidate cytidyl transferase* to form CDP-diglyceride (cytidine diphosphate-diacyl glycerol).
2. In the next step, this compound reacts with inositol, catalysed by the enzyme, *CDP-diacyl glycerol inositol transferase* to form phosphatidyl inositol.

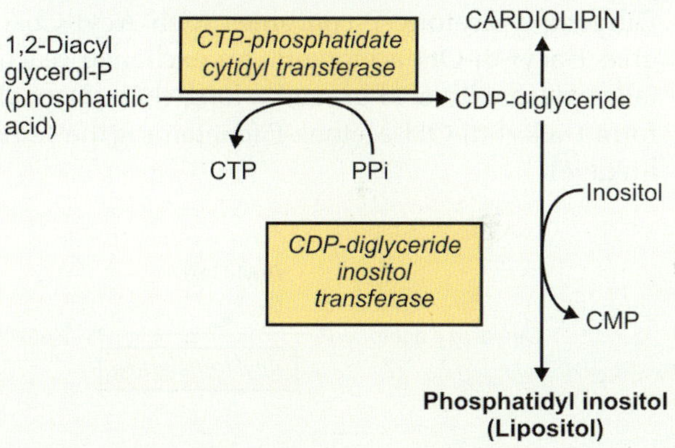

By successive phosphorylations, phosphatidyl inositol is transformed first to "Phosphatidyl-inositol-4-P", and then to "phosphatidyl-inositol, 4,5 - biphosphate". The latter compound is broken down to diacyl glycerol and Inositol-tri-P by certain protein hormones that increase Ca^{++} inside the cell which acts as second messenger in hormone action (Refer to function of PL).

D. Synthesis of Cardiolipin

Cardiolipin is a phospholipid present in mitochondria. Chemically it is "Di-phosphatidyl glycerol".

Substrates required are:

- **α-Glycero-P**
- **CDP-diglyceride** (formed in synthesis of phosphatidyl inositol—see above).

Both of the above two reacts to produce phosphatidyl glycerol which forms cardiolipin, with the help of another molecule of CDP-diglyceride.

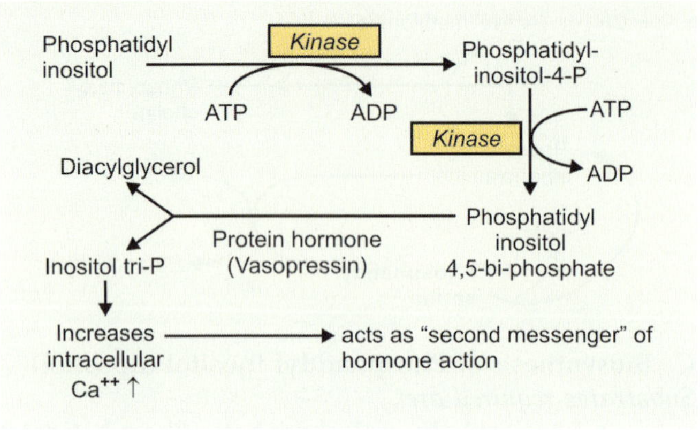

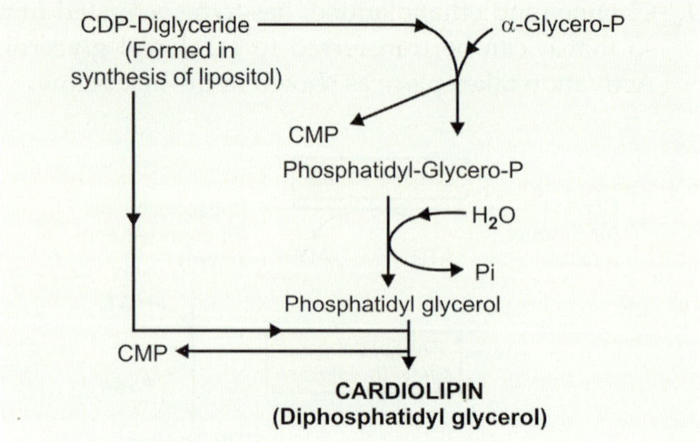

E. Biosynthesis of plasmalogens: A plasmalogen is the PL in which 1 or 2 position has an alkenyl residue containing the vinyl ether aldehydogenic linkage ($-CH_2-O-CH = CH-R'$). Plasmalogens are present in mitochondria.

Substrate required are:

1. **di-OH-acetone-P,** and
2. **Acyl-CoA**

Steps

- Dihydroxy-acetone-P combines with Acyl-CoA to give 1-acyl-di-OH-acetone-P. An exchange reaction takes place with acyl-gr. and a long chain alcohol to form 1-alkyl-di-OH-acetone-P (containing the "ether" linkage).
- In next step, with the help of NADPH, it is reduced to form 1-alkyl-glycerol-3-P.
- The compound undergoes further acylation in 2-position resulting to formation of '1-alkyl-2-acyl-glycerol-3-P (similar to phosphatidic acid).
- It is further hydrolysed to give the free glycerol derivative.
- Finally plasmalogens are formed by desaturation of the analogous 3-phospho-ethanol amine derivative. Formation is shown schematically in Figure 25.7.

Plasma activating factor (PAF): Plasma activating factor (PAF) is synthesised from the corresponding 3-phospho-choline derivative and has been identified chemically as "1-alkyl-2- acetyl-Sn-glycerol-3-phosphocholine".

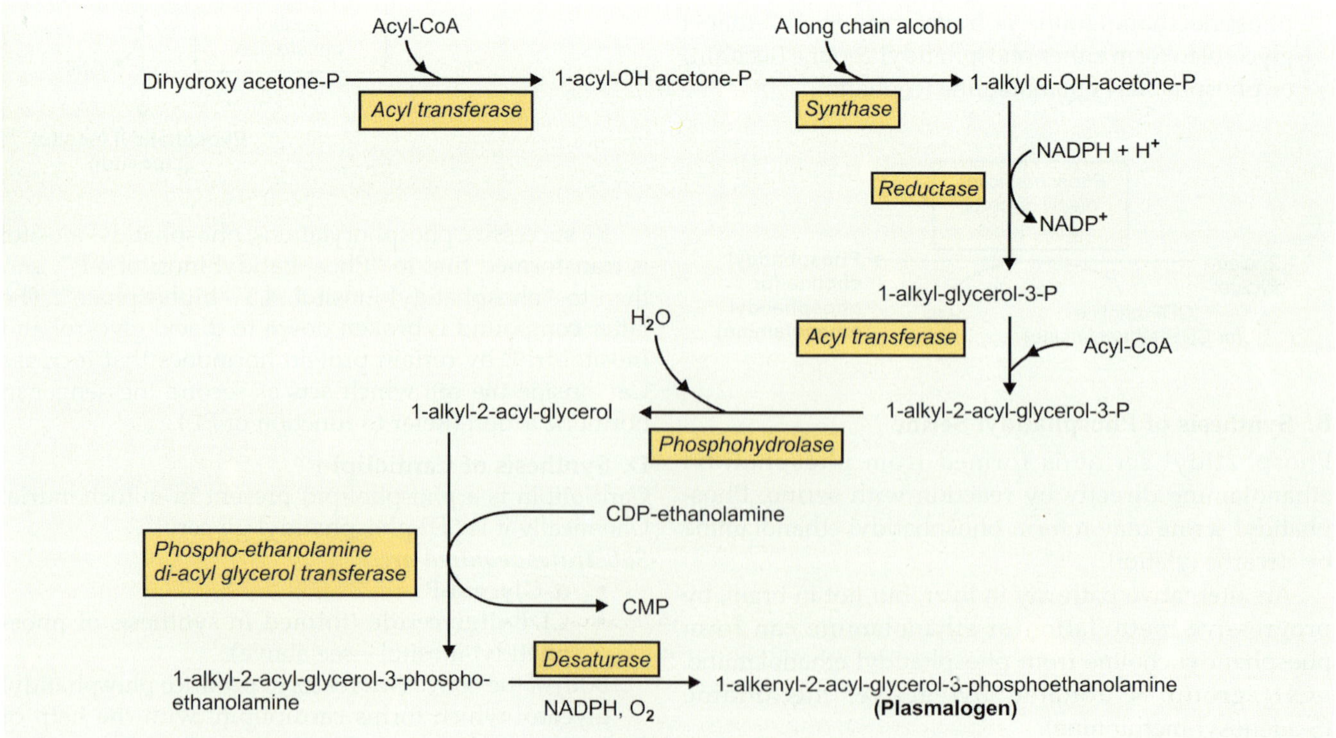

Fig. 25.7: Biosynthesis of plasmalogen

SECTION FOUR

Functions

- It is formed by many blood cells and other tissues and it brings about aggregation of platelets.
- It has also hypotensive action, i.e. lowers BP
- Also possesses ulcerogenic property.

F. Biosynthesis of Sphingomyelin

Steps of Synthesis

- First the aminoalcohol "Sphingosine" (Sphingol) is synthesised
- Next 'ceramide' is formed.
- Finally from ceramide → the PL sphingomyelin is synthesised.

1. ***Synthesis of sphingosine:*** Sphingosine is synthesised in endoplasmic reticulum.

Substrates required are:

- ***Palmitoyl-CoA,*** and
- ***Amino acid serine.***

Coenzymes/cofactors required are:

- Pyridoxal-P (B_6-PO_4)
- Mn^{++}
- NADPH and
- Fp (Flavo-Protein).

Steps

- Following activation by combination with B_6-PO_4 and Mn^{++}, the amino acid serine combines with palmitoyl-CoA to form ***3-keto-sphinganine***, after loss of CO_2 and CoA-SH is liberated.
- 3-ketosphinganine is reduced by the enzyme *3-keto-sphinganine reductase* in presence of NADPH to form ***Dihydrosphingosine***.
- Finally this is followed by an oxidative step which is catalysed by another reductase, *sphinganine reductase* which is a FP-enzyme, forming finally *sphingosine*.

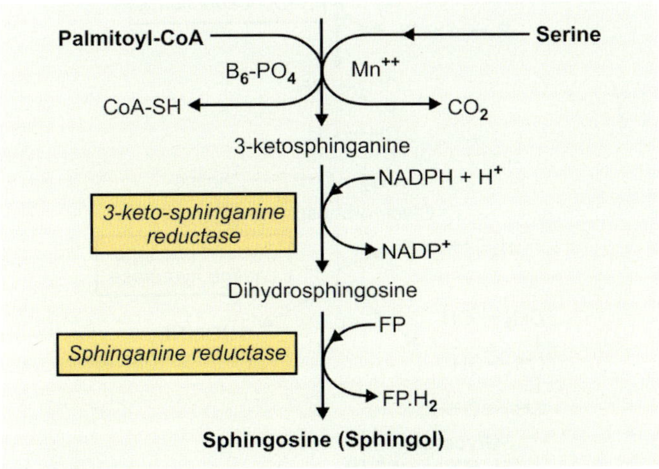

2. ***Synthesis of ceramide: (N-acyl sphingosine):*** Ceramide is formed by combination of sphingosine and acyl-CoA. Acyl group is usually represented by long-chain saturated FA or a monoethenoid.

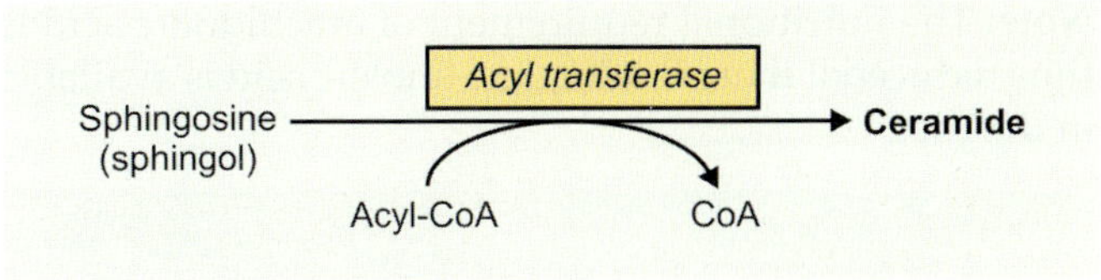

3. ***Synthesis of Sphingomyelin***

Finally sphingomyelin is formed either ***of the following two ways:***

- Ceramide reacts with a molecule of CDP-choline
- Alternatively, ceramide can react with a molecule of phosphatidyl choline (Lecithin) (See box below).

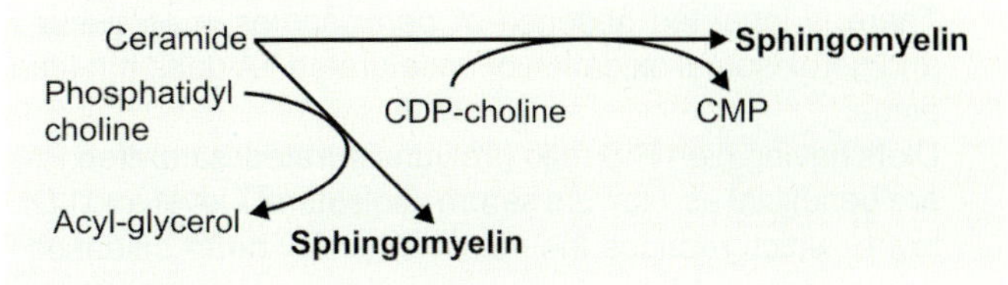

G. Synthesis of Arachidonic Acid

Linoleic acid, if available, in the body and provided in the diet, can be converted to arachidonic acid by a **3-stage reaction.**

- After the linoleic acid is activated to form 'Linoleyl-CoA' it is first dehydrogenated to form "γ-linolenyl-CoA".
- γ-linolenyl-CoA is converted to form 'Dihomo-γ-linolenyl-CoA' ("eicosatrienoyl-CoA"), by addition of a 2-carbon unit (acetyl-CoA) in the microsomal system of chain elongation in presence of NADPH.
- The latter forms arachidonic acid by further dehydrogenation.

Synthesis of Arachidonic Acid

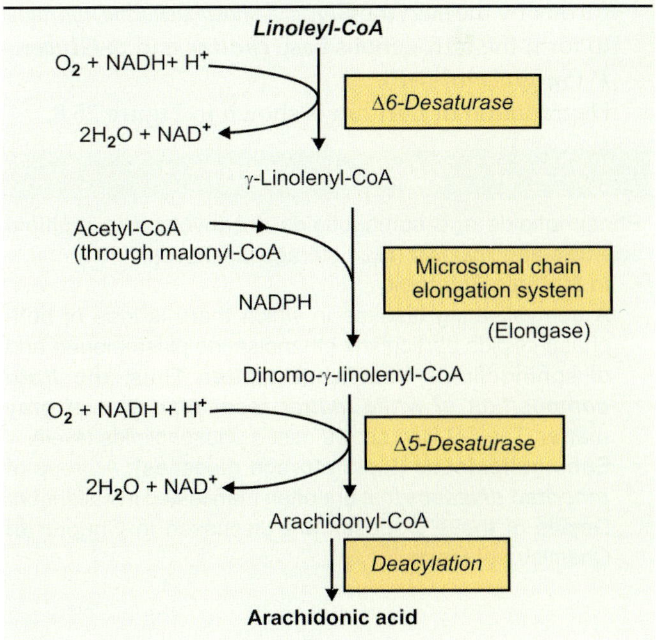

Note: The nutritional requirement of arachidonic acid is thus reduced if there is adequate linoleic acid is available in the diet.

CLINICAL ASPECT

- Abnormal metabolism of essential fatty acids has been correlated with various clinical disorders, viz. cystic fibrosis, hepatorenal syndrome, Sjögren syndrome, multisystem neuronal degeneration, alcoholism and cirrhosis liver, acrodermatitis enteropathica, Crohn's disease and Reye's syndrome.
- High levels of very long chain polyenoic acids have been found in the brains of patients with Zellweger's syndrome. There is inherited absence of *peroxisomes* in all tissues and peroxisomal oxidation of unsaturated FA does not take place.
- Diets having high P:S ratio (Polyunsaturated: saturated FA) are beneficial as it lowers serum cholesterol↓ level and LDL level↓ which reduces the risk of coronary heart diseases.

CATABOLISM OF LECITHIN (PHOSPHATIDYL CHOLINE)

Steps

- Lecithin is degraded in the body by the enzyme *Phospholipase A_2*, which catalyses the hydrolysis of the ester bond in Position-2 (β-position) to form a free fatty acid and Lysolecithin (Lysophosphatidyl choline).
- Lysolecithin is attacked by the enzyme *Lysophospholipase (Phospholipase B)*, which hydrolyses the ester bond at α-position (1-position), liberating another molecule of free fatty acid and forms "Glycerol-phosphoryl choline" (Glyceryl phosphocholine).
- Finally, Glyceryl-phosphocholine is hydrolysed further by the enzyme *Glyceryl phosphocholine hydrolase* to form the nitrogenous base ***choline*** and ***α-Glycero-P (Sn-glycerol-3-P).***

  Degradation of Lecithin is shown in Figure 25.8.

CLINICAL ASPECT

Phospholipids and sphingolipids are involved in multiple sclerosis and lipidosis (Lipid storage diseases).

- **In Multiple Sclerosis**

  A demyelinating disease in which there is loss of both phospholipids particularly ethanolamine plasmalogen and of sphingolipids from white matter. Thus, the ***lipid composition of white matter resembles that of gray matter.*** The CS fluid shows raised phospholipids levels.
- **Sphingolipidoses (Lipid storage diseases):** A group of ***inherited diseases*** that are often manifested in childhood. Details of these diseases are discussed in Chapter of Chemistry of Lipids.

Treatments: There is no effective treatment for many of these diseases.

Recently some success has been achieved with enzymes that have been ***chemically modified to ensure binding to receptors of target cells,*** e.g. to macrophages in the liver in order to deliver β-glucosidase (glucocerebrosidase) in the treatment of Gaucher's disease.

- A recent promising approach is **substrate reduction therapy** to inhibit the synthesis of sphingolipids.
- **Gene therapy** for lysosomal disorders is currently under investigation.

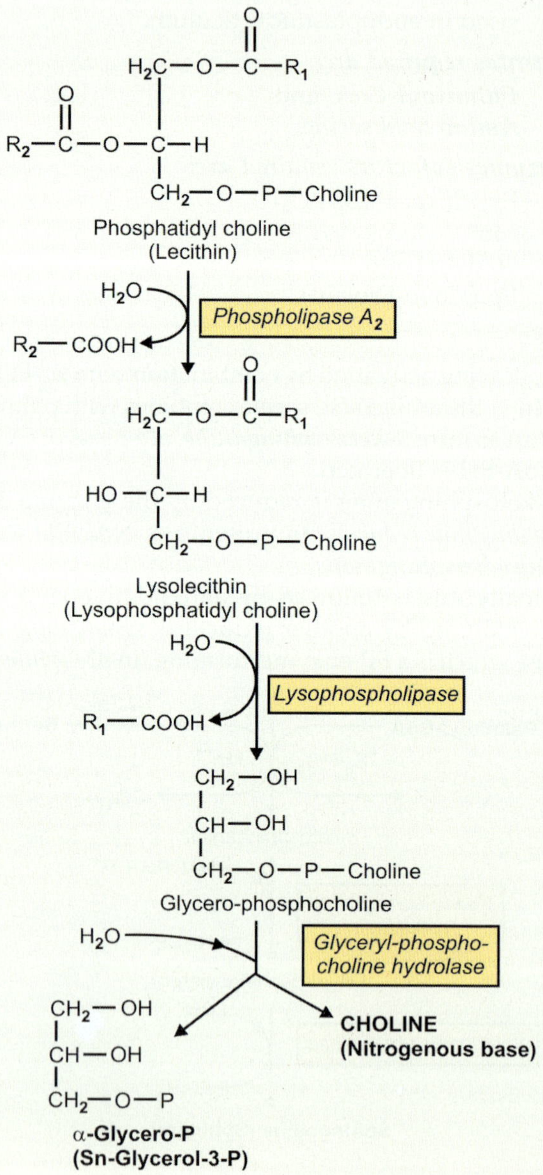

Fig. 25.8: Degradation of lecithin

KETOSIS

Under certain metabolic conditions associated with a high rate of fatty acid oxidation, liver produces considerable quantities of compounds like **acetoacetate** and **β-OH butyric acid**, which pass by diffusion into the blood. Acetoacetate continually undergoes spontaneous decarboxylation to produce **acetone**. ***These three substances are collectively known as "ketone bodies" (or "acetone bodies").*** Sometimes also called as "ketones", which is rather a misnomer.

Inter-relationship of these three substances are shown below:

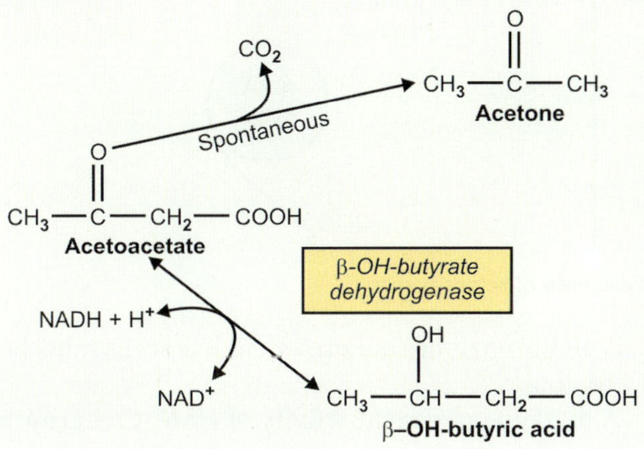

Concentration of Ketone Bodies

Concentration of total ketone bodies in the blood of well-fed individuals does not normally exceed 1 mg/100 ml (as acetone equivalents).

Urine: Loss via urine is usually less than 1 mg/ 24 hrs in humans.

Ketoacidosis: Acetoacetic acid and β-OH-butyric acid are moderately strong acids. They are buffered when present in blood and tissues, entailing some loss of buffer cations, which progressively depletes the **alkali reserve↓** causing ketoacidosis.

Note: This may be fatal in uncontrolled diabetes mellitus.

> ***Certain Terminologies***
>
> ***Ketonaemia:*** Rise of ketone bodies in blood above normal level is known as ***ketonaemia.***
>
> ***Ketonuria:*** When the blood level of ketone bodies rises above the renal threshold, they are excreted in urine and is called as ***ketonuria.***
>
> ***Ketosis:*** Accumulation of abnormal amount of ketone bodies in tissues and body fluids is termed as ketosis, where the urinary excretion of β-OH butyric acid exceeds 200 mg daily (normal 5 to 10 mg). The ***overall pattern is called ketosis.***

Causes

1. ***Starvation:*** Simplest form of ketosis occurs in starvation. ***Mechanism:*** Involves depletion of available carbohydrate reserve, coupled with mobilisation of FFA and oxidation to produce energy.
2. ***In Pathologic States***
   - In ***Diabetes mellitus:*** Clinical and experimental.
   - In some types of ***alkalosis:*** Ketosis may develop.
   - Pregnancy toxaemia in sheep and in lactating cattle.
3. In prolonged ***ether anaesthesia.***
4. Other non-pathologic forms of ketosis are found under conditions of:
   - **High fat feeding.**
   - **After severe exercise** in the postabsorptive state.
5. **Injection of anterior pituitary extracts.**

Site of Formation and Fate

Liver appears to be the only organ which produces ketone bodies and add to the blood. Extrahepatic tissues can pick up ketone bodies from the circulating blood and utilise them as respiratory substrates. Net flow of ketone bodies from the liver to extrahepatic tissues results from an active enzymatic mechanism in the liver which exists, for the production of ketone bodies, coupled with very low activity of the enzymes rather than their absence, responsible for their degradation or utilisation. The reverse situation exists in extrahepatic tissues (Fig. 25.9).

KETONE BODY FORMATION IN LIVER (KETOGENESIS)

Enzymes are mitochondrial:

Steps

1. ***Aceto-acetyl-CoA: Aceto-acetyl-CoA is the starting material for ketogenesis.*** This can arise in two ways:
   (a) Directly during the course of β-oxidation of fatty acids, or
   (b) As a result of condensation of two C-2 units, i.e. 'active acetate' (acetyl-CoA) by reversal of *thiolase* reaction.
2. ***Formation of Acetoacetate: Acetoacetate is the first ketone body to be formed.*** This can occur in two ways:

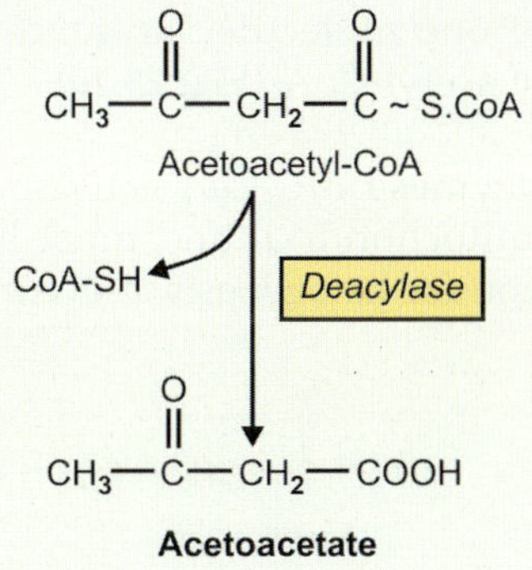

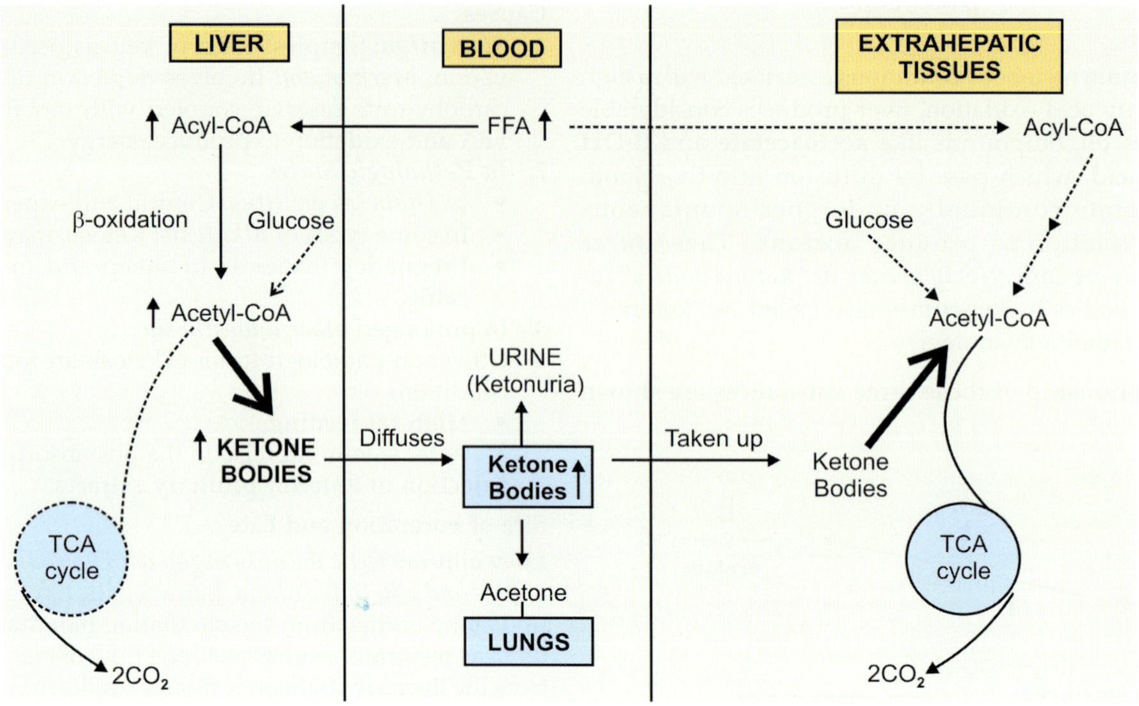

Fig. 25.9: Formation, utilisation and excretion of ketone bodies

(a) ***By deacylation:*** Acetoacetate can be formed from aceto-acetyl-CoA by simple deacylation catalysed by the enzyme *aceto-acetyl-CoA deacylase.*

The above does not seem to be the major pathway when excessive amount of ketone bodies are formed, the deacylation reaction is not enough to cope up.

(b) ***Second pathway:*** Formation of acetoacetate via intermediate production of **"β-OH-β-methyl glutaryl CoA" (HMG-CoA).** ***Present opinion favours the HMG-CoA pathway as the major route of ketone body formation.***

Steps

Involves two steps:

- Condensation of aceto-acetyl-CoA with another molecule of acetyl-CoA ***to form β-OH-β methyl glutaryl-CoA (HMG-CoA)*** catalysed by the enzyme *HMG-CoA synthase* (mitochondrial enzyme).
- HMG-CoA is then acted upon by an another enzyme, *HMG-CoA Lyase,* which is also mitochondrial enzyme, to produce one molecule "acetoacetate" and one molecule of acetyl-CoA (Fig. 25.10).

Note

- Both the enzymes *HMG-CoA synthase* and *HMG-CoA Lyase* are mitochondrial and must be available in mitocondrion for ketogenesis to occur.
- Both the enzymes are present in liver cells mitochondria only.
- A marked increase in activity of *HMG-CoA Lyase* has been noted in fasting.
- HMG-CoA is a ***committed step***. Cholesterol also can be formed by *"HMG-CoA reductase".*

3. ***Formation of Acetone:*** As stated earlier, acetone is formed from acetoacetate by spontaneous decarboxylation (Non-enzymatic).

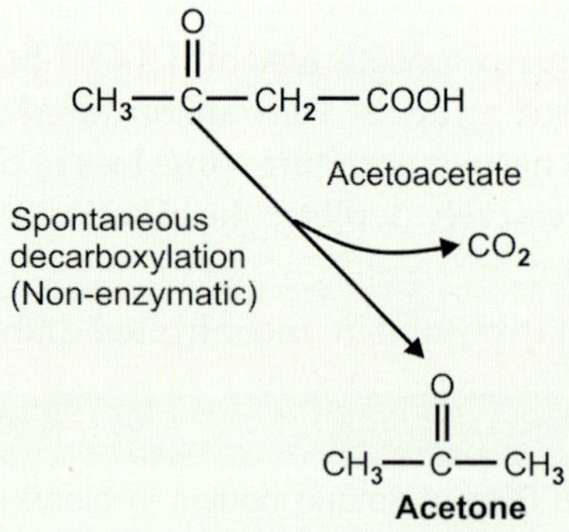

4. ***Formation of β-OH Butyrate:*** Acetoacetate once formed is converted to β-OH-butyric acid; the reaction

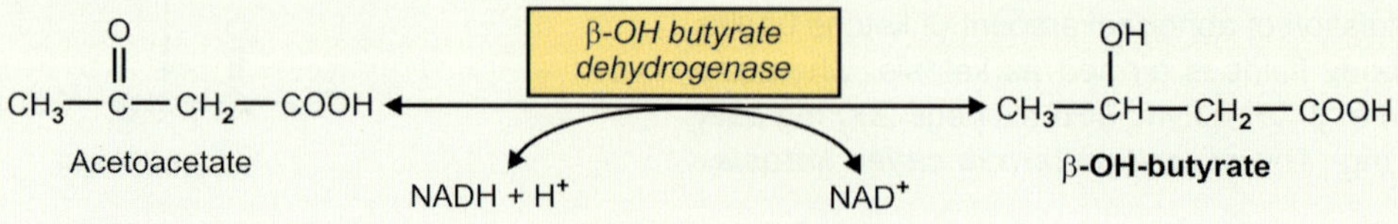

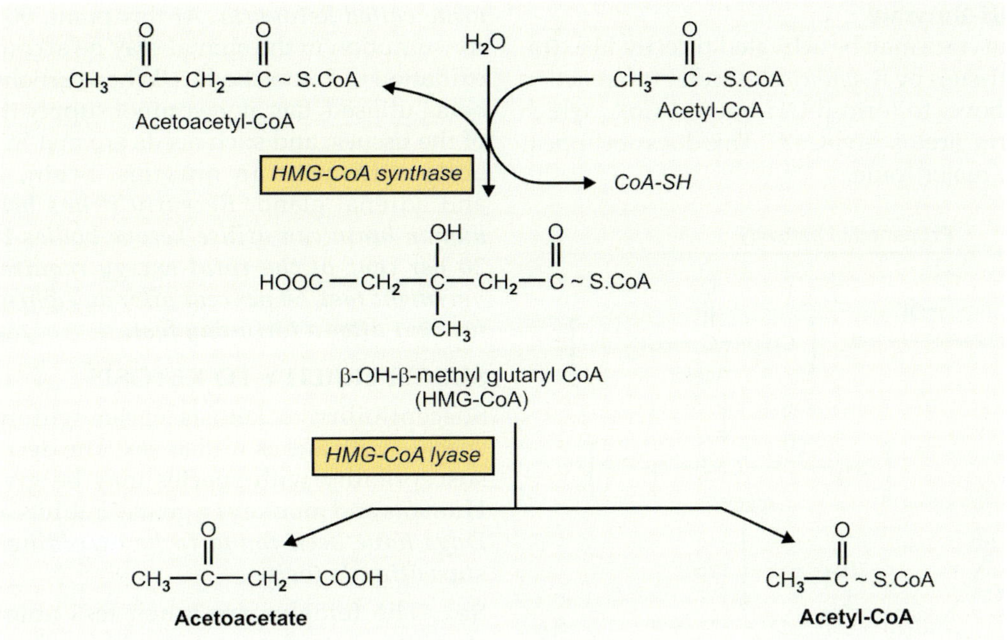

Fig. 25.10: Ketogenesis

is catalysed by the enzyme β-*OH-butyrate dehydrogenase*, which is present in liver and also found in many other tissues. ***β-OH-butyrate is quantitatively the predominant ketone body present in blood and urine in Ketosis*** (Refer to the box on bottom of previous page).

UTILISATION OF KETONE BODIES

Ketolysis

Ketone bodies are utilised by extrahepatic tissues as "fuel". Acetoacetate cannot be reactivated directly in Liver, it can be in cytosol, which is a much less active pathway, and there acetyl-CoA serves as precursor in cholesterol synthesis. Thus, Liver though is equipped with enzymatic machinery for production of acetoacetate, it cannot utilise it further as explained. Hence, it accounts for the net production of the Ketone bodies by the Liver, which diffuses out into the blood.

1. **Activation of Acetoacetate**

   Two reactions take place in extrahepatic tissues which activate acetoacetate to form aceto-acetyl-CoA, which is further utilised.

   (a) Action of acetoacetate with succinyl-CoA: Major Pathway by which acetoacetate is activated in extrahepatic tissues. Acetoacetate reacts with one molecule of succinyl-CoA (intermediate of TCA cycle), catalysed by the enzyme ***CoA transferase*** (also called ***thiophorase***), which transfers CoA from succinyl-CoA to acetoacetate, thus forming acetoacetyl-CoA and succinate.

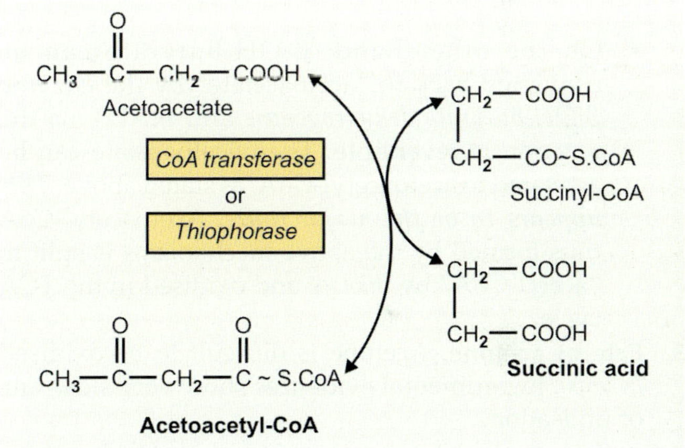

   (b) Second mechanism: Activation of acetoacetate with ATP and CoA-SH, catalysed by the enzyme *Acetoacetyl-CoA synthetase*. This is probably not a major pathway, can occur to some extent.

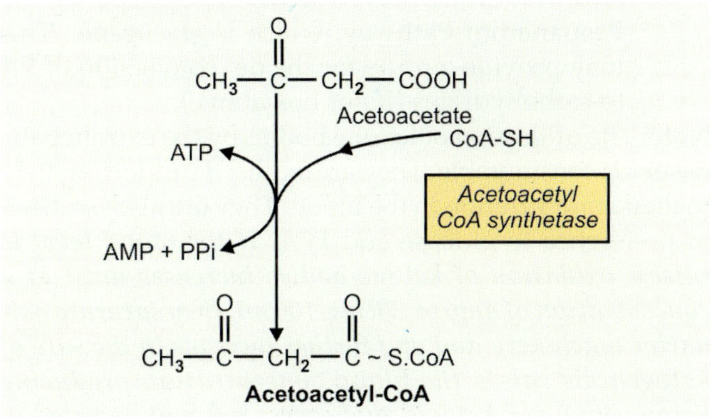

SECTION FOUR

2. **Fate of β-OH-Butyrate**
   - β-OH-butyrate may be activated directly in extrahepatic tissues by a *synthetase*, similar to a reaction stated above, to form β-OH-butyryl-CoA, which can reform 'acetoacetyl-CoA'. This does not appear to be the major route.

Propanediol pathway

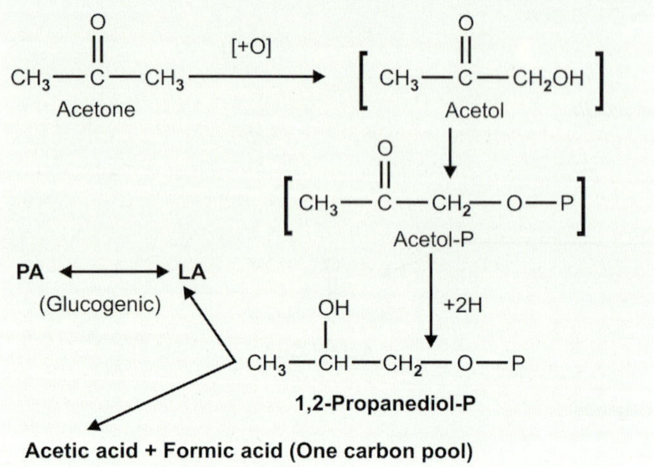

   - On the other hand, β-OH-butyrate can be converted back to "acetoacetate" by the enzyme *β-OH-butyrate dehydrogenase* and NAD^+, as the reaction is **reversible.** Then acetoacetate can be activated to acetoacetyl-CoA, as stated above. ***This appears to be the major route.*** Acetoacetyl-CoA thus formed by the above mechanisms is split to "acetyl-CoA" by *thiolase* and oxidised in the TCA cycle.
3. **Fate of acetone:** Acetone is difficult to be oxidised *in vivo*. Experimental evidences show very slow rate of utilisation.
   - Some authorities propose a reversal of the reaction, in which acetone is converted back to acetoacetate.
   - Excess of acetone can be breathed out and also excreted in urine. This gives a **fruity smell,** in the breath and in urine
   - Another possible pathway proposed is the **Propanediol Pathway**, which is *glucogenic*. This may provide a route for the net conversion of FA to carbohydrates (Refer box above).

Note: The utilisation of ketone bodies by the extrahepatic tissues is considerable. They are oxidised proportionately to their concentration in the blood. They are also oxidised in preference to glucose and FFA. ***If the blood level is raised, oxidation of ketone bodies increases until at a concentration of approx. 70 mg/100 ml, they saturate oxidative machinery and any further increase in the rate of ketogenesis raises the blood concentration producing ketonaemia and ketosis, and excess amount excreted in urine called ketonuria.*** At this point, 90 per cent of O_2 consumption in the animal may be accounted for by the oxidation of ketone bodies. When carbohydrates are not being utilised, fats alone cannot supply the "fuel" needs of the tissues, and such needs are met in part by ketone body utilisation in muscles, brain, kidney, heart and adrenal gland. ***Recently it has been shown that human brain can utilise ketone bodies to the extent of 20 per cent of the total energy requirement after an overnight fast, 60 per cent after an eight-day fast and 80 per cent after a forty-day fast.***

SUSCEPTIBILITY TO KETOSIS

Susceptibility to ketosis varies widely with animal species, with age as well as sex. The decreasing order of susceptibility with species may be given as follows: **Humans and monkeys > goats > rabbits and rats > dogs.** ***Dogs have been found to be exceedingly resistant to starvation ketosis.***

Sex: The females are much less able to withstand starvation ketosis as compared to males.

Age: Infants and young children are more susceptible than adults.

FACTORS DETERMINING MAGNITUDE OF KETOGENESIS

Though ketone bodies are being formed constantly and being utilised, ***in vivo*** ketosis does not occur unless there is a concomitant rise in the level of circulating FFA. ***Severe ketosis is accompanied invariably by very high concentration of plasma FFA.*** Numerous experiments *in vivo* have demonstrated that ***fatty acids are the precursors of ketone bodies and liver is the main site of ketone body formation.***

Liver in both fed and fasting conditions, is capable of extracting 30 per cent or more of FFA passing through it. So when FFA concentration in plasma is very high, substantial amount of FFA passes through the liver.

Two fates await the FFA, taken up by the liver cells after activation to Acyl-CoA.

Either,

- They are esterified to form TG, phospholipids and cholesterol esters, or
- They undergo β-oxidation to form acetyl-CoA.

Acetyl-CoA in turn is oxidised in TCA cycle to form CO_2 and water. But if in excess, they are used for ketone bodies formation.

The two fates are shown schematically in Figure 25.11.

Note

1. Control is excercised initially in adipose tissue. As stated above, ketosis does not occur *in vivo* unless there is a concomitant rise in free fatty acid (FFA) level arising from lipolysis in adipose tissue.

SECTION FOUR

2. Hence, factors regulating mobilisation of FFA from adipose tissue are important in controlling ketogenesis.

ANTIKETOGENIC MECHANISMS

1. Esterification to Form TG

Once FFA, derived from lipolysis of TG in adipose tissue, are esterified in the liver, they become negligible source of ketone bodies. Hence, ***esterification of FFA can be regarded as a significant antiketogenic mechanism in the liver.***

From Figure 25.11 it will be seen that if pathway (a) predominates, there will be less acyl-CoA to be diverted to Pathway (b) and (c). Precursor substance for esterification is "α-Glycero-P". Hence, ***capacity for esterification by Liver depends on the availability of the precursor substance "α-Glycero-P".*** During conditions of ketosis, viz. in starvation, strenuous exercise, etc., there is little glycogen in liver and to maintain the blood glucose to meet "basal needs", liver forms glucose by gluconeogenesis from glucogenic amino acids, lactates and pyruvates and glycerol. Hence in these conditions, insufficient carbohydrate is available to supply enough 'α-Glycero-P' to esterify all increased influx of FFA. As quantity of acyl-CoA available for oxidation increases ↑, the oxidative machinery TCA cycle gets saturated and more acetyl-CoA is diverted for ketone bodies formation [Pathway (b) and (c) operate more].

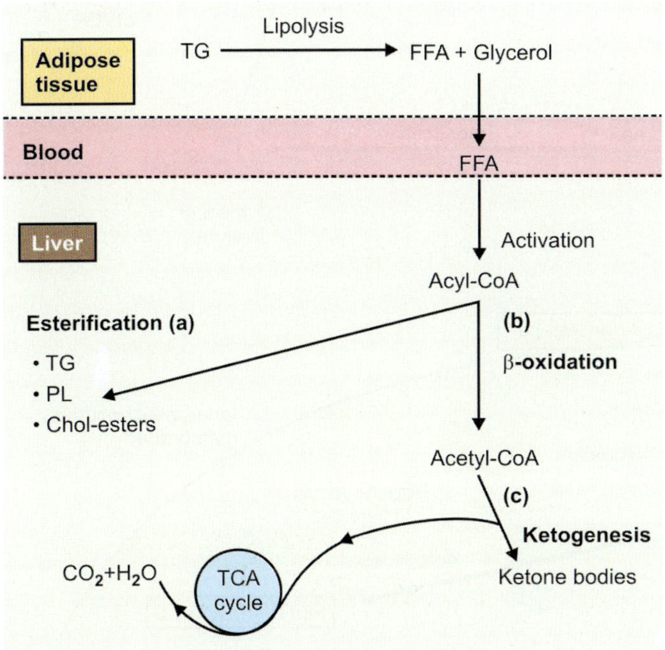

Fig. 25.11: Fate of FFA in liver

2. Role of Oxaloacetate

Theoretically a fall in concentration of 'oxaloacetate'↓ particulary within the mitochondria could cause impairment of TCA cycle to metabolise acetyl-CoA. Since oxaloacetate is in the main pathway of gluconeogenesis, enhanced gluconeogenesis leading to a fall in concentration of OAA may account for severe forms of ketosis. Role of OAA is shown schematically in Figure 25.12.

Oxaloacetate (OAA) plays an important key role in metabolism in that by combining with acetyl-CoA it forms citrate to start the TCA cycle. If OAA is not present in adequate amounts, which can be due to enhanced gluconeogenesis, the acetyl-CoA instead of entering TCA cycle, is diverted to formation of ketone bodies. Thus, ***OAA can prevent ketosis by taking up acetyl-CoA to form citrate, in its absence, ketone bodies accumulate.***

3. Role of Malonyl-CoA

Malonyl-CoA is the first compound formed in extramitochondrial *de novo* FA synthesis which regulates the activity of the enzyme, *carnitine palmitoyl-transferase I.*

- In "fed state", FA synthesis is more and thus concentration of Malonyl-CoA ↑ increases which inhibits the enzyme, thus **"switching off" β-oxidation.** ***Thus in fed condition there is active lipogenesis, FFA are mostly esterified to TG and transported out of liver as VLDL.***
- **With starvation and in DM,** as the concentration of FFA increases, the long acyl-CoA inhibits *acetyl-CoA carboxylase,* the first enzyme in FA synthesis, thus ***decreasing the concentration of Malonyl-CoA↓*** significantly, which releases the inhibition of the enzyme *carnitine palmitoyl-transferase I,* allowing more acyl-CoA to be oxidised by β-oxidation.

These events are aggravated by the decrease in [Insulin]/[Glucagon] ratio, which increases lipolysis in adipose

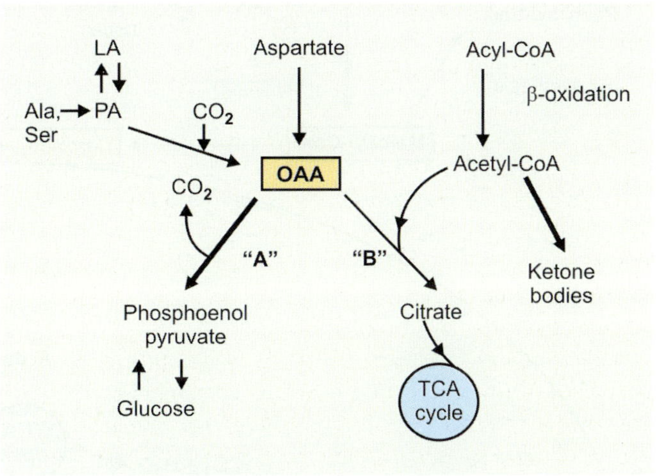

Fig. 25.12: Role of OAA in ketone bodies formation

SECTION FOUR

tissue, releasing more FFA and inhibiting *acetyl-CoA carboxylase*. ***Increased acetyl-CoA is diverted to ketone bodies formation.*** Thus, ***increased malonyl-CoA concentration can prevent ketosis,*** by decreasing β-oxidation of FFA and producing less acetyl-CoA. Role of malonyl-CoA and "Carnitine-Palmitoyl *Transferase I*" is shown in Figure 25.13.

GLUCOSE-FA CYCLE OF RANDLE

The suppression of glucose oxidation during periods of carbohydrate deprivation leading to release of NEFA (non-esterified fatty acid)/or FFA, from the fat depots into the blood and their subsequent oxidation has been termed by Randle as the **Glucose-FA cycle**.

Summary

1. Ketosis arises as a result of deficiency in available carbohydrates.
2. Two principal actions in fostering ketogenesis are:
   - An imbalance between esterification and lipolysis in adipose tissue with consequent release of FFA in circulation.
   - FFA are principal substrates for ketone bodies formation in Liver.

   Hence, all factors, metobolic or endocrine affecting the release of FFA from adipose tissue influence ketogenesis.
3. Availability of carbohydrates in Liver provides α-Glycero-P, which determines the extent to which the large influx of FFA into the liver is esterified. FFA which remains unesterified is oxidised to CO_2 and forms ketone bodies.
4. As quantity of FFA undergoing oxidation increases in starvation and DM, more form ketone bodies and less is oxidised in TCA cycle to form CO_2 and H_2O. This is regulated in such a manner that the total energy production remains constant.
5. Ketone bodies produced in liver are not oxidised by liver and they diffuse into the circulation from where they are extracted by extrahepatic tissues preferentially to other tissues.

KETOGENIC/ANTIKETOGENIC RATIO IN DIET

While prescribing diets the proportion of the ketogenic and antiketogenic substances should be so regulated that ketosis may be avoided.

It is found that ***if the ratio between the molecules of the ketogenic substances and the molecules of the antiketogenic substances exceeds 2, ketone bodies appear in urine. The clinical rule is that the total fat (F) content of the diet must not exceed the sum of twice the carbohydrates (C) and ½ of the Protein (P), i.e.***

$$F \leq (2\,C + 1/2\,P)$$

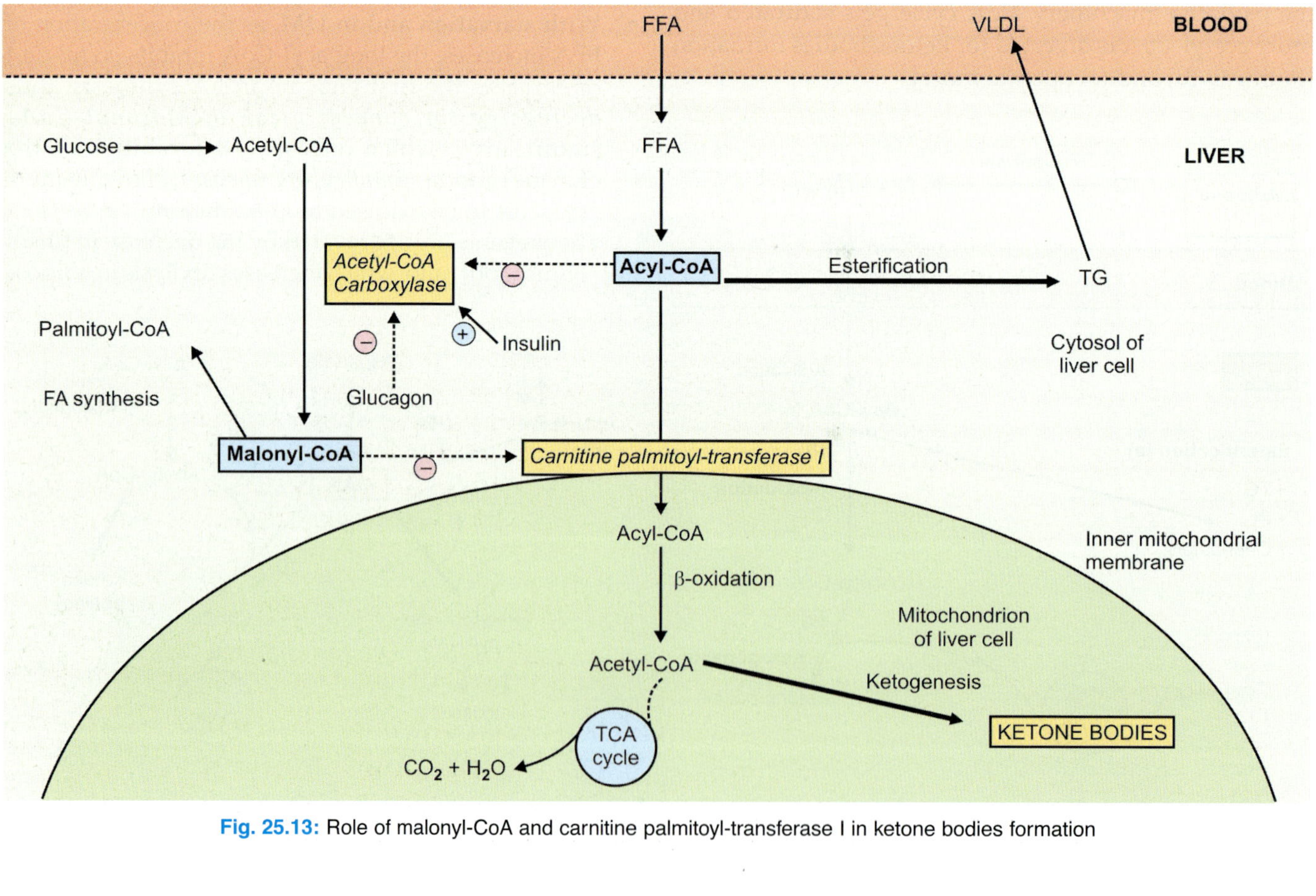

Fig. 25.13: Role of malonyl-CoA and carnitine palmitoyl-transferase I in ketone bodies formation

SECTION FOUR

Ketosis is abolished by increasing the metobolism of carbohydrates in the liver. In diabetes mellitus, this is achieved by giving insulin, and in ketosis due to carbohydrate deprivation by giving glucose or substances readily convertible to glucose or glycogen.

Ketogenic substances: The ketogenic substances are:

- **All FFA** (i.e. 90% of food fats)

Note: Glycerol part, the product of hydrolysis of TG, is glucogenic and hence, it is antiketogenic.

- **Proteins:** Ketogenic amino acids (40%).

The above are the sources from which ketone bodies are formed.

Antiketogenic substances: These are substances which prevent the formation of ketone bodies. They provide glucose, which in turn can provide α-Glycero-P, required for esterification.

- ***All carbohydrates***
- ***Insulin***
- ***60 per cent of proteins:*** Glucogenic amino acids
- ***10 per cent of dietary fats:*** Glycerol part which is glucogenic.

METABOLISM OF CHOLESTEROL

For chemistry of cholesterol, its properties and occurrence/distribution—refer to Chemistry of Lipids.

For absorption of cholesterol—refer to chapter on Digestion and Absorption of Lipids.

BIOSYNTHESIS OF CHOLESTEROL

A number of established facts regarding cholesterol biosynthesis are as follows:

- **Site of Synthesis**

  Essentially all tissues form cholesterol. ***Liver is the major site of cholesterol biosynthesis;*** also other tissues are active in this regard, e.g. adrenal cortex, gonads, skin, and intestine are most active. Low order of synthesis: Adipose tissue, muscle, aorta and neural tissues. Brain of newborn can synthesise cholesterol while ***the adult brain cannot synthesise cholesterol.***

  Efficiency of formation of cholesterol from labelled C_{14} acetate:

Tissues	*Efficiency of cholesterol formation (Liver = 100)*
• Liver	100
• Adult skin	90
• Small intestine	60
• Gonads	31
• Kidney	4
• Adult brain	0
• Newborn brain	185

- **Enzymes:** Enzyme system involved in cholesterol biosynthesis are associated with:
  - Cytoplasmic particles "microsomes"
  - Soluble fraction—cytosol.
- **Acetate**

  'Active' acetate (acetyl-CoA) is the starting material and principal precursor. The entire carbon skeleton, all 27 C of cholesterol in humans can be synthesised from active acetate.

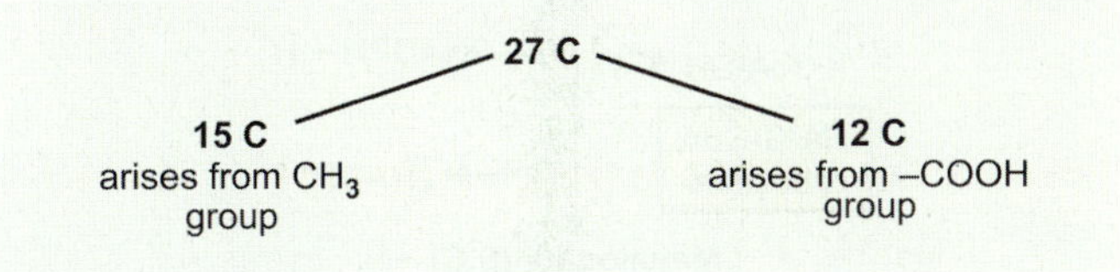

Steps of biosynthesis: Cholesterol biosynthesis can be thought of as occurring in **Five groups of reactions.** They are:

I. **Synthesis of mevalonate:** A 6-C compound from acetyl-CoA.

II. **Formation of "Iso-Prenoid units"** (C-5) from Mevalonate: By successive phosphorylations and followed by loss of CO_2.

Note: The isoprenoid units are regarded as the building blocks of the steroid nucleus.

III. **Formation of Squalene:** A 30-carbon aliphatic chain, formed by condensation of six isoprenoid units.

IV. **Cyclisation of Squalene** to form **Lanosterol.**

V. **Conversion of Lanosterol** → to form **cholesterol.**

I. Synthesis of Mevalonate from Acetyl-CoA

Consists of two steps:

- ***Formation of HMG-CoA:* (β-OH-β-methyl glutaryl-CoA):** HMG-CoA can be formed in the cytosol from acetyl-CoA in two steps catalysed by the enzyme ***thiolase*** and ***HMG-CoA synthase.***

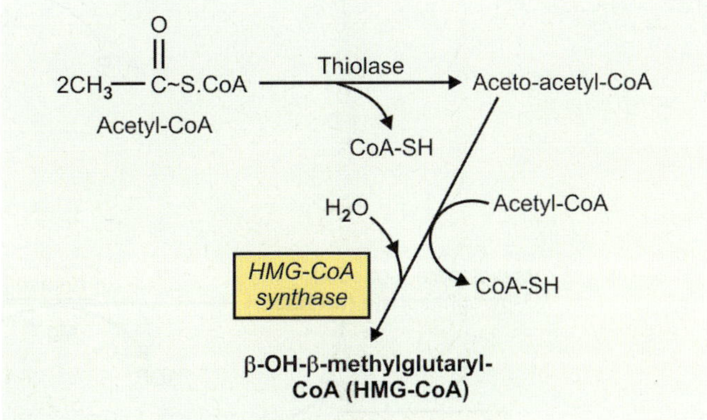

Note

1. HMG-CoA may also be produced as an intermediate in the metabolic degradation of amino acid L-Leucine.
2. There are two pools of HMG-CoA:

i. Mitochondrial: Concerned with ketogenesis
ii. Extramitochondrial (cytosolic): Concerned with synthesis of Mevalonate and iso-Prenoid units.

- In the next step, which is the ***rate-limiting step***, HMG-CoA is converted to Mevalonic acid (Mevalonate) catalysed by the enzyme *HMG-CoA reductase*.

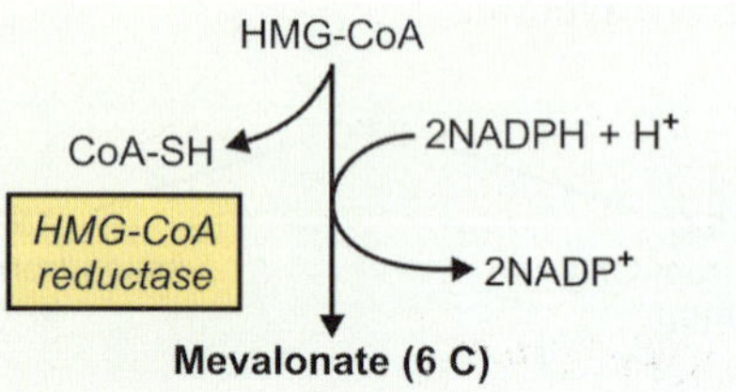

Characteristics of this reaction:

- Most important and ***rate limiting step*** and ***irreversible*** reaction
- Enzyme contains—SH group
- NADPH required as cofactor—supplied by HMP-Pathway.
- Enzyme activity not affected in Diabetic rats; Fasting animals show decreased activity of the enzyme; Dietary cholesterol and endogenously synthesised cholesterol inhibits this 'rate-limiting' step ("feedback" inhibition).
- **Hormones:** Insulin and thyroid hormones—increases reductase activity; Glucagon and gluco-corticoids—reduces the activity.
- Also inhibited by cyclic AMP.

II. Formation of Isoprenoid Units

- Mevalonate is phosphorylated by ATP to form several 'active' phosphorylated intermediates.
- Three such phosphorylated compounds are formed and it is followed by decarboxylation to form first "active" iso-prenoid unit: ***"Iso-pentenyl pyrophosphate" (5 C).*** One of intermediate phosphorylated compound is "Mevalonate-3-phospho-5-Pyrophosphate" which is unstable.
- Iso-pentenyl pyrophosphate undergoes isomerisation to form another 5 C iso-prenoid unit, called ***"3-3′-Dimethyl allyl pyrophosphate".***

III. Formation of squalene:

The pyrophosphorylated isoprenoid units condense to form ultimately a 30-carbon aliphatic chain called **Squalene**. The condensation occurs in **three steps:**

- One molecule of iso-pentenyl-pyrophosphate first condenses with one molecule of 3,3-dimethyl allyl pyrophosphate to form a **10-C** compound called ***geranyl pyrophosphate,*** the reaction is catalysed by the enzyme *geranyl pyrophosphate synthase.*

  Characteristics of this reaction

  Iso-pentenyl pyrophosphate by virtue of terminal double bond methylene carbon is a "nucleophilic" agent; and 3-3′-dimethyl allyl pyrophosphate by virtue of its double bond esterification with a strong acid is "electrophilic" agent, hence both of them are ideal agents for condensation.
- Another molecule of iso-pentenyl pyrophosphate reacts with geranyl pyrophosphate to form the **15C**

Synthesis of Isoprenoid Units

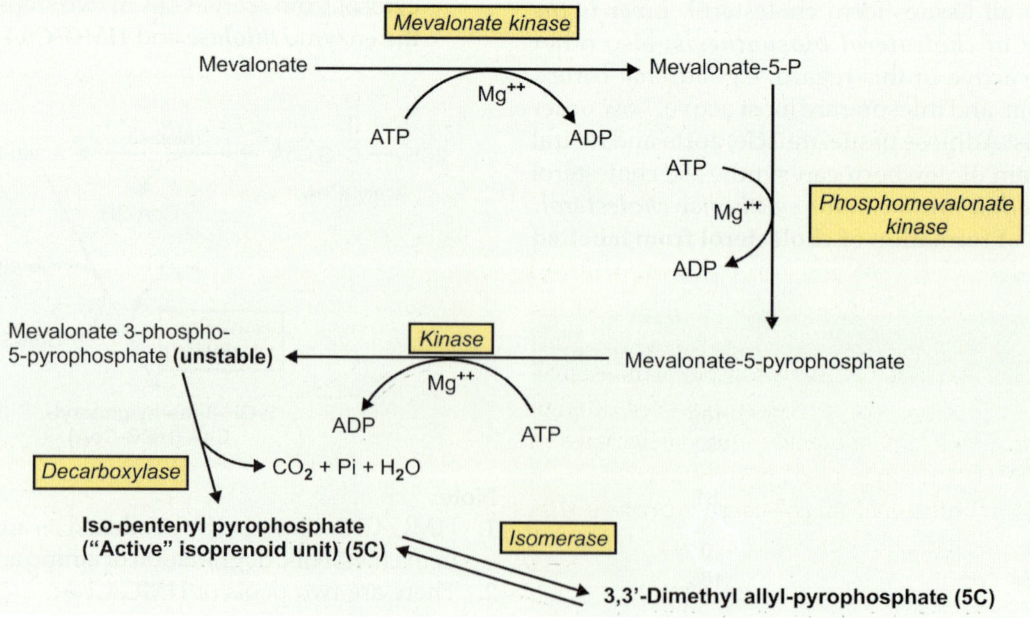

Synthesis of Squalene

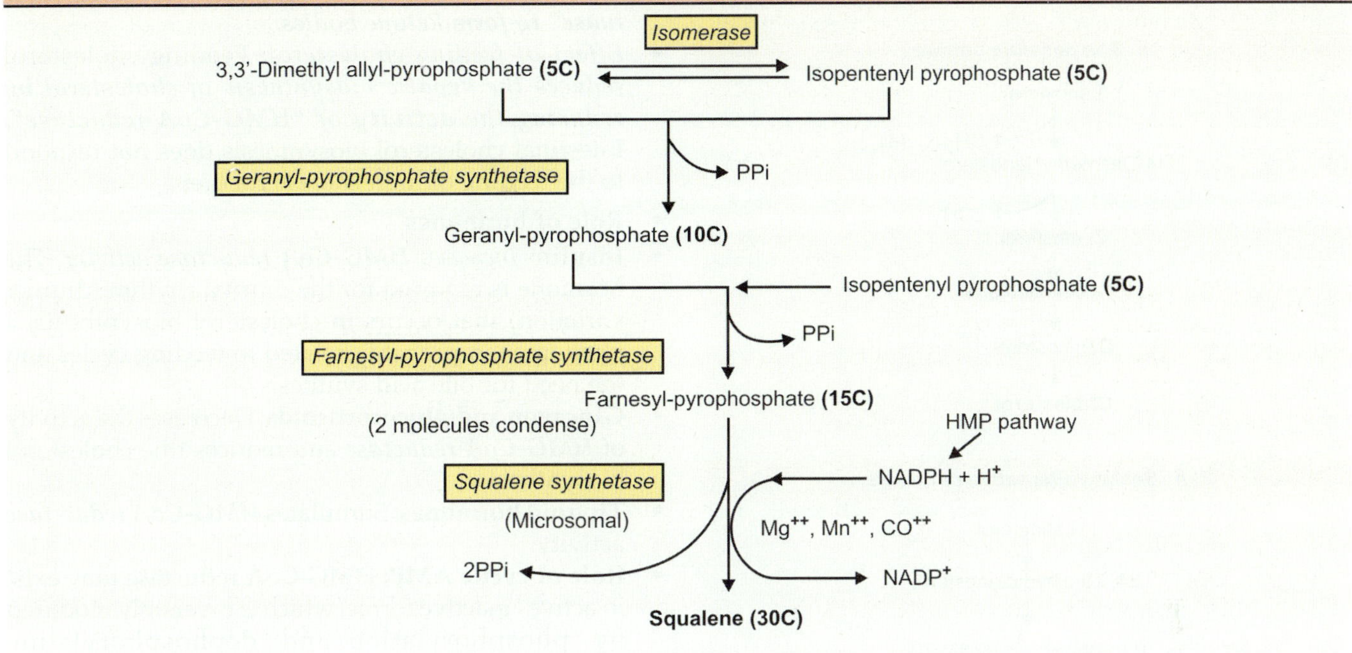

compound ***farnesyl pyrophosphate***, the reaction is catalysed by the enzyme *"farnesyl pyrophosphate synthetase"*.

- Finally, two molecules of farnesyl pyrophosphate condenses at "Pyrophosphate end" to form **30C** aliphatic compound called **squalene,** reaction is catalysed by the enzyme *squalene synthetase*.

Characteristics of this reaction:

- The enzyme *squalene synthetase* is microsomal—liver enzyme is firmly bound to microsomes.
- NADPH is required as a coenzyme as electron donor, provided by HMP pathway.
- Cofactors required: Mg^{++}, Mn^{++} and CO^{++}.

Note: Squalene or sterol carrier protein:

1. It is suggested that intermediates from squalene to cholesterol may be attached to a special "carrier protein", known as ***"squalene or sterol carrier protein".*** This protein binds sterols and other insoluble lipids, allowing them to react in an 'aqueous phase' of the cell.
2. It is also envisaged that it is in the form of "cholesterol-sterol carrier protein" that cholesterol is converted to steroid hormones and bile acids and participates in formation of membranes and lipoproteins.
3. It is also as cholesterol-sterol carrier protein that cholesterol might affect the activity of *β-HMG-CoA-reductase* activity.

IV. Cyclisation of squalene to form lanosterol: The formation of lanosterol from squalene takes place in two steps:

- In the first step squalene-2,3-epoxide is formed catalysed by the enzyme *squalene mono-oxygenase;* which requires NADPH and molecular O_2.
- In the next step, an enzyme ***cyclase*** brings about the cyclisation of squalene to form lanosterol.

V. Conversion of lanosterol to cholesterol: Main changes that are brought about are:

- ***Removal of three angular –CH_3 groups.*** This involves a series of reactions, mechanism of demethylation is not properly known. CH_3 group at C_{14} is first eliminated.
- Shift of double bond between C_8 and C_9 to C_5 and C_6, and
- Saturation of double bond in side chain.

Formation of Lanosterol

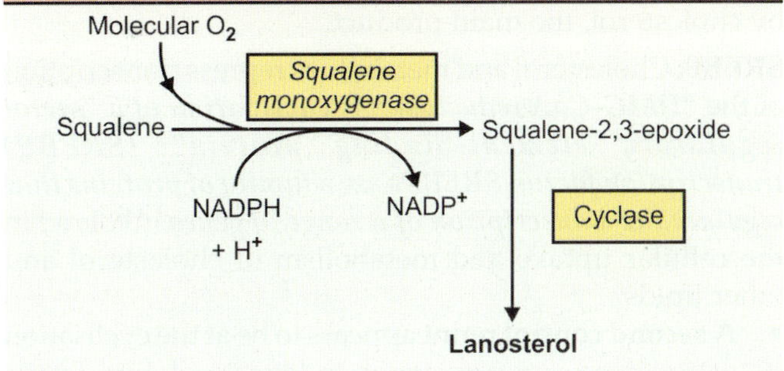

Two possible pathways have been suggested: (shown in box).

I. One possible pathway

Lanosterol
↓ → CH_3 (C-14)
14-Desmethyl lanosterol
↓ → $2CH_3$
Zymosterol
↓
$\Delta^{7,24}$-cholestadienol
↓
Desmosterol
↓
Cholesterol

II. Second alternative pathways

Lanosterol
↓
24, 25-dihydrolanosterol
↓
α 4-methyl-Δ^8-cholesterol
↓
α 4-methyl-Δ^7-cholesterol
↓
Δ^7-cholesterol
↓
7-dehydrocholesterol
↓
Cholesterol

Control of Cholesterol Biosynthesis

- **SREBP Pathway**

The steps in the biosynthesis of cholesterol to HMG-CoA are **reversible.** The formation of mevalonate, however, in the next step is **irreversible** and is the **committed** step. Regulation of cholesterol biosynthesis is exerted near the beginning of the pathway at the **HMG-CoA reductase step, the rate limiting step.** The reduced synthesis of cholesterol in starving animals is accompanied by a decrease in the activity of the enzyme. However, it is only hepatic synthesis that is inhibted by dietary cholesterol. "HMG-CoA reductase" in liver is inhibited byy mevalonate, the immediate product of the pathway and by cholesterol, the main product.

SREBP: Cholesterol and metabolites repress transcription of the ***"HMG-CoA reductase"*** via ***activation of a "sterol regulatory element-binding protein" (SREBP) transcription factor. SREBPs*** are a ***family of proteins that regulate the transcription of a range of genes*** involved in the cellular uptake and metabolism of cholesterol and other lipids.

- A **second control point** appears to be at the cyclisation of squalene and conversion to lanosterol, but details of the regulation at this step is not clear.
- ***Effect of starrving/fasting:*** Fasting and/starvation also inhibt the enzyme and ***activate "HMG-CoA lyase" to form ketone bodies.***
- ***Effect of feeding cholesterol:*** Feeding cholesterol ***reduces the hepatic biosynthesis of cholesterol by reducing the activity of "HMG-CoA reductase".*** Intestinal cholesterol biosynthesis does not respond to the feeding of high cholesterol diets.
- **Role of hormones**
- **Insulin:** *Increases **HMG-CoA reductase** activity.* The hormone is required for the diurnal rhythm (diurnal variation) that occurs in cholesterol biosynthesis, a phenomenon probably related to feeding cycles and the need for bile acid synthesis.
- **Glucagon and/glucocorticoids:** Decreases the activity of ***HMG-CoA reductase*** and reduces tthe cholesterol biosynthesis.
- **Thyroid hormones:** Stimulates ***HMG-CoA reduc-tase*** activity.
- **Role of cyclic AMP:** HMG-CoA reductase may exist in active/inactive forms, which is reversibly modified by phosphorylation/and dephosphorylation mechanisms, whhich may be mediated by c-AMP dependent protein kinases. ***Cyclic AMP inhibits cholesterol biosynthesis by converting "HMG-CoA reductase" to inactive form.***

TRANSPORT OF CHOLESTEROL

Cholesterol in the diet is absorbed from the intestine, and in company with other lipids, are incorporated into "chylomicrons" and also to some extent 'VLDL'. Of the cholesterol absorbed, 80 to 90 per cent in the lymph is esterified with long-chain FA. Esterification may occur in the intestinal mucosa. In man, the total plasma cholesterol varies from 150 to 250 mg% (average 200 mg%), rising with age, although there are wide variations between individuals. The greater part is found in the 'esterified' form and is transported as "lipoproteins" in plasma. ***Highest proportion of circulating cholesterol is found in LDL (β-lipoproteins)*** which carry cholesterol to tissues and also in HDL, which takes cholesterol to liver from tissues for degradation ***(scavenging action).*** However, under conditions, where the VLDL are quantitatively predominant, an increased proportion of plasma cholesterol will reside in this fraction. Dietary cholesterol takes several days to equilibrate with cholesterol in plasma, and several weeks to equilibrate with cholesterol in tissues. Free cholesterol in plasma and liver equilibrates in matter of hours. In general, free cholesterol exchanges readily between tissues and lipoproteins, whereas cholesterol esters do not exchange freely. Some plasma cholesterol ester may be formed in HDL as a result of transesterification reaction in plasma between cholesterol and FA in position-2 of lecithin which

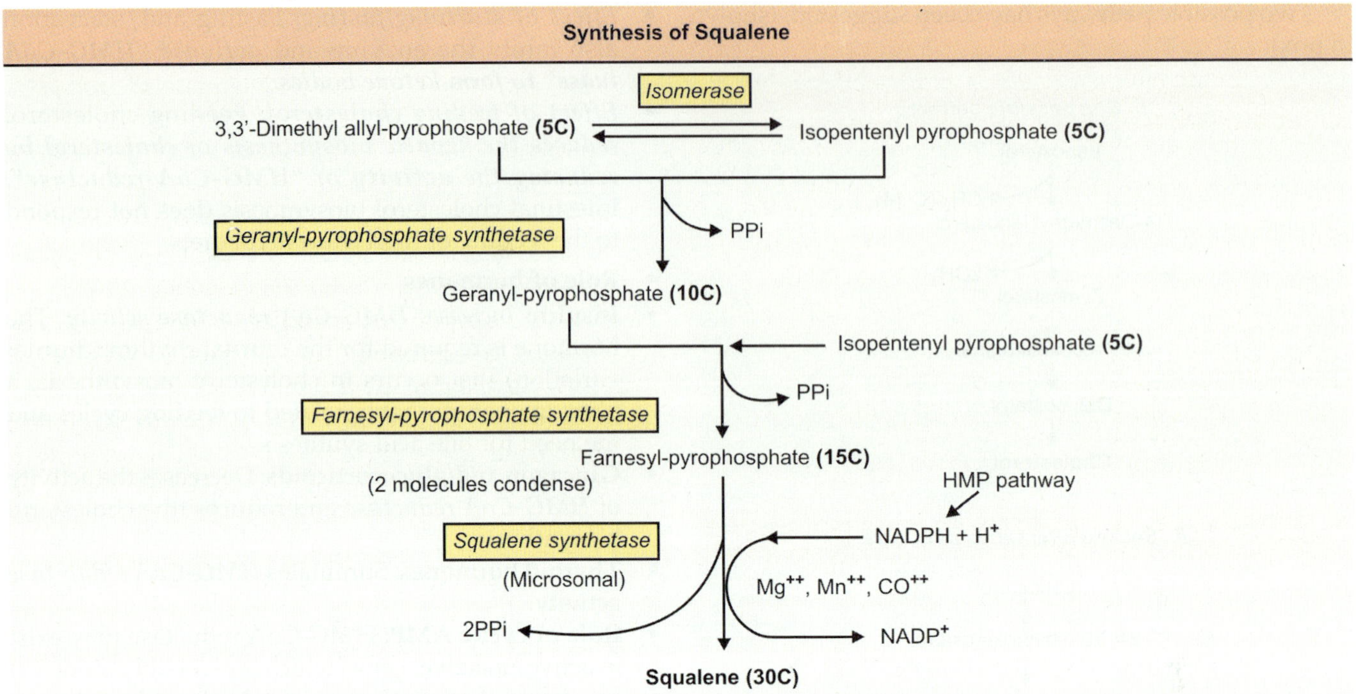

compound ***farnesyl pyrophosphate***, the reaction is catalysed by the enzyme *"farnesyl pyrophosphate synthetase"*.

- Finally, two molecules of farnesyl pyrophosphate condenses at "Pyrophosphate end" to form **30C** aliphatic compound called **squalene,** reaction is catalysed by the enzyme *squalene synthetase*.

Characteristics of this reaction:

- The enzyme *squalene synthetase* is microsomal—liver enzyme is firmly bound to microsomes.
- NADPH is required as a coenzyme as electron donor, provided by HMP pathway.
- Cofactors required: Mg^{++}, Mn^{++} and CO^{++}.

Note: Squalene or sterol carrier protein:

1. It is suggested that intermediates from squalene to cholesterol may be attached to a special "carrier protein", known as ***"squalene or sterol carrier protein"***. This protein binds sterols and other insoluble lipids, allowing them to react in an 'aqueous phase' of the cell.
2. It is also envisaged that it is in the form of "cholesterol-sterol carrier protein" that cholesterol is converted to steroid hormones and bile acids and participates in formation of membranes and lipoproteins.
3. It is also as cholesterol-sterol carrier protein that cholesterol might affect the activity of *β-HMG-CoA-reductase* activity.

IV. Cyclisation of squalene to form lanosterol: The formation of lanosterol from squalene takes place in two steps:

- In the first step squalene-2,3-epoxide is formed catalysed by the enzyme *squalene mono-oxygenase;* which requires NADPH and molecular O_2.
- In the next step, an enzyme ***cyclase*** brings about the cyclisation of squalene to form lanosterol.

V. Conversion of lanosterol to cholesterol: Main changes that are brought about are:

- ***Removal of three angular $-CH_3$ groups.*** This involves a series of reactions, mechanism of demethylation is not properly known. CH_3 group at C_{14} is first eliminated.
- Shift of double bond between C_8 and C_9 to C_5 and C_6, and
- Saturation of double bond in side chain.

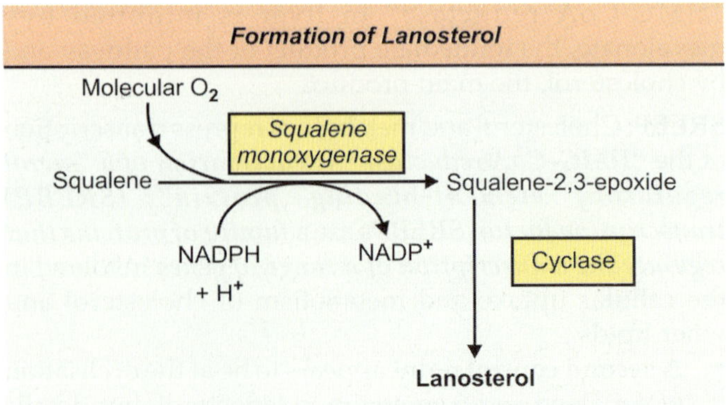

SECTION FOUR

Two possible pathways have been suggested: (shown in box).

I. One possible pathway

Lanosterol
↓ → CH_3 (C-14)
14-Desmethyl lanosterol
↓ → $2CH_3$
Zymosterol
↓
$\Delta^{7,24}$-cholestadienol
↓
Desmosterol
↓
Cholesterol

II. Second alternative pathways

Lanosterol
↓
24, 25-dihydrolanosterol
↓
α 4-methyl-Δ^8-cholesterol
↓
α 4-methyl-Δ^7-cholesterol
↓
Δ^7-cholesterol
↓
7-dehydrocholesterol
↓
Cholesterol

Control of Cholesterol Biosynthesis

- **SREBP Pathway**

The steps in the biosynthesis of cholesterol to HMG-CoA are **reversible.** The formation of mevalonate, however, in the next step is **irreversible** and is the **committed** step. Regulation of cholesterol biosynthesis is exerted near the beginning of the pathway at the **HMG-CoA reductase step, the rate limiting step.** The reduced synthesis of cholesterol in starving animals is accompanied by a decrease in the activity of the enzyme. However, it is only hepatic synthesis that is inhibted by dietary cholesterol. "HMG-CoA reductase" in liver is inhibited byy mevalonate, the immediate product of the pathway and by cholesterol, the main product.

SREBP: Cholesterol and metabolites repress transcription of the ***"HMG-CoA reductase"*** via ***activation of a "sterol regulatory element-binding protein" (SREBP) transcription factor. SREBPs*** are a ***family of proteins that regulate the transcription of a range of genes*** involved in the cellular uptake and metabolism of cholesterol and other lipids.

- A **second control point** appears to be at the cyclisation of squalene and conversion to lanosterol, but details of the regulation at this step is not clear.
- ***Effect of starrving/fasting:*** Fasting and/starvation also inhibt the enzyme and ***activate "HMG-CoA lyase" to form ketone bodies.***
- ***Effect of feeding cholesterol:*** Feeding cholesterol ***reduces the hepatic biosynthesis of cholesterol by reducing the activity of "HMG-CoA reductase".*** Intestinal cholesterol biosynthesis does not respond to the feeding of high cholesterol diets.
- **Role of hormones**
- **Insulin:** *Increases* ***HMG-CoA reductase*** *activity.* The hormone is required for the diurnal rhythm (diurnal variation) that occurs in cholesterol biosynthesis, a phenomenon probably related to feeding cycles and the need for bile acid synthesis.
- **Glucagon and/glucocorticoids:** Decreases the activity of ***HMG-CoA reductase*** and reduces tthe cholesterol biosynthesis.
- **Thyroid hormones:** Stimulates ***HMG-CoA reduc-tase*** activity.
- **Role of cyclic AMP:** HMG-CoA reductase may exist in active/inactive forms, which is reversibly modified by phosphorylation/and dephosphorylation mechanisms, whhich may be mediated by c-AMP dependent protein kinases. ***Cyclic AMP inhibits cholesterol biosynthesis by converting "HMG-CoA reductase" to inactive form.***

TRANSPORT OF CHOLESTEROL

Cholesterol in the diet is absorbed from the intestine, and in company with other lipids, are incorporated into "chylomicrons" and also to some extent 'VLDL'. Of the cholesterol absorbed, 80 to 90 per cent in the lymph is esterified with long-chain FA. Esterification may occur in the intestinal mucosa. In man, the total plasma cholesterol varies from 150 to 250 mg% (average 200 mg%), rising with age, although there are wide variations between individuals. The greater part is found in the 'esterified' form and is transported as "lipoproteins" in plasma. ***Highest proportion of circulating cholesterol is found in LDL (β-lipoproteins)*** which carry cholesterol to tissues and also in HDL, which takes cholesterol to liver from tissues for degradation ***(scavenging action).*** However, under conditions, where the VLDL are quantitatively predominant, an increased proportion of plasma cholesterol will reside in this fraction. Dietary cholesterol takes several days to equilibrate with cholesterol in plasma, and several weeks to equilibrate with cholesterol in tissues. Free cholesterol in plasma and liver equilibrates in matter of hours. In general, free cholesterol exchanges readily between tissues and lipoproteins, whereas cholesterol esters do not exchange freely. Some plasma cholesterol ester may be formed in HDL as a result of transesterification reaction in plasma between cholesterol and FA in position-2 of lecithin which

is catalysed by the enzyme *lecithincholesterol acyl transferase* (LCAT).

Cholesterol balance in tissues: Many factors will determine the cholesterol balance at the cellular level.

(a) *Increase of cholesterol in cells:*
- Increased synthesis of cholesterol.
- Hydrolysis of cholesterol ester by the enzyme *"cholesterol ester hydrolyase"*.
- Uptake and delivery of cholesterol in cells by circulating LDL (uptake by specific receptors).
- Uptake of cholesterol containing lipoproteins by 'non-receptor' mediated pathway.
- Uptake of free cholesterol by cell membranes.

(b) Decrease of cholesterol in cells:
- Efflux of cholesterol from cells to HDL ***(scavenging action).***
- Esterification of cholesterol by the enzyme *Acyl-CoA-cholesterol acyl transferase* (ACAT).
- Utilisation of cholesterol for synthesis of steroid hormones, viz. glucocorticoids, mineralo-corticoids, Gonadal hormones.
- In liver cells: formation of cholic acid.
- Formation of vit D_3.

CONSIDERATION OF OTHER FACTORS THAT INFLUENCE CHOLESTEROL LEVEL IN BLOOD

1. ***Dietary fats:*** Increased intake of fats in the diet increases level of cholesterol by increased synthesis. Greater amount of saturated fatty acids increases cholesterol level. ***Substitution in the diet of saturated FA by polyunsaturated FA has beneficial effect and lowers cholesterol level.***

Fats rich in saturated FA	*Oils that are rich in polyunsaturated FA*
Butter fat, ghee, dalda vanaspati, beef fat, coconut oil are rich in saturated FA	Sunflower oil, cottonseed oil, mustard oil, soyabean oil, groundnut oil

***Mechanism by which polyunsaturated FA lowers cholesterol level is not known exactly but possibilities are* (Tables 25.6 to 25.8):**

Table 25.6: Percentage of cholesterol in some common food substances

Food items	*Cholesterol content in mg%*
• Butter	280
• Fresh whole egg	468
• Fresh yolk of egg	2000
• Hen meat	70
• Lamb	70
• Pork	60

- It stimulates oxidation of cholesterol to bile acids
- Stimulation of cholesterol excretion in intestine
- May be a shift of cholesterol from plasma to tissues
- Cholesterol esters of polyunsaturated FA are more rapidly metabolised by liver and other tissues.

2. ***Dietary cholesterol: Increased feeding of cholesterol in diet decreases endogenous synthesis and reduces cholesterol level.*** It is difficult to lower the normal blood cholesterol level by taking food of low cholesterol. ***Restricted dietary intake of cholesterol is usually balanced by increased biosynthesis.***
3. ***Dietary carbohydrates:*** Increased consumption of carbohydrates increases cholesterol level. Consumption of excessive amount of sucrose and fructose cause increase in plasma lipids particularly TG and also cholesterol.

A diet providing 50 per cent carbohydrates, if ratio of starch: Sucrose is 4:1, Plasma cholesterol level is not much affected. When ratio between starch: sucrose is 1:4, an increase of plasma cholesterol is observed.

4. ***Heredity:*** Hereditary factors play the greatest role in determining individual blood cholesterol concentrations.
5. ***Blood groups:*** Cholesterol level found to be slightly higher in persons belonging to blood groups 'A' and 'AB', than those belonging to 'O' and 'B' groups.
6. ***Calorie Intake:*** Intake of excess calories increases cholesterol level.
7. ***Vitamin B-Complex***
   - **Nicotinic acid:** ***In large doses has cholesterol lowering effect.***
   - ***Pyridoxine deficiency:*** Produces increase in blood cholesterol level and atherosclerosis in monkeys.
8. ***Minerals***
   - ***In vitro acetate to cholesterol conversion in tissue cell cultures depressed↓ by addition of Vanadium and increased by chromium and Manganese salts.***
   - Conversion of mevalonate to cholesterol was inhibited by "Vanadyl SO_4".

Table 25.7: Cholesterol rich and poor diets

Cholesterol rich food items	*Cholesterol poor food items*
• Milk, cream, cream soups	• Skimmed milk
• Egg yolk	• Butter milk without fat
• Liver, brain, heart and kidney	• White of egg
• Animal fats: Pork, bacon, lard, etc.	• Lean meats, lean fish
	• Cooked and raw vegetables and dry cereals – Fruits, lemon juice – Vinegar, tea, coffee – Tomato juice/soup – Vegetable fats/oils, margarine contain practically no cholesterol

9. ***Dietary fibres:*** Increased fibres in the diet, caused an increased excretion of cholesterol and bile acids in faeces in experimental animals and produced significant reduction in serum cholesterol.
10. ***Physical exercise:*** Studies on human volunteers showed hard physical exercise brought about lowering in serum cholesterol↓ level and increased level of HDL ↑.
11. ***Life style of the individual:*** Life style of the individual also affects serum cholesterol level. Additional factors which play a part in coronary heart disease include:
    - Obesity
    - Lack of exercise and sedentary habits
    - Smoking
    - High blood pressure
    - Associated diabetes mellitus
    - Drinking of soft as opposed to hard water.
12. ***Elevation of Plasma FFA:*** Elevation of plasma FFA due to any cause will enhance increased VLDL secretion by the liver by enhancing endogenous TG synthesis, involving extra-TG and cholesterol output to the circulation.

Factors leading to higher or fluctuating levels of FFA include:

- Emotional stress
- Nicotine from cigarette smoking
- Coffee drinking and
- Partaking a few large meals rather than more continuous feeding.

Pre-menopausal women appear to be protected against these deleterious factors, probably due to the hormones, estrogens and high HDL, as compared to men and postmenopausal women.

Hypolipidaemic drugs: Several drugs are known to block the formation of cholesterol at various stages in the biosynthetic pathway. Some may increase the catabolism / excretion of cholesterol also; many of the drugs have

Table 25.8: Some of important factors which increase/decrease cholesterol level and synthesis

Increase	*Decrease*
1. Dietary cholesterol: A reduction in dietary cholesterol enhances synthesis and increase in cholesterol	1. Cholesterol feeding: "feedback" inhibition: Inhibits *HMG-CoA reductase*
2. Dietary fats: Feeding of more saturated fatty acids increases cholesterol	2. Fastings/starvation: Inhibits *HMG-CoA reductase* activity. Increases *HMG-CoA Lyase* activity and formation of ketone bodies ↑
3. High carbohydrate diet: Increase in dietary sucrose and fructose	3. Administration of analogues of mevalonate or squalene
4. Loss of bile: Drainage of bile by fistula increases synthesis, major factor is bile acid concentration in liver	4. Administration of cholates: Increase bile acid concentration in liver will decrease synthesis
5. Administration of plant sterols—sitosterol: Competes with esterification, decreases absorption, lowering cholesterol level which enhances endogenous synthesis	5. Presence of fats and bile acids in intestinal lumen increases intestinal absorption which inturn decrease synthesis
6. Lack of dietary fibres	6. Feeding of polyunsaturated fatty acid decreases synthesis.
7. Pyridoxal deficiency	7. Cyclic AMP: Increased cyclic AMP inhibits synthesis by converting *"HMG-CoA reductase"* to inactive form
8. Hormones: • Insulin • Thyroid hormones. Both increase *HMG-CoA reductase* activity.	8. Hormones: • Glucagon and glucocorticoids decreases ↓ synthesis
	9. Hypolipidaemic drugs: Lower the cholesterol level by inhibiting synthetic pathway/or catabolism.

Hypolipidaemic drugs used by clinicians

Drug	*Effect*	*Dosage*
• Niacin (Nicotinic acid)	Lowers cholesterol and TG	1/2 to 1.0 gm three times a day
• Colestipol	Lowers cholesterol and TG	10 gm three times a day (sachets)
• Lovastatin	Lowers cholesterol markedly	20 mg once or twice a day
• Gerfibrozil	Lowers TG mainly, lowers LDL also, Increases HDL slightly	30 mg cap. Two cap, twice before meals
• Garlic and onion extracts	Lowers TG and cholesterol	To be used in dietary preparation
• Guggul	Lowers TG and cholesterol	One cap three times a day

Table 25.9: Hypolipidaemic drugs and possible mechanism of action

Drugs	*Mechanism of action*
1. Aromatically substituted carboxylic acids, e.g. p-phenyl butyrate, p-bi-phenyl butyrate	• Inhibits acetate incorporation. Experiments in human volunteers—disappointing
2. **Triparanol:** Compounds related to non-steroidal estrogens and estrogens antagonist (Not used)	• Inhibits reduction of desmosterol. After widespread use, the drug was withdrawn due to side effects like: Cataract, alopecia, changes in hair colour, leucopenia.
3. **Pro-adifen HCl**	• Blocks pathway between mevalonate to squalene (Not used-due to side effects)
4. **Nicotinic acid:** In large doses has hypocholesterolaemic effect	• Reduces the flux of FFA by inhibiting adipose tissue lipolysis, thereby inhibiting VLDL production in liver. In large doses, may produce fatty liver
5. **Estrogen**	• Lower cholesterol level and increases HDL
6. **Sitosterol**	• Acts by blocking esterification of cholesterol in gut thus reducing cholesterol absorption (Synthesis may be increased later on)
7. **Dextrothyroxine (cholaxin), and Neomycin**	• Increases faecal excretion of cholesterol and bile acids
8. **Clofibrate** (Atromid S), **Gemfibrozil** CPI B (Ethyl-p-chlorophenoxy isobutyrate) (A commonly used drug)	• Acts by various ways: – Inhibits secretion of VLDL by liver – Inhibiting hepatic cholesterol synthesis – Probably also increases faecal excretion – They facilitate hydrolysis of VLDL triacyl glycerol by *lipoprotein* lipase
9. **Certain Resins, e.g.** • **Colestipol** • **Cholestyramine (Questran)**	• Prevent the reabsorption of bile salts by combining with them, increasing their faecal loss
10. **Probucol**	• Increases catabolism of LDL by receptor independent pathway
11. **Statins** Statins currently in use include: • ***Atorvastatin*** • ***Simvastatin*** • ***Pravastatin***	• All statins inhibit **HMG-CoA reductase**-enzyme, thus **up-regulating LDL receptors** reduces LDL-cholesterol level

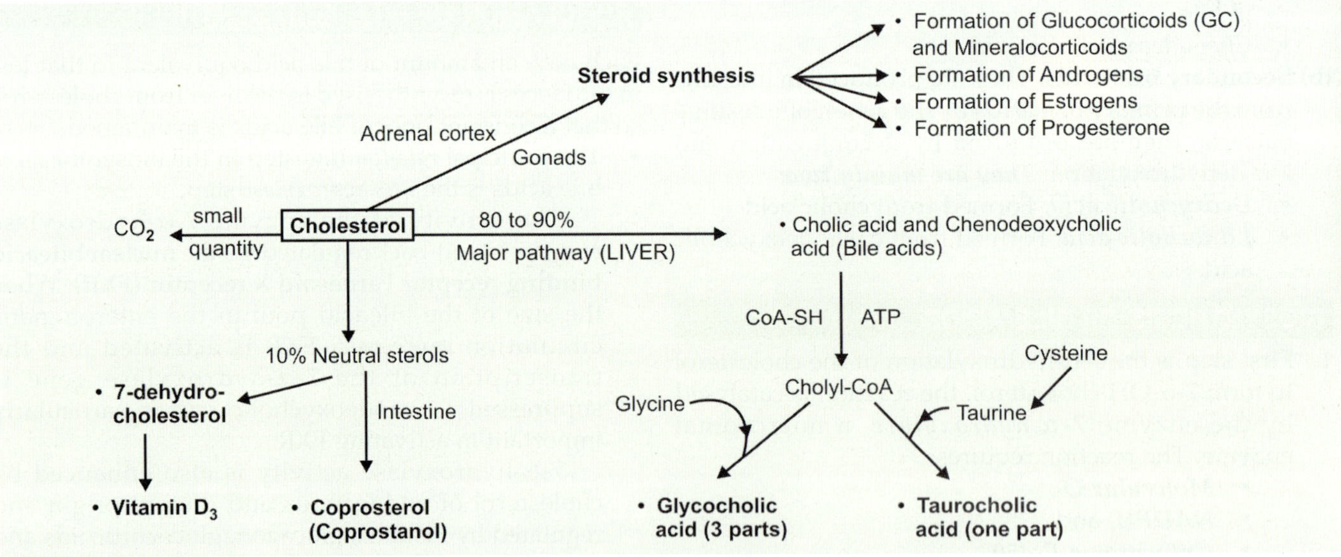

Fig. 25.14: Fate of cholesterol in the body

also harmful side effects. Some hypolipidaemic drugs and their mechanism of action are shown in Table 25.9.

FATE OF CHOLESTEROL

Fate of cholesterol in body is shown schematically in Figure 25.14.

About 1.0 gm of cholesterol is eliminated from the body per day. Fate of cholesterol has been studied in rats by giving labelled C^{14} cholesterol and H_3 cholesterol. Ring-labelled cholesterol has been shown to be transformed to various compounds (Refer to box—Fate of Cholesterol).

Fate of Cholesterol

- ***Degradation to CO_2:*** In human tissues—conversion to CO_2 does not occur.
- ***Conversion to Bile Acids***: Major pathway, more than 50 per cent is converted to bile acids and excreted in faeces (See below, Bile acid formation).
- ***Conversion to Neutral Sterols:*** 10 per cent of cholesterol is converted to neutral sterols, called as **coprosterol (coprostanol),** which is formed in lower part of intestine by the bacterial flora and excreted in faeces.
- ***Conversion to 7-Dehydrocholesterol:*** In skin, by UV light of Sun's rays, 7-dehydrocholesterol is converted to vit. D_3 (cholecalciferol).
- ***Formation of Adrenocortical Hormones:*** Glucocorticoids and mineralocorticoids are formed from cholesterol in adrenal cortex.
- ***Formation of Androgens***
- ***Formation of Estrogens***
- ***Formation of Progesterone***

BILE ACID

Bile acids are formed from cholesterol.

Types

(a) **Primary bile acids:** They are synthesised in the liver from cholesterol. They are mainly two:
- *Cholic acid:* Quantitatively the largest in amount in bile.
- *Chenodeoxy cholic acid.*

(b) **Secondary bile acids:** They are produced in intestine from the primary bile acids by the action of intestinal bacteria, they are produced by deconjugation and 7-α-dehydroxylation. ***They are mainly two:***
- *Deoxycholic acid:* Formed from cholic acid
- *Lithocholic acid:* Formed from chenodeoxycholic acid.

BIOSYNTHESIS OF BILE ACIDS

1. First step is the 7-α-hydroxylation of the cholesterol to form 7-α-OH cholesterol, the reaction is catalysed by the enzyme ***7-α-hydroxylase***, a microsomal enzyme. The reaction requires:
   - *Molecular O_2*
   - *NADPH*, and
   - *Cytochrome P-450.*

   The enzyme is a typical mono-oxygenase. The enzyme also requires vit. C as a coenzyme.

Note

- It is the ***rate-limiting*** reaction and controls the synthesis of bile acids.
- Vit. C deficiency interferes with bile acid formation and leads to cholesterol accumulation and athero-sclerosis in scorbutic animals.

2. Pathway of bile acid biosynthesis bifurcates from 7-α-OH-cholesterol to two directions:
   - Cholic acid formation, and
   - Chenodeoxy cholic acid formation.

   The conversion of 7-α-OH-cholesterol to formation of cholic acid/or chenodeoxycholic acid is catalysed by *12-α-hydroxylase* and involves several steps. The enzyme requires:
   - Molecular O_2
   - NADPH and
   - CoA-SH.
   - Propionyl-CoA is split off the side chain leaving 'cholyl CoA'/"chenodeoxy cholyl-CoA" respectively.
3. A second enzyme catalyses the conjugation of the CoA-derivatives with glycine or taurine, to form the Primary bile acids:
   - ***Glycocholic acid*** and/or ***glycochenodeoxycholic acid***, and
   - ***Taurocholic and/or taurochenodeoxycholic acid.*** In humans, the ratio of the glycine to taurine conjugates is usually 3:1.

Since bile has alkaline pH and it has sodium and K, it is assumed that the bile acids exist in bile as corresponding sodium salts usually as **sodium glycocholate** and **sodium taurocholate**. Hence they are called as **bile salts**.

Formation of Bile acids are shown schematically in Figure 25.15.

REGULATION OF BILE ACID SYNTHESIS

Each day, an amount of bile acid equivalent to that lost in the faeces is resynthesised in the liver from cholesterol, so that a constant pool of bile acids is maintained.

- The principal *rate-limiting* step in the biosynthesis of bile acids is the *7-α-hydroxylase* step.

  The activity of the enzyme 7-α-hydroxylase enzyme is feed-back regulated via the **nuclear bileacid binding receptor Farnesoid X receptor (FXR).** When the size of the bileacid pool in the enterohepatic circulation increases, FXR is activated and the transcription of the 7-α-hydroxylase gene is suppressed. Chenodeoxycholic acid is particularly important in activating FXR.

  7-α-hydroxylase activity is also enhanced by cholesterol of endogenous and dietary origin and regulated by insulins, glucagon, glucocorticoids and thyroid hormones.
- The controlling enzyme for cholesterol biosynthesis is *HMG-CoA reductase*.
- Activities of both these probably change in parallel and undergo similar 'diurnal' variation. Both the enzymes may exist in 'active' and 'inactive' form which may be regulated by phosphorylation/dephosphorylation mechanisms.
- Cholesterol feeding exerts a stimulatory effect on *7-α-hydroxylase*. Bile acids, on the other hand, exert a *"feedback" inhibition* on the enzyme *7-α-hydroxylase*.

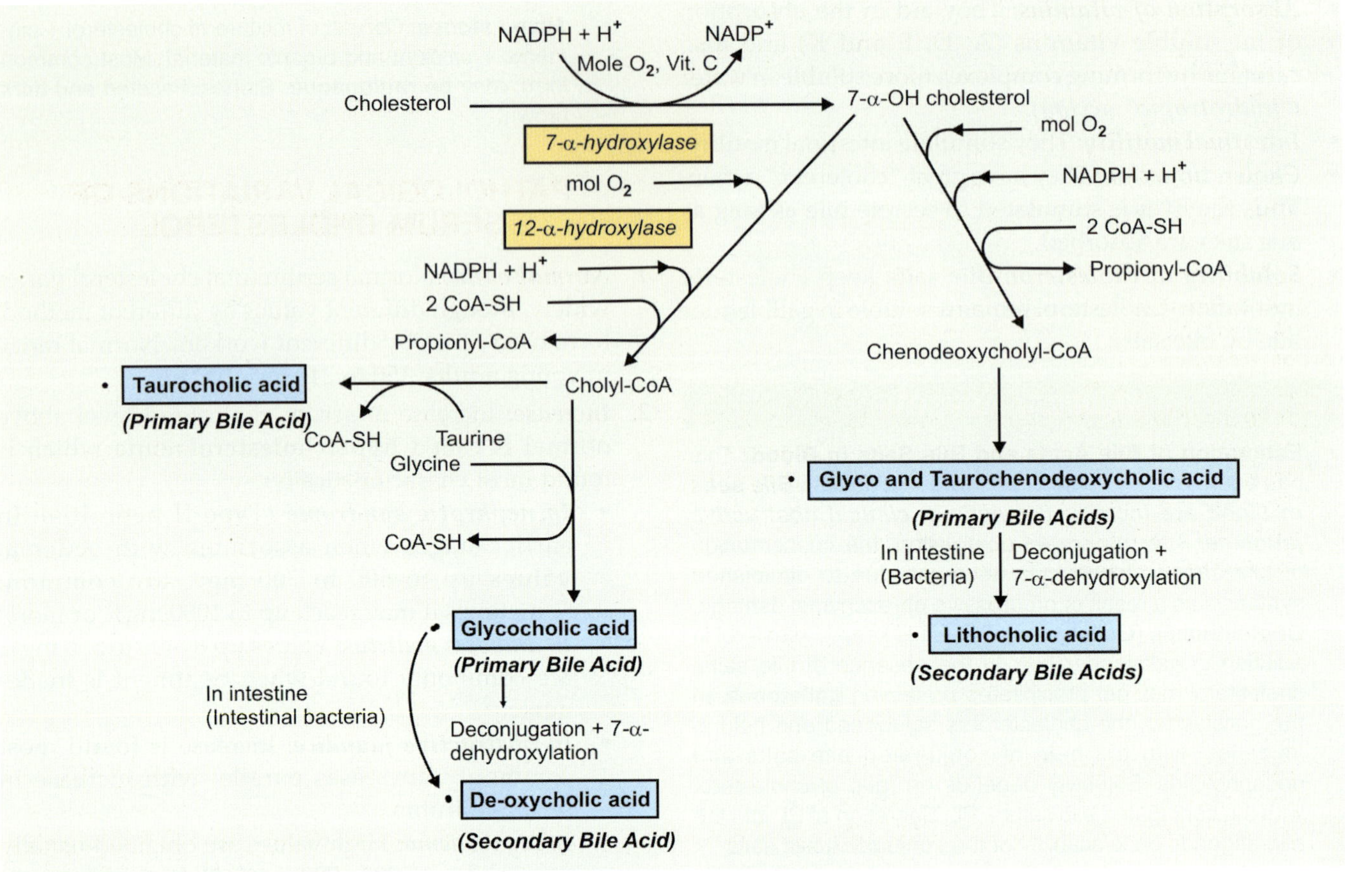

Fig. 25.15: Formation of bile acids

BILE

Bile is a viscous fluid produced by the liver cells. Strictly speaking it is not a digestive secretion as it ***does not contain any digestive enzymes*** but it helps in digestion and absorption of lipids. It is secreted continuously by the liver and through bile canaliculi and bile duct it accumulates in the gallbladder, where it is stored.

In gallbladder, certain changes take place by reabsorption of large amount of water leading to concentration of bile. Water is absorbed along with inorganic components as isotonic solution. ***Mucin is added and bicarbonate and chlorides are reabsorbed.*** Hence organic constituents like cholesterol, bile pigments—bilirubin get concentrated in gallbladder bile. During digestion, gallbladder contracts by the stimulation of the GI hormone **cholecystokinin** which is produced by small intestine and release bile rapidly to the intestine by the way of common bile duct. ***Approximately 500 to 1000 ml of bile is secreted by liver in a day.***

Functions of Bile Salts

- ***Lowering of surface tension:*** Because of their power of lowering surface tension, they aid in the emulsification of fats and tend to stabilise such emulsions. The ***emulsification is a prerequisite for action of pancreatic lipase on fats.***

Differences in composition of hepatic bile and gallbladder bile

	Hepatic bile	*Gallbladder bile*
pH	7.0 to 8.2	6.0 to 7.0
Specific gravity	1.010	1.040
Water	97.2%	88.0%
Solids	2.7%	12.0%
Bile acids	1.2%	6.2%
Mucin and bile pigments	0.58%	3.2%
Total lipids	0.3%	2.5%
TG	0.1%	0.4%
Phospholipids	< 0.1%	0.2%
Cholesterol	0.08%	0.5%
Inorganic salts	0.84%	0.75%

- ***Bile salts accelerate the action of pancreatic lipase:*** In the presence of bile salts, a ***colipase*** (molecular wt = 10,000) binds to lipase and shifts the optimal pH of the enzyme from 9.0 to 6.0.
- ***Micelles formation:*** *Bile salts form 'micelles'* with fatty acids, mono and diacyl glycerols and also TG which are made water soluble and helps absorption.

- ***Absorption of vitamins:*** They aid in the absorption of fat soluble vitamins (A, D, E and K) and also carotene by forming complexes more soluble in water *("hydrotropic" action).*
- ***Intestinal motility:*** They stimulate intestinal motility.
- ***Choleretic action:*** They have great "choleretic" action. Thus the liver is stimulated to secrete bile as long as bile salts are absorbed.
- ***Solubility of cholesterol:*** Bile salts keep cholesterol in solution. Cholesterol remains soluble in gallbladder bile by bile salts.

CLINICAL ASPECT

- **Estimation of Bile Acids and Bile Salts in Blood:** This has been recently used for as liver function test. ***Bile salts in blood are increased greatly in clinical obstructive jaundice.*** After prolonged obstruction, the concentration of bile salts in blood may decrease due to diminished synthesis as a result of progressive parenchymal damage.
- **Cholelithiasis (Gallstones):** Bile salts keep cholesterol in solution in gallbladder bile. In the absence of bile salts, cholesterol may get precipitated producing **gallstones**. In the gallbladder, the cholesterol is solubilised and held in 'micelles' with the help of conjugated bile salts and phospholipids. Solubility depends on ratio of cholesterol with the conjugated bile salts + PL. Secretion of PL into the bile depends on availability of the conjugated bile salts.

  If bile salts content is decreased due to any cause, the phospholipid also decreases leading to an imbalance of the ratio. The solubility of cholesterol is hampered as a result it crystallises out. The crystals grow to form the stones.

A. Conditions which Favour Stone Formation

1. ***Infection:*** Favours stone formation.
   Infection causes:
   - Deconjugation of bile acids leading to decrease in solubility.
   - Production of phospholipase, which converts Lecithin to Lysolecithin.

   Thus, the ratio is disturbed leading to precipitation of cholesterol.
2. ***Decreased availability of bile salts (Reduction in bile salt pool):***
   - Defect in enterohepatic circulation.
   - Disease of terminal ileum.
   - In patients with cirrhosis liver.

B. Types of Gallstones: Gallstones can be of mainly ***three types:***

- ***Cholesterol stones:*** They are single or multiple, mainly formed of cholesterol, mulberry-shaped and are **not radiopaque.**
- ***Pigment stones:*** Consist of bile pigments and calcium with other organic substances. Small multiple stones, dark green or black, **not radiopaque.**
- ***Mixed stones:*** Consist of mixture of cholesterol + pigments + calcium and organic material. Most common form, **may be radiopaque.** Stones: Faceted and dark brown.

PATHOLOGICAL VARIATIONS OF SERUM CHOLESTEROL

1. ***Normal value:*** Normal serum total cholesterol varies widely, though different values by different method, have been given by different workers. Normal range in young adults: 150 to 240 mg/100 ml.
2. ***Increase:*** Increase of serum cholesterol level above normal is called **hypercholesterolaemia** which is found most characteristically:
   - ***In nephrotic syndrome*** (Type II nephritis): In earlier stages when associated with oedema, values up to 600 to 700 mg% are common. Sometimes it may reach up to 1000 mg% or more.
   - ***In diabetes mellitus:*** Values up to 400 to 550 mg% are commonly found when treatment is inadequate.
   - ***In obstructive jaundice:*** Increase is found most commonly. Increases parallel with increase in serum bilirubin.
   - ***In myxoedema:*** High values are obtained usually ranging from 500 to 700 mg%. ***Helps in diagnosis.***
   - ***In xanthomatous biliary cirrhosis:*** Very high values are seen.
   - ***In hypopituitarism:*** Small increases ranging from 250 to 350 mg% may be seen.
   - ***In xanthomatosis:*** Frequently found to be associated with high cholesterol values.
   - ***In coronary thrombosis and in angina pectoris:*** Value between 300 to 400 mg% are rather of frequent finding.
   - ***Idiopathic*** hypercholesterolaemia has also been described.
3. ***Decrease:*** Decrease in blood cholesterol below normal is called **hypocholesterolaemia**. Hypocholesterolaemia is characteristically seen:
   - ***In thyrotoxicosis:*** Values as low as 80 to 100 mg% may be seen. But quite a number of hyperthyroidism cases may have a serum cholesterol within normal range.
   - In pernicious anaemia and in other anaemias.
   - In haemolytic jaundice.
   - In malabsorption syndrome.
   - In wasting diseases.
   - In acute infections and in a number of terminal states.

RELATION OF CHOLESTEROL AND OTHER LIPIDS AS RISK FACTOR IN CORONARY HEART DISEASE (CHD)

Of the serum lipids, cholesterol has been the one most often incriminated as the risk factor. However, other parameters such as serum TG, VLDL and LDL have been incriminated. Patients with CHD can have any one of following abnormalities:

- *Elevated concentrations of VLDL with normal concentrations of LDL*
- *Elevated LDL with normal VLDL*
- *Elevation of both VLDL and LDL.*

1. **Role of Cholesterol**
   An elevation of the total cholesterol in plasma is considered to be a 'Prime risk factor' for CHD. The **Framingham Study** has demonstrated a linear increase in coronary "risk" with increment of total plasma cholesterol level from 180 mg% upwards. The Lipid Research Clinics Coronary Primary Prevention Trial had presented firm proof that in humans, a lowering of plasma cholesterol level reduces the coronary thrombosis and myocardial infarction and mortality. One conclusion deduced from this pioneering work is: ***A 1 per cent fall in cholesterol predicts a 2 per cent reduction in CHD risk.***

2. **Role of LDL and HDL**
   Recent studies have shown that atherogenic significance of the total cholesterol concentration must be viewed with restrictions.

   From numerous studies, it is now concluded that ***LDL is the carrier of 70 per cent of total cholesterol and it transports cholesterol to tissues and thus is most potential atherogenic agent.*** On the other hand, an increase of second cholesterol rich class HDL is not associated with 'risk' at all. ***An inverse relation between CHD and HDL concentration has been found. A raised HDL concentration is beneficial and protective against CHD.*** This protective mechanism is explained by the following two mechanisms operating in parallel:
   - "Reverse transport" of cholesterol from peripheral tissues into the Liver by way of HDL which thus reduces the intracellular cholesterol content ***(scavenging action of HDL).***
   - Control of catabolism of TG rich lipoproteins. High HDL concentrations are associated with a faster elimination from the plasma of TG rich lipoproteins and their atherogenic intermediate.

3. **Role of TG and VLDL**
   Elevated VLDL and hypertriglyceridaemia may also be considered a primary 'risk' factor because it is associated in specific cases, with an increased atherogenic risk.
   - A low blood TG level is suggestive of efficient intravascular lipolysis and thus of enhanced formation of HDL by this route.
   - Hypertriglyceridaemia, on the other hand, indicates less effective intravascular lipolysis and hence a reduced formation of HDL which is in turn associated with a higher atherogenic risk.

4. **Role of Apolipoproteins (Apoproteins)**
   Refer to role of apoproteins and plasma lipoproteins and atherosclerosis in chapter on Lipoproteins ahead.

FORMATION AND FATE OF "ACTIVE" ACETATE (ACETYL-CoA) (TWO CARBON METABOLISM)

'Active' acetate or acetyl-CoA is the C-2 compound, a key substance in third phase of metabolism. It is produced from various sources, viz. metabolism of carbohydrates, lipids and proteins and is metabolised to CO_2 and water. It also produces large number of "biologically" important compounds (Fig. 25.16).

Formation and fate of active acetate is shown in Table 25.10.

SECTION FOUR

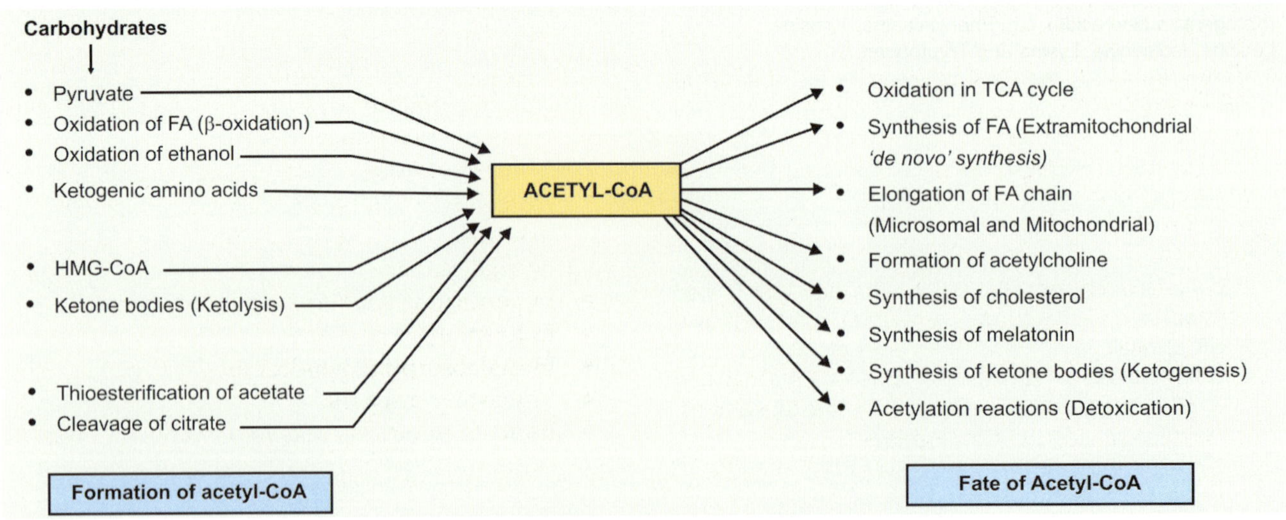

Fig. 25.16: Different sources and fate of acetyl-CoA schematically

Table 25.10: Sources of acetyl-CoA and its fate

Formation	Fate
1. ***From metabolism of glucose:*** Glucose forms pyruvate by glycolysis. Pyruvate is oxidatively decarboxylated in mitochondria by *pyruvate dehydrogenase complex* to form acetyl-CoA.	1. ***Principal fate is oxidation in TCA cycle:*** Most of acetyl-CoA combines with OAA to form citrate and further oxidised to CO_2 and H_2O in TCA cycle.
2. ***β-oxidation of FA:*** Acetyl-CoA is produced in mitochondria from β-oxidation of FA.	2. ***In cholesterol biosynthesis:*** Acetyl-CoA is the starting material for cholesterol biosynthesis. All the 27 carbon atoms are derived from acetyl-CoA.
3. ***Cleavage of citrate:*** In the cytosol, acetyl-CoA is produced from citrate, which is cleaved by *ATP-citrate lyase* in presence of ATP and CoA to form OAA and acetyl-CoA.	3. ***Ketogenesis:*** 'Acetyl-CoA' is the starting material required for the formation of first ketone body 'aceto-acetate' in liver.
4. ***Oxidation of ethanol:*** Alcohol is oxidised by the enzyme *Alcohol dehydrogenase* to form acetyl-CoA.	4. ***Fatty acid synthesis:*** • Cytoplasmic *de novo* fatty acid synthesis (Extramitochondrial): Acetyl-CoA is the starting material for synthesis of palmitic acid. • Microsomal elongation system: Uses Malonyl-CoA synthesised from acetyl-CoA by carboxylation reaction for elongation of pre-existing acyl-CoA by addition of C-2 units. • Mitochondrial elongase system: Uses acetyl-CoA itself in incorporating C-2 units into acyl-CoA.
5. ***Thio-esterification of acetate:*** Acetate can be activated to acetyl-CoA by the enzyme *Acetyl-CoA-Synthase* in presence of ATP and CoA-SH. Acetate can be formed in ruminants from cellulose. In humans, small amount of acetate may be obtained from oxidation of ethanol, hydrolysis of aspirin, and catabolism of amino acid threonine.	5. ***Acetylation reactions (Detoxication):*** Acetyl-CoA is used in detoxication of many substances by 'acetylation' reaction. *Acetyl transferases (acetylases)* transfer the acetyl group from acetyl-CoA to many substrates, e.g. • Sulfanilamide is detoxicated to N-acetyl sulfanilamide in the liver by the enzyme *sulfanilamide acetylase* with acetyl-CoA, and excreted in urine. • Bromobenzene is detoxicated by cysteine and acetyl-CoA to form p-bromophenyl mercapturic acid and excreted in urine.
6. ***By ketolysis:*** Acetoacetyl-CoA is formed from acetoacetate in extrahepatic tissues which is further split to form acetyl-CoA by thiolase.	6. ***Formation of acetylcholine:*** From choline in cholinergic neurons. The enzyme *choline acetylase* transfers acetyl group of acetyl-CoA to choline
7. ***HMG-CoA:*** Forms acetyl-CoA by the action of the enzyme HMG-CoA lyase.	7. ***In melatonin synthesis:*** Formation of N-acetyl serotonin from serotonin.
8. ***From metabolism of certain amino acids:*** Catabolism of certain amino acids produces 'acetyl-CoA' (Ketogenic amino acids). e.g. Phenylalanine, Tyrosine, Leucine, Isoleucine, Lysine and Tryptophan.	

PART III

Major Concepts

A. To study the chemistry and metabolism of lipoproteins and the clinical disorders associated with them.
B. To learn what is fatty liver and how it is formed.

Specific Objectives

A. 1. What are lipoproteins? Study the structure of a lipoprotein complex.
2. Study how various lipoproteins are classified:
   - Depending on hydrated density
   - Depending on electrophoretic mobility
   - Classification based on apolipoproteins content
3. Learn the types of apoproteins present in various Lipoprotein fraction.
   - ***How does "nascent" chylomicrons and VLDL differ from the "circulating" chylomicrons and VLDL?***
4. Study how chylomicrons and VLDL are synthesised in the intestinal mucosal cells and the liver cells respectively and secreted in the blood.
5. Learn the fate of Chylomicrons and VLDL
   - What is *Lipoprotein lipase*? Study the location of lipoprotein lipase, its mode of action on chylomicrons and VLDL.
   - How does circulating chylomicrons and VLDL interact with HDL.
   - Study how LDL is produced from VLDL via IDL.
   - Study the formation of "chylomicrons remnants" and its fate.
6. - Study the metabolic fate of LDL.
   - Study the mechanism how LDL interacts with cell membrane, learn about LDL receptors and increases concentration of cellular cholesterol and its regulation.
   - Study how LDL is destroyed in the body.
7. Learn the synthesis and metabolism of HDL. Learn about HDL receptor protcins and HDL cycle.
   - ***How does "nascent" intestinal HDL differ from "nascent" hepatic HDL.***
   - ***Learn about the "Scavenging" action of HDL.***
8. Learn the major functions of Lipoproteins in the body.
9. Study the clinical disorders associated with Lipoprotein metabolism.
   (a) Hyperlipoproteinaemias: It may be primary and secondary.
   - Study the Frederickson's "five" types of primary hyperlipoproteinaemias, their inheritance and clinical/biochemical features in brief.

   (b) Learn briefly the types of hypolipoproteinaemias.
10. Study about atherosclerosis and the role of lipoproteins in atherosclerosis.

B. 1. Learn the amount of fats present in liver in a normal individual and the types of the lipids.
2. Study the factors which tend to increase/and decrease the fat content of liver.
   The amount of lipids in the liver at any given time will be the resultant of these influences, some acting in conjunction with and some in opposition to others.
3. Theoretically fatty liver can be of following types:
   - Type 1: due to 'overfeeding" of fat.
   - Type 2: due to "oversynthesis" of fat from excessive carbohydrate intake.
   - Type 3: due to 'over mobilisation' of fats from depot to liver: "physiological" fatty liver.
   - Type 4: due to "under mobilisation" from liver to depot: "Pathological" fatty liver.
   - Type 5: due to "under utilisation" of fats in Liver.
4. - Study the biochemical mechanisms by which the different types of fatty liver can be produced.
   - Learn the biochemical mechanism by which the following can produce fatty liver: (a) Carbon tetrachloride (CCl_4) (b) Ethionine (c) Orotic acid (d) Ethyl alcohol.

SECTION FOUR

PLASMA LIPOPROTEINS AND METABOLISM

INTRODUCTION

Hypercholesterolaemia, hypertension, cigarette smoking and **obesity** have been identified as major independent **risk factors** for the development of premature cardiovascular diseases. The clear delineation of hypercholesterolaemia, in the begining, as a risk factor has stimulated active investigation into the metabolism of cholesterol and TG in normal healthy man and in patients with disorders of lipid metabolism and atherosclerosis.

What are Lipoproteins?

In plasma, cholesterol and TG form integral components of macromolecular complexes called as "lipoproteins"

which are conjugated proteins; Lipid part is the prosthetic group and lipid-free proteins are designated as ***apo-lipoproteins*** or ***apoproteins.***

Structure of a lipoprotein complex: Extraction of plasma lipids with a suitable lipid solvent and subsequent separation of the extract into various classes of lipids, shows the presence of:

• ***TG (triacyl glycerol),*** • ***Phospholipids (PL),*** • ***Cholesterol*** and • ***Cholesterol-esters*** and in addition, the existence of a much smaller fraction of • ***"unesterified long chain FA"*** (Free FA/FFA) that accounts for less than 5 per cent of total FA present in the plasma. Free fatty acid is also called as **"unesterified FA (UFA)" or "non-esterified FA (NEFA)"**. ***FFA is now considered to be metabolically most active of plasma Lipids.*** Lipoproteins serve as 'carrier' of lipids in plasma. Since lipids account for much of the energy expenditure of the body, the problem is presented of transporting a large quantity of ***hydrophobic*** material, the lipids, in an aqueous environment, i.e. plasma. This is achieved by associating the more insoluble lipids with more ***polar*** ones, such as PL and then combining with cholesterol/and cholesterolesters and a specific protein called "apo-protein" to form the so-called ***hydrophilic lipoprotein complexes.*** It is in this way that TG formed in intestinal epithelial cells (exogenous TG) and TG formed in Liver by synthesis ("endogenous" TG) are carried as Lipoprotein complexes 'chylomicrons' and VLDL (very low density lipoprotein) respectively. Thus, **"Chylomicrons are the chief carrier of "exogenous" TG, Hepatic VLDL Chief carrier of endogenous hepatic TG"**. (Fig. 25.17).

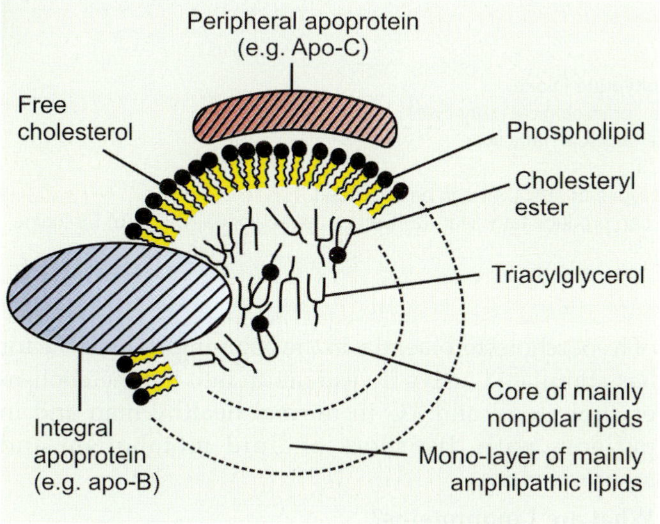

Fig. 25.17: Structure of lipoprotein molecule

CLASSIFICATION OF LIPOPROTEINS

Lipoproteins can be classified according to their • ***hydrated density***, to • ***electrophoretic mobility*** and based on • ***'apo-lipoprotein'*** content.

1. **Classification as per hydrated density:** Pure fat is less dense than water. As the proportion of lipid to protein in lipoprotein complexes increases ↑, the density of the macro-molecule decreases↓. Use of the above property has been made in separating various lipoproteins in plasma by ultracentrifugation.

Svedberg units of floatation (Sf unit): The rate at which each lipoprotein floats through a solution of NaCl (Sp. gr. 1.063) is expressed as "Svedberg" units of floatation (Sf).

Sf unit: One Sf unit is equal to 10^{-13} cm/S/dyne/g at 26°C.

Gofman and colleagues (1954) separated lipoproteins by ultracentrifugation into **four** major density classes:

- *Chylomicrons: Density lowest-floats*
- *Very low density lipoproteins (VLDL or VLDLP)*
- *Low density lipoproteins (LDL),* and
- *High Density lipoproteins (HDL):* settles below.

LDL: It has been further divided into LDL 1, IDL (intermediate density lipoprotein) and LDL-2

HDL: It has been further separated into HDL-1 (this fraction is quantitatively insignificant), HDL-2 and HDL-3. Recently HDL c has been described.

2. **Classification Based on Electrophoretic Mobility (Frederickson and colleagues, 1967):** The most widely used and simplest classification for lipoproteins is based on the separation of major four classes by electrophoresis. The most frequently employed electrophoretic media are "paper" and 'agarose'. Plasma lipoproteins separated by this technique are classified in relation to comparable migration of serum proteins.

   On electrophoresis, the different fractions according to mobility appear at:

- The origin is ***chylomicrons,***
- Migrating into β-globulin region is called ***β-lipoproteins*** **(LDL)**
- Migrating into Pre-β-globulin region, called as ***pre-β-lipoproteins*** **(VLDL).**
- Migrating to "α_1-globulin region called ***α-lipoproteins*** **(HDL)**

  Migration is shown diagrammatically in Figure 25.18.

3. **Classification Based on Apo-lipoproteins [Alaupovic and colleagues (1972)]:** In this classification, Lipoproteins are designated by their "apo-lipoprotein" composition. At present, **five major** lipoprotein families have been identified and they are shown in Table 25.11.

SECTION FOUR

Table 25.11: Lipoproteins classes as per apoprotein contents				
Families	***Apolipoproteins***	***Density class***	***Mol. wt. range***	***Function***
LP A	A-I and A-II	HDL	17000 to 28000	• LCAT activator • 'Scavenger'
LP B	Apo-B (B_{48} and B_{100})	LDL and VLDL	250,000	• Cholesterol carrier to tissues
LP C	APO-C I, C II, C III	VLDL, LDL and HDL	6500 to 10,000	• *"Lipoprotein Lipase"* activator, CIII is lipase inhibitor
LP D	Apo-D	HDL_3	~ 20,000	• LCAT activator 3
LP E	Apo-E (Arginine rich)	VLDL, LDL, and HDL	32,000 to 39,000	• Cholesterol transport

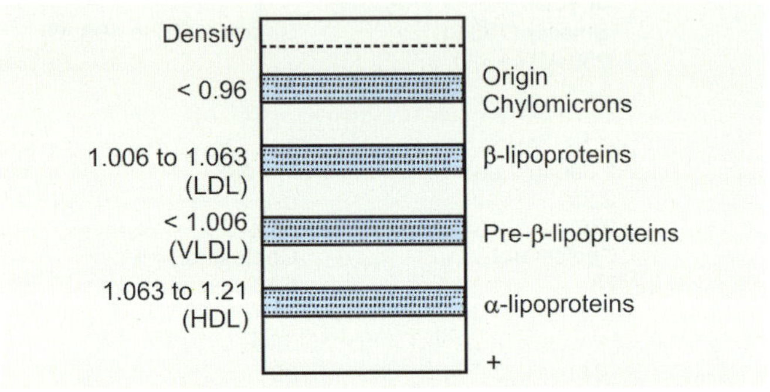

Fig. 25.18: Electrophoretic separation of plasma lipoproteins

TYPES OF APOPROTEINS PRESENT IN VARIOUS LIPOPROTEIN FRACTIONS (CHEMISTRY OF APOPROTEINS)

As stated above, lipoproteins are characterised by the presence of one or more proteins or polypeptides known as apoproteins. **According to ABC nomenclature:**

1. **HDL:** Two major apoproteins of HDL are designated as **apo-A-I** and **apo-A-II**. In addition to above, HDL also contains apo-C-I, C-II and C-III. HDL-3 is characterised by having apo-D and HDL may also acquire arginine-rich apo-E.
2. **LDL:** The main apoprotein of LDL is **apo-B_{100}**, which is also present in VLDL.
3. **Chylomicrons:** Principal apoprotein of chylomicrons is **apo-B_{48}** (Mol.wt = 200 Kd). In addition, chylomicrons also contain apo-A (AI and AII) and apo-C (C-II and C-III), also arginine rich apo-E (34 Kd). ***Apo-C seems to be freely transferable between chylomicrons and VLDL on one hand and HDL on the other.***
4. **VLDL and LDL:** Principal apoproteins of VLDL, IDL and LDL is **apo-B_{100}** (350 Kd). They also contain apo-C (C-I, C-II and C-III), and apo-E. ***IDL carries some apo-E apoprotein.***

Apo-E: ***Arginine rich apo-E***, isolated from VLDL. It contains arginine to the extent of 10 per cent of the total amino acids and accounts for 5 to 10 per cent of total VLDL apo-proteins in normal subjects but is present in excess in the "broad" β-VLDL of patients of type III hyperlipoproteinaemia.

Carbohydrate content: Carbohydrates account for approx. 5 per cent Apo-B and include mannose, galactose, fucose, glucose, glucosamine and sialic acid. So, some of lipoproteins are glycoproteins.

Difference of "Nascent" Chylomicrons and VLDL from "Circulating" Chylomicrons and VLDL

"Nascent" chylomicrons and VLDL contains the principal apoprotein B_{48} and B_{100} respectively. During circulation they acquire apo-C principally by interaction with HDL and the other apoproteins like apo-E. 'Nascent' chylomicrons may have also apo-A.

Distribution of Apoproteins in the Molecule (Fig. 25.17)

1. Apoproteins like **apo-B** are **integral** proteins and are embedded deep into the surface PL layer of the droplet.
   - They cannot be extracted easily from the lipoprotein droplets except by drastic treatment with chaotropic agents, and
   - They are not freely transferable to other lipoprotein particles.
2. Other apoproteins such as **apo-CI, CII and CIII, apo-D** and arginine-rich **apo-E** are **Peripheral** Proteins, located over the surface of the lipoprotein particle.
   - They are easily extractable from lipoprotein particle, and
   - They are freely transferable between the particles. Thus they can interact during circulation and transferred from one to other.

FUNCTIONS OF APOLIPOPROTEINS

Apolipoproteins carry out several roles:

- By entering into the "polar" surface layer they make the lipoprotein molecules **water miscible (hydrophilic)**.

Electrophoresis

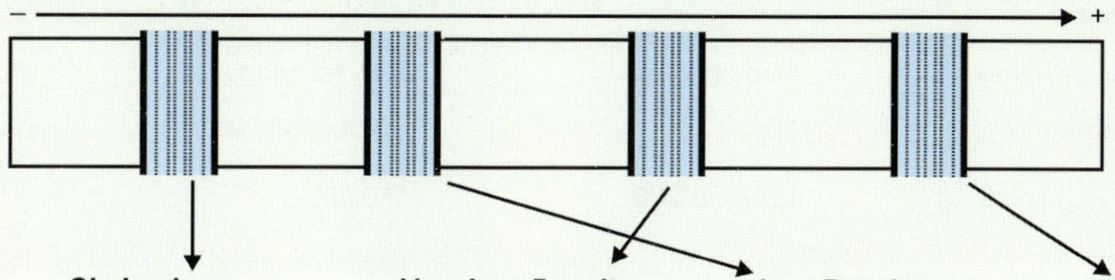

I. Classification:				
• Hydrated Density	***Chylomicrons***	***Very Low Density lipoproteins (VLDL)***	***Low Density lipoprotins (LDL)***	***High Density lipoproteins (HDL)***
• Electrophoresis	origin	Pre-β (α_2)	β	α
• Apoproteins	Apo-B_{48} (also apo-C and E)	Apo-B_{100} (also apo-C and E)	Apo-B_{100} (apo-C,E)	Apo-A-I and A-II (apo-E, and apo-C and D)
II. Origin (where synthesised):	Intestine	Liver and intestine	In blood by degradation of VLDL	Liver (Intestine ?)
III. Comparative size:	very large (1000 to 10,000 A°)	Large (300 to 700 A°)	Smaller (150 to 250 A°)	Smallest (75 to 100 A°)
IV. Composition and other properties:				
1. Density	< 0.96	0.96 to 1.006	1.006 to 1.063	> 1.063
2. Sf	> 400	20-400	2-20	0-2
3. Protein (%)	0.5 to 2	12%	25%	50%
4. Major Lipid	T.G	T.G	Cholesterol	Cholesterol = PL
5. Lipids (%)	98-99	88	75	50
Other fractions: (wt %)				
• PL	4	19	24	30
• Cholesterol	6	19	45	18
• TG	90	50	10	5
V. Chief Function:	• Transports exogenous TG	• Transports endogenous TG	• Metabolic end-product of VLDL • Transports cholesterol to peripheral tissues	• Scavenging action • Transports cholesterol from peripheral tissues to liver for degradation

Fig. 25.19: Characteristics of human plasma lipoproteins

- They can form part of the structure of the lipoprotein itself, e.g. Apo-B.
- They are enzyme cofactors and can act as activator or inhibitor of the enzyme, e.g.
  - **Activators**—C II for lipoprotein lipase
    A-1 for lecithin—cholesterol acyl transferase (LCAT)
  - **Inhibitors**—Apo-A-II and Apo-C III for lipoprotein lipase
    Apo-C1 for cholesteryl ester transferase protein.
- They act as **ligands** for interaction with lipoprotein receptors in tissues, e.g.
  Apo B-100 and apo-E for the LDL receptor
  - Apo-E for the LDL receptor related protein (LRP), which has been identified as the **remnant receptor** and apo-Al for the HDL receptor.
- Apo-D probably functions as "cholesteryl ester transfer protein" for transferring cholesteryl esters between different lipoproteins.

Though the functions of apo-A IV and apo-D are yet not clearly defined, ***recently apo-D has been incriminated to be an important factor in*** '**human neurodegenerative disorders**".

Figure 25.19 shows the characteristics of Human Plasma lipoproteins.

SYNTHESIS OF CHYLOMICRONS AND VLDL

Chylomicrons are found in "chyle" formed only by the Lymphatic system draining the intestine. However, it is now realised that a smaller and denser particle similar to VLDL is now known to be synthesised in small quantity in intestinal cells also.

Chylomicrons formation fluctuates with the load of TG absorbed, whereas VLDL formation is quantitatively less, but is more constant and occurs even in the fasting state. However, bulk of plasma VLDL is of hepatic origin, being the vehicle for transport of TG from Liver to extrahepatic tissues (carrier of "endogenous" TG). There are many similarities in the mechanism of formation of chylomicrons by intestinal epithelial cells and VLDL by hepatic parenchymal cells.

Salient Features

- Chylomicrons and VLDL are synthesised in intestinal mucosal cells and Liver cells respectively.

- Polysomes on rough endoplasmic reticulum (ER) of these tissues synthesise apo-B_{48} and apo-B_{100} respectively.
- TG is synthesised in smooth endoplasmic reticulum (ER) of both tissues.
- Microsomal and cytoplasmic enzymes participate in cholesterol synthesis.
- In the smooth endoplasmic reticulum (ER) of intestinal mucosal cells and Liver cells, the lipids are incorporated with apo-B_{48} and apo-B_{100} to form chylomicrons and VLDL respectively.
- B-apoproteins are finally glycosylated in the Golgi-complex, which then packs the lipoproteins into secretory vesicles and budded off from Golgi cisternae.

Release: Chylomicrons and VLDL are released from either the intestinal or hepatic cell by fusion of the "secretory vacuole" with the cell membrane, i.e. by ***reverse Pinocytosis***. Chylomicrons pass into the spaces between the intestinal cells, eventually making their way into the Lymphatic system ("Lacteals") draining the intestine. VLDL are secreted by hepatic parenchymal cells into the "Space of Disse" and then into the hepatic sinusoids.

Note: After entry into the blood, these "nascent" lipoproteins gradually acquire the full contingent of their apoproteins by the transfer of 'apo-C' and 'apo-E' apoproteins from circulating HDL particles.

CATABOLISM OF CHYLOMICRONS AND VLDL

Chylomicrons: Fate of Labelled Chylomicrons

1. ***Clearance*** of labelled chylomicrons from the ***blood is rapid.***
2. ***Half time of disappearance*** in humans is ***less than one hour.***
3. Larger particles are catabolised more quickly than smaller ones.
4. When chylomicrons "labelled" in FA of TG are given IV, 80 per cent of label is found in adipose tissue, heart and muscles and approx 20 per cent in Liver.

Fate of labelled VLDL: When I^{125}-VLDL were injected into humans, labelled apo-C was found in HDL as it became distributed between VLDL and HDL.

On the other hand, labelled apo-B_{100} disappeared from VLDL and appeared in a lipoprotein of **"intermediate density"** (1.006 to 1.019), called as **IDL.** Finally, radioactivity was found in apo-B_{100} of LDL, showing that the apo-B of VLDL is the precursor of apo-B of LDL. ***Only one IDL particle is formed from each VLDL particle, which in turn produces one LDL particle. In humans, virtually all of the VLDL is converted to LDL.***

Note: ***LDL is not formed in liver.*** It is a degradation product of VLDL through intermediate IDL.

(a) Role of Lipoprotein Lipase in Degradation of Chylomicrons and VLDL

- ***Lipoprotein lipase*** hydrolyses TG of circulating chylomicrons and VLDL, through di and monoacyl glycerols to release FFA and glycerol. Chylomicrons and VLDL provide the enzyme with both the substrates and cofactors. Hydrolysis takes place while the lipoproteins are attached to the enzyme on the endothelium.
- Most of the released FFA are taken up by tissue cells while smaller amounts circulate in plasma as ***Albumin-FFA complex.***
- As *lipoprotein lipase* hydrolytically removes TG from circulating chylomicrons and VLDL, the diameters and TG content of these get decreased and the percentage amounts of cholesteryl esters and cholesterol are almost doubled in them. Simultaneously apo-A and apo-C are transferred from them back to circulating HDL. Chylomicrons and VLDL are thereby converted to **"chylomicrons-remnants"** and **"IDL"** in circulating blood respectively. ***IDL thus represents the end of degradation of VLDL by lipoprotein lipase and thus corresponds to Chylomicron-remnant.***

(b) Role of Liver

1. The "E-apoproteins" of "chylomicron remnants" and some IDL particles may bind with specific "apo-E receptors" on the hepatic cell membranes, chylomicron 'remnants' and some IDL thus get concentrated on the hepatic cells and are eventually internalised into them by "pinocytosis".
2. Hepatic cells hydrolyse TG and cholesteryl esters of these lipoproteins into FA, cholesterol and glycerol which are further metabolised by Liver Cells.
3. Some of the "chylomicrons remnants" may be incorporated mainly into PL and are resecreted from the Liver as a "PL-rich" lipoprotein particles of density less than 1.006, called as ***remnant-remnants,*** whose metabolic fate is not definitely known.

CLINICAL ASPECT

Apo-B_{48} and apo-B_{100} are most essential for chylomicrons and VLDL formation. In "abeta-lipoproteinaemia", a rare disease, apo-B is not synthesised and hence chylomicrons and VLDL cannot form and Lipids accumulate in intestinal mucosal cells and hepatic cells (fatty infiltration).

Regulation of Lipoprotein Lipase Activity

- *Lipoprotein lipase* activity declines in adipocytes on starvation and rises after feeding. Hence starvation reduces and feeding enhances the uptake and storage of Fat by adipose tissue.

SECTION FOUR

- On the other hand, starvation enhances *lipoprotein lipase* activity in cardiac and striated muscles enabling them to take up and oxidise more of FA.

METABOLIC FATE OF LDL

- As stated above, LDL is not synthesised/or secreted by Liver or intestine. It is formed principally by degradation of circulating VLDL, which initially forms IDL.
- ***Most of IDL particles change into LDL particles, by losing their apo-E and some of TG.*** This makes LDL richer in cholesteryl esters, and cholesterol, and poorer in TG and total lipids. ***Thus they become smaller in diameter and higher in density than IDL.*** Cholesterol Linoleate containing polyunsaturated linoleic acid is the principal cholesteryl ester present in LDL. The 1/2 time of disappearance of apo-B_{100} of LDL from circulation is approximately 2½ days.
- ***Site of catabolism of LDL was initially thought to be solely by liver. But recent studies on partially hepatectomised dogs have indicated that catabolism of LDL in addition to liver, principally occurs in peripheral tissues viz.***
  - ***Fibroblasts,***
  - ***Lymphocytes*** and
  - ***Arterial smooth muscle cells.*** Studies on cultured fibroblasts, lymphocytes and arterial smooth muscle cells have shown the existence of "specific LDL-receptors" (Apo B-100, E).

LDL RECEPTORS AND LDL CATABOLISM

The liver and many extrahepatic tissues, as stated above express the **LDL receptor (apo-B-100, E)**

It is so designated because it is ***specific for apo B-100 but not apo B-48***, which lacks carboxy terminal domain of B-100 containing the LDL receptor ligand, ***and it also takes up lipoproteins rich in apo-E.*** This ceceptor is defective in familial hypercholesterolemia. Approximately 30 per cent LDL is degraded in extrahepatic tissues and 70 per cent in the liver.

A positive correlation exists between the incidence of coronary atherosclerosis and the plasma concentration of LDL cholesterol (bad cholesterol).

Interaction with cell membrane:

- LDL (apo B-100, E) is highly regulated.
- **LDL receptors** ***occur on the cell surface in*** **"pits"** that are coated on the cytoplasmic side of the cell membrane with a protein called **clathrin**. The glycoprotein receptor spans the membrane, the B-100 binding region being at the exposed amino terminal end. ***After binding LDL is taken up intact by endocytosis.***
- The apoprotein and cholesteryl ester are then hydrolysed in the lysosomes and cholesterol is translocated into the cell. ***The receptors are recycled to the cell surface.***
- This **influx of cholesterol inhibits the transcription of the genes encoding *HmG-CoA synthase/HmG-CoA reductase*** and other enzymes involved in cholesterol synthesis as well as the LDL receptor itself **via SREBP** *(sterol regulating element binding protein)* pathway and thus coordinately suppresses cholesterol synthesis and uptake.
- In addition ACAT ***(Acyl-CoA cholesterol acyl transferase) activity is stimulated promoting cholesterol esterification.***

In this way, **LDL receptor activity on the cell surface is regulated by the cholesterol requirement for membranes, steroid hormones or bile acid synthesis.**

CETP (Cholesteryl Ester Transfer Protein)

CETP is an enzyme protein, synthesised in the liver. If facilitates the transfer of cholesteryl esters from HDL to VLDL or LDL, in exchange of TG.

CLINICAL ASPECT

In Japan, a group of people have a genetic mutuation that causes high levels of HDL, and have a low incidence of heart disease. Later on, it was shown that these people ***lack the enzyme protein CETP due to genetic mutation,*** which increased HDL↑. Scientists are now trying to ***make drugs which can block CETP.*** Pfizer Inc has developed an oral drug, **Torcetrapib which blocks CETP** producing increased level of HDL. This drug has been found to increase BP search is on for CETP blockers.

Lipoprotein (a) [LP(a)]

It is a special type of lipoprotein ***not present in all people.*** In normal individuals it may be present only in 20 per cent of the population. In 40 per cent to 80 per cent of the population, normal LP (a) is not present in the serum in detectable amounts.

In 20 per cent of normal individuals, if present, it is found to be more than 30 mg/dl. When present, ***it is attached to apo-B100 by a disulphide bond.*** Indians have a higher level of LP (a) than in the western population.

Clinical Importance

The persons having LP (a) are ***more susceptible to heart attack at the younger age group of 30 to 40 years.*** LP (a) ***inhibits fibrinolysis.*** Levels more than 30 mg/dl increases the risk of myocardial infarction by 3 times and when

the increased LP (a) level is ***associated with increased LDL***, the risk for myocardial infarction increases further.

METABOLISM OF HDL

Synthesis: HDL is synthesised in Liver cells and also in intestinal mucosal cells.

(a) Hepatic HDL

- Apo-A and apo-C are synthesised by polysomes on the rough endoplasmic reticulum (ER).
- They are assembled with lipids to form the nascent HDL which is released in the circulation.

(b) Intestinal HDL

- In a similar manner, apo-A is synthesised by polysomes on the rough endoplasmic reticulum (ER).
- It is assembled with lipids to form the "nascent"-HDL which is released in circulation from intestinal mucosal cells.

Difference of nascent intestinal HDL from "nascent" hepatic HDL

- "Nascent" intestinal HDL contains only apo-A, when it circulates, it acquires apo-C and apo-E.
- "Nascent"-Hepatic HDL on the other hand contains both apo-A and apo-C.

Note: ***Apo-C and apo-E are only synthesised in liver and not in intestinal mucosal cells.***

Scavenging action of HDL: Glomset (1968) has suggested that HDL plays a major role in the removal of cholesterol from peripheral extrahepatic tissues and transport of this cholesterol to the liver where it is further metabolized. This has been called as ***scavenging action of HDL (reverse cholesterol transport).***

METABOLISM OF HDL

HDL Receptor Proteins and HDL Cycle

Two types of HDL receptor proteins haven been described involved in HDL metabolism. They are:

1. **HDL receptor protein—SR-B_1**
2. **ATP-binding cassette transporter A_1 Protein (ABCA1)**

1. **HDL receptor protein SR-B1**

The **class B scavenger receptor B1 (SR-B1)** has been identified recently as HDL receptor with a ***dual role in HDL metabolism.***

- **In the liver, and steroidogenic tissues,** it binds HDL via apo-A1 and ***cholesteryl ester is selectively delivered to the cells,*** although ***the particle itself, including apo-A1 is not taken up.***
- **In the tissues:** On the other hands, SR-B_1 mediates the acceptance of cholesterol from the cells by HDL, which then transports it to the liver for excretion via the bile (either as cholesterol or after conversion to bile acids), the process known as ***reverse cholesterol transport.***

Nascent HDL: Nascent HDL is made up of a bilayer of PL and free cholesterol arranged in disc-like form called as **discoid HDL containing apoproteins.**

Formation of HDL_3 by LCAT: Lecithin-cholesterol acyl transferase of plasma binds with nascent HDL discs and gets activated by A-1 and C-II apoproteins of HDL itself and transfers acyl groups from HDL-PL to the free cholesterol thus producing **Lysophospholipids** and **cholesteryl-esters.**

Lysophospholipids thus formed are released from HDL to plasma where they bind to albumin and cholesteryl esters and transferred into the 'central core' of the HDL particles.

Thus by the above reactions 'Discoid HDL' is converted to 'spherical' HDL which is called as *HDL-3 particle,* which is bounded by a lipid layer with apoproteins and cholesteryl esters embedded in its core.

Nascent HDL (discoidal) → HDL-3 particle (spherical)

CLINICAL ASPECTS

The lipoprotein particles, similar to nascent discoid HDL are found in plasma of patients with LCAT deficiency and also in plasma of patients with obstructive jaundice.

HDL Cycle

HDL_3 generated from discoidal HDL by the action of LCAT, accepts cholesterol from the tissues via the SR-B1 and cholesterol is then esterified by LCAT, increasing the size of the particles to **form the less dense HDL_2.**

HDL_3 is then reformed either after selective delivery of cholesteryl ester to the liver via the SR-B1 or by hydrolysis of HDL_2 phospholipids and triacyl glycerol by "hepatic lipase". *This interchange of HDL_2 and HDL_3 is called the HDL cycle.*

Formation of Pre-β HDL

Free apo-A1 is released by these processes and forms pre β-HDL after associating with a minimum amount of phospholipids and cholesterol. **Surplus apo-A_1 is destroyed in the kidney.**

SECTION FOUR

Apolipoprotein J (Apo-J)

It is a glycoprotein, a dimer found in association with HDL-2. Its molecular weight is approximately 50,000. Two monomeric units are α and β. α-subunit consists of 205 amino acids and β-subunit has 222 amino acids. It is found in atheromatous plaques. *Apo J has been found to inhibit macrophage mediated cell damage and thus, it is antiatherogenic and offers protection to endothelial and smooth muscle cells from injury (protective).*

2. ATP-binding Cassette Transporter A_1 (ABCA1)

Recently another second important mechanism for reverse cholesterol transport has been found involving ABCA1.

ABCA1 is a member of a family of transporter proteins that couple the hydrolysis of ATP to the binding of a substrate, enabling it to be transported across the membrane.

ABCA1 preferably transfer cholesterol from cells to poorly lipidated particles such as pre-β-HDL or apo-A1, which are then converted to HDL_3 via discoidal HDL.

Pre-β-HDL is the most potent form of HDL inducing cholesterol efflux from the tissues.

Regulation of HDL Concentration

HDL concentration varies reciprocally with plasma TG concentrations, and indirectly with the activity of lipopoprotein lipase. This may be due to surplus surface constituents, e.g. PL and apo-A1 being released during hydrolysis of chylomicrons and VLDL and contributing toward the formation of pre-β-HDL and discoidal HDL.

CLINICAL ASPECT

Role of HDL-2 and HDLc in Atherosclerosis

- **HDL-2:** Concentrations are found to be inversely related to the incidence of coronary atherosclerosis, possibly because they reflect the efficiency of cholesterol scavenging from the tissues.
- **HDLc:** Recently HDLc has been described which is found in the blood of diet-induced hypercholesterolaemia. HDLc is rich in cholesterol and its sole apo-protein is apo-E. It is taken up by the liver via the apo-E 'remnant' receptor and also by LDL- receptors.

In atherosclerosis, the atherosclerotic plaques contain 'scavenger cells' (macrophages) that have taken up so much cholesterol that they are converted into "cholesteryl-ester" laden ***foam cells***. Most of the "foam cells" arise from macrophages that ingest the more abnormal cholesterol rich lipoproteins as chemically modified LDL or β-VLDL. The macrophages secrete both cholesterol and apo-E to a suitable recipient as HDL. This apo-E after suitable processing in presence of LCAT may be the probable source of cholesterol-rich HDLc. Hence, HDLc may be an important component in the movement of cholesterol from the tissues to the liver.

SECTION FOUR

MAJOR FUNCTIONS OF LIPOPROTEINS

1. ***Chylomicrons:*** Acts as carrier of ***exogenous TG*** chylomicrons transport mainly TG, smaller amounts of PL, cholesterol-esters and fat soluble vitamins from intestine to liver and adipose tissue. Thus, the lipids carried by chylomicrons are principally dietary lipids.
2. ***VLDL:*** Acts as carrier of ***endogenous TG*** VLDL transports mainly TG synthesised in hepatic cells from the Liver to the extrahepatic tissues for storage.

   Note: High carbohydrate intake, high insulin/glucagon ratio, high plasma FFA and alcohol intake increase the hepatic synthesis of TG and thus VLDL.
3. ***LDL***
   - LDL rich in cholesterol and cholesterol-esters ***bad cholesterol*** transport and delivers cholesterol to extrahepatic tissues.
   - LDL also regulates cholesterol synthesis in extra-hepatic tissues, as cholesterol delivered by LDL to cells inhibits *HMG-CoA reductase*, the ratelimiting enzyme for cholesterol synthesis.
4. ***HDL***
   - ***Scavenging action:*** HDL scavenges the body cholesterol and blood vessel wall cholesterol by *"reverse cholesterol transport"*. With the help of apo-A, it strips off the cellular cholesterol from peripheral cells and smooth muscle cells of arteries and apo-A, activates the enzyme "LCAT", which helps in esterification of cholesterol. The cholesterol ester formed at the surface, being hydrophobic, moves into the interior and makes the ***"discoid" HDL to spherical.*** The mature HDL ***(good cholesterol)*** moves to Liver where cholesterol is catabolised.
   - HDL, with the help of apo-E, competes for the common ***BE binding site*** on the membranes and prevents internalisation of LDL cholesterol in the smooth muscle cells of the arterial wall.
   - HDL contributes its apo-C and apo-E to "nascent" VLDL and "nascent" chylomicrons. After receiving the apo-C and apo-E from HDL, both VLDL and chylomicrons become substrates to be acted upon by the enzyme *lipoprotein lipase.*
   - HDL stimulates prostacyclin synthesis by the endothelial cells. Prostacyclins inhibits platelets aggregation and thus ***HDL prevents thrombus formation.***
   - HDL also helps in removal of macrophages from the arterial wall.
   - HDL-3 contains apo-D. It functions as the *cholesteryl ester transfer protein* and transfers some "cholesteryl esters" from HDL to VLDL, LDL and chylomicrons in

the plasma. These lipoproteins then transfers these "cholesteryl esters" to Liver for degradation.

5. ***Albumin-FFA complex:*** Albumin-FFA complexes transport mainly free FA, released by adipose tissue lipolysis and small amounts of Lysophospholipids from extrahepatic tissues to the Liver.

CLINICAL ASPECT

CLINICAL DISORDERS ASSOCIATED WITH LIPOPROTEIN METABOLISM

Clinical disorders may be:

A. Hyperlipoproteinaemias

B. Hypolipoproteinaemias

Hyperlipidaemias may be further divided into:

(a) ***Primary:*** They are genetic disorders characterised by distinct clinical syndromes.

(b) ***Secondary:*** Due to underlying disease process usually thyroid, Liver and renal diseases.

INHERITED DISORDERS

A. Primary Hyperlipoproteinaemias:

Frederickson *et al.* (1967) proposed ***five types*** based on changes in plasma lipoproteins.

1. Type-I: Familial Lipoprotein Lipase Deficiency:

- A rare disorder and is characterised by Hypertriglyceridaemia (TG ↑), and hyperchylomicronaemia. Chylomicrons grossly increased ↑↑ and there is slow clearing of chylomicrons.
- VLDL (Pre-β-Lipoproteins) also increased, more so in, increased carbohydrate intake.
- Decrease in α-lipoprotein (HDL↓) and β lipoproteins (LDL↓).

Inheritance: Autosomal recessive.

Enzyme deficiency:

- Deficiency of the enzyme ***lipoprotein lipase.***
- A variant of the disease can be produced by deficiency of apo-C II.

Clinical feature: Presents in early childhood and is characterised by

- Eruptive Xanthomas
- Recurrent abdominal pain.

Note

1. Disease is fat induced. Patient may be effectively treated by low dietary fat.
2. High carbohydrate diet can raise Pre-β-lipoprotein levels as TG synthesis in Liver increases.
3. Premature cardiovascular disease is not encountered.

Refrigeration test:

If serum of suspected patient is taken in a narrow small tube and kept in refrigerator temperature for 24 hrs undisturbed, a clear zone of chylomicrons is seen to float on the top and make a distinct separate layer **(Figs 25.20 and 25.21)**.

2. Type II: Familial Hypercholesterolaemia (FHC): A common disorder which has been extensively investigated. The disease is characterised by:

- Hyper β-lipoproteinaemia (LDL ↑)
- Associated with increased total cholesterol ↑
- VLDL may be raised, hence total TG may be high. But plasma usually remains clear.

Inheritance: Autosomal dominant; Frequency—0.2 per cent

Enzyme deficiency (metabolic defect): There is no enzyme deficiency. Metabolic defects are:

- An increased synthesis of apo-B ↑
- Defective catabolism of LDL. Deficiency of LDL receptors in fibroblasts demonstrated.

Clinical features: (a) Xanthomas of tendinous and tuberous type have been described, (b) Corneal arcus, (c) Occasionally xanthelesma. Clinically most important is the ***increased incidence of atherosclerosis and premature cardiovascular diseases.***

Note: Type-II Pattern can develop as a result of ***hypothyroidism*** (secondary hyper or lipoproteinaemia).

3. Type-III: Familial Dys-Beta Lipoproteinaemia:

Synonyms: Broad Beta disease, 'Remnant' removal disease

The disease is characterised by:

- Increase in β-lipoproteins (LDL ↑)
- Increase in Pre-β-lipoproteins (VLDL ↑);
- Actually rise is in IDL (VLDL remnant). This appears as **broad β-band *("floating" β-band)***-β-VLDL on electrophoresis. Also there is hypercholesterolaemia and hypertriglyceridaemia.

Inheritance: Autosomal dominant.

Metabolic defects:

- Increased concentration of apo-E ↑.
- Increased synthesis of apo-B ↑.
- Conversion of normal VLDL to β-VLDL (IDL) and its degradation without conversion to LDL. Defect is in "Remnant" metabolism.

Clinical feature: Presents with Xanthomastuberous and palmar Xanthomas. Premature cardiovascular diseases and atherosclerosis common. Also prone to develop peripheral vascular diseases.

SECTION FOUR

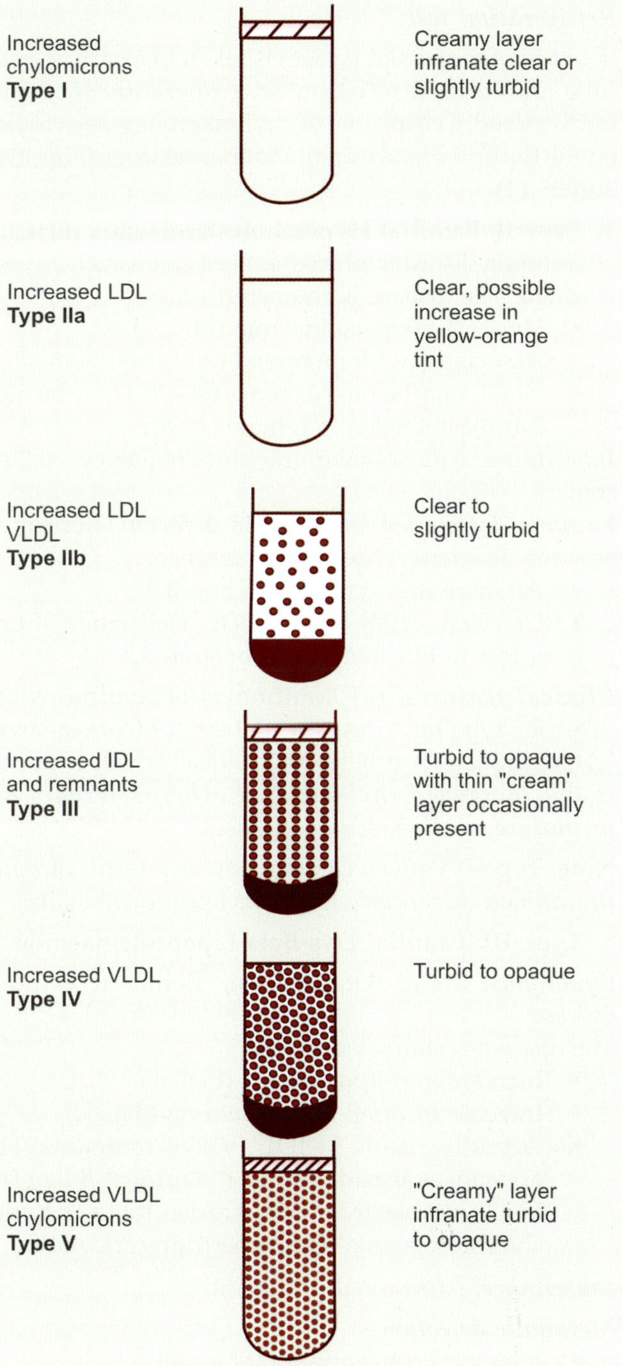

Fig. 25.20: Refrigeration test

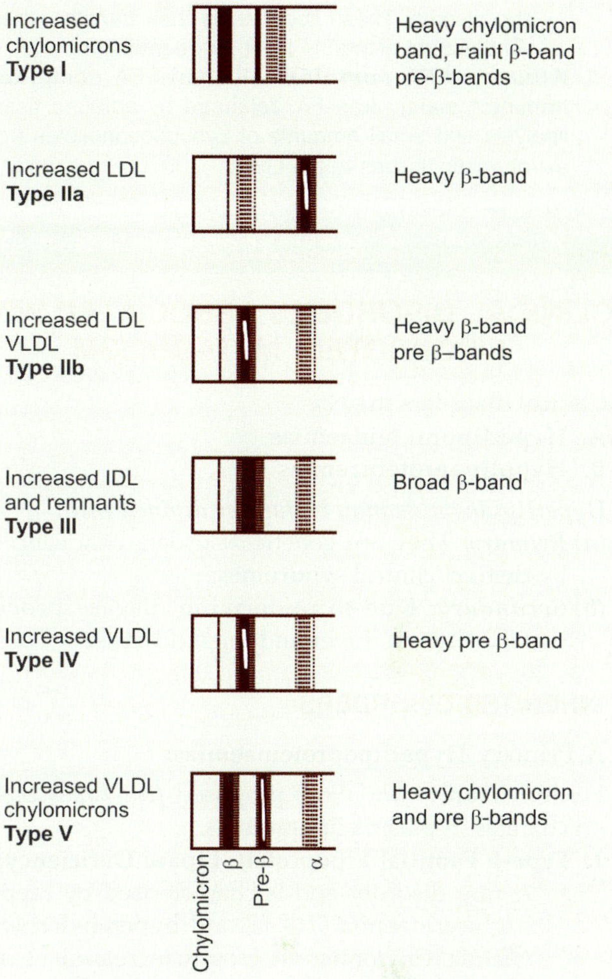

Fig. 25.21: Lipoprotein electrophoresis pattern

Treatment:

- By weight reduction
- Low carbohydrate diets, containing unsaturated fats and little cholesterol.

4. Type-IV: Familial Hypertriglyceridaemia (FHTG):

The disease is characterised by:

- Hyper Pre-β-lipoproteinaemia (VLDL ↑)
- With increase in endogenous synthesis of TG ↑
- Cholesterol level may be normal or increased
- α and β-lipoproteins are subnormal (HDL↓, LDL↓)

Inheritance: Autosomal dominant.

Metabolic Defects:

- Increased endogenous synthesis of TG ↑
- Decreased catabolism of both
- Glucose intolerance is frequent.

Clinical features: Usually presents in early adulthood.

The above-mentioned lipoprotein pattern is also found:

- In association with coronary heart disease (CHD)
- With maturity onset Diabetes mellitus
- In obesity
- With chronic alcoholism, and
- With taking of progestational hormones.

Treatment:
- Weight reduction,
- Decrease of carbohydrate intake in diets with unsaturated fats
- Low cholesterol diet
- Use of hypolipidaemic agents.

5. **Type-V: Combined Hyperlipidaemias:** In this disease, the lipoprotein pattern is complex. Increase in both chylomicrons and Pre-β-lipoproteins (VLDL) are seen. Hypertriglyceridaemia (TG ↑), Hypercholesterolaemia? Concentration of α-lipoproteins (HDL) and β-lipoproteins (LDL) are decreased.

Inheritance: Autosomal dominant.

Clinical feature: The disorder is manifested only in adulthood. Xanthomas are frequently present; incidence of atherosclerosis is not striking. Glucose tolerance is usually abnormal. Frequently found associated with obesity and Diabetes mellitus. The reason for the condition being familial is not known.

Treatment:
- Weight reduction
- Followed by diet not too high in carbohydrates and fats.

WOLMAN'S DISEASE: Also called cholesteryl ester storage disease. Increased levels of cholesterol is seen (hypercholesterolaemia).

Enzyme deficiency: Due to a deficiency of *cholesteryl ester hydrolase* in Lysosomes; Such deficiency in cells of fibroblasts have been demonstrated.

Table 25.12 shows five types of Hyperlipoproteinaemias in a tabular form.

B. Hypolipoproteinaemias

1. **Abeta Lipoproteinaemia:** A rare inherited disorder. The disease is characterised by decreased plasma cholesterol↓ due to absence of β-lipoproteins (LDL). Most lipids are present in low concentration specially TG, which is virtually absent. No chylomicrons and/or pre-β lipoproteins (VLDL) are formed.

Clinical feature: Associated with above lipid changes. There may be:
- ***Atypical relinitis pigmentosa***
- Red blood cells abnormalities like **acanthosis**
- Malabsorption of fats
- Both small intestine mucosal cells and in Liver cells, accumulation of fats occur ***(fatty infiltration).***

Metabolic defect: Principal metabolic defect is in "synthesis of apo-B" leading to gross deficiency of apo-B resulting to deficiency of lipoproteins containing apo-B viz. chylomicrons, VLDL and LDL. Classic form of this disease is called as **Bassen-Kornzweig syndrome.**

2. **Familial α-Lipoprotein Deficiency (Tangier's disease):** This disease is characterised by deficiency of α-lipoprotein (HDL↓). In homozygous patient: ***plasma HDL may be nearly completely absent.***

Inheritance: Autosomal recessive

Metabolic defect:
- Reduction in apo-A I and apo-A II
- Leading to accumulation of cholesteryl esters in different tissues.

Clinical features: Clinically cardinal feature of this disease is:
- ***Hyperplastic orange yellow tonsils,*** and
- ***Adenoids.***

There is no impairment of chylomicrons formation or secretion of endogenous TG by the Liver. However, on electrophoresis, there is no pre-β-lipoprotein, but a ***broad β-band*** is found containing the endogenous TG.

Note: ***The presence of low plasma cholesterol levels, associated with normal or elevated TG levels is often diagnostic of this disease.*** Due to HDL deficiency, clearance of TG from Plasma is slow tending to elevated TG levels (hypertriglyceridaemia), probably as a result of absence of apo-C-II which is an activator of *lipoprotein lipase*.

ROLE OF LIVER IN LIPID TRANSPORT AND METABOLISM

Liver plays a central role in lipid transport and metabolism. The liver carries out the following major functions in lipid metabolism:
- **It facilitates the digestion and absorption of lipids by the production of bile**, which contains cholesterol and bile salts synthesised within the livber *denovo* or from uptake of lipoprotein cholesterol.
- ***It actively synthesises fatty acids.***
- It ***oxidises fatty acids*** for provision of energy.
- Also synthesises triacylglycerols and phospholipids.
- It converts fatty acids to ketone bodies ***(ketogenesis)***
- Liver plays an integral part in the ***synthesis and metabolism of plasma lipoproteins.***

ATHEROSCLEROSIS

Atherosclerosis is a slowly progressive disease of large to medium-sized muscular arteries and large elastic arteries characterised by elevated focal intimal *fibrofatty Plaques.*

Principal larger vessels affected are the abdominal aorta, descending thoracic aorta, internal carotid arteries

Table 25.12: Five types of hyperlipoproteinaemias

Type	Genetic classification	Electrophoretic classification	Inheritance	Plasma lipids	Plasma lipoproteins	Clinical features	Treatment
I.	**Familial Lipoprotein Lipase deficiency**	Hyperchylomic-ronaemia	Autosomal recessive	TG ↑, may be cholesterol increased	Chylomicrons ++, Pre-β-Lipoproteins may be ↑ and α and β-Lipoproteins↓	• Rare • Early childhood • Eruptive xanthomas • Recurrent abdominal pain	Fat induced, Diet low in fat
Note: 1. Slow clearing of chylomicrons 2. Premature CV disease does not occur.							
II.	**Familial hypercholestero-laemia (FHC)**	Hyper-β-lipopro-teinaemia	Autosomal Dominant	Total Cholesterol ↑, TG ↑ or N	LDL↑, VLDL may be↑	Common occurrence, Associated with Xanthomas-tendinous and tuberous	Reduction of dietary cholesterol and saturated fats
Note: 1. Increased incidence of Premature CV diseases and atherosclerosis. 2. Metabolic defects: (a) Increased synthesis of apo-B ↑ (b) Defective catabolism of LDL ***(Defective LDL receptor)***							
III.	**Familial dysbetalipopro-teinaemia (Broad β-disease) ('Rem-nant-remnant' disease)**	Broad-β-Lipopro-teinaemia (floating β-band)	Autosomal Dominant	Cholesterol ↑ TG ↑	VLDL ↑, IDL ↑ LDL ↑	Rare, xanthomas, tuberous and palmar	• Weight reduction • Low carbohydrate diets • Unsaturated fats with little cholesterol
Note: 1. Metabolic defects: (i) Increased synthesis of apo-B↑ (ii) Increased synthesis of apo-E↑ 2. Premature CV disease and Peripheral Vascular disease.							
IV.	**Familial hyper-triglyceridaemia (FHTG)** **Note:** This lipoprotein pattern is associated with coronary heart disease, obesity, Maturity onset DM (Type-I), alcoholism and taking progesterone hormones.	Hyper Pre-β-lipoproteinaemia	Autosomal dominant	TG and choles-terol ↑ or N	VLDL and α and β lipoproteins subnormal	• Present in early adulthood • Synthesis of lipids from carbohydrates ↑	• Weight reduction • Replacement of much carbohydrates with unsaturated fats • Low cholesterol diet • Hypolipidaemic drugs
V.	**Combined Hyperlipidaemia**	Variable	Autosomal Dominant	Both TG and chylomicrons-	Both VLDL and chylomicrons ↑ α- and β-lipo-proteins ↑	• uncommon occurrence • xanthomas present • abnormal glucose tolerance	• Weight reduction followed by a diet not too high either in carb or fats
N.B. Associated with ketotic DM, Incidence of atherosclerosis less.							

and medium to smaller sized vessels affected are popliteal arteries, coronary arteries and circle of Willis in brain.

The atheroma may be preceeded by ***fatty streaks*** that are intimal collection of lipid-laden macrophages and smooth muscle cells, occurring in persons as young as one year of age.

The disease typically manifests in later life as the vessel lumen is compromised, predisposing to thrombosis and the underlying media is thinned, predisposing to aneurysm formation.

It is number one killer disease, 50 per cent of all deaths in USA are attributed to atherosclerosis and half of these are due to acute myocardial infarctions. The remainder include cerebrovascular accidents ("stroke"), aneurysm rupture, mesenteric occlusion and gangrene of the extremities.

Etiological Factors

Major risk factors in CHD have been discussed earlier.

Risk of developing atherosclerosis increases with age, a positive family history, cigarette smoking, diabetes mellitus, hypertension and hypercholesterolaemia. ***The risk is correlated with elevated LDL and inversely related to the HDL level.***

Hereditary defects, e.g. familial hypercholesterolaemia involving the LDL receptor or the LDL apoproteins cause elevated LDL, hypercholesterolaemia and accelerated atherosclerosis.

Lesser influences on the risk of atherosclerosis include sedentary, or high-stress lifestyle, obesity and oral contraceptives.

Pathology

- The characteristic atheromatous plaque, called as ***atheroma*** is a white-yellow internal lesion upto 1.5 cm in diameter protruding into the vessel lumen.
- Microscopically by studying a section—it is found to be composed of a (i) • ***Superficial fibrous cap:*** Containing smooth muscle cells, scattered leucocytes and dense connective tissue.
- ***Below is a "cellular zone"*** with smooth muscle cells, macrophages, and T lymphocytes.
- ***A central necrotic zone:*** Containing dead cells, lipid, cholesterol clefts, and lipid laden ***foam cells*** (macrophages and smooth muscle cells) and plasma proteins.

  Complicated "Plaques" are ***calcified*** and ***fissured*** or ulcerated predisposing to thrombosis and to cholesterol microemboli. There may be haemorrhage inside the lesion, medial thinning and aneurysmal dilatation.

Pathogenesis: Most theories postulate some damage to endothelium or underlying smooth muscle cells, with consequent proliferation of smooth muscle cells and fibroblasts as an inflammatory response to such damage accordingly **two views** expressed:

(a) Endothelial injury: Mechanisms of injury include gross physical or chemical trauma or more subtle insults such as hyperlipidaemia, hypertension and diabetic angiopathy.
- Injury induces increased permeability to plasma constituents, including lipids and permits platelets and monocytes to adhere to endothelium.
- Factors from activated platelets and monocytes, e.g. PGDF (platelets derived growth factor) induce smooth muscle migration from media to intima, followed by proliferation.
- Smooth muscle cells also synthesise extracellular matrix like collagen, elastic fibres and proteoglycans, and along with macrophages from monocytes emigration which accumulate lipids to become **foam cells.**
- Macrophages bring about further injury by contributing of ***enzymes***, ***cytokines*** like IL-1 and TNF and oxidants that oxidise LDL to propagate further injury, repeated insults lead to atheromatosis plaque formation.

(b) Smooth muscle injury: More accepted view and it proposes smooth muscle proliferation as the initial events. Smooth muscle injury is brought about by ***lipid peroxidation***. Endothelial activation is subsequent feature. This view is supported in part by the finding of monoclonal/oligoclonal smooth muscle proliferation within individual human plaque.

Role of LDL in Atherosclerosis

- ***Modification of apo-LDL by lipid peroxidation attracts macrophages which are subsequently converted to "foam cells".***
- LDL is the principal factor in promoting atherosclerosis. They are deposited under the endothelial cells. LDL undergoes oxidation by ***free radicals*** released from endothelial cells. Local elevation of lipid hydroperoxides leads to a decrease in the prostacyclin (PI) and thromboxane (Tx) ratio, which can lead to thrombosis.

Consequences of Atherosclerosis

Atherosclerosis is a slowly progressive disease. It remains asymptomatic for decades until it causes disease.

1. Sudden ***occlusion*** of narrowed vascular lumen by superimposed thrombus, e.g.
   - Myocardial infarction precipitated by thrombotic occlusion of coronary artery.
   - Thrombosis of cerebral vessels.
2. Providing a site for thrombosis and then ***embolism***, e.g. renal infarction from a mural thrombus overlying an ulcerated atherosclerotic aortic plaque.
3. Weakening the wall of a vessel followed by ***aneurysm*** formation and may be rupture, e.g. an abdominal aortic aneurysm.
4. Narrowing vascular lumina, e.g. *gangrene* of the lower leg because of stenosis of atherosclerosis in the popliteal artery.

PLASMA LIPOPROTEINS AND ATHEROSCLEROSIS

Over the last several years, an intensive investigation has focussed on the identification of "risk factors" for the development of premature atherosclerosis. The high incidence of cardiovascular diseases in the western world has necessitated a major scientific effort to elucidate the etiology of this disease.

Hypercholesterolaemia or more accurately, hyper-β-lipoproteinaemia (LDL ↑) has clearly been identified as a major risk factor. Of recent interest, has been the retrospective analysis of epidemiological data which has suggested ***a negative correlation between HDL cholesterol in human plasma and risk of premature heart disease***. In a series of patient with any given LDL cholesterol level, the probability of CV disease increases as HDL-cholesterol level decreases. ***These studies are consistent with the hypothesis that HDL-deficiency is an independant risk factor for premature CV disease.***

Two separate mechanisms have been postulated to explain the role of HDL in the regulation of intracellular concentrations of cholesterol:

1. **Glomset (1968)** proposed that HDL played a role in cholesterol metabolism ***by facilitating removal of cholesterol from peripheral cells and transporting to Liver for degradation.*** This has been termed as **reverse cholesterol transport.** As discussed above, VLDL are metabolised to LDL, which enter the peripheral cells for degradation. The cellular concentration of cholesterol is controlled both by:
   - Regulation of endogenous cholesterol synthesis and
   - Egress of cholesterol from the cell, which is facilitated by HDL.

Hence, lowered levels of HDL in the plasma would be less effective in the removal of cholesterol from peripheral cells.

Note: Role of HDL-2 and HDLc has already been discussed above.

2. An additional mechanism has recently been proposed to explain the inter-relationship between LDL and HDL. In these studies, HDL was demonstrated to influence the binding and uptake of LDL by the peripheral cells. In *in vitro* experiments, in fibroblast, endothelial cells, Lymphocytes and arterial smooth muscle cells in tissue culture ***HDL was shown to competitively inhibit LDL binding and uptake.***

Thus, decreased plasma levels of HDL could be postulated to increase LDL uptake, while increased HDL levels would decrease cellular uptake of LDL. ***A ratio of LDL cholesterol and HDL cholesterol is important. If it is high the risk is more and if the ratio is low risk is less.***

Recently, ***estimation of apolipoproteins*** are being done in clinical biochemistry. ***Apo-A-I apoprotein of HDL and apo-B apoprotein of LDL have been considered as the better determinants of the "risk" and are regarded as better indicators for myocardial infarction.*** They can be estimated by various immunoassay techniques, viz. RIA, immunonephelometric, immunoturbidimetric and radial immunodiffusion.

Boosting HDL Level in Blood

Studies suggest that reducing LDL by 1 mg/dl cuts cardiovascular 'risk' by 1 per cent, but raising (boosting) HDL by 1 mg/dl reduces risk by 2 to 3 per cent. Statins are good drugs that lower LDL by 30 to 35 per cent, but it is more difficult to raise HDL which is more effective in reducing c.v. risk.

The following are some ways that have been proved to improve HDL level:

- **Exercise:** Aerobic exercises for 30 minutes several times a week can raise HDL level by 3 to 9 per cent in sedentary healthy people.

  There is ***little evidence that walking increases HDL.***
- **Weight control:** Every one kg (2-2 pounds) of weight loss raises HDL by an average of 0.35 mg/dl.
- **Alcohol consumption:** Mild to moderate drinking (one or two drinks a day) can raise HDL by an average of 4 mg/dl.
- **Stopping of smoking:** Quiting of smoking increases on an average about 4 mg/dl.
- **Diet:** A diet low in trans fatty acids and high in mono unsaturated and poly unsaturated fatty acids (PUFA) can raise HDL.

  Select oils like olive oil, flax seed oil and canola; nuts; and cold water fish and shell fish.

  Limit high-glycaemic-load foods such as pasta and white bread made with refined flour which can lower HDL.
- **Niacin therapy:** Niacin is the most effective drug, increases of 20 to 35 per cent seen. But use is limited due to side effects like intense itching and facial and upper body flushing.

RECENT ADVANCES: AN EXCERPT ARTIFICIAL HDL

Recently scientists have created tiny particles in the laboratory by nanotechnology that mimic the natural HDL (High density Lipoprotein) good cholesterol, and they help in scooping up the cholesterol before it can grow up into dangerous deposits of plaque.

The surfaces of these numoparticles are coated with fats are proteins as they can bind tightly with the sticky cholesterol to transport it through block stream to liver for degradation (Fig 25.22).

Researchers have demonstrated that these synthetic nanoparticles (synthetic version of natural HDL) binds cholesterol very tightly not only in the laboratory but in animals

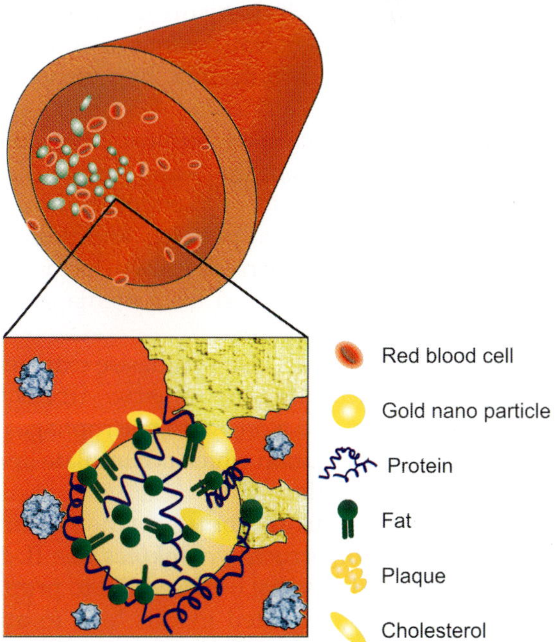

Fig. 25.22: The surface of a synthetic particle (right) is designed to bind with sticky cholesterol and sweep it from arterial walls

SECTION FOUR

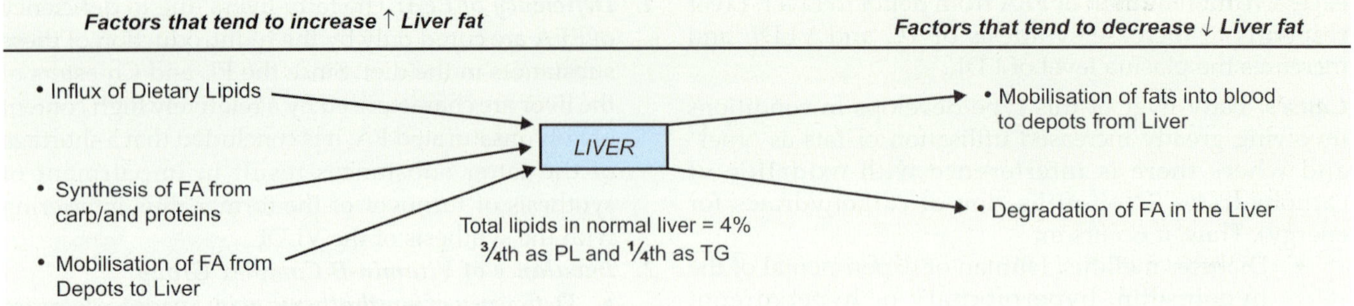

also Dr Andre Nel, chief and Director of nanotechnology at the university of California, Los Angles, said that these particles may some day be important in treating cardiovasculor disease.

He said, the particles may be useful not only for cardiovascular therapy, but also in diagnosis. The researchers have put gold and other metal cores at the centre of the particles, so that they show up well in medical imaging. Such imaging could be used, to monitor plaques as they build up in blood vessels.

FATTY LIVER

The amount of lipids in the liver at any given time is the resultant of several influences, some acting in conjunction with and some in opposition to other. ***Normal liver contains about 4 per cent as total lipids,*** three-fourths of which is phospholipids (PL) and one-fourth as neutral fats (TG).

Factors that Regulate Fat Content of Liver

(a) Factors that tend to increase ↑ the fat content of liver are:

- Influx of dietary lipids,
- Synthesis of FA from carbohydrates and proteins,
- Mobilisation of FA from depots to liver.

(b) Factors that tend to decrease↓ the liver fats are:

- Mobilisation of fats into the blood and then to the depots from the liver.
- Degradation of FA within the liver itself.

Normal levels of Lipids in the Liver are the result of maintenance of a proper balance between the above mentioned factors. ***A relative increase or decrease in the rate of one or other of these processes can result in accumulation of abnormal quantity of lipids in the Liver, producing fatty Liver.***

TYPES OF FATTY LIVER

Biochemically, theoretically fatty liver can be of five types:

- **Type-1** overfeeding of fat
- **Type-2** oversynthesis of fats from carbohydrates
- **Type-3** over mobilisation from depots to liver
- **Type-4** under mobilisation from liver to depots
- **Type-5** under utilisation in the liver.

Type 1: Overfeeding of Fats

Overfeeding of fats produce increase in circulating chylomicrons:

- Liver can take up by pinocytosis, leading to increased TG in Liver cells.
- Circulating chylomicrons are acted by *Lipoprotein lipase,* which produces increase in FFA by hydrolysis of TG of chylomicrons. Leads to influx of FFA in liver, synthesis of TG is enhanced and formation and secretion of more VLDL.
- ***Lipids deposited in type-1, reflects the composition of dietary Lipids.***

Type 2: Oversynthesis of Fats from Carbohydrates

Ingestion of carbohydrates in excess of caloric requirement, overloads the capacity of the cells which normally store glycogen. Surplus carbohydrates are channeled to synthesis of FA and TG (lipogenesis) in Liver and adipose tissue.

Plasma lipids: Increase in plasma TG ↑ and VLDL ↑, which in turn increases LDL level.

Causes: Oversynthesis from carbohydrates can result from:

1. Forced overfeeding of carbohydrates,
2. Experimentally,
   - By administration of excessive amounts of certain B-vitamins viz. thiamine, riboflavin and biotin, and
   - Administration of amino acid cystine.

Explanation

- Probably stimulates and increases appetite.
- Increase in general metabolic activity, leading to greater demand of Inositol and choline, producing relative deficiency of them.
- Deficiency of amino acids Threonine and Isoleucine also increases FA synthesis.

Type 3: Overmobilisation from Depots to Liver

This type of fatty liver is referred as **physiological** fatty liver. This represents an exaggeration of normal process,

excessive mobilisation of FFA from depot to Liver. Liver responds to increase synthesis of TG and VLDL and increases the plasma level of LDL.

Causes: Fatty liver of this type develops in conditions involving greatly increased utilisation of fats as "fuel" and where there is interference with oxidation of carbohydrates. (Non-utilisation of carbohydrates for energy). Thus, it occurs in:

- Diabetes mellitus: Human or experimental of the hypoinsulin, hyperpituitary or hyperadreno-cortical type,
- Starvation and
- Carbohydrate deprivation.

Owing to the non-utilisation of carbohydrates, adipose tissue cannot esterify FFA due to lack of α-glycero-P, thus aggravating the hyperlipaemia. In addition to fatty liver and hyperlipaemia, this condition is characterised by ketosis and in advanced cases acidosis.

Note: Hydrazine fatty liver is partially caused by over-mobilisation of FFA.

Type 4: Under Mobilisation from Liver to Depot

Fatty Liver of this type has been differentiated from the preceding type by being designated as ***Pathological fatty liver.*** It is accompanied by a decrease↓ in plasma lipids (hypolipaemia), which affects mainly PL and Ch-esters. The pattern of liver lipids is also abnormal, ***Fatty Livers of this type, if not treated, eventuates in cirrhosis liver and there may be associated haemorrhagic lesions in the kidneys.***

Causes: They appear to be caused by agents or conditions, which produce either absolute or a relative deficiency in certain of the ingredients used by the Liver for synthesis of VLDL. Such as the:

- Protein: Apoprotein itself or
- The building blocks of its structural lipid moieties, such as cholesterol esters and PL viz. Inositol phosphatides, choline and the polyunsaturated FA
- Factors interfering with secretory mechanism.

LIPOTROPIC AGENTS

Agents such as **choline, methionine, Betaine, inositol,** etc. which have the apparent effect of facilitating the removal of fat from liver, and thus prevents accumulation of fat in Liver cells. ***Such substances which prevent accumulation of fat in liver are said to be 'Lipotropic' ('Lipotropic agents' or 'Lipotropins').*** The phenomenon itself is called "Lipotropism". Antagonistic agents and the converse condition are "anti-lipotropic" and "anti-lipotropism" respectively.

Causative agents: Fatty livers of Type-4 may be roughly classified according to the causative agent or phenomenon as follows:

1. *Deficiency of EFA:* The fatty livers due to deficiency of EFA are cured only by the re-introduction of these substances in the diet. Since the PL and Ch-esters of the liver are characterised by a relatively high content of Polyunsaturated FA, it is concluded that a shortage of the latter substances result in impairment of synthesis or turnover of the former thus interfering with the synthesis of the VLDL.
2. *Imbalance of Vitamin-B Complex Group*
   - *Deficiency of panthothenic acid:* Leads to decrease in CoA-SH. Thus, activation of FFA and its oxidation suffers.
   - *Lack of pyridoxine and an excess of biotin:* They are said to elicit a greater demand upon the supply of inositol, thus supposedly interfering with the synthesis of inositol-containing phosphatides.
   - *Excess biotin:* This provides more of malonyl-CoA and there may be oversynthesis of FA.
   - *Choline deficiency:* Certain fatty livers of type-4 are due, more or less directly, to a deficiency of choline. Since choline is synthesised by successive methylations of ethanolamine, it is true that ***induced deficiencies of the –CH_3 group produce fatty livers,*** solely, however, due to resulting shortage of choline.

A relative deficiency of choline can be induced also by inclusion in the diet of compounds which will compete with ethanolamine for available –CH_3 groups. Nicotinic acid or its amide and guanidoacetic acid (glycocyamine) are methylated in the body to form N'-methyl nicotinamide and creatine respectively. Administration of excessive quantities of these compounds deplete the supply of –CH_3 groups available for synthesis of choline, producing choline deficiency leading to fatty liver. Lecithin is also lipotropic as it contributes to availability of choline in the body.

Mechanism of Action of Choline as a Lipotropic Agent

- Choline deficiency results in depression of oxidation of long-chain FA, due to decreased levels of 'carnitine' (as carnitine formation requires –CH_3 group). ***Reduced oxidation of long-chain FA enhances TG synthesis.***
- Deficiency of PL containing choline impairs synthesis of intracellular membranes concerned in lipoprotein synthesis.
- Choline deficiency may impair availability of phosphoryl choline which stimulates incorporation of glucosamine into glycolipoproteins.

Other Lipotropic Agents

1. *Lipocaic:* Pancreatectomised animals maintained adequately with insulin, develop fatty livers, not of

SECTION FOUR

overmobilisation type characteristic of Diabetes mellitus, but of Type 4. The condition is alleviated by the administration of raw pancreas.

Explanation

- According to some authorities, "lipocaic", an internal secretion of pancreas, is the lipotropic agent responsible for the alleviation of the condition.
- Majority believe that an external secretion of the pancreas, is involved; probably a proteolytic enzyme which specifically facilitates the liberation of methionine from the food.

2. Other substances that exert a lipotropic action but the mechanism of which has not been explained are oestrogens, certain *androgens* and *growth hormone*. They have been found to be 'Lipotropic' in the case of experimental fatty liver induced by feeding Ethionine.
3. *Casein:* is also a lipotropic agent, probably due to high content of methionine.

Type 5: Underutilisation in the Liver

1. It is possible that the fatty livers of pantothenic acid deficiency are of this type, i.e underutilisation. Deficiency of pantothenic acid leads to decrease↓ in availability of CoA-SH. Hence, activation of FA and its oxidation suffers.
2. Poisoning by salts of rare earth elements (e.g. Cerium) also appears to cause underutilisation, by inhibition of the mitochondrial system which oxidises FA.

Note

Though we have discussed above, 5 types of biochemical mechanisms which can cause fatty liver, in **practice clinically** type 1 and type 5 are rather rare.

Causes of fatty liver seen in clinical practice are:

- ***Alcohol abuse:*** Most common cause in India (Mechanism of production of fatty liver by alcohol is discussed below).
- ***Malnutrition:*** Deficiency of protein, EFA and lipotropic agents.
- ***Diabetes mellitus***
- ***Obesity***
- ***Hepatotoxins and Drugs.***

BIOCHEMICAL MECHANISMS OF SOME AGENTS

The following agents are involved in production of fatty liver.

1. *Carbon Tetrachloride (CCl_4)*

   CCl_4 produces the fatty liver by the following mechanisms:

   - Interferes with synthesis of apoprotein required to be incorporated in lipoprotein complex in Liver.
   - Also affects the secretory mechanism itself, or
   - Can interfere with conjugation of the lipid moiety with lipoprotein apoprotein.
   - Also mobilises FA through release of catecholamines.

2. *Ethionine*

   Ethionine is chemically α-amino-γ-ethylmercaptobutyric acid. Ethionine produces fatty liver, probably ***due to decline in m-RNA↓*** and ***protein synthesis↓ caused by a reduction in availability of ATP.***

Mechanism: The above happens when ethionine replacing methionine in "S-Adenosyl methionine" traps available adenine and prevents synthesis of ATP.

3. *Orotic Acid*

   Administration of orotic acid causes fatty liver.

Mechanism of Action

- Probably blocks specifically the synthesis of apo-VLDL (apo-B_{100}).
- Also probably interferes with the inclusion of glucosamine in VLDL-apoprotein.

4. *Ethyl Alcohol*

   Chronic alcoholism leads to fat accumulation in liver, which leads to cirrhosis liver. Plasma shows hyperlipidaemia.

Lipid Changes

- Increased FFA level ↑. Extra FFA mobilisation plays some part or not is not clear. Experimental studies in rats, after a single intoxicating dose of ethanol shows elevated level of FFA ↑ (increased synthesis of FA?)
- Increased TG synthesis ↑ occurs
- Decreased FA oxidation↓ and inhibition of TCA cycle↓
- Increased cholesterol synthesis ↑
- Depresses transport of fats from liver
- In chronic alcoholics, there is also associated nutritional deficiencies:
  - Deficiency of vitamins
  - Deficiency of proteins/amino acids like threonine, isoleucine, glycine, tryptophan, etc.

The above occurs due to lack of appetite, and associated gastritis and thus can aggravate.

Metabolism of Ethanol

Metabolism takes place exclusively in liver. Ethanol oxidation is catalysed by the enzyme ***alcohol dehydrogenase,*** which is a zinc-containing metallo-enzyme and requires NAD^+ as acceptor of H^+.

Metabolism of Ethanol

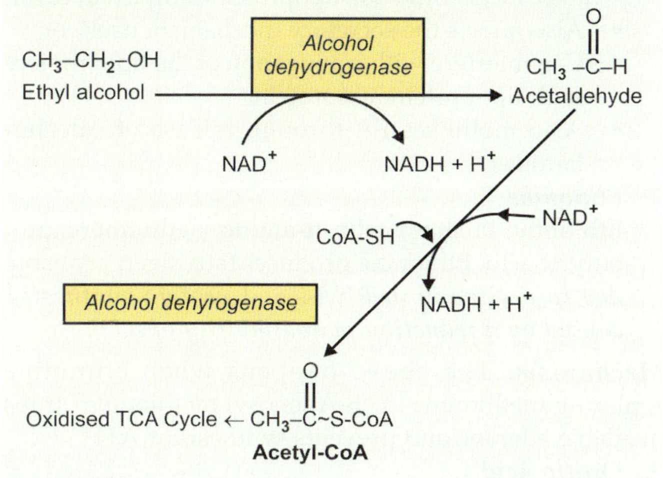

Biochemical mechanism:

Due to ethanol oxidation, ratio of $\frac{NADH + H^+}{NAD^+}$ ↑

This leads to following alterations:

(a) **Shift to the right** of the following reaction:

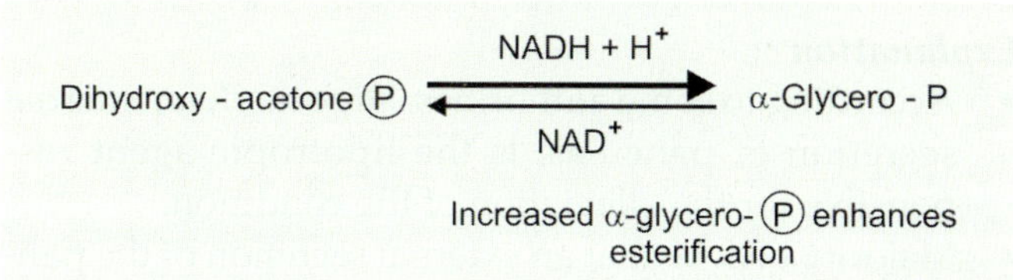

(b) **Shift to Left:**

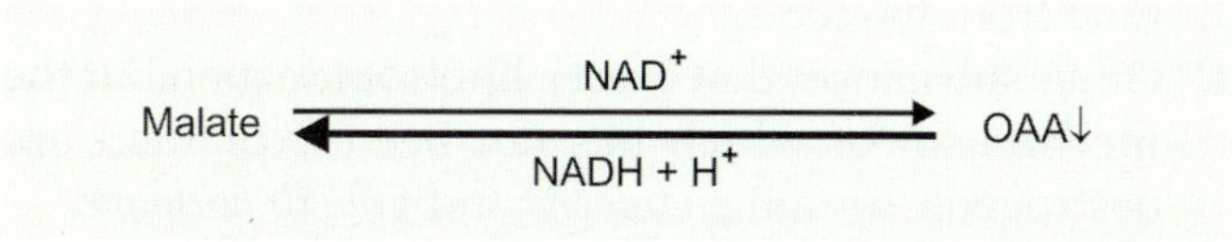

Produces relative deficiency of OAA and thus reduces activity of TCA cycle.

I. Questions (Essay type and short notes)

1. Describe beta-oxidation of fatty acids.
2. What is beta oxidation? Enumerate the pathway for fatty acid degradation.
3. Describe biosynthesis of fatty acids and their role in metabolism during exercise.
4. Enumerate the ketone bodies. Describe the formation and utilisation of ketone bodies in the body.
5. Describe the role of liver in fat metabolism.
6. Describe the formation and fate of "active acetate" (acetyl-CoA) in the body.
7. Enumerate the ketone bodies. How are they formed? Describe the conditions in whcih ketosis develops. How can it be prevented?
8. What is a fatty liver? Describe the biochemical mechanisms of production of fatty liver.
9. Describe briefly the pathways for fatty acid synthesis in mammalian cells.

Short Notes

a. Carnitine
b. Lipoproteins
c. Fatty liver
d. Bile salts
e. Biosynthesis of cholesterol
f. Role of carnitine
g. Ketosis
h. β-oxidation
i. Cholesterol
j. Lipogenesis
k. Atherosclerosis
l. Ketone bodies
m. Lipotropic factors
n. Mechanism of fatty Liver formation by carbon tetrachloride and ethanol

Differentiate

Mitochondrial and extramitochondrial FA synthesis.

II. MCQs (Give one correct answer)

1. A pathway that requires NADPH as a cofactor is:
   a. Fatty acid oxidation
   b. Extramitochondrial *de novo* fatty acid synthesis
   c. Ketone bodies formation
   d. Glycogenesis
   e. Gluconeogenesis
2. LCAT activity is associated with which of the lipoprotein complex?
   a. VLDL
   b. Chylomicrons
   c. IDL
   d. LDL
   e. HDL
3. The 'committed step' in the biosynthesis of cholesterol from acetyl-CoA is:
   a. Formation of acetoacetyl-CoA from acetyl-CoA
   b. Formation of mevalonate from HMG-CoA
   c. Formation of HMG-CoA from acetyl-CoA and acetoacetyl-CoA
   d. Formation of squalene by squalene synthetase
   e. Formation of lanosterol from squalene by cyclisation.
4. In β-oxidation of fatty acids, which of the following are utilised as coenzymes?
   a. NAD^+ and $NADP^+$
   b. FAD H_2 and NADH + H^+
   c. FAD and FMN

d. FAD and NAD^+
e. FAD and $NADP^+$
5. The most important source of reducing equivalents for FA synthesis in the Liver is:
a. Glycolysis b. HMP-shunt
c. TCA cycle d. Uronic acid pathway
e. Gluconeogenesis
6. The lipoproteins with the fastest electrophoretic mobility and the lowest TG content are:
a. VLDL b. IDL
c. LDL d. HDL
e. Chylomicrons
7. All of the following tissues are capable of using ketone bodies, except:
a. Brain b. Renal cortex
c. Red blood cells d. Cardiac muscle
e. Skeletal muscle
8. The major source of cholesterol in arterial smooth muscle cells is from:
a. IDL b. LDL
c. HDL d. Chylomicrons
e. VLDL
9. Ketone bodies are synthesised from fatty acid oxidation products by which of the following organs?
a. Skeletal muscles b. Kidney
c. Erythrocytes d. Brain
e. Liver
10. Chain elongation of fatty acids occurring in mammalian Liver takes place in which of the following subcellular fractions of the cell?
a. Nucleus b. Ribosomes
c. Lysosomes d. Microsomes
e. Golgi apparatus
11. Which of the following cofactors of their derivatives must be present for the conversion of acetyl-CoA to Malonyl-CoA in extramitochondrial FA synthesis?
a. FAD b. FMN
c. ACP d. NAD+
e. Biotin
12. Match the following:
a. Lipoprotein lipase deficiency
b. Obesity
c. Juvenile Diabetes mellitus—Type-A
d. α-lipoprotein deficiency
e. Deficiency of choline.
1. Increase FA acid esterification________
2. Increased fat accumulation in liver ________
3. High levels of free FA-albumin complex ______
4. Low level or absent high density lipoproteins______
5. High level of chylomicrons in plasma following a 2 hr fast
13. Which of the following statement regarding β-oxidation is true:
a. Requires β-ketoacyl-CoA as a substrate
b. Forms CoA thioesters
c. Requires GTP for its activity
d. Yields acetyl-CoA as a product
e. None of the above
14. All statements regarding 3-OH-3 methly glutaryl CoA are true, except:
a. It is formed in the cytoplasm
b. Required in ketogenesis
c. Involved in synthesis of FA
d. An intermediate in cholesterol biosynthesis
e. Enzyme involved in HMG-CoA synthase
15. Which of the following lipoproteins would contribute to a measurement of plasma cholesterol in a normal individual following a 12 hour fast?
a. Chylomicrons
b. VLDL
c. Both VLDL and LDL
d. LDL
e. Both chylomicrons and LDL
16. All statements regarding ketone bodies are tue ***except:***
a. They may result from starvation
b. They are formed in kidneys
c. They include acetoacetic acid and acetone
d. They may be excreted in urine
e. They are present in high concentration in uncontrolled Diabetes mellitus
17. In synthesis of TG from α-Glycero-P and acyl-CoA, the first intermediate formed is:
a. α, β-diacyl glycerol
b. Acyl carnitine
c. β-monoacyl glycerol
d. Phosphatidic acid
e. Lysophosphatide
18. During each cycle of β-oxidation of FA, all the following compounds are generated, except:
a. NADH b. H_2O
c. $FAD.H_2$ d. Acyl-CoA
e. Acetyl-CoA
19. The energy yield from complete oxidation of products generated by second reaction cycle of β-oxidation of palmitoyl-CoA will be:
a. 5 ATP b. 12 ATP
c. 17 ATP d. 34 ATP
e. 42 ATP
20. β-oxidation of odd-carbon fatty acid chain produces:
a. Succinyl-CoA b. Propionyl-CoA
c. acetyl-CoA d. Malonyl-CoA
e. acetoacetyl-CoA
21. Match the following:
a. Mevalonic acid b. Cholic acid
c. Zymosterol d. Acyl synthase
e. Coprosterol
1. Cholesterol precursor produced by removal of $-CH_3$ groups ______
2. Brings about activation of FA ______
3. Precursor of isoprenoid units in cholesterol biosynthesis ______
4. Product of microbial reduction of cholesterol _____
5. Main metabolic end-product of cholesterol ______.

SECTION FOUR

22. Brown adipose tissue is characterised by the following:
    a. Present in large quantities in adult humans
    b. Mitochondrial content higher than white adipose tissue
    c. Oxidation and phosphorylation are tightly coupled
    d. Absent in hibernating animals
    e. All of the above
23. Ketosis is partly ascribed to:
    a. Over production of glucose
    b. Under production of glucose
    c. Increased carbohydrate utilisation
    d. Increased fat utilisation
    e. Increased pyruvate in liver
24. The free fatty acids in blood are:
    a. Stored in fat depots
    b. Mainly bound to β-lipoproteins
    c. Mainly bound to serum albumin
    d. Metabolically mast inacive
    e. All of the above
25. Triacyl glycerol (TG) Present in VLDL is hydrolysed by:
    a. Intestinal lipase
    b. Lipoprotein lipase
    c. Hormone sensitive TG lipase
    d. Pancreatic lipase
    e. Co-lipase
26. Carnitine is synthesised from:
    a. Lysine b. Serine
    c. Choline d. Arginine
    e. Threonine
27. A metabolite which is common to pathways of cholesterol biosynthesis from acetyl-CoA and cholecaliferol formation from cholesterol is:
    a. Zymosterol
    b. Lumisterol
    c. Ergosterol
    d. 7-Dehydrocholesterol
    e. Pregnenolone
28. Acetyl-CoA required for extramitochondrial FA synthesis is produced by:
    a. Pyruvate dehydrogenase complete
    b. Thiolase
    c. Acetyl-CoA synthase
    d. Carnitine-acyl tranferase
    e. Citrate Lyase
29. Biosynthesis of TG and lecithin both require an intermediate
    a. Monoacyl glycero-(P)
    b. Phosphatidic acid
    c. Phosphatidyl ethanolamine
    d. Phosphatidly cytidylate
    e. Phosphatidyl adenylate
30. The rate limiting step in cholesterol biosynthesis is:
    a. Squalene synthetase
    b. Mevalonate kinase
    c. HMG-CoA synthetase
    d. HMG-CoA reductase
    e. Thiolase

Answers to MCQs

1. b 2. e 3. b 4. d 5. b 6. d
7. c 8. b 9. e 10. d 11. e
12. 1-b, 2-e, 3-c, 4-d, 5-a 13. b 14. c 15. d
16. b 17. d 18. b 19. d 20. b
21. 1-c, 2-d, 3-a, 4-e, 5-b 22. b 23. d 24. c
25. b 26. a 27. d 28. e 29. b 30. d.

CHAPTER 26

DIGESTION AND ABSORPTION OF PROTEINS

Major Concepts

A. To study how the proteins present in foodstuffs are broken down to simpler forms in the GI tract.
B. To study how the simpler forms are absorbed from GI tract.

Specific Objectives

A. 1. Study the principal proteins present in the foodstuffs which we take normally in the diet. List them.
   2. Digestion of dietary proteins in stomach:
      - Revise your knowledge about gastric juice.
      - List and name the proteolytic enzymes present in gastric juice.
      - Study in detail the activation of the enzymes, pH range of action, substrates on which they act and products formed.
   3. Digestion in duodenum and small intestine:
      (a) • Revise your knowledge about pancreatic juice.
      - List and name the *proteolytic enzymes* present in pancreatic juice.
      - Study in detail the activation of the enzymes, their pH range of activity, substrates on which they act and products formed.
      (b) • Revise your knowledge about composition of intestinal juice.
      - List and name the proteolytic enzymes present in intestinal juice.
      - Study the nature of the enzymes, their mode of action, substrates for action and products formed.

B. Absorption of amino acids:
Under normal circumstances the dietary proteins are almost completely digested by the proteolytic enzymes present in various juices to their constituent amino acids, some peptides, oligopeptides, di and tripeptides.The products are rapidly absorbed.
- Learn the site of absorption of amino acids and oligopeptides like tri- and dipeptides.
- Study how the absorbed products are carried to Liver.
- Study whether there is any difference in the rate of absorption of two isomers of amino acids.
- Study in detail the process of "active" transport of *L-amino* acids and try to find out evidences to support this hypothesis.
- Study the role of glutathione in the absorption of *L-amino* acids, Meister's hypothesis: Gammaglutamyl cycle (γ-Glutamyl cycle)
- Sensitivity to Proteins: Some individuals show sensitivity to dietary proteins. Protein is antigenic and evokes immune response. Study the mechanism.

Examples of protein antigens:
- Antibody to 'colostrum' in some infants.
- Antibody to wheat "gluten" in nontropical sprue.

DIGESTION OF PROTEINS

INTRODUCTION

Dietary proteins: proteins which we take in our diet are either from animal source or vegetable source.
- ***Principal animal sources:*** Milk and dairy products, meat, fish, liver, eggs.
- ***Principal vegetable sources:*** Cereals, pulses, peas, beans and nuts.

Some food materials contain enzyme inhibitors or certain enzymes which can destroy certain vitamins. Egg-white and soybeans contain *trypsin inhibitors*. Raw clams contain *thiaminase* which destroys thiamine (Vit B_1). Such inhibitors or enzymes are also destroyed on cooking. Cooking also destroys harmful bacteria and other pathogenic microorganisms existing in the raw food materials.

DIGESTION IN MOUTH

There are no proteolytic enzymes in mouth. After mastication and chewing, the bolus of food reaches stomach where it meets the gastric juice.

DIGESTION IN STOMACH

Gastric juice contains a number of proteolytic enzymes. They are:

- ***Pepsin***
- ***Rennin***
- ***Gastriscin***
- ***Gelatinase.***

1. **Pepsin:** It is a potent proteolytic enzyme and is present in gastric juices of different species including the mammals. It is secreted as ***inactive zymogen*** form, ***pepsinogen,*** having a mol. wt. of 42,500 approx. It is synthesised in "chief cells" of stomach and 99 per cent is poured in gastric juice as ***pepsinogen***. Remaining 1 per cent is secreted in the blood stream from where it is ultimately excreted in the urine. urinary pepsin is known as *uropepsin*.

Pepsinogen is hydrolysed in the stomach with the help of HCl or pepsin itself ***(autocatalytically)*** to form the "active" *pepsin* (mol. wt = 34,500). In the process of activation (i) An inactive peptide called as ***"pepsin inhibitor"***. (mol. wt. 3242) and (ii) 5 smaller peptides are liberated.

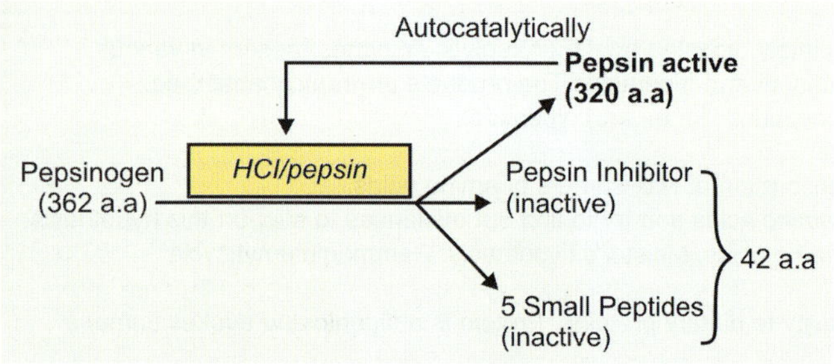

HCl maintains the gastric pH at about 1 to 2 and ensures maximum pepsin activity. ***Optimum pH for pepsin is 1.6 to 2.5*** and pepsin gets denatured if the pH is greater than 5.

Pepsin is a ***proteinase, a non-specific endopeptidase,*** and it hydrolyses peptide bonds well inside the protein molecule and produces ***proteoses*** and ***peptones***.

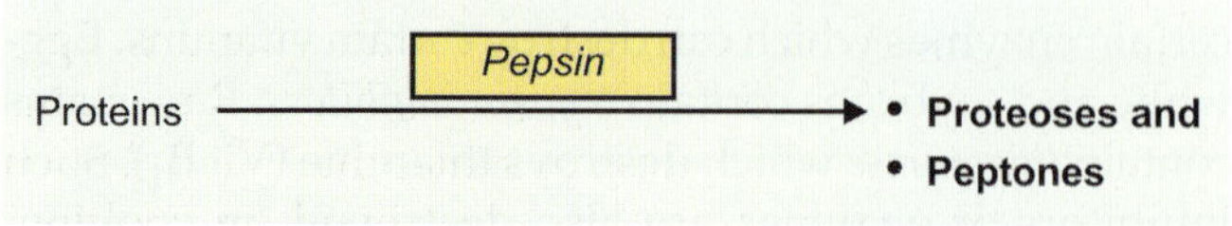

It is particularly active on a peptide bond, which connects the –COOH group of an aromatic amino acid like *Phe, Tyr,* and *Tryp* with the amino group of either a dicarboxylic acid or an aromatic a.a. It can also hydrolyse the peptide bonds of:

- COOH group of methionine and leucine
- Leucine and glutamic acid
- Glutamic acid and asparagine
- Leucine-valine
- Valine and cysteine.

Pepsin cannot act on proteins like keratins, Silk-fibroins, mucoproteins, mucoids and protamines.

Action on Milk: Pepsin can act on milk. It hydrolyses the soluble phosphoprotein **casein** of milk to produce **paracasein** and a proteose, the latter is the whey protein. Paracasein is then precipitated as **Ca-paracaseinate**, which is further digested by **pepsin to peptones.**

2. **Action of Rennin:** ***Rennin*** is absent in adult humans, and many non-ruminants. Certain amount of rennin activity is seen in babies, in infancy. In the calf, it is secreted in zymogen form as ***prorennin,*** which is activated in the stomach to form ***active rennin*** (mol. wt. 40,000) and in the process of activation an inactive peptide is split off. ***Optimum pH for activity is 4.0*** and specificity of action is very similar to pepsin, in that it hydrolyses peptide bonds connected with *L*-aromatic amino acids. Like pepsin, it also acts on casein of milk to form paracasein which is immediately precipitated by Ca^{++}. ***Thus it also coagulates milk like pepsin.***
3. **Gastriscin:** The enzyme is secreted in the gastric juice of humans as inactive zymogen form, which is activated in presence of HCl. Optimum pH is 3 to 4. It acts as a *Proteinase* and requires an acidic medium for its activity.
4. **Gelatinase:** Gelatin is hydrolysed by the enzyme *Gelatinase* present in gastric juice to form polypeptides. It acts in an acidic medium.

DIGESTION IN DUODENUM

The bolus of food after leaving stomach reaches duodenum, where it meets with pancreatic juice. A number of proteolytic enzymes are present in Pancreatic juice to act on proteins and partly digested products. Chief enzymes are:

- ***Trypsin***
- ***Chymotrypsin***
- ***Carboxy peptidases***
- ***Elastases***
- ***Collagenases.***

1. **Trypsin**

Trypsin, ***a proteinase,*** is secreted as an inactive zymogen form ***trypsinogen,*** which is activated to form *active Trypsin,* which has strong proteolytic activity and an inactive hexapeptide which is produced and

Action of Proteolytic Enzymes on Milk Protein

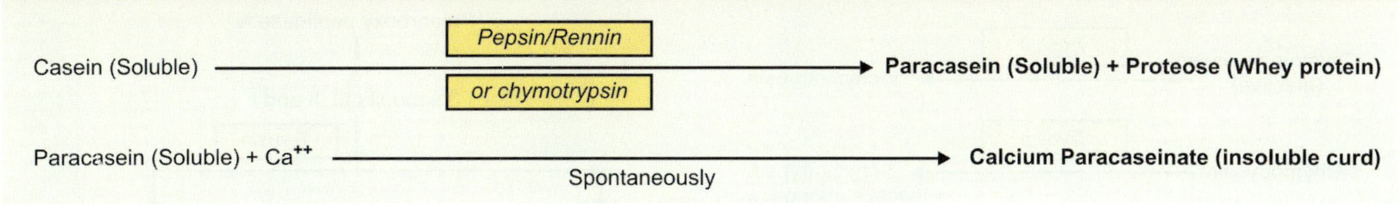

Activation of Trypsinogen

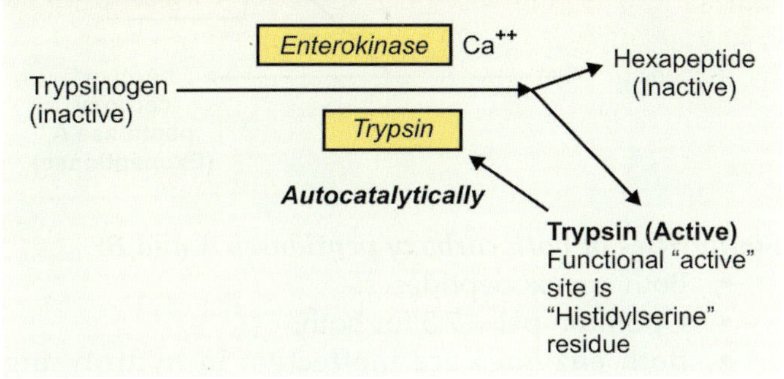

liberated during the process of activation. Activation is brought about by:

- A glycoprotein enzyme called as ***enterokinase*** of the intestinal juice at a pH of 5.5
- Also by *trypsin* itself once it is formed, ***autocatalytically***, at a pH of 7.9.
- Ca^{++} also is required for the activation.

In the process of activation, the "active site" of the enzyme trypsin, which is ***histidylserine*** residue is unmasked. Hence trypsin belongs to the group of *serine proteases*. ***Trypsin acts in an alkaline medium*** pH 8 to 9 (optimum pH-7.9) and has low Michaelis constant.

Proteolytic Functions of Trypsin

- As stated above, Trypsin can hydrolyse a peptide bond formed by the carbonyl gr. of a Lysine residue on trypsinogen, converting the latter to trypsin and inactive hexapeptide.
- Trypsin can also hydrolyse the peptide bond connected to the carbonyl group of an arginine residue in chymotrypsin which is converted to "π-chymotrypsin" (See under "Chymotrypsin").

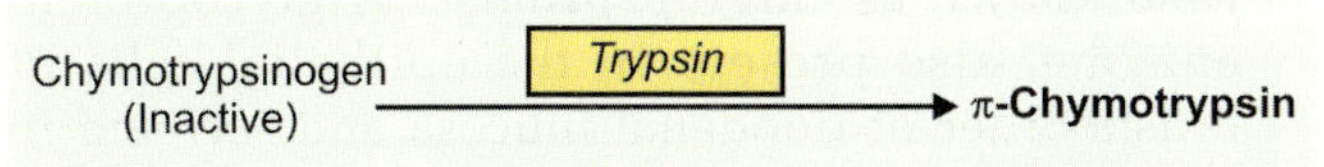

- Trypsin hydrolyses the native proteins particularly the basic proteins, by splitting the peptide bonds connected with carbonyl groups of basic a.a. viz arginine and Lysine and forms various polypeptides, proteoses, peptones, tri and dipeptides.
- It can activate zymogen *proelastase* to *elastase* by hydrolysing a peptide bond.

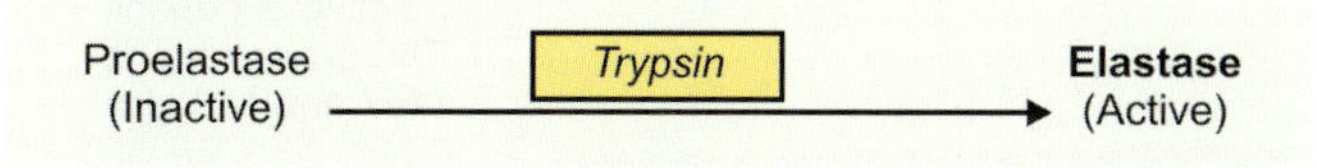

- It has a very weak action on milk protein casein (cf. chymotrypsin)
- It can convert fibrinogen to fibrin and thus can coagulate blood.

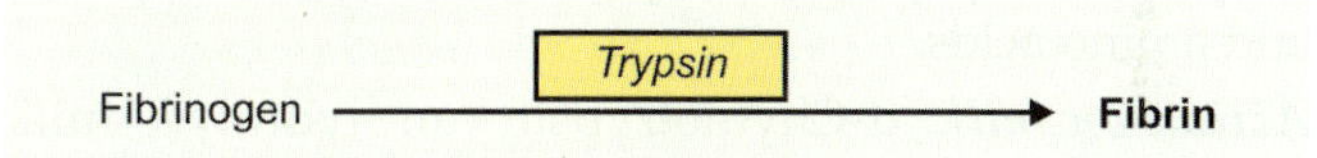

Note

(a) Though trypsin is a strong proteolytic enzyme, ***it cannot hydrolyse any peptide bond with proline residue.***

(b) ***Trypsin-inhibitors:*** Our food may contain trypsin-inhibitors. For example:

- Egg–white contains water soluble mucoprotein, a very potent trypsin inhibitor; and
- Human and bovine colostrum and raw soyabeans have also been shown to contain trypsin inhibitors.
- Trypsin inhibitors have also been reported recently from lung tissues and blood.
- Experimentally, trypsin can be inhibited by di-isopropylfluorophosphate (DFP).

2. Chymotrypsin

Chymotrypsin, a ***proteinase*** is secreted as inactive zymogen ***chymotrypsinogen***, which is activated by trypsin and completed by chymotrypsin, which acts autocatalytically.

Three chymotrypsinogens A, B and C are found in pancreatic juice of vertebrates. During its activation two inactive peptides are liberated in two stages:

- Seryl-arginine dipeptide (a.a 14 and 15), and
- Threonyl-asparagine (a.a 147 and 148)

α-Chymotrypsin is the active form, it is included under ***serine proteases*** like trypsin. ***Optimum pH = 7 to 8.*** It has a high Michaelis constant. It hydrolyses peptide bonds which are connected with carbonyl groups of

SECTION FOUR

Activation of Chymotrypsinogen

- Chymotrypsiongen (Inactive) —*Trypsin*→ π-chymotrypsin
- π-Chymotrypsin —*Trypsin*→ δ-chymotrypsin + inactive di-peptide Ser-Arg (14. and 15 a.a)
- δ-Chymotrypsin —*π-Chymotrypsin*→ **α-Chymotrypsin (active) +** Inactive di-peptide Threo-Asparagine (147 and 148 a.a)

aromatic a.a. like *Tryp, Tyr,* and *Phe.* To some extent, it can also attack peptide bonds connected with *Met, His, Leu* and *Asparagine* residues. α-Chymotrypsin converts the proteoses, peptones and peptides to smaller peptides and amino acids.

Action on milk: α-Chymotrypsin can hydrolyse milk protein casein to paracasein and a proteose (Whey protein). Paracasein is then precipitated by a spontaneous reaction with Ca^{++} and forms Ca-paracaseinate.

3. Carboxy Peptidases

Two types of carboxy peptidases are:
- ***Carboxy peptidase A***
- ***Carboxy peptidase B***

(a) *Carboxy Peptidase A*

It is a metalloenzyme, contains zinc *(Zn-protein).* Secreted as inactive zymogen ***procarboxy peptidase** A*, which has three subunits, III, II and I.
- Trypsin converts subunit II to a proteinase and subunit III is degraded.
- Then both trypsin and proteinase formed from subunit II, changes subunit I of procarboxy peptidase A to ***active carboxy peptidase A.*** (Refer box next column). It is an ***exopeptidase*** and cannot act on peptide bonds well inside the protein molecule. The enzyme hydrolyses the terminal peptide bond connected to an end a.a bearing free α-COOH group, particularly if the end a.a is *Tyr, Phe* or *Trypt.* It liberates the end a.a as "free" form, so that the peptide becomes shorter by one a.a.

(b) *Carboxy Peptidase B*

It is also an **"exopeptidase"**. Also hydrolyses terminal peptide bonds, which are connected with "basic" amino acids e.g., *Arg, Lysine* bearing free-COOH gr. (cf. carboxy peptidase A).

Activation of Procarboxy Peptidase A

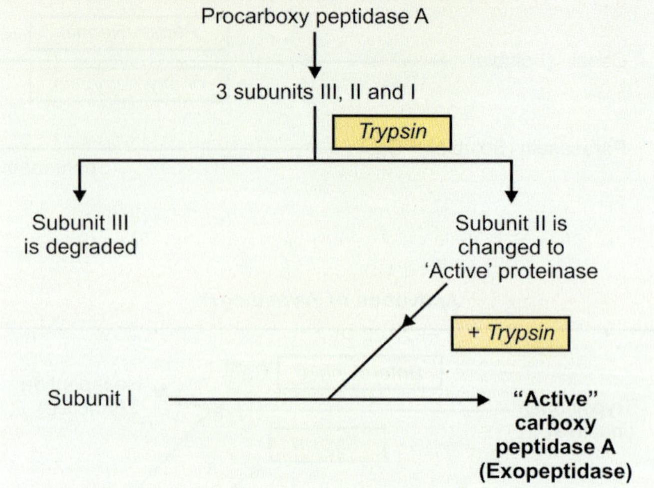

Similarities of both carboxy peptidases A and B:
- Both are exopeptidases.
- Optimum pH = 7.5 for both.
- Both enzymes are ineffective in hydrolysing dipeptides.

4. Elastase and Collagenase

Elastase: A serine protease, secreted as inactive zymogen ***proelastase,*** activated by trypsin to *active elastase.*

The enzyme has maximum activity on peptide bonds connected to carbonyl groups of neutral aliphatic a.a.

Collagenase: An enzyme which can act on proteins present in collagen.

Both the enzymes can digest yellow and white connective tissue fibres respectively to yield peptides.

DIGESTION IN SMALL INTESTINE

Proteolytic enzymes present in intestinal juice are:

- ***Enterokinase***
- ***Amino peptidases***
- ***Prolidase, and***
- ***Tri and Di-peptidases***

1. Enterokinase

Also known as ***enteropeptidase***. A glycoprotein enzyme, also present in the epithelial cells of brushborder of duodenal mucosa and secreted in duodenum.

Action: It hydrolyses a peptide bond connected to a lysine residue in pancreatic trypsinogen, so that an inactive hexapeptide is split off from the N-terminal end and "active" trypsin is produced.

$$\text{Trypsinogen} \xrightarrow[\text{Ca}^{++}]{\text{Enterokinase}} \text{'Active' trypsin}$$

Ca^{++} is required for this activation. *Bile salts* help in liberation of *enterokinase* from the brush border membrane of intestinal epithelial cells to intestinal lumen.

2. Amino Peptidases

Best example is LAP (Leucine amino peptidase):
- Can hydrolyse peptides to tripeptides.
- Cannot hydrolyse a dipeptide.
- Requires presence of Zn^{++}, Mn^{++} and Mg^{++} which help in formation of a metal-enz-substrate coordination complex for the catalysis.
- Can hydrolyse a terminal peptide bond connected to an end a.a bearing a free-α NH_2 group and thus splits off the end a.a. from the N-terminal end of a peptide, changing the latter gradually stepwise to a "tripeptide".
- Cannot hydrolyse a peptide bond connected with a proline residue.

3. Prolidase
- An ***exopeptidase*** and can hydrolyse a proline peptide of collagen molecule, acts on terminal peptide bond connected to proline as end a.a, liberating a proline molecule.

4. Tri and Di-peptidases

These enzymes hydrolyse the peptides at either of two places:
- In microvillus membrane of intestinal epithelial cells, or inside the epithelial cells after the peptides have been absorbed inside the cell.
- *Tri-peptidase* acts on a tri-peptide and produces a di-peptide and free a.a.
- A di-peptidase hydrolyses a di-peptide to produce two molecules of amino acids.
- They require the presence of Mn^{++}, Co^{++} or Zn^{++} as cofactors for their activity.

ABSORPTION OF AMINO ACIDS

Under normal circumstances, the dietary proteins are almost completely digested to their constituent amino acids. But some amounts of oligopeptides like tri and dipeptides may remain as such. The products of digestion are rapidly absorbed.

Site of Absorption: ***Amino acids are absorbed from ileum and distal jejunum.*** Oligopeptides like di and tri-peptides are absorbed from duodenum and proximal jejunum.

How Do They Reach Liver? Amino acids and other products of digestion like di and tripeptides, if any, after absorption are carried by portal blood to Liver. There is a marked rise in amino acid level in portal blood after a protein meal.

Rate of Absorption: There is difference in rate of absorption from the intestine of the two isomers.
- ***L-amino acids and L-peptides are absorbed more rapidly than D-isomers and they have been shown to be absorbed by'active' transport process.*** *L*-amino acids are 'actively' transported across the intestine from mucosa to serosal surface. Pyridoxal-(P) (B_6-PO_4) is probably involved in this process.
- *D*-amino acids are absorbed slowly and they are absorbed by ***simple passive diffusion.***

Mechanism of Absorption L-Amino Acids

Ion gradient hypothesis: *L*-amino acids are absorbed from small intestine by sodium (Na^+) dependant, carrier-mediated process. This transport is energy dependant and energy is provided by ATP (similar to absorption of glucose and galactose).
- *L*-amino acids and Na^+ combine with a common "carrier protein" molecule present on the outer or mucosal surface of the microvillous membrane to form a "Carrier- a.a.-Na^+" complex. The complex passes to the inner or cytoplasmic surface of the same membrane. There it dissociates to liberate free a.a and Na^+.
- Na^+ is actively carried out through the cell membrane by a "Sodium pump" mechanism, with the help of transport *ATP-ase,* so that intracellular Na^+ concentration is always maintained low.
- Carrier-protein molecule comes back to the brush border again.
- The amino acid ultimately passes out through the serosal membrane of the cell by diffusion down an outward concentration gradient of a.a and taken by portal blood to liver.

Note: Different classes of *L*-amino acids viz., diamino-acids, small neutral a.a, imino acids, and large neutral a.a are believed to be absorbed by different "carrier" protein molecules present in the microvillus membrane of intestinal cells.

Model for study: Synthetic amino acids utilised as model for study are:
- **α-amino *iso*-butyric acid**
- **1-amino-cyclopentane-1-carboxylic acid.**

Evidences for "Active" Absorption
1. Rate and extent of absorption of *L*-amino acids is considerably higher than D-isomers.
2. If Na^+ is replaced by lithium and/or K^+ in bathing fluid, rate of absorption is depressed and may be practically nil.
3. Inhibitors like dinitrophenol (DNP) or cyanide depress *L*-amino acid absorption. DNP acts as an uncoupler in oxidative phosphorylation and thus interferes with ATP formation.

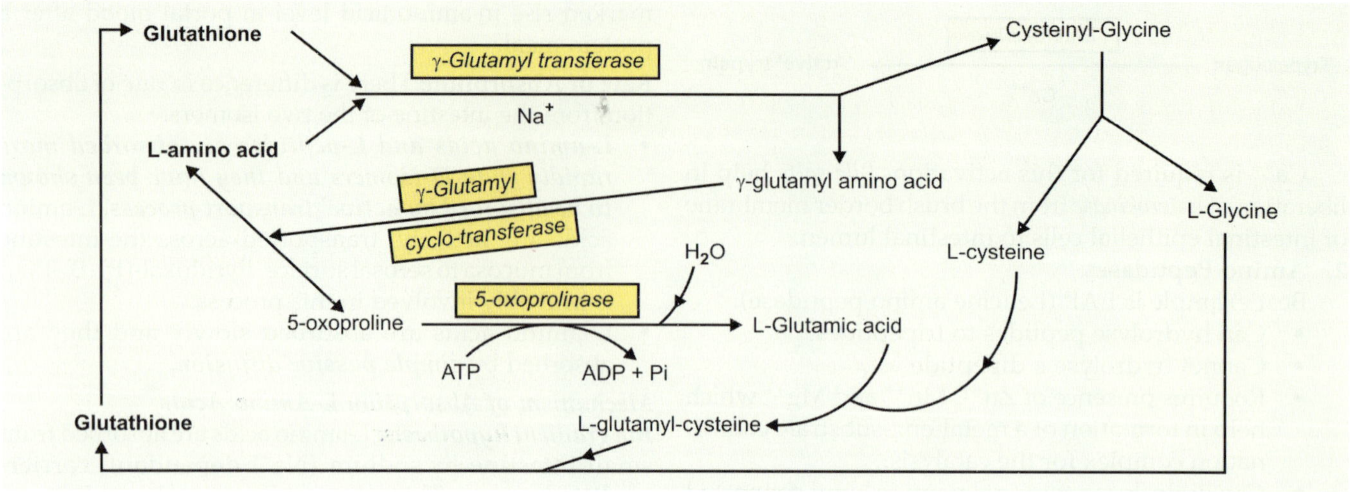

Fig. 26.1: γ-glutamyl cycle

4. Rates of absorption of different *L*-amino acids are different from each other and are independent of their diffusibilities and concentration gradients.
5. High concentration of one *L*-amino acid sometimes reduces the rate of absorption of some other *L*-amino acids, indicating several *L*-amino acids ***may share a common "carrier" molecule and may compete with each other.***
6. Basic and dicarboxylic amino acids are generally more slowly absorbed than neutral amino acids.

Absorption of *L*-Oligo Peptides: *L*-oligo peptides are also actively transported. Intracellular peptidases hydrolyse them into a.a. This hydrolysis within the intestinal epithelial cells is rapid enough to keep peptide concentration low in these cells. Transport mechanisms for *L*-peptides appear to be independant of *L*-amino acids.

Role of Glutathione in Amino Acid Absorption

Meister has proposed that glutathione participates in an ***active group translocation*** of *L*-amino acids ***(except L-proline)*** into the cells of small intestine, kidneys, seminal vesicles, epididymis and brain. He proposed a "cyclic" pathway, ***in which the Glutathione is regenerated again, and it is called as γ-glutamyl cycle*** (Fig. 26.1).

- Glutathione combines with *L*-amino acid in presence of the enzyme *γ-glutamyl transferase* and Na^+ and forms cysteinyl-Glycine and γ-glutamyl amino acid complex.
- γ-glutamyl amino acid in presence of the enzyme *γ-glutamyl cyclotransferase* forms 5-oxoproline and L-amino acid, which is absorbed.
- 5-oxoproline in presence of the enzyme *5-oxoprolinase* and ATP form L-Glutamic acid.
- Cysteinyl-glycine formed in first step is hydrolysed to form *L*-cysteine and *L*-Glycine.
- *L*-Glutamic acid and *L*-cysteine joins together to form ***"L-glutamyl-cysteine"***, which combines with *L*-glycine to form ***'Glutathione'*** again and the cycle is repeated.

Sensitivity to Dietary Proteins (immunologic response)

Some individuals show sensitivity to dietary proteins. Proteins to be antigenic, so that can stimulate an immunologic response, it should be relatively large molecule. Normally proteins are digested in GI tract to small peptides and amino acids. Digestion of proteins to small peptides and amino acids stage destroys the antigenicity.

How dietary proteins can be antigenic is a puzzling feature? Hence it is suggested that in some individuals, there must be absorption of some "unhydrolysed proteins", probably by *pinocytosis*. ***It has been shown that to be antigenic, the protein must be a polypeptide containing 6 to 7 amino acids, having mol. wt. 820-928 daltons, and presence of glutamine and proline in the protein molecule is a must.***

Examples

- Antibodies to "colostrum" are known to be formed in infants and has been demonstrated.
- In "non-tropical sprue", there is basic enzymatic defect that 'gluten' the wheat protein is not digested completely and polypeptides of wheat gluten may be absorbed, which exerts local harmful effects and produce antibodies to gluten polypeptide.

I. Questions (Essay type and short notes)

1. Describe the constituents of pancreatic secretion, their stimulation and functions during digestion.
2. Describe the composition and functions of pancreatic juice.
3. Describe the mechanism of digestion and absorption of proteins.

Short Notes

a. "Active" transport
b. Trypsin inhibitors
c. γ-glutamyl cycle
d. Proteolytic enzymes of pancreatic juice
e. Pepsin

II. MCQs (Give one correct answer)

1. Pancreatic juice contains all of the following, except:
   a. Trypsinogen b. Lipase
   c. Cholecystokinin d. Chymotrypsinogen
   e. Amylase
2. The milk protein in the stomach in an adult is digested by:
   a. Pepsin b. Rennin
   c. HCI d. Trypsin
   e. Chymotrypsinogen
3. Carboxypeptidase A enzyme of pancreatic juice contains:
   a. Copper b. Zinc
   c. Magnesium d. Manganese
   e. Cobalt
4. The enzyme collagenase of pancreatic juice hydrolyses collagen present in:
   a. Milk b. Eggs
   c. Cereals d. Meat
   e. Soyabeans
5. The zymogen form trypsinogen of pancreatic juice is converted to 'active' trypsin by:
   a. Pepsin b. Enterocrinin
   c. Enterokinase d. Rennin
   e. Gastrin
6. Carboxypeptidase B acts as an exopeptidase and hydrolyses the terminal peptide bonds containing:
   a. Lysine and Arginine b. Leucine and Arginine
   c. Glycine and leucine d. Glycine and Valine
   e. Leucine and Isoleucine
7. Inactive zymogens (proenzymes) are precursors of all the following gastrointestinal enzymes, except:
   a. Trypsin b. Carboxypeptidase
   c. Pepsin d. Aminopeptidase
   e. Chymotrypsin
8. Which of the following regarding digestion of proteins are correct, except:
   a. Pepsinogen is activated by autoactivation (pH-2) or autocatalysis
   b. The major products or protein hydrolysis are large petides and some free amino acids
   c. Most of the digested proteins are absorbed in the intestine in the form of polypeptides
   d. Trypsin and chymotrypsin are secreted by the pancreas as inactive zymogens
   e. Enterokinase activates pancreatic trypsinogen
9. Rennin acts on casein of milk in infants in presence of:
   a. Mg^{++} b. Zn^{++}
   c. Co^{++} d. Ca^{++}
   e. Mn^{++}
10. Active chymotrypsin is:
    a. π-chymotrypsin b. α-chymotrypsin
    c. δ-chymotrypsin d. γ-chymotrypsin
    e. All of the above
11. During intestinal proteolysis:
    a. Enterokinase converts trypsinogen to trypsin
    b. Involves conformational change, but no change in covalent structure
    c. Trypsin activates procarboxypeptidase A and proelastase
    d. a + b true
    e. a + c true
12. Pepsin contains large amounts of which of the following amino acids:
    a. Aromatic amino acids
    b. Acidic amino acids
    c. Basic amino acids
    d. Sulphur containing amino acids
    e. Neutral amino acids
13. *D*-amino acids are absorbed by:
    a. Active transport
    b. In presence of a 'carrier' molecule
    c. In presence of sodium ions
    d. Passive diffusion
    e. None of the above
14. Chymotrypsin hydrolyses the peptide linkages containing
    a. Phenyl alanine and tyrosine:
    b. Phenyl alanine and lysine
    c. Tyrosine and lysine
    d. Tyrosine and leucine
    e. Phenylalanine and leucine
15. Match the following:
    a. Trypsin b. Pepsin
    c. Chymotrypsin d. Endopeptidases
    e. Exopeptidases
    1. Splits peptide bonds well inside the protein molecule ________
    2. Converts proteins to proteoses and peptones _____
    3. Splits peptide bonds involving carboxyl groups of Arginine and Lysine ________
    4. Splits only terminal peptide bonds ________
    5. Splits Peptide bonds involving carboxyl groups of Phenylalanine and tyrosine ________.

Answers to MCQs

1. c 2. a 3. b 4. d 5. c 6. a
7. d 8. c 9. d 10. b 11. e 12. b
13. d 14. a 15. 1-d, 2-b, 3-a, 4-e, 5-c.

SECTION FOUR

CHAPTER 27

METABOLISM OF PROTEINS AND AMINO ACIDS

PART I

Major Concepts

A. To study the "General amino acid pool" and utilisation of amino acids.
B. To study the dissimilation of amino acids, learn how nitrogen part is removed as ammonia (NH_3).
C. To study the fate of ammonia and fate of carbon skeleton in the body.

Specific Objectives

A. 1. Study the concept of **General amino acid pool,** how it is formed?

2. Learn about plasma amino acids level and "circadian changes" of plasma amino acid levels. Note the total amino acid level in blood.

   Study the mechanism of uptake of amino acids by various tissues: Role of pyrixodal-(P) and the hormones in the process. Study the utilisation of amino acids by various tissues.

3. Learn what is meant by **state of dynamic equilibrium** and positive and negative nitrogen balance (nitrogen balance and related experiments).
   - Study the **Essential amino acids:** Define and list them. What is meant by "Semiessential" amino acids? Name them.
   - Study about the value of essential amino acids in nutrition in terms of "administration of complete group" and "optimal ratio".
   - Learn about the minimal requirement of essential amino acids for normal adults (Try to study experiments carried out by Rose).

B. Dissimilation of amino acids (N-Catabolism of amino acids).

In mammalian tissues, the α-NH_2 group of amino acids are removed as ammonia and finally converted to urea. Ammonia Formation involves the following processes:

- Transamination
- Oxidative and Non-oxidative deamination
- Transdeamination

(a) • Study in detail the process of transamination and clinical importance of transaminases.
- Note the limitation of transamination reaction.

(b) • Study in detail the process of oxidative deamination—enzymes and coenzymes required for the reaction. The enzymes are **autooxidisable.**
- Study with examples of processes of non-oxidative deamination.

(c) • L-Glutamate produced by specific transamination can be deaminated to produce NH_3 by specific enzyme. The combination of transamination and deamination of Glutamate is called *transdeamination.*
- Study in detail the "transdeamination" reaction, enzyme and coenzymes required and its regulation.

(d) • Study the transport of ammonia, and its fate.

The ammonia is toxic substance hence, it is removed from the body quickly and converted into nontoxic substances.
- Study the causes of ammonia toxicity and learn the clinical manifestations.

C. Ammonia can undergo the following fates:
- Principal fate: Conversion to urea
- Formation of glutamine
- Amination of α-Ketoacid to form amino acid

1. (a) Study in detail the urea formation, organs involved, site of synthesis and steps of synthesis alongwith enzymes and coenzymes required.
   - Study the normal level of blood urea and urinary urea and clinical significance.
   - Learn the formula of urea, and try to find out the sources of two nitrogen atoms.

- Study the regulation of urea synthesis, how ammonia is diverted to urea formation by linkage of Glutamate dehydrogenase with Carbamoyl-P-synthetase 1.
- Learn about "Glucose-alanine cycle".
- Carbamoyl-(P) is a "committed step" and a branch point compound, study the different fates of carbamoyl-(P) and its regulation.
- Study briefly the metabolic disorders associated with urea cycle, inherited disorders.

(b)
- Learn what is glutamine and its chemical formula.
- Study how glutamine is synthesised, the enzyme and coenzymes involved. Learn how glutamine is hydrolysed and tissues in which it takes place.
- Study the important functions of glutamine in the body.

(c) Amination of α-ketoacid to form α-amino acid.

2. Study the fate of C-Skeletons.

INTRODUCTION

AMINO ACID POOL

Amino acids, on absorption from intestine are carried to Liver through portal blood. They are taken up by Liver cells to some extent and remainder enters the systemic circulation and diffuse throughout the body fluids and taken up by tissue cells. At the same time, most of the tissue proteins both "structural" proteins and functional proteins, (including plasma proteins) are continually undergoing disintegration to release amino acids which likewise enter the circulation. There is also a continuous synthesis of amino acids (except the "essential" amino acids).

Amino acids from all these sources get mixed up to constitute what is known as **"general amino acid pool"** of the body. ***"Amino acid pool" has no anatomical reality but represents an availability of amino acid building units. No functional distinction can be drawn between the fate of the amino acids derived from the dietary source and those derived from the tissue breakdown.*** All tissues including exocrine and endocrine glands draw freely from the amino acid pool to synthesise the tissue proteins, enzymes and protein hormones. ***Amino acid is taken up by each cell according to its own specific needs, to be built into the cell structure and materials as required.***

If a cell takes up as much amino acids as it loses, it is in a state of ***"dynamic equilibrium"***, if the loss is greater, the cell wastes, and if the gain is greater the cell grows. In experimental animals, this protein "turnover" is greatest in intestinal mucosa, followed by kidneys, liver, brain and muscles in that order. Each protein species is constantly lost and resynthesised at a characteristic rate, in man, the protein turnover involves the breakdown and resynthesis of 80 to 100 gm of tissue protein per day, about ½ of it occurring in liver. ***On an average, plasma proteins are completely replaced every 15 days.***

The 'pool' is constantly undergoing depletion because,

- Large scale deamination of presumably surplus amino acids take place.
- Amino acids and their derivatives viz urea, creatinine are lost in the urine and other excretions.
- Amino acids are continually being built up into those proteins, e.g. hair, collagen proteins, which are not part of dynamic systems.

  On the other hand, amino acid pool is being always re-established by amino acids, derived from the following:
- Reamination of certain non-nitrogenous residues.
- Amination of appropriate fragments which are present in the common metabolic pool (and therefore derived from carbohydrate and fat breakdown).
- Amino acids split off from dietary proteins and absorbed from the intestine into the blood.

This state is called ***continuing metabolism*** of the amino acids.

Amino Acids in Blood

All the amino acids occur in blood in varying concentrations and make a total of 30 to 50 mg/100 ml in the "post-absorptive" state. In terms of amino acid N_2, it is 4 to 5 mg/100 ml. Following a protein containing meal, the amino acid levels rise to 45 to 100 mg per 100 ml (amino acid N_2 6 to 10 mg/100 ml).

"Circadian" Changes in Plasma Amino Acid Levels

The plasma levels of most amino acids do not remain constant throughout 24-hour period; but rather change by varying in a ***circadian rhythm*** about a "mean" value. This was first noted for the amino acid 'Tyrosine' and later on confirmed for most other amino acids. In general, plasma amino acid levels are lowest at early morning (4 AM) and rise 15 to 35 per cent by noon to early afternoon.

Tissue Amino acids

The amino acids are transported into tissues **actively**. Pyridoxal-P (B_6-P) is one of the requirement for this active transport. Tissue uptake is also favoured by hormones:

- Insulin, growth hormone and testosterone favour the uptake of amino acids by tissues **(anabolic hormones)**.
- **Oestradiol** stimulates selectively their uptake by uterus.

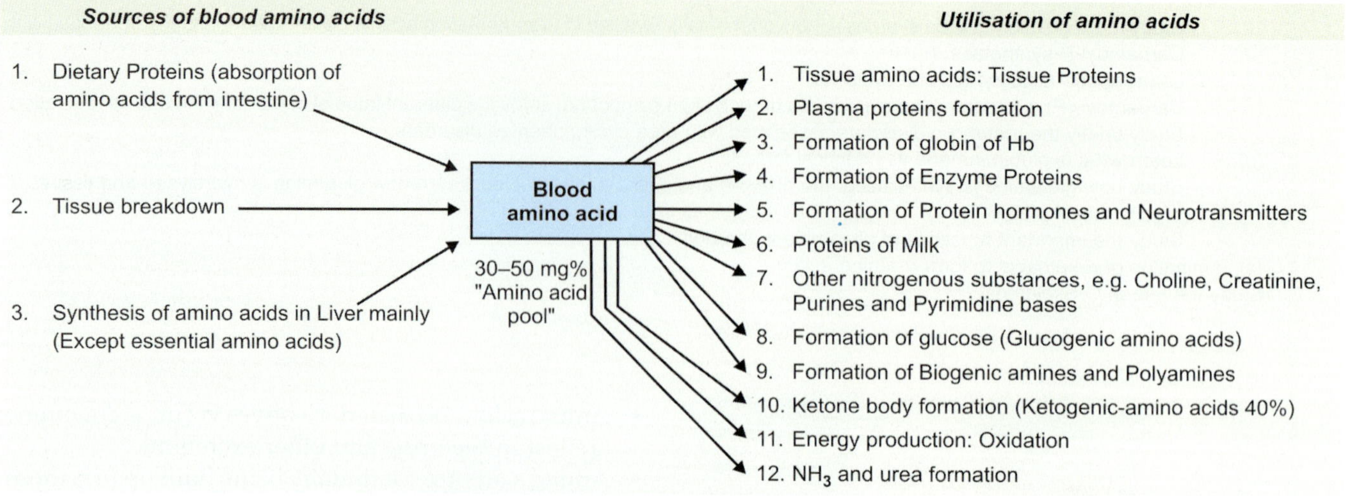

Fig. 27.1: Sources and utilisation of amino acids

- **Epinephrine and glucocorticoids:** Stimulate the uptake of amino acids by the Liver.
  Sources and utilisation of blood amino acids are shown schematically in Figure 27.1.

NITROGEN BALANCE

In an adult healthy individual maintaining constant weight, the amount of intake of N in food (mainly as dietary proteins) will be balanced by an excretion of an equal amount of N in urine (in form of urea mainly, uric acid, creatinine/and creatine, and amino acids contribute to a minor extent) and in faeces (mainly as unabsorbed N). The individual is then said to be in ***nitrogen balance or nitrogenous equilibrium.***

Experiments measuring N intake and excretion under specified conditions are called "nitrogen balance experiments".

- A subject in nitrogenous equilibrium is said to be in ***Nitrogen balance,*** i.e. intake of N equalises the output.
- A subject whose intake of N is greater than the output, e.g. in growth, is said to have a ***+ve nitrogen balance.*** In the growing period and also during convalescence from illness or when anabolic hormones are given, the body puts on weight and N-intake will be more than N-output, since some of the N is retained as tissue proteins.
- A subject whose intake of N is less than the output of N, (e.g. in losing weight), is said to have a ***–ve nitrogen balance.*** In old age and during illness and starvation weight is lost and results in –ve nitrogen balance.

Lability of proteins: There is no special storage form for proteins like glycogen for carbohydrates or fats in adipose tissue. ***Protein storage is always accompanied by tissue growth.*** During starvation when protein is not available from dietary sources, it is the Liver, which loses the largest proportions of its proteins compared to other proteins. Additions of Proteins in food similarly cause an appreciable increase in liver weight first. ***Thus the liver proteins appear to be more labile than the proteins of other tissues. Kidney and blood proteins come next in degree of lability.***

Types of proteins required for N-balance: To establish N-balance, certain minimum amounts of proteins or equivalent amino acids must be provided to replace the inevitable losses from the dynamic equilibrium and metabolic utilisation of amino acids. This minimum replacement requires amino acids of ***specific type*** in adequate amounts and in appropriate ratios. They are ***essential amino acids*** which must be provided in the diet simultaneously together and they cannot be synthesized in the body. ***It is impossible to maintain N-equilibrium on diets which are deficient in anyone or more of these essential amino acids, no matter how much protein is consumed.***

Examples of some incomplete proteins:

- **Gelatin:** Which lacks tryptophan
- **Zein** of corn/maize: Low in both tryptophan and Lysine.

ESSENTIAL AMINO ACIDS

Definition: An essential or indispensable amino acid is defined as one which cannot be synthesised by the organism from substances ordinarily present in the diet at a rate commensurate with certain physiological requirements and they must, therefore, be supplied in the diet usually combined in proteins.

Non-essential amino acids: The non-essential or dispensable amino acids can be synthesised in the body either:

- By the amination of appropriate non-nitrogenous fragments derived from other sources.
- In special cases, directly from the essential amino acids, e.g. formation of Tyrosine from phenyl alanine or formation of cysteine from methionine.
- But it should be emphasised that the body is "spared" the trouble of this synthesis if the dispensable amino acids are also available in the diet *(sparing effect).*

Features of Essential Amino Acids

1. Animals given a basal diet which contains no proteins or amino acids, but which is otherwise complete in all respects, will rapidly die; if, however, right type of protein supplements are added, then normal health and reproductive power are maintained in adult animals and growth occurs in young animals.
2. By measuring the N-balance on various amino acid supplements, it has been found that following **eight amino acids** are indispensable for human adults under normal conditions. ***Exclusion of any one of these essential amino acids leads to a –ve N-balance*** manifesting as loss of weight, fatigue, loss of appetite and nervous irritability. When missing essential amino acid is supplemented in the diet, perfect health is promptly restored.
3. **The eight essential amino acids are:** ***valine; leucine; isoleucine; threonine; methionine; phenylalanine; tryptophan and lysine.*** To remember one may use the formula **Mattvillphly**.

Note

- Presence of tyrosine in the diet can spare 70 to 75 per cent of phenyl alanine requirement in humans *(sparing action).*
- Similarly, Presence of cystine / and cysteine in the diet can spare 80 to 90 per cent of methionine requirement in humans.

4. *Administration of "complete group":*
   - A curious fact concerning essential amino acids to be noted is that ***the complete group of eight amino acids must be administered to the organism simultaneously and together.***
   - If a single essential amino acid is omitted from the group and fed separately several hours later, the nutritional effectiveness of the entire group is impaired. The "excess" amino acids not utilised, in absence of a missing one, are almost completely oxidised during the elapsed period and not utilised for tissue protein synthesis.
5. *Optimal ratio of essential amino acids* in the diet should approximate that found in carcass of the animal concerned. Significant deviations from the "optimal ratio" result in certain adverse effects:
   - Certain amino acids are toxic to the experimental animal when fed at a high level.
   - Vitamin deficiencies may be aggravated by addition to the diet of increased quantities of a single amino acid.

   Those effects may be due to an increase in the rate of amino acid metabolism, leading to increased utilisation and breakdown of certain vitamins.
6. *Quantitative aspect: How much to take?*
   For normal adult, the minimum amount of each essential amino acid which must be supplied/day, when all other amino acids are present has been set as 0.3 to 1.0 gm of the natural L-form. **Rose** after performing experiments on animals suggested that a "safe" intake would be double of the amount mentioned. The requirement for EAA varies with the amount of "strain" placed on the protein-synthesising capacity of the body and during growth, pregnancy or lactation, a higher intake (as mixed protein) is required.
7. *"Semiessential" amino acids:* In addition, animal growth experiments indicate that dietary supplies of two other amino acids **Histidine** and **Arginine** may be required under conditions of growth or equivalent physiological stress as pregnancy and Lactation. The capacity of the body to synthesise Histidine and Arginine, though adequate for protein maintenance, may not suffice for the more extensive calls of protein accumulation.

DISSIMILATION OF AMINO ACIDS (N-catabolism of amino acids)

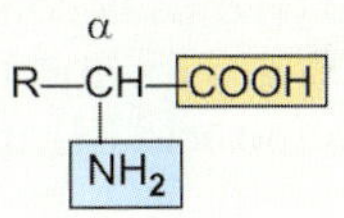

Amino acid

In mammalian tissues, α-NH_2 group of amino acids, derived either from the diet or breakdown of tissue proteins, ultimately is converted first to NH_3 and then to urea and is excreted in the urine.

α-NH_2 group → NH_3 → Urea (Excreted in urine)

The formation of urea involves the action of several enzymes.

Formation of NH_3 and urea can be discussed under the following heads:

- Transamination
- Deamination → Oxidative deamination; → Non-oxidative deamination
- Transdeamination
- NH_3 transport, and
- Formation of urea

Verebrates other than mammals share all features of the above scheme except urea formation.

Urea is the characteristic end-product of amino acid N-catabolism in human beings and **ureotelic** organisms.

Urea synthesis is replaced,

- By uric acid formation in **uricotelic** organism, e.g. Reptiles and birds,
- By NH_3 in **ammonotelic** organism, e.g. bony fish.

TRANSAMINATION

It was first discovered by **Braunstein and Kritzmann** (1947). It is a process of combined deamination and amination.

Definition: Transamination is a reversible reaction in which α-NH_2 group of one amino acid is transferred to a α-ketoacid resulting in formation of a new amino acid and a new ketoacid.

The general process of transamination may be represented as follows:

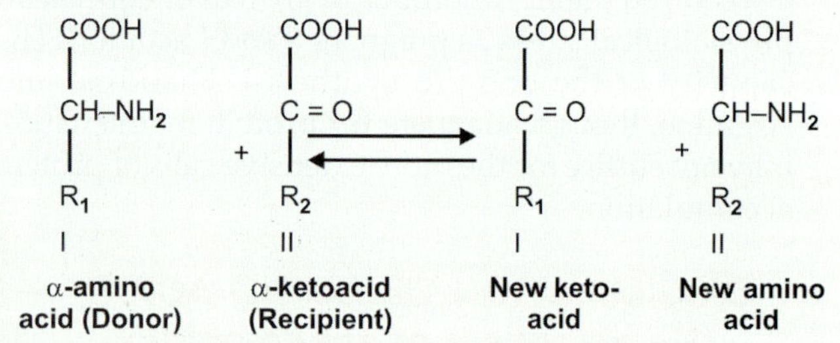

Donor amino acid (I) thus becomes a new ketoacid (I) after losing the α-NH_2 group, and the recipient ketoacid (II) becomes a new amino acid (II) after receiving the NH_2 group.

Note

The process represents only an **intermolecular transfer of NH_2 group** without the splitting out of NH_3. ***Ammonia formation does not take place by transamination reaction.***

SALIENT FEATURES: POINTS TO REMEMBER

1. ***Reversible reaction:*** The reaction is **reversible** and is catalysed by enzymes.
2. ***Site of transamination:*** Transamination takes place principally in liver, kidney, heart and brain. But the enzyme is present in almost all mammalian tissues and transamination can be carried out in all tissues to some extent.
3. ***Enzymes:*** The enzymes concerned in transamination are called ***transaminases*** (better called as ***amino transferases***).
4. ***Coenzyme for the reaction:*** The ***coenzyme required for the reaction is pyridoxal-P (B_6P).*** In the process of transamination, the amino acid reacts with enzyme-bound pyridoxal-P to form an ***enzyme-bound complex of the Schiff-base type,*** which then yields ketoacid and pyridoxamine-P. The pyridoxamine-P then reacts with a second ketoacid to produce a similar enzyme bound Schiff-base complex, which then decomposes forming a second new amino acid and regenerates the pyridoxal-P. The role of pyridoxal-P is shown ahead in next page in box.

Limitations of Transamination Reactions

There are certain limitations to the reaction of transamination:

I. *Group transaminases:* While most amino acids may act as Donor I, ***the recipient ketoacids may be either-α-keto-(oxo) glutarate,*** or ***oxaloacetate or pyruvate.*** It is to be noted that all of these recipient ketoacids are "components of TCA cycle" and hence they are common metabolites of the cell and are easily available. The amino acids formed from these recipient ketoacids are respectively glutamic acid, aspartic acid and alanine. Those transaminases, which can use any of these three ketoacids are called ***"Group" transaminases.***

II. *Specific transaminases:* But there are two transaminases of clinical importance in the body in that ***they use specific amino acid and specific ketoacid.*** These two specific transaminases are:

- *Aspartate transaminase (or Aspartate amino transferase)*

  Previously used to be called as S-GOT *(serum Glutamate oxaloacetate transaminase).* In this Aspartic acid is the donor amino acid and α-oxoglutarate is the recipient ketoacid. ***New amino acid formed is always glutamic acid.***

  Asparate + α-oxoglutarate ⇌ Oxaloacetate + Glutamate

- *Alanine transaminase (or Alanine amino-transferase)*

  previously used to be called as S-GPT *(Serum Glutamate pyruvate transaminase).* In this alanine is the donor amino acid and α-oxoglutarate is the recipient ketoacid. ***New amino acid formed is again always glutamic acid.***

  Alanine + α-oxoglutarate ⇌ Pyruvate + Glutamic acid

III. *Amino acids which do not Take Part in Transamination:* Though most of the amino acids can act as substrates for transamination, there are certain exceptions. Exceptions include: ***Lysine, Threonine, the cyclic iminoacids, Proline and OH-Proline.***

IV. There are a few transaminases which do not utilise the three ketoacid recipient mentioned above.

V. Transamination is not restricted to α-NH_2 groups. The δ-NH_2 group of ornithine is readily transaminated forming Glutamate γ-semialdehyde.

Role of Pyridoxal-P in Transamination

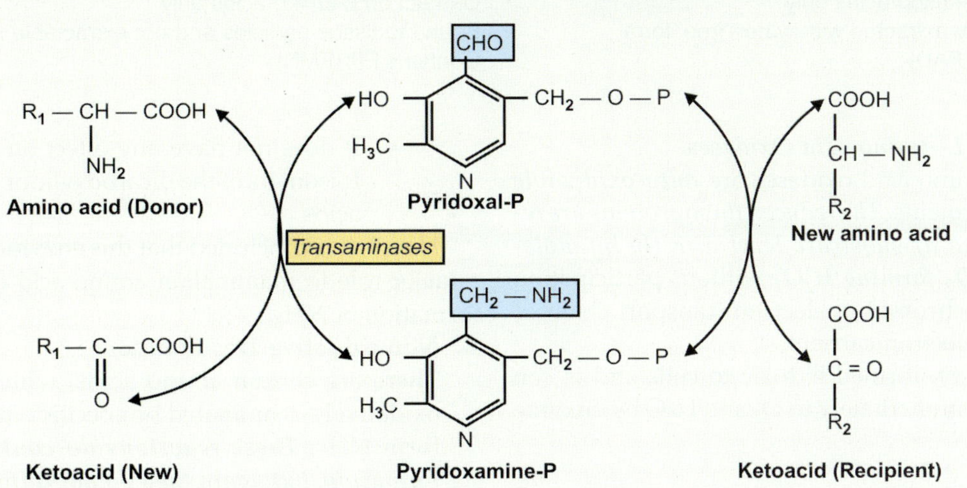

CLINICAL SIGNIFICANCE

As mentioned above two specific transaminases are of clinical importance.

1. ***Aspartate transaminase (Aspartate amino transferase/ or aminoferase), S-GOT (old nomenclature):***

   Normal serum activity is 4 to 17 I.U/L (7 to 35 Karmen units/ml). **Concentration of the enzyme is very high in myocardium and also in liver cells.** The enzyme is also distributed in other tissues, viz. muscles, pancreas, kidney, etc. The enzyme is ***cytoplasmic*** and also ***mitochondrial. Helpful in acute myocardial infarction*** (Refer to Chapter on Enzymes and Isoenzymes of Clinical Importance).

Other extracardiac factors:

- Increases in Liver diseases, but it is less than Alanine transaminase (S-GPT).
- Increase in muscular dystrophies—myositis. No increases in muscular disease of nervous origin.
- Increased activity seen in acute pancreatitis, leukaemias, in acute haemolytic anaemia.
- In normal persons, after prolonged severe exercise.
- A rise has been seen in therapy with erythromycin.

2. ***Alanine transaminase (Alanine amino transferase/or aminoferase)-S-GPT (old nomenclature)***

   The enzyme is found mainly in Liver, Liver cells are rich. It is **entirely cytoplasmic** (cf. S-GOT).

   Normal enzyme activity is 3 to 15 IU/L (6-32 Karmen units/ ml).

Most helpful in liver diseases: Increases in both transaminases are common finding in hepatic diseases but always S-GPT > than S-GOT, ***though in normal healthy persons, S-GOT is slightly more than S-GPT.***

It is most useful in assessing severity and progress of the disease in acute viral hepatitis. Serial estimations are most useful. Highest values of enzyme activity seen in acute viral hepatitis, peak values 250 to 1500 IU/L or more seen at the time of maximum illness. It is useful for ***'screening'*** in outbreaks of acute viral hepatitis, useful for segregating the contacts (Refer to Chapter on Liver Function Tests).

DEAMINATION

Deamination is the process by which N– of amino acid is removed as NH_3.

Types: It can be of **2 types:**

A. Oxidative deamination

B. Non-oxidative deamination.

A. Oxidative Deamination

1. ***Site of oxidative deamination:*** **Krebs (1935)** studied deamination of amino acids in various tissue slices and found that liver and kidney to be very active.
2. ***Enzymes:*** He also demonstrated the presence of *D and L-amino acid oxidases* enzymes in these tissues, which can act on D-amino acids and L-amino acids respectively and can oxidatively liberate NH_3 from these amino acids. Essential differences between these two enzymes are tabulated ahead in the box.

Note

It is to be noted that it is rather peculiar that despite the absence of D-amino acids in tissues, the D-amino acid oxidases is generally much greater than that of the L-amino acid oxidases. The function of the D-amino acid oxidases is not clear.

D-amino acid oxidase	L-amino acid oxidase
• Can act on D-amino acids only	• Can act on L-amino acids only
• Can be readily extracted with water-'free' form	• Bound to tissue particles and not extractable with water
• Contains FP (FAD)	• Contains FP (FMN)

3. ***Nature of L-Amino acid Oxidases***
   - The amino acid oxidases are ***auto-oxidizable*** flavoproteins. ***The reduced flavoproteins are re-oxidised at substrate level directly by molecular O_2 forming H_2O_2*** without participation of cytochromes of electron transport chain or other electron carriers.
   - The H_2O_2 formed is toxic to cells and is converted immediately to O_2 and H_2O by enzyme *catalase*.

Note

It is to be noted, if *catalase* is absent genetically, the α-Ketoacid produced by oxidative deamination is decarboxylated non-enzymatically by H_2O_2, forming a carboxylic acid with one less carbon atom.

4. ***Process of Oxidative Deamination***

   This takes place in **two steps:**

   (i) The Amino acid is first dehydrogenated by the flavoprotein (FP) of the enzyme, ***L-amino acid oxidase***, forming an **α-Imino acid.**

   (ii) In the next step, water molecule is added spontaneously, and decomposes to the corresponding α-ketoacid, with loss of the α-imino nitrogen as NH_3.

The process of oxidative deamination is shown schematically below in the box.

5. ***Remarks and Conclusion:***
   - Mammalian L-amino acid oxidase, an FMN flavoprotein is restricted to liver and kidney only.
   - Activity of the enzyme in these tissues is quite low.

Oxidative Deamination

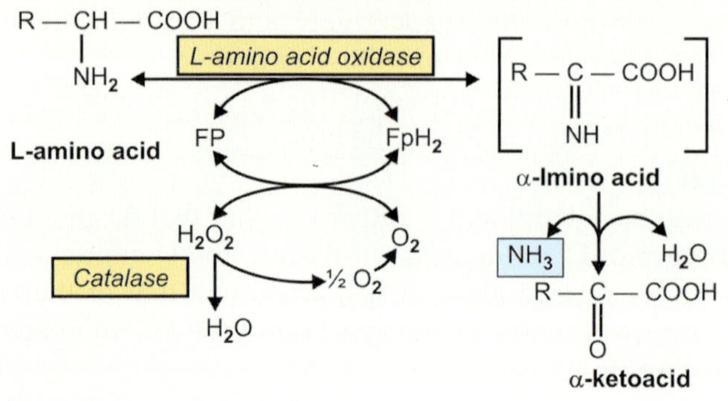

   - It does not have any effect on glycine, or the L-isomers of the dicarboxylic or β-OH-α-amino acids.

Hence, it is concluded that this enzyme does not fulfil a major role in mammalian amino acid catabolism and formation of NH_3.

B. Non-oxidative Deamination:

There are certain amino acids, which can be non-oxidatively deaminated by specific enzymes, and can form NH_3. ***These reactions do contribute to NH_3 formation, but again they do not fulfill, a major role in NH_3 formation.***

Examples: Only three types of non-oxidative deamination will be discussed:

1. ***Amino acid dehydrases:*** The hydroxy amino acids viz serine, threonine and homoserine are deaminated by specific enzymes, called *amino acid dehydrases* which requires Pyridoxal-P (B_6-P) as coenzyme. The enzymes catalyze a primary dehydration followed by spontaneous deamination.

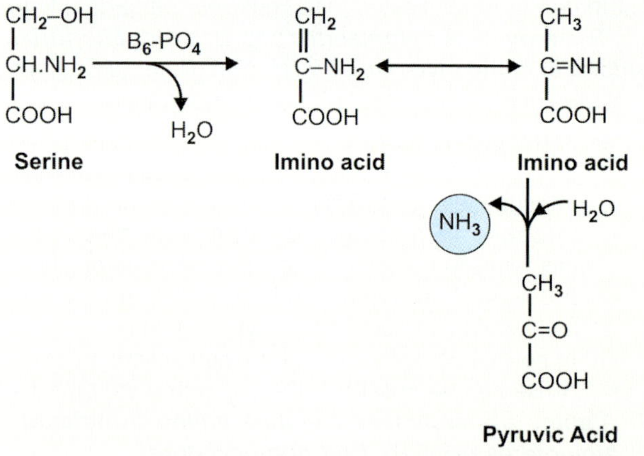

2. ***Deamination of histidine:*** Histidine is non-oxidatively deaminated by the specific enzyme *Histidase* to form NH_3 and urocanic acid.

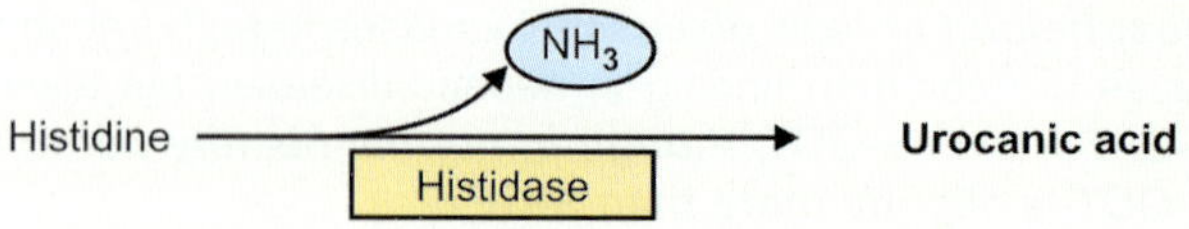

3. ***Amino acid desulfhhydrases:*** S-containing amino acids, e.g. cysteine, and homocysteine are deaminated

SECTION FOUR

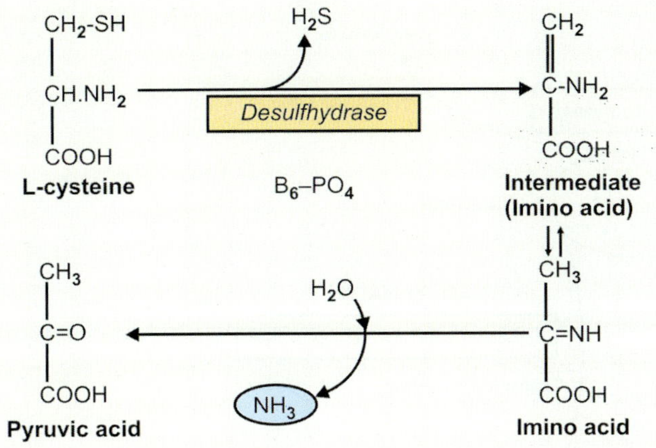

by a primary desulfhydration (removal as H_2S), forming an imino acid, which is then spontaneously hydrolysed.

TRANSDEAMINATION (Deamination of L-Glutamic Acid)

It is to be noted that L-glutamic acid is not deaminated by *L-amino acid oxidase* but by a specific enzyme called ***L-glutamate dehydrogenase.***

Characteristics of the enzyme L-Glutamate dehydrogenase:

- The enzyme has four polypeptide chains (a ***tetramer***)
- A Zn^{++}-containing ***metalloenzyme,*** one atom of Zn^{++} present in each peptide chain
- It is widely distributed in tissues in humans and has high activity, and is ***specific for L-Glutamate***. It requires NAD^+ or $NADP^+$ as coenzymes
- It is a regulated enzyme whose activity is affected by allosteric modifiers as ATP, GTP and NADH which inhibit the enzyme, and ADP activates the enzyme
- Certain hormones appear to influence the enzyme activity *in vitro*.

Reaction

The enzyme *L-Glutamate dehydrogenase* catalyzes the deamination of L-Glutamate to form **α-Iminoglutaric acid**, which on addition of a molecule of water forms NH_3 and α-Ketoglutarate. It is to be noted that the reaction is ***reversible,*** and the equilibrium constant favours glutamate formation, but the quick removal of NH_3 to form urea in urea cycle and α-Ketoglutarate to TCA cycle favours onward reaction, i.e. NH_3 formation (Fig. 27.2).

Remarks and Conclusion: The amino groups of most amino acids are transferred to α-ketoglutarate by group transaminases and specific transaminases by the process of transamination ***forming L-glutamate as end-product.*** Release of this N_2 as NH_3 from L-Glutamate is catalysed by *L-Glutamate dehydrogenase*, an enzyme of high activity and wide distribution in mammalian tissues. ***It is the coupled action of an amino acid-α-oxoglutarate transaminase and L-Glutamate dehydrogenase might explain the oxidative deamination of L-amino acids.*** As it involves first transamination and coupled with oxidative deamination, the process is called as ***Transdeamination. This mechanism seems to be the major pathway for removal of NH_2 group from an L-amino acid and formation of NH_3.***

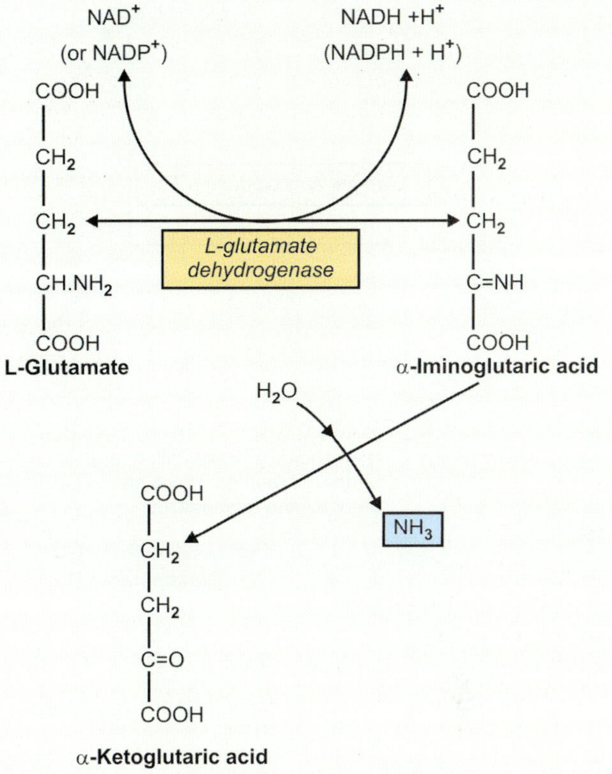

Fig. 27.2: Transdeamination

Other sources of NH_3

- Absorption from gut produced by intestinal bacteria, which can be a major source in intestinal obstruction.
- Pyrimidine catabolism.

NH_3 TRANSPORT

Formation of NH_3 has been discussed above and different sources from which NH_3 is formed shown in Figure 27.3. It is stressed that in addition to NH_3 formed in the tissues, a considerable quantity of NH_3 is produced in the gut by intestinal bacterial flora, both

- ***From dietary proteins,*** and
- ***From urea present in fluids secreted into the GI tract.***

This NH_3 is absorbed from the intestine into portal venous blood which contains relatively high

SECTION FOUR

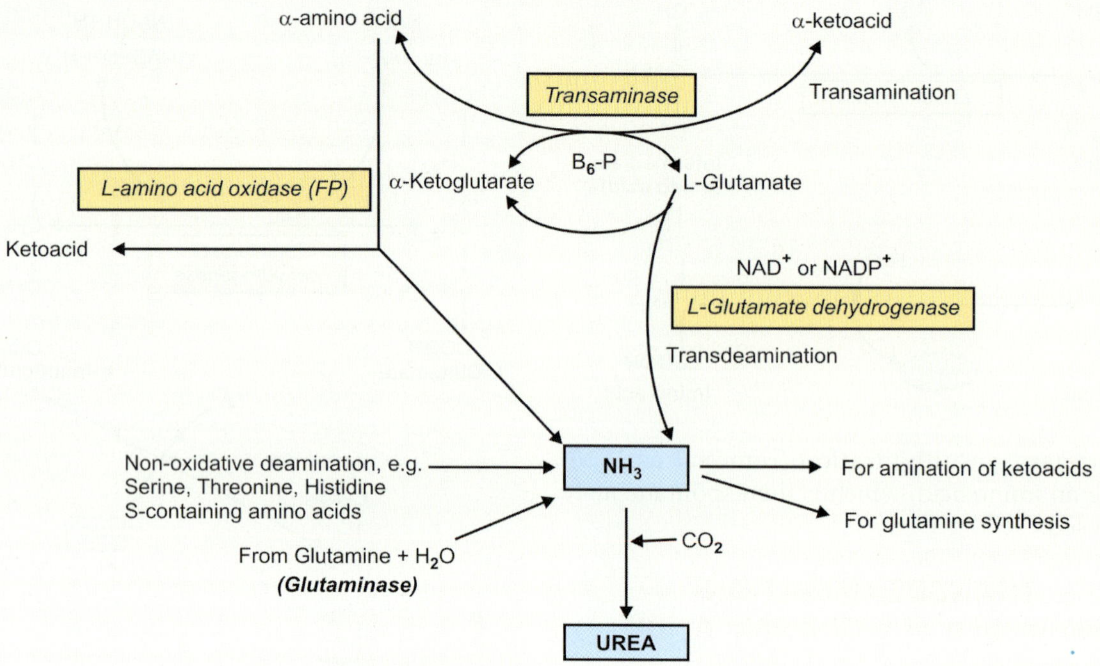

Fig. 27.3: Overall pattern of N-removal from an L-amino acid

concentration of NH_3 as compared to systemic blood. Under normal conditions of health, ***Liver promptly removes the NH_3 from the portal blood, so that blood leaving the liver is virtually NH_3-free. This is essential since even small quantities of NH_3 are toxic to CNS.***

CLINICAL SIGNIFICANCE

With severely impaired hepatic function or the development of collateral communications between portal and systemic veins as may occur **in cirrhosis Liver,** the portal blood may bypass the liver. Surgically produced shunting procedures so-called **Eck-fistula** or other forms of **portocaval shunts** are conducive to NH_3 intoxication, particularly after ingestion of large quantities of proteins or after haemorrhage into GI tract.

Normal Blood Ammonia Level

In man, normal blood level of NH_3 varies from 40 to 70 μg/100 ml. Free NH^+_4 (ammonium ion) concentration of fresh plasma is less than 20 μg per 100 ml. Such low concentrations suggest that the mechanism for removal for this highly toxic substance is extremely efficient.

CLINICAL ASPECT

Hyperammonaemia: Hyperammonaemia is associated with comatose states such as may occur in hepatic failure. May be of **2 types**

1. ***Acquired hyperammonaemia:*** It is usually the ***result of cirrhosis of the Liver*** with the development of a collateral circulation, which shunts the portal blood around the organ, thereby severely reducing the synthesis of urea.
2. ***Inherited hyperammonaemia:*** Results from genetic defects in the urea cycle enzymes.

Features of NH_3 intoxication: The symptoms of NH_3 intoxication include:

- A peculiar flapping tremor
- Slurring of speech
- Blurring of vision
- In severe cases follows to coma and death.

These resemble those of syndrome of hepatic coma, where blood and brain NH_3 levels are elevated.

Why NH_3 is Toxic?

The cause of NH_3 toxicity is not definitely known. Following associated biochemical changes are important.

- Increased NH_3 concentration enhances amination of α-ketoglutarate, an intermediate in TCA cycle to form Glutamate in brain. This ***reduces mitochondrial pool of α-ketoglutarate*** ↓ consequently depressing the TCA cycle, affecting the cellular respiration.
- Increased NH_3 concentration **enhances glutamine** formation from Glutamate and thus reduces 'brain-cell' pool of glutamic acid. Hence there is ***decreased formation of inhibitory neurotransmitter GABA (γ-aminobutyric acid)*** ↓.

- Rise in brain glutamine level enhances the outflow of glutamine from brain cells. Glutamine is carried 'out' by the same "transporter" which allows the entry of 'tryptophan' into brain cells. Hence ***'tryptophan' concentration in brain cells increases which leads to abnormal increases in synthesis of "serotonin", a neurotransmitter.***

Metabolic fate of NH_3 in the body: Three main fates:
1. Mainly NH_3 is converted to urea (urea cycle)
2. Formation of Glutamine
3. Amination of α-ketoacid to form α-amino acid

UREA FORMATION (KREBS-HENSELEIT CYCLE)

The removal of excess of NH_3 derived from amino acid catabolism in the tissues or from bacterial action in the gut is accomplished by the production of urea which is excreted in the urine. Steps of urea synthesis have been elucidated by **Krebs** and **Henseleit** (1932) (Fig. 27.4).

Characteristic Features

- It is a *cyclic process,* **five reactions** which involves ***ornithine, citrulline, arginine and aspartic acid.***
- *Site of synthesis:* urea formation takes place in liver in mammals and all of the enzymes involved have been isolated from Liver tissue.

Note

Other Organs
- **Kidneys:** Urea cycle operates in a limited extent. Kidney can form up to ariginine but cannot form urea, ***as enzyme arginase is absent in kidney*** tissues.
- **Brain:** Brain can synthesise urea from citrulline, but lacks the enzyme for forming citrulline from ornithine.

Thus, neither the kidneys nor the brain can form urea in significant amounts.

- *Location of enzymes:* It is partly mitochondrial and partly cytosolic.
  - One mol. of NH_3 and one mol. of CO_2 are converted to one mol. of urea for each turn of the cycle and ***orinithine is regenerated at the end, which acts as a catalytic agent.***
  - The overall process in each turn of cycle ***requires 3 mols of ATP.***

Evidences to show that the synthesis occurs only in Liver:
1. If kidneys are removed in an experimental animal, there is sharp rise in blood urea level.
2. The above can be prevented if Liver is also removed.
3. The enzyme *Arginase* has only been isolated from Liver. The enzyme is absent in kidneys, Brain (activity is very low) and other tissues.
4. In cirrhosis liver, functioning of liver is much below normal, blood urea levels decrease with a simultaneous increase in NH_3.

SECTION FOUR

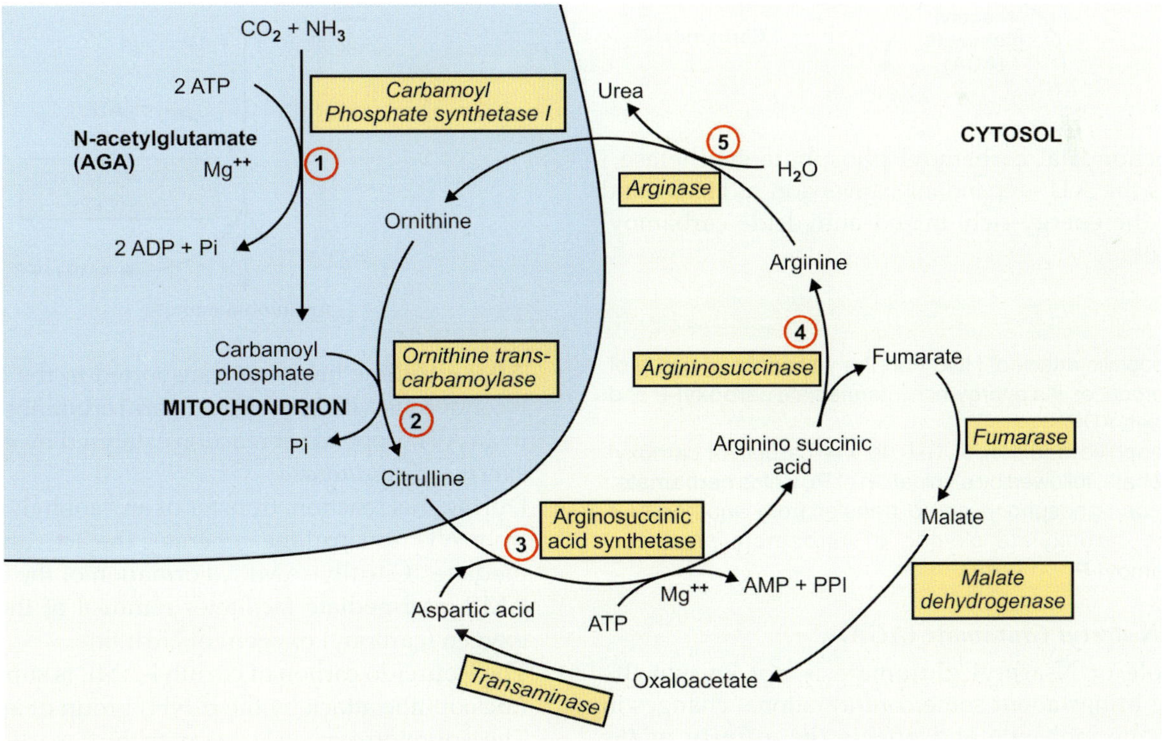

Fig. 27.4: Biosynthesis of urea or ornithine—urea cycle

5. Similar results are seen where the liver is excluded from circulation by an anastomosis between portal vein and vena cava (portocaval shunt).

Stages

The reactions of urea cycle can be studied in **five** sequential enzymatic reactions.

- *Reaction 1:* Synthesis of carbamoyl-phosphate
- *Reaction 2:* Synthesis of citrulline
- *Reaction 3:* Synthesis of argininosuccinate
- *Reaction 4:* Cleavage of argininosuccinate
- *Reaction 5:* Cleavage of arginine to form ornithine and urea

Reaction 1: Synthesis of Carbamoyl-P (Mitochondrial)

In this reaction, HCO_3^-, NH_4^+ and phosphate derived from ATP reacts to form **carbamoyl-P** (also called **Carbamyl-P**). The reaction is catalysed by the mitochondrial-enzyme ***Carbamoyl phosphate synthetase 1.***

There are **2 types** of the enzyme:

- ***Carbamoyl synthetase I:*** Occurs in mitochondria of Liver cells. It is involved in urea synthesis.
- ***Carbamoyl synthetase II:*** Present in cytosol of liver cells which is involved in pyrimidine synthesis (Refer to metabolic fate of carbamoyl-P).

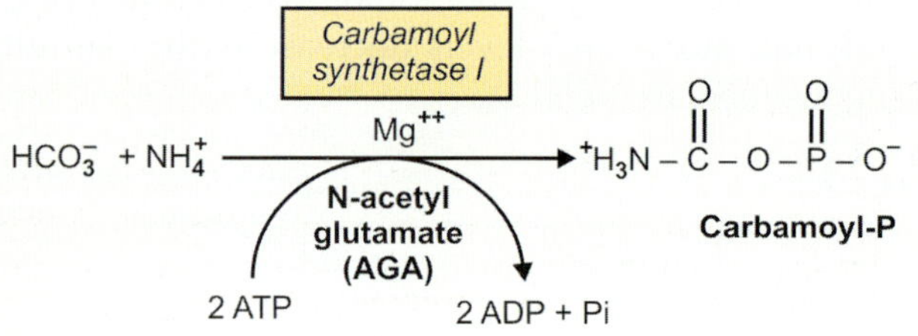

Mitochondrial carbamoyl phosphate synthetase I catalyses the ATP-dependant conversion of HCO_3^- and NH_4^+ to the energy-rich, mixed anhydride carbamoyl phosphate.

Mechanism of Formation of Carbamoyl-P

- Nucleophilic attack of HCO_3^- on the γ-phosphoryl group of ATP produces the energy-rich intermediate **carboxyl-P** and releases ADP.
- Nucleophilic addition of NH_3 to the carbon of carboxyl phosphate, followed by elimination of Pi, yields **carbamate**.
- A second-phosphoryl group transfer from another ATP to the carboxylate oxygen of carbamate produces **carbamoyl-P**.

Role of N-acetyl Glutamate (AGA)

Exact role of N-acetyl glutamate is not known. Its presence brings about some conformational changes in the enzyme molecule and ***affects the affinity of the enzyme for ATP.***

Reaction 2: Synthesis of Citrulline: (Mitochondrial)

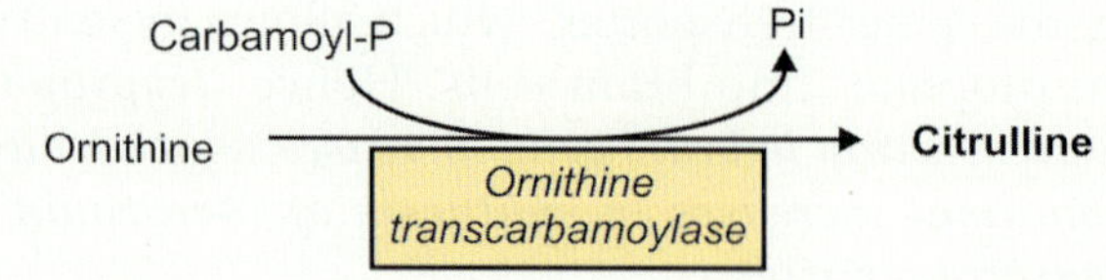

Ornithine transcarbamoylase enzyme, also called as ***ornithine carbamoyl transferase*** is found associated with carbamoylphosphate synthetase I in the mitochondrial matrix.

- It catalyses the nucleophilic addition of ornithine to the carbonyl group of carbamoyl-P to produce Citrulline.
- During this reaction, the δ-NH_2 group of ornithine attaches to the carbonyl group of carbamoyl-P and the phosphate group (Pi) is released.

Note

1. Ornithine which is regenerated in cytosol in the 5th reaction, is transported into the mitochondrial matrix by a specific ***transport protein*** in the inner mitochondrial membrane.
2. Similarly Citrulline which is produced in mitochondrial matrix is transported across the inner mitochondrial membrane to the cytosol by a specific ***transport protein.***

Reaction 3: Synthesis of Argininosuccinate: (cytosolic):

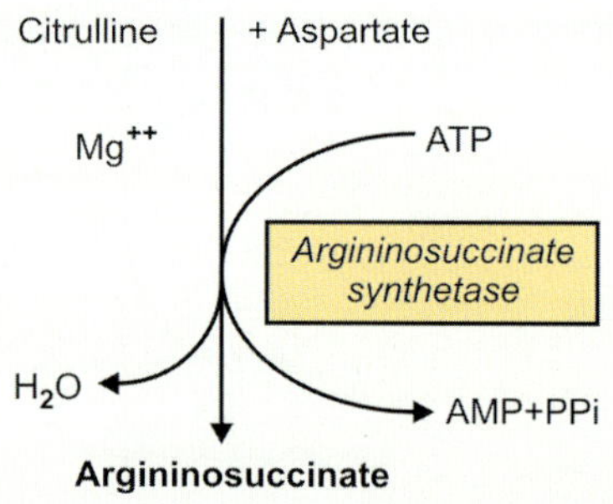

- After citrulline has been transported to the cystosol, it condenses with Aspartate to form argininosuccinate in an ATP-dependant reaction catalysed by ***argininosuccinate synthetase.***
- During this reaction, transfer of an "adenylyl" group from ATP to citrulline generates the activated intermediate "Citrullyl-AMP". Formation of the citrullyl-AMP intermediate facilitates removal of the ureido oxygen (carbonyl oxygen) of citrulline.
- The isoureido carbon of citrullyl-AMP is subjected to nucleophilic attack by the α-NH_2 group of aspartate. The isoureido oxygen leaves with the departing AMP, and **'argininosuccinate' is formed.**

Reaction 4: Cleavage of Argininosuccinate: (Cytosolic)

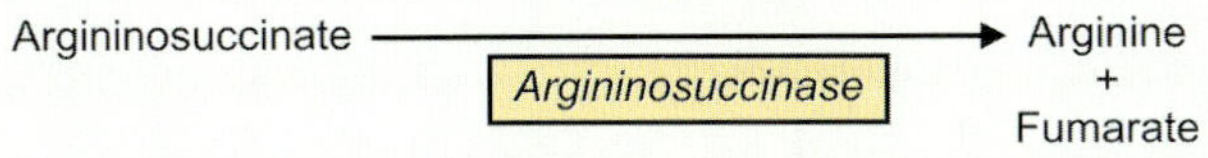

- In this reaction of urea cycle, the enzyme ***arginino-succinase*** also known as ***Argininosuccinate Lyase*** catalyses conversion of Argininosuccinate to arginine and fumarate. ***The urea cycle is linked to the TCA cycle through the production of fumarate. Amino acid catabolism, is therefore directly coupled to energy production.***
- *Argininosuccinase* is ***cold-labile enzyme*** of mammalian liver and kidney tissues. Loss of activity in the cold is associated with dissociation into two-protein components. This dissociation is prevented by Pi, arginine, and argininosuccinate.

Fate of fumarate: The fumarate is converted to oxaloacetate (OAA) via the *fumarase* and *malate dehydrogenase* reactions and then transaminated to regenerate aspartate to participate in the cycle.

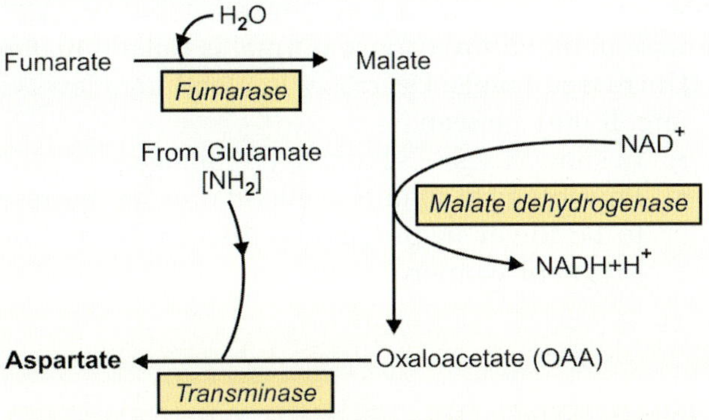

Reaction 5: Cleavage of Arginine to Ornithine and Urea

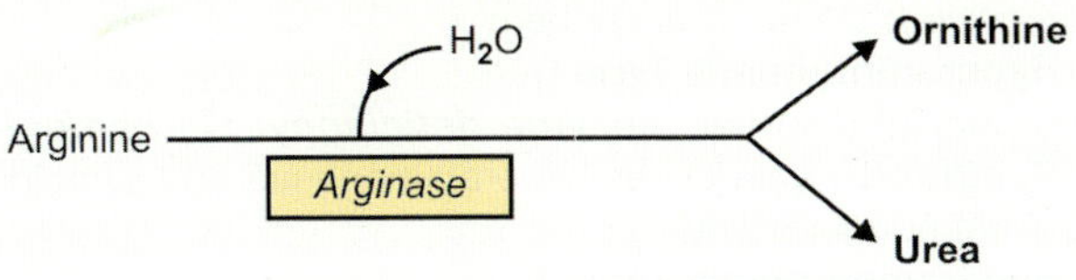

The last reaction of the urea cycle completes the cycle. It is catalysed by the enzyme ***arginase,*** which is found only in the liver cells. Arginase catalyses hydrolysis of the guanidine group of arginine, ***releasing urea*** and ***regenerating ornithine.*** Ornithine now enters mitochondrion through inner mitochondrial membrane by a specific transport protein. Ornithine and lysine are potent inhibitors competitive with arginine. Highly purified *arginase* from mammalian liver cells is activated by CO^{++} and Mn^{++}.

Significance of Urea Cycle

1. ***Detoxification of NH_3:*** Major biological role of this pathway is the detoxication of NH_3. Toxic ammonia is converted into a nontoxic substance urea and excreted in urine.
2. ***Biosynthesis of arginine:*** The urea cycle also serves for the biosynthesis of arginine from ornithine in liver, kidney and intestinal mucosa. Kidney and intestinal mucosa probably contribute most of the body arginine because they possess all the urea cycle enzymes ***except arginase***. Hence they can form upto arginine and cannot form urea. The arginine is used for protein synthesis.

Source of C and N of Urea

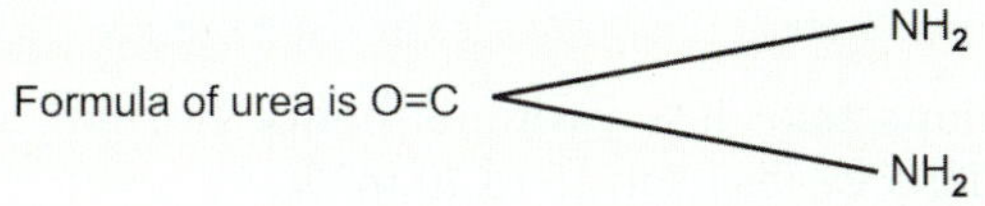

- One Nitrogen of NH_2 group is derived from the NH^+_4 ion (Reaction 1).
- Other Nitrogen of NH_2 group is provided by Aspartate (Reaction 3).
- Bicarbonate, HCO^-_3 ion, provides the carbon atom of urea.

Regulation of Urea Synthesis

1. Achieved by linkage of mitochondrial glutamate dehydrogenase with carbamoyl-P-synthetase I.
2. Carbamoyl-P-synthetase I is thought to act in conjunction with mitochondrial glutamate dehydrogenase to channelise nitrogen from glutamate and, therefore, from all amino acids as NH_3 and then through carbamoyl-P and thus finally to urea.
3. Though the equilibrium constant of the glutamate dehydrogenase reaction favours glutamate formation rather than formation of NH_3, but removal of NH_3 by the carbamoyl-P-synthetase I reaction and oxidation of α-ketoglutarate by TCA cycle favours the glutamate catabolism.
4. The above effect is favoured by the presence of ATP, which in addition to being a requirement for carbamoyl-P-synthetase I reaction, it also stimulates *Glutamate dehydrogenase* activity unidirectionally in the direction of NH_3 formation (Refer box on next page).

SECTION FOUR

CLINICAL SIGNIFICANCE OF UREA

A moderately active man consuming about 300 gm carbohydrates, 100 gm of fats and 100 gm of proteins daily must excrete about 16.5 gm of N daily. 95 per cent is eliminated by the kidneys and the remaining 5 per cent, for the most part as N, in the faeces.

1. **Normal Level**
   The concentration of urea in normal blood plasma from a healthy fasting adult ranges from 20 to 40 mg%.

Regulation of urea synthesis

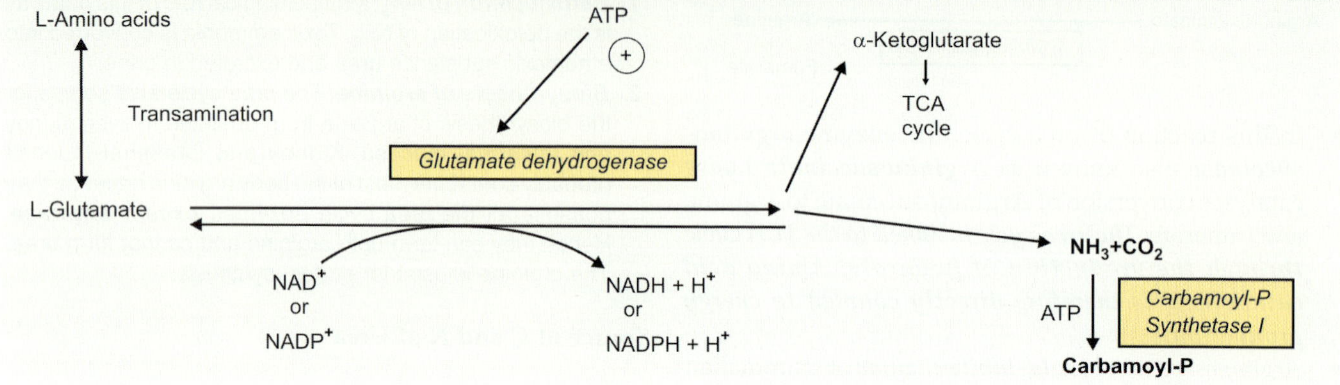

Indians take less proteins hence normal level in Indians varies from 15 to 40 mg%.

2. **Increase of Levels**

Increases in blood urea may occur in a number of diseases in addition to those in which the kidneys are primarily involved. The causes can be classified as:

- **Prerenal,**
- **Renal,** and
- **Postrenal**

(a) **Prerenal**

Most important are conditions in which plasma vol/body-fluids are reduced:

- Salt and water depletion,
- Severe and protracted vomiting as in pyloric and intestinal obstruction,
- Severe and prolonged diarrhoea,
- Pyloric stenosis with severe vomiting,
- Haematemesis,
- Haemorrhage and shock; shock due to severe burns,
- Ulcerative colitis with severe chloride loss,
- In crisis of Addison's disease (hypoadrenalism).

(b) **Renal**

The blood urea can be increased in all forms of kidney diseases:

- In acute glomerulonephritis.
- In early stages of type II nephritis (nephrosis) the blood urea may not be increased, but in later stages with renal failure, blood urea rises.
- Other conditions are malignant nephrosclerosis, chronic pyelonephritis and mercurial poisoning.
- In diseases such as hydronephrosis, renal tuberculosis; small increases are seen but depends on extent of kidney damage.

(c) **Postrenal Diseases**

These lead to increase in blood urea, when there is obstruction to urine flow. This causes retention of urine and so reduces the effective filtration pressure at the glomeruli; when prolonged, produces irreversible kidney damage.

Causes:

- Enlargement of prostate,
- Stones in urinary tract,
- Stricture of the urethra,
- Tumours of the bladder affecting urinary flow.

Note

Increase in blood urea above normal is called ***uraemia.***

3. **Decreased levels:** Decreases in blood urea levels are rare. It may be seen:
   - In some cases of severe liver damage,
   - Physiological condition: Blood urea has been seen to be lower in pregnancy than in normal non-pregnant women.

CLINICAL ASPECT

INHERITED DISORDERS ASSOCIATED WITH UREA CYCLE

Inherited disorders due to inherited deficiency of enzymes of urea cycle have been described:

1. **Hyperammonemia Type I:**
   A familial disorder, ***enzyme deficiency: Carbamoyl-P-Synthetase I,*** produces hyperammonemia and symptoms of ammonia toxicity.
2. **Hyperammonemia Type II:**
   (Also called ***Ornithinemia***)
   - ***Inheritance:*** X-chromosome linked
   - ***Enzyme deficiency: Ornithine transcarbamoylase***
   - Produces hyperammonemia and symptoms of NH_3 toxicity.
   - ***Blood, urine and CS Fluid:*** shows characteristically increased level of glutamine ↑ (glutamine synthesis is enhanced). Also increase in blood NH_3 level and ornithine.
   - Mother of affected child also exhibit hyperammonemia and aversion to protein foods.

3. Citrullinemia

A rare disorder

- ***Inheritance:*** Autosomal recessive
- ***Enzyme deficiency: Argininosuccinate synthetase.***
- **Two types** of deficiency:
  - **(a)** One type: Mutation in regulator gene: Enzyme is absent in Liver, normal Km for citrulline.
  - **(b)** Other type: Mutation in structural gene: Amount of enzyme normal in liver, affects catalytic site and has abnormally high Km for citrulline.
- ***Clinically:*** Presents with hyperammonemia and NH_3 toxicity, and mental retardation
- ***Biochemically:*** Blood and CSF: Increased level of NH_3 and marked increase in citrulline
- ***Urine:*** Large quantities of citrulline are excreted in urine (1 to 2 g/d). Feeding arginine in these patients enhances citrulline excretion.
- Feeding benzoate diverts ammonia N to form hippuric acid with glycine and increased hippuric acid excretion occurs.

4. Argininosuccinic Aciduria

A rare inherited disorder, ***usually fatal.***

- ***Inheritence:*** Autosomal recessive
- ***Enzyme deficiency: Argininosuccinase.*** (Also called ***Argininosuccinatelyase***)
- ***Age:*** Usually manifest before 2 years of age and terminates fatally in early life.
- ***Clinically:*** Hyperammonemia and NH_3 toxicity, mental retardation, occurrence of friable, tufted hairs called ***tricorrhexis nodosa***.
- ***Biochemically: Blood and CS Fluid:*** Shows elevated levels of argininosuccinate.
- ***Urine:*** Increased excretion of argininosuccinate.

The enzyme deficiency has been demonstrated in brain, Liver, kidney and RB Cells of the patient. Early diagnosis can be made by demonstrating enzyme deficiency in erythrocytes from cord blood and deficiency in amniotic fluid (by amniocentesis).

Feeding of arginine and benzoate promotes nitrogen waste excretion.

5. Hyperargininemia

- ***Enzyme deficiency: Arginase.*** Deficient in Liver and RB Cells.
- ***Clinically:*** Manifests as hyperammonemia
- **Biochemically: Blood and CSF** elevated levels of arginine.
- ***Urine:*** Increased urinary excretion of Lysine, cystine, ornithine and arginine. ***Low Protein diet resulted in lowering of plasma NH_3 levels and disappearance of urinary Lysine-cystinuria Pattern.***

GLUCOSE-ALANINE CYCLE

1. Skeletal muscle transports NH_3 to the Liver in the form of the amino acid 'alanine'. The alanine is formed in the muscle tissue by a transamination reaction between pyruvate (PA) and glutamate.
2. The alanine is transported through the bloodstream to the liver, where it reacts with α-ketoglutarate to reform Pyruvate and glutamate. This reaction is catalysed by *alanine transaminase*.
3. The nitrogen originating from the glutamate is processed by the urea cycle.
4. When the blood glucose is low, the Pyruvate resulting from alanine transamination is used to make glucose via the gluconeogenesis pathway.
5. The glucose can be returned to the skeletal muscle to supply quick energy.

Thus, ***the transport of alanine from muscle to Liver results in a reciprocal transfer of glucose to muscle. The entire cyclical process is referred to as the "glucose-alanine cycle"*** (Fig. 27.5).

Its importance is proportional to the muscular activity of the organism. It is to be noted that active muscle tissue operates anaerobically, producing large quantities of Pyruvate (PA) and consuming large quantities of glucose.

FATE OF CARBAMOYL-P

- Carbamoyl-P is a ***committed step*** and a branch point compound in metabolic pathway and exerts independent regulation at the level of these two enzymes.
- Urea formation, a distinctive process, occurs only in Liver is an example of biochemically ***differentiated*** system.
- Pyrimidine synthesis, necessary for all tissue cells and all cell division, may be considered as an ***undifferentiated*** system (Refer box next page).

In a regenerating liver tissue

1. *Ornithine transcarbamoylase* activity decreases ↓, while *aspartate transcarbamoylase* level and cytosolic *carbamoyl synthetase II* activity increases, ↑
2. When regeneration is complete, biochemical differentiation occurs, and *aspartate transcarbamoylase* level ↓ decreases, at the same time *ornithine transcarbamoylase* level increases ↑ and activity of mitochondrial *carbamoyl synthetase I* increases. ↑

The above is an example of ***biochemical dedifferentiation.***

GLUTAMINE FORMATION AND FUNCTIONS

1. *Chemically* glutamine is ***δ-amide of α-aminoglutaric acid.*** Glutamine formation is a ***manoeuvre to remove*** NH_3

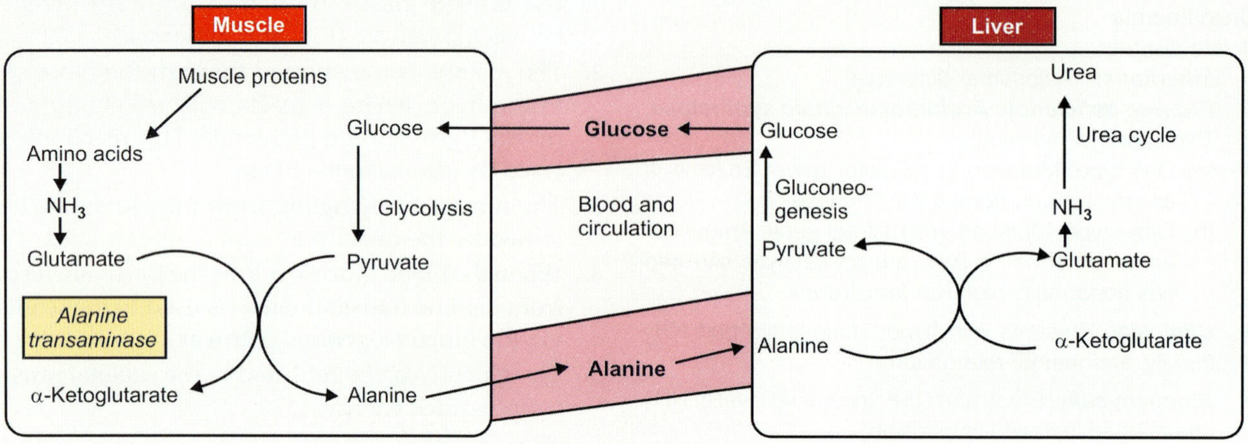

Fig. 27.5: Glucose-alanine cycle

Metabolic Fate of Carbamoyl (P)

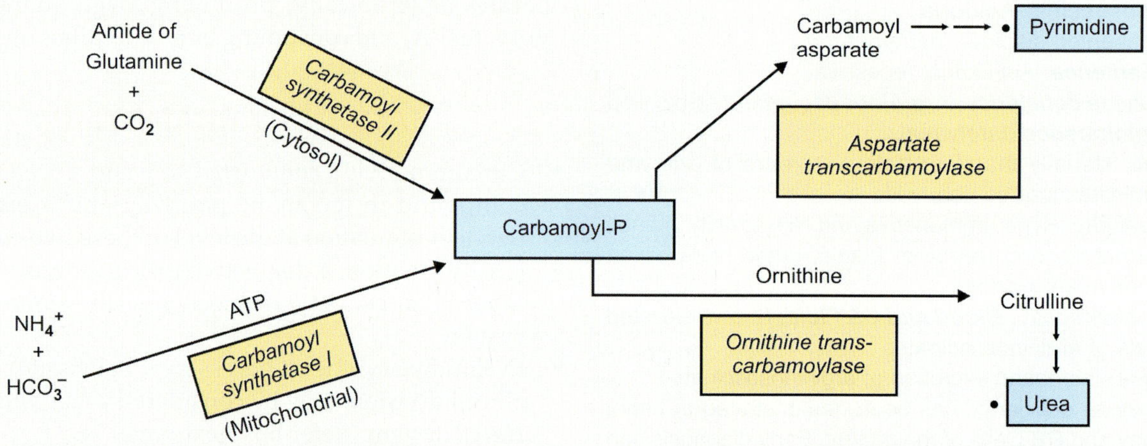

from blood, another alternative mechanism of ammonia detoxication in many tissues. ***Glutamine serves as an important reservoir of NH_3 nitrogen in tissues which can be drawn upon for various synthetic processes.***

2. *Synthesis of Glutamine*
   - Glutamine is synthesised in tissues from glutamic acid and NH_3 by the action of the enzyme ***glutamine synthetase,*** a mitochondrial enzyme.
   - The reaction is ***irreversible.***
   - It requires ATP. Enzyme first binds to ATP and then reacts to glutamate.
   - *Site:* Synthesis takes place in tissues such as Liver, kidney, brain and retina.

The reaction is shown in the box:

Synthesis of Glutamine

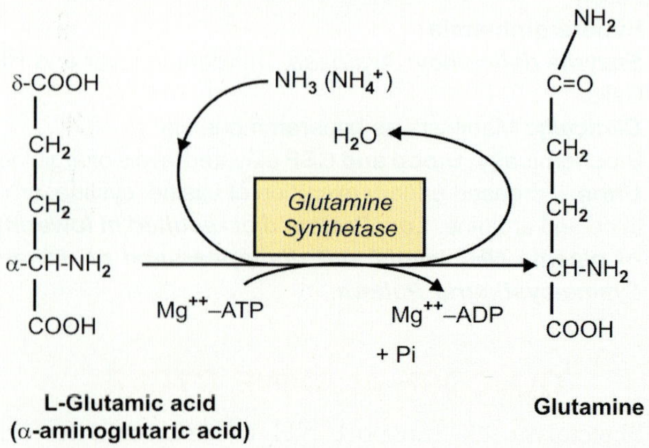

Reaction probably takes place as follows:

ATP + Enzyme ⇌ Enz-ATP

Enz-ATP + Glutamate ⇌ Enz-ATP-Glutamate **(Complex 1)**

Complex-1 (Rearranges) ⇌ Enz-ADP-γ-Glutamyl-P **(Complex-2)**

Complex-2 + NH_3 ⇌ Enz-ADP- Glutamine + Pi

Enz-ADP—Glutamine ⇌ Glutamine + ADP + Enz

3. ***Regulation of Glutamine Synthetase Activity***
   - *Glutamine synthetase* regulation has been studied in *E. Coli*. It is largely regulated by reversible *adenylation* and *deadenylation*.
   - ***Glutamine synthetase*** occurs as **active** and **inactive** form.
     - "Active" form is GSa ("deadenylated" form).
     - "Inactive" form is GSb ("adenylated" form).

(a) Conversion of active to inactive:

GSa (active) ⟶ **GSb (inactive)**

Active form is converted to inactive form **by adenylation** by a Mg^{++} dependant GSa *adenyl transferase* and 12 ATP molecules (multiple adenylation)

(b) Conversion of inactive to active:

GSb (inactive) ⟶ **GSa (active)**

Inactive form is converted to active form **by deadenylation** by a Mn^{++}-dependant enzyme *deadenylase* which removes 12 adenylate groups from GSb by phosphorolysis.

- Glutamine itself inhibits the enzyme *deadenylase* and activates the *adenyl transferase* to reduce its further synthesis.

4. ***Feedback inhibition:*** Glutamine synthesis is also reduced by the "Feedback" inhibition of *glutamine synthetase* by:
   - ***Carbamoyl-(P)***
   - ***Glucosamine-6-P***
   - ***AMP*** and
   - ***CTP***
5. ***Blood level of glutamine:*** Blood glutamine is probably mostly synthesised in Liver. ***Normal blood plasma level in humans ranges from 6 to 12 mg%***. It represents approx. 18 to 25 per cent of total free amino nitrogen of the plasma.
6. ***Hydrolysis of glutamine:*** A ready source of NH_3. Glutamine is hydrolysed by a specific enzyme *Glutaminase.* The reaction is ***irreversible.*** The reaction can occur in various tissues specially liver, kidney, brain and retina.

Note: The reaction is specially important in renal distal tubular epithelial cells where it is a source of NH_3, which is used for exchange of Na^+, thus conservation of base (Refer to Chapter on Acid-base Balance and Imbalance).

Glutamine + H_2O —(*Glutaminase*)⟶ L-Glutamic acid + NH_3

7. ***Transamination:*** Glutamine is capable of transamination with a variety of α-ketoacids to produce α-ketoglutaramic acid which can be hydrolysed by a specific *deamidase* to produce α-ketoglutaric acid and NH_3.

Glutamine + Pyruvic acid (PA) ⇌ (*Transaminase*, B_6–PO_4) α-ketoglutaramic acid + Alanine

α-ketoglutaramic acid + H_2O —(*Deamidase*)⟶ α-ketoglutaric acid + NH_3

8. ***Effect of GH on plasma glutamine level:*** Administration of GH (growth hormone) to dogs causes a marked increase in the plasma glutamine along with N_2-storage and weight gain.

FUNCTIONS OF GLUTAMINE

I. ***Transamidation***

(a) Formation of glucosamine-6-P: The amide-N of glutamine can be transferred to a ketogroup.

- If transferred to an α-ketoacid, an amino acid can be formed.
- If transferred to a Keto group of fructose-6-P, it forms Glucosamine-6-P by a transamidation reaction from Glutamine. The catalysing enzyme is ***L-glutamine-D-fructose-6-P transamidase.***

Fructose-6-P + Glutamine —(*Transamidase* or *(Amidotransferase)*)⟶ Glucosamine-6-P

This *transamidase* reaction is unique in that the ***necessary energy is derived solely from the cleavage of the amide bond of the glutamine*** in contrast to the usual reactions involving transfer of amide-N_2 of Glutamine to an acceptor other than water, all of which require an additional source of energy such as ATP.

SECTION FOUR

(b) Other Examples of Transamidation

Other examples of transfer of amide-N_2 of glutamine for synthesis of various biomedical compounds:

1. ***Formation of guanylate (GMP):*** The amide group of glutamine is transferred to C_2 of Xanthylate (XMP), by the enzyme *GMP-synthetase*, forming Guanylate (GMP).

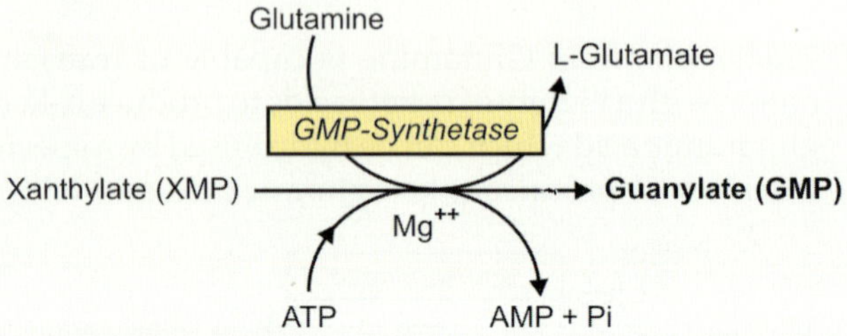

2. ***Formation of 5-phosphoribosyl-1-amine:*** Amide group of glutamine is transferred to 5-phosphoribosyl-1-pyrophosphate (PRPP) to form 5-phosphoribosyl-1-amine. It is catalysed by the enzyme *Glutamine PRPP-amidotransferase. **It is the first rate-limiting step of purine synthesis.***

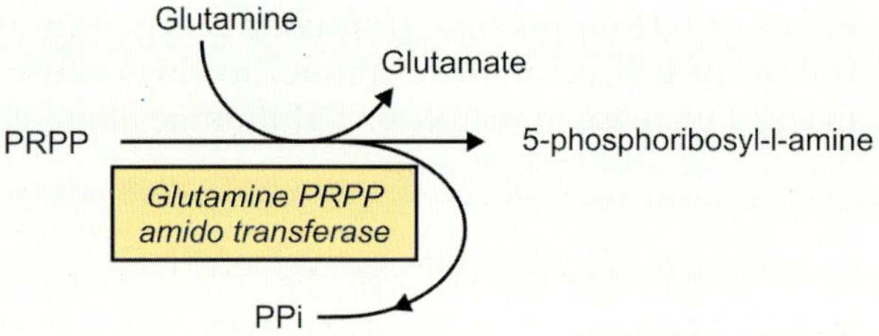

3. ***Formation of N-formyl glycinamidine ribonucleotide:*** An intermediate compound in purine synthesis. N-formyl-glycinamide ribonucleotide undergoes trans-amidination with glutamine amide group to form "N-formyl glycinamidine ribonucleotide".

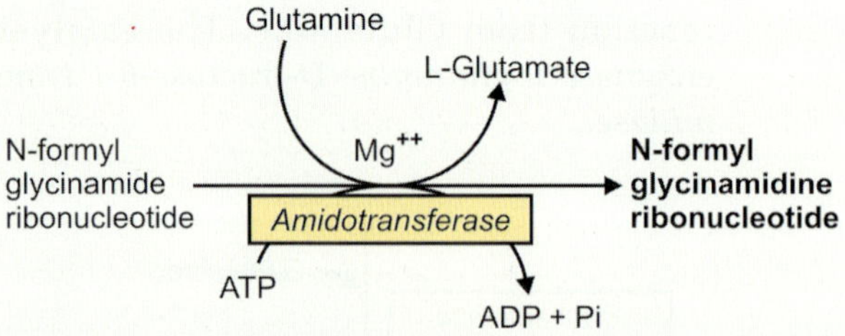

4. ***Synthesis of asparagine:*** Amide group of glutamine is transferred to the β-COOH group of aspartic acid

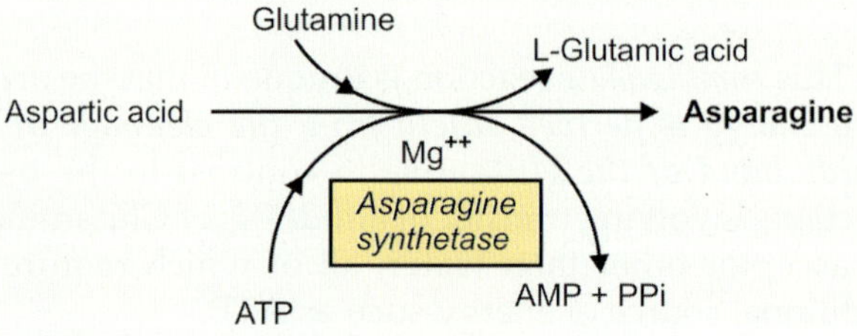

forming asparagine. The reaction is catalysed by the enzyme *Asparagine synthetase*, which require ATP and Mg^{++}.

II. ***Carbamoyl-P synthesis in cytosol for pyrimidine synthesis:*** Carbamoyl-P is also synthesised in cytosol from amide-N of Glutamine, and HCO_3^-. It is catalysed by the enzyme *carbamoyl-P-synthetase II*, which requires two molecules of ATP.

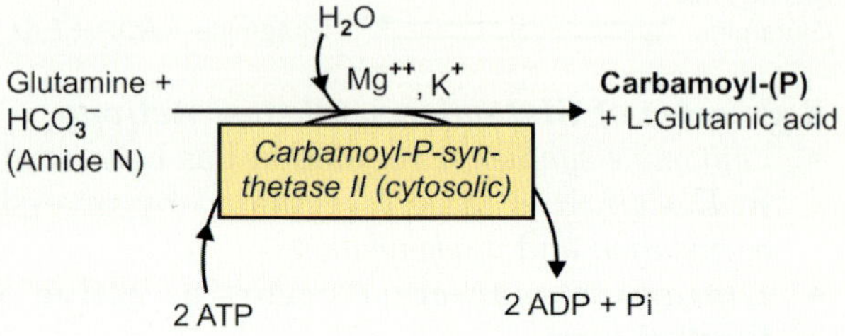

Cytosolic carbamoyl-(P) is used for Pyrimidine syn thesis. Through the carbamoyl-group of carbamoyl-p, the amide-N of glutamine goes to form N_3 of the Pyrimidine ring.

Note: For differentiation of two carbamoyl-P-synthetase I and II, refer to box next page.

III. ***Role of glutamine in kidney—Conservation of Na^+:*** (Refer Chapter on Acid-base Balance and Imbalance).

IV. ***Role of glutamine in brain: The role of glutamate in detoxifying NH_3 in the brain by formation of glutamine is extremely important.*** Glutamate is a major acceptor of NH_3 produced either in the metabolism or delivered to the brain when arterial blood NH_3 is elevated ↑. In this latter reaction, glutamic acid accepts one molecule of NH_3 and is thus converted to glutamine. ***Formation of urea does not play a significant role in the removal of NH_3 in the brain.*** When the levels of NH_3 in brain increases, as **in hepatic failure,** the supply of glutamic acid from the blood may be insufficient to form the additional amounts of glutamine required to detoxify the NH_3 in the brain. Under these circumstances, glutamic acid is synthesized in the brain by amination of α-ketoglutarate produced by TCA cycle within the brain itself. However, continuous utilisation of α-ketoglutarate for this purpose would rapidly deplete the TCA cycle of its intermediates. Repletion is achieved by CO_2^- fixation, involving PA to form OAA, which enters the TCA cycle and proceeds to the formation of α-ketoglutarate.

CLINICAL ASPECT OF CSF GLUTAMINE

- Estimation of CS fluid glutamine level has been taken as ***indirect evidence of hepatic function test.***
- Normal range of CSF glutamine in health ranges from 6 to 14 mg%.

Differentiation of two 'Carbamoyl-P-Synthetase' Enzymes	
Carbamoyl-P-synthetase-I	***Carbamoyl-P-synthetase-II***
1. ***Present in mitochondrial matrix***	1. ***Present in cytosol***
2. ***Required for urea synthesis*** and is rate limiting enzyme.	2. ***Required for Pyrimidine synthesis***
3. Is closely associated with *ornithine transcarbamoylase* in mitochondrial matrix	3. Is closely associated with *Asparate transcarbamoylase and dihydro-orotase* as multienzyme complex in cytosol
4. Uses HCO_3^-, free NH_4^+ and a phosphate from ATP to produce carbamoyl-(P)	4. Utilises HCO_3^-, a phosphate from ATP and the amide N of glutamine to produce carbamoyl-(P)
5. 'Carbamoyl' group of carbamoyl-P is transferred by *Ornithine transcarbamoylase* to ornithine	5. 'Carbamoyl-group from carbamoyl-(P) is transferred by *Aspartate transcarbamoylase* to aspartate
6. 'Carbamoyl group' is the source of carbon of urea	6. ***Carbamoyl-group is the source of C-2 of pyrimidine ring***
7. Requires N-acetyl glutamate (AGA) as allosteric activator of the enzyme	7. Does not require N-acetyl glutamate. It is allosterically activated by PRPP

- Studies have shown, in cirrhosis Liver, CSF glutamine level is increased ↑ and ranges from 16 to 31 mg%.
- In hepatic failure and hepatic coma very much increased values are obtained, ranging from 30 to 54 mg% or more.
- In coma due to various other causes, normal CSF glutamine level obtained. It has been observed that high CSF glutamine level has a bad prognosis. **40 mg% has been kept as the dividing line; values greater than 40 mg% in hepatic failure cases usually show bad prognosis and ends fatally.**

V. ***Role of glutamine in conjugation reaction:*** In man and chimpanzee, phenyl acetic acid is conjugated with glutamine to form **Phenyl acetyl glutamine**, which is excreted in urine, which gives ***"mousy odour"*** (Refer to Phenyl alanine metabolism).

VI. ***Role in cancer:*** Recently *glutaminase* and *asparaginase* have both been investigated as **anti-tumor agents**, since certain tumors exihibit abnormally high requirements for glutamine and asparagine for their growth.

C. Amination of α-ketoacids to form amino acids: NH_3 is also used to aminate certain α-ketoacids to form corresponding α-amino acids.

Examples are

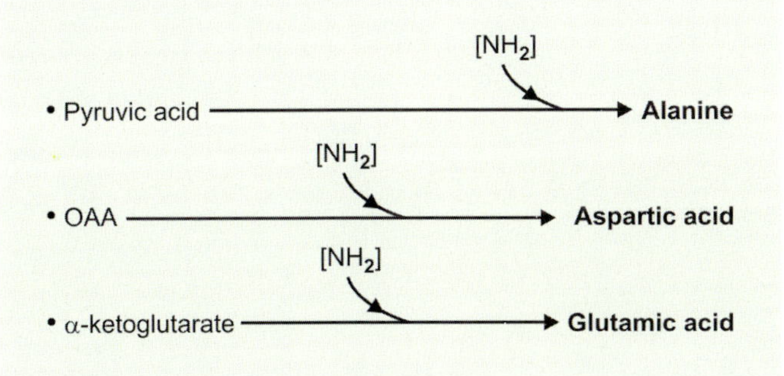

Fate of C-skeletons: The fate of the ketoacids yielded by removal of the NH_2 group of the amino acids are now to be considered. The ketoacids formed after removal of NH_3 can undergo one of **3 fates:**

- ***Reamination to form original amino acids.***
- ***Formation of CO_2 and H_2O and energy, after entering TCA cycle.***
- ***Formation of glucose (glucogenic) or 'ketone bodies' (ketogenic) or both.***

If amino acids are fed one at a time to a "starving phloridzinised" dog, it is found that some fed amino acids give rise to glucose, in the urine; while others give rise to aceto acetic acid. A few give rise to neither. In such an animal, ***about 60 gm. (approx. 58 G.) of glucose are formed and excreted in the urine for 100 gm of proteins metabolized.*** In other words, **60 per cent of protein is potentially glucogenic**. When the liver is removed, however, no such glucose formation takes place. Amino acids are, therefore, **divided into 3 groups:**

1. ***Glucogenic (antiketogenic):*** They form on deamination ***amphibolic*** intermediates, ketoacids, which enter the TCA cycle and either can be oxidised to CO_2 and H_2O or they can go reverse pathway and form glucose or glycogen.
2. ***Ketogenic:*** After deamination, they yield ketoacids, which during subsequent oxidation to H_2O and CO_2, pass through the stage of "aceto acetate".
3. ***Both glucogenic and ketogenic:*** The amino acids of this group give rise to both glucose and ketone bodies.

List of amino acids of the above three groups is given in the box on next page.

Glucose: Nitrogen ratio (G:N ratio) (Dextrose: Nitrogen ratio D:N ratio)

It is ratio of glucose and nitrogen excreted in urine. G:N ratio (also called D:N ratio) is important, as it reflects the conversion of protein into glucose.

It is experimentally proved by measuring the glucose and Nitrogen excreted in urine in:

Glycogen (Glucogenic amino acids)	Fat (Ketogenic amino acids)	Both glycogen and fat (Glucogenic and ketogenic)
Alanine		
Arginine		Isoleucine
Asparate	L-Leucine	Lysine
Cysteine/Cystine		Phenylalanine
Glycine		Tyrosine
Glutamate		Tryptophan
Histidine		
OH-Proline		
Methionine		
Proline		
Serine		
Threonine		
Valine		

(a) "Phloridzinidied" animal (when glucose is not reabsorbed in renal tubules but is excreted).

(b) "Starving" animal (as amino acids contribute to formation of glucose by gluconeogenesis, as glycogen store is depleted).

The nitrogen excreted in such an animal comes from catabolism of proteins and the urinary glucose is thought to be formed from proteins.

It is assumed that one gram of urinary nitrogen represents 6.25 gm of proteins.

Normally, the G:N ratio = 3.65: 1, i.e. 3.65 gm of glucose has come from 6.25 gm of proteins.

$$\therefore \frac{3.65 \times 100}{6.25} = 58 \text{ per cent is the average conversion rate.}$$

PART II

Major Concepts

A. To study the decarboxylation reactions and the functions of various biogenic amines.
B. To study the metabolic fate and metabolic role of aromatic amino acids, phenyl alanine and tyrosine and the inherited disorders associated with them.
C. To study the metabolic fate and metabolic role of sulphur containing amino acid and the inherited disorders associated with them.
D. To study the metabolic role of other amino acids, viz. glycine, serine, histidine, tryptophan and other amino acids.
E. To study the metabolism of creatine.

Specific Objectives

A. 1. Study the decarboxylation reaction, enzyme and coenzymes involved in the reaction and tissues involved.
2. List in tabular form the biogenic amines formed from various amino acids and their biologic importance.
3. Study about histamine, how it is formed and its actions-through histamine receptors. Study how histamine is catabolised.
4. Study about γ-amino butyric acid (GABA): Its formation, action as neurotransmitter and its degradation in the body. Learn what is ***'GABA-shunt' (a by-pass in TCA cycle).***
5. Study what are ***polyamines***, list them. How they are synthesised and degraded in body? List the important functions, learn the normal range of excretion in urine of different polyamines and clinical importance if any, learn about polyamine synthesis inhibitor; DFMO

B. 1. Write the structural formula of phenyl alanine and tyrosine. Note the difference.
2. Study how phenyl alanine is converted to tyrosine in the body, enzyme and coenzymes required, mechanism of hydroxylation of phenyl alanine at p-position.
3. Learn in detail the metabolic fate of tyrosine. Study all the reactions involved with enzyme and co-enzymes required for each reaction. Note the end products produced after complete degradation. What are the nature of the products-glucogenic and/or ketogenic or both.
4. Study the metabolic role of tyrosine: Formation of thyroid hormones, synthesis of catecholamines, and melanin, formation of tyramine, phenol and cresol. Formation of tyrosine-O-sulphate. What is the function of this substance?
5. Study the inherited disorders associated with phenyl alanine and tyrosine metabolism. Study the inherited disorders in brief, specific enzyme which is deficient and clinical presentation.

C. 1. Revise your knowledge about S-containing amino acids. Name the essential/non-essential amino acids. What are the other sources of sulphur? Write the structure of methionine and cysteine. Note the difference between cysteine and cystine.
2. Study the metabolic fate of L-methionine. Learn how L-methionine is activated before it enters into catabolic pathway. What are the end products of methionine degradation, note their fate. Study how methyl mercaptan is formed from methionine, what is its clinical importance?
NB: ***Note cysteine is produced during degradation of methionine,*** the carbon skeleton is contributed by hydroxy amino acid serine and 'S' is directly transferred from methionine.
3. Study the pathway for degradation of cysteine-enzyme and coenzymes required.
Note: Pyruvate is the end product in all the pathways, hence cysteine is glucogenic. Study how cysteine and cystine are interconvertible.
4. Study the metabolic role of cysteine.
5. Study what is glutathione. Learn how glutathione is synthesised and list the important functions of glutathione in the body.
6. Learn what is transmethylation. Study how methionine is activated. List about ten examples of transmethylation, in which $\sim CH_3$ gr. is transferred from "active" methionine.
7. Study the inherited disorders associated with metabolism of S-containing amino acids, pin point the specific enzyme deficiency in each and study the clinical features in brief.

D. Study briefly the metabolic role of following amino acids:
1. Glycine (amino acetic acid).
2. Serine.
3. Histidine:
   - Histamine formation (refer to biogenic amines)
   - Other Histidine compounds and their importance if any, like Ergothioneine, Carnosine, Anserine
   - Learn the basis of "Figlu Test"
   - Inherited disorder: Histidinaemia.
4. Tryptophan:
   - Study the chemical structure.
   - Formation of 5-HT (serotonin). Study how serotonin is formed from tryptophan. Functions of serotonin in the body. How serotonin is destroyed? Role of drugs like iproniazid and reserpine.
   - Clinical disorder: 'carcinoid syndrome': in which excessive serotonin is formed and its features.
   - Nicotinic acid pathway: ***60 mg of tryptophan gives rise to 1 mg of nicotinic acid.***
   - Study briefly inherited disorder "Hartnup's disease".
5. Study the metabolic role of other amino acids viz: Arginine, threonine, glutamic acid and aspartic acid, proline and hydroxyproline, lysine and branched chain amino acids, viz., valine, leucine and isoleucine.
   - Study how ***Nitric oxide (NO) is formed from the amino acid arginine*** and its metabolic role in the body.

E. • Learn the difference between creatine and creatinine.
• Study the occurrence and distribution of these two compounds.

SECTION FOUR

- Study the biosynthesis of creatine and its regulation. Name the starting material and learn the site of synthesis, enzymes, and coenzymes required.
- Learn the causes of creatinuria.
- Study the role of creatine in muscle. What is "Löhman reaction" and "myokinase" reaction.
- Learn the principles of creatinine estimation (Jaffe's reaction). Learn what is meant by "creatinine coefficient" and its significance.

DECARBOXYLATION REACTION AND BIOGENIC AMINES

Decarboxylation

Decarboxylation is the reaction by which CO_2 is removed from the COOH group of an amino acid as a result ***an amine is formed.*** The reaction is catalysed by the enzyme ***decarboxylase,*** which requires pyridoxal-P (B_6-PO_4) as coenzyme. Tissues like liver, kidney, brain possess the enzyme *decarboxylase* and also by microorganisms of intestinal tract. The enzyme removes CO_2 from COOH gr. and converts the amino acid to corresponding amine. This is mostly a process confined to putrefaction in intestines and produces amines. Biogenic amines formed from various amino acids and their biologic importance are listed in **Table 27.1**.

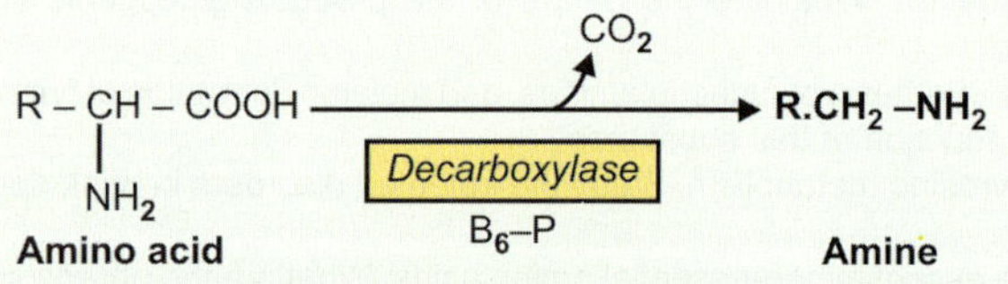

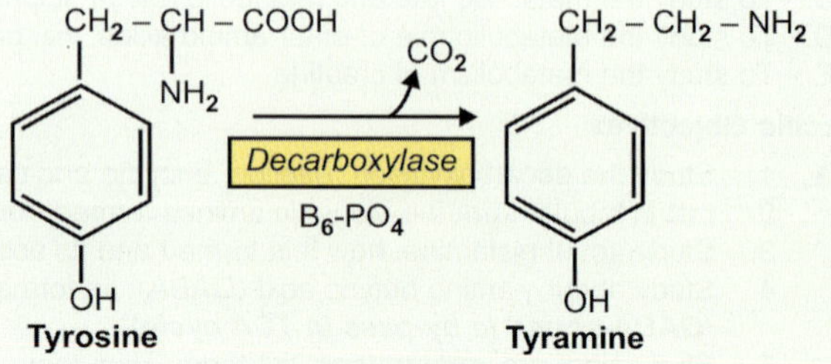

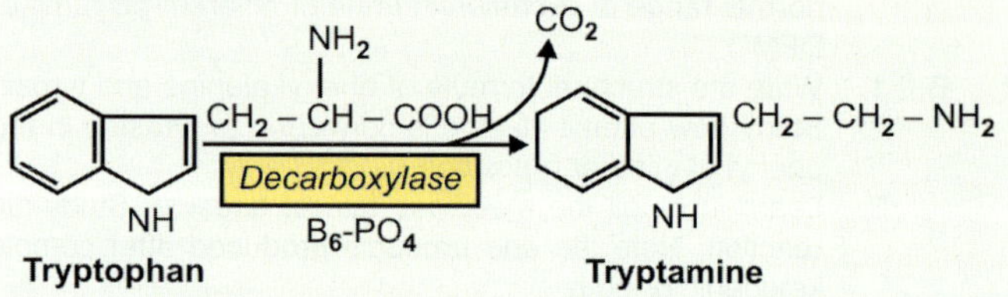

Table 27.1: Biogenic amines and their functions

Nos	Amino acids	Amine	Biologic importance
1.	Tyrosine	• Tyramine	• Increases blood pressure (Vasoconstriction) • Contracts uterus
2.	Tryptophan	• Tryptamine	• Tissue hormone: A derivative 5-OH Tryptamine (Serotonin) • Vasoconstriction • BP ↑
		• 5-methoxy Tryptamine (Melatonin)	• Hormone of pineal gland
3.	Histidine	• Histamine	• Vasodilator, Bl. pr ↓ • HCl ↑ • Pepsin ↑
4.	Serine	• Ethanolamine	• Forms choline by three methylations • Constituent of Phospholipid like cephalin
5.	Threonine	• Propanol amine	• Constituent of Vit B_{12}
6.	Cysteine	• β-mercaptoethanolamine	• Constituent of coenzyme A
7.	Aspartic acid	• β-alanine	• Constituent of pantothenic acid (coenz. A) • As a constituent of dipeptide carnosine and Anserine
8.	Glutamic acid	• γ-amino butyric acid (GABA)	• Presynaptic inhibitor in brain. • Forms a bypass in TCA cycle (GABA-shunt)
9.	3,4,-di-OH-phenylalanine (DOPA)	• Dopamine	• Precursor of Epinephrine and Nor-epinephrine
10.	Cysteic acid	• Taurine	• Constituent of Bile acid taurocholic acid
11.	Lysine	• Cadaverine	• Product of Putrefaction in the gut
12.	Ornithine	• Putrescine	• Product of Putrefaction in the gut
13.	Arginine	• Agmatine	• Product of Putrefaction in the gut

SOME OF THE IMPORTANT BIOGENIC AMINES

1. **Tyramine**
   Decarboxylation of tyrosine forms tyramine. This occurs in the gut as a result of bacterial action. Also this reaction takes place in kidney. The ***reaction is favoured by O_2-deficiency.*** In the presence of sufficient O_2, tissue deaminates tyrosine. Tyramine elevates blood pressure.
2. **Tryptamine**
   Mammalian kidney, liver and bacteria of gut can decarboxylate the amino acid, tryptophan to form the amine ***tryptamine***. Tryptamine also elevates blood pressure. Hydroxylation at 5-position produces 5-OH-tryptamine-5-HT (Serotonin).
3. **Decarboxylation of amino acids lysine and arginine**
   Amino acids Lysine and Arginine may undergo decarboxylation to corresponding "diamines" called ***cadaverine*** and ***Putrescine*** respectively, which are largely excreted in faeces, but are essentially non-toxic in amounts ordinarily formed.

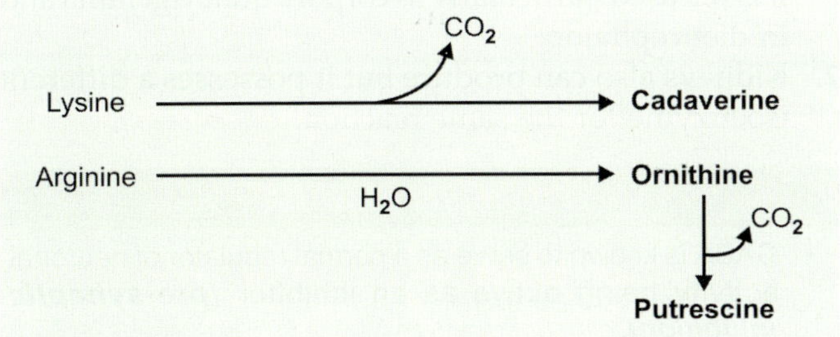

4. **Histamine**
   Histamine is formed by decarboxylation of amino acid "Histidine" by the enzyme ***Histidine decarboxylase*** or aromatic L-amino acid decarboxylase in presence of B_6-PO_4.

Site of Formation

- Mast cells are the chief source of histamine in the tissues and histamine constitutes about 10 per cent of the weight of mast cell granules.

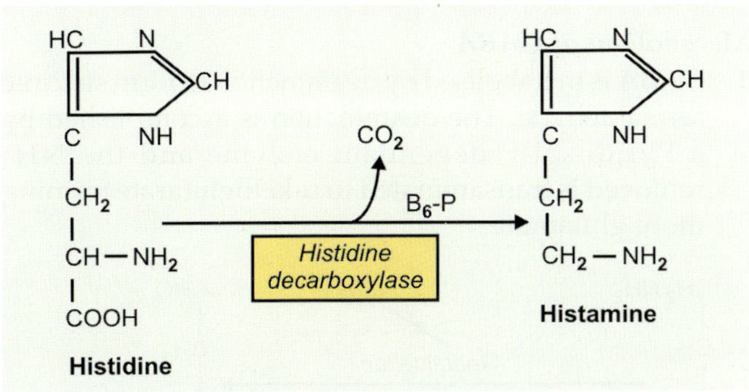

- Also produced by gastric mucosa cells and histaminergic neurones of the central nervous system.
- ***Basophils are the chief source of histamine in the circulating cells.***
- Also produced in the gut by bacterial decarboxylation of Histidine.

Storage: Other than in the enterochromaffin cells of the gastric mucosa virtually no histamine is stored in the tissues except for that found in mast cells.

Release

When histamine is packaged in the low pH environment of the granules, it is ionically bound to acidic groups on a heparin-protease matrix. Upon release of the granule contents and exposure to the higher extracellular pH, the histamine becomes freely soluble, because of its small size, it would be expected to diffuse rapidly from the site of release.

Mechanisms of action and effects: ***Histamine acts as a neurotransmitter, particularly in the hypothalamus.*** It acts as an anaphylactic and inflammatory agent on being released from mast cells in response to antigens. Effects of released histamine are ***mediated through 2 types of receptors designated as H_1 and H_2 receptors.*** Similarities and Differences between H_1 and H_2 receptor action are given in the box below.

Actions through H_1 receptors	*Actions through H_2 receptors*
• Contracts smooth muscle including airways and the GI tract	• Produces bronchodilation
• Increases venular permeability	• Increases vasopermeability and dilation
• Induces nasal mucus production	• Induces airway mucus production
• Causes pruritus, with cutaneous vasodilation	• Also causes pruritus with H_2 receptor, stimulates gastric acid secretion. HCl ↑ and pepsin ↑

SECTION FOUR

APPLIED CLINICAL ASPECT

Elevated plasma levels of histamine have been demonstrated in:

- Patients with anaphylaxis, provoked by exercise or antigen. Such reactions are related to the explosive liberation of histamine caused by entrance of the sensitising substances in the tissues.
- During spontaneous episodes of increased symptoms in patients with "mastocytosis", mast cells tumor.
- During experimentally induced angio-oedema in patients with cold urticaria.
- In patients with antigen-induced bronchial asthma.
- Also formed in injured tissues. Excessive liberation of histamine may be related to traumatic shock.
- Histamine markedly depresses blood pressure ↓ and large doses may cause extreme vascular collapse.
- After challenge by specific antigens in patients with 'atopy', histamine demonstrated in nasal lavage fluid and skin blister fluid.

Local Action of Histamine

Upon SC injection of hsistamine, it causes (i) pruritus, • erythema, (ii) circumferential flare and a central raised (iii) wheal ***(wheal and flare).***

Blockers of Histamine (Antihistaminics)

- ***Blockers of H_1 receptors:*** The anaphylactic reaction can be minimised by pharmacological agents, e.g. Promethazine and Mepyramine which block H_1 receptors.
- ***Blockers of H_2 receptors:*** 'Cimetidine' is used to reduce the gastric acidity in peptic ulcer patients, it is blocker of H_2 receptor.

Metabolism of Histamine

- Histamine is oxidatively deaminated to "β-imidazole-acetaldehyde" in kidneys by the enzyme ***Histaminase (Diamine oxidase)*** using B_6-PO_4 as a coenzyme.
- β-imidazole-acetaldehyde is further oxidised in Liver by ***aldehyde oxidase (aldehyde dehydrogenase)*** to form β-imidazole acetic acid which is excreted in urine after conjugation with ribose.

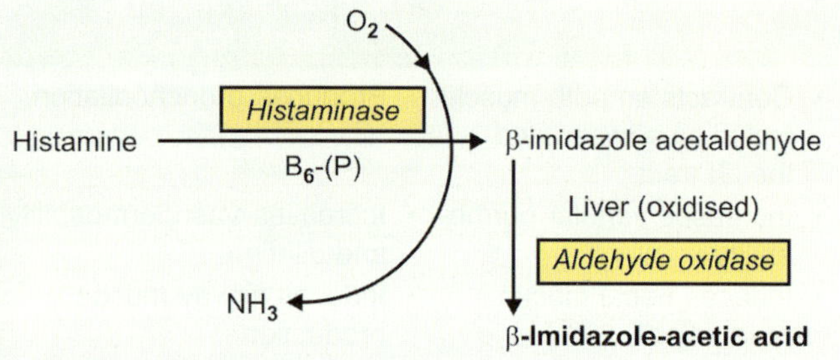

- Alternatively, Histamine can be methylated in the Liver to form "N-methyl histamine", ***–CH_3 group donor is "active" methionine*** and the enzyme catalysing is ***methylferase.***
- N-methyl histamine undergoes same fate as stated above and excreted in urine as "N-methyl-β-imidazole acetate". (Refer to box below).

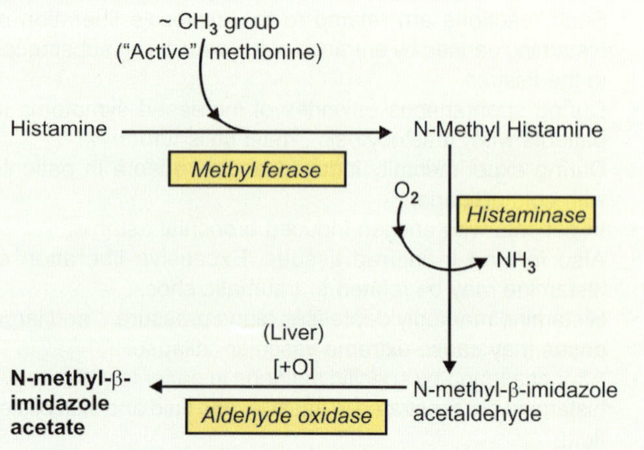

5. γ-aminobutyric Acid (GABA) formation:

Decarboxylation of glutamic acid produces γ-aminobutyric acid (GABA).

- Reaction is ***irreversible***
- glutamate *α-decarboxylase* is the enzyme which catalyses the reaction.
- It requires B_6-PO_4 as coenzyme and Mg^{++} as cofactor.

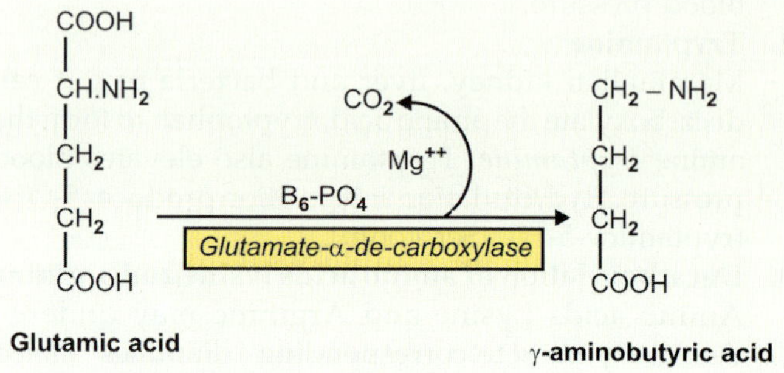

Site of formation

1. ***Principally formed in CN system in the gray matter.*** It is released particularly in corpora quadrigemina and in diencephalon.
2. Kidneys also can produce but it possesses a different isoenzyme for the same reaction.

Functions of GABA

- GABA is known to serve as a normal regulator of neuronal activity being active as an inhibitor ***(pre-synaptic inhibition).***
- It is released at the axonterminals of neurons in grey matter and acts as inhibitory neurotransmitter by enhancing K^+ permeability of postsynaptic membranes.

CLINICAL ASPECT

Vit B_6 deficiency in children may be responsible for some of the cases of infantile convulsions. B_6-deficiency causes less formation of GABA leading to neuronal hyperexcitability and convulsions.

Metabolism of GABA

1. GABA is metabolised by deamination to form succinic semialdehyde. The deamination is accomplished by a Pyridoxal-P dependant enzyme and the NH_3 removed is transaminated to α-ketoglutarate forming more glutamate.

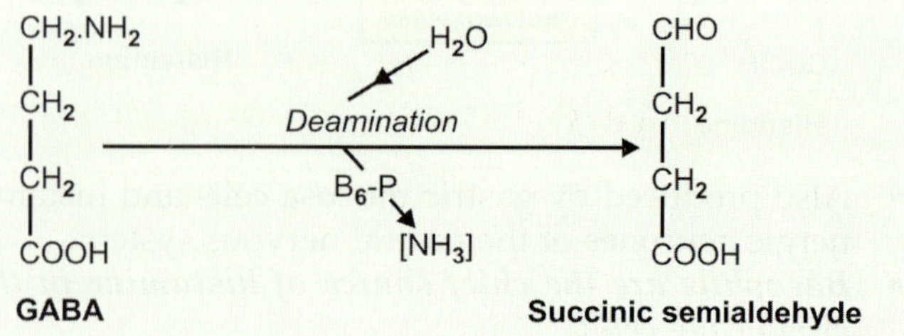

2. ***Succinic semialdehyde*** thus formed has **two fates:**
   - Either it is oxidised to succinate, the reaction is catalysed by the enzyme *Succinic semialdehyde dehydrogenase* using NAD^+ as H-acceptor, or
   - It is reduced to γ-OH butyrate by the enzyme *lactate dehydrogenase* (LDH) using NADH as H-donor (Refer to box below).

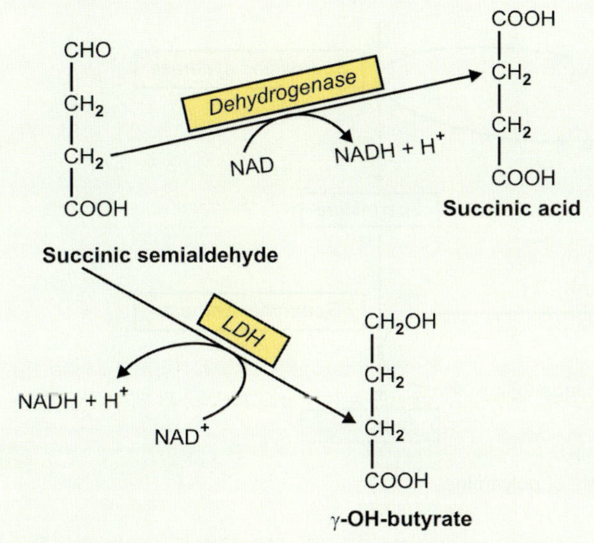

GABA SHUNT

GABA by its conversion to succinic acid **can form a "by-pass" in TCA cycle** and this is called as GABA-shunt (shown in the box below).

6. Polyamines

Types of Polyamines are:

- **Spermidine**
- **Spermine**

Ornithine in addition to its role in urea cycle, serves as the precursor of ubiquitous mammalian and bacterial polyamines, spermidine and spermine. It requires 'active' methionine. Normal human can synthesise approx 0.5 n mol of spermine/day.

Synthesis of Polyamines

Steps in synthesis are as follows:

1. Ornithine on decarboxylation forms **Putrescine** by the enzyme *ornithine decarboxylase.*
2. 'Active' methionine on decarboxylation by "S-adenosyl *methionine decarboxylase"* produces decarboxylated S-adenosyl methionine, which reacts with Putrescine in presence of *Spermidine synthase* forming **spermidine.**
3. The enzyme *spermine Synthase* in presence of another molecule of decarboxylated S-adenosyl methionine then forms **spermine.** Synthesis is shown schematically in **Figure 27.6** (next page).

Regulation: Stimulation/Inhibition of Enzymes

1. Ornithine decarboxylase and S-Adenosyl methionine decarboxylase both are ***inducible*** enzymes with short half-lives.
2. Spermidine synthase and spermine synthase are not inducible. Also not labile.
3. Hormones like growth hormone (GH), corticosteroids, testosterone and epidermal growth factor (EGF) increases *ornithine decarboxylase* activity.
4. The activity of *S-adenosyl methionine decarboxylase* is inhibited by decarboxylated S-adenosyl methionine and activated by Putrescine.

Degradation of Polyamines

1. The enzyme ***Polyamine oxidase*** present in liver peroxisomes, oxidises spermine to spermidine.

Degradation of Polyamines

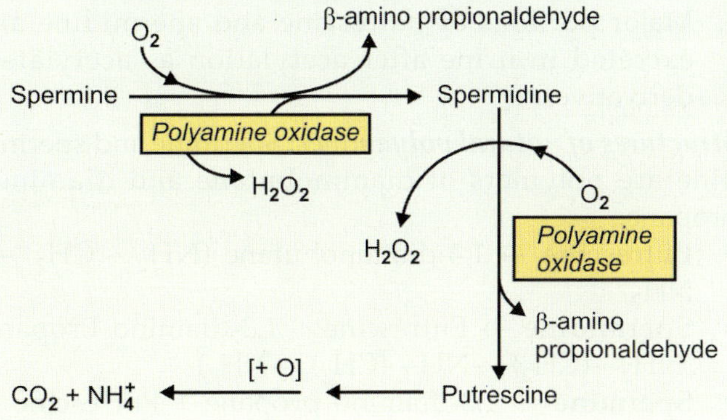

GABA Shunt

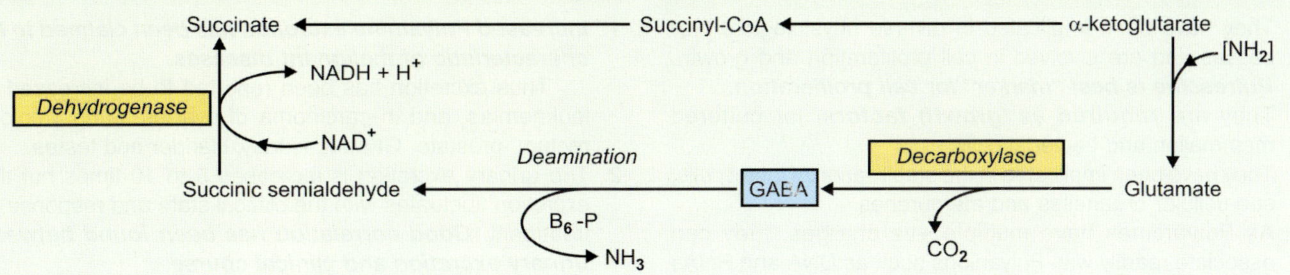

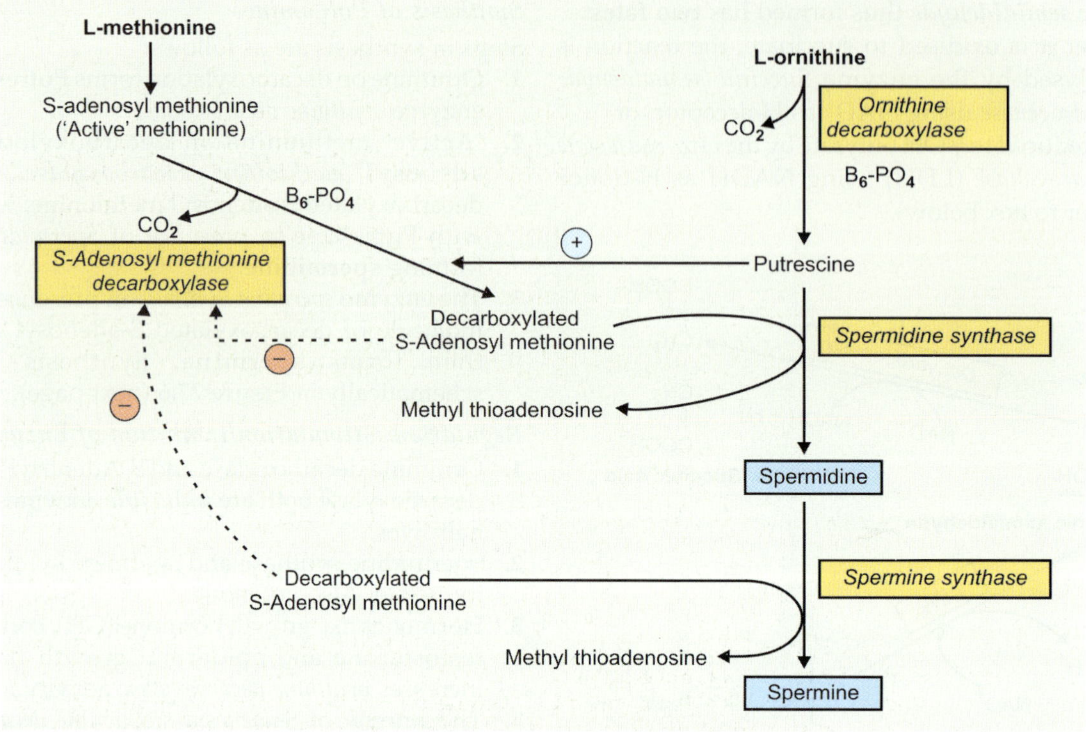

Fig. 27.6: Showing synthesis of polyamine

2. Spermidine is then again undergoes oxidation by the same enzyme to form putrescine.
   Both "di-amino propane" moieties are converted to "β-amino Propionaldehyde".
3. ***Putrescine is finally oxidised to NH_4^+ and CO_2 by mechanisms not known exactly.***
4. Major portions of putrescine and spermidine are excreted in urine after acetylation as acetylated derivatives.

Structures of natural polyamines: Spermine and spermidine are polymers of diaminobutane and diaminopropane.

- **Putrescine** → 1,4-diaminobutane (NH_2—$(CH_2)_4$—NH_2)
- **Spermidine** → Putrescine + 1,3-diamino Propane (NH_2—$(CH_2)_3$—NH—$(CH_2)_4$—NH_2)
- **Spermine** → 1,3-diamino propane + Putrescine + 1,3-diaminopropane. (NH_2-$(CH_2)_3$-NH-$(CH_2)_3$-NH_2)

FUNCTIONS OF POLYAMINES

- They have been implicated in diverse physiological processes and are involved in cell proliferation and growth. ***Putrescine is best "marker" for cell proliferation.***
- They are ***required as 'growth factors'*** for cultured mammalian and bacterial cells.
- They have been implicated in the stabilisation of intact cells, sub-cellular organelles and membranes.
- As Polyamines have multiple +ve charges, they can associate readily with Polyanions such as DNA and RNAs and have been implicated in such fundamental processes as stimulation of DNA and RNA biosynthesis, DNA stabilisation and packaging of DNA in bacteriophages.
- Polyamines also exert diverse effects on protein synthesis.
- They act as inhibitors of enzymes that include *Protein kinases.*
- Polyamines added to cultured cells induce synthesis of a ***protein antienzyme*** that binds to ornithine decarboxylase and inhibits putrescine formation.
- ***Spermidine has been claimed to be best "marker" of tumor cell destruction.***
- In Pharmacologic dosage Polyamines have been found to be **hypothermic** and **hypotensive**.

Ranges of Normal Excretion of Polyamines: The polyamines are excreted normally in urine as conjugates.

Normal values:

- **Putrescine** : 2.7 ± 0.5 mg
- **Spermine** : 3.4 ± 0.7 mg
- **Spermidine** : 3.1 ± 0.6 mg

CLINICAL SIGNIFICANCE

1. ***Increased Polyamine excretion has been claimed to be characteristic of malignant diseases.***
   Thus excretion has been reported to be increased in leukaemias, and in carcinoma of ovaries, Lungs, colon, rectum, prostate, GI tract, kidney, bladder and testes.
2. The urinary excretion is increased 5 to 10 times but the excretion fluctuates with the clinical state and response to treatment. ***Good correlation has been found between urinary excretion and clinical course.***

3. However, not all patients with malignant diseases exhibit increased excretion of Polyamines.
4. **Russell *et al*** have postulated that ***spermidine is the best "marker" of tumour cell destruction, whereas putrescine is the best 'marker' for cell proliferation.***

POLYAMINE SYNTHESIS INHIBITOR

DFMO

Chemically it is ***α-difluoromethyl ornithine.*** Also known as ***Eflornithine***. It is a strong inhibitor of polyamine synthesis-an example of ***suicide inhibition***.

CLINICAL ASPECTS

- **Treatment of cancer:** DFMO was initially developed for cancer treatment, but ***failed to have malignancies.*** A recent study (2008) showed that when the drug is combined with an anti-inflammatory drug, it significantly reduces the risk of recurring colorectal polyps.
- DFMO is found to be ***highly effective in African trypanosomiasis ("sleeping sickness"). DFMO appears to kill Tripanosomas by acting as a "suicide inhibitor" of the enzyme "ornithine decarboxylase".*** This enzyme regulates cell division by catalysing the first step in polyamine synthesis. As the inhibitor has a low half life in humans, it is broken down quickly while the parasite cannot metabolise it quickly enough. This means ***that is preferentially harms the parasite.*** DFMO's effects against *Trypanosoma gambiense* were discovered by chance and because of its ability to bring patients back from come, it became known as ***resurrection drug.***
- DFMO has also been found to be an effective ***hair growth inhibiting agent***. As a topical application the drug has been shown to be an effective hair growth retardant in some patients. ***It may be particularly effective in postmenopausal women.***

METABOLISM OF INDIVIDUAL AMINO ACIDS

Two groups will be discussed in detail:

A. Metabolism of Aromatic amino acids, viz. phenyl alanine and Tyrosine.

B. Metabolism of Sulphur-containing amino acids, viz. methionine, cysteine.

The other amino acids will be discussed in brief but stress will be given on metabolic role in the body.

A. AROMATIC AMINO ACIDS

Structures of Phenylalanine and tyrosine are given below:

Difference in structure:

Tyrosine possesses an additional –OH group at para position of benzene ring.

Phenyl alanine (Phe)

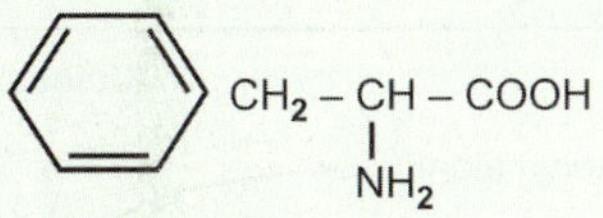

α-amino-β-Phenyl Propionic acid

Tyrosine (Tyr)

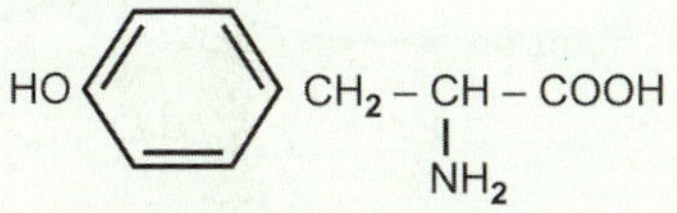

α-amino-β-(p-OH) Phenyl Propionic acid

POINTS TO REMEMBER

- Phenyl alanine is nutritionally an ***essential amino acid.*** It cannot be synthesised in humans, hence must be provided in diet.
- ***Tyrosine is not essential,*** as it can be formed in the body from Phenyl alanine.
- Phenyl alanine is readily converted to Tyrosine, **but the reaction is NOT reversible.**
- The feeding of tyrosine decreases the need of phenyl alanine in the diet **(sparing action).**
- In phenyl ketonuric patient, where phenyl alanine cannot be converted to tyrosine in the body due to inherited deficiency of the enzyme, ***tyrosine becomes essential amino acid to the patient.***
- Both amino acids are ***glucogenic*** and ***ketogenic.***
- Both can participate in transamination reaction.

A. Metabolic fate

This can be discussed as:

(a) Conversion of phenyl alanine to tyrosine and

(b) Metabolic fate of tyrosine.

(a) Conversion of phenyl alanine to tyrosine: The reaction involves hydroxylation of phenyl alanine at p-position in benzene ring.

Enzyme: Phenyl alanine hydroxylase. Present in liver and the conversion occurs in Liver.

Coenzymes and cofactors: The enzyme requires the following for its activity:

- **Molecular oxygen**
- **NADPH**
- **Fe^{++} and**
- **Pteridine (folic acid) coenzyme: Tetrahydrobiopterin-FH_4**

The reaction is complex and takes place in ***two stages*** as shown below:

I. Reduction of O_2 to H_2O and conversion of phenyl-alanine to tyrosine. Reduced form of pteridine, FH_4 acts as H-donor to the molecular O_2.

II. Reduction of dihydrobiopterin, FH_2 by NADPH, catalysed by the enzyme *Dihydrobiopterin reductase.*

Conversion of Phenyl Alanine to Tyrosine

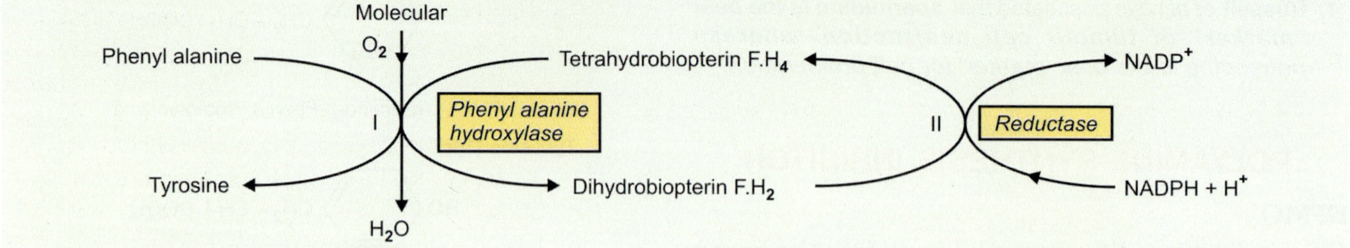

The overall reaction in reaction I involves incorporation of one atom of molecular O_2 into the p-position of phenyl alanine while the other atom of O_2 is reduced to form H_2O.

(b) Metabolic Fate of Tyrosine:

1. Tyrosine is degraded to produce as end products 'Fumarate' and 'acetoacetate'.
2. ***Fumarate is glucogenic***, whereas ***acetoacetate is ketogenic***.
3. Phenyl alanine is catabolised via tyrosine. ***Hence both Phenyl alanine and tyrosine are glucogenic and ketogenic.***

Reaction sequences: Five sequential enzymatic reactions are necessary for complete degradation.

1. ***Conversion of tyrosine to p-OH phenyl pyruvic acid (PHPPA):*** Tyrosine undergoes transamination and forms p-OH-phenyl Pyruvate. The reaction is catalysed by the enzyme ***Tyrosine-α-ketoglutarate transaminase*** which requires B_6-PO_4 and ascorbic acid.

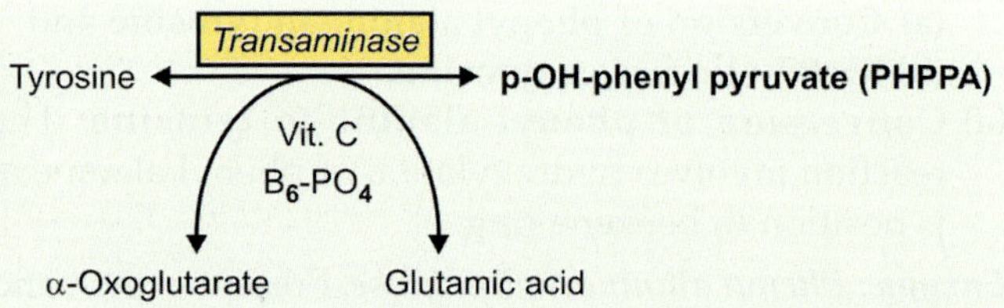

2. ***Conversion of p-OH phenyl pyruvate to homogentisic acid:*** The reaction is catalysed by the enzyme ***p-OH-phenyl pyruvate oxidase,*** which is a Cu-containing metalloenzyme. It requires ascorbic acid (Vit C) and vit B_{12}.

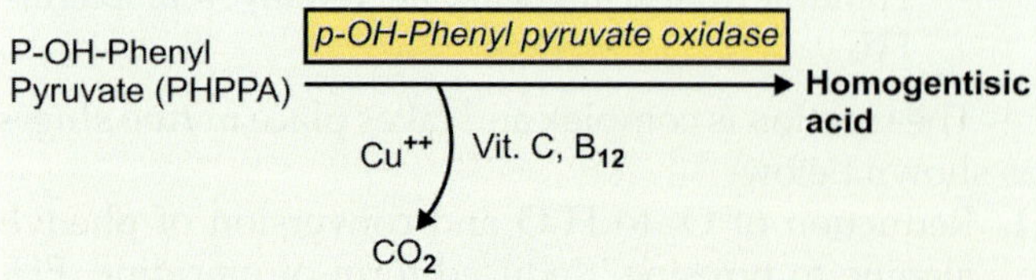

Although the reaction appears to involve hydroxylation of PHPPA at ortho-position accompanied by oxidative loss of the carboxyl carbon, ***it actually involves migration of the side chain. Ring hydroxylation and side chain migration occur in a concerted manner.***

3. ***Conversion of homogentisate to maleyl acetoacetate:*** The oxidative reaction is catalysed by the enzyme ***Homogentisate oxidase,*** an Fe-containing metalloenzyme present in Liver. The benzene ring of Homogentistic acid is ruptured forming **"Maleyl acetoacetate"**.

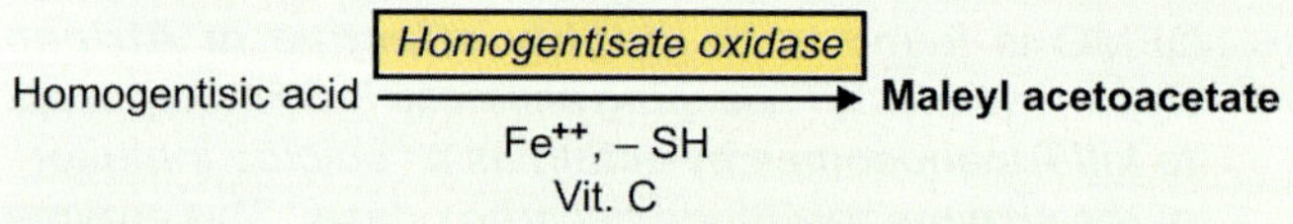

Inhibitor

α-α′-dipyridil: A chelating agent can strongly bind to Fe^{++} of the enzyme and inhibits the reaction. Thus homogentisic acid accumulates and appears in urine ***(Experimental alkaptonuria).***

4. ***Conversion of maleyl acetoacetate to fumaryl acetoacetate:*** A *cis-trans* isomerisation about the double bond, is catalysed by ***Maleyl acetoacetate cis-trans isomerase*** an –SH enzyme present in liver.

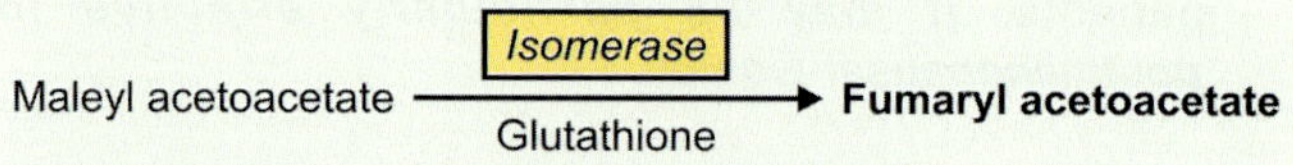

5. ***Hydrolysis of fumaryl acetoacetate:*** Hydrolysis of Fumaryl acetoacetate is catalysed by the enzyme ***Fumaryl acetoacetate hydrolase*** and forms fumarate

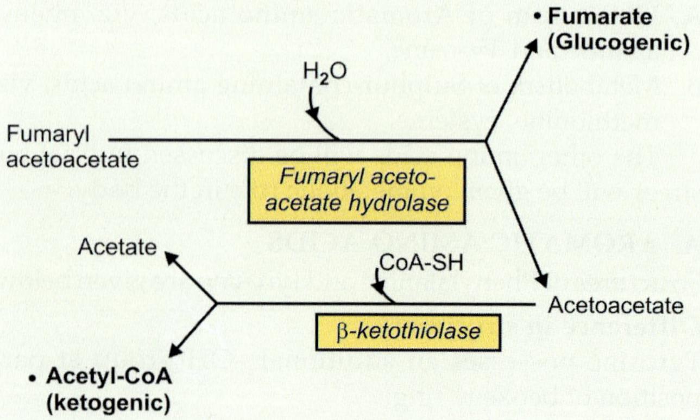

and acetoacetate. Acetoacetate can then be converted to acetyl-CoA and acetate by the β-*ketothiolase* reaction.

B. Metabolic Role of Tyrosine

Tyrosine though it is 'dispensible' (non-essential amino acid), but it is of great importance in human body. ***Many biological compounds of importance are synthesised from tyrosine.*** They are shown in the box below.

Metabolic role of Tyrosine

- ***Synthesis of thyroid hormones: Thyroxine (T_4) and triiodothyronine (T_3)***
- ***Synthesis of catecholamines***
- ***Synthesis of melanin pigment***
- ***Formation of tyramine***
- ***Formation of phenol and cresol***
- ***Formation of tyrosine-O-sulphate***

1. ***Synthesis of thyroid hormones:*** Refer to Chapter on Hormones and Metabolic Role.
2. ***Biosynthesis of catecholamines:*** Catecholamines are: Epinephrine (adrenaline), norepinephrine (noradrenaline) and Dopamine. ***All three are synthesised from tyrosine.*** Epinephrine and norepinephrine are hormones produced by adrenal medulla and dopamine is precursor. ***All three act as "neurotransmitters"***. Epinephrine and norepinephrine are released at axon terminals of adrenergic sympathetic fibres. Dopamine is produced in nerve terminals particularly in hypothalamus and diencephalon.

Steps of synthesis

Tyrosine is **taken up actively** by cells of adrenal medulla pheochromocytes and neuroglial cells.

(a) Conversion of tyrosine to DOPA (In mitochondrion)

Tyrosine enters the mitochondrion where it is hydroxylated to form 3,4-dihydroxy phenylalanine (DOPA) catalysed by the enzyme ***Tyrosine hydroxylase.*** It requires tetrahydrobiopterin (F. H_4) as hydrogen donor and Fe^{++}. Tyrosine hydroxylase, also called tyrosinase is a Cu^{++}-containing metallo-enzyme. It is the ***rate-limiting*** enzyme in biosynthetic pathway. It also requires NADPH. By donating hydrogen to one atom of O_2 molecule to produce H_2O, the F.H_4 is converted to F. H_2, which is immediately reduced to F. H_4 by NADPH in presence of the enzyme *Dihydrobiopterin reductase*.

Note: The reaction is similar to conversion of phenylalanine to tyrosine.

Conversion of Tyrosine to DOPA

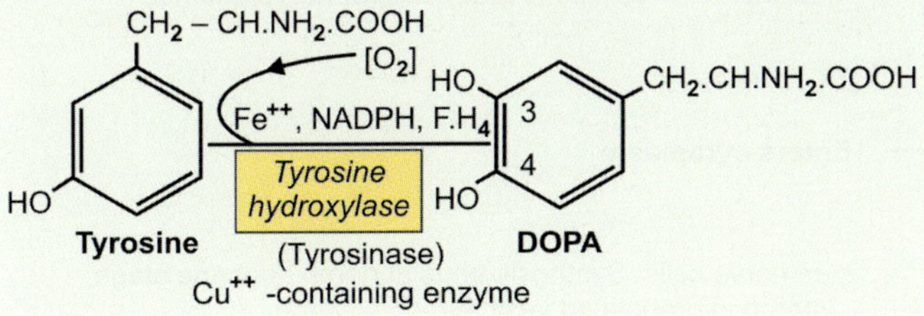

(b) Conversion of DOPA to dopamine (In cytoplasm)

DOPA comes out of mitochondrion into cytosol, where it is ***decarboxylated to Dopamine.*** The enzyme catalyzing the reaction is *Dopa-decarboxylase (Aromatic-L-amino acid decarboxylase)* and requires B_6-PO_4.

(c) Conversion of dopamine to norepinephrine (In granules/vesicles)

Dopamine from cytosol enters Chromaffin granules of Pheochromocytes or granulated vesicles of brain cells or nerve endings. Dopamine is hydroxylated to Norepinephrine by the enzyme *Dopamine-β-oxidase,* a Cu^{++}-containing enzyme. Vit C is requuired for the reaction. The Cu^{+} ion of the enzyme is oxidised to Cu^{++} during the reaction and must be reduced back to Cu^{+} by vit C.

(d) Conversion of Nor-epinephrine to epinephrine (In cytosol)

Nor-epinephrine comes out of the chromaffin granules into cytosol, where it is ***methylated. CH_3 group is donated by "active" methionine (S-adenosyl methionine)*** and the enzyme catalysing the reaction is ***phenylethanolamine N-methyl transferase. This reaction does not take place in nerve cells, where synthesis stops at Nor-epinephrine stage.*** Epinephrine after synthesis in cytosol moves back to chromaffin granules, where it is stored.

Synthesis of catecholamines is shown schematically in Figure 27.7.

Conversion of DOPA to Dopamine

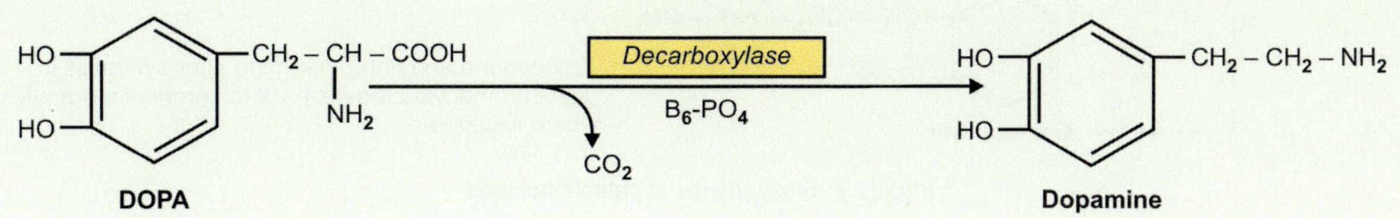

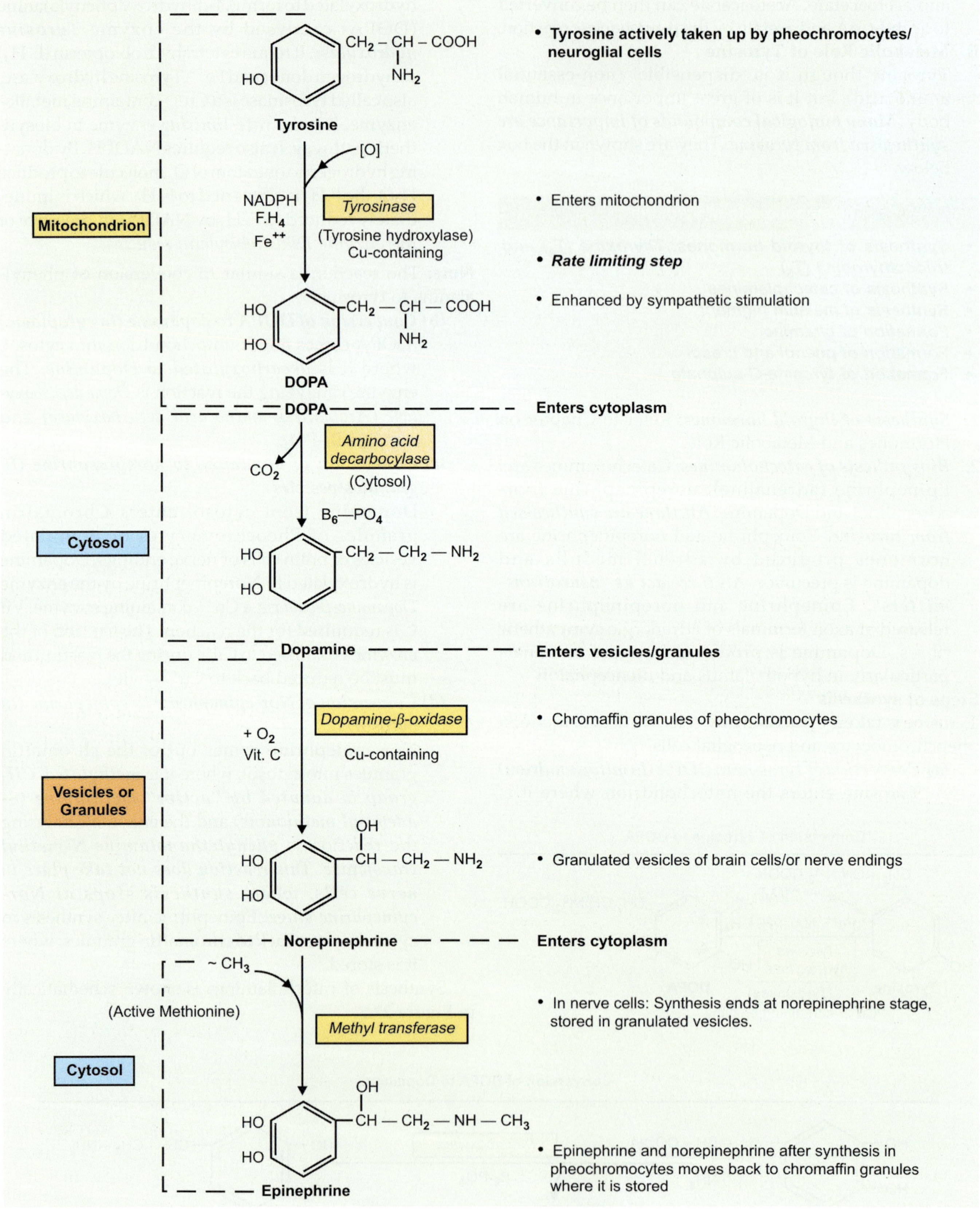

Fig. 27.7: Biosynthesis of catecholamines

Conversion of Dopamine to Norepinephrine

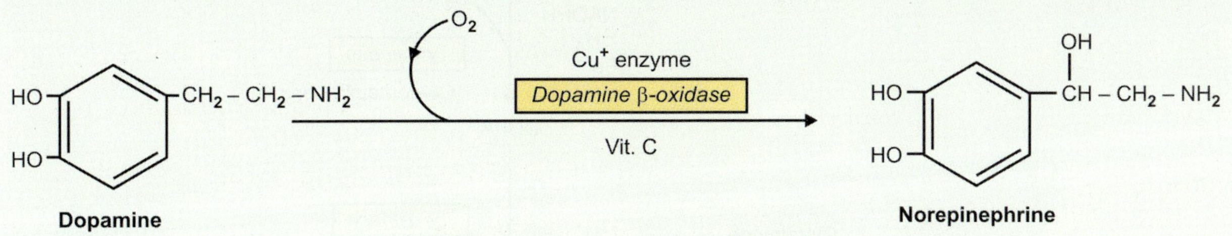

Conversion of Norepinephrine to Epinephrine

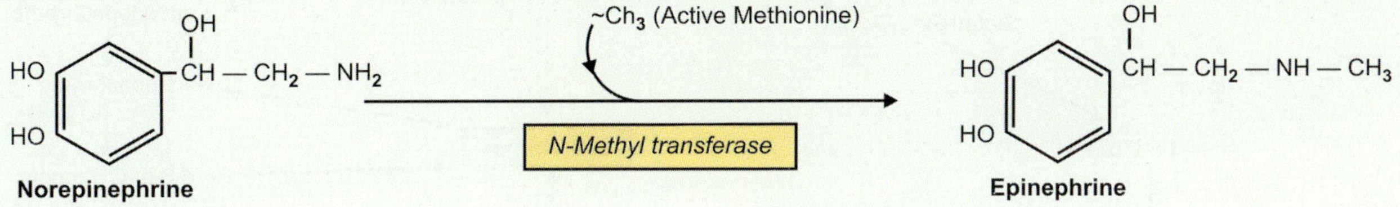

3. ***Synthesis of Melanin***

   Melanins are synthesised from tyrosine in **melanosomes**, membrane bound particles within melanocytes in skin which are cells of neural crest origin.

Types of melanins

- ***Eumelanins:*** These are insoluble, heterogenous, high molecular weight, black to brown **heteropolymers** of 5, 6-dihydroxy indole and several of its biosynthetic precursors, viz. Leucodopachrome and Dopachrome.
- ***Pheomelanins:*** These are yellow to reddish-brown polymers, though of high molecular weight are soluble in dilute alkali. ***They contain sulphur.***
- ***Trichochromes:*** These low molecular weight compounds, ***contains sulphur*** and are related to pheomelanins.

Both pheomelanins and Trichochromes are derived from cysteine and Dopaquinone. Pheomelanins and trichochromes are present primarily in hairs and feathers. Melanins have complex chemical structures and they are heteropolymers. They complex with proteins of the melanosomal matrix, forming melanoprotein.

STEPS OF BIOSYNTHESIS

Biosynthesis is complex and not fully understood. The biosynthetic pathway is shown diagrammatically in Figure 27.8.

1. Initial reactions in melanin biosynthesis involve oxidations by *Tyrosinase,* a phenol mono-oxygenase. It is Cu^+-containing mixed function oxidase, which produces first DOPA, the same enzyme converts it to **DOPA quinone.**
2. **Dopa-Quinone is starting compound from which:**
   - Non-sulphur containing **Eumelanin** is formed. Dopa quinone undergoes a series of fast spontaneous reactions in which an "indole" ring is formed and decarboxylation of COOH group occurs producing "5, 6-di-OH indole" and "indole-5,6 Quinone". An intermediate compound is ***Dopachrome*** (or ***Hallochrome***) a red Pigment. ***The Quinones Polymerise to form Eumelanin.***
   - Dopa-Quinone can react with Glutathione and after a series of changes form sulphur containing **Trichochromes** and **Pheomelanins**.
4. ***Formation of Tyramine:*** (Refer to biogenic amines).
5. ***Formation of Phenol and Cresol***

   Phenylalanine (through tyrosine) and Tyrosine are acted upon by intestinal bacteria in the gut to form p-cresol and Phenol. These are absorbed from the gut and conjugated in Liver with H_2SO_4 and D-Glucuronic acids and are excreted in urine.

   Phenol and p-cresol are formed from tyrosine in 2 ways:

   (i) Tyrosine is decarboxylated to form tyramine, which is reduced to p-cresol which is further demethylated to form phenol.

   (ii) Alternatively, tyrosine is deaminated first to form p-OH phenyl Propionic acid, which is oxidised to form "p-OH phenyl acetic acid", the latter is decarboxylated to form p-cresol and then demethylated to form phenol.

SECTION FOUR

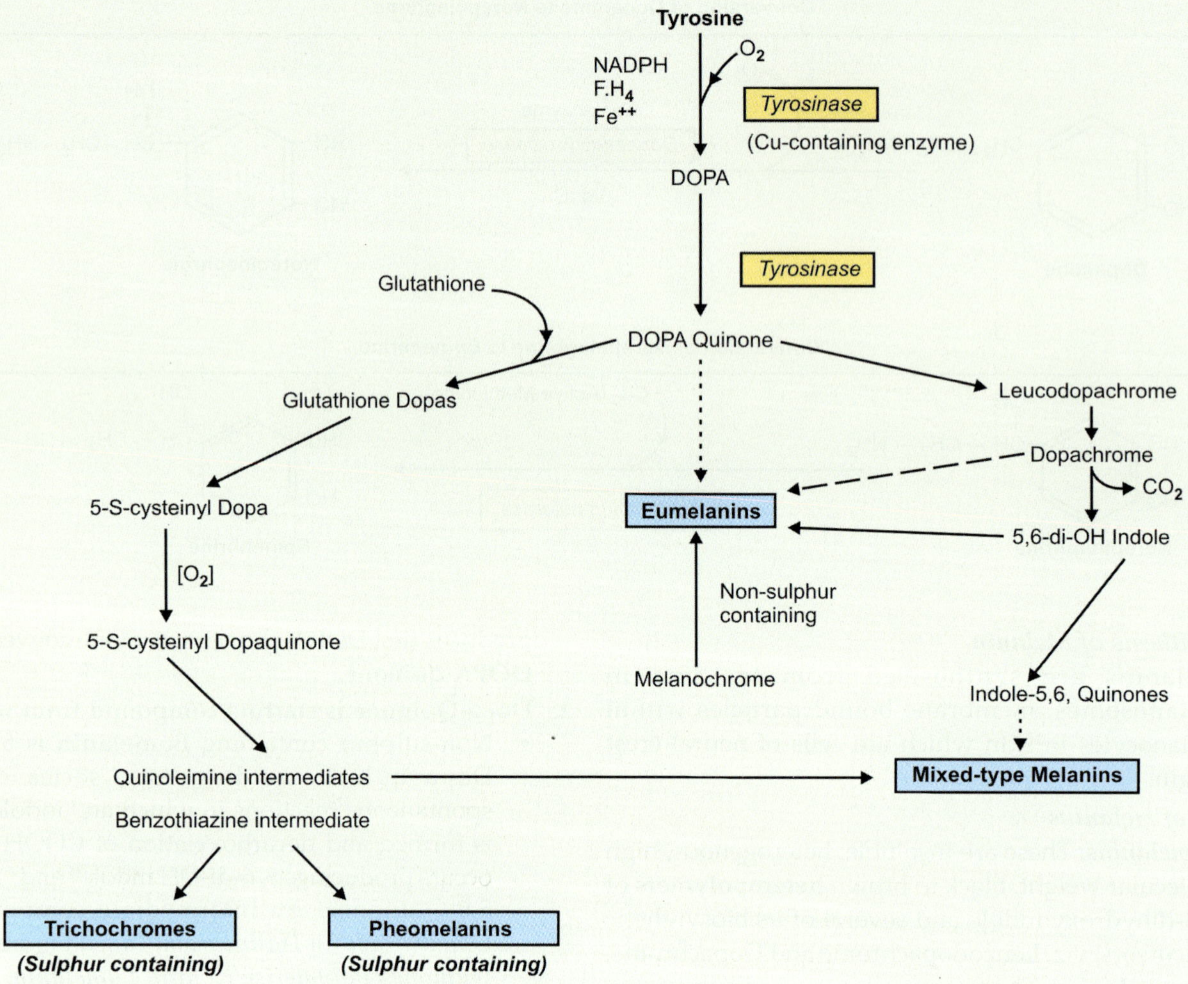

Fig. 27.8: Synthesis of melanins

Formation of p-cresol and Phenol is shown on next page in the box.

6. ***Formation of tyrosine-O-sulphate:*** Tyrosine-O-Sulphate is formed by sulphation by 'active' sulphate. This is present in fibrinogen molecule. During conversion of fibrinogen to fibrin two peptides are liberated—one of the peptides contains tyrosine-O-sulphate.

CLINICAL ASPECT

INHERITED DISORDERS

Following disorders are associated with Phenyl alanine and tyrosine metabolism.

1. Phenyl Ketonuria

Five types of hyperphenylalaninaemias have been described. The **five types** of hyperphenylalaninaemias along with the corresponding enzyme defects are shown in Table 27.2.

Classical type of phenyl ketonuria (PKU): An inherited disorder with incidence of 1 in 10,000 live births.

Enzyme deficiency

Phenyl alanine hydroxylase is absent.

Metabolic changes due to absence of phenylalanine hydrodylase:

Phenyl alanine cannot be converted to tyrosine, as a result alternative catabolites are produced. Phenyl alanine accumulates in the blood; phenyl alanine undergoes transamination to form phenyl pyruvic acid and its products as phenyl lactic acid and phenyl acetic acid are produced.

Phenyl acetic acid is conjugated with glutamine and excreted as phenyl acetyl glutamine in urine (responsible for "mousy odour" of urine).

Accumulation of phenylalanine leads to

- Defective "serotonin" formation.
- Also impairs melanin synthesis, Children with the defect tend to have flair skin and fair hair.

Formation of Phenol and p-cresol

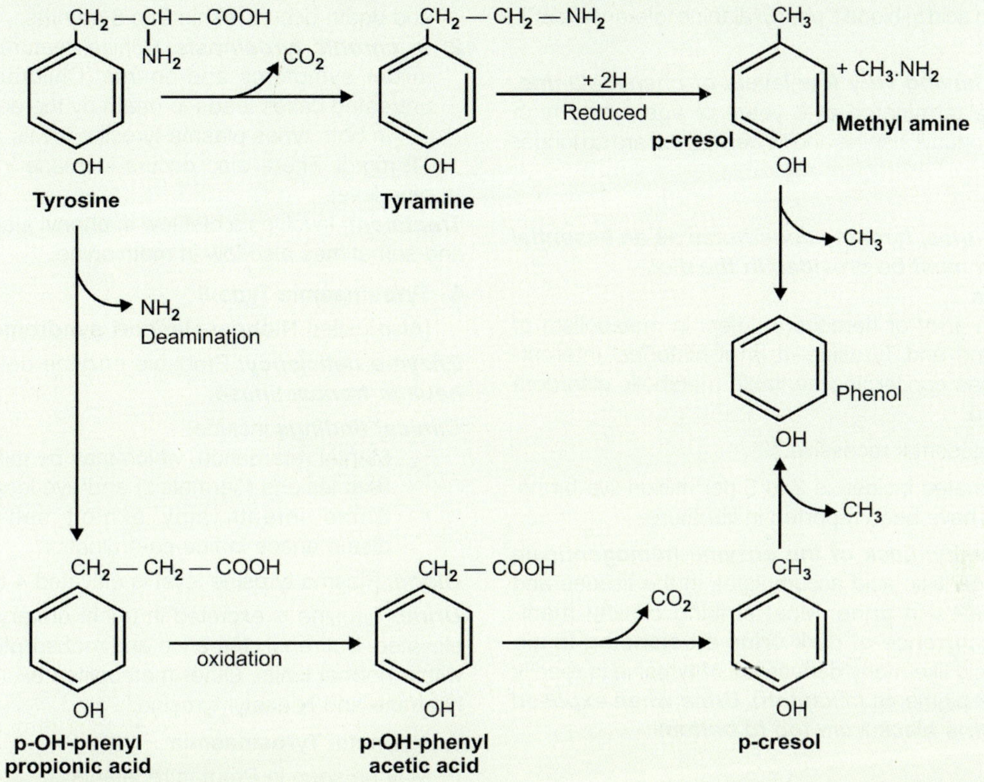

Table 27.2: Hyperphenylalaninaemias

Type	Condition	Probable enzyme defect	Treatment
I.	**Classical type of phenyl ketonuria (PKU)**	Phenyl alanine hydroxylase enzyme absent	Low phenyl alanine diet
II.	**Persistent hyperphenylalaninaemia**	Decreased Phenyl alanine hydroxylase enzyme	None but temporary dietary therapy
III.	**Transient mild hyperphenylalaninaemia**	Maturational delay of phenyl alanine hydroxylase enzyme	Same as Type II
IV.	**Dihydropteridine reductase deficiency**	Deficient or absent *dihydropteridine reductase*	Dopa, 5-OH tryptophan, carbi Dopa
V.	**Abnormal dihydrobiopterin function**	Dihydrobiopterin synthesis defect	Same as Type IV

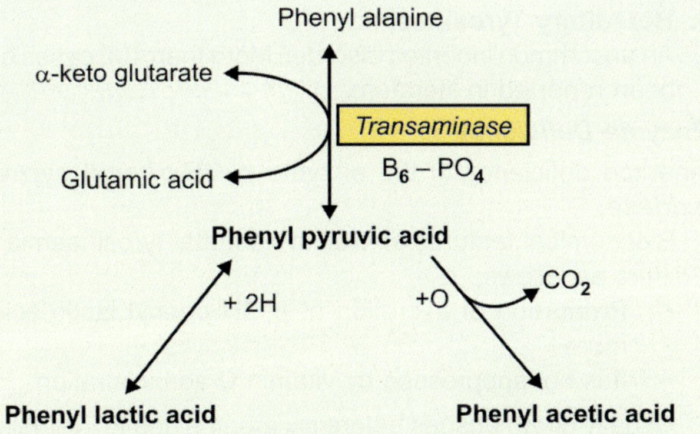

- Excess of phenyl alanine in blood leads to excretion of this amino acid into the intestine. Here it competes with tryptophan for absorption. Tryptophan becomes subject to action of intestinal bacteria resulting in formation of indole derivatives which are absorbed and excreted in urine.

Clinical features: Child is ***mentally retarded***, other features include seizures, Psychoses and eczema.

Blood: Increased levels of phenyl alanine. Normal level in blood is 1 to 2 mg%. It increases to 15 to 65 mg%.

Urine

- Excretion of phenyl alanine, and its catabolites: Phenyl pyruvic acid, and phenyl lactic acid.
- Phenyl acetic acid excreted as phenyl acetyl glutamine which produces ***'mousy' odour.***
- Also abnormal O-hydroxy derivative is formed, whose metabolites may also be found in urine.

Diagnosis

- By estimation of plasma Phenyl alanine level.
- ***"Screening test":*** for presence of phenyl pyruvate with $FeCl_3$ (In urine).

- Administration of phenyl alanine to a phenylketonuric patient should result in prolonged elevation of the level of this amino acid in blood ("phenyl alanine tolerance test").

Treatment

By giving diet having very low levels of phenyl alanine. The diet can be terminated at 6 years of age, when high concentration of phenyl alanine and its derivatives are no longer injurious to brain.

Note

In phenylketonurias, tyrosine constitutes as an essential amino acid and must be provided in the diet.

2. Alkaptonuria

A rare inborn error or hereditary defect in metabolism of Phenyl alanine and Tyrosine. It is of historical interest- **Garrod's** ideas concerning heritable metabolic disorders was proposed.

Inheritance: Autosomal recessive.

Incidence: Estimated incidence 2 to 5 per million live births. Over 600 cases have been reported in literature.

Enzyme deficiency: ***Lack of the enzyme homogentisate oxidase.*** Homogentisic acid accumulates in the tissues and blood and appears in urine. Most striking clinical manifestation—**is occurrence of dark urine on standing in air.** Homogentisic acid like many derivatives of tyrosine is readily oxidised to black pigments ***(alkapton). Urine when exposed to air slowly turns black from top to bottom.***

Ochronosis

In long-standing cases, deposition of homogentisic acid derivatives in cartilages of ears and other exposed parts leads to generalised pigmentation of connective tissues and deposition in joints leads to arthritis, a condition called ***Ochronosis***.

Mechanism of Ochronosis

The precise mechanism of ochronosis is not known. It probably involves oxidation of homogentisate by ***Polyphenol oxidase,*** forming benzoquinone acetate, which polymerizes and binds to connective tissue macromolecules.

Note

- Alkaptonuria may also occur in premature infants on account of vitamin C deficiency. Condition is improved in premature infants by administration of vit. C.
- Experimental alkaptonuria may be produced by "α-α' dipyridyl"-chelating agent of Fe.

3. Tyrosinaemia Type I

(Also called ***Tyrosinosis***) A rare inherited disorder. Tyrosinosis is characterised by accumulation of metabolites that adversely affect the activities of several enzymes and transport systems. Pathophysiology of this disorder is complex.

Enzyme deficiency: There is lack of the enzyme *Fumaryl acetoacetate hydrolase* and possibly also *Maleyl acetoacetate isomerase.*

Types: Both acute and chronic forms are known.

1. ***In acute tyrosinosis:*** Infants exhibit diarrhoea, vomiting, a "cabbage"-like odour. They do not thrive well, and there is usually associated Liver damage. ***Infants die from liver failure.*** Untreated acute tyrosinosis cases do not survive and death occurs within 6 to 8 months.
2. ***In chronic tyrosinosis:*** Clinical features are similar but milder symptoms and course. Children survive and in untreated cases leads to death by the age of 10 years.

   In both types plasma tyrosine levels are elevated: 6 to 12 mg/dl. There also occurs increase in plasma methionine level.

Treatment: Involves a diet low in phenyl alanine and tyrosine and sometimes also low in methionine.

4. Tyrosinaemia Type II

(Also called **Richnar-Hanhart syndrome**)

Enzyme deficiency: Probable enzyme defect is the lack of ***hepatic transaminase.***

Clinical findings include:

- Mental retardation, which may be mild to moderate.
- Skin lesions (dermatitis) and eye lesions.
- Some infants may exhibit self-mutilation and disturbances in fine co-ordination.

Blood: Plasma tyrosine level is elevated 4 to 5 mg/dl.

Urine: Tyrosine is excreted in urine, urinary concentration is elevated. But renal clearance and reabsorption of tyrosine fall within normal limits. Other metabolites excreted in urine are tyramine and N-acetyl tyrosine.

5. Neonatal Tyrosinaemia

May be seen in Premature infants.

Enzyme deficiency: Relative deficiency of p-OH-*Phenyl pyruvate hydroxylase.* There may be transient deficiency also, which may be the delay in maturation of the enzyme.

- Blood levels of tyrosine and phenyl alanine are elevated.
- Urinary excretion of the following increases: Increased excretion of tyrosine, tyramine, p-OH-Phenyl acetate, and N-acetyl tyrosine.

  Treatment: Involves feeding a diet low in protein, specially with low phenyl alanine and tyrosine.

  In premature infants: Administration of vitamin C may improve the condition.

6. Hereditary Tyrosinaemia

An uncommon inherited disorder. More than 100 cases have been reported in literature.

Enzyme Deficiency

Inherited deficiency of the enzyme ***p-OH-phenyl pyruvate oxidase.***

Biochemical features similar to neonatal tyrosinaemia but it differs as follows:

- Proportion of excretion of p-OH phenyl lactic acid is more
- It is not suppressed by vitamin C administration.

Usually death ensues before six months due to liver failure. Those who survive develops later on:

- ***Nodular cirrhosis,***
- Multiple defects in renal tubular reabsorption like proteinuria, aminoaciduria, and hyperphosphaturia,
- Tendency to develop hypoglycaemia,

- A number of patients develop hypophosphataemic rickets,
- Hepatic carcinoma may develop in some cases as a late life complication.

7. Albinism

It includes a spectrum of clinical syndromes characterized by **hypomelanosis**, arising from inherited defects in the pigment cells (melanocytes) of eye and skin. There are various forms of the disease. But can be divided into **two major groups:**

(a) Oculocutaneous albinism: More than ten forms are known. In this there is decreased pigmentation of skin and eyes. They can be differentiated by clinical presentation and biochemical and other features. Such ***'albinos'*** can be biochemically of two types:

(i) 'Tyrosinase' negative albinos

(ii) 'Tyrosinase' positive albinos

Differentiating feature of these two types are presented in Table 27.3.

(b) Ocular albinism: Affects only eye and not the skin. Occurs both as autosomal recessive and as an X-linked trait.

Figure 27.9 shows a flow chart of Metabolic fate and Metabolic role of Phenylalanine and tyrosine.

Table 27.3: Differentiating features of tyrosinase-negative albinos and tyrosinase-positive albinos

"Tyrosinase"-negative albinos	*"Tyrosinase"-positive albinos*
1. Completely lacks visual pigment	1. They have some visual pigment. Hair colour ranges from white yellow to light tan Lightly pigmented naevi may be present.
2. Hair bulbs from these patients fail to convert added tyrosine to pigment *in vitro*	2. Hair bulbs from these patients may be able to convert added tyrosine to pigment "Eumelanin" *in vitro*
3. Melanocytes in these patients contain unpigmented melanosomes	3. Melanocytes in these patients contain lightly pigmented melanosomes

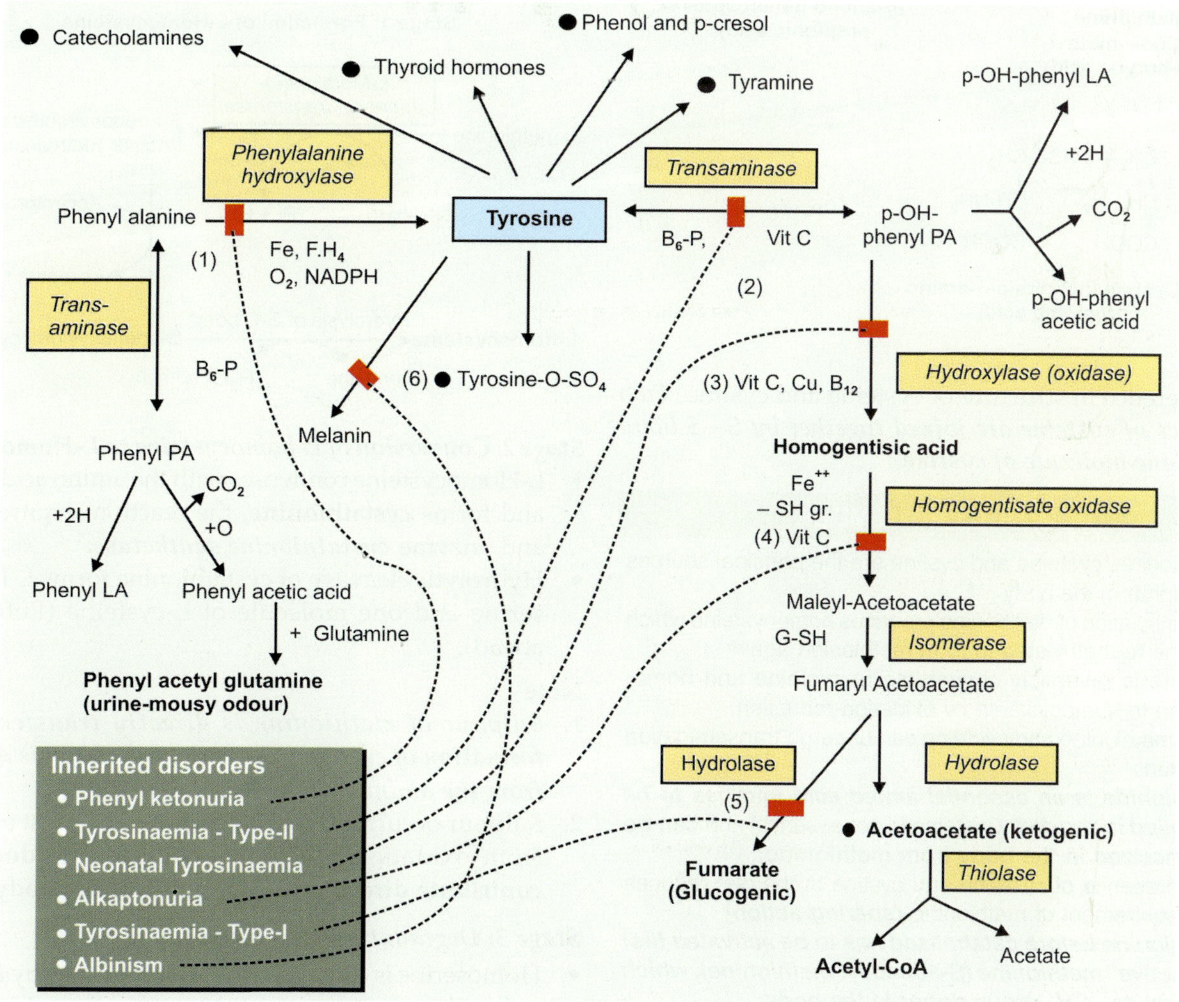

Fig. 27.9: Flow chart of metabolic fate and metabolic role of phenylalanine and tyrosine

B. METABOLISM OF SULPHUR-CONTAINING AMINO ACIDS

Sulphur containing amino acids are three:

(i) L-Methionine: essential amino acid,

(ii) L-Cysteine } **Non-essential amino acid (both)**
(iii) L-Cystine

Note

Other sources of sulphur in the body are the sulphur-containing vitamins: thiamine (vit B_1), lipoic acid and biotin.

Structure of S-containing Amino acids

Structure of S-containing Amino acids:

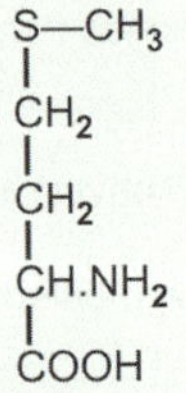

L-Methionine
(α-amino-γ-methyl thio-n-butyric acid)

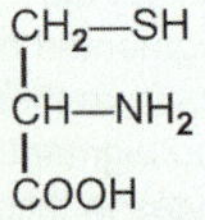

L-Cysteine
(α-amino-β-mercapto-propionic acid)

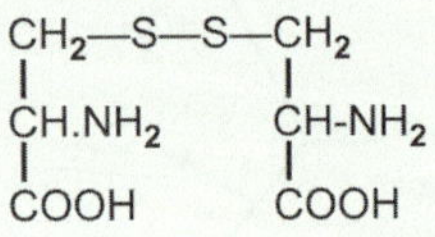

L-Cystine (β-β-dithio-α-amino propionic acid)

Note

The difference in structure of cysteine and cystine. *Two molecules of cysteine are joined together by S—S bond to form one molecule of cystine.*

POINTS TO REMEMBER

- Methionine, cysteine and cystine are the principal sources of sulphur in the body.
- Demethylation of methionine produces homocysteine which may be re-methylated to form methionine again.
- Cystine is reversibly convertible to cysteine and homocystine to homocysteine by oxidation-reduction.
- Both methionine and cysteine can undergo transamination reaction.
- ***Methionine is an essential amino acid and has to be supplied in the diet.*** Cysteine is not essential and **can be synthesized in the body from methionine.**
- The presence of cysteine and cystine in the diet reduces the requirement of methionine ***(sparing action).***
- ***Methionine before catabolised has to be activated first to "Active" methionine (S-adenosyl methionine), which can act as $-CH_3$ group donor in the body.***

Metabolic Fate of L-Methionine

Metabolic fate of L-methionine can be discussed in **three stages:**

Stage 1: ***Activation of methionine and its demethylation to form L-Homocysteine.***

Stage 2: ***Conversion of L-homocysteine to L-homoserine.***

Stage 3: ***Degradation of L-homoserine to end products L-propionyl-CoA and α-amino butyrate.***

Stage 1: Activation of L-methionine and its demethylation to form L-homocysteine:

- L-methionine condenses with ATP in presence of an activating enzyme in Liver to form "active" methionine (S-adenosyl methionine). It requires Mg^{++} and reduced glutathione.
- "Activated" S-methyl group may donate the CH_3 group to an "acceptor" forming S-adenosyl homocysteine.
- Hydrolysis of the S-C bond yields L-homocysteine and adenosine.

Stage 1: Formation of L-Homocysteine

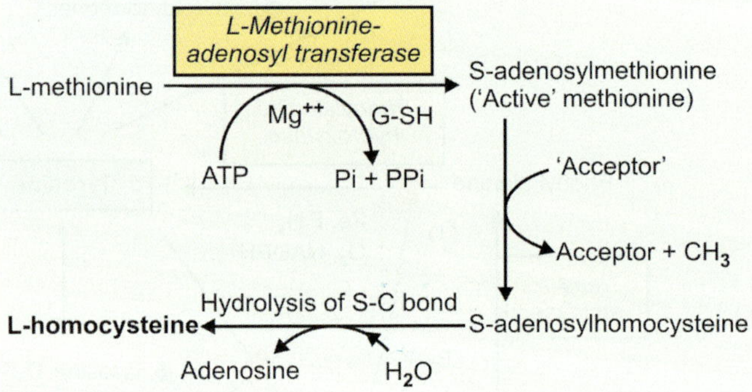

Stage 2: Conversion of L-homocysteine to L-Homoserine:

- L-Homocysteine condenses with the amino acid serine and forms **cystathionine,** the reaction requires B_6-P and enzyme ***cystathionine synthetase.***
- Hydrolytic cleavage of cystathionine forms L-Homoserine and one molecule of L-cysteine (Refer box ahead).

Note

1. ***Sulphur of methionine is directly transferred in formation of cysteine, the carbon-skeleton is derived from the amino acid serine.***
2. Sulphur occurring in urine is derived almost entirely from oxidation of cystine. **Methionine does not contribute directly to "SO_4-pool" of the body.**

Stage 3: Degradation of L-Homoserine

- Homoserine is converted to α-Ketobutyrate by ***Homoserine deaminase*** via formation of imino acid.

Stage 2: Formation of L-Homoserine

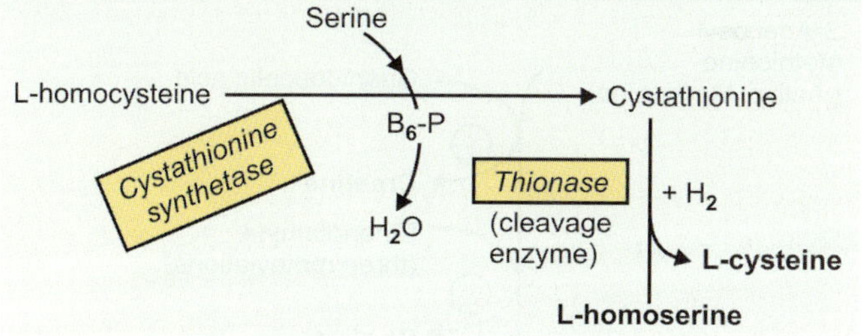

- α-ketobutyrate is converted to **Propionyl-CoA** by oxidative decarboxylation or it can be aminated by transamination to α-amino butyrate.

Thus end-products of Methionine degradation are
- **Propionyl-CoA**
- **α-amino butyrate**

Fate of End Products

- Propionyl CoA is converted to succinyl-CoA through formation of methyl malonyl-CoA and thus it is ***glucogenic.***
- α-amino butyrate is excreted in urine. Excretion in urine increases after giving "loading dose" of methionine.

Stage 3: Catabolism of L-Homoserine

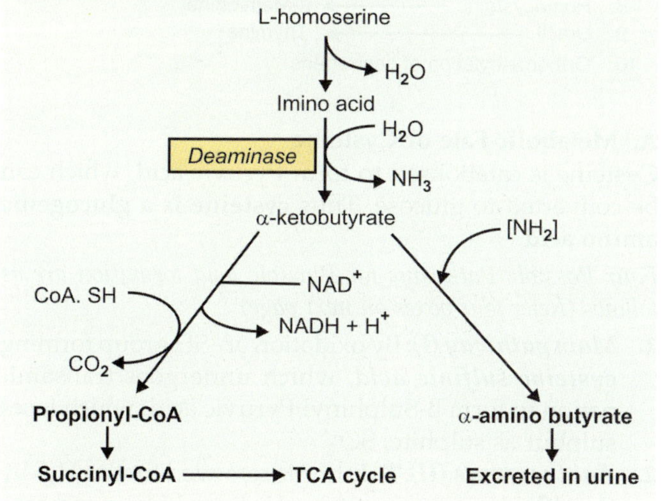

METABOLIC ROLE OF METHIONINE

- ***Methionine is "glucogenic"***: Propionyl-CoA the endproduct is glucogenic.
- ***Cysteine formation:*** (see stage 2)
- ***Lipotropic function:*** "Active" methionine can donate "methyl group" and can form choline from ethanolamine. Choline is lipotropic and prevents accumulation of fat in Liver.
- ***Polyamine synthesis:*** 'Active' methionine after decarboxylation combines with putrescine to form first polyamine **Spermidine** (Refer, biogenic amines).

- ***Formation of methyl mercaptan and its clinical significance:***
  Patients with severe Liver diseases, exhibit foul odour in breath called as ***Foetor hepaticus***. It has been attributed to methyl mercaptan, which appears to be formed from methionine. Methyl mercaptan has been found in urine of these patients. Methionine on transamination produces corresponding ketoacid which on hydrolysis produces methyl mercaptan.

Formation of Methyl Mercaptan

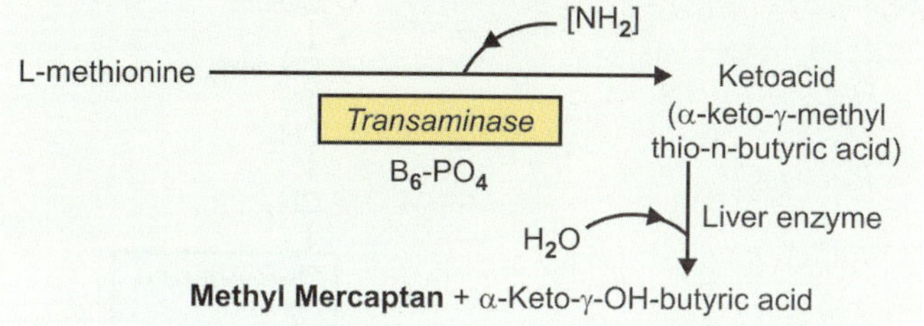

- ***Transmethylation***
  Certain compounds of the body, with structures containing ***CH_3 group attached to an atom other than carbon*** can take part in enzymic reactions, whereby these $-CH_3$ groups are transferred to a suitable "acceptor", which have no $-CH_3$ group. Such reactions are termed as ***"transmethylation reaction"***, and the substrate, i.e. the $-CH_3$ donor is said to possess ***biologically labile $-CH_3$ group"***. The most important compounds with biologically labile methyl group are:

- "Active" methionine containing $-S \sim CH_3$ group.
- Choline-containing N+ (CH_3, CH_3, CH_3)
- Betaine — an oxidation derivative of choline.

"ACTIVE" METHIONINE

Chemically called as ***"S-adenosyl methionine"***. Activation of methionine occurs in presence of ATP, catalysed by an enzyme, called ***L-methionine adenosyl transferase.*** It requires presence of Mg^{++} and G-SH. In the process of activation, ATP donates the entire adenosine moiety to methionine and loses 3 molecules of PO_4, one as orthophosphate and two as Pyrophosphate. CH_3 group forms a 'high energy' bond with sulphur ($-S \sim CH_3$), this attributes to lability of methyl group.

Transmethylation reaction is highly ***"exergonic"***. In most cases, a hydrogen ion is released in the reaction, contributing to the liberation of free energy at physiological pH. The free energy of transmethylation approximates that of hydrolysis of a high energy bond. ***Examples of transmethylation reactions*** are shown schematically ahead in box.

Formation of 'active' methionine

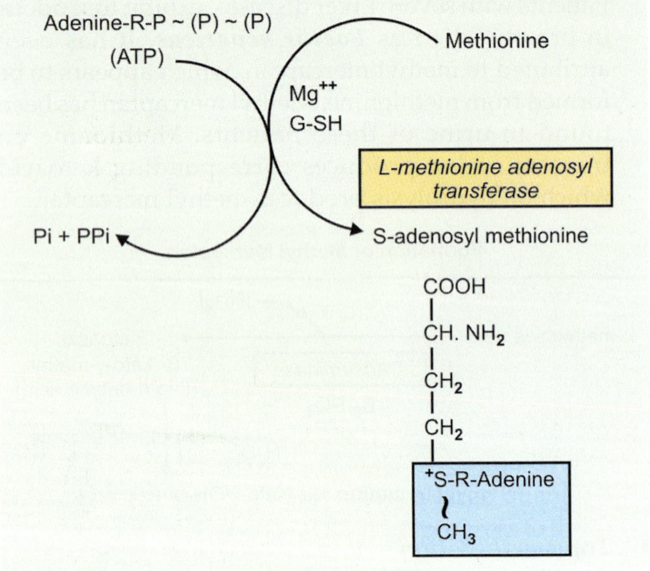

Transmethylation reactions

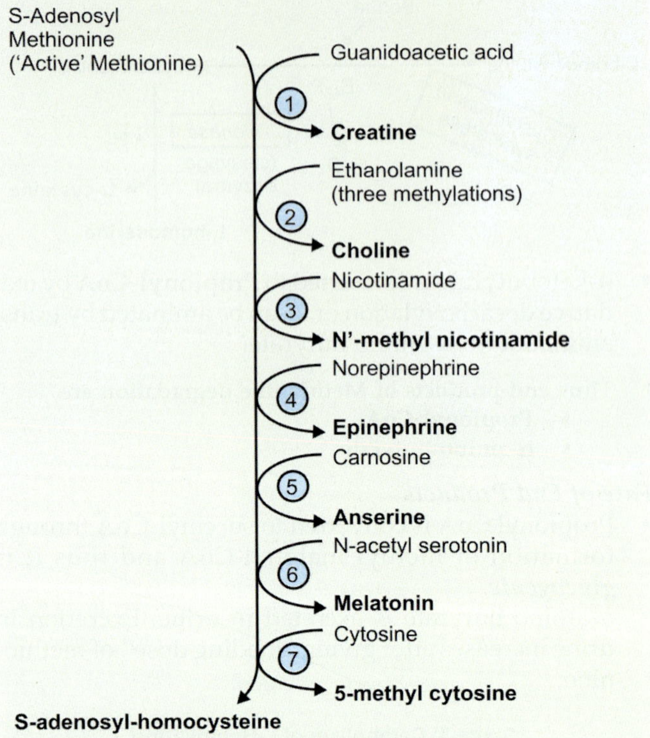

Other examples are:

8. Homocysteine → L-Methionine
9. Uracil → Thymine
10. Orthomethylation of oestrogens.

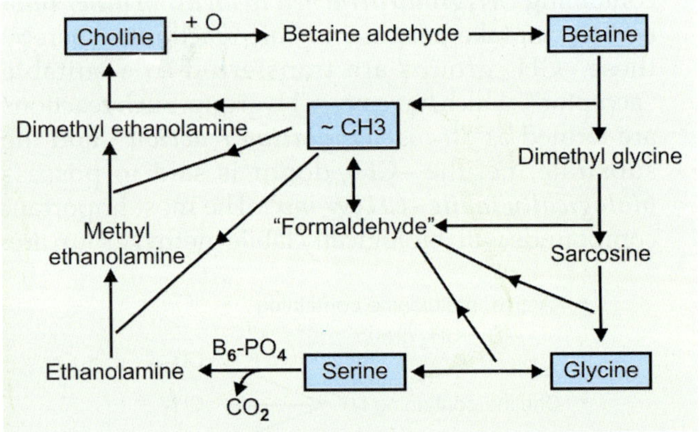

Fig. 27.10: Interrelation of glycine and choline derivatives in transmethylation

Interrelationship of glycine and choline derivatives in transmethylation is shown diagrammatically in Figure 27.10.

Metabolism of Cystine

Cystine metabolism proceeds through cysteine, to which it is readily converted. The reaction is ***reversible*** and catalyzed by NADH-dependant *oxidoreductase.*

One molecule of cystine is reduced to form two molecules of cysteine and vice-versa.

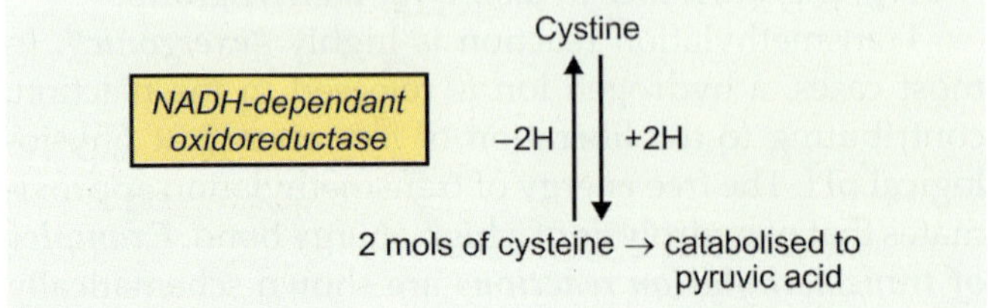

A. Metabolic Fate of Cysteine

Cysteine is catabolised to form Pyruvic acid, which can be converted to glucose. Thus **cysteine is a glucogenic amino acid**.

Four Possible Pathways for Pyruvic acid formation are as follows (Refer four boxes on next page)

1. ***Main pathway (I):*** By oxidation of –SH group forming ***cysteine sulfinic acid***, which undergoes transamination to form β-Sulphinyl Pyruvic acid, which loses sulphur as sulphite, SO_2^-.
2. ***Another route (II):*** Sulphur is removed first as H_2S by ***desulfhydrase*** enzyme, found in liver, kidney and pancreas, which requires B_6-PO_4. An imino acid is formed, which is hydrolysed to form NH_3 and pyruvic acid.
3. ***Another possible route (III):*** L-Cysteine first undergoes transamination, which is followed by loss of sulphur as H_2S, and formation of pyruvic acid.
4. ***Another possible alternative route (IV):*** It is a bifurcation from main pathway. L-cysteine-sulphinic acid, instead of undergoing transamination, is further oxidised to form **cysteic acid,** which then undergoes transamination to form ***β-sulfonyl-pyruvate. Sulphur is then removed as SO_4^- and Pyruvic acid is formed.***

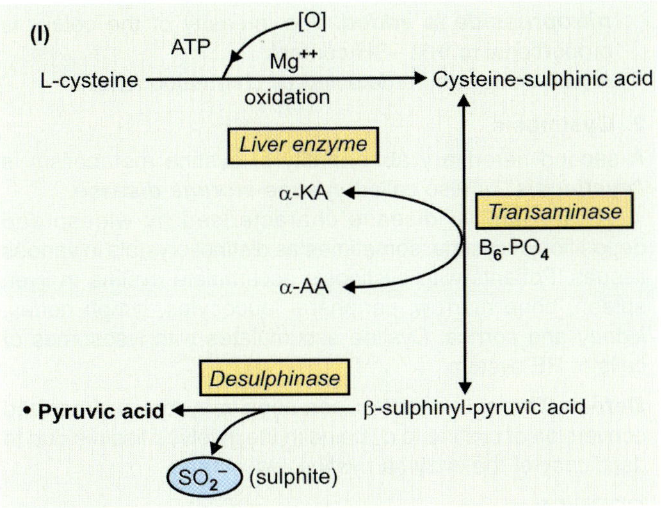

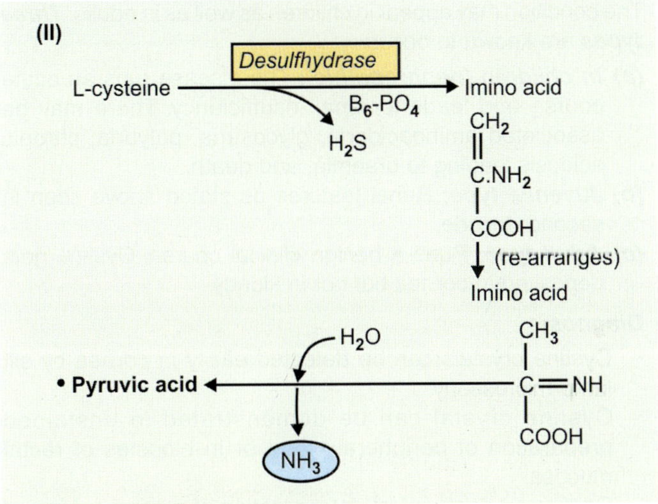

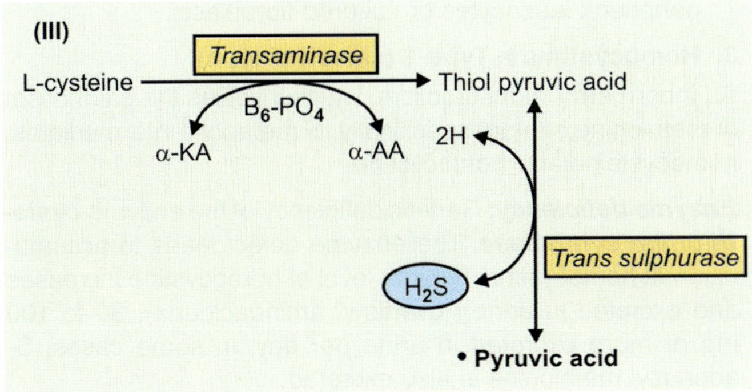

Conversion of SO_2^- to $SO_4^=$

SO_2^- formed in main pathway is oxidised to form $SO_4^=$, catalysed by *sulphite oxidase*, which requires Lipoic acid and hypoxanthine for its activity.

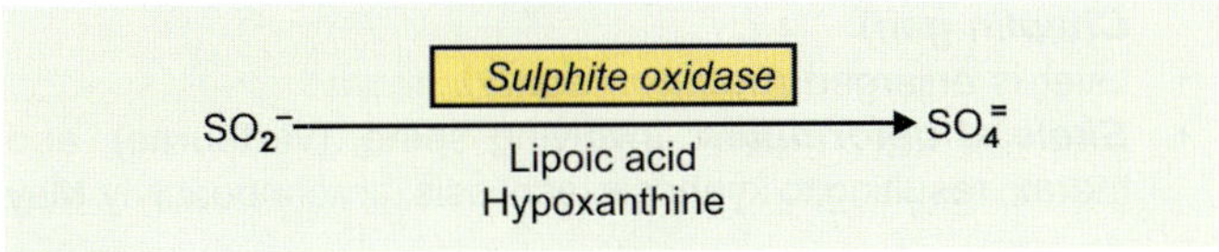

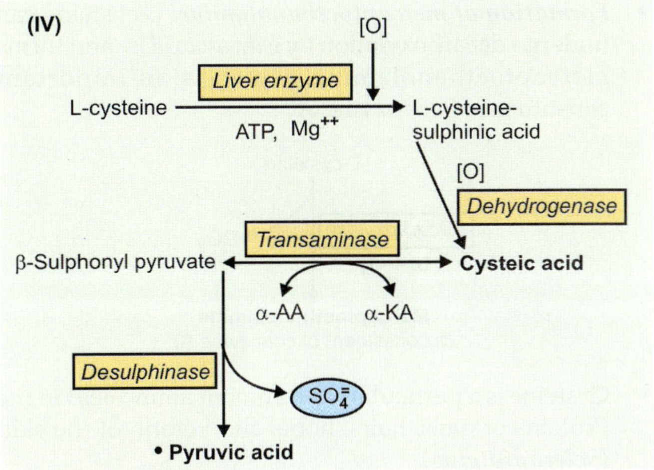

B. Metabolic Role of Cysteine

- ***Glucogenic:*** Cysteine is catabolised to Pyruvic acid which is glucogenic.
- ***Formation of glutathione:*** Cysteine is required for synthesis of glutathione. G-SH is the reduced form, active group is SH group. G-S-S-G is the oxidised form.
- ***Formation of taurine:*** Cysteine is utilised in the formation of 'taurine', which combines with cholic acid (obtained from degradation of cholesterol in Liver,) to form Bile acid 'taurocholic acid'.

Formation of Taurine

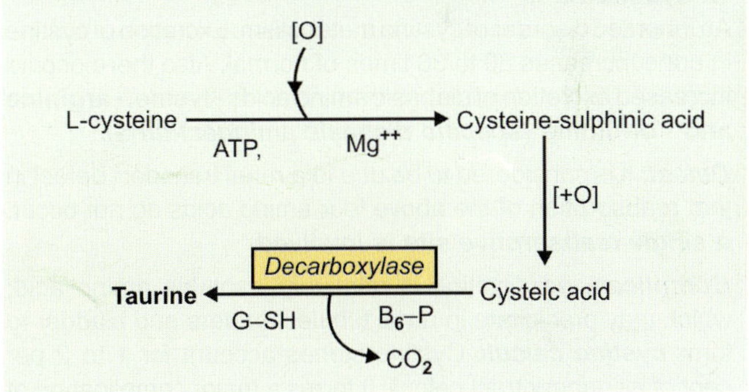

Formation of Taurocholic acid

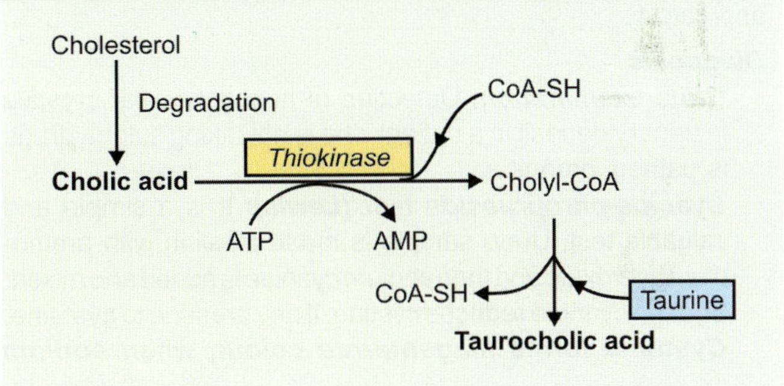

- ***Formation of mercaptoethanolamine:*** L-cysteine can undergo decarboxylation by a *decarboxylase* and forms mercaptoethanolamine, which is an important constituent of coenzyme A.

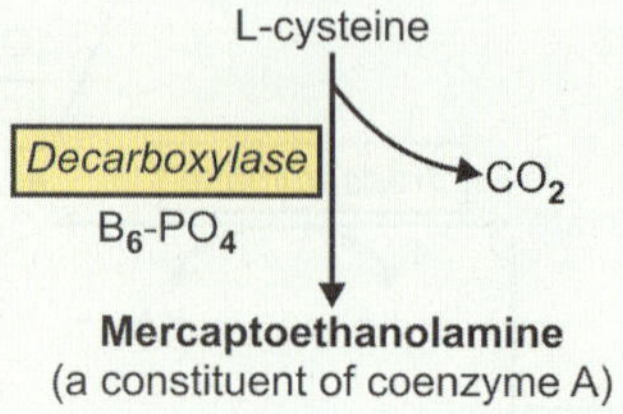

- Cysteine is a particularly prominent amino acid in the Proteins of nails, hairs, hoofs and keratin of the skin *(Scleroproteins).*
- Cysteine is also a constituent of many other proteins, including certain protein hormones like Insulin, Vasopressin, etc. where it is of great importance in maintaining secondary and tertiary structures of the proteins.
- ***Role of cysteine in detoxication:*** Cysteine is involved in detoxication reactions (Refer to chapter on detoxication).

CLINICAL ASPECT

INHERITED DISORDERS OF S-CONTAINING AMINO ACIDS

1. Cystinuria

An inherited disorder of cystine metabolism. Excretion of cystine in urine increases 20 to 30 times of normal. Also there occurs increased excretion of diabasic amino acids • **lysine,** • **arginine and** • **ornithine** ***(specific diabasic aminoaciduria).***

Defect: It is considered to be due to a renal transport defect in that reabsorption of the above four amino acids do not occur, ***a single reabsorptive site is involved.***

Complications: Cystine is relatively insoluble amino acid, which may precipitate in renal tubules, ureters and bladder to form ***cystine calculi***. Cystine stones account for 1 to 2 per cent of all urinary tract calculi. It forms a major complication of the disease. A **mixed disulphide** consisting of L-cysteine and L-homocysteine has been found in urine. This is more soluble and thus reduces the tendency to formation of cystine crystals/ and calculi.

Diagnosis

- ***Urine examination:*** Detection of hexagonal, flat crystals in urinary deposit in a patient who is not taking sulpha drugs is pathognomonic.
- ***Cyanide-nitroprusside test (Lewis):*** It is a simple and valuable test. Urine sample is made alkaline with ammonium hydroxide and then sodium cyanide is added and mixed. Sodium cyanide reduces cystine, if any present, to cysteine. ***Cysteine forms magenta-red colour, when sodium nitroprusside is added.*** The intensity of the colour is proportional to free –SH content.
- Amino acids can be detected by chromatography.

2. Cystinosis

A second hereditary abnormality of cystine metabolism is ***"cystinosis"*** or also called ***cystine-storage disease***.

A rare familial disease characterised by widespread deposition of cystine, sometimes as distinct crystals in various tissues. Patients with cystinosis accumulate cystine in liver, spleen, bone marrow, peripheral leucocytes, lymph nodes, kidney and cornea. Cystine accumulates with lysosomes of cells of RE system.

Defect: The cause of the condition may be an impaired conversion of cystine to cysteine in the involved tissues due to deficiency of the enzyme *cystine reductase*.

Clinical features

The condition may appear in children as well as in adults. ***Three types*** are known to occur:

(a) In children (nephropathic): The disease runs an acute course and leads to renal insufficiency. There may be associated aminoaciduria, glycosuria, polyuria, chronic acidosis leading to uraemia, and death.

(b) Juvenile type: Renal features as stated above seen in second decade.

(c) Adult type: Runs a benign clinical course. Cystine gets deposited in cornea but not in kidney.

Diagnosis

- Cystine crystals can be detected easily in cornea by slit lamp microscopy.
- Cystine crystal can be demonstrated in unstained preparation of peripheral blood or in biopsies of rectal mucosa.
- Confirmed by chemical determination of cystine content of peripheral leucocytes or cultured fibroblasts.

3. Homocystinuria Type-1 (classical type)

An inborn error of metabolism, which involves the catabolism of methionine or more specifically its metabolic intermediates, homocysteine/and homocystine.

Enzyme deficiency: Genetic deficiency of the enzyme ***cystathionine synthetase.*** The enzyme defect leads to accumulation of homocystine. Plasma level of homocystine increases and excreted in urine ("overflow" aminoaciduria), 50 to 100 mg or more excreted in urine per day. In some cases, S-adenosyl methionine is also excreted.

Incidence: 1 in 60,000 live births.

Clinical features

- **Mental retardation:** In children and surviving adults.
- Some affected individuals, are extraordinarily tall, with long extremities, frequently with flat feet with toes out ***(Charlie-Chaplin gait).***
- Liver is enlarged ***(hepatomegaly).***
- ***Skeletal deformities:*** Involving spine, (vertebrae), and thorax, resulting to kyphosis, scoliosis, arachnodactyly. May

be premature osteoporosis which also accounts to above deformities. X-ray spine shows ***cod fish*** Vertebrae.

- ***Ectopia lentis:*** Curious dislocation of lens of the eye. Not seen at birth, may show at the age of 2 to 3 years.
- ***Life-threatening arterial/venous thrombosis.***
- Most of the patients show abnormal EEG.

Urine: Sodium cyanide-nitroprusside test is positive and helps in diagnosis.

The classical type of homocystinuria is described above. In addition to above classical type, two more types of homocystinurias have been described.

a. Homocystinuria Type-2

- ***Inheritance:*** Autosomal recessive
- ***Enzyme deficiency:*** N^5-methyl-Tetrahydrofolate-homocysteine methyl transferase.
- ***Clinical feature***
  - Mental retardation +
  - No ectopia lentis or thrombotic episodes seen.

Blood: Shows increased level of homocysteine.

Urine: Homocysteine is excreted in urine. Nitroprusside test +ve.

b. Homocystinuria Type-3

- ***Inheritance:*** Autosomal recessive.
- ***Enzyme deficiency:*** N^5, N^{10}-methylene tetra-hydrofolate reductase deficiency.
- ***Clinical features***
  - Mental retardation +
  - No ectopia lentis or thrombotic episodes

Blood: Shows increase homocysteine.

Urine: Excretion of homocystine, nitroprusside test +ve.

Note: Both type 2 and type 3 show response to folic acid administration.

Relation of B-complex Vitamins, Homocysteine and Heart Attack

Recently it has been shown that some cases of CHD where there are no obvious risk factors like increased cholesterol/TG/LDL, obesity, hypertension, etc. are ***due to increased level of homocysteine in the blood***. Individuals with the highest homocysteine levels have three times the risk of precipitating heart attack, even if all other risk factors are under control.

How elevated homocysteine contributes CHD remains unclear, but it is postulated that ***it increases the possibility of thrombosis***, and promotes also plaque formation thus damaging the arteries.

It is also shown that an insufficient concentration of three B-vitamins viz. folic acid, vit B_{12} and pyridoxine (B_6) cause increase in the homocysteine level in the blood.

Sufficient amounts of folic acid, B_{12} and B_6 reduce the blood levels of homocysteine. The higher the concentration of folic acid in the diet, the lower the homocysteine levels in the blood.

Figure 27.11 shows flow chart for metabolic fate and role of S-containing amino acids (next page).

GLUTATHIONE—CHEMISTRY AND FUNCTIONS

Glutathione is a **tripeptide** of three amino acids, **(i) glutamic acid**, **(ii) cysteine** and **(iii) glycine**. Chemically it is α-L-Glutamyl-cystinyl-glycine.

Synthesis

It is synthesised in cytosol outside the ribosomes from the above three amino acids. Synthesis requires ATP, but does not require the participation of RNA.

Steps

- L-Glutamate first condenses with L-cysteine to form γ-L-glutamyl-L-cysteine. The reaction is catalysed by the enzyme *γ-glutamyl-cysteinyl-synthetase* in presence of ATP, K^+ and Mg^{++}. Probably there is formation of enzyme bound γ-glutamyl-(P) as an intermediate.

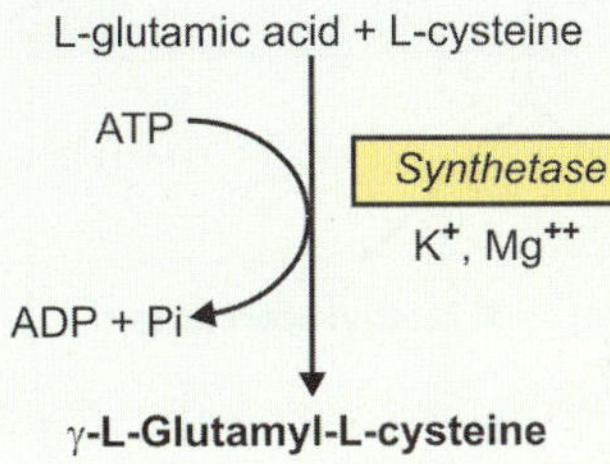

- γ-glutamyl-cysteine then becomes phosphorylated by the enzyme *"Glutathione synthetase"* in presence of ATP to form γ-glutamyl-cysteinyl-(P), which remains enzyme-bound.
- Enzyme bound γ-glutamyl-cysteinyl-(P) then reacts with glycine to liberate free glutathione and Pi.

FUNCTIONS OF GLUTATHIONE

Glutathione is an ***important reducing agent in the tissues.*** Oxidised glutathione G-S-S-G is harmful to the tissues, specially to RB Cells and Lens protein and is converted to reduced glutathione G-SH, which is required for the integrity of RB Cells membrane and Lens proteins.

- By donating H_2, it ***helps to destroy H_2O_2*** and other peroxides in cells. The reaction is catalysed by the Selenium-containing enzyme *glutathione peroxidase.*
- It acts as a ***coenzyme with Liver enzyme Glutathione-insulin transhydrogenase*** which helps in the catabolism and degradation of protein hormone insulin.
- Glutathione and the enzyme *glutathione transhydrogenase* help to cause reductive cleavage of S-S linkages in thyroglobulin glycoprotein.
- Many –SH group containing enzymes are also protected by glutathione against the oxidation of their –SH groups, e.g. *Glyceraldehyde-3-P-dehydrogenase* enzyme.

SECTION FOUR

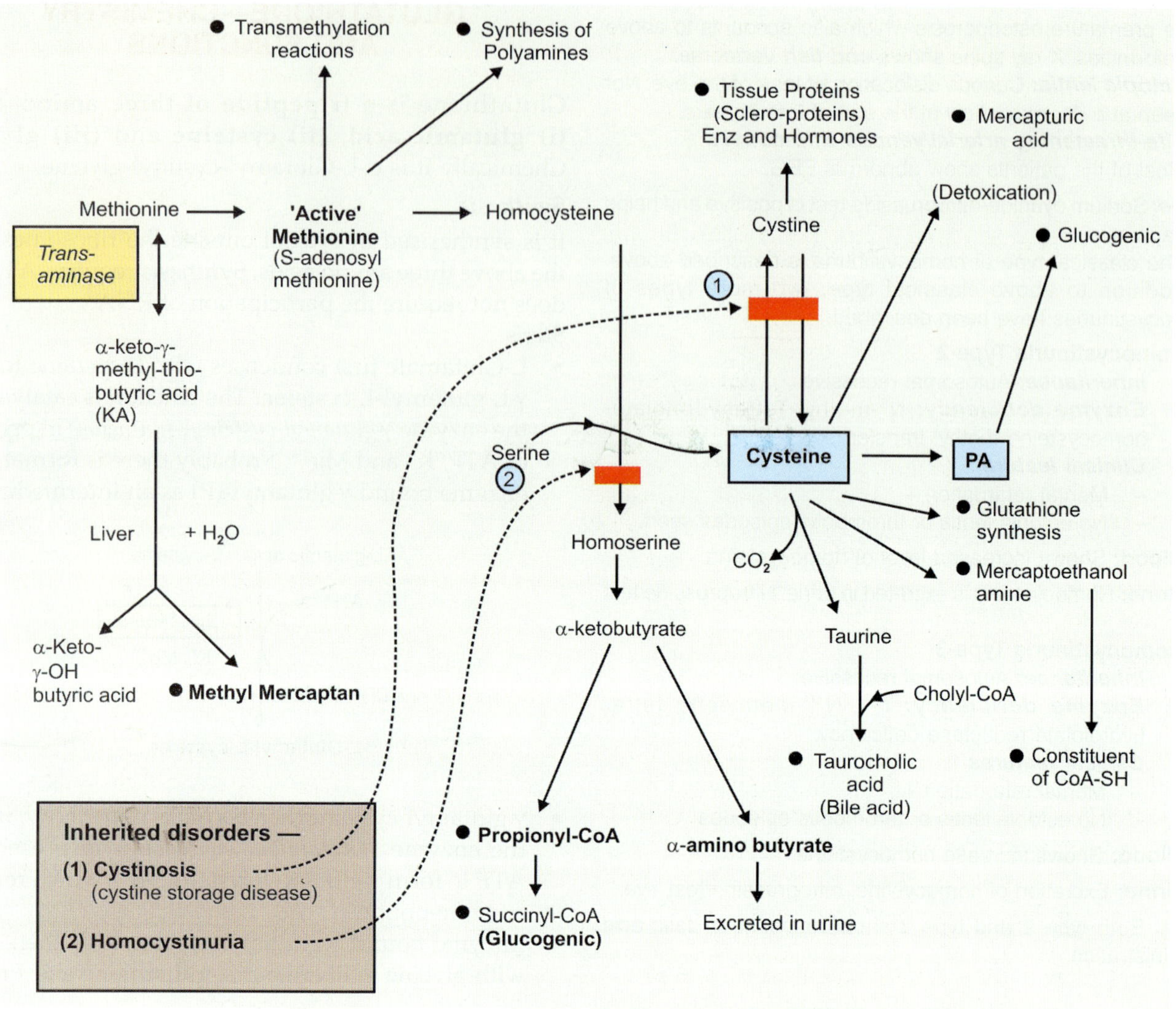

Fig. 27.11: Showing flow chart—metabolic fate and metabolic role of S-containing amino acids

- Glutathione is required as coenzyme with *formaldehyde dehydrogenase* which catalyses the oxidation of formaldehyde to formic acid.
- It is also ***required as a 'coenzyme' for the enzyme Glyoxylase,*** which converts "methyl glyoxal" to Lactic acid through intramolecular oxidation-reduction.
- It also acts as coenzyme with *Maleyl-acetoacetate isomerase,* which catalyses "Cis-trans" isomerisation of Maleyl-acetoacetate to fumaryl-acetoacetate.
- Glutathione takes part in ***γ-glutamyl cycle*** for absorption of amino acids from gut.
- Oxidised glutathione is harmful for Red cell membranes and lens proteins. Oxidised glutathione formed during coenzyme actions is reduced back to glutathione by an NADPH-dependant flavoprotein called *Glutathione reductase* which contains FAD as prosthetic group.

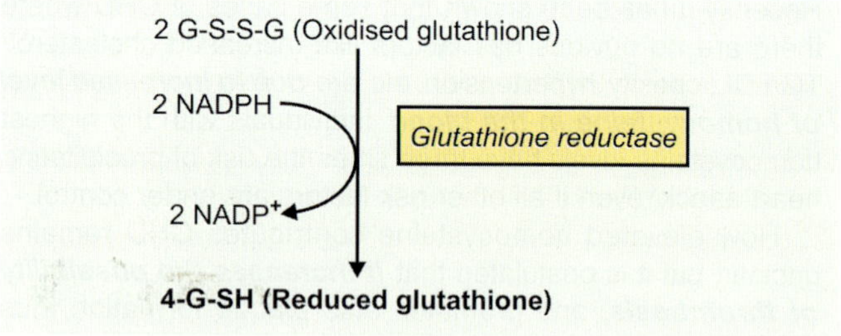

- Reduced glutathione is required for conversion of *Dopa-quinone* to *Glutathione*-Dopas synthesis of pheomelanins and trichochromes.
- G-SH is required as coenzyme/cofactor with *PG-Synthetase system (Cyclo-oxygenase)* required for formation of endoperoxides from arachidonate in PG synthesis.

- G-SH is required as a coenzyme for Liver enzyme *for activation of methionine to form "S-adenosyl methionine" (Active methionine).*

METABOLISM OF OTHER AMINO ACIDS

GLYCINE

Glycine is the simplest of amino acids. Chemically it is *amino acetic acid*

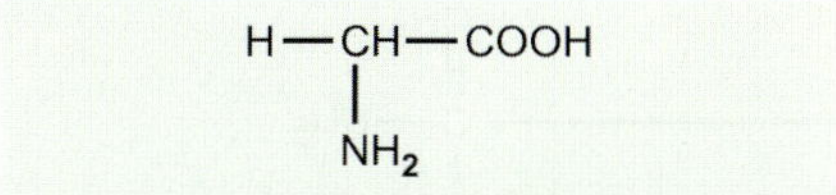

Amino acetic acid (glycine)

It does not possess asymmetric carbon atom, hence does not exist in isomeric forms. It is dispensible or nonessential amino acid and can be synthesised in animal tissues. ***Though it is non-essential but it is an important amino acid as it forms many biologically important compounds in the body.***

A. Metabolic Fate

1. ***Deamination***

   Glycine is deaminated by a specific enzyme ***glycine oxidase*** (a flavoprotein enzyme) present in Liver and kidney to produce ***glyoxylic acid*** (glyoxylate). Glyoxylate can further be converted to either oxalic acid or formic acid and thus enters "one-carbon pool".

Note

Amino acid glycine is a source of oxalic acid in the body.

2. ***Glycine Cleavage***

   Major pathway for glycine catabolism in vertebrates involves conversion to CO_2, NH_4^+, and N^5, N^{10}-methylene-F.H_4 catalysed by the enzyme ***Glycine Synthase complex.*** The ***reaction is reversible,*** occurs in Liver of most vertebrates including humans and the enzyme complex is found as aggregates in mitochondrion.

3. ***Conversion to Serine***

   Glycine can be converted to serine which by non-oxidative deamination can form pyruvic acid, thus glycine may be ***glucogenic*** (Fig. 27.12).

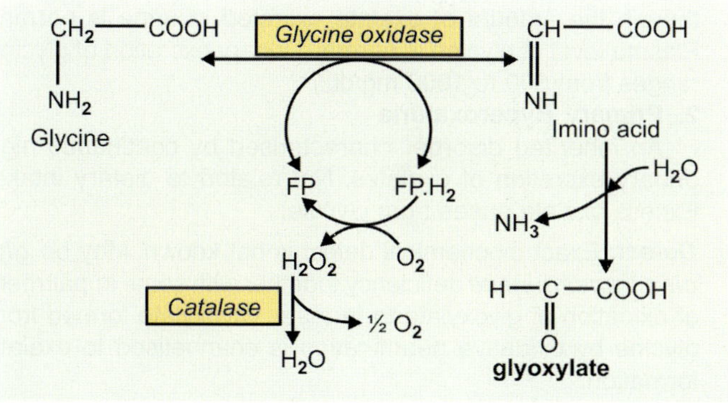

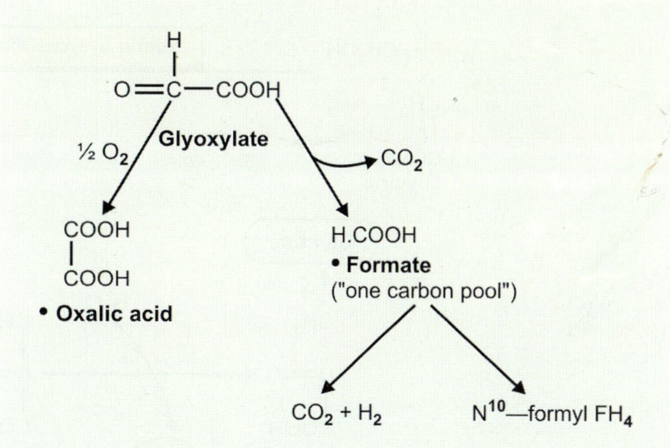

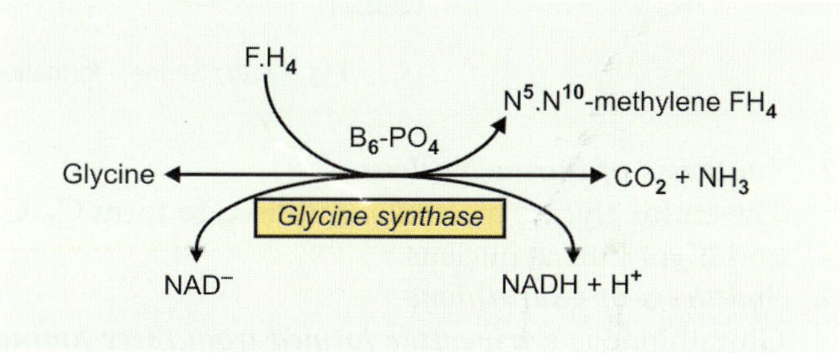

4. ***Oxidation to form Aminoacetone***

   An alternative pathway for oxidation of glycine. Glycine may be oxidised by conversion to **amino acetone** through the action of ***aminoacetone synthetase*** in presence of acetyl-CoA. The aminoacetone may further be metabolised through *methyl glyoxal* to Lactic acid and Pyruvic acid (see box below).

B. Metabolic Role of Glycine

1. ***Synthesis of Haem***

   Glycine is necessary in the first reaction of haem synthesis (Refer Chapter on Synthesis of Haem).

Formation of Amino Acetone

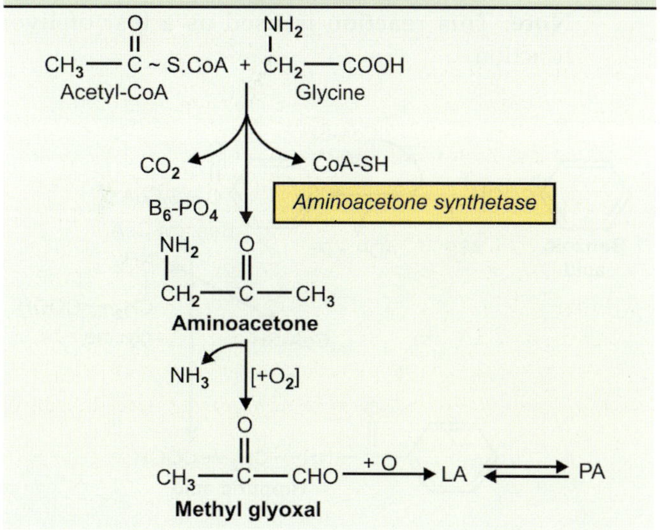

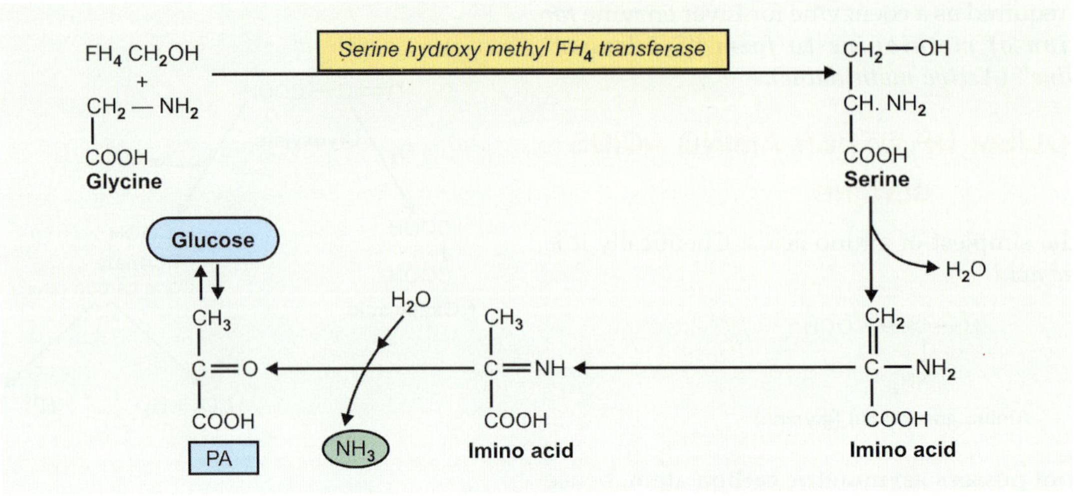

Fig. 27.12: Serine—formation from glycine and metabolic fate

2. ***Synthesis of Purine Nucleus***
   The entire glycine molecule is utilised to form C_4, C_5 and N_7 of Purine nucleus.
3. ***Synthesis of Glutathione***
   Glutathione is a ***tripeptide formed from three amino acids—Glutamic acid, cysteine*** and ***glycine.***
4. ***Synthesis of Creatine***
   Arginine and glycine, in presence of the enzyme *transamidinase* in kidney, reacts to form ***Glycocyamine (Guanidoacetic acid),*** which is further converted to creatine-P in liver.
5. ***Conjugation***
   Glycine also acts as a conjugating agent.
   - Benzoic acid and other aromatic acids absorbed from the alimentary tract or arising during the processes of metabolism are conjugated with glycine before excreted in urine. Thus, benzoic acid is converted to **hippuric acid** and excreted in urine (Refer to box below).

     Note: This reaction is used as a test of liver function.

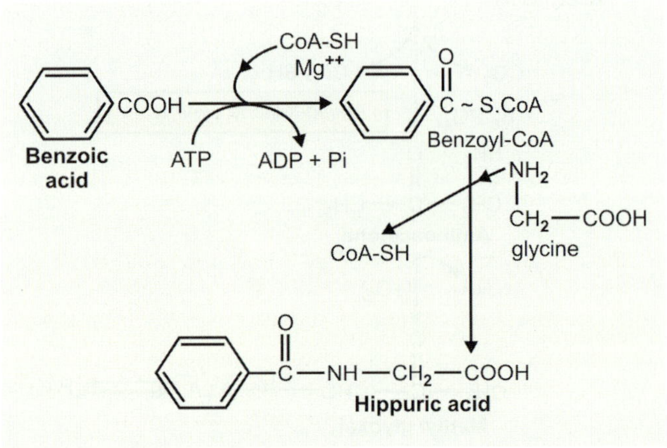

   - In a similar way, cholic acid formed from degradation of cholesterol in Liver is conjugated with glycine to form **glycocholic acid**, a bile acid which is excreted in bile as sodium salt, **sodium glycocholate** (bile salt).
6. ***Glycine is Glucogenic***
   Glycine is converted to Serine which is converted to PA (glucogenic).
7. ***Source of Formate (one carbon pool) and oxalate*** (See next page).

Figure 27.13 is flow chart of glycine, showing metabolic fate and metabolic role.

CLINICAL ASPECT

INHERITED DISORDERS OF GLYCINE METABOLISM

Two disorders are associated with glycine metabolism:

1. Glycinuria:

The disease is characterised by excess urinary excretion of glycine.

Inheritance: Autosomal dominant may be X-linked trait.

Defect: There is no enzyme deficiency. Defect is attributed to renal tubular reabsorption of glycine.

Clinically: Tendency to formation of oxalate stones in kidney though the amount of oxalate excreted in urine is normal. Plasma level of glycine is normal. Urinary excretion of glycine ranges from 600 to 1000 mg/dl.

2. Primary Hyperoxaluria

An inherited disorder characterised by continuous high urinary excretion of oxalates. Not related to dietary intake. Excess oxalate arises from glycine.

Defect: Exact biochemical defect is not known. May be ***glycine transaminase*** deficiency together with some impairment of oxidation of glyoxylate to formate. Glyoxylate formed from glycine by oxidative deamination is channelised to oxalate formation.

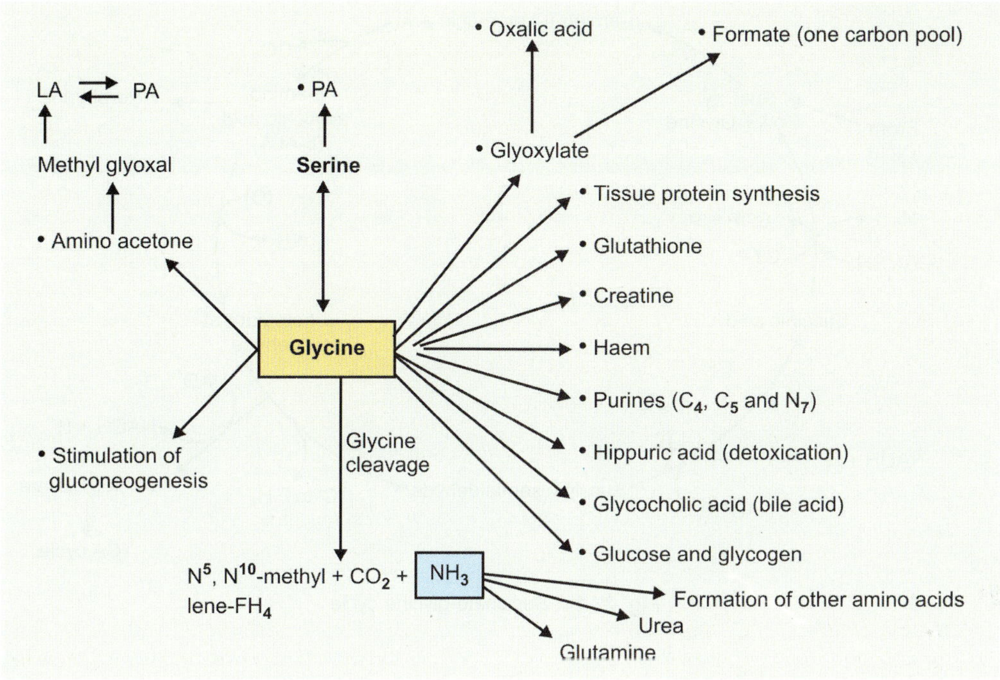

Fig. 27.13: Glycine, showing metabolic fate and metabolic role

Clinical features: Progressive bilateral calcium oxalate urolithiasis, oxalate stone formation in genitourinary tract, also may be nephrocalcinosis, and recurrent infection of the urinary tract.

Prognosis: Death occurs in childhood or early adult life from renal failure or hypertension.

Succinate-Glycine Cycle

Succinate-glycine cycle is shown diagrammatically in Figure 27.14.

- Succinyl-CoA produced in TCA cycle condenses on the α-carbon atom of glycine to form α-amino-β-keto-adipic acid.
- α-amino-β-ketoadipic acid is decarboxylated to form δ-aminolaevulinic acid (δ-ALA), which is precursor for porphyrin synthesis.
- δ-ALA by oxidative deamination forms α-Keto-glutaraldehyde which forms α-ketoglutarate and joins TCA Cycle (an intermediate).
- α-ketoglutaraldehyde is converted to succinic semi-aldehyde the carbon removed forms "one carbon pool" (FH_4).
- Succinic semialdehyde is oxidised to succinic acid, which is converted to succinyl-CoA. Both joins TCA cycle as intermediates.

SERINE

- Serine is a hydroxy amino acid.
- It is ***non-essential (dispensable)*** and can be synthesised in the body.
- Chemically it is ***α-amino-β-OH-propionic acid.***

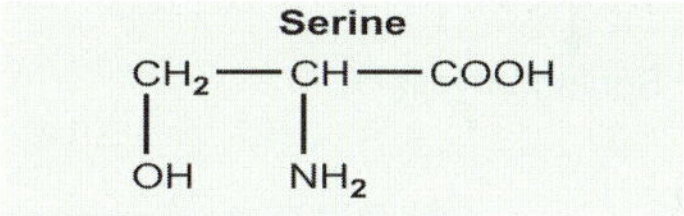

A. Metabolic Fate (Fig. 27.15)

It is deaminated by ***L-serine-dehydrase*** in Liver to form **Pyruvic acid** (non-oxidative deamination).

B. Metabolic Role (Fig. 27.15)

- As serine produces pyruvic acid, it is ***glucogenic.***
- Serine can be utilised, like all amino acids for formation of tissue proteins.
- Serine is a ***"carrier" of PO_4 group*** in phosphoproteins.
- ***Serine contributes the carbon-skeleton to form cysteine. Sulphur of cysteine comes from methionine.***
- Serine undergoes decarboxylation to form ***Ethanolamine*** by the enzyme *decarboxylase* in presence of B_6-PO_4. This is very important reaction as 'ethanolamine' is the precursor for:

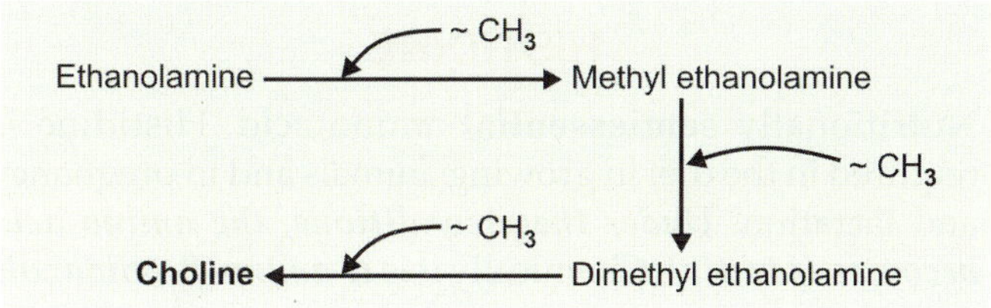

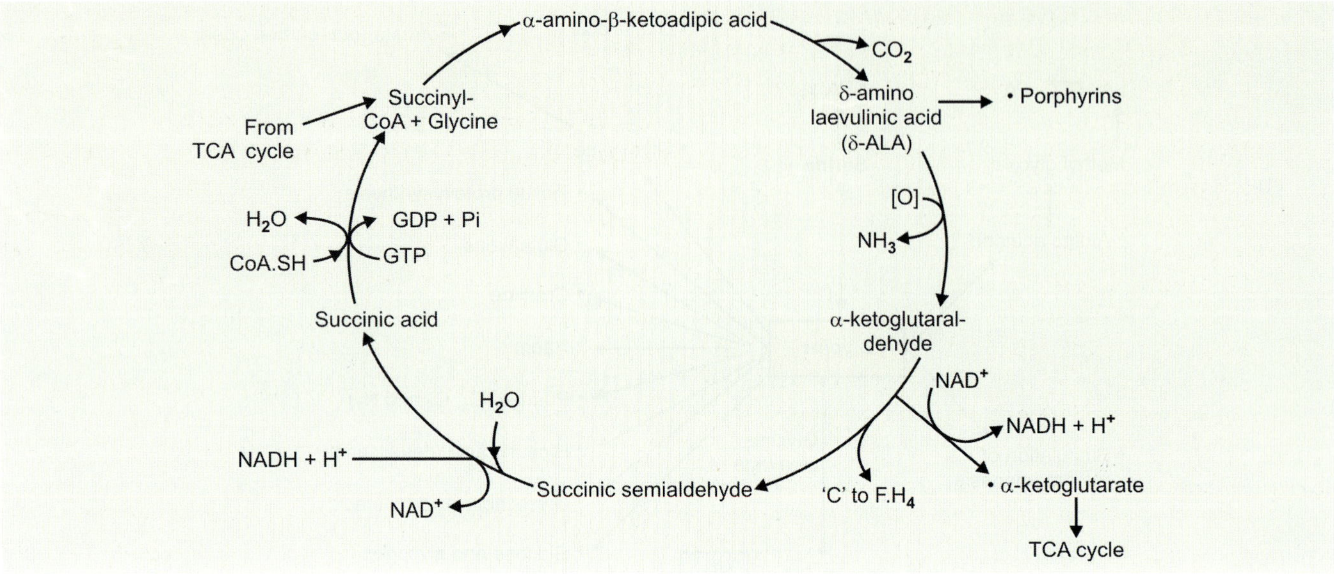

Fig. 27.14: Succinate-glycine cycle

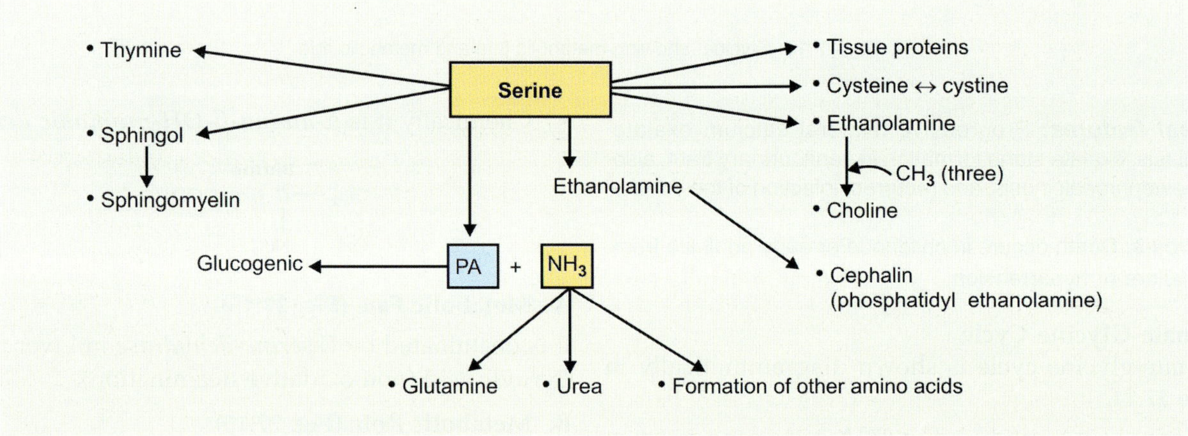

Fig. 27.15: Serine—showing metabolic fate and metabolic role

- – Formation of phosphatidyl ethanolamine (cephalin).
- – Formation of 'choline' (a lipotropic factor) by three successive methylations, ~ CH_3 group is donated by S-adenosyl methionine ("Active" methionine).
- Serine is used for synthesis of sphingol.
- β-Carbon of serine used for thymine formation.
- ***Hydroxyl group of serine in an enzyme protein is phosphorylated/dephosphorylated to form active/inactive forms of the enzyme.***

HISTIDINE

Nutritionally **semiessential** amino acid. Histidine is required in the diet in growing animals and in pregnancy and lactation. ***Under these conditions, the amino acid becomes essential.*** Chemically it is ***α-amino-β-imidazole propionic acid.***

A. Metabolic Fate

- Histidine on deamination produces urocanic acid, which is converted to 4-imidazolone-5-propionate by the enzyme *urocanase.*

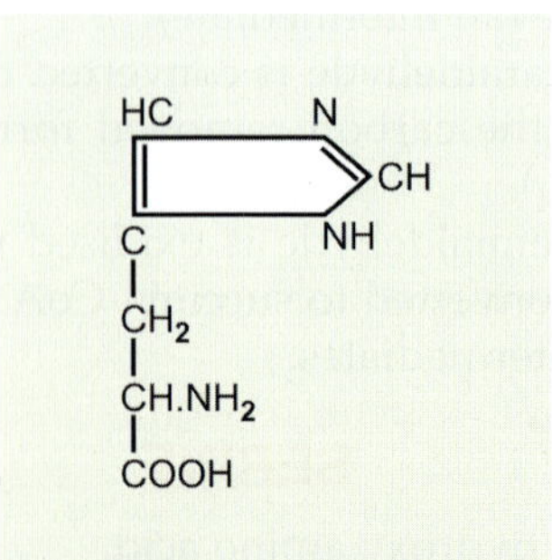

Structure of Histidine

- This product on addition of water produces ***formiminoglutamic acid (Figlu)***, which is converted to glutamate, the latter is transaminated to α-ketoglurate, which is an intermediate of TCA cycle. Metabolic fate is shown in next page.

B. Metabolic Role (Fig. 27.16)

- It is **glucogenic** through formation of glutamate to α-ketoglutarate.
- **Histamine** formation: Decarboxylation of histidine produces histamine.
- ***Formate*** can serve as one carbon moiety. The 'one-carbon' fragment of histidine is taken up by folic acid and metabolised by transformylation reaction normally.

In deficiency of folic acid, the histidine derivative, formiminoglutamic acid, (figlu) accumulates and excreted in urine, used as a test for folic acid deficiency (Refer folic acid).

- ***Other histidine compounds***
  - ***Ergothioneine***: Present in RB cells and liver. It is a reducing substance and reduces alkaline copper sulphate solution. Hence its liberation from haemolyzed red blood cells can give false high blood glucose when estimation of blood glucose is done by Folin and Wu's method.
  - ***Carnosine:*** A dipeptide of histidine with β-alanine (β-alanyl histidine).
  - ***Anserine:*** A methyl derivative of carnosine "1-methyl carnosine". It is formed by methylation of carnosine; methyl group is donated by "active" methionine.

Both carnosine and anserine occur in muscles. In myopathies: Methyl derivatives of histidine is found in urine.

Distribution of Carnosine and Anserine

- Major fraction of β-alanine in human tissues is its presence as dipeptide carnosine in skeletal muscle.
- Anserine is not present in human muscle but present in skeletal muscles, in rapidly contractile muscles like pectoral muscles of birds and rabbit limbs.

Functions of Carnosine and Anserine

Incompletely understood. But they may serve:

- To buffer the pH of anaerobically contracting skeletal muscle.
- Carnosine and anserine both activate *myosin-ATPase* activity *in vitro*.
- Both dipeptides also chelate copper (Cu) and enhance copper uptake. Thus they may participate in pathologic process in Wilson's disease.

Catabolism of Carnosine: Carnosine is hydrolysed to β-alanine and L-histidine by the zinc-containing metalloenzyme *carnosinase*.

β-ALANINE

Chemistry and Functions

Little free β-alanine is present in tissues. It is found in combination as:

- **β-alanyl dipeptides, e.g. carnosine and anserine;**
- **As a constituent of coenzyme A.**

Source: In mammalian tissues: β-alanine arises principally from catabolism of uracil, carnosine and anserine.

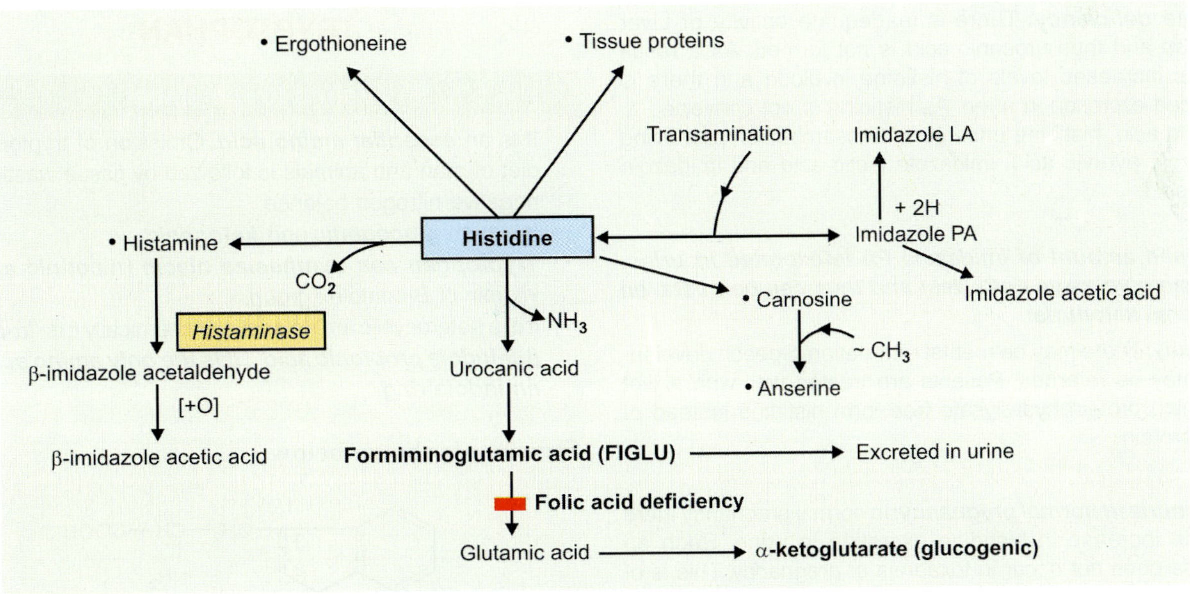

Fig. 27.16: Histidine showing metabolic role

Metabolic fate of Histidine

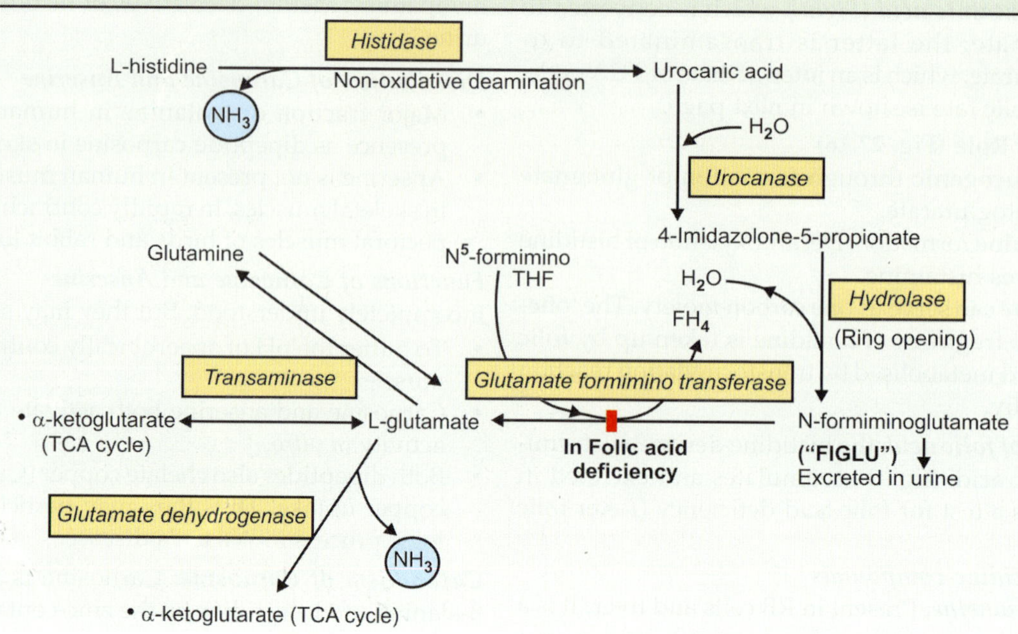

Catabolism: Catabolism of β-alanine in mammals involves transamination to form ***malonate semialdehyde,*** which is oxidized to acetate and thence to CO_2.

CLINICAL ASPECT

INHERITED DISORDERS

1. Histidinaemia

It is an inherited disorder.

Enzyme deficiency: There is inadequate activity of Liver *histidase* and thus urocanic acid is not formed. As a result there is increased levels of histidine in blood and there is increased excretion in urine. As histidine is not converted to urocanic acid, histidine undergoes transamination producing imida-zole pyruvic acid, imidazole lactic acid and imidazole acetic acid.

Note

Increased amount of imidazole PA is excreted in urine. This can give a +ve $FeCl_3$ test and thus can be mistaken as phenyl ketonuria.

Clinically: There may be mental retardation. Speech develop-ment may be retarded. Patients are treated well with a diet containing protein hydrolysate free form histidine instead of intact protein.

Note

Histidinuria in normal pregnancy: In normal pregnancy there is some increase in histidine excretion in urine. Such an increase does not occur in toxaemia of pregnancy. This is of diagnostic and clinical importance. Histidinuria in normal pregnancy is not a metabolic defect, probably due to changes in renal function.

2. Hyper-β-alaninaemia

A rare metabolic disorder. Free β-alanine levels are elevated in body fluids viz. Plasma, CS fluid and urine. Free β-alanine also increases in various tissues like Liver, kidney, brain and skeletal muscle. There is also increase in levels of taurine and β-aminoisobutyric acid.

TRYPTOPHAN

Points to remember

- It is an ***essential amino acid.*** Omission of tryptophan in diet of man and animals is followed by tissue wasting and negative nitrogen balance.
- It is both ***glucogenic*** and ***ketogenic.***
- ***Tryptophan can synthesize niacin*** **(nicotinic acid)**, a vitamin of B-complex group.
- It is a heterocyclic amino acid and chemically it is ***"α-amino-β-3-indole propionic acid". It is the only amino acid with an indole ring.***

Structure is shown below:

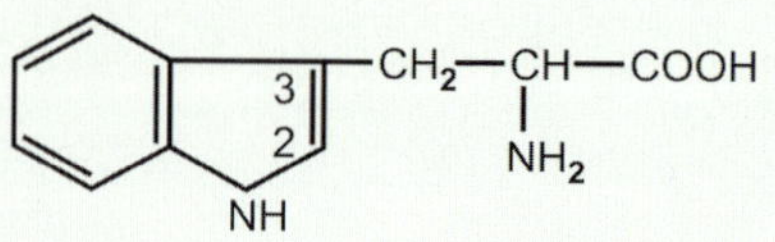

A. Metabolic Fate (Fig. 27.17)

1. Main pathway of its catabolism is by way of anthranilic acid called as **"Kynurenine-anthranilate" pathway**. In this pathway, tryptophan is finally converted to glutaric acid, which in turn gives two molecules of acetyl-CoA (thus it is ***ketogenic***) from acetoacetyl-CoA. It also produces alanine which on transamination can form ***Pyruvic acid*** (thus it is ***glucogenic***).

Main Pathway of catabolism is shown schematically below:

Characteristics of Tryptophan Pyrrolase

- The first enzyme in catabolic pathway is ***tryptophan pyrrolase*** (also called ***tryptophan oxygenase***).
- It catalyses cleavage of indole ring with incorporation of 2 atoms of molecular O_2 forming N-Formyl Kynurenine.
- The oxygenase is an Fe-porphyrin metalloprotein.
- It is ***inducible*** enzyme in Liver-increased synthesis may be induced by adrenal glucocorticoids (cortisol) and also by tryptophan itself.
- A considerable portion of newly synthesised enzyme is in a 'latent' form that requires further activation.
- The enzyme is "feedback" inhibited by nicotinic acid derivatives including NADPH.

B. Metabolic Role (Fig. 27.18)

- The amino acid tryptophan is both **glucogenic** and **ketogenic** (See below).
- ***Nicotinic acid formation***

  Amino acid tryptophan has been shown to synthesise nicotinic acid in the body and normally contributes the niacin supply of body. Many of the diets causing Pellagra are low in good quality protein as well as vitamins. ***Staple maize eaters suffer from pellagra as maize protein is deficient/lacks in tryptophan.*** Pellagra is usually due to a combined deficiency of tryptophan and the vitamin niacin.

Tryptophan rich diet has "sparing effect" on niacin requirement in diet. ***60 mg of tryptophan can give rise to 1 mg of niacin.***

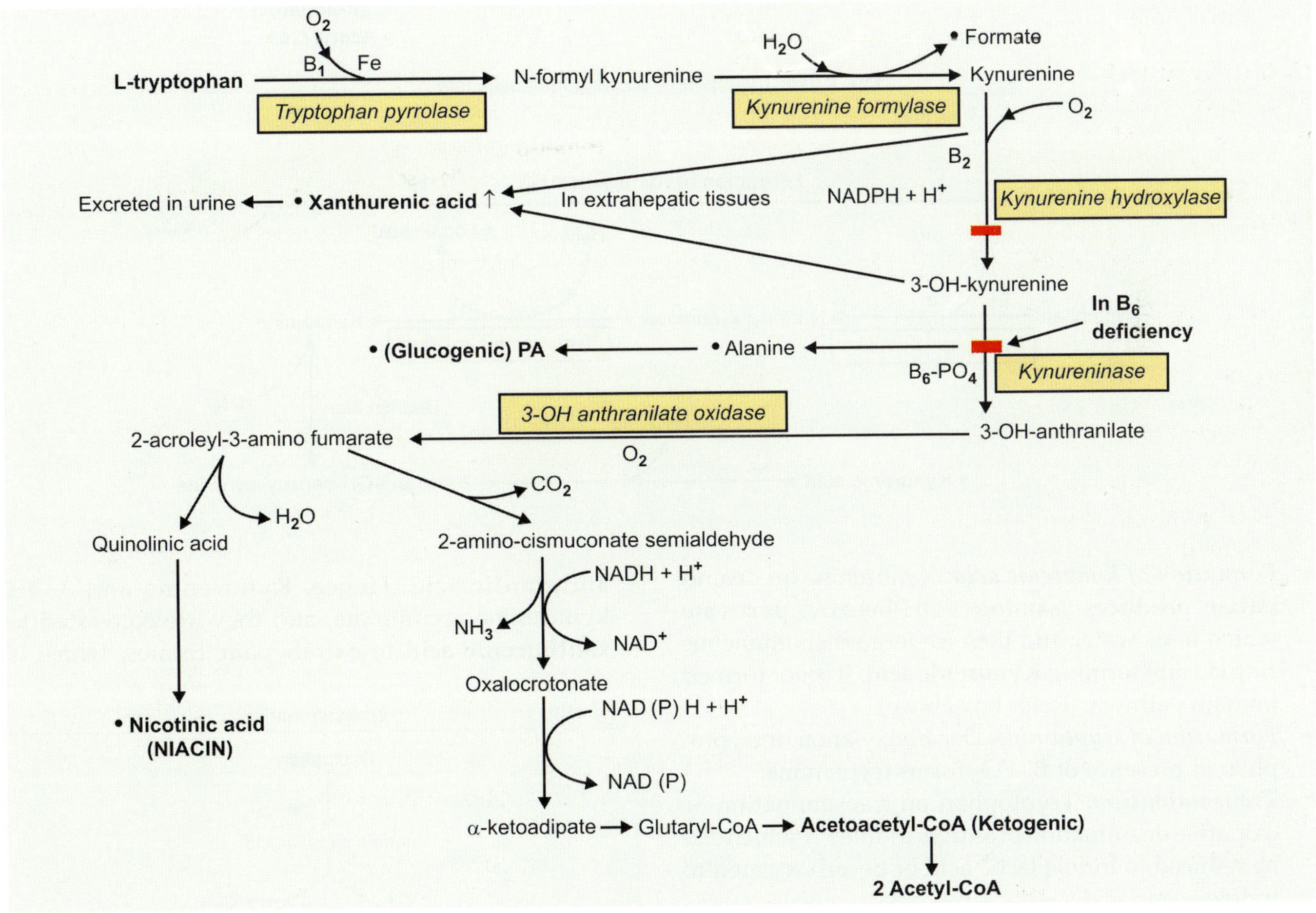

Fig. 27.17: Metabolic fate of tryptophan

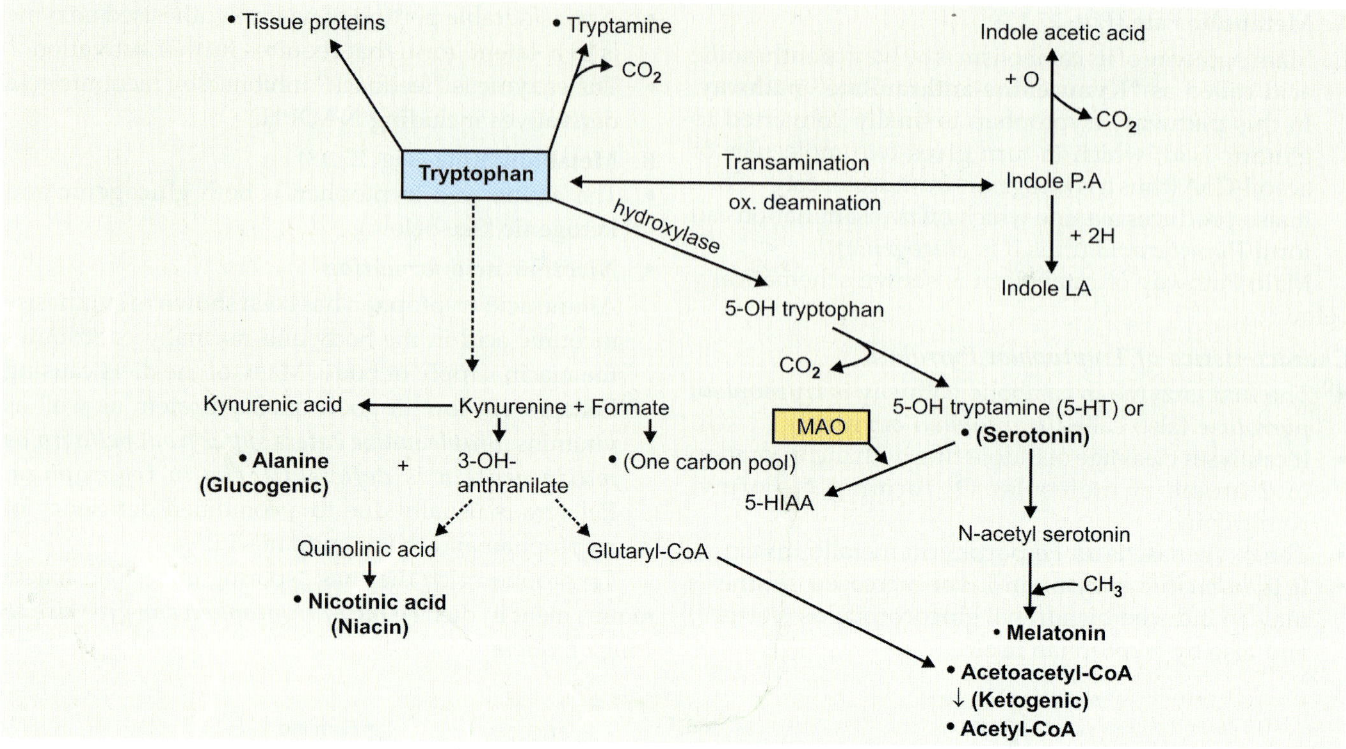

FIG. 27.18: Tryptophan—showing metabolic role

Formation of Kynurenic Acid

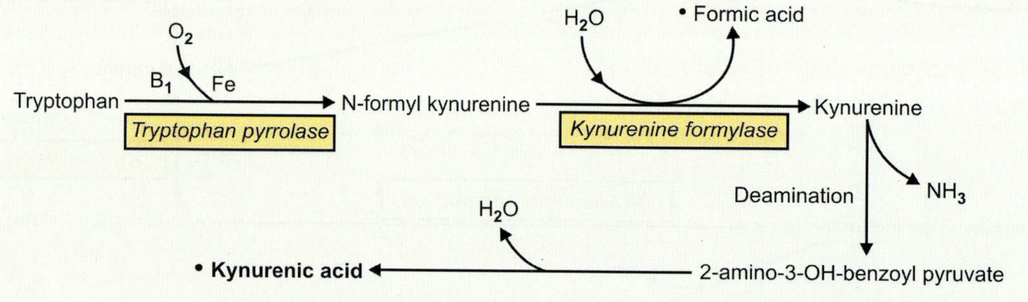

- ***Formation of kynurenic acid:*** Kynurenine on deamination produces 2-amino-3-OH-benzoyl pyruvate which loses water and then undergoes spontaneous ring closure forming **Kynurenic acid**. It is not formed in main pathway (Refer box above).
- ***Formation of tryptamine:*** Decarboxylation of tryptophan in presence of B_6-PO_4 forms tryptamine.
- ***Transamination:*** Tryptophan on transamination or oxidative deamination produces Indole PA which can be reduced to Indole lactic acid or decarboxylated to Indole acetic acid.
- ***Formation of xanthurenic acid: In B_6-deficiency,*** 3-OH kynurenine cannot be converted to 3-OH-anthranilic acid. Hence, Kynurenine and 3-OH kynurenine accumulate, and they are converted to **xanthurenic acid** in extrahepatic tissues, which is

Transamination

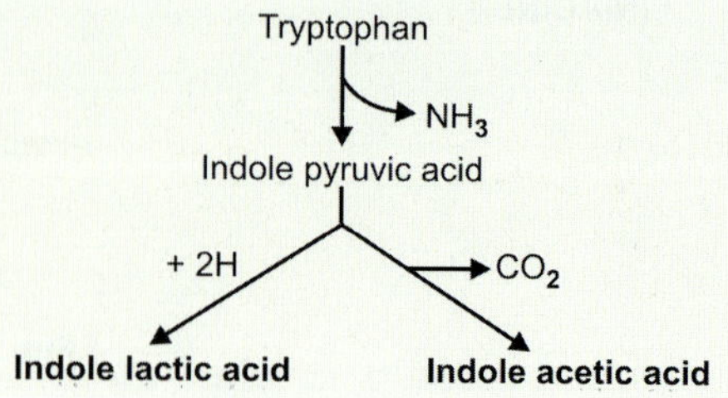

excreted in urine. ***Xanthurenic acid excretion in urine is an index for B_6-deficiency*** (Refer to box below).

Formation of Xanthurenic Acid

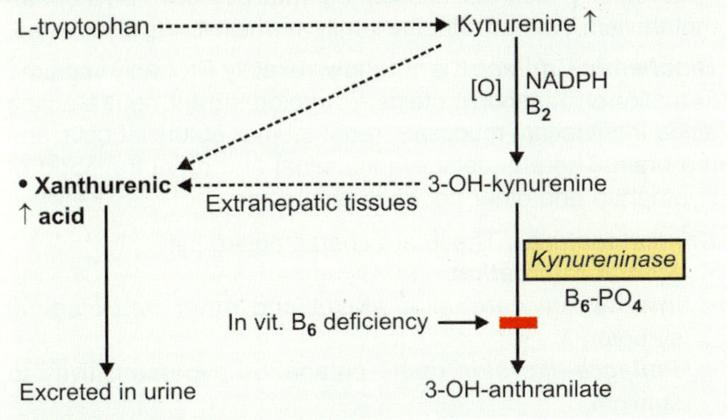

CLINICAL INTERPRETATION

1. Normal persons excrete only 1 to 3 mg of xanthurenic acid per day and 2 to 11 mg in 24 hours urine after taking a 2 gm "load" of tryptophan (loading test).
2. ***In Pyridoxine deficiency,*** much greater amount of Xanthurenic acid is excreted in urine. Following a "loading" test up to 60 mg be excreted in 24 hours urine.

- *Formation of serotonin: Another major pathway. Synonyms:* other names of serotonin are "enteramine" or "thrombocytin" (See below).

SEROTONIN

Chemistry

It is 5-OH tryptamine (5-HT). Serotonin is a vasoconstrictor substance. It is present in the blood and is produced in tissues like gastric mucosa, intestine, brain, mast cells and platelets (?), probably stored in platelets. It is produced by special cells called as ***serotonin-producing*** cells. These cells take up silver staining hence also called **Argentaffin cells**. The cells are also known as **Kultchitsky's cells**.

Synthesis of Serotonin

- Tryptophan is first hydroxylated to form 5-OH-tryptophan in liver. The reaction is analogous to conversion of Phe → to tyrosine. Liver phenyl alanine hydroxylase also can catalyse hydroxylation of tryptophan.

  In the next step, 5-OH-tryptophan is decarboxylated, by the enzyme ***5-OH-tryptophan decarboxylase,*** in presence of B_6-PO_4 to form **5-hydroxy tryptamine (5-HT), also called serotonin.** The enzyme is present in kidney, liver and stomach. Aromatic-L-amino acid decarboxylase, widely distributed in tissues can also catalyse this reaction.

Formation of Serotonin

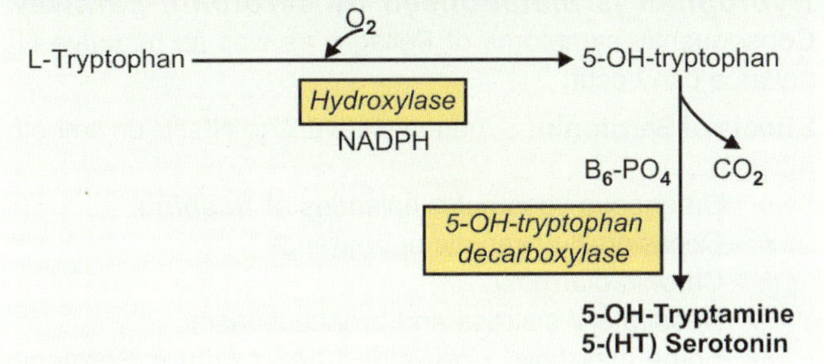

Functions

- It is a ***potent vasoconstrictor***
- Produces ***contraction of smooth muscles***
- ***Stimulator of cerebral activity.***

Note: Serotonin is actually synthesised in tissues where it is found rather than produced in one organ and carried by blood to other organs.

Effects of Serotonin on Brain

Serotonin does not pass *blood-brain barrier* to any significant amount. For action it has to be produced locally from the amino acid.

- Excess of serotonin in brain tissues produces stimulation of cerebral activity (excitation).
- Deficiency of serotonin produces depressant effect.

Catabolism of Serotonin: The enzyme which catalyses the conversion of serotonin to 5-HIAA (5-OH-Indole acetic acid) is called ***Monoamine oxidase (MAO).*** 5-HIAA is excreted in urine. Normal adults excrete about 7 mg HIAA per day.

Effects of drugs on enzyme MAO

- Drugs which inhibit the enzyme, e.g. iproniazide, will prolong serotonin action on the brain and produce a psychic stimulation due to increased cerebral activity.
- Serotonin of the brain is in a bound form. Drugs like reserpine, a common anti-hypertensive drug, acts by releasing the serotonin from its bound form and thus making it readily available to MAO action. Hence reserpine produces a depression of cerebral activity.

CLINICAL ASPECT: CARCINOIDS

A malignant tumor of serotonin producing cells is called **carcinoids (or argentaffinoma)** and the clinical features associated with is called as **carcinoid syndrome**.

Clinical Features: Symptoms are mainly due to presence of excessive amount of serotonin produced by malignant cells. ***Normal persons utilise 1 per cent of tryptophan in sero-***

tonin production; in this condition 60 per cent of tryptophan is metabolised by serotonin pathway. Consequently symptoms of Pellagra as well as negative N_2 balance can occur.

Effects of Serotonin: Symptoms are due to effects on smooth muscles:

- Cutaneous vasomotor episodes of ***flushing***
- Occasionally cyanotic appearance
- Chronic diarrhoea
- Respiratory distress and bronchospasm
- Some may have ***right-sided heart failure***. Serotonin passing through lungs is destroyed by MAO hence left side is not affected.

Urine: In malignant carcinoids, urinary excretion of 5-HIAA may increase to as much as 400 mg per day.

MELATONIN

Melatonin is a hormone of Pineal body and peripheral nerves of man.

Biosynthesis: The hormone is synthesised from serotonin:

- by N-acetylation, in which acetyl-CoA serves as acetyl donor, and
- followed by methylation of the 5-OH group, in which "S-adenosyl methionine" ("Active" methionine) serves as methyl donor. The reaction of methylation of –OH group is localised in Pineal body tissue. *Serotonin-N-acetylase* is the ***rate-limiting*** enzyme. Both synthesis and secretion of melatonin by the pineal gland is regulated by light (Refer box below).

Melatonin synthesis

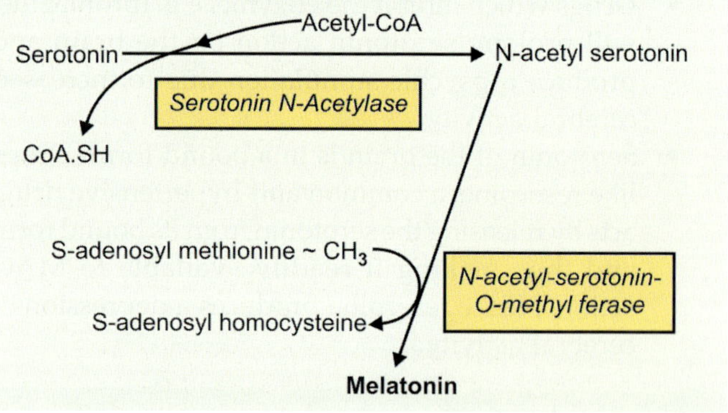

Functions of Melatonin

- Melatonin may mediate in the effect of light on seasonal reproductive cycles.
- It participates in diurnal biological rhythms.

CLINICAL ASPECT

INHERITED DISORDER

Hartnup Disease

A hereditary disorder associated with defective tryptophan metabolism. Named after the family in which it was discovered.

Biochemical defect: It is not known exactly. Probably impaired formation of "transport proteins" for tryptophan and neutral amino acids in intestinal mucosal, renal tubular epithelial cells and the brain. There is defective intestinal and renal transport of tryptophan and other neutral amino acids.

Clinical features: These are characterised by:

- ***Mental retardation***.
- Intermittent cerebellar ataxia and other neurological symptoms.
- ***Pellagra-like skin*** rash—cutaneous hypersensitivity to sunlight.

Blood: Plasma level of tryptophan and other neutral amino acids are reduced.

Faeces and Urine: The neutral amino acids, including tryptophan are excreted in urine and faeces, at least 5 to 10 times of normal average. *Faecal* excretion of tryptophan is specially marked after a "loading" dose of tryptophan given orally.

Urine: Also shows greatly increased amounts of Indoleacetic acid.

Note

There is decreased synthesis of serotonin and nicotinic acid, which accounts for neurological symptoms and pellagra like rash respectively.

METABOLIC ROLE OF OTHER AMINO ACIDS

Chemical and metabolic role in brief of other amino acids viz. Arginine, ornithine and citrulline, threonine, glutamic acid and aspartic acid, proline and OH-proline, lysine and branched chain amino acids are given in Table 27.4.

ROLE OF NITRIC OXIDE

Nitric oxide (NO) is formed in the body from amino acid arginine

It is a wonder molecule having diverse biological functions like PGs. Endothelium derived relaxing factor (EDRF) which produces vasodilatation is now proved to be nitric oxide.

Formation of NO

Arginine is acted upon by an enzyme called ***nitrogen oxide synthase***, a cytosolic enzyme and converts arginine to citrulline and nitric oxide (NO).

Table 27.4: Showing important metabolic role of other amino acids

S.N.	*Name of the amino acid*	*Nature and properties*	*Metabolic role*
1.	**Arginine**	• Basic amino acid • Semiessential • Becomes essential in growing children, in pregnancy and lactation	• Tissue protein formation • ***Required in creatine synthesis*** • Required ***in urea formation*** • Glucogenic amino acid • ***Forms nitric oxide***
2.	**Ornithine and citrulline**	• Both are basic amino acids • Both are non-essential	• Both are intermediates in formation of urea (urea cycle) • Ornithine is regenerated in urea cycle and acts as catalytic agent to continue the cycle • Ornithine is ***required for synthesis of polyamines*** • Ornithine is glucogenic
3.	**Threonine** **Note:** Metabolic fate—Figure 27.19	• Essential amino acid • Cannot be formed by transamination of the ketoacids	• Glucogenic amino acid • By non-oxidative deamination forms α-ketobutyric acid, which on oxidative decarboxylation ***gives Propionyl-CoA*** (glucogenic) • Cleaved by the enzyme threonine aldolase to form glycine and acetaldehyde **(Refer Fig. 27.19)** • Threonine can also be converted to pyruvic acid (thus glucogenic) • Like serine, it is a hydroxy amino acid acts as PO_4 carrier
4.	**Glutamic acid and aspartic acid**	• Both are acidic amino acids • Both are non-essential • Both can participate in transamination reactions	• Both glucogenic—on deamination forms OAA and α-ketoglutarate (intermediates of TCA cycle) • Glutamic acid on decarboxylation ***forms GABA*** • Glutamate is a ***constituent of glutathione*** • Aspartate participates in synthesis of purines and pyrimidines • Glutamate is a constituent of folic acid • ***Glutamine is formed from glutamate*** • Glutamic acid helps in transport of K^+ ions in brain tissue • N-acetyl glutamate (N-AGA) and asparate both take part in urea cycle • Proline and OH-proline can be formed from glutamic acid • Amide of glutamic acid, glutamine takes part in conjugation reactions • Amide of aspartic acid, asparagine, is an important NH_3 donor
5.	**Proline and OH-proline** **Note:** Inherited disorder—hyperprolinaemia (Refer box page 523)	• Both are non-essential amino acids	• OH-proline forms structural part of collagen tissues and elastin • OH-proline is synthesised in collagen tissue from proline for which vitamin C is required • OH-proline is catabolised to form pyruvic acid and glyoxylate (glucogenic)
6.	**Lysine** **Note:** ***This amino acid is not present in adequate amounts in cereal proteins. Hence its deficiency may occur in strict vegetarians***	• Essential amino acid	• It is both glucogenic and ketogenic • Cannot take part in transamination reactions • OH-lysine, like OH-proline is required for collagen tissues • OH-lysine is synthesised in collagen tissues from lysine for which vitamin C is required
7.	**Branched-chain amino acids valine, leucine and isoleucine** **Note:** For metabolic fate—refer text.	• All the three amino acids are essential	• Valine is glucogenic. On deamination forms methyl malonyl-CoA which is converted to succinyl-CoA (glucogenic) • Leucine is potent ketogenic amino acid • Isoleucine, after oxidative removal of one carbon forms propionyl-CoA (glucogenic) and one molecule of acetate (ketogenic).

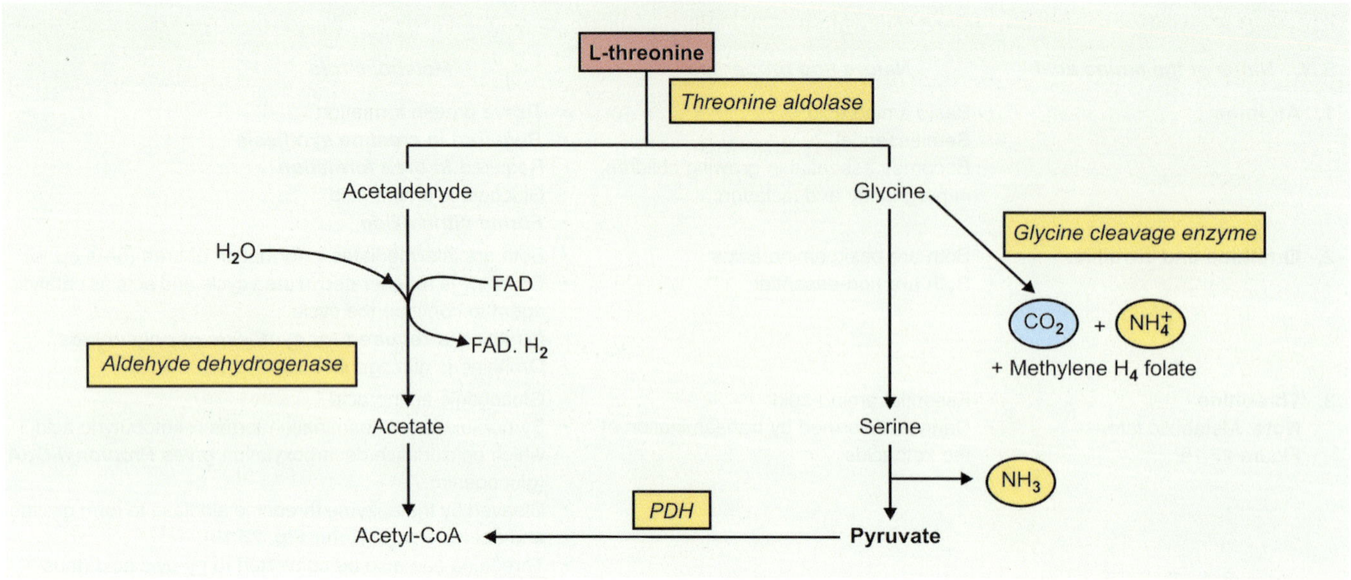

FIG. 27.19: Metabolic fate of L-threonine

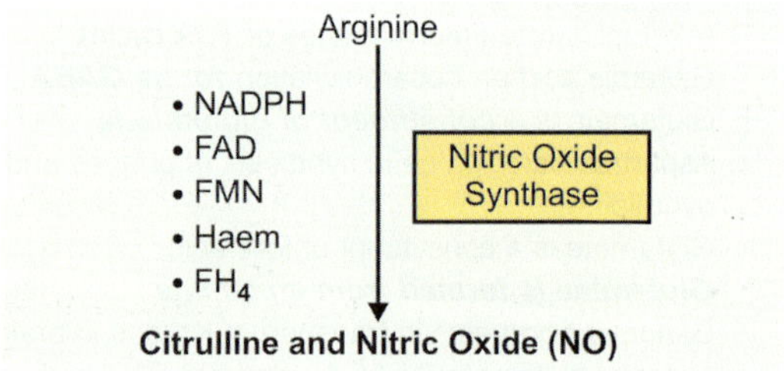

Mechanism of Enzyme Action

Nitric oxide (NO) synthase catalyses a five-electron oxidation of an amidine nitrogen of arginine. L-OH-arginine is an intermediate in the reaction that remains tightly bound to the enzyme. Nitric oxide synthase (NOS) is a very complex cytosolic enzyme which requires five redox cofactors: **NADPH, FAD, FMN, haem and tetrahydrobiopterin** (FH_4).

(a) Endothelial cell: Shows formation of nitric oxide (NO) from argonine catalysed by the enzyme ***nitric oxide synthase***. Interaction of an agonist, e.g. (acetylcholine) with a receptor (R), leads to ***release of intracellular*** Ca^{++} via inositol triphosphate generated by the phosphoinositide pathway resulting in activation of no synthase.

(b) Smooth muscle cell: The nitric oxide (NO) subsequently diffuses into **adjacent smooth muscle** cell, where it leads to activation of guanylyl cyclase, **forming cGMP**, which stimulates cGMP-protein kinases and **subsequent relaxation**.

The vasodilator nitroglycerine (glyceryl trinitrate) enter the smooth muscle cell, where its metabolism also leads to formation of nitric oxide (Fig. 27.20).

Isoenzymes of Nitric Oxide Synthase (NOS)

Three isoenzymes have been isolated and described:

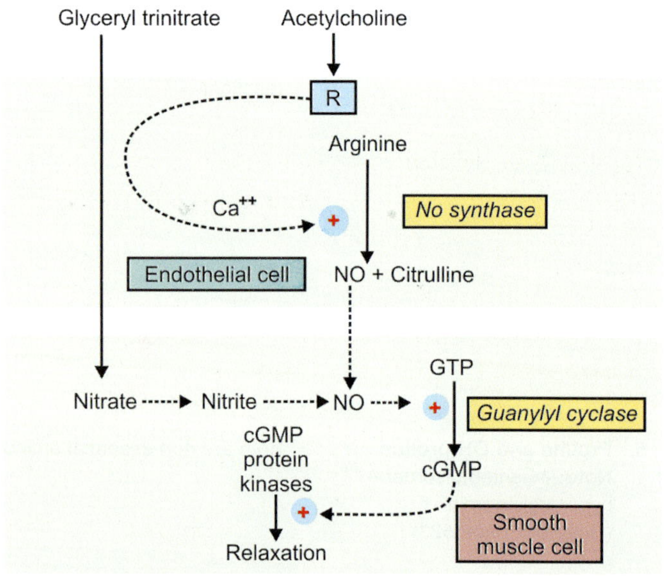

Fig. 27.20: Showing formation of nitric oxide in endothelial cell and relaxation of smooth muscle cell

- ***Endothelial NOS (eNOS):*** Identified first in endothelial cells. Also later on found in myocardium, endocardium, and platelets. In these sites, nitric oxide is constantly produced and released so as to have arterial relaxation. Activity of eNOS depends on elevated Ca^{2+} ions.
- ***Neuronal NOS (nNOS):*** Identified in central and peripheral neurones. Nitrogen oxide producing neurones are seen specially in cerebellum. Activity of this isoenzyme also depends on elevated Ca^{2+} ions.
- ***Macrophage NOS (iNOS):*** Found in macrophages. Responsible for bactericidal actions of macrophages. The action is independent of elevated Ca^{2+}.

Duration of action of NO: Nitric oxide formed in the tissues has a very short half-life, approximately 3 to 4 seconds because it reacts with oxygen and superoxide. The product of the reaction with superoxide is Peroxynitrite ($ONOO^-$), which decomposes to form the highly reactive $OH^{\cdot}$ radical.

Functions of Nitric Oxide (NO)

Functions of nitric oxide are summarised and given in the box below:

Functions of Nitric Oxide

- It acts as a **vasodilator** and causes relaxation of smooth muscles.
- It has important role in the regulation of blood flow and maintaining blood pressure.
- It is involved in penile erection.
- Acts as a **neurotransmitter** in the brain and peripheral autonomic nervous system.
- May have also role in relaxation of skeletal muscles.
- Inhibits adhesion, activation and aggregation of platelets.
- May constitute part of a primitive immune system and may mediate bactericidal actions of macrophages.
- Low level of nitric oxide may be involved in causation of pylorospasm of infantile hypertrophic pyloric stenosis.

Mechanism of Action: Vasodilatation effect of nitric oxide and inhibition of platelet aggregation are mediated through cyclic GMP and protein kinase.

Inhibitors

- Nitric oxide (NO) is inhibited by Haemoglobin and other haem proteins which bind it tightly.
- Chemical inhibitors of *NO synthase* are now available that causes marked decrease formation of NO.
- ***Endogenous inhibitor:*** Asymmetric dimethyl arginine (ADMA), an endogenous arginine analogue may function as a competitive inhibitor of NO synthase. ADMA has been found to be increased in pre-eclampsia.

CLINICAL ASPECT

- ***Nitroglycerine:*** The important coronary artery vasodilator used in Angina Pectoris acts to increase intracellular release of EDRF (now proved to be NO) and cGMP↑.
- ***In septic shock:*** Bacterial lipopolysaccharide present in blood causes uncontrolled production of NO leading to dilatation of blood vessels and lowering of BP.
- ***In eclampsia and pre-eclampsia:*** The hypertension is due to decreased production of nitric oxide (NO) due to probably formation of ADMA (asymmetric dimethyl arginine).
- ***Iron supplements:*** Iron supplements can dramatically reduce dry cough symptoms in heart patients. Cardiac patients using ACE inhibitors, widely prescribed for hypertension, heart failure and other cardiac conditions often suffer from a dry cough. It is the biggest reason for people stopping taking their medication. Iron supplements act by decreasing the production of Nitric oxide, which is linked to inflammation of the bronchial cells in the lungs.

Inherited Disorders: Hyperprolinaemias

Two types of hyperprolinaemias:

Type I: Enzyme deficiency is *proline dehydrogenase.*

Type II: Metabolic defect is the *dehydrogenase* that is involved in conversion of glutamate-γ-semialdehyde to glutamic acid.

Inheritance: Both are autosomal recessive traits.

Though mental retardation may be present in 50 per cent of cases, both the conditions are harmless.

Metabolic Fate of Branched Chain Amino Acids

- Due to their structural similarities, catabolism of leucine, valine and isoleucine initially involves the same reactions.
- The above common pathway then diverges, and each amino acid then follows a unique pathway to yield amphibolic intermediates. The nature of these amphibolic intermediates determine whether the amino acid is glucogenic/or ketogenic or both.

A. Initial Reactions Common to All Three Amino Acids

- ***Reversible transamination,*** probably by a single transaminase.
- ***Oxidative decarboxylation*** to Acyl-CoA thioesters. This reaction is analogous to oxidative decarboxylation of PA to form acetyl-CoA. The branched chain *α-ketoacid dehydrogenase complex* is an intramitochondrial multienzyme complex that catalyzes the oxidative decarboxylation of corresponding ketoacid of the three amino acids.
- ***Dehydrogenation*** to form α-β-unsaturated acyl-CoA thioesters. These three reactions common to all three-branched chain amino acids are shown schematically in box on next page.

B. Subsequent Catabolism: Unique to each of three amino acids are shown on next page:

(a) Specific reactions of leucine catabolism: This follows the following steps:

- ***Carboxylation*** of β-methyl-crotonyl-CoA, which forms β-methyl-glutaconyl-CoA.
- ***Hydration*** of β-methyl-glutaconyl-CoA to form HMG-CoA. This is precursor not only of ketone bodies, but also precursor for Mevalonic acid so that Polyisoprenoids and cholesterol can be formed.

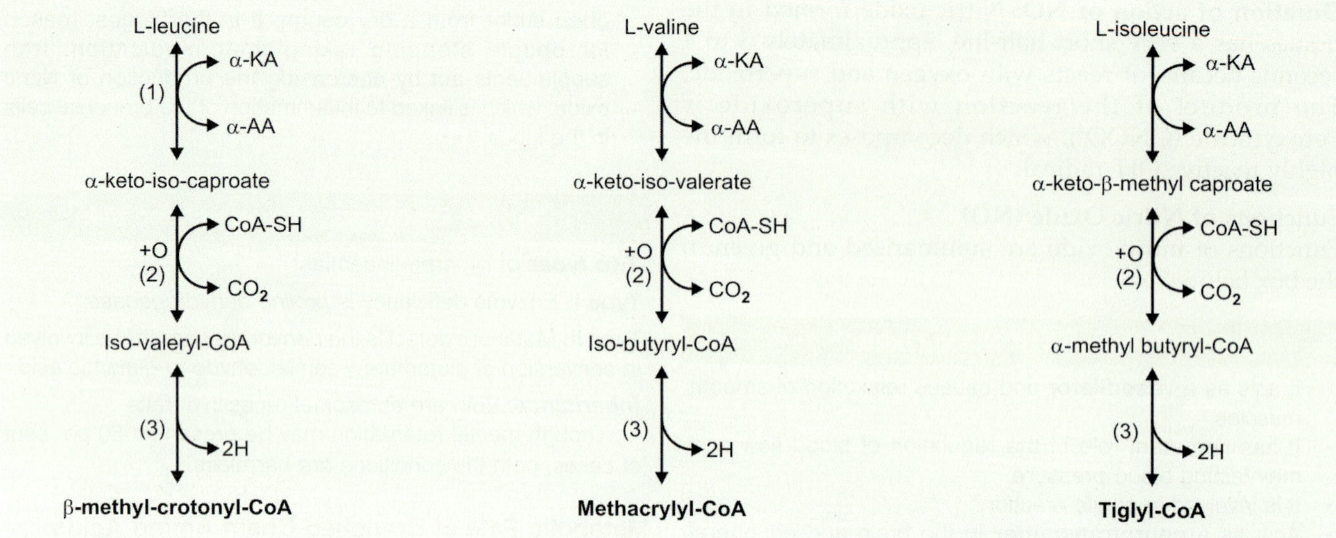

- ***Cleavage of HMG-CoA:*** Cleavage of HMG-CoA to acetyl-CoA and acetoacetate occurs in Liver, kidney and Heart muscles in mitochondria.

Thus, ***Leucine is strongly ketogenic,*** not only one molecule of acetoacetate is formed per mole of Leucine catabolised but another ½ mole of ketone bodies may be formed indirectly from the remaining product acetyl-CoA.

(b) Specific reactions of valine catabolism: This follows the following steps:

- ***Hydration*** of metha-acrylyl-CoA to form β-OH-isobutyryl-CoA. This reaction which occurs non-enzymatically at a relatively rapid rate, is catalysed by ***crotonase,*** a hydrolase of broad specificity.
- ***Deacylation*** of β-OH-isobutyryl-CoA to form β-OH-isobutyrate, the reaction is catalysed by a *deacylase* present in mammalian tissues.

- **Oxidation** of β-OH-isobutyrate to form methyl malonate semialdehyde, the reaction is rapidly reversible.
- ***Fate of methylmalonate semialdehyde***

  Two possible fates:
  - It can be transaminated to form β-amino-isobutyrate.
  - More commonly it is ***converted to succinyl-***CoA (***Major fate***): Involves oxidation to form methyl malonate, which is acylated to form methyl malonyl-CoA, subsequently isomerised to form succinyl-CoA (thus **glucogenic**).

(c) Specific reactions of isoleucine catabolism: This follows the following steps:

- **Hydration** of Tiglyl-CoA to form α-methyl-β-OH-butyryl-CoA, the reaction is catalysed by *crotonase.*
- **Dehydrogenation** of α-methyl-β-OH-butyryl-CoA to form α-methyl acetoacetyl-CoA.
- **Thiolysis** of α-methyl acetoacetyl-CoA: Reaction is catalysed by *β-ketothiolase* and products formed are acetyl CoA **(Ketogenic)** and Propionyl-CoA **(Glucogenic)**. ***Hence this amino acid is both glucogenic and ketogenic.***

CLINICAL ASPECT

INHERITED DISORDERS

1. Maple Syrup Urine Disease

An inherited disorder of branched chain amino acids.

Enzyme defect: Absence of ***α-ketoacid decarboxylase*** or greatly reduced activity of the enzyme. As a result the conversion of all three branched chain α-ketoacids to CO_2 and acyl CoA-thioesters is interferred with.

Clinical features: The disease is evident by the end of first week of extrauterine life. Infant does not take feed and may vomit, poor muscle tone. The patient may exhibit lethargy and

Catabolism of Leucine

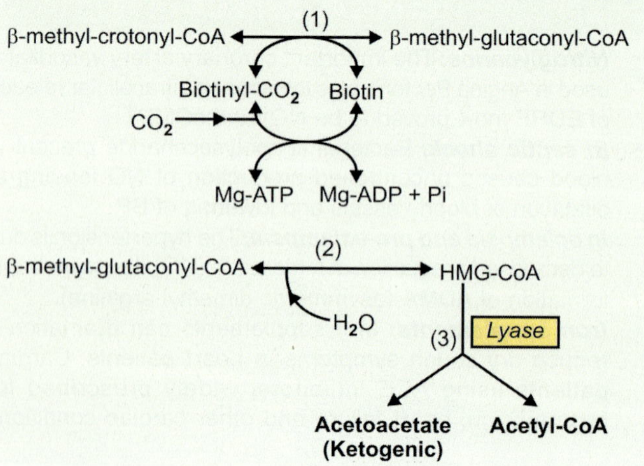

Catabolism of Valine

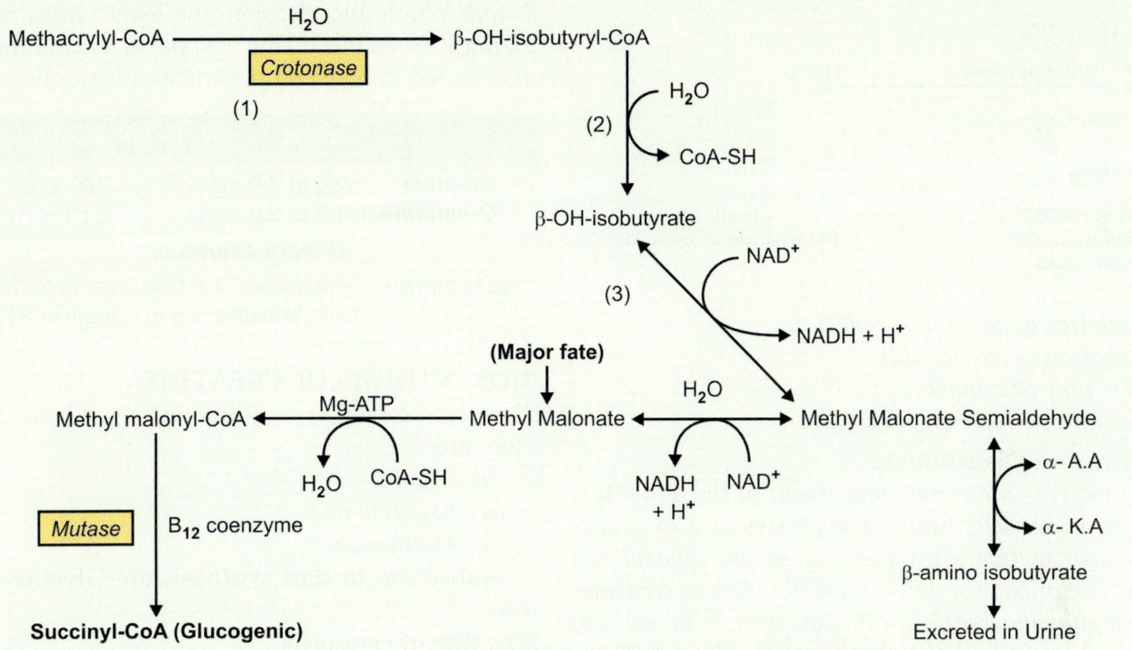

Catabolism of Isoleucine

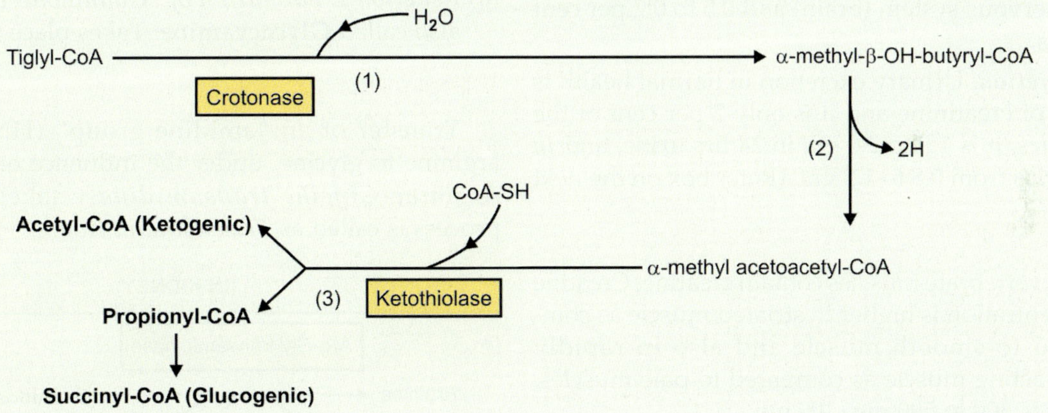

convulsive seizures. Extensive brain damage can occur in surviving children and mental retardation. Without treatment, death usually occurs by the end of the first year of life.

Blood: Plasma levels of the branched chain amino acids leucine, isoleucine, valine and their corresponding α-ketoacids are greatly elevated.

Urine: Branched chain amino acids leucine, isoleucine, valine and their corresponding α-ketoacids are excreted. Hence it is also called as ***Branched-chain ketonuria.*** Small amounts of branched-chain α-OH-acids, formed by reduction of α-keto-acids are also excreted in urine. The urine has characteristic odour, which resembles that of ***maple syrup or burnt sugar,*** hence the name.

Treatment: Infant is fed a diet in which protein is replaced by a mixture of purified amino acids from which branched chain amino acids leucine, valine and isoleucine are omitted.

METABOLISM OF CREATINE

Two closely related nitrogenous compounds which are connected with protein metabolism are:

- **Creatine** and
- **Creatinine**.

Structure and relationship of these two compounds are shown in the box:

SECTION FOUR

Formation of creatinine

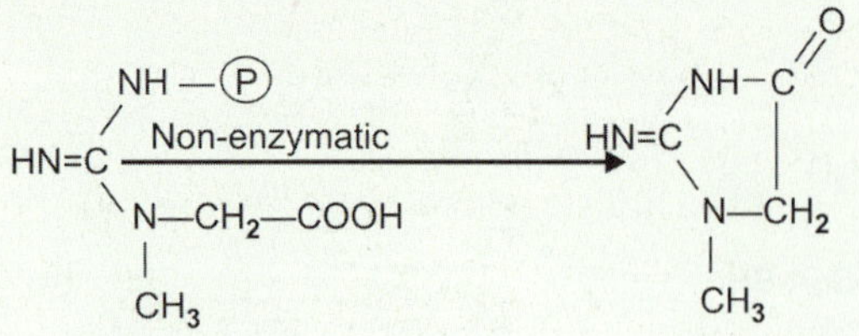

Creatine-(P) Methyl guanido-acetic acid

Creatinine (Anhydride of creatine)

Characteristics of the above reaction

- Reaction is ***irreversible***
- It is *non-enzymatic*
- Creatinine has ***ring structure.***

Occurrence and Distribution:

A. *Creatine:* ***It is a normal constituent of the body.*** It is present in muscle, brain, liver, testes and in blood. Can occur in *free* form and also as ***phosphorylated*** form. The phosphorylated form is called as ***creatine-*** PO_4 or ***phosphocreatine*** or ***Phosphagen.*** Total amount in adult human body is approximately 120 gm. 98 per cent of total amount is present in muscles, of which 80 per cent occurs in phosphorylated form, 1.3 per cent in nervous system (brain) and 0.5 to 0.7 per cent in tissues.

Urinary excretion: Urinary excretion in normal health is in the form of creatinine and it is only 2 per cent of the total. ***In males,*** it is 1.5 to 2.0 gm in 24 hrs urine, and ***in females,*** varies from 0.8 to 1.5 gm. (Refer box on the next page).

Note

- Only vertebrate muscles contain creatine. Creatine concentration is higher in striated muscle as compared to smooth muscle and also in rapidly contracting muscle as compared to pale muscles. Total is 300 to 500 mg/100 gm.
- In invertebrates: Arginine replaces creatine in muscles.

Blood and plasma level

- **In whole blood:** Creatine level varies from 2 to 7 mg%.
- **In plasma:** It is less than 1 mg%.
  In male: It varies from 0.2 to 0.6 mg%.
  In females: 0.35 to 0.9 mg%.
  (Refer box on the next page).

B. *Creatinine:* Creatinine is the anhydride of creatine, and ***it is in this form that creatine is excreted in normal health.*** Removal of one molecule of H_2O is *non-enzymatic* and ***irreversible.*** Formation of creatinine is a preliminary step and prerequisite for excretion of most of creatine. Total creatinine in muscle is only 0.01 per cent (10 mg).

Blood: Whole blood creatinine level varies from 1.0 to 2.0 mg% (Refer box on the next page). Creatinine is evenly distributed in between plasma and RB Cells.

	Whole blood	Muscles
• *Creatine*	2.0 to 7.0 mg%	300 to 500 mg
• *Creatinine*	1.0 to 2.0 mg%	0.01% (10 mg)
	Urinary excretion	
• *Creatinine*	• In males: 1.5 to 2.0 gm in 24 hr • In females: 0.8 to 1.5 gm in 24 hr	

BIOSYNTHESIS OF CREATINE

Three amino acids are required in biosynthesis of creatine. They are:

(i) Glycine
(ii) Arginine and
(iii) Methionine

Substrates to start synthesis are Glycine and Arginine.

Site: **Site of synthesis**

- ***In kidney***
- ***In liver***

(a) Reaction 1: Formation of "Guanidoacetic acid": It is also called **Glycocyamine.** Takes place in Kidney.

Transfer of an "amidine group" (HN = C with NH_2) from arginine to glycine, under the influence of the enzyme ***Arginine-Glycine transamidinase*** takes place. The process is called as ***Transamidination.***

(IN KIDNEY)

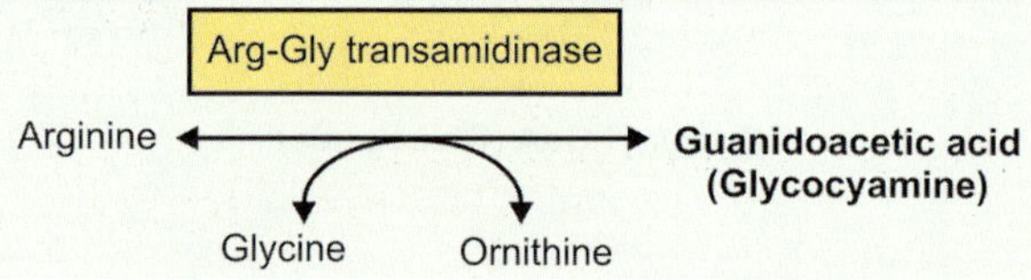

(b) Reaction 2: Formation of Creatine-(P)

- Enzyme catalysing the reaction is ***Guanidoacetate methylferase.*** S-adenosyl methionine (Active methionine) is the "methyl" donor, for methylation.
- ATP is required for the synthesis which donates the PO_4. ***Also*** O_2 ***is required for the reaction (aerobic).***
- Reaction is ***irreversible,*** and occurs in liver.

Once creatine-(P) is formed in liver, it goes to muscles, and stored. ***Creatinine is formed from creatine-(P) in muscles by non-enzymatic*** and ***irreversible reaction.***

(IN LIVER)

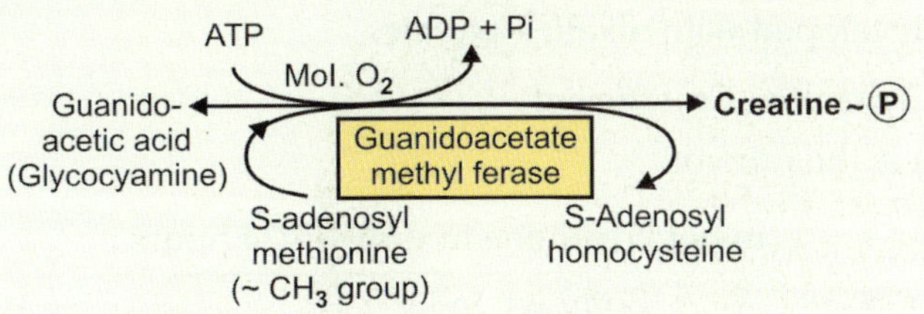

Other methyl donors: Betaine and choline (after oxidation to Betaine) may also serve indirectly by producing methionine through the methylation of homocysteine.

Other sites of creatine synthesis: Recently it has been shown that **pancreas** can synthesise glycocyamine and creatine-(P). An enzyme preparation has been isolated from pancreatic tissues of beef as well as dogs, which can catalyse the synthesis of creatine from glycocyamine and "active" methionine. It has also been shown recently that pancreatic tissue (in contrast to liver) can synthesise guanidoacetic acid (glycocyamine). Thus it has been concluded that ***pancreas singly may play a unique role in the synthesis of creatine within the body of mammals.***

Regulation of Creatine Synthesis

1. Dietary creatine has effect on creatine synthesis. In rats, fed a complete diet containing 3 per cent creatine, *transamidinase* activity of the kidney was markedly lower as compared to control animals.
2. But dietary creatine or a high blood creatine has no effect on rate of synthesis of creatine in liver.
3. It is also shown that hepatic synthesis of creatine is related to the blood glycocyamine levels and that this compound is produced in kidney, suggests that the ***rate of creatine biosynthesis is actually dependent on kidney transamidinase activity.*** Thus, the activity of transamidinase in kidney is the key regulating enzyme for creatine biosynthesis and is affected by creatine level as a "feedback" mechanism.

Excretion

1. When creatinine is ingested, most of it is rapidly eliminated in urine. It can be quantitatively recovered.
2. But when creatine is taken, some is retained in the body. It has been seen by giving labelled creatine that 20 to 30 per cent is excreted as creatinine and some is retained in the body whose fate is not known.

Urine of normal healthy adult male contains creatinine but no creatine. Amount of creatinine excreted as discussed above is approximately 1.0 to 1.5 gm/day and this is:

- Independant of amount of proteins taken in the diet.
- Excretion is greater in muscular persons and appears to be related to muscular development and muscular activity.
- After severe exercise, it may increase, but total amount remains constant from day-to-day.

Creatinuria

Excretion of creatine in urine is called creatinuria. Creatine excretion occurs:

- In children: Reason probably lack of ability to convert creatine to creatinine.
- In adult females in pregnancy and maximum after parturition (2 to 3 weeks).
- *In febrile conditions*
- *In thyrotoxicosis*, probably due to associated myopathies.
- In muscular dystrophies, myositis, and myasthenia gravis.
- Lack of carbohydrate in diets and in diabetes mellitus.
- In wasting diseases, e.g. in malignancies.
- In starvation.

Role of Creatine in Muscles

1. Creatine is the reservoir of energy in muscles. When muscles contract, energy is derived from breakdown of ATP to ADP and Pi. ***ATP must be reformed quickly,*** to supply the energy, which initially comes from creatine ~ (P), subsequently from glycolysis (contracting muscle).

Löhmann Reaction

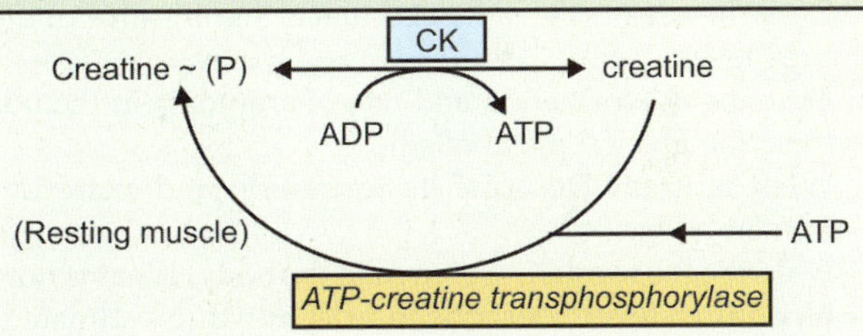

From the above reaction, ATP is formed from creatine ~ (P). The high energy phosphate is transferred to ADP and ATP is formed. This reaction is called **Löhmann reaction** and it takes place during activity of the muscles. In the resting condition, creatine ~ (P) is reformed, the enzyme that catalyses the reaction is ***ATP-creatine transphosphorylase.***

2. A further source of ATP in muscle is by the **Myokinase reaction.** Two ADP molecules react to produce one molecule of ATP and AMP, the reaction is catalysed by the enzyme ***myokinase (Adenylate kinase).***

Myokinase Reaction

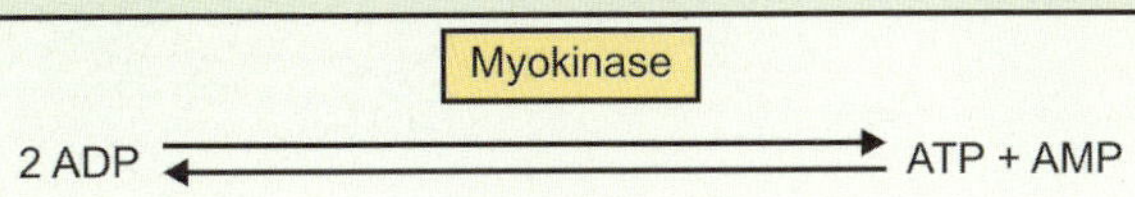

In this reaction, one high energy phosphate is transferred from one ADP to another ADP molecule to form one ATP.

Estimation of Creatinine

Jaffe's reaction: Serum is treated with alkaline picrate solution when a red colour develops **(Jaffe's reaction).** The colour is read against a 'standard' similarly treated in a colorimeter

Estimation of Creatine

When heated with acid solution, creatine is converted to creatinine, which can be measured in a similar way as stated above.

Value after boiling with acid solution –value before boiling = creatine content.

1.0 gm of creatinine is formed from 1.16 gm of creatine. Hence, substract the pre-formed creatinine from the total creatinine × multiply by 1.16.

"True" Creatinine

Serum creatinine estimation by Jaffe's reaction does not give "true" creatinine. ***It measures also certain non-creatinine chromogens,*** upto 20 per cent in blood and up to 5 per cent in urine. For excluding the chromogens and to get 'true' creatinine, after precipitating the proteins, creatinine is adsorbed on to **Lloyd's reagent (Fuller's earth)**, a hydrated aluminium silicate, and then colour developed with alkaline picrate.

Creatinine Co-efficient

It is the ratio of

$$= \frac{\text{mg of creatinine in urine in 24 hours}}{\text{Body wt. in kg}}$$

The value is 20 to 26 for males and 14 to 22 in females.

Significance

- It depends on muscular development and remains fairly constant.
- As the rate is so constant in a given individual the creatinine co-efficient may serve as a reliable index of the adequacy of a 24-hr urine collection.

Creatinine Clearance

Endogenous creatinine clearance is used as renal function test. At normal levels of creatinine in the blood, this metabolite is filtered at the glomerulus but neither secreted nor re-absorbed by the tubules. Hence its clearance measures the glomerular filtrate rate (GFR) (Refer Chapter on Renal Function Tests).

SECTION FOUR

I. Questions (Essay type and short notes)

1. Describe urea cycle. Discuss clinical significance of blood urea level.
2. Describe the formation and fate of ammonia in the body.
3. Describe glycine metabolism.
4. What is urea? Describe its formation and excretion in human body.
5. Describe the formation of urea in the body. What is normal blood urea level? Describe in brief how it is estimated.
6. Describe the metabolic fate and role of tyrosine in the body.
7. Describe the metabolism of sulphur-containing amino acids.
8. Describe the metabolic fate and role of methionine in the body.
9. Describe the fate of amino acids after their absorption.
10. Comment on the concept of "Amino acid Pool". Give an account of fate of amino acids in the body after its absorption.
11. Describe the metabolic fate and role of cysteine in the body.
12. Describe the formation and metabolic importance of the following:
    a. Glutamine b. 'Active' methionine
    c. Carbamoly-(P)
13. Give an outline of urea cycle reactions. Also mention other fates of NH_3 in the body.

Short Notes

a. Alkaptonuria
b. Phenyl ketonuria
c. Glutathione
d. Transamination
e. Transmethylation
f. Biosynthesis of creating
g. Serotonin
h. Nitrogen balance
i. Homocystinuria
j. Urea cycle
k. Creatinine coefficient
l. Albinism
m. Essential amino acids
n. Biosynthesis of catecholamines (adrenaline)
o. Creatine
p. Histamine
q. Decarboxylation
r. Urea
s. Catecholamine-Biosynthesis and degradation
t. Oxidative deamination
u. Glycine
v. Abnormalities of phenylalanine metabolism
w. Polyamines.

II. MCQ's (Give one correct answer)

1. Quantitatively the most important enzyme involved in formation of NH_3 from amino acids in humans is:
   a. L-amino acid oxidase
   b. Serine dehydratase
   c. Glutamate dehydrogenase
   d. Histidase
   e. Desulfhydrase
2. The two nitrogen atoms in urea arise from:
   a. Ammonia and glutamine
   b. Glutamine and aspartic acid
   c. Glutamine and glutamic acid
   d. Ammonia and aspartic acid
   e. Glutamine and alanine
3. Ochronosis is a feature of:
   a. Phenyl ketonuria
   b. Hereditary tyrosinaemia
   c. Albinism
   d. Hypervitaminosis D
   e. Alkaptonuria
4. Which of the following substance is responsible for GI disturbances and respiratory distress seen in "carcinoid syndrome"?
   a. Tryptamine b. 5-OH tryptamine
   c. Tyramine d. Histamine
   e. GABA
5. A product of the series of reactions that converts carbamoyl-(P) to urea is:
   a. Arginine b. Citrulline
   c. Fumarate d. Aspartate
   e. Adenosine triphosphate
6. Ornithine has all of the following properties except:
   a. It is an intermediate of urea cycle
   b. It is not found in proteins
   c. It is glucogenic
   d. It can be formed from arginine
   e. It is a mono-amino mono-carboxylic acid
7. Epinephrine is formed from norepinephrine by:
   a. Hydroxylation b. Oxidative deamination
   c. Decarboxylation d. N-methylation
   e. O-methylation
8. The rate limiting step in the biosynthetic pathway of catecholamines is:
   a. The hydroxylation of phenylalanine
   b. The hydroxylation of tyrosine
   c. The reduction of biopterine
   d. The formation of dopamine
   e. None of the above
9. Which of the following compounds serve as a primary link between the TCA cycle and the urea cycle?
   a. Malate b. Citrate
   c. Oxaloacetate d. Fumarate
   e. Succinate
10. Serine can be a percussor of each of the following, except:
    a. Cysteine b. Methionine
    c. Ethanolamine d. Glycine
    e. Choline
11. Homocystinuria is an inherited disorder characterised by:
    a. Mental retardation b. Frequent renal stones
    c. Aortic aneurysm d. Ectopia lentis
    e. Paralysis of limbs
12. All are true about phenylketonuria, except:
    a. Deficiency of phenylalanine hydroxylase
    b. Mental retardation
    c. Increased urinary excretion of p-OH phenyl pyruvic acid
    d. Decrease serotonin formation
    e. Increase serum phenyl alanine
13. The rate of excretion of certain substances in 24-hr sample of urine can be used to estimate the rate of muscle protein loss. Which of the following substances falls into this category?
    a. Ammonia b. Urea
    c. Uric acid d. 3-Methyl histidine
    e. Ergothioneine
14. Degradation and excretion of biogenic amines involve all of the following reactions, except:
    a. Conjugation to asparagine
    b. Methylation
    c. Conjugation to sulphates
    d. Conjugation to glucuronic acid
    e. All of the above
15. Tryptophan is best described by which of the following statement?
    a. It produces thyroid hormones
    b. Is a precursor for melanin
    c. It is a precursor of the pineal hormone melatonin
    d. It produces catecholamines
    e. It is glucogenic only
16. Which of the following amino acids are required for synthesis of creatine?
    a. Arginine, Asp, 'active' methionine
    b. Arginine, glycine 'active' methionine
    c. Arginine, lysine, methionine
    d. Glycine, Asp, ornithine
    e. Arginine, glutamic acid, 'active' methionine
17. Deficiency of which of the following vitamins cause creatinuria?
    a. Vitamin A b. Vitamin K
    c. Vitamin E d. Vitamin D
    e. Vitamin B_6
18. Match the following:
    a. Hereditary tyrosinaemia
    b. Homocystinuria
    c. Hartnup disease
    d. Carcinoid syndrome
    e. Alkaptonuria
    1. Deficiency of tryptophan pyrrolase ________
    2. Malignancy of Argentaffin cells ________
    3. Urine turns black on exposure to air ________
    4. Deficiency of cystathionine synthase ________
    5. Excretion of p-OH-phenyl pyruvic acid ________.

SECTION FOUR

19. Match the following:
    a. Alanine b. Glutamate
    c. Spermidine d. Fumarate
    1. Formed from putrescine ________
    2. Product of tryptophan catabolism ________
    3. Product of phenylalanine catabolism ________
    4. Product of histidine catabolism ________.
20. Which of the following enzyme catalyses reactions in the biosynthesis of both catecholamines and 5-OH tryptamine (serotonin)?
    a. Tryptophan hydroxylase
    b. Dopamine-β-hydroxylase
    c. Phenyl ethanolamine N-methyl transferase
    d. Tyrosine hydroxylase
    e. Aromatic amino acid decarboxylase
21. The following statements regarding creatinine are correct except:
    a. Blood levels of creatinine is a good indicator of kidney function
    b. Although 24-hr excretion rate is very constant, there is a marked diurnal variation
    c. Urinary creatinine levels give a measure of muscle mass
    d. Excretion of creatinine is urine of adults is very constant from day-to-day
    e. Cratinine is an anhydride of creatine-(P)
22. Phenylketonuria is an inherited disorder due to deficiency of the enzyme:
    a. Transaminase
    b. Homogentisate oxidase
    c. Phenylalanine hydroxylase
    d. Isomerase
    e. None of the above
23. Which of the following amino acids on degradation produces a glucogenic intermediate of TCA cycle and ketone body?
    a. Glycin b. Phenylalanine
    c. Alanine d. Cysteine
    e. Serine
24. Reactions of urea cycle take place in liver cells
    a. In cytosol
    b. Only in lysosomes
    c. In mitochondrial matrix
    d. Both cytosol and mitochondrial matrix.
    e. Both cytosol and lysosomes
25. Depletion of α-ketoglutarate during increased NH_3 influx leads to the formatin of :
    a. Arginine b. Proline
    c. Glutamine d. Histamine
    e. Serine
26. Which of the following amino acids produce a vasodilator on decarboxylation:
    a. Glutamic acid b. Histidine
    c. Ornithine d. Cysteine
    e. Arginine
27. Breakdown of amino acid tryptophan begins with formation of:
    a. Urocanic acid b. Nicotinic acid
    c. Anthranilic acid d. N-formyl kynurenine
    e. None of the above
28. During degradation of amino acid leucine to acetyl-CoA, it produces which compound that help in cholesterol biosynthesis:
    a. β-methylglutaconyl-CoA
    b. Isopentenyl pyrophosphate
    c. β-OH-β-methyl glutaryl-CoA
    d. Isovaleryl-CoA
    e. All of the above.
29. Urinary oxalates in small quantities may arise from:
    a. Tyrosine b. Glycine
    c. Serine d. Citrates
    e. Oxaloacetate
30. Match the following:
    a. Glutamate b. Pyruvate
    c. Serotonin d. Formate
    1. Produced from catabolism of cysteine ________
    2. Catabolism of histidine ________
    3. Catabolism of glycine ________
    4. From tryptophan ________.
31. In humans the sulphur of methionine and cysteine is excreted mainly as:
    a. Ethereal SO_4 b. Inorganic SO_4
    c. Sulphites d. Taurocholates
    e. Cysteic acid
32. Match the following:
    a. Histamine b. Serotonin
    c. γ-amino butyrate d. Encephalin
    e. Reserpine
    1. A brain substance with opiate properties ________
    2. A neurotransmitter ________
    3. Produced by argentaffin cells ________
    4. Prevents storage of catecholamines ________
    5. A vasodilator ________.

Answers to MCQs

1. c	2. d	3. e	4. b	5. c	6. b
7. d	8. b	9. d	10. b	11. d	12. c
13. d	14. a	15. c	16. b	17. c	

18. 1. c, 2. d, 3. e, 4. b, 5. a,
19. 1. c, 2. a, 3. d, 4. b,

20. e	21. b	22. c	23. b	24. d	25. c
26. b	27. d	28. c	29. b		

30. 1. b, 2. a, 3. d, 4. c,
31. b 32. 1. d, 2. c, 3. b, 4. e, 5. a.

CHAPTER 28

INTEGRATION OF METABOLISM OF CARBOHYDRATES, LIPIDS AND PROTEINS

Major Concept

To understand various points at which various metabolic pathways of carbohydrates, proteins and lipids are interlinked and the significance of the same.

Specific Objectives

1. Know the three stages of energy production from nutrients.
2. Study the details of processes taking place at each stage.
3. Study the interconversion of carbohydrates and lipids.
4. Study the details of conversion of fatty acids into amino acids.
5. Study the regulation and control of interconversions.

INTRODUCTION

Though metabolism of each of major food nutrients, viz. carbohydrates, lipids and proteins have been considered separately for the sake of convenience, ***it actually takes place simultaneously*** in the intact animal and are closely interrelated to one another. The metabolic processes involving these three major food nutrients and their interrelationship can be broadly divided into **three stages** (Fig. 28.1):

1st stage: *Stage of hydrolysis to simpler units*

2nd stage: *Preparatory stage*

3rd stage: *Oxidative stage–Aerobic final (TCA Cycle).*

1st Stage

Stage of Hydrolysis to Simpler Units

- The complex polysaccharides, starch/glycogen are broken down to glucose; and disaccharides are hydrolysed to monosaccharides in GI tract by various carbohydrate-splitting enzymes present in digestive juices.
- Similarly, principal lipids, triacylglycerol (TG) is hydrolysed to form FFA and glycerol.
- Proteins are hydrolysed by proteolytic enzymes to amino acids.

The above is the prelude to either further synthesis of new substances or for their oxidation. ***Very little of energy is produced in this hydrolytic phase and it is dissipated away as heat. There is no storage of energy at this stage.***

2nd Stage

Preparatory Stage

- The monosaccharide glucose runs through the glycolytic reactions to produce the 3-C keto acid pyruvic acid (PA) in the cytosol, which in turn is transported to mitochondrion where it undergoes oxidative decarboxylation to produce 2-C compound **"acetyl-CoA" ("active" acetate).**
- The glycerol of fat, either goes into formation of glucose (gluconeogenesis) or by entering the same glycolytic pathway through the triose-P, forms PA and then finally 2-C compound "acetyl-CoA"
- The fatty acids undergo principally β-oxidation and form several molecules of "acetyl-CoA".
- The amino acids are deaminated/and/or transaminated first and the C-skeleton is metabolised differently from amino acid to amino acid
  - In the case of amino acids, viz. Glycine, Alanine, Serine, Cysteine/Cystine and threonine when catabolised form pyruvic acid (PA) similar to carbohydrates and is finally converted to 'Acetyl CoA'
  - In the case of amino acids, viz. Glutamic acid, Histidine, Proline and OH-proline, Arginine and Ornithine produces α-ketoglutaric acid when catabolised and thus they enter the TCA cycle.
  - Yet a few others like Leucine, Phenyl alanine, Tyrosine and Isoleucine yield acetate or acetoacetate, the latter can be converted to "acetyl-CoA".

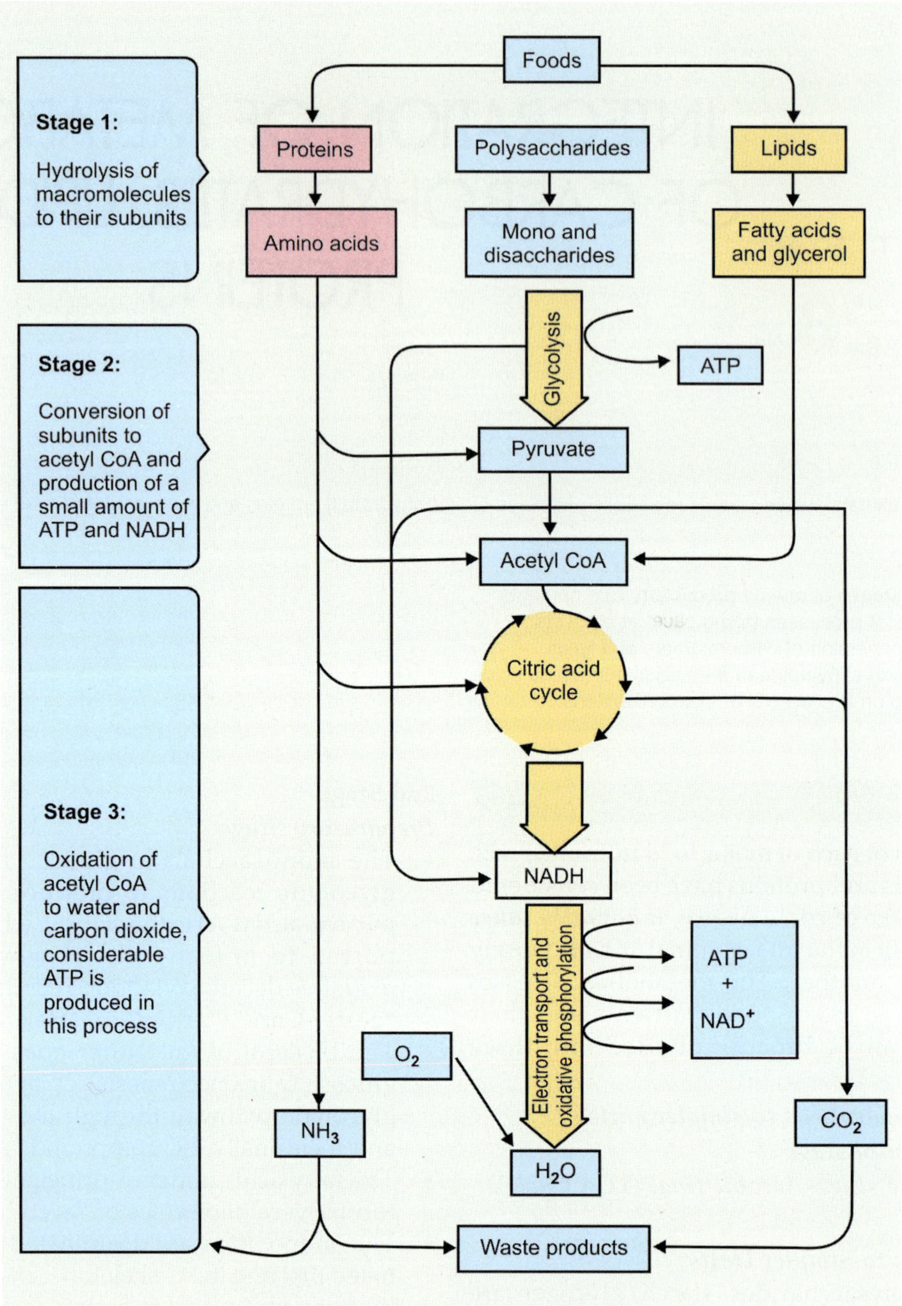

Fig. 28.1: Three stages of metabolism

During the second stage (glycolysis, β-oxidation, etc.) relatively small amount of energy is produced and this is stored as ATP.

3rd Stage

Oxidative Stage: Aerobic Final (TCA Cycle)

In presence of oxygen, acetyl-CoA is oxidised to CO_2 and H_2O by common final pathway TCA cycle.

The carbohydrates, lipids and proteins all form acetate or some other intermediates like oxaloacetate (OAA), α-ketoglutarate, succinyl-CoA, or fumarate, which are all intermediates of TCA cycle. Having gained entry into the TCA cycle at any site, two of carbons of "citrate" constituting an acetate moiety are oxidised finally to CO_2 and H_2O and the energy of oxidation by the electron transport chain is captured as energy-rich PO_4^- ATP mostly. ***This stage yields the largest amount of energy*** of all three stages. Thus, the pathways are similar to a large extent and identical in the final stage of oxidation of the metabolites, whether derived from carbohydrates, lipids or proteins.

This is schematically represented in Figure 28.2, alongwith the entry of various amino acids.

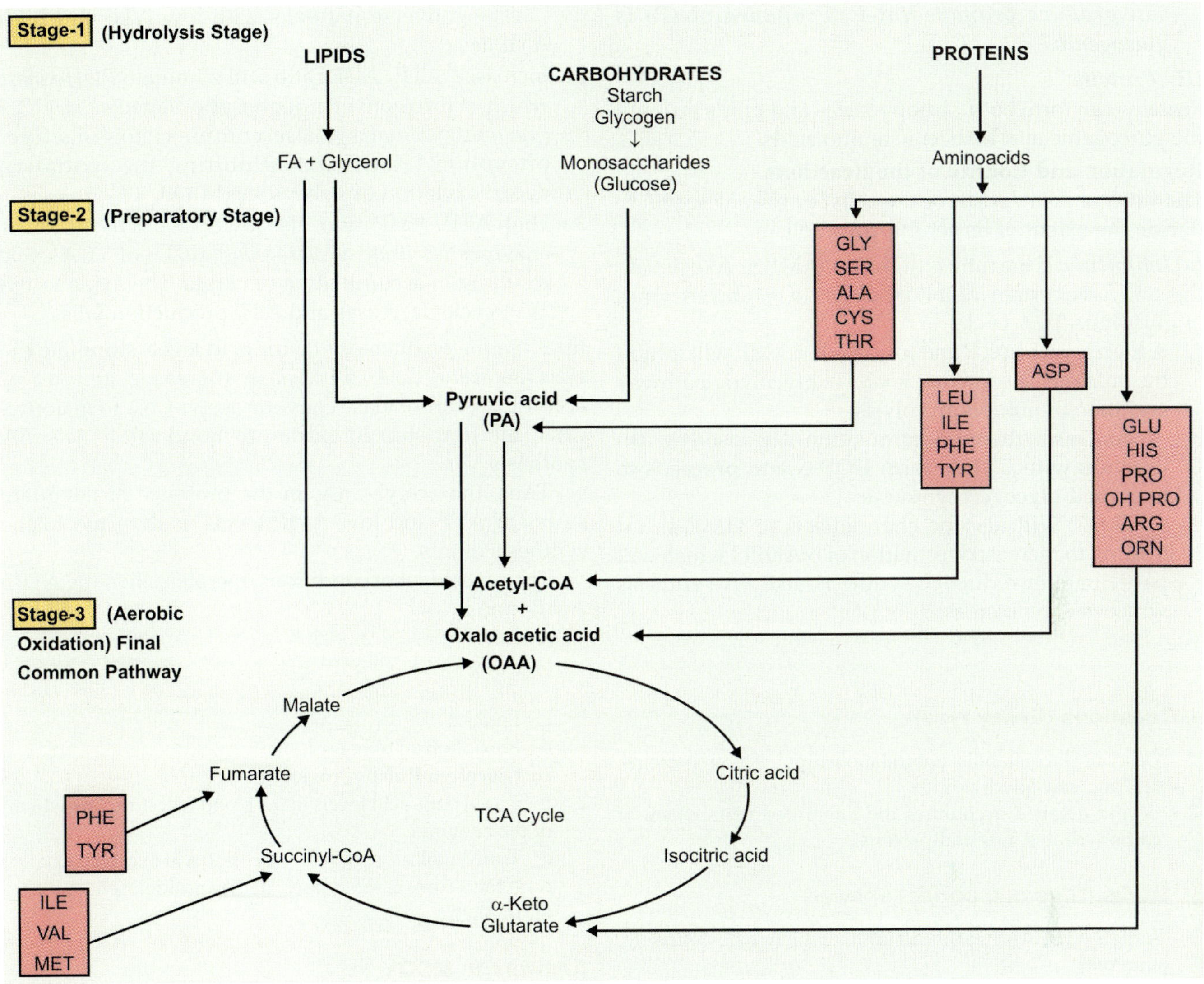

Fig. 28.2: Three stages and their relationship with various amino acids

INTERCONVERSION BETWEEN THE THREE PRINCIPAL COMPONENTS

I. Carbohydrates

1. **Carbohydrates can form lipids:** Through formation of: (a) α-glycero-P from glycerol or di-hydroxy acetone-P (from glycolysis) which is necessary for Triacyl glycerol (TG) and (b) FA from acetyl-CoA-extramitochondrial *de novo* synthesis.
2. **Carbohydrates can form non-essential amino acids:** Through amination of α-ketoacids, viz. pyruvic acid (PA), oxaloacetic acid (OAA) and α-ketoglutarate to form amino acids alanine, aspartate and glutamate respectively.

II. Fats

1. Fatty acids can be converted to some amino acids by forming the dicarboxylic acids like malic acid, oxalo acetic acids and α-ketoglutarate.
2. Fatty acid carbon may theoretically be incorporated into carbohydrates by the acetate running through TCA cycle. But there is no net gain in carbohydrates, since two carbons, equivalent of acetate are oxidised in the cycle.
3. However acetate can form glucose by running through the glyoxylate cycle.
4. Acetone, one of the ketone bodies may be glucogenic. Acetone can be converted to acetol-P which in turn

can ***produce propanediol-P. Propanediol-(P) is glucogenic.***

III. Proteins

Proteins can form both carbohydrates and lipids through the glucogenic and ketogenic amino acids.

Regulation and Control of the Reactions

The ratio of ATP/AMP of the cells/or tissues seems to decide the extent of its aerobic metabolism.

(a) Inhibition: If the ratio is high (low AMP or ADP level), this will have certain inhibitory effects of certain enzymes of glycolytic-TCA cycle.

1. A high level of ATP and low level of AMP will inhibit the enzyme *phosphofructokinase* of glycolytic pathway and thereby inhibit glycolysis.

   As a result there is accumulation of hexose-P which interacts with UTP to form UDP-G and proceeds to increased glycogen synthesis.

   G-6-P will also be channelised to HMP-shunt leading to increased formation of NADPH which will participate in reductive synthesis, like FA synthesis which will be increased.

   The converse happens with low ATP and high AMP levels.
2. Increased ATP/ADP ratio will stimulate *PDH-kinase* which in turn converts dephosphorylated *active PDH (pyruvate dehydrogenase* complex) to 'inactive' phosphorylated PDH inhibiting the oxidative decarboxylation of pyruvic acid (PA).
3. High ATP/AMP ratio, also lowers the activity of the enzymes *Isocitrate dehydrogenase* (ICD) of TCA cycle resulting in accumulation of citrate. The oxidation in TCA cycle decreases and ATP production falls.

(b) Stimulation: Increased citric acid levels stimulate the enzyme *acetyl-CoA carboxylase.* Increased activity of acetyl-CoA carboxylase converts acetyl-CoA to malonyl-CoA, the first step in extramitochondrial *de novo* FA synthesis.

Thus, the acetyl-CoA, in the presence of adequate stores of ATP and low AMP levels, is diverted to the synthesis of fats.

The reverse set of conditions operates when the ATP/AMP ratio is low.

I. Questions (Essay type)

1. Describe the metabolic interrelationships between proteins, fats and carbohydrates.
2. Give a detailed account of the integrated metabolism of carbohydrates, fats and proteins.

II. MCQs (Give one correct answer)

1. A high ATP/AMP ratio will inhibit which of the following enzyme?
   a. Glyceraldehyde-3-P dehydrogenase
   b. Pyruvate kinase
   c. Enolase
   d. Phosphofructokinase
   e. Glucose-6-P dehydrogenase
2. Increased citric acid levels in the blood will stimulate which of the enzyme?
   a. Transketolase
   b. Pyruvate carboxylase
   c. Acetyl-CoA carboxylase
   d. Transaldolase
   e. Enolase

Answers to MCQs

1. d 2. c

SECTION FOUR

CHAPTER

29

METABOLISM IN STARVATION

Major Concept

Study the general effects and metabolic changes in starvation in the body.

Specific Objectives

1. Note the general physiological changes that occur as a result of starvation.
2. Study the effects of starvation associated with carbohydrate metabolism.
3. Study in detail metabolic changes in lipid metabolism associated with starvation, including starvation ketosis and its aggravating factors.
4. Study feedback mechanism and carbohydrate and lipid cycles.
5. Study metabolic changes of protein metabolism in starvation.
6. Study how water and mineral metabolism is affected in starvation.

INTRODUCTION

Total starvation includes complete deprivation of foods, salts and water. It results in death of the animal in the shortest possible time. Deprivation of water only follows death of the animal in approximately 7 to 10 days time, that of salts in two weeks time, whereas food starvation in 3 to 4 weeks or even longer depending on the reserve of fats in the body. The length of time, a man can survive, depends upon his fat stores, but the longest period of survival never exceeds 9 to 10 weeks.

Starvation induces a number of metabolic changes, some occurring within a few days and others occurring late. In addition to metabolic changes, a progressive fall in BMR occurs but pulse rate and blood pressure are affected much later. Ketosis develops and some retention of salt and water occurs and in prolonged starvation fatty liver develops.

I. EXPERIMENTAL OBSERVATIONS

The effects of food starvation have been mostly observed in animals. But some direct observations have been carried out on volunteers and on professional fasting men.

General considerations: The following general effects are seen:

(a) General Condition

- During the first few days, there is a craving for foods, particularly at meal times. But later on, this craving subsides, provided water and salts are freely allowed. Gradually, desire for food vanishes.
- "Weakness" gradually increases and a strong dislike to undertake any physical or mental effort develops. At about this time the subject falls into a state of semiconsciousness.
- The pulse rate and body temperature remain almost normal till before death, these are affected very late. The sleep increases and respiration becomes slower. Temperature falls before death.
- The amount of urine as well as its urea content falls.

(b) Body Weight

- The body weight is steadily lost. The daily loss in man during the first 10 days, amounts to between 1 to 1.5 per cent of the original body weight. At the onset of the fast, after depletion of glycogen stores, the fat depots and subcutaneous tissues bear the brunt. The extracellular fluid in large quantities is also lost.
- Dissolution of the muscular tissues and protoplasmic structures occurs much later. The muscle

fibres are much reduced in size and many of the fibres are degenerated.

- Organs and tissues of the body are not affected alike. The more vital organs lose the least weight, whereas the less vital ones lose the most.

II. EFFECTS ON METABOLISM

During starvation, the body has to depend upon its own tissue materials to get energy in absence of foods. Even if no physical work is being done, 2000 K. cal (C) are needed daily approximately.

For discussion, metabolic changes can be arbitrarily divided into the following three stages:

(a) ***First Stage***

- Out of the principal three foodstuffs—glycogen, fat, and proteins, the liver ***glycogen is first mobilised.*** But due to its limited storage, 6 per cent, i.e. approx. 108 grams, it cannot last long.
- Glucose is synthesised in the body by the process of *gluconeogenesis* principally in the beginning from glucogenic amino acids from the mobilisation of stored proteins from tissues (later on gluconeogenosis is maintained and supplemented by other substrates also). This initial stage lasts for not more than 2 to 3 days.

(b) ***Second Stage***

- ***80 to 90 per cent of energy requirement will be derived from fats and the remainder 10 to 20 per cent from proteins.*** Utilisation of fats will depend on the "fat storage" of the body. Since the adipose tissue represents the largest amount of stored food, the second stage will last for longest period, usually over two weeks.

(c) ***Third Stage***

- In the third stage when the fat stores are almost exhausted, ***energy requirement is obtained from the breakdown of tissue proteins.*** The cell substance will break up with a consequent dislocation of cell metabolism and cell life. This stage of affairs, if continued, leads to death. The stage lasts for less than one week.
- The determination of total RQ and of non-protein RQ will indicate the extent to which these three foodstuffs are burning at the three stages. The RQ is highest at first stage and diminishes later on.

A. CARBOHYDRATE METABOLISM

- ***Tissue glycogen is utilised initially for energy production and for maintaining blood sugar level,*** consequently, the liver glycogen and the blood sugar fall on fasting, the blood sugar may become less than 40-60 mg % and total carbohydrate reserves may decline by 40 per cent.
- ***Hypoglycaemia,*** in turn depresses insulin secretion ↓ and thus ***increases gluconeogenesis,*** the latter may maintain the blood sugar and carbohydrates reserves in liver and muscles, though at a subnormal level.
- Antagonist hormone glucagon activity increases, which increases glycogenolysis. Insulin/glucagon ratio decreases. ***Supply of glucose thus balances the "obligatory demands" for glucose oxidation and utilisation.*** The glycogen content of cardiac muscle may, however, still be normal and not exhausted.
- Activities of key gluconeogenic enzymes, viz. *pyruvate carboxylase, fructose 1,6-bi-phosphatase (FDPase), PEP carboxykinase* and *glucose 6-P-ase* are increased manifold in starvation or carbohydrate deprivation, enhancing both gluconeogenesis and glycogenolysis.
- But the hepatic removal of blood glucose and glycogenesis are depressed due to reduction of *glucokinase* activity caused by hypoglycaemia.
- Starvation simultaneously depresses activities of *glucose-6-phosphate dehydrogenase (G-6-PD)* and citrate cleavage enzyme, thus reducing HMP shunt and lipogenesis from carbohydrates.

Summary

- All the above factors try to provide for a continuous supply of blood sugar to the tissues including the brain and red blood cells.
- In the first few days of starvation, considerable amounts of gluconeogenesis occur from proteins in liver and kidney to supply the "basal needs"; subsequently however, gluconeogenesis continues mainly in kidneys, though at a reduced state, while still smaller amounts of carbohydrates are formed in the liver from catabolic products of carbohydrates and glycerol.

B. LIPID METABOLISM

- Carbohydrate reserves being insufficient to fulfil the calories requirements for more than a couple of days, ***fats of adipose tissue are largely mobilised to the liver as FFA and oxidised for energy purposes.*** Fat utilised in starvation in the beginning is only triacylglycerol (TG), from fat depots, i.e. the "element variable". ***The burning of fat is reflected by a lower RQ (about 0.73)*** ↓.
- Starvation increases the activities of the "hormone-sensitive" *TG lipase* of the adipose tissue and also increases the enzymes of β-oxidation in Liver. With increased activity of TG lipase there is increase in lipolysis in the adipose tissue, increases FFA↑ in plasma due to increased entry into the blood stream. Liver takes up FFA and there is increased FA oxidation in Liver.
- Amounts of FFA progressively rises in the plasma from the very first day of starvation and this rise is

SECTION FOUR

enhanced by muscular work, probably because of increased secretion of catecholamines from adrenal medulla and sympathetic nerve endings which induces adipose tissue lipolysis by activating *hormone-sensitive lipase.*

- Glycerol component released from lipolysis acts as a substrate for gluconeogenesis and joins the carbohydrate pool after activation to α-glycero-P.
- On the other hand, *de novo* extramitochondrial synthesis of FA is reduced during starvation as activities of *acetyl-CoA carboxylase* and *citrate synthetase,* as well as translocation of citrate from the mitochondrion to the cytosol are ***inhibited by the rise of plasma level of long-chain FA,*** liberated from adipose tissue by lipolysis. Long-chain acyl-CoA is inhibitor of the two enzymes stated above.
- All the above cause a loss of body fat, to the extent of 75 per cent in several weeks; adipose tissues thin out and finally disappear; visceral organs get displaced due to loss of adipose tissue (viz. perirenal fat, etc.) around them and even the skin hangs loose with folds, owing to loss of subcutaneous fats.
- In prolonged starvation the activity of *lipoprotein lipase* declines in the adipose tissue; but rises in cardiac and skeletal muscle fibres. This leads to a fall in the fat storage in adipose tissues and a rise in the availability of FFA in cardiac and skeletal muscles for energy production.
- ***Starvation ketosis*** owing to gradual decline in carbohydrate store and lack of carbohydrate, energy is obtained from fat burning. ***As FFA level increases, ketogenesis is stimulated,*** and excess of ketone bodies formed which pass into blood faster than they are utilised by tissues, resulting in *ketosis* and *ketonuria.* ***Starvation ketosis is more pronounced in females than in males and in children than in adults.***
- Recently it has been shown that human brain can utilise ketone bodies for energy to the extent of 20 per cent of total energy requirement after 48 hrs fast, 60 per cent after a week's fast and 80 per cent after 40 days fast. It has been suggested ***that brain adaptation is such that it can utilise preferably ketone bodies than glucose and spare glucose for energy.***

Aggravating Factors for Ketosis

- Relative lack of oxaloacetic acid (OAA) is produced when gluconeogenesis is more as OAA is utilised for gluconeogenic pathway and less available for TCA cycle. This aggravates ketogenesis due to availability of more acetyl-CoA.
- With starvation and increased concentration of FFA, *acetyl-CoA carboxylase* is inhibited, and malonyl-CoA concentration decreases, releasing the inhibition of *carnitine palmitoyl-transferase I* (i.e. enzyme is stimulated) and allowing more acyl-CoA to be oxidised. These events are reinforced in starvation by the [insulin]/[glucagon] ratio, which decreases, causing increased lipolysis in adipose tissue, the release of free FFA and inhibition of *acetyl-CoA carboxylase* in the liver. Availability of more acetyl-CoA increases *ketogenesis.*
- ***Starvation acidosis:*** Ketosis and ketonuria cause **acidosis**. There occurs:
  - Reduction of $[HCO_3^-]$↓ due to buffering of the strong and non-volatile acids.
  - Increased pulmonary ventilation and fall in alveolar CO_2 tension.
  - Increased acidity of urine (titratable acidity)↑.
  - Increased NH_3 secretion and excretion of NH_4Cl.
- ***Fatty liver:*** Due to lack of carbohydrate, esterification will be less↓ α-glycero-P, VLDL formation and elimination will decrease leading to accumulation of fat producing fatty liver.

Feedback Mechanism

- A feedback mechanism has been suggested for controlling FFA output from adipose tissue. In starvation it may operate as a result of the action of ketone bodies and FFA to directly stimulate the pancreas to produce insulin. Under most conditions, FFA are mobilised in excess of oxidative requirements, since a large proportions are esterified even during fasting in the beginning.
- As the liver takes up and esterifies a considerable proportion of the FFA output, it plays a regulatory role in removing excess FFA from circulation. Till the time when carbohydrate supplies are maintained and adequate, most of the FFA influx is esterified and ultimately re-transported from the liver as VLDL to be utilised by other tissues.
- However, when the capacity of the liver to esterify is not sufficient and diminishes in face of an increased influx of FFA, an alternative route ketogenesis which enables the liver to continue to combat much of influx of FFA by converting it to a suitable form, i.e. ketone bodies, which can be readily utilised by extrahepatic tissues, under all nutritional conditions.

Thus, ***in starvation, two cycles operate:***

- A ***carbohydrate cycle,*** involving release of glycerol from adipose tissue and its conversion in liver to glucose, followed by its transport to adipose tissue to complete the cycle.
- The other is a ***lipid cycle*** and involves release of FFA by adipose tissue, its transport to and esterification in liver and retransport as VLDL back to adipose tissue.

C. PROTEIN METABOLISM

- ***In starvation, tissue proteins are treated as 'food proteins".*** They are hydrolysed to amino acids, but to a larger extent and at an increased scale than that occurs normally and constitutes the "general amino acid pool". For catabolism of tissue proteins, tissues are not uniformly used. Brain and heart are spared and only lose 3 per cent of the bulk. Muscles lose 30 per cent, liver 55 per cent, and spleen 70 per cent. The breakdown of tissue proteins in starvation is controlled possibly through the action of adrenal cortex (Glucocorticoids)
- The released amino acids from amino acid pool are utilised as follows:
  - *First call:* Utilised for the maintenance of the structural and functional efficiency of vital organs. Also utilised for formation of enzymes and hormones which are protein/amino acid derivatives.
  - *Second call:* The amino acids also undergo de-amination in the liver and the non-nitrogenous part helps in the maintenance of blood sugar level by gluconeogenesis. In prolonged starvation, in man, gluconeogenesis from protein is diminished gradually due to reduced release of amino acids, particularly alanine from muscles, the principal protein stores.
- *N_2 excretion in urine:* ***It is an index of tissue protein consumption,*** when fat stores are depleted and exhausted, proteins alone are available for energy purposes and death rapidly results.
  - The amount of N_2 excretion in urine during first few days is directly proportional to the amount of protein intake before starvation.
  - The average daily excretion in the first week of starvation is about 10 grams daily. On a normal mixed diet, an adult's daily excretion of urea is 15 to 49 gm depending on dietary intake of proteins (= 7 to 18 gm of urea N_2).
  - During second and third weeks of starvation, there is a gradual decline in N_2 excretion and the values are very low.
  - But just before death, when tissue proteins are being catabolised for energy, the urinary N_2 excretion again rises (***"pre-mortal" rise***).
- The end-products of endogenous protein metabolism, i.e. creatine/creatinine, neutral sulphur compounds and uric acid are the main other nitrogenous products. Creatine excretion gradually falls as the weight of muscle diminishes. Increased protein catabolism lowers the secretion of insulin, thyroxine, gonadotrophins and several other hormones and may lead to reproductive failure.

D. WATER AND MINERAL METABOLISM

- The ECF is reduced during the first few days, due to stoppage of water intake and continued ***obligatory*** losses
- On prolonged starvation, the ICF volume may also fall by 25 per cent or more, because of cellular breakdown. ECF shows a relative expansion due to the subsequent shrinkage of cell mass, and fluid accumulates, to restore ECF volume and may produce oedema
- Due to cellular disintegration, there is loss of intracellular K^+ thus reducing the total body K^+. Na^+ in the ECF may be maintained in the normal range. There may be a reduction of 10 per cent or more of total mineral content of the body
- The need for drinking water is reduced due to increase in "metabolic water" and relative expansion of ECF by reduced glomerular filtration
- ***Oedema:*** May appear due to relative expansion of EC compartment and 'decline' in serum albumin level ***(starvation oedema).***

I. Questions (Essay type and short notes)

1. Describe the metabolic changes in body during starvation.

Short Notes

a. Starvation ketosis
b. Maintenance of blood sugar level in starvation.

II. MCQs (Give one correct answer)

1. During starvation increased serum alanine represents:
   a. Decreased utilisation of non-essential aminoacids
   b. Leakage of aminoacids from cells due to cell damage
   c. Decreased utilisation of aminoacids for gluconeogenesis
   d. Net breakdown of muscle proteins
   e. Impaired renal function
2. In starvation, all the following enzyme activities are decreased, except:
   a. Lipoprotein lipase
   b. Acetyl-CoA carboxylase
   c. Fatty acid synthetase
   d. Citrate cleavage enzyme
   e. Carnitine palmitoyl-transferase I
3. In starvation, in the beginning, the activities of the following enzymes are increased except:
   a. Glucose-6-phosphatase
   b. Glucose-6-phosphate dehydrogenase
   c. Pyruvate carboxylase
   d. Fructose-1, 6-diphosphatase
   e. PEP-carboxykinase

SECTION FOUR

4. During the stages of starvation, blood or tissue levels of all the following would be expected to be elevated except:
   a. Epinephrine
   b. Norepinephrine
   c. Glucagon
   d. Ketone bodies
   e. Glycogen
5. During starvation, the non-protein RQ value is close to:
   a. 1.50
   b. 0.75
   c. 0.50
   d. 0.25
   e. 0.20
6. The main type of lipids found in tissues in starving individual are in end of first week of fasting:
   a. Triacyl glycerol
   b. Phospholipids
   c. Cholesterol
   d. Cholesterol esters
   e. Glycolipids
7. The source of glucose for glucose oxidation (glycolysis) by renal medullary cells in the fasting state is believed to be:
   a. Blood glucose
   b. Conversion of fructose to glucose
   c. Glycogen in medullary cells
   d. Gluconeogenesis by the renal cortex
   e. Gluconeogenesis by medullary cells
8. During prolonged fasting, the major control of rate of gluconeogenesis is the:
   a. Insulin/glucagon ratio
   b. Availability of FFA to liver
   c. ATP level in liver
   d. C-AMP level in liver
   e. Availability of alanine in the liver

Answers to MCQs

1. d 2. e 3. b 4. e 5. b 6. b
7. d 8. e.

CHAPTER 30

PORPHYRINS AND PORPHYRIAS (SYNTHESIS OF HAEM)

Major Concept

To study how haem is synthesised and learn about different types of porphyrias.

Specific Objectives

1. Learn the various steps of haem synthesis—can be divided into the following stages:
   (a) **Stage I:** Synthesis of δ-ALA (intramitochondrial)
   (b) **Stage II:** Synthesis of coproporphyrinogen III (cytosol)
      - Learn how 2 mols of δ-ALA combine to form first 'pyrrole' ring porphobilinogen
      - Study the various reactions involved to form the tetrapyrrole, coproporphyrinogen III in "major series" and coproporphyrinogen I in 'minor series'.
   (c) **Stage III:** Formation of protoporphyrin IX (Intramitochondrial) from coproporphyrinogen III from the major series.
   (d) **Stage IV:** Formation of haem from Protoporphyrin IX. Haem thus produced may be coupled to various proteins to produce conjugated haemoproteins, like Hb, myoglobin, cytochromes, catalase, peroxidase.
2. Study how haem synthesis is regulated (Regulation of haem synthesis)
   - Learn the "rate limiting" enzyme—*δ-ALA synthetase.* Study the factors that stimulate the enzyme and how 'feedback inhibition' occurs by haem.
   - Learn about second "rate limiting" enzyme ***δ-ALA dehydratase***—which is Cu-dependent and also inhibited by 'feedback' inhibition by haem.
   - Study the effect of O_2 on haemsynthesis.
   - Study the effects of certain insecticides, carcinogens and drugs that require cyt P_{450} for their metabolism.
3. **Porphyrias:** Condition in which there is increased excretion of both coproporphyrin and uroporphyrin.
   - List the various types of porphyrias and classify them.
   - Note the enzyme deficiency and clinical manifestations in hereditary porphyrias.
   I. Congenital erythropoietic porphyria.
   II. Hepatic Porphyrias: Which can be of 3 types
      - Acute intermittent porphyria ("paroxysmal")
      - Porphyria cutanea tarda
      - Variegate porphyria (mixed or combined type).

PORPHYRINS

INTRODUCTION

- The porphyrins are complex structures consisting of 4 pyrrole rings, united by *"methyne" bridges (or methylidene bridges)*

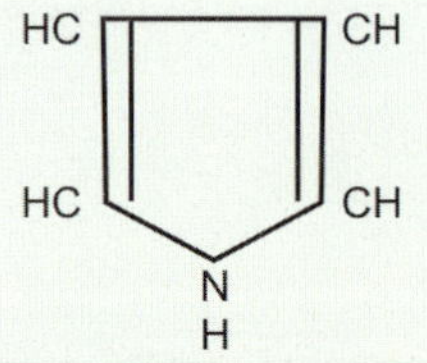

Pyrrole ring

- The nitrogen of 4 pyrrole rings can form complex with metallic ions such as Fe^{++} and Mg^{++}.
- They form the prosthetic groups of conjugated proteins, viz.
   – Haemoglobin of mammalian erythrocytes
   – Myoglobin of muscle
   – Erythrocruorins of some of the invertebrates, which occur in blood and tissue fluids
   – Cytochromes: Respiratory enzymes in electron transport chain
   – *Catalase* and *peroxidase* enzymes and
   – Oxidative enzyme like *tryptophan pyrrolase.*
   All the above contain Fe-porphyrins as prosthetic groups.
   – Chlorophyll, occurring in plants, contain Mg-porphyrin as the prosthetic group.

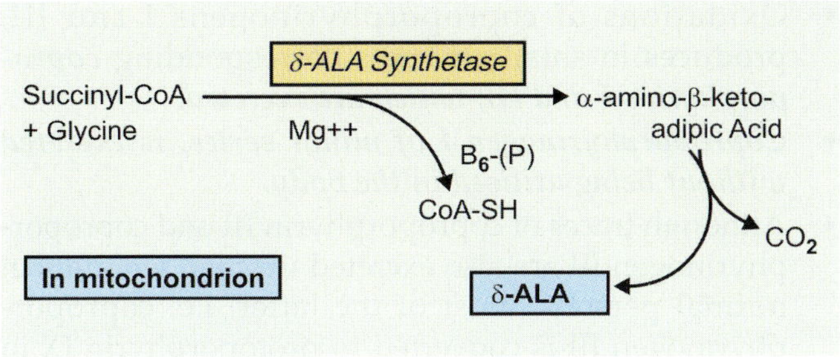

Fig. 30.1: Formation of δ-ALA

BIOSYNTHESIS OF PORPHYRINS

Porphyrins are synthesised partly in the ***mitochondrion*** and partly in ***cytosol*** of **aerobic cells like developing erythrocytes and hepatic cells.**

Stages of Biosynthesis

Arbitrarily the synthesis of porphyrins can be divided into **three stages** for understanding.

Stage I: Synthesis of δ-Amino Laevulinic acid (δ-ALA), which ***occurs in mitochondria***.

Stage II: Synthesis of coproporphyrinogen III ***(major series)*** and coproporphyrinogen I ***(minor series)*** which ***occurs in cytosol***.

Stage III: Synthesis of protoporphyrin IX, – which ***occurs in mitochondria again.***

Stage I: Synthesis of δ-Amino Laevulinic Acid (δ-ALA) (Intramitochondrial) (Fig. 30.1)

- Biosynthesis begins with the ***condensation of 'succinyl CoA' ("active" succinate)*** and ***glycine*** to form 'α-amino-β-Ketoadipic acid".
- α-amino-β-ketoadipic acid then undergoes decarboxylation to produce δ-ALA.

Both the reactions are catalysed by the enzyme ***δ-ALA-synthetase,*** which requires pyridoxal-P (B_6-P) and Mg^{++} as coenzymes. In liver cells, the synthesis occurs in the mitochondrion. Panthothenic acid is also required at this stage being a constituent of CoA-SH.

Mechanism of Action

- Glycine first combines with "Enz – B_6 – complex" to form enzyme bound "schiff base".
- The above then condenses with Succinyl-CoA forming a "Ternary complex", α-amino-β-ketoadipic acid + B_6P + Enz and CoA-SH is liberated.
- α-amino-β-ketoadipic acid then loses a mol. of CO_2, liberating δ-ALA in free form from the complex.

δ-ALA Synthetase Enzyme and its Regulation

δ-ALA synthetase enzyme is very unstable, low in concentration in tissues, ***Main rate-limiting enzyme*** in the synthetic pathway.

Regulation

- Many erythropoietic substances including hormones stimulate haem synthesis by inducing the production of the enzyme.
- ***End product 'haem' inhibits the enzyme by "feed-back" inhibition.***
- Haem also causes a repression of the synthesis of the enzyme, "end-product repression".

Stage II: Synthesis of coproporphyrinogen III and I (cytosolic):

1. *Formation of Porphobilinogen*
   - ***δ-ALA comes out of mitochondrion into the cytosol.*** Two molecules of δ-ALA condense further to form a molecule of **porphobilinogen,** ***which is the precursor of 'pyrrole' ring.***
   - The reaction is catalysed by the enzyme ***δ-ALA dehydratase,*** for which Cu^{++} is required as a cofactor. It is a Zn-containing enzyme.

Regulation

This is a ***second rate-limiting enzyme,*** which is inhibited by 'feedback' inhibition by endproduct haem.

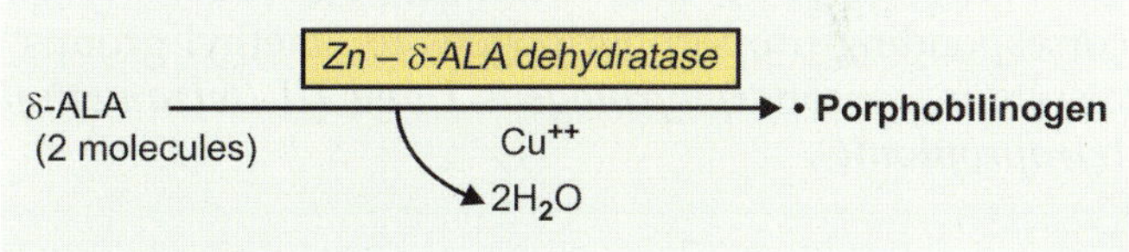

2. *Formation of Uroporphyrinogen I and III*

I. *Uroporphyrinogen I (minor series)*
   - In presence of a porphobilinogen *deaminase* (also called *uroporphyrinogen*-I synthetase), 4 mol. of porphobilinogens condense, losing 4 mol. of NH_3 and forms ***uroporphyrinogen I*** **(minor series),** in which the acetic acid and propionic acid side chains alternate.

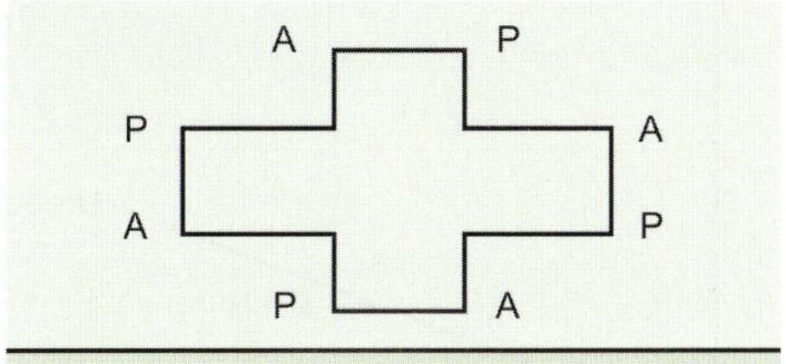

Uroporphyrinogen I

- In formation of uroporphyrinogen I, as above, "Di-pyrroles" and "tetrapyrroles" may be formed as intermediates.
- Oxidation of uroporphyrinogen-I, produces ***uroporphyrin I,*** which ***may be excreted in urine*** in small amounts normally.

II. *Uroporphyrinogen III (Major series)*
- Concomitant operation of an ***isomerase*** (also called as *uroporphyrinogen III cosynthetase*) with *deaminase*, results in reversal of one porphobilinogen residue, so

that the cyclisation results in the formation of ***uroporphyrinogen III* (major series).** In this, in IV pyrrole ring, ***acetic acid and propionic acid side chains are "reversed" (cf. uroporphyrinogen I).***

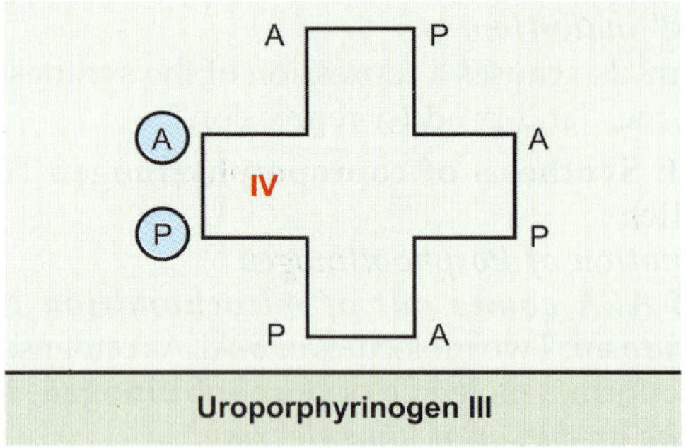

- Oxidation of uroporphyrinogen III produces **uroporphyrin III**, minute amounts of which may be ***excreted in urine*** in normal healthy individuals.

3. *Formation of Coproporphyrinogen I and III*

Decarboxylation, catalysed by ***uroporphyrinogen decarboxylase*** of the four acetic acid side chains of the corresponding uroporphyrinogens to "methyl groups" results in ***coproporphyrinogens I and III (tetramethyl tetrapropionic).***

- Oxidations of coproporphyrinogens I and III, produces in small amounts corresponding ***coproporphyrins I and III, which are excreted.***
- ***Coproporphyrinogen I of minor series, is excreted without being utilised in the body.***
- Although traces of coproporphyrin III and coproporphyrinogen III are also excreted in small amounts in normal persons, most of the latter, i.e. coproporphyrinogen III is converted to protoporphyrin IX in human beings (Fig. 30.2).

Stage III: Formation of Protoporphyrin IX (Intramitochondrial) (Fig. 30.3)

- Coproporphyrinogen III **enters mitochondrion**.
- Steps between coproporphyrinogen III (tetramethyl, tetrapropionic) and protoporphyrin IX (tetramethyl, divinyl, dipropionic acid) are obscure. An ***oxidative decarboxylase system*** containing flavins as coenzyme (probably the enzyme system consists of more than one enzyme) converts coproporphyrinogen III to protoporphyrinogen IX.
- Protoporphyrinogen IX is converted to protoporphyrin IX by another *oxidase* enzyme.
- The ***above steps require the presence of molecular O_2.***

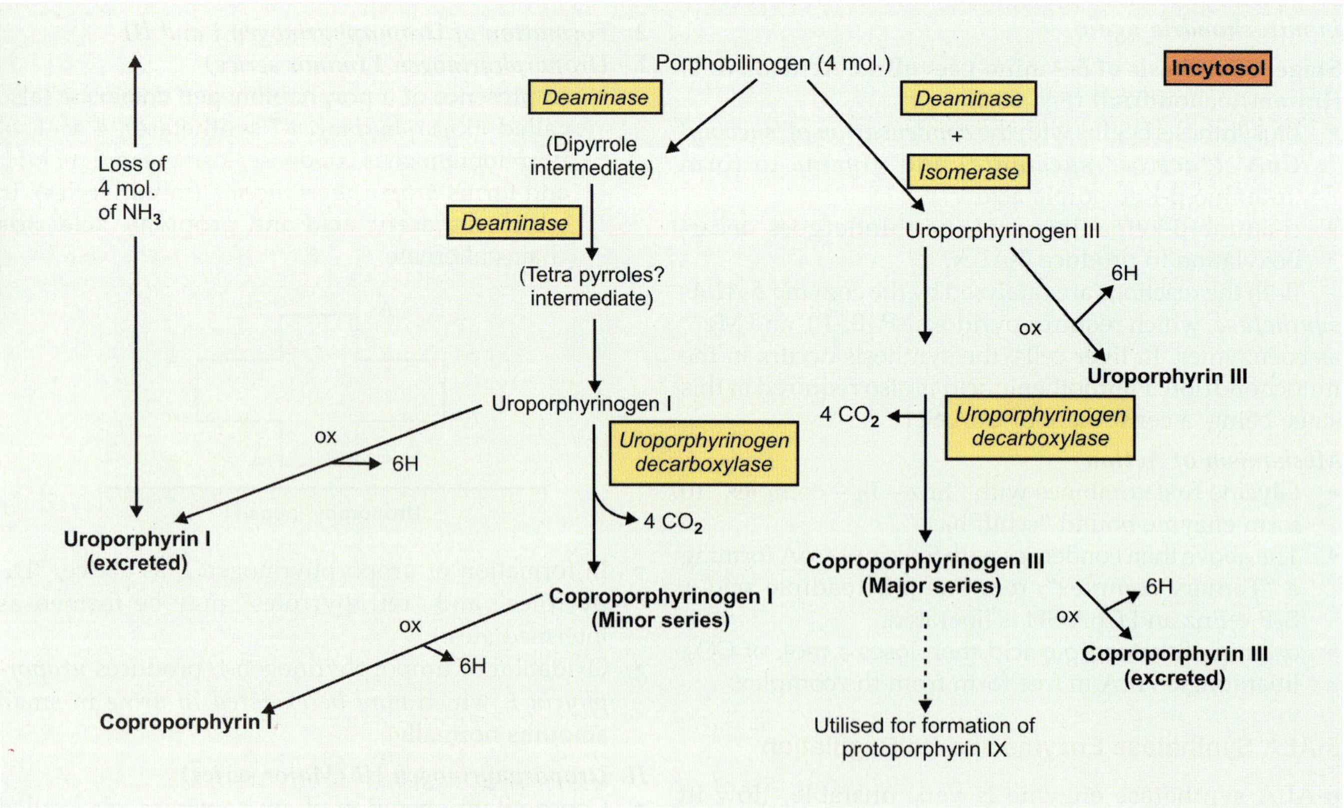

Fig. 30.2: Formation of coproporphyrinogens I and III from porphobilinogen

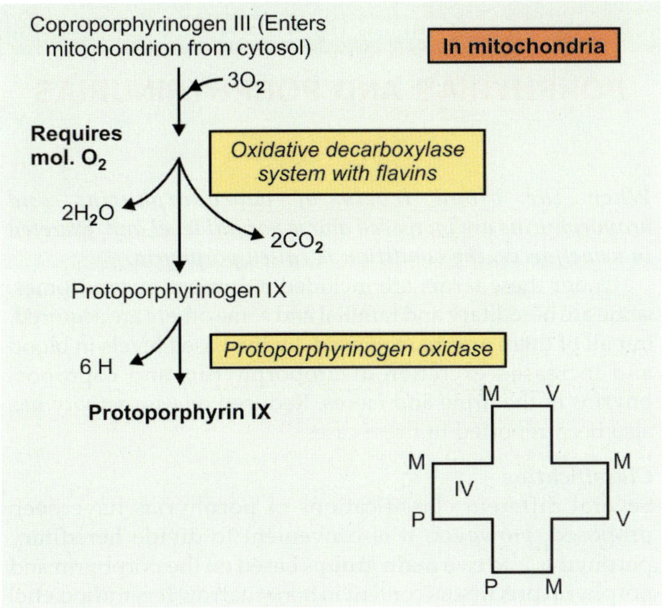

Fig. 30.3: Formation of protoporphyrin IX

Note

1. Though protoporphyrin is derived from type III (major series), it is called as "protoporphyrin IX". (Probably discovered in ninth experiment.)
2. Also note formation of protoporphyrin IX from coproporphyrinogen III is an **"aerobic" process** and requires presence of molecular O_2.

Formation of Haem and Haemoproteins (Intramitochondrial)

- Insertion of an atom of Fe^{++} into central position of protoporphyrin IX is catalysed by ***haem synthetase (ferrochelatase)*** which for optimal function requires:
  - ***Anaerobiosis,*** and
  - ***Reducing agents such as glutathione.***
- The "haem" which is produced is then coupled to various proteins and thus form the conjugated proteins, viz. haemoglobin, myoglobin, cytochrome C, catalases and peroxidases.
- This pathway operates inside mitochondrion (Fig. 30.4).

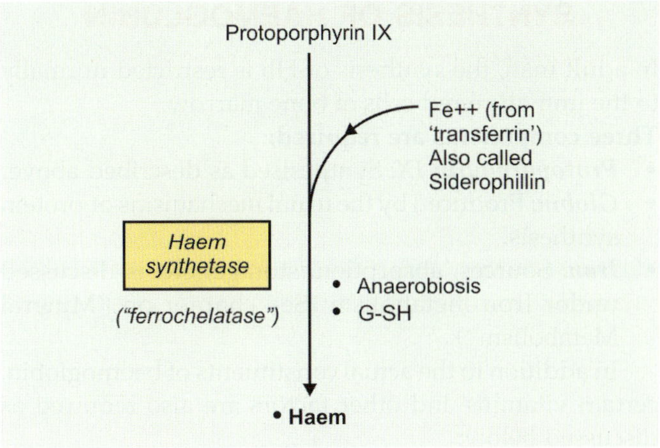

Fig. 30.4: Formation of haem

REGULATORY INFLUENCES AND EFFECTS OF INHIBITORS

1. **Effect of O_2:** Effect of O_2 on haem synthesis is rather complex.
   - ***In vivo:*** Stimulated by low O_2 tension (e.g. living at high attitudes)
   - ***In vitro:*** Conversion of porphobilinogen to uroporphyrinogen and protoporphyrin to haem, are both inhibited by O_2
   - But O_2 is required for decarboxylation of coproporphyrinogen and oxidation of protoporphyrinogen.
2. **Enzyme Inhibition:** Enzymes which catalyse the synthesis and utilisation of δ-ALA are important sites of regulation. **Haem, the end-product of the metabolic sequence, inhibits the activity of synthetase.**
3. **Drugs:** Many compounds of diverse structures, viz. certain insecticides, carcinogens and others, when administered to human beings can result in marked increase in hepatic δ-ALA synthetase, leading to increased porphyrins.

Explanation: Most of these drugs are metabolised by a system in the liver that utilises a specific cytochrome called as ***cytochrome P_{450}***. During metabolism of the drugs by cytochrome P_{450}, consumption of 'haem' by cytochrome P_{450} is greatly increased, which in turn diminishes the cellular concentration of haem, leading to 'derepression' of δ-ALA synthetase with a corresponding increase in rate of haem synthesis.

Other Factors

- ***Lead:*** It is known to produce profound abnormalities in porphyrin metabolism. ***It inhibits δ-ALA synthetase, δ-ALA dehydratase*** and ***haem synthetase 'in vitro'.***
- ***Glucose:*** It can prevent induction of δ-ALA.
- ***Hypoxia:*** It increases *δ-ALA synthetase* activity in erythropoietic tissues.
- ***Steroids:*** These play a permissive role in the drug-meditated 'derepression' of *δ-ALA synthetase 'in vivo'.*
- ***Iron:*** In chelated form it exerts a synergistic effect on induction of *δ-ALA synthetase.*
- ***Haematin:*** Administration of haematin *in vivo* can prevent the drug-mediated 'derepression' of *δ-ALA synthetase.*

SECTION FOUR

SYNTHESIS OF HAEMOGLOBIN

In adult man, the synthesis of Hb is restricted normally to the immature red cells of bone marrow.

Three components are required:

- ***Protoporphyrin IX:*** Synthesised as described above.
- ***Globin:*** Produced by the usual mechanisms of protein synthesis.
- ***Iron:*** Sources, absorption, storage, etc. is discussed under Iron metabolism (See chapter on "Mineral Metabolism").

In addition to the actual constituents of haemoglobin, certain vitamins and other factors are also required as discussed below:

- ***Pantothenic acid:*** It is necessary for synthesis of CoA-SH, required for the substrate succinyl-CoA.
- ***Pyridoxal-P:*** It is required as a coenzyme for the activity of *δ-ALA synthetase.*
- ***Copper:*** No satisfactory explanation of the role of Cu^{++} has been offered, except its participation as a co-factor with *δ-ALA dehydratase* is established.
- ***Vit C (ascorbic acid):*** It is essential as (*a*) it helps in absorption of Fe from the gut, converts Fe^{+++} to Fe^{++} and (*b*) it helps in mobilisation of Fe from ferritin.
- ***Intrinsic factor:*** It is necessary for vit B_{12} absorption.
- ***Vit B_{12} and folic acid:*** The requirement of vit B_{12} and folate may not be concerned directly for the formation of porphyrins. It is required indirectly as follows: the rapid rate of cell growth and division occurring during erythropoiesis results in a correspondingly rapid rate of nucleic acid synthesis, which require B_{12} and folate, to ensure adequate supplies of formate for purine synthesis.
- ***Erythropoietin:*** A low molecular wt. glycoprotein originating in the kidneys, stimulates production, maturation and release of red blood cells from the blood forming tissues.

Synthesis of Hb appears to proceed concurrently with the maturation of erythrocytes. The primitive red cells contain free porphyrins rather than Hb. As the red blood cells mature, the content of free porphyrin decreases and that of Hb rises. ***These biochemical changes are correlated with the alterations in the staining properties of the cells.*** Regulatory mechanisms exist which coordinate the synthesis of haem with that of globin. Haem has been shown to stimulate the synthesis of 'globin' on the ribosomal level. ***In an adult human of 70 kg body wt, approx. 6.25 gm of Hb (90 mg/kg) is synthesised and degraded per day, corresponding to approx. 300 mg of porphyrin (porphyrin rings are about 4 per cent by wt. of Hb molecule).***

CLINICAL ASPECT

PORPHYRIAS AND PORPHYRINURIAS

PORPHYRIA

When the blood levels of coproporphyrins and uroporphyrins are increased above normal level and excreted in urine/faeces, the condition is called porphyria.

Under these terms are included a number of syndromes, some are hereditary and familial and some others are acquired, but all of them are characterised by increased levels in blood and increased excretion of uroporphyrins and coproporphyrins in the urine and faeces. Reduced *catalase* activity has also been reported in these cases.

Classification

Several different classifications of porphyrias have been proposed. However, it is convenient to divide hereditary porphyrias into **two main groups** based on the porphyrin and porphyrin precursors content in bone marrow (erythropoietic) and in liver (hepatic). Hepatic porphyrias may further be subdivided into 3 groups depending on the clinical presentation and enzyme deficiency:

Classification of Porphyrias

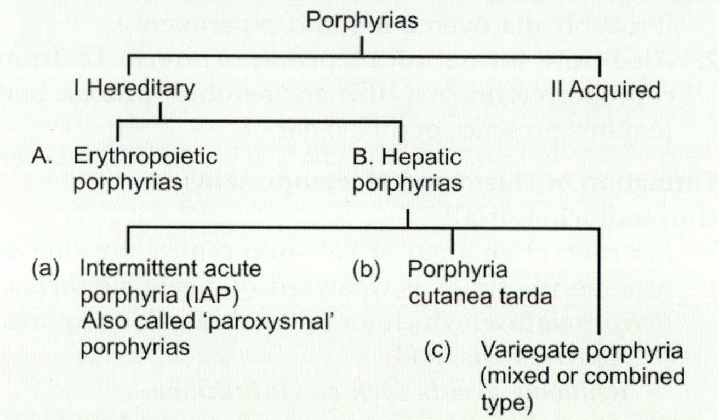

I. Hereditary (Inherited) Porphyrias

A. Congenital Erythropoietic Porphyrias

Rare inherited disorder

Site of Lesion

Genetic defect is present in all cells but peculiarly due to unknown reasons, expressed in erythropoietic tissues and affects red bone marrow.

Inheritance

Autosomal recessive.

Enzyme Deficiency

Preponderance of type I porphyrins, both uroporphyrin type I and coproporphyrin type I (minor series) suggest:

(a) Deficiency of isomerase enzyme, or

(b) Relative preponderance of deaminase activity with isomerase deficiency (low in activity).

Clinically

Affected individuals exhibit abnormal sensitivity to light ***(photosensitivity)*** and develop skin lesions. Porphyrins accumulate under the skin, normoblasts and erythrocytes. Teeth and bones may be brownish or pink due to porphyrin deposition. There may be tendency to haemolysis and defective erythropoiesis.

Biochemically

Blood contains ***increased amounts of porphyrins of type I (minor series)*** which are excreted in urine, they are excreted as free form. Also an increased production and excretion of porphobilinogen and δ-ALA.

Urinary findings

Urine is usually ***'portwine' or 'red'*** coloured. ***Urine contains type I isomers, oxidised to uroporphyrin I and coproporphyrin I (both red pigments).***

Explanation: As uroporphyrinogen III is less formed or absent, haem formation suffers. Relative deficiency of haem produces induction of δ-ALA *synthetase* (an inducible enzyme) leading to massive overproduction of Type I.

B. Hepatic Porphyrias

In these conditions, organ responsible for the dysfunction is the Liver, in which there occurs abnormal and excessive production of porphyrins (chiefly type III), their precursors δ-ALA and porphobilinogen. ***The hepatic porphyrias are divided into 3 groups*** depending on their clinical presentation and enzyme deficiency.

(a) Intermittent acute porphyria (IAP)

Also called as ***paroxysmal*** porphyria.

Inheritance: Is autosomal dominant.

Enzyme deficiency: Partial deficiency of deaminase (uroporphyrinogen I synthetase). In heterozygous—50 per cent activity is present; enzyme defect is present in other cells also, but expressed in hepatic cells only.

Race: Found in Swedish family.

Clinical features: Mainly presents with GI symptoms: acute attacks of abdominal pain, nausea and vomiting, constipation; CV abnormalities and neuropsychiatric signs and symptoms. The above correlates with increased production of porphobilinogen and δ-ALA.

Do not have abnormal sensitivity to light ***(photosensitivity is absent), there may be increase in serum protein bound iodine (PBI), some degree of hypercholesterolaemia and diabetic type of Glucose tolerance curve.***

Urinary findings: Freshly passed urine is often normal in colour ***but on standing in sunlight turns to red wine colour.*** Excretes massive quantities of porphobilinogen and δ-ALA (precursors). Both are colourless compounds but on exposure to air, polymerises to form two coloured red compounds porphobilin and porphyrin.

Biochemically: There is always associated *catalase* deficiency.

Exacerbating factors: Drugs and steroids requiring Cyt. P_{450} can precipitate an acute exacerbation. Reason being excessive utilisation of Cyt. P_{450}, for which haem is utilised. This produces relative diminution of haem in the cells, which produces derepression of *δ-ALA synthetase*.

(b) Porphyria Cutanea Tarda

Most common form, seen in ***South African whites.***

Inheritance: Autosomal dominant.

Usually associated with some form of hepatic injury, particularly alcohol or iron overload.

Enzyme deficiency: Not well defined. ***Probably partial deficiency of uroporphyrinogen decarboxylase.***

Clinical features: Characterised principally by ***skin photosensitivity***

Urinary findings: Contains increased quantities of uroporphyrins and coproporphyrins of both type I and type III, as zinc complexes ***(Not free forms).***

Elevated urinary excretion of d-ALA and porphobilinogen rarely occurs.

Biochemically: Frequent rise in serum iron.

(c) Variegate Porphyria

Also called mixed or combined type in which abdominal, neurological as well as cutaneous symptoms are seen.

Inheritance: Autosomal dominant.

Enzyme deficiency: Not known precisely. Probably partial block of enzymatic conversion of protoporphyrin IX to haem, as a result haem deficiency occurs. ***Enzymes involved are:***

- ***Protoporphyrinogen oxidase,*** and
- ***Ferrochelatase (haemsynthetase).*** Both are mitochondrial enzymes.

Explanation: Relative haem deficiency produced under stressful conditions leads to *derepression* of hepatic *δ-ALA synthetase*, which results in increased activity of δ-ALA synthetase leading to overproduction of all the intermediates of haem synthesis (precursors over production).

Clinical features: Mixed presentation.

Abdominal: Acute attacks of pain in abdomen, nausea and vomiting, constipation.

+ Neuropsychiatric signs and symptoms and cutaneous photosensitivity.

Urinary findings: Excretes excessive quantities of δ-ALA, porphobilinogen and Type I and III isomers.

PORPHYRINURIAS

Urine: Normally only small amounts of coproporphyrins of type I and III, 60 to 280 μg in the ratio of 70 per cent type I and 30 per cent type III are excreted in urine per day in normal health. Uroporphyrins are excreted only in negligible amounts of 15 to 30 μg per day.

Porphyrins excreted in urine in hepatic porphyrias are present as zinc complexes, whereas those of "erythropoietic" porphyrias are in the "free" state.

Faecal porphyrins: The faeces normally contain 300 to 1000 μg of coproporphyrins per day, mostly of the type-I.

Acquired Porphyrias

(a) ***Coproporphyrin type III:*** Coproporphyrin type III is excreted in excessive quantities as the result of:

- Exposure to certain ***toxic chemicals*** and ***heavy metals***, e.g. lead (Pb), arsenic.

SECTION FOUR

- *Acute alcoholism:* Temporary increase in output.
- *Cirrhosis of the liver* in chronic alcoholics (persistent porphyrinurias can occur)
- In certain miscellaneous conditions like poliomyelitis, aplastic anaemia, Hodgkin's disease.

(b) *Coproporphyrin type I:* Abnormally large quantities of coproporphyrin type I found in the urine in:

- Obstructive jaundice (deviation from biliary obstruction)
- Certain liver diseases (including cirrhosis in nonalcoholics)
- Certain blood dyschrasias, viz. haemolytic anaemia, pernicious anaemia and leukaemias.

Experimental Porphyrinuria: A type of porphyrinuria resembling acute intermittent porphyria may be induced in experimental animals by administration of ***Sedormid***, a hypnotic, which ***is chemically "allyl isopropyl acetyl carbamide".*** In this uroporphyrin and coproporphyrins mostly type-III, appear in the urine, along with porphobilinogen and concentrations of these substances are greatly increased in liver. Similar results are obtained by administration of certain barbiturates containing allyl groups. Other chemicals and drugs which can produce porphyrinurias are tolbutamide, sulfonamides, chloroquine and sex hormones.

Laboratory test: Watson and Swartz Test for Porphobilinogen

1. Mix equal volumes of *fresh urine* and Ehrlich's reagent (0.7 gm of p-dimethyl amino benzaldehyde + 150 ml of conc. HCl and 100 ml DW).
2. Allow to stand for 3 minutes.
3. To this add saturated aqueous sodium acetate solution (2 volumes) and leave for 3 minutes.
4. Add a few ml of chloroform and shake thoroughly.

Interpretation

- Porphobilinogen forms a red aldehyde compound, which unlike that formed by urobilinogen, is insoluble in chloroform.
- Hence any red colour remaining in the aqueous phase after a second extraction with chloroform, constitutes a +ve test for porphobilinogen and highly suggestive of acute intermittent porphyria.

Fluorescence of Porphyrins

When porphyrins are dissolved in strong mineral acids or in organic solvents and ***illuminated by UV light, they emit a strong red fluorescence.*** The fluorescence is so characteristic it is used to detect small amounts of free porphyrins.

The double bonds joining the pyrrole rings in the porphyrins are responsible for the characteristic absorption and fluorescence of these compounds.

Note: The double bonds are absent **in the porphyrinogens** hence they **do not give fluorescence.**

CLINICAL ASPECT

Cancer Phototherapy

An interesting application of the photodynamic properties of porphyrins is their possible use in the treatment of certain types of cancer, a procedure called **cancer phototherapy.**

Tumours often take up more porphyrins than do normal tissues. Thus, haematoporphyrins or other related compounds are administered to a patient with appropriate tumour. The tumour is then ***exposed to an argon laser***, which excites the porphyrins producing ***cytotoxic effects***.

SECTION FOUR

I. Questions (Essay type and notes)

1. Describe the synthesis of haemoglobin.
2. Give diagrammatically the biosynthesis of haem depicting all essential details. What are the starting materials for the synthesis?

Short Notes

a. Synthesis of haemoglobin
b. Functions of Hb
c. Porphyrias

II. MCQs (Give one correct answer)

1. Which of the following steps in the biosynthesis of porphyrins is the rate-controlling?
 a. Uroporphyrinogen I synthetase
 b. Uroporphyrinogen decarboxylase
 c. δ-amino laevulinate synthetase
 d. Protoporphyrinogen oxidase
 e. None of the above
2. Substrates required for haemoglobin biosynthesis are:
 a. 'Active acetate' and glycine
 b. Glycine and 'formate'
 c. 'Active succinate' and lysine
 d. 'Active succinate and glycine
 e. Formate and lysine
3. Which of the porphyrins go into the formation of protoporphyrin IX?
 a. Type I series b. Type II series
 c. Type III series d. None of the above
 e. All of the above
4. In the biosynthesis of porphyrins, which of the coenzyme is required for δ-ALA formation:
 a. FAD b. NAD^+
 c. $NADP^+$ d. $B_6 - PO_4$
 e. FMN
5. δ-ALA dehydratase enzyme requires for its activity:
 a. Fe^{++} b. Cu^{++}
 c. Mn d. Co
 e. Mg^{++}

6. Oxidative decarboxylase system which converts coproporphyrinogen III to protoporphyrinogen IX requires which of the coenzyme:
   a. NAD^+ b. $B_6 - PO_4$
   c. TPP d. Flavins
   e. CoA-SH
7. The enzyme which catalyses the synthesis of Haem from protoporphyrin IX is:
   a. Ferroreductase b. Ferrochelatase
   c. Ferro oxidase d. None of the above
   e. All of the above
8. In mammalian liver, conversion of coproporphyrinogen III to protoporphyrinogen IX requires the presence of:
   a. ATP b. Molecular oxygen
   c. Mg^{++} d. B_6 – (P)
   e. NAD^+
9. Uroporphyrin and coproporphyrin of which series is excreted in urine in congenital erythropoietic porphyria:
   a. Type I series b. Type II series
   c. Type III series d. None of the above
   e. All of the above

Answers to MCQS

1. c 2. d 3. c 4. d 5. b 6. d
7. b 8. b 9. a.

CHAPTER

31

HAEM CATABOLISM

Major Concepts

- To study the catabolism of haem, and formation of bile pigments and their metabolism.
- To learn the different types of jaundice that can develop in case of partial or complete failure in metabolic pathway.

Specific Objectives

Haem catabolism (bile pigment metabolism)

The formation of bilirubin, chief bile pigment produced principally by haem catabolism, is a waste product which requires elimination. It involves a series of metabolic alterations and transport processes. Partial or complete failure of any point in this sequence can result in ***jaundice.***

1. Learn the sources of bilirubin.
   - From 'Haem' of 'effete' senescent erythrocytes (85%)
   - From other sources (15%), List the possible sources.
2. Learn the principal sites of Hb breakdown. Study the steps, enzymes involved, and the different views.
   - ***Note that the first product is 'biliverdin'***
   - Study how biliverdin is converted to bilirubin, enzyme and coenzymes required
   - Study what is meant by term **"shunt" hyperbilirubinaemia.**
3. Bilirubin produced in RE cells, called as "unconjugated bilirubin" are released, is **highly lipid soluble** and has limited aqueous solubility. Hence it is transported in blood in combination with albumin
   - Study the binding site of albumin for bilirubin and its capacity
   - Learn the factors that can alter the binding capacity of albumin for bilirubin, physical and chemicals and note clinical importance
   - Study about **bilirubin "encephalopathy"** ***(kernicterus).***
4. Transfer of bilirubin from plasma to liver, study how bilirubin is taken up by liver cells
   - Selective affinity of liver cells to bilirubin
   - Two nonalbumin proteins Y and Z, "Ligandins" acting as acceptors.
5. Conjugation of bilirubin with glucuronic acid
   - Learn the steps of conjugation and role of enzyme *Glucuronyl transferase*
   - Study how bilirubin mono and di-glucuronides are formed (conjugated bilirubin)
   - Note how conjugated bilirubin differs from unconjugated bilirubin: Regarding size, solubility and glomerular filtration
6. Study about *Glucuronyl transferase* activity, inhibition and stimulation of Glucuronyl transferase activity.
   Inhibition: Drugs like novobiocin and steroids (unusual isomer of pregnanediol.)
   - Study about Lucey-Driscoll syndrome, and transient familial neonatal hyperbilirubinaemia.

   Stimulation: List the drugs which can stimulate or increase the enzyme activity. Note the clinical importance of phenobarbitone as a drug.
7. Inherited disorders associated with the 'Glucuronyl transferase' activity, delayed development, partial/complete deficiency. When bilirubin in blood exceeds 1 mg/dl hyperbilirubinaemia occurs and produces jaundice. Clinically visible **jaundice is detected above 2.5 mg/dl.**
   - Learn about different types of unconjugated hyperbilirubinaemias producing clinical jaundice. Study the following unconjugated hyperbilirubinaemias:
     - Transient "Neonatal" physiological jaundice.
     - Crigler-Najjar syndrome: Type I and Type II
     - Gilbert's disease or syndrome.
8. Conjugated bilirubin is secreted in bile canaliculi and from there stored in Gallbladder and excreted in GI tract.
   Study the inherited disorder "Dubin-Johnson syndrome" in which the defect is hepatic secretion of conjugated bilirubin in bile.
9. Study in detail the changes that take place in lower gut, and how L-Stercobilinogen is formed. Learn about "Enterohepatic circulation".

INTRODUCTION

Under physiologic conditions in the human adult, 1 to 2 × 10^8 erythrocytes are destroyed per hour, thus ***in one day, a 70 kg man turns over approx. 6.0 gm of Hb.***

When Hb is destroyed:

- Protein portion globin is reutilised or constituent amino acids are reutilised after proteolysis.
- **Fe^{++} of haem enters "iron pool"** for reutilisation or stored as "ferritin", and
- Fe-free porphyrin portion of haem is degraded to bile pigments, ***Biliverdin*** and ***Bilirubin***, in RE cells.

The formation of Bilirubin, the chief bile pigment in humans, and its elimination from the body as a waste product of haem catabolism requires a series of metabolic alterations and transport processes. ***Partial or complete failure at any point in this sequence can result in clinical condition, Jaundice*** (Fig. 31.1).

SOURCES OF BILIRUBIN

Mainly Two:

(a) From 'haem' of "effete" erythrocytes and (b) From sources other than effete erythrocytes.

(a) From 'Haem' of erythrocytes

Approximately 85 per cent of bilirubin is derived from senescent erythrocytes by conversion of 'haem' of Hb to biliverdin within RE cells.

Site

The principal sites are ***bone marrow, spleen*** and ***liver***. The bone marrow appears to be the most active site.

Amount: Approximately 210 to 250 mg of bile pigments are excreted daily by the liver in the bile. ***One gram of Hb yields approx. 35 mg bilirubin.***

Nature of bile pigments: Principal bile pigments are ***biliverdin*** and ***bilirubin***. The colour of the bile is due primarily to these and to derivatives of them. Normally,

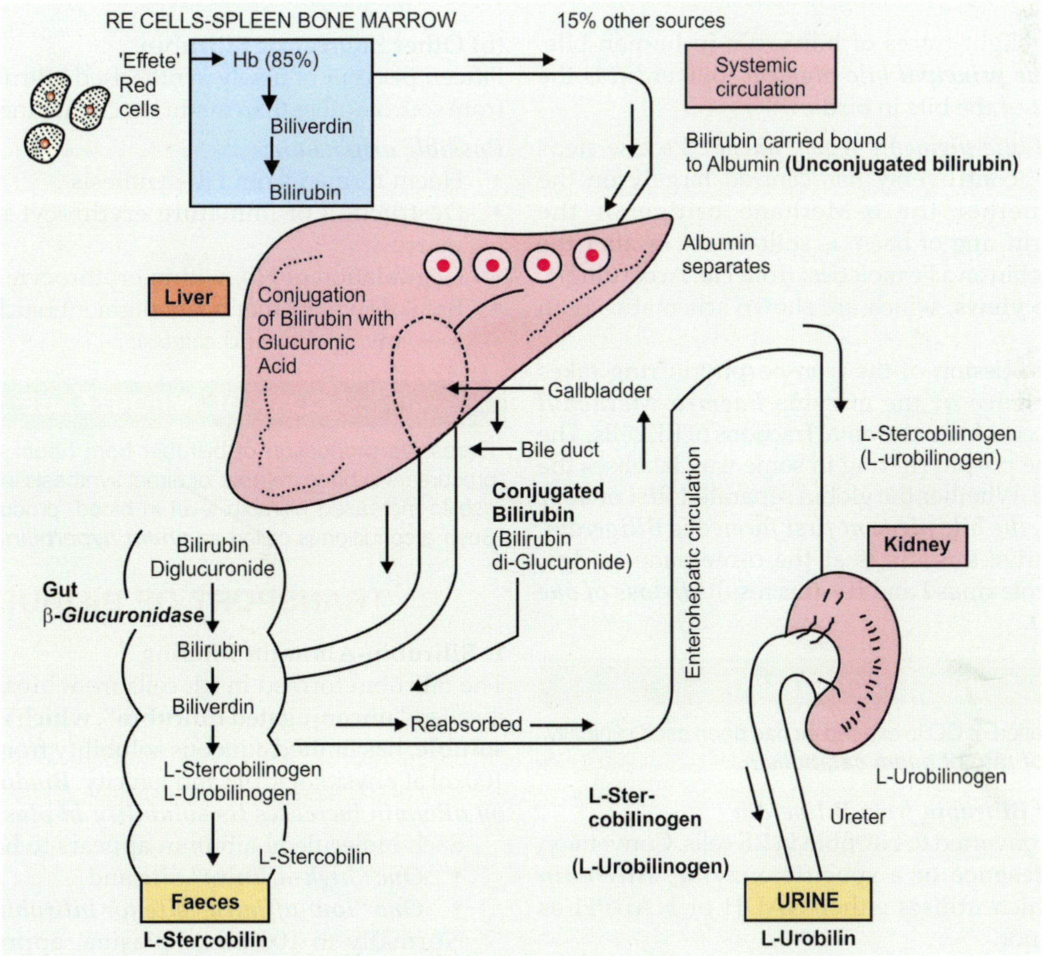

Fig. 31.1: Diagrammatic representation of bile pigment metabolism

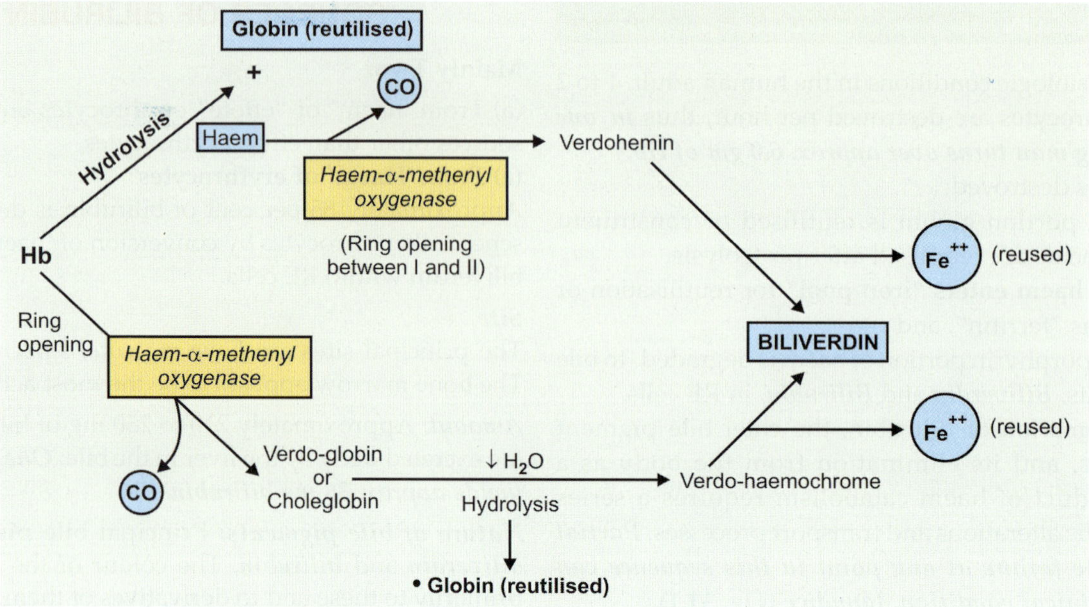

Fig. 31.2: Breakdown of Hb

there is only slight traces of biliverdin in human bile, ***bilirubin is the principal bile pigment.*** Biliverdin is the chief pigment of the bile in birds.

Formation of bile pigments from "haem": Precise steps are not clear, controversy has centred largely on the question whether the α-Methane bridge of the protoporphyrin ring of haem is split before or after the removal of globin and Fe moieties from Hb. Accordingly, there are **two views,** which are shown schematically in (Fig. 31.2).

Oxidative scission of the iron-porphyrin ring takes place in presence of the enzyme ***haem-α-Methenyl oxygenase,*** occurs in microsomal fractions of RE cells. The opening of the porphyrin ring in some way labelises the removal of Fe. Whether the globin separated first or after ring opening, ***the bile pigment first formed is Biliverdin.*** Site of oxidative scission is at the α-methane bridge, between pyrrole rings I and II—this results ***in loss of one carbon as CO.***

CLINICAL IMPORTANCE

Rate of elimination of CO in expired air has been used clinically ***as an index of rate of haem catabolism.***

Formation of Bilirubin from Biliverdin

Biliverdin is converted to bilirubin in RE cells. Conversion occurs in presence of a specific enzyme, ***Bilirubin reductase*** which utilises either NADH or NADPH as hydrogen donor.

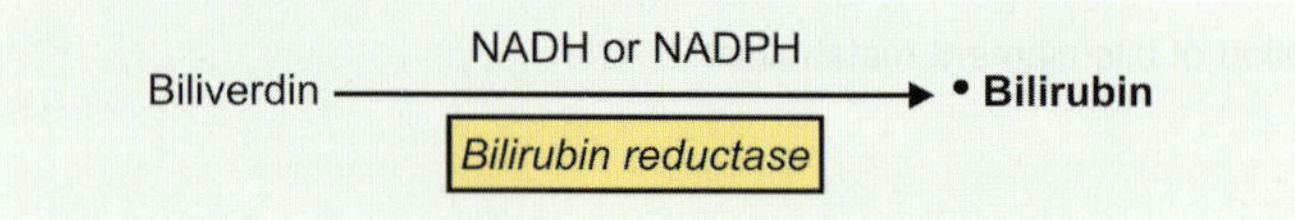

(b) Other Sources of Bilirubin

Fifteen per cent of newly synthesised bilirubin is derived from sources other than maturing circulating erythrocytes.

Possible origins are:

- Haem formed from Hb-synthesis
- Destruction of immature erythrocytes in the bone marrow
- Degradation of Hb, within erythrocyte precursors
- Breakdown of other haem pigments such as cytochromes, myoglobin and *catalase.*

CLINICAL IMPORTANCE

Excessive production of bilirubin from haem or erythrocyte precursors in bone marrow, or direct synthesis in marrow, gives rise to increased bilirubin level in blood, producing jaundice. Such a condition is called as ***shunt hyperbilirubinaemia.***

TRANSPORT OF BILIRUBIN

1. Bilirubin-Albumin Binding

The bilirubin formed in RE cells from breakdown of Hb is called **"unconjugated bilirubin", which is highly lipid soluble,** has limited aqueous solubility from 0.1 to 5 mg/100 ml at physiologic pH and tonicity. ***Binding of bilirubin by albumin increases its solubility in plasma.***

Each molecule of albumin appears to have:

- ***One "high-affinity" site*** and
- ***One "low-affinity" site for bilirubin.***

Normally in 100 ml of plasma, approx. 25 mg of bilirubin can be tightly bound to albumin to its high affinity site. Bilirubin in excess of this quantity can be bound only loosely and can thus be easily detached and can diffuse into the tissues.

> **CLINICAL IMPORTANCE**
>
> Clinical interest in binding of bilirubin by albumin has primarily related to the development of **bilirubin encephalopathy, (kernicterus).** In this condition, seen only rarely outside the newborn period, ***unconjugated bilirubin enters the neurons of the basal ganglia, hippocampus, cerebellum, and medulla, causing necrosis of nerve cells, probably by interfering with cellular respiration.***

Alterations of Albumin-Bilirubin Binding and its Biomedical Significance

The binding capacity of albumin for bilirubin can be modified by a variety of physical and chemical alterations:

- Several "anionic drugs" such as sulphonamides. ***Administration of sulphonamides to pregnant women and neonates increases the risk of kernicterus in the jaundiced infants.***
- Increased free fatty acids behave similarly.
- Asphyxia, hypoxia and acidosis are also associated with increased risks:
  - By interfering with bilirubin-albumin binding, and
  - May also increase the permeability of brain for unconjugated bilirubin.

2. Transfer of Bilirubin from Plasma to Liver Cells

Liver appears to have a ***selective affinity to remove unconjugated bilirubin.***

Two views are prevalent:

- Lateral extension of plasma membrane of Liver cells facing hepatic sinusoids has specific *receptor sites* for bilirubin, and
- Plasma membrane permeable to "non-polar molecules" like dissociated unconjugated bilirubin and an "intracellular" protein (or proteins) which act as an acceptor and facilitates the transfer of bilirubin to liver cells. Earlier two ***non-albumin proteins,*** designated as **'Y'** and **'Z'** have been isolated from liver cytoplasm and account for most of intracellular binding of bilirubin. ***Recent studies have shown that the proteins are same single one and has been named as ligandins.***

CONJUGATION OF BILIRUBIN WITH D-GLUCURONIC ACID IN LIVER CELLS

Mammalian liver cells contain an enzyme referred to as ***glucuronyl transferase.*** The enzyme catalyses the transfer of Glucuronic acid from UDP-GA to various phenolic, carboxylic and amine receptors. The process is called ***conjugation*** reaction and it is carried out in the smooth endoplasmic reticulum of liver cells.

Two glucuronyl groups are transferred from "active-glucuronide" (UDPGA) by the catalytic action of glucuronyl transferase. Glucuronic acid is attached through "ester-linkage" to the propionic acid carboxyl group of bilirubin to form the glucuronide. In the conjugation reaction, ***monoglucuronide*** is formed first, followed by formation of ***bilirubin di-glucuronide*** (Fig. 31.3).

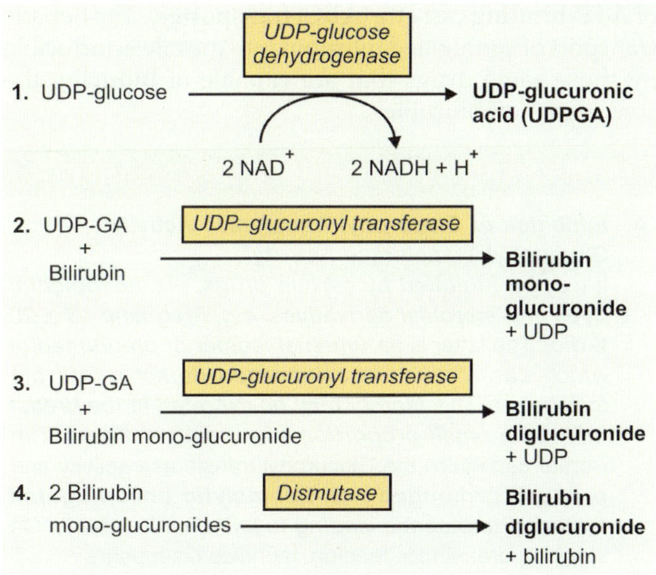

Fig. 31.3: Steps of conjugation of bilirubin in liver cells

Two Bilirubin monoglucuronides can form one molecule of Bilirubin diglucuronide and one molecule free 'bilirubin' by the action of the enzyme ***dismutase.***

Mono and diglucuronides of bilirubin are called as **conjugated bilirubins.** Conjugated bilirubins, unlike unconjugated bilirubins are:

- **Water soluble** and
- **Smaller in molecular size** as they are not bound to albumin.

Hence, conjugated bilirubin can pass through glomerular filter and can appear in urine (bilirubinuria). Unconjugated bilirubin cannot pass through glomerular filter and does not appear in urine.

Glucuronyl Transferase Activity in Extrahepatic Tissues

Recently Glucuronyl transferase activity has also been detected in certain extrahepatic tissues, viz. skin, kidneys, adrenal glands, ovary, testes, intestinal mucosa and synovial membrane. But the role of the enzyme in these extrahepatic tissues is uncertain. ***Probably Bilirubin monoglucuronide may be formed but not the diglucuronide in these tissues.***

SECRETION OF BILIRUBIN IN THE BILE

Secretion of conjugated bilirubin into the bile occurs by an "**active transport**" **mechanism,** which is **rate limiting** for the entire process of hepatic bilirubin metabolism.

MRP-2: The protein involved is ***MRP-2*** (***multi drug resistance protein*** 2) also called *"**multispecific organic anion transporter**"* (**MOAT**). It is located in the plasma membrane of the bile canalicular membrane and handles a number of organic anions. It is a member of the family of **ATP-binding cassette (ABC) transporters**. The hepatic transport of conjugated bilirubin into the bile is inducible by those same drugs that are capable of inducing the conjugation of bilirubin.

BIOMEDICAL IMPORTANCE: CLINICAL ASPECT

A. *Inhibition of Glucuronyl Transferase Activity*

1. **Glucuronyl Transferase Activity**

   It may be ***inhibited by certain drugs, viz. novobiocin, dyes and steroidal derivatives, e.g. Pregnane –3 α-20 β-diol.*** The latter is an ***unusual isomer of pregnanediol*** which can form due to *inherited defect* in steroid metabolism. This ***isomer may be excreted in the breast milk by a small proportion of nursing mothers.*** The isomer can inhibit the Glucuronyl transferase activity and ***produce prolonged non-haemolytic unconjugated hyperbilirubinaemia leading to jaundice in infants.*** On stopping breast milk feeding, jaundice disappears.

2. **Lucey-Driscoll Syndrome**

   This is a transient familial neonatal ***nonhaemolytic unconjugated hyperbilirubinaemia.*** Healthy looking women can give birth to infants with severe nonhaemolytic unconjugated hyperbilirubinaemia ***with risk of Kernicterus***. An ***unidentified factor, probably progestational steroid***, has been isolated from serum of the mother, which inhibits the glucuronyl transferase activity producing the condition.

3. **Transient Neonatal "Physiological" Jaundice**

   Most common cause of ***neonatal unconjugated hyperbilirubinaemia.*** It results from an accelerated haemolysis and due to an immature hepatic system for uptake, conjugation and secretion of bilirubin. Glucuronyl transferase activity is delayed and reduced. Probably also reduced synthesis of substrate, i.e. UDP-Glucuronic acid. ***Risk of kernicterus is present.***

   Administration of phenobarbital, which stimulates the enzyme activity is useful.

   Exposure to visible light (phototherapy) is helpful, it promotes hepatic excretion of unconjugated bilirubin by converting some of bilirubin to other derivatives:
   - ***Maleimide fragments,*** and
   - ***Geometric isomers,*** which are excreted in bile.

4. **Crigler-Najjar Syndrome**

   (a) Type I: A rare autosomal recessive disorder. Primary metabolic defect is inherited absence of ***glucuronyl transferase*** activity.

   Genetic defect: It is due to ***mutations*** in the ***gene encoding bilirubin-UGT*** activity in hepatic tissues. The gene encoding human bilirubin-UGT is part of a large UGT-gene complex ***situated on chromosome*** 2, characterised by severe congenital ***nonhaemolytic unconjugated hyperbilirubinaemia and jaundice.*** Usually fatal within the first 15 months of life. When untreated, serum bilirubin usually exceeds 20 mg/dl leading to risk of kernicterus. ***Phototherapy has been found to be useful.***

   (b) Type II: A rare inherited disorder. Milder defect in the bilirubin conjugating system and has a more benign course.

   Genetic defect: Also results from mutations in the gene encoding bilirubin-UGT, but ***some activity of the enzyme*** is retained and thus the ***condition has more benign course*** than Type-1. Unconjugated hyperbilirubinaemia, serum bilirubin usually do not exceed 20 mg/dl. ***No risk of kernicterus.*** Bile of these patients have been found to contain "bilirubin monoglucuronide" only. Patients respond to treatment with large doses of phenobarbitone.

5. **Gilbert's Syndrome**

   A heterogenous group of diseases, many of which are now recognised to be:
   - Due to a compensated haemolysis associated with unconjugated hyperbilirubinaemia
   - Due to a defect in hepatic clearance of bilirubin, possibly due to defect in uptake of bilirubin by liver cells
   - Due to reduced *glucuronyl transferase* activity
   - Due to mutations in the gene encoding bilirubin UGT.

   Gilbert and his colleagues described the syndrome to be characterised by low grade chronic unconjugated hyperbilirubinaemia and jaundice. ***Bilirubin level in 85 per cent cases is usually less <3 mg/dl.***

   Age: 18 to 25 years, more common in males, detected suddenly during examination—a mild icterus of sclera of eye. Patient usually complaints of fatigue, weakness, and abdominal pain.

6. **Toxic Hyperbilirubinaemia**

   Unconjugated hyperbilirubinaemia can result from toxin induced liver dysfunctions such as that caused by chloroform, arsphenamines, carbon tetrachloride, acetaminophen, hepatitis virus, cirrhosis, and Amanita mushroom poisoning. These acquired disorders are due to **hepatic parenchymal cell damage** which **impairs conjugation**.

7. **Dubin-Johnson Syndrome**

   An autosomal recessive disorder. Characterised by ***conjugated hyperbilirubinaemia*** and jaundice in childhood and during adult life.

   Genetic defect: The hyperbilirubinaemia is caused by ***mutations*** in the ***gene encoding MRP-2, the protein involved in the secretion of conjugated bilirubin into bile.***

   BSP Test
   - When performed **shows a secondary rise** in plasma concentration due to reflux of the conjugated BSP ***(pathognomonic)***. Dyes, viz. Indocyanine Green and Rose Bengal, do not require conjugation hence secondary rise do not occur.
   - Another interesting feature is 80 to 90 per cent of coproporphyrins excreted in urine are of type I, reasons not known. No abnormalities in porphyrin synthesis seen.
   - Hepatocytes in centrilobular area have been found to **contain an abnormal** pigment in this disease that may be derived from epinephrine.

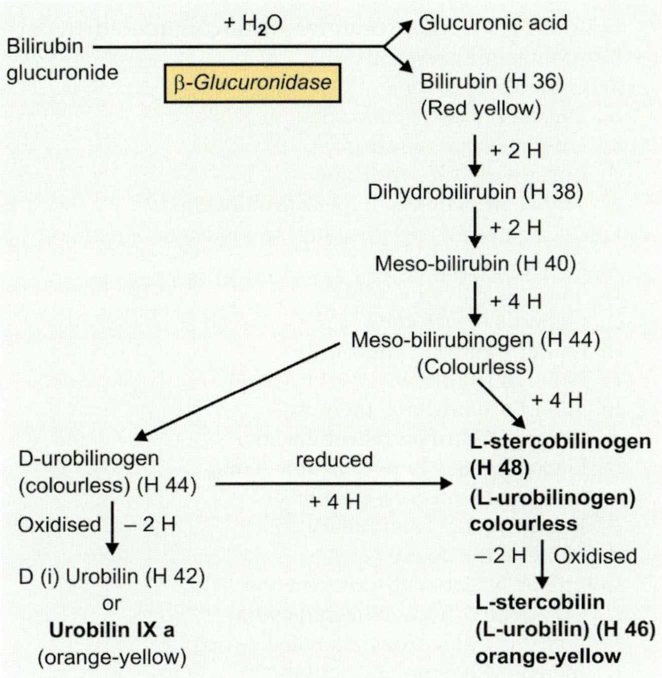

Fig. 31.4: Reductive changes of bilirubin in intestine

8. **Rotor Syndrome**
   A rare benign condition characterised by chronic conjugated hyperbilirubinaemia and normal liver histology. Its precise cause has not been identified.
9. **Obstruction of the Biliary Tree**
   Conjugated hyperbilirubinaemia commonly results from blockage of the hepatic or common bile ducts, often due to:
   - A gall stone or
   - Cancer of the head of the pancreas.

   Because of the obstruction, bilirubin diglucuronide cannot be excreted. It thus regurgitates into the hepatic veins and lymphatics and conjugated bilirubin appears in blood and urine (**choleric jaundice**).

B. *Increased Activity of Glucuronyl Transferase*
Hepatic ***glucuronyl transferase*** activity is increased after administration of certain drugs, viz. Benzpyrene, aminoquinolines, chlorcyclizine and phenobarbitones to normal adults and neonates. Administration of these drugs results in proliferation of smooth endoplasmic reticulum and increases the synthesis of the enzyme.

EXCRETION OF BILE PIGMENTS

Conjugated bilirubins are secreted in the GI tract in bile. In the lower portion of the intestinal tract, specially in the caecum and the colon, the bilirubin is released from the glucuronides with the help of the enzyme ***β-glucuronidase*** produced by bacteria, and then the released bilirubin is subjected to series of reductive action of enzyme systems present in the intestinal tract, mainly derived from the anaerobic bacteria in the caecum.

Experimental Evidence: Faecal flora as well as a pure strain of a *Clostridium* derived from the rat colon have been demonstrated ***in vitro*** to be able to complete the reduction of bilirubin to L-stercobilinogen, the normal end product of bilirubin metabolism in the colon. The series of reductive changes that take place is shown in Figure 31.4.

In the intestine, Progressive hydrogenation (reduction) occurs, as shown above, to produce a series of intermediary compounds which, beginning with **meso-bilirubinogen** comprise a number of colourless urobilinoids, which may be oxidised, with loss of hydrogen, to coloured compounds.

The ***end product is colourless "L-stercobilinogen" (L-urobilinogen).*** Auto-oxidation in the presence of air, produces ***L-stercobilin (L-urobilin),*** an orange-yellow pigment which contributes to the normal colour of the faeces and urine. Stercobilin is strongly *Laevo-rotatory*.

Urobilin IX-D(1) urobilin or inactive (1) urobilin, is an optically inactive urobilinoid that has been identified in the faeces. It is less stable than stercobilin, is oxidised in air to form violet or blue-green pigments.

Enterohepatic Circulation of Bile Pigments: The various products derived from the progressive reduction of bilirubin may in part be absorbed from the intestine and returned to the liver for its re-excretion, called as ***entero-hepatic circulation*** of bile pigments. A small part escapes enterohepatic circulation and excreted in urine, which normally contains traces of "urobilinogen" and urobilin as well as mesobilirubinogen and perhaps other intermediary products. The great majority of the metabolites of bilirubin are, however, excreted with faeces.

SECTION FOUR

CLINICAL SIGNIFICANCE

Alteration of Intestinal Flora with Bile Pigments Metabolism
If the intestinal flora is modified or diminished, as by the administration of ***orally effective broad spectrum antibiotics,*** which are capable of producing partial sterilisation of the intestinal tract, bilirubin may not be further reduced and may later be auto-oxidised, in contact with air, to **biliverdin**. Thus, ***the faeces acquire a greenish tinge under the above circumstances. Similar condition may develop in premature babies/or in infants where the bacterial flora develop late.***

In the patients whose intestinal flora is altered by oral administration of oxytetracyclines/or chlortetracyclines, as stated above, a dextrorotatory urobilinoid, ***D-urobilin***, has been identified. It is believed to be derived from dihydrobilirubin by way of ***D-urobilinogen.***

Note: For significance of these pigments in different types of jaundice, see chapter on "Liver Function Tests".

I. Questions (Essay type and short notes)

1. Describe briefly the fate of Hb in the body.
2. What are bile pigments? How are they formed? Describe briefly the metabolism of bile pigments.

Short Notes

a. Enterohepatic circulation
b. Catabolism of haem
c. Breakdown of haem
d. Breakdown of Hb
e. Bile pigments
f. Bile salts and bile pigments
g. Bilirubin metabolism and the measurement of serum bilirubin
h. Bilirubin.

II. MCQs (Give one correct answer)

1. Bile pigments are:
   a. Cholic acid
   b. Deoxycholic acid
   c. Bilirubin
   d. Chenodeoxycholic acid
   e. Lithocholic acid
2. Degradation of Hb takes place in:
   a. Mitochondrion
   b. Erythrocytes
   c. Cytosol of cell
   d. Reticuloendothelial cells
   e. All of the above
3. In haem catabolism, the first bile pigment formed is:
   a. Cholic acid
   b. Bilirubin
   c. Lithocholic acid
   d. Deoxycholic acid
   e. Biliverdin
4. Biliverdin is converted to bilirubin by the process of:
   a. Oxidation
   b. Reduction
   c. Conjugation
   d. Decarboxylation
   e. Dehydrogenation
5. Bilirubin formed in reticuloendothelial cells is transported to liver in combination with:
   a. Transferrin
   b. Albumin
   c. Globulin
   d. Haptoglobin
   e. Ceruloplasmin
6. In liver cells, bilirubin is conjugated with:
   a. Glucose
   b. Cholic acid
   c. Glycine
   d. Glucuronic acid
   e. Iduronic acid
7. Bilirubin is derived from all of the following except:
   a. Destroyed effete red blood cells
   b. Cytochromes
   c. Haemoglobin
   d. Catalase
   e. Coenzymes
8. The enzyme responsible for conjugation of bilirubin is:
   a. Bilirubin esterase
   b. Haemoglobin reductase
   c. Bilirubin conjugase
   d. Glucuronyl transferase
   e. Glutamyl-bilirubin esterase
9. Following are the examples of unconjugated hyperbilirubinaemias, except:
   a. Gilbert's syndrome
   b. Dubin-Johnson syndrome
   c. Criglar-Najjar syndrome
   d. Lucey-Driscoll syndrome
   e. Transient neonatal hyperbilirubinaemia
10. Unconjugated hyperbilirubinaemia is characteristic of:
    a. Rotor syndrome
    b. Chronic active hepatitis
    c. Gilbert's syndrome
    d. Dubin-Johnson syndrome
    e. None of the above
11. In Gilbert's syndrome, there is:
    a. Conjugated hyperbilirubinaemia
    b. Unconjugated hyperbilirubinaemia
    c. Increased transferase activity
    d. Frequent attacks of biliary colic
    e. None of the above
12. In Criglar-Najjar syndrome, there is:
    a. Conjugated hyperbilirubinaemia
    b. Increased glucuronyl transferase activity
    c. Increased haemolytic activity
    d. Unconjugated hyperbilirubinaemia
    e. None of the above
13. About L-stercobilinogen all are true, except:
    a. L-stercobilinogen and L-urobilinogen are structurally similar
    b. Is formed from unconjugated bilirubin in large intestine by action of bacteria
    c. Stercobilin is the oxidation product
    d. Daily excretion in faeces is about 250 mg
    e. Small amount is reabsorbed from gut and undergoes enterohepatic circulation
14. All are true about Gilbert's syndromes except:
    a. Serum bilirubin is usually < than 5 mg/dl
    b. Liver function tests are normal
    c. Sclera of the eye usually shows mild icterus
    d. Liver biopsy shows dark pigmentation of hepatic cells
    e. There is unconjugated hyperbilirubinaemia
15. All of the following statements regarding haem catabolism are correct, except:
    a. The reduction of biliverdin yields bilirubin
    b. The formation of biliverdin is the first step in catabolism of haem
    c. Bilirubin is conjugated with glucuronic acid in liver
    d. Failure to conjugate bilirubin with glucuronic acid leads to jaundice
    e. Free unconjugated bilirubin is reduced to urobilinogen in liver

Answers to MCQs

1. c	2. d	3. e	4. b	5. b	6. d
7. e	8. d	9. b	10. c	11. b	12. d
13. b	14. d	15. e.			

SECTION FOUR

CHAPTER

32

DETOXICATION

Major Concepts

A. Learn various theories of detoxication.
B. Learn the details of each process of detoxication.

Specific Objectives

A. 1. Define detoxication.
2. Know the theories of Sherwin, Berczeller and Quick.
Understand the 2-phase concept of Xenobiotics.

B. 1. Know the mechanism of detoxication as found in
• Oxidation • Reduction • Hydrolysis • Conjugation.
2. Define conjugation.
3. Learn in detail various types of conjugation reactions.
4. Study the mechanism of detoxication of drugs.
5. Learn the biomedical importance of mono-oxygenase cyt P_{450} system in detoxication.

WHAT IS DETOXICATION?

INTRODUCTION

By the term "detoxication" it is meant, all the biochemical processes, whereby noxious substances are rendered less harmful and are more easily excreted in urine.

Several theories have been put forward from time to time to explain detoxication:

- *Theory of Sherwin:* Detoxication mechanisms render so called toxic compounds less toxic by transforming them into more soluble derivatives, which are then more easily excreted.
- *Theory of Berczeller:* Proposed that toxic compounds are made less toxic by transformation into compounds having a surface tension nearly like water than the parent compound. In this way toxic compounds are prevented from accumulating at the surface of cells, since the non-toxic forms are swept into body fluids and excreted.
- *Theory of Quick:* Proposed that the important factor is conversion of a weakly acidic substance to a strongly acidic one. Kidney can excrete stronger acids and their salts more readily than weaker acids.

Definition

The term detoxication covers all those biochemical changes proceeding in the body, which convert foreign molecules, generally toxic, ***but not always so***, to generally non-toxic or less toxic but not always so, and more soluble so that they can be easily excreted.

Foreign molecules may be ***exogenous***, which include those substances which are not ordinarily ingested or utilised by the organism. Those may enter the body through the dietary foodstuffs or in the form of certain medicines / or drugs which are administered to the body.

- Some of them may be ***endogenous*** and may be produced in the body by synthesis or as metabolites of various processes in the body.

Note: Detoxification or detoxication is not now considered always as an appropriate term.

Xenobiotics: Foreign molecules, which enter the body are called ***xenobiotics.*** Humans are now subjected increasingly to exposure to various foreign chemicals (xenobiotics) whether they may be **drugs, food additives** or **pollutants** or **carcinogens**. In some cases, the reactions to which xenobiotics are subjected may increase their biologic activity and even toxicity, instead of making less toxic.

MECHANISM OF DETOXICATION

They are mainly of **four** types:

- **Oxidation**
- **Reduction**
- **Hydrolysis** and
- **Conjugation**

Sometimes they may occur independently and in others there may be combination of these processes. As for example, in many cases, in humans, oxidation and other reactions may be followed by conjugation. ***In man, detoxication is principally carried out in liver, but to some extent it can be carried out in kidneys also.*** Present concept is that the reactions of xenobiotics occur in **two phases:**

Phase 1: This phase involves the hydroxylation, the major reaction, catalysed by *mono-oxygenases* or *cytochrome* P_{450} species. Other types of reactions in Phase 1 include oxidation, reduction and hydroxylation.

Phase 2: The hydroxylated or other compounds produced in phase 1 are converted by specific enzymes to various water soluble polar metabolites by conjugation with various conjugating agents, viz. Glucuronic acid, "active" sulfate, methylation, acetylation, etc.

The overall purpose of these two phases is to increase their water solubility and thus facilitate their excretion from the body.

A. OXIDATION

A large number of foreign substances are destroyed in the body by oxidation. Aliphatic as well as aromatic alcohols may be oxidised to corresponding acids, probably via aldehyde formation. In addition certain amines, anilides and drugs also can undergo oxidation.

Examples are:

1. ***Methyl groups:*** These groups can be oxidised to form – COOH group through formation of aldehyde.

$$—CH_3 \rightarrow —CH_2OH \rightarrow —CHO \rightarrow —COOH$$

2. ***Primary aliphatic and aromatic alcohols:*** They are oxidised to corresponding acids, e.g.

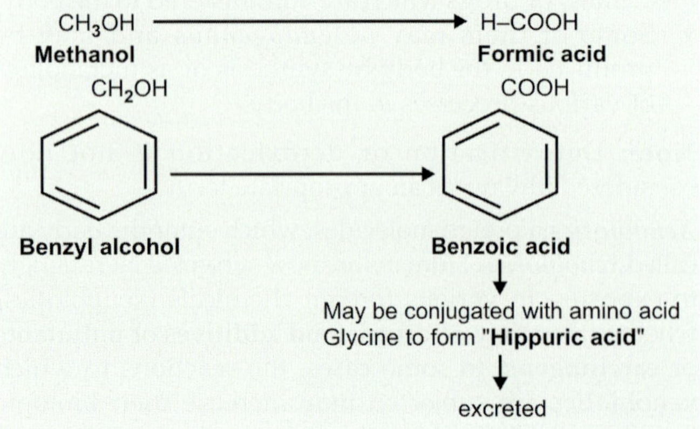

3. ***Aromatic hydrocarbons:*** Aromatic hydrocarbons are oxidised to Phenol and other phenolic compounds. Again they are conjugated with glucuronic acid or sulphuric acid and excreted as corresponding *glucuronides* and *sulphates*.

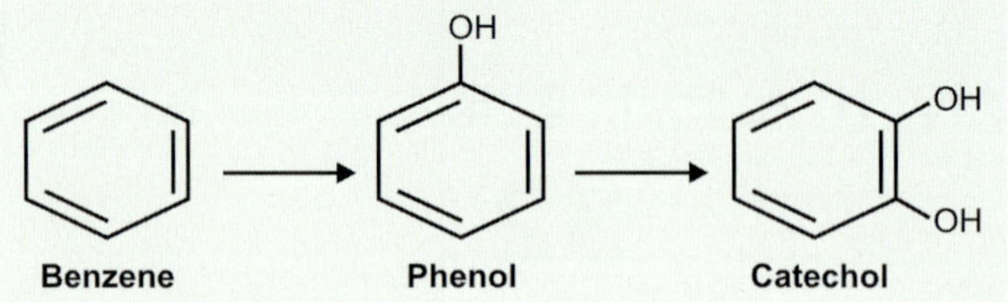

In rare cases, aromatic ring may open, but only to a slight extent, for example:

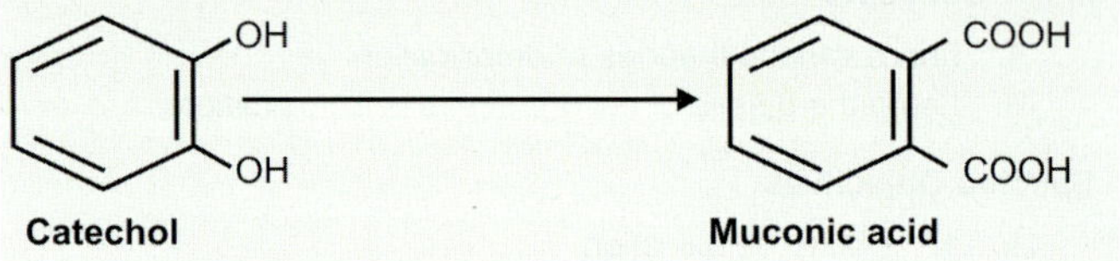

4. ***Aldehydes:*** Aldehydes are oxidised to form the corresponding acids.

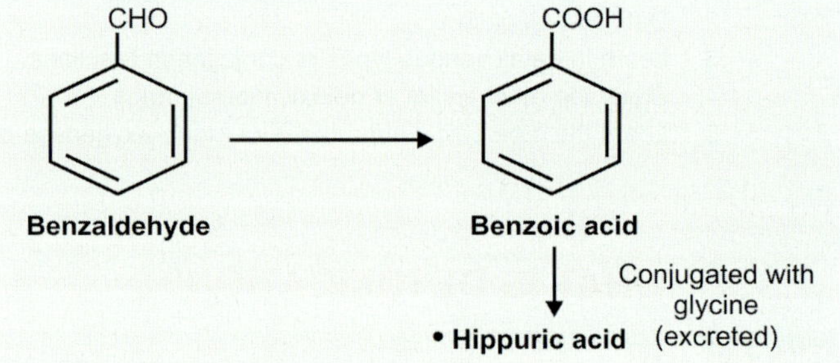

5. ***Anilides:*** Anilides are oxidised to the corresponding Phenols, e.g. Acetanilide is present as a constituent of analgesic drugs, which relieves pain. It is oxidised in the body to form p-acetyl amino phenol.

Acetanilide ⟶ **p-acetyl amino phenol**

6. ***Amines:*** Many primary aliphatic amines undergo oxidation to the corresponding acids and N is converted to urea.

Benzylamine ⟶ **Benzoic acid + urea**

- Aromatic amines like aniline is oxidised to corresponding phenol.

Aniline ⟶ **p-amino phenol**

7. ***Sulphur compounds:*** The sulphur present in organic sulphur compounds is oxidised to SO_4^- which in turn may be excreted in inorganic or organic form or as neutral (unoxidised) sulphur.

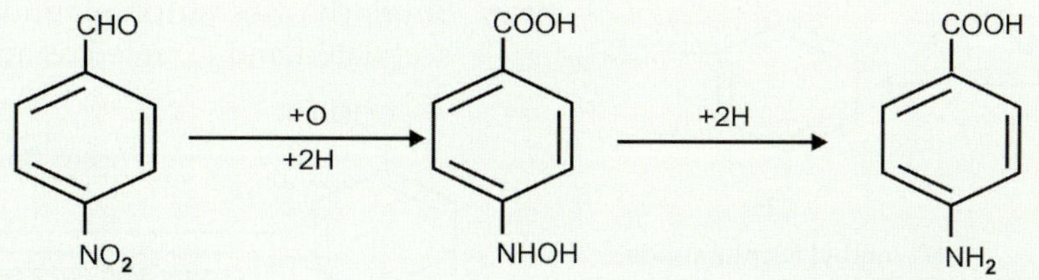

p-nitrobenzaldehyde → **p-amino benzoic acid** excreted after conjugation

8. ***Drugs:*** Certain drugs can be oxidised in the body and are excreted as hydroxy derivative or salts.

Examples are:

- ***Meprobamate:*** A tranquilizer used in psychiatric disorders is excreted largely as the oxidation product hydroxy meprobamate.

Meprobamate ⟶ OH–meprobamate

- ***Chloral:*** Used as a hypnotic. Most of the chloral undergoes reduction and conjugation; but partly it can be oxidised to form trichloroacetic acid which is excreted as its salt.

Chloral ⟶ Trichloroacetic acid

(See hydroxylation of drugs and role of Cyt. P_{450} species below)

B. REDUCTION

Reduction usually does not occur extensively in man.

Examples are:

- ***Certain aldehydes,*** e.g. chloral, a hypnotic, principally undergoes reduction in the body to form corresponding alcohol, which is then conjugated with D-glucuronic acid and excreted as corresponding glucuronides.

CCl_3CHO (**Chloral**) ⟶ CCl_3CH_2OH (**Trichlorethanol**)
+ D-Glucuronic acid
excreted as corresponding glucuronides↓

- ***Aromatic nitrocompounds,*** e.g. p-nitrobenzaldehyde is reduced to corresponding amines and excreted after conjugation (Refer box – top).

Note: Certain of the reduced metabolites, instead of being less toxic, may be more toxic.

C. HYDROLYSIS

There are quite a number of therapeutic compounds, used as drugs, which undergo hydrolysis, usually in liver.

Examples are:

- Acetyl salicylic acid (Aspirin) ⟶ Salicylic acid + Acetic acid
  Salicylic acid can reduce Benedict's Qualitative reagent.
- Atropine (Tropyl tropate) ⟶ Tropic acid + tropin
- Digitalis (A cardiac glycoside) ⟶ Sugar + Aglycone ↓ (Nonsugar component)
- Procaine ⟶ p-aminobenzoic acid + Diethylamino ethanol

D. CONJUGATION

Definition

A process by which the foreign molecules or its metabolites are coupled with a conjugating agent and converted to soluble, nontoxic derivatives which are easily excreted in urine.

Features

- Various conjugating agents are available in the body and some of them are synthesised in the body, e.g. *D*-glucuronic acid formed from glucose by uronic acid pathway. Certain amino acids as glycine, cysteine, can be available from dietary proteins/or breakdown of tissue proteins or synthesised.
- Conjugation reaction principally occurs in liver and to some extent it can occur in kidneys also.
- Conjugation produces less toxic, more soluble compounds which are excreted.
- Conjugation can occur independently or it can follow oxidation, reduction or hydroxylation of a compound.

Types of Conjugation

1. ***Methylation***
   - Methylation as a detoxication process though limited in the body, at the same time is quite important. ***Usual methyl donor is "S-adenosyl methionine" ("active" methionine).***
   - Methylation of heterocyclic N-atom of compounds of the Pyrimidine and Quinoline types, e.g. Nicotinamide. This occurs also with other heterocyclic aromatic compounds, e.g. Histamine.

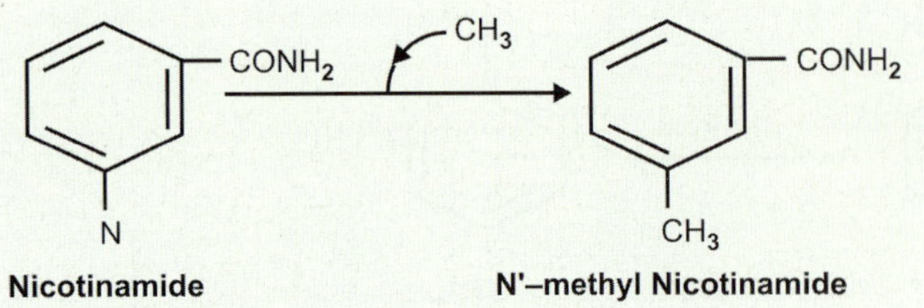

- Methylation of p-aminomethyl amino azo benzene to p-dimethyl amino azo benzene (butter yellow), a potential hepatic carcinogen.
- O-methylation of certain naturally occurring amines (with phenolic hydroxyl group), e.g. epinephrine and norepinephrine and their metabolites are methylated at the phenolic hydroxyl group.
- O-methylation of natural estrogens.

2. ***Acetylation Reactions***

In detoxication reactions, conjugation with acetic acid occurs only with aromatic NH_2 group.

Exception: Acetylation of –OH group takes place in physiological compounds like formation of acetyl choline.

- Acetic acid helps in conjugation of aromatic compounds alongwith cysteine to form corresponding mercapturic acids (see below in next page—conjugation with cysteine).
- In humans, certain drugs, e.g. sulpha drugs are conjugated by acetylation. As much as 50 per cent of excreted sulpha drugs may be acetylated and excreted as acetylated derivatives.

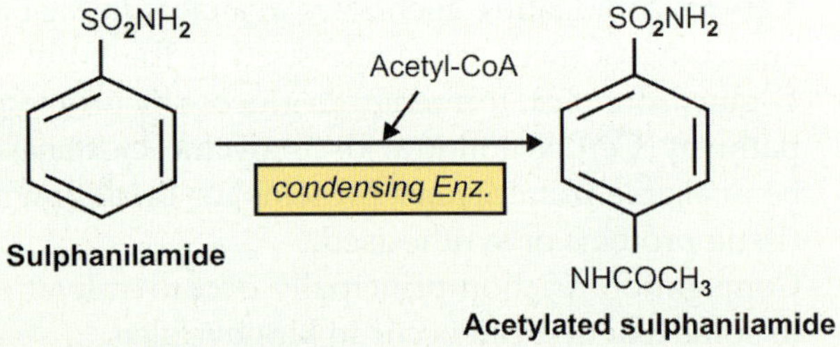

- Similarly, like sulphonamide drugs, PABA is also acetylated and excreted as acetyl derivative.

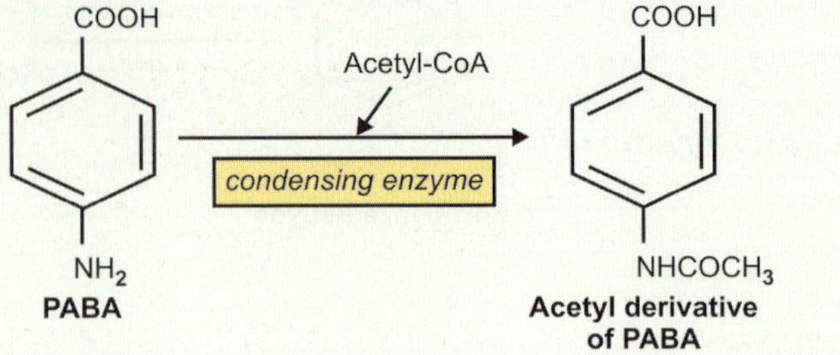

Acetylation is done by active acetate (Acetyl-CoA): It is catalysed by the enzyme ***acetyl transferase*** present in the cytosol of various tissues.

Biomedical importance

The drug "Isoniazid", used in treatment of TB is detoxicated by acetylation.

Polymorphic types of *acetyl transferases* exist, resulting in individuals who are classified as:

- ***'Slow' acetylator*** and
- ***'Fast' acetylator.***
  - **'Slow' acetylators:** More prone to certain toxic effects of isoniazid because the drug persists longer in these individuals due to slow acetylation.
  - **'Fast' acetylators:** Removes the drug by acetylation in a faster rate.

3. ***Conjugation with Sulphuric Acid***

- Sulphuric acid is used by human beings for detoxication of various compounds having phenolic or hydroxyl groups.
- Substances like ***phenol, cresol, indole and skatole*** formed in the gut by the action of intestinal bacteria are absorbed and transported to liver, where they are conjugated with sulphate to form ***Ethereal sulphates,*** which are excreted in urine, being less toxic and more acidic.
- "Active" sulphate acts as the donor, chemically it is **3′-phosphoadenosine-5′-phosphosulfate (PAPS).**

Formation of Indican

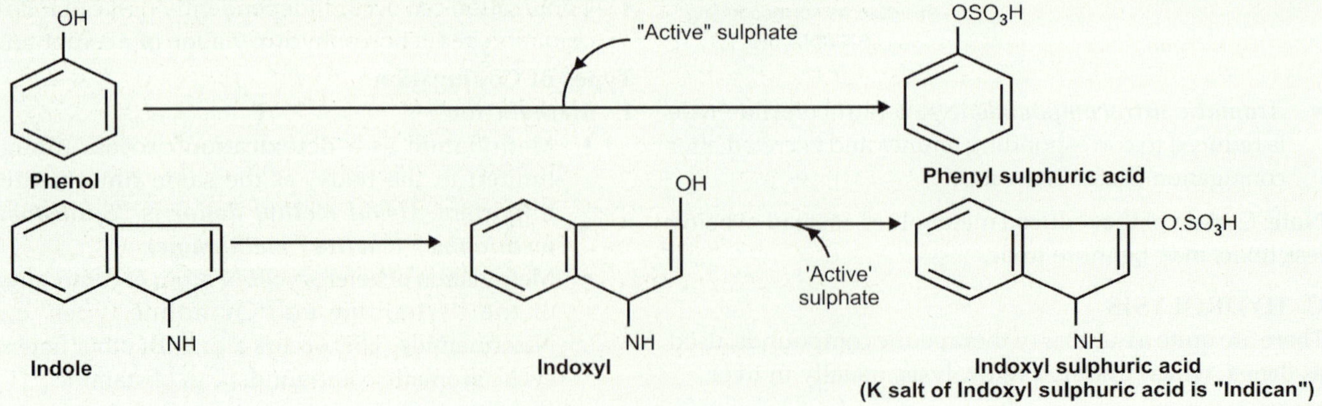

- Other compounds which are conjugated in the body to form corresponding esters are tyrosine to form tyrosine-O-SO_4 required for fibrinogen molecule, the amino sugars, certain hormones like oestrogens and androgens.

4. *Conjugation with D-Glucuronic Acid*

Most important and commonest detoxication reaction. D-Glucuronic acid participates in its detoxication reaction as its active form "UDP-glucuronic acid" which is formed in "uronic acid" pathway of glucose oxidation. Enzyme required is *Glucuronyl transferase*.

Linkages: In the process of conjugation, the glucuronic acid can form **two types of linkages:**

- An "ether" (glucosidic) linkage, e.g. in phenyl glucuronide.
- An "ester" linkage, e.g. in benzoyl glucuronide (Fig. 32.1).

Various compounds that are conjugated with glucuronic acid are:

- Bilirubin to form Bilirubin diglucuronide.
- Aromatic acids, e.g. benzoic acid.
- Phenols and other secondary and tertiary aliphatic alcohols.
- Certain drugs like morphine, menthol, pyramidon, acetanilide, sulfa pyridine, etc.
- Antibiotics like chloramphenicol
- Hormone like thyroid hormones; derivatives of steroids, e.g. tetrahydroderivatives of cortisol, sex hormones metabolites.

Thus, formation of glucuronides play an important role in detoxication mechanisms of exogenous and endogenous compound and their excretion as corresponding glucuronides.

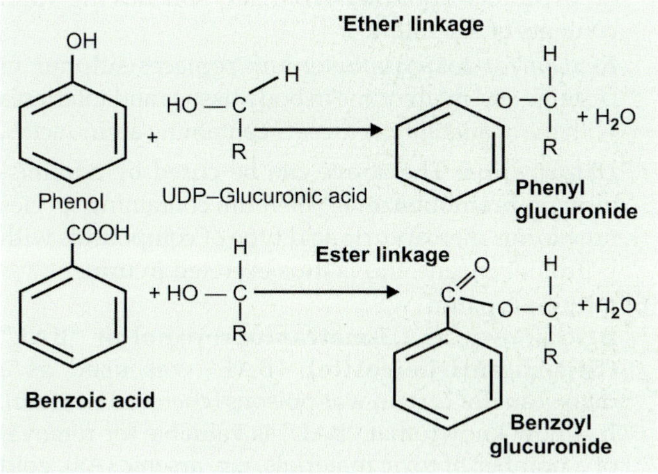

Fig. 32.1: Linkages in glucuronides

5. *Conjugation with Amino Acids*

I. *Glycine*

(a) Glycine combines with potentially harmful substances, mainly aromatic carboxylic acids in the body to form harmless derivatives which are excreted in urine.

Examples are:

(i) Reaction with "Nuclear" carboxyl group (Fig. 32.2).

(ii) Reaction with aromatic –COOH group separated from the aromatic ring by a "Vinyl" group, e.g. Cinnamic acid.

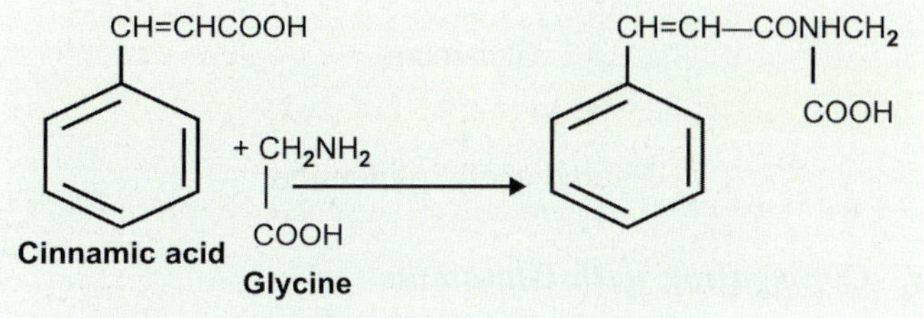

(b) Other aromatic carboxylic acids that can be conjugated with glycine and excreted are: Naphthoic acid, furoic acid, thiophene carboxylic acid, etc.

(c) Aliphatic carboxyl group do not combine as a rule except endogenous cholic acids produced in the liver by catabolism of cholesterol. Thus cholic acid and deoxycholic acid are conjugated with glycine to form glycocholic acid and glycodeoxy cholic acid.

II. *L-cysteine*

In man, a few aromatic compounds are conjugated with *L*-cysteine in presence of acetic acid to form **mercapturic acids**. Coupling of cysteine with aromatic compounds is linked with acetylation reaction (see box on next page).

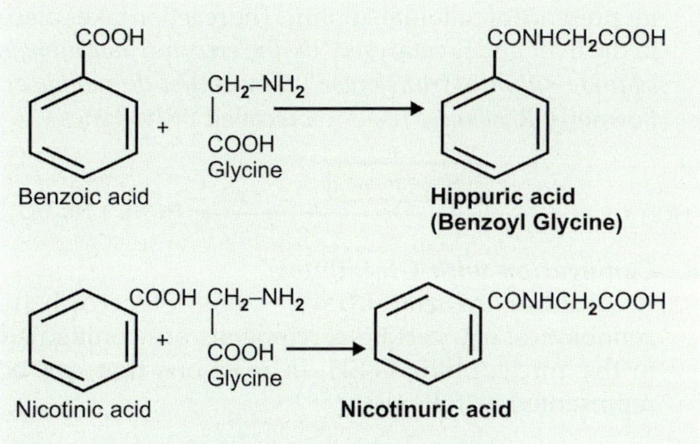

Fig. 32.2: Conjugation with glycine (reaction with nuclear-COOH GR)

Conjugation with Cysteine

- Bromobenzene + L-cysteine + acetic acid
  ↓
  Bromophenyl mercapturic acid
- Naphthalene + L-cysteine + Acetic acid
  ↓
  Naphthyl mercapturic acid

Conjugation with Glutamine

Phenyl acetic acid
+
Glutamine
↓
Phenyl acetyl glutamine

6. ***Conjugation with Glutamine***
   In man and in primates (chimpanzee), glutamine conjugates phenyl acetic acid to form 'phenylacetyl glutamine' and excreted in urine. **This accounts for "mousy" odour of urine in phenyl ketonurics** (see box above).

Note: Different mechanisms operate in different species:

- **In dogs,** phenylacetic acid is conjugated with glycine forming ***phenyl aceturic acid.***
- **In fowls,** it conjugates with ornithine to form "di-phenyl acetyl ornithine".

7. ***Conjugation with Thiosulphates (thiocyanate formation)***
   Animal organism normally excretes thiocyanates which is non-toxic. Human saliva contains an average of 0.01 per cent. Normal human blood contains about 1.31 mg KCNS (potassium thiocyanates) per 100 ml. Highly toxic cyanides are derived in body in small amounts from fruits, proteins and tobacco smoke. Cyanides are conjugated by "thiosulphates" or even in presence of colloidal sulphur. The reaction takes place in the liver and is catalysed by the enzyme *thiosulphate cyanide sulphur transferase* (also called *Rhodanase*). Formerly *Rhodanase* used to be called "Rhodanese".

$$HCN + S \xrightarrow{\text{Rhodanase}} \mathbf{HCNS + Na_2SO_3}$$

8. ***Conjugation with Glutathione***
   A number of potentially toxic electrophilic xenobiotics, e.g. certain carcinogens are conjugated to the nucleophilic G-SH, in reactions that can be represented as follows:

$$R + G - SH \longrightarrow R\text{–}S\text{–}G$$

Where R represents an electrophilic xenobiotic

The reaction is catalysed by the enzyme ***glutathione-S-Transferases***. The enzyme is present in high amounts in liver cytosol and in lower amounts in other tissues.

Note: If the potentially toxic xenobiotics are not conjugated with G-SH, they would be free to combine covalently with DNA, RNA or cell proteins and can produce serious cell damage.

G-SH is thus an important defence mechanism against certain toxic compounds, such as some drugs and carcinogens.

9. ***Detoxication of Drugs***
   Most of the drugs more than 50 per cent are detoxicated by hydroxylation. Enzymes concerned are ***mono-oxygenases*** or ***cytochrome P_{450}*** species. The reaction can be represented as follows:

$$AH + O_2 + NADPH + H^+ \rightarrow A - OH + H_2O + NADP^+$$

AH represents the drugs, which can be of wide variety: carcinogens, chemicals, pollutants, and certain endogenous compounds, such as steroids or its metabolites.

Characteristic features of cytochrome-P_{450} system

- Present in endoplasmic reticulum (ER), microsomal fraction of liver (highest concentration).
- **At least six closely related species** of Cyt-P_{450} in ER of liver described.
- Chemically they are "haemoproteins."
- Enzyme is "NADPH-Cyt *P_{450} reductase.*"
- The enzyme requires NADPH.
- Cyt-P_{450} system contains lipids and most common lipid is phosphatidyl choline (lecithin).
- It is an inducible enzyme.

(a) Selenium Poisoning

Selenium poisoning develops due to high feeding of products obtained from the soil having high contents of selenium.

Reason for toxicity: Selenium replaces sulphur in cysteine and methionine in body tissues and interferes with the availability of these S-containing amino acids.

Detoxication: The above can be cured by administering **p-bromobenzene**. Selenium containing species now forms mercapturic acid type of compounds with p-Bromobenzene and is thus excreted in urine.

(b) Dithiopropanol

Also known as **2,3-mercaptopropanol or "BAL" (British-anti-Lewesite).** 'BAL' was used as a detoxicant for certain war poisons (chemical warfare). It is now known that "BAL" is valuable for removal of a number of toxic materials, e.g. arsenic (As), gold (Au), mercury (Hg), cadmium (Cd).

Mechanism of action: Exact mechanism of action is not known. Toxic metal ions combine with –SH groups of body enzymes or other important –SH groups containing molecules and thus inactivate them. 'BAL' having a greater affinity for certain metals, when administered pulls the metal ions from their enzyme combinations and forms a similar complex which is rather readily excreted.

Biomedical Importance of Mono-oxygenases: Cyt-P_{450} System

Recently another *cytochrome P_{450}* species has been found called as ***cytochrome P_{448}***. This species has been found to be specific for metabolism of *polycyclic aromatic hydrocarbons (PAHs)*. Hence this species has also been named as ***aromatic hydrocarbon hydroxylase (AHH).***

Importance: It has been found to be very important enzyme for metabolism of PAHs and in carcinogenesis produced by these agents.

Studies have shown that:

- In the lungs of cigarette smokers, the enzyme may be involved in conversion of inactive PAHs *(procarcinogens)* present in cigarette smoke to active carcinogens by hydroxylation reactions. Smokers were found to have higher levels of this enzyme in cells and tissues than nonsmokers.
- Some reports have shown that activity of this enzyme is increased (induced) in placentae of pregnant women who are cigarette smokers and thus foetus is exposed to potentially harmful metabolites (carcinogens?).

I. Questions (Essay type and short notes)

1. What are various important conjugation compounds utilised by human body for detoxication of foreign substances?
2. Write about various mechanisms of detoxication in the human body.
3. Give important conjugation reactions taking place in our body.
4. Discuss the mechanisms of detoxication.

Short Notes

a. Conjugation reaction in detoxication
b. Detoxication by conjugation
c. Detoxication
d. Detoxication mechanisms
e. Cytochrome-P-450 system

II. MCQs (Give one correct answer)

1. In phenyl ketonurics, the phenyl acetic acid is conjugated with:
   a. Glycine b. D-glucuronic acid
   c. Acetylation d. Glutamine
   e. Methylation
2. The water-soluble vitamin, nicotinic acid, is detoxicated in the body with:
   a. Glucuronic acid b. Glycine
   c. Glutamine d. Cysteine
   e. 'Active' sulphate
3. The drug "isoniazid" (isonicotinic acid hydrazide) used in the treatment of TB is detoxicated by:
   a. Methylation b. 'Active' sulphate
   c. By acetylation d. By glutathione
   e. Cytochrome P-450 species
4. Indican excreted in urine is formed from detoxication of indoxyl in liver by:
   a. Acetylation b. PAPS
   c. Glutathion d. Methylation
   e. Glycine
5. The enzyme "cytochrome p-450 reductase" which catalyses hydroxylation of drugs requires the coenzyme:
   a. NAD^+ b. $NADH + H^+$
   c. $NADP^+$ d. $NADPH + H^+$
   e. FAD
6. Selenium poisoning can be cured by the administration of:
   a. Benzyl amine b. Acetanilide
   c. P-bromobenzene d. Aniline
   e. P-nitrobenzaldehyde
7. Nicotinamide is detoxicated by:
   a. 'Active' sulphate
   b. By methylation
   c. By acetylation
   d. Glucuronic acid
   e. Glycine
8. Ingestion of sodium benzoate in man results in an increase in urinary excretion of:
   a. Phenyl pyruvic acid
   b. Glyoxalic acid
   c. Oxalic acid
   d. Hippuric acid
   e. Ethereal sulphates

Answers to MCQs

1. d 2. b 3. c 4. b 5. d 6. c
7. b 8. d.

CHAPTER

33

HORMONES—CHEMISTRY, MECHANISM OF ACTION AND METABOLIC ROLE

Major Concepts

A. To study what are hormones, how do they differ from enzymes. Learn how hormones are classified, their general characteristics, mechanism of action.

B. To study the chemistry, mechanism of action, and metabolic role of various hormones secreted by various endocrine glands.

Specific Objectives

A. 1. Define hormone. List in a tabular form the similarities and differences between hormones and enzymes.
   2. Classify hormones.
   3. Study briefly the various ways a hormone acts, i.e. mechanism of action.
   4. Learn the various hypothalamic releasing factors that control the secretion.

B. 1. List the hormones produced by pituitary gland anterior, intermediate and posterior pituitary.
   2. Study the chemistry, and metabolic role of growth hormone.
   3. Define 'tropins'? List the tropic hormones produced by the anterior pituitary gland. Learn the chemistry and metabolic role of the various tropins.
   4. Study the chemistry and metabolic role of melanocyte-stimulating hormone (MSH) produced by middle (intermediate) lobe.
   5. List the hormones produced by the posterior pituitary. Learn the chemistry, mechanism of action and metabolic role of vasopressin and oxytocin.
   6. Thyroid gland:
      - List the hormones produced by the thyroid gland.
      - Study the chemistry, biosynthesis, mechanism of action and metabolic role of thyroid hormones.
      - List the different antithyroid drugs and study their mechanism of action.
      - What are natural goitrogens?
   7. Parathyroid glands:
      - Study the chemistry, biosynthesis, mechanism of action and metabolic role of parathormone.
      - What is calcitonin? Study the chemistry, mechanism of action, and metabolic role of calcitonin.
      - Study the role of parathormone and calcitonin in calcium homeostasis.
   8. Pancreas (Islet of Langerhans):
      - List the hormones produced by different cells of islet of Langerhans of pancreas.
      - Learn the chemistry, biosynthesis and metabolic role of Insulin. Study how Insulin is degraded?
      - What are 'Insulin receptors'? Learn the chemistry of insulin receptors. Study the possible mechanism of action.
      - Study the chemistry of "Insulin-like Growth factors" (IGF). In a tabular form differentiate: Insulin, IGF-I and IGF-II.
      - Learn the chemistry, biosynthesis, mechanism of action and metabolic role of glucagon. Study the clinical and therapeutic uses of glucagon.
      - What is somatostatin? Learn the different sources of somatostatin and their metabolic role.
   9. Adrenal glands
      (a) Adrenal cortex:
         - List the important steroid hormones produced by adrenal cortex. Classify steroid hormones, according to structure and according to function.
         - Learn the chemistry, biosynthesis, mechanism of action and metabolic role of glucocorticoids.
         - What are mineralocorticoids? Which layer of adrenal cortex produces mineralocorticoids and why? List the mineralocorticoids.
         - Study the synthesis and mechanism of action and metabolic role of aldosterone.
         - Learn how aldosterone secretion is regulated. Study 'Renin-angiotensin system'. Learn the actions of angiotensin-II.
      (b) Adrenal medulla:
         - Study the chemistry, biosynthesis, mechanism of action and metabolic role of catecholamines.

Hypothalamic Releasing Factors: Control of hormone secretion from the pituitary is in part modulated by regulating factors or hormones from the hypothalamus. The median eminence of the hypothalamus is connected directly to the pituitary stalk. Within this stalk is a portal system of blood vessels required to maintain normal secretory activity of the pituitary gland. The activities of the cells of the anterior lobe are controlled by the nerve cells of the hypothalamus which send axons to the capillary beds. The nerve endings liberate chemical substances, hypothalamic releasing factors or hormones. At present **10 discrete regulatory factors** have been described that may affect the synthesis as well as secretion of specific pituitary hormone. They are listed as follows:

Hypothalamic Hormone or Factor	Abbreviation
• Corticotropin (ACTH) releasing hormone	CRH or CRF
• Thyrotropin (TSH) releasing hormone	TRH or TRF
• Follicle stimulating hormone (FSH) releasing hormone	FSH-RH or FSH-RF
• Luteinizing Hormone (LH) releasing hormone	LH-RH or LH-RF
• Growth-hormone (GH) releasing hormone	GH-RH or GH-RF
• Growth-hormone release inhibiting hormone	GH-RIH or GIF
• Prolactin (PL) release inhibiting hormone	PL-RIH or PL-RIF
• Prolactin (PL) releasing hormone	PRH or PRF
• Melanocyte stimulating hormone (MSH) release inhibiting hormone	MSH-RIH or MSH-RIF
• Melanocyte stimulating hormone (MSH) releasing hormone	MSH-RH or MSH-RF

HORMONES OF THE ANTERIOR PITUITARY

The hormones secreted by the anterior lobe of the pituitary gland are:

- **Growth hormone,** and
- **Pituitary tropic hormones** such as prolactin, gonadotropins FSH and LH, Thyrotropic hormones (TSH) and Adrenocorticotropic Hormone (ACTH).

GROWTH HORMONE (SOMATOTROPIN)

Growth hormone (GH) or somatotropin (STH) was first isolated in sufficient quantity from cattle, now it has been prepared in crystalline form from several species including man.

Chemistry: Growth hormone from all mammalian species consists of a single polypeptide with a molecular weight of about 21500. ***It consists of 191 amino acids.*** There are two disulfide bridges between the adjacent cysteine residues (53 and 165 and 182 and 189). Although there is a high degree of similarity in the amino acid sequences of human, bovine and porcine GH; only human GH or that of other primates is active in man. GH can bring about some of the actions of prolactin and human placental lactogen (HPL) due to amino acid sequence homology.

Metabolic role: Growth hormone has a variety of effects on different tissues. ***The hormone acts slowly requiring from 1-2 hours to several days before its biological effects are detectable.*** This slow action and its stimulatory effects on RNA synthesis suggest that it is involved in protein synthesis. The hormone acts by binding to specific membrane receptors on its target cells. But its exact mechanism of action and the second messenger are not yet known.

1. *Protein synthesis:* Growth hormone brings about **positive nitrogen balance** by retaining nitrogen. ***It stimulates overall protein synthesis*** with an associated retention of phosphorus probably by increasing tubular reabsorption. Blood amino acid and urea level are decreased. ***It facilitates the entry of amino acids into the cell.*** In addition, growth hormone facilitates protein synthesis in muscle tissue by a mechanism independent of its ability to provide amino acids. Thus protein synthesis carries on even if the amino acid transport is blocked.
   - Growth hormone increases DNA and RNA synthesis.
   - It increases the synthesis of collagen.
2. *Lipid metabolism:* ***Growth hormone brings about lipolysis*** in a mild way by mobilising fatty acids from adipose tissue by activating the hormone sensitive *triacylglycerol lipase*. Thus it increases circulating free fatty acids.
3. *Carbohydrate metabolism:* Growth hormone is a diabetogenic hormone, ***antagonises the effect of insulin***. Hypersecretion of GH can result in ***hyperglycaemia***, poor sugar tolerance and glycosuria. Growth hormone produces:
   - Hyperglycaemia by increasing gluconeogenesis.
   - It reduces insulin sensitivity and thereby decreases the hypoglycaemic effect of insulin.
   - It brings about glycostatic effect, i.e. increases liver glycogen. It can also increase muscle and cardiac glycogen level probably by reducing glycolysis.
4. *Effect on growth of bones and cartilages:* Growth hormone when secreted in abnormally high concentration prolongs the growth of epiphyseal cartilages to cause overgrowth of long bones.

SECTION FOUR

Acromegaly is found in adults. Hyposecretion causes 7stunted stature due to premature cessation of growth of the epiphyseal cartilages and consequently of long bones.

- The effect of growth hormone partly depends upon its calcium anabolic action. It promotes the retention of calcium and phosphate which helps in ossification and osteogenesis.
- It enhances the incorporation and hydroxylation of proline in the matrix collagen, incorporation of amines into glycosaminoglycans of cartilage, incorporation of sulphate into matrix proteoglycans like chondroitin sulphates, the synthesis of DNA and RNA in chondrocytes.
- The growth effects are mediated by a peptide called insulin-like growth factor I (IGF-I or somatomedin-C).

5. ***Prolactin action:*** Growth hormone has a sequence homology with prolactin. Growth hormone binds to membrane receptors for prolactin and stimulates the growth and enlargement of mammary glands.
6. ***Ion or mineral metabolism:*** It is observed that the intestinal absorption of calcium is increased by GH, since the bone growth and development is stimulated by growth hormone. Growth hormone retains Na, Ca, K, Mg and PO_4^{-3}.

PITUITARY TROPIC HORMONES

In addition to GH, anterior pituitary gland secretes some tropic hormones usually called as pituitary tropins.

What are "tropins"?

A tropin or tropic hormone is the one which influences the activities of other endocrine gland, principally those involved in stress and reproduction. These are carried by the blood to other target gland. The pituitary tropins are under the positive and negative control of peptide factors from hypothalamus. Further the tropic hormones are usually subject to feedback inhibition at the pituitary or hypothalamic level by hormone product of the final target gland. Prolactin (mammotropin), TSH (or thyrotropin), FSH and LH (gonadotropins), ACTH (corticotropin) are the tropic hormones secreted by the pituitary gland.

A. Prolactin: PRL or Leuteotropic Hormone (LTH)

This is a monomeric simple protein (MW 23,000). It contains 199 amino acids with three –S–S–linkages. It is ***secreted by*** lactotroph **α-cells** of anterior pituitary and as already mentioned has sequence homology with growth hormone.

Metabolic role

- The main function of PRL is ***to stimulate mammary growth and the secretion of milk.*** By acting through specific glycoprotein receptors on plasma membrane of mammary gland cells, it stimulates mRNA synthesis. This ultimately leads to enlargement of breasts during pregnancy. This is called as ***mammotropic action.***
- The synthesis of milk proteins such as lactalbumin, and casein takes place after parturition such an effect is called as ***lactogenic action.***
- Estrogens, thyroid hormones and glucocorticoids increase the number of prolactin receptors on the mammary cell membrane.
- Progesterone has the opposite effect.

B. Thyrotropic Hormone or Thyroid Stimulating Hormone (TSH)

This is ***produced by basophil cells*** of anterior pituitary and is ***glycoprotein*** in nature. Its molecular weight is approximately 30,000. This ***consists of a and b subunits.***

- The ***α-subunit of TSH, LH, HCG and FSH are nearly identical.***
- ***The biological specificity of thyrotropin must therefore be in β-subunit.*** The **α-subunit** consists of 92 amino acids while **β-subunit** has 112 amino acids. Both α and β have several disulfide bridges. Its carbohydrate content is 21 per cent and its α and β chains bear two and one oligosaccharide chains linked by N-glycosidic linkages to specific asparagine residues. The chains are synthesised separately by separate structural genes and later undergo posttranslation modifications and glycosylations separately.

Metabolic role

There are glycoprotein receptors on the thyroid cell membrane which bind to the receptor binding site on β-subunit of TSH. The complex then activates *adenylate cyclase* which catalyses the formation of c-AMP which acts as the second messenger for most TSH actions as follows:

- ***The TSH stimulates the synthesis of thyroid hormones at all stages*** such as Iodine uptake, organification and coupling.
- It enhances the release of stored thyroid hormones.
- It increases DNA content, RNA and translation of proteins, cell size.
- It stimulates glycolysis, TCA cycle, HMP and phospholipid synthesis. Stimulation of last two does not involve c-AMP.
- It activates adipose tissue *lipase* to enhance the release of fatty acids (lipolysis).

C. Adrenocorticotropic Hormone (ACTH) or Corticotropin

It is a single polypeptide containing 39 amino acids in its structure with a molecular weight of 4500. **Two forms** have been isolated, ***α-corticotropin*** and ***β-corticotropin.***

Biological activity of ACTH resides in the first 23 amino acids from N-terminal end. The sequence of these 23 amino acids in the peptide chain is the same in all species tested. The remaining biologically inactive 16 amino acid residues vary according to sources. ACTH is ***synthesised as a part of precursor peptide*** of mol. wt. of 31500 with 260 amino acids. ***ACTH contains sequences of amino acids common for LPH, MSH and the endorphins.*** The precursor molecule is synthesised as a glycoprotein called **Pro-opiomelano cortin peptide (POMC)**. Various proteolytic enzyme hydrolyse POMC to give different peptides. ***Thus POMC is broken down into ACTH, β-lipotropin (LPH). β-LPH is further cleaved into γ-LPH and endorphins.***

Metabolic role

- The principal actions of corticotropin are exerted on the adrenal cortex and extraadrenal tissue. ***ACTH increases the synthesis of corticosteroids by the adrenal cortex and also stimulates their release from the gland.*** Profound changes in the adrenal structure, chemical composition and enzymatic activity are observed as a response to ACTH. Total protein synthesis is found to be increased. Thus, ACTH produces both a tropic effect on steroid production and tropic effect on adrenal tissue. It is observed that ACTH has specific receptors on cells of fasciculata which increases c-AMP levels in the cell. This activation is calcium dependent. This results in DNA content and RNA is transcribed. This leads to proliferation of fasciculata cells and growth of adrenal cortex.
- ACTH also stimulates the synthesis and secretion of glucocorticoids.
    - ACTH is found to increase the transfer of cholesterol from plasma lipoproteins into the fasciculata cells.
    - The ACTH induces rise in c-AMP, brings about phosphorylation and activation of cholesterol esterase. The enzyme action ultimately makes a large pool of free cholesterol.
    - Corticotropin promotes the binding of cholesterol to mitochondrial cytochrome P_{450} required for hydroxylating cholesterol.
    - It activates the rate limiting enzyme for conversion of cholesterol to pregnenolone.
    - It activates dehydrogenases of HMP to increase the conc. of NADPH required for hydroxylation.
- By activating *adenylate cyclase* of adipose tissue it increases intracellular c-AMP which in turn activates hormone sensitive lipase. This enzyme is involved in lipolysis which increases the level of free fatty acids.
- It leads to increased ketogenesis and decreased RQ.
- Direct effects on carbohydrate metabolism include:
    - **Lowering of blood glucose ↓;**
    - Increase in glucose tolerance;
    - Deposition of glycogen in adipose tissue is increased, regarded as due to stimulation of insulin secretion.
- ***It has MSH activity*** due to homology in amino acid sequence.

D. Pituitary Gonadotropins

These tropic hormones influence the function and maturation of the testes and ovary and are of **two types:**

- *Follicle Stimulating Hormone (FSH)*
- *Luteinising Hormone (LH)*

Both of them are glycoproteins with sialic acid, hexose and hexosamine as the carbohydrate moiety (16%). Molecular weight of FSH is 25000 and that of LH is 40000. As already mentioned, FSH and LH are dimers of α and β-chains linked noncovalently. ***The α-chain is identical for TSH, FSH and LH of the same species.*** The β-chain of human FSH and LH have respectively 118 and 112 amino acid residues. Each chain has several disulfide bridges. A large precursor protein molecule for α and β chains is synthesised separately in gonadotroph β-cells.

Metabolic Role of FSH

It brings about its action by specific receptor binding and c-AMP.

In females:

- It promotes follicular growth
- Prepares the Graafian follicle for the action of LH and
- ***Enhances the release of estrogen*** induced by LH.

In males:

- It stimulates seminal tubule and testicular growth, and
- Plays an important ***role in maturation of spermatozoa.***

Role of FSH in Spermatogenesis

The conversion of primary spermatocytes into secondary spermatocytes in the seminiferous tubules is stimulated by FSH. ***In absence of FSH, spermatogenesis cannot proceed. However, FSH by itself cannot cause complete formation of spermatozoa. For its completion, testosterone is also required.*** Thus, FSH seems to initiate the proliferation process of spermatogenesis, and testosterone is apparently ***necessary for final maturation of spermatozoa.*** Since the testosterone is secreted under the influence of LH, ***both FSH and LH must be secreted for normal spermatogenesis.***

Metabolic Role of LH

This hormone is **also known as interstitial cells stimulating hormone (ICSH).**

In females:
- It causes the ***final maturation of Graafian follicle*** and ***stimulates ovulation.***
- Stimulates secretion of estrogen by the theca and granulosa cells.
- It helps in the formation and development of corpus luteum for luteinisation of cells.
- In conjunction with luteotropic hormone (LTH), it is concerned with the production of estrogen and progesterone by the corpus luteum.
- In the ovary it can stimulate the nongerminal elements, which contain the interstitial cells to produce the androgens, androstenedione, DHEA and testosterone.

Action of LH in Ovulation : Ovulatory surge for LH

It is necessary for final follicular growth and ovulation. Without this hormone, even though large quantities of FSH are available, the follicle will not progress to the stage of ovulation. ***LH acts synergistically with FSH to cause rapid swelling of the follicle shortly before ovulation.*** It is worth noting that especially large amount of LH called ***ovulatory surge*** is secreted by the pituitary during the day immediately preceding ovulation.

Regulation of Testosterone Secretion by LH

Testosterone is produced by the interstitial cells of Leydig only when the testes are stimulated by LH from the pituitary gland, and the quantity of testosterone secreted varies approximately in proportion to the amount of LH available. Thus ***in males, LH stimulates the development and functional activity of Leydig cells (interstitial) and consequently testicular androgen.***

ENDORPHINS AND ENCEPHALINS

Endorphins are a group of polypeptides which influence the transmission of nerve impulses. They are also known as **opioides** because they bind to those receptors which bind opiates like morphine and plays a role in pain perception.

The opioides first discovered were two pentapeptides in the brain and were named ***"encephalins"***. They are of **2 types:**

1. ***Methionine-encephalin*** and
2. ***Leucine-encephalin.***

Formation of Endorphins

β-lipoprotein (β-LPH) is the precursor for endorphins, all the three types α, β and γ and also for β-MSH.

β-Lipoprotein (β-LPH) is derived from the precursor molecule ***Pro-opiomelanocortin Peptide (POMC).*** It is a single chain polypeptide containing 93 amino acids. γ-LPH containing 60 amino acids is a part of β-LPH.

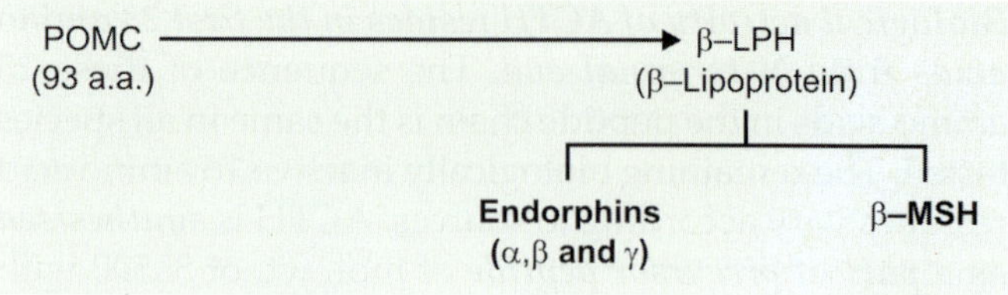

Types of Endorphins

There are **three types** of endorphins, as mentioned above, α, β and γ.

- The sequence of 31 amino acids at the C-terminus of β-LPH, (obtained from POMC) i.e. aa 104 to 134 gives ***β-endorphin.***
- ***α-endorphin*** (104 to 117) containing 17 aa less than the β from the C terminus.
- ***γ-endorphin*** (104 to 118) containing 16 aa less than the β from C-terminal end.

Function: Endorphins bind to the same CNS receptors like the morphine opiates and ***they play a role in the endogenous control of pain perception. They have higher analgesic potency than morphine.***

HORMONE OF MIDDLE LOBE OF PITUITARY

Melanocyte Stimulating Hormones: The hormones secreted by intermediate lobe or middle lobe of pituitary gland are called melanocyte-stimulating hormones or MSH. POMC is the precursor molecule which is cleaved by *proteases* to give ACTH and β-lipotropin. The ACTH is further cleaved to β-MSH which has 13 amino acids. There is also α-MSH which is present in larger quantities. Amino acids 11-17 of β-MSH are common to both α-MSH and ACTH. ***MSH darkens the skin and is involved in skin pigmentation by deposition of melanin by melanocytes.***

HORMONES OF POSTERIOR PITUITARY LOBE

The hormones have been isolated and characterised from extracts of posterior pituitary gland. They are:

1. ***Vasopressin (Pitressin) or Arginine Vasopressin (ADH)*** and
2. ***Oxytocin.***

Both are small peptides containing nine amino acids. Oxytocin differs from Vasopressin with respect to 3rd and 8th amino acid residues. Their biological activities depend on C-terminal glycinamide, the side chain amide groups of glutamine and asparagine, the hydroxyphenyl group of tyrosine, and the intra-chain –S–S–linkage between cysteine of Ist and 6th amino acid.

Posterior pituitary hormones are synthesised in neurosecretory neurons. They are ***stored in the pituitary***

in association with two proteins neurophysin I and II with molecular weights of 19000 and 21000 respectively. The release of these two hormones is independent of each other.

A. Metabolic Role of Vasopressin

1. ***Antidiuretic action:*** Antidiuretic effect is its main function. It reabsorbs water from the kidneys by distal tubules and collecting ducts. It is found to be mediated through formation of c-AMP. It is released due to rise in plasma osmolarity. This leads to formation of hypertonic urine having low volume, high sp. gr. and high conc. of Na^+, Cl^-, phosphate and urea. Halothane, colchicine and vinblastine inhibit antidiuretic effect of vasopressin.

CLINICAL IMPORTANCE

Condition of **diabetes insipidus** is described due to failure in secretion or action of vasopressin. It is characterised by very high volumes of urine output, up to 20-30 litres per day with a low specific gravity and excessive thirst.

- ***In primary, central or neurohypophyseal diabetes insipidus***, vasopressin secretion is poor.
- ***In nephrogenic diabetes insipidus***, kidneys cannot respond to vasopressin due to renal damage. The damage is common in psychiatric patients on Lithium therapy.

Inappropriate vasopressin secretion is characterised by a persistently hypertonic urine, progressive renal loss of Na^+ with low plasma levels of Na^+, symptoms of water intoxication like drowsiness, irritability, nausea, vomiting, convulsions, stupor and coma. It could be due to pulmonary infection and ectopic ADH secretions from lung tumours.

2. ***Urea-retention effect:*** Permeability of medullary collecting ducts to urea is increased by vasopressin. This leads to retention of urea and subsequently contributes to hypertonicity of the medullary interstitium. Urea retention effect can be reversed by phloretin.
3. ***Pressor effect:*** It stimulates the contraction of smooth muscles and thus causes vasoconstriction by increasing cytosolic Ca^{+2} concentration.
4. ***Glycogenolytic effect:*** By increasing intracellular calcium concentration.

B. Metabolic Role of Oxytocin

Contraction of smooth muscle is the primary function of oxytocin. There are basically two effects, one on mammary glands called as ***galactobolic effect*** and the other on uterus called as ***uterine effect.***

1. ***Galactobolic effect:*** This is released due to neuro-endocrinal reflex such as suckling of nipples. By doing so it causes the contraction of myoepithelial cells around mammary aleveoli and ducts and the smooth muscles surrounding the mammary milk sinuses; estrogen increases the number of oxytocin receptors during pregnancy while progesterone decreases the same and also inhibits the secretion of oxytocin.
2. ***Uterine effect:*** It is found to be elevated at full term pregnancy. It causes contraction of uterine muscle for child-birth. Estrogens enhance while progesterone decreases oxytocin receptors as well as its secretion. Oxytocin is also secreted during coitus by the female uterus which promotes the aspiration of semen into the uterus. This is also augmented by rise in estrogen in the follicular phase of menstrual cycle.

THYROID GLAND AND ITS HORMONES

Hormones Produced by Thyroid Gland

1. ***Follicular cells:*** Produces T_4, T_3 and "reverse" T_3
2. ***Parafollicular C-cells:*** Produces calcitonin (hence also called thyrocalcitonin).

THYROID HORMONES

The principal hormones secreted by the follicular cells of thyroid are:

1. **Thyroxine (T_4)**
2. **Tri-iodo thyronine (T_3)** and
3. **'Reverse' T_3**

Chemistry of Thyroid Hormones: The hormones T_4, T_3 and "reverse" T_3 are ***iodinated amino acid tyrosine.*** The iodine in thyroxine accounts for 80 per cent of the organically bound iodine in thyroid venous blood. Small amounts of 'reverse' tri-iodo thyronine, Monoiodotyrosine (MIT) and other compound are also liberated. The chemical name and structures of thyroid hormones are shown in the box on next page:

BIOSYNTHESIS OF THYROID HORMONES

Two raw materials (substrates) required by thyroid gland to synthesise the thyroid hormones are:

1. **Thyroglobulin**
2. **Iodine**

A. Thyroglobulin

Thyroid hormones are synthesised by the iodination of tyrosine residues of a large protein called "thyroglobulin".

Chemistry of thyroglobulin

- Thyroglobulin is a dimeric glycoprotein, 19S in type (a macroglobulin) with a molecular weight of 660,000.
- The receptor tyrosine molecules are present in this macroglobulin protein, each molecule containing ***115 tyrosine residues.***
- Carbohydrates account for 8 to 10 per cent of the weight of thyroglobulin and iodide for about 0.2 to 1 per cent, depending on the iodine content of the diet. The carbohydrates are N-acetyl glucosamine, mannose, glucose, galactose, fucose and sialic acid.
- About 70 per cent of the iodide in thyroglobulin exists as inactive precursors ***'mono-iodo-tyrosine' (MIT)***

Structure of Thyroid Hormones

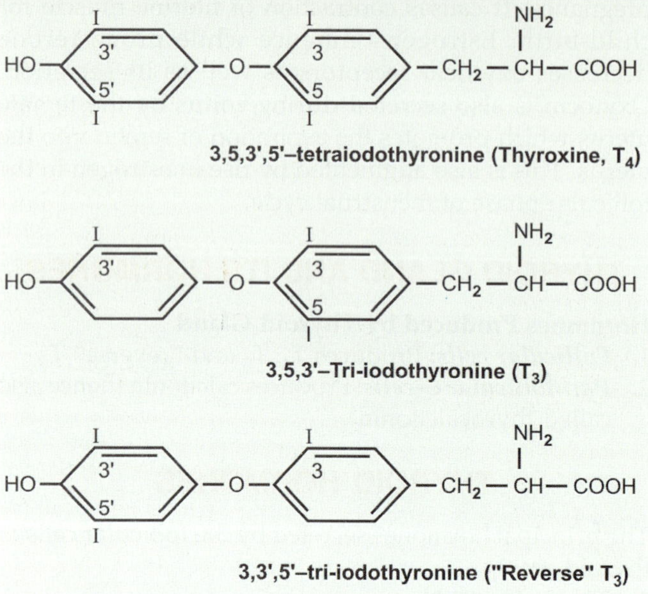

and *'di-iodotyrosine' (DIT)* while 30 per cent is in the iodothyronyl residues T_4 and T_3.

When Iodine supplies is sufficient, $T_4 : T_3$ ratio is about 7:1. ***In iodine deficiency, this ratio decreases,*** including the MIT/DIT ratio. T_3 and T_4 after being synthesised, remains in the bound form until it is secreted. When they are secreted, the peptide bonds are hydrolysed and free T_3 and T_4 enter the thyroid cells, cross them and are discharged into the capillaries.

Thyroid aciner cells have three functions:

1. They synthesise thyroglobulin and stores as colloid in follicles.
2. They collect and transport I_2 for synthesis of the hormones in the colloid (see below), and
- They remove T_3 and T_4 from thyroglobulin secreting the hormones into the circulation.

Proteins other than thyroglobulin: Besides thyroglobulin, some other albumin-like and hormonally inactive two other 4S iodoproteins of uncertain function are also found in thyroid.

Note: The above can appear in circulating blood and can contribute to the measured protein bound iodine (PBI).

Synthesis of thyroglobulin: Thyroglobulin is synthesised in acinar cells and stored in colloid.

Steps:

1. Thyroglobulin is translated by the polysomes on the granular endoplasmic reticulum (ER) of thyroid acinar cells. The polypeptide portion is synthesised as two subunits and later aggregated to form the dimer.
2. The glycosylation of the molecule starts in the smooth endoplasmic reticulum (SER) with the incorporation of mannose and is completed in the Golgi cisternae where the other sugars as N-acetyl glucosamine, galactose, fucose and sialic acid are added in the oligosaccharide chains (Studied by EM auto-radiography).
3. The glycoproteins formed as above are packaged into small vesicles. Electron-lucent membrane bound vesicles containing thyroglobulin are then pinched off from the Golgi cisternae.
4. The vesicles then move towards the apical plasma membrane and fuse with it, releasing their contents into the colloid of thyroid follicles.

B. Iodine

The other substrate required for thyroid hormone synthesis is iodine.

Iodine metabolism: Vegetables and fruits grown and obtained from sea-shore and also sea fishes are rich in iodine. Vegetables and fruits in hilly regions lack iodine (for people residing in hilly regions, table salt should be iodinated).

Ingested dietary iodine is converted to iodide and absorbed from the gut. Of a total of 50 mg of iodine in the body about 10 to 15 mg are in thyroids. The normal daily intake of iodide is 100 to 200 μg. Minimum requirement is 25 μg. This iodide is absorbed mainly from small intestine and is transported in plasma in loose attachment to protein; can also be absorbed from lungs, other mucous membranes and skin. Small amounts of iodide are secreted by the salivary glands, stomach, and small intestine and traces in milk. About 2/3 (40-80%) of the ingested iodide is excreted by the kidneys, the remaining 1/3 is taken up by the thyroid glands for synthesis of thyroid hormones. ***Thyroid-stimulating hormone (TSH) of anterior pituitary gland stimulates iodide uptake by the thyroid gland.***

Inorganic iodide in plasma and cells varies from 0.3 to 1.0 μg per cent. Part of the "circulating pool" is iodide liberated from thyroid hormones broken down in the tissues and to a minor extent in the thyroid itself.

In the kidneys, 97 per cent of the filtered iodide is reabsorbed, so that the loss from the body by this route, at normal plasma iodide levels is about 15 μg/day.

Steps of Iodine Incorporation in Thyroid Gland for Synthesis of Thyroid Hormones

1. Iodine trapping

The thyroid concentrates iodide by ***"actively"*** and ***"selectively"*** transporting it from the circulation to the colloid. The transport mechanism is called as **"iodide-trapping"** mechanism or **"iodide-pump"**.

The I_2 trapping is done:

- *Against electrical gradient*

- ***Against concentration gradient,***
- As it is 'actively' taken in against electrochemical gradient, it is ***energy dependent*** and ***requires energy.***

The thyroid acinar cell is about 50 mV–ve to the interstitial area and the colloid, has a resting membrane potential of–50 mV. Hence, iodide is presumably pumped in the acinar cell at its base against the electrical gradient. The ratio of thyroid to plasma/or serum free iodide (T/S ratio) ranges from 10: 1 to over 100: 1. ***Hence, iodine is taken up against 'concentration gradient'***. For this active transport it requires energy. The iodide "transporter" (Pump) is located in the basal plasma membrane in association with ***"Na^+-K^+ dependent ATPase"*** and requires a simultaneous activity of the "sodium-pump" (Na^+ pump). The iodide-pump requires ATP and shows substrate specificity for iodides (I^-).

Note:

- Thiocyanates/or perchlorates compete with iodide for uptake mechanism and causes the rapid discharge of the "exchangeable iodide" from the thyroid gland.
- ***The iodide pump activity is stimulated by TSH.***
- Hypophysectomy decreases the iodide concentrating ability.
- As uptake mechanism is energy dependent it can be inhibited by **Cyanide** or **Dinitrophenol (DNP).**
- It can also be inhibited by **'ouabain'** which will inhibit 'sodium-pump'.
- Intracellular iodide in acinar cells exists in two pools:
    - One pool is freely exchangeable with blood iodide and
    - The other pool arising from de-iodination of unused iodotyrosines (shown later).

Extrathyroidal iodide uptake and iodide concentration: The salivary glands, gastric mucosa, placenta, ciliary body of eye, choroid plexus and the mammary glands also transport iodide against a concentration gradient but this uptake is not affected by TSH.

Note:

- The ***mammary gland is the only one extrathyroidal tissue which binds the iodine. It has been shown that "Di-iodotyrosine" (DIT) may be formed in mammary gland, but thyroxine (T_4) and tri-iodothyronine (T_3) are not formed.***
- There is also a minor uptake by the posterior pituitary and adrenal cortex, but the function is not known.

The physiological significance of all these extrathyroidal iodide concentrating mechanisms is obscure.

2. *Oxidation of iodide*

Oxidation of iodide and other steps as mentioned below in thyroid hormone synthesis are catalysed by a ***"haem-containing"*** particulate-bound ***"peroxidase"*** called ***thyroperoxidase*** which requires H_2O_2 for its activity. Thyroperoxidase is a tetramer, having a molecular wt = 90,000. H_2O_2 is produced by an NADPH-dependent enzyme system similar to cytochrome-c-reductase. "Inactive" form of the enzyme *"Thyroperoxidase"* is synthesised on endoplasmic reticulum (ER) and packaged by the Golgi cisternae into vesicles pinched off from them. ***The enzyme is 'activated' during the fusion of the vesicle with apical plasma membrane by H_2O_2.*** At the colloid-membrane interface, *"thyroperoxidase"* binds iodide (I^-) and thyroglobulin at distinct sites of its molecule and then in presence of H_2O_2, the enzyme oxidises the enzyme bound I^- to form "active" iodine, which may be, **"iodinium" ion (I^+)** or **"hypoiodite" (HIO)** or both or **"free" iodine** radical (I)

$$I_2 + H_2O \rightleftharpoons HIO + I^- + H^+$$

Oxidation of iodide to 'active' iodine formation thus **involves two distinct steps:**

- ***Production of H_2O_2*** and
- ***Oxidation of iodide by the "peroxidase" enzyme in presence of H_2O_2. TSH is active in stimulating this process.***

3. *Iodination of tyrosine*

'Active' iodine transfers iodine from its iodidebinding site to a tyrosine residue of the enzyme bound thyroglobulin under the influence of *thyroperoxidase* enzyme. Thyroperoxidase probably uses H_2O_2 to oxidise the tyrosine residue to a free tyrosine radical before its interaction with "active iodine". Iodination of the tyrosine residues in thyroglobulin occurs first in "3 position" of the aromatic nucleus forming ***Mono-iodotyrosine (MIT).*** Monoiodotyrosine is next iodinated in the "position 5" to form ***"Di-iodotyrosine" (DIT).***

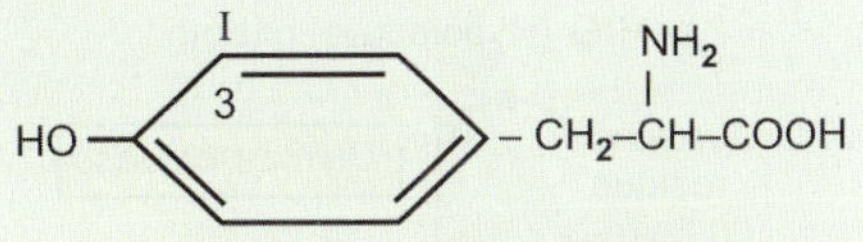

3–Mono–iodotyrosine (MIT)

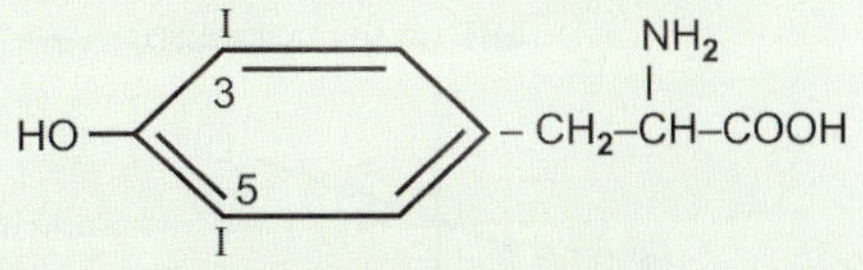

3,5–Di–iodotyrosine (DIT)

Normally, the two are present in approximately equal concentrations, but with I_2-deficiency, more MIT is formed. This process of iodination, also called as ***"organification"*** occurs within seconds in luminal thyroglobulin. Once iodination occurs, the iodine does not readily leave the thyroid.

4. Coupling of iodotyrosines

- **Two molecules of DIT when undergo an oxidative condensation,** under the influence of the enzyme *thyroperoxidase,* forms Thyroxine (T_4) molecule still in peptide linkage. In the process ***an "alanine" residue is liberated,*** which ultimately forms pyruvate and NH_3.

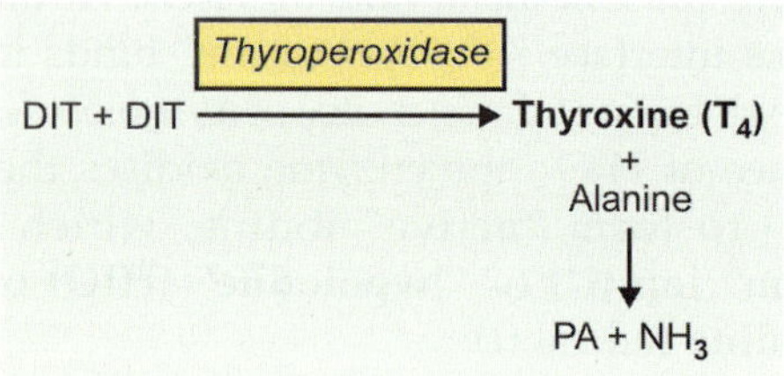

- Similarly, Tri-iodothyronine (T_3) is probably formed by condensation of MIT with a molecule of DIT and 'reverse' tri-iodothyronine ('reverse' T_3) by condensation of DIT with MIT.

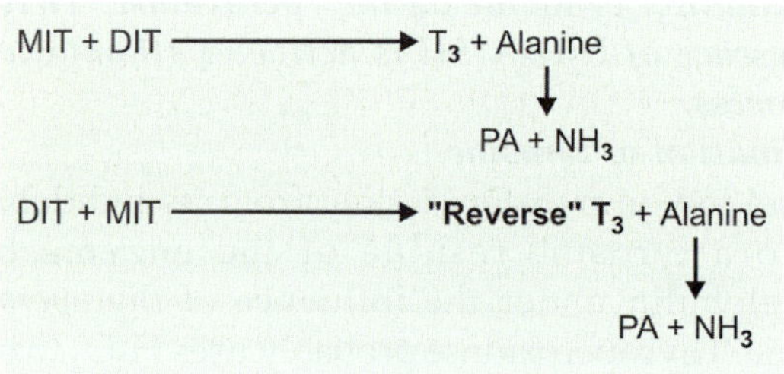

The condensation reaction is an ***aerobic,*** and ***energy requiring reaction.*** Tri-iodothyronine (T_3) may possibly be formed also by "partial de-iodination" of T_4.

Approximately 20 per cent of the thyroglobulin tyrosine residues are iodinated. Only 8 to 10 per cent of the total tyrosine bound iodine in the thyroid is in the form of thyroxine (T_4), most is found as DIT or MIT, only traces of "reverse" T_3 and other components are present. **TSH stimulates the synthesis of thyroglobulin and all the steps from oxidation to coupling reactions for forming thyroid hormones.**

Formation of thyroid hormones are shown schematically in Figure 33.4. ***About 80-95 μg of thyroxine is secreted daily under normal physiological conditions.***

Recycling of Iodine in the Gland

The hydrolysis of thyroglobulin also liberates MIT and DIT. If these iodotyrosines are lost from the gland, considerable amounts of iodide would be biologically unavailable for the synthesis of active hormones. Thyroid cells have microsomal enzyme ***de-iodinase (dehalogenase),*** which uses NADPH, and they rapidly de-iodinate the iodotyrosines, and the removed iodides are recycled in the gland and utilised for thyroid hormones synthesis. ***Approximately 1/3 of the total iodides in the thyroid gland is recycled in this manner.***

Transport

Within the plasma, T_4 and T_3 are mostly transported almost entirely in association with two proteins, the so-called **"thyroxine-binding proteins"** which act as specific carrier agents for the hormones.

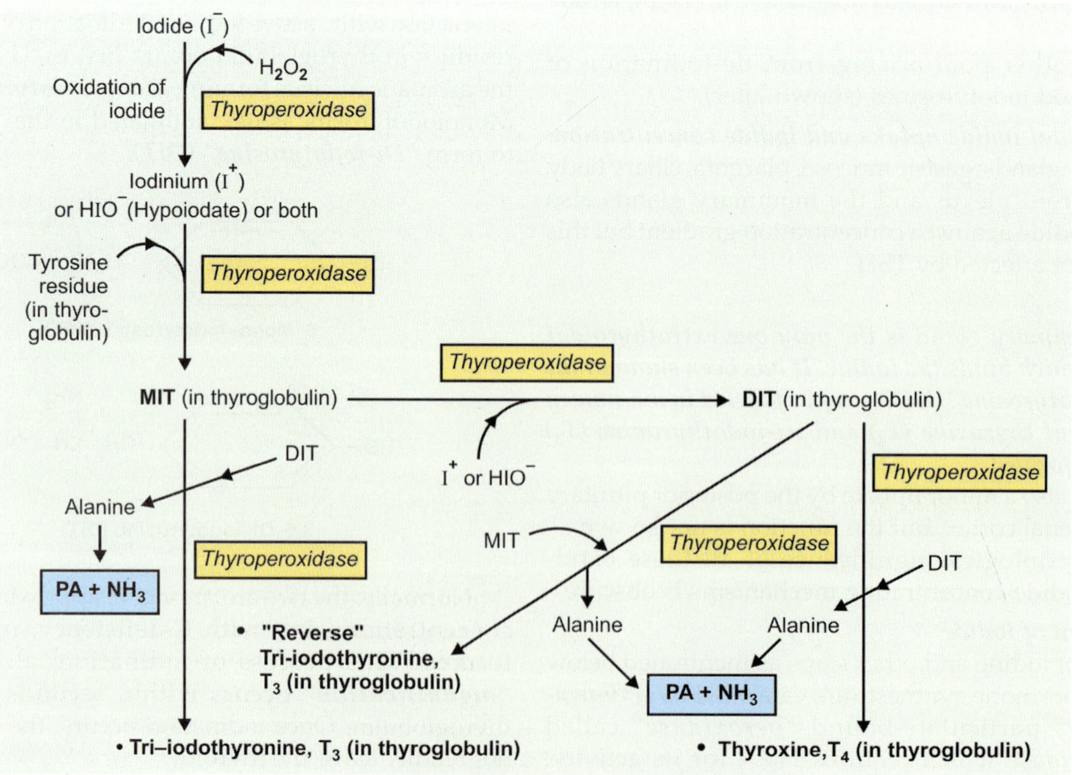

Fig. 33.4: Biosynthesis of thyroid hormones

Two main carrier proteins are:
- Thyroxine-binding globulin (TBG)
- **Thyroxine binding prealbumin (TBPA).**

When large amounts of T_4 and T_3 are present and the binding capacities of the above two specific carrier proteins are saturated, ***the hormones can be bound to "Serum albumin"***. Approximately, 0.05 per cent of the circulating thyroxine is in the 'free', unbound form. ***"Free" T_3 and T_4 are the metabolically "active" hormones in the plasma.***

Characteristics of the two specific carrier proteins: These are given in Table 33.1.

Table 33.1: Showing characteristics of two specific carrier proteins

TBG	*TBPA*
• A glycoprotein; High carbohydrate content (32%)	Poor in carbohydrate Tryptophan occurs in high concentration
• Molecular weight = 59,000	Molecular weight = 73,000
• Electrophoretically moves between α_1 and α_2 globulins	Moves ahead of albumin
• A single binding site/per molecule	A single binding site/per molecule
• Average TBG concentration is 1.5 mg/100 ml and average thyroxine binding capacity is 20 μg/100 ml	Average TBPA concentration is 23-35 mg%

The comparative binding affinities for T_4 and T_3, by the proteins are as follows:

Proteins	*Relative binding*	*Affinity*	*Capacity*
• **TBG**	$T_4 > T_3$ *Tetrac = 0	High	Low
• **TBPA**	T_4 $T_3 = 0$ Tetrac > T_4	Moderate	High
• **Serum albumin**	Same for all	Low	Very High

**Tetrac* = Tetra iodo thyroacetic acid.

In normal subject: Only about one-third of the maximum binding capacity is utilised.

In a hyperthyroid patient: This may increase to 1/2 or more.

Abnormal TBG Level: In certain circumstances, TBG levels may become abnormal and it may affect binding of hormones.

An increase occurs in:
- Pregnancy
- After administration of estrogens
- Women taking contraceptive pills.

Decrease levels seen in:
- In nephrosis
- After treatment with androgenic or anabolic steroids.
- Hypoproteinaemic states, viz. liver diseases, cirrhosis liver.

T_3 versus T_4: Although the circulating levels of T_3 are much lower than the corresponding T_4 levels, ***T_3 appears to be the major thyroid hormone metabolically.***

Characteristics of T_3

- Extrathyroidal de-iodination converts T_4 to T_3
- T_3 binds to the "thyroid receptor" in target cells with 10 times the affinity of T_4
- About 80 per cent of circulating T_4 is converted to T_3 or reverse T_3 (r T_3) in the periphery
- T_3 is loosely bound to serum proteins
- ***T_3 is 3 to 5 times more active than T_4***
- Has a more rapid onset of action
- It is also more rapidly degraded in the body.

Note:
- 'Reverse' T_3 (r T_3) is a very weak agonist that is made relatively in larger amounts in chronic diseases, in carbohydrate starvation and in foetus.
- Propylthiouracil and propranolol decrease the conversion of T_4 to T_3.

Chemical Hyperthyroidism: In rare subjects with chemical hyperthyroidism, in whom circulating level of bound and free T_4 is normal, the ***T_3 concentration is elevated and accounts for the thyrotoxic state.*** In these patients, significant amount of T_3 probably arises by de-iodination of T_4 at peripheral level.

MECHANISM OF ACTION OF THYROID HORMONES

Thyroid hormones are transported into their target cells by a "carrier-mediated" active transport system of the cell membrane. Target organs include: Liver, kidneys, adipose tissue, cardiac, neurons, lymphocytes, etc.

1. *Nuclear Action:* T_4 and T_3 pass into the nucleus and bind directly to specific high affinity "nuclear receptors", which are non-histone chromatin proteins of specific genes. This receptor hormone binding increases the action of nuclear *DNA-dependent RNA polymerase* increasing gene transcription, which inturn enhances m-RNA synthesis ↑ and induces synthesis of specific proteins and enzymes.
2. *Na^+ K^+ ATPase Pump:* Thyroid hormones exerts its most of metabolic effects by increasing O_2 – consumption. It has been suggested that much of the energy utilised by a cell is for driving the *Na^+ K^+ ATP-ase Pump*. Thyroid hormones enhance the function of this pump by increasing the number of pump units, almost in all cells.
3. *Translation of Proteins:* Besides its direct 'nuclear action', as stated above, thyroid hormones may stimulate translation of proteins by directly increasing the binding of amino acid–t RNA complex to ribosomes or by increasing the activity of *peptidyl transferase* or *translocase* enzymes.

METABOLIC ROLE OF THYROID HORMONES

1. *Effects on Protein Metabolism:*
   - In hypothyroid children and in physiological doses, thyroid hormones when given in small doses, favour protein anabolism, leading to N-retention (+ve N-balance), because they stimulate growth.
   - Large, unphysiological doses of thyroxine, cause protein catabolism, leading to –ve N-balance.

CLINICAL SIGNIFICANCE

The catabolic response in skeletal muscle, in cases of hyperthyroidism, is sometimes so severe that muscle weakness is a prominent symptom and creatinuria is marked, called ***thyrotoxic myopathy***. The K^+ liberated during protein catabolism appears in urine and there is an increase in urinary hexosamine and uric acid excretion.

Effects on bone proteins: Mobilisation of bone proteins leads to hypercalcaemia and hypercalciuria, with some degree of osteoporosis.

Effect on skin: The skin normally contains a variety of proteins combined with polysaccharides, hyaluronic acid and chondroitin sulphuric acid.

CLINICAL SIGNIFICANCE

In hypothyroidism, these complexes accumulate promoting water retention, which produces characteristic puffiness of the skin; when thyroxine is administered, the proteins are mobilised and diuresis continues until the puffiness (myxoedema) is cleared.

SECTION FOUR

2. *Effects on Carbohydrate Metabolism:* Net effect on carbohydrate metabolism:
   - ***Increase in blood sugar ↑ (hyperglycaemia),*** and ***glycosuria***
   - Increased glucose utilisation, and decreased glucose tolerance. ***Thyroid hormones are, therefore, antagonistic to insulin.***
   - Thyroid hormones increase the rate of absorption of glucose from intestine.
   - Decreased glucose tolerance may be contributed to also by acceleration of degradation of insulin.

Note: Diabetes mellitus is aggravated by coexisting thyrotoxicosis or by administration of thyroid hormones.

- Increased hepatic glycogenolysis ↑, because they enhance the activity of *Glucose-6-phosphatase.*
- In addition there is increased sensitivity to catecholamines, they potentiate the glycogenolytic effect of epinephrine by increasing the β-adrenergic receptors on hepatic cell membrane.
- ***Stimulate glycolysis*** as well as oxidative metabolism of glucose via TCA cycle and also increasing Hexose-monophosphate pathway (HMP-Shunt). Thyroxine increases the activity of "G-6-PD" enzyme in liver.
- Thyroid hormones cause a decrease of glycogen store ↓ in liver and to a lesser extent, in the myocardium and skeletal muscle.
- At the same time, thyroid hormones increase hepatic gluconeogenesis ↑ by increasing the activities of *pyruvate carboxylase* and *PEP carboxykinase.*

3. *Effects on Lipid Metabolism:*
   - ***Increases lipolysis*** ↑ in adipose tissue thus ***increasing plasma FFA*** ↑. This effect is rather indirect in the sense it increases sensitivity to catecholamines, potentiates the lipolytic effect of epinephrine, by increasing the β-adrenergic receptors on adipocyte cell membrane.
   - They may stimulate, at the same time, lipogenesis ↑ by increasing the activities of malic enzyme, ATP citrate lyase and G-6-PD.
   - ***Cholesterol:*** Despite the fact that hepatic synthesis of cholesterol and PL is depressed following thyroidectomy and is increased in thyrotoxicosis, the concentration of cholesterol and to a lesser extent PL in plasma is increased in hypothyroidism and decreased in hyperthyroidism. Decreased value in hyperthyroidism is explained as follows:

   Although thyroid hormones increase the rate of biosynthesis of cholesterol, they increase:
     - The rate of degradation
     - Increases the formation of bile acids (cholic acid/deoxycholic acid) and
     - Increases biliary excretion, to a greater extent accounting for the lowered blood concentration.
   - ***Lipoproteins:*** The concentration of plasma lipoproteins of Sf 10 to 20 class (LDL) is frequently increased in hypothyroidism and decreased in thyrotoxicosis, or following administration of thyroid hormones to normal subjects.

4. *Calorigenic Action:* Thyroid hormones increases considerably O_2 consumption and oxygen coefficient of almost all metabolically active tissues. ***Exceptions are:*** Brain, testes, uterus, lymph nodes, spleen and anterior pituitary. There is increase in heat production and BMR. This effect is due to:
   - Induction of *glycerol-3-P-dehydrogenase* and other enzymes involved in mitochondrial oxidation.
   - More important is increased activity and increased units of *Na^+-K^+ ATP-ase pump*. It hydrolyses ATP for transmembrane extrusion of Na^+, leading to enhanced heat production, O_2 consumption and oxidative phosphorylation.

5. *Vitamins:*
   - Administration of large amounts of thyroid hormones increase the requirement of certain members of vitamin B-complex (thiamine, pyridoxine, pantothenic acid) and for vitamin C. These are presumably related to the stimulation of oxidative and catabolic processes.
   - Thyroxine is necessary for hepatic conversion of carotene to vitamin A and the accumulation of carotene in the bloodstream ("carotinaemia") in hypothyroidism is responsible for yellowish tint of the skin.

Antithyroid Drugs

Most of the drugs which inhibit thyroid function act either by:

- Interfering with 'iodide trapping'
- Inhibiting iodination and coupling
- Inhibiting hormone release
- Inhibiting conversion of T_4 to T_3 at target tissues.

Important antithyroid drugs with examples and mechanism of action are given in Table 33.2.

Effect of Iodine Excess on Thyroid Gland

Another substance which inhibits thyroid under certain conditions is "iodides" itself. The position of iodide in thyroid physiology is unique in that a minimal amount is necessary for functioning of thyroid normally, while a large amount is inhibitory, when the gland is hyperplastic. At high serum concentrations, exceeding 30 μg/dl, iodide exerts antithyroid effects and sometimes produces even a transient goitre. This is called as ***Wolff-Chaikoff effect.***

Mechanism of action

- Initially, iodination of tyrosine is inhibited due to formation of "inactive" iodine I^-_3, in presence of excess iodine.

$$I^- + I_2 \rightleftharpoons I^-_3$$

- Later effect is decrease in TSH; induced rise in c-AMP in thyroid cells with a consequent inhibition of TSH action.
- Also inhibits 'Pinocytosis' of colloid droplets by acinar cells.
- Decrease in proteolysis and release of hormones.

CLINICAL IMPORTANCE

Iodide therapy is sometimes done by surgeons to hyperthyroid patients for a short interval to prepare the patient for surgery (subtotal thyroidectomy).

Advantages

- Colloid accumulates and enhances firmness to the gland
- Vascularity of the gland is decreased
- Decreases the blood thyroid hormone level
- Reduces the chance of acute postoperative hyperthyroidism.

Table 33.2: Showing antithyroid drugs with mechanism of action and examples

Important antithyroid drugs		
Type of drug	***Examples***	***Mechanism of action***
1. Monovalent anions	Chlorate, hypochlorite, periodate, nitrate, perchlorate pertechnate, etc.	• Compete with iodide for transport into thyroid and inhibit iodine uptake (Iodine trapping)
2. Thiocarbamides	Thiourea, thiouracil, Propyl thiouracil, Methimazole (tapazole), carbimazole, etc.	• Block oxidation of iodide to active iodine • Inhibits iodination of MIT • Block the coupling reaction • May inhibit synthesis of thyroglobulin
3. Aminobenzenes	Sulfonamides, PABA, Sulfonyl urea, tolbutamide, carbutamide, etc.	• Inhibit the conversion of iodide to "active" iodine • Inhibit thyroperoxidase • Reduces iodination and coupling reactions.
4. Drugs that inhibit release of thyroid hormones	Colchicine, vinblastin, vincristine, cytochalasin	• Inhibit the formation of microtubules and microfilaments in apical thyroidal cells, which are required for pinocytosis of colloid droplets
5. Drugs that inhibit $T_4 \rightarrow T_3$ conversion	Propylthiouracil, Propanolol	• Inhibit conversion of $T_4 \rightarrow T_3$ in target cells.

PARATHYROID GLANDS AND THEIR HORMONES

INTRODUCTION

The parathyroid glands are intimately concerned with regulation of the concentration of Ca and PO_4 ions in the blood plasma. This is accomplished by secretion of a hormone, **parathormone (PTH)** by the chief cells, the net effect of which is:

- To increase the concentration of Ca ↑ and
- Decrease the PO_4 ↓.

In addition to its effects on plasma ionised Ca via its action on bone, parathormone controls renal excretion of Ca and PO_4.

PARATHORMONE (PTH)

Chemistry: Parathormone is ***a linear polypeptide consisting of 84 amino acids.*** N-terminal amino acid is alanine and C-terminal is glutamine. Bovine PTH has molecular wt. of 9500. Parathormone from different species differ only slightly in structure.

Core of Activity: Studies on the synthetic PTH indicate that the amino acid sequence 1 to 29 or possibly 1 to 34 from N-terminal end is essential for the physiologic actions of this hormone on both skeletal and renal tissues. ***Methionine*** is important amino acid and necessary for calcium mobilising effect. The N-terminal end up to 34 amino acids possesses the "receptor-binding" ability.

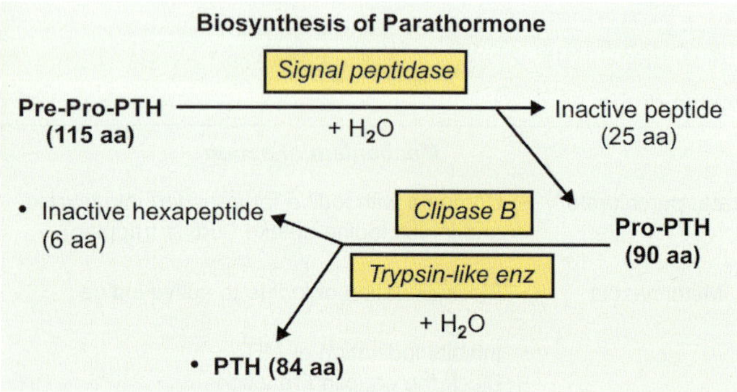

Biosynthesis: PTH is initially synthesised in chief cells as a pro-hormone.

- *Pre-Pro-PTH:* Consisting of 115 amino acids is first formed in polysomes, adhering on the rough ER membrane.
- *Pro-PTH:* Before the formation of Pre-Pro-PTH is completed, its N-terminal end protrudes into the lumen of rough ER and a *signal peptidase* of rough ER membrane hydrolyses the molecule to split off 25 aa and thus pre-pro-PTH is changed to pro-PTH having 90 amino acids.
- *PTH:* Pro-PTH is transferred to rough ER lumen and moves to Golgi cisternae. A "trypsin-like" enzyme, called ***Clipase B*** hydrolyses its N terminal end amino acids and ***remove 6 amino acids, (hexapeptide)*** rich in basic amino acids, thus converting Pro-PTH to PTH.

PTH thus formed is packaged and stored in secretory vesicles. Increased c-AMP concentration and a low Ca^{++} level stimulates its release from secretory vesicles. On the other hand, a high concentration of Ca^{++} stimulates the degradation of the stored PTH in secretory vesicles instead of its release (Refer box LHS).

Mechanism of Action:

PTH increases serum Ca^{++} level by acting on bones, kidney and intestines.

(a) *Increasing Cyclic AMP Level:* PTH binds to "specific receptor" on the plasma membrane of bone cells, renal tubule cells, it activates the *adenyl cyclase* to form c-AMP in the cells. c-AMP acts as the "second messenger" which activates specific c-AMP dependent *protein kinases,* which phosphorylate and thereby modulate the activities of specific proteins in the bone cells and kidney cells.

(b) *Role of Ca^{++}:* c-AMP also increases the Ca^{++} concentration in these cells, which in turn may act as a "messenger" to modulate the activities of some intracellular proteins.

(c) *pH change in Tissues:* The hormone increases the amounts of both LA and citric acid in the tissues and both of these acids may act to aid bone resorption.

METABOLIC ROLE OF PTH

The actions of PTH are reflected in the consequences of:

- ***Its administration*** and
- ***Removal of the parathyroid glands.***

A. ***The most conspicous metabolic consequences of administration of PTH are:***

- Increase in serum Ca^{++} concentration ↑.
- Decrease in serum inorganic PO_4 ↓ concentration.
- Increased urinary Ca^{++} ↑ following an initial decrease.
- Increased urinary PO_4 ↑.
- Removes Ca from bones, particularly if dietary intake of Ca is inadequate.
- Increase in 'citrate' ↑ content of blood plasma, kidney and bones.
- Activates vit D in renal tissue by increasing the rate of conversion of 25-OH-cholecalciferol to 1,25-di-OH-cholecalciferol, ***by stimulating α-1-hydroxylase enzyme.***
- *Effect on Mg metabolism:* PTH has been reported to exert an influence on Mg metabolism. Primary

hyperparathyroidism has been found to be associated with excessive urinary excretion of Mg and –ve Mg balance.

B. **Actions on Different Organs:**

(a) *Action on Kidneys:* ***PTH acts through by increasing c-AMP.*** PTH binds to specific 'receptors' on plasma membrane of renal cortical cells of both proximal and distal tubules and stimulates *adenyl cyclase* to produce c-AMP ↑. c-AMP then is transported to apical/luminal part of the cell where it activates c-AMP dependent *protein kinase*, which phosphorylates specific proteins of the apical membrane to affect the several mineral transport, across the membrane.

- PTH decreases the transmembrane transport and reabsorption of filtered Pi in both proximal and distal tubular cells and increases the urinary excretion of inorganic phosphate ↑ ***(phosphaturic effect)***
- Fall in serum inorganic PO_4 level leads to mobilisation of PO_4 from bones, which also mobilises Ca^{++} along with, ***resulting to hypercalcaemia.***
- PTH ***stimulates α-1-hydroxylase*** enzyme located in mitochondria of proximal convoluted tubule cells, which converts 25-OH-cholecalciferol, to 1-25, di-OH-cholecalciferol which in turn ***increases the intestinal and renal absorption of Ca^{++} resulting to hypercalcaemia.***
- PTH inhibits the transmembrane transport of K^+ and HCO_3^- to decrease their reabsorption by renal tubules.
- PTH increases the transmembrane transport and reabsorption of filtered Ca^{++} in the distal tubules resulting initially to decrease urinary excretion of Ca^{++}. But later on, PTH-induced hypercalcaemia enhances the amount of filtered Ca^{++} which increases the renal excretion.

(b) *Action on Bones:* PTH binds to specific 'receptors' present on membranes of osteoclasts, osteoblasts and osteocytes and increases cyclic AMP level in these cells, which acts through c-AMP dependent *protein kinases*.

Following actions are seen:

- *Osteoclastic activity:* It stimulates the differentiation and maturation of precursors cells of osteoclasts to mature osteoclasts.
- *Osteoclastic osteolysis:* PTH stimulates the osteoclasts through "second messenger" c-AMP to increase the resorption of bones which enhances mobilisation of Ca and P from bones.
- *Osteocytic osteolysis:* PTH also stimulates osteocytes which increases bone resorption thus mobilising Ca^{++} and Pi. There occurs enlargement of bone lacunae.
- *Action on alkaline phosphatase:* Alkaline phosphatase activity varies as per PTH concentration. At low concentrations, PTH stimulates the sulfation of cartilages and increases the number of osteoblasts and alkaline phosphatase activity of bone osteoblasts. At higher levels of physiological concentrations, PTH inhibits alkaline phosphatase activity and collagen synthesis in osteoblasts and decreases the Ca^{++} retaining capacity of bones. PTH induced rise in intracellular c-AMP in osteoclasts/and osteocytes leads to secretion of lysosomal hydrolases/and collagenases which increase breakdown of collagen and MPS in bones matrices.

(c) *Action on Intestinal Mucosa:* PTH does not act directly on intestinal mucosal cells as the cells do not possess the specific 'receptors' for PTH. But it increases the absorption of Ca^{++} and PO_4 through production, 1-25, di-OH-cholecalciferol (calcitriol).

PTHrP (Parathormone-related Peptide):

- Also called as ***Humoral hypercalcaemic factor of malignancy*** **(HHFM)**
- It is a peptide containing 141 amino acids. Amino acid sequence on first 13 same (8 of 13 are homologous) from N-terminus
- ***Produced by a number of tumours specially squamous cells carcinomas of lungs, oesophagus, cervix and head and neck.***
- ***Also produced by renal carcinoma (hypernephroma), carcinoma of pancreas, breast carcinoma, etc.***
- PTHrP ***can bind to parathormone receptor and can mimic the action of parathormone (PTH).***
- Target tissues are bones and kidneys and produces hypercalcaemia, hypophosphataemia like PTH and also increases urinary cyclic AMP.
- PTHrP is produced by a gene on chromosome 12 which is distinct from PTH gene which is located on chromosome 11.

CLINICAL IMPORTANCE

- Serum level of PTHrP are ***low*** or ***absent*** in normal healthy persons and in patients with primary hyperparathyroidism ***but it is high in majority of patients in malignancy*** and is responsible for HHM (humoral hypercalcaemia of malignancy).
- Determination of serum PTHrP is becoming an important diagnostic tool in evaluation of hypercalcaemia.

CALCITONIN

Calcitonin is a calcium regulating hormone. It is proved that calcitonin originates from special cells, called **"C-cells"**, parafollicular cells. ***C-cells constitute an endocrine system which,*** are derived from "neural crest" and are found in thyroid, parathyroids and in thymus.

Chemistry: Calcitonin is a single chain lipophilic polypeptide, having a mol. wt. of 3600. As many as four separate active fractions have been isolated and they have been designated as α, β, γ and δ-calcitonin. Amino acid sequence of calcitonin has now been established. ***It contains 32 amino acids,*** N-terminal amino acid is cysteine, and C-terminal prolinamide. An interchain disulfide bridge joins two cystine residues between position 1 and 7. Low number of ionisable groups are present, 5 of 6 possible –COOH groups being amidated. Isoleucine and lysine are absent conspicuously from the molecule. There is high content of aspartic acid and threonine (Fig. 33.5).

Mechanism of Action

1. *Role of Cyclic AMP:* Calcitonin binds to specific calcitonin receptors on the plasma membrane of bone osteoclasts and renal tubular epithelial cells, activates *adenyl cyclase* which increases c-AMP level ↑ which mediates the cellular effects of the hormone. This is the principal method by which calcitonin acts.
2. *Cellular Shift:* It has been suggested that calcitonin may directly affect the relative distribution of bone cells. The hormone both *in vitro* and *in vivo* produced a cellular shift, in which the number of osteoclasts decreased.
3. *pH Change:* Calcitonin may regulate pH at cellular level producing more alkaline medium which diminishes resorption.

METABOLIC ROLE

Calcitonin acts both on (a) **bone** and (b) **kidneys.**

Indirectly, the effects of these two organ systems account for:

- **Hypocalcaemia** and
- **Hypophosphataemia.**

(a) Action on Bones:

- Calcitonin inhibits the resorption of bones by osteoclasts and thereby reduced mobilisation of Ca and inorganic PO_4 from bones into the blood.
- It also stimulates influx of phosphates in bones.
- There is decrease in activities of lysosomal hydrolases, pyrophosphatase and alkaline phosphatase in bones.
- Decrease in collagen metabolism and decreased excretion of urinary OH-proline.

```
1         7
Cys – S – S – Cys – Met – Leu – Gly – Thr – Tyr – Thr
|           |                                        |
Gly         Thr   Phe – Lys – Asn – Phe –Asp –Gln
|           |     |
Asn – Leu – Ser   His – Thr – Phe – Pro – Gln –Thr
                                              |
       H2N – Pro – Ala – Gly – Val – Gly – Ile – Ala
             32
```

Fig. 33.5: Amino acid sequence of human calcitonin

- Whether or not calcitonin promotes bone formation is uncertain and controversial. But it has been established that the hormone in addition to causing a decrease in number of osteoclasts, it increases osteoblasts cells, which are thought to be involved in bone laying.

(b) Action on Kidneys:

- The hormone acts on the distal tubule and ascending limb of Loop of Henle and decreases tubular reabsorption of both calcium and inorganic PO_4 ***thus producing calcinuria and phosphaturia.***
- The hormone inhibits *α-1-hydroxylase* and inhibits synthesis of 1,25-di-OH-D_3 thus ***decreasing calcium absorption from intestine.***

Both the above effects account for hypocalcaemia.

CLINICAL ASPECT

Abnormal calcitonin secretion is now proved in case of **"medullary carcinoma of thyroid".** Histologically, medullary carcinoma is composed of a single cell and it is now established that this tumour arises from para-follicular C-cells of thyroid. Frequently patients with this tumour have been found to have ***associated cutaneous neuromas, adrenal tumours*** and ***parathyroid enlargement.*** Patient also suffers from severe diarrhoea. Though the tumour produces very high serum calcitonin level, hypocalcaemia and bone demineralisation is not marked.

Reason: Probably high calcitonin concentration may regulate the number of calcitonin specific receptors on bone cells.

Therapeutic Uses of Calcitonin

Calcitonin has been used:

- In Paget's disease
- Idiopathic hypercalcaemia of infancy
- Hypercalcaemia secondary to malignancies, hyperparathyroidism and vit. D intoxication.

Figure 33.6 shows schematically role of PTH and calcitonin on blood calcium.

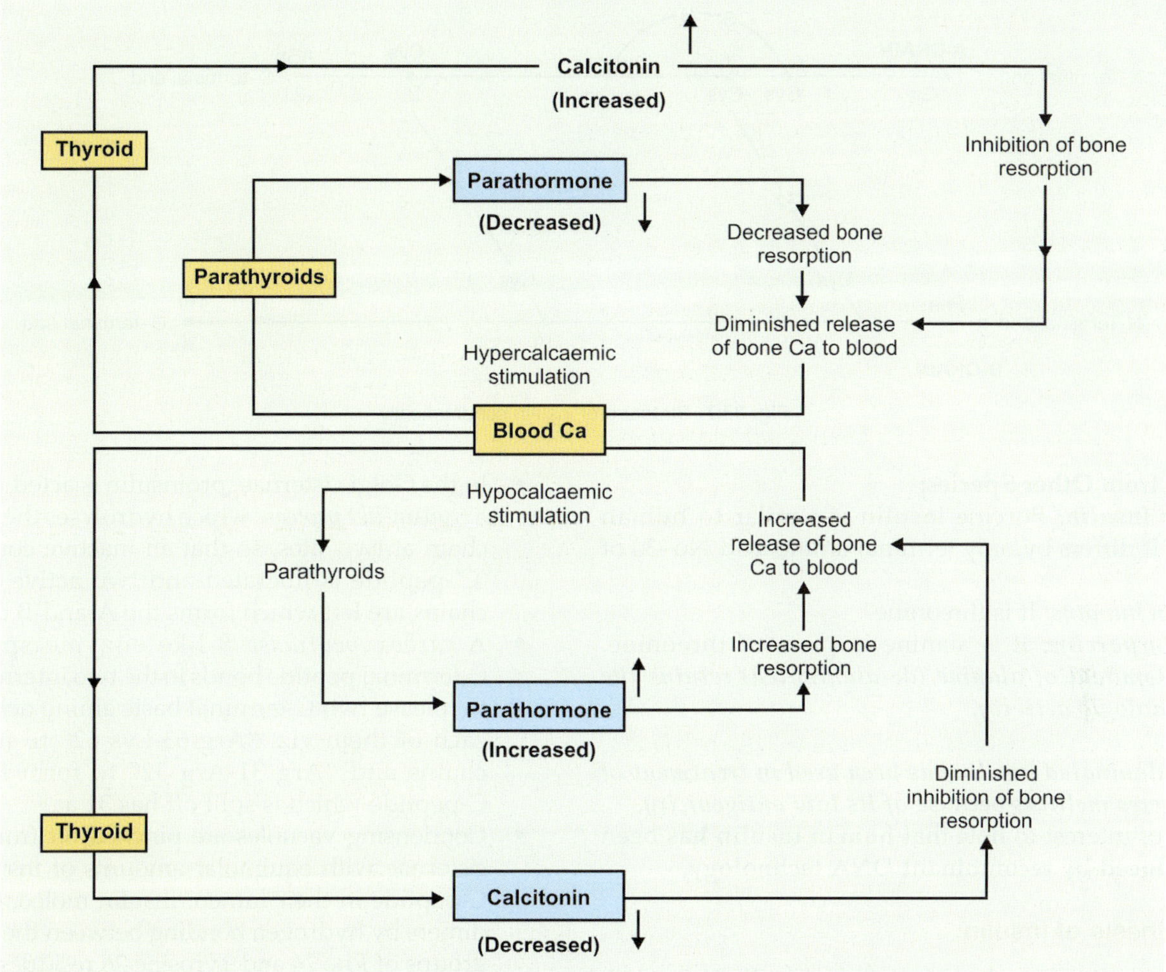

Fig. 33.6: Control of blood calcium by PTH and calcitonin

INSULIN

Insulin is ***a protein hormone***, secreted by β-cells of islets of Langerhans of pancreas. It plays an important role in metabolism causing increased carbohydrate metabolism, glycogenesis and glycogen storage; FA synthesis/TG storage and amino acid uptake/protein synthesis. **Thus insulin is an important anabolic hormone** which act on variety of tissues. Major target tissues of insulin are the muscles, liver, adipose tissue and heart.

Note: RB cells, GI tract epithelial cells and renal tubular epithelial cells are rather generally unresponsive to insulin.

Chemistry: Insulin is a heterodimeric protein; has been isolated from pancreas and prepared in crystalline form. For crystallisation, it requires Zn^{++}. Zinc is also a constituent of stored insulin and normal pancreatic tissue is relatively rich in Zn. Insulin molecule is ***composed of two polypeptide chains, called 'A'-chain and 'B'-chain, containing total of 51 amino acids***. **A-chain** contains 21 amino acids and **B-chain** contains 30 amino acids. In A-chain, N-terminal amino acid is glycine and C-terminal is asparagine. In B-chain, N-terminal amino acid is phenyl alanine and C- terminal is threonine.

- ***Disulfide bridges:*** Both the chains are held together by two S–S linkages. Cys 7 and Cys 20 of A-chain are joined to Cys 7 and Cys 19 of B-chain respectively. In addition, the 'A' chain carries an "intra-chain" S-S linkage between Cys 6 and Cys 11 (Fig. 33.7).
- ***Importance of S–S bridges:*** Breaking of the disulfide bonds with alkali or reducing agents inactivate insulin. Digestion of insulin protein with proteolytic enzymes also inactivates the hormone.

Note: *The above is the reason why insulin cannot be given orally.*

- ***Molecular Weight of Insulin:*** Minimum calculated molecular wt. is 5734. Most estimates of mol. wt. by physical measurements range from 12,000 to 48,000. Insulin can exist in different ***'polymeric' forms*** (dimers, trimers, etc.) depending on pH, temperature and concentration.

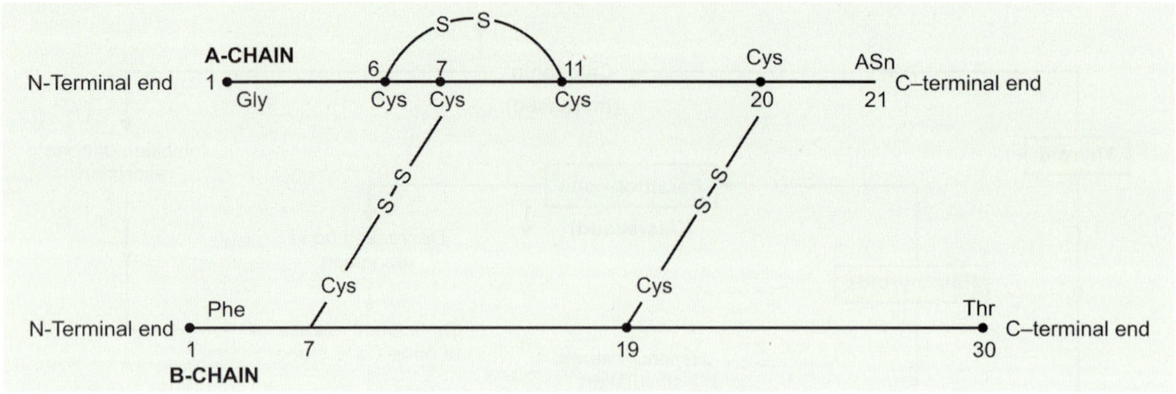

Fig. 33.7: Structure of insulin schematically

Insulin from Other Species:

Porcine insulin: Porcine insulin is similar to human insulin. It differs by only terminal amino acid No.-30 of B-chain.

- ***In humans:*** It is threonine.
- ***In porcine:*** It is alanine in place of threonine. ***Removal of alanine (de-alaninated) retains the biologic activity.***

Note:

- ***De-alaninated insulin has been used in treatment of diabetes mellitus because of its low antigenicity.***
- It is of interest to note that **human insulin has been produced by recombinant DNA technology.**

Biosynthesis of Insulin:

In biosynthesis of insulin, first **pre-proinsulin is formed, which is converted to pro-insulin.** The latter is finally converted to insulin.

1. ***Synthesis of Pre-Proinsulin:*** Pre-proinsulin is synthesised in polysomes, attached to the membrane of rough endoplasmic reticulum (ER) in β-cells of islet of Langerhans.
   - It is a polypeptide consisting of 109 amino acids. Mol. wt = 11,500
2. ***Conversion of Pre-Proinsulin to Proinsulin:***
   - Pre-proinsulin after synthesis is transferred to lumen of rough endoplasmic reticulum cisternae.
   - A peptide chain consisting 23 amino acids in its N-terminal, called ***"leader sequence"*** is split by an enzyme called ***signal peptidase*** present in the membrane of rough endoplasmic reticulum (RER) and pro-insulin is formed.
   - ***Proinsulin has 86 amino acids.*** Molecular wt = 9000
3. ***Conversion of Pro-Insulin to Insulin:***
   - Pro-insulin containing small vesicles are detatched from ER and fuses with the cisternae of Golgi apparatus.
   - In the Golgi cisternae, proinsulin is acted upon by a *trypsin-like protease* which hydrolyses the peptide chain at two sites, so that an inactive connective 'C'-peptide is liberated and two active peptide chains are left which forms the A-and-B chain.
   - A *carboxypeptidase B* like enzyme splits the C-terminal peptide bonds in the two intermediates to release two C-terminal basic amino acids from each of them viz. "Arg 63-Lys 62" to form 'A' chains and "Arg 31-Arg 32" to form B-chain. C-peptide which is split off has 31 aa.
   - Condensing vacuoles are pinched off from Golgi cisternae with equimolar amounts of insulin and C-peptide in their lumen. Insulin molecules form dimers by hydrogen bonding between the peptide groups of Phe 24 and Tyrosine 26 residues of their B-chains. Gradually with increasing concentrations, condensing vacuoles change into secretory granules. In them, insulin forms crystalloid form of hexamers with two Zn^{++}. C-peptides remain in the fluid spaces surrounding the crystalloid granules.

Biosynthesis of Insulin

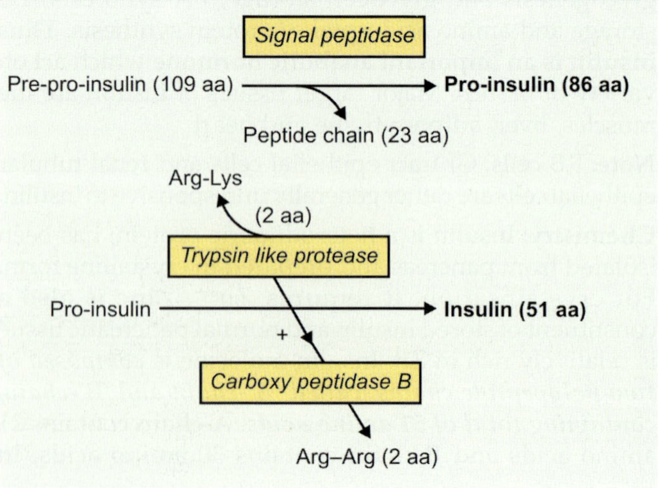

Note

- Pro-insulin is comparatively inactive biologically, but it can cross-react with anti-sera prepared against insulin.
- Plasma pro-insulin is not elevated in human diabetes or in normals after glucose stimulation, but it may be the predominant circulating form in some subjects with islet-cell tumours.

CATABOLISM OF INSULIN

Insulin is very rapidly catabolised. ***Its plasma 1/2 life is less than 3 to 5 minutes under normal conditions.*** Major organs where insulin is catabolised are liver, kidneys and placenta. About 50 per cent of insulin is degraded in its single passage through the liver.

Mechanism: **Two enzyme systems** are involved for degradation of insulin.

- ***Protease:*** An insulin-specific *protease* has been found in many tissues with highest concentration in liver and kidneys. The protease is –SH dependent and active at physiological pH.
- Second mechanism is more important. The enzyme is ***glutathione-insulin transhydrogenase*** (also called ***insulinase***). This enzyme is found in highest concentration in liver and kidneys. Also present in skeletal muscles and placenta. This brings about reductive cleavage of the "S-S bond" which connects the A and B-chains of insulin molecule. *Reduced glutathione* (G-SH), acting as a coenzyme for the trans-hydrogenase, donates the H-atoms for the reduction and is itself thus converted to oxidised glutathione (G-S-S-G).
- After insulin is reductively cleaved, the A-chains and B-chains are further hydrolysed by *proteolysis*.

INSULIN RECEPTORS

Insulin acts on target tissues by binding to specific **"insulin receptors"**, which are **'glycoproteins'**. The ***human insulin receptor gene is found on chromosome 19.*** The insulin receptors are being constantly synthesised and degraded. ***Their 1/2 life is 6 to 12 hours only.*** It is synthesised as a single-chain polypeptide, **"Pro-receptor"** in the rough endoplasmic reticulum (RER) and is rapidly glycosylated in Golgi region. The 'pro-receptor' has 1382 amino acids and mol. wt. of 190,000.

The pro-receptor is cleaved to form mature 'α' and 'β' subunits (α_2 β_2), which is a heterodimer, linked by S-S bonds. Both subunits are extensively glycosylated and removal of sialic acid and galactose decreases insulin binding and insulin action. Insulin receptors are found in target cell membrane, up to 20,000 per cell.

Chief differences in structure and function of α and β subunits are tabulated below:

α-Subunit	*β-Subunit*
• The heavy chain	• The lighter chain
• Entirely extracellular	• Transmembrane protein
• Mol. wt. = 135,000	• Mol. wt. = 95,000
• Binds insulin via 'cysteine rich' domain (insulin binding site)	• Functions as "Signal transducers"
• Inter-linked by S-S linkage to a b-chain	• Its N-terminal third on outer surface of the membranes, a narrow 23 aa domain in the membrane and C-terminal 2/3 possesses *tyrosine kinase* activity and ATP-binding activity.

Binding of insulin to the receptor, stimulates its *tyrosine kinase* activity. *Tyrosine kinase* enzyme phosphorylates the phenolic-OH group of tyrosine residues in specific protein including that of a tyrosine in the β-chain of insulin receptor itself to modulate their activities, ATP + tyrosine protein → ADP + phospho-tyrosine protein.

Regulation of Insulin Receptors

A high blood insulin level decreases the number of insulin receptors on target cell membrane, probably through internalisation of the insulin-receptor complex into the cell and thus decreases the insulin sensitivity of the target tissues.

MECHANISM OF ACTION OF INSULIN

When insulin binds to the specific receptor, several events of actions take place:

- A conformational change of the receptor
- The receptors cross-link and form "microaggregates".
- The receptor complex is internalised, and
- One or more signals are generated.

But nature of the intracellular signal and intracellular "second messenger" remains still uncertain and vague. Various mechanisms have been proposed:

1. ***Role of Cyclic AMP:*** It is proposed that insulin promotes the phosphorylation of cyclic AMP *phosphodiesterase*. The "active" phosphodiesterase hydrolyses c-AMP and ***lowers the c-AMP level ↓ in the cells.*** The consequent fall in activities of c-AMP dependent *protein kinase* reduce phosphorylation of specific enzymes.
2. ***Role of Cyclic - GMP:*** The insulin-receptor binding may activate *guanylate cyclase* which forms cyclic-

GMP. Increased concentration of cyclic GMP act as "second messenger" to activate cyclic-GMP dependent *protein kinase*. These may phosphorylate some enzymes to modulate their activities.

Note: ***Cyclic AMP and cyclic GMP function in a reciprocal relationship*** which has been called the ***"yin-yang hypothesis".***

3. ***Role of Protein-phosphatases:*** Insulin may act through the *protein-phosphatase 1* which may dephosphorylate certain key enzymes thereby activating them. Best examples are the key enzyme *glycogen synthase* and PDH (*pyruvate dehydrogenase* complex). On the other hand, it inhibits *phosphorylase* enzyme and *triacyl glycerol lipase*.
4. ***Action through 'Tyrosine Kinase' Activity of β-Subunit of Receptor:*** The binding of insulin to its receptor enhances *tyrosine kinase* activity. Tyrosine kinase inturn phosphorylates phenolic-OH group of tyrosine residues of specific proteins leading to changes in enzyme activities.
5. ***Role in m-RNA Translation:*** Insulin is known to affect the activity or amount of at least more than 50 proteins in a variety of tissues and many of these effects involve covalent modification. A role of insulin in the translation of m-RNA has been proposed largely based on studies of ribosomal protein 6S, a component of the 40S ribosomal unit. Such a mechanism accounts for the general effect of insulin on protein synthesis in liver, heart muscle and skeletal muscle.
6. ***Role on Gene Expression (Nuclear Action):*** Insulin also affects the rate of transcription of specific genes, thereby regulates the synthesis of specific m-RNAs and thus changing the rate of synthesis of specific proteins coded by them.

Example: Insulin decreases the transcription of gene involved in synthesis of the enzyme *phosphoenol-pyruvate carboxy kinase* (PEPCK), the key enzyme for gluconeogenesis. On the other hand, insulin induces the synthesis of *phosphofructokinase* and *pyruvate kinase* required for glycolysis, by increasing the transcription of these genes.

SECTION FOUR

METABOLIC ROLE OF INSULIN

ACTION ON CARBOHYDRATE METABOLISM

Net effect is (1) lowering of blood glucose ↓ level and (2) increased glycogen store ↑.

The above is achieved by several mechanisms:

1. ***Increases glucose uptake:***
   - Insulin increases glucose uptake from EC fluid by the various tissues, viz. muscles, adipose tissue, mammary glands, lens, etc.
   - In adipose tissue and probably other extrahepatic tissues, insulin stimulates translocation of glucose ***'transporters'*** from their intracellular pool in Golgi cisternae to the plasma membrane where they participate as "carriers" in transportation of D-glucose and D-galactose across the membrane.
   - Also in hepatocytes, insulin increases hepatic uptake of glucose (freely permeable to liver cells). It induces the synthesis of the enzyme *glucokinase*, which simultaneously phosphorylates glucose, thereby lowering intracellular concentration.
2. ***Increases glycolysis***
   Increases utilisation of glucose for providing energy which takes place in muscles, liver and many other tissues. Insulin enhances glycolysis ↑ because it induces the synthesis of key enzyme *phosphofructokinase* and also *"pyruvate kinase"*.
3. ***Increases conversion of pyruvate to acetyl-CoA:***
   Insulin increases aerobic oxidative decarboxylation of pyruvate to acetyl-CoA ↑, because it causes dephosphorylation of *pyruvate dehydrogenase complex* (PDH) which is thus converted to 'active' form.
4. ***Stimulates glycogenesis:***
   Insulin stimulates glycogenesis ↑ in the liver and muscles by increasing dephosphorylation of the key and rate limiting enzyme *"glycogen synthase"*, thus converting to its 'active' form. Insulin stimulates the *protein-phosphatase-1* directly, which brings about dephosphorylation.
5. ***Decreases gluconeogenesis:***
   Insulin reduces gluconeogenesis ↓ :
   - By repressing the synthesis of the key rate limiting enzyme *PEP-carboxykinase* (PEPCK), by decreasing the transcription rate of the gene.
   - Also inhibits allosterically *fructose-1, 6-bi-phosphatase*, another key enzyme for gluconeogenesis.
   - Insulin dephosphorylates *fructose-2, 6-bi-phosphatase* so that it is converted to 'inactive' form, which increases the concentration of "fructose-2, 6-bi-P" in the cell, which in turn allosterically inhibit *fructose-1, 6-bi-phosphatase*.
6. ***Decreases glycogenolysis:***
   Insulin decreases glycogenolysis ↓
   - By dephosphorylating the key and rate limiting enzyme glycogen *phosphorylase* thus converting it to "inactive form".
   - Also represses the enzyme *glucose-6-phosphatase*.
7. ***Increasing HMP-shunt:***
   Insulin stimulates HMP-shunt producing more NADPH (required for FA synthesis), by inducing the synthesis of *glucose-6-P-dehydrogenase* (G-6-PD) and *6-phosphogluconate dehydrogenase*.

ACTION ON LIPID METABOLISM

Net effects are: (1) lowering of FFA level ↓ and (2) increase in TG store ↑.

The above is achieved as follows:

1. ***Decreases Lipolysis:***
   Insulin decreases lipolysis ↓ in adipose tissue cells and ***consequently lowers plasma FFA ↓.***
   Lipolysis is reduced due to:
   - Insulin activates *phosphoprotein phosphatase* which dephosphorylates the *triacyl glycerol lipase* and thus is converted to 'inactive' form.
   - At the same time, insulin activates *phosphodiesterase* which degrades cyclic-AMP and prevents phosphorylation and reactivation of *TG lipase*.
2. ***Increases FA Synthesis:***
   Insulin increases the extramitochondrial de novo FA synthesis ↑, by making available of more substrate acetyl-CoA and ***also increasing the activity of acetyl-CoA carboxylase.***
   The above is done as follows:
   - Insulin promotes dephosphorylation of *pyruvate dehydrogenase complex* and converts into "active form" so that more acetyl-CoA is available from pyruvate.
   - Insulin induces the synthesis of *ATP-citrate lyase* to increase cleavage of citrate, so that more acetyl-CoA is available in cytosol.
   - Insulin lowers the plasma FFA level, so prevents long-chain acyl-CoA from inhibiting *acetyl-CoA carboxylase*.
   - It induces the synthesis of *acetyl-CoA carboxylase* and *fatty acid synthase*, the cytosolic enzymes required for FA synthesis.
   - Insulin activates *acetyl-CoA carboxylase* by dephosphorylation of the enzyme (converting to "active" form).
   - Provides more NADPH for the reductive steps in FA synthesis by stimulating HMP-shunt pathway.
3. ***Increases Synthesis of TG:***
   Insulin enhances TG synthesis ↑ in adipose tissue by:
   - ***Providing more α-glycero-p, as glucose uptake and utilisation is enhanced in adipocytes.***
   - Increased synthesis of FA provides the acyl-CoA (FFA pool 1), required for TG synthesis.
   - Insulin also induces the synthesis of *lipoprotein lipase*. This enzyme hydrolyses TG of circulating chylomicrons and VLDL and releases FFA (FFA pool-2), which are taken up by adipocytes and used for TG synthesis.
4. ***Decreases Ketogenesis:***
   As plasma FFA level is decreased, less is oxidised by β-oxidation and less acetyl-CoA will be available for cholesterol synthesis and ketogenesis.

ACTION ON PROTEIN METABOLISM

Net effect is that insulin promotes protein synthesis ↑. This is achieved as follows:

- Insulin ***increases amino acids uptake*** by the tissues, by enhancing the rate of synthesis of membrane ***"transporters"*** for amino acids.
- Adequate supply of insulin is necessary for protein anabolic effect of GH ***(permissive effect)***.
- Insulin increases protein synthesis by providing more amino acids in cells, by affecting gene transcription (nuclear level), by regulating specific m-RNA synthesis and affecting translation at ribosomal level.
- ***Regulation of ribosomal translation is done by two ways:***
  - ***Increases the synthesis of polyamines,*** required for ribosomal RNA synthesis, by increasing the synthesis of key and ratelimiting enzyme *ornithine decarboxylase*.
  - Secondly, insulin modulates ribosomal activity by causing phosphorylation of 6S ribosome (a component of 40S).

Action on Mineral Metabolism

Decrease in concentration of K^+ ↓ and inorganic P ↓ in blood due to enhanced glycogenesis and phosphorylation of glucose.

Actions on Growth and Cell Replication

Insulin stimulates growth *in vivo* and also cell proliferation *in vitro*. Cultured fibroblasts have been used most frequently in studies of cell proliferation. It has been found that ***insulin potentiates the ability of fibroblast growth factor (FGF), platelet-derived growth factor (PDGF) and epidermal growth factor (EGF), etc.*** The effects on growth and cell proliferation are seen in many tissues such as liver, mammary glands and adrenals and also in embryogenesis and tissue differentiation. These effects are largely due to stimulation of DNA replication, gene transcription, protein synthesis and modulation of various enzyme activities through phosphorylation dephosphorylation.

RELATION OF TYROSINE KINASE ACTIVITY WITH GROWTH FACTORS

Insulin receptors along with receptors of many other growth promoting peptides including those of EGF, PDGF, IGF-I, etc. have *tyrosine kinase* activity. It is of interest to note that at least ten or more "oncogene" products, many of which are suspected to be involved in stimulating malignant cell replication are also *tyrosine kinases*. Tyrosine kinase activity is now thought to be an essential factor in the action of a number of viral oncogene

Table 33.3: Differences between insulin, IGF-I and IGF-II are shown below in a tabular form

	Insulin	IGF-I	IGF-II
• Synonyms	—	Somato-medin C	Multiplication stimulating activity (MSA)
• Source	Produced by β-cells of islet of Langerhans of pancreas	Liver and other tissues	Diverse tissues
• Structure	Heterodimer, Two polypeptide chains A- and B-chain. Total amino acids = 51	A single chain polypeptide having 70 amino acids	A single chain polypeptide having 67 amino acids
• Plasma level	0.3 to 2 ng/ml	—	—
• Carrier protein in plasma	No carrier protein	Carried by plasma proteins—albumin chiefly	Carried by plasma proteins
• Regulation of plasma level	Chiefly by glucose level	• Growth hormone • Nutritional status	—
• Receptors	Insulin receptor hetero-dimer, $\alpha_2\beta_2$ β-subunit has tyrosine kinase activity	Dimer, $\alpha_2\beta_2$, also have tyrosine kinase	Single chain polypeptide, no tyrosine kinase
• Affinity of receptor to hormones:			
• Insulin receptors	High	Moderate	Negligible
• IGF-I receptor	Low	High	Low
• IGF-II receptor	Negligible	Moderate	High
• Major physiological function	Control of metabolism	Skeletal and cartilage growth	Not known, probably role in embryogenesis.

products. Mammalian cells contain analogous of these 'oncogenes', as proto-oncogenes which may be involved in the replication of normal cells.

SECTION FOUR

Insulin-Like Growth Factors:

- ***Two insulin like growth factors, IGF-I and IGF-II have recently been found.***
- They are ***not of pancreatic origin but produced by liver and other tissues.***
- It is difficult to separate the effects of insulin on cell growth and replication from similar actions exerted by IGF-I and IGF-II. ***Insulin and IGFs may interact in this process.***
- Insulin is more potent metabolic hormone and IGFs are involved more in stimulating growth.
- Each hormone has unique receptor to act (Table 33.3).

INSULIN ANALOGUES

Recently major breakthrough of synthesising insulin analogues with pharmacologic advantages have been made possible by ***"computer modelling"***. The programme furnishes appropriate modifications in the molecular structure necessary for altering the stability, self association and pharmacologic activity of insulin. Synthesis of these altered molecules is then done by ***"Re-combinant DNA technology"***. Using technique, **three types** of insulin analogues have been prepared. They are:

- *Short (fast) acting analogues: monomeric insulins*
- *Intermediate acting analogues* and
- *Long acting analogues.*

Note: It is not possible to discuss all the analogues, a few prototype examples are given:

1. **Short (Fast) acting analogues:**
   - ***B9 Asp B27 Glu:*** In normal subjects this analogue is absorbed 2 to 3 times faster than soluble human insulin after SC injection and it is accompanied by more rapid rise in plasma insulin and more rapid onset of hypoglycaemia.
   - ***Lys (B28), Pro (B29): (Lispro Insulin):*** Studies have shown onset of hypoglycaemic activity seen within 15 minutes of administration, peak serum insulin reached in approx. one hour, and duration of action was shorter (3.5 to 4.5 hrs).
2. **Intermediate acting insulin analogues:**
   - ***Diarginyl insulin:*** This is an interesting product of recombinant DNA technology, an intermediary metabolite in the bioconversion of proinsulin to insulin. It behaves like an intermediate acting insulin preparation.
   - ***DES 64, 65 HPI (D PRO):*** This is a normal metabolite of pro-insulin formed by split between positions 65 and 66 and removal of Arg and Lys at position 64 and 65 of human proinsulin. Data generated in humans are very preliminary.
3. **Long acting insulin analogues:**
   - ***Novosol basal:*** In this, substitution of threonine in position B27 with arginine, and amidation of the 'C' terminal of the B-chain adds two positive charges and increases the isoelectric point (pI) from 5.4 to 6.8. A further substitution of asparagine in A21 with glycine renders the molecule stable in acid solution. This preparation is soluble in its formulation of pH 3.0. After injection, when the pH rises to about 7.4, it crystallises (crystal less

```
      1                                           10
NH2— His – Ser – Gln – Gly – Thr – Phe – Thr – Ser – Asp – Tyr —
      11                                          20
      Ser – Lys – Tyr – Leu – Asp – Ser – Arg – Arg – Ala – Gln —
      21
      Asp – Phe – Val – Gln – Trp – Leu – Met – Asn – Thr – COOH
```

Fig. 33.8: Amino acid sequences of glucagon

than <5.0 micrometers in diameter) ***acts as a subcutaneous depot*** from which insulin is slowly absorbed, thus acts as long acting insulin analogue.

GLUCAGON HYPERGLYCAEMIC-GLYCOGENOLYTIC FACTOR (HGF)

INTRODUCTION

Glucagon is a hormone **produced by α-cells of islet of Langerhans of pancreas** and is an important hormone involved in:

- Rapid mobilisation of hepatic glycogen to give glucose by glycogenolysis, and
- To a lesser extent FA from adipose tissue.

Thus, ***it acts as a hormone required to mobilise metabolic substrates from storage depots.***

Chemistry

Glucagon has been purified and crystallised from pancreatic extracts and also the hormone has been synthesised. ***It is a polypeptide containing 29 amino acids.*** There are only 15 different amino acids in the molecule. Amino acid sequence has been determined, histidine is the N-terminal amino acid and threonine is the C-terminal (Fig. 33.8). Molecular wt. is approx. 3485.

Unlike Insulin:

- ***It does not require Zinc*** or other metals for its crystallisation.
- Glucagon contains no cystine, proline or isoleucine, but contains tyrosine, methionine and tryptophan.

SYNTHESIS

It is synthesised first as a pro-hormone, ***"proglucagon"*** in α-cells. Lysosomal enzymes peptidases like *carboxypeptidase B* and *trypsin-like peptidases* in α-cells hydrolyse pro-glucagon from both its N-terminal end and C-terminal end to yield glucagon and inactive peptides.

Note:

Entero-glucagon or glucagon-like immune reactive factor (GLI).

A glucagon-like immuno-reactive factor (GLI) has been identified in gastric and duodenal mucosa. GLI is immunologically similar though not identical to the pancreatic hormone. Moreover, it is less active than pancreatic glucagon in stimulating *adenyl cyclase* and therefore cannot duplicate many of the functions of pancreatic hormone. GLI is stimulated by absorbed glucose causing an apparent elevation of circulating pancreatic glucagon.

Recently, two different molecular fractions have been isolated:

- One having mol. wt. = 3500, has hyperglycaemic, and glycogenolytic activity but far less potent than pancreatic glucagon.
- The other fraction, mol. wt. = 7000, is devoid of the above activity.

But both have insulin releasing activity.

MECHANISM OF ACTION

Glucagon binds to specific receptors on the plasma membranes of hepatocytes and adipocytes and activates *adenyl cyclase* to produce c-AMP in these cells, which is the principal "second messenger" and duplicates the functions of the hormone. c-AMP in turn activates c-AMP dependent *protein kinases* which further phosphorylates specific enzymes to increase/decrease their activities. c-AMP also induces synthesis of certain specific enzymes like *glucose-6-phosphatase* by increasing the transcription of their genes.

METABOLIC ROLE

1. Action on Carbohydrate Metabolism:

Net effect of the hormone is to increase the blood sugar level (hyperglycaemia) ↑. Hyperglucaemic effect is due to various causes:

- ***Glycogenolysis:*** Glucagon increases glycogenolysis in liver. In muscles, ***it cannot bring about glycogenolysis as muscle cell membrane lacks the glucagon specific receptors.*** Glucagon also induces the synthesis of *glucose-6-phosphatase* enzyme.
- ***By increasing gluconeogenesis in liver:*** Glucagon stimulates the conversion of LA and glucogenic amino acids to form glucose.
  - The increased hepatic cyclic AMP produced after glucagon action has been shown to increase *protein kinases* that catalyse nuclear histone phosphorylation in liver cell nucleus. This reaction inhibits the repressive effect normally exerted by histones on DNA and allows the initiation of a sequence of events leading to the synthesis of new enzyme proteins involved in gluconeogenesis. Thus, ***glucagon induces the synthesis of PEP-carboxykinase, pyruvate carboxylase*** and ***fructose-1, 6-biphosphatase enzyme, all key enzymes for gluconeogenesis.***
- Also glucagon increases the pool of glucogenic amino acids in liver, so that they can be used for gluconeogenesis. This is achieved by increasing protein breakdown in liver and by reducing hepatic protein synthesis.

2. **On Lipid Metabolism:**
   - ***Lipolysis:*** In adipose tissue and also possibly in liver, glucagon increases the breakdown of TG to produce FFA ↑ and glycerol ↑. FA undergo β-oxidation, increased breakdown may lead to Ketone bodies formation and Ketosis. ***Thyroid hormones help in the lipolytic action of glucagon, probably the hormones increase the number of glucagon specific receptors on adipocytes.***
   - ***Anti-lipogenic Action:*** ***Glucagon reduces FA synthesis.*** This is achieved in **2 ways:**
     - Increased lipolysis raises the concentration of FFA in blood. Long-chain acyl-CoA inhibits the rate-limiting enzyme *acetyl-CoA carboxylase*.
     - Increased c-AMP level in cells activates the c-AMP-dependent protein kinase which phosphorylates acetyl-CoA carboxylase. Phosphorylated form of the enzyme is "inactive".
3. **On Protein Metabolism:**
   - ***Glucagon reduces protein synthesis by depressing incorporation of amino acids into peptide chains.*** This may be due to the inactivation of some ribosomal component by a *protein kinase* whose activity is enhanced by glucagon-induced rise in c-AMP.
   - Glucagon also stimulates protein catabolism ↑ specially in liver thus increases the hepatic amino acid pool which is utilised for gluconeogenesis. Also increases urinary NPN and urea.
4. **Action on Heart:**
   Glucagon ***exerts a +ve ionotropic effect on heart without producing increased myocardial irritability. Hence, use of glucagon in treatment of heart disease, viz. in cardiac failure and in cardiogenic shock.***
   Advantage over nor-epinephrine: Glucagon increases the force of contraction, but does not produce any arrhythmias, tachycardia or increase in O_2 consumption.
5. **Calorigenic Action:**
   Glucagon ***increases heat production and rise in BMR.*** The calorigenic action is not due to hyperglycaemia *Perse* but is probably due to increased hepatic deamination of amino acids, with thyroid hormones stimulating the utilisation of deaminated residues. The calorigenic action requires the presence of thyroid and adrenocortical hormones and fails to occur in their absence.
6. **On Mineral Metabolism:**
   - ***Potassium:*** Glucagon increases K^+ release from the liver, an action which may be related to its glycogenolytic activity.
   - ***Calcium:*** Recently it has been shown that glucagon can increase the release of calcitonin from the thyroid, thus have calcium lowering affect.

CLINICAL ASPECTS

Clinical and Therapeutic Uses

- ***Most important use is in treatment of severe insulin induced hypoglycaemia.***
- Long acting Zinc glucagon has been used in inoperable pancreatic cell tumours.
- Has been used in ***heart failure and cardiogenic shock*** due to its direct ionotropic effect on cardiac muscle.
- It also improves the renal perfusion by decreasing renal vascular resistance. It is mainly indicated in low output failure and in toxicity of β-blockers (as its action is not blocked by α-blockers).
- Recently it has been used also in treatment of ***Acute Pancreatitis*** due to inhibitory effect on exocrine secretions of pancreas.

SOMATOSTATIN

The peptide somatostatin (also called as **"GH release inhibiting factor"**) was first isolated from the hypothalamus and was implicated as a regulator of GH secretion.

Chemistry: ***It is a peptide consisting of 14 amino acids.*** There is an intrachain S–S linkage joining cysteine 3 and cysteine at position 14.

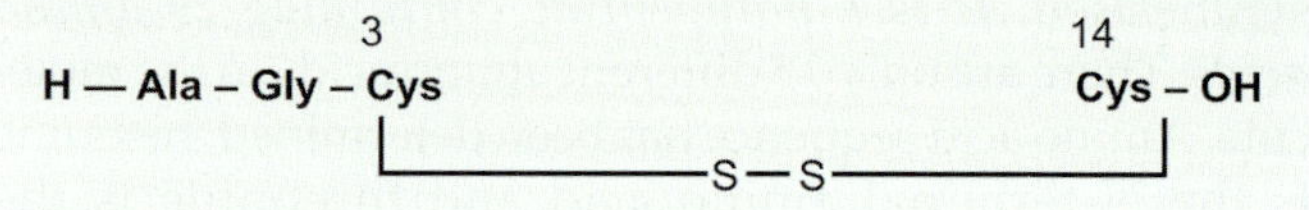

Source: There are ***three sources:***
1. ***Hypothalamus*** as stated above.
2. ***Pancreas:*** Somatostatin is also ***secreted by δ-cells of islet of Langerhans of pancreas.***
3. ***GI tract:*** It is also ***produced by D-cells of antral mucosa of stomach*** and ***also duodenal mucosa.***

The above suggest that the hypothalamic releasing hormones may actually be more widely distributed.

METABOLIC ROLE

In contrast to 'telecrine' action of hypothalamic somatostatin on anterior pituitary, the GI somatostain has local "paracrine" actions limited to GI mucosa and pancreas.

(a) Hypothalamic Somatostatin:
- Acts as a regulator of growth hormone secretion
- It inhibits GH release
- It may also serve as a neurotransmitter substance in the brain.

(b) Pancreatic Somatostatin:

- It inhibits both insulin and glucagon secretion and thus may serve as an "intra islet" (paracrine) regulator of secretion of these hormones. Thus acts as intraorgan "synaptic transmitters" or neuromodulators.
- Somatostatin is secreted into the portal vein blood as a result of glucose or amino acid stimulus indicating extra-islet role.
- Also directly inhibit secretion of both HCO_3^- and enzymes in pancreatic juice.

(c) GI Somatostatin:

- Inhibits the secretions of gastrin, CCK, GIP and motilin.
- Also inhibits gastric acid secretion, secretion of Brunner's glands, pancreatic HCO_3^- and enzyme secretions, gastric emptying and gall bladder contraction.

Since somatostatin can inhibit a variety of GI functions (gastric emptying, GI motility), its major function may be to regulate nutritional influx at the level of GI tract.

ADRENAL STEROID HORMONES

Steroid Hormones Produced by Adrenal Cortex:

About 50 steroids have been isolated from the adrenal cortex. But out of them **only 7 (seven) are important** and known to possess physiologic activity. They are all derived from cholesterol which ***can be synthesised from "active" acetate,*** and they contain the steroid nucleus, called ***"cyclo-pentano perhydro phenanthrene"*** nucleus.

Seven important hormones are:

- 11-dehydro corticosterone (DOC) (Earlier called as compound A)
- Corticosterone (Compound B)
- Cortisone (Compound E)
- Cortisol (17-OH corticosterone, Compound F)
- Aldosterone (mineralo-corticoid)
- Androstenedione } Two androgens
- Dehydroepiandrosterone (DHE) } Two androgens

Cortisol is the major free-circulating adrenocortical hormone (gluco-corticoid) in human plasma.

CLASSIFICATION

I. According to structure: Adrenocortical hormones are mainly of two structural types:

1. ***C-21 steroids:*** Those which have a two carbon side chain at position 17 of the 'D' ring and contain total 21 carbon atoms.
2. ***C-19 steroids:*** Those which have an O_2 atom or – OH group at position 17 and contain 19 carbon atoms. Most of the C-19 steroids have a = 0 group at position 17 and are, therefore, called as ***"17-oxosteroids" (17-ketosteroids).***

Note: The C-21 steroids which have a –OH GR at the position 17, in addition to the side chain are often called '17-OH corticoids' or "17-OH-corticosteroids".

In general:

- ***C-19 steroids have androgenic activity*** and
- ***C-21 steroids have glucocorticoids and mineralo-corticoids activity.***

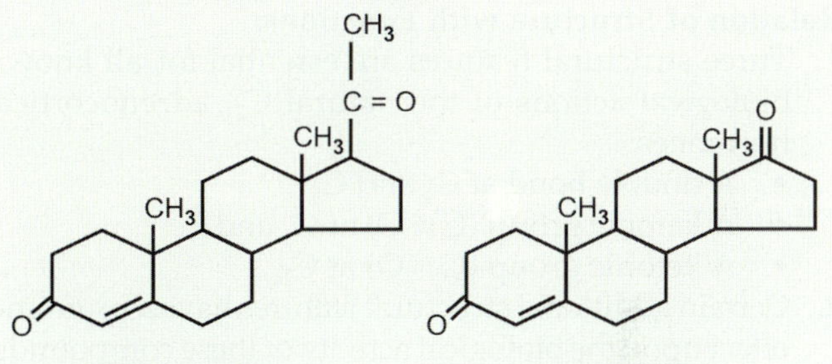

(C-21 steroids) **(C-19 steroids)**

II. According to function

Steroids are divided into ***three types*** according to function (using ***Seyle's terminology***):

1. ***Glucocorticoids:*** Which primarily affect metabolism of carbohydrates, proteins and lipids and relatively minor effects on electrolytes and water metabolism, e.g. ***cortisol, cortisone*** and ***corticosterone.***
2. ***Mineralo-corticoids:*** Mineralo-corticoids are those which primarily affect the reabsorption of Na^+ and excretion of K^+ (Mineral metabolism) and distribution of water in tissues, e.g. ***Aldosterone*** (chief mineralocorticoid). Others are ***corticosterone, 11-deoxycortisol*** and ***11-deoxycorticosterone (DOC).***

Structures of corticosteroids

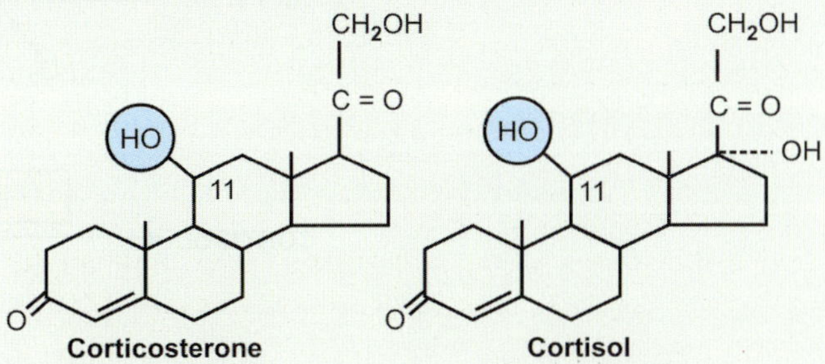

Corticosterone (Compound B) **Cortisol (Compound F)**

SECTION FOUR

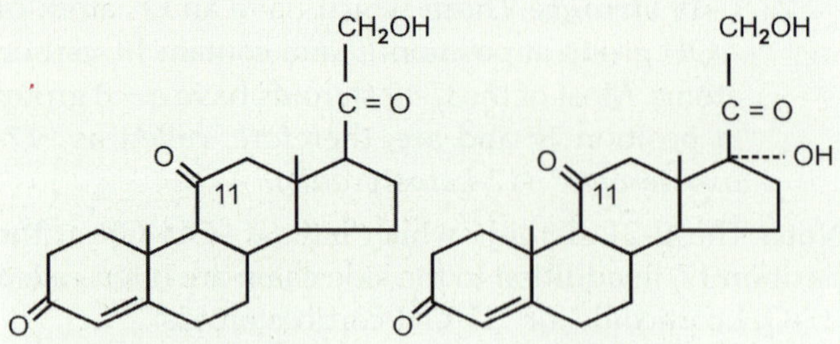

11–dehydro corticosterone (Compound A)   Cortisone (Compound E)

3. *Cortical sex hormones (Androgens and estrogens):* Primarily affect secondary sex characters.

Relation of Structure with Functions:

1. Three structural features are essential for all known biological actions of the natural C_{21} adrenocortical hormones:
   - A double bond at C_4 and C_5
   - A ketonic group (C = O) at C_3 and
   - A ketonic group (C = O) at C_{20}.
2. Certain additional structural features have a profound effect upon the biological activity of these compounds:
   - ***An –OH group at C-21 enhances Na-retention*** and ***is required for activity in carbohydrate metabolism.***
   - The presence of 'O' either as –OH GR or as = O group, i.e. hydroxyl or ketonic group of C_{11} is necessary for carbohydrate activity and decreases Na^+ retention.
   - ***An –OH GR at C_{17} increases carbohydrate activity.***
   - ***A –CHO group at C_{18} necessary for mineralo-corticoid activity.***

GLUCOCORTICOIDS

Biosynthesis of Glucocorticoids: (Fig. 33.9).

I. *Common Pathway for all Corticosteroids:*

Corticosteroids are synthesised by a common pathway from cholesterol in the adrenal cortex.

In all the three zones of adrenal cortex:

- ***Cholesterol is first changed to form pregnenolone (common pathway).*** For this free cholesterol is released in the cytosol from cholesteryl esters of cytoplasmic lipid droplets and transferred into mitochondria. An enzyme called ***"cytochrome-P-450-sidechain cleavage"*** enzyme (P_{450} sce) present in inner mitochondrial membrane hydroxylates cholesterol at C_{22} and C_{20} (also called ***"20, 22-desmolase"***) and then cleaves the side chain to form ***pregnenolone*** and ***isocaproic aldehyde.*** The enzyme requires molecular O_2 and NADPH like all mono-oxygenases and also require FAD containing Fp, an Fe_2S_2 protein (called ***adreno-doxin***).

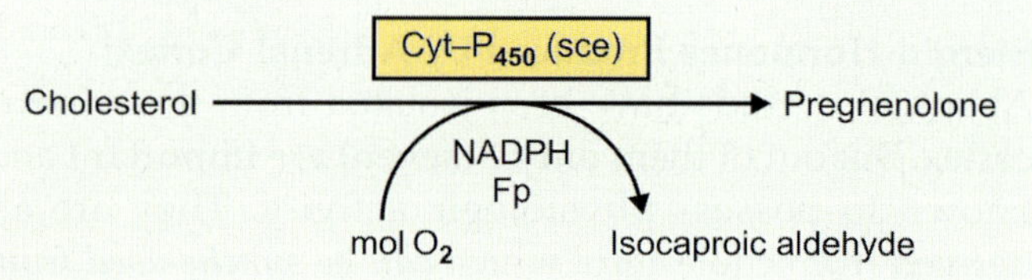

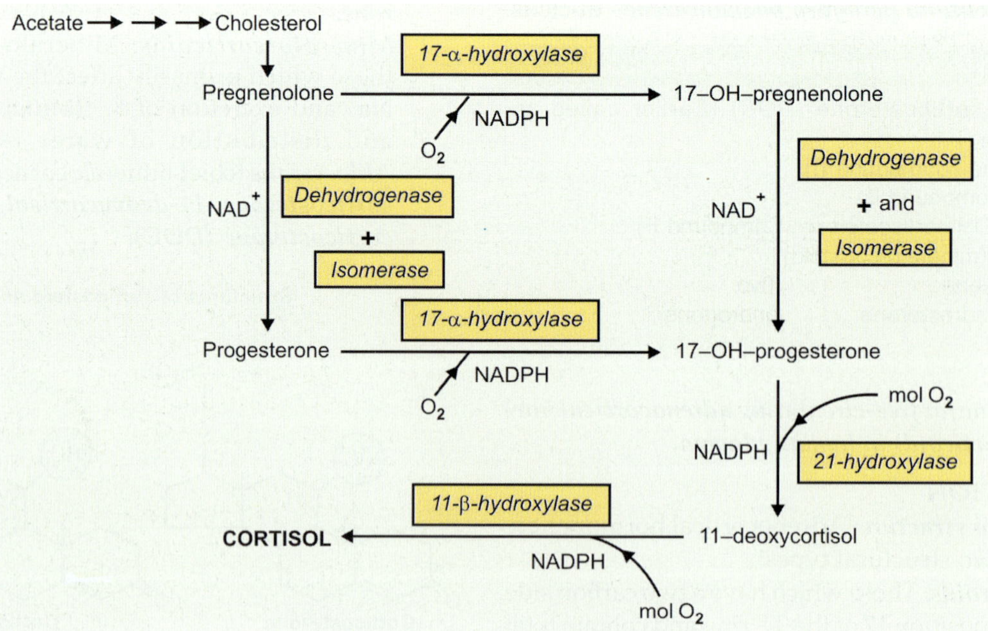

Fig. 33.9: Synthesis of glucocorticoids

II. Glucocorticoid synthesis

Glucocorticoids, as mentioned above, are synthesised in zona fasciculata cells. The steps are:

- ***Pregnenolone to 17-OH pregnenolone:*** Pregnenolone is transferred to smooth endoplasmic reticulum (ER), where it is converted to 17-OH-pregnenolone catalysed by the enzyme ***17-α-hydroxylase.***
- ***Conversion of 17-OH pregnenolone to 17-OH progesterone:*** This is achieved by two enzymes, one is NAD$^+$ dependent ***3 β-OH-steroid dehydrogenase*** and the other is ***$\Delta^{4,5}$-isomerase.*** Alternatively, the same pregnenolone may be first converted to **'progesterone'** by the action of the two enzymes *dehydrogenase* and *isomerase* and it is acted upon by the enzyme *17-α-hydroxylase* to form 17-OH progesterone.
- ***17-OH-progesterone to 11-deoxycortisol:*** Catalysed by the enzyme *21-hydroxylase* present in endoplasmic reticulum (ER).
- Finally 11-deoxycortisol is transferred to mitochondrion where it is acted by the enzyme *11-β-hydroxylase* and is converted to **'cortisol'.**

Note:

- All the three hydroxylases are mono-oxygenases containing Cyt-P_{450} and they require NADPH and molecular O_2 for their action.
- There are evidences that pregnenolone/and/or progesterone can be synthesised from "acetate" by a pathway other than through cholesterol, possibly from 24-dehydrocholesterol.

ACTION OF ACTH ON CORTISOL FORMATION

ACTH stimulates the synthesis and secretion of glucocorticoids. It acts in several ways:

- ***Increases the availability of "free cholesterol" in fasciculata cells.*** This is achieved in two ways:
  - Through cyclic-AMP, activates the enzyme *cholesteryl esterase,* which hydrolyses cholesterol esters and increases free cholesterol in cells.
  - ***Increases transfer of free cholesterol from plasma lipoproteins into fasciculata cells,*** probably by increasing "lipo-protein" receptors on plasma membrane of fasciculata cells.
- ***ACTH increases the conversion of cholesterol to pregnenolone, the "rate-limiting" step.***
- ACTH also stimulates the HMP-shunt pathway by increasing the activity of *G-6-PD and phosphogluconate dehydrogenase*. So that more NADPH is provided which is required for hydroxylation reactions.
- ACTH also increases the binding of cholesterol to mitochondrial Cyt-P_{450} necessary for hydroxylation reactions.

MECHANISM OF ACTION

All of the steroids act primarily at the level of cell nucleus ***('nuclear' action)*** to increase m-RNA synthesis and increased protein synthesis.

- The first step occurs within minutes, which involves the binding of the steroids to a corresponding **specific receptor protein** present in cytosol.
- Glucocorticoids pass into target cells through plasma membrane and binds to specific ***glucocorticoid receptor protein*** present in **cytosol.** The receptors occur in a wide variety of target tissues, viz. liver, muscles, adipose tissue, lymphoid tissue, skin, bone, fibroblasts, etc.

Types of receptors

In humans, there are **two types of receptor proteins:**

1. **α form:** Containing approx 777 amino acids
2. **β form:** Having 742 amino acids.

Both differ in amino acid sequence in the C- terminal end. The **receptor molecule has three distinct domains:**

i. A ***steroid binding domain*** near c-terminus
ii. A ***DNA binding domain*** near the middle of the molecule in c-terminal half, and
iii. A ***transcription-activating domain*** near the N-terminal side.

A heat-shock protein, **hap 90**, binds to the receptor in the absence of hormone and prevents folding into the active conformation of the receptor protein.

Glucocorticoids binds to the specific receptor in cytosol to steroid-binding site. This binding causes dissociation of the "hsp 90" stabiliser and permits conversion to the active configuration.

The steroid-receptor complex enters the nucleus, and bind by DNA-binding site to the **Hormone responsive element (HRE)** of specific nuclear genes. This modulates the transcription rate of those genes, leading to increased synthesis of many proteins and enzymes and also to decreased synthesis of some proteins like corticotrophin **(Fig. 33.10)**.

Note: The levels of specific 'receptors' vary in different cells and can decrease in clinical states causing decreased sensitivity to steroids, e.g. steroids "resistance" occurring in fibroblasts and lymphoid cells.

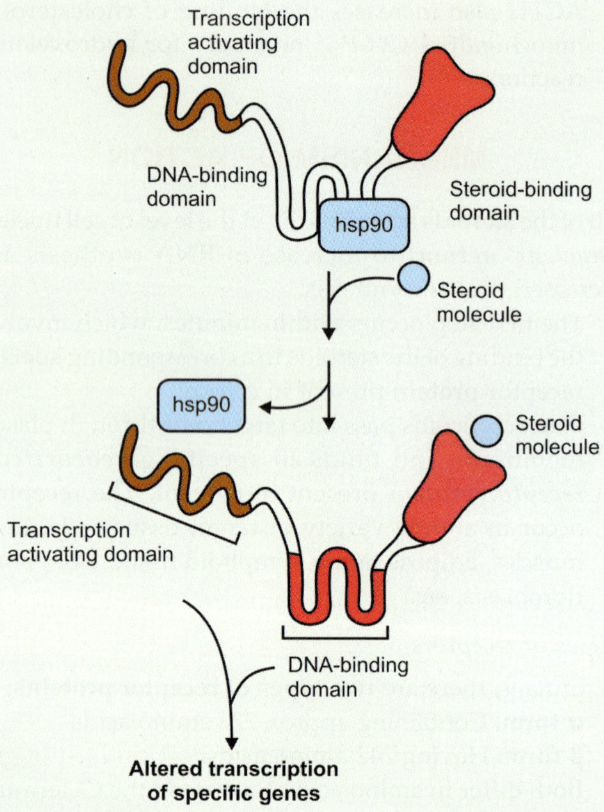

Fig. 33.10: Mechanism of glucocorticoid action. The glucocorticoid receptor polypeptide is schematically depicted as a protein with three distinct domains. A heat-shock protein, hsp90, binds to the receptor in the absence of hormone and prevents folding into the active conformation of the receptor. Binding of a hormone ligand causes dissociation of the hsp90 stabiliser and permits conversion to the active configuration

CLINICAL ASPECT

Steroids in disease states:

- Hepatic inactivation of corticoids decline during prolonged malnutrition and in hepatic diseases like cirrhosis liver.
- Decreased excretion may also occur in renal insufficiency. Under both the states as above → the levels of corticosteroids in blood is raised ↑ .
- Similarly, in chronic liver diseases like cirrhosis liver and congestive cardiac failure, liver does not completely inactivate "mineralocorticoids", ***leading to excessive "salt retention"***, which account for oedema and ascites found in certain stages of these diseases.

ADRENAL ANDROGENS

Main hormones are:

- **Dehydroepiandrosterone (DHEA)**
- **DHEA-SO_4**
- **Δ^4-androstenedione**
- **11-β-OH-androstenedione.**

Note:

- DHEA and DHEA-SO_4 are freely interconvertible
- Δ^4-androstenedione is derived from DHEA.
- **Androstenedione is the precursor of testosterone,** more potent male hormone.
- A proportion of androstenedione may be converted to 11-β-OH androstenedione, a much weaker androgen.

In liver: Some of DHEA and all Δ^4-androstenedione are converted to:

- Androsterone and its isomer and
- **Etiocholanolone,** which is conjugated to form 'glucuronides' or 'sulphates' and excreted in urine. 11-β-OH-androstenedione is also converted to 11-oxygenated derivatives of "androsterone" and 'etiocholanolone'. Both of which are conjugated to 'sulfates' or 'glucuronides' water soluble compounds and excreted in urine as 17-oxosteroids.

METABOLIC ROLE OF GLUCOCORTICOIDS

1. Metabolic Actions

Points to note

- In general, glucocorticoids have ***anti-insulin effect.***
- ***Glucocorticoids are catabolic to peripheral tissues and anabolic to liver.***

(a) Effects on carbohydrate metabolism: Over-all effect increases blood glucose ↑ level ***(Hyperglycaemia).***

Mechanism of hyperglycaemia:

1. *Decreases glucose uptake* ↓ and utilisation in muscles, in adipocytes and lymphoid cells by inhibiting the membrane transport of glucose into these cells.
2. *Enhancing gluconeogenesis ↑ in liver.*
   - Induces the synthesis of key gluconeogenic enzymes such as *pyruvate carboxylase, PEP carboxykinase, fructose 1-6-bi-phosphatase* and also *Glucose-6-phosphatase*.
   - By making available more of substrates required for gluconeogenesis. This is achieved by:
     - Increasing protein catabolism in extrahepatic tissues.
     - Decreasing incorporation of amino acids in protein in peripheral tissues.
     - Also increasing synthesis of some key enzymes required for amino acid catabolism like, *alanine transaminase, tyrosine transaminase, tryptophan pyrrolase, etc.*
3. *Decreases glycolysis ↓ in peripheral tissues:*
   - **In liver: Glucocorticoids are anabolic.** It increases the glycogen store ↑ in liver. This is due to:
     - *Increase in gluconeogenesis ↑ from amino acids and glycerol.*

– Activates *protein-phosphatase-1*, which dephosphorylates and activates *glycogen synthase*, the key enzyme for glycogen synthesis.
– Stimulates the synthesis of *glycogen synthase* also.

(b) Effects on lipid metabolism: Net effect increases FFA ↑ in plasma and also glycerol. Glycerol is utilised for gluconeogenesis in liver.

In adipocytes:

- Glucocorticoids ***increases 'lipolysis' and liberates FFA*** and glycerol by activating hormone sensitive *TG lipase*.
- As glucocorticoids decrease the uptake of glucose in adipose tissue, there will be reduction in α-glycero-P, as a result esterification suffers, hence net flow of FFA in plasma increases.

(c) Effects on protein metabolism:

- In peripheral extrahepatic tissues, ***cortisol is catabolic and increases protein breakdown, leading to increased 'amino acids' availability in plasma.***
  Reasons of increased catabolism:
  – Enhances synthesis of key enzymes of amino acid catabolism like *transaminases, tyrosine transaminase, tryptophan pyrrolase*, etc.
  – Also there is decreased incorporation of amino acids in protein molecule.
- **In liver:** ***Cortisol is anabolic, it increases protein synthesis ↑. It increases:***
  – ***Hepatic uptake of amino acids ↑.***
  – ***Incorporation of amino acids into ribosomal proteins.***
  – Increased m-RNA formation and synthesis of proteins including plasma proteins.
  – In liver, cortisol also enhances urea synthesis ↑ from amino acids. There is increased synthesis of enzymes necessary for urea cycle, e.g. *arginino succinate synthetase, arginase*, etc.

Over-all effect on protein metabolism by cortisol is "Negative Nitrogen Balance".

Summary

1. **Action on 'Peripheral' tissues** like muscles, adipose tissue and lymphoid tissue is ***'catabolic'*** ("spares" glucose)
   - Glucose uptake ↓ and glycolysis ↓
   - Lipolysis ↑, FFA in plasma ↑, glycerol in plasma ↑, Esterification, i.e. TG formation ↓, α-glycero-p ↓.
   - Protein synthesis ↓, protein breakdown ↑, plasma amino acids ↑.
2. **Action on liver:** It is ***anabolic***
   - Gluconeogenesis ↑, from amino acids and glycerol. Glycogen in liver ↑ increased.
   - Protein synthesis in liver cells ↑ enhanced.

II. Other Actions

1. ***Permissive Action:*** Small amount of glucocorticoid is required for a number of metabolic reactions to occur:
   - Required for adipokinetic activity of GH.
   - Required for calorigenic action of glucagon and catecholamines.
2. ***Anti-inflammatory Action:*** Normal cortisol level does not affect; but therapeutic doses exert an anti-inflammatory effect.

Mechanism: **Three** basic mechanisms by which glucocorticoids exerts anti-inflammatory effects are:

i. ***Action on Lysosomes:*** Stabilises cell membrane of lysosomes and thus block the release of lysosomal hydrolases.
ii. ***Action on Kinin Formation:*** Prevents formation of bradykinin which is produced by action of Kallikrein, a proteolytic enzyme on α-globulin.
iii. ***Action on Capillaries:*** Decreases permeability of capillary walls and prevent protein leakage.

Other effects which help in anti-inflammatory action are:

- Decreases the formation of PGs, $PG\text{-}I_2$, Tx and leukotrienes by inhibiting *phospholipase* A_2.
- Prevents the release of histamine from mast cells.
- Reduces fibroblastic proliferation and collagen synthesis.
- Inhibits the release of ***"interleukin-I"*** from granulocytes.
- Decreases the number of circulating lymphocytes (lymphopenia), eosinophils and monocytes.

CLINICAL ASPECTS

Based on above, steroids have been used as anti-inflammatory drugs in treatment of rheumatoid arthritis, rheumatic fever and acute glomerulonephritis.

3. ***Immuno-suppressive Effect:*** Cortisol decreases immune response associated with infections and allergic states. Also used for purpose of repressing antibody formation.
   Mechanism: High doses of glucocorticoids decrease gene transcription and protein synthesis and antigen-induced proliferation of lymphocytes in thymus, spleen and lymph glands and also shifts lymphocytes from circulation to lymphoid tissues. The lymphoid tissue mass, circulating lymphocytes in blood, T-lymphocyte mediated cellular immunity are all reduced, ***causing suppression of immune response.***

CLINICAL ASPECT

- Based on above, glucocorticoids have been ***used in organ transplantation*** as it prevents graft rejection.
- As it is anti-allergic also ***used for treatment of bronchial asthma and status asthmaticus.***

4. ***Effect on Exocrine Secretion:*** Chronic and prolonged treatment with glucocorticoids causes:
   - Increased secretion of HCl ↑ and
   - Increased secretion of pepsinogen ↑ in stomach
   - Also increases trypsinogen ↑ secretion in pancreatic juice.

CLINICAL ASPECT

In prolonged treatment with glucocorticoids, the patient may develop GI ulcers.

5. ***Effect on Bones:*** Glucocorticoids reduce the osteoid matrix of bone, thus favouring ***osteoporosis*** and there may be excessive loss of calcium from the body.

CLINICAL ASPECT

Osteoporosis is a major complication of prolonged glucocorticoid therapy.

6. ***Haematological Changes:*** Large doses of glucocorticoids and in hypertrophy of adrenal cortex, the hormone brings about destruction of lymphocytes and also shift of lymphocytes to lymphoid tissues producing ***"Lymphopaenia". Also there is reduction in circulating monocytes and eosinophils.*** Hypofunctioning of adrenal cortex, on the other hand, results in ***'Lymphocytosis'***.

MINERALO-CORTICOIDS

Mineralo-corticoids are C_{21} steroids, which influence mainly the metabolism of Na^+ and K^+. ***The chief mineralo-corticoid is aldosterone.*** It is ***produced by zona glomerulosa of the adrenal cortex. Structurally, it bears a –OH GR. at C_{11} and an aldehyde (–CHO) group at C_{18}.***

Other corticosteroids which have mineralo-corticoid activity are:
- ***Corticosterone***
- ***11-deoxycortisol***
- ***11-deoxycorticosterone (DOC):*** It is secreted in minute quantities and has almost the same effects as aldosterone, but a potency only 1/30th that of aldosterone.

Biosynthesis

Mineralocorticoids are synthesised in Zona glomerulosa cells only. They cannot be synthesised in other two layers of adrenal cortex. Only zonaglomerulosa cells have the enzymes ***18-hydroxylase*** and ***18-hydroxysteroid dehydrogenase, which are lacking in other layers.***

Steps in Synthesis **(Fig. 33.11).**

1. Cholesterol is converted to pregnenolone. Pregnenolone is then converted to progesterone in smooth ER catalysed by the enzymes *3-β-OH-steroid dehydrogenase* and *$\Delta^{4,5}$-isomerase.*
2. Progesterone is then directly hydroxylated by the enzyme *21-hydroxylase* and forms 11-deoxy-corticosterone (DOC).

   Note: Enzyme *17-α-hydroxylase* is absent in zona glomerulosa.
3. 11-deoxycorticosterone is next translocated to mitochondrion, where it is converted to corticosterone, the reaction is catalysed by the enzyme *11-β-hydroxylase.*
4. In the next step, by the enzyme *18-hydroxylase,* corticosterone is converted to 18-OH-corticosterone, which

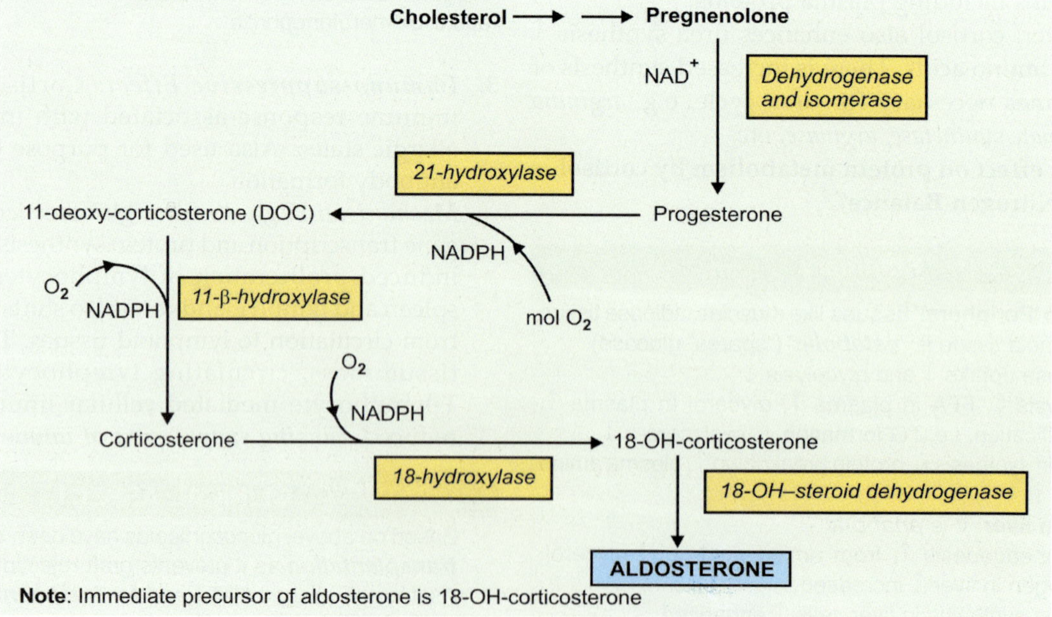

Fig. 33.11: Biosynthesis of aldosterone

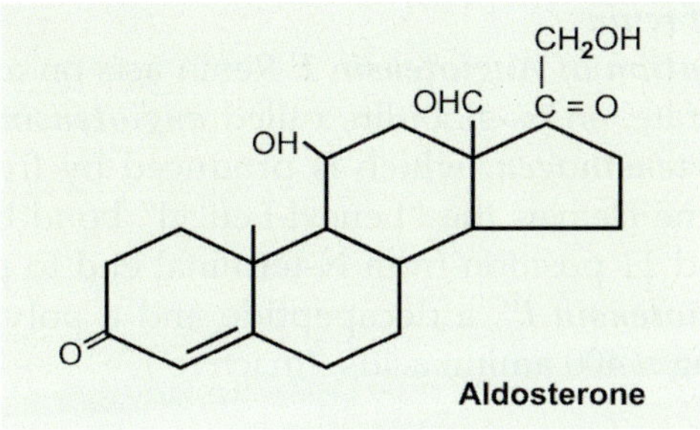

is then acted upon by a dehydrogenase, ***18-hydroxy steroid dehydrogenase to form aldosterone.*** Both the reactions take place in mitochondria, the dehydrogenase oxidises the 18-CH_2OH GR to an aldehyde (–CHO) group.

5. *In Circulation:* ***Aldosterone exists both in 'aldehyde' and 'hemi-acetal' form*** as shown in the box below. It is believed that the hormone exists in solution in hemi-acetal form.

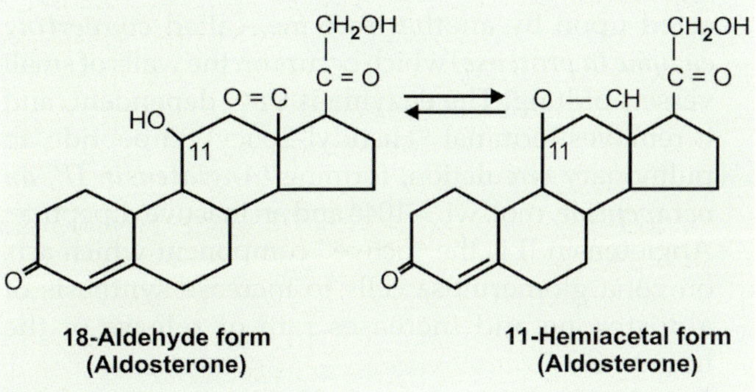

Mechanism of Action

Mineralocorticoids enter the target cells through the plasma membranes and binds to a specific protein present in cytosol, and nucleoplasm, called ***'Mineralocorticoid receptors'.*** Both high-affinity and low-affinity mineralocorticoid receptors have been described. They are present in epithelial cells of renal distal tubular cells and collecting ducts and also in gastrointestinal mucosa, salivary gland ducts and sweat ducts. The ***steroid-receptor complex*** then enters the nucleus and binds to ***hormone responsive element (HRE)*** of specific nuclear genes and increases the transcription rates of genes. Thus, aldosterone initiates an increase in m-RNA synthesis ↑, at the level of transcription of DNA. The induced m-RNA stimulates protein synthesis at the ribosomal level.

METABOLIC ROLE OF ALDOSTERONE

(a) Renal Effects of Aldosterone:

1. *Effect on Tubular Reabsorption of Sodium:* By far the most important effect of aldosterone and other mineralocorticoids is ***to increase the rate of tubular reabsorption of Na***. Sodium is reabsorbed from the renal tubules along their entire extent. Aldosterone has a specially potent effect in the distal tubule, collecting tubule and at least a part of Loop of Henle.

Note: Total lack of aldosterone secretion can cause loss of as much as 12 grams of Na in the urine in a day, an amount equal to 1/7th of all the sodium in the body.

2. *Effect on tubular reabsorption of chlorides:* Aldosterone also ***increases the reabsorption of Cl^- ions from the tubules***. This probably occurs secondarily to the increased Na reabsorption. Absorption of +vely charged Na ions causes an electrical potential gradient to develop between the lumen and outside of the tubules with positivity on the out side. This positivity in turn attracts –vely charged diffusible anions through the membrane. Since Cl^- ions are by far the most prevalent anions in the tubular fluids, the absorption of Cl^- increases.
3. *Increased renal excretion of K^+:* As aldosterone causes increased tubular reabsorption of Na^+, at the same time ***it also increases loss of K^+ in the urine by the renal distal tubules and collecting ducts***. This may result from the elimination of K^+ in exchange of the reabsorbed Na^+.

CLINICAL SIGNIFICANCE

Hypokalaemia and muscle paralysis: The loss of K^+ in urine decreases K^+ in ECF resulting to ***'hypokalaemia'***. Thus at the same time that Na^+ and Cl^- ions become increased in ECF, there will be gross decrease in K^+ ions. The low K^+ concentration sometimes ***leads to muscle paralysis***; this is caused by hyperpolarisation of the nerve and muscle fibre membrane which prevents transmission of action potentials.

4. *Effect on acid-base balance (Alkalosis):* A large proportion of Na^+ reabsorption from the tubules results from an exchange reaction in which H^+ ions are secreted into the tubules to take the place of Na^+ that is reabsorbed. Hence, when the rate of Na^+ reabsorption is enhanced, in response to aldosterone, the H^+ concentration in the body fluids is reduced. For each Na^+ ion reabsorption by H^+ exchange, one HCO_3^- ion enters the ECF which shifts the reaction to alkaline side. Thus, ***increased secretion of aldosterone promotes alkalosis,*** whereas decreased secretion produces acidosis.

(b) Effects of Aldosterone on Fluid Volume

1. *Effect on ECF volume:* Mineralocorticoids greatly increase the quantities of Na^+, Cl^-, and HCO_3^- ions in the ECF, increasing the electrolyte concentration in ECF. This in turn increase water reabsorption from the tubules by:

SECTION FOUR

- Stimulating the hypothalamic ADH system, and
- Creating an osmotic gradient across the tubular membrane. When the electrolytes are absorbed, carries water through the membrane in the wake of electrolytes absorption.
- Also increased electrolyte concentration of ECF causes thirst, thereby making the person to drink excessive amount of water.

Hence, ***the final result is an increase in ECF volume ↑, sometimes enough to cause generalised oedema.***

Note: ECF volume must increase about 30 per cent before frank oedema appears.

2. ***Effect on blood volume:*** The plasma volume ↑ increases almost proportionally during the early part of increase in ECF volume. Hence, one of the effects of increased aldosterone secretion is a mild to moderate increase in blood volume.

(c) **Effects of Aldosterone on Sweat Glands, Salivary Glands and Gastric Mucosa:** The mineralocorticoids have almost the same effect on the sweat glands, salivary glands, intestinal glands as on the renal tubules, greatly reducing the loss of Na^+ and Cl^- in the glandular secretions. The effect on the sweat glands is important to conserve body salt in hot environments, whereas, the effect on intestinal glands is probably of importance to prevent salt loss in the gastrointestinal excretory products.

HEPATIC ALDOSTERONE

Recently it has been reported that ***large amounts of aldosterone are produced from 'androstenedione' in the liver.*** But this may have little peripheral biologic activity since it could be inactivated by reduction and conjugation before leaving the liver. Hepatic aldosterone would, however, contribute to the conjugates of this hormone measured in the urine.

RENIN-ANGIOTENSIN SYSTEM

JG Cells

Afferent arteriole of nephron ***(vascular pole?)*** show cytoplasmic granules which contain an enzyme called **Renin** (granules are secretory vesicles for renin). ***A fall in sodium concentration, hypovolaemia, hypotension and a fall in intracellular Ca^{++} stimulate release of "renin" from JG cells to blood.*** Bradykinin and glucagon also stimulate release of renin.

Chemistry

Renin is a proteolytic enzyme, Mol. wt. = 35,000. Recently, renin isoenzymes or renin like enzymes have been detected in brain, placenta, sub-maxillary duct and at the junction of uterine endometrium and myometrium. Exact function not known.

Action of renin

- ***Formation of Angiotensin I:*** Renin acts on a plasma substrate, an α_2-globulin, called ***angiotensinogen*** or ***hypertensinogen***, which is produced by liver. The enzyme cleaves the "Leucyl-Leucyl" bond between 10 and 11 position from N-terminal end to produce ***"Angiotensin I"***, a decapeptide and a polypeptide having > 400 amino acids ('inactive').

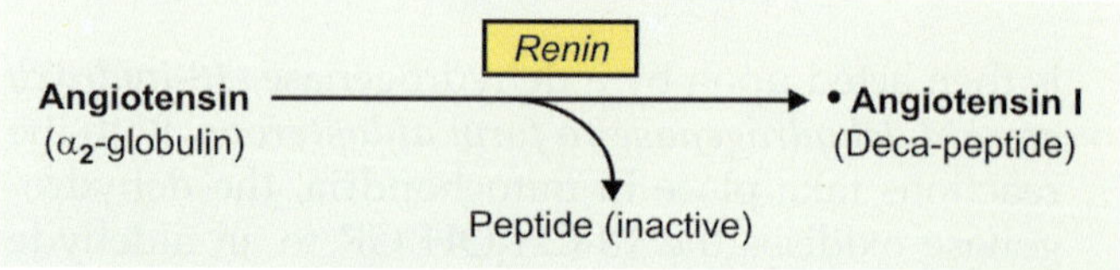

This is the ***rate-limiting*** step. Cortisol and β-estradiol enhances this reaction, probably by increasing hepatic synthesis of "angio-tensinogen".

- ***Formation of Angiotensin II: Angiotensin I, a decapeptide,*** having a mol. wt. of 1296, while circulating, is acted upon by another enzyme, called ***converting enzyme (a protease)*** which occurs on the walls of small vessels of lung. The enzyme is Ca^{++} dependent, and it removes terminal "Histidyl-Leucyl" dipeptide, in pulmonary circulation, forming ***'Angiotensin II', an octapeptide,*** mol. wt. = 1046 and an inactive dipeptide. Angiotensin II is the "active" component which acts on zona glomerulosa cells to increase synthesis of aldosterone and increases rate of release of the hormone.

  Inactivation of Angiotensin II: An enzyme *angiotensinase*, an amino-peptidase, present in kidney, intestine and blood inactivates angiotensin II by hydrolysis.
- ***Angiotensin III:*** Recently in rats, ***heptapeptide angiotensin III*** has been isolated. It is claimed to be also present in humans. Both heptapeptide (angiotensin III) and octapeptide (angiotensin II) are claimed to be equipotent in stimulating aldosterone secretion (Fig. 33.12). Aldosterone can 'inhibit' the enzyme 'renin' by "feed-back" inhibition so that angiotensin II formation is decreased.

Inactivation of Renin: In addition to "feed-back" inhibition by aldosterone:

- Renin is also destroyed by a cephalin derivative in plasma and
- Also inhibited by a lysophospholipid, liberated by the action of *phospholipase A_2*.

Actions of Angiotensin II

Principal action is angiotensin II stimulates aldosterone synthesis in zona glomerulosa cells and increases rate of secretion of aldosterone.

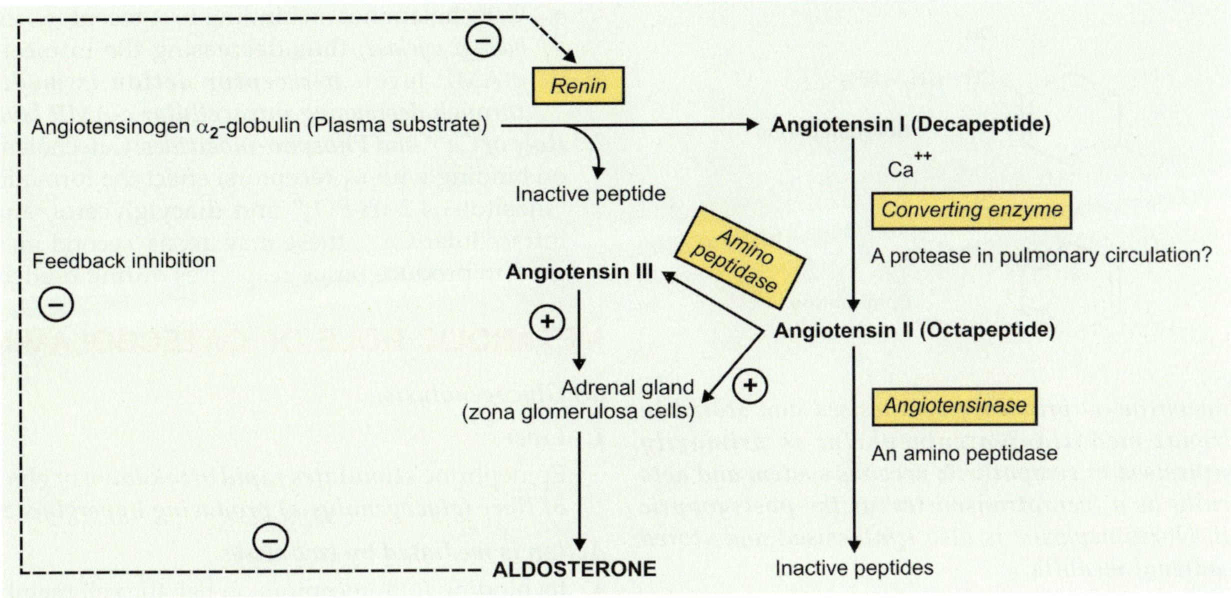

Fig. 33.12: Formation of angiotensin II

Mechanism of action and effects: Angiotensin II binds with specific "receptor" on membrane of zona glomerulosa cells and

- Enhances cytosolic concentration of Ca^{++} ions ↑ in cells, and
- Formation of "inositol-1, 4, 5-triphosphate".

The above act as "second messenger" and inturn:

(a) Enhances conversion of cholesterol to pregnenolone

(b) And corticosterone to aldosterone by increasing the activity of *18-hydroxylase*

Aldosterone thus formed and secreted:

- Increases the active tubular reabsorption of Na^+, and
- Consequently, 'Passive' reabsorption of Cl^- and water.

Renal retention of water restores the falling ECF volume and helps in long-term increase of arterial BP.

Other Actions of Angiotensin II

- Stimulates contraction of smooth muscles on the walls of alimentary canal, uterus, arteries/arterioles, but unlike catecholamines, it cannot relax the sphincter muscles. Probably, it reacts with a specific "receptor" in the cell membrane of smooth muscles, leading to a rise in intracellular Ca^{++}, which then promotes contraction of smooth muscle fibres.
- May raise arterial Bl. Pr. ↑ by causing arteriolar constriction and is thought to be responsible for hypertension associated with ischaemic kidney.
- Also stimulates VM centre in the hindbrain leading to reflex rise in cardiac output, reflex arteriolar constriction and a rise in BP. ↑
- May stimulate vasopressin ↑ secretion, indirectly causing water retention.
- May also stimulate synthesis and release of PGs in renal medulla.

On the other hand, PGs, particularly PGE_2 may act against angiotensin II and reduce the renal vasoconstrictor and antidiuretic effect of angiotensin II.

ADRENAL MEDULLARY HORMONES

Chemistry

1. Two biologically active compounds have been isolated from the adrenal medulla and synthesised. They are:
   - ***Epinephrine (Adrenaline or Adrenine)***
   - ***Norepinephrine (Nor-adrenaline or Arterenol)***
2. The naturally occurring forms are laevorotatory, the synthetic are racemic, the former being almost twice as active as the latter.
3. The above two hormones are called **catecholamines** and are closely related to tyrosine and ***synthesised in body from tyrosine.***
4. Their structures are shown ahead in the box.

Epinephrine differs from tyrosine in following respects:

- Contains an additional phenolic –OH GR in meta-position to benzene ring.
- Additional –OH GR attached to β carbon of the sidechain
- Has no –COOH GR
- Has a $-CH_3$ GR attached to N-atom in side chain.

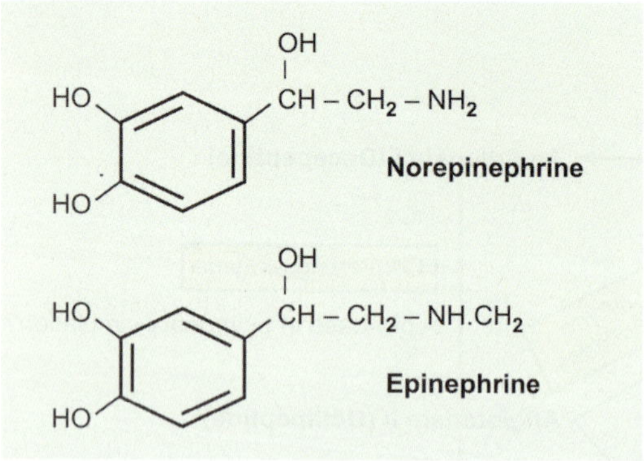

5. ***Epinephrine is primarily synthesised and stored in adrenal medulla. Norepinephrine is primarily synthesised in sympathetic nervous system and acts locally as a 'neurotransmitter' at the postsynaptic cell. Norepinephrine is also synthesised and stored in adrenal medulla.***

Biosynthesis

In adrenal pheochromacytes and neuronal cells, the synthesis of catecholamines is essentially same. Both are produced from the amino acid tyrosine. (For details of synthesis refer to Metabolic Role of Tyrosine in Protein Metabolism).

Storage

- Epinephrine, nor-epinephrine and dopamine are stored in the form of granules, 0.1 to 0.5 μ in diameter, in the pheochromocytes of adrenal medulla.
- Norepinephrine only occurs in adrenergic nerve terminals as granules/or vesicles 400 to 500 Å in diameter, and some is probably free in the cytoplasm.

Both the hormones are stored in the granules in the adrenal medulla and in adrenergic neurons ***as a complex containing ATP in the ratio, about 4 molecules of hormones: one molecule of ATP*** and in combination with several incompletely characterised proteins like *chromogenin A* and *chromomembrin B*.

CLINICAL IMPORTANCE

As catecholamines cannot penetrate blood-brain barrier, the norepinephrine in the brain must be synthesised within that tissue. **'L-DOPA', the precursors of catecholamines does penetrate the barrier. It is, hence, *used to increase brain catecholamine synthesis in Parkinson's disease.***

Mechanism of Action

1. *Role of Cyclic AMP:* Catecholamines on binding to β-receptors (β_1 and β_2) activate *adenyl cyclase* which increases cyclic AMP level in the cells. Increased cyclic AMP activates c-AMP-dependent *protein kinases* which phosphorylates specific protein/or enzymes and activate/inactivate them. ***β-receptor action is mediated through increased intracellular c-AMP level.***
    - Catecholamines on binding to α-receptors, inhibit *adenyl cyclase*, thus decreasing the intracellular c-AMP level. ***α-receptor action is mediated through decreasing intracellular c-AMP level.***
2. *Role of Ca^{++} and Phospho-inositides:* Catecholamines on binding with α_1 receptors, effect the formation of "inositol-1,4,5-tri-PO_4" and diacylglycerol, and/or intracellular Ca^{++}, these may act as 'second messengers' to produce tissue responses during α-effects.

METABOLIC ROLE OF CATECHOLAMINES

(a) Glycogenolysis:

1. *Liver*

   Epinephrine ***stimulates rapid breakdown of glycogen of liver (glycogenolysis) producing hyperglycaemia.***

Action is mediated by two ways

- Its binding to β_2 receptors on hepatic cell membrane by increasing cyclic AMP level.
- Also exerts its effect by binding to α_1 receptors on hepatic cell membrane, which increases intracellular Ca^{++} level which acts as second messenger.

The effect of cyclic AMP increase in hepatic cell is similar to glucagon. But measurements of cyclic AMP levels after epinephrine and/or glucagon indicate that ***glucagon is by far the more active hormone in liver tissue.*** Norepinephrine has very little effect on blood glucose.

2. *Muscle: In muscle,* epinephrine also causes breakdown of glycogen (glycogenolysis) by increasing cyclic AMP level (β-effect), but in this tissue it is more active than glucagon. ***Glucagon has very little effect or no effect due to lack of specific receptors.*** In exercising muscle, this can result in increased LA formation, which passes to blood.
3. *Heart muscle:* Increases in cyclic AMP after epinephrine administration is seen in 2 to 4 seconds, the effect of epinephrine on cardiac output (ionotropic effect) is seen shortly afterwards, whereas activation of phosphorylase is not detectable for 45 seconds.
    - **Heart Glycogen:** *In vivo,* actually epinephrine action can result in an increase in heart glycogen. This is probably secondary to the hormone action on adipose tissue causing adipolysis and increased FFA. ***Fatty acids are utilised as fuel. Increased glycogen is due to gluconeogenesis, the glucose is not utilised for energy and diverted to glycogen formation.***

(b) Lipolytic Action: Both epinephrine and norepinephrine ***increases breakdown of TG in adipose tissue*** by increasing cyclic-AMP level (**β_1 effect**). Net effect of lipolysis is rapid release of FFA and glycerol from adipose tissue to blood.

(c) ***Gluconeogenic Action:*** Epinephrine increases hepatic gluconeogenesis **(β_2 effect).** Epinephrine increases cyclic AMP which induces the synthesis of key enzymes *pyruvate carboxylase, PEP carboxy kinase,* and *fructose-1-6-bi-phosphatase*. Increased FFA level in blood produced by lipolytic action can also activate hepatic gluconeogenesis.

(d) ***Action on Glycolysis:*** Epinephrine increases blood LA level by promoting muscle glycolysis, nor-epinephrine has very little effect on blood lactic acid (LA).

(e) ***Action on Insulin Release:*** Epinephrine has a direct inhibitory action on insulin release from β-cells of pancreas (α_2-effect). Thus, in pancreas, the α-adrenergic response to epinephrine predominates, cyclic AMP decreases ↓ and insulin release is inhibited. However, in the presence of an α-blocker, such as "phentolamine" (Regitine), the β-effect predominates and epinephrine causes increased cyclic AMP and increased insulin release.

(f) ***Calorigenic Action:*** Norepinephrine and epinephrine are almost equally potent in their calorigenic action. They produce:

- A prompt rise in the metabolic rate which is independent of the liver, and
- A smaller delayed rise which is abolished by hepatectomy and coincides with rise in blood LA.

The calorigenic action does not occur in the absence of the thyroid and adrenal cortex.

Explanation

- The cause of the initial rise in metabolic rate is not clearly understood. It may be due to cutaneous vaso-constriction which decreases heat loss and leads to a rise in body temperature, or to increased muscular activity or to both.
- The second rise is probably due to oxidation of LA in the liver.

GONADAL HORMONES

The sex hormones or the gonadal hormones are elaborated by the testes, ovary and corpus luteum mostly, and also in small quantities by the placenta and adrenal cortex. They are all steroid compounds related to cholesterol and are synthesised from that precursor. Sex hormones are also related to the adrenal cortical hormones both in the chemical nature and in the common biosynthetic pathway and interconversions.

Types

Sex hormones are of 3 types:

1. **Androgens or male hormones**
2. **Oestrogens or female hormones**
3. **Gestogens or progestational hormones.**

ANDROGENS

Androgens are hormones capable of producing certain characteristic masculinizing effects, i.e they maintain the normal structure and function of the prostate and seminal vesicles and influence the development of secondary male sex characteristics, such as hair distribution and voice.

Chemistry: The naturally occurring androgens in man are:

1. **Testosterone**
2. **Epiandrosterone (3 β-androsterone)**
3. **Androsterone** and
4. **Dehydroepiandrosterone (DHEA).**

All have: –CH_3 GR at C_{10} and C_{13} and contain 19 C atoms.

Structure of testosterone

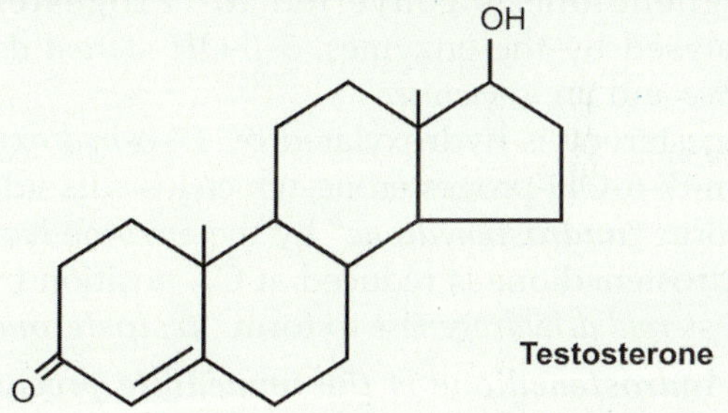

Structure of immediate precursor—androstenedione

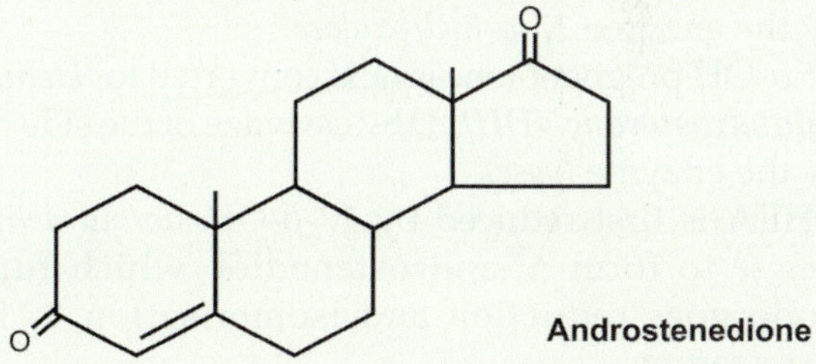

Δ^4-pathway

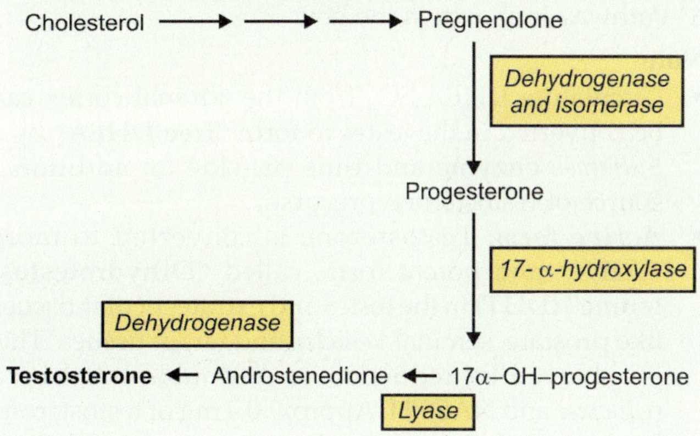

Biosynthesis

Androgens are produced in testes (Leydig cells), adrenal cortex, ovary and placenta. They may be formed from either (1) acetate ('active' acetate) or (2) cholesterol, and ***pregnenolone being an important intermediate.***

Steps

- Cholesterol is converted to pregnenolone in the Leydig cells mitochondria (pathway similar to adrenal cortex). This is the ***"rate limiting" step.***
- Next pregnenolone is translocated to smooth endoplasmic reticulum where testosterone is synthesised.

There are two pathways

1. **Δ^4-Pathway** and
2. **Δ^5-Pathway**

(a) In humans, Δ^4–pathway predominates.

In this pathway:

- Pregnenolone is converted to **"Progesterone"**, catalysed by the enzymes, *3-β-OH steroid dehydrogenase* and an *isomerase.*
- Progesterone is hydroxylated by *17-α-hydroxylase* to form 17-α-OH-progesterone, which loses its side chain to form ***"androstenedione"*** by the enzyme *lyase.*
- Androstenedione is reduced at C_{17} position by *17-β-OH steroid dehydrogenase* to form ***"testosterone".***

Note: ***Androstenedione is the immediate precursor of testosterone.***

Δ^4-pathway is shown in the box.

(b) Δ^5-Pathway: In this pathway:

- Pregnenolone is converted to 17-α-OH-pregnenolone by the enzyme *17-α-hydroxylase*
- 17-α-OH pregnenolone is next converted to ***'Dehydroepiandrosterone' (DHEA)*** by cleavage of the side chain by the enzyme *lyase.*
- DHEA is first reduced by *17-β-OH steroid dehydrogenase* to form Δ^5-androstenediol, which further undergoes reduction and isomerisation to form testosterone.
- Alternatively, DHEA may be converted to 'Androstenedione' by *dehydrogenase* + and *isomerase* and androstenedione is reduced to form testosterone.

Δ^5-Pathway is shown in the box.

Note

- Circulating DHEA-SO_4 from the adrenal cortex can be converted in the testes to form **"free DHEA"** by a *Sulfatase* enzyme and thus provide an additional source of testosterone precursor.
- ***Active form:*** Testosterone is converted to more "active" and potent form, called **"Dihydrotestosterone" (DHT)** in the testes and extratesticular tissues like prostate, seminal vesicles and target tissues. This is achieved by reduction of Δ^5-double bond by a *reductase* and NADPH. Approx. 0.4 mg of testosterone is reduced daily to 'dihydrotestosterone' (DHT) in testes and extratesticular tissues.

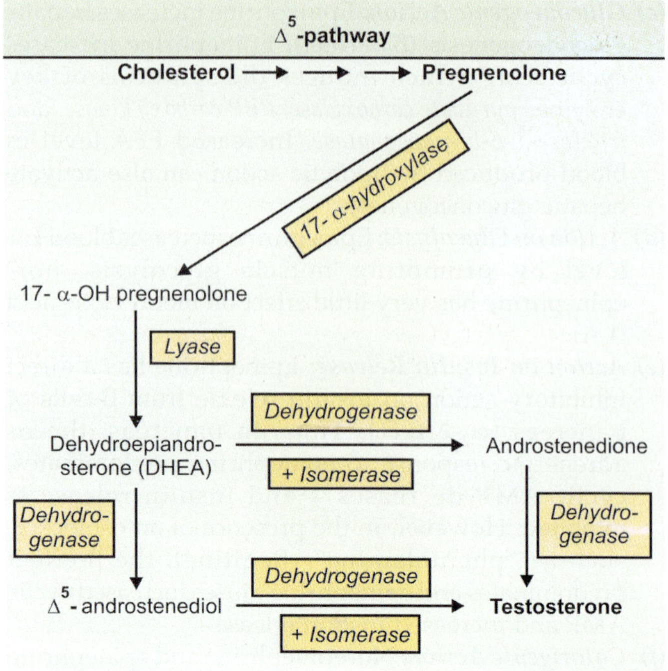

- ***The small amount of testosterone in females, results mainly from peripheral conversion of androstenedione to testosterone by the ovary.***

Mechanism of Action: Free testosterone enters the target cells by simple diffusion or facilitated diffusion. In the cytoplasm of target cells, the testosterone is ***converted to 'active' form dihydrotestosterone.*** There are specific 'androgen receptors' present in the cytosol. Dihydrotestosterone has greater affinity than testosterone for the specific receptor. The hormone is tightly bound to the receptor and ***"hormone-receptor complex"*** then binds to the ***'hormone-responsive element'*** **(HRE)** present with specific nuclear genes. This induces the transcription of those genes leading to increased synthesis of respective proteins which produce cellular effects (Receptor structure is similar to receptor described under mechanism of action of glucocorticoids).

17-Ketosteroids (17-oxo-steroids)

The androgens excreted in urine are classed as 17-ketosteroids (17-oxo-steroids). In the case of females, it gives an idea about the condition of the adrenal cortex and its functions.

- **In males:** 17-ketosteroids arise from testes (1/3 of the total), while the major amount arises from the adrenal cortex (2/3 of total).
- **In females:** The 17-ketosteroids are almost entirely from adrenal cortical origin.
- **Normal value:** In 24 hr excretion of urine, ***normal adult males*** excrete 9 to 24 mg of neutral 17-ketosteroids while ***normal adult females*** excrete 5 to 17 mg.

METABOLIC ROLE

Both testosterone and dihydrotestosterone are protein anabolic and growth promoting hormone.

1. *Protein Metabolism*

   Dominant general metabolic effect is stimulation of protein anabolism. This is reflected in:
   - A decrease in urinary N_2 (urea) ↓ without an increase in blood NPN. Decrease in hepatic *arginine synthetase* activity ↓, the enzyme that catalyses conversion of citrulline to arginine.
   - *Creatine metabolism:* Creatine is virtually absent from the urine of normal men, increases after castration. This increase is abolished by administration of testosterone, owing to increased storage of creatine in the muscles.
   - There is increase in body weight, due chiefly to an increase in skeletal muscle. There is evidence that the associated increase in O_2 consumption by muscle tissue is due to increased activity of *NADH-cytochrome-C reductase*.
2. *Protein Synthesis:* Androgens promote protein synthesis in male accessory glands. It causes increased RNA and *RNA polymerase* in the nucleus, and increased *amino acyl transferase* at the ribosomal level. Androgens also act at the mitochondrial level to increase the respiratory rate, the number of mitochondria and the synthesis of mitochondrial membrane.
3. *Carbohydrate Metabolism:* Androgens **increase the fructose production ↑ by seminal vesicles** and utilisation of this sugar by the seminal plasma by enhancing the activity of both *aldose reductase* as well as *ketoreductase*.
4. *Skeletal Growth:* In the growing organism, a growth spurt is induced with increase in bone matrix and skeletal length. Androgens stimulate the growth of bones before the closure of epiphyseal cartilage. Mineralisation of added skeletal tissue is accompanied by decreased excretion of Ca and PO_4, i.e. a more +ve balance in respect of these.
5. *Renotropic Action:* Androgens cause a rather selective increase in size and weight of the kidneys **("renotropic" action).** This is accompanied by a decrease in renal *alkaline phosphatase* activity and by increase in activity of *D-amino acid oxidase*, *arginase* and *acid phosphatase* in the kidney. The relation of these phenomena to metabolic activity of testosterone is not clear.
6. *Mineral Metabolism:* The decreased excretion of urinary N_2 (chiefly urea) that follows administration of androgens is accompanied by a lower urine volume and diminished excretion of Na, Cl, K, SO_4 and PO_4, with no increase in their concentration in blood plasma. The tissue retention of K, SO_4, PO_4 is probably related to the increased storage of proteins. The retention of Na, Cl and water are due to increased tubular reabsorption.
7. *Citrate Excretion:* Androgen reduces (and oestrogen increases) the excretion of citrates in the urine. This is due to increased reabsorption of citrate by renal tubular epithelium.

FEMALE SEX HORMONES

Two main types of female hormones are secreted by the ovary:

1. **The follicular or estrogenic hormones:** Produced by cells of developing Graffian follicles; and
2. **The progestational hormone:** Derived from the corpus luteum that is formed in the ovary from the ruptured follicles.

ESTROGENS

Estrogens are hormones capable of producing certain biological effects, the most characteristic of which are the changes which occur in mammals of estrus. They include:

- Growth of female genital organs,
- The appearance of female secondary sex characteristics
- Growth of the mammary duct system and numerous other phenomena which vary somewhat in different species.

Chemistry

The naturally occurring estrogens in humans are:

- **β-Estradiol**
- **Estrone and**
- **Estriol**

The principal estrogenic hormone in circulation and the most active form of the estrogen is β-estradiol, which is in metabolic equilibrium with estrone. Estriol is the principal estrogen found in the urine of pregnant women and in the placenta.

Essential Features:

- Aromatic character of ring A (three double bonds)
- Absence of $-CH_3$ group at C_{10}
- OH GR at C_3 possesses the properties of a phenolic –OH group (weakly acid)

  All naturally occurring estrogens are C_{18} steroids.

Note

- Estriol is produced from estrone by hydroxylation of estrone at C_{16} and reduction of the ketone group at C_{17}. ***It is the principal metabolite found in urine.***
- Estrone is the hormone produced in the follicles but released into the blood as β-estradiol. In the liver, it is converted to estriol. β-***estradiol and estrone are interconvertible.***

- ***Potency:*** β-Estradiol is 10 times more potent than estrone and 300 times more potent than estriol.
- **Site of Formation**
  In the ovary estrogens are produced by the maturing Graffian follicles, both thecal cells and granulosa cells are involved, and also in Corpus luteum. All the three pituitary gonadotropins—FSH, LH and LTH are involved in stimulation of estrogen secretion. Estrogens are also formed in the adrenal cortex, placenta and testes in small amounts.

Structure of estrogens

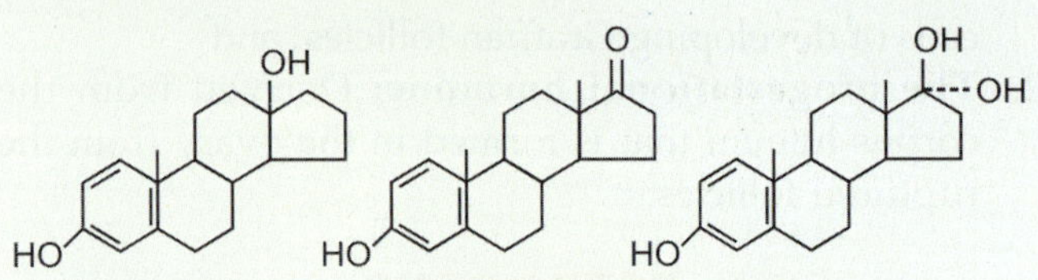

β–Estradiol   Estrone   Estriol

BIOSYNTHESIS

Androgenic steroids—testosterone and androstenedione—are precursors for the synthesis of estrogens in testes, ovaries, adrenal cortex and placenta. Transformation of the natural C_{19} steroids, e.g. testosterone and androstenedione to β-estradiol and estrone involves:

- Removal of the angular $-CH_3$ group at C_{10}, and
- Aromatisation of ring A.

The above is achieved by:

- Enzymes: *Aromatase, hydroxylases (mono-oxygenases)* and *dehydrogenases*.
- *Hydroxylase:* Requires mol O_2 and NADPH.

Steps (Fig. 33.13):

- Cholesterol is converted to pregnenolone and progesterone by the ovarian steroidogenic cells similar to pathway operating in adrenal cortex. Luteal cells also provide certain amount of progesterone.
- Theca interna cells of graafian follicles converts both pregnenolone and progesterone to testosterone and androstenedione, which are the precursors.
- ***In granulosa cells of the follicle → testosterone is converted to β-estradiol,*** catalysed by a microsomal enzyme *aromatase*, which brings about aromatisation of ring A. The enzyme contains a *mono-oxygenase* as a component, it contains Cyt. P_{450} and requires molecular O_2 and NADPH. It brings about three successive hydroxylations of testosterone and the final hydroxylated product loses C_{18} nonenzymatically and converted to β-estradiol.
- Androstenedione can be converted to estrone by the enzyme *aromatase*, with molecular O_2 and NADPH.
- β-Estradiol and estrone are interconvertible by the enzyme *β-estradiol dehydrogenase*.
- Estrone is converted to 16-α-OH-estrone by the enzyme *16-α-hydroxylase*, also a mono-oxygenase which has Cyt. P_{450} and requires molecular O_2. Subsequently 16-α-OH-estrone is reduced to form estriol by the enzyme *reductase*.

Note

- β-Estradiol and estrone can be produced in extra-ovarian tissues, viz. liver, muscles, adipose tissue, etc. from DHEA, androstenedione and testosterone.
- β-Estradiol and estrone are interconvertible in extra-ovarian tissues like liver and placenta, catalysed by the specific enzyme *17-β-estradiol dehydrogenase*.

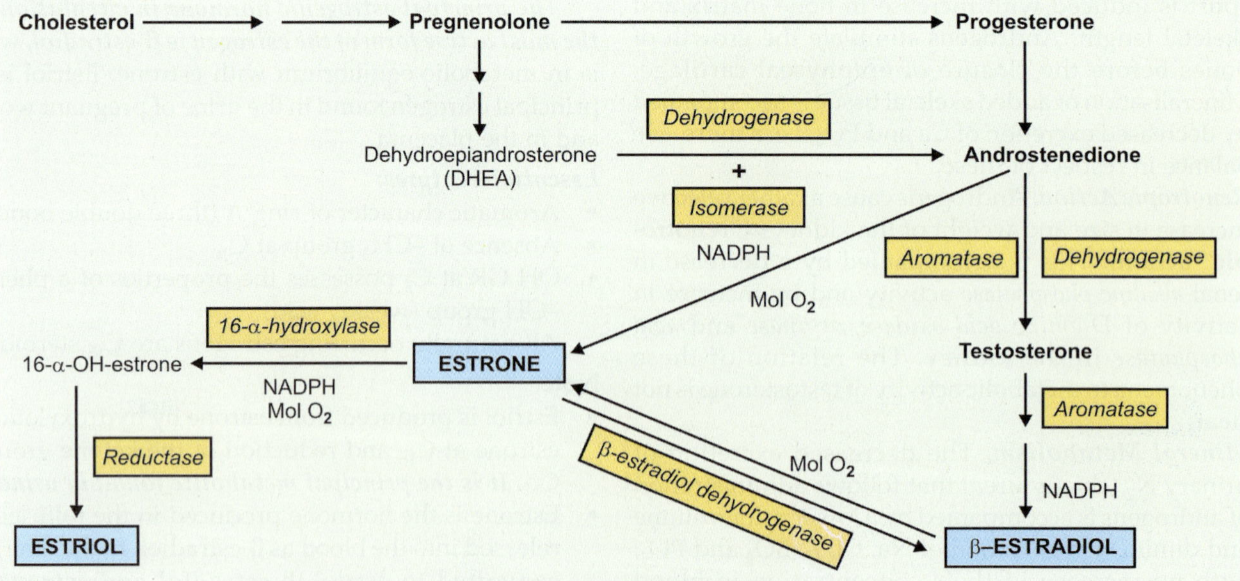

Fig. 33.13: Synthesis of estrogens

Mechanism of Action (Similar to Androgens): After entering the target cells, it binds to a specific 'receptor' present in cytosol, "**receptor-steroid complex**" then binds to hormone-responsive element (HRE) associated with specific nuclear genes, which translates for synthesis of specific proteins and enzymes (Refer to mechanism of action of glucocorticoids).

METABOLIC ROLE

I. *After Administration of Estrogens:* The following biochemical changes are observed to occur:

- Proliferation of vaginal epithelium and endometrium, an increase in glycogen ↑ in the cells, and increase in alkaline phosphatase activity ↑ in endometrium. Glycogen also increases in vaginal epithelial cells.
- There is increased rate of glycolysis with accumulation of lactic acid (LA). The vaginal glycogen is probably the source of LA, which by increasing the acidity of the vaginal secretion (pH 4.0 to 5.0), favours a homogeneous flora of acid bacteria.
- Acceleration of incorporation of amino acids into proteins of uterus, increased protein synthesis ↑, which is preceded by an increase in *RNA polymerase* activity and RNA synthesis.
- Favours retention and elevation of Ca and P and skeletal deposition of Ca producing hypercalcaemia and hyperphosphataemia and calcification and ossification of bones.
- Estrogens also stimulate the closure of bone epiphyses.
- **β-Estradiol prevents osteoporosis,** which is frequently seen in menopausal women, when estrogens decrease. Menopausal women are liable to get fractures due to weakness of bones from osteoporosis.
- An increase in O_2 - consumption in endometrium, placenta, mammary gland and adenohypophysis, owing to a specific effect on estradiol-sensitive *Isocitrate dehydrogenase* enzyme (ICD).
- Estrogens also produce an effect on mineral metabolism. β-Estradiol particularly causes a slight retention of Na, Cl and water.
- In certain mammalian species, estrogens may exert a 'lipotropic effect', i.e tendency to prevent accumulation of fats in the liver.
- Estrogens also have a ***cholesterol lowering effect*** ↓ and reduces plasma cholesterol level and a fall in the level of β-lipoproteins ↓ (LDL).

CLINICAL ASPECT

Young women are protected against myocardial infarction whereas women in menopause, with decline in estrogenic activity are more susceptible to myocardial infarction. ***Estrogenic activity in pre-menopause is associated with increased HDL.***

II. *Transhydrogenation Reaction*

Estrogens may also act as cofactor in transhydrogenation reaction in which H^+ ions and electrons are transferred from reduced $NADP^+$ to NAD^+.

Transhydrogenation

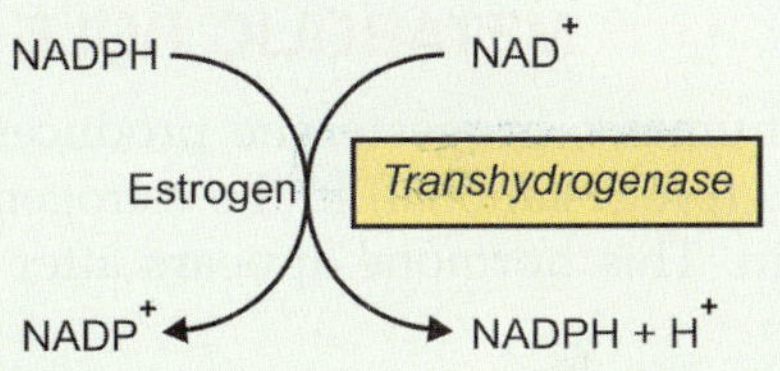

BIOMEDICAL IMPORTANCE

The estrogen dependent ***transhydrogenase*** which catalyses the transfer of H^+ from NADPH to NAD^+ may bring about an increased rate of biologically useful energy in **two ways:**

- Considering that concentration of $NADP^+$ in the cells is ordinarily very low, increased oxidation of NADPH would maintain availability of $NADP^+$ for the activity of those dehydrogenases which require the oxidised form of this cofactor.
- Because direct oxidation of NADP yields no high energy PO_4, a method of deriving high energy PO_4 would be made available by transfer of H^+ to NAD^+, thus forming NADH which can be oxidised in electron transport chain.

PROGESTATIONAL HORMONES: (LUTEAL HORMONES)

PROGESTERONE

- Progesterone is the hormone of the corpus luteum, the structure which develops in the ovary from the ruptured graafian follicle. It is also formed by the placenta, which secretes progesterone, during the later part of pregnancy. Progesterone is also formed in the adrenal cortex, as a precursor of both C_{19} and C_{21} corticosteroids. It is also formed in the testes.

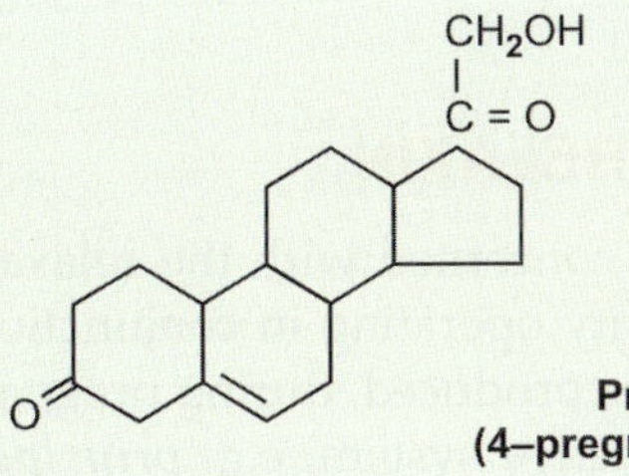

Progesterone
(4–pregnane–3, 20–dione)

Chemistry: Progesterone may be regarded as a derivative of "pregnane" and is designated chemically as "4-pregnane-3, 20-dione". It is a C_{21} steroid and has a $-CH_3$ group at C_{10} and C_{13}.

Biosynthesis: Progesterone has a role as an intermediate in the biogenesis of adrenocortical hormones and of androgens. Indirectly via androstenedione and testosterone, it also serves as precursors for estrogens also. Progesterone is formed from acetate via cholesterol, ***'Pregnenolone' is the immediate precursor.***

Mechanism of action: It is similar to estrogen.

METABOLIC ROLE

In the humans, progesterone produces characteristic changes (progestational) in the estrogen primed endometrium. This hormone appears after ovulation and causes:

- Extensive development of the endometrium preparing the uterus for the embedding of the embryo and for its nutrition.
- It causes an increase in glycogen ↑, mucin ↑ and fat ↑ in the lining epithelial cells. *Alkaline phosphatase* ↓ decreases in activity.
- The hormone also suppresses estrus, ovulation and the production of pituitary luteinizing hormone (LH). Progesterone modifies the action of estrogen on the vaginal epithelium during the menstrual cycle, causing desquamation and basophilia of the superficial layer of cells and leucocytic infiltration.
- Hormone also stimulates the mammary glands. In conjunction with estrogen, progesterone causes development of the alveolar system of the breasts and sensitises them for the action of lactogenic hormone.
- Progesterone is responsible for the rise in basal temperature ↑, which occurs during the corpus luteum phase of the normal menstrual cycle. This is due to increase in basal metabolic rate (BMR ↑).
- In large doses, progesterone exerts androgenic effects, perhaps by conversion to androgenic metabolites.
- The hormonal effects on electrolyte and water metabolism vary in different species. In dogs and rodents, it appears to favour retention of Na, Cl and water. In humans, there is evidence that it exerts an opposite effect.

RELAXIN

Relaxin is a hormone concerned with the relaxation of pelvic tissues and cavity operating in conjunction with other factors. Relaxin is produced, during pregnancy, in tissues of the reproductive system, e.g. ***principally by corpus luteum and also by placenta***. Its production is stimulated by progesterone, pregnenolone and related adrenocortical steroids, e.g. deoxy corticosterone.

Chemistry: Porcine relaxin is made up of two peptide chains, consisting of 22 and 26 amino acid residues. It has two S–S bonds linking its two chains and one intra-chain S–S bond in the A-chain. Approximate mol. wt. = 9000. It is inactivated by proteolytic enzymes or by reagents which breaks the S–S bond (reduction) to form –SH groups.

METABOLIC ROLE

The specific effect of relaxin consists of:

- Increased vascularity of the connective tissue of the symphysis.
- Followed by imbibition of water, dissolution and splitting of collagen fibres, and disorganization of the fibrous structures. There is depolymerization of MPS of ground substance.
- The above makes separation of the symphysis pubis and, a dilatation and softening of uterine cervix, facilitating child birth.

PLACENTAL HORMONES

Pregnancy activates the placental hormones. The implanted blastocyst forms the trophoblast which is subsequently organised into the placenta. The placenta provides the nutritional connection between the embryo and the maternal circulation.

Human placenta produces and secretes:

(a) Peptide hormones

Mainly two:

1. ***Human chorionic gonadotropin hormone (hCG)*** and
2. ***Chorionic somatomammotropin (CS) (also called placental lactogen).***

(b) Ovarian steroid hormones

1. ***Progestins***
2. ***Estrogens, chiefly Estriol.***

1. Human Chorionic Gonadotropin (hCG):

Chemistry

It is a glycoprotein, a heterodimer consisting of two subunits α and β.

- **α-chain** is made up of 92 amino acids and is identical to human FSH, LH and TSH.
- **β-chain** is made up of 145 amino acids.
- Carbohydrate moieties present are as follows:
  - α-chain carries two asparagine linked oligosaccharides
  - β-chain has more carbohydrate and contains two asparagine-linked oligosaccharides and four serine-linked oligosaccharides.

SECTION FOUR

Origin: It is formed by the syncytiotrophoblast of chorionic villi within 12 to 14 days of fertilization.

Mechanism of Action: The hormone binds to "specific receptor" on the cell membrane of target tissues like ovaries and testes, activates *adenyl cyclase*, which in turn increases cyclic AMP level ↑. Cyclic AMP acts as "second messenger" to produce the biological effects.

METABOLIC ROLE

- *Luteotrophic effect*: The hormone produces enlargement of corpus luteum and stimulates its secretion. It maintains a secretory corpus luteum in first three months of pregnancy.
- *Testosterone secretion:* Like LH, the hormone stimulates the growth of interstitial cells (Leydig cells) of embryonic testes and produces testosterone. This helps in virilisation of the reproductive system of male embryo.

2. **Chorionic Somatomammotropin (CS): (Placental Lactogen)**

The hormone has biologic properties of prolactin and growth hormone of anterior pituitary. It is a peptide hormone and amino acid sequences are similar to GH and prolactin (85% homology). ***The hormone is secreted by the syncytiotrophoblast from about the second week of pregnancy,*** rises slowly and reaches a peak approximately by 36 weeks of pregnancy.

METABOLIC ROLE

The exact role of this hormone is not clear, because pregnant women lacking this hormone have normal pregnancies and deliver normal babies.

But as the hormone has similar structure to anterior pituitary GH and prolactin, it exerts similar effects:

- *Somatotrophic effect:* May promote growth of maternal tissues.
- *Luteotrophic effect:* Stimulates the enlargement, growth and secretion of corpus luteum and helps to maintain a secretory corpus luteum.
- *Mammotrophic effect:* Stimulates alveolar growth of mammary glands during pregnancy.
- *Lactogenic effect:* Also stimulates lactation.
- *Anabolic effect:* Stimulates foetal and maternal tissue growth. Promotes retention of N, Ca^{++} and inorganic P.
- *Anti-insulin effect:* May decrease glucose utilisation, decreased carbohydrate tolerance and hyperglycaemic effect.

3. **Ovarian Steroids**

   A. **Progestins:** The corpus luteum is the major source of progesterone for the first 6 to 8 weeks of pregnancy and then placenta takes over this function. The corpus luteum though continues to function, but in third trimester onwards, the placenta produces 30 to 40 times more progesterone than the corpus luteum. ***Placenta cannot synthesize cholesterol from 'active' acetate,*** hence for cholesterol it has to depend on maternal supply.

   B. **Estrogens:** Plasma concentrations of estradiol, estrone and estriol gradually increase throughout pregnancy. Estriol is produced in the largest amount. Adrenal cortex of foetus produces DHEA and DHEA SO_4, which are converted to 16 α-OH derivatives by the foetal liver, and these are subsequently converted to estriol by the placenta. After its formation, travels via the placental circulation to the maternal liver, where they are conjugated to glucuronides, and then are excreted in the urine.

CLINICAL SIGNIFICANCE

1. **Feto-placental function:** As estriol is formed by placenta, ***the measurement of urinary estriol levels has been used as a test of feto-placental function.*** Failure of urinary estriol or total estrogens to rise in late pregnancy reflects the integrity of the feto-placental unit and ***may indicate imminent fetal death or placental insufficiency,*** e.g pre-eclamptic toxaemia.

   Note: Urinary pregnanediol (or plasma progesterone) reflects placental function only as does estimation of plasma placental lactogen. ***A falling titre of urinary estrogens or plasma placental lactogen is serious.***

2. **Pregnancy tests:** Increased urinary excretion of HCG which occurs in early pregnancy (as early as 10th day of gestation) forms the basis for pregnancy tests.

SECTION FOUR

I. Questions (Essay type and short notes)

1. Describe synthesis of epinephrine and its role in carbohydrate metabolism.
2. Describe the chemistry and functions of steroids.
3. Describe the chemistry and functions of thyroxine.
4. What is insulin? Describe the chemistry and role of insulin in metabolism.
5. Enumerate the hormones of anterior pituitary gland. Describe the actions of any three of them.
6. Classify hormones secreted by adrenal cortex and describe the functions of mineralocorticoids.
7. Name the hormones secreted by the ovary. Describe their functions.

8. Classify hormones with examples. What are the mechanisms of action of different hormones? What are the various methods for the study of functions of hormones?
9. What is the role of insulin in metabolism.
10. Write in brief, the role of any three of the following:
    a. Enterogastrone b. Cholecystokinin
    c. Villikinin d. Secretin
11. Enumerate the female sex hormones. Give an account of chemistry, mode of action and functions of progesterone.
12. What are the physiological and metabolic functions of thyroxine in the body?
13. Give an account of the regulation and actions of posterior pituitary hormones.
14. Describe the endocrine functions of pancreas.
15. Describe the chemistry and physiological actions of testosterone. How the secretion of testosterone is controlled?
16. Classify the hormones of adrenal cortex. Give an account of the chemistry, regulation of secretion and functions of aldosterone.
17. Enumerate the female sex hormones. Give an account of the chemisry, mode of action and functions of estrogens.
18. What are pituitary 'Tropins'? Describe the chemistry, regulation of secretion and functions of:
    a. Thyrotropic hormone (TSH) and
    b. Adrenocorticotropic hormone (ACTH)
19. Mention the steroid and non-steroid hormones and describe their mode of action common to each group. What are the metabolic effects of glucocorticoids and their significance?

Short Notes

a. Somatostatin
b. Thyroxine
c. Glucocorticoids
d. Role of insulin in carbohydrate metabolism
e. Epinephrine
f. Renin-angiotensin system
g. Parathormone
h. Steroids of physiological importance
i. Biosynthesis of catecholamines
j. Human chorionic gonadotropins
k. Calcitroil
l. Aldosterone
m. Vasopressin
n. Biosynthesis of thyroid hormones
o. Mechanism of action of steroid hormones
p. Antidiuretic hormone
q. Hypothalamic releasing factors
r. Cushing's syndrome
s. Use of hormones in contraception
t. Conn's syndrome
u. Gonadotropins
v. Catecholamines-biosynthesis and degradation
w. Glucagon
x. Calcitonin.

SECTION FOUR

II. MCQs (Give one correct answer)

1. The rate limiting step in the biosynthetic pathway of catecholamines is:
   a. The hydroxylation of phenylalanine to tyrosine
   b. The decarboxylation of DOPA
   c. The reduction of biopterin
   d. The hydroxylation of tyrosine
   e. Methylation of norepinephrine
2. Which of the following substances is present in high concentration in the urine of patients with pheochromocytomas?
   a. Epinephrine
   b. Metanephrine
   c. Norepinephrine
   d. Dopamine
   e. 3-methyoxy-4-OH-Mandelic acid
3. Epinephrine is formed from norepinephrine by:
   a. N-methylation b. O-methylation
   c. Decarboxylation d. Hydroxylation
   e. Transamination
4. Which of the following is required for crystallisation and storage of the hormone insulin:
   a. Mn^{++} b. Mg^{++}
   c. Ca^{++} d. Zn^{++}
   e. None of the above
5. All of the following hormones use cyclic-AMP as a second messenger except:
   a. FSH b. LH
   c. Epinephrine d. Glucagon
   e. Estrogen
6. Following release of epinephrine from pheochromocytes and norepinephrine from sympathetic nerve-endings, all of the following metabolic processes are enhanced *except:*
   a. Gluconeogenesis b. Glycolysis
   c. Glycogenolysis d. Ketone bodies formation
   e. Lipolysis
7. In patients with primary hypothyroidism, which of the following laboratory test will be useful:
   a. Estimation of T_3 b. Estimation of T_4
   c. Estimation of TBG d. Estimation of iodine
   e. Detection of auto-antibodies
8. Which of the following enzymes catalyses reactions in the synthesis of both catecholamines and serotonin?
   a. Tyrosine hydroxylase
   b. Dopamine-β-hydroxylase
   c. Tryptophan pyrrolase
   d. Aromatic amino acid decarboxylase
   e. N-methyl transferase
9. Which of the following hormone is involved in increased reabsorption of water from renal tubular epithelial cells?
   a. Insulin b. Glucagon
   c. Vasopressin d. Epinephrine
   e. Cortisol
10. In hyperparathyroidism, which of the following is correct?
    a. Low serum calcium
    b. High serum phosphorus

c. Low serum calcium and high serum phosphorus
d. High serum calcium and low serum phosphorus
e. None of the above

11. Insulin causes all of the following *except:*
a. Increased glucose uptake and utilisation
b. Increased amino acids uptake
c. Shift of K^+ from ECF to cells
d. Increased ketogeneisis
e. Increased glycogenesis

12. Following statements regarding insulin are all correct except:
a. Induces glucokinase activity
b. Inhibition of acetyl-CoA carboxylase
c. Converts glycogen phosphorylase to inactive form
d. Stimulaes FA synthase activity
e. Increases activity of glycogen synthase

13. Match the following:
A. Cushing's disease B. Myxoedema
C. Pheochromocytoma D. Conn's syndrome
E. Hyperthyroidism
1. Periodic elevation of BP associated with sweating ______
2. Tachycardia, tremor and exophthalmos ______
3. Retention of Na^+ and increased excretion of K^+ ______
4. Obesity, hypertension, glycosuria, hirsutism ______
5. Bradycardia, falling of hair, and thickening of skin ______.

14. What is untrue of a steroid hormone?
a. Synthesised from cholesterol
b. Binds to specific receptor protein in cytosol
c. Steroid receptor complex enters nucleus
d. Stimulates adenyl cyclase activity
e. Binds to HRE of the gene and increases gene transcription.

15. Which of the following statement is true of glucagon?
a. Accelerates incorporation of amino acids into liver protein
b. Stimulates muscle glycogenolysis
c. Increases gluconeogenesis in liver
d. Inhibits lipolysis in adipocytes
e. Inhibits secretion of insulin from β-cells of islet of Langerhans in pancreas.

16. Thiocyanates and perchlorates interfere with:
a. Iodination of tyrosine b. Release of T_3
c. Release of T_4 d. Trapping of I_2 by thyroid
e. All of the above

17. All of the following are true about the hormone parathormone except:
a. Occurs as a single chain polypeptide
b. Synthesised initially as prohormone
c. Decreases serum calcium level and increases serum inorganic phosphates
d. Acts on kidneys and bones
e. Stimulates 1-α-hydroxylase in kidney tubules

18. Match the following:
A. Tyrosine
B. 3-Methoxy epinephrine
C. "Active" methionine
D. Epinephrine
E. 3-methoxy-4-OH Mandelic acid
1. Stored in chromatin granules ______
2. Major urinary excretory product ______
3. Intermediate in degradation of catecholamines ______
4. Required for formation of epinephrine from norepinephrine ______
5. Starting material for catecholamine synthesis ______.

19. All of the following hormones are produced by anterior pituitary gland except:
a. TSH b. Gonadotropic hormone
c. Growth hormone d. MSH
e. Prolactin

20. All of the following are expected to be elevated in blood and tissues during starvation except:
a. Glucagon b. Glycogen
c. Ketone bodies d. Epinephrine
e. Norepinephrine

21. Which of the following defects in adrenal cortex is responsible for lack of glucocorticoids and mineralocorticoids?
a. Androstenedione deficiency
b. 17-α-OH progesterone deficiency
c. C_2 hydroxylase deficiency
d. Testosterone deficiency
e. Estrone deficiency

22. All of the following statements regarding hormones are true except:
a. May influence the synthesis of enzymes
b. May act as enzymes or coenzymes
c. May act by regulating pre-existing processes
d. May affect the rate of enzymatic catalysis
e. May alter the permeability of plasma membranes

23. Match the following:
A. Amino acid derivative B. Cholesterol derivative
C. Glycoprotein D. Polypeptide
E. Tri-peptide
1. Glucagon ______
2. Glucocorticoids ______
3. Thyroxine ______
4. Insulin receptors ______
5. Thyrotropin releasing hormone (TRH) ______.

Answers to MCQs

1. d 2. e 3. a 4. d 5. e 6. b
7. e 8. d 9. c 10. d 11. d 12. b
13. 1-C, 2-E, 3-D, 4-A, 5-B, 14. d 15. c
16. d 17. c 18. 1-D, 2-E, 3-B, 4-C, 5-A 19. d
20. b 21. c 22. b 23. 1-D, 2-B, 3-A, 4-C, 5-E.

CHAPTER

34

METABOLISM OF MINERALS AND TRACE ELEMENTS

Major Concepts

A. Study and classify elements present in the body.
B. Study in detail the minerals or principal elements.
C. Study in detail the trace elements.

Specific Objectives

A. 1. Learn the five groups of elements and elemental content of each group.
B. 1. Study in detail the metabolism of minerals such as Na, K, P, Cl, S and Ca. Study the homeostasis of each and clinical conditions associated with their excess and deficiencies in diet.
   2. Learn their dietary sources, normal blood levels and daily requirement.
   3. Learn the functions of each of them.
C. 1. Define "trace elements" and clearly learn how they are different from minerals.
   2. Learn the body content of various trace elements.
   3. Study what are their dietary sources and daily requirement, specially Fe, Cu, Mg.
   4. Study the general metabolism, homeostasis and clinical conditions associated with their excess or deficiency in diet.

INTRODUCTION

It is observed that there are at least 29 different types of elements in our body. Organic components such as carbohydrates, proteins, and lipids form about 90 per cent of the solid matter and mainly consist of C, H, O and N.

The elements of the body are divided in **five major groups:**

Gr. I: C, H, O, N. Components of macromolecules such as carbohydrates, proteins, lipids, etc.

Gr. II: Nutritionally important minerals or principal elements. The daily requirement of these is > 100 mg. The deficiency of these can prove fatal. These include Na, K, Cl, Ca, P, Mg and S. They are also called ***macroelements.***

Gr. III: "Trace elements" which are essential. The requirement is less than 100 mg per day. Deficiency can lead to serious disorders. They include Cr, Co, Cu, I, Fe, Mn, Mo, Se, Zn.

Gr. IV: These are additional trace elements which may be possibly essential. The exact role is not known. They include Cd, Ni, Si, Sn, Vn.

Gr. V: These are not essential elements and may be toxic. They have no known function in the body and may enter the body through polluted air, water, soil or food substances, e.g. As, CN^-, Hg, etc.

SODIUM

As already mentioned in the chapter on Water Metabolism, body water is found mainly in intracellular and extracellular compartments with a small amount in interstitial fluid compartment. ***Sodium is the chief electrolyte which is found in large conc. in extracellular fluid compartment.***

Approx. body distribution of sodium is as follows:

	Total mmol	*Conc. mmol/L*
• Total body	3150	—
• Intracellular	250	10
• Extracellular	2900	140
• Plasma	400	140

The sodium is found in the body mainly associated with chloride as NaCl and $NaHCO_3$.

Sources: Sodium is widely distributed in food material; more in animal sources than plants. However, major source is table-salt used in cooking or seasoning. It is also found in cheese, butter, khoa.

Daily Requirement:

- 1-3.5 g of Na is required daily for adults.
- Infants need 0.1-0.5 g and
- Children 0.3-2.5 g daily.

Absorption of Sodium: Sodium is absorbed by sodium pump situated in basal and lateral plasma membrane of intestinal and renal cells. Na-pump actively transports Na into extracellular fluid.

SODIUM PUMP

This is ***also called as Na^+-K^+ ATPase.*** It requires ATP and Mg^{++}. Intracellular Na^+ conc. is around 10 mM while that of extracellular is 150 mM. This high inward conc. gradient is contrary to what could be expected from the Gibbs-Donnan effect. There is high conc. of proteins and phosphate anions inside the cell than outside it. Conc. of K^+ inside the cell is 100 mM while outside it is 5 mM. This observation is also unexplainable following the Gibbs-Donnan effect. Na-pump is found to maintain both magnitudes and direction of transmembrane concentration gradients of those ions.

Na-pump is an enzyme, Na^+-K^+-ATPase. It is a glycoprotein composed of 2 α and 2 β chains. Its activity depends on presence of Na^+ and K^+ and requires ATP and Mg^{++} ions as cofactor. The enzyme hydrolyses a high energy phosphate bond of ATP and uses the energy thus released to transport ***three Na^+ ions outside and simultaneously two K^+ ions inside across the cell membrane.*** In this way, each Na^+-K^+ pump transfers 9000 Na^+ ions outside and 6000 K^+ ions inside the cell in one minute. The Na-pump is very active in those cells where activities depend largely on transmembrane Na^+ fluxes, e.g. nervous, muscle fibres, renal tubules cells, intestinal mucosal cells (Fig. 34.1).

Forms of Sodium Pump and Mechanism: Na^+-K^+ ATPase exists in ***two forms: E_1 and E_2.***

- **The E_1 form:** Presents its ion binding and phosphate-binding sites on the cytoplasmic surface of the membrane. Three sodium ions from cytoplasm bind with the ion binding sites of E_1. This leads to the phosphorylation of aspartate residue of E_1 with the help of ATP and Mg^{++}. This results in conformational change and E_1 becomes E_2.

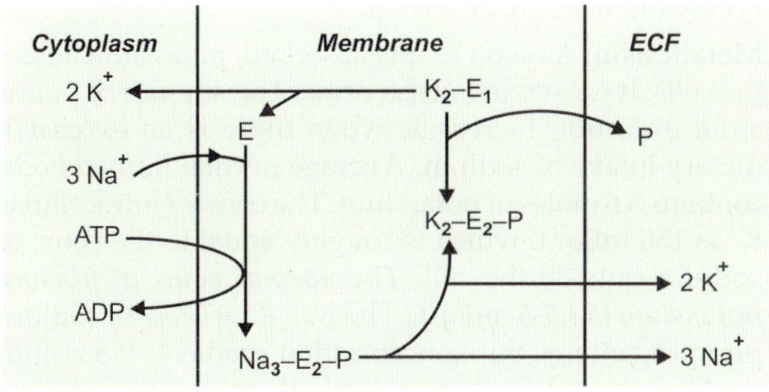

Fig. 34.1: Action of Na^+-K^+ pump

- **Now E_2 exposes** both ion binding and phosphate binding sites on the extracellular surface of the membrane, lowers the affinity of the ATPase for Na^+ and releases it into the ECF. On the contrary, now the K^+ ions from ECF bind to the respective ion binding site of the pump. This lowers the affinity of E_2 for phosphate. This dephosphorylation changes the conformation of E_2 to E_1 again and lowers its affinity for K^+ ions. This leads to release of the K^+ ions from ATPase into the cell.

Inhibitors

- ***Ouabain:*** A glycoside of a steroid and ***digitalis*** is the cardiotonic drug which inhibits the Na^+-K^+ pump by blocking the step of dephosphorylation.
- ***Vanadate:*** Inhibits the pump when present inside the cell.

Note: Thus, the sodium transported actively by Na-pump diffuses into the cell across its luminal or microvillus membrane from the lumen. Active absorption of Na^+ is coupled with glucose absorption or amino acid absorption. This carrier mediated transport is explained in chapter on digestion and absorption.

Excretion of Na: Every 24 hours approximately, 25000 mmol of sodium are filtered by the kidneys. However, due to tubular reabsorption less than 1 per cent of this sodium appears in the urine (100-200 mM/day). Approximately 70 per cent of the filtered sodium is reabsorbed in proximal tubule. Further 20–30 per cent of filtered Na^+ is reabsorbed by ascending loop of Henle.

FUNCTIONS

- ***Fluid balance:*** Sodium maintains crystalloid osmotic pressure of extracellular fluids and helps in retaining water in ECF.
- ***Neuromuscular excitability:*** Alongwith other cations, Na^+ is also involved in neuromuscular irritability which is given:

$$\text{Neuromuscular irritability } \alpha = \frac{[K^+]\,[Na^+]}{[Ca^{++}] + [Mg^{++}] + [H^+]}$$

- ***Acid-base balance:*** Na^+-H^+ exchange in renal tubule to acidify urine. (Refer to chapter on Acid-base Balance and Imbalance).
- ***Maintenance of viscosity of blood:*** The salts of Na with globulins are soluble and further Na^+ and K^+ both regulate in maintaining the degree of hydration of the plasma proteins.
- ***Role in resting membrane potential:*** Plasma membrane has a poor Na^+ permeability and passive Na^+ inflow through it. Na-pump keeps Na^+ conc. far higher outside than inside. This separation of charges is called

polarisation of the membrane. It creates a potential difference of –70 to –95 millivolts across the membrane and is called as *resting membrane potential.*

- ***Role in Action Potential:*** A local depolarisation of nerve or muscle fibre is observed in stimulation. This rapidly increases its permeability to Na^+ causing considerable transmembrane influx of Na^+ down its inward conc. gradient.

CLINICAL ASPECT

Clinical conditions are of ***two major types:***

I. **Hypernatraemia** and
II. **Hyponatraemia.**

I. Hypernatraemia

A high plasma sodium concentration does not necessarily mean that the total body sodium content is increased, but infers that the extracellular sodium is excessive relative to water. Decrease in body water and increase in body sodium.

Specific conditions in which hypernatraemia occurs are as follows:

- ***Simple dehydration:*** This occurs as a result of excessive sweating with inadequate or no water replacement. (Refer to chapter on Water and Electrolytes Balance and Imbalance).
- ***Diabetes insipidus:*** A special type of water loss occurs in DI. These diseases are characterised by lack of antidiuretic hormone (ADH) or failure of the hormone to act on its target cells. It occurs usually as a complication of pituitary surgery, when hormone is not produced in adequate amount. In nephrogenic *Diabetes insipidus*, the kidney cells are unable to respond to the hormone.
- ***Osmotic loading:*** If the kidney is required to excrete large quantities of very soluble substances such as glucose, urea, amino acids, osmotic effect of these substances on the urine causes the co-excretion of large amounts of water. Relatively little sodium is excreted. So the plasma level rises. In case of very ill patients on high protein diet, urea is formed which is then excreted in very large amounts along with large volumes of water.
- ***Excess sodium intake:*** Normally in clinical medicine, 0.9 per cent NaCl is administered intravenously –154 mEq/L. Excessive use of isotonic saline particularly in children leads to hypernatraemia. Hypernatraemia may also occur following the administration of $NaHCO_3$ in treatment of acidosis.
- ***Steroid therapy:*** Certain adrenal steroids, the mineralocorticoids, control the metabolism of sodium. Among other effects, mineralocorticoids cause the kidney to absorb sodium from the glomerular filtrate, which results in increased plasma concentration. In certain tumours of adrenal gland, large amounts of most potent mineralocorticoid, aldosterone is produced **(Conn's syndrome).**

II. Hyponatraemia

- ***Diuretic medication:*** One of the most commonly encountered causes of hyponatraemia today is due to the use of diuretics. Many of the standard diuretic medications act by promoting excretion of Na by kidney. The object is to lower the total body sodium and thus reduce the total extracellular water. There are many diseases in which this objective is desired, viz. congestive heart failure, chronic kidney disease and hypertension. Reduced total body sodium is achieved but as the extracellular volume reaches critical dimension, there is a counter effort to retain water which then dilutes the sodium and hyponatraemia results.
- ***Excessive sweating:*** Loss of fluids of high Na^+ and Cl^- content (like sweating) but are replaced by salt deficient fluids such as water by mouth or glucose solution by IV.
- ***Kidney diseases:*** Kidneys are impaired, glomerulus is the one in which blood is filtered and Na is reabsorbed by the renal tubules. Due to kidney dysfunction, Na^+ is not reabsorbed and is thus excreted in the urine. There is also a progressive failure to excrete water.
- ***Congestive heart failure:*** Hyponatraemia is common in heart failure for two reasons:
  - Diuretics administration
  - In advance stages, congestive heart failure cause low Na^+ conc. It is because the low cardiac output is sensed incorrectly by the brain as low blood volume, calling forth increased secretion of ADH.
- ***Gastrointestinal loss:*** Diarrhoea, particularly if prolonged and severe, will result in reduced sodium/chloride levels in plasma and extracellular fluid.

POTASSIUM

Potassium is the major intracellular cation. It is widely distributed in the body fluids and tissues as follows:

• Whole blood	:	200 mg/dl
• Plasma	:	20 mg/dl
• Cells	:	440 mg/100 g
• Muscle tissue	:	250-400 mg/100 g
• Nerve-tissue	:	530 mg/100 g

It is widely distributed in the vegetable foods. An average amount of 4 g of potassium is present in the diet. Potassium is easily absorbed.

Metabolism: As soon as it is absorbed, potassium enters the cells. It is excreted in the urine. The amount of potassium excretion increases, when there is an excessive dietary intake of sodium. Average normal human body contains 3.6 moles of potassium. The conc. of intracellular K^+ is 150 mEq/L which is roughly equal to the conc. of sodium outside the cell. ***The normal conc. of plasma potassium is 3.5-5 mEq/L.*** The *Na^+-K^+ ATPase* or sodium pump maintains this concentration gradient. Potassium is also excreted in gastrointestinal tract, saliva, gastric juice, bile, pancreatic and intestinal juices. This fact

SECTION FOUR

becomes clinically important if these secretions are lost in large amounts. Potassium is continuously filtered by the glomeruli of the kidney and reabsorbed by the cells of proximal convoluted tubules. Potassium (and hydrogen) ions are also secreted in distal tubule in exchange for sodium.

FUNCTIONS

Many functions of potassium and sodium are carried out in coordination with each other and are common. These functions have already been described under sodium. Briefly:

- It influences the muscular activity
- Involved in acid-base balance
- It has an important role in cardiac function
- Certain enzymes such as *pyruvate kinase* require K^+ as cofactor
- Involved in neuromuscular irritability and nerve conduction process.

CLINICAL IMPORTANCE

Extracellular levels of potassium are measured on a sample of serum. ***Since RBCs contain a large amount of potassium, care must be taken that sample is not hemolysed.*** On standing, potassium value changes, so the plasma potassium must be measured as soon as possible on fresh sample. Both high values and low values are clinically important.

CLINICAL ASPECT

I. Hyperkalaemia

The mechanisms for excretion of potassium in normal persons are so effective that it is difficult to produce hyperkalaemia by simply increasing the oral intake. Increases however may occur after rapid intravenous infusion of potassium salts. In clinical practice, most cases of hyperkalaemia are due either to:

- Kidney failure with decreased excretion of potassium or
- To the sudden release of potassium from the intracellular compartment which may happen in a variety of diseases.
- ***Anuria:*** Complete shut-down of kidney function regardless of cause results in increasing conc. of K^+. The rise may be particularly rapid if the kidney failure is associated with sudden release of intracellular potassium from any organ.
- ***Tissue damage:*** Damage to body cells from any cause results in release of cell contents including K^+ into ECF. Crush injuries, with damages to large volumes of muscle tissue, massive hemolysis are examples. In both of these conditions, there is often reduced kidney function which adds to the hyperkalaemia.
- ***Violent muscle contraction:*** Vigorous exercise produces a release of potassium from muscle cells into the extracellular space and may cause a temporary elevation in plasma potassium. The same mechanism is responsible for the increase seen in status epilepticus.
- ***Addison's disease:*** In the absence of aldosterone, the exchange of sodium for potassium in the kidney is reduced, with increased loss of sodium and retention of K^+ in body. Low serum sodium and high serum K^+ are characteristic of this disease.
- ***Diabetes mellitus:*** In ketoacidosis, there is substantial loss of intracellular K^+ to the ECF. This is partly due to increased activity of *Na^+-K^+ ATPase* which results from impaired glucose metabolism. If ketoacidosis presents for a long time, there will be major depletion of total body K^+. Treatment with insulin allows resumption of sodium pump activity and movement of K^+ back into the cells. This in turn causes an abrupt fall in plasma K^+. This must be treated by administration of K^+ to restore that which has been lost in the urine during the period of acidosis. Frequent monitoring of K^+ is vital to DM with ketoacidosis.

II. Hypokalaemia

Low serum K^+ usually results from the depletion of total body K^+. Since nearly all food contains large quantities of K^+, dietary deficiency by itself is uncommon. Dietary supplements of K^+ are required only in patients with some disease or drug use which causes potassium depletion. Some of the more common causes of hypokalaemia are:

- ***Loss of K^+ in GI secretions***
  - Both prolonged vomiting and severe diarrhoea cause depletion of total body potassium and causes hypokalaemia.
  - The intestinal fluid which drains from a recent ileostomy is rich in K^+. Excessive loss of fluid from this site may rapidly produce hypokalaemia.
  - Habitual users of laxative eventually develop a state of chronic mild diarrhoea which may cause low K^+-levels.
  - A special instance of K^+ loss through GIT is a **mucous secreting tumour called a cillous adenoma.** This tumour secretes large amounts of K^+ into lumen of colon.
- ***Loss of K^+ in urine***
  - Many medications which are used to decrease total body sodium such as some diuretics also cause loss of K^+. This is especially true of thiazides, acetazolamide, and the organic mercurial diuretics. In clinical practice, the use of thiazide diuretics is one of the most common causes of low plasma K^+. The degree of hypokalaemia is usually not severe. It can be corrected by increasing K^+ intake.
  - Rarely a tumour **(Conn's tumour)** develops in the adrenal gland which produces excess amounts of aldosterone. The resulting syndrome of primary hyperaldosteronism is characterised by increased loss of K^+ in urine.
  - A compound found in *licorice*, glycyrrhizinic acid has aldosterone like activity and people who eat large amounts of licorice may develop hypokalaemia.

- In Cushing syndrome (adrenocortical hyperfunction), hypokalaemia is the rule. It is particularly severe if the process is due to production of ACTH by an ectopic tumour.

- ***Loss of extracellular potassium into the intracellular space:*** As already mentioned above, treatment of diabetes ketoacidosis causes a rapid fall in plasma K^+ from a state of hyperkalaemia to normal and then to hypokalaemia as the acidosis is brought under control.
- ***Other causes of hypokalaemia:***
  - There is an inherited disorder called **familial periodic paralysis** in which there is sudden shift of K^+ into ICF causing paralysis.
  - ***In thyrotoxic periodic paralysis (TPP)*** similar thing happens. In both cases, it normally happens after heavy exercise or large carbohydrate meal. Thyroid hormone excess is also observed in TPP patients.
  - ***Renal tubular acidosis:*** Low serum K^+ is seen ***in renal tubular acidosis, in Bartter's syndrome*** and ***after administration of steroids*** such as deoxy-corticosterone, cortisone and testosterone.

CHLORINE

Chloride is taken in diet as sodium chloride. Many vegetables and meats have small proportions of chloride. It is also available in the 'chlorinated water' which is normally supplied as a process of purification of water for drinking purpose.

Daily Requirement: About 100-200 mmol is taken in diet as sodium chloride (table salt).

Distribution

- Whole blood : 250 mg/dl
- Plasma : 375 mg/dl
- CSF : 440 mg/dl
- Cells : 190 mg/100 g
- Muscles : 40 mg/100 g

Absorption and Excretion

Absorption: It takes place in small intestines. The mechanism of chloride uptake is unclear, but it appears to depend on an exchange process with the HCO_3^-, whilst the accompanying sodium exchange for a hydrogen ion.

Excretion

- ***Sweat:*** 5 mMol/day depends on weather
- ***Faeces:*** 5 mMol/day
- ***Renal:*** 100-200 mM/day. 99 per cent of the Cl^- in the glomerular filtrate is reabsorbed by renal tubules mainly in proximal tubule (60-70%) and then in ascending loop of Henle (20-25%) followed by distal tubule, collecting duct (10-15%).

Regulations: Control of absorption and excretion of chloride appears to be similar to that of sodium. Increase in blood volume decreases reabsorption of chloride and vice-versa. Plasma levels of chloride vary with and to a great extent depend on the plasma conc. of Na and HCO_3^-.

$\downarrow$ Na associated with Cl^- $\downarrow$
$\uparrow$ Na usually associated with Cl^- $\uparrow$
$\uparrow$ HCO_3^- associated with Cl^- $\downarrow$
$\downarrow$ HCO_3^- associated with Cl^- $\uparrow$

FUNCTIONS

- It is important in the production of HCl in the gastric juice.
- It is important in chloride shift (Refer to chapter on Chemistry of Respiration).

CALCIUM

Calcium is an important mineral mainly found in bone and teeth.

Dietary sources: It is widely distributed in food substances such as milk, cheese, egg-yolk, beans, lentils, nuts, figs, cabbage.

Body distribution: The total calcium of the body is 25-35 mols (100 g-170 g). About 99 per cent of it is found in bones. It exists as carbonate or phosphate of calcium. About 0.5 per cent in soft tissue and 0.1 per cent in ECF. ***The normal level of plasma calcium is 9-11 mg/dl.*** The ***calcium in plasma is of 3 types,*** namely:

1. **Ionised calcium (diffusible),**
2. **Protein bound calcium and**
3. Complexed calcium, it is probably complexed with organic acids. ***About 40 per cent of total calcium is in ionised form.*** Albumin is the major protein with which calcium is bound. All the three forms of calcium in plasma remain in equilibrium with each other. ***Ionised calcium is physiologically active form of calcium.***

Absorption: Calcium is taken in the diet principally as calcium phosphate, carbonate and tartarate. Unlike Na and K which are readily absorbed, the absorption of Ca is rather incomplete. About 40 per cent of average daily dietary intake of Ca is absorbed from the gut. Calcium is absorbed mainly from the duodenum and first half of jejunum against electrical and concentration gradients.

MECHANISM

Two mechanisms have been proposed for absorption of calcium by gut mucosa.

1. ***Simple diffusion***
2. ***An "active" transport*** process involving energy and *Ca^{++} pump*. Both the processes require 1, 25-di hydroxy-D_3 (calcitriol) which regulates the synthesis of Ca-binding proteins and transport and also a Ca^{++}-dependent ATPase.

SECTION FOUR

FACTORS AFFECTING ABSORPTION

Various factors which influence the absorption of calcium are discussed below:

1. ***pH of intestinal milieu: An acidic pH favours calcium absorption*** because the Ca-salts, particularly PO_4 and carbonates are quite soluble in acid solutions.

- In an alkaline medium, the absorption of calcium is lowered due to the formation of insoluble tricalcium PO_4.

2. ***Composition of the diet***

- ***High protein diet:*** *A high protein diet favours absorption,* 15 per cent of dietary Ca is absorbed. If the protein content is low, only 5 per cent may be absorbed.

  Reason: Amino acids increase the solubility of Ca-salts and thus its absorption. *Lysine* and *arginine* obtained from basic proteins cause maximal absorption of Ca.

- ***Fatty acids:*** In malabsorption syndrome, fatty acids are not absorbed properly. Fatty acids produce insoluble calcium soaps which are excreted in faeces, thus, decreasing the Ca absorption.
- ***Sugars and organic acids:*** Organic acids produced by microbial fermentation of sugars in the gut, increases the solubility of Ca-salts and increases their absorption. Citric acid also may increase the absorption of calcium.
- ***Phytic acid:*** Cereals contain phytic acid (inositol hexaphosphate) which forms insoluble Ca-salts and decreases the absorption of Ca.
- ***Oxalates:*** Oxalates present in vegetable like cabbage and spinach forms insoluble calcium oxalates which are excreted in the faeces, thus lowering the calcium absorption.
- ***Fibres:*** Presence of excess of fibres in the diet interferes with the absorption of calcium.
- ***Minerals***
  - ***Phosphates:*** Excess of phosphates lower calcium absorption.
  - ***Magnesium:*** High content of magnesium in the diet decreases absorption of calcium.
  - ***Ca:P ratio:*** A ratio of food Ca to P not more than 2:1 and not less than 1:2 (ideal 1:1) is necessary for optimal absorption of calcium.
  - ***Fe in diet:*** Food Fe may form insoluble ferric phosphates. These indirectly increases the Ca:P ratio in the gut beyond the range of optimal absorption.
- ***Vitamin D:*** Promotes Ca absorption.

3. ***State of health of the individual and aging***

- A healthy adult absorbs about 40 per cent of dietary calcium.
- Above the age of 60 years, there is a gradual decline in the intestinal absorption of Ca.
- In sprue syndrome, the intestinal absorption of calcium suffers due to formation of Ca-soap with FA which are excreted in faeces.

4. ***Hormonal***

- ***PTH (Parathormone):*** PTH directly cannot increase the calcium absorption. But PTH stimulates *"1, α-hydroxylase"* enzyme in the kidney and increases the synthesis of 1, 25-$(OH)_2$-D_3 (calcitriol) which enhances calcium absorption (refer to Vitamin D).
- ***Calcitonin:*** Calcitonin directly cannot affect Ca absorption. Increased calcitonin level inhibits *"1-α-hydroxylase"* enzyme, thus decreasing synthesis of calcitriol and Ca-absorption.
- ***Glucocorticoids:*** Diminishes intestinal transport of calcium.

Calcium is secreted into the gut as a normal constituent of bile and intestinal fluids. Faecal output of calcium could exceed intestinal absorption on situations where the diet contains high levels of phytates or other sequestrating substances. Under normal circumstances, the faeces are not an important excretion route for calcium.

Regulation: Kidneys filter about 250 mmol of Ca^{++} every day, some 95 per cent of which is reabsorbed by the tubules. The major portion of this filtered Ca^{+2} is taken up by proximal tubule without hormonal regulation. A fine adjustment to the amount reabsorbed occurs in distal tubules under the influence of PTH (PTH-uptake). ***Plasma level of ionised calcium conc. is the principal regulator of PTH secretion by a simple negative feedback mechanism.*** A threshold level of magnesium is required for PTH release. Hypermagnesaemia inhibits PTH secretion. PTH secretion is also subject to negative feedback by the vit D metabolite 1,25 $(OH)_2D_3$. PTH rapidly stimulates osteoclast activity, the increased bone resorption causing an increase in plasma Ca^{+2} and PO_4; Vit D_3 plays a permissive role for this effect. PTH stimulates more slowly (days) osteoblast activity. PTH via c-AMP increases the distal nephron reabsorption of calcium and decreases that of PO_4 in the proximal tubule. In doing so, PTH increases the tubular synthesis and excretion of c-AMP. PTH also stimulates the enzyme complex that converts 25, OH D_3 to 1, 25 $(OH)_2$ D_3, thereby increasing calcium uptake from the gut. (For details see chapter on Vit D, PTH and Calcitonin). Hypercalcaemia stimulates calcitonin and katacalcin release while hypocalcaemia has inhibitory effect. Calcitonin strongly inhibits osteoblastic bone resorption. However the role of calcitonin in calcium regulation is controversial. Thyroid hormones, ACTH, prostaglandins have some effect on Ca level of plasma.

FUNCTIONS

- ***Calcification of Bones and Teeth:*** The process of bone formation and teeth formation is known as calcification which is a continuous process for bones. Osteoblasts secrete an enzyme *alkaline phosphatase* which can hydrolyse certain phosphoric esters.
- Calcium plays ***a role in blood coagulation*** by producing substances for thromboplastic activity of blood.
- Calcium has a role in neuromuscular transmission.
- Calcium ions are needed for excitability of nerves.
- Calcium plays role in muscle contraction.
- Normal excitability of heart is Ca ion dependent.
- It plays role as secondary or tertiary messenger in hormone action.
- It plays role in permeability of gap junctions.

CLINICAL IMPORTANCE

The two conditions namely Hypercalcaemia and Hypocalcaemia occur.

I. Hypercalcaemia

When the serum calcium level ***exceeds 11.0 mg/dl, it is called as hypercalcaemia*** (Normal serum calcium level is 9 to 11 mg/dl).

Causes

1. Primary hyperparathyroidism: Most common cause for outpatients (OPD cases).

May be due to:

Familial

Hyperplasia-chief cells: Hyperplasia involving all four parathyroid glands (15% cases)

- ***Tumours***
  - Solitary adenoma (80 to 85% cases)
  - Multiple adenomas (2% cases)
  - Parathyroid carcinoma (< 1% cases).
- ***"Ectopic" hyperparathyroidism:***
  - Multiple endocrine neoplasia type I (**MEN I**) with pituitary and pancreatic tumours.
  - Multiple endocrine neoplasia type II (**MEN II**)
    - Medullary carcinoma of thyroid and
    - Pheochromocytoma.

2. Malignancy

Most important cause for hospital in-patients. Hypercalcaemia in malignancies may be due to:

- ***Humoral*** factors ***(HHM-Humoral hypercalcaemia of malignancy):*** No direct skeletal involvement. May be:
  - PTH related protein (PTHγP)
  - Growth factors, e.g. Tumor growth factor (TGF), Epidermal growth factor (EGF), Platelet derived growth factor (PDGF).
- ***Direct skeletal involvement by the tumours, viz.***
  - Direct erosion of bone by tumour
  - Production of PGE_2 by the tumour which can produce bone resorption.
- ***Haematological malignancies:*** Production of
  - Cytokines: Interleukin-1, Tumor necrosis factor (TNF), Lymphotoxin
  - 1,25-di (OH)-D_3 (Calcitriol) by lymphomas.

3. Other Endocrine Causes: Hyperthyroidism, Hypothyroidism, Acromegaly, Acute adrenal insufficiency.

4. Granulomatous Diseases: Tuberculosis, sarcoidosis, berylliosis, coccidioidomycosis.

5. Overdosage of Vitamins: Vitamin A intoxication, and Hypervitaminosis D.

6. Drug-induced Hypercalcaemia (Iatrogenic): Thiazide diuretics, Spironolactone, Milk-alkali syndrome.

7. Miscellaneous Other Causes

- Idiopathic hypercalcaemia of infancy ***(William syndrome)***
- Familial hypocalcinuric hypercalcaemia
- Prolonged immobilisation
- ***Increased serum proteins***
  - Hyperalbuminaemia—due to haemoconcentration
  - Hyperglobulinaemia—due to multiple myeloma
- ***Renal failure:*** Acute renal failure—diuretic phase, chronic renal failure, postrenal transplantation.

II. Hypocalcaemia

Hypocalcaemia is said to ***exist when serum calcium is less than 8.5 mg/dl*** as determined by a standard method.

Causes: The commonest cause of hypocalcaemia is hypoalbuminaemia, closely followed by renal failure. The other ***most common cause of hypocalcaemia is surgically-induced hypoparathyroidism.***

The various causes of hypocalcaemia can be grouped as follows:

1. Reduction in serum albumin: (Hypoalbuminaemia)

- Malnutrition, malabsorption states, nephrotic syndrome, chronic liver disease and liver failure, protein losing enteropathy.

2. Hypoparathyroidism

- May be surgical-induced partial or complete (90% cases)
- Idiopathic—may be autoimmune (10% cases)
- Bio-inactive parathyroid hormone (PTH)
- Transient hypoparathyroidism of infancy, may be partial.

3. Renal Diseases and Renal Failure

- Renal tubular dysfunction, acute tubular necrosis
- Chronic renal failure contributing factors for low calcium values are:
  - Hyperphosphataemia
  - Impaired synthesis of 1,25-diOH D_3 (calcitriol) due to inadequate renal mass due to disease process and tubular damage.

Note: ***Hypocalcaemia in chronic renal failure may not be associated with tetany.*** Renal failure is associated with acidosis, in which ionisation of calcium is not suppressed ***thus "ionic" calcium is not lowered. Hence there may not be any tetany.***

4. Pseudohypoparathyroidism

5. Hypoparathyroidism in association with other disease states which may be familial, viz.

- Addison's disease
- Pernicious anaemia
- Fungal disease like candidiasis

6. Other Miscellaneous Causes
- ***Acute pancreatitis:*** Haemorrhagic or oedematous
- ***Osteomalacia and rickets*** due to vitamin D deficiency or resistance.
- ***Medullary carcinoma of thyroid:*** With or without associated endocrinopathies.
- ***"Healing phase" of bone disease*** of treated hyperparathyroidism, hyperthyroidism and haematological malignancies ***("Hungry bone" syndrome).***
- ***Magnesium deficiency***
- ***Iatrogenic:***
  - **Administration of Foscarnate:** A drug given for the therapy of cytomegalovirus retinitis in patients with "Acquired immune deficiency syndrome" (AIDS) has been reported to result in hypocalcaemia.
  - Mithramycin, glucocorticoids, calcitonin.

7. Neonatal hypocalcaemia
- ***Prematurity:*** Mg^{++} deficiency, immature parathyroid gland, immature vit. D metabolism.
- *Poor feeding:* Oral or IV.

PHOSPHORUS

Food Sources: Foods rich in phosphorus content are cheese, milk, nuts, organ meats, egg.

Body Distribution: Total body phosphate is about 25 mol (700 g). More than 85 per cent (600 g) is found in bones, 15 per cent in soft tissues and 1 per cent is found in ECF. About 5 g in brain and 2 g in blood is found. About 1.5 g of phosphate is required to be taken in the diet daily.

Absorption: 90 per cent of daily dietary phosphate is absorbed. The absorption is stimulated by both PTH and Vit. D_3. The Ca:P ratio in diet affects the absorption and excretion of phosphorus. If one is in excess in diet, the excretion of the other is increased.

Regulation: Regulation of Ca and P is under the similar control mechanisms by kidney with respect to PTH and vit. D.

Role of Kidneys

1. Phosphate uptake is sodium dependent, about 85 per cent of filtered PO_4 is reabsorbed by the proximal tubules. Phosphate reabsorption is increased when dietary intake is reduced by a PTH-dependent mechanism.
2. Plasma inorganic phosphate is a major regulator of 25-OH-D_3. Increase of↑ 1,25 $(OH)_2$ D_3 activity increases↑ PO_4 absorption. Decrease↓ of 1,25 $(OH)_2$ D_3 decrease↓ PO_4 absorption.

FUNCTIONS

- Phosphate is the constituent of bone and teeth.
- *Energy transfer:* The free energy produced by metabolic reactions may be stored as high energy phosphate ATP, creatine phosphate.
- *Acid-base balance:* The buffer which is effectively handled by kidneys is a phosphate buffer. It is a mixture of dibasic and monobasic phosphates.
- Phosphorylation/dephosphorylation and phosphorolysis reaction involves phosphate.
- *Enzyme action:* Phosphate of several coenzymes such as NADP, TPP, is involved in enzymatic reactions.
- Constituent of phospholipids, nucleotides/nucleic acids, lipoproteins, phosphoproteins is phosphate.

CLINICAL IMPORTANCE

Rickets and osteomalacia are important dietary deficiency disorders of calcium, phosphorus or vit-D. Plasma levels of adult 0.6-1.2 mmol/L are lower compared to childhood 1.3-2.8 mmol/L. There is often a slight fall in PO_4 after a meal rich in carbohydrates. Plasma Ca and phosphate together are normally measured.

↑ Ca + ↓ PO_4	Primary hyperparathyroidism
↑ Ca + ↑ PO_4	Malignancy (1° or 2°) tumour deposits in bone, post-dialysis in renal failure.
↓ Ca + ↑ PO_4	Hypoparathyroidism
↓ Ca + ↓ PO_4	Vit. D deficiency

I. Hypophosphataemia: It may be due to:
- ***Decreased intake:***
  Starvation, Malabsorption, Vomiting.
- ***Increased cell uptake:***
  High dietary carbohydrate, Liver disease.
- ***Increased Excretion:***
  Diuretics, Hypomagnesaemia, ↑PTH.

II. Hyperphosphataemia: It may be due to:
- ***Factitious hemolysis, prolonged contact of plasma with red cells 7-8 hours.***
- ***Increased intake:***
  Diet, Vit. D.
- ***Increased release from cells:***
  Diabetes mellitus, Acidaemia, Starvation.
- ***Increased release from bone.***
  Malignancy, Renal failure, ↑PTH.
- ***Decreased excretion:***
  Renal failure, Hypoparathyroidism, ↑growth hormone.

SULPHUR

Sources

Sulphur is an essential element. The sulphur is made available to the body by the proteins containing methionine, cystine or cysteine. These amino acids contain sulphur. Also available from S-containing B-vitamins, viz. Thiamine (TPP), coenzyme A, Lipoic acid and biotin. Certain sulpholipids and glycoproteins (mucoitin and chondroitin sulphuric acid) also provide sulphur (Sulphur as free element cannot be utilised).

SECTION FOUR

Sulphur is thus available in meat, fish, legumes, egg, liver, cereals. Adequate protein in diet fulfills sulphur requirement.

Absorption: Sulphur is ingested as organic sulphates as in proteins or as inorganic sulphate. Inorganic sulphate is absorbed as such from the intestines, while sulphur containing amino acids are absorbed by active transport.

Distribution: About 0.25 per cent (150-200 g) of the total body wt. is sulphur. It is mainly found as organic compounds such as Met, Cys, heparin, glutathione, thiamine, biotin, CoA, lipoic acid, taurocholic acid, etc. Many proteins, hormones, keratin of hair contain sulphur. Small amounts of inorganic sulphates occur in tissues and body fluids. 100 ml of blood contains 0.1 to 1.0 mg sulphur as organic compounds.

Fate: Catabolism of S-containing amino acids yields inorganic sulphates. Liver converts inorganic sulphate to ethereal sulphate by conjugation.

METABOLIC FUNCTIONS

- ***Formations of 'active sulphate' (PAPS):*** Active sulphate participates in several transulfuration reactions.
- Sulphur is involved in the formation of proteins such as keratin, chondroproteins, sulpholipids.
- It is also involved in the formation of –SH groups which act as active centres of enzymes such as Acyl carrier protein (ACP) and multienzyme complex of fatty acid synthesis.
- It forms –S–S–linkages between two –SH groups of cysteine to form a secondary and tertiary structure of proteins.
- Iron-sulphur proteins are found in electron transport chain.
- ***S-adenosylmethionine*** is a co-substrate for methylferases (transmethylation).
- S-adenosylmethionine also acts as the initiator in initiation process of protein synthesis.
- Sulphur containing vitamin such as biotin, pantothenic acid, thiamine, lipoic acid are involved as coenzymes.
- Sulphates hexosamines and hexuronic acids are important constituents of mucopolysaccharides, sulphated galactose occurs in sulpholipids.
- Phenol, skatole, indole and steroids may be detoxicated in the liver with sulphate ions.
- Acyl complexes of CoA, S-adenosyl methionine, 'active' sulphate are high energy sulphur compounds.

IRON

Iron is one of the ***most essential trace elements in the body.*** In spite of the fact that iron is the fourth most abundant element in the earth's crust, iron deficiency is one of the most important prevalent nutritional deficiencies in India. The reasons for this are numerous and, in many cases, not well understood. ***Total iron content in a human of 70 kg body weight varies approximately from 2.3 gm to 3.8 gm.*** Average iron content of adult males is about 3.8 gm and of females about 2.3 gm.

Types of Iron Present in Body: There are ***two broad categories*** that are used to describe iron in the body. They are:

A. Essential (or functional) iron

B. Storage iron.

A. Essential Iron

Essential or functional iron is one which is involved in the normal metabolism of the cells.

They are mainly divided into **three** groups:

1. Haem Proteins

- ***Haemoglobin and Myoglobin:*** These are haem proteins, which are proteins with an iron-porphyrin prosthetic group attached to the protein globin. They are most abundant of the essential (or functional) iron compounds in the body (Refer chapter on Chemistry of Haemoglobin).

Other haem proteins are:

- ***Catalases:*** A haem containing enzyme. It contains four haem groups and is found in blood, bone marrow, mucous membrane, liver and kidney. Its molecular weight is 225, 000. It destroys hydrogen peroxide, H_2O_2, formed in the tissues and molecular O_2 is evolved in the reaction (Refer to chapter on Biologic Oxidation).

$$2H_2O_2 \xrightarrow{\text{Catalase}} 2H_2O + O_2$$

The reaction is inhibited by CN^-, F^- and H_2S. Peroxisomes are found in liver and they are rich in aerobic dehydrogenases and catalases.

- ***Peroxidases:*** It is typically a plant enzyme, but it is also found in milk, erythrocytes, leucocytes and lens fibres. Its molecular weight is 44,100 and its prosthetic group is protohaem which is only loosely bound to apoprotein.

Example: Typical example is glutathione peroxidase which also catalyses destruction of H_2O_2, but it works in conjunction with reduced glutathione, G-SH.

$$\underset{\text{(Reduced glutathione)}}{H_2O_2 + 2\ \text{G-SH}} \xrightarrow[\text{(Se-containing)}]{\text{Glutathione peroxidase}} \underset{\text{(Oxidised glutathione)}}{H_2O + \textbf{G–S–S–G}}$$

Note:
- Both catalase and glutathione peroxidase catalyses the destruction of H_2O_2 and forms H_2O and O_2.
- **Catalase can directly act on H_2O_2, but glutathione peroxidase cannot, it requires reduced glutathione. The reason is Km of catalase for H_2O_2 is much greater than that of glutathione peroxidase.**
- To scavenge small amounts of H_2O_2 formed in cells like RB cells and lens fibres, glutathione peroxidase becomes the active enzyme.

2. ***Cytochromes:*** A second group of organo-iron compounds in the body are the cytochromes. Cytochromes are chiefly found in mitochondria (Refer to chapter on Biologic Oxidation).
3. ***Iron Requiring Enzymes:*** A third group of iron containing compounds is the iron-requiring enzymes.
   a. This group contains enzymes that also use riboflavin as coenzyme.

      Examples are: ***Xanthine oxidase, Cytochrome C reductase, Acyl-CoA dehydrogenase, NADH-reductase.***
   b. Other enzymes in this group that require the metal only as cofactor:
      Succinate dehydrogenase, Aconitase, Ribonucleotide reductase
   c. Fe^{++} is required for conversion of O_2^-, superoxide radical to free OH^- radical. This is called as **Haber's reaction**. (Refer to chapter on Chemistry of Respiration).

B. Storage Iron

Storage iron is present in ***two major compounds.*** They are:

1. **Ferritin**
2. **Haemosiderin.**

1. Ferritin

Free iron is toxic and catalyses the conversion of O_2^- to hydroxy OH^- oxy radicals. Iron bound to ferritin is non-toxic. It is the storage protein of iron and found in blood, liver, spleen, bone marrow and intestine (mucosal cells).

Apoferritin is the apoprotein with a molecular weight of 550,000. Apoferritin is an interesting compound in that it is ***composed of 24 monomeric units,*** each having molecular weight of 18,000, that form a spherical shell. (Figs 34.2A and B). There are ***six pores in the shell*** that allow molecules of a certain size to enter and exit the shell. The pores have been ***shown to have catalytic activity,*** most notably the binding of ferrous iron (Fe^{++}) and its subsequent oxidation to ***"ferric oxy hydroxide"*** (FeO.OH). Bound form of iron with ferritin is more soluble and iron is present as ***"ferric oxyhydroxy phosphate"*** complex in ferritin and it is reddish brown in colour. ***Up to 4500 Fe^{+++} atoms are found stored in a ferritin complex.***

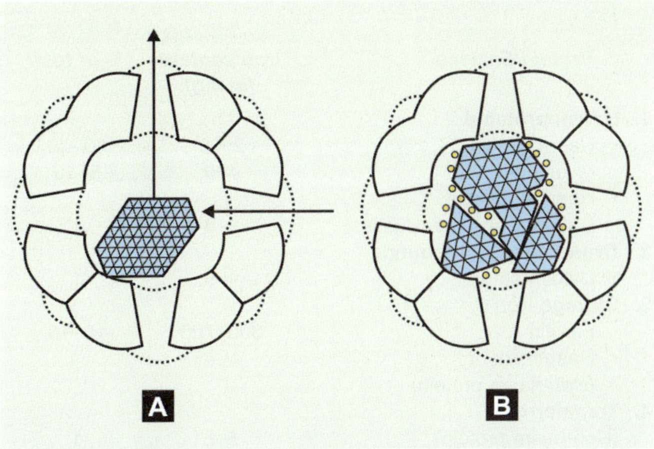

Figs 34.2A and B: Diagrammatic representation of ferritin molecules. (A) A partially filled ferritin molecule with a hydrous Fe (III) oxide microcrystal growing from a nucleation centre inside the apoferritin shell: the arrows indicate addition and release of iron (or phosphate) at the microcrystal surface. For the single perfect microcrystal shown, addition and release would be expected to follow a 'last-in first-out' principle. (B) A nearly full ferritin molecule containing three microcrystals not perfectly aligned: added ions (such as phosphate), represented by full circles, might be expected to bind both at crystallite surfaces and within the 'imperfections' ions bound within these faults would be released after those at surface positions

2. Haemosiderin

Evidence suggests that ***haemosiderin is derived from ferritin*** and is ferritin with partially stripped shell. Haemosiderin contains a larger fraction of its mass as Fe than does ferritin and exists as microscopically visible Fe-staining particles. ***Haemosiderin is usually seen in states of iron overload or when Fe is in excess,*** when the synthesis of apoferritin and its uptake of Fe are maximum. Haemosiderin is rather insoluble. Fe in haemosiderin is available for formation of Hb, but mobilisation of iron is much slower from haemosiderin than ferritin.

Distribution of iron in the body is shown in the Table 34.1.

TRANSFERRIN

Transferrin is a non-haem iron binding glycoproteins. ***Apotransferrin is the apoenzyme and Fe is its prosthetic group.*** It has a molecular weight of 70,000 and ***it can bind with two atoms of iron in the ferric state (Fe^{+++})*** synergistically in presence of HCO_3^- ion. It exists in plasma as β_1-globulin and **is the true carrier of iron.** In plasma, transferrin is saturated only to the extent of 30 per cent to 33 per cent with iron. Prior to binding to transferrin, Fe^{++} (ous) iron has to be oxidised to Fe^{+++}(ic) form. **Ceruloplasmin and ferroxidase II are required for this conversion.**

Note: In copper deficiency, this conversion cannot occur and haemopoietic system will not be getting required amount of Fe for inclusion in Hb synthesis.

Table 34.1: Distribution of iron in the body		
Protein/Enzyme	***Iron content (in mg)***	***% of total***
1. Haemoproteins		
• Haemoglobin	2500	60–70
• Myoglobin	400	5–10
• Haem enzymes: Catalase and peroxidase	2–3	1
2. Organo-iron compounds		
• Cytochromes	4–5	1
3. Storage iron		
• Ferritin	300–700	10–15
• Haemosiderin (Non-haem protein)		
4. Transferrin (Non-haem protein)	6–8	1
5. Iron requiring enzymes		
• Fp, Fe-S Non-haem enzymes	–	1
• Other-dehydrogenases	–	1
• Non-haem enzymes		

Function of Transferrin

Major function of transferrin is transport of iron to RE cells, bone marrow to reach the immature red blood cells. Specific receptors are available on cells surface. Transferrin is internalised by receptor mediated endocytosis. Within the target cells, iron is released and apotransferrin is recycled to form new transferrin molecules **(Fig. 34.3).**

DIETARY SOURCES OF IRON

*I. **Exogenous:*** Foods rich in iron include:

a. ***Animal Sources:*** Meat, fish, liver, spleen, red marrow are very rich sources (2.0 to 6.0 mg/100 gm). Also found in shellfish.

b. ***Vegetable Sources:*** Cereals (2.0 to 8.0 mg/100 gm) are the major rich source. Legumes, molasses, nuts, amaranth leaves. ***Dates are other good sources.***

*II. **Endogenous:*** Fe is utilised from ferritin of RE system and intestinal mucosal cells. Fe obtained from "effete" red cells are also reutilised.

SECTION FOUR

ABSORPTION OF IRON AND FACTORS REGULATING ABSORPTION

Normally, the loss of iron from the body of a man is limited to 1 mg per day. Menstruating women lose iron with menstrual blood. Around 10 to 20 mg of Fe is taken in the diet and only about 10 per cent is absorbed. The greatest need of iron is during infancy and adolescence.

The only mechanism by which total body stores of iron is regulated is at the level of absorption. ***Garnick*** proposed a ***"mucosal block theory"*** for iron absorption.

Mucosal Block Theory

1. Soluble inorganic salts of iron are easily absorbed from the small intestine. HCl present in gastric juice liberates free Fe^{3+} from non-haem proteins. Vitamin C and glutathione in diet reduce Fe^{3+} to Fe^{2+}, which is less polymerisable and more soluble form of iron. Vitamin C and amino acids can form iron-ascorbate and iron-amino acid chelates which are readily absorbed. Haem is absorbed as such.
2. ***Gastroferrin,*** a glycoprotein in gastric juice is believed to bind iron and facilitate its uptake in duodenum and jejunum.
3. The absorption of iron from intestinal lumen into mucosal cells takes place as Fe^{2+}.
4. ***Events in intestinal mucosal cells (Enterocyte):*** **(Fig. 34.4)**
   - **Enterocytes** in the ***proximal duodenum are responsible for absorption of iron.***
   - Incoming iron in the **Fe^{3+} state is reduced to Fe^{2+}** by an enzyme ***"Ferrireductase"*** present on the surface of enterocytes, it is helped by **vitamin C** present in the foods.
   - The transfer of iron (Fe^{2+}) from the apical surfaces of enterocytes into their interiors is performed by a **proton-coupled divalent metal transporter (DMT1).** This protein is **not specific** for iron as it can transport a wide variety of divalent cations.

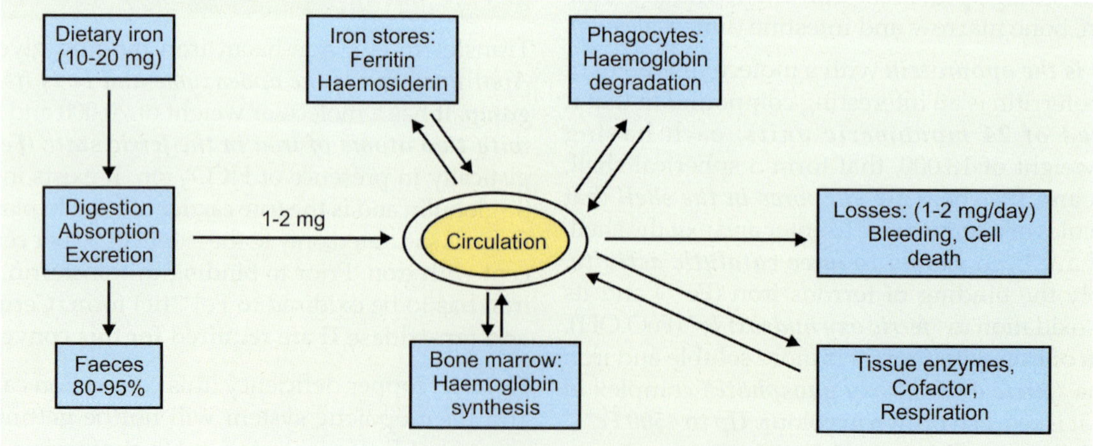

Fig. 34.3: Iron distribution and transport in humans

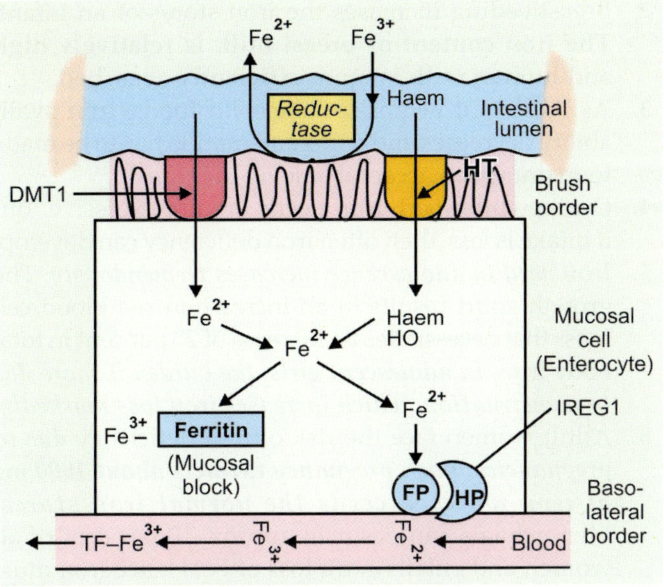

Fig. 34.4: Absorption of iron

- Once it is inside, it can either be **stored as "ferritin"** or it can be transferred across the busolateral membrane into the plasma where it is carried bound to **transferrin.**
- Passage of Fe^{2+} across the basolateral membrane is carried out by another protein called **iron regulatory protein 1 (IREG 1).**
- Most of Fe^{2+} required to be absorbed is transferred to plasma by a **Fe^{2+} transporter (FP).**
- Fe^{2+} in the enterocytes also come from **"haem"** by the action of ***"haem oxidase"*** enzyme on haem.
- **IREG1** may interact with the copper containing protein called **"hephaestin"**, a protein similar to caeruloplasmin. Hephaestin is thought to have a **'ferroxidase' activity** which is important in the release of iron from cells ***as Fe^{3+}, the form in which it is transported in the plasma by transferrin.***

Overall regulation of iron absorption is complex and not well understood mechanistically. It is exerted at the level of the enterocyte where further absorption of iron is blocked if sufficient amount taken up, for body need—so called dietary regulation exerted by **"mucosal block" (Garnick's hypothesis).**

Key

- **HT** – Haem transporter
- **DMT-1** – Iron transporter
- **HO** – Haem oxidase
- **FP** – Iron transporter present in basolateral border
- **HP** – Hephaestin
- **IREG1** – Iron regulatory protein 1

Other Factors

(a) ***Source of Fe has marked effect on absorption:***
- ***Haem iron*** which comes mainly from animal products and is from Hb and myoglobin, is efficiently absorbed (about 20 to 30%).
- ***Non-haem iron,*** which is present in plants, though ingested in larger amount than haem iron, are inefficiently absorbed (only 1 to 5%).

(b) ***The absorption of non-haem iron is influenced by the:***
- ***Composition of the diet***
- ***pH of the intestinal milieu,*** and
- ***State of health of the individual.***

1. ***Composition of the Diet:*** The composition of the diet exerts a profound effect on non-haem iron absorption.
   - Dietary factors that increase iron absorption are the presence of vitamin C (ascorbic acid), glutathione, and some form of meat, fish or poultry ***(all contain an unknown "meat factor").***
   - Foods that inhibit non-haem iron absorption to some extent are:
     – Tea (diminishes absorption by > 60%)
     – Coffee (reduced absorption by > 35%)
     – Phytates, found in corn, soya products, grains, and bran (producing insoluble complex)
     – Oxalates found in spinach and chocolates
     – Some dietary fibres may also bind the iron or decrease gastrointestinal transit time.
2. ***pH of Intestinal Milieu***
   - HCl secreted in gastric juice liberates Fe^{3+} from non-haem iron and serves to increase solubility of dietary non-haem iron.
   - pH of duodenum is most conducive for absorption. Rate of absorption further decreases down the intestines as the pH becomes more alkaline.
   - At high alkaline pH, the ingested iron is precipitated.
3. ***State of Health of the Individual***
   - Healthy adults absorb about 5 to 10 per cent of dietary iron, which is approx. 1 to 2 mg of iron.
   - Iron-deficient adults absorb 10 to 20 per cent of the dietary iron equivalent to 3 to 6 mg of Fe.

Note

- Individuals having achlorhydria or achylia gastrica and in persons having resection of gut, partial or total gastrectomies, iron absorption is diminished and they are at risk of iron deficiency.
- Intestinal epithelia normally are desquamated and again regenerated. ***Significant quantity of iron as ferritin is lost during desquamation.***
- ***Parasitic infection,*** viz. Ankylostoma duodenale produces Fe loss as the parasites suck blood and thrive on it.

SECTION FOUR

Iron Transport and Utilisation: Transport of Fe throughout the body is accomplished with a specific protein called **transferrin** (Shown earlier).

Transferrin transports Fe from the GI tract to the bone-marrow for Hb synthesis and to all other cells as required. Transferrin can transport a maximum of two atoms of iron as Fe^{3+} per molecule. ***Normally, in plasma/serum transferrin is about 33 per cent saturated with Fe.*** As discussed above, cell surface specific receptors are available for the iron-transferrin complex. Tissues having high uptake, e.g. liver, have a larger number of receptors present. ***The number of receptors decreases when a person is replete with iron and increases with depletion.*** Iron is transported to bone marrow where it is required for Hb synthesis. Fe^{2+} is incorporated in protoporphyrin IX with the help of the enzyme ***"ferrochelatase".***

Iron is also transported into cells where it is used for both oxidative phosphorylation and as an enzyme cofactor.

A small amount of Fe is released each day from 'effete' red cells, which are destroyed by phagocytes, but this released Fe^{2+} is recycled into new Hb in the erythroblasts. A small amount of released Fe is also stored as ferritin. ***The turnover of iron in an adult in 24 hours has been calculated to be 35 to 40 mg.*** Plasma "transferrin iron pool" is in equilibrium with the iron in storage forms ferritin and haemosiderin. Ferritin in storage form of Fe occurs in reticuloendothelial system (RES), viz. liver, spleen and bone marrow and also in intestinal mucosal cells.

When Fe is mobilised from ferritin, the storage form, the sequence is as follows:

- **First call:** From ferritin of RE system (liver, spleen and bone marrow)
- **Second call:** From ferritin of intestinal mucosal cells.
- **Thirdly:** Absorbed iron from intestines.

Before Fe is released from ferritin to blood, Fe^{3+} of ferritin is first reduced to Fe^{2+}.

Iron Requirements

Requirement of iron varies according to age, sex, weight and state of health. An adult male requires approx. 10 mg/day and adult female 20 mg/day. Pregnancy and lactation demands more: Pregnant women require 10 mg/day and lactating mothers 25 to 30 mg/day. Children require 10 to 15 mg/day.

Note

1. In the last trimester of pregnancy, the foetus sequesters Fe from the mother at an average rate of 3 to 4 mg/day. At birth, the infant's iron stores average 75 mg/kg body wt. A premature infant will not have had time to build its stores sufficiently.
2. Breastfeeding increases the iron stores of an infant. **The iron content of breast milk is relatively high** and **human milk is more efficiently absorbed.**
3. As the infant begins to take solid foods, iron availability decreases and supplementation has to be made to maintain an acceptable level in infants.
4. During childhood, iron needs are not excessive. But if intake is less, then often iron deficiency can develop.
5. ***Iron need in adolescence increases tremendously.*** The growth spurt results in an increase in red blood cell mass that necessitates an increase of 25 per cent in total body iron. ***In adolescent girls, the burden is more due to menstruation, which increases iron loss markedly.***
6. Adult women face the risk of iron deficiency ***due to pregnancies. Each pregnancy extracts about 1000 mg of iron which exceeds the normal iron stores.*** Menorrhagia can occur in approx. 10 per cent of all women and entails extra loss of Fe. Hence iron must be supplemented in such cases to avoid iron deficiency anaemia.
7. Iron status of adult males and postmenopausal women tends to increase throughout life and normally iron deficiency should not be a problem. ***But certain chronic diseases at this age group, if not treated, can lead to iron deficiency. They are:***
   - *Chronic infections including tuberculosis*
   - *Malignancies*
   - *Chronic aspirin ingestion*
   - *Peptic ulcers*
   - *Rheumatoid arthritis* and
   - *Parasitic infections, etc.*

CLINICAL ASPECT

A. IRON DEFICIENCY: *Three stages* of iron deficiency are:

1. ***Iron storage depletion***
2. ***Iron deficiency***
3. ***Iron deficiency anaemia.***

1. ***Iron Storage Depletion:*** This phase is not usually recognisable by the patient and normally does not elicit a medical examination. Serum ferritin decreases during this phase and is the only good indication of possible iron deficiency. ***Many women of child bearing age remain in this phase for years without being identified.***
2. ***Iron Deficiency:*** In this phase, iron stores are almost exhausted. Biochemically the ***serum ferritin is low↓ and transferrin saturation is low↓.*** Erythrocyte protoporphyrin↑ increases, as erythropoiesis is slowed down due to nonavailability of Fe which cannot be incorporated in protoporphyrin IX. Haemoglobin concentration falls↓ to the lowest limit of normal.
3. ***Iron Deficiency Anaemia:*** Iron deficiency anaemia is manifested as ***hypochromic microcytic anaemia.*** At this phase:

- ***Hb concentration continues to fall*↓**
- Serum ferritin level shows slow decline↓
- ***Transferrin saturation continues to fall*↓** and
- Erythrocyte protoporphyrin increases to upper limit of normal↑.

To classify as iron deficiency anaemia, a low Hb plus a documented abnormal serum ferritin or other iron test must be present.

Note: It is imperative to determine iron levels in all patients with anaemias, since there are disorders such as thalassaemias that may be present and misdiagnosed as iron deficiency.

B. IRON OVERLOAD

Iron overload can also be an important clinical concern. Iron stores may increase due to:

- ***Excessive absorption,*** or
- ***Parental iron therapy*** or
- ***Repeated transfusions.***

Cells start to fill with excess of haemosiderin↑. Both reticuloendothelial cells and parenchymal cells sequester iron.

Types: ***Two broad types*** of iron overload are seen:

1. **Haemochromatosis:** When iron overload is ***associated with injury to cells.***
2. **Haemosiderosis:** Iron overload ***without cell damage*** is called haemosiderosis.

1. HAEMOCHROMATOSIS

Haemochromatosis can be of ***two types:***

1. ***Primary hereditary haemochromatosis (Idiopathic)***
2. ***Secondary haemochromatosis.***

Note:

- ***Early diagnosis of haemochromatosis is essential.*** Untreated haemochromatosis can lead to liver, pancreatic and cardiac impairment, diabetes mellitus, and hepatic carcinoma.
- ***A helpful screening test is:*** Serum transferrin saturation. Patients with serum transferrin saturation > 62 per cent may have haemochromatosis.

a. Primary (Idiopathic) Haemochromatosis:

- Inherited disorder
- Autosomal recessive
- Massive accumulation of iron, mainly as ferritin and haemosiderin, in visceral organs, principally liver and skin.
- ***Classic "triad" for diagnosis***
  1. ***Micronodular cirrhosis*** with marked brown pigmentation.
  2. ***Diabetes mellitus*** and
  3. Skin pigmentation called as **"Bronze diabetes".**
- ***Deposition of iron in myocardium can cause cardiomyopathy and heart failure.***

  Excess iron accumulates in the cytoplasm of parenchymal cells with possible "free" radical generation, leading to lysosomal disruption and cell damage.
- ***Increased risk of hepatocellular carcinoma***
- Haemochromatosis "gene" is linked to HLA-A_3, with a frequency of 1:200.

Postulated Mechanisms for the Disease:

- Hereditary defect in regulation of iron absorption by the duodenum and jejunum.
- A defect in immediate postabsorptive excretion of iron.
- A genetic inability of phagocytes to take up iron with loss of their regulatory control signals over iron absorption.

Defect at Molecular level

Feder and colleagues isolated a gene called as **HFE, located on chromosome 6 close** to the major histocompability complex genes. **The encoded protein HFE was found to be related to MHC class I antigens.**

Two different missense mutations can take place in HFE in individuals with primary haemochromatosis:

1. **More common and frequent mutation** is that **changes cysteinyl** residue ***282 to a tyrosyl residue (CY 282 Y)*** which disrupts the structure of HFE protein.
2. **The other mutation** changes **histidine residue 63** to an **aspartyl residue (H 63 D).**

In some patients, neither of the above mutations occur, perhaps some other mutations in HFE gene or because one or more other genes may be involved in its causation.

Effect of HFE

HFE has been shown to be **located in cells in the crypts of the small intestine,** the site of iron absorption. There is evidence that **it is associated with β_2-microglobulin,** an association necessary for its stability, intracellular processing and cell surface expression. The complex interacts with the transferrin receptor (Tfr); how this leads to excessive storage of iron when HFE is altered by mutation is not very clear.

Summary

Suggested scheme of events that take place in primary idiopathic haemochromatosis.

- Mutations in HFE gene located on chromosome 6p21.3

↓

- Leading to abnormalities of structure of encoded HFE protein

↓

- Loss of regulation of iron in small intestine

↓

- Results to accumulation of iron in various tissues particularly in pancreatic islets, liver, heart, muscle, and skin.

↓

- Excessive iron damages the tissues leading to cirrhosis liver, diabetes mellitus, damage to cardiac muscles, skin pigmentation (bronzed diabetes).

b. Secondary Haemochromatosis:

- Due to ineffective erythropoiesis as in thalassaemia, erythrogenesis imperfecta or
  - With repeated blood transfusion/transfusion overload or
  - Patients with haemodialysis.
- It exhibits more even distribution of iron between macrophages and hepatocytes. Long-term accumulation of Fe in hepatocytes leads to hepatocellular necrosis and secondary scarring and rarely a micronodular cirrhosis develop.

2. SIDEROSIS OR HAEMOSIDEROSIS

- ***Bantu siderosis:*** Bantus in Africa cook their food in iron pots. This causes enhanced absorption of Fe leading to Bantu siderosis. Their PO_4 intake is usually low as they consume plenty of corn. ***Low PO_4 aggravates increased Fe absorption.***

  Note: ***Fe deficiency anaemia is not found in pregnant Bantu women.***
- Repeated blood transfusions, thalassaemia and hereditary haemolytic anaemias may also result in this condition.
- ***Idiopathic pulmonary haemosiderosis:*** Chronic episodic haemorrhages of the lungs of unknown etiology result in prominent haemosiderin deposition and fibrosis.

SECTION FOUR

COPPER

Adult humans contain 100 to 150 mg of copper, out of which approximately 65 mg is found in muscles, 23 mg in bones and 18 mg in liver. Foetal liver contains approximately ten times more copper than adult liver.

It occurs as:

1. *Erythrocuprein* (in red blood cells),
2. *Hepatocuprein* (in liver) and
3. *Cerebrocuprein* (in brain).

Erythrocuprein is a colourless protein containing 2 atoms of Cu per molecule. Molecular weight approx. 33,000.

Source: Average diet provides 2 to 4 mg/day in the form of meat, shellfish, legumes, nuts and cereals. Milk and milk-products are poor sources.

Absorption: ***Primarily absorbed from the duodenum.*** About 32% of the dietary Cu can be absorbed. Phytates, Zinc, Mo, Cd, Ag, Hg and high amount of Vit. C inhibit Cu absorption.

Absorption of Cu from GI tract requires a specific mechanism because of highly insoluble nature of Cu^{++} ions. ***An unidentified low molecular weight substance from human saliva and gastric juice complexes with Cu^{++} to keep it soluble at pH of intestinal fluid.*** In the intestinal mucosal cells, Cu is associated with low molecular weight metal binding protein called as ***metallothionein.***

Plasma: After absorption, Cu enters plasma, where it is bound to amino acids, particularly histidine and to serum albumin at a single strong binding site. In less than an hour, the recently absorbed Cu is removed from the circulation by liver.

Role of Liver in Copper Absorption

Liver processes absorbed Cu through **two routes:**

1. Cu is excreted in the bile into the GI tract from which it is not reabsorbed. In fact, copper homeostasis is maintained almost exclusively by biliary excretion, the higher the dose of the Cu, more it is excreted in faeces. Normally, human urine contains only traces of Cu.
2. Second route: Incorporation as an integral part of *Caeruloplasmin*, a glycoprotein synthesised exclusively by liver.

Serum Copper

Serum Cu level is approximately 90 μg% (average). In red blood cells: 93 to 115 μg/100 ml. During pregnancy, the serum levels of Cu rises steadily and reaches peak level at the time of parturition. Cu of red blood cells, however, remains almost constant.

Serum Cu is present in **two distinct forms:**

1. *Direct reacting Cu:* Which is loosely bound to albumin. Approximately 4 per cent present in this form. So-called as it reacts directly with diethyl dithiocarbamate.
2. *Bound form:* Which remains bound to α-globulin fraction of the serum, called "Caeruloplasmin" (as stated above). About 96 per cent of serum Cu is found in combination with caeruloplasmin (For Caeruloplasmin, refer chapter on Plasma Proteins—Chemistry and Functions).

Excretion: Under normal conditions, 85 to 99 per cent of the ingested Cu is excreted in the faeces via the bile, and remaining 1 to 15 per cent in the urine. The amount retained in the body is probably dependent mainly on the Cu status of the tissues and is influenced relatively slightly by the intake.

Requirements

- **Infants and children:** 0.05 mg Cu/kg body wt. per day
- **Adult requirement:** Is approximately 2.5 mg/day. Ordinary diets consumed daily contain about 2.5 to 5.0 mg Cu.

FUNCTIONS

1. *Role in Enzyme Action:* Cu forms integral part of certain enzymes, e.g. some of cytochromes, *cytochrome oxidase, tyrosinase, Monoamine oxidase (MAO), Lysyl oxidase, Catalase, Ascorbic acid oxidase, uricase* and

superoxide dismutase. These enzymes contain about 550 μg of Cu per gram of enzyme protein.

Superoxide dismutase: A colourless dimeric enzyme, having mol. wt. = 32,000, present in cytosol of mammalian liver, nerve and red cells and contain 2 Cu^{++} and 2 Zn^{++} per molecule.

Function of Superoxide dismutase: Changes superoxide radicals, formed by univalent reduction of O_2 in tissues, to hydrogen peroxide (H_2O_2)

$$O_2' + O_2' + 2H^+ \rightarrow O_2 + H_2O_2$$

There is another **"mitochondrial" form** of 'superoxide dismutase' enzyme which is a different protein with Mn^{++} instead of Cu^{++} and Zn^{++} as its prosthetic group.

2. ***Role of Cu^{++} in Fe Metabolism***
   - Cu helps in the utilisation of Fe for Hb synthesis in the body. It is believed that 'caeruloplasmin', a blue-Cu protein complex of blood plasma, ***functions as serum Ferro-oxidase,*** catalyses the oxidation of Fe^{++} to Fe^{+++}. ***This helps in the incorporation of Fe in "transferrin" to facilitate mobilisation and utilisation of Fe.***
   - Facilitatory role of Cu^{++} in iron absorption also.
   - A yellow copper-protein called serum *Ferro-oxidase II* or non-caeruloplasmin ferro-oxidase may also participate in the oxidation of Fe^{++} in human plasma.
3. ***Role in Maturation of Elastin:*** Copper helps to form insoluble elastin fibres by cross-linking soluble proelastin chains through the oxidation of some lysine side chains of the latter into aldehydes. Proelastin rises significantly in copper deficient animals.
4. ***Role in Bone and Myelin Sheath of Nerves:*** Copper has been reported to help in the formation of bones and maintenance of myelin sheaths of nerve-fibres.
5. ***Role in Haemocyanin:*** One of the copper protein "haemocyanin" found in blood of certain invertebrates functions as Hb in the storage and transport of O_2.

Copper Deficiency Manifestations

1. ***Loss of weight:*** A diet deficient in copper causes loss of weight which sometimes may prove fatal.
2. ***Bone disorder:*** A bone disorder has been reported in the dogs on a copper-deficient diet which is characterised by abnormally thin cortices, deficient trabeculae, and wide epiphyses.
3. ***Anaemia:*** Copper deficiency produces ***microcytic hypochromic anaemia,*** due to impairment of erythropoiesis and decreases in erythrocyte survival time, which cannot be corrected by administration of iron.
4. Copper deficiency turns ***hair grey,*** which however, can be controlled by administration of Cu.
5. Copper deficiency has been reported to involve ***atrophy of myocardium***. The elastic tissue of aorta, coronary arteries and pulmonary artery is deranged. These vessels may rupture and cause death.
6. Depletion of brain Cu stores has been observed to cause non-coordinated movements and demyelination of the nerves in the animals. Histopathological changes may occur in the cerebrum, brainstem, and spinal cord.

CLINICAL ASPECT

INHERITED DISORDERS

1. WILSON'S DISEASE (hepatolenticular degeneration):

Inheritance: It is inherited as autosomal recessive.

Metabolic defects: Mainly **two defects:**

1. Defect in incorporation of Cu into newly synthesised "apo-caeruloplasmin" to form caeruloplasmin. It is not clear whether the genetic defect is in the structural gene for caeruloplasmin or in the process of incorporating Cu into caeruloplasmin.
2. In addition to above, the patients have impaired ability of the liver to excrete Cu into bile.

Defect at Molecular level

In 1993, it was reported that a variety of mutations in a gene, encoding a ***copper binding P type ATPase*** was responsible for Wilson's disease. The gene is estimated to encode a **protein of 1411 amino acids.**

In a manner not yet fully explained, a **non-functional ATPase causes (1) defective excretion of copper into the bile, (2) a reduction of incorporation of copper into apocaeruloplasmin and (3) the accumulation of copper in liver, brain, kidney and RB cells.** It can be regarded as inability to maintain a near-zero copper balance, resulting in **copper toxicity**.

Clinical features: Total body retention of Cu is increased particularly in organs like ***liver, brain, kidney*** and ***cornea.***

- ***Liver:*** Produces progressive ***hepatic cirrhosis*** of a coarse nodular type which leads to hepatic failure.
- ***Brain:*** There is dysfunction of lenticular region of the brain, necrosis and sclerosis occurs.
- ***Kidneys:*** Defects in renal tubular reabsorption producing aminoaciduria.
- ***Eyes:*** Copper deposition in "Descemet's membrane" of the eye causes a golden brown, yellow or green ring round the cornea, called as ***"Kayser-Fleischer ring".***

Blood: The ***serum Cu is low***, as caeruloplasmin in patient's plasma contains no Cu.

Urine: Urinary excretion of Cu is markedly increased↑.

Treatment: Improvement can be achieved by removing the excess of tissue Cu by administering Cu-chelating agent like **"penicillamine"**.

2. MENKE'S DISEASE

***Synonym:* Kinky or Steel hair syndrome**

Inheritance: It is an ***X-linked disorder*** of intestinal copper absorption.

Metabolic defect: The first phase of Cu absorption its uptake into the mucosal cells and the second phase of its intracellular transport within the mucosal cells are both normal in patient's with Menke's disease. ***The third phase: Transport across the serosal aspect of the mucosal cell membrane is defective.***

Defect at molecular level

In 1993, it has been reported that ***defect in Menke's disease is mutations in the gene for a "copper binding P type ATPase".*** This ATPase is thought to be responsible for directing the efflux of copper from cells.

When altered by mutation, copper is not mobilised normally from the intestine in which it accumulates, as it does in a variety of other cells and tissues from which it cannot exit.

Clinical Features: IV administered Cu is handled normally by these children but unless therapy is commenced promptly at birth, many of severe signs of this disease like mental retardation, temperature instability, abnormal bone formation and susceptibility to infection are not prevented.

3. ACAERULOPLASMINAEMIA:

Recently another condition involving caeruloplasmin "**Acaeruloplasminaemia**" has been described.

In this genetic disorder, ***levels of caeruloplasmin is very low*** and hence its ***ferroxidase activity is markedly deficient.***

This leads to ***failure of release of iron*** from cells and iron accumulates in certain brain cells, hepatocytes, and pancreatic islet cells. Affected individuals exhibit **severe neurologic signs** and have **diabetes mellitus.**

Use of a chelating agent or administration of plasma/or caeruloplasmin concentrate may be beneficial.

SECTION FOUR

MAGNESIUM

Magnesium is the fourth most abundant and important cation in humans. It is extremely essential for life and is present as ***intracellular ion in all living cells and tissues.***

Sources: Magnesium is widely distributed in vegetables, found in porphyrin group of chlorophyll of vegetable cells and also found in almost all animal tissues. Other important sources are cereals, beans, green vegetables, potatoes, almonds and dairy products, e.g. cheese.

Distribution: Total body magnesium is approximately 2400 mEq. Approximately 2/3 occurs in bones, 1 per cent in EC fluid and remainder in soft tissues.

Plasma level: 1.5 to 1.8 mEq/L, which is rigorously maintained within normal limits. 15 per cent of total body Magnesium is exchangeable with tissues but there is wide variations. Muscles contain 20 per cent of exchangeable Mg and bones only 2 per cent. Hyperthyroidism markedly increases the amount of exchangeable Mg, whereas it is reverse in hypothyroidism.

Blood: Magnesium exists in blood partly bound to proteins. Under conditions of physiological pH roughly 1/3 is 'protein-bound', the remainder 2/3 is ionic.

CS Fluid: Concentration of Mg in CS fluid is ½ as high as in plasma.

ABSORPTION

Average daily intake in humans is 250–300 mg, much of which is obtained from green vegetables where Mg is found in porphyrin group of chlorophyll. Roughly 1/3 of dietary Mg is absorbed; the remainder is passively excreted in faeces. ***Absorption takes place primarily in small bowel,*** beginning within hour after ingestion and continues at a steady rate for 2 to 8 hours, by that time 80 per cent of total absorption has taken place.

FACTORS AFFECTING ABSORPTION

- ***Size of Mg load:*** Absorption is doubled when normal dietary Mg requirement is doubled and vice versa.
- ***Dietary calcium:*** Increased absorption in calcium deficient diets. Decreased absorption occurs in presence of excess of Ca. A common transport mechanism from intestinal tract for both Ca and Mg suggested.
- ***Motility and mucosal state:*** This also affects absorption. In hurried bowel, absorption is decreased. Absorption decreases in damaged mucosal state.
- ***Vit. D:*** Helps in increased absorption.
- ***Parathormone:*** Increases absorption.
- ***Growth hormone:*** Increases absorption.
- ***Other factors***
  - High protein intake and neomycin therapy increases absorption.
  - Fatty acids, phytates and phosphates decrease absorption.

Excretion: Magnesium is lost from the body in faeces, sweat and urine. 60 to 80 per cent of orally taken Mg is lost in faeces.

Sweat loss: Currently it is drawing attention; 0.75 mEq of Mg is lost daily in perspiration in normal health with normal diet. Loss is much increased with visible frank sweating.

Urine: Regulation of Mg balance is principally dependent on renal handling of the ion. In a normal healthy adult with normal diet 3 to 17 mEq are excreted daily.

Factors Affecting Renal Excretion

- ***Calcium intake:*** Increased dietary calcium produced increased excretion of Mg.
- ***Parathormone (PTH):*** Diminishes excretion.

- *Antidiuretic hormone (ADH):* Increases Mg excretion
- *Growth hormone (GH):* Also increases excretion of Mg.
- *Aldosterone:* Increases excretion
- *Thyroid hormones:* 80 per cent greater excretion in hyperthyroidism.
- *Alcohol ingestion:* Oral ingestion of as little as 1.0 ml of 95 per cent alcohol per kg, increases urinary excretion 2 to 3-fold. The increased excretion partially accounts for Mg-deficiency in chronic alcoholics with Delirium tremens.
- Administration of acidifying substances (NH_4Cl) is followed by increased urinary elimination of Mg.

FUNCTIONS

1. *Role in Enzyme Action:* Mg is involved as a cofactor and as an activator to wide spectrum of enzyme actions. It is essential for *peptidases, ribonucleases, glycolytic enzymes* and co-carboxylation reactions.
2. *Neuromuscular Irritability:* Mg exerts an effect on neuromuscular irritability similar to that of Ca^{++}, high levels depress nerve conduction and low levels may produce tetany **(hypomagnesaemic tetany).**
3. *As Constituent of Bones and Teeth:* About 70 per cent of body magnesium is present as apatites in bones, dental enamel and dentin.

CLINICAL ASPECT

Plasma Mg in Diseases

a. **Hypermagnesaemia:** Raised values have been reported in:
 • Uncontrolled *Diabetes mellitus* • Adrenocortical insufficiency • Hypothyroidism • Advanced renal failure and • Acute renal failure.

b. **Hypomagnesaemia:** Low values are observed in:
 • Malabsorption syndrome and Kwashiorkor • Prolonged gastric suction • Hyperthyroidism • Portal cirrhosis • Prolonged use of diuretics • Chronic alcoholism • Delirium tremens • Renal diseases • Primary aldosteronism.

Magnesium Deficiency

In man, 'overt' magnesium deficiency rarely occurs.

1. ***In Animals:*** In cattles, two types:
   a. Unsupplemented whole milk (in calves)
   b. Endemic disease: called as ***Grass staggers (or Grass Tetany).*** Cattles grazing in fields fertilised with Nitrates. Condition occurs due to high NH_3 content of diet. Absorption of Mg is impaired by the formation of insoluble ammonium-Mg-phosphates.

Clinical features: In both similar: Restlessness and convulsions followed by death.

2. ***In Humans:*** Experimentally induced prolonged Mg-depletion reported in two patients (reported by **Shils**). Both were fed Mg-deficient synthetic diets: One for 274 days and another for 414 days. In both, plasma Mg fell slowly over several months.

Clinical features: Personality changes, GI disturbances, gross tremors, hyporeflexia, abnormal electromyograph, +ve Chvostek's sign, epileptiform convulsions. Both cases, despite adequate Ca and K intake, developed hypocalcaemia and hypokalaemia.

FLUORINE

Sources: Fluoride is solely derived in human from drinking water.

Other sources: Tea, salmon, sardine and Mackerel contain small amounts of fluoride.

Requirements: About one part of fluorine in one million parts of drinking water (one PPM) seems to serve the daily requirement of fluorine in human adults and children. ***Daily intake of fluoride should not exceed 3 mg as it is a toxic element. For an adult individual, the lethal dose is 2.5 gm.***

Absorption and Excretion: Dietary soluble fluorides are absorbed by diffusion from the intestine. About 10 to 20 μg of fluoride, mostly ionised are present in 100 ml of blood. Fluorides are present in significant amounts in calcified tissues like bones and teeth. Fluorides are excreted mainly in the urine.

FUNCTIONS

1. *Role in Tooth Development and Dental Health:* Fluorine is present in human tooth in trace amounts and helps in tooth development, normal maintenance and hardening of dental enamel and ***prevention of "dental caries"***. Destruction of dental enamel and incidence of dental caries are widespread among both adults and children in areas where drinking water contains less than 0.5 PPM of fluorine. Cariostatic effect of fluorine is due to its entry into the apatite salts of dental enamel. Greater than 1.2 PPM of fluorine in drinking water of infants/children may increase fluoride contents of the enamel and dentine, may reduce Ca deposition in those tissues and may cause ***mottling of enamel,*** in newly erupted permanent teeth: Discoloration, corrosion and stratification of enamel including formation of pits are observed.
2. *Role in Bone Development:* Fluorine is present in human bones in trace amounts. Very small amount of fluorine in food and drinking water promote normal bone development, increases retention of Ca^{++} and PO_4 and prevent old age osteoporosis.
   - High fluoride intakes may raise the fluoride content of bone, stimulate osteoblast activity and

cause an abnormal rise in calcium deposition and increased density of bone. Catalytic amounts of fluorine are required for the conversion of the phosphates of calcium to 'apatite salts' of bones and teeth, these may be the basis of role of fluorine in teeth and bone development.

CLINICAL ASPECT

Fluoride Toxicity: Fluorosis

Fluorosis: Excess of fluoride in water or diet or its inhalation is harmful and is considered to be the main cause of the crippling disease known as ***'fluorosis'.***

There is still dearth of information on the precise manner in which fluoride ions act upon the body tissues which cause such a derangement. The fluoride levels in GI tract, blood and urine have been reported to be high in fluoride toxicity.

Biochemical Changes in Fluorosis:

- ***Mitochondrial damage:*** Administration of sodium fluoride in dosage of 50 mg/kg body weight for 45 days in rabbits resulted in mitochondrial damage and release of CPK. The fluoride ions adversely affect the sarcolemma by enhancing its permeability↑ and CPK level in blood was raised.
- ***Action on enzymes:*** High fluoride concentration inhibits particularly Mg^{++} dependent enzymes. High fluoride consumption also led to reduction in *succinic dehydrogenase* activity of rabbit diaphragm as well as reduction in the diameter of the fibres. The structural and biochemical changes led to muscle wasting and impairment of energy metabolism.
- ***Effects of protein synthesis:*** Fluoride intake in higher amounts resulted 10 to 46 per cent reduction, in protein contents of various organs, e.g. adrenal gland, cardiac muscles, kidneys, lungs, pancreas, skeletal muscles, spleen, stomach, testes and spinal cord, ***suggestive of inhibition of protein synthesis by fluorides.***
- ***Steroid synthesis:*** Steroid production was impaired because of depletion of δ-5-3-β-*hydroxysteroid dehydrogenase* enzyme.
- ***Effect on collagen synthesis:*** Collagen content is found to be reduced and its biosynthesis adversely affected due to reduced proline uptake↓. ***Formation of deficient collagen fibres with abnormal biochemical sites, provide an impetus for pathological calcification which occurs during fluoride intoxication.***

ZINC

Sources

(a) ***Animal sources:*** Good sources of zinc are liver, milk and dairy products, eggs.

(b) ***Vegetable sources:*** Good vegetable sources are unmilled cereals, legumes, pulses, oil seeds, yeast cells, and vegetables (spinach, lettuce).

Distribution: An adult man weighing 70 kg contains approximately 1.4 to 2.3 gm of zinc in the body. It is distributed in different parts of the body as follows:

- High (70 to 86 mg/100 gm): In skin, and prostate
- Average (15 to 25 mg per 100 gm): In bones and teeth
- Low (2.3 to 5.5 mg/100 gm): In kidneys, muscles, heart, pancreas and spleen and
- Very low (1.4 to 1.5 mg/100 gm): In brain and lungs.

Absorption and Excretion

- Only a small percentage of dietary zinc is absorbed and the absorption occurs mainly from duodenum and ileum.
- ***Zinc-binding factor:*** It has been reported and claimed that a ***low molecular weight zinc-binding factor is secreted by the pancreas, which forms complex with zinc and helps in its absorption.***
- High amounts of dietary calcium, phosphates and phytic acid have been found to interfere with zinc absorption.

Loss in faeces and urine: In a normal healthy adult human, approximately 9.0 mg of zinc is lost in the faeces and about 0.5 mg is lost in the urine and 0.5 mg is retained in the body.

Sweat: Trace amount is lost in sweat.

Blood level: Whole blood contains about 650 to 680 μg of zinc per 100 ml. It is present in red blood cells mainly in *carbonic anhydrase* enzyme molecules and WB cells as other zinc-protein complexes.

Plasma: Plasma contains approximately 120 to 140 μg/100 ml of plasma. ***Approximately 10 per cent of zinc in plasma is transported by*** α_2***-macroglobulin and the remainder being transported by albumin.***

Requirements: As a result of balance studies, the requirement for normal health has been recommended as 0.3 mg zinc/per kg body wt. ***Adult men and women require about 15 to 20 mg.*** For pregnant and lactating women, the requirement is 25 mg and for infants and children, it is 3 to 15 mg.

FUNCTIONS

1. ***Role in Enzyme Action:*** Zinc forms an integral part of several enzymes (metallo-enzymes) in the body. Important zinc containing enzymes are:
   - ***Superoxide dismutase:*** The enzyme is present in cytosol of brain cells, liver cells and blood cells. It is a Cu-Zn protein complex with two Zn^{++} per molecule of the enzyme.
   - ***Carbonic anhydrase:*** Molecular weight is ≈ 30,000. It is present in red blood cells, parietal cells and renal tubular epithelial cells and contains one Zn^{++} per molecule of the enzyme.
   - ***Leucine amino peptidase (LAP):*** Of intestinal juice.
   - ***Carboxy peptidase 'A':*** Of pancreatic juice.

SECTION FOUR

Examples of other zinc containing enzymes are:

- *Alcohol dehydrogenase* of mammalian liver and yeast cells
- *Retinine reductase* of retina
- *Alkaline phosphatase* enzyme
- *Glutamate dehydrogenase* involved in transdeamination
- *Lactate dehydrogenase* which brings about reversible reaction PA to LA.
- *DNA and RNA polymerase*
- *δ-ALA dehydratase*

2. ***Role in Vitamin A Metabolism:*** Zn^{++} has been claimed to stimulate the release of vitamin A from liver into the blood and thus increases its plasma level and its utilisation in *rhodopsin synthesis*. In addition, Zn^{++} containing metallo-enzyme *"retinine reductase"* participates in the regeneration of rhodopsin in the eye during dark adaptation after illumination with light.
3. ***Role in Insulin Secretion:*** Protamine zinc-insulin and globin zinc insulin contain Zn^{++} for its functioning. Rise of blood glucose after glucose administration to a normal animal increases the release of insulin and simultaneously lowers the zinc content of pancreas specially of β-cells of islets of Langerhans. Zinc content of pancreas also have been found to diminish in *Diabetes mellitus*. The above facts indicate the participation of zinc in storage and secretion of Insulin.
4. ***Role in Growth and Reproduction:***
   - **Prasad *et al*** have shown that zinc deficiency may lead to ***'dwarfism'*** and ***'hypogonadism'***. In such dwarfs, zinc concentration in plasma, red blood cells, hair, urine and faeces was found to be less than control subjects. Pubic hair disappear and do not grow. However, zinc supplementation caused improved growth and appearance of pubic hair. Growth retardation and gonadal hypofunction in these subjects were also related to zinc deficiency.
   - Zinc deficiency also lowers spermatogenesis in males and menstrual cycles are disturbed in females.
   - Zinc deficiency due to phytate-rich diet may cause poor body growth, failure of full reproductive maturity and hypogonadism in humans.
5. ***Role in Wound Healing:*** Zinc is necessary for wound healing. Zinc has been found to accumulate in granulation tissues and in and around the healing wounds. ***Zinc deficiency delays wound healing.*** Thus, ***zinc plays a vital role in wound healing.***
6. ***Role in Biosynthesis of Mononucleotides:*** The biosynthesis of mononucleotides and their incorporation into the nucleic acids has been found to be impaired in zinc deficiency. *Ribonuclease* activity has been reported to be higher in zinc deficiency.

CLINICAL ASPECT

ZINC DEFICIENCY DISEASE

Acrodermatitis Enteropathica:

- A rare inherited disorder in which primary defect is in zinc absorption.
- ***Inheritance:*** Autosomal recessive.
- ***Clinically:*** The disease is characterized by dermatologic, ophthalmologic, gastrointestinal and neuropsychiatric features alongwith growth retardation and hypogonadism.

CLINICAL SIGNIFICANCE

1. ***In diabetes mellitus:*** Total amount of zinc in pancreas has been reported to be reduced to half. Deficiency of zinc may interfere with storage and secretion of insulin.
2. ***Leukaemias:*** Normally leucocytes also contain zinc as stated above. In Leukaemias, zinc content is almost reduced to 10 per cent of the normal amount. The functional role played by zinc in leucocytes is not clearly delineated, probably it has immunological function.
3. ***Malignancies:*** It has been reported that the liver zinc concentration is significantly higher in subjects dying due to malignant diseases than in subjects who are nonmalignant. The increase was found to be confined to the parts of the liver which showed no macro or microscopic evidence of carcinomatous invasion. ***The zinc content of the malignant deposits themselves was found to be lower than that of the normal liver.*** It has been suggested that the liver zinc concentration rises probably as a part of the normal tissue biochemical defence reaction against infection by the malignant cells.
4. ***Atherosclerosis:*** Zinc therapy has been found to be useful in some cases of atherosclerosis.

   Experimental evidence: Administration of 3.4 mg of elemental zinc per rabbit per day significantly prevents rise in serum cholesterol, LDL and aortic wall cholesterol content (which are produced by cholesterol feeding). It also prevents platelet adhesiveness and increases the fibrinolytic activity.
5. ***Wound healing:*** (Given earlier)
6. ***Hepatic diseases: Serum zinc level decreases in cirrhosis liver.*** Low plasma level of zinc has been noted in acute viral hepatitis which returns to normal with recovery.
7. ***Acute myocardial infarction: Decreased plasma zinc level has been observed in acute myocardial infarction. Maximum fall found on 3rd day after the attack.*** It has been suggested the decrease in serum zinc in acute myocardial infarction may be mediated by a humoral factor released from polymorphonuclear leucocytes called ***"Leucocytes endogenous mediator (LEM)."***
8. ***Sickle cell anaemia:*** Recently in sickle cell anaemia, decreased zinc level (hypozincaemia) with hyperzincuria have been noted.
9. ***Dermatitis:*** Zinc deficiency produces skin lesions, scaly parakeratotic plaques with acanthosis. Zinc therapy is found to help healing intractable chronic leg ulcers.

Role of Zinc in binding of regulatory proteins to DNA: The specificity involved in the control of transcription requires that regulatory proteins bind with high affinity to the correct region of DNA.

Three Unique Motifs:

- Helix-turn helix
- Zinc finger motif; and
- Leucine-Zipper, account for many of these specific protein-DNA interactions.

Zinc-finger motif: Protein TF III A, which is positive regulator of 5S RNA transcription, requires zinc for its activity. It has been shown that each TF III A molecule contains 9 zinc ions in a repeating co-ordination complex. Two types of zinc finger are:

- Cys-Cys zinc finger and
- Cyst-His zinc finger.

CLINICAL IMPORTANCE

It has been claimed that a single amino acid mutation in either of the two zinc-fingers of the calcitriol receptor protein results in resistance to the action of this hormone and the clinical syndrome rickets may occur.

MANGANESE

Sources: Manganese is also an essential trace element and is required by the body. The average diet can provide approximately 3 to 4 mg of manganese, which is obtained principally from cereals, vegetables, fruits, nuts and tea.

From ***animal source:*** Liver and kidneys are rich source and can supply sufficient Mn^{++} to meet the daily requirement.

Absorption: About 3 to 4 per cent of dietary Mn^{++} is absorbed. Dietary Ca and P have been found to reduce Mn^{++} absorption.

Distribution: The total amount of this trace element in our body has been estimated to be approximately 15 mg average (range 10 to 18 mg) and is found concentrated mainly in the kidneys and liver.

Blood Level: Blood contains about 4 to 20 µg manganese per 100 ml. It is present mainly in red blood cells in combination with several porphyrins and is transported in the plasma in combination with a β_1-globulin called as ***transmanganin***. Because of its presence in plasma in protein bound form very little of it is excreted in urine.

FUNCTIONS

1. ***Role in Enzyme Action:***
   - Acts either as a 'cofactor' or as an activator of many enzymes like *arginase, isocitrate dehydrogenase* (ICD), *cholinesterase, lipoprotein lipase, enolase, leucine amino peptidase* in intestine, phosphotransferases and 5-*oxo-prolinase* of kidneys, and small intestine and many others.
   - Manganese and magnesium may replace one another in case of some of the enzymes.
   - ***Mitochondrial form of Superoxide dismutase contains Mn^{++}*** in its prosthetic group unlike the cytosol form of the enzyme which contains Cu and Zn.
   - Another mitochondrial enzyme, the ATP-dependent tetrameric *ligase* called *pyruvate carboxylase* contains Mn^{++}, and also the vitamin Biotin in its prosthetic group, which is involved in "CO_2-fixation reaction". A similar enzyme is *Acetyl-CoA carboxylase* which also contains Mn^{++} and biotin.
   - Mn^{++} may be associated with mitochondrial respiratory chain enzymes.
   - Manganese also acts as a cofactor of all *hydrolases* and *decarboxylases*.
2. ***Role in Animal Reproduction:*** In animals, Mn^{++} deficiency has been shown to produce sterility in cattles, disturbances of estrous cycles, resorption of foetus and sterility in cows, and degeneration of testes as well as inability to feed the offspring in rats.
3. ***Role in Bone Formation: Mn^{++} plays a part in the synthesis or deposition of Mucopolysaccharides (MPS) in the cartilaginous matrices of long bones.*** Mn-deficiency causes significant lowering in the content of chondroitin SO_4. Abnormal bone formation due to Mn deficiency may lead to ***perosis*** (slipping of gastrocnemius tendon) and bone deformities in chicks.
4. ***Role in Carbohydrate Metabolism:*** Mn^{++} is reported to influence carbohydrate metabolism by affecting the peripheral utilisation and their conversion to MPS. In Mn deficiency, pancreatic hypoplasia, associated with 'diabetic type' of GTT has been reported.
5. ***Role in Porphyrin Synthesis:*** Some porphyrins of RB cells contain Mn^{++}. Manganese also helps in porphyrin synthesis by participating in *δ-ALA synthetase* activity. Hb-synthesis appears to be depressed in Mn-deficient rats.
6. ***Role in Fat Metabolism:*** Manganese has been reported also to exhibit ***"lipotropic effect"*** and it stimulates FA synthesis and cholesterol synthesis.
7. ***Role in Proteoglycan Synthesis:*** Manganese also participates in glycoprotein and proteoglycan synthesis.

Manganese Toxicity

Inhalation poisoning produces psychotic symptoms and Parkinsonism like symptoms.

CHROMIUM

Distribution and Source: Chromium is widely distributed throughout the body. Infants have a higher chromium concentration than adults. Brewer's yeast is rich in chromium and most grains and cereal products contain significant quantities. Significant amount of chromium is obtained in the diet by cooking foods in stainless steel cookwares.

Absorption and Excretion: Chromium is absorbed poorly in the diet. It is absorbed mainly in the small intestine by a pathway it appears to share with zinc. It is transported to tissues, ***bound to 'transferrin'*** and appears in liver mitochondria, microsomes and the cytosol.

Blood level: Serum level of chromium in normal healthy adult is about 6 to 20 μg/100 ml.

Requirements: Recommended daily requirement for an adult is approximately 50–200 μg/day.

FUNCTIONS

Chromium plays an important role in carbohydrate, lipid and protein metabolism.

1. *Role in Carbohydrate Metabolism:* Chromium is ***a true potentiator of insulin*** and is known as ***Glucose tolerance factor (GTF).*** Trivalent chromium Cr^{3+} has been claimed to be a constituent of Glucose tolerance factor.
2. *Role in Lipid Metabolism:* Chromium supplementation in deficient diets decreases serum cholesterol levels and prevents/decreases atheromatous plaque formation in aorta.
3. *Role in Protein Metabolism:* When given with insulin in chromium deficiency state, it improves amino acid incorporation mainly α-amino isobutyric acid, glycine, serine and methionine. In protein-energy malnutrition (PEM) states, chromium supplementation is beneficial for weight gain.

CLINICAL SIGNIFICANCE

- ***Risk of Cancer:*** Hexavalent chromium is much more toxic than the trivalent chromium. Chronic occupational exposure to chromate is associated with increased **risk of lung cancer** (occupational hazard).
- Serum chromium levels decrease during pregnancy and acute infections.

Deficiency manifestations: Impaired glucose tolerance, secondary to parental nutrition.

NICKEL

Source and Distribution: Nickel occurs in trace amount in human and animal tissues. Nickel does not accumulate with age in any of human tissues other than lungs. Dietary Nickel is poorly absorbed from intestine to the extent of about 1 to 10 per cent. It is excreted mainly in faeces.

Blood level: Blood contains 1.1 to 3.6 μg nickel per litre of blood, average being 2.6 μg/litre.

Requirements: The minimum daily requirement of nickel is approximately 20 μg per adults.

FUNCTIONS

1. *Role in enzyme action:* Nickel activates several enzyme systems, viz. *arginase, carboxylase, trypsin* and *acetyl-CoA synthetase.*
2. *Role in growth and reproduction:* Nickel is required in trace amount for growth and reproduction.

Deficiency Manifestations: Nickel deficiency in humans is unknown. In experimental animals, impaired growth and reproduction, and anaemia occur. Ultrastructural changes in liver have been observed.

TOXICITY AND CLINICAL SIGNIFICANCE

Nickel is relatively non-toxic metal, but ***prolonged exposure results in respiratory tract neoplasia and dermatitis*** as observed in workers in nickel refineries (occupational hazard).

(a) Decreased Level: Diminished values have been reported in cirrhosis liver and in cases of chronic uraemia.

(b) Increased Level

- Nickel levels are markedly increased in blood to about twice the normal values within 12 to 36 hours in acute myocardial infarction.
- An abnormally high nickel concentration also occurs in cases of acute 'stroke' and in severe burns.

COBALT

Cobalt forms an integral part of vitamin B_{12} and is required as a constituent of this vitamin.

Sources and Requirements: Normal average diet contains about 5 to 8 μg of cobalt which is far more than the recommended daily allowance (1 to 2 μg of vitamin B_{12} contains approximately 0.045 to 0.09 μg of cobalt).

Main source: Foods from animal source. Not present in vegetables.

Absorption and Excretion: About 70 to 80 per cent of the dietary cobalt is absorbed readily from the intestine. Isotopic studies have shown that about 65 per cent of the ingested cobalt is excreted almost completely through the kidney. Cobalt is stored mainly in the liver, it being the principal storage site, is only trace amount; present in other tissues.

FUNCTIONS

1. *Role in Formation of Cobamide Enzyme:* In formation of *cobamide coenzyme (Adenosyl co-enzyme),* cobalt of B_{12} undergoes successive reduction in a series of steps catalysed by the enzyme *"B_{12} reductase",* which requires NADH and FAD (Refer to chapter on Vitamins).
2. *Bone Marrow Function*
   - Cobalt is required to maintain normal bone marrow function and required for development

and maturation of red blood cells. A deficiency of cobalt results in decreased B_{12} supply which produces nutritional macrocytic anaemia.
- Excess of cobalt results in overproduction of red blood cells causing polycythaemia. The ***polycythaemic*** effect may be due to inhibition of certain respiratory enzymes, viz. ***cytochrome oxidase, succinate dehydrogenase,*** etc. leading to relative anoxia.

3. ***Role as Cofactor:*** Cobalt may act as a cofactor for enzyme like *glycyl-glycine dipeptidase* of intestinal juice.

Cobalt Deficiency: In ruminants, but not in other species, cobalt deficiency results in anorexia, fatty liver, macrocytic anaemia, wasting and haemosiderosis of spleen.

MOLYBDENUM

Source and Requirement: Cereals and dry legumes supply more than 50 µg per day. Requirement for:
- *Adults:* Approximately 0.5 mg per day
- *Children:* 0.05 to 0.3 mg per day.

Absorption, Distribution and Excretion: It is absorbed from the intestine and excreted mainly in the urine and to a small extent in faeces probably via the bile. Tissue storage is very little, mainly in the bones and to a smaller extent in liver and kidneys.

SECTION FOUR

FUNCTIONS

1. ***Role in Enzyme Action***
   - *Xanthine oxidase:* It occurs in several non-haem iron flavoproteins, e.g. xanthine oxidase of liver. Xanthine oxidase enzyme contains Molybdenum, S, Fe and flavine. It oxidises xanthine to uric acid.
   - Molybdenum also occurs in some haemoflavoproteins, e.g. *NADH-nitrate reductase.*
   - *Sulfite oxidase:* Of human liver contains molybdenum and haem, and oxidises inorganic sulphite to SO_4^-.

   Molybdenum of all these enzymes participates in internal electron transfers during oxidoreductions.
2. ***Relation with Copper Utilisation:*** Presence of small amounts of Molybdenum help in the utilisation of copper. On the other hand, high molybdenum intake produces copper deficiency, producing microcytic anaemia and low tissue copper in cattle and sheep.

SELENIUM

Selenium was found to prevent liver cell necrosis, which was discovered by **Schwartz** and **Flotz** in 1957. Since then a wide variety of animal diseases have been shown to respond to selenium.

Current evidences indicate selenium as an essential trace element for all species including humans. A positive role of selenium in human health has been suggested. On the other hand, excess selenium is harmful and produces toxic manifestations.

Occurrence and Distribution: Biological forms of selenium which occur in animal body are selenium analogues of S-containing amino acids, viz. ***selenomethionine, selenocysteine and selenocystine*** found at a mean concentration of 0.2 µg/g. It is widely distributed in all the tissues, highest concentrations are found in liver, kidneys and fingernails. Muscles, bones, blood and adipose tissues show a low concentration of selenium.

Selenium in cereal ranges from less than 0.1 µg/g to 1.0 µg/g wet weight; whereas dairy products, fruits, and vegetables are relatively poor sources of selenium. ***Principal source of selenium for the food is plant material,*** selenium uptake in plant tissue is passive and is influenced by its concentration in soil.

Absorption and Excretion

Food constitutes the major route of human exposure to environmental selenium.

Intake: It is in the range of 20 to 300 µg/day. Infants get their selenium through breast milk. ***Total body selenium has been estimated to be approx. 4 to 10 mg (average 6 mg).*** A good correlation between selenium intake in food and blood levels has been shown.

There is also evidence to show that:
- Selenium is assimilated more effectively from plant food than animal products.
- Nature of diet plays a major role in determining the forms of selenium consumed.
- Other dietary constituents, e.g. vit. A, C and E may also affect its absorption.

Selenium is ***absorbed mainly from the duodenum*** and is transported actively across the intestinal brush border particularly in the form of methionine analogue.

Selenium after absorption is ***transported bound to plasma proteins particularly β-lipoproteins*** in humans. Selenomethionine can be deposited directly in tissues and taken up also by myoglobin, cytochrome C, myosin, aldolase and nucleoproteins. Selenocysteine is not directly incorporated into proteins but is catabolised, releasing selenium for utilisation.

Main route of excretion of selenium appears to be through urine. Also small amount is excreted through faeces and expired air.

Blood and tissue levels: Selenium levels in blood and tissues are very much influenced by dietary selenium intake. Blood level varies 0.05 to 0.34 µg/ml. Selenium levels are very low at 0.05 to 0.08 µg/ml in people of New Zealand, where the dietary intake is approx. 20 to 30 µg per day. In selenium deficient areas of China, blood levels may be as low as 0.009 µg/ml.

METABOLIC ROLE

1. ***The only metabolic role of selenium which has been established is as the prosthetic group of selenium enzyme Glutathione peroxidase*** which is present in cell cytosol and mitochondria and ***functions to reduce hydroperoxide.***

$$R.OOH + 2\,GSH \xrightarrow[\text{(Se containing)}]{\text{Glutathione peroxidase}} R\text{–}OH + H_2O + G\text{–}S\text{–}S\text{–}G$$

The reaction has special significance in the protection of polyunsaturated FA located within the cell membranes, where the enzyme ***functions in the cytosol as part of a multicomponent antioxidant defence system within the cell.*** It is supplementary to vitamin E and acts as primary antioxidant by scavenging reactive oxygen species and free radical intermediates of polyunsaturated lipid peroxidation.

2. *Selenide containing NHI proteins:* Selenium probably occurs as selenide at the active site of some nonhaem iron proteins, located as integral proteins in microsomal and other cellular membranes. Probably they are associated with the mixed function oxidase system of membranes.
3. *Relation with Vitamin E:* **Selenium has sparing effect on vitamin E** and it reduces the vit. E requirements at least in **3 ways:**
   - Selenium is required for normal pancreatic function and thus the digestion and absorption of lipids including vit. E.
   - As a component of *glutathione peroxidase*, selenium helps to destroy peroxides and thereby reduces the peroxidation of polyunsaturated acids of lipid membranes (discussed above). This diminished peroxidation greatly reduces the vit. E requirement for the maintenance of membrane integrity.
   - In some unknown way, selenium helps in retention of vit. E in the blood plasma lipoproteins.

   Conversely, ***vit. E appears to reduce the selenium requirement,*** at least, in experimental animals, by preventing loss of selenium from the body or maintaining it in an active form. However, there are certain symptoms which cannot be reversed by vit. E in selenium deficient states. Neither can vit. E overcome poor growth of animals on selenium deficient diets.
4. *Relation with Heavy Metals:* Selenium, in common with sulphur, shares an affinity with heavy metals such as cadmium, mercury (Hg) and silver (Ag). ***Supplements of selenium probably protect against toxic effects of these heavy metals.***

SELENIUM TOXICITY

Numerous reports on toxicity in animals are on record. Ruminant farm animals grazing on soils which are selenium rich or high levels of selenium in plants develop toxicity.

1. *Acute poisoning:* Manifests as diarrhoea, elevated pulse rate and temperature, tetanic spasms, laboured breathing and respiratory failure.

   Pathological changes include: Haemorrhage, necrosis, congestion and oedema of various tissues. However, such acute poisoning is rare, as plants containing high selenium are not palatable.
2. *Chronic Poisoning:* Occurs when plants containing selenium are consumed over long periods. The animals develop impaired vision and movement disorders ***(Blind staggers)***, which ultimately result in paralysis and death. Chronic poisoning also results in so-called ***Alkali disease,*** which manifests as anorexia, loss of hair and vitality, with myocardial atrophy and liver necrosis.

TOXICITY IN HUMANS

Reports of toxicity in humans are available, which manifests as chronic dermatitis, loss of hair and brittle nails. No hepatotoxicity has been observed in humans. ***An early hall-mark of selenium toxicity is a garlicky breath, caused by exhalation of dimethyl selenide.*** Likely cause is occupation exposure in electronics, glass and paint industries.

Selenium Deficiency: Specific features of severe selenium deficiency have been reported in a number of species. These are:

Liver cell necrosis, exudative diathesis, pancreatic degeneration, muscular dystrophies, myopathy, infertility, failure of growth and dilatation of the heart resulting in congestive cardiac failure. Evidences exist to implicate selenium in muscular discomfort following parenteral nutrition.

CLINICAL ASPECT

Naturally Occurring Selenium Deficiency in Humans

1. **Keshan disease: *Manifesting principally as cardiomyopathy.*** Has been reported from Keshan country of north-eastern China. Dietary intake of selenium average around 7 to 11 μg/day and blood levels are in the range of 0.001 to 0.01 μg/ml. Affects mainly children and younger women.

   Clinically: Manifests as acute or chronic cardiac enlargement, arrhythmia and ECG changes. Cardiomyopathy is endemic. Remarkable improvement seen clinically with sodium selenite. Prophylaxis with sodium selenite is highly effective.

2. **Kaschinbeck disease:** ***It manifests as endemic human osteopathy (as osteoarthritis).*** Seen in several parts of the eastern Asia and is characterised by degenerative osteoarthrosis particularly affecting children between 5 and 13 years of age. The disease shortens fingers and long bones with severe enlargement and dysfunction of the joints resulting in retardation of growth.
Etiological factors are not well established, the disease is endemic in low selenium zones and selenium administration has been reported to have both therapeutic and prophylactic effects.

Recommended Dietary Allowances: Since definite description of biological conditions in humans that can be totally attributed to selenium deficiency is lacking, **requirements have been fixed between 0.1 to 0.2 mg/kg diet as allowances** for most species of animals.

- For an adult of 70 kg, the required daily selenium is approx. 50 to 100 µg.
- For children it is between 20 to 120 µg.

CLINICAL SIGNIFICANCE

1. ***Parenteral nutrition:*** Several reports have linked total parenteral nutrition to selenium deficiency. Selenium responsive muscular discomfort and cardiomyopathy have been reported.
2. ***Kwashiorkor:*** Low levels of selenium have been reported in the blood of kwashiorkor children and selenium supplementation in kwashiorkor children has been documented to stimulate growth.
3. ***GI cancers:*** Low selenium levels are observed in cancer patients specially those with ***GI cancers, specially oesophageal cancer*** and ***oropharyngeal cancers.***
4. ***Cardiovascular disorders:*** Recently selenium has been implicated in cardiovascular disorders also. Inverse epidemiological relationships have been observed between selenium distribution in plants and cardiovascular and cerebrovascular diseases in humans.
5. ***Thrombosis:*** Low selenium intakes have been found to increase the risk of thrombotic episodes.
6. ***Role of Selenium in Cancer:*** Evidences have shown that selenium may be ***a cancer protective agent.*** Animal experimentations have shown that increased selenium intake can decrease the incidence of several viral and chemically induced tumours. ***Areas having high selenium content in foods and high selenium levels in blood have lower incidence of cancers.*** In a recent study it has been shown that with low selenium intakes, the risk of human cancer is higher.
Mechanism of Anti-cancer Activity: Several mechanisms are considered for antitumour activity:
   - Probably selenium brings about changes in carcinogen metabolism
   - Protects from carcinogen induced oxidant damage and
   - Toxicity of selenium metabolites to tumour cells.
7. ***Role of Selenium in HIV infection:*** Recently, the researchers in a new study have shown that giving selenium, an antioxidant trace element as a dietary supplement to HIV patients reduces the amount of virus in their blood. They observed that patients taking 200 µg (micrograms) of high selenium yeast daily, produces on an average ***12 per cent drop in blood virus levels.*** The leading author ***likened "selenium to a lion tamer in a circus".*** What selenium appears to do is ***make the virus more docile, less virulent*** and ***less likely to replicate.*** Earlier ***"in vitro"*** studies have also shown that the trace element suppresses HIV infection in the laboratory also.

SECTION FOUR

I. Questions (Essay type and short notes)

1. Describe the metabolism of calcium and phosphorus.
2. Discuss the maintenance of serum calcium level.
3. Describe the factors which favour the absorption of iron from the intestine.
4. Describe the metabolism of calcium.
5. Describe iron absorption in the intestinal tract.
6. Give an account of the dietary sources, absorption and utilisation of iron.
7. Describe iodine metabolism.
8. Describe the absorption, transport and metabolism of iron.
9. Describe metabolic functions of zinc, selenium and phosphorus.
10. Enumerate the factors that influence the absorption of calcium. Give briefly how blood calcium levels are regulated.

Short Notes

a. Trace elements
b. Calciostatic mechanism of the body
c. Iron absorption
d. Ferritin
e. Transferrin
f. Factors affecting serum calcium level
g. Iron metabolism
h. Metabolic role of copper and zinc
i. Metabolic function of Ca, Na and Fe in human body
j. Regulation of level of serum Ca
k. Metabolic role of Zn and selenium
l. Iron containing compounds in human body and their functions.

II. MCQs (Give one correct answer)

1. A hypochromic microcytic anaemia with increased Fe stores in the bone marrow may be:
   a. Folic acid responsive b. Vitamin B_{12} responsive
   c. Pyridoxine responsive d. Vitamin C responsive
   e. Iron responsive
2. A deficiency of copper affects the formation of normal collagen by reducing the activity of which of the following enzyme?
   a. Prolyl hydroxylase b. Lysyl oxidase
   c. Lysyl hydroxylase d. Glucosyl transferase
   e. Galactosyl transferase
3. Molecular iron (Fe), is:
   a. Stored primarily in spleen
   b. Absorbed in the intestine by transferrin
   c. Absorbed in the ferric, Fe^{+++} form
   d. Stored in the body in combination with ferritin
   e. Excreted in the urine as Fe^{++}
4. All the following statements regarding calcium are correct except:
   a. It diffuses as a divalent cation
   b. It freely diffuses across the endoplamic reticulum of muscle cells
   c. It can exist in the blood as ionic form and also protein bound
   d. It is found in high concentration in bones
   e. Required as a cofactor in some of reactions
5. Iron is absorbed from:
   a. Stomach b. Duodenum and jejunum
   c. Ileum d. Caecum
   e. Transverse colon
6. The normal route of calcium excretion is:
   a. Kidney
   b. Kidney and liver
   c. Kidney and intestine
   d. Kidney, intestine and pancreas
   e. Pancreas, liver and kidney.
7. Hypocalcaemia affects:
   a. Skeletal muscles
   b. Smooth muscles
   c. Cardiac muscles
   d. Skeletal muscles + smooth muscles
   e. Skeletal muscles + smooth muscles + cardiac muscles
8. Transferrin is a type of:
   a. Albumin b. α-globulin
   c. β_1-globulin d. γ-globulin
   e. A lipoprotein complex
9. In case of Wilson's disease (hepatolenticular degeneration), the features include all of the following, *except*:
   a. Progressive hepatic cirrhosis
   b. Kayser-Fleischer ring
   c. Aminoaciduria
   d. Urinary excretion of Cu is decreased
   e. Low levels of caeruloplasmin
10. In vitamin D poisoning (hypervitaminosis):
    a. Both serum and urinary Ca are high
    b. The serum Ca is low and urinary calcium high
    c. The serum Ca is increased and urinary Ca is normal
    d. Both serum and urinary Ca are low
    e. None of the above
11. The % of K in extracellular fluid is about:
    a. 1% b. 2 to 3%
    c. 10% d. 15%
    e. 20%
12. The Fe containing pigments are:
    a. Haematoidin b. Bilirubin
    c. Haemosiderin d. Urobilinogen
    e. None of the above
13. All of the following are true of Wilson's disease, ***except***:
    a. Low total plasma Cu b. Elevated urinary copper
    c. Arthritis d. Aminoaciduria
    e. Low caeruloplasmin
14. An increased serum iron and decreased Fe-binding capacity are found in:
    a. Fe-deficiency anaemia
    b. Sideroblastic anaemia
    c. Thalassaemia
    d. Anaemia of chronic disorders
    e. All of the above
15. Iron therapy is ineffective in which of the following conditions:
    a. Chronic blood loss
    b. Inadequate Fe-intake in diet
    c. Hypochromic anaemia of pregnancy
    d. Thalassaemia minor (α-chain thalassaemia)
    e. Impaired absorption of Fe
16. In haemochromatosis, the liver is infiltrated with:
    a. Copper b. Iron
    c. Manganese d. Chromium
    e. Lipids
17. All of the following are true of hypochromic anaemias not due to iron deficiency, ***except***:
    a. Serum Fe is high
    b. Normal or low transferrin
    c. Stainable iron in bone marrow
    d. Vitamin B_6 deficiency
    e. Iron therapy is effective
18. Match the following:
    A. Molybdenum B. Copper
    C. Zinc D. Trivalent chromium
    E. Manganese
    1. Cytochrome oxidase _______
    2. Glutamate dehydrogenase __________
    3. Glucose tolerance factor (GTF) _________
    4. Mitochondrial superoxide dismutase _______
    5. Xanthine oxidase _________
19. Cytosolic superoxide dismutase contains:
    a. Zn only b. Copper only
    c. Zn and Cu both d. Manganese
    e. Cu and Mn both
20. Match the following:
    A. Calcium B. Fluoride
    C. Magnesium D. Selenium
    E. Potassium

SECTION FOUR

1. Alkalosis __________
2. Muscular tremors __________
3. Dental caries ____________
4. Rickets __________
5. Cardiomyopathy (Keshan disease) __________

21. A rise in blood Ca may indicate:
    a. Paget's disease
    b. Vitamin D deficiency
    c. Cushing disease
    d. Hypervitaminosis D
    e. All of the above
22. The essential trace element which catalyses the formation of Hb in the body is:
    a. Mn
    b. Se
    c. Mg
    d. Cu
    e. Fe
23. Match the following:
    A. Copper
    B. Iron
    C. Selenium
    D. Cobalt
    E. Caeruloplasmi
    1. Required for the enzyme glutathione peroxidase _______
    2. Component of vitamin B_{12} ____________
    3. Cofactor for an enzyme MAO ________
    4. Acts as a ferroxidase _________
    5. Approx 90% of the intake is excreted in stool ________
24. Zinc is a constituent of the enzyme:
    a. Succinate dehydrogenase
    b. Carbonic anhydrase
    c. Mitochondrial superoxide dismutase
    d. Aldolase
    e. Amylase.
25. The active transport of Ca is regulated by _______, which is synthesised in kidneys:
    a. Cholecalciferol
    b. Ergosterol
    c. 25-OH cholecalciferol
    d. 1,25-diOH-cholecalciferol
    e. All of the above
26. Caeruloplasmin shows the activity of:
    a. As ferroxidase
    b. As reductase
    c. As ligase
    d. As transferase
    e. As hydrolase

Answers to MCQs

1. c	2. b	3. d	4. b	5. b	6. c
7. e	8. b	9. d	10. a	11. b	12. c
13. c	14. b	15. d	16. b	17. e	
18. 1-B; 2-C; 3-D; 4-E; 5-A			19. c		
20. 1-E; 2-C, 3-B; 4-A; 5-D			21. d	22. d	
23. 1-C; 2-D; 3-A; 4-E; 5-b			24. b	25. d	26. a.

SECTION FOUR

SECTION

5

Clinical Biochemistry

CHAPTER 35

ENZYMES AND ISOENZYMES OF CLINICAL IMPORTANCE

Major Concepts

A. Learn the rational of clinical enzymology.
B. List and study various enzymes that are used in diagnosis.
C. Study how enzymes and isoenzymes can be used in differential diagnosis.

Specific Objectives

A. 1. Learn the sources of plasma enzymes.
  a. Plasma derived
  b. Cell derived
 2. Study various possible mechanisms responsible for abnormal levels of plasma enzymes.
 3. Study the reasons for decreased levels of plasma enzymes.
 4. Learn the value of enzyme levels in diagnosis.

B. 1. Study different enzymes useful in the diagnosis of myocardial infarction.
 2. Study the enzymes of liver diseases.
 3. Study the enzymes of GI disorders.
 4. Study the enzymes of muscle diseases.
 5. Study the enzymes of bone diseases.
 6. Study the enzymes of malignancies.

C. 1. Define isoenzyme.
 2. Give examples of isoenzymes and study them briefly.
 3. Study how isoenzymes of LDH and CPK are useful in diagnosis of MI.
 4. Study isoenzymes of ALP and their importance.

ENZYMES

INTRODUCTION

The investigation and interpretation of changes in serum enzymes in diseases is one of the most rapidly expanding fields in clinical biochemistry. **Wröblewski** and his coworkers in 1956 published their first papers on serum Glutamate-oxaloacetate transaminase (S-GOT) and followed by serum lactate dehydrogenase (LDH) and brought the possibilities of these enzyme assays in general notice. Thus began the present efflorescence of clinical enzymology and large number of enzymes have been used for diagnosis and prognosis of various diseases.

GENERAL CONSIDERATIONS

Sources of Plasma Enzymes: They can be:

- **Plasma derived,** and
- **Cell derived.**

(a) ***Plasma derived enzymes:*** These act on substrates in plasma, and their activity is higher in plasma than in cells, e.g. coagulation enzymes. This group will not be further considered.

(b) ***Cell-derived enzymes:*** These have a high activity in cells and overflow into the plasma. They are further subdivided into:

- ***Secretory:*** These are mainly derived from digestive glands and function in the extracellular space, and
- ***Metabolic:*** These are concerned with intermediary metabolism and function in the cells and those enzymes found in the plasma are mainly derived from the soluble and microsomal fractions of the cells.

The cell-derived enzymes enter the plasma in small amounts as a result of:

- ***Continuous normal ageing of the cells,*** or
- ***Owing to diffusion through undamaged cell membranes.***

They leave the plasma through:
- Inactivation.
- *Catabolism in general protein pool.*
- *Rarely excretion in bile and urine.*

Possible Mechanisms Responsible for Abnormal Levels: Serum level of a particular enzyme may be increased by diseases that provoke: *(a)* an increase in its rate of release, or *(b)* a decrease in rate of disposition or excretion.

1. Increase Serum Level

(a) Increased Release
- *Necrosis of cells:* Due to damage to cells of the tissue. The resultant pattern will depend on:
  - Normal enzyme content of the tissue/organ.
  - On the extent and type of necrosis.
- ***Increased permeability of cell membrane*** without necrosis of cells: Increased permeability without gross cellular damage/necrosis can increase the enzyme level, e.g.
- In early stage of viral hepatitis, before jaundice appears. There is ***"ballooning" degeneration*** of liver cells, leading to elevated levels of transaminases (S-GPT).
- Progressive muscular dystrophy-elevated levels of Aldolase, GOT and CPK.
- ***Increased production of the enzyme within cell:*** Such a situation may be seen in treatment of which patients with protein anabolic drugs, results in increased synthesis of liver cell *transaminases* and *serum transaminases* will increase by overflow.
- ***An increase in tissue source of enzymes*** due to either:
  - Increased rate of production in cells as mentioned above, or
  - Increase in the number of cells/and cell mass, as seen in malignancies, e.g. *alkaline phosphatase* increase in patients with osteoblastic bone lesions, or *acid phosphatase* increase in patients with carcinoma prostate.

(b) Impaired Disposition/excretion
- Increased levels of serum LAP and ALP seen in patients with obstructive jaundice
- Certain increased enzyme levels in cases of renal failure.

2. Decreased Serum Levels

(a) Decreased formation of the enzyme which may be:
1. *Genetic*
   - Hypophosphatasia, with decreased ALP level in serum,
   - Wilson's disease with decrease in serum ceruloplasmin.
2. *Acquired*
   - ***In hepatitis:*** Decreased serum level of pseudocholinesterase due to decreased production.
   - Decreased serum amylase in patients with chronic hepatic, or pancreatic diseases, or those who are severely malnourished.

(b) Enzyme inhibition: Decreased serum pseudocholinesterase in insecticide poisoning.

(c) Lack of cofactors: Decreased serum GOT level in pregnancy and cirrhosis.

Unit of Serum Enzyme Activity

Various workers have used various units. It is better to have uniformity, the serum enzyme activity is expressed in ***International units (IU).***

Definition
- One IU is defined as the activity of the enzyme which transforms one μ mole of substrate per minute under optimal conditions and at defined temperature, and expressed as IU/ml.
- When milli-micromole of the substrate is transformed/mt, it is IU/L or m-IU/ml.
- Katal—Another unit is katal which stands for catalytic unit.

Definition

One katal unit is defined as the number of mole of substrate transformed per second per litre of sample.

Katal is abbreviated as Kat or K.

Value of Serum Enzyme Assay in Clinical Practice

Single or serial assay of the serum activity of a selected enzyme or enzymes may provide information on the nature and extent of a disease process.

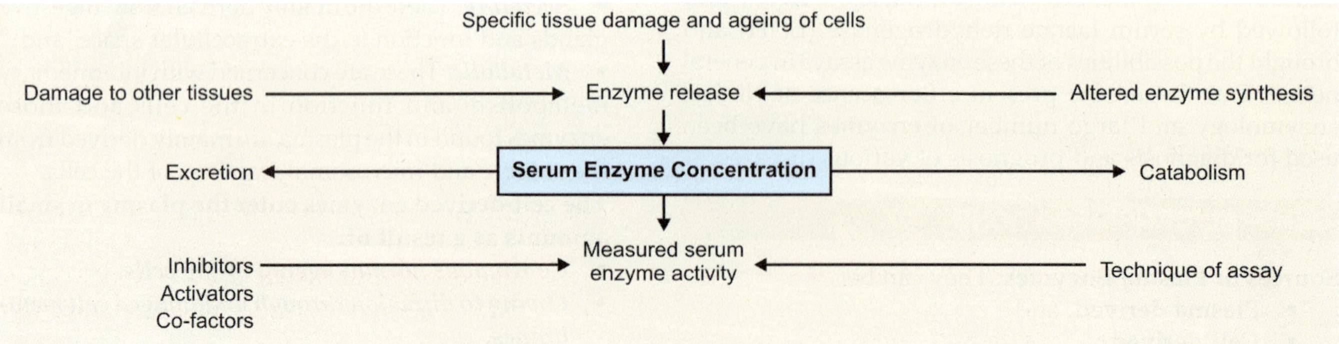

1. ***Value in diagnosis:*** An enzyme assay of serum CK on the day of a suspected case of myocardial infarction will be helpful for diagnosis if ECG changes are doubtful.
2. ***In differential diagnosis:*** When the differential diagnosis lies between a disease that is known to cause a particular pattern of serum enzyme change and one that does not, e.g. as an aid in differentiating myocardial infarction and pulmonary embolism both presenting with chest pain.

	Serum GOT	Serum LDH
• Myocardial infarction	↑	↑
• Pulmonary embolism	Normal	↑

3. ***In ascertaining prognosis:*** Serial enzyme assays are required:
   - To ascertain progress in viral hepatitis: Serial enzyme assays of S-GPT are of great help.
   - Response to endocrine therapy of carcinoma of prostate is shown by degree of reduction of the elevated serum acid phosphatase.
4. ***Early detection of a disease:*** When damage to a tissue is suspected which is so slight that it cannot be detected otherwise.

Example
- Minimal hepatotoxic effects of antidepressant drugs can be detected by a raised serum ICD / or OCT before the patient is clinically ill.
- Increased S-GPT in early stage of viral hepatitis when jaundice has not appeared ***(subclinical stage).***

CLINICAL SIGNIFICANCE OF ENZYME ASSAYS

Value of enzyme assays will be discussed organwise.

SERUM ENZYMES IN HEART DISEASES

Before the introduction of serum GOT assay for the investigation of myocardial infarction ***the heart had been a "biochemically inaccessible" organ.*** In cases of suspected myocardial infarction when clinical and ECG evidence was equivocal, there was no other means of specifically investigating possible injury to cardiac muscle.

Why Enzyme Diagnosis?
1. 25 to 30 per cent of myocardial infarctions are not diagnosed "antemortem" sometimes.
2. Clinical diagnosis and angiographic studies do not correlate in 25 to 33 per cent of patients.
3. ECG findings may not be helpful if:
   - Prior left bundle branch block is present.
   - Old changes exist that may obscure current ECG interpretation.
   - Intramural infarctions may not change ECG pattern.
   - Diaphragmatic infarctions often missed on ECG.

ENZYMES

Enzyme Assays that are Carried out in Myocardial Infarction

(a) Commonly done:
- *Creatine phosphokinase (CK)*
- *Aspartate transaminase (G-OT)/or (AST)*
- *Lactate dehydrogenase (LDH)*

(b) Other enzymes which have been studied but not commonly done:
- γ-Glutamyl transpeptidase (GGTP)
- *Histaminase*
- *Pseudocholinesterase*

1. CREATINE PHOSPHOKINASE: (CPK OR CK)

This enzyme catalyses the following reaction:

$$\text{Creatine} \sim (\text{P}) + \text{ADP} \rightarrow \text{Creatine} + \text{ATP}$$

The enzyme is also called as ***creatine kinase.***

Site: Found in high concentration in skeletal muscle, myocardium and brain but not found at all in liver and kidney. Small amounts are found in lung, thyroid and adrenal gland. ***NOT found in RB cells and its level is not affected by haemolysis.***

Normal value: Serum activity varies from 4 to 60 IU/L (at 37°C).

Behaviour in Acute Myocardial Infarction

After myocardial infarction, serum value is found to increase after about 6 hours, reaches a peak level in 24 to 30 hours, and returns to normal level in 2 to 4 days (usually in 72 hours).

Remarks
- Studies suggested that serum CK activity is a more sensitive indicator in early stage of myocardial ischaemia.
- Potentially more useful in subendocardial infarction.
- No increase in activity noted in heart failure and coronary insufficiency.
- Magnitude of elevation was found to be greater than that observed with G-OT or LDH.

Note
- ***Storage:*** There is ***50 per cent loss of serum CK activity after 6 hours at room temperature*** and 24 hours at refrigerated temperature. Hence, ***all determinations of serum CK activity should be done on fresh blood samples.***

- The above can be circumvented by adding to the reaction mixture cysteine or other compounds containing–SH group *(cysteine stimulated CPK assay).*

2. SERUM GLUTAMATE OXALOACETATE TRANSAMINASE (S-GOT)

Also called as ***Aspartate transaminase*** or ***amino-transferase (AST).***

Site: Concentration of the enzyme is very high, in myocardium.

Normal value: Serum activity of S-GOT varies from 4 to 17 IU/L (25°C) (10-35 of original Karmen spectrophotometric units/ml.)

Behaviour in Acute Myocardial Infarction

In acute myocardial infarction, serum activity rises sharply within the first 12 hours, with a peak level at 24 hours or over and returns to normal within 3 to 5 days.

Remarks

- Level of serum enzyme has been correlated well with prognosis.
  - Levels > 350 IU/L usually fatal, (due to massive infarction)
  - Levels > 150 IU/L associated with high mortality and
  - Levels < 50 IU/L are associated with low mortality.
- Elevation has been noted in absence of any ECG change.
- ***Highest incidence of abnormal levels occurs on second day of infarction.***
- Rise depends on size of the infarction
- ***Extracardiac factors:*** Elevation seen in other diseases e.g. Muscle disease and hepatic diseases. But these can be differentiated clinically and simultaneous determination of S-GPT. ***There is no rise of S-GPT in myocardial infarction.***
- Reinfarction results in a secondary rise of S-GOT.

3. LACTATE DEHYDROGENASE (LDH)

LDH catalyses the reversible conversion of pyruvic acid (PA) and lactic acid (LA).

Normal value: Normal serum LDH activity ranges from 60 to 250 IU/L (120-500 units/ml, original Karmen spectrophotometric method).

Behaviour in Acute Myocardial Infarction

In acute myocardial infarction, serum activity rises within 12 to 24 hours, attains peak at 48 hours (2 to 4 days) reaching about 1000 IU/L and then return gradually to normal from 8th to 14th day.

Remarks

- The peak rises in S-LDH is roughly proportional to the extent of injury to the myocardial tissue.
- ***S-LDH elevation may persist for more than a week after CPK and S-GOT levels have returned to normal levels.***
- S-LDH level > 1500 IU/L in acute myocardial infarction suggests a grave prognosis.

Disadvantage

The enzyme is relatively ***non-specific*** for myocardial tissue. It is so widespread in body cells that coexistent disease processes in other organs may cause elevations. Thus, S-LDH levels are raised in: carcinomatosis, acute leukaemias, granulocytic leukaemia, pulmonary infarction, renal necrosis, muscle diseases, etc. Less pronounced S-LDH increases are seen in inflammatory hepatic disorders.

Precaution: Red blood cells are rich in LDH, hence avoid haemolysis. ***Haemolysed samples should not be assayed.***

Other Enzymes

4. γ-GLUTAMYL TRANSPEPTIDASE (G-GTP)

Also called γ-Glutamyl transferase (γ-GT). γ-Glutamyl transpeptidase catalyses the transfer of the γ-glutamyl group from one peptide to another peptide or to an amino acid.

Normal value: Normal serum activity has been shown to be:

- **Men:** 10 to 47 IU/L
- **Women:** 7 to 30 IU/L

Site: Highest tissue activity of this enzyme is found in kidneys, but activity is relatively high in liver, lungs, pancreas and prostate. Some activity is present in intestinal mucosa, thyroid gland and spleen. ***Normal heart contains very little γ-GT.***

Behaviour in Acute Myocardial Infarction

Several investigators recently have demonstrated increases in serum γ-GT in acute myocardial infarction. Increase serum activity is found to be late, peak activity between 7th and 11th day and lasts as long as a month.

Conclusion: Hence, it has been proposed as a ***useful test for myocardial infarction in later stages.*** The enzyme does not come from heart muscle, it is increased tissue levels develop with the repair process. ***Source of the enzyme is from vascular endothelium from angioblastic proliferation.***

Remarks: Elevated levels of serum activity are found also in many other conditions:

- In hepatobilliary disorders: It is useful in detecting obstructive jaundice, cholangitis and cholecystitis, with primary and secondary neoplasms of liver.

- Also found elevated in alcoholics and also in alcoholic cirrhosis. ***γ-GT is the most sensitive indicator in alcoholics.***
- Increase serum activity seen with pancreatic diseases.
- Elevated levels seen in epileptic patients with drug therapy with anticonvulsants, probably due to enzyme induction.
- Since serum γ-GT levels are not elevated in any form of bone disorders ***it is a valuable parameter in differentiating between skeletal (bone) and hepatic dysfunction associated with increased serum-ALP.***

5. HISTAMINASE

The enzyme histaminase occurs in different organs in various species. In man, however, normal plasma contains either very small amount of histaminase or none at all, but considerable amount of histaminase has been found in human heart muscle.

Estimation of serum histaminase has been done by volumetric method of Kapeller-Adler.

Normal value: 0.12 to 0.76 PU/ml. (mean 0.41 ± 0.17)

Interpretation

- Raised histaminase found > 0.8 PU/ml in 97.14 per cent of ECG proved cases of myocardial infarction.

Pattern of Rise in Acute Myocardial Infarction

Serum enzyme activity rises within 6 hours of myocardial infarction and persists for whole of first week.

Remarks

- It helps in early diagnosis of myocardial infarction even when ECG failed to reveal.
- It also has a prognostic value as higher serum histaminase levels were found to be associated with worse prognosis. (Mean value in fatal cases > 3.48 ± 0.97)

6. CHOLINESTERASES

Cholinesterases are enzymes which hydrolyse esters of choline to give choline and acid.

Types: There are **two types:**

(i) ***True cholinesterase:*** Responsible for destruction of acetyl choline at the neuromuscular junction and is found in nerve tissues and RB cells.

(ii) ***Pseudocholinesterase:*** Found in various tissues such as liver, heart muscle and intestine and it is this type which is present in plasma.

Normal value: Normal value of serum pseudocholinesterase is 2.17 to 5.17 IU/ml (De la Huerga *et al.*). By Michaelis method it is 1.05 to 2.45 units/ml (mean 1.5 ± 0.33).

Pattern in Acute Myocardial Infarction

Elevation of plasma pseudocholinesterase was observed in 90.5 per cent cases of acute myocardial infarction.

Raised serum activity is found within 12 hours (or even as early as 3 hours found in some cases). Serum enzyme activity has been ***considered as a sensitive index for determination of cellular necrosis in myocardium.***

Remarks

Serum enzyme activity in other conditions:

- Serum enzyme activity is decreased in acute hepatitis. Its regular low level in chronic hepatitis is more often found valuable (owing to diminished synthesis by hepatic cells).
- ***In organophosphorous poisoning:*** Cholinesterase is also inhibited by organophosphorous compound (insecticides) and serum cholinesterase assay is, therefore, useful to detect poisoning by these compounds in agriculture and industries ("Diazinon" poisoning).

1. Cardiac Troponins (CT1 and CTT)

Troponins are **of 3 types:**

(i) Troponin-C (Calcium binding) non cardiac

(ii) Troponin I (Actomycin ATPase inhibitory element) Cardiac-T_1 (CT 1)

(iii) Troponin T (Tropomycin binding element) - Cardiac T (CT)

Behaviour in Acute Myocardial Infarction

Troponin 1 (CT 1) is released into the blood within 4 hours after the onset of MI. Peaks reaches at 14 to 24 hours and remains elevated for 5 to 7 days.

Methods—Assayed by RIA technique or Elisa method

Normal value—Less than 1.5 mg/L

Remarks

- Specific for cardiac muscle
- Extremely useful for early diagnosis of MI
- Sensitive index is > 75 per cent
- No increase in muscle injury
- CT 1 level greater than 1-5 mg/L is indicative of myocardial infarction.

Cardiac Troponin T (CTT)

CTT is also present in adult cardiac muscle. Two isoforms described T_nT_1 and T_nT_2.

Behaviour in Acute Myocardial Infarction

CTT level increases within 6 hours of myocardial infarction, reaches peak level in 3 days and then remains elevated upto 8 to 10 days.

Table 35.1: Important cardiac markers useful for diagnosis of AMI

Cardiac Markers	*Pattern of increase*			*Remarks*
	Early rise	*Peak level*	*Time to return to normal*	
A. Enzymes:				
• **Creatine phosphokinase (CPK or CK)**	Rises after 6 hours	24 to 30 hours	2 to 4 days usually 72 hours	Early marker and cardiac specific
• **Aspartate transaminase (AST)/(S-GOT)**	Rises within 12 hours	24 hours	3 to 5 days	Relatively late cardiac specific but can rise in hepatitis
• **Lactate dehydrogenase (LDH)**	Rises within 12 to 24 hours	2 to 4 days (48 hours)	8 to 14 days	Relatively late marker and nonspecific
• **γ-glutamyl transpeptidase (G-GTP)**	Rises very late	Peak between 7 to 10 days	Lasts longer about 4 weeks	Helpful in late diagnosis. Serum rise due to repair process—source vascular endothelium from angioblastic proliferation
B. Non-enzyme markers				
• **Cardiac Troponins – CT 1 and CTT**				
– CT 1	within 4 hours	14 to 24 hours	5 to 7 days	Earliest marker and cardiac specific
– CTT	within 6 hours	3 days	8 to 10 days	Also earlier marker and cardiac specific
• Myoglobin	within 2 to 3 hours	4 to 6 hours	20 to 24 hours	Earliest marker and **not cardiac specific**

2. Myoglobin

Recently estimation of myoglobin has also been used in acute myocardial infarction. There is early rise of the protein, hence can be used in early diagnosis of MI, unfortunately it is ***not specific for myocardial injury,*** hence not of much use. **Table 35.1** shows the important cardiac 'markers' useful for diagnosis of AMI.

SERUM ENZYMES IN LIVER DISEASES

Refer to Chapter on Liver Function Tests.

SERUM ENZYMES IN GI TRACT DISEASES

Assays in serum of proteolytic enzymes and their precursors or their inhibitors have not yet found a place in the normal investigation of diseases of GI tract. The only enzyme of GI origin which is regularly assayed is ***serum amylase.*** The other enzyme is ***serum lipase,*** but it is not routinely done in most of the laboratories.

1. SERUM AMYLASE

There are problems in methodology and numerical, results of assay by one procedure cannot easily be converted by a factor to those obtained by another procedure and IU are difficult to apply.

Many laboratories use:

1. A very quick and rapid ***amyloclastic*** method for rapid diagnosis.
2. ***Saccharogenic*** method: Which is more accurate (Somogy's method).

Normal value: By Somogy method: 80 to 180 Somogy units/100 ml.

Interpretation

- ***Acute pancreatitis:*** Serum amylase assay is the investigation of choice in the diagnosis of **Acute pancreatitis.** Serum enzyme activity > 1000 units seen within 24 hours returns to normal within 3 days. Also urinary amylase increases and persists a little longer than serum activity.
- ***In other diseases:*** As amylase is secreted in the parotid glands, raised serum values not exceeding 1000 units, are usually found in **Mumps** and other forms of parotitis and also when there is salivary duct stone. This may be of value occasionally in differential diagnosis of
  - Meningoencephalitis and
  - In facial swellings of other causes.
- A raised serum amylase though not usually exceeding 500 units is often found in other acute abdominal catastrophes like
  - Perforated peptic ulcer,
  - Intestinal obstruction.
- ***After administration of opiates:*** Raised values may be seen.
- ***Macroamylasemia:*** In some individuals, a form of amylase with a high molecular weight occurs in the circulation. It cannot pass the glomerular filter and consequently accumulates in the bloodstream. Macroamylasemia should be suspected when there is (a) an increase in serum amylase and (b) no increase in urinary amylase output. ***Macroamylase can be formed by combination of ordinary serum amylase with an antibody.*** It can probably result from polymerisation of the enzyme molecule.

2. SERUM LIPASE

Serum Lipase assay is more specific in pancreatic disorders and remains raised for longer periods. But it is not valuable in practice because of the absence of quick assay methods. The lipolytic activity of the serum may be determined by the amount of "olive-oil emulsion" hydrolysed by a given quantity of serum in a given time at 37°C. Values for Lipase can be expressed as the amount of 0.05 M NaOH required to neutralise the FA produced by one ml of serum **(Cherry-Crandall).** ***A colorimetric assay has also been designed (Seligman and Nachlas).***

Normal Value

- ***By titrimetric method:*** 0.06 to 1.02 ml of 0.05 (N) Sodium hydroxide.
- ***By colorimetric assays:*** 9.0 to 20 m IU. (Seligman and Nachlas).

Remarks: Increase in serum lipase is a reflection of pancreatic disorders. ***In acute pancreatitis*** serum lipase activity increases promptly at the time of onset of symptoms, values as high as 2800 U/L having been reported. The subsequent fall is more gradual than in the case of amylase. ***Elevated levels persist in some cases 10 to 14 days or longer*** (less rapid removal from circulation).

- Elevated serum lipase levels also reported in perforated duodenal and peptic ulcers and in intestinal obstruction.
- Moderate increases of serum lipase were found in about 1/3 patients with cirrhosis.
- **Provocative Tests** with secretin and PZ, have been reported:
  - In *normal* cases duodenal juice amylase is increased and serum level is unaltered.
  - In *chronic pancreatitis* duodenal juice value is unaltered but serum level increases.

SERUM ENZYMES IN MUSCLE DISEASES

Enzyme assays used in muscle diseases are:

- **S-GOT/S-GPT**
- **Aldolase**
- **CPK**

1. S-GOT/S-GPT: These are not used now.

2. SERUM ALDOLASE

Aldolase was until recent years the enzyme of choice in the investigation of diseases of muscles being more sensitive than GOT. This enzyme catalyses the interconversion of fructose-1,6-bi-phosphate and triose phosphate.

Site: It has a wide tissue distribution specially found in high concentration in liver, skeletal muscle, brain. Also found abundant in neoplastic tissues.

Normal Value: 2 to 6 m-IU

Remarks

- Normal values of serum aldolase found in neurogenic muscular weakness, e.g. in poliomyelitis, peripheral neuritis; Moderate increase in dermatomyositis, and muscular dystrophies; highest values are seen in Deuchenne type of muscular dystrophy.
- Other conditions: Increase in serum aldolase activity has also been reported in liver diseases specially viral hepatitis, and myocardial infarction.

3. SERUM CPK

Assay of serum aldolase has been replaced by serum CPK which is more sensitive and more specific.

Remarks

- Serum CPK is slightly elevated occasionally in neurogenic muscular atrophy.
- Raised values are seen in most cases of muscular dystrophies and dermatomyositis, usually 1000 IU/L.
- Highest values are found in Deuchenne type of muscular dystrophies (10,000 IU/L)

The increase is most marked in acute phase, in early childhood; the actual value depend on both:
  - Severity of the disease, or
  - On mass of diseased muscle. The rise occurs before the clinical manifestations and serum enzyme levels are used to detect "carriers".
- Raised values are also found in hypothyroidism owing to secondary muscle disease.

SERUM ENZYMES IN BONE DISEASES

Serum alkaline phosphatase remains the only useful enzyme assay for investigation.

ALP is a most valuable index of osteoblastic activity.

- Increases in serum ALP activity seen in rickets, osteomalacia, hyperparathyroidism and particularly in Paget's disease.
- In primary and secondary malignancies of bone, the level depends on the severity and degree of new bone formation. When the lesion is purely destructive as myelomatosis, the value is normal.
- In hypophosphatasia, where there is defective calcification, low tissue and serum ALP activity is observed.

VALUE OF ENZYMES IN MALIGNANCIES

Chief enzyme assays useful in malignancies are listed in Table 35.2.

1. ACID PHOSPHATASE

There are **two types** depending on activity in different pH.

Table 35.2: Chief enzyme assays useful in malignancies

Enzymes assayed	Diseases
1. Serum acid phosphatase (AP)	• Cancer of prostate with/without metastasis
2. Serum alkaline phosphatase (ALP)	• Metastasis in liver • Osteoblastic metastasis in bone. • Jaundice due to carcinoma of head of pancreas.
3. Serum LDH, aldolase, phosphohexose isomerase	• Widespread malignancies • Advanced leukaemias.
4. β-glucuronidase in urine and serum	• Cancer of urinary bladder • Cancer head of pancreas
5. LDH in effusion fluids	• Local malignancies.
6. LAP	• Liver cell carcinoma • Primary or secondary hepatoma superimposed on cirrhosis liver.

	Tissues
• A type of acid phosphatase (pH = 6.0)	found in erythrocytes
• A type of acid phosphatase (pH = 5.0)	Prostate epithelium, spleen, kidney, plasma, liver, pancreas.

Normal value: Normal serum contains small amount of acid phosphatase 0.6 to 3.1 K.A. units/100 ml.

Precautions

1. It is ***extremely labile enzyme***. **Enzyme assays should be done on fresh samples immediately**.
2. ***Avoid haemolysis*** (due to presence of acid phosphatase in RB cells. Haemolysed sample gives high results).

Remarks

- Main value in relation to diagnosis of metastasising prostate carcinoma. Enzyme is formed from mature prostatic epithelial cells. Not formed by immature prostatic epithelial cells.
- ***Highly anaplastic carcinoma may not produce the enzyme.***
- Acid phosphatase of prostate is inhibited by L-tartarate. **"Tartarate-liable" AP is more specific**, Normal value: 0.0 to 0.5 K.A. units %.
- Acid phosphatase of RB cells are inactivated by 20 per cent neutral formaldehyde.
- *Sullivan test:* It can be used in cases of highly anaplastic carcinoma to stimulate AP production. Injection of 25 mg of testosterone propionate is given daily for 5 days. This may stimulate enzyme production by anaplastic cancer cells.

Other conditions: Acid phosphatase activity in serum may also rise in certain other diseases:

- ***Marked rise is seen in Gaucher's disease*** and it is characteristic of that disorder.
- Occasional rise is seen in Paget's disease, hyperparathyroidism, and osteolytic metastasis from breast and other carcinomas.
- Marked rise seen with thrombocytosis, chronic granulocytic leukaemia, myeloproliferative disorders, etc.

Note: No rise occurs in lymphocytic leukaemias, or Lymphomas.

- A rise is seen in haemolytic anaemia.
- Small elevations with thromboembolic disorders, e.g. Pulmonary embolism.

2. β-GLUCURONIDASE

Though it is not routinely done, it is useful in certain malignancies. β-glucuronidase catalyses *glucuronotransferase* reactions as well as the hydrolysis of β-D-glucopyranosides by means of which its activity is usually estimated.

Site: The enzyme is widespread in human tissues but is most abundant in liver, spleen, endometrium, breast and adrenals. Human RB cells contain little or no β-glucuronidase but leucocytes have a high enzyme content.

Normal value: Serum β-glucuronidase activity ranges from 210 to 550 m-IU for males, and 90 to 400 m-IU in females.

Remarks

- Serum and urinary β-glucuronidase activity is ***increased markedly in cancer of urinary bladder.***
- Very high serum activity reported in carcinoma of head of pancreas and in 50 per cent cases of cancer breast and cervix without liver metastasis.
- Serum β-glucuronidase activity increases in last trimester of pregnancy, and then falls to normal values by about the 5th of post-partum day.
- Assay of β-glucuronidase activity of vaginal fluid has been suggested as useful in diagnosis of malignancies of female genital tract.
- β-glucuronidase activity in ascitic fluid has been found to increase twice in malignant diseases compared to ascitic fluid of non-malignant origin.

Table 35.3 shows increase/decrease of different enzymes in various diseases.

ISOENZYMES

Definition: Isoenzymes (or Isozymes) are the physically distinct forms of the same enzyme but catalyse the same chemical reaction or reactions, and differ from each other structurally, electrophoretically and immunologically.

Table 35.3: Increase/decrease of different enzymes in diseases

Serum enzymes	Normal value	Concentrations increased in	Concentrations decreased in
1. ***Aspartate transaminase (AST) (SGOT)***	4-17 IU/L	***Myocardial infarction,*** elevation slight to moderate in muscle diseases, acute liver disease, toxic liver cells necrosis, haemolytic anaemia.	—
2. ***Alanine transaminase (ALT) (S-GPT)***	3-15 IU/L	• ***Marked increase: Viral hepatitis,*** • Slight-to-moderate-obstructive jaundice, cirrhosis of the liver, toxic liver cells necrosis, skeletal muscle disease	—
3. ***Lactate dehydrogenase (LDH)***	60 to 250 IU/L	Acute myocardial infarction, acute hepatitis, also raised in muscle diseases, leukaemias, renal tubular necrosis, carcinomatosis, cerebral infarction, pernicious anaemia.	
4. ***Alkaline phosphatase (ALP)***	3 to 13 K.A. units% (23-92 IU/L) Infants and growing children 12-30 K.A. units per 100 ml	• ***Marked increase: Obstructive jaundice*** (> 35 K.A. units%), bone diseases—rickets, Paget's disease, hyperparathyroidism • Slight-to-moderate increase: Acute liver diseases, metastatic carcinoma, ***space-occupying*** lesions of liver, kidney disease, osteoblastic sarcoma.	
5. ***Creatine kinase (CK or CPK)***	4-60 IU/L	• ***Marked increase: Acute myocardial infarction,*** muscular dystrophies; • Mild-to-moderate rise: Muscle injury, severe physical exertion, hypothyroidism.	
6. ***Aldolase***	2 to 6 m-IU	***Muscular dystrophies,*** acute liver diseases, myocardial infarction, diabetes mellitus, leukaemias, etc.	
7. ***Amylase***	80 to 180 Somogyi units %	***Acute pancreatitis,*** acute parotitis (mumps), perforated peptic ulcer, intestinal obstruction, macroamylasemia, renal failure	Acute liver diseases, D. mellitus
8. ***Lipase***	• Colorimetric assay 9.0 to 20 m-IU (Seligman and Nachlas) • Titrimetric method 0.06 to 1.02 ml of 0.05 (N) NaOH. • Chery-Crandal units. 1.0 to 1.5 units %	***Acute pancreatitis,*** perforated peptic ulcer, cirrhosis of the liver, pancreatic carcinoma	acute liver diseases, D. mellitus, vitamin A deficiency
9. ***Cholinesterase***	2.17 to 5.17 IU/ml, 130-310 units (dela Huerga)	Nephrotic syndrome, acute myocardial infarction	Acute liver diseases, malnutrition, acute infectious diseases, organophosphorous poisoning (diazinon poisoning)
10. ***Acid phosphatase (ACP)***	0.6 to 3.1 KA units/ 100 ml; Tartaratelabile ACP: 0 to 0.8 KA units %	***Metastasizing prostatic carcinoma, marked rise seen in Gaucher's disease*** • Slight-to-moderate rise seen in Paget's disease, hyperparathyroidism, osteolytic lesions from breast carcinoma, thrombocytosis • ***Slight increase after rectal examination (PR),*** chronic granulocytic leukaemia, myeloproliferative lesions.	
11. ***Ceruloplasmin (Ferroxidase)***	3 to 58 mg%	Cirrhosis, bacterial infections, pregnancy	Wilson's disease (hepatolenticular degeneration)
12. ***Isocitrate dehydrogenase (ICD)***	0.9 to 4.0 IU/L	• Marked increase seen in viral hepatitis, • Slight-to-moderate rise-cirrhosis of the liver	
13. ***Ornithine carbamoyl transferase (OCT)***	8 to 20 m-IU	• Marked elevation in viral hepatitis; • Slight elevation-cirrhosis of the liver, obstructive jaundice, metastatic carcinoma	—
14. ***Leucine aminopeptidase (LAP)***	15 to 56 m-IU	• ***Marked rise in-liver cell carcinoma,*** • Slight increase-cirrhosis of the liver, marked rise in superimposed hepatoma in cirrhosis of the liver, • Moderate rise: Viral hepatitis	—

Table 35.3 contd....

SECTION FIVE

Table 35.3 contd...

Serum enzymes	Normal value	Concentrations increased in	Concentrations decreased in
15. γ-Glutamyl transpeptidase (γ-GT)	10 to 47 IU/L	Acute hepatobiliary diseases, ***alcohol abuse marked rise characteristic,*** alcoholic cirrhosis, slight-to-moderate increase seen in epileptic patients with drug therapy with anticonvulsants, pancreatic diseases	
16. 5'-Nucleotidase	2 to 17 IU/L	Acute liver diseases, obstructive jaundice, tumours	

VALUE AND SIGNIFICANCE OF DIFFERENT ISOENZYMES

1. LDH ISOENZYMES

1. LDH catalyses the reversible oxidation of lactate to pyruvate.

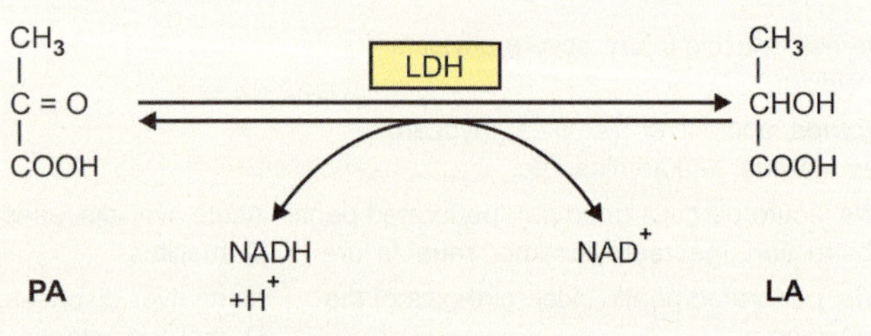

2. In serum as many as 5 (five) physically distinct isoenzymes of this enzyme exist and are known as **LDH-1, LDH-2, LDH-3, LDH-4 and LDH-5.**
3. All these isoenzymes though different physically, they catalyse the same reaction of oxidation of LA to PA.
4. The different forms can be separated by electrophoresis. The difference in electrophoretic mobilities is due to different electric charges on the isoenzymes due to difference in contents of acidic and basic amino acids.

   Example
   - **LDH-1** has the highest negative charge and hence, moves fastest during electrophoresis. It contains a higher proportions of Asp and glutamate than the other forms.
   - **LDH-5** is the slowest moving fraction.
5. Though the same chemical reaction is catalysed, the different isoenzymes may catalyse the same reaction at different rates.

   Example: Rate of oxidation of –OH- butyrate is greater by LDH-1 and LDH-2, when compared with rate of oxidation of LDH-4 and LDH-5.
6. The isoenzymes may have different physical properties also.

   Example: LDH-4 and LDH-5 are easily destroyed by heat, whereas LDH-1 and LDH-2 are not, if heated up to about 60°C ("Heat-resistant")
7. The isoenzymes have different pH optima and Km values.
8. It has been shown that the existence of different isoenzymes is due to difference in the quarternary structure of the enzyme protein.

Structure of LDH Isoenzymes

1. In man, there are **5 (five)** principal isoenzymes of LDH as mentioned above.
2. A sixth, atypical isoenzyme LDH has been found in male genital tissues, called LDH_x.
3. Each isoenzyme protein is made up of four polypeptide subunits, thus each is a ***tetramer***. Each subunit may be one of two types termed H and M and the different isoenzymes contain H and M in different proportions.
4. Thus five possible combinations occur as shown in the box.

LDH Isoenzymes

Type	Polypeptide chains	Electrophoretic mobility	Tissue rich in isoenzyme type
• **LDH-1**	(H_4) H H H H	Fast moving (fastest)	Found in myocardium
• **LDH-2**	(H_3M) H H H M		
• **LDH-3**	(H_2M_2) H H M M		
• **LDH-4**	(HM_3) H M M M		
• **LDH-5**	(M_4) M M M M	Slowest moving	Found in liver (Hepatic)

CLINICAL SIGNICANCE

- After damage to either of these tissues, viz. myocardium or liver, total serum LDH is increased and it may be useful to know the origin of this enzyme increase.
- ***In normal serum, LDH_2 (H_3M) is the most prominent isoenzyme*** and the slowest peak of LDH-5 is rarely seen.
- After myocardial infarction, the faster isoenzymes LDH-1 and LDH-2 predominate.
- In acute viral hepatitis, the slowest isoenzymes LDH-5 and LDH-4 (HM_3) predominate.

Chemical Differentiation of Isoenzymes of LDH

Attempts have been made at simpler chemical identification of these patterns by:

- **Heat stability,**
- **Inhibition with urea,**
- **Reaction with changed "substrate".**

Remarks

- *Myocardial LDH* (LDH-1) is found to be more "heat stable" than that of hepatic LDH (LDH-5)
- *Hepatic* LDH (LDH-5) is inhibited by urea. The above two properties have been utilised to differentiate these two isoenzymes by chemical methods.
- Reaction with changed substrate: *Cardiac LDH* (LDH-1 and 2) utilises oxobutyrate preferentially to pyruvate as a substrate, whereas liver LDH (LDH-5 and 4) has relatively less activity with oxobutyrate.

LDH Isoenzymes in Malignancy: Total serum LDH is frequently elevated in neoplastic diseases. In malignancies, isoenzyme pattern shifts towards slower migrating zone, there is usually an increase in LDH-3, LDH-4 and LDH-5.

- An increase in LDH-5 seen in breast carcinoma, malignancies of CNS, prostatic carcinoma.
- In leukaemias, rise is more in LDH-2, and LDH-3.
- Malignant tumors of testes and ovary show rise of LDH-2, LDH-3, and LDH-4.

2. ISOENZYMES OF CPK

In human tissues, CPK exists as **three** different isoenzymes. Each isoenzyme is a ***dimer,*** composed of two protomers 'M' (for muscle) and 'B' (for brain).

Types: Thus three possible isoenzymes are:

CPK Isoenzymes

Type	Polypeptide chains	Electrophoretic mobility	Tissues found
• **CPK-1**	BB	Fast moving (more –ve charge)	Brain
• **CPK-2**	MB		Myocardium
• **CPK-3**	MM	Slow moving	Skeletal muscle

CK isoenzymes can be separated by:

1. Electrophoresis.
2. Ion exchange chromatography techniques.

Remarks

- ***Normally:*** CK-2 (MB) isoenzymes is very small, (accounts for about 2 per cent of total CK activity of plasma), and almost undetectable.
- ***In myocardial infarction:*** Increase of CK-2 (MB) occurs within 4 hours, maximum in 24 hrs, then falls rapidly. MB accounts for 4.5 to 20 per cent of the total CK activity in plasma of patients with a recent myocardial infarction and the total MB isoenzyme level is elevated up to 20-fold above normal.

Atypical CPK Isoenzymes

In addition to above three distinct forms CK-BB, CK-MB and CK-MM, as stated above, ***two atypical isoenzymes of CPK*** have been reported. They are:

- **Macro-CK (CK-macro)**
- **Mitochondrial CK (CK-Mi)**

(a) Macro-CK (CK-macro)

- ***Formation:*** It is formed by aggregation CK-BB with immunoglobulin usually with IgG but sometimes IgA. It may also be formed by complexing CK-MM with lipoproteins.
- ***Electrophoresis:*** Electrophoretically migrates between CK-MB and CK-MM.
- ***Incidence:*** 0.8 to 1.6 per cent
- ***Age and sex:*** Occurs frequently in women above 50 years of age.
- ***Clinical significance:*** No specific disease has been found to be associated with this isoenzyme.

(b) CK-Mi (Mitochondrial CK-isoenzyme)

- ***Formation:*** It is present bound to the exterior surface of inner mitochondrial membrane of muscle, liver and brain. It can exist in dimeric form or as an oligomeric aggregates having high molecular weight of approximately 35,000.
- ***Electrophoresis:*** Electrophoretically, it migrates towards cathode and is behind CK-MM band.
- ***Incidence:*** It is not present in normal serum. Incidence is from 0.8 to 1.7 per cent.
- ***Clinical significance:*** It is only present in serum when there is extensive tissue damage causing breakdown of mitochondrial and cell wall. Thus, its presence in serum indicate severe illness and cellular damage.
  It is not related with any specific disease states, but it has been detected in cases of malignant tumours.

3. ISOENZYMES OF ALKALINE PHOSPHATASE (ALP)

ALP exists as a number of isoenzymes, the major isoenzymes found in serum are derived from liver, bone, intestine and placenta.

Assay: The techniques used most frequently for separating the isoenzymes are:

- Electrophoresis.
- Chemical inhibition.
- Heat inactivation.

Electrophoresis is considered the most useful single technique for ALP isoenzyme analysis. By starch gel electrophoresis at pH 8.6, at least *six* **isoenzyme bands** have been delineated.

- ***Hepatic isoenzyme:*** Travels fastest towards the anode and occupies the same position as the fast α_2-globulin.

- ***Bone isoenzyme:*** The hepatic isoenzyme is closely followed by bone isoenzyme in β-globulin region.
- ***Placental isoenzyme:*** Follows bone isoenzyme.
- ***Intestinal isoenzyme:*** Slow moving and follows the placental isoenzyme.

Remarks

- ***The major ALP isoenzyme in normal serum of adult healthy person is derived from liver, and it shows main liver band.*** In growing child, bone isoenzyme predominates.

The presence of intestinal isoenzyme in serum depends on blood group and secretor status. ***Individuals who have B or O blood group and are secretors are more likely to have intestinal isoenzyme.***

- Nearly all tissues show a ***subsidiary band*** near the point of insertion, this approximates in position of serum β-lipoproteins.
- ***Liver isoenzyme*** can actually be divided into ***two fractions:***
  - The major liver band.
  - A subsidiary smaller fraction, called 'fast' liver or α_1-liver, which migrates anodal to the major band and corresponds to α_1-globulin.

When total ALP levels are increased, it is the major liver fraction that is most frequently elevated.

CLINICAL SIGNIFICANCE

- The major liver band increased in many hepatobiliary diseases.
- 'Fast' liver band is found in many hepatobiliary diseases and in metastatic carcinoma of liver. The two subsidiary bands form a ***doublet*** which ***is of diagnostic significance in extrahepatic obstructive jaundice.***
- ***Bone isoenzyme:*** Increases due to osteoblastic activity and is normally elevated in children during periods of growth and in adults over the age of 50. In these cases, an elevated ALP level may cause difficulty in interpretation.
- ***In pregnancy:*** During last six weeks of pregnancy, placental isoenzyme of ALP increases. Placental isoenzyme is ***heat stable*** and resists heat denaturation at 65°C for ½ hour. ***It is inhibited by L-phenylalanine.***
- Increases of intestinal isoenzyme occurs after consumption of fatty meal. It may increase in several disorders of GI tract and cirrhosis of liver. Increased levels are also found in patients undergoing chronic haemodialysis.

Characteristics of intestinal isoenzyme are given below:

- Slow moving in electrophoresis.
- Inhibited by L-phenyl alanine.
- Resistant to neuraminidase.

Atypical ALP-isoenzymes-"oncogenic markers": In addition four major ALP isoenzymes, ***two more abnormal fractions are seen associated with tumors.*** They are:

- ***Regan isoenzyme.***
- ***Nagao isoenzyme.***

They have been called as "carcinoplacental ALP isoenzymes" as they resemble placental isoenzyme. Frequency of occurrence in cancer patients is 3 to 15 per cent.

Properties

- ***Regan isoenzyme***
  Electrophoretically migrates to same position as bone fraction. It is extremely 'heat-stable' and resists heat denaturation of 65°C for ½ hour. It is inhibited by L-phenyl alanine.
- ***Nagao isoenzyme***
  May be considered as a variant of Regan isoenzyme. Other properties and electrophoretic mobility are similar to Regan isoenzyme. It can be inhibited by L-leucine.

CLINICAL SIGNIFICANCE

- Regan isoenzyme is produced by malignant tissues. It has been detected in various carcinomas of breast, lungs, colon and ovary. ***Highest incidence of positivity found in cancers of ovary and uterus.***
- Nagao isoenzyme has been detected in metastatic carcinoma of pleural surfaces and adenocarcinoma of pancreas and bile duct. Both have prognostic significance. They disappear on successful treatment.

I. Questions (Essay type and short notes)

1. Describe critically the interpretations of the following enzymes in serum and their merits and demerits:
   a. Acid phosphatase b. Lactate dehydrogenase
   c. Creatine phosphokinase d. Amylase
2. Discuss the diagnostic value of isoenzyme measurement.
3. Describe the clinical significance of various serum enzyme assays.

Short Notes

a. Amylase in serum and urine
b. Use of enzyme estimation as LFT
c. Clinical importance of enzymes
d. LDH isoenzymes
e. Isoenzymes
f. Biochemical properties and clinical significane of isoenzyme

g. Transaminases
h. Creatine kinase
I. Lactate dehydrogenase

II. MCQs (Give one correct answer)

1. The most useful test for the diagnosis of acute hemorrhagic pancreatitis during the first few days is:
   a. Urinary lipase test
   b. Serum calcium
   c. Urinary amylase
   d. Serum amylase
   e. Serum inorganic
2. The best test for acute pancreatitis in the presence of mumps is:
   a. A serological test for mumps
   b. Virus isolation
   c. Serum amylase
   d. Urinary amylase
   e. Serum lipase
3. The slow moving fraction of LDH is typically increased in patients with:
   a. Cerebrovascular accidents
   b. Acute myocardial infarction
   c. Acute pancreatitis
   d. Acute viral hepatitis
   e. Peptic ulcer
4. Which of the following enzyme typically elevated in alcoholism:
   a. Serum ALP
   b. Serum GOT
   c. Serum γ-GT
   d. Serum acid phosphatase
   e. Serum LDH
5. Patients with hepatocellular jaundice, as compared to those with purely obstructive jaundice, tend to have:
   a. Lower serum ALP, LDH and AST activity
   b. Lower serum ALP, higher LDH and AST activity
   c. Higher serum ALP, LDH and AST activity
   d. Higher serum ALP, lower LDH and AST activity
   e. None of the above
6. If results of the serum bilirubin, serum ALP, LDH and AST determinations suggest obstructive jaundice, the best confirmatory test would be the estimation of:
   a. Serum ALT
   b. Serum 5'-nucleotidase
   c. Serum protein electrophoresis
   d. Serum pseudocholinesterase
   e. None of the above
7. Which enzyme estimation will be helpful in differentiating the elevated serum ALP found in obstructive jaundice as well as bone disorders?
   a. Serum AST
   b. Serum ALT
   c. Serum LDH
   d. Serum γ-GT
   e. All of the above
8. Cardiac muscle contains which of the following CK-isoenzyme?
   a. BB only
   b. MM and BB only
   c. MM, BB and MB all three
   d. MM and MB only
   e. MB only
9. Liver and skeletal muscle disorder are characterised by a disproportionate increase in which of the LDH isoenzyme fraction?
   a. LDH-1
   b. LDH-1 and LDH-2
   c. LDH-3 and LDH-4
   d. LDH-2 and LDH-3
   e. LDH-5
10. On the third day following onset of acute myocardial infarction which enzyme estimation will have the best predictive value?
    a. Serum AST
    b. Serum CK
    c. Serum ALT
    d. Serum LDH
    e. Serum ALP
11. Serum ASt activity is not characteristically elevated as the result of:
    a. Myocardial infarction
    b. Passive congestion of liver
    c. Muscular dystrophies
    d. Peptic ulcer
    e. Infectious hepatitis
12. On which day following acute myocardial infarction the estimation of serum AST will be of greatest significance?
    a. First day
    b. Second day
    c. Third day
    d. Fourth day
    e. Fifth day
13. In which diseases of the following organs, isoenzymes LDH-1 and LDH-2 will be released in plasma?
    a. Kidney, red blood cells, liver
    b. Heart, kidney, red cells,
    c. Heart, kidney, liver
    d. Heart, lungs, brain
    e. Liver, redcells, brain
14. Rise of which serum enzyme activity 4-8 hours after acute myocardial infarction is characteristically seen?
    a. AST
    b. ALT
    c. LDH
    d. CK
    e. γ-GT
15. Which serum enzyme activity will be useful to establish myocardial infarction if the patient is seen after three weeks of suspected attack?
    a. AST
    b. LDH
    c. γ-GT
    d. CK
    e. GOT
16. An increase in LDH-5 isoenzyme is seen in the following except:
    a. Acute hepatitis,
    b. Muscular dystrophies
    c. Breast carcinoma
    d. Pulmonary embolism
    e. Prostatic carcinoma

Answers to MCQs

1. d	2. e	3. d	4. c	5. b	6. b
7. d	8. d	9. e	10. d	11. d	12. b
13. b	14. d	15. c	16. d		

CHAPTER

36

RENAL FUNCTION TESTS

Major Concepts

A. To review the physiological aspect of kidney.

B. To study the various function tests that can be employed to assess the renal function.

Specific Objectives

A. Anatomy and details of physiology of kidney have not been included in the text. However, students are encouraged to learn them before they study the renal function tests.

B. 1. Classify the various function tests based on Glomerular Filtration Rate (GFR), Renal Plasma Flow (RPF) and tubular function.

2. • Define "clearance"
   • Learn in details the procedure, and interpretation of various clearance tests.
   (a) Urea clearance test
   (b) Endogenous creatinine clearance test
   (c) Inulin clearance test.

3. Learn the tests based on Renal Plasma Flow:
   • Study PAH clearance and its value • Learn about filtration fraction (FF) and its normal range.
   • Study the significance of FF in various renal diseases.

4. Learn the tests based on tubular function:
   • Study the two common bedside tests used for tubular function based on sp. gr. of urine.
   (a) Water concentration test (b) Water dilution test. Learn the precautions, procedure and interpretation of the tests.
   • Study phenol sulphthalein (PSP) test, its procedure and interpretation.

5. Learn other miscellaneous tests, viz. IV pyelography, radioactive scanning/renogram in brief, which are helpful in assessing renal function as well as size, shape, etc. of the kidneys.

INTRODUCTION

The body has a considerable factor of safety in renal as well as hepatic tissues. ***One healthy normal kidney can do the work of two,*** and if all other organs are functioning properly, less than a whole kidney can suffice. On the other hand, there are certain extrarenal factors which can interfere with kidney function, specially circulatory disturbances. Hence, methods that appraise the functional capacity of the kidneys are very important. Such tests have been devised and are available, ***but it is stressed that no single test can measure all the kidney functions. Consequently, more than one test is indicated to assess the kidney function.***

Preliminary Investigations to Renal Function Tests

Assessment of renal function begins with the appreciation of:

- ***Patient's history:*** A proper history taking is important, particularly in respect of oliguria, polyuria, nocturia, ratio of frequency of urination in day time and night time. Appearance of oedema is important.
- ***Physical examination:*** This is followed by side room analysis of the urine specially for presence/or absence of albumin, and microscopic examination of urinary deposits specially for pus cells, RB cells and casts.
- ***Biochemical parameters:*** Certain biochemical parameters also help in assessing kidney function. ***A stepwise increase in three nitrogenous constituents of blood is believed to reflect a deteriorating kidney function. Some authorities claim that serum uric acid normally rises first, followed by urea and finally increase in creatinine.*** By determining all the above three parameters a rough estimate of kidney function can be made. However, other causes of uric acid rise should be kept in mind.

Other biochemical parameters which help are determination of total plasma proteins, albumin, globulins and total cholesterol. In nephrosis there is marked fall in albumin and rise in serum cholesterol level.

PHYSIOLOGICAL ASPECT

Main functions of the kidney are:

- To get rid of the body waste products of metabolism,
- To get rid of foreign and non-endogenous substances,
- To maintain salt and water balance, and
- To maintain acid-base balance of the body.

Glomerular Function: The glomeruli act as **filters,** and the fluid which passes from the blood in the glomerular capillaries into Bowman's capsule is of the same composition of **protein-free plasma.** The effective filtration pressure which forces fluid through the filters is the result of (a) the blood pressure in the glomerular capillaries and (b) the opposing osmotic pressure of plasma proteins, renal interstitial pressure and intratubular pressure. Thus,

- Capillary pressure = 75 mmHg
- Osmotic pressure of plasma proteins = 30 mmHg
- Renal interstitial pressure = 10 mmHg
- Renal intratubular pressure = 10 mmHg

Hence, ***net effective filtration pressure:***

= 75–(30 + 10 + 10)

= **25 mmHg**

Rate of filtration is influenced by:

- Variations in BP in glomerular capillary
- Concentration of plasma proteins
- Factors altering intratubular pressure, viz.
  - Rise with ureteral obstruction
  - During osmotic diuresis
- State of blood vessels.

If the efferent glomerular arteriole is constricted, the pressure in the glomerulus rises and the effective filtration pressure is increased. On the other hand, if the affarent glomerular arteriole is constricted, the filtration pressure is reduced.

The volume of glomerular filtrate formed depends on:

- The number of glomeruli functioning at a time,
- The volume of blood passing through the glomeruli per minute and
- The effective glomerular filtration pressure.

Under normal circumstances, about 700 ml of plasma (contained in 1300 ml of blood or approximately 25 per cent of entire cardiac output at rest) flow through the kidneys per minute and 120 ml of fluid are filtered into Bowman's capsule.

The volume of the filtrate is reduced in extrarenal conditions, such as dehydration, oligaemic shock and cardiac failure which diminish the volume of blood passing through the glomeruli, or lower the glomerular filtration pressure, and when there is constriction of the afferent glomerular arterioles or, changes in the glomeruli such as occurring in glomerulonephritis. If the volume of glomerular filtrate is lowered below a certain point, the kidneys are unable to eliminate waste products which accumulate in blood.

Tubular function: Whereas, the glomerular cells act only as a passive semipermeable membrane, ***the tubular epithelial cells are a highly specialised tissue able to reabsorb selectively some substances and secrete others.***

- About 170 litres of water are filtered through the glomeruli in 24 hours, and only 1.5 litre is excreted in the urine. Thus ***nearly 99 per cent of the glomerular filtrate is reabsorbed in the tubules.***
- Glucose is present in the glomerular filtrate in the same concentration as in the blood but practically none is excreted normally in health in detectable amount in urine and the tubules reabsorb about 170 GM/day. At an arterial plasma level of 100 mg/100 ml and a GFR of 120 ml/mt, approximately 120 mg of glucose are delivered in the glomerular filtrate in each minute. Maximum rate at which glucose can be reabsorbed is **about 350 mg/mt (Tm G),** which is an 'active' process.
- About 50 gm of urea are filtered through the glomeruli in 24 hours, but only 30 gram are excreted in the urine, this is a passive diffusion.
- Certain substances foreign to the body, e.g. diodrast, para-amino hippuric acid (PAH) and phenol red are: **(1)** filtered through the glomeruli and in addition are **(2)** secreted by the tubules. Thus the amount of these substances excreted per minute in the urine is greater than that filtered through the glomeruli per minute. At low blood levels, the tubular capacity for excreting these compounds is so great that the plasma passing through the kidneys is almost completely cleared of them.
- Another group of substances, e.g. inulin, thiosulphate, and mannitol are eliminated exclusively by the glomeruli and are neither reabsorbed nor secreted by the tubules. Hence, amount of these substances excreted per minute in the urine is the same as the amount filtered through the glomeruli per minute, thus they give the glomerular filtration rate (GFR).

RENAL FUNCTION TESTS

Based on the above functions, the renal function tests can be classified as follows.

CLASSIFICATION OF RENAL FUNCTION TESTS

I. *Tests based on Glomerular filtration:*
- Urea clearance test.
- Endogenous creatinine clearance test.
- Inulin clearance test.
- Cr^{51}-EDTA clearance test.

II. *Tests to measure Renal Plasma Flow (RPF):*
- Para-amino hippurate test (PAH).
- Filtration fraction

III. *Tests based on tubular function:*
- Concentration and Dilution tests.
- 15 minute-PSP excretion test.
- Measurement of tubular secretory mass.

IV. *Certain Miscellaneous tests:*
Can determine size, shape, asymmetry, obstruction, tumour, infarct, etc.

GLOMERULAR FILTRATION TESTS

Three clearance tests are:
- *Urea clearance,*
- *Endogenous creatinine clearance,*
- *Inulin clearance tests:* These are used to examine for impairment of glomerular filtration and recently.
- *Cr^{51}- EDTA clearance* has been described.

What is Meant by Clearance Test?

As a means of expressing quantitatively the rate of excretion of a given substance by the kidney, its "clearance" is frequently measured.

This is defined as a volume of blood or plasma which contains the amount of the substance which is excreted in the urine in one minute.

Or alternatively, the clearance of a substance may be defined as that volume of blood or plasma cleared of the amount of the substance found in one minute excretion of urine.

1. Urea Clearance Test

Ambard was the first to study the concentration of urea in blood and relate it to the rate of excretion in the urine, and *Ambard's coefficient* was, for a while, the subject of much clinical study. At present, the blood/plasma urea clearance test of van Slyke is widely used.

Blood urea clearance is an expression of the number of ml of blood/plasma which are completely cleared of urea by the kidney per minute.

As a matter of fact, the plasma is not completely cleared of urea. Only about 10 per cent of the urea is removed. Consequently, 750 ml of plasma pass through the kidney per minute and 10 per cent of the urea is removed, this is equivalent to completely clearing of 75 ml of plasma per minute.

(a) Maximum clearance: **If the urine volume exceeds 2 ml/mt,** the rate of urea elimination is at a maximum and is directly proportional to the concentration of urea in the blood. Thus, provided the blood urea remains unchanged, urea is excreted at the same rate whether the urinary output is 4 ml or 8 ml/mt. Volume of blood cleared of urea per minute can be calculated from the formula:

$$\frac{U \times V}{B}$$

where,

U = Concentration of urea in urine (in mg/100 ml)
V = Volume of urine in ml/mt and
B = The concentration of urea in blood (in mg/100 ml)

Substituting average values, the number of ml of blood cleared of urea per minute =

$$\frac{1000 \times 2.1}{28} = 75$$

A urea clearance of 75 does not mean that 75 ml of blood has passed through the kidneys in one minute and was completely cleared of urea. But it means that the amount of urea excreted in the urine in one minute is equal to the amount found in 75 ml of blood.

Maximum urea clearance: The clearance which occurs when the urinary volume exceeds 2 ml/mt is termed as *Maximum urea clearance (Cm)* and average normal value is 75.

Cm = 75 ml (normal range 75 ±10)

*(b) Standard clearance: **When the urinary volume is less than 2 ml/mt,*** the rate of urea elimination is reduced, because relatively more urea is reabsorbed in the tubules, and is proportional to the square root of the urinary volume. Such clearance is termed as ***standard clearance of urea (Cs)*** and average normal value is 54.

$$Cs = \frac{u \times \sqrt{V}}{B} = 54 \text{ ml (Normal range} = 54 \pm 10)$$

Note: Provided no prerenal factors are temporarily reducing the clearance of urea, the volume of blood cleared of urea per minute is an index of renal function.
- If a larger volume than normal is cleared/mt, renal function is satisfactory.
- If a smaller volume is cleared, renal function is impaired.

Expression of result as %

Sometimes the result of a urea clearance test is expressed as a per cent of the normal maximum or of the normal standard urea clearance depending on whether the urinary output is greater or lesser than 2 ml/mt.

Expressed as % of normal

$$Cm = \frac{u \times v}{B} \times \frac{100}{75} \% = 1.33$$

$$Cs = \frac{u \times \sqrt{v}}{B} \times \frac{100}{54} \% = 1.85$$

Relation with Body Surface

The urea clearance is proportional to the surface area of the body and if the result is to be expressed as a % of normal, a correction must be made in the case of children and those of abnormal stature.

- The Cm is directly proportional to the body surface and if any correction is required the result should be multiplied by $\frac{1.73}{BS}$, where BS = the patient's body surface derived from the height and weight.
- In the case of Cs, the correction factor is $\frac{\sqrt{1.73}}{BS}$

Procedure

The test should be performed between breakfast and lunch, as excretion is more uniform during this time.

1. The patient, who is kept at rest throughout the test, is given a light breakfast and 2 to 3 glasses of water.
2. The bladder is emptied and the urine is discarded, the exact time of urination is noted.
3. One hour later, urine is collected and a specimen of blood is withdrawn for determining urea content.
4. A second specimen of urine is obtained at the end of another hour.

The volume of each specimen of urine is measured accurately and the concentration of urea in the specimen of blood and urine is determined.

The average value of the two specimens of urine is used for assessing the quantity and urea content of urine.

Interpretation of the Test

- Urea clearance of 70 per cent or more of average normal function indicates that the kidneys are excreting satisfactorily.
  Values between 40 to 70 per cent indicate mild impairment, between 20 to 40 per cent moderate impairment and below 20 per cent indicates severe impairment of renal function.
- *In acute renal failure:* The urea clearance Cm or Cs is lowered, usually less than 1/2 the normal and increases again with clinical improvement.
- *In chronic nephritis:* The urea clearance falls progressively and reaches a value 1/2 or less of the normal before the blood urea concentration begins to rise. With values below 20 per cent of normal, prognosis is bad, the survival time rarely exceeds two years and death occurs within a year in more than 50 per cent cases.
- *In terminal uraemia:* The urea clearance falls to about 5 per cent of the normal values.
- *In nephrotic syndrome:* The urea clearance is usually normal until the onset of renal insufficiency sets in and produces the same changes as in chronic nephritis.
- *In benign hypertension:* A normal urea clearance is usually maintained indefinitely except in few cases which assume a terminal malignant phase when it falls rapidly.

Note: A very low protein diet can lead to low clearance value even in normal persons and in patients with mild renal disease.

2. Endogenous Creatinine Clearance Test

At normal levels of creatinine, this metabolite is filtered at the glomerulus, but is neither secreted nor reabsorbed by the tubules. Hence, its clearance gives the GFR. This is a convenient method for estimation of GFR since:

- It is a ***normal metabolite in the body.***
- It does not require the intravenous administration of any test material.
- Estimation of creatinine is simple.

Measurement of 24-hour excretion of endogenous creatinine is convenient. This longer collection period minimizes the timing error.

Procedure of the Test

1. An accurate 24-hour urine specimen is collected ending at 7 AM and its total volume is measured.
2. Collect a blood sample for serum creatinine determination.
3. Estimate the serum and urinary creatinine concentration.

Result:

$$C\,cr = \frac{U \times V}{P}$$

where,

U = Urine creatinine concentration in mg/dl
P = Serum creatinine in mg/dl
V = Volume of urine in ml/mt.

Normal values: for creatinine clearance varies from 95 to 105 ml/mt.

3. Inulin Clearance Test

Inulin, a homopolysaccharide, polymer of fructose is an ideal substance as:

- Not metabolised in the body,
- Following IV administration, it is excreted entirely through glomerular filtration, being neither excreted nor reabsorbed by renal tubules.

Hence the number of ml of plasma which is cleared of inulin in one minute is equivalent to the volume of glomerular filtrate formed in one minute.

Procedure of the Test

1. Preferably performed in the morning. Patient should be hospitalised overnight and kept reclining during the test.
2. A light breakfast is given consisting of 1/2 glass milk, one slice toast can be given at 7.30 AM
3. At 8 AM 10 gm of inulin dissolved in 100 ml of saline, at body temperature, is injected IV at a rate of 10 ml per minute.
4. One hour after (9 AM) the injection, the bladder is emptied and this urine is discarded.
5. Note the time and collect urine after one and two hours. Volume of urine is measured and analysed for inulin content.
6. At the mid-point of each collection of urine, 30 and 90 minutes after the initial emptying of bladder, 10 to 15 ml of blood is withdrawn (in oxalated bottle), plasma is separated and analysed for inulin concentration.

Calculation and result: Values obtained of two samples of blood is averaged.

$$C_{In} = \frac{U \times V}{P}$$

where

U = mg of inulin/100 ml of urine

V = ml of urine/mt

P = mg of inulin/dl of plasma (average of two samples).

Normal average: Inulin clearance in an adult (1.73 sq m) = 125 ml of plasma cleared of inulin/mt Range = 100 to 150 ml.

Note

- To promote a free flow of urine, one glass of water is given at 06.30 AM and repeated every ½ hour until the test is completed. This step may be eliminated if administration of fluid is contraindicated.
- Inulin clearance test is definitely superior for determination of GFR but requires tedious and intricate chemical procedure for determination.

4. Determination of ^{51}Cr-EDTA Clearance

Currently simplified single injection method for determination of ^{51}Cr-EDTA plasma clearance is widely used, for routine assessment of glomerular filtration rate (GFR) in adults as well as in children. It is particularly convenient in children where it is not easy to collect 24 hr. urine sample. It has been used for children younger than 1-year-old.

TESTS FOR RENAL BLOOD FLOW

1. Measurement of Renal Plasma Flow (RPF)

Para-amino hippurate (PAH) is filtered at the glomeruli and secreted by the tubules. At low blood concentrations (2 mg or less/100 ml) of plasma, PAH is removed completely during a single circulation of the blood through the kidneys. Tubular capacity for excreting PAH of low blood levels is great. Thus, the amount of PAH in the urine becomes a measure for the value of plasma cleared of PAH in a unit time, i.e. ***PAH clearance at low blood levels measures renal plasma flow (RPF). RPF (for a surface area of 1.73 sq. m) = 574 ml/mt.***

2. Filtration Fraction (FF): The filtration fraction (FF) is the fraction of plasma passing through the kidney which is filtered at the glomerulus and is obtained by dividing the inulin clearance by the PAH clearance.

$$FF = \frac{C\ In}{C_{PAH}} = \frac{GFR}{RPF}$$

If we take, GFR = 125 and RPF = 594, then the

$$FF = \frac{125}{574} = 0.217\ (21.7\%)$$

Normal range: 0.16 to 0.21 in an adult.

Interpretations

- The FF tends to be normal in early ***essential hypertension,*** but as the disease progresses, the decrease in RPF is greater than the decrease in the GFR. ***This produces an increase in FF.***
- ***In the malignant phase of hypertension:*** These changes are much greater, consequently the ***FF rises considerably.***
- ***In glomerulonephritis:*** The reverse situation prevails. In all stages of this disease, a progressive decrease in the FF is characteristic because of much greater decline in the glomerular filtration rate (GFR), than the renal plasma flow (RPF).
- A rise in FF is also observed early ***in congestive cardiac failure.***

TESTS OF TUBULAR FUNCTION

Pathophysiological aspect: Alterations in renal tubular function may be brought about by:

- ***Ischaemia*** with reduction in blood flow through the peritubular capillaries,
- ***Direct action of toxic substances*** on the renal tubular cells, and
- ***Biochemical defects,*** impairing transfer of substances across the tubular cells.

Adequate renal tubular function requires adequate renal blood flow, a significant reduction in the latter is reflected in impaired tubular function. Hence arteriolar nephrosclerosis and other diseases diminishing blood flow, causes inability to concentrate or dilute the urine with resulting ***"isosthenuria" ("fixation" of sp. gr. at 1.010).***

A. Concentration Tests

Principle: Based on the ability of the kidneys to concentrate urine, and based on measuring sp. gr. of urine. They are simple bedside procedures, easy to carry out and extremely important.

The tests are conducted either:

- Under conditions of restricted fluid intake, or
- By inhibiting diuresis by injection of ADH.

1. **Fishberg concentration test:** This test imposes less strenuous curtailment of fluid intake and may be completed in a shorter period of time. Most commonly used simple bedside concentration test.

Procedure

Patient is allowed ***no fluids from 8 PM until 10 AM*** next morning. The evening meal is given at 7 PM. It should be high protein and must have a fluid content of less than 200 ml. Urine passed in the night is discarded. Nothing by mouth next morning. Collect urine specimens next morning at 8 AM, 9 AM and 10 AM and determine the specific gravity of each.

Result and Interpretation

- ***If tubular function is normal, the sp.gr. of at least one of the specimens should be greater than 1.025,*** after appropriate correction made for temperature, albumin and glucose.
- Impaired tubular function is shown by a sp.gr. of 1.020 or less and may be fixed at 1.010 in cases of severe renal damage.

Note: A false result may be obtained, if the patient has:

- Congestive cardiac failure because elimination of oedema fluid in night and will simulate inability to concentrate.
- Inability to concentrate is also characteristic of Diabetes insipidus.

2. **Lashmet and Newburg concentration test:** This test imposes:
   - ***Severe fluid intake*** restriction over a period of 38 hours, and
   - Involves the use of a special dry diet for one day.
3. **Concentration test with posterior pituitary extract:** The subcutaneous inj. of 10 pressor units of posterior pituitary extract (0.5 ml of vasopressin injection) in a normal person will inhibit the diuresis produced by the ingestion of 1600 ml of water in 15 minutes.

   The test has the advantage of short performance time, and minimising the necessity of preparation of the patient.

   Posterior pituitary extract will also inhibit the diuresis seen in congestive heart failure under active treatment as well as that of *D. Insipidus*, allowing sufficient concentration to determine degree of tubular function in these conditions.

Interpretation: Under the conditions of the test, individual with normal kidney function, excrete urine with sp. gr. 1.020 or higher. Failure to concentrate to this degree indicates renal damage.

B. Water Dilution/Elimination Test

Principle: The ability of the kidneys to eliminate water is tested by measuring the urinary output after ingesting a large volume of water.

Note: Water excretion is not only a renal function but also depends on extrarenal factors and prerenal deviation will reduce the ability of the kidneys to excrete urine.

Procedure

The patient remains in bed throughout the test because elimination of water is maximal in the horizontal position. On the day before the test, the patient has an evening meal but takes nothing by mouth after 8 PM. On the morning of the test, he empties his bladder at 8 AM which is discarded, and then drinks 1200 ml of water within half hour. The bladder is emptied at 9, 10, 11 and 12 noon and ***the volume and the sp. gr. of the four specimens are measured.***

Interpretations

- **If renal function is normal, more than 80 per cent (1000 ml) of water is voided in 4 hours, the larger part being** excreted in the first 2 hours. **The sp.gr. of at least one specimen should be 1.003 or less.**
- ***If renal function is impaired,*** less than 80 per cent (1000 ml) of water is excreted in 4 hours, and the ***sp.gr. does not fall to 1.003 and remains fixed at 1.010 in cases of severe renal damage.***

C. Tests of Tubular Excretion and Reabsorption

Principle: The reserve function of secretion of foreign non-endogenous materials by the tubular epithelium is most conveniently tested for by the use of certain dyes and measuring their rate of excretion.

1. Phenol Sulphthalein (PSP) Excretion Test

Use of PSP (Phenol red) to measure renal function was first introduced by **Rowntree** and **Geraghty** in 1912. Later on, **Smith** has shown that with the amount of dye employed, 94 per cent excreted by tubular action and only 6 per cent by glomerular filtration. Thus the test measures primarily tubular activity as well as being a measure of renal blood flow.

15-MINUTE PSP TEST: It has been shown that the test is reliable and sensitive if the amount of dye excreted in the first 15 minutes is taken as the criterion of renal function.

Test and Interpretation

- When 1.0 ml of PSP (6 mg) is injected IV, normal kidneys will excrete 30 to 50 per cent of the dye during the first 15 minutes.
- ***Excretion of less than 23 per cent of the dye during this period regardless of the amount excreted in 2 hours indicates impaired renal function.***
- It is also used to determine the function of each kidney separately. Here the appearance time as well as the rate of excretion of the dye is of importance. After IV injection, the normal appearance time of the dye at the tip of the catheters is 2 minutes or less and rate of excretion from each kidney is greater than 1 to 1.5 per cent of the injected dye per ml. ***Increase in appearance time and decrease in excretion rate indicate impaired function.***

D. Tests to Measure Tubular Secretory Mass

Principle: If Diodone/or PAH concentration in the plasma is gradually raised above the level at which it is wholly excreted whilst traversing the kidney on a single occasion, the amount of Diodone/PAH actually excreted per minute increases, but the removal of the presented Diodone is no longer complete.

Eventually, a plasma concentration will be reached at which the tubules are excreting the "Maximum" amount possible, they are said to be "saturated" and since they are working at their utmost capacity, further elevation of plasma Diodone level produces no increase in the tubular excretion. Hence the total excretion/mt under these conditions is the

- Amount excreted by glomerular filtration +
- The amount excreted by the tubules.

$$\text{Total excretion/mt} = U_D \times V$$

The glomerular contribution is the glomerular volume/mt (C_{In}) and diodone concentration in the glomerular filtrate (P_D). Since filtrate and plasma contain the same concentration the.

Maximum contribution by tubules is

$$= UD \times V - C_{In} \times P_D$$

The above represents the **tubular excretory capacity or mass** for Diodone expressed in mg/mt. and represented by the symbol **"T_{mD}"**.

Normally: T_{mD} lies in the range of 36 to 72 in adults.

OTHER MISCELLANEOUS TESTS TO ASSESS RENAL FUNCTION

1. **Test of renal ability to excrete acid:** A number of workers have studied the excretion of acid by the kidneys following stimulation by giving NH_4Cl.

Procedure (Davies and Wrong, 1957)

Give NH_4Cl, 0.1 gm/kg in grams or half gram gelatin coated capsules over a period of an hour, e.g. from 10 to 11 AM empty the bladder an hour later and discard the specimen. Collect all urine specimens passed during the next 6 hours and empty the bladder at the end of that period. Make sure that the urine is collected in specially cleaned vessels preferably under oil. A crystal of thymol can be placed in the vessel. Measure the pH of the urine specimens and determine the NH_3 content of the combined urine specimens.

Interpretation

- Normal persons pass urine during the 6-hour period with pH = 5.3, and have an NH_3 excretion between 30 and 90 microequivalents/mt.
- In most forms of renal failure, the pH falls in the same way, but the NH_3 excretion is low.
- In renal tubular acidosis, pH remains between 5.7 and 7.0 and NH_3 excretion is also low.

2. **Intravenous pyelography:** When injected IV, certain radio-opaque organic compounds of iodine are excreted by the kidneys in sufficient concentrations to cast a shadow of the renal calyces, renal pelvis, ureters and the bladder on an X-ray film and ***gives lot of informations regarding size, shape, and functioning of the kidneys.***

The most commonly used substances are:

- *Iodoxyl:* Available as "Pyelectan" (Glaxo), Uropac (M and B), Uroselectan B, etc.
- *Diodone 30 per cent:* More recently introduced, which gives better results. Available as Perabrodil (Bayer), Pyelosil (Glaxo), etc.

Indications: IV pyelography is widely used in the investigation of diseases of urinary tract and should be a routine procedure for investigation:

- Patients of renal calculi, • repeated urinary infections, • renal pain, • haematuria, • prostatic enlargement, • suspected tumours, and • congenital abnormalities

By pyelography the relationship of the renal tract to calcified abdominal shadows and masses can be demonstrated.

- The excretion and concentration of diodone may be used as a rough indication of renal function.

- If the calyces and pelvis of one kidney are outlined, while the other remains invisible, it can be assumed that the function of the invisible side is impaired.

Contraindications: IV pyelography should not be done in patients with:

- Acute nephritis, • congestive cardiac failure, • severely impaired liver function, • in frank uraemia, • in hypersensitive patients and sensitivity to organic iodine compounds.

Note: Sensitivity test should be done before injecting the drug.

3. **Radioactive renogram:** I^{131}-labelled Hippuran is given IV and simultaneously the radioactivity from each kidney is recorded graphically in a stripchart recorder by electronic device. Hippuran-I^{131} is actively secreted by the kidney tubules and it is not concentrated in the liver.

Dose: 15 to 60 μci of Hippuran-I^{131} is given IV slowly in a single dose.

Interpretation: With the limitations and complexities of the interpretation of the results, the investigation is of great practical clinical use. The following information is obtained:

- ***Whether any major asymmetry*** in function between the two kidneys is present.
- A reasonable assessment of overall renal function, given ***by the ratio of bladder activity/heart activity in 10 minutes time.***
- The ***presence of obstruction to urine flow*** in renal pelvis or ureters.

No other means exist for obtaining so much information in a short time about the differential function of the kidneys.

4. **Radioactive Scanning**

A recent development is the renal scintiscan. This has the theoretical advantage over the renogram of being able ***to detect segmental lesions.*** In this technique, Hg^{203}-labelled chlormerodrin or Hg^{197}-labelled chlormerodrin is injected intravenously and a renal scan can be obtained by a scintillation counter over the lumbar regions.

- Renal scanning is helpful for detection of abnormalities in size, shape and position of the kidneys.
- Renal tumours and renal infarcts are shown in scintiscan which may be missed in pyelography.

I. Questions (Essay type and short notes)

1. Classify kidney function tests. Describe the urea clearance test in detail.
2. Classify, enumerate kidney function tests. Describe any one test in detail.
3. Give an account of creatinine clearance test and discuss its significance.
4. Describe the renal function tests.
5. What is meant by the term "clearance"? Describe the urea clearance test and its significance.
6. Outline the function of the distal tubule and describe the tests that assess the same.
7. Enumerate the kidney function tests. Describe "endogenous creatinine clearance test" and give its significance.
8. Name the tests performed for the localisation of kidney dysfunction and discuss the merits of the tests performed for assessment of the glomerular filtration rate.

Short Notes

a. Urea clearance test
b. Creatinine clearance
c. Glomerular filtration rate (GFR)
d. Filtration fraction
e. Endogenous
f. Tubular maximum for Glucose (TmG)
g. Specific gravity of urine and renal function
h. Inulin clearance test
i. Renal clearance test.

II. MCQs (Give one correct answer)

1. Normal renal plasma flow in healthy adults averages about:
   a. 125 ml/mt b. 200 ml/mt
   c. 450 ml/mt d. 574 ml/mt
   e. 750 ml/mt
2. All of the following substances have been used to estimate GFR *except:*
   a. Inulin b. Creatinine
   c. Phenol red d. Mannitol
   e. Urea Nitrogen
3. Relationship between GFR and serum creatinine concentration is:
   a. Non-existent b. Inverse
   c. Direct d. Indirect
   e. Haphazard
4. Normal maximum urea clearance (Cm) in an adult averages about
   a. 60 b. 65
   c. 70 d. 75
   e. 85
5. In healthy adult the upper limit of normal serum creatinine concentration is
   a. 0.75 mg/dl b. 1.2 mg/dl
   c. 2.1 mg/dl d. 2.1 mg/dl
   e. 3.5 mg/dl
6. In patients with renal failure all of the following are typically elevated in serum, *except*
   a. Urea nitrogen b. Phosphate
   c. Uric acid d. Albumin
   e. Creatinine

7. Excretion of PSP primarily reflects:
   a. Liver function
   b. Kupffer cell activity
   c. Maximal tubular excretory capacity
   d. Glomerular filtration rate (GFR)
   e. None of the above
8. The nephrotic syndrome includes, all of the following *except*
   a. Protein loss of 3 to 3.5 gm/24 hrs or more in urine,
   b. Dehydration
   c. Elevated serum cholesterol level
   d. Low serum albumin level
   e. Elevated levels of α_2-globulin in serum
9. Of the following, the first laboratory evidence of chronic pyelonephritis is often
   a. Raised serum uric acid level
   b. Raised serum urea nitrogen level
   c. Decreased creatinine clearance
   d. Loss of concentrating ability
   e. None of the above

III. Fill in the blanks

1. Stepwise increase in______ , _______ and _______ is believed to reflect kidney dysfunction.
2. Net effective filtration pressure which forces the fluid through the glomerular filter is due to ______ and _____ and is____.
3. Normally _______ ml of fluid is filtered into Bowman's capsule.
4. About _____ litres are filtered through the glomeruli per day and only ______litres are excreted in urine.
5. Renal tubules reabsorb almost ________ gm of glucose per day and about _____ gm of urea are filtered in 24 hours but only____ gm are excreted in the urine.
6. Substances like_____, _______, and _______gm are excreted without being reabsorbed or secreted by the tubules.
7. Classification of renal function tests is based on _______, _______ and _______.
8. When urine volume excretes 2 ml/mt, the rate of urea elimination is called as_______clearance and it is directly proportional to the_______in the blood and its normal average value is_______.
9. Standard clearance of urea, the rate of urea elimination is proportional to_______ of urine volume and its average normal volume is_______.
10. The urea clearance is proportional to the _______ of the body.
11. At normal blood level, _______is the endogenous metabolite which is filtered at the glomerulus but neither secreted nor reabsorbed by the tubules.
12. Normal average inulin clearance in an adult is_______.
13. Normally _______/kg body weight of ^{15}Cr-EDTA is injected for its clearance and in a single sample procedure it is given as_______.
14. The average normal value or renal plasma flow is _______ and of filtration fraction is _______.
15. In early essential hypertension the filtration fraction tends to be_______ but as the disease progresses it is found to be_______.
16. In glomerulonephritis the filtration fraction is_______.
17. Concentration and dilution tests are employed to assess_______.
18. In arteriolar nephrosclerosis, there is inability to_______ or_______ the urine resulting in_______.
19. Phenolsulphthalein (PSP) excretion test is employed to assess_______.
20. Normally_______of PSP is excreted by tubular action and _______by glomerular filtration.

Answers to MCQs

1. d	2. c	3. b	4. d	5. c	6. d
7. c	8. b	9. d			

Answers to fill in the blanks

1. Uric acid, urea, creatinine
2. Blood pressure and osmotic pressure and it 25 mmHg
3. 120 ml
4. 170 litres, 1.5 litres
5. 170 gm 50,30
6. Inulin, Thiosulphate Mannitol
7. GFR, Renal plasma Flow and Tubular function
8. Maximum, Conc of urea, 75
9. Square root, 54
10. Surface area
11. Creatinine
12. 125 ml/mt
13. 4.5 μCi (or 0.17 mBq), Cl = E Cv/t
14. 574 ml/minute, 21.7%
15. Normal, Increased
16. Decreased
17. Tubular function
18. Concentrate, dilute, isosthenuria
19. Tubular excretion and reabsorption
20. 94%, 6%

CHAPTER

37

LIVER FUNCTION TESTS

Major Concepts

A. To know various functions of liver.
B. To classify the liver function tests and their interpretations.
C. To learn how to use various tests collectively to evaluate the liver dysfunction.

Specific Objectives

A. 1. Review anatomy and physiology of liver. (Not included in the text, but students are advised to do so before studying this chapter.)
   2. Learn in detail, various functions of liver such as metabolic functions, secretory functions, excretory functions, hematologic functions, protective functions, storage functions.

B. 1. Learn the classification of various liver function tests based on different functions of the liver.
   2.
   - Learn the details of van den Bergh test; and total direct bilirubin estimation and it's interpretation.
   - Classify jaundice and learn the liver function tests associated with it and their usefulness.
   - Study glucose, galactose, fructose and epinephrine tolerance tests and the implication of these tests in assessing liver dysfunction.
   - Learn the tests of protein, albumin estimation, fibrinogen estimation, and flocculation and turbidity tests.
   - Study the tests based on lipid metabolism such as cholesterol-cholesteryl ester ratio and the interpretation.
   - Learn hippuric acid test and its importance and detoxication functions of the liver.
   - Learn details of excretory function tests such as BSP retention test, Rose-Bengal dye test, bilirubin-tolerance test and their interpretations.
   - Study the details of tests based on hematological functions of liver such as, prothrombin time, prothrombin index and its interpretation.
   - Learn the details of tests based on amino acid catabolism such as estimation of ammonia, ammonia tolerance test, glutamine estimation in CSF and their importance in assessment of liver function.
   - Activities of several enzymes are elevated in liver dysfunction, such as ALP, GPT, OCT, GGT, etc. How these can be useful in differential diagnosis.

C. Normally more than one test is indicated to assess the liver function. Learn how different tests can be employed and used collectively for assessment of liver function.

INTRODUCTION

Numerous laboratory investigations have been proposed in the assessment of liver diseases. From among these host of tests, the following battery of blood tests: Total bilirubin and VD Bergh test, total and differential proteins and A: G ratio and certain enzyme assays as aminotransferases, alkaline phosphatase and γ-GGT have become widely known as ***Standard Liver Function Tests (LFTs).***

Urine tests for bilirubin and its metabolites and the prothrombin time (PT) and index (PI) are also often included under these headings; but tests such as Turbidity/flocculation test, Icteric index, etc. are now becoming outdated. ***Second generation*** LFTs attempt to improve on this battery of tests and to gain a genuine measurement of liver function, i.e. quantitative assessment of functional hepatic mass. These include the capacity of the liver to eliminate exogenous compounds such as aminopyrine or caffeine or endogenous compounds such as bile acids which have gained much importance recently. However, such investigations are not yet routinely or widely used due to lack of facilities and are useful for research purpose only. Hence in our discussion we will confine to "Standard LFTs" which are routinely done and possible in any standard laboratory. It is stressed that with the advent of more sophisticated techniques for the diagnosis of liver diseases, ***particularly ultrasound and CT scanning together with percutaneous and endoscopic cholangiography and liver biopsy, routine use of standard LFTs is being questioned now.***

- ***In haemolytic jaundice***
  There is an increase in unconjugated bilirubin, hence indirect reaction is obtained, occasionally it may be a delayed direct reaction.
- ***In obstructive jaundice***
  Conjugated bilirubin is increased, hence an immediate direct reaction is obtained.
- ***In hepatocellular jaundice***
  Either or both may be present. In viral hepatitis, direct reaction is the rule, because it is associated with intrahepatic obstruction.
- An immediate direct reaction is also observed in ***cholestatic jaundice***. In low-grade jaundice present in some cases of cirrhosis of the liver, results are variable, but an indirect reaction is usually seen.

An immediate direct reaction is obtained whether the obstruction is intrahepatic or extrahepatic. This does not, therefore, differentiate between an infectious hepatitis or toxic jaundice on one hand and posthepatic (obstructive jaundice) on the other. ***Hence a direct VD Bergh reaction is only of limited value.***

(b) *Serum bilirubin*

It gives a ***measure of the intensity*** of jaundice. Higher values are found in obstructive jaundice than in haemolytic jaundice.

Usefulness of quantitative estimation of serum bilirubin:

- ***In subclinical jaundice:*** Where the demonstration of small increases in serum bilirubin 1.0 to 3.0 mg/dl is of diagnostic value.
- ***In clinical jaundice:*** Useful to follow the development and course of the jaundice.

3. Bile Pigments in Urine/Faeces (Bilirubinuria)

Principle: Most of the tests used for detection of bile pigments depend on the oxidation of bilirubin to differently coloured compounds such biliverdin (green) and bilicyanin (blue).

Interpretations

- Bilirubin is found in the urine in obstructive jaundice due to various causes and in "cholestasis". Conjugated bilirubin can pass through the glomerular filter.
- Bilirubin is not present in urine in most cases of haemolytic jaundice, as unconjugated bilirubin is carried in plasma attached to albumin, hence it cannot pass through the glomerular filter.
- ***Bilirubinuria is always accompanied with direct VD Bergh reaction.***

Note: Bilirubin in the urine may be detected even before clinical jaundice is noted.

Bile Pigments in Faeces

- Bilirubin is not normally present in faeces since bacteria in the intestine reduce it to urobilinogen.
- Some may be found if there is very rapid passage of materials along the intestine.
- Sometimes it is found in faeces of very young infants, if bacterial flora in the gut is not developed.
- It is regularly found in faeces of patients who are being treated with gut sterilising antibiotics such as neomycin.
- Biliverdin is found in meconium, the material excreted during the first day or two of life.

4. Urinary and Faecal Urobilinogen

Faecal Urobilinogen

Normal quantity of urobilinogen excreted in the faeces per day is from 50-250 mg. Since urobilinogen is formed in the intestine by the reduction of bilirubin, the amount of faecal urobilinogen depends primarily on the amount of bilirubin entering the intestine.

- Faecal urobilinogen is increased in haemolytic jaundice, in which ***dark-coloured faeces*** is passed.
- Faecal urobilinogen is decreased or absent if there is obstruction to the flow of bile in obstructive jaundice, in which ***clay-coloured faeces*** is passed. Complete degree of obstruction is found in tumours, whereas obstruction due to gallstones is intermittent. **A complete absence of faecal urobilinogen is strongly suggestive of malignant obstruction.** ***Thus, it may be useful in differentiating a non-malignant from a malignant obstruction.***
- A decrease may also occur in extreme cases of diseases affecting hepatic parenchyma.

Urine Urobilinogen

- Normally there are mere traces of urobilinogen in the urine. Average is 0.64 mg, maximum normal 4 mg/24 hours.
- ***In obstructive jaundice:*** In case of complete obstruction, no urobilinogen is found in the urine. Since bilirubin is unable to get into the intestine to form it. ***The presence of bilirubin in the urine, without urobilinogen is strongly suggestive of obstructive jaundice either intrahepatic or posthepatic.***
- ***In haemolytic jaundice:*** Increased production of bilirubin leads to increased production of urobilinogen which appears in urine in large amounts. Thus, ***increased urobilinogen in urine and absence of bilirubin in urine are strongly suggestive of haemolytic jaundice.***
- Increased urinary urobilinogen may be seen in ***damage to the hepatic parenchyma***, because of inability of the liver to re-excrete into the stool by way of the bile and urobilinogen absorbed from the intestine ***Enterohepatic circulation*** suffers.

SECTION FIVE

TESTS BASED ON LIVER'S PART IN CARBOHYDRATE METABOLISM

The tests are based on tolerance to various sugars since liver is involved in removal of these sugars by glycogenesis or in conversion of other monosaccharides to glucose.

1. Glucose Tolerance Test

- Not of much value in liver diseases
- Although glucose tolerance is sometimes diminished, it is often difficult to separate the part played by the liver from other factors influencing glucose metabolism.

2. Galactose Tolerance Test

Basis: The normal liver is able to convert galactose into glucose; but this function is impaired in intrahepatic diseases and the amount of blood galactose and galactose in urine is excessive.

Advantages of this test

- It is used primarily ***to detect liver cell injury.***
- It can be performed in presence of jaundice.
- As it measures an intrinsic hepatic function, it ***may be used to distinguish obstructive and non-obstructive jaundice.***

Note: In prolonged obstruction, if untreated, secondary involvement of liver leads to abnormality in the galactose tolerance.

Methods: This can be of **two types:**

(a) ***Oral galactose tolerance test (Maclagan)*** and

(b) ***IV galactose tolerance test.***

(a) *Oral galactose tolerance test (Maclagan):* The test is performed in the morning after a night's fast. A fasting blood sample is collected which serves as "control". 40 gm of galactose dissolved in a cup-full of water is given orally. Further blood samples are collected at 1/2 hourly intervals for two hours (similar to GTT).

Interpretations

- ***Normally or in obstructive jaundice:*** 3 gm or less of galactose are excreted in the urine within 3 to 5 hours and the blood galactose returns to normal within one hour.
- ***In intrahepatic (Parenchymatous) jaundice:*** The excretion amounts to 4 to 5 gm or more during the first five hours.

Galactose index (Maclagan)

It is obtained by adding the four blood galactose levels.

Interpretations

- Upper normal limit of normal was taken as 160.
- In healthy medical students, range varied from 0 to 110 and in hospital patients not suffering from liver disease the value ranged from 0 to 160.
- ***In liver diseases*** very high values are obtained.
- In infective and toxic hepatitis values up to about 500 are seen, decreasing slowly as the clinical condition improves.
- ***In cirrhosis of the liver*** increased values may be obtained up to 500, depending on the severity of the disease.

(b) *IV galactose tolerance test (King):* The test is performed in the morning after a night's fast. A fasting blood sample is collected which serves as "control". An IV injection of galactose, equivalent to 0.5 gm/kg body weight is given as a sterile 50 per cent solution. Blood samples are collected after five minutes, ½ hour, 1 hour, 1½ hours, 2 hours and 2½ hours after IV injection and blood galactose level is estimated.

Interpretations

- ***A normal response*** should have a curve beginning on the average at about 200 mg galactose/100 dl, falling steeply during the one hour and reaching a figure between 0 to 10 mg% by end of 2 hours.
- In most cases of ***obstructive jaundice*** similar results are obtained, unless there is parenchymal damage.
- ***In parenchymatous diseases*** with liver cell damage, ***the fall in blood galactose takes place more slowly.*** Normally no galactose is detected in 2½ hours sample, but ***in parenchymatous disease, value is greater than 20 mg/dl.***

3. Fructose Tolerance Test

Method

50 gm of fructose given to the fasting patient as for GTT. Fasting blood sugar is estimated and blood sugar is estimated in samples is taken at ½ hourly intervals for 2½ hours after taking the oral fructose. The usual methods for estimation of blood sugar measures both the glucose and fructose present.

Interpretations

- ***Normal response*** shows little or no rise in the blood sugar level. The highest blood sugar value reached during the test should not exceed the fasting level by more than 30 mg%.
- Similar result is obtained in most cases of ***obstructive jaundice*** cases (provided no parenchymal damage).
- ***In infectious hepatitis*** and parenchymatous liver cells damage: Rise in blood sugar is greater than above, but the increases obtained are never very great.

4. Epinephrine Tolerance Test (Storage Function)

Principle: The response to epinephrine as evidenced by elevation of blood sugar is a manifestation of glycogenolysis and is directly influenced by glycogen stores of liver.

Method

The patient is kept on a high carbohydrate diet for three days before the test. After an overnight fast, the fasting blood sugar is determined. 0.01 ml of a 1 in 1000 solution of epinephrine per kg body weight in injected. The blood sugar is then determined in samples collected at 15 minutes intervals up to one hour.

Interpretations

- ***Normally*** in the course of an hour, the rise in blood sugar over the fasting level exceeds by 40 mg% or more.
- ***In parenchymal hepatic diseases*** the rise is less.
- It is of much use for diagnosis of ***glycogen storage diseases***, specially in ***von Gierke's disease***, in which blood glucose rise is not seen due to lack of glucose-6-phosphatase.

TESTS BASED ON CHANGES IN PLASMA PROTEINS

1. **Determination of Total Plasma Proteins and Albumin and Globulin and A:G Ratio:**
   This yields most useful information in chronic liver diseases. Liver is the site of albumin synthesis and also possibly of some of α-and β-globulins.

Interpretations

- ***In infectious hepatitis*** quantitative estimations of albumin and globulin may give normal results in the early stages. ***Qualitative changes may be present, in early stage rise in β-globulins and in later stages γ-globulins show rise.***
- ***In obstructive jaundice*** normal values are the rule, as long as the obstructive jaundice is not associated with accompanying liver cell damage.
- ***In advanced parenchymal liver diseases and in cirrhosis liver*** the albumin is grossly decreased and the globulins are often increased, so ***that A:G ratio is reversed, such a pattern is characteristically seen in cirrhosis of the liver***. The albumin may fall below 2.5 gm% and may be a contributory factor in causing oedema in such cases. Fractionation of globulins, reveals that the increase is usually in the γ-globulin fraction, but in some cases there is a smaller increase in β-globulins.

Note:

- ***The severity of hypoalbuminaemia in chronic liver diseases is of diagnostic importance and may serve as a criterion of the degree of damage.***
- A low serum albumin which fails to increase during treatment is usually a poor prognostic sign.

2. **Estimation of Plasma Fibrinogen**
   Fibrinogen is formed in the liver and likely to be affected if considerable liver damage is present.

- ***Normal value*** is 200-400 mg%.
- Values below 100 mg% have been reported in ***severe parenchymal liver damage***. Such a situation is found in severe acute insufficiency such as may occur in:
  - ***Acute hepatic necrosis***
  - ***Poisoning from carbon tetrachloride***
  - ***In advanced stages of liver of the cirrhosis.***

3. **Flocculation Tests**

The tests have become outdated and not routinely carried out.

Principle: Flocculation tests depend on an alteration in the type of proteins present in the plasma. The alteration may be either quantitative or qualitative and most frequently involves one or more of the globulin fractions.

(a) Thymol turbidity and flocculation test: The degree of turbidity produced when serum is mixed with a buffered solution of thymol is measured. Turbidity produced is compared with a set of protein standards, or turbidity is read in a colorimeter against a $BaSO_4$ standard.

Maclagan unit: Maclagan expressed the results in units, so that a turbidity equivalent to that of 10 mg/100 ml protein standard is one unit.

Basis of the reaction: The thymol turbidity test requires lipids (phospholipids). ***The turbidity/and flocculation in this test is a complex of "lipothymoprotein".*** The thymol seems to decrease the dispersion and solubility of the lipids, and the proteins involved is mainly β-globulin, though some γ-globulin is also precipitated.

Interpretations

- ***Normal range*** is 0 to 4 units.
- It measures only an acute process in the liver, but the degree of turbidity is not proportional to the severity of the disease.
- ***In infectious hepatitis,*** it is highest soon after the onset of the jaundice, but frequently remains raised for several weeks.
- Sera with high β-and γ-globulin fractions, due to other causes may give a positive test.
- ***A negative thymol test in the presence of jaundice is very useful for distinguishing between hepatic and extrahepatic jaundice.***

Thymol flocculation: After the turbidity has been measured, the tubes are kept in the dark for overnight and read the degree of flocculation if any. Flocculation is graded as –ve no flocculation, +ve flocculation as +, ++, +++ and ++++.

(b) Zinc sulphate turbidity test: When a serum having an ***abnormally high content of γ-globulin*** is diluted with a solution containing buffered $ZnSO_4$ solution,

a turbidity develops. The amount of turbidity is proportional to concentration of γ-globulin. Turbidity is measured as discussed in thymol turbidity test.

Interpretations
- *Normal range:* Varies from 2 to 8 units.
- All cases of cirrhosis of liver gives +ve results.
- *In infectious hepatitis:* γ-globulin is increased in later stage. ***$ZnSO_4$ turbidity becomes +ve later as compared to thymol turbidity which becomes +ve early.***
- It may be +ve in other cases where there is increase in γ-globulin.

Other turbidity/and flocculation test viz, cephalin-cholesterol flocculation test, Takata-Ara test, etc. have become obsolete.

4. **Amino acids in urine (Aminoaciduria):** The daily excretion of amino acid nitrogen in normal health varies from 80-300 mg. Aminoaciduria found in severe liver diseases is of **overflow type,** with accompanying increase in plasma amino acids level.

CLINICAL IMPORTANCE

In severe liver diseases like acute yellow atrophy and sometimes in advanced cirrhosis of liver crystals of certain amino acids may be found in urinary deposits microscopically.
- ***Tyrosine crystals:*** Tyrosine crystallises in sheaves or tufts of fine needles.
- ***Leucine crystals:*** Leucine has spherical shaped crystals, yellowish in colour, with radial and circular striations.

Solubility: Both are insoluble in acetone and ether but soluble in acids/and alkalies. Tyrosine is only slightly soluble in acetic acid and insoluble in ethanol, whereas leucine is soluble in the former and slightly soluble in the latter.

TESTS BASED ON ABNORMALITIES OF LIPIDS

- **Cholesterol-Cholesteryl Ester Ratio**
  The liver plays an active and important role in the metabolism of cholesterol including its synthesis, esterification, oxidation and excretion.

Interpretations
- Normal total blood cholesterol ranges from 150-250 mg/dl and approx. 60 to 70 per cent of this is in esterified form.
- *In obstructive jaundice,* an increase in total blood cholesterol is common, but the ester fraction is also raised, so that % esterified does not change.
- *In parenchymatous liver diseases,* there is either no rise or even decrease in total cholesterol and the ***ester fraction is always definitely reduced.*** The degree of reduction roughly parallels the degree of liver damage.
- *In severe acute hepatic necrosis,* the total serum cholesterol is usually low and may fall below 100 mg/dl, whilst there is marked reduction in the % present as esters.

TESTS BASED ON THE DETOXICATING FUNCTION OF THE LIVER

HIPPURIC ACID TEST OF QUICK

Principle
1. Best known test for the detoxicating function of liver.
2. Liver removes benzoic acid, administered as sodium benzoate, either orally or IV, and combines with amino acid glycine to form hippuric acid. The amount of hippuric acid excreted in urine in a fixed time is determined. (Refer chapter on detoxication).
3. The test thus depends on *two factors*
   (i) The ability of liver cells to produce and provide sufficient glycine and
   (ii) The capacity of liver cells to conjugate it with the benzoic acid.
4. ***For reliable result, renal function must be normal.*** If there is any reason to suspect renal impairment, a urea clearance test should be done simultaneously.

Methods: Both oral and IV forms of the hippuric acid test are in use.

1. *Oral hippuric acid test:* Dissolve 6.0 gm of sodium benzoate in approx 200 ml of water. The test may be started 3 hours after a light breakfast of toast and tea. Food should not be given until late in the test. The patient empties the bladder, the urine being discarded. The patient is allowed to drink the sodium benzoate solution and **time is noted.** The bladder is again emptied 4 hr later. Any urine passed during this 4 hours is kept and added to that passed at the end of 4 hours. The amount of hippuric acid excreted in this 4 hours period is estimated.

Interpretations
- *Normally,* at least 3.0 gm of hippuric acid, expressed as Benzoic acid or 3.5 gm of sodium benzoate should be excreted in health.
- Smaller amounts are found when there is either acute or chronic liver damage. Amounts lower than 1.0 gm may be excreted by patients with infectious hepatitis.

2. *IV Hippuric Acid Test*

Indications: Normally oral test is preferred. An IV test is indicated:
- When there is impairment of absorption due to absorption defects.
- If there is accompanying nausea/vomiting.

Procedure

1.77 gm of sodium benozate dissolved in 20 ml of DW as a sterile solution given IV. Shortly before the injection, the patient empties the bladder, which is discarded. The bladder is emptied after ***one hour*** and ***two hours after the injection.***

Interpretations

- ***In normal health,*** hippuric acid equivalent to at least 0.85 gm of sodium benzoate, or to 0.7 gm of benzoic acid should be excreted in the one hour, or equivalent to 1.15 gm of benzoic acid in the first two hours.
- Excretion of smaller amounts more than above indicate the presence of liver damage.

TESTS BASED ON EXCRETORY FUNCTION OF LIVER

1. BSP Retention Test (Bromsulphthalein Test)

Principle

1. The ability of the liver to excrete certain dyes, e.g. BSP is utilised in this test.
2. In normal healthy individual, a constant proportion (10 to 15% of the dye) is removed per minute. In hepatic damage and insufficiency, BSP removal is impaired by cellular failure, as damaged liver cells fail to conjugate the dye or due to decrease blood flow.
3. ***Removal of BSP by the liver involves conjugation of the dye as a mercaptide with the cysteine component of glutathione.*** The reaction of conjugation of BSP with glutathione is ***rate-limiting***, and thus it exerts a controlling influence on the rate of removal of the dye.

Procedure

With the patient fasting, inject IV slowly, an amount of 5 per cent BSP solution, which contains 5 mg of BSP/Kg, body weight. Withdraw 5 to 10 ml of blood, 25 and 45 minutes after the injection and allow the specimens to clot. Separate the sera and estimate the amount of the dye in each sample.

Interpretations

- *In normal* healthy individual not more than 5 per cent of the dye should remain in the blood at the end of 45 minutes. The bulk of the dye is removed in 25 minutes and less than 15 per cent is left at the end of 25 minutes.
- *In Parenchymatous liver diseases* removal proceeds more slowly. In advanced cirrhosis removal is very slow and ***40 to 50 per cent of the dye is retained in 45 minutes sample.***

Contraindication: Since the dye is removed in bile after conjugation, this test can only be used in cases in which there is no obstruction to the flow of bile. Hence ***the test is of no value if obstruction of biliary tree exists (obstructive jaundice).***

CLINICAL SIGNIFICANCE

- BSP-excretion test is ***a useful index of liver damage,*** particularly when the damage is diffuse and extensive.
- **The test is most useful:**
  - **In liver cell damage without jaundice**
  - **In cirrhosis of the liver**
  - **In chronic hepatitis.**

2. Rose-Bengal Dye Test

Rose-Bengal is another dye which can be used to assess excretory function. 10 ml of a 1 per cent solution of the dye is injected IV slowly.

Interpretation: Normally 50 per cent or more of the dye disappears within 8 minutes.

I^{131}-labelled Rose-Bengal: I^{131}-Rose-Bengal has been used where isotope laboratory is present. I^{131}-labelled Rose-Bengal is administered IV. Then count is taken over the neck and abdomen. Initially, count is more in neck and practically nil over abdomen. ***As the dye is excreted through liver, neck count goes down and count over abdomen increases.***

Interpretation

- *In parenchymal liver diseases* high count in the neck persists and there is hardly a rise in count over abdomen, as the dye is retained.

3. **Bilirubin tolerance test:** 1 mg/kg body weight of bilirubin is injected IV. If more than 5 per cent of the injected bilirubin is retained after 4 hours, the excretory and conjugating function of the liver is considered abnormal. The bilirubin excretion test has been recommended by some authorities as a better test of excretory function of the liver as compared to dye tests as bilirubin is a normal physiologic substance and the dyes are foreign to the body. But the test is not used routinely and extensively due to its high cost.

Note: The three substances listed above, with the exception of BSP, are excreted almost entirely by the liver. No significant amounts are taken up by RE cells.

FORMATION OF PROTHROMBIN BY LIVER

Determination prothrombin time: Prothrombin is formed in the liver from ***inactive "preprothrombin"*** in presence of vitamin K. ***Prothrombin activity is measured as prothrombin time (PT).*** The term prothrombin time was given to time required for clotting to take place in citrated plasma to which optimum amounts of "thromboplastin" and Ca^{++} have been added. The "one-stage" technique introduced by Quick, the prothrombin time is related inversely to the concentration not only to prothrombin, but also of factors V, VII and X and it can be more sensitive

SECTION FIVE

to a lack of VII and X than to prothrombin alone. In spite of above restriction, as it is simple and quick in performance, it is still much used.

Interpretations

- ***Normal value:*** Normal levels of prothrombin in control give prothrombin time of approx 14 seconds. (Range: 10-16 Sec). Results are always expressed as patient's prothrombin time in seconds to normal control value.
- ***In parenchymatous liver diseases:*** Depending on the degree of liver cells damage plasma prothrombin time may be increased from 22 to as much as 150 secs.
- ***In obstructive jaundice:*** Due to absence of bile salts, there may be defective absorption of vitamin K, hence PT is increased, as prothrombin formation suffers.

Note

- From above, it is observed that PT is increased both in obstructive jaundice and in diseases of liver cells damage. Hence ***PT cannot be used to differentiate between them.***
- ***However, if adequate vitamin K is administered parenterally, the PT returns rapidly to normal in uncomplicated obstructive jaundice, whereas in liver damage the response is less marked.***

Other Clinical Uses

- PT is used mostly in controlling anticoagulant therapy.
- Determination of PT is also used to decide whether there is danger of bleeding at operation in biliary tract diseases.

Prothrombin index

Prothrombin activity is also sometimes expressed as ***"prothrombin index" in %,*** which is the ratio of prothrombin time of the normal control to the patient's prothrombin time multiplied by 100. Thus,

- Prothrombin index $= \dfrac{\text{PT of normal control}}{\text{PT of patient}} \times 100$

Normally index is 70 to 100 per cent. The "critical level" below which bleeding may occur is not fixed one, but there is always a possibility of this occurring if prothrombin index is below 60 per cent.

TESTS BASED ON AMINO ACID CATABOLISM

1. **Determination of blood NH_3:** Nitrogen part of amino acid is converted to NH_3 in the liver mainly by transamination and deamination (transdeamination) and it is converted to urea in liver only (Refer Chapter on Protein Metabolism).

> ***Other sources of NH_3***
>
> Following are the other sources of ammonia:
>
> - NH_3 is formed from nitrogenous material by bacterial action in the gut.
> - In kidneys, by hydrolysis of glutamine by ***glutaminase.***
> - From pyrimidines catabolism: A small amount of NH_3 is formed from catabolism of Pyrimidines.

Interpretations

- ***Normal range:*** Blood ammonia varies from 40 to 75 μg ammonia nitrogen per 100 ml of blood.
- ***In parenchymal liver diseases:*** The ability to remove NH_3 coming to liver from intestine and other sources may be impaired. Increases in NH_3 can be found in more advanced cases of cirrhosis of the liver, particularly when there are associated neurological complications. In such cases blood levels may be over 200 μg/100 ml. Very high values may be obtained in hepatic coma.

2. **Ammonia tolerance test:** An ammonia tolerance test has been devised to test the ability of the liver to deal with NH_3 coming to it from the intestine.

Procedure

The patient should come for the test after overnight 12 hours fast, only small amounts of fluids can be taken during that time. Take fasting specimen of blood for NH_3 determination. After that, give by mouth 10 gm of ammonium citrate dissolved in water and flavoured with fruit juice/lemon. Take blood samples after 30, 60, 120 and 180 minutes and determine blood NH_3.

Note: In patients with increased initial levels, give smaller doses, e.g. only 5 grams.

Interpretations

- ***In normal healthy persons:*** Little increase is found; blood NH_3 levels remaining within normal range.
- ***In advanced cirrhosis of the liver:*** Marked rise to twice the initial level or more, exceeding 200 to 300 μg% are seen.
- Considerable increases are also seen when there is a collateral circulation and in patients who have a portocaval anastomosis.

3. **Determination of glutamine in CS Fluid (An Indirect Liver Function Test)**

Glutamine, the amide of glutamic acid, is formed by *glutamine synthetase* by glutamic acid and NH_3.

Glutamine in CS fluid can be estimated by the method of **Whittaker** (1955). The glutamine is hydrolysed to glutamic acid and NH_3 by the action of dilute acid at 100 degree centigrade. A correction is made for a small amount of NH_3 produced from urea. No other substances present in CS fluid were found to form NH_3 under above conditions.

Interpretations

- ***The normal range*** found to be 6.0 to 14.0 mg%.
- ***In infectious hepatit is*** found is range from 16 to 28 mg%, but is usually less than 30 mg%.
- ***In cirrhosis of the liver*** the increase is more; depending on the severity. It varies from 22 to 36 mg% or more.
- ***In hepatic coma*** increase is very high, ranging from 30 to 60 mg% or more.
- ***In other types of coma, normal values are obtained.***

Note: ***Some authorities put 40 mg% as a critical level. Prognosis of the case is fatal if CS fluid glutamine level is more than 40 mg%, in case of cirrhosis of the liver and hepatic coma.***

TESTS BASED ON DRUG METABOLISM

The capacity of the liver to metabolise certain drugs can successfully be used as a measure of hepatic function. Tests based on drug clearance or metabolite formation kinetics reflect the actual functional state of the liver and are therefore called ***dynamic liver function tests***. Tests based on the rate of metabolite formation are of particular interest.

1. MEGX Test

Principle: ***Lidocaine*** is rapidly converted to its primary metabolite ***monoethyl glycine xylidine*** (MEGX) by the hepatic microsomal cytochrome P_{450} system. A loss of hepatic cytochrome P_{450} activity or major changes in hepatic blood flow (due to portosystemic shunting) result in decreased MEGX formation. ***Lidocaine metabolite formation has been used as an index of hepatic function.***

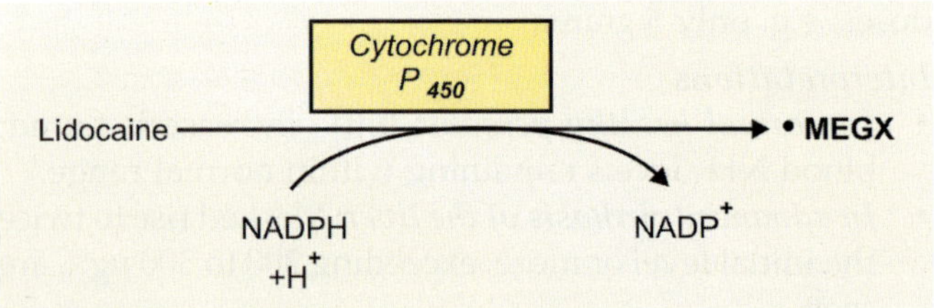

Procedure

An IV bolus injection of a small lidocaine test dose, 1 mg/kg, is given. Blood sample is taken before the injection. Another blood sample is taken 15 or 30 minutes after the injection. MEGX is determined in the serum by use of an automated fluorescence polarisation immuno-assay within about 20 minutes in both the samples.

Interpretations

- The highest MEGX test results are observed in liver donors with unimpaired organ function and in normal healthy subjects.
- Liver recipients with uncomplicated postoperative course show somewhat lower test results.
- In patients with ***cirrhosis of the liver,*** the increase of MEGX concentration in serum is much less marked and decrease value is dependant on disease severity.

Note

- The test is rapid and easy to function
- Potential sources of extrahepatic MEGX formation seem to have no noticeable influence on the test.

2. Antipyrine breath test: ***Antipyrine*** like lidocaine is also metabolised by cytochrome P_{450} system. When given orally it is absorbed from intestine completely, not bound to plasma proteins and metabolised by liver only.

Procedure: C^{14}-labelled aminopyrine (dimethyl amino antipyrine) is given orally in dosage of 1 to 2 microcurie. Breath samples are collected for 2 and 24 hours and analysed.

Interpretations

- Normal subjects excrete 5 to 8 per cent of the administered dose in 2 hours.
- Patients with hepatitis and cirrhosis excretes only 2 to 3 per cent.

VALUE OF SERUM ENZYMES IN LIVER DISEASES

Quite a large number of enzyme estimations are available which are used to ascertain liver function.

They can be divided into **2 groups**

I. Most commonly and routinely done in the laboratory.
II. Not routinely done in the laboratory.

Most commonly and routinely employed in laboratories are two:

A. Serum transaminases (amino transferases), and
B. Serum alkaline phosphatase.

A. Serum Transaminases (Amino transferases)

Interpretations

- Normal ranges for these enzymes are as follows:
  - **SGOT** (aspartate transaminase): 4 to 17 IU/L (7 to 35 units/ml)
  - **SGPT** (alanine transaminase): 3 to 15 IU/ L (6 to 32 units/ml)
- Both these enzymes are found in most tissues, but the relative amounts vary. ***Heart muscles are richer in SGOT, whereas liver contains both but more of SGPT.***
- Increases in both transaminases are found in liver diseases, with SGPT much higher than SGOT.
- Their determination is of limited value in differential diagnosis of jaundice because of considerable overlapping.
- ***But their determination is of extreme use in assessing the severity and prognosis of parenchymal liver***

diseases specially acute infectious hepatitis and serum hepatitis. In these two conditions, highest values in thousand units are seen.

- *Screening test:* Also useful as a screening test in outbreak of infectious hepatitis (viral hepatitis), it is the most sensitive diagnostic index. The increase can be seen in prodromal stage, when jaundice has not appeared clinically. ***Such cases can be isolated and segregated from others,*** so that spread of the disease can be checked.
- Very high values are also obtained in toxic hepatitis, due to carbon tetrachloride poisoning. Increases are comparatively less in drug hepatitis (cholestatic) like chloropromazine.
- In obstructive jaundice (extrahepatic) also increases occur, but usually do not exceed 200 to 300 IU/L.

B. Serum Alkaline Phosphatase

Alkaline phosphatase enzyme is found in a number of organs, mostly in bones and liver, then in small intestine, kidney and placenta. ***Placental isoenzyme of alkaline phosphatase is heat-stable.***

Interpretations

- *Normal range* for serum ALP as per King-Armström method is 3 to 13 KA Units/100 ml (23 to 92 IU/L).
- ***It is used for many years in differential diagnosis of jaundice.*** It is increased in both infectious hepatitis (viral hepatitis) and posthepatic jaundice (extrahepatic obstruction) but ***the rise is usually much greater in cases of obstructive jaundice.*** **Dividing Line which has been suggested is 35 KA units/100 ml.** ***A value higher than 35 KA units/100 ml is strongly suggestive of diagnosis of obstructive jaundice,*** in which very high figures even up to 200 units or more may be found. There is certain amount of overlapping mostly in the range of 30 to 45 KA Units/100 ml.
- Very high values are occasionally found in certain liver diseases, e.g. xanthomatous biliary cirrhosis in which there is no extrahepatic obstruction.
- Higher values are also obtained in *space-occupying lesions of liver,* e.g.
  - *Abscess*
  - *Primary carcinoma (hepatoma)*
  - *Metastatic carcinoma*
  - *Infiltrative lesions like lymphoma*
  - *Granuloma and amyloidosis.*

A diagnostic triad suggests:
- *High serum ALP*
- *Impaired BSP-retension* and
- *Normal/or almost normal serum bilirubin.*

- Serum ALP is found to be normal in haemolytic jaundice.

Mechanism of increase in ALP in liver diseases

Increase in the activity of ALP in liver diseases is not due to hepatic cell disruption, nor to a failure of clearance, but rather **to increased synthesis of hepatic ALP.** The stimulus for this increased synthesis in patients with liver diseases has been attributed to bile duct obstruction either extrahepatically by stones, tumours, strictures or intrahepatically by infiltrative disorders or ***space-occupying lesions.***

Note

- The relation of the amino transferase to ALP level may provide better evidence than either test alone, as to whether or not the jaundice is cholestatic.
- **High ALP with low amino transferase activity is usual in cholestasis and the converse occurs in non-cholestatic jaundice.**

  It is, however, stressed that there are several intrahepatic causes of cholestasis such as primary biliary cirrhosis, acute alcoholic hepatitis and sclerosing cholangitis in which laparotomy is inappropriate. ***Hence, even after a confident diagnosis of cholestatic jaundice based on the LFTs, further investigation to define the site of obstruction is imperative.***

II. Other Enzymes (not done routinely)

Other enzymes which have been found to be useful but not routinely done in the laboratory are discussed below briefly:

1. **Serum 5′-Nucleotidase**

   This enzyme hydrolyses nucleotides with a phosphate group on carbon atom 5′ of the ribose, e.g. adenosine 5′-P. On hydrolysis it produces adenosine and inorganic PO_4. These nucleotides are also hydrolysed by non-specific phosphatases such as alkaline phosphatase present in the serum. However, 5′-nucleotidase is inactivated by nickel, hence if hydrolysis is carried out with and without added nickel, the difference gives the 5′-nucleotidase activity.

Interpretations

- *Normal range* is 2 to 17 IU/L
- *Liver diseases*
  - ***Serum 5′ nucleotidase is raised along with serum ALP in diseases of liver and biliary tract in a roughly parallel manner.*** It is thus highest in post-hepatic obstructive jaundice frequently over 100 units. **It has added advantage over serum ALP in that enzyme is not affected in bone diseases.**
  - Smaller increases are found in hepatic jaundice, e.g. in infectious hepatitis, in some cases of which normal results are obtained.
- *Bone diseases*

  5′ nucleotidase is normal in patients with increased serum ALP in bone diseases such as Paget's disease.

2. Serum Lactate Dehydrogenase (LDH)

LDH enzyme is widely distributed, found in all cells in man, but is specially plentiful in cardiac and skeletal muscle, liver, kidney and the red blood cells.

Interpretations

- *Normal range* is 70-240 IU/L
- ***In liver diseases*** an increased activity is found particularly in infectious hepatitis, but the increase is not so great as that of the transaminases and its behaviour is less predictable.
- The enzyme is ***less specific*** and as it is widespread, increase of the enzyme activity is also seen in many other diseases like leukaemias, pernicious anaemia, megaloblastic and haemolytic anaemias, in renal diseases and in generalised carcinomatosis.
- ***In cirrhosis of the liver*** and posthepatic jaundice (obstructive jaundice) normal results are often found.

Isoenzymes LDH

Refer to Chapter on Enzymes and Isoenzymes of Clinical Importance.

3. Serum Isocitrate Dehydrogenase (ICD)

A specific enzyme found in liver only.

- ***Normal range*** is 0.9 to 4.0 IU/L
- ***In liver diseases*** a marked increase in ICD activity seen whether it is inflammatory like infectious hepatitis, malignancy or from taking drugs. Large increases are seen in infectious hepatitis; serum activity almost returns to normal by the 3rd week after the onset of jaundice.
- ***In obstructive jaundice*** normal values are the rule.
- In most cases of cirrhosis of the liver, serum enzyme activity is either normal or slightly raised.

4. Serum Cholinesterases

Cholinesterases are enzymes which hydrolyse esters of choline to give choline and acid. **Two types** have been distinguished:

(a) *"True"*, and

(b) *"Pseudo"*.

- ***'True' cholinesterase:*** It is thought to be responsible for the destruction of acetylcholine at the neuromuscular junction and is found in nerve tissues and RB cells.
- ***'Pseudo' cholinesterases:*** These are found in various tissues such as liver, heart muscle and intestine and it is this type which is present in plasma.

Interpretations

- ***Normal range*** is 2.17 to 5.17 IU/ml. (130 to 310 units of de la Huerga)
- ***In liver diseases***
  - The enzyme is formed in liver and serum activity is reduced in liver cells damage. Hence determination has been used for recognising liver damage. (Protein synthesis)
  - Low values are also obtained in advanced cases of cirrhosis of the liver.
- Normal serum activity seen in obstructive jaundice cases.
- Serial estimations has been found to be of value in prognosis of infectious hepatitis and cirrhosis of the liver.

5. Serum ornithine carbamoyl transferase (OCT): This enzyme catalyses the following reaction.

$$\text{Ornithine} + \text{Carbamoyl–P} \xrightleftharpoons{\text{OCT}} \text{Citrulline} + PO_4$$

It is involved in urea synthesis.

Note

This enzyme is exclusively found in liver and virtually no activity in other tissues.

Interpretations

- Serum enzyme activity in normal healthy individuals are usually very low and ranges from 8 to 20 m-IU.
- ***In liver diseases***
  - The enzyme level is markedly elevated 10 to 200-fold in patients with acute viral hepatitis depending on the severity and also those with other forms of hepatic necrosis.
  - Relatively slight elevations occur in obstructive jaundice, cirrhosis of the liver, metastatic carcinoma, etc.
- **Serum OCT appears to be a specific and sensitive measure for hepatocellular injury.**

6. Serum Leucine Amino Peptidase (LAP)

It is a proteolytic enzyme which splits off N-terminal residues from certain L-peptides and amides having a free NH_2 group, especially when the N-terminal residue is leucine or related amino acid.

Interpretations

- ***Normal range*** is between 15 to 56 m-IU.
- ***In viral hepatitis*** shows mild-to-moderate increase and ranges from 30.0 to 130.0 m-IU.
- Increases is also seen ***in cirrhosis of the liver,*** but rise is less. It has been observed by some workers and corroborated by PM studies that ***marked increase in cirrhosis of the liver is usually associated with superimposed hepatoma.***
- ***In obstructive jaundice:*** Marked increase is seen like alkaline phosphatase. ***Increase is more in malignant obstruction than that of benign obstruction.*** In one series benign obstruction showed 75.0 to 184.0 m-IU (average 101.25 m-IU), whereas malignant obstruction showed 67.0 to 340 m-IU (average 105.0 m-IU). ***Advantage over serum ALP is that LAP does not rise in osseous involvement.***
- Marked rise has been seen ***in Liver cell carcinoma (Hepatoma).***

7. **Serum SHBD (Hydroxy Butyrate Dehydrogenase):** An enzyme acting on α-OH butyric acid has been identified in the serum and studied as a diagnostic aid in liver diseases.

Interpretations

- ***Normal*** serum HBD between 56 to 125 IU/L
- ***In liver diseases*** elevated levels of this enzyme is observed in acute viral hepatitis. Also elevated level is seen in myocardial infarction.

Ratio of LDH/SHBD

To differentiate the liver diseases and acute myocardial infarction:

Ratio of $\frac{\text{LDH}}{\text{SHBD}}$ has been found more useful

- Normal ratio of $\frac{\text{LDH}}{\text{SHBD}}$ = 1.18 to 1.60
- ***Less than 1.18 is observed in most cases of myocardial infarction. Greater than 1.60 is observed in liver diseases.***
- In infectious hepatitis, the ratio is frequently > 2.0. In chronic hepatitis and obstructive jaundice the ratio ranged from 1.6 to 2.0.

8. **Serum Aldolase and Phosphohexose Isomerase:** These are both markedly increased in serum of patients with acute hepatitis. No increase is found in cirrhosis, latent hepatitis or biliary obstruction.
9. **Serum Amylase**
   Liver is a major, if not the only source of amylase found in the serum under normal physiologic conditions. Studies have shown low serum amylase levels in liver diseases like acute infectious hepatitis.
10. **Serum Sorbitol Dehydrogenase (SDH)**
   The enzyme catalyses the following reaction:

$$\text{Sorbitol} + \text{NAD}^+ \xrightleftharpoons{\text{SDH}} \text{Fructose} + \text{NADH}$$

Interpretations

- ***Normal value:*** Normal values for serum found to be less than 0.2 m-IU.
- Striking elevation seen in ***acute viral hepatitis*** and carbon tetrachloride poisoning up to 17 m-IU. In viral hepatitis, values of SDH return to normal before transaminases.
- ***In chronic hepatitis and in obstructive jaundice:*** Serum levels of SDH are normal or only slightly elevated.
- Myocardial and other extrahepatic diseases do not lead to elevated levels.

Advantages

- Like OCT, it is a ***hepatospecific enzyme.***
- The serial estimation is of immense value in diagnosis and follow up for prognosis of infectious hepatitis.
- Also of immense value in differential diagnosis of jaundice.
- Enzyme has been recently demonstrated in small amount in kidney and prostate but no increase in activity in the diseases of these organs noted.

11. **Serum γ-Glutamyl Transferase (γ-GT)**
   - *Normal range:* 10 to 47 IU/L
   - **Recently, the importance of this enzyme in alcohol abuse has been stressed.**

The activity of this microsomal enzyme has been found to increase in most of hepatobiliary diseases but, largely because of the enzyme's wide tissue distribution, the specificity of a high value is very low. Unlike the amino transferases, the elevated levels do not necessarily indicate liver cell disruption but may be due to enzyme induction by drugs such as, phenobarbitone, phenytoin, warfarin and alcohol.

- These severe limitations have meant that this test has now *only two, practical uses*

(a) *An elevated γ-GT implies that an elevated ALP is of hepatic origin,* and

(b) Secondly, it may be useful in screening for alcohol abuse. ***Sudden increase in γ-GT in chronic alcoholics suggests recent bout of drinking of alcohols.***

Table 37.1 shows the enzyme assays as per priorities useful in detecting alterations in liver diseases.

Table 37.2 shows the differentiating features of three types of jaundice.

Table 37.1: Enzyme assays as per priorities useful in detecting alterations in liver diseases

Alterations detected	Principal enzyme assays
• Hepatocellular damage/or increased permeability of liver cells.	1. Transaminases (amino transferases) (AST and ALT) 2. Ornithine carbamoyl transferase (OCT) 3. Sorbitol dehydrogenase (SDH)
• Extrahepatic or intrahepatic obstruction	1. Alkaline phosphatase (ALP) 2. 5'-Nucleotidase 3. γ-GT 4. LAP
• Protein synthesis	1. Pseudocholinesterase
• Alcohol abuse	1. γ-GT

SECTION FIVE

Table 37.2: Differentiation of three types of jaundice

	Haemolytic or prehepatic jaundice	*Hepatic or parenchymatous jaundice*	*Obstructive or posthepatic jaundice*
I. Causes:	Due to excessive haemolysis (a) Intrinsic defects in RB cells (b) Extrinsic causes external to RB Cells	Disease of parenchymal cells of liver viz. Viral hepatitis, toxic jaundice Cirrhosis of the liver	Due to obstruction of biliary passage (a) Extrahepatic gallstones, tumors, enlarged lymph nodes, etc (b) Intrahepatic cholestasis.
II. Clinical findings:			
(a) Degree of jaundice	Usually low +	Marked jaundice ++ to +++	Marked Jaundice ++ to +++
(b) Faeces	Dark coloured	Variable, usually pale	Clay coloured
III. Biochemical findings:			
Based on bile pigment metabolism:			
1. VD Bergh reaction	Indirect, may be delayed positive	Biphasic	Direct
2. Type of bile pigment in circulation	Unconjugated bilirubin	Mixture of conjugated and unconjugated bilirubin	Conjugated bilirubin
3. Serum bilirubin	Usually low 3 mg to 5 mg%	High, up to 20 mg%	Very high, may be up to 50 mg%
4. Bile Pigments in urine:			
· Bilirubin	Not detected	Present	Present ++
· Urobilinogen	Increased ++	May be increased + or normal	Decreased or Absent
5. Faecal Stercobilinogen	Increased ++	Decreased	Decreased or Absent
IV. Steatorrhoea:	Not present	Present	Present
V. Other biochemical features:			
1. Prothrombin time (PT)	Normal	Increased	Increased, After parental vit K becomes normal
2. Turbidity and flocculation tests			
· Thymol turbidity	Negative	++ to +++	Negative
3. Enzyme assays:			
· Amino transferase activity ALT (S-GPT)	Usually normal	Marked increase +++ to ++++ (goes in thousand units). Usually 500 to 1500 IU/L or may be more	Increased to ++. Usually 100 to 300 IU/L Do not exceed 300 IU/L
· Alkaline Phosphatase (ALP)	Normal	Increased slightly (+) ***usually less than 30 KA Units %***	Marked increase 30 to 100 KA units %, ***more than 35 KA units % suggests obstructive jaundice.***
· 5'-Nucleotidase	Normal	Increased (+) Slight	Marked increase ++ to +++

I. Question (Essay type and short notes)

1. Enumerate various functions of the liver. Describe any one test based on its detoxication function in details.
2. What are the various liver function tests? Describe any one test based on conjugation mechanism.
3. Enumerate various functions of liver. Describe the liver function tests based on excretion of bilirubin.
4. Classify liver function tests. Describe the tests based on bile pigment metabolism.
5. Classify the liver function tests. Describe the liver function test based on either
   a. Excretory function or
   b. Detoxicating function
6. Describe briefly the liver function tests.
7. What are the conjugation tests performed for the assessment of liver function? Give details of the oral and the IV procedure.
8. Enumerate the diverse metabolic tests performed for the assessment of liver function.
9. Name liver function tests and give an account of certain enzyme estimations used as liver function tests.
10. Enumerate the tests performed to assess liver function and discuss the excretory tests.

Short Notes

a. Liver function test based on dextoxication.
b. Hippuric acid test
c. Vanden Bergh test
d. Direct and indirect Van den Bergh test
e. Prothrombin time and vit K.
f. Hippuric and synthesis test
g. Galactose tolerance test
h. BSP retention test
i. Liver function tests based on protein metabolism
j. Biochemical tests for obstructive liver disease
k. Liver function tests, related to carbohydrate metabolism
l. Bilirubin metabolism and the measurement of serum bilirubin
m. Use of enzyme estimations as liver functions tests.

II. MCQs (Give one correct answer)

1. In a case of jaundice, there is no trace of bile pigments in urine, the most probable diagnosis is:
   a. Infections hepatitis b. Obstructive jaundice
   c. Serum hepatitis d. Haemolytic jaundice
   e. None of the above
2. The following enzyme activities are often increased along with the diseases as mentioned in each:
   a. Hepatic bilirubin, UDP—Glucuronyl transferase and Gilbert's syndrome
   b. Gammaglutamyl oxidase and viral hepatitis
   c. Alkaline phosphatase and cholestatic and obstructive jaundice
   d. Gammaglutamyl transferase and obstructive jaundice
   e. Serum amylase and obstructive jaundice
3. Jaundice is clinically detected in sclerae when serum bilirubin concentration reaches above:
   a. 0.5 to 1 mg/100 ml b. 1 to 2 mg/100 ml
   c. 2 to 3 mg/100 ml d. 3 to 4 mg/100 ml
   e. >5 mg/100 ml
4. When jaundice results from hepatitis, the unconjugated fraction of total serum bilirubin is usually:
   a. At least 50% b. Less than 50%
   c. 50 to 85% d. 85 to 90%
   e. 90–100%
5. Patients with hepatocellular jaundice as compared to those with purely obstructive jaundice, tend to have:
   a. Higher serum alkaline phosphatase, LDH and ALT activity
   b. Higher serum alkaline phosphatase activity, lower LDH and ALT activity
   c. Lower serum alkaline phosphatase activity, LDH and ALT activity,
   d. Lower serum alkaline phosphatase activity, higher LDH and ALT activity
   e. None of the above
6. If results of the serum bilirubin, alkaline phosphatase, LDH and ALT determination suggest obstructive jaundice, the ideal confirmatory test would be:
   a. Serum ICD b. Protein electrophoresis
   c. Serum 5-Nucleotidase d. Serum OCT
   e. None of the above
7. The slow moving fraction of LDH isoenzyme LD5 (M4) is typically elevated in patients with:
   a. Myocardial infarction
   b. Cerebrovascular accident (stroke)
   c. Pancreatitis
   d. Hepatitis
   e. In all of the above
8. Drugs can cause jaundice by all of the following mechanisms except:
   a. By causing haemolysis
   b. By selective toxic effects on kupffer cells of liver
   c. By causing intrahepatic cholestasis
   d. By competition with bilirubin for albumin binding site
   e. By direct toxic effect on hepatic cells
9. Prothrombin time in obstructive jaundice:
   a. Normal
   b. Decreases
   c. Becomes normal after vit K injection
   d. Increases after vit K injection,
   e. None of the above.
10. Prothrombin time in parenchymal disease of the liver:
    a. Normal
    b. Decreases
    c. Increases
    d. Becomes normal after parenteral vit K injection
    e. None of the above.
11. Overflow aminoaciduria is one of the abnormalities observed in patients with:
    a. Acute hepatic necrosis b. Obstructive jaundice
    c. Cholecystitis d. Nephrotic syndrome
    e. Hepatoma
12. In which of the following conditions the plasma activities of both ALP and GGT are likely to be increased:
    a. Carcinoma of prostrate
    b. In trimester of normal pregnancy
    c. Osteomalacia
    d. Alcoholic cirrhosis
    e. Infectious hepatitis.
13. Increased unconjugated bilirubin is found in all of the following except:
    a. hemolytic anemia
    b. Crigglar Najjar syndrome
    c. Gilbert's syndrome
    d. Dubin-Johnson syndrome
    e. Physiological jaundice in infants.
14. In the liver, a substantial proportion of the activity of the following enzyme is membrance bound:
    a. Aspartate Aminotransferase
    b. Alanine Aminotransferase
    c. Lactate dehydrogenase
    d. Alkaline phosphatase
    e. Ornithine Transcarbamoylase

Fill in the blanks

1. The purple colored complex formed in Vanden Bergh test in called_________.
2. Certain drugs such as chlorpromazine cause_________ jaundic while gall stones may cause_________ jaundice.
3. Presence of bilirubin in urine without the presence of____________ is strongly suggestive of obstructive jaundice.
4. In advanced parenchymal liver diseases and cirrhosis of liver albumin is_________ and globulins are_________.
5. A complete absence of faecal____________is strongly suggestive of malignant obstruction obstruction.
6. In parenchymal liver, diseases in fall in blood galactose in galactose tolerance test takes place_________.
7. Epinephrime tolerance test is of much use in glycogen storage disease specially_________.
8. In advanced parenchymal liver diseases and in cirrhosis liver the albumin is grossly_________.

9. Amino acid involved in hippuric acid test is _________.
10. The enzyme which is used for differential diagnosis of jaundice is_________.

Answers to MCQs

1. d. 2. c. 3. c 4. b 5. d 6. c
7. d 8. b 9. c 10. c 11. a 12. d
13. d 14. d.

Answers to the fill in blanks

1. Azobilirubin
2. Cholestatic/obstructive
3. Urobilinogen
4. Decreased/increased
5. Urobilinogen
6. Slowly
7. Von-Gierke's disease
8. Decreased
9. Glycine
10. Alkaline phosphatase

CHAPTER 38

GASTRIC FUNCTION TESTS

Major Concepts

A. Study the chemistry of gastric juice.
B. Study various gastric function tests and learn their interpretations.

Specific Objectives

A. 1. Study what are the indications for gastric function tests.
   2. Learn the details of procedure how gastric juice is obtained.
   3. Study the constituents of gastric juice and learn their significance.

B. 1. Classify gastric function tests.
   2. Study normal response of fractional test meal analysis (FTM).
   3. Study abnormal responses of FTM.
   4. Learn about Hyperchlorhydria, hypochlorhydria and achylia gastrica.
   5. Study various stimulation tests with interpretations.
   6. Learn about tubeless gastric analysis.

INTRODUCTION

In diseases of the stomach and duodenum alterations of gastric secretion often occur. Chemical examination of gastric contents has a limited but specific value in the diagnosis and assessment of disorders of the upper gastrointestinal tract, e.g. peptic ulcer, cancer of the stomach, etc. In order to obtain complete data regarding gastric function, the contents of the stomach should be examined

(a) During the resting period
(b) During the period of digestion after giving a meal
(c) After stimulation.

In 24 hours the normal healthy stomach secretes about 1000 ml of gastric juice when the subject is fasting. But the stomach of a person taking a normal diet secretes 2000 to 3000 ml of juice per 24 hrs.

The chief constituents of gastric juice are:

- *HCl* secreted by the parietal cells,
- *Pepsinogen:* Secreted by zymogen cells or "chief" cells.
- *Rennin:* Not found in adult gastric juice. Only found in infants/babies.
- *Intrinsic factor:* Required for absorption of vitamin B_{12}, and
- Other cells produce an alkaline mucus.

Indications of gastric function tests: Gastric analysis may be of value in the following:

- Diagnosis of gastric ulcer.
- Exclusion of diagnosis of pernicious anaemia and of peptic ulcer in a patient with gastric ulceration.
- Presumptive diagnosis of Zollinger-Ellison syndrome.
- Determination of the completeness of surgical vagotomy.

The above are the only situations in which gastric analysis has significant clinical value.

Note: Cytologic examination of gastric juice fluid has not been included as part of gastric analysis.

CLASSIFICATION

Tests commonly employed for assessing gastric function are:

A. Examination of resting contents in resting juice (gastric residuum).
B. Fractional gastric analysis using a test 'meal'.
C. Examination of the contents after stimulation:
- "Alcohol" stimulation.
- Caffeine stimulation.

- Histamine stimulation.
- Augmented histamine test.
- Insulin stimulation.
- Pentagastrin test.

D. Tubless gastric analysis.

Collection of Contents of Stomach

1. The stomach contents are collected after introducing a stomach tube by nasogastric route into the stomach and removing the contents by aspiration. The resting gastric contents are completely removed for examination.
2. Gastric contents are removed after a "test meal" to see the response of stomach. In this, small samples 5 to 6 ml of the gastric contents are removed after every 15 minutes and the samples are collected in small sterile clean penicillin bottles.

Types of Stomach Tubes

1. The stomach tube is made of rubber or plastic and has an external diameter of 4 mm.
2. **Two types** of tubes are in use:
   (a) *Rehfuss tube:* This has an uncovered metal end with openings about the size of the bore of the tube
   (b) *Ryle's tube:* This is commonly used. It has a covered end containing a small weight of lead, the holes being in the tube a short distance from the end.
3. *Markings on the tubes:* Both tubes have markings to indicate how far the tube has been swallowed by the patient. The markings are in the form of black rings.
   - When the single ring reaches the lips, sufficient tube has been swallowed so that tip reaches the cardiac end.
   - When the double ring reaches the lips, the tube should be in body of the stomach, sometimes almost to pylorus (about a distance of 50 cm).

Precaution: The tube should be boiled in water and before passing it should be lubricated with liquid Paraffin or glycerol.

Note

- Ryle's tube is easier to swallow and less likely to cause trauma.
- But disadvantage is that the Ryle's tube tends to block more easily.

Errors in Collection of Samples

Common errors are as follows:

- Tube may be blocked with mucus or food residues, so that the stomach is wrongly assumed to be empty.
- Tube may not be placed properly in the stomach so that either no specimen is obtained or if saliva is being swallowed, a series of samples containing saliva may be sent for analysis and a wrong diagnosis of achlorhydria may be made.
- Too much tubing may be swallowed resulting to aspiration of heavily bile stained duodenal contents.

EXAMINATION OF RESTING CONTENTS

The tube is passed after a night's fast and the stomach contents are removed completely. Valuable informations can be obtained by the examination of resting stomach contents. The following physical and chemical characteristics are important from point of view of diagnosis of diseases of stomach.

1. *Volume:* ***In most normal cases after a night's fast only 20 to 50 ml of resting contents is obtained. Volume greater than 100 to 120 ml is considered abnormal.*** An increase in volume of resting contents may be due to:
   - Hypersecretion of gastric juice
   - Retention of gastric contents due to delayed emptying of the stomach
   - Due to regurgitation of the duodenal contents.
2. *Consistency:* The normal resting gastric juice is fluid in consistency and does not contain food residues. It may contain small amounts of mucus. ***Food residues are present in carcinoma of the stomach.***
3. *Colour*
   - In more than 50 per cent of normal individuals, the gastric residuum is clear or colourless, or it may be slightly yellow or greenish due to regurgitation of bile from duodenum.
   - A bright red or dark red or brown colour in the residuum is due to presence of blood-fresh/or altered blood.
4. *Bile:* Bile may be found occasionally but is not usually of any particular significance.

A small amount may be regurgitated from the duodenum as stated above, as a result of nausea which some people may experience in swallowing the tube.

Increase quantities of bile is abnormal which may result from intestinal obstruction or ileal stasis.

5. *Blood*
   - ***Normally blood should not be present.*** A small amount of fresh bright blood may be traumatic.
   - *Pathologically*
     – Blood which has stayed for sometime in stomach is usually brown or reddish-brown in colour. In the presence of HCl, red blood cells are haemolyzed and dark brown acid haematin is formed. This can occur in gastric ulcer (bleeding) and occasionally in gastric carcinoma.
     – When bleeding is associated with delayed emptying of stomach, the blood is usually mixed with food residues giving dark brown colour, called as ***coffee-grounds*** appearance. ***This is characteristically seen in gastric carcinoma.***

– Occasionally bleeding can occur from *gastritis*.
– ***Sudden bleeding from swallowing aspirin tablets,*** due to irritation of mucous membrane of stomach and erosion of small capillaries.

Note: Possibility of blood arising from a lesion of upper or lower respiratory tract which may be swallowed, appear as altered blood in gastric contents.

6. *Mucus*
   - Normally mucus is present in only small amounts.
   - Increased mucus is found in gastritis and in gastric carcinoma. Presence of mucus is inversely proportional to the amount of HCl present.

Note: Swallowed saliva may account for excess of mucus.

7. *Free and Total Acidity*
   Determined by titrating a portion of the filtered specimen with a standard solution of NaOH.

Two indicators are used in succession. The indicators most commonly used are:

- ***Methyl orange 0.1%*** aqueous solution or ***Topfer's reagent*** (0.5% solution of dimethyl amino azobenzene in absolute ethanol). Measures pH 2.9 to 4.4 (change from red to yellow colour)
- ***Phenolphthalein 1%*** solution in 50% ethanol. This indicator measures, pH 8.3 to 10.0, colour change yellow to red again.

Inferences: The following inferences should be drawn:

- *Free acidity:* The first titration to about pH 4.0 measures the amount of free HCl present, i.e. free acidity.
- *Total acidity:* The complete titration is said to give the total acidity. Some protein hydrochloride and any organic acids present are titrated. Proteins present include mucin in the gastric secretion and protein in meal (this will be in the juice obtained after test meal).
- *Combined acid:* The difference between the two titrations gives the combined acid.

Results: Result of titration is expressed as ml of 0.1 N HCl per 100 ml of gastric contents. This is same as mEq/litre. To get this figure multiply the above titration by 10.

Normal values:

- **Free acid**: 0 to 30 mEq/L
- **Total acid**: 10 mEq/L higher (10 to 40 mEq/L)

Note:

- ***Thymol blue can be used as indicator.*** It has the ***advantage of having two colour changes.*** One red to yellow at pH 1.2 to 2.8, and the other from yellow to blue at pH 9.0 to 9.5. Titration to the first colour change has been used for free acid and second titration colour change for total acid.
- Concentration of free acid above 50 mEq/L indicate hyperacidity.

8. *Organic acids:* Lactic acid and butyric acid may be present in large amounts in cases where there is achlorhydria and hypochlorhydria and residual foods must remain in stomach. In absence of HCl, the micro-organisms can thrive well and ferment the food residues to produce the organic acids, lactic acid and butyric acid. ***Achlorhydria associated with retention of food materials is exclusively found in carcinoma stomach.***

FRACTIONAL GASTRIC ANALYSIS: USING TEST MEALS

Fractional Gastric analysis: *Also called Fractional Test Meal (FTM)*

It consists of the following steps:

1. Introduction of Ryle's tube in stomach of a fasting patient (overnight).
2. Removal of residual gastric contents and its analysis. The above two have already been discussed above.
3. Ingestion of "test meal"
4. Removal of 5 to 6 ml of gastric contents after meal by aspiration using a syringe and analysis of the samples.

Test Meals

Several types of test meals have been used:

- *"Ewald" test meal:* It consists of two pieces (35 gm) of toast and approx 8 oz (250 ml) of light tea.
- *"Oatmeal" porridge:* This is prepared by adding 2 tablespoonfuls of oatmeal to one quart of boiling water and straining the porridge through fine thin muslin.
- *"Riegel" meal:* It consists of 200 ml of beef broth, 150 to 200 gm of broiled beef steep and 100 gm of smashed potatoes. This meal is not used normally.

Ewald meal has to be consumed by the patient before the introduction of Ryle's tube and the tube is introduced after one hour. This is a little disadvantageous. In the case of oat porridge, it can be taken by the patient with tube *in situ* after clipping the tube.

Collection of samples: At intervals of exactly 15 minutes, about 10 ml of gastric contents are removed by means of syringe attached to the tube. If the stomach is not empty at the end of 3 hours, the remaining stomach contents are removed and the volume noted.

Analysis of the samples: Each sample is strained through a fine mesh cheese cloth. The residue on the cloth is examined for mucus, bile, blood and starch. The strained samples are analyzed for free and total acidity.

Results and Interpretation of the Tests

A. *Normal Response*

In normal health: After taking the meal, free acid is again found after 15 to 45 minutes (Fig. 38.1). The free acid then

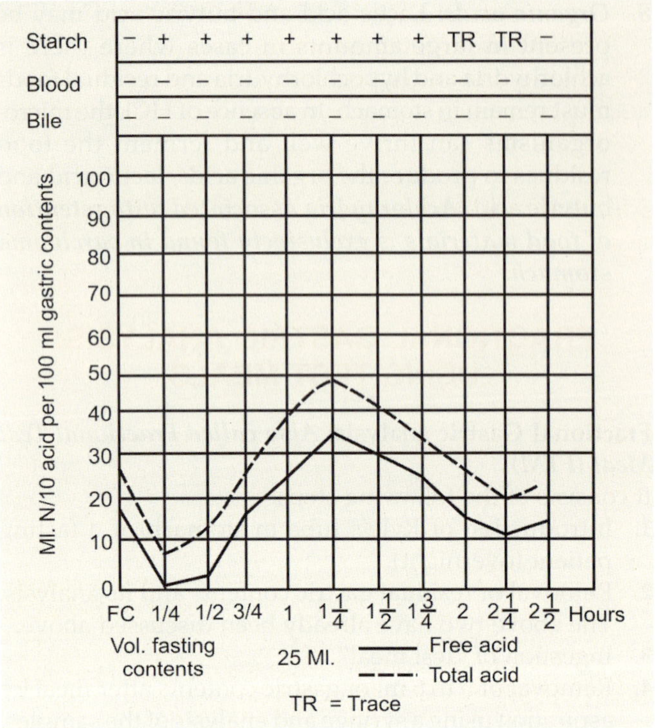

Fig. 38.1: Fractional test meal—normal result

rises steadily to reach a maximum at about 15 mts to ½ hour, after which the concentration of free acid begins to decrease. Free acid ranges from 15 to 45 mEq/litre at the maximum with total acid at about 10 units higher. About 80 per cent of normal people fall within these limits. Blood should not be present and there should not be any appreciable amount of bile.

B. Abnormal Responses

Three types of abnormal responses:

1. *Hyperacidity (hyperchlorhydria):* In which free acid reaches a higher concentration than in normal persons.
2. *Hypoacidity (hypochlorhydria):* In which though free acid is present, it is present in a concentration below the normal range.
3. *Achlorhydria:* In which there is no secretion of free acid at all.

1. Hyperchlorhydria

This occurs when the maximum ***free acidity exceeds 45 mEq/L***, some prefer to keep at 50 mEq/L, combined acid remains the same as in normal persons.

Causes: Hyperacidity is found in the following:

- *In duodenal ulcer* a climbing type of curve is seen.
- *In gastric ulcer:* Though hyperacidity is common, 50 per cent cases may give normal results, whilst in some chronic cases, due to associated gastritis, hypoacidity may be found. Blood may be present in gastric contents. ***Blood together + hyperchlorhydria is suggestive of gastric ulcer.***
- *Gastric carcinoma:* Small percentage of cases show hyperacidity and blood.
- Jejunal and gastrojejunal ulcers occur as sequelae to ***gastroenterostomy;*** they are often found associated with hyperacidity after operation.
- Other disorders where hyperacidity may be found are ***gastric neurosis, hyperirritability and pylorospasm, pyloric stenosis, chronic cholecystitis, chronic appendicitis, etc.***

2. Hypoacidity (Hypochlorhydria)

It is difficult to define this zone. Low acidities are found in carcinoma of stomach and in atonic dyspepsia. In pernicious anaemia, free HCl is absent in gastric secretion. In gastroenterostomy hypoacidity seen.

3. Achlorhydria

This term is used when there is no secretion of HCl, but enzyme like pepsin is present. ***Achlorhydria can be differentiated from hypochlorhydria by stimulation test with histamine.*** In hypochlorhydria, histamine stimulation shows rise in free HCl but in achlorhydria-histamine stimulation does not show any response.

Causes

- Found in some normal people increasing with age at about 60 to 75 years.
- High incidence in ***carcinoma of stomach.***
- ***In chronic gastritis,*** there is tendency of gastric acidity to be reduced. As the disease progresses, increasing incidence of achlorhydria seen.
- Partial gastrectomy leads to reduction of gastric acidity often and to achlorhydria in a considerable number of cases.
- In pernicious anaemia.
- Other diseases viz., ***Microcytic hypochromic anaemia* (in 80% cases), *hyperthyroidism* and *myxoedema*** may be associated with achlorhydria.

ACHYLIA GASTRICA

The term is used when ***both enzymes and acids are absent*** indicating there is a complete absence of gastric secretion.

Causes

- In advanced cases of cancer of stomach
- Advanced cases of gastritis
- Typically found in pernicious anaemia and of sub-acute combined degeneration of the spinal cord (100% cases).

STIMULATION TESTS

A. Alcohol Stimulation

7% ethyl alcohol has been used as a stimulant of gastric secretion.

Procedure

1. After overnight fast, the Ryle's tube is passed into the stomach and resting contents are removed for analysis.
2. ***One hundred ml (100 ml) of 7% ethyl alcohol is administered.***

   Note: A little of methylene blue can be added in alcohol meal so that it gives an indication of emptying time of the stomach.
3. Samples of gastric contents are removed every 15 minutes.
4. All the collected samples are analysed for free and total acidity, peptic activity, presence of blood, bile and mucus.

Advantages: Advantages of alcohol test meal over "oat-meal" porridge are:

- More easily administered and prepared.
- It is consumed better than porridge,
- Specimens are clear and easily analysed.
- The gastric response is more rapid and more intense.
- The stomach empties more quickly as compared to porridge meal

Disadvantages

- Stimulus with alcohol is not so strictly physiological as with oatmeal porridge.
- Stimulus is more vigorous as compared to oatmeal
- Rather higher levels of free acidity are obtained and the limits of normal are wider.

B. Caffeine Stimulation

Caffeine can be used as a stimulus instead of alcohol. Procedure remains same as above.

Procedure

1. Ryle's tube is introduced after an overnight's fast and the resting gastric contents are removed and analysed.
2. Caffeine sodium benzoate, 500 mg dissolved in 200 ml of water is given to the patient orally.
3. Samples of stomach contents are removed every 15 minutes and analysed for free and total acidity, peptic activity, blood, bile and mucus.

Advantages of caffeine stimulation is similar to alcohol stimulation.

C. Histamine Stimulation Test

Histamine is a powerful stimulant for the secretion of HCl in the normal stomach. It acts on receptors on the oxyntic cells, increasing the cyclic AMP level, which causes secretion of an increased volume of highly acidic gastric juice with low pepsin content.

Indications: *To differentiate "true" achlorhydria from "false" achlorhydria due to various causes.* "True" achlorhydria which is histamine-resistant is seen in achylia gastrica. Demonstration of such an achlorhydria is useful in the diagnosis of subacute combined degeneration of the cord and pernicious anaemia.

Types of histamine test:

I. Standard histamine test and

II. Augmented histamine test.

I. *Standard Histamine Test*

Procedure

1. After an overnight's fast, Ryle's tube is passed into the stomach and stomach contents are removed for analysis.
2. Patient is given a subcutaneous injection of histamine, 0.01 mg/kg body wt.
3. After the injection, 10 ml of stomach contents are removed every 10 minutes for one hour. The samples are analyzed for free and total acidity, peptic activity, and for presence of blood, bile and mucus.

CLINICAL SIGNIFICANCE

- ***Absence of free HCl in the secretions after histamine indicate "achylia gastrica" ("true" achlorhydria).***
- ***In duodenal ulcer:*** More juice may be secreted and a higher concentration of acid may be found in the specimen obtained after histamine administration than in normal cases.

Note: Standard histamine test may be combined with the FTM. If no free acid is found in the resting contents in FTM by the end of an hour after giving the gruel meal, histamine can be given and standard test can be carried out.

II. *Augmented Histamine Test (Kay)*

It is a more powerful stimulus than the original standard test used, and provides a more reliable proof of an inability to secrete acid.

Disadvantage: Larger doses of histamine used in this test causes sometimes untoward severe reactions and hence an antihistaminic will have to be given side by side to prevent any such reactions.

Note: The antihistamine does not interfere in gastric stimulation action of histamine.

Indications: The test has been used for two purposes:

- To show an inability to secrete acid which is present with pernicious anaemia and subacute combined degeneration of the cord.
- To assess the maximum possible acid secretion as in the diagnosis and surgical treatment of duodenal ulcer.

Procedure

1. After an overnight fast, pass a Ryle's tube and remove the residual gastric contents for analysis.

2. Collect resting contents every 20 minutes for an hour.
3. Halfway through this period, give 4 ml of anthisan (100 mg of mepyramine maleate) intramuscularly (IM).
4. At the end of the hour, give histamine (0.04 mg histamine acid phosphate per kg body wt) subcutaneously (SC) and remove gastric contents every 15 minutes for one hour (4 specimens) or three 20-minute interval specimens.

Specimens obtained are: Resting contents, an hour pre-histamine specimen and three 20-minute post-histamine specimens.

CLINICAL SIGNIFICANCE

- ***In normal persons:*** Up to 10 mEq/hour acid is present in the prehistamine specimen, with 10 to 25 mEq in the combined post-histamine ones.
- ***In pernicious anaemia:*** No free HCl is secreted after augmented histamine stimulation (achylia gastrica), but in other forms of achlorhydria (false achlorhydria), some amount of free HCl is secreted after histamine stimulation.
- ***In duodenal ulcers:*** Higher values are obtained sometimes reaching even exceeding 100 mEq. The maximum acidity, reached in the second 20-minute specimen has been used by some workers for duodenal ulcers.

Note: Recently, a histamine analogue, called **'histalog'** (3β-aminoethylpyrazole) has been used in place of histamine.

Dose: Recommended dose is 10 to 50 mg.

Advantages

- No side effects like histamine hence no anti-histaminic is required to be administered along with.
- It is highly effective in stimulating gastric secretion.

D. Insulin Stimulation Test (Hollander's Test)

Hypoglycaemia produced by administration of insulin is a potent stimulus of gastric acid secretion. **Hollander** suggested that to be effective blood sugar must be brought below 50 mg per cent, whereas other workers have recommended a level below 45 mg per cent is a necessity for a reliable test.

Indication

To ascertain the effectiveness of vagotomy (vagal resection) in patients with duodenal ulcer. Insulin test meal was suggested by Hollander to determine whether the section of vagus has been successfully performed.

Procedure

1. After an overnight fast, pass a Ryle's tube and empty the stomach.
2. Then give 15 units of soluble Insulin intravenously (IV)
3. After injecting the insulin, withdraw approximately 10 ml samples of gastric contents every 15 minutes for 2½ hours.
4. Samples to be analysed for free and total acidity, peptic activity and presence of blood, bile and starch. ***No starch should be present.***

Note

- The test is not without hazard as blood sugar may go down to dangerously low level in some, which may require glucose treatment and should be readily available.
- Blood sugar, may be determined at least one ½ hour after giving insulin in order to make sure a sufficiently low value 45 to 50 mg per cent has reached.

CLINICAL SIGNIFICANCE

- In patients suffering from duodenal ulcer, before operation, there is a marked and prolonged output of acid in response to insulin. The concentration of free acid may rise well over 100 mEq/litre.
- After a successful vagotomy there is no response to Insulin and the gastric acidity remains at a low level of 15 to 20 mEq/L, before and after insulin injection (Fig. 38.2).

Note

- Some surgeons prefer to have the test done preoperatively and then soon after the operation and once more several months later.

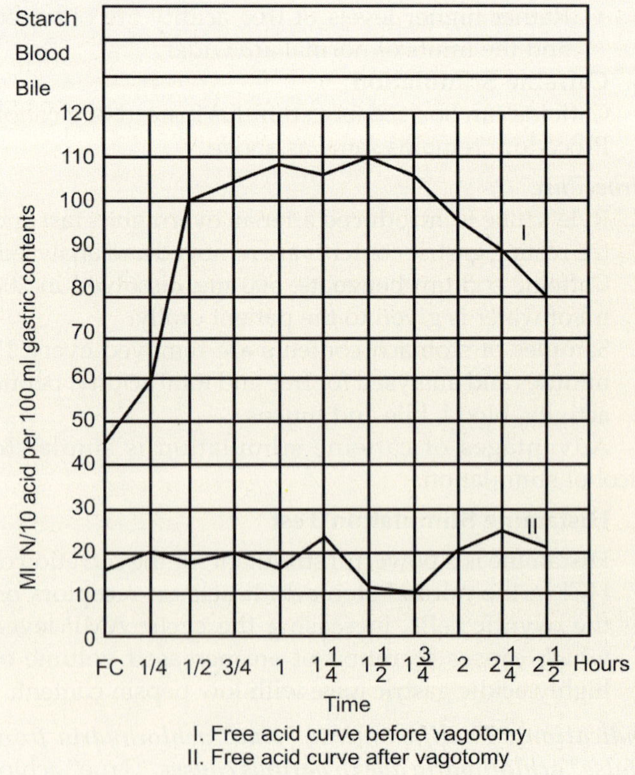

Fig. 38.2: Insulin "test meal"

Others have suggested that it is sufficient and quite satisfactory only to do it once, at least six months after the operation.

- The degree of stimulation of acid secretion is related to the degree of hypoglycaemia obtained and hence indirectly to the dose of insulin given.

E. **Pentagastrin Test**

Pentagastrin is a synthetic peptide in which N-terminal end is blocked by butyloxycarbonyl-β-alanine. Thus, it is "butyloxycarbonyl-β-alanine." **Trp-Met-Asp-phe ($CONH_2$),** the four c-terminal amino acids form the "active" part of the molecule. Pentagastrin is a potent stimulator, and involves the maximal stimulation of stomach after a period of assessment of the basal secretion rate. ***This is thus a measure of the total parietal mass.***

Indications

- Useful in investigation of patients with "active" duodenal ulcer, which may suggest appropriate surgical measures.
- In pernicious anaemia.
- Useful in suspected cases of Zollinger-Ellison syndrome.

Procedure

- After an overnight fast, stomach tube (Ryle's tube) is passed into the stomach and the resting contents completely removed.
- After emptying the stomach of resting contents, collect two 15-minute specimens to have the **basal secretion**.
- Then injection of pentagastrin, 6 μg/kg body wt is given subcutaneously (SC) and collect four specimens, accurately timed, at 15 minutes intervals.
- All the specimens are analysed.

CLINICAL SIGNIFICANCE

- ***Normal basal secretion rate*** is 1 to 2.5 mEq/hour. After pentagastrin stimulus, maximal secretion in normal persons roughly varies from 20 to 40 mEq/hr.
- ***In duodenal ulcer:*** The range was 15 to 83 mEq/hour with a mean of 43. Values above 40 mEq/hour has been kept which is suggestive of duodenal ulcer.
- ***Zollinger-Ellison syndrome:*** It is characterised by a high basal secretion usually above 10 mEq/hr; if, it is maximal then, there will be no further rise after giving pentagastrin, otherwise only a small to moderate increase is seen.
- ***In gastric ulcer:*** The test is of little value.
- ***In cancer of the stomach:*** "True" achlorhydria is found in about 50 per cent of cases, and hypochlorhydria in about 25 per cent.
- Output of acid is also reduced transiently in acute gastritis, and permanently in chronic gastritis.
- ***In pernicious anaemia:*** The basic pathology is gastric mucosal atrophy with lack of intrinsic factor and in great majority of cases "true" achlorhydria. Some occasional young persons with pernicious anaemia have been found to have acid secretion.

Note

- **Zollinger-Ellison syndrome:** Zollinger-Ellison syndrome is characterised by a peptic ulcer, intractable to medical treatment, gastric hypersecretion and diarrhoea in patient with "gastrin" secreting pancreatic islet cell (δ-cells) adenoma. It is sometimes accompanied by other endocrine adenomas or hyperplasias, especially parathyroid adenomas with hyperparathyroidism.

SERUM PEPSINOGEN

Pepsinogen determination has been used to investigate the gastric secretion of this enzyme.

A convenient method using the digestion of dried serum has been used.

Interpretations

- *Normal value:* Ranges from 30 to 160 units/ml.
- *In pernicious anaemia:* Serum pepsinogen is absent or very low.
- *In duodenal ulcer:* An increase is often found upto and above twice the upper limit of normal. If the serum pepsinogen is less than < 80 units/ml; it is considered that an ulcer is not present.

TUBELESS GASTRIC ANALYSIS

Swallowing a stomach tube (Ryle's tube) is an unpleasant and cumbersome procedure and sometimes inadvisable hence attempts have been made to devise tests which can be done without using a stomach tube.

Initially **Segal and coworkers** used a quininium resin indicator given orally, from which H^+ ions if present in stomach could liberate quinine ions (QH^+ cation) at a pH less than 3.0. The quinine liberated forms quinine HCl which is absorbed in small intestine and then excreted in the urine from which quinine is extracted and determined fluorimetrically. Thus, it gives indirect measure for acid secretion.

Modification: Subsequently the test was simplified. They introduced **Diagnex Blue** prepared by reacting carbacrylic cation exchange resin with **Azure A,** an indicator. The hydrogen ions of the resin exchanged with 'Azure A' ions, the reaction is reversed in the stomach when acid, if present, in a concentration giving a pH less than 3.0. By the action of acid, the indicator 'Azure A' is released, which is absorbed in the small intestine and excreted in the urine, the colour of which can then be matched with known standards.

CLINICAL SIGNIFICANCE

The test is of value if it is used as a "screening test" only.

- A positive result, provided no other cations such as K^+, Ba^{++}, Fe^{++}, etc. are present, indicates that acid is being secreted by the stomach.
- A negative result is an unreliable indicator of "true" achlorhydria since 50 per cent of these cases secrete acid in response to pentagastrin.
- The test is not reliable in patients suffering from renal diseases, urinary retention, malabsorption, pyloric obstruction and after gastrectomy and gastroenterostomy.

Note: Vitamin preparations should not be taken on the day preceding the test or medicaments which might contain substances decolourised by ascorbic acid.

I. Questions (Essay type and short notes)

1. Describe gastric function tests.
2. Enumerate gastric function tests. Give details of alcohol fractional test meal.

Shorts Notes

a. FTM
b. Insulin test meal (Hollander's test)
c. Pentagastrin test
d. Tubeless gastric analysis
e. Alcohol fraction analysis
f. Gastric analysis
g. Alcohol fractional test meal

II. MCQs (Give one correct answer)

1. The stimulation routinely and commonly used for the gastric analysis procedure is:
   a. Caffeine b. Alcohol test meal
   c. Epinephrine d. Strychnine
   e. Amphetamine
2. The stimulation test of gastric analysis used to differentiate "true" achlorhydria from 'false' achlorhydria is:
   a. Alcohol b. Caffeine
   c. Pentagastrin d. Histamine
   e. Epinephrine
3. All of the following are associated with Zollinger-Ellison syndrome except:
   a. Peptic ulcer
   b. Interstitial thyroid cell adenoma
   c. Gastric hypersecretion
   d. Diarrhoea
   e. Intractability to medical treatment
4. The Hollander's gastric stimulation test routinely requires use of:
   a. Alcohol b. Gastrin
   c. Glucose d. Insulin
   e. Epinephrine
5. Vagal nerves are intact if basal gastric test results indicate:
   a. Hypochlorhydria
   b. Presence of pepsin
   c. Normal hydrogen ion
   d. Presence of gastrin secretion
   e. Achlorhydria
6. Presence of free acidity in gastric aspirate makes it possible to:
   a. Rule out pernicious anaemia
   b. Rule out pulmonary tuberculosis
   c. Diagnose Zollinger-Ellison syndrome
   d. To confirm that a bilateral vagotomy was complete
   e. None of the above
7. Presence of stimulant-fast gastric achlorhydria:
   a. Confirms the diagnosis of gastric ulcer
   b. Is uniformly found in patients with gastric carcinoma
   c. Rules out gastric carcinoma
   d. Rule out a peptic ulcer
   e. Confirms the diagnosis of peptic ulcer
8. While performing Hollander's test the patient becomes nervous and sweats, develops tremors and complains of "feeling unwell" the probable diagnosis is:
   a. Psychosomatic problem
   b. Hysterical
   c. Hyperglycaemia
   d. Anxiety neurosis
   e. Hypoglycaemia
9. If the patient's gastric analysis shows total absence of free acidity and pepsin, the probable condition is:
   a. Achlorhydria
   b. Hypochlorhydria
   c. Zollinger-Ellison
   d. Achylia gastrica syndrome
   e. None of the above.

Answers to MCQs

1. b 2. d 3. b 4. d 5. c 6. a
7. d 8. e 9. d.

CHAPTER 39

THYROID FUNCTION TESTS

Major Concepts

A. To review the anatomy and physiology of the thyroid gland.
B. To describe various laboratory tests to assess the thyroid function.
C. To learn the use of various tests under specific thyroid disease or dysfunction.

Specific Objectives

A. To study the anatomy and physiology of thyroid gland (The details are not included in the text; Students are encouraged to review them before studying this chapter).
B. Classify thyroid function tests based on various functions of the gland.
   1. Describe the tests based on primary function of thyroid such as
      - Radioactive uptake studies
      - Serum PBI^{131}
      - T_3 suppression test
      - TSH stimulation test
      - TRH stimulation test and learn their interpretations.
   2. Describe the tests based on the measurement of blood levels of thyroid hormones. Such as serum
      - PBI or BEI levels,
      - T_4, T_3 and TSH
      - *In vitro*^{131} I-T_3
      - Tyrosine and their interpretations.
   3. Describe the test based on metabolic effects of thyroid hormones such as:
      - BMR
      - Cholesterol level in plasma
      - Serum creatine level
      - Uric acid level
      - Serum CK level and learn their interpretations.
   4. Learn briefly about thyroid scanning with a particular reference to hot or warm areas, cold nodules and use of 99 technetium and its usefulness.
   5. Describe briefly the immunological tests for thyroid function such as:
      - Precipitation test
      - Tanned red cell haemagglutination test
      - Complement fixation test and their usefulness in autoimmune thyroid diseases.
C. Evaluate critically each of the above tests with respect to its application under various thyroidal disorders. Learn how to interpret the results.

INTRODUCTION

The main objectives for the laboratory procedures in evaluation of thyroid diseases are:

- To assess the functional status of the gland
- To characterise the anatomical features of the thyroid gland, and
- To possibly evaluate the cause for the thyroid dysfunction.

With the advent of "tracers" especially I^{131}, (a) uptake studies: Reflecting substrate input in hormone synthesis; and (b) 'scanning' characterising benign and malignant lesions, localising "ectopic" thyroid tissue or functioning metastasis, have contributed a great deal in improving the thyroid diagnostic acumen.

This was followed by the development of 'radioimmunoassays' (RIA) and the prospect of determining the actual minute circulating quantities of thyroid hormones, viz. T_4, T_3 and TSH, further augmented the precision in diagnosis of thyroid diseases.

It is emphasised that a single thyroid function test is not absolute in diagnostic accuracy and it must be thus a careful selection of such tests so that their combination can give comprehensive data that would enhance the diagnostic accuracy.

CLASSIFICATION

Classification of various tests can be made on the basis of the functions of the gland.

N.B. Students should review the knowledge of physiology of thyroid gland.

I. Tests based on primary function of thyroid, viz. substrate input and hormone synthesis
- Radioiodine "uptake" studies and 'turnover' (RAI or RIU) studies
- PBI^{131} in serum
- T_3-suppression test
- TSH-stimulation test
- TRH-stimulation test.

II. Tests measuring blood levels of thyroid hormones
- Serum PBI and BEI
- Circulating T_3 and T_4 level
- Circulating TSH level
- *In vitro* resin-uptake of T_3
- Plasma tyrosine level

III. Tests based on metabolic effects of thyroid hormones
- BMR
- Serum cholesterol level
- Serum creatine level
- Serum uric acid
- Serum CK enzyme.

IV. "Scanning" of thyroid gland

V. Immunological tests to detect autoimmune diseases of thyroid gland
- Agar gel diffusion test (precipitation test)
- TRCH test: Tanned red cells haemagglutination test.
- Complement fixation test.

SECTION FIVE

TESTS BASED ON PRIMARY FUNCTION OF THYROID

(a) Radioactive "Uptake" Studies

Iodine plays a key role in the metabolism of the thyroid gland. I^{131} "tracer" is most commonly used for thyroid function studies because of low cost, easy availability, and convenient shelf life. Short lived isotopes of iodine like I^{132} and I^{123} are preferred for use in paediatric practice and in pregnant and lactating women. Recently, ^{99m}Tc has also been used as it behaves like iodine and has added advantage of lower radiation dose to the patient.

Procedure

Dose of I^{131} = 10 μci given orally. Thyroid accumulation of radio-I_2 is measured externally over the gland. Radioiodine uptake of the gland reflects the iodine-"trapping" ability. Thyroid uptake of I^{131} is routinely measured 24-hours after the administration of oral dose, although 4-hour uptake or 48-hour uptake are also measured when rapid turnover or delayed uptake situation is expected. "Turnover" is faster in 'active' and hyperfunctioning gland and slower in underactive hypofunctioning gland.

- *Normal range:* **20 to 40 per cent**

In Indian subjects a value of 15 to 35 per cent has been found. The range varies from one population to another depending on dietary iodine intake.

Interpretations
- An abnormally high RAI uptake is usually consistent with hyperthyroid state.
- In endemic goitre and some cases of non-toxic sporadic goitre also may be high.
- Abnormally low thyroid uptake is characteristic of hypothyroidism, but not specific since subacute thyroiditis and administration of large doses of I_2 and thyroid hormones may also lower the I^{131} uptake of the gland.

Urinary excretion of I^{131} and "T" Index: Renal excretion of I^{131} is an indirect evidence of thyroid function. Proportion of the administered dose excreted is inversely proportional to thyroid uptake. If uptake is "more", less of I^{131} will be excreted and vice versa. 24 hours urine is collected accurately and radioactivity is measured.

- *Normal range:* It is 30 to 60 per cent of the administered dose.

"T"-index

Activity is measured in urine sample after 0 to 8 hours, 0 to 24 hours and 0 to 48 hours.

'T'-index is calculated as follows:

$$T = \frac{\text{0-8 hrs excretion expressed as per cent} \times 100}{\text{(0-24 hrs excretion expressed as \%)} \times \text{(0-48 hrs excretion expressed as \%)}}$$

- *Normal value of "T"* = 2.5 to 12

Interpretations
- A 'T'-index > 17 indicates hyperfunctioning of the gland.
- A "T"-index < 2.5 indicates hypothyroidism.

Thyroid "Clearance" Rate

The amount of I^{131} that is accumulated in thyroid over a fixed interval, in relation to the mean plasma concentration of I^{131} mid-way in that time period provides the index of rate at which the thyroid gland is handling I^{131}. (Rationale is similar to the concept of renal clearance.)

Hence, Thyroid Clearance rate =

$$\frac{\text{Thyroid } I^{131} \text{ accumulation rate}}{\text{Plasma } I^{131} \text{ concentration. (Midway between the time period)}}$$

This gives a direct index of thyroid activity with regard to I_2 accumulation.

- *Normal value:* 60 ml/mt.

Interpretations

- Clearance rate is high with thyroid hyperfunction, the value has been distinctly high with no overlap.
- The value is also high when "intrathyroidal iodine pool" is small.
- Lower values are indicative of hypothyroid status.

(b) Serum PBI^{131}

Administered I^{131} accumulates in the thyroid gland and appears as "labelled" hormone bound to proteins. Normally it is a slow process, but in hyperthyroidism, level of protein-bound radioactivity increases in plasma, which can be measured accurately by a scintillation counter. The result is conveniently expressed as "conversion ratio", which indicates the proportion of the total plasma radioactivity at 24 hrs.

- *Normal value:* 35 per cent.

Interpretations

- *In hyperthyroidism:* It is usually greater than 50 per cent.
- It is of no value in the assessment of patients who have been treated for hyperthyroidism, either surgically or with radioactive I_2, as high values may persist for long time after such treatments.
- PBI^{131} is found to be elevated in 50 per cent of the patients with ***Hashimoto's thyroiditis***, when the thyroid uptake is usually normal or low, ***a combination of findings which is very suggestive of this condition.*** The reason for these discrepancies is that PBI^{131} is not a measure of plasma thyroxine concentration.

Factors Determining Serum PBI^{131} Level

The level of serum PBI^{131} is dependent on several factors:

- The initial proportion of the "tracer" dose accumulated by the thyroid.
- The rate of secretion of the thyroid hormones and
- The size of the "intrathyroidal iodine pool".

In primary hyperthyroidism, the intrathyroidal iodine pool is similar to that of the normal thyroid gland so that in untreated hyperthyroidism, the elevated PBI^{131} is largely a reflection of increased secretion rate.

On the other hand, the "intrathyroidal iodine pool" is markedly reduced after treatment either surgically or with radioiodine, and ***also a striking feature in Hashimoto's thyroiditis,*** so that under these circumstances an elevated PBI^{131} is mainly due to markedly reduced intrathyroidal iodine pool, the secretion rate of the thyroid hormones being normal or even reduced.

(c) T_3-Suppression Test

1. After a 24 hrs RIU studies and obtaining the basal value and serum T_4 values, 20 μg of T_3 four times daily is given for 7 to 10 days (or alternatively 25 μg three times a day for 7 days).
2. RIU is repeated after T_3 administration and serum T_4 values are also determined.

Interpretations

- A suppression is indicated by the 24 hrs RIU falling to < 50 per cent of the "initial" uptake (as exogenous T_3 suppresses TSH) and total T_4 to approx 2 μg/100 ml or less.
- ***Non-suppression indicates autonomous thyroid function.*** In Graves' disease, no change seen as the action is due to LATS (long-acting thyroid stimulator) and is not under control of hypothalamopituitary axis.

Use

To differentiate borderline high normal from primary hyperthyroidism (Graves' disease).

(d) TSH-Stimulation Test

1. Following completion of 24 hrs RIU studies, 3 injections of TSH, each 5 USP units are given at 24 hrs intervals.
2. 24-hour thyroidal RIU is measured after 42 hours after the final TSH dose.

Interpretations

- *In primary hypothyroidism:* There is failure of stimulation of the gland.
- *In secondary hypothyroidism:* There is stimulation of the gland showing increase RIU.

Use

The test is useful in differentiating primary hypothyroidism from secondary hypothyroidism.

(e) TRH-Stimulation Test

With the availability of synthetic TRH, which is a tripeptide, suitable for human use, it is now possible to assess the functional integrity of thyrotropic cells or the factors that influence the secretory response.

Procedure

200 to 400 μg of TRH is administered IV and blood samples at 0, 20, 40 and 60 minutes are analysed for TSH content.

Interpretations

- Peak response in ***normal*** is about 4 times elevation of TSH levels at 20 and 40 minutes sample as compared to basal TSH level.
- *In primary hypothyroidism:* The response will be exaggerated and prolonged.
- *In secondary hypothyroidism:* The response will be blunted.
- *In tertiary hypothyroidism,* i.e. hypothalamic in origin, the increase in TSH is delayed.

Use

Currently this test is used to locate the site of pathological lesion for hypothyroid states.

TESTS MEASURING BLOOD LEVELS OF THYROID HORMONES

(a) Serum PBI and BEI Levels: Chemical estimation of Protein bound I_2 is used for long time as a test for thyroid function.

- It is indirect measure of thyroid hormones.
- It is useful where isotope techniques are not available.

Disadvantage

- Technically time consuming lengthy procedure
- Also measures non-hormonal I_2 and iodotyrosines.

Normal value: Ranges from 4.0 to 8.0 μg%

Interpretations

- More than 95 per cent of hyperthyroidism cases show greater than 8.0 μg%
- 87 per cent of hypothyroidism cases show value below 3 μg%
- Care should be taken to interpret values between 4.0 and 5.0 μg%.

Precautions and Limitations

1. Easily affected by I_2 contamination both exogenous and endogenous.
   - ***Exogenous:*** To eliminate exogenous contamination, all glass wares and syringes should be iodine free.
   - ***Endogenous:*** Iodides, iodine containing drugs and I_2-containing radiological contrast media can give false high results.
2. The test is also affected by "trace" elements and chemicals that interfere iodine-reduction reaction.
3. Values are also affected by alterations in serum TBG level. ***Increased serum TBG gives higher values whereas decreased TBG gives lower values.***

Serum TBG may be increased in:

- Pregnancy
- Oestrogen therapy
- On oral contraceptive pills.

Serum TBG may be decreased in:

- Hypoproteinaemic states
- Nephrotic syndrome
- Androgen therapy and anabolic drugs like danazol
- Dicoumarol therapy
- Inherited TBG deficiency.

4. Certain drugs may give misleading results by competing with T_4 for Protein binding sites, e.g. phenytoin sodium, salicylates, etc.

Serum BEI: Butanol extractable iodine involves extraction of serum with n-butanol and subsequent washing of the extracts with alkaline solution. This removes the inorganic iodine and iodotyrosines.

Interpretations

- ***In normal:*** Value ranges from 3.5 to 7.0 μg%.
- ***In hyperthyroidism:*** Values are more than 10 μg%.

(b) Serum T_4 Levels

Most commonly used methods are:

- Competitive protein binding assay (CPBA)
- Radioimmunoassay (RIA)
- ELISA technique.

Interpretations

- ***Normal range*** of serum T_4 is 4.0 to 11.0 μg%.
- ***In hyperthyroidism:*** The value is usually more than 12.0 μg%
- ***In hypothyroidism:*** Less than 2.5 μg%.

(c) Effective Thyroxine Ratio (ETR)

This integrates into a single procedure the measurement of total serum thyroxine and also binding capacity of thyroid hormone proteins. At the present time, the ETR provides the most reliable single test of thyroid function available which can be readily carried out on a sample of serum and only requires radioisotope laboratory.

Advantage: It is not affected by oral contraceptives, pregnancy, excess iodine or any other drugs.

(d) Serum T_3 Level

Radioimmune assay is the method of choice for measurement of serum T_3 level. CPBA is not good and accurate as T_3 has very low affinity for TBG.

Normal Range and Interpretations

- ***Normal value:*** 100 to 250 ng% (μg%).
- Values in females tend to be slightly on higher side than compared to males.
- ***In hyperthyroidism:*** It is usually more than 350 ng%
- ***In hypothyroidism:*** Less than 100 ng%. It may be useful test for hyperthyroidism, but it is less useful for diagnosis of hypothyroidism.

(e) Serum TSH Level

Measurement of serum TSH also provides a very sensitive index of thyroid function. By radioimmunoassay, the normal range is 0 to 3 μu/ml average being 1.6 μu/ml. It is of particular value in the diagnosis of primary hypothyroidism.

(f) *In vitro* I^{131}-T_3 uptake by resin/red cells (Hamolsky *et al* 1957):

Method

1. A known amount of I^{131}-T_3 is added to a standard volume of serum from a patient.

SECTION FIVE

2. The amount of I^{31}-T_3 which binds to the serum proteins varies inversely with the endogenous thyroid hormones already bound to serum proteins (TBG).
3. Residual free I^{131}-T_3 is then adsorbed by resin/sponge/sephadex/red cells, which is removed from the sample and then the adsorbed/bound I^{131} is measured.

This method thus gives the measure of T_4 binding in the serum and not the actual level of thyroid hormones.

Interpretations

- ***In normal subjects:*** The value is 21 to 35 per cent.
- ***In hyperthyroidism:*** Saturation of binding of TBG with endogenous T_4 and T_3 is greater than normal, hence little of tracer I^{131}-T_3 can bind to TBG and more I^{131}-T_3 will be free to be adsorbed by resin/sponge. ***The resin uptake in hyperthyroidism will be more, greater than 35 per cent.***
- ***In hypothyroidism:*** The reverse will occur. The proportion of I^{131}-T_3 taken up by the resin is inversely reduced and less than 21 per cent.
- Resin uptake of I^{131}-T_3 also gets influenced by drugs, hormones, pregnancy, etc. Thus ***false high result*** may occur in hypoproteinaemic states, Nephrotic syndrome and androgen therapy as TBG is decreased. Similarly ***false low result*** may occur where TBG is increased as in pregnancy, estrogen therapy and women on oral contraceptive pills.

(g) Plasma Tyrosine Level

Rivlin *et al* (1965) studied plasma tyrosine level in normal subjects and in thyroid disorders.

Interpretations

- ***Normal level:*** It was found to be from 11.8 ± 0.4 μg/ml.
- ***In hyperthyroidism:*** Plasma tyrosine level was found to be elevated in more than 70 per cent cases.

Increased tyrosine level in hyperthyroidism: Its Mechanism: It is suggested that excess thyroid hormones has inhibitory effect on hepatic and tissue *tyrosine transaminase,* as a result tyrosine catabolism is reduced and thus increasing plasma tyrosine level. The authors proposed the use of *tyrosine loading test* for hyperthyroidism and claimed that it is not influenced by age, sex, pregnancy or by previous iodides/radioisotope administration. Using "tyrosine loading test" the authors observed markedly increased plasma tyrosine level in cases of hyperthyroidism.

- ***In hypothyroidism:*** The decreased level of plasma tyrosine was observed (average 9.8 μg/ml).

TESTS BASED ON METABOLIC EFFECTS OF THYROID HORMONES

These tests are of much use where facilities for isotope techniques are not available.

(a) BMR: The test is helpful in diagnosis and is of particular value in assessing the severity and prognosis. At least two estimations consecutively after proper sedation and physical/mental rest will be helpful.

Interpretations

- A BMR between: 5 per cent and +20 per cent is considered as **normal.**
- ***In euthyroid states:*** –10 to +10 per cent of normal
- ***In hyperthyroidism:*** +50 per cent to +75 per cent is usually found.
- ***In hypothyroidism:*** Value below –20 per cent is suggestive (usually –30 per cent to –60 per cent seen in hypothyroid states).

(b) Serum Cholesterol Level

It is useful in assessment of hypothyroidism, where it is usually high. Not of much value in hyperthyroidism, though it is usually low. **Baron** has shown that 90 per cent of hypothyroidism cases have serum cholesterol greater than 260 mg%. He found poor correlation with severity as judged by BMR. In hypothyroidism, the synthesis of cholesterol is impaired, but its catabolism is reduced more, leading to high cholesterol level.

(c) Serum Creatine Level

Griffiths advocated the estimation of serum creatine level for diagnosis of hyperthyroidism, who considered a serum level greater than 0.6 mg% is diagnostic.

He compared serum creatine with BMR. A raised serum creatine, between 0.6 and 1.6 mg% may or may not be accompanied by increased BMR. ***He considered a normal serum creatine and normal BMR excludes thyroid dysfunction*** and held that when symptoms of thyroid disorders is present, a raised serum creatine is highly significant even though BMR is normal.

(d) Serum Uric Acid Level

Serum uric acid has been found to be increased in myxoedematous males and postmenopausal women, ranging from 6.5 to 11.0 mg%.

(e) Serum CK Level

Serum CK level are often raised in hypothyroidism but the estimation does not help in diagnosis. CK levels are also raised in thyrotoxic myopathy.

(f) Hypercalcaemia

It is very rarely found in severe thyrotoxicosis; there is an increased turnover of bone, probably due to direct action of thyroin hormones.

THYROID SCANNING

Scintiscans provide visualisation of the distribution of radioactive I_2 in the gland and also permits characterization of its anatomical features.

Advantages/Uses of Scintiscan

- Readily distinguishes the diffuse glandular activity from the patchy pattern seen in nodular goitres.
- The scan also permits functional classification of nodules as:
  - *'Hot' or 'warm':* Areas of increased uptake. Hot nodules suggest-increased thickness of the gland in those regions / or due to functioning adenoma or carcinoma.
  - *'Cold' nodules:* Which are due to reduced / or absent uptake. It may be due to cysts, haemorrhagic nodules, degeneration in an adenoma or carcinoma.
- In association with thyroid suppression regimes, helps to determine the TSH dependant or autonomous nature of the 'hot' / warm nodules.
- Scanning also provides useful information regarding size, shape, position of the gland.
- Facilitates identification and localisation of functioning thyroid tissues in "ectopic" or 'Metastatic' sites, e.g. in lungs and bones.

Use of 99^m technetium pertechnate: Recently, 99^m technetium pertechnate has been used. It has similar properties as I_2. Thyroid follicles 'trap' pertechnate ions, similar to I_2.

Advantages

- Radiation effect is low
- Has very short half-life of 6 hours
- Virtual absence of Particulate radiations.

Limitations

- Remains unaltered in the gland
- Cannot demonstrate retrosternal extension of thyroid, if any, due to attenuation of low energy γ-radiations passing through sternum.
- Fails to identify functioning metastasis from differentiated carcinomas of thyroid due to short ½ life and lack of fixation of ^{99m}Tc by the functioning metastasis.

IMMUNOLOGICAL TESTS FOR THYROID FUNCTIONS

I. Determination of Antithyroid Autoantibodies

Antithyroid autoantibodies are found in a variety of thyroid disorders, as well as, in other autoimmune diseases and certain malignancies. These autoantibodies are directed against several thyroid components and thyroid hormone antigens. They are:

- **Thyroglobulin (Tg)**
- **Thyroid microsomal antigen**
- **TSH receptor**
- **A non-thyroglobulin (non-Tg) colloid antigen**
- **Thyroid stimulating hormone (TSH)** and
- **Thyroxine (T_4).**

Of these antibodies, only anti-Tg (antithyroglobulin) and antimicrosomal autoantibodies are commonly used in evaluating thyroid status and function.

Anti-Tg autoantibodies are directed against thyroglobulin (Tg), a major constituent of thyroid colloid. Several different techniques are available and used in clinical laboratory to detect and quantify Tg-autoantibodies in blood.

They are mainly:

- **Agargel diffusion precipitation** (Fig. 39.1)
- **Tanned red cells haemagglutination test (TRCH Test)**
- **Enzyme-linked immunoabsorbent assay (ELISA)**
- **Immunofluorescence of tissue sections**
- **Radioimmunoassay (RIA) method.**

Most widely used method is based on haemagglutination.

- *Tanned Red Cells Haemagglutination Test (TRCH Test)*

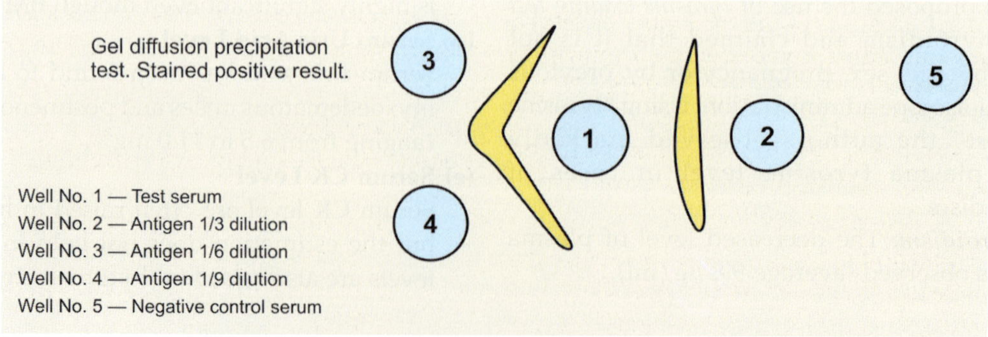

Fig. 39.1: Thyroid antibodies in thyroid diseases by gel diffusion

Principle: In TRCH test, an aliquot of patient's serum is mixed with erythrocytes that have been treated/coated with tannic acid and then quoted with purified human Tg-antigen.

When antibodies, if present in patient's serum, combine with tanned red cells coated with antigen, agglutination occurs which is visible as a 'carpet' at the bottom. Lack of agglutination is indicated by setting of the cells at the bottom as a compact button or ring.

Note: Use of Tg-coated erythrocytes makes the agglutination reaction much more sensitive than a simple antigen-antibody reaction.

Procedure

- Prior to testing, patient's serum is inactivated at 56°C × for ½ hour.

Note: Heating is important for inactivation of complement and thyroid binding globulin (TBG), which otherwise would interfere with the assay.

- A dried perspex tray with wells is taken. Serial double dilutions of the patient's inactivated serum is made to establish Tg-antibody titre.
- A suspension of tanned-red cells coated with Tg-antigen is put in each well.
- Tray is shaken and then kept in 4°C undisturbed for overnight.
- Reading is taken next morning.

Interpretation

- Titres are usually considered negative at less than 1 in 10 dilution ratio.
- The reported result is the highest dilution that causes agglutination *(carpet of red cells at bottom of the well).*
- The test is not highly specific and about 5 to 10 per cent of the normal population may have a low titre of Tg-autoantibodies with no symptoms of the disease.
- Reactivity occurs more frequently in Hashimoto's thyroiditis. It is positive in very high titre in more than 85 per cent of the patients.
- In Grave's disease (thyrotoxicosis) a high titre even greater than 1600 are common in more than 30 per cent of patients.
- Positive responses with high titre also observed in spontaneous adult myxoedema (primary) in more than 45 per cent of cases. In another 30 per cent cases titres may be low but positive.
- Weakly positive and low titres may also be found in patients with non-toxic goitre, thyroid carcinoma and pernicious anaemia.

ELISA and RIA methods: These methods have been developed for measuring anti-Tg antibodies. Correlate well with agglutination tests but are generally more sensitive and specific for thyroid autoimmune diseases. Some assays also allow identification of subclasses of Tg-antibodies. The clinical significance of these subclasses is still not clear.

II. Determination of Antimicrosomal Antibodies

Antimicrosomal antibodies are directed against a protein component of thyroid cells microsomes. These antibodies can be measured using:

- *Complement fixation test (CFT)*
- *Immunofluorescence of tissue sections*
- *Passive haemagglutination test similar to TRCH*
- *ELISA techniques*
- *Radioimmunoassays (RIA) method.*

(a) *Tanned Red Cells Haemagglutination Test—Using Microsomal Antigen*

Tanned erythrocytes agglutination method uses red cells coated with tannic acid and with microsomal antigen isolated from human hyperplastic thyroid glands. The procedure is simple and is easily carried out in clinical laboratory.

Interpretation

- Positive reactivity occurs in nearly all adult patients with Hashimoto's thyroiditis and in nearly 85 per cent of patients with Grave's disease.
- Low titres may, however, be seen in 5 to 10 per cent of normal asymptomatic individuals.
- When compared with TRCH test of Tg-antibody (as described above), the result of microsomal antibody is more frequently *positive* for thyroid autoimmune diseases and usually titres are much higher.

(b) *Complement Fixation Test (CFT)*

CFT is used also in clinical laboratory but not routinely as compared to TRCH test.

Limitations of anti-microsomal assays:

- Limited availability of human thyroid tissue
- Contamination of microsomal preparations with Tg.
- Presence of irrelevant thyroid antigens and auto-antibodies.

Approximate positivity reactions of TRCH (Tg) and CFT in normal and various thyroid disorders and other autoimmune disorders as reported in a study group are shown in the box on next page.

III. Newer Tests

Recently the following newer techniques have been put forward:

- ***Determination of antithyroid peroxidase antibody (anti-TPo antibodies)***
- ***Determination of thyrotropin-receptor antibodies (TRab)***

Comparison of positivity of TRCH and CFT in normal and disease states			
Group	**TRCH**	**CFT**	**Remarks**
• Normal (control group)	< 10%	< 10%	% may increase with age and more often in females
• Thyrotoxicosis	50%	80%	
• Myxoedema (primary)	43%	35%	
• Autoimmune thyroiditis	71%	92%	
• Non-toxic goitres and carcinoma of thyroid	< 10%	< 10%	
• Collagen diseases and other autoimmune disorders	< 10%	< 10%	

Note: It is important to realize that autoantibody presence only in high titre should be taken indicative of autoimmune thyroiditis.

(a) ***Determination of Antithyroid Peroxidase Antibody (Anti-TPo Antibody)***

In recent years, TPo has been identified and claimed as the main and possibly the only autoimmune component of microsomes. Its purification by using affinity chromatography and its production by recombinant technology has led to the development of ELISA and RIA methods for measuring anti-TPo antibodies.

Methods are easy to perform, provide greater sensitivity and specificity as compared to TRCH Tests, and can be used for "screening".

A suitable "immunometric assay" has been developed.

Immunometric Assay

Principle: Immunometric assay is based on competitive inhibition of the binding of radioiodinated TPo to an anti-TPo monoclonal antibody coated onto plastic tubes.

Advantages

- Easy to perform
- Assay is rapid (only 2 hours incubation period is required).

Result: The antibody concentration is expressed as units/ml.

Interpretation

- ***In normal healthy persons:*** The mean anti-TPo activity in serum is 69 $\pm$ 15 units/ml.
- Detectable concentration of anti-TPo antibodies are observed in nearly all patients with Hashimoto's thyroiditis, spontaneous adult myxoedema (idiopathic primary type) and in a majority of patients with Grave's disease.
- The frequency of detectable anti-TPo autoantibodies found in normals and nonthyroid cases is similar.

(b) ***Determination of Thyrotropin-Receptor Antibodies (TRAb)***

- The first indication that autoantibodies to TSH receptor plays a role in the pathogenesis of Grave's disease came with the discovery of LATS (long acting thyroid stimulator) in serum of some patients.
- Thyrotropin-receptor antibodies (TRAb) are group of related immunoglobulins (Igs) that bind to thyroid cell membranes at or near the "TSH receptor" site.
- These antibodies have recently been demonstrated frequently in patients with Grave's disease specially and also in other thyroid autoimmune disorders.

Note

- These antibodies show substantial heterogeneity.
- Some cause thyroid stimulation.
- Some others may have no effect or decrease thyroid secretion by blocking/inhibiting action of TSH.

Types of receptor antibodies: **Two types** have been described:

1. **Thyrotropin binding inhibitory immunoglobulins (TBI)**
2. **Thyroid stimulating immunoglobulins (TSIgs).**

Methodology: At present these abnormal antibodies, Igs cannot be differentiated by chemical or immunological methods. Their presence is determined by either: (i) radioreceptor assays; (ii) bioassays.

1. ***Thyrotropin-binding Inhibitory Immunoglobulins (TBI)***

- Determined by direct radioreceptor assay.
- The method assesses the capacity of Igs to inhibit the binding of radioisotope labelled TSH to its receptors in human or animal thyroid membrane preparations.
- In this method, detergent-solubilised porcine TSH-receptors and ^{125}I-labelled TSH are used.
- The ability of a purified fraction of serum Igs to displace ^{125}I-labelled TSH from the receptors is measured.

Interpretation

- Normal immunoglobulin G (IgG) concentrates do not produce significant displacement, and produces only less than 10 per cent inhibition.
- This method detects over 85 per cent of patients with Grave's disease.

2. *Thyroid Stimulating Immunoglobulins (TSIgs)*
   - *In vitro* bioassay utilised. The method assesses the capacity of the Igs (antibodies) to stimulate a functional activity of the thyroid gland such as adenylcyclase stimulation leading to increase in cyclic-AMP formation.
   - Measurement of increase in cyclic-AMP level can be done using human thyroid slices, frozen human thyroid cells culture or a cloned line of thyroid follicular cells.

Interpretations

- The effect of stimulation is expressed as a percentage of basal activity.
  In normal: Range is 70 to 130 per cent.
- *TSIgs* have been detected in 95 per cent of patients with untreated Grave's disease. It has been claimed to be highly *sensitive* and *specific* technique in diagnosing Grave's disease.
- TSIgs measurement has also been found to be useful for predicting relapse or remission in hyperthyroid patients.
- Also found useful for predicting the development of neonatal hyperthyroidism.

CLINICAL ASPECT

Practical Implications of Immunological Tests

Thyroid autoantibodies detection is of importance in diagnosis of the following conditions:

- In nodular goitres, detection of thyroid autoantibodies in high titres make the possibility of goitres being due to carcinoma less likely.
- Primary hypothyroidism can be differentiated from obesity and other hypometabolic states.
- Autoimmune thyroiditis diagnosis is confirmed.
- In differential diagnosis of endocrine exophthalmos other ocular lesions can be excluded.
- Serological tests may provide choice of line of treatment in patients with Grave's disease.

I. Questions (Essay type and short notes)

1. Describe the functions of thyroid gland. How would you investigate them?
2. Enumerate thyorid function tests. Write any one of them in detail.
3. Classify the various thyroid function tests an discuss briefly the tests used in thyroid disorders.

Short Notes

a. Thyroxine
b. Biosynthesis of thyroid hormones
c. T_3-suppression test
d. TSH-stimulation test
e. *In vitro* I^{131}-resin uptake test

II. MCQs (Give one correct answer)

1. Thyroid status mast accurately be mirrored by:
   a. Serum TBG concentration
   b. Serum T_4 level
   c. T_3 uptake
   d. Free thyroxine index
   e. Serum TSH level
2. The TSH does all the following except:
   a. Stimulate production of thyroid hormones.
   b. Promote iodide uptake by the thyroid gland.
   c. Promote irreversible bonding of hormones to globulin
   d. Promote conversion of iodide to iodine
   e. Promote coupling of iodinated tyrosine molecules
3. The production of T_3 and T_4 is normally influenced by all of the following except:
   a. Plasma thyroglobulin
   b. Plasma T_3 concentration
   c. Plasma somatostatin
   d. Stress
   e. Plasma T_4 concentration
4. Pregnancy, oestrogen therapy and oral contraceptives characteristically:
   a. Cause hyperthyroidism
   b. Cause hypothyroidism
   c. Have no effect on circulating T_4 concentration
   d. Change levels of circulating free T_4
   e. Increase circulating TBG
5. Most circulating T_3 and T_4 is bound to:
   a. Thyroxine-binding pre-albumin
   b. Thyroxine-binding globulin
   c. Albumin
   d. γ-globulins
   e. Thyroglobulin
6. Elevated T_4, lowered T_3 uptake together suggest the likelihood of:
   a. Hyperthyroidism
   b. Hypothyroidism
   c. Secondary hyperthyroidism d. Increased TBG
   e. Decreased TBG
7. Decreased T_4 and decreased T_3 uptake together suggest the likelihood of:
   a. Hyperthyroidism b. Hypothyroidism
   c. Increased TBG d. Decreased TBG
   e. Secondary hyperthyroidism
8. Hypothyroidism is characteristically associated with high serum:

a. Glucose b. Phosphate
c. Calcium d. Cholesterol
e. Osmolality.

9. A serum TSH level five times the upper limit of normal in the presence of a low T_4 and low T_3 upake:
a. Establishes the presence of primary hyperthyroidism
b. Rules out primary hypothyroidism
c. Establishes the presence of secondary hyperthyroidism
d. Establishes the diagnosis of primary hypothyroidism
e. Fails to do any of the above.

10. Hyperthyroidism is characteristically associated with:
a. High serum calcium b. Low serum phosphate
c. High serum creatine d. High serum glucose
e. High serum cholesterol.

11. High iodine uptake by the thyroid gland are caused by all of the following except:
a. Iodine deficiency
b. Prior tri-iodothyronine treatment
c. Prior intake of antithyroid drugs
d. Increased TSH
e. Thyroid enzyme defects.

12. Circulating thyroid autoantibodies are usually detected in good titre in patients with all of the following, except:
a. Hashimoto's thyroiditis
b. Chronic autoimmune thyroiditis
c. Spontaneous adult primary myxoedema
d. Thyrotoxicosis
e. Thyroid carcinoma

13. The resin-T_3 uptake may be elevated in presence of all except:
a. Anticoagulants b. Acidosis
c. Liver disease d. Pregnancy
e. Androgens.

Answers to MCQs

1. d 2. c 3. a 4. e 5. b 6. d
7. b 8. d 9. d 10. c 11. b 12. e
13. d.

CHAPTER

40

WATER AND ELECTROLYTE BALANCE AND IMBALANCE

Major Concepts

A. To learn the distribution of water and electrolytes in the body and their exchanges.
B. To learn the mechanism of normal water and electrolyte balance in health.
C. To study the regulatory mechanisms by which the water and electrolytes balance is maintained.
D. To study the abnormalities of water and electrolyte metabolism:
- Dehydration and
- Water intoxication.

Specific Objectives

A. 1. Learn the different body compartments.
2. Study the distribution of water in these compartments expressed as per cent of 'lean' body mass (fat-free mass).

N.B. Note 7.5 per cent of total body water (4.0 L) present in dense connective tissue, cartilage and bone does not exchange fluid and electrolytes readily.

3. Study what is meant by "transcellular" fluid.
4. Study briefly the methods (principles only) by which body water can be measured.
5. Define non-electrolytes, electrolytes and law of electrical neutrality.
6. Study the types of solutes present in body fluids.
7. Study the electrolyte composition of ECF (plasma and tissue fluids) and ICF

N.B. Note that:
- Total electrolytes concentrations is higher in ICF as compared to ECF.
- Chief cations in ECF is Na^+ and chief cations in ICF are K^+ and Mg^{++}.

8. Study the movement or exchanges of water and electrolytes from one compartment to other in health.
   - Learn the properties of vessel wall (called as "rapid membrane" by Darrow) and the cell wall (called as "slow membrane" by Darrow) as regards water and electrolytes movements.
   - Study the fluid exchange that takes place at the capillary beds ("Starling" hypothesis).
   - Study the mechanism of movement of electrolytes mainly sodium and potassium into the cells and out of the cells.

B. Body water is constantly carrying out exchanges with external environment.
1. Study the various sources of Intake of water.

   Note
   What is "metabolic" water?
2. Study the various processes by which water is lost from the body, i.e. output of water.
3. Note what is meant by "obligatory losses" and how much is this loss?
4. Note the difference between "insensible perspiration" and "sensible" perspiration (sweating).
   Normally intake of water greater than the loss and the surplus is excreted by the kidneys. Thus, ***in health, the urinary volume largely depends on intake of water.***
5. Note what is meant by "minimum excretory volume" and study the factors that determine this volume.
6. Study how the electrolyte balance is maintained in health, the organs principally involved.
   - Learn what is meant by "internal circulation of salts" by GI tracts and kidneys.

C. 1. Learn the various regulatory mechanisms that operate to maintain the homeostasis.
   - Neural mechanisms: "thirst" mechanisms
   - Humoral mechanisms • Role of antidiuretic hormone (ADH) • Role of aldosterone: The mineralocorticoid • Study how aldosterone secretion is regulated.
   - Study the role of kinins and prostaglandins in water and electrolyte metabolism.
   - Study what is "atrial natriuretic peptide" (ANP)

D. Abnormalities of water and electrolyte metabolism can produce: (a) dehydration (b) water intoxication.
(a) 1. Study what is meant by dehydration.

2. Study the types of dehydration as per classification given by Marriott:
   - Pure water depletion (primary dehydration)
   - Pure salt depletion (secondary dehydration)
   - Mixed type
3. Learn the causes, pathophysiology, clinical and biochemical findings of each type.
4. Differentiate primary dehydration from secondary dehydration in a tabular form.
5. Learn how from clinical findings an approximate estimate of water and salt loss can be made.
6. Note the type of fluids to be administered in each type of dehydration.
7. Learn what is "Fantus" test.

N.B. Pathologically water and electrolyte imbalance can be of **six types:** Hypotonic expansion, isotonic expansion, hypertonic expansion, hypotonic contraction, isotonic contraction and hypertonic contraction.

(b) Study what is water intoxication, its causes and clinical feature.

DISTRIBUTION OF BODY WATER AND ELECTROLYTES

DISTRIBUTION OF BODY WATER

Total body water in an adult of 70 kg varies from 60 to 70 per cent (36-49 litres) of total body weight, when expressed as percentage of ***lean body mass***, i.e., sum of the "fat-free tissue." The body water can be visualised to be distributed mainly in two "compartments", viz.

(a) ***Intracellular fluid (ICF):*** The fluid present in the cells which is approx. 50 per cent (35 L), and

(b) ***Extracellular fluid ECF:*** The fluid present outside the cells which constitutes approx 20 per cent (14 L). The extracellular fluid (ECF) is considered to be present in the two compartments as follows:

1. ***Plasma:*** The fluid present in heart and blood vessels, approx, 5 per cent (3 L) and
2. ***Interstitial tissue fluid (ITF):*** 15 per cent (11 L).

Distribution of body water is shown in Figure 40.1.

More recently, it has become clear that although the concept of a single homogenous compartment of the body water is still useful, the extracellular compartment must be recognised as more "heterogenous" and should be subdivided into **four main subdivisions.**

1. ***Plasma,***
2. ***Interstitial and lymph fluid,***
3. ***Fluid of dense connective tissue, cartilage and bones,*** and
4. ***Transcellular fluid***

1. ***Plasma volume:*** This comprises in general the fluid within the heart and blood vessels.
2. ***Interstitial and lymph fluid:*** This is considered to represent an approximation of actual fluid environment outside the cells.
3. ***Fluid of dense connective tissue, cartilage and bones:*** Due to differences in structure and relative avascularity, the fluid present in these tissues does not exchange fluid and electrolytes readily with remainder of body water. ***This includes approx. 4.0 L (7.5 per cent of total body water) and should be considered a "distinct subdivision of ECF".***
4. ***"Transcellular" fluid:*** A variety of extracellular fluid collections formed by the "transport" or "secretory activity" of cells. **Examples are:**
   - Fluids found in salivary glands, pancreas, liver and biliary tract, skin, mucous membrane of Respiratory and GI tracts; and
   - The fluids present in "spaces" within the eyes (aqueous humour), cerebrospinal fluid (CSF) in spinal canal and ventricles of brain, and that within the lumen of GI tract (mostly reabsorbed and not lost).

Thus excepting the fluid of dense connective tissue, cartilage and bone (4.0 L) remaining part of ITF (7.0 L) and plasma (3.0 L) constitute the ***active and mobile part of ECF.***

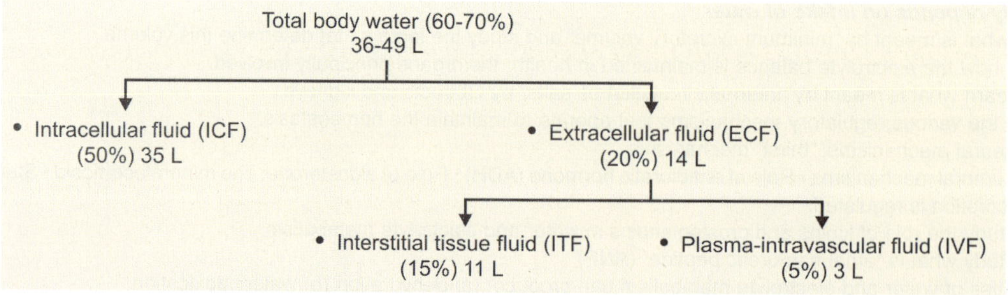

Fig. 40.1: Distribution of body water in an adult of 70 kg

MEASUREMENT OF BODY WATER

(a) **Total body water**: This has been estimated by various methods:
- Has been determined in animals by desiccation,
- Recently, the distribution of heavy water, deuterium oxide (D_2O) or tritium oxide (^3H_2O) has been used in the Living animal and in human subjects as a method of measuring total body water.
- A drug called ***antipyrine*** has also been used.

Considerable variations are obtained when different subjects are compared even by the same analytical method. This is due mainly to variations in the amount of fat in the body. The higher the fat content of the subject, the smaller the % of water that subject will contain in his/her body. If a correction for the fat content of the subject is made, the total body water in various subjects is relatively constant, 60 to 70 per cent of body weight, when expressed as % of "lean body mass", i.e. the sum of fat free tissue.

(b) **Plasma volume:** This may be measured by the following methods:
- *Evans Blue dye (T-1824):* In this procedure a carefully measured quantity of the dye is injected IV After a lapse of time, to allow for mixing, a blood sample is withdrawn and the concentration of the dye in plasma is determined colorimetrically.
- *Other methods:* For plasma or blood volume measurements are based on IV injection of radio-phosphorus P^{32}-labelled RB cells or radio-iodinated labelled I^{131}-human serum albumin (RIHSA). These substances distribute themselves in blood/plasma and after a mixing period of ten or more minutes, their volume of distribution may be calculated from their concentration in an aliquot of blood or plasma. **Normal plasma volume is 40 to 50 ml/kg body weight.**

(c) **ECF volume:** The volume of ECF may be measured by dilution of a substance:
- Which does not penetrate into the cells,
- Which is distributed rapidly and evenly in all of the plasma as well as the remainder of the EC fluid. No such ideal substance has yet been found. However, the volume of distribution of certain 'saccharides' such as ***mannitol*** or ***inulin*** has been found to give reasonably accurate measurement of the volume of interstitial and lymph fluid.

(d) **ICF volume:** Intracellular water volume is calculated simply as the difference between the total body water and extracellular water.

DISTRIBUTION OF ELECTROLYTES IN THE BODY

Non-electrolytes, such as glucose, urea, etc. do not dissociate in solution. While substances like NaCl, KCl in solution dissociate into sodium (Na^+), potassium (K^+) and chloride (Cl^-) ions, they are called as ***electrolytes***. Water molecules completely surround these dissociated ions and prevent union of +vely charged particles with –vely charged ones. ***The +ve ions are called cations and negatively (–vely) charged ions are called anions.***

Law of electrical neutrality: Fluid in any body compartment will contain equal number of +vely charged and –vely charged ions.

Solutes in body fluids: Solutes in body fluids are mainly of *three categories.*

1. *"Organic" compounds of small molecular size* like glucose, urea, uric acid, etc. They are "nonelectrolytes" as they do not dissociate or ionise in solution. Since these substances diffuse relatively freely across cell membrane they are not important in the distribution of water. If it is present in large quantities, however, they aid in retaining water and thus do influence total body water.
2. *"Organic" substances of large molecular size,* mainly the ***proteins.*** Effect of protein fractions of the plasma and tissues is mainly on the transfer of fluid from one compartment to another and not on the total body water.
3. *Inorganic "electrolytes":* Because of the relatively large quantities of these materials in the body, ***they are by far the most important both in distribution and retention of body water.***

Electrolytes composition of ECF: Both plasma (IVF) and tissue fluid (ITF) may be considered as ***one single compartment*** for all practical purposes as both resemble each other and both differ grossly from ICF. ***Electrolytes composition of TF is similar to plasma except that Cl^- largely 'replaces' proteins as anion. Predominant cation is Na^+*** (Table 40.1).

Table 40.1: Electrolytes of plasma and tissue fluid

Cations mEq/L			*Anions mEq/L*		
(a) Plasma					
Na^+	=	143	Cl^-	=	103
K^+	=	5	HCO_3^-	=	27
Ca^{++}	=	5	$HPO_4^=$	=	2
Mg^{++}	=	2	$SO_4^=$	=	1
Total	=	**155**	$Proteins^-$	=	16
			Organic $acids^-$	=	6
			Total	=	**155**
(b) Tissue fluid					
Na^+	=	145	Cl^-	=	116
K^+	=	5	HCO_3^-	=	27
Ca^{++}	=	3	$HPO_4^=$	=	3
Mg^{++}	=	2	$SO_4^=$	=	2
Total	=	**155**	$Proteins^-$	=	1
			Organic $acids^-$	=	6
			Total	=	**155**

Electrolytes composition of ICF: ICF contains 195 mEq of cations and anions. Values of different electrolytes in

ICF differs in different tissues. But chief cations are K^+ and then Mg^{++}. These are balanced by the chief anions $PO_4^=$ and next by Pr^-. About two-thirds of K^+ within cells is "protein-bound", while remaining one-third is 'free' which exchanges with ECF. Cl^-, HCO_3^- and Na^+ are present in intracellular fluid only in minimal amounts. Total electrolytes concentration is higher than that of the ECF (Table 40.2). The phosphates of the cells are phosphoric esters of hexoses, creatine phosphate, ATP and inorganic phosphates.

Table 40.2: Electrolytes of ICF

Cations mEq/L			*Anions mEq/L*		
K^+	=	150	$HPO_4^=$	=	110
Mg^{++}	=	40	Protein⁻	=	50
Na^+	=	5	$SO_4^=$	=	20
Total	=	**195**	HCO_3^-	=	10
			Cl^-	=	5
			Total	=	**195**

Normal Fluid and Electrolytes Exchanges in the Body

Under normal conditions in health, the relative volumes of water in above three compartments is kept constant (Fig. 40.2).

Water can pass freely through the membrane which divide plasma from tissue fluid; and tissue fluid from intracellular fluid; ***but distribution of water is controlled by the osmotic pressure exerted by substances present in each compartment, i.e. the electrolytes and protein molecules.***

- Membrane separating ICF from tissue fluid is ***Semi-permeable*** called as ***slow*** membrane by **Darrow**, and it allows only passage of water but not electrolytes and protein molecules in health. Normally, there is osmotic equilibrium between these two compartments, but if this is disturbed, water is drawn from the compartment with lower osmotic pressure into that with higher osmotic pressure until equilibrium is restored. ***The osmotic imbalance between these two compartments results in water being either sucked out of the cells (producing cellular dehydration) or water is drawn into the cells (producing cellular oedema) to restore the balance.***
- Membrane separating the vascular compartment from tissue fluid is more ***permeable***, called as ***Rapid*** membrane by **Darrow,** to water and electrolytes but not to protein molecules.

Starling hypothesis: There is also fluid exchange taking place at the 'capillary beds'. The mechanism Starling hypothesis, provides for a continual circulation of fluid between the capillaries and the tissue spaces, a balance being maintained between the quantity of water filtered and that reabsorbed. ***Net filtration pressure that drives fluid out at the arterial end is 7 mmHg,*** (22 mmHg Hydrostatic pressure–15 mmHg osmotic pressure), hydrostatic pressure being greater than osmotic pressure. On the other hand, at the venous end, ***net absorption pressure which absorbs fluid is 8 mmHg,*** (15 mmHg osmotic pressure –7 mmHg hydrostatic pressure), osmotic pressure is greater than hydrostatic pressure (Refer chapter on Plasma Proteins).

Electrolytes Movement In and Out of Cells

- Much higher concentration of Na^+ and Cl^- in interstitial fluid and K^+ in intracellular fluid are accompanied by a difference in electrical potential. The resting skeletal muscle cells being about 90 mv –ve to the interstitial fluid. It is believed that the Lipid-protein membrane plays an important role in determining and maintaining these differences in concentration and potential.
- K^+ ions tend to diffuse out of and the Cl^- ions into the cells because of their concentration gradients, but this is almost exactly counter-balanced by a tendency to diffuse in the opposite direction due to the difference in electrical potential, i.e. the relative negativity on the inside of the cells tend to keep Cl^-out and K^+ in.

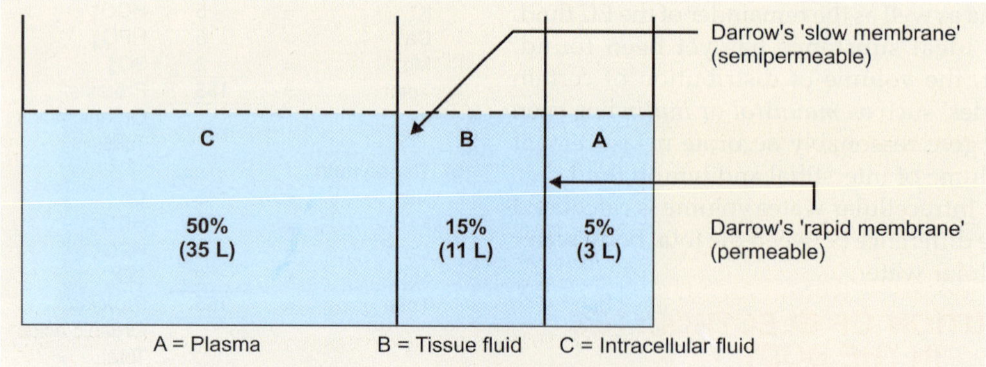

Fig. 40.2: Distribution of water in three compartments schematically

- ***In the case of Na^+, however, diffusion into the cells is favoured by both the concentration gradient and electrical potential.*** Cells do not allow accumulation of Na^+, hence under normal healthy conditions, there must be some mechanism for removing Na^+ from the cell, virtually as rapidly as it enters. Since this has to be accomplished in opposition to forces of concentration and electrical potential, it involves expenditure of energy, derived from cellular metabolism. This process of "Active transport" of Na^+out of cells (Pumping out) is done by the **Sodium Pump,** which effectively extrudes Na^+ from the intracellular fluid. This extrusion of Na^+ from the cell is associated with splitting of ATP by "Na^+ – K^+ *ATPase*" located at the inner surface of the cell membrane. The enzyme is activated by Mg^{++}, has a molecular weight of 2,50,000 to 3,00,000. It consists of two large subunits viz. one with mol. weight 1,00,000 to 1,30,000 called **α-subunit** and another smaller subunit, mol. wt. 55,000, a glycoprotein called **β-subunit.** ***It is the α subunit which is catalytically active.*** The energy of hydrolysis of ATP is used by the transport mechanism for the coupled-exchange of Na^+ for K^+ ions between the intracellular and tissue fluids.

NORMAL WATER BALANCE

Body water is constantly exchanged with external environment.

INTAKE OF WATER

- Water is normally absorbed into the body from the bowel (taken by mouth as water and beverages)
- Water taken in cooking in cooked foods,
- ***Metabolic water: Formed from oxidation of food stuffs.*** Each gram of carbohydrates, fats and proteins yield 0.55 gm, 1.06 gm and 0.45 gm of water respectively on complete oxidation. In ml., on oxidation of 1 gm of carbohydrates, fats and proteins produces 0.56 ml, 1.07 ml, and 0.34 ml of water respectively. ***In general 10 to 15 ml of water are produced per 100 calories of energy produced.***

OUTPUT OF WATER

Water is lost from the body constantly from various routes, they are as follows:

- ***Via kidney as urine:*** 1000 to 1500 ml in 24 hours.
- ***Via skin as "insensible perspiration":*** 600 to 800 ml of water in 24 hrs.

  Note: Frank sweating is abnormal. Sweat is a "hypotonic" solution, 30 to 90 mEq/litre of NaCl are lost in sweating. In "insensible perspiration" there is no loss of salt, it is equivalent to distilled water.
- ***Via lungs in the expired air:*** Approx 400 to 600 ml of water is lost in 24 hours.
- To a minor degree approx 100 to 150 ml of water is lost in 24 hrs from large intestine ***in faeces*** (Table 40.3).

Normally in health, the intake of water is more than the loss via skin, lungs and faeces and the surplus is excreted by the kidneys. Thus, **in health, the urinary volume largely depends on intake of water.**

If the intake of water is low or excessive amounts are lost via extrarenal channels, the excretion of urine is diminished until only sufficient is excreted to eliminate the "waste products" (metabolic loads) of metabolism. Urinary volume may be reduced to 500 to 600 ml in 24 hours and this is called as ***minimum excretory volume. The exact quantity will depend on:***

- ***"Concentrating" power of the kidneys*** and
- ***The quantity of "waste materials" required to be eliminated (solute load).***

Obligatory Loss

The loss through expired air (minimum 400 ml), by insensible perspiration through skin (minimum 600 ml), loss through faeces (minimum 100 ml) and the minimum excretory volume of kidney to eliminate waste products, i.e. 500 ml is called as ***obligatory losses*** (Approx 1600 ml). This loss will continue as long as the individual is surviving.

NORMAL ELECTROLYTE BALANCE

Though human systems consume fluids and food which vary markedly both in quality and quantity, electrolyte

Table 40.3: Average water intake and output in an adult

Intake		Output	
• Fluid by mouth as water and beverages	1000-1500 ml	• Urine (via kidney)	1000-1500 ml
• Water in cooking	700 ml	• Lungs	400 ml
• ***Metabolic water***	400 ml	• Skin (insensible perspiration)	600 ml
		• Faeces	100 ml
	Total = 2100-2600 ml		**Total = 2100-2600 ml**

levels in subjects from any two widely located regions of the world are within narrow normal ranges. The organs which are constantly regulating the electrolyte levels are the:

(i) Intestine, and (ii) the kidneys, process is termed as the ***internal circulation of salts.*** Principal ECF ions enter the lumen of GI tract and renal tubules and their near complete reabsorption regulate electrolyte levels.

(a) GI Tract: About 8 litres of fluid of different electrolytes enter GI tract every day and are reabsorbed almost completely with fluid loss approx 100 to 150 ml, and electrolyte loss of Na^+ approx 10 to 30 mEq and of K^+ approx 10 mEq (Table 40.4).

Table 40.4: Electrolyte composition of various secretions of GI tract

Secretions	*Volume ml/day*	*Electrolytes in mEq/L*			
		Na^+	*K^+*	*Cl^-*	*HCO_3^-*
• **Saliva**	1500 ml	33	20	34	—
• **Gastric juice**	2500 ml	70	10	90	10
• **Bile**	500 ml	145	5	100	40
• **Pancreatic juice**	700 ml	145	5	70	115
• **Intestinal juice**	3000 ml	140	5	110	25
	8200 ml	**533**	**45**	**404**	**190**

Loss through Faeces : Fluid : 100 to 150 ml
Na^+ : 10 to 30 mEq
K^+ : 10 mEq

(b) Kidneys: *Internal circulation of salts* constantly occurring in kidneys is at a much faster rate than that observed in GI tract. In kidneys, a volume of plasma equal to ECF (12- 15 L) is filtered and reabsorbed every 2 hours and about 25,000 mEq of Na^+ are filtered and reabsorbed everyday (Table 40.5).

Table 40.5: Glomerular filtration and tubular reabsorption of water and Na^+

Substances	*Filtered per day*	*Excreted per day*	*Reabsorption*
• **Water**	180 L	1 L	99.4%
• **Na^+**	180 × 140 mEq (25,000 mEq approx)	100 mEq	99.6%

Na^+ is reabsorbed from the renal tubules in exchange with H^+ and NH_4^+ in proximal and distal tubules respectively by the following mechanisms:

- H^+ exchange against bicarbonate (bicarbonate system)
- H^+ exchange against Na_2HPO_4 (phosphate system) and
- Ammonia mechanism against NaCl (Refer Chapter on Acid Base Balance and Imbalance).

REGULATORY MECHANISMS

1. In health, the volume and composition of various body fluid compartments are maintained within physiological limits even in the face of wide variations in intake of water and solutes.
2. Osmolarity of ICF is determined mainly by its K^+ concentration, while that of ECF by Na^+ concentration. If the volumes of these compartments are to be maintained at constant levels, a mechanism must be provided for adjustments in excretions of not only of water but also of Na^+ and K^+ in response to variations in amounts of each supplied to the organism. These adjustments are accomplished mainly by the kidneys. ***The kidneys respond promptly to deviations in osmolarity or individual ions concentration in ECF.***
3. ***Homeostasis of body fluids:*** Therefore, involves mechanisms that:
   - Responds to fluctuations in volume, as well as,
   - To changes in concentration of total solutes or of individual ions.
4. Current concepts of the nature of the regulatory mechanisms include the existence of ***receptors*** sensitive to variations in:
   - Osmolar concentration ***(osmoreceptors)***
   - Or individual ions ***(chemoreceptors)*** concentration in ECF and,
   - To local or general variations in intravascular pressures ***(baroreceptors)***
   - And plasma/or ECF volume ***(volume receptors or stretch receptors)***

The intrarenal mechanisms concerned with excretion of water and solutes may be influenced by stimuli initiated in these receptors either:

- By direct neural connections, or
- Through the medium of "humoral factors", i.e. alterations in production and release of certain hormones, these are **mainly two:**
  1. ***Antidiuretic hormone (ADH) or Vasopressin,***
  2. ***Aldosterone,***

The former regulating the excretion of water and the latter Na^+ and K^+.

A. Thirst mechanism (neural mechanism): The intake of fluid is regulated by the mechanism of "thirst". ***A thirst centre is located in III ventricle*** which regulates the amount of water consumed as water or beverages. A deficient intake of water with continuing "obligatory losses" leads to concentration of body fluids with respect to solutes and a rise in osmotic pressure. This tends to draw water from ICF, the dehydration of the cells seem to be the main stimulus for thirst mechanisms through osmor eceptors as well as sensory nerves of mouth and pharynx (IX and X), which respond to dryness of the mouth and pharynx.

B. **Antidiuretic hormone (vasopressin):** A protein hormone, an octapeptide produced by the supraoptic nuclei in hypothalamus, one-sixth amount can be produced by paraventricular nuclei also. The hormone is stored in posterior pituitary as granules in combination with two proteins **neurophysin I and II**. The hormone exerts an antidiuretic effect (Refer to Chapter on Chemistry of Hormones).

Role of prostaglandins: Antidiuretic action of vasopressin is modulated by PG's. Vasopressin stimulates PG synthesis in kidney and produces PG-E_2, which in turn decreases the cyclic-AMP level in tubular epithelial cells thereby opposing the cyclic-AMP mediated anti-diuretic effect of vasopressin.

C. **Aldosterone:** Aldosterone, a steroid hormone, classified as Mineralocorticoid, produced by zona glomerulosa of Adrenal cortex has the most important effect on mineral metabolism. The hormone increases the rate of tubular reabsorption of Na^+. (Refer to Chapter on Chemistry of Hormones).

D. **Role of kinins:** Kinins also take part in homeostatic control of blood pressure, ECF volume and metabolism of salt and water. Chief kinins are:
- ***Bradykinin:*** A nona peptide and
- ***Kallidin:*** A decapeptide, which is *lysyl bradykinin*. Kinins are inactivated in a few minutes by plasma and tissue *Kininases* and plasma carboxy *peptidases*. Kallidin can also be hydrolysed by a plasma *amino-peptidase* into Bradykinin and Lysine. Bradykinin can also be inactivated by Angiotensin I and converting enzyme.

Formation of kinins: Kinins are formed from an 'inactive' pre-enzyme called *pre-kallikreins*. It is activated when blood is diluted or comes in contact with foreign surfaces or get its cells damaged. On activation pre-kallikreins is converted to an active proteolytic enzyme called *kallikreins*, in blood/tissue fluids. This in turn, acts on serum α_2-globulin, called 'kininogens' (kallidinogens) present in plasma/tissue fluids and converted to kinins: **Bradykinin** and **kallidin** (Fig. 40.3).

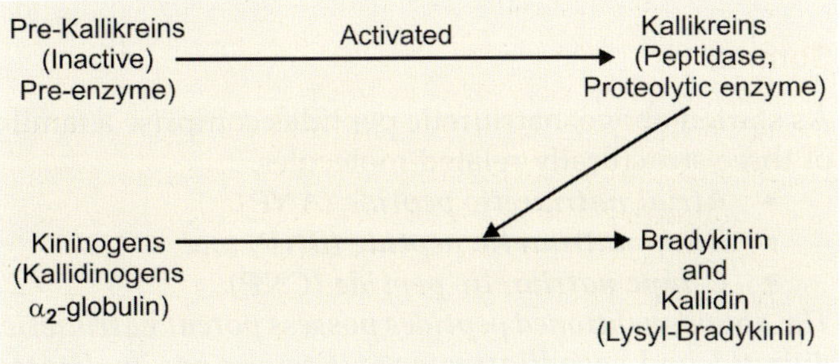

Fig. 40.3: Formation of kinins

Kinins produced in the kidney increase salt and water excretion. (Natriuretic effect and diuretic effect). action of kinins is just opposite of A D H and aldosterone.

Role of PGs in Kinin Formation in Kidney

Prostaglandins modulate the activity of Kinins in kidney; distal tubular cells of kidney secrete "kallikreins" into the tubular lumen. The enzyme hydrolyses 'kininogens', present in tubular filtrate to 'kinins'. Considerable concentration of kinins are found in tubular filtrate in distal tubules and collecting tubules. Kinins stimulate the activity of *phospholipase A_2* and releases PGE-2, which in turn mediate the renal effects of kinins, viz. natriuresis and diuresis. Kinins also participate in conversion of PG-E_2 to PG-$F_2\alpha$ in renal medulla by stimulating the enzyme *PGE-9 keto-α-reductase* and thus lower the PG-E_2/PGF-2 α ratio. ***PGF-2 α has no renal effects—thus kinins modulate and reduce their own vasodilator, diuretic and natriuretic effects.*** State of Na^+ balance determines the capacity of kinins to activate the enzyme reductase.
- ***High Salt intake*** inhibits *reductase* activity and PGE-2 action is more
- ***Low salt diet*** and a fall in ECF volume stimulates *reductase* activity producing more PGF-2α and this, in turn decreases the diuretic and natriuretic effects.

E. Natriuretic Peptides

Three natriuretic peptides have been found. They are:
1. **Atrial natriuretic peptide (ANP)**
2. **Brain natriuretic peptide (BNP)**
3. **C-type natriuretic peptide (CNP)**

1. Atrial Natriuretic Peptide (ANP)
- Also called—**Atriopeptin**
- **Atrial natriuretic hormone (ANH)**
- Atrial natriuretic peptide (ANP) is ***a peptide of 28 amino acids.*** It has a 17-amino acid ring in the middle of the molecule formed by S-S bond between two cysteine molecules at positions 7 and 23.
- ANP was discovered in **1983 by Adolfo J de Bold in Canada.**
- ANP is produced, stored and released by the atria of the heart by **atrial myocytes).**
- ANP is **secreted in response to:**
  1. Atrial distension and stretching.
  2. Raised sodium concentration (*hypernatraemia*)
  3. Sympathetic stimulation of β-adrenoceptors.
  4. Angiotensin II and
  5. ***Endothelin***—a potent vasoconstrictor.
- **Receptors:** ***ANP is closely related to BNP and CNP. Three call surface Receptors have been identified*** on which ANP act. They are: **ANP_A,**

SECTION FIVE

ANP_B and **ANP_C**. ANP_A and ANP_B Receptors have **guanylate cyclase** activity and ***mediate the biological effects by producing C-GMP.*** ANP and BNP act through the ANP_A Receptor and CNP acts through ANP_B receptor. ***All natriuretic peptides are bound by the ANP_C receptor.*** ANP_C receptor functions mainly as a clearance receptor by bonding and sequestering ANP from the circulation.

Functions of ANP

1. **Renal:** Main action is renal; **it dilates the afferent glomerular arteriole,** constricts the efferent glomerular arteriole and relaxes the mesangial cells. This increases pressure in the glomerular capillaries, thus **increasing glomerular filtration rate** (GFR↑) resulting in greater excretion of Na **(*natriuresis*)** and water **(diueresis** GFR↑**)**
2. *Decreases sodium reabsorption in the distal convoluted tubules.*
3. Inhibits renin secretion thus inhibiting renin-angiotensin system
4. Reduces aldosterone secretion by adrenal cortex.
5. **Vascular:** A **powerful Vasodilator,** relaxes vascular smooth muscle in arterioles and venules by: 1) membrane receptor-mediated elevation for vascular smooth muscle C-GMP. 2) and inhibition of the effects of catecholamines.
6. **Adipose tissue:** Increase the release of free fatty acids (FFAs) from adipose tissue **(*lipolysis increases*)**. Plasma concentration of glycerol and non-esterified fatty acids (NEFAs) are increased by IV infusion of ANP in humans.
   - **Degradation**
     Regulation of the effects of ANP is achieved through gradual degradation of the peptide by the enzyme ***neutral endopeptidase (NEP)***

Note

Recently NEP inhibitors have been developed which may be useful in future ***for treating congestive heart disease***

2. **Brain Natriuretic Peptide (BNP)**
   Also called now as B-type natriuretic peptide.
   - BNP is a peptide consisting of **32 amino acids, secreted by the ventricles of the heart in response to excessive stretching of heart muscle cells. BNP is so named because it was originally identified in extracts of porcine brain.**
   - It is produced in humans by cardiac ventricles as a *precursor 76 amino acid N-terminal fragement (NT-ProBNP) which is biologically inactive.*
   - **Receptors:** BNP binds to and activates the same atrial antinatriuretic factor receptors (NPRA) and to a lesser extent NPRB. Affinity of BNP to the receptor is 10 times lesser than that or ANP.
   - **Action:** Physiological action of BNP is similar to ANP and includes:
   - Decrease in systemic vascular resistance and cenntral venous pressure.
   - **An increase in natriuresis↑:** Thus the net effects of BNP and ANP is an increase in cardiac output and decrease in blood volume.
   - Excretion: BNP is cleared by binding to natriuretic peptide receptors (NPRs) and neutral peptidase (NEP)
   - Less than 5 per cent of BNP is cleared renally. BNP can be elevated in renal failure

CLINICAL ASPECTS

Both BNP and NT-ProBNP levels in the blood are used for:
- Screening and diagnosis of acute congestive heart failure (CHF)
- Also used to establish prognosis in heart failure as both markers are typically higher in patients with worst outcome.
- Plasma concentrations of both BNP And NT Pro-BNP are typically increased in patients with asymptomatic/ symptomatic left ventricular dysfunction.
- BNP plays a very important role in prognostication in diabetic cases.
- Also has an important role in prognostication of heart surgery patients.

3. **Natriuretic peptide C:** Also called ***C-type natriuretic peptide-CNP***
   - **Precursor molecule** is natriuretic peptide precursor C, also called **NPPC**—a protein that in humans is **encoded by the NPCC gene.**
   - NPCC precursor molecule is cleaved to form the C-type natriuretic peptide (CNP) having 22 amino acids.

Note

Unlike ANP and BNP, CNP does not have direct natriuretic activity. This is because CNP is a selective agonist for the B-type natriuretic receptor (NPRB), whereas ANP and BNP are selective for NPRA.

Summary

As started above, natriuretic peptides comprise a family of three structurally related molecules.
- ***Atrial natriuretic peptide (ANP)***
- ***Brain natriuretic peptide (BNP) and***
- ***C-type natriuretic peptide (CNP)***

The above-mentioned peptides possess potent natriuretic, diuretic, and vasodilating activities and are implicated in body fluid homeostasis and blood pressure control.

SECTION FIVE

ABNORMAL WATER AND ELECTROLYTE METABOLISM

Abnormalities can be **two types:**

1. *Dehydration:* Due to loss of water, or electrolytes or both and
2. *Water intoxication.*

DEHYDRATION

Dehydration is a disturbance of water balance in which the output exceeds the intake, causing a reduction of body water below the normal level. Although the term implies loss of body fluid, the clinical state to which it refers is more than this; because, characteristically there is an accompanying disturbance of electrolytes.

Dehydration may be the result of:

- **Pure water depletion,**
- **Pure salt depletion or**
- **Mixed type in which both occur.**

Marriott called pure water depletion as ***primary dehydration***, and pure salt depletion as ***secondary dehydration***.

The following discussion will stress the two forms of depletion as separate entities in order to understand and emphasise certain basic pathophysiologic principles. However, it should be noted that most patients with dehydration have mixed type of depletion, with one or the other predominating.

A. Pure Water Depletion (Primary Dehydration)

Definition: Pure water depletion occurs when water intake is stopped or water intake is inadequate and there is **no parallel loss of salt** in the secretions from the body.

Causes

- When a patient is too weak or too ill to satisfy his/her water needs,
- In mental patients who refuse to drink,
- In cases of coma, dysphagia (difficulty in swallowing),
- In individuals lost in desert or shipwrecked.

Pathophysiology

Water depletion occurs almost always because of lack of intake, rather than because of losses from the body. When a person stops intake of water, body water stores become depleted, because of the ***continuing obligatory losses*** and later on, supplemented by the continued excretion of "minimal volume" of urine required for excretion of "metabolic loads". Only source of water supply to the body in the complete absence of intake becomes the water obtained from oxidation of food stuffs (metabolic water). As obligatory water loss continues, the concentration of the electrolytes rises in the ECF, which becomes ***hypertonic (or hyperosmolar).*** Water flows from ICC to ECC to correct this imbalance and to maintain uniform osmotic pressure throughout the body. ***Thus the volume of ECF is maintained almost to normal at the expense of ICF which is grossly reduced in volume causing intracellular dehydration*** (Fig. 40.4).

Clinical and Biochemical Findings

- *Thirst:* ***It is the earliest symptom*** due to intracellular dehydration. Dehydration is shown by a dry tongue and 'pinched' facies.
- *Oliguria:* Hyperosmolarity stimulates the release of ADH, which causes reabsorption of water from kidney tubules, causing a gradual diminution of urine volume.
- A normal or slightly increased blood urea, a normal or slightly reduced plasma volume, may occur.
- There is usually no circulatory collapse or fall in BP seen as the plasma volume is maintained.
- *Urinary chlorides:* It is important to note that ***in this type of dehydration urine will contain NaCl, rather it is to the higher side.***

Death: Occurs when water loss amounts to approx 15 per cent of body wt. (about 22% of total body water), which happens on about the 7th to 10th day of complete water deprivation and if not treated.

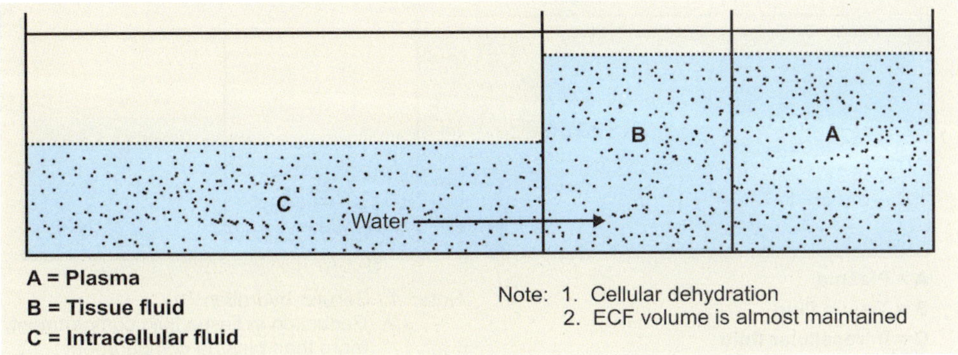

Fig. 40.4: Schematically distribution of water in three compartments in primary dehydration (cf Fig. 40.2)

Table 40.6: Estimation of water loss in primary dehydration from clinical manifestations

Clinical phases	*Clinical features*	*Estimated deficit*
1. **Early**	Thirst +	2% of body wt (approx 1.5 L)
2. **Moderately severe (72 to 96 hrs without water)**	Thirst ++ Dry mouth pinched facies, oliguria, weakness, seriously ill-early personality changes.	6% of body wt (approx 4.2 L)
3. **Very severe**	Above features + Diminution of physical and Mental capabilities, Hallucinations, and Delirium	7 to 14% of body wt (approx. 5 to 10 L)

Estimation of Water Loss

Marriott divided states of water depletion into three clinical phases along with fluid deficit (Table 40.6).

B. Pure Salt Depletion

Definition: Pure salt depletion occurs when fluids of high Na^+ or Cl^- content are lost from the body and are **replaced by salt-deficient fluids** such as water by mouth or glucose solution IV.

The term "sodium depletion" is now used rather than salt depletion, to lay stress on the fact that Na^+ is the significant ion concerned with the maintenance of ECF volume, further, it is preferable because there is not a loss of sodium chloride (salt) as such, but rather a mixture of ions along with Na^+.

Note

Lack of dietary intake in a healthy subject does not cause serious sodium depletion, because the kidneys can effectively conserve this ion in such a situation.

Causes

- Loss of Na^+ can occur by ***excessive sweating,*** when only water is taken in as replacement.
- Another important means of sodium depletion is ***loss of GI fluids*** as in vomiting, diarrhoea, pancreatic/or biliary fistulaes, cholera, and continuous aspirations through intubation (suction).
- Urinary losses of Na^+ are perhaps not as common, but can occur in such clinical states as ***Addison's disease, diabetic acidosis, cerebral saltwasting syndrome*** and certain instances of chronic renal diseases. In these cases, loss of Na^+ may be aggravated by accompanying vomiting.
- Vigorous ***use of diuretics*** and low sodium or salt-free diets in the management of congestive heart failure may induce sodium depletion.

Pathophysiology

With sodium depletion, ***the ECF becomes hypotonic.*** The lowered osmotic pressure inhibits the release of ADH and the kidneys excrete water in an attempt to maintain normal extracellular Na^+ concentration. Because of the above, plasma and interstitial fluid volume are decreased. Also the extracellular hypotonicity, allows water to flow into the cells where the concentration is greater, thus further reducing the volume of the ECF, ***the cellular hydration is in contrast to cellular dehydration noted in pure water depletion.*** The reduction of volume of the interstitial fluid exceeds that of the plasma. Probably ***because of two factors:***

1. ***A higher osmotic pressure of plasma*** due to the proteins, and
2. ***Diminished filtration*** caused by decline of hydrostatic pressure within the circulation, a consequence of lowered plasma volume (Fig. 40.5).

Clinical and biochemical features: Clinical and biochemical findings mainly attributed to reduction in volume of ECF and sodium deficiency:

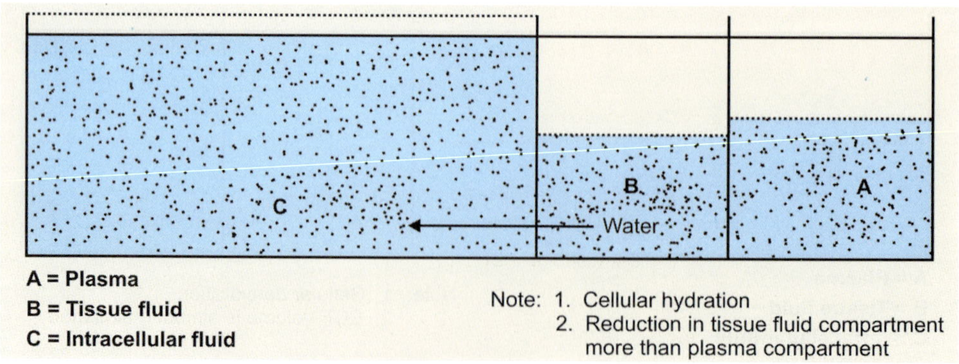

Fig. 40.5: Schematically distribution of water in three compartments in secondary dehydration (Ref. Figs 40.2 and 40.4)

- Because of hypotonicity, **thirst is not a striking** feature and ***absence of thirst*** is an important negative finding.
- The patient appears apathetic and listless, mental changes are common and hallucinations, and confusions are common and sometimes delirious.
- Anorexia and Nausea / vomiting often aggravates the "vicious circle" of sodium depletion. Thus

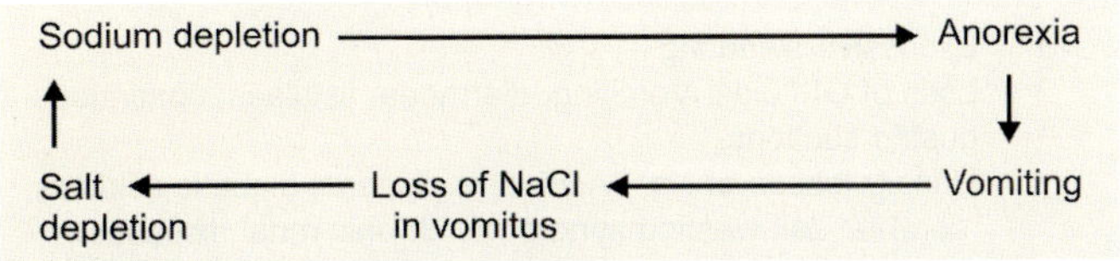

- ***Cramps*** are common and may occur in thigh, abdominal and respiratory muscles.
- Loss of interstitial fluid is manifested clinically by sunken eyes and inelastic skin.
- ***Reduced plasma volume*** leads to haemoconcentration. As a result of lowered blood volume there are decreased cardiac output, lowering of BP and a tendency to orthostatic fainting.
- Decreased glomerular filtration leads to N_2 retention with increase urea concentration.
- ***Urine analysis:*** Patients drinking freely maintain a normal or slightly increased urinary volume, but ***there is no salt present in urine*** (except in Addison's disease).

Death: Due to oligaemic shock.

Estimation of Loss of Fluids and Electrolytes

Marriott divided states of water and salt depletion into three clinical phases along with fluid and salt deficit **(Table 40.7)**.

Essential differences between two types of dehydration: Pure water depletion (primary dehydration) and pure salt depletion (secondary dehydration) are given in **Table 40.8**.

C. Mixed Water and Salt (Sodium) Depletion

In clinical practise, depletion of both water and salt (sodium) is more common than depletion of either alone.

Definition: Mixed depletion occurs when there is loss of fluids containing high concentration of Na and Cl without a free intake of water.

Pathophysiology

Initially the ECF is hypotonic. Later water loss ***outstrips*** the salt loss and ECF becomes ***hypertonic.***

Clinical and Biochemical Features

The clinical picture is a mixture of pure salt depletion and pure water depletion. The volume of fluid in both ECF and ICF is reduced. The patient appears dehydrated and complaints of thirst. The BP may be lowered, blood urea is raised and there is haemoconcentration; urinary output is diminished and the excretion of salt is reduced.

Types of Fluids to Administer

1. ***For pure water depletion:*** Water by mouth or per rectum or *5 per cent* ***Glucose by IV/SC*** or intraperitoneal routes depending on the case.

Note: ***Never give Isotonic Saline,*** which will increase hypertonicity.

2. ***Pure sodium depletion:*** Corrected by Isotonic saline solution.
3. ***Mixed water and sodium depletion:*** Treated with a mixture of saline and 5 per cent Glucose, usually,
   - In the proportion 1 : 1 (half-normal saline or
   - In 1 to 2 (one-third normal saline)

Note

Urinary chloride level to be kept as a general guide:

- Normal saline to be given when chloride is absent from urine.
- Half normal saline when urinary chlorides are between 2.0 and 5.0 G and
- Not more than one-third normal saline for maintenance therapy, when urinary chloride excretion is greater than 5.0 G per litre.

Estimation of Chloride Content of Urine

A rough estimate of chloride excretion can be obtained in bedside of patient by a simple test devised by **Fantus.**

Fantus's Test

- Ten drops of urine are taken by a dropper / pipette into a clean test tube.
- Dropper / pipette is rinsed thoroughly in distilled water.
- One or two drops of 20 per cent solution of potassium chromate is added as an indicator.
- The dropper / pipette is rinsed thoroughly again in distilled water.
- Take 2.9 per cent solution of silver Nitrate ($AgNO_3$) by the dropper / pipette and add drop by drop, the test tube being shaken after addition of each drop. ***Count the drops.***

Table 40.7: Estimation of water and salt loss in secondary dehydration from clinical manifestations

Clinical phases	*Clinical features*	*Estimated deficit*
• **Early (slight to moderate)**	Lassitude, indifference/apathy, syncope, urine Cl^--reduced	0.5 gm NaCl/Kg = 4 L
• **Moderate to severe**	Above features + Nausea/vomiting, cramps, BP $^-$ but > 90 mmHg, urinary Cl^- = Absent	0.5 to 0.75 gm/Kg (4.0-6.0 L)
• **Severe to very severe**	Above features + BP< 90 mmHg, urinary Cl^- = absent	0.75 to 1.25 gm/Kg (6.0-12.0 L)

Table 40.8: Essential differences between pure water depletion (primary dehydration) and pure salt depletion (secondary dehydration)

Pure water depletion (Primary dehydration)		Pure salt depletion (Secondary dehydration)
1. Definition Pure water depletion occurs when water intake stops or inadequate and there is ***no paralleled loss of salt in the secretions.***		• Pure salt depletion occurs when fluids of high Na or Cl content are lost from the body and are ***replaced by salt deficient fluids*** such as water by mouth or glucose solution by IV
2. Causes		
• Patient too weak or too ill to satisfy water needs.		• Excessive sweating
• Mental patients		• Loss of GI fluids: Vomiting, diarrhoea, fistulaes, continuous gastric suctions.
• Comatose patients		• Urinary losses of Na^+, Addison's disease, diabetic acidosis, cerebral salt wasting syndrome, chronic renal diseases.
• Patients with dysphagia		• Diuretics and low salt diet in congestive heart failure.
• Individuals lost in desert or shipwrecked.		
3. Nature of Dehydration +++ Primary or simple, due to loss of ICF (cellular dehydration)		• +++ Secondary or extracellular, due to loss of ECF
4. Clinical features and findings		
• Thirst	+++ (marked)	Absent
• Lassitude	+ (slight)	++ to +++ (marked)
• Orthostatic fainting	: Absent till late	++ to +++ (marked)
• Nausea and vomiting	: Absent	may be + to ++
• Cramps	: Absent	may be + to ++
• Pulse	: Normal till late stage	Rapid and thready
• BP	: Normal till late	Fall in BP ++ to +++
5. Urinary findings		
• Urine volume	: Scanty (oliguria)	Normal and colourless and increased volume
• Sp. gr.	: High	usually low
• NaCl in urine	: Often + (usually to higher side)	Always absent except Addison's disease.
6. Biochemical findings		
• Tonicity of ECF	: ***Hypertonic***	***Hypotonic***
• Plasma volume	: Normal till late	Decreased ++ to +++
• Haemoconcentration	: Not till late stage and slight	Increased ++ to +++
• Blood viscosity	: Normal till late	Increased ++ to +++
• Plasma (Na^+)	: Normal or slight increase in late stage	Decreased ++ to +++
• Blood urea	: + in late stage	Usually increased ++ to +++
• Water absorption	: Rapid	Slow
7. Mode of death Due to rise in os. pr and cellular dehydration		Oligaemic shock and peripheral circulatory failure.

End Point

End point is shown by a sharp colour change from canary yellow to brown-brick red due to formation of silver chromate.

Result

The number of drops of silver nitrate required to produce the change gives Grams of sodium chloride per litre of urine.

Precautions

- Same dropper/pipette should be used throughout, as the whole test depends on the volume contained in the drops.
- Test should be repeated with distilled water instead of urine to ensure that potassium chromate solution is not contaminated with chlorides.

Interpretation

- Urine normally contains 6.0 to 16.0 gm of NaCl per litre.
- Chloride may be regarded as absent if the colour changes with the first drop of $AgNO_3$. This may be normal finding in very dilute urine, which is unlikely to be encountered in dehydration.
- *If urine of a sp. gr. 1020 or more contains less than 3.0 gms NaCl/litre, salt depletion is present.*
- If the urine contains more than 5.0 gm/litre, chloride deficiency is unlikely unless the patient is suffering from Addison's disease or saline is being given intravenously.

PATHOLOGICAL VARIATIONS OF WATER AND ELECTROLYTES

The variations seen pathologically in body fluid and electrolytes are not so simple as stated above. There may be ***three types of expansion:*** Hypotonic, isotonic, and hypertonic. Similarly, there may be ***three types of contraction:*** Hypotonic, isotonic and hypertonic. Salient features of these six types will be discussed briefly

1. *Hypotonic expansion:* Accumulation of water without an equivalent amount of salt. It is occasionally encountered when copious quantities of salt free fluids viz. 5 per cent glucose solution is given to persons with inadequate renal function. The accumulated water distributes osmotically among all the fluid compartments. The cells of CNS also share in this process, which may lead to convulsions and even death ***(water intoxication).***

Changes
- Volume of ICF ↑
- Volume of ECF ↑
- Plasma (Na^+) ↓, Haematocrit and plasma proteins ↓
- Urinary excretion: Na^+ ↓ and H_2O ↑ (lower osmotic pressure inhibits ADH)

2. *Isotonic Expansion*
Accumulation of water and salt in isotonic amounts, i.e. accumulation of water with equivalent amount of salt. This expands the ECF, with no alteration of intracellular volume or composition. The water distributes between interstitial fluid (TF) and plasma, thereby lowering the concentration of plasma proteins and haematocrit. Clinically may manifest as palpable oedema of extremities or pulmonary oedema. Such a condition may ***arise as a serious complication of parenteral fluid therapy.***

Changes
- Volume ICF—
- Volume of ECF ↑
- Plasma (Na^+) —
- Haematocrit and plasma proteins ↓
- Urinary excretion:
  Na^+ ↑
  H_2O ↑

3. *Hypertonic expansion:* Accumulation or retention of sodium leads to an increase in extracellular fluid volume (ECF). If somehow, this sodium is not accompanied by an equivalent amount of water, the resultant ECF becomes "hypertonic" and water is transferred from cells to ECF until osmotic equilibrium is attained. ***Thus ECF expands at the cost of cells, producing Intracellular dehydration.*** If this state is allowed to continue, death may occur because CNS is damaged under such circumstances.

Changes
- Volume ICF ↓
  ECF ↑
- Plasma (Na^+) ↑
- Haematocrit and plasma proteins ↓
- Urinary excretion: Na^+ ↑
  H_2O ↑

4. *Hypotonic contraction:* This results when salt is lost in excess from the body, unaccompanied by an equivalent amount of water, i.e. without simultaneous loss of equivalent amount of water. Excess water distributes itself in all the compartments. However, the serious aspects are those due to diminution in plasma volume. Such a condition can be seen in ***'Adrenal cortical insufficiency'.***

Changes
- Volume ICF ↑
  ECF ↓
- Plasma (Na^+) ↓
- Haematocrit and plasma proteins ↑
- Urinary excretion: Na^+ ↑, later ↓
  H_2O ↑

5. *Isotonic contraction:* Most frequently encountered condition since there is no normal obligatory sodium loss from the body, isotonic contraction can occur by abnormal losses of Na^+ from the body, most commonly in one or more of the secretions of GI tract. These secretions are virtually isotonic with plasma.

The total daily production of these secretions is equal to 65 per cent of the volume of entire ECF and continued loss of these secretions, if not treated, would soon be serious. As these fluids are all isotonic, their loss does not result in a change in ICF volume, and the entire loss must be from ECF, which contracts to an equivalent amount the water of interstitial fluid (TF) is drained, which results to dehydration, CV disturbances, oliguria and finally anuria. The patient may become unconscious and dies of circulatory collapse. Such a condition might arise in ***severe prolonged untreated diarrhoea.***

Changes
- Volume ICF —
  ECF↓
- Plasma (Na^+) —
- Haematocrit and plasma proteins ↑
- Urinary excretion: Na^+ ↓
  H_2O ↓

Table 40.9: Changes in six types of contractions/expansions

Types	Volume ICF	Volume ECF	Plasma (Na^+)	Haematocrit and plasma proteins	Urinary excretion Na^+	Urinary excretion H_2O
1. Hypotonic expansion	↑	↑	↓	↓	↓	↑
2. Isotonic expansion	—	↑	—	↓	↑	↑
3. Hypertonic expansion	↓	↑	↑	↓	↑	↑
4. Hypotonic contraction	↑	↓	↓	↑	↑ late ↓	↑
5. Isotonic contraction	—	↓	—	↑	↓	↓
6. Hypertonic contraction	↓	↓	↑	↑	↑	↓

6. *Hypertonic Contraction*
The situation arises when there occurs excessive loss of water without simultaneous loss of Na^+. This results in contraction of both extracellular and intracellular compartments, such a condition is known as "hypertonic contraction" and it may arise in following conditions:
- In patients with diabetes insipidus.
- In persons who on account of debility are unable to feed themselves and remain unattended.
- Those persons to whom water is unavailable, or
- In persons who lose unusual amount of fluid in perspiration without getting it compensated.

Osmotic pressure of both compartments increases.

Changes
- Volume ICF ↓
  ECF ↓
- Plasma (Na^+) ↑
- Haematocrit and plasma proteins ↑
- Urinary excretion: Na^+ ↑
  H_2O ↓

Note
(In practise), pure examples of these six situations are rarely encountered. Thus, although diarrhoea/or vomiting may produce *isotonic contraction,* the individual may fail to ingest water in sufficient quantity to meet the 'obligatory' water losses, thus converting the situation into *hypertonic contraction.*

Summary of the changes in six types is given in Table 40.9.

WATER INTOXICATION

This condition is caused by excess of water retention in the body and can occur due to the following causes:
- Renal failure
- Excessive administration of fluids parenterally.
- Hyper secretion of ADH following the administration of an anaesthetic for surgery, administration of narcotic drugs or in stress (including any surgery)
- Excess of aldosterone may lead to an overhydration of the body and subsequent water intoxication (Conn's Syndrome).

Clinically: Headache, nausea, incoordination of movements, muscular weakness and delirium are the main symptoms of water intoxication.

Changes:
- PCV, Hb concentration and plasma proteins concentration are all decreased.
- Plasma electrolytes are lowered
- Urinary volume is usually increased and is of low sp. gravity.

Treatment: Withholding fluids by mouth and administering 3 to 5 per cent hypertonic saline IV.

I. Questions (Essay type and short notes)

1. Write in brief about various factors maintaining electrolyte balance in the body.
2. Describe the role of kidneys in maintenance of the composition of extracellular fluid.
3. Describe the distribution of water in the body. How fluid volume is regulated?
4. Give an account of the distribution of fluids in the human body. Describe the normal metabolism of water and its regulation.
5. Give an account of the distribution of fluids in the human body. Describe how water metabolism is regulated?
6. State the distribution of the water in the body. Mention how extracellular fluid differs from intracellular fluid in

composition. Describe the mechanism of exchange of fluids in the body.

7. Describe the electrolytes composition of intracellular and extracellular compartments of the body.
8. Describe briefly regulatory mechanisms of water balance and abnormalities related to it.
9. Describe water balance. Give its regulatory mechanisms. Also describe briefly its abnormalities.

Short Notes

a. Fluid compartments of the body
b. Dehydration
c. Metabolic water
d. Water balance
e. Aldosterone
f. Antidiuretic hormone
g. Extracellular fluid volume
h. Water intoxication
i. Antinatriuretic peptide (ANP)
J. Role of Kinins in water and electrolytes metabolism

Differentiate

Pure water depletion and pure salt depletion.

II. MCQs (Give one correct answer)

1. Quantity of fluid present in dense connective tissue, cartilage and bones which do not exchange readily is:
   a. 2.0 L b. 3.0 L
   c. 4.0 L d. 5.0 L
   e. 6.0 L
2. Predominant cation of plasma is Na^+ and it is balanced mostly by which of the anion
   a. HCO_3^- b. $HPO_4^=$
   c. $SO_4^=$ d. Cl^-
   e. Organic acids
3. In the arterial end of capillary loop, net filtration pressure which drives the fluid out is:
   a. 5 mmHg b. 7 mmHg
   c. 8 mmHg d. 9 mmHg
   e. 10 mmHg
4. Chief cation K^+ of ICF is mostly balanced by which of the following anions:
   a. Pr^- b. HCO_3^-
   c. Cl^- d. $HPO_4^=$.
   e. $SO_4^=$
5. Net absorption pressure which draws fluid in at the venous end of capillary loop is:
   a. 4 mmHg b. 6 mmHg
   c. 8 mmHg d. 10 mmHg
   e. 15 mmHg
6. Loss of water by insensible perspiration through the skin in 24 hours is:
   a. 200–400 ml b. 400–600 ml
   c. 600–800 ml d. 800–1000 ml
   e. None
7. Minimum excretory volume to eliminate waste-products from the body in dehydration is:
   a. 200–400 ml b. 500–600 ml
   c. 600–800 ml d. Less than 200 ml
   e. None of the above
8. Aldosterone increases:
   a. Reabsorption of Na^+ b. Excretion of Na^+
   c. Reabsorption of K^+ d. Excretion of water
   e. All of the above
9. Aldosterone is produced by:
   a. Zona reticularis b. Zona glomerulosa
   c. Zona fasciculata d. All of the above
   e. None of the above
10. Atrial natriuretic peptide (ANP) is produced by the atrial wall in response to:
    a. Aldosterone b. Decreased BP
    c. Kinins d. Increased Intravascular volume
    e. None of the above
11. In primary dehydration ECF becomes:
    a. Isotonic b. Hypotonic
    c. Hypertonic d. None of the above
12. In pure salt depletion (secondary dehydration) ECF becomes:
    a. Hypertonic b. Hypotonic
    c. Isotonic d. None of the above
13. In pure water depletion, the fluid to be administered IV is:
    a. Isotonic saline b. $^1/_2$ Normal saline
    c. $^1/_3$ Normal Saline d. 5 per cent Glucose solution
    e. None of the above
14. Urine normally contains NaCl.
    a. 2 to 6 gm/litre b. 3–8 gm/litre
    c. 4–8 gm/litre d. 6 to 16 gm/litre
    e. None of the above

Answers to MCQs

1. c 2. d 3. b 4. d 5. c 6. c
7. b 8. a 9. b 10. d 11. c 12. b
13. d 14. d.

CHAPTER

41

ACID BASE BALANCE AND IMBALANCE

Major Concepts

A. Learn what are acids, bases, and buffers and study the mechanisms of normal regulation of pH in the blood.

B. Learn the abnormalities that can take place in acid-base balance and study the different types of acidosis and alkalosis.

Specific Objectives

A. 1. Define pH. Revise your concept of pH and pK, Hasselbalch-Henderson equation.

2. Study what are acids and bases with suitable examples. What is meant by strong acids/and bases and weak acids/and bases.

3. Define buffer. Study how a buffer acts in the body.

4. List the major sources of acids in the body which tend to decrease the pH.

5. Learn the different mechanisms which regulate the pH of blood.

• Buffer systems in the body-first line of defence. • Respiratory mechanisms. • Renal mechanisms-second line of defence.
• Dilution factor

(a) Buffer systems in the blood

• List the various buffer systems in plasma and the erythrocytes. • Study how the bicarbonate buffer system works and explain how it is linked with respiration. • What is meant by the term "alkali reserve" • Study how the phosphate buffer system works and explain how it is linked with kidneys. • Study how plasma proteins help in buffering action. • Learn how haemoglobin acts as a buffering agent. • Oxygenated Hb is more acidic than deoxygenated Hb. Explain. • Draw a diagram and show the buffering action of Hb in tissues and in lungs. • What is meant by the term "isohydric transport of CO_2"?

(b) Study the role of respiration in pH regulation.

(c) Study the renal mechanisms for regulation of pH of blood.

Following three renal mechanisms to be studied:

• Bicarbonate mechanism: Which operates in proximal tubule.
• Study the factors that affect bicarbonate reabsorption.
• Study the action of CA inhibitor on excretion of bicarbonate.
• Phosphate mechanism: Which operates in distal tubule.
• Ammonia mechanism: Which operates in distal tubular epithelial cells. Draw neat diagrams of above three mechanisms and label them.

(d) Study the relation of K^+ excretion to acid-base equilibrium.

• Explain what is ***paradoxic aciduria***.
• Learn what is meant by anion gap and its significance.

B. 1. Study the types of acid-base imbalances that can occur in human body.

2. List the types of imbalances. • Acidosis: Which can be 'metabolic' acidosis and/or "respiratory" acidosis. • Alkalosis: Which can be metabolic' alkalosis and/or 'respiratory' alkalosis. They can be in "compensated phase" or "uncompensated phase".

3. Study the synonyms of each type, their causes, mechanism, biochemical characteristics in "uncompensated" phase and when fully 'compensated'.

4. In tabular form differentiate

• Metabolic acidosis and respiratory acidosis.
• Metabolic alkalosis and respiratory alkalosis.

INTRODUCTION

Under normal conditions, ***the pH of ECF usually does not vary beyond the range 7.35 to 7.5*** and is maintained approximately at 7.4, (pH of arterial blood is approx. 7.43 and venous blood is 7.4). ***Maintenance of this constant blood reaction is one of prime requisites of life and any material variation on either side, seriously disturbs the vital process and may lead to death.*** pH < 7.3 leads to acidosis and pH > 7.5 leads to alkalosis.

Large amounts of H^+ are continually contributed to these fluids from intracellular metabolic reactions, hence to maintain a constancy it is necessary and imperative that they are removed from the fluids effectively and promptly. ***The mechanisms of neutrality regulations are concerned, therefore, with maintaining a state of equilibrium between production, i.e. introduction of H^+ ions and removal of the same.***

pH of interstitial fluid: This is probably somewhat lower than that of blood plasma, since it occupies an intermediate position between the plasma and the site of production of acids within the cells.

Intracellular pH: Very little precise information is available regarding intracellular pH. **Robin *et al*** has shown clinically and with basic experimental studies that intracellular pH is more acidic than plasma, averaging approximately 7.05. But it is not same in all tissues and differ widely according to the functional activity. It may be higher in osteoblasts, pH 8.0 or more, for facilitating the action of alkaline phosphatase; similarly, it may be quite low, pH below < 5.0 in prostatic cells, which is required for the action of the enzyme acid phosphatase.

*Mitochondrial pH: **Robin et al (1960)*** in an ingenious experiment measured the internal pH of mitochondrion. They showed that the internal pH of mitochondria, under the conditions of their experiment reached 6.6, at a plasma pH of 7.4 and an intracellular pH of approximately 7.0. They described ***mitochondria appear to be small islands of acidity in the relatively alkaline sea of intracellular water.***

ACID-BASE BALANCE IN NORMAL HEALTH

According to the modern concept of **Brönsted-Lowry:**

- An ***acid*** is defined as a substance, ion, molecule or particle, that yields H^+ ions (protons) in solution, and
- A ***base*** is anything that combines with H^+ ions (protons). Accordingly, whereas H_2CO_3 is an acid, dissociating into H^+ and HCO_3^- ions, its anionic component HCO_3^- is a base.

Other examples are

Acid		Base
HSO_4^-	$\Longleftrightarrow$	$H^+ + SO_4^-$
CH_3COOH	$\Longleftrightarrow$	$H^+ + CH_3COO^-$
H_2PO_4	$\Longleftrightarrow$	$H^+ + HPO_4^-$

$NaHCO_3$ acts as a base because it yields HCO_3^- ions, which can combine with H^+ ions.

H_2O **(Acid)**	$\Longleftrightarrow$	$H^+ + OH^-$
$H^+ + H_2O$ **(base)**	$\Longleftrightarrow$	H_3O^+ **Hydroxonium ion (strong acid)**

Water is a substance that can act either as acid or base. HCl is a strong acid by virtue of its extensive dissociation into H^+ and Cl^- ions, Cl^- ions is an extremely weak base, because it has very little capacity for combining firmly with H^+ ions. On the other hand, such anions as HCO_3^-, HPO_4^{--}, $H_2PO_4^-$ and protein are comparatively strong bases, because they have a relatively strong affinity for H^+ ions, forming weak acids (i.e. relatively slight dissociation). **The stronger the acid, the weaker the base,** which results from its dissociation and vice versa. ***Such pairs have been termed conjugate acid base pairs.*** Thus the conjugate base of the acid HCl is the Cl^- ions. Several such pairs important in the body, arranged in order of descending strength of acid and hence of increasing strength of base are shown below in the box.

Note: It is to be noted that cations such as Na^+, K^+, Ca^{++}, Mg^{++} cannot donate or accept protons and so are neither acids nor bases. Such substances have been termed ***aprotes.***

BUFFERS

Definition: A buffer may be ***defined*** as a solution which resists the change in pH which might be expected to occur upon the addition of acid or base to the solution.

Buffers consist of mixtures of weak acids and their corresponding salts, alternatively, weak bases and their salts. The former type is the more important and common in human body.

Mechanism of action: Its action against added acid or base may be illustrated as follows:

The sketchy diagram in the box next page shows the weak acid HA and its completely ionised salt B^+A^-.

- *Added H^+ ions,* in the form of strong acid, combine with anions A^- (largely from the salt component of the buffer), to form the weakly dissociable HA, so that

	Acids			Bases	
Decreasing strength ↓	HCl	very strong	very weak	Cl^-	↓ ***Increasing strength***
	H_3O^+			H_2O	
	H_2CO_3	rather weak	rather strong	HCO_3^-	
	$H_2PO_4^-$			HPO_4^{--}	
	$Hpr^{(n-1)-}$			$Proteinate^{n-}$	
	NH_4^+	very weak	very strong	NH_3	
	H_2O			OH^-	

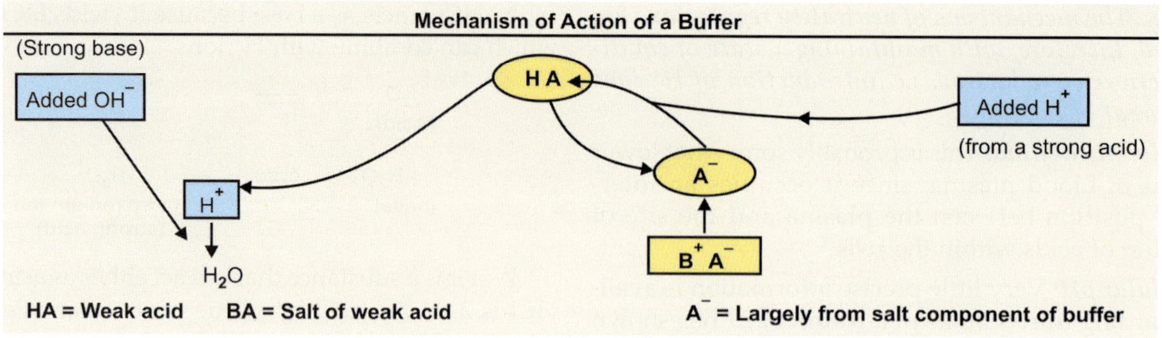

pH does not become as acid as it would be in the absence of the buffer. The capacity to combine with added acid remains so long as there is a supply of the buffer salt in the medium.

- ***Added OH^- ions,*** in the form of a strong base, combine with H^+ ions derived from the acid HA and form the weakly dissociable H_2O molecules. Hence pH does not become as alkaline as would happen in absence of the buffer. OH^- ions can be buffered as long as some of the acid HA remains to supply the H^+ ions.

ACIDS PRODUCED IN THE BODY

The following are the major sources of H^+ (protons) production in the human body.

- ***Carbonic acid (H_2CO_3):*** It is the chief acid produced in the body in the course of oxidation in the cells. Oxidation of C-compounds resulting in CO_2 production. About 10 to 20 or more moles being produced daily from oxidation of food stuffs in the body. ***Approx. 300 litres of CO_2 are produced and eliminated daily in the*** body of an adult.
- ***Sulphuric acid (H_2SO_4):*** A strong dissociable acid produced during oxidation of S-containing amino acids, e.g. cysteine/cystine and methionine.
- ***Phosphoric acid:*** Products of metabolism of dietary phosphoproteins, nucleoproteins, phosphatides and hydrolysis of phosphoesters.
- ***Organic acids:*** Abnormal production and accumulation of certain intermediary organic acids from oxidation of carbohydrates, fats and proteins, under certain circumstances, e.g. pyruvic acid, lactic acid, acetoacetic acid, β-OH-butyric acid, etc. Under ordinary conditions, PA/and LA and β-OH butyric acid are produced in quantities of about 80 to 120 millimoles daily, which may increase considerably under certain abnormal circumstances.
- ***Iatrogenic:*** Certain medicines like NH_4Cl, mandelic acid, etc. may increase H^+ concentration of blood when they are used as treatment, when administered in excess.

Note: Although certain foodstuffs may provide a certain amount of potentially "basic" substances, this is far exceeded by their potential acid content. Both the H^+ ions and the anions produced by these acids must be disposed of, i.e. ultimately excreted from the body, in such a manner that their temporary sojourn in the EC fluids does not unduly affect the pH under normal health. ***The means whereby these ends are accomplished comprise the mechanism of regulation of acid-base balance.***

MECHANISMS OF REGULATION OF pH

The mechanisms of regulation of blood pH involves the following factors:

(a) ***"Front-line" defence:*** They are mainly:
- ***Buffer systems*** in the blood: Which restricts pH change in body fluids.
- ***Respiratory mechanisms:*** Regulation of excretion of CO_2 and hence, regulation of H_2CO_3 concentration in EC fluid.

(b) ***"Second-line" defence:*** This is achieved by kidneys ***(Renal mechanisms).*** Ultimate excretion of excess of acid or base and thus ultimate regulation of concentration of H^+ and HCO_3^- ions in EC fluid.

(c) ***Dilution factor:*** The acids introduced into and formed in the body are distributed throughout the ECF volume. Although this may not properly be regarded as a regulatory mechanism, entrance of a given amount of acid into a smaller volume of fluid, as in conditions of severe dehydration, results in relatively greater rise in H^+ ion concentration and decrease in effective buffer base.

Physiological Buffer Systems

The capacity of the EC fluids for transporting acids from the site of their formation (cells) to the site of their excretion (e.g. lungs and kidneys), without undue change in pH is dependant chiefly on the presence of efficient buffer systems in these fluids and in the erythrocytes.

Blood Buffers

Each of the buffer system consists of a mixture of a weak acid, HA, and its salt BA, which gives the mixture the

SECTION FIVE

ability to resist change in the H^+ ion concentration and thus prevents any change of pH of the medium.

Most important buffer systems of blood are as follows:

Plasma buffers

- $\frac{NaHCO_3}{H_2CO_3}$
- $\frac{Na_2HPO_4 \text{ (Alk } PO_4)}{NaH_2PO_4 \text{ (Acid } PO_4)}$
- $\frac{Na - Pr}{H\ Pr}$
- $\frac{\text{Na organic acid}}{\text{H organic acid}}$

Buffers of RB Cells

- $\frac{K\ HCO_3}{H_2CO_3}$
- $\frac{K_2H\ PO_4}{K\ H_2\ PO_4}$
- $\frac{K\ Hb}{H\ Hb}$
- $\frac{K\ Hb\ O_2}{H\ Hb\ O_2}$
- $\frac{\text{K organic acid}}{\text{H organic acid}}$

Note

- The buffer systems in the interstitial fluids and lymph are much the same as in the blood plasma, except that proteins are generally present in much smaller quantities.
- The buffer systems in intracellular fluids are also qualitatively same as in the plasma, but the cell fluids contain much higher concentration of proteins.

ROLE OF DIFFERENT BUFFER SYSTEMS

1. Bicarbonate Buffer System

($NaHCO_3/H_2CO_3$ = [Salt]/[Acid])

This consists of weak acid "Carbonic acid" (H_2CO_3) and its corresponding salt with strong base (HCO_3^-), $NaHCO_3$ (Sodium bicarbonate).

Normal ratio in blood $\frac{NaHCO_3}{H_2CO_3} = \frac{20}{1}$

They are the chief buffers of blood and constitute the so called **alkali reserve**.

Neutralisation of strong and non-volatile acids entering the ECF is achieved by the bicarbonate buffers. Such acids, e.g. HCl, H_2SO_4, Lactic acid, etc. which are strong and non-volatile react with $NaHCO_3$ component. Thus, Lactic acid will be buffered as follows:

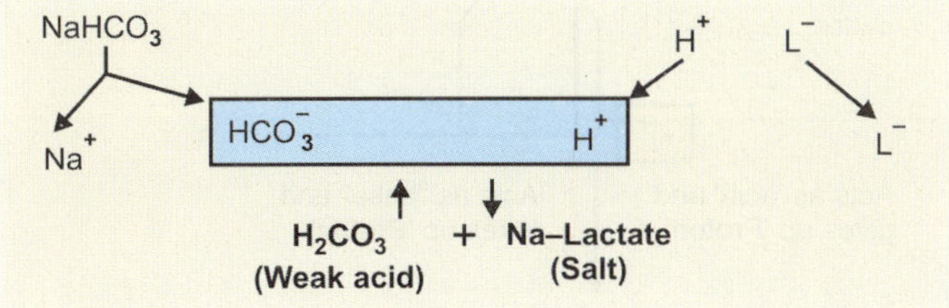

- ***A strong and non-volatile acid is converted into weak (less dissociable) and volatile acid at the expense of $NaHCO_3$ (salt component of the buffer).***
- H_2CO_3 thus formed, as it is volatile, is eliminated by diffusion of CO_2 through alveoli of lungs.

Note: Proper lung functioning is important.

$$H_2CO_3 \xrightarrow[\text{Low } CO_2 \text{ tension}]{CA} H_2O + CO_2\uparrow$$

CA = *Carbonic anhydrase*

Hence, **bicarbonate buffer system is directly linked up with respiration.**

Similarly, when alkaline substance, e.g. NaOH enters the ECF, it reacts with the acid component, i.e. H_2CO_3 of the buffer system.

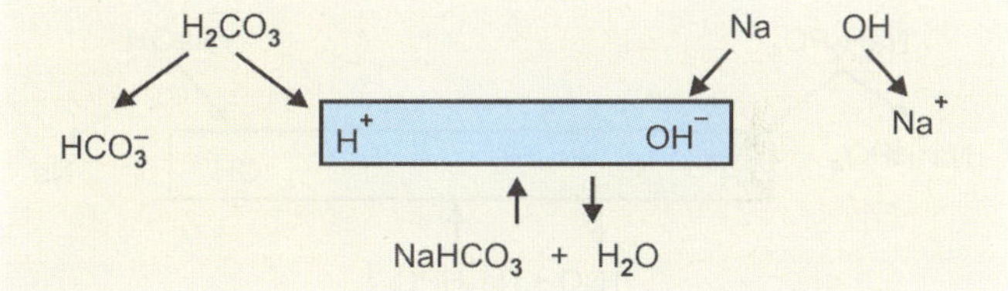

Alkali reserve: It is represented by the $NaHCO_3$ concentration in the blood that has not yet combined with strong and non-volatile acid.

Normally, all acids except carbonic acid reacts with bicarbonate to liberate CO_2.

Strong non-volatile acids, e.g. LA, H_2SO_4 HCl, etc. $+ NaHCO_3 \rightarrow H_2CO_3 +$ Salt of acid

$$H_2CO_3 \Leftrightarrow H_2O + CO_2\uparrow$$

Advantages of bicarbonate buffer system: Bicarbonate buffer system is efficient as compared to other buffer systems as:

- It is *present in very high concentration* than other buffer systems. (26 to 28 millimole per litre)
- *Produces H_2CO_3*, which is a *weak acid* and *volatile* and *CO_2 is exhaled out.*
- Hence, *it is a very good physiological buffer* and acts as a front line defence.

Disadvantage: As a chemical buffer, it is rather weak, pKa is further away from the physiological pH.

2. Phosphate Buffer System

(Na_2HPO_4 / NaH_2PO_4 = [Alk PO_4]/[Acid PO_4])

Normal ratio in plasma is 4:1. This ratio is kept constant with the help of the kidneys. Thus, **phosphate buffer system is directly linked up with the kidneys.**

(a) When a strong acid enters the blood, it is fixed up by alkaline PO_4 (Na_2HPO_4) which is converted to acid PO_4 as follows:

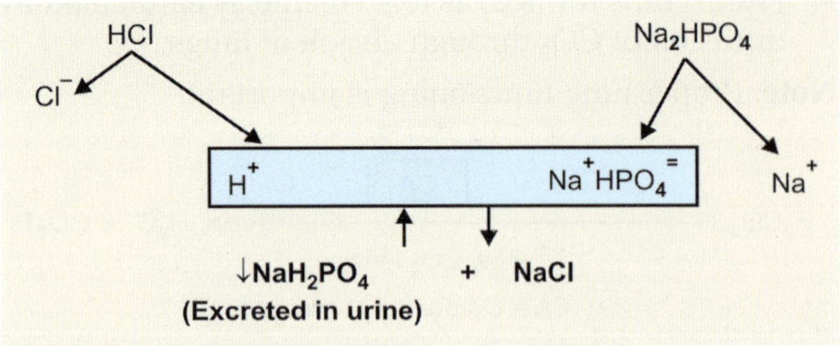

The acid PO_4 (NaH_2PO_4) thus produced are excreted by the kidneys, hence urine becomes more ***acidic.***

(b) When an alkali enters, it is buffered by the acid PO_4, which is converted to alkaline PO_4 and is excreted in urine, producing increased alkalinity of urine.

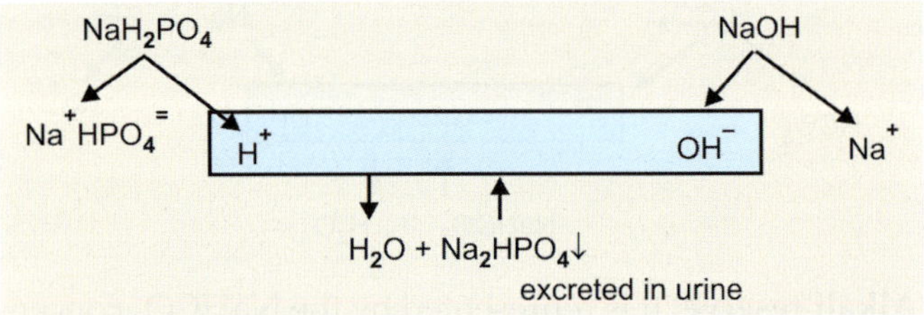

Thus, **phosphate buffer system works in conjunction with the kidneys.** A normal healthy kidney is necessary for proper functioning.

Disadvantage

- Concentration in blood is low (1.0 millimole/litre),
- As a physiological buffer it is less efficient.

Advantage: As a chemical buffer it is very effective and better, as pk_a approaches physiological pH.

3. Protein Buffer System

($Na^+ Pr^-/H^+ Pr^-$ = [Salt]/[Acid])

Buffering capacity of plasma proteins is much less than Hb. The latter operates only in erythrocytes.

Example:

1. One gm. of Hb binds 0.183 mEq of H^+. On the other hand, one gram of plasma proteins binds 0.110 mEq of H^+, when titrated between pH 7.5 and 6.5.
2. Hb of one litre of blood as buffer can bind 27.5 mEq of H^+. But plasma proteins present in one litre of blood can buffer 4.24 mEq of H^+ only between pH 7.5 and 6.5.

From the above examples, it is clear that **Hb has more buffering capacity than plasma proteins.**

Buffering action of proteins

- *In acidic medium:* protein acts as a base, NH_2 group takes up H^+ ions from the medium forming NH_3^+, Proteins become +vely charged.
- *In alkaline medium:* Proteins act as an acid. Acidic COOH gr dissociates and gives H^+, forming COO^-. H^+ combines with OH^- to produce a molecule of water, proteins become –vely charged.
- *Na^+ proteinate:* Salt component can combine with strong acids and thus produces weak acid H^+ pr^-.

$$\underset{\text{(Strong acid)}}{Na^+Pr^- + \underset{LA}{H^+L^-}} \longrightarrow \underset{\text{(Salt)}}{\textbf{Na L}} + \underset{\text{(Weak acid)}}{\textbf{H}^+\textbf{Pr}^-}$$

- Other factor that contributes to the removal of CO_2 is by formation of ***Carbamino-compounds,*** thus directly fixing CO_2.

$$PrNH_2 + CO_2 \Longleftrightarrow \underset{\text{(Carbamino compound)}}{\textbf{PrNHCOOH}}$$

This reaction achieves the binding of CO_2 without passing through the carbonic acid stage.

4. Hemoglobin as a Buffering Agent

The buffering capacity of Hb, as of any protein, depends on the number of dissociable buffering groups, viz. acidic-COOH gr, basic-NH_2 gr, Guanidino group and most important is ***imidazole group,*** which varies with the pH of the medium. ***With the pH range of 7.0 to 7.8, most of the physiological buffering action of Hb is due to the "imidazole" group of amino acid "histidine".*** Each molecule of Hb contains 38 mols of Histidine. In α-chain, histidine at 87 position, and in β-chain, histidine at 92 position is directly linked with Fe^{++} of "haem" (Refer chapter on Chemistry of Haemoglobin).

Imidazole contains two groups

1. ***Fe^{++} containing group*** which is concerned with carriage of O_2, and
2. ***Imidazole N_2 group,*** which can give up H^+ (proton) and accept H^+ depending on the pH of the medium.

Thus, buffering capacity of Hb is due to the presence of "Imidazole" nitrogen group which remains dissociated in acidic medium and conjugate base forms.

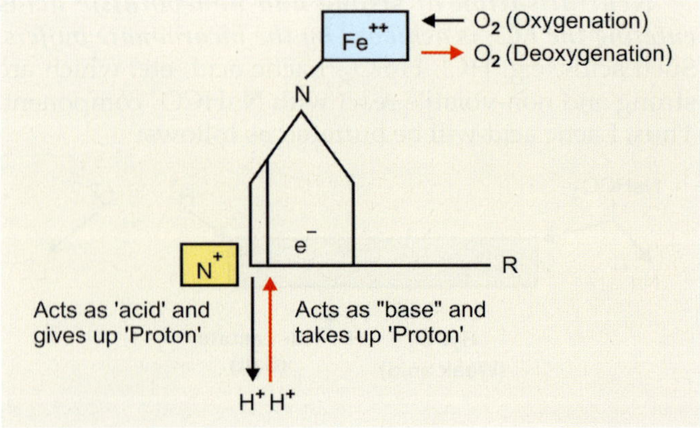

- **Oxygenated Hb is a stronger acid than deoxygenated Hb.** 21On oxygenation, the imidazole N_2 group acts as acid and donates protons in the medium.
- Deoxygenated Hb is less acidic, less dissociable and imidazole N_2 group acts as 'base' and takes up protons from the medium.
- **Acidity of the medium favours delivery of O_2**
- **Alkalinity of the medium favours oxygenation of Hb.**

In the case of "Imidazole" group of histidine, which is intimately associated with the Fe^{++} of Hb, its strength as a buffer is affected by changes in degree of oxygenation of Hb. When O_2 is removed (in deoxygenated Hb), the imidazole group is rendered "less acidic", consequently less dissociated, removing a H^+ from solution and becoming electrically +ve. This effect is reversed with increased oxygenation of Hb.

Sequence of events that occur in lungs and tissues is shown schematically in Figure 41.1.

1. In the Lungs

The formation of oxy-Hb (HbO_2) from deoxygenated Hb (HHb), must release H^+ ions, which will react with HCO_3^- to form H_2CO_3. Because of Low CO_2 tension in the lungs, the equilibrium then shifts towards the production of CO_2, which is continually eliminated in the expired air.

2. In tissues

Due to reduced O_2 tension, local acidity, and aided by CO_2 Böhr effect, Oxy-Hb (HbO_2) dissociates delivering O_2 to the cells and deoxygenated Hb (HHb) is formed. At the same time, CO_2 produced as a result of metabolism in the cells, is hydrated to form H_2CO_3, which ionizes to form H^+ and HCO_3^-. ***Deoxygenated Hb (HHb) acting as an anion, accepts the H^+ ions,*** forming so-called acid-reduced Hb (HHb). Very little change in pH occurs because the newly arrived H^+ ions are buffered by formation of a very weak acid.

Isohydric Transport of CO_2

At a pH of 7.25, one mol. of oxy-Hb donates 1.88 mEq of H^+; on the other hand, one mol. of reduced Hb, because it is less ionized, donates only 1.28 mEq H^+. It may, therefore, be calculated that at the tissues a change of one mol. of oxy-Hb to reduced Hb allows 0.6 mEq H^+ to be bound (buffered), so that these newly formed H^+ ions do not bring about a change in pH. This circumstance as it relates to the role of Hb buffers is sometimes referred to as ***isohydric transport of CO_2***.

ROLE OF RESPIRATION IN ACID-BASE REGULATION

Participation of the respiratory mechanism, in the regulation of acid-base balance is depend upon:

- The ***sensitivity*** of the respiratory centre (RC) in medulla oblongata to very slight changes in pH and pCO_2, and
- The ready diffusibility of CO_2 from the blood, across the pulmonary alveolar membrane, into the alveolar air. Hence, the lungs should be healthy so that diffusion of CO_2 take place properly.
  - An increase in blood pCO_2 and of only 1.5 mmHg (0.2 per cent increase in CO_2) results in 100 per cent

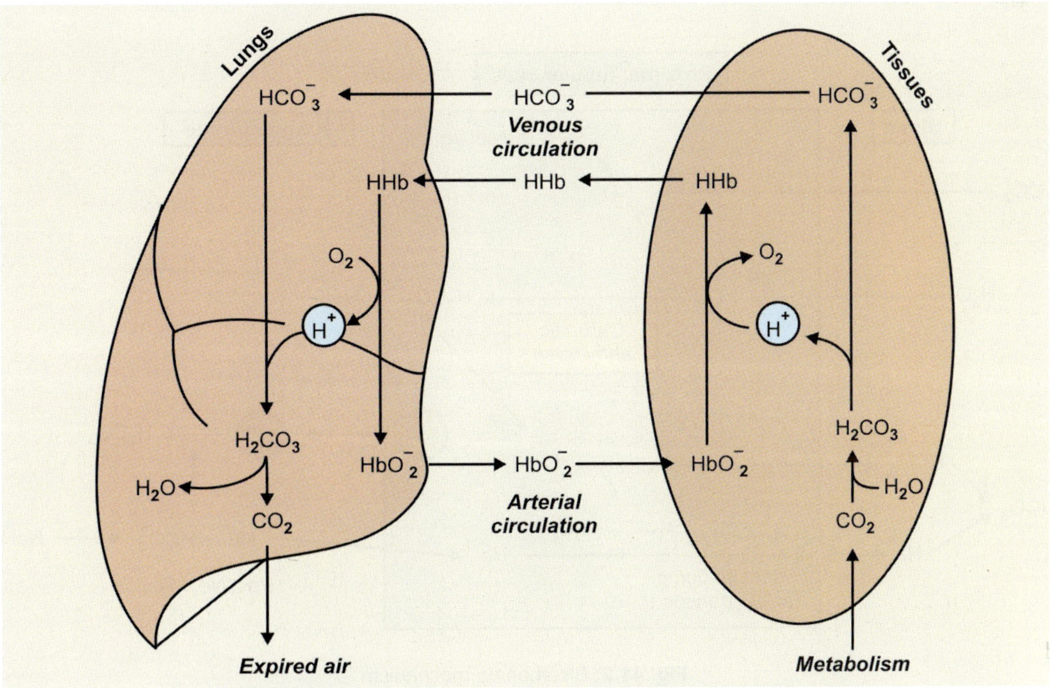

Fig. 41.1: Showing sequence of events that occur in lungs and tissues

increase in pulmonary ventilation (stimulation of respiratory centre), which increases also with slight increases in H^+ ion concentration of the blood (acidosis). The excess CO_2 is thereby promptly removed from the ECF in the expired air.
– A decrease in blood $pCO_2\downarrow$ or H^+ ion concentration (alkalosis), causes depression of respiratory centre, with consequent slow and shallow respiration *(hypoventilation)* resulting to retention of CO_2 in the blood until the normal pCO_2 and pH are restored.

This respiratory mechanism, therefore, tends to maintain the normal B-H CO_3/H_2CO_3 ratio in the EC fluids.

RENAL MECHANISMS FOR REGULATION OF ACID-BASE BALANCE

Kidneys also affect acid-base equilibrium:

- By providing for elimination of non-volatile acids viz. Lactic acid, H_2SO_4, ketonebodies, etc. after being buffered with cations (principally Na^+) are first removed by glomerular filtration.
- ***Body cannot afford to lose Na^+, being extremely important. It is recovered in the renal tubules by reabsorption in exchange of H^+ ions*** which are secreted. It is recovered as $NaHCO_3$ ("alkali reserve").

There are **three mechanisms** by which the above is achieved.

A. Bicarbonate mechanism
B. Phosphate mechanism
C. Ammonia mechanism

A. Bicarbonate Mechanism

- Mobilisation of H^+ ions for tubular secretion is accomplished by ionisation of carbonic acid (H_2CO_3) which itself is formed from metabolic CO_2 and H_2O.
- This reaction $CO_2 + H_2O \Leftrightarrow H_2CO_3$ is catalysed by the enzyme (Zn-containing enzyme) *carbonic anhydrase*, present in renal tubular epithelial cells.

Site: In proximal tubular epithelial cells, the exchange of H^+ ions proceed first against sodium bicarbonate.

Sequence of events that take place is shown in Figure 41.2.

Under normal conditions, the rate of H^+ secretion is about 3.50 millimoles per minute, and the rate of filtration of HCO_3^- ions is about 3.49 millimoles per minute. ***Hence, all the HCO_3^- ions are normally reabsorbed, while a slight excess of H^+ ions remains in the tubules to react with other substances and to be excreted in urine.***

Above mechanism provides:

- For complete reabsorption of all of $NaHCO_3$,
- Reduction of H^+ ion load of plasma with little change in pH of urine.

Note: ***HCO_3^- moiety filtered is not that which is reabsorbed into the blood.***

Carbonic Anhydrase (CA)

It is a Zn-containing metalloenzyme. It specifically catalyses the removal of CO_2 from H_2CO_3, however, ***the reaction is reversible.*** At the tissues, the formation of H_2CO_3 from CO_2 and H_2O is also accelerated by CA.

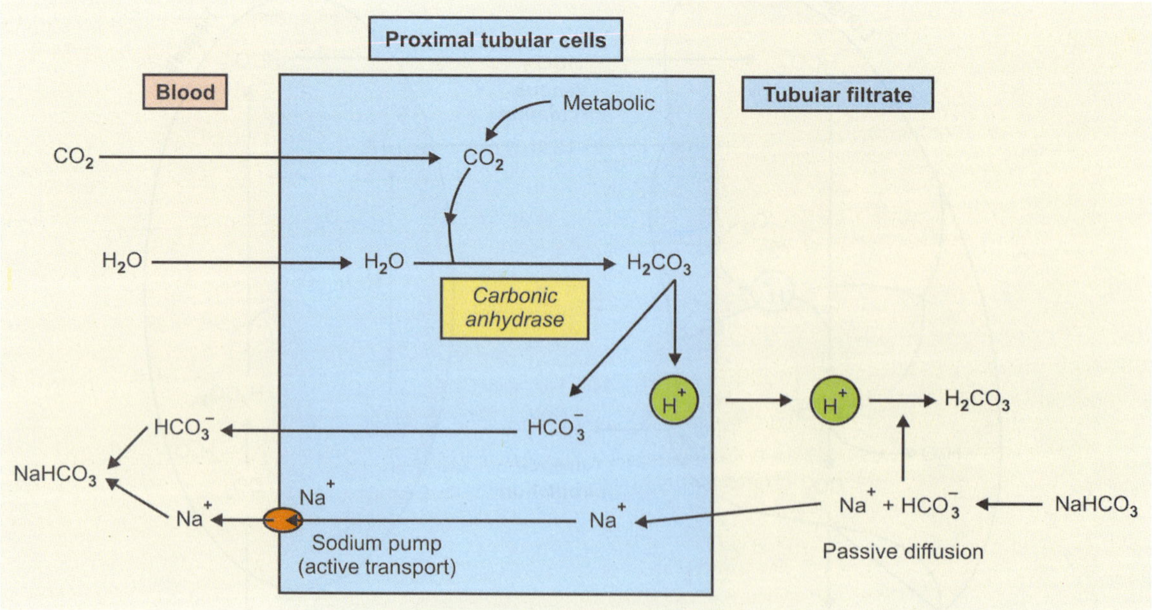

Fig. 41.2: Bicarbonate mechanism

Source: The enzyme is present:
- ***RB cells:*** Where associated with Hb; ***never found in plasma.***
- In most of the tissues, where catalyses formation of H_2CO_3 from H_2O and metabolic CO_2.
- ***In parietal cells of stomach:*** Where the enzyme is involved in secretion of HCl.
- ***In renal tubular epithelial cells***

Recently, it has been demonstrated in small quantities in:
- Muscle tissue,
- Pancreas and
- Spermatozoa.

Factors Affecting Bicarbonate Reabsorption:

1. ***Influence of CO_2 tension (pCO_2):***
   - Increase in pCO_2 ↑, accelerates formation of H_2CO_3 and thus increases the H^+ secretion by renal tubular epithelial cells, which facilitate reabsorption of HCO_3^-.
   - Decrease in pCO_2 ↓ will do the reverse.
2. ***Variations in the body store of K^+:*** The intracellular concentration of K^+ and NOT of plasma is the factor that controls the HCO_3^- absorption.

When K^+ is administered (in excess), K^+ rapidly enters into the cell in exchange of H^+ ions. H^+ ions leaving the cells are buffered by bicarbonate of ECF as a result plasma bicarbonate content is reduced. Hence, renal tubular epithelium, secretes less H^+ ions and thus less HCO_3^- is absorbed. ***Bicarbonate excretion in urine becomes more and urine becomes alkaline, though ECF is acidic.***

CLINICAL IMPORTANCE

In K^+ deficiency, K^+ ions leave the cell. H^+ ions enter the cells ***producing intracellular acidosis***. ECF becomes alkaline. More H^+ secretion from tubular epithelial cells and increased excretion of H^+, NaH_2PO_4 and NH_4Cl, increasing the titratable acidity. Though **the ECF is alkaline, highly acidic urine is excreted. This condition is called as *paradoxic aciduria.*** Such a situation may occur:
- Patients treated for long with cortisone or corticotropin (ACTH) or
- In persons with hypercorticism (Cushing's syndrome)
- In postoperative patients with K-free fluids, in whom depletion may occur due to continued loss in urine and GI fluids. Although alkalosis in these surgical cases usually accompanied by depletion of chloride ***(Hypochloraemic alkalosis).*** Correction cannot be achieved only with NaCl alone but by administration of K-salts along with. When adequate repletion of K^+ is done, a fall in serum HCO_3^- ↓ and a rise in serum Cl^- ↑ together with elevation of urinary pH to normal levels will then occur.

3. ***Variations in plasma level of Cl^-:*** Increase or decrease of Cl^- concentration in plasma causes decrease or increase of HCO_3^- concentration respectively. Increase in (Cl^-) ↑ causes decrease in (HCO_3^-) ↓ and vice versa.
4. ***Variations in secretion of adrenocortical hormones:*** In Cushing's syndrome, plasma [HCO_3^-] increases with augmented HCO_3^- reabsorption (see K^+ deficiency above).

CLINICAL SIGNIFICANCE

Action of carbonic anhydrase inhibitor: The enzyme carbonic anhydrase which, catalyses the important reaction in renal tubules by which H^+ ions are produced for secretion, may be inhibited by sulphonamide derivatives, the most potent inhibitor yet found is **Acetazolamide ('Diamox').** When the drug is administered, the urine becomes alkaline, increased amount of $NaHCO_3$ appears in urine. There is also reduction in titratable acidity ↓ and in the NH_3 ↓ excretion in the urine, with an increase in K^+ excretion ↑ from distal tubular epithelial cells. All of the above can be explained by inhibiting action on CA due to less H^+-Na^+ exchange. Drug is used clinically as a diuretic to induce a loss of Na^+ and H_2O in patients who will be benefitted by such a regime, e.g. congestive cardiac failure and in hypertensive heart diseases.

B. Phosphate Mechanisms

Both disodium hydrogen phosphate (Na_2HPO_4, alkaline PO_4) and monosodium dihydrogen PO_4 (NaH_2PO_4, acid phosphate) are present in the plasma. The pH of the urine is determined by the ratio of these two phosphates. In plasma, concentration of Na_2HPO_4 exceeds that of NaH_2PO_4 and the ratio is maintained to 4:1. ***But in urine, the concentration of NaH_2PO_4 exceeds that of Na_2HPO_4 and the ratio becomes 9:1.*** Glomerular filtrate of pH 7.4 is converted to a urine having pH –6.0 or even as low as 4.8. After all the HCO_3^- has been reabsorbed by the mechanism stated above, H^+ secretion proceeds against Na_2HPO_4. The exchange of Na^+ ion for secreted H^+ ion changes Na_2HPO_4 to NaH_2PO_4 with consequent increase in acidity of urine, resulting in decrease in pH.

Site: Operates in "distal tubule" of kidney **(Fig. 41.3)**.

C. Ammonia Mechanism

A third mechanism operates in the ***distal renal tubule cells,*** for the elimination of H^+ ions and the conservation of Na^+, by production of NH_3 by the renal tubular epithelial cells.

Source of NH_3 in Distal Tubular Epithelial Cells

1. NH_3 is produced by the hydrolysis of Glutamine by the enzyme *Glutaminase* which is present in these cells.

$$\text{Glutamine} \xrightarrow[\text{Glutaminase}]{H_2O} \text{Glutamine acid} + NH_3$$

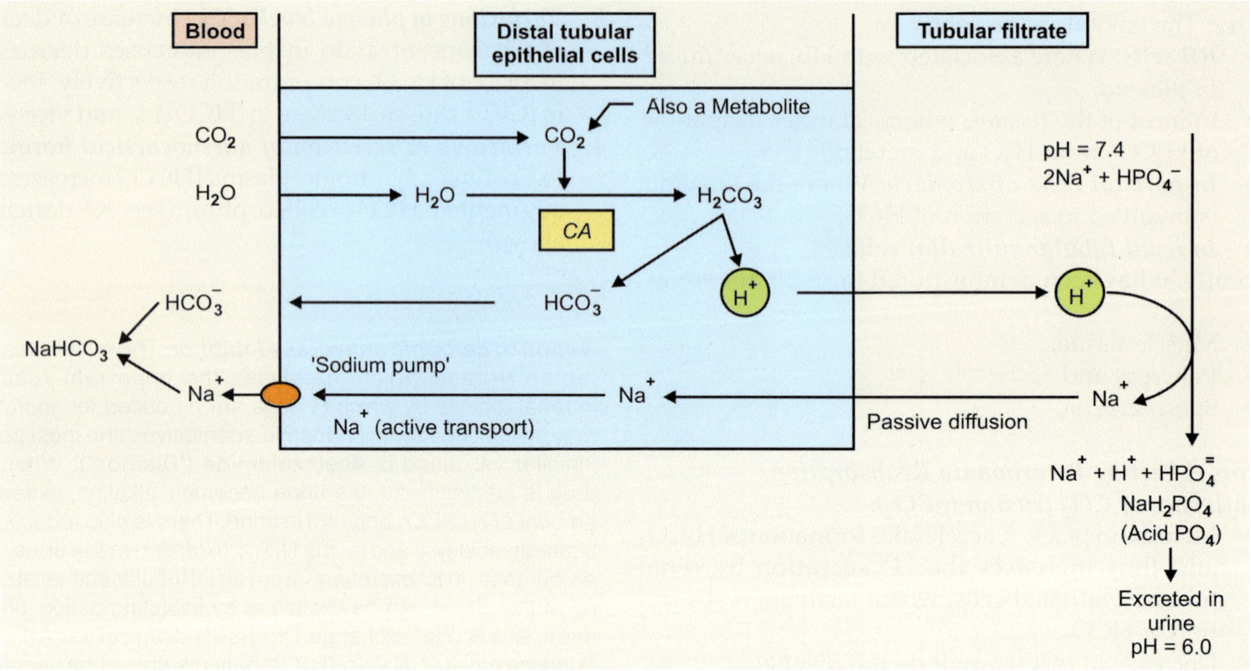

Fig. 41.3: Phosphate mechanism

2. In addition to above, if the cells require NH_3 more
   - NH_3 can also be formed from other amino acids by oxidative deamination by *L-amino acid oxidase*.
   - NH_3 can also be formed from glycine by *glycine-oxidase.*

The NH_3 thus formed forms NH_4^+ ions by combining with H^+ ions and NH_4^+ ions can exchange Na^+ ion from NaCl.

NH_4^+ ions formation

- NH_3 can diffuse into the tubular filtrate and there forms NH_4^+ ions in combination with H^+ ions.
- NH_3 can combine with H^+ ions inside the cells and then NH_4^+ ions come into tubular filtrate. This probably is not the principal mechanism as NH_4^+ ions are less readily permeable to tubular epithelial cells (Fig. 41.4).

The NH_3 production is greatly increased in metabolic acidosis and negligible in alkalosis. It is also observed that activity of renal *glutaminase* is enhanced in acidosis. The NH_3 mechanism is a valuable device for the conservation of fixed base. Under normal conditions, 30 to 50 mEq of H^+ ions are eliminated per day, by combination with NH_3 and about 10 to 30 mEq, as titratable acid, i.e. buffered with PO_4.

ANION GAP

The "anion gap" is a mathematical approximation of the difference between the anions and cations routinely measured in serum.

Routine electrolyte measurements include Na^+, K^+, Cl^- and HCO_3^- (as total CO_2). The unmeasured cations, i.e. Ca^{++}, Mg^{++} average 7 mmol/L and the unmeasured anions, i.e. $PO_4^=$, $SO_4^=$, proteins$^-$ and organic acids average 24 mmol/L.

If the Cl^- and the total CO_2 concentrations are summed and subtracted from the total of Na^+ and K^+ concentrations, the difference should be less than 17 mEq/L (mmol/L).

(a) ***If the anion gap exceeds 17 mmol/L this usually indicates significantly increased concentrations of unmeasured anions.***

Causes

- Uraemia with retention of "fixed" acids.
- Ketotic states, e.g. diabetes mellitus, alcoholism, starvation.
- Lactic acidosis, e.g. shock.
- Toxin ingestion, e.g. methanol, salicylate, ethylene glycol.
- Increased plasma proteins, e.g. in dehydration.

(b) ***An increased "anion gap" occurs occasionally in metabolic alkalosis.*** This is felt to be due to filtration of plasma proteins, resulting in loss of H^+ and consequent increase in the proteins net negative charge. In addition, plasma protein concentration may increase from the ECF deficit that occurs in metabolic alkalosis.

(c) ***Decreased anion gap less than 10 mmol/L can result from either:***

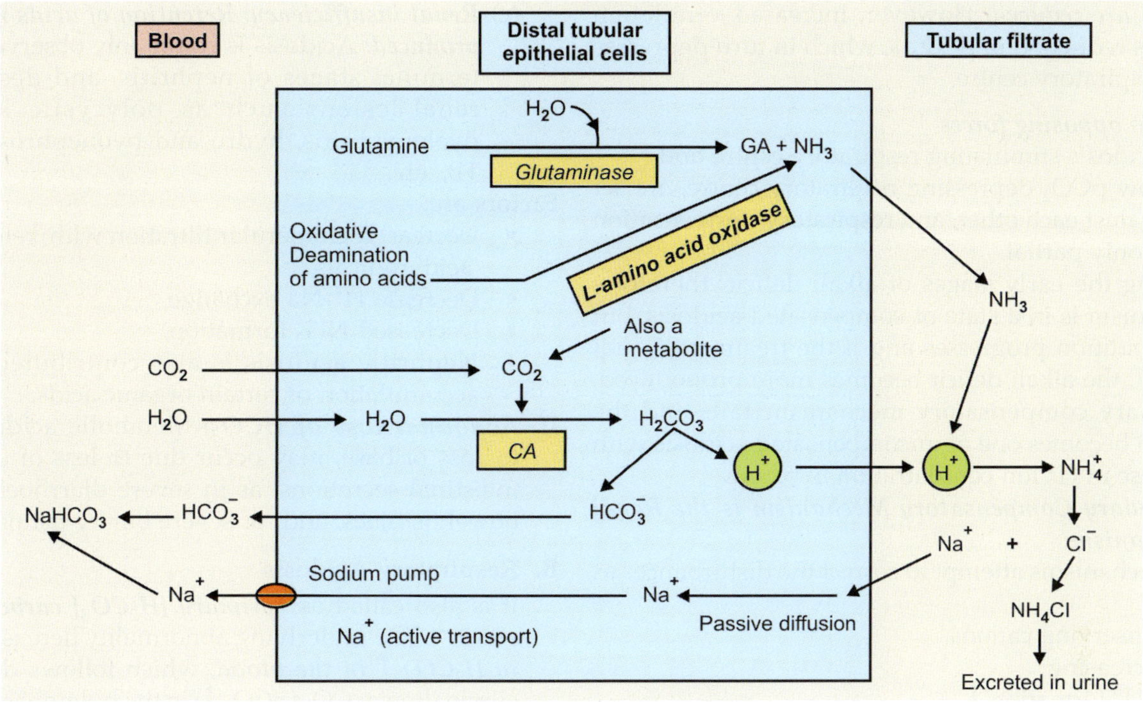

Fig. 41.4: Ammonia mechanism

- An increase in the unmeasured cations, or
- A decrease in unmeasured anions.

1. ***An increase in unmeasured cations:*** This can be seen in:
   - Lithium intoxication,
   - Hypermagnaesemia,
   - Multiple myeloma,
   - Polyclonal gammopathy and
   - Polymyxin B therapy, since the drug is polycationic.

The reason for the decreased gap due to the presence of increased γ-globulins is the fact that these proteins may have net +ve charge at physiologic pH.

2. ***Decrease in unmeasured anions occurs in***
   - Hypoalbuminaemia and
   - Hyponatraemia, with normal or increased EC fluid (e.g., SIADH). This is postulated to result from the selective renal excretion of unmeasured anions in this condition. Finally a spurious increase in measured Cl^- caused by Bromide intoxication can cause a spurious increase in the calculated gap.

CLINICAL USE

The "anion gap" is useful also for Quality control of laboratory results for Na^+, K^+, Cl^- and total CO_2. If an increased or decreased anion gap is calculated for a set of electrolytes from a healthy individual, this would indicate that one or more of the laboratory results are erroneous. Another possible explanation is that a mixed acid-base disturbance is present.

ACID-BASE IMBALANCE

Acid-Base imbalance can manifest as ***acidosis*** and ***alkalosis.***

A. Acidosis: Which can be **(1)** ***metabolic acidosis*** and **(2)** ***respiratory acidosis.***

B. Alkalosis: Which can be **(1)** ***metabolic alkalosis*** and **(2)** ***respiratory alkalosis.***

All of the above may be in *compensated* phase and *uncompensated* phase.

ACIDOSIS

A. Metabolic Acidosis

Also called as ***primary alkali deficit***. It is the commonest disturbance of acid-base balance observed clinically. It is caused when there is a reduction in the plasma HCO_3^- ↓ (B. HCO_3 ↓) with either no or little change in the H_2CO_3 fraction.

Mechanisms: If primary deficit of HCO_3^- occurs the ratio $[HCO_3^-]/[H_2CO_3] = 20/1$, is decreased, i.e. pH is decreased resulting in metabolic acidosis (primary bicarbonate deficit).

(a) ***Primary compensatory mechanism:*** The respiratory centre is stimulated by acidosis causing deep and rapid ***(Kausmaul)*** breathing. This increased ventilation will result in CO_2 loss and reduction in $[H_2CO_3]$ ↓ (carbonic acid). As a result, the ratio of $[HCO_3^-]/[H_2CO_3]$ is restored towards 20:1, ***as levels of both in***

blood are reduced. However, increased ventilation causes reduction in $pCO_2 \downarrow$, which in turn depresses the respiratory centre.

Thus, two opposing forces

1. Acidosis stimulating respiratory centre and
2. Low pCO_2 depressing respiratory centre, are set against each other, and respiratory compensation is only partial.

During the early stages of alkali deficit, therefore, the organism is in a state of compensated acidosis, but as the condition progresses and if the treatment is not instituted, the alkali deficit becomes more pronounced, the primary compensatory mechanism fails and the condition becomes one of uncompensated acidosis with an increase in H^+ ion concentration in blood.

(b) *Secondary Compensatory Mechanism is the Renal Mechanism*

Renal mechanisms attempt to correct the disturbances as follows:

1. By conserving cations
2. By increasing
   - NH_3 formation $\uparrow$
   - H^+ excretion compared to K^+ excretion in distal tubule, and
   - HCO_3^- reabsorption

Biochemical Characteristics

(a) ***Uncompensated:*** If uncompensated, it is characterised biochemically in plasma or blood as follows:
- Disproportionate decrease in $[HCO_3^-] \downarrow$
- Decrease in $[H_2CO_3] \downarrow$ and $pCO_2 \downarrow$
- ***Decrease in total CO_2 content*** $[HCO_3^-] + [H_2CO_3]$
- Decrease in $[HCO_3^-] : [H_2CO_3]$ ratio $\downarrow$
- Decrease in pH $\downarrow$

(b) ***Fully compensated:*** If fully compensated the CO_2 content is low, but the decrease in $[HCO_3^-]$ and $[H_2CO_3]$ is proportionate, the $[HCO_3^-] : [H_2CO_3]$ ratio and pH remain within normal limits.

Urinary findings: The urinary $NH_3 \uparrow$ and titratable acidity $\uparrow$ are increased (if kidneys are functioning normally).

Causes

I. ***Abnormal increase in "anions",*** other than HCO^-_3 ("acid-gain" acidosis) resulting from:

(a) ***Endogenous production of acid ions when excessive*** as occurring in:
• Diabetic acidosis, • Lactic acidosis, • Starvation, • High fever, • Violent exercise, and shock, • Haemorrhage and anoxia.

(b) ***Ingestion of acidifying salts,*** dietary or iatrogenic: May be produced by the administration of excessive quantities of acids, e.g. acetyl salicylic acid, phosphoric acid, HCl, NH_4Cl and NH_4NO_3, Mandelic acid, etc.

(c) ***Renal insufficiency: Retention of acids normally produced:*** Acidosis is commonly observed in the terminal stages of nephritis, and destructive renal lesions such as polycystic kidneys, pyelonephritis, hydro and pyonephrosis, renal TB, etc.

Factors are:
- Decreased glomerular filtration with retention of 'acid' radicals,
- Decrease H^+-Na exchange,
- Decreased NH_3 formation,
- Nephritic acidosis is also contributed by the accumulation of certain organic acids.

II. ***Abnormal loss of HCO_3^-:*** Metabolic acidosis due to loss of base, may occur due to loss of excessive intestinal secretions, as in severe diarrhoeas, small bowel fistulaes, and/or severe biliary fistulaes.

B. Respiratory Acidosis

It is also called as ***"primary $[H_2CO_3]$ carbonic acid excess".*** The underlying abnormality here is ***increase in $H_2CO_3 \uparrow$ in the blood,*** which follows decreased elimination of CO_2 ($pCO_2 \uparrow$) in the pulmonary alveoli. This may result from:
- Breathing air containing abnormally high percentage of CO_2, and
- Conditions in which elimination of CO_2 through lungs is retarded.

Mechanism: If excretion of CO_2 through lungs is impaired (e.g. emphysema or depression of respiratory centre), more CO_2 will accumulate in blood, resulting in excess H_2CO_3 formation $[H_2CO_3] \uparrow$. This results in lowering the ratio of $[HCO_3^-]/[H_2CO_3]$, resulting lowering in pH $\downarrow$ and is described as ***"Respiratory acidosis" (carbonic acid excess).***

Compensatory mechanism: In this condition, the respiratory mechanism becomes secondary and renal mechanism becomes of prime importance.

(a) Respiratory mechanism: Increased stimulation to respiratory centre (RC) by the increased CO_2 tension ($pCO_2 \uparrow$) results in increased depth and rate of respiration with consequent increased ventilation. This mechanism becomes secondary in importance as the defect may be with the RC, its depression/or some pathology in the Lungs. **As a result this compensatory mechanism becomes less effective.**

(b) Renal mechanism: It is of prime importance. More HCO_3^- are reabsorbed from tubules in response to raised pCO_2 in blood and ratio of $[HCO_3^-]/[H_2CO_3]$ is restored towards 20:1 as the levels of both in blood are increased.

Biochemical characteristics

(a) If *uncompensated*, it is characterised biochemically (plasma or blood) as follows:

- Disproportionate increase in $[H_2CO_3]$ ↑ (pCO_2) ↑
- Increase in $[HCO_3^-]$ ↑
- ***Increase in total CO_2 content*** ↑
- Decrease in $[HCO_3^-]$: $[H_2CO_3]$ ratio ↓
- Decrease in pH ↓

(b) If fully ***compensated,*** the CO_2-content is high, but the increase in $[H_2CO_3]$ and $[HCO_3^-]$ are proportionate, the $[HCO_3^-]$: $[H_2CO_3]$ ratio and pH remaining within normal limits.

Urinary findings: The urinary NH_3 ↑ and titratable acidity are increased ↑ (if kidneys are functioning normally).

Causes

I. Conditions in which there is depression or suppression of respiration

(a) Damage to CNS:

- *Brain damage:* Trauma, inflammation, or compression and convulsive disorders.
- *Drug poisoning:* Like morphine and barbiturates.
- Excessive anaesthesia, and bulbar polio.

(b) Loss of "ventilatory functions" due to increased intrathoracic pressure or loss of elasticity.

- Tension cyst/and tension pneumothorax, pulmonary and mediastinal tumours, emphysema.

(c) Effects of pain, e.g. Pleurisy

II. Conditions causing impairment of diffusion of CO_2 across alveolar membrane ***(Reduced alveolorespiratory function).*** Reduction of respiratory surface:

- Emphysema
- Pulmonary oedema, and congenital alveolar dysplasia (they also cause thickening of alveolar membrane or exudates).

III. Conditions in which there is an obstacle to the escape of CO_2 from the alveoli:

- Obstruction to respiratory tract
  - Laryngeal obstruction
  - Asthma.
- Rebreathing from a closed space.

IV. Conditions in which pulmonary blood flow is insufficient:

- Certain congenital heart diseases.
- Ayerza's disease.

For differentiation of metabolic and respiratory acidosis, (Table 41.1).

ALKALOSIS

A. Metabolic Alkalosis

Also called as ***primary alkali excess.*** This condition results from an absolute or relative increase in $[HCO_3^-]$. Primary alkali excess or increase in the "alkali reserve" is the most frequent cause of clinically observed alkalosis.

Mechanism

1. Excess of HCO_3^- accumulation (soluble alkali ingestion) causes an increase in the ratio of $[HCO_3^-]/[H_2CO_3]$ (i.e. pH is increased ↑) and it is known as ***"Metabolic alkalosis" ("bicarbonate excess").***
2. The respiratory centre (RC) is inhibited by alkalosis causing shallow, irregular breathing. This reduced ventilation will result in CO_2 retention and increases in carbonic acid level $[H_2CO_3]$ ↑.
3. The ratio of $[HCO_3^-]/[H_2CO_3]$ will be restored towards 20:1 as the levels of both in blood are increased.
4. However, decreased ventilation raises pCO_2, which tends to stimulate the RC.

Again, opposing forces

- Alkalosis depressing, and
- Raised pCO_2 stimulating the RC are working simultaneously and the respiratory compensation is incomplete

Renal mechanisms: Increases the excretion of:

- Cations ↑
- HCO_3^- ↑ (replacing Cl^- in urine). Both are due to decrease H^+-Na^+ exchange.
- K^+ excretion increases in the distal tubules instead of H^+.
- There is reduced NH_3 ↓ formation and excretion of non-volatile acids, viz. lactic acid and ketoacids.

Following compensatory mechanisms operate

- Decreased pulmonary respiration ↓
- Increased alkali excretion ↑
- Decreased acid excretion ↓
- Decreased NH_3 formation ↓
- Retention of acid metabolites

Urinary findings: Urinary acidity decreases ↓ and decrease NH_3 formation. Decrease in titratable acidity ↓.

Other Biochemical Changes and Clinical Manifestations

The following may accompany alkalosis:

- *Tetany:* In both types of alkalosis, respiratory/metabolic, tetany may occur due to decreased ionisation of calcium salts in the EC fluid. ***The total serum calcium may remain within normal limits, but ionic calcium may decrease to produce tetany.***
- *Hypokalaemia:* Increased excretion of K^+ from distal tubules can produce K^+ depletion (decrease serum K^+ concentration—hypokalaemia).
- *Kidney damage:* Mainly degenerative changes in the tubules (nephrosis) occurs frequently with oliguria and N_2-retention.

Table 41.1: Differentiation of metabolic acidosis and respiratory acidosis

Metabolic acidosis	***Respiratory acidosis***
1. ***Primary bicarbonate, HCO_3^- deficit***	1. ***Primary carbonic acid excess***
2. $\frac{B.\ HCO_3 \downarrow}{H_2CO_3} = \frac{20}{1} \downarrow = pH \downarrow$	2. $\frac{B.\ HCO_3}{H_2CO_3 \uparrow} = \frac{20}{1} \downarrow = pH \downarrow$
3. (a) In uncompensated phase • Disproportionate decrease in [HCO_3^-] ↓ • [H_2CO_3] ↓ • pCO_2 ↓ • **Total CO_2 ↓ decreased** • Ratio ↓ (b) In fully compensated phase: Total CO_2 is low ↓, but the decrease in [HCO_3^-] and [H_2CO_3] is proportionate and ratio 20:1 and pH maintained.	3. (a) In uncompensated phase • Disproportionate increase in [H_2CO_3] ↑ • increase in [HCO_3^-] ↑ • pCO_2 ↑ • **Total CO_2↑ increased** • Ratio ↓ (b) In fully compensated phase: Total CO_2 content is high↑, but the increase in [H_2CO_3] is proportionate and ratio 20:1 and pH maintained.
4. ***Compensatory mechanism*** (a) ***Primary:*** Respiratory-low pH stimulates respiratory centre producing hyperventilation and decrease in [H_2CO_3] ↓ (b) ***Secondary:*** Renal • H^+ -Na exchange ↑ increased • HCO_3^- reabsorption ↑ increased • NH_3 formation ↑ increased (c) ***Urinary findings*** • pH: Acidic • Increase excretion of NH_4Cl and NaH_2PO_4 • Increase in titratable acidity	4. ***Compensatory mechanism*** (a) ***Primary:*** Renal: Most important • Increase in H^+: Na exchange • More HCO_3^- reabsorption • Increase NH_3 formation (b) ***Secondary:*** Respiratory—partial Low pH and high CO_2 induces hyperventilation But CO_2 elimination is partial as the pathogenesis involves Lung disorders/or depression of respiratory centre. (c) ***Urinary findings*** • pH: Acidic • Increase in excretion of NH_4Cl and NaH_2PO_4 • Increase in titratable acidity
5. ***Causes*** 1. Abnormal increase in anions other than HCO_3^- (acid gain acidosis) • Endogenous production of acid ions when excessive – Diabetic acidosis – Starvation – High fever – Violent exercise (LA) – Lactic acidosis due to other causes like shock and haemorrhage • Ingestion of acidifying salts • Renal insufficiency: Retention of acids normally produced 2. Abnormal loss of HCO_3^-, e.g. in severe diarrhoea, fistulas	5. ***Causes*** 1. Conditions in which there is depression/or suppression of respiration (a) Damage to CNS • Brain damage (trauma, inflammation, compression), convulsions • Drug poisoning like morphine or barbiturates • Excessive anaesthesia • Bulbar polio (b) Loss of "ventilatory functions" due to increased intra-thoracic pressure or loss of elasticity, e.g. tension cyst, tesion pneumothorax, pulmonry and Mediastinal tumours, Emphysema. (c) Effects of pain like pleurisy 2. Conditions causing impairment of diffusion of CO_2 across alveolar membrane, "Reduced alveolar-respiratory function" • Reduction of respiratory surface: Emphysema, pneumonia, Pulmonary fibrosis, pulmonary oedema, etc. 3. Condition in which there is 'obstruction' to escape of CO_2 from the alveoli: • Obstruction to respiratory tract, • Rebreathing from a closed space 4. Conditions in which pulmonary blood flow is insufficient, e.g. certain congenital heart diseases, Ayerza's disease

- ***Ketosis and ketonuria:*** Ketosis and ketonuria may occur frequently in alkalosis, due to excessive vomiting because of inadequate carbohydrates intake.

Biochemical characteristics

(a) If ***uncompensated*** phase, it is characterised biochemically (plasma or blood) as follows:

- Disproportionate increase in [HCO_3^-] ↑
- Increase in [H_2CO_3] ↑, pCO_2 ↑
- ***Increase in total CO_2 content***
- Increase in [HCO_3^-]:[H_2CO_3] ratio ↑
- Increase in pH ↑

(b) If ***fully compensated*** the CO_2 content is high, but the increase in $[HCO_3^-]$ and $[H_2CO_3]$ are proportionate, and the $[HCO_3^-]$: $[H_2CO_3]$ ratio and pH remaining within normal limits.

Urinary findings: The urinary $NH_3\downarrow$ and titratable acidity $\downarrow$, both are decreased (if kidneys are functioning normally).

Causes

1. ***Excessive loss of HCl from stomach:*** The loss of excessive quantities of HCl from the stomach is encountered most frequently in individuals with:
   - Pyloric obstruction,
   - High intestinal obstruction,
   - Protracted gastric lavage without proper provision of acid replacement, in infants with pylorospasm,
   - Sometimes in patients with generalised peritonitis.

   As a result of the loss of Cl^- ions from the blood there is present in the body an excess of base, chiefly Na^+ and K^+, which is retained in the form of bicarbonate. ***In this way, a neutral salt (NaCl) is replaced by an alkaline salt ($NaHCO_3$).***
2. ***Alkali administration:*** Excessive intake of bases like $NaHCO_3$, Na and K acetates, lactates or citrates. **Lactates and citrates are converted into** HCO_3^-.
3. ***Potassium deficiency:*** Produces alkalosis (see above).
4. ***Roentgen ray, UV irradiation and radium therapy:*** A decrease in the H^+ ion concentration of the blood plasma (i.e. increased pH) has been observed following deep X-ray therapy, radium therapy and prolonged exposure to UV rays. The mechanism is not clear. In some cases (radiation sickness), it may be due to excessive voming.

B. Respiratory Alkalosis

Also called as ***primary H_2CO_3 deficit***. This condition occurs when there is a decrease in $[H_2CO_3]\downarrow$ fraction with no corresponding change in HCO_3^- in plasma.

$$\frac{B\,HCO_3}{\downarrow H_2CO_3} = \frac{20}{1} \uparrow = pH\uparrow$$

Excessive quantities of CO_2 may be washed out of the blood by hyperventilation.

Mechanism: Increased loss of CO_2 (due to hyperventilation), results in diminution of $[H_2CO_3]\downarrow$. The ratio of $[HCO_3^-]/[H_2CO_3]$ is increased $\uparrow$ (i.e. pH is increased) and is termed ***"respiratory alkalosis" (carbonic acid deficit)***. In this condition, due to increased CO_2 loss, pCO_2 is low, which leads to less H^+–Na exchange and less bicarbonate is reabsorbed (i.e. more HCO_3^- is excreted) by the renal tubules and the ratio of $[HCO_3^-]/[H_2CO_3]$ returns towards normal, i.e. 20:1, as levels of both in blood are decreased. Alkalosis and low pCO_2 depress respiratory centre and excretion of CO_2 is reduced.

Compensatory mechanisms

In this condition, main compensatory mechanism is 'renal'.

- Excretion of alkali in the form of HCO_3^-
- Decreased excretion of acid
- Decreased excretion of NH_3 in the urine
- Retention of Cl^- in the blood.

In view of the pathogenesis of this condition, the task of compensating for this defect falls on the kidneys. Other features are similar to metabolic alkalosis

Biochemical Characteristics

(a) If ***uncompensated,*** it is characterised biochemically (plasma or blood) as follows:
- Disproportionate decrease in $[H_2CO_3]\downarrow$ and $pCO_2\downarrow$
- Decrease in $[HCO_3^-]\downarrow$
- ***Decrease in CO_2 content*** $\downarrow$
- Increase in $[HCO_3^-]$: $[H_2CO_3]$ ratio $\uparrow$
- Increase in pH $\uparrow$

(b) If ***fully compensated,*** the CO_2 content is low, but the decrease in $[HCO_3^-]$ and $[H_2CO_3]$ is proportionate, the $[HCO_3^-]$: $[H_2CO_3]$ ratio and pH remaining within normal limits.

Urinary findings: The urinary $NH_3\downarrow$ and titratable acidity $\downarrow$ are both decreased (if kidneys are functioning normally).

Causes

1. *Stimulation of Respiratory Centre (RC)*
   - *In CNS diseases*, e.g. meningitis, encephalitis. Alkalosis due to hyperventilation has been observed in some cases of meningitis/encephalitis, etc. manifesting hyperpnoea, over prolonged periods of time.
   - *Salicylate poisoning:* Large doses of salicylates, such as are sometimes given in the treatment of acute rheumatic fever, produce stimulation of Respiratory centre (RC) with consequent hyperventilation and tendency towards alkalosis.
   - *Hyperpyrexia:* Hyperventilation may occur as a result of the increased respiratory rate associated with increase in body temperature.
2. *Other Causes*
   - *Hysteria:* Hyperventilation during hysterical attacks.
   - *Apprehensive blood donors:* Hyperventilation tetany with alkalosis has been observed in apprehensive and hyperexcitable donors.
   - *High altitude effects:* Hyperpnoea occurring in untrained individuals ascending to high altitudes where the atmospheric O_2-tension is low (anoxic anoxaemia) commonly results in primary H_2CO_3 deficit and alkalosis.

- ***Injudicious use of respirators.***
- Sometimes in ***hepatic coma.***

For differentiation of metabolic and respiratory alkalosis (Table 41.2).

Table 41.2: Differentiation of metabolic alkalosis and respiratory alkalosis

Metabolic alkalosis	***Respiratory alkalosis***
1. ***Primary HCO_3^- excess***	1. ***Primary H_2CO_3 deficit***
2. $\frac{B\,HCO_3\uparrow}{H_2CO_3} = \frac{20}{1}\uparrow = pH\uparrow$	2. $\frac{B.HCO_3}{H_2CO_3\downarrow} = \frac{20}{1}\uparrow = pH\uparrow$
3. (a) In uncompensated phase: Disproportionate increase • $[HCO_3^-]\uparrow$ • $[H_2CO_3]\uparrow$ or N • $pCO_2\uparrow$ or N • Ratio ↑ • **Total $CO_2\uparrow$** • pH ↑ (b) In fully compensated phase: Total CO_2 is high ↑ but increase in $[HCO_3^-]$ and $[H_2CO_3]$ are proportionate and ratio 20:1 and pH maintained.	3. (a) In uncompensated phase: Disproportionate decrease • $[H_2CO_3]\downarrow$ • $[HCO_3^-]\downarrow$ or N • $pCO_2\downarrow$ or N • Ratio ↑ • **Total $CO_2\downarrow$** • pH ↑ (b) In fully compensated phase: Total CO_2 content is low↓ decrease in $[HCO_3^-]$ and $[H_2CO_3]$ are proportionate and ratio 20:1 and pH maintained.
4. ***Compensatory mechanisms*** (a) ***Primary:*** Respiratory Depression of RC and hypoventilation leading to retention of CO_2 (b) ***Secondary:*** Renal • H^+ –Na exchange ↓ • NH_3 formation ↓ • bicarbonate (HCO_3^-) reabsorption ↓ • K^+ excretion ↑ • Cl^- retention (c) ***Urinary findings*** • pH of urine: Alkaline • Decrease $NH_3\downarrow$ • Decrease titratable acidity ↓	4. ***Compensatory mechanisms*** (a) ***Primary:*** Renal • Decreased H^+ –Na^+ exchange ↓ • Decreased excretion of acid ↓ • Increased excretion of HCO_3^- ↑ • Decreased excretion of NH_3 ↓ • K^+ excretion ↑ • Cl^- retention (b) ***Secondary:*** Respiratory-High pH and low pCO_2 produces hypoventilation and increase in H_2CO_3 (c) ***Urinary findings*** • pH of urine: Alkaline • Decrease in $NH_3\downarrow$ • Decrease titratable acidity↓
5. ***Other findings*** In both the conditions in association with alkalosis both may have: • Low ionic Ca^{++} leading of tetany • K^+ depletion: Hypokalaemia • Ketosis and ketonuria may develop • Kidney damage degenerative changes in tubules leading to N_2 retention and oliguria may occur	5. ***Other findings*** • Low ionic Ca^{++} leading to tetany • K^+ depletion: Leading to hypokalaemia • Ketosis and ketonuria may develop • Kidney damage may occur leading to oliguria N_2-retention
6. ***Causes*** • Excessive loss of HCl: – Protracted gastric lavage – Pyloric obstruction – High intestinal obstruction – Pylorospasm • Alkali ingestion and alkali administration • Excessive loss of K^+ leading to K^+ deficiency • X-ray therapy, UV radiation and Radiation therapy.	6. ***Causes*** 1. Stimulation of respiratory centre (RC) • CNS disease: Meningitis, encephalitis, etc. • Salicylate poisoning • Hyperpyrexia 2. Other causes • Hysteria • Apprehensive blood donors • High altitude ascending • Injudicious use of respirator • Some cases of hepatic coma

I. Questions (Essay type and short notes)

1. What is meant by acid-base balance in the body? Describe the role of blood buffers and the kidneys in maintaining the same.
2. Describe in detail the renal and respiratory mechanisms for regulation of acid-base balance.
3. What is normal blood pH? What are buffers? Describe their role in maintaining the normal blood pH.
4. Describe in brief, the role of kidney and respiration in the maintenance of acid-base balance.
5. What is meant by acid-base balance in the body? Describe the role of lungs and kidneys in maintaining the blood pH.
6. Explain the following:
   a. Bicarbonate buffer system is linked with respiration
   b. Phosphate buffer system is linked with kidneys.
7. Describe the role of blood buffers and kidney in maintenance of acid-base balance.
8. Define buffer. Name the important buffer systems in the body. Describe in brief the role of kidney in maintenance of acid-base balance.
9. What are buffers? Describe their role in maintaining the normal blood pH.
10. Describe the role of kidney in the regulation of acid-base balance.
11. What is respiratory acidosis? Describe the role of lungs in maintenance of acid-base balance.

Short Notes

a. Respiratory control of acid-base balance.
b. Alkali reserve
c. Blood buffers
d. Titratable acidity of urine and urinary NH_3
e. Plasma buffers
f. Role of haemoglobin as a buffering agent
g. Paradoxic aciduria
h. Respiratory alkalosis
i. Metabolic alkalosis
j. Respiratory acidosis
k. Respiratory alkalosis and its compensation
l. Metabolic acidosis
m. pH and blood buffers
n. Buffer systems of blood.

Differentiate

a. Respiratory and metabolic alkalosis
b. Respiratory and metabolic acidosis
c. Acidosis and alkalosis.

II. MCQs (Give one correct answer)

1. The most important physiological buffer in the blood is:
   a. Haemoglobin buffer system
   b. Acetate buffer
   c. Phosphate buffer
   d. Bicarbonate buffer
   e. Proteinate buffer
2. In the reaction $HA \rightleftharpoons H^+ + A^-$, the conjugate acid and conjugate base are:
   a. HA and H^+ b. HA and A^-
   c. H^+ and A^- d. A^- and H^+
   e. A^- and HA
3. If the pH of blood is 7.4, then the ratio of [$NaHCO_3$] and [H_2CO_3] will be:
   a. 5 : 1 b. 10 : 1
   c. 20 : 1 d. 25 : 1
   e. 30 : 1
4. The ratio of Alkaline PO_4 (Na_2HPO_4) to acid PO_4 (NaH_2PO_4) in blood is:
   a. 2 : 1 b. 4 : 1
   c. 8 : 1 d. 6 : 1
   e. 10 : 1
5. The ratio of Alkaline PO_4 (Na_2HPO_4) to acid PO_4 (NAH_2PO_4) in urine is:
   a. 1 : 2 b. 1 : 4
   c. 1 : 6 d. 1 : 9
   e. 1 : 12
6. All of the following are associated with metabolic alkalosis, except:
   a. Ingestion of NH_4CI
   b. Loss of HCI
   c. Prolonged steroid therapy
   d. Diuree therapy
   e. Pyloric obstruction
7. Metabolic alkalosis associated with prolonged vomiting or gastric aspirations is due to loss of which of the following:
   a. Hydrogen ion b. Chloride ion
   c. Sodium ion d. Potassium ion
   e. Bicarbonate ion
8. In which of the following enzyme, zinc constitutes an integral part of the enzyme:
   a. Pepsin b. Trypsin
   c. Cytochrome oxidase d. Carbonic anhydrase
   e. Xanthine oxidase
9. Which of the following biochemical characteristic is true in uncompensated phase of metabolic alkalosis.
   a. Disproportionate decrease in [HCO_3^-]$\downarrow$
   b. Increase in total CO_2 $\uparrow$ content
   c. Decrease in pCO_2 $\downarrow$
   d. Decrease in ratio of [HCO_3^-]:[H_2CO_3] $\downarrow$
   e. Decrease in [H_2CO_3] $\downarrow$
10. Which of the following biochemical parameter is true in a case of Respiratory acidosis in uncompensated phase:
    a. Disproportionate increase in [H_2CO_3] $\uparrow$
    b. Decrease in total CO_2 content $\downarrow$
    c. Increase in ratio of [HCO_3^-]: [H_2CO_3]
    d. Decrease in [HCO_3^-] $\downarrow$
    e. None of the above
11. Paradoxic aciduria may be seen in following condition, except:
    a. K^+ deficiency
    b. Prolonged treatment with steroids

c. Na^+ deficiency
d. In Cushing's syndrome (hypercorticism)
e. Administration of K^+ free fluids in post-operative patients.

12. In prolonged administration of acetazolamide (Diamox) as diuretic in a patient of congestive cardiac failure produces:
a. Acidic urine
b. Increase in titratable acidity of urine
c. Alkaline urine
d. Decreased bicardonate excretion
e. Increase in NH_4CL excretion.

Answers to MCQs

1. d	2. b	3. c	4. b	5. d	6. d
7. b	8. d	9. b	10. a	11. c	12. c.

CHAPTER 42

CEREBROSPINAL FLUID (CSF)–CHEMISTRY AND CLINICAL SIGNIFICANCE

Major Concepts

A. To study what is CSF? Learn its formation and circulation. To know the composition of normal CS fluid in health, its physical and chemical characteristics.

B. To study the changes that can occur in physical appearance and chemical constituents in diseases and how these changes help in diagnosis of CNS diseases.

Specific Objectives

A. 1. Cerebrospinal fluid is not just the ultrafiltrate of plasma. Study its formation and circulation.
   2. Learn briefly how a sample of CSF is collected for analysis.
   3. Study the composition both physical and chemical of normal CSF.
   4. Study the various chemical constituents present in normal CSF and their normal value in health.

B. 1. Learn the various changes in physical characteristics like appearance, colour, turbidity, coagulum formation, pressure, etc. that can occur and their significance in various central nervous system (CNS) disorders.
   2. Study the chemical constituents that are routinely done in the laboratory and their alterations and significance in various CNS disorders.
   3. Learn also the other chemical constituents which are not routinely done and their alterations and significance in various CNS diseases.
   4. Make in a tabular form the characteristic alterations physical and chemical that take place in some of the common CNS diseases.

INTRODUCTION

The surface of the central nervous system is covered by the meninges, three layers called as the piamater, arachnoid mater and duramater. The last is the outermost layer. CSF is found between the pia and the arachnoid, i.e. sub-arachnoid space and is formed by 'active' secretion from the cells of the choroid plexuses, the vascular structures lying within the ventricles of the brain. ***It is not just a plasma ultrafiltrate.***

In normal healthy adults, the rate of formation of CSF is 100 to 250 ml per 24 hours and total volume of CSF is approx 100 to 200 ml. In addition to the cell count, microbiological and serological tests, the chemical tests that are commonly and routinely carried out on CSF are protein and glucose estimations, chloride tests are rarely done now.

Other biochemical tests, viz. urea, calcium, bicarbonates, enzymes, etc. are rarely done. The concentration in the CSF of certain drugs may sometimes be required.

Site of withdrawal: The chemical composition of CSF from a normal subject depends on the site of withdrawal, ventricular and lumbar CSF differ from each other in certain respects, which will be discussed under the various constituents. Generally only lumbar CSF is examined. The spinal cord ends near the 1st lumbar vertebra and the accumulation of fluid below this, called as lumbar fluid, is the portion of CSF commonly subjected for analysis. It is obtained by passing a lumbar puncture needle, aseptically, between the 3rd and 4th lumbar vertebrae into the subarachnoid space.

"Blood-Brain" Barrier

"A blood CSF barrier" exists for many substances including the blood constituents, drugs, enzymes, etc. and their concentration in CSF is lower than in plasma. In inflammatory states and in cerebrovascular accidents, the blood-CSF barrier may be impaired and the differences may be less marked.

Collection of sample: A sample of CSF submitted for chemical analysis, should if possible, be ***fresh and free***

from blood. The fluid should be collected in sterile containers and to be sent to the laboratory at the earliest. If the fluid cannot be analysed for glucose within half an hour of withdrawal, then to prevent glycolysis by any cells or bacteria present, the CSF should be collected in a bottle containing sodium fluoride.

COMPOSITION OF NORMAL CSF (LUMBAR FLUID)

Colour and appearance	:	Clear, colourless, no coagulum or deposit.
Pressure	:	60 to 150 mm CSF
Specific gravity	:	1.006 to 1.007
Cells	:	0 to 4 mononuclear cells per C. mm
pH	:	7.3 (anaerobically)
Protein content	:	10 to 45 mg per 100 ml
Globulins (Qualitative)	:	Not increased. Pandy's test and Nonne-Apelt tests-negative
Glucose	:	45 to 100 mg per 100 ml
Chlorides	:	700 to 760 mg per 100 ml as NaCl (120 to 130 mEq per litre)
Urea	:	20 to 40 mg per 100 ml
Calcium	:	5.5 to 6 mg per 100 ml

APPEARANCE OF CEREBROSPINAL FLUID

Normal CSF is clear and colourless and gives no coagulum or sediment on standing, if it is not contaminated (sterile). Abnormalities in appearance may arise in regard to:

(a) *Colour*
(b) *Turbidity*
(c) *Coagulum*

(a) Colour

The presence of blood is the main cause of an abnormal colour. ***Normally no RB cells should be present.***

1. *Trauma:* Some blood may be introduced as a result of trauma, while doing the LP In such cases, the first few drops are most mixed up with blood, so that if the first one or two ml is collected separately, the subsequent fluid should be clear or nearly clear (use two to three consecutive bottles for collection). The supernatent fluid after centrifugation would also be clear.
2. *Pathological:* Haemorrhagic fluid obtained in subarachnoid haemorrhage, haemorrhage into the ventricles, and following neurosurgical operations.

Note

Differentiation from Traumatic Blood

- Blood obtained in pathological conditions is more homogeneously mixed with CSF than is that introduced during collection so that there is not the clearance in subsequent collections.
- ***The RB cells of haemorrhagic fluid from pathological conditions have a crenated appearance,*** whereas blood obtained by puncture have a normal shape when seen microscopically.

3. *Xanthochromia:* This is the yellow colouration of CSF. This can be due either to Hb or to other pigments, usually bilirubin/or carotenoids.

After haemorrhage, xanthochromia is initially due to oxy-Hb which is then converted into bilirubin. Bilirubin can be detected in CSF after 6 hours after the haemorrhage, reaches a maximum concentration in approx 10 days time and after that the RB cells usually disappear. In such situations, the supernatant fluid obtained after centrifugation shows yellow colouration.

4. *Froin's syndrome:* This term used to denote to xanthochromic fluid obtained from lumbar region in cases in which there is complete block (spinal block) due to a tumour. The fluid has high protein concentration which can coagulate spontaneously. The cause of xanthochromia is capillary haemorrhage.
5. **Other Causes of Yellow Colouration of CSF**
   - The yellow colouration can be due to high CSF bilirubin in conditions like cholestatic jaundice and in icterus neonatorum.

   Cause: Both unconjugated bilirubin and conjugated bilirubin can pass across to CSF due to alteration in blood-brain barrier in jaundice. In premature infants also there is increased permeability of the blood-brain barrier.
   - Carotenoids can also pass from plasma into CSF when the permeability to blood-brain barrier is altered which may occur in inflammatory disorders or in space below a spinal block.

(b) Turbidity: Turbidity is seen when there is marked increase in the number of cells or when organisms are present and hence ***found in meningitis specially in coccal type. At least 400 to 500 polymorphs per c.mm are needed to give a visible turbidity.***

The CSF obtained from viral meningitis or tubercular meningitis is usually not turbid as the cell response in these cases are lymphocytic.

Note

- Small numbers of red cells may also give CSF an opalascent appearance.
- If traces of substances such as alcohol are mixed with the fluid during its collection some opalascence may result.

(c) Coagulum: Normal CSF does not form a fibrin clot on standing.

SECTION FIVE

Causes of Fibrin Clot

- In pathological haemorrhagic CSF, fibrinogen present in the blood of CSF may be sufficient to form a clot.
- Fibrin clot is formed readily on standing, when the protein content of CSF is high. ***Such fibrin clot can occur, when the CSF protein is above 2 g/l.***

CSF obtained from below a spinal block usually contains a high concentration of fibrinogen.

- *In tuberculous meningitis:* Often a fine delicate ***cob-web like coagulum*** forms if the fluid is allowed to stand overnight. Such a 'web' may take up *Myco-bacterium tuberculosis* along with, which can be easily demonstrated when a smear is made, stained and seen microscopically.
- Such fine clot on standing may develop in cases of neurosyphilis or poliomeningitis.

Note

CSF must be examined for "clot formation" within 24 hours of withdrawal, otherwise autolysis destroys the fibrin.

PRESSURE OF CSF

When collecting sample of CSF one can measure its pressure within the lumbar sac. With the patient lying quietly on his side, the pressure is normally 60 to 150 mm CSF (40 to 90 in children), increasing to 200 to 250 mm CSF on sitting.

Causes of Alterations in Intracranial Pressure

- Raised intracranial pressure can occur from a variety of causes, e.g. meningitis, and in tumours (intracranial).
- Low pressure may be seen below a block-tumours compressing spinal cord.

Note

In raised intracranial pressure, one should be careful in doing LP. ***If too much of fluid is removed there is danger of 'coning' of the brain and sudden death.***

BIOCHEMICAL CHANGES IN CSF

A. pH Value

pH of CSF collected anaerobically is 7.31 approx and is mainly dependant on the pCO_2 content. The pH may be a major regulator of the activity of the respiratory centre, while CO_2 equilibrates between plasma and CSF relatively quickly, changes in HCO_3 concentration are slower. Hence too rapid replacement of a HCO_3-deficit in plasma may adversely affect the CSF pH.

B. Chemical Constituents

Following are the chemical constituents which are routinely and commonly done in the laboratory:

I. Estimation of Glucose in CSF

This can be done by any of the usual blood glucose methods.

Precautions

Since the glucose concentration in CSF is normally somewhat lower than in blood, it is recommended that a larger volume of CSF may be used, when it is anticipated that there may be gross reduction of CSF glucose in certain diseases

- Streptomycin given intrathecally shortly before CSF sampling may interfere with routine method of copper reduction used.
- CSF, if it is contaminated during laboratory sampling may show a fall in glucose content if kept at room temperature.

CLINICAL SIGNIFICANCE

(a) Normal value: **Varies from 50 to 80 mg per 100 ml,** though a range of 45 to 100 is often allowed. CSF glucose level is slightly lower than the blood glucose. The ventricular CSF glucose is rather higher than the lumbar CSF glucose and approximates to blood glucose.

Note

- The concentration of glucose in CSF largely depends on blood glucose, and normally stays at about 60 per cent of its concentration in blood because of incomplete penetration of blood-brain barrier.
- The CSF glucose changes slowly when the blood glucose changes, and there is even a slight postprandial rise.
- If the blood-brain barrier is damaged, it becomes more permeable to glucose and CSF glucose approaches the blood glucose level.

(b) Decrease in CSF Glucose: Decrease in CSF glucose is the most important pathological change seen.

- *In coccal meningitis:* Due to meningococci, staphylococci, pneumococci, etc. glucose often disappears completely in CSF and it may be totally absent.

  Causes
  - In these CSF shows polymorphonuclear leucocytosis. Glucose level is lowered because the leucocytes and the cocci are glycolytic and thus use up the glucose.
  - Moreover, in chronic cases, there is also decreased permeability.
- *In tuberculous meningitis:* Glucose content may be reduced but it is rarely absent completely, usually it varies from 10 to 40 mg per 100 ml.
- *In viral meningitis:* The glucose concentration is often normal, but it is occasionally as low as 20 mg%.
- *In neurosyphilis:* It is almost always within normal limits.
- *In cryptococcal meningitis:* Usually low values are seen.
- Low values have occurred following carcinomatous infiltration of meninges, also in leukaemic cells infiltrations in leukaemias.

- Low CSF glucose also seen in hypoglycaemia due to various causes.

(c) Increase in CSF glucose

- Small increases are found in some cases of encephalitis, poliomyelitis, and cerebral abscess and values between 150 to 180 mg% have been found in some cases, but this is of little diagnostic value.
- Considerable increases are seen to occur in diabetic hyperglycaemia but the CSF value remains lower than in the blood.

Note: In the majority of cases of other CNS diseases, glucose content of CSF is usually normal.

II. Estimation of Chlorides in CSF

This is rarely done now.

Determination of Chlorides in CSF

- It is possible to titrate most CSF directly with an appropriate $AgNO_3$ solution using potassium dichromate as an indicator.
- When CSF is titrated with mercuric nitrate, unionised mercuric chloride is formed. When all the chloride ions are used up, the excess mercuric ions will react with indicator diphenyl carbazone to give a purple-colour complex, which gives the end point (Schales and Schales titrimetric method).

CLINICAL SIGNIFICANCE

(a) Normal value: The chloride content of normal CSF whether lumbar, cisternal or ventricular, lies between 700 to 760 mg NaCl/100 ml (120 to 130 mEq/L) and thus is appreciably higher than the plasma chloride.

Note

- It is mainly a consequence of "active" secretion at the choroid plexuses, only a small proportion being attributable to the Donnan effect of the difference in protein concentration between plasma and CSF.
- CSF chloride is affected by the plasma concentration and the difference between CSF and plasma becomes less if the meninges are inflamed.

(b) Decrease in CSF chloride in meningitis: A decrease in CSF chloride is seen in meningitis, ranging between 700 to 600 mg NaCl% (120 to 102 mEq/L). Occasionally the decrease may be between 600 to 500 mg NaCl% (102 to 85 mEq/L). ***The reduction is generally more marked in tubercular meningitis as compared to coccal meningitis.*** It is very unusual for the chloride content to be below 600 mg NaCl% (102 mEq/L) except in tubercular meningitis and found usually late in the disease.

Very occasionally, even in tuberculous meningitis, the chlorides may be almost normal. Earlier the low values in tuberculous meningitis used to be taken as diganostic for the disease. The fall is more due to associated vomiting in the disease.

(c) Increase in CSF chloride: An increase is sometimes found in the hyperchloraemic acidosis in chronic renal failure.

Note

- Normal figures are the rule in viral meningitis and in all other CNS disorders.
- CSF chloride estimation is now rarely undertaken as it gives no additional diagnostic information as are provided by CSF glucose and proteins.

III. Estimation of CSF Proteins

Determination of CSF proteins is usually done by turbidimetric methods but some of the colourimeteric methods for estimation of proteins have also been used.

Methodology (Turbidimetric Method)

Reagents

1. Sulphosalicylic acid 3 per cent solution
2. Proteinometer standards—one set

Procedure

- One ml of CS fluid is mixed with 3 ml of sulphosalicylic acid reagent in a small test tube
- Mix and allow to stand for 5 minutes
- The turbidity developed is compared against the tubes of proteinometer set.

Note: For values above 90 mg per 100 ml, dilute the CS fluid suitably with water before mixing with sulphosalicylic acid reagent.

CLINICAL SIGNIFICANCE

(a) Normal values: Protein content of normal lumbar CSF ranges between 15 to 45 mg per 100 ml, most containing 25 to 30 mg%. Protein content is almost entirely albumin, with small amounts of globulins.

Cisternal CSF: It has slightly lower protein, being most often in the region of 20 mg%.

Ventricular CSF: Usually still lower, approx. 10 mg%.

(b) Increase of CSF protein: An increase in the total protein is the commonest abnormality seen. The protein in such cases is a mixture of albumin and globulins. Increase in globulins can be roughly ascertained by qualitative test. Quantitative analysis can be done by electrophoresis and immunological methods.

Cause

The increase in proteins results from a breakdown of "blood-CSF and brain-CSF barriers" which may be due to: **(a)** An inflammatory reaction,

(b) ***Occasionally may be due to obstruction.***

- *Acute meningitis:* Marked increase in CSF protein is seen in various forms of acute meningitis. In many cases, the concentration does not exceed usually 1 g/L though increases up to 4 g/L are not infrequent in acute meningitis.

SECTION FIVE

- Also marked increase seen in polyneuritis and in presence of such tumours as acoustic neuroma and meningiomas.
- In chronic conditions, such as multiple sclerosis and general paresis there may be slight rise in CSF protein, which is principally due to an increase in γ-globulins.
- Very high CSF protein concentrations up to approx. 10g/L are found in CSF ***below a "spinal block"***, and this is mainly due to albumin which has leaked from the plasma. A spinal block is usually due to a tumour, but arachnoid adhesions in pyogenic meningitis can also cause loculation of fluid.
- CSF proteins also alter following marked changes in plasma proteins as occurs in multiple myelomatosis.
- Also increase in CSF protein is found occasionally in general toxic states such as uraemia or after a myelogram.

Relation of Increase in CSF Protein with Cellular Changes

The increase in CSF protein content is not always accompanied by an increase in cells.

- ***Both cells increase and proteins increase occur in inflammatory lesions*** and found in all types of meningitis. Also found in poliomyelitis, in general paralysis of insane, and in tabes dorsalis.
- Small increases in cells may occur in polyneuritis and in multiple sclerosis, not always with a protein increase except later one.
- A disproportionate increase in protein compared to cells ***albuminocytologic dissociation*** most commonly found in tumours, particularly spinal ones, also after cerebral arteriosclerosis and cerebral infarction, and in acute post-infective polyneuritis ***(Guillain-Barre Syndrome).***

GLOBULINS

1. Qualitative Tests

Simple chemical tests for excess of CSF globulins, e.g.

- **Pandy's test** using saturated phenol solution or
- **Nonne-Apelt test** using half-saturated ammonium sulphate are in common use which demonstrate the presence of an increase of globulins in CSF. Qualitative test allows a rapid approximate comparison to be made between the γ-globulin increase and that of total proteins.

Pandy's Test

One drop of CS fluid is added to one ml of Pandy's reagent (a clear 7% solution of phenol in water)

- A turbidity indicates increased globulins in CS fluid.

Nonne-Apelt Test

One ml of CS fluid is carefully layered over one ml of saturated Ammonium sulphate solution.

- A white ring at the junction of the two liquids indicates increased globulins.

2. **Electrophoresis:** Electrophoresis of CSF shows the presence of many different globulins. When electrophoresis is done on normal CSF in cellulose-acetate/ or agargel, the proteins are separated into different fractions. When expressed as percentage of total proteins of CSF the typical findings of normal CSF are as follows:

Percentage of Total Proteins	
• Pre-albumin	5.0
• Albumin	60.0
• α_1-Globulin	6.0
• α_2-Globulin	5.5
• β_1-Globulin	10.0
• β_2-Globulin	5.5
• γ-Globulin	8.0

Although some changes in globulins other than γ-globulin have been associated with neurological diseases, the results have not been consistent and are not used routinely for diagnosis.

3. **Immunoglobulins in CSF:** Predominant γ-globulin is IgG with a concentration of 30 mg/L and small amounts of IgA-4 mg/L and IgM varying from 0 to 6 mg/L.

Sources of Protein in CSF and its Relation with Diseases

(a) When CSF contains proteins derived from plasma as occurs in acute inflammatory lesions like acute meningitis, due to increased capillary permeability albumin and γ-globulins are increased. In coccal meningitis there is a particular increase in IgM.

(b) When CSF contains proteins derived from brain tissue and spinal cord, then mainly the local immunoglobulins are increased.

Recently considerable interest has been focussed on the CSF changes of γ-globulins in multiple sclerosis in particular. ***γ-globulins increase are disproportionately greater in multiple sclerosis and in neurosyphilis.*** In multiple sclerosis there is a predominant increase in IgG and "oligoclonal" γ-globulin bands derived from plasma cells and lymphocytes are often present. ***In multiple Sclerosis, the IgG antibodies may be "anti-myelin" antibodies, which accumulate in the plaques of demyelination.***

4. **"Index" of γ-globulin:** Attempts have been made to define an index which would give good discrimination between multiple sclerosis and neurosyphilis in one hand and other neurological conditions on the other. After estimation of γ-globulins by standard method it is expressed as % of CSF proteins. ***A figure of above 29 has been suggested as indicating multiple sclerosis and neurosyphilis.***

OTHER CHEMICAL CONSTITUENTS

Discussed below are the other chemical constituents which are ***not routinely done in the laboratory.***

1. **Urea:** Urea concentration in CSF is similar to that of plasma, unless the latter is changing rapidly, when the CSF response is delayed.

   When there is nitrogen retention due to chronic renal failure the CSF urea concentration rises in parallel with the plasma urea.

Value of CSF urea in postmortem diagnosis: It may be useful when deciding the cause of death at postmortem to know what the patients blood urea was during his final illness. It is known that an appreciable rise in blood urea may occur during the period immediately preceding the death **("agonal" period)** and also as a postmortem change. But urea content of CSF does not alter significantly during the "agonal" period and postmortem and that postmortem CSF urea levels correlate closely with antemortem blood levels before the 'agonal' period. Thus assay of CSF urea performed at autopsy may be valuable for retrospective diagnosis.

2. **CSF glutamine level:** Can be used as an 'indirect' Liver function test. (Refer chapter on Liver Function Tests).
3. **Calcium:** CSF calcium level is 1.2 to 1.4 mmol/L of which 95 per cent is ionised calcium. However, because CSF is an active secretion this does not strictly follow the changes in plasma ionised calcium in diseases. Calcium is lower than the serum value because of lower protein concentration. The level decreases if the ionic calcium value in serum is reduced as in tetany.
4. **CSF phosphate:** CSF phosphate level is somewhat lower than plasma but it is not helpful in diagnosis.
5. **Uric acid:** CSF uric acid is present in only low concentration 0.20 mg/100 ml. ***It has been claimed to be increased in progressive cerebral atrophy.***
6. **Bicarbonate:** The CSF bicarbonate concentration is 18 mmol/L, lower than in plasma to compensate for the increased chloride value and normal sodium and potassium concentration.
7. **Phenyl alanine:** In phenylketonuria, the CSF phenylalanine is raised with the plasma phenylalanine increased.
8. **CSF copper:** The copper concentration in CSF is increased in Wilson's disease.
9. **Lipids:** Lipids are also found in low concentration and it has been claimed recently that ***the cephalin fractions are increased in demyelinating disorders and desmosterol increases in gliomas.***
10. **Vitamins**
    - Vitamins B_{12} concentration in CSF is almost same as that of plasma. It is low in pernicious anaemia but the estimation does not aid in detection of cases with neurological involvement.
    - CSF folic acid is approximately five times the plasma level and it does not give any extra diagnostic information.
11. **Enzymes of CSF**

Various CSF enzymes have been investigated in CNS disorders.

(a) Aspartate Transaminase (AST)

- *Normal* 5 to 12 units/L.
- AST increase up to ten times of normal are found due to release of the enzyme from damaged brain tissue when there has been an abscess, cerebral haemorrhage and infarction and in primary or metastatic malignant disease, (even with a normal CSF protein).
- Rise has also been noted in some cases of multiple sclerosis.

(b) Lactate Dehydrogenase (LDH)

- Normal CSF LDH is 5 to 40 U/L. Similar to AST, increase of CSF LDH observed in abscess, cerebral haemorrhage/infarction and metastatic carcinoma.
- LDH isoenzyme pattern is similar to serum. ***Increase of LDH_4 isoenzyme of CSF has been reported in tubercular meningitis.***

(c) Creatine Kinase (CK): Creatine kinase is present in high concentration in brain tissue.

- CSF CK-enzyme is found to be increased in Duechene muscular dystrophy and in many other neurological disorders including post-ictal stage. The increase is due to BB isoenzyme.
- CSF-CK also rises in cerebral haemorrhage/ and cerebral infarction. CK also rises in plasma after cerebral infarction and the rise is mainly due to muscle CK-MM and not from brain (CK-BB) and has not been shown to have diagnostic value.
- Also increase in CSF-CK has been observed in pyogenic meningitis and tuberculous meningitis. ***A dividing line of 30 units/ml has been suggested, >30 is suggestive of tubercular meningitis and <30 pyogenic meningitis.***

LANGE COLLOIDAL GOLD REACTION

Serial dilutions of CSF are mixed with a colloidal gold solution and any colour change is noted. Colour changes are graded as 0 to 5.

SECTION FIVE

Table 42.1: Cerebrospinal fluid findings in CNS diseases

Disease	*Appearance*	*Cells*	*Proteins*		*Glucose mg%*	*Chlorides mg%*	*Lange's colloidal reaction*	*Remarks*
			Qualitative	*Quantitative mg%*				
Normal	Clear, colourless no clot	0-4 mononuclears per C. mm	0	10-15	45 to 100	700-760	0	—
(a) Acute Meningitis								
1. Pyogenic meningitis	Turbid may be thick clot	100 to 6000 95% polymorphs	++ to ++++	Markedly increased (1.0 to 10.0 g/l)	0 to 15 or absent	630 to 680	Meningitic	Cocci can be seen in smear or culture
2. Tuberculous meningitis	Clear, opalascent or white 'cob-web' clot on standing	Children early 10-100 Late: 100 to 1000 70 to 90% Lymphocytes, poly ±	± to +++	30 to 400 Highest shortly before death	15 to 20	Early 680-700 Late: 500-650	Weak Meningitic	*M. tuberculosis* from clot
3. Pneumococcal meningitis	Turbid to yellow clots	Acute cases slight increase. Less acute cases 100 to 5000 95% polymorphs	++ to ++++	100 to 200 mg% or higher	0 to 10	600 to 650	Meningitic	Pneumococci in smear and culture
4. Syphilitic meningitis	Clear to turbid may be fibrin clot	10-500 90% mononuclears (Lymphocytes)	±	25 to 60 mg%	Usually Normal	650 to 720	Luetic or paretic	—
Viral Meningitis	Clear or sometimes opaque	Early: 10-100 Late: 100-500 Lymphocytes ++	+	45-100	Normal	Normal	Normal	—
Meningism in acute fever	Clear	Slight increase Lymphocytes +	—	15-50	Normal	Normal	Normal	—
(b) Epidemic Encephalitis	Clear Occasional fibrin clot	10 to 200 All monocytes Less than 10 in 30 to 50% cases	±	25-60	Above Normal 65-120	Normal or increased	Paretic or Meningitic	—
(c) Acute Poliomyelitis	Clear to milky Occasional fibrin clot	Pre-paralytic 15 to 2000 polys Paralytic 10 to 100 mononuclears	± to ++	Pre-paralytic 25 to 60 paralytic 60-300	Normal	Normal	Variable	—
(d) Cerebral Tumour	Normal or Xanthochromic	Normal or 10-80	± to ++	50-200	40-100	Normal	Variable	High protein in acoustic neuroma and meningiomas
(e) Spinal Block (Froin's syndrome)	Opalascent with clot ++	Normal or slight increase	+ to +++	50-100	Normal	Normal	Variable or Meningitic	—
(f) Multiple Sclerosis	Normal	70 to 90% cases normal, others 5-50 (Lympho)	0 to +	30 to 90 mg% 10 to 40% of cases above normal Increase in γ-globulins	Normal	Normal	50-60% paretic curve	'Oligoclonal' γ-globulin (weak paretic)

1. Normal CSF usually gives zero colour change at all dilutions.
2. **Three types** of colour changes and abnormal curves have been recognised.

(a) Paretic curve: Colour change at low dilutions, known as "first zone" or paretic curve.

Example: **5 5 5 5 4 3 2 1 0 0 (strong reaction)**
2 3 3 3 3 1 0 0 0 0 (weak reaction)

Basis of the test: This is shown when total proteins of CSF is within normal limits or slightly increased and there is marked increase in γ-globulins.

(b) Luetic or tabetic curve: Colour change at medium dilutions, known as "Mid-zone" or Luetic (tabetic) curve.

Example: **0 1 2 3 3 3 2 1 0 0.**

Basis: This type of curve is seen when the total protein is raised but γ-globulin is raised more than albumin.

(c) Meningitic curve: Colour changes are seen at high dilutions, known as the "end-zone" or meningitic curve.

Example: **0 0 0 1 2 3 4 5 4 4 (strong reaction)**
0 0 0 1 2 3 3 2 1 0 (weak reaction)

Basis: This type of curve is seen when total protein is greatly raised due to rise in both albumin and γ-globulin, the albumin is 'diluted' out in higher dilutions so that γ-globulins have precipitating action.

Note

The Lange's colloidal test was principally used earlier for the investigation of neurosyphilis, and multiple sclerosis. But now with advent of electrophoresis and Immunodiagnostic methods it has become out dated.

Table 42.1 shows the CSF changes in different diseases.

I. Questions (Essay type and short notes)

1. Discuss the value of CSF examination in various CNS disorders.

Short Notes

a. Lange's colloidal gold test
b. Xanthochromia
c. Froin's syndrome
d. Albuminocytologic dissociation
e. Value of CSF enzymes

II. MCQs (Give one correct answer)

1. Xanthochromia may be caused by all of the following except:
   a. Pus b. Traumatic tap
   c. Recent haemorrhage d. Recent infarct
   e. Previous traumatic tap
2. Lumbar spinal glucose concentration is characteristically low in patients with:
   a. Meningioma b. Infarction
   c. Cerebral haemorrhage d. Bacterial infection
   e. Multiple sclerosis
3. CSF electrophoresis done in a suspected case of multiple sclerosis which fraction of CSF protein is typically elevated?
   a. α_1-Globulin, b. β_1-Globulin
   c. γ-Globulins d. α_2-Globulin
   e. Pre-albumin
4. Albumino-cytologic dissociation may occur in patients with all of the following *except:*
   a. Cerebral tumours b. Guillain-Barre syndrome
   c. Thrombosis d. Multiple sclerosis
   e. Coccal meningitis
5. Lumbar spinal fluid protein increase may be characteristically associated with all of the following *except:*
   a. Infections b. Idiopathic epilepsy
   c. Haemorrhages d. Tumours
   e. Drug toxicities
6. The "triad" of CS fluid findings including protein concentration 500 mg% or higher, xanthochromia and spontaneous clotting is termed:
   a. Hypercoagulability b. Spinal proteinosis
   c. Pott's disease d. Froin's syndrome
   e. Albuminocytologic dissociation
7. On standing CS fluid overnight a 'cob-web' clot is formed in:
   a. Meningism due to acute fever
   b. Multiple sclerosis
   c. Staphylococcal meningitis
   d. Tuberculous meningitis
   e. Pneumococcal meningitis
8. pH of CSF in normal health collected anaerobically is:
   a. 6.8 b. 7.0
   c. 7.3 d. 7.5
   e. 7.8

III. Fill in the blanks

1. Chemical composition of CSF depends on the ________ of the withdrawal.
2. Presence of ________ is the main cause of abnormal colour of CSF.
3. Glucose value of CSF is considerably ________ in bacterial infection.
4. In tuberculous meningitis on standing CSF forms ________ clot.
5. Normal pH of CSF is________.

6. Immunoglobulin that is raised predominantly in multiple sclerosis is__________.
7. High CSF protein with significantly low or absent sugar level is indicative of__________.
8. Index of γ-globulins of a figure above________ is suggestive of multiple sclerosis and neurosyphilis.

Answers to MCQs

1. b 2. d 3. c 4. e 5. b 6. d
7. d 8. c.

Answers to fill in the blanks

1. site
2. blood
3. decreased
4. cob-web
5. 7.3
6. IgG
7. coccal meningitis
8. 29

CHAPTER 43

RADIOACTIVITY: RADIOISOTOPES IN MEDICINE

Major Concepts

A. Learn the phenomenon of radioactivity, its hazards and protection.
B. Study the diagnostic and therapeutic uses of radioisotopes in medicine.

Specific Objectives

A. 1. Define radioactivity and learn the different series of radioactive elements.
2. Radioactive emissions consist of three rays: α, β, and γ-rays. Learn the properties of each.
3. Learn radioactive and non-radioactive stable isotopes. Differentiate them in a tabular form.
4. Study radiation hazards, radiation safety and protection.
5. List and define various units of radioactivity measurement.
6. Prepare a table of radioactive elements with their properties that are used in medicine and biology.

B. 1. Study the details of diagnostic applications of radioisotopes in
(a) Dilution studies
• Blood volume, • Protein loss, • Blood loss, • RBC life span study.
(b) Dynamic function tests such as
• Thyroid uptake of I^{131} • Regional blood flow • Anaemia studies • Pulmonary function studies • Scanning of organs.
2. (a) Study the therapeutic uses of radioisotope as external sources:
• Teletherapy • Beads, needles and applicators • Heavy particles • Extracorporeal irradiation of blood • $Boron^{10}$ Neutron irradiation.
(b) Therapeutic uses as internal sources:
• Regional applications • IV applications
• Intralymphatic applications.
(c) Study systemic uses of radioisotopes.

RADIOACTIVITY

INTRODUCTION

An unstable nucleus atom of some elements spontaneously emit (a) Accelerated particles or (b) Photons of radiation. This phenomenon is termed as ***radioactivity***. During this process the nuclei lose energy and change into other specific elements. This property of spontaneous nuclear changes is called ***radioactive decay***.

Natural radioactive elements belong to one of the three different series: (a) uranium series (b) thorium series, and (c) actinium series. Radioactive elements can be produced artificially.

Atomic nucleus holds an equal number of protons and neutrons with the help of nuclear forces also called as nuclear energy. Radioactive element has an unstable combination of neutrons and protons in its nucleus. The initial radioactive precursor of each of the three series changes by successive radioactive decay into other radioactive members of that series and ultimately gives a non-radioactive end product with a stable nucleus.

RADIOACTIVE EMISSIONS

Radioactive emissions consist of α, β and γ-rays as shown below:

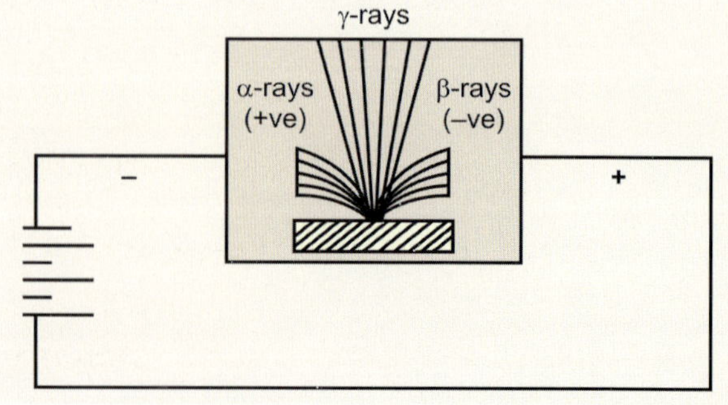

The α, β and γ-rays interact with matter to produce ion pairs and are called as ionising radiations.

- **Properties of α-Rays**

  These are streams of ***positively charged particles*** called α-particles. The initial velocity is 1.4×10^9 to 2.2×10^9 cm/sec and fixed initial kinetic energies of 3 to 9 million electron volts or MeV (1 MeV = 1.602×10^6 erg). They are readily absorbed by matter, ***cannot penetrate even a sheet of thick paper*** and are deflected towards cathode. Each α-particle is a tight cluster of two neutrons and two protons, corresponds to a mass number of 4, carries two positive charges and is identical with the Helium nucleus (He^{+2}).

- **Properties of β-Rays**

  β-rays are streams of very light, ***negatively charged*** and accelerated β-particles. They ***resemble electrons*** but they originate from the atomic nucleus and not electron shells. β-particles are deflected towards anode and possess variable initial kinetic energies (near 0 to 2 MeV) and initial velocities like light. They are less easily absorbed and generally ***have greater penetrability.*** The penetrating power of β-particle depends upon its energy content. For example, tritium ^3H has 0.018 MeV and travels only a few μm and cannot penetrate even the glass wall of the container while P^{32} atom has (1.71 MeV) and travels across a room and has a far higher penetrability.

- **Properties of γ-Rays**

  The γ-rays are electromagnetic radiations which unlike X-rays originate from nuclear and not electronic parts. They have higher frequencies and shorter wave lengths 10^{-4} to 10^{-1} nm. They ***carry no charge*** hence remain undeflected in an electric field, little absorbed and ***capable of penetrating very thick and dense material.*** An unstable nucleus attains an excited state on emission of α or β-particles or positrons β^+ with less than the maximum possible kinetic energy; the excess of the energy is then immediately emitted by the resulting nucleus as γ-rays. Sometimes the mutual annihilation of a positron and an electron generates γ-rays ***(annihilation radiation).*** The energy of γ-rays amounts to about 0.5, 2 and 0.5 MeV according as their emission follows α-decay, β-decay or positron electron annihilation respectively.

RADIOISOTOPES

Isotopes are the elements having the same atomic number but differing in mass number. Thus, in isotopes the number of neutrons is different. Identity of atomic number places the isotopes in the same position in periodic table. Radioactive isotopes differ from each other in radioactivity and mass number and are formed during the decay of a radioactive element. Emission of β-particle raises the atomic number by 1 without altering the mass number, thus changing the element into its isobar. But whenever a radioactive element decays by emitting one α-particle and two β-particles in successive steps, the final product is its isotope with the mass number lowered by 4. For example, in the uranium series, uranium I (U) and uranium II (UII) are radioisotopes, each with an atomic number of 92, while uranium XI (UxI), Ux2 and UII are isobars each with a mass number of 234. Radioactive decay of uranium is shown below in the box.

Radioactive Decay

Radioactive elements continuously lose radioactivity. Thus ^{24}Na is reduced from 100 mci to 50 mci after 15 hours. This loss of radioactivity takes place in a well-defined manner in a logarithmic fashion. ***The time required for a given isotope to disappear to half its original value is called the physical half-life of that isotope.*** In general, the radioactivity of an isotope at a given time can be calculated by the following formula:

$$\text{Present Activity} = \frac{Ao}{2^m}$$

where Ao = Activity at zero hour

m = Number of half-life.

The time required by an isotope to reduce its body concentration to half of that administered is known as the biological half-life.

Stable non-radioactive isotopes: They exist in nature, e.g. $^{14}{}_7N$ and $^{15}{}_7N$, $^{206}{}_{82}Pb$, $^{207}{}_{82}Pb$, etc. Some elements occur naturally as both radioactive and non-radioactive isotopes, e.g. nature carries mainly non-radioactive carbon as ^{12}C, also small amounts of non-radioactive ^{13}C and radioactive ^{14}C. Hydrogen exists as non-radioactive ^1H and ^2H Deuterium of heavy hydrogen, etc.

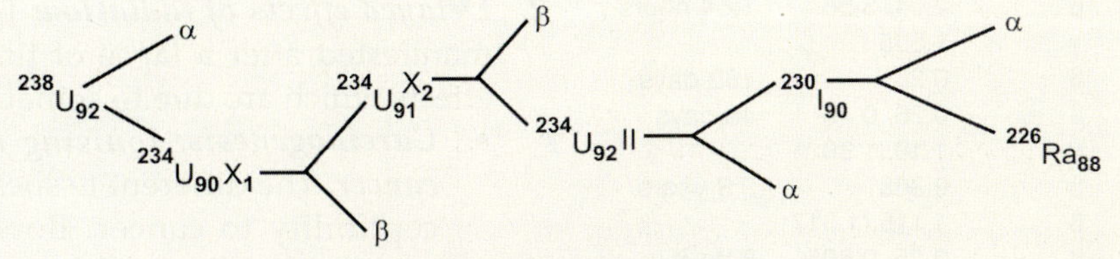

Radioactive decay of uranium

SECTION FIVE

Artificial radioisotopes: These can be produced by bombarding the stable nuclei of some natural non-radioactive elements, e.g. C, I, Al, B, Br, P, etc. with α-particles, neutrons, γ-rays or accelerated particles like protons, electrons, tritons and deuterons. Artificial radioisotopes obey the law of radioactive decay and possess specific half-life.

MEASUREMENT OF RADIOACTIVITY

1. ***Curie (Ci):*** It is the international unit of radioactivity based on the radioactivity of 1 g of radium. One curie equals 3.7×10^{10} nuclear disintegrations per second. 1 mCi = 10^{-3} and 1 μCi = 10^{6} Ci.
2. ***Roentgen (r):*** It is the amount of γ or X-rays which deposits 8.33×10^{-6} J of energy per g of air to produce 1.61×10^{12} ion pairs/g.
3. ***Radiation absorbed dose (rad):*** It is the quantity of γ or X-rays that can deposit 100 ergs or 1×10^{-5} J/g in matter including biological tissues.
4. ***Relative biological equivalence (RBE):*** It is the ratio between the doses (rad) of the radiation under 'investigation' and of γ-rays producing identical biological effects. RBE is 1 for γ-rays.
5. ***Roentgen equivalent mammals (REM):*** It measures a radiation in terms of its effects on mammalian tissues. Rem = rad × RBE.

RADIOISOTOPES IN MEDICINE

Radioisotopes commonly used in medicine and biology emit β-particles which can be most easily detected and located by autoradiography or estimated by liquid scintillation counter, spectrometry or autoradiography. The list of radioisotopes commonly used in medicine is shown in Table 43.1.

Table 43.1: List of radioisotopes commonly used in medicine and biology with their half-lives

Symbol	*At No*	*Mass No*	*Type of Emission*	*Energy (MeV)*	*Half-life*
^{3}H	1	3	β	0.018	12.3 years
^{24}Na	11	24	β	1.39	
			γ	1.37	5.1 hours
				2.75	
^{32}P	15	32	β	1.71	14.3 days
^{35}S	16	35	β	0.167	87.1 days
^{42}K	19	42	β	2.04, 3.58	12.4 hours
			γ	1.395	
^{45}Ca	20	45	β	0.26	152 days
^{59}Fe	26	59	β	0.26, 0.46	45 days
			γ	1.10, 1.30	
^{60}Co	27	60	β	0.308	5.3 years
			β	1.115, 1.317	
^{131}I	53	131	β	0.25, 0.605	8 days
			γ	0.164	

RADIATION HAZARDS

Exposure to radioactivity is hazardous to living organisms. The understanding regarding the biological effects of radiation has come mainly from somatic and hereditary damages sustained by the atomic explosions.

- **Stochastic effects:** These effects of radiation are those for which probability of a certain kind of damage is directly proportional to the administered dose.
- **Non-Stochastic effects**: These effects are those which have a threshold exposure beyond which can bring about the effects. For these, the severity of the effect is directly proportional to the administered dose. Examples of such effects are erythema, epilation, cataract, bone marrow depletion. These type of effects are cumulative.

In general the effects of radiation can be divided into *three categories:*

1. **Immediate effects**
2. **Delayed effects** and
3. **Genetic effects.**

1. ***Immediate effects of radiation:*** Immediate effects are obvious only at very high doses of exposure such as nuclear explosions, nuclear accidents or high doses of radiation therapy. ***Three types of syndromes*** are associated with such high doses:
   (i) ***Bone marrow syndrome:*** Whole body exposure of 200-1000 rads results in severe damage to hematopoietic system due to higher sensitivity of proliferating cells. Depletion of leucocytes and myelocytes leads to gross immunosuppression and increased susceptibility to infection. Death occurs within 10 to 20 days.
   (ii) ***Gastrointestinal tract syndrome:*** The radiation exposure of 1000 to 5000 rads causes severe damage to mucosal epithelium, fluid loss, electrolyte imbalance and haemorrhage in GIT. There is nausea, vomiting, acute diarrhoea. Death occurs within 3 to 5 days.
   (iii) ***Central nervous system syndrome:*** In case of nuclear holocast or nuclear reactor accidents, the exposure is as high as 5000-10,000 rads. Such doses are lethal for even non-dividing cells like neurons. ***The blood-brain barrier is lost leading to cerebral vasculitis, meningitis and choroid plexitis.*** Death occurs within 8 to 48 hours.
2. ***Delayed effects of radiation:*** The effects of radiation manifested after a lapse of time are called delayed effects which are due to somatic mutations.
   - ***Carcinogenesis: Ionising radiations can cause cancer.*** The different tissues have a different susceptibility to cancer. Bone marrow and other rapidly dividing cells are particularly prone to cancer, e.g. leukemia. Persons exposed to < 10 rads

SECTION FIVE

at Hiroshima and Nagasaki in 1945 are still showing higher leukemia mortality rate. Thyroid cancer is found in children whose scalps were irradiated for ring worms. Other types include polycythemia vera, breast cancer and bone cancer.

- ***In Utero radiation exposure:*** Embryonic stage is the most susceptible in the life of an organism to the radioactivity. *In utero* exposure leads to ***three types*** of damages:
  - (i) Growth retardation,
  - (ii) Congenital malformation, and
  - (iii) Fetal or neonatal death.
- The probability of three types depends on the radiation dose, the dose rate and stage of development of the embryo. The exposure of pre-implantation embryo to radiation can be lethal. Whereas after implantation growth of embryo will be retarded. Exposure at the time of organogenesis or later can cause congenital malformations and growth retardation.
- ***Shortening of life span:*** It has been observed in animals (mice and rats) that the general life expectancy of experimentally irradiated animals with moderate doses of X-ray, γ-rays, the life span was reduced.
- ***Miscellaneous effects:*** Other effects of radiation exposure are found to be endocrine imbalance, nephrosclerosis, decreased fertility or sterility, cataract, etc.

3. ***Genetic effects of radiation:*** According to ***target theory of Lea,*** each interaction of radiation with a target molecule can lead to molecular damage of target molecule. If the target molecule is DNA, it will undergo a damage-mutagenesis. However, the manifestation of mutation will depend upon the efficiency of DNA repair mechanism. Even the small doses of radiation are capable of inducing mutation. The magnitude of the genetic damage depends on:
   - The stage of germ cell development
   - Dose rate
   - Dose fractionation
   - Interval between exposure and conception.

RADIATION SAFETY AND PROTECTION

Eventhough it is risky to have an exposure to radiation, it is definitely has many diagnostic and therapeutic uses. The International Commission on Radiological Protection (ICRP) and the United Nations Scientific Committee of Effects of Atomic Radiation (UNSCEAR) are continuously busy in evolving a rational balance between "risk" and "benefits" of radioactivity to the human beings.

At present the recommended level of maximum individual radiation exposure for the occupational workers is 5.0 rem per year. It is 0.5 rem for general public. There is no totally safe dose of radiation. Therefore reducing the use of radiation to the minimum is desirable. The general use should be based on ALARA, i.e. ***As Low As Reasonably Achievable.***

The most popular "triad" of radiation protection is time, distance and shield (TDS).

- ***Minimum possible time should be spent near the radiation zone.***
- ***Handling of radioactive material should be done from a maximum possible distance.***
- ***Person should be shielded by lead.***

Further, continuous monitoring of the dose received by the occupational worker must be done.

Table 43.2 gives essential differences between radioactive isotopes and stable isotopes.

DIAGNOSTIC AND THERAPEUTIC USES OF RADIOISOTOPES

A. Diagnostic Uses

Diagnostic applications of radioisotopes can be described ***under two major headings:***

1. Procedures in which radioactivity is administered to the patient.
2. Procedures in which no radioactivity is administered to the patient, but samples obtained from the patient are analysed by techniques involving use of radioisotopes *in vitro.*

I. First Group

First group can be further divided under at least ***four categories:***

1. ***Dilution studies:*** Volume and space, 'turnover' rates and loss of 'tracers' from the body as in protein losing enteropathy.
2. ***Dynamic function tests:*** For example, absorption tests, "uptake" tests, clearance tests, blood flow to organs, etc.
3. ***Organ 'scanning':*** Pictorial representation of distribution of radioactivity in the organ.
4. ***Autoradiography:*** Essentially a research procedure.

1. Dilution Studies

(a) ***Volume and Space***

Principle: When a known amount of a radioactive "tracer" is introduced into an unknown volume, and if after thorough mixing, the concentration of the radioactive 'tracer' is estimated, then the total volume in which the "tracer" has been diluted is given by the formula.

$$V = \frac{N}{n}$$

where

V = Volume to be measured
N = Total number of counts injected
n = Number of counts per ml

Table 43.2: Essential differences between stable isotope and radioactive isotope

	Radioactive isotope	Stable isotope
1.	Natural abundance of radioactive isotope is less than that of stable ones.	Most abundantly found in nature.
2.	The atomic number and/or atomic mass are continuously changing.	Atomic number and atomic mass are constant.
3.	N/P ratio is generally high.	Generally the neutron to proton (N/P) ratio is not very high.
4.	Spontaneous emission of radiation (α, β, γ-rays) or neutrons.	Do not involve emission of any kind of radiation.
5.	They continuously disintegrate with a specific decay constant and half-life.	Do not disintegrate spontaneously. No energy spectrum.
6.	New elements are produced upon the disintegration or decay of radio-isotopes.	No new elements are produced.
7.	Cannot be distinguished from the stable isotopes by chemical and spectroscopic methods.	Detection by chemical or spectroscopic methods.
8.	Are detected by external detectors like gas chamber of scintillation counters.	Cannot be detected by an external detectors.
9.	Emits radiations and can produce deleterious effects on biological tissues.	Are not hazardous unless in the form of toxic chemicals.
10.	Special precautions are required while handling radioactive elements.	Special handling precautions are not required unless in the form of chemicals that are explosives, strong acids or carcinogens.
11.	Are used for generation of power (energy) in nuclear reactors.	Not used.
12.	Have special applications in medical diagnosis (RIA/Gamma ray scanning) and therapy (treatment of cancer and thyrotoxicosis), etc.	No such use.
13.	Used in research to induce mutations and in special techniques like autoradiography.	No such research applications.
14.	Used for sterilisation of medical instruments and utility items.	Cannot be used.
15.	Can be used for destructive purposes also, i.e. for manufacture of nuclear bombs.	No such use.

Tests: Following tests utilise dilution principle for estimating volume and space:

- ***Plasma volume:*** I^{131} labelled human serum albumin IV ("RIHSA" method)
- ***RB cells volume:*** Injection of Cr^{51}-labelled RB cells from the patient itself IV.
- ***Total body water:*** H^3-tritiated water given orally.
- ***ECF Volume:*** Br^{82} orally or IV.
- ***Total exchangeable sodium*** and 'sodium' space: Na^{24} orally or IV.

  There are many other tests also.

(b) Estimating GI Protein Loss in Protein-Losing Enteropathy

GI protein loss can be estimated by giving IV injection of I^{131}-labelled human serum albumin. "Tracer" will remain normally in the circulating compartment and will not appear in the stools in significant quantities unless there is GI protein loss.

Interpretation

- ***In normal healthy persons:*** Less than 0.1 to 1 per cent of injected radioactivity appear in 24-hours stool collection.
- More than 2 per cent of injected radioactivity in stools would indicate excessive protein loss.

(c) To Measure GI Blood Loss

Cr^{51}-labelled RB cells when given IV will not appear in the stools except a loss through GI tract. By measuring faecal radioactivity over a 72-hour period and by comparing it with blood radioactivity one can quantitate the faecal blood loss.

Precaution: Care has to be taken that stool samples are not contaminated with patient's urine, which will contain some radioactivity due to elution of Cr^{51}.

Interpretation: By this method as little as 5 ml of blood loss in GI tract can be estimated.

(d) Study of RB Cells Life Span

After an IV injection of Cr^{51}-labelled RB cells and estimation of the RB cells volume, if blood samples are drawn daily or on alternate days, there will be a progressive reduction in the blood activity. This reduction in radioactivity is due to several factors.

- Mostly due to destruction of *senescent* red blood cells.
- Partly also due to elution of labelled Cr.
- If there is any blood loss in GI tract.

Interpretation

- ***In normal healthy subjects:*** 50 per cent of the radioactivity will be still present in the circulating blood at the end of a month.
- ***In haemolytic anaemias:*** Half chromium time is reduced to below 24 days in mild to moderate cases and in severe cases it may be as low as 8 to 10 days.

2. Dynamic Function Tests

(a) Thyroid "uptake" of Radioactive Iodine: (see thyroid function tests)

(b) Study of Regional Blood Flow: Central, regional, and peripheral blood flows have been studied using a number of isotopes as "tracer" indicator.

Estimation of regional blood flow has also been made where organ uptake is proportional to blood flow. This principle finds its application in the estimation of coronary blood flow by a single injection of Rb^{86} (Rubidum). Radioisotopes have also been employed recently in the study of 'Left to right' and 'Right to left' shunts in congenital heart diseases.

(c) Radioactive iron studies: With the help of Fe^{59} we can determine:

1. The rate of absorption from the GI tract.
2. The rate of disappearance of Fe^{59} from the plasma.
3. The rate of ***turnover*** of plasma Fe indicates the extent of activity of the erythropoietic marrow, and
4. The rate of incorporation of Fe^{59} into the red blood cells which indicates the extent of effective erythropoiesis.

Interpretations

1. *In Normal Healthy Subjects*
   - Fe rapidly disappears from the circulation with the 1/2 time of 60 to 120 minutes.
   - Most of the Fe accumulates in erythropoietic bone marrow as shown ***by a rapid rise of sacral counts*** (counts taken over the sacrum). The counts of the liver and spleen rise only slightly initially and then remain steady.
2. *In aplastic anaemia:* T 1/2 is greater than 120 minutes. ***Sacral counts do not show rise.*** If at all there is any rise, it is very slow. ***Liver and spleen counts rise very rapidly showing that Fe is going to the reticuloendothelial cells rather than being used for erythropoiesis.***

(d) Absorption Tests

The clinician is interested in knowing the patient's capabilities in absorbing various nutrients, e.g. Fats, Calcium, Iron, Vitamins like B_{12}, folic acid, etc.

1. Vitamin B_{12} Absorption Studies

Oral dose of 0.5 to 1.0 μg of labelled B_{12} is used.

Methods: Several methods are used to assess the B_{12} absorption:

- Measurement of faecal radioactivity.
- Measurement of rate of accumulation of B_{12} in Liver (storage organ), by surface counting.
- Measurement of 8-hour plasma activity.
- **Schilling test:** Most commonly used, in which measurement of the urinary excretion of the labelled B_{12}, following a saturating dose of non-labelled stable B_{12} ***("flushing out" dose)*** is done.

For labelling B_{12}, following isotopes may be used:
Co^{60}, Co^{58} (1/2 life = 71 days)
Co^{57} (1/2 life = 270 days)

Method

- 1.0 μg of labelled-B_{12} is given orally.
- 1000 μg of non-radioactive stable B_{12} is given IM ***("flushing out" dose)***
- Urine is collected over a 24 hour period, (the urine passed before giving the labelled B_{12} is discarded).
- Radioactivity of this urine and of a "standard" is measured.

Calculation: The percentage of dose excreted in the urine is calculated as follows:

$$\frac{\text{Total counts/min. in 24 hr urine sample}}{\text{counts/min. in standard (test dose)}} \times 100$$

Interpretation

- ***Normal urinary excretion*** is greater than 15 per cent of the test dose.
- In patients with ***pernicious anaemia*** or with vitamin B_{12} deficiency associated with intestinal malabsorption or other causes, the excretion is less than 5 per cent.

Note: Low results may be obtained in patients with associated renal diseases.

2. *Fat Absorption:* Use of radio-iodine I^{131} labelled tri-olein and oleic acids and measurement of blood radioactivity 3 to 4 hours after oral ingestion has been tried.

(e) Study of Pulmonary Function

Radioisotope technique is perhaps the only method to study pulmonary function on a regional basis. Radioisotopes used for this purpose are radioactive Xenon (Xe^{133}), CO_2 (cyclotron produced). Such studies enable one to detect regional abnormalities in lung before total lung function is impaired or one can locate areas of maximum involvement in generalised pulmonary diseases.

Usefulness of such studies: Such informations are not only of diagnostic value but also helps the surgeon in deciding which part of the lung requires resection.

(f) Renal clearance-study: Recently Cr^{51}-labelled EDTA has been used for study of renal clearance and GFR.

(g) Isotope Renogram

(Refer to Chapter on Renal Function Tests)

(h) Study of Hepatic Function by I^{131} ***Labelled Rose Bengal Dye***

(Refer Chapter on Liver Function Tests).

3. *Scanning of Organs:* If an organ selectively concentrates a γ-emitting radioisotope, an image of the

distribution can be obtained by using a scanner over the organ, which will provide:

- Position of the organ, any "ectopic site" like thyroid,
- Size and boundaries of the organ,
- Shape of the organ,
- Presence or absence of any lesion.

A localised impairment of function like cyst, abscess, etc. may be noticed as a 'void' or so-called **cold area.** On the other hand, abnormal accumulation of radioactivity much in excess of the surrounding tissue, which may occur with tumours, appears as "dense", or so-called **Hot area.**

*(a) **Thyroid scanning:*** (Refer to Chapter on Thyroid Function Tests)

*(b) **Renal scan:*** (Refer to Chapter on Renal Function Tests)

*(c) **Brain scanning:*** Brain is difficult to approach due to its position inside the cranium. Radioisotope scanning is a safe and informative procedure in neurological investigations. Technetium (^{99m}Tc) and Indium (^{133m}In) are currently the radioisotopes of choice used for brain scanning.

Interpretations

- ***Brain tumors:*** Overall detection rate of brain tumors have been 70 to 80 per cent. Vascular tumors like meningioma and Astrocytoma grade-III show very well.
- ***Cerebral infarction:*** Immediately after a cerebral infarction the scan is normal. Later it becomes +ve (a "hot" spot scan) and subsequently the 'hot' spot disappears.

Thus, serial scanning may help to differentiate an infarct from a tumor.

(d) Lung Scanning

Isotope used: Human serum albumin is tagged with I^{131} and 'macro-aggregates' are prepared by heating, to an optimal size of 10 to 50 μ.

Principle: When such I^{131} human serum albumin macro-aggregates are injected IV, because of their particle size, they will be trapped in the first capillary bed which they come across viz. pulmonary capillary bed, which will temporarily hold the particles for a few hours and their distribution will give a lung perfusion scan. If there is no blood flow to a region, particles will not go there for being trapped, and the scan will show a "void" or 'cold' area.

Usefulness

- Scanning thus provides a safe and rapid screening test ***for detection of pulmonary infarction, especially in early stages when X-ray may be quite normal.***
- Many other organ scanning has been introduced like liver scan, spleen scan, placental scan, myocardial scan and bone scanning, which have been found to be quite informative.

SECTION FIVE

II. Second Group

In this group of procedures no radioactivity is administered to the patient but *in vitro* activity is assessed:

T_3-Resin/sponge uptake studies: (Refer to Chapter on Thyroid Function Tests)

B. Therapeutic Uses of Isotopes

General Principles: Radioisotopes have a useful role in the treatment of diseases, particularly malignancies. The crux of the problem is to get the ionising radiation dose to concentrate selectively in tumour tissue without harming the adjacent healthy tissues. The tumour tissues may be attacked by a beam of radiation.

- From outside the patient's body ***(external sources),*** or
- From within the body ***(internal sources).***
  The radioactive material may also be applied locally over the surface of malignant tissue, or it may be implanted into the tumour tissue, or it may be instilled into serous cavities or in the bladder. It may be injected.
- Intra-arterially as microspheres, or
- Intralymphatically during performing lymph-angiography. Most commonly it is given by mouth or IV injection taking advantage of selective localization, e.g. P^{32} is concentrated more in malignant cells than in normal cells. Tritiated (H^3)-thymidine is taken up by rapidly proliferating cells for DNA synthesis and radio-iodine (I^{131}) is selectively concentrated in thyroid gland.

1. External Sources

(a) Teletherapy

Co^{60} offers a powerful source of radiation for the treatment of various malignant disorders. When compared to radium, Co^{60} sources are cheap. Co^{60} has certain advantages over the conventional 200 KeV X-ray therapy. Its energetic γ-rays can penetrate deep into the tissue to enable treatment of internal cancers without causing severe skin reactions.

(b) Beads, Needles, and Applicators

- Radium was initially first used for radiation treatment of various cancerous conditions. Radium was very expensive and very longlived and they were used in temporary implants.
- Co^{60} now has successfully replaced radium, in the treatment of cancer of cervix. Co^{60} can be encapsulated in gold or silver needles, wires, rods or cylinders.
- ***In dermatology:*** P^{32} is applied to paper or polythene sheets for treatment of squamous cell carcinomas, superficial angiomas, mycosis fungoides, and senile keratosis.
- ***In ophthalmology:*** Special applicators or Sr^{90} are used for lesions of cornea, conjunctiva and sclera.

(c) Heavy Particles

Heavy particles have a great capacity to produce dense ionisation in tissues, which give them important therapeutic advantages over electromagnetic radiations (like γ-rays and X-rays). ***Recently, heavy particle proton irradiation of the pituitary gland has been used in diabetic retinopathy in an attempt to improve vision or slow down the rate of deterioration of vision.***

(d) Extracorporeal Irradiation of Blood

One serious limitation of radiation therapy of "radiomimetic" alkylating drugs is the depression of the normal haemopoietic stem cells in the bone marrow by the dose required to kill the malignant cells.

Recently, a new approach to this problem has been ***extracorporeal irradiation of blood*** of patients like a case of chronic leukaemia. The blood is taken out from the patient via forearm artery, then it is circulated around a cesium^{137} source which emits powerful γ-rays, and then the irradiated blood is returned to the same patient via a forearm vein.

(e) Boron^{10} Neutron Irradiation

Boron^{10} has been used recently in the treatment of the inoperable and rapidly fatal brain tumour like glioblastoma multiforme.

- When injected IV into the patient, ***Boron^{10} will be rapidly taken up by the tumour tissue in brain.***
- After 10 minutes, the head of the patient is placed in a beam of slow neutrons. ***Boron^{10} in tumour tissue absorbs neutrons rapidly and becomes transformed to Boron^{11}, which disintegrates almost immediately into α-particles and Lithium isotope.***
- α-particles have an extremely limited range and produces dense ionisation. Hence all their powerful ionising property is utilised in destroying the tumour cells in tumour tissue, without virtually producing any harm to the adjacent brain cells.

2. Internal Sources

(a) Regional Applications

- Gold^{48} (Au^{48}) is a very useful radioisotope, since it is chemically inert in the body, it is non-toxic and it needs no protective coating. If it is introduced into pleural or peritoneal cavities in the form of colloidal suspensions, it will remain within the cavities without getting absorbed. Thus ***it has been used in treatment of malignant pleural/peritoneal effusions.*** Its β and γ-radiations are almost entirely absorbed by the tissue in the immediate vicinity.
- Other isotopes which have been used in treatment of pleural/and peritoneal effusion are Gold^{198} chromic PO_4 labelled with P^{32} and Yttrium^{90}.
- **Yttrium^{90} Synovectomy**

  Recently yttrium radioisotope has been used to produce radiation synovectomy in arthritis of haemophiliac patients. Joint deterioration and recurrent arthritis is a major problem for haemophiliac patients. Synovium is a contributory factor. Bleeding in joints in haemophilia patients irritates the synovium, causing synovial swelling, and increased bleeding and irritation. Thus vicious cycle set up by a joint bleed; the destructive cycle can ultimately lead to joint destruction.

Yttrium synovectomy is a relatively new treatment for haemophiliacs who have severe joint pain and bleeding. It is simple and safe and over long-term decreases bleeding episodes, gives pain relief and may decrease development of severe arthritis.

Method

- The affected joint is injected with yttrium^{90} radioisotope under X-ray control and joint is immobilised for 36 hours.
- Yttrium isotope ***emits β-particles locally.*** It has low penetration and rapid decay which make it an ideal isotope for use in joint spaces to dissolve and destroy the inflamed synovium.

(b) IV applications: Tiny ceramic microspheres of radioactive gold (Au^{198}) or Yttrium^{90} are stopped at the arterial end of the capillary because of their size and deliver local radiations to the tumour cells. Tumours of the lung, prostate, and bone have been treated by this method. Injection into the mesenteric vein is given for hepatic tumours.

Disadvantage: Many tumours in advanced stages become necrotic and blood supply is poor hence erratic dose/and result.

(c) Intralymphatic applications have also been used.

(d) Systemic Uses

1. **Use of P^{32}:** P^{32} has been used in the treatment of following diseases:
   - Chronic myeloid leukaemia.
   - Polycythemia vera.
   - Multiple myeloma.
   - Primary haemorrhagic thrombocytosis.
   - Carcinoma of the breast.
   - Carcinoma of prostate.
2. ***Use of I^{131} for thyroid cancers:*** I^{131} has been used for treatment of thyroid cancers. Compared to normal thyroid, the uptake of I^{131} by thyroid cancer tissue is much less.
3. ***Radio-iodine therapy for non-malignant diseases:***

a. ***Thyrotoxicosis:*** I^{131} has been used also for treatment of primary thyrotoxicosis.

Present trend is that young patients below the age group of 25 are not given radioiodine. Pregnancy is absolute contraindication for I^{131} in women.

Greatest problem of radio-iodine therapy in thyrotoxicosis is the incidence of subsequent development of permanent hypothyroidism. Studies have shown at the termination of therapy, the incidence is about 7 per cent and in 10 years it rises to 30 per cent.

b. *I^{131} therapy of cardiac disease in euthyroid patients:* A deliberate production of hypothyroid state in euthyroid patients suffering from intractable angina pectoris or intractable congestive cardiac failure has been tried.

c. *I^{131} for control of ectopic arrhythmias:* When usual therapy fails to control troublesome ectopic rhythms, I^{131} has been found useful in some of the cases.

I. Questions (Essay type and short notes)

1. What are radioisotopes? Describe the clinical and therapeutic uses of radioisotopes.

Short Notes

a. Radioisotopes of medical significance
b. Hazards of radiation
c. Radioisotopes and their hazards
d. Isotopes
e. Hazards of radioactivity
f. Diagnostic and therapeutic uses of radioisotopes
g. Radioisotopes
h. Uses of radioactive isotopes in medicine

Compare and contrast

Stable and radioactive isotopes.

II. MCQs (Give one correct answer)

1. Natural radioactive elements belong to one of the series containing:
 a. Uranium b. Thorium
 c. Actinium d. All the above (a + b + c)
 e. None of the above (a + b + c)
2. The most penetrating rays are:
 a. α-rays b. β-rays
 c. γ-rays d. X-rays
 e. δ-rays
3. Which radioisotope is used for schilling test?
 a. Labelled iodine b. Labelled folic acid
 c. Labelled chromium d. Labelled B_{12}
 e. None of the above
4. Elevated T_4, lowered T_3 uptake together suggest the likelihood of:
 a. Increased TBG b. Decreased TBG
 c. Hyperthyroidism d. Hypothyroidism
 e. Secondary hyperthyroidism
5. A 4-month-old child exhibits failure of thrive and develop mentally at a normal rate. The thyroid gland is palpably enlarged to three times the normal size. The best sequence for laboratory tests would be:
 a. T_3 uptake T_4, TSH and T_3
 b. T_4 only, followed by T_3 uptake if T_4 is low
 c. T_4 only, followed by T_3 uptake and TSH if T_4 is abnormal
 d. T_3 and T_4 only
 e. None of the above
6. Primary and secondary hypothyroidism, in a patient with an established diagnosis of hypothyroidism, are most directly differentiated by:
 a. Serum TSH b. T_4
 c. I^{131} uptake d. T_3 uptake
 e. Free thyroxine index

III. Fill in the blanks

1. Radioactive element has unstable combination of________and________.
2. Radioactive emission consist of________,________and ________rays.
3. The α-rays carry positive charge β-rays carry________ charge while γ-rays ________ carry.
4. Radioisotopes are the elements having the same________ but differing in________.
5. When the probability of damage is directly proportional to the administered dose, it is called________.
6. Exposure of________can damage non-dividing cells such as________.
7. In GI protein loss, more than________of the injected radioactivity appears in stool.
8. Effective erythropoiesis can be assessed by rate of incorporation of________ in red cells.
9. The isotope compound used for renal clearance is________.

Answers to MCQs

1. d 2. c. 3. d 4. a 5. c 6. a

Answers to fill in the blanks

1. Neutrons/protons
2. α, β, γ
3. Negative/No.
4. atomic number/mass number
5. stochastic effect
6. 5000-10000 rads/Neurons
7. 2 per cent
8. Fe^{59}
9. Cr^{57}-EDTA

SECTION

6

Miscellaneous

CHAPTER 44

DIET AND NUTRITION

PART I
ENERGY METABOLISM

Major Concepts

A. Study the concept of energy as regards to diet.
B. Study the details of BMR and RQ.

Specific Objectives

A. 1. Study what is free energy and exergonic and endergonic reaction.
   2. Define caloric value of food and how it is calculated.
   3. Define unit of energy—kilocalorie.
B. 1. Study BMR in detail.
   2. Study factors influencing BMR.
   3. Study pathological variations of BMR.
   4. Define Respiratory Quotient (RQ).
   5. Study various factors affecting RQ.
   6. Study in detail, specific dynamic action (SDA).

INTRODUCTION

Throughout the discussions of the metabolism of organic foodstuffs, attention has been focussed particularly on the mechanisms and nature of their chemical transformations, various components of the body tissues are undergoing degradation (catabolism) and resynthesis (anabolism) continually. Certain of the chemical reactions involved in these metabolic processes are *exergonic,* i.e. they are accompanied by liberation of energy, whereas others are *endergonic,* i.e. they require the introduction of energy. The manner in which energy is produced, stored, transferred and utilised has already been considered.

Energy expenditure by the body consists of **two parts:**

1. Energy utilised for doing physical works and exercises.
2. Energy utilised for doing involuntary works.
   - **First category** includes: expenditure of energy in movement of body, lifting of any object, doing day-to-day works and muscular exercises. The extent of energy expenditure depends upon the extent of physical work done.
   - **Second category** includes: the energy expenditure in doing osmotic work, absorption, transport of food materials, excretion, contraction of involuntary muscles, active transports, etc. ***Energy is continuously expended in such involuntary works throughout the life period for which we are not conscious.*** This part of expenditure is relatively constant and ***expenditure in such involuntary works occurs at a basal rate.***

CALORIC VALUE OF FOODS

Different foodstuffs on burning give different amounts of energy. How much heat will be obtained by burning a particular foodstuff is expressed by the term ***caloric value.***

Definition: Caloric value is defined as amount of heat-energy obtained by burning 1.0 gm of the food stuff completely in the presence of O_2. Caloric value of different foodstuffs is determined *in vitro* in a special apparatus called ***bomb calorimeter.***

Principle and Procedure

1. A weighed amount of the sample is burnt in an atmosphere of O_2 by an electrically heated platinum wire.
2. The heat evolved is absorbed in a weighed amount of water which surrounds the burning chamber.
3. The rise of temperature is recorded with the help of a sensitive thermometer.

Calculation: From the above data, heat evolved by 1.0 gm of food can be calculated with the following formula:

$$H = \frac{W (T_2 - T_1)}{M (1000)} \text{ Kcal/gm}$$

where,

W = Water equivalent of calorimeter and its water.
T_2–T_1 = Rise in temperature in centigrade.
M = Denotes amount of food burnt in gm.

Unit of Energy: The unit of energy is calorie (c).

Definition of Calorie: It is defined as the amount of heat required to raise the temperature of 1.0 gm of water by 1°C (specifically from 15°C to 16°C).

This is the ordinary calorie and is found too small a unit for measuring the energy value of foods.

A unit thousand times of the ordinary calorie is called **kilo-calorie** or simply calorie **(by capital 'C')** is used for this purpose. ***Calorie in biological science always means a "kilocalorie" ("C").***

Food materials undergo combustion in the animal body and liberate energy in the same way as in a bomb calorimeter, but in a graded and continuous stepwise manner instead of in an explosive way.

The average values in calories obtained per gm have been found as follows:

- **Glucose** = 3.6
- **Starch** = 4.2
- **Animal fats** = 9.5
- **Animal proteins** = 5.6
  (vegetable proteins slightly lower)

In the body, proteins do not undergo complete oxidation, a portion of its amino groups being converted to and excreted as NH_3 and urea (*in vitro* it is further oxidisable). This involves a loss of about 1.3 cal/gm, leaving 4.3 C produced/gram of protein metabolised.

Taking into consideration, the variations in caloric value of individual carbohydrate/fat/protein, their average energy value when metabolised may be represented as follows in C/gm:

- **Carbohydrates** = 4.1
- **Fats** = 9.3
- **Proteins** = 4.1

When these corrected figures are applied, it is found that the amount of energy (calories) produced by a given quantity of these foodstuffs in the body is the same as that produced by their combustion outside the body. On accounts of losses in digestion and absorption and other unaccountable factors, the caloric value are usually rounded off and said to be 4.0 calories/gm of carb and proteins and 9.0 calories/gm for fats (Table 44.1).

BASAL METABOLISM AND BMR

The amount of energy required for any individual varies directly with the degree of activity and environmental conditions, but the rate of energy production in an individual by its over-all cellular metabolism is more or less constant under some standard conditions ***basal conditions*** and is known as ***basal metabolism.***

The basal conditions are as follows:

1. Person should be awake but at complete rest—both physical and mental.
2. Person should be without food at least 12 to 18 hrs, i.e. in the "postabsorptive state".
   Postabsorptive state: This period is allowed to pass for avoiding:
   - Effects of digestion and absorption
   - The effects of SDA of foodstuffs, and
   - Also to prevent any chance of starvation.
3. Should be in recumbent/reclining position in bed.
4. Person should remain in normal condition of environment, i.e. at normal temperature, pressure and humidity (environmental temperature of between 20° to 25°C).

Under above conditions, energy output of the individual is to maintain respiration, circulation, muscle tone (skeletal and smooth muscles), functions of visceras like the kidney, liver and brain for the maintenance of the body temperature.

Table 44.1: Caloric value, O_2 and CO_2 equivalent of carb, fats, and proteins

	Calories per gm	*Litres of CO_2/ gm*	*Litres of O_2 per gm*	*RQ*	*Caloric value per litre O_2*
Carbohydrates	3.7–4.3 (4.1)	0.75–0.83 (0.8)	0.75–0.83 (0.8)	1.0	5.0
Fats	9.5	1.43	2.03	0.707	4.7
Proteins	4.3	0.78	0.97	0.801	4.5

BMR

The rate of energy production under such basal conditions per unit time (one hour) and per sq. metre of body surface is known as ***basal metabolic rate (BMR).***

Definition: The BMR may be **defined** as the amount of heat given out by a subject who though awake is lying in a state of maximum physical and mental rest under comfortable conditions of temperature, pressure and humidity, 12 to 18 hours (postabsorptive) after meal.

A constant ratio of endogenous carbohydrates, lipids and proteins are metabolised under such basal conditions. Under such conditions, RQ is 0.82 and each litre of O_2 consumed represents 4.825 C of energy output.

Determination of BMR

BMR can be determined by the following methods:

(a) Open-circuit system: In which both O_2 consumption and CO_2 output are measured. It requires a high degree of technical skill and more cumbersome apparatus and is less rapid but is more accurate. **Tissot method** and **Douglas method** are both open-circuit methods.

(b) Closed-circuit method: In clinical practice, the BMR is estimated with sufficient accuracy merely by measuring O_2 consumption of the patient for 2 to 6 minutes period under "basal" conditions. The O_2 consumption is measured in a ***closed-circuit system***. The apparatus commonly used is the **"Benedict-Roth metabolism apparatus"** (Fig. 44.1).

The apparatus consists of:

A cylindrical spirometer vessel closed at upper end. Floating on water-jacket below, to make an air-tight water-seal. The spirometer is filled with O_2 and is connected by passages regulated by valves. The outlet passage directly to the mouth piece and the inlet passage through a sodalime container, to absorb the CO_2 of the expired air. The subject kept under basal conditions will be made to breathe by mouth by closing the nostrils and inserting the mouth piece into his mouth. The respiratory excursions are transmitted to the floating respirometer which moves up and down in the water jacket surrounding its lower portion. The movements of the respirometer, in turn, are transmitted to a pen connected to the top of the respirometer by a pulley and chain. The movements are recorded by the pen on a recording drum which is rotated by a mechanical or electrical clock work. The test is usually run for a period of 6 minutes and the volume of O_2 consumed in that period is obtained from the tracing on the recording drum. This is corrected to standard conditions of temperature and barometric pressure.

Calculation: The average O_2 consumption for the two periods is multiplied by 10 to convert it to an hourly basis, and then multiplied by 4.825 C, the heat production represented by each litre of O_2 consumed. This gives the heat production in C/hour. Since BMR is to be expressed as C/sq. metre/hour, the energy output per hour obtained above has to be divided by the surface area of the individual.

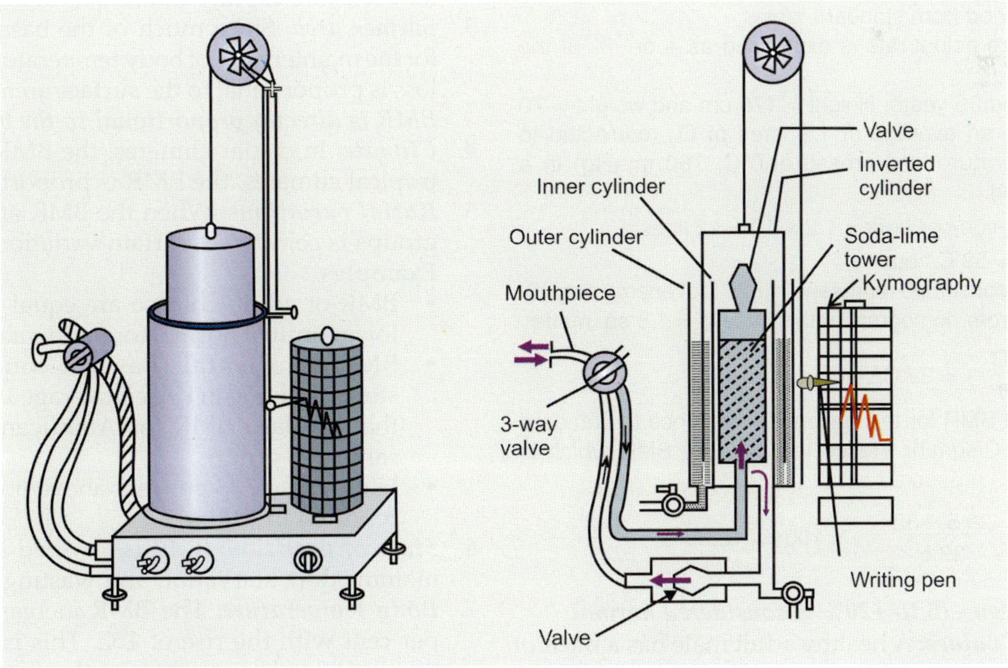

Fig. 44.1: Benedict-Roth spirometer

Calculation of surface area: The surface area of an average adult is about 1.8 square metres.

1. A simple formula for calculating the surface area is as follows:

   Oee of mid-thigh × 2 × Height = surface area
   (in cm) (in cm) (in sq. cm)

2. The classical formula is that of **Du Bois,** as follows:

   Du Bois' surface area formula

$$A = H^{0.725} \times W^{0.425} \times 71.84$$

Where,

A = Surface area in sq. cm
H = Height in cm
and W = Weight in Kg.

Surface area thus obtained in sq. cm has to be divided by 10,000 to get surface area in sq. metre.

To have the value directly in sq. metre, the formula can be modified as:

$$A \text{ (in sq. metre)} = H^{0.725} \times W^{0.425} \times 0.007184$$

3. **By using nomogram:** More conveniently, in practice, the nomogram prepared by **Boothby** and **Sandiford** can be used for finding surface area for an individual if heights and weights are known.
   - Height: Given in feet/cm
   - Surface area: In square metres
   - Weight: Given in pounds/kg.

Example of calculation of BMR

1. The normal BMR for an individual of the patient's age and sex is obtained from standard tables.
2. The patient's actual rate is expressed as + or –% of the normal.

A male aged 35 years, Height = 170 cm and weight = 70 kg, consumed an average of 1.2 litres of O_2 (corrected to normal temperature and pressure: 0°C, 760 mmHg) in a 6-minute period.

∴ O_2 consumption per hour = 1.2 x 10 = 12 litres.

∴ 12 x 4.825 = 58 C/hour

(one litre of O_2 consumed represents 4.825 C of energy output).

Surface area from nomogram in this case is = 1.8 sq. metre

$$\therefore \text{BMR} = \frac{58\text{C}}{1.8} = 32 \text{ C/sqm/hr.}$$

The normal BMR for this patient by reference to standard tables is 39.5 C/sqm/hr. Hence the patient's BMR which is below normal is presented as:

$$\frac{39.5 - 32}{39.5} \times 100 = -18.98\%$$

A BMR between –15 to +20% is considered normal:

Normal BMR values: A healthy adult male has a BMR of about 40 C/sqm/hr and adult female about 37 C/sqm/hr.

Determination of BMR by Read's formula: This formula gives a rough estimate of BMR and is often used at the bedside.

The formula given is as follows:

$$\text{BMR} = 0.75\ (\text{PR} + 0.74 \times \text{PP}) - 72$$

where:

- PR = Pulse rate
- PP = Pulse pressure

The result obtained as the % of the normal and is correct within a range of ± 10 per cent, viz. if above 10 per cent, the BMR is higher, and if below 10 per cent, the BMR is lower than normal.

Factors Influencing BMR

The rate of metabolism at "basal" conditions has been found to vary in different individuals and therefore BMR varies with different factors.

1. *Age:* The BMR of children is much higher than the adults. Roughly speaking it is inversely proportional to the age.
   - At the age of 6, it is 57.5 C
   - At 12, it is 50.4 C
   - Between 20 and 30 yrs, it is 40 C and
   - Between 40 and 70 yrs, it varies between 38.5 and 35.5 C.

   In other words, ***with advancing age, BMR gradually falls***. This is due to the fact that children possess a greater surface area in proportion to their body weight (Exception, in newly born babies it is low, about 25 C/sqm/hr. In premature infants, it is still lower).
2. *Sex:* ***Women normally have a lower BMR than men.*** The BMR of females decline between the ages of 5 and 17 more rapidly than those of males.
3. *Surface area:* Since much of the basal metabolism is for the maintenance of body temperature and since heat loss is proportional to the surface area of the body, ***the BMR is directly proportional to the body surface.***
4. *Climate:* In colder climates, the BMR is high and in tropical climates, the BMR is proportionally low.
5. *Racial variations:* When the BMR of different racial groups is compared, certain variations are noted.

   Examples
   - BMR of adult Chinese are equal to or below the lower limit of normal for occidentals (Westerners),
   - BMR of oriental (Eastern countries) female students living in USA is average 10 per cent below the standard BMR for American women of the same age-groups.
   - High values, 33 per cent above normal have been reported in Eskimos.
6. *State of nutrition:* BMR is lowered in conditions of malnutrition, starvation and wasting diseases.
7. *Body temperature:* The BMR increases by about 12 per cent with the rise of 1°C. This is due to the fact that increased temperature stimulates the chemical processes of the body and thereby increases the BMR.

8. *Barometric pressure:* Moderate reduction of atmospheric pressure does not affect the BMR, but a fall of pressure to ½ an atmosphere (viz. O_2 tension 75 mmHg) as occurs in mountain climbing increases BMR, but increased pressure of O_2 does not raise BMR.
9. *Habits:* Trained athletes and manual workers have a slightly higher BMR than persons leading a sedentary life.
10. *Drugs:* Quite a number of drugs like Caffeine, Benzedrine, Epinephrine, Nicotine, Alcohol, etc. increase the BMR. On the other hand, reverse is observed with most of the anaesthetics.
11. *Hormones:* Circulating levels of hormones secreted by thyroid, adrenal medulla and anterior pituitary increase BMR. ***One mg of thyroxine increases BMR by about 1000 calories (1 C).***

- *In thyrotoxicosis:* ***BMR may increase by 50 to 100 per cent above normal,*** but RQ remains unaltered since both O_2 consumption and CO_2 production increase proportionately in such cases.
- *In myxoedema:* BMR is ***diminished to 30 per cent or even 45 per cent below normal.***

Anterior pituitary through its TSH affects BMR; GH also causes about 20 per cent rise in BMR, catecholamines increase BMR by about 20 per cent of the resting value. Male sex hormones cause a 10 per cent increase in BMR.

12. *Pregnancy:* The BMR of pregnant mother after six months of gestation rises. It may be noted in pregnancy, the BMR of the mother is the sum total of:
    - Her own metabolism as in her nonpregnant state and
    - Combined with that of the foetus.

Hence, pregnancy exerts no specific effect upon BMR.

CLINICAL ASPECT

Pathological Variations in BMR

1. ***Fever:*** Infections and febrile diseases elevate the BMR, usually in proportion to the increase in temperature.
2. ***Diseases:*** They are characterised by increased activity of cells, also increased heat production due to increased cellular activity. Thus, BMR may increase in such diseases as:
   - Leukaemias (21 to 80%)
   - Polycythemia (10 to 40%)
   - Sometypes of anaemias, cardiac failure, hypertension and dyspnoea (25-80%).

   All of which involve increased cellular activity.
3. ***Perforation of an eardrum:*** This causes falsely high readings.
4. ***Endocrine diseases:*** The most important factor which alters BMR is the state of function of thyroid. In fact, determination of BMR is mainly used for the assessment of thyroid function.
   - ***In hyperthyroidism:*** BMR is increased to + 75 per cent or more.
   - ***In hypothyroidism (myxoedema):*** BMR is reduced to 40 per cent or more.
   - BMR is also increased in Cushing's disease and Cushing's syndrome and also in Acromegaly.
   - BMR is decreased in Addison's disease (Hypofunction of Adrenal cortex).

IMPORTANCE OF BMR

1. *As a diagnostic aid:* For the diagnosis of various pathological conditions specially assessing the thyroid function (specially useful where hormone assays and isotope laboratory are not available).
2. *For calculation of caloric requirements:* Essential in the calculation of calorie requirements of an individual for prescribing a diet of adequate calorific value and planning nutrition for individuals as well as communities and populations at large.
3. *Effect of foods and drugs:* To note the effect of different types of foods and drugs on basal metabolic rate.

RESPIRATORY QUOTIENT (RQ)

Definition: ***RQ is the ratio of the volume of CO_2 produced by the volume of O_2 consumed (i.e. CO_2/O_2) during a given time.***

Note

- ***RQ is simply a ratio.*** It gives no idea as to the absolute quantity of gaseous exchange.
- Proportional increase or diminution of CO_2 produced and O_2 utilised, will keep the ratio unchanged. But any disproportionate variation will be reflected by a corresponding change in the RQ.

Normal RQ: ***In a healthy adult, on a mixed diet, it is 0.85.***

Method of determination: It is done by measuring the volume of O_2 consumed and CO_2 produced during a given time with the help of Douglas bag and other similar instruments.

FACTORS AFFECTING RQ

1. *Role of Diet*

(a) *Carbohydrate:* ***In case of carbohydrate diet, RQ is 1 (one).*** Because in carbohydrate diet, the volume of CO_2 produced is the same as the volume of O_2 consumed.

Explanation: This is due to the fact that in carbohydrate molecule, the amount of O_2 present is just sufficient to oxidise the H present in the same molecule. Hence, external O_2 is necessary only to convert the C of the molecule to CO_2 so that the volume of O_2 consumed and the volume of CO_2 produced will be same.

$$C_6H_{12}O_6 + \mathbf{6\,O_2} = \mathbf{6\,CO_2} + 6\,H_2O$$

$\therefore$ RQ for carbohydrate =

$$\frac{CO_2 \text{ produced}}{O_2 \text{ consumed}} = \frac{6}{6} = 1$$

(b) *Fats:* ***In case of fats, the RQ will be lowest and is about 0.7, because fat is an oxygen poor compound.*** The oxygen present in it cannot fully oxidise the H of the molecule so that O_2 consumed from outside is used for two purposes:

- Firstly, for oxidising C and producing CO_2, and
- Secondly, for oxidising H and giving H_2O.

Consequently, **the volume of CO_2 produced will be less than the volume of O_2 consumed. Hence, RQ will be low.**

Example: Oxidation of tristearin will be used to exemplify the RQ for fats

$$2\,C_{15}H_{110}O_6 + 163\,O_2 \rightarrow 114\,CO_2 + 110\,H_2O$$

$$\therefore \text{RQ for tristearin} = \frac{114}{163} = 0.70$$

(c) *Proteins:* The oxidation of proteins cannot be so readily expressed. By indirect methods, ***the RQ for proteins has been calculated to be about 0.8.***

Example: RQ for Alanine is:

$$2\,C_3H_7O_2N + 6\,O_2 \rightarrow (NH_2)_2CO + 5\,CO_2 + 5\,H_2O$$

$$\therefore RQ = \frac{5}{6} = 0.8$$

(d) *RQ of mixed diets under varying conditions:* In mixed diets, containing varying proportions of proteins, fats and carbohydrates, the **RQ is about 0.85. As the proportion of carbohydrates metabolised is increased, the RQ approaches closer to 1.**

2. *Effect of Interconversion in the Body*

When carbohydrates are converted into fats in the body, RQ will rise.

Explanation: In this process, an O_2-rich substance is converted into an O_2-poor compound, so that some amount of O_2 liberated from carbohydrates will be utilised for purposes of oxidation. Consequently, less O_2 will be needed from outside. *Hence, the amount of CO_2 produced will be more than the amount of O_2 consumed.* So that RQ will rise, and will be considerably elevated. A reversal of the above process, i.e. conversion of fats to carbohydrate, would lower the RQ below 0.7. This has been reported but has not been generally confirmed. It is, therefore, evident that RQ value will indicate the following:

- *The type of foodstuffs burning in the body, or*
- *The nature of conversion of one foodstuff into another in the body.*

3. *Muscular Exercise*

- With moderate exercise (with a normal mixed diet), the RQ remains almost unaltered. Because in exercise, the body uses different foodstuffs in the same proportion as at rest.
- *With violent exercise:* Lactic acid enters blood and produces acidosis, as a result pulmonary ventilation will be increased, washing out more CO_2. Consequently, RQ rises, and may go above 2 even.
- During recovery from violent exercise: RQ falls, because less CO_2 is evolved. Gradually it goes back to normal.

CLINICAL ASPECT

1. ***In acidosis:*** During acidosis, CO_2 output is greater than O_2 consumption, hence ***RQ increases in acidosis.***
2. ***In alkalosis:*** In this ***RQ will fall***, because respiration is depressed and CO_2 will be retained in the body, i.e. less CO_2 is produced.
3. ***In febrile conditions: It may increase RQ.*** Rise of body temperature such as in high fever, will cause increased breathing and thereby will wash out more CO_2, hence CO_2 production increases.
4. ***In diabetes mellitus:***
   - In advanced cases of Diabetes mellitus, when little carbohydrate is burning, energy is supplied mainly by oxidation of fats. Hence RQ will fall.
   - In such cases, when Insulin is administered, carbohydrates will start burning and RQ will rise.
5. ***In starvation:*** Here the subject has to live on its own body tissues.
   - In the first stages (initial 1 to 2 days), energy is derived mainly from the stored glycogen so that the RQ although it falls below normal 0.85, is proportionately high 0.78.
   - But later on, when energy is derived chiefly from the combustion of fats, RQ will fall still further, and will be about 0.7.

Value and Significance of Determination of RQ

- RQ acts as ***a guide as to the type of food burning*** or to the nature of synthesis taking place in the whole body as well as in a particular organ.
- RQ is ***very helpful in determining metabolic rate.***
- Non-protein RQ can be used (indirect method) for calculating the total energy output and the proportions of various foodstuffs being burnt.
- Determination of RQ helps in the diagnosis of various pathological conditions such as acidosis, alkalosis, diabetes mellitus, etc.

CALORIC REQUIREMENTS

To maintain caloric balance in an adult, it is necessary to supply enough foods to replace the calories expended per day.

These include the following:

- A supply of "basal" requirements (BMR)
- Supplies of calories to meet the extra requirement caused by *specific dynamic action (SDA),* also called "calorigenic action" of foods.

- Supply for physical activity over and above the basal requirements (most 'Variable' under normal conditions).
- During periods of growth and / or convalescence, extra provision has to be made to meet the synthesis of tissues and weight gain.

1. BMR

It is quite constant in health, for any given individual and is influenced by various factors as stated above.

For an adult man of 70 kg body wt. and surface area 1.7 square metre, the basal requirement will be @ 40 C / sq. metre / hr

$= 1.7 \times 40 \times 24 =$ **1632 C (approx 1600 C/day)**

2. Specific Dynamic Action (SDA)

In an adult individual, whose BMR is 1600 C, is fed with just enough food to provide 1600 C and is kept under basal conditions (except that he is not under post absorptive state), it is found that his energy output has increased beyond the basal output of 1600. The increase varies with the type of food that has supplied the calories. ***This stimulant action of foods on the metabolism is known as the "specific dynamic action" (SDA)*** or ***calorigenic action of food.***

Definition: SDA may be **defined** as "extra heat" production, over and above the actual heat ought to be produced outside from a given amount of food, when this food is metabolised inside the body. The mechanism of stimulation is not clear.

- ***Proteins have the greatest SDA,*** amounting to about 30 per cent above its caloric value.
- Carbohydrates cause an increase of about 5 per cent or 6 per cent, and
- Fats cause about 4 per cent
- Ordinarily, the ***SDA of all together amounts to about 6 per cent of the BMR.***

  Explanation: The explanation for SDA is not clear. It cannot be due to a production of heat as a result of digestion as used to be thought earlier, because feeding of the products of digestion is as effective as undigested substances. In fact, intravenous administration of amino acids or glucose give rise to a SDA of the same order as results from feeding.

Several studies indicate that the SDA of the various amino acids is best correlated with the metabolisable energy of the individual amino acid, i.e. it is not related to the N_2 but rather to the non-nitrogenous fraction. This fraction undergoes oxidative and synthetic changes that liberate heat. In other words, heat is evolved during the intermediary metabolism of carbon chains. The SDA of glucose is increased if thiamine is administered at the same time. Since thiamine stimulates the formation of fat from glucose, the SDA of glucose has been suggested as being due to the energy required to prepare it for deposition of fat. Possibly this is the explanation for the SDA of all foodstuffs, i.e. the energy required to prepare the nonnitrogenous parts of the molecule for storage. To provide for this increase in metabolism, an extra provision 5 to 10% of basal requirement (usually take 6%) has to be made. ***Thus for a 70 kg man with BMR of 1600 C, another 80 to 160 C (96 C) have to be added on this account.***

3. Physical Activity

Most variable element in the calculation of energy requirements.

a. ***Influence of Muscular Work on Total Metabolism***
   Muscular work is accomplished by the body at the expense of increased metabolism. The potential energy of the foodstuffs is transformed to the free energy of work and the energy of heat. A man sitting quietly has a total metabolism, on the average, of about 100 C / hr. When he stands up, his metabolism increases by about 10 per cent because of the greater tonus of the muscles. If he engages in active works, it may increase 300 C or more / hr. ***The type of work or exercise influences the total amount of energy output,*** heavy work requiring more energy than light works. Many tables are available which give the total energy expenditure of various types of activity.

b. ***Influence of Mental Work on Total Metabolism***
   Mental work results in very little increase in total metabolism. **Benedict** found, that the effort involved in solving mathematical problems increases metabolism by only 3 per cent or 4 per cent. Brain tissue has a high "basal" metabolism, amounting to about 1 / 10th of that for the entire body but the additional work it performs in thinking does not result in much of an increase over this high basal figure.

c. ***Influence of sleep:*** During normal sleep, the muscles are relaxed and the total metabolism is correspondingly low. It is usually 10% below the BMR.

Measurement of Total Heat Production

Example

- A carpenter of 70 kg body wt. having 1.7 sqm of body surface. He does carpentary works for 8 hrs, and 8 hours he does sedentary works and sleeps for 8 hrs.

Calculation

BMR for 24 hrs	=	1600 C
8 hrs sleep	=	–53 C
8 hrs carpentary (164 extra C / hr)	=	1312 C
8 hrs sedentary (74 extra C / hr)	=	592 C
SDA	=	96 C
Total	=	3547 C

- ***Alternatively,*** using the data of Tables 44.2 and 44.3, calculation for the same carpenter will be:

8 hrs carpenter work at 240 C	=	1920 C
8 hrs Light work (sedentary) at 170 C	=	1360 C
8 hrs sleep 65 C/hr.	=	520 C
SDA	=	96 C
Total	=	3896 C

The above two hypothetical cases, have about the same total caloric output.

Note

- In first case, we began with basal metabolic rate and added and subtracted additional energy factors + the SDA.
- In the second case, the total caloric output per hour was tabulated + the SDA.

Table 44.2: Energy expenditure per hour under different conditions of muscular activity (includes BMR)

Forms of physical activity	***C/hour per 70 kg***	***Per kg***
Sleeping	65	0.93
Awake but lying still	77	1.10
Sitting at rest	100	1.43
Standing relaxed	105	1.50
Singing	122	1.74
Tailoring	135	1.93
Carpentry	240	3.43
Typewriting rapidly	140	2.0
Ironing (with 5 tb iron)	144	2.06
Walking slowly (2.6 miles/hr)	200	2.86
Walking moderately fast (3.75 miles/hr)	300	4.28
Walking downstairs	364	5.20
Swimming	500	7.14
Running (5.3 miles/hr)	570	8.14
Walking upstairs	1100	15.8
Walking very fast (5.3 miles/hr)	650	9.28

Table 44.3: Total calorie requirements for 24 hrs

	Calories	
• ***Men***		
Shoe maker	2000–2400	(Sedentary work)
Carpenter or Mason	2700–3200	(Light work)
Farmer	3200–4000	(Moderate work)
Lumber man	4000 or more	(Heavy work)
• ***Women***		
House hold work	2300–2900	
Seamstress (needle)	1800	
Seamstress (machine)	2300–2900	

4. ***Metabolism in Children***
   - The total metabolism in childhood is relatively much greater than, in adult life. There is, in the first place, ***the high BMR of childhood.***
   - The ***physical activity of children is usually greater,*** despite the fact that their period of sleep is longer than that of an adult.
   - Their games and play involve tremendous amount of muscular exercise. The food intake, therefore, must cover the caloric needs in addition to the extra food required for growth.
   - A child of 12 years, consequently, needs about the same amount of food as an adult, whereas an active boy of 16 years may require 3600 C or more per day.
   - Caloric requirement of infants 0 to 1 year are determined on the weight basis.

Infants up to 2 months of age	=	120 C/kg
Infants from 2 to 6 months	=	110 C/kg
Infants from ½ to 1 year	=	100 C/kg
Children: 1 to 10 years	=	110 to 2200 C
Boys: 10 to 18 years	=	2500 to 3000 C
Girls: 10 to 18 years	=	2250 to 2300 C

DETERMINATION OF CALORIC REQUIREMENTS FOR A "FAMILY"

Caloric requirement of a family can be calculated from the formula:

caloric requirement = "man value" of the family × energy required for each "man unit".

Man Unit

A ***man unit*** is defined as a normal man who has attained puberty and belongs to the age group of 15 to 45 years.

Man Value

The "man value" of a family can be obtained by "adding the man units, of all the members which can be obtained with the help of the Table 44.4.

Example: Let us consider a family consisting of an adult man, a pregnant wife, and four children of the age 2 yrs, 5 yrs, 7 yrs and 9 yrs.

∴ "Man value" of the family

= 1.0 + 1.10 + 0.35 + 0.50 + 0.60 + 0.70

= 4.25 man units.

If the energy requirements of "man unit" is about 3000 C. Then caloric requirements of the family

= 3000 × 4.25

= 12,750 C.

Table 44.4: Man value of different age groups

Age groups	***Man value***	***Age groups***	***Man value***	***Age groups***	***Man value***
1–2 yrs.	0.35	9–10 yrs.	0.70	Adult man and woman	1.0
3–4 yrs.	0.42	11 yrs.	0.80		
5–6 yrs.	0.50	12 yrs.	0.90	Pregnant lady	1.10
7–8 yrs.	0.60	13 yrs. and above	1.0	Lactating lady	1.30

PART II
NUTRITIONAL ASPECTS

Major Concepts

A. Study the role of various nutritional factors in the diet.
B. Study the nutrient requirement under normal conditions.
C. Study the clinical conditions of malnutrition.
D. Study the nutrient requirement under special conditions.
E. Study composition and nutritive value of common foodstuffs.

Specific Objectives

A. 1. Study role of proteins in the nutrition.
 2. Study factors influencing biological value of proteins.
 3. Study the quantitative aspects of proteins in the diet.
 4. Study role of carbohydrates in diet, about its requirements and other aspects, study about glycaemic index (GI)
 5. Study the role of lipids in diet, learn about diet or value of trans fatty acids

B. 1. Study in detail what is balanced diet and how it can be formulated.

C. 1. Study the conditions of protein-energy malnutrition (PEM).
 2. Differentiate in a tabular form—marasmus and kwashiorkor.
 3. Study about obesity—definition, causes, clinical features, learn about hormones of adipose tissue and its role in obesity

D. 1. Study the nutrient requirements during pregnancy.
 2. Study the dietary requirements of lactating mothers.

E. Study the composition and nutritive value of certain common food substances.

PROTEIN FACTOR IN NUTRITION

The large amount of informations which are available on nutritive value of dietary proteins has been obtained mainly from experimental studies carried out on albino rats and to a limited extent studies carried out on the dogs, specially plasma proteins and Hb. The relatively few observations made on human volunteers support the view, that with occasional exceptions, the result of studies in these experimental animals may be generally applied to man.

Functions of Dietary Proteins

- *Proteins are primarily NOT meant for energy,* their principal function is to synthesise tissue proteins of the body. Taken in excess, they may be utilised for the production of energy and may be converted to carbohydrates and fats.
- **Dietary Proteins and their Influence on Growth**
  Growth is manifestated by formation of tissue proteins at a rate exceeding that of their degradation, i.e. +ve N-balance. Proteins in diet is much more critically concerned in growth and tissue repair than are carbohydrates and lipids. ***In this respect, quality of proteins taken in diet is more important than the quantity consumed.*** One may take large amount of protein in the diet, but if it is not of good quality, i.e. lacks essential amino acids, it will not be utilised for tissue protein synthesis and repair. There is evidence that protein intake determines significantly the adult size of the individual, over and above, the genetic influences. Adult heights and weights of people in various geographic areas also vary directly with the amount and quality of proteins in their diets.

Example

- Members of certain African tribes subsisting largely on vegetables are small in stature, whereas
- Neighbouring tribes of similar origin, subsisting mainly on meat and milk are tall. Similar correlations have been obtained in studies of genetically comparable groups of individuals of other countries.

Protein factors in nutrition can be studied under two headings:

A. Quality of Proteins

Earlier stress was given more on how much proteins to be consumed but now with development of more satisfactory analytical methods for amino acids and accumulation of information on "essential amino acids", ***it is now known that quality of proteins consumed is also of equal, rather more important than the quantity.***

Quality of proteins can be discussed under the following heads:

(a) *Biological value of proteins*
(b) *Amino acid composition of the dietary proteins*
(c) *Balance of dietary amino acids*
(d) *Availability of amino acids from foods*
(e) *Supplementary relationship of amino acids.*

(a) Biological Value

Food proteins differ considerably in the efficiency of their utilisation for synthesis of body proteins.

Definition: Biological value is **defined** as that % of absorbed Nitrogen which is retained in the body.

Procedures

There are several procedures which may be employed for evaluation of the biological value, i.e. quality of the protein.

1. ***Measurement of weight increase:*** Measurement of its influence on the weight increase of weanling animals. This can be expressed by ***protein efficiency ratio (PER).***

$$\text{Protein efficiency ratio (PER)} = \frac{\text{Weight increase (in gm)}}{\text{Gm of proteins consumed}}$$

2. ***On retention of absorbed N_2:*** Determination of the biological value (BV) in terms of % of the absorbed N_2 retained by the organism. To estimate "biological value" (BV) of a protein:
   - The animal is first kept on protein free diet for a couple of days and the faecal N_2 and urinary N_2 are estimated to obtain the amounts of the metabolic faecal N_2 and endogenous urinary N_2 respectively.
   - Then the animal is fed with a measured amount of the test Protein and the faecal and urinary N_2 are determined again.

Calculation: Biological value (BV) is then calculated as follows:

$$\mathbf{BV} = 100 \times \frac{\text{Food } N_2 - (\text{Faecal } N_2 - \text{Metabolic faecal } N_2) - (\text{Urinary } N_2 - \text{Endogenous urinary } N_2)}{\text{Food } N_2 - (\text{Faecal } N_2 - \text{Metabolic Faecal } N_2)}$$

"Biological value" (BV) of some common proteins is given in Table 44.5.

Table 44.5: Biological values (BV) of certain common food proteins

Animal protein	*BV*	*Vegetable protein*	*BV*
Egg, whole	94	Rice	86
Milk (cow)	85	Barley	71
Milk powder	83	Wheat	67
Egg white	83	Maize	60
Fish	70 to 80	Bengal gram	76
Liver	77	Green gram	51
Pork	77	Soya bean	64
Beef	69	Cashew nut	72
Mutton	60	Ground nut	54
		Peas	56
		Cotton seed	63

Net protein utilisation (NPU)

It is ***defined*** as % of food N_2 that is retained in the body. This depends on both:

- Content of essential amino acids and
- Digestibility and absorbability of the protein.

$$\text{NPU} = \frac{\text{Digestibility coefficient} \times \text{Biological value}}{\text{Protein intake (gm)}}$$

Its value for some dietary proteins are as follows:

Proteins	NPU	Proteins	NPU
Egg proteins	91	Maize	36
Milk	75	Wheat	47
Meat	76	Peas	45
Fish	72	Ground nut	45
Liver	65	Soyabean	54
Rice	57		

3. Measurement of its influence on the rate of regain of body weight or of liver protein by previously depleted animals.
4. Measurement of its influence on the rate of restoration of plasma proteins or of Hb in animals previously depleted of these specific proteins (e.g. by plasmapheresis).

CLASSIFICATION OF QUALITY OF PROTEINS

Evaluated on the basis of above criteria, animal proteins generally are of "higher" quality ("first class" proteins/or **complete** proteins) as compared to those of vegetable proteins (**Incomplete** proteins).

- Whole egg and milk proteins, specially Lactalbumin rank highest in this respect and they contain the highest percentages of the "essential amino acids".
- Meat, fish, poultry and glandular tissues occupy next position in the scale. In the same class, are also yeast and soyabean.
- Cereals, legumes (peas, beans, etc) and nuts are generally poor because they lack some essential amino acids and thus they are incomplete proteins and are of poor quality.

Gelatin: Commonly used in desserts and in preparation of ice cream. It does not occur naturally as such, but is prepared from collagen (cartilage, bone, tendon and skin) by boiling in water. The protein is palatable, tasty and easily digested and used in diets of children and invalids/convalescing patients. Though the protein is animal origin, it is wholly an "inadequate protein", as it ***lacks essential amino acid tryptophan, also low in tyrosine and cystine.***

Factors Influencing Biological Value of Proteins

Certain factors are known to affect the biological value of proteins:

- The amounts and relative proportions of their constituent amino acids (essential amino acids).
- ***The nutritional availability:*** The rates of liberation and absorption of their constituent amino acids under conditions of digestions in GI tract. This may be influenced beneficially or adversely by various methods of processing or preparation (cooking) of foodstuffs.

CHEMICAL SCORE

Since egg proteins contain all essential amino acids in adequate amounts and possess the highest nutritive value, it has been assigned a chemical score of 100. (reference protein). Chemical scores of different proteins have been calculated in terms of egg proteins. ***It will depend on the essential amino acid most limiting in a particular dietary protein and thus it serves of an index of nutritive value of the particular protein.***

Definition

The chemical score is ***defined*** as the ratio between the content of the most limiting amino acid in the test protein to the content of the same amino acid in egg protein expressed as a percentage.

Example

1. *Milk protein:* The limiting amino acid is S-containing amino acids. Chemical score

$$= \frac{3.4}{5.5} \times 100 = 65$$

2. Chemical score of Gelatin and Zein are 0 (zero), chemical score of some of the common proteins are given below:

Chemical score of proteins

Proteins	Chemical score	Proteins	Chemical score
Egg proteins	100	Peas	42
Milk proteins	65	Bengal gram	44
Meat	70	Soya bean	57
Fish	60	Groundnut	44
Liver	66	Gelatin	0
Rice	60	Zein	0
Wheat	42		

(b) Amino Acid Composition of Dietary Proteins

The role of essential amino acids have already been stressed. To be a 'complete' protein and of high biologic value, the protein must have all the essential amino acids and ***they must be available to the organism together and simultaneously,*** so that they can be utilised for protein synthesis. It has been observed, ***if one of the essential amino acid is lacking and there is an interval of two hours or more, the amino acids are not utilised for protein synthesis.***

Proportionality Relationship of Essential Amino Acids

The biological value of protein is also related to proportionality relationship of its constituent essential amino acids. Studies in rats have shown that for optimal growth, the following proportional relationship of EAA is necessary, taking tryptophan as unity.

Proportional relationship of EAA

Tryptophan	1
Threonine	2.5
Isoleucine	2.5
Methionine	3.0
Phenylalanine	3.5
Valine	3.5
Leucine	4.0
Lysine	5.0

There is evidence that above proportionality also is applicable approximately to human beings also.

Based on above, the following daily intake values in gm have been recommended tentatively for human beings, which is approximately double of the minimum requirement. This should provide an adequate excess from which cell can select the proper mixture for its specific protein synthesis.

	Minimal requirement	Recommended daily intake
Tryptophan	0.25	0.5
Threonine	0.5	1.0
Isoleucine	0.7	1.4
Valine	0.8	1.6
Lysine	0.8	1.6
Leucine	1.1	2.2
Phenyl alanine	1.1	2.2
Methionine	1.1	2.2

"Sparing" Action

Nonessential amino acids tyrosine and cysteine can be synthesised in the body from essential amino acids phenyl alanine and methionine respectively. ***Presence in diet of these two nonessential amino acids reduce the necessity of their corresponding essential amino acids ("Sparing" action).*** On the other hand, if nonessential amino acids tyrosine and cysteine are not provided in diet, requirement of phenyl alanine and methionine increases and the necessity for synthesis constitute an additional excessive burden on the metabolic activities of the cell. This will be more under conditions of rapid growth, e.g. growing child, convalescing patient, pregnancy/and lactation.

(c) **Balance of dietary amino acids:** When the protein intake is quantitatively inadequate and also in terms of deficiency of certain essential amino acids, proper balance of dietary amino acids is particularly important.

Limiting Amino Acid

Essential amino acid which is most deficient in a protein is called limiting amino acid of that protein. When protein intake is inadequate, the addition of the most limiting amino acid may precipitate deficiency in the next most limiting amino acid. Conversely, addition of the latter may induce deficiency of the former.

Examples: Experimental studies in rats have shown:

- Addition of methionine to a low protein diet deficient in this amino acid may cause threonine deficiency.
- Similarly, addition of lysine to a diet low in histidine increases the requirement of histidine and urinary excretion of cystine.

Vitamin requirement: Provision of excessive amounts of single amino acids may alter the requirement of certain vitamins.

Examples

- Addition of glycine and leucine in excessive amount to a sucrose diet increases the requirement for niacin.
- An excess of methionine increases the pyridoxine requirement.

(d) **Availability of amino acids of foods:** Digestibility of various proteins and the rates of release of amino acids may be affected by the way the foods are prepared before eating, e.g. heating during cooking, in some it may be beneficial and in some cases it may affect adversely.

Coefficient of digestibility: Coefficient of digestibility means that the amino acids are readily split off by the enzymes and are readily available for absorption.

Definition: It is the % of food N_2 that is absorbed from the alimentary canal after digestion.

For its determination:

- Animal is first kept on a protein free diet for a couple of days and its metabolic faecal N_2 is determined.
- Then the animal is fed with the test protein and its faecal N_2 is again estimated, while on the diet.

Calculation

Digestibility co-efficient can be calculated as follows:

$$100 \times \frac{\text{Food } N_2 - (\text{Faecal } N_2 - \text{metabolic faecal } N_2)}{\text{Food } N_2}$$

Note

- Usually vegetable proteins from cereals, legumes and seeds have lower digestibility coefficient than what animal proteins have. Meat proteins have a high "co-efficient of digestibility". In this respect, kidney is superior to other meats, liver is next and muscle meats third.
- Different proteins with similar amino acids compositions may not liberate a given amino acid at the same rate during the course of digestion and the rates at which different amino acids are liberated may vary widely. ***The determining factor may be the nature of "linkage" of amino acids in the protein molecule.***

(e) **Supplementary relationships of amino acids and time factor:** This is a very important factor from nutritional point of view. As discussed above, most of vegetable proteins are "incomplete" proteins and of poor quality, lacking in one or other of essential amino acid. ***A vegetable protein lacking a particular essential amino acid, if supplemented by another vegetable protein possessing that amino acid and taken simultaneously, the 'biological value' of each of these two low-quality proteins is enhanced and the constituent amino acids can be utilised for protein synthesis.***

Some illustrative examples are given below:

- Wheat, an important dietary protein of Indians, is deficient in lysine. All 'cereals' lack lysine.
- Rice, another staple food for Indians, not only lacks lysine, but is also poor in threonine.
- Maize/corn lacks in tryptophan.
- Legumes (peas and beans), usually deficient in methionine and also tryptophan.
- Roots/tubers are generally deficient in methionine.

Supplementary action is manifested as follows:

- The combination of wheat, low in lysine, but adequate in methionine, when taken together either with potatoes or peas, which is adequate in lysine but low in methionine, enhances the biological value of each of these low quality proteins to a considerable degree and the amino acids can be utilised for protein synthesis.
- Similar enhancement of the biological value of relatively low quality vegetable proteins is accomplished by the addition of high-quality animal proteins.

Time factor: As stressed earlier, a protein molecule can be synthesised only if all the constituent 'essential' amino acids are available simultaneously and together in proper amounts. If an interval of one hour is allowed to elapse between ingestion of an "incomplete" amino acid mixture and administration of the missing essential amino acid, the supplementary action is not seen, and the amino acids are not utilised for tissue protein synthesis resulting to 'Negative' N-balance.

B. Quantitative Aspect

1. If the intake of protein is reduced gradually, urinary N-excretion also diminishes correspondingly and the organism may remain in N-equilibrium until a "critical intake" level above 0.25 to 0.33 gm/kg body wt. is reached, below which the N-balance becomes negative. Continuation of such low-protein diet for prolonged periods may endanger health and is harmful for body.

2. ***What should be the optimum quantity of protein to be consumed by an individual?*** This question has been discussed for years.
3. **Voit** in Germany, in latter part of 19th century, studied protein intake of many healthy adults in Germany and found their average consumption 118 gram/day. He suggested "optimum" standard for protein intake as **118 gram.** Only evidences for such an assumption are:
   - That it was the amount usually consumed.
   - Such persons were healthy.

   Disadvantage: No N-balance studies were carried out.
4. **Chittenden** in 1904 carried out N-balance studies on students, instructors, soldiers and others and came to a different conclusion. He found that optimum protein requirement was much less than given by Voit. Chittenden's subjects could be kept in N-equilibrium on from **45 to 53 grams/day** and this was suggested as the "Optimum" requirement.
5. Many investigators, after that have even shown that the minimum protein requirement compatible with N-equilibrium is much lower even than the figures set by Chittenden. But it was suggested that in order to attain N-equilibrium with a low protein intake, enough carbohydrates and fats must be consumed ***("protein-sparing" effect),*** to provide sufficient energy, for the individual's needs so that proteins are spared for giving energy.
6. The question was raised as to whether an individual could continue on such a low protein diet for considerable length of time, even if he was in N-equilibrium. The experiments, which have been conducted, are restricted for weeks/or months. Could a person live for years, and live comfortably on a minimum or near minimum protein ration?

Nutritionists now doubt if that could be done. They argue that with the modern concept of "essential amino acids", a person on a highly restricted diet with minimum protein, might easily experience a lack of one or more of these essential amino acids, the vital tissue building material.

7. Hence, an amount of protein greater than Chittenden, and yet lower than Voit would be optimum and ideal for good nutritive condition. This would constitute desirable *factor of safety* and would provide an excess of indispensible and semidispensible amino acids.
8. Studies of various investigators have now established that ***an intake of about 1 gram of protein per kg of body weight is adequate to maintain N-equilibrium.*** In growing children, convalescents, and pregnancy and lactation, this must be increased considerably to permit growth of new tissues. In elderly persons, the protein requirement is usually somewhat higher even though there is a somewhat lower calorie need. Negative N-balances are more common in this group. ***Hence it is recommended that elderly aged persons should consume more than 1 gm of "high-quality" proteins per kg of body wt. daily.***

The "recommended daily allowance" (RDA) value for proteins is given below

Category	Age in years	Protein in gm
• Men	18–35	70
	35–55	70
	55–75	70
• Women	18–35	58
	35–55	58
	55–75	58
• Pregnancy (2nd and 3rd trimesters)	—	+ 20
• Lactation	—	+ 40
• Infants	—	kg × 2.5 ± 0.5
• Children	1 to 3	32
	3 to 6	40
	6 to 9	52
• Boys	9 to 12	60
	12 to 15	75
	15 to 18	85
• Girls	9 to 12	55
	12 to 15	62
	15 to 18	58

The daily requirement of course will be influenced by the quality and biologic value of the proteins.

CONSEQUENCES OF PROTEIN DEFICIENCY

If the protein intake does not meet the immediate requirements, protein anabolism cannot be maintained at the required rate.

- In the child, ***growth is retarded.***
- In the adult, ***weight is lost.***
- Haemoglobin formation is impaired with consequent ***anaemia.***
- ***Wound healing is delayed.***
- If the deficiency is marked, excessive amount of fats may accumulate in liver producing ***fatty liver*** due to (1) choline deficiency, a consequence of methionine deficiency, and (2) impaired 'apoprotein' synthesis, thus lipoproteins formation suffers. ***Fatty liver later on may lead to fibrosis (cirrhosis liver).***
- Prolonged deficiency may result to ***inadequate synthesis of plasma proteins***—specially albumin and fibrinogen. **Fibrinogen deficiency may lead to bleeding disorders.**
- If the deficiency progresses to the point of significant decrease in plasma albumin concentration, ***oedema may develop***, and also increased susceptibility to shock.
- ***Resistance to infections may be diminished*** as a result of impaired capacity for forming γ-globulins antibodies (IgGs).
- In severe protein restrictions, certain hormones, protein in nature, such as those of anterior pituitary may not be synthesised in adequate amounts and ***endocrine abnormalities may appear,*** viz. amenorrhoea (gonadotropin deficiency).

- Since enzymes are proteins and must be synthesised in the body, the enzyme content of certain tissues {viz. liver cholinesterase, ornithine carbamoyl transferase (OCT), etc.} and enzymes in secretions (viz. pepsinogen in gastric juice, etc) falls in advanced deficiency stage. This may result in disturbances of functions of organs affected.

ROLE OF CARBOHYDRATES IN DIET

1. Glucose, fructose, galactose and to a minor degree, mannose, as well as those carbohydrates that yield them on digestion, are available to the body as energy producers. The pentoses in foods seem to be of limited value nutritionally. Ribose, deoxy ribose required for nucleic acid synthesis, is obtained from HMP-shunt. Moreover, pentoses form a very small fraction of the total carbohydrate intake of diet.
2. Dietary polysaccharides or disaccharides cannot be utilised until digested to the monosaccharide stage. When introduced directly into blood stream, they act as foreign bodies and are excreted, chiefly by the kidneys.
3. *Requirement of carbohydrates in diet:* ***Normally 55 to 65 per cent of the total food calories should come from carbohydrates.*** A moderately active man requiring 3000 C/day, ***should take about 450 gm carbo hydrates daily.*** But in India, poorer sections of the population derive more than 85 per cent of the food calories from carbohydrates.
4. Undue restriction of dietary carbohydrates influences both fat and protein metabolism adversely, even if the calorie intake is adequate. Fat mobilisation from the depots and utilisation are exaggerated, ketogenesis is increased and ketosis may develop. The effect on protein metabolism is apparently of a specific nature, not shared by other substances, e.g. fats, alcohol and not related to its calorigenic action. This is referred to as ***protein sparing action of carbohydrates.***
5. *Protein sparing action of carbohydrates:* ***Adequate amount of carbohydrates and fats in the diet may reduce the protein requirement.*** This may be due to:
   - Metabolic products of carbohydrates, e.g. oxaloacetate (OAA), pyruvates (PA) and α-oxoglutarates provide the C-skeletons for the formation of non-essential amino acids through transamination
   - Carbohydrates reduce the need for gluconeogenesis from amino acids, and
   - Both carbohydrates and fats are catabolised for energy and thus spare the proteins from being used for this purpose. Dietary fats may also depress the SDA of proteins.
6. *Effect on N-Excretion*
   - Isocaloric substitution of fat for carbohydrate in the diet is followed by an increase in N-excretion.
   - If the protein and carbohydrate components of an adequate diet are ingested separately, at wide time intervals, there is a transitory increase in N-excretion. No such effect is exhibited by fats.
   - On an exclusively fat diet, the N-output is the same as during starvation, whereas administration of carbohydrates reduced the nitrogen output.
7. *Action of carbohydrates on plasma lipids:* Replacement of a 'low' or moderate carbohydrate diet by a high carbohydrate diet may produce temporary rise in plasma TG and VLDL and temporary reduction in blood cholesterol. Substitution of starch by fructose or sucrose in the diet may also increase plasma TG by increasing lipogenesis from fructose.
8. *Relation with B-vitamins:* With diets rich in carbohydrates, the requirements for B-vitamins, particularly thiamine (vit B_1) increases because of the essential role of these in carbohydrate metabolism.
9. *Role of cellulose:* Celluloses are polysaccharides found in plants. They are indigestible by human beings as there is no enzyme in our GI tract which can split β-1 → 4 linkage. But, celluloses in diet, contribute bulk to the intestinal contents, and therefore, in normal amounts promote intestinal motility, i.e. increases peristalsis **("Roughage" action)** and removes constipation. When present in excess, they may be irritating to intestinal mucous membrane, producing diarrhoea or a spastic type of constipation.
10. *Excessive Intake of Carbohydrates in Diet*
    - Ingestion of excessive amounts, specially in infants, may occasionally produce intestinal disturbances due to irritation induced by products of bacterial fermentation.
    - There is some evidence in experimental animal (rats) that continued ingestion of super tolerance amounts of galactose, may result in the **formation of cataracts.** A similar phenomenon has been suggested as possibly occurring in man, especially in children, in the presence of hepatic functional impairment, in which condition the conversion of galactose to glucose (or glycogen) may be impaired, with consequent elevation of blood galactose level.
    - It is to be noted that the incidence of cataract is relatively high also in uncontrolled diabetes mellitus, a phenomenon presumably related to persistently elevated to blood glucose concentration and altered metabolism.

Glycaemic Index (GI)

The glycaemic index or GI is a measure of effects of carbohydrates on blood sugar levels.

Definition: The glycaemic index (GI) of a food is defined as the area under the two hour blood glucose response curve (AUC) following the ingestion of a fixed portion of carbohydrates (usually 50 gm)

$$GI = \frac{\text{AUC of the test food}}{\text{AUC of the standard}} \times 100$$

Standard is either 50 gm glucose or white bread equivalent to 50 gm.

Calculation: The average GI value is calculated from data collected from 10 human subjects. Both the standard and test food must contain an equal amount of carbohydrates. The results give a relative ranking for each tested food.

Interpretation

- Carbohydrates that break down quickly during digestion releasing glucose ***rapidly*** into the circulation have a ***high GI***.
- Carbohydrates that break down more ***slowly***, releasing glucose more gradually into the bloodstream have a ***low GI***.

Food with Low GI has Significant Health Benefits

The concept was developed by Dr Jenkins and colleagues (1981) in their research to find out which food items would be best for diabetic patients. A lower glycaemic response usually equates to a lower insulin demand and may improve long-term blood glucose control and blood lipids.

Classification of Food According to GI

Food can be arbitrarily classified into three groups according to GI values (Table 44.6).

Factors Modifying GI Value of Food

The glycaemic effects of food depends on a number of factors such as:

- ***Type of starch*** (amylase vs amylopectin)
- Physical entrapment of starch molecules within the food
- ***Fat and protein content*** of the food and organic acids or their salts in the meals, e.g. ***adding vinegar lowers the GI.***
- ***The presence of soluble dietary fibres*** can slow the gastric emptying rates, thus lowering the GI
- ***Unrefined bread*** with higher amounts of fibre generally have a lower GI value than white bread
- Many ***brown breads*** are often ***treated with enzymes to soften the crust,*** which ***makes the starch more accessible thus increasing the GI.***
- ***Alcoholic beverages*** have been reported to have low GI values.

Recent studies (2005) have shown that the consumption of an alcoholic drink prior to a meal reduces the GI of the meal by approximately 15 per cent.

CLINICAL ASPECTS

- Several studies have shown that individuals who followed a low GI diet over many years were at a ***significantly lower risk for developing type 2 diabetes and coronary artery disease.***
- High blood glucose levels or repeated glycaemic "spikes" following a meal may promote these diseases by increasing the oxidative stress to the vasculature and also by the direct increase in insulin levels.
- ***Post prandial hyperglycaemia was*** considered as a ***risk factor*** associated with diabetes but ***recent studies have shown*** that *also presents an increased risk for atherosclerosis, in the non-diabetic populations.*
- A recent study from the University of Sydney, Australia (2008) suggests that having a breakfast of white bread and sugar

Table 44.6: Foods with their GI values

Classification	*GI range*	*Examples of foods*
Low GI	55 or less	• Most fruits (except water melon, bananas) • Vegetables (except potatoes) • Legumes, peanuts and pulses • Products extremely low in carbohydrates—fish, eggs, meat and some cheeses • Brown rice • Ice cream
Medium GI	56–69	• Whole wheat products • Basmati rice • Table sugar • Most of the white rice products • Sweet potatoes • Bananas
High GI	70 and above	• White rice (polished) • Cornflakes • White bread • Baked potatoes (potato chips) • Water melon

rich cereals overtime, may ***make a person susceptible to diabetes, heart disease and even cancer.***
- Recent researches provide evidence that ***high GI carbohydrate is associated with increased risk of obesity.***

ROLE OF LIPIDS IN THE DIET

1. A wide variety of lipids is provided in a balanced diet. In as much as, under normal circumstances, all components of biologically significant lipids, with the exception of essential fatty acids, can be synthesised in the body from non-lipid precursors. The main function of dietary lipids, like that of carbohydrates, is ***to provide energy***, largely through oxidation of their constituent free fatty acids.
2. The dietary lipids serve another indirect function, serving ***as "carriers" of certain fat soluble vitamins*** (vit A, D, E and K) and provitamins like carotenes, which because of their solubility in fats, occur in nature mainly in association with these substances.
3. Lipids may also exert a relatively ***minor "protein-sparing effect"***, apart from their calorie contribution.
4. *Requirement:* Neutral fats (TG), comprising the largest fraction of food lipids are quantitatively the most important of these substances. Under usual conditions, fats provide 20 to 35% of the calories of the diet, i.e. 1 to 2 gm/kg of body weight in the average moderately active adult.
5. Dietary fat has a ***high "satiety value"***, i.e. the ability to satisfy hunger.
6. ***Supply of polyunsaturated FA:*** Food fats should contain adequate amounts of polyunsaturated FA to supply at least 1 per cent of total calories in adult man and 4 per cent of same in children. The nutritional significance of the polyunsaturated "EFA" have already been discussed. They are provided in ample amounts by a balanced diet that contains an adequate quantity of naturally occurring lipids (Linoleic acid mainly in plant and seed oils and arachidonic acid of animal origin). Not all vegetable fats are rich in linoleic acid (low content in coconut oil), nor are all animal fats deficient in this substance (high content in chicken fat). It is generally agreed that in man elevated plasma cholesterol levels of certain types may be lowered by:
   - Restriction of fat intake and
   - Substitution of polyunsaturated FA for saturated FA.
7. ***Quality of fat:*** Chain length and saturation of FA and MP of TG influence the nutritive quality of food fats. TG of short chain, medium chain, or polyunsaturated FA are more easily digested by *Lipases* in the intestine. While the TG of caprylic acid (C_8) has a digestibility of more than 97 per cent, that of palmitic acid (C_{16}) has a digestibility of about 70 per cent only. Unsaturated TG are also more readily absorbed than saturated ones. Oils of vegetable and seeds (e.g. sunflower oil, groundnut oil, soyabean oil, mustard oil) contain mainly unsaturated FA and are preferable.
8. ***Medium chain TG has been used in treatment of chyluria and chylothorax as they are absorbed directly in portal blood.***
9. ***Excess of fats in diet:*** An excessive high fat intake inhibits gastric secretion and motility, producing anorexia and gastric discomfort. Intestinal irritation and diarrhoea may result from excessive amounts of FA in the intestine. Excess of fats, particularly saturated fats, in the diet may reduce the gastric digestion of proteins, because fat digestion starts mainly in the intestine, thus preventing exposure of food proteins to pepsins. Excess fat intake can cause excessive production of ketone bodies, due to high FA oxidation and also can lead to Type 1 fatty liver.
10. Delay or failure of fat absorption may also reduce Ca^{++} absorption as calcium forms insoluble soaps with higher FA in intestine.
11. Cotton seed oil contains a pigment **gossypol,** which has anti-oxidant and antitryptic activity, diminishes appetite and interfere with protein digestion.

Dietary Role of Trans Fatty Acids (TFAs) or Trans Fats

Trans fatty acids are **principally artificial fats** but a small amount of TFAs occur naturally in meat and dairy products.

- Trans fatty acids are **formed artificially during the hydrogenation of vegetable oils.** In this process, the vegetable oils which are liquid turn solid. **Hydrogenation improves the shelf-life and palatability of oils and thus they are used in food industry.** Examples of hydrogenated fats include vanaspati sold under the brand names of Dalda, Rath, Sarathi, etc. Foods cooked in partially hydrogenated fats contain high levels of TFAs.
- **Naturally occurring TFAs:** Some TFA is found naturally in small amounts in meat and dairy products. The health effects of these naturally occurring TFAs are being studied.
- **Effects of TFAs in our body:** TFAs are found to be a greater risk for chronic degenerative diseases. ***TFAs are more atherogenic than saturated fatty acids.***
  1. **TFAs *alter secretion and composition of apo-B 100 containing lipoproteins LDL and VLDL.***

2. TFA is *involved in altering metabolism, viz.:*
   - **Increasing catabolism of apo-A_1**
   - **Increasing LDL levels (bad cholesterol)**
   - **Decreasing HDL (good cholesterol) level**
   - **Increasing the abnormal clotting of blood. Studies have also reported a positive correlation of trans fatty acids with diabetes type-2 and some cancers.**

Some ways to avoid TFA intake in the diet:

- Avoid using vanaspati and hydrogenated margarines
- Do not heat the oil for long while cooking or frying at home
- Any oil left over after frying should not be reused.
- Avoid processed foods like cookies, chips, cakes, and patties
- Cook and back using cold pressed oils like mustard oil, olive oil, seasame oil, etc.
- Avoid reheating oil
- Check the nutrition facts label on packaged food items for TFA content
- Check the ingredient list on packaged foods for words like "shortening", "partially hydrogenated vegetable oil" or "hydrogenated vegetable oil". These are nothing but trans fats.

ROLE OF MINERALS IN DIET

Refer to chapter on Metabolism of Minerals and Trace Elements.

ACID-FORMING AND BASE-FORMING PROPERTIES OF FOODS

- When a food is incinerated in a crucible, the ash remaining will have an acid, alkaline or neutral reaction depending upon the proportion and type of anions and cations present and the effect of heat upon them. When the same food is consumed by an individual, its final product will sometimes have the same reaction as the ash, but there are other factors, which modify the "ash" left by vital processes.
- Proteins, phospholipids, and neucleoproteins yield sulfuric, phosphoric and uric acids respectively. These acids are neutralised by basic elements before excretion and thus tend to diminish the alkali reserve and alkaline factors in blood and urine.
- Fruits and vegetables usually have enough positive (+ve) radicals, such as Ca, Mg, Na and K, to combine with the acids produced by the proteins or with other acids. Organic acids such as citric acid, malic acid, tartaric acid, and lactic acid present in fruits and vegetables are oxidised to CO_2. Most of this is lost by way of lungs, whereas K-salts of the above acids, also occurring in fruits, are oxidized to $KHCO_3$, which, if present in excess, is excreted in the urine. Thus vegetables, even acid fruits, usually have an alkaline effect *(alkalinisers)*.

But there are some exceptions, for example:

- Benzoic acid present in "cranberries" is not oxidised by the body and is excreted as ***hippuric acid,*** after conjugation with glycine and thus has an acidic effect on urine.
- Oxalic acid found in Rhubarb, beet leaves, cocoa, and tea is also an exception. It is oxidised very poorly and is neutralised and excreted as oxalates.

BALANCED DIET

Definition: A diet is said to be a balanced one, when it includes proportionate quantities of food items selected from the different basic food groups so as to supply the essential nutrients in complete fulfilment of the requirement of the body.

Basis

A balanced diet should be based on:
- Locally available foods
- Should be within the economic means of the people
- Should fit with the local food habits
- Diet should be easily digestible and palatable
- Should contain enough roughage materials.

Such a diet containing the required quantities of the different essential nutrients would perform the basic functions of food.

BASIC FOOD GROUPS

All essential nutrients including accessory food factors like vitamins and 'trace' elements required for the proper functioning of the body are distributed in varying quantities in different natural foods, constituting the ***basic food groups.*** Basic food groups have been initially divided into seven groups, but now they have been put mainly into 4 **(four)** basic food groups. They are:

- *Milk group:* Including dairy products
- *Meat group:* Including meat, fish, eggs and pulses/ beans/nuts, etc.
- *Green leafy vegetables* and fruits group.
- *Cereal groups:* Bread, rice, wheat, barley, etc.

A balanced diet should be an intelligent assortment of items from each of these four basic food groups so that different foods, rich in different nutrients can contribute to the total nutritive value of the diet (Fig. 44.2).

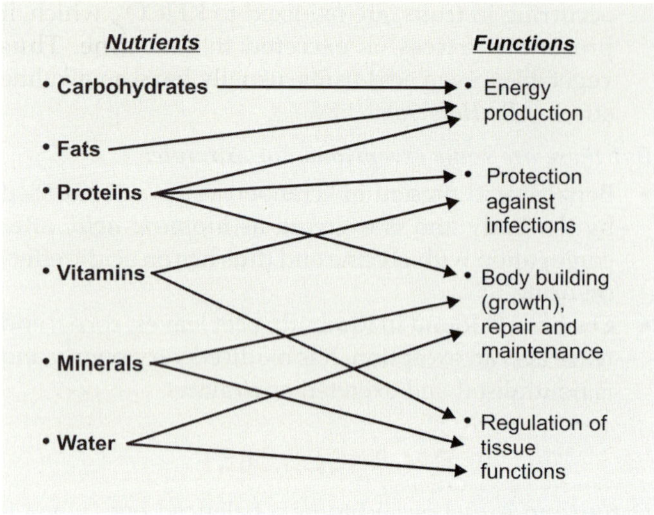

Fig. 44.2: Basic functions of foods (nutrients)

How to Plan a Balanced Diet?

1. ***Age, sex and calorie requirements:*** While planning diet for any individual, his/her age and sex, physical activity involved and special nutritional needs, viz. a growing child or a pregnant/lactating lady, if any, must be taken into account in determining the total calorie requirements and total daily requirement of nutrients.
2. ***Selection of Nutrients from 'Basic Food Groups'*** The required quantities of food items are to be selected from the four basic food groups in such a way that their total nutritive values satisfy the estimated requirements. The selection should be based on varieties of foods and if the requirements for calories and proteins are met, the vitamins and mineral requirements are automatically fulfilled.
3. ***Economic status of the individual:*** In formulating balanced diet, it is imperative and necessary that the economic status of the individual is taken into account so that the ***diet is within the purchasing capacity of the individual.***
4. ***High-cost and low-cost diet:*** For a rich person planning of a balanced diet does not pose any problem, because he can afford to purchase the foods recommended. In case of a person, belonging to low-income group, it poses the problem. Cheaper items have to be selected and at the same time care should be taken that nutritional and calorie requirements are fulfilled. A balanced diet formulated loses relevance if it does not conform to his or her purchasing capacity.

Example of a balanced diet: A typical balanced diet (non-veg.) for an adult man doing moderate work is shown in Table 44.7.

Table 44.7: Composition of a typical balanced diet for an adult male doing moderate work (non-veg.)—'high-cost' diet

Food items (Nutrients)	*Quantity gm/day*
Cereals	475
Pulses	65
Green leafy vegetables	125
Other vegetables	75
Roots and tubers	100
Fruits	30
Milk	100
Meat/fish	30
Fats and oils	40
Eggs	30
Sugars	40

Formation of any balanced diet must be followed by an approx. costing of the diet based on the local market values of the different commodities. The above diet provides about **2800 C** and about **75 gm** proteins of high quality.

Need for "low cost" balanced diet: Need for formulating 'low-cost' balanced diets is necessary to bring a compromise between the nutritive values and the cost. But now even the cheapest sources of nutrients like pulses, nuts, etc. have become costlier and so as long as the diet is made a balanced one, the cost cannot be low. Hence, the term 'low-cost' diet should be replaced by **'minimum cost'** diet. An example of a "minimum cost" balanced diet for an adult male doing moderate work is given in Table 44.8.

Such a diet will supply approx. **2700 C** and approx. **80 gm of vegetable proteins.**

Note: Following to be noted:

- Costlier items like meat/fish/milk and eggs have been replaced by comparatively cheaper foods of vegetable origins. Their cost is less and nutritionally considered adequate.
- ***Limitation:*** Such a diet based exclusively on vegetarian foods there will be ***a possibility of developing deficiency of vit. B_{12} in future. The above can be prevented by taking small fishes once or twice a week or***

Table 44.8: Composition of a "minimum cost" balanced diet for an adult male doing moderate work

Food items (Nutrients)	*Quantity gm/day*
Cereals	285
Pulses	100
Green leafy vegetables	200
Potatoes	200
Colocasia	100
Ground nut (Kernel)	50
Germinated bengal grams	50
Oils	35
Sugar	35

supplemented by an egg. As the requirement of vit. B_{12} is very small and as the storage capacity in liver very high, the chance of developing such a deficiency is rare.

Use of certain unconventional food items: Certain food items are discarded due to ignorance of their food values. Many of them are often found to be rich sources of proteins and calories and if consumed can effectively protect the body against nutritional disorders.

Examples include: Jack fruit seeds, pumpkin seeds, water-lilly seeds, water-melon seeds, etc. The kernel portions if suitably cooked can be good sources of calories and proteins. Similarly, small snails, oysters and crab meat are also delicious, highly nutritive and can be consumed. Various types of mushrooms, available in the rural areas are good sources of proteins and B-vitamins and are extremely popular in developed countries.

Guidelines for Certain Dietary Foodstuffs

- ***Nature of lipids:*** Avoid too much of fat in the diet. ***Saturated fats to be avoided*** and include oils having polyunsaturated fatty acids, use egg, meat, butter, ghee, cream, etc., in moderation, excess to be avoided.
- ***Nature of carbohydrates:*** Eat foods with complex carbohydrates like starches, e.g. cereals, whole grain, breads, etc. than simple sugars. ***Avoid too much sucrose.***
- ***Dietary fibres:*** Should be included, include vegetables, fruits, peas, beans and nuts.
- ***Avoid excess calorie:*** Maintain ideal body weight according to height, obesity predisposes hypertension, diabetes mellitus and cardiovascular diseases. Variety of foods should be taken to ensure an adequate intake of all essential nutrients. Choose food from basic food groups.
- ***Salt:*** ***Avoid too much of salt.***
- ***Alcohol:*** If the individual drinks alcohol, it must be in a limited quantity. Heavy drinking is detrimental for health and can lead to fatty liver, cirrhosis of liver, and certain neurological disorders. Alcohols are high in calorie value but low in nutrients.

PROTEIN-ENERGY MALNUTRITION (PEM)

Synonym: Earlier used to be called as ***protein calorie malnutrition (PCM).***

What is Malnutrition?

Malnutrition is a state arising from:

1. An insufficient calorie intake causing undernutrition or inanition and/or
2. Insufficient intake of one or more of the essential nutrients, specially proteins causing deficiency.
   The above two are ***primary*** causes and responsible for Marasmus and Kwashiorkor respectively.

Other causes lead to ***secondary*** malnutrition:

- Due to inadequate absorption or utilisation of essential nutrients (malabsorption syndrome) or
- Due to increase in their requirement, destruction or excretion generally secondary to diseases with special features superadded by the pre-existing disease.

Types: Protein-energy malnutrition (PEM) are mainly **of 3 types:**

1. ***Marasmus***
2. ***Kwashiorkor, and***
3. ***Marasmic-kwashiorkor.***

The above are quite common diseases in children and often met within India, Bangladesh, SE Asian countries, West Africa and Arab countries.

The two main types are discussed below and given in a tabular form to differentiate the two diseases **(Table 44.9).**

- *Marasmic-Kwashiorkor:* Symptoms of both marasmus and Kwashiorkor are sometimes produced in a mixed way depending on the relative degrees of protein and calorie deficiencies. Marasmus and Kwashiorkor may follow each other in some patients. Besides the above, symptoms of deficiencies of vitamins and minerals may also be found in these diseases, e.g. hypoprothrombinaemia (vit K-deficiency), Pellagra (Niacin deficiency), etc. and anaemia.

Note: Protein energy malnutrition cases should be treated early with suitable diets and antibiotics and other ancillary measures, otherwise permanent stunting of growth, hepatic cirrhosis, permanent mental retardation, low IQ and even death may result.

OBESITY

INTRODUCTION

It is difficult to define obesity—various definitions have been given.

"Anyone who is more than: 20 per cent above the 'Standard' weight for people of the same age, sex and race must generally be considered to be at least overweight".

Alternatively, "Obesity is that physical state in which the amount of fat stored in the body is excessive".

"Obesity is due to excess of adipose tissue and is defined as that body weight over 20 per cent above mean ideal body weight".

It is still not clear whether obesity represents a disease process or a symptom, a common clinical manifestation of a group of disorders, like diabetes, hypertension and certain endocrine disorders. But though it may be a symptom, it commands the medical attention and accorded as

Table 44.9: Differentiation of marasmus from kwashiorkor

Marasmus	*Kwashiorkor*
I. **Marasmus** of primary or dietary origin is most common in tropics, result of starvation in small children.	I. **Kwashiorkor is *primarily due to diet very low in proteins.***
II. ***Nature of diet and calories:*** Diet may be adequate in qualitative term but insufficient in calories for the rapidly growing ***child. Calories become the most limiting factor. Infant survives on utilising its own tissues.***	II. It is the result of a diet very low in proteins but provides enough calories to satisfy the need of the child. ***Proteins become the most limiting factor and not calorie.*** Protein lack is quantitative and also qualitatively of low biological value.
III. ***Causes:*** • Exclusively breast fed infant of a malnourished mother. Milk supply is grossly reduced. • Prolonged breastfeeding with inadequate supplementation of other foods. • Artificial feeds inadequate, less nutritive in proteins and calories. • Fear of diarrhoea, less feeding.	III. Seen in artificially fed and weaned children. Occurs weeks or months after weaning. Proteins of low quality are fed, e.g. cereal grains, starchy foods and roots. No milk, eggs, etc.
IV. ***Predisposing factors:*** • Premature babies • Local disease and/or • Malformations of mouth and nose, interfere with adequate feeding.	IV. Kwashiorkor seldom occurs as a consequence of an improper diet alone. In almost all cases, an infectious disease acts as the precipitating factor. These may be: • Acute diarrhoea • A respiratory infection • Measles. They add deterioration of diet, poorer utilisation, and higher requirement.
V. Age: Usually seen in infants less than one year	V. Seen in older children: In second/third year of life.
VI. ***Clinical features:*** • Growth: Retarded growth, child is grossly emaciated, and underweight. Depleted of subcutaneous fats and muscles. • Infant is very hungry and cries continuously. • ***Diarrhoea and vomiting:*** The infant may present frequent small dark green mucous stools of ***hungry diarrhoea.*** It aggravates the disease further. Vomiting is more common. • Diarrhoea is almost always present. It becomes chronic and remittant.	VI. • Retardation in growth and development has been described. Child is not emaciated like marasmus, rather looks blown up due to oedema. • Apathy and anorexia are early manifestations. The child is less lively and refuses to eat.
• ***Skin and mucous membranes:*** Skin is thin, attached to bone, flaccid and wrinkled. Bony prominence are marked. Mucous membranes of mouth are usually reddish.	• Skin lesions although not always present are very characteristic, patches of hyperpigmentation, exfoliation, desquamation and ulceration are seen in skin of legs, buttocks and perineum. Unlike marasmus, some subcutaneous fats are present.
• ***Oedema:*** No oedema is present	• ***Pitting oedema*** is the main clinical characteristic on which the diagnosis is made. It is soft, painless. Usually, first affects legs and then spreads to upper extrem,ities and face.
• ***Hairs:*** Usually thin and lustreless Face: All the above features make up a typical face of marasmic child which has been compared to a **"little monkey".**	• Hairs usually dry and thin. Black hairs become brown or reddish-yellow and sometimes even white. Depigmented hairs alternate with more pigmented hairs, called as ***Flag sign.***
• ***Dehydration and electrolytes imbalance*** May be present in both if complicated with excessive diarrhoea and vomiting.	
VII. ***Haematological and biochemical alterations:*** • ***Anaemia:*** Hb and haematocrit values are slightly reduced.	VII. • Some degree of anaemia is always found, it is mild to moderate. Type of anaemia varies, usually serum Fe and Cu are low.
• ***BMR:*** Usually subnormal	• May be low

Table 44.9 contd...

Table 44.9 contd...

Marasmus	Kwashiorkor
• ***Serum proteins:*** Total and differential and A:G ratio Total serum proteins and their fractions are reduced but not to that extent as Kwashiorkor. Total serum protein usually ranges 5 to 6 gms% and albumin approx. 3.0 g% A:G ratio: maintained.	• Total serum proteins are always reduced. Albumin↓ Hypoalbuminaemia is a characteristic feature. α_1 and α_2-globulins found to be relatively increased; β-globulins frequently decreased↓ γ-globulins variable, but usually high. A:G ratio: frequently reversed.
• ***Plasma Lipids*** Not much affected	• Unlike marasmus fall in plasma levels of cholesterol, TG and b-lipoproteins seen.
• ***Fatty Liver not common***	• ***Fatty liver may be seen***
• ***Carbohydrate metabolism*** Hypoglycaemia not a constant feature.	• Hypoglycaemia frequently found.
• ***Other electrolytes*** Not much alterations.	• A marked K-depletion, primary at intracellular level is a constant and important alteration. Mg ↓ and Pi↓ depletion also described.
• ***Serum enzymes*** Not much alterations.	• The serum activity of various enzymes viz. Amylase, alkaline phosphatase, and others are frequently reduced.
Note: Histological, functional and metabolic alterations are less marked in Marasmus than Kwashiorkor. This is due to the fact that the child developing marasmus is forced in view of his very limited calorie intake to consume its own tissues.	
VIII. ***Prognosis:*** Is Good, unless severe complications like dehydration and infections are present. Recover well with adequate dietary treatment. A complete and balanced diet adequate for his apparent or biological age but much higher in calories for normal child, 200 calories/kg or more may be required.	**VIII.** Not so good Even under the best conditions, the mortality of children admitted is still relatively high in the order of 10 to 20%.

the status of a serious condition due to its implications and associations with certain diseases.

1. IMPORTANCE OF OBESITY

Obese persons are more prone than the average populations to certain disease processes. They are:

- *Diabetes mellitus* type II (maturity-onset).
- *Cardiovascular disorders:* Hypertension, angina of efforts, widespread atherosclerosis, varicose veins and thromboembolism.
- *Liver diseases:* Prone to develop fatty liver, cholelithiasis and cholecystitis.
- *Physical consequences of too much fat:*
  - Bronchitis
  - Alveolar hypoventilation associated with massive obesity eventually leading to CO_2 retention (obesity hypoventilation syndrome or **Pickwickian syndrome**)
  - Backache, arthritis of hips and knee joints, flat feet; and
  - Hernias, ventral and diaphragmatic.
- *Metabolic diseases:* Like gout (hyperuricaemia).
- *Skin disorders:* Intertriginous dermatitis. Intertrigo is quite common in the folds below the breasts and in the inguinal regions.
- *Gynaecological disorders:*
  - Amenorrhoea, oligomenorrhoea
  - Toxaemia of pregnancy; and
  - Endometrial carcinoma.
- *Surgical postoperative complication:* Surgical "risks" in general is greater in obesity.
- *Industrial, household and street accidents:* Obese persons are susceptible to these accidents.

2. TYPES OF OBESITY

A. Immediate cause of obesity is always a positive energy balance, but there are many ways in which the balance may be tilted towards the positive side. Thus obesity is often divided into **2 types:**

a. ***Exogenous obesity.***
b. ***Endogenous obesity.***

a. Exogenous Obesity

Overfeeding and gluttony with less physical activity. Many people overeat than the calorie requirements either because they are too fond of their foods which is a pleasure, or quite often because they are unhappy, foods give them solace.

b. Endogenous Obesity

There may be one or more endogenous factors: endocrinal, metabolic, hypothalamic lesion.

B. ***Pathologically,*** **the types of obesity are:**

a. **Hyperplastic type.**
b. **Hypertrophic type.**

a. *Hyperplastic type:* This type is a life long obesity characterised by an increase in adipose cell number as well as increase in adipose cell size. ***Fat distribution is usually peripheral as well as central.*** Long-term response to treatment is not good. After weight reduction, ***adipose cell size may shrink but the increased number of cells persist.***

b. *Hypertrophic type:* It is seen in adults after twenty years of age ***(adult onset type).*** It is characterised by hypertrophy of adipose tissue cells without increase in adipose cells number. There is ***increase in cell size only. Fat distribution is usually central.*** The energy requirements of the body diminish with the advancing age and if there is no corresponding reduction in eating habits, a ***middle-aged spread*** is the natural result. Long term response to treatment is fairly good.

3. CAUSES

Obesity is most commonly due to overeating than the caloric requirement. Obesity can be encountered with other diseases, viz. certain metabolic disorders, and endocrine disorders. Thus, the causes of obesity as listed below, though may not be all complete, but encompasses the more common and certain uncommon syndromes which have been reported.

a. *Genetic influences.*

b. *Physiological:*
   - Overeating than caloric requirement.
   - Pregnancy.
   - Postmenopausal women.
   - Use of oral contraceptives for prolonged periods.

c. *Metabolic:*
   - Diabetes mellitus maturity onset (Type II).
   - Hyperlipidaemic states specially, type IV and type V.

d. *Hypothalamic injuries or abnormalities* **(e.g. Prader-Willi syndrome).**

e. *Miscellaneous and endocrine disorders:*
   - Hypothyroidism, Cushing's disease and Cushing's syndrome, pseudohypoparathyroidism, islet cell tumour (insulinoma), polycystic ovary syndrome, Laurence-Moon-Biedl syndrome, Fröhlich syndrome, Acromegaly.

4. PATHOGENESIS

Genetic and Other Factors in Obesity

Age: Immoderate accumulation of adipose tissue may occur at any age, but is more common in middle life. Minor degrees of corpulence, 10 to 15 per cent above optimal weight are the rule rather than the exception after the age of 30 years.

Sex: Adult women are more prone to obesity as compared to men. The normal fat content of an average young woman, approximately 15 per cent of body weight, is twice that of young men of comparable age. Women in menopausal period become usually obese. Obesity is also more frequent in pregnancy and women on oral contraceptives.

- *Genetic factor:* Obesity occurs much more frequently among the members of certain families than among others. A genetic factor may be identified in many cases but its mode of transmission and operation is still not known.
- *Psychological factors:* Psychological factor also plays an important role. Obese persons are often psychologically imbalanced. People who are suffering from anxieties, worries, and under constant tension or are frustated, they eat more to compensate.
- *Hypothalamic factor:* **Two mechanisms** within the hypothalamus appear to regulate food intake:
  - If certain **lateral centres** are bilaterally destroyed, aphagia results.
  - When the *medially controlled centres* are bilaterally destroyed, the lateral "feeding" areas are freed of their usual regulatory checking action and hyperphagia occurs. The individual eats more than requirements and obesity results.

The exact site of the hunger sensation accompanying hypoglycaemia is not well understood. Persons with so-called *pituitary obesity* presumably suffer from a hypothalamic disturbance. It has been established that experimental pituitary destruction does not cause obesity unless the hypothalamus is also injured.

- *Epidemic encephalitis:* May be followed by the development of obesity, and in such cases hypothalamic lesions have been found which resemble those known to cause experimental obesity.
- *Endocrine factors:* Certain endocrinal disorders may predispose to obesity:
  - *Fröhlich's syndrome:* It is characterised by hypogonadism and obesity, has been considered the result of hypopituitarism. In adiposogenital dystrophy, the excessive fat accumulation may result from hypothalamic disturbance, but its typical distribution is characteristic of hypogonadism, which may result from pituitary insufficiency.
  - *Cushing's syndrome:* (Adrenocortical hyperfunction): It is often associated with an increase in body fat mainly confined to the head, neck and trunk (turncal obesity and ***buffalo hump***), but spares the limb. It is often associated with a gain in weight. Although a low BMR cannot explain the usual type of obesity, hypothyroidism may be associated with gain in weight, partly due to water retention in tissues and partly to fat storage; which is evident in particular sites stated above.
  - *Functional or organic hypoglycaemia (hyperinsulinism):* It is frequently associated with abnormal

hunger leading to excessive food intake and obesity. Hyperinsulinism may aggravate the disability by promoting lipogenesis and inhibiting lipolysis.

- *In pregnancy:* Endocrine factors play part in increasing weight and producing obesity.
- *Hypothyroidism:* Diminished BMR and energy expenditure may be associated with gain in weight and obesity.
- *Hypogonadism:* In man as well as in animals, removal or destruction of the gonads by diseases predisposes to obesity. Many women show such changes and gain in weight after the menopause. The adiposity characteristic of hypogonadism involves chiefly the breast, abdomen, hips and thighs.

The endocrine disorders do not cause the obesity as such, but may favour its development by increasing food intake or decreasing energy expenditure or both. Localisation of fat deposits is, however, specifically influenced by certain abnormalities of the internal secretions.

5. METABOLIC CHANGES IN OBESITY

Various metabolic abnormalities observed in obesity are not permanent in nature. They are induced with weight gain and are reversible with weight reduction.

1. *Changes in Fat Metabolism*
   - *Serum triglyceride level:* Increased TG level (hypertriacylglycerolaemia) is seen characteristically in obesity. This may be explained partly due to associated hyperinsulinism seen in obese patients. Studies have shown a good correlation between hypertriacylglycerolaemia and hyperinsulinism.
   - *Serum cholesterol level:* In obesity associated with type IV and type V hyperlipoproteinaemias, alongwith hypertriglyceridaemia, there may be slight to moderate hypercholesterolaemia. As such, serum cholesterol levels are less closely related with obesity, but statistically significant relationship do exist. It may be explained partly ***by the increased cholesterol production rate in*** relationship of degree of obesity. It is supported by the fact that cholesterol gallstones are more common in obese individuals.
   - *Mobilisation of FFA:* As obesity is usually associated with hyperinsulinaemia, it is expected to play a part in lipogenesis. Fatty acid mobilisation from adipose tissue appears to be less affected and is considered to be normal in obesity.
   - *Lipoprotein lipase activity:* Lipoprotein lipase brings about the delipidation of TG of circulating chylomicrons and VLDL. It appears to be sensitive to the availability of insulin and ***its activity has been found to be increased in adult-onset type of obesity (hypertrophic type).*** Increased activity of the enzyme would lead to increased FFA assimilation in adipose tissue and thus it can lead to increased fat deposition, in adipose tissue.
2. *Changes in carbohydrate metabolism:* Obesity is ***associated with hyperinsulinaemia.*** The β-cells of islet of Langerhans of pancreas are stimulated to produce more insulin. The nature of the stimulus is not known which may be hormonal or neuronal or by some specific amino acids or fatty acids. ***Hyperinsulinism may aggravate obesity by promoting lipogenesis and inhibiting lipolysis.*** Prolonged hyperinsulinism in obesity might lead to the exhaustion of β-cells in those individuals who are genetically susceptible to diabetes mellitus.

   Insulin resistance is associated with obesity. The obesity has been found to be associated with fewer numbers of insulin "receptors", on adipose tissue, liver and muscle. A high blood insulin level (hyperinsulinaemia) decreases the number of insulin receptors on target cell membrane, probably through internalisation of the ***insulin-receptor complex*** into the cell and thus decreases the insulin sensitivity of the target tissues, thus contributing, to insulin resistance and impaired glucose utilisation by the cells.
3. *Changes in acid-base status:* Massive obesity may be ***associated with alveolar hypoventilation leading to CO_2 retention.*** PCO_2 may be high ↑ and this can bring about certain personality changes, fatiguability, dyspnoea and somnolence, called as ***"obesity-hypoventilation syndrome" (Pickwickian syndrome).***
4. *Energy metabolism in obesity:* BMR as ordinarily determined, is usually normal in obese subjects. Their energy expenditure per unit mass is the same as in normal people. It appears that since BMR of an obese person is normal and his surface area large, his total O_2 consumption must be greater than normal. It may be as much as 25 per cent more than that of normal persons of the same age. The individual uses more oxygen, burns fuel and yet continues to store fat.

6. CLINICAL FEATURES

Most of the obese patients are asymptomatic. When obesity is marked, exertional dyspnoea, depression, somnolence and easy fatiguability are likely to occur. Marked obesity may be associated with alveolar hypoventilation leading to CO_2 retention (PCO_2 ↑) which may account for above features. Many of the symptoms attributed to obesity actually result from an associated disorder like DM or endocrinopathy, rather than from obesity *per se*.

Symptoms: The more common symptoms seen in obese individuals are as follows:

- Fatigue/tiredness on exertion.
- Exertional dyspnoea, weakness, malaise.
- *Symptoms of reactive hypoglycaemia* like weakness, palpitation, sweating, often seen in obese and adult-onset diabetics about 3 to 5 hours after meals.
- Excessive weight gain in spite of normal or reduced calorie intake, frequent steroid therapy, and in Cushing's disease or syndrome.
- Excessive hunger found in obesity associated with pregnancy, women taking oral contraceptives, steroid therapy, adult onset DM, etc.

Signs: Obesity *per se* may produce some physical findings, but most of the signs seen in obese individuals are primarily related to associated underlying disorders like endocrinopathy.

- *Pink striae* are commonly seen over abdomen, thighs, buttocks, breasts, particularly in young women, pink colour usually disappears leaving shiny and white striae.
- When obesity is massive, *exertional dyspnoea* and tachypnoea may be seen.
- *Intertrigo* is quite common in the folds below the breast and in the inguinal regions.
- *Plethora* involving the cheeks and neck is not unusual.
- *Blood pressure* is usually normal. Sometimes systemic hypertension may be present due to associated disorders like DM.
- Occasionally *ankle oedema* may be noted.
- With certain endocrinopathies associated findings may be of help in diagnosis.

Hormones of Adipose Tissue and Its Role in Obesity

Two peptide hormones are produced by adipose tissue cells (adipocytes). They are:
1. **Leptin**
2. **Adiponectin**

1. Leptin

It is a 16 kDa *protein hormone* produced by adipose tissue cells (adipocytes).

- Leptin was discovered in 1994 by **Jeffrey M Friedman and colleagues** at the Rockefeller University
- **Gene:** The *"ob (lep)" gene* is located in humans on *chromosome* 7.
- Leptin is a *hormone because it is produced by adipose tissue cells abut acts through brain receptors.*
- Leptin *circulates at levels proportional to body fat.* It enters the CNS in proportion to its plasma concentration. *Receptors for leptin are found in brain neurons* involved in regulating energy intake and expenditure. It also controls food intake (appetite) and energy expenditure by acting on receptors in medio-basal hypothalamus.
- **Mechanism of action:** *Leptin receptors*—Leptin interacts *with six types of receptors (ob-Ra-ob-Rf,* or *LepRa-LepRf)* which in turn are encoded by single gene LEPR. Ob-rb is the receptor isoform that *can signal intracellularly and is present in hypothalamic nuclei.* Once Leptin is bound to the ob-rb receptors, it activates the stat-3, which is phosphorylated and travels to the nucleus, to presumably effect changes in gene expression.
  One of the main effects on gene expression is the down regulation of the expression of "endocannabinoids", responsible for increasing appetite. There are other intracellular pathways activated by leptin, but less is known about how they function in this system.
- **Other sources of leptins:** Recently it has been shown that in addition to white adipose tissue adipocytes being the major source, leptins *can also be produced in small amounts by other tissues, viz.* brown adipose tissue, placenta (syncytio trophoblasts), ovaries, mammary epithelial cells, bone marrow pituitary and liver.
- **Paracrine action:** Recently it has been shown that in addition to its endocrine action at a distance (from adipose tissue to brain); leptin also acts paracrine mediator.
- **Leptin resistance and obesity:** Although leptin is a circulating signal that reduces appetite, in general, obese people have an unusually high circulating concentration of leptin. These obese people *are considered to be resistant to the effects of leptin,* in much the same way that people with type 2 diabetes are resistant to the effects of insulin.

2. Adiponectin

Adiponectin is another protein hormone, a polypeptide consisting of 244 amino acids, produced and secreted by adipose tissue cells (adipocytes).

- Adiponectin in humans is encoded by the **ADIPOQ gene.**
- *Polymeric forms: Adiponectin automatically self-associates into larger polymeric forms,* viz. homotrimers, hexamers, etc. Females have increased proportions of high molecular weight forms.
- **Mechanism of action:** Adiponectin though produced by tissue cells (adipocytes), *it exerts its effects through brain. This is similar to the action of leptin,* but the two hormones perform complementary actions and can have additive effects.
- **Receptors:** Adiponectin binds to number of receptors. So far *three receptors* have been identified. These receptors are:
  1. **ADIPOR 1**—Adiponection receptor 1.
  2. **ADIPOR 2**—Adiponectin receptor 2.
  3. **T-Cad**—T-Cadherin.

SECTION SIX

These have distinct tissue specificities within the body and have different affinities to the various forms of adiponectin. The receptors affect the downstream target "AMP kinase", an important cellular metabolite rate control point.

- **Metabolic effects and functions of adiponectin:**
  - Increases glucose by glucose uptake and gluconeogenesis (glucose flux).
  - Lipid catabolism—β-oxidation increases and triglyceride clearance.
  - Protection from endothelial dysfunction (important facet in atherosclerosis formation).
  - Insulin sensitivity.
  - Weight loss—Adiponectin exerts some of its weight reduction effects via the brain.
  - Control of energy metabolism.

Levels of the hormone are inversely correlated with body fat percentage in adults. The hormone plays a role in the suppression of the metabolic derangements that may result in type-2 diabetes, obesity, atherosclerosis, nonalcoholic fatty liver disease (NAFLD).

Structure of Adiponectin—Structurally it has 4 distinct regions:

- First is a short signal sequence that targets the hormone for secretion outside the cell
- Next is a short region that varies between species
- Third is a 65 amino acid region with similarity to collagenous proteins and
- Last is a globular domain.

However, when the 3-dimensional structure of the globular region was determined, a striking similarity to TNFα was observed.

Antiobesity Vaccine

- **Scripp's vaccine:** Recently, scientists at the Scripp's Research Institute claim they have developed an antiobesity vaccine that significantly slows weight gain in experimental mice, by tackling the **"ghrelin", a naturally occurring hormone that helps regulate energy balance** in the body.
  Experimental mice given shots of the vaccine ate just as much as the untreated controlled group of mice, and ***showed about 20 to 30 per cent reduction in weight gain*** as compared to the control. Scripp's vaccine has yet to be tried in humans.
- **Cytos antiobesity vaccine:** Cytos, a swiss based biotechnology company is already testing different vaccines on humans. The cytos vaccine works in a different way than the Scripp's vaccine, ***preventing the uptake of ghrelin by the brain.***

DIET IN PREGNANCY AND LACTATION

Even though pregnancy and lactation are normal physiological processes, they increase considerably the nutritional requirements of the mother:

1. Due to nausea, vomiting and loss of appetite in early months of pregnancy, the food intake is generally reduced.
2. Further, additional nutrients are required for the growth of the foetus. A baby weighing about 3.2 kg at birth, will contain about 500 gm of proteins, 30 gms of calcium and 0.4 gm of Fe and varying quantities of vitamins. This amount has to be supplied by the mother during whole of pregnancy.
3. The increased nutritional requirements during lactation are due to the milk secreted by mother for feeding the baby.

A. Nutrient Requirements During Pregnancy

(a) Calorie requirement: The total calorie cost of supplying and maintaining the foetus has been estimated to be approx. 40,000 C. Since the greater part of calories will be required mostly during 2nd and 3rd trimesters, the additional work out as 200 C/ day. ICMR Nutrition Expert Group recommends ***extra allowance of 300 C/ day.***

Reasons for this additional requirement are:

- Increased energy requirements for the growth and development of the foetus, placenta and maternal reproductive organs like uterus and mammary glands.
- The rise in BMR up to even 25 per cent.
- Enhanced needs for increased haemopoiesis in mother.
- Increased muscular work in carrying the rising weight of the foetus during each movement and activity. However, restriction of physical work in third trimester may somewhat reduce this additional calorie need.

(b) Protein requirement: Available evidence indicates about 910 gm of proteins are deposited in the foetus and maternal tissues during the pregnancy period. The average daily increment is estimated to be about 5 gm as tissue proteins during the last six months of pregnancy. This works out to 10 gm of additional proteins in terms of dietary protein of NPU 50. ICMR nutrition expert group recommends an ***extra allowance of 10 gm/day.***

Reasons for this additional requirement:

- Synthesis of haemoglobin
- Synthesis of plasma proteins in the increasing blood volume
- Formation of tissue proteins in growing tissues of foetus and in maternal reproductive organs.

(c) ***Iron requirement:*** It has been estimated that approx. 540 mg of Fe are found in the foetus and maternal tissues. This works out to about 2 to 3 mg of Fe per day during last six months of pregnancy. Since the Fe from diet is utilised only to an extent of 20% in pregnant women, the extra dietary Fe requirements will be approx. 10 to 15 mg/day. ICMR expert group ***recommends 10 mg/day.*** Iron should be supplied by iron-rich foods or by supplementation with $FeSO_4$ so that haemopoiesis at enhanced rate may continue, otherwise hypochromic microcytic anaemia results.

(d) ***Calcium, phosphorus and vit D requirements:*** It has been estimated that about 30 gram of Ca is deposited in the foetus during pregnancy. This works out to approx. 150 mg of extra calcium during last six months of pregnancy. Since dietary Ca is utilised to the extent of about 25 per cent in pregnancy, additional requirements work out to 600 mg in terms of dietary Ca. ICMR expert group recommends an ***extra allowance of 500 to 600 mg/day.*** Some authorities recommend also additional phosphorus and 400 IU of vit. D daily in later half of pregnancy.

Reasons for extra allowance:

- To provide minerals for the growing foetal bones for calcification (mineralisation of bones).
- If this additional requirement is not met, maternal bones may get decalcified to supply the minerals to the foetus and *osteomalacia* may result.

(e) ***Vit A requirement:*** Quantity of vit. A found in liver of infant is about 5400 to 7200 μg of retinol. This works out to about 25 to 35 μg additional retinol/day. Since the additional requirements are small as compared with the daily requirement of 750 μg for a normal woman; ICMR nutrition expert group did not recommend any additional allowance for vitamin A.

(f) ***Vit K requirement:*** Vit K should be provided in the last trimester of pregnancy as a protection against excessive bleeding during parturition and also to build up body stores of vit K in the foetus. Dark green leafy vegetables should be taken in the diet for vit K.

(g) ***Water soluble B-vitamins***

- ***Thiamine, Riboflavine and Nicotinic acid:*** Small quantities of these vitamins are present in the tissues of newborn infants. The extra calorie allowance of 300 C/day for pregnant women needs an increase in the requirements of these vitamins as they act as coenzymes. ICMR expert group recommended an additional allowance of:
  - 0.2 mg thiamine
  - 0.2 mg riboflavine and
  - 2.0 mg of nicotinic acid per day.
- ***Folic acid, vit B_{12} and vit C:*** Small amounts of these vitamins are present in the tissues of newborn infants. ICMR recommends an additional daily allowance of:
  - Free folic acid 50 to 200 μg
  - Vit B_{12} 0.5 μg
  - Vit C 50 mg.

Reasons for additional allowance:

- Increased folic acid and vit B_{12} are needed for helping in nucleic acid synthesis in proliferating tissue cells and maturing RB cells both in mother and foetus.
- Enhanced need for vit C is due to increasing synthesis of collagen in growing tissues.

B. Nutrients Requirement in Lactation:

Milk output varies in lactating mother. The WHO expert committee assumed the average output to be 850 ml/day, while the ICMR expert group assumed an output of 600 ml/day in Indian women. The nutrients contained in 600 ml of milk are given below:

Nutrients and calories	Nutrients present in 600 ml of human milk
Calories (C)	420
Proteins (gm)	7.2
Calcium (mg)	205
Fe (mg)	0.75
Vit A (μg)	300
Vit C (mg)	15 to 30
Thiamine (mg)	0.09
Riboflavine (mg)	0.37
Nicotinic acid (mg)	1.2
Folic acid (μg)	6.0
Vit B_{12} (μg)	0.14

(a) ***Calorie requirement:*** It has been estimated that ***the efficiency of conversion of dietary calories to human milk calories is only 60 per cent.*** On this basis, production of 420 milk calories will need 700 diet calories approx. ICMR recommends ***an additional 700 C/day*** in diet during lactation.

Reasons

- Required in lactating mother to synthesise and provide milk constituents.
- Also for energy required for milk secretion.

A part of extra calories, however, may be met from the fats deposited during pregnancy.

(b) ***Proteins requirement:*** Average NPU of the diets consumed in India is about 50. For the production of 7.2 gm of milk proteins, about 14.4 gm of additional dietary proteins will be required. ICMR recommended an ***extra allowance of 20 gm of proteins/day.***

Reasons for extra allowance:

- For supplying amino acids for synthesis of milk proteins

- Also for maintaining secreting mammary glands, for synthesis of protein hormones required for lactation.

(c) ***Calcium requirement:*** Quantity of calcium present in 600 ml of human milk is 205 mg. Since the retention of dietary Ca in lactating women is about 30 per cent, the additional dietary Ca requirement will be approx. 700 mg/day. ICMR recommends ***500 to 600 mg/day.***

(d) ***Iron requirement:*** Fe content of 600 ml of milk is 0.75 mg. Since only approx. 20 per cent of dietary Fe is retained in lactating women, it will be essential to provide about 3.6 mg of extra Fe in the diet. ICMR expert group did not recommend an extra allowance for Fe in lactating women.

(e) ***Vit A Requirement:*** Quantity of vit A present in 600 ml of human milk is 300 µg. ICMR recommended an additional allowance of 400 µg of vit A.

(f) ***Water Soluble B-Vitamins***

- ***Thiamine (B_1), Riboflavine (B_2) and Nicotinic acid***
  Additional calorie requirements during lactation are 700 C/day. The extra requirements of B_1, B_2 and Nicotinic acid to meet the needs of extra calories will be:
  - B_1 = 0.3 mg/day
  - B_2 = 0.4 mg/day
  - Nicotinic acid = 4.6 mg/day.

ICMR recommended as follows:

- Vit. B_1 = 0.4 mg/day
- Vit. B_2 = 0.4 mg/day
- Nicotinic acid = 5.0 mg/day

- ***Folic Acid, Vit. B_{12} and Vit. C***

	Quantities present in 600 ml of human milk	ICMR recommendation of extra daily allowance
• Folic acid	6 µg	50 µg
• Vit. B_{12}	0.14 µg	0.5 µg
• Vit. C	15 to 30 mg	30 mg

COMPOSITION AND NUTRITIVE VALUE OF COMMON FOODSTUFFS

1. NUTRITIVE VALUE OF MILK

No other single food has as many nutritional virtues as that of milk. Milk supplies proteins of high biological value, easily digestible fats, lactose (milk sugar) and calcium, phosphorus, vit A and certain B-vitamins in sufficient amounts. It is an ideal food for the infant, but it must be supplemented with other foods as the child grows. However, it is not a perfect food since **it lacks Fe, Cu and vitamin C.**

Chemistry

(a) Proteins

Chief proteins are:

1. **Caseinogen**
2. **Lactalbumin and**
3. **Lactoglobulin.**

Caseinogen

Is a ***phosphoprotein*** (0.7% P) and carries Ca^{++} bound with it, is more in amount in cow's milk (2.8%) as compared to human milk (0.5%). Boiling of milk increases digestibility of casein. Bovine caseins are more difficult to digest than human milk casein, as bovine caseins form harder calcium paracaseinate during milk digestion due to higher Ca^{++}: casein ratios.

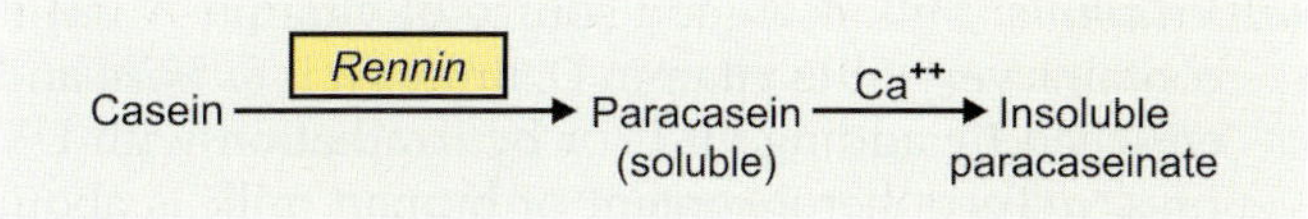

Caseinogen is insoluble at its isoelectric pH (pI = 4.6). Usual pH of fresh milk is 6.6 to 6.9.

Lactalbumin and Lactoglobulin

Have also very high biologic values and both are good proteins. ***Lactalbumin is heat-coagulable and is most easily digested.***

Note: Milk proteins have sufficient amount of tryptophan and this compensates for the low niacin content of milk.

(b) ***Lipids:*** Milk fats are in the form of very fine and stable emulsion and are the most palatable and digestible fats known. It differs from other fats in containing all saturated even carbon FA from butyric (C_4) to lignoceric acid (C_{24}) as well as a variety of unsaturated FA, viz. oleicacid, linoleic, linolenic and arachidonic acid. About 30 per cent of FA in milk TG contains polyunsaturated fatty acids. Human milk differs from cow's milk in FA composition. Oleic acid predominates in both 30 to 35 per cent of total and FA C_{12} to C_{18} consists 80 to 90 per cent of total. 10 per cent of human milk FA are highly unsaturated (linoleic, linolenic, etc.) as opposed to 0.5% of cow's milk. Total fat content in human milk is 4.0 gm% as compared to cow's milk 5.0 g%; other lipids include cholesterol and phospholipids in small quantities. Boiling reduces the fat content of milk as some of the fat separates along with some of the coagulated lactalbumin as a floating layer of clotted cream.

(c) ***Carbohydrates:*** Principal carbohydrate present is the disaccharide **Lactose** (milk sugar). Human milk contains 7.0 g% as compared to cow's milk 5.0 g%.

(d) ***Minerals:*** Milk is rich in mineral elements specially calcium, phosphorus, potassium, sodium, chloride and zinc. ***Milk is poor in iron and copper.*** Both

humans and experimental animals develop a dietary anaemia, ***hypochromic microcytic type*** on an exclusive milk diet.

Calcium and Phosphorus: About 120 mg Ca and 90 mg of P are present in 100 ml of cow's milk; the amounts are higher in buffalo's milk but much lower (Ca = 40 mg% and P = 30 mg%) in human milk. Calcium is present in combination with casein, as free Ca^{++} and also as inorganic phosphates. Phosphorus is present as phosphoprotein casein, as inorganic phosphates and also non-protein organic PO_4 esters. The ratio of Ca: P of milk facilitates the formation of soluble calcium phosphates in the intestine. Lactic acid produced by bacterial fermentation of lactose increases calcium absorption.

(e) *Vitamins:* Milk is a good source of vitamin A but it contains very little vitamin D, unless it is *fortified* and enriched by adding vit D or by irradiation with UV rays. α-Tocopherol content of human milk is about twice that of bovine milk. ***Vitamin C content of milk is very low*** and pasteurisation destroys ½ of the original content. Milk has rather low concentrations of soluble B-vitamin group, but comparatively rich in riboflavine (B_2) and good in thiamine (B_1).

Human milk: Human milk differs markedly from cow's milk in a number of ways:

- The protein content of human milk is far lower, while,
- The lactose content is much higher.

	Cow's milk	Human milk
• Caseinogen: Lactalbumin	3 : 1	1 : 2
• Protein: Non-protein N_2 ratio	11 : 1	3 : 1

Differentiation in composition of human milk and cow's milk is given in tabular form in Table 44.10.

Humanisation of cow's milk: Cow's milk, when fed to newborn babies has to be diluted with water to lower its protein content to that of human milk and lactose, glucose, maltose or sucrose has to be added to it to raise its sugar content to the level of human milk. This process is known as ***humanisation of cow's milk.***

Note: Bovine milk normally contains far less linoleic acid, α-Tocopherol, vit. C and niacin than those in human milk and hence humanised cow's milk will have still less of them due to the dilution.

Colostrum: The secretion of the lactating mammary glands during the first few days of lactation (first 4 to 5 days) after parturition is called **colostrum.** It is thick, viscous yellow liquid and is heat coagulable. Heat coagulability is due to presence of increased amounts of globulins and lactalbumins. ***Colostrum is richer than mature milk in proteins, vit. A and D, α-tocopherols and calcium, but comparatively poorer than mature milk in casein, fats and lactose.*** Though it has less casein, but the total protein content is twice as much. The proteins have a high percentage of globulins and next is lactalbumins. Globulins which are the highest include some lactoglobulins and ***various Igs (immunoglobulins) coming from the maternal blood.*** These immunoglobulins may be absorbed from the small intestine of newborns probably by ***'pinocytosis'*** and ***confer temporary immunity. A trypsin-like inhibitor present in colostrum may help to preserve the Igs in the alimentary canal of newborns by preventing hydrolysis.*** Colostrum also contains larger amounts of B-vitamins like thiamine, riboflavine and folic acid.

Certain Milk Preparations

1. *Channa (coagulated casein or cottage cheese):* It is prepared by treating milk with lemon juice or alum or by the enzyme rennin. By this treatment casein, lactalbumin and fats are coagulated to separate as "channa" and a faint greenish yellow fluid separates called as "whey".
   - **Channa:** Contains about 15 to 25 per cent of proteins, 20 per cent fats and 3 per cent lactose and some minerals, specially calcium and vitamins. It is easily digestible.

Table 44.10: Difference between the composition of human milk and cow's milk

	Human milk	Cow's milk
Solids	12.5	13.5
Water	87.5	87.0
Proteins	1.0 to 2.5	2.0 to 6.0
(gm%)	(1.5)	(4.0)
Carbohydrates	4.5 to 8.0	2.0 to 6.0
(gm%)	(7.0)	(5.0)
Fats (gm%)	1.0 to 8.0	1.5 to 6.5
	(4.0)	(5.0)
Calcium (mg%)	40	120
Phosphorus (mg%)	30	90
Magnesium (mg%)	5	20
Sodium (mg%)	15	50
Potassium (mg%)	60	140
Chloride (mg%)	40	110
Vit A (μg%)	50	35
Vit D (IU)	5.0	2.5
Vit C (mg%)	4.5	2.0
Vit B_1 (μg%)	15	45
Riboflavin (μg%)	45	200
Niacin (μg%)	180	80
Vit B_6 (μg%)	10	50
Pantothenic acid (μg%)	200	350
Calories/100 g	67	69

- **Whey:** It is faint greenish yellow fluid which separates. Proteins in whey are lactalbumin and lactoglobulins to some extent + proteose like proteins derived from casein. In addition to proteins, it contains lactose, some fats, minerals and B-complex vitamins specially riboflavine which accounts for the colour.

2. *Dahi (curd):* It is a common preparation in our daily foods. Dahi contains about 4 per cent fats, 3 per cent proteins and 3 per cent lactose. Certain amount of lactose is converted to lactic acid which causes the sour taste.
   - **Proteins:** They are usually coagulated casein, peptones and small peptides which are easily digested and absorbed.
   - **Minerals:** Mainly calcium, 100 ml of Dahi contains approx. 145 to 150 mg Ca. Calcium is easily absorbed due to acidity caused by lactic acid.
   - **Vitamins:** Thiamine and riboflavine are present in high quantities.

2. NUTRITIVE VALUE OF EGG

Egg is a delicious, palatable and enjoyable food item liked by adults as well as children. It can be taken in the diet in various forms and preparations. It is highly nutritive. From a nutritional stand point, the egg stands with dairy products and meat.

Composition of an Average Hen's Egg

It consists of approx. 30 per cent yolk (yellow portion), 59 per cent white and 11 per cent shell. In the edible part there is 15 per cent proteins, 10.5 per cent fats and 1 per cent ash (Table 44.11).

I. *White part:* It is essentially a solution of proteins and salts. Proteins of eggs are high quality proteins and are used as reference standard for assessing other proteins.

Proteins in white part: The main proteins are:
- *Ovalbumin:* A typical albumin; other proteins are:
- *Conalbumin:* Another albumin
- *Ovoglobulin:* A globulin and
- *Ovo-mucoid:* A glycoprotein

Note: The pale-yellow colour of egg white is partly due to riboflavine.

II. *Yolk:* It is more concentrated, containing only 51 per cent of water. The chief constituents are proteins and fats. Also there is about 1 per cent of mineral matters and vitamins.

1. *Proteins:* Proteins present in yolk are mainly two:
   - *Vitellin:* A phosphoprotein, resembling caseinogen of milk
   - *Livetin:* A globulin. Vitellin predominates and the ratio of Vitellin/Livetin = 3.6/1.

The vitellin and lecithin present in yolk are associated as lipoprotein complex called Lecitho-vitellin.

2. *Fats:* Yolk contains nearly 30 per cent of phospholipids—cephalin and lecithin mainly and to a small extent other phospholipids. ***Yolk is rich in cholesterol.*** An average hen's egg weighing 2 oz contains approximately 250 mg cholesterol. Fats present in egg-yolk are readily and thoroughly digested and absorbed. It contains appreciable amounts of polyunsaturated FA like linoleic acid.
3. *Minerals:* The yolk is well supplied with mineral matters, specially Ca, Fe and PO_4. ***It is very good source of Fe which is easily assimilable.*** Few foods supply as much available Fe as egg. It appears to be present practically all in inorganic forms. Much of the phosphates are present in PL and vitellin.
4. *Vitamins:* The yolk is rich in vitamins A and D. Also in B-vitamins like thiamine, riboflavine but ***not vit C***. The other vitamins of B-group and vit E are also present (Table 44.12).
   - *Lutein and Zeaxanthin:* Two carotenoids in egg yolk that protect against degeneration of the retina. New studies say they have a role in prevention of heart disease and stroke. Carotenoids in eggs are better absorbed than those from plant sources.

Table 44.11: Composition of different nutrients of egg

	Water	Proteins (in gm%)	Lipids	Carb	Calories per 100 g
• Whole egg	73%	12.5	12.0	1.6	162
• Yolk	51%	16.0	33.0	2.1	381
• White	85%	11.5	trace (0.1)	1.4	51

3. OTHER BASIC FOOD GROUPS

1. Meat and Fish

Meat and fish, due to their flavour and taste, are used by most as a dietary constituent, but due to their high price all sections of people cannot afford to buy. Principal nutrient supplied is the proteins although from the quantity standpoint many meat cuts contain a higher percentage of fats. ***Proteins of both meat and fish are of high biological value.*** Muscle meat excluding cartilages and bones, in itself is not a complete food as it is deficient in Ca and has high P content and this makes Ca:P ratio out of balance. ***Small fishes, though comparatively of***

Table 44.12: Vitamin content of egg (fresh) per 100 gm of edible portion

Vit. A	B_1	B_2	Niacin	Vit C	Vit D
(IU)	(mg)	(mg)	(mg)	(mg)	(IU)
1000	0.15	0.40	0.1	0	60

cheaper cost, are good sources of Ca and P because the whole fish is eaten. Fish and Meat supplies adequate Fe but vit. C, the fat soluble vitamins and certain B-vitamins are slightly deficient.

2. *Pulses and Legumes*

Pulses and legumes are commonly known ***as poor man's meat*** and are rich sources of protein. Now the pulses price has gone so high that it is not within the reach of poor men even. Chief differences between dried pulses (peas, beans and lentils, etc,) and cereals is the high proportion of proteins in the former. ***Chief protein is a globulin, called "Legumin"***. The protein content of dried pulses is from 20 to 25 per cent, i.e. double that of cereals but they have the same calorific values as those of cereals (350 calories per 100 gm). They are rich sources of B-vitamins specially thiamine, riboflavine and niacin. Pulses and grams, if germinated, become rich sources of B-group vitamins and vit. C.

3. *Nuts*

Nuts are characterised by a high fat and low carbohydrate content. The protein content is slightly lower than that of dried pulses. Raw nuts are not easily digested. This difficulty is overcome largely by grinding and cooking the nuts.

4. *Cereals*

General composition of crude cereals (oats, wheat, barley, rice, rye, etc.) approximately:

- Protein 11 per cent,
- Carbohydrates 70 per cent
- Mineral matter 2 per cent
- Fats varying from 0.5 to 8 per cent and
- Water 11 per cent.

In flours made from cereals there is usually not more than 3 per cent of cellulose. Oat meal is richest in proteins and fats and rice the poorest.

- *Proteins:* Chief proteins of cereals are ***glutelins*** and ***gliadins;*** usually they are not complete proteins as they may be deficient in certain amino acids. Some amounts of albumins and globulins are also present. Though protein content of cereals is not high, but in view of large quantities consumed per day, the total intake is good.
- *Carbohydrates:* Mainly in the form of starch, covered by a thin membrane of insoluble carbohydrate mainly cellulose, starch grains are insoluble and indigestible. Cooking by bursting the cellulose covering of the grains, renders starch soluble and digestible.
- *Fats:* Contain sufficient olein.
- *Minerals:* Most abundant mineral constituents are Ca and PO_4, the latter is partly in the form of phytic acid (inositol hexaphosphate), which is not utilisable by humans and hinders the absorption of Ca. (anticalcifying effect). Some cereals, wheat and rye contain a *phytase* which hydrolyses phytic acid and diminishes the anticalcifying effect.

Note: Rice and wheat are the commonly consumed cereals in our country. Cereals constitute the major bulk of the diet, satisfy hunger and are the main suppliers of calories (350 Cal/100 gm). In a good balanced diet, cereals contribute about 50 per cent of the total calories.

5. *Roots and Tubers*

The most important tuber is potato, which is common food used daily. Sweet potatoes, colocasia, tapioca, carrots, etc. are other belonging to this group.

- *Carbohydrates:* They are rich sources of carbohydrates and the most important substance present is starch. Main source of energy and gives 100 calories/100 gm, because of higher moisture content.
- *Proteins:* Only about ½ the total N_2 is present as protein chiefly as the globulin, ***Tuberin.*** The remaining nitrogen is in the form of simple soluble nitrogenous compounds such as ***Asparagine.***
- *Minerals and vitamins:* ***Potatoes contain good quantities of vit C and Fe.*** vit C content varies with the time of the year. Newly raised potatoes contain about 28 mg vit C per 100 gm, but on storage this amount gradually falls and the old potatoes contain less than 5 mg/100 gm. Carrots which are richest in sugars (10%) are valuable source of carotene, precursor of vit A.

Nutrients in Leafy Vegetables

- Carotene (Precursor for vit A)
- B-vitamins like riboflavine, folic acid
- Vit C
- Fe and
- Ca and salts.

6. *Green Leafy Vegetables*

This is an important basic food group and includes fresh leafy vegetables of all kinds, e.g. spinach, amaranth, lettuce, cabbage, etc. ***Leafy vegetables are good sources of at least six essential nutrients,*** e.g. see box above.

The cost is quite cheap. If these nutrients, in similar quantities are to be obtained from alternate animal sources, the cost will be comparatively much higher. Excess of cabbage should be avoided as it is goitrogenic. Cooking causes loss of B-vitamins and Vit. C, hence uncooked salads are, therefore more valuable. Cellulose present is not digested and absorbed, but add bulk to intestinal contents which stimulate peristalsis. Cellulose provides ***roughage*** value. Green leafy vegetables are essentially ***protective foods*** due to their vitamin and mineral contents, they prevent development of deficiency diseases.

Table 44.13: Nutritional value of some of common foodstuffs

Items	Calories	Proteins gm%	Carbohydrates gm%	Fats gms%	Fe mg%	Ca mg%	NaCl mg%	Vit. A IU	B_1 mg%	Riboflavine mg%	Niacin mg%	Vit C mg%
Atta	353	12.0	72.3	1.8	7.4	39	116	60	451	120	4.9	0
Rice (Raw, undermilled)	349	7.1	79.0	0.4	2.8	11	88	0	198	53	1.8	0
Butter	744	0.4	0	82.5	0	14	1764	2716	0	0	0	0
Cow's milk	65	3.2	4.9	3.5	0	120	123	180	46	200	80	0
Buffalo's milk	116	4.2	5.3	8.8	0	21.2	—	162	—	—	—	—
Hen's Egg	138	10.9	6.7	10.2	2.5	53	141	882	trace	353	0	0
Mutton (Goat) with Bones	155	14.8	0	10.9	1.8	123	141	28	49	219	5.3	0
Onion	49	1.1	11.6	0	0.7	169	—	0	35	11	0.4	11
French beans	14	1.1	2.5	0	0.7	46	—	120	71	46	0.4	11
Dal Musoor	346	26.1	60.1	0.7	2.1	134	88	250	444	—	1.4	0
Apple	46	0.4	10.9	0	1.4	8.8	—	trace	116	28	0.4	4
Banana	102	1.1	24.0	0	0.4	7	282	trace	78	32	0.4	4

7. *Fruits*

Fresh fruits are also essential ***protective foods,*** although the energy value is twice that of green vegetables. This is due to sugars and starch. Proteins and fats usually amounting to less than 0.5 per cent. Many fruits contain pentoses and pectins. Neither pentoses nor pectins are utilised by the body. They are good source of B-vitamins, Vit A and Vit C. Fruits like orange, lemons and guava are good sources of Vit C. ***Amla is the richest source of Vit C.*** Ripe yellow fruits, e.g. mango, or papaya, contain carotene (precursor for vit. A). Fruits are good 'alkalinisers' of blood and hence useful in pyrexia.

Banana: Cheaper among the fruits and found plenty in our country. Ripe banana mainly contains carbohydrates and have definite energy value, contains starch as well as other sugars. It also has a higher protein content. Starch present in banana is said to be easily digested.

Table 44.13 shows nutritional value of some common foodstuffs.

ROLE OF DIETARY FIBRES

1. Dietary fibres denote all plant cell wall components that cannot be digested by an animal's own digestive enzymes, e.g. cellulose, hemicellulose, pectins, gums, lignins and pentosans.
2. *Beneficial effects of high fibre diets:*
   - Helps in aiding water retention during passage of foods along the gut, and thereby producing larger, softer faeces.
   - They help in increasing bulk of the faeces, which induces peristalsis ***("Roughage" action)*** and removes constipation.
   - More insoluble fibres such as cellulose and lignin found in wheat bran helps in colonic function.
   - Whereas more soluble fibres found in legumes and fruits, e.g. gums and pectins can lower blood cholesterol, probably:
     - by binding bile acids and
     - by binding dietary cholesterol thus preventing absorption.
   - Soluble fibres also slow emptying of stomach and attenuate the post-prandial rise in blood glucose, with consequent reduction in insulin secretion. This effect is beneficial to diabetic patients and to dieters because it reduces the rebound fall in blood glucose that stimulates appetite.
3. **Clinical importance:** A fibre diet is associated with reduced incidence of:
   - ***Diverticulosis***
   - ***Cancer of colon***
   - ***Atherosclerosis and cardiovascular diseases***
   - ***Diabetes mellitus.***

TEA, COFFEE AND COCOA

A. **Tea:** Types of tea leaves:
   1. The "black tea" ordinarily sold consists of the leaves of young shoots of tea plants which have been fermented and dried by heat.
   2. In "green tea", the fermentation process is omitted. Green tea leaves contain a compound called Epigallocatectin gallate (EGCg) which has anti-cancer activity (Refer the chapter on Biochemistry of Cancer).

Properties of tea (infusion produced with tea leaves):
1. It has negligible calorie value.
2. *Caffeine:* It contains caffeine which is a stimulant and diuretic. It is present to the extent of 2 to 4 per cent in the dry tea, is readily soluble and is quickly extracted when tea is made.

3. *Tannic acid:* Other constituent present. It is an astringent. It is present to the extent of 5 to 15 per cent, is less soluble and only passes into the infusion slowly.

The infusion which has stood over the tea leaves for longer time contains more of tannic acid and that accounts for the increased bitterness of tea. ***A cup of strong tea contains about 0.1 gm of caffeine.*** Strong tea owing to increased concentration of tannic acid can retard gastric digestion.

B. **Coffee:** Coffee is the roasted seeds of the cherry-like fruit of "coffea arabica". The aroma or smell is due to an oil, called *caffeol,* formed when the beans are roasted. Like tea, the infusion, although containing more solids, is of little calorific value and contains caffeine and tannic acid in amounts of the same order as in tea infusion. Neither tea nor coffee, unless taken with milk and sugar can be regarded as foods. Their value is largely due to the pharmacological properties of caffeine.

CLINICAL ASPECT

Recently it has been claimed by scientists that coffee has many beneficial effects as a beverage and it protects from certain illnesses, like:
- Diabetes
- Parkinson's disease
- Reduces the risk of cirrhosis liver by 60 to 80 per cent. Regular intake of coffee, four to six cups daily protect from digestive or heart diseases. It has been widely recognised that the research about coffee focussed on caffeine only. But recently the French researcher **Astrid Nehlig** claimed that coffee bean contains:
  - **Clorogenic** and
  - **Melanic acids.**

  Both of them have been found to be ***potent antioxidant.*** Some scientists have claimed coffee is more efficient than fruits and vegetables to counteract DNA oxidation which causes several serious diseases including neoplasia.

C. **Cocoa:**
1. The seeds obtained from the pods of the cocoa tree, *Theobroma cocoa,* after fermentation and roasting are known as "cocoa-nibs".
2. The "nibs" contain about 50 per cent of fats, part of which is removed by pressure "cocoa-butter".
3. The finely ground residue is known as "cocoa" and contains from 25 to 30 per cent fats. Sugar and starch are added in some marketed brands.
4. Constituents:
   (a) *Caffeine:* Present up to 0.5 per cent in cocoa.
   (b) *Theobromine:* "3,7-dimethylxanthine", which has similar pharmacological properties as caffeine.
5. *Nutritive value:* Although its analysis suggests a high nutritive value, cocoa as a beverage is of little importance, since so little is consumed. The milk and sugar taken with it provides most of the nourishment.

D. **Chocolate:** Consists of ground cocoa nibs mixed with sugar. Starch and flavouring materials are frequently added. The powder owing to its high fat content, melts easily and can be cast into bars, etc. (Table 44.14).

Table 44.14: Composition % of cocoa and chocolate

	Protein	Fat	Carb	Calories per 100 gm
• Cocoa	18.1	26.8	40.3	489
• Chocolate	4.8	31.1	59.9	555

E. **Alcohol**

Alcohol has an energy value of 7 calories (C) per gram, i.e. greater than carbohydrates. As it yields energy in the body, when consumed, it must be admitted as a food. But its use is somewhat restricted because of its side effects and other pharmacological actions.

Types of alcohol: It is available in different forms and consumed in different forms.
1. *Beer:* Formed by fermentation of malt and contains from 4 to 8 per cent alcohol by volume. Fermentation is not complete and there is usually some sugar (up to 2%) and dextrins (up to 4%) left, in addition to small amounts of protein.
2. *Wines:* Are the product of fermentation of fruit juices (usually grapes). The alcohol content is from 10 to 20 per cent by volume, and sugar from 0.1 to 4 per cent. Chief chemical characteristic is the presence of a large number of organic acids (e.g. malic, tartaric, succinic acid, etc.) as well as several alcohols and esters.
3. *Cider and Perry:* Considered as wines of low alcohol content (3-8%) and higher sugar content.
4. *Port:* Contains sugar up to 7 per cent.
5. *Spirits:* Formed by distillation of various fermented products.
   Whiskey and Gin: They are regarded as distilled Beer.
   Brandy: Distilled wine. They are, therefore, practically free from sugar or solid matter. The alcoholic content is from 30 to 50 per cent by volume.
6. *Liqueurs* are essentially alcohol sweetened with cane sugar and flavoured with aromatic herbs or essences. Sugar content is usually in the region of 30 per cent and alcohol 35 to 55 per cent by volume (Table 44.15).

Table 44.15: Average % composition of alcoholic liqueurs

	Alcohol (By volume)	Alcohol (By weight)	Carbohydrate gm%	Calories per 100 ml.
• Beer	5	4	6	52
• Wines	12	10	2	78
• Spirits	40	33	0	224
• Liqueur	45	38	30	386

Coconut Water

Dietary Value as a Beverage

Young green coconut water is a delicious, refreshing and hygienically very safe drink. Coconut water is ***rich in minerals*** especially ***potassium, calcium*** and ***magnesium*** making it an ***excellent remedy for replacing electrolytes,*** rejuvenating the body on a hot day. Being a naturally isotonic beverage, it has a similar level of electrolyte balance as our body; often referred to as ***fluid of life.***

Composition

Composition of coconut water is given below:

I. Minerals

Minerals	*Value per 100 ml.*
• Potassium	250 mg
• Sodium	150 mg
• Magnesium	25 mg
• Calcium	24 mg
• Phosphorus	20 mg

II. Nutrients

Nutrients	*Value per 100 ml*
• Protein	1.4 gm
• Carbohydrates	4.4 gm
• Fats	0.1 gm
• Cholesterol	Nil
• Water	93.8 gm
• Energy	24 C

Clinical Importance

- Its quick absorption ***helps to replace fluids and salts during diarrhoea and intestinal infections*** both in infants and adults. In cases of severe diarrhoea it may not be a substitute for "oral rehydration solution (ORS) which is a must in such a situation, but certainly it ***can be a useful adjunct.***
- Coconut water contains organic compounds that ***possess growth promoting properties.***
- The minerals and albumin in coconut water makes it ***a good drink for urinary infections and kidney stones.***
- Its ***cooling effect*** also prevents development of summer rashes and boils.
- **A source of cytokinins:** Cytokinins protect the cells as they undergo cell division. Cytokinins help cellular DNA to replicate perfectly during cell division.
- **Sports drink:** Coconut water has all the functional characteristics required in a "sports drink". For an athlete or sportsman it is a healthy drink which quenches thirst, corrects dehydration if any and improves neuromuscular control. ***It has been considered as "a natural health drink for jogger's and athletes".***

EFFECT OF COOKING

Foods undergo considerable change in the process of cooking and preparation.

1. Inedible portions are removed.
2. Harmful bacteria and organisms are destroyed.
3. Effect on raw meat:
   - The chief difference between raw and cooked meat is that, the latter, even boiled, has less water, so that 4.0 gm of cooked meat have the nutritive value of approx. 5.0 gm of raw meat. The soluble portion is coagulated.
   - Some fats and extractives are lost, and collagen fibres are converted to gelatin, thus loosening the muscle fibres.
   - *Effect on digestibility:* Cooking does not necessarily increase the digestibility. Clinical practice, in fact, suggests that well-disintegrated raw or underdone meat is the most easily digested. But cooking, by breaking down connective fibres, makes meat easier to masticate and so assists digestion, it also increases the palatability of the meat. Overcooking, by causing shrinkage of coagulated proteins, decreases the digestibility.
4. *Vegetables:* Cooking usually increases both the water content and digestibility of vegetables. The chief effect of cooking is the loosening of the cellulose framework and the liberation of starch from starch grains (granules). Raw starch is practically indigestible.
5. Fats are little changed in the process of cooking.
6. *Taste and flavour:* Cooking enhances the taste and flavour of food by the addition of seasoning and in dry cooking (like roasting and baking) by the formation of caramel from sugar and from partial decomposition products from fats and proteins. Many of these substances stimulate the secretion of digestive juices.
7. *Loss of vitamins:* Cooking involves loss in nearly all vitamins of foods, especially of soluble substances in boiling processes. Vitamins B_1 and C are specially liable to destruction when vegetables are cooked.
8. *Effect of cooking on green vegetables:* May be summarised as follows:
   - Vit. A is unlikely to suffer damage.
   - Water soluble B vitamins and vit. C are likely to be lost by diffusion into the soaking or cooking water.
   - Raw vegetables contain enzymes which destroy vitamins and become "active", if the vegetables are kept after bruising or cutting up.
   - They act more rapidly if the temperature is raised and are only themselves destroyed at about 80°C. It is therefore, better to cook vegetable by plunging them into boiling water or hot fat/oil rather than

by raising to boil from cold. In the latter method, there may be considerable destruction of vitamins before the enzyme is destroyed.

- Water soluble vitamins are destroyed by prolonged heating and the vitamin content of cooked food diminishes if they are kept. These losses are reduced by adding salt or sugar before cooking.
- Vit. B_1 and C are more stable in acid solution. Alkali, e.g. sodium bicarbonate hastens the destruction of vit. B_1.

I. Questions (Essay types and notes)

1. Discuss the role of proteins and carbohydrates both from qualitative and quantitative point of view in human nutrition.
2. What is respiratory quotient? Discuss various factors affecting respiratory quotient.
3. What is BMR (basal metabolic rate)? How is it determined? What are the factors which influence the BMR?
4. Describe the composition of milk. Explain the role of milk in human nutrition.

Short Notes

a. Essential amino acids
b. Respiratory quotient and its significance
c. Kwashiorkor
d. Nutritive value of proteins in the diet
e. Adequate protein
f. Marasmus
g. Basal metabolic rate (BMR)
h. Protein-caloric malnutrition
i. Indirect calorimetry methods
j. Role of minerals in the diet
k. Specific dynamic action (SDA) of food
l. Green vegetables as an article of diet
m. Protein sparers
n. Composition and nutritive value of egg
o. Requirements of phosphorus
p. Balanced diet
q. Caloric value of foods
r. Role of dietary calcium
s. Factors affecting BMR
t. Composition and nutritive value of milk
u. Diet of a pregnant lady
v. Protein sparing action of carbohydrates
w. Role of vit. B_{12} in nutrition.

Differentiate

a. Marasmus and kwashiorkor

II. MCQs (Give one correct answer)

1. Oxidation of which substance in the body yields the most calories per gram?
   a. Glucose b. Glycogen
   c. Animal protein d. Plant protein
   e. Lipids
2. Milk is notoriously deficient of which vitamin?
   a. Vitamin B_1 b. Vitamin C
   c. Vitamin A d. Pantothenic acid
   e. Vitamin B_2
3. Cellulose cannot be digested by the humans because of the inability to hydrolyse:
   a. α-Glycosidic bonds b. β-Glycosidic bonds
   c. Xylans d. Pectins
   e. α-Amyloses
4. Milk is notoriously deficient in which of the minerals?
   a. Calcium b. Iron
   c. Phosphorus d. Sodium
   e. Potassium
5. The biological activity of vit E has been attributed, in part, to its action as:
   a. A carrier in electron transport chain
   b. An antidote for selenium poisoning
   c. An anti-coagulant
   d. An anti-oxidant
   e. A reducing substance
6. Dietary fibre:
   a. Includes limit dextrins
   b. Affects digestion of other dietary components
   c. Is defined as nondigestible and water insoluble polysaccharides
   d. Is composed of undigested proteins
   e. Serves as carrier of micronutrients
7. Which of the following essential dietary factors is a precursor for a compound that can act as carrier of "one carbon" fragments at different levels of oxidation?
   a. Methionine b. Folic acid
   c. Biotin d. Pyridoxine
   e. Thiamine
8. Rats fed a fat free diet from birth would be deficient in:
   a. Sphingolipids b. Phospholipids
   c. Triacylglycerols d. Prostaglandins
   e. Cholesterol
9. A cup of strong coffee would be expected to:
   a. Decrease the effects of glucagons
   b. Enhance the effects of epinephrine
   c. Provide the vitamin nicotinic acid
   d. Interfere with the synthesis of prostaglandins
   e. None of the above
10. Humans most easily tolerate a lack of which of the following nutrients?
    a. Protein b. Lipids
    c. Carbohydrates d. Calcium
    e. Iodine
11. In patients fed an equal amount (on a molar basis) of carbohydrates and fats, the RQ is:
    a. 2.70 b. 2.00
    c. 1.00 d. 0.86
    e. 0.72

12. The intake of which food stuffs result in the greatest SDA:
    a. Carbohydrates b. Fats
    c. Proteins d. Vitamins
    e. Minerals
13. An important etiological factor in kwashiorkor is:
    a. Excess dietary ethionine
    b. Steatorrhoea
    c. Anaemia
    d. Dietary protein deficiency
    e. Dietary mineral deficiency
14. Oxidation of which substance yields the most calories per gram:
    a. Animal proteins b. Plant proteins
    c. Glucose d. Vitamins
    e. Lipids
15. Whole wheat is an excellent source of:
    a. Vitamin A b. Vitamin D
    c. Ascorbic acid d. Thiamine
    e. Riboflavine
16. Feeding of raw white of egg in the diet may result in deficiency of:
    a. Riboflavine b. Biotin
    c. Choline d. Thiamine
    e. Cyanocobalamine
17. The RQ of a mixed diet is:
    a. 0.70 b. 0.75
    c. 0.80 d. 0.85
    e. 1.0
18. When the NPU (Net protein utilisation) is low, the requirements for proteins should be:
    a. High b. Moderate
    c. Low d. None of the above
19. The semiessential amino acid is:
    a. Alanine b. Arginine
    c. Glycine d. Cysteine
    e. Serine

Answers to MCQs

1. e	2. b	3. b	4. b	5. d	6. b
7. b	8. d	9. b	10. c	11. d	12. c
13. d	14. e	15. d	16. b	17. d	18. a
19. b.					

CHAPTER 45

ENVIRONMENTAL BIOCHEMISTRY*

Major Concepts

A. Definition.
B. Importance of pollution free and ecofriendly environment.
C. Sources of environmental changes.

Specific Objectives

I. **Alteration in atmospheric temperature**
   A. ***Exposure to cold stress***
      - Learn about the ability to survive cold stress by metabolic adjustments.
   B. ***Exposure to heat***
      - Learn about different disorders related to heat stress and their management in brief.

II. **Chemical stress and pollution**
   A. ***Air pollution***
      1. Definition.
      2. Learn about sources of air pollution
         - Industrial
         - Combustion
         - Vehicular
         - Miscellaneous.
      3. Indicators of air pollution.
      4. Health effects of air pollution.
   B. ***Water pollution***
      1. Importance of pure water.
      2. Water pollution
         - Organic: Microorganisms, synthetic organic chemicals and pesticides and petroleum oil
         - Inorganic: Heavy metals, fluorine
         - Sediments
         - Radioactive materials
         - Thermal.
   C. ***Food pollution***
      1. Factors leading to food pollution.
         a. Processing of food
         b. Natural toxins
         c. Changes occurring during storage
         d. Adulteration
            a. ***Processing of food***
               - Defective operation in freezing: Typhoid
               - Defective packing techniques: Botulism
               - Food additives—carcinoma.
            b. ***Natural toxins***
               - Lathyrism
               - Favism
               - Alkaloids

* Contributed by Dr (Mrs) Alka Sontakke MBBS MD (Biochemistry), Professor and Head of the Department of Biochemistry, Padmashree Dr DY Patil Medical College for Women, Pimpri, Pune-411 018 (Ex-Reader, Department of Biochemistry, Armed Forces Medical College, Pune 411040)

- Pressor amines
- Goitrogens.

c. ***Storage contamination***
- Aflatoxins
- Ergot poisoning.

d. ***Adulteration***
- Epidemic dropsy
- Endemic ascites.

INTRODUCTION

The habitat of *Homo sapiens* is but a small planet (the earth) in the bound less universe. Life, with its complexities of the physicochemical process, has developed, modified and evolved under changing and variable environmental situations over millions of years of existence of earth. ***Hippocrates*** and ***Darwin*** first recognised the interaction and interdependence of life and environmental and the organism is at the mercy of the environment.

Life requires very narrow limits of environmental conditions. The world population is concentrated in a small percentage of the area of earth with favourable environmental conditions. But now it is becoming necessary under the related threats of population growth and environmental degradation for man to contend with a somewhat inhospitable environment which includes the following parameters: Variation in atmospheric temperature, i.e. cold and heat, variation in atmospheric pressure, electromagnetic variations, humidity, toxic gases and particulates in the atmosphere, noise, gravitation, chemical toxins in food and water and altered rhythms. Each stress factor is tolerated within limits and the organism tends to adapt to the altered situation.

Definition: Environmental biochemistry is defined as the metabolic responses to the environmental factors and the limits of changes there of, leading to adaptation.

As we develop more understanding of the earth, its ecosystems, and the pathways of chemical exposure, problems once thought to be limited to natural environment have been shown to pose risks to human health, e.g. contamination of drinking water by hazardous water dump. The most substantial long-range threat to human health posed by environmental chemicals is the alteration of the Earth's atmosphere. Chlorofluorocarbons and other compounds are producing a decrease in the level of stratospheric ozone which prevents shorter range ultraviolet light rays from penetrating to the Earth's surface, a protective mantle which has been present through much of evolution.

The demographic growth and fast urbanisation all over the world are bringing profound social and environmental changes. Therefore, the attainment of a healthy environment is becoming more and more complex. This chapter emphasises mainly the effects of variation in atmospheric temperature, i.e. cold and heat exposure and chemical stress in form of air, water and food pollution.

I. Alteration in Atmospheric Temperature

The temperature of the air varies in different parts of the day and also in the different seasons. The factors which influence the temperature are, latitude of the plane, altitude, direction of wind and proximity to sea.

A. ***Exposure to cold stress***: Injury due to cold may be general or local. In general cold injury (hypothermia) the individual is said to be suffering from exposure to cold. This is characterised by numbness, loss of sensation, muscular weakness, desire to sleep, coma and death.

Local cold injury may occur at temperatures above freezing (wet-cold conditions) as in immersion or ***trench foot*** or at temperature below freezing (dry-cold conditions) ***frostbite***, where the tissues freeze and ice crystals form in between the cells. Frostbite is common at high altitudes.

The ability to survive the stress of environmental cold depends upon the capacity to produce extra heat to compensate for the loss due to increased temperature differential between the body and the surrounding.

(i) ***Shivering phase***: During short temperature exposure, the heat loss is minimised by mechanisms of heat conservation such as incubative, behavioural, homodynamical and neurophysiological in the initial stages, the skeletal muscle plays a key role in the production of the extra heat by the process of shivering which is the first response.

(ii) ***Non-shivering phase***: In chronic exposure, metabolic adjustments take place. Heat generation is increased by a process called non-shivering thermogenesis or chemical thermogenesis.

Adaptations thereof seems to be initiated by increased food intake followed by adjustment in metabolism diverted towards maintaining the thermal state of the body at the expense of less vital function.

(a) ***Nutritional modification*:** The body develops capacity to use all the components of the diet for combustion. The increased calories can be provided by carbohydrates and fats. Proteins requirement is less. The two catabolic enzymes, *tyrosine transaminase* and *tryptophan pyrrolase* are raised within hours as a response to cold stress for serum transamination reactions are activated, amino acids are broken down and carbon skeletons are used for oxidation. Increased *arginase* facilitates excretion of urea.

(b) ***Lipid metabolism:*** Mobilisation of fat is an important feature of acclimatisation to cold. Free fatty acids in blood and liver increase, due to mobilization of adipose tissue. The total free fatty acids decrease on acclimatisation possibly due to their increased oxidation for heat production. The ketone body production is also reduced. Brown adipose tissue has role in non-shivering thermogenesis in cold exposure, as in normal life. The tissue increases in size and the synthesis of fat is enhanced.

(c) ***Calorigenic shunts*:** Following increased intake of food, rate of glycolysis is enhanced by the way of activation of the enzymes. Cold exposure favours gluconeogenesis and this is in line with the utilization of non-carbohydrate sources for calorigenic purposes. Increased activity of TCA cycle enzymes and electron transport chain components is also seen during cold exposure. Alternate pathway of electron transport not restricted by coupled phosphorylation in order to provide calorigenic shunts are activated. It appears that some components like ubiquinone and biogenic amines play significant roles in switching in favor of thermogenesis.

(d) ***Hormonal effects*:** Thyroid hormones are associated with adaptation to cold and they are increased during cold exposure. This appears to be a secondary response. The same is true for corticosteroids.

(e) ***Sympathetic nervous systems*:** The output of noradrenaline as well as tissue sensitivity to it are increased.

(f) ***Hypothermia*:** The body temperature has certain small variations during a day, low in the morning and rising by the evening. But the body can tolerate further cooling for short periods. This fact is used by surgeons to reduce body temperature by as much as 10°C, thereby reducing the metabolic needs of the tissues to permit stopping blood supply for the valuable minutes required for open heart surgery.

B. *Exposure to heat:* Many areas experience hot spells during summer resulting in a number of deaths due to heat exhaustion and stroke. This is caused by the mobility to keep the thermal equilibrium between the organism and the environment.

(i) **Heat Balance:** The 'metabolic heat' generated as a byproduct of ATP synthesis, is exchanged with the environment to maintain the body temperature.

Heat stress is the burden or load of heat that must be dissipated if the body is to remain in thermal equilibrium. The factors which influence heat stress are metabolic rate, air temperature, humidity, air movement and radiant temperature.

(ii) **Effects of heat stress:** A number of disorders resulting from exposure to heat have been recognised and documented.

(a) ***Heat stroke*:** This is attributed to failure of the heat regulating mechanism. It is characterised by very high body temperature which may rise to 110°F and profound disturbances including delirium, convulsions and partial or complete loss of consciousness. The skin is dry and hot. Death is often sudden and may be due to hyperkalaemia. It may be due to release of potassium from red blood cells. The treatment involves rapidly cooling the body, i.e. water bath till the rectal temperature falls below 103°F. Active cooling should then be stopped. Confinement of the patient to bed until the temperature control becomes stable.

(b) ***Heat hyperpyrexia*:** Impaired functioning of heat regulating mechanism without characteristic features of heat stroke. It may proceed to heat stroke.

(c) ***Heat exhaustion*:** Salt deficiency heat exhaustion occurs due to replenishment of water without salt supplements. It leads to circulatory failure. This can also give rise to heat cramps due to sodium deficiency.

(d) ***Heat syncope*:** This is common ill effect of heat. In its milder form, the person standing in the sun becomes pale, his blood pressure falls and he collapses suddenly. There is no rise in body temperature. The condition results from pooling of the blood in lower limbs due to dilatation of blood vessels leading to decreased venous return and lack of blood supply to brain. It occurs quite commonly in parades. The treatment is simple. Patient is made to lie down in shade with head low position. Recovery is seen within 5 to 10 minutes.

II. Chemical stress and pollution

Man has been disturbing the ecological equilibrium on the earth by his excessive use of the natural resources.

The depletion of resources without regeneration and the concentration of waste products without possible neutralisation, leads to pollution of air, water and foodstuffs.

(A) Air Pollution: It is defined as excessive concentration of foreign matter in the outdoor atmosphere which is harmful to man or his environment.

Air is a mechanical mixture of gases. The normal composition of external air is approximately as follows:

Nitrogen 78.1 per cent, Oxygen 20.93 per cent, Carbon dioxide 0.03 per cent. In addition argon, neon, krypton, Xenon and helium occur in traces, with water vapor, ammonia, and suspended matter such as dust, bacteria, spores and vegetable debris. With rapid growth of industries and urbanisation, the primary pollutants which together contribute more than 90 per cent of global pollutants ***include carbon monoxide (CO), Nitrogen oxides, hydrocarbons, sulphur oxides and particulates.***

Sources of air pollution

(1) *Industrial processes*: Chemical, metallurgical, oil refineries, fertilizer factories, etc.

(2) *Combustion*: Industrial and domestic combustion of coal, oil and other fuel is another sources of smoke, dust and sulphur dioxide.

(3) *Motor vehicles*: Motor vehicles are a major source of air pollution throughout the urban areas. Motor vehicles, trucks, trains, aircraft and other forms of transport contribute to air pollution by emitting hydrocarbons, carbon monoxide, lead, nitrogen oxides and particulates. In strong sunlight, certain of these hydrocarbons and oxides of nitrogen may be converted in the atmosphere into a "photo-chemical" pollutant of oxidising nature.

(4) *Miscellaneous*: Burning of refuse, agricultural activities and nuclear energy programmers also contribute to air pollution.

Pollutants: More than 100 contaminants have been identified. The important ones are CO_2, CO, sulphur dioxide, hydrogen sulphide and organic sulphides, phluorine compounds, oxides of nitrogen and ammonia, aldehydes, beryllium, carcinogenic agents.

Meteorological condition: Concentration of air pollutants depends upon meteorological conditions. Wind and temperature play a major role in the dissemination of air pollutants.

Indicators of air pollution: ***The best indicators of the general level of air pollution are sulphur dioxide, smoke and suspended particles.***

(1) *Sulphur dioxide*: This gas is a ***major contaminant*** in many urban and industrial areas. It is produced by the burning of coal and fuel oil. Its concentration is estimated in all air pollution surveys. CO, nitrogen oxides, hydrocarbon (HC), sulphur oxides (SO_2), particulates, (part) are principal air pollutants.

(2) *Smoke index or soiling index*: This varies daily, weekly or seasonally.

(3) *Suspended particles*: Dust and soot from domestic heating and industry is another useful indicator of air pollution.

Health effects of air pollution

1. *Oxides of sulphur* are poisonous and corrosive gases. They are produced chiefly by burning of oil and coal containing sulphur as impurity. These cause injury to the respiratory system. The damage to lung tissue is chiefly due to the acidic pH. Dipalmitoyl lecithin (DPL) which acts as lung surfactant is affected. Continued exposure for a few days to SO_2 concentration above 0.1 ppm concentration result in bronchitis and lung cancer. Plants are very sensitive to SO_2, fluoro-compounds, smog, etc. Spotting and burning of leaves, destruction of crops and retarded growth of plants have been observed.

2. *Oxides of Carbon*: The changes in CO_2 in the cells due to variations in its removal can be tolerated within limits. High concentration of CO_2 up to 7.5 per cent can be tolerated for short periods. Adaptation to CO_2 (1.5%) on prolonged exposure involves changes in acid-base equilibrium, increase in sodium and decrease in potassium in RBCs.

   The increase in CO levels in the atmosphere is increasing due to incomplete burning of the fuels in internal combustion engines. It displaces oxygen in the blood and creates indirectly a condition equivalent to hypoxia. At the levels of 10 ppm or higher, it can lead to decreased mental performance and cause severe effects in those suffering from diseases of heart, lung and anemia. (For source of carbon monoxide (CO), reaction with Hb and toxic effects refer to Chapter on Chemistry of Haemoglobin and Haemoglobino-pathies).

3. *Photochemical oxidants*: Under the influence of sunlight, nitrogen oxides, produced by burning fuels at high temperature and hydrocarbons produced by incomplete combustion, form complex secondary products. These together with ozone, also a photo-chemical oxidant, constitute potential health hazard, and cause eye and lung irritation, attacks of asthma and also damage to vegetation.

4. *Particulates*: Dust is common occurrence and its effects on health cannot be ignored. Particulates of solid and liquid substances of varying sizes of 1 to 5 microns, from visible type of soot and smoke to tiny invisible particles of a variety of chemical composition

are produced by industrial processes and fuel burning. Pulmonary abnormalities resulting from the inhalation of dust particles known as pneumoconiosis is the serious occupational health problem, siderosis (iron), anthracosis (soot and carbon smoke), silicosis (silica dust), asbestosis (asbestos, μgm silicate), byssinosis (cotton dust) and bagassosis (bagasse, sugar cane rice dust). Recent studies have shown that the particulate load indoor is directly proportional to the number of cigarette smokers. Increased prevalence of respiratory illnesses and higher risk of lung cancer are seen in nonsmokers exposed to passive smoking.

Other Effect of Air Pollution

1. *Acid rain*

Acid rain is the precipitation that contains high levels of sulphuric acid (H_2SO_4) and nitric acid (HNO_3).

Sulphur dioxide (SO_2) produced during burning of coal and oil is oxidised to sulphur trioxide (SO_3) by atmospheric oxygen (O_2). This dissolves in the moisture present in the clouds or in rain water forming sulphuric acid. Similarly, nitrogen dioxide produced and released by factories and automobiles forms nitric acid with moisture/rain water.

Acids formed by above manner pollute rivers, streams and soil. Acid rain pollutants also decrease the fertility of the soil.

The acid rain pollutants often travel long distances and even pollute neighbouring towns and countries.

2. *Increased CO_2 concentration: Green house effect*

The burning of coal and oils release CO_2 into the air. Increased population and felling down of trees and deforestation has upset the ecological balance leading to gradual increase in concentration of CO_2 in the atmosphere.

Normally, the heat received by the sun is radiated into the atmosphere. But, if the CO_2 concentration increases, the excessive CO_2 accumulating near the Earth's surface acts as a mantle and traps the heat and prevents the heat from escaping resulting in warming up of the Earth. This is similar to a ***'green house'*** which lets in sunlight through its glass panes but traps it when released as heat. Hence, the action of increased CO_2 accumulating near the Earth's surface which traps the sunlight has been called as ***"Green house effect".***

Future Consequences: This global warming by even a few degrees produced by increased CO_2 concentration could cause the "Polar" ice to melt. This inturn would raise the level of seas all over the world. Many coastlines would be flooded and low lying towns and cities would be at risk.

(B) Water Pollution: Water is the matrix of life and forms the bulk of the weight of the living cells. The resources of usable water have been diminishing and are unable to meet the variety of needs of modern civilisation. Pollution by municipal, industrial and agricultural wastes disposal by "infinite dilution" aggravates the situation.

A large number of water pollutants may be broadly classified as:

1. **Organic pollutants.**
2. **Inorganic pollutants.**
3. **Sediments pollutants.**
4. **Radioactive materials.**
5. **Thermal pollutants.**

1. Organic Pollutants

This group includes oxygen demanding wastes, disease carrying agents, plant nutrients, sewage, synthetic organic compounds and oil.

Dissolved oxygen (DO) is an essential requirement of aquatic life. The optimum DO in natural water is 4 to 6 ppm. ***Decrease in this DO value is an index of pollution mainly due to organic matter,*** e.g. sewage (domestic and animal), industrial waters from food-processing plants, paper mills and tanneries, waters from slaughter houses and meat packing centres, run off from agricultural lands, etc.

(a) ***Water*** is the carrier of pathogenic microorganisms and can cause immense harm to public health. The water borne diseases are typhoid and paratyphoid fever, dysentery and cholera, polio and infectious hepatitis. The responsible organisms occur in the faeces or urine of infected people and are finally discharged into a water body. Historically, the first step in water pollution control was the disinfection technique for the prevention of water borne diseases, which are still in use.

Management and treatment of domestic sewage becomes increasingly important in highly populated cities. Sewage and run off from agricultural land provide plant nutrients in natural settings, in the natural biological process called ***"eutrophication"*** (Greek well nourished). Algal blooms and large amounts of other aquatic weeds cause serious problems. The excessive plant growth prevents an unaesthetic scene and disturbs recreational area of water. The water body, in the process of eutrophication loses all its DO in the long run and ends up in a dead pool of water.

(b) ***Production of synthetic organic chemicals*** which includes fuels, plastics, plasticisers, fibres, elastomers, solvent detergents, paints, insecticides, food additives and pharmaceuticals has multiplied about 10 times since 1950. Their presence in water imparts objectionable and offensive taste, odour and colour.

(c) ***Pesticides***: Include insecticides, rodenticides, molluscides, herbicide and fungicides. The negative aspects of these is that their residues have moved through ecosystems and are threatening to destroy the food chain. The positive aspect is eradication of disease like malaria and typhus.

Classification

a. ***Chlorinated hydrocarbons***, e.g. DDT, aldrin, dieldrin, etc. Persistent stay in environment for 20 years.

b. ***Chlorophenoxy acids***, e.g. Atrazine, 2,4-D. Non persistent, last only for few days.

c. ***Organophosphorus (ORP)*** e.g. Malathion, diazinon.

d. ***Organocarbamates (ORC)*** Baygon, sevin.

Nature and mode of action of the insecticides

- **DDT:** It is chemically "dichlorodiphenyl-trichloro ethane." It is widely used in agricultural practice.

It is fat soluble, hence it is deposited in the adipose tissue and it is difficult to be excreted. Hence its concentration in the body gradually builds up. Though DDT is banned in many countries, it is still used in our country and is a source of environmental pollutants.

- **Aldrin and Dieldrin:** It is a cyclic halogenated hydrocarbon. Like DDT, it is also fat soluble and gets deposited in the adipose tissue. It is a very toxic insecticide. It is metabolised by microsomal monooxygenase system and is converted to ***"epoxides"*** which is also toxic.
- ***Organophosphorus (ORP) and organo-carbamates (ORC)***: They are powerful neurotoxic agents. They enter through skin, mucous membranes and respiratory tract. They inhibit the enzyme ***"cholinesterase (cholinesterase inhibitor)*** by phosphorylation of the active site of the enzyme. Acetylcholine accumulate in the nerve endings and prevent the transmission of nerve impulses across synapses.

(d) Another form of organic contamination of growing concern is petroleum oil, released into water either by accidents of oil tanker or as wastes, which results in the destruction of algae and invertebrate larvae.

The overall effects of oil on marine organism are (a) direct lethal toxicity, (b) disruption of physiological or behavioural activities, (c) changes in biological habitat.

2. Inorganic Pollutants

a. Heavy metal pollutants, either directly or indirectly entering the food chain, are becoming an increasing threat to health. Several trace elements (few ppm or less) are found in polluted water. The most dangerous among them are the heavy metals, e.g. Pb, Cd, Hg and metalloids, e.g As, Se, Sb, etc. ***The heavy metals have a great affinity for sulphur and attack sulphur bonds in enzymes, thus immobilising the latter.*** Other vulnerable sites are protein carboxylic acid and amino groups. Heavy metals bind to cell membrane, affecting transport processes through the cell wall. They also tend to precipitate phosphate biocompounds or catalyse their decomposition.

Organic mercurials used as bactericides or fungicides can cause neurological disorders and death. Lead (Lead tetraethyl) used in gasoline causes blocking of spindle fiber mechanism in cell division, uncoupling oxidative phosphorylation and altered rapid eye movement phase of sleep, leading to insomnia.

Cadmium—Potent uncoupler of oxidative phosphorylation

Beryllium—A powerful phosphate inhibitor.

Strontium—A competitor for calcium in bone.

The most disheartening feature of metal toxicity is the early effect on behaviour and intelligence. Before appearance of any outward sign of disability or diseased condition, areas of the brain are damaged, out of which arise many neurological and behavioural abnormalities. Pathetically, children suffer most from this malady as the neuronal functions are more readily affected during the early phase of development.

A. Lead

Amongst the heavy metal pollutants ***lead deserves special mention.*** Lead is one of mankind's oldest environmental and occupational toxins. Although the symptoms are acute and chronic. Lead poisoning was known to the ancient Greeks, it is only in the 20th century that substantial progress has been made in reducing the incidence of lead poisoning. **The United States Centres for Disease Control (CDC) recently issued new guidelines on preventing lead poisoning, specially in young children.**

CDC guidelines for lead poisoning is given as follows:

≤ 10 μg/dL: Not considered to indicate lead poisoning.

10 to 14 μg/dL: A high prevalence of persons with levels >10 μg/dL should trigger community wide lead poisoning prevention. Individual should be retested more frequently.

15-19 μg/dL: Should receive education on preventing lead poisoning and on nutrition. Individuals should be retested more frequently; if levels persist sources of lead exposure to be investigated.

20-44 μg/dL: Full medical evaluation is indicated. Sources of lead exposure should be removed from the environment. Drug therapy may be required.

45-69 μg/dL: Chelation therapy is indicated. The individual should be removed to a lead-free environment.

≥70 μg/dL: A medical emergency requiring immediate hospitalisation and chelation therapy.

The new guidelines recommend universal lead testing for all children under six years of age, and expand the definition of lead poisoning to include any child with a blood lead level of 10 μg/dL or more. Screening should be done by direct measurement of blood lead rather than by protoporphyrin levels.

Recognition of subclinical lead poisoning, in which various biochemical and biological functions are impaired at levels below those required to produce symptoms is also a relatively recent development. The ability of the laboratory to measure blood lead levels and biochemical ***'markers'*** of subclinical toxicity has contributed greatly to this progress.

Sources

Young children are particularly susceptible to lead poisoning. The "hand to mouth" activity of the toddler greatly increases the likelihood of ingesting lead-contaminated dust and dirts. Paint is the major source of exposure in the children as they bite painted toys and suck lead pencils.

Lead is ubiquitous in the environment. Increased content of lead is found in air, water and vegetables in cities and near highways which is due to tetraethyl lead from exhaust of vehicles. Other sources are lead pipes, soldering by lead, newspapers and xeroxed copies contain lead. Lead chromate is used as an adulterant and all yellow powders.

Clinical features

While adults absorb only above 10 per cent of ingested lead, young children may absorb upto 50 per cent.

The CNS is principally involved, the principal target of lead toxicity. In children, the CNS is still in developing stage and ***hence is more susceptible to toxic insults.*** In children, mental retardation, poor concentration skills, behavioural changes, learning disabilities are seen.

A number of recent studies have shown impairment of childhood development and intelligence by relatively low levels of lead. Meta-analysis of the conjoint studies provided convincing evidence, a loss of about 4 to 5 IQ points for each 10 μg/dL of blood lead.

In adults, headache, irritability, confusion and anaemia are seen. If the blood level is more than 60 μg/dL acute toxicity is seen which is manifested as encephalopathy, neuropathy, abdominal colic and severe anaemia. There may be associated porphyrinuria. ***Discolouration and blue line along the gums are characteristic features of acute lead poisoning.***

Pathogenesis of anaemia:

- Lead inhibits almost all the enzymatic steps of haem-synthesis. Lead particularly inhibits *"delta aminolaevulinic acid synthase (δ-ALA)"* and *"ALA dehydratase"* (Refer to Chapter on Haem Synthesis).
- Uptake of Fe by the reticulocytes from transferrin is inhibited.
- Also inhibits the enzyme *"ferrochelatase"*, causing an increased level of free erythrocyte protoporphyrines.

Diagnosis: Made by measuring the level of lead in blood and urine. CDC guidelines for childhood lead poisoning is 10 μg/dL or above.

B. Arsenic

Sources

- Oxides of arsenic are commonly used as fruit sprays, rat poisons, pesticides, etc.
- Water pollution—in certain areas water has been found to have high content of arsenic (recently high arsenic content of water reported from Malda district of West Bengal).

Toxic effects: Produces dizziness, chills, cramps and paralysis leading to death.

- Arsenic inhibits the sulfhydryl enzymes and interferes with cellular metabolism.
- It may also cause intravascular haemolysis leading to haemoglobinaemia and haemoglobinuria.

(a) Other Chemical Pollutants

- Fluorine in excessive amounts produces body injury probably because of its inhibition of a number of enzymes particularly those in carbohydrate metabolism. But in low concentration fluorine of drinking water is beneficial in the prevention of dental caries (Refer Chapter on Minerals and Trace Elements).

(b) Chlorofluorocarbons (CFCs) and Halons

- **CFCs:** Invented in 1928, are not found in nature. It was considered as pathbreaking invention being inert, cheap, non-explosive, non-toxic and easy to store. ***Hence CFCs found its use in refrigerators, airconditioners, perfumes, room fresheners, making foam for mattresses and cushions, etc.***

Halons: Used as fire extinguishing agents, do not harm people but are more dangerous to ozone than CFCs.

Both CFCs and halons when leaked in the atmosphere during manufacture, testing and repair of products are carried by air currents and slowly diffuse into the stratosphere.

CFCs do not destroy ozone directly, they are photodissociated when they reach the upper atmosphere and the chlorine atoms are released which destroy the ozone layer and produces ozone depletion. Hence CFCs and Halons are called ***"Ozone-Depleting Substances" (ODS).***

Ozone is a bluish coloured gas having a pungent smell. The ozone layer is a thin invisible layer of ozone gas. The stratosphere contains about 90 per cent of all the ozone in the atmosphere spread thinly and unevenly. ***Ozone in the stratosphere acts as a protective layer which shields the earth and its inhabitants from harmful ultraviolet radiations of the sun (UV rays).***

Ozone is also found in the lower level of the atmosphere, i.e. Troposphere. **In the troposphere, ozone acts as a harmful pollutant and, more than a trace of this gas can damage human lungs and tissues** and also human plants. Human activities are increasing the ozone in the troposphere.

Harmful effects of UV radiation due to ODS:
Exposure to too much UV radiation may lead to widespread damage to all life forms on Earth. The continual exposure could cause:

- ***The skin to freckle and age faster***
- ***It may increase the occurrence of skin cancer***
- ***It may increase eye infections and cataracts in humans and animals.***

Plant life: Increased UV radiation also damages plant.

- Leaf size will be reduced and germination time increased. This could decrease crop yield.
- Too much UV radiations kill phytoplanktons—the base of the aquatic food chain. Thus the entire aquatic food chain is disturbed.

3. **Sediments:** The natural process of soil erosion gives rise to sediments in water. It represents the most extensive pollutants of surface water. Bottom sediments are important sources of inorganic and organic metals in streams.
4. **Radioactive material:** Four human activities are responsible for radioactive pollution
   - Mining and processing of ores to produce usable radioactive substances.
   - Radioactive material in nuclear weapons.
   - Radioactive material in nuclear power plants.
   - Radioactive isotopes in medical, industrial and research.

   The nuclear power plants generate the following types of pollutants: (a) Low-level radioactive liquid water, (b) Liquid and gaseous wastes from fuel elements of fission products, and (c) Heat.

   Exposure of high-levels of radiation adversely affects human health. It is decidedly connected with the incidence of leukemia and other forms of cancer (Refer to Chapter on Radioactivity: Radioisotopes in Medicine).
5. **Thermal pollutants:** Coal fired or nuclear fuel fired steam power plants are associated with problem of thermal pollution. Dissolved oxygen (DO) of water is decreased. Affects aquatic life. It is worthwhile exploring the possibility of transferring the water heat to atmosphere rather than to the surrounding water bodies through cooling towers.

Food Pollution

It may occur due to various factors like:
(a) Processing of food.
(b) Natural toxins present in plants.
(c) Changes occurring during storage.
(d) Adulteration of food.

(a) Processing of food. This includes:

(i) **Defective operation in freezing,** e.g. milk may be contaminated by ***Salmonella*** and ***Staphylococcus aureus***.

(ii) **Defective packing techniques:** This can lead to Botulism especially with marine products and cooked soups.

(iii) **Food additives** for preservation or enhancing the flavour or aroma, e.g. aniline dyes to enhance colour are carcinogenic. Cyclamate used as sweetening agent may cause carcinoma of bladder, monosodium glutamate used for enhancing aroma in Chinese food is supposed to be toxic for children below five. It causes transient symptoms like numbness, general weakness and palpitations. Benzoate, sulphite and polyphenols are comparatively safer preservatives.

(b) Natural toxins

- ***Lathyrism***

  It is a crippling disease leading to paralysis of lower limbs due to excessive consumption of Kesari dal (*Lathyrus sativus*) seen commonly in Madhya Pradesh, UP and Bihar. Neurotoxins present in lathyrus damage the upper motor neurons. **The toxin is identified as betaoxalyl amino alanine (BOAA).** The signs and symptoms include exaggerated knee jerk, ankle clonus, scissor gait and in extreme cases complete spastic paralysis. Thorough cooking and decanting the supernatant two or three times will remove these toxins. Table 45.1 shows the various toxic factors found in lathyrus and vicia species.

Table 45.1: Toxic factors in lathyrus and vicia species

Lathyrus species		*Chemical nature of toxic factors*
• *Lathyrus sativus*	—	β-oxalyl-amino-alanine (BOAA)
• *Lathyrus odoratus*	—	γ-glutamyl-β-amino-propionitrile
• *Lathyrus latifolius and L. sylvestris*	—	α-γ,-diaminobutyric acid
• *Vicia sativa*	—	β-cyanoalanine

- **Favism:** This is a condition seen in individuals having glucose-6-phosphate dehydrogenase deficiency on ingestion of uncooked broad bean (*Vicia faba*). Manifested in form of hemolytic anaemia. Cooking and decanting minimises the toxicity (Refer to Chapter on Metabolism of Carbohydrates).
- **Alkaloids:** Alkaloids like muscarine found in toxic variety of mushroom may produce nausea, vomiting, diarrhoea and in large quantity can cause acute necrosis of liver and death.
- **Pressor amines:** Food rich in amines such as histamine, tyramine, tryptamine and serotonin include plantains, bananas, cheese, etc. Normally, on consumption they are catabolised by MAO. But they may lead to hypertension in patients on MAO inhibitors.
- **Goitrogens:** Many foodstuffs of Brassicaceae family contain organic compounds which have goitrogenic properties. These compounds prevent iodine uptake or utilisation of iodine by thyroid gland.

  The goitrogens present in foods may be broadly grouped under the following heads:

  1. *Thio-oxazolidone derivatives*

     Vegetables of the Brassicaceae family particularly rutabagas, cabbage, mustard seeds, turnips, etc. contain thioglycosides called as **"Progoitrin"** chemically it is **"1-5-vinyl-2-thio-oxazolidone"**.

     Progoitrin can be converted to "goitrin", an active antithyroid agent.

     $$\text{Progoitrin} \xrightarrow{\text{Activator}} \textbf{Goitrin}$$

     Progoitrin "activator" present in vegetables is "heat-labile", but their activators present in intestine (colonic bacteria), goitrin is formed in the intestine by bacterial action even if the vegetables are cooked.

     The goitrin intake on a normal mixed diet is usually not great enough to be harmful, but ***in vegetarians and in food faddists "cabbage goitres" may occur.***
  2. *Thiocyanates and Isothiocyanates:* Certain oil seeds, e.g. rapeseed, mustard, etc. contain thioglycosides which yield on hydrolysis "isothiocyanate".
  3. *Thiocyanogen:* Thiocyanogen is present in some foods or is formed from cyanoglycosides present in some foods. These act as goitrogens.
  4. *Indolyl acetonitrile:* Indolyl acetonitrile. present in "brassicae" species has been shown to possess goitrogenic properties. Another sulphur containing compound possessing strongly goitrogenic properties has been identified as ***1, 2-dithiocyclopentane, 1-4, ene-3-thione"***.
  5. *Polyphenolic glycosides:* Polyphenolic glycosides present in red skin of groundnuts and almonds have been shown to possess goitrogenic properties.
- *Protease inhibitors:* Many cereals (corn), tubers (Potato and sweet potato), peanut, legumes, (Soyabeans) contain ***"trypsin inhibitors"***. Usually they are ***heat-labile*** and are destroyed in cooking, hence not harmful. But partially cooked foods or raw foods may have these and may inhibit the digestion and absorption of amino acids (Refer to Chapter on Digestion and Absorption of Proteins and Amino Acids).
- *Cyanogenic glycosides:* Cyanogenic glycosides are present in certain cereals like sorghum, legumes like lima bean and in tubers like tapioca. ***These compounds on hydrolysis produce hydrocyanic acid. Hence they are highly toxic when taken raw.***

Ackee fruit poisoning (Jamaican vomiting)

Ackee fruit grown in Jamaica and also in Nigeria is extremely popular; when the unriped fruit is taken raw it produces toxic manifestations (Refer to Chapter on Lipid Metabolism).

(c) Storage contamination: This mainly occurs due to fungus growing on stored grains.

(i) *Aflatoxins* produced by ***Aspergillus flavus*** which grow in moist conditions on groundnut, coconut, etc. are ***hepatotoxic and carcinogenic,*** maximum permissible limit of contamination is 0.05 ppm.

(ii) ***Ergotamine, ergotoxin and ergometrine*** found in ergot that grows on moist food grains like rye, millet, wheat, bajra may produce peripheral vascular contraction leading to cramps, gangrene in extremities and convulsions. This is known as **ergotism.** The fungal contamination may be prevented by proper storage condition in non humid and dry atmosphere.

(iii) ***Contamination of rice*** can take place during storage with a fungus called ***"Penicillium islandicum"***. It produces yellow discolouration of rice. ***The fungus has got toxic effects on liver and kidney, i.e. it is hepatotoxic and nephrotoxic.***

Table 45.2 shows food borne diseases caused by pathogenic organisms (due to defective processing and storage).

(d) Adulteration of food: As the list of adulterants of food is long, only common ones are considered.

Table 45.2: Shows food borne diseases caused by pathogenic microorganisms (due to defective processing and storage)

	Microorganisms	Foods commonly involved	Ill effects and diseases
a. Bacterial	*Salmonella*	Defectively processed meat, fish and egg products, raw vegetables grown on sewage	Salmonellosis (vomiting, diarrhoea and fever)
	Cl. botulinum toxins	Defectively processed meat and fish	Botulism (muscular paralysis, death due to respiratory failure)
	Cl. welchii	Defectively processed. Precooked meat	Nausea, abdominal pain, and diarrhoea
	Bacillus cereus	Cereal products	Nausea, vomiting, abdominal pain
	Shigella sonnei	Foods kept exposed for sale in unhygienic surroundings	Bacillary dysentery
	Staphylococcus aureus	Foods exposed for sale in unhygienic surroundings	Increased salivation, vomiting, abdominal pain and diarrhoea
b. Fungal	*Aspergillus flavus (aflatoxin)*	Corns and groundnuts infected with *Aspergillus flavus*	Liver damage and cancer liver
	Penicillium islandicum	Contamination of rice during storage producing yellow discolouration	Liver damage and kidney damage
	Claviceps purpurea (Ergot)	Rye and pearl millet infected with ergot	Ergotism (Burning sensation in extremities, cramps, peripheral gangrene)
	Fusarium sporotrichioides	Cereals and millets infected with fusarium	Alimentary toxic aleukia
c. Parasitic	*Trichinella spiralis*	Pork and Pork products	Trichinosis—nausea, vomiting, diarrhoea, colic and muscular pains
	Ascaris lumbricoides	Raw vegetables grown on sewage farms	Ascariasis
	Entamoeba histolytica	Raw vegetables grown on sewage farms	Amoebic dysentery

(i) **Epidemic dropsy:** Mustard oil may be adulterated with argemone oil which is derived from a wild plant **Argemone mexicana**. Argemone oil contains the alkaloid, ***"sanguinarine"*** which causes vomiting, diarrhoea and congestive cardiac failure leading to edema. The condition is referred to as **epidemic dropsy**. Sanguinarine interferes with oxidation of pyruvic acid leading to its accumulation in the blood. From time to time, outbreaks of epidemic dropsy are reported in India, the latest and lethal being in 1999 in India.

(ii) **Endemic ascites:** This is seen in areas where local population subsist on millet (known locally as Gondhi) which may be adulterated with weed seeds of crotalaria (locally known as Jhunjhunia). On chemical analysis these ***seeds were found to contain pyrrolizidine alkaloids which are hepatotoxic leading to ascites and jaundice.***

CHAPTER

46

BIOCHEMISTRY OF CANCER

Major Concepts

A. Learn what is cancer and study the mechanisms of carcinogenesis, learn about role of apoptosis in carcinogenesis.
B. Learn how cancer cells spread (Biochemistry of metastasis).
C. Study the different "oncogenic markers" (Tumour markers).

Specific Objectives

I. ***Chemistry of Cancer and Carcinogenesis***
   1. Define cancer
      - Study the properties of cancer cells—morphological and biochemical alterations.
   2. Study carcinogenesis
      - Study the predisposing factors
      - Learn about different carcinogenic agents:
        - Physical: Radiations
        - Chemicals: As carcinogens
      - Viral oncogenesis.
   3. Study experimental carcinogenesis
      - Initiation, and
      - Promotion.
   4. Learn mechanisms of chemical carcinogenesis
      - Direct acting chemical carcinogens
      - Procarcinogens.
   5. Learn in detail about viral oncogenesis:
      - Study different DNA viruses that cause cancer and their mechanism of action
      - Study the different RNA viruses and their mechanism of action
      - Learn about acute transforming retroviruses and slow transforming retroviruses.
   6. Study the biochemical mechanisms of viral oncogenesis: Rous sarcoma virus as a model.
   7. Learn in detail how 'Proto-oncogenes' are converted to 'oncogenes': Its biochemical mechanisms:
      - Study the **five** mechanisms postulated:
        - Single point mutation
        - Gene amplification
        - Promoter insertion
        - Enhancer insertion
        - Translocations.
   8. Study about cancer-suppressor genes (growth suppressor genes)—antioncogenesis.
   9. Learn about different growth factors:
      - Types of growth factors
      - Mechanism of action of growth factors
      - Action of growth factors at molecular level.
   10. Learn about interferons (IFN) and its use in clinical medicine as therapeutic agent, learn about role of telomeres and telomerase in cancer and in ageing.

II. ***Learn what is apoptosis.*** *Study about the role of apoptosis in carcinogenesis.*

III. ***Biochemistry of Metastasis of Cancer***
   1. Define metastasis.

2. Learn in detail about the biochemistry of Metastasis:
   - Composition of extracellular matrix (ECM) and basement membrane (BM)
   - Learn how cancer cells interact with basement membrane (BM). Study 3 steps:
     – Step 1: Attachment of metastatic cancer cell to BM
     – Step 2: Dissolution of the BM
     – Step 3: Migration of cancer cells.
   - Study the various enzymes which help in spread of cancer cells.

IV. ***Oncogenic Markers (Tumour Markers)***

1. Define oncogenic marker.
2. Study the different types of tumour markers.
3. Learn the characteristics of an ideal tumour marker.
4. Study how the tumour markers can be classified.
5. Learn the commonly used tumour markers: Their chemistry, clinical uses, and limitations
   - Carcinoembryonic antigen (CEA)
   - Human chorionic gonadotrophin (β-HCG)
   - α-fetoprotein (AFP).
6. Learn about other tumour markers which are not used commonly.

I. Biochemistry of Cancer Cells and Carcinogenesis

What is Cancer? Cancer is a cellular tumour that, unlike benign tumour cells, can metastasize and invade the surrounding and distant tissues.

INTRODUCTION

Cancer has been a major cause of death in the USA for the past few decades, being second only to cardiac diseases. Approximately 20 per cent of all deaths in America are due to cancer.

There are at least fifty different types of malignant tumours being identified. More than 50 per cent of the newly diagnosed cancers occur in five major organs: **(i) lungs, (ii) colon/rectum, (iii) breast, (iv) prostate and (v) uterus.**

Cancers of the lungs, colon/rectum and prostate are the principal leading causes of deaths in males and in females, breast, colorectal and uterine cancers are the most common.

Environmental factors play a very important part. In Japan, death rate from cancer of stomach is about seven times more than that in the USA. Other examples are:

- Increased risk of certain cancers with occupational exposures to asbestos, naphthylamine, etc.
- Association of cancers of oropharynx, larynx, oesophagus and lungs with tobacco chewing and cigarette smoking.

PROPERTIES OF CANCER CELLS

Cancer cells are characterised by **three important properties:**

1. *Diminished or unrestricted control of growth.*
2. *Capability of invasion of local tissues, and*
3. *Capable of spreading to distant parts of body by metastasis.*

Characteristics of Cancer Cells: Morphological and Biochemical

Changes shown by cultured cells undergoing malignant transformation *in vitro* have been studied.

(a) Morphological Changes

- Have usually rounded shape, larger than normal cells.
- Cells show nuclear and cellular pleomorphism, hyperchromatism, ***altered nuclear: cytoplasmic ratio,*** abundant mitosis, sometimes tumour giant cells.
- Transformed cells often grow over one another and form multilayers.
- Can grow without attachment to the surface *in vitro*, diminished adhesion.

(b) Biochemical Changes

- ***Increased synthesis of DNA and RNA.***
- Show increased rate of glycolysis both aerobic and anaerobic.
- ***Show alterations of permeability and surface charge.***
- Changes in composition of glycoproteins and glycosphingolipids on cell surfaces.
- Alterations of the oligosaccharide chains.
- ***Increased activity of ribonucleotide reductase*** and decreased catabolism of pyrimidines.
- Secretion of certain *proteases* and *protein kinases.*
- ***Alterations of isoenzyme patterns often to a foetal pattern and synthesis of foetal proteins,*** e.g. carcino-embryonic antigen (CEA), α-fetoprotein (AFP), etc.
- Appearance of new antigens and loss of certain antigens.

- Inappropriate synthesis of certain hormones and growth factors. Often there may be increased secretion of certain growth factors into the surrounding medium.

ETIOLOGY OF CANCER (Carcinogenesis)

(a) Predisposing Factors

1. *Age:* Cancer can develop in any age, though it is most common in those over 55 years of age.
   Certain cancers are particularly common in children below 15 years of age, viz.
   - ***Retinoblastomas***
   - ***Neuroblastomas***
   - ***Wilms' tumours***
   - Certain tumours of haemopoietic tissues as ***lymphomas*** and ***leukaemias.***
   - ***Sarcomas of bones and skeletal muscles.***
2. *Heredity:* ***Heredity plays an important role in carcinogenesis.*** Certain precancerous conditions are inherited.
   Examples are:
   - Susceptibility to childhood retinoblastomas is inherited as an autosomal dominant trait and approximately 40 per cent of retinoblastomas are familial.
   - Susceptibility to multiple colonic polyposis is inherited as autosomal dominant trait and almost all cases develop into ***adenocarcinomas*** in later life.
   - Chromosomal DNA instability may be inherited as an autosomal recessive trait. Conditions are characterised by some defect in DNA repair.
   - ***In xeroderma pigmentosa,*** a skin condition, the affected individuals develop carcinomas of skin in areas exposed to UV rays of sunlight.
3. *Environmental factors:* Statistically it has been shown that ***80 per cent of human cancers are caused by environmental factors,*** principally chemicals, viz.
   - *Lifestyle:* Cigarette smoking, tobacco chewing.
   - *Dietary:* Groundnuts and other foodstuffs infected with fungus like *Aspergillus* produce **aflatoxin B_1** which is carcinogenic.
   - *Occupational:* Asbestos, benzene, naphthylamines, beryllium, etc.
   - *Iatrogenic:* Certain therapeutic drugs may be carcinogenic.
4. *Acquired precancerous disorders:* Certain clinical conditions are associated with increased risk of developing cancers.
   Examples are:
   - *Leukoplakia* of oral mucosa and genital mucosa developing into squamous cell carcinomas.
   - *Cirrhosis of liver:* A few cases can develop hepatoma (hepatocellular carcinoma).
   - *Ulcerative colitis:* Can produce adenocarcinoma of colon.
   - *Carcinoma in situ of cervix:* Can produce squamous cell carcinoma of cervix.

(b) Carcinogenic Agents (Agents Causing Cancer): Carcinogens that cause cancer can be divided into **three main broad groups:**

1. **Physical:** Radiant energy
2. **Chemicals:** Variety of chemical compounds can cause cancer. Some of these can act directly and others can act as procarcinogens
3. **Biological:** Oncogenic viruses.

I. RADIANT ENERGY (RADIATIONS): MECHANISM OF CARCINOGENESIS

Radiations can cause cancer mainly in **two ways:**

1. *Direct Effect*
   By producing damage to DNA, which appears to be the basic mechanism but the details are not clear. Radiations like X-rays, γ-rays or UV rays are harmful to DNA of cells and they can be mutagenic and carcinogenic.
   Damages to DNA brought about by radiations may be as follows:
   - ***Single or double strand breaks.***
   - ***Elimination of purine/pyrimidine bases.***
   - ***Cross-linking of strands.***
   - ***Formation of pyrimidine dimers.***
2. *Indirect Effects*
   In addition to direct effects on DNA as stated above, radiations like γ-rays and X-rays produce **free radicals**, viz. OH^-, superoxide and others which may interact subsequently with DNA and other macromolecules leading to molecular damage.

UV rays: Natural UV rays from sun can cause skin cancer. Fair-skinned people living in places where sunshine is plenty are at greatest risk. Carcinomas and melanomas of exposed skin are particularly common in Australia and New Zealand.

UV rays produce:
- ***Damage to DNA by formation of pyrimidine dimers.***
- ***Secondly by immunosuppression.***

Ionising Radiations

The ability of ionising radiations to cause cancer lies in their ***ability to produce mutations*** (mechanisms discussed above). ***Particulate radiations such as α-particles and neutrons are more carcinogenic than electromagnetic radiations like X-rays and γ-rays.***

Evidences in favour of carcinogenicity of ionising radiations:

- Incidence of leukaemias increased in Japan after atom bomb explosion.
- Development of thyroid cancer in later life in children exposed to therapeutic radiation in neck.
- Lung cancer is more in miners who work in radioactive ores.

II. CHEMICALS AS CARCINOGENS

A large number of chemicals have been incriminated as carcinogenic. Some of these are direct reacting and majority occur as procarcinogens which are converted in the body to ultimate carcinogenic chemicals. Many of the chemicals have been tested on animals (experimental carcinogenesis).

A list of carcinogenic chemicals is given below

Class	***Nature of Chemicals Compound***
1. Polycyclic aromatic hydrocarbons	• Benzpyrene • Dimethyl-benzanthracene

Note: Aromatic hydrocarbons are present in cigarette smoke and they are thus relevant in pathogenesis of lung cancer.

2. Azo dyes (Aromatic amines)	• β-Naphthylamine • N-methyl-4-amino-azobenzene • 2-acetylaminofluorine

Note: β-naphthylamine, an aniline azo dye used in the rubber industries has been held responsible for bladder cancers in exposed workers.

3. Nitrosamines and amides	• Dimethylnitrosamine • Diethylnitrosamine

Note: Nitrosamines and amides can be synthesised in GI tract from ingested nitrites or derived from digested proteins and may contribute to induction of gastric cancer.

4. Naturally occurring compounds	• Aflatoxin B_1 produced by the fungus, *Aspergillus flavus.*

Note: The fungus grows on groundnuts, peanuts and other grains in congenial environmental conditions. It ***produces "aflatoxin B_1" which is a potent hepatocarcinogen.*** This is believed to be responsible for high incidence of liver cell carcinoma in Africa, where the contaminated foods are eaten.

5. Various Drugs	• Alkylating and acylating agents, e.g. cyclophosphamide and busulfan.

Note: The drugs are used in cancer treatment and also as immunosuppressants. Patients receiving such therapy are at a higher risk for developing cancer.

	• Diethylstilbestrol, oestrogen. • Nitrogen mustard. • β-propiolactone
6. Miscellaneous agents:	• Beryllium, cadmium, nickel, chromium, arsenic • Asbestos • Vinyl chloride • Saccharin and cyclamates

EXPERIMENTAL CARCINOGENESIS

Carcinogenesis has been studied in experimental animals. **Neoplastic transformation produced by chemicals is a dynamic multi-step process.**

Experimental development of skin tumours in mice have been studied. **Two definite stages** can be observed:

1. When the skin of mice is painted with a chemical **benzpyrene** only and no other subsequent treatment is done, no skin tumours develop.
2. But if the area of benzpyrene application is followed by several applications of **croton oil**, subsequently, tumours develop. On the other hand, application of croton oil alone without pretreatment with benzpyrene does not result in development of skin tumours.

Conclusions: From above, it is clear that two definite stages can be observed in this experimental study.

I. Initiation

The stage of carcinogenesis caused by application of benzpyrene is called initiation. Benzpyrene is thus called as initiating agent.

This stage appears to be:

- **Rapid and irreversible**
- It involves in the induction of certain irreversible changes in the genome of the cells, i.e. brings about modifications in DNA perhaps resulting in one or more mutations.

Characteristics of initiated cells

- They are ***not transformed cells***
- They ***do not have growth autonomy***
- Unlike normal cells, they ***give rise to tumours when stimulated by the promoting agents.***

II. Promotion

The second stage is the stage of carcinogenesis resulting from application of *croton oil* and is called promotion. Croton oil is thus a **promoter** or **promoting agent**. Promoters cannot produce initiation. Promoters bring about tumour induction in a previously initiated cell.

Note

Most chemical carcinogens are capable of acting as both initiators and "promoter".

Biochemical mechanism of action of croton oil as promoter:

- ***Active agent present in croton oil is a mixture of phorbol esters.*** The most active one is **TPA** (chemically 12-O-tetradecanoylphorbol-13-acetate).

- In cells, '*protein kinase C*' can act as a receptor for TPA.
- Interaction of TPA with protein kinase C brings about phosphorylation of a number of membrane proteins leading to effects on transport and other functions.
- The above allows in transmitting a message across the plasma membrane to the interior of the cell called as **transmembrane signal transduction** which produces alterations in gene expression.

Thus many tumour promoters appear to act by alterations of gene expression but the precise mechanisms by which promoters transform the initiated cell to cancer cell still remains obscure.

Mechanisms of Chemical Carcinogenesis

As discussed above, chemical carcinogens may be:

- **Direct acting**
- **Procarcinogens**

(a) Direct Acting

A few chemical carcinogens like alkylating agents, e.g. cyclophosphamide, busulfan, etc. can interact directly with target molecules.

(b) Procarcinogens

Vast majority of the chemicals act as **procarcinogens.** Procarcinogens are not chemically reactive. In the body, after metabolism they are converted to **"ultimate carcinogens"** which are highly carcinogenic.

Procarcinogen → Proximate Carcinogen → Ultimate Carcinogen (highly carcinogenic)

Most of the ultimate carcinogens are *"electrophiles"*, i.e. the molecules are deficient in electrons and thus they can readily react with "nucleophilic electron rich" groups in DNA, RNA and various proteins.

Metabolic activation: The process by which a procarcinogen is converted in the body to highly active ultimate carcinogen by one or more enzyme catalysed reactions is called as metabolic activation.

Enzymes involved: The enzyme systems involved in metabolic activation are ***cytochrome P_{450} species*** present in the endoplasmic reticulum of cells. Recently ***a particular mono-oxygenase species cytochrome P_{498} (AHH-Aromatic hydrocarbon hydroxylase) has been incriminated in the metabolism of polycyclic aromatic hydrocarbons.***

Molecular Targets of Chemical Carcinogens

DNA is the primary and most important target of chemical carcinogens. Hence ***chemical carcinogens are mutagens.***

Damage to DNA can be:

- Binding covalently with DNA, (also to RNA and proteins).
- Interaction with the purine, pyrimidine and phosphodiester groups of DNA.
- Most common site of attack is guanine and addition of various carcinogens to the N_2, N_3, N_7, O_6 and O_8 atoms of this base has been observed.

Note: Critical to carcinogenesis are alterations in two types of genes: proto-oncogenes and recessive cancer genes. The sequences encoded by those families of genes regulate normal growth and differentiation.

III. VIRAL ONCOGENESIS

Oncogenic viruses may be either DNA viruses or RNA viruses. A variety of them are now known to cause cancer in animals and some have been implicated in human cancer.

Some important oncogenic viruses are listed in Table 46.1.

Oncogenes and Proto-oncogenes

- ***Oncogenes***

  They are genes whose products are associated with neoplastic transformation **(V-onc)**.
- ***Proto-oncogenes***

  They are normal cellular genes that affect growth and differentiation **(V-onc proto-oncogene)**.

Proto-oncogenes are converted into oncogenes before they can be carcinogenic by:

- Transduction into retroviruses (V-oncs) or
- Changes *in situ* that affect their expression and function thereby converting them into cellular oncogenes (c-oncs).

A. DNA VIRUSES

Many of the DNA viruses cause tumours in animals. ***Three DNA viruses have been established as causing human cancers, viz. EBV, HBV and HPV.***

1. Epstein-Barr Virus (EBV)

(a) Burkitt's Lymphoma

- Belongs to herpes family and produces Burkitt's lymphoma.
- It is a tumour of B lymphocytes that is consistently associated with a [t 8:14] translocation.
- It is endemic in Africa and patient's tumour cells carry EBV genome.
- EBV alone cannot cause the tumour. In patients with subtle or overt immune dysregulation, EBV causes sustained B-cells proliferation; they acquire additional mutations and sometimes translocation [t 8:14] and becomes tumourogenic.

(b) Nasopharyngeal Carcinoma

- Is endemic in southern China.
- EBV genome is found in all such tumour cells.

Table 46.1: Some important oncogenic viruses

Class	*Members*	*Associated tumours in humans*
(a) DNA viruses		
• *Papovavirus*	Polyomavirus, SV 40 virus, Human Papillomavirus (HPV)	—
		• Warts leading to skin cancers • Carinoma-*in-situ* of cervix leading to cancer cervix
• *Adenovirus*	Adenovirus 12, 18 and 31	—
• *Hepadnavirus*	Hepatitis B virus	• Liver carcinoma
• *Herpes virus*	Epstein-Barr virus (EBV)	• Burkitt's lymphoma • Immunoblastic lymphoma
	Herpes simplex type 2	• Nasopharyngeal carcinoma • Cancer of cervix
(b) RNA viruses		
• *Retrovirus (type C)*	Murine sarcoma virus Murine leukaemia virus Avian sarcoma and Leukaemia viruses	—
	Human T-cell leukaemia viruses I and II	Adult T-cells Leukaemia and lymphoma

2. **Hepatitis B Virus (HBV):** Hepatitis B virus infection is found to be closely associated with formation of liver cancer.

3. **Human Papillomavirus (HPV):**

   (a) *Multiple Warts*
   - Give rise to multiple warts (benign squamous papillomas).
   - In 30 per cent cases, some of the warts undergo malignant transformation.
   - Usually associated with depressed cell-mediated immunity.
   - Several types of HPV identified but types 1, 2, 4 and 7 are important.

   (b) *Cervical Cancer:*
   - Carcinoma *in situ* (precancerous condition) and squamous cell carcinoma of cervix have been found to be associated in HPV specially types 16 and 18 (in more than 90% cases).

Mechanism of Action of DNA Viruses

- DNA viruses form stable associations with host cell genome.
- Integrated virus is not able to complete its replicative cycle.
- Early genes, i.e. those viral genes that are transcribed early in viral lifecycle are important for transformation.
- Early genes produce specific proteins which can act on nucleus and derange normal growth regulation.
- ***Examples of Specific Proteins are:***
  - SV 40 and polyoma virus produce proteins called as **'T' antigens.**

    SV 40 produces 'T' and 't'. Polyoma virus produces 'T', mid-T, and small 't'.
- Adenoviruses produce proteins called EIA and EIB.
- Papillomaviruses produce E_5 and E_6 proteins.

How these proteins cause malignant transformation is still under investigation. T antigens are known to bind tightly to DNA and bring about alterations in gene expression.

B. RNA VIRUSES

- ***All oncogenic RNA viruses are retroviruses***
- They are of **two types** as given below

(a) *Acute Transforming Retroviruses*

These include type C viruses and cause rapid induction of tumours in animals. Transforming sequences of these viruses are 'Viral Oncogenes' (V-oncs).

(b) *Slow Transforming Retroviruses*

- These do not contain V-oncs and are replication-competent and cause transformation of the cells slowly.
- Mechanism of transformation is insertional mutagenesis (see below).

Human T-Cell Leukaemia Virus (HTLV)

- HTLV-1, found associated with human leukaemia/lymphoma.
- It is endemic in parts of Japan. Sporadic cases seen in other parts.
- Mechanism of HTLV-1 induced transformation is not clear. It neither has V-oncs, nor is it found integrated near a proto-oncogene.
- HTLV-1 contains a segment in its genome called **"tat"**. ***The proteins encoded by 'tat' gene are believed to be responsible for transformation.*** They affect the transcription of certain growth factors and receptors like IL-2 and IL-2R.

Recently researchers from university of Hong Kong identified **a new human protein called TAX IBP2 which ensures proper cell division of white blood cells.**

They claimed that abnormal division of white blood cells took place when a **foreign protein called TAX in the Human T-cell Leukaemia Virus (HTLV)** bound itself to the newly discovered protein TAX IBP2.

When the **foreign protein TAX merges with the human protein the function of TAX IBP2 is disrupted,** and leads to the **generation of abnormal number of chromosomes in daughter cells** and considered to be the driving force in the development of Leukaemia.

The findings of the research group might also help the way for the **design of new drugs** to stop the foreign protein TAX from disrupting the normal cell division. They are considering means to **block or inhibit the binding of TAX to the human protein TAX BP2.** ***If this is achieved then it will stop this foreign protein TAX from causing blood cancer leukaemia.***

Rous Sarcoma Virus—Biochemical Mechanisms of Oncogenesis

The genome of Rous sarcoma virus has been analysed and its oncogenic products along with mechanism of oncogenesis has been studied in detail.

Genome of this retrovirus contains 4 genes. The genes and their functions are shown below in the box.

genes	functions
• **Gag** :	Codes for group specific antigens of the virus
• **Pol** :	Codes for reverse transcriptase
• **Env** :	Codes for certain glycoproteins for viral envelope
• **Src** :	Sarcoma causing gene

The product of 'src' gene is a "protein tyrosine kinase", which has been found to be responsible for the cell transformation. The protein is produced in cytoplasm by inner cell membrane and called **PP60 src.** The specific biochemical mechanism involved is abnormal phosphorylations of a number of proteins.

The proteins that are phosphorylated by "protein tyrosine kinase" are:

1. *Certain glycolytic enzyme proteins*: They appear to be target proteins for phosphorylation. It is supported by the fact that transformed malignant cells show increased glycolysis as compared to normal cells.
2. *Vinculin:* A critical protein. Abnormal phosphorylation of this critical protein leads to transformation of the cells.

Phosphatidyl-inositol: Can bring about phosphorylation of phosphatidylinositol to phosphatidyl-inositol-mono and biphosphates. When phosphatidyl 4,5-bi-P is hydrolysed by phospholipase C, it produces two "second" messengers:

- ***Inositol-tri-P, and***
- ***Di-acyl glycerol (DAG)***
  - **(a) Inositol-tri-P:** Releases Ca^{++} from intracellular storage calcium from endoplasmic reticulum increasing the intracellular cytosolic Ca^{++} ions.
  - **(b) Di-acyl glycerol (DAG):** stimulates the activity of plasma membrane bound *protein-kinase C*, which in turn phosphorylates a number of proteins, some of which may be components of 'ion-pumps'.

Other Oncogenes of Retroviruses: In addition to Rous sarcoma virus, approximately 20 oncogenes of other retroviruses have been identified. Some of the oncogenes of retroviruses and their products are listed next page top (Table 46.2).

ONCOGENIC PRODUCTS

More than 50 per cent of these oncogenes produce ***protein tyrosine kinases***. Remainders produce various other proteins, viz, truncated PDGF, truncated EGF receptor, DNA binding protein, CSF-1, and GTP binding proteins, etc.

1. ***Protein tyrosine kinases:*** Several proto-oncogenes code for this enzyme which catalyses transfer of phosphate groups to the tyrosine residue of proteins. Proto-oncogenes with protein tyrosine kinase activity may be associated with cell membrane receptor or cytoplasmic proteins.
   - *(a) Receptor associated tyrosine kinases:* Receptors for EGF, PDGF, and CSF-1 are transmembrane proteins with tyrosine kinase activity associated with their cytoplasmic domain.

   At least **three oncogenes** codes for members of this receptor family, e.g.
   - (i) ***Erb-B*** codes for EGF receptor
   - (ii) ***Neu gene*** codes for a protein similar to EGF receptor
   - (iii) ***Fms*** codes for CSF-1 (myeloid growth factor)
   - *(b) Cytoplasmic tyrosine kinases:* C: Src is associated with inner cell membrane and encodes a phosphoprotein with tyrosine kinase activity, called PP60src. Several proteins like glycolytic enzymes, cytoskeletal protein (vinculin) are phosphorylated (see above).
2. ***GTP binding proteins:*** Ras oncogenes, most commonly found in human cancers, produce GTP binding proteins (see in "single point mutation").
3. ***Nuclear proteins:*** Products of myc, fos and myb oncogenes are "nuclear proteins". How these proteins modify cell growth in transformed cells is not very clear.

Table 46.2: Oncogenes and oncogenic products of retroviruses

Retroviruses	Corresponding Oncogenes	Oncogenic Product	Associated Human Cancers
• *Rous sarcoma virus*	src	Protein tyrosine kinase	—
• *Murine sarcoma virus*	ras	GTP-binding proteins with GTPase activity	Large variety of human cancers, e.g. lung, bladder, colon, leukaemias, neuroblastomas
• *Simian sarcoma virus*	sis	Truncated PDGF (B chain)	—
• *Feline sarcoma virus*	fms	CSF-1 receptor protein-tyrosine kinase	—
• *Avian myelocytoma virus*	myc	DNA-binding protein	• Burkitt's lymphoma • Small-cell carcinoma of lung • Neuroblastoma
• *Avian erythroblastosis virus*	erb-B	Truncated EGF-receptor	—
• *Abelson murine*	abl	Protein-tyrosine kinase	—

ACTIVATION OF PROTO-ONCOGENES TO ONCOGENES: BIOCHEMICAL MECHANISMS

At least **five mechanisms** are known that alter the structure/expression of proto-oncogenes and convert them to "oncogenes" which produces cancer.

1. **Single Point Mutation**
   - Proto-oncogene can be converted to oncogene by a single-point mutation, in that ***it differs from each other in one base only*** resulting in an amino acid substitution.
   - Murine retrovirus oncogene V-ras produces a protein called P_{21}, having molecular wt of 21,000. Functionally it is related to G-protein, which is acted upon by the enzyme *'GTPase'*, and modulates the activity of adenylate cyclase and thus the cyclic-AMP level in the cell.
   - Proto-oncogene of C-ras from normal human cells and C-ras of oncogene from a cancer differ only in one base, ***resulting to an amino acid substitution at position 12 of P_{21} protein.***
   - ***This mutational change brings about diminished activity of the enzyme GTPase.*** Lowered GTP-ase activity can result in chronic stimulation of adenylate cyclase; which normally is decreased when GDP is formed from GTP. The resulting increased activity of adenylate cyclase increases cyclic AMP level, which produces a number of effects on cellular metabolism. It affects the activities of various cyclic AMP-dependent protein kinases which brings about phosphorylations of various proteins. Such altered cellular metabolism favours transformation.
2. **Gene Amplification**
   - Gene amplifications can be seen in certain tumour cells. The amplified genes may be detected as a "homogeneously stained regions" on a specific chromosome.
   - Activation of certain "C-onc" is brought about by amplification.
   - Increased amount of products formed from certain oncogenes may help in progression of a tumour cell to malignant (cancerous) form.
   - ***N-myc amplification, 3 to 300 copies, seen in neuroblastomas.*** A strong correlation observed amongst 'N-myc' amplification, advanced stage and poor prognosis.
3. **Promoter Insertion**
   - It has been seen that certain retroviruses, e.g. Avian leukaemia virus do not have 'V-oncs'. They can cause cancer over a longer period of time as compared to retroviruses which possess V-oncs, called slow-transforming retroviruses (see above).
   - When such a retrovirus infects cells, a DNA copy, called cDNA is synthesised from the RNA genome of the virus by ***reverse transcriptase***. The cDNA, thus formed, is integrated in the host genome. ***The integrated double-stranded cDNA is called as a "provirus".***
   - ***The cDNA copies thus formed are usually flanked at both ends by 'long terminal repeats' (Jumping genes) which act as a 'promoter' resulting in transcription.***

   Example
   - Normal chicken B lymphocytes have "myc" gene on its chromosome which is inactive.
   - Infection of such a cell with Avian leukaemia virus (ALV) produces cDNA copy (Provirus) which becomes integrated.
   - The 'myc' gene is activated in an upstream, by its right hand long terminal repeat (LTR), which is a strong promoter and results in transcription of myc mRNA.
4. **Enhancer Insertion**

   In this mechanism, ***the cDNA copy (Provirus) gets inserted 'downstream' from the gene and hence cannot***

act as a promoter. Instead, a certain portion of proviral sequence acts as an "enhancer" element leading to activation of upstream gene and its transcription.

5. **Translocations**

Most of the tumour cells show chromosomal abnormalities. ***Translocation is the most commonly seen chromosomal abnormality, in which a piece of one chromosome is split off and then is joined to another chromosome (direct translocation).***

Reciprocal translocation: In this the second chromosome donates the piece to the first.

Examples

- ***Burkitt's lymphoma***, a fast growing cancer of human B lymphocytes is an example of "reciprocal translocation".
- Usually chromosomes 8 and 14, (t 8:14) are involved. ***In 90 per cent of the patients with Burkitt's lymphoma, there is an exchange of genetic material between chromosomes 8 and 14.*** Chromosome 14, in humans, bears the genes of the H-chain locus.
- In 10 per cent of the patients with Burkitt's lymphoma, there is exchange between chromosomes 8 and 2 or 22 (t 8:2 or t 8:22), which in the human genome carry the "kappa" (κ) and "lambda" (λ) immunoglobulin genes respectively.
- *Chronic granulocytic leukaemia:* In chronic granulocytic leukaemia, the ***Philadelphia chromosome*** involves chromosomes 9 and 22. It is an example of "chromosomal direct translocation". There is fusion of the gene with new genetic sequences (t 9:22). Translocation relocates the 'C-abl' gene from chromosome 9 to the 'bcr' locus on chromosome 22. ***The "C-abl-bcr" hybrid gene (Philadelphia chromosome) codes for a chimeric protein that exhibits the 'tyrosine kinase' activity.***

CANCER SUPPRESSOR GENES (GROWTH SUPPRESSOR GENES)

- *Antioncogenesis*

Recently it has been shown that genes other than oncogenes can play a role in etiology of certain types of cancer. These are called ***cancer suppressor genes (or growth suppressor genes or antioncogenes).*** Mechanism of action is quite different here. A loss or inactivation of such genes removes certain mechanisms of growth control.

Example

An important model for understanding is the tumour known as **retinoblastoma** which occurs in children. Retinoblasts are precursor cells of cones, the photoreceptor cells in retina. 40 per cent of retinoblastomas are familial; and the remaining are sporadic. ***Rb gene is located on chromosome 13q14 which is the cancer suppressor gene and exerts an inhibitory effect.*** Hence both copies of normal Rb gene has to be inactivated for the tumour to develop.

Two-hit hypothesis has been proposed as follows:

- Both normal alleles of the Rb locus must be inactivated (two hits) for the development of retinoblastoma.
- *In "familial" cases,* the children inherit one defective Rb gene and the other is normal. ***Retinoblastoma develops only when the normal Rb gene is lost in the retinoblasts as a result of mutation.***
- In sporadic cases, both normal Rb alleles are lost by somatic mutation.

Other examples: Loss of antioncogenes producing cancer has been incriminated also in the following:

- ***Wilms' tumour of kidney***
- ***Small cell carcinoma of lung (Oat-cell carcinoma)***
- ***One type of breast cancer.***

Note

Patients with familial retinoblastomas have been found to have increased risk of developing osteosarcomas and soft tissue tumours.

Mechanism of action of anti-oncogenes: Mechanism of action of Rb gene and other antioncogenes is not very clear. Their products are nuclear proteins and they may act as repressors of DNA synthesis and modulate gene expression.

A Cancer Causing Tumour Suppressor Gene Discovered

Recently two research teams have simultaneously discovered a gene that plays an important role in aggressive brain tumour like *Glioblastoma multiforme.* The gene is also incriminated in other types of cancers viz., breast, prostate and renal cancers and highly malignant skin tumour melanoma.

- The gene has been designated:
  - **P-TEN** *(Phosphatase and tension homolog)* by one group and
  - **MMAC 1** *(Mutated multiple advanced cancers)* by other group
- The gene appears to be located on **chromosome 10** and appears to be an important ***"tumour suppressor"*** and ***when mutated it allows the cells to grow out of control and become malignant.***
- Studies have shown that an area of genetic code corresponding to gene on chromosome 10 is lost when a glioma, a benign tumour of brain progresses into glioblastoma multiforme, a more aggressive and spreading form of the disease.

- It has been suggested that mutations in "MMAC 1"/ "P-TEN" gene play an important role in the aggressiveness of the tumour.

SUMMARY

Pathogenesis of Cancer and Mechanisms of *Action of Oncogenes*

- Cancer is a genetic disease; changes in the genome of somatic cells may be brought about by genetic factors.
- It can be acquired environmental factors, viz. radiations, chemicals, and viral oncogenes.
- Tumour cells can gain growth autonomy by either.
  1. Activation of growth-promoting antigens, or
  2. Loss of growth inhibitory 'cancer suppressor genes' (antioncogenes).
- Activated oncogenes may promote growth by various mechanisms:
  1. May encode membrane proteins critical to signal transduction and may act on key intracellular pathways involved in growth control.
  2. May encode growth factors, or
  3. May encode for growth factor receptors that are either defective or amplified.

GROWTH FACTORS

A variety of growth factors have recently been isolated and identified:

- Are usually polypeptides.
- Can initiate cell migration, differentiation and tissue remodelling.
- May be involved in various stages of wound healing.
- Can affect many different types of cells, viz. haemopoietic cells, epithelial tissues, nervous system and mesenchymal tissues.
- Play a major role in regulating differentiation of stem cells to form various types of mature cells.
- Are mitogenic to target cells.
- Products of several oncogenes are either growth factors or parts of receptors for growth factors.
- Some growth factors may be inhibitory to growth of certain cells, e.g. TGF-β (see below).
- Chronic exposure of target cells to increased amounts of a growth factor or to decreased amounts of a growth inhibitory factor may alter the balance of cellular growth (growth autonomy).

Types of Growth Factors

Some of the important polypeptide growth factors identified so far, along with their sources and functions are listed in Table 46.3.

Mechanism of Action of Growth Factors

Growth factors may act in the ***following 3 ways:***

*(a) **Endocrine action:*** Similar to hormone action. May be synthesised in one place and then carried by bloodstream to target cells where they exert their effects.

*(b) **Autocrine action:*** Synthesised and act on the same cells.

*(c) **Paracrine action:*** Synthesised in certain cells and secreted from them to affect the neighbouring cells.

Action of Growth Factors at Molecular Level

1. Most of the growth factors have high affinity protein receptors on the membranes of target cells.
2. Genes for receptors of EGF and IGF have been extensively studied. Structurally they are found to have short membrane spanning segments and external and cytoplasmic domains of varying lengths.
3. Most of the receptors, e.g. of PDGF, EGF, etc. have been found to exhibit *protein tyrosine kinase* activity which is located in cytoplasmic domain.
4. The kinase activity brings about autophosphorylation of the receptor protein and also phosphorylation of other proteins of target cells.
5. Growth factors interact with the specific membrane receptor and transmits the message across the plasma membrane to the interior of the cells by ***trans-membrane signal transduction,*** which finally affect one or more processes involved in mitosis of the cells.
6. Molecular event is similar to 'src' gene and has been studied with PDGF. Phospholipase c is stimulated in cells exposed to PDGF resulting in hydrolysis to form "second messengers", Inositol-tri-P (ITP) and Di-acyl-glycerol (DAG). IT-P releases intracellular bound calcium increasing cellular Ca^{++} ions. DAG stimulates *protein kinase c*; these events in turn phosphorylates number of proteins in target cells, affecting cell metabolism.
7. In addition, subsequent hydrolysis of Diacylglycerol (DAG) by *Phospholipase* A_2 releases arachidonic acid which results in increased production of PG's and leukotrienes which in turn exert their biological effects.
8. Growth factors like PDGF can bring about rapid activation of certain cellular "proto-oncogenes".

INTERFERONS (IFN)

Chemistry

Interferons are cellular glycoproteins produced as soon as virus infects the cells. Other inducers are inactivated viruses, microorganisms, fungal extracts, synthetic polynucleotides, etc. Molecular weight approximately 20,000 to 25,000.

Types

There are **three types** of interferons IFN-α, IFN-β and IFN-γ.

Mechanism of Action

- ***Direct inhibition of translation of viral mRNA:*** IFN induces an enzyme, called ***oligo-2′,5′-adenylate***

Table 46.3: Growth factors: Sources and functions

Name of growth factor	Source	Functions
1. Platelet derived growth factor (PDGF) *Nature* *• A highly cationic polypeptide* *• Composed of A and B chains (Dimer)*	• α-granules of platelets • Activated macrophages, endothelium and smooth muscles • Certain tumour cells	• Stimulates growth of mesenchymal and glial cells. • Causes migration and proliferation of fibroblasts, smooth muscle cells and monocytes. • Plays a role in normal wound healing *in vivo.*
2. Epidermal growth factor (EGF)	• Mouse salivary gland	• Stimulates growth of many epidermal and epithelial cells • Mitogenic for epithelial cells and fibroblasts.
3. Fibroblast growth factors (FGFs)	• Basic FGF is present in many organs and secreted by activated macrophages • Acidic FGF is found in neural tissue	• Binds to heparin and cause fibroblast proliferation and neovascularisation.
4. Nerve growth factor (NGF)	• Mouse salivary gland	• Tropic effect on sympathetic and certain sensory neurons,
5. Insulin-like growth factors (IGF-I and IGF-II) *Syn* *Somatomedin C and somatomedin A* *Note* *For differentiation refer Insulin*	• Serum	• Exert insulin-like effect on many cells, • Stimulates sulfation of cartilages, • Mitogenic for chondrocytes.
6. G-CSF	• Monocytes • Fibroblasts	• Stimulates growth of neutrophil colony formation, • Activates neutrophils.
7. GM-CSF	• T lymphocytes • Fibroblasts • Endothelial cells • Keratinocytes	• Activates granulocytes, monocytes, NK cells, • Stimulates haemopoietic progenitor cells of multiple lineages.
8. M-CSF	• Monocytes • Fibroblasts • Endothelial cells	• Activates monocytes, • Stimulates macrophage colony formation.
9. Interleukin-1 (IL-1)	• Conditioned media virtually any cell including monocytes, B-cells, keratinocytes, macrophages, endothelium and glial cells	• Stimulates T lymphocytes to produce IL-2 • Stimulates mesenchymal cells to produce CSFs • Chemotactic for fibroblasts and increases collagen synthesis • Stimulates systemic acute phase responses.
10. Interleukin-2 (IL-2)	• Activated T lymphocytes	• Stimulates growth of T lymphocytes, • Co-stimulates B-cells and Monocyte differentiation, • Activates and promotes growth of NK cells.
11. Interleukin-3 (IL-3)	• T lymphocytes • NK cells	• Multipotential haemopoietic cell growth factor.
12. Interleukin-4 (IL-4)	• T lymphocytes	• Mast cell growth factor, • Stimulate IgG and IgE production, • Inhibits some macrophage cell lines, • Stimulates fibroblasts to produce CSFs. • Co-stimulates proliferation of T and B lymphocytes and multiple lineage haemopoietic progenitor cells.
13. Interleukin-5 (IL-5)	• T lymphocytes	• Stimulates T lymphocytes, • Stimulates IgM and IgA production, • Stimulates eosinophil colony formation and enhances differentiation
14. Interleukin-6 (IL-6)	• Monocytes • T and B lymphocytes	• B and T lymphocytes stimulating activities, • Synergizes with IL-3 to stimulate very primitive haemopoietic blast cells, • Hepatocyte stimulating factor, • Activates IL-2 production and enhances cytotoxic T lymphocyte killing, • Supports of growth of pre-B lymphocytes.

Contd....

Table 46.3 contd....

Name of growth factor	Source	Functions
15. Interleukin-7 (IL-7)	• B lymphocytes • Bone marrow • Fibroblasts	• Supports growth of pre-B lymphocytes, • Co-stimulating activity on T lymphocytes.
16. Interleukin-8 (IL-8)	• Monocytes	• Neutrophil activating and chemotactic factor.
17. Interleukin-9 (IL-9)	• T lymphocytes	• Helper T lymphocyte growth factor, • Stimulates proliferation of erythromegakaryoblastic leukaemic cell line, • Stimulates proliferation and differentiation of early erythroid progenitors.
18. Interleukin-10 (IL-10)	• B lymphocytes	• Cytokine synthesis inhibitory factor (IL-2, IFN α-Interferon α).
19. Interleukin-11 (IL-11)	• Bone marrow • Fibroblasts	• Increases number of plaques forming B lymphocytes, • Synergizes with IL-3 to stimulate murine megakaryocytic production.
20. Interleukin-12 (IL-12)	—	• Stimulates NK cells.
21. EPO	• Renal non-tubular mesangial cells	• Stimulates erythroid precursors to proliferate and mature into erythrocytes.
22. Transforming growth factor α (TGF-α)	• Conditioned media or transformed or tumour cells	• Similar to EGF causes fibroblast proliferation.
*23. Transforming growth factor β *(TGF-β)*	• Kidneys • Platelets • T-cells • Endothelium and macrophages	• Inhibits growth of most cells except fibroblasts, • Inhibits collagen degradation, • Deactivates macrophages, • Favours fibrogenesis in that stimulates fibroblast chemotaxis and collagen production.

* **Note**
Originally thought to be a +ve growth factor, now it is known as "inhibitory factor"

synthetase, which activates a latent "endonuclease" (RNAse L) which degrades the viral mRNA.

- ***Inhibition of protein synthesis***: IFN can inactivate IF-2 required for protein synthesis and removes cyclic AMP required for genetic expression for protein synthesis.

Functions

- ***Antiviral:*** Inhibits the multiplication of viruses. They are not virus-specific but cell-specific.
- ***Anticancerous:*** Anti-proliferative action.

CLINICAL ASPECT

Various Interferons (IFN), ***IFN-α has been used recently most extensively*** in clinical practice. Utilising recombinant DNA technology, this agent is now produced in large scale.

1. ***Used in Haematological Malignancies, viz.***
   - Chronic myeloid leukaemia (CML)
   - Hairy cell leukaemia (HCL)
   - Multiple myeloma
   - Low grade non-Hodgkin's lymphomas.
2. ***In Myeloproliferative Disorders (MPD)***
   Other than CML, IFN-α appears to be highly effective.
3. It has also been found to be highly effective in:
   - Regressing haemangiomas of infants and young adults by its anti-angiogenic effect.
4. Among the solid tumours:
   - Kaposi's sarcoma,
   - Renal cell carcinoma, and
   - Breast cancer have been found to be sensitive to IFN-α.
5. Also it has been used in connective tissue disorders and chronic neurological diseases.

Naturally Occurring Anticancer Substances

It is worthwhile to note that there are number of substances occurring in nature have anticancer activity. Some of them are:

1. ***Vitamin A and β-carotene:*** (antioxidant action, mops up free radicals)
2. ***Vitamin E*** (Tocopherols)—antioxidant
3. ***Lycopene:*** A carotenoid present in ripe red tomatoes (antioxidant action)
4. ***Ascorbic acid*** vitamin C: (antioxidant)
5. ***Selenium:*** A trace element has anticancer activity (Refer Chapter on Metabolism of Minerals and Trace Elements)
6. ***Zinc*** (a trace element)
7. ***Quercetin*** a flavonoid present in apple which is anticancer, shown to prevent growth of prostatic cancer cells.
8. ***Glucosinolates:*** Present in bitter **Brussel sprouts.** It contains:
   - ***Sinigrin:*** Suppresses development of precancerous cells
   - ***Glucoraphanin:*** It breaks down into a compound called **sulphoraphane** which neutralises carcinogens.

9. *Epigallocatectin Gallate* (EGCg)

EGCg is a compound ***found in green tea leaves*** also to a lesser extent in black tea leaves. This compound has been shown to inhibit the activity of an enzyme called **tumor-associated quinol oxidase (tNox),** which is an overactive form of an enzyme known as **NOX:** The 'NOX' enzyme is found on the surface of cells, and plays a key role in growth of both normal and cancerous cells.

Normal cells produce the "NOX" enzyme only when hormonal signals prompt the cells to divide. But cancerous cells appear to be able to produce the abnormal form "tNOX", all the time.

The "tNOX" protein has been found on many types of cancer cells including the breasts, prostate and colonic cancer cells.

Drugs that inhibit 'tNOX' have been shown to block cancer cell growth in the laboratory.

The compound 'EGCg' is present in ten to hundred times more in green tea leaves as compared to black tea leaves. Thus ***tea has been found to have anticancer activity.***

Some Drugs that are Used in Cancer Chemotherapy

Various drugs that have been used in chemotherapy of cancer are classified as follows:

- *Polyfunctional alkylating agents:*
  - Nitrogen mustard, busulfan, chlorambucil, triethylene-melamine.
- *Hormones*
  - Sex hormones: Oestrogens
  - Corticosteroids: Prednisone
- *Antimetabolites*
  - Folate antagonist: Methotrexate
  - Purine antagonist: Mercaptopurine
  - Pyrimidine antagonist: 5-fluorouracil
- *Antibiotics*
  - Actinomycin D
  - Doxorubicin
- *Miscellaneous agents*
  - Vinblastine
  - Vincristine
  - Cisplatin
  - Retine—a nontoxic anticancer agent, a natural constituent of body cells.
- *Enzyme:* L-Asparaginase

Resistance of cancer cells to drugs and multidrug resistance (MDR)

The drugs are effective initially to cancer chemotherapy, but after several months of treatment, the drugs become ineffective by the mechanisms developed by the tumour cells producing acquired resistance to drugs. This is called drug resistance.

Multidrug Resistance (MDR)

In multidrug resistance (MDR), the resistance is not only to a particular drug but also to other structurally unrelated anti-cancer agents.

Role of P-glycoprotein in MDR: Phosphorylated glycoprotein (P-glycoprotein) present in the plasma membrane. It contains 1280 amino acids and has a molecular weight of 12,000. It acts as energy-dependent efflux pump expelling a variety of drugs and thus mediating MDR.

Telomeres and Telomerase: Role in Ageing and Cancer

A. Telomeres

The ends of each chromosome contain structures called telomeres. They are composed of DNA and proteins that are located at the ends of chromosomes of lower and higher eukaryotes. They consist of short repeat TG-rich sequences.

In humans, telomeres consist of many (1000 or more) arrays of **TTAGGG** repeats at the terminals of the 3' ending strands. These repeats are maintained in germ-line cells by the action of a special enzyme called *Telomerase* (see below).

Significance and Importance of Telomeres

Telomeres confer stability on the ends of chromosomes and are necessary because the RNA Primers at the 5'-end of a completed lagging strand cannot be replaced with DNA since the primer would have no place to bind. This would inevitably ***lead to shortening of the chromosomes*** at each replication of DNA with resultant ***loss of important genes.***

B. Telomerase

Telomerase or called as ***"Telomere terminal transferase" a 'reverse transcriptase' an enzyme responsible for telomere synthesis and thus for maintaining the length of telomere.*** Telomerase is a ribonucleoprotein and in humans contains an RNA molecule that has one segment that is complementary to the **"TTAGGG"** repeat. This is used as a template for the replication of the telomeric sequences using the appropriate deoxynucleotide triphosphates.

C. Clinical Significance

1. Role in Ageing or Senescence

Telomeres were observed to shorten during senescence (ageing) and lack "telomerase" activity, which leads to chromosome instability and cell death.

The above is supported by the fact by ***in vitro*** studies of cell culture, Telomerase were observed to shorten in their length during ageing of primary culture of cells and also it was

SECTION SIX

found that ageing cultured primary cells lack "Telomerase" activity. At the same time, studies have shown when cells are "immortalised" in culture, the lengths of their 'telomeres' are stabilised and they exhibited "telomerase" activity.

2. Role in Cancer

It was found that ***many tumour cells had shorter lengths of 'telomeres' than normal*** and that "telomerase" activity was present in most tumour cells but lacking in most normal cells. These findings suggested that telomeres and telomerase might prove to be useful targets for anticancer therapy.

Recent studies have shown that "telomerase" enzyme is active in cells normal or malignant with ***high Proliferation rate.*** Telomerase null mice (i.e. mice in which the gene encoding telomerase is knocked out) were able to form tumours indicating that in mice the enzyme is *not* necessary for cancer production.

There is now evidence that telomerase may be active in some tumours and *not* in others. Thus it may prove to be useful therapeutic target in a spectrum of carefully selected tumours with telomerase activity. ***Measurement of telomerase activity may help in the staging of certain cancers for which no good "oncogenic markers" are available.***

II. Apoptosis: Biochemistry and Role in Carcinogenesis

The term apoptosis was not known earlier. The term apoptosis was first coined in 1972. Earlier apoptotic cells were known by a variety of names found in various diseases, e.g. ***council man bodies*** in viral hepatitis, ***civatte bodies*** in epidermis, and ***tingible bodies*** in macrophages of reactive lymph nodes.

Apoptosis is a co-ordinated and apparently internally programmed process that mediates the death of cells in a variety of biologically significant situations viz. cytotoxic T-cell killing, atrophy in response to withdrawal of endocrine and other stimuli, etc. ***Identical changes are observed in normal and tumour tissues exposed to low or moderate doses of chemotherapeutic agents or to ionising radiations.***

Structural and Morphological Changes in Apoptosis

- Show dramatic shrinkage of cell volume, accompanied by dilatation of endoplasmic reticulum and convolution of the plasma membrane.
- The cells break up into membrane bound spherical bodies containing structurally normal but compacted organelles.
- Nucleus undergoes profound chromatin condensation round the nuclear periphery.
- Nucleolus falls apart, nuclear pores appear to aggregate in those areas of membrane which are free from condensed chromatin.

How does Apoptosis Differ from Necrosis of Cell?

The changes of apoptosis grossly contrast with those of necrosis the disruption of cells observed in severe hypoxia, or in acute high dose toxicological exposures. In necrosis, the injured cells swell and their plasma membranes rupture ***to release proinflammatory materials*** from interior of cells to the extracellular spaces.

Fate of Apoptotic Cells

The ultimate fate of apoptotic cells in tissues varies:

- Some of the apoptotic cells are rapidly phagocytosed.
- Some lose contact with their neighbours and basement membrane and hence float in adjacent spaces.
- ***There is no acute inflammation associated with apoptosis,*** hence the speed with which apoptotic cells are removed, major cell losses can be effected within tissues with minimal disturbance of over all tissue structure.

Biochemical and Molecular Changes in Apoptosis

Many of the biochemical processes responsible for the structural and morphological changes of apoptosis are now understood.

(a) Biochemical Changes

- Several cytoskeletal proteins and the nuclear lamins undergo site specific proteolysis. Activation of ***trans-glutaminase*** results in protein-protein cross-linking.
- ***Nuclease*** activation cleaves first large and then small oligonucleosomal fragments of chromatin.
- On the cell membrane, ***phosphatidyl serine*** appears on the outer as well as the inner surface, and previously hidden residues including charged amino sugars become exposed on the cell surface.
- Mitochondrial membrane potential alters sometime before the structural changes begin.
- No satisfactory explanation is yet available for the rapid cell shrinkage.

(b) The Effector Pathway of Apoptosis

1. *Role of ICE Family Proteases*

   A family of cysteine proteases exhibit preference for cleavage adjacent to aspartate residue, mammalian cells contain about 10 (ten) distinct members of which the earliest one found is ***Interleukin-1-β - converting enzyme (ICE).*** This being an extracellular cytokine is unlikely ***to play a major role in activation of apoptosis.*** Inhibitors of ICE family proteases do inhibit the chromatin cleavage of apoptosis.

   The nuclear protein ***Poly ADP-ribose Polymerase,*** a high abundance nuclear protein that contributes to recognition and repair of DNA double stranded breaks—is consistently cleaved and inactivated early in apoptosis by the **ICE family protease CPP 32,** perhaps preventing wasteful and ultimately abortive DNA repair reactions.

   Most cells appear to be endowed with at least some members of the *"ICE-family proteases"*. The initiation of apoptosis must, therefore, be critically dependant upon the initial activation of the proteolytic cascade and the set point of intracellular balance between lethal and survival factors.

2. *Role of Cytokines*

   Cytokines play an important role in the signalling of cell death; it triggers the ICE family protease and activation and setting of proteolytic cascade.

   - **"CD-95/apo-1/fas"** is a transmembrane receptor belonging to the receptor family **"CD-40/TNF"** recep-tor. Historically, it was the first receptor discovered to be associated with apoptosis.
   - "Fas" signalling is now considered to be one of the key events in CTL killing. On finding the "fas" ligand, a cytokine of the tumour necrosis factor (TNF) family trimerises and initiates a chain of protein-protein interactions:

i. First between its own cytosolic domain and a cytosolic protein called **FADD** ***(Protein containing fas activated death domain)***.
ii. Secondly, between **FADD** and a newly discovered *"ICE family protease"*, called **FLICE** ***(FADD - like ICE)*** so triggering the ICE family protease cascade.

3. ***Role of Ceramide***
Ceramide has been incriminated as signal for activation of apoptosis that is released from membrane lipids on digestion by "sphingomye linases". The enzyme sphingomyelinase is activated within seconds of plasma membrane damage, specially in cases of *ionising radiations.*

It has been shown that ceramide production is negatively regulated by ***Protein kinase C***, suggesting that cells with high ***Protein kinase C*** activity might be preferentially protected against some types of injury.

Ceramide production and sphingomyelinase activation have been shown to be one result of activation of CD-95/fas and the TNF receptor.

4. ***Role of Cytotoxic T-cells***
Cytotoxic T-cells also possess granules that release a variety of effector proteins, e.g.
i. ***Perforin:*** It permeabilises the target cell membrane
ii. ***Granzyme B:*** It penetrates the target cell and is activated by perforin attack.

Granzyme B is not a cysteine protease, although it shares the property of effecting cleavage after Asp. residues, but it has the capacity to activate ICE and CPP 32, and thus can trigger the effector events dependant on the cysteine protease cascade in the target cell.

5. ***Role of P^{53} and interferon-response factor-1 (IRF-1)***
P^{53} and IRF-1 are both ***oncosuppressor genes*** that signal DNA injury. ***Both oncosuppressors can induce apoptosis in the appropriate circumstances.*** Induction of apoptosis by P^{53} is observed most consistently in cells committed to the cell cycle. Thus, there is a complex interrelationship between P^{53} changes, apoptosis, and the expression of genes (such as rb-1) that remove cells from cycle. The effects of P^{53} and IRF-1 also appear to be interdependent.

Role of Apoptosis in Carcinogenesis

Studies have suggested that abrogation of apoptosis is a general mechanism in carcinogenesis.

Many non-viral carcinogenic agents also work through survival and replication of cells bearing mutations. In at least a proportion of cases, these mutations arise through double-strand breaks, e.g. after ionising radiation or single-strand break (after UV rays) or bulky adducts after exposure to many chemical carcinogens. Although mechanisms exist to repair all these lesions, but there is at least the theoretical possibility that the ***damaged DNA may be a sufficient stimulus for the initiation of apoptosis. Failure in apoptosis may permit clonal out growth of cells with inappropriate recombination events resulting to tumour.***

The following points favour the above hypothesis:

- There is interesting circumstantial evidence that the silencing of P^{53} oncosuppressor gene, leads to the production of tumours in cell populations in which there is a high level of apoptosis through deficiency of "rb". Thus, in tissues of genetically modified animals bearing tissue specific inactivation of "rb" there is a high level of apoptosis, but in those animals ***simultaneous inactivation of P^{53} in these tissues renders them prone to cancinogenesis.***
- In a skin carcinogenesis model in mice, in which chemicals were applied directly to the skin of mice producing both benign and malignant tumours, the animals without P^{53} showed accelerated development of carcinomas relative to wild type mice.
- In another animal model, blockade of apoptosis in the Lymphoid system produced Lymphomas of low-grade malignancy, whilst concurrent blockade of P^{53}, in the absence of exposure to carcinogens, produced much more aggressive tumours.

There is now good evidence that in the lymphoid lineage exposure to ionising radiation in the *absence* of P^{53} permits the survival of increased numbers of mutated cells, whereas exposure to similar doses of ionising radiation in the *presence* of P^{53} permits the survival of an immeasurably *small* proportion of the population.

Mutation incidence in pre-B cells derived from the bone marrow of P^{53} null animals measured by the production of hprt-deficient colonies *in vitro* was found to be at least 10-fold higher than in the un-irradiated controls.

The above entirely compatible with the high incidence of Lymphoid tumours in P^{53} null animals in general.

Thus it is seen that DNA damage, if not repaired, may stimulate and initiate apoptosis, the degree of which vary from tissue to tissue. If oncogenic suppressor gene P^{53} is simultaneously absent the mutagenic effect progresses to cancer.

III. Biochemistry of Metastasis

Definition of metastasis: Metastasis is the spread of cancer cells from the primary site of origin to other tissues, both neighbouring and distant, where they grow as the secondary tumours.

INTRODUCTION

1. Benign tumours can grow very rapidly and attain big sizes and may be sometimes life-threatening but they do not metastasize.
2. It is the malignant tumours, cancerous ones, invade surrounding tissues and send out cells to begin new tumours at distant sites. The spread may be blood-borne/or through lymphatics.
3. This ***colonisation at distant sites is metastasis*** and is the major cause of death from human malignancies.
4. Tumour cells must attach to degrade and penetrate the "extracellular matrix" (ECM) at several steps of metastasis. **Thus metastasis biochemically is a multi-step process.**
5. Approximately 50 per cent of patients who develop malignant tumours can be cured with various therapies, viz. surgical removal, radiation therapy and chemotherapy. Of the remaining 50 per cent , majority die because of metastasis. **Hence, in a real sense, if metastasis could be controlled, cancer could be controlled and for the most part, cured.**

Biochemical Basis of Metastasis

Knowledge regarding biochemical basis of metastasis was quite limited. But the past few years have brought a dramatic increase in biochemical understanding of how metastasis occurs. Current consensus opinion is that metastasis is an **active process of invasion**.

Metastasis has been shown to require:

- *Specific surface receptors*
- *Requires enzymes*
- *The process uses energy and*
- *Requires protein synthesis.*

Note

Theoretically, if any of the above is blocked, invasion can be prevented.

As stated above, metastasis is not a passive phenomenon. ***It is an active process.*** The phenomenon of metastasis involves:

- A metastatic cell has to penetrate the extracellular matrix (ECM) that surrounds the tumour.
- Travels through the tissue till it reaches a blood vessel/or a lymphatic.
- In case of blood borne metastasis, the tumour cell then attaches to the blood vessel wall, dissolves a portion of the wall and propels itself through into the circulating blood.
- Metastatic cells often travel in the circulation as small clumps of cells, called ***emboli.***
- At a distant site, the tumour cell again re-attaches to the blood vessel wall and repeats the process, travelling as much as two or three cell diameters into the invaded tissue before it settles down and begins to form a new tumour.

INTERACTION OF INVADING CANCER CELL WITH EXTRACELLULAR MATRIX

Composition of Extracellular Matrix

Extracellular matrix (ECM) can be divided into **two major categories:**

- ***Basement membrane (BM)***
- ***Interstitial connective tissue (ICT)***

Important Constituents of ECM

- ***Collagen:*** Basement membrane contains type IV collagen and interstitial connective tissues type I and type III collagen.
- ***Adhesion-promoting proteins:*** Basement membrane contains **laminin** and interstitial connective tissue **fibronectin**.
  - Both laminin and fibronectin are large multi-functional molecules that can bind to other ECM components such as collagen, proteoglycans and to cells.
  - Attachment of cells to laminin and fibronectin is brought about by distinct and specific cell surface receptors.

Interaction with basement membrane: Basement membrane (BM) is the first tough elastic barrier that surrounds both tissues and blood vessels. Hence, an invading cancer cell must pass this barrier several times, in order to establish metastatic colonies in distant tissues.

Stages

The above interaction of cancer cell with BM can be considered arbitrarily under **three steps.**

I. ***Step 1:*** Attachment of the invading metastatic cell to basement membrane (BM).

II. ***Step 2:*** Dissolution of the basement membrane (BM), so that the cell can pass through it.

III. ***Step 3:*** Migration of tumour cells. It has been shown that specific biochemicals are required for the tumour cell to complete the process.

Note

To the histopathologist an intact BM is an important criterion to differentiate benign and malignant tumours.

Benign tumours are always surrounded by an intact BM whereas in invading malignant tumours, the membrane becomes thin and broken and often entirely lacking.

I. Step 1

- The tumour cell binds to one of the membrane's glycoproteins, a cross-shaped molecule called as **laminin** (see above).
- Binding sites for laminin on the surfaces of certain types of cancer cells have been demonstrated.
- Laminin appears to serve as a bridge between receptors on the surface of the invading cancer cell and the BM itself.

Note

Monoclonal antibodies, if prepared against the receptor protein, can be used to block these receptors on metastatic cells. When the antibodies bind to receptors of metastatic cells, the cells will not be able to bind to laminin and metastasis can be prevented.

II. Step 2

Once the tumour cell is attached to laminin, ***the invasive tumour cell secretes certain proteolytic enzymes that degrade the BM.***

Several such proteolytic enzymes have been incriminated:

1. **Collagenases:** A collagen degrading enzyme that acts specifically on type IV collagen, the principal structural component of membrane has been isolated from highly metastatic cells.

Properties

- It is a ***metalloenzyme,***
- Secreted as zymogen latent form and is clipped to form the "active" enzyme by a second enzyme ***cathepsin B.***
- Active enzyme has a molecular wt of 60,000.

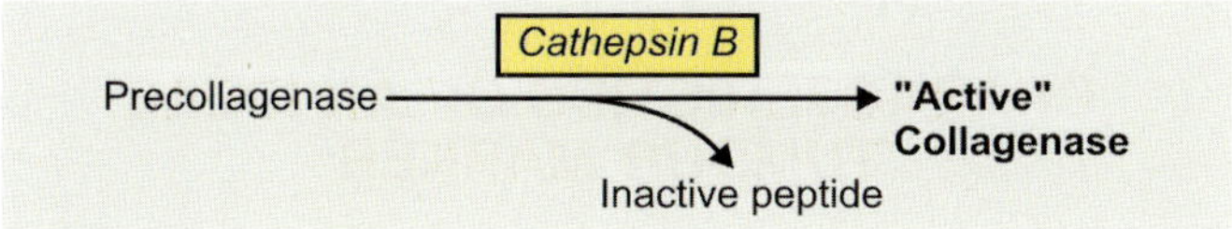

Note: Type IV collagenases cleave BM collagen, whereas "interstitial collagenases" cleave type I and type III collagen.

2. **Heparanase**

   Many metastatic cells also produce an enzyme called ***"heparanase" that degrades heparan SO_4, the predominant proteoglycan of the basement membrane.*** Many other components of the membrane attach to heparan SO_4, including laminin and fibronectin.

Note

- Heparanase enzyme is also produced by monocytes. But it has been claimed that ***heparanase produced by malignant metastatic cells is slightly different than the comparable enzyme produced by monocytes and even normal cells.***
- The enzyme produced by tumour cells appears to clip the heparan SO_4 chain at a slightly different position and produce different fragments than the enzyme produced by non-malignant cells.

3. **Cathepsin B:** A *lysosomal protease.*
   - ***Cathepsin B activates "Latent" collagenases.*** It clips type IV collagenase to its 60,000 molecular weight 'active' form.
   - This enzyme has been found in metastatic cancer cells in high concentration.
4. **Plasmin:** Degrades several non-collagenous extracellular matrix proteins.

III. Step 3

Factors that favour migration of tumour cells in the passage created by the degradation of EC matrix including BM are not well understood.

Implicated in this process are:

- ***Autocrine motility factors called as migration factors.***
- Tumour cells induced degradation of the interstitial matrix produces fragments of ECM that are attractive to the tumour cells and cause it to move forward.

Summary

1. Process of metastasis is ***selective.***
2. It is not a passive diffusion, but an ***active process*** and requires specific receptors, enzymes, protein synthesis and energy.
3. Not every cell in a malignant tumour is capable of forming metastasis.
4. The classic oncogenes cause tumours to grow, but they do not necessarily cause metastasis and invasion.
5. It has been postulated that there can be new special types of genes different from the classic oncogenes.
6. Ability to metastasise can be transformed genetically to cells.
7. The vast majority of metastatic cells that enter the bloodstream are destroyed "en route". ***Natural killer cells (NKC), a part of host's immune system have been claimed to be reasonably efficient in destroying metastatic cells in bloodstream.***
8. Natural killer cells (NKC) are also present in tissues, but they are much less efficient at destroying tumour cells outside the bloodstream.
9. It is claimed that cells populating a metastatic colony are more capable of metastasising than the cells populating the parent tumour.
10. **Monoclonal antibodies if produced can be used to block the receptors on metastatic cells so that they can not bind to "Laminin" and "fibronectin" and thus can prevent metastasis.**

IV. Oncogenic Markers or Tumour Markers

WHAT ARE TUMOUR MARKERS?

Definition: Tumour "markers" are defined as a biochemical substance (e.g. hormone, enzymes, or proteins) synthesized and released by cancer cells or produced by the host in response to cancerous substance and are used to monitor or identify the presence of a cancerous growth.

Sites: Tumour markers may be present in

- Blood circulation
- Body cavity fluids
- Cell membranes
- Cell cytoplasm

Tumour markers are different from substances produced by normal cells, in quantity and quality.

Methods for Detection:

1. ***Immunohistological and immunocytological tests are used*** to detect those tumour markers which are present only on cell-membranes and cytoplasm of cells and ***not in blood circulation.***

Examples

- Immunofluorescence
- Immunoperoxidase
- Monoclonal antibody technology

2. ***Biochemical methods are used for measuring tumour markers found in the blood circulation.***

Examples

- Radioimmunoassay (RIA)
- Enzyme-immune assay
- Immunochemical reactions

Clinical Uses of Tumour Markers

Ideally tumour markers have following **six** potential uses in cancer patient care. They are:

- For screening specially in asymptomatic population.
- For diagnosis in asymptomatic patients.
- As a ***prognostic predictor.***
- As an adjunct in clinical staging of the cancerous condition.
- ***For monitoring during treatment*** of the patients.
- And for early detection for relapse/recurrence of the cancerous process.

Although it seems unlikely that an ideal tumour marker will be identified for every cancer, but several workable 'markers' are available and can be used. Increasing knowledge about the capabilities and limitations of existing tumour markers will enable the oncologist to use them judiciously in cancer patient care and treatment.

TYPES OF TUMOUR MARKERS

Two types of tumour antigens have been described:

1. ***Tumour-specific antigens***
2. ***Tumour-associated antigens***

1. Tumour-Specific Antigens

- These are a direct product of oncogenesis induced by an oncogene (viral), radiation, chemical carcinogen or an unknown risk factor.
- Oncogenesis causes abnormalities of genetic information available to the cancer cells, which then subsequently synthesises **neoantigens** specific to cancer cells.
- They play an important role in clinical oncology.

2. Tumour-Associated Antigens

- Also called as **oncofoetal** proteins/antigens.
- Shown to exist in both in embryo-foetal tissues and cancer cells.
- These are produced in large quantities in foetal life and released in foetal circulation. ***After birth,*** these oncofoetal antigens disappear from blood circulation and may be present in trace amounts in normal healthy adults.
- ***With the onset of malignancy in adult life, the synthesis of oncofoetal antigens in foetal life which was suppressed in adult life, is again*** **reactivated** ***with malignant transformation of cells and reappears in cancer cells and in blood circulation (retrogenetic expression theory).***

Examples of such oncofoetal antigens are:

- **CEA (carcinoembryonic antigen)**
- **AFP (Alpha-fetoprotein)**

CHARACTERISTICS OF AN IDEAL TUMOUR MARKER

An ideal tumour marker is yet to be found out. ***Ideal tumour marker must satisfy the following criteria:***

(a) Analytical Criteria

- Should have high ***sensitivity.*** The test method should be sensitive to measure very low concentrations.
- Should have high ***specificity.*** It should measure a particular tumour marker only and no other substances should interfere.
- Should have high ***accuracy.***
- Should have high ***precision.***
- Method should be ***simple*** and easy to measure.
- ***Should not be very costly.***

(b) Clinical Criteria

- Should be ***disease-sensitive.***
  - It must be positive in all patients with particular cancer.
  - No false-negative results.
  - Should show increased level in presence of micro-metastasis.
  - Should be able to detect relapse/and recurrences.
- Should have high ***disease-specificity.***
  - Should not be detectable in normal healthy people.
  - Should be associated only with a particular cancer.
  - Should not show increase with other tumours.
  - Should not show rise in benign tumours and other diseases.
  - No false-positives.
- Should be stable and should not show wide fluctuations.
- Should correlate well with the cancerous process, i.e. its extent and the volume of the tumour.
- Should correlate well with cure rate.
- Should prognosticate the 'high risk' cancer patients from "lower risk".
- Should be able to detect relapse/recurrence of the cancer.

CLASSIFICATION OF TUMOUR MARKERS

Quite a large number of 'tumour markers' have been used in cancer. Their list is quite big and ***no universally accepted classification exists.***

Tumour markers which have been found clinically useful in cancer patients may be grouped as follows:

I. Tumour Associated Antigens (Oncofoetal Antigens)

(a)
- Carcinoembryonic antigen (CEA)
- α-fetoprotein (AFP)

(b) Other antigens
- Tissue polypeptide antigen (TPA)
- Pancreatic oncofoetal antigen (POA)
- Colonspecific antigen
- Beta oncofoetal antigen
- Tennessee antigen (TENAGEN)

II. Carbohydrate Antigens
- Detected by monoclonal antibody.
- More organ and tumour-specific

  (a)
  - CA-125
  - CA-15-3
  - CA-19-9
  - CA 72-4 (TAG-72)
  - CA 50

  (b) Others
  - Mammary serum antigen (MSA)
  - MAM-6
  - Mucin like carcinoma associated antigen (MCA)

III. Pregnancy Associated Antigens

(a)
- Human chorionic gonadotropin-β-subunits (β-HCG).
- Placental like alkaline phosphatase (Regan isoenzyme-PLAP)

(b) Other antigens
- Pregnancy specific glycoprotein (SP-1).
- Human placental lactogen (HPL).
- α_2 pregnancy associated globulin (PAG).
- Other placental proteins.
- Sex hormone binding globulin (SHBG).
- Steroid binding β-globulin (SBβG).

IV. Hormones Used as Tumour Markers

(a)
- Parathormone (PTH)
- Calcitonin
- Growth hormone (GH)
- ACTH

(b) Other hormones: Insulin, Glucagon, Catecholamines, Serotonin

V. Enzymes and Isoenzymes—Used as Tumour Markers:

(a) Enzymes synthesised by tumour tissue:
- Prostatic Acid Phosphatase (PAP) and Prostate specific antigen (PSA).
- Neuron specific enolase (NSE).
- LDH isoenzymes
- CK isoenzymes
- Glycosyl transferases II isoenzymes (GT II).

(b) Enzymes derived from organ tissues with metastatic carcinoma:

1. From Bone
- Bone isoenzymes of ALP (osteoblastic)

2. From Liver
- Liver isoenzyme of ALP (from liver cells)
- Gamma-glutamyl transferase (GGT)
- 5'-Nucleotidase

3. From Prostate
- Acid phosphatase (ACP)

VI. Miscellaneous Tumour Markers
- Sialic acid concentration in serum
- Polyamines
- Monoclonal immunoglobulins
- Steroid receptors
  - Oestrogen receptors
  - Progesterone receptors
- Cellular markers
  - T lymphocytes
  - B lymphocytes
- α_1-antitrypsin (α_1-AT)

CLINICAL USEFULNESS OF DIFFERENT TUMOUR MARKERS

It is not possible to go into indepth discussion for all-tumour markers noted above. Discussion will be done on those biochemical tumour markers which are commonly done in hospitals, viz. CEA, AFP, and β-HCG. Other tumour markers will be discussed briefly.

I. COMMONLY USED TUMOUR MARKERS

A. Carcinoembryonic Antigen (CEA)

CEA is one of the oncofoetal antigens used most frequently and widely as a tumour marker in clinical oncology. It was originally described by **Gold** and **Freedman** as a tumour specific antigen present only in cancer cells, in the circulation of patients with gastrointestinal malignancy and in the normal epithelial cells of foetal GI tract, hence it was named as CEA ***because of its presence in both carcinoma and embryonic tissue.*** It was discovered in 1965 by raising antiserum against a colon cancer.

Properties of CEA and Chemical Composition

- It is a ***glycoprotein.***
- Molecular weight varies from 150,000 to 300,000 (average 185,000).

- *Protein Part*
  - ***A single polypeptide chain (monomeric unit)*** consisting of 30 aa. with lysine at N-terminus.
  - By EM, it appears as a twisted rod.
  - Protein content is 46 to 75 per cent .
- *Carbohydrate Component*
  - Carbohydrates surround the protein and constitutes 45 to 57 per cent . On analysis of carbohydrates, it is found to contain fucose, mannose and galactose.
  - N-acetyl galactosamine is low whereas large amount of N-acetyl glucosamine is present.
  - Sialic acid varies significantly.

PHYSIOLOGY AND METABOLISM

1. *Sites*: CEA is chiefly present in:
   - ***Endodermally derived tissues, viz. GI mucosa, lungs and pancreas.***
   - Also may be in nonendodermally derived tissues (?), conclusive evidences lacking.

   It has been detected in GI tract of foetuses as early as three months of gestation. Also found in embryonic liver, pancreas and lungs. CEA has been detected in free brush border of normal mucosal cells and also in cytoplasm of colonic carcinoma cells.
2. ***Metabolism:*** Not known exactly. CEA is probably broken down in liver. It disappears from circulation in 3 to 4 weeks after removal of CEA-producing tumour.

CLINICAL USES AND REMARKS

1. CEA has been reported to be ***most useful as tumour marker in colorectal Cancer.***
2. It is elevated also in other malignancies. Found to be useful in:
   - Breast cancer
   - Bronchogenic carcinoma of lung specially small cell carcinoma of lungs (SCCL)

- Other malignancies where the value is raised are:
  - Pancreatic carcinoma
  - Gastric carcinoma
  - Cancer of urinary bladder
  - Prostatic cancer, neuroblastomas, ovarian cancer and carcinoma of thyroids.

3. *Value in Colorectal Cancer*
   - Most valuable, has been ***used as an aid in diagnosis. Value of CEA as a tumour marker is greatest in colorectal cancer.***
   - Has been ***useful in staging***. Found to be elevated in 28 per cent of patients with stage A colorectal cancer and in 45 per cent of patients with stage B colorectal cancer.
   - Most important use of CEA has been ***monitoring*** the response of colorectal cancer to treatment.
   - Patients with colorectal cancer who initially had elevated CEA show return of CEA values to normal after complete and successful surgical removal. ***Values of CEA remain normal as long as remission persists (serial assays are helpful). A rise again in postsurgical patients is definite indication of relapse/recurrence.***
   - ***Prognostic usefulness:*** Patients of colorectal carcinoma with near normal pretreatment CEA levels had a lower incidence of metastasis. On the other hand, majority of patients having high CEA pretreatment levels developed metastasis.

Note

CEA is considered as best available noninvasive tumour marker for the postoperative monitoring of surgically treated patients with colorectal cancer.

4. *Value in Other Malignant Tumours*
   - Reports are conflicting.
   - A role of CEA in monitoring the therapeutic response in patients with gastric carcinoma and lung carcinoma is not proven.
   - But in small cells lung cancer (SCLC) it is claimed that chemotherapy may show dramatic, short lived responses, monitoring the CEA level may be of value.

Limitations of CEA Assay

Though CEA is the most widely investigated and most frequently used tumour marker in clinical oncology, ***it has certain limitations.***

1. *High false –ve results:* A significant number of patients with adenocarcinoma of the GI tract will not have an elevated CEA level. ***Hence, a normal CEA value would not rule out the presence of cancer.***
2. *False +ve results*: ***Abnormally elevated CEA values have been reported in certain benign diseases,*** viz. ulcerative colitis, in benign breast conditions, in emphysema, in rectal polyps and even in heavy smokers. In these, increase in CEA levels usually does not exceed 3 to 4 times of the upper limit of reference range.

B. Human Chorionic Gonadotropin (β-HCG)

HCG is a ***placental hormone. It is synthesised by the syncytiotrophoblastic cells of placental villi.***

Normally

- It is present in the serum of nonpregnant women in very trace amounts or not at all.
- But it is markedly elevated in pregnancy. Maximum peak level is reached by 12 weeks of pregnancy, then it declines slowly, reaching 1/5th of peak by the end of 20th week and then continues at a very low level for a few days even after parturition.
- Measurement of elevated HCG in serum and urine has been used to diagnose pregnancy.

Chemistry

- It is a ***glycoprotein***. Molecular weight averages 45,000.
- Protein is present as a central core with branched carbohydrate side chains, which terminate with sialic acid.
  - It is ***dimer*** and has two dissimilar subunits:
    - ***α-subunit*** and
    - ***β-subunit***

α-subunit:

- Molecular wt 15,000 to 20,000
- Consists of 92 amino acids
- Is identical with α-subunit of FSH, LH and TSH

β-subunit

- Molecular wt 25,000 to 30,000
- β-subunit or c-terminal part of the β-subunit is specific immunologically.

CLINICAL USES AND REMARKS FOR β-HCG

1. The β-subunit of HCG is typically measured because of its increased specificity and because some tumours secrete only β-subunit.
2. β-HCG is ***an ideal tumour marker for diagnosing and monitoring gestational trophoblastic tumours and germ cell tumours of testes and ovary.***
3. Frequency of elevated β-HCG has been observed to be as follows:

• **Seminomas**	15%
• **Embryonal carcinomas**	50%
• **Teratocarcinoma**	42%
• **Choriocarcinoma**	100%

Specificity increases when AFP and LDH isoenzymes are done simultaneously. Both LDH and LDH-1 isoenzyme show increased levels in 50 to 80 per cent of patients of testicular cancers.

4. **β-HCG in CS fluid:** Recently, measurement of β-HCG in cerebrospinal fluid (CSF) has aided in diagnosis of brain metastases. **A serum/CS fluid ratio of less than 60:1 points to central nervous system (CNS) metastasis.** The response of therapy in patients with CNS metastases can be monitored using HCG levels.

Limitations

Elevated β-HCG levels have also been reported in other tumours, viz. lung cancer, breast cancer, GI cancer, ovarian cancer, hypernephroma, etc.

Conclusion

In spite of limitations, **serum β-HCG level is an ideal and superb 'tumour marker' in patients with gestational trophoblastic tumours.** Serial assays of β-HCG levels would be the best tool for monitoring the clinical course of those patients.

C. Alpha-Fetoprotein (AFP)

Like CEA, α-Fetoprotein (AFP) is another ***oncofoetal antigen.*** AFP is synthesised in the liver, yolk sac and GI tract in foetal life and is released into the serum of foetus.

It is a normal component of serum protein in human foetus. The concentration is highest during embryonic and foetal life. At birth, the serum AFP declines to 1/100th of AFP value at the highest foetal concentration. At one year of life, the value decreases further and in normal adults it is negligible, less than 20 ng/ml.

Chemistry

- It is a ***glycoprotein.*** Protein constitutes 95 per cent and carbohydrate moiety 5 per cent. Having molecular weight 61,000 to 70,000.
- Physically and chemically it is related to albumin, pI is similar to albumin (4.8).
- It is a single polypeptide chain (monomeric unit) with regions in the interior being similar to human serum albumin.

CLINICAL USES AND REMARKS

1. **AFP is the most specific and ideal tumour marker for primary carcinoma of the liver (hepatocellular carcinoma).** Serum level of AFP level is elevated markedly. Hepatoma cells are analogous to foetal hepatocytes and are capable of synthesising AFP.
2. AFP assay has been used in case of hepatic mass:
   - In suspected hepatoma.
   - In patients with cirrhosis liver suspected to have superimposed hepatoma.
   - Also serial assay in established case of hepatoma to follow the effect of therapy.
3. AFP as tumour marker has been found to be also most useful in germ cell tumours of the testes and ovary. **Serum AFP and β-HCG are the best available tumour markers for germ cell type of tumours.**

Note

- AFP is not elevated in seminomatous testicular tumours. ***It is found to be elevated in majority of the patients with non-seminomatous tumours, e.g. choriocarcinoma, embryonal carcinoma, yolk sac tumour, teratomas and teratocarcinomas.***
- AFP is specifically elevated in embryonal carcinoma and yolk sac tumour.
- Combined use of AFP, β-HCG and LDH isoenzyme has proved clinically useful.

II. TUMOUR MARKERS NOT USED COMMONLY

A brief discussion with clinical usefulness of the other 'tumour markers' will be done.

1. **Tissue polypeptide antigen (TPA):** It was first evaluated as tumour marker in 1978.
   - *Not tumour specific.*
   - Elevated levels are found in various malignancies, viz. in colonic cancer, breast cancer, prostatic cancer and metastatic cancer.
   - Can be used in combination with other tumour markers which improves the sensitivity but diminishes the specificity of the other marker.
2. **Tennesee Antigen (Tenagen)**
   - Not in use routinely
   - Clinical uses in colorectal cancer, pancreatic cancer, lung cancer and stomach cancer.
3. **CA 125:** CA 125 is an antigenic determinant expressed by epithelial ovarian carcinomas that can be detected by a monoclonal antibody.
   - **Has been used for screening and diagnosis of ovarian carcinoma.** CA 125 is not specific for ovarian carcinoma. It is also elevated in breast carcinoma and colorectal cancers. ***An elevated CA 125 level is observed in 80 to 90 per cent patients with ovarian cancer.***
4. **CA 15-3**
   - **Has been found useful as tumour marker in breast carcinoma.**
   - CA 15-3 is an antigen that is detected by a monoclonal antibody generated against extracts of metastatic human breast cancer.
   - Most studies have shown that CA 15-3 level is a more sensitive marker than the CEA level. Elevated CA 15-3 levels are found in 70 to 80 per cent of patients with metastatic or recurrent breast cancer.
   - A more important role of CA 15-3 is early detection of recurrence.

Limitations: Though CA 15-3, is useful in breast cancer as a 'marker', ***its specificity is low***; because elevated levels have been observed in other malignancies and also in patients with benign breast lesions and liver diseases.

5. **CA 19-9**
   - **Has been found useful as tumour marker in pancreatic cancer.** ***It is an asialylated Lewis blood group antigen*** that is detected by monoclonal antibody. Like CEA, it was first detected in a colorectal carcinoma.
   - Found to be ***elevated markedly in patients with pancreatic cancer (80 to 100% cases).*** Sensitivity of CA 19-9 level in patients with pancreatic cancer is relatively high.
   - Excellent correlations have been reported between CA 19-9 assay and relapse/recurrence in post-surgical resection cases.

Limitations: CA 19-9 specificity is lowered as:
   - It is found to be elevated in other malignancies also, viz. colorectal cancer (20%), hepatomas (20 to 50%) and gastric cancer.
   - Elevated levels have also been found in benign disorders, e.g. pancreatitis, and liver diseases.

6. **CA 72-4**
   - It is a radioimmunoassay that detects a tumour-associated glycoprotein termed TAG 72, which has been found by immunohistochemistry in tissue sections from more than 90 per cent of patients with colo-rectal, gastric and ovarian cancers and from 70 per cent of patients with breast cancers.
7. **CA 50:** A carbohydrate antigen detected by monoclonal antibody.
   - It has been found useful in lung cancer, gastrointestinal cancer including colorectal carcinoma.
   - It is claimed that CA 50 level can be used to identify postoperative recurrences of colorectal cancer which are not identified by the CEA level.
8. **Mammary Antigens**
   Several new antigens have been recognised by monoclonal antibodies. Have been identified in patients with breast cancer. They have been proposed as 'tumour markers'.

- **MCA** (Mucin-like carcinoma associated antigen) It is a mucin glycoprotein that is detected by monoclonal antibody against human breast cancer cell lines. The antigen is not sensitive in early stage of breast cancer.
- **MAM 6**
  An epithelial membrane antigen present on ductal and alveolar epithelial cells that is detected by monoclonal antibody raised against human milk-fat globule membranes.
- **MSA** (Mammary serum antigen)
  It is detected by an antibody raised against a whole cell suspension of intraductal breast cancer. ***Studies have shown MSA to be superior than CA 15-3 as a tumour marker in breast cancer.*** It has better sensitivity and specificity. Incidence of elevated level in patients with stage I and stage II breast cancer has been observed to vary from 68 to 90 per cent and specificity from 82 to 95 per cent .
- **MAP (Mitogen Activated Protein) Kinase**
  Recently a new breast cancer *"marker"* found, called as ***mitogen activated protein (MAP) kinase.*** Normally this enzyme helps cells to divide. But breast cancer cells were found to contain up to 20 times of the enzyme level found in normal. This enzyme in excess appears to be the *trigger* for breast cells to proliferate wildly, a hallmark of cancer.

  Researchers claim to be hopeful to use the "MAP-kinase marker" to distinguish cancerous breast lumps from harmless benign one. The test is performed on biopsy tissues taken after a woman has had a suspicious looking mammogram. ***MAP-kinase levels were found to be 5 to 20 times higher in Breast Cancer as compared to normal breast tissue.*** More studies are needed to repeat the results and to determine if inhibiting MAP-kinase can influence the growth of breast cancer.

9. **Prostatic Acid Phosphatase (PAP) and Prostate specific antigen (PSA)**
   - **PAP** has been recognised as a tumour marker for prostatic cancer since 1938. But it is not an ideal marker. Its use has been hampered by poor sensitivity and specificity.
   - **PSA** is an organ-specific, localised to prostatic ductal cells. As a tumour marker it is nearly an ideal marker and better than PAP. Levels are undetectable in women, in normal men (below 2.0 ng/ml), but it is elevated in benign or malignant prostatic disease. The clinical recurrence/relapse of prostate cancer is associated with raised serum PSA levels in 90 to 100 per cent of patients. PSA has been found to be very sensitive in detecting disease recurrence.

10. **Neuron-specific Enolase (NSE):** Recently the use of NSE as a specific tumour marker has been advocated for tumours of neuroendocrine origin.
    - Although NSE has been detected with various tumours, studies of NSE as serum tumour marker have been found to be extremely useful in:

Tumour markers	Tumour types and associated abnormalities
A. Hormones	
• **ACTH**	Lung cancer, carcinoids, pancreatic carcinoma, medullary carcinoma of thyroid
• **Catecholamines**	Pheochromocytoma
• **Insulin**	Insulinoma
• **Glucagon**	Glucagonoma
• **Gastrin**	Gastrinoma
• **Serotonin**	Carcinoids
• **Calcitonin**	Medullary carcinoma of thyroid, lung cancer
B. Enzymes	
• **Total LDH**	Lymphomas, leukaemia, germ cell tumours, breast and lung cancer and others
• **LDH1**	Germ cell tumours, ovarian carcinoma, osteosarcoma ***Other abnormalities:*** Acute myocardial infarction, renal infarct, haemolytic disease
• **LDH 2,3,4**	Leukaemias, Lymphomas
• **LDH 5**	Hepatoma, breast cancer, prostatic cancer, colorectal cancer ***Other abnormalities:*** Hepatitis and other benign Liver diseases, Skeletal muscle injury
• ***Alkaline Phosphatase (ALP)***	
– ***Fast liver isoenzyme***	Metastatic liver cancer
– ***Bone isoenzyme***	Metastatic bone disease, Benign bone disease
– ***Regan isoenzyme***	Lung cancer, ovarian cancer, breast cancer, colonic cancer, uterine cancer
– ***Nagao isoenzyme***	• Metastatic carcinoma of pleural surfaces • Adenocarcinoma of pancreas and bile duct
– ***Prostatic acid Phosphatase (PAP)***	Prostatic cancer
• ***Creatine kinase (CK/CPK)***	
– **Isoenzyme CK-BB useful as a "tumour marker"**	Adenocarcinomas, prostatic carcinoma
– **α_1-antitrypsin (α_1-AT)**	Germ cell tumours of testes and ovary

 - **Neuroblastomas,** and
 - **Lung cancer.**
- Immunoreactive NSE is found in the tumours of most patients with "small-cell carcinoma of lung" (SCLC). ***NSE level is found to be elevated in serum in 80 to 87 per cent of cases of SCLC.***

Limitations: Specificity is rather low. In 10 per cent patients of non-SCLC and non-malignant lung diseases elevations are seen.

11. Other tumour markers and their clinical uses are tabulated an previous page.

CONCLUSION

During the past two decades, there has been remarkable efflorescence in search of 'tumour-markers'. Since the introduction of CEA, numerous tumour-markers have been identified and has been used by oncologists in cancer patient care.

As seen from above discussion, as diagnostic tools, tumour markers have certain limitations. Quite a number of them lack in specificity. Most of the markers can be elevated in benign disorders also. A few markers are not elevated in early stages of malignancy. Also, the lack of tumour markers does exist in patients with tumours, and hence, ***a negative result does not exclude the diagnosis of cancer.***

Further research for more specific and more sensitive tumour markers is in progress. ***Monoclonal antibody technology has opened a new era and has sparked the search for tumour markers that are more organ and tumour specific.***

CHAPTER 47

BIOPHYSICS: PRINCIPLES AND BIOMEDICAL IMPORTANCE

Major Concepts

A. Define, pH, Acids, bases, buffers and learn their biomedical importance.

B. Learn biophysical properties of fluids such as diffusion, osmosis and osmotic pressure, dialysis, Donnan equilibrium, surface tension, viscosity, colloids and their importance in living organisms as applied to body fluids.

Specific Objectives

A. 1. Define acids and bases.

2. • Define pH • What is Henderson-Hasselbalch equation, derive it. • Learn applications of HH equation and significance of pK.
• What is conjugate base-pairs • Define buffer • Learn the mechanism of action of buffers. • List the common buffers and enumerate base-pairs as found in blood. • Study biomedical importance of buffer

B. 1. • Define diffusion • Study the characteristics of diffusion • Learn about Fick's law and physiological role of diffusion.

2. • Define osmosis and osmotic pressure • Learn the factors affecting osmotic pressure. • Learn what is meant by isotonic, hypotonic and hypertonic solutions. • Study the physiological importance and application of osmosis in biological system.

3. • Learn what is dialysis? List the names of some of dialysis membranes used in the laboratory. • Study what is Donnan equilibrium and learn its clinical and physiological importance.

4. Differentiate between diffusion and osmosis, and differentiate dialysis and Donnan equilibrium.

5. • Define surface tension • Learn how surface tension is determined and the factors that affect surface tension. • Study the Gibbs-Thompson principle in relation to surface tension. • Learn what are surfactants and their biomedical importance.

6. • Define viscosity • Learn how viscosity is determined and the factors affecting viscosity. • Study the biomedical importance of viscosity.

7. • Define colloids • Learn the properties of colloidal particles • Study the different types of colloidal solutions, viz. emulsoids and suspensoids, differentiate the two • What is meant by "protective colloids"?

INTRODUCTION

Most properties and laws of matter are applicable to biological systems. Some of such important properties are discussed in this chapter.

HYDROGEN ION CONCENTRATION (pH)

It is defined as the negative logarithm of hydrogen ion concentration to the base 10. The hydrogen ion concentration $[H^+]$ is expressed in moles/litre.

$$pH = -\log [H^+] = \log \frac{1}{[H^+]}$$

Pure water or neutral aqueous solutions have $[H^+] = 1 \times 10^{-7}$ mol/litre. Therefore their pH according to definition can be calculated to be equal to 7. pH can also be expressed as the index of the exponential term obtained by writing the molar conc. of H^+ as a power of 10, omitting the negative sign of the term. According to this the pH scale ranges from 0 to 14. It was introduced by **Sorensen.** ***pH = 7 is neutral, while pH > 7 is alkaline and pH < 7 is acidic.*** A rise or fall in pH by 1 signifies a tenfold fall or rise in the H^+ conc. respectively. pOH is the negative logarithm or $[OH^-]$ to the base 10. Ionisation exponent of water (PKw) is the negative logarithm of Kw to the base 10 and equals the sum of pH and pOH values. PKw = pH + pOH. pKw of water at room temp. is 14.

Henderson-Hasselbalch Equation

Let us consider HA a weak acid that ionises as follows:

$$HA = H^+ + A^-$$

Equilibrium constant K can be written as,

$$K = \frac{[H^+] + [A^-]}{[HA]}$$

Simplifying we get

$$K \times [HA] = [H^+][A^-]$$

$$\text{or } [H^+] = \frac{K \times [HA]}{[A^-]}$$

Take log of both sides,

$$\log [H^+] = \log K + \log \frac{[HA]}{[A^-]}$$

Change sign of both sides,

$$-\log [H^+] = -\log K - \log \frac{[HA]}{[A^-]}$$

Now $-\log [H^+] = pH$

$-\log K = pK$

Substituting, we get

$$\mathbf{pH = pK + \log \frac{[A^-]}{[HA]}}$$

The above is known as ***Henderson-Hasselbalch equation***.

Significance of pK: pK of an acid group is that pH at which the protonated and unprotonated species are present at equal concentrations.

Applications of Henderson-Hasselbalch Equation

1. *Can be used to determine pH of blood,* if the concentration of salt, i.e. bicarbonate and acid (carbonic acid) is known.

Example

- Approx. concentration of bicarbonate in normal health = 0.025 M
- Approx. concentration of carbonic acid = 0.00125 M.
- pKa of carbonic acid = 6.1

  Then applying HH equation

$$pH = 6.10 + \log \frac{0.025}{0.00125}$$
$$= 6.10 + \log 20$$
$$= 6.1 + 1.3$$
$$= 7.4$$

2. *Determination of pH of an unknown buffer solution*

By using HH equation, one may directly calculate the pH of a buffer solution if pKa of the buffer acid and the molar ratio of salt to acid in the solution are known.

Examples

(a) A solution containing 0.05 M Na-acetate and 0.1 N Acetic acid. Value of pKa of acetic acid is 4.73 at room temperature.

∴ pH of the solution

$$= 4.73 + \log \frac{0.05}{0.1}$$
$$= 4.73 + \log 0.5$$
$$= 4.73 - 0.30$$
$$= 4.43$$

(b) In the above example, if the solution contained 0.1 M Na-acetate and 0.1 N acetic acid.

$$pH = 4.73 + \log \frac{0.1}{0.1}$$
$$= 4.73 + \log 1$$
$$= 4.73 + 0 = 4.73$$

This shows that ***when the molar ratio of salt to acid in a buffer solution is unity, the pH of the solution becomes equal to the value of pKa for the buffer acid.***

Biological Importance of pH

pH plays an extremely important role in biological systems. It is already mentioned in the chapter on enzymes that enzymes have an optimum pH. Various body fluids and cell organelles, etc. maintain a specific pH in order to allow the activity of the enzymes located there. Gastric juice is highly acidic with pH = 1.2. The gastric juice contains pepsin whose optimum pH is around 2. The other functions of HCl are mentioned in gastric function tests. Osteoblasts possess a highly alkaline pH. The values of pH lies between 9 and 10. The alkaline phosphatase located there has the optimum pH in that range.

Prostate has acidic pH (3-5) in order to allow acid phosphatase activity. pH of arterial blood is 7.4 and is effectively regulated as described in Chapter on "Acid-base Balance and Imbalance".

- *Tautomeric forms of purines and pyrimidines:* Keto/enol (lactam/lactim) forms of nucleic acid bases depend on pH. Specific tautomeric form exist at pH 7.4. This helps in proper hydrogen bonding between the complementary base pairs.
- *Isoelectric pH and optimum pH of proteins:* Amino acids and proteins exist as Zwitterions at isoelectric pH. The magnitude of the charge depends on the pH. By influencing the ionised states of proteins, pH considerably affects the formation and maintenance of ionic and hydrogen bonds. The solubility and biological activity of protein depends on their 3-D structure. Hence protein has an optimum pH where it best can maintain its 3 dimensional conformation befitting its biological activity.
- *pH and Gibbs-Donnan effect:* The physiological pH 7.4 renders body proteins to exist as anions. Since they are large molecules, they do not cross semipermeable biological membrane resulting in an unequal distribution of diffusible cations and anions on the two sides of membranes. The diffusible cations are more concentrated, on the side containing a higher concentration of the protein anions, while diffusible anions have a higher concentration on the opposite side of the membrane (Gibbs-Donnan membrane equilibrium). pH is responsible for this.

- ***Ionic states of nucleic acids, lipids and mucopolysaccharides:*** It is observed that nucleic acids, MPS, phosphoglycerides, sphingolipids, etc. exist as ionised form either as cations or anions depending on the pH. At pH 7, lysylphophatidylglycerols bear net positive charges, gangliosides carry negative charge. By influencing the electrical charges of the polar head groups of membrane lipids, pH influences the membrane structure.
- ***pH and K_{eq}:*** pH influences the K_{eq} product yield and spontaneity of metabolic oxidation-reduction and some nonenzymatic acid-base catalysis.

BUFFERS

A buffer solution is a solution that resists any change in its pH value on addition of strong acid or alkali. The process by which the added H^+ or OH^- are removed is called as buffering action. The extent of resistance of change in pH by buffer is called its buffering capacity.

1. Principal buffers of extracellular fluids are:
   - Bicarbonate buffer
   - Phosphate buffer
   - Protein buffer.
2. Principal intracellular buffers are:
   - Phosphate buffer,
   - Protein buffers (Hb).

The details of the mechanisms of action of these buffers are discussed in Chapter on Acid-Base Balance and Imbalance.

DIFFUSION

Definition

It is *defined* as a process in which solute particles move from a more conc. environment to a less concentrated one in order to bring a uniform concentration throughout.

Although particles (solutes) move at random in all directions, a greater member of them move from a region of higher concentration to one of lower concentration. ***Thus the diffusion of solute particles takes place down the concentration gradient (pressure gradient with respect to partial pressure of gases), until uniform concentration is achieved.*** In equilibrium condition, equal numbers of particles move in all directions so the net diffusion is zero. Diffusion of gases, liquids and solute particles in solution can occur even across the permeable membrane whose pores allow a free passage of diffusing particles.

Factors affecting diffusion: Following factors affect the diffusion:

- ***Mean molecular velocity and mean free path:*** Diffusion directly depends on mean molecular velocity and mean free path.
- ***Pressure of concentration gradient:*** A gas or solute diffuse down their own pressure or concentration gradient.
- ***Conc. gradient of solute:*** Diffusion of a solute is directly proportional to its conc. gradient.
- ***Electrical gradient of ions:*** When a potential difference exists between two sides of a membrane, anions and cations diffuse to electropositive and electronegative sides.
- ***Solubility:*** Diffusion of a gas or of a solute through a medium is directly proportional to its solubility.
- ***Partition coefficient:*** The diffusion of a solute from one aqueous phase to another across an intermediate lipid phase is directly proportional to its partition coefficient which is the ratio of its solubilities in the lipid and water respectively.
- ***Temperature:*** Diffusion of a gas or solute is directly proportional to temperature.
- ***Pore size of membrane:*** It also depends on pore size of the membrane that separates the two.
- ***Surface area of the membrane:*** Diffusion varies directly with the cross-sectional area across which it occurs.
- ***Thickness of membrane:*** It is inversely proportional to the square of the distance to be traversed.
- ***Size and shape of solute particles:*** Diffusion is inversely related to the diameter of diffusing particles.
- ***Viscosity of solvent:*** More viscous solvent or solution resist diffusion.

Determination in a non-homogeneous solution: Diffusion between two planes X and Y in a non-homogeneous solution can be expressed quantitatively by **Fick's law** as follows:

$$\frac{ds}{dt} = DA\frac{dc}{dx}$$

Where, $\frac{ds}{dt}$ = Rate of movement of solute

D = Diffusion constant

A = The area of the planes

$\frac{dc}{dx}$ = the concentration gradient, i.e. the difference in concentration between X and Y/distance between X and Y.

BIOMEDICAL IMPORTANCE OF DIFFUSION

The process of diffusion is a commonly observed phenomenon in biological system.

- ***Respiratory exchange of gases:*** There is a pressure gradient of gases such as O_2, CO_2 which brings about transport of these gases following diffusion. Thus the whole process of respiration involves diffusion at various sites.
- Increased thickness of inflamed alveolar membrane reduces the diffusion of O_2 in pneumonia.
- Keratinised epidermis minimises loss of water.

- Intestinal absorption of pentoses, minerals, water-soluble vitamins and renal absorption of urea are carried out by diffusion.
- Water, ions and small molecules pass largely by diffusion through plasma membrane.
- Protein cations and Ca^{+2} present in the plasma membrane, electrostatically impede the diffusion of cations and enhance the diffusion of anions, the diffusion rate of Cl^- far exceeds that of K^+ in spite of their similar effective diameters.
- Diffusion of ions across cell membranes influences polarisation of membranes and membrane potentials.

What is Diffusion Trapping? Weak acids such as salicylic acid and phenobarbital are excreted passively in distal tubule and collecting duct of kidney by the ***diffusion trapping*** mechanism, so long as the urine is alkaline. They diffuse from the tubule cells to tubular lumen down their concentration gradient, as the tubular membrane is permeable to them, though not to their anions. In the alkaline urine, they ionise into anions, hence they cannot diffuse back into the tubule cells but lower the concentrations of the respective non-ionised weak acids, enabling their diffusion to continue.

OSMOSIS AND OSMOTIC PRESSURE

Definition

It is ***defined*** as the process of net diffusion of water molecules from a dilute solution or pure water (solvent) itself to a more concentrated solution, when both are separated by a semipermeable membrane.

This membrane allows the water to diffuse but not the solute. Thus, separated from water by semipermeable membrane, water flows into concentrated solution across the membrane. Water molecules diffuse in both directions across the semipermeable membrane, but a net diffusion or osmosis of water from the dilute to the concentrated solution results from a larger number of water molecules diffusing in that direction than in the reverse direction. Water continues to flow into the more conc. solution across the membrane in this way until the hydrostatic pressure rises so high on the concentrated side of the membrane to cause a transmembrane diffusion of water in the opposite direction at the rate as the osmotic inflow. ***This hydrostatic pressure which exactly balances the osmotic influx of water from pure water to conc. solution is called the osmotic pressure of that solution.*** Thus, ***osmotic pressure (π)*** can also be defined as the pressure which has to be exerted on the concentrated solution, separated from pure water by a semipermeable membrane, in order to counteract and stop the osmotic inflow into the solution. It equals the difference between the hydrostatic pressures on the two sides of membrane. Osmotic pressure is a colligative property of a solution. A rise in the number of solute particles in the solution increases the number of solvent particles bound by solute particles in complexes called ***solvates.*** Osmotic pressure may be considered to be caused by the higher partial pressure of solvent molecules on that side of the semipermeable membrane as has a higher concentration of the solvent.

- Osmotic pressure of a dilute solution is directly proportional to other colligative properties like the fall in freezing point, lowering of vapour pressure, etc.
- Osmotic pressure is inversely proportional to the MW of the solute.

Osmol units (Osm) give the number of osmotically active particles per mole of a solute, each mole of a nonionised solute is equivalent to 1 Osm. ***Osmolarity*** of a solution is its solute concentration in Osmols / litre or dm^2. A solution of 1 m Osm of any solute in 1 litre of solution gives an osmotic pressure of 19.3 mmHg at 38°C.

Osmolality of a solution is its solute concentration in Osm / Kg of solvent. Human plasma has an osmolarity of 303.7 Osm / dm^2.

Osmotic pressure is measured by:

(a) Osmometer or

(b) Berkeley-Hartley method or

(c) Freezing point method. The phenomenon of osmosis and osmotic pressure is widely observed in human beings and other living organisms.

BIOMEDICAL IMPORTANCE OF OSMOSIS

- ***Hemolysis, crenation and plasmolysis:*** Plasma membrane is permeable to water and few solutes but not to proteins and some electrolytes. Thus plasma membrane behaves like a semipermeable membrane. Osmosis takes place through plasma membrane.
- Solutions with identical osmotic pressure are called ***isosmotic*** solutions. All the colligative properties of such solutions are also identical. A solution is called ***hyperosmotic*** **or** ***hyposmotic*** in comparison to another when the former has respectively a higher or lower osmotic pressure than the other.
- The extracellular fluid exerts some osmotic pressure called as tonicity of the solution and is proportional to sum of molar concentrations of only such solutes for which it is not permeable. Two solutions of identical tonicity are called ***isotonic solutions*** though they may not be isosmotic. "Tonic" solutions are isotonic, hypotonic or hypertonic. When the cells are kept in hypotonic solutions the cells rupture. This is called as ***osmolysis.*** Physiological or isotonic solution is 0.9 per cent of NaCl. If red cells are kept in 0.3 per cent of NaCl they rupture. This kind of osmolysis of RBC is called ***haemolysis.*** It is due to endosmosis, i.e. water entering cells. Osmotic fragility test is an important test in haemolytic anaemia.

On the other hand if the cells are kept in hypertonic saline, say 1.5 per cent NaCl the cells get shrunken due to osmotic outflow of water (exosmosis). This is called ***crenation.*** Plant cells also lose water by exosmosis in a hypertonic solution, making the cell membrane collapse and withdraw from the cell wall ***(plasmolysis).***

- ***Normal osmotic distension of RBCs:*** Due to intracellular and plasma levels of electrolytes, etc. the RBCs remain well distended in fluid in blood. However, even slightest distension can cause hemolysis as in malformation of RBCs, genetic disorders of enzyme deficiencies such as G-6-P.D, vit E and selenium deficiency.
- ***Colloid osmotic pressure of blood and plasma:*** Colloid osmotic pressure of plasma proteins partly counteracts the filtering effect of blood pressure and retains water in the plasma. (For details see Starling Hypothesis in Chapter on Plasma Proteins). In Kwashiorkor, hepatic cirrhosis and nephrosis, a fall in plasma concentration of albumin, reduces the colloid osmotic pressure of blood and lowers the retention of water in circulation leading to oedema.
- ***Osmotic work:*** In moving *n* molecules in dilute solution (with conc. C_1) to conc. solution (with conc. C_2) cell has to perform osmotic work. Such osmotic work is performed by intestinal mucosa, renal tubules for Na^+.
- ***Water absorption and loss by animal cell:*** Absorption of water from intestines and renal tubules is due to osmosis caused by Na^+ and glucose. Polyuria is caused in diabetes by a fall in water reabsorption due to osmotic effect of urinary glucose. Facultative renal absorption of water also depends on the osmotic effects of the more conc. intestinal fluid in the renal medullary tissue.
- ***Exchange of fluids:*** Exchange of fluids at capillary end depends on osmotic pressure (See plasma proteins: Starling Hypothesis for details).

DIALYSIS

The disperse phase of a colloidal system have such large particles as cannot diffuse through the pores of a membrane made of cellophane, parchment, collodion, or inert cellulose esters like cellulose nitrate or acetate. However, such membranes are freely permeable to water and small molecules or ions in true solution.

Electrodialysis: This is another form of dialysis. Here the semipermeability and migration of electrolyte ions in an electric field are utilised in separating colloid particles from electrolyte ions.

BIOMEDICAL IMPORTANCE OF DIALYSIS

Dialysis is observed and made use of widely in medicine.

- ***Separation of proteins from small solutes:*** Dialysis is used to separate proteins in pure form from the mixture with salts. It is used to separate out macromolecules cell extract. Dialysis can also be used for stopping enzymatic or metabolic reactions by removing small cofactor molecules from the cell extract.
- ***Biological ultrafiltrates:*** Many extracellular fluids like interstitial fluids, CSF, glomerular filtrate of kidney are formed by ultrafiltration. Proteins do not appear in ultrafiltrate.
- ***Dialysis by artificial kidney:*** This is the important application of dialysis in the field of medicine. Patients with acute kidney failure and uremia, blood is dialysed in artificial kidneys to eliminate nonelectrolyte waste products as well as the excess of electrolytes. However, the blood cells are proteins, which are retained in the plasma. Blood of patient is heparinised to prevent coagulation and then passed through these membrane-bound channels, separated by the cellophane membrane from a dialysing fluid.

GIBBS-DONNAN MEMBRANE EQUILIBRIUM

When two solutions containing diffusible and nondiffusible ions are separated by a semipermeable membrane, the nondiffusible ions enhance the diffusion of oppositely charged diffusible ions. The diffusion takes place towards nondiffusible ion containing side. This also reduces the diffusion of like charged ions to that side. As a result, on the side which contains nondiffusible ions, diffusible counterions are more concentrated while the like charged diffusible ions concentrate more on the opposite side. This is called as *Gibbs-Donnan effect.* However, the total number of cations and anions is equal on both sides at equilibrium.

Example: Let us assume there are two compartments A and B, which are separated by a semipermeable membrane. (Refer box).

- **The compartment A** contains a solution of sodium proteinate in which Na exists as Na^+ and protein as Pr^- and to maintain electrical neutrality Na^+ balances Pr^-. Sodium salt of protein is a colloidal solution, the proteinate Pr^- is colloidal and is not diffusible through the membrane.
- **The compartment B** contains a solution of NaCl, in which both Na^+ and Cl^- are diffusible.

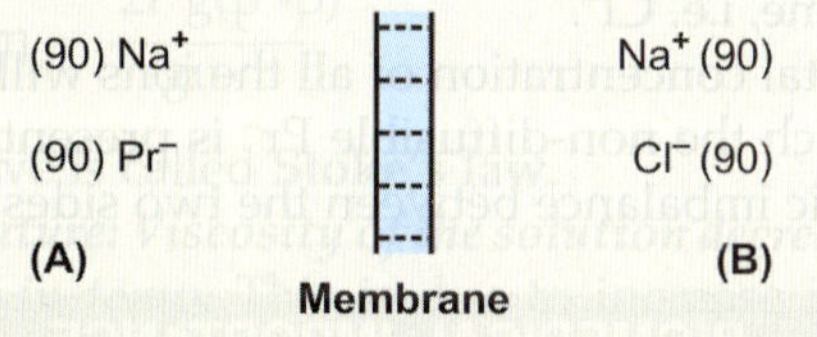

- According to Donnan effect, the non-diffusible ion or ions on one side of the membrane influences the diffusion of diffusible ions, and both quality and quantity of diffusible ions will be influenced.
- In above situation, Na^+ ions can diffuse either way, but Cl^- ions can diffuse only to the left, i.e. to

infection. Among the factors associated with false-positive EIA tests are antibodies to class II antigens, autoantibodies, hepatic disease, recent influenza vaccination, and acute viral infections. For these reasons, anyone suspected of having HIV infection based upon a positive or inconclusive EIA rsult must have the result confirmed with a more specific assay.

- The **most commonly used confirmatory test is the Western blot**. This assay takes advantage of the fact that **multiple HIV antigens of different, well-characterised molecular weights elicit the production of specific antibodies**. These antigens can be separated on the basis of molecular weight, and antibodies to each component can be detected as discrete bands on the Western blot. A negative Western blot is one in which no bands are present at molecular weights corresponding to HIV gene products. In a patient with a positive or indeterminate EIA and a negative Western blot, one can conclude with certainty that the EIA reactivity was a false positive. On the other hand, a **Western blot demonstrating antibodies to products of all three of the major genes of HIV (*gag, pol*, and *env*) is conclusive evidence of infection with HIV.** By definition, Western blot patterns of reactivity that do not fall into the positive or negative categories are considered "indeterminate". There are two possible explanations for an indeterminate Western blot result. The most likely explanation in a low-risk individual is that the patient being tested has antibodies that cross-react with one of the proteins of HIV. The most common patterns of cross-reactivity are antibodies that react with p24 and/or p55. The least likely explanation in this setting is that the individual is infected with HIV and is in the process of mounting a classic antibody response. In either instance, the western blot should be repeated in 1 month to determine whether or not the indeterminate pattern is a pattern in evolution. In addition, one may attempt to confirm a diagnosis of HIV infection with the p24 antigen capture assay or one of the tests for HIV RNA. While the Western blot is an excellent confirmatory test for HIV infection in patients with a positive or indeterminate EIA, it is a poor screening test. Among individuals with a negative EIA and PCR for HIV, 20 to 30 per cent may show one or more bands on Western blot. While these bands are usually faint and represent cross-reactivity, their presence creates a situation in which other diagnostic modalities **[such as DNA PCR, RNA PCR, the (b) DNA assay, or p24 antigen capture]** must be employed to ensure that the bands do not indicate early HIV infection.

A **variety of laboratory tests are available for the direct detection of HIV or its components.** These tests may be of considerable help in making a diagnosis of HIV infection when the Western blot results are indeterminate. In addition, the tests detecting levels of HIV RNA can be used to determine prognosis and to assess the response to antiretroviral therapies. The simplest of the direct detection tests is the ***p24 antigen capture assay***. This is an EIA-type assay in which the solid phase consists of antibodies to the p24 antigen of HIV. **It detects the viral protein p24 in the blood of HIV-infected individuals where it exists either as free antigen or complexed to anti-p24 antibodies.** Overall, approximately 30 per cent of individuals with untreated HIV infection have detectable levels of free p24 antigen. This increases to about 50 per cent when samples are treated with a weak acid to dissociate antigen-antibody complexes. Throughout the course of HIV infection, an equilibrium exists between p24 antigen and anti-p24 antibodies. **During the first few weeks of infection, before an immune response develops, there is a brisk rise in p24 antigen levels. After the development of anti-p24 antibodies, these levels decline.** Late in the course of infection, when circulating levels of virus are high, p24 antigen levels also increase, particularly when detected by techniques involving dissociation of antigen-antibody complexes. **This assay has its greatest use as a screening test for HIV infection in patients suspected of having the acute HIV syndrome, as high levels of p24 antigen are present prior to the development of antibodies.** In addition, it is currently routinely used along with the HIV EIA assay to screen blood donors in the United States for evidence of HIV infection. **Its utility as an assay is decreasing with the increased use of the reverse transcriptase PCR (RT-PCR) and bDNA technique for direct detection of HIV RNA.**

The ability to measure and monitor levels of HIV RNA in the plasma of patients with HIV infection has been of extraordinary value in furthering our understanding of the pathogenesis of HIV infection and in providing a diagnostic tool in settings where measurements of anti-HIV antibodies may be misleading, such as in acute infection and neonatal infection. **Two assays are predominantly used** for this purpose. They are the **RT-PCR (Amplicor) and the bDNA (Quantiplex).** It should be pointed out that the only test approved by the FDA at this time for the measurement of HIV RNA levels is the RT-PCR test. While this approval is limited to the use of the test for determining prognosis, it is the general consensus that these tests as well as the bDNA test are also of value for monitoring the effects of therapy and in making a diagnosis of HIV infection. **In addition to these two commercially available tests,** the ***DNA PCR*** **is also employed by research laboratories for making a diagnosis of HIV infection by amplifying HIV proviral DNA from peripheral blood mononuclear cells.** The commercially available RNA detection tests have a sensitivity of 40 to 50 copies of HIV RNA per millilitre of plasma, while the DNA PCR tests can detect proviral DNA at a frequency of one copy per 10,000 to 100,000 cells. Thus, these tests are extremely sensitive. One frequent consequence of a high degree of sensitivity is some loss of specificity, and false-positive results have been reported with each of these techniques. For this reason, **a positive EIA with a confirmatory Western blot remains the "gold standard" for a diagnosis of HIV infection,** and the interpretation of other test results must be done with this in mind.

In the **RT-PCR technique,** following DNAase treatment, a cDNA copy is made of all RNA species present in plasma. Insofar as HIV is an RNA virus, this will result in the production of DNA copies of the HIV genome in amounts proportional to the amount of HIV RNA present in plasma. This proviral DNA is then amplified and characterised using standard PCR techniques, employing primer pairs that can distinguish genomic cDNA from messenger cDNA. The bDNA assay involves the use of solid-phase nucleic acid capture system and signal amplification through successive nucleic acid hybridisations to detect small quantities of HIV RNA. Both tests can achieve a tenfold increase in sensitivity to 40 to 50 copies of HIV RNA per millilitre with a preconcentration step in which plasma undergoes ultracentrifugation to pellet the viral particles. **In addition to being a diagnostic and prognostic tool, RT-PCR is also useful for amplifying defined areas of the HIV genome for sequence analysis and has become an important technique for studies of sequence diversity and microbial resistance to antiretroviral agents. In patients with a positive or indeterminate EIA test and an indeterminate Western blot, and in patients in whom serologic testing may be unreliable (such as patients with hypogammaglobulinemia or advanced HIV disease), these tests provide valuable tools for making a diagnosis of HIV infection. They should only be used for diagnosis when standard serologic testing has failed to provide a definitive result.**

ANTIRETROVIRAL THERAPY (ART)

There has been reduction in number of new AIDS cases in the developed countries with the advent of **Highly Active Anti-Retroviral Therapy [HAART].** When potent combination therapy is administered effectively, the levels of RNA in plasma and infected cells in lymphoid tissue clear rapidly. Virtually all the compounds that are currently used, or under advanced clinical trial, for the treatment of HIV infections, belong to one of the following classes:

- **Nucleoside/nucleotide reverse transcriptase inhibitors (NRTIs):** For example zidovudine (AZT), didanosine (ddI), zalcitabine (ddC), stavudine (d4T), lamivudine (3TC), abacavir (ABC), emtricitabine [(-) FTC], tenofovir (PMPA) disoproxil fumarate;
- **Non-nucleoside reverse transcriptase inhibitors (NNRTIs):** For example nevirapine, delavirdine, efavirenz, emivirine (MKC-442); and
- **Protease inhibitors (PIs):** For example saquinavir, ritonavir, indinavir, nelfinavir, amprenavir, and lopinavir.

In addition to the reverse transcriptase and protease step, various other events in the HIV replicative cycle are potential targets for chemotherapeutic intervention:

- Viral adsorption, through binding to the viral envelope glycoprotein gp120 (polysulphates, polysulphonates, polyoxometalates, zintevir, negatively charged albumins, cosalane analogues);
- Viral entry, through blockade of the viral coreceptors CXCR4 and CCR5 [bicyclams (i.e. AMD3100), polyphemusins (T22), TAK-779, MIP-1 alpha LD78 beta isoform];
- Virus-cell fusion, through binding to the viral glycoprotein gp41 [T-20 (DP-178), T-1249 (DP-107), siamycins, betulinic acid derivatives];
- Viral assembly and disassembly, through NCp7 zinc finger-targeted agents [2,2'-dithiobisbenzamides (DIBAs), azadicarbonamide (ADA) and NCp7 peptide mimics];
- Proviral DNA integration, through integrase inhibitors such as L-chicoric acid and diketoacids (i.e. L-731, 988);
- Viral mRNA transcription, through inhibitors of the transcription (transactivation) process (fluoroquinolone K-12, Streptomyces product EM2487, temacrazine, CGP64222).

Also, in recent years, new NRTIs, NNRTIs and PIs have been developed that possess respectively improved metabolic characteristics. ***Although, a multitude of anti-HIV agents are being pursued actively, it has not been possible to eradicate HIV completely in an infected individual.*** The viral suppression is inadequate (failure to reduce viral copy number to less than 50 copies/ml) and unsustainable. Over a period of time, expansion of resistant variants takes place and these viral populations overtake the immune system leading eventually to AIDS. According to the guidelines laid down, **it is advised to administer a multi-drug regimen consisting of non-nucleoside reverse transcriptase inhibitors, nucleoside**

reverse transcriptase inhibitors and protease inhibitors to avoid faster emergence of resistant viruses. The use of existing therapies in the developing world, where more than two-thirds of the total HIV infection prevails, is limited owing to their high cost. Apart from the high cost and emergence of resistant mutants, another limiting factor is low patient compliance owing to the cumbersome drug regimens and side effects.

HIV VACCINE

Identifying the epitopes of HIV that are most critical in establishing infection or, conversely, which epitopes should be targeted for the development of cell-mediated and humoral immune responses to control HIV, is a major concern in vaccine development. **HIV vaccine can be either preventive vaccine**, which can be given to healthy individuals who are HIV negative, or it can be a **therapeutic vaccine,** which can be given to people who are already ill with the goal of curing them or improving their health.

The **goals for an HIV vaccine** should include:

- Protection against HIV infection, i.e. against all routes of transmission, against intravenous transmission only, against mucosal transmission only;
- Protection against progression to disease, i.e. reduction of the viral load;
- Reduction of transmission, i.e. vaccines likely to have lower viral load or lower transmission rate.

An Ideal HIV Vaccine

The ***ideal characteristic of an AIDS vaccine would include:***

- Efficacy in preventing transmission by the mucosal and parenteral route
- Excellent safety profile
- Single dose administration
- Long lived effect resulting in protection many years after vaccination
- Low cost
- Stability under field conditions
- Ease of transportation and administration and
- Ability to induce protection against infection with diverse viral isolates preventing the need for many isolate specific vaccines.

Although the overall strategy is to achieve sterilising immunity, a more realistic goal is to develop a vaccine, which could control viral replication, delay the onset of the disease and to reduce viral transmission. Modeling studies have revealed that even a partially efficacious vaccine would still have a major medical and socio-economic impact particularly in the developing countries.

Barriers to HIV Vaccine Development

Obstacles to the development of an effective HIV vaccine include factors related to the biology of HIV-1 infection and practical realities of developing and testing an AIDS vaccine are as follows:

Sequence variation: The rapid replication of HIV-1 *in vivo* produces 10^{10} new virions/day which facilitates rapid generation of sequence variants. Because a significant proportion of HIV specific neutralising antibodies and CTL are subtype specific, this sequence diversity has fostered efforts to induce broadly reactive immune responses or to utilise multivalent HIV vaccine.

Protective immunity: Another fundamental barrier is the lack of information regarding the type of immune response that may protect against HIV infection. CTL responses may be important to induce vaccine mediated protective immunity. HIV vaccine should be able to induce both HIV specific CTL and neutralising antibody responses.

Latency: Like other retroviruses, HIV integrates into the host genome where it can remain in a latent form that does not express HIV structural proteins and is thus less likely to be eliminated by the host cellular and humoral immune responses.

Transmission: HIV-1 is predominantly transmitted by mucosal route. Yet our knowledge of the event occurring during mucosal infection and immune responses responsible for defending against mucosal infection are quite limited. In addition, HIV transmission may occur by both cell free and cell associated virus particles. Cell associated virus is thought to be resistant to neutralising antibodies and will not be recognised by the host CTL responses, unless there is a fortuitous match between the HLA molecules between the host and the donor.

HIV Vaccine Concepts

Several different HIV vaccine concepts have been used in the animal model to elicit HIV specific immune response as follows:

- **Recombinant subunit vaccine:** A vaccine produced by genetic engineering simulating a part of the outer surface envelope or other part of HIV. gp 120 is the most well studied candidate for HIV-1 vaccine. **VaxGen**, a San Francisco based company, initiated the first phase III-efficacy trial of an HIV vaccine in 1998 using ***its gp120-subunit vaccine*** known as **AIDS VAX**.
- **Synthetic peptide vaccines:** Synthetic peptides of HIV are small epitopes of HIV proteins. Peptide based

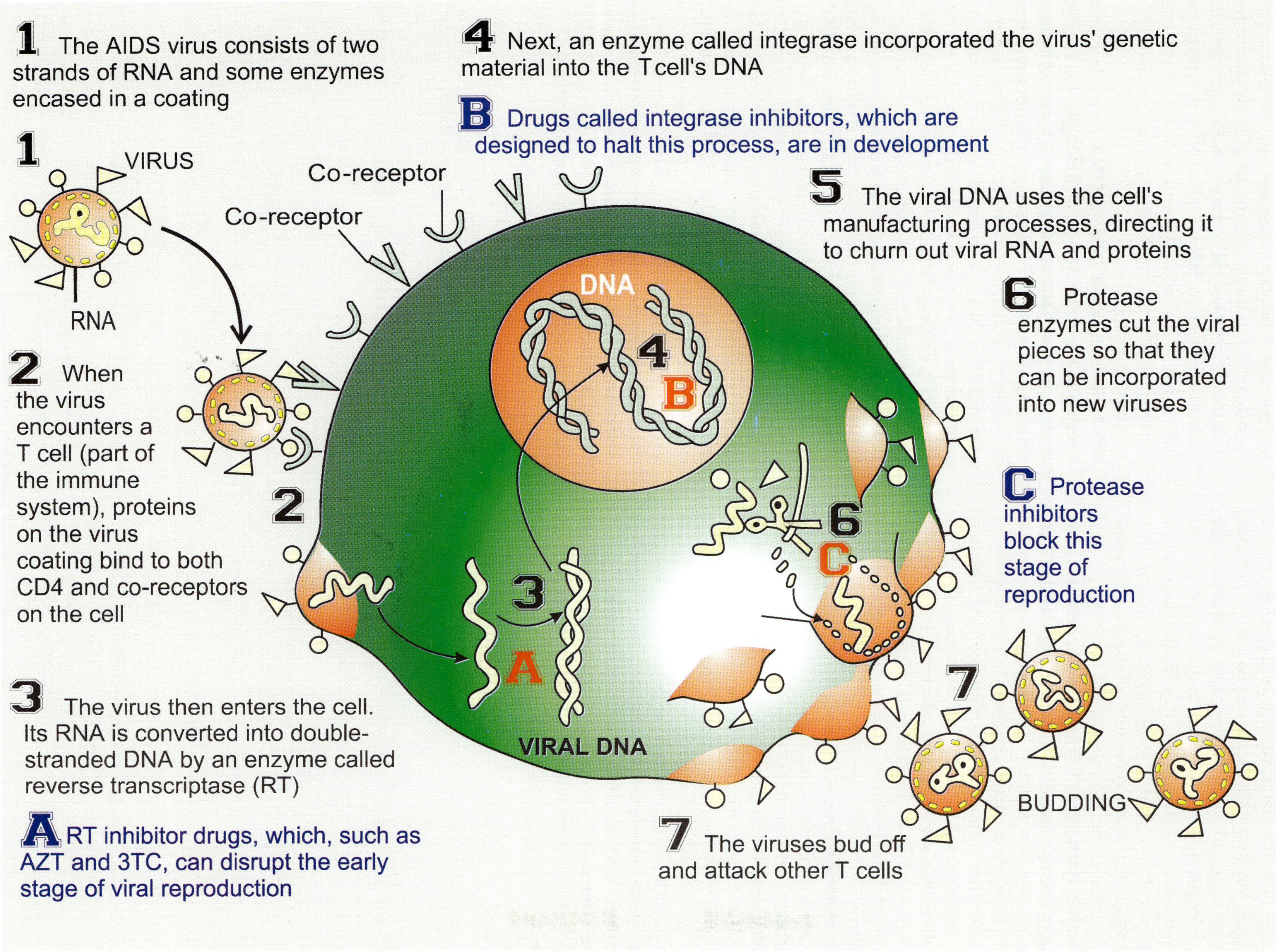

Fig. 49.6: Diagrammatic representation of life cycle of AIDS virus and possible methods to stop the disease

approaches offer the advantage of targeting specific epitopes that lie within the conserved area of the virus. Synthetic peptides can be linked to lipid molecules (e.g. lipopeptides) to facilitate induction of cellular immune responses such as CTLs. Finally the peptide can be combined as a multipeptide vaccine in a strategy to include diverse subtypes so as to increase the breadth of the vaccine induced response.

- **Virus like particles (VLP) and pseudovirions:** These are non-infectious particles resembling HIV that has one or more HIV proteins. Pseudovirions are replication incompetent viruses produced in mammalian cell cultures that contain all the viral proteins required for viral assembly, but do not contain the RNA genome, thus making it non-infectious. For example, core particles of hepatitis B virus have been engineered and evaluated preclinically to present HIV antigens.
- **Live vector vaccine:** A live bacteria or a virus that is harmless to humans and is used to transport a gene that makes HIV proteins. These include live attenuated bacterial vectors such as bacille Calmette-Guerin (BCG) and *Salmonella*. These vectors are safe and can establish infection via a mucosal route and can elicit strong mucosal immune responses.
- **Whole killed or inactivated vaccine:** In this, the entire virus particle is presented to the immune system but it cannot infect or replicate and is thus safer.
- **Live-attenuated vaccines:** Live attenuated virus vaccines have been successfully used to protect against a great number of diseases including polio and measles. *Nef*-deleted strains of simian immunodeficiency viruses (SIV) have shown promising immune protection from challenge with infectious SIV. Safety is a serious concern with this vaccine as the chances of reverting back to a more virulent HIV strain is quite high.
- **DNA immunisation (Naked DNA or nucleic acid vaccine):** One of the newest technologies for vaccine design offer significant advantages in ease of manufacturing. Pieces of HIV DNA are incorporated into harmless plasmid DNA from bacteria. These bacterial plasmids that have been genetically engineered to contain viral genes are injected into the muscle or skin. HIV DNA vaccines have been developed using HIV antigen from the *env* and core region of the virus.
- **Prime boost protocols/combination vaccines:** Recognising that protection from HIV may require a broad spectrum of immune responses including humoral, cellular and mucosal immunity, scientists have designed **combination regimens** in attempts to **elicit such broad-spectrum immunity**. Prime boost refers to a vaccination regimen involving a primary vaccination with one vaccine generating CMI response followed by a boost with another vaccine, often a subunit protein to elicit humoral response. Combination vaccine approaches elicit the most potent immune responses in non-human primates and in humans. Fig. 49.6 shows diagrammatically life cycle of AIDS virus and possible methods to stop the disease.

CHAPTER

50

BIOCHEMISTRY OF AGEING*

Ageing is characteristic of all multicellular organisms. Old age followed by death is accepted fact of life, and no serious effort were made by the scientists in this field. The tremendous efforts made by the biological and medical research has succeeded in controlling several disease, this has led to decline in death rate and at the same time pushed large number of people in old age. This has called for intense research to keep people physically fit and useful to society.

Definition of Ageing

- Ageing has been defined as a biological process, which causes increased susceptibility of organism to disease. (Comfort)
- Ageing has been defined as gradual decline in adaptation of the organism to its normal environment, following the onset of reproductive maturity.

Life Span and Life Expectancy

Life expectancy is the number of years an individual is expected to live, it is based on average of life span of maximum members of that species. **Average life span and life expectancy in our country has dramatically increased from 44 years in 1900 to 64 years in 1990.** Now the dream of extending life has shifted from fabled fountains of youth and biblical tales of long-lived patriarchs, to laboratory as gerontologists explore the genes and organs involved in ageing, and are uncovering more and more secrets of longevity.

So far **three stages** of problems have been identified in ageing:

- **Biological:** Includes study of molecular, physiological, and structural aspects of cell during ageing process
- **Clinical:** Includes study and cure of diseases, common during old age. (Geriatrics)
- **Sociopsychological:** Study of problems of old people. (Gerontology).

Ageing Theories

There are two main theories:

1. **Programmed theories**
2. **Error theories**

1. Programmed Theories

Ageing is the result of sequential switching on and off of certain genes, with senescence. Programmed senescence is defined as the time when age associated deficits are manifested.

- **Endocrine theory:** Biological clock acts through hormones to control the pace of ageing.
- **Immunological theory:** A programmed decline on immune functions leading to increased vulnerability to infectious diseases, ageing and death.

2. Error Theories

In laboratories around the world, scientists are isolating specific genes, cloning them, mapping them and studying their products to find what and when they do to influence the ageing and longevity. Following theories are proposed:

- ***Wear and tear***
- ***Crosslinking***
- ***Free radicals***
- ***Somatic mutations***
- ***Errors in proteins***
- ***The genetic connection***

---

* Contributed by Lt Col HS Batra, MD (Path); MD (Biochemistry), Associate Professor of Biochemistry, Armed Forces Medical College, Pune 411040.

- **Wear and tear pigment or age pigment:** One of the most distinguishable change is **accumulation of pigment lipofuscin in cytoplasm of neurons, skeletal muscles, and cardiac muscles.** Such pigment is also seen in connective tissues also.

  This pigment is acid insoluble and is rich in peroxidation products of polyunsaturated fatty acids and has high concentration of neutral and acidic polymerases. The cause and origin of this pigment is not clear but if large quantity of such inert substances will accumulate then it will adversely affect the metabolism and functions of the cells.
- **Cross linking:** It has been proposed that ageing is **due to cross linking of macro molecules, nucleic acids** and **proteins,** which are vital for the functioning of the cell. Several compounds like pyruvic acid, succinic acid, acetaldehyde, silicon, lead, manganese, calcium, and zinc are found in our body and are **potential cross linkers.** These accumulate with age and may not **only cross link with two or more enzymes and inactivate them but also act at the level of genes.**

  Increased amount of calcium is known to cross-link as it is a divalent ion and can bind with proteins and nucleic acid having negative charge.
- **Free radicals,** atoms and molecules having one unpaired electron can cross link two molecules. These compounds are produced during oxidation of organic compounds by molecular oxygen as follows:

$$RH + O_2 \rightarrow R^{\cdot} + H_2O_2$$
$$R^{-} + O_2 \rightarrow RO_2$$
$$RO_2 + RH \rightarrow R + ROOH$$
$$R^{-} + R \rightarrow R{:}R$$

- **Somatic mutations:** It has been proposed that ageing is **due to random mutations**, which destroys genes and causes loss of chromosomes of somatic cells like that of cardiac muscles, skeletal muscles and brain. These cells do not divide after certain stage of growth, due to mutations these genes either do not produce any protein or if they do so, the proteins are defective. With age mutations accumulate and after reaching certain level inactivate the cell and causes death of cell. Therefore number of functional somatic cells decreases with age. The organism dies when number of functional cells decreases below certain level.
- **Errors in proteins:** According to error theory of ageing, the proteins are of **two types:**
  A. Proteins needed for structure (Collagen and keratin) and for metabolism (Enzymes).
  B. Those needed for protein synthesis, e.g. RNA polymerase, Amino acyl tRNA synthetase, etc.

  Errors in amino acid sequence of enzymes involved in metabolism, results in decreased or complete loss of enzyme activity. This effect may not accumulate; once faulty proteins are degraded, their effects are lost.

  In second case change in sequence of amino acids of RNA polymerase results in synthesis of wrong mRNA and protein. This effect will continue till whole cell is filled with wrong protein, resulting in deterioration of all functions.

  There is no proof that errors occur in functional molecule during transcription and translation after adulthood, if they do so then how rate of errors are regulated to account for life span of the species. Is error the primary cause of ageing or is it secondary effect of primary protein, question still remained unanswered.
- **The genetic connection:** The process by which various cells arises from single cell is called ***cellular differentiation.*** It is believed that due to gene activity these differentiated cells differ in the capacity to synthesise the different proteins. The method by which genes are activated or inactivated in micro-organisms if not different may operate in the higher organisms for control of gene activity.

  The proteins expressed by genes carry out a multitude of functions in each cell and tissue of the body, and some of the functions are related to the ageing. We can say, that it is the protein product of these genes which are responsible for longevity or ageing. Some products like antioxidants prevent damage to the cells while other products prevent damage to the DNA and some products control cell senescence.

  Proliferating genes found to trigger cell proliferation. C-fos gene encodes a short lived protein thus regulates the expression of other genes involved in gene expression. This may be countered by anti-proliferating genes like, the anti-proliferating gene from retinoblastoma (RB gene), when RB gene become inactive the cells divide indefinitely to produce tumour and when gene product is activated then cells stop dividing.

In Search of the Secrets of Ageing

Proteins, in their myriad forms and functions, are the substances most responsible for the day-to-day functioning of living organisms. some of these proteins seem to affect the way we age and how long we live. Tracherous oxygen molecules, protective enzymes, hormones that seem to turn back the clock, and proteins that may speed it up: The biochemistry of ageing is a rich territory with an expanding frontier. Major areas of exploration include oxygen radicals and glucose cross linking of proteins, both of which damage cells; the substances that help prevent and repair damage; and the role of specific proteins, particularly heat shock proteins, hormones, and growth factors.

Oxygen Radicals

Demolishing proteins and damaging nucleic acids, oxygen radicals are thought to be the day-to-day life of cells. The free radical theory of ageing, first proposed by **Denham Harman** at the University of Nebraska, holds that damage caused by oxygen radicals is responsible for many of the bodily changes that come with ageing. Free radicals have been implicated not only in ageing but also in degenerative disorders, including cancer, atherosclerosis, cataracts, and neurodegeneration.

A free radical is a molecule with an unpaired, highly reactive electron. An oxygen-free radical is a by product of normal metabolism, produced as cells turn food and oxygen into energy.

In need of a mate for its lone electron, the free radical takes an electron from another molecule, which in turn becomes unstable and combines readily with other molecules. A **chain reaction can ensue, resulting in a series of compounds, some of which are harmful. They damage proteins, membranes, and nucleic acids, particularly DNA,** including the DNA in mitochondria, the organelles within the cell that produce energy. But free radicals do not go unchecked. Mounted against them is a Multilayer defense system manned by antioxidants include nutrients like vitamins C and E and beta carotene as well as enzymes such as superoxide dismutase (SOD), catalase, and glutathione peroxidase. They prevent most, but not all, oxidative damage. Little by little the damage mounts and contributes, so the theory goes, to deteriorating tissues and organs.

Support for the free radical theory comes from studies of antioxidants, particularly SOD. SOD converts oxygen radicals into the also harmful hydrogen peroxide, Which is then degraded by another enzyme, catalase, to oxygen and water.

Antioxidants and Ageing Gerbils

A boost for the hypothesis that high levels of antioxidants can slow the ageing process comes from a study of N-tert-butyl-alpha-phenylnitrone or PBN in gerbils. Although it does not occur naturally in the body. PBN works in much the same way as beta-carotene and other antioxidants by increased levels of oxidised protein in their brains.

While it is only one study and more are needed, this investigation supports the idea that **maintaining antioxidant defense levels may be critical during ageing.** It also suggests that an intervention such as PBN may some day provide the means.

At the National Institute on Ageing (NIA), **Richard Cutler** has found that SOD levels are directly related to life span in 20 different species; **longer-lived animals have higher levels of SOD, suggesting that the ability to fight free radicals has something to do with longer life spans.**

Levels of other antioxidants vitamin E and beta-carotene, for example have also been correlated with life span.

Other studies have shown that **inserting extra copies of the SOD gene into fruit flies extends their average life span.** In three different laboratories researchers have reported that transgenic fruit flies, carrying extra copies of the gene for SOD, ***live 5 to 10 per cent longer than average.***

Other experimental evidence lends support to the free radical hypothesis. For example, higher levels of SOD and catalase have been found in long-lived nematodes. And in another important study, giving gerbils a synthetic antioxidant has reduced high levels of oxidised protein, a sign of ageing, in their brains.

The discovery of antioxidants raised hopes that people could retard ageing simply by adding them to the diet. Unfortunately taking SOD tablets has no effect on cellular ageing; the enzyme is simply broken down in the body during digestion. And when antioxidant vitamins are added to cells. They compensated by halting production of their own antioxidants, leaving free radical levels unchanged.

Researchers have not abandoned all hope for dietary antioxidants, however. Current studies, are exploring the possibility that vitamin C can reduce heart disease by blocking oxidation of low-density lipoproteins. Oxidation of these cholesterol-carrying proteins is thought to be a key element in hardening of the arteries. In addition, there is evidence that vitamin E in the diet may be linked to heart attacks, with low vitamin E intake appearing to increase the risk.

Glucose Crosslinking

Glucose, the fundamental source of energy, **reacts with and crosslinks essential molecules causing cellular deterioration,** process called **non-enzymatic glycosylation** or **glycation,** glucose molecules attach themselves to proteins, setting in motion a chain of chemical reactions that ends in the proteins binding together or crosslinking, thus altering their biological and structural roles. The process is slow but increase with time.

Crosslinks, which have been termed "**advanced glycosylation end products" (AGEs), seem to toughen tissues and may cause some of the deterioration associated with ageing. AGEs have been linked to stiffening connective tissue (collagen), hardened arteries, clouded eyes, loss of nerve function, and less efficient kidneys.**

These are deficiencies that often accompany ageing. They also appear at younger ages in people with diabetes, who have high glucose levels. Diabetes, infact, is sometimes considered an accelerated model of ageing. Not only do its complications mimic the physiologic changes that can accompany old age, but it victims have shorter-than-average life expectancies. As a result, much research on crosslinking has focussed on its relationship to diabetes as well as ageing.

One happy finding is that the body has its own defense system against crosslinking. Just as it has antioxidants to fight free-radical damage, it has other guardians, **immune system cells called that macrophages, that combat glycation. Macrophages with special receptors for AGEs seek them out, engulf them, break them down, and eject them into the blood stream where they are filtered out by the kidneys and eliminated in urine.**

The only apparent drawback to this defense system is that it is not complete and ***levels of AGEs increase steadily with age.*** One reason is that kidney function tends to decline with advancing age. Another is that macrophages, like certain other components of the immune system, become less active. Why is not known, but immunologists are beginning to learn more about how the immune system affects and is affected by ageing. And in the meantime, diabetes researchers are investigating drugs that could supplement the body's natural defenses by blocking AGE formation.

Crosslinking interests gerontologists for several reasons. It is associated with disorders that are common among older people, such as diabetes; it progresses with age; and AGEs are potential targets for anti-ageing drugs. In addition, crosslinking may play a role in damage to DNA, which has become another important focus for research on ageing.

DNA Repair

In the normal wear and tear of cellular life, DNA undergoes continual damage. Attacked by oxygen radicals, ultraviolet light, and other toxic agents, it suffers damage in the form of deletions, or destroyed sections, and mutations, or changes in the sequence of DNA bases that make up the genetic code.

DNA is damaged throughout life; the repair process may be a major factor in ageing, health, and longevity. DNA damage, which gradually accumulates, leads to malfunctioning genes, proteins, cells, and, as the years go by, deteriorating tissues and organs.

Not surprisingly, numerous enzyme systems in the cell have evolved to detect and repair damaged DNA. The repair process interests gerontologists. ***It is known that an animal's ability to repair certain types of DNA damage is directly related to the life span of its species. Humans repair DNA, more quickly and efficiently than mice or other animals with shorter life spans.*** This suggests that DNA damage and repair are in some way part of the ageing puzzle.

In addition, researchers have ***found defects in DNA repair in people with a genetic or familial susceptibility to cancer.*** If DNA repair processes decline with age while damage accumulates, as scientists hypothesise, it could help explain why cancer is so much more common among older people. Gerontologists who study DNA damage and repair have begun to uncover numerous complexities. Even within a single organism, repair rates can vary among cells, with the most efficient repair going on in terms (sperm and egg) cell. Moreover, certain genes are repaired more quickly than others, including those that regulate cell proliferation.

Research on sunlight may help explain what happens to skin as we age. As anyone who reads beauty magazines knows, sunlight damages skin in ways that seem similar to ageing. It causes wrinkles, to begin with. And in both normal ageing and photoageing – the process initiated by sunlight – the skin becomes drier and loses elasticity. Although gerontologists think that the normal or intrinsic ageing process is probably not the same as photoageing, there are enough similarities to make this a tantalising field of study.

The process of photoageing may hold clues to normal ageing because many of the same cells are affected. **Photoageing**, for example, **damages collagen and elastin, the two proteins that give skin its elasticity**. These proteins decline as we age, along with the fibroblast cells that manufacture them. In addition, the enzymes that break down collagen and elastin increase. Other changes occur in keratinocytes, upper-layer skin cells that are shed and renewed regularly. **In the normal ageing process the turnover of keratinocytes slows down and in photoageing, they are damaged.** Still other skin cells, called **melanocytes,** are also **affected by both processes: they decline with normal ageing, are killed in photoageing.**

What we don't known yet is exactly how photoageing damages cells. Ultraviolet light can damage DNA and could be the culprit. Free radicals could be involved in some way. Researchers continue to explore these and other factors in the effort to understand photoageing.

Especially intriguing is repair to a kind of DNA that resides not in the cell's nucleus but in its mitochondria. These small organelles are the principal sites of metabolism and energy production, and cells can have hundreds of them. **Mitochondrial DNA is thought to be injured at a much greater rate than nuclear DNA,** possibly because the mitochondria produce a stream of damaging oxygen radicals during metabolism. Adding to its vulnerability, mitochondrial DNA is unprotected by the protein coat that helps shield DNA in the nucleus from damage.

Research has shown that **mitochondrial DNA damage increases exponentially with age**, and several diseases that appear late in life, including late-onset diabetes, have been traced to defects in mitochondria. While such disorders seem to be linked to metabolism, it is not yet known whether age-associated damage also impairs metabolism.

Heat Shock Proteins

Despite their name, heat shock proteins (HSPs) are produced when cells are exposed to various stresses, not only heat. Their expression can be triggered by exposure to toxic substances such as heavy metals and chemicals and even by behavioural and psychological stress. Produced in response to stress, HSPs decline with age. What attracts ageing researchers to HSPs is the finding that the levels at which they are produced depend on age. ***Old animals placed under stress – physical restraint, for example – have lower levels of a heat shock protein designated HSP-70 than young animals under similar stress.*** Moreover, in laboratory cultures of cells, researchers **have found a striking decline in HSP-70 production as cells approach senescence.**

Exactly what role HSPs play in the ageing process is not yet clear. They are known to help the cell disassemble and dispose of damaged proteins and to facilitate the making and transport of new proteins. But what proteins are involved and how they relate to ageing is still the subject of speculation and study.

Hormones

In 1989, at Veterans Administration hospitals in Milwaukee and Chicago, a small group of men aged 60 and over began receiving injections three times a week that dramatically reversed some signs of ageing. The injections increased their lean body (and presumably muscle) mass, reduced excess fat, and thickened skin. When the injections stopped, the men's new strength ebbed and signs of ageing returned. **Declining levels of these chemical messengers may trigger some ageing processes.**

What the men were taking was ***recombinant human growth hormone (GH),*** a synthetic version of the hormone that is produced in the pituitary gland and plays a critical part in normal childhood growth and development. Now researchers are learning that **GH, or the decline of GH, seems also to play a role in the ageing process in at least some individuals.**

The idea that hormones are linked to ageing is not new. We have long known that some hormones decline with age. **Human growth hormone levels decrease in about half of all adults with the passage of time. Production of the sex hormones estrogen and testosterone tends to fall off.** Hormones with less familiar names, like melatonin and thymosin, are also not as abundant in older as in younger adults.

Hormones and Research on Ageing

Produced by glands, organs, and tissues, hormones are the body's chemical messengers, flowing through the blood stream and searching out cells fitted with special receptors. Each receptor, like a lock, can be opened by the specific hormone that fits it and also, to a lesser extent, by closely related hormones. Here are some of the hormones and other growth factors of special interest to gerontologists.

- **Estrogen:** The female hormone, estrogen is **used in hormone replacement therapy to relieve discomforts of menopause**. Produced mainly by the ovaries, it *slows the bone thinning that accompanies ageing* and may help prevent frailty and disability. After menopause, fat tissue is the major source of a weaker form of estrogen than that produced by the ovaries.
- **Growth hormone:** This product of the pituitary gland appears to play a role in body composition and muscle and bone strength. It is released through the action of another tropic factor called growth hormone releasing hormone, which is produced in the brain. It works by stimulating the production of insulin-like growth factor, which comes mainly from the liver. All three are being studied for their potential to strengthen muscle and bones and prevent frailty among older people.
- **Melatonin:** This hormone from the pineal gland responds to light an seems to regulate various seasonal changes in the body. As it declines during

ageing, it may trigger changes throughout the endocrine system.

- **Testosterone:** The male hormone, testosterone is produced in the testes and may decline with age, though less frequently or significantly than estrogen in women. Researchers are investigating its ability to strengthen muscles and prevent frailty and disability in older men when administered as testosterone therapy. They are also looking at its side effects, which may include an increased risk of certain cancers, particularly prostate cancer.
- **DHEA:** (Dehydroepiandrosterone), DHEA is produced in the adrenal glands. It is a weak male hormone and a precursor to some other hormones, including testosterone and estrogen. **DHEA is being studied for its possible effects on selected aspects of ageing, including immune system decline, and its potential to prevent certain chronic diseases, like cancer and multiple sclerosis.**

Hormone Replacement

We also know that when some declining hormones are replaced, various signs of ageing diminish. Most, like growth hormone, are still in the experimental stage, but one, estrogen, is used in medical practice to alleviate the discomforts of menopause. ***Estrogen replacement therapy also lessens the accelerated bone loss that comes with menopause and may help prevent cardiovascular disease.*** Preliminary studies suggest that testosterone replacement may likewise have benefits for ageing men, by increasing bone and muscle mass and strength. However, questions about cancer and other risks surround both estrogen and testosterone replacement therapy and have not yet been resolved.

A hormone that has attracted the interest of many researchers is DHEA (dehydroepiandrosterone), which is abundant in youth but begins to decline in humans at about age 30. Very low levels of DHEA have been linked to cardiovascular disease in men, some cancers, trauma, and stress; low levels are also associated with old age, particularly in the unwell, institutionalised elderly. **In animal studies, replacing DHEA has had Startling anti-ageing effects. Large doses of the hormone have restored older animals' strength and vigor.**

How DHEA works is still not clear. Circulating through the blood stream in an inactive form, called DHEA sulphate, this hormone becomes active when it comes in contact with a specific cell or tissue that "needs" it. When this happens, the sulfate is removed.

DHEA seems to be needed, for example, to assist in the function and proliferation of immune cells. In experiments with mice, DHEA sulphate boosted the older animals' levels of a **substance called interleukin-2, important in the immune response.**

Growth Factors

Hormones are aided and abetted by an arsenal of other substances that also stimulate or modulate cell activities. Known collectively as growth or trophic factors, these include substances such as insulin-like growth factor (IGF-1), which mediates many of the actions of GH. Another trophic factor of interest to gerontologists is growth hormone releasing hormone, which stimulates the release of GH.

The mechanisms – how hormones and growth factors produce their effects – are still a matter of intense speculation and study. Scientists know that these chemical messengers selectively stimulate cell activities which in turn affect critical events, such as the size and functioning of skeletal muscle. However, the pathway from hormone to muscle is complex and still unclear.

Consider growth hormone. It begins to stimulating production of insulin-like growth factor. Produced primarily in the liver, IGF-1 enters and flows through the blood stream, seeking out special IGF-1 receptors on the surface of various cells, including muscle cells. Through these receptors it signals the muscle cells to increase in size and number, perhaps by stimulating their genes to produce more of special, muscle-specific proteins. Also involved at some point in this process are one or more of the six known proteins that bind with IGF-1; their regulatory roles are still a mystery.

As if the cellular complexities were not enough, the action of growth hormone also may be intertwined with a cluster of other factors – exercise, for example, which stimulates a certain amount of GH secretion on its own, and obesity, which depresses production of GH. Even the way fat is distributed in the body may make a difference; lower levels of GH have been linked to excess abdominal fat but not to lower body fat.

Role of Dopamine Receptors in Ageing

Some of the changes associated with ageing appear due to the loss of a particular type of a chemical receptor in the brain.

Researchers investigated the effects of ageing on dopamine, which is a chemical in the brain that modulates the communications between areas in the brain that are involved with movement, cognition, motivation and reward. Cognition can be defined as memory and learning ability.

By using an imaging technique called positron emission tomography (PET) scanning to visualise the brain, the investigators measured both the levels of dopamine D_2 receptors as well as regional brain glucose metabolism which serves as an index of brain function.

After analysis of the datas, the researchers found that ***ageing correlated with a loss of dopamine receptors and that these losses were associated with a decline in activity in the frontal regions of the brain.***

The researchers interpret these results as an indication that the age-related losses in brain dopamine activity result in dysfunction of brain regions that are known to be involved with cognition, attention and mood.

They concluded that the decline in cognitive abilities and the higher propensity for depression in the elderly with ageing may in part be due to the losses of brain dopamine receptors.

Macular Degeneration of Eye

Recently a team of international researchers have **identified a protein called VEGFR-1** and found that when levels are low or absent blood vessels begin to grow in the cornea of eye impairing vision.

Normally cornea is transparent and lacks blood vessels. Cornea contains a chemical called **VEGF-A,** which **Promotes blood vessel growth**. But the new protein discovered by the researchers **acts like a mop, absorbing VEGF-A**, which would otherwise make blood vessels to grow. **The new protein discovered blocks the blood vessel growth.**

The researchers are now focussing on understanding how the body produces the protein and how to deliver a synthesised version of the protein into the eye through eyedrops.

BIBLIOGRAPHY

1. Alberti KG MN (Ed). *Recent Advances in Clinical Biochemistry*, Churchill Livingstone, 1978.
2. Astwood EB. *Recent Progress in Hormone Research*, Vol 24, Academic Press, New York, 1968.
3. Baron DN. *A Short Textbook of Chemical Pathology*, 4th edn, 1982.
4. Bell GH, Davidson JN, Scarborough. *Textbook of Physiology and Biochemistry*, E & S Livingstone, 1965.
5. Bloom SR, Polak JM. *Gut Hormones*, 2nd edn, Churchill Livingstone, 1981.
6. Bondy PK, Rosenberg LE. *Duncan's Diseases of Metabolism*, 7th edn, WB Saunders, Philadelphia, London, 1974.
7. Bowen HJA. *Trace Elements in Biochemistry*, Academic Press, New York, 1966.
8. Brewer HB, Bronzert TJ. *Human Plasma Lipoproteins*, Fractions No-I, 1977.
9. Cantarow A, Schepartz B. *Biochemistry*, 4th edn, WB Saunders, Philadelphia, London, 1970 (reprint).
10. Conn EE, Stumpf PK. *Outlines of Biochemistry,* 2nd edn, Wiley Eastern, New Delhi, 1969.
11. Coodley EL. *Diagnostic Enzymology*, Lea & Febiger, Philadelphia, 1970.
12. Davidson AN (Ed.). *Biochemistry and Neurological Disease*, Blackwell Scientific Publications, Oxford, 1976.
13. Davidson JN. *Biochemistry of Nucleic Acids*, 5th edn, Wiley, New York, 1965.
14. Daven Port HW. *ABC of Acid-base Chemistry*, 6th edn, University of Chicago Press, 1974.
15. Das D. *Biophysics and Biophysical Chemistry*, 2nd edn, Academic Publishers, 1987.
16. De Luca HF, Schnoes HK. Vitamin D: Recent Advances, *Ann Rev Biochem*, 1983.
17. De AK. *Environmental Chemistry*, 3rd edition.
18. Dixon M, Webb EC. *Enzymes*, 2nd edn, Academic Press, New York, 1964.
19. Elkeles RS, Javill AS. *Biochemical Aspects of Human Disease*, Blackwell Publications, 1983.
20. Fersht A. *Enzyme Structure and Mechanism*, 2nd edn, Freeman, 1985.
21. Fourgerean M, Dausset J (Ed). *Progress in Immunology* (Vol IV), Academic Press, 1981.
22. Frisell WR. *Acid-base Chemistry in Medicine,* Macmillan, New York, 1968.
23. Fruton JS, Simmonds SS. *General Biochemistry*, 2nd edn, John Wiley & Sons, New York, 1965.
24. Ganong WF. *Review of Medical Physiology*, 6th edn, Lange Medical Publications, 1973.
25. Goldberger, Emanuel. *A Primer of Water, Electrolytes and Acid-base Syndromes*, 4th edn, Lea and Febiger, Philadelphia, 1971.
26. Goodhart RS, Maurice E Shils. *Modern Nutrition in Health and Disease*, 5th edn, Lea and Febiger, 1973.
27. Gopalan C, Rao, Nara Singa BS. *Dietary Allowances for Indians*, Indian Council of Medical Research, New Delhi, 1980.
28. Halkerston Ian DK. *Biochemistry*, 2nd edn, John Wiley and Sons, 1990.
29. Hoffman WS. *The Biochemistry of Clinical Medicine*, 4th edn, Year Book Medical Publishers, Chicago, 1970.
30. Harper HA. *Review of Physiological Chemistry*, 17th edn, Lange Medical Publication, 1979.
31. *Harper's Biochemistry: A Lange Medical Book*, 25th edn, 1999.
32. *Harper's Illustrated Biochemistry, A Lange Medical Book,* International Edition, 2003, 26th Edition.
33. *Harper's Illustrated Biochemistry A Lange Medical Book*, International Edition, 2006, 27th Edition.
34. Harrison A. *Chemical Methods in Clinical Medicine*, 4th edn, J and A Churchill, 1957.
35. Hawk's. *Physiological Chemistry* (Ed Oser BL), 14th edn, Blackiston Division, McGraw-Hill, New York, 1965.
36. Hobbs JR. Immunoglobulins in Clinical Chemistry, *Advances in Clinical Chemistry,* 1971.
37. Heftman E (Ed). *Chromatography*, 3rd edn, Reinhold, 1975.
38. Hsia DY. *Inborn Errors of Metabolism*, 2nd edn, Year Book Medical Publications, Chicago, 1966.
39. King EJ. *Practical Clinical Enzymology*, D Von Nostrand, London, 1965.

40. Kleiner IS, Orten JM. *Biochemistry*, 7th edn, CV Mosby, St Louis, 1966.
41. Khan RH, Lands WEM. *Prostaglandins and Cyclic AMP*, Academic Press, New York, 1973.
42. Kornberg A. *DNA Replication*, Freeman, 1980.
43. Krishna Swamy K. Selenium in Human Health, *ICMR Bulletin*, 1990.
44. Lands WEM. The biosynthesis and metabolism of prostaglandins, *Ann Rev Physiol*, 1979.
45. Latner AL, Cantarow, Trumper. *Clinical Biochemistry*, 7th edn, Saunders, Philadelphia, 1975.
46. Lehninger AL. *Biochemistry*, 2nd edn, (Reprint) Kalyani Publishers, Ludhiana, New Delhi, 1984.
47. Levinsky NG. Renal Kallikrien-kinin system, *Clin Res*, 1979.
48. Mazur A, Harrow B. *Textbook of Biochemistry*, 10th edn, Saunders, Philadelphia, 1971.
49. Moncada S (Ed). Prostacyclin, thromboxane and leukotrienes, *Brit Med Bull*, 1983.
50. McGilvery RW. *Biochemistry—A Functional Approach*, 3rd edn, Saunders, Philadelphia, 1983.
51. Murray. *Harpers Biochemistry*, Harper & Row, 1990.
52. Orten JM, Neuhaus W. *Human Biochemistry*, 10th edn, CV Mosby, BI Publications Ltd, New Delhi.
53. Prasad AS. *Trace Elements and Iron in Human Metabolism*, Plenum Press, 1978.
54. *Park's Textbook of Preventive and Social Medicine*, 15th edn.
55. Putman FW (Ed). *The Plasma Proteins—Structure, Function and Genetic Control*, 2nd edn, Academic Press, New York, 1977.
56. Ramakrishnan S, Swamy J. *Textbook of Clinical (Medical) Biochemistry and Immunology*, 1st edn, TR Publications, 1995.
57. Rawn JD. *Biochemistry*, Neil Patterson Publishers, Burlington, North Carolina, 1989.
58. *Samson Wright's Applied Physiology*: The English Language Book Society and Oxford University Press, London, 12th edn, 1971.
59. Sittes DP. *Basic and Clinical Immunology*, 4th edn, Lange Medical Publication, 1982.
60. Smith LC. Plasma Lipoproteins: Structure and metabolism. *Ann Rev Biochem*, 1978.
61. Smith EL, Hill RL, Lehman IR, *et al*. *Principles of Biochemistry*, 7th edn, McGraw-Hill International, 1983.
62. Stryer L. *Biochemistry*, 3rd edn, WH Freeman, 1975.
63. Sunderman FW, Sunderman FW Jr. *Serum Proteins and the Dysproteinaemias*, Pitman Medical Publication, Philadelphia, 1964.
64. Suttie John W. *Introduction to Biochemistry*, Holt Rinehart and Winston, New York, 1977.
65. Swaminathan M. *Biochemistry for Medical Students*, 1st edn, Geetha Book House Publishers, Mysore, 1981.
66. Tanaka N, Ishihara M, Lamphier MS. Cooperation of the tumour suppressors IRF-1 and P53 in response to DNA Damage, *Nature* 382; 816: 1996.
67. Thompson G. Plasma lipoproteins and their disorders, *Medicine*, 3rd series, 1978.
68. Thompson RHS, Wotton IDP. *Biochemical Disorders in Human Diseases*, 3rd edn, J and A Churchill Ltd, London, 1970.
69. Thorpe WB, Bray HG, James HP. *Biochemistry for Medical Students*, 9th edn, Churchill, London, 1970.
70. Underwood EJ. *Trace Elements in Human and Animal Nutrition*, 4th edn, Academic Press, New York, 1977.
71. Varley H. *Practical Clinical Biochemistry*, William Heinemann Medical Books Ltd, London, 1969.
72. von Euler, Eliasson R. *Prostaglandins*, Academic Press, New York, 1967.
73. Wasserman RH (Ed). *Calcium Binding Proteins and Calcium Function*, Elsevier, 1977.
74. Weisberg HF. Water, *Electrolytes and Acid-Base Balance*, 2nd edn, Williams and Wilkins, Baltimore, 1962.
75. West ES, Todd WR, Mason HS, Van Bruggen JT. *Textbook of Biochemistry*, Macmillan, New York, 1966.
76. Wikipedia, the free encyclopedia.
77. Wilkinson R. *Isoenzymes*, 2nd edn, Chapman and Hall, London, 1970.
78. Wilkinson JH (Ed). *Principles and Practice of Diagnostic Enzymology*, Edward Arnold, London, 1976.
79. Williams RH. *Textbook of Endocrinology*, WB Saunders, Philadelphia, Indian Reprint, 1970.
80. Wootton IDP. *Microanalysis in Medical Biochemistry*, 6th edn, J and A. Churchill Ltd., London, 1982.
81. Wyllie AH. Apoptosis, *Recent Advances in Histopathology*.
82. Wyllie AH, Carder PJ, Clarke AR. Apoptosis in carcinogenesis; the role of p53, Gold Spring Harbor Symposia on Quantitative Biology, 403-09, 1994.
83. Yudkin M, Offord K. *Comprehensive Biochemistry*, Longman (England), 1973.
84. Zubay Geoffrey. *Biochemistry*, 2nd edn, Maxwell Macmillan (International edn), 1989.

INDEX

*Page numbers followed by **f** refer to figures and **t** refer to tables*

B

C

INDEX

INDEX

E

F

G

H

J

K

L

M

N

O

INDEX

S

T

U

V

INDEX